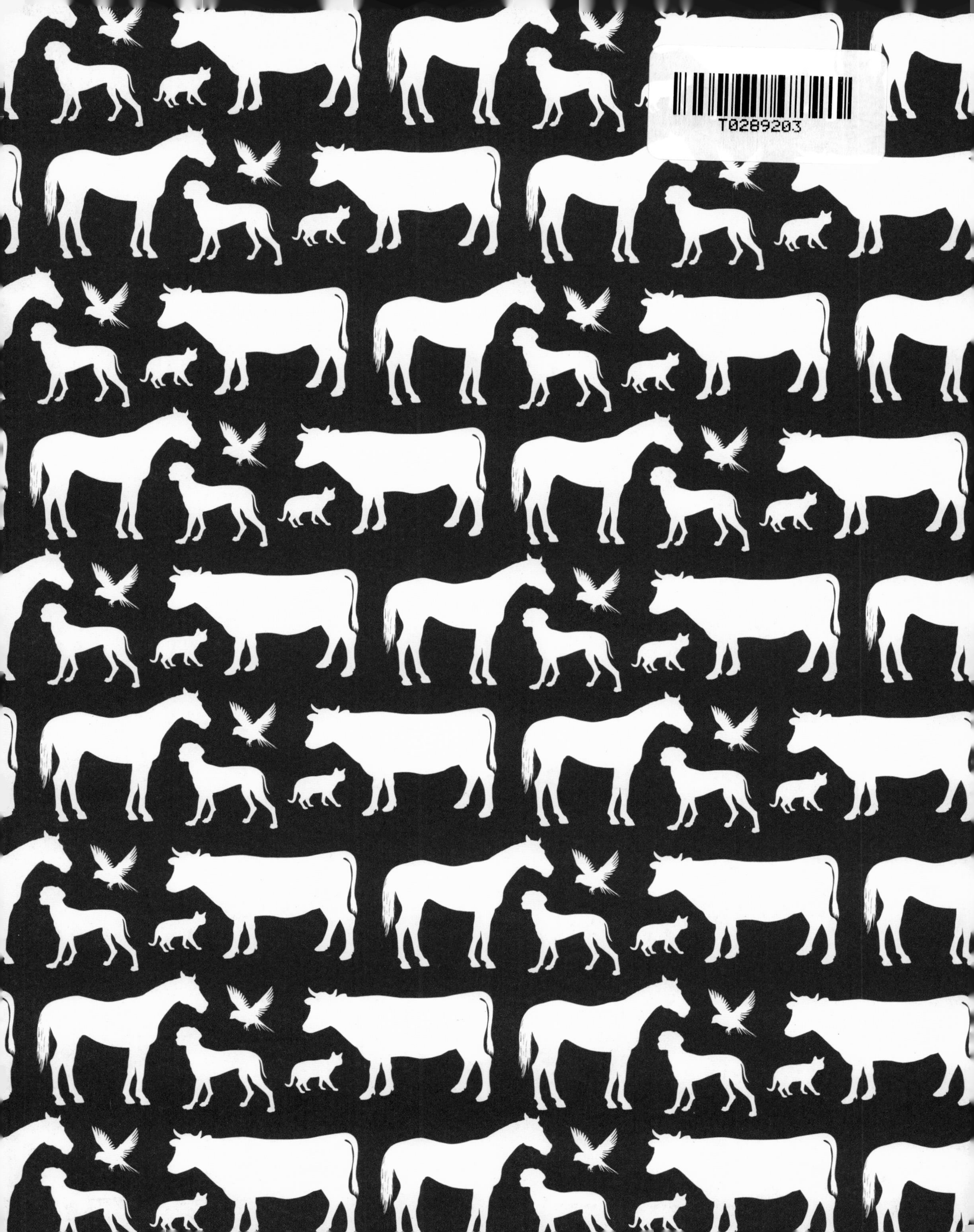
T0289203

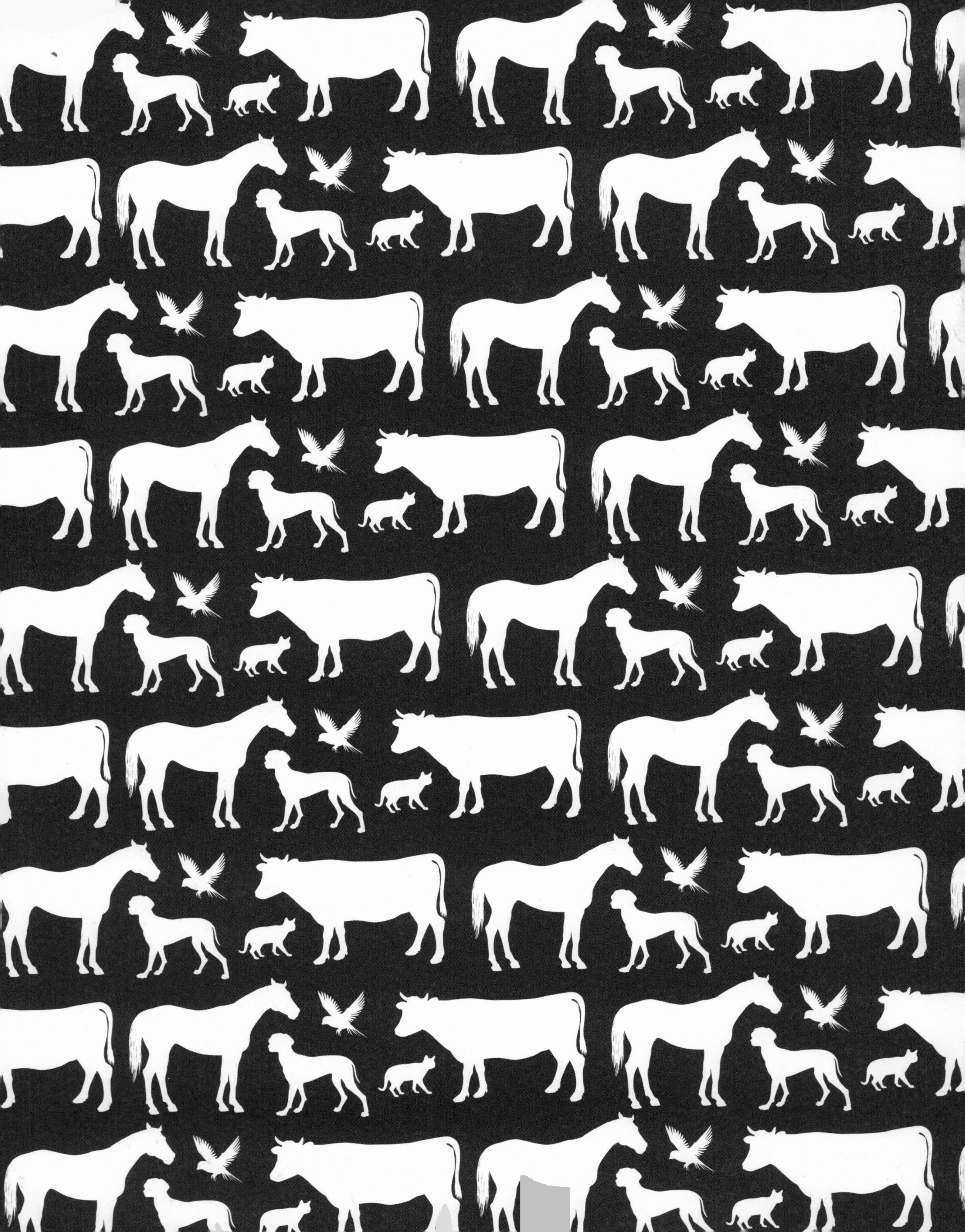

Comparative Veterinary Anatomy: A Clinical Approach

Comparative Veterinary Anatomy: A Clinical Approach

Edited by:

James A. Orsini
Associate Professor of Surgery
Department of Clinical Studies - New Bolton Center
University of Pennsylvania School of Veterinary Medicine
Kennett Square, Pennsylvania, US

Nora S. Grenager
Director of Internal Medicine
Steinbeck Peninsula Equine Clinics
Menlo Park, California, US

Alexander de Lahunta
Emeritus James Law Professor of Anatomy
Department of Biomedical Sciences
Cornell University College of Veterinary Medicine
Ithaca, New York, US

Academic Press is an imprint of Elsevier
125 London Wall, London EC2Y 5AS, United Kingdom
525 B Street, Suite 1650, San Diego, CA 92101, United States
50 Hampshire Street, 5th Floor, Cambridge, MA 02139, United States
The Boulevard, Langford Lane, Kidlington, Oxford OX5 1GB, United Kingdom

Notices
Knowledge and best practice in this field are constantly changing. As new research and experience broaden our understanding, changes in research methods, professional practices, or medical treatment may become necessary.

Practitioners and researchers must always rely on their own experience and knowledge in evaluating and using any information, methods, compounds, or experiments described herein. In using such information or methods they should be mindful of their own safety and the safety of others, including parties for whom they have a professional responsibility.

To the fullest extent of the law, neither the Publisher nor the authors, contributors, or editors, assume any liability for any injury and/or damage to persons or property as a matter of products liability, negligence or otherwise, or from any use or operation of any methods, products, instructions, or ideas contained in the material herein.

Library of Congress Cataloging-in-Publication Data
A catalog record for this book is available from the Library of Congress

British Library Cataloguing-in-Publication Data
A catalogue record for this book is available from the British Library

ISBN **978-0-323-91015-6**

For information on all Academic Press publications
visit our website at https://www.elsevier.com/books-and-journals

Publisher: Charlotte Cockle
Acquisitions Editor: Anna Valutkevich
Editorial Project Manager: Pat Gonzalez
Production Project Manager: Kiruthika Govindaraju
Cover Designer: Matthew Limbert

Typeset by STRAIVE, India

Printed in India

Last digit is the print number: 9 8 7 6 5 4 3

MARY ALICE D. MALONE

As a great woman you have touched every element of the human spirit with your huge heart and tremendous kindness.

It was your trust, enduring integrity, and unwavering dedication to a shared vision that allowed this book to be published.

Thank you for inspiring us to bring together the priceless knowledge of so many to advance education and research—forever!

We are deeply grateful for your passion for excellence and joyfully dedicate this book to you.

PHOTO CREDIT: SUSAN SEXTON

We Remember *and* Celebrate our colleague

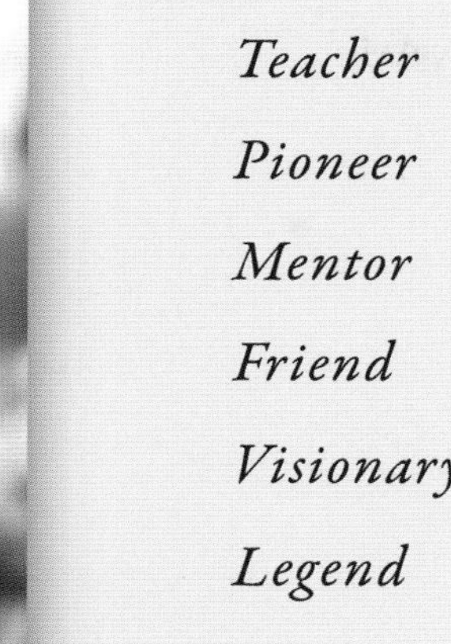

Teacher

Pioneer

Mentor

Friend

Visionary

Legend

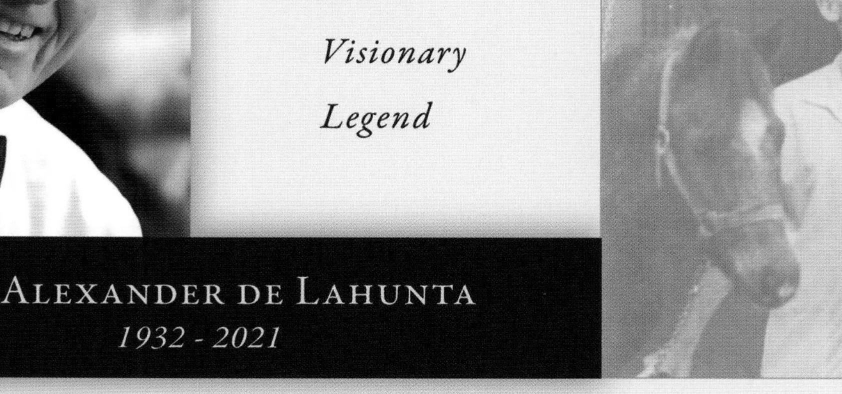

Dr. Alexander de Lahunta

1932 - 2021

"Sandy was a unique individual who almost single handedly developed clinical veterinary neurology. The esteem and respect that he evoked amongst his colleagues, and the dedication and love from his students, is beyond anything that I have experienced. He embodied humility and brilliance and was the model physician scientist."

Michael Kotlikoff, VMD, PhD, ScD (hc) Provost, Cornell University

"On top of his extraordinary contributions to neuroscience and the practice of clinical neurology, Dr. de Lahunta was above all else an inspiring educator. The clarity and enthusiasm of his presentations were an example for all of us who knew him as faculty colleagues, and he will be sorely missed. His influence on veterinary students was immense, whether through his teaching and mentorship in neuroanatomy, applied anatomy, and clinical interactions at Cornell University, or through foundational textbooks used throughout the world. The veterinary profession benefitted from his work in numerous ways and his legacy will continue to impact our work far into the future."

Lorin D. Warnick, DVM, PhD • Austin O. Hooey Dean of Veterinary Medicine
Cornell University College of Veterinary Medicine

Dr. Alexander (Sandy) de Lahunta was a once-in-a-lifetime professor and mentor, and a genuinely remarkable educator who embodied the role of scientist, clinician, and teacher. His energy, passion, and kindness were captivating and infectious, giving tirelessly to veterinary medicine for more than 40 years. We were all blessed and benefited, directly or indirectly, from his ability to explain the most complex processes methodically and, most importantly, systemically in the basic and clinical sciences in a manner so that everyone understood the information. Dr. de Lahunta was indeed a true "giant" amongst educators, beloved by generations of students, and we are enormously thankful for his generosity of spirit as a professor, mentor, and friend. And the greatest professional compliment is we have modeled our careers after Dr. D.!

Eric N. Glass, DVM, DACVIM (Neurology) Redbank Veterinary Hospital
Amy L. Johnson, DVM, DACVIM (LAIM & Neurology) University of Pennsylvania
Marc Kent, DVM, DACVIM (Internal Medicine and Neurology) University of Georgia

Alexander de Lahunta- teacher extraordinaire, colleague, and mentor. Thank you for believing in us by honoring us with this gift of a lifetime as our co-editor.

Nora S. Grenager *and* James A. Orsini Editors

Photo credits: Cornell University College of Veterinary Medicine

CONTENTS

CONTRIBUTORS xv
PREFACE xxi
ACKNOWLEDGMENTS xxiii

SECTION I: INTRODUCTION 1

CHAPTER 1: Clinical Anatomy Nomenclature
N.S. Grenager, J.A. Orsini, and A. de Lahunta 3

SECTION II: DIAGNOSTIC IMAGING 15

CHAPTER 2: Introduction to Imaging Modalities 17
Current Imaging Technologies
2.1 Endoscopy – Nora S. Grenager and James A. Orsini 18
2.2 Radiography – Aitor Gallastegui 23
2.3 Ultrasonography (US) – Aitor Gallastegui 35
2.4 Computed tomography (CT) – Aitor Gallastegui 47
2.5 Magnetic resonance imaging (MRI) – Aitor Gallastegui 57
2.6 Nuclear medicine (including PET) – Aitor Gallastegui 67
Novel Technologies
2.7 Advances in diagnostic imaging 74
- Files in diagnostic imaging – Aitor Gallastegui
- 3-D printing – Aitor Gallastegui
- Cone beam computed tomography – Aitor Gallastegui
- Total or whole-body PET scanner – A. Gallastegui, N.S. Grenager, and J.A. Orsini

SECTION III: CANINE AND FELINE CLINICAL CASES 81

3.0/Canine Landscape Figures (1–9)
N.S. Grenager, J.A. Orsini, J.F. Randolph, H.M.S. Davies, and A. de Lahunta. 83

3.1/Feline Landscape Figures (1–9)
N.S. Grenager, J.A. Orsini, J.F. Randolph, H.M.S. Davies, and A. de Lahunta. 93

CHAPTER 3: Axial Skeleton: Head, Neck, and Vertebral Column 103
Nasal Cavity, Pharynx, and Paranasal Sinuses
3.1 Nasopharyngeal polyp – Meredith Miller 104
3.2 Brachycephalic airway syndrome – David Holt 112
Mouth
3.3 Odontogenic disease – Nadine Fiani and Santiago Peralta 119
Eye
3.4 Retrobulbar mass – Eric Ledbetter 126
Ear
3.5 Otitis interna/media – Adalberto Merighi and Laura Lossi 133
Neck
3.6 Thyroid adenocarcinoma – Takanori Sugiyama and Helen M.S. Davies 146
3.7 Hyperthyroidism – Mark E. Peterson and John F. Randolph 164

Vestibular System, Brain and Vertebral Column
3.8 Cervical intervertebral disc disease – Takanori Sugiyama and Helen M.S. Davies ... 170
3.9 Vestibular dysfunction – Marc Kent and Eric N. Glass ... 181
3.10 Glioma – Marc Kent and Eric N. Glass ... 188
3.11 Meningioma – Fred Wininger ... 196
3.12 Lumbar intervertebral disc disease – Marc Kent and Eric N. Glass ... 207

CHAPTER 4: Thorax ... 219
Pleura, Mediastinum, and Lungs
4.1 Aspiration pneumonia – N. Bamford, C. Beck, and H.M.S. Davies ... 220
4.2 Pyothorax – L. Dooley, C. Beck, and H.M.S. Davies ... 230
4.3 Diaphragmatic rupture – Cathy Beck and Helen M.S. Davies ... 236
Mediastinal Organs
4.4 Feline cardiomyopathy – Mark Oyama and Simon Bailey ... 243
4.5 Persistent right fourth aortic arch – David Holt ... 251
4.6 Patent ductus arteriosus – Mark Oyama ... 258
4.7 Mitral valve disease – L. Dooley, C. Beck, and S. Bailey ... 265
4.8 Esophageal foreign body – David Holt ... 272

CHAPTER 5: Abdomen ... 277
Liver, Pancreas, Spleen, and Adrenal Glands
5.1 Portosystemic vascular anomaly – Sharon A. Center and John F. Randolph ... 278
5.2 Extrahepatic bile duct obstruction secondary to acute pancreatitis – Shannon M. Palermo and Mark P. Rondeau ... 287
5.3 Hyperadrenocorticism – N.S. Grenager, R.S. Hess, and J.A. Orsini ... 292
5.4 Splenic torsion – D. Holt, N.S. Grenager, and J.A. Orsini ... 298
Stomach
5.5 Gastric dilatation and volvulus – Carol A. Carberry ... 304
Small Intestine
5.6 Small intestine obstruction – D. Holt, N.S. Grenager, and J.A. Orsini ... 310
Large Intestine, Anal Canal, and Rectum
5.7 Megacolon – Takanori Sugiyama and Helen M.S. Davies ... 316
Body Wall
5.8 Abdominal wall hernia – Takanori Sugiyama and Helen M.S. Davies ... 325

CHAPTER 6: Pelvic Organs ... 337
Female Urogenital System
6.1 Ectopic ureters – James Flanders ... 338
6.2 Pyometra – Natali Krekeler and Helen M.S. Davies ... 346
6.3 Dystocia and the mammary gland – N. Krekeler, H.M.S. Davies, and C. Beck ... 360
Male Urogenital System
6.4 Benign prostatic hyperplasia – M. Schrank, S. Romagnoli, and N. Krekeler ... 373
6.5 Congenital phimosis – M. Schrank, N. Krekeler, H.M.S. Davies, and S. Romagnoli ... 385

CHAPTER 7: Thoracic Limb ... 391
Proximal Thoracic Limb (shoulder, brachium, and antebrachium)
7.1 Osteochondritis dissecans of the shoulder – Kimberly A. Agnello ... 392
7.2 Incomplete ossification of the humeral condyle – T. Sugiyama, H.M.S. Davies, and C. Beck ... 400
Distal Thoracic Limb (carpus and manus)
7.3 Carpal valgus deformity – L.A. Wallett, C. Beck, and H.M.S. Davies ... 416
7.4 Phalangeal fracture – Ray G. Ferguson and Helen M.S. Davies ... 430
Innervation of the Thoracic Limb
7.5 Nerve sheath neoplasm – Eric N. Glass and Marc Kent ... 446

CHAPTER 8: Pelvic Limb 455

Proximal Pelvic Limb (hip, stifle, crus)

8.1 Hip dysplasia – Christina Murray and Cathy Beck 456

8.2 Femoral fracture – Christina Murray and Cathy Beck 469

8.3 Cranial cruciate ligament tear – Kimberly A. Agnello 486

8.4 Tibial fracture – L.A. Wallett, C. Beck, and H.M.S. Davies 494

Distal Pelvic Limb (tarsus and pes)

8.5 Fracture of the tarsal bones – C. Boemo, O. Al-Juhaishi, Z. Akbar, and H.M.S. Davies 506

Innervation of the Pelvic Limb

8.6 Degenerative lumbosacral stenosis – Marc Kent and Eric N. Glass 520

8.7 Calcaneal tendon injury – M. Kent, E.N. Glass, and A. de Lahunta 531

CHAPTER 9: Integument and Mammary Gland 539

9.1 Sebaceous adenitis – Karen Trainor and Brian Palmeiro 540

(Mammary Glands – see Chapter 6.3)

SECTION IV: EQUINE CLINICAL CASES **551**

4.0/Equine Landscape Figures (1–11)

N.S. Grenager, M. Gerard, J.A. Orsini, and A. de Lahunta 553

CHAPTER 10: Axial Skeleton: Head, Neck, and Vertebral Column 565

Eye

10.1 Squamous cell carcinoma – Amanda Davis and Brian Gilger 566

Mouth

10.2 Septic sialoadenitis – Timo Prange and Mathew Gerard 576

Paranasal Sinuses

10.3 Paranasal sinus cyst – Ferenc Toth and James Schumacher 586

10.4 Dental disease and sinusitis – Callie Fogle and Mathew Gerard 599

Pharynx

10.5 Guttural pouch disease – Olivier M. Lepage 611

10.6 Dorsal displacement of the soft palate – Tara R. Shearer and Susan J. Holcombe 624

10.7 Laryngeal hemiplegia – Eric Parente 636

Cranial Nerves

10.8 Vestibular disease – William Gilsenan 646

Ear

10.9 Ear sarcoid – Annette M. McCoy 655

Poll

10.10 Nuchal bursitis – José García-López 663

Neck

10.11 Esophageal obstruction – Judith Koenig and Shune Kimura 670

Central Nervous System

10.12 Cervical vertebral osteoarthritis – Amy L. Johnson 679

10.13 Congenital cerebellar disorder – Monica Aleman 688

CHAPTER 11: Thorax 693

Heart

11.1 Tetralogy of Fallot – Cristobal Navas de Solis 694

11.2 Mitral regurgitation– Kari Bevevino and Cristobal Navas de Solis 704

Pleura, Mediastinum, and Lungs

11.3 Pleuropneumonia– Michelle Coleman 714

CHAPTER 12: Abdomen ... 723

Stomach, Spleen, and Small Intestine

12.1 Gastric ulcer disease – C. Xue, K. Withowski, A. St. Pierre, and K. Epstein ... 724
12.2 Duodenitis-proximal jejunitis – Katherine Christie and Jarred Williams ... 732
12.3 Epiploic foramen entrapment – J. Tyma, J. Goni, and J. Williams ... 737

Cecum and Colon

12.4 Cecal impaction – Euan Murray and Kira Epstein ... 744
12.5 Large colon volvulus – Jessica Bramski and Kira Epstein ... 750
12.6 Small colon enterolith – Phillip Kieffer and Kira Epstein ... 756

Organs

12.7 Cholangiocarcinoma – Brina Lopez and Kira Epstein ... 763

CHAPTER 13: Pelvic Organs ... 771

Female Urogenital System

13.1 Urovagina – Jennifer Linton ... 772
13.2 Second-degree perineal laceration – Candace Lyman and G. Reed Holyoak ... 779
13.3 Uterine artery rupture – Maria Ferrer ... 786
13.4 Granulosa cell tumor – Dirk Vanderwall ... 793
13.5 Endometrial cysts – Carlos Pinto and Luis Henrique de Aguiar ... 801
13.6 Oviduct/uterine tube obstruction – Candace Lyman and Patricia Sertich ... 808

Male Urogenital System

13.7 Squamous cell carcinoma of the penis – David Levine and Carrie Jacobs ... 816
13.8 Seminal vesiculitis – Malgorzata Pozor ... 825
13.9 Inguinal hernia – N.S. Grenager, M. Gerard, and J.A. Orsini ... 834

Urinary Bladder and Urachal Remnant

13.10 Omphalitis and bladder rupture – Singen Elliott and Jarred Williams ... 845
13.11 Cystic calculus –Tamara Dobbie ... 851

CHAPTER 14: Thoracic Limb ... 861

Proximal Thoracic Limb (shoulder, brachium, and antebrachium)

14.1 Radial neuropathy – Laura Johnstone ... 862
14.2 Supraglenoid tubercle fracture – Nick Carlson ... 873
14.3 Ulnar fracture – Nick Carlson ... 882
14.4 Radial fracture – Liberty Getman ... 891

Distal Thoracic Limb (carpus and manus)

14.5 Superficial digital flexor tendonitis – Nick Carlson ... 898
14.6 Osteochondral fragment of the metacarpophalangeal joint – Nick Carlson ... 905
14.7 Fracture of the 2nd phalanx – Nick Carlson ... 912
14.8 Foreign body penetration of the hoof – Nick Carlson ... 918
14.9 Laminitis – Nick Carlson ... 925

CHAPTER 15: Pelvic Limb ... 933

Proximal Pelvic Limb (hip, stifle, crus)

15.1 Coxofemoral joint luxation – Nick Carlson ... 934
15.2 Osteochondritis dissecans – Sarah DeSante ... 942
15.3 Disruption of the fibularis (peroneus) tertius – Sarah DeSante and Nick Carlson ... 952

Distal Pelvic Limb (tarsus and pes)

15.4 Gastrocnemius tendonitis – Sarah DeSante and Nick Carlson ... 958
15.5 Fracture of the 4th metatarsal bone – Nick Carlson ... 969

CHAPTER 16: Integument and Mammary Gland ... 977

16.1 Hereditary equine regional dermal asthenia – Daniela Luethy ... 978
16.2 Mastitis – Maria Ferrer ... 984

SECTION V: FARM ANIMAL CLINICAL CASES 989

5.0/Bovine Landscape Figures (1–9)
Nora S. Grenager, James A. Orsini, André Desrochers, and Alexander de Lahunta 991

CHAPTER 17: Axial Skeleton: Head, Neck, and Vertebral Column 1001
Head and Neck
17.1 Maxillary sinusitis – Caroline Constant 1002
17.2 Tooth root abscess – Andrew Niehaus 1013
17.3 Dehorning – Marjolaine Rousseau 1020
Vertebral Column
17.4 Spinal lymphoma – André Desrochers and Gilles Fecteau 1030

CHAPTER 18: Thorax 1041
Heart and Lungs
18.1 Pericarditis – Marie-Eve Fecteau 1042
18.2 Endocarditis and atrial lymphoma – Marie-Eve Fecteau and Gilles Fecteau 1049

CHAPTER 19: Abdomen 1055
Forestomachs (rumen, reticulum, omasum, and abomasum)
19.1 Traumatic reticuloperitonitis – Emma Marchionatti 1056
19.2 Left displacement of the abomasum – Brenton C. Credille and Susan Fubini 1063
Small and Large Intestine
19.3 Intestinal volvulus – David E. Anderson 1069
Cecum
19.4 Cecal dilatation/volvulus – Emma Marchionatti 1077
Liver
19.5 Caudal vena cava syndrome – Julie Berman 1082
19.6 Hepatic lipidosis – Julie Berman 1090
Kidney
19.7 Hydronephrosis – André Desrochers 1097

CHAPTER 20: Pelvic Organs 1105
Female Urogenital System
20.1 Perineal laceration – Michael Pesato and Billy I. Smith 1106
20.2 Dystocia with cesarean section – Andrew Niehaus 1112
Male Urogenital System
20.3 Urolithiasis – Marie-Eve Fecteau 1127
20.4 Penile hematoma – David E. Anderson 1135

CHAPTER 21: Thoracic Limb 1145
Thoracic Limb
21.1 Shoulder luxation – Marjolaine Rousseau 1146
21.2 Septic arthritis of the distal interphalangeal joint – Karl Nuss 1161
21.3 Sole ulcer – Karl Nuss 1169
21.4 Metacarpal fracture – André Desrochers 1181

CHAPTER 22: Pelvic Limb 1193
Pelvic Limb
22.1 Coxofemoral luxation – Emma Marchionatti 1194
22.2 Patellar luxation – David E. Anderson 1198
22.3 Cranial cruciate ligament tear – Emma Marchionatti and Caroline Constant 1205
22.4 Gastrocnemius rupture – André Desrochers 1211

CHAPTER 23: Integument and Mammary Gland/Udder 1221
23.1 Contagious ecthyma – Cynthia M. Faux and Luise King 1222
23.2 Chronic udder abscess – Sylvain Nichols 1236
23.3 Teat obstruction – Sylvain Nichols 1242

SECTION VI: AVIAN CLINICAL CASES 1249

6.0/Avian Landscape Figures (1–7)
C.M. Faux, M.L. Logsdon, N.S. Grenager, J.A. Orsini, A. de Lahunta 1251

CHAPTER 24: Adaptations to Flight: *Cynthia M. Faux and Marcie L. Logsdon* 1260

CHAPTER 25: Head and Neck 1263
25.1 Infraorbital sinusitis – Cynthia M. Faux and Marcie L. Logsdon 1264
25.2 Crop impaction – Cynthia M. Faux and Marcie L. Logsdon 1271
25.3 Syringeal obstruction – Cynthia M. Faux and Marcie L. Logsdon 1276
25.4 Beak fracture – C.M. Faux, M.L. Logsdon, and L. Lossi 1285
25.5 Obstruction of external ear canal – Cynthia M. Faux and Marcie L. Logsdon 1299
25.6 Ocular trauma – Cynthia M. Faux and Marcie L. Logsdon 1305

CHAPTER 26: Thoraco-abdominal Cavity 1315
26.1 Ischemic stroke – Cynthia M. Faux and Marcie L. Logsdon 1316
26.2 Egg-yolk peritonitis – Cynthia M. Faux and Marcie L. Logsdon 1326
26.3 Air sacculitis – Cynthia M. Faux and Marcie L. Logsdon 1334
26.4 Ventricular foreign body – Cynthia M. Faux and Marcie L. Logsdon 1343
26.5 Sertoli cell tumor – Cynthia M. Faux and Marcie L. Logsdon 1349
26.6 Marek's disease – Ricardo de Matos and James K. Morrisey 1355

CHAPTER 27: Thoracic and Pelvic Limb. 1365
27.1 Humeral fracture – Cynthia M. Faux and Marcie L. Logsdon 1366
27.2 Vertebral column trauma – Cynthia M. Faux and Marcie L. Logsdon 1377
27.3 Pododermatitis (bumblefoot) – Cynthia M. Faux and Marcie L. Logsdon 1384

CHAPTER 28: Integument/Feathers 1399
28.1 Impacted uropygial (preen) gland – *Cynthia M. Faux and Marcie L. Logsdon* 1400

SECTION VII: APPENDICES 1417
Appendix 1: Standard abbreviations 1419
Appendix 2: Normal respiratory rate, heart rate, and temperature reference ranges 1423
Appendix 3: Hematology reference intervals 1425
Appendix 4: Biochemistry reference intervals 1427

Bibliography 1429

List of Illustrations under Editor Copyright 1431

SECTION VIII: INDEX 1433

CONTRIBUTORS

Kimberly A. Agnello, DVM, MS, DACVS, DACVSMR
Associate Professor of Small Animal Surgery,
Department of Clinical Sciences &
Advanced Medicine,
University of Pennsylvania School of Veterinary Medicine,
Philadelphia, Pennsylvania, US

Zeeshan Akbar, DVM
Veterinary and Agricultural Sciences,
The University of Melbourne,
Melbourne, Victoria, AU

Oday Alawi Al-Juhaishi, DVM
Veterinary and Agricultural Sciences, The University of Melbourne, Melbourne, Victoria, AU

Monica Aleman, MVZ, PhD, DACVIM
Professor, Medicine and Epidemiology, University of California-Davis School of Veterinary Medicine, Davis, California, US

David E. Anderson, DVM, MS, DACVS
Associate Dean for Research and Graduate Studies,
Department of Large Animal Clinical Sciences,
University of Tennessee College of Veterinary Medicine,
Knoxville, Tennessee, US

Simon Bailey, BVMS, PhD, FHEA, DECVPT, MRCVS
Professor, Veterinary Biosciences,
The University of Melbourne, Parkville, Victoria, AU

Nicholas Bamford, BVSc(Hons), PhD, DACVIM
Lecturer, Veterinary Biosciences,
The University of Melbourne, Parkville,
Victoria, AU

Cathy Beck, BVSc(Hons), DipVetClinStud, MVS, GradCertUniTeach, FANZCVS
Senior Lecturer, Veterinary Hospital, The University of Melbourne, Melbourne, Victoria, AU

Julie Berman, DVM, DACVIM
DrVet, Clinician, Centre Hospitalier Universitaire Veterinaire (CHUV), Université de Montréal,
Montréal, Québec, CA

Kari E. Bevevino, DVM, DACVIM
Roaring Fork Equine Medical Center, Glenwood Springs, Colorado, US

Christopher M. Boemo, BVSc (Hons)
Principal Veterinarian, Keysborough Veterinary Practice, Keysborough, Victoria, AU

Jessica Bramski, DVM
Resident, Department of Large Animal Medicine,
University of Georgia College of Veterinary Medicine,
Athens, Georgia, US

Carol Carberry, DVM, DACVS
Department of Surgery, Oradell Animal Hospital,
Paramus, New Jersey, US

Nick Carlson, DVM, DACVS
Director of Surgery, Steinbeck Peninsula Equine Clinics, Salinas, California, US

Sharon A. Center, DVM, DACVIM
Professor, Internal Medicine, Department of Clinical Sciences, Cornell University College of Veterinary Medicine, Ithaca, New York, US

Katherine Christie, DVM, MSc, DACVIM
Internal Medicine, Rood and Riddle Equine Hospital,
Lexington, Kentucky, US

Michelle Coleman, DVM, PhD, DACVIM
Assistant Professor, Large Animal Clinical Sciences
Texas A&M University College of Veterinary Medicine,
College Station, Texas, US

Caroline Constant, DMV, MSc, MENG, DACVS
Junior Project Leader, Preclinical Surgeon,
AO Research Institute, Davos Platz, CH

Brenton C. Credille, DVM, PhD
Associate Professor, Food Animal Health and Management Program, Department of Population Health, University of Georgia College of Veterinary Medicine, Athens, Georgia, US

Helen M.S. Davies, BAgSci, MAgrSc, BVSc, PhD
Associate Professor, Veterinary Anatomy, Veterinary Biosciences, The University of Melbourne, Melbourne, Victoria, AU

Amanda Davis, DVM, DACVO
Red Bank Veterinary Hospital, Ophthalmology, Mount Laurel, New Jersey, US

Luis Henrique de Aguiar, Med. Vet., DACT
Department of Veterinary Clinical Sciences, Louisiana State University School of Veterinary Medicine, Baton Rouge, Louisiana, US

Alexander de Lahunta, DVM, PhD
Professor Emeritus, Biomedical Sciences, Cornell University College of Veterinary Medicine, Ithaca, New York, US

Christobal Navas de Solis, LV, MS, PhD, DACVIM
Assistant Professor, Cardiology and Internal Medicine, Department of Clinical Studies - New Bolton Center, University of Pennsylvania School of Veterinary Medicine, Kennett Square, Pennsylvania, US

Ricardo de Matos, LMV, MSc, DABVP, DECZM
Senior Lecturer, Department of Clinical Sciences, Cornell University College of Veterinary Medicine, Ithaca, New York, US

Sarah DeSante, DVM, DABVP
Veterinarian, Steinbeck Peninsula Equine Clinics, Salinas, California, US

André Desrochers, DMV, MS, DACVS, Dip. ECBHM
Department of Clinical Sciences, Université de Montréal, Faculty of Veterinary Medicine, St-Hyacinthe, Québec, CA

Tamara Dobbie, DVM, DACT
Associate Professor of Clinical Reproduction Department of Clinical Studies - New Bolton Center, University of Pennsylvania School of Veterinary Medicine, Kennett Square, Pennsylvania, US

Laura M. Dooley, BVSc(Hons), GCert(UniTeach), PhD
Senior Lecturer, Veterinary Biosciences, The University of Melbourne Melbourne, Victoria, AU

Singen Elliott, DVM, DACVS
Mid-Atlantic Equine Medical Center, Ringoes, New Jersey, US

Kira Lyn Epstein, DVM, DACVS, DACVECC
Professor, Department of Large Animal Medicine, University of Georgia College of Veterinary Medicine, Athens, Georgia, US

Cynthia M. Faux, DVM, PhD
Professor, University of Arizona, College of Veterinary Medicine, Oro Valley, Arizona, US

Gilles Fecteau, DMV, DACVIM
Department of Clinical Sciences, Université de Montréal, Faculty of Medicine, Saint Hyacinthe, Québec, CA

Marie-Eve Fecteau, DVM, DACVIM
Associate Professor, Food Animal Medicine and Surgery, Department of Clinical Studies - New Bolton Center, University of Pennsylvania School of Veterinary Medicine, Kennett Square, Pennsylvania, US

Ray Ferguson, BVSc, OAM
Consultant, Monash Veterinary Clinic, Oakleigh East, Victoria, AU

Maria Ferrer, DVM, MS, DACT
Associate Professor, Large Animal Medicine, University of Georgia College of Veterinary Medicine, Athens, Georgia, US

Nadine Fiani, BVSc (Hons), DAVDC
Assistant Clinical Professor, Dentistry and Oral Surgery, Department of Clinical Sciences, Cornell University College of Veterinary Medicine, Ithaca, New York, US

James Flanders, DVM, DACVS
Emeritus Professor of Surgery, Department of Clinical Sciences, Cornell University College of Veterinary Medicine, Ithaca, New York, US

Callie Fogle, DVM, DACVS
Clinical Associate Professor, Equine Surgery, Department of Clinical Sciences, North Carolina State University College of Veterinary Medicine, Raleigh, North Carolina, US

Susan Fubini, DVM, DACVS
Professor, Large Animal Surgery, Cornell University College of Veterinary Medicine, Ithaca, New York, US

Aitor Gallastegui, LV, MSc, DACVR
Clinical Assistant Professor, Small Animal Clinical Sciences, University of Florida College of Veterinary Medicine, Gainesville, Florida, US

José M. García-López, VMD, DACVS, DACVSMR
Associate Professor, Large Animal Surgery, Clinical Sciences, Tufts University Cummings School of Veterinary Medicine, North Grafton, Massachusetts, US

Mathew Gerard, BVSc, PhD, DACVS
Teaching Professor, Veterinary Anatomy, Department of Molecular Biomedical Sciences, North Carolina State University College of Veterinary Medicine, Raleigh, North Carolina, US

Liberty M. Getman, DVM, DACVS
Surgeon, Tennessee Equine Hospital, Thompson's Station, Tennessee, US

Brian Gilger, DVM, MS, DACVO
Professor of Ophthalmology, Department of Clinical Sciences, North Carolina State University College of Veterinary Medicine, Raleigh, North Carolina, US

William F. Gilsenan, VMD, DACVIM
Internal Medicine, Rood and Riddle Equine Hospital, Lexington, Kentucky, US

Eric N. Glass, MS, DVM, DACVIM
Section Head, Neurology and Neurosurgery, Red Bank Veterinary Hospital, Tinton Falls, New Jersey, US

Jose Goni, DVM
Resident, Department of Veterinary Clinical Sciences, Large Animal Medicine, Purdue University College of Veterinary Medicine, West Lafayette, Indiana, US

Nora S. Grenager, VMD, DACVIM
Director of Internal Medicine, Steinbeck Peninsula Equine Clinics, Menlo Park, California, US

Rebecka S. Hess, DVM, DACVIM
Professor, Internal Medicine, Department of Clinical Sciences and Advanced Medicine, University of Pennsylvania School of Veterinary Medicine, Philaphelphia, Pennsylvania, US

Susan J. Holcombe, VMD, MS, PhD, DACVS, DACVECC
Professor Emeritus, Department of Large Animal Clinical Sciences, Michigan State University College of Veterinary Medicine, East Lansing, Michigan, US

David Holt, BVSc, DACVS
Professor of Surgery, Department of Clinical Sciences and Advanced Medicine, University of Pennsylvania School of Veterinary Medicine, Philadelphia, Pennsylvania, US

G. Reed Holyoak, DVM, PhD, DACT
Professor, Oklahoma State University College of Veterinary Medicine, Stillwater, Oklahoma, US

Carrie Jacobs, DVM, DACVS
Assistant Clinical Professor, Equine Orthopedic Surgery, North Carolina State University College of Veterinary Medicine, Raleigh, North Carolina, US

Amy L. Johnson, BA, DVM, DACVIM
Associate Professor, Large Animal Medicine and Neurology, Department of Clinical Studies - New Bolton Center, University of Pennsylvania School of Veterinary Medicine, Kennett Square, Pennsylvania, US

Laura Johnstone, BVSc, MVSc, DACVIM-LA
Equine Veterinary Clinic, Massey University, Palmerston North, NZ

Marc Kent, DVM, DACVIM
Professor, Department of Small Animal Medicine and Surgery, University of Georgia College of Veterinary Medicine, Athens, Georgia, US

Phillip Kieffer, DVM
Evidensia Specialisthastsjukhus Helsingborg, Helsingborg, SE

Shune Kimura, DVM
Department of Clinical Sciences, Auburn University College of Veterinary Medicine, Auburn, Alabama, US

Titia Luise King, DVM, PhD
Assistant Professor, University of Arizona College of Veterinary Medicine, Oro Valley, Arizona, US

Judith Koenig, DMV, DVSc, DACVS, DACVSMR
Associate Professor, University of Guelph Ontario Veterinary College, Guelph, Ontario, CA

Natali Krekeler, Dr. med.vet., PhD, DACT
Senior Lecturer in Veterinary Reproduction, Veterinary Biosciences, The University of Melbourne, Melbourne, Victoria, AU

Eric Ledbetter, DVM, DACVO
Professor of Ophthalmology, Department of Clinical Sciences, Cornell University College of Veterinary Medicine, Ithaca, New York, US

Olivier M. Lepage, DMV, MSc, DES, PD, HDR, DECVS
Professor and Head, Equine Health Center, Veterinary School of Lyon, VetAgro Sup, Marcy l'Etoile, FR

David Levine, DVM, DACVS, DACVSMR
Associate Professor of Surgery, Department of Clinical Studies - New Bolton Center, University of Pennsylvania School of Veterinary Medicine, Kennett Square, Pennsylvania, US

Jennifer Linton, VMD, DACT
B.W. Furlong & Associates, Oldwick, New Jersey, US

Marcie L. Logsdon, DVM
Clinical Instructor, Exotics & Wildlife Department, Washington State University College of Veterinary Medicine, Pullman, Washington, US

Brina Lopez, DVM, PhD, DACVIM
Assistant Professor, Midwestern University College of Veterinary Medicine, Glendale, Arizona, US

Laura Lossi, DVM, PhD
Professor, Department of Veterinary Science, University of Turin, Turin, IT

Daniela Luethy, DVM, DACVIM
Large Animal Internal Medicine, Large Animal Clinical Sciences, University of Florida College of Veterinary Medicine, Gainesville, Florida, US

Candace Lyman, DVM, DACT
Associate Professor, Equine and Small Animal Auburn University College of Veterinary Medicine, Auburn, Alabama, US

Emma Marchionatti, DMV, MSc, DACVS
Senior Lecturer, Clinic for Ruminants, Vetsuisse Faculty University of Bern, Bern, CH

Annette M. McCoy, DVM, MS, PhD, DACVS
Assistant Professor, Equine Surgery, Veterinary Clinical Medicine, University of Illinois Urbana-Champaign College of Veterinary Medicine, Urbana, Illinois, US

Adalberto Merighi, DVM, PhD
Professor, Department of Veterinary Science, University of Turin, Turin, IT

Meredith Miller, DVM, DACVIM
Lecturer, Department of Clinical Sciences, Cornell University College of Veterinary Medicine, Ithaca, New York, US

James K. Morrisey, DVM, DABVP
Department of Clinical Sciences, Cornell University College of Veterinary Medicine, Ithaca, New York, US

Christina Murray, BVSc (Hons), GCUT, MScAgr
Senior Lecturer, Veterinary Biosciences, The University of Melbourne, Melbourne, Victoria, AU

Euan Murray, DVM
Ophthalmology Intern, BluePearl Veterinary Partners, Louisville, Kentucky, US

Sylvain Nichols, DMV, MS, DACVS
Associate Professor, Department of Clinical Sciences, Université de Montréal, St-Hyacinthe, Québec, CA

Andrew J. Niehaus, BS, DVM, MS, DACVS
Professor, Veterinary Clinical Sciences, The Ohio State University College of Veterinary Medicine, Columbus, Ohio, US

Karl Nuss, Prof., DMV, DECVS
Farm Animal Surgery Section, Farm Animals, Vetsuisse-Faculty, University of Zürich, Zürich, CH

James A. Orsini, DVM, DACVS
Associate Professor of Surgery, Department of Clinical Studies - New Bolton Center, University of Pennsylvania School of Veterinary Medicine, Kennett Square, Pennsylvania, US

Mark A. Oyama, DVM, MSCE, DACVIM
Professor, Department of Clinical Sciences and Advanced Medicine, University of Pennsylvania School of Veterinary Medicine, Philadelphia, Pennsylvania, US

Shannon M. Palermo, VMD, DACVIM
Department of Internal Medicine, Veterinary Specialists and Emergency Services, Rochester, New York, US

Brian Palmeiro, VMD, DACVD
Lehigh Valley Veterinary Dermatology & Fish Hospital, Allentown, Pennsylvania, US

Eric Parente, DVM, DACVS
Professor of Surgery, Department of Clinical Studies - New Bolton Center, University of Pennsylvania School of Veterinary Medicine, Kennett Square, Pennsylvania, US

Santiago Peralta, DVM, DAVDC, FF-AVDC-OMFS
Assistant Professor, Department of Clinical Sciences, Cornell University College of Veterinary Medicine, Ithaca, New York, US

Michael Pesato, DVM, DABVP
Department of Pathobiology and Population Medicine, Mississippi State University College of Veterinary Medicine, Mississippi, US

Mark E. Peterson, DVM, DACVIM
Director, Animal Endocrine Center, New York, New York, US

Carlos Pinto, MedVet, PhD, DACT
Professor of Theriogenology, Veterinary Clinical Sciences, Louisiana State University School of Veterinary Medicine, Baton Rouge, Louisiana, US

Malgorzata Pozor, DVM, PhD, DACT
Clinical Assistant Professor, Large Animal Clinical Sciences, University of Florida College of Veterinary Medicine, Gainesville, Florida, US

Timo Prange, DMV, MS, DACVS
Equine Surgery, Department of Clinical Studies, North Carolina State University College of Veterinary Medicine, Raleigh, North Carolina, US

John F. Randolph, DVM, DACVIM
Professor of Medicine, Department of Clinical Sciences, Cornell University College of Veterinary Medicine, Ithaca, New York, US

Sarah M. Reuss, VMD, DACVIM
Technical Manager, Equine Veterinary Services, Boehringer Ingelheim Animal Health, Duluth, Georgia, US

Stefano Romagnoli, DVM, MS, PhD, DECAR
Department of Animal Medicine, Production and Health, University of Padua, Padova, IT

Mark P. Rondeau, BS, DVM, DACVIM
Department of Clinical Sciences and Advanced Medicine, University of Pennsylvania School of Veterinary Medicine, Philadelphia, Pennsylvania, US

Marjolaine Rousseau, DMV, MS, DACVS
Assistant Professor, Department of Clinical Sciences, Université de Montréal, Faculty of Veterinary Medicine, Saint-Hyacinthe, Québec, CA

Magdalena Schrank, DVM, PhD
Department of Animal Medicine, Production and Health, University of Padua, Padova, IT

James Schumacher, DVM, MS, DACVS
Professor, Department of Large Animal Clinical Sciences, University of Tennessee College of Veterinary Medicine, Knoxville, Tennessee, US

Patricia L. Sertich, VMD, DACT
Associate Professor, Department of Clinical Studies - New Bolton Center, University of Pennsylvania School of Veterinary Medicine, Kennett Square, Pennsylvania, US

Tara R. Shearer, DVM, DACVS
Department of Large Animal Clinical Sciences, Michigan State University College of Veterinary Medicine, East Lansing, Michigan, US

Billy I. Smith, BS, DVM, MS DABVP
Associate Professor of Medicine, Department of Clinical Studies - New Bolton Center, University of Pennsylvania School of Veterinary Medicine, Kennett Square, Pennsylvania, US

Alexandra St. Pierre, DVM
Hamilton Wenham Veterinary Clinic, South Hamilton, Massachusetts, US

Takanori Sugiyama, MS, MVDr/DVM, MANZCVS, MVS, MVSc, GCCT, DACVS
Animalius Vet, Bayswater, WA, AU

Ferenc Toth, DVM, PhD, DACVS
Assistant Professor, Veterinary Clinical Sciences, University of Minnesota College of Veterinary Medicine, St. Paul, Minnesota, US

Karen Trainor, DVM, MS, DACVP
Director and Founder, Dermatopathology Service, Innovative Vet Path, Leawood, Kansas, US

Jesse Tyma, DVM, DACVS
Mid-Atlantic Equine Medical Center, Ringoes, New Jersey, US

Dirk K. Vanderwall, DVM, PhD
Department Head and Professor, Animal, Dairy and Veterinary Sciences, Utah State University School of Veterinary Medicine, Logan, Utah, US

Lane A. Wallett, DVM, PhD
Scholarly Assistant Professor, Integrative Physiology and Neuroscience, Washington State University College of Veterinary Medicine, Pullman, Washington, US

Jarred Williams, MS, DVM, PhD, DACVS, DACVECC
Clinical Associate Professor, Large Animal Medicine, University of Georgia College of Veterinary Medicine, Athens, Georgia, US

Fred Wininger, VMD, MS, DACVIM
Neurologist, Neurology Department, Charlotte Animal Referral and Emergency, Charlotte, North Carolina, US

Katie Withowski, DVM
Resident, North Caroline State University, College of Veterinary Medicine, Raleigh, North Carolina, US

Cynthia Xue, DVM, DACVIM - LA
Department of Clinical Sciences, Ross University School of Veterinary Medicine, St. Kitts and Nevis, WI

PREFACE

The creation of *Comparative Veterinary Anatomy: A Clinical Approach* has been a long journey that began with a single step—a vision and dream to publish a book that bridged basic anatomy and the clinical arena. The original concept for this clinical anatomy reference book came from a compilation of the experiences of three generations of clinicians who recognized the importance and value of understanding the critical anatomy when presented with a patient—be it a cat, dog, horse, cow, or bird. We felt we could reinforce the anatomy that is truly useful in the clinical arena by going through real clinical cases that highlight these features. Our combined experience over the years as students, clinicians, and teachers supported this concept, and from this, we realized that such a book was truly needed in veterinary medicine for students, colleagues, and other health professionals.

We strive to be the change that we want to see in our profession and perfection has been our top priority. We feel we realized this goal in the beautifully created, anatomically precise, and clearly presented figures and descriptions that unambiguously highlight each clinically relevant anatomical feature, integrated with clinically relevant cases. The first step in our strategy for success was to find a team of professionals committed to the level of excellence required for this impactful work. We were fortunate to find the right people, with the right resources, in the right professional culture, to make this book a reality.

One of the many goals we established during the developmental stage of the book was to maintain consistency throughout in the presentation of each case, the anatomical descriptions, the terminology (Clinical Anatomy Nomenclature—Chapter 1), and the supportive descriptive figures. To this end, our team looked to make each case and its relevant anatomy memorable and adaptable to the clinical setting. The imaging chapter at the beginning of the book aims to provide a foundation and understanding of the relationship of these diagnostic techniques and their importance in interpreting anatomy in health and disease. The landscape figures, placed at the beginning of each species section, were meticulously researched, edited at multiple levels, and provide a readily available resource of anatomical relationships from integument to skeleton. In every case throughout the book, the reader will notice what we refer to as "side boxes" that have pearls of comparative anatomy, surgery highlights, and other interesting facts that apply to the case to help cement the pertinent anatomy. There are two icons that you will find in the clinical cases: (1) symbolizes additional or complementary information as relates to the clinical anatomy and (2) symbolizes surgery-related pearls associated with the clinical anatomy. There are four appendices in the back of the book with common abbreviations, vital signs, and clinical laboratory values to be used for the clinical cases.

And finally, because of space and page limitations, we wanted to list the anatomical and clinical textbooks referenced throughout the book here:

- Budras KD, Habel RE, Müllerig CKW, Greenough PR, Weinche A, and Budras S. *Bovine anatomy*. 2nd ed. Germany: Schlütersche; 2011.
- Budras KD, Sack WO, Röck S, Horwitz A, Berg R. *Anatomy of the horse*. 6th ed. Germany: Schlütersche; 2011.
- de Lahunta A. *Applied veterinary anatomy*. Philadelphia: W.B. Saunders; 1986.

- *Dorland's Illustrated Medical Dictionary*. 32nd ed. Philadelphia: Elsevier/W.B. Saunders; 2012.
- Evans HE, de Lahunta A. *Miller's dissection guide for the dog*. 8th ed. Philadelphia: Elsevier/W.B. Saunders; 2017.
- Getty R. *Sisson and Grossman's – The anatomy of the domestic animals*. 5th ed. Philadelphia: W.B. Saunders; 1975.
- Hermanson JW, de Lahunta, Evans HEA. *Miller and Evan's anatomy of the dog*. 5th ed. Philadelphia: Elsevier/W.B. Saunders; 2020.
- Popesko P. *Atlas of topographical anatomy of the domestic animals*. Vols. 1, 2, 3. Philadelphia: W.B. Saunders; 1978.
- McIlwraith CW, Nixon A, Wright IM. *Diagnostic and surgical arthroscopy in the horse*. 4th ed. St. Louis: Elsevier; 2015.
- Pollitt CC. *The illustrated horse's foot: a comprehensive guide*. St. Louis: Elsevier; 2016.
- Schummer A, Nickel R, Sack WO. In: Nickel R, Schummer A, Seiferle E. editors. 2nd ed. *The viscera of the domestic mammals*. New York: Springer-Verlag; 1979.
- Shively MJ. *Veterinary anatomy basic, comparative, and clinical*. College Station: Texas A&M Press; 1984.
- Singh B. *Dyce, Sack, and Wensing—Textbook of veterinary anatomy*. 5th ed. St. Louis: Elsevier; 2018.

Our hope is that your mind, stretched with this new approach to learning and truly knowing clinical anatomy, will never go back to its original dimensions and—with this—you will feel confident and succeed in meeting your professional goals.

James A. Orsini

Nora S. Grenager

Alexander de Lahunta

ACKNOWLEDGMENTS

Thanking the many individuals who have contributed to the publishing of the first edition of *Comparative Veterinary Anatomy: A Clinical Approach* brings the editors great joy and professional gratification. We could never have realized our goals for this edition without the colleagues who contributed to this book's development during its five-year production. The original idea for this edition was conceived when we were students and first introduced to "applied anatomy" by numerous outstanding professors at Cornell and Penn who demonstrated the value of "learning anatomy again"—but in a different way—to make us better clinicians in the end. Dr. Paul Orsini—anatomist, clinician, and surgery and dentistry specialist—deserves special thanks for his expertise in the early stages of the book's development by underscoring the importance of linking basic and clinical anatomy using a relevant clinical case as the opening scene in every lesson to learn and retain important anatomy specific to the area of study. Paul, unfortunately, was unable to continue as an editor because of many professional and personal commitments. We are incredibly grateful that Dr. Alexander de Lahunta, who first published *Applied Veterinary Anatomy* in 1986, was so gracious in signing on to round out our editorial team.

For starters, we want to recognize members of our support team who played a vital role in this book's development from its start, especially Kate Shanaghan—Production Coordinator, and Jeanne Robertson—Chief Medical Illustrator, whose special talents, further emphasizes and illustrates the complexity of the many features of this book. Other individuals who contributed much to this book's success include Libby Wagner, Jason Mc Alexander, and Stefan Németh—medical illustrators; Dee Crandall—library resources; and Zoe Papas and Belinda Norris—cover design.

The section and chapter editors are an extraordinary group of veterinary colleagues who served as an important layer of checks and balances in many facets—coordinating with contributing authors, editing each clinical case in their chapter, serving as outstanding anatomy experts, and helping to create and edit each figure in their section. Special gratitude goes to Aitor Gallastegui—Diagnositic Imaging; John F. Randolph and Helen M.S. Davies—Canine and Feline Clinical Cases; Mathew Gerard, Amy L. Johnson, Sarah Reuss, Kira Epstein, Jarred Williams, Dirk Vanderwall, and Nick Carlson—Equine Clinical Cases; André Desrochers—Farm Animal Clinical Cases; and Cynthia M. Faux and Marcie Logsdon—Avian Clinical Cases.

We would like to especially acknowledge Dr. Roy V. H. Pollock, a well-known editor, author, academician, veterinarian, and leader in the field of medical informatics, product development, and learning, and who has held multiple executive positions in the animal health industry. Roy was a terrific resource who contributed to the book's development by providing his honest and skilled recommendations on how to make each clinical case clear, memorable, and outstanding for learning. Thanks, Roy—you are a special friend and colleague.

With more than 100 contributing authors, we are unable to use this format to recognize and acknowledge each author who generously contributed his or her expertise, creativity, clinical experiences, and nuggets of wisdom in the preparation and presentation of each clinical case and its applied anatomy. However, please take a moment to note their names at the beginning of each case. The figures and illustrations included by each author attempt to achieve a happy medium of not being too complex, thus causing confusion and difficulty in understanding, while also not being overly simplified, inadequate, or misrepresenting the clinical anatomy central to the understanding and retention of the information. Thank you, authors, for your excellence.

Throughout our careers, we are inspired and mentored by professors, colleagues, friends, and family. We would like to acknowledge some of the many that changed our careers and lives: A big thank you to all of you!

(JAO) thank Drs. Willard Daniels, Nate Hale, Wayne Schwark, Larry Kramer, Don Lein, Jay Georgi, Robert Hillman, Jack Lowe, Eric Trotter, William Kay, William Donawick, Charles Ramberg, Robert Whitlock, William Boucher, Bernard Shapiro, Francis Fox, Charles Short, Tom Divers, Rustin Moore, James (Jeanne) Geer, Ed Kanara, Joe Mankowski, John Lee, Malcolm Kram, Tom McGrath, and Brian Harpster; dear friends Marianne and John Castle, Mary Alice Malone, Margaret and Bob Duprey, Marian and J. Gibson McIlvain, Michael and Meredith Rotko, Candace and Kent Humber, Jennifer and Mike Wrigley, Lisa Gaudio, Jimmy Kazanjian, John Garafalo, Vicki and John Price, Robin Bernstein, Herb and Ellen Moelis, and Mark and Carol Zebrowski. And the many wonderful interns, residents, referring veterinarians, students, clients, faculty, and colleagues, in challenging me to be the best I could be as a clinician, teacher, researcher, and mentor.

Of course, I (JAO) was fortunate to have 2 very bright, creative, and professional coeditors, making this book unsurpassed in content and creation. Thank you, Dr. D. for both mentoring and acting as a role model for not just me but every student whom you taught and trained and Nora, as one of the best students ever, reassuring me that the profession and world are in exceptional hands as the next generation of clinicians and leaders.

(NSG) I would not be where I am in my career today without the mentorship and support of Jim Orsini, Tim Eastman, Alex Eastman, Eric Davis, and Tom Divers. I am also grateful to the clients over the years who have become like an extended family and have entrusted me with the care of their horses.

(ADL) Owes his opportunity to teach at Cornell University to Dr. Robert Habel, Professor of Applied Anatomy at the New York State College of Veterinary Medicine at Cornell University, who convinced him to leave general practice in Concord, NH and return to Cornell to pursue a PhD in anatomy. Dr. Habel asked Dr. D to teach a new course in neuroanatomy. To make this course more practical and useful to the students, Dr. D. established a neurology clinical service in the Teaching Hospital. This provided Dr. D. with a wealth of practical teaching material that he could incorporate in the neuroanatomy course. He felt indebted to the clinical faculty and pathologists for his education in this specialty.

The team at Elsevier was always professional, supportive, flexible, and receptive to new ideas in making this book the best possible publication. Thank you to Charlotte Cockle—publisher; Anna Valutkevich—acquisitions editor; Pat Gonzalez—editorial project manager; Kiruthika Govindaraju—production project manager; and Alan Studholme—cover designer. Thanks to the administrative assistant team at PENN VET—Bethany Healy, Cindy Stafford, and Karen McAvoy—helped in keeping us organized and focused from day two of book development—thank you!

And Finally, most notably, we are thankful to our loving and supportive families who cheered us on, kept our spirits high, and never grew impatient.

(JAO) I am grateful to Toni, Colin, Angela, Marco, and Joanna, and my parents Anne and Sal, who taught me to be the best at everything I undertake, the importance of grit, honesty, responsibility, kindness, and generosity; and my personal role models during the basic developmental stages of life.

(NSG) I am grateful for the tireless support and love of my incredible husband, David, and our beloved daughter, Sally; they mean the world to me. To my father, Trond Grenager, who passed away in the fall of 2020, I attribute my love for animals along with my work ethic and perseverance; thank you to him and my mother, Suzanne, for supporting me at each twist and turn of life and teaching me to believe in myself.

James A. Orsini (JAO)

Nora S. Grenager (NSG)

Alexander de Lahunta (ADL)

The editors would like to thank those individuals who provided so much time and work to our book.

Chief Medical Illustrator: Jeanne Robertson

Production Coordinator: Kate Shanaghan

SECTION I

INTRODUCTION

CHAPTER 1: CLINICAL ANATOMY NOMENCLATURE 3
Nora S. Grenager, James A. Orsini, and Alexander de Lahunta

CHAPTER 1

CLINICAL ANATOMY NOMENCLATURE

Nora S. Grenager[a], James A. Orsini[b], and Alexander de Lahunta[c]
[a]Steinbeck Peninsula Equine Clinics, Menlo Park, California, US
[b]Clinical Studies - New Bolton Center, University of Pennsylvania School of Veterinary Medicine, Kennett Square, Pennsylvania, US
[c]Professor Emeritus, Biomedical Sciences, Cornell University College of Veterinary Medicine, Ithaca, New York, US

Effective communication in clinical medicine and surgery is dependent on speaking the language—i.e., understanding and correctly using the [clinical anatomical] terminology. Medical etymology (Gr. *étuacon*, sense of truth + *-logia*, the study of) comes from the actual Greek term *etymon*, meaning the original form of the word or morpheme (form and structure of any part of a word as it relates to language and meaning), and it has an extensive and rich history in Latin (L.) and Greek (Gr.). Frank H. Netter, MD (1906–1991)—the basic and clinical human anatomist—is quoted: *"Anatomy, of course, does not change, but our understanding of anatomy and its clinical significance does."* This is applicable to veterinary as well as human anatomy, and it remains the foundation of the One Health enterprise in human and animal health care.

Throughout the book, we adhered to the anatomical terms published in the 6th edition of the *Nomina Anatomica Veterinaria* (NAV) in 2017. The nomenclature used in the clinical cases in this book defines relationships between organs, anatomical regions, and distinct body parts as they relate to one another, a body plane, or an axis, even though the majority of medical language in the text is anatomical terminology describing distinct parts of the body. These directional and positional terms are especially helpful when performing or interpreting diagnostic imaging—endoscopy, radiography, ultrasonography (US), computed tomography (CT), magnetic resonance imaging (MRI), nuclear scintigraphy, positron emission tomography (PET) imaging, and even in 3D printing used for surgical planning.

The following figures will help you think 3-dimensionally when learning and interpreting veterinary clinical anatomy. Fig. 1-1A depicts the four major body planes in quadrupeds, while Fig. 1-1B shows the equivalent terminology and major body planes in humans for comparison; the terminology applied to the head and—more specifically—the mouth/dental anatomy is shown in Fig. 1-2; the terminology used for the eye is shown in Fig. 1-3; the terminology relating to the distal limb in quadrupeds is represented in Fig. 1-4A (horse) and B (dog); Fig. 1-5 shows different nerve fibers, and Fig. 1-6 characterizes the change in abnormal limb movement seen with neurologic and musculoskeletal disorders. Tables 1-1 and 1-2 summarize the important terms included in Figs. 1-1–1-6, which are critical for learning and understanding clinical anatomy. Each term is linked with its opposite or antonym (in red) wherever appropriate.

For ease in use and understanding, each table is arranged alphabetically. The information focuses on directions, relationships, and word roots that simplify the use of clinical anatomical terminology as a noun (n.) or adjective (adj.) form. Reading the tables from left to right, the following information for each term is given: (1) medical term (green); (2) antonym (red); (3) derivation; (4) definition; (5) clinical anatomy significance; and (6) example in clinical anatomy. The tables are not comprehensive in incorporating all anatomical terms; however, they represent the most important words you will find helpful in maximizing your retention of clinical anatomy.

The Latin (L.), Greek (Gr.), and [occasionally] Old French roots are given under "derivation" wherever possible.

Comparative Veterinary Anatomy: A Clinical Approach. https://doi.org/10.1016/B978-0-323-91015-6.00168-0

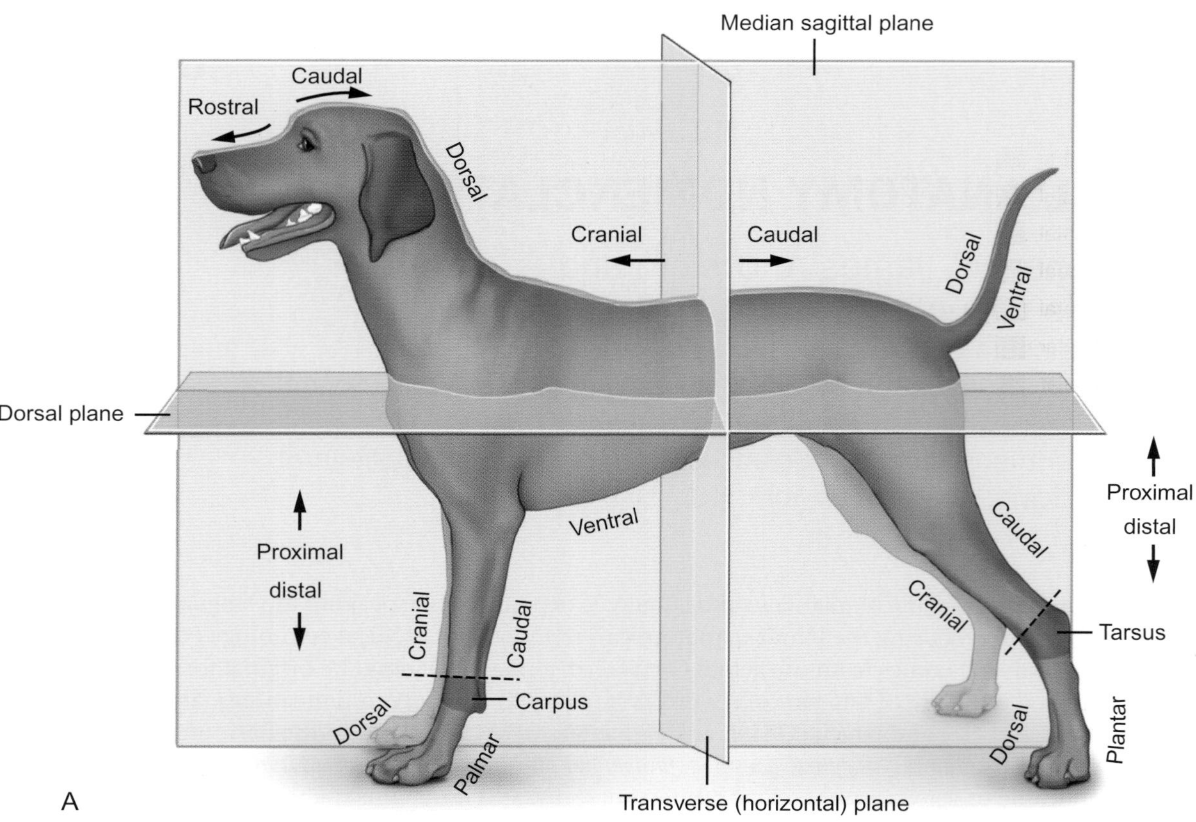

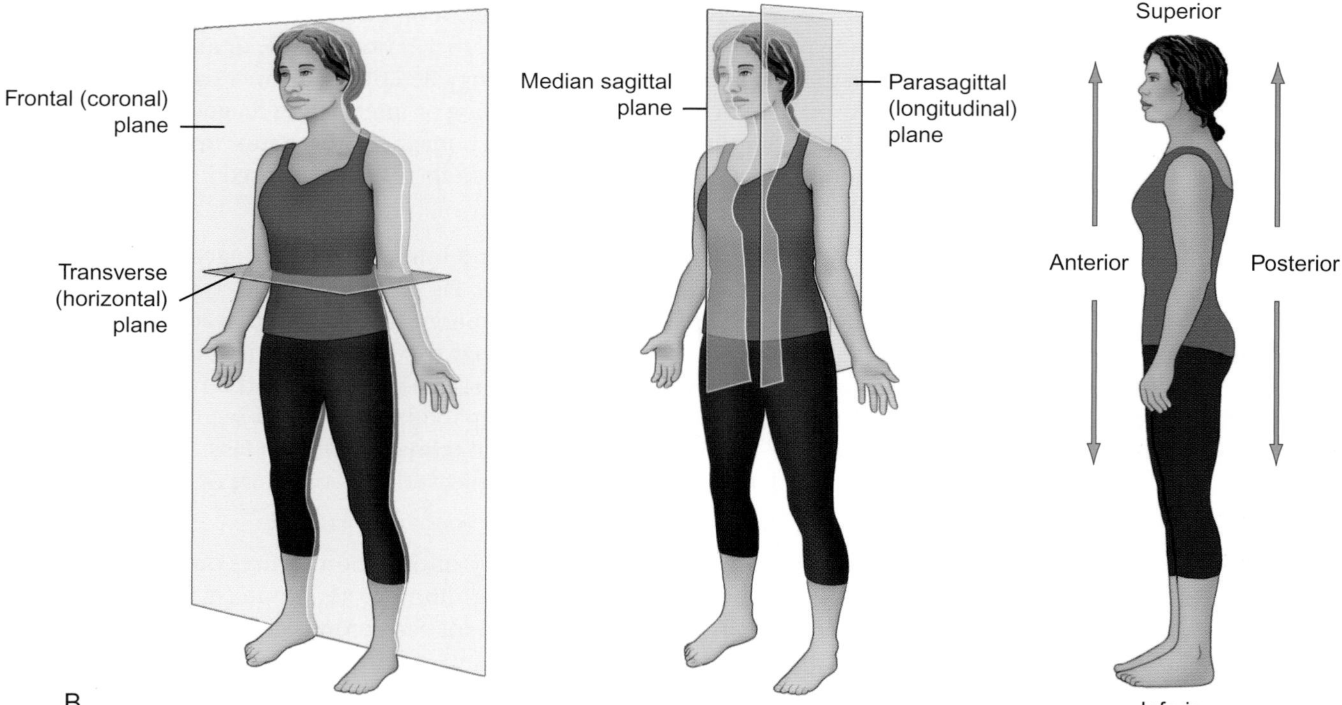

FIGURE 1-1 Anatomical nomenclature: (A) Directional terms and planes used in describing the anatomy of animal species. (B) Directional terms and planes used in describing the anatomy of humans.

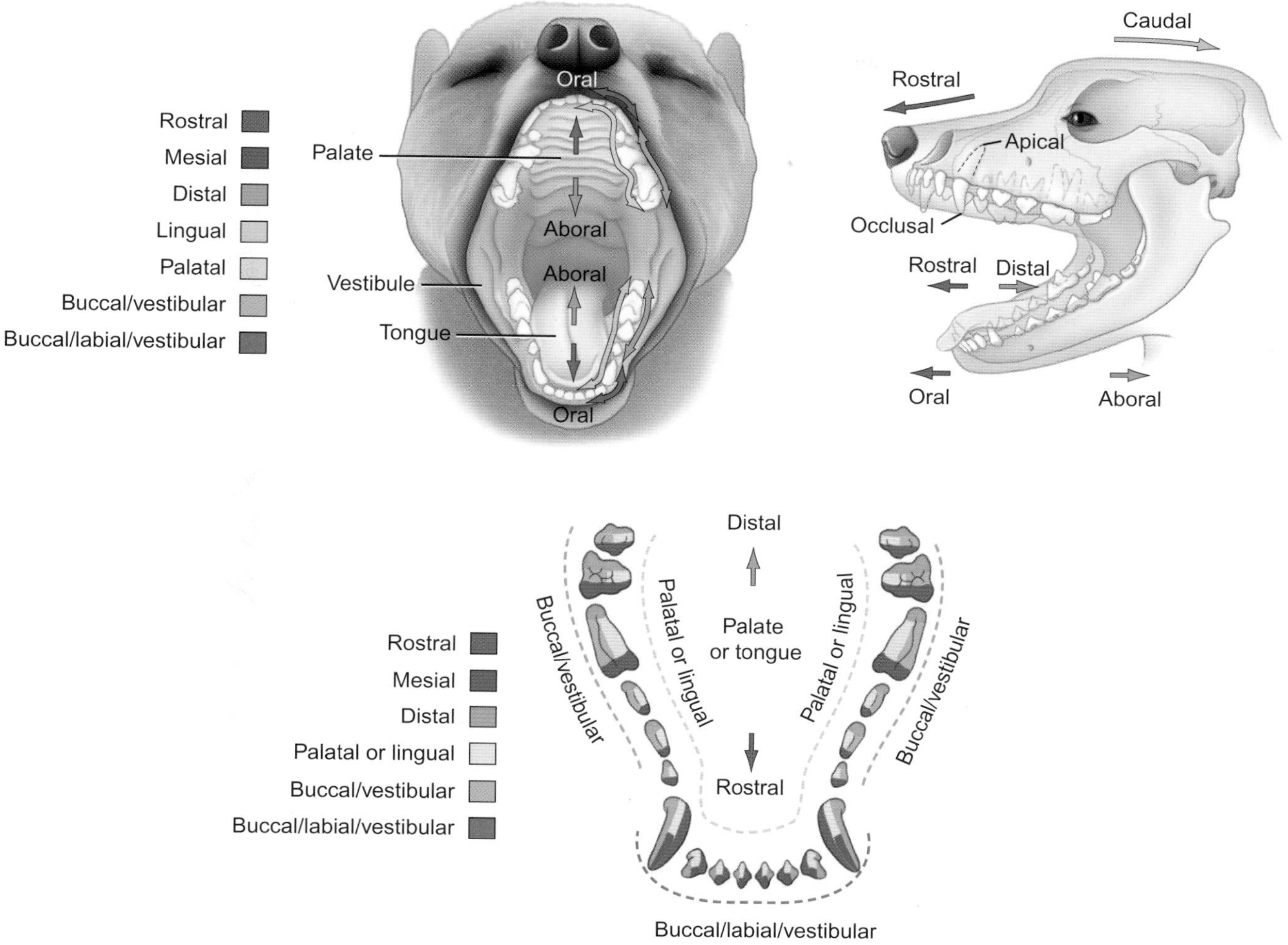

FIGURE 1-2 Dental: Directional terms and planes used in describing the anatomy of the head and, specifically, dental anatomy. Shown using a canine head, but terms are used across veterinary species.

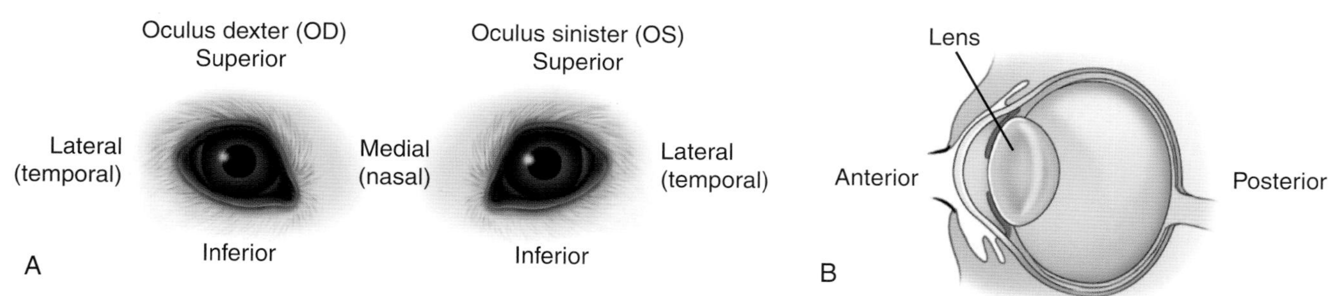

FIGURE 1-3 Eye: Directional terms and planes used in describing the anatomy of the head as relates to the eyes. Depicted using canine eyes, but terms are used across veterinary species.

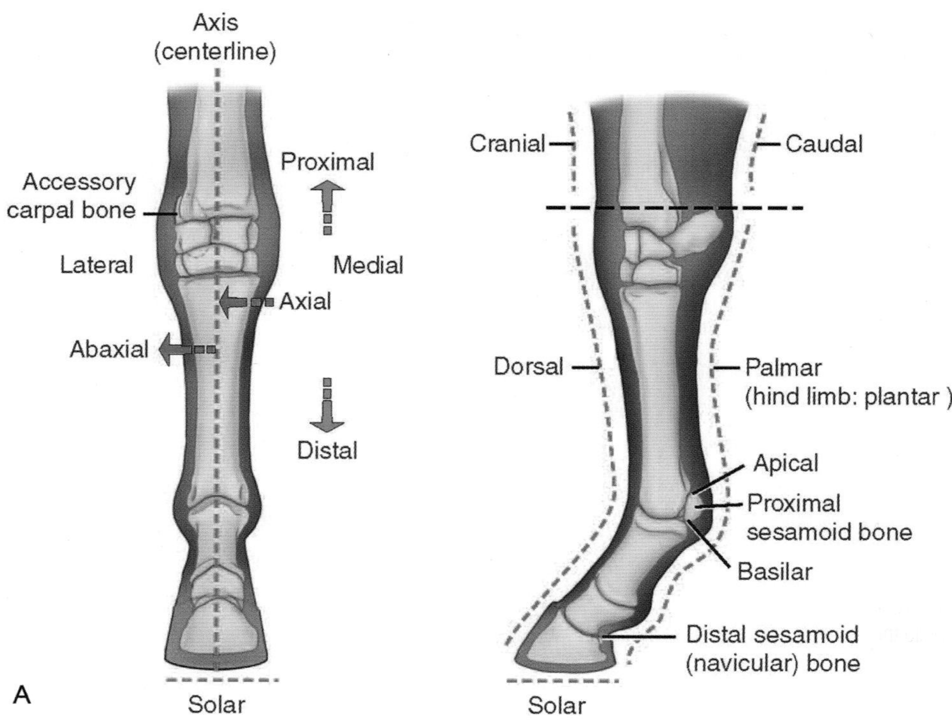

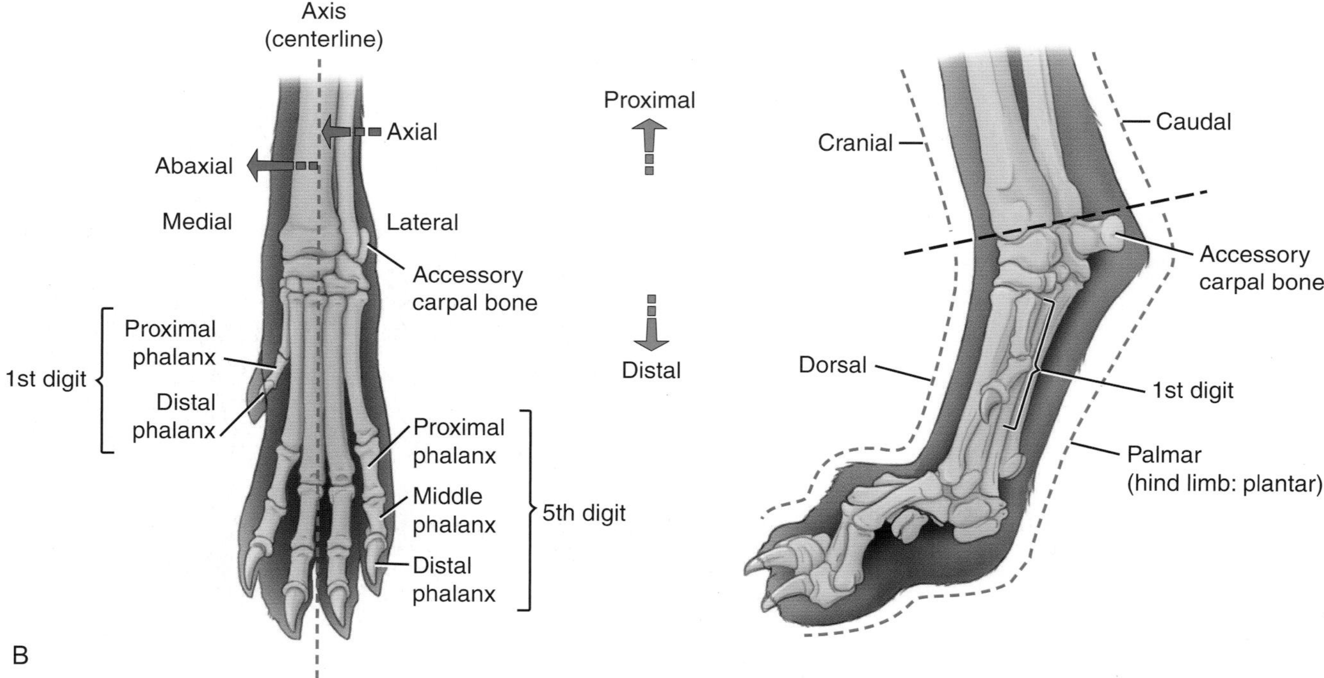

FIGURE 1-4 Distal limb: Directional terms and planes used in describing the anatomy of the distal limb. (A) Depicted using equine forelimb, but terms are used across quadruped veterinary species. (B) Depicts the directional terms and planes in the dog of the forelimb.

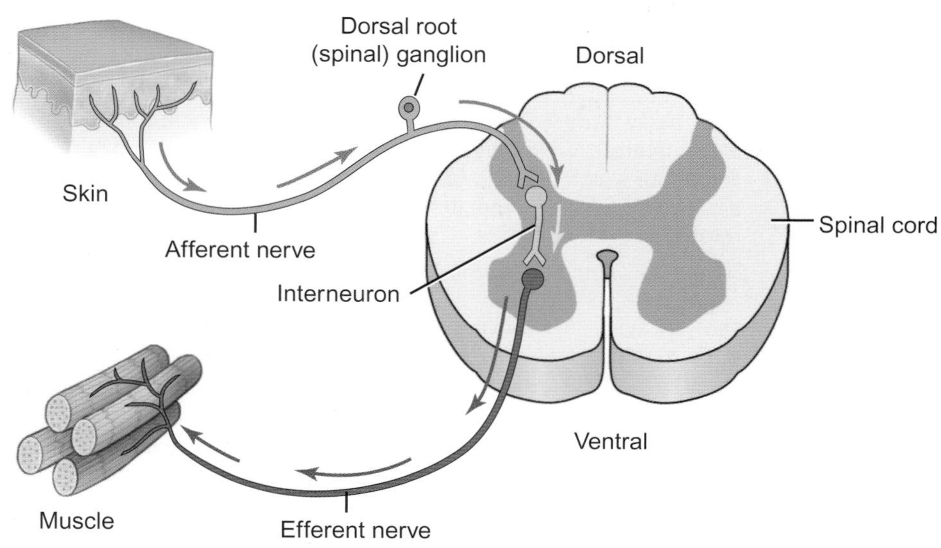

FIGURE 1-5 Afferent and efferent nerves: Nerves providing sensory (afferent) and motor (efferent) function. Afferent nerves travel toward the central nervous system (CNS), while the efferent nerves travel away from the CNS.

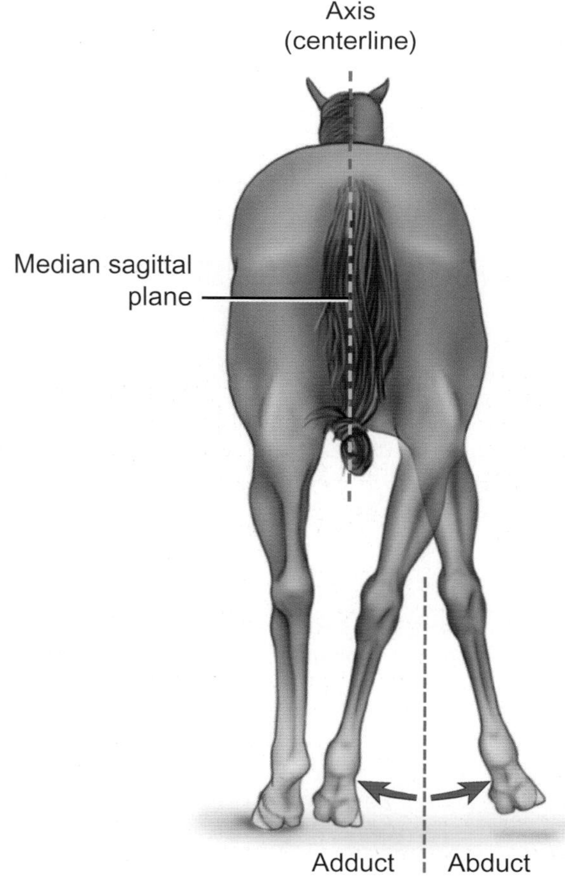

FIGURE 1-6 Limb movement in abduction and adduction: Abducting the limb moves the limb away from the median plane toward the outside vs. adducting the limb, which moves the limb in the direction of the median plane toward the inside.

TABLE 1-1 Medical nomenclature for the term abaxial → efferent.

Medical term	Antonym	Derivation	Definition	Clinical anatomy significance	Example in clinical anatomy
Abaxial	Axial	L. *ab-*, away from; *axis-*, central line about which a specific part rotates	Defines a lesion or part of an organ or limb farthest away from its center	A boundary furthest from the center of a structure—e.g., bone; position of a finding/ lesion relative to the long axis of a limb	There are axial and **abaxial** palmar/ plantar ligaments of the proximal interphalangeal joint in the distal limb of a horse (see Fig. 1-4)
Abduct	Adduct	L. *ab-*, away + *ducere,* to draw away	Moving away from the median plane	Important in assessing abnormal limb movement	Horses and dogs with neurologic diseases may **abduct** a limb (see Fig. 1-6)
Adduct	Abduct	L. *adducer*, to draw toward	Movement of a limb toward the median plane	Orthopedic and neurologic problems of the pelvic limb commonly affect limb movement	In dogs with intervertebral disc disease, the pelvic limbs frequently cross the midline (median plane) thereby **adducting** the limbs when walking (see Fig. 1-6)
Aboral	Oral	L. *ab-*, away from; *oralis*, mouth	Away from the mouth	Moving away from the mouth, as it relates to ingesta	A small intestinal obstruction prevents food from moving **aborally** into the large intestine
Afferent	Efferent	L. *ad-,* to + *ferre,* to carry	Neurons transmitting nerve impulses to the CNS	Nerve fibers from the periphery to CNS	Sensory information from muscles/limbs/ skin travels via **afferent** nerves to CNS (see Fig. 1-5)
Anterior	Posterior	L. *anterius* or *ante,* before	Forward part of an organ or body part; toward the head of the body; applied to anatomy of the eye and ear in veterinary medicine	Anterior and posterior are generally not applied to quadrupeds because of the confusion arising from their meaning in human anatomy; in veterinary anatomy, used only for structures of the head—e.g., eye and ear	More commonly used in human anatomy vs. in quadrupeds (see Fig. 1-1B); **anterior** semicircular canal; **anterior** ampulla is in the inner ear of quadrupeds
Apical	Basilar	L. *apices,* tip, point	Adj. for apex or the highest point	Refers to the superior aspect of the body or organ	Apices or extremities/ tips of the heart, lungs, bones—e.g., **apical** part of the proximal sesamoid bones
Apical	Occlusal	As mentioned previously	As mentioned previously	Used in dentistry/dental examinations; refers to the tip of the tooth root	Dogs may get a root abscess leading to **apical** periodontitis

TABLE 1-1 Medical nomenclature for the term abaxial → efferent—cont'd

Medical term	Antonym	Derivation	Definition	Clinical anatomy significance	Example in clinical anatomy
Axial	Abaxial	Gr. *axon,* axle	Defines a lesion or part of an organ or limb closest to its center	A boundary closest to the center of a structure—e.g., bones; position of a finding/ lesion relative to the long axis of a limb	There are **axial** and abaxial palmar/ plantar ligaments of the proximal interphalangeal joint in the distal limb of the horse (see Fig. 1-4)
Basilar	Apical	L. *basilaris*, from base	Adj. for base and serves as a foundation for a structure	Refers to the base of an organ or structure—e.g., the skull	The base of the skull, heart, and bone—e.g., bottom or **basilar** aspect of the proximal sesamoid bones in the horse
Buccal/ Vestibular	Lingual/ Palatal	L. *buccalis*, from *bucca*, cheek	Adj. used when referring to the mouth	Denotes the cheek or the side of the teeth facing the cheek	In dentistry, the **buccal** side of a tooth is next to the cheek, and the opposite is next to the tongue
Caudal	Cranial	L. *cauda*, tail	Adj. denoting the posterior part of a structure or tail end of the body	Used when referring to the tail area or the posterior/back part of the proximal limb	Cranial and **caudal** describe the front and back of a limb above the carpus and tarsus in the dog and horse (see Fig. 1-4)
Caudal	Rostral	As mentioned previously	Adj. as mentioned previously and back portion of the head	Describes relationships of structures in the head	The brain is **caudal** to the paranasal sinuses
Cranial	Caudal	L. *cranialis*, head or anterior	Pertains to the cranium or the anterior (in animals and superior in humans) part of a structure	Used when referring to the area closest to the head or the anterior/ front part of the proximal limb	More common to use **cranial** when referring to the limbs of the dog and horse—cranial is the front of the limb above the carpus and tarsus (see Fig. 1-4)
Dexter	Lev(o)/Sinister	L. *dexter*, right	The right side of the body; dextr(o)- combining form denoting the right	"Oculus dexter" (OD) is Latin for the right eye	A medication may be administered **OD** every (q) 12 hours (h)—i.e., in the right eye every 12 h
Dextr(o)	Lev(o)	L. *dexter*, right	Adj. Combining form signifying "to the right"	Describes the position of similar structures or those situated on the right side of the body	**Dextro**rotation is a clockwise rotation; **dextro**rotation/ volvulus of the small intestinal mesentery in a calf or foal causes an obstruction of the bowel

Continued

TABLE 1-1 Medical nomenclature for the term abaxial → efferent—cont'd

Medical term	Antonym	Derivation	Definition	Clinical anatomy significance	Example in clinical anatomy
Distal	Proximal	L. *distans,* distant	Further away from a point of reference	Common positioning term in dentistry, imaging, and orthopedics	Fracture description—e.g., the fracture includes the **distal** growth plate of the tibia
Dorsal	Ventral	L. *dorsalis*, from; L. *dorsum,* back	Located closer to the back	Commonly used in quadrupeds to describe the front side of a limb from the carpus and tarsus distal	Stress-related bone injury of MCIII and MTIII includes fractures of the **dorsal** cortex
Efferent	Afferent	L. *ex-,* out of, away from + *ferre*, to bear	Away from the center	Defines the conducting nerve leaving the CNS and affecting an action—e.g., motor function of muscles	The radial nerve has sensory/afferent nerve fibers and motor/**efferent** nerve fibers important in extensor function of the thoracic limb

TABLE 1-2 Medical nomenclature for the term inferior → vestibular.

Medical term	Antonym	Derivation	Definition	Clinical anatomy significance	Example in clinical anatomy
Inferior	Superior	L. *inferius*, lower	The lower surface of an organ or the lower part of two similar structures	Superior and inferior terms are generally not applied to quadrupeds because of the confusion arising from their meaning in human anatomy; use in veterinary anatomy is restricted to structures of the head—e.g., eyes	**Inferior** or ventral eyelid (see Fig. 1-3)
Labial	N/A	L. *labialis*, pertains to a lip or fleshy border—e.g., labium	A lip, a fleshy edge, boundary, or surface of a tooth	The adj. form of the term is used in relation to the lip	In dentistry—the **labial** tooth surface faces the lip (see Figure 1-2)
Lateral	Medial	L. *lateralis*, away from the median plane	Point or region furthest from the median plane or midline of a structure	Use this term when referring to a structure/tissue away from the median plane—e.g., eyes and limbs	The MC/MTIV is also referred to as the **lateral** splint bone in horses or the **lateral** (temporal) side of the eye in all species (see Figs. 1-3 and 1-4)

TABLE 1-2 Medical nomenclature for the term inferior → vestibular—cont'd

Medical term	*Antonym*	*Derivation*	*Definition*	*Clinical anatomy significance*	*Example in clinical anatomy*
Lev(o)	Dextro	L. *laevus*, left	Combining form adj.—to the left or on the left side; sinister (synonym)—on the left	Signifies on the left side; also describes a structure rotating left or in a counterclockwise direction	**Levo**rotation—counterclockwise rotation/volvulus of the abomasum in dairy cows (viewed from the right side); canine gastric dilatation and volvulus is a counterclockwise or **levo**rotation of the stomach
Lingual	Buccal/ Vestibular	L. *lingualis*, from *lingua*, tongue	The surface of the tooth facing the tongue	The adj. form is used when referring to the tongue itself or the tooth surface facing the tongue	The inside surface of the mandibular teeth is called the **lingual** surface (see Fig. 1-2)
Medial	Lateral	L. *medialis*, relating to the middle	Point or region closest to the median plane or midline of the body	The adj. form is used when referring to a structure/tissue toward the middle—e.g., eyes and limbs	MC/MTII is commonly referred to as the **medial** splint bone in horses or the **medial** (nasal) side of the eye in all species (see Figs. 1-3 and 1-4)
Mesial	Distal	Gr. *mesos*, in the middle	Tooth surface facing forward or rostral	The adj. form refers to the surface of the tooth facing forward	The **mesial** surface of the first molar borders the distal surface of the 4th premolar tooth (see Fig. 1-2)
Occlusal	Apical	L. *occlūsiō*, occluding, obstruction	Contact or grinding surfaces of opposing teeth	The adj. form denotes the contact surfaces of the incisive, maxillary, and mandibular teeth	The cup in the infundibulum of the incisors is on the **occlusal** surface and is used in estimating the age of a horse (see Fig. 1-2)
Oral	Aboral	L. *oralis*, mouth	Anything relating to the mouth or in the direction of the mouth	The adj. & n. denote the mouth; ingested food moves in an oral → aboral direction	An **oral** examination is the foundation of clinical dentistry (see Fig. 1-2)
Palatal	Buccal	L. *palatum*, roof of the mouth	Pertaining to the palate, roof of the mouth, or surface of a tooth facing the palate	The adj. form refers to the roof of the mouth; this term is also used when stipulating a surface of a tooth	The **palatal** surface of the maxillary teeth facing the hard palate (see Fig. 1-2)

Continued

TABLE 1-2 Medical nomenclature for the term inferior → vestibular—cont'd

Medical term	*Antonym*	*Derivation*	*Definition*	*Clinical anatomy significance*	*Example in clinical anatomy*
Palmar	Plantar	L. *palmaris*, palm	Relating to the palm; in quadrupeds, the posterior surface of the thoracic limb distal to the radius	The adj. refers to the surface opposite to the dorsum of the manus and pes, respectively ("volar" is no longer used)	The **palmar** surface in the dog and cat includes the carpus, metacarpus, and digits, and the area where the flexor tendons are located (see Fig. 1-4)
Parasagittal/ Paramedian	Transverse	Gr. *para-*, at or from the side of + L. *sagittalis, sagitta,* arrow	Parallel to the median plane and at right angles to the transverse/horizontal plane	Useful in reporting the location of an abnormal finding when using imaging technologies—e.g., radiography, ultrasonography, CT, and MRI	A subchondral bone cyst in MCIII is best depicted in the **parasagittal/ paramedian** plane on CT
Plantar	Palmar	L. *plantaris*, sole of the foot	Referring to the sole of the foot; in quadrupeds, the posterior surface of the pelvic limb distal to the tibia	The pelvic limb structures—e.g., ligaments, tendons, vessels, and nerves on the back part of the leg	The flexor tendons are found on the **plantar** surface of the pelvic limb (see Fig. 1-4)
Posterior	Anterior	L. *posterius*, behind	The back of, or in the back, or affecting the back part of a structure (used commonly in human medicine)	Used in referencing direction or the back area in humans and the eye and ear in quadrupeds	More commonly used in human anatomy (see Fig. 1-1B); **posterior** semicircular canal; the **posterior** ampulla is in the inner ear in quadrupeds
Proximal	Distal	L. *proximus*, next	Near or next to a point of reference in direction	Describes the location of a structure(s)—such as bones in the appendicular skeleton	**Proximal** sesamoid bones vs. distal sesamoid (navicular) bones in the limb of the horse
Rostral	Distal	L. *rostralis*, originating from *rostrum*, beak	Situated toward the oral and nasal region; nearer to the tip of the nose	This term normally refers to the head region in quadrupeds	The nose is **rostral** to the ears
Sinister	Dexter	L. *sinister*, left; Old French *sinistre*	Structure situated on the left side of the body	"Oculus sinister" (OS) is Latin for the left eye	A dog with a cataract **OS** may have unilateral, left-sided blindness
Solar	Vertex	L. *solaris,* from *solea*; *planta*, the bottom of the foot	As an adj., in veterinary medicine, it refers to the underside of the foot	In ungulates, horses, cattle, small ruminants, and camelids, it is the surface of the foot in contact with the ground	The frog is located on the **solar** surface of the hoof

TABLE 1-2 Medical nomenclature for the term inferior → vestibular—cont'd

Medical term	*Antonym*	*Derivation*	*Definition*	*Clinical anatomy significance*	*Example in clinical anatomy*
Superior	Inferior	L. *superius*, upper, above; Old French *superiour,* which is above	Directed upward, above or near the top	The term is commonly used in relation to the eye in veterinary medicine	Ptosis is drooping of the **superior** (upper) eyelid (see Figure 1-3)
Transverse/ Horizontal plane	Median plane	L. *transversus*, crosswise	The horizontal plane of the body, or at right angles to the parasagittal and median planes	Used in human medicine dividing the body into superior and inferior parts	The **transverse plane** is at right angles to the long axis of the body (see Fig. 1-1A and B)
Ventral	Dorsal	L. *ventralis, venter,* belly, abdominal	Situated toward the lower abdominal plane of the body or belly	The term is applied to the tail, trunk, neck, and head, but never to the limbs	The sternum is the **ventral**-most aspect of the rib cage in quadrupeds
Vestibular	Lingual	L. *vestibularis*, a space, cavity	In dentistry, the term denotes the surface of the tooth facing the oral cavity or "vestibule" of the mouth	Buccal and vestibular (adj.) are used interchangeably when annotating findings during a dental examination	The canine parotid gland duct opens into the **vestibule** opposite the 4th upper premolar tooth (see Fig. 1-2)

Selected references

[1] Dorland and Dorland. Dorland's illustrated medical dictionary. 32nd ed. W.B. Saunders Co.; 2011.
[2] Getty R. Sisson and Grossman's, the anatomy of the domestic animals. 5th ed. Philadelphia: W.B. Saunders; 1975.
[3] International Committee on Veterinary Gross Anatomical Nomenclature: Nomina Anatomica Veterinaria (NAV). 6th ed; 2017.
[4] Singh B. Dyce, Sack, and Wensing—Textbook of veterinary anatomy. 5th ed. St. Louis: Elsevier; 2018.

SECTION II

DIAGNOSTIC IMAGING

CHAPTER 2: INTRODUCTION TO IMAGING MODALITIES 17
Aitor Gallastegui, Nora S. Grenager, and James A. Orsini

CHAPTER 2

INTRODUCTION TO IMAGING MODALITIES

Aitor Gallastegui, Chapter editor

Current Imaging Technologies

2.1 Endoscopy—*Nora S. Grenager and James A. Orsini* 18
2.2 Radiography—*Aitor Gallastegui* 23
2.3 Ultrasonography (US)—*Aitor Gallastegui* 35
2.4 Computed tomography (CT)—*Aitor Gallastegui* 47
2.5 Magnetic resonance imaging (MRI)—*Aitor Gallastegui* 57
2.6 Nuclear medicine (Including PET)—*Aitor Gallastegui* 67

Novel Technologies

2.7 Advances in diagnostic imaging
- Files in diagnostic imaging—*Aitor Gallastegui* 74
- 3-D printing—*Aitor Gallastegui* 75
- Cone beam computed tomography—*Aitor Gallastegui* 77
- Total or whole-body PET scanner—*A. Gallastegui, N.S. Grenager, and J.A. Orsini* 78

CHAPTER 2

DIAGNOSTIC IMAGING 2.1

Endoscopy

Nora S. Grenager[a] and James A. Orsini[b]
[a]Steinbeck Peninsula Equine Clinics, Menlo Park, California, US
[b]Department of Clinical Studies - New Bolton Center, University of Pennsylvania School of Veterinary Medicine, Kennett Square, Pennsylvania, US

Introduction

Endoscopy (Gr. *endo-* within; *skopein*, to look at or examine) is an imaging modality used in veterinary medicine to examine the interior of a hollow viscus (internal organ) or body cavity. Numerous endoscopic images are included in this book, so a basic understanding of clinical endoscopy is useful.

Endoscopy provides a minimally invasive look into a hollow viscus or body cavity. Endoscopy is often used to supplement the information obtained through other imaging modalities such as radiography and ultrasonography but, in many cases, it is used as the primary or sole means of diagnostic examination. For example, endoscopy may be the sole imaging modality used to explore abnormal respiratory noises heard during exercise in horses. The upper airway may be examined while the horse is at rest or during high-speed stress testing on a treadmill or racetrack—the latter is a procedure known as dynamic endoscopy.

However, from an anatomical viewpoint, it is important to understand that the field of view with endoscopy is limited by several factors, including the features of the particular endoscope (type, diameter, length, lens angle, etc.), the experience and anatomical knowledge of the operator, and the size and shape of the cavity being examined. Endoscopy is like looking through a camera lens: what is seen within the static field of view is only part of what there is to be seen. Even so, with enough training and experience, endoscopy provides unsurpassed visual and physical access to structures that would not otherwise be reached except through an alternate invasive procedure.

Many different types of endoscopes (i.e., "scopes") are used in human and veterinary medicine and are generally divided into two basic types: rigid and flexible. Rigid scopes (Fig. 2.1-1A and B) consist of a metal tube with a fixed-angle lens at the tip (ranging from 0 to 70 degrees, depending on the scope), so altering the field of view requires altering the angle of the entire scope. Otoscopes, used for examining the external ear canal, are another common example of a specific type of rigid endoscope. Flexible scopes (Figs. 2.1-2 and 2.1-3) consist of a flexible tube that can conform somewhat to the shape of the body part being examined or traversed. In addition, most flexible scopes have an adjustable tip that can be controlled remotely by the clinician to alter the field of view without moving the rest of the scope. The scopes used to examine the upper gastrointestinal tract (esophagus, stomach, and duodenum) and the upper airways are examples of a flexible endoscope.

In (Fig. 2.1-3), the image resolution and anatomical detail are excellent with a deflection up/down of 180°/110° and left to right of 110°/110° and a field of view of 120°.

Comparative Veterinary Anatomy: A Clinical Approach. https://doi.org/10.1016/B978-0-323-91015-6.00002-9

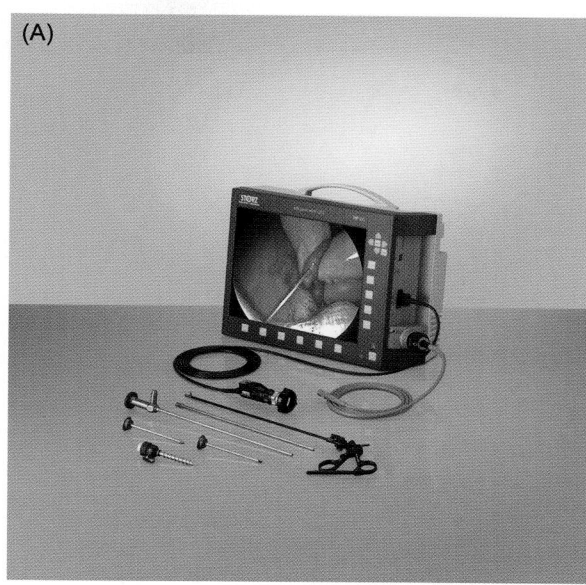
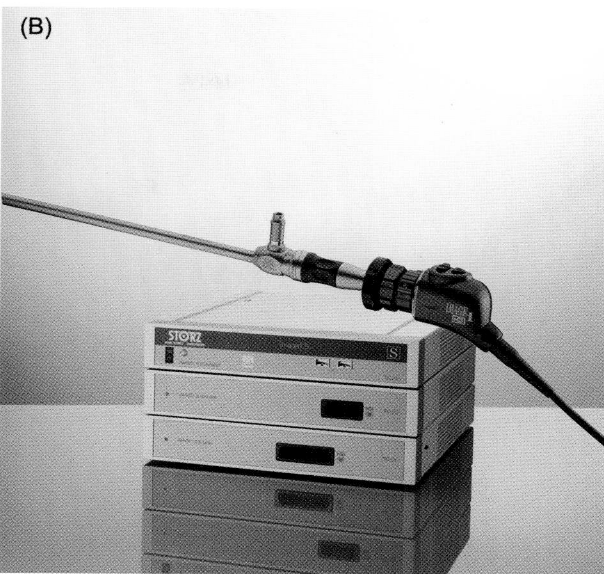

FIGURE 2.1-1 (A) Arthroscope and (B) laparoscope are examples of rigid endoscopes used in veterinary and human medicine. (Photo courtesy of Karl © KARL STORZ SE & Co., Germany.)

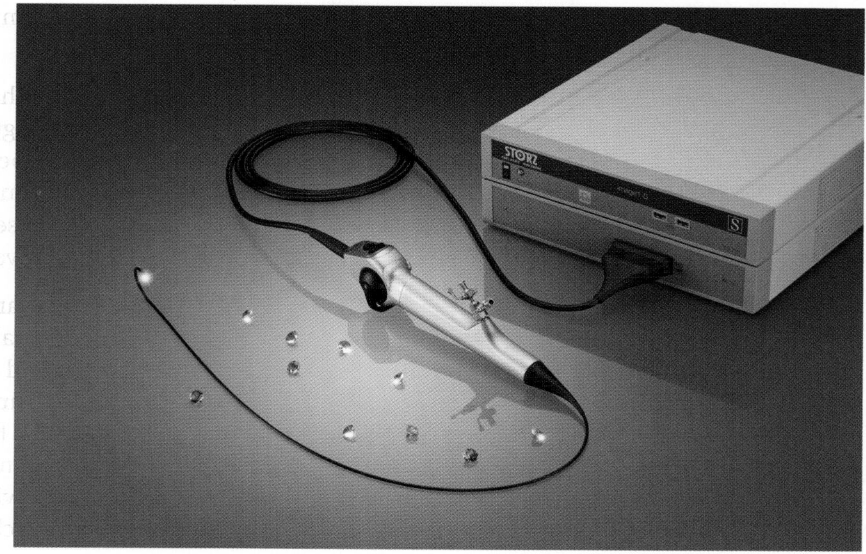

FIGURE 2.1-2 Small video endoscope with outside diameter of 2.8–5.2 mm (0.1–0.2 in.) used in veterinary and human medicine. (Photo courtesy of © KARL STORZ SE & Co., Germany.)

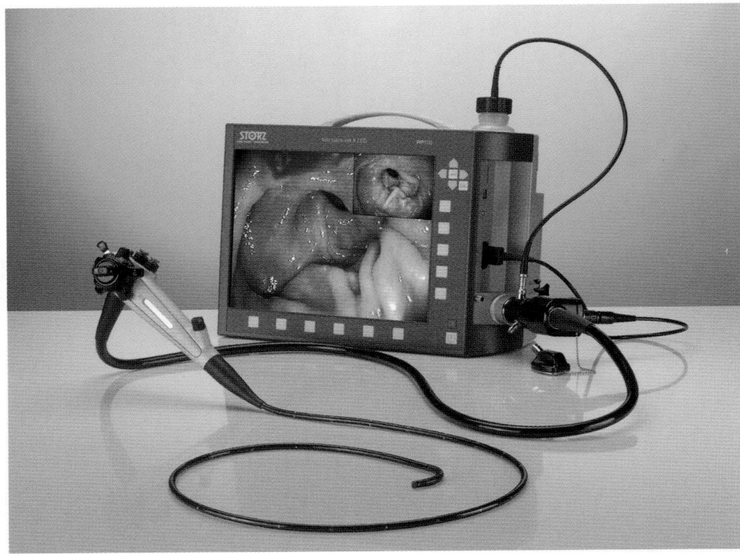

FIGURE 2.1-3 Flexible video endoscope for examination of the gastrointestinal, urogenital, and respiratory systems in veterinary and human medicine. (Photo courtesy of © KARL STORZ SE & Co., Germany.)

Endoscopic procedures

Endoscopy may be used diagnostically and therapeutically, and both uses may be combined during the same examination. **Diagnostic uses** include the following:

- Visual examination
- Fluid aspiration or mucosal swabbing/brushing for biochemical, cytologic, and/or microbiologic examination (e.g., Gram staining, bacterial culture, and antimicrobial sensitivity testing)
- Tissue biopsy for microbiologic and/or histopathologic examination

Therapeutic uses include the following:

- Lavage (flushing)
- Retrieval of foreign material (e.g., aspirated material that is causing airway obstruction)
- Debridement of damaged or devitalized tissue
- Debulking or removal of abnormal tissue masses (using either endoscopic instruments or laser)
- Selected surgical procedures, such as ovariectomy, with the help of specially designed surgical instruments and separate portals

The ability to record still digital images and videos of the endoscopic procedures permits the clinician to preserve the images as part of a patient's permanent medical record. Several such images appear throughout the book.

When describing endoscopic procedures, a prefix that denotes the target organ or cavity is used before the suffix "-oscopy." For example, gastroscopy indicates an endoscopic examination of the stomach, specifically of the gastric lumen. Table 2.1-1 summarizes the endoscopic terms used in the book.

TABLE 2.1-1 Endoscopic terms used in the book.

Region and term	*Target organ/cavity*	*Access route*
Gastrointestinal tract		
Esophagoscopy	Esophagus	Via the oral cavity in most mammals and birds; via the nares (nostrils) in horses[a]
Gastroscopy	Stomach	As mentioned previously
Duodenoscopy	Duodenum	As mentioned previously
Esophagogastroduodenoscopy (EGD)	Esophagus, stomach, and duodenum	As mentioned previously; this is a common combination in clinical practice
Endoscopic retrograde cholangiopancreatography (ERCP)	Common bile duct and pancreatic duct	Performed during duodenoscopy
Colonoscopy	Colon	Via the anus and rectum
Proctoscopy	Rectum	Via the anus
Cloacoscopy (birds)	Cloaca (terminal cavity shared by the digestive and urogenital tracts in birds)	Via the vent (external opening of the cloaca)
Respiratory tract		
Rhinoscopy	Nasal passages	Mammals: via the nares (nostrils); except in horses, the caudal aspect of the nasal cavity may also be accessed via the nasopharynx and choanae (caudal openings of the nasal cavity) by the retroflexion of the endoscope around the caudal (free) margin of the soft palate from the oral cavity Birds: via the nares or the choanal opening in the roof of the mouth
Sinoscopy	Paranasal sinuses	Generally, via a small opening made in the skin and bone overlying the frontomaxillary sinus; the nasal route may be used if an opening is created between the sinus and the nasal passage
Pharyngoscopy	Pharynx (most often the nasopharynx in mammals); may also include the guttural pouches in horses[a]	Via the nares in horses[a], via the oral cavity in most other mammals and in birds (note: birds lack a soft palate, so they have a simple pharynx)
Laryngoscopy	Larynx	Via the nares in horses[a], via the oral cavity in most other mammals and in birds
Tracheobronchoscopy	Trachea and main bronchi	Via the nares in horses[a], via the oral cavity in most other mammals and in birds
Syringoscopy (birds)	Syrinx (the vocal organ in birds)	Via the oral cavity; in most avian species, the syrinx is located at the tracheobronchial bifurcation, so this procedure is like tracheobronchoscopy
Ear		
Otoscopy	External ear canal	Via the external opening of the canal; this is a very common procedure in companion animal practice
Urogenital tract		
Urethroscopy	Urethra	Via the external urethral orifice
Cystoscopy	Urinary bladder	Via the external urethral orifice
Colposcopy	Vagina and cervix	Via the vulva
Hysteroscopy	Uterus	Via the vulva

TABLE 2.1-1 Endoscopic terms used in the book—cont'd

Region and term	*Target organ/cavity*	*Access route*
Body cavities		
Laparoscopy	Peritoneal cavity	Via a small incision in the abdominal wall
Thoracoscopy	Pleural cavity	Via a small incision in the thoracic wall
Coelomoscopy (birds)	Coelomic cavity	Via a small incision in the body wall
Musculoskeletal system		
Arthroscopy	Joint space	Via a small incision in the skin and joint capsule
Tenoscopy	Tendon sheath	Via a small incision in the skin and tendon sheath
Bursoscopy	Bursa	Via a small incision in the skin and bursal wall

[a] In horses, the caudal (free) margin of the soft palate normally rests under the epiglottis, forming an anatomic separation between the nasopharynx and oropharynx; therefore, the upper gastrointestinal tract and the airways are generally accessed via the nares (nostrils) to the nasopharynx and then into either the esophageal opening or the glottis. The guttural pouches (auditory tube diverticula) in horses are also accessed via the nasopharynx.

Selected references

[1] Abutarbush S, Carmalt J. Endoscopy and arthroscopy for the equine practitioner. New York: Teton Newsmedia; 2008.
[2] Dallap Schaer B, Aldrich E, Orsini JA. Emergency diagnostic endoscopy. In: Orsini JA, Divers TJ, editors. Equine emergencies: treatment and procedures. 4th ed; 2014. p. 61–9.
[3] Tams TR, Rawlings CA. Small animal endoscopy. 3rd ed. St. Louis: Mosby-Elsevier; 2010.
[4] Lierz M. Diagnostic value of endoscopy and biopsy. In: Harrison GJ, Lightfoot TL, editors. Clinical avian medicine. Ithaca, NY: International Veterinary Information Service; 2020. e-book Available at: http://www.ivis.org/advances/harrison/chap24/chapter.asp?LA=1.

CHAPTER 2

DIAGNOSTIC IMAGING 2.2

Radiography

Aitor Gallastegui
Small Animal Clinical Sciences, University of Florida, Gainesville, Florida, US

Introduction

Radiology is the use of x-rays for medical diagnostic purposes. X-rays were discovered by Wilhelm Roentgen in 1895 (the first known radiograph is of his wife's hand), for which he was awarded the Nobel Prize in physics in 1901. The first publications reporting the use of radiology in the veterinary medical field go back to 1896, just a year after its discovery. Radiology is the most widely used diagnostic imaging modality worldwide due to its affordability and reliability. Screen-film radiography was the first-generation radiography and has largely been replaced by digital radiography (DR), which provides rapid results and improved image quality compared to screen-film radiography.

Understanding the imaging technology/mode of action

X-rays are produced within an x-ray tube. The x-ray tube consists of a vacuum chamber that contains a cathode and an anode. At the cathode, an electron cloud is produced and then electrons within the cloud are projected against a target in the anode at high velocities. The vacuum environment of the tube and electric focusing cups help control the direction of the electrons as they travel from the cathode to the anode. Any break in the tube impairs its function and increases undesirable exposure to high energy radiation. The anode and the cathode are typically made of tungsten, a material with ideal properties for x-ray production, including its ability to withstand high temperatures without melting. The invention of the rotating anode, which increases the target surface for electrons and allows for higher energy x-rays, also helps with heat dissipation. This is particularly useful when radiographing dense body regions such as the trunk and proximal limbs in horses (Fig. 2.2-1).

To minimize radiation exposure to staff and patients, the tube is enclosed by—or contained in—a lead chamber, except for a small window that permits x-rays used for diagnostic purposes to exit. As the x-ray beam travels from the target to the patient, various filters and collimation are used to improve the quality of the x-ray beam. All these procedures aim to optimize the energy of the x-ray beam for diagnostic purposes and reduce unnecessary radiation exposure.

When selecting the parameters for taking a radiograph, the following need to be considered: kilovoltage, milliamperage, time, and distance. **Kilovoltage (kVp)** is the peak voltage applied to the tube, which is applied between the cathode and the anode. The kVp determines the speed of the electrons impacting the target in the anode and, therefore, the energy the x-rays generated. Following the interaction of the electrons with the target, a large variety of x-ray energies are created with a maximum x-ray energy (KeV) been equal to the KVp. **Milliamperage** (mA) determines the heat achieved at the cathode, the quantity of electrons available in the electron cloud, and therefore, the number of x-rays generated. **Time** (sec.) refers to the exposure time; the longer the exposure time, the more x-rays are generated per exposure. Because these two factors (mA and sec.) affect the number of x-rays generated and therefore the film density, mA and time are often expressed as a combined factor: **mAs**. The distance from the x-ray source to the detector, depending on the setting, ranges between 80 and 100 cm (31.5 and 39.4 in.). In many x-ray units, and particularly in small animal radiology suites, the distance is fixed and determined by the manufacturer. All these parameters must be adjusted on the x-ray unit's control panel depending on the size of the animal and the body region radiographed; Tables 2.2-1–2.2-4 show examples of radiographic technique charts regularly used for small and large animal settings.

Comparative Veterinary Anatomy: A Clinical Approach. https://doi.org/10.1016/B978-0-323-91015-6.00003-0

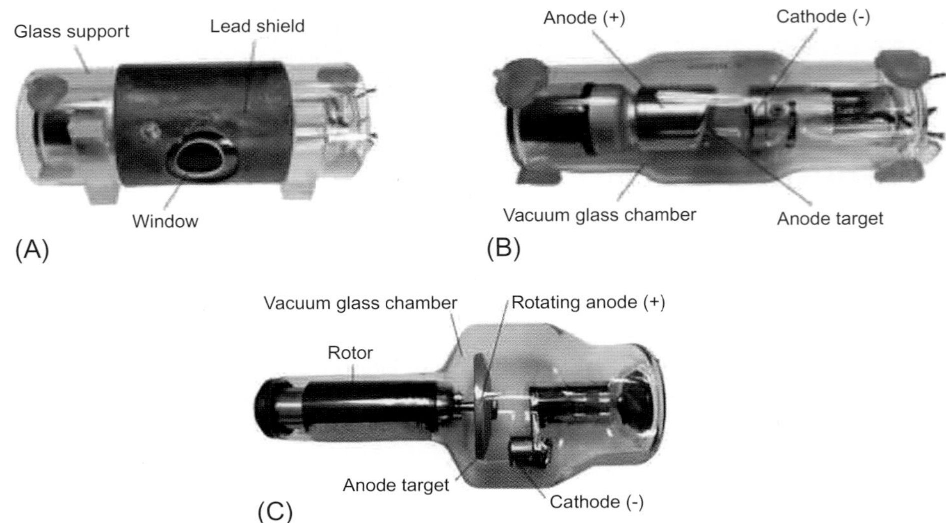

FIGURE 2.2-1 (A) A fixed x-ray tube in glass support with a lead shield (Note the window at the site where the x-ray beam exits the vacuum glass chamber. (B) Same fixed x-ray tube as in A without the glass support and the shield (note the oblique angle of the anode, which determines the width of the x-ray beam cone; also, note the brown-colored glass due to repetitive radiation exposure demonstrating that generated x-rays are multidirectional, which justifies the use of shields. (C) In rotating x-ray tubes, the anode is a disk; as the disk rotates, it increases the surface area of the target zone resulting in improved heat dissipation, providing longer exposures and higher energy x-rays to be created.

TABLE 2.2-1 Strengths and weaknesses of film and digital radiographic receptors.

	Film-screen systems	*Digital systems*
Strengths	• Detector, viewer, and storage in one • Simple and reliable (used > 100 years) • Low-investment cost • No need for expensive PACS	• Allow for a wider range in kVp and mAs settings to result in diagnostic quality radiographs • Thick and thin body regions can be displayed in a single radiograph with diagnostic quality throughout • Due to the ability to postprocess, fewer radiographs are taken per patient, resulting in reduced absolute radiation dose to the patient • Images are processed faster, allowing more patients to be radiographed • Digital radiographic images are highly portable and can be shared via the internet, allowing consultation with specialists via teleradiology • Digital radiographic images can be postprocessed on a computer, providing more details and increasing their diagnostic value (e.g., changing brightness (WW/LL), zoom, rotate) • Digital receptors are reusable, resulting in the use of few chemicals and reduced environmental contamination

TABLE 2.2-1 Strengths and weaknesses of film and digital radiographic receptors—cont'd

	Film-screen systems	*Digital systems*
Weaknesses	• The radiographic technique must be exact • Increased overall radiation dose to the patient due to: ◦ Multiple radiographs often needed for a diagnostic quality image ◦ Multiple radiographs needed in regions of marked difference in the denseness of anatomy in the field of view—e.g., hip and stifle • The film can only be viewed in one location at a time • Cannot visualize immediately • Patients waiting for confirmation • Throughput suffers • Film processor problems cause artifacts • Lack of image manipulation capabilities • Not a reusable format • Environmental hazards from chemicals used • Physical storage limitations	• Potential loss of spatial resolution due to the pixel size • Overexposure leads to saturation of the detector and the loss of stored information • The tendency to increase dose per radiographic exposure because it guarantees a better image quality ("dose creep") • Expensive PACS required

TABLE 2.2-2 Example of a radiographic technique chart used in small animal practice.

Body region (thickness)	*kVp*	*mA*	*Milliseconds*	*Grid*
Thorax (1–14 cm or 0.4–5.5 in.)	75–85	250	10	No
Thorax (15–30 cm or 5.9–11.8 in.)	95–120	500–800	10	Yes
Abdomen (1–14 cm or 0.4–5.5 in.)	75–85	500	10	No
Abdomen (15–30 cm or 5.9–11.8 in.)	85	500–800	25–40	Yes
Pelvis (1–14 cm or 0.4–5.5 in.)	70–85	320	20	No
Pelvis (15–30 cm or 5.9–11.8 in.)	85–90	500–800	20–25	Yes
Extremity (1–6 cm or 0.4–2.4 in.)	70–74	320	10	No
Extremity (7–13 cm or 2.8–5.1 in.)	76	320	10–16	No
Vertebral column (1–14 cm or 0.4–5.5 in.)	70–80	320	25	No
Vertebral column (15–30 cm or 5.9–11.8 in.)	85–90	630–800	25–40	Yes

TABLE 2.2-3 Example of a radiographic technique chart used in equine practice (portable machine).

Body region (views)	*kVp*	*mA*	*Milliseconds*	*Grid*
Foot (DP and LM)	70	320	0.08	No
Fetlock (DP, LM, and obliques)	70	320	0.08	No
Tarsus (DP, LM, obliques)	70	320	0.08	No
Carpus (DP, LM, obliques)	70	320	0.08	No

Key: DP, dorsopalmar or dorsoplantar; LM, lateromedial; obliques, dorsolateral-palmaromedial or dorsomedial-palmarolateral oblique or dorsolateral-plantaromedial or dorsomedial-plantarolateral oblique.

TABLE 2.2-4 Example of a radiographic technique chart used in equine practice (overhead machine).

Body region (view)	*kVp*	*mA*	*Milliseconds*	*Grid*
Thorax (latero-lateral)	100–120	640–800	25–32	Yes
Thorax of foal (latero-lateral)	100	800	32	Yes
Abdomen (latero-lateral)	100	800	32	Yes
Neck	85–120	800	64–400	Yes
Stifle (caudo-cranial)	80	800	50	No
Stifle (latero-medial)	75	800	8	No
Stifle (caudolateral-craniomedial oblique)	75	800	8	No

Similar to the radiance of torchlight on a foggy night, the x-ray beam divides as it fans out from its source, resulting in a cone shape affecting the radiographic image generated (Fig. 2.2-2). As the distance between the patient and the receptor increases, the image created undergoes **magnification** (Fig. 2.2-3). If the x-ray beam is not perpendicular to the receptor, the image created is subject to **distortion** (Fig. 2.2-3). Another characteristic of the x-ray beam is that one edge of the x-ray beam transmits more x-rays than the other, resulting in differential opacification of the film or receptor (the **heel effect**), due to differential absorption of newly produced x-rays at the target (Fig. 2.2-2).

Once the x-ray beam reaches the patient, it interacts in several ways with the atoms that constitute the different body tissues. Generally, the photons that form the x-ray beam are absorbed, scattered, or transmitted.

Scatter occurs when an x-ray photon loses some energy and alters its direction upon tissue interaction, causing the atoms in the tissues to release an electron. The scattered x-ray photon can go in any direction and (1) interact with another atom; (2) reach the receptor at a site out of its original path, creating a signal in the incorrect spot and resulting in a "noise" on the image; or (3) reach the imaging staff in the room. The amount of scatter increases with increased x-ray beam kVp, collimation up to 30 × 30 cm (11.8 × 11.8 in.), and patient width/depth. In addition, loss of an electron ionizes the atom and can damage the DNA, leading to reparable or irreparable cell injury. The latter results in cell mutation and cancer development, or cellular death.

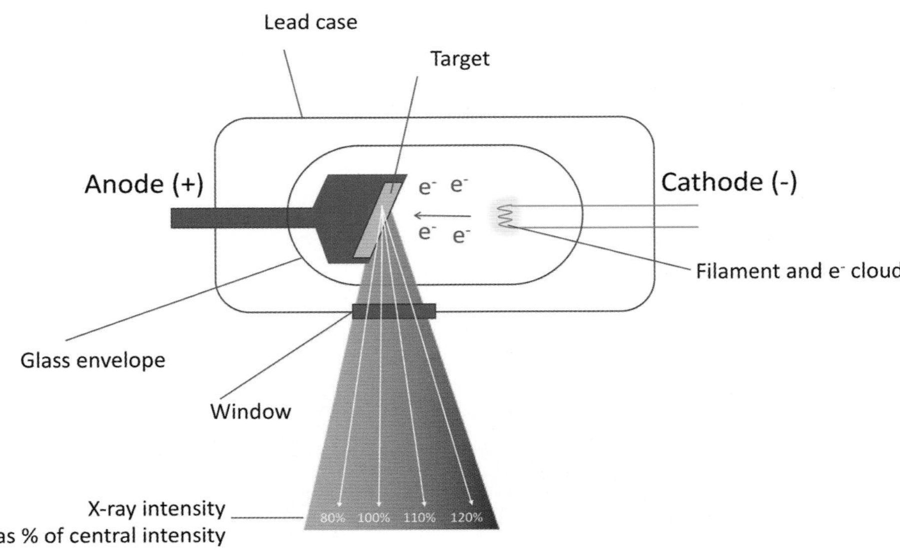

FIGURE 2.2-2 Configuration of a fixed x-ray tube and diverging configuration of the created x-ray beam. The "heel effect" consists of differential x-ray beam intensity, with higher intensity toward the cathode due to partial x-ray absorption by the anode.

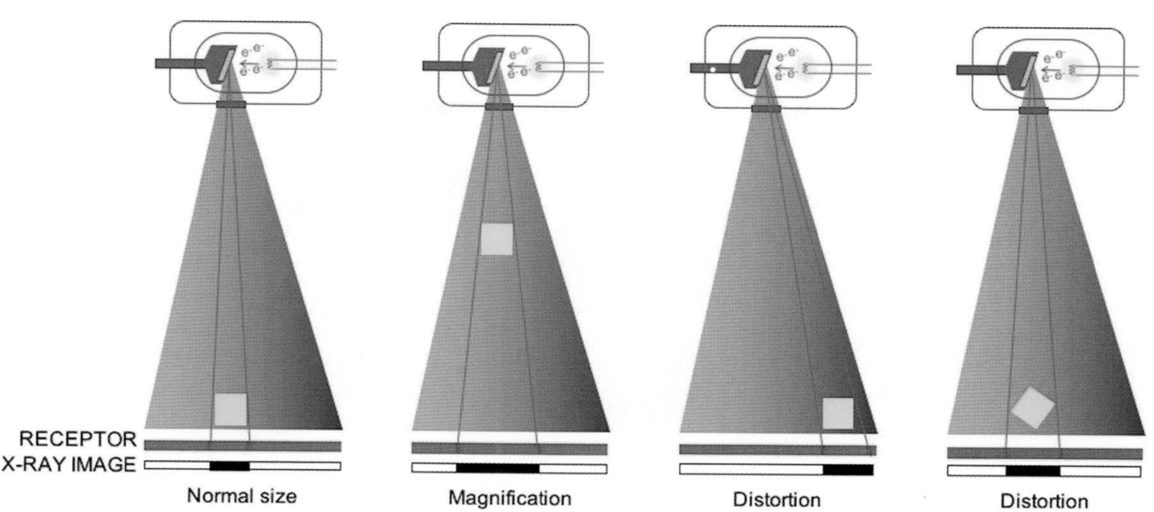

FIGURE 2.2-3 Radiographic magnification and distortion. An increase in patient-to-receptor distance leads to magnification of the patient or body region imaged. Obliquity of the patient, obliquity of the x-ray beam, and displacement of a body region lateral from the center of the x-ray beam leads to distortion of the region imaged.

Attenuation is the phenomenon in which x-rays lose energy and/or intensity as they pass through the body of the patient, depending on the types of tissue with which the x-rays interact. The amount of attenuation affects the quantity and energy of the x-rays that reach the x-ray receptor. The more radiation a region of the receptor receives, the more black or radiolucent this region appears on the generated image. Conversely, the less radiation received, the more white or radiopaque it appears on the image generated. Based on the x-ray attenuation experienced, five **radiographic opacities** can be differentiated, from the least to the most opaque: (1) gas or air opacity; (2) fat opacity; (3) soft tissue or fluid opacity; (4) mineral or bone opacity; and (5) metal opacity (Fig. 2.2-4).

It is important to remember that the final image generated is not solely a factor of the type of tissue or x-ray energy, but the density and volume of the tissue through which the x-ray beam passes. An x-ray beam of certain energy is more attenuated by a 1-cm (0.4-in.) dense soft tissue structure than by a 1-cm (0.4-in.) dense fatty structure. However, as the thickness of the fatty structure increases, the resultant attenuation also increases and, at some point, causes enough attenuation to result in an opacity similar to that of soft tissue or fluid. Similarly, as the x-ray beam travels through the body, it is attenuated by different organs, each of whose opacities summates into a final opacity. As an example, radiographing a stomach containing fluid and gas with an x-ray beam horizontal to the ground differentiates

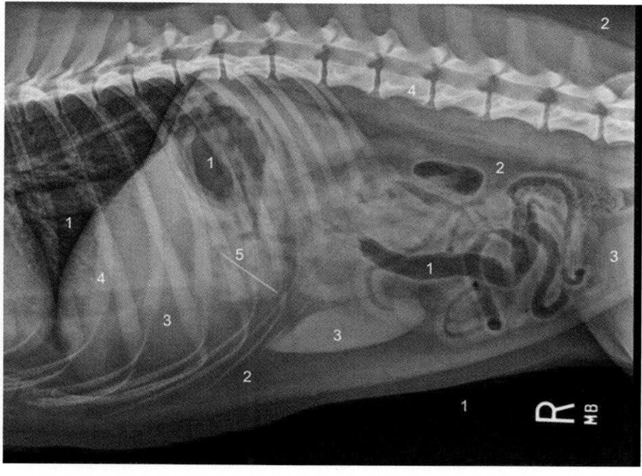

FIGURE 2.2-4 Abdominal radiographs of a dog with a metallic (needle) foreign body in the pyloric antrum of the stomach. (1) Gas opacity, (2) fat opacity, (3) soft tissue or fluid opacity, (4) mineral opacity, and (5) metal opacity.

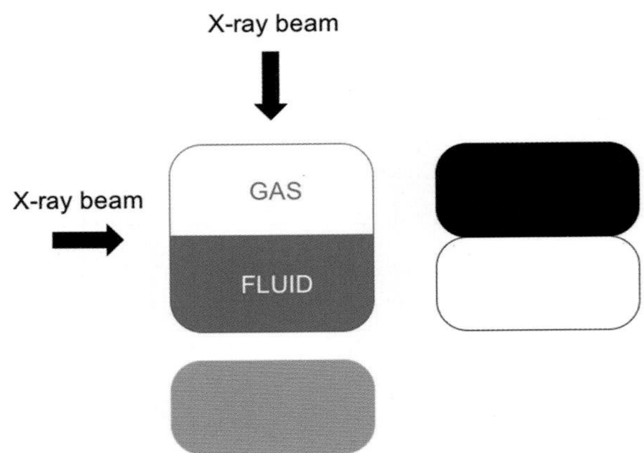

FIGURE 2.2-5 Resulting opacities of summating tissues or materials such as gas and fluid, e.g., gastric contents, results in intermediate opacities, e.g., mild soft tissue opacity and fat opacity.

both opacities with a clear fluid–gas line interphase; whereas, an x-ray beam perpendicular to the ground results in an intermediate opacity (Fig. 2.2-5). In a similar way, when the x-ray beam undergoes attenuation from two organs of the same opacity, those opacities summate—termed **summation** (Fig. 2.2-6). If organs of the same opacity touch each other, **border effacement** occurs, in which their individual margins are lost, and the opacity appears continuous (Fig. 2.2-6).

After having interacted with the tissues, the resulting x-ray beam interacts with the receptor to create an image. **Film** or **screen-film** has been the only receptor element used for radiographic image recording and viewing since its invention in 1918 until the early 2000s. In addition to a support layer, screen-film contains one (single emulsion) or two (double emulsion) gelatin layers. Each gelatin layer contains innumerable crystals such as silver halides that form a latent image of the body tissues following interaction with x-rays. The film is processed and—by means of various chemical interactions—the crystals exposed to x-rays remain in the film, and the ones that did not interact with x-rays fall off the film during processing. Therefore, tissues or materials that allow x-rays to pass are called **radiolucent** and create a black image, and tissues or materials that do not allow x-rays to pass through are called

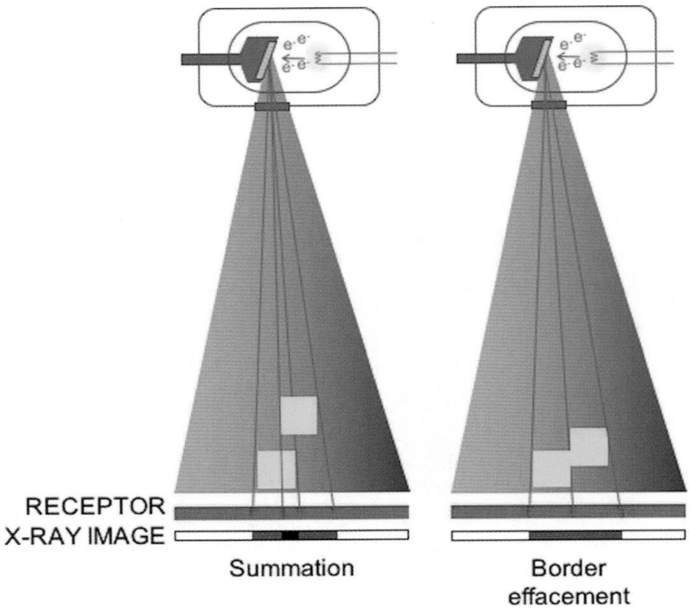

FIGURE 2.2-6 Summation and border effacement. The attenuation of structures in the path of the same x-rays sums their opacities. When organs, tissues, or materials of similar opacity are in contact in a plane perpendicular to the x-ray beam, the touching borders are lost or effaced.

radiopaque and create a white or transparent image. The silver halide crystals are much more sensitive to light than they are to x-rays, which is why films need to be housed in light-tight cassettes.

Intensifying screens generate light as a response to interaction with x-rays. The addition of intensifying screens to the cassettes, located between the case and the film, significantly increases the speed of image generation and dramatically reduces the amount of radiation needed to create a diagnostic image. The use of intensifying screens does, however, carry some loss in image quality, so its use is contraindicated for detailed radiographic studies.

Currently available digital receptors include **computed radiography** (CR) and **digital radiography** (DR). In CR, a photostimulable phosphor contained within a cassette is used as a receptor element, which stores the latent image. The cassette is then introduced in a specifically built reader in which the x-ray image is released by laser light and converted into an electronic signal or image file. One of the benefits of this system is that a single receptor is used to acquire thousands of radiographs. In DR, flat-panel detectors or charge-coupled devices are used to record the x-ray image. The **flat-panel detectors** convert the x-ray image directly into an electronic signal or image by direct or indirect mechanisms. Initially, these panels were connected to the computer by a cable, but most recent versions have wireless connections. The spatial resolution is determined by the size of the 6–7 million tiny electronic elements or pixels that make up each panel. The **charge-coupled devices** are receptors of a much smaller size and require focusing lenses to fit the x-ray image in. Because of the need for focusing lenses, the distance between the x-ray tube and the receptor must be constant and stable. Therefore, these receptors are built into x-ray tables and cannot be used for mobile radiography. Despite the small size, the receptor is also composed of millions of pixel elements. This type of receptor is often used in human mammography, dental radiography, and in many video camcorders and digital cameras. Digital radiography systems (CR and DR) have replaced film-based systems in most countries for many reasons: (1) it is easier to store and share images via the use of PACS (picture archiving and communication systems); (2) even poor techniques create good quality images compared to traditional radiography because the images can be manipulated (Fig. 2.2-7); (3) postprocessing includes capabilities such as window and leveling or use of zoom and rotation; (4) there is reduced overall patient exposure to radiation; and (5) there is reduced environmental contamination since no chemicals are used during "processing" of digital radiographs. Other strengths and weaknesses of these systems are listed in Table 2.2-1. For additional information on PACS, refer to section 2.7 in this chapter.

Digital radiographs are viewed on computer monitor screens, which are comprised of arrays of pixel elements. Average monitors have less than 1,000,000-pixel (or 1MP) elements and display fewer colors or shades of gray. Medical grade monitors normally display 2–10 MP and many more colors or shades of gray, which significantly increases diagnostic accuracy. Using nonmedical grade monitors can lead to misdiagnosis, with small lesions such as pulmonary or bone metastasis often missed.

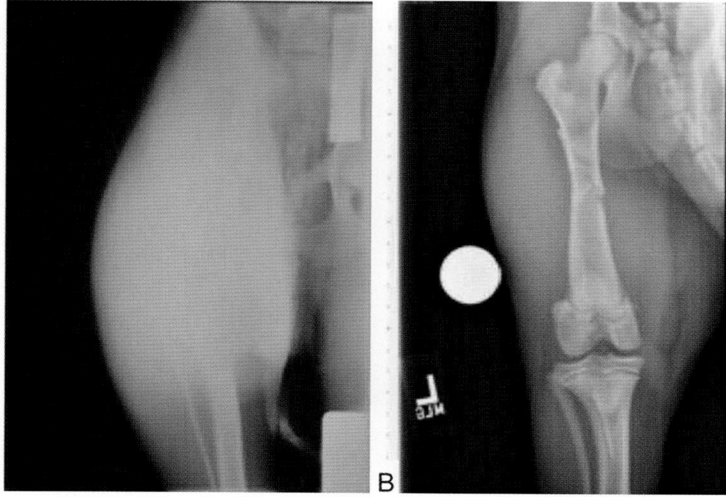

FIGURE 2.2-7 Film vs. digital radiography (DR). These radiographs of the left thigh in a canine patient demonstrate how DR (B) allows for quality radiographic images of thin and thick regions to display in a single image, while film radiography (A) is more affected by tissue thickening due to its smaller dynamic range.

Improving radiographic image quality

Scatter radiation is caused by random x-rays that cause secondary interaction with tissues resulting in artifactual regions of false exposure or increased "noise" on the film. The effects of scatter radiation on image quality can be reduced by using a **grid** between the patient and the receptor. Grids are comprised of parallel strips of a radiopaque material such as lead on a radiolucent material such as aluminum (Fig. 2.2-8). The parallel lines of the grid are aligned with the trajectory of the primary x-ray beam as it originates from the x-ray tube, allowing most of the beam to pass through while absorbing most of the scatter radiation. Because the grid absorbs a large portion of the primary beam radiation, the radiographic exposure—kVp and mAs, and therefore, the dose to the patient—must be increased to maintain image quality. The total increase in exposure needed to maintain image quality is determined by the "grid ratio."

Additional radiographic imaging techniques

Contrast media are used primarily to improve the imaging of certain abdominal organs. Routinely, two types of contrast media are used in radiology: positive or negative. **Positive contrast media** are substances that cause strong attenuation of x-rays and, therefore, create images with more intense opacities (Fig. 2.2-9). Barium-based contrast media are used primarily for GI studies; however, this is contraindicated if there is a possible rupture of the GI tract because contrast media outside of the GI tract cause severe—often fatal—mediastinitis and peritonitis. Aspiration of barium-based contrast media are not harmful to the patient's health, but contrast media often remain in the lower airways and the tracheobronchial lymph nodes indefinitely, potentially confusing future radiographic interpretation. Nonionic, low-osmolarity iodinated contrast media such as iohexol are primarily used for IV, intrathecal, and retrograde urogenital administration but can be administered by any route due to their increased safety. Although some patients may have temporary reactions, severe adverse events such as acute renal failure are extremely rare and are prevented by maintaining good patient hydration. Some ionic-iodinated contrast media are commercially available but are not currently recommended given their increased risk—e.g., neurotoxicity, nephrotoxicity, and allergic and anaphylactic reactions; in addition, these contrast media cannot be administered intrathecally (into the subarachnoid space).

Negative contrast media are gases—e.g., room air or carbon dioxide—that cause reduced attenuation of x-rays and, therefore, cause images of gas opacity (Fig. 2.2-10). These contrast media are normally used to distend hollow viscera such as the GI tract and lower urogenital system. The main risk in using negative contrast media is the development of a secondary air embolism, which is more likely to occur when the blood-mucosal barrier is injured, as in cases

FIGURE 2.2-8 Radiographic grids are composed of hundreds of tightly packed radiopaque laminae that absorb scatter radiation and radiolucent laminae that mostly allow transmission of only x-rays with diagnostic information, as seen in this broken grid.

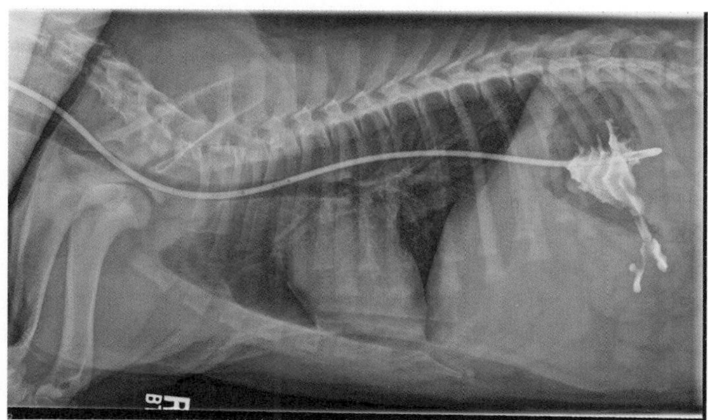

FIGURE 2.2-9 Use of positive contrast media. Lateral thoracic radiograph in a dog with a gastroesophageal tube; barium sulfate was administered via the nasogastric feeding tube, partially filling the stomach and "contrasting" the rugae folds. Positive contrast media exhibit variable metallic opacity depending on their concentration.

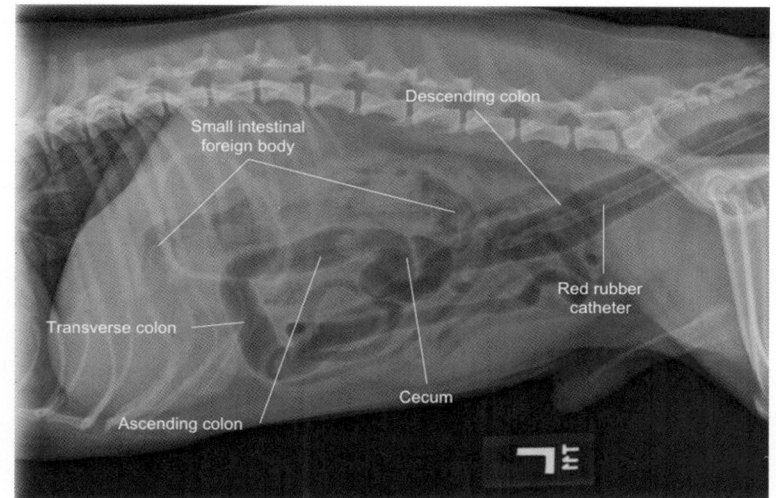

FIGURE 2.2-10 Use of negative contrast media. Left lateral abdominal radiograph in a dog; a red rubber catheter was inserted rectally and room air was introduced, distending the colon and a small segment of the small intestine. Distension of the colon with gas provided localization of a foreign body in a small intestinal segment located in the mid-craniodorsal abdomen.

of mucous membrane ulceration. When performing negative contrast studies, the patient should be positioned in left lateral recumbency, so that if there is an air embolus, the gas accumulates in the apex of the right ventricle and slowly mixes in the blood before embolization of large regions of the lung with potentially deadly consequences. Negative contrast media are often administered in conjunction with positive contrast media, termed **double-contrast studies,** to highlight luminal mural masses or defects.

Radiographic safety

When x-rays were first discovered and used in medical imaging, the exposure times for a radiograph of a hand and full body were 20 and 30 min, respectively. This discovery was an enormous advancement in medicine, but the long exposure times and doses of radiation resulted in long-term adverse events, such as cancer ascribed to radiation exposure. Since the original introduction of x-rays, the scientific community has gained a better understanding of the effects and proper medical use of radiation in patients and workers. Currently, radiation exposure limits are determined by international, federal, and state laws. Pregnant individuals and those younger than 18 years old are generally subject to stricter regulations.

The principle of "*as low as reasonably achievable*" (ALARA) as defined in the Code of Federal Regulations should be followed. Simply, the reduction of exposure from external radiation is achieved by (1) reducing the time of exposure to a source of radiation; (2) increasing the distance from the source of radiation; and (3) by using appropriate shielding. Reducing exposure by increasing the distance from the source does not only refer to the primary beam but also to the scatter radiation originating from the patient and equipment. The highest source of radiation contamination for radiology employees comes from exposure to scatter radiation. Shielding used to protect from scatter radiation includes lead aprons, gloves, neck collars, glasses, and windows. However, it is important to remember that these defenses do not protect against primary beam radiation. Therefore, radiology technicians should avoid holding patients during radiographic acquisition whenever possible (Fig. 2.2-11).

Fluoroscopy

The first fluoroscopy system became available in 1896, just a year after the discovery of x-rays. Fluoroscopy uses a constant x-ray beam to display connecting x-ray images and creates a video file. Fluoroscopy is an extremely valuable diagnostic imaging tool because it acquires real-time images of internal organs and their function in patients—e.g., cardiac pumping, swallowing, and tracheal collapse. Since the 1950s, fluoroscopy has used intensifying screens as receptors. Later models use flat-panel detectors with higher sensitivity to x-rays and better temporal and contrast resolution. This results in (1) dose reduction to the patient; (2) reduced motion blurring of the image; and (3) an improved contrast ratio. To further enhance fluoroscopic contrast resolution, a positive contrast medium is often used. The positive contrast medium is administered at different locations and in different preparations depending

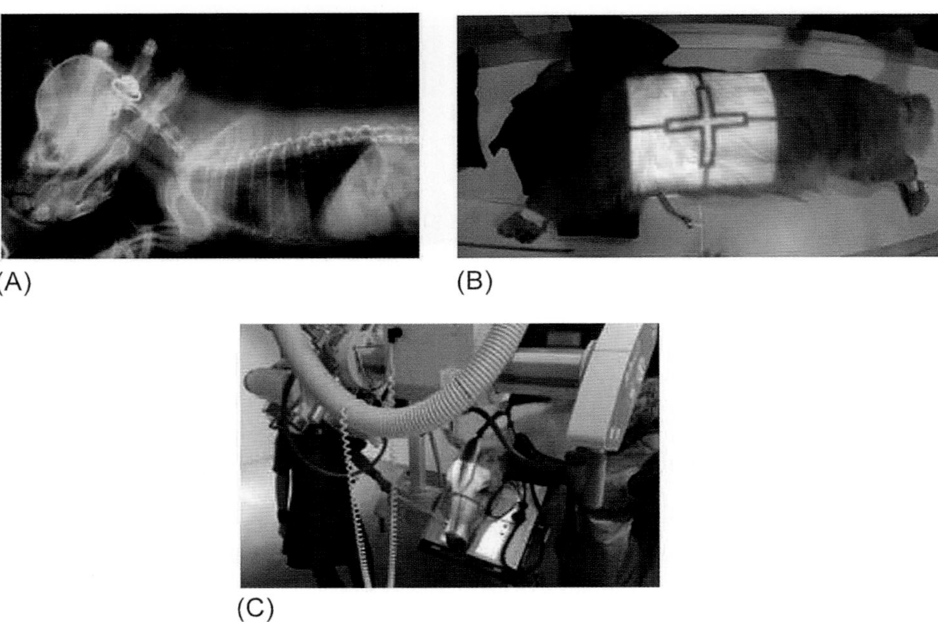

FIGURE 2.2-11 Radiation safety. Holding patients (A) should be avoided, and the use of sedation and support systems (such as sandbags and tape in figures B and C) should be always enforced. If a patient needs to be held during the radiographic examination, lead gloves should always be worn, and hands should always be kept out of the collimation, as depicted by the last two images. The reader must be aware of the fact that lead shields ONLY protect from scatter radiation, not from the radiation of the primary x-ray beam.

on the study performed. Fluoroscopy is commonly used intra-operatively—especially in cardiac, vascular, and orthopedic surgeries—leading to the specialty of "interventional radiology."

Fluoroscopic units are available in numerous configurations, optimized for different clinical needs. As opposed to radiology systems, **conventional fluoroscopy** systems consist of a patient table with the x-ray tube positioned *under* the table-top and the receptor *above* the table and patient. These systems are used for most diagnostic procedures performed with the patient in various positions and states of consciousness.

The **C-arm** configuration facilitates intra-operative uses of fluoroscopy, primarily in orthopedic surgery, as well as cardiac and vascular procedures (Fig. 2.2-12). These systems include advanced features such as digital subtraction, which allows for clear imaging of vascular structures that are superimposed over a bony or dense soft tissue structure. Owing to its configuration, the C-arm results in increased scatter radiation to the operator and OR staff, who should always wear protective shielding and monitoring devices.

The newest fluoroscopic systems incorporate 3D imaging capabilities achieved by combining a C-arm that spins around the patient, and tomographic reconstruction that results in a volumetric data set—referred to as **cone-beam CT**. Cone-beam CT is also known as cone-beam computed tomography or CBCT, C-arm CT, cone-beam volume CT, or flat-panel CT.

Indications for radiography

- Acute bone trauma—fracture rule-out
- Thoracic evaluation for pneumonia, metastases, and diaphragmatic hernia
- Suspicion of stone formation—e.g., uroliths in small animals, enteroliths in large animals, and sialoliths
- Identification of metallic foreign bodies

Contradictions for radiography

- Unstable patient—e.g., a small animal that cannot be sedated to facilitate images without motion artifact

Strengths of radiography

- See Table 2.2-1

Weaknesses of radiography

- See Table 2.2-2

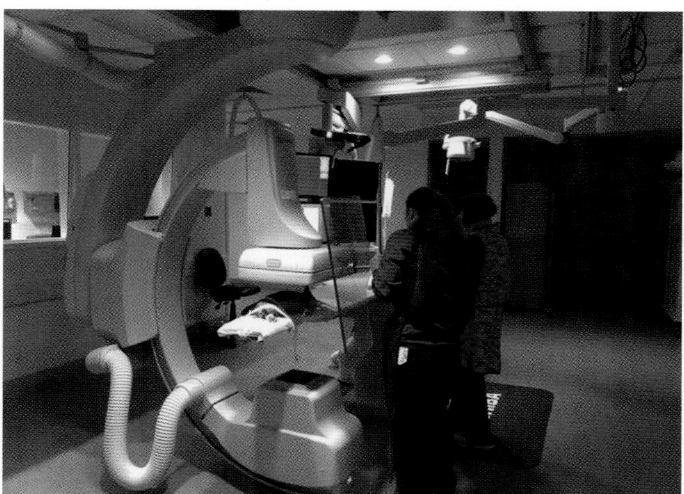

FIGURE 2.2-12 C-arm fluoroscopy in an interventional radiology suite used in the depicted case for assessment of tracheal collapse in a chihuahua.

Selected references

[1] Buttler JA, Colles CM, Dyson SJ, Kold SE, Poulos PW. Clinical radiology of the horse. 4th ed. Wiley-Blackwell; 2017.
[2] Thrall DE. Textbook of veterinary diagnostic radiology. 7th ed. Elsevier; 2017.
[3] Kirberger R, McEvoy F. BSAVA manual of canine and feline musculoskeletal imaging. 2nd ed. British Small Animal Veterinary Association; 2016.
[4] Thrall DE, Robertson ID. Atlas of normal radiographic anatomy and anatomic variants in the dog and cat. 2nd ed. Elsevier; 2015.
[5] Holloway A, McConnell F. BSAVA manual of canine and feline radiography and radiology: a foundation manual. 1st ed. British Small Animal Veterinary Association; 2014.
[6] Coulson A, Lewis N. An Atlas of interpretative radiographic anatomy of the dog and cat. 2nd ed. Wiley-Blackwell; 2011.
[7] Schwarz T, Johnson V. BSAVA manual of canine and feline thoracic imaging. British Small Animal Veterinary Association; 2008.

CHAPTER 2

DIAGNOSTIC IMAGING 2.3

Ultrasonography (US)

Aitor Gallastegui
Small Animal Clinical Sciences, University of Florida, Gainesville, Florida, US

Introduction

Ultrasonography (US) was first developed from Italian physiologist Lazzaro Spallanzani's 1794 work on echolocation in bats. Later, neurologist Karl Dussik was credited with being the first to use ultrasonic waves in 1942 as a diagnostic tool, which he used for the detection of brain tumors through the human skull. The first documented use of ultrasonography in veterinary medicine was in 1966, subsequent to its rise in the human medical field. For many decades, US has been—and remains—the second-most widely used imaging modality, following radiography.

Understanding the imaging technology/mode of action

Ultrasonography uses vibrations or high-frequency sound waves transmitted through tissues. The human ear can perceive sounds of up to 20 kHz (20–20,000 sound wave cycles per second); sound above this level is called "ultrasound." Interactions of US waves with tissues create "echoes" that return to the US probe or transducer, generating a grayscale image of the internal body anatomy. Ultrasonography machines come in multiple sizes and shapes ranging from large carts with multiple probes to small portable units with a single probe. Many portable units allow the use of handheld devices such as tablets or smartphones to view the images; some display the image on a headset, which is wireless in the newer versions, permitting a full freehand use (Fig. 2.3-1).

Ultrasound waves are created within the US probe. A majority of the currently available US probes contain piezoelectric crystal elements such as PZT. Piezoelectric crystals are fascinating because they can deform, and these deformations convert mechanical or kinetic energy into electric charge energy, and vice versa. Applying bursts of electric charge on the piezoelectric elements causes them to deform and create US waves. Conversely, the kinetic energy from the returning echoes causes the piezoelectric elements to release electric charge bursts that are used by the US machine to generate a diagnostic image. The piezoelectric elements are arranged along the surface of the US probe and covered by materials that protect and facilitate the transmission of US through the tissue. Because air impedes the propagation of US from the probe to the body and vice versa, a coupling gel or a wetting substance such as alcohol must be applied to the body surface.

Ultrasound waves are characterized by frequency (f), wavelength (λ), and velocity (v), as depicted by the following formula: $f = v/\lambda$. For a particular US probe, the frequency of the US that it generates is determined by the size of the piezoelectric elements it contains. Newer technologies allow US probes to work on a range of frequencies—e.g., 5–8 MHz for ultrasonography of the abdomen in feline patients—allowing a single probe to be used for multiple types of imaging studies. The latest technologies use ultrasound-on-chip technology instead of piezoelectric elements, which allows a single probe to produce US of any frequency, making these probes suitable for all diagnostic purposes (Fig. 2.3-1B).

Ultrasound probes are classified as high-frequency or low-frequency probes, according to the US wave frequency generated. **High-frequency probes** have short wavelengths that permit the waves to interact with smaller structures and, therefore, provide better spatial resolution. Because this results in increased interactions among the tissues being scanned, the US wave is reflected, refracted, scattered, or absorbed in more superficial tissues, and consequently, there

Comparative Veterinary Anatomy: A Clinical Approach. https://doi.org/10.1016/B978-0-323-91015-6.00004-2

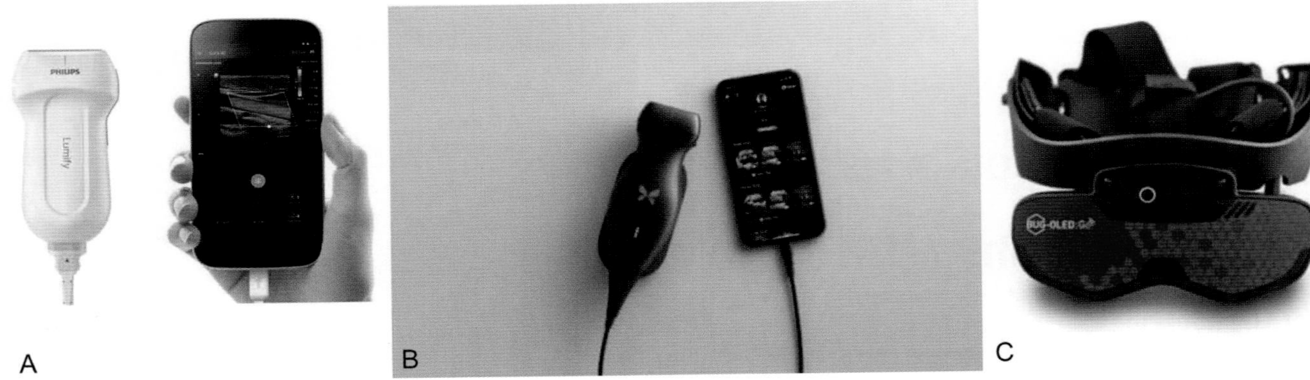

FIGURE 2.3-1 Philips Ultrasound, Bothell, WA USA (A) and Butterfly Network (B) are handheld US systems with US probes that connect directly to phones and tablets, increasing portability. IMV imaging (C) displays US images on goggles, allowing increased mobility of the ultrasonographer due to the use of Wi-Fi technology, which connects the probe and the goggles; this feature is very convenient, especially during large animal reproductive examinations.

is lower tissue penetration. **Low-frequency probes** have longer wavelengths that result in fewer interactions with small structures and inferior spatial resolution. However, because there are fewer tissue interactions, the energy of the US wave is maintained for a longer period and results in deeper tissue penetration. In summary, high-frequency probes provide better image quality of superficial structures but cannot penetrate to deeper tissues, whereas low-frequency probes can penetrate to deeper tissues but with poorer image quality (Table 2.3-1). Table 2.3-2 depicts the recommended uses of different frequency probes depending on the desired penetration and spatial resolution requirements.

Ultrasound probes or transducers are available in different sizes, shapes, and configurations. **Linear-array probes** (Fig. 2.3-2A) have a flat surface, and the US waves are emitted in a perpendicular direction creating a squared field of view that has the width of the probe, e.g., trans-rectal US probes. **Curved-array probes** (Fig. 2.3-2B) have a curved surface, and the US waves are emitted in a perpendicular direction creating a diverging cone-shaped or sectorial

TABLE 2.3-1 Effect of ultrasound wave frequency on spatial resolution and penetration.

	Wavelength	*Spatial resolution*	*Penetration*
High-frequency probe	Short	Best	Superficial
Low-frequency probe	Long	Reduced	Deep

TABLE 2.3-2 Relationship between ultrasound probe frequency, spatial resolution, and penetration.

Frequency (MHz)	*Spatial resolution*	*Penetration*	*Uses*
10–15	Very high	Very low	Superficial structures: thyroids, parathyroids, eyes, tendons, masses, etc.
7.5–10	High	Low	Superficial structures Small dogs, cats Equine reproductive exams
5	Medium	Medium	Medium-sized dogs Equine abdomen
3	Low	High	Large-breed dogs Large animal cardiology

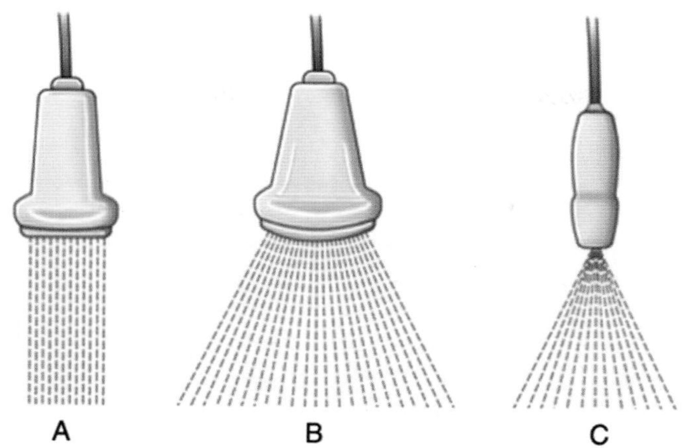

FIGURE 2.3-2 (A) Linear-array probes. (B) Curved-array probes. (C) Phased-array probes.

field of view that is larger than the size of the probe's contact surface—e.g., abdominal probes in horses. **Phased-array probes** (Fig. 2.3-2C) allow the piezoelectric elements to be excited precisely, such that the user can shape and control the US beam in a particular way—e.g., echocardiography probes.

When an US wave with a certain wave frequency travels through a specific tissue, it has a constant velocity and, therefore, a constant wavelength. When this US wave moves into a different type of tissue along its path, it changes its velocity and, therefore, its wavelength according to the earlier mentioned formula: $f = v/\lambda$. The velocity of sound in a tissue is a factor of the density and the acoustic impedance of such a tissue (Table 2.3-3). **Acoustic impedance** defines the acoustic characteristics of tissue, dictating its level of reflection and thus its **echogenicity**. The magnitude of reflection of the US wave is dictated by the difference in the acoustic impedance between the tissues. The greater the difference in acoustic impedance between the adjacent tissues, the stronger the reflection of the US wave. Most of the soft tissues within the body have similar acoustic impedance, allowing the US waves to travel through them. These slight changes in the acoustic impedance and US wave velocity are desirable because at each interphase between tissues of similar acoustic impedance, a small reflection is produced—e.g., the liver parenchyma to the liver lobule and the liver lobule to the central vein. The summation of these small reflections results in variable echogenicity and echotexture of the organs being examined. When US waves encounter tissue interphase between tissues with a large difference in acoustic impedance, such as a bone or gas, the transmission of most—if not all—of the US waves is reflected back to the US probe. Consequently, an image deep to that interphase is not produced.

TABLE 2.3-3 Velocity of sound and acoustic impedance in materials and biological tissues.

Tissue/medium	*Velocity (m/s)*	*Acoustic impedance (Rayls, kg/(m²s))*
Air	**331**	**0.0004 × 106**
Fat	1450	1.34×10^6
Water	1480	1.48×10^6
Soft tissue average	**1540**	**1.62 × 106**
Kidney	1560	1.63×10^6
Blood	1570	1.65×10^6
Muscle	1580	1.71×10^6
Bone	**4080**	**7.8 × 106**

Source: Bushberg et al. (2011) and Middleton et al. (2004).

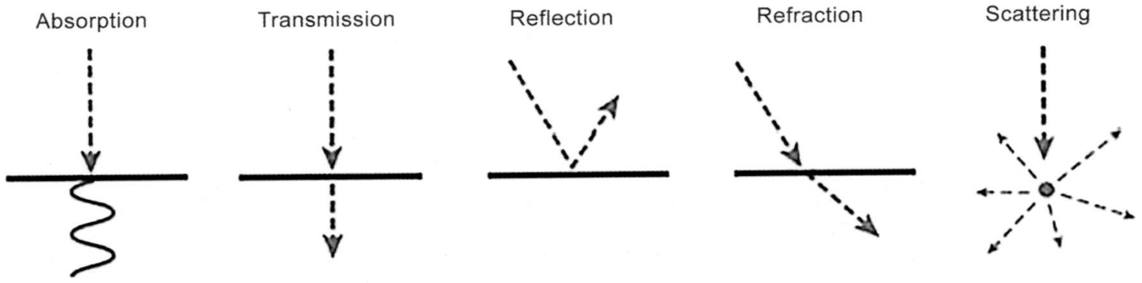

Absorption Transmission Reflection Refraction Scattering

FIGURE 2.3-3 The most common interactions of US waves at different tissue interphases.

Because of the difference in acoustic impedance of tissues, the US wave undergoes an interaction at the level of each tissue interphase. Depending upon this interaction, the US wave is transmitted through, reflected back to the probe, refracted, absorbed, or scattered (Fig. 2.3-3). These interactions determine (1) what echoes return to the probe, (2) the strength or signal intensity of the returning echoes, (3) the time for an echo to return to the probe from the time the US wave was produced, and (4) the direction of return of the echo with respect to the direction of the US wave from which it originated. The echoes received at the probe create an US image in a grayscale color scheme. Within the US image, each bright point represents an anatomical landmark, and (1) its location is based on the direction of the initial US beam and the returning echo, (2) its depth in the field of view is determined by the return time of the echoes from the time the generating US wave was produced, and (3) its intensity is determined by the amplitude or strength and the amount of the returning echoes. All these characteristics confer an echogenicity and echotexture for each of the different tissues.

In medical imaging, several US modes are used:

- **A-mode (Amplitude mode)**: The simplest mode uses a single transducer or crystal and displays the focal image as a function of depth; it is used in therapeutic ultrasonography such as tumor or calculus destruction
- **B-mode (Brightness mode) or 2D mode**: It uses an array of transducers or crystals, scanning a plane of the body that is displayed in a 2D image
- **M-mode (Motion mode)**: Ultrasonography waves are emitted in quick succession creating a video image file; this is the most frequently used US mode in clinical settings

Multiple settings can be adjusted in the US machine to better optimize the image for a particular purpose. The most frequently used settings are the gain, time gain control (TGC), depth, and focus. The **gain** is used to increase or decrease the overall amplitude (or "brightness") of the image. The **TGC** is a zonal control of gain based on time or depth and is used to adjust the intensity of the echoes originating from deeper tissue regions. The **depth** controls the tissue depth of the US image displayed on the monitor. The **focus** controls the location of the focal zone within the image and determines the area of best lateral spatial resolution within the US beam; typically, the focus is placed at the depth of the tissue of interest (Fig. 2.3-4).

When tissue interphase causes a strong reflection of the US wave, a bright image or "signal" is created, and the tissue is referred to as **hyperechoic**—e.g., gas and bone. When tissue interphase causes a mild reflection of the US wave, a dark gray image or "signal" is created, and the tissue is referred to as **hypoechoic**—e.g., renal medulla. When a tissue or material has a homogeneous architecture resulting in a lack of interphases and constant acoustic impedance, a strong black image or "signal" is created, and the tissue is referred to as **anechoic**—e.g., urine and bile. The echogenicity of an organ or region can also be referred to as hyper- or hypo-echoic when compared to other organs or the known ideal echogenicity of that organ or tissue. Either increased or decreased echogenicity is possible depending on the disease process (Fig. 2.3-5). As a rule, decreased echogenicity reflects increased water content and/or reduced cellularity within the tissues, and increased echogenicity reflects decreased water content and/or an increase in cellularity, fibrosis, etc.

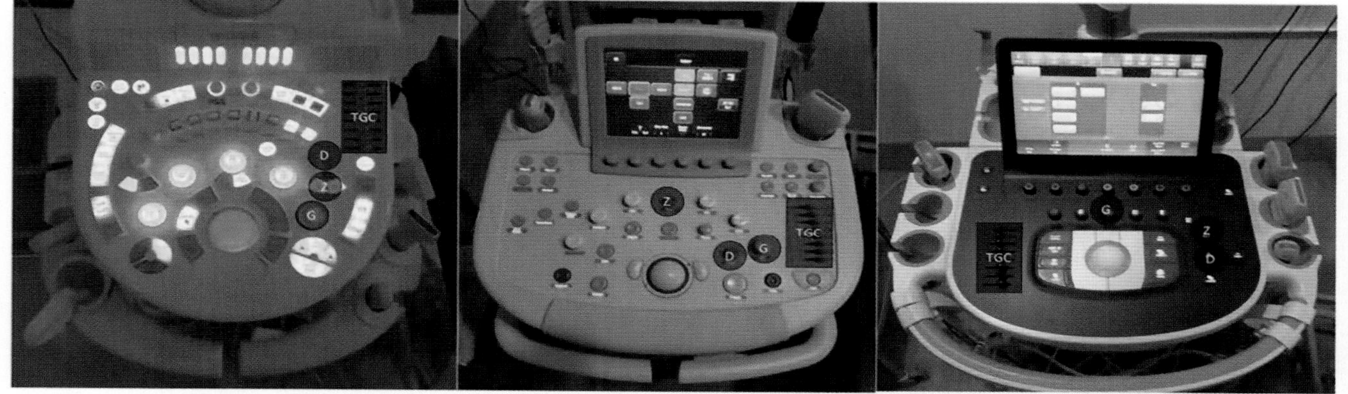

FIGURE 2.3-4 Ultrasonography control panels have multiple buttons and knobs; the newest units incorporate tactile screens. This permits easier management of the various settings that control the US image; some of which are user-dependent. Regardless of the manufacturer, all US systems include controls for basic features—i.e., general gain (G), time-dependent gain control (TGC), depth (D), and zoom (Z). The location of these features varies with the manufacturer and model as depicted in the images, from left to right: Hitachi Prierus, Philips Epiq 5, and Philips iU22.

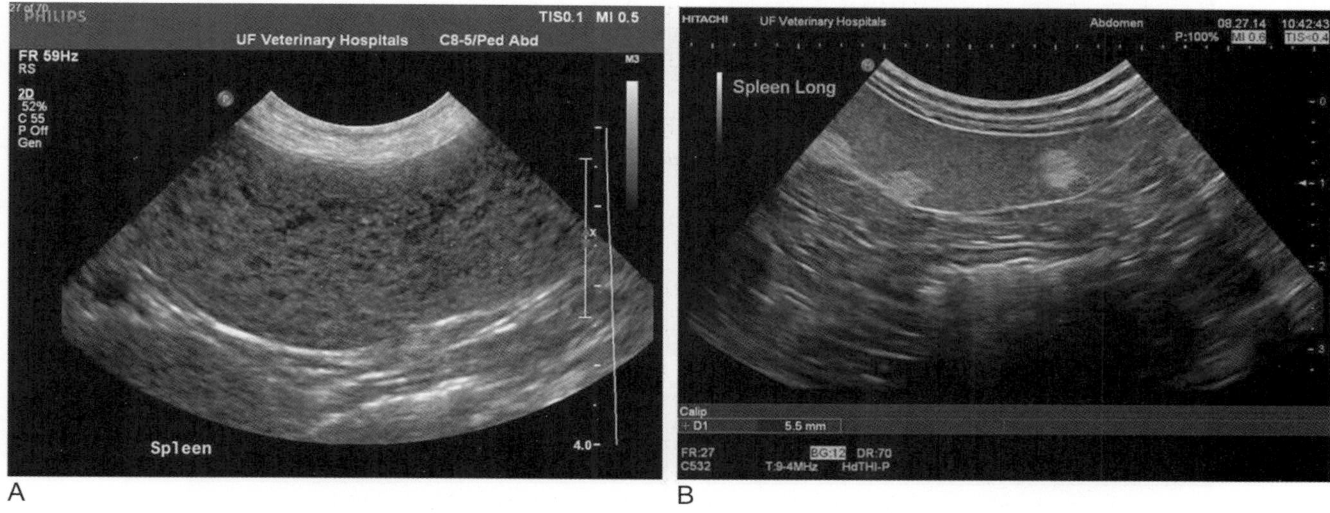

FIGURE 2.3-5 The splenic parenchyma is homogenous, has a fine echotexture, and is covered by a thin, hyperechoic capsule. In a dog with lymphoma, multiple tiny to small hypoechoic nodules give the organ a leopard-spotted or mottled appearance (A). In a dog with benign myelolipomas, small- and medium-sized hyperechoic, well-defined round and irregular nodules can be seen in the spleen (B).

Additional US imaging techniques

As previously discussed, when an US wave is generated, it has a frequency that remains constant. The returning echo has the same frequency as the US wave from which it originated. When US is used to examine moving objects, the frequency of the returning echo is different. This is attributed to the **Doppler effect** and is primarily used to detect and characterize tissue motion and blood flow. Like hearing a pitch change from a moving ambulance, the frequency of the returning echo varies depending on the direction of the movement or blood flow. Doppler information can be presented in audible, color Doppler, and spectral Doppler forms. Audible Doppler is primarily used in echocardiography; color Doppler is primarily used in the determination of presence and direction of blood flow; and spectral Doppler is primarily used in characterizing blood flow velocity, resistive index (RI), etc.

Color Doppler ultrasonography is the most frequently used Doppler tool, except for echocardiography that uses both color and spectral Doppler. On color Doppler, a red and blue color scale is used to indicate the direction of the blood flow. However, it is important to note that these *colors do not necessarily reflect venous or arterial flow*. Changes in the orientation of the probe change the interpretation of the flow direction and, therefore, assign a different color to the vessels (Fig. 2.3-6). In addition, most US machines permit the individualization of the colors or color patterns used. Doppler is a useful tool in the detection of normal anatomic landmarks, masses, thrombi, infarcts, shunts, ischemic lesions, and other vascular abnormalities (Fig. 2.3-7).

The contrast of US images can be improved by using **contrast-enhanced ultrasonography (CEUS)**, which uses microbubble contrast agents as contrast enhancers. When the US wave reaches the microbubbles, they resonate with a specific frequency depending on the diameter of the microbubbles. The resonating frequencies are known

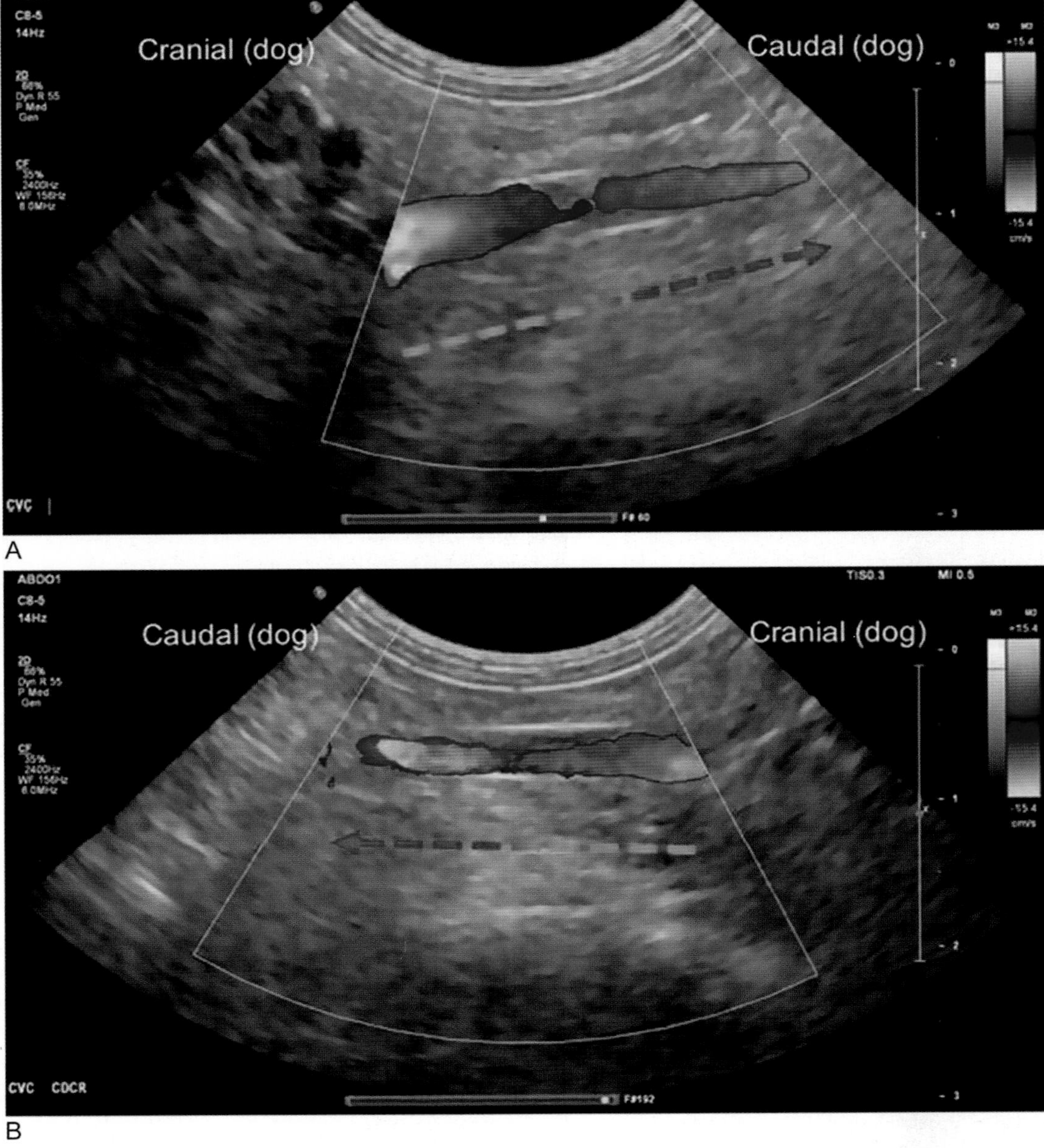

FIGURE 2.3-6 Color Doppler detects flow toward the probe (red) and away from the probe (blue). A change in the orientation of the probe over the abdominal caudal vena cava (CVC) from cranial-to-caudal (A) to caudal-to-cranial (B) changes the color pattern.

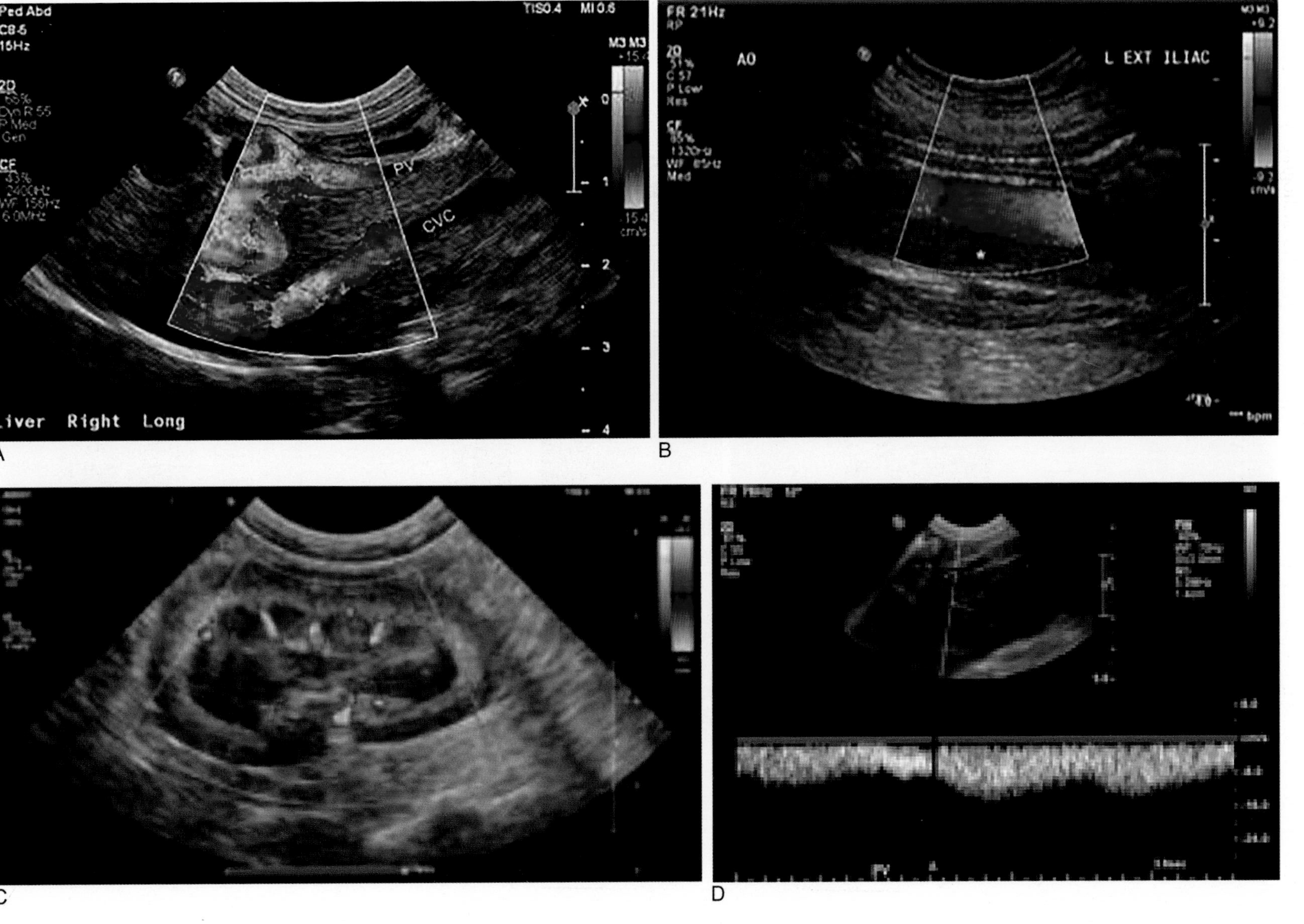

FIGURE 2.3-7 Use of color Doppler. (A) Right divisional intrahepatic shunt; notice the large tortuous vessel in the liver connecting the portal vein (PV) to the caudal vena cava (CVC). (B) Caudal aortic thrombus (*) in a longitudinal plane, causing nearly 50% occlusion of the lumen. (C) Normal interlobar vessels in a canine kidney. (D) Spectral Doppler of the portal vein is often used to detect and characterize congenital extrahepatic portosystemic shunts; in this case, portal flow exhibits normal hepatoportal flow but is wavering because of the effect of respiratory motion.

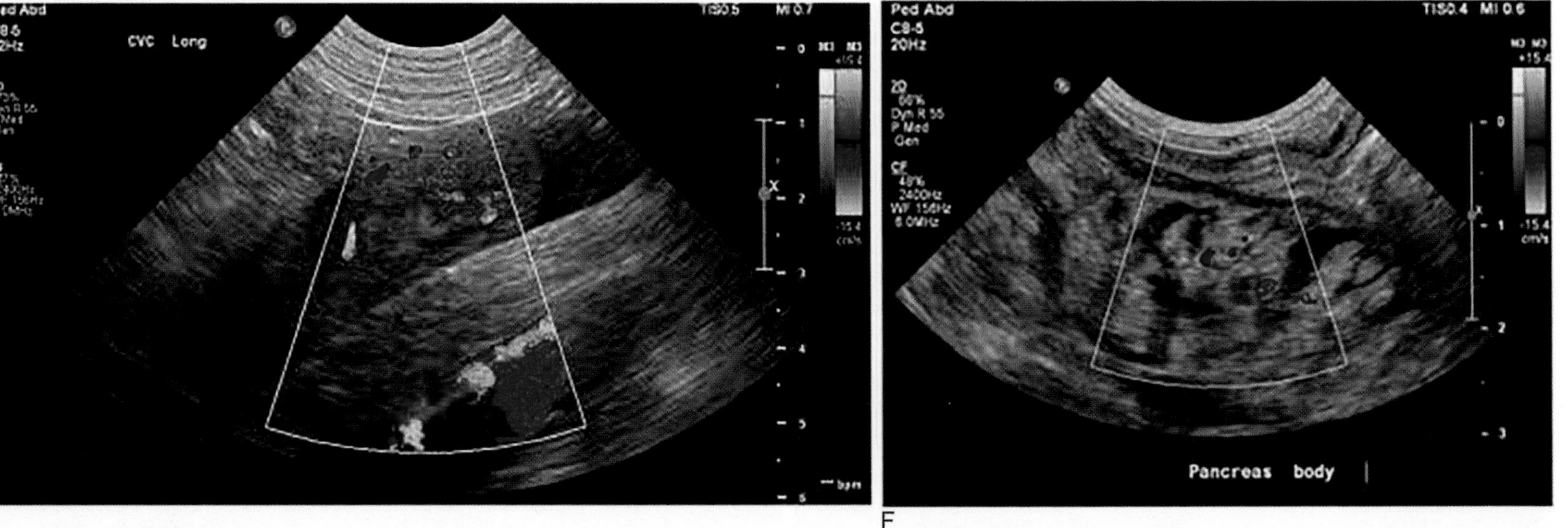

FIGURE 2.3-7, CONT'D (E) Within the CVC there is a medium echogenic structure exhibiting multifocal regions of blood flow, consistent with vascular infiltration of an adjacent adrenal malignant neoplastic mass. (F) Confirmation of blood flow in an enlarged and edematous pancreas supports the clinical diagnosis of nonnecrotizing acute pancreatitis in a dog.

as harmonic frequencies. Out of the harmonic frequencies, the second harmonic is strong enough to be used for diagnostic purposes. Using an adequate US machine and appropriate settings, users can target the signal from the echoes in the second harmonic frequency alone, subtracting the information from the surrounding tissues. This substantially improves the contrast resolution of the US image. Unfortunately, in the USA, just a few US contrast agents are approved by the FDA for clinical use in humans, and none are currently licensed for clinical use in veterinary medicine. In Europe, US contrast agents are frequently used in the clinical setting. **Harmonic ultrasonography** substantially reduces noise and artifacts secondary to reverberation and aberration and improves image quality even without the use of an US contrast medium.

A recent focus of research in the veterinary US is elastography. **Elastography** is a noninvasive technique used to assess the elasticity or stiffness of a tissue, in conjunction with traditional B-mode ultrasonography (Fig. 2.3-8). To date, most available publications in veterinary medicine have established normal values for different body regions, with a focus on equine tendons and multiple organs in small animals. Some of these publications have identified associations with pathologic processes. Unfortunately, elastography can be user-dependent, and further research is needed before it becomes a tool that is used routinely in clinical settings.

Artifacts and maximizing image acquisition

The same tissue interactions that allow the creation of different echotextures and differentiation of dissimilar structures within the body can create artifacts (Fig. 2.3-9). **Distal acoustic shadowing** is a hypoechoic or anechoic area distal to an object, caused by tissues or objects with high attenuation or reflection of the incident US beam—often seen with bones, mineral stones, and many gastrointestinal foreign bodies. **Distal acoustic enhancement** is a hyperechoic area distal to a normal anatomical structure or lesion, which occurs when the US beam passes through an area of lower attenuation than the adjacent tissues, with a resulting greater amplitude or energy when it reaches deep tissues—often seen with fluid-filled structures such as cysts, gallbladders, and urinary bladders. **Edge shadowing** is the presence of narrow shadows distal to the lateral edges of a curved surface, which occurs because part of the US beam is reflected and another part is refracted, and the refracted portion of the signal does not return to the US probe—often seen at the edges of fluid-filled structures, kidneys, and margins of tendons. **Reverberation**

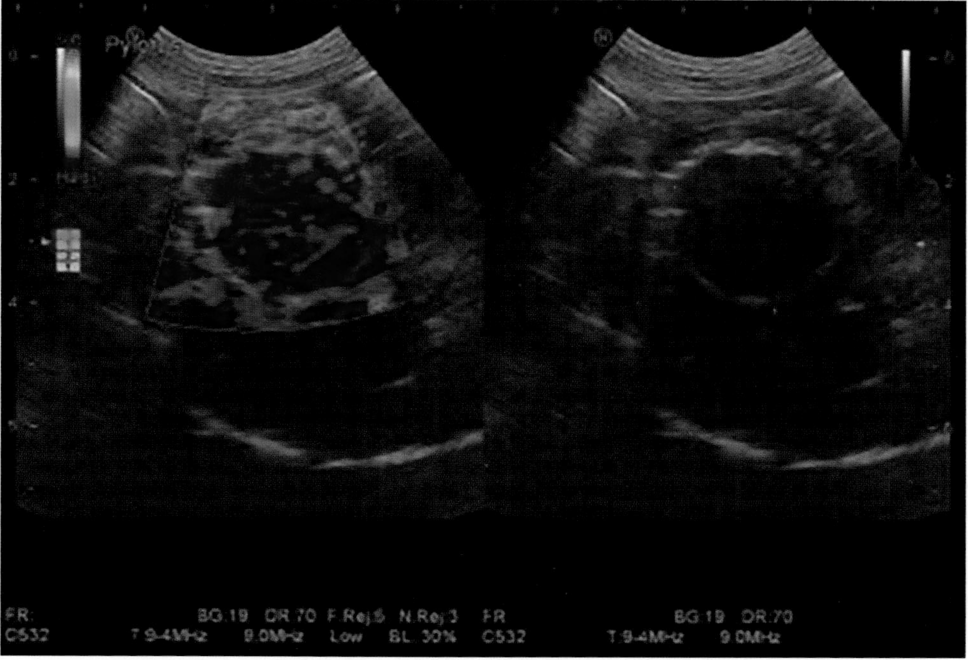

FIGURE 2.3-8 Elastography used in an 8-year-old mixed-breed dog with a history of vomiting, characterizing a round hypoechoic structure located in the pyloric antrum of the stomach; the structure was primarily blue (firm) and diagnosed as a nonobstructing foreign body.

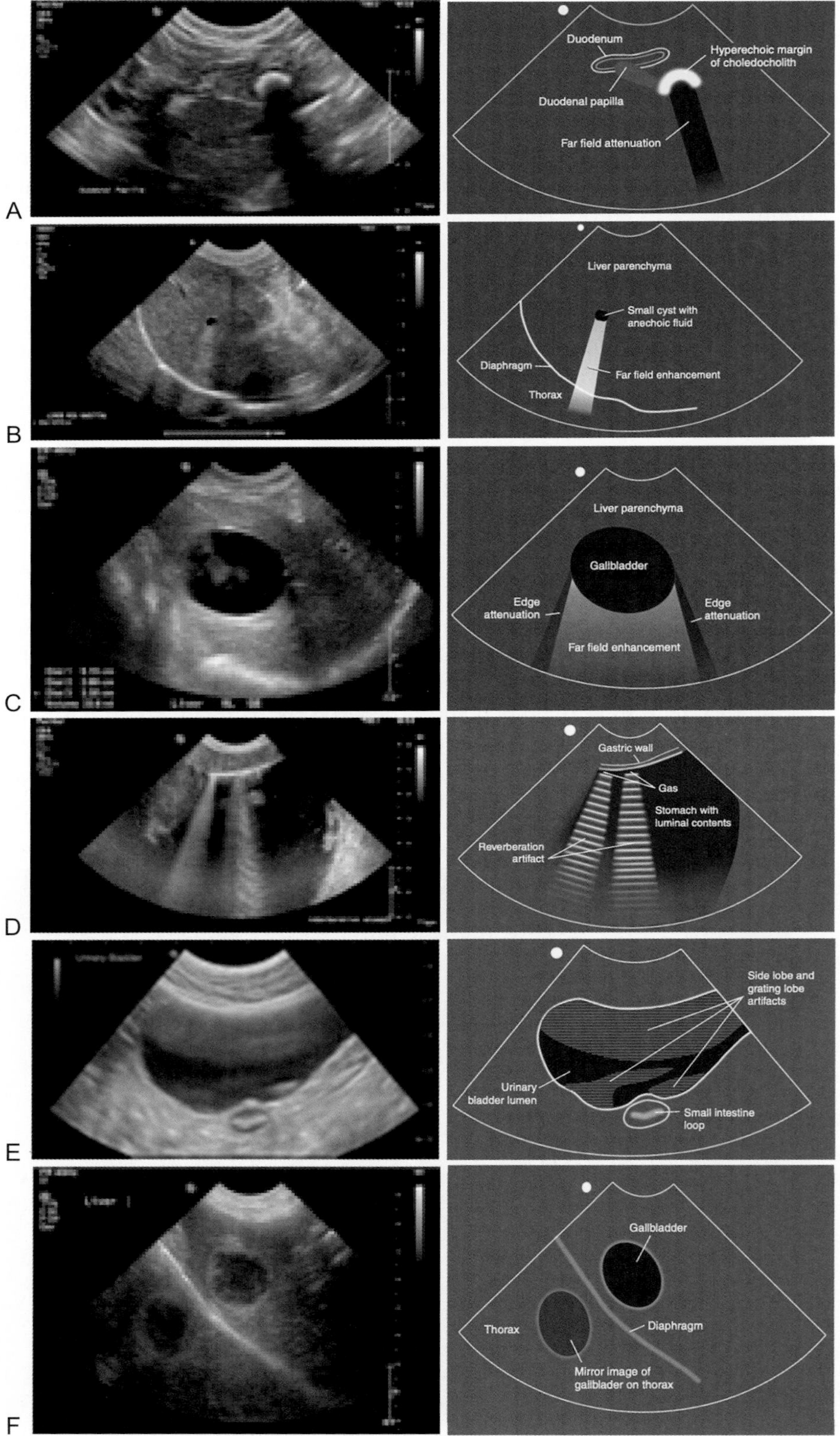

FIGURE 2.3-9 Common artifacts with ultrasound. (A) Distal acoustic shadowing from a choledocholith in the common bile duct by the duodenal papilla of the duodenum. (B) Distal acoustic enhancement in a small hepatic cyst. (C) Edge shadowing from the lateral margins of the gallbladder in a transverse plane. (D) Reverberation artifact from gas in the gastric lumen. (E) Side lobe and grating lobe artifacts causing a mirror image. (F) Gallbladder and diaphragm resulting in a mirror image artifact in thorax.

is the presence of distally attenuating equidistant echogenicities, which occurs when multiple echoes are generated between two closely spaced tissue interphases that reflect US energy back and forth—often seen with gas. **Side lobe** and **grating lobe artifacts** are due to the emission of US beams in directions different from the main US beam of the probe that generate echoes and are mapped in the path of the main US beam, creating the appearance of a noise signal—frequently giving the appearance of sludge in the urinary bladder. **Mirror image** is the creation of an anatomic ultrasonographic image mirroring a real ultrasonographic image and occurs due to multiple reflections and refractions that eventually find their way to the transducer, creating the illusion of misplaced anatomy—often seen near highly reflective surfaces such as the diaphragm.

It is important to always perform the ultrasonographic examination in a systematic way to prevent missing an organ. In addition, the probe should always be oriented consistently to facilitate pattern recognition over time. To assist with recognizing the orientation of the probe and the image, all US probes have a palpable knob on one side; this side is marked on the US image with a small dot displayed at the top of the image, either on the right or on the left side. Although a standard for orientation has not been defined in veterinary medicine, all organs and body regions of interest should be imaged in long- and short-axes, and the orientation of the images displayed ideally in cranial to caudal, right to left, proximal to distal, and medial to lateral planes. Currently, the Ultrasound Society of the American College of Veterinary Radiology (ACVR) is working on a consensus statement on the standardization of the abdominal US exam. It is expected that additional consensus statements will follow on the standardization of additional ultrasonographic examinations in veterinary medicine.

Indications for US

- Rapid assessment of abdominal and thoracic cavities for fluid accumulation focused assessment with sonography for trauma (FAST)
- A quick evaluation of the abdominal cavity in cases with gastrointestinal signs (e.g., horses with acute abdomen/colic)—fast localized abdominal sonography (FLASH)
- Evaluation of cardiac function or anatomy—echocardiography
- Evaluation of the surface of the lungs and pleural space (e.g., in horses with pleuropneumonia)
- In-depth evaluation of the abdominal cavity and the ability to guide aspirates/biopsies
- A detailed evaluation of soft tissues of the musculoskeletal system (e.g., especially in horses)
- Assistance in centesis/injection of synovial structures and the cerebrospinal canal (e.g., atlantoaxial aspiration)
- Trace a wound tract
- Evaluating dermal or subcutaneous masses/swellings

Contraindications of US

- Very few clinical contraindications because the technique is noninvasive
- Might induce biological effects by heat production secondary to absorption of the US beam energy and nonthermal or mechanical mechanisms secondary to radiation force, streaming, and cavitation

Strengths of US

- A minimally invasive method of obtaining diagnostic information
- Performed in the nonsedated or minimally sedated patient
- Equipment is portable and affordable
- Biopsy and aspirate guidance for lesions viewed at the time of examination
- Technology applicable in all species
- The risk from an US is minimal; therefore, follow the ALARA (as low as reasonably achievable) principle of prudent scanning implemented to minimize the risk while obtaining the necessary information for a diagnosis

Weaknesses of US

- Distal acoustic shadowing prevents imaging structures deep to the bone/air/gas
- Training needed to operate and maintain equipment for proper function, implementation, and diagnosis
- Comprehensive knowledge of applied anatomy needed to interpret US findings and to be able to differentiate normal from abnormal lesions
- Requires a reduced light environment/area to perform the examination—e.g., difficult for the ambulatory clinician

Selected references

[1] Reef VB. Equine diagnostic ultrasound. 2nd ed. Elsevier; 2019.
[2] Penninck D, d'Anjou MA. Atlas of small animal ultrasonography. 2nd ed. Wiley-Blackwell, John Wiley & Sons, Inc.; 2015.
[3] Kidd JA, Lu KG, Frazer ML. Atlas of equine ultrasonography. Wiley-Blackwell, John Wiley & Sons, Inc.; 2014.
[4] Mattoon JS, Nyland TG. Small animal diagnostic ultrasound. 3rd ed. Elsevier; 2014.
[5] Barr FJ, Gaschen L. BSAVA manual of canine and feline ultrasonography. British Small Animal Veterinary Association; 2012.

DIAGNOSTIC IMAGING 2.4

Computed Tomography (CT)

Aitor Gallastegui
Small Animal Clinical Sciences, University of Florida, Gainesville, Florida, US

Introduction

The first computed tomography (CT) scanner was developed by Sir Godfrey Hounsfield in 1973, and the first uses of CT in veterinary medicine date back to the 1980s. The technology has evolved dramatically and the currently available multidetector scanners (also known as MDCT or multi-slice scanners) are very fast and provide excellent submillimeter anatomical spatial resolution. With an increase in the rotating speed of the gantry and advanced software technologies, most body regions are scanned within seconds. Because of these improved technologies, CT has established itself as an indispensable diagnostic tool in human and veterinary medicine with applications in oncology, orthopedic and soft tissue surgery; emergency and internal medicine, and virtually every medical specialty.

Computed tomography is a cross-sectional diagnostic imaging modality that uses x-ray energy to create diagnostic-quality images. The main difference from "regular" x-ray imaging is that CT allows visualization of all the body structures without superimposition—making interpretations of complex anatomical areas such as the skull and the pelvis much easier to evaluate. It is the current gold standard for imaging bones and lungs. This modality has excellent spatial resolution and fair to good contrast resolution for assessment of the soft tissues.

There are multiple types of CT scanners available, with the main differences among them being the configuration and number of detectors in the gantry, and the rotation modes of the elements in the gantry. Earlier CT scanners contained a single detector element, and the gantries could not complete a full rotation, requiring all the moving parts to return to starting position before acquiring each new cross-sectional image slice of the patient. This caused slow acquisition times and images with limited spatial resolution. The use of contactless or cordless connections now allows contiguous 360° rotations of the gantry and increased image acquisition speed compared with earlier CT scanners.

Today, 3rd or 4th generation scanners are the most commonly used CT scanners. The 3rd generation CT scanners are composed of a vertically positioned round gantry, with an x-ray tube on one side of the gantry focusing onto an assembly of detectors located on the opposite side of the gantry. During image acquisition, the x-ray tube and the detector array rotate at the same time (Fig. 2.4-1A). The 4th generation CT scanners have fixed detectors along the entire surface of the gantry, which permit the x-ray tube to rotate along the gantry during image acquisition (Fig. 2.4-1B). Besides the gantry, CT scanners have a table that positions the patient at the isocenter and moves along the long axis in a cranial to caudal or caudal to cranial direction.

In addition to these configurations, gantries have one or multiple rows of detectors in single-slice scanners and multidetector computed tomography (MDCT), respectively (Fig. 2.4-2). Depending on the CT system, the minimal collimated slice width is 0.5, 0.6, or 0.625 mm (0.019, 0.023, or 0.024 in.), which represents the size of the detector unit and the limit of the spatial resolution for the system. Currently, 4-, 8-, 12-, 16-, and 64-row scanners are commercially available, with a maximum collimated detector width of 38.4 mm (1.5 in.) (64 detectors × 0.6 mm (0.023 in.) detector width). The more rows of detectors in the scanner, the larger the body area that can be imaged in the long axis per full rotation of the gantry elements. Therefore, MDCT scanners are much faster than earlier single-slice scanners.

Comparative Veterinary Anatomy: A Clinical Approach. https://doi.org/10.1016/B978-0-323-91015-6.00005-4

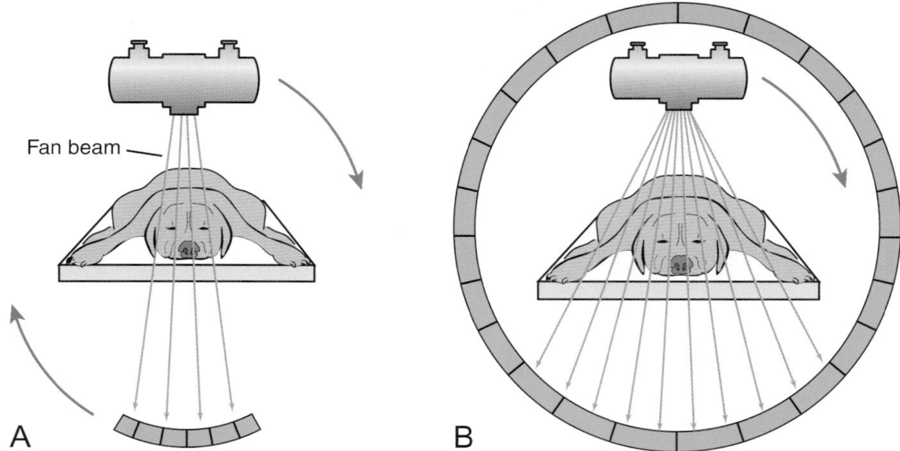

FIGURE 2.4-1 Computed tomography scanner generations. In the 3rd generation CT scanner (A), a limited number of detectors are placed in a position of the gantry opposite to the x-ray tube; in this system, both the x-ray tube and the detector array rotate together. In 4th generation scanners (B), the whole perimeter of the gantry is lined by detectors; during image acquisition, only the tube rotates.

As the CT gantry rotates around the patient during image acquisition, the CT table moves in the long axis. This is known as "helical" or "spiral" scanning and is the predominant acquisition mode. In helical scanning mode, the speed of the table movement in the axial plane can be varied. Because of the cone shape of the x-ray beam, changes in the speed of the CT table movement cause varying overlaps of the radiographic image profile at the detector. The amount of overlap of the radiographic image profile on the detector is controlled by the pitch (Fig. 2.4-3). By modifying the pitch, we can under-sample or over-sample a body region. Under-sampling (high pitch) results in faster speed with poor image resolution and over-sampling (low pitch) results in a slower speed and better image resolution. Under-sampling is desired in CT scans of the thorax to reduce the effect of breathing motion artifact and over-sampling is preferred in CT scans of the head to improve spatial resolution.

During image acquisition, as the gantry (in 3rd generation CT scanners) or the x-ray tube (in 4th generation scanners) rotates around the patient, radiographic images of the body region of interest are generated in 360° projections that overlap variably depending on the pitch selected. Specific built-in software combines the information from all the x-ray projections that have been acquired into a volume data set from which 2D and 3D images of diagnostic value are generated. The 2D images are known as multiplanar reconstructions (MPR). By standard, the scanner displays MPRs in a transverse plane, but the user can create images in additional planes including, but not limited to, dorsal and sagittal MPRs. Generation of MPRs in any plane is possible because the CT data are acquired as a volume data set, as opposed to radiographs that are acquired in a 2D format. The 3D images generated from CT data are known as surface rendering (SR) and volume rendering (VR) and are used to create more realistic final images of bones, blood vessels, and virtual endoscopic recreations among others (Fig. 2.4-4).

Like MRI, a limitation of CT comes from the size of the bore of the gantry, and the weight the CT table can support. Since most of the CT scanners used in veterinary medicine are developed for human use, the diameter of regular bores is in the range of 70 cm (27.6 in.), and the weight limit of the CT tables is in the range of 400–425 lbs (181–193 kg). Thus, CT scanners are readily available for the companion animal patient (Fig. 2.4-5A). However, problems arise when larger animals—e.g., horses, pigs, wildlife—need to be scanned. Specific tables built into the units have allowed the use of standard CT scanners in large animal species, but the size of the bores still essentially restrict the use of CT to scanning heads and distal extremities in horses, pigs, and wildlife (Fig. 2.4-5B).

Another limitation of using standard CT scanners in large animals is that these patients typically need to be placed under general anesthesia for good imaging quality, along with increasing the safety of the patient, staff, and equipment. Because of the increased use of CT imaging in large animal medicine, several companies are developing technologies that overcome some of the previously mentioned problems. These advancements include the use of regular CT scanners in standing sedated horses for imaging the head (Fig. 2.4-6A), horizontal gantries for imaging the distal extremities of standing sedated patients (Fig. 2.4-6A), wide bore gantries for anesthetized or standing

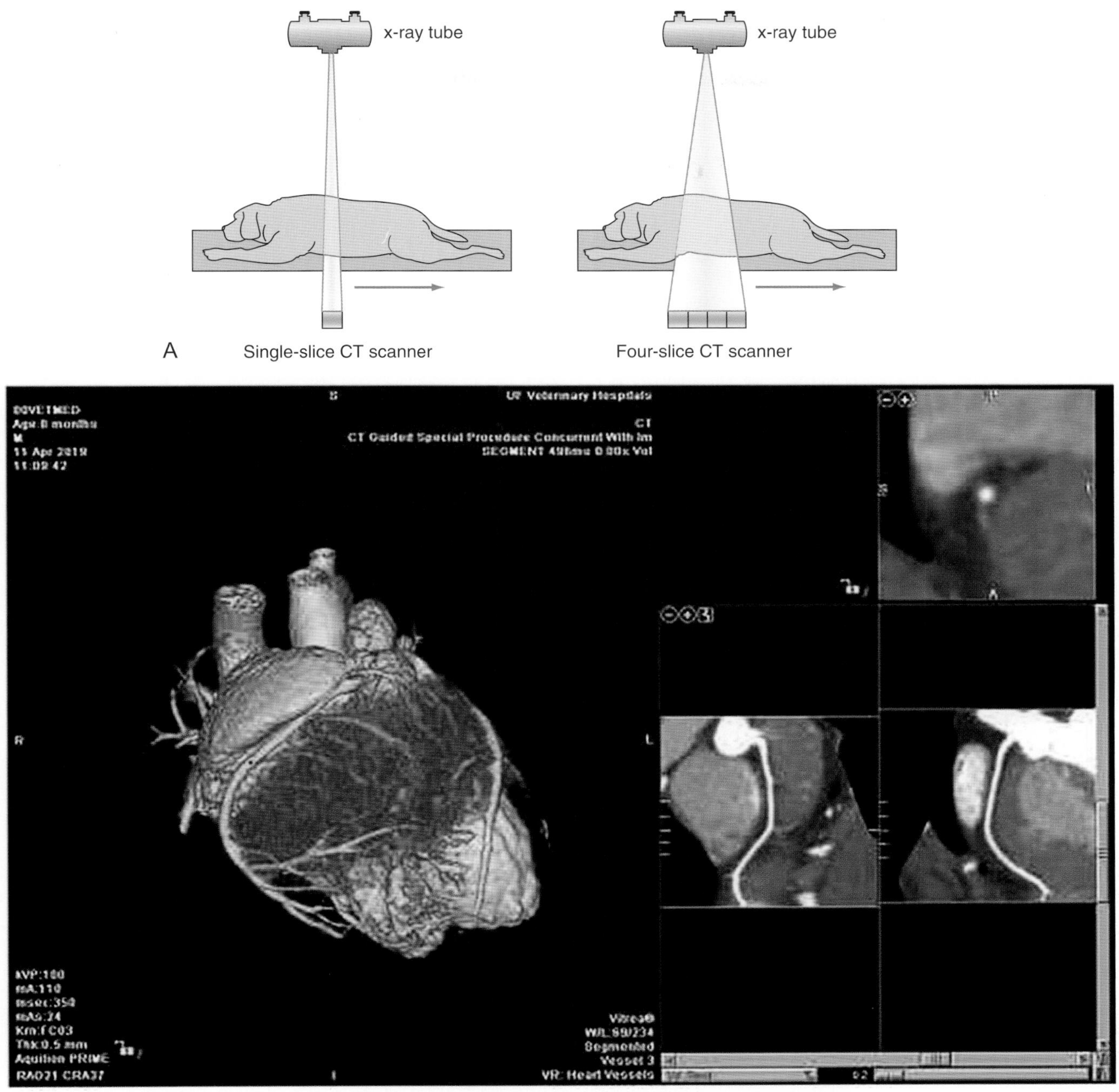

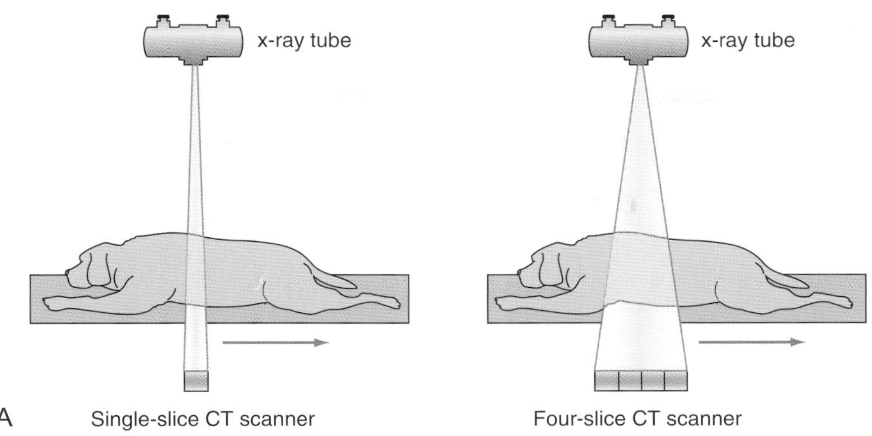

B

FIGURE 2.4-2 Single-slice scanners and multidetector computed tomography (MDCT) multi-slice scanners. (A) Multidetector computed tomography scanners allow for more than one slice to be acquired in the *z*-axis per gantry rotation, resulting in faster image acquisition and increased capability for performing dynamic studies. (B) Multiplanar and 3D reconstructions of a cardiac CT angiography study in an 8-month-old Bulldog puppy with pulmonic stenosis, acquired with a 160-slice CT scanner.

sedated patients (Fig. 2.4-5), and flat panel-based robotic CT scanners with the goal to produce diagnostic CT with the radiographic image quality of horses in motion (Fig. 2.4-6B).

Digital radiography and CT scanners use x-ray energy for image production. However, in CT, the term "x-ray attenuation" is used instead of the term "opacity," and is measured in Hounsfield Units (HU). As compared to the five opacities detected radiographically, CT provides a much larger range of different attenuations; this

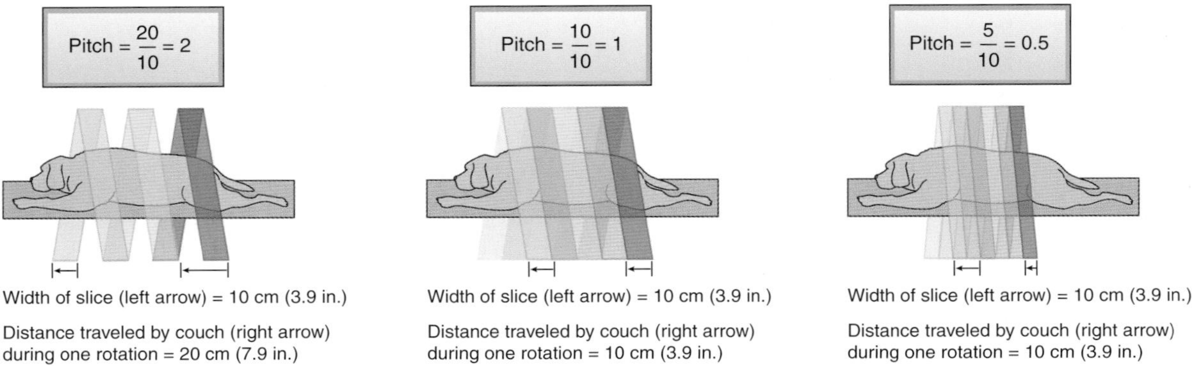

FIGURE 2.4-3 Pitch in CT scanners. The pitch indicates the movement of the table in the *z*-axis per full gantry rotation (360°). Increasing the pitch results in under-sampling of the patient and decreasing the pitch results in over-sampling the patient.

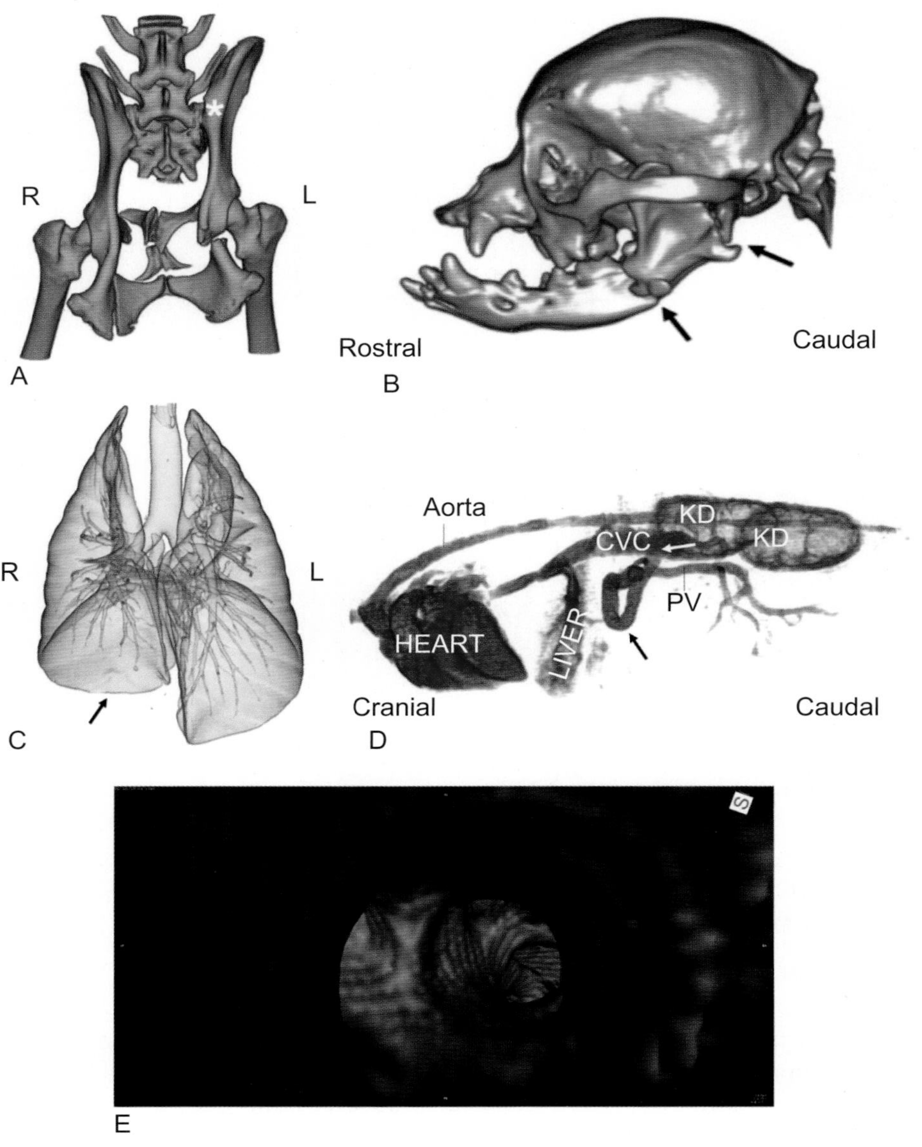

FIGURE 2.4-4 Examples of 3D images generated from CT data. (A) Multiple fractures involving the left ischium and the pelvic floor and a left sacroiliac luxation (*). (B) Segmental fracture of the right mandible (*arrows*). (C) Lower airways with atelectasis of the right caudal lung lobe (*arrow*). (D) Targeted CT angiogram demonstrating a portocaval shunt (*black arrow*) communicating the portal vein with the caudal vena cava (*white arrow*). (E) Virtual bronchoscopy allows noninvasive examination of the tracheal wall, carina, and branching of the principal bronchi.

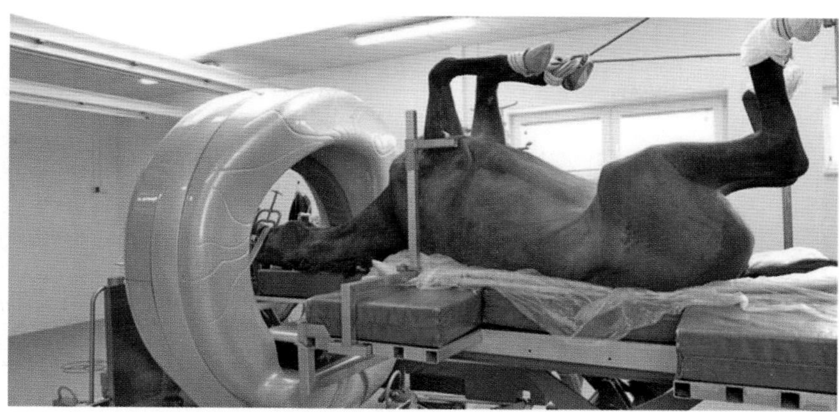

FIGURE 2.4-5 Most large and small animal patients undergoing CT are under general anesthesia. Patients are positioned on the table with the help of support devices. Standard CT scanners can be used in large animal patients with the use of purposely built CT tables. The large animal table is controlled or rolled over the CT table. Due to bore size limitations, only head and distal extremities can be imaged with these systems.

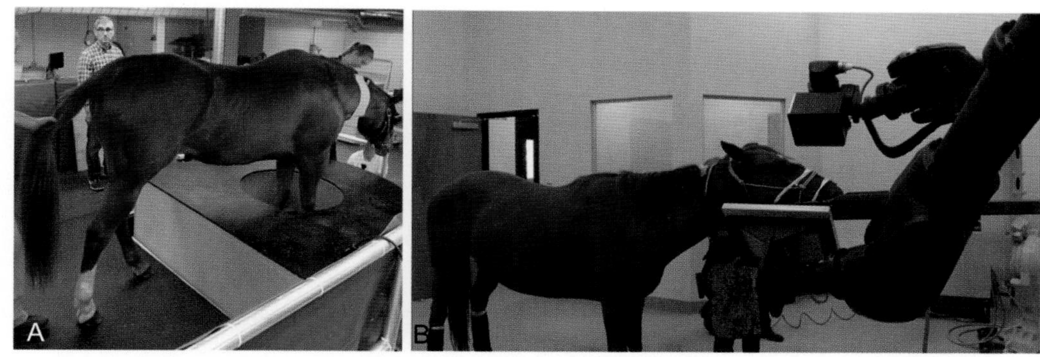

FIGURE 2.4-6 Computed tomography scanners for the imaging of large animal patients. (A) Standing CT at the University of Auburn, courtesy of Dr. Ana Velloso Alvarez. (B) Standing robotic CT.

range extends from − 1000 HU for gas to + 3000 HU for dense bones (Table 2.4-1). When a body region causes a strong attenuation of the x-ray energy, a bright image is created and is referred to as hyperattenuating—e.g., bones. When a body region causes low attenuation of the x-ray energy, a dark image is created and is referred to as hypoattenuating—e.g., gas. When tissues cause similar x-ray attenuation to that of other or surrounding tissues, they are referred to as isoattenuating—e.g., the liver and pancreas.

Even if CT can distinguish between some 4000 different shades of gray, standard computer monitors with 8-bit pixel depths can display only 256 different shades of gray, and the human eye is only capable of recognizing about 30 different shades of gray. To accommodate the large amount of information provided by CT scans to our visual capabilities, images can be postprocessed at the image viewing workstation. Among other postprocessing tools, the CT images can be leveled and windowed at a particular level of the x-ray attenuation or HU scale. By windowing and leveling the images, the operator can generate images that are optimized for a better assessment of bones ("bone algorithm"), lungs and gas ("lung algorithm"), soft tissues ("soft tissue algorithm"), and neural tissues ("brain algorithm") (Fig. 2.4-7).

The contrast resolution of CT images can be further improved using positive contrast media, with iodinated contrast solutions being the most widely used clinically. Of all the available iodinated solutions, the nonionic iodinated contrast media solutions such as iohexol are preferred for their increased safety and lower cost. These positive contrast media are generally administered IV and provide a temporary increase in contrast resolution of all the structures in the body, facilitating the identification and further characterization of masses (Fig. 2.4-8A and B), inflammation, infection, metastasis, etc. Contrast can also be tracked as it travels along a vascular structure of interest

TABLE 2.4-1 Hounsfield unit chart.

Tissue or substance	*HU*
Air (gas)	−1000
Lung	−500
Fat	−100 to −50
Water	0
Blood	+30 to +45
Muscle	+10 to +40
White matter	+20 to +30
Gray matter	+37 to +45
Liver	+40 to +60
Soft tissues with contrast	+100 to +300
Cancellous bone	+700
Compact bone	+3000

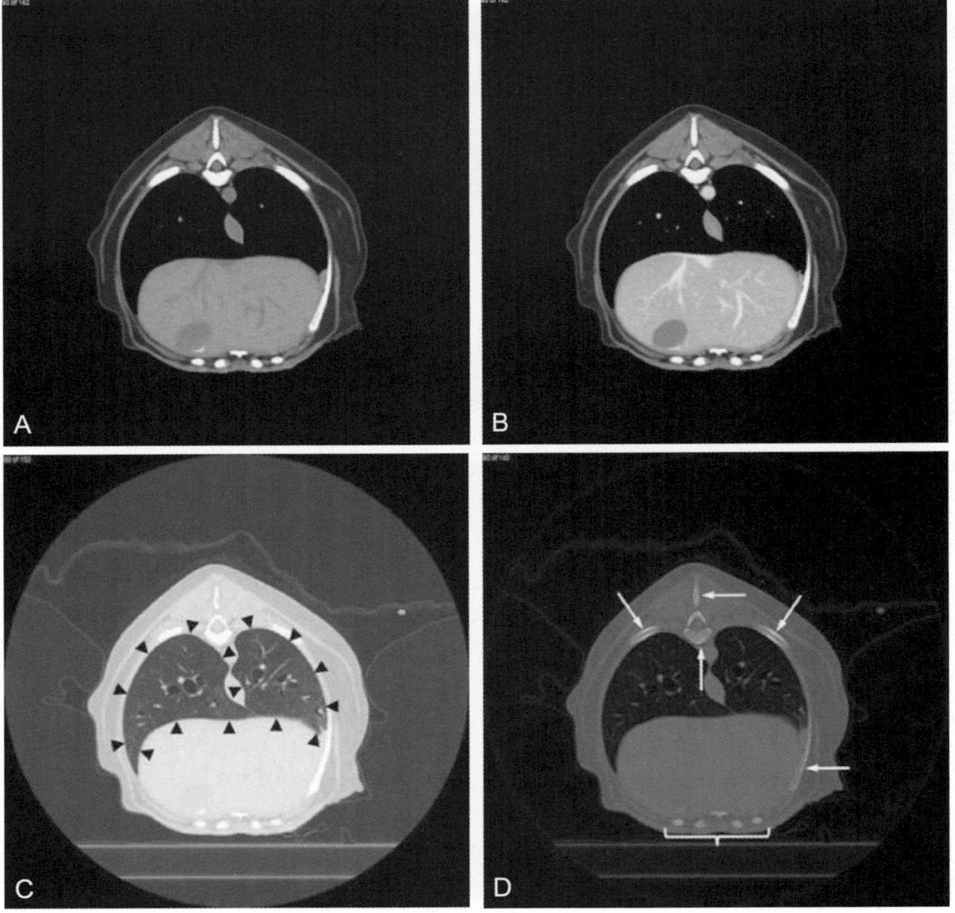

FIGURE 2.4-7 Windowing and leveling of images allow optimization of CT data to improve different tissue viewing. (A) Soft tissue algorithm. (B) Soft tissue algorithm enhanced by intravenous contrast. (C) Lung algorithm (*arrowheads* mark the lung margins). (D) Bone algorithm (*arrows* and *bracket* mark the bony structures).

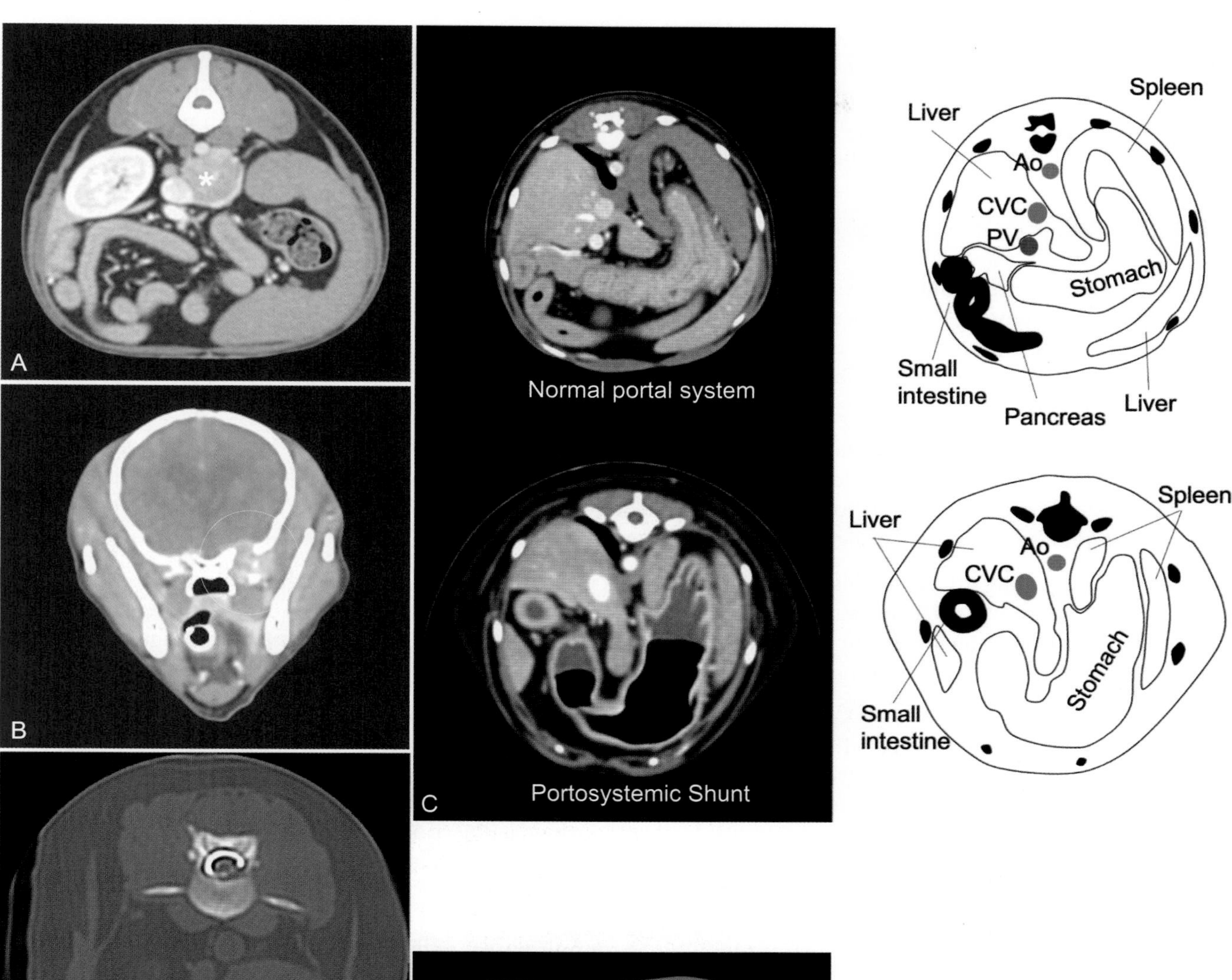

FIGURE 2.4-8 The use of IV contrast media improves the contrast of soft tissue structures, highlighting a left adrenal mass (*) causing ventral displacement of the phrenicoabdominal vein and caudal vena cava (CVC), without evidence of intraluminal invasion (A) and a left retrobulbar mass with associated calvarial lysis and extension into the intracranial cavity (B). The use of targeted CT angiography can be used to detect congenital extrahepatic portosystemic shunts (C), often characterized by enlarged CVC and an absent or small portal vein (PV). Positive contrast CT myelography can be used to identify the site of intervertebral disk herniation and the distribution of herniated disk material for surgical planning, characterized by focal filling defects of the normal contrast column (D). Administration of positive contrast medium into a popliteal lymph node highlights its tract along the lymphatic system in popliteal lymphangiography; in this image, a maximum intensity projection (MIP) reconstruction highlights the abdominal lymphatic system, cisterna chyli, and thoracic duct in a dog (E).

(termed "targeted CT angiography") to detect and characterize shunts (Fig. 2.4-8C), arteriovenous fistulae, ischemic lesions, internal bleeds, and thrombi, primarily. In addition, contrast can be injected into the subarachnoid space for a CT myelogram (Fig. 2.4-8D); into body cavities—e.g., CT retrograde urethrocystography, colon enema; lymph nodes—e.g., popliteal lymphangiogram in patients with chylothorax (Fig. 2.4-8E), sentinel lymph nodes; and wounds—e.g., CT fistulogram, CT sinogram.

Risks and limitations of CT

Iodinated, nonionic contrast media solutions are safe with the most frequent adverse events being transient tachycardia or bradycardia. Acute renal failure and cardiac arrest are extremely rare adverse events associated with CT contrast studies. The risk of these complications can be minimized by using nonionic low-osmolarity contrast solutions, maintaining good patient hydration before and after contrast administration, and slowly administering the contrast medium.

During image acquisition, the gantry elements rotate, and the patient moves along the long axis, with image quality affected by motion artifact, primarily due to breathing or movement of the patient on the table. Patient movement can be minimized using general anesthesia; however, diagnostic-quality images can also be obtained in sedated and nonsedated patients. Complex generating image algorithms are one means to adjust for motion artifact in 3D image reconstructions.

Computed tomography images are also affected by the presence of metallic implants, high-density foreign objects, and superimposition of thick bony structures, creating artifacts that cause variable degrees of shadowing and can affect and compromise the diagnostic value of the study (Fig. 2.4-9). Planning ahead with the radiologist and radiology technician before beginning the CT study results in the best diagnostic images.

Most importantly, CT uses ionizing X-ray energy and increases the potential for health hazards if improperly used for the patients and staff. Acute health effects such as skin burns and acute radiation syndrome can occur when doses of radiation exceed recommended levels. Lower doses of ionizing radiation can increase the risk of long-term effects such as cancer and heritable mutations. Because of the increase in radiation dose associated with the use of CT scanners, shielding is an important factor to consider, both during facility construction and when determining the personnel in the room while CT images are being acquired, as depicted by scatter maps (Fig. 2.4-10).

Indications for CT

- Primary or metastatic neoplasia (superior to radiology for pulmonary nodules)
- Staging of cancer patients
- Sinonasal disease
- Evaluation of the middle and inner ear
- Acute intracranial hemorrhage
- Gross identification of intracranial masses
- Multifocal trauma injuries—e.g., vehicular trauma, dog fight
- Mediastinal, thoracic, and abdominal masses
- Assessment of renal function and ureters
- Congenital malformations of the musculoskeletal and vascular systems
- Intervertebral disk extrusion in young chondrodystrophic breeds
- Vascular incidents—e.g., ischemia, thrombosis
- Morphological assessment of the coronary arteries

Contradictions for CT

- Severe azotemia or renal insufficiency/failure

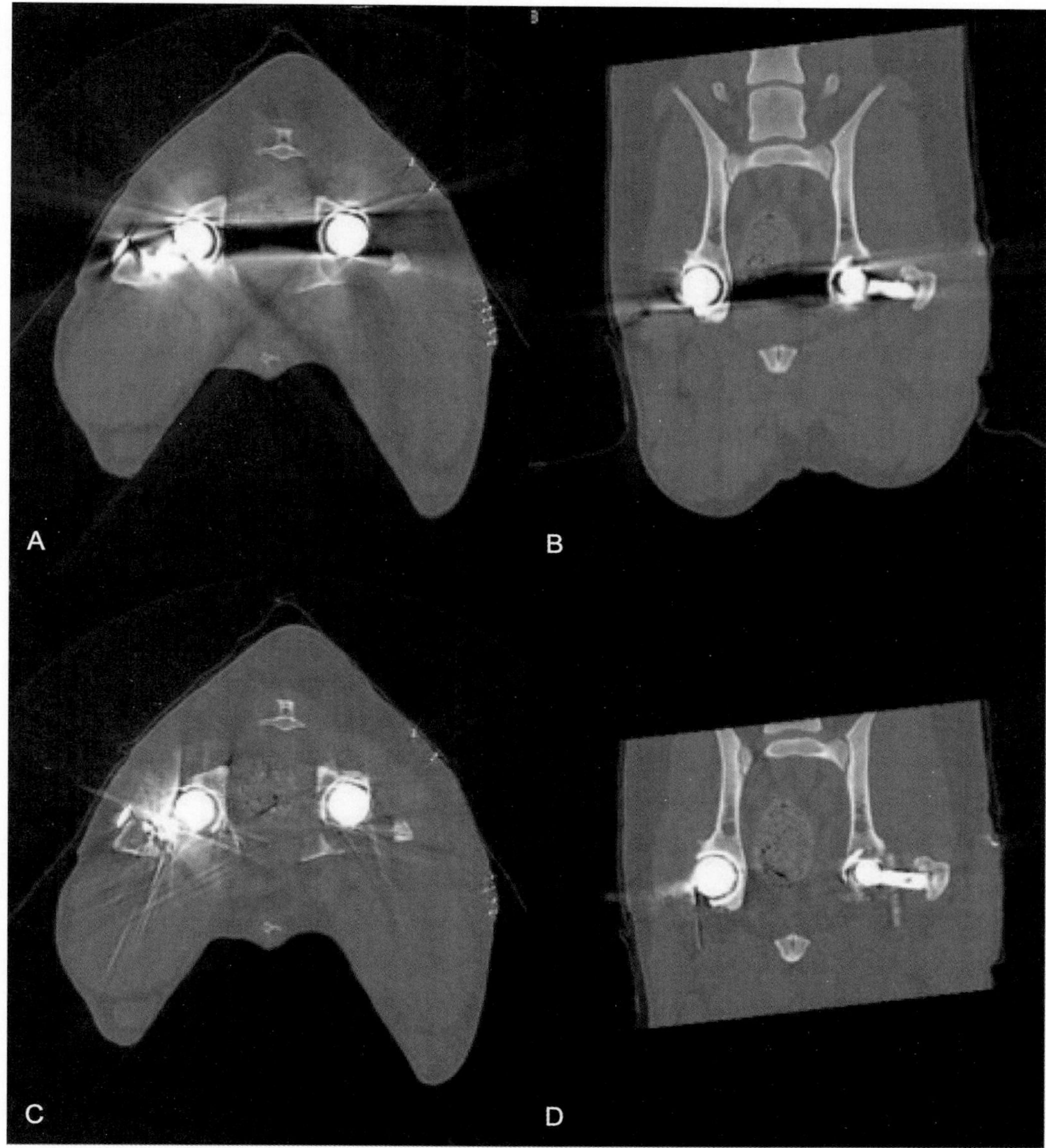

FIGURE 2.4-9 Artifacts with CT. Metallic implants create artifacts that create hyperattenuating and hypoattenuating bands. In this case, bilateral total hip replacement implants cause severe striking artifacts and loss of signal within the pelvic canal (A and B). The use of postprocessing technologies can reduce the effect of some artifacts, and newer vendor-dependent techniques such as the single energy metal artifact reduction (SEMAR) algorithm by Canon Medical (C and D) can substantially improve image quality.

- Known allergy to IV contrast media
- Unstable patients
- Pregnant patients—particularly in early gestational stages

Strengths of CT

- Fast acquisition times allow studies in awake, restrained, and sedated patients
- Superior to MRI in detecting acute hemorrhage
- Superior to MRI and radiography in detecting bone lesions
- Superior to radiographs in detecting pulmonary metastasis

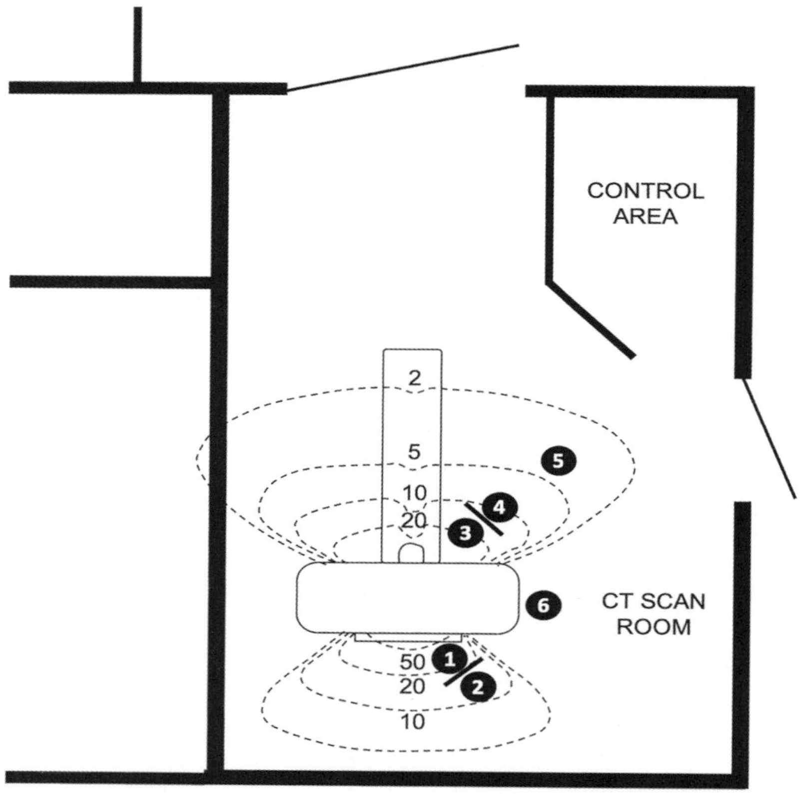

FIGURE 2.4-10 Computed tomography scatter maps display the extent and severity of radiation exposure (measured in mR) the staff is exposed to within the surroundings of the CT scanner and help design the CT room and appropriate shielding. (1) Behind the gantry with no shield (9.6 mR). (2) Behind the gantry with a shield (0.1 mR). (3) Front of the gantry with no shield (18.8 mR). (4) Front of the gantry with a shield (0.2 mR). (5) Front of the gantry with a shield and 3 feet back (0.1 mR). (6) Side of the gantry with no shield (0.1 mR).

- Superior to ultrasound in detecting abdominal lesions in dogs larger than 25 kg (55 lbs)
- Raw data are acquired as a volume data set
- Allows image postprocessing (any plane and optimization for various tissues)

Weaknesses of CT

- Uses large doses of ionizing radiation energy (x-rays)
- Subject to motion—though less than MRI—and metal artifacts (among others)
- Inferior soft-tissue resolution compared to MRI
- Inferior to MRI in assessing the nervous system
- Inferior to MRI in assessing inflammatory processes

Selected references

[1] Bertolini G. Body MDCT in small animals. Springer International Publishing AG; 2017.
[2] Kimberlin L, zur Linden A, Ruoff L. Atlas of clinical imaging and anatomy of the equine head. Wiley-Blackwell, John Wiley & Sons, Inc.; 2017.
[3] Wisner E, Zwingenberger A. Atlas of small animal CT and MRI. Wiley-Blackwell, John Wiley & Sons, Inc.; 2015.
[4] Schwarz T, Saunders J. Veterinary computed tomography. Wiley-Blackwell, John Wiley & Sons, Inc.; 2011.

DIAGNOSTIC IMAGING 2.5

Magnetic Resonance Imaging (MRI)

Aitor Gallastegui
Small Animal Clinical Sciences, University of Florida, Gainesville, Florida, US

Introduction

The use of magnetic resonance imaging (MRI) in veterinary medicine started almost 30 years ago and it is currently the gold standard for imaging the CNS in dogs and cats, and for evaluating musculoskeletal disease in horses.

Varying MRI magnets create magnetic fields of different strengths and are measured in Tesla (T) units. Currently, low-gradient field strength magnets (≤ 0.3T) and high-gradient field strength magnets or superconducting magnets (1.5–3T) are used in veterinary medicine (Fig. 2.5-1). Higher-strength magnets are available but are currently only used in human medicine or for research purposes. Low-gradient field strength magnets are less costly and require minimal installation; however, the image quality is poorer with limited capability to run various MRI sequences. High-gradient field strength or superconducting magnets are more costly to purchase and maintain and importantly require a cooling system to operate. All magnets require magnetic and radiofrequency shielding adequate for the strength of the magnet, both for safety and image quality purposes.

Magnetic resonance imaging units are available in various configurations allowing imaging of different body regions and types and sizes of patients. Close bore MRI units create stronger and more stable magnetic fields, and this results in better image quality; however, the size of the patient or the body region being imaged is limited by the size of the bore. Open bore MRI units facilitate imaging of larger body regions and allow improved patient access for monitoring during the procedure. Most MRI units use a horizontal configuration requiring the patient to lay on a table whereas others have a vertical configuration allowing acquisition of images in the standing patient. The latter system is used in the equine patient for imaging soft tissues of the distal extremity. Horizontal configurations are primarily used in imaging small animals and—with a purposely built table—the extremity and head of large animal species (Fig. 2.5-2).

Understanding the imaging technology/mode of action

Magnetic resonance imaging uses magnetic field energy to align protons in a certain direction, combined with radiofrequency pulses to alter the position of these protons. When the frequency pulses are interrupted and the protons return to their original relaxed state position, electromagnetic energy is released. This electromagnetic energy is then read by the MRI receptor and transformed into a diagnostic image. When the energy released is *high energy*, a bright image is created and referred to as a **hyperintense** image or "signal." When the energy released is *low energy*, a dark image is created and referred to as a **hypointense** image or "signal." When the energy signal from a particular region is canceled or nulled, this is an image or signal **void**.

Within a body, hydrogen is the main proton that takes part in image formation. Depending on the tissue in which the hydrogen protons are located—e.g., the muscle, fat, and CSF—they are affected by the magnetic field and the radiofrequency pulses differently, and thus release different energies (Tables 2.5-1 and 2.5-2). The difference in energy released by specific tissues allows differentiation of each tissue in the image, and in a similar way, a disease process alters the normal signal intensity of the tissues. Most disease processes cause hyperintense lesions because of the increase in water content associated with inflammation, but not all (Figs. 2.5-3B, 2.5-4A, 2.5-5, 2.5-6B, 2.5-7, and 2.5-8).

Comparative Veterinary Anatomy: A Clinical Approach. https://doi.org/10.1016/B978-0-323-91015-6.00006-6

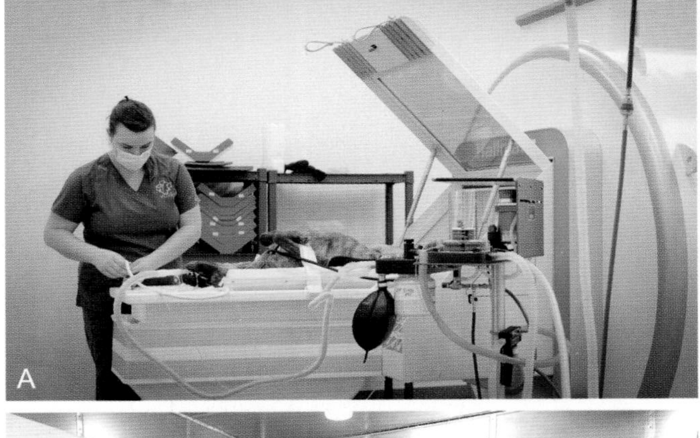

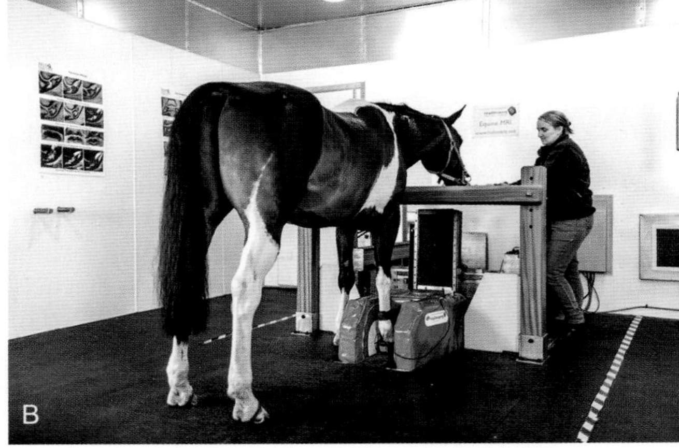

FIGURE 2.5-1 Examples of MRI systems used for companion animals (A) and horses (B). (Courtesy: ©Hallmarq Veterinary Imaging 2021.)

Advanced software allows the creation of complex magnetic field sequences highlighting and invaliding different tissue characteristics—primarily differentiating fat, water, blood products, and calcium. Thus, this imaging modality permits achieving functional, vascular, and metabolic analyses of MRI changes. These techniques are used to further characterize ischemic lesions, blood vessels, nerve tracts, and infectious and neoplastic processes, as examples.

There are two basic characteristics of MRI image acquisition that are important to understand—T1 time and T2 time. The physics involved with these two concepts is beyond the scope of this book, but simplistically, in a T2 weighted (T2w) image, water is hyperintense and fat is hypointense; whereas in a T1 weighted (T1w) image, fat is hyperintense and water is hypointense. The newer "spin-echo" sequencing techniques used in an MRI for the acquisition of T2w images increase the speed of the image acquisition and the quality of the image generated, at the expense of making water and fat hyperintense. To differentiate water from fat in these T2w spin-echo sequences, additional sequences have been developed that cause cancellation of the fat signal and highlight the water signal—i.e., fat saturation and STIR (short T1 inversion recovery) sequences, respectively (Figs. 2.5-4C and 2.5-5B). Another variation of the T2w sequence is the FLAIR (T2-weighted fluid-attenuated inversion recovery) sequence, which cancels the water signal. The FLAIR sequence is used to cancel the signal from CSF, allowing better visualization of lesions in the brain parenchyma (Fig. 2.5-6B) and canceling the signal from the synovial fluid. This allows better imaging of ligaments, joint capsules, cartilages, and subchondral bone lesions.

In addition to using different sequences to characterize various tissues and disease processes, contrast media are used to acquire more information from an MRI. The contrast media used in an MRI is gadolinium (Gad), a nontoxic paramagnetic contrast enhancement agent. Although contrast media have an effect on both T1 and T2 times, its effect is most noticeable on T1w, with contrast enhancement being characterized by increased T1 hyperintensity. Gadolinium-enhanced images are particularly useful in imaging vascular structures and failure of the blood-brain barrier—e.g., tumors, inflammation, and infection (Figs. 2.5-3B, 2.5-4B and C, and 2.5-7D).

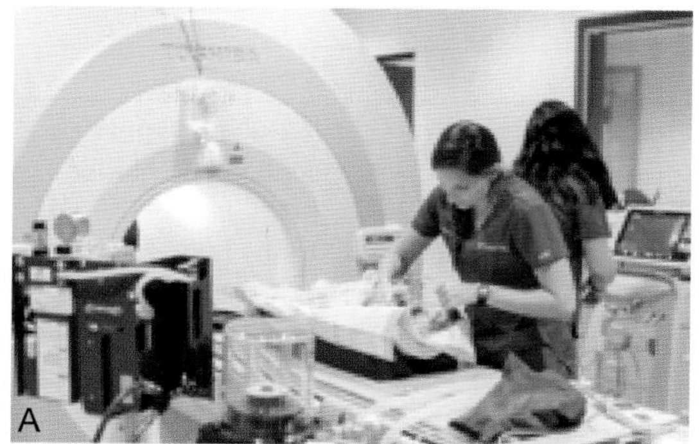

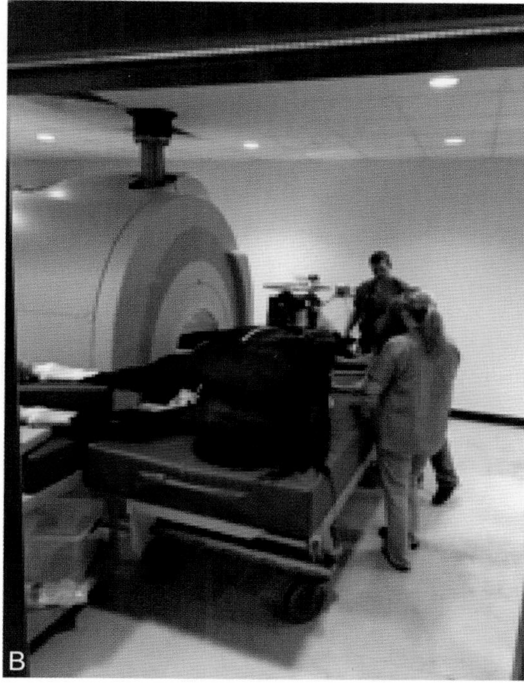

FIGURE 2.5-2 (A) A small canine patient with acute onset of paraplegia is being prepared for an MRI of the thoracolumbar region in a 1.5T magnet for evaluation of a suspected acute intervertebral disk herniation. (B) With the use of a purposely built MRI-safe table, high-gradient field strength magnets can be used to image the distal limbs of horses. In the photo, a horse is being positioned for an MRI of the thoracic fetlocks and feet.

TABLE 2.5-1 T1 and T2 signal intensities of various body tissues.

Type of tissue	***T1 weighted signal***	***T2 weighted signal***
CSF	Hypointense (dark)	Strongly hyperintense (bright)
Muscle	Isointense (gray)	Hypo- to isointense (dark gray)
Spinal cord	Isointense (gray)	Isointense (light gray)
Fat	Strongly hyperintense (bright)	Slightly hyperintense (light)
Nucleus pulposus	Iso- to hypointense (gray)	Strongly hyperintense (bright)
Air	Strongly hypointense (very dark)	Strongly hypointense (very dark)
Inflammation	Hypointense (dark)	Hyperintense (bright)

TABLE 2.5-2 T1, T2, and FLAIR intensities of intracranial tissues.

Tissue type	*T1 weighted*	*T2 weighted*	*FLAIR*
CSF	Hypointense (dark)	Strongly hyperintense (bright)	Hypointense (dark)
White matter	Hyperintense (light)	Slightly hypointense (dark gray)	Slightly hypointense (dark gray)
Cortex	Isointense (gray)	Slightly hyperintense (light gray)	Slightly hyperintense (light gray)
Fat (in bone marrow)	Strongly hyperintense (bright)	Hyperintense (light)	Hyperintense (light)
Inflammation	Hypointense (dark)	Hyperintense (bright)	Hyperintense (bright)

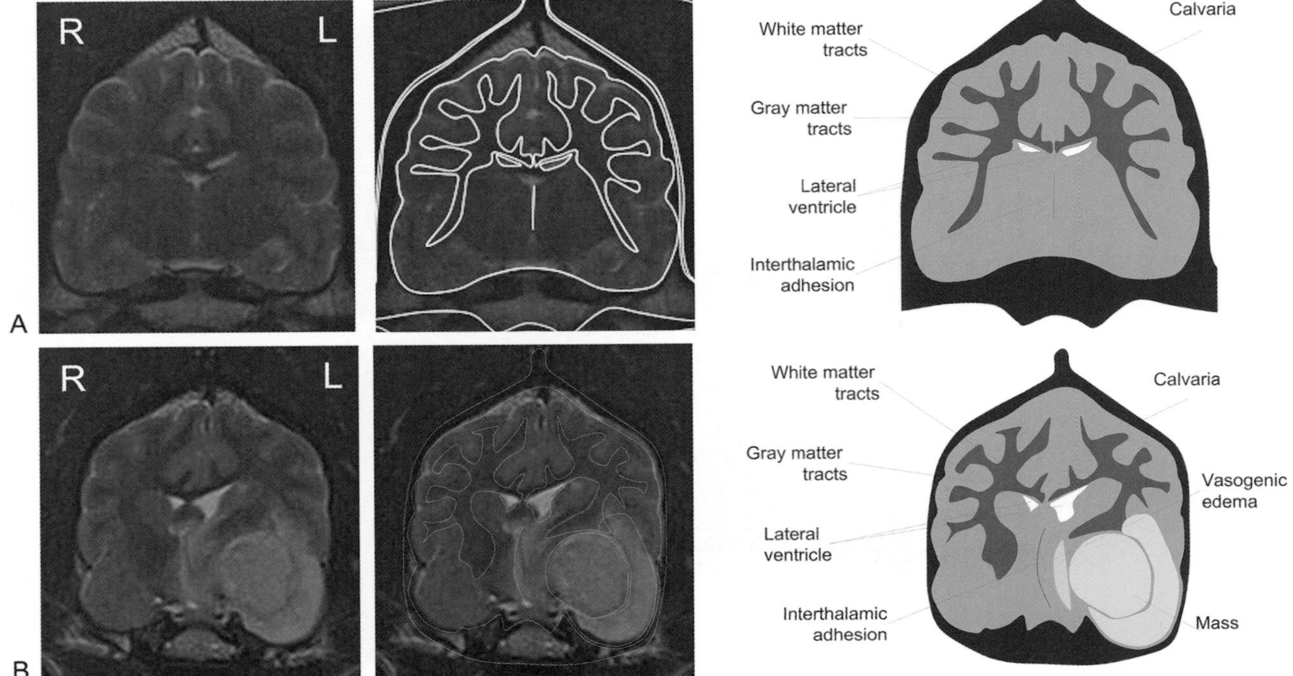

FIGURE 2.5-3 Transverse images in T2 sequences of the brain at the level of the interthalamic adhesion in two Labrador Retriever dogs. (A) Normal brain. (B) Mass in the left cerebral hemisphere centered on the region of the thalamus and the piriform lobe. Note the hyperintensity of the mass, the hyperintense edema axial to the mass, and the mass effect causing displacement of the interthalamic adhesion to the right and asymmetry of the lateral ventricles.

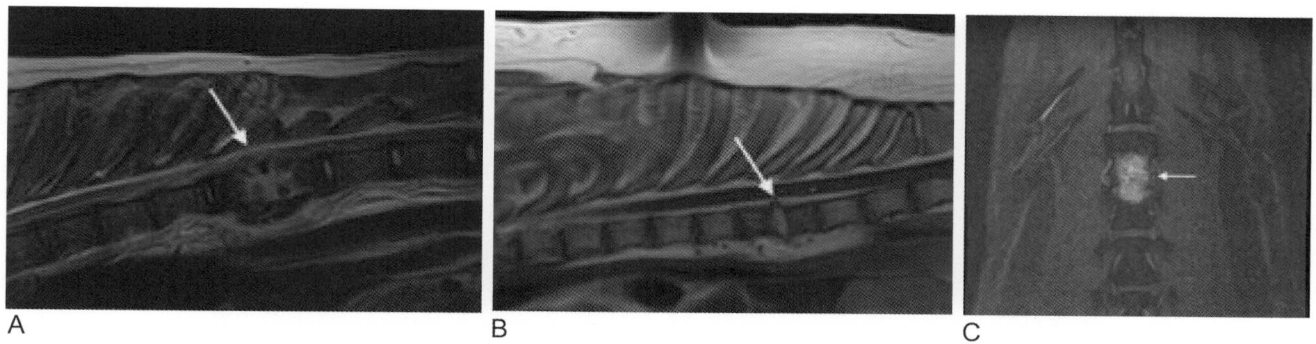

FIGURE 2.5-4 Discospondylitis at T6-7 in a canine patient with T3-L3 deficits. (A) T2 sagittal image; notice the hyperintense signal at T6-7 (*arrow*) extending into the adjacent vertebral endplates and associated ventral and dorsal soft-tissue reaction. (B) T1 + contrast sagittal image; notice the hyperintense contrast enhancement at T6-7 (*arrow*) and associated decreased intensity in the adjacent vertebral bodies compared to the remaining vertebral bodies consistent with bone sclerosis. (C) T1 + contrast dorsoventral image with fat saturation, nulling the fat signal of the subcutaneous tissues, and deep fascial planes improves visualization of contrast enhancement (*arrow*) compared to image B. Due to time constraints, different sequences are often acquired in different planes to obtain as much information in as short a time as possible.

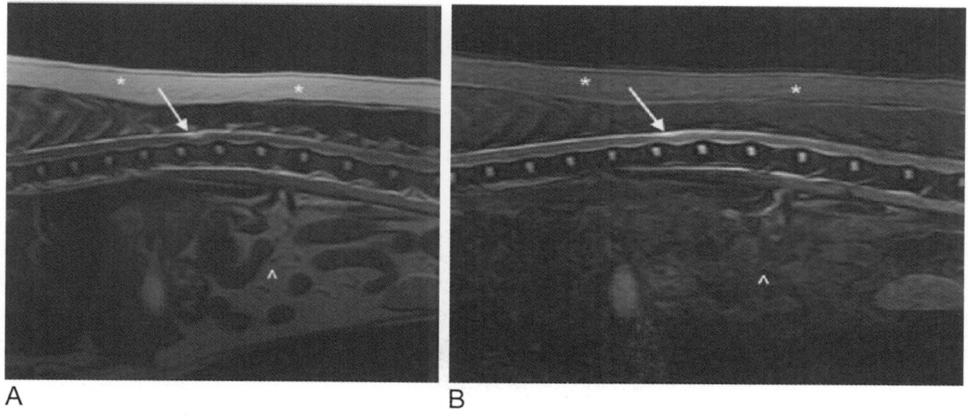

FIGURE 2.5-5 Subarachnoid diverticulum (*arrow*) in a 2-year-old Maltese with left-sided monoparesis. (A) T2 sagittal image; notice the hyperintensity of the lesion (*arrow*) and subarachnoid space along the vertebral canal and the hyperintensity of the subcutaneous fat (*) and the mesenteric fat (^). (B) STIR sagittal image of the same patient; notice the decreased intensity of the subcutaneous fat (*), mesenteric fat (^), and fat in the deep fascial muscular planes, while the hyperintensity of the subarachnoid space (*arrow*) is maintained along the vertebral canal.

Because the MRI localizes the energy received from a specifically magnetized and excited region in the body, it is very sensitive to motion artifact. This is important because the acquisition of most MRI images takes several minutes, with some sequences taking > 10 min. Motion artifact, therefore, is a major concern in the standing MRI. In horizontal configurations, motion artifact is prevented by anesthetizing the patient. However, even under general anesthesia, motion from respiration, along with cardiac motion and the rhythmic expansion of an artery pulse causes enough artifact to affect image quality and diagnostic capabilities.

Magnetic resonance imaging safety considerations are important because the MRI uses strong magnetic fields that not only affect hydrogen protons in the body's region of interest but all protons within the created magnetic field,

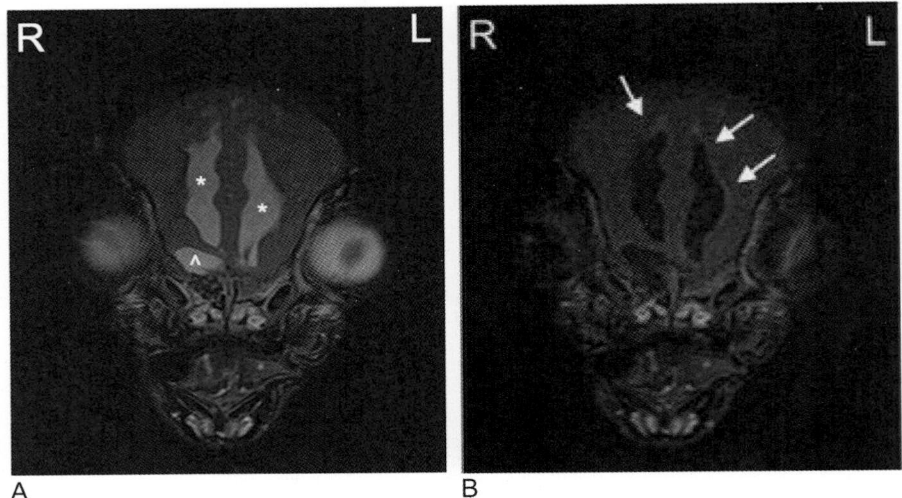

FIGURE 2.5-6 Transverse images of the rostral aspect of the cranium in a 9-month-old Yorkshire Terrier with congenital obstructive hydrocephalus. (A) T2 transverse image with markedly dilated lateral ventricles (*) and olfactory peduncle (^) with homogeneous hyperintense CSF. (B) FLAIR transverse image with a nulled signal from the CSF signal highlighting periventricular hyperintensities consistent with periventricular edema (*arrows*).

especially objects that are ferromagnetic; the created magnetic field extends beyond the MRI system (Fig. 2.5-9). The magnetic fields created by MRI units are so powerful that they can cause ferromagnetic objects to move suddenly and with great force to the isocenter of the magnetic field—i.e., where the patient is positioned for image acquisition. This poses a potentially fatal risk to the patient and anyone in the flight path of the object (Fig. 2.5-10), along with the destruction of the expensive magnet. Therefore, MRI units are housed in a magnetic field shield, and access to the MRI room is restricted and monitored. Additionally, all equipment used around the MRI unit must be made of nonferromagnetic materials, including the anesthesia machine, oxygen tanks, and clothing.

Metallic surgical implants, surgical clips, and bone plates also suffer the effects of the magnetic field (Fig. 2.5-11); however, even though implants are subject to torque under a strong magnetic field, most are protected within a few months after the surgical procedure because the surrounding fibrosis and soft tissues provide resistance and support. Implants are also exposed to the heat produced from the torque created by the magnetic field; however, the heat produced normally diffuses to the surrounding soft tissues with minimal adverse effects. The primary negative effect from metallic surgical implants comes from what is termed the **susceptibility artifact**, causing strong focal distortion of the magnetic field that leads to a strong signal void on the areas surrounding the metallic implants. The magnitude of the signal void depends on the strength of the magnetic field, the ferromagnetic properties of the metal, and the size and shape of the implant.

Similar to ferromagnetic metals, paramagnetic materials can also cause susceptibility artifacts. To prevent expending time and money on nondiagnostic studies, patients are screened for the presence of metal before performing certain MRI studies. In the horse, for example, shoes are removed, and radiographs are used to look for nail fragments and other metallic debris before performing an MRI study of the foot.

Indications for MRI studies

- Primary or metastatic neoplasia to the CNS
- Infectious processes with central or peripheral neurologic deficits
- Vascular accidents—e.g., infarct or thrombosis of the brain or spinal cord
- Intervertebral disk extrusion and/or protrusion with neurologic deficits
- Subacute brain trauma
- Congenital malformations of the CNS

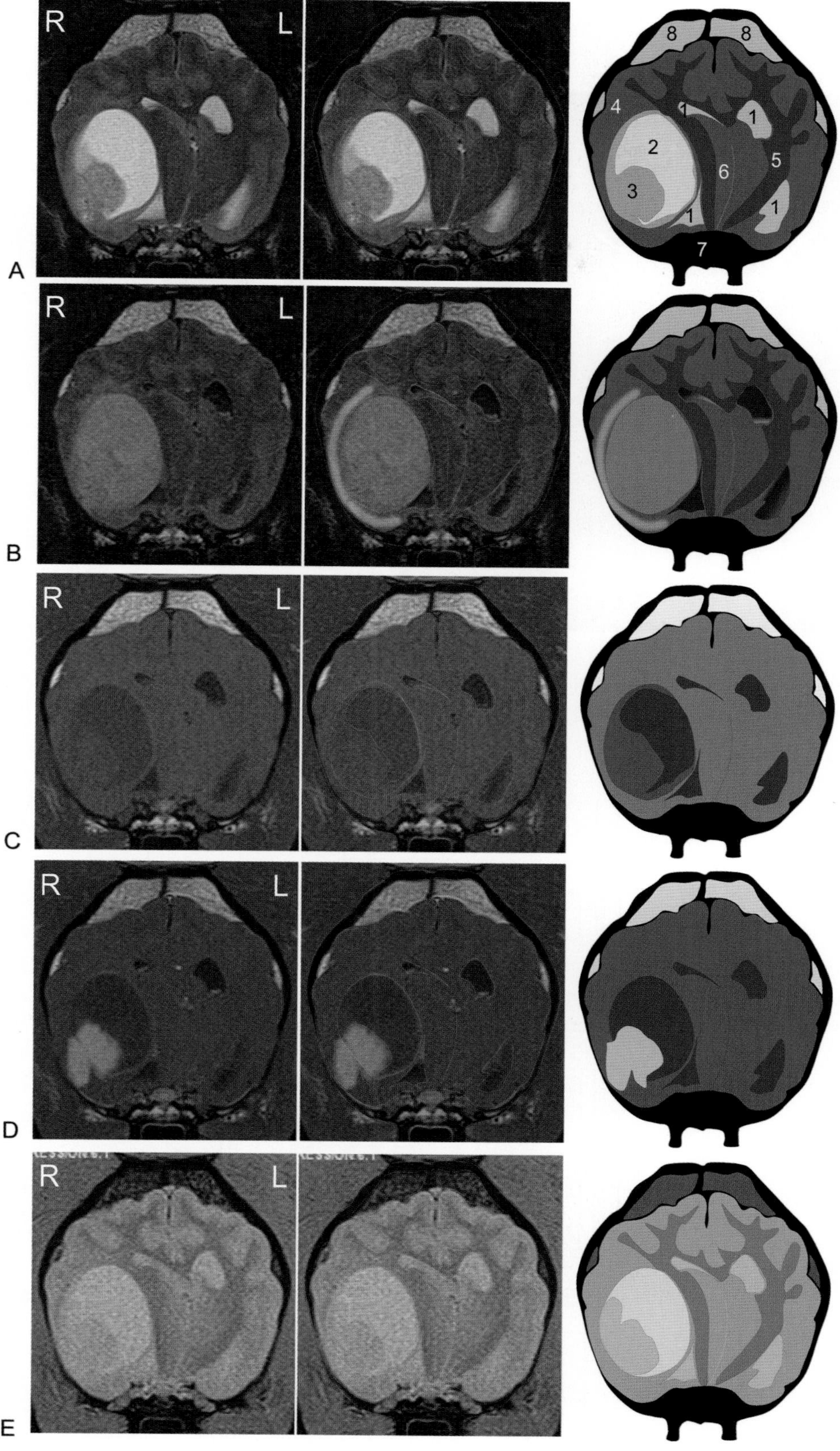

FIGURE 2.5-7 Transverse images at the level of the thalamus in an 11-year-old male castrated Boston Terrier with a right-sided cystic meningioma. (A) T2 sequence, the right and left lateral ventricles (1) and the large cyst (2) in the right side have strong hyperintensity, and a small lobular mass (3) at the right ventrolateral margin of the cyst is mildly hyperintense. (B) FLAIR sequence, the right and left lateral ventricles are now severely hypointense (T2 hyperintensity has been nulled) and the large cyst is moderately hyperintense and isointense to the right-sided mass. (C) T1 sequence, the brain parenchyma, and the mass have medium intensity, and the fluid in the lateral ventricle and the cyst are slightly hypointense. (D) T1 sequence postcontrast, the mass exhibits strong contrast enhancement, and the cyst remains slightly hypointense. (E) T2* sequence, allows better definition of bony structures and increases white/gray matter layer distinction. (4) Gray matter, (5) white matter, (6) interthalamic adhesion, (7) calvaria, and (8) diploe.

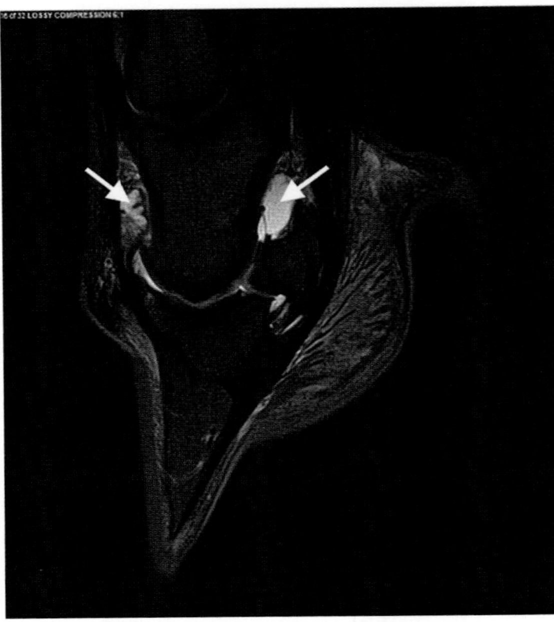

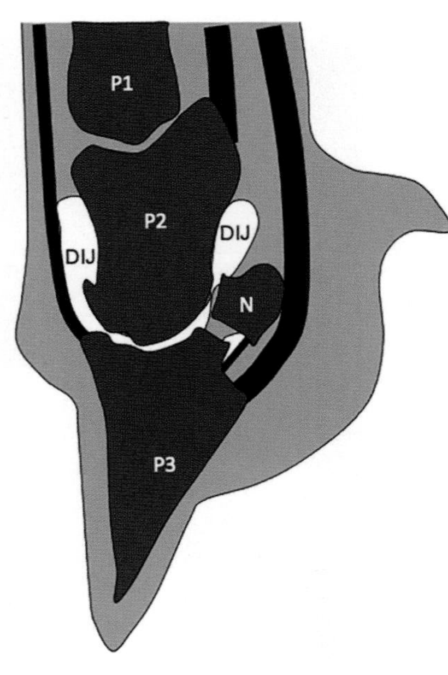

FIGURE 2.5-8 PD sagittal image of the distal interphalangeal joint (DIJ) in a horse with thoracic limb lameness with articular erosions, osteophytes, and associated intracapsular effusion (notice the hyperintense synovial fluid) consistent with synovitis. P1, proximal phalanx; P2, middle phalanx; P3, distal phalanx; N, navicular bone.

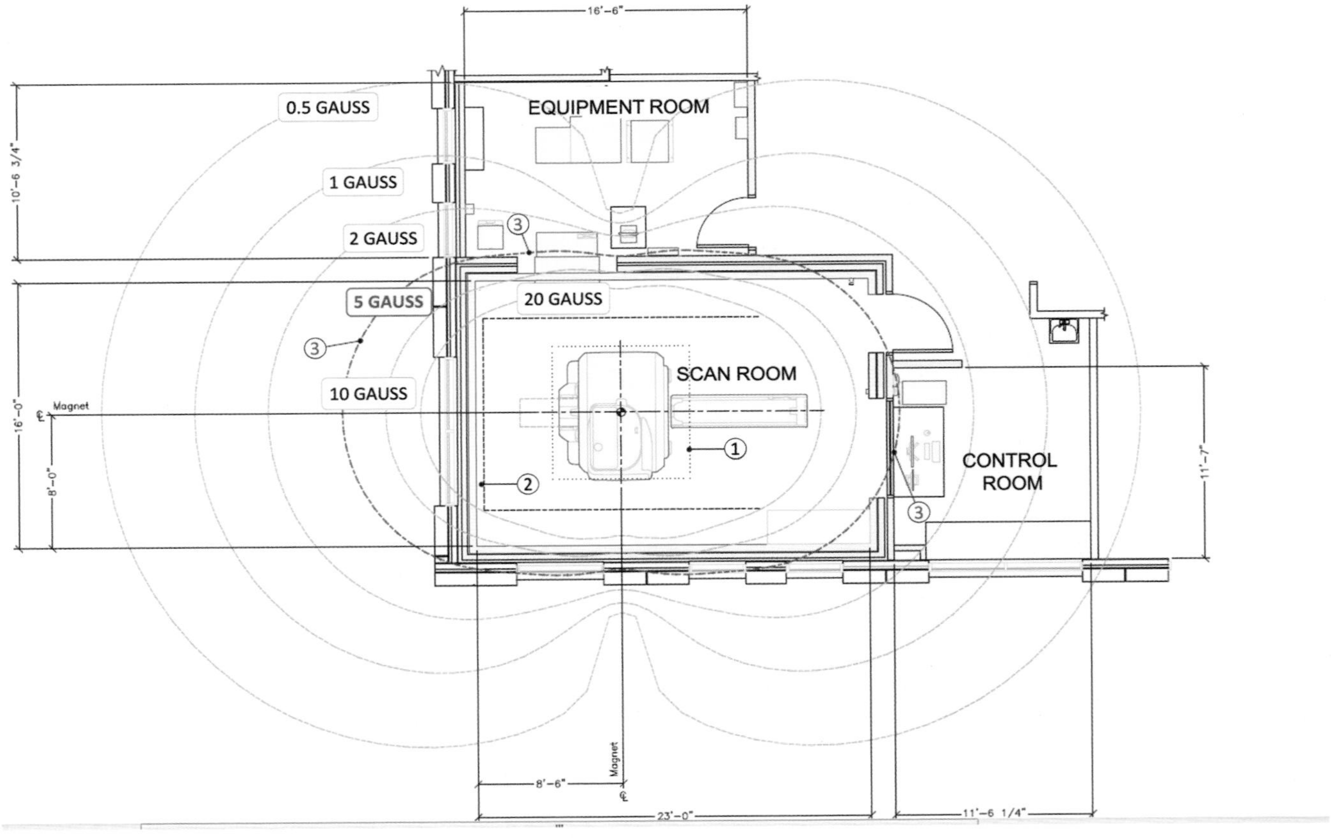

FIGURE 2.5-9 Example of room design for an MRI magnet (1). The five Gauss line (3) defines the limit beyond which ferromagnetic objects are strictly prohibited. Five gauss (*red line*) and below are considered "safe" levels of static magnetic field exposure for the general public. The MRI room design (2) is important since gauss lines of higher energies can affect nearby environments. (Image courtesy of Canon Medical Systems USA, Inc.)

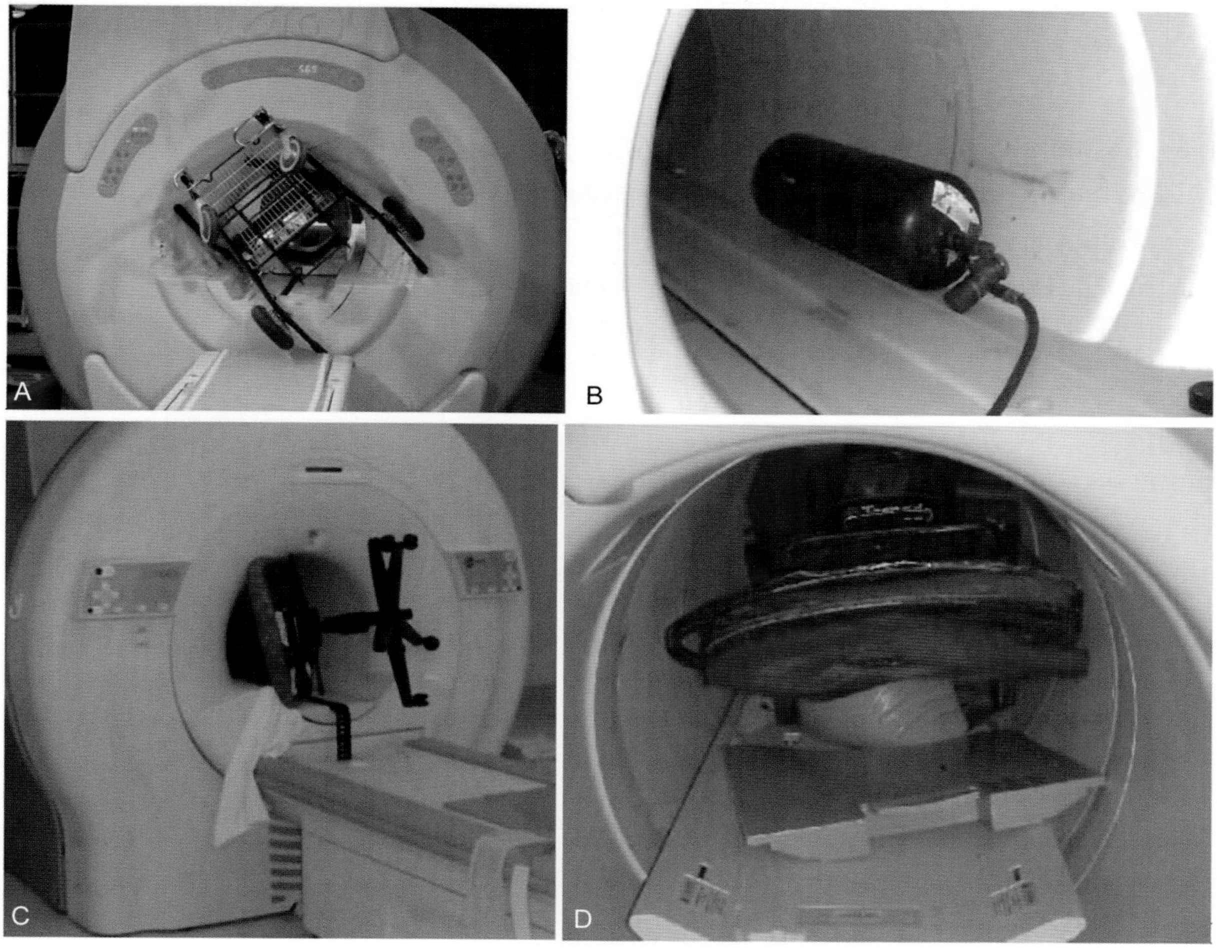

FIGURE 2.5-10 Magnetic resonance imaging missile-effect accident pictures that demonstrate the power of MRI magnets and should make the reader aware of the risk that the ferromagnetic equipment can pose to patients if these were to happen during the examination. (A) A wheelchair. (B) A welding tank. (C) An office chair. (D) An industrial polisher.

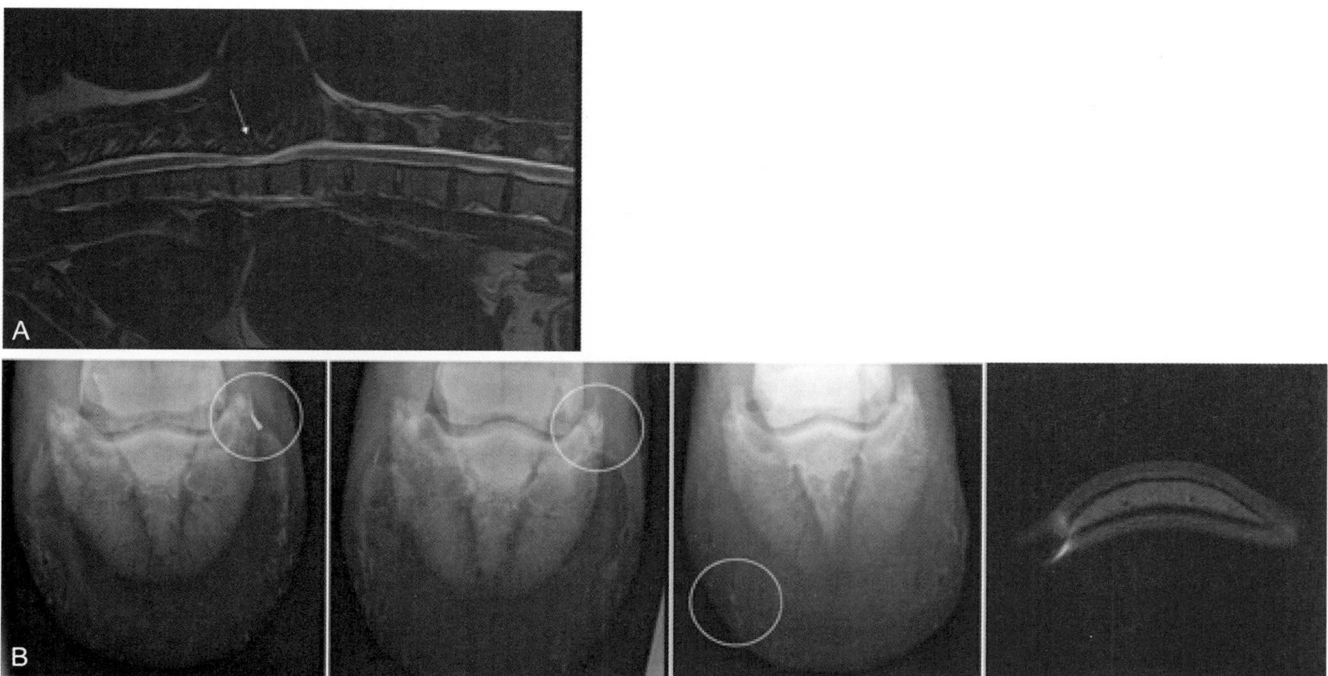

FIGURE 2.5-11 Magnetic resonance imaging artifacts caused by ferromagnetic and paramagnetic elements. (A) A microchip is causing severe focal distortion of the magnetic field and secondarily, the vertebral column and the spinal cord in the image (T2 sagittal), hampering its diagnostic value. (B) Metallic nail fragment by the palmar process of the distal phalanx (*circle*), the metallic fragment has been removed before the MRI; small metallic debris remains in nail tract (*circle*), and metal artifact damaging image quality on an axial PD image of the distal phalanx.

Indications for MRI studies, continued

- Musculoskeletal disease—e.g., inflammation and bone edema
- Staging of cancer patients
- Morphological assessment and quantification of the ventricular function of the heart
- Soft-tissue lesions of the extremities—e.g., tendons and ligaments in horses

Contraindications for an MRI

- Presence of certain metallic implants and other metallic devices—e.g., pacemakers, some artificial heart valves, aneurysm clips, and cochlear implants
- Intraocular metallic implants
- Patients that have health risks for general anesthesia
- Pregnant patients (unknown effects to the fetus)

Strengths of an MRI

- Does not expose the patient and imaging personnel to x-ray ionizing radiation
- Superior to CT in detecting inflammation (e.g., discospondylitis, Fig. 2.5-4)
- Superior soft-tissue contrast resolution compared to CT and radiography
- Images can be acquired in multiple planes
- Ability for 3-D reconstruction

Weaknesses of an MRI

- Subject to motion and susceptibility artifacts (among others)
- Inferior to CT at detecting acute hemorrhage
- Inferior to CT at detecting bone fractures
- Long imaging acquisition times
- General anesthesia required in most cases, except standing equine MRI

Selected references

[1] Wilfrey M. Diagnostic MRI in dogs and cats. CRC Press; 2018.
[2] Wisner E, Zwingenberger A. Atlas of small animal CT and MRI. Wiley-Blackwell, John Wiley & Sons, Inc.; 2015.
[3] Murray RC. Equine MRI. Blackwell Publishing Ltd., John Wiley & Sons, Inc.; 2011.
[4] Gavin PR, Bagley RS. Practical small animal MRI. Wiley-Blackwell, John Wiley & Sons, Inc.; 2009.

CHAPTER 2

DIAGNOSTIC IMAGING 2.6

Nuclear Medicine (Including PET)

Aitor Gallastegui
Small Animal Clinical Sciences, University of Florida, Gainesville, Florida, US

Introduction

Nuclear medicine was developed by John Lawrence in 1930 after the invention of the cyclotron by his brother Ernest Lawrence at the University of Berkeley. Nuclear medicine has been used for more than 50 years in veterinary medicine. It is an especially important imaging modality in equine practice for the musculoskeletal system (bone primarily, Fig. 2.6-1) and in companion animals for metastatic disease (oncology, Fig. 2.6-2), endocrinology, and nephrology (Fig. 2.6-3). The use of nuclear medicine is limited to approved referral centers because of the expense of

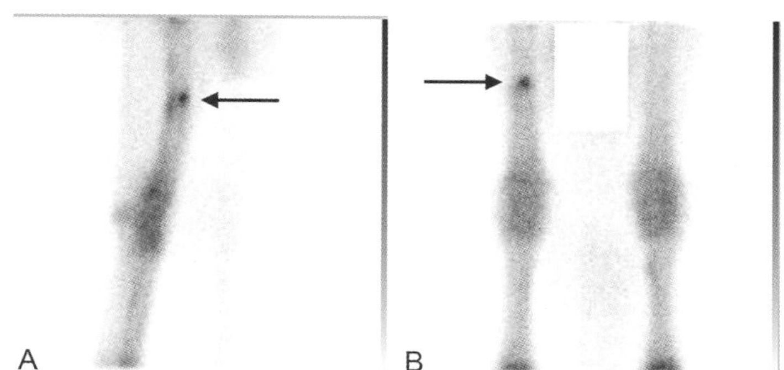

FIGURE 2.6-1 Right lateral (A) and cranial (B) views of the right radius in a lame horse with a focal area of markedly IMA (*arrows*) attributed to an enostosis-like lesion or a stress fracture.

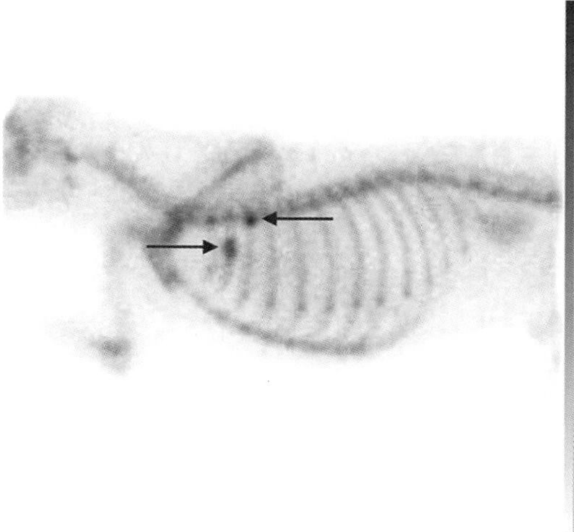

FIGURE 2.6-2 Focal areas of markedly increased metabolic activity (IMA) in two ribs (*arrows*) consistent with metastatic lesions from a previously diagnosed and treated thoracic limb osteosarcoma.

Comparative Veterinary Anatomy: A Clinical Approach. https://doi.org/10.1016/B978-0-323-91015-6.00008-X

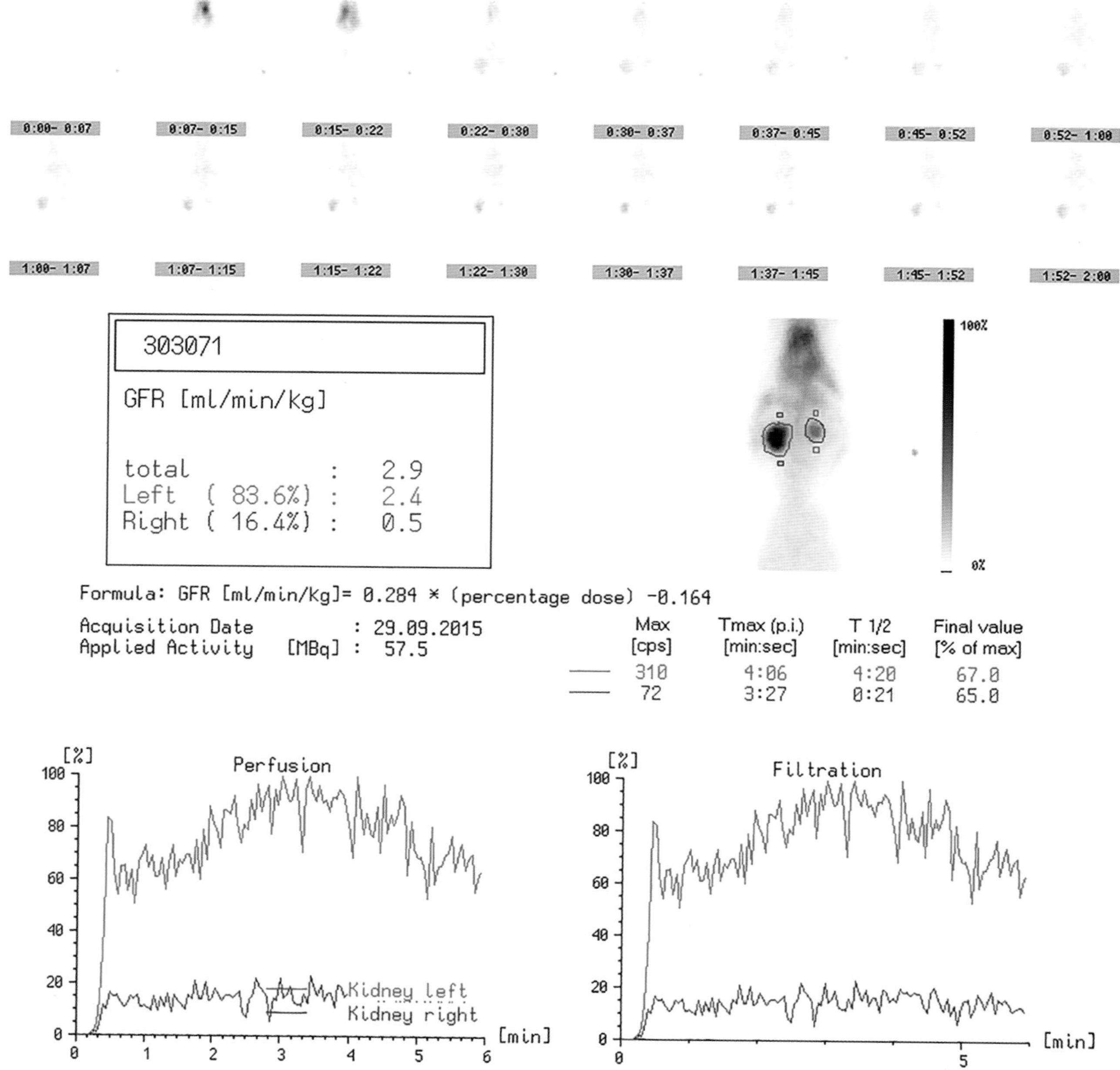

FIGURE 2.6-3 Renal scintigraphy allows the clinician to assess renal function individually for each kidney. In this canine patient, renal scintigraphy identified the right kidney (green) to be small with reduced radiopharmaceutical uptake and clearance, accounting for 16.4% of the total GFR, which was within normal limits (2.9 mL/min/kg).

supplies (radionuclides—e.g., radiopharmaceuticals) and the rigorous regulations regarding the imaging system and procedures. The main advantage of nuclear medicine is that it offers a *functional* diagnosis rather than an *anatomical* diagnosis. In addition, nuclear medicine is used to treat specific diseases such as hyperthyroidism in cats (Fig. 2.6-4).

Multiple terms are used synonymously with nuclear medicine and include nuclear scintigraphy, bone scintigraphy, bone scan, and gamma scintigraphy. Recognizing the nuances among the different terms, most equine clinicians commonly refer to nuclear medicine as bone scintigraphy or "bone scan" in the clinical setting, whereas companion animal clinicians refer to nuclear medicine as "scintigraphy." Bone scintigraphy is highly sensitive and specific

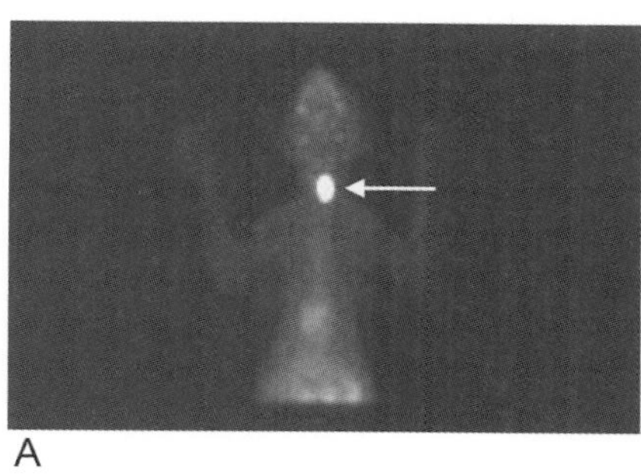
A

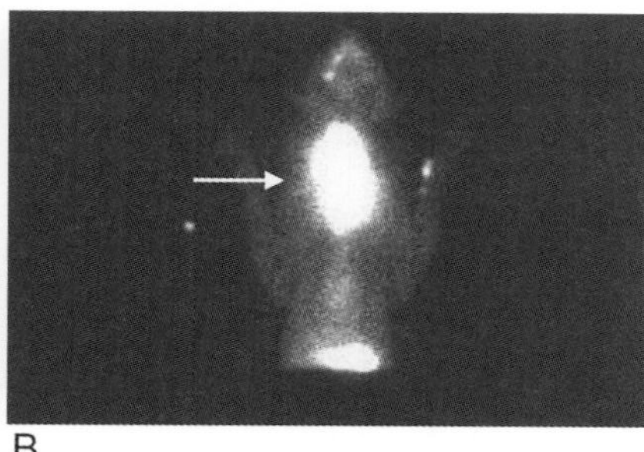
B

FIGURE 2.6-4 (A) Focally, markedly IMA at the level of the left thyroid gland (*arrow*) consistent with unilateral thyroid hyperplasia in a hyperthyroid cat. (B) A large region of markedly IMA (*arrow*) in the neck and cranial mediastinum consistent with thyroid carcinoma in a cat.

compared to digital radiography and detects as little as 10^{-13} g of radionuclide in a bone before a change in digital radiography abnormalities would be detected. Nuclear medicine is commonly referred to as a "physiological imaging modality" because it measures tissue activity at the point of interest at a point in time. Lesions in nuclear medicine are most frequently identified as focal areas of increased radiopharmaceutical uptake (IRU) that represent increased metabolic activity (IMA), which is currently the preferred terminology. Occasionally, lesions are also associated with regions of decreased metabolic activity.

Understanding the imaging technology/mode of action

Nuclear medicine uses **radionuclides**, which are unstable atoms that emit radiation as they decay to a more stable particle. Different radionuclides emit different types and energies of radiation—e.g., x-ray, gamma radiation, annihilation radiation—which makes them ideal for different medical diagnostics. The radionuclides used for medical purposes are generated in either a nuclear reactor—e.g., technetium-99m (Tc-99m), iodine-131 (I-131)—or a cyclotron—e.g., fludeoxyglucose (F-18).

Radionuclides are frequently bound to other substances that enable targeting the effect of the emitted radiation to a particular body region or tissue type. The labeled radionuclides are called **radionucleotides,** and the group of pharmaceutical drugs that have radioactivity is called **radiopharmaceuticals**. These substances are administered by various routes (generally IV) and, based on their distribution, can be used for the diagnosis or treatment of disease. In veterinary medicine, Tc-99m is the most widely used radionuclide for diagnostic purposes with applications in the detection of hyperthyroidism (primarily in cats), identification of portosystemic shunts, assessment of bone pathology, and evaluation of renal function (Table 2.6-1). Other radionuclides less frequently used in veterinary medicine but of increasing clinical importance are listed in Table 2.6-1.

As a radionuclide decays to a more stable state, it emits radiation, which is then sensed by the detector unit (a gamma camera) to generate an image of diagnostic value. The period required to reduce the radioactivity level to one-half of its original value due to radioactive decay is known as the **physical half-life (t1/2)**. Some radionuclides decay within yoctoseconds ($\times 10^{-24}$ seconds), and others take centuries to decay (hydrogen-7—$t_{1/2}$: 23×10^{-24} seconds; polonium-210—$t_{1/2}$: 138 days; uranium-232—$t_{1/2}$: 68.9 years; and carbon-14—$t_{1/2}$: 5730 years). The physical half-life determines how long a dose is radioactive, which in turn determines (1) the diagnostic or treatment interval and (2) how long the patient must be isolated to prevent radioactive contamination of other patients, clinical staff, and/or owners. Based on these factors, for most of the clinically used radionuclides, the protocols for each standard procedure are well established. Additionally, each country or state regulates when patients can be safely released from the imaging center after administration of a specific radionuclide. The regulations also specify the handling, transportation, and disposal of the radionuclide; the patient management; the specific training of imaging staff; and the authorizations and radiation exposure surveillance measures. In the USA, the Nuclear Regulatory Commission (NRC) regulates the production, use, and disposal of nuclear products. The International Commission on Radiological Protection (ICRP) is a nongovernmental independent institution with the mission to provide recommendations and guidance on radiological protection concerning ionizing radiation.

TABLE 2.6-1 Radionuclides most commonly used in veterinary medicine.

Radionuclide	*Radiopharmaceutical*	*Use*	*Modality*	*Administration*	*Excretion and sources of contamination*	*Half-life ($t_{1/2}$)*
Tc-99m	+ Pertechnetate	Thyroid scintigraphy	Gamma camera	Subcutaneous	Urine, salivary glands, and feces	6.02 h
		Portal scintigraphy		Per-rectal, trans-splenic		
	+ Methyldiphosphonate (MDP) or Hydrometyldiphosphonate (HDP)	Bone scintigraphy		Intravenous		
	+ Diethylenetriamine-pentaacetic acid (DTPA)	Renal scintigraphy		Intravenous		
I-131		Treatment of hyperthyroidism and thyroid carcinoma		Subcutaneous preferred to intravenous	Urine, salivary glands, and feces	8.02 days
F-18	Fluoxydiglucose (FDG)	Brain function, metastatic disease, and cardiac glucose metabolism	PET	Intravenous	Urine	110 min
	+ Sodium fluoride (NaF)	Bone activity				

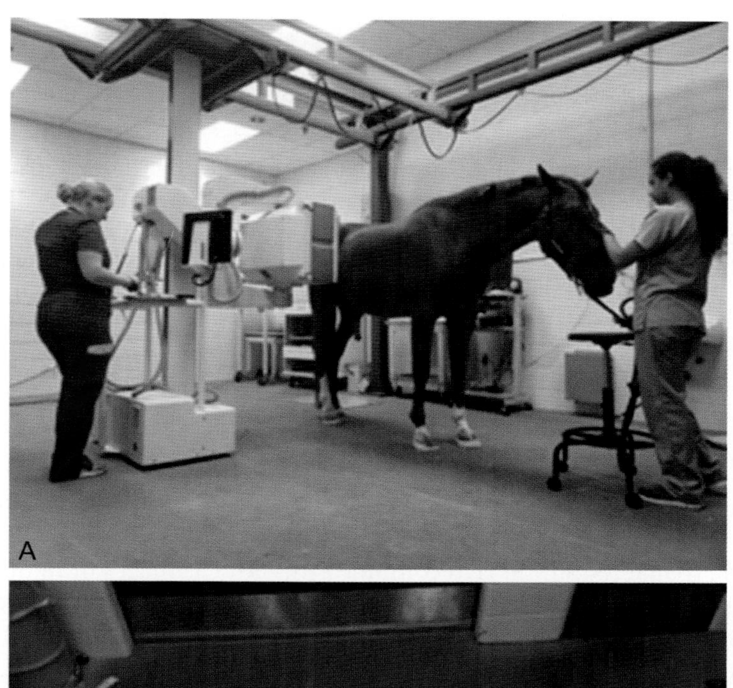

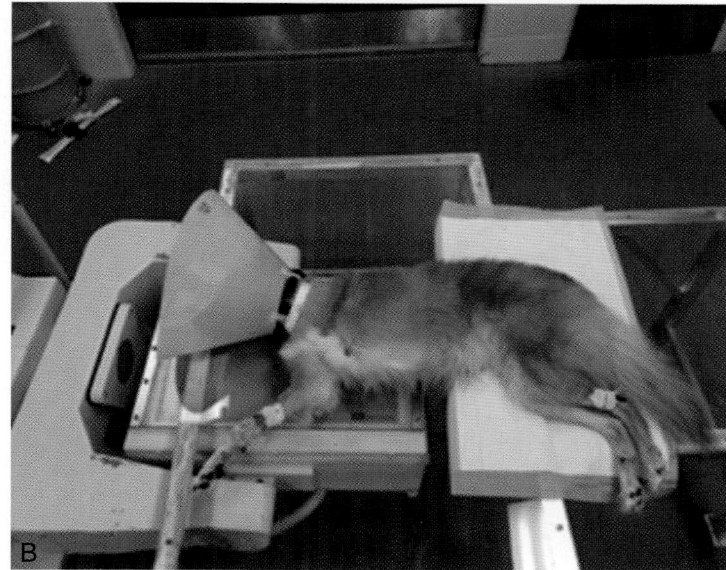

FIGURE 2.6-5 (A) Equine patient undergoing bone scintigraphy, with images of the lumbar vertebral column. (B) Canine patient undergoing full-body scintigraphy for the staging of metastatic osteosarcoma; diapers and urinary catheters are frequently used to reduce contamination and artifacts. Patients are frequently sedated to reduce motion artifacts during image acquisition.

The detector unit used in nuclear medicine is the **gamma camera**, which comes in various setups for use in small and large animal medicine (Fig. 2.6-5). When using radionuclides that emit x-rays or gamma-radiation, a collimator—which defines the dimensions and direction of the radiation beam—is always used to improve image quality and spatial resolution at the expense of absorbing most of the radiation emitted by the patient. Different types of collimators are used depending on the study type and region of interest. The most frequently used collimators and the purpose of each include (1) parallel collimators that maintain proportion, (2) pin-hole collimators that magnify and invert the image, (3) diverging collimators that magnify the image, and (4) converging collimators that minimize the image. The physics of nuclear medicine image processing are beyond the scope of this section (please read more in the selected references). The reader should be aware that the image created by a nuclear medicine system lacks adequate spatial resolution, although newer systems have significantly improved upon this deficiency.

Additional nuclear medicine techniques

Positron emission tomography (PET) uses annihilation radiation (the physics of converting a particle and an antiparticle into radiation) to localize the origin of the radiation. The key characteristics of annihilation radiation are that it allows the identification of the exact point from which it originated because it creates two photons of known energy that are emitted in opposite directions—i.e., 180° apart—and it demonstrates real-time metabolic activity. In veterinary medicine, F-18 is the most frequently used radionuclide for PET. Because F-18 is a glucose analog, it marks regions with IMA such as the brain, heart, and sites of metastatic disease. Limitations to the increased clinical use of PET include:

- F-18 has a short half-life, requiring a commercial source to be available within a few hours from the imaging center
- Active digestion or muscular activity can create transient areas of IMA, leading to artifactual images that can mimic metastatic lesions
- It has a poor spatial resolution

One solution to improve the spatial resolution is to combine PET with CT images from the same body region. This can be done by "postprocessing" the PET and CT images with compatible software; however, the final images have certain limitations. The best way to nullify these limitations is to combine the PET and CT scanner under the same gantry, which is known as a **PET/CT scanner**; the patient undergoes each separate study in the same session, and then the images are combined into a single superimposed image. The advantage of a PET/CT scanner is that since the patient position and other technical components are the same for both studies, the purposely built-in software analyzes the combined images with better resolution. Therefore, PET/CT allows a more precise alignment of the depicted spatial distribution of abnormal metabolic activity from the PET scan with the images obtained by CT scanning.

Indications for nuclear medicine studies

- Metastatic disease in oncology
- Lameness diagnosis for anatomical locations that are unspecified and/or difficult to radiograph in horses
- Increased IRU comparisons in bone activity—baseline and follow-up studies
- Specific radiopharmaceuticals for special studies—e.g., gallium for bone infections and iridium-131 for improved sensitivity
- Radiopharmaceutical uptake in determining growth plate closure

Contraindications for nuclear medicine

- Unlikely to reflect the changes in a bone that occurred more than 3–4 months before the imaging study ***unless*** there is continued:
    - ↑ Osteoblastic activity
    - ↑ Blood flow
    - ↑ Surface area
    - ↑ Capillary permeability
    - ↑ Extraction efficiency—i.e., young vs. old
    - Exposed hydroxyapatite crystals
- A lag phase of 24–48 h is best if muscle injury is likely in acute trauma
- Osteoclastic activity predominates during bone resorption, with osteoblastic activity dominating during bone remodeling; therefore, delay nuclear medicine study for 1–2 weeks to maximize IRU after an acute injury

Strengths of nuclear medicine

- Highly sensitive to binding sites in actively remodeling bone
- Highly sensitive to binding sites in soft tissue undergoing mineralization
- Increased blood flow does not significantly affect a bone scan; however, adequate blood flow is needed to deliver radiopharmaceuticals to binding sites in bones
- Bone scintigraphy is highly sensitive compared to digital radiography, detecting as little as 10^{-13} g of radiopharmaceutical uptake as compared to grams of bone change before a lesion would be diagnosed using digital radiography
- Assessment of metastatic disease in companion animals
- Three phases of nuclear medicine studies—flow (vascular), pool (soft tissue), and delayed (bone)
- Maximum IRU occurs 8–12 days after bone injury

Weaknesses of nuclear medicine

- False-negative results affected by distance, shielding, high background activity, and motion
- Radiopharmaceuticals and study procedures subject to local and regional regulations
- Photopenia (Gr. *photos* light + Gr. *penia* poverty, need) occurs if blood supply is inadequate during flow- and pool-phases, resulting in the inability to differentiate the normal from the abnormal bone
- Injured muscle behaves like an injured bone, confounding interpretation
- Equipment requirements—e.g., gamma camera and computer software
- Isolation of nuclear medicine patients during radiopharmaceutical decay

Selected references

[1] Gregory BD, Clifford RB. Veterinary nuclear medicine. Am Coll Vet Radiol 2006.
[2] Dyson SJ, Pilsworth RC, Twardock AR, Martinelli MJ. Equine scintigraphy. Equine Vet J 2003.

CHAPTER 2

DIAGNOSTIC IMAGING 2.7

Advances in Diagnostic Imaging

Aitor Gallastegui[a], Nora S. Grenager[b], James A. Orsini[c]
[a]Small Animal Clinical Sciences, University of Florida, Gainesville, Florida, US
[b]Steinbeck Peninsula Equine Clinics, Menlo Park, California, US
[c]Department of Clinical Studies - New Bolton Center, University of Pennsylvania School of Veterinary Medicine, Kennett Square, Pennsylvania, US

Files in diagnostic imaging

Introduction

Digital images made using the various diagnostic imaging modalities discussed in this section become part of the patient's medical record, which is a legal document. In addition, the possibility of sharing these files online through the internet or various portable disk formats makes them vulnerable to corruption or loss of information. Damage to these files can have unintended consequences—for example, if a patient was euthanized based on a radiographic diagnosis made on radiographs from another patient due to an error or file corruption. In addition, given the space limitations of hard drives, cables, and storage units, creating compressed versions of the medical diagnostic images, such as.jpg, results in significant loss of spatial and contrast resolution that can secondarily lead to misdiagnosis.

To maintain the quality and the integrity of the medical diagnostic images, an open-source image format was created—digital imaging and communications in medicine (DICOM). This is the standard for the communication and management of medical imaging information and its related data. It is most commonly used for storing and transmitting medical images, enabling the integration of medical imaging devices such as scanners, servers, workstations, printers, network hardware, and picture archiving and communication systems (PACS) from multiple manufacturers. Because other image files such as .jpg, .jpeg, .png, .tiff, etc., do not fit these criteria, these image files should never be used for diagnostic purposes (Fig. 2.7-1).

Understanding the technology/mode of action

Picture archiving and communication systems is a medical imaging technology that provides economical storage and convenient access to images from all the imaging modalities. This allows electronic transmission of DICOM images and imaging study reports. Other file formats such as PDF can be included once captured in DICOM. A PACS is composed of four major components: (1) the imaging modalities (e.g., radiology, fluoroscopy, CT, MRI, and US); (2) a secured network for transmission of sensitive and private patient/owner information; (3) workstations for image viewing and interpretation; and (4) archives for storage and retrieval of images and reports. These PACS overcome some of the limitations of the film-based systems regarding the sharing and storage of the imaging diagnostic medical information by replacing hard copies, facilitating remote access, facilitating electronic image integration with other hospital platforms [e.g., the hospital information system (HIS), the radiology information system (RIS), and the electronic medical report (EMR)], and improving the radiology workflow management. Basic PACS software can be found free of cost on the internet.

Indications for files in diagnostic imaging

- Improved storage and retrieval of a large number of medical records
- Minimizes storage space requirements

Comparative Veterinary Anatomy: A Clinical Approach. https://doi.org/10.1016/B978-0-323-91015-6.00011-X

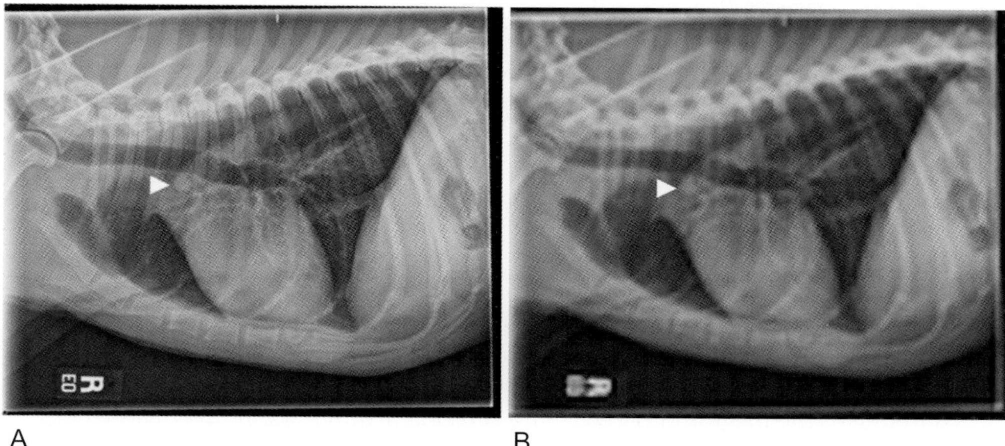

FIGURE 2.7-1 Digital imaging and communications in medicine (DICOM) is the standard image format used in medical imaging (A). Other image formats do not contain the same amount of information, so they are less precise (B). In these images, a nodule is apparent on both radiographs superimposed with the aortic arch but on the nonDICOM image (B) the margins of the nodule (*arrowhead*), the bones, and the pulmonary blood vessels are poorly defined, which could lead to radiographic misdiagnosis; therefore, this image (B) should never be used for diagnostic purposes.

- Simplifies sharing of diagnostic imaging studies with other clinicians and clients
- Combines various diagnostic studies for improved follow-up comparisons
- Comingling of multiple imaging technologies in one file

Contraindications of files in diagnostic imaging

- No major contraindications in implementing the technology

Strengths of files in diagnostic imaging

- No limit on the amount of data that can be saved for future retrieval
- Multiple sources can access the diagnostic images
- Reduced physical facilities for storage of imaging files
- Provides an excellent, accessible teaching and research resource

Weaknesses of files in diagnostic imaging

- Information technology support required
- Investment in software and computer hardware
- Requires periodic upgrade and software updates
- Requires cloud storage availability and access
- Needs excellent cybersecurity for health record confidentiality

3-D printing

Introduction and understanding the technology

A more recent method of displaying information about body regions in studies performed by tomographic modalities such as CT and—to a lesser extent—MRI is the printing of 3-D models.

Indications and strengths of 3-D printing

- Surgical planning of complex or newly developed procedures
- Surgeons can practice the surgical approach and technique multiple times
- Improved surgical outcome
- Surgeons can create patient-specific surgical guides to improve accuracy and reduce surgery time (Fig. 2.7-2)

Contraindications of 3-D printing

- No known contraindications reported in the literature

Strengths of 3-D printing

- Improved understanding of the magnitude of the clinical problem and assist in the treatment protocol
- Interpreting the precise location of the disease process (e.g., fracture), selecting the appropriately sized implants (e.g., plates and screws in fracture repair), and identifying potential problems
- Reduced operative time and improved pre-operative surgical planning
- An anatomic replica helps predict whether a certain clinical strategy is likely to be successful or even beneficial in a clinical case
- 3D-printed models are valuable in the education of residents and fellows
- Improved owner education in explaining the recommended procedure

Weaknesses of 3-D printing

- Technology is new
- Not yet expanded into daily clinical work
- Equipment and materials are expensive
- Printing times are relatively long

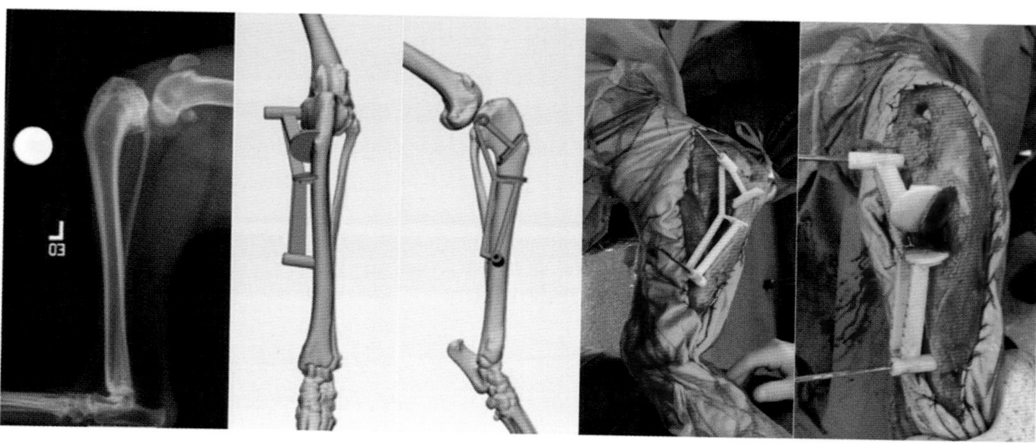

FIGURE 2.7-2 3-D printing of patient-specific surgical guides improves surgical outcomes and reduces surgical and anesthetic times. In this case, a 3-D printed guide was used to perform a cranial wedge ostectomy to correct a proximal tibial deformity (*left image*) in a dog. (Images courtesy of Dr. Christina De Armond from the University of Florida.)

Cone beam computed tomography

Introduction

Cone beam computed tomography (CBCT) is also known as C-arm CT, cone beam volume CT, or flat panel CT. The technology was first introduced in Europe in 1996 and in the United States in 2001. Divergent x-rays create a cone in combination with CT to produce multiple images that can be anatomically reconstructed.

This modality is used in veterinary medicine in the sedated standing horse for multiple diagnostic procedures in orthopedics, sports medicine, dentistry, neurology, and upper airway (respiratory and sinuses). In human medicine, the technology has been important in treatment planning and diagnosis in dentistry (oral surgery, endodontics, and orthodontics), otolaryngology, orthopedics and interventional radiology (IR), and whole organ perfusion imaging. Another clinical use of integrated CBCT in oncology is to assist with patient positioning when performing image-guided radiation therapy (IGRT).

Understanding the imaging technology/mode of action

The CBCT scanner acquires defined anatomical volume images by rotating around the region of interest in the patient—e.g., the limb or head—without concurrent motion of the patient in a plane perpendicular to the rotation. Approximately 400–600 different images are created within several minutes. A 180- to 360-degree rotation around the region of interest produces a volume data set that the software collects and reconstructs, producing 3D voxels (each defined volume unit of the anatomy being scanned) of anatomical data. Using specialized software, this volumetric anatomical data can be manipulated and visualized. Traditional CT—also known as "slice CT" or **fan beam computed tomography (FBCT)**—shares several similarities to CBCT, but there are important differences that affect the ability of the system to reconstruct the anatomical region of interest; in FBCT, the images are acquired slice-by-slice in 360-degree rotations as the patient moves in a plane perpendicular to the rotation, and that the whole region of interest is imaged within seconds. Because of these differences, CBCT is much more sensitive to motion artifacts and has reduced contrast resolution (Fig. 2.7-3).

Indications for cone beam CT

- Image acquisition in the nonanesthetized patient
- Rapid image study
- Imaging the distal limb and head in the horse

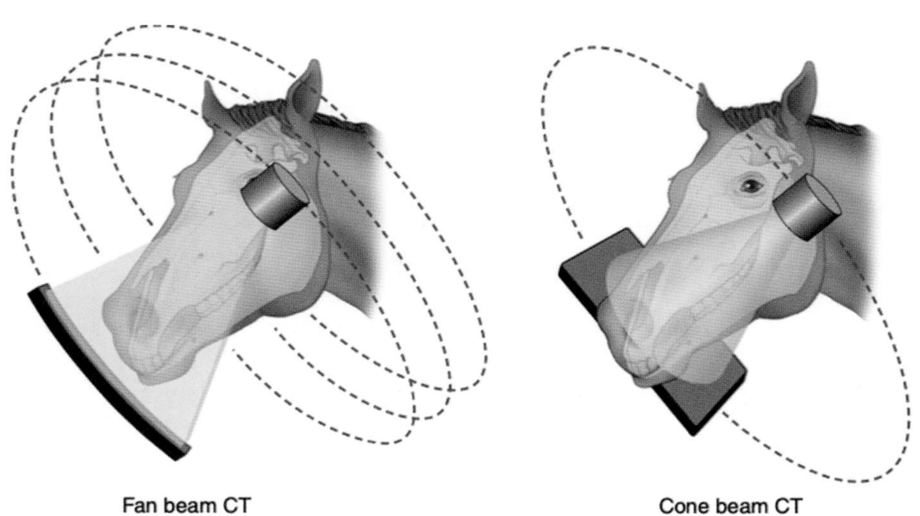

FIGURE 2.7-3 Standard slice CT or fan beam CT and cone beam CT.

TABLE 2.7-1 Comparison of uses in conventional (FBCT) and cone beam (CBCT) computed tomography.

	Indicated studies due to adequate image quality	Contraindicated studies due to poor image quality
Conventional CT (FBCT)	• Abdomen and thorax imaging studies where fast gantry rotations are needed to avoid motion artifacts and maximize the contrast resolution • Soft tissue studies requiring a superior contrast resolution	Small anatomical regions or patients where slice thickness does not provide enough spatial resolution: • Some small mammals • Small birds • Small exotic species
Cone beam CT (CBCT)	Situations in which scatter from the cone beam conformation is reduced: • Small anatomical regions (e.g., the head, distal extremities, and patients <20 pounds) • Bony structures (e.g., the skull, elbow, and distal extremities) where there are fewer soft tissues superimposed resulting in reduced scatter	• Soft-tissue applications requiring a minimum contrast resolution • Abdomen and thorax due to motion artifacts and a reduced contrast resolution caused by increased scatter

- Bone imaging is the primary area of interest
- Confirm questionable findings from digital radiographic studies

Contraindications of cone beam CT

- Technology is still evolving; thus, fan beam CT is currently superior (see Table 2.7-1)

Strengths of cone beam CT

- Multiple hundreds of images acquired in a short time frame
- Avoids the need for general anesthesia
- 3D image reconstruction
- Application in multiple species
- Contrast studies possible for soft-tissue lesions

Weaknesses of cone beam CT

- Requires special equipment and physical facilities
- The image quality is heavily software-dependent
- Subject to motion and artifacts at the imaging site
- Specialized training of the imaging personnel

Total or whole-body PET scanner

Introduction and understanding the imaging technology

Positron emission tomography (PET) is a safe medical imaging technology that can map the location or track the movement of radioactive-tagged compounds called radiotracers that have been injected into the body. Scanners are being developed that have a sensitivity 40-fold higher than what is currently available and therefore provide new ways this technology can be used in clinical practice.

The clinical applications are multiple and include (1) reduced anesthesia requirements; (2) oncology—better detection of cancer; (3) monitoring cell-based therapies for cancer treatment; (4) better studies for metabolic and autoimmune diseases; (5) research applications in toxicology, endocrine, and immune system signaling; and (6) opportunities in improved orthopedic, neurologic, and cardiologic diagnoses.

Besides the anatomic and physiologic information acquired with PET scans, there is a superior resolution and sensitivity for enhanced care and treatment of companion animals and humans (Fig. 2.7-4).

Indications for whole-body PET scanner

- Lower-radiation exposure compared to radiography or CT
- Rapid acquisition of scans, some in < 1 min
- Track injected radiotracers for a longer period of time
- Reduced dose of radiopharmaceutical needed to obtain the image

Contraindications of whole-body PET scanner

- General anesthesia is a risk to the patient
- Expertise and training of the imaging personnel
- Technical personnel for support of the PET scanner currently confined to a few imaging centers

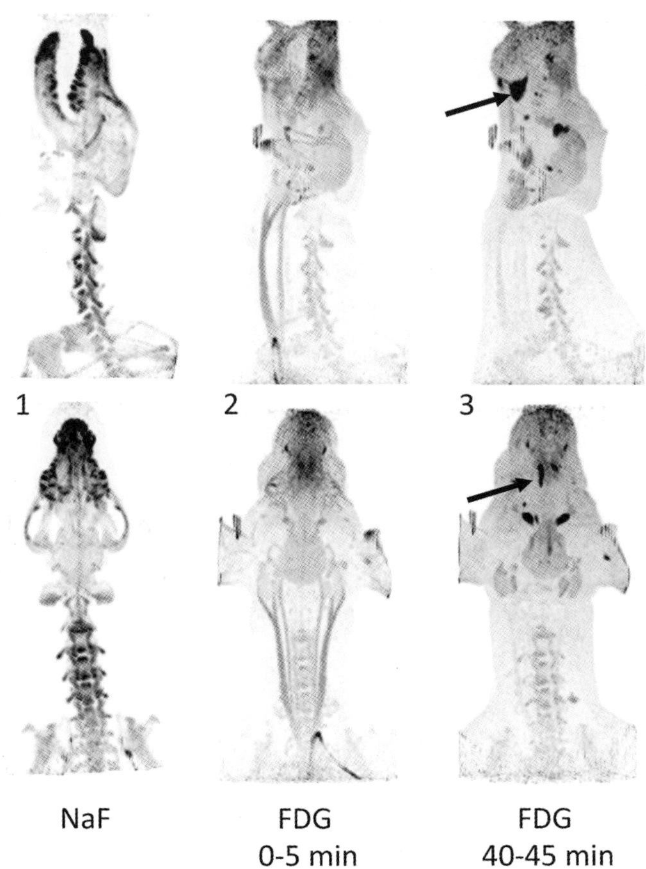

FIGURE 2.7-4 Lateral (*top row*) and dorsal (*bottom row*) 3-D maximal intensity positron emission tomography images (mini-Explorer, United Imaging) of the head and neck of a 5-year-old Labrador presenting for oral pain. The figure on the left (image 1) was obtained 1 h after the injection of 1.5 Mbq/kg of 18F-Sodium Fluoride (NaF) (a marker of bone turnover). The image demonstrates the remodeling of the alveolar bone surrounding the teeth; there is no significant abnormality. 3 MBq/kg of 18F-Fluorodeoxyglucose was then injected. Image 2 was obtained immediately after 18F-FDG injection, and image 3 was obtained 40 min later. Note the visualization of the radiotracer within the vessels on Image 2, before specific distribution demonstrating metabolic activity in Image 3. There is an abnormal uptake in the tongue (*arrow*), which is due to the presence of inflammation surrounding a foreign body (grass awn), found at surgery. (Images courtesy of M Spriet, UC Davis.)

Strengths of whole-body PET scanner

- Reduced anesthesia time for study
- More sensitive in cancer detection
- Can assess the success of cancer treatment(s)
- Useful in metabolic and autoimmune diseases
- Applications in orthopedics, neurology, and cardiovascular diseases
- Multiple research applications
- Real-time physiological study

Weaknesses of whole-body PET scanner

- General anesthesia is usually required
- Limited clinical sites where the technology is currently available
- Applications remain more research-focused with evolving translation into the clinical arena

Selected references

[1] Anon. Digital radiography. J Vet Radiol Ultrasound (VRU) 2008;49(s1).[2]Journal of 3D Printing in Medicine. ISSN: 2365-6271.Journal of 3D Printing in Medicine. ISSN: 2365-6271.

SECTION III

CANINE AND FELINE CLINICAL CASES

3.0 CANINE LANDSCAPE FIGURES (1–9) 83
N.S. Grenager, J.A. Orsini, J.F. Randolph, H.M.S. Davies, and A. de Lahunta

3.1 FELINE LANDSCAPE FIGURES (1–9) 93
N.S. Grenager, J.A. Orsini, J.F. Randolph, H.M.S. Davies, and A. de Lahunta

CHAPTER 3: AXIAL SKELETON: HEAD, NECK, AND VERTEBRAL COLUMN 103
John F. Randolph, Chapter editor

CHAPTER 4: THORAX 219
Helen M.S. Davies, Chapter editor

CHAPTER 5: ABDOMEN 277
John F. Randolph, Chapter editor

CHAPTER 6: PELVIC ORGANS 337
Helen M.S. Davies, Chapter editor

CHAPTER 7: THORACIC LIMB 391
Helen M.S. Davies, Chapter editor

CHAPTER 8: PELVIC LIMB 455
Helen M.S. Davies, Chapter editor

CHAPTER 9: INTEGUMENT AND MAMMARY GLAND 539
Helen M.S. Davies, Chapter editor

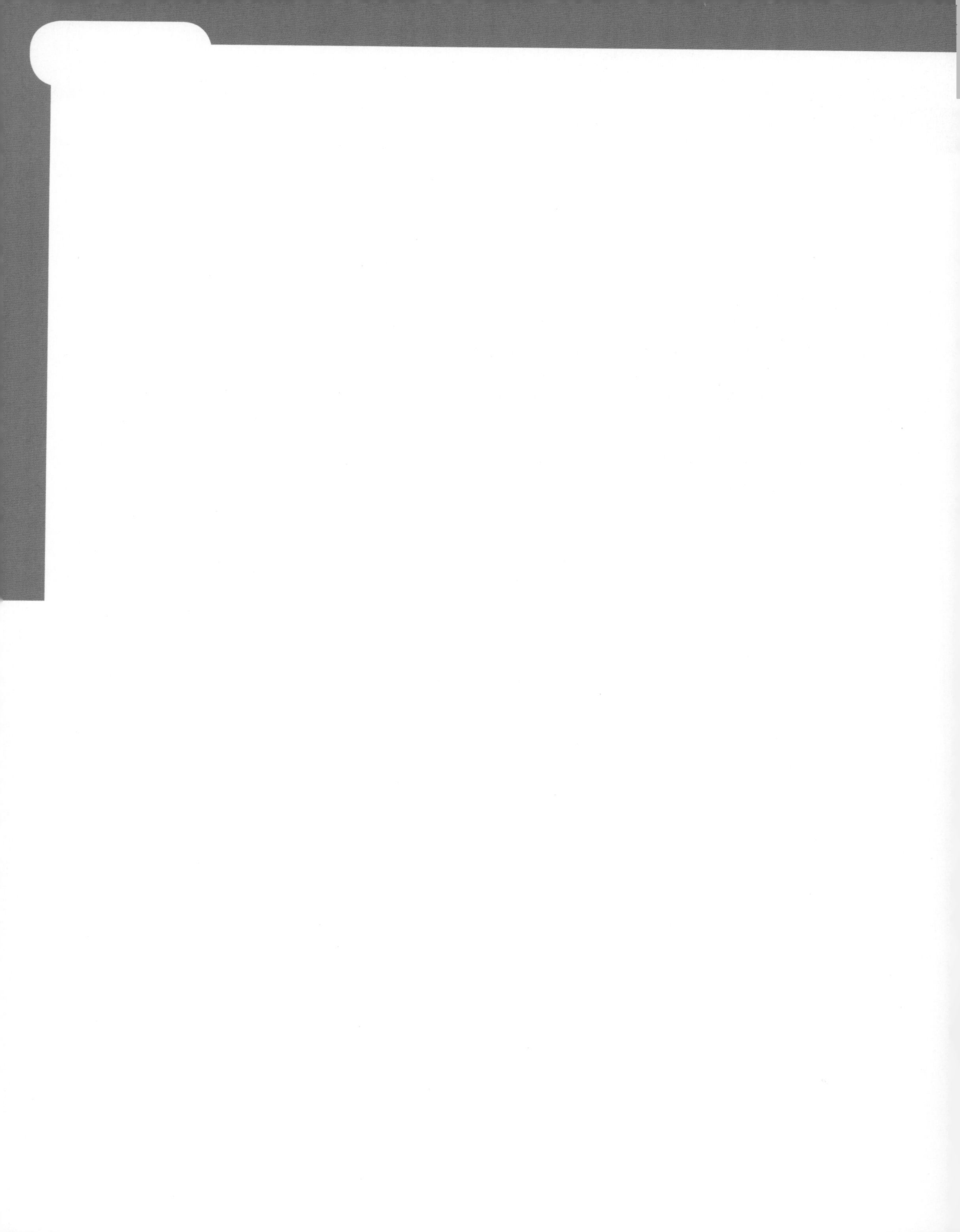

SECTION III

3.0 CANINE LANDSCAPE FIGURES (1–9)

N.S. Grenager[a], J.A. Orsini[b], John F. Randoph[c], Helen M.S. Davies[d], and A. de Lahunta[e]

[a]Steinbeck Peninsula Equine Clinics, Menlo Park, California, US

[b]Department of Clinical Studies - New Bolton Center, University of Pennsylvania School of Veterinary Medicine, Kennett Square, Pennsylvania, US

[c]Department of Clinical Sciences, Cornell University College of Veterinary Medicine, Ithaca, New York, US

[d]Veterinary Biosciences, University of Melbourne, Melbourne, AU

[e]Professor Emeritus, Biomedical Sciences, Cornell University College of Veterinary Medicine, Ithaca, New York, US

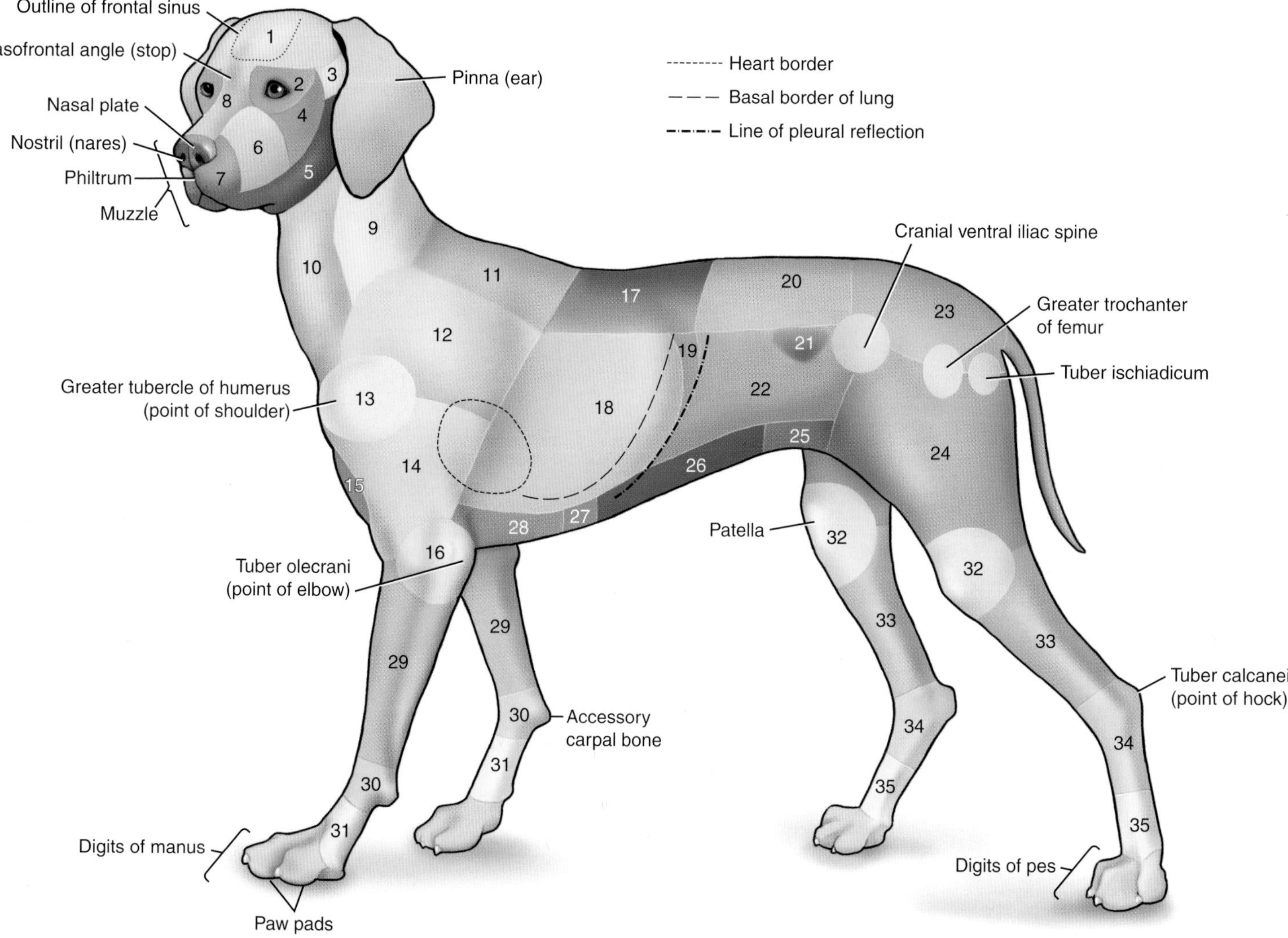

FIGURE 3.0-1 Regional canine anatomy.
Canine anatomical regions: 1, frontal; 2, orbital; 3, temporal; 4, zygomatic; 5, mandibular; 6, incisive (premaxillary); 7, maxillary; 8, nasal; 9, ventral cervical; 10, dorsal cervical; 11, interscapular; 12, thoracic appendage; 13, shoulder; 14, brachial; 15, presternal; 16, elbow; 17, dorsum (back); 18, costal; 19, hypochondriae; 20, lumbar (loin); 21, paralumbar fossa; 22, lateral abdominal; 23, sacral (croup); 24, femoral; 25, prepubic; 26, umbilical; 27, xiphoid; 28, sternal; 29, antebrachial (forearm); 30, carpal; 31, metacarpal; 32, stifle (knee); 33, crural (leg/gaskin); 34, tarsal (hock); 35, metatarsal.

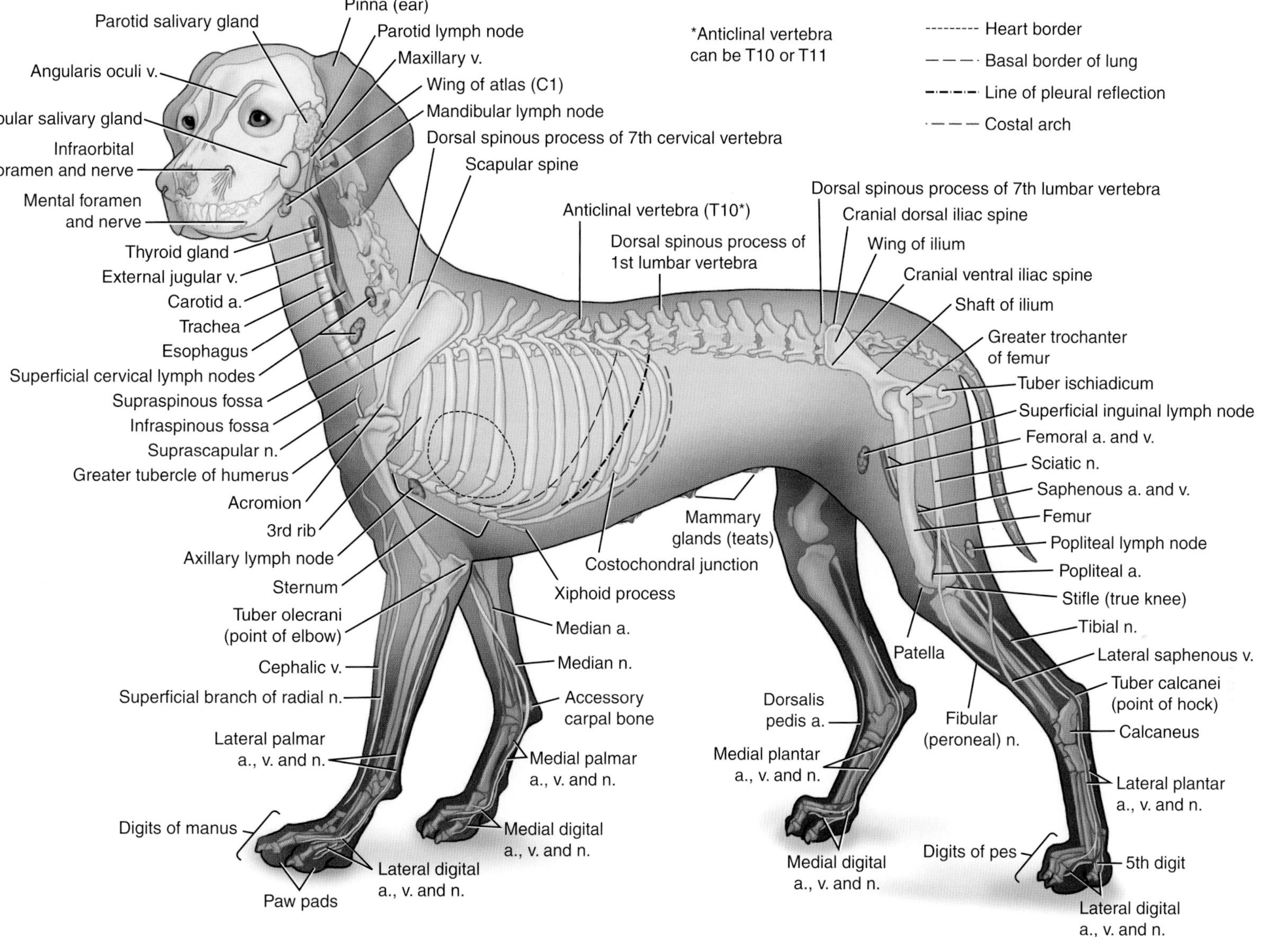

FIGURE 3.0-2 Topographical canine anatomy.

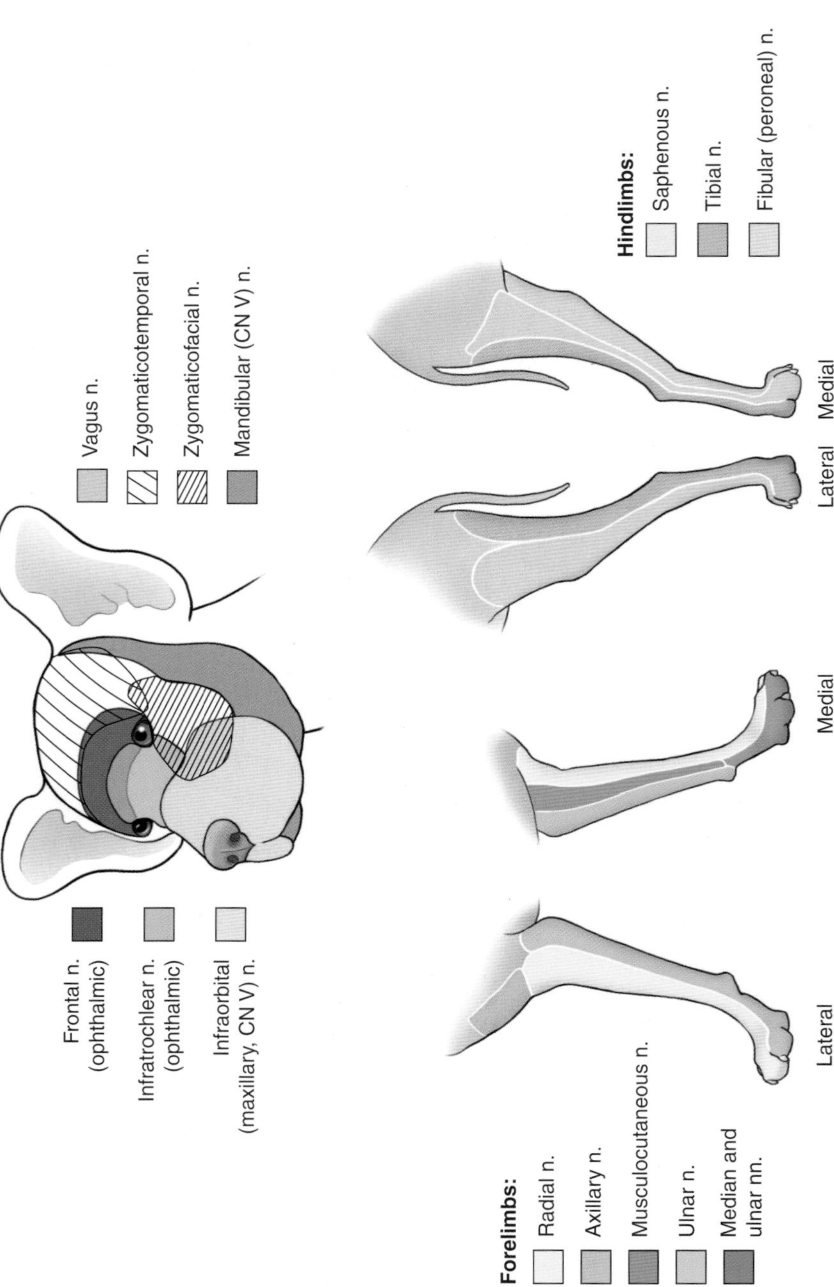

FIGURE 3.0-3 Dermatomes of the canine head and thoracic and pelvic limbs.

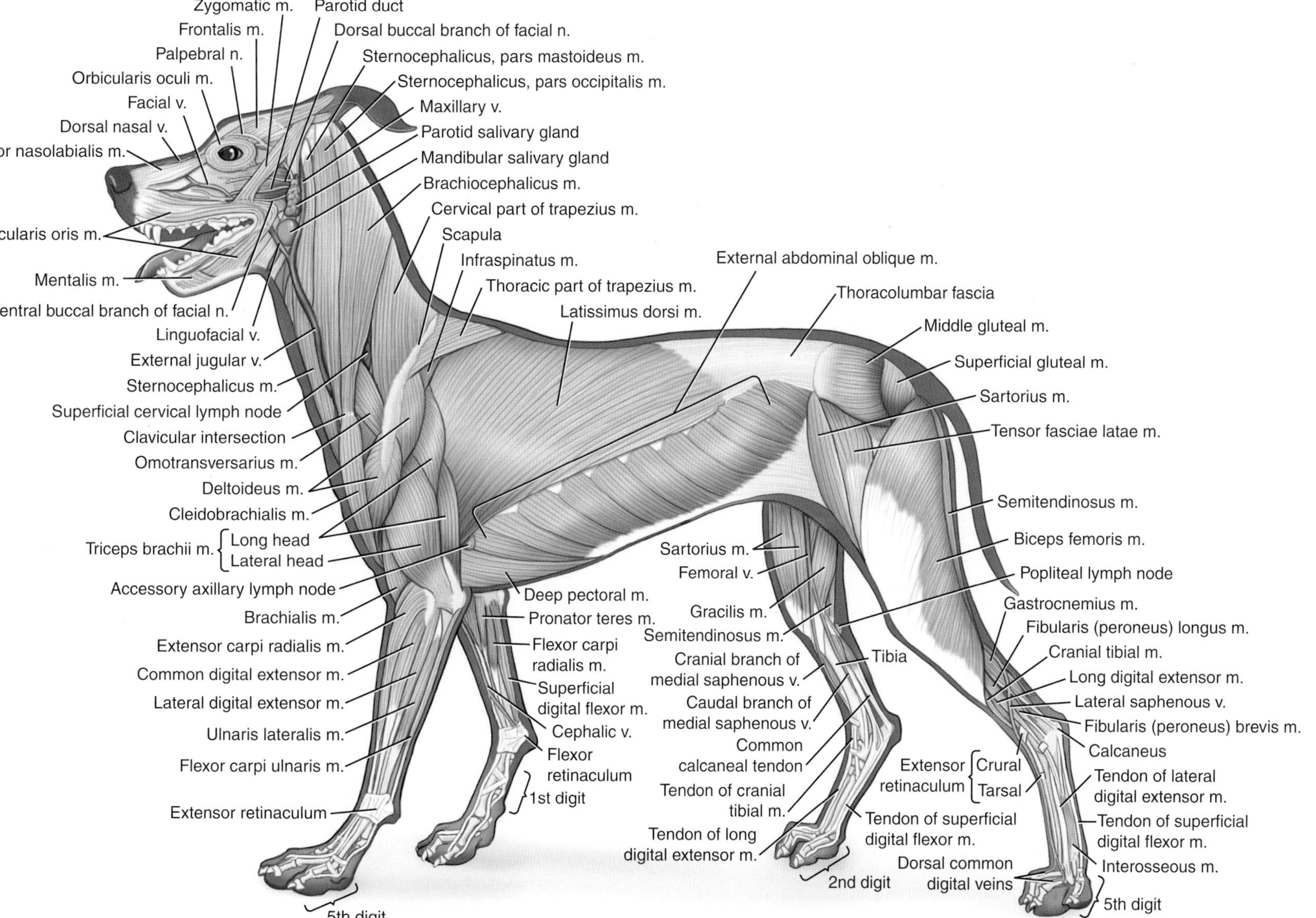

FIGURE 3.0-4 Superficial canine muscle layer.

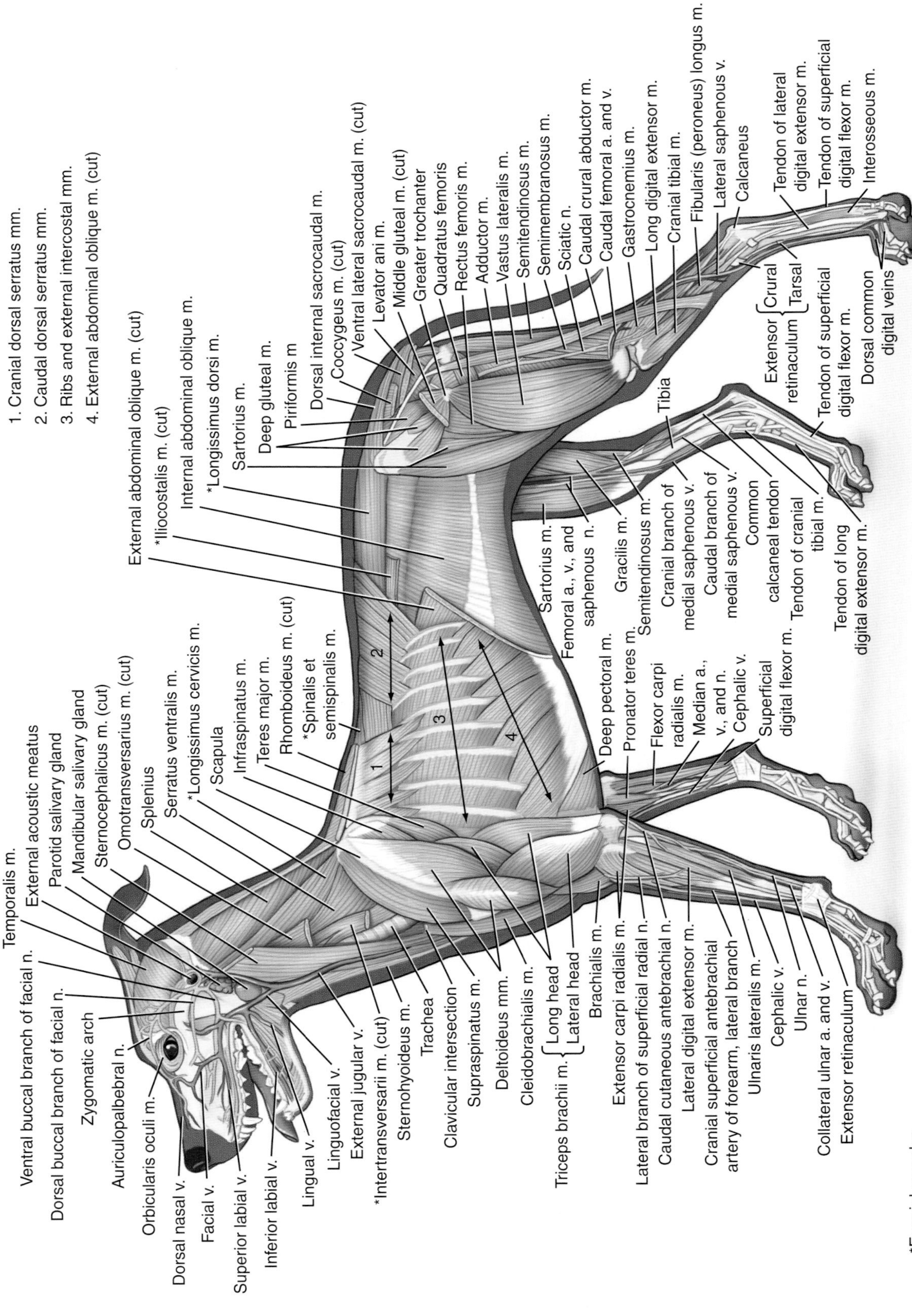

Temporalis m.
Ventral buccal branch of facial n.
External acoustic meatus
Dorsal buccal branch of facial n.
Parotid salivary gland
Mandibular salivary gland
Zygomatic arch
Sternocephalicus m. (cut)
Auriculopalbebral n.
Omotransversarius m. (cut)
Orbicularis oculi m.
Splenius
Dorsal nasal v.
Serratus ventralis m.
Facial v.
*Longissimus cervicis m.
Scapula
Superior labial v.
Infraspinatus m.
Inferior labial v.
Teres major m.
Rhomboideus m. (cut)
Lingual v.
*Spinalis et semispinalis m.
Linguofacial v.
External jugular v.
*Intertransversarii m. (cut)
Sternohyoideus m.
Trachea
Clavicular intersection
Supraspinatus m.
Deltoideus mm.
Cleidobrachialis m.
Triceps brachii m. {Long head, Lateral head}
Brachialis m.
Extensor carpi radialis m.
Lateral branch of superficial radial n.
Caudal cutaneous antebrachial n.
Lateral digital extensor m.
Cranial superficial antebrachial artery of forearm, lateral branch
Ulnaris lateralis m.
Cephalic v.
Ulnar n.
Collateral ulnar a. and v.
Extensor retinaculum

Deep pectoral m.
Pronator teres m.
Flexor carpi radialis m.
Median a., v., and n.
Cephalic v.
Superficial digital flexor m.

1. Cranial dorsal serratus mm.
2. Caudal dorsal serratus mm.
3. Ribs and external intercostal mm.
4. External abdominal oblique m. (cut)

External abdominal oblique m. (cut)
*Iliocostalis m. (cut)
Internal abdominal oblique m.
*Longissimus dorsi m.
Sartorius m.
Deep gluteal m.
Piriformis m
Dorsal internal sacrocaudal m.
Coccygeus m. (cut)
Ventral lateral sacrocaudal m. (cut)
Levator ani m.
Middle gluteal m. (cut)
Greater trochanter
Quadratus femoris
Rectus femoris m.
Adductor m.
Vastus lateralis m.
Semitendinosus m.
Semimembranosus m.
Sciatic n.
Caudal crural abductor m.
Caudal femoral a. and v.
Gastrocnemius m.
Long digital extensor m.
Cranial tibial m.
Fibularis (peroneus) longus m.
Lateral saphenous v.
Calcaneus
Tendon of lateral digital extensor m.
Tendon of superficial digital flexor m.
Interosseous m.

Sartorius m.
Femoral a., v., and saphenous n.
Gracilis m.
Semitendinosus m.
Tibia
Cranial branch of medial saphenous v.
Caudal branch of medial saphenous v.
Common calcaneal tendon
Tendon of cranial tibial m.
Tendon of long digital extensor m.
Extensor retinaculum {Crural, Tarsal}
Tendon of superficial digital flexor m.
Dorsal common digital veins

*Epaxial muscles

FIGURE 3.0-5 Deep canine muscle layer.

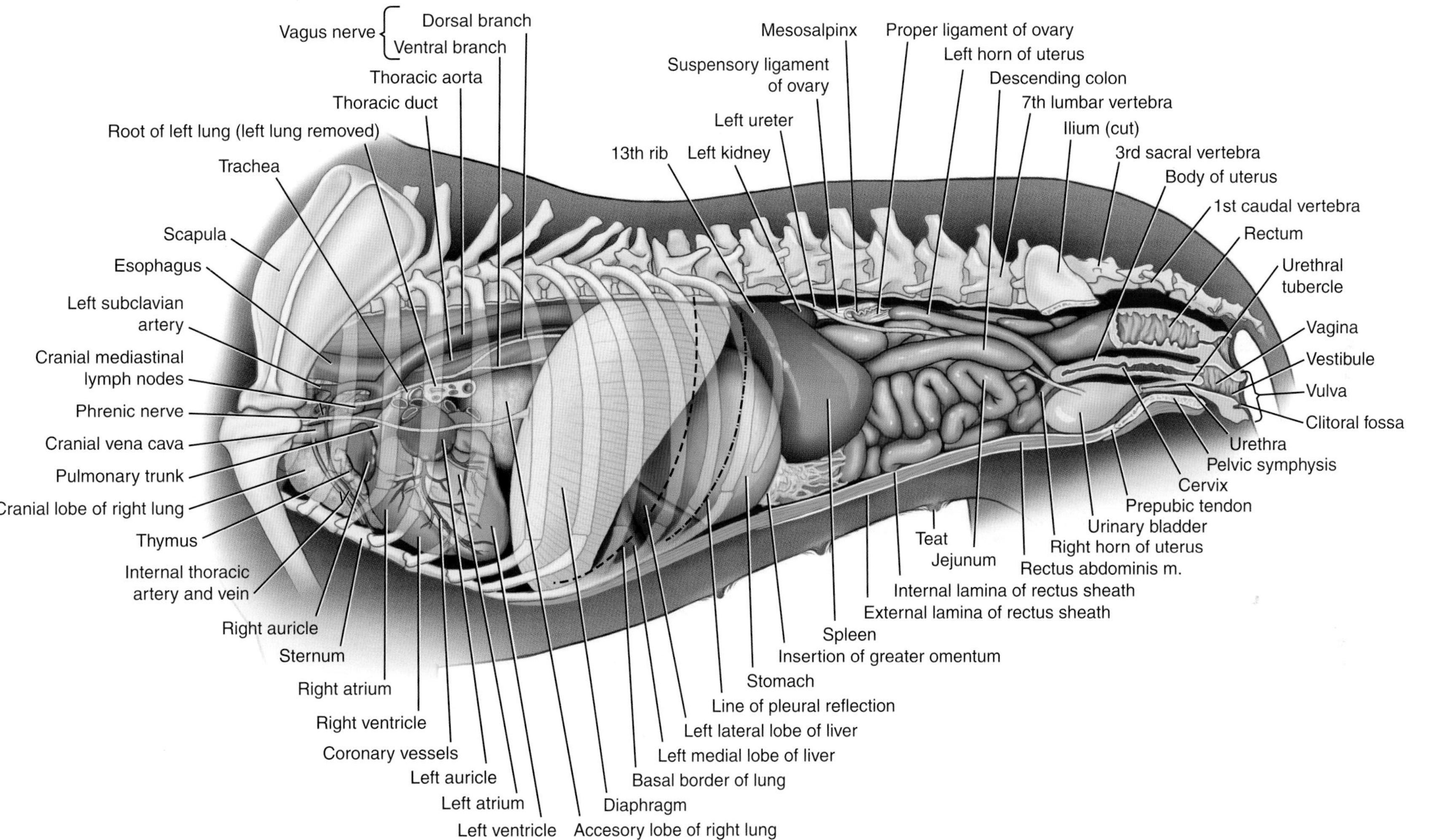

FIGURE 3.0-6 Left view of the canine thoracic and abdominal cavities and pelvis (female).

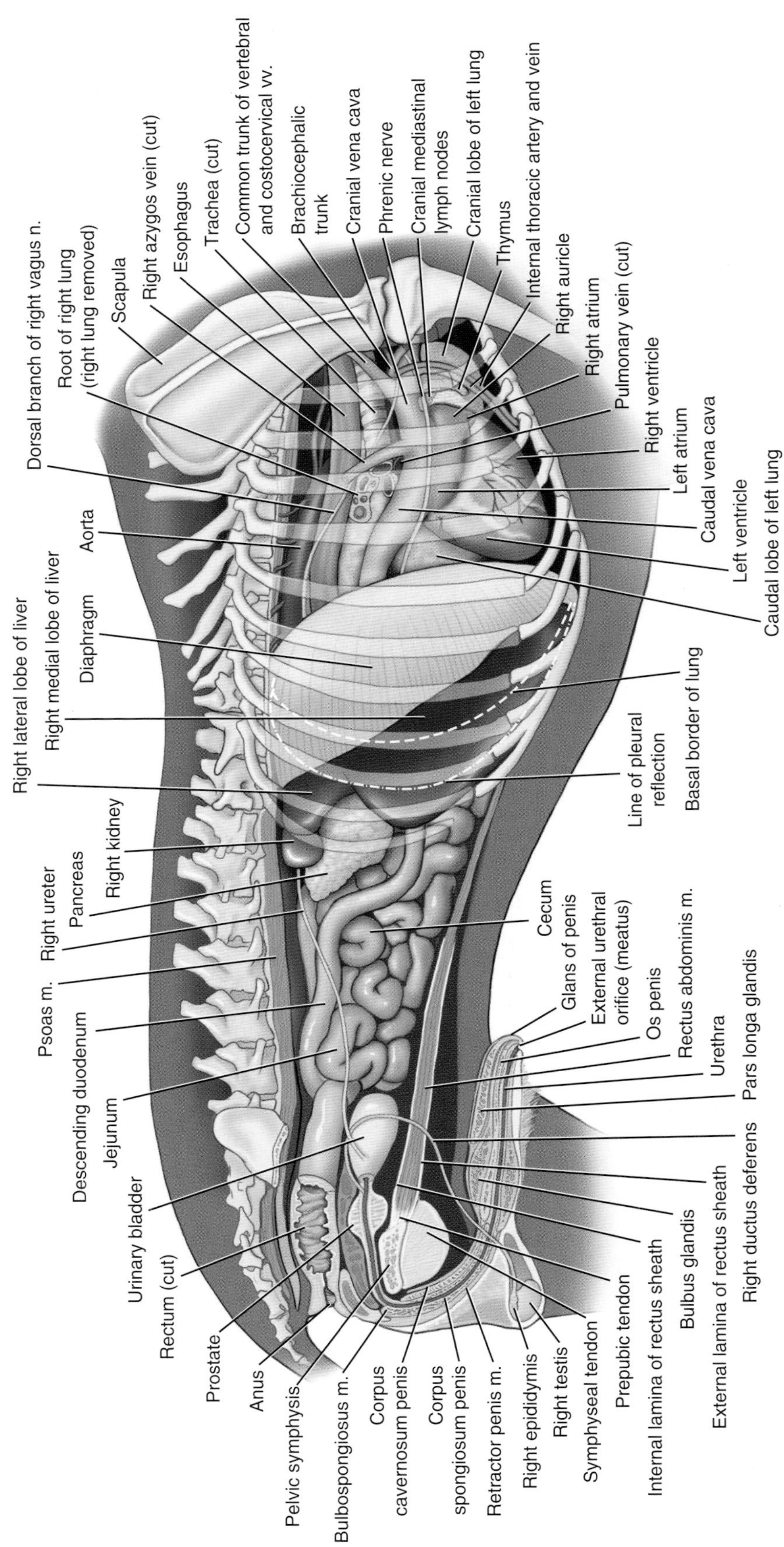

FIGURE 3.0-7 Right view of the canine thoracic and abdominal cavities and pelvis (male).

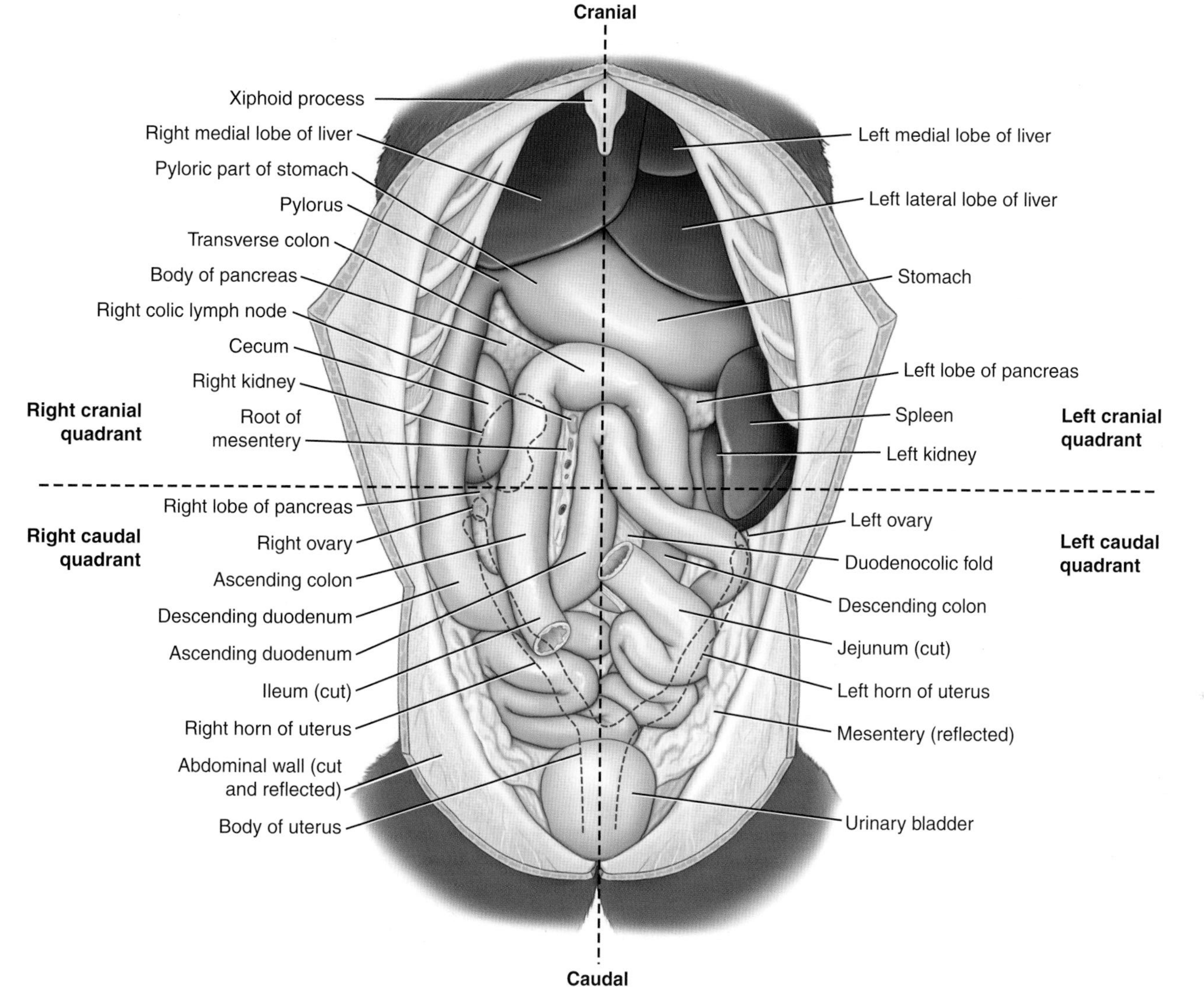

FIGURE 3.0-8 Ventral view of the canine abdominal cavity.

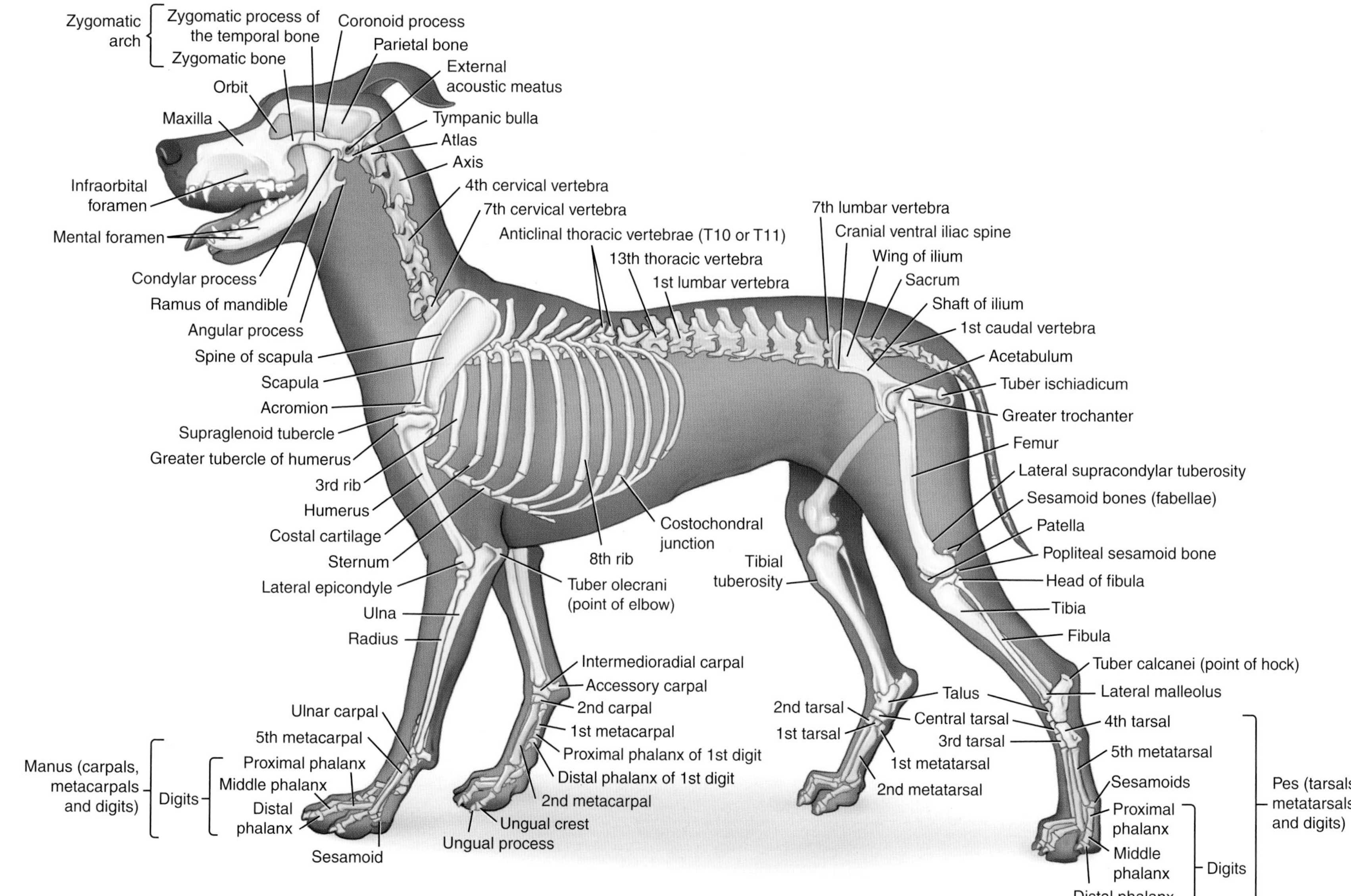

FIGURE 3.0-9 Lateral view of the canine skeleton. *Vertebral column formula*: C_7T_{13} L_7 S_3 $Cd_{20\text{-}23}$.

SECTION III

3.1 FELINE LANDSCAPE FIGURES (1–9)

N.S. Grenager[a], J.A. Orsini[b], J.F. Randolph[c], H.M.S. Davies[d], and A. de Lahunta[e]

[a]Steinbeck Peninsula Equine Clinics, Menlo Park, California, US

[b]Department of Clinical Studies - New Bolton Center, University of Pennsylvania School of Veterinary Medicine, Kennett Square, Pennsylvania, US

[c]Department of Clinical Sciences, Cornell University College of Veterinary Medicine, Ithaca, New York, US

[d]Veterinary Biosciences, University of Melbourne, Melbourne, AU

[e]Professor Emeritus, Biomedical Sciences, Cornell University College of Veterinary Medicine, Ithaca, New York, US

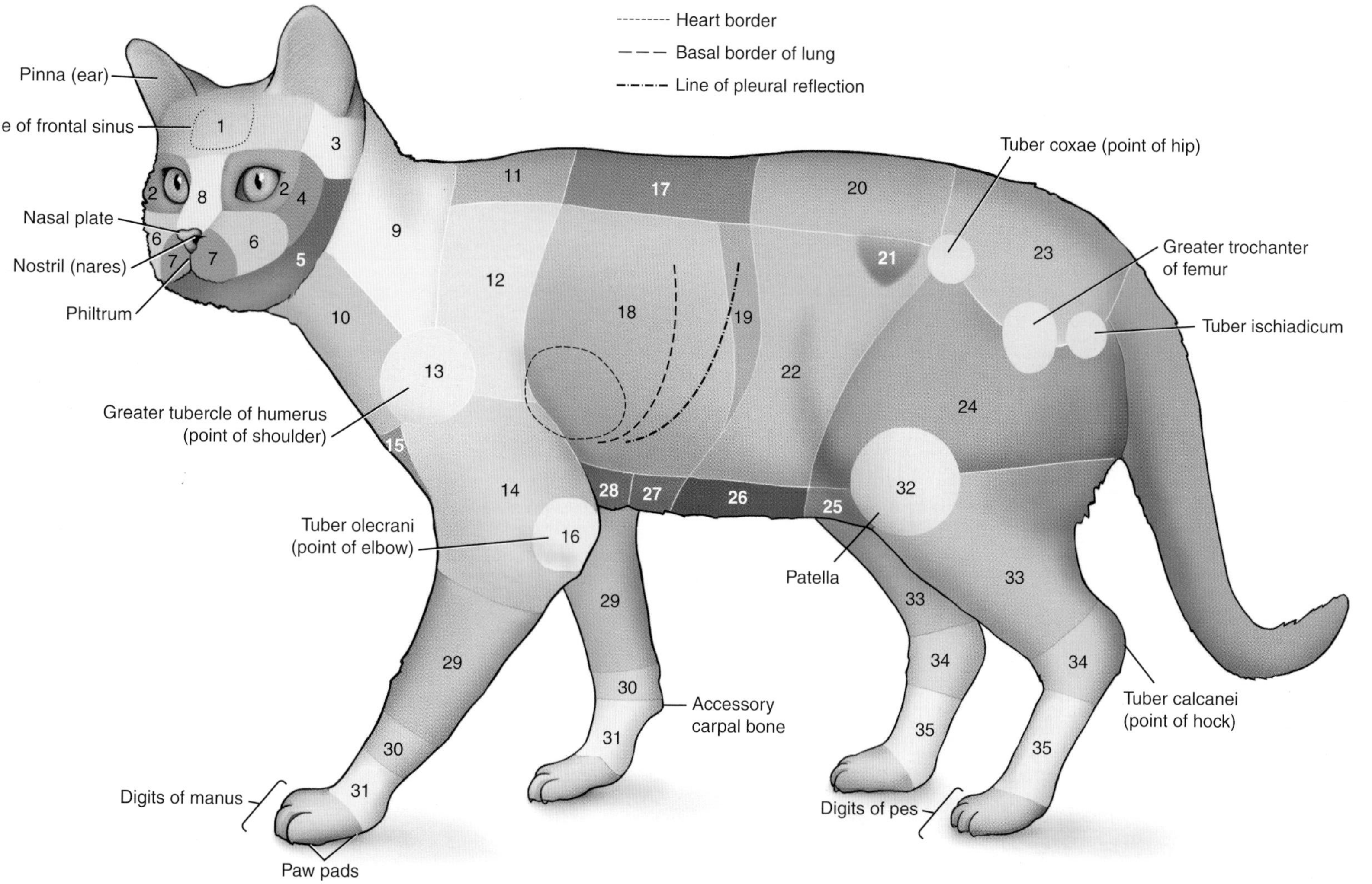

FIGURE 3.1-1 Regional feline anatomy.

Feline anatomical regions: 1, frontal; 2, orbital; 3, temporal; 4, zygomatic; 5, mandibular; 6, maxillary; 7, incisive (premaxillary); 8, nasal; 9, dorsal cervical; 10, ventral cervical; 11, interscapular; 12, scapular; 13, shoulder; 14, brachial; 15, presternal; 16, elbow; 17, dorsum (back); 18, costal; 19, hypochondriae; 20, lumbar (loin); 21, paralumbar fossa; 22, lateral abdominal; 23, sacral (croup); 24, femoral; 25, prepubic; 26, umbilical; 27, xiphoid; 28, sternal; 29, antebrachial (forearm); 30, carpal; 31, metacarpal; 32, stifle (true knee); 33, crural (leg/gaskin); 34, tarsal (hock); 35, metatarsal.

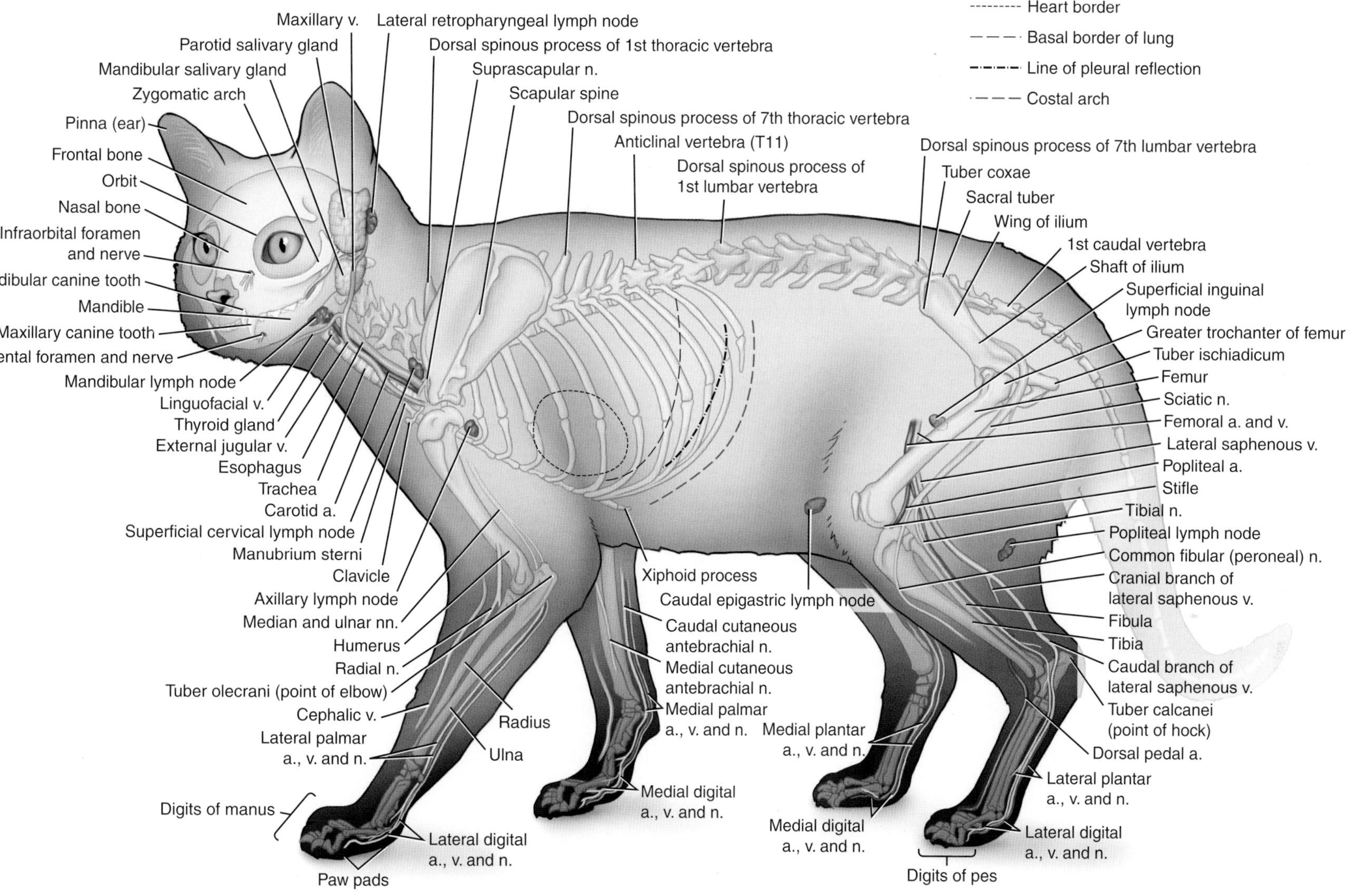

FIGURE 3.1-2 Topographical feline anatomy.

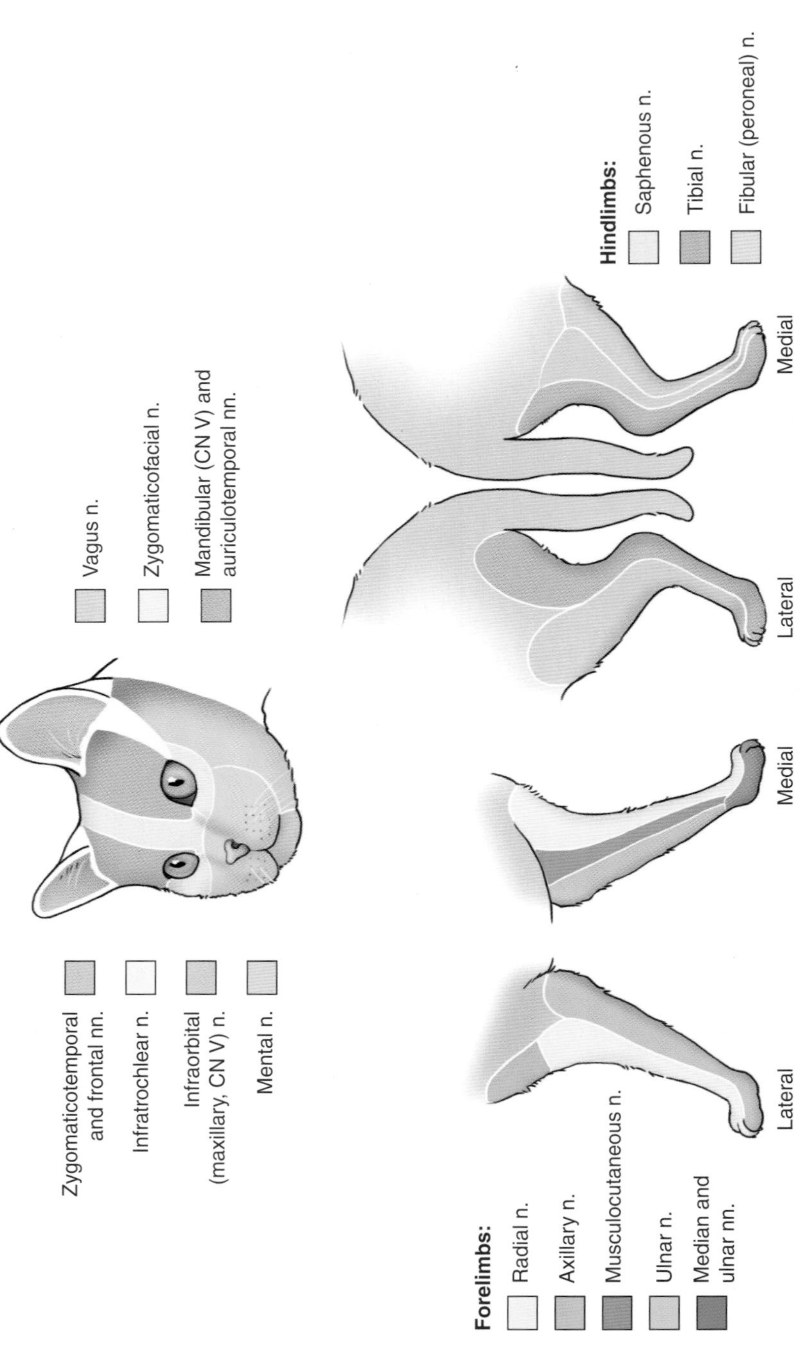

FIGURE 3.1-3 Dermatomes of feline head and thoracic and pelvic limbs.

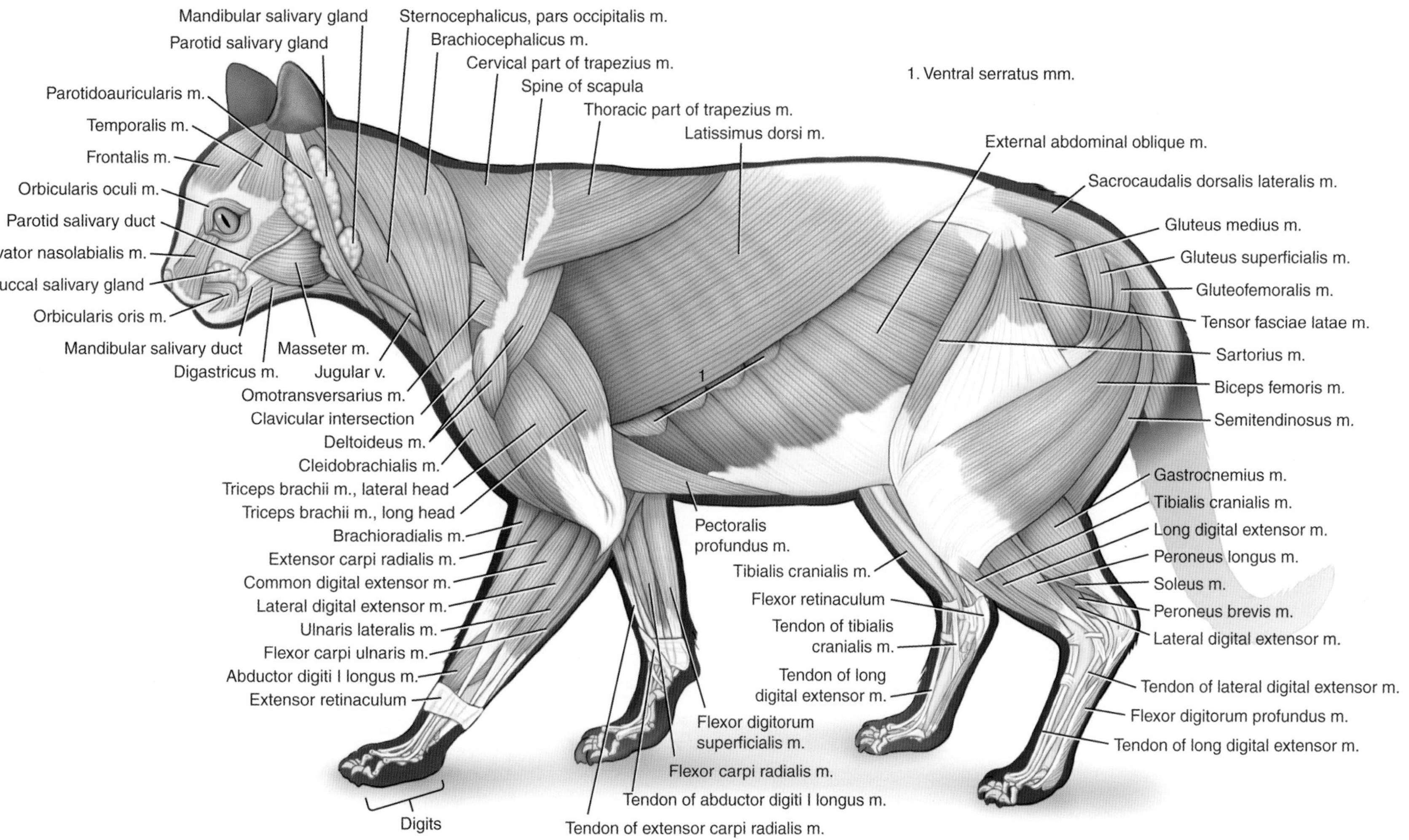

FIGURE 3.1-4 Superficial feline muscle layer.

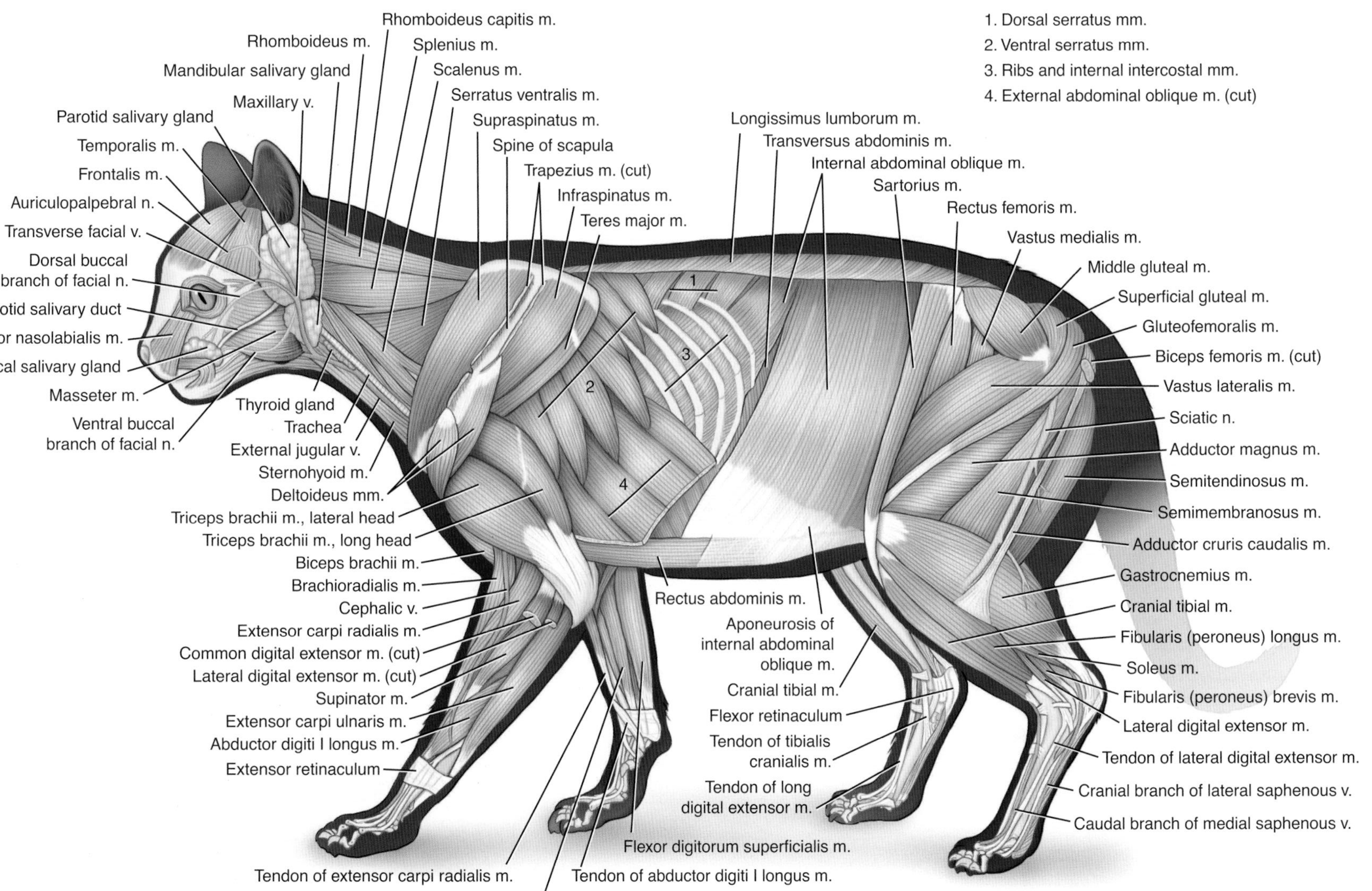

FIGURE 3.1-5 Deep feline muscle layer.

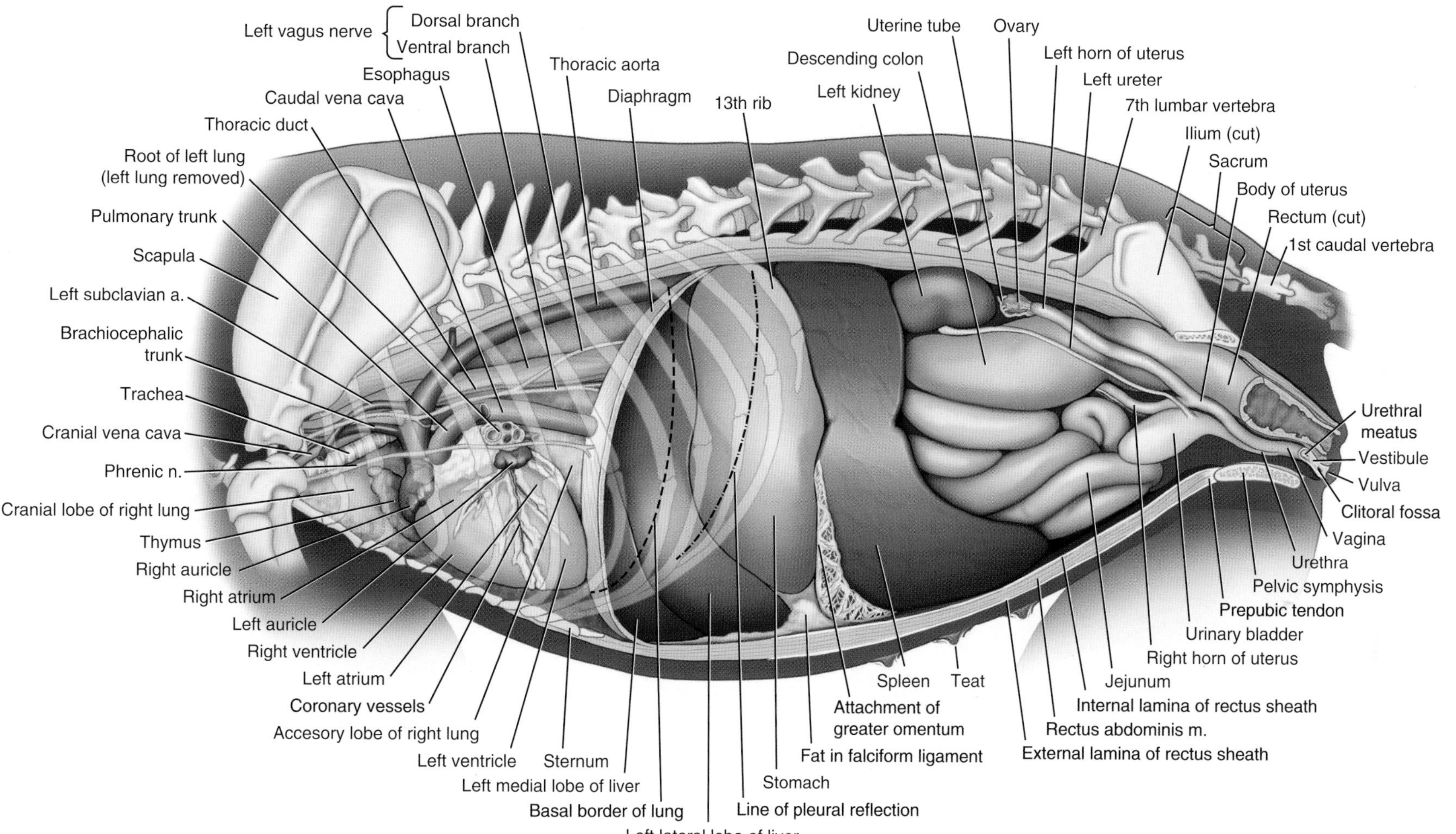

FIGURE 3.1-6 Left view of the feline thoracic and abdominal cavities and pelvis (female).

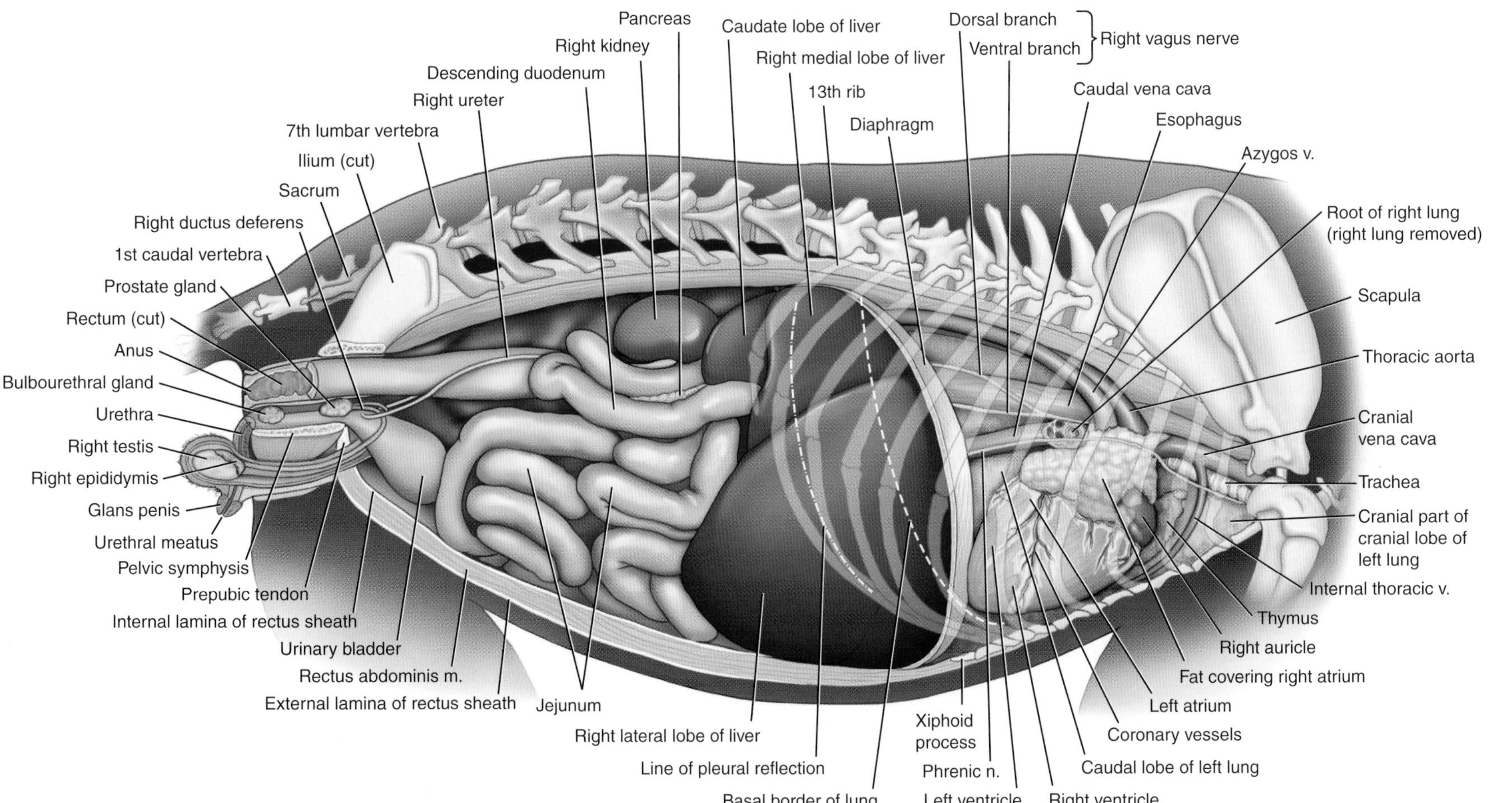

FIGURE 3.1-7 Right view of the feline thoracic and abdominal cavities and pelvis (male).

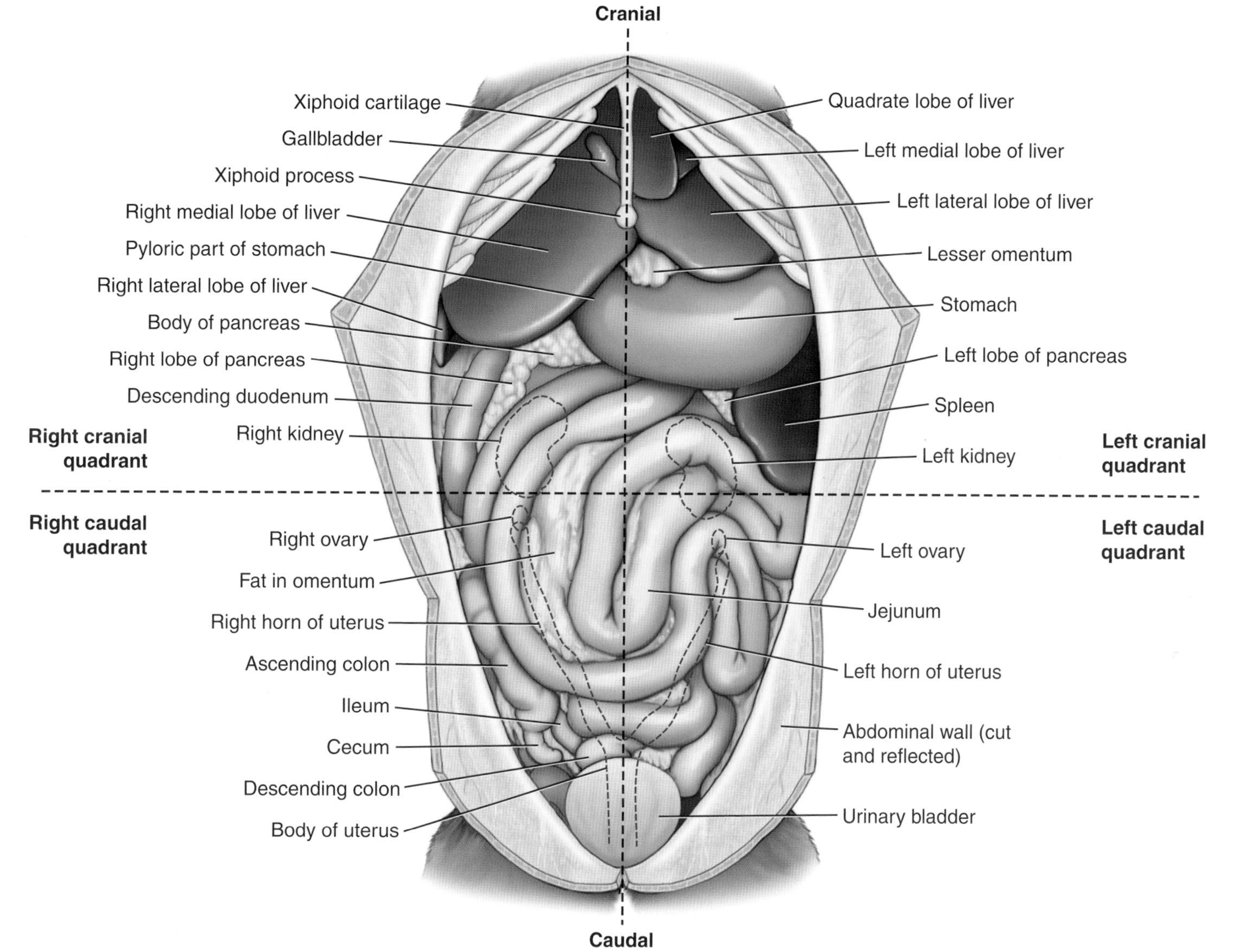

FIGURE 3.1-8 Ventral view of the feline abdominal cavity.

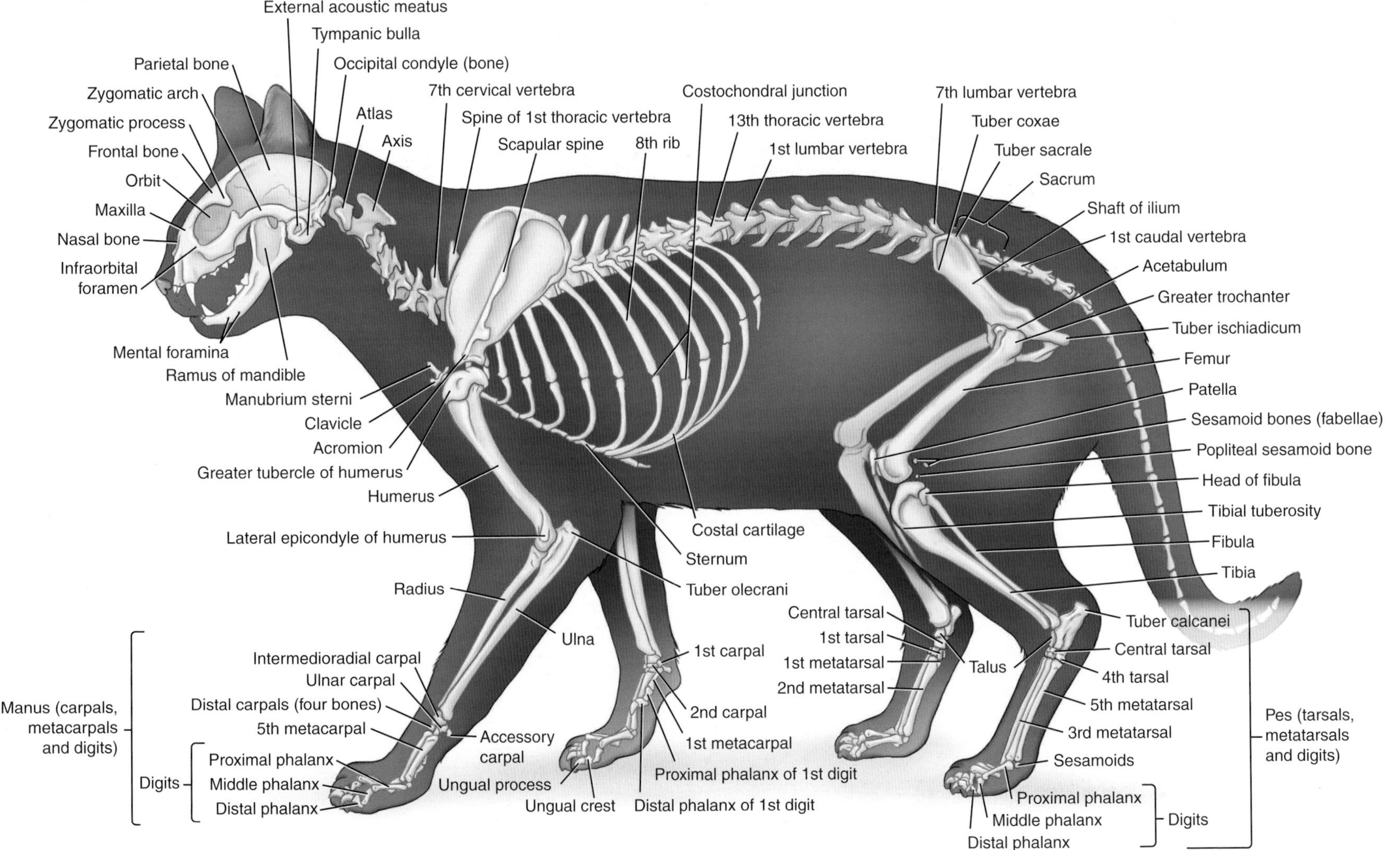

FIGURE 3.1-9 Lateral view of the feline skeleton. *Vertebral column formula:* C_7T_{13} L_7 S_3 Cd_{5-23}.

CHAPTER 3

AXIAL SKELETON: HEAD, NECK, AND VERTEBRAL COLUMN

John F. Randolph, Chapter editor

Nasal Cavity, Pharynx, and Paranasal Sinuses

3.1 Nasopharyngeal polyp—*Meredith Miller* 104
3.2 Brachycephalic airway syndrome—*David Holt* 112

Mouth

3.3 Odontogenic disease—*Nadine Fiani and Santiago Peralta* 119

Eye

3.4 Retrobulbar mass—*Eric Ledbetter* 126

Ear

3.5 Otitis interna/media—*Adalberto Merighi and Laura Lossi* 133

Neck

3.6 Thyroid adenocarcinoma—*Takanori Sugiyama and Helen M.S. Davies* 146
3.7 Hyperthyroidism—*Mark E. Peterson and John F. Randolph* 164

Vestibular System, Brain, and Vertebral Column

3.8 Cervical intervertebral disc disease—*Takanori Sugiyama and Helen M.S. Davies* 170
3.9 Vestibular dysfunction—*Marc Kent and Eric N. Glass* 181
3.10 Glioma—*Marc Kent and Eric N. Glass* 188
3.11 Meningioma—*Fred Wininger* 196
3.12 Lumbar intervertebral disc disease—*Marc Kent and Eric N. Glass* 207

CHAPTER 3

CASE 3.1

Nasopharyngeal Polyp

Meredith Miller
Department of Clinical Sciences, Cornell University College of Veterinary Medicine, Ithaca, New York, US

Clinical case

History

A 1-year-old female spayed Persian cat presented with a 3-month history of progressive bilateral, whitish-yellow nasal discharge, noisy breathing, sneezing, and occasional open-mouth breathing after play. Her appetite and activity levels were normal, and no other abnormalities were reported. She was an indoor-only cat, adopted from an animal shelter at 4 months of age, at which time she tested negative for feline leukemia virus (FeLV) and feline immunodeficiency virus (FIV).

"Referred" sounds are those that are heard in a location distant and distinct from their origin. In this case, the stertor arising from the upper airways was heard as abnormal lung sounds on thoracic auscultation. Stertor often causes referred sounds that can be confused for abnormal lung sounds. When such sounds are heard loudest over the trachea, it indicates that they originate in the upper airways.

Physical examination findings

The cat was bright, alert, and active. She had the typical stenotic nares of a brachycephalic breed, along with decreased airflow through both nostrils, prolonged inspiratory phase of respiration, and pronounced stertor. In addition, she had bilateral mucopurulent nasal discharge and sneezed frequently during the examination. On thoracic auscultation, the lung sounds were bilaterally harsh and increased in intensity, but they were heard loudest over the trachea, consistent with referred upper airway sounds. No other abnormalities were found on physical examination.

UPPER RESPIRATORY NOISES

Brachycephalic breeds such as the Persian cat are prone to partial airway obstruction caused by stenotic nares (narrowed nostrils) and/or distorted nasal passages. Stertor (L. *stertō* snore) refers to a snoring or snorting sound caused by soft tissue obstruction of the nasal cavity or nasopharynx (Fig. 3.1-V1 in the online version at https://doi.org/10.1016/B978-0-323-91015-6.00012-1). It is clearly audible without the aid of a stethoscope and may be alleviated by opening the mouth (open-mouth breathing), which creates an unobstructed pathway for airflow.

Stridor is a typically high-pitched upper respiratory noise associated with obstruction of the larynx or trachea.

Comparative Veterinary Anatomy: A Clinical Approach. https://doi.org/10.1016/B978-0-323-91015-6.00012-1

Differential diagnoses

Feline herpesvirus-1 (feline viral rhinotracheitis) or feline calicivirus infection, *Cryptococcus neoformans* rhinitis, chronic rhinosinusitis, allergic rhinitis, cuterebriasis (nasal bots—*Cuterebra* spp.), brachycephalic conformation, nasopharyngeal polyp, nasal or nasopharyngeal neoplasia, nasal foreign body, oronasal fistula, and nasopharyngeal stenosis

Diagnostics

The cat was placed under general anesthesia and the oral cavity and pharynx were visually and digitally examined. During palpation of the palate, a firm mass was felt in the nasopharynx dorsal to the soft palate. A spay hook was used to retract the caudal margin of the soft palate, which exposed a smooth, grayish-pink mass in the nasopharynx, most consistent with an inflammatory polyp. (A similar nasopharyngeal polyp in an older, domestic longhair cat is shown in Fig. 3.1-1.)

Although unnecessary in this case, other diagnostic options include radiography or CT of the skull and retrograde endoscopy of the nasopharynx and nasal cavity (retrograde rhinoscopy (Gr. *rhinos* nose + Gr. *skopein* to examine)). When a polyp is too small or located too far rostrally to appreciate during oropharyngeal examination, radiography may reveal a soft tissue mass within the nasopharynx (nasopharyngeal polyp), loss of air contrast in the external ear canal (middle ear polyp), and/or increased radiodensity of the tympanic bulla. Computed tomography requires specialized equipment and expertise, but it provides greater detail because it eliminates superimposition of the bony structures in the area of interest. Use of contrast-enhanced CT highlights the margins or "rim" of the polyp, more clearly defining its body and stalk (Fig. 3.1-2). In addition, rim enhancement helps distinguish an inflammatory polyp from fluid accumulation in the tympanic cavity or a neoplastic mass.

Retrograde rhinoscopy can be performed using a flexible endoscope, inserted into the oral cavity, and retroflexed around the caudal margin of the soft palate. It allows visualization of the nasopharynx, nasal choanae, and caudal nasal passages, and if indicated, biopsy of suspected lesions.

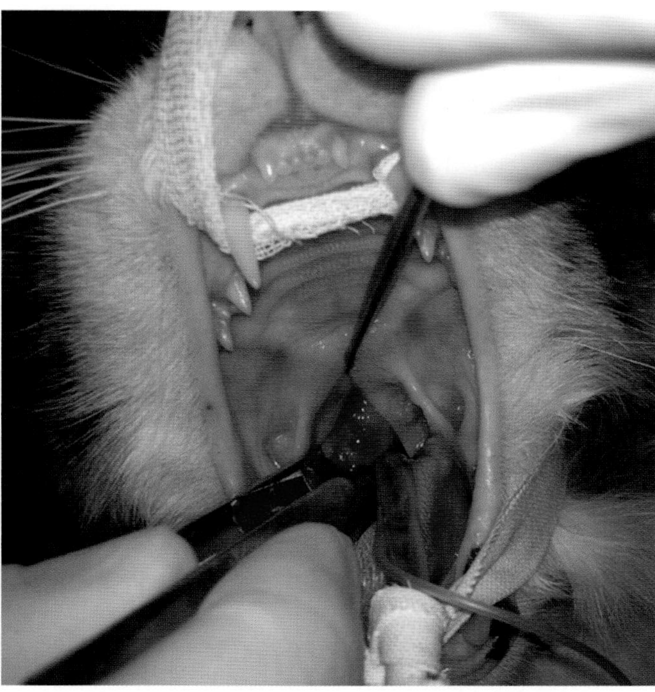

FIGURE 3.1-1 A nasopharyngeal polyp in a cat. With the cat under general anesthesia, retraction of the soft palate using a spay hook *(upper right)* has exposed the polyp in the nasopharynx. A forceps handle *(lower left)* is used to draw the body of the polyp forward for better visualization. The endotracheal tube *(lower right)* is seen lying against the tongue and running under the polyp. (Courtesy of James Flanders, DVM, DACVS; Cornell University.)

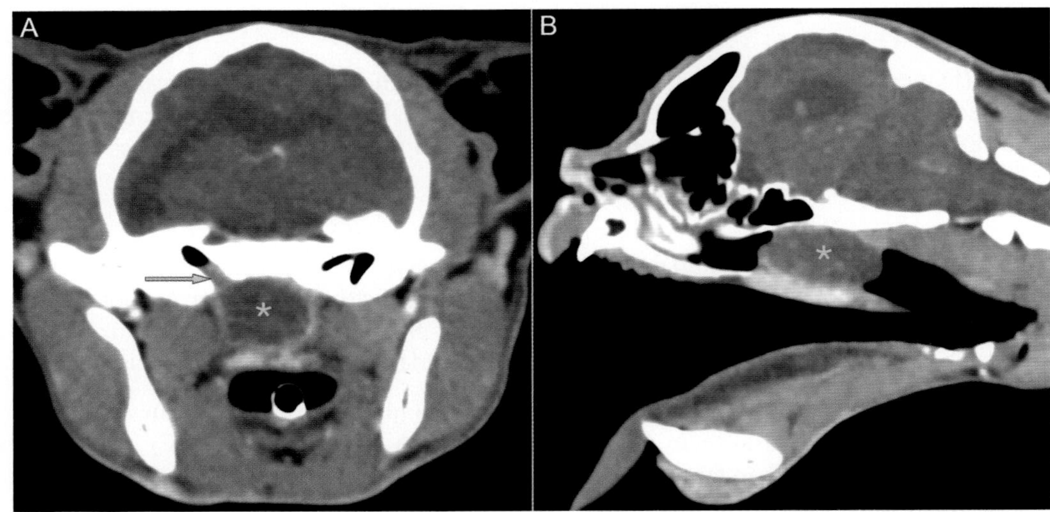

FIGURE 3.1-2 Contrast-enhanced CT images of a nasopharyngeal polyp (*) in the transverse (A) and sagittal (B) planes. Most obvious in image B, this large polyp lies dorsal to the soft palate in the rostral nasopharynx, just caudal to the hard palate. Rim enhancement outlines the polyp, clearly differentiating it in image A from the surrounding soft tissues and revealing its origin in the auditory tube *(arrow)* on that side.

Diagnosis

Nasopharyngeal polyp

Treatment

The polyp was removed under general anesthesia. After retracting the caudal margin of the soft palate with a spay hook, curved hemostats were used to grasp the polyp; then firm traction and rotation were applied to detach the polyp and its stalk—a surgical procedure termed "traction-avulsion" (Fig. 3.1-V2 in the online version at https://doi.org/10.1016/B978-0-323-91015-6.00012-1). In this case, staphylotomy (incision of the soft palate to access the rostral nasopharynx) was not necessary. Post-operatively, the cat received broad-spectrum oral antimicrobial therapy and a tapering anti-inflammatory regimen of corticosteroids.

FELINE INFLAMMATORY POLYPS

Polyps are the most common nonneoplastic growths found in the nasopharynx or external ear canal in cats. They arise from the epithelial lining of the tympanic cavity or auditory tube and may be unilateral (most common) or bilateral. Histopathology reveals a loosely arranged fibrovascular core, covered by stratified squamous or ciliated columnar epithelium.

The polyp may grow through the auditory tube into the nasopharynx (nasopharyngeal polyp), through the tympanic membrane into the external ear canal (middle ear polyp), or occasionally in both directions. The tympanic bulla is generally filled with tissue and fluid.

This condition is more common in young cats, but it occurs in cats of all ages. The cause is unknown, but proposed causes include a congenital abnormality of the branchial arch, chronic viral infection, and chronic middle ear infection (otitis media). Recurrence months or even years after removal is common (33–57% of cases).

Clinical anatomy in canids and felids

Introduction

This section describes the anatomy of the nasopharynx and ear as they relate to one another, along with regional blood supply, lymphatics, and innervation. Additional information on the ear of small animals is found in Case 3.5.

Function

The nasal cavity provides for the passive transit of air and communicates with the middle ear via the auditory (Eustachian) tubes, serving as a pressure equilibration system between the two sides of the eardrum. The ear, comprised of the three main parts—external, middle, and internal compartments—is the hearing mechanism for animals and man and in addition plays a role in the "sense of balance" and therefore is also referred to as the vestibulocochlear organ (L. *vestibularis* a space or cavity + L. *cochlea* snail shell). The vestibulocochlear n. is (CN VIII) comprised of the vestibular n., important for balance, and the cochlear n., important for hearing.

Anatomy of the nasopharynx

The relationship between the nasopharynx and the middle ear is shown in Fig. 3.1-3. The **nasopharynx** extends from the caudal openings of the nasal cavity, the **choanae** (singular, **choana**, left and right), to the caudal margin of the soft palate and the paired **palatopharyngeal arches** where the soft palate meets the wall of the pharynx. The nasopharynx is bounded dorsally by the base of the skull and ventrally by the soft palate.

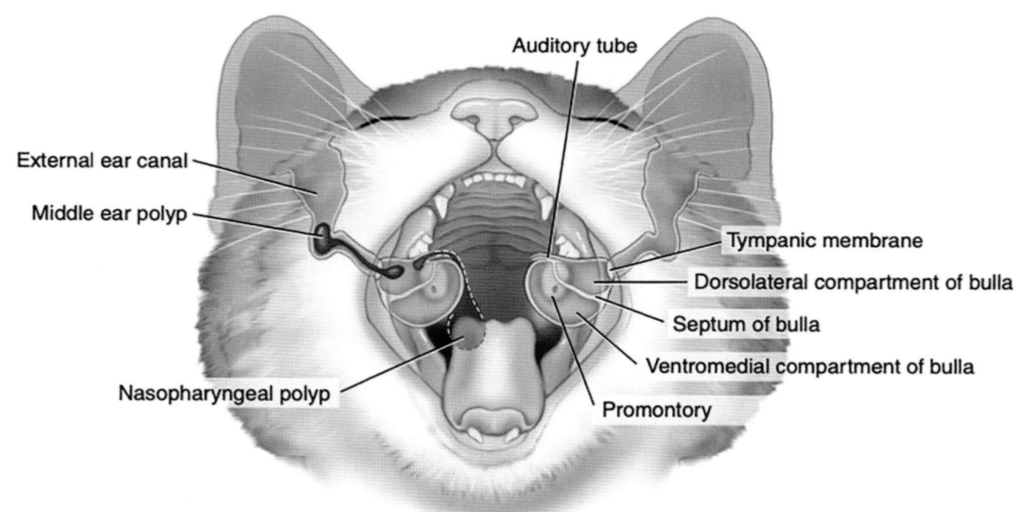

FIGURE 3.1-3 Relationship of the nasopharynx, auditory tube, and tympanic cavity. Inflammatory polyps originating in the tympanic cavity or auditory tube may extend into the nasopharynx (nasopharyngeal polyp) or expand within the middle ear (middle ear polyp), eventually breaching the tympanic membrane and invading the external ear canal.

VENTRAL BULLA OSTEOTOMY (VBO)

This procedure is recommended by some clinicians as part of the treatment for nasopharyngeal polyps, as middle ear involvement is common and such polyps have a high recurrence rate (up to 50%) when removed via traction-avulsion. Surgical access to the tympanic cavity via its ventral bony wall allows complete exploration of both chambers of the feline "double-chambered" tympanic bulla (described below). The polyp is removed, and then, the epithelial lining and accumulated material within the bulla are debrided using a curette (a sharp, spoon-shaped surgical instrument). Although this procedure reduces the recurrence rate (to less than 33%), it requires surgical expertise; is expensive and time-consuming; and has a higher rate of complications than traction-avulsion, so it is usually reserved for cases in which polyps recurred after less invasive procedures.

On each side of the nasopharynx, the small, slit-like pharyngeal opening of the **auditory tube** (Eustachian tube equivalent in humans) is located adjacent to the **tonsil**. The auditory tube extends from the tympanic cavity of the middle ear (described later) to the nasopharynx. It functions as a pressure-equalizer across the tympanic membrane (eardrum) between the middle ear and the external acoustic meatus (external ear canal).

Blood supply, lymphatics, and innervation of the nasopharynx

The blood supply in general to the nasal cavity, including the nasopharynx, is provided by several arteries and their corresponding veins: the **sphenopalatine**, **ethmoidal**, and **greater palatine arteries** and **veins** Lymphatic vessels empty into the **mandibular** and **retropharyngeal lymph nodes**. Innervation is primarily provided by the **olfactory** and **trigeminal nerves**.

Anatomy of the ear (see also Case 3.5)

The ear is divided into three parts: the inner ear, middle ear, and external ear.

The **inner ear** consists of a fluid-filled, **membranous labyrinth** within a **bony labyrinth**, all contained within the **petrous temporal bone**. It houses the end-organs of the **vestibulocochlear nerve** (CN VIII): the vestibular apparatus and the cochlear apparatus.

The **vestibular apparatus** is responsible for the sense of orientation in space, and thus equilibrium or balance. It principally consists of the **semicircular canals** and is supplied by the vestibular branch of CN VIII. The **cochlear apparatus** is responsible for the sense of hearing. It principally consists of the **cochlea** and is supplied by the **cochlear branch** of CN VIII, also called the **auditory** or **acoustic nerve**. These two components are connected by the **vestibule** of the inner ear; combined, these structures comprise the labyrinth (Figs. 3.1-4 and 3.1-5).

The vestibulocochlear nerve exits the cranial cavity via the **internal acoustic meatus**, along with the **facial nerve** (CN VII). It then divides into the vestibular and cochlear nerves, which innervate their respective components of the inner ear.

> Because of their location, CN VII and VIII are vulnerable to injury in diseases involving the middle/inner ear. (Further description of these nerves is found in Case 3.9).

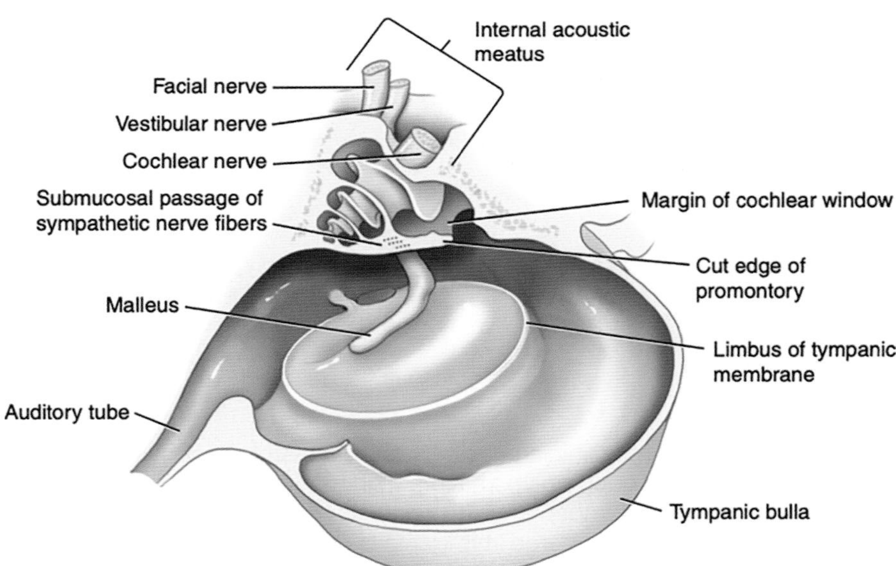

FIGURE 3.1-4 Middle ear, with the promontory cut to show the cochlea. Note the proximity of the facial nerve (CN VII) and the vestibular nerve (vestibular branch of CN VIII), and the sympathetic innervation to the eye along the promontory of the temporal bone that protrudes into the tympanic cavity. These structures are vulnerable to injury from inflammation, infection, or surgical trauma in patients with middle ear disease. (Image adapted from Miller and Evans' Anatomy of the dog. 5th ed.)

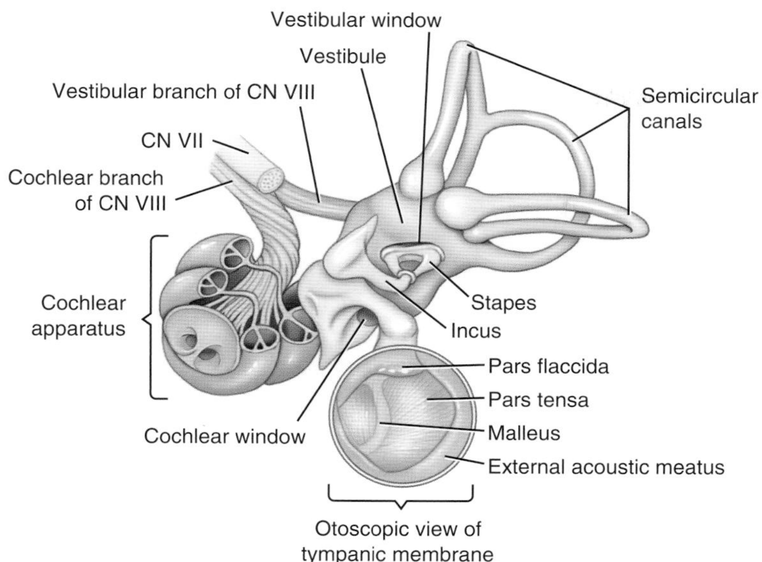

FIGURE 3.1-5 Relationship of middle and inner ear. The external acoustic meatus is separated from the tympanic cavity by the tympanic membrane which is divided into two parts: the pars flaccida and pars tensa. The pars tensa transmits sound vibrations to the cochlea through the auditory ossicles. The manubrium of the malleus is attached to the pars tensa and can be visualized on otoscopic examination. The inner ear is comprised of the semicircular canals, cochlear apparatus, and connecting vestibule; the latter contains the vestibular window (occupied by the stapes) and the cochlear window. The osseus labyrinth is shown here with a section of osseous cochlea removed to demonstrate the membranous cochlear duct within.

The **middle ear** is the normally air-filled compartment between the external acoustic meatus and the inner ear. It transmits sound waves to the inner ear via the **cochlear window** and the three small **auditory ossicles—malleus**, **incus**, and **stapes**. The malleus is attached to the center of the tympanic membrane and transmits vibrations to the incus and stapes, the latter of which occupies the **vestibular window** of the inner ear (Fig. 3.1-5).

The central feature of the middle ear is the **tympanic cavity**. It consists of the epitympanic recess dorsally, the tympanic cavity proper, and the ventral or fundic portion within the tympanic bulla. The **epitympanic recess** is the relatively narrow part of the tympanic cavity dorsomedial to the external acoustic meatus. It is the smallest part and mostly contains the malleus and incus.

The **tympanic cavity proper** is the middle portion, opposite the tympanic membrane. An important feature is the **promontory**, a bony eminence of the temporal bone that houses the cochlear apparatus and protrudes into the tympanic cavity opposite the tympanic membrane. The cochlear and vestibular windows are located on its lateral surface. The **sympathetic nerves of the eye** course along the promontory.

These sympathetic nerves of the eye are also vulnerable in patients with middle/inner ear disease. (See box on Horner syndrome.)

MIDDLE EAR POLYPS IN CATS

These polyps also arise from the tympanic cavity or auditory tube but extend into the external ear canal rather than into the nasopharynx. Clinical signs include accumulation of earwax, ear scratching, and head shaking. In more severe cases, the cat may exhibit Horner syndrome, facial nerve paralysis, or vestibular dysfunction. The diagnosis is made by otoscopic examination, which reveals a fleshy, pink mass in the external ear canal. Treatment involves polyp removal via traction-avulsion, as described for nasopharyngeal polyps. If the polyp recurs, more invasive surgical intervention may be required, either a VBO or a lateral approach to the tympanic bulla.

This bony septum is often perforated for complete surgical drainage of the middle ear. Leaving the dorsal portion of the septum intact reduces the risk of damage to the sympathetic nerves in this vicinity. Overzealous curettage of the tympanic cavity during VBO may damage the cochlear or vestibular window and even the vestibular apparatus itself, resulting in vestibular dysfunction. Peripheral vestibular dysfunction typically causes ipsilateral head tilt, circling, falling, or rolling (e.g., head tilt to the right with a right-sided lesion). Ocular abnormalities include horizontal or rotary nystagmus (uncontrolled, rapid eye movements) and/or positional ventrolateral strabismus on the affected side. Patients may also demonstrate imbalance or ataxia. Peripheral vestibular dysfunction may even be a presenting complaint in cats with middle ear polyps.

Inflammatory processes involving the tympanic cavity (otitis media) typically result in fluid accumulation in the bulla due to the presence of these goblet cells.

Inadvertent injury to the facial nerve or carotid artery can occur during surgery of the middle ear. Transient facial nerve paralysis may be caused by stretching or retraction of the nerve during surgery involving the middle ear. It presents clinically as ipsilateral drooping of the ear and muzzle (lips and nostril). As facial muscle tone is normal on the unaffected side, the muzzle may appear drawn to that side. Other signs include ipsilateral decreased or absent palpebral reflex and menace response (inability to blink caused by paralysis of the eyelids) along with excessive salivation or dropping of food from the affected side. Given the intimate contact between the facial nerve and the peripheral vestibular system, facial nerve paralysis is commonly seen with peripheral vestibular dysfunction; however, it may occur independently following ventral bulla osteotomy in cats. Although uncommon, facial nerve paralysis may also be a presenting complaint in cats with middle ear polyps.

The **tympanic bulla** is the part of the temporal bone that houses the fundus, which is by far the largest part of the tympanic cavity and is generally referred to simply as the tympanic bulla. In cats, the tympanic bulla is divided by a thin **bony septum** that separates the bulla into a larger ventromedial and a smaller dorsolateral compartment.

The **auditory tube** originates in the rostrodorsal wall of the tympanic bulla and extends in a rostroventromedial direction to open into the lateral wall of the nasopharynx, as described earlier. Both the tympanic cavity and the auditory tube are lined with pseudostratified ciliated epithelium that contains goblet cells.

As mentioned, the **facial nerve** leaves the cranial cavity with the vestibulocochlear nerve via the internal acoustic meatus. It then exits the skull through the **stylomastoid foramen** dorsal to the tympanic bulla. The **carotid artery** runs ventromedial to the tympanic bulla.

The **external ear** extends from the tympanic membrane at the base of the external acoustic meatus to the apex of the auricle or pinna. The anatomy of the external ear is discussed in Case 3.5.

Blood supply, lymphatics, and innervation of the ear

The blood supply of the ear varies with each part of the ear with most branches originating from the **external carotid**, **superficial temporal**, and **occipital arteries** and their corresponding veins.

The external ear is primarily supplied by the **posterior auricular artery**, with the

HORNER SYNDROME

Injury to the sympathetic nerves of the eye may result in Horner syndrome, characterized by ptosis (drooping of the upper eyelid), miosis (pupillary constriction), enophthalmos (posterior displacement of the eye within the orbit), and protrusion of the nictitating membrane (3rd eyelid) on the affected side. Injury may occur at any point along the sympathetic pathway from the preganglionic nuclei (1st to 3rd thoracic vertebrae) to the orbital fissure—including the promontory. Horner syndrome is a common complication of inflammatory polyps in cats, particularly in those with concurrent otitis media/interna (inflammation of the middle/inner ear) or secondary to surgical trauma (traction-avulsion or bulla osteotomy).

anterior auricular artery providing some supply to the outer edge of the ear; the middle ear is supplied by branches of the **occipital**, **posterior auricular**, and **deep auricular arteries**; and the inner ear is supplied by branches of the **maxillary**, **posterior auricular**, and **middle meningeal arteries**.

Lymphatic drainage is to the **parotid lymph nodes**, along with the **superficial** and **deep cervical lymph nodes**.

Innervation of the ear, like the blood supply, differs depending on the part of the ear innervated with branches originating from cranial nerves—trigeminal (CN V), facial (CN VII), vestibulocochlear (CN VIII), glossopharyngeal (CN IX), and vagus (CN X) (see Landscape Fig. 3.1-3).

Selected references

[1] Donnelly KE, Tillson DM. Feline inflammatory polyps and ventral bulla osteotomy. Compendium 2004;446–54.
[2] Degner DA. Surgical removal of feline inflammatory polyps. Clinician's Brief 2012;90–4.
[3] Greci V, Mortellaro CM. Management of otic and nasopharyngeal, and nasal polyps in cats and dogs. Vet Clin N Am Small Anim Pract 2016;46:643–61.

CHAPTER 3

CASE 3.2

Brachycephalic Airway Syndrome

David Holt
Department of Clinical Sciences and Advanced Medicine, University of Pennsylvania School of Veterinary Medicine, Philadelphia, Pennsylvania, US

Clinical case

Stertor (L. snoring; L. sonorus resonant breathing) is a low-pitched, snoring, or snuffling sound heard on inhalation. It is caused by vibration of the soft tissues of the upper airway rostral to the larynx (i.e., nasopharynx or soft palate). Stridor, in contrast, is a high-pitched, wheezing, or rasping sound, heard loudest on inhalation. It is caused by abnormal airflow in the larynx or trachea. Both abnormal respiratory sounds indicate partial obstruction at some location in the upper airway.

Brachycephalic airway syndrome occurs in breeds that have a relatively short, broad skull (G. *brachy* short + *cephalo* head). In these breeds, the bones of the skull are shortened in relation to the soft tissues, such as the soft palate and pharynx (Fig. 3.2-1A and B). Common clinical findings include stenotic nares, increased contact between the nasal turbinates, an overlong soft palate, a hypoplastic trachea, and everted laryngeal saccules (secondary to the other abnormalities). Alone or in combination, these abnormalities cause partial airway obstruction that may be severe; the snuffling or snoring sound is typical of these breeds.

History

A 2-year-old male pug presented with intermittent stertor when awake and snoring when asleep. The dog's noisy breathing worsened with exercise as the weather warmed during the spring.

Physical examination findings

The dog was slightly overweight with a body condition score of 6 on a scale from 1 (emaciated) to 9 (obese). He was panting at rest, with slight to moderate stertor that was audible without the help of a stethoscope and was loudest on auscultation over the larynx. There was airflow through both nostrils, but the nostrils were somewhat narrowed, as is a common finding in brachycephalic breeds. The rest of the physical examination was unremarkable.

Differential diagnoses

Elongated soft palate as a component of brachycephalic airway syndrome; other less likely differentials included pharyngeal or nasopharyngeal foreign body or mass

Comparative Veterinary Anatomy: A Clinical Approach. https://doi.org/10.1016/B978-0-323-91015-6.00013-3

Diagnostics

Radiographs of the neck and thorax failed to reveal evidence of hypoplastic trachea and aspiration pneumonia. The dog was placed under general anesthesia and the soft palate and larynx were visually examined. The soft palate was observed to extend through the glottis into the larynx rather than ending rostral to the glottis (Fig. 3.2-2). There was no evidence of everted laryngeal saccules, a condition where the mucosa lining the lateral ventricles (saccules) of the larynx everts and protrudes into the laryngeal lumen.

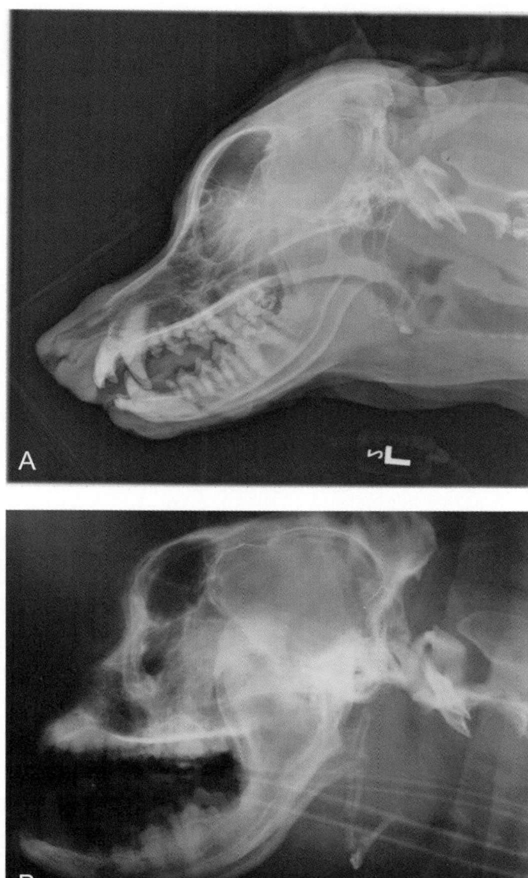

FIGURE 3.2-1 (A) Radiograph of a mesocephalic (normal) dog skull. (B) Radiograph of a brachiocephalic dog skull.

TRACHEOSTOMY

Tracheotomy (L. *trachealis/trachea* pertaining or relating to the trachea + Gr. *tomē* a cutting) is an incision in the trachea, usually made to create an opening called a tracheostomy (*tracheo-* + Gr. *stomoun* to furnish with an opening). This surgery is often performed as an emergency procedure to bypass a life-threatening obstruction of the upper respiratory tract in dogs, cats, and horses. The tracheostomy is typically located in the proximal ventral trachea and allows insertion of a tube to facilitate ventilation with upper airway obstruction. When a tracheostomy is no longer needed, it is allowed to heal shut or is surgically closed. For some animals with irreparable laryngotracheal disease, tracheostomy may be permanent.

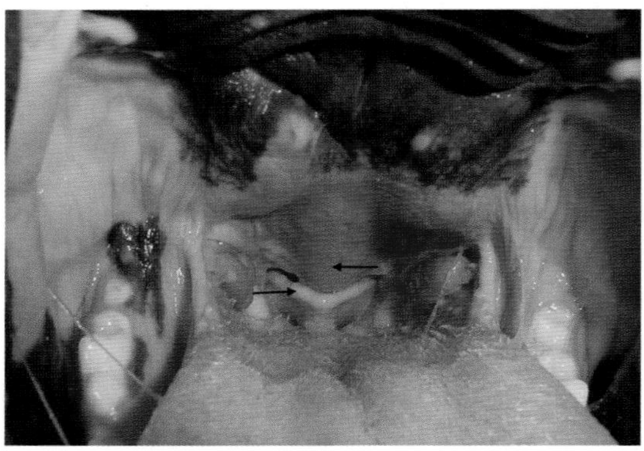

FIGURE 3.2-2 Overlong soft palate extending beyond the epiglottis.

Staphylectomy involves trimming the caudal free margin of the soft palate to the appropriate length to alleviate clinical signs. Through the oral cavity, a "stay" suture is placed in the caudal aspect of the soft palate so that it can be retracted into the oral cavity for easier access. The free margin of the palate is then resected as planned (in this case, to the caudal border of the tonsils), and the oral and nasal surfaces of the soft palate are apposed with a fine, absorbable suture to minimize bleeding (Fig. 3.2-3A–E). Widening stenotic nares surgically requires removing a wedge of tissue from the dorsolateral aspect of the nostril (including nasal cartilage and overlying epithelium); the incised edges of the dorsal lateral nasal cartilages and the overlying pigmented epithelium are apposed with fine interrupted suture to permanently widen the nostrils.

Diagnosis

Elongated soft palate

Treatment

Under general anesthesia, the soft palate was shortened (staphylectomy) and the stenotic nostrils were widened. The dog made an uneventful recovery from anesthesia. During the post-operative period, the dog was carefully monitored for difficulty breathing or swallowing associated with the staphylectomy; no such problems arose. Surgery resolved the stertorous breathing.

Anatomical features in canids and felids

Introduction and function

The hard and soft palates separate the nasal passages and nasopharynx dorsally from the oral cavity and oropharynx ventrally in dogs and cats. Both structures are readily observable on oral examination (Fig. 3.2-4). This section covers the hard and soft palates and their innervation and vasculature.

BRACHYCEPHALIC BREEDS

Muzzle length—and therefore respiratory compromise—varies by breed and individual.

Dogs: Affenpinscher, Boston Terrier, Boxer, Brussels Griffon, Bulldog, Bullmastiff, Cavalier King Charles Spaniel, Dogue de Bordeaux, French Bulldog, Japanese Chin, Lhasa Apso, Pekingese, Pug, Shar Pei, Shih Tzu, and others

Cats: British Shorthair, Burmese, Exotic Shorthair, Himalayan, Persian, Scottish Fold

Rabbits: Jersey Wooly, Lionhead, Lop

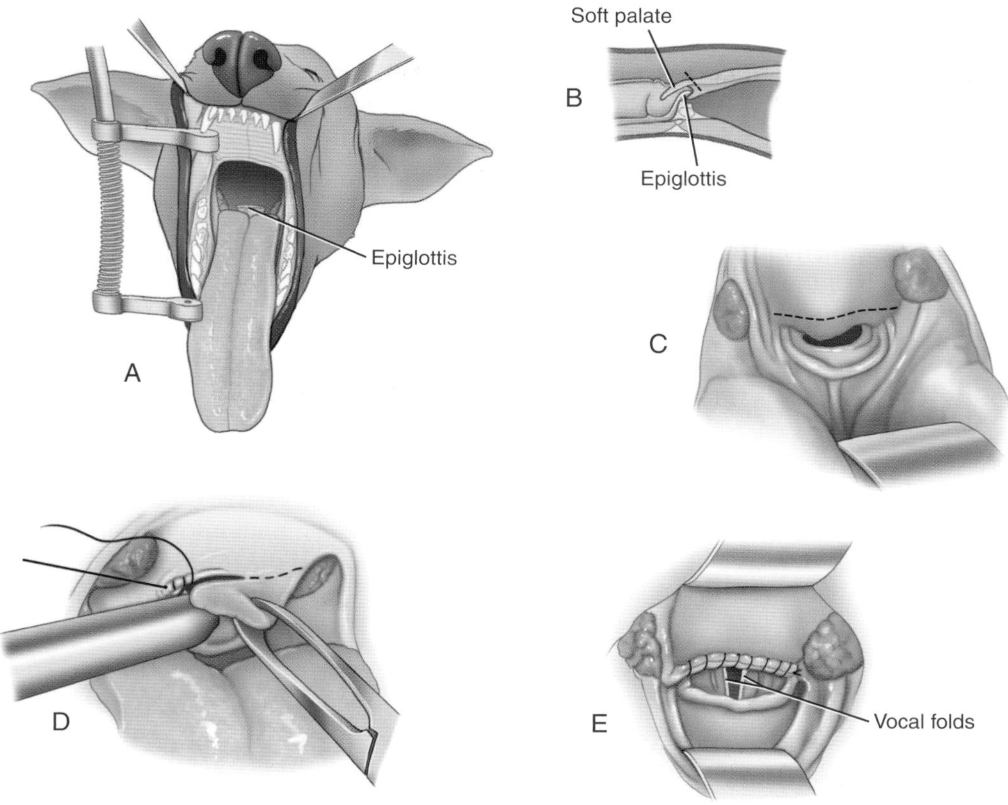

FIGURE 3.2-3 (A) A dog is positioned for surgery in sternal recumbency with the mouth opened. (B) Lateral view. (C) Intraoral view showing proposed line of soft palate resection. (D) A stay suture or instrument is placed in the palate to help in manipulation and the incision into the palate is started. The nasal and oral mucosal edges of the resected palate are sutured together to minimize bleeding. (E) The shortened palate with suturing complete, showing the now visible vocal folds.

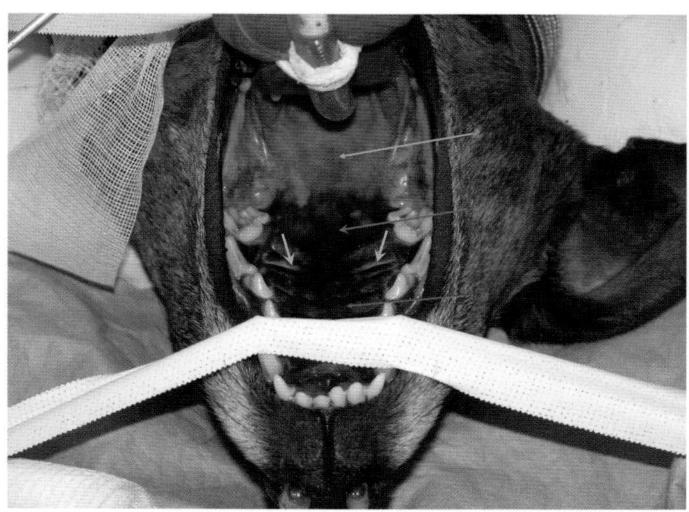

FIGURE 3.2-4 Hard and soft palates are visible. The dog in this image is in dorsal recumbency with the mouth opened for a surgical procedure. *Red arrow*: hard palate; *blue arrow*: junction between hard and soft palate; *green arrow*: soft palate; *yellow arrows*: approximate location of major palatine arteries.

Hard palate

The bony **hard palate** is comprised of the **palatine**, **maxillary**, and **incisive bones**. It is covered by a thick, stratified squamous **mucoperiosteum**, which is folded into several curved, transverse ridges in roughly parallel rows from rostral to caudal. Cats have papillae in the grooves between the ridges, whereas dogs do not. The palatal mucosa may be pigmented.

The paired **major palatine arteries** supply the palatine mucoperiosteum and exit the palatine bone through the **major palatine foramina**, which are located medial to the maxillary 4th premolar, or carnassial, teeth (Fig. 3.2-5). The **major palatine branch** of the **maxillary division** of the **trigeminal nerve** (CN V) also runs through this foramen and supplies sensation to the hard palate and adjacent gingiva of the maxillary teeth.

Soft palate

The **soft palate** is the caudal, soft tissue extension of the hard palate. It extends into the **pharynx**, incompletely dividing the chamber into the **nasopharynx** dorsally and the **oropharynx** ventrally. Because the soft palate does not form an airtight

Horses, are obligate nasal breathers due to their long soft palate over which the epiglottis lies. This arrangement makes it impossible for a horse to normally move air from the mouth to the trachea.

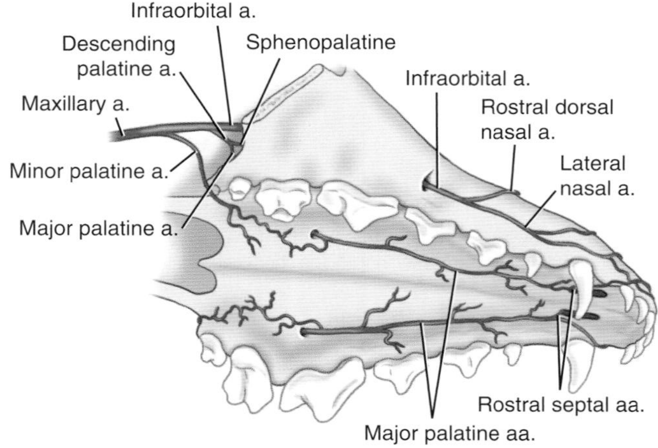

FIGURE 3.2-5 Blood supply to the palate. The major palatine arteries exit the major palatine foramina medial to the fourth premolar teeth.

"STICKS AND STONES"

The hard palate is widest between the maxillary 4th premolar teeth. It is common for dogs to get a stick or bone wedged across the hard palate between these teeth.

CLEFT PALATES

Congenital oronasal fistulas can involve the lip/premaxilla (termed "cleft lip") or the hard/soft palate (termed "cleft palate"). These occur during fetal development (day 25–28 in dogs) when the two palatine shelves fail to fuse. They are more common in brachycephalic dogs and causes include genetic, nutritional, toxic, mechanical, and hormonal factors. Clinical signs range from nasal discharge to aspiration pneumonia to death, depending on the severity. They are typically seen on oral examination though skull or thoracic imaging may be indicated to assess the middle ear (can have concurrent disease) and pneumonia, respectively. Following medical stabilization, most palatoplasty surgeries (a complicated surgery to recreate a complete hard and soft palate) are not performed until at least 16 weeks of age to allow for growth. Prognosis depends on the extent of the defect, but is good with small clefts with appropriate surgical repair.

seal with the glottis in dogs and cats, they can breathe through the nostrils or the mouth. This flexible breathing pattern is enabled by the **intrapharyngeal opening**, which is the caudal orifice that connects the oropharynx and the nasopharynx, formed by the space between the caudal free margin of the soft palate rostrally, the **larynx** caudally, and the **palatopharyngeal arches** laterally.

In nonbrachycephalic breeds, the soft palate begins just caudal to the last maxillary molar teeth and extends as far caudally as the epiglottis. On its ventral or oral surface, the soft palate is composed of stratified squamous epithelium that is rich in palatine glands. On its dorsal or nasal surface, it is covered by nasal epithelium.

The soft palate is easily deflected dorsally during endotracheal intubation. With the help of an illuminated laryngoscope, the base of the tongue and epiglottis are depressed, and the endotracheal tube is passed through the oral cavity and the now-visible glottis (between the vocal cords of the larynx) and into the trachea (Fig. 3.2-6). In the procedure, the endotracheal tube lifts the soft palate up out of the way.

In between these two mucosal layers are the palatine muscles (Fig. 3.2-7). The **palatinus muscle** runs from the **palatine process** of the palatine bone to the caudal border of the soft palate; it shortens the soft palate rostrocaudally when it contracts. The **levator veli palatini muscle** originates from the skull rostral to the tympanic bulla and then courses caudally and ventrally to insert on the caudal half of the soft palate. It elevates the caudal aspect of the soft palate, thereby helping to seal off the nasopharynx during swallowing and vomiting. Both muscles are innervated by **pharyngeal branches** of the **glossopharyngeal** and **vagus nerves**.

The **tensor veli palatini muscle** also arises from the skull rostral to the tympanic bulla and passes ventrally over the pterygoid bone before inserting onto the **palatine aponeurosis**—a dense, fibrous sheet at the caudal margin of the hard palate that connects and supports the soft palate. As its name suggests, the tensor veli palatini muscle stretches and tenses the soft palate between the pterygoid bones. It is innervated by the **mandibular branch** of the **trigeminal nerve**. The **minor palatine arteries** supply blood to the soft palate. The **minor palatine branch** of the **maxillary branch** of the trigeminal nerve provides sensory innervation to the soft palate.

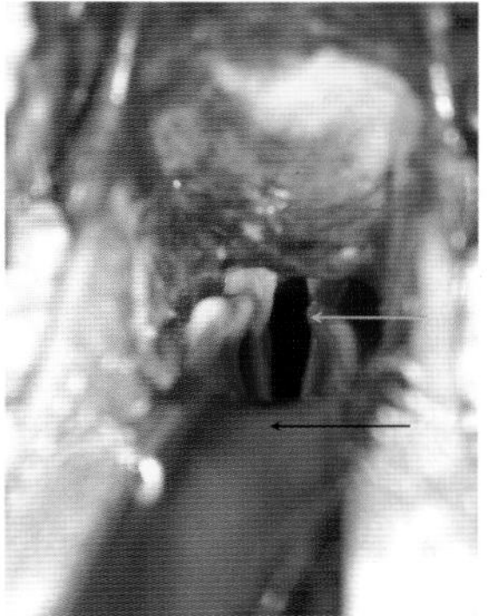

FIGURE 3.2-6 Larynx visible for intubation with the tongue depressed by a laryngoscope. *Blue arrow*: soft palate; *green arrow*: larynx; and *black arrow*: laryngoscope. The epiglottis is ventral to the laryngoscope.

SOFT PALATE FUNCTION

The soft palate is important in protecting the airway during swallowing. Excessive shortening of the soft palate, such as an overly aggressive staphylectomy, allows water and food to enter the nasopharynx and then the nasal passages and trachea. Aspiration of water or food into the lungs may lead to pneumonia, appropriately termed "aspiration pneumonia."

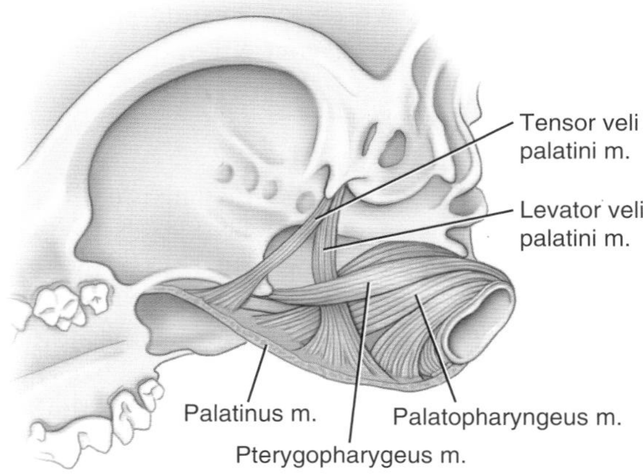

FIGURE 3.2-7 Muscles of the soft palate.

The sensory nerves that supply the soft palate are an important part of the mechanism that triggers the swallowing reflex. During swallowing, the caudal soft palate is elevated dorsally by the paired levator veli palatini muscles to close the intrapharyngeal opening, preventing liquid and food from entering the nasopharynx and subsequently being aspirated. The **palatopharyngeus muscles**, which originate on the palate and insert onto the dorsal pharyngeal fascia, draw the caudal soft palate, palatopharyngeal arches, and caudal pharynx together during swallowing, thereby closing the intrapharyngeal opening.

Regional blood supply and lymphatics

The arterial supply (see Fig. 3.2-5) and innervation are discussed in the sections entitled "Hard palate" and "Soft palate." The mucosa of the hard and soft palates is drained by a plexus of veins that flow into the **maxillary vein**. Lymphatic vessels from the canine palate generally join the **retropharyngeal lymph nodes**.

SOFT PALATE FUNCTION IN BRACHYCEPHALIC DOGS

Brachycephalic airway syndrome (BAS) is not solely dependent on the dog's anatomy. The function of the palate and nasopharynx is also important. Many dogs tense their palatal muscles, effectively shortening the soft palate and reducing resistance to air flow. However, other dogs have clear evidence of nasopharyngeal collapse, similar to sleep apnea in humans.

Selected references

[1] Harvey CE. Inherited and congenital airway conditions. J Small Anim Pract 1989;30:184–7.
[2] Harvey CE. Upper airway obstruction surgery. 1. Stenotic nares surgery in brachycephalic dogs. J Am Anim Hosp Assoc 1982;18(4):535–7.
[3] Harvey CE. Upper airway obstruction surgery. 2. Soft palate resection in brachycephalic dogs. J Am Anim Hosp Assoc 1982;18(4):538–44.
[4] Harvey CE. Tracheal diameter: Analysis of radiographic measurements in brachycephalic and nonbrachycephalic dogs. J Am Anim Hosp Assoc 1982;18:570–6.
[5] Oechtering GU, Pohl S, Schlueter C, Schuenemann R. A novel approach to brachycephalic syndrome. 2. Laser-assisted turbinectomy (LATE). Vet Surg 2016;45:173–81.

CHAPTER 3

CASE 3.3

Odontogenic Disease

Nadine Fiani and Santiago Peralta
Department of Clinical Sciences, Cornell University College of Veterinary Medicine, Ithaca, New York, US

Clinical case

History

A 4-year-old female spayed Labrador Retriever was presented for a nonhealing wound below the right eye. The wound was first noted 5 months before presentation. The dog received several courses of systemic antibiotics over that period. The wound was noted to resolve during treatment but recurred soon after antibiotics were discontinued. The patient otherwise appeared well, had a normal appetite, showed no obvious signs of discomfort or pain, and enjoyed chewing on toys and bones.

Physical examination findings

Extraoral examination revealed a soft tissue swelling in the right infraorbital area. On close inspection, a draining tract was found in the center of the swollen tissue (Fig. 3.3-1). The patient showed some discomfort when the area was palpated. The right mandibular lymph node was slightly larger than the left. No other abnormalities were noted on extraoral examination. Oral examination revealed minimal generalized plaque and calculus accumulation.

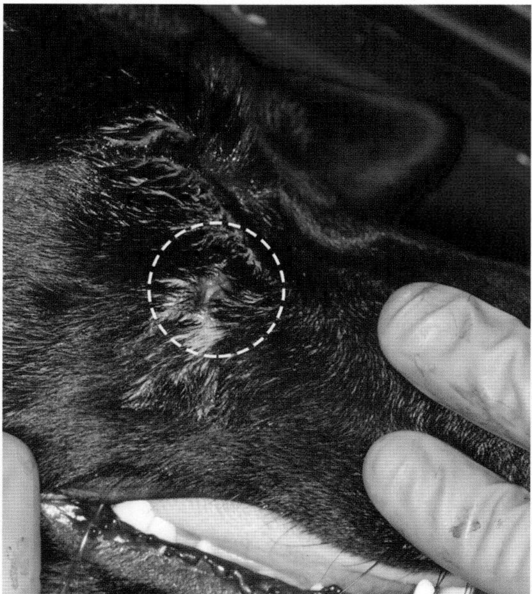

FIGURE 3.3-1 Clinical photograph of the soft tissue swelling and draining tract *(circle)* below the right eye.

Comparative Veterinary Anatomy: A Clinical Approach. https://doi.org/10.1016/B978-0-323-91015-6.00014-5

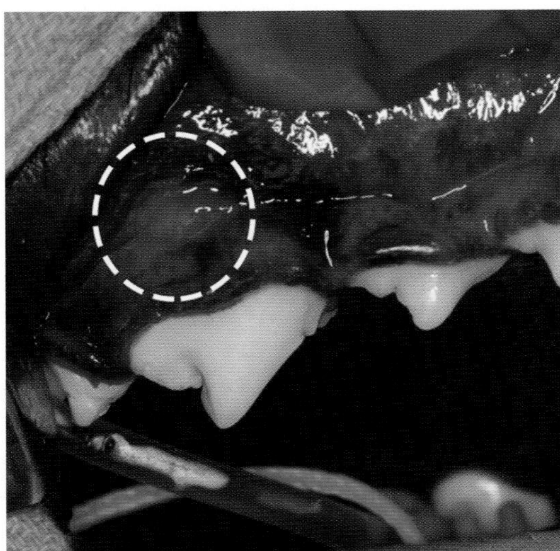

FIGURE 3.3-2 Intraoral clinical photograph of the caudal right maxilla. Soft tissue swelling and a parulis *(circle)* are present at the mucogingival junction at the level of the right maxillary 4th premolar tooth.

Some soft tissue swelling and a draining tract (parulis; Gr. *para* to, at, or from the side of + Gr. *oulon* gum, gingiva) were also noted at the mucogingival junction at the level of the right maxillary 4th premolar tooth (Fig. 3.3-2). No obvious tooth fractures or discoloration were noted during the conscious oral examination.

Differential diagnoses

Odontogenic infection; foreign body with secondary infection; neoplasia with secondary infection

Diagnostics

On anesthetized oral examination of the soft tissues, the only abnormality detected was the swelling and parulis noted at the mucogingival junction of the right maxillary 4th premolar tooth. The use of an explorer/probe helped to identify a crown fracture at the distal aspect of the right maxillary 4th premolar tooth. Further exploration of the fracture revealed that the pulp was exposed. This fracture type is termed a complicated crown fracture. On periodontal probing, the right maxillary 4th premolar tooth had a deep (> 12 mm or >0.5 in.) pocket buccally and a stage 2 furcation (anatomical point where two roots meet in a multirooted tooth) was noted. All other teeth had mild gingivitis with no abnormal periodontal probing. Intraoral radiographs were obtained under anesthesia and the findings are described in the figure legend (Fig. 3.3-3).

The crown of the right maxillary 4th premolar tooth was fractured, exposing the pulp. The pulp then went on to become nonvital and infected. Inflammatory mediators and infectious pathogens had exited at the apical delta, resulting in inflammation of the periapical structures (periodontal ligament, cementum, and alveolar bone). With chronic inflammation, the alveolar bone was resorbed leading to a periapical lucency radiographically. The infection in this case was not limited to the periapex but also extended coronally and destroyed the periodontium, resulting in periodontitis. The combined effect is termed an endodontic-periodontal lesion (Fig. 3.3-4).

Diagnosis

Odontogenic infection

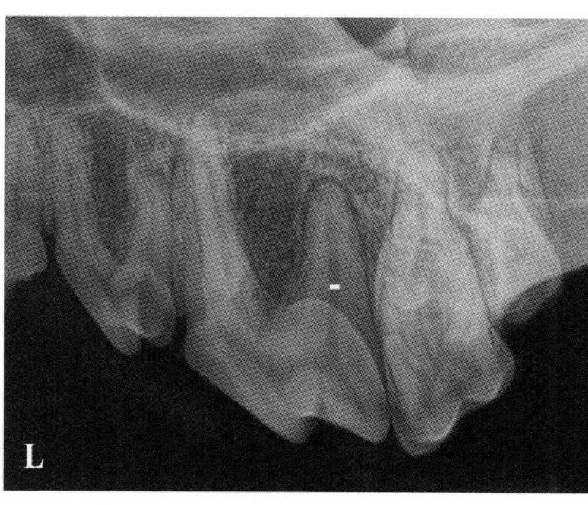

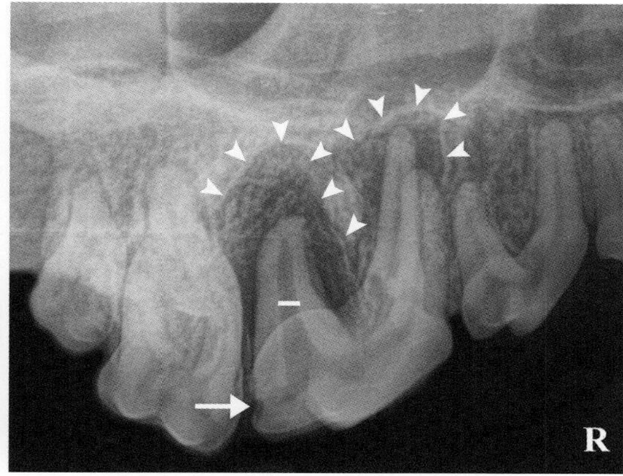

FIGURE 3.3-3 Intraoral radiographs of the caudal maxilla of the patient. The left (L) caudal maxillary view depicts normal radiographic anatomy of the 3rd and 4th premolar and 1st and 2nd molar teeth. The right (R) caudal maxillary view shows the 4th premolar tooth affected by endodontic-periodontal disease. The crown is fractured at the distal aspect *(white arrow)*. The pulp cavity has failed to narrow *(lines)*. This is evident by comparing the right and left 4th premolar teeth *(white line)*. Periapical lucencies and extensive periodontitis *(white arrowheads)* including furcation (anatomical area where a multirooted tooth divides) involvement affect the tooth.

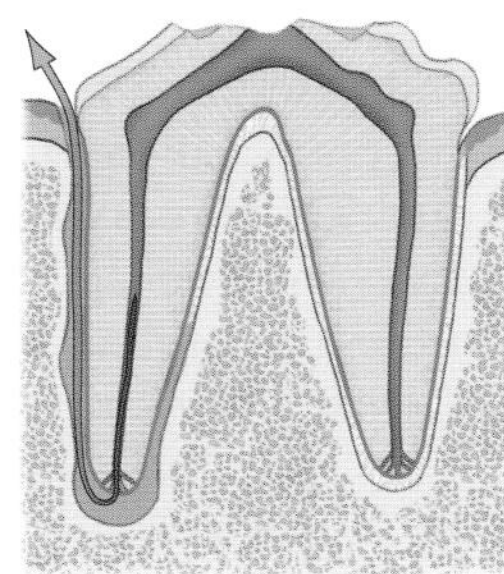

FIGURE 3.3-4 Graphic demonstrating the pathogenesis of endodontic disease that extends coronally (*green arrow*), destroying the periodontium. Collectively the process is referred to as "endodontic-periodontal disease."

Treatment

Routine surgical extraction of the right maxillary 4th premolar tooth was performed. The alveolus was debrided and lavaged/irrigated with saline before apposition of the mucogingival flap. The patient was discharged; broad-spectrum oral antimicrobials and analgesics were prescribed for a week. A 2-week follow-up examination showed complete resolution of the swelling and draining tracts, with healing of the extraction site.

Anatomical features in canids and felids

Introduction

The teeth of the dog are described as brachydont, anelodont, and diphyodont, meaning that they have a short crown, form a complete apex, and have two generations—deciduous and permanent, respectively. Adult dogs have a total of 42 teeth. The teeth can be divided into four groups: incisors, canines, premolars, and molars. Relevant anatomical differences regarding feline teeth are in the side box entitled "Feline dental anatomy."

Function

The incisor teeth are used primarily for prehension. The canines are designed to bring down and hold prey. The premolars play an important role in tearing and ripping food. The molars are used for grinding down food so that it can be swallowed more easily.

Incisors

There are three **incisors** in each maxillary and mandibular quadrant. They are single-rooted and increase in size from the 1st to the 3rd. The maxillary incisors are in the **incisive bone**. The 3rd incisor is large, and its apex is adjacent to the nasal cavity.

Canines

There is a single **canine** in each maxillary and mandibular quadrant. They are single-rooted and are the largest teeth in the oral cavity. A fully developed canine in the dog has a crown-to-root ratio of 2:3 or approximately a 40 (crown) to 60 (root) percent ratio. The apex of the mandibular canine tooth is lingual to the middle mental foramen. The maxillary canine tooth apex is dorsal to the 2nd premolar tooth. It is separated from the nasal cavity by a thin bone plate.

Premolars

There are four **premolars** in each of the maxillary and mandibular quadrants. The 1st premolar teeth are small and single-rooted. The 2nd, 3rd, and mandibular 4th premolar teeth are 2-rooted. The maxillary 4th premolar tooth is the largest premolar and has 3 roots. The junction of roots of multirooted teeth is called the **furcation**. The roots of the maxillary 1st and 2nd premolar teeth are short, and their apices are separated from the nasal cavity by thin bone. The 3rd and 4th premolar teeth are near the infraorbital canal and maxillary recess. All the mandibular premolar teeth are in the body of the mandible and their apices are dorsal to the mandibular canal.

ENDODONTIC DISEASE

Endodontic disease manifests at the periapical region of the teeth. Being familiar with the number of roots that each tooth has, as well as the location of their apices, can be very helpful in localizing endodontic disease because endodontic disease does not always result in a draining tract. However, when it does, the tract can develop intraorally, extraorally, intranasally, and/or in the periocular region, depending on the tooth and its location.

FELINE DENTAL ANATOMY

Adult cats have a total of 30 teeth. As with many mammalian species, cats have evolved to require less teeth. The typical location of tooth reduction is at the beginning of the premolar teeth and the end of the molar teeth.

Incisors—there are three single-rooted incisor teeth in each of the maxillary and mandibular quadrants.

Canines—There is a single canine tooth in each quadrant.

Premolars—There are three premolars in the maxilla. These are the 2nd, 3rd, and 4th premolar teeth. The maxillary 2nd premolar tooth is single-rooted, the 3rd is 2-rooted, and the 4th is 3-rooted. The mandible has only two premolar teeth. These are the 3rd and 4th premolar teeth. All the mandibular premolar teeth are 2-rooted.

Molars—There is a single molar in each of the maxillary and mandibular quadrants. The maxillary molar tooth is a small single-rooted vestigial tooth. In contrast, the mandibular molar tooth is a large 2-rooted tooth.

The maxillary 4th premolar and the mandibular 1st molar teeth are the carnassials in the cat.

Molars

The dog has two maxillary and three mandibular **molars** in each quadrant. Both maxillary molar teeth are 3-rooted with relatively flat crowns. The apices of these teeth are located close to the maxillary foramen and the pterygopalatine fossa, ventral to the globe. The mandibular 1st molar tooth is large and 2-rooted. It has a large triangular central cusp and a flattened distal cusp. Its apices can be located within the mandibular canal or dorsal to it, depending on the size of the dog. The 2nd mandibular molar tooth is also 2-rooted with a flat crown; however, it is much smaller. The mandibular 3rd molar tooth is very small and single-rooted. The 2nd and 3rd molar teeth apices are dorsal to the mandibular canal.

Carnassial teeth—the maxillary 4th premolar and mandibular 1st molar teeth—act together in a powerful shearing action.

Microstructure of the tooth and periodontium (Fig. 3.3-5)

Enamel is the outermost layer covering the crowns of all teeth. It is the hardest tissue in the body, composed of 96% mineral. Enamel is formed by ameloblasts and is completed before the eruption of a tooth. The cells are exfoliated or regress to remnant forms at the time of eruption so enamel does not have the ability to repair itself.

Dentin-pulp complex

Dentin composes the bulk of the crown and root of a tooth. It is softer than enamel, consisting of 70% inorganic material. Dentin is produced throughout the life of a tooth by a cell called the **odontoblast**. These cells lie at the periphery of the pulp and gradually move toward the center of the tooth as they produce dentin (thickening of the dentinal walls) (Fig. 3.3-6).

Odontoblasts are part of the pulp and they secrete dentin throughout the life of a tooth. If the pulp becomes nonvital, so do the odontoblasts and therefore no more dentin can be secreted.

Three types of dentin are recognized: primary, secondary, and tertiary. **Primary dentin** is produced until the external form of the tooth is complete. **Secondary dentin** forms more slowly after tooth eruption and is continuous with the primary dentin. **Tertiary dentin** is produced as a result of various noxious stimuli such as abrasion, caries, or

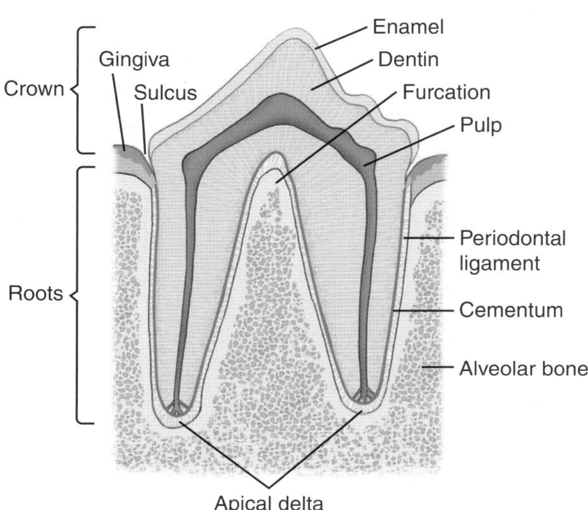

FIGURE 3.3-5 Graphic demonstrating the normal dental and periodontal anatomy of a tooth of a dog.

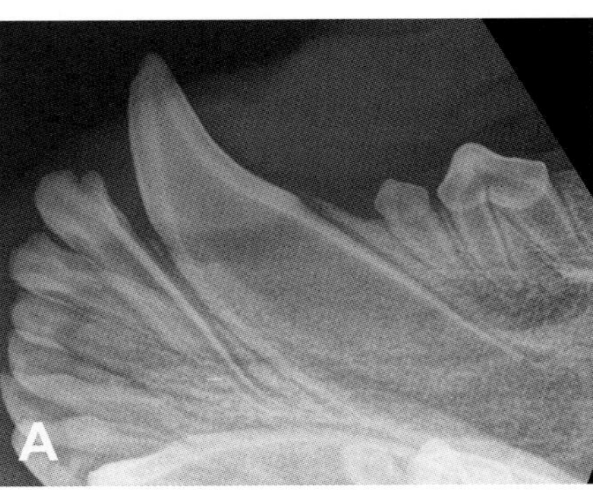

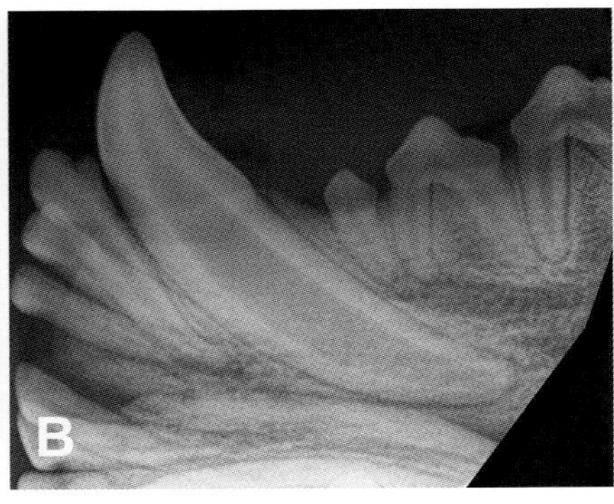

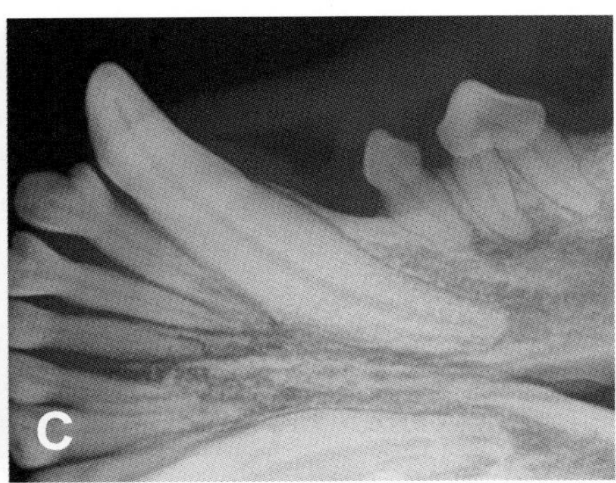

FIGURE 3.3-6 Normal aging of a left mandibular canine tooth. (A) The canine tooth of a 6-month-old dog. The dentinal wall is very thin, and the apex of the tooth has not completely formed yet. (B) The canine tooth of an 18-month-old dog. The dentin has thickened, and the apex is fully formed. (C) The canine tooth of a 6-year-old dog. The dentin has continued to thicken and therefore the pulp cavity is gradually narrowing.

These odontoblasts produce dentin at a more rapid rate than those unaffected by the trauma. Depending on the intensity and duration of the stimulus, the dentin may have a histologically haphazard appearance. As this exposed dentin is rougher and more porous than enamel, it tends to become easily stained.

restorative procedures. Tertiary dentin is only produced by the odontoblasts that are directly affected by the stimulus.

The **pulp** is in the **pulp cavity** and contains blood vessels, nerves, loose connective tissue, and the previously mentioned odontoblasts. The pulp cavity in the crown is referred to as the **pulp chamber** and in the root, the **root canal**. The neurovascular supply to the tooth enters the pulp cavity at the apex. In the dog, the apex consists of numerous fenestrations, collectively referred to as the **apical delta**.

Periodontium

The **periodontium** is a group of four structures that retain the tooth within the jaw: cementum, periodontal ligament, alveolus, and gingiva.

Cementum is a thin layer of mineralized tissue that has a similar structure to bone. It covers the outer surface of the tooth root. Cementum is produced by cementoblasts and is largely avascular. Cementum slowly thickens with time, especially apically.

The **periodontal ligament** is a highly vascular, cellular, and well-innervated fibrous connective tissue that contains collagen fibers. The ligament embeds into cementum and alveolar bone, thereby suspending the tooth within the alveolus.

The tooth-bearing bones of the jaw are the mandibles, maxillae, and incisive bones. All of them have **alveolar processes**, which are composed of two layers of cortical bone on either side of cancellous bone. The alveolar processes bear the **alveoli,** "sockets," for each tooth root. The alveoli are lined with a thin layer of perforated bone, the **cribriform plate**, confluent with the surrounding bone.

Radiographically, the well-defined cribriform plate of the alveoli is referred to as the lamina dura. It appears as a radiopaque line around the roots of the tooth.

The **gingiva** (L. gum of the mouth) is a cuff of keratinized soft tissue around the **cementoenamel junction** and coronal (L. *coronalis* a crown) to the alveolar bone. It consists of two portions—the **free gingiva** is the unattached portion of the gingiva and should have a sharp margin. The space created between the free gingiva and the tooth is called the **sulcus**. The sulcus is lined by a thin nonkeratinized oral sulcular epithelium. In health, the sulcus should measure 0.5–2 mm (0.01–0.07 in.) in the dog. There is constant flow of neutrophils and transudate that emerge from the venous plexus deep to the sulcular epithelium, making their way to the gingival margin. The neutrophils are part of the normal local defense system. The **attached gingiva** is tightly adhered to the underlying periosteum. The width of the attached gingiva varies and ends when it meets the alveolar mucosa at the **mucogingival junction**.

Under normal circumstances, the gingiva is the only part of the periodontium that should be seen on oral examination.

Blood supply, lymphatics, and innervation to the dentition

Blood supply to the dentition originates from the **external carotid artery** that becomes the **maxillary artery** as it tracks rostrally. The maxillary artery becomes the **infraorbital artery** before entering the infraorbital canal, where it supplies the maxillary teeth via the **dental branches**. The **mandibular artery**, which emerges from the maxillary artery, becomes the **inferior alveolar artery** before entering the mandibular canal and supplying the mandibular dentition via the **dental branches**.

Venous return from the maxillary dentition occurs via the **infraorbital vein** that courses caudally via the infraorbital canal and emerges into the pterygopalatine fossa via the maxillary foramen where it joins the **deep facial vein**. Venous return from the mandibular dentition occurs via the **inferior alveolar vein** which, upon exiting the mandibular canal via the mandibular foramen, joins the **maxillary vein**.

Nerve supply to the dentition ultimately derives from the **maxillary** and **mandibular branches of the trigeminal nerve**. As the maxillary nerve travels rostrally, it is known as the **infraorbital nerve** before entering the infraorbital canal via the maxillary foramen. During its course, the infraorbital nerve supplies the maxillary dentition via the **caudal**, **middle**, and **rostral superior alveolar branches**. The mandibular nerve becomes the **inferior alveolar nerve** before entering the mandibular canal via the mandibular foramen on the medial aspect of the ramus. As the nerve courses rostrally, it supplies the mandibular dentition via the **caudal**, **middle**, and **rostral branches of the inferior alveolar nerve**. Once the infraorbital and inferior alveolar nerves emerge from their respective canals, they no longer innervate teeth.

Although pulp tissue in the dog lacks lymphatic vessels, other oral tissues drain to the **mandibular lymph nodes** located caudal to the angle of the mandible. The superficial lymphatic vessels randomly cross midline and drain to either the ipsilateral or contralateral mandibular lymph nodes. The efferent drainage from the mandibular lymph nodes primarily goes to the ipsilateral **medial retropharyngeal lymph node**.

Selected references

[1] Hermanson JW, Miller ME, DeLahunta A, Evans HE. Miller and Evans' anatomy of the dog. 5th ed. St. Louis, MO: Elsevier; 2020.
[2] British Small Animal Veterinary Association. In: Reiter AM, Gracis M, editors. BSAVA manual of canine and feline dentistry and oral surgery. 4th ed. Quedgeley, Gloucester: British Small Animal Veterinary Association; 2018.
[3] Lobprise HB, Dodd JR, editors. Wiggs's veterinary dentistry: principles and practice. 2nd ed. Hoboken, NJ: Wiley-Blackwell; 2018. Retrieved from: https://onlinelibrary.wiley.com/doi/book/10.1002/9781118816219.

CHAPTER 3

CASE 3.4

Retrobulbar Mass

Eric Ledbetter
Department of Clinical Sciences, Cornell University College of Veterinary Medicine, Ithaca, New York, US

Clinical case

History

A 7-year-old female spayed mixed breed dog was presented for evaluation of a swelling above the right eye, first noted 5 months previously. The dog was initially treated with a course of oral antibiotics and nonsteroidal anti-inflammatories without noticeable improvement. The mass had slowly and progressively enlarged since it was first observed. The dog did not appear painful and the dog's vision, activity level, and appetite were unchanged. The dog had no history of ocular or systemic health problems.

Ophthalmic and physical examination findings

Complete ocular examination was performed, including slit-lamp biomicroscopy, indirect ophthalmoscopy, Schirmer I tear tests, ocular surface fluorescein staining, and applanation tonometry. The palpebral reflex, menace response, and dazzle reflex were present OU (L. *o' culus uter' que* each eye). Direct and consensual pupillary light responses were present OU. The Schirmer I tear test results were 21 mm (0.8 in.)/min OD (L. *o' oculus dex' ter* right eye) and 18 mm (0.7 in.)/min OS (L. *o' oculus sinis' ter* left eye). The cornea was fluorescein stain negative OU. Intraocular pressure was 18 mmHg OD and 14 mmHg OS. All these results were within normal limits.

> The slit-lamp biomicroscope is a high-magnification binocular viewing system with an adjustable, high-intensity light source. By changing and pivoting the light source, different angles of illumination are achieved, which permit detailed viewing of the ocular tissues.

A large, semifluctuant mass was visible and palpable superior, temporal, and inferior to the globe OD (Fig. 3.4-1). The overlying conjunctiva was moderately hyperemic and chemotic. There was mild mucoid ocular discharge OD. Ocular motility assessed by the oculocephalic reflex—moving the dog's head both side-to-side and up-and-down—was mildly reduced in all directions OD. The globe was mildly exophthalmic OD, and manual globe retropulsion through closed eyelids was moderately reduced OD.

> Exophthalmos and resistance to retropulsion most commonly indicate increased orbital contents, as occurs with the presence of space-occupying orbital masses.

Comparative Veterinary Anatomy: A Clinical Approach. https://doi.org/10.1016/B978-0-323-91015-6.00015-7

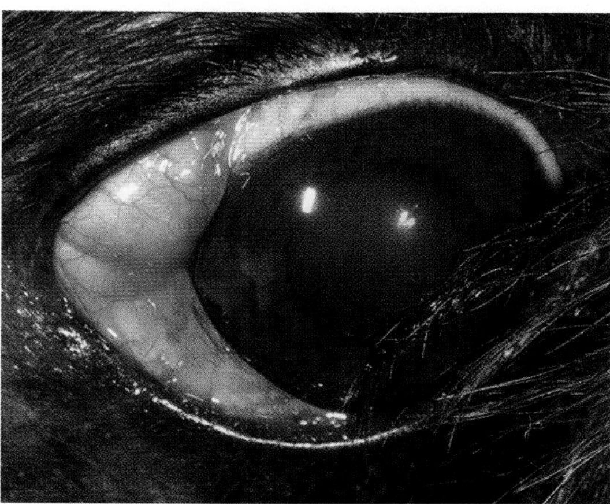

FIGURE 3.4-1 Photograph of a dog with a large, semifluctuant subconjunctival mass superior, temporal, and inferior to the right globe.

Globe retropulsion and palpation of the orbital and periorbital structures, including the bony orbital margin, appeared nonpainful.

Pain during orbital palpation is expected with inflammatory conditions such as orbital abscesses or cellulitis, and the absence of discomfort during palpation is most consistent with orbital neoplasms and cysts. Palpation of the orbital margin is a routine component of the complete ophthalmic examination that is especially important when evaluating dogs for suspected orbital disease.

The cornea, anterior chamber, iris, lens, retina, and optic nerve appeared normal OD. Complete ophthalmic examination appeared normal OS. Complete physical examination was normal, including during forced opening of the mouth.

When a dog's mouth is opened, the vertical ramus of the mandible is displaced into the orbit and compresses orbital contents. This compression is often associated with pain in the presence of inflammatory orbital conditions and these dogs generally vocalize or struggle with this manipulation.

Differential diagnoses

Orbital neoplasia, orbital abscess or cellulitis, orbital cysts (e.g., zygomatic mucocele, inclusion cysts), extraocular myositis, masticatory muscle myositis, orbital fractures and hematoma, ocular onchocerciasis, orbital vascular anomalies (e.g., varices or arteriovenous fistula)

Diagnostics

Complete blood count, serum biochemistry panel, and urinalysis were performed; the results were normal. Thoracic radiographs (standard 3 views) and complete abdominal ultrasound were performed and were normal with no evidence of metastatic disease.

Orbital ultrasound revealed a large mass of moderate and mixed echogenicity surrounding the temporal aspect of the right globe. The globe itself appeared ultrasonographically normal. Contrast CT of the head showed that the right globe was mildly exophthalmic and was being displaced anteriorly and slightly nasally by a large homogenous mass with pronounced contrast enhancement (Fig. 3.4-2). The mass circumscribed the globe superiorly, temporally, and inferiorly, and extended posteriorly into the orbit behind the globe. Globe integrity was conserved. The mass was confined to the orbit with no evidence of orbital bone involvement or destruction. Retropharyngeal and mandibular lymph nodes were within normal limits on the CT scan. Magnetic resonance imaging was performed to further characterize the soft tissues of the orbit and the mass (Fig. 3.4-3). On MRI, the mass was large, lobulated, septate, and avidly contrast-enhancing. Like what was observed on the CT scan, the mass surrounded the superior, temporal,

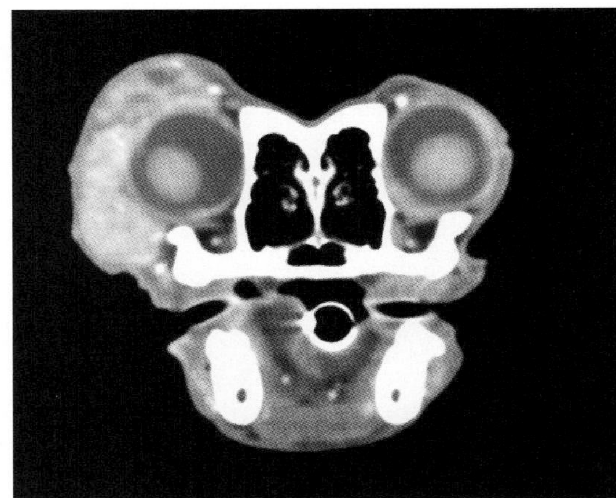

FIGURE 3.4-2 Contrast CT image of the head showing a large homogenous mass with pronounced contrast enhancement surrounding the right globe superiorly, temporally, and inferiorly. Globe integrity is conserved, and the mass is confined to the orbit with no evidence of orbital bone involvement or destruction.

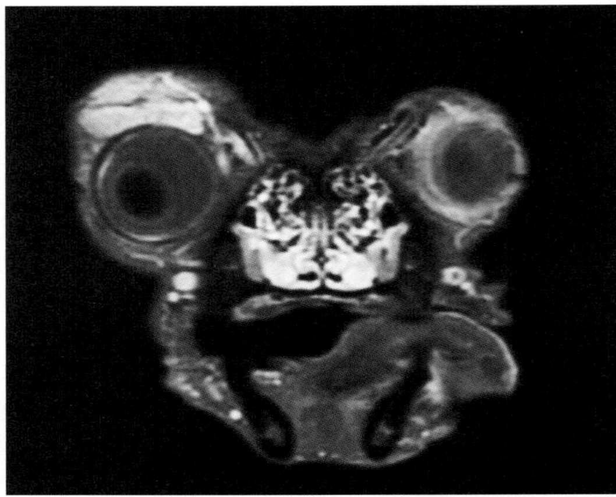

FIGURE 3.4-3 An MRI image of a contrast-enhancing, large mass surrounding the superior, temporal, and inferior aspects of the right globe. The globe appears normal and no nasal cavity, sinus, or intracranial abnormalities are present.

and inferior aspects of the globe OD. The globe appeared normal and no nasal cavity, sinus, or intracranial abnormalities were noted.

A transconjunctival, ultrasound-guided biopsy of the temporal aspect of orbital mass was performed under general anesthesia. Histopathologic examination of the biopsy revealed a densely cellular, well-demarcated neoplasm with histologic features consistent with a benign neoplasm arising from the lacrimal gland.

Diagnosis

Lacrimal gland adenoma

Treatment

The mass was surgically excised through a modified lateral orbitotomy. Starting at the level of the lateral canthus, a curvilinear skin incision was made along the dorsal rim of the zygomatic arch and extended posteriorly to the base of the ear. The subcutaneous tissues were bluntly dissected to expose the surface of the zygomatic arch. The orbital ligament and temporal muscle aponeurosis were sharply transected from the dorsal zygomatic arch. An osteotomy of the central zygomatic arch was then performed, and the bone deflected ventrally leaving it attached to the masseter muscle. The exposed mass in the dorsal and posterior orbit was then completely excised by a combination of blunt and sharp dissection with sparing of the globe. The zygomatic arch was repaired with cerclage wires and the orbital ligament sutured back into place. The subcutaneous tissues and skin were routinely closed.

Anatomical features in canids and felids

Introduction

The orbit is the conical bony fossa of the skull that surrounds and protects the globe and its associated tissue structures. Like other carnivores, dogs possess an open or incomplete orbit. Open orbits are only partially surrounded by bone at their external margins in contrast to the enclosed orbit of many herbivorous species that is entirely bordered by bone. The feline orbital and periorbital tissues are very similar in their arrangement to the dog with only minor anatomical differences.

Function

The orbit in the dog and cat houses, supports, and protects the globe and its associated tissue structures. The borders of the orbit serve as a physical barrier and anchor point for ligament attachment.

Bony orbit

Six bones constitute the osseous portion of the canine **orbit** (80%) and the **orbital ligament** forms the soft tissue of the dorsolateral external border of the orbit. Although there is some variation between different canine skull conformations, the representative bones constituting the canine orbit are the **frontal**, **maxillary**, **lacrimal**, **zygomatic**, **sphenoid**, and **palatine bones** (Fig. 3.4-4). In dogs, the orbital ligament is a thick collagenous tissue that transverses from the **zygomatic process** of the frontal bone to the **frontal process** of the zygomatic bone. The **orbital margin** is the external opening of the conical orbital cavity and it is directed anterolaterally in dogs. The frontal bone comprises the dorsal and medial segments of the orbital margin. The lacrimal bone forms a small section of the ventromedial orbital margin in most dogs; however, the lacrimal bone does not contribute to the formation of the orbital margin in brachycephalic breeds.

> The lateral aspects and floor of the canine orbit are almost entirely composed of soft tissues, permitting access to some regions of the orbit for diagnostic sample collection (e.g., aspirates and biopsies) and surgical approaches without osteotomy.

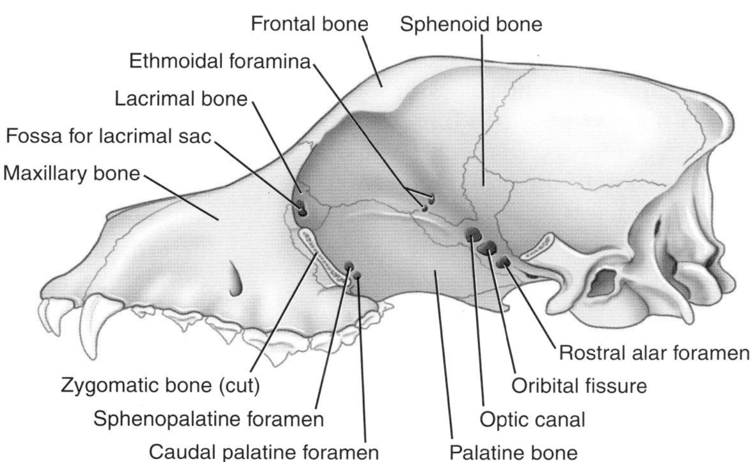

FIGURE 3.4-4 The canine skull displaying the major bones, foramina, and fissures located in the medial wall of the orbit. The foramina and fissures in the bones permit vessels, nerves, and ducts to pass through the bone and enter or exit the orbital space.

> The structural importance of these muscles and their contributions to orbital anatomy are clinically appreciated with masticatory myositis (which can result in exophthalmos) and with conditions resulting in masticatory muscle atrophy (such as denervation or postinflammatory conditions) where permanent enophthalmos can be observed.

> Inflammation, cysts, and neoplasms of the zygomatic salivary gland are relatively common causes of orbital disease in dogs and frequently produce exophthalmos with a dorsal globe deviation.

Orbital soft tissues

The lateral aspects and floor of the canine orbit are almost entirely composed of soft tissues. In addition to the bones and orbital ligament, the masticatory muscles (including the **masseter**, **temporalis**, and **medial pterygoid muscles**) are present as portions of the lateral wall, dorsal wall, and floor of the orbit. The **zygomatic salivary gland** lines the lateral two-thirds of the orbital floor where it is located on the dorsal aspect of the **pterygoid muscle**. The zygomatic gland extends from the ventral orbital margin posteriorly to the **optic canal**.

All structures within the orbit are enclosed within a thin sheet of tough connective tissue referred to as the **orbital fascia**. The orbital fascia of the dog can be further subdivided into the **periorbita**, **Tenon's capsule**, and **muscular fasciae.** The periorbita is the conically shaped layer of fascial tissue that lines all the orbital bones and the orbital ligament. The periorbita surrounds the globe and its extraocular muscles, nerves, and vessels. The canine periorbita is a double-layered structure with superficial and deep layers. At its posterior aspect, the apex of the periorbita is firmly attached to the borders of the orbital fissure and optic canal. Here the nerves and vessels for the eye enter the periorbital cone and the periorbita is continuous with the dura mater of the optic nerve at this location. The base of the periorbita splits at the anterior aspect of the orbital margin and becomes continuous with the periosteum of the facial bones and the tarsal plates of the eyelids. The periorbita is reflected over the extraocular muscles (i.e., **dorsal rectus**, **ventral rectus**, **medial rectus**, **lateral rectus**, **dorsal oblique**, **ventral oblique**, **and retractor bulbi muscles)** forming the muscular fasciae. The periorbita also envelopes the anterior portion of the globe to become Tenon's capsule, which inserts at the limbus beneath the conjunctiva.

> Loss of orbital fat is observed in a variety of conditions (e.g., age-related, masticatory muscle myositis, extraocular myositis) and can result in enophthalmos. Prolapse of orbital fat into the subconjunctival space can occur due to congenital or acquired defects in the orbital fascia. This condition appears as a nonpainful, soft, subconjunctival mass or swelling. Cytology of a fine needle aspirate is usually adequate for diagnosis. Treatment, when required, includes surgical excision of the prolapsed fat and repair of the orbital fascial defect.

The **lacrimal gland** is located on the dorsolateral aspect of the globe, just under the periorbita, and adjacent to the orbital ligament and zygomatic process of the frontal bone. The fossa of the lacrimal gland is the shallow depression on the ventral surface of the zygomatic process of the frontal bone surrounding the lacrimal gland. The **orbital fat body** is found within the periorbita lining the extraocular muscles and other ocular structures, and outside the periorbita (between the periorbita and the orbital bones). Orbital fat serves to cushion and support the globe.

Eyelids

The canine **nictitating membrane** is well developed and is situated in the anterior orbit between the globe and inferior eyelid. The nictitating membrane is composed of connective tissue lined on both sides by conjunctiva. It contains a T-shaped hyaline cartilage that provides structural strength. The periorbita attaches to the base of the nictitating membrane. The **nictitans gland** is an accessory lacrimal gland that surrounds the base of the nictitating membrane. Dogs do not have active muscular control over the nictitating membrane. The position of the nictitating membrane is passive in the dog and its elevation over the ocular surface occurs secondarily when the globe is retracted and displaces the nictitans membrane and its gland anteriorly. In contrast to dogs, small smooth muscle bundles are present within the feline nictitating membrane that appears to facilitate some degree of active control over the membrane's position.

Prolapse of the nictitans gland is a relatively common problem in young dogs that is believed to result from abnormalities in the connective tissue structures attaching the gland to the periorbita. Surgical treatments for this condition include a variety of "anchoring" surgical techniques (in which the nictitans gland is repositioned and anchored to another anatomical structure) or "pocket" surgical techniques (in which the nictitans gland is repositioned and held in place by creation of a pocket of tissue) for returning the untraumatized gland to its normal anatomic position.

The canine eyelids represent modified, mobile folds of skin. In dogs, the eyelids are composed of an external skin layer, a muscular layer, a dense connective tissue layer referred to as the **tarsus**, and the **palpebral conjunctiva** that lines the internal layer sitting adjacent to the globe. Relative to many other species, the canine tarsus is relatively poorly developed. Closing of the eyelids is mediated by the **orbicularis oculi** muscle that is in the eyelid tissues surrounding the palpebral fissure. The **levator palpebrae superioris** is the primary muscle that opens the eyelids through contraction and insertion on the upper tarsus.

Elevation of the nictitating membrane is also a frequent finding in orbital disease that either increases or decreases orbital content volume.

The **eyelid margin** is where the skin and palpebral conjunctiva meet on the edge of the eyelid. This mucocutaneous junction contains both **cilia** and the orifices of the **meibomian glands**. Meibomian glands are modified sebaceous glands located in the tarsus. Dogs typically possess 20–40 meibomian glands per eyelid. The meibomian glands produce an oily secretion that forms the lipid layer of the precorneal tear film. The **superior** and **inferior lacrimal puncta** are located on the inner surface of the eyelid margins adjacent to the **medial canthus**. These puncta are the round to oval openings of the **lacrimal canaliculi** that meet to form the **lacrimal sac**. In the dog, the lacrimal sac is merely the dilated origin of the nasolacrimal duct and is poorly developed compared to some other species.

The most common clinical issue of the canine meibomian glands is the formation of benign neoplasms, including adenomas and epitheliomas. Although typically benign, these meibomian gland tumors can ulcerate or cause frictional irritation of the ocular surface. Treatment is best performed by complete surgical excision of the tumor and associated gland.

The eyelids are connected to the globe by the **conjunctiva**. Conjunctiva is the mucous membrane that lines the posterior surface of the eyelids (**palpebral conjunctiva**), anterior and posterior surface of the nictitating membrane (**nictitans conjunctiva**), and the anterior portion of the globe (**bulbar conjunctiva**). The conjunctiva is composed of nonkeratinized, stratified squamous epithelium and the underlying loose connective layer called the **substantia propria**. The conjunctival epithelium contains mucin-producing goblet cells that contribute to the precorneal tear film. The substantia propria contains vessels, nerves, and resident lymphoid tissue. The conjunctiva is highly vascular and very mobile, permitting ocular and eyelid motion.

Regional foramina, fissures, and fossae

Foramina and fissures located in the medial wall of the orbit serve as openings for vessels, nerves, and ducts to pass through the bone and enter or exit the orbital space (Fig. 3.4-4). The largest of these in the dog (and the anatomic structures they contain) include the **optic canal** (optic nerve, internal ophthalmic artery), **orbital fissure** (oculomotor nerve, trochlear nerve, abducens nerve, ophthalmic nerve, anastomotic branch of the external ophthalmic artery, orbital venous plexus), and the **rostral alar foramen** (maxillary branch of the trigeminal nerve, maxillary artery). Smaller foramina in the medial orbital wall of the dog include the two **ethmoidal foramina** (external ethmoid artery and ethmoidal nerve), the **sphenopalatine foramen** (caudal nasal nerve, and sphenopalatine artery and vein), and the **caudal palatine foramen** (major palatine nerve, artery, and vein).

The fossa for the lacrimal sac is in the rostral orbital margin and, as the name implies, is a small bony depression in the lacrimal bone that contains the underdeveloped canine lacrimal sac. The **lacrimal canal**, containing the **nasolacrimal duct**, exits the fossa of the lacrimal sac and is directed rostroventrally toward the nasal cavity. On the ventral surface of the zygomatic process of the frontal bone is the fossa of the lacrimal gland. This fossa marks the dorsal origin of the orbital ligament and the location of the lacrimal gland.

Selected references

[1] Dorbandt DM, Joslyn SK, Hamor RE. Three-dimensional printing of orbital and peri-orbital masses in three dogs and its potential applications in veterinary ophthalmology. Vet Ophthalmol 2017;20:58–64.
[2] McDonald JE, Knollinger AM, Dees DD. Ventral transpalpebral anterior orbitotomy: surgical description and report of 3 cases. Vet Ophthalmol 2016;19:81–9.
[3] Lederer K, Ludewig E, Hechinger H, et al. Differentiation between inflammatory and neoplastic orbital conditions based on computed tomographic signs. Vet Ophthalmol 2015;18:271–5.
[4] van der Woerdt A. Orbital inflammatory disease and pseudotumor in dogs and cats. Vet Clin North Am Small Anim Pract 2008;38:389–401.
[5] Wang FI, Ting CT, Liu YS. Orbital adenocarcinoma of lacrimal gland origin in a dog. J Vet Diagn Invest 2001;13:159–61.

CHAPTER 3

CASE 3.5

Otitis Interna/Media

Adalberto Merighi and Laura Lossi
Department of Veterinary Science, University of Turin, Turin, IT

Clinical case

History

A 6-year-old male castrated domestic shorthair cat was presented with signs of a right-sided head tilt that was acute in onset. The owner had been treating the ears for what was believed to be ear mites with an undisclosed product. The cat preferred a soft diet and any dry food fed was left untouched in the feed dish.

Physical examination findings

The cat had a right-sided head tilt, ataxia, horizontal nystagmus with fast phase to the left, and fell toward the right side when walking. Postural reactions and spinal reflexes were normal. Reluctance was noted when opening the cat's mouth for an oral examination.

Differential diagnoses

Otitis media due to iatrogenic rupture of the eardrum during ear cleaning, foreign body, or microbial infection; middle ear polyp or neoplasm; neurologic disease involving the vestibulocochlear nerve (CN VIII)

Diagnostics

The cat was sedated for otoscopic examination and digital radiographs of the head. Otoscopic examination revealed redness and swelling of the skin lining the ear canals consistent with otitis externa, and intact tympanic membranes bilaterally. Digital radiographs of the head were within normal limits. Magnetic resonance imaging performed under anesthesia revealed an effusion within the right tympanic bulla with abnormal fluid opacity. Myringotomy for culture and sensitivity of the right middle ear isolated *Staphylococcal* spp.

Diagnosis

Bilateral otitis externa and right-sided otitis interna/media (due to bacterial infection with secondary peripheral vestibular dysfunction)

Treatment

Gentle irrigation and lavage was performed with suction of the external ear canal. The cat was placed on a 6-week course of antimicrobial therapy. Bulla osteotomy (Gr. *osteon* bone + Gr. *temnein* to cut) was not needed to confirm the infection or diagnosis in this case. The cat recovered from the ear infection though continued to have a mild right-sided head tilt.

Comparative Veterinary Anatomy: A Clinical Approach. https://doi.org/10.1016/B978-0-323-91015-6.00016-9

Anatomical features in canids and felids

Introduction

The ear is the organ of hearing and balance. It is divided into 3 parts: outer, middle, and inner. The outer ear includes the auricle (pinna) and the external acoustic meatus (ear canal). The middle ear is contained within the temporal bone and includes the tympanic cavity that accommodates the 3 auditory ossicles: malleus (hammer), incus (anvil), and stapes (stirrup). Finally, the inner ear, also confined within the temporal bone, consists of the bony and membranous labyrinths, where acoustic receptors and statokinetic receptors are located.

Function

The outer ear collects sounds from the environment and transmits the sound waves to the tympanic membrane. The middle ear operates as the first transducer of sound waves, which impact onto the tympanic membrane and move the lever system consisting of the chain of auditory ossicles (Fig. 3.5-1). Movements of the stapes (stirrup) lead to the formation of pressure waves in the inner ear corresponding to the vibrations of the tympanic membrane. Receptors

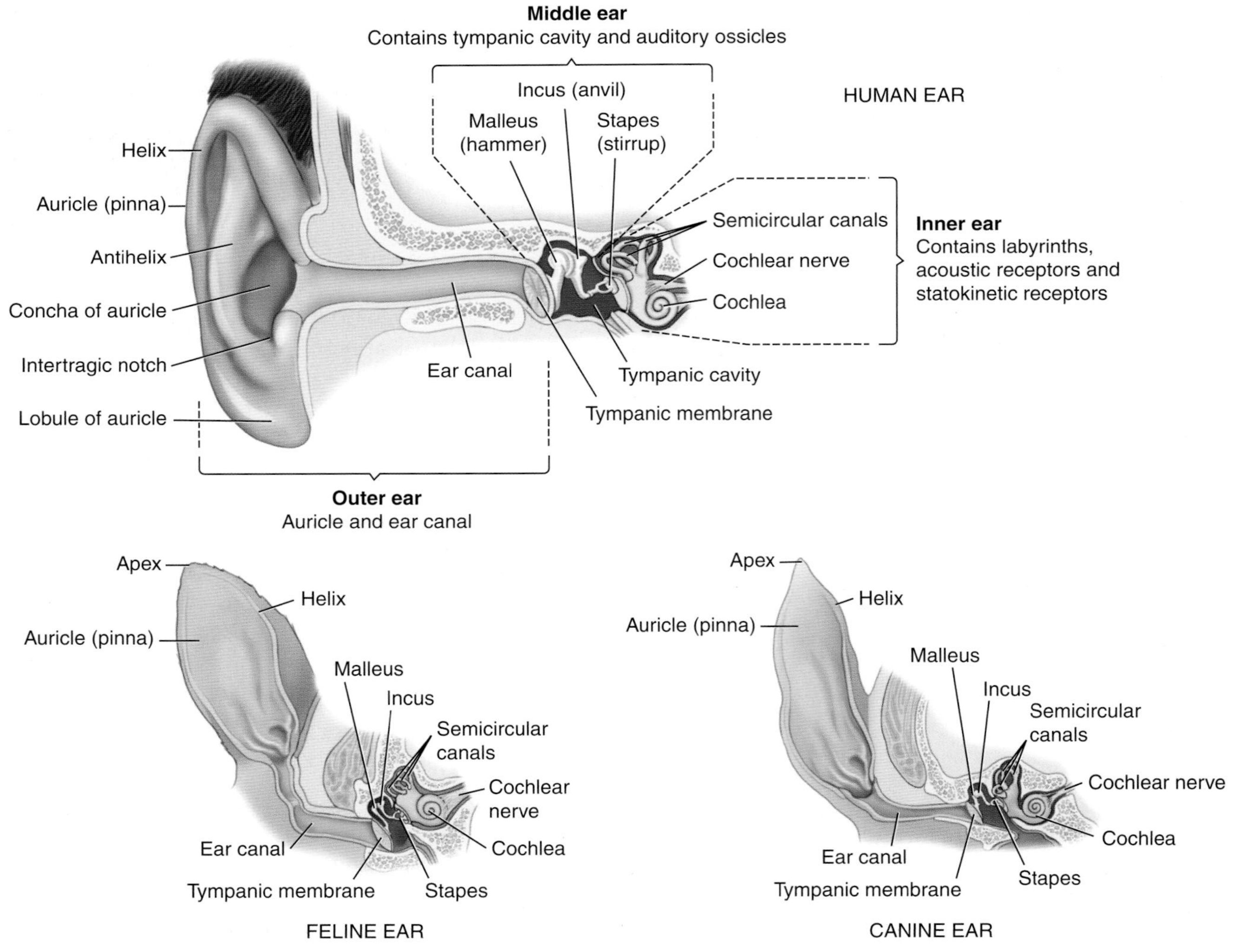

FIGURE 3.5-1 Schematic drawings of the ear in humans, cats, and dogs. Note, in particular, the differences in size of the pinna, as well as in the length and direction of the ear canal.

in the inner ear, referred to as hairy cells, assemble information about sounds and balance that is transmitted to the brainstem. These receptors are modified epithelial cells with shared histology (Fig. 3.5-2).

In clinical veterinary medicine, the function of balance regulation is especially important, since deafness constitutes a relatively less severe impairment in companion animals than in man.

Acoustic receptors perform a second transduction of sound waves collected from the environment, whereby pressure waves in endolymph are encoded into electrical signals (nerve impulses) to be sent to the brainstem.

Statokinetic receptors, instead, respond to mechanical stimuli to adjust balance of the animal's body in space. These receptors transduce mechanical stresses of the otoliths of the macules and movements of endolymph in the semicircular canals into electrical signals (Fig. 3.5-3).

External anatomy of the pinna

The **pinna** (**auricle**) is a skin fold that collects sound waves from the environment to convey them into the external acoustic meatus. Its presence allows the animal, in a general sense, to perceive sounds at a greater distance than man. The size and shape affect the optimal frequencies of perceived sounds: the larger the auricle, the more readily sensed are low frequency, longer waves. The shape is variable between different animal species and breeds and can be a characteristic feature of specific breeds.

The auricle has several landmarks that are important in surgery (Fig. 3.5-4).

The **helix** is the border of the auricle. It has 2 margins with different orientations in different species/breeds according to how the ears are carried. In animals with erect ears, one of the margins is medial and the other lateral, whereas in those with floppy ears, one is cranial and the other caudal. The 2 edges of the helix meet at the top (apex) of the pinna. The **scapha** is a triangular portion of the auricle bordered by the margins of the helix, extending from the dorsal apex to the ventrally located **antihelix**. The latter is the transverse margin found at the base of the auricle on its medial surface, bordering the external acoustic meatus.

The enlarged initial portion of the external acoustic meatus is called the **concha**. It is continuous with the vertical portion of the meatus. The combined configuration of the concha and external acoustic meatus allows for selective

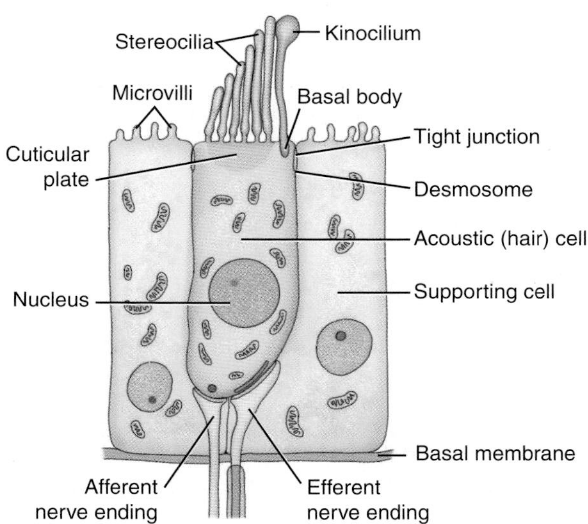

FIGURE 3.5-2 A schematic drawing of a hairy cell of the inner ear. Hairy cells are columnar or flask-shaped and have an array of stereocilia at the apical end. Additionally, there is one true cilium (the kinocilium) at one flank of each tuft of stereocilia. The position of the kinocilium "polarizes" the cell. Research has shown that hairy cells also receive efferent innervation from olivocochlear fibers of the cranial olivary complex. These efferent fibers inhibit the auditory end-organs and have a more variegated effect on the vestibular end-organs.

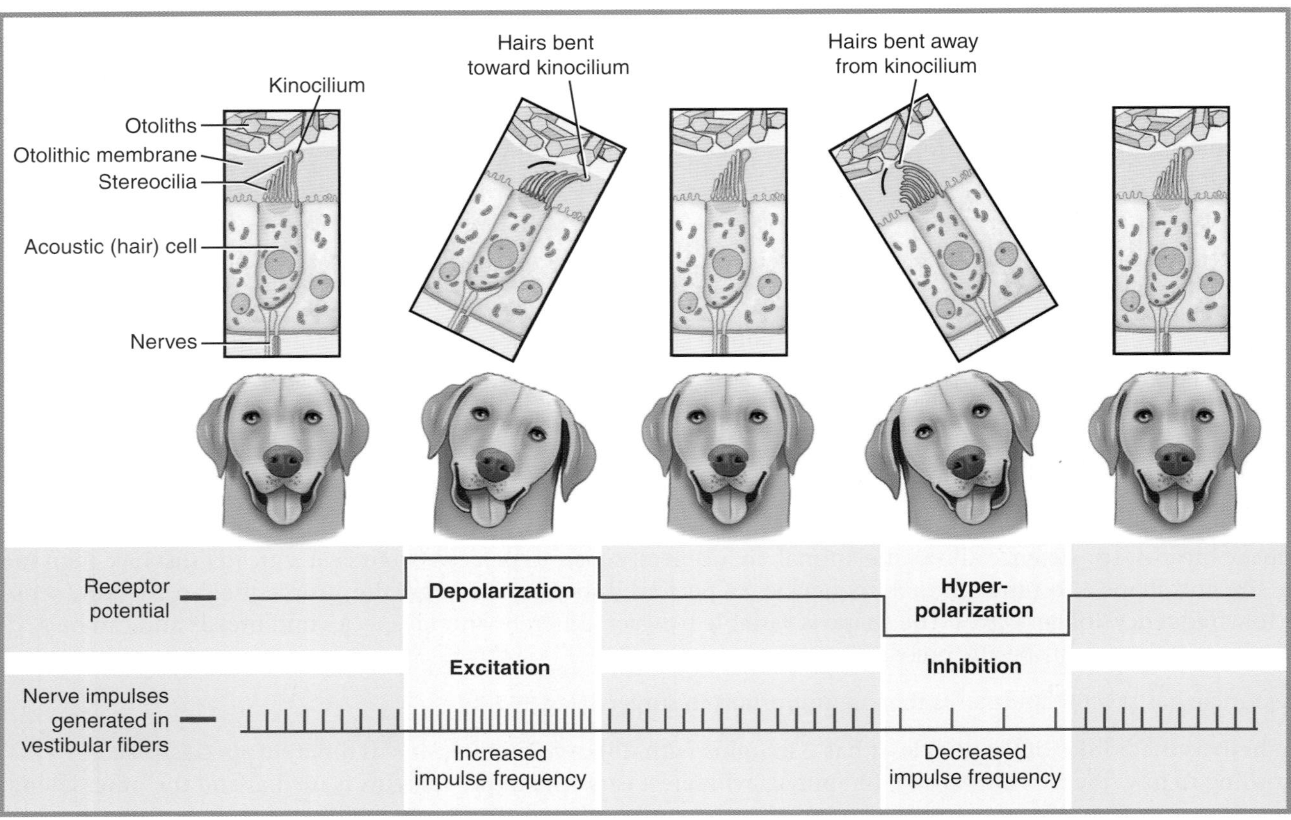

FIGURE 3.5-3 Schematic drawing of the mechanism of activation of statokinetic receptors. As the dog moves the head, there are variations in the receptors' membrane potential according to the direction of the head's movement. This causes depolarization (activation) when the stereocilia move toward the kinocilium, or hyperpolarization (inhibition) when they move opposite to it. The nerve impulses are then transferred to the vestibular nerve dendrites at the cytoneural synapses located at the basal pole of the hairy cells.

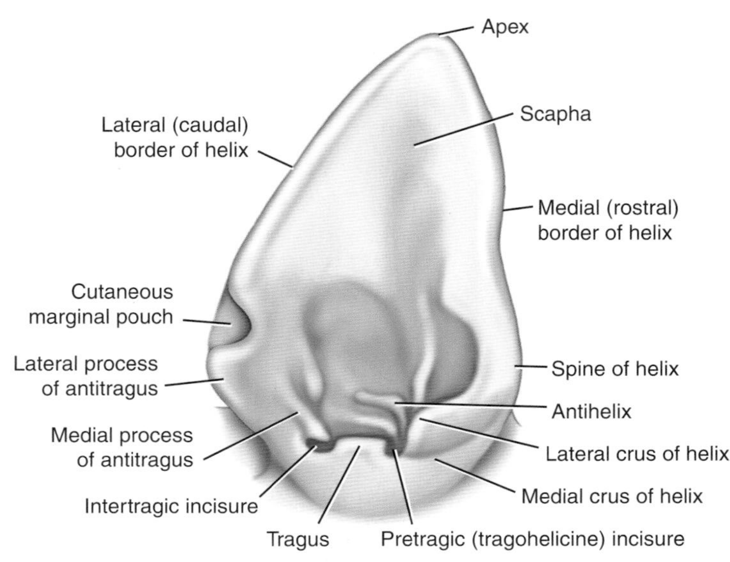

FIGURE 3.5-4 Schematic drawing of the right pinna. The main reference points useful for surgery are indicated. (Modified from Miller and Evans' Anatomy of the dog. 5th ed.)

amplification of certain sound frequencies. The **tragus** is the quadrangular side edge of the external acoustic meatus, opposite to the antihelix. The **antitragus** is the caudolateral margin of the external acoustic meatus. It extends to a small lateral **incisure** (marginal cutaneous pocket). Rostrally, the **pretragic (tragohelicine) incisure** divides the tragus from the helix, and the **intertragic incisure** separates the caudal part of the tragus from the antitragus.

The antitragus is an important point of reference for the insertion of the otoscope's speculum. The ear canal is straightened by gently applying traction on the pinna. The straighter the ear canal, the easier the visualization of the tympanic membrane and more comfortable the examination is for the patient. Otoscopy of the external acoustic meatus and the tympanic membrane is very important to properly diagnose external otitis. The skin of the external acoustic meatus should be examined to detect the presence of parasites or changes due to hyperemia and inflammation. The tympanic membrane should be scanned for integrity to exclude a perforation.

Muscles of the pinna

The position and movements of the pinna depend on **auricular muscles**, whose development varies widely in different species. The cat has greater mobility of the auricle and a total of 32 auricular muscles, 20 of which directly connect to the pinna, allowing 180° rotations. The dog has approximately 20 auricular muscles. These numbers are very high compared to humans' 6 auricular muscles. The auricular muscles in domestic carnivores are divided into 3 main groups: (1) the **rostral auricular muscles** rotate the pinna medially and reduce the amplitude of the scapha by their contraction, (2) the **caudal auricular muscles** are elevators and lateral rotators of the pinna, and (3) the **ventral auricular muscles** move it downward.

Most of the auricular muscles are innervated by branches of the **facial nerve**; however, some of the caudal muscles are innervated by the **first cervical nerves**.

External acoustic meatus

The external acoustic meatus is a canal that runs in a ventromedial direction and continues the cavity of the concha to its termination at the **tympanic membrane** (Fig. 3.5-5).

In dogs and cats, the external acoustic meatus has **vertical** and **horizontal portions** forming an angle of 90°–110° between them. It is lined internally by skin throughout its course and, for most of the length, it has 2 elastic support cartilages, the **auricular** and **annular cartilages**. A short terminal portion of the horizontal canal has a bony skeleton in the form of a thick osseous ridge, called the **bony external acoustic meatus**. The skin has **hair**, which generally decreases in quantity as it nears the tympanic membrane.

The disappearance or loss of hair near the tympanic membrane is a useful landmark indicating the proximity of the tympanic membrane. Because the external acoustic meatus is not straight in domestic carnivores, it is not always easy to see the tympanic membrane. Therefore, as the veterinarian moves the cone of the otoscope toward the membrane and/or inserts tubing to irrigate the external acoustic meatus, care must be taken to avoid iatrogenic injury to the tympanum. Location of hair disappearance thus becomes very useful. On the contrary, if an excessive quantity of hair is present, it supports stagnation of glandular secretions and predisposes to otitis externa, as these secretions provide a very good environment supporting parasites and/or bacteria.

Sebaceous glands are more abundant and numerous in the first part of the ear canal. They are superficial and consist of 6–10 acini in the shape of a club surrounded by connective tissue. The **ceruminous glands** are tubular, wrapped in a spiral pattern, and found in the deeper part of the dermis under the sebaceous glands; they increase in number in the third terminal section of the external acoustic meatus. Their excretory channels open both in the hair follicles and at the cutaneous surface and consist of a layer of secretory epithelial cells surrounded by myoepithelial cells. The number

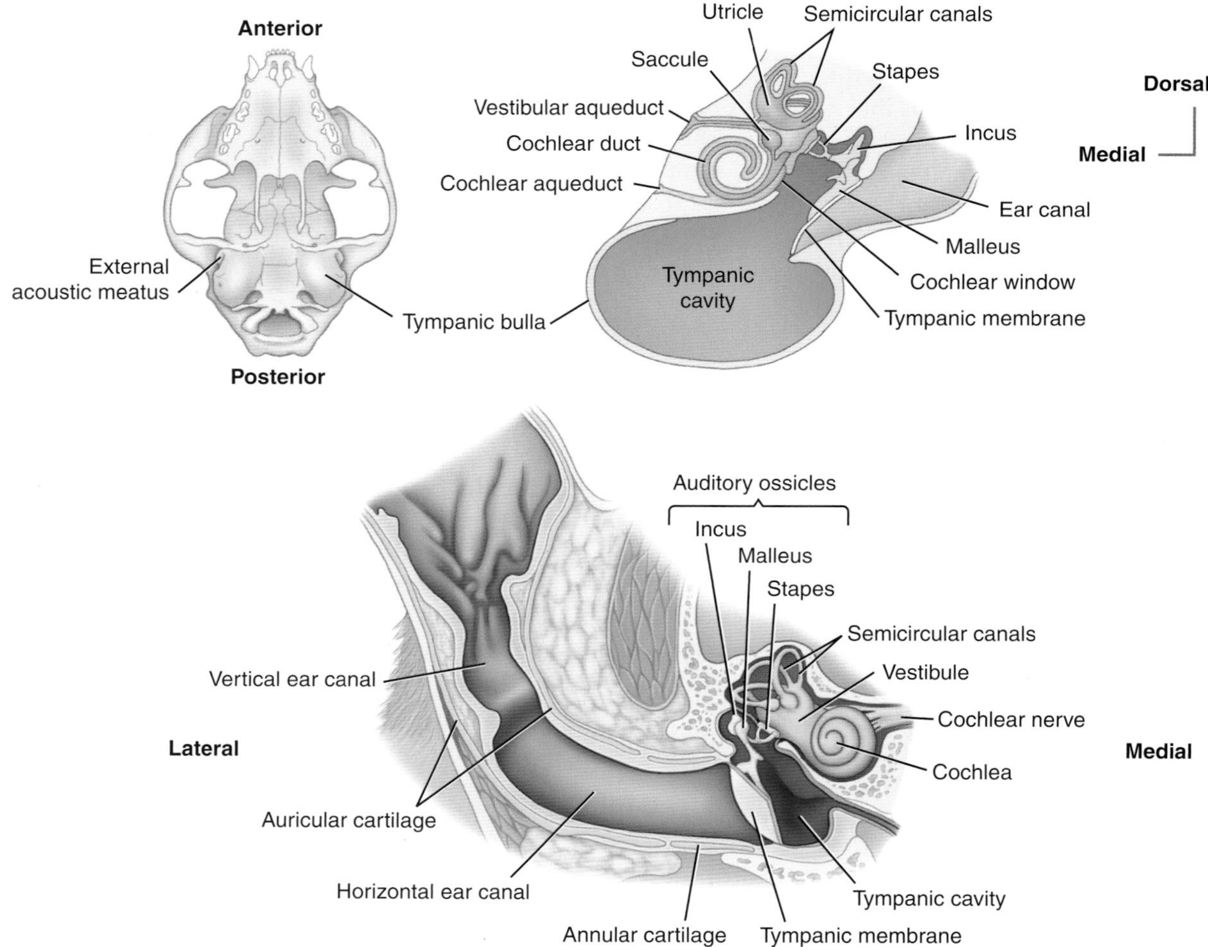

FIGURE 3.5-5 Schematic drawing of the canine ear. At *upper left* is the position of the middle and inner ear in reference to the skull. The image at the *bottom* shows the course of the ear canal. Note the existence of an approximately 110° angle between its vertical and horizontal portions. The figure at *top right* shows the middle and inner ear. Note the large size of the tympanic bulla.

Long-haired dog breeds—e.g., Spaniels or Irish Setters—have more glandular tissue than short-haired breeds—e.g., Boxers. These differences contribute to the predisposition of different breeds to otitis externa.

This cartilaginous crest makes it difficult to perform an otoscopic examination of the horizontal portion of the canal if the pinna is not gently elevated to straighten the canal.

of glands associated with the external acoustic meatus varies in different species and breeds.

The **auricular cartilage** is the continuation of the cartilage of the pinna and narrows to a funnel to form the vertical portion of the external acoustic meatus. The canal deviates medially immediately dorsal to the eardrum to form the horizontal portion of the canal. There is a prominent cartilaginous crest that separates the 2 parts of the external acoustic meatus.

The **annular cartilage** supports the horizontal part of the external acoustic meatus and extends to the bony external acoustic meatus of the temporal bone where it is fixed by fibrous connective tissue, allowing some flexibility in movement. There are significant differences between breeds and individuals in the joining of the external acoustic meatus with the surrounding bone, primarily with the **retroarticular process** of the mandible.

Differences in the attachment of the external acoustic meatus to bone affect ease of access to the most distal part of the external acoustic meatus during otoscopic examination of the tympanic membrane.

Tympanic membrane

The **tympanic membrane** (Fig. 3.5-6) is a thin semitransparent membrane that attaches to the **tympanic groove** of the temporal bone's **tympanic ring**. It is divided into a smaller dorsal flaccid part (**pars flaccida**) and wider ventral stretched part (**pars tensa**).

The tympanic membrane has an external face that concludes the acoustic meatus and an internal, **tympanic face** directed medially toward the cavity of the middle ear. In most dogs and cats, the flaccid part is flat on the lateral side when viewed otoscopically. However, in dogs, it can protrude laterally and appear convex, even under normal conditions.

The pars flaccida frequently protrudes toward the external acoustic meatus in the case of otitis media.

The pars tensa is thin, uniform, and strong because of radial thickenings of fibrous connective fibers. On the inner face, in the medial part of its perimeter, the membrane protrudes into the middle ear toward the tympanic cavity. The flat surface of the **manubrium** (handle) **of the malleus (hammer)** faces craniolaterally and adheres to the inner face of the tympanic membrane. When viewed from the outside, the elongated portion appears concave due to traction applied on the inner face at the attachment to the manubrium. During otoendoscopic examination, the manubrium is observed on the outer face forming a transparent strip called the **stria mallearis** (luminous triangle). The point of greatest depression, on the side opposite to the distal end of the manubrium, is called the **umbo**.

The presence of the taut and flaccid portions of the tympanic membrane depends on structural differences. In the taut part, the membrane consists of 3 layers. The first (**external layer**) is cutaneous and thin, the second (**intermediate layer**) is fibrous, and the third (**internal mucous**) layer is formed by a single layer of cuboidal cells. In the flaccid part, however, the intermediate layer is comprised of loose connective tissue, which is responsible for the flaccidity.

A unique feature of the tympanic membrane is its ability to be a thin and resilient structure in spite of continuous secretions of the skin glands attached to the external acoustic meatus. The slender nature of the tympanum relies on an epithelial migration process. This is associated with a self-cleaning process of the acoustic meatus, which includes a means by which ear wax is conveyed from the tympanic membrane to the distal opening of the ear canal. Failure of this process and accumulation of debris can lead to reduced—or complete loss of—hearing. Another important characteristic of the tympanic membrane associated with epithelial migration is its regenerative capacity. If a myringotomy (L. *myringa* combining form denoting the tympanic membrane) is performed in dogs or cats, the membrane begins to regenerate on the 14th day and is healed in 21–35 days.

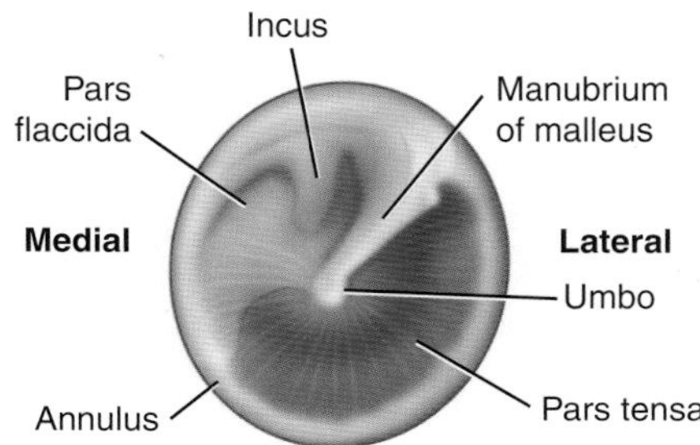

FIGURE 3.5-6 Schematic drawing showing the tympanic membrane.

As a whole, the average volume of the middle ear cavity measured by CT is 1.5 mL in mesomorphic (heavily muscled) dogs and increases nonlinearly with body weight. This is important with regard to the volume of liquids infused when irrigating the tympanic cavity.

Tympanic cavity

The **tympanic cavity** is the main cavity of the middle ear, corresponding to the intermediate compartment of the middle ear in a dorsoventral direction. It has the shape of a flattened fissure, in which lateral (**membranous**) and medial (**labyrinthic**) walls are recognized. The lateral wall is formed by the medial face of the tympanic membrane. This has a convexity—equivalent to the umbo. The medial wall is bony and has a raised part, the **promontory**, with 2 openings: the window of the vestibule (**oval window**) dorsally, providing access to the vestibule of the inner ear, which is normally closed by a membrane on which the **stapes** inserts; and the window of the cochlea (**round window**) more ventrally, which, instead, provides access to the cochlea and is closed by the **secondary tympanic membrane**.

Epitympanic recess

The **epitympanic recess** is the most dorsal and smallest compartment of the middle ear. It contains part of the malleus (hammer) and incus (anvil).

Tympanic bulla

The **tympanic bulla**, the most ventral cavity of the middle ear, is a blind-ended diverticulum of the tympanic cavity, with variable development depending on the species. In dogs and cats, the bulla is divided by the **septum of the tympanic bulla** (Fig. 3.5-7). In dogs, the septum is a small incomplete crest that rostrally contacts the **petrous**

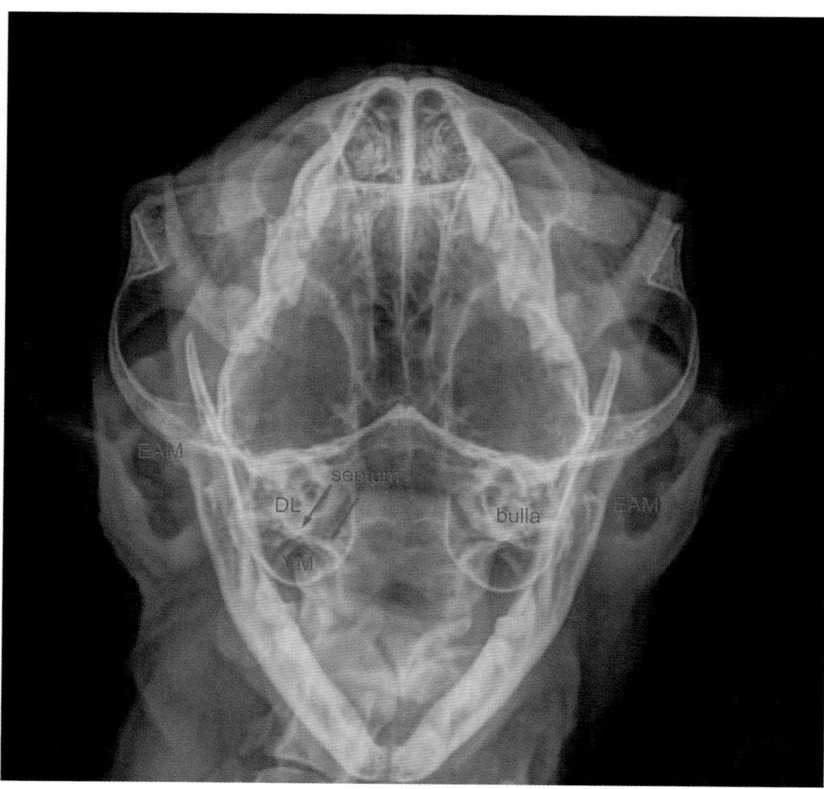

FIGURE 3.5-7 Ventrodorsal radiograph of a feline skull depicting the position and subdivision of the tympanic bulla. *DL*, dorsolateral compartment of the bulla; *EAM*, external acoustic meatus; *VM*, ventromedial compartment of the bulla.

portion of the temporal bone. In cats, the septum attaches to the petrosal bone and divides the tympanic cavity into 2 compartments: the **lateral epitympanic cavity (pars tympanica)** and the **ventromedial tympanic cavity (pars endotympanica)**. This separation is almost complete and the communication between the 2 compartments only occurs through 2 small openings: the first is situated between the septum and the petrous bone, and the other is located caudally, just to the side of the round window.

The difference in size and development of the septum between dogs and cats is of clinical importance in the treatment and management of diseases of the middle ear. The septum is smaller in dogs and leaves a large communication between the tympanic cavity proper and that of the bulla. This allows for complete emptying after an osteotomy. In cats, both compartments of the tympanic bulla can be accessed by perforation of the bony septum during ventral bulla osteotomy, but sympathetic fibers in the middle ear on the dorsal aspect of the bulla are easily traumatized. Furthermore, caution is advised in cats when administering drugs in cases where there has been any trauma or tearing of the tympanic membrane because medications can easily pass into the large ventromedial compartment of the bulla and remain sequestered in this compartment.

Nerves and vessels in proximity of the tympanic bulla

The tympanic bulla has shared associations with important regional vessels and nerves. The larger arteries in contact with the ventrolateral surface of the bulla are the **facial** and **maxillary arteries**, both arising from the **external carotid artery**. The **facial nerve** (CN VII) supplies motor innervation to the muscles of the lips and cheeks. It is the largest nerve associated with the ventrolateral surface of the tympanic bulla. This association occurs immediately after the nerve exits the **stylomastoid foramen** of the skull. The stylomastoid foramen is the terminal opening of the facial canal, lying deep in the petrous bone. In its course, the canal forms a double curve. The channel originates at the level of the **internal acoustic meatus** through the **internal acoustic pore,** an opening on the medial face of the auricular portion of the temporal bone. Emerging from the stylomastoid foramen, the facial nerve only contains somatic motor fibers as it goes a short distance in the parotid region before giving rise to the **internal auricular nerve** (2 nerves in dogs) and the **auriculopalpebral nerve**.

It is important to identify and safeguard these branches of the facial nerve at surgery because of their innervation of the external ear and eyelids, respectively. Injury to these nerve structures leads to partial/total paralysis of the pinna. It also affects the orbicularis oculi muscle with the possible sequela (L. lesion or affection caused by a disease) of making it difficult or impossible to close the eyelids.

There is a group of nerves that course inside the bulla or the cavity of the tympanum: the **internal carotid** (caroticotympanic) **nerves**, which contain sympathetic fibers from the cranial cervical ganglion; the **tympanic nerve**; and the **tympanic plexus**. The tympanic nerve is a branch of the **glossopharyngeal nerve** (CN IX). Above the promontory of the medial wall of the tympanic

HORNER SYNDROME IN DOGS AND CATS

Horner syndrome is a common neurological disorder of the eye and facial muscles caused by a dysfunction of the sympathetic nervous system. The condition usually occurs suddenly and without warning. The most common clinical signs of Horner syndrome in dogs and cats are ptosis, miosis, enophthalmos, and protrusion of the nictitating membrane on the affected side.

In cats, Horner syndrome is often caused by trauma—e.g., vehicular injury or a bite wound from another animal. Other causes include conditions affecting the retrobulbar structures, the middle ear, or tumors of the chest, neck, or brain.

In dogs, approximately 50% of cases of Horner syndrome are idiopathic.

Primarily in cats, the sympathetic branches of the tympanic nerve entering the plexus are situated immediately below the lining mucosa of the tympanic cavity. Here they can be injured during curettage of the middle ear, resulting in Horner syndrome.

Inflammatory exudate can injure the facial nerve in cases of otitis media/interna, causing ipsilateral facial drooping as well as decreased or absent palpebral reflex and menace response on the affected side. If the vestibulocochlear nerve (CN VIII) is also involved, then signs of ipsilateral peripheral vestibular dysfunction may develop (see Case 3.9).

cavity, it branches repeatedly, joining the sympathetic fibers of the internal carotid nerves and forming the tympanic plexus. From this plexus originate the sensory fibers for the mucosa of the middle ear (tympanic cavity and auditory tube) and the visceral motor fibers for the parotid gland via the **lesser petrosal nerve**, which contains parasympathetic postganglionic fibers. These preganglionic fibers synapse onto the otic ganglion neurons, from which the parasympathetic postganglionic fibers originate.

A small part of the bony canal protecting the facial nerve is incomplete dorsal to the caudal extension of the stapes. The stapedius muscle emerges here, inserting on the head of the stapes with the tympanic nerve in close contact with the mucosa.

Auditory tube

The **auditory tube** is a connecting channel between the tympanic and pharyngeal cavities. It has important interspecies differences.

In Carnivora, the auditory tube is a short channel originating from the nasopharynx at the **pharyngeal ostium** and ending in the rostral part of the tympanic cavity through the **tympanic ostium**. The lumen of the auditory tube is covered by a mucosa. The wall has different histological characteristics in its 3 parts starting from the nasopharynx to reach the middle ear: (1) the proximal cartilaginous part; (2) the junctional part or isthmus, which connects the cartilaginous and bony portions; and (3) the distal bony part, which runs through the musculotubal canal of the temporal bone. The lumen of the **bony portion** is always open, while the **cartilaginous portion** is closed at rest and opens during swallowing, following synergistic contraction of the **levator** and **tensor muscles** of the **palatal veil** that act together in keeping the lumen open. The pharyngeal ostium is concealed by the soft palate and is located halfway between the caudal limit of the nasal cavity and the palate. Contrast CT scans show that the auditory tube extends from the rostral and dorsomedial part of the tympanic bulla to the dorsolateral part of the nasopharynx, just caudal to the **pterygoid hamulus**.

The auditory tube has various functions and is best known for pressure regulation in the tympanic cavity—i.e., balance between the internal and external environment. Therefore, both sides of the tympanic membrane have equal pressures, allowing the membrane to vibrate normally when contacted by sound waves. In addition, the auditory tube can control the size of its lumen via the tensor and levator muscles of the palatal veil to protect the middle ear from noises and secretions from the nasopharynx. It also allows for physiological drainage of secretions from the mucosa of the middle ear into the nasopharynx.

Bony labyrinth

The **bony (osseous) labyrinth** of the inner ear is a complex set of narrow spaces inside the petrous part of the temporal bone (**petrous bone**), which contains an equally complex set of membranous structures (the **membranous labyrinth**). The bony labyrinth, in turn, consists of the **bony vestibule**, the 3 **osseous semicircular canals**, and the **bony cochlea**.

The **petrous bone** is composed of the densest bone in the body. It is grossly yellow in color compared to other cranial bones and lacks spongy tissue. Its shape is conical, with rounded corners and a rostroventrally directed apex. Its lateral face, constituting the medial wall of the tympanic cavity, has a projection that restricts the basal coil of the bony cochlea, called the **promontory**. In dogs and cats, there is a second projection that corresponds with the other parts of the cochlea. The **oval** and **round windows** flank the promontory on opposite sides.

Membranous labyrinth

The **membranous labyrinth** is also made up of a **vestibule**, formed by the **saccule** and the **utricle**; 3 **membranous semicircular canals**; and the **membranous cochlea**. It is separated from the bony labyrinth by a space containing a small amount of **perilymph**, a liquid with a composition like that of extracellular fluid. The membranous labyrinth contains **endolymph**, whose features are similar to that of cytosol (liquid part of cytoplasm). The individual elements of the membranous labyrinth cannot be visualized in vivo except with sophisticated CT or MRI techniques.

Vestibular ganglion and cochlear ganglion

The **vestibular** and **cochlear ganglia** are comprised of clusters of bipolar primary sensory neurons that convey special information regarding balance and hearing, respectively, to the CNS. The vestibular ganglion lies in the internal acoustic meatus, whereas the cochlear ganglion sits within the **modiolus** (central axis) of the **bony cochlea**. The neurons of these 2 ganglia give rise to the **vestibular nerve** and **cochlear nerve**, the 2 subdivisions of the acoustic or vestibulocochlear nerve (CN VIII).

Vestibular labyrinth

The vestibular system regulates the balance of an animal in space, preventing it from falling to the ground under the pull of gravity. This is done by monitoring and adjusting the position of the eyes, head, and body with respect to gravity. The system consists of **hairy cells**, a specialized class of mechanoreceptors; the **vestibular nerve**; and 4 **vestibular sensory nuclei** on each side of the 4th ventricle floor, extending within the gray matter from the rostral part of the medulla oblongata to the caudal third of the pons. The vestibular system intervenes in the regulation of vestibulospinal reflexes (static labyrinth) and vestibuloocular reflexes (dynamic labyrinth).

The hairy cells (Fig. 3.5-2) clustered at the maculae of the saccule and the utricle detect linear acceleration and deceleration of the head in space and, functionally, constitute the **static labyrinth** of the inner ear. The **maculae** have an oval shape, are present in the walls of the utriculus and sacculus, and monitor the position of the head with respect to gravity. Their major axis is vertical in the saccule and horizontal in the utricle. Each macula is covered with a layer of polysaccharides adhering to small crystals of calcium carbonate called **otoliths** (Fig. 3.5-3). The otoliths have a high specific mass and a higher density than endolymph. Therefore, when the head moves, the force of gravity makes them passively displace the stereocilia of the apical pole of the hairy cells. When the head bends to one side, some hairy cells are depolarized, others hyperpolarized, and still others show no changes. The complex signal that is thus generated is interpreted centrally.

The **semicircular canals** detect the angular acceleration and deceleration of the head and functionally constitute the **dynamic labyrinth** of the inner ear. In this case, the endolymph movement causes passive inclination of the stereocilia on the apex of the hairy cells of the **ampullary crests**. The stereocilia thus approach or move away from the **kinocilium**, another special type of cilium on the apex of hairy cells. The endolymph flows in a relatively opposite direction compared to that of the rotation of the head. Its movements involve the semicircular canals according to the relationship between the plane of rotation of the head and the plane on which the canal lies.

For example, if the head rotates in a counter-clockwise direction in the horizontal plane, the endolymph flows in a clockwise direction in the left horizontal semicircular canal, causing a folding of the stereocilia of the hairy cells of the ampullary crest toward the kinocilium and the depolarization of the receptors. On the contrary, in the right horizontal semicircular canal the endolymph flows in a counterclockwise direction, such that the stereocilia move away from the kinocilium and the receptors are hyperpolarized. In this example, a depolarization and increased frequency of action potentials in the left vestibular nerve follow, together with hyperpolarization and decreased discharge frequency in the right vestibular nerve.

Vestibular nerve

The **vestibular nerve** can be subdivided into 2 parts: the **superior**, which innervates the macula of the utricle and the ampullary crests of the anterior and lateral semicircular canals; and the **inferior**, which innervates the macula of the saccule and the ampullary crest of the posterior semicircular canal. The nerve then joins with the **cochlear nerve** to give rise to the **vestibulocochlear nerve** that enters the brainstem.

The **pontine vestibular nuclei** receive signals from the axons of the vestibular nerve and intervene to keep the animal in a vertical position by facilitating an effect on the ipsilateral extensor muscles of the limbs via the vestibulospinal tracts—i.e., the vestibulospinal reflexes. The vestibular nuclei also send axons through the medial longitudinal fasciculus to the motor nuclei of the CN III, IV, and VI that control the extrinsic ocular muscles to modify the position of the eyes in relation to the position of the head. In this way, the visual system can compensate for head movements and maintain stable images on the retina—i.e., the vestibuloocular reflexes.

The strict anatomical relationship between the vestibulocochlear nerve and the facial nerve explains why facial nerve palsy is one of the most common neurological complications following a total ear canal ablation (TECA) intervention and osteotomy of the tympanic bulla. The palsy, if not resolved, can lead to corneal alterations because the tear film is no longer evenly distributed on the corneal surface due to the lack of palpebral movements.

The vestibulocochlear nerve is enclosed in a common sheath of the dura mater with the facial nerve at the passage through the internal acoustic meatus to reach the pons.

Cochlea

The **cochlea** is the part of the inner ear where the **acoustic cells** are found. These specialized receptors are responsible for the perception of sounds. The cochlea is a 3D spiral formation divided into separate compartments; it has a membranous wall and is filled with liquid (Fig. 3.5-8).

The 2 main compartments of the cochlea are the **scala vestibuli (vestibular canal)** and the **scala tympani (tympanic canal)**, both of which contain perilymph derived from filtration of cerebrospinal fluid. At the apex of the cochlea, the two canals communicate through the **helicotrema**. Between the vestibular and tympanic canals lies the **scala media (cochlear duct)**, which is smaller and filled with endolymph. The cochlear duct is separated from the vestibular canal by **Reissner's vestibular membrane** and from the tympanic canal by the **basilar membrane**. The 3 compartments form 2.5 turns of coils around the bony axis of the cochlea.

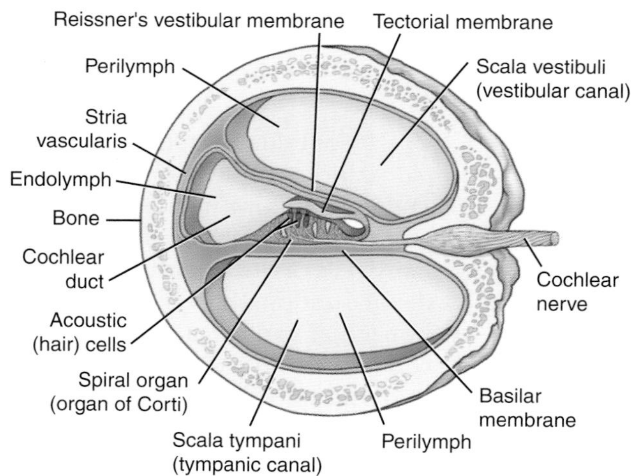

FIGURE 3.5-8 Schematic drawing of a transverse section through the cochlea.

Above the basal membrane is located the **spiral organ (organ of Corti)**, containing the acoustic cells. These cells are divided into **inner** and **outer acoustic cells**, using the modiolus (L. "hub" central pillar of the cochlea) as a reference point. The acoustic cells are related to different types of supporting cells that contribute to demarcate a spiral tunnel which contains the **cortilymph**, which is a specialized endolymph. Cortilymph contains high concentrations of sodium ions necessary to depolarize the cochlear fibers—i.e., the dendrites of the cochlear ganglion neurons, which cross the spiral tunnel to reach the acoustic cells.

A gelatinous substance containing collagen forms the **tectorial membrane,** which is suspended above the organ of Corti. The tectorial membrane overlies the sensory inner acoustic cells and electrically motile outer acoustic cells of the organ of Corti. During acoustic stimulation, the tectorial membrane stimulates the inner acoustic cells through fluid coupling, and the outer acoustic cells via direct connection to their stereocilia and kinocilia. The pressure waves of the perilymph of the scala vestibuli (vestibular canal), caused by the movements of the stapes (stirrup), give rise to a deflection of the basilar membrane of the cochlear duct. This causes movement of the organ of Corti and, thus, of the tectorial membrane resting on the acoustic cells, causing their stereocilia to bend. The bending direction determines the depolarization or hyperpolarization of the cells. The depolarized acoustic cells release glutamate at cytoneural junctions with the dendrites of the neurons of the cochlear nerve, while neurotransmitter release is inhibited in the hyperpolarized cells. Along its length, the basilar membrane varies in diameter and thickness—allowing it to detect the different wave frequencies perceived by the ear.

Selected references

[1] Anders BB, Hoelzler MG, Scavelli T, et al. Analysis of auditory and neurologic effects associated with ventral bulla osteotomy for removal of inflammatory polyps or nasopharyngeal masses in cats. J Am Vet Med Assoc 2008;233:580–5.
[2] Njaa BL, Cole L, Tabacca N. Practical otic anatomy and physiology of the dog and cat. Vet Clin North Am Small Anim Pract 2012;42:1109–26.
[3] Strain GM. Canine deafness. Vet Clin North Am Small Anim Pract 2012;42:1209–24.
[4] Ryugo DK, Menotti-Raymond M. Feline deafness. Vet Clin North Am Small Anim Pract 2012;42:1179–207.

CHAPTER 3

CASE 3.6

Thyroid Adenocarcinoma

Takanori Sugiyama[a], Helen M.S. Davies[b]
[a]Animalius Vet, Bayswater, WA, AU
[b]Veterinary BioSciences, The University of Melbourne, Melbourne, Victoria, AU

Clinical case

History

A 10-year-old male neutered Australian cattle dog was presented for a cervical mass located in the right ventral region of the neck. The mass was first noticed about 1 month before veterinary examination. The owner reported no other signs of illness. The referring veterinarian had performed a fine-needle aspirate, but it was nondiagnostic, showing only an abundance of red blood cells.

Physical examination findings

Other than the mass, no abnormalities were found on physical examination. The ovoid mass measured approximately 4.5 × 5 × 5 cm (1.75 × 2 × 2 in.) and was located in the middle of the neck, on the right ventral aspect. The mass was mobile beneath the skin and subcutis and attached cranially to a deeper structure. The dog showed no signs of pain on palpation of the mass, nor any other part of the neck.

Differential diagnoses

Thyroid tumor (adenoma or adenocarcinoma), benign thyroid hyperplasia, localized lymphadenopathy, hematoma

Diagnostics

Routine hematology and serum biochemistry, including indices of thyroid function, were normal. The dog was anesthetized for contrast-enhanced CT of the neck and thorax. The mass was located on the right side of the neck, lateral to the trachea and ventrolateral to the esophagus, displacing the trachea to the left but not compressing it (Figs. 3.6-1–3.6-3). Contrast enhancement revealed the mass to be vascular. Its location, size, and vascularity were strongly suggestive of thyroid adenocarcinoma. In addition, contrast enhancement showed enlargement of the right medial retropharyngeal lymph node (Fig. 3.6-4). No masses suggestive of metastasis were identified in the thorax and the left lobe of the thyroid gland appeared normal (Fig. 3.6-3).

> Thyroid adenocarcinomas are extremely vascular tumors, so it is not surprising that fine-needle aspiration yielded only red blood cells and no neoplastic cells. Incisional biopsy is not recommended when this tumor is suspected because the procedure carries too great a risk of hemorrhage. Rather, definitive diagnosis is based on histopathology following surgical excision (hemithyroidectomy or thyroid lobectomy), during which bleeding can be surgically controlled.

Comparative Veterinary Anatomy: A Clinical Approach. https://doi.org/10.1016/B978-0-323-91015-6.00017-0

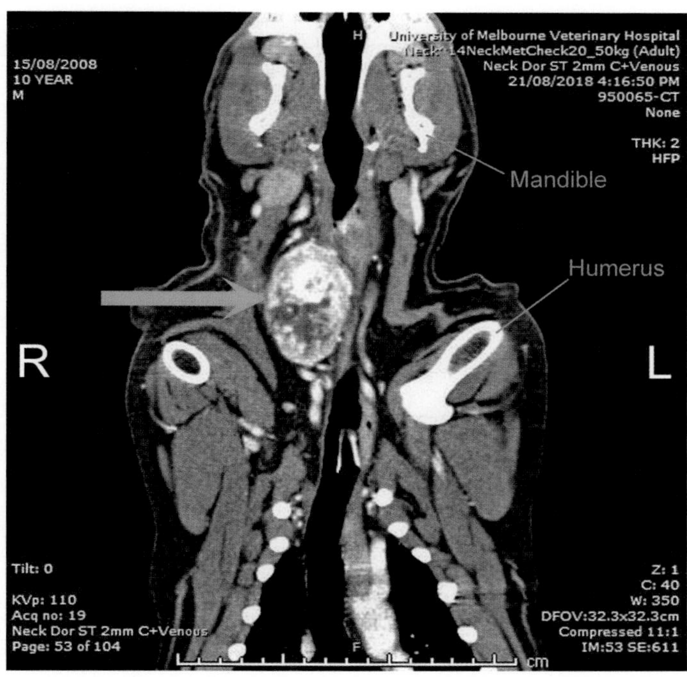

FIGURE 3.6-1 Contrast-enhanced CT image of the head, neck, and cranial thorax, dorsal view. A large mass, approximately 6 × 4 cm (2.4 × 1.6 in.), on the right side of the neck *(arrow)* extends across the midline. Contrast enhancement shows a mass of mixed radiodensity, indicating high but nonuniform vascularity.

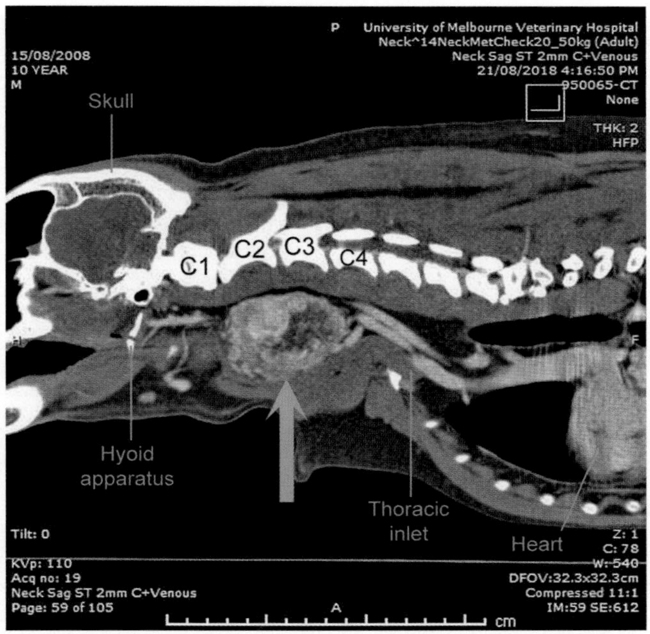

FIGURE 3.6-2 Contrast-enhanced CT image of the head, neck, and cranial thorax, sagittal view, showing the location, size, and radiodensity of the mass *(arrow)* from a lateral aspect. The mass extends from the caudal margin of the 1st cervical vertebra (C1) to the body of the 4th (C4).

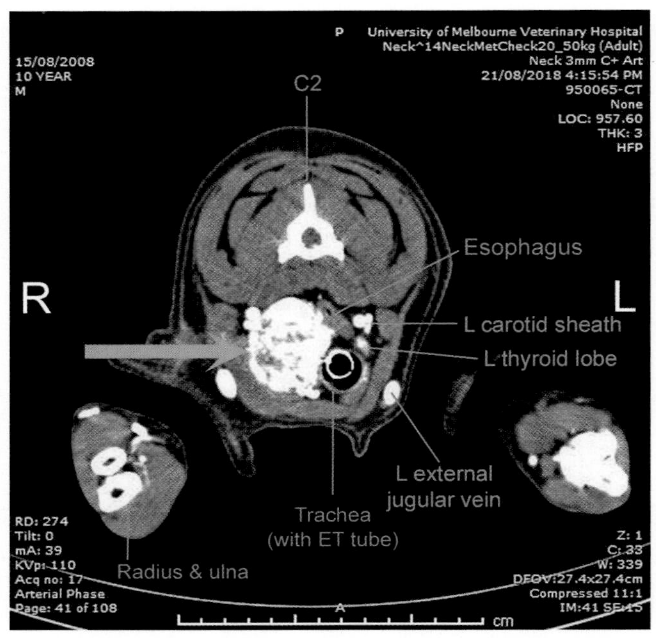

FIGURE 3.6-3 Contrast-enhanced CT image of the neck, transverse view at the level of the second cervical vertebra (C2). The mass *(arrow)* on the right side of the neck is displacing the trachea to the left but not compressing it. (The endotracheal [ET] tube can be seen in cross section within the tracheal lumen.) Contrast enhancement shows the high vascularity of the mass—important for presurgical planning. The left lobe of the thyroid gland appears normal in size, position, and vascularity. (Only the left [L] carotid sheath and external jugular vein are labeled. These structures are also present on the right side.)

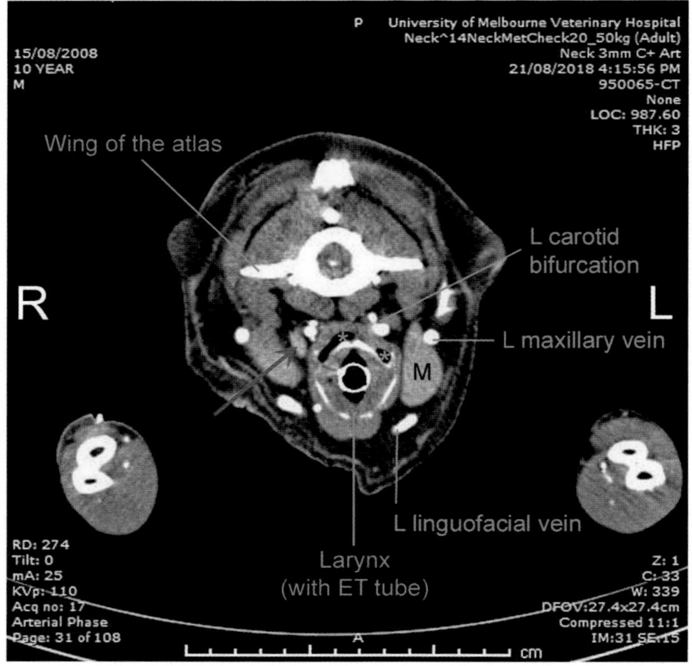

FIGURE 3.6-4 Contrast-enhanced CT image of the neck, transverse view at the level of the first cervical vertebra (the atlas), cranial to the mass. Contrast enhancement shows an enlarged right medial retropharyngeal lymph node *(arrow)*. Also note the elliptical shape of the laryngeal lumen compared with that of the trachea in Fig. 3.6-3. Parts of the air-filled caudal pharynx (*) are seen as black areas dorsal to the larynx. Key: *ET*, endotracheal; *M*, mandibular salivary gland. (Only the left [L] carotid bifurcation, internal maxillary vein, linguofacial vein, and mandibular gland are labeled. These structures are also present on the right side.)

Diagnosis

Thyroid adenocarcinoma (presumptive diagnosis, to be confirmed by histopathology)

Treatment

Due to the high likelihood of thyroid adenocarcinoma, the owner elected the recommended surgical excision of the mass, with wide surgical margins as the goal. With the patient under general anesthesia and in dorsal recumbency, a skin incision was made along the ventral midline of the neck adjacent to the mass. The left and right sternohyoideus and sternothyroideus muscles were separated to expose the trachea and the adjacent mass. The mass was adhered to the esophagus, trachea, and right carotid sheath. Meticulous dissection along the numerous tissue planes was performed to separate the mass from the surrounding structures, using electrocautery for hemostasis. The mass was removed in total, without the need for a blood transfusion. However, owing to its proximity to several vital structures, the surgical margins were limited. The patient recovered from surgery without problems.

Because thyroid adenocarcinomas are so vascular and are situated close to the carotid sheath (containing the common carotid artery, internal jugular vein, and vagosympathetic trunk), presurgical planning for a whole blood transfusion is important for these cases.

Histopathologic examination confirmed the diagnosis of thyroid adenocarcinoma. Although no obvious metastases were found on CT, this neoplasm is very aggressive. Because wide surgical margins could not be achieved, surgery was followed by chemotherapy and radiotherapy using the iodine radioisotope I^{131}.

About 90% of thyroid adenocarcinomas in dogs are malignant, with a propensity for local, regional, and even distant metastasis. In this case, the tumor may already have spread to the right medial retropharyngeal lymph node. Hormonally, however, most of these tumors are nonfunctional, so the majority of affected dogs are not hyperthyroid. Treatment options include surgical excision, radiotherapy (I^{131}), and chemotherapy. If the mass is fully resectable, surgery is the best single-treatment option.

Anatomical features in canids and felids

Introduction

The cervical spine, the principal nerves and muscles of the neck, and the hyoid apparatus are described in Case 3.8. The thyroid and parathyroid glands of cats are covered in detail in Case 3.7, with a brief mention herein. Following are the descriptions of the soft tissue structures in the neck.

Function

The anatomy of the neck is a complex network of arteries, veins, and nerves, intermixed with the trachea, esophagus, spinal cord, salivary glands, and muscles functioning as the connection between the thoracic and abdominal cavities and the head. The neck area assists in balance and locomotion during movement at different gaits.

Salivary glands (Fig. 3.6-5)

The salivary glands include the parotid, mandibular, sublingual, and zygomatic glands. They anatomically belong to the head, because they all empty into the oral cavity. However, these glands are briefly described and illustrated here because the **parotid** and **mandibular glands** are located at the junction of the head and neck. Additionally, these two glands partially or completely overlie the mandibular, parotid, and medial retropharyngeal lymph nodes that are described later.

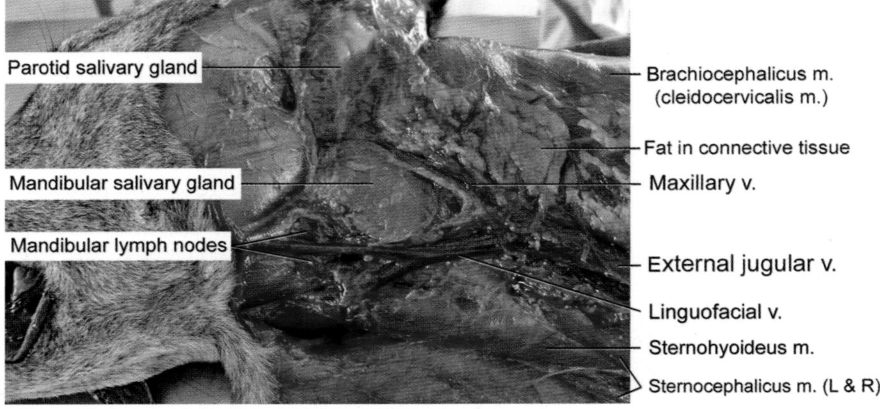

FIGURE 3.6-5 Superficial structures in the cranial portion of the canine neck. Cranial is to the left. Key: *L*, left; *m.*, muscle; *R*, right; *v.*, vein.

Caudal pharynx (Figs. 3.6-6 and 3.6-7)

The **pharynx** is a cavity that lies caudal to the oral and nasal cavities and is common to both the digestive and respiratory systems. Its rostral extent is the junction of the **hard** and **soft palates**, and for most of its length, the pharynx is divided by the soft palate into the **nasopharynx** dorsally and the **oropharynx** ventrally. The soft palate and nasopharynx are described further in Cases 3.2 and 3.1, respectively. The caudal extent of the pharynx is marked by the **esophageal opening** dorsally and the **larynx** ventrally. The pharynx therefore extends a short distance into the cervical region, to about the level of the 2nd cervical vertebra (C2).

The intrinsic and extrinsic muscles of the pharynx, larynx, soft palate, and tongue coordinate the flow of air and food/water from the nasal and oral cavities into their respective orifices (larynx and esophagus). This process is complex, as airflow from the nasal cavity to the larynx takes a caudoventral route through the pharynx, whereas the flow of food and water from the oral cavity to the esophagus takes a caudodorsal route. These flows are sometimes illustrated as 2 curved lines that cross in the pharynx, but in reality, respiration and deglutition (swallowing) are mutually exclusive functions within the pharynx.

CRICOPHARYNGEAL ACHALASIA

This condition is an uncommon cause of dysphagia in dogs. It involves incoordination of the pharyngeal muscles, particularly failure of the cricopharyngeal muscle to relax and allow food to enter the esophagus.

Cricopharyngeal achalasia (failure of smooth muscle to relax) is primarily seen in puppies soon after weaning onto solid food. Affected puppies are often thin or small for their age (the result of malnutrition) and they may gag or cough when eating. Aspiration pneumonia is a common complication.

The diagnosis is confirmed using fluoroscopy: the pharynx is observed radiographically while the puppy eats food mixed with barium (a radiopaque material). In these cases, the pharynx becomes distended with food as the caudal pharynx fails to relax and allow food to enter the esophagus, except in intermittent small amounts.

Treatment involves myectomy (surgical transection) of the cricopharyngeal muscle and sometimes a portion of the thyropharyngeal muscle rostral to it. These small, paired muscles originate on the lateral surface of the larynx and insert onto a median raphe on the dorsal aspect of the caudal pharynx. The caudal aspect of the cricopharyngeal muscle also blends with some of the muscle fibers of the esophagus. Both muscles thus contribute to the upper esophageal sphincter.

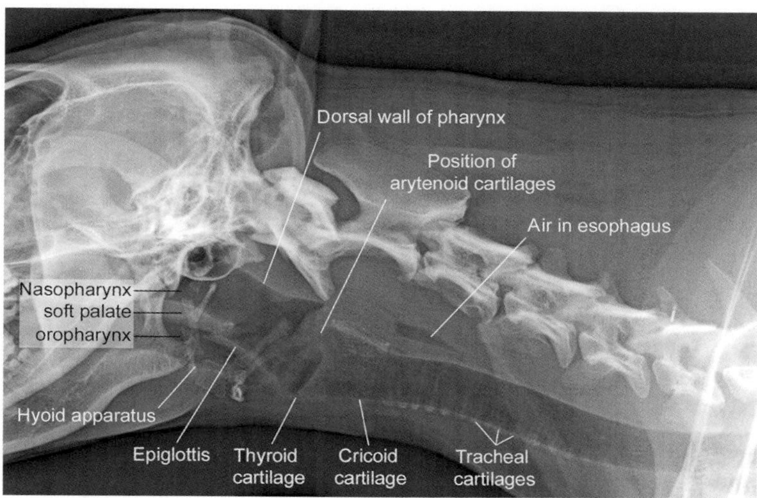

FIGURE 3.6-6 Lateral radiograph of the head and neck of a dog, highlighting the structures of the pharynx and larynx.

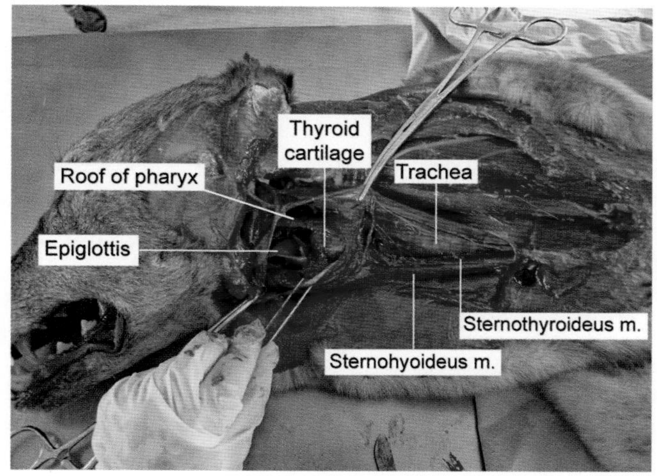

FIGURE 3.6-7 Lateral view of the canine pharynx and larynx. The lateral wall of the pharynx has been incised to expose the larynx. The dorsal clamp is retracting the left carotid sheath; the ventral clamp is retracting the incised lateral wall of the pharynx and overlying tissues; and the forceps are retracting the left thyrohyoid bone. Key: *m.*, muscle.

With the **epiglottis** at rest, as shown in Figs. 3.6-6 and 3.6-7, air moves freely from the nasal or oral cavity into the larynx, and thus, into the trachea. During deglutition, the epiglottis elevates to seal the glottis, and food/water is moved into the esophagus by the coordinated action of the muscles at the base of the tongue, the wall of the pharynx, and the soft palate.

> The glottis is somewhat like the pupil, in that it is the opening circumscribed by the surrounding structure(s). In the larynx, the glottis is defined by the arytenoid cartilages dorsally and the vocal folds laterally and ventrally (both described later).

The paired **palatine tonsil** (left and right) is a long, thin lymph node that is recessed in the lateral wall of the pharynx, just caudal to the **palatoglossal arch** (the junction of the soft palate and the pharyngeal wall at the base of

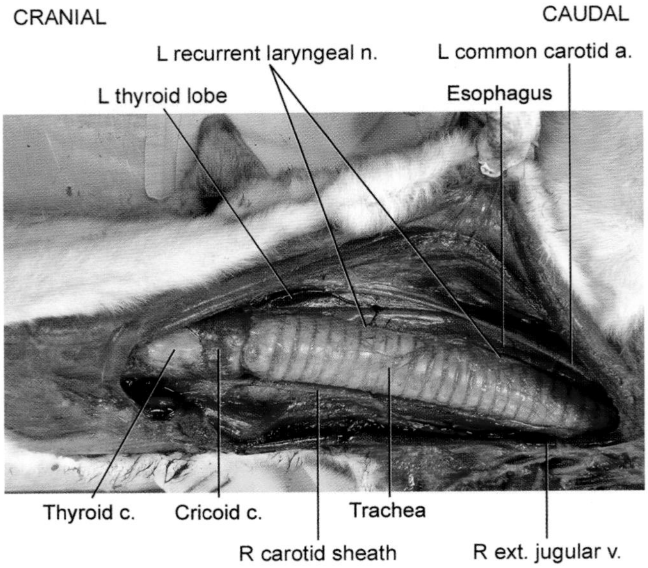

CRANIAL CAUDAL
L recurrent laryngeal n. L common carotid a.
L thyroid lobe Esophagus
Thyroid c. Cricoid c. Trachea
R carotid sheath R ext. jugular v.

FIGURE 3.6-8 Dissection of the ventral aspect of the canine neck. The thoracic inlet is on the right side of the photograph and the throat is on the left. Key: *a.*, artery; *c.*, cartilage (laryngeal); *ext.*, external; *L*, left; *n.*, nerve; *R*, right; *v.*, vein.

As the palatine tonsils are part of the first line of defense against ingested or inhaled pathogens, inflammation and infection of the tonsils (tonsillitis and tonsillar abscess, respectively) occur in dogs, cats, and humans. The tonsils may also be the site of neoplasia, such as squamous cell carcinoma (SCC). In dogs and cats, tonsillar SCC tends to be more aggressive than oral SCC.

the tongue). The recess forms a shallow well or moat, called the **tonsillar sinus**, around the body of the tonsil. The palatine tonsil drains into the mandibular and medial retropharyngeal lymph nodes on the same side (described later).

Esophagus and its vasculature and innervation (Fig. 3.6-8)

The **esophagus** is a soft, distensible, tubular structure that connects the pharynx (and thus, the oral cavity) with the stomach. As it extends from the pharynx to the stomach, it has cervical and thoracic portions, plus a very short abdominal portion.

The cervical portion of the esophagus originates in the caudodorsal wall of the pharynx, dorsal, and slightly caudal to the larynx. (Notice, that in Fig. 3.6-4 the pharynx drapes the dorsal surface of the larynx.) Beginning on midline dorsal to the trachea, in most dogs and cats, the esophagus shifts to the left of midline on its course toward the thoracic inlet, lying dorsolateral to the trachea through the midcervical region. At the thoracic inlet, the esophagus typically lies to the left of the trachea, but it may be found either dorsolateral, or ventrolateral to the trachea. In some individuals, the esophagus lies to the right of midline for much of its course through the cervical region—this is an uncommon, yet normal variant seen in a small percentage of horses.

The caudal pharynx and the cervical portion of the esophagus are supplied by branches of the **cranial thyroid artery**, which is the main branch of the common carotid artery. When present, the **caudal thyroid artery** in dogs also contributes to the esophageal blood supply. The caudal pharynx and cervical esophagus are drained by satellite veins of these arteries that empty into the **internal** and **external jugular veins**. Cranially, the lymphatic vessels drain into the **medial retropharyngeal lymph nodes**; in the middle and caudal portions of the neck, the esophageal lymph vessels drain into the **deep cervical lymph nodes**. These structures are all described in detail later. The caudal pharynx and cervical esophagus are supplied by branches of the **glossopharyngeal** and **vagus** nerves, and by the **recurrent laryngeal nerve**, both described in Case 3.8.

USE OF A LARYNGOSCOPE FOR ENDOTRACHEAL INTUBATION

A laryngoscope is an instrument that is used to facilitate endotracheal (ET) intubation in dogs and cats under general anesthesia. It has a slightly curved, blunt "blade" for depressing the base of the tongue and a light for illuminating the larynx. One side of the blade has a raised edge to help guide the ET tube into the glottis.

With the patient anesthetized and the head extended, the mouth is held open and the tongue is gently drawn rostrally to better see the larynx and lower the epiglottis. The laryngoscope is introduced into the oral cavity and oropharynx, and gentle pressure is applied to depress the base of the tongue. The ET tube is then carefully passed between the vocal folds and into the trachea (Figs. 3.6-10 and 3.6-11).

Cats are prone to laryngospasm (reflex closure of the glottis) during intubation, because the feline larynx is very sensitive to mechanical stimulation. To avoid this problem, a small amount of lidocaine (local anesthetic agent) is applied to the larynx a minute or so before intubation, with extra care taken when passing the ET tube through the glottis. If necessary, a smaller diameter ET tube is used than initially indicated based on the size of the cat and the diameter of the trachea.

Larynx and trachea (Figs. 3.6-9–3.6-13)

The **larynx** is a short, tubular, and relatively rigid structure that opens in the caudoventral aspect of the pharynx and acts as the "gatekeeper" for the trachea. Its functions are to prevent food and water from entering the trachea during deglutition (L. *deglutitio* the act of swallowing), modulate airflow to/from the trachea, and enable vocalization (hence, its colloquial name, the "voicebox").

The larynx is basically composed of a series of small cartilages, a set of intrinsic muscles, and a pair of vertically oriented mucosal folds (the vocal folds), which together create a structure that resists collapse under external and inspiratory pressures, yet is able to respond to changes in respiratory demand and to generate vocal sounds by altering the diameter of its rostral opening, the **glottis**.

Widening of the glottis during exercise and other situations of increased respiratory demand is an obvious function of the larynx. Closure of the glottis is a less appreciated function that is required when increased intraabdominal pressure is needed, such as during parturition (giving birth). Tenesmus (straining to defecate) may also elicit partial or complete closure of the glottis, depending on the intensity of abdominal "press" needed to expel the feces.

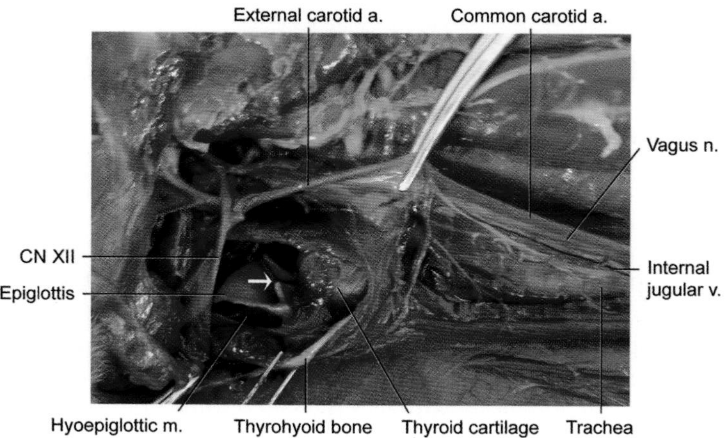

FIGURE 3.6-9 Close-up of the dissection shown in Fig. 3.6-7. The *arrow* overlying the epiglottis is pointing toward the glottis (not seen on this lateral view). Key: *a.*, artery; *CN XII*, cranial nerve 12 (hypoglossal nerve); *m.*, muscle; *n.*, nerve; *v.*, vein.

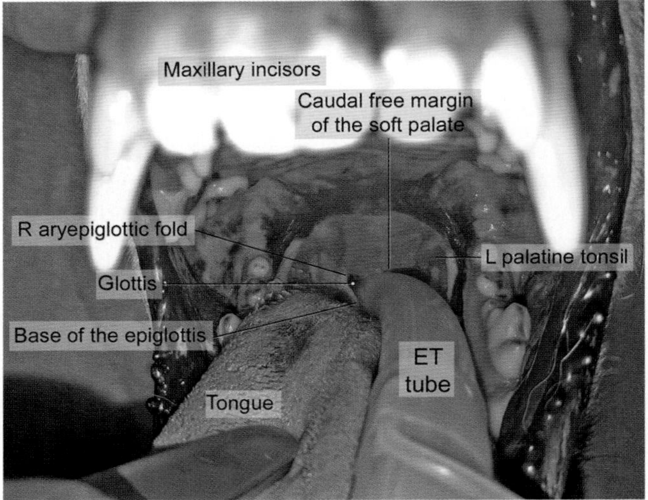

FIGURE 3.6-10 Oral view of an endotracheal (ET) tube, correctly passing through the glottis into the trachea of a dog. In this photograph, the caudal free margin of the soft palate is covering the dorsal aspect of the larynx and the ET tube is overlying most of the epiglottis, so only the right base of the epiglottis and the right (R) aryepiglottic fold are visible in the caudal pharynx. Also note that the left (L) palatine tonsil is everted from its mucosal recess, whereas the right tonsil is not visible in this photograph.

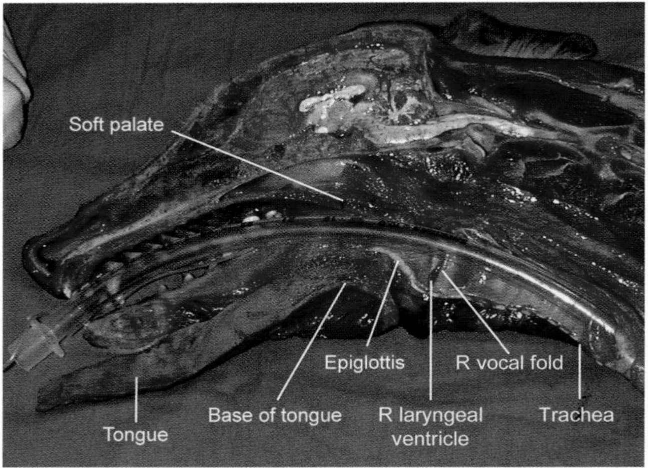

FIGURE 3.6-11 Sagittal section of the head and neck of a dog, showing an endotracheal (ET) tube in situ. The tube is passed through the oral cavity, over the base of the tongue and the epiglottis, between the left and right vocal folds (only the right [R] is shown), and into the trachea. The inflatable cuff near the tip of the ET tube is partially visible where the trachea has been incised (in the lower-right corner of the image).

The individual cartilages that comprise the larynx are all unpaired, except for the arytenoid cartilages (described later). The **epiglottis** is the most rostral of the laryngeal cartilages. In its resting position, it overlies the caudal floor of the pharynx. From a dorsal perspective, the epiglottis is shaped roughly like a spear tip, the apex directed rostrally toward the oral cavity, and the base attached to the thyroid cartilage immediately caudal to it. Laterally, the base of the epiglottis is attached to the bilateral arytenoid cartilages by paired folds of mucosa and connective tissue, the **aryepiglottic folds** (left and right).

On the ventral surface of the epiglottis, about midway between base and apex, the **hyoepiglottic muscle** ties the epiglottis to the **basihyoid bone** rostroventral to it and functions to draw the epiglottis horizontally, like the

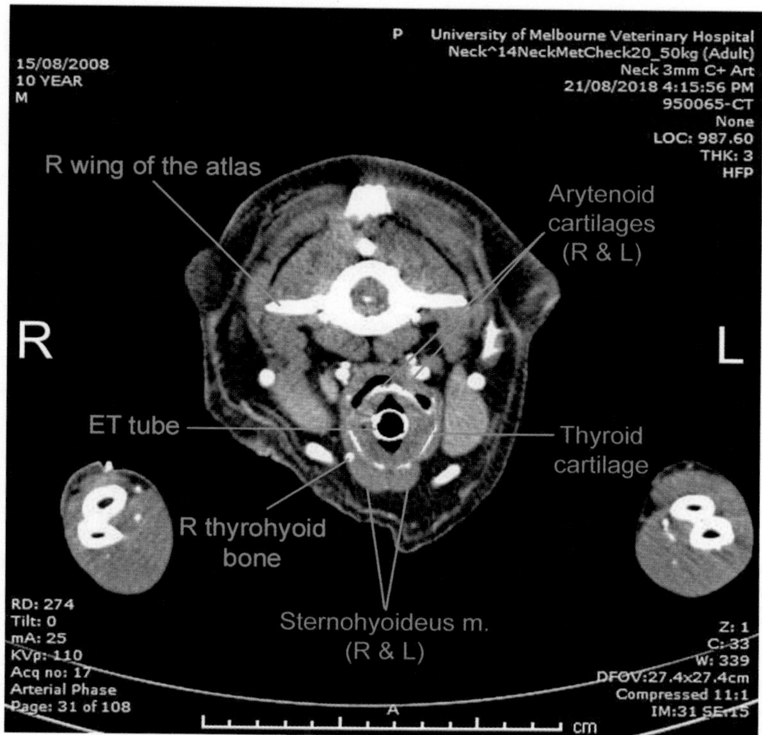

FIGURE 3.6-12 The CT image shown in Fig. 3.6-4, relabeled to highlight some features of the larynx. The "trough" shape of the thyroid cartilage (which appears incomplete in this "slice") is illustrated by the dashed green line. Key: *ET*, endotracheal; *L*, left; *m.*, muscle; *R*, right.

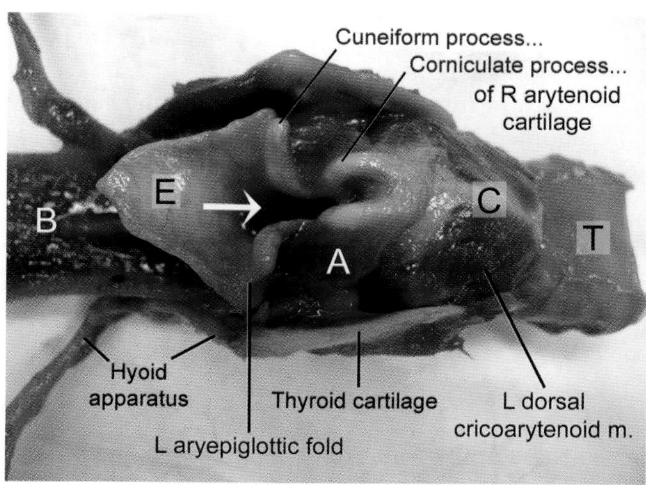

FIGURE 3.6-13 Dorsal view of the canine larynx. The *arrow* overlying the base of the epiglottis is pointing toward the glottis (obscured in this image by the arytenoid cartilages). Key: *A*, arytenoid cartilage (the letter A is over the left arytenoid cartilage); *B*, base of the tongue; *C*, cricoid cartilage; *E*, epiglottis; *L*, left; *m.*, muscle; *R*, right; *T*, trachea.

The short hyoepiglottic muscle also plays a passive role when the tongue and the basihyoid bone are drawn rostrally. For example, when a dog pants, protrusion of the tongue draws the basihyoid bone rostrally, which pulls the epiglottis down, thereby minimizing resistance to airflow between the oral cavity and the glottis. Drawing the tongue out during endotracheal intubation is another way in which the hyoepiglottic muscle passively lowers the epiglottis.

lowering of a drawbridge. The position of this muscle can be seen in Fig. 3.6-9, although the muscle itself is covered by pharyngeal mucosa. Conversely, when the animal swallows, the epiglottis is both passively pushed and actively drawn up by the base of the tongue and the muscles of the pharynx and larynx, such that it completely covers the glottis and prevents food or water from entering the larynx.

Immediately caudal to the epiglottis is the **thyroid cartilage**. One of the largest of the laryngeal cartilages, it is shaped like a deep trough that is open dorsally. The small, paired **arytenoid cartilages** (left and right) lie on the dorsomedial surfaces of the thyroid cartilage, toward its cranial aspect. As mentioned earlier, the aryepiglottic fold connects the arytenoid cartilages to the base of the epiglottis. The "roof" of the larynx, dorsal to the thyroid cartilage, is enclosed by the dorsal laryngeal muscles, notably the paired **dorsal cricoarytenoid muscle** that abducts the arytenoid cartilages (see later).

Caudal to the thyroid cartilage and forming the connection between the larynx and the trachea is the **cricoid cartilage**. It is the only one of the laryngeal cartilages that form a complete ring, although the cricoid ring is much narrower ventrally than it is dorsally. Ventrally, the wedge-shaped gap between the cranial aspect of the cricoid cartilage and the caudal aspect of the thyroid cartilage may be seen on lateral radiographs of the laryngeal region. This gap is filled by a fibrous membrane and is spanned ventrally by the **cricothyroid ligament**.

The epiglottic, thyroid, and cricoid cartilages are each apparent radiographically (Fig. 3.6-6). The arytenoid cartilages generally are not, as they are small, irregularly shaped, and covered by the thyroid cartilage and laryngeal muscles.

When viewed from the rostral aspect, the glottis at rest is a vertically elongated, triangular, or diamond-shaped opening between the arytenoid cartilages dorsally and the paired **vocal folds** laterally and ventrally. Abduction of the arytenoid cartilages by the dorsal cricoarytenoid muscles, innervated by the recurrent laryngeal nerves, expands the glottis and decreases resistance to airflow through the larynx and the trachea. Between the aryepiglottic fold craniolaterally and the vocal fold caudomedially is a mucosa-lined recess—the paired **laryngeal ventricles** or **saccules** (left and right).

The **trachea** begins at the larynx and extends to the cranial thoracic cavity, where it bifurcates into the **primary bronchi** at the base of the heart. The cervical portion begins immediately caudal to the cricoid cartilage of the larynx, to which it is attached by a narrow membrane. The trachea consists of a long series of cartilage rings, each connected to adjacent rings by a membrane that is about one-quarter the width of the cartilage rings. The **tracheal rings** are

LARYNGEAL COMPONENT OF BRACHYCEPHALIC AIRWAY SYNDROME

Brachycephalic airway syndrome is a complex of structural and functional abnormalities of the upper airway that cause varying degrees of airway obstruction in brachycephalic ("short skull" or short-nosed) dogs and cats, such as Pug dogs and Persian cats. This syndrome may eventually lead to laryngeal collapse.

Eversion of the laryngeal ventricles or saccules represents the mildest form (grade 1) of laryngeal collapse. With greater degrees of upper airway obstruction and, therefore, inspiratory pressure, the cuneiform (grade 2) and corniculate (grade 3) processes of the arytenoid cartilages also collapse into the laryngeal lumen during inspiration. Brachycephalic airway syndrome is described further in Case 3.2.

incomplete dorsally—the left and right sides of the rings connected by a short, transverse membrane (Fig. 3.6-13). The lumen of the trachea is lined by respiratory mucosa characterized as ciliated columnar epithelium, which forms the "mucociliary escalator" for the cranial (ascending, antigravitational) movement of mucus and debris from the lower airways to the pharynx.

Care is needed when inflating the cuff on an endotracheal tube. If the cuff is underinflated, it cannot serve its functions of (a) creating a "closed" system for gas delivery/retrieval during gaseous anesthesia and (b) protecting the lower airways from aspiration of fluid in the oropharynx, such as saliva and regurgitated gastric contents. If the cuff is overinflated, it can cause a ring of pressure necrosis that injures the tracheal mucosa and, if severe and prolonged, the tracheal rings can also be affected.

Thyroid gland and its vasculature and innervation (Fig. 3.6-14)

The thyroid (Gr. *thyreoeidēs, thyreos* shield) and parathyroid glands are described as they relate to the other structures in the cranial part of the neck.

The **thyroid gland** is a bilobed structure that, in most dogs and cats, is located lateral to the trachea, immediately caudal to the larynx. It is a common misconception that there are 2 thyroid glands. There is one thyroid gland, comprising 2 **lobes** (left and right) that are connected near their caudal poles by a thin, narrow **isthmus** (Gr. *isthmos* a narrow connection between two larger bodies or parts) of either glandular thyroid tissue or fibrous connective tissue that traverses the trachea ventrally. Because the isthmus embryologically, and postnatally in some animals (cattle), is composed of glandular tissue, the thyroid gland is considered a single structure comprising 2 interconnected lobes. The isthmus is most likely glandular in young animals and in brachycephalic breeds. In some adults, there is no longer any trace of an isthmus.

In the mature dog described in this case, the isthmus was composed entirely of connective tissue, so the disease was confined to just the right thyroid lobe, and hemithyroidectomy—removal of just the affected lobe (lobectomy)—was performed.

Each thyroid lobe is a flattened, elongated, oval or elliptical structure that narrows at its cranial and caudal poles. Both lobes are symmetrical in size and shape, but it is relatively common for one to be larger than the other

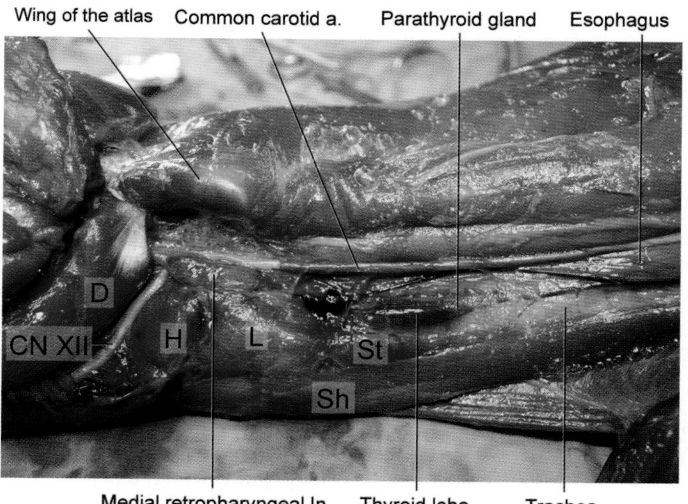

FIGURE 3.6-14 Deep dissection of the cranial cervical region of a dog, left side. (Note that in this dog, the left external parathyroid gland is located near the caudal pole of the left thyroid lobe.) Key: *a.*, artery; *CN XII*, cranial nerve 12 (hypoglossal nerve); *D*, digastricus muscle; *H*, hyoid bone; *L*, larynx; *ln*, lymph node; *Sh*, sternohyoideus muscle; *St*, sternothyroideus muscle.

(by up to 50%), and for the left lobe to be situated slightly caudal to the right lobe. In general, the thyroid gland tends to be relatively larger in young animals than in adults, in small (particularly brachycephalic) breeds than in medium and large breeds, and in females than in males, especially during estrus and pregnancy—periods of relatively greater metabolic activity.

The thyroid lobes are covered by the paired sternocephalicus muscles laterally and by the paired sternothyroideus and sternohyoideus muscles ventrally. (These muscles are described in Case 3.7.) The fibrous capsule of the thyroid gland is loosely attached to these overlying muscles by connective tissue. Medially, each lobe lies alongside the trachea, to which it is loosely attached by connective tissue. Dorsally, each lobe is in close association with its respective carotid sheath and recurrent laryngeal nerve. (See Case 3.7.) As described later, the **parathyroid glands** are intimately associated with the thyroid gland.

The thyroid gland is a highly vascular tissue, and its arterial supply requires special surgical consideration. The main blood supply to the thyroid gland is the paired **cranial thyroid artery**, which is the main branch of the **common carotid artery**. As each thyroid lobe is closely associated with the carotid sheath, the cranial thyroid artery is short; it arises at the level of the larynx and courses caudally for a short distance to enter the thyroid lobe on that side. Along its short course, the cranial thyroid artery sends off pharyngeal, caudal laryngeal, and muscular branches, and it also supplies the parathyroid glands (described later).

> Because the cranial thyroid artery also supplies structures other than the thyroid lobe, ligation during thyroid lobectomy should be performed as close to the lobe as possible.

In dogs, the paired **caudal thyroid artery** is the smaller of the 2 main arteries of the thyroid gland and is not always present. Cats do not have a caudal thyroid artery. When present, the caudal thyroid artery may arise from any of the large arteries within the thoracic inlet—most often from the brachiocephalic artery. It courses cranially along the trachea and esophagus, with the recurrent laryngeal nerve, to supply the caudal portion of the thyroid lobe. Occasionally, the caudal thyroid artery is the principal arterial supply for the thyroid gland.

Caudal to the thyroid lobe, a large branch of the cranial thyroid artery (usually a ventral branch) courses caudally along the trachea toward the thoracic inlet, in close association with the recurrent laryngeal nerve. When the caudal thyroid artery is present, both cranial and caudal thyroid arteries anastomose in the midcervical region. Together, they supply the cervical portions of the trachea and esophagus.

Venous drainage from the thyroid gland is primarily via the paired **cranial** and **caudal thyroid veins**. Although venous drainage is variable, the cranial thyroid vein typically empties into the internal jugular vein and the caudal thyroid vein usually joins the brachiocephalic vein. The caudal thyroid vein is usually the larger of the 2, exiting the thyroid lobe at its caudal pole. Lymph vessels from the thyroid gland drain into the **cervical lymph nodes**, either the cranial deep cervical lymph node or, when absent, the nearest lymph node, such as the medial retropharyngeal lymph node (described later).

THYROID COMPARTMENT

The thyroid gland lies in a fascial compartment between the trachea and the superficial cervical muscles which extends from the sternum to the head. This compartment communicates readily with the mediastinal compartment in the thoracic cavity, so surgical access, foreign body penetration, tracheal perforation, or other disease processes involving this fascial compartment may result in significant cervical and even thoracic extension.

Surgical access to the thyroid gland, such as for the removal of a neoplastic lobe, is generally easy using a ventral midline approach over the trachea just caudal to the larynx. Normally, the mobility of the thyroid lobe allows it to be easily separated from the surrounding structures and its blood supply identified and ligated. Even so, lobectomy must be undertaken with care—not only to avoid bleeding from this vascular organ, but to avoid extension of bleeding, subcutaneous emphysema (accumulation of air beneath the skin), or possible infection throughout and beyond this fascial compartment. Care is also needed to identify and preserve the external parathyroid glands with their blood supply and drainage.

The thyroid gland receives sympathetic innervation from the **cranial cervical ganglion** (described in Case 3.8) and the cranial laryngeal nerve (a branch of the vagus nerve), which usually merge to form the **thyroid nerve**. In addition, the **middle cervical ganglion** may supply the thyroid gland via its vascular plexuses.

In dogs, it is common for **accessory** or **ectopic thyroid glands**—small, embryological remnants of thyroid tissue—to be found in various locations, including the connective tissues of the neck (anywhere from the larynx to the thoracic inlet), the periaortic fat at the base of the aorta, the mediastinum and/or pericardium, and even the abdomen. In the neck, these small masses may number from 2 to > 10. Accessory thyroid glands are uncommon in cats.

While not universal, accessory thyroid glands are normal structures. Typically, they are too small to appreciate clinically or on routine diagnostic imaging. However, they may enlarge to several times their normal size following total thyroidectomy and in about 4% of cats with hyperthyroidism. Accessory thyroid glands should not be mistaken for metastases in patients with thyroid tumors.

Parathyroid glands and their blood supply and innervation

The **parathyroid glands** are so-named because of their anatomical association with the thyroid gland (G. *para*-beside, alongside). It was once thought that they were embryological remnants of thyroid tissue, but other than sharing a blood supply, the 2 glands are embryologically, structurally, and functionally distinct.

The parathyroid glands in dogs and cats consist of multiple, small, ovoid or disc-shaped glands that are located adjacent to, and within, the thyroid gland (see later). Parathyroid gland size is variable in normal dogs, averaging 3.7 × 2.6 × 1.6 mm (0.15 × 0.10 × 0.06 in.). The parathyroid glands are usually some variation of yellow, red, or brown.

Unless enlarged, the parathyroid glands are difficult to distinguish from surrounding structures, even with CT, MRI, or ultrasonography.

In dogs, about half of the parathyroid glands are "external," located exterior to the thyroid capsule, and the other half are "internal," located within the thyroid capsule and even within the parenchyma of the thyroid lobe. In most dogs, the **external parathyroid glands** are situated over the cranial half of their respective thyroid lobe, usually lateral to the lobe, but occasionally at its cranial pole or over its dorsal aspect. Uncommonly, as in the prosection shown in Fig. 3.6-14, the external parathyroid gland may be found over the caudal aspect or at the caudal pole of its thyroid lobe.

The **internal parathyroid glands** are typically found on the medial or "tracheal" surface of their respective thyroid lobe about midway between the cranial and caudal poles. Although similar in shape and color to the external parathyroid glands, the internal parathyroid glands tend to be smaller.

The parathyroid glands vary in size and number among animals, but typically each dog or cat has 4 principal parathyroid glands—2 on each side (left and right), one internal and one external to the thyroid lobe. Other smaller glands may also be present, as well as **accessory** or **ectopic parathyroid glands** in other locations, such as in the connective tissue surrounding the larynx, carotid sheath, cranial mediastinum, and thymus. Accessory parathyroid glands are common in cats but uncommon in dogs.

When present, hypertrophy of accessory parathyroid glands can enable a patient to restore or maintain normal calcium homeostasis following inadvertent parathyroidectomy during surgical removal of both thyroid lobes.

The parathyroid glands, like the thyroid gland, are abundantly vascular. In most cases, the external parathyroid glands have their own blood supply—a separate branch of the cranial thyroid artery. However, the internal

It is impossible to avoid removing the internal parathyroid gland(s) along with the thyroid lobe or entire thyroid gland during hemi- or total thyroidectomy. However, if the external parathyroid glands have their own blood supply and they, along with their vasculature, are identified and spared, the external parathyroid glands may be left in situ, preventing the development of hypoparathyroidism, post-operatively.

parathyroid glands, and in some dogs the external parathyroid glands as well, are supplied by small branches of the thyroid arteries. The venous and lymphatic drainage for both the external and internal parathyroid glands are as described for the thyroid gland.

Arteries of the head and neck

Two separate arterial systems supply the head and neck, although both originate from the **brachiocephalic trunk**: the common carotid artery and 3 branches of the subclavian artery (superficial cervical, costocervical, and vertebral arteries). All are paired.

The **common carotid artery** is the main blood supply to the head and to the deeper structures of the neck. It courses cranially along the neck within a deep layer of fascia, the **carotid sheath**, along with the **internal jugular vein** (described later) and the **vagosympathetic trunk** (described in Case 3.8). Owing to the importance of its contents, the carotid sheath is well-protected beneath the cervical vertebrae and adjacent musculature (Figs. 3.6-3 and 3.6-4). The left carotid sheath lies along the esophagus or between the esophagus and trachea on the left side of the neck and is loosely attached to the esophagus by the deep cervical fascia (Figs. 3.6-8 and 3.6-9). The right carotid sheath lies on the right dorsolateral aspect of the trachea.

The principal branch of the common carotid artery is the **cranial thyroid artery**, which exits the carotid artery just caudal to the larynx. It branches extensively to supply the thyroid gland, parathyroid glands, pharynx, larynx, trachea, and esophagus, as described earlier. Each common carotid artery terminates at the skull as it bifurcates into the external and internal carotid arteries. The much larger **external carotid artery** branches extensively as it supplies most of the structures of the head. The small **internal carotid artery** enters the cranial cavity and supplies the brain and meninges.

The **subclavian artery** sends off 3 branches before continuing as the axillary artery. The **superficial cervical** and **costocervical arteries** supply the caudal aspect of the neck, most notably the cervical muscles and skin. The **vertebral artery** is a deeper branch that courses cranially along the **longus colli muscle** immediately ventral to the cervical spine traversing the transverse foramina of the cervical vertebra each side from the 6th cervical vertebra onwards (see Case 3.8) to supply the cervical portion of the spinal cord and its meninges.

Veins of the head and neck

The veins of the head and neck drain into the small internal jugular vein and the much larger external jugular vein. As mentioned previously, for most of its length, the **internal jugular vein** courses toward the thoracic inlet within the carotid sheath. It joins the external jugular vein within the thoracic cavity, along with veins from the forelimb, all of which converge to form the **cranial vena cava**.

LYMPHADENOPATHY

Lymphadenopathy is an abnormal enlargement of one or more lymph nodes. It may involve a single lymph node (solitary or localized), multiple regional lymph nodes, or many lymph nodes in different parts of the body (generalized).

Any condition that triggers an immune response, particularly involving the recruitment and activation of neutrophils and/or T-lymphocytes, may result in lymphadenopathy. The many possible causes can broadly be grouped into infectious (bacterial, fungal, viral), noninfectious immune-mediated (allergic, autoimmune), and neoplastic (primary lymphoid tumors or metastasis of other tumor types to the draining nodes).

Lymphoma is a common cause of regional or generalized lymphadenopathy in dogs and cats, and the cervical lymph nodes are often involved.

The **external jugular vein** is about 1 cm (0.4 in.) in diameter in the average-size dog, as it is the main venous drainage for the head. It arises at the union of the **maxillary** and **linguofacial veins** just caudal to the **mandibular gland** (Fig. 3.6-5). For its entire course toward the thoracic inlet, the external jugular vein is superficial, lying in the fascia just beneath the skin, within a discernible **jugular groove** formed on the ventrolateral side of the neck (left and right). In the caudal third of the neck, the external jugular vein may appear to dip beneath the **brachiocephalicus muscle**, but it remains superficial, especially from the ventral aspect (Fig. 3.6-8).

The external jugular vein—commonly referred to as the "jugular vein" in this context—is a common site for blood collection in dogs and cats. By extending the patient's head and neck, the loose skin over the jugular vein is stretched and the lower jaw is raised for improved access (Fig. 3.6-15). To further aid identifying and sample collection, the vein is gently occluded with the thumb just cranial to the thoracic inlet. The needle is then inserted through the skin, at an angle of about 45 degrees to the skin surface and directed cranially into the now-distended jugular vein. If necessary, the hair overlying the vein can be clipped or dampened with an alcohol-soaked swab to aid identification of the vein.

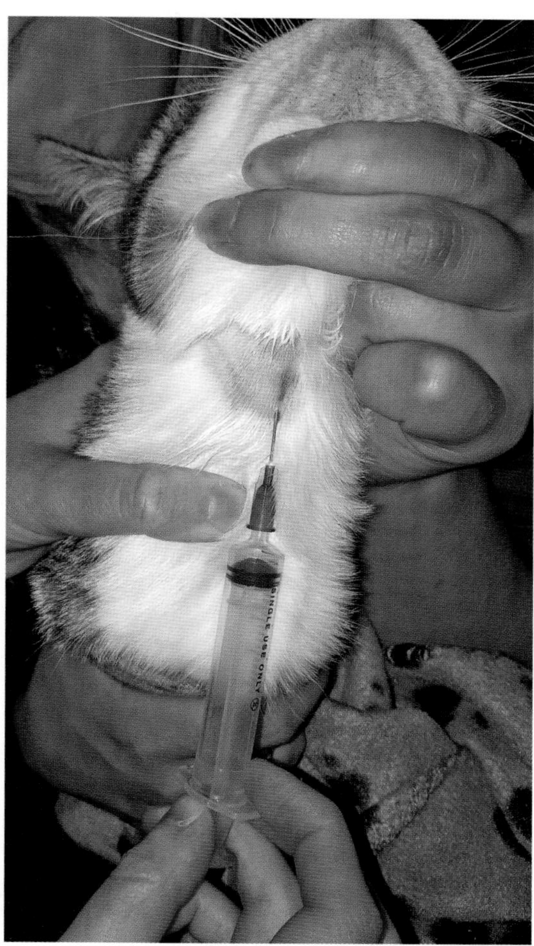

FIGURE 3.6-15 Performing jugular venipuncture in a cat. The patient's neck is extended, the chin is raised, and the external jugular vein is occluded with a thumb just cranial to the thoracic inlet.

On palpation, the cervical lymph nodes have a firmer texture than adjacent glandular tissue such as the salivary glands or the fatty connective tissue in which they may be embedded, and their discrete outline helps distinguish them from adjacent muscle tissue of similar texture. However, while their smooth surface can also distinguish them from lobulated glandular tissue, a cluster of small lymph nodes may palpate like a lobulated structure, so it is important to be familiar with the normal position, size, and number (single or multiple/cluster) of the palpable lymph nodes.

Cervical lymph nodes (Fig. 3.6-16)

The principal lymph nodes of the head and neck are all paired (left and right) and include the mandibular, parotid, medial retropharyngeal, and superficial cervical. For the most part, the cervical lymph nodes are ovoid or elliptical in shape, although some are flattened.

The **mandibular lymph nodes** are located superficially, caudolateral or caudoventral to the angle of the mandible, at the confluence of the maxillary and linguofacial veins into the external jugular vein (Fig. 3.6-5). They usually comprise a small cluster of 2 or 3—and occasionally up to 5—individual lymph nodes that range in size from <1 to >5 cm (<0.4 to >1.9 in.) long in dogs and from 2.5 mm to 2.5 cam (0.1 to 0.9 in.) long in cats, depending on the size and age of the animal. Cats may also have a few tiny "accessory" mandibular lymph nodes caudal to the main nodes, caudal to where the maxillary vein empties into the external jugular vein. In both species, the mandibular lymph nodes drain much of the head and they empty primarily into the ipsilateral medial retropharyngeal lymph node.

The **parotid lymph nodes** are located near the temporomandibular joint and the base of the ear, in close association with the dorsal portion of the parotid gland (Fig. 3.6-5). These nodes are superficial, but they are partially or completely covered by the parotid gland, so they are not as readily palpable as the mandibular lymph nodes. Dogs usually have a single parotid lymph node, 1–3 cm (0.4–1.2 in.) long, and occasionally there is a small string of 2 or 3 nodes.

Cats most often have 2 or more parotid lymph nodes that form a loose "V" configuration around the base of the ear, along the superficial temporal vein cranially and the caudal auricular vein caudally. The more cranial nodes are very small (1–9 mm or 0.03–0.35 in.) and are often embedded within the parotid gland. The usually single, caudal parotid lymph node is much larger, reaching up to 3 cm (1.2 in.) in length. It is embedded in the fatty connective tissue caudal to the parotid gland, so it is often palpable caudoventral to the base of the ear.

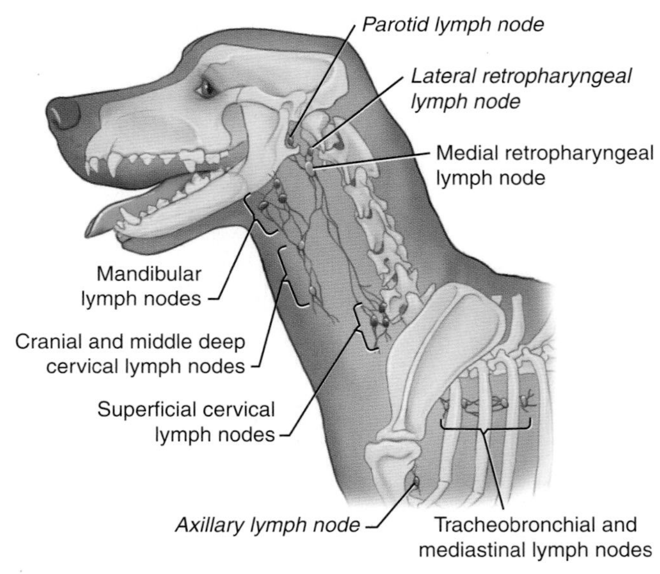

FIGURE 3.6-16 Principal lymph nodes of the head and neck in the dog (excluding the palatine tonsils). All of these lymph nodes are paired (left and right). Except where noted in the text, the same lymph nodes are present in the same locations in cats. Those in *green* are superficial enough to be palpable, although those with *italics* are only rarely identifiable. Those in *orange* are the deeper lymph nodes; they are not palpable. (The deep cervical lymph nodes are inconsistently present in dogs and cats.) Key: *ln*, lymph node (single); *lnn*, lymph nodes (multiple).

In both species, the parotid lymph node drains the more dorsal and lateral aspects of the head and empties primarily into the ipsilateral medial retropharyngeal lymph node.

The **medial retropharyngeal lymph node** is a large lymph node situated immediately dorsocaudal to the pharynx, dorsolateral to the larynx and cranial trachea, lateral to the carotid sheath, ventral to the wing of the atlas, and deep to the brachiocephalicus (specifically, cleidomastoideus) and sternocephalicus muscles (Figs. 3.6-4 and 3.6-14). The medial retropharyngeal lymph node in the average-size dog is about 5 cm (2.0 in.) long and 2 cm (0.8 in.) wide and in cats is up to 2.25 cm (0.9 in.) long; but owing to its depth, it is not normally palpable. It is named the *medial* retropharyngeal lymph node because some dogs also have a lateral retropharyngeal lymph node (briefly described later).

This important lymph node drains the deep structures of the head and the cranial portion of the neck, including the mandibular, parotid, and (when present) lateral retropharyngeal lymph nodes. Its efferent (outflow) vessels form the start of the **tracheal duct**, which is the paired lymphatic duct (left and right) that runs caudally toward the thoracic inlet, dorsolateral to the trachea, and lateral to the carotid sheath, returning lymph from the head and neck to a systemic circulation. It typically empties into the thoracic duct on the left side and the external jugular or brachiocephalic vein on the right side.

The **superficial cervical lymph nodes** usually comprise a linear arrangement of 2 nodes (range, 1–4) in dogs and 3 or more nodes in cats. They are oriented obliquely within fatty connective tissue along the cranial border of the scapula and, in cats, the scapulohumeral (shoulder) joint, deep to the superficial muscles in this region (cleidocervicalis, omotransversarius, and ventral portion of trapezius). Because of their location, these nodes are sometimes referred to as "prescapular."

In dogs, these lymph nodes are grouped close together to form a short string. Individual nodes may reach up to 7.5 cm (3.0 in.) long in large dogs, but usually they are collectively only about 3 cm (1.2 in.) long and 1 cm (0.4 in.) wide. In cats, the nodes are more widely distributed and are separated into dorsal and ventral. The dorsal nodes are in a similar position to those in dogs, and individual nodes may be up to 3 cm (1.2 in.) long. The ventral nodes are much smaller (<1–15 mm or <0.04–0.59 in.) and are located near the external jugular vein, just cranial to the thoracic inlet.

In both species, the superficial cervical lymph nodes are readily palpable when enlarged, but normally they may barely be differentiated from surrounding and overlying muscle, fascia, and superficial fat. These nodes drain a wide area that includes skin; subcutis; and muscles of the head (caudally), neck, thoracic wall (cranially), and most of the ipsilateral thoracic limb. They empty directly into systemic circulation at the **thoracic inlet**; and variably into the tracheal duct (right side), the thoracic duct (left side), or the external jugular vein (either side).

> Although uncommon, unilateral enlargement of the superficial cervical lymph nodes may occur following ipsilateral subcutaneous or intramuscular injection of an irritating or antigenic substance. Neck wounds or abscesses—common in outdoor cats, particularly intact males—may also cause regional lymphadenopathy involving these nodes.

The other cervical lymph nodes variably include the lateral retropharyngeal lymph nodes (absent in cats and many dogs) and the deep cervical lymph nodes (cranial, middle, and caudal), which are located along the tracheal duct. These are much smaller than the principal lymph nodes described already, and none are consistently present or palpable.

Thymus

The **thymus** is a lymphoid structure that is contained within the thoracic cavity, specifically within the cranioventral mediastinum. However, in young animals, < 6 months of age, the thymus may extend cranially, for up to 1 cm (0.4 in.), through the thoracic inlet into the caudal cervical region, where it lies ventral to the trachea.

Selected references

[1] Deitz K, Gilmour L, Wilke V, et al. Computed tomographic appearance of canine thyroid tumours. J Small Anim Pract 2014;55(6):323–9.
[2] Liptak JM. Canine thyroid carcinoma. Clin Tech Small Anim Pract 2007;22(2):75–81.
[3] Turrel JM, McEntee MC, Burke BP, et al. Sodium iodide I^{131} treatment of dogs with nonresectable thyroid tumors: 39 cases (1990–2003). J Am Vet Med Assoc 2006;229(4):542–8.

CHAPTER 3

CASE 3.7

Hyperthyroidism

Mark E. Peterson[a], **John F. Randolph**[b]
[a]Animal Endocrine Clinic, New York, New York, US
[b]Department of Clinical Sciences, Cornell University College of Veterinary Medicine, Ithaca, New York, US

Clinical case

History

A 13-year-old female spayed domestic long-haired cat was presented with signs of weight loss, polyphagia, nervousness, and unkempt hair coat. The cat had lost almost half of her body weight in the previous 2 years. The cat had vomited daily, usually shortly after eating, for the past month. There was no coughing, sneezing, diarrhea, or polyuria/polydipsia.

Physical examination findings

On physical examination, the cat was easily agitated and resented handling. The cat weighed 2.8 kg (6.2 lbs) with marked muscle wasting, and all bony prominences were easily palpable. The hair coat along the trunk and neck had been trimmed by the owners due to severe matting and slight greasiness. The body temperature was mildly increased at 102.7 °F (39.3 °C). The cat was tachycardic at 260 bpm with a grade 3/6 systolic murmur. Abdominal palpation was normal. A left-sided movable mass (approximately 3 × 2 cm or 1.2 × 0.8 in.) was palpable in the ventral cervical area adjacent to the trachea (Fig. 3.7-V1 in the online version at https://doi.org/10.1016/B978-0-323-91015-6.00023-6).

Differential diagnoses

Unilateral lymphadenopathy, thyroid tumor or cyst, parathyroid tumor or cyst, salivary tumor or mucocele, carotid body tumor, branchial cleft cyst, thyroglossal duct cyst, dermoid cyst, or teratoma

Diagnostics

Diagnostic testing included CBC, urinalysis, serum biochemical profile, and serum thyroxine (T4) concentration. Laboratory abnormalities included increased serum ALP activity of 191 IU/L and serum ALT activity of 320 IU/L. Serum T4 concentration was increased at 18.2 µg/dL.

Electrocardiography (ECG/EKG) revealed sinus tachycardia (260 bpm) and occasional ventricular premature contractions. Radiographs of the thorax and abdomen were normal except for mild cardiac enlargement.

Scintigraphic imaging of this cat's thyroid gland with technetium-99m pertechnetate revealed bilaterally enlarged thyroid lobes (left lobe larger than right lobe) with a homogeneous pattern of increased thyroid uptake of the injected radionuclide compared to a normal cat (Fig. 3.7-1). Pertechnetate was also taken up by the salivary glands and gastric mucosa. The thyroid lobes of this cat had more intense uptake of pertechnetate compared to the salivary glands (Fig. 3.7-1), as is seen in hyperthyroid cats.

Comparative Veterinary Anatomy: A Clinical Approach. https://doi.org/10.1016/B978-0-323-91015-6.00023-6

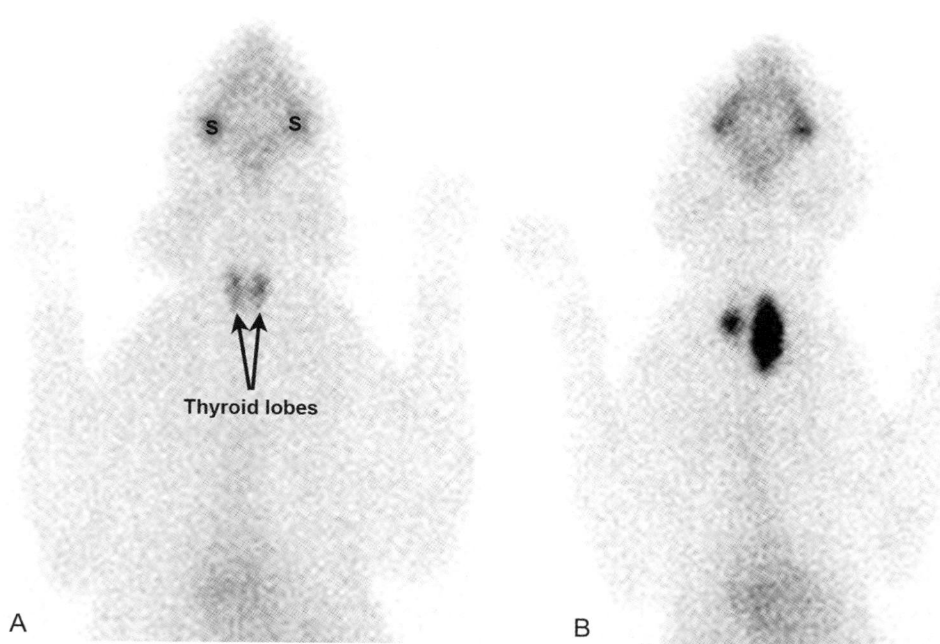

FIGURE 3.7-1 Thyroid scintigraphy illustrating technetium-99m pertechnetate scans of a clinically normal cat (A) and this cat with hyperthyroidism (B). In the normal cat, note the similar uptake of radionuclide in thyroid lobes and salivary glands (S). In the hyperthyroid cat, the thyroid lobes are asymmetrically enlarged with increased thyroid uptake of the injected radionuclide compared to the salivary glands and compared to the normal cat's thyroid lobes.

Diagnosis

Bilateral thyroid neoplasia causing hyperthyroidism

Treatment

Although there are many treatment options for feline hyperthyroidism (see side box entitled "Treatment options for feline hyperthyroidism"), the caretakers elected bilateral thyroidectomy for this cat. In order to minimize anesthetic complications from uncontrolled thyrotoxicosis, short-term administration of methimazole was instituted before surgery to reduce circulating concentrations of T4. The cat was placed under general anesthesia in dorsal recumbency and a ventral midline incision was made from the larynx to the manubrium. The sternohyoideus and sternothyroideus muscles were bluntly separated on the midline and retracted. The left thyroid gland was grossly enlarged. The caudal thyroid vein was ligated and transected. Thyroidectomy was performed using a modified extracapsular technique (Fig. 3.7-2). The thyroid capsule was resected together with the thyroid mass to minimize any retention of thyroid tissue remnants that may lead to relapse of hyperthyroidism. Careful attention was paid to ensure that the external parathyroid gland was not injured, devascularized, or removed in the process. To that end, the thyroid capsule was cauterized approximately 2 mm (0.08 in.) from the edge of the external parathyroid gland (Fig. 3.7-2A),

TREATMENT OPTIONS FOR FELINE HYPERTHYROIDISM

Options include chronic administration of oral antithyroid drugs (e.g., methimazole), exclusive feeding of a commercially available iodine-restricted diet, injection of radioactive iodine (I^{131}), or surgical removal of the affected thyroid lobe(s). Although methimazole and iodine-restricted food can block thyroid hormone synthesis, reduce circulating concentrations of T4, and resolve clinical signs of hyperthyroidism, those therapies do not eliminate the thyroid tumors. Definitive treatment requires either administration of radioactive iodine or thyroidectomy.

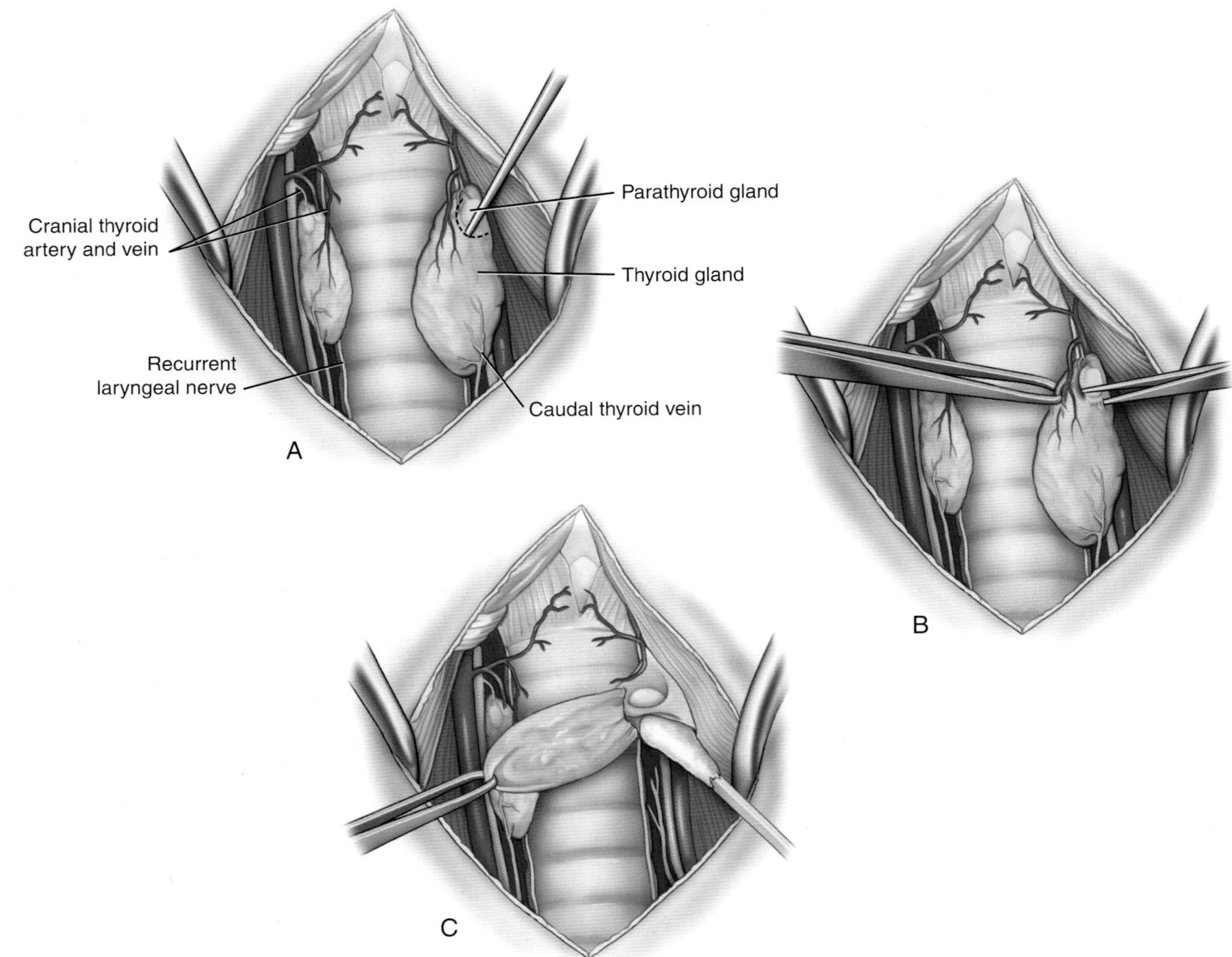

FIGURE 3.7-2 Modified extracapsular thyroidectomy technique in a cat with bilateral thyroid adenomas exposed by a ventral midline cervical approach. The thyroid capsule was cauterized approximately 2 mm (0.08 in.) from the junction of the external parathyroid gland and the left thyroid lobe (A). The cauterized thyroid capsule was incised (B). The external parathyroid gland was bluntly separated from the thyroid lobe (C).

and the cauterized thyroid capsule was then incised (Fig. 3.7-2B). The external parathyroid gland was carefully separated from the thyroid tissue (Fig. 3.7-2C) and the encapsulated thyroid tumor removed. The removal of the smaller right thyroid lobe was completed in the same manner. Both excised thyroid lobes were fixed in formalin and histopathologic examination confirmed bilateral thyroid adenomas.

The cat was observed closely over the next 48 hours for signs of hypocalcemia, which can develop if the parathyroid glands are inadvertently injured or removed during bilateral thyroidectomy. Blood collected from the cat 3 days after surgery demonstrated normal calcium concentration, so the cat was discharged from the hospital.

Fourteen days after thyroidectomy, the cat returned for suture removal. The cat had an improved hair coat and normalized ALP and ALT activities. The serum T4 concentration was subnormal at 0.6 µg/dL, as expected for iatrogenic hypothyroidism from bilateral thyroidectomy. Thyroid hormone replacement was initiated. When rechecked 3 weeks later, the cat appeared calmer. She had gained 0.3 kg and repeat T4 was normal at 2.6 µg/dL.

Anatomical features in canids and felids

Introduction

The endocrine structures of the neck include the thyroid lobes and parathyroid glands. The anatomic relationship of these endocrine organs is illustrated in Fig. 3.7-3.

Function of thyroid and parathyroid glands

The thyroid gland secretes the thyroid hormones thyroxine (T4) and triiodothyronine (T3), which act on nearly every cell in the body. Thyroid hormones increase basal metabolic rate and regulate protein, fat, and carbohydrate metabolism.

The major function of the parathyroid glands is to secrete parathyroid hormone. Parathyroid hormone, in turn, acts on the bone, kidney, and intestinal tract to maintain the body's circulating calcium and phosphorus concentrations within narrow ranges, thus supporting proper function of the nervous and muscular systems.

Thyroid gland

The **thyroid gland** is comprised of 2 separate paired lobes, which lie on the lateral aspects of the cranial trachea (Fig. 3.7-3). Each lobe measures approximately 1 cm (0.4 in.) long and 3–5 mm (0.1–0.2 in.) wide in the normal cat. The thyroid lobes are closely adhered but loosely attached to the trachea and lie deep to the **sternohyoideus** and **sternothyroideus** muscles. The right lobe is closely associated with the structures of the ipsilateral **carotid sheath** (containing the carotid artery, internal jugular vein, and vagosympathetic trunk). The left thyroid lobe is closely associated with the esophagus, which lies dorsolateral to the gland and separates it from the carotid sheath. The **recurrent laryngeal nerves** pass dorsal to the thyroid lobes.

> To palpate an enlarged thyroid gland, extend the cat's neck and tilt its head backward. Gently pass your thumb and index finger of one hand along both sides of the trachea in the jugular furrows from the larynx to the thoracic inlet. Thyroid enlargement is typically felt as a movable nodule, varying in size from a pea to a grape.

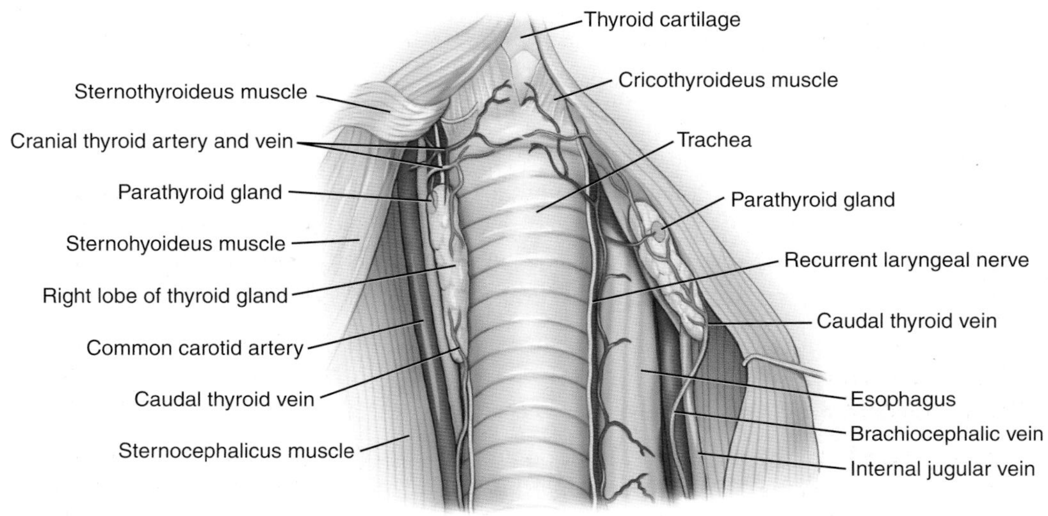

FIGURE 3.7-3 Normal anatomy of the neck of a cat. (Modified from Miller and Evan's Anatomy of the dog. 5th ed.)

Parathyroid glands

There are a total of 4 **parathyroid glands**, with one external parathyroid gland and one internal parathyroid gland intimately associated with each thyroid lobe. Although variable in location, the external parathyroid gland is typically located on the ventral surface of the cranial aspect of each thyroid lobe. The internal parathyroid gland is usually embedded within the parenchyma of the caudal aspect of each thyroid lobe, but its location also may vary. Each parathyroid gland measures approximately 3 mm (0.1 in.) in diameter in normal cats and dogs.

Several techniques for thyroidectomy in cats have been described, but the modified extracapsular thyroidectomy procedure minimizes the risk of external parathyroid gland injury or removal. The internal parathyroid gland embedded in the thyroid lobe is removed with any thyroidectomy technique. However, preservation of at least one external parathyroid gland is necessary for normal maintenance of calcium balance with bilateral thyroidectomy.

Damage to adjacent structures (e.g., carotid artery, internal jugular vein, vagosympathetic trunk, and recurrent laryngeal nerve) can also complicate the thyroidectomy surgical procedure if the thyroid mass is large or invasive. Injury to the recurrent laryngeal nerve may lead to ipsilateral vocal cord paralysis, whereas damage to the sympathetic trunk can result in ipsilateral Horner syndrome (i.e., enophthalmos, miosis, ptosis, and protrusion of third eyelid [nictitating membrane]).

Blood supply and lymphatic drainage

In the cat, blood is primarily supplied to the thyroid and parathyroid glands by the **cranial thyroid artery**, which arises from the **common carotid artery** (Fig. 3.7-3). Venous drainage of the thyroid and parathyroid glands is variable but is predominantly provided by the **cranial** and **caudal thyroid veins**. The cranial thyroid vein drains into the **internal jugular vein**, whereas the caudal thyroid vein joins the **brachiocephalic vein**.

Lymphatic drainage from the thyroid gland is to the **cranial** and **caudal deep cervical lymph nodes**. Efferent lymphatics then reach the venous system by way of the **right lymphatic duct** and **left tracheal duct**. Although the neck anatomy of dogs and cats is similar, dogs also have a **caudal thyroid artery**, which usually arises from the brachiocephalic artery between the common carotids, but its origin is variable.

PATHOPHYSIOLOGY OF FELINE HYPERTHYROIDISM

Hyperthyroidism, typically resulting from functional benign thyroid adenomas/adenomatous hyperplasia, is the most common endocrine disorder of cats. These thyroid tumors can be either unilateral (30%) or bilateral (70%). Rarely (1–2% of cats with hyperthyroidism), the tumors are malignant carcinomas. The thyroid tumors produce excessive amounts of thyroid hormones (T4 and T3) causing the cats to develop the clinical syndrome of hyperthyroidism.

Canine hyperthyroidism from thyroid tumors is rare (see side box). Instead, hypothyroidism due to lymphocytic thyroiditis or degenerative atrophy is more commonly encountered in dogs.

THYROID TUMORS IN DOGS

In contrast to cats, most (90%) thyroid tumors in dogs are malignant carcinomas. And, also unlike cats, only 10–25% of canine thyroid tumors produce excess amounts of thyroid hormone. Thyroid tumors in dogs are often invasive and highly vascular.

Selected references

[1] Birchard SJ. Thyroidectomy in the cat. Clin Tech Small Anim Pract 2006;21:29–33.
[2] Flanders JA. Treatment of hyperthyroidism: surgical thyroidectomy. In: Feldman EC, Fracassi F, Peterson ME, editors. Feline endocrinology Milan. EDRA; 2019. p. 21–226.
[3] Mooney CT, Peterson ME. Feline hyperthyroidism. In: Mooney CT, Peterson ME, editors. Manual of canine and feline endocrinology. 4th ed. Quedgeley, Gloucester: British Small Animal Veterinary Association; 2012. p. 199–203.
[4] Peterson ME. Hyperthyroidism in cats. In: Rand J, Behrend E, Gunn-Moore D, Campbell-Ward M, editors. Clinical endocrinology of companion animals. Ames, Iowa: Wiley-Blackwell; 2013. p. 295–310.
[5] Peterson ME. Hyperthyroidism. In: Greco D, Davidson A, editors. Blackwell's five-minute veterinary consult clinical companion: small animal endocrinology and reproduction. Ames, Iowa: Wiley-Blackwell; 2017. p. 245–62.
[6] Radlinsky MG. Thyroid surgery in dogs and cats. Vet Clin North Am Small Anim Pract 2007;37:789–98.

CHAPTER 3

CASE 3.8

Cervical Intervertebral Disc Disease

Takanori Sugiyama[a] and Helen M.S. Davies[b]
[a]Animalius Vet, Bayswater, WA, AU
[b]Veterinary BioSciences, The University of Melbourne, Melbourne, Victoria, AU

Clinical case

History

A 6-year-old male neutered Pekingese dog was presented for acute, nonambulatory tetraparesis. There was no history or evidence of trauma.

Physical examination findings

Abnormal findings were limited to the neurologic examination. The dog was nonambulatory and tetraparetic. Mentation and cranial nerve reflexes and responses were normal. Motor reflexes in thoracic and pelvic limbs were normal to exaggerated with increased muscle tone and normal nociception (pain perception). The dog was painful when pressure was applied to the cranial cervical vertebrae.

Differential diagnoses

Cervical (C1–C5) myelopathy (e.g., intervertebral disc disease [IVDD], infection [e.g., bacterial meningitis], inflammation [e.g., granulomatous meningoencephalitis], and neoplasia)

Diagnostics

Routine bloodwork was within normal limits. Magnetic resonance imaging was performed under general anesthesia and showed compression of the cervical spinal cord between the 2nd and 3rd cervical vertebrae (C2–C3). Intervertebral disc material appeared to be protruding into the vertebral canal at this site (Fig. 3.8-1).

Diagnosis

Cervical (C2–C3) IVDD

Treatment

The dog was placed under general anesthesia in dorsal recumbency and an incision was made along the ventral midline of the neck, through the skin, subcutis, and thin fascial division between the left and right sternohyoideus and sternothyroideus muscles to expose the trachea deep to these ventral neck muscles. The trachea, esophagus, and both carotid sheaths (left and right) were gently retracted to the left to expose the paired longus colli muscles ventral to the vertebral column. The muscles were elevated and retracted to expose the ventral surface of the intervertebral articulation at C2–C3.

After creating a slot in the ventral aspect of the intervertebral disc annulus (fibrous outer ring), disc material was meticulously retrieved from the joint. The longus colli muscles and then the trachea and associated structures were returned to their normal anatomical positions. The incision separating the left and right sternothyroideus and sternohyoideus muscles was sutured, followed by closure of the subcutaneous tissue and the skin.

Comparative Veterinary Anatomy: A Clinical Approach. https://doi.org/10.1016/B978-0-323-91015-6.00018-2

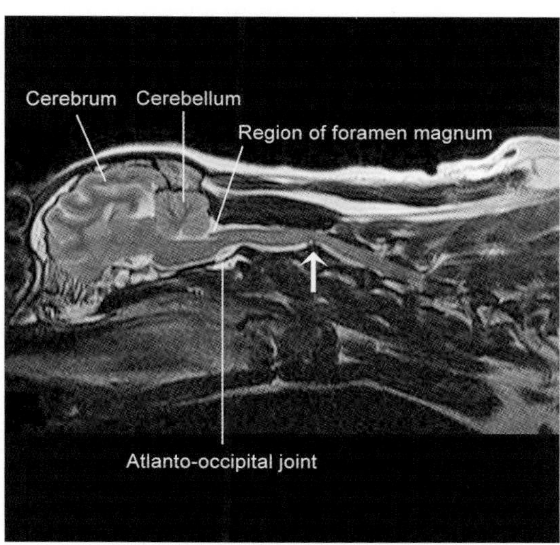

FIGURE 3.8-1 Sagittal MRI of the dog in this case, showing compression of the cervical spinal cord between the 2nd and 3rd cervical vertebrae *(arrow)*. Intervertebral disc material appears to be protruding into the vertebral canal at this site. The mild deviation at C1–C2 is within normal limits for the dens, as there is no intervertebral disc at this space to impinge on the cord.

On recovery from general anesthesia, the dog was tetraplegic but gradually regained motor function during his post-operative recovery in the intensive care unit, and was walking by post-operative day 5.

Anatomical features in canids and felids

Introduction

The neck is a complex anatomical region that contains parts of many different systems, including the musculoskeletal, nervous, digestive, respiratory, cardiovascular, endocrine, immune, and integumentary systems. That is also true of the head, the thorax, and—except the respiratory system—the abdomen; but unlike these other regions, the various structures of the neck must adapt to the large range of motion characteristic of the cervical spine.

WOBBLER SYNDROME IN DOGS

Wobbler syndrome is the common name for caudal cervical spondylomyelopathy (CCSM). It is a developmental abnormality of the caudal cervical spine in which dynamic compression of the spinal cord results in ataxia and paresis. Neck pain may also be present as a result of compression of the spinal nerve roots. Wobbler syndrome occurs most often in large- and giant-breed dogs, although it also occurs in chondrodystrophoid breeds such as Bassett Hounds.

In medical terms, wobbler syndrome is a spondylomyelopathy: pathology involving the spine (Gr. *spondulos* spine, vertebrae) and spinal cord (Gr. *myelos* marrow, spinal cord). The osseous form primarily involves abnormalities of vertebral development and is most common in young dogs. The disc-associated, distraction-responsive form primarily involves soft tissues, including intervertebral discs and connective tissues such as the ligamentum flavum. It is responsive to the distraction of the intervertebral joints, such as flexion/extension of the neck, and is more often diagnosed in mature dogs.

Owing to abnormalities involving the vertebrae and/or the discs and connective tissues, the spinal canal is narrowed when the neck is dorsiflexed or extended, which impinges on the spinal cord, exacerbating the clinical signs. To reduce cord compression, these patients tend to carry the head lower with the neck ventroflexed.

Diagnosis is confirmed using MRI or cervical myelography. Many affected dogs respond to anti-inflammatory therapy, their clinical signs either improving or stabilizing. Surgical options—and even the advisability of surgery—vary with lesion type, location, and severity.

This section describes the cervical spine and associated neural, muscular, and connective tissues. The central and peripheral components of the companion animal nervous system are described in other parts of this textbook, including Cases 3.9–3.12. The following is an abbreviated discussion of some characteristics of the peripheral nervous system within the neck region that are of clinical importance. The other soft tissues of the neck are described in Cases 3.6 and 3.7.

Function

As a suite of movements, dorsiflexion and ventroflexion, left and right lateral bending, and axial rotation are greatest in the cervical region of the spine, so the anatomical structures within the neck must either facilitate or tolerate this considerable mobility.

The main function of the cervical spine is to protect the cervical portion of the spinal cord and the spinal nerve roots.

Cervical spine (Figs. 3.8-2–3.8-4)

Intervertebral disc disease is not as common in the cervical spine as in the thoracolumbar spine. Less than 20% of IVDD cases involve the cervical spine. Spondylosis, a form of degenerative joint disease with partial or complete bony bridging of the affected intervertebral joints, also occurs in the cervical spine, although less frequently than in the thoracolumbar spine and lumbosacral joint.

In Greek mythology, Atlas supported the heavens on his shoulders. Hence, C1 is called the atlas because in human anatomy, it "supports" the head.

The cervical spine consists of 7 **vertebrae**, designated C1–C7. The first 2 and the last 2 cervical vertebrae are unique and distinctly different from the middle 3. Except for the 1st intervertebral joint (C1–C2), all the adjacent vertebral bodies in the cervical spine are separated by intervertebral discs that are essentially the same as those in the thoracolumbar spine (see Case 3.12).

The **1st** cervical vertebra (C1) is also called the **atlas**. It articulates with the pair of occipital condyles at the base of the skull, forming the atlantooccipital (AO) joint. A distinctive clinical feature of the atlas is its pair of broad, flat transverse processes, called the **wings of the atlas** (left and right). The lateral margins of the wings are palpable just caudal to the skull, about level with the base of the ear and dorsal to the larynx.

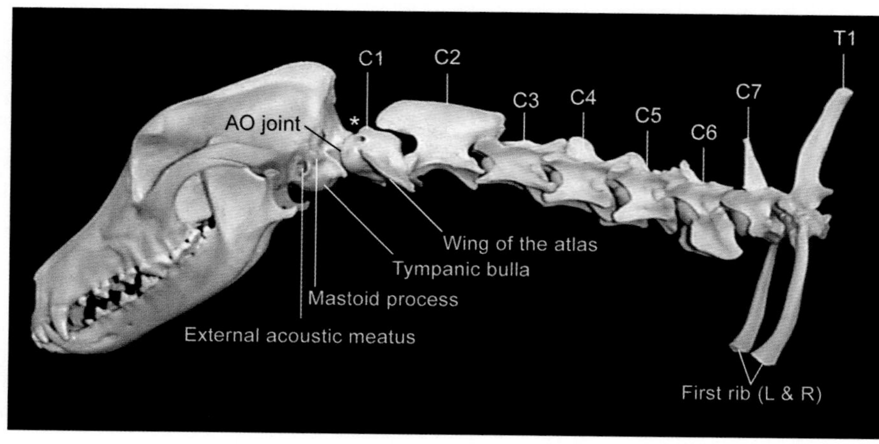

FIGURE 3.8-2 Skull and cervical spine of a dog. The *asterisk* is placed directly over the site used to access the cervicomedullary cistern for myelography or collection of cerebrospinal fluid (see Fig. 3.8-4). Key: *AO*, atlantooccipital; *C*, cervical vertebra (numbered 1–7); *L*, left; *R*, right; *T1*, 1st thoracic vertebra.

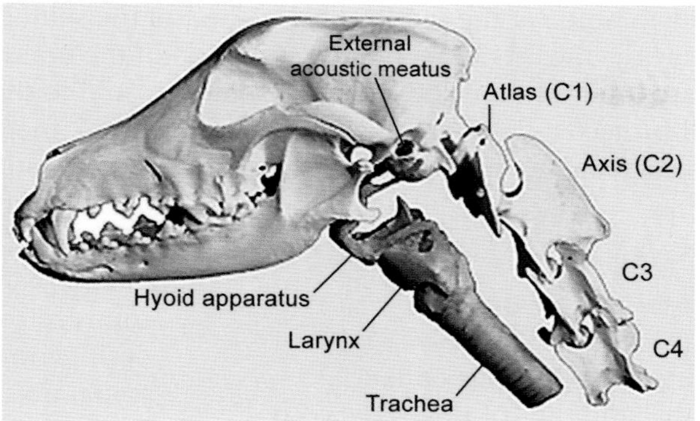

FIGURE 3.8-3 Canine skull and first 4 cervical (C) vertebrae, showing the relative positions of the hyoid apparatus, larynx, and trachea.

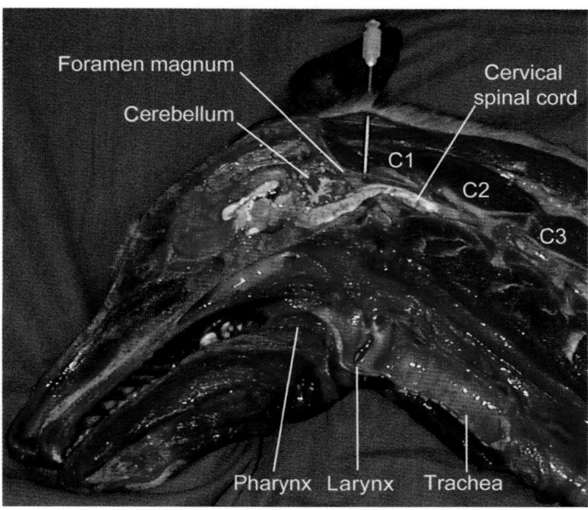

FIGURE 3.8-4 Sagittal section of the head and neck of a dog, showing placement of a spinal needle into the cervicomedullary cistern between the foramen magnum and the 1st cervical vertebra (C1).

CAUDAL OCCIPITAL MALFORMATION SYNDROME IN DOGS

Caudal occipital malformation syndrome (COMS) is a congenital abnormality that primarily affects small breeds of dog, most notably Cavalier King Charles Spaniels. Malformation of the occiput involving the foramen magnum results in compression of the cerebellum and cervicomedullary junction and impedes the normal flow of CSF through this area. Commonly, CSF accumulates within the central canal of the cervical spinal cord and causes cord compression from within, a condition known as "syringohydromyelia." Hence, many dog breeders refer to this condition as "SM" (for syringomyelia).

The most common and distinctive clinical sign is persistent scratching of the shoulder, neck, or head. Many affected dogs also show signs of neck pain which may be chronic and of moderate intensity—e.g., neck stiffness—but more severely affected dogs commonly have episodes of acute, severe pain characterized by yelping, hiding, and not wanting to be touched.

Definitive diagnosis requires MRI, although a presumptive diagnosis is often made based on age, breed, and clinical signs. Age at diagnosis varies widely from young adult to senior dog, but mean age is 4–6 years in most studies.

Surgical treatment involves relieving the compression at the cervicomedullary junction. However, many owners opt for medical management with gabapentin (an analgesic drug used for neuropathic pain) and nonsteroidal anti-inflammatory drugs and muscle relaxants as needed.

The **foramen magnum** is the bony foramen at the caudoventral aspect of the skull through which the spinal cord exits the cranial cavity and continues as the cervical spine. The junction of the brainstem and the origin of the spinal cord is the **cervicomedullary junction**; it is located immediately rostral to the foramen magnum. Although the brain and spinal cord are well-protected by the skull and vertebral column, there are a couple of anatomical sites at which the subarachnoid space can be accessed for diagnostic and therapeutic purposes. One is at the dorsal aspect of the **atlantooccipital (AO) junction**. Cranially, the dorsal arch of C1 has a broad, transverse notch that creates a rounded or elliptical space between the occiput and the atlas, through which a spinal needle may be placed into the cervicomedullary "cistern" (subarachnoid space surrounding the most cranial part of the spinal cord).

The cervicomedullary cistern is one of the 2 principal sites for CSF collection and/or injection of contrast material during myelography (contrast-enhanced radiography of the vertebral canal and thus of the spinal cord and spinal nerve roots). The other site is the lumbar cistern, usually accessed between lumbar vertebrae L5 and L6.

The convex **occipital condyles** sit in the concave **condylar fossae** at the cranial aspect of C1, forming a pair of shallow ball-and-socket joints that allow movement in all planes: dorsiflexion/ventroflexion (nodding), left and right lateral bending, and axial rotation. However, the myofascial structures that support and activate the AO joint primarily allow dorsiflexion/ventroflexion.

The **2nd cervical vertebra** (C2) is also known as the **axis**, so C1–C2 is the atlantoaxial joint. The axis has 2 notable features. The first is the vastly elongated **dorsal spinous process** that projects the length of C2 like a blade or fin. In dogs, the nuchal ligament (described later) attaches here, particularly to the caudal aspect, which broadens laterally. The second feature is the **dens** or odontoid process (both terms referring to a "tooth-like" structure), which is a pointed, cranial projection of the dorsal aspect of the vertebral body. The dens lies in a depression in the floor of the vertebral canal in the caudal part of C1. It is held in a place within the canal by a short, strong transverse ligament. Owing to the shape and position of the dens, movement at the atlantoaxial joint is restricted to axial rotation. However, this arrangement allows considerable rotation; in fact, much of the animal's ability to rotate the head and cranial portion of the neck in the longitudinal plane is attributable to the dens, and it is why C2 is called the "axis" (L. axle or pivot).

The **3rd** to **5th cervical vertebrae** (C3–C5) are similar to one another. They all have small dorsal spinous and transverse processes, and they articulate with one another in a way that provides stability to the cervical spine while still allowing mobility in all planes (dorsoventral, lateral, and axial). The **6th cervical vertebra** (C6) is distinctive for its expanded, flap-like transverse processes that project ventrolaterally. The **7th cervical vertebra** (C7) is somewhat transitional, in that it has a much taller dorsal spinous process than the other six cervical vertebrae, being more like the 1st thoracic vertebra (T1) with which it articulates. Furthermore, C7 articulates with the cranial part of the head of the first rib on either side in most mammalian species, including the dog and cat. Although the range of motion allowed at each of the cervical joints caudal to C2 is somewhat limited, as a unit the cervical spine has an extensive range of motion.

Dorsally supporting the cervical spine, and indirectly the skull, in dogs is the **nuchal ligament**, a strong fibroelastic band that extends like a "suspension bridge" from the dorsal spinous processes of the first 2 thoracic vertebrae to the dorsal spinous process of C2. The **supraspinous ligament** is the caudal continuation of this fascial band. It extends caudally along the tops of the dorsal spinous processes to the 3rd caudal vertebra at the base of the tail. Cats do not have a nuchal ligament but do have a supraspinous ligament.

The cervical spine is supported ventrally by a pair of **ventral longitudinal ligaments** that run along the ventral surfaces of the vertebral bodies, from C2 to the sacrum. The **dorsal longitudinal ligaments** run along the dorsal surfaces of the vertebral bodies, on the floor of the vertebral canal.

One other set of intervertebral ligaments are of clinical importance: the **ligamentum flavum** (plural, **ligamenta flava**) or interarcuate ligaments. These thin, loose, elastic connective tissue sheets span successive vertebral arches. They blend with the capsules of the small intervertebral joints on either side of the dorsal arches (left and right), and with the periosteum of the vertebrae cranial and caudal to each joint.

> The ligamentum flavum is so-named because of its high elastin content, which gives it a yellowish color (L. *flavum* yellow, blonde). These ligaments may become hypertrophied or abnormally thickened in dogs with caudal cervical spondylomyelopathy (wobbler syndrome), contributing to cord compression in the dorsolateral portion of the spinal canal.

Principal nerves of the neck (Fig. 3.8-5)

All of the structures discussed in this section are paired.

The **vagosympathetic trunk** tracks the length of the neck alongside the **common carotid artery** and **internal jugular vein** within the **carotid sheath**. The vagosympathetic trunk comprises 2 functionally distinct elements of the autonomic nervous system: the vagus nerve (parasympathetic) and the cervical portion of the sympathetic trunk.

The **vagus nerve** (cranial nerve X) exits the cranial cavity and proceeds caudally, ventrolateral to the spine, to eventually innervate the abdominal viscera, with major branches to the heart and lungs along the way. In the cranial portion of the neck, it sends branches to the pharynx and most of the larynx.

The **recurrent laryngeal nerve** (RLN) is a small but important branch of the vagus nerve, so-named because it exits the vagus nerve within the thorax, near the base of the heart, and then "returns" cranially along the neck, separate from the vagus nerve (i.e., external to the carotid sheath), to innervate the larynx, trachea, and

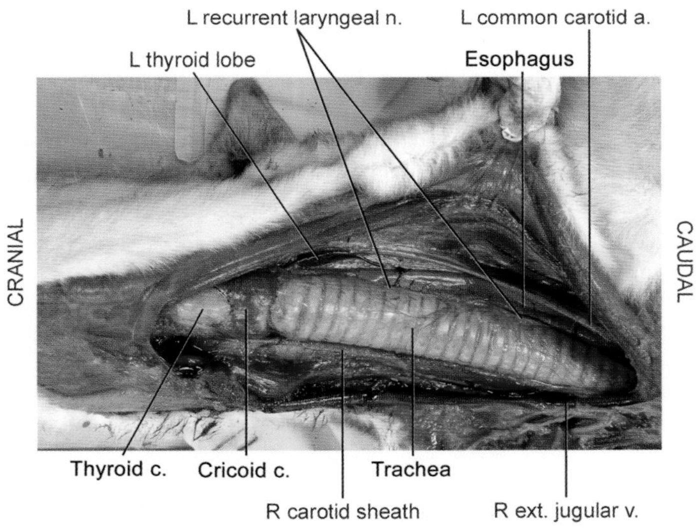

FIGURE 3.8-5 Prosection of the ventral aspect of the neck in a dog. The thoracic inlet is on the right side of the figure and the throat is on the left. Key: *a.*, artery; *c.*, cartilage (laryngeal); *ext.*, external; *L*, left; *n.*, nerve; *R*, right; *v.*, vein.

Injury or neuropathy involving the recurrent laryngeal nerve anywhere along its path results in laryngeal paresis or paralysis. When only one RLN is involved, the laryngeal paresis/paralysis is unilateral and may be inconsequential other than affecting vocal sounds. However, when both left and right RLNs are involved, bilateral laryngeal paresis/paralysis causes potentially life-threatening airway occlusion. Comparatively, left RLN is a common cause for exercise intolerance and an inspiratory noise in the horse, frequently termed "roaring" (see Case 10.7).

esophagus. The left and right RLNs follow slightly different paths initially. The right RLN leaves the right vagus nerve and curves around the right subclavian artery before continuing cranially into the cervical region along the right dorsolateral surface of the trachea, deep to the right carotid sheath. The left RLN leaves the left vagus nerve at the level of the aortic arch; it curves around the aorta before continuing cranially into the neck between the trachea and esophagus. The terminal portions of the RLNs innervate the dorsolateral muscles of the larynx, notably the paired dorsal cricoarytenoid muscle, responsible for abducting the arytenoid cartilages and thus expanding the diameter of the glottis. (The larynx is described in Case 3.6.)

Nerve fibers of the cervical portion of the **sympathetic trunk** arise from the first few segments of the thoracic spinal cord, converging at the **cervicothoracic (stellate) ganglion** adjacent to the 2nd or 3rd thoracic vertebra, and then continuing cranioventrally a short distance around the subclavian artery to the **middle cervical ganglion**. After sending fibers to the heart and lungs, the cervical portion of the sympathetic trunk tracks cranially alongside the common carotid artery, terminating deep to the tympanic bulla as the **cranial cervical ganglion**. Its sympathetic nerve fibers supply the eye, various glands of the head and neck (salivary, sweat, mucosal), arteries of the head and neck, and erector pili muscles at the base of the hair follicles in the skin.

Hyoid apparatus (Figs. 3.8-3, 3.8-6, and 3.8-7)

The **hyoid apparatus** is a fine bony structure that is suspended from the ventrolateral aspect of the skull and supports the base of the tongue and larynx. Although it is located between the mandibles, it is not palpable because it consists of a series of small, thin bones, the structure is quite mobile, and it is embedded in—or covered by—regional muscles.

The basic structure of the hyoid apparatus is shaped somewhat like a swing. The single horizontal **basihyoid** bone forms the "seat" of the swing; it is embedded in the muscles at the base of the **tongue**. The paired **thyrohyoid** bone (left and right) extends caudodorsally to articulate with the **thyroid cartilage** of the larynx, thus supporting the larynx. The 2 "chains" of the swing (left and right) are the paired series of bones that, from ventral to dorsal, consist of the **ceratohyoid**, **epihyoid**, and **stylohyoid** bones. Each side sweeps first rostrodorsally then caudodorsally from the basihyoid bone to the **mastoid process of the temporal bone**, just dorsal to the tympanic bulla (Fig. 3.8-2). A short **tympanohyoid cartilage** forms the articulation between the stylohyoid bone and the skull.

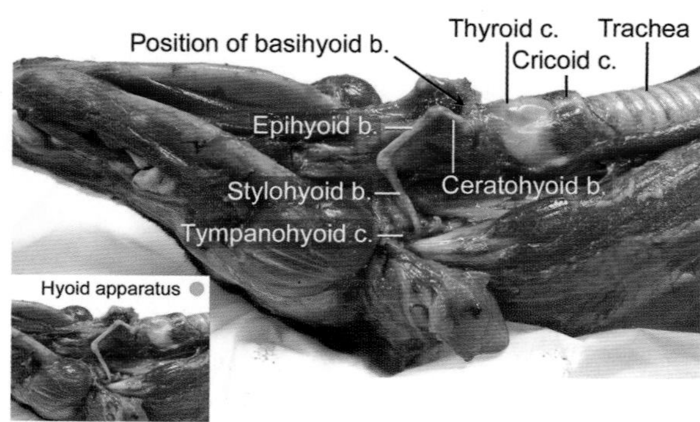

FIGURE 3.8-6 Deep prosection of the head and neck of a Greyhound, ventrolateral aspect, showing the components of the hyoid apparatus in situ. Key: *b.*, bone; *c.*, cartilage (laryngeal).

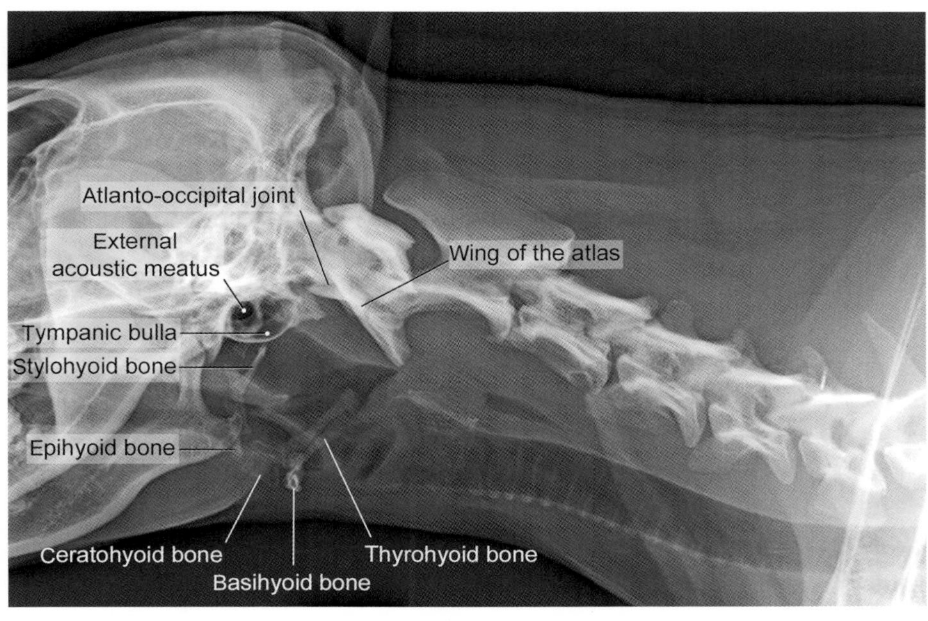

FIGURE 3.8-7 Lateral radiograph of the head and neck of a dog, highlighting the hyoid apparatus. Except for the basihyoid bone, all of the hyoid bones are paired.

Cervical muscles (Figs. 3.8-8–3.8-10)

Many muscles of the neck are arranged in multiple layers from deep to superficial, and various segmental lengths from single intervertebral joints (deep) to the entire span of trunk-to-head (superficial). Those of greatest clinical relevance are briefly described here. The largest muscles in the neck, in terms of length/span and thus action, are those that originate on the sternum or humerus and insert on or near the head. All of the following muscles are paired.

The **brachiocephalicus muscle** is a composite of 3 muscles, all connected by the remnant of the clavicle, the **clavicular tendon**. The caudal portion, the **cleidobrachialis**, originates on the crest of the humerus and crosses the craniolateral aspect of the shoulder joint to insert onto the clavicular tendon just cranial to the thoracic inlet on the same side. The 2 cranial portions, the **cleidomastoideus** and **cleidocervicalis**, originate on the clavicular tendon and course craniodorsally over the lateral and dorsolateral aspects of the neck to insert, respectively, onto the mastoid part of the temporal bone and the fascia on the dorsum of the neck just caudal to the skull.

> The term clavicle is derived from the Latin word *clavicula*, meaning "little key." However, the prefix denoting the clavicle in anatomical terminology is *cleido-*, from the Greek word *kleis*, meaning bar or bolt. Generally, the clavicular remnant is not visible radiographically in dogs and cats. Occasionally, a small and mostly cartilaginous bone is present.

The **sternocephalicus muscle** arises on the manubrium of the sternum. It courses cranially along the ventrolateral aspect of the neck and inserts onto the skull at the mastoid part of the temporal bone and the dorsolateral part of the occipital bone. From their origins on the sternum, the left and right muscles are united along the ventral midline of the neck, but by the middle of the neck, they have diverged as they course toward their separate insertions. Thus, except for the **cutaneous colli muscles** attached to the skin, the sternocephalicus is the most superficial of the ventral neck muscles in the caudal third to half of the neck, where it covers the ventral and lateral aspects of the trachea as well as the origins of the following 2 muscles.

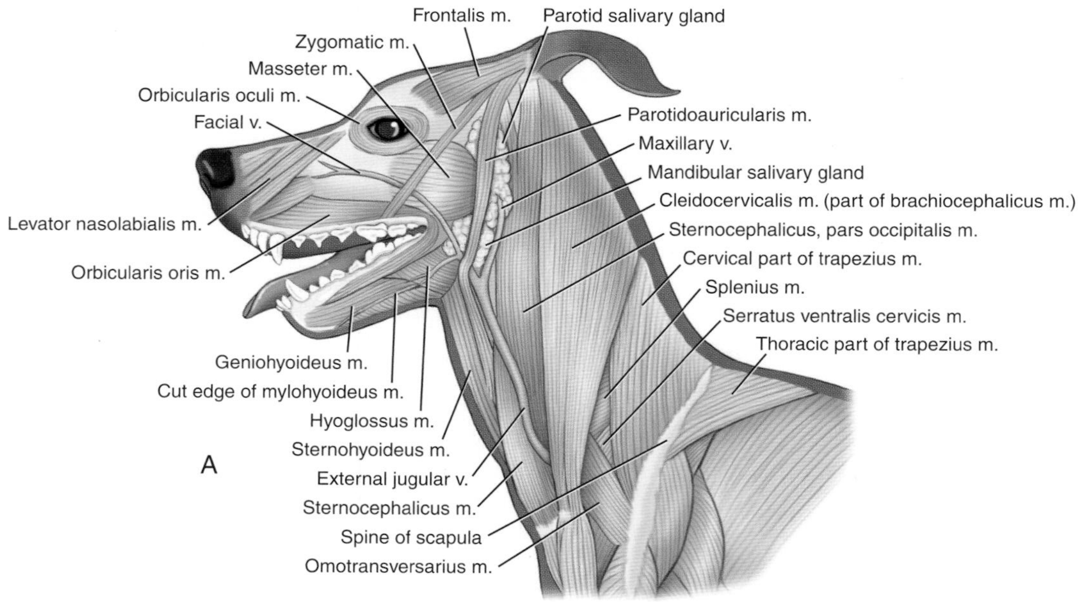

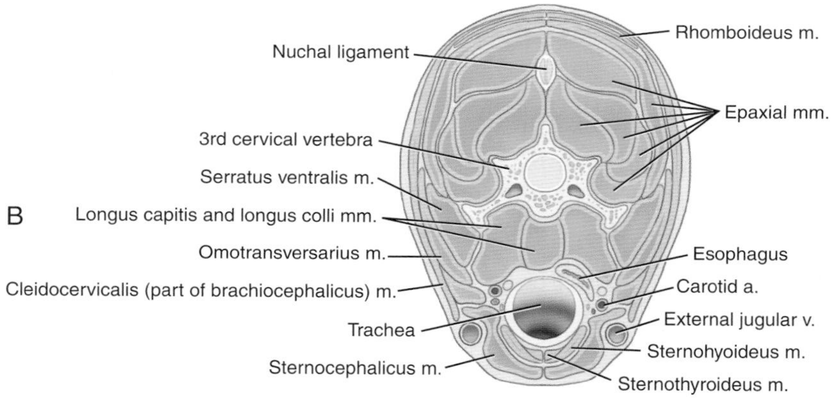

FIGURE 3.8-8 (A) Lateral view of the superficial muscles of the head and neck with veins and salivary glands. (B) Transverse section of the neck at the level of the 3rd cervical vertebra. Key: *m.*, muscle; *mm.*, muscles; *a.*, artery; *v.*, vein.

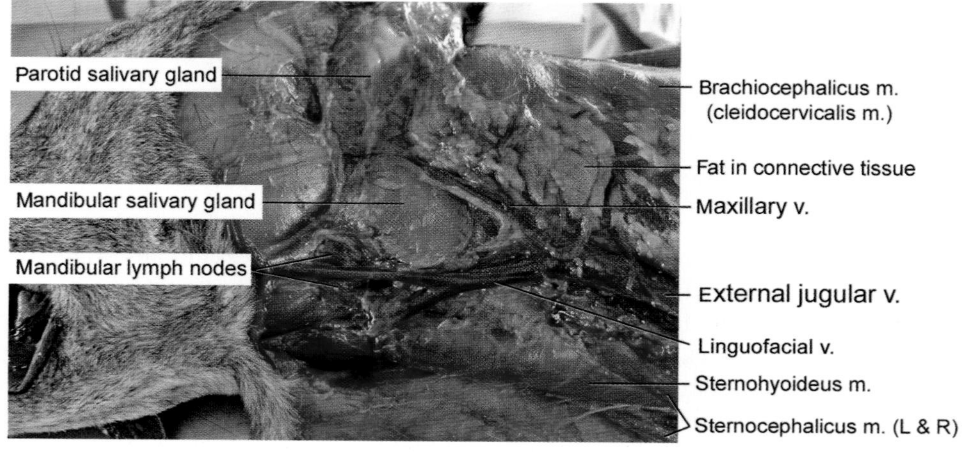

FIGURE 3.8-9 Superficial structures in the cranial portion of the canine neck. Cranial is to the left. Key: *L*, left; *m.*, muscle; *R*, right; *v.*, vein.

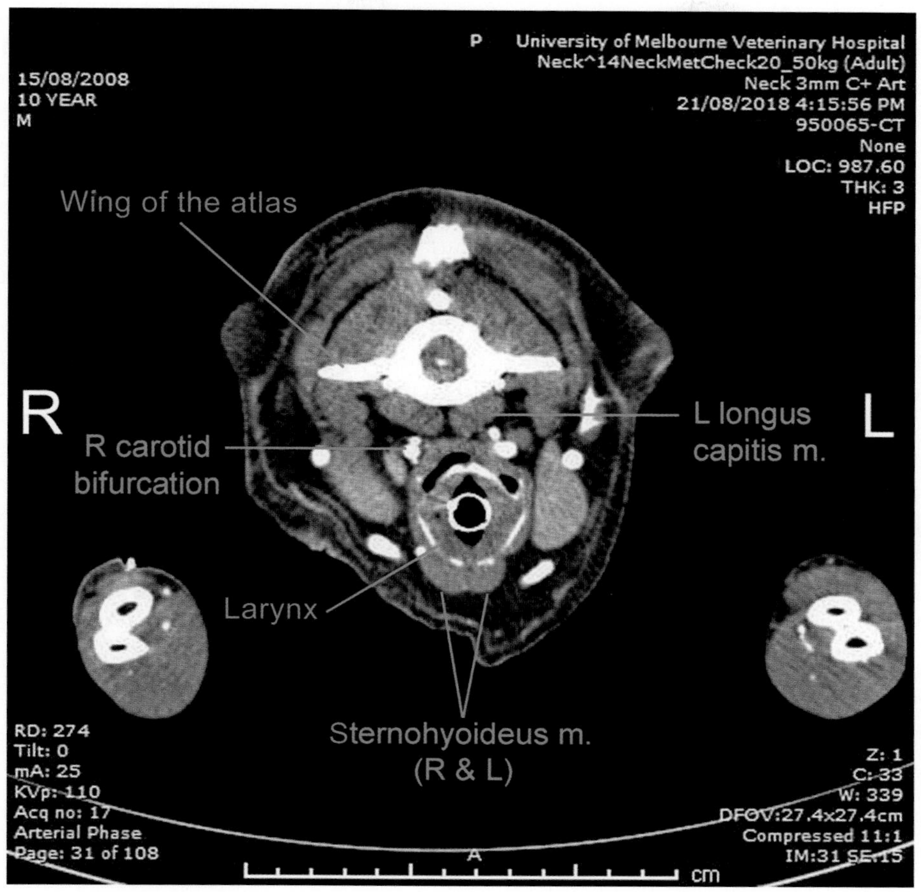

FIGURE 3.8-10 Transverse CT image at the level of the 1st cervical vertebra (atlas) showing the position of the paired sternohyoideus and longus capitis muscles. Key: *L*, left; *m.*, muscle; *R*, right.

The **sternohyoideus** and the smaller, deeper **sternothyroideus muscles** both originate on the manubrium of the sternum and/or the 1st costal cartilage (cartilage of the 1st rib). Both are strap-like muscles that course cranially with their pair (left and right) along the ventral and ventrolateral surfaces of the trachea, inserting onto the basihyoid bone and the thyroid cartilage of the larynx, respectively.

Surgical access to the ventral surface of the trachea or deeper neck structures, such as the ventral aspect of the spine in this case, requires incision through the thin fascial connection between left and right pairs of sternohyoideus and sternothyroideus mm.

The paired **external jugular vein** lies superficial to all the aforementioned muscles (except cutaneous colli) in the **jugular groove**. This groove runs from the cranial neck to thoracic inlet and is a depression between the margins of various structures along its path: dorsally, the transverse processes of the cervical vertebrae and associated cervical musculature, and ventromedially the trachea. The sternohyoideus and sternothyroideus muscles lie ventral to the jugular groove along their whole length. The sternocephalicus muscle passes obliquely under the external jugular vein (Fig. 3.8-9), and the ventral margin of brachiocephalicus (specifically, cleidomastoideus) forms part of the dorsal boundary of the jugular groove.

The left and right pairs of longus colli join on midline, so they must be separated and then elevated in order to access the ventral aspect of the vertebral bodies and the intervertebral joints, as in this case.

Two deep muscles of the neck relevant to this case of cervical IVDD are the paired longus colli and longus capitis muscles. Both track cranially along the ventral surfaces of the cervical vertebrae, in close contact with the vertebral bodies. **Longus colli** is a long series of short muscle bundles that extend from the sixth thoracic vertebra to the atlas. In the cervical region, each muscle bundle originates on the ventral surface of a transverse process and inserts onto the ventral midline of a vertebra cranial to it; thus, each muscle bundle crosses 1 or 2 intervertebral joints.

Longus capitis is a long muscle that lies lateral to longus colli on the ventrolateral aspect of the vertebral bodies (Fig. 3.8-10). It extends from the transverse processes of C6 to the base of the skull, inserting onto the basioccipital bone between the tympanic bullae. Collectively, longus colli and capitis ventroflex the neck; longus capitis also ventroflexes the AO joint.

Selected references

[1] Böttcher P, Böttcher IC, Truar K, et al. Effect of ventral slot procedure on spinal cord compression in dogs with single static intervertebral disc disease: preliminary findings while evaluating a semiquantitative computed tomographic myelographic score of spinal cord compression. Vet Surg 2013;42(4):383–91.

[2] Hakozaki T, Iwata M, Kanno N, et al. Cervical intervertebral disk herniation in chondrodystrophoid and nonchondrodystrophoid small-breed dogs: 187 cases (1993–2013). J Am Vet Med Assoc 2015;247(12):1408–11.

CHAPTER 3

CASE 3.9

Vestibular Dysfunction

Marc Kent[a] and Eric N. Glass[b]
[a]Department of Small Animal Medicine and Surgery, University of Georgia College of Veterinary Medicine, Athens, Georgia, US
[b]Neurology and Neurosurgery, Red Bank Veterinary Hospital, Tinton Falls, New Jersey, US

Clinical case

History

A 4-year-old male castrated Cocker Spaniel was presented for ataxia, falling, and poor appetite. Two days before presentation, the dog acutely started vomiting, developed a right-sided head tilt, and had nystagmus after falling onto his right side. The remainder of the history was unremarkable aside from a history of chronic otitis externa. The owner administered ear drops (unknown type) approximately 4–6 times per year to prevent worsening of clinical signs of ear infections. The owner tried to clean the dog's ears with an otic cleaning solution, but he typically resisted and would bite her.

Physical examination findings

The physical examination was normal except for abnormal pinnae and external ear canals, bilaterally. Lichenification and erythema of the epithelium were visible in the external orifice of the external ear canals bilaterally with purulent, malodorous exudate. The pinnae were painful, eliciting a biting reaction when touched.

Neurological examination findings

Mentation

- Normal

Gait/posture

- Right-sided vestibular ataxia; as the dog walked, he leaned to the right and occasionally fell onto his right side; in the examination room, the dog leaned against the wall with the right side of his body; when he shook his head, he almost fell over onto his right side; there was a right-sided head tilt

Postural reactions

- Normal hopping and proprioceptive placing in all four limbs

Spinal reflexes, muscle tone, and muscle mass

- Normal in all four limbs

Cranial nerves

- There was spontaneous horizontal nystagmus with the fast phase directed to the left; there was no change in the direction of the nystagmus with changes in the position of the head; there was a head tilt to the right
- With extension of the head and neck, ventrolateral strabismus in the right eye developed
- Normal menace response and palpebral reflex OU

Comparative Veterinary Anatomy: A Clinical Approach. https://doi.org/10.1016/B978-0-323-91015-6.00019-4

- Normal pupil size, symmetrical pupils, and pupillary light reflex OU
- Normal response to stimulation of the nasal septum with the closed end of a hemostat bilaterally
- There was no atrophy of the muscles of mastication
- There was a normal gag response
- The tongue was symmetric, was normal in size, and had normal movement
- Abnormal movements of the superior eyelids occurred simultaneously with the fast phase (jerk) of nystagmus to the left

Sensory

- Normal

Causes of otitis externa may be divided into primary, secondary, and predisposing causes. Common primary causes include atopy; food allergy; systemic disease—e.g., hypothyroidism or hyperadrenocorticism; foreign material—e.g., plant material; neoplasms arising in the external ear canal, tympanic cavity, or tympanic bulla; and generalized skin disease. Common secondary causes include bacterial and fungal/yeast infections. Common predisposing causes include ear pinnae conformation and stenosis of the external ear canals.

Neuroanatomic diagnosis

Right peripheral vestibular dysfunction

Differential diagnoses

Otitis externa with secondary otitis media/interna; idiopathic peripheral vestibular disease (though affected dogs are typically > 7 years old); hypothyroidism; neuritis involving the vestibulocochlear nerve

Diagnostics

Thorough otic examination revealed proliferation, lichenification, and erythema of the skin lining the external ear canals and stenosis of the external ear canals, bilaterally. Routine bloodwork (including testing of thyroid function) and 3-view radiographs of the thorax (to rule out metastatic neoplasia) were within normal limits. The dog was anesthetized for an MRI of the head, which showed effusion in both tympanic cavities with complete filling of the right tympanic cavity (Figs. 3.9-1 and 3.9-2). The right tympanic cavity appeared enlarged compared to the left. There was abnormal enhancement of the mucoperiosteum of the tympanic cavity. The external ear canals appeared thickened and stenotic. The normal high T2 signal intensity of the fluid-filled portions of the vestibule and cochlea of the inner ear was absent on the right side. The intracranial structures were normal. On CT, the tympanic bullae were abnormal bilaterally. On the right, the tympanic bulla was expanded, was extensively remodeled from its normal shape, and had areas of thickening as well as punctate lysis. On the left side, the tympanic bulla was slightly thickened. There were focal areas of ossification of the scutiform cartilage. Material was present in both tympanic cavities and the external ear canals bilaterally consistent with exudate (Fig. 3.9-3).

Diagnosis

Otitis media/interna (presumptive bacterial) with secondary peripheral vestibular dysfunction

Treatment

The treatment of otitis externa and otitis media/interna requires topical and systemic therapy, respectively. Choice of topical antimicrobials for the external ear canal should be based on cytologic and microbiological testing. Anti-inflammatories and analgesics also may be used if necessary. Otitis media/interna requires long-term parenteral antimicrobial therapy. Ideally, antimicrobials are chosen based on cytology and microbiological testing of effusion/material from the tympanic cavity. The effusion/material from the tympanic cavity may be obtained via myringotomy. Often, empirical parenteral antimicrobial therapy is chosen based on results of cytology and microbiological testing of exudate from the external ear canals. With severe disease as in the present case, surgical intervention (total ear canal ablation and bulla osteotomy) should be considered.

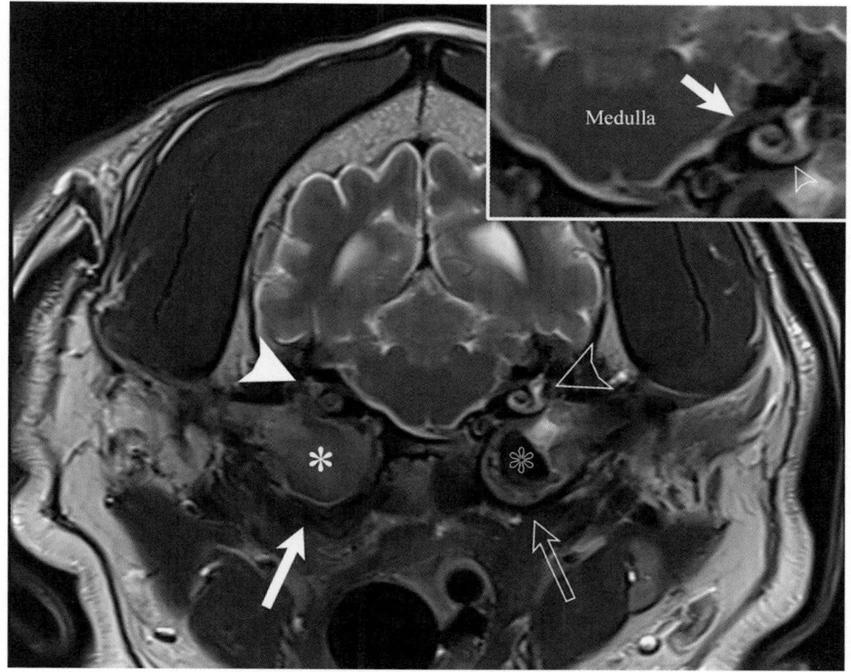

FIGURE 3.9-1 Transverse T2-weighted MRI image at the level of the tympanic cavities and rostral medulla. The right tympanic bulla *(white arrow)* is thickened and effusion fills the right tympanic cavity *(asterisk)*. There is a loss of the normal high T2 signal intensity in the fluid-filled cochlea and vestibule on the right side *(white arrowhead)* in comparison with the normal-appearing fluid-filled cochlea and vestibule on the left *(open arrowhead)*. The left tympanic bulla *(open arrow)* is normal but the left tympanic cavity is partially filled by effusion and a thickened mucoperiosteum *(open asterisk)*. Inset: magnified view of the medulla and fluid-filled cochlea and vestibule on the left. Note CN VII and CN VIII passing through the internal acoustic meatus *(white arrow)*. The fluid-filled cochlea and vestibule are seen ventrolaterally *(open arrowhead)*.

The dog was treated with oral antimicrobials (cephalexin) along with a topical antimicrobial and anti-inflammatory (Mometamax® Otic Suspension). The vestibular function resolved in the typical way: the abnormal nystagmus improved within 3–5 days, the vestibular ataxia within 3 weeks, and the head tilt over a couple of months.

Anatomical features in canids and felids

Introduction

Vestibular dysfunction is commonly encountered in small animal practice. Establishing an accurate neuroanatomic diagnosis is critical in order to define appropriate differential diagnoses and choose appropriate diagnostic tests. Ultimately, this enables one to establish a correct etiological diagnosis, prescribe proper therapy, and accurately communicate a prognosis to the owner.

The vestibular system can be divided into peripheral (vestibular component of CN VIII [vestibulocochlear nerve] and the receptors in the 3 semicircular canals and vestibule of the inner ear) and central (vestibular nuclei and cerebellum [fastigial nucleus or flocculonodular lobes of the cerebellum]) components. Based on the neurological examination, a neuroanatomic diagnosis can be defined as involving the components of the peripheral (peripheral vestibular dysfunction) or central (central vestibular dysfunction) nervous system. Knowledge of the anatomy of the vestibular system, along with nervous system structures near the vestibular system, is important to establish a correct neuroanatomic diagnosis. For a complete discussion of the anatomy of the ear, please see Case 3.5.

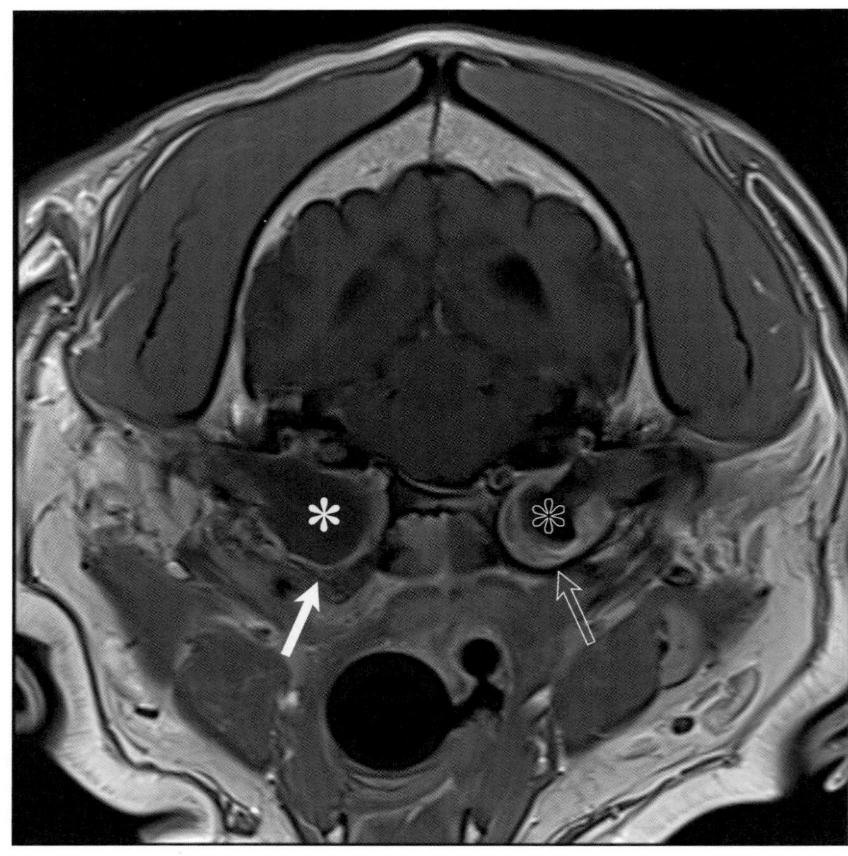

FIGURE 3.9-2 A corresponding transverse T1-weighted postcontrast image to Fig. 3.9-1. On T1-weighted images, the effusion in the right tympanic cavity *(asterisk)* is hypointense and the mucoperiosteum lining the tympanic cavity is enhanced. The right tympanic bulla is thickened *(arrow)*. The tympanic bulla and adjacent tissues enhance along with the right rectus capitis ventralis muscle (medial to the tympanic bulla). The left tympanic bulla is normal *(open arrow)*; however, there is thickened mucoperiosteum that is enhanced *(open asterisk)*.

Function

The vestibular system provides for special proprioception (balance). Special proprioception is the ability to maintain balance and proper orientation of the head and body in relation to gravity—i.e., knowing which way is up. The vestibular system maintains balance when the patient is static (standing still) as well as during rotation, acceleration, or deceleration. Finally, the vestibular system has diffuse projections throughout the CNS, which enables coordination of the head, neck, limbs, trunk, and eye movements in relation to changes in the position of the head.

Peripheral vestibular system

The peripheral vestibular system is contained within the inner ear. The inner ear is made up of the bony labyrinth of the petrosal portion of the temporal bone. The bony labyrinth is comprised of 3 fluid-filled compartments—the three semicircular canals, a vestibule, and a cochlea. Receptors involved in vestibular function (special proprioception) are in the 3 semicircular canals and vestibule. The 3 **semicircular canals** are oriented along planes that are at right angles to each other. This orientation provides for the determination of the spatial orientation and movement (acceleration or deceleration) of the head. Within the **vestibule** are the **utricle** and **saccule**, which determine gravitational forces. Receptors in the semicircular canals and vestibule are in synaptic contact with the dendritic processes of the **vestibular nerve**.

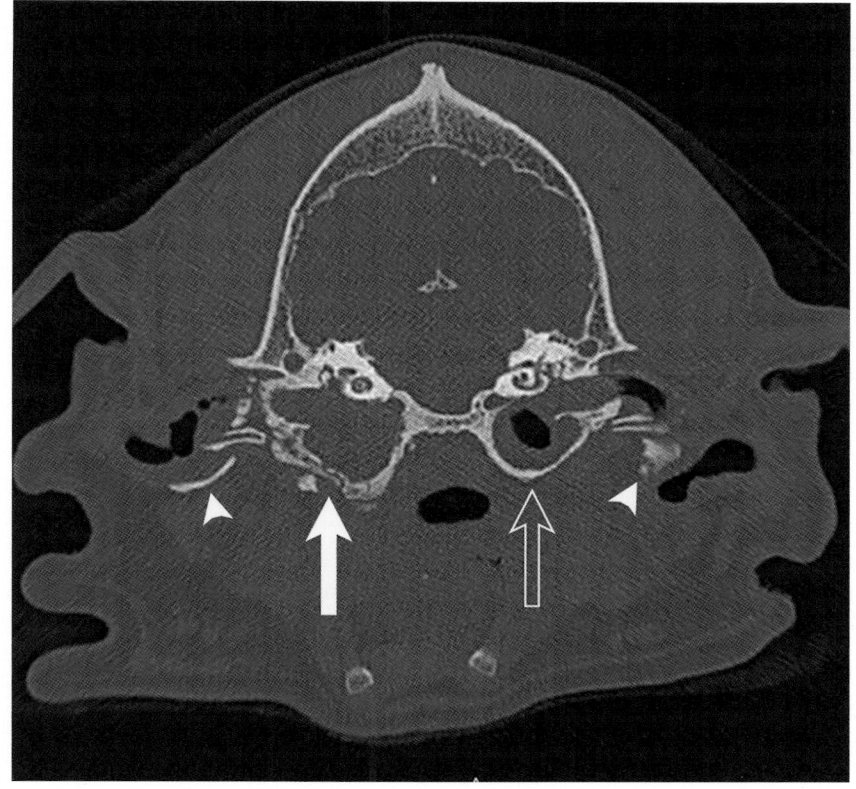

FIGURE 3.9-3 A corresponding CT image to Figs. 3.9-1 and 3.9-2. The right tympanic bulla is enlarged and thickened with an irregular surface and areas of lysis *(arrow)* in comparison with the normal-appearing left tympanic bulla *(open arrow)*. Mineralization of the external ear canals is visualized *(arrowheads)*.

Nervous system structures near the components of the peripheral vestibular system include CN VII (facial nerve) and the postganglionic sympathetic axons that track through the tympanic cavity (Figs. 3.9-4 and 3.9-5).

Therefore, the finding of ipsilateral deficits related to the facial nerve and/or Horner syndrome (loss of sympathetic innervation to the eye) along with vestibular dysfunction establishes a neuroanatomic diagnosis of peripheral vestibular dysfunction (Fig. 3.9-4).

IS IT CENTRAL OR PERIPHERAL VESTIBULAR DYSFUNCTION?

Lesions that affect either the peripheral or the central components of the vestibular system result in the same signs of vestibular dysfunction—i.e., head tilt, vestibular ataxia, abnormal nystagmus, and vestibular strabismus (ventral deviation of the eye with the extension of the head and neck).

Therefore, differentiation between peripheral and central vestibular dysfunction is made by identifying neurological deficits other than those related to vestibular dysfunction that can be ascribed to involvement of either the inner/middle ear or the medulla, respectively.

In summary, the finding of vestibular dysfunction along with abnormal mentation, ipsilateral GP ataxia and UMN paresis, and/or dysfunction implicating CN V–XII establishes a neuroanatomic diagnosis of central vestibular dysfunction (Fig. 3.9-5). The finding of vestibular dysfunction along with facial paresis/paralysis and Horner syndrome establishes a neuroanatomic diagnosis of peripheral vestibular dysfunction (Fig. 3.9-4). Likewise, the finding of vestibular dysfunction alone (without evidence of involvement of the medulla) also establishes a neuroanatomic diagnosis of peripheral vestibular dysfunction.

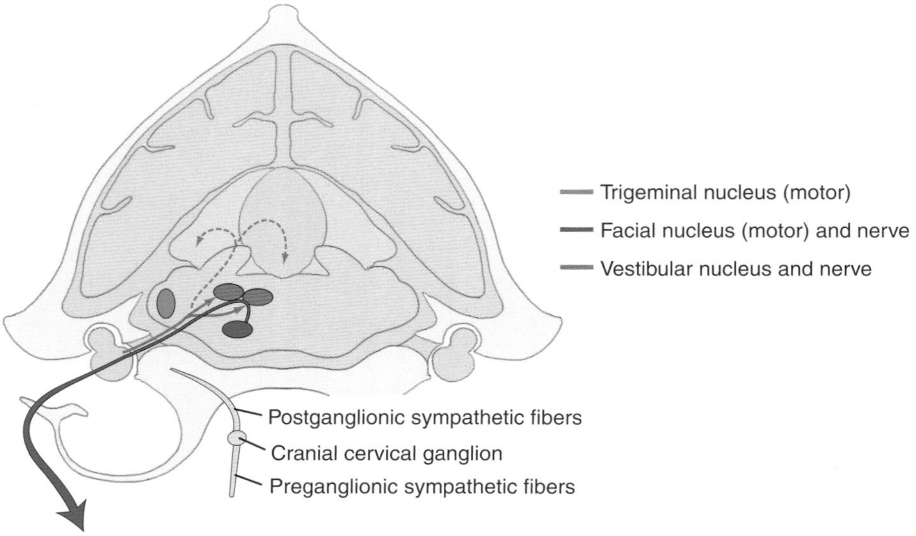

Trigeminal nucleus (motor)
Facial nucleus (motor) and nerve
Vestibular nucleus and nerve
Postganglionic sympathetic fibers
Cranial cervical ganglion
Preganglionic sympathetic fibers

FIGURE 3.9-4 A transverse section of a canine brain at the junction between the pons and rostral medulla depicting the location of the vestibular nuclei and nerve, facial nucleus (motor) and nerve, trigeminal nucleus (motor), and postganglionic sympathetic axons and their relationship to the tympanic cavity.

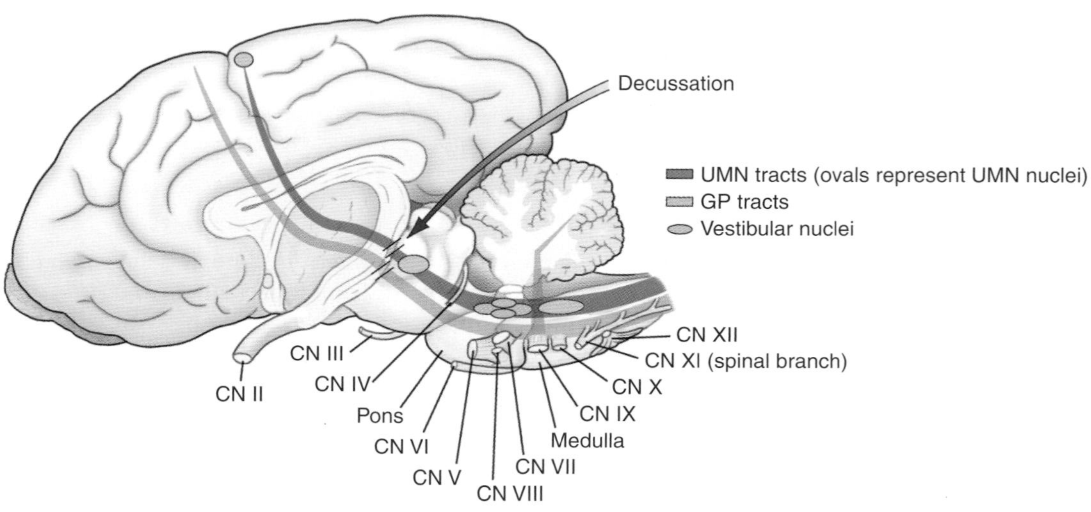

FIGURE 3.9-5 A sagittal view of the canine brain depicting the location of the vestibular nuclei and their proximity to cranial nerves V–XII.

Central vestibular system

The axons of the vestibular nerve pass through the **internal acoustic meatus** to enter the cranial cavity. The axons enter the **medulla** at the **cerebellopontine medullary angle** where the majority of axons synapse on the **vestibular nuclei** in the medulla. Some axons pass via the **caudal cerebellar peduncle** to synapse in the **cerebellum** (fastigial nucleus and flocculonodular lobes).

The vestibular nuclei have widespread projections to the entire CNS. From the nuclei, projections track via the medial longitudinal fasciculus to CN nuclei III, IV, and VI that provide general somatic efferent innervation to the extraocular muscles involved in movement of the eye. Projections caudally in the spinal cord form the **vestibulospinal tracts**, which are involved in upper motor neuron (UMN) innervation to the head, neck, limbs, and trunk. Axons from the vestibular nuclei also project to the vomiting center in the medulla.

Rhythmic dorsolateral movement of the superior eyelids may be present that move in concert with the nystagmus (as was seen in this case). As the eye deviates laterally, the superior eyelid elevates dorsolaterally. The muscle that is likely involved in this eyelid movement is the **levator anguli oculi medialis muscle**, which is innervated by CN VII. There are probably projections from the vestibular nuclei to the facial nuclei which mediate this movement. On rare occasions, patients with vestibular dysfunction may demonstrate rhythmic oscillations of the head. Like the movement of the superior eyelids, the oscillations of the head are in concert with the nystagmus. The vestibular nuclei likely project via the vestibulospinal tracts to activate neck musculature. Often, dogs and cats with vestibular dysfunction and concurrent head oscillations are blind (the cause of blindness is unrelated to vestibular dysfunction; the blindness is typically due to a prior disease process). The role that blindness plays is undetermined.

Nervous system structures near the components of the central vestibular system include the cranial nerve VII (facial nerve) and cranial nerve V (trigeminal nerve), which are located at the cerebellopontine medullary angle. The facial nerve exits the medulla adjacent to the vestibulocochlear nerve and, together, they pass to the internal acoustic meatus in a common dural sheath. Additionally, the motor nucleus of the facial nerve and the vestibular nuclei are located in the rostral medulla, while immediately rostral to the medulla the pontine sensory and motor nucleus of the trigeminal nerve is located in the pons.

A lesion located at the cerebellopontine medullary angle or unilaterally affecting the medulla often results in ipsilateral deficits involving vestibular dysfunction, facial paresis/paralysis, and deficits involving the sensory and motor function of the trigeminal nerve (Fig. 3.9-5).

The cranial nerve nuclei VI–XII, which are in the medulla, also are near the vestibular nuclei (Fig. 3.9-5).

Lesions that affect the vestibular nuclei may cause deficits involving CN V–XII.

The general proprioceptive (GP) tracts projecting to the cerebrum and cerebellum, along with the descending UMN tracts, track through the medulla (Fig. 3.9-5).

Lesions affecting the vestibular nuclei may result in ipsilateral GP ataxia and UMN paresis.

Some of the nuclei that make up the ascending reticular activating system reside in the medulla.

Lesions resulting in central vestibular dysfunction may affect mentation in the dog and cat.

Selected references

[1] de Lahunta A, Glass EN, Kent M. Vestibular system: special proprioception. In: Veterinary neuroanatomy and clinical neurology. Elsevier Health Sciences; 2014. p. 338–67.
[2] Kent M, Platt SR, Schatzberg SJ. The neurology of balance: function and dysfunction of the vestibular system in dogs and cats. Vet J (London, England: 1997) 2010;185:247–58.
[3] Rossmeisl Jr JH. Vestibular disease in dogs and cats. Vet Clin North Am Small Anim Pract 2010;40:81–100.
[4] Lowrie M. Vestibular disease: anatomy, physiology, and clinical signs. Compendium (Yardley, PA) 2012;34:E1.
[5] Garosi LS, Dennis R, Penderis J, et al. Results of magnetic resonance imaging in dogs with vestibular disorders: 85 cases (1996-1999). J Am Vet Med Assoc 2001;218:385–91.

CHAPTER 3

CASE 3.10

Glioma

Marc Kent[a] and Eric N. Glass[b]
[a]Department of Small Animal Medicine and Surgery, University of Georgia College of Veterinary Medicine, Athens, Georgia, US
[b]Neurology and Neurosurgery, Red Bank Veterinary Hospital, Tinton Falls, New Jersey, US

Clinical case

History

A 9-year-old female spayed Boxer was presented for a 3-week progressive history of being quiet and less responsive to her owner, and walking in circles to the right. The day before presentation, the dog experienced a generalized seizure.

Physical examination findings

Physical examination was normal.

Neurological examination findings

Mentation

- Abnormal, obtunded

Gait/posture

- Normal gait and posture
- Compulsively circling to the right (constantly walking in a circle)
- Tendency to turn the head to the right and gaze to the right

Postural reactions

- Delayed hopping and placing in the left thoracic and pelvic limbs

Spinal reflexes/muscle tone/muscle mass/nociception

- Thoracic limbs—Normal withdrawal reflexes, muscle tone, and muscle mass
- Pelvic limbs—Normal patellar and withdrawal reflexes, muscle tone, and muscle mass
- Normal cutaneous trunci reflex, perineal reflex, and external anal sphincter tone
- Hypalgesia—Reduced but not absent response to noxious stimulus (pinching skin with a hemostat) applied to the left side of the body; on the right side of the body, patient turned around and looked at area being pinched; on the left side of the body, the patient did not look toward stimulus or would walk away

Comparative Veterinary Anatomy: A Clinical Approach. https://doi.org/10.1016/B978-0-323-91015-6.00020-0

Cranial nerves

- Absent menace response OS with normal-sized pupils and normal pupillary light reflexes OU
- Left-sided facial hypalgesia (reduced response to stimulation of mucosa of the nasal septum on the left side)
- Remaining cranial nerves were normal

Sensory

- Normal with exception of hypalgesia to the left side of the face and body

Neuroanatomic diagnosis

Right prosencephalon—cerebrum and/or diencephalon (primarily thalamus) (see Table 3.10-1)

Differential diagnoses

Neoplasia—primary (e.g., glioma, meningioma) vs. secondary (e.g., metastatic or secondary extension of a neoplasm from adjacent anatomy, such as pituitary neoplasms, sinonasal neoplasms, or neoplasms arising from the skull); infectious meningoencephalitis (bacterial; fungal such as *Cryptococcus* spp., *Blastomyces dermatitidis*, or *Aspergillus* spp.; protozoal such as *Neospora caninum*, *Toxoplasma gondii*; viral such as canine distemper or rabies), or immune-mediated meningoencephalitis (e.g., granulomatous meningoencephalitis)

Diagnostics

A CBC, serum chemistry profile, and urinalysis were assessed to identify concurrent disease that may impact or preclude general anesthesia. Three-view thoracic radiographs were performed to screen for the evidence of metastatic neoplasia. The results of these diagnostics were within normal limits.

The dog was placed under general anesthesia and MRI identified a single, round to oval, moderately well-circumscribed, intraparenchymal mass in the right piriform lobe of the cerebrum (Fig. 3.10-1). The mass caused compression of the lateral ventricle and deviation of the adjacent thalamus to the left side. The lesion was hyperintense on T2-weighted and hypointense on T1-weighted sequences. The white matter adjacent to the mass and extending into the internal capsule and adjacent to the thalamus were hyperintense on T2-weighted and T2-weighted fluid-attenuated inversion recovery sequences, consistent with vasogenic edema. Following IV administration of contrast medium, the mass displayed mild heterogeneous enhancement.

Diagnosis

Glioma (presumptive diagnosis based on signalment and MRI findings)

TABLE 3.10-1 Key clinical neurologic signs of prosencephalic lesions.

Abnormal mentation
Normal gait with no ataxia (incoordination) or paresis (weakness)
Compulsive circling toward the side of the lesion—*adversive syndrome*
Contralateral deficits:
Postural reactions
Menace response
Response to stimulation of the nasal mucosa—nasal hypalgesia

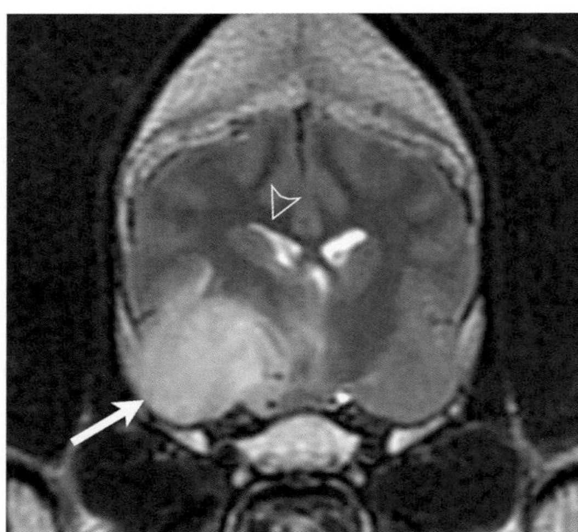

FIGURE 3.10-1 Transverse T2-weighted MRI at the level of the rostral cerebrum of a 9-year-old female spayed Boxer with signs referable to the right prosencephalic lesion. In the right piriform lobe, there is a round to oval T2-hyperintense mass (*arrow*) compressing the adjacent thalamus and the right lateral ventricle (*arrowhead*). The patient's right side is on the left side of the image.

Treatment

The dog underwent 5 treatments of stereotactic radiation therapy for the presumed glioma. The dog also was treated with an anti-inflammatory regimen of prednisone (0.5–1.0 mg/kg q 24 h) and phenobarbital (2.2 mg/kg orally q 12 h) for seizures. Following radiation therapy, the dog lived approximately 11 months before clinical signs returned and the dog was euthanized. The prognosis for dogs with intracranial neoplasms treated by radiation therapy varies widely.

Anatomical features in canids and felids

Introduction

The term prosencephalon is used to define the neuroanatomic diagnosis and is an embryological term. The prosencephalon is composed of the telencephalon (cerebrum) and the diencephalon (primarily the thalamus, but also the epithalamus and hypothalamus). Clinically, a lesion affecting the cerebrum results in deficits that are indistinguishable from a lesion affecting the diencephalon. Therefore, the term prosencephalon is used to succinctly denote that the lesion responsible for the observed deficits may involve either the cerebrum and/or the diencephalon.

Function

The prosencephalon functions in the maintenance of a normal mentation—establishing the awake state and consciousness of the patient, general proprioception (GP), vision, and perception of noxious stimuli. An important, overarching clinical concept of prosencephalon function is that all information (sensory and motor functions) that projects into or out of the prosencephalon crosses before entering into or exiting out of the prosencephalon, respectively. The clinical consequence of this routing of sensory and motor pathways means that unilateral lesions of the prosencephalon result in contralateral deficits. This is in contradistinction to all other anatomic regions of the CNS where a lesion results in ipsilateral deficits.

Ascending reticular activating system

Mental state (wakefulness and consciousness) is maintained by the **cerebrum** and the **ascending reticular activating system** (ARAS). The ARAS is a collection of nuclei located throughout the brain stem from the thalamus through the medulla. All sensory information destined for the cerebrum—conscious projection pathways involved in touch, nociception, and proprioception, as well as sensory information involved in assessment of the internal environment of the body (temperature, blood pressure, blood gases, and viscera), smell, and taste—synapses directly or indirectly via collaterals on the ARAS. These ARAS nuclei project rostrally to thalamic nuclei, synapse, and ultimately send diffuse projections to the cerebrum. Therefore, the awake state and level of consciousness may be affected with lesions affecting either the cerebrum or the ARAS (Fig. 3.10-2). In general, lesions of the diencephalon result in a more severe alteration of mentation than other parts of the ARAS.

Seizures are a clinical manifestation of dysfunction of the prosencephalon. Therefore, the observation of a seizure mandates a prosencephalic neuroanatomic diagnosis. It is important to note that a neuroanatomic diagnosis should not be confused with an etiological diagnosis. For example, seizures may occur secondary to an intracranial pathology—e.g., a primary brain neoplasm—or an extracranial disorder—e.g., hypoglycemia—or idiopathic epilepsy. In all cases, the neuroanatomic diagnosis is the same—prosencephalon.

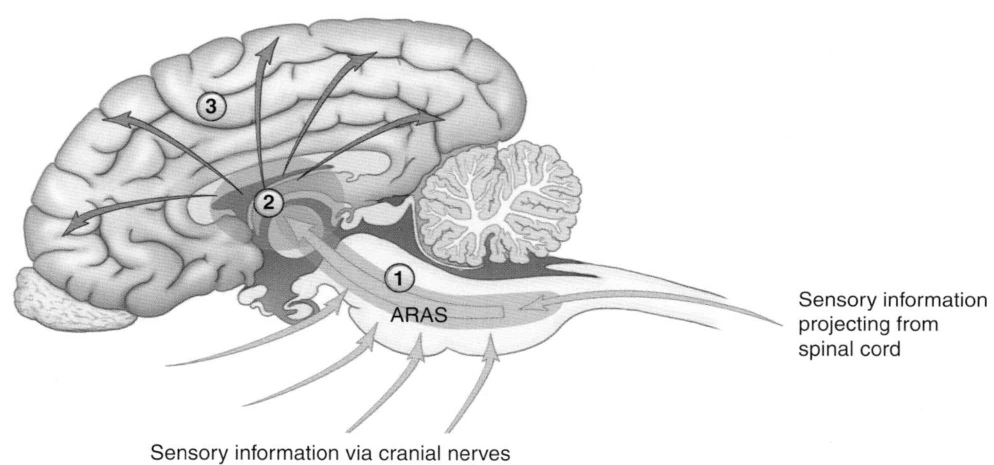

FIGURE 3.10-2 Ascending reticular activating system. (1) Neurons that comprise the ascending reticular activating system. (2) Thalamic nuclei (thalamic reticular system). (3) Diffuse cerebral cortical projections.

GAIT WITH PROSENCEPHALON LESIONS

As a general rule, the gait remains normal with lesions affecting the prosencephalon in veterinary patients. Lesions affecting the prosencephalon do not result in ataxia (general proprioceptive, vestibular, or cerebellar) or paresis. There are several notable exceptions to the generalization that prosencephalic lesions do not affect the gait (e.g., see next paragraph). Unilateral prosencephalic lesions may result in compulsive circling where the patient consistently circles toward the side of the prosencephalic lesion. This is referred to as the "adversive syndrome." The observation of a normal gait yet compulsive circling toward one side is unique to lesions affecting the prosencephalon.

With peracute or acute lesions such as an ischemic stroke or a traumatic brain injury that involves the prosencephalon, patients may experience paresis/paralysis of the limbs contralateral to the side of the lesion. Typically, the gait gradually returns to normal over 3–4 days. Also, patients with metabolic or toxic encephalopathies that affect the prosencephalon may be unable to generate a gait (i.e., be unable to walk). In metabolic or toxic encephalopathies, this may reflect diffuse involvement of the brain. Likewise, patients with severe obtundation, stupor, or coma will not walk.

General proprioception

Anatomically, general proprioception (GP) (position sense of the limbs and trunk) is composed of conscious proprioception (CP) and unconscious proprioception (UCP). Conscious proprioception is the proprioceptive information destined for the contralateral somesthetic cerebral cortex, whereas UCP is the proprioceptive information destined for the ipsilateral cerebellum. The ascending GP tracts are located in close proximity to the descending, upper motor neuron (UMN) tracts. Therefore, a lesion affecting the GP tracts will also affect the UMN tracts that result in postural reaction deficits—i.e., delayed or absent hopping and proprioceptive placing/knuckling. Additionally, the contributions the 2 GP pathways make in position sense are indistinguishable, despite being anatomically distinct pathways projecting to different brain structures.

The observation of postural reaction deficits with lesions affecting the GP pathways should not be termed a "CP deficit." The term "CP deficit" is a misnomer because the contribution of the CP pathways cannot be separated from that of the UCP pathways.

Of importance, despite the location of the *anatomic* crossover being located at the level of the caudal medulla, the GP pathways destined for the contralateral somesthetic cerebral cortex *clinically* behave as if the crossover is located rostrally at the level of the midbrain. The midbrain is the rostral most site of the brain stem in which all the ascending GP pathways have crossed to the contralateral side.

Consequently, animals with lesions affecting the prosencephalon display postural reaction deficits on the side contralateral to the lesion whereas animals with lesions caudal to the midbrain display postural reaction deficits on the side ipsilateral to the lesion.

Anatomic pathway for conscious proprioception (Fig. 3.10-3)

Proprioceptors—sensory receptors concerned with position sense of the limbs and trunk—are invested in muscles (intrafusal fibers), tendons, and joints. Proprioceptors are in synaptic contact with **sensory nerves**. Impulses are conducted proximally along sensory nerves to enter the spinal cord along the **dorsal spinal nerve roots** to synapse in the dorsal horn. Dorsal horn neurons send axons that project cranially in the **dorsal funiculus**. The GP axons from the pelvic limbs (and caudal aspect of the trunk) course in the **fasciculus gracilis** within the medial aspect of the dorsal funiculus, whereas GP axons from the thoracic limb (and cranial aspect of the trunk) track in the **fasciculus cuneatus** within the lateral aspect of the dorsal funiculus. Axons in the fasciculus gracilis and cuneatus project ipsilaterally in the spinal cord to synapse in the ipsilateral nucleus gracilis and medial cuneate nucleus in

CRANIAL NERVE EVALUATION WITH PROSENCEPHALON LESIONS

The menace response and response to stimulation of the nasal mucosa may be abnormal in animals with prosencephalic lesions. In general, the function of the cranial nerves is assessed through the evaluation of reflexes analogous to how spinal reflexes are used to assess the spinal cord segments, their roots and spinal nerves, and the named nerves of the limbs. Therefore, evaluation of cranial nerve reflexes only assesses specific areas of the brain stem. However, there are 2 exceptions to this generality: the menace response and response to stimulation of the nasal mucosa. These 2 responses necessitate normal function of the prosencephalon. As with other information projecting to the prosencephalon, the visual pathway (assessed via the menace response) and the conscious projection pathways for sensation of the face (assessed via stimulation of the nasal mucosa) project to the contralateral prosencephalon. Consequently, unilateral lesions affecting the prosencephalon may result in a contralateral menace response deficit and contralateral facial hypalgesia.

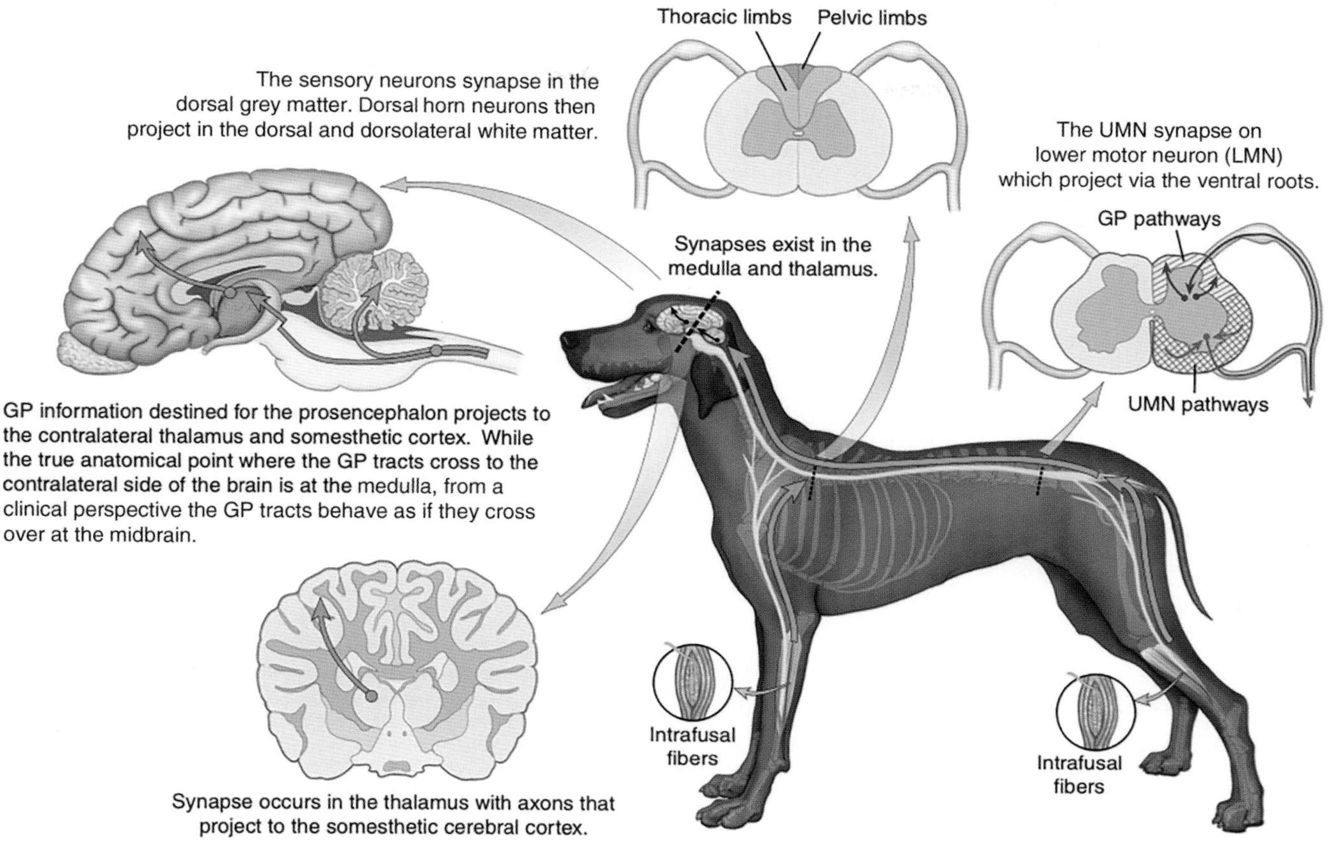

FIGURE 3.10-3 The general proprioceptive pathways.

the caudal medulla, respectively. From the nucleus gracilis and medial cuneate nucleus, axons decussate via the deep arcuate fibers in the medulla and project rostrally in the contralateral medial lemniscus to synapse in the contralateral thalamus. From the thalamus, axons project in the internal capsule to the somesthetic cerebral cortex.

Visual and quintothalamic pathways

The menace response is elicited by gesturing one's hand as if to strike the face. To assess each eye individually, the eye contralateral to that being assessed should be covered (Fig. 3.10-4). Facial hypalgesia is assessed through stimulation of the nasal mucosa by touching the nasal mucosa of the nasal septum with the tip of a closed hemostat (Fig. 3.10-5). The reason why the nasal mucosa is assessed is that there is almost always a robust, vigorous resentment by the patient to touching the nasal mucosa with a hemostat, whereas the response to stimulation elsewhere on the face is less predictable. Likewise, pinching of the skin on the lateral side of the body may be used to assess hypalgesia along the trunk. In patients with a prosencephalic lesion that results in facial hypalgesia, there also should be hypalgesia along the body on the same side as the facial hypalgesia. However, hypalgesia along the body may be difficult to appreciate in a practical sense. Often, normal patients do not demonstrate a vigorous response to pinching the skin along the body. Therefore, detecting an observable difference between the sides of the body may be challenging. In the end, observing hypalgesia on one side of the body in a patient with unilateral facial hypalgesia does not contribute more to the neuroanatomic diagnosis.

The use of the term "response" denotes a difference from a reflex activity. Normal reflex activity needs only the function of an afferent limb, a defined portion of the CNS, and an efferent limb. For example, the patellar reflex needs only the femoral nerve and the L4–L6 spinal cord segments to function normally. Lesions cranial or caudal

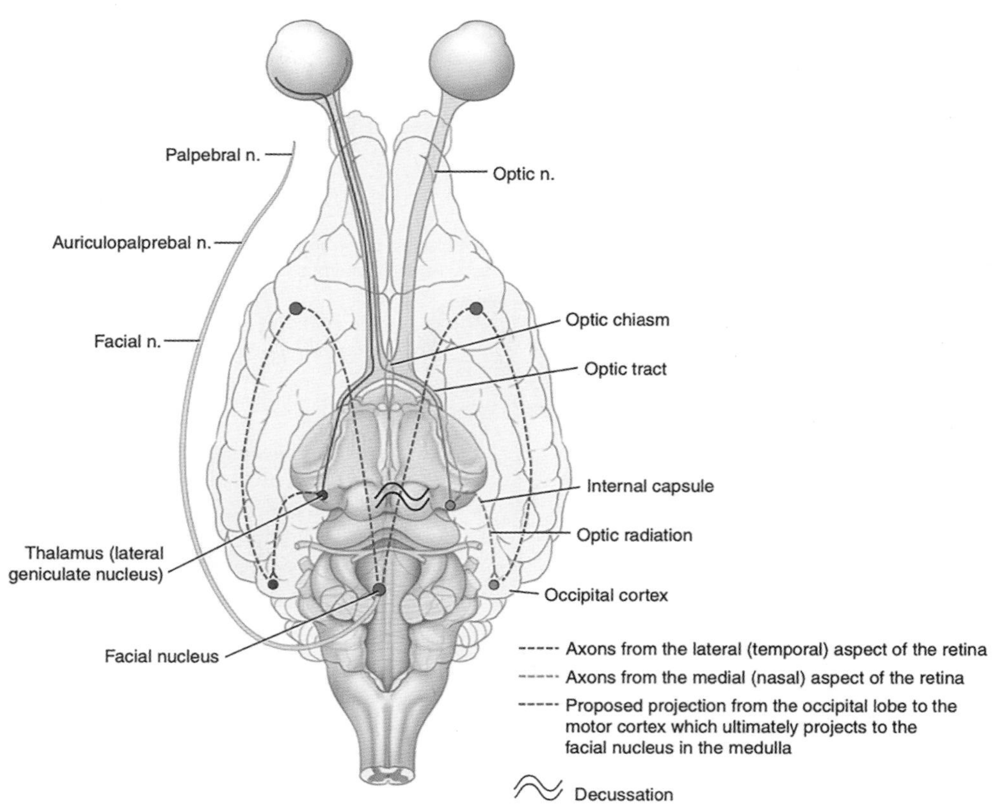

FIGURE 3.10-4 The menace response.

to the L4–L6 spinal cord segments do not impair the patellar reflex. Likewise, cranial nerve reflexes assess specific segments of the brain stem without influence from other areas of the brain. In contrast, the use of the word response (menace response and response to stimulation of the nasal mucosa) signifies a more complex reaction necessitating normal prosencephalic function. As with all sensory information, the pathways involved in these responses cross over to project to the contralateral prosencephalon. For the menace response, this crossover occurs at the optic chiasm. Conscious projection of sensation of the face projects in a pathway, the quintothalamic tracts, similar to the pathway for CP projections (discussed previously).

HYPALGESIA

Hypalgesia is defined as a reduced or less vigorous response. While the majority of general somatic afferent (sensation) pathways of the face project to the contralateral somesthetic cerebral cortex, some pathways remain ipsilateral. Therefore, a lesion affecting only one side of the prosencephalon results in a reduced, but not an absent response, as some pathways for sensation remain unaffected. This is in contrast to analgesia, which is defined as an absent response to a noxious stimulus. In order to cause analgesia of the face, there has to be a complete loss of the general somatic afferent information of the face. This would likely necessitate a lesion affecting the trigeminal nerve or its main branches—the maxillary, ophthalmic, and mandibular nerves.

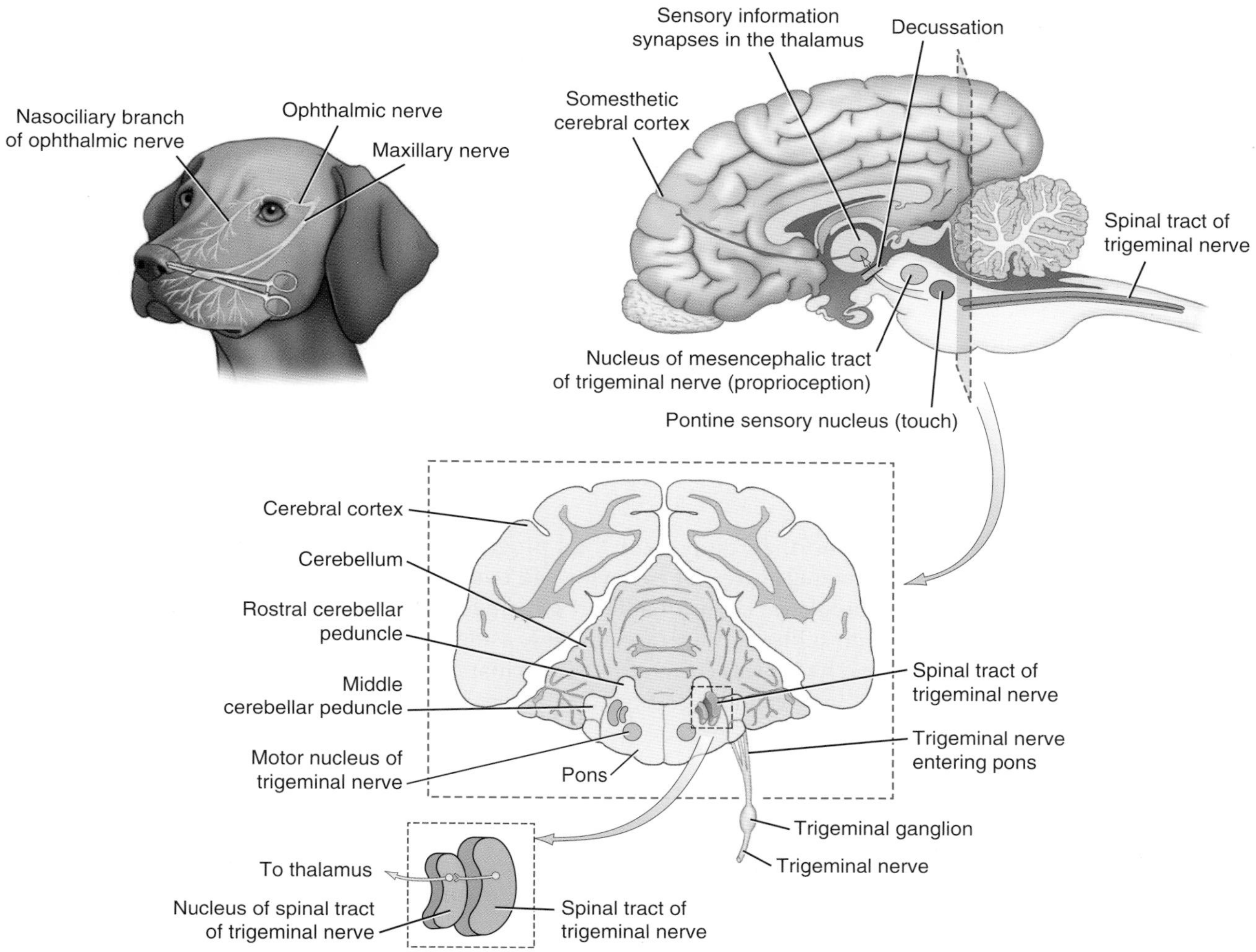

FIGURE 3.10-5 Assessing facial sensation.

Selected references

[1] de Lahunta A, Glass E, Kent M. Veterinary neuroanatomy and clinical neurology. St. Louis, MO: Elsevier Saunders; 2014.
[2] Lorenz MD, Coates JR, Kent M. Stupor or coma. In: Handbook of veterinary neurology. Elsevier/Saunders; 2011.

CHAPTER 3

CASE 3.11

Meningioma

Fred Wininger
Neurology Department, Charlotte Animal Referral and Emergency, Charlotte, North Carolina, US

Clinical case

History

A 10-year-old female spayed domestic shorthair cat presented for progressive behavior changes over the last month. The previously affectionate cat was showing aggressive behavior and had recently bitten the owner. The owner described the cat as confused and compulsive. Occasional head-pressing was noted along with tripping on the left thoracic limb. Previous pertinent medical history was unremarkable aside from a subcutaneous abscess on the left flank surgically treated 6 years before. The cat was current on all vaccinations and had negative FIV/FeLV status as per a recent SNAP ELISA test. The cat was on a balanced commercial diet and received no medications.

Physical examination findings

On presentation, the patient was bright, alert, and responsive. There was mild dental calculus and gingivitis. The general physical examination was otherwise considered normal.

A full neurologic examination was performed, including gait evaluation, proprioceptive placing, segmental reflex testing, cranial nerve assessment, and spinal hyperesthetic manipulations. Though the gait was considered normal, the cat compulsively circled to the right. Postural reaction testing showed a proprioceptive deficit in both the left thoracic and left pelvic limbs. Proprioception was considered normal in the right limbs. Segmental reflexes were normal in all limbs including patellar reflexes in the pelvic limbs and withdrawal reflex in all limbs. Cranial nerve examination revealed a decreased menace response OS (left eye), though the palpebral reflex and pupillary light reflex (PLR) were normal. Left facial sensation was decreased based on only a twitching of the whiskers when a hemostat touched the left nasal planum compared to the right side of the face where the same stimulus caused the cat to pull away its face and growl. The remainder of the cranial nerve examination was considered normal. There was no evidence of spinal hyperesthesia (Gr. *hyper-* above +Gr. *aisthēsis* perception = increased sensitivity).

REFLEX VS. RESPONSE

When assessing cranial nerves, it is important to recognize the distinction between reflexes and responses. Responses are often learned and require cortical input. This is different from reflexes that are "hardwired" and require brain stem function without "conscious awareness." The absent menace response is distinguished from the palpebral and pupillary light reflexes. Similarly, movement away from a stimulus to the nasal planum is a response to facial sensation as opposed to the flickering of whiskers or licking, which are reflexes.

Comparative Veterinary Anatomy: A Clinical Approach. https://doi.org/10.1016/B978-0-323-91015-6.00021-2

Differential diagnoses

Right forebrain lesion —neoplasia (primary extra-axial tumors such as meningioma, vs. primary intraaxial tumor such as a glioma, vs. infiltrative neoplasms such as lymphoma), encephalitis (infectious vs. noninfectious); less likely—cerebrovascular event; unlikely—metabolic or toxic lesions, anomalous diseases (hydrocephalus)

The patient localized to a lesion of the right forebrain based on the compulsive circling to the right, left appendicular paresis, absent menace response OS, and poor left facial sensation. The circling was a consequence of a hemineglect (Gr. *hĕmi-* half + L. *neglegere* to disregard = inability to process and perceive stimuli) syndrome in which a right forebrain lesion causes "lack of awareness or attention" to the left side of the body. The left-sided postural reaction deficits were caused by dysfunction of the sensory and motor cortices on the contralateral side of the body. The ascending proprioceptive and descending motor pathways decussate at the level of the midbrain. The menace deficit was also caused by a lesion of the contralateral visual and/or motor cortices. The lack of facial sensation on the left side of the face was also caused by a lesion of the contralateral somatosensory cortex, like the proprioceptive loss.

Diagnostics

Following an unremarkable preanesthetic screen of CBC, blood chemistry, and 3-view thoracic radiographs, the patient was placed under general anesthesia for a complete MRI study of the brain with a 1.5T unit (see Chapter 2, Section 2.5). Included in the study were T1- and T2-weighted (T1W and T2W, respectively) sequences, FLAIR, hemorrhage sequences (T2*GRE and SWI), and contrast imaging in multiple planes (Fig. 3.11-V1 in the online version at https://doi.org/10.1016/B978-0-323-91015-6.00021-2). There was a large, broad-based, right extra-axial lesion in the temporoparietal lobe at the level of the thalamic adhesion that was T2W isointense with the adjacent gray matter. The lesion was homogeneously contrast enhancing on T1W sequences. The lesion was causing a substantial mass effect with falcine and caudal transtentorial herniation, as well as obstructive hydrocephalus of the ipsilateral olfactory recess of the lateral ventricle. The mass was associated with a "dural tail" extending caudally and hyperostosis of the overlying calvarium.

INS AND OUTS—ASSESSMENT OF AFFERENT VS. EFFERENT CRANIAL NERVES

Most cranial nerve (CN) assessment involves an "in and out" (afferent vs. efferent or sensory vs. motor) principle.

For example, the menace response assesses CN II as the "in" (afferent) and CN VII as the "out" (efferent). To discover which is affected, the PLR can also assess the "in" of II and the palpebral reflex can assess the "out" of VII. It is the combination of cranial nerve testing that allows the clinician to sift out which component is dysfunctional.

CAUSES OF FOREBRAIN DISEASE IN CATS

- Acute onset and nonprogressive:
  1. Cerebrovascular
  2. Traumatic
- Acute onset and progressive:
  1. Infectious (FIP, Toxoplasma, Cryptococcus)
  2. Neoplastic
- Chronic onset and progressive:
  1. Nutritional (thiamine deficiency)
  2. Degenerative (storage disorders)
  3. Anomalous (hydrocephalus)

Imaging diagnosis

Meningioma

Additional diagnostics

Following the MRI study, CT of the head was performed with the goal of 3D printing the skull (please see Chapter 2, Section 2.7; see also Table 3.11-1). An 8-slice Toshiba CT acquired a 0.5-mm (0.002 in.) slice thickness data set in both a bone and a soft tissue window. Notably, the previously identified hyperostosis (Gr. hypertrophy of bone, exostosis) on the MRI was more conspicuous on the CT. Proprietary software was used to create a fusion of the CT and MRI data sets (Fig. 3.11-V2 in the online version at https://doi.org/10.1016/B978-0-323-91015-6.00021-2A and B). This fusion facilitated the printing of a to scale skull with the tumor included in the model. A secondary 3D print was made as a craniotomy jig, delineating the extent of the tumor on the inner skull surface with a 2-mm (0.08-in.) addition to the identified base (Fig. 3.11-1). The jig would facilitate complete and precise craniotomy borders to remove the meningioma en bloc (Fr. as a whole).

TABLE 3.11-1 CT vs. MRI.

Both CT and MRI (see Chapter 2, Sections 2.4 and 2.5) are tomographic (cross-sectional) imaging modalities that allow for the discerning of structures that would otherwise be superimposed on each other. This is especially important when it comes to the CNS as it is highly segmented despite its tissue being of similar density inside of a bony skull. Like traditional radiography, CT is a modality based on density. For this reason, it has excellent bony resolution but suffers in differentiating the similar tissues of the brain. Magnetic resonance imaging is based on water content and is much more capable of differentiating parts of brain tissue because of its soft tissue resolution. Though MRI has largely replaced CT for brain imaging, there are appropriate applications for both modalities.

Advantages of CT:

1. Fast
2. Less expensive
3. Excellent bone resolution
4. Capable of detecting acute-subacute hemorrhage
5. Very thin isotropic (equal size) voxels for 3D reconstructions

Disadvantages of CT:

1. Ionizing radiation
2. Poor nervous tissue resolution
3. Beam hardening artifact (the dense bone in the caudal aspect of the skull [specifically the petrous temporal bones] can block some of the photons passing through this region); the consequence is an artifact appearing as dark lines across the image set that disrupts brainstem visualization

Applications of CT brain imaging:

1. Fast and simple imaging
2. Head trauma
3. Brain hemorrhage
4. 3D printing applications

Advantages of MRI:

1. Excellent soft tissue imaging
2. More easily picks up pathologic lesions because it is water-based
3. Multiple sequences allow the users to select sequences for evaluation of certain tissues

Disadvantages of MRI:

1. Poor imaging of bone and air
2. Susceptible to metal artifacts (they disrupt the magnetic field causing image distortion)
3. Can be more complex to operate

Applications of MRI:

1. All brain imaging in general

In the case of a meningioma, MRI is preferable to CT. This tumor can be identified on both modalities because this tumor takes up contrast agents due to its high vascularity and because in cats it is often mineralized. However, MRI also allows for imaging the other sequelae of such a lesion including perilesional edema, obstructive hydrocephalus, brain shifting/herniation, and extent of meningeal involvement.

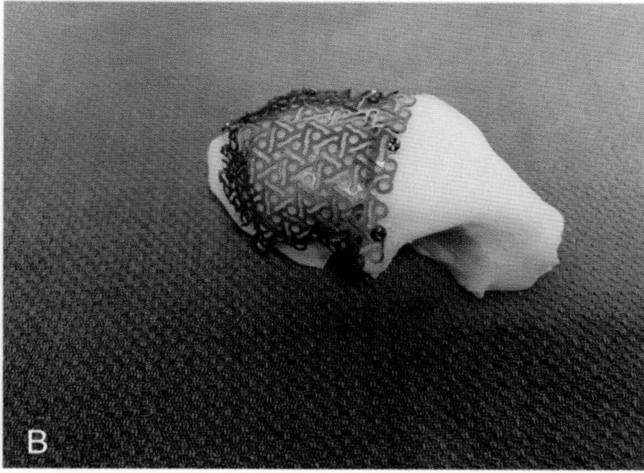

FIGURE 3.11-1 (A) The printed 3D jig is placed on a practice 3D print of the actual patient's skull. The internal contour of the jig interdigitates with the skull in a novel way for precise location placement. (B) The area the jig touches has been outlined in a *red surgical marker*. A titanium mesh is prefitted to cover the craniotomy site.

The following day, the patient was prepared for excisional biopsy of the mass under general anesthesia with the addition of mannitol IV as a presurgical osmotic agent. A 6-cm (2.4-in.) curvilinear incision was made extending 0.5 cm (0.2 in.) caudal to the zygomatic process of the frontal bone and extending to the level of the nuchal crest. The overlying platysma muscle was incised, exposing the temporalis muscle fascia. The temporalis muscle was incised at its midline insertion and reflected laterally with a periosteal elevator and Gelpi retractors. The 3D craniotomy jig was placed and articulated with the underlying calvarial curvature (Fig. 3.11-2). A sterile surgical marker outlined the borders of the craniotomy. The craniotomy was brought to calculated depth with a pneumatic burr drill and then laterally reflected with 2 periosteal elevators. Removal of the bone revealed a firm, tan,

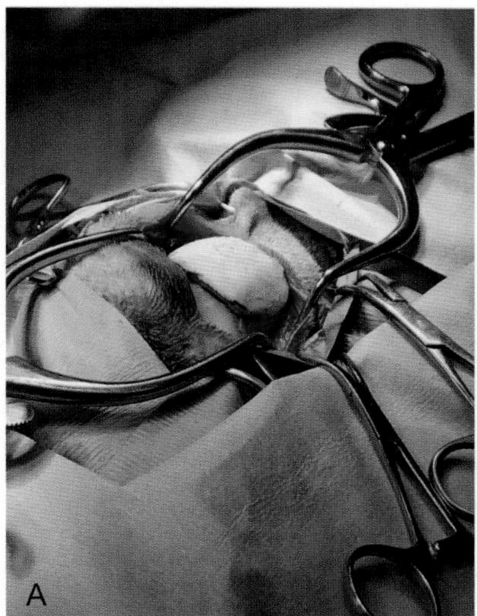

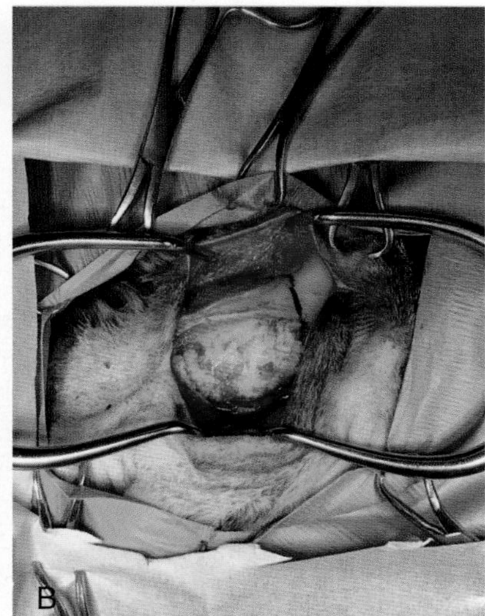

FIGURE 3.11-2 (A) The jig is placed on the skull and the craniotomy site drawn with a surgical marker. (B) The craniotomy defect is created by a pneumatic drill fitted with burring drill bits.

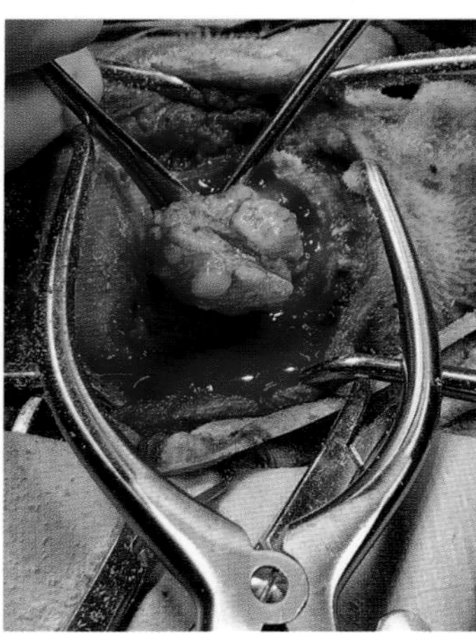

FIGURE 3.11-3 The meningioma is removed through the craniotomy defect. Most feline meningiomas are dense and can be removed en bloc with minimal bleeding or neuroparenchymal dissection.

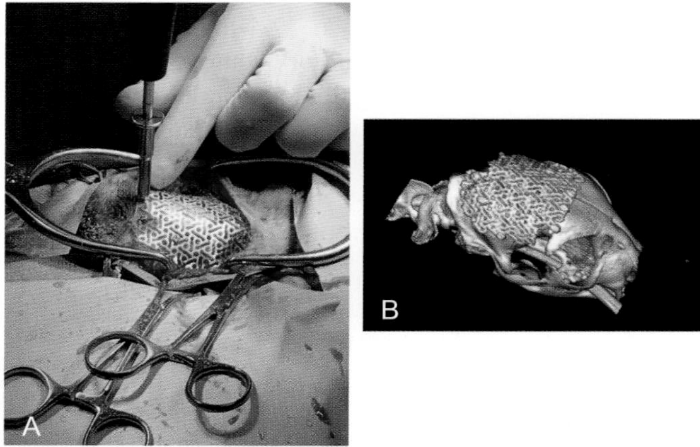

FIGURE 3.11-4 (A) Placement and securing of the titanium mesh posttumor removal. The mesh and screws are made of titanium to facilitate later MRI with minimal distortion artifacts. (B) Post-operative CT showing the titanium mesh in place.

vascularized mass that grossly interdigitated with the adjacent dura overlying the normal brain (Fig. 3.11-3). The dura had been partially torn during bone removal. A dural incision was created approximately 0.3 cm (0.1 in.) around the visualized mass. The dura and mass were easily removed en bloc via blunt dissection. The minimal bleeding was controlled with gel foam. The premolded titanium mesh was placed over the defect and maintained in position by 8 self-drilling titanium 2.0 mm × 6 mm (0.08 in. × 0.2 in.) screws (Fig. 3.11-4). The muscle, subcuticular layer, and skin were closed anatomically in a routine manner.

The following morning the cat had an overall improved neurologic examination and the compulsive circling had improved (Fig. 3.11-5). The paresis and menace deficit persisted but subjectively appeared less intense. The patient was discharged 36 hours post-operatively. At the 14-day recheck and suture removal, the owner reported that the cat's behavior had returned to normal, including the previous level of affection. The neurologic examination was normal at that time. Follow-up MRI 8 months later showed no evidence of tumor recurrence.

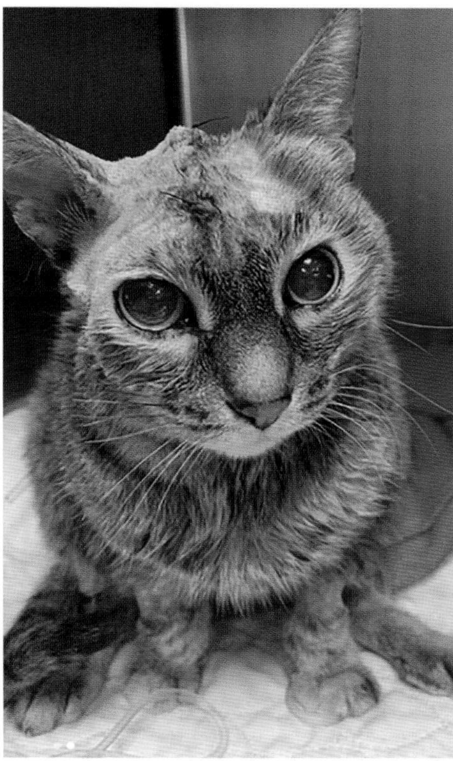

FIGURE 3.11-5 Post-operative image of the patient.

Anatomical features in canids and felids

Introduction

The embryologically characterized diencephalon and telencephalon together form the clinical unit, the prosencephalon, also known as the "forebrain" due to its general location (or "supratentorial," a term used more frequently by neurosurgeons). The prosencephalon is generally thought of as a clinical unit as lesions in any area of this region cause similar clinical signs.

The diencephalon is composed predominantly of the thalamus, pineal gland, and hypothalamus.

This section primarily covers the gross and microscopic anatomy of the prosencephalon and its related functions, with a focus on the visual pathways and postural reactions.

Function

The prosencephalon serves many functions, which are discussed in the following sections as relevant to their anatomy.

The thalamus is a multifunctional relay system that organizes brain stem signals for higher processing. The hypothalamus also serves multiple functions, including control of satiety, thirst, thermoregulation, hormonal release, and pituitary function.

Telencephalon anatomy (Fig. 3.11-6)

The **telencephalon** is comprised of the paired **cerebral hemispheres**. To maximize surface area, many gross undulations make up the protruding **gyri** and deeper **sulci**. The 4 lobes of the brain are the **frontal**, **parietal**, **occipital,** and **temporal lobes**. Separated from the frontal lobe by the **cruciate sulcus**, the parietal lobe is the site of the "sensory homunculus"—a map of the body. The correlate motor homunculus (L. a little man) is in the adjacent

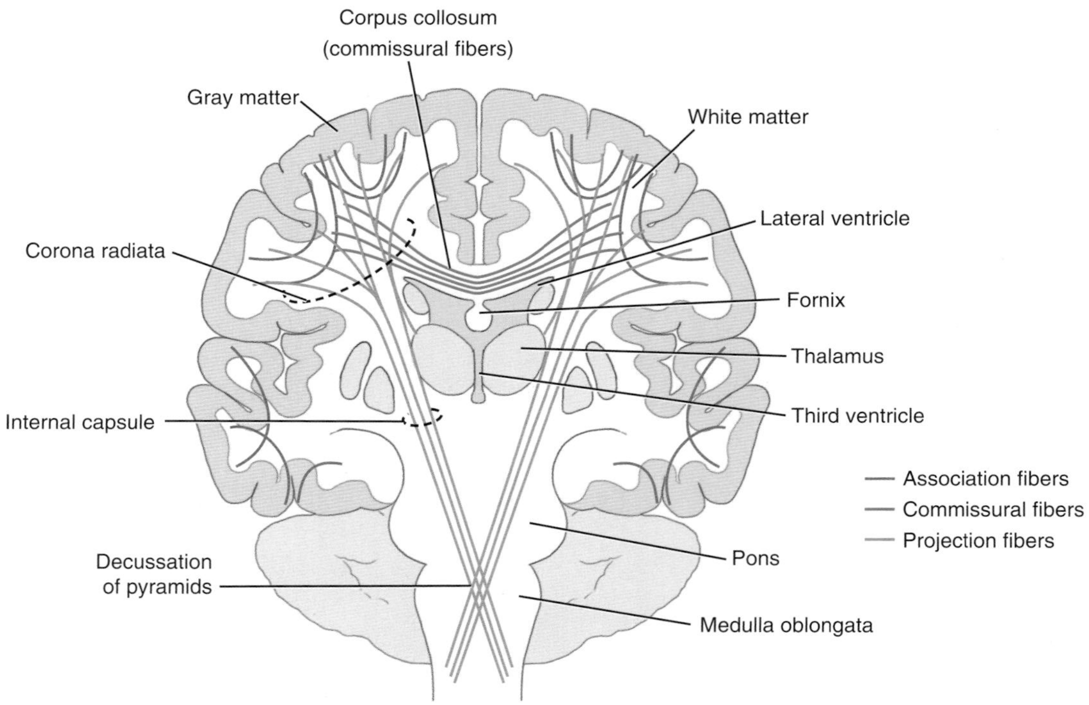

FIGURE 3.11-6 Gross anatomy of the telencephalon.

frontal lobe. The occipital lobe is largely the location of the visual cortex, and the temporal lobe is the primary site of the auditory cortex, which plays a large role in memory and emotion.

On cross section, the **gray matter** (superficial) of the forebrain surface forms the **cortical ribbon**. The ribbon is histologically made up of 6 layers or **laminae** with alternating large pyramidal neurons and smaller granular neurons.

The **white matter** of the telencephalon lies deep into the gray matter with the **corona radiata** mirroring the undulations of the cortical ribbon. Deep to it, the **centrum semiovale** and **internal capsule** comprise the largest area of myelinated white matter axons. This white matter is divided into 3 predominant fiber types. First, the **association fibers** are neurons that synapse with other neurons within the same hemisphere. Second, **commissural fibers** cross midline and connect neurons of both hemispheres. Most commissural fibers pass through the **corpus callosum** with fewer crossing through the **rostral** and **caudal commissures**. Third, **projection fibers** are neurons that send fibers caudal to the brain stem and spinal cord.

Pathway of vision—menace response (Fig. 3.11-7)

The visual pathway starts with light as its stimulus. As light enters the eye, **photoreceptors** in the retina convert the stimulus into an electrical potential. There, the potential is transmitted from retinal photoreceptors to bipolar neurons, which form the large **optic nerve** (cranial nerve II). The optic nerve enters the skull through the **optic canal** and joins with the contralateral CN II to form the **optic chiasm**. At the optic chiasm, neurons either remain ipsilateral or decussate (cross) to relay through the contralateral pathway.

> The degree of decussation is species-specific. Species requiring greater field of view, such as prey animals—e.g., cattle and horses—generally have their eyes more lateral on their face and a greater degree of decussation. Species requiring more depth perception or stereopsis—i.e., predators such as cats—have eyes closer to the centerline of their face and less decussation, facilitating their binocular vision.

Caudal to the chiasm, axons project through the **optic tracts** and synapse on neurons within the **lateral geniculate nucleus** of the thalamus. Neurons from these nuclei then project to the **visual cortex** in the occipital and temporal

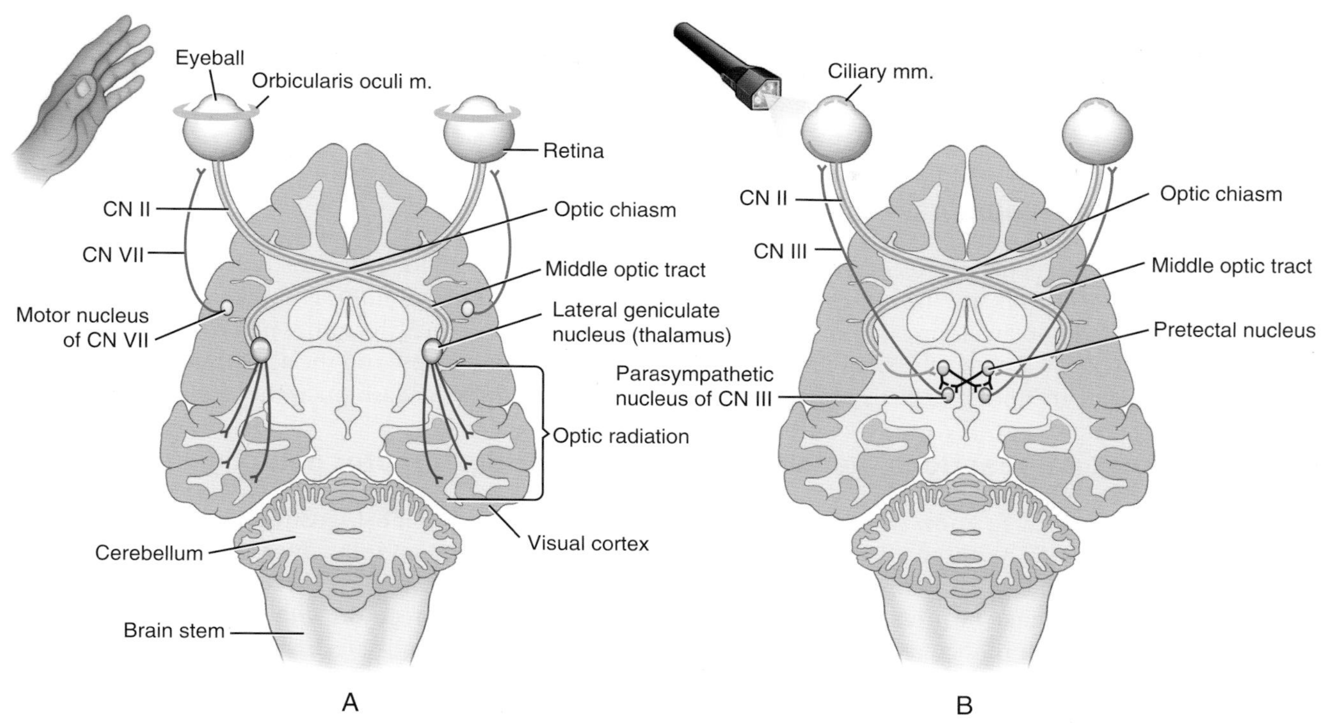

FIGURE 3.11-7 The pathway for menace (vision) (A) and the pupillary light reflex (PLR) (B) share many similarities, specifically the "prechiasmic components." They diverge at this point, with the PLR afferent continuing to the pretectal sensory nucleus. Connections to the motor component of the reflex are in the parasympathetic nucleus of CN III. Cranial nerve III innervates the ciliary ganglion adjacent to the eye, which synapses on the postganglionic neurons that innervate the iris, constricting the pupil. The menace response instead has the afferent neurons synapse on nuclei within the lateral geniculate nucleus of the thalamus. Higher order neurons reach the visual cortex via optic radiations of the white mater. From the visual cortex, association fibers innervate the motor cortex and descending fibers from it synapse on the motor nucleus of cranial nerve VII. Cranial nerve VII innervates the periocular muscles, facilitating eyelid closure. If both the menace and PLR are affected, the lesion is generally considered "prechiasmic." If the menace is affected but the PLR is normal, the lesion is often termed "postchiasmic."

lobes. Association fibers innervate the motor cortex with projection fibers then synapsing on the motor nucleus of the **facial nerve** (cranial nerve VII). Cranial nerve VII forms at the level of the pons and exits the skull through the **stylomastoid foramen.** It innervates the **orbicularis oculi muscles**, facilitating closure of the palpebral fissure.

The menace response also incorporates a relay through the **cerebellum** contralateral to the visual cortex from which it is perceived.

> Therefore, a menace deficit OS could be caused by a right forebrain lesion or left cerebellar lesion. The distinction between the 2 is regarding visual function. The cat with the cerebral lesion has no menace because of absent vision as opposed to the cerebellar lesion, which is related to the blink response itself. Therefore, if the patient is blindfolded over the right eye, it is unable to navigate an obstacle course with a cerebral but not a cerebellar lesion.

Pathway of vision—pupillary light reflex (Fig. 3.11-7)

The **optic chiasm** is considered a clinically relevant structure in that abnormalities of vision and pupils are often labeled as "pre-" or "postchiasmic." This is because the optic chiasm is the neuroanatomic site at which alternate nonvisual pathways deviate from the previously described pathway of the menace response. An example of this is the pathway of the PLR. Instead of following the pathway of

> Cases that have a dysfunctional PLR and menace deficits are considered prechiasmatic, whereas cases that have an intact PLR with a dysfunctional menace are postchiasmatic.

the optic tract to the lateral geniculate nucleus, the fibers synapse on the ipsi- and contralateral **sensory pretectal nucleus**. This sensory nucleus has multiple connections with the parasympathetic nucleus of the **oculomotor nerve** (cranial nerve III). Cranial nerve III exits the brain at the level of the **rostral colliculus** through the orbital fissure of the skull. There, it synapses on the **ciliary ganglion** and then the second-order neuron synapses on the ciliary muscles, which cause the pupil to constrict.

Pathway of vision—palpebral reflex

To ensure that an absent menace is not caused by peripheral motor dysfunction, the palpebral reflex is used to confirm the proper function of CN VII and the muscles it innervates. When a mechanical stimulus is applied to the medial canthus, it is converted into an electrical impulse by the ophthalmic branch of the **trigeminal nerve** (CN V). The ophthalmic branch of CN V enters the skull through the **orbital fissure** and synapses on the **sensory nucleus of CN V**. Interneurons stimulate the motor nucleus of CN VII and cause a blink in the same fashion as the menace response.

Pathway of postural reactions/general proprioception/upper motor neurons

Proprioception, the "ability to know where the legs are in space," is perhaps the most sensitive and sophisticated sensory system in the body. Without visual confirmation, the degree of tension and stretch of the muscles and joints send constant relays to the somatosensory cortex, which in turn facilitates descending motor signals from the motor cortex to the muscles to make near-instantaneous movements. The speed and precision with which this occurs is extremely rapid and is seen in the grace with which animals move. As proprioceptive testing requires both sensory perception and motor corrections, both the ascending and descending tracts are assessed at once.

To assess proprioception, postural reaction testing (specifically knuckling and hopping testing) is performed. The impulses originate when a limb is placed in an abnormal knuckled position, stretching the muscle spindles and Golgi tendon organs. The stimulates the dorsal and ventral spinocerebellar tracts of the pelvic limbs and the cuneocerebellar and cranial spinocerebellar tracts in the thoracic limbs.

Proprioceptive receptors-neuromuscular spindles (Fig. 3.11-8)

The proprioceptive mechanoreceptors have dendritic zones that lie within the muscle spindles, Golgi tendon organs, and joints. The afferent axons are large, myelinated axons that rapidly transmit action potentials to the cell body located in the dorsal root ganglion. (**Ganglions** are accumulations of cells outside the CNS as opposed to **nuclei,** which are groups of neurons within CNS tissue.)

Pelvic limbs—dorsal spinocerebellar tracts (DSCT)

A first-order neuron carrying proprioceptive information from the dorsal root ganglion synapses on a cell body in the dorsal horn of the gray matter (thoracic or Clarke's nucleus). A second-order neuron projects through the DSCT in the dorsal funiculus on the ipsilateral side. Fibers can project to the cerebellum through the caudal (inferior) cerebellar peduncle or the nucleus gracilis.

Pelvic limbs—ventral spinocerebellar tracts (VSCT)

Tension on the Golgi tendon organ is transduced into an electrical signal that ascends the peripheral nerve into its cell body in the dorsal root ganglion. Like the DSCT, the neuron synapses on a cell body in the dorsal horn and then immediately decussates to run in the VSCT in the ventral funiculus. It enters the rostral (superior) cerebellar peduncle where it decussates for a second time.

Thoracic limbs

The **cuneocerebellar tract** is identical to the DSCT except that it originates from the thoracic limb and the first-order neuron synapses on the lateral cuneate nucleus.

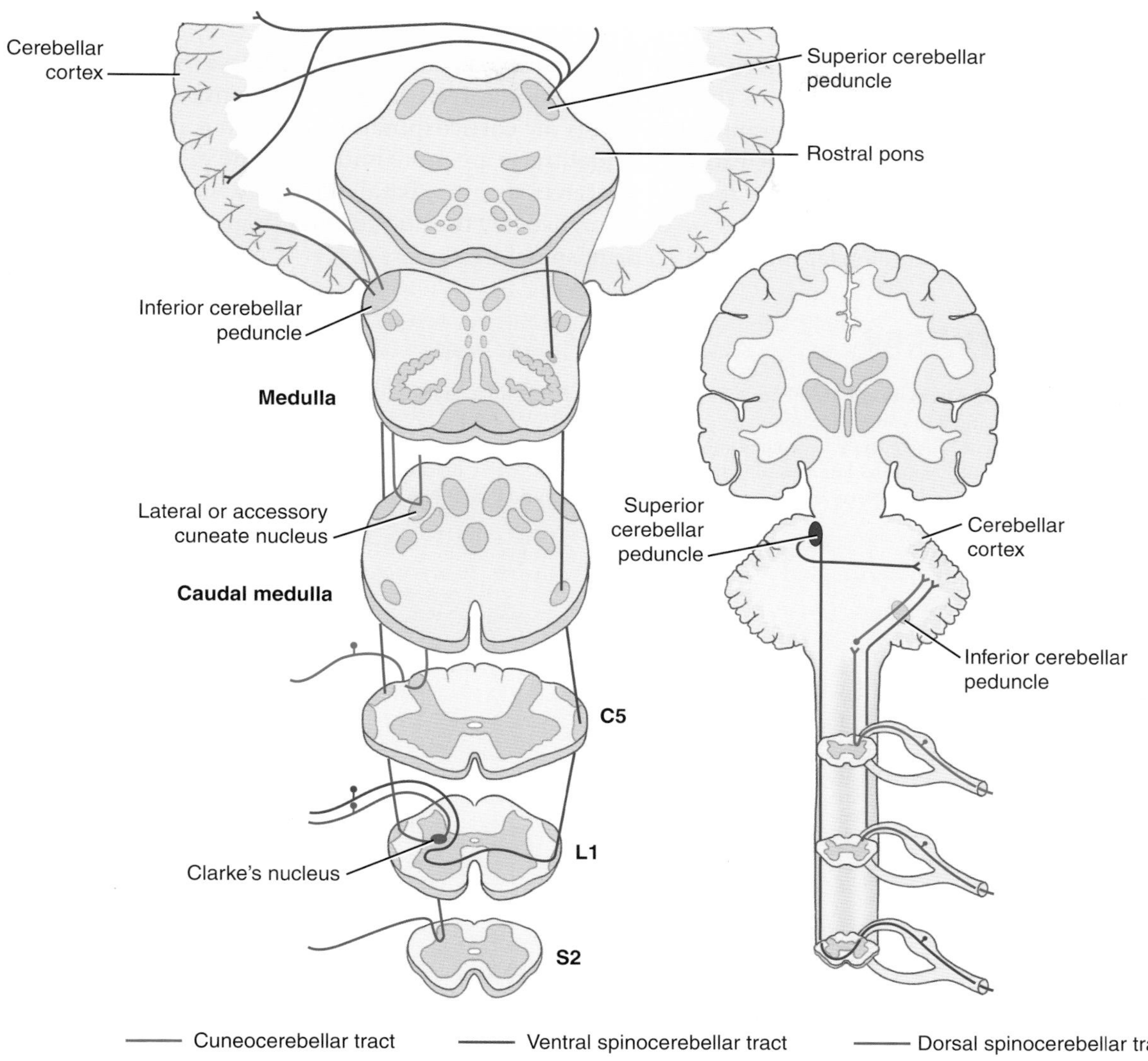

FIGURE 3.11-8 Proprioceptive receptors are found in all muscles and joints of the limbs. Afferent action potentials reach the cell bodies of the dorsal root ganglion found at each spinal segment via the peripheral nerves. Axons from the dorsal root ganglion enter the spinal cord and ascend through the different proprioceptive pathways. The dorsal and ventral spinocerebellar tracts are specific to the pelvic limbs and the cuneocerebellar and cranial spinocerebellar tracts are specific to the thoracic limbs. Proprioceptive signals reach both the cerebellum and cerebral cortex via interneurons that join the dorsal column-medial lemniscal tract.

The **cranial spinocerebellar tract** is more like the VSCT in that its primary receptor is the Golgi tendon organ and it runs in the ventral funiculus. Upon entering the spinal canal, however, it does not decussate like the VSCT does and it enters the cerebellum through the rostral (superior) and caudal (inferior) peduncles.

Though these proprioceptive tracts primarily project to the cerebellum as a subcortical sensory system, there are direct projections to the cerebrum. Collaterals from the DSCT and cuneocerebellar tracts synapse on the nucleus gracilis and cuneatus, respectively, along with tracts from the dorsal column. Neurons from these nuclei enter the medial lemniscus, projecting to the ventral thalamic nuclei and subsequently the somatosensory cortex. Anatomically, most proprioception destined for the somatosensory cortex decussates at the level of medulla, accounting for the contralateral deficits with prosencephalic lesions (see also Cases 3.10 and 3.12).

Motor pathway

Once the proprioceptive information has been perceived by the somatosensory cortex, relays to the motor cortex are necessary for the correction of limb position. The descending motor tracts are often differentiated into pyramidal vs. extrapyramidal systems. The primary **pyramidal system** is the **corticospinal tract**, which is thought to be responsible for finer motor movements. The **extrapyramidal tracts** are more numerous and arise from several nuclei throughout the entire brain, including those in the cerebral cortex, basal nuclei, red nucleus, reticular nuclei, and vestibular nuclei. The names of these tracts are often prefixed by the nucleus they originate from—e.g., red nucleus = rubrospinal tract and reticular formation = reticulospinal tract.

Selected references

[1] Bagley R. Fundamentals of veterinary clinical neuroanatomy. Blackwell Publishing; 2005. p. 100–4.
[2] Platt S, Olby N. BSAVA manual of canine and feline neurology. 4th ed. BSAVA; 2014. p. 27–8.
[3] De Lahunta A, Glass R, Kent M. Veterinary neuroanatomy and clinical neurology. 4th ed. Elsevier; 2015. p. 189, 233–234.
[4] Dewey C, da Costa R. Practical guide to canine and feline neurology. 3rd ed. Wiley Blackwell; 2016. p. 55–57 and 101–102.
[5] Lorenz M, Coates J, Kent M. Handbook of veterinary neurology. 5th ed. Saunders; 2011. p. 88–92.

CASE 3.12

Lumbar Intervertebral Disc Disease

Marc Kent[a] and Eric N. Glass[b]
[a]Department of Small Animal Medicine and Surgery, University of Georgia College of Veterinary Medicine, Athens, Georgia, US
[b]Neurology and Neurosurgery, Red Bank Veterinary Hospital, Tinton Falls, New Jersey, US

Clinical case

History

A 5-year-old male castrated Dachshund was presented with a chief complaint of pelvic limb weakness. Two days before presentation, the owner noticed that the dog was reluctant to jump onto furniture. When picked up, he would cry out. The following day, the owner noticed that the dog's pelvic limbs were weak and appeared wobbly and uncoordinated as he walked. The owner also could hear him scuffing the toenails of his pelvic limbs.

Physical examination findings

Physical examination was normal.

Neurological examination findings

Mentation

- Normal

Gait/posture

- The gait was characterized by a general proprioceptive (GP) ataxia and upper motor neuron (UMN) paraparesis—i.e., pelvic limb weakness
- The dog displayed kyphosis (Gr. *kyphōsis* humpback) at the thoracolumbar vertebral junction
- The dog had difficulty standing up on his pelvic limbs from a sitting or lying position
- When walking, there was a delay in the onset of protraction (swing phase of the gait) of the pelvic limbs; during protraction, the dog scuffed his nails, occasionally the foot knuckled over, and he stood on the dorsum of the digits; his pelvic limbs frequently crossed midline, which made the dog stumble and fall; the length of the strides in the pelvic limbs was slightly longer than the length of the strides of the thoracic limbs; the pelvic limbs appeared stiff as the dog walked

Postural reactions

- Normal hopping and proprioceptive placing in the thoracic limbs; delayed to absent hopping and proprioceptive placing in the pelvic limbs; the right pelvic limb was more affected than the left

Spinal reflexes/muscle tone/muscle mass/nociception

- Thoracic limbs—normal withdrawal reflex, muscle tone, and muscle mass

Comparative Veterinary Anatomy: A Clinical Approach. https://doi.org/10.1016/B978-0-323-91015-6.00022-4

- Pelvic limbs—hyper-reflexic (exaggerated) patellar reflex, normal withdrawal reflex, increased muscle tone, and normal muscle mass, bilaterally
- Normal perineal reflex, normal tone of the external anal sphincter, normal tail tone
- The cutaneous trunci reflex was absent caudal to the L1 vertebra, bilaterally
- Nociception was not assessed in the pelvic limbs given that the dog retained voluntary motor function

Cranial nerves

- All cranial nerve reflexes and responses were normal

Sensory

- The dog was painful when pressure was applied to the spinous processes of the L1 and L2 vertebrae

Neuroanatomic diagnosis

A T3–L3 myelopathy; likely secondary to a focal lesion near the L1–L2 vertebrae given the region of discomfort and where the cutaneous trunci reflex stopped

Differential diagnosis

Intervertebral disc herniation; meningomyelitis (either infectious or immune-mediated); discospondylitis with secondary epidural empyema; neoplasia (primary or metastatic); traumatic injury (vertebral fracture, subluxation, or luxation)

Diagnostics

Complete blood count, serum biochemical profile, and urinalysis were performed to screen for a concurrent disorder that would either impact or preclude general anesthesia. These results were within normal limits and therefore, an MRI of the thoracolumbar vertebral column from the T1 vertebra through to the sacrum was performed under general anesthesia.

On MRI, there was a single, right-sided, well-circumscribed, extradural lesion centered over the L1–L2 intervertebral disc space resulting in severe spinal cord compression (Figs. 3.12-1 and 3.12-2). The lesion was hypointense (signal void) on T1-weighted and T2-weighted sequences. Abnormal contrast enhancement was not observed.

Diagnosis

L1–L2 intervertebral disc herniation with secondary spinal cord compression

SPINAL CORD IMAGING MODALITIES

The gold standard imaging modality for evaluation of the vertebral column and spinal cord is MRI. Alternative imaging modalities include CT, CT combined with myelography (iodinated contrast medium injected into the subarachnoid space), myelography alone, and plain radiographs. While CT has excellent spatial resolution (the ability to discern 2 objects positioned close together as separate), it has less contrast resolution (ability to discern separate tissue types) than MRI. Therefore, it may be difficult to identify lesions that affect the spinal cord unless the lesion has a different density (attenuation) than the spinal cord.

In chondrodysplastic dogs, there is frequently mineralization of the herniated, degenerative nucleus pulposus. Consequently, mineralized disc material herniated into the vertebral canal can be identified with CT. Combined CT/myelography improves the ability to identify spinal cord compression.

Myelography and plain radiographs suffer from contrast resolution and may be challenging to perform and interpret. A definitive diagnosis of intervertebral disc herniation is rarely made with plain radiography; additionally, plain radiography cannot detect spinal cord compression. Plain radiography is generally discouraged by neurologists unless boney lysis or trauma to the vertebral column is strongly suspected, because it is rarely useful in making a definitive diagnosis in cases of intervertebral disc disease, requires further restraint of the patient, and adds additional expense for the owner.

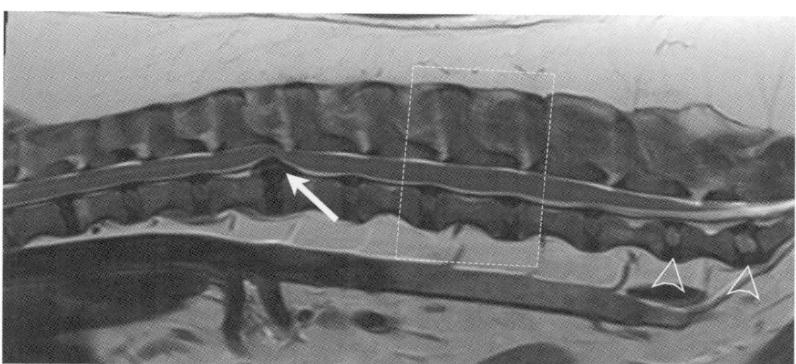

FIGURE 3.12-1 A sagittal T2-weighted MRI of the vertebral column from the T11 vertebra to the cranial aspect of the sacrum (cranial is to the left). The L1–L2 intervertebral disc is herniated *(arrow)* into the ventral epidural space resulting in dorsal deviation of the ventral subarachnoid space and severe compression of the spinal cord. With the exception of the L6–L7 and lumbosacral intervertebral discs, there is degeneration of the nucleus pulposus of the other intervertebral discs as evidenced by the loss of their normal high signal *(white)* intensity. Compare the normal high signal intensity of the nucleus pulposus of the L6–L7 and lumbosacral intervertebral discs *(arrowheads)* with the loss of signal intensity of the nucleus pulposus of the other intervertebral discs. The box outlines a section of normal anatomy that is annotated in Fig. 3.12-2.

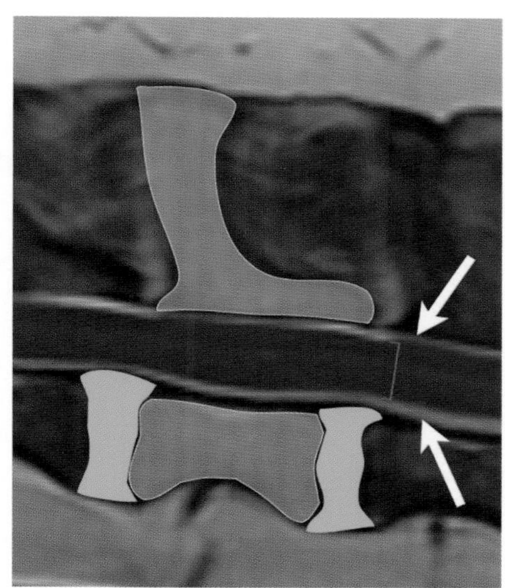

FIGURE 3.12-2 Outlined in *yellow* is the body of the vertebra (L4) ventrally and the spinous process and lamina dorsally. The *red bracket* denotes the dorsoventral dimension of the spinal cord. In T2-weighted images, the signal intensity of the epidural fat is the same as that of the cerebrospinal fluid. Therefore, the high signal intensity *(white)* dorsal and ventral to the spinal cord represents the subarachnoid space and epidural fat *(arrows)*. The intervertebral disc cranial and caudal to the highlighted vertebra is outlined in *blue*. These intervertebral discs show evidence of degeneration of their nucleus pulposus given the loss of signal intensity. Compare these intervertebral discs with normal L6–L7 intervertebral disc in Fig. 3.12-1.

Treatment

The dog underwent a right-sided hemilaminectomy (Gr. *hēmi-* half + L. *lamina* layer + Gr. *ektomē* excision) under general anesthesia to remove the herniated intervertebral disc material that was compressing the spinal cord. The prognosis for return of function following hemilaminectomy was excellent. The dog recovered from surgery with improved strength and gait over the course of the subsequent 5 days. Two weeks post-operatively, the dog had regained normal strength and gait.

Anatomical features in canids and felids

Introduction

The goal of the neurological examination is to determine the anatomical location of a lesion in the nervous system (central or peripheral nervous system) that would result in the deficits identified on examination. The spinal cord has 3 main functions that are assessed in order to make a neuroanatomic diagnosis: strength, coordination, and spinal reflexes. These functions are evaluated through visual assessment of the gait—i.e., can the patient walk; evaluation of postural reactions; and assessment of spinal reflexes, muscular tone, and presence of muscle atrophy. The ability to walk with normal strength and coordination is provided by 2 systems: (1) voluntary motor—i.e., upper motor neuron (UMN) and lower motor neuron (LMN) and (2) general proprioception (GP). There are no significant anatomical differences between the dog and the cat.

Function

An important function of the spinal cord involves the motor system, which is involved in both the generation of the gait and strength. Paresis (weakness) can be defined as an inability to generate a gait as well as strength—i.e., the ability to support the body's weight when standing. Paralysis is defined as an absence of voluntary movement. The suffix -plegia (Gr. *plēgē* paralysis, a blow, stroke) denotes an absence of voluntary movement.

The important UMN tracts in nonprimate species involved in walking include the rubrospinal tracts (red nucleus), reticulospinal tracts, and vestibulospinal tracts. The corticospinal tracts (originating in the motor cortex of the cerebrum and projecting caudally through the brain via the internal capsule, crus cerebri [midbrain], longitudinal fibers of the pons, and pyramids of the medulla) are referred to as the pyramidal (L. *pyramidalis* shaped like a pyramid) system and have little to no clinical influence on the generation and maintenance of a normal gait in nonprimates. Therefore, with both experimentally and naturally occurring diseases of the motor cortex (e.g., such as neoplasms or granulomas), the nonprimate has a normal gait. The exception to this is with peracute insults to the cerebrum or thalamus—e.g., that which occurs with strokes (hemorrhagic or ischemic) and traumatic injuries where an obvious paresis is observed initially. In peracute insults, the gait returns to normal within 2–3 days. In contrast, primates will have an abnormal gait with lesions of the motor cortex of the cerebrum.

Upper and lower motor neurons

Upper motor neurons are a collection of neurons distributed throughout the brain—i.e., motor cortex of the cerebrum and brainstem nuclei. The caudally projecting axons from these nuclei comprise the UMN pathways that include the corticospinal, rubrospinal, reticulospinal, and vestibulospinal tracts, all of which project in the white matter of the lateral and ventral funiculi of the spinal cord.

LOWER MOTOR NEURON DYSFUNCTION

A lesion affecting the LMN (also known as an alpha motor neuron) results in LMN quality paresis. The same LMN quality paresis is present if any part of the LMN unit is affected. Said another way, the same examination findings of LMN paresis are present whether the lesion affects the LMN cell bodies, roots, nerves, neuromuscular junctions, or the muscles themselves. The clinical relevance is that it is often challenging to determine what part of the LMN unit is affected in a patient with LMN quality weakness using the clinical examination alone. Electrophysiological testing such as electromyography and nerve conduction studies, along with measurement of serum creatinine kinase and other specialized blood tests, may aid in determining what part of the LMN unit is affected.

Lower motor neurons (LMN, or **general somatic efferent [GSE] neurons**) are located in the ventral gray column of the spinal cord. The axons of these GSE neurons form the named nerves, which innervate striated skeletal muscles. The size of the ventral gray matter at each spinal cord segment depends on the number of muscles innervated by that segment. Consequently, the size of the ventral gray column that gives rise to the nerves that innervate the thoracic and pelvic limbs is large relative to other spinal cord segments, which results in 2 grossly visible enlargements of the spinal cord called the cervical and lumbar intumescence, respectively. The **cervical intumescence** (L. *intumescentia* a normal or abnormal swelling) is located from the C6 through the T2 spinal cord segments, whereas the **lumbar intumescence** is located from the L4 through S1 spinal cord segments. It is important to note that the specific spinal cord segments do not necessarily reside in the exact same numbered vertebra along the vertebral column (see "Spinal segments" section later in this case). At each spinal cord segment, GSE axons emerge ventrolaterally to form the **ventral spinal roots**. A ventral spinal root joins with its paired **dorsal root** to form a **spinal nerve**. At the cervical and lumbar intumescences, spinal nerves exit their respective **intervertebral foramina** and coalesce to form the **brachial** and **lumbosacral plexus**, respectively. The named nerves of the limbs—e.g., radial nerve and femoral nerve—are formed out of these plexi.

A lesion within the cervical or lumbar intumescence results in paresis or paralysis of the thoracic or pelvic limb(s), respectively. The quality of the resultant weakness is termed LMN paresis/paralysis (also referred to as neuromuscular paresis), which is characterized by the patient having difficulty or being unable to support its weight when standing and walking. The gait and posture are characterized by a shortened stride length and a crouched stance, reduced or absent spinal reflexes, and flaccid (reduced) muscular tone to the affected limb. In short time, severe muscular atrophy develops in the affected limb due to a loss of innervation (denervation atrophy). For more information, see Cases 7.5 and 8.6.

Voluntary movement as it relates to clinical anatomy

The initiation of voluntary movement is complex and requires an integration of multiple areas of the brain including the cerebral cortex, basal nuclei, brainstem nuclei, and the cerebellum. However, simplistically, voluntary movement is initiated by UMNs that send impulses via descending UMN tracts to directly or indirectly, via interneurons, synapse on GSE neurons. In turn, impulses initiated by GSE neurons project via nerves to synapse on a skeletal muscle via a **neuromuscular junction**. Acetylcholine is thus released at the nerve terminus, diffuses across the neuromuscular junction to bind to postsynaptic acetylcholine receptors, and results in depolarization of the sarcolemma of the myofiber, culminating in the activation of actin and myosin and ultimately a muscle contraction.

Proprioception as it relates to clinical anatomy

Changes in posture—i.e., limb position—are detected by **proprioceptors**—receptors invested in muscles, tendons, and joint capsules. Proprioceptors detect the position of the limbs and body in space. General proprioceptive (GP) information is conducted along sensory nerves proximally to spinal nerves and ultimately enters the dorsolateral sulcus of the spinal cord along dorsal roots and rootlets. Within the spinal cord, GP information is conducted cranially in the ipsilateral dorsal funiculus (fasciculus gracilis [pelvic limbs] and fasciculus cuneatus [thoracic limbs]) and dorsal aspect of the lateral funiculus (spinocerebellar tracts) to ultimately project to the contralateral somesthetic cerebral cortex and ipsilateral cerebellum, respectively.

Spinal segments

Anatomically, the spinal cord can be divided into segments. Each segment is named according to a correspondingly numbered vertebra with the exception of the cervical spinal cord segments. The C1 and C2 spinal cord segments reside within the vertebral canal of the C1 vertebra. The C1 spinal nerve exits the lateral vertebral foramen of the C1 vertebra. The remaining cervical spinal nerves exit via the intervertebral foramina just *cranial* to the vertebrae of the same number. Consequently, there are 8 cervical spinal cord segments and 7 cervical vertebrae. The C8 spinal nerve exits the C7–T1 intervertebral foramen. The thoracic and lumbar spinal nerves exit the intervertebral foramina just *caudal* to the vertebra of the same number.

Two generalizations arise from the grouping of spinal cord segments into these functional divisions. First, a lesion affecting one spinal cord segment results in deficits that are indistinguishable from a lesion in another spinal cord segment within the same functional division. Likewise, in general, a focal lesion within a functional division results in deficits indistinguishable from a diffuse lesion affecting all the segments of that functional division.

In contrast to a lesion affecting the intumescence, a lesion in the C1–C5 or T3–L3 spinal cord segments disrupts the descending (caudally tracking) UMNs that project to the LMNs at the intumescence(s). Similarly, there is a disruption of the ascending (cranially tracking) GP information. Clinically, GP ataxia and UMN paresis/paralysis are observed.

Clinically, the spinal cord segments can be grouped into 5 functional divisions. The spinal segments are grouped as follows: C1 through C5 spinal cord segments (cervical), C6 through T2 spinal cord segments (the cervical intumescence), T3 through L3 spinal cord segments (thoracolumbar), L4 through S1 spinal cord segments (lumbar intumescence), and S1 through the caudal (Cd) segments (sometimes called sacrocaudal). With a few exceptions, these 5 functional divisions are the most precise neuroanatomic localizations that can be made in the spinal cord in dogs and cats.

Neurological assessment as relates to clinical anatomy

Examination of patients with a suspected or known lesion involving the spinal cord begins with the observation of the gait. Some affected patients may not be able to walk despite being able to move their limbs. Similarly, severely affected patients may be paralyzed—i.e., unable to voluntarily move their limbs. In patients capable of walking, the patient should be observed walking from a side view as well as walking toward and away from the examiner. The gait should be evaluated for lameness, paresis (weakness), and GP ataxia (incoordination).

In some patients, it may be challenging to discern whether neurological dysfunction is present. In these patients, the evaluation of the postural reactions helps determine the presence of a neurological deficit. Postural reactions are tests designed to make neurological dysfunction obvious in patients with subtle gait deficits or in those with a normal gait. In patients in which there is obvious neurological dysfunction (paresis or paralysis), postural reactions may not contribute more in determining an anatomic diagnosis; in some instances, postural reaction testing may help define asymmetrical deficits.

Implied in the name, postural reactions assess the patient's ability to react or correct for changes in posture. For the patient to perform the postural reaction tests normally, both the sensory and motor components of the peripheral and central nervous system—as well as the muscles of the limbs—must be normal (see side box entitled "Assessing postural reactions").

In part, postural reactions assess general proprioception. A common misnomer is to refer to abnormal postural reactions as conscious proprioception deficits or "CP deficits" (conscious proprioception is the GP information that projects to the contralateral somesthetic cortex, whereas unconscious proprioception projects to the ipsilateral cerebellum). Postural reactions not only assess sensory information projecting to the cerebrum (conscious) and cerebellum (unconscious), but also require normal UMN and LMN function as well as normal muscle activity to move the limb.

General proprioception involves knowledge of the position of the limbs and body in space. General proprioceptive information is conveyed to both the ipsilateral cerebellum (unconscious proprioception) and the contralateral somesthetic cerebral cortex (conscious proprioception) (Fig 3.12-3). Clinically, it is impossible to discern the contribution of unconscious vs. conscious proprioception. No clinical test exists that can assess these pathways alone nor is it important to attempt to discern these functions separately. The purpose of the neurological examination is not to identify a deficiency in the specific pathway (conscious vs. unconscious proprioception) but rather to define a location along these pathways where a lesion could exist to cause the observable deficits. Therefore, all that is necessary is to be able to identify deficits in GP function.

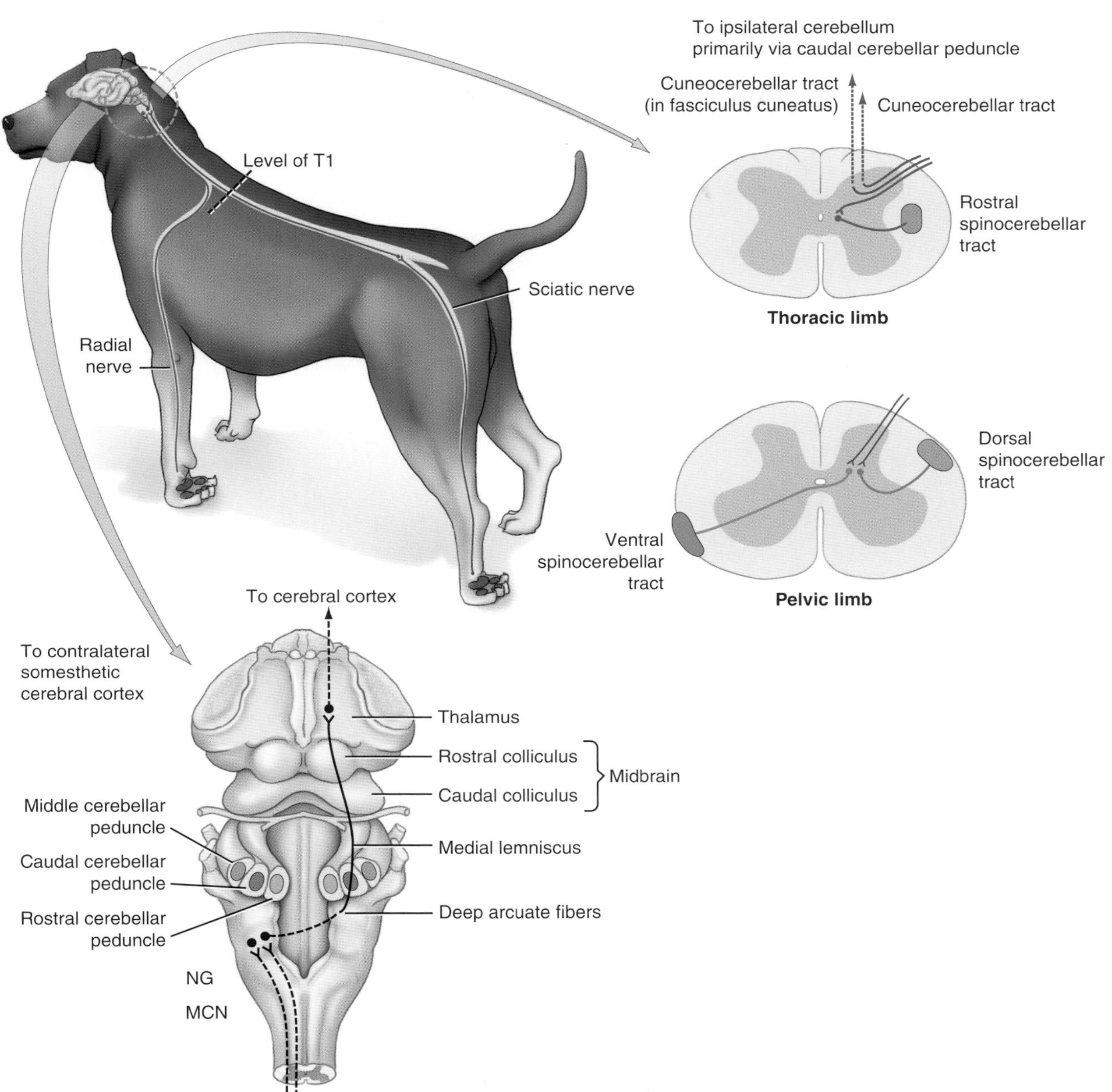

FIGURE 3.12-3 Illustration depicting the general proprioception (GP) tracts. To the right of the dog are the pathways for unconscious proprioception (spinocerebellar and cuneocerebellar tracts). Below the dog are the pathways for conscious proprioception that are conveyed via the fasciculus gracilis (from pelvic limbs) and fasciculus cuneatus (from thoracic limbs) to the nucleus gracilis (NG) and medial cuneate nucleus (MCN), respectively, in the medulla.

ASSESSING POSTURAL REACTIONS (FIG. 3.12-4)

Commonly used postural reaction tests are hopping and proprioceptive placing (also called knuckling or paw placement). Hopping starts with holding the patient in a standing position. With one limb flexed, the patient is slowly pushed/leaned laterally over the limb being tested (the limb being tested is left standing on the ground—i.e., the nonflexed limb). In the normal patient, as the shoulder or hip moves laterally over the paw, the patient should "hop" laterally to keep from falling. A deficit consists of a delay or lack of moving the limb to "catch" the patient from falling.

Proprioceptive placing consists of turning the paw over so that the patient stands on the dorsal surface of the digits. The normal dog or cat rapidly replaces the paw into a normal standing position, whereas a deficit results in the patient displaying a delay or absence in replacing the paw in a normal position.

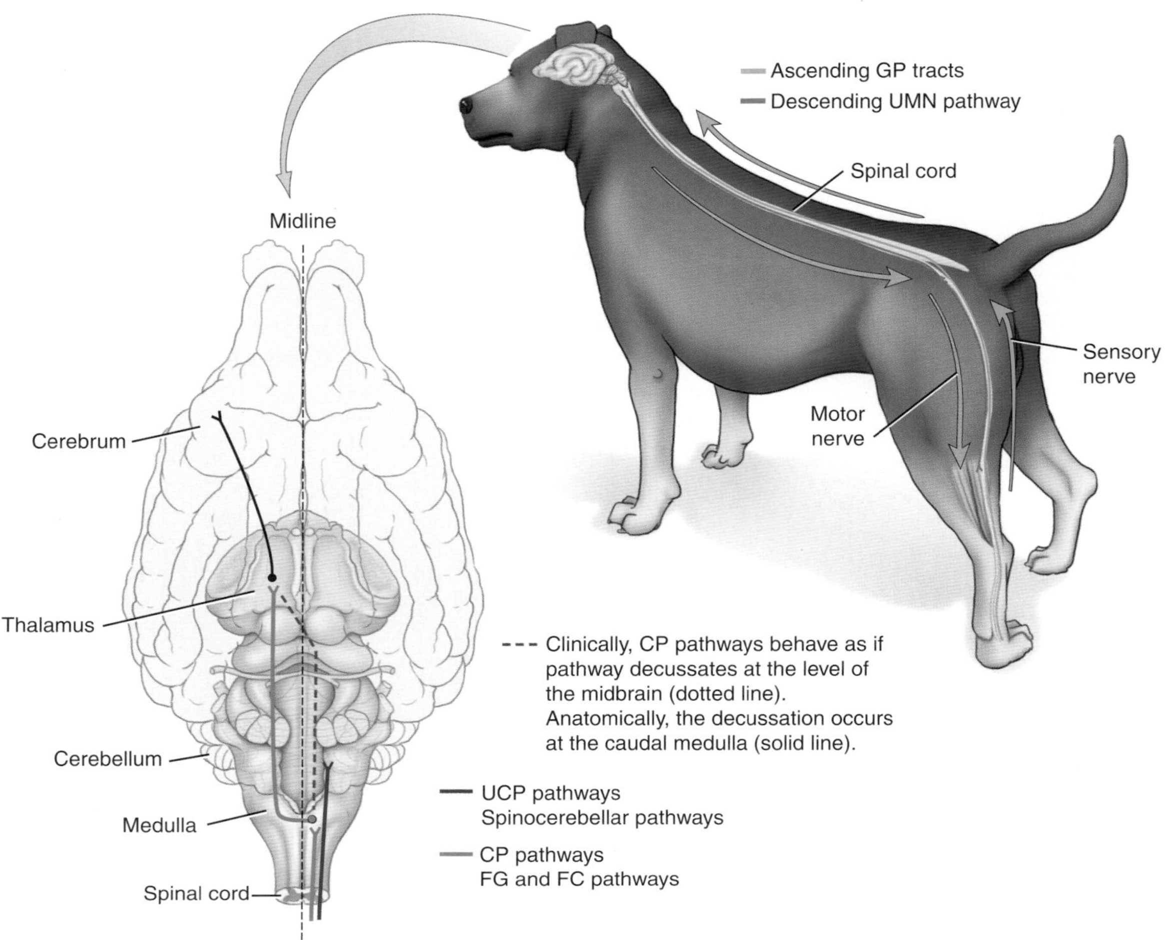

FIGURE 3.12-4 Assessing postural reactions. *GP*, general proprioception; *CP*, conscious proprioception; *UCP*, unconscious proprioception; *UMN*, upper motor neuron; *FG*, fasciculus gracilis; *FC*, fasciculus cuneatus.

The ability to perform the postural reaction tests normally also requires normal UMN and LMN function. Descending UMN tracts from the cerebrum and UMN nuclei in the midbrain, pons, and medulla course in the lateral and ventral funiculi to influence the LMN axons that give rise to the nerves of the limb (Fig. 3.12-5). Therefore, postural reaction testing is a means to evaluate both the GP systems and the motor systems (UMN and LMN).

Conscious proprioception pathways (Figs. 3.12-3 and 3.12-4)

Anatomically, the **fasciculus gracilis** (pelvic limbs) and **fasciculus cuneatus** (thoracic limbs) synapse on the nucleus gracilis and medial cuneate nucleus in the caudal medulla, respectively. From these nuclei, axons decussate (L. *decussare* to cross in the form of an X) via the deep arcuate fibers to track in the contralateral medial lemniscus to synapse in the thalamus. From the thalamus, axons project to the contralateral somesthetic cortex via the internal capsule, the centrum semiovale, and corona radiata. Therefore, from an anatomy perspective, most proprioception destined for the somesthetic cortex decussates at the level of the medulla.

> Clinically, patients display neurological deficits as though the decussation occurs at the level of the midbrain. This translates into the clinical observation that lesions at or caudal to the midbrain result in ipsilateral deficits, whereas lesions affecting the prosencephalon (cerebrum and thalamus) result in contralateral deficits.

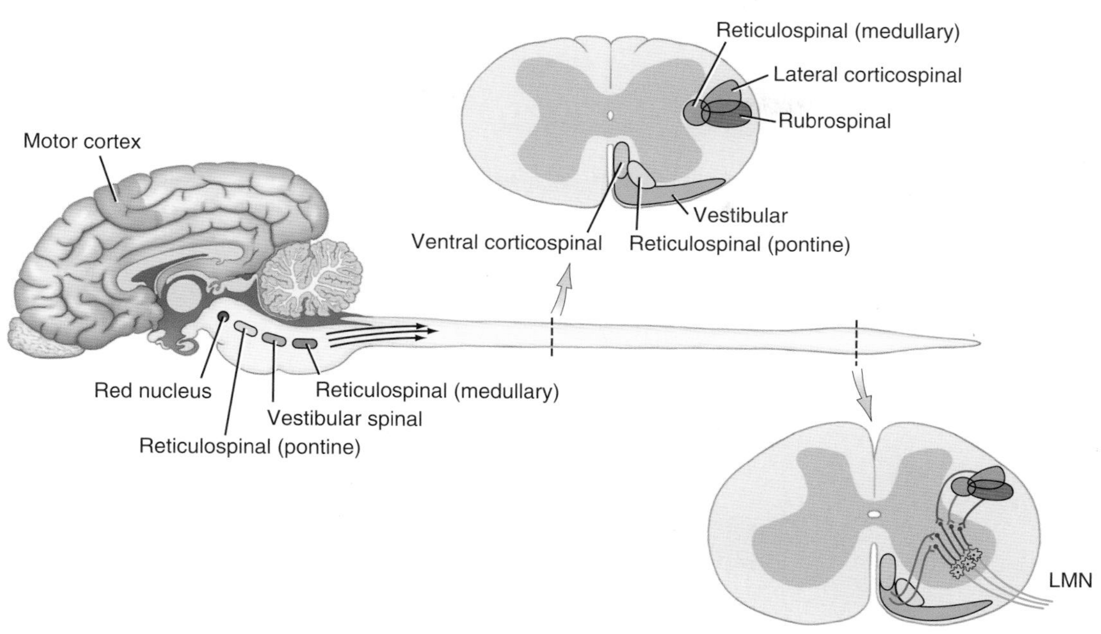

FIGURE 3.12-5 Descending UMN tracts from the cerebrum and UMN nuclei in the midbrain, pons, and medulla course in the lateral and ventral funiculi to influence the LMN axons that give rise to the nerves of the limb. The corticospinal and rubrospinal (from red nucleus) tracts project contralaterally, whereas the remaining tracts function as though ipsilateral to their nuclei.

CUTANEOUS TRUNCI REFLEX (CTR)

The CTR is elicited by pinching the skin approximately 1 cm (0.4 in.) lateral to the dorsal midline on both sides, starting around the iliac wings and continuing along each vertebra until the cranial thoracic vertebrae. The anticipated reaction is for the cutaneous trunci muscle to contract, which causes an observable twitch to the skin on both sides of the trunk. The CTR can reliably be elicited in most normal dogs and cats. The skin along the trunk is segmentally innervated at each spinal cord segment by the dorsal branch of the corresponding spinal nerve. Therefore, at each spinal cord level, the afferent limb of the reflex arc is provided by a dorsal branch of a spinal nerve. The afferent impulses generated by pinching the skin track along the spinal nerve to enter the vertebral column at or just caudal to where the skin is pinched. The impulse travels along the spinal nerve to enter the spinal cord via the dorsal root and synapse on an interneuron in the dorsal horn. These interneurons send their axons cranially in the fasciculus proprius (primarily ipsilaterally but also in the contralateral fasciculus) to the C8 and T1 spinal cord segments. The efferent limb of the arc is provided by the lateral thoracic nerve, which innervates the cutaneous trunci muscle. The lateral thoracic nerve has its cell bodies in the ventral gray column of the C8 and T1 spinal cord segments (Figs. 3.12-6 and 3.12-7). The CTR is sometimes incorrectly referred to as the panniculus reflex. This should be discouraged as the panniculus layer (subcutaneous) is not involved in the movement of the skin.

In most dogs and cats, the CTR can be elicited as caudal as the wings of the ilia. The examination should begin at the wings of the ilia and continued cranially to the cranial thoracic vertebrae. If present at the level of the ilia, there is not a need to examine the CTR more cranial as the entire pathway is normal. If not present, the examiner should progressively attempt to elicit the CTR more cranial until the CTR becomes observable. The point at which the CTR can be elicited marks the approximate level of the lesion.

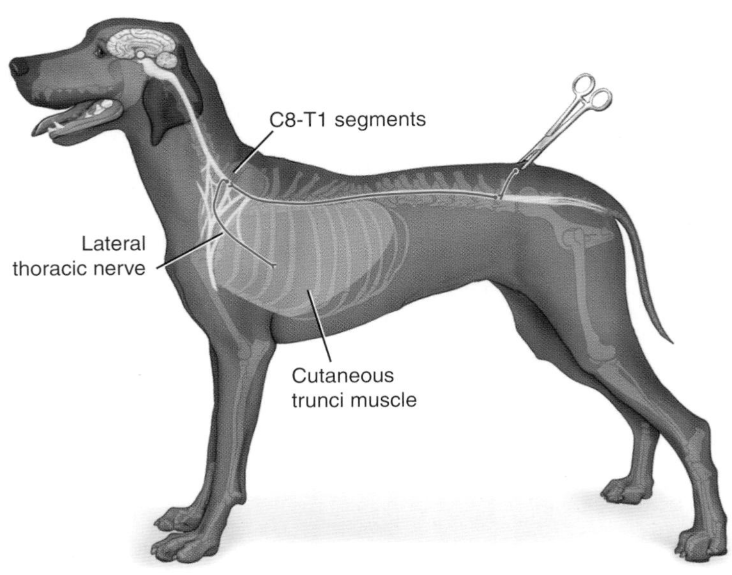

FIGURE 3.12-6 Pathway for cutaneous trunci reflex.

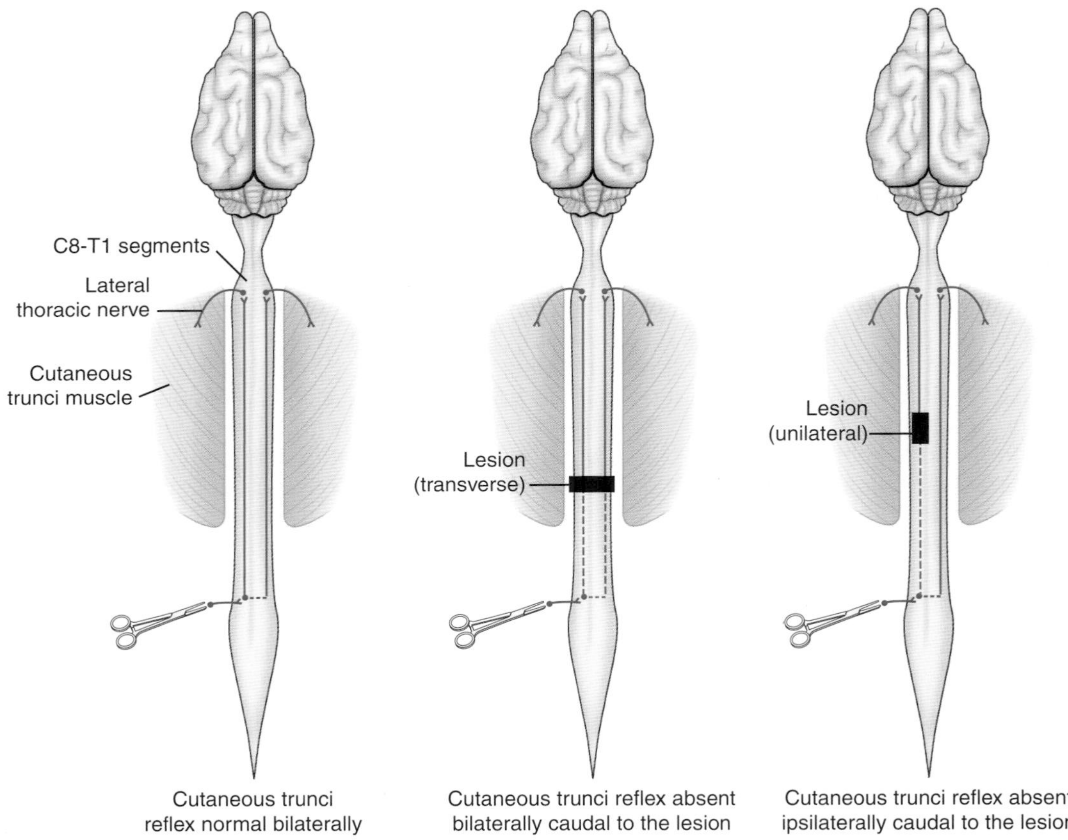

FIGURE 3.12-7 The cutaneous trunci reflex for localizing lesions.

Unconscious proprioceptive pathways (Figs. 3.12-3 and 3.12-4)

For the pelvic limbs, GP axons entering the spinal cord synapse in the dorsal gray column. From the dorsal gray column, some axons project via the ipsilateral dorsal spinocerebellar tracts in the dorsal aspect of the lateral funiculus, whereas others decussate in the ventral commissure of the spinal cord to project in the contralateral ventral spinocerebellar tract. These spinocerebellar tracts ultimately project to the ipsilateral cerebellum. Axons projecting via the ventral spinocerebellar tract ultimately recross within the cerebellum.

For the thoracic limbs, some GP axons enter the spinal cord to pass directly into the ipsilateral fasciculus cuneatus (the cuneocerebellar tract) in the dorsal funiculus, without synapsing, to project to the medulla. At the medulla, the synapse occurs in the lateral cuneate nucleus. Axons from the lateral cuneate nucleus project to the ipsilateral cerebellum. Additionally, some GP axons enter the spinal cord and synapse in the dorsal gray column. From the dorsal gray column, axons enter the ipsilateral funiculus to project in the ipsilateral rostral spinocerebellar tract to the ipsilateral cerebellum.

General proprioceptive ataxia is characterized by a delay in the protraction phase (advancing the limb) of the gait, hypermetria, incoordination, and overreaching as the limb is advanced, which in part contributes to a longer than normal stride length. In part, the long-strided gait also reflects dysfunction of the UMN system. In addition, there also may be scuffing of the toes, knuckling over of the paw, or even standing on the dorsum of the digits. The limbs may cross midline and the patient may frequently stumble and fall. In addition to GP loss, there also is UMN paresis/paralysis, which is characterized by normal to increased spinal reflexes and increased (spastic) muscular tone. The spastic tone gives the appearance of a stiff, long-strided gait.

Neuroanatomical localization as relates to clinical anatomy

Ultimately, defining a neuroanatomic diagnosis for a lesion affecting the spinal cord is determined by identifying LMN paresis or GP ataxia/UMN paresis. Lesions affecting the C1–C5 or C6–T2 spinal cord segments result in tetraparesis/paralysis. Lesions affecting the C1–C5 spinal cord segments result in GP ataxia/UMN paresis or paralysis in both the thoracic and pelvic limbs. Lesions affecting the C6–T2 spinal cord segments result in LMN deficits to the thoracic limbs and GP ataxia/UMN paresis or paralysis to the pelvic limbs.

Lesions affecting the T3–L3 spinal cord segments or the L4–S1 spinal cord segments result in paraparesis. Lesions affecting the T3–L3 spinal cord segments result in GP ataxia/UMN paresis in pelvic limbs. Lesions affecting the L4–S1 spinal cord segments result in LMN deficits to the pelvic limbs. Unilateral lesions of the spinal cord affect the limbs ipsilateral to the lesion.

Finally, lesions affecting S1 through the Cd segments affect bowel and bladder function, as well as the ability to move the tail. The S1 through the S3 spinal cord segments contain the LMN cell bodies that give rise to the pudendal and pelvic nerves. Lesions affecting the S1–S3 spinal cord segments, the S1–S3 spinal nerves, or the pudendal and pelvic nerves result in LMN paresis of the external urethral sphincter and external anal sphincter resulting in decreased sphincter tone and, therefore, LMN urinary and fecal incontinence. Lesions affecting the caudal spinal cord segments, spinal nerves, or caudal nerves result in LMN paresis of the tail.

Within the T3–L3 spinal cord segments, it is sometimes possible to narrow the site of the lesion within a few spinal cord segments. Intuitively, the location where the dog or cat is painful when pressure is applied over a spinous process is likely to be the site of the lesion. In patients with severe spinal cord lesions that result in a lack of nociception (sensation) to the limbs, the site along the thoracolumbar vertebral column where normal sensation to the skin changes to analgesia marks the approximate location of the lesion. In less severely affected dogs and cats, the cutaneous trunci reflex may be used similarly (see side box entitled "Cutaneous trunci reflex").

Determining lesion severity as it relates to clinical anatomy

Separate from localizing a lesion to a specific neuroanatomic region of the spinal cord, the neurological examination findings can help define the severity of the lesion. As mentioned, the primary goal of the neurological examination is to define where in the nervous system a lesion is located, and not which tracts are affected. However,

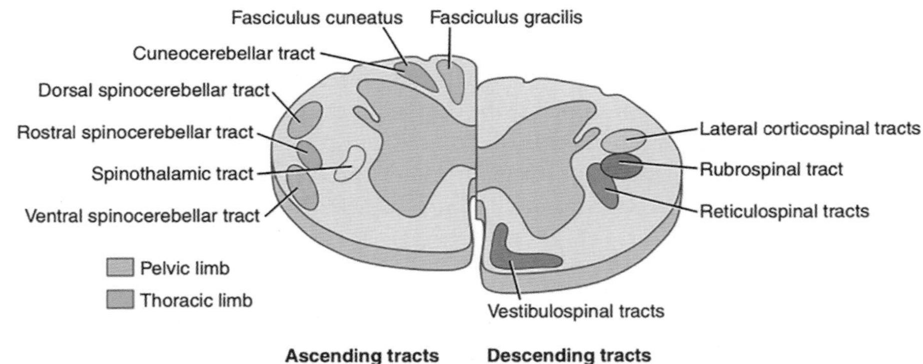

FIGURE 3.12-8 A cross section of the spinal cord showing the location of the tracts.

understanding the anatomic location of the tracts helps explain the severity of the spinal cord dysfunction. As a rule, with increasing compression of the spinal cord by an extradural lesion—i.e., lesions located external to the spinal cord parenchyma and meninges, such as a herniated intervertebral disc—there is a predictable pattern of increasing loss of neurologic function based on the locations of tracts in the spinal cord. Intuitively, patients unable to walk are more severely affected than those that can walk. Likewise, the first axons to experience compression by an extradural lesion are those located superficially in the spinal cord. Additionally, large-diameter, heavily myelinated axons are more susceptible to compression than smaller diameter, less myelinated axons. Small-diameter, unmyelinated axons are most resistant to compression.

With the knowledge of the location of GP tracts, UMN tracts, and pathways for nociception, the amount of spinal cord compression can be estimated based on neurological deficits (Fig. 3.12-8). The most superficial tracts in the spinal cord are those that conduct GP information. These axons are in the dorsal funiculus and dorsal aspect of the lateral funiculus. These also are heavily myelinated, large axons that are most susceptible to compression. Therefore, with mild extradural spinal cord compression, patients may only display GP ataxia. In these cases, there is also a degree of UMN paresis given the close anatomic relationship of the GP tracts with the UMN tracts. However, clinically the UMN paresis may not be perceptible. The descending UMN axons are less myelinated, smaller diameter axons that are located deeper in the spinal cord. The descending UMN axons reside in the lateral and ventral funiculi. With greater compression, the UMN tracts become affected and UMN weakness is clinically more obvious. With increasing compression, there is a greater loss of voluntary motor function from ambulatory paresis, to nonambulatory paresis, to paralysis.

Finally, the axons conducting nociceptive information (spinothalamic tracts) are unmyelinated, small-diameter axons that are located deep within the spinal cord parenchyma. With severe compression, there is a loss of nociceptive function. The spinothalamic tracts, which conduct nociceptive information, are a multisynaptic pathway in which axons are distributed bilaterally in all funiculi of the spinal cord. This has a significant clinical implication. In order to affect nociception, the lesion must involve the entire cross-sectional areas of the spinal cord. Such a lesion is referred to as a transverse myelopathy and is of greatest severity and consequence. Depending on the etiology, the loss of nociception often carries a guarded prognosis for return of spinal cord function.

Ultimately, in the context of a compressive extradural lesion, the neurological examination finding may suggest the degree of spinal cord compression. However, exception frequently occurs based on chronicity (dogs and cats with chronic lesions may exhibit less severe clinical signs yet have severe compression of the spinal cord) and etiologies other than extradural compressions, such as with intraparenchymal pathologies. Finally, clinically, it is almost never necessary to distinguish GP ataxia from UMN paresis. It is accepted that both systems are affected with naturally occurring lesions.

Selected references

[1] Langerhuus L, Miles J. Proportion recovery and times to ambulation for non-ambulatory dogs with thoracolumbar disc extrusions treated with hemilaminectomy or conservative treatment: a systematic review and meta-analysis of case-series studies. Vet J 2017;220:7–16.

[2] de Lahunta A, Glass EN, Kent M. Small animal spinal cord disease. In: Veterinary neuroanatomy and clinical neurology. St. Louis, MO: Elsevier Health Sciences; 2014. p. 257–303.

[3] Evans HE, de Lahunta A. Spinal cord and meninges. In: Miller's anatomy of the dog. St. Louis, MO: Elsevier Health Science; 2013. p. 589–610.

CHAPTER 4

THORAX

Helen M.S. Davies, Chapter editor

Pleura, Mediastinum, and Lungs

4.1 Aspiration pneumonia—*Nicholas Bamford, Cathy Beck, and Helen M.S. Davies* 220
4.2 Pyothorax—*Laura Dooley, Cathy Beck, and Helen M.S. Davies* 230
4.3 Diaphragmatic rupture—*Cathy Beck and Helen M.S. Davies* 236

Mediastinal Organs

4.4 Feline cardiomyopathy—*Mark Oyama and Simon Bailey* 243
4.5 Persistent right fourth aortic arch—*David Holt* 251
4.6 Patent ductus arteriosus—*Mark Oyama* 258
4.7 Mitral valve disease—*Laura Dooley, Cathy Beck, and Simon Bailey* 265
4.8 Esophageal foreign body—*David Holt* 272

CHAPTER 4

CASE 4.1

Aspiration Pneumonia

Nicholas Bamford[a], Cathy Beck[b], and Helen M.S. Davies[c]
[a]Veterinary Biosciences, The University of Melbourne, Parkville, Victoria, AU
[b]Veterinary Hospital, The University of Melbourne, Melbourne, Victoria, AU
[c]Veterinary Biosciences, The University of Melbourne, Melbourne, Victoria, AU

Clinical case

History

A 12-year-old male castrated, mixed-breed terrier presented with a 2-day history of lethargy, reduced appetite, and intermittent cough. Three days before the onset of these signs, he had experienced several episodes of vomiting and diarrhea, which had then resolved.

Physical examination findings

The dog was febrile (rectal temperature 104 °F [40 °C]) and exhibited a moist cough on several occasions during the examination. His respiratory rate was increased, 32 brpm, although the depth and rhythm of ventilation were normal. Thoracic auscultation revealed a focal area of abnormal ("adventitious") lung sounds in the region of the right 6th intercostal space, with soft crackles heard during the inspiratory phase. Thoracic percussion over the area produced sounds that were duller than normal, and it induced a short episode of coughing. Auscultation of the remainder of the thorax, including the heart, was normal. Thoracic auscultation and percussion are described at the end of the section on the relevant clinical anatomy of the lower airways.

Differential diagnoses

Pneumonia, pulmonary hemorrhage, pulmonary edema, and pulmonary neoplasia

PNEUMONIA

Pulmo- and pneumo- both refer to the lungs. While "pneumonia" is generally taken to mean infection of the lungs, it simply means inflammation of the lung parenchyma, the cause of which may or may not be a pathogen. "Pulmonary infection" is a more precise term for infectious processes involving the lungs, but "pneumonia" persists as the term used in clinical practice.

"Bronchopneumonia" or "pleuropneumonia" is used when the bronchi (lower airways) or pleura (serosal coverings of the lungs and inner chest wall), respectively, are also involved.

Comparative Veterinary Anatomy: A Clinical Approach. https://doi.org/10.1016/B978-0-323-91015-6.00036-4

Diagnostics

Thoracic radiography revealed a well-demarcated area of homogeneous soft tissue opacity in the region of the right middle lung lobe (Figs. 4.1-1 and 4.1-2). A transtracheal wash (TTW) was performed to characterize the pathologic process. Cytologic analysis of the fluid revealed many degenerate neutrophils, indicative of a purulent exudate. No bacteria were seen, but culture of the sample yielded a mixed growth of bacteria. The recent history of repeated vomiting was consistent with aspiration of stomach contents as the presumptive source of the bacterial infection.

Under light sedation and local anesthesia, a needle was introduced percutaneously into the tracheal lumen on the caudoventral aspect of the neck, just cranial to the thoracic inlet. A catheter was passed through the needle, and a small volume of sterile fluid was instilled into the trachea and then aspirated and submitted for cytologic and microbiologic examination. Alternatives to TTW include peroral bronchoalveolar lavage (BAL) and percutaneous fine-needle aspiration (FNA) of the affected lung lobe. With BAL, an endoscope or a dedicated BAL tube is passed via the oral cavity into the trachea, and passed caudally until it lodges in a primary or secondary bronchus (depending on patient size). As with TTW, sterile fluid is instilled and then aspirated for examination. When an endoscope is used for BAL, the endoscope may be directed into a specific lung lobe, whereas the BAL tube must be passed "blindly." Ultrasound-guided FNA permits precise placement of the needle into the affected lung lobe.

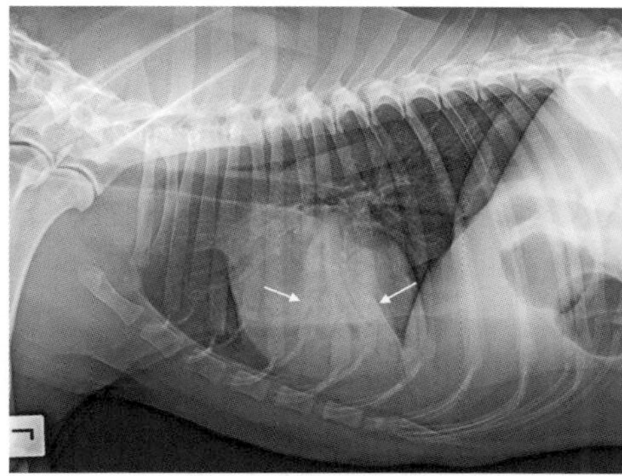

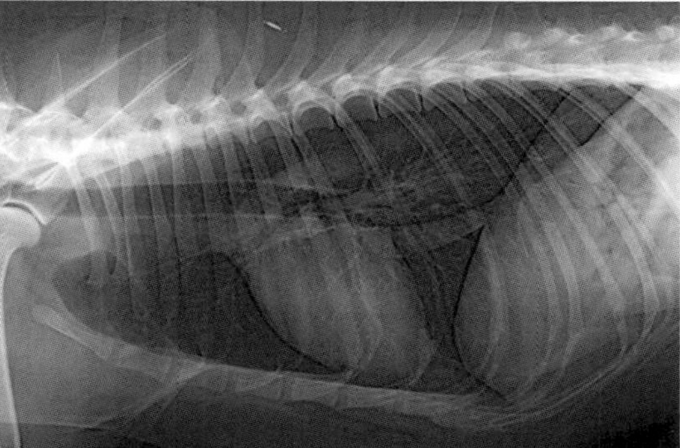

FIGURE 4.1-1 Lateral thoracic radiographs of the dog with pneumonia (left) and, for comparison, a dog with normal lungs (right). In the image on the left, there is opacity within the right middle lung lobe (between the *arrows*) (best seen in the ventrodorsal view in Fig. 4.1-2).

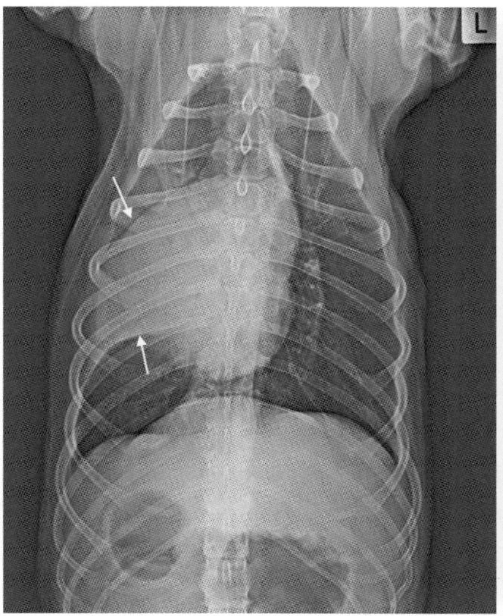

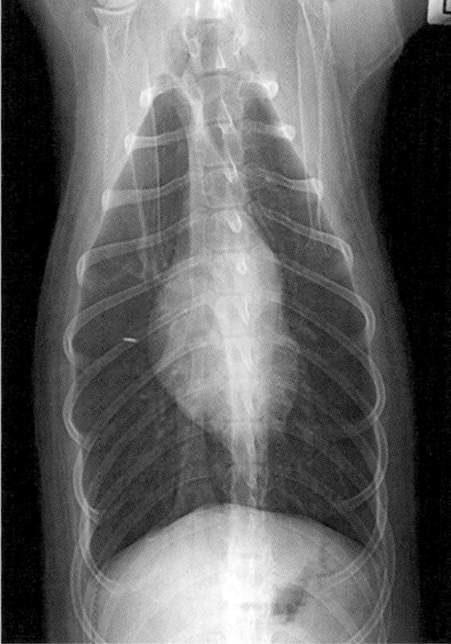

FIGURE 4.1-2 Ventrodorsal thoracic radiographs of the dog with pneumonia (left) and, for comparison, a dog with normal lungs (right). In the image on the left, the opacity within the right middle lung lobe is clearly seen (between the *arrows*).

Diagnosis

Aspiration pneumonia affecting the right middle lung lobe

Treatment

The dog was hospitalized and treated with IV fluid therapy and broad-spectrum antimicrobials. He was discharged from the hospital after 3 days, with normal appetite and demeanor, on prescribed oral antimicrobial therapy. Thoracic radiographs repeated 3 weeks later showed that the soft tissue opacity had resolved. Antimicrobial therapy was continued for an additional 2 weeks, and no relapse was observed after ending therapy.

Anatomical features in canids and felids

Introduction

This section describes the lower respiratory system: the thoracic portion of the trachea and the lungs. The pleura, pleural cavity, and mediastinum are described in Case 4.2 and the diaphragm in Case 4.3.

Because radiography is an important component of clinical assessment of the lower respiratory system, this section focuses on the radiographic appearance of the trachea, lungs (bronchi and parenchyma), and pulmonary vessels in health and disease. The various normal and abnormal respiratory sounds that may be heard on thoracic auscultation are also described, as are normal and abnormal findings with thoracic percussion.

Function

The trachea serves to conduct inhaled and exhaled air to and from the lungs. A critical part of its function is to maintain a patent (clear or unobstructed) airway despite changing pressures within the thorax. The lungs are the organs of respiration, comprised of a fine network of airways and blood vessels that facilitate gas exchange, with oxygen moving from the airways into the systemic circulation, and carbon dioxide moving in the opposite direction.

Trachea (Figs. 4.1-3 and 4.1-4)

The cervical portion of the trachea is described in Case 3.6. The **trachea** enters the thoracic cavity at the **thoracic inlet** before dividing at its terminal **bifurcation**, situated over the heart base, into the paired **primary bronchi** (left and right). Each primary bronchus enters its respective lung, extensively branching and narrowing to ultimately supply air to the alveoli. A ridge of cartilage, the **carina**, separates the left and right primary bronchi at the bifurcation. Hence, "tracheal bifurcation" and "carina" are sometimes used interchangeably in clinical terminology.

Radiographically, the air-filled trachea is easily seen on lateral projections of the thorax (e.g., Fig. 4.1-1), but on the ventrodorsal and dorsoventral projections, it is obscured by the sternum and thoracic vertebrae (e.g., Fig. 4.1-2). Intrathoracic tears in the tracheal wall allow air to escape into the mediastinum—a condition known as pneumomediastinum—surrounding and thus outlining the outer wall of the trachea in that location.

Lungs (Figs. 4.1-5–4.1-7)

The paired **lungs** (left and right) are attached to the **mediastinum** at the **root**, a structure that includes the primary bronchus and the pulmonary vessels, nerves, and lymphatics of their respective lung, enclosed within a reflection of the mediastinal pleura that is continuous with the pulmonary pleura. The window through which these structures enter the lung is the **hilus** (adjective, "hilar") (*L. hilus* a depression or fissure). The fold of pleura that extends caudal to the root of the lungs, between the caudal lobe of the lung and the mediastinum, is the **pulmonary ligament**.

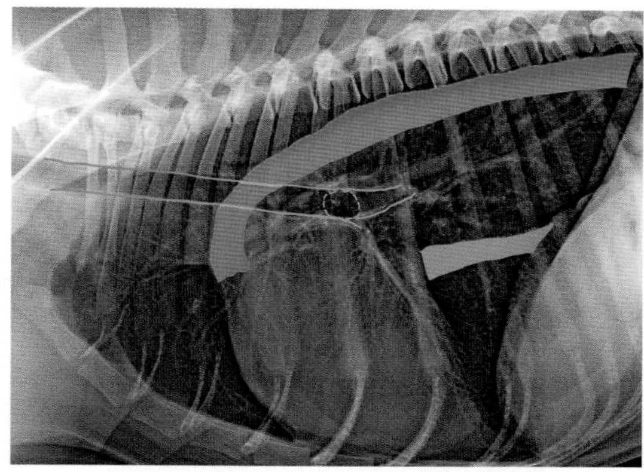

FIGURE 4.1-3 Normal lateral radiograph of the canine thorax (right laterally recumbent projection). The trachea is outlined in *green*; the tracheal bifurcation is seen as a radiolucent (*dark gray*) circle dorsal to the heart. The aorta is shown in *red* and the caudal vena cava in *blue*.

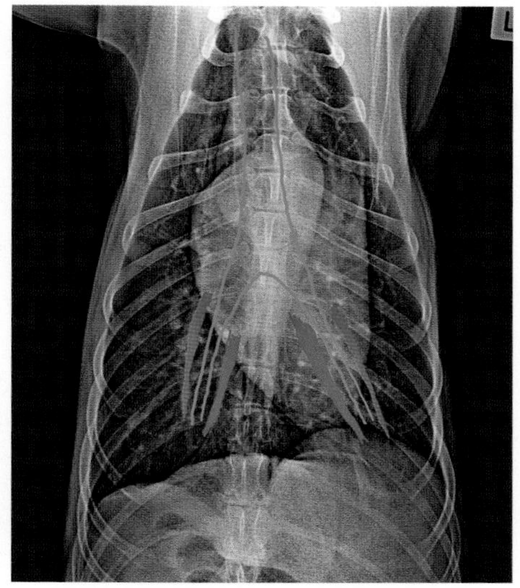

FIGURE 4.1-4 Normal ventrodorsal radiograph of the canine thorax. The trachea (outlined in *green*) is seen tracking through the thorax from the thoracic inlet to its bifurcation over the heart into the left and right primary bronchi. The caudal lobar arteries (*red*) and veins (*blue*) are highlighted adjacent to their respective primary bronchus. The normal trachea often sits slightly to the right side of the thorax, but it can be difficult to see the trachea along its entire length on this projection due to superimposition of the sternum and spine.

As described below, each lung is divided into separate lobes. But taken together, each lung has a roughly conical shape, with a pointed **apex** adjacent to the thoracic inlet, a concave **base** adjacent to the diaphragm, a convex **costal surface** that is in contact with the lateral thoracic wall, and a **medial surface** that is indented by the mediastinal contents, including the heart. The lungs separate the heart from the thoracic wall, except at the **cardiac notch**, which is a gap in the ventral margin of the lung at approximately the level of the 3rd to 5th intercostal space. The cardiac notch is larger in the left lung than in the right.

The cardiac notch provides a useful "acoustic window" for echocardiography (see Case 4.4).

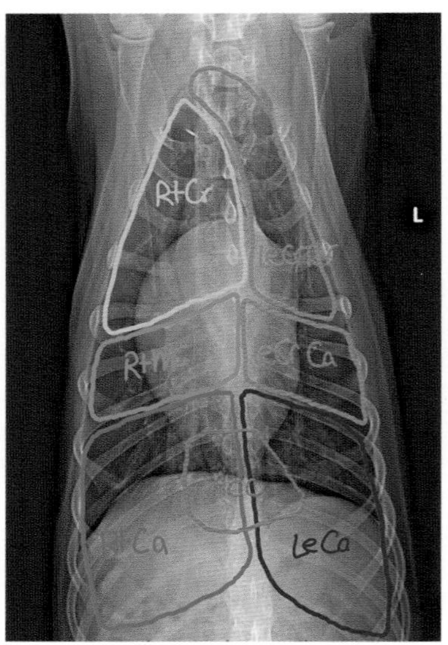

FIGURE 4.1-5 Normal ventrodorsal radiograph of the canine thorax, with the position of all lobes of the left and right lungs outlined. Note that the cranial tip of the left cranial lung lobe extends across midline to the right side. Key: *RtCr,* right cranial lung lobe; *RtM,* right middle lung lobe; *RtCa,* right caudal lung lobe; *Acc,* accessory lobe of the right lung; *LeCrCr,* cranial part of the left cranial lung lobe; *LeCrCa,* caudal part of the left cranial lung lobe; *LeCa,* left caudal lung lobe.

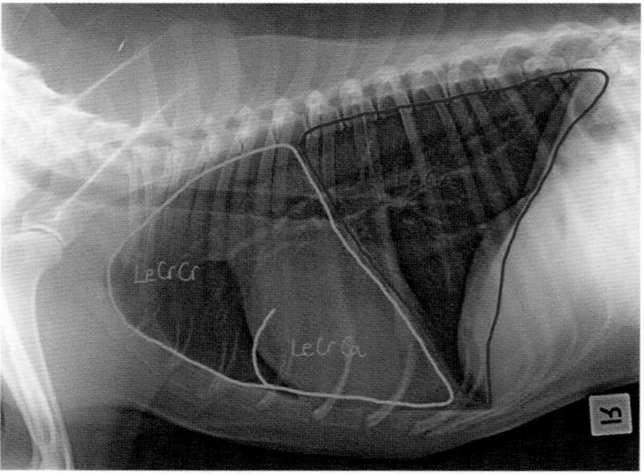

FIGURE 4.1-6 Normal lateral radiograph of the canine thorax, right laterally recumbent ("right lateral") projection, with the lobes of the left lung outlined. Key: *LeCrCr,* cranial part of the left cranial lung lobe; *LeCrCa,* caudal part of the left cranial lung lobe; *LeCa,* left caudal lung lobe.

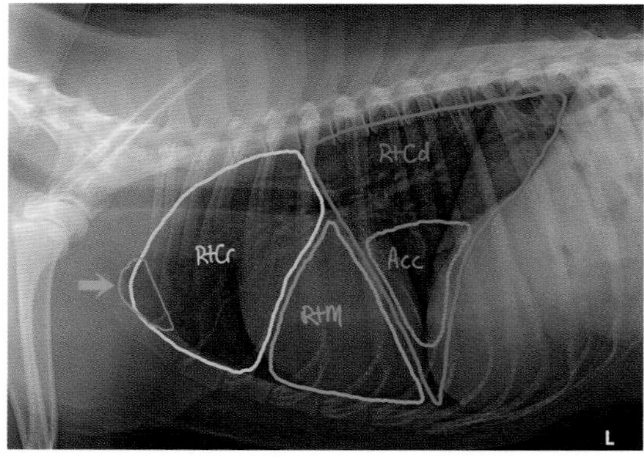

FIGURE 4.1-7 Normal lateral radiograph of the canine thorax, left laterally recumbent ("left lateral") projection, with the lobes of the right lung outlined. As shown in Fig. 4.1-5, the cranial tip of the left cranial lung lobe extends a short distance to the right of midline, wrapping part of the way around the cranial aspect of the right cranial lung lobe. This portion of the left cranial lung lobe (*arrow*) is often superimposed on the right cranial lung lobe in the left lateral view. It appears as a *rounded, dark gray* region overlapping the cranial aspect of the right cranial lung lobe. Key: *RtCr,* right cranial lung lobe; *RtM,* right middle lung lobe; *RtCd,* right caudal lung lobe; *Acc,* accessory lobe of the right lung.

The lungs are each divided into **lobes** by deep fissures. Lung lobes can be defined as regions supplied by **secondary bronchi**, which are lobar branches of the primary bronchus.

Because the lungs are divided into lobes, it is possible—although rare—for a lung lobe to rotate around its long axis, a condition known as "lung lobe torsion." It results in occlusion of the bronchus and pulmonary vessels to/from the affected lobe, which can present a life-threatening emergency, with peracute onset of respiratory distress. Treatment involves immediate surgical removal of the affected lung lobe (lobectomy).

The **right lung** is the larger of the two lungs and has **cranial**, **middle**, and **caudal lobes**, named for their relative positions. A fourth lobe on the right side is the **accessory lobe**, which lies in the space between the caudal mediastinum and a fold of pleura associated with the vena cava (the plica venae cavae), caudal to the heart. The **left lung** has **cranial** and **caudal lobes**, with the cranial lobe being incompletely divided into cranial and caudal parts by a division of the secondary bronchus.

The right middle lung lobe is most commonly affected in cases of aspiration pneumonia, presumably because of the ventral orientation of the secondary bronchus to this lobe where it branches from the right primary bronchus (discussed later). As a consequence, gravity-dependent pooling of aspirated material can occur when the body is in an upright position (standing, sitting, or in sternal recumbency). However, the number and location of affected lung lobes likely varies in each case.

Pulmonary vessels

The large **pulmonary arteries** leave the right side of the heart, branching immediately and tracking alongside the lobar bronchi (see Fig. 4.1-4) to supply each lung lobe with deoxygenated blood expelled from the right ventricle of the heart. A separate supply of oxygenated arterial blood supplies the bronchi and connective tissues of the lungs; these **bronchial arteries** arise directly from the aorta and vary in number and origin. The freshly oxygenated **pulmonary veins** leave each lung lobe and join together as they enter the left atrium of the heart. The superficial and deep lymphatic networks of the lungs drain into the **tracheobronchial lymph nodes** at the root of the lungs.

The pulmonary arteries and veins are paradoxical, in that the pulmonary arteries carry deoxygenated blood (from the heart to the lungs) and the pulmonary veins carry oxygenated blood (from the lungs to the heart). Based on the oxygen and carbon dioxide levels of their contents, the pulmonary arteries should be called the pulmonary veins, and vice versa, but the convention is to call vessels leaving the heart "arteries" and those returning to the heart "veins," so these paradoxical names persist for these pulmonary vessels.

Clinical assessment: thoracic auscultation

On auscultation of the thorax, **normal** breath sounds, termed **vesicular sounds**, are the soft, low-pitched, rustling sounds generated by turbulent air moving through the thoracic portion of the trachea and the primary and secondary bronchi. Inspiratory sounds are normally a little louder than expiratory sounds.

Abnormal breath sounds, termed **adventitious sounds**, are generated when the usual flow of air is disrupted, such as by changes in airway diameter or the accumulation of fluid within the airways. Examples of adventitious sounds include the following:

- **Crackles**: popping or bubbling sounds, generated by the opening of small airways or alveoli that are otherwise collapsed or fluid-filled; causes include pulmonary edema and pneumonia
- **Wheezes**: high-pitched, shrill sounds with a musical quality, generated by air moving through narrowed small airways; causes include asthma and bronchitis

- **Rhonchi**: low-pitched rattling or gurgling sounds, generated by air moving through narrowed large airways; causes include intraluminal mucus or exudate accumulation
- **Pleural rubs**: rustling or squeaking sounds, generated when dry pleural surfaces move against each other, creating friction; causes include pneumonia and pleuritis

Clinical assessment: thoracic percussion

Thoracic percussion is performed by lightly pinging or tapping a finger against the chest wall over several different areas, listening for the resonance it produces. Percussion over normal (aerated) lungs produces a relatively hollow sound, whereas percussion over a more dense structure, such as the heart or an area of pleural effusion or lung consolidation, produces a relatively dull sound. As the vibration of percussion is transmitted through the chest wall into the adjacent lung, it can induce coughing in patients with bronchial inflammation.

Clinical assessment: thoracic radiography

Thoracic radiography is commonly performed in veterinary practice. At least 2 radiographic projections are required for the evaluation of any body part, but 3 projections of the thorax are commonly made, and occasionally 4. Routinely, the patient is placed in right lateral recumbency ("right lateral"), left lateral recumbency ("left lateral"), and then dorsal recumbency (ventrodorsal or VD) to produce a series of 3 projections of the thorax. The patient may also be placed in sternal recumbency for a dorsoventral (DV) projection.

On the right lateral projection, the left lung lobes are projected (see Fig. 4.1-6) while the right lung lobes are projected on the left lateral projection (see Fig. 4.1-7). This seeming paradox is due to the increased aeration of the uppermost lung lobes (e.g., the left lung lobes when the dog is lying on his right side) when the patient is in lateral recumbency.

On both lateral projections, the **aorta** is seen tracking dorsally from the heart, then caudally into the abdomen (see Fig. 4.1-3). The **caudal vena cava** runs from the diaphragm to the heart. In most dogs, the caudal vena cava runs slightly ventrally from the diaphragm to the heart, but it may also run horizontally.

The **pulmonary vessels** are projected as soft tissue opacity (white) tubular structures that taper toward the lung periphery. The vessels are able to be visualized when they are surrounded by air. Thus, thoracic radiographs are routinely made while the patient is in the inspiratory phase of respiration to maximize the amount of air within the lungs.

On the lateral projections, the **cranial lobar arteries** and **veins** can be seen as separate structures (Fig. 4.1-8). The artery tracks dorsal to the **bronchus**, and the vein lies ventral to the bronchus. Within the caudal thorax, the arteries and veins are superimposed and are seen as multiple soft-tissue-opacity tubular structures, but the arteries and veins cannot be identified separately.

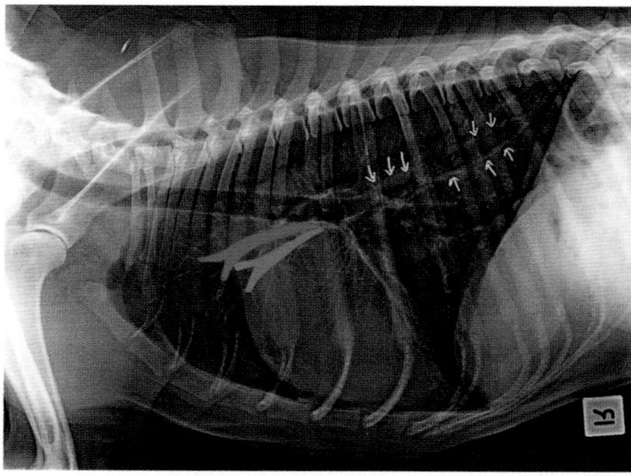

FIGURE 4.1-8 Normal lateral radiograph of the canine thorax, highlighting the cranial lobar vessels. The cranial lobar artery (*red*) is dorsal to the vein (*blue*). The bronchus lies between the artery and vein, but the normal bronchial wall is often not visible. Within the caudal thorax, the pulmonary vessels (*yellow arrows*) can be seen extending to the periphery of the lung. Because of superimposition, it is not possible to distinguish between the arteries and veins in the caudal thorax on the lateral projections.

On the VD and DV projections, the **caudal lobar arteries** and **veins** can be seen on either side of the **caudal lobar bronchus** (see Fig. 4.1-4) The artery is abaxial (lateral) to the bronchus and the vein is axial (medial). One way to remember it is with the saying, "the veins are ventral and central." That is, on the lateral projection, the veins are ventral to the artery and bronchus, and on the VD/DV projection, the veins are central (medial) to the bronchus and artery. The vessels are easier to see in some patients than in others because of variations in the shape of the thorax among animals.

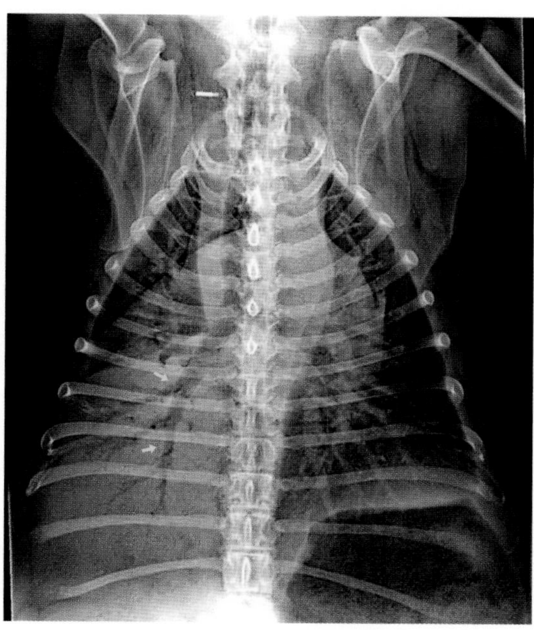

FIGURE 4.1-9 Dorsoventral thoracic radiograph of a dog with cardiogenic pulmonary edema. There is increased opacity (*whiteness*) within the right caudal lung lobe, and air bronchograms (*yellow arrows*) can be seen. In comparison, the left caudal lung lobe is unaffected, so its pulmonary vessels are clearly visible.

COMMON RADIOGRAPHIC PATTERNS OF DISEASE

The ability to see the pulmonary vessels within the pulmonary parenchyma is due to the soft tissue/fluid opacity of the vessels when they are surrounded by air. In pulmonary parenchymal disease, the vessels are obscured if the air within the alveoli is replaced by fluid, such as in cardiogenic pulmonary edema (Fig. 4.1-9), or by exudate such as in pneumonia (see Figs. 4.1-1 and 4.1-2). When the air is replaced the margins of the vessels can no longer be seen the lung has an increased opacity (i.e., it is whiter than normal).

In severe disease, all of the alveolar air is replaced by the disease process and the only air remaining in the lung is that within the bronchi. When air-filled bronchi, visible as grey tubes, are seen within the diseased (white) lung, they are referred to as "air bronchograms." The presence of air bronchograms is consistent with an "alveolar" pulmonary pattern (Fig. 4.1-9).

In some disease processes, not all of the air within the alveoli is replaced, or the disease process (e.g., neoplasia) may be within the supporting interstitium of the lung rather than the air spaces. When this situation occurs, the vessels may still be seen, but they are hard to see (Fig. 4.1-10). This presentation is called an "interstitial" pulmonary pattern.

In most dogs and cats, especially young patients, the walls of the bronchi are not able to be seen on thoracic radiographs. Some older dogs, and dogs with hyperadrenocorticism, have mineralization of the bronchial walls. The mineralized bronchial walls are seen as thin, white lines within the pulmonary parenchyma. In diseases, such as bronchitis in dogs and asthma in cats, the bronchial walls may become thickened and are seen as thick, white rings ("donuts") and thick, white lines ("tram tracks") within the pulmonary parenchyma (Fig. 4.1-11). This finding indicates a "bronchial" pulmonary pattern.

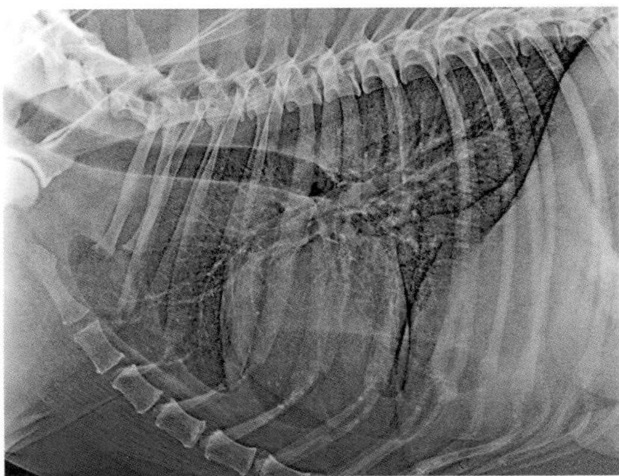

FIGURE 4.1-10 Right lateral thoracic radiograph of a dog with lymphoma. There is increased opacity within the lung (i.e., it is whiter than normal), but the pulmonary vessels can still be seen, although they are harder to see than in the normal lung.

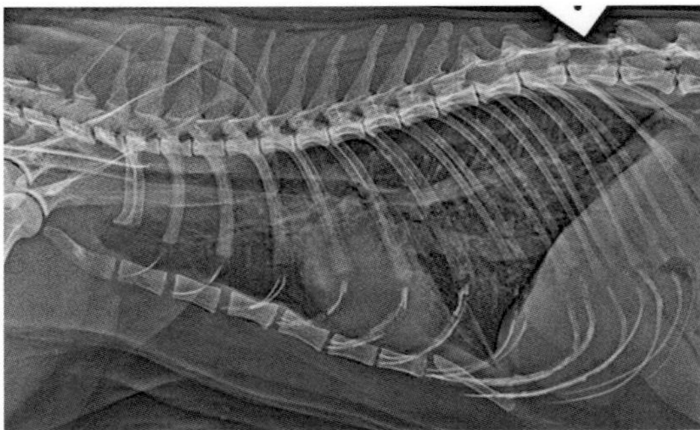

FIGURE 4.1-11 Left lateral thoracic radiograph of a cat with feline lower airway disease (feline asthma). There is increased opacity within the pulmonary parenchyma. The pulmonary vessels can be seen, but they are harder to see than in a normal patient. There are multiple, small ring-like, and linear opacities within the pulmonary parenchyma consistent with thickened bronchial walls.

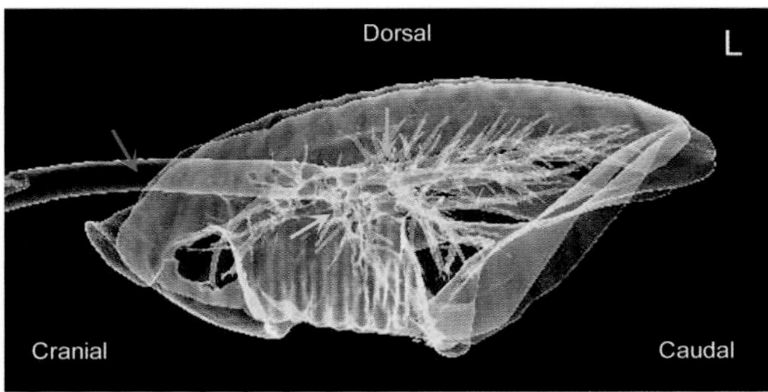

FIGURE 4.1-12 A volume-rendered CT image of the lungs of a normal dog. The trachea (*red arrow*) is seen at the left of the image, extending into the lungs. The bronchi can be seen arising from the trachea. The cranial lobar bronchi arise from the ventral aspect of the principal bronchi (*yellow arrow*), whereas the caudal lobar bronchi arise slightly dorsally from the principal bronchi (*blue arrow*). This cranioventral origin of the cranial lobar bronchi makes the cranial lung lobes (right cranial, right middle, and left cranial) more susceptible to aspiration pneumonia than the caudal lung lobes.

WHY IS ASPIRATION PNEUMONIA COMMON IN THE CRANIOVENTRAL LUNG FIELDS?

The trachea divides into the left and right principle lobar bronchi at the tracheal bifurcation. As shown in the CT image of the trachea and lungs in Fig. 4.1-12, the bronchi of the cranial lung lobes arise from the cranioventral aspects of the principle lobar bronchi, whereas the caudal lobar bronchi arise from the principle bronchi more caudodorsally. As the aspiration of liquids is gravity-dependent, the cranioventral origin of the bronchi supplying the right middle, right cranial, and left cranial lung lobes make these lobes more likely to be affected in aspiration pneumonia.

Selected references

[1] Kogan DA, Johnson LR, Jandrey KE, et al. Clinical, clinicopathologic, and radiographic findings in dogs with aspiration pneumonia: 88 cases (2004–2006). J Am Vet Med Assoc 2008;233:1742–7.

[2] Kogan DA, Johnson LR, Sturges BK, et al. Etiology and clinical outcome in dogs with aspiration pneumonia: 88 cases (2004–2006). J Am Vet Med Assoc 2008;233:1748–55.

[3] Goggs RAN, Boag AK. Aspiration pneumonitis and pneumonia. In: Silverstein DC, Hopper K, editors. Small animal critical care medicine. 2nd ed. St. Louis: WB Saunders; 2015. p. 127–33.

[4] Neath PJ, Brockman DJ, King LG. Lung lobe torsion in dogs: 22 cases (1981–1999). J Am Vet Med Assoc 2000;217:1041–4.

CASE 4.2

Pyothorax

Laura Dooley[a], Cathy Beck[b], and Helen M.S. Davies[a]
[a]Veterinary Biosciences, The University of Melbourne, Melbourne, Victoria, AU
[b]Veterinary Hospital, The University of Melbourne, Werribee, Victoria, AU

Clinical case

History

A 5-year-old male castrated Siamese cat was presented with a 1-week history of lethargy and decreased appetite. On the morning of presentation, he was inappetant and acutely dyspneic.

Physical examination findings

On examination, the cat assumed a sternal position with abducted elbows. He was tachycardic with a respiratory rate of 60 brpm, with a restrictive (rapid, shallow) respiratory pattern and increased inspiratory effort. On thoracic auscultation, lung sounds were decreased bilaterally and were absent over the ventral thorax; however, no crackles or wheezes were heard in the dorsal thorax. Although his heart rate was normal and there was no audible murmur, the cardiac sounds were muffled on auscultation. In addition, the cat was febrile (rectal temperature 104.2 °F [40.1 °C]).

This posture (sternal position, elbows abducted) is common in cats with respiratory distress and should alert the clinician to focus on the thorax and on those disease processes that limit lung expansion.

PLEURAL EFFUSION IN CATS

Radiography is useful for identifying pleural effusion, but it is neither able to determine the type of fluid nor, in most cases, the cause. Differential diagnoses include:

- pyothorax—accumulation of purulent material ("pus") in the pleural space
- left-sided congestive heart failure—accumulation of modified transudate in the pleural space
- chylothorax—leakage of chyle (lipid-rich lymphatic fluid) from the thoracic duct into the pleural space
- neoplastic effusion—exudation secondary to intrathoracic neoplasia
- hemothorax—free blood in the pleural space

Rounding of the lung margins in this case suggests chronicity—the presence of pleural effusion for days or weeks. Cats are good at "masking" disease. The increased opacity within the lung lobes may have been secondary to atelectasis (lung collapse) caused by pressure from the pleural fluid, or reflect pulmonary disease, such as pneumonia, cardiogenic pulmonary edema, or neoplasia. A pleural fluid sample would be required for added information.

Comparative Veterinary Anatomy: A Clinical Approach. https://doi.org/10.1016/B978-0-323-91015-6.00024-8

Differential diagnoses

Pleural effusion, pneumothorax, diaphragmatic rupture (with displacement of abdominal contents into the thoracic cavity), other intrathoracic mass

Diagnostics

Thoracic radiography revealed the presence of pleural effusion (Figs. 4.2-1 and 4.2-2). With the cat positioned in sternal recumbency following local anesthesia of the skin and subcutaneous tissues, a butterfly needle connected to extension tubing and a 3-way stopcock was introduced into the left pleural space cranial to the costochondral junction at the 8th intercostal space (i.e., thoracocentesis). Pleural fluid was incrementally aspirated using a 20-mL syringe until no

> This combination of syringe and 3-way stopcock prevents aspiration of air into the pleural space during thoracocentesis, and thus the development of pneumothorax and lung collapse. During inhalation, negative pressure is generated within the thoracic cavity to pull air into the lungs. Any disruption in the thoracic wall, including insertion of a needle for fluid aspiration, allows air to be drawn into the pleural space via the thoracic wall unless the opening is sealed in some way (e.g., syringe and 3-way stopcock).

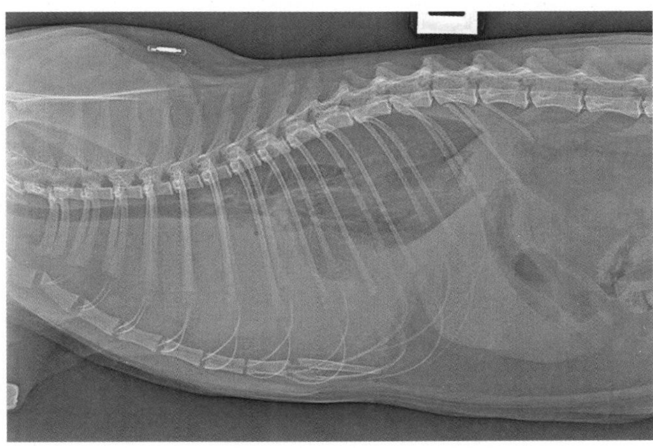

FIGURE 4.2-1 Lateral radiograph of a cat with pleural effusion. A diffuse soft-tissue opacity fills the pleural cavity ventrally, forcing the trachea and lungs dorsally and obscuring the cardiac silhouette. In comparison, there is good serosal detail in the abdomen. (Note: this cat has a microchip implanted behind the right scapula.)

PLEURA VS. THORAX

"Pleura" and "thorax" are terms that are often used interchangeably, but they are not synonymous. Thorax (n.) is the broader term, as it describes the entire area bounded by the chest wall laterally, the thoracic (adj.) inlet cranially, the diaphragm caudally, the sternum ventrally, and the spine and heads of the ribs dorsally. The thoracic cavity encompasses the pleura, mediastinum, and all their contents.

Pleura (n.) is specific to the area bounded by the parietal and visceral pleura, which is the serous membrane that lines the chest wall (parietal pleura) and covers the thoracic organs (visceral pleura). For example, with pleural (adj.) effusion the fluid occupies the area between the outer surface of the heart and lungs and the inner surface of the chest wall. While it is an intrathoracic process, it is more precisely described as "pleural." The fluid can be a transudate (low WBC and total protein concentration) or an exudate (increased WBC and total protein concentration).

However, the "rules" do not apply for a number of clinical conditions that primarily or exclusively involve the pleural space, yet are described as thoracic, such as pneumothorax, pyothorax, and chylothorax.

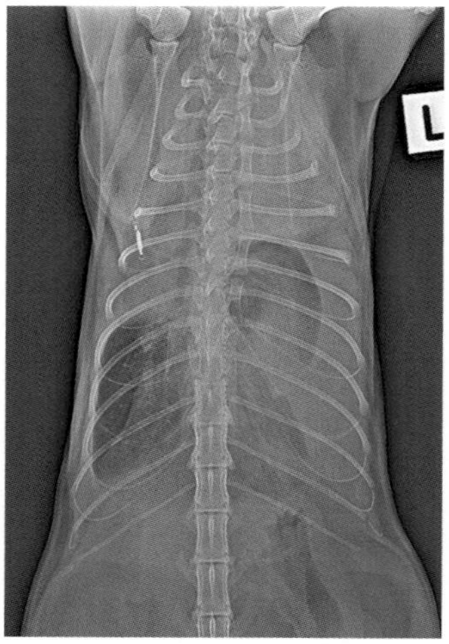

FIGURE 4.2-2 Ventrodorsal (VD) radiograph of the cat in Fig. 4.2-1 with pleural effusion. (This VD projection is slightly obliqued, so the sternum is seen lateral to the spine rather than being superimposed on the spine.) The cardiac silhouette is indistinct. The lung lobes are retracted from the thoracic wall, most evident on the left side, and their margins are rounded. A diffuse soft-tissue opacity fills the space between the lung lobes and the thoracic wall. In addition, there is increased opacity within the lung lobes, most severe in the right middle and right cranial lobes. The cranial part of the left cranial lung lobe is completely collapsed.

more fluid could be obtained. Aspiration was then repeated on the right side. This yielded a net of approximately 150 mL of malodorous, pink, turbid fluid. Repeat radiographs showed that only a small volume of pleural fluid remained after drainage, and no other abnormalities were evident.

Analysis of the pleural fluid confirmed it to be a purulent exudate based on elevated total protein concentration and total nucleated cell count, with a predominance of degenerated neutrophils. Microbial culture yielded a mixed growth of anaerobic bacteria.

In most dogs and cats, the mediastinum (median wall dividing left and right sides of the thoracic cavity) is fenestrated (L. *fenestrare* window, opening), so both pleural spaces (left and right) normally communicate. However, in patients with pyothorax, fibrinous clots may block these channels, so it is generally best to place a thoracostomy tube in each side of the chest rather than relying on a single tube to drain both sides. In performing a thoracotomy incision remember that the intercostal vessels track along the caudal border of the ribs and should be avoided to minimize unintended bleeding and formation of a dissecting hematoma between tissue planes.

Diagnosis

Pyothorax

Treatment

Under general anesthesia, indwelling thoracostomy tubes ("chest tubes" or "chest drains") were inserted bilaterally. A small skin incision was made over the 10th intercostal space and the tube was tunneled under the skin for a short distance, entering the pleural cavity at the level of the costochondral junction in the 7th intercostal space (Fig. 4.2-3). This approach minimized the risk of pneumothorax developing from aspiration of air around the tube. The pleural space was lavaged (flushed) with warm, sterile, physiologic saline solution and subsequently drained. Systemic broad-spectrum antimicrobial therapy was also initiated.

During hospitalization, regular intermittent suction of the chest drains was performed, and the fluid volume recorded. After 6 days, pleural fluid production had decreased to less than 10 mL per day, so the chest drains were removed, and the cat was discharged on oral antimicrobial therapy. Thoracic radiographs repeated 2 weeks after discharge showed complete resolution of the pleural effusion.

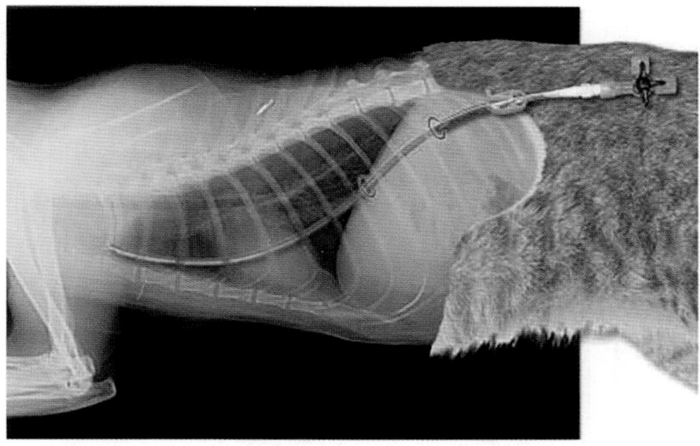

FIGURE 4.2-3 Positioning of a thoracostomy tube in a cat. (Image courtesy of Drs. Vanessa Barrs and Julia Anne Beatty.)

Anatomical features in canids and felids

Introduction

This section describes the mediastinum (and its contents) and the pleura. The lungs are described separately in Case 4.1 and the diaphragm in Case 4.3. The heart and its specialized membrane, the pericardium, are described in Cases 4.4 and 4.7.

Function

The function of the mediastinum and pleura is to support the organs of the thoracic cavity. The pleura provides a frictionless surface that allows the lung lobes to move frictionless along side each other and effortlessly opposite the diaphragm and chest wall.

Mediastinum

The thoracic cavity is divided into left and right sides by a median sheet of serosa-covered fascia, the **mediastinum** (L. *mediastīnus* midway). The left and right sides of the thoracic cavity contain the left and right lungs, respectively, and the mediastinum contains most of the other thoracic structures of clinical interest. In dogs and cats, the mediastinum is usually fenestrated by many small channels through which pleural fluid and pressure may be exchanged or equalized between left and right sides (Fig. 4.2-4).

Although the mediastinum is a single structure that extends from the cranial to the caudal part of the thoracic cavity, it is often divided into separate regions for the purpose of discussion. The cranial mediastinum is the part that lies cranial to the heart; the middle mediastinum contains the heart; and the caudal mediastinum lies caudal to the heart.

The **cranial mediastinum** widens dorsally, and it contains the **trachea** and **esophagus** as they enter the thorax at the thoracic inlet. It also contains the **cranial vena cava** and other large vessels, as well as the **thymus**, **cranial mediastinal lymph nodes**, and **thoracic duct**.

The **thymus** is a bilobed, lobulated, laterally compressed organ within the cranial mediastinum that is located ventral

Tears in the thoracic portion of the trachea or esophagus can allow air to enter the mediastinum, a condition called "pneumomediastinum." Because the mediastinum normally does not communicate with the pleural cavity, air within the mediastinum does not enter the pleural cavity unless the mediastinum is also ruptured or incomplete. Pneumomediastinum is typically diagnosed using thoracic radiography or ultrasonography.

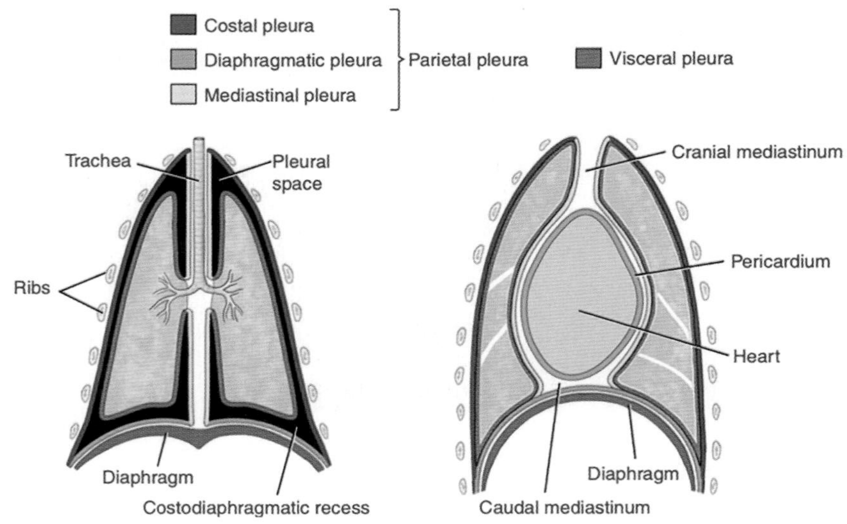

FIGURE 4.2-4 Schematic diagram of the pleural membranes. The left and right pleural cavities in the drawing on the left are greatly expanded for illustration purposes. In healthy animals, each pleural cavity is merely a potential space. Only a thin film of serous fluid normally separates the parietal and visceral pleura.

to the trachea. It is large in young animals (maximum size at 4–5 months of age), but it gradually atrophies and is replaced by fat with maturity, such that only remnants remain in adulthood. When fully developed, the thymus extends caudally from just cranial to the thoracic inlet to the level of the 5th costal cartilage, where it lies against the pericardium.

The **thoracic duct** is the major route for the return of lymph to the systemic circulation. It begins in the sublumbar region between the crura of the diaphragm as a cranial continuation of the **cisterna chyli** (the "cistern" into which lymph from the intestinal tract and lumbar region flows). The thoracic duct runs cranially within the mediastinum on the right side, between the aorta and the azygous vein, to about the level of the 6th thoracic vertebra. It then passes to the left within the cranial mediastinum and empties into the large veins cranial to the heart. The thoracic duct is thin-walled (only a few millimeters in diameter) and usually covered by fat, so it is difficult to identify unless it is obstructed, in which case numerous random dilations may be visible along its length.

Any disruption in the drainage of chyle through the thoracic duct can increase the pressure in the duct, leading to leakage into the pleural space—a condition known as "chylothorax."

The **middle mediastinum** contains the **heart** and pericardium, the **aortic arch**, and the roots of the lungs.

The dorsal part of the **caudal mediastinum** contains the **descending aorta**, **azygous vein**, and **esophagus**. The ventral part extends from the caudal surface of the pericardium to the diaphragm. It deviates to the left of the thoracic cavity, owing to the greater size of the right lung in relation to the left and to the leftward position of the heart.

CHYLOTHORAX

Chylothorax is the term used to describe chylous effusion in the pleural space. It is one of the differential diagnoses in any case of pleural effusion. Chyle contains high levels of triglycerides, so chylous effusion has a milky appearance. Chylothorax may be caused by any disease process that disrupts or obstructs the thoracic duct, but in many cases no underlying cause can be identified, so the "working" diagnosis is idiopathic chylothorax.

Pleura (Fig. 4.2-4)

The **pleura** is the serous membrane that covers the thoracic organs and lines the internal wall of the thoracic cavity. The part that covers the lungs is the **pulmonary** or **visceral** pleura. The part that lines the thoracic wall is the **parietal** pleura.

The parietal pleura is **costal** where it covers the internal surface of the rib cage, **diaphragmatic** where it covers the thoracic surface of the diaphragm, **mediastinal** where it covers the mediastinum, and **pericardial** where it covers the pericardium.

The **mediastinal pleura** is continuous with the **pulmonary pleura** at the root of the lung, and the fold of pleura that extends caudal to this junction, between the caudal lobe of the lung and the mediastinum at the level of the esophagus, is called the **pulmonary ligament**. The **plica vena cava** is the fold of pleura that contains the caudal vena cava. It extends from the caval foramen of the diaphragm to the heart on the right side of the thoracic cavity.

In a healthy animal (i.e., one with an intact thorax), the **pleural cavity** is a potential rather than a real space, and negative pressure between the parietal and visceral pleura maintains the position of the lung lobes within the thoracic cavity. The normal relationship between the lung lobes and the ribs is lost the moment the seal between the pleural space and the exterior of the body is broken.

The negative pressure in the pleural cavity causes an inflow of air whenever there is a surgical or traumatic opening in the pleural cavity. This abnormal inflow of air, termed pneumothorax, causes the lungs to collapse. Pneumothorax is usually bilateral in dogs and cats due to their typically fenestrated mediastinum.

Regional blood supply and innervation

The mediastinal arterial supply derives from a series of branches from the **internal thoracic arteries** that themselves branch from the **subclavian arteries** either side. Two to 4 branches directly supply the cranial mediastinum while further branches perforate the overlying transverse thoracic muscle to supply the cardiac and caudal mediastinum as well as the diaphragm (phrenic arteries) and, via extensions of the **phrenic branches**, the **plica vena cava**. The arteries to the visceral pleura, and minor branches to the bronchi and bronchial lymph nodes and tissues around the root of the lungs, are also derived from the internal thoracic arteries. The parietal pleura is supplied by many small branches from local arteries, such as the **intercostal arteries** that themselves are mostly paired branches from the aorta.

The venous drainage of the mediastinum is via the satellite **mediastinal veins** that drain into the **internal thoracic veins**, which may join into a single vein before draining on the ventral surface of the cranial vena cava.

Sensation to the thoracic wall and the parietal pleura shows similarity via branches from the **intercostal nerves**, which branch from the ventral branches of the corresponding thoracic spinal nerves and course ventrally with the intercostal veins and arteries along the caudal border of each successive rib.

Selected references

[1] Beatty J, Barrs V. Pleural effusion in the cat: a practical approach to determining aetiology. J Feline Med Surg 2010;12:693–707.
[2] Stillion JR, Letendre JA. A clinical review of the pathophysiology, diagnosis, and treatment of pyothorax in dogs and cats. J Vet Emerg Crit Care 2015;25:113–29.
[3] Barrs VR, Beatty JA. Feline pyothorax—new insights into an old problem. Part 2. Treatment recommendations and prophylaxis. Vet J 2009;179:171–8.

CHAPTER 4

CASE 4.3

Diaphragmatic Rupture

Cathy Beck[a] and Helen M.S. Davies[b]
[a]Veterinary Hospital, The University of Melbourne, Melbourne, Victoria, AU
[b]Veterinary Biosciences, The University of Melbourne, Melbourne, Victoria, AU

Clinical case

History

A 10-year-old female spayed Miniature Fox Terrier was presented to the emergency center for dyspnea after having been hit by a car approximately 4 hours earlier.

Physical examination findings

The dog's heart and respiratory rates were elevated and her mucous membranes were tacky, but her rectal temperature was normal. On thoracic auscultation, lung sounds were decreased ("dull") on her right side, and it was difficult to hear the heart sounds on the right. No other clinical abnormalities were found.

Differential diagnoses

Right-sided pneumothorax or hemothorax (free air or blood, respectively, in the right pleural cavity); diaphragmatic rupture

DIAPHRAGMATIC DEFECTS

Displacement of abdominal contents into the thoracic cavity through a defect in the diaphragm may be congenital or acquired. While uncommon overall, the most common type of congenital defect is a peritoneal-pericardial diaphragmatic hernia (described later).

Traumatic rupture of the diaphragm is the most common type of acquired defect in dogs and cats. Often, these are the result of trauma—e.g., hit-by-car—and accompanied by other injuries. This condition is referred to in several ways—rupture, tear, or hernia. By definition, a hernia is simply an abnormal passage of an organ or tissue through a defect in a cavity where it normally resides. Frequently, the tissue exiting through the defect is contained in a sac referred to as a "hernial sac" and is thus referred to as a true hernia. Nonetheless, diaphragmatic rupture is commonly referred to as a diaphragmatic hernia.

Depending on the cause and size of the defect, some diaphragmatic defects are life-threatening emergencies, while others remain clinically silent for months or years. As displacement of abdominal organs into the thoracic cavity interferes with lung expansion and function, traumatic diaphragmatic rupture typically causes tachypnea (increased respiratory rate), often with a shallow respiratory pattern and/or dyspnea. Signs of shock, pain, and upper gastrointestinal disorders—e.g., vomiting, inappetence—may also be part of the clinical picture.

Comparative Veterinary Anatomy: A Clinical Approach. https://doi.org/10.1016/B978-0-323-91015-6.00025-X

Diagnostics

Thoracic radiography revealed the right side of the diaphragm to be incomplete, with the displacement of the stomach, some small intestine, and likely a portion of the liver into the right hemithorax (Figs. 4.3-1 and 4.3-2). In the ventrodorsal view, the heart was displaced to the left.

Normally, the "domed" cranial margin of the diaphragm is clearly delineated on both the lateral and ventrodorsal radiographic views, as the diaphragm is bordered cranially by aerated lung and caudally by the liver (Fig. 4.3-3). Traumatic rupture usually occurs in the ventral portion of the diaphragm, and abdominal contents are maybe seen in the ventral part of the thoracic cavity. It is common for a rupture or tear to occur on only one side of the diaphragm, so only one hemithorax may contain abdominal organs, although their presence may displace the heart and compress the lung on the other side.

Diagnosis

Acute, traumatic diaphragmatic rupture

Treatment

Under general anesthesia and with the dog in dorsal recumbency, a ventral midline incision was made through the abdominal wall and the displaced abdominal organs were returned to their normal position within the peritoneal cavity, while being carefully examined for signs of injury. The tear in the diaphragm was sutured closed, followed by closure of the body wall, as described in Case 5.8. The patient made a full recovery.

A traumatic rupture of the diaphragm is usually a surgical emergency. However, if the stomach is displaced cranially and distended with gas, percutaneous needle decompression may be needed before the patient is stable enough for general anesthesia. Reducing the gastric distension in this way allows the lungs sufficient room to expand, reducing the risks of general anesthesia in these patients.

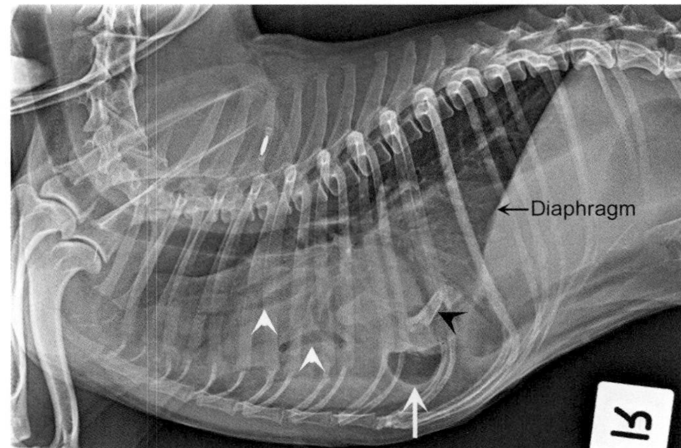

FIGURE 4.3-1 Lateral radiograph of the dog in this case. The diaphragm is intact dorsally, but its ventral margin is indistinct. (Compare this image with the normal dog in Fig. 4.3-3A.) Part of the stomach (*white arrow*) and loops of small intestine (*white arrowheads*) are located cranial to the diaphragm, making it difficult to see the heart. These findings are indicative of diaphragmatic rupture and displacement of abdominal contents into the thoracic cavity. The piece of bone in the dog's stomach (*black arrowhead*) is an incidental finding.

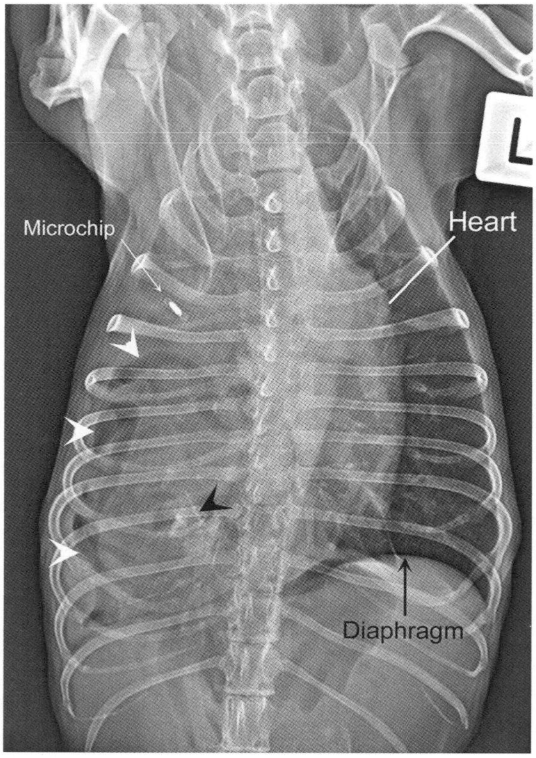

FIGURE 4.3-2 Ventrodorsal radiograph of the dog in this case. The cranial margin of the diaphragm is clearly identifiable on the dog's left side, but not on the right. (Compare this image with the normal dog in Fig. 4.3-3B.) Loops of small intestine (*white arrowheads*) are present on the right side of the thoracic cavity, and the heart is displaced to the left. The incidental piece of bone in the stomach can be seen in the caudal right hemithorax (*black arrowhead*).

Anatomical features in canids and felids

Introduction

This section describes the diaphragm and its relation to the thoracic organs and to respiration. The pleura and mediastinum are described in Case 4.2.

Function

The diaphragm is a respiratory muscle which also serves to divide the coelom (Gr. *coelom* cavity) into separate thoracic and abdominal compartments in mammals.

RESPIRATION

The diaphragm is actively involved in inhalation. Its muscular contraction reduces (flattens) its curvature, causing an increase in intrathoracic volume and a decrease in intrathoracic pressure, which together draw air into the lungs. Exhalation involves relaxation of the diaphragm. At the same time, the muscles of the abdominal wall contract, pushing the abdominal contents against the diaphragm and thereby assist in exhalation.

In its relaxed state, at the end of exhalation and before the next breath, the most cranial portion of the diaphragm extends cranially to the level of the 6th rib. From there, the diaphragm curves caudally, dorsally, and laterally to attach to the costal arch and the bodies of the first 4 lumbar vertebrae. During inhalation, the central part of the diaphragm moves caudally.

Body of the diaphragm

The **diaphragm** is a dome-shaped sheet of muscle and fascia. The convex side of the "dome" faces cranially toward the thoracic cavity, and the concave side is largely filled by the liver, whose cranial aspect is similarly domed. The diaphragm consists of an outer ring of muscle and a central tendinous portion (Fig. 4.3-4).

Dorsally, the muscular portion of the diaphragm is comprised of the **left** and **right crura** (L. *crus* leg; plural, **crura**). Each originates as a narrow tendon on the ventral surfaces of the bodies of the **lumbar vertebrae**, from L4 cranially, in intimate association with the **psoas muscles** (psoas major and minor). Ventrally, the diaphragm attaches to the caudal part of the **sternum**.

> The left and right crura of the diaphragm are visible separately on lateral radiographs of the thorax, as shown in Fig. 4.3-3A. When a patient is in lateral recumbency, the abdominal contents push the dependent crus (e.g., the right crus in right lateral recumbency) cranially, leading to separation of the crura on the lateral projection. However, superimposition of the crura is not abnormal.

Because of the somewhat conical shape of the thorax (expanding caudally), the costal attachments of the diaphragm extend from the 9th to the 13th ribs. The diaphragm is attached to the inner surfaces of the 9th and 10th ribs, the

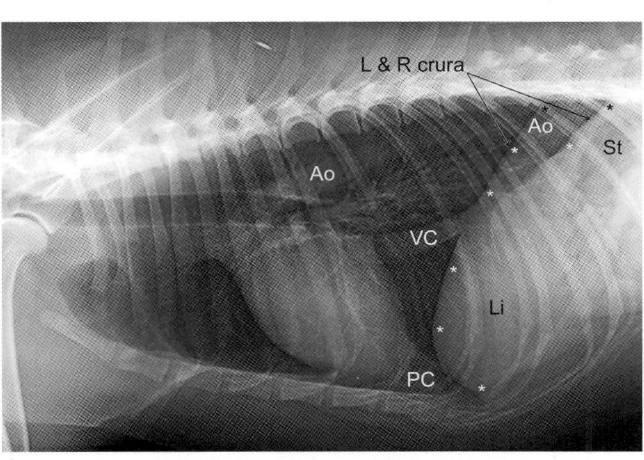

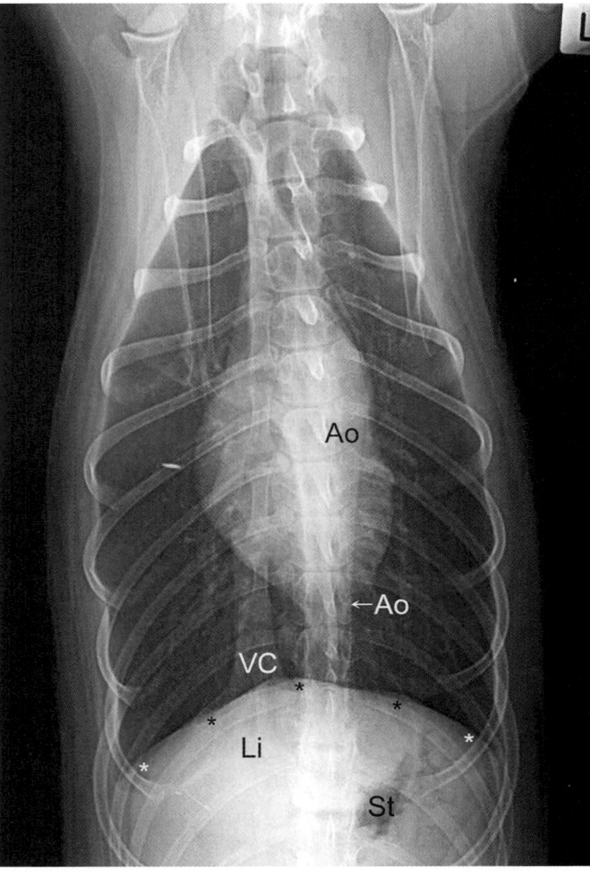

FIGURE 4.3-3 Normal thoracic radiographs of a dog. (A) Right laterally recumbent view. The domed shape of the diaphragm (outlined by *asterisks*) is clearly visible from spine to sternum. Dorsally, both crura of the diaphragm can be seen separately, and the descending aorta is seen passing caudally between them. The stomach in this dog contains a mix of gas and more radiopaque food and fluid. Key: *Ao*, descending aorta; *L*, left; *Li*, liver; *R*, right; *St*, stomach; *VC*, caudal vena cava. (B) Ventrodorsal view (slightly skewed). The domed shape of the diaphragm (outlined by *asterisks*) is clearly visible from left to right sides of the ribcage. Key: *Ao*, descending aorta; *Li*, liver; *St*, stomach; *VC*, caudal vena cava.

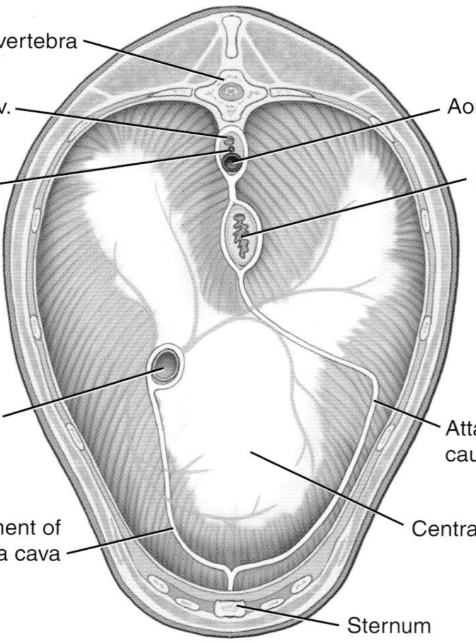

FIGURE 4.3-4 The canine diaphragm viewed from the thoracic side.

more distal parts of the 11th and 12th ribs, and the proximal part of the 13th rib. Hence, both 13th ribs can project beyond the diaphragm caudally, and may often be palpated against the abdominal wall.

The muscular portion of the diaphragm is the weakest part, particularly in the ventrolateral region (the costal parts). Therefore, traumatic rupture typically occurs in the ventrolateral region of the diaphragm and results in displacement of abdominal contents into just one side of the thoracic cavity. The most common cause is a sudden, dramatic increase in intra-abdominal pressure—for example, when a dog is hit by a car. The liver is suddenly forced against the diaphragm, which may then fail/tear, allowing part of the liver and the organs immediately caudal to it (stomach, small intestine, omentum, spleen) to enter the thoracic cavity.

The central **tendinous portion** of the diaphragm consists of connective tissue that radiates medially from the crural and costal muscular portions.

The **thoracic** surface of the diaphragm is covered by **pleura**, and the **abdominal**

PERITONEAL-PERICARDIAL DIAPHRAGMATIC HERNIA (PPDH)

With this type of congenital hernia, abdominal contents prolapse into the pericardial sac through a defect in the ventral diaphragm. Other abnormalities that may also be present—such as umbilical hernia, ventral abdominal hernia cranial to the umbilicus, and/or sternal defects—illustrate the fundamental defect in embryonic development of the ventral body wall. Clinical signs are indicative of cardiorespiratory compromise and may also include upper gastrointestinal signs (e.g., anorexia, vomiting).

Among dog breeds, this uncommon abnormality is seen most often in Weimaraners. Diagnosis is usually made with radiography and ultrasonography, although more advanced diagnostic imaging such as CT provides a more complete picture of the hernia and any associated defects. In some cases, however, PPDH is simply an incidental finding at necropsy, having caused the animal no clinical problems.

surface is covered by **peritoneum**. As described in Case 4.4, the pericardial sac which surrounds the heart is firmly attached to the ventral portion of the diaphragm by the **phrenico-pericardiac ligament**.

The word "phrenic" has its origins in the Greek word phren, which means the diaphragm. The phrenico-pericardiac ligament is a short, strong band of connective tissue, covered by pleura, that firmly attaches the pericardium to the ventral portion of the diaphragm in dogs and cats.

Diaphragmatic foramina (Fig. 4.3-4)

The diaphragm contains 3 "breaks" in the muscular or tendinous portion through which the aorta, the esophagus and vagal trunk, and the caudal vena cava pass between the thoracic and abdominal cavities.

The **caval foramen** is the most ventral of these foramina. It lies on the right side, at the junction of the muscular and tendinous parts of the diaphragm. Here, the caudal vena cava enters the thoracic cavity from the abdomen on its way to the right atrium.

The esophagus and its associated vessels, along with the vagal nerve trunks (right and left) pass through the **esophageal hiatus**, which is more centrally located in the muscular part, just dorsal and medial to the caval foramen. It is surrounded by the muscle extending from the crura and acts as a partial sphincter which helps prevent reflux from the stomach into the esophagus during respiration. The crural part of the diaphragm is preferentially inhibited during swallowing to allow the passage of a food bolus through the esophagus into the stomach.

Dorsal to the esophageal hiatus, between the left and right crura of the diaphragm, is the **aortic hiatus** through which the descending aorta enters the abdomen (see Fig. 4.3-3A).

Blood supply and innervation

The diaphragm is supplied ventrally by terminal branches of the **internal thoracic arteries**, and dorsally and laterally by the **caudal intercostal arteries**, which branch on either side from the descending aorta. Venous drainage is via **intercostal veins** into the **azygous vein** dorsally or the **internal thoracic veins** ventrally, which drain into the **cranial vena cava**, or via the **phrenic veins** into the **phrenicoabdominal veins**, which enter the **caudal vena cava**.

The diaphragm is supplied by the **phrenic nerve**, which originates from the ventral branches of the **cervical spinal nerves** at C5–C7. The phrenic nerve tracks over the base of the heart as it traverses the thoracic cavity on its way to innervate the diaphragm. Caudal to the heart, the right phrenic nerve lies next to the caudal vena cava, within a slight fold of pleura associated with the vena cava (the **plica [L. a ridge or fold] venae cavae**). At the caval foramen, the right phrenic nerve branches to supply the muscle of the diaphragm on the right side. The left phrenic nerve lies to the left of midline in its own small fold of pleura. It enters the diaphragm just to the left of the tendinous center to supply the muscle of the diaphragm on the left side.

If the depolarization threshold in the phrenic nerve is reduced for any reason (e.g., the nerve is irritated or the animal is hypocalcemic, hypomagnesemic, or both), then it may depolarize every time the myocardium is depolarized during the normal cardiac cycle. In such a situation, the diaphragm contracts with every heartbeat. This synchronized contraction of the diaphragm with the heart causes the conditions commonly known as hiccups and thumps. Hiccups (or hiccoughs) are an audible short, sharp intake of breath with every contraction of the diaphragm. Thumps are a visible movement of the abdominal wall with every contraction of the diaphragm (also called "synchronous diaphragmatic flutter"). The latter condition is not uncommon in exhausted performance horses with a metabolic alkalosis and hypocalcemia.

Selected references

[1] Schmiedt CW, Tobias KM, Stevenson MA. Traumatic diaphragmatic hernia in cats: 34 cases (1991–2001). J Am Vet Med Assoc 2003;222:1237–40.
[2] Valentine BA, Cooper BJ, Dietze AE, et al. Canine congenital diaphragmatic hernia. J Vet Intern Med 1988;2:109–12.
[3] Legallet C, Thieman Mankin K, Selmic LE. Prognostic indicators for perioperative survival after diaphragmatic herniorrhaphy in cats and dogs: 96 cases (2001–2013). BMC Vet Res 2017;13(1):16.

CASE 4.4

Feline Cardiomyopathy

Mark Oyama[a] and Simon Bailey[b]
[a]Department of Clinical Sciences and Advanced Medicine, University of Pennsylvania School of Veterinary Medicine, Philadelphia, Pennsylvania, US
[b]Veterinary Biosciences, The University of Melbourne, Parkville, Victoria, AU

Clinical case

History

An 8-year-old male neutered domestic long-haired cat presented with a 4-day history of hiding, decreased appetite, and mildly elevated respiratory rate and effort.

Physical examination findings

The cat was bright, alert, and responsive, with a mild increase in respiratory effort evidenced by mild flaring of the nostrils on inspiration. His heart rate was normal for a clinic setting (200 bpm) and the rhythm was regular, but a grade 3 out of 6 left parasternal systolic murmur was heard on auscultation. His femoral pulses were strong and synchronous with the heart rate—i.e., no pulse deficit. However, his respiratory rate was elevated (45 brpm) and diffuse bronchovesicular sounds were heard over the lung fields bilaterally.

Heart murmurs are graded by their intensity on a scale from 1 (barely audible) to 6 (very loud, barely requiring a stethoscope, and palpable through the chest wall). A grade 3 left parasternal systolic murmur is of moderate intensity that is heard best over the left ventrolateral thoracic wall during systole. The fact that a point of maximum intensity (PMI) was not described suggests more "global" cardiac disease than typically occurs with a single valve defect.

Bronchovesicular sounds simply reflect audible airflow through the small airways of the lung. These soft sounds are normally heard during both inhalation and exhalation; they are loudest centrally—i.e., closest to the tracheobronchial bifurcation—and diminish in intensity toward the lung periphery. An increase in their intensity or audible area may indicate increased ventilation or consolidation of the lung parenchyma.

Differential diagnoses

Heart disease, lung disease (bronchial or parenchymal), thoracic space disease—e.g., pleural effusion

Diagnostics

Thoracic radiography revealed severe cardiomegaly (Figs. 4.4-1 and 4.4-2). On the dorsoventral projection, the cardiac silhouette was grossly widened and had the "valentine" shape of marked atrial enlargement. In addition, the lungs showed a mixed alveolar-interstitial pattern and enlarged pulmonary vessels, consistent with pulmonary congestion. Small pleural fissure lines and retraction of the lung lobes from the thoracic wall indicated mild pleural effusion.

Comparative Veterinary Anatomy: A Clinical Approach. https://doi.org/10.1016/B978-0-323-91015-6.00026-1

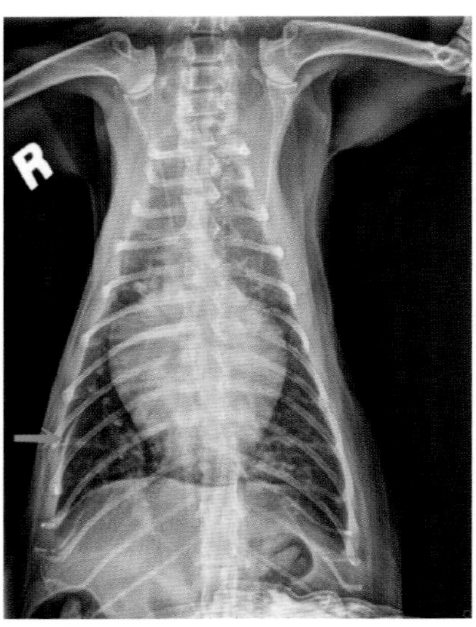

FIGURE 4.4-1 Dorsoventral thoracic radiograph of the cat in this case. The cardiac silhouette is widened and has a "valentine" shape. The lungs show a mixed alveolar-interstitial pattern with enlarged pulmonary vessels. Small pleural fissure lines (*green arrow*) and retraction of the lung lobes from the thoracic wall indicate mild pleural effusion.

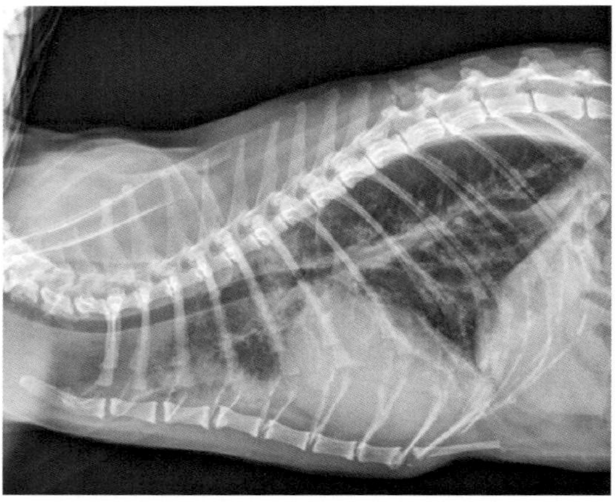

FIGURE 4.4-2 Lateral thoracic radiograph of the cat in this case, showing generalized cardiomegaly, a mixed alveolar-interstitial pattern in the lung fields, and retraction of the lung lobes from the thoracic wall.

Echocardiography identified severe concentric thickening of the interventricular septum and mild pericardial effusion (Figs. 4.4-3 and 4.4-V1 in the online version at https://doi.org/10.1016/B978-0-323-91015-6.00026-1). The left atrium was markedly enlarged (Figs. 4.4-4 and 4.4-V2 in the online version at https://doi.org/10.1016/B978-0-323-91015-6.00026-1) and the left ventricular posterior wall was abnormally thin, with poor systolic contractile function (Figs. 4.4-5 and 4.4-V3 in the online version at https://doi.org/10.1016/B978-0-323-91015-6.00026-1).

UNCLASSIFIED CARDIOMYOPATHY IN CATS

Unclassified cardiomyopathy is the term used in cats for a broad range of echocardiographic abnormalities involving the myocardium, such as regional hypertrophy or thinning of the wall, systolic and diastolic dysfunction, and left atrial enlargement. Cats with this group of findings defy classification into one of the more defined categories of cardiomyopathy—e.g., hypertrophic, dilated, and restrictive. The clinical sequelae of unclassified cardiomyopathy—congestive heart failure, thromboembolism, and sudden death—are like those associated with the other forms of cardiomyopathy.

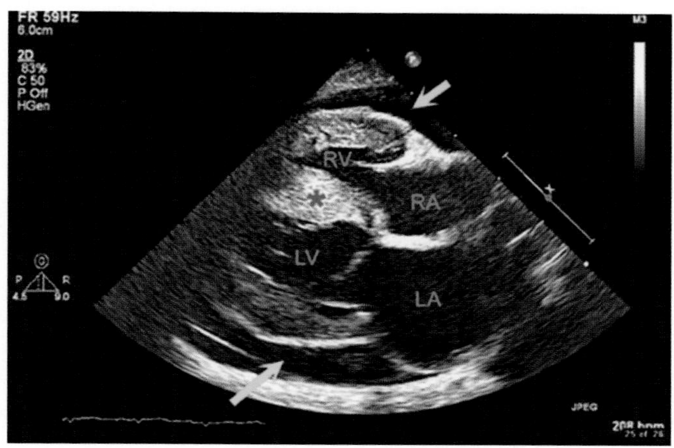

FIGURE 4.4-3 Echocardiogram, long-axis view, showing severe thickening of the interventricular septum (*red asterisk*) and mild pericardial effusion (*yellow arrows*). Key: *LA*, left atrium; *LV*, left ventricle; *RA*, right atrium; *RV*, right ventricle. See Fig. 4.4-V1 in the online version at https://doi.org/10.1016/B978-0-323-91015-6.00026-1 for a short video clip of the heart in motion.

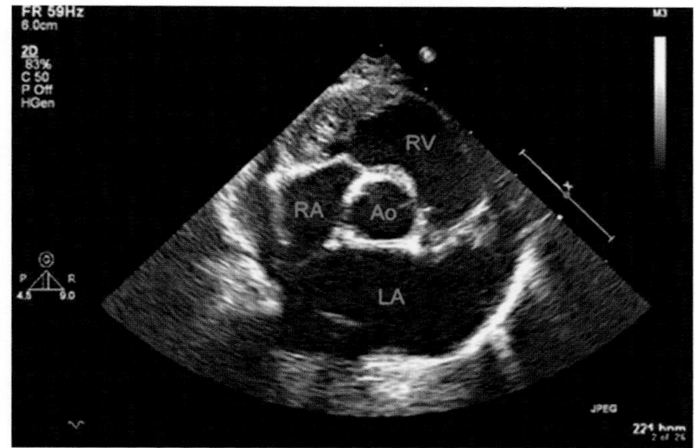

FIGURE 4.4-4 Echocardiogram, short-axis view, showing marked enlargement of the left atrium. Key: *Ao*, aorta; *LA*, left atrium; *RA*, right atrium; *RV*, right ventricle/ventricular outflow tract. See Fig. 4.4-V2 in the online version at https://doi.org/10.1016/B978-0-323-91015-6.00026-1 for a short video clip of the heart in motion.

Diagnosis

Left-sided congestive heart failure secondary to unclassified cardiomyopathy

Treatment

Treatment included supplemental oxygen, a diuretic agent (IV furosemide), and a positive inotrope (oral pimobendan). After 18 hours of treatment, the cat's respiratory rate and effort had improved, so he was weaned off supplemental oxygen and switched from IV to oral furosemide. The cat was sent home 36 hours after presentation, with instructions for the owner to monitor his respiratory rate and effort, appetite,

A diuretic is used to increase urine output and thereby relieve pleural effusion and pulmonary congestion as fluid shifts from the pleural compartment and lung parenchyma into the intravascular space to replace that lost in the urine. Positive inotropes improve myocardial contractility, so the heart pumps more effectively and efficiently. This drug—pimobendan—is also a peripheral vasodilator, so it further reduces the cardiac workload by lowering the peripheral vascular resistance against which the left ventricle must pump.

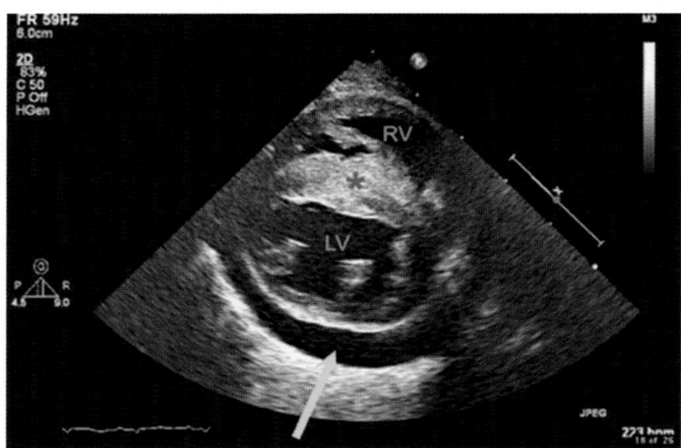

FIGURE 4.4-5 Echocardiogram, short-axis view through the ventricles, showing the abnormally thin posterior or "free" wall of the left ventricle (it should be at least twice the thickness of the right ventricular free wall), along with severe thickening of the interventricular septum (*red asterisk*), and mild pericardial effusion (*yellow arrow*). Key: *LV,* left ventricle; *RV,* right ventricle. See Fig. 4.4-V3 in the online version at https://doi.org/10.1016/B978-0-323-91015-6.00026-1 for a short video clip of the heart in motion.

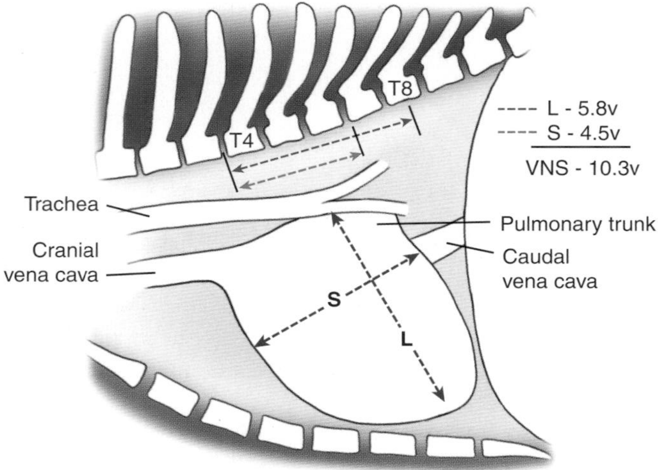

FIGURE 4.4-6 Measurement of vertebral-heart size (VHS) in dogs and cats. Key: *L,* long axis of the heart; *S,* short axis of the heart; *T4,* body of the 4th thoracic vertebra.

VERTEBRAL-HEART SIZE

Vertebral-heart size (VHS) is an index that relates ("normalizes") heart size to body size by using the length of the vertebral bodies in the mid-thoracic area as the unit of measure (Fig. 4.4-6). It is a standard indicator of radiographic heart size in dogs and cats, and is particularly useful in identifying the presence and extent of cardiomegaly and monitoring progressive changes in heart size.

On a lateral thoracic radiograph, the cardiac dimensions in the long axis (L) and short axis (S) are transposed onto the vertebral column and then recorded as the number of vertebral bodies, beginning at the cranial margin of the 4th thoracic vertebra (T4), and estimated to one decimal place—i.e., each vertebral body is divided into 10. The values for L and S are added to arrive at the VHS.

The normal range in dogs is 8.5–11 and the normal range in cats is 6.8–7.8.

and activity. The cat's long-term prognosis was guarded, owing to the severity of his cardiomegaly and underlying myocardial disease. However, his quality of life was expected to be acceptable if the medications were able to control the consequent pulmonary congestion.

Anatomical features in canids and felids

Introduction

This section covers the pericardium and the heart's location, muscular walls, and external features. The mediastinum and its contents are described further in Case 4.2. The cardiac chambers and valves are described further in Case 4.7.

The heart accounts for about 0.5% of the total body weight in a cat, which is like most other domestic animals. However, the feline heart is less clearly conical in shape than it is in dogs and other domestic animals, appearing almost globular.

Exceptions include Greyhound dogs, Thoroughbred horses, and pigs. Through generations of selective breeding for racing, Greyhounds and Thoroughbreds have relatively large hearts, which become even larger with training. Compared with other domestic animals, pigs have relatively small hearts.

Function

The pericardium in toto forms a protective sac around the heart, helps maintain the heart in its normal position within the thorax, helps minimize friction between the heart and surrounding structures, and may also prevent the heart from becoming over-distended.

Heart position

The **heart** lies within the **mediastinum** (the partition between left and right pleural cavities), enclosed within the **pericardial sac**. In most normal dogs and cats, the heart is situated between the 3rd and 6th ribs. The dorsal boundary of the heart lies on a horizontal plane through the center of the 1st rib; its ventral boundary is the **sternum**; its caudal boundary is the dome of the **diaphragm**.

The long axis of the heart is obliquely oriented in dogs and cats, the heart lying in a craniodorsal-caudoventral orientation (Fig. 4.4-7). Unlike in horses and ruminants, whose hearts are more vertically oriented, the apex of the heart almost touches the diaphragm in dogs and cats, and the **phrenicopericardial ligament** anchors the pericardium to the diaphragm rather than to the sternum.

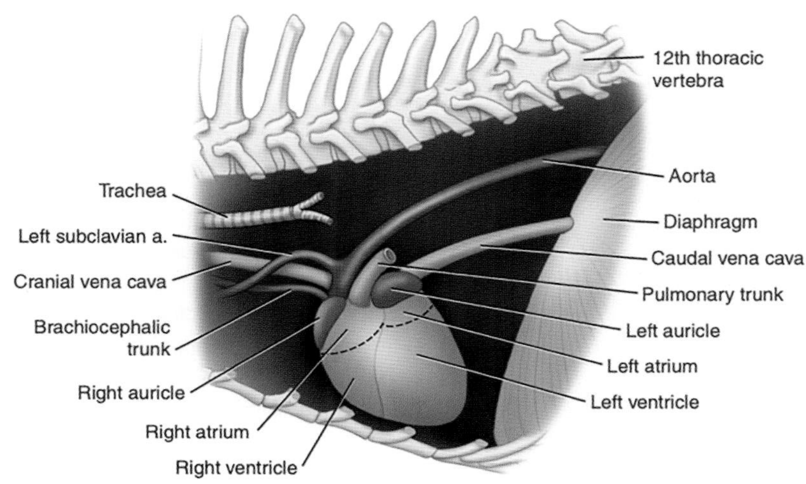

FIGURE 4.4-7 The size, shape, and orientation of the heart in a normal cat as viewed from the left. Note that the pericardial sac has been removed.

The cardiac notch provides a useful "acoustic window" for echocardiography.

As in other domestic animals, most of the canine/feline heart lies to the left of midline, so the left surface of the heart is closer to the thoracic wall than is the right surface. The **cardiac notch** is the gap in the ventral border of left and right **lungs** (larger on left than right), which allows the pericardium to contact the thoracic wall.

Pericardium

The **pericardium** (Fig. 4.4-8) is a fibroserous sac that surrounds the heart. Embryologically, it begins as a deep invagination of the ventral mediastinum around the developing heart. These origins are reflected postnatally by the phrenico-pericardiac ligament—a short, strong band of connective tissue covered by pleura which firmly attaches the pericardium to the caudoventral thoracic wall (in dogs and cats, to the ventral portion of the diaphragm).

These origins are further reflected in the structure of the pericardial cavity, which is like that of the pleural cavity: parietal and visceral layers of mesothelium are separated by a thin film of serous fluid, which lubricates the 2 surfaces and allows them to glide over each other during the cardiac cycle.

The **parietal** layer of the pericardium has a strong, fibroelastic outer coating—which itself is covered by the visceral pleura—giving the parietal pericardium a whitish, semi-opaque appearance. This outer part of the pericardial sac is generally referred to as the **fibrous pericardium** or simply as the pericardium. Its fibrous tissue is continuous with that of the phrenico-pericardiac ligament. The **visceral** layer of the pericardium is thin and firmly adhered to the underlying myocardium, so it is generally referred to as the **epicardium**.

Excess fluid in the pericardial sac—whether serous fluid (pericardial effusion), blood (hemopericardium), or inflammatory exudate (exudative pericarditis)—may result in compression of the heart, a condition known as "cardiac tamponade" (Fr. *tampon* a plug). Excessive pericardial fluid prevents normal dilation and filling of cardiac chambers, impeding return of blood to the heart and resulting in distention of peripheral veins such as the external jugular vein. When severe, the combination of pericardial fluid accumulation and reduced cardiac filling may muffle the heart sounds on thoracic auscultation. (Note: Heart sounds may also be muffled with pleural effusion and in obese patients, it is not specific for pericardial pathology.)

The 2 layers of pericardium (parietal and visceral) fuse where the ascending aorta and the pulmonary trunk exit the heart base. Thus, the pericardial sac completely encloses the heart—i.e., the pericardial space does not communicate with the mediastinal space or the pleural cavity—and only at its base is the heart attached to the fibrous pericardium.

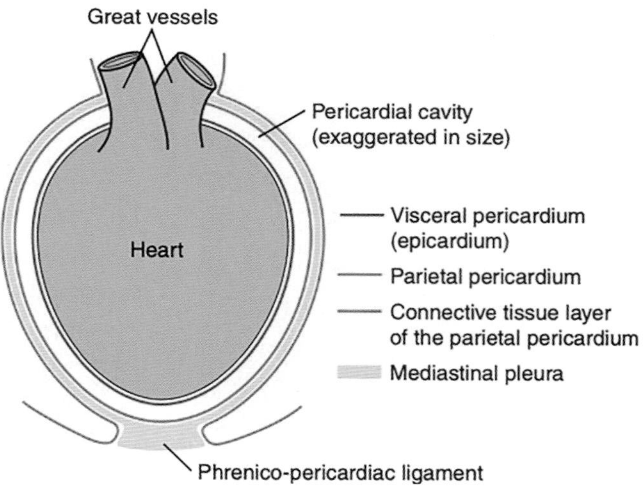

FIGURE 4.4-8 Diagram of the pericardium.

External features of the heart

The base of the heart (Fig. 4.4-9) is formed by the thin-walled **atria** (left and right), which are separated from the ventricles by the encircling **coronary groove**. Each atrium has a blind diverticulum or free appendage, the **auricle**, which wraps part of the way around its respective atrium. The thicker-walled **ventricles** (left and right) meet and form a rounded cone at the cardiac **apex**. The position of the **interventricular septum** (the wall dividing the internal chambers of the left and right ventricles) is marked externally by the **left interventricular**, or **paraconal**, **groove** on left side of the heart and the **right interventricular**, or **subsinuosal**, **groove** on the right side of the heart.

The left (or "auricular") surface of the heart is formed mainly by the left atrium and left ventricle, but the right ventricle and the auricle of the right atrium extend around the cranial border of the heart and contribute substantially to the left surface (as shown in Fig. 4.4-7). The right ventricle thus lies both cranial to, and to the right of, the left ventricle. The right (or "atrial") surface of the heart is formed mainly by the right atrium and right ventricle, but the left ventricle extends around the caudal border of the heart, thereby contributing somewhat to the right surface.

This orientation explains why the pulmonary valve (in the right ventricular outflow tract) is auscultated best on the left side of the patient's thorax (see Case 4.7).

Heart wall

The thick middle layer of the heart wall (Fig. 4.4-10)—the **myocardium**—is composed of cardiac muscle, which is a type of striated muscle that is specific to the heart (Fig. 4.4-10). As mentioned above, the exterior of the myocardium is covered by the epicardium (visceral pericardium). The interior of the myocardium is covered by the **endocardium**, which is a smooth endothelial layer that is continuous with the endothelium of the great vessels that enter and exit the heart. The interior of the heart is described in detail in Case 4.7.

The myocardia of the atria and ventricles are completely separated by a fibrous "skeleton." This connective tissue framework is essential for preventing indiscriminate electrical conduction between the atria and ventricles. **Atrial muscle** is thin and is arranged in superficial and deep bundles. Some of the superficial bundles are common to both atria, but the remainder of the superficial and the entire deep bundles are confined to the separate atria. The **ventricular muscle** is much thicker, but it too is arranged in superficial and deep bundles. Some superficial bundles coil around both chambers, using the interventricular septum to complete a figure-of-8 course. Others, like the deeper bundles, encircle only one ventricle.

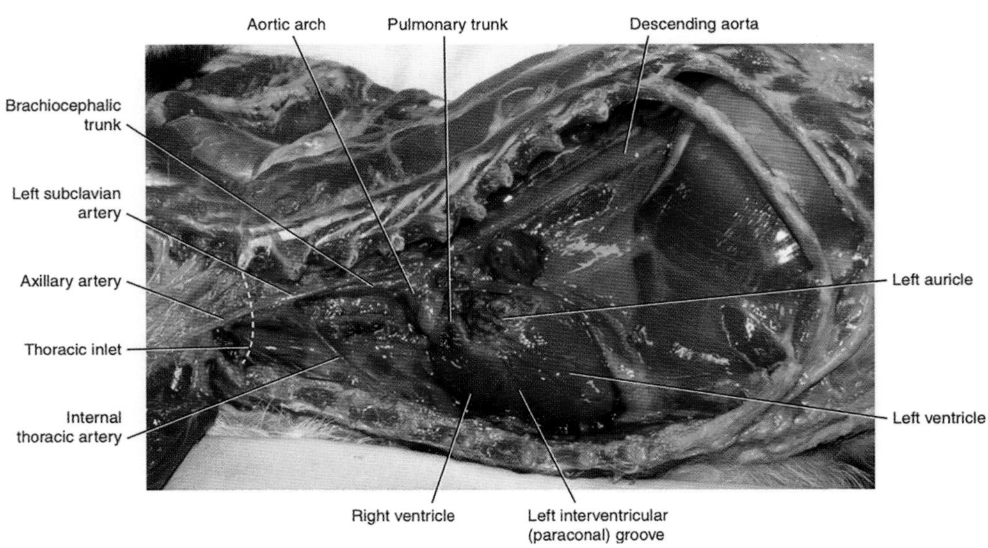

FIGURE 4.4-9 External features of the canine/feline heart.

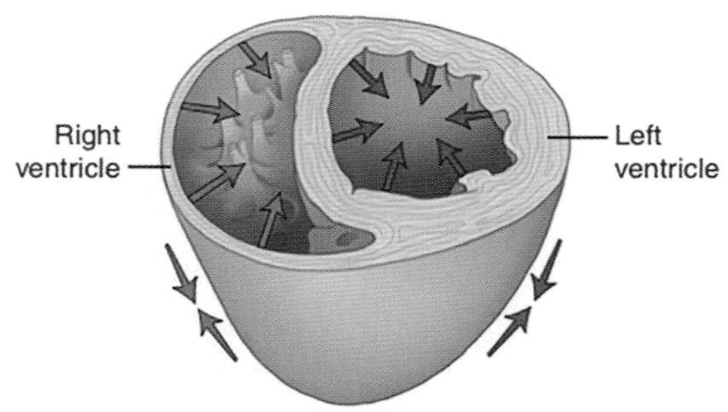

FIGURE 4.4-10 Mode of contraction of the left and right ventricles. The right ventricular muscle squeezes the lumen in a "bellows" action, whereas the left ventricular wall contracts radially.

The thick, muscular **interventricular septum** has 2 components. The larger **muscular** part is thick myocardium, formed by the combined walls of the 2 ventricles. The surface facing the lumen of the left ventricle is concave, whereas the surface facing the lumen of the right ventricle is convex. The thin, **collagenous** or membranous part is a small, inconspicuous area in the extreme dorsal part of the septum.

The membranous septum marks the site of final closure of the embryonic interventricular foramen. It is the most common site of congenital interventricular (ventricular septal) defects.

Selected references

[1] https://biosphera.org/international/ 3D cat anatomy.
[2] Dyce, Sack, and Wensing. Textbook of veterinary anatomy, 5th ed. Elsevier.
[3] VIN.com James Buchanan Cardiology Library: https://www.vin.com/apputil/content/defaultadv1.aspx?pId=84&id=4253805.

CASE 4.5

Persistent Right Fourth Aortic Arch

David Holt
Department of Clinical Sciences and Advanced Medicine, University of Pennsylvania School of Veterinary Medicine, Philadelphia, Pennsylvania, US

Clinical case

History

A 4-month-old female intact Labrador Retriever puppy was presented for vomiting after eating, according to the owner. The problem began when the puppy was weaned onto solid food at around 6 weeks of age. The puppy seemed otherwise normal; she was active and growing well, with a good appetite.

Physical examination findings

No significant clinical abnormalities were found on physical examination.

Differential diagnoses

Esophageal foreign body, vascular ring anomaly, congenital esophageal stricture, another congenital defect

Diagnostics

Thoracic radiographs with a barium swallow (Fig. 4.5-1) revealed marked dilation of the esophagus just cranial to the heart base. In addition, the trachea was deviated to the left just cranial to the heart base on the DV projection (Fig. 4.5-2). No other radiographic abnormalities were noted in the thorax.

> Diagnostic procedures that may be used in these cases include contrast radiography (contrast esophagram), in which a barium suspension is given by mouth just before thoracic radiography and CT with contrast enhancement to visualize the major arteries. Conclusive findings include focal narrowing of the esophageal lumen just cranial to the heart base. An additional reason for obtaining thoracic radiographs in these cases is because aspiration pneumonia is common in animals experiencing frequent regurgitation.

Diagnosis

Probable persistent right aortic arch (PRAA), causing partial but persistent esophageal obstruction. (Fig. 4.5-3)

> The most common vascular ring anomaly in domestic animals involves persistence of the right fourth aortic arch during embryonic development (called PRAA, see side box entitled "Embryonic development"). The esophagus becomes trapped between the right fourth aortic arch, the base of the heart, and the ligamentum arteriosum (a vestige of the ductus arteriosus). The diagnosis is suspected based on the clinical signs of regurgitation of solid food and confirmed by radiographs. Clinicians must have an index of suspicion for regurgitation in any young animal presented for "vomiting." Successful treatment involves identifying, ligating, and transecting the ligamentum arteriosum. In rare cases, the ligamentum can be a patent ductus arteriosus. In these cases, a continuous or "machinery" murmur is heard on thoracic auscultation.

Comparative Veterinary Anatomy: A Clinical Approach. https://doi.org/10.1016/B978-0-323-91015-6.00027-3

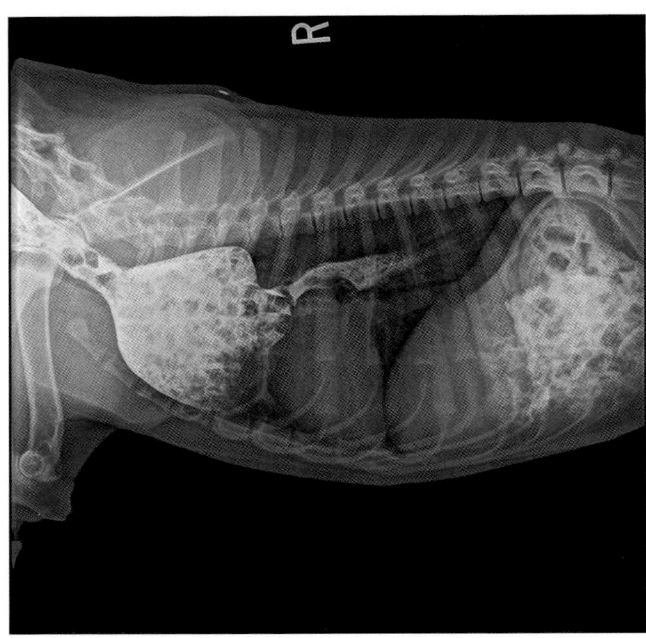

FIGURE 4.5-1 Lateral radiograph with a barium swallow indicating marked dilation of the esophagus cranial to the heart base with subsequent narrowing (*arrow*).

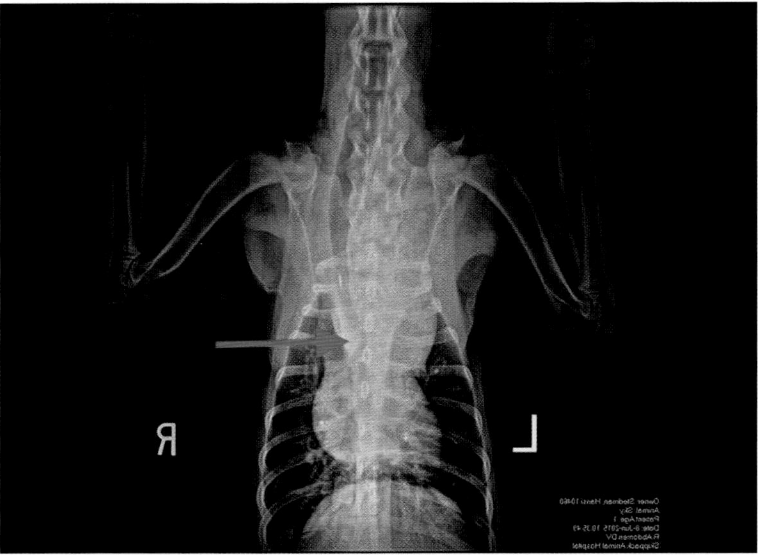

FIGURE 4.5-2 Dorsoventral radiograph showing leftward deviation of the trachea at the level of the aorta, indicating a persistent right aortic arch (*arrow*).

THORACOTOMY VS. THORACOSCOPY: THE DIFFERENCES

Thoracoscopy or video-assisted thoracoscopic surgery (VATS) is a minimally invasive surgery using a camera attached to a surgical endoscope and instruments (see Section 2.1) introduced into the thoracic cavity through small incisions (ports or portals), allowing the surgeon to operate the instruments while viewing the anatomy on which the surgeon is operating. The advantages are faster post-operative recovery, improved wound healing, and less post-operative pain.

Exploration of the thorax in cases of pyothorax and pericardectomy (removal of the pericardium surrounding the heart) are examples of surgical procedures that can be performed via thoracoscopy and thoracotomy.

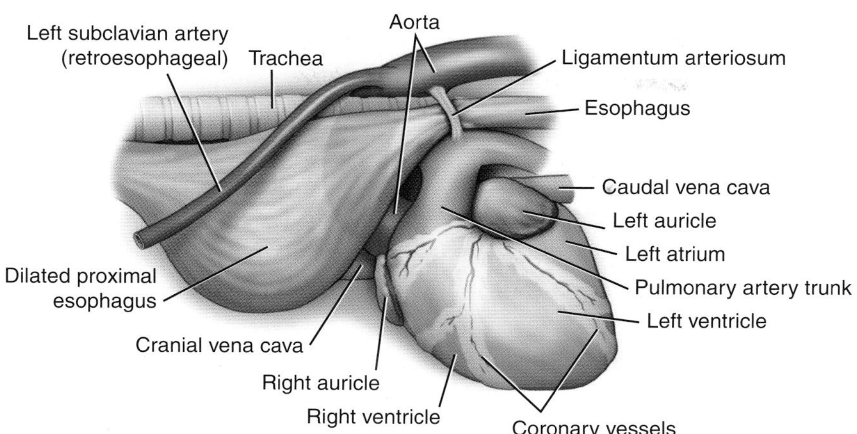

FIGURE 4.5-3 Persistent right aortic arch. The esophagus is trapped between the right aortic arch on the right, the ligamentum arteriosum on the left, and the heart base and pulmonary arteries ventrally.

Treatment

The dog was anesthetized and prepared for thoracotomy for identification and surgical correction of the suspected vascular ring anomaly/PRAA. A left lateral 4th intercostal space thoracotomy was performed. A vertical incision was made just caudal to the scapula. The subcutaneous tissues and ventral aspect of the latissimus dorsi muscle were incised. The intercostal spaces were palpated and counted and the serratus ventralis muscle separated over the fourth intercostal space. The intercostal muscles and pleura were incised to enter the thoracic cavity and expose the left heart base. The left ligamentum arteriosum was identified, ligated proximally and distally, and transected to relieve the annular constriction of the esophagus. The dog made an uneventful recovery from surgery and continued to thrive and grow normally.

Chronic dilation of the cranial thoracic esophagus is a common consequence of vascular ring anomalies such as PRAA. In many cases, normal esophageal dimensions return after surgery, but motility may remain abnormal. In more severe or chronic cases, the esophageal dilation can persist, and these animals must be fed small meals frequently from a raised platform to avoid regurgitation after eating and chronic malnutrition.

Anatomical features in canids and felids

Introduction

This section covers the aorta and its thoracic branches along with a brief discussion of normal embryonic development of the aorta.

Function

The aorta is the chief and largest artery in the dog and cat, originating from the left ventricle of the heart, separated by the aortic valve, and tracks through the abdomen separating into the paired internal iliac and single median sacral arteries at its termination. The aorta distributes oxygenated blood to the entire body via the systemic circulation except for the respiratory region of the lung. Specifically, branches from the ascending aorta supply the heart; branches from the aortic arch supply the head, neck, and thoracic limbs; branches from the thoracic part of the descending aorta supply the chest, except the heart and the respiratory region of the lung; and branches from the abdominal aorta supply the abdomen. The pelvis and pelvic limbs receive blood from the external and internal iliac arteries.

Aorta

The normal **aorta** consists of an ascending portion, an arch, and a descending portion. The **ascending aorta** is attached to the fibrous base of the heart and is largely contained within the pericardial sac. The first branches of the aorta are the **coronary arteries**; they arise from a dilation (the "bulb") of the ascending aorta immediately distal to—i.e., downstream of—the **aortic valve**. The coronary arteries supply the heart muscle (**myocardium**) with blood.

The **aortic arch** connects the ascending aorta with the descending aorta. As the name suggests, the aortic arch curves or "arches" in a broad, almost 180-degree bend to direct the aorta caudally. The blood supply to the head, neck, and thoracic limbs is supplied by 2 branches from the aortic arch: the **brachiocephalic trunk** and the **left subclavian artery**.

Thoracic branches of the aorta

The **brachiocephalic trunk** is the first large branch of the aorta. It runs cranially, ventral to the trachea, and gives rise to the left and right **common carotid arteries** and the **right subclavian artery**. The **left subclavian artery** is the second largest branch of the aorta. Together, the left and right subclavian arteries give rise to the bilateral **vertebral arteries**, **costocervical trunk**, **superficial cervical arteries**, **axillary arteries**, and **internal thoracic arteries**. The aortic arch continues as the **descending aorta**, supplying arterial blood to the rest of the body (Fig. 4.5-4).

HUMAN ORIGINS OF SOME TERMS USED IN VETERINARY ANATOMY

The ascending aorta is so-named because in humans, this short initial portion of the aorta "ascends" from the heart in the standing or seated person, being directed toward the head. The same anatomical term is used in quadrupeds, although the ascending aorta ascends only in the sense that it likewise flows in a cranial direction (toward the head).

The descending aorta in humans "descends" from the aortic arch in the standing or seated person as it takes blood away from the heart and into the lower portion of the thorax and the abdomen. The same term is used in quadrupeds for the portion of the aorta which takes blood into the caudal thorax and the abdomen, even though the descending aorta travels horizontally, parallel to the thoracolumbar spine.

FATE OF THE 6 PAIRS OF AORTIC ARCHES

Arches 1 and 2. The first and second pairs of arches involute very early.

Arch 3. The ventral roots of the third arches persist and become the left and right common carotid arteries.

Arch 4. The left fourth arch expands and, with the left dorsal root and part of the left ventral root, forms the definitive aortic arch. The right fourth arch combines with part of the right dorsal aorta to form the right subclavian artery.

Arch 5. The fifth arches involute bilaterally.

Arch 6. The sixth arches expand to become the pulmonary arteries, and the left sixth arch also connects the pulmonary arteries to the left dorsal aorta as the ductus arteriosus. (The ductus arteriosus is discussed in greater detail in Case 4.6).

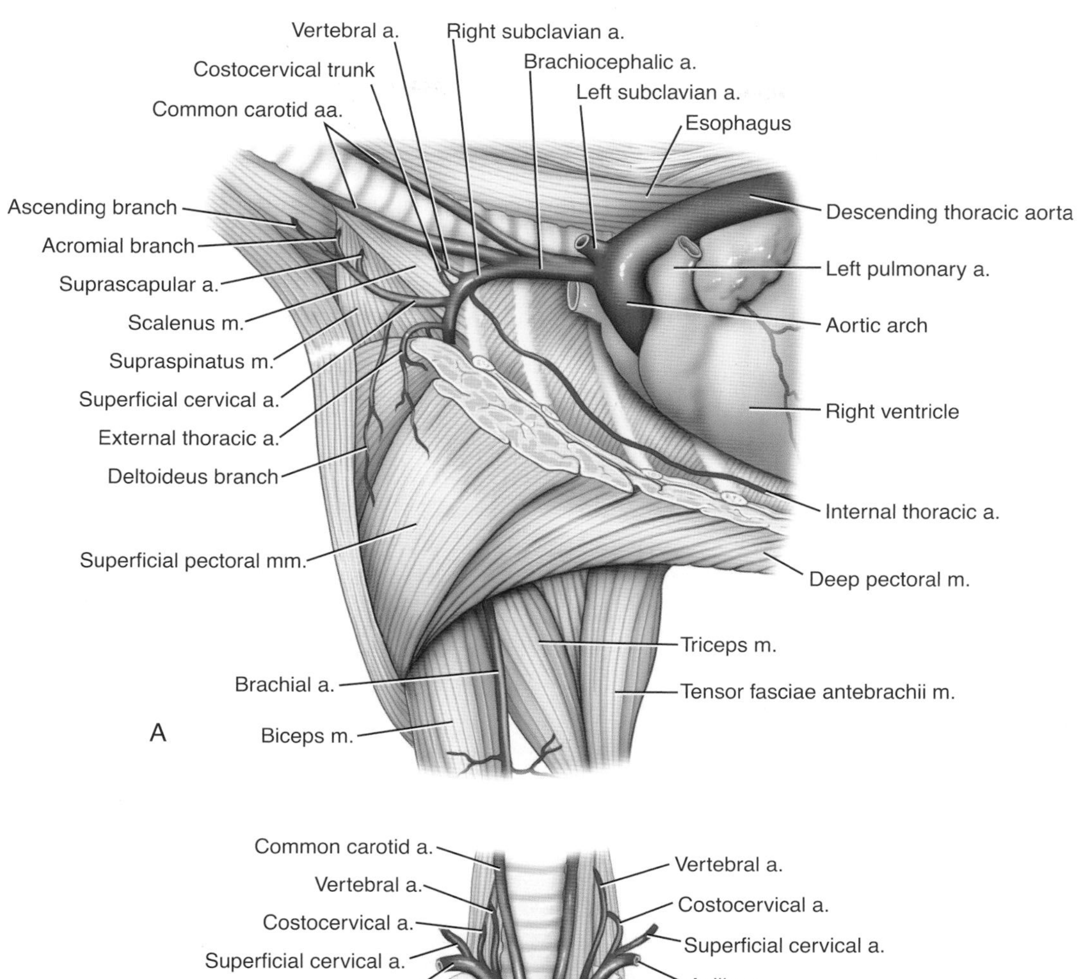

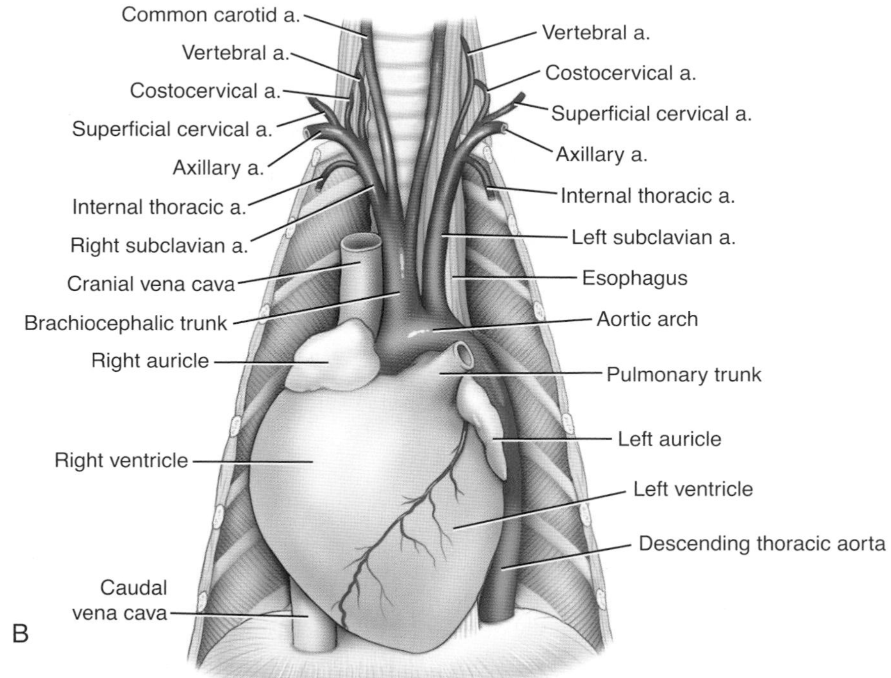

FIGURE 4.5-4 Normal anatomy and branches of the thoracic aorta.

Embryonic development

In the normal embryo, the aorta begins as a set of 4 vessels, the **left** and **right dorsal** and **ventral aortas**. During embryonic development, the left and right dorsal aortas fuse to form the **descending aorta**, and the left and right ventral aortas fuse to form the **heart**.

Initially, the dorsal and ventral aortas are connected by 6 pairs of vessels (left and right) called the **aortic arches**, which form around the embryonic foregut and associated pulmonary bud (Fig. 4.5-5). The aortic arches eventually become the major arteries of the neck and head (Fig. 4.5-6).

The aortic arches develop and regress or transform sequentially, so they are not present at the same time; and although they begin as symmetrical pairs, normally their transformation during embryonic development is somewhat asymmetrical.

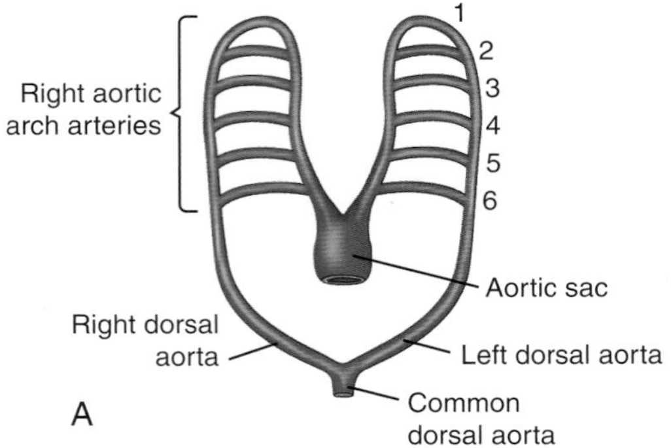

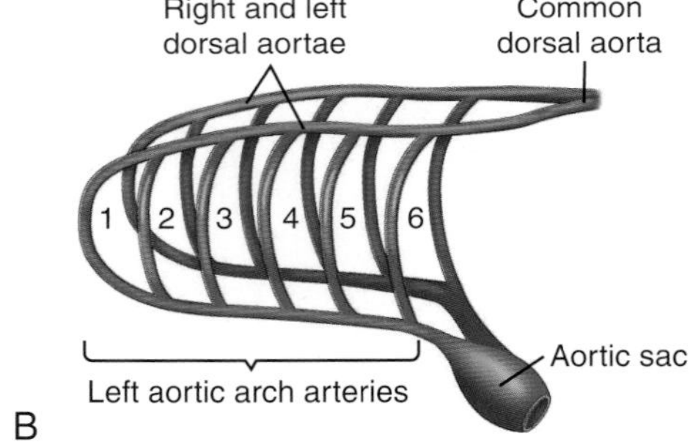

FIGURE 4.5-5 Dorsoventral and lateral views of the paired dorsal and ventral aortas in the embryo connected by the 6 pairs of aortic arches.

VASCULAR RING ANOMALIES

Most developmental anomalies involving the major arteries represent persistence of a right aortic arch, or less commonly a right ligamentum arteriosum, which should have regressed during embryonic development or involuted at birth. The aortic arches initially connect the embryonic dorsal and ventral aortas, forming vascular rings, so these developmental defects are called vascular ring anomalies. Vascular ring anomalies include PRAA with a left ligamentum arteriosum, left aortic arch with a right ligamentum arteriosum, and double aortic arches. Patent ductus arteriosus occurs as a separate condition where the ductus, connecting the aorta to the pulmonary artery in the embryo, fails to involute, resulting in shunting of blood from the aorta to the pulmonary artery, left heart volume overload, and heart failure.

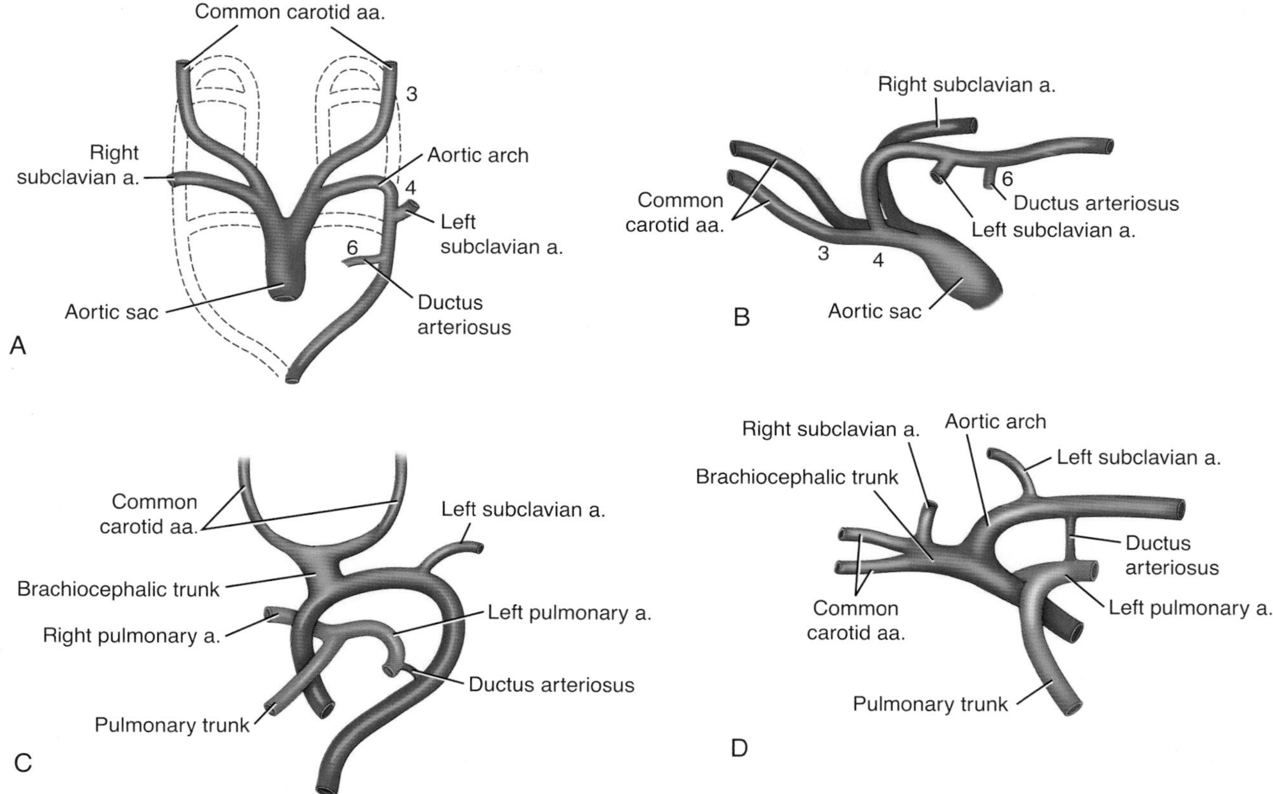

FIGURE 4.5-6 The fate of the embryologic arches.

DOUBLE AORTA

Double aorta is a rare vascular ring anomaly in which both the left and right fourth embryonic arches persist as fully formed aortic arches, whereas normally just the left fourth arch remains and becomes the definitive left aortic arch. Both the esophagus and the trachea may be narrowed at the base of the heart with clinical signs of esophageal obstruction. Thoracic radiographs may show focal narrowing of the esophageal lumen.

Treatment involves performing a thoracotomy. The arches are dissected and the smaller of the 2 arches is divided between vascular clamps. The ends are then oversewn. The left ligamentum arteriosum is ligated and divided. Thus, the esophageal and tracheal compressions are relieved, and the dog has one fully functional aortic arch.

Selected references

[1] Aultman SH, Chambers JN, Vestre WA. Double aortic arch and persistent right aortic arch in two littermates: surgical treatment. J Am Anim Hosp Assoc 1980;16:533–6.
[2] Buchanan JW. Tracheal signs and associated vascular anomalies in dogs with persistent right aortic arch. J Vet Intern Med 2004;18:510–4.
[3] Buchanan JW. Patent ductus arteriosus and persistent right aortic arch surgery in dogs. J Small Anim Pract 1968;9:409–28.
[4] Shires P, Liu W. Persistent right aortic arch in dogs: a long term follow-up after surgical correction. J Am Anim Hosp Assoc 1981;17:773–6.

CHAPTER 4

CASE 4.6

Patent Ductus Arteriosus

Mark Oyama
Department of Clinical Sciences and Advanced Medicine, University of Pennsylvania School of Veterinary Medicine, Philadelphia, Pennsylvania, US

Clinical case

A hyperkinetic or "bounding" peripheral pulse is felt when there is a greater-than-normal difference in pulse pressures between systole and diastole—i.e., greater-than-normal amplitude in the arterial pressure wave. The pulse not only feels stronger than normal, it feels more intense than expected for the animal's physical condition and mental/emotional state. It is an important clinical finding and, with the presence of a continuous murmur, it is also highly suggestive of a congenital heart defect that represents a failure to complete the transition from fetal to neonatal circulation.

A heart murmur is indicative of turbulent and higher than normal velocity blood flow. A "thrill" in this case refers to palpable vibration emanating from a vascular structure. For example, a thrill may be palpated in large arteries experiencing unusual flow—e.g., the uterine arteries in cows during the second half of pregnancy (commonly called fremitus [L. a vibration detectable on palpation]). A thrill may also be palpated over the heart or great vessels when a structural abnormality is present that alters the normal course or intensity of blood flow through the cardiopulmonary circulation. A continuous murmur describes abnormal heart sounds that are heard during both systole and diastole. As it is heard throughout the entire cardiac cycle, it is also called a "machinery" murmur. Together with the dog's young age, this murmur is highly suggestive of a congenital heart defect.

History

A 5-week-old female Poodle was presented for lethargy and failure to thrive. Compared with her littermates, the puppy was nursing less vigorously and was underweight. Occasional panting was also reported.

Physical examination findings

The puppy was quiet but responsive to handling. Palpation of both femoral arteries revealed hyperkinetic peripheral pulses. The mucous membranes of the mouth were pink, but the capillary refill time was 3 seconds (normal < 2 seconds). A cardiac "thrill" was palpated on the left side of the chest over the base of the heart, and a continuous murmur was heard on auscultation of the same area. The murmur was particularly prominent over the main pulmonary artery (craniodorsally at the left heart base), and it radiated cranially to the thoracic inlet and to the right side of the chest (to the right heart base).

Comparative Veterinary Anatomy: A Clinical Approach. https://doi.org/10.1016/B978-0-323-91015-6.00028-5

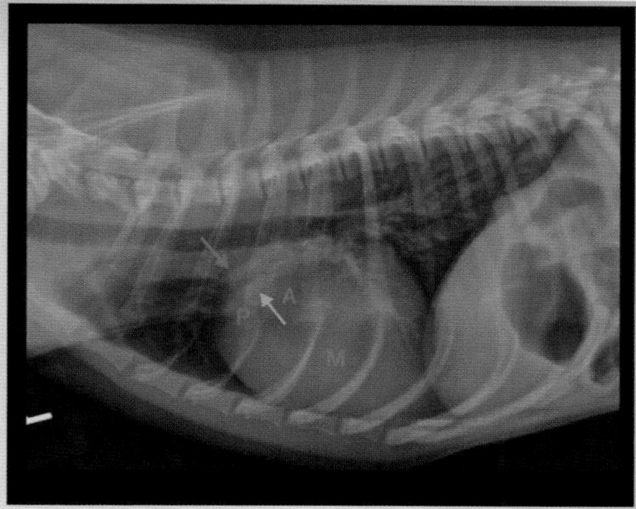

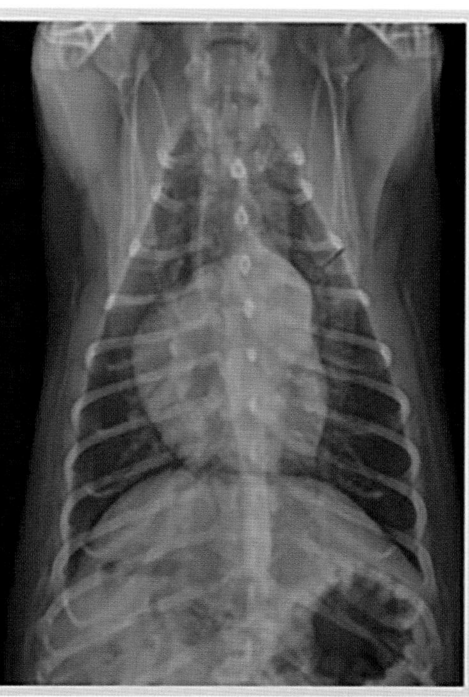

FIGURE 4.6-1 Radiographs of a dog with a patent ductus arteriosus. Left: Lateral projection showing the approximate locations on the chest wall of the pulmonic valve *(P)*, aortic valve *(A)*, and mitral valve *(M)*; the dilated pulmonary artery *(blue arrow)* and pulmonary vein *(orange arrow)* indicate pulmonary overcirculation. Right: Dorsoventral view (right radiograph) showing a bulge in the descending aorta *(blue arrow)* consistent with the presence of a patent ductus arteriosus.

Differential diagnoses

Aortic-pulmonary shunt, coronary arteriovenous fistula, aorticopulmonary window, patent ductus arteriosus (PDA), various other arteriovenous shunts

> Each of these differential diagnoses is characterized by (1) continuous flow through 2 vessels (either an artery and vein, or systemic artery and pulmonary artery) that bypass the usual capillary bed and (2) decreased systemic diastolic blood pressure that increases the difference in systolic and diastolic pulse pressures.

Diagnostics

Thoracic radiographs showed left-sided cardiac enlargement and pulmonary overcirculation, consistent with left-to-right shunting of blood. In the dorsoventral view, the aortic arch, left auricle, and main pulmonary artery were abnormally prominent. The most significant radiographic finding was an aortic bulge, also called a ductus bump, which is a dilation of the descending aorta at the level of the ductus arteriosus (Fig. 4.6-1).

> Pulmonary overcirculation is indicated by an increase in the size of both the pulmonary arteries and veins visible on routine thoracic radiographs. Not only are the normally visible pulmonary vessels increased in width, but pulmonary vessels can also be visible further into the periphery of the lung fields than normal.

High-velocity, continuous flow through the ductus arteriosus from the aorta to the pulmonary artery occurs when there is incomplete smooth muscle development in the wall of the fetal ductus, which prevents constriction and closure of the ductus shortly after birth. Typically, the ductus has little or no smooth muscle at the aortic end, but sufficient smooth muscle at the pulmonary artery end to effect partial constriction, resulting in a funnel-shaped ductus with a wide aortic opening and narrower pulmonary artery opening. The failure of the ductus to completely close creates a connection between the high-pressure aorta and lower pressure pulmonary artery which allows blood flow to bypass the systemic capillary bed. The murmur heard on physical examination is continuous because pressure in the aorta is always greater than the pressure in the pulmonary artery, even during diastole, so blood flows continuously from the aorta to the pulmonary artery through the PDA. The high-velocity, continuous ductal flow from the aorta into the main pulmonary artery explains the continuous murmur and cardiac thrill. During diastole, blood flow in the aorta continues to flow into the pulmonary artery, resulting in a lowered diastolic blood pressure, widened pulse pressure, and hyperkinetic femoral artery pulses. The echocardiographic study helped confirm that this puppy failed to complete the transition from fetal to neonatal circulation.

Electrocardiography (ECG) indicated left ventricular enlargement: increased R wave amplitudes in leads II, III, and aVF, and in the left precordial leads V2 and V3. In addition, the P waves appeared wider than normal, indicating left atrial enlargement.

Echocardiography revealed eccentric left ventricular hypertrophy and dilation of the left atrium and main pulmonary artery (Fig 4.6-2). Using the color-flow Doppler, high-velocity and continuous blood flow was seen flowing from the aorta into the main pulmonary artery through a PDA (Figs. 4.6-3 and 4.6-V1 in the online version at https://doi.org/10.1016/B978-0-323-91015-6.00028-5). In addition, there was mildly increased left ventricular outflow velocity and moderate secondary mitral valve insufficiency due to dilation of the mitral valve annulus.

Cardiac catheterization and contrast angiocardiography can also be used to confirm the diagnosis of a PDA, but the echocardiographic findings in this case were sufficient.

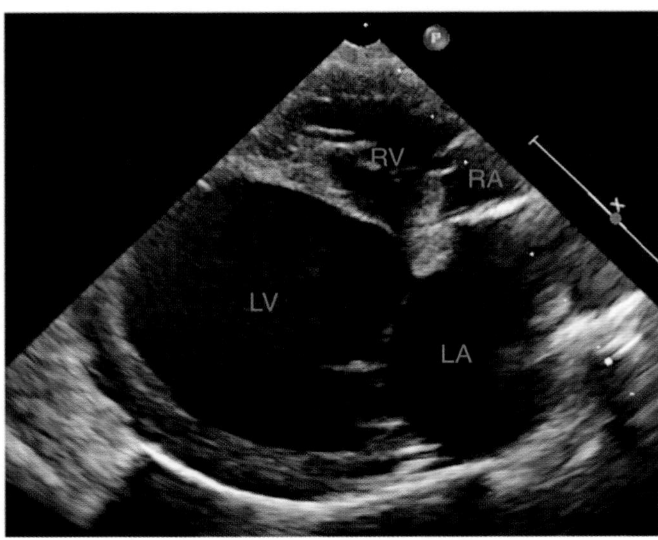

FIGURE 4.6-2 Echocardiogram of a dog with a PDA. The left ventricle *(LV)* and left atrium *(LA)* are dilated and volume overloaded as compared to the right ventricle *(RV)* and right atrium *(RA)*. This finding in association with a continuous heart murmur is suggestive of a left-to-right extracardiac shunting defect, such as a PDA. Blood flow in the aorta is diverted into the PDA and pulmonary vessels causing pulmonary overcirculation. The extra blood flow through the lungs then returns to the left side of the heart causing dilation before it is once again pumped back out into the aorta.

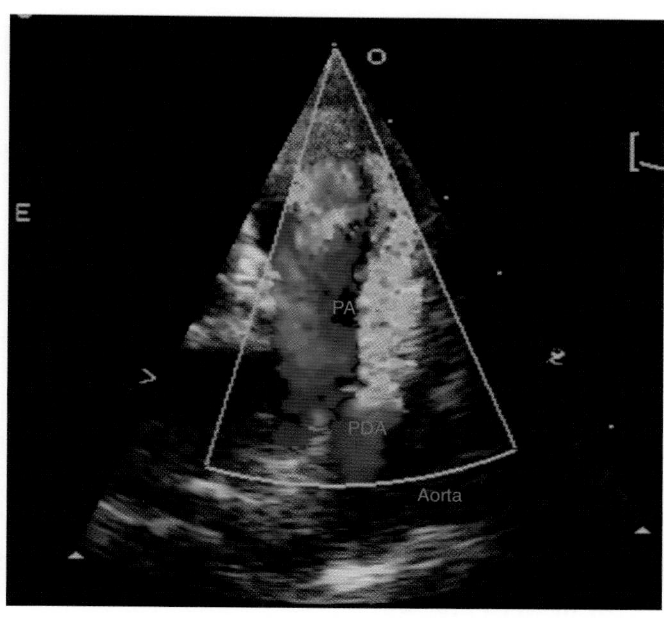

FIGURE 4.6-3 Close-up echocardiogram of a dog with a PDA. Color flow Doppler demonstrates the presence of a high-velocity and turbulent jet of blood originating from the high-pressure aorta, through the PDA, and into the lower pressure pulmonary artery *(PA)*. The high velocity and turbulent blood flow produce a continuous heart murmur.

Diagnosis

Patent ductus arteriosus (PDA)

Treatment

The PDA was closed using a minimally invasive transcatheter approach under general anesthesia. Recovery from anesthesia was uneventful, and the continuous murmur was no longer present. Follow-up examination revealed no problems, and the puppy thrived and grew normally.

In cats and very small dog breeds, the PDA is surgically ligated via thoracotomy. Minimally invasive transcatheter procedures are now widely used in most dog breeds. With the puppy under general anesthesia, a catheter is inserted into a femoral artery and, under fluoroscopic guidance, a thrombogenic coil or medical duct occluder is inserted into the ductus arteriosus to block flow through the PDA. Unless the PDA is complicated by other congenital heart defects or treatment delay has resulted in congestive heart failure, ductus closure is generally curative.

PATENT DUCTUS ARTERIOSUS (PDA)

Patent ductus arteriosus is the most common type of congenital heart disease seen in dogs. It is less common in cats. The dog breeds most often affected include Bichon Frise, Chihuahua, German Shepherd dog, Keeshond, Maltese, Pomeranian, and Poodle. At least in Miniature Poodles, it appears to be caused by a lack of smooth muscle in the wall of the ductus arteriosus, preventing its normal postnatal closure. Females are more likely to have PDA than males. A common presenting complaint is failure to thrive or slowed growth, so puppies with PDA are often described as “the runt of the litter” or may have respiratory signs due to congestive heart failure.

Anatomical features in canids and felids

Introduction

Understanding fetal circulation and the changes that normally take place shortly after birth are useful in the diagnosis and treatment of congenital heart disease in mammals.

Blood flow to the fetal heart (Fig. 4.6-4)

During gestation, blood from the **placenta** enters the fetus via the **umbilical vein**. In all mammalian species, there is one umbilical vein within the fetal abdomen at birth. Carnivores have 2 umbilical veins in the umbilical cord, but these unite at the umbilicus.

The umbilical vein enters the liver between the quadrate and left lobes, and it joins the left end of the left intrahepatic branch of the portal vein. This junction is connected by a large shunt, the **ductus venosus,** to the caudal vena cava near the beginning of the left hepatic vein. As a result, a small portion (about 10%) of the umbilical blood bypasses the liver. The ductus venosus is still present at birth in carnivores. After birth, the ductus venosus closes and the intrahepatic part of the umbilical vein becomes the left extremity of the left branch of the portal vein.

Blood flow through the fetal heart (Fig. 4.6-4)

Blood from the caudal vena cava enters the right atrium, but approximately 60% of the blood is directed through the **foramen ovale** into the left atrium, and from there into the left ventricle. The remaining 40% from the caudal vena cava mixes with blood from the cranial vena cava and flows into the right ventricle.

The foramen ovale is an exclusively fetal aperture between right and left atria. Its closure at birth is the result of several contemporaneous phenomena: (1) increased blood flow into the left atrium from the now-functional lungs and pulmonary circulation, resulting in a pressure gradient that resists flow from right to left atrium, (2) decreased pressure in the right heart, owing to the now-functional lungs and closure of the umbilical vein, and (3) persistence of the ductus arteriosus, which remains patent for a few hours or days after birth and thereby allows some extracardiac left-to-right shunting of blood.

During fetal life, the stroke volume of the left and right ventricles is similar, as is the thickness of the ventricular walls. After birth, the right ventricle gradually atrophies and the left hypertrophies as pressures in the left heart exceed those in the right heart.

COMPARATIVE FETAL CIRCULATION

Like carnivores, ruminants have 2 umbilical veins in the umbilical cord, but these unite at the umbilicus. The pig and the horse have only one vein in the umbilical cord, the other disappearing during gestation.

Like carnivores, the ductus venosus is still present at birth in ruminants, but it disappears during gestation in the pig and the horse.

Unlike carnivores, ruminants and horses have an additional sphincter in the cord at the umbilicus, which facilitates the transition from fetal to neonatal circulation.

ATRIAL SEPTAL DEFECT

An atrial septal defect (ASD) is distinct from persistent foramen ovale and is a relatively rare developmental failure of atrial septal construction. An ASD permits blood flow from left to right atrium. This increases the volume of blood that the right side of the heart must pump through the tricuspid and pulmonic valves and into the pulmonary circulation, so pulmonary overcirculation is a shared feature of large ASDs.

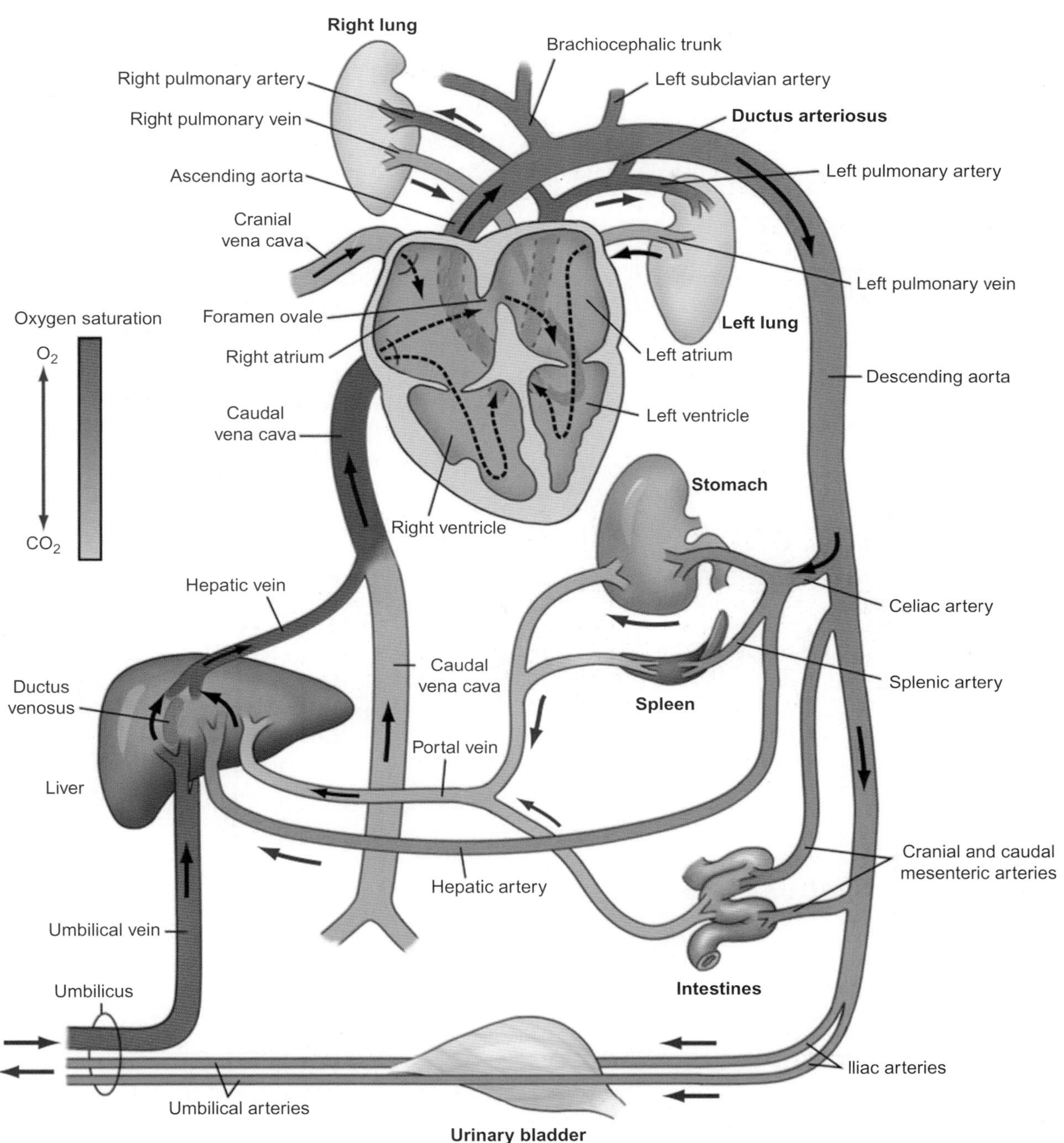

FIGURE 4.6-4 Fetal circulation and the normal transition from fetal to neonatal life.

Blood flow from the fetal heart (Fig. 4.6-4)

Noting that umbilical cord blood has been oxygenated by the mother's lungs and then filtered through the endometrium and placenta, blood leaving the fetal right ventricle into the pulmonary trunk is about 50% saturated with oxygen. Approximately 20% of it goes to the fetal lungs, and the rest passes through the **ductus arteriosus** into the aorta, which eventually returns it to the placenta via the **umbilical arteries**. All mammalian species have 2 umbilical arteries, which in the fetus are the terminal branches of the aorta.

Transition from fetal to neonatal circulation

A complex but well-orchestrated series of events occur at birth as the umbilical cord is ruptured, bitten, or otherwise severed and the neonate takes its first breaths. These changes are crucial for the pivotal transition from reliance on the mother's cardiorespiratory system to self-reliance.

The umbilical arteries and veins are muscular vessels, difficult to distinguish grossly in the cord, and the spasm of their musculature is sufficient to occlude them when the cord is transected at birth. The ductus arteriosus is also a muscular vessel (unlike the elastic arteries it connects), so it too is designed to constrict at birth.

> After birth, the umbilical arteries retract into the lateral ligament of the bladder and become the round ligaments of the bladder.

Closure of the umbilical vessels at birth contributes to the pressure changes in the right and left chambers of the heart which characterize the transition from fetal circulation, with its approximately equal left and right ventricular pressures, to neonatal circulation, with its disparate ventricular pressures (left much greater than right). Closure of the umbilical arteries increases pressure in the aorta, while expansion of the neonatal lungs and their circulation lower pressure in the pulmonary trunk. These factors result in reversal of the flow of oxygenated blood through the ductus arteriosus, which in fetal life is from pulmonary artery to aorta. However, closure is not completed for some hours or days, during which the reverse flow—now from the high-pressure aorta to the low-pressure pulmonary artery—continues. After birth, the ductus arteriosus becomes the **ligamentum arteriosum**, unless it persists as a PDA.

> This case of PDA is an example of an instance in which the transition from fetal to neonatal life is incomplete or imperfect, resulting in a recognizable clinical entity. The reasons for persistence of a PDA beyond the first few days of life are not completely understood, although lack of smooth muscle in the wall of this normally muscular vessel appears to be a primary factor in dogs. As described in this case, PDA typically causes significant cardiorespiratory compromise and failure to thrive in young animals. Without corrective treatment, it may result in congestive heart failure and premature death.

Selected references

[1] Broaddus K, Tilson M. Patent ductus arteriosus in dogs. Compend Contin Educ Vet 2010;32, E9.

[2] Saunders AB, Gordon SG, Boggess MM, Miller MW. Long-term outcome in dogs with patent ductus arteriosus: 520 cases (1994–2009). J Vet Intern Med 2014;28:401–10.

[3] Beijerink NJ, Oyama MA, Bonagura JD. Congenital heart disease. In: Ettinger SJ, Feldman EC, Cote E, editors. Textbook of veterinary internal medicine. 8th ed. Elsevier Inc.; 2017. p. 1207–48.

CASE 4.7

Mitral Valve Disease

Laura Dooley[a], Cathy Beck[b], and Simon Bailey[a]
[a]Veterinary Biosciences, The University of Melbourne, Melbourne, Victoria, AU
[b]Veterinary Hospital, The University of Melbourne, Werribee, Victoria, AU

Clinical case

History

A 14-year-old female neutered Maltese cross dog presented with a history of exercise intolerance. She had recently become reluctant to go on walks, and the owners reported that she seemed to be breathing more rapidly when sleeping.

Physical examination findings

The dog was alert and responsive, although she was panting during the examination. On thoracic auscultation, the lung sounds were increased in intensity bilaterally, particularly over the caudodorsal lung fields. Cardiac auscultation revealed tachycardia (heart rate, 150 bpm) with a regular rhythm and a grade 5/6 systolic heart murmur with a point of maximal intensity over the left apical region. No other abnormalities were found.

> Heart murmurs are graded by their intensity on a scale from 1 (barely audible) to 6 (very loud, barely requiring a stethoscope and palpable through the chest wall). Pathologic murmurs are most commonly attributable to regurgitant flow of blood through the heart, such as backflow through an abnormal/diseased valve. A grade 5 systolic murmur is an obvious murmur that is heard between the first (S1) and second (S2) heart sounds. The point of maximal intensity (often abbreviated PMI) is the location at which the murmur is loudest. The left apical region is the location of the mitral valve (Fig. 4.7-1).

Differential diagnoses

Myxomatous mitral valve degeneration, mitral valve endocarditis, and congenital mitral valve dysplasia

> In this case, the murmur was most consistent with regurgitant flow through the mitral valve. While the dog's breed and age made myxomatous mitral valve degeneration most likely, the other differential diagnoses could not be ruled out at that point.

Diagnostics

Thoracic radiography revealed generalized cardiomegaly with left atrial enlargement (left atrial "wedge") and an alveolar pattern in the caudodorsal lung fields (Figs. 4.7-2 and 4.7-3). Echocardiography revealed the left atrium to be grossly dilated when compared to the aorta (Fig. 4.7-4). The mitral valve was thickened and prolapsed into the left atrium. The ratio between the diameter of

> Left atrial "wedge" is a term used when the enlarged left atrium creates a triangular shape just caudal to the tracheal bifurcation on the lateral radiographic projection. Alveolar opacity (an alveolar pattern) in the presence of cardiomegaly and left atrial enlargement is most consistent with cardiogenic pulmonary edema.

Comparative Veterinary Anatomy: A Clinical Approach. https://doi.org/10.1016/B978-0-323-91015-6.00029-7

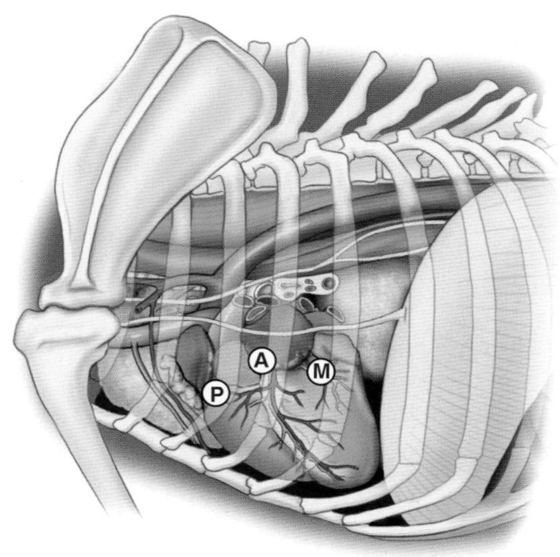

FIGURE 4.7-1 Location of heart valves on auscultation of the left side of the thorax in a dog or cat. A, aortic valve; M, mitral valve; P, pulmonary valve.

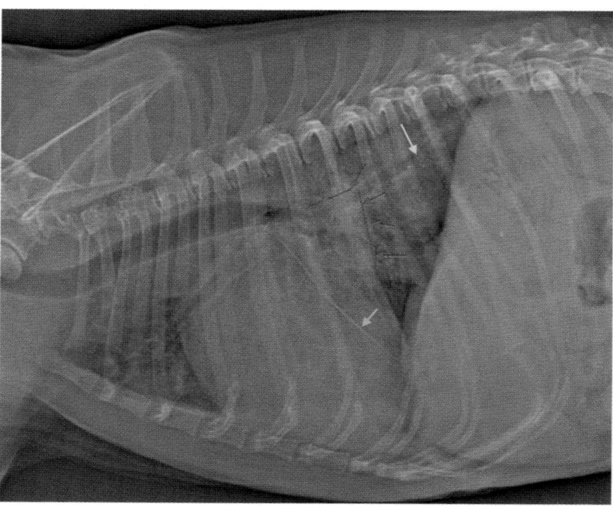

FIGURE 4.7-2 Lateral radiograph of the thorax with the patient in right lateral recumbency. The entire cardiac silhouette is enlarged (generalized cardiomegaly). In addition, a left atrial "wedge" (triangular shape outlined by the *red arrows*) is evident just caudal to the tracheal bifurcation, indicating significant left atrial enlargement. There is also increased opacity within the dorsocaudal lung fields consistent with cardiogenic pulmonary edema (*yellow arrow*). The white line (*blue arrow*) is a pleural fissure line, in this case most likely caused by fat within the pleural space. This is a common finding in older dogs.

the left atrium and the aorta on the short-axis view shown in Fig. 4.7-4 was 2.95, indicating that the left atrium was grossly enlarged (normal is < 1.3).

Cardiogenic pulmonary edema is typically evident on radiographs in the caudodorsal lung region in dogs vs. a patchy, diffuse, and lobar pattern in the cat.

Diagnosis

Myxomatous mitral valve disease, resulting in left-sided congestive heart failure and cardiogenic pulmonary edema

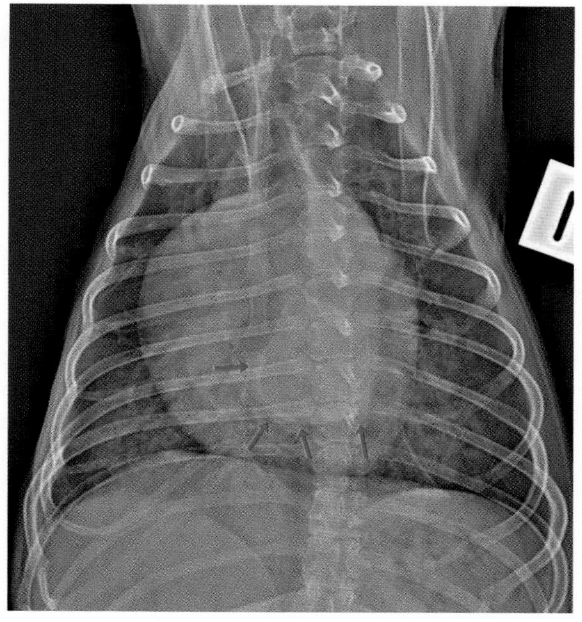

FIGURE 4.7-3 Ventrodorsal radiograph of the thorax with the patient in dorsal recumbency. Generalized cardiomegaly, left atrial enlargement (*red arrows*) and left auricular enlargement (*purple arrows*), and an alveolar pattern in the caudal lung fields are also apparent on this projection.

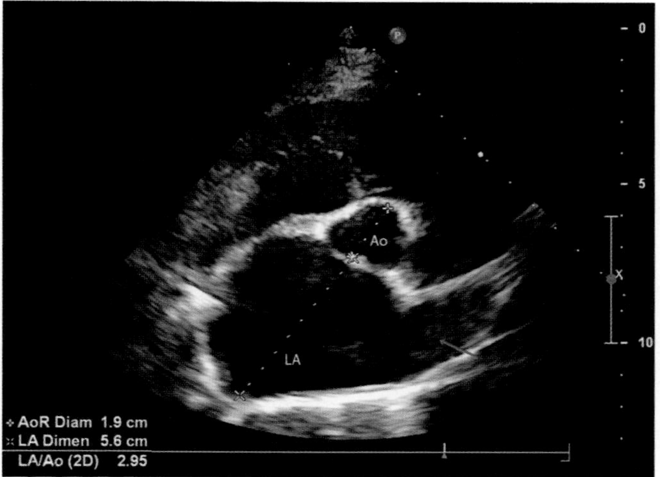

FIGURE 4.7-4 Echocardiogram, short-axis view, showing the left atrium (*LA*) and aorta (*Ao*). The left atrium is grossly enlarged, with a LA:Ao of 2.95:1. The left auricle (*red arrow*) is seen projecting cranially.

Treatment

The dog was treated with a diuretic agent (furosemide), a peripheral vasodilator (benazepril, which is an angiotensin-converting enzyme inhibitor), and a positive inotrope that also has vasodilatory properties (pimobendan). When the dog returned for re-examination after 2 weeks of therapy, her owners reported improved appetite and decreased respiratory rate at rest.

Anatomical features in canids and felids

Introduction

This section describes the interior of the canine and feline heart. The pericardium, exterior of the heart, and the myocardium are described in Case 4.4.

Function

The heart (weighing approximately 0.7–2.2% of body weight in companion animals) consists of 4 chambers—2 atria (upper chambers) and 2 ventricles (lower chambers)—each with a valve through which blood passes before exiting the chambers, preventing retrograde flow of blood. These valves consist of cusps or leaflets and are the right and left atrioventricular, aortic, and pulmonary valves, acting as ingress for blood coming into a ventricle and egress outlets for blood leaving a ventricle. The valves open and close as the heart muscle (myocardium) contracts and relaxes. The right atrioventricular (tricuspid) valve may have anywhere from 2 to 5 cusps in dogs and is of little clinical significance, while the left atrioventricular (bicuspid or mitral) valve has 2 major cusps.

Right atrium (Fig. 4.7-5A; Landscape Figs. 3.0-6 and 3.0-7)

The **right atrium** is the cardiac chamber into which the principal systemic veins enter. In all domestic species, the right atrium has 4 main openings: (1) the **cranial vena cava**, which drains the head, neck, thoracic limbs, ventral thoracic wall, and adjacent part of the abdominal wall; (2) the **caudal vena cava**, which drains the abdominal viscera, part of the abdominal wall, and the pelvic limbs; (3) the **coronary sinus**, which returns venous blood from

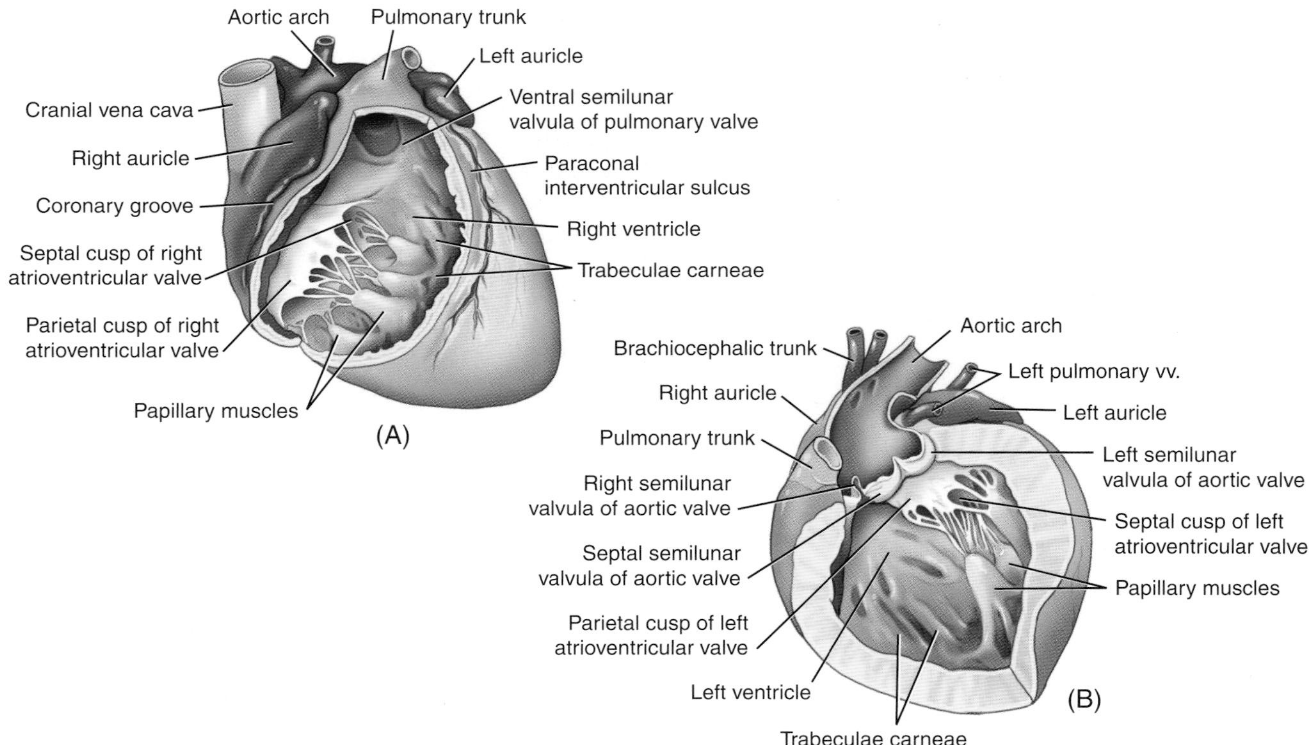

FIGURE 4.7-5 (A) Right side of the canine heart, highlighting the right ventricular outflow tract. (B) Left side of the canine heart, highlighting the left ventricular outflow tract.

the heart itself, opening into the right atrium immediately ventral to the opening of the caudal vena cava; and (4) the **right atrioventricular (AV) orifice**—protected by the **right AV (tricuspid) valve**—through which blood flows between the right atrium to the right ventricle. (The right AV valve is described below, along with the right ventricle.)

Several other anatomical structures worth noting as they relate to the right atrium are: (1) the **azygous vein** (a single vein on the right side in carnivores) drains blood from part of the lumbar region, the caudal three-quarters of the thoracic wall, the bronchial circulation, and the esophagus, and enters the cranial vena cava near the right atrium; (2) small **coronary veins** open directly into the right atrium through numerous small foramina in the atrial wall; and (3) the **right auricle**—a blind diverticulum of the right atrium—opens from the cranial end of the right atrium and winds around the cranial aspect of the heart. The internal wall of the right atrium is smooth, but the auricle is interlaced with muscular ridges, the **pectinate muscles**. Several vestiges of fetal circulation are also present in the right atrium, such as the **fossa ovalis**, an ovoid depression on the interatrial septum near the opening of the caudal vena cava. (The fetal circulation is described in Case 4.6.)

Right ventricle (Fig. 4.7-5)

The **right ventricle** receives blood from the **right atrium** and pumps blood into the **pulmonary trunk**. Because the pulmonary circulation is a low-pressure system, the wall of the right ventricle is approximately half as thick as that of the left ventricle (described later). The lumen of the right ventricle is crescent-shaped in transverse section (see Fig. 4.4-10) and its opening is spanned by the **right AV**, or **tricuspid**, **valve**, so-named because it has three cusps.

The **valve cusps** consist of a collagen fiber layer sandwiched between two sheets of endothelium. The collagen fibers are continuous with those of the fibrous ring that surrounds the right AV opening. Each valve cusp has a slightly thickened free border.

Chordae tendineae (tendinous cords) arise from **papillary muscles** that project from the internal surface of the ventricular wall and fan out to attach to the cusps of the AV valve. Each papillary muscle modulates the movements of two cusps. The right ventricle has three papillary muscles: two arise from the interventricular septum; a third, larger, papillary muscle (the **great papillary muscle**) arises from the outer or lateral wall of the ventricle.

The right ventricular cavity is divided into two functional components: an inflow channel and an outflow channel. The **inflow channel** extends from the **right AV orifice** to about the level of the septomarginal trabeculae (described below). The inner surface of the right ventricle is rendered irregular by the **trabeculae carneae**, which are subendocardial ridges of myocardium that protrude into the lumen, and by the presence of the papillary muscles. The right **septomarginal trabeculum**, or "moderator band," is a rounded bundle of tissue that crosses the lumen of the ventricle from the interventricular septum to the free wall. Several smaller bundles pass from the interventricular septum to the base of the papillary muscles and distribute conducting fibers to the papillary muscles.

CARDIAC AUSCULTATION

Fig. 4.7-1 shows the normal positions of the heart valves that may be ausculted over the left thoracic wall. As the right ventricular outflow courses toward the left side of the body, the pulmonary valve (P) is best heard between the left second and fourth intercostal spaces. The aortic valve (A) is best auscultated over the left fourth intercostal space, just dorsal to the costochondral junction, so slightly caudal and dorsal to the pulmonary valve. The mitral valve (M) is located within the left fifth intercostal space, around the costochondral junction—caudal and ventral to the pulmonary and aortic valves.

The tricuspid valve is auscultated over the right thoracic wall, between the right third and fifth intercostal spaces, near the costochondral junction. It is important to adjust the position of the stethoscope to locate the point of maximum intensity over each valve area.

The right ventricular **outflow channel** begins at the septomarginal trabeculae and consists mainly of the funnel-shaped **conus arteriosus** that directs blood into the **pulmonary trunk**. The conus arteriosus has smooth walls with no papillary muscles or trabeculae carneae. The root of the pulmonary trunk slightly increases in diameter in the region of the **pulmonary valve**, comprised of three semilunar ("half-moon" shaped) valvules; each valvule has a slightly thickened free border, with a small nodule on the midpoint, like the AV valves. There are three bulges in the wall of the pulmonary trunk corresponding to the three semilunar valvules; these are the three **sinuses** of the pulmonary trunk. They resemble the sinuses of the aortic bulb (described below) but are less distinct.

Left atrium (Fig. 4.7-5B; Landscape Figs. 3.0-6 and 3.0-7)

The **left atrium** receives arterial blood from the lungs via the **pulmonary veins**. As in the right atrium, small coronary veins also empty into the left atrium. Like the right atrium, the **left auricle** is a blind diverticulum of the left atrium with pectinate muscles on its internal surface, while the remainder of the left atrium has a smooth internal wall.

Left ventricle (Fig. 4.7-5B)

The **left ventricle** communicates with the left atrium via the **left AV orifice**, spanned by the **left AV**, or **mitral**, **valve**. The mitral valve is also known as the "bicuspid valve" because it has two cusps. The general structure and relationships between the cusps, chordae tendineae, and papillary muscles of the left AV valve are the same in principle as described above for the right AV valve. As in the right ventricle, there are the same number of cusps as papillary muscles—2—and chordae tendineae arising from each muscle attach to both cusps. Both papillary muscles in the left ventricle originate from the outer wall of the ventricle.

Like the right ventricle, 1 large **septomarginal trabeculum** and several smaller trabeculae traverse the ventricular lumen from the interventricular septum to the base of the papillary muscles on the free wall. Similarly, in the right ventricle, the outflow channel of the left ventricle is smooth. However, the walls of the inflow channel in the left ventricle are very irregular because of the presence of strong trabeculae carneae, as well as prominent papillary muscles. The systemic circulation is a high-pressure system, and therefore, the wall of the left ventricle is normally 2–3 times as thick as that of the right ventricle. Correspondingly, the cusps of the left AV valve and the semilunar valvules of the aortic valve (described below) are much stronger and thicker than those in the right side of the heart.

Like the pulmonary valve, the **aortic valve** consists of three semilunar valvules. The **root of the aorta,** the site of the aortic valve, is expanded into the **aortic bulb** by three bulges, the **aortic sinuses**, which correspond to the position of the three aortic valvules. The **left coronary artery** originates from the left aortic sinus and the **right coronary artery** originates from the right aortic sinus; the remaining sinus is referred to as the **septal** aortic sinus.

The relative position of the cardiac valves, along with the number of valve cusps or valvules for each, is depicted in Fig. 4.7-6.

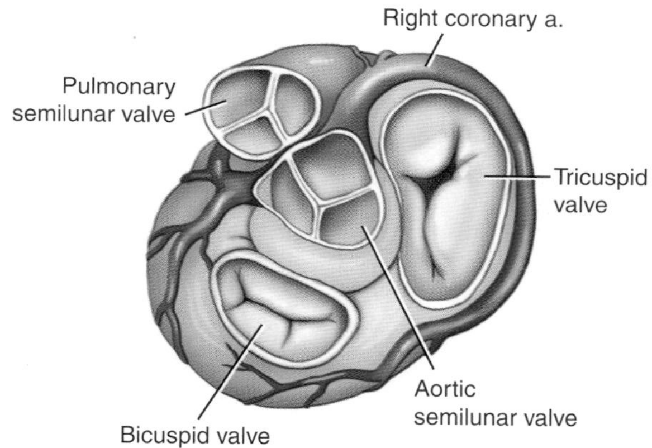

FIGURE 4.7-6 Cross-sectional (short-axis) view of heart with the atria removed to show the relative positions of the mitral, tricuspid, aortic, and pulmonary valves. (This view is relevant to echocardiographic imaging.)

Aorta and its major branches (see Landscape Figs. 3.0-6 and 3.0-7)

The **aorta** is the systemic arterial trunk, distributing oxygenated blood to the various regions of the body. It arises from the left ventricle and is divided into three main segments: ascending aorta, aortic arch, and descending aorta.

The **ascending aorta** is short, approximately 2 cm (0.8 in.) long in the average dog. It arises from the left ventricle and passes craniodorsally between the pulmonary trunk on its left and the right atrium on its right. It supplies blood to the myocardium via the **coronary arteries**.

The **aortic arch** begins where the aorta bends dorsocaudally toward its eventual caudal path. Here is where the aorta penetrates the **pericardium**, tracking dorsally within the **mediastinum** until it reaches the left ventral aspect of the seventh thoracic vertebra. (The pericardium is described in Case 4.4 and the mediastinum in Case 4.2.)

The aortic arch supplies blood to the head, neck, shoulder, thoracic limbs, and thoracic wall. Two major vessels arise from the aortic arch: the brachiocephalic trunk and the left subclavian artery. These vessels give rise to the various branches that supply the head, neck, thorax, and thoracic limbs. The short **brachiocephalic trunk** passes obliquely to the right and cranially across the ventral surface of the trachea and gives rise to the left and right **common carotid arteries** and the **right subclavian artery**.

The **left and right subclavian arteries** are the origin of four main branches: (1) the **vertebral artery** supplies the cervical muscles, spinal cord, and brain (via the circle of Willis); (2) the **costocervical artery** extends dorsally as far as the vertebral end of the first rib and supplies the first to third intercostal spaces, the muscles at the base of the neck, and the muscles of first 3 or 4 thoracic vertebrae; (3) the **internal thoracic artery** supplies the thoracic wall; and (4) the **superficial cervical artery** (or "omocervical artery") supplies the base ofthe neck and adjacent scapular region. The subclavian artery then exits the thorax, bends around the cranial border of the first rib, and continues as the **axillary artery**, which passes through the axilla and down into the thoracic limb.

Distal to the aortic arch, the **descending aorta** extends caudally along the ventral aspect of the vertebral column. In the thoracic region, the aorta supplies blood to the thoracic wall and internal organs of the thorax, and in the abdominal region, it supplies blood to the abdominal wall, abdominal organs, and the pelvic limbs.

Selected references

[1] Atkins CE, Häggström J. Pharmacologic management of myxomatous mitral valve disease in dogs. J Vet Cardiol 2012;14(1):165–84.
[2] Menciotti G, Borgarelli M. Review of diagnostic and therapeutic approach to canine myxomatous mitral valve disease. Vet Sci 2017;4(4):47.
[3] Keene BW, Atkins CE, Bonagura JD, Fox PR, Häggström J, Fuentes VL, et al. ACVIM consensus guidelines for the diagnosis and treatment of myxomatous mitral valve disease in dogs. J Vet Intern Med 2019;33:1127–40.

CHAPTER 4

CASE 4.8

Esophageal Foreign Body

David Holt
Department of Clinical Sciences and Advanced Medicine, University of Pennsylvania School of Veterinary Medicine, Philadelphia, Pennsylvania, US

Clinical case

History

A 6-year-old male neutered Labrador Retriever was presented with a 3-day history of vomiting. The dog had a history of dietary indiscretion. The dog had been somewhat lethargic since the vomiting began.

Physical examination findings

On physical examination, the dog was depressed and hypersalivating. There were no abnormal sounds on thoracic auscultation or abnormal findings on abdominal palpation; the abdomen was not painful. The remainder of the physical examination parameters were normal.

Differential diagnoses

GI foreign body, pancreatitis, dietary indiscretion, toxin ingestion, hepatic disease, renal disease

Diagnostics

Abdominal radiographs were normal and routine laboratory testing—a CBC and serum biochemistry profile—revealed only a slightly elevated serum albumin level suggesting mild dehydration. On further questioning of the owner, the dog was more likely regurgitating rather than vomiting. Thoracic radiographs revealed a radiodense structure in the esophagus at the level of the heart base. The lungs appeared normal with no evidence of aspiration pneumonia (Fig. 4.8-1).

Diagnosis

Esophageal foreign body causing esophageal obstruction

Advanced imaging (including CT) is generally not necessary when a radiodense foreign body is seen in the esophagus. However, radiographs should be evaluated carefully for any evidence of mediastinal effusion (presenting as widening of the mediastinum), pleural effusion, or alveolar lung disease. Mediastinal or pleural effusions are suggestive of esophageal perforation; alveolar lung disease suggests aspiration pneumonia.

Esophageal foreign bodies tend to lodge at areas of anatomical narrowing. In cats, needles often lodge at the level of the cranial esophageal sphincter, formed by the cricopharyngeus muscle. Other areas of narrowing are the thoracic inlet, heart base, and caudal esophageal sphincter, just cranial to the diaphragm.

Comparative Veterinary Anatomy: A Clinical Approach. https://doi.org/10.1016/B978-0-323-91015-6.00030-3

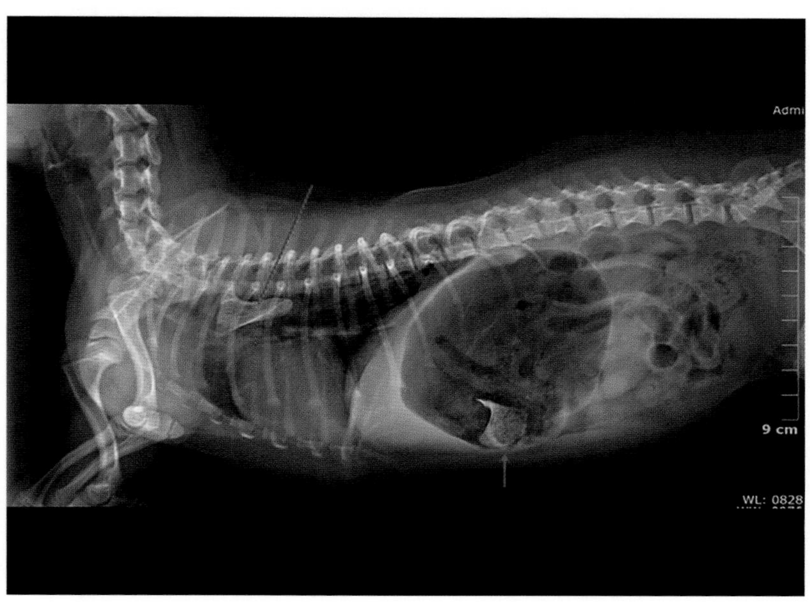

FIGURE 4.8-1 A foreign body (bone) is visible in the thoracic esophagus dorsal to the heart base (*long red arrow*). Note the gastric dilation and the presence of an additional bone foreign body in the stomach (*short red arrow*).

Treatment

The dog was anesthetized for upper gastrointestinal endoscopy. The foreign body could not be retrieved with forceps and a basket catheter and could not be advanced into the stomach. Thus, the right thorax was aseptically prepared for surgery; an incision was made through the skin, subcutaneous tissue, and ventral aspect of the latissimus dorsi muscle. The intercostal spaces were palpated and counted to verify the location for the thoracotomy, the serratus ventralis muscle separated over the 4th intercostal space, and the intercostal muscles and pleura were incised to expose the esophagus at the level of the heart base. The foreign body was lodged just caudal to the azygos vein, which crosses the esophagus in a dorsoventral direction before entering the cranial vena cava (Fig. 4.8-2). The esophagus was incised longitudinally, and the foreign body removed. The esophageal wall was judged to be healthy; the incision was closed with a single row of interrupted, appositional sutures. A chest tube was placed and air removed from the pleural space after the closure of the thoracic incision. The dog made an uneventful recovery from surgery.

A right lateral thoracotomy was chosen for access to the esophagus because the aorta prevents access to an esophageal foreign body at this level of the esophagus from a left-sided approach. For foreign bodies lodged in the caudal thoracic esophagus, a left 8th or 9th space intercostal thoracotomy is indicated (Fig. 4.8-3).

Anatomical features in canids and felids

Introduction

The esophagus is a strong, distensible, muscular tube that connects the pharynx to the stomach. This section covers the esophagus along with its innervation and blood supply with relevant adjacent anatomical structures—e.g., the vagus nerve.

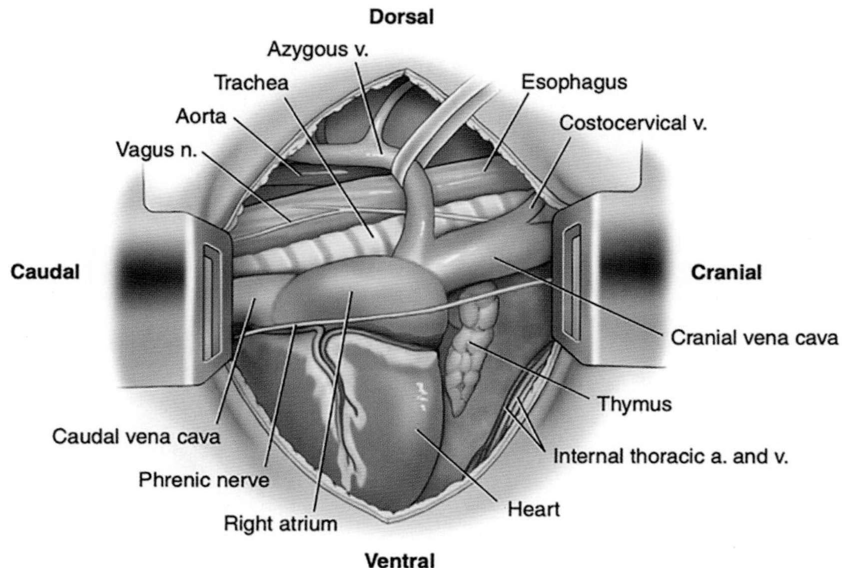

FIGURE 4.8-2 Mid-thoracic esophagus and related structures. The esophagus courses dorsal to the heart and on the right side of aorta in the mid-thorax. The azygos vein crosses over the thoracic esophagus in this location before entering the cranial vena cava.

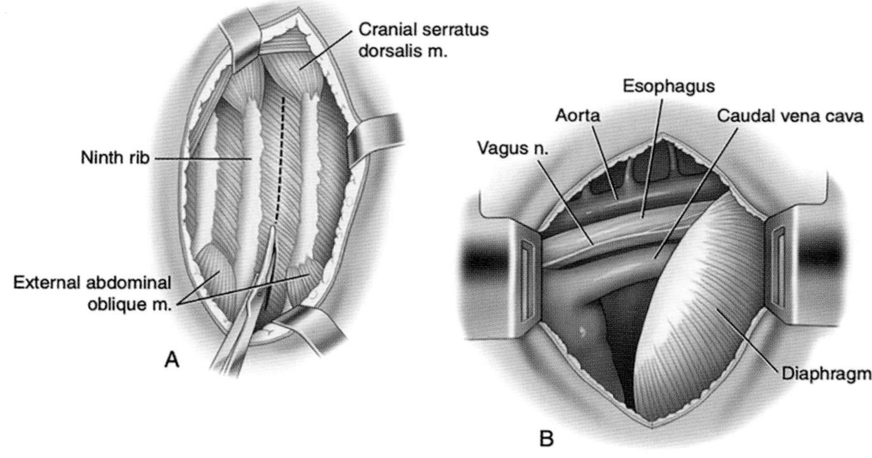

FIGURE 4.8-3 Approach to the caudal esophagus through a left lateral thoracotomy. The aorta is dorsal to the esophagus at this level of the thorax.

Function

The esophagus functions to move boluses of food and liquid from the pharynx to the stomach. During swallowing, a wave of primary peristalsis (Gr. *peri-* around + Gr. *stalsis* contraction), a coordinated muscular contraction, moves ingested food and liquid from the proximal esophagus into the stomach. Any food remaining in the esophagus stimulates a wave of secondary peristalsis to empty the esophagus.

Gross anatomy of the esophagus

The proximal **esophagus** begins dorsal to the **cricoid cartilage** of the **larynx**. At this level, it is surrounded by the **cricopharyngeus muscle**, which functions as the cranial esophageal sphincter. The esophagus is generally located on the left side of the trachea in the cervical region and through the **thoracic inlet**. In the thorax, it is positioned in the **mediastinum** dorsal to the trachea at the level of the heart base where it passes between the **aorta** on the left side and the **azygos vein** on the right side. In the caudal thorax, it passes slightly to the left side of the mediastinum to enter the abdomen through the **esophageal hiatus** (opening) of the **diaphragm**. The abdominal section of the esophagus is short and extends from the diaphragm to the **cardia** of the stomach. There is not a distinct caudal esophageal sphincter. The striated muscular coats of the esophagus and the esophageal hiatus of the diaphragm provide a functional caudal sphincter.

Histologic anatomy of the esophagus

The **mucosa** is a resilient stratified squamous epithelium with a **submucosa** containing substantial collagen and elastin. The **muscular layers** of the esophagus are composed of 2 oblique layers of striated muscle. These are continuous throughout the esophagus in the dog, but transition to smooth muscle in the distal one-third of the feline thoracic esophagus. The outer layer of the esophagus is a thin fibrous coat termed the **adventitia** (Fig. 4.8-4).

The submucosa is the strongest suture-holding layer and should always be included in esophageal closure.

The submucosa is flexible because of the many elastin fibers facilitating distention. The esophagus must accommodate large boluses of food (including bones) in carnivores that often eat and swallow rapidly.

Blood supply, lymphatics, and innervation of the esophagus

The esophagus is supplied with blood from the **cranial** and **caudal thyroid arteries** in the cervical region. The thoracic esophagus blood supply is by the **bronchoesophageal arteries** in the cranial two-thirds of the thorax;

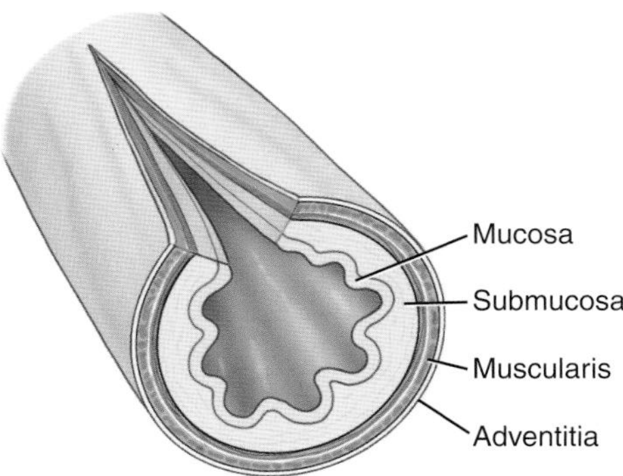

FIGURE 4.8-4 The canine esophagus is composed of a tough mucosa, a submucosa containing many elastic fibers that allow distension during swallowing large boluses of food, two layers of striated muscle, and a thin adventitia.

ESOPHAGEAL FOREIGN BODIES

In dogs, esophageal foreign bodies are most commonly bones, but rawhides, other chew treats, sticks, fishhooks, and small toys have also been reported. In cats, needles (often with thread attached), trichobezoars (hair balls), and string are most common. Foreign bodies can often be removed endoscopically by dilating the esophagus around the foreign body with air. This must be done carefully because insufflation of a perforated thoracic esophagus can lead to an immediately life-threatening tension pneumothorax. Once the esophagus is dilated, the foreign body is removed with forceps or an endoscopic basket, or gently advanced into the stomach. Surgery is indicated in cases of suspected perforation, when the foreign body is immovable, or if esophageal wall necrosis is found upon endoscopy after foreign body retrieval.

the caudal thoracic esophagus blood supply is by branches from the aorta and **dorsal intercostal arteries**, while the abdominal esophagus blood supply is by a branch of the **left gastric artery**. Venous drainage is via the **jugular veins** in the cervical part of the esophagus. Veins from the thoracic portion of the esophagus drain into the **azygos vein**. Lymphatic vessels draining the cervical esophagus join the **deep cervical lymph nodes**. Lymphatics from the thoracic esophagus drain to the **cranial mediastinal** and **bronchial lymph nodes**.

The esophagus is innervated by the **pharyngoesophageal branches** of the **vagus nerve**, the **recurrent laryngeal** and **pararecurrent laryngeal nerves** in the neck, and by the dorsal and ventral vagus nerves in the thorax.

Adjacent structures

The left and right branches of the **vagus nerve** course along the caudal thoracic esophagus. The dorsal left and right branches of the vagus unite on the dorsal esophagus cranial to the esophageal hiatus and the ventral branches unite more cranially. Both the dorsal and ventral vagal trunks pass through the esophageal hiatus of the diaphragm with the esophagus.

Selected references

[1] Dallman MJ. Functional suture-holding layer of the esophagus in the dog. J Am Vet Med Assoc 1988;192:638–40.
[2] Thompson HC, Cortes Y, Gannon K, et al. Esophageal foreign bodies in dogs: 34 cases. J Vet Emerg Crit Care 2012;22:253–61.
[3] Deroy C, Benoit Corcuff J, Hamaide A. Removal of oesophageal foreign bodies: comparison between oesophagoscopy and oesophagotomy in 39 dogs. J Small Anim Pract 2015;56:613–7.

CHAPTER 5

ABDOMEN

John F. Randolph, Chapter editor

Liver, Pancreas, Spleen, and Adrenal Glands

5.1 Portosystemic vascular anomaly—*Sharon A. Center and John F. Randolph* 278
5.2 Extrahepatic bile duct obstruction secondary to acute pancreatitis—*Shannon M. Palermo and Mark P. Rondeau* 287
5.3 Hyperadrenocorticism—*Nora S. Grenager, Rebecka S. Hess, and James A. Orsini* 292
5.4 Splenic torsion—*David Holt, Nora S. Grenager, James A. Orsini* 298

Stomach

5.5 Gastric dilatation and volvulus—*Carol A. Carberry* 304

Small Intestine

5.6 Small intestine obstruction—*David Holt, Nora S. Grenager, and James A. Orsini* 310

Large Intestine, Anal Canal, and Rectum

5.7 Megacolon—*Takanori Sugiyama and Helen M.S. Davies* 316

Body Wall

5.8 Abdominal wall hernia—*Takanori Sugiyama and Helen M.S. Davies* 325

CHAPTER 5

CASE 5.1

Portosystemic Vascular Anomaly

Sharon A. Center and John F. Randolph
Department of Clinical Sciences, Cornell University College of Veterinary Medicine, Ithaca, New York, US

Clinical case

History

A 1.5-year-old female spayed Pug was presented for intermittent hematuria and stranguria. Adopted at 8 weeks of age, the dog had a small body stature compared to her littermates. She had a historically finicky appetite with occasional lethargy and mental dullness 30 minutes after eating. The breeder denied behavioral abnormalities or inappetence until the dog was weaned onto commercial puppy chow 2 weeks before adoption. Since adoption, there was occasional bilious vomiting. Stools were usually well-formed and there was no coughing or sneezing. The dog was reportedly polyuric and polydipsic. The dog had been routinely dewormed. During a recent bout of stranguria, the dog became disoriented, displayed ptyalism, and aimlessly circled while repeatedly attempting to urinate.

Physical examination findings

Physical assessment revealed small body stature for this breed (bodyweight = 7 kg [15.5 lbs]), normal heart rate, normal respiratory effort and rate, a body condition score of 4/9, and slightly dull mentation. The dog dribbled blood-tinged urine during abdominal palpation. An oral examination revealed modest dental plaque with normal tonsils. Menace and pupillary light responses were intact. Thoracic auscultation was within normal limits with some referred airway sounds attributed to the brachycephalic breed. Abdominal palpation could not detect a liver margin, but kidneys were easily palpated. The urinary bladder was the size of a hen's egg and its palpation seemed to elicit a painful response.

A combination of stunted body stature and intermittent dull behavior associated with the initiation of puppy chow (i.e., a higher protein intake) raised suspicion of growth retardation and neurobehavioral signs associated with PSVA.

Differential diagnoses

The episodic neurobehavioral features were suggestive of hepatic encephalopathy likely secondary to a congenital portosystemic vascular anomaly (PSVA); the hematuric dysuria was consistent with urinary tract infection or calculi (possible ammonium biurate urolithiasis associated with PSVA)

Comparative Veterinary Anatomy: A Clinical Approach. https://doi.org/10.1016/B978-0-323-91015-6.00031-5

Diagnostics

A CBC revealed nonanemic microcytosis (56 fL) and mild leukocytosis ($16.6 \times 10^3/\mu L$) with mature neutrophilia ($11.3 \times 10^3/\mu L$). Serum biochemical profile revealed mild hypoalbuminemia (2.7 g/dL) and hypoglobulinemia (1.6 g/dL) and mildly increased ALP (195 U/L) and ALT (175 U/L), along with low BUN (6 mg/dL), creatinine (0.3 mg/dL), and cholesterol (82 mg/dL). Urinalysis revealed a specific gravity of 1.016, moderate hematuria, mild pyuria, and numerous ammonium biurate crystals (Fig. 5.1-1). Urine culture was negative. Preprandial and 2-hour postprandial total serum bile acid concentrations were increased: 79 μM/L and 329 μM/L, respectively. Upon receipt of the high serum bile acid concentrations, a protein C activity was submitted (54% activity, reference interval > 70%).

Abdominal radiographs (Fig. 5.1-2) revealed microhepatia (Gr. *mikros* small + Gr. *hēpar* liver = smallness of the liver), prominent kidneys, and no evidence of radiodense urinary calculi (renal, ureteral, urethral, or cystic). Abdominal ultrasound confirmed a subjectively small liver with normal echogenicity and an impression of reduced portal venous perfusion with concurrently enhanced arterial flow. The size of the portal vein at the porta hepatis compared to the hepatic artery was subnormal (0.57, reference measurement > 0.91). A thorough examination of the portal vein using color flow Doppler could not definitively identify vasculature shunting (hepatofugal portal flow [flow away from rather than to the liver]). Ultrasound revealed uroliths in the urinary bladder and left renal pelvis.

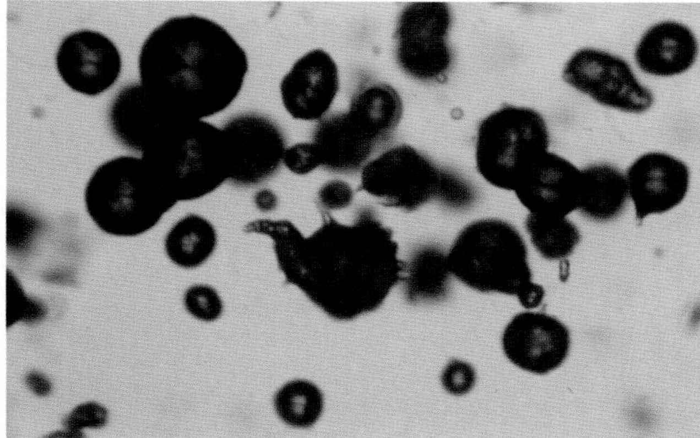

FIGURE 5.1-1 Ammonium biurate crystalluria as observed on 400× microscopic magnification in the urine of a dog with PSVA. Crystals are golden brown and may appear spherical with "fuzzy" striations but also frequently have crooked pointed extensions classically described as "thorn apple" in appearance. The urine of PSVA dogs often appears golden pigmented because of these crystals.

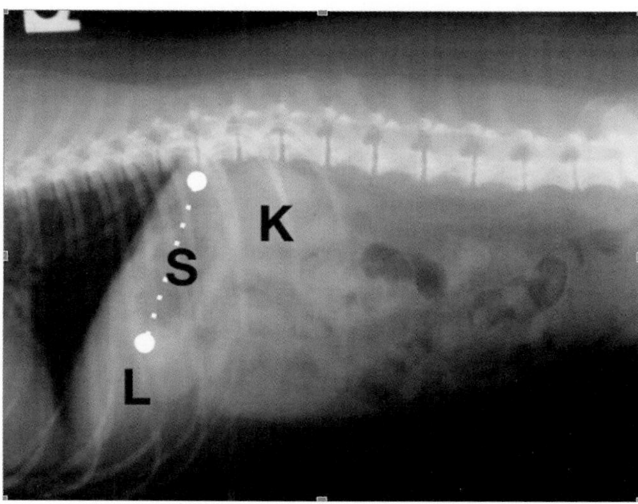

FIGURE 5.1-2 Lateral abdominal radiograph of a dog with PSVA showing small liver (*L*) and prominent kidneys (*K*) under the last 2 ribs; also note the abnormal gastric axis (not parallel with the ribs; *S*, *dotted line*) that further supports microhepatia.

Colorectal scintigraphy can be used to definitively confirm macroscopic portosystemic shunting, and is most often used to differentiate dogs with MVD (microscopic portal shunting of blood as the cause of increased bile acid concentrations) from dogs with PSVA. With CR scintigraphy, a technetium[99] enema is delivered deep into the rectum (so isotope is placed in the colon) followed by immediate gamma camera scanning of the isotope circulation. The isotope has rapid absorption into colonic veins that then empty into the portal vein. Avoid shallow isotope administration into the rectum, as absorption into rectal veins that interconnect directly with the caudal vena cava will result in isotope bypassing the portal vein.

Colorectal (CR) scintigraphy confirmed macroscopic shunting (Fig. 5.1-3). Results for this dog and a dog with microvascular dysplasia (MVD) are illustrated in the figure. In normal dogs or dogs with only MVD, isotope circulates to the liver before the heart. In dogs with macroscopic shunting—e.g., PSVA or acquired portosystemic shunts (discussed later), the isotope is first circulated to the heart and then to the liver.

A CT angiogram (CTA) was performed under general anesthesia to illustrate the vessels involved in this PSVA (Figs. 5.1-4 and 5.1-V1 in the online version at https://doi.org/10.1016/B978-0-323-91015-6.00031-5).

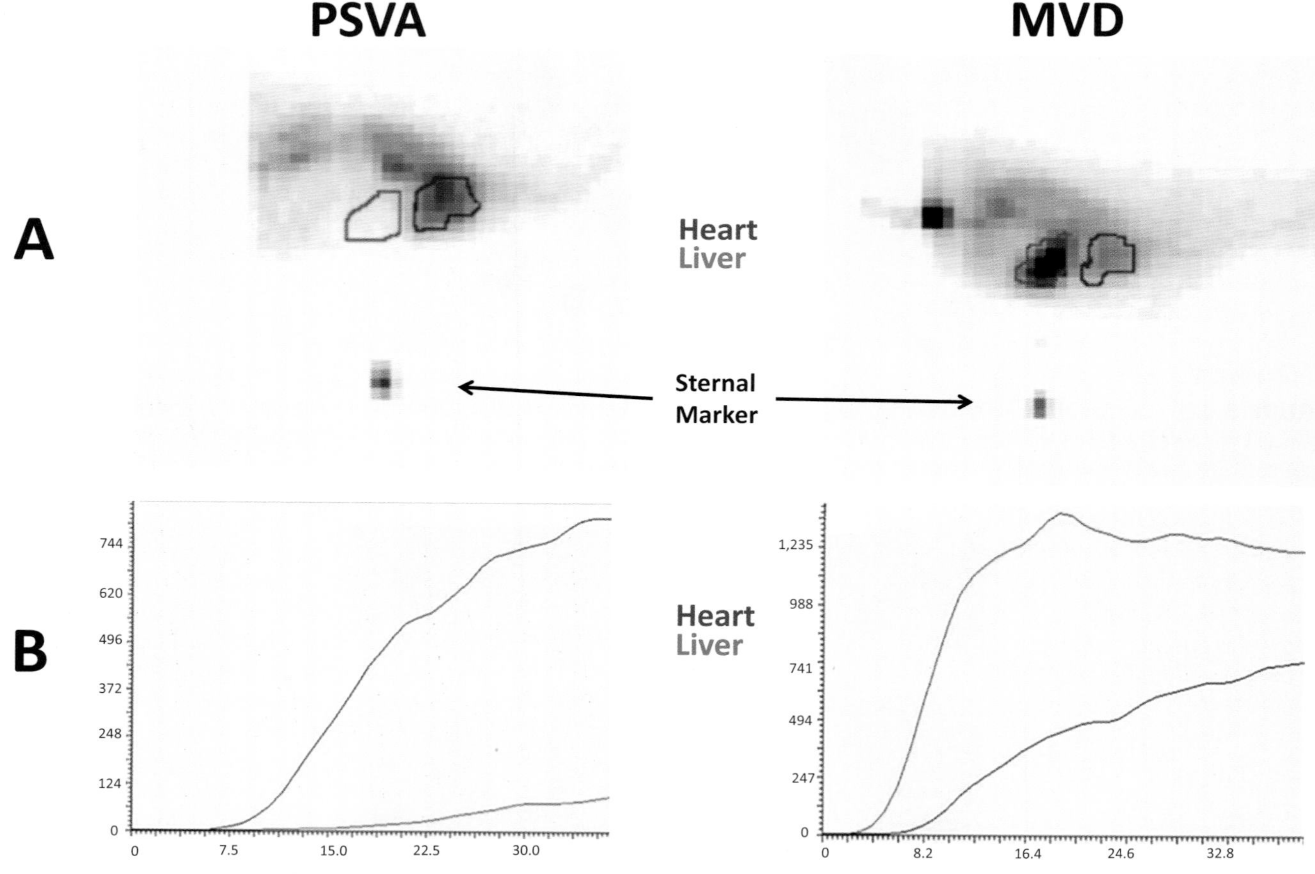

FIGURE 5.1-3 Colorectal (CR) scintigraphy depicted as collated images (A) and graphic plots of isotope distribution time activity curves (B) from a dog with PSVA *(left)* and a dog with MVD *(right)*. For the images in A, note the sternal marker placed on the scan to determine the location of heart and liver.

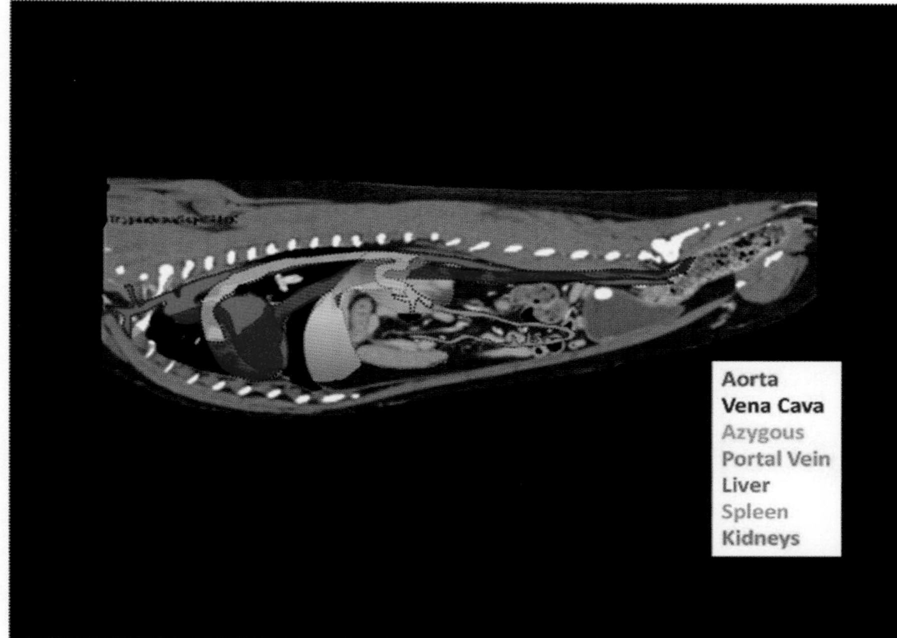

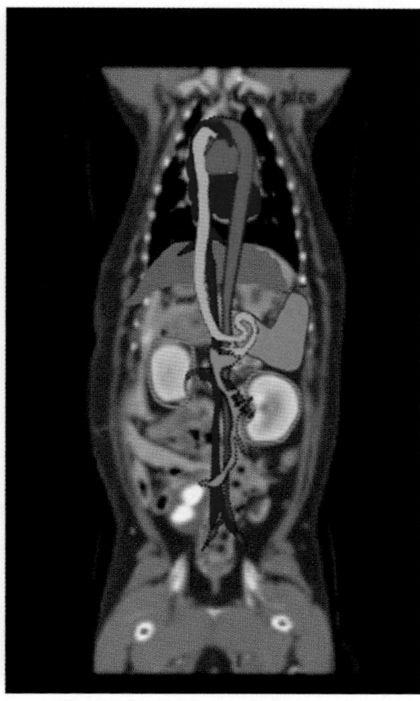

FIGURE 5.1-4 Sagittal and ventrodorsal images from CT angiogram (CTA) of a dog with PSVA providing a legend of the vascular anatomy. Video clips of the CTA for this PSVA are available as 5.1-V1 in the online version at https://doi.org/10.1016/B978-0-323-91015-6.00031-5.

Diagnosis

Congenital portoazygous vascular anomaly with left renal pelvic and cystic uroliths consistent with ammonium biurate urolithiasis

Treatment

Definitive management of PSVA requires shunt attenuation. Unfortunately, a cure cannot be guaranteed, as some dogs cannot tolerate complete shunt closure.

Medical management of PSVA is initiated as a first step to improve patient health before attempting shunt attenuation. The cornerstone of medical management is nutritional support as well as avoidance of conditions that provoke hepatic encephalopathy—e.g., azotemia, infection, constipation, enteric bleeding, or swallowing blood (e.g., associated with tooth eruption). Dietary management includes restriction of red meat or fish-derived protein sources and replacement with alternative dairy and vegetable-based protein. Carbohydrates associated with these protein sources beneficially modify the enteric biome (production of organic acids that diminish ammonia production and absorption) that mitigates hepatic encephalopathy. Treatment with lactulose, a fermentable synthetic disaccharide not absorbed or metabolized by mammals, is usually combined with low-dose metronidazole; together these also beneficially manipulate the enteric biome, and fermentation of lactulose further attenuates ammonia formation and uptake in the colon.

Under general anesthesia, after visual identification of the vascular anomaly, a venous branch of a jejunal mesenteric vein was catheterized for monitoring intra-operative portal pressures and for injection of iodinated contrast for intra-operative pre- and postligation portograms (Fig. 5.1-V2 in the online version at https://doi.org/10.1016/B978-0-323-91015-6.00031-5). This imaging procedure was used to ascertain successful attenuation of the shunting vessel. Surgical ligation with silk achieved complete shunt closure in this case. The renal and cystic uroliths were routinely

It is important to avoid taking a liver biopsy from the caudate lobe in patients with suspected PSVA or MVD because this lobe is generally the best perfused liver lobe in dogs with these conditions. The caudate lobe receives its blood supply from the first branch of the portal vein and therefore a biopsy from this site is not representative of the overall severity of hepatic portal venous hypoperfusion.

removed and submitted for stone analysis and bacterial culture. Liver biopsies were collected from 2 sites. The dog was monitored for 3 days in the intensive care unit and discharged on continued medical management until reassessment in 6 weeks. The stone analysis confirmed ammonium biurate calculi. The bacterial culture of the urinary stones was negative.

Six weeks after surgery, with the dog on continued medical management, there was notable clinical improvement. There were no neurobehavioral episodes; there was complete resolution of hematuric dysuria, polyuria, and polydipsia.

Anatomical features in canids and felids

Introduction and function

The liver performs many functions essential for life, including the metabolism of proteins, carbohydrates, and fats; the production and secretion of bile; and the filtration of the splanchnic-derived portal circulation (i.e., assimilating nutrients and removing bacteria, endotoxins, and numerous noxious products). Importantly, the liver is central to drug biometabolism, detoxification, and elimination. Compared to other organs, the liver is unique in receiving a dual blood supply—the portal venous and hepatic arterial circulations. The following section reviews hepatic circulation and the embryology of hepatic vascular development.

Hepatic circulation

The **portal vein** delivers ~80% of the hepatic circulation, 50% of the oxygen essential to aerobic metabolism, absorbed enteric products (nutrients, bacteria, endotoxins, and other toxins), and hepatotrophic factors (insulin and nutrients) necessary for normal liver size. The portal vein is best categorized as a "capacitance vessel" and maintains a low circulatory pressure (~5–8 mmHg). In health, the pressure in the vena cava is ~3 mmHg. Because the main trunk of the portal vein is devoid of valves, portal flow follows a path of least resistance into the vena cava (hepatofugal or retrograde flow) in the presence of PSVA or in the circumstance of portal hypertension (through acquired portosystemic shunts [APSS]).

Portal hypertension is encountered when the forward flow of blood to the liver is thwarted by (1) resistance within the liver—e.g., fibrosis, hepatic venule/vein attenuation or occlusion), or (2) flow obstruction in the prehepatic portal vasculature—e.g., thrombus, vascular compression by neoplasia, or phlebitis. In these circumstances, hepatofugal portal circulation is accommodated by development of acquired portosystemic shunts (APSS) anastomosing the portal vein with the systemic circulation (vena cava). Schematic and ultrasound images of APSS are shown in Figs. 5.1-5 and 5.1-6.

The **hepatic artery** provides ~20–25% of liver circulation, 50% of its oxygen supply, and nutrient vasculature (via a peribiliary arterial plexus) essential for the intrahepatic bile ducts and portal vein structures. The hepatic artery is a high-pressure system (80–90 mmHg) that can dynamically compensate for diminished portal circulation that occurs in PSVA. Although the pathomechanism remains unclear, the hepatic arterial buffer response (HABR) is associated with vascular exposure to adenosine, a vasodilating agent normally washed from the portal tract by portal venous circulation. The HABR increases arterial perfusion by up to 200% (experimental studies); this response causes arterial tortuosity (vascular coiling) and angiogenesis (formation of arterial "twigs"). Microscopic features characteristic of portal venous hypoperfusion (as seen in PSVA or MVD) include lobular atrophy and increased arterial cross sections with thick muscular walls and serpiginous arterial profiles in portal tracts.

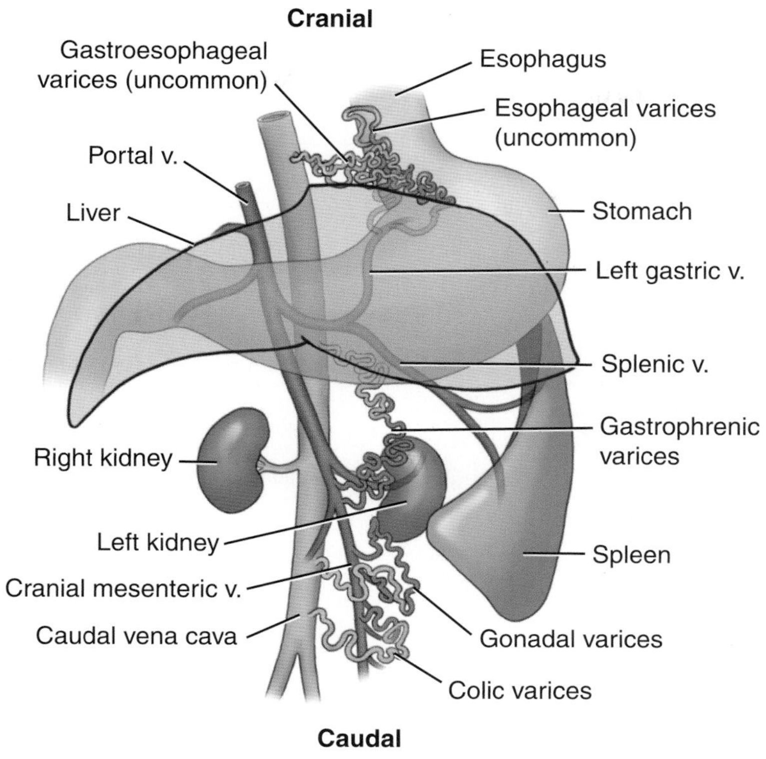

FIGURE 5.1-5 Schematic diagram of different types of acquired portosystemic shunts (APSS; varices) in dogs.

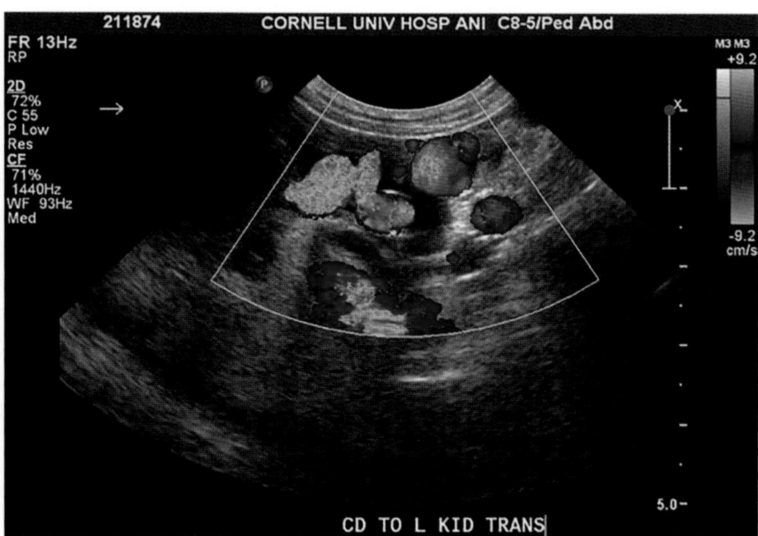

FIGURE 5.1-6 Detection of acquired portosystemic shunts using ultrasound color flow Doppler in a dog with portal hypertension due to chronic liver disease. Turbulent blood flow (chaotic kaleidoscope of color) and numerous vascular profiles are imaged where there should be none, caudal to the left renal vein.

The anatomic morphology of the hepatic vein has a different microscopic appearance in the dog vs. the cat. In the dog, medium- and large-sized hepatic veins have a distinctive perivenular spiral throttling muscle, associated with a resident mast cell population, and a meshwork of perivenular lymphatics. The canine hepatic vein structure is nourished by **vasa vasorum** derived from small branches of the internal thoracic and phrenic arteries. In the domestic cat, the hepatic vein is comparatively simple, without throttling musculature or resident mast cells but has a similar smaller perivenular lymphatic plexus.

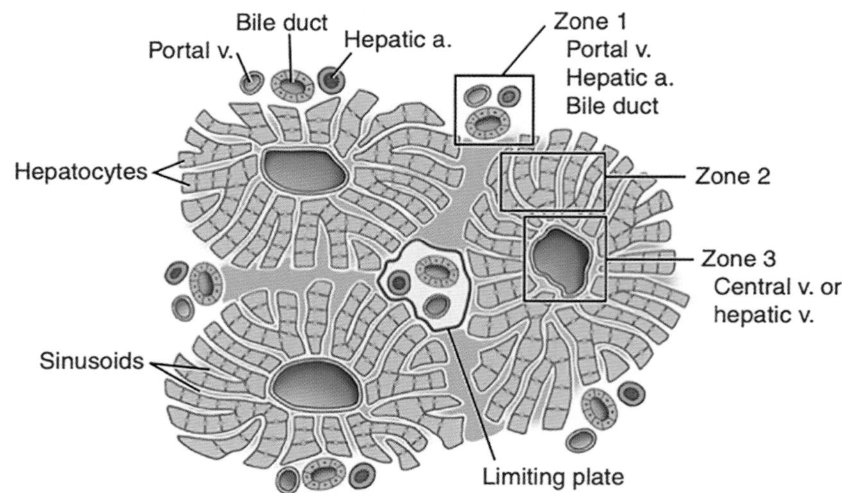

FIGURE 5.1-7 The drawing depicts normal hepatic architecture. Hepatic anatomy is described relative to the directional flow of blood across the hepatic sinusoids: portal region (zone 1), midzonal region (zone 2), and centrilobular region (zone 3).

In health, **portal tracts** (Fig. 5.1-7) have a single portal vein (thin nonmuscular vessel), 1–2 arteries (thicker-walled muscular vessels), and 1–2 bile ducts (with cuboidal epithelium). Approximately 80% of portal tracts are triads (all 3 anatomic structures present), with a subset termed "dyads" (i.e., 2 components present) or "monads" (one component present; usually a lone arteriole.).

Hepatic lymphatics

Hepatic lymphatic capillaries, composed of a single layer of lymphatic endothelium lacking a continuous basement membrane, have patency maintained by "wispy" connective tissue attached to surrounding adventitia. These lymphatic spaces are particularly prominent surrounding hepatic veins (in the dog) and in the subcapsular regions of the liver, and become apparent when there is stasis of circulatory flow. In health, the liver produces ~25–50% of lymph coursing through the **thoracic duct**. Hepatic lymphatic channels fall into 3 categories, classified by location: portal, sublobular, and superficial (or capsular). Hepatic lymph primarily derives from highly permeable **hepatic sinusoids** (ultralymph filtered into the **space of Disse**). Smaller contributions derive from the **peribiliary lymphatic plexus** (largely derived from hepatic arterials) and the **perivenular plexus** (surrounding branches of the hepatic vein). Ultralymph initially accumulates in the space of Disse and then follows sinusoids (or vascular channels in cirrhotic livers) to the perivenular plexus or the interstitial plexus beneath the liver capsule (**Glisson's capsule**). In health, the directional flow of hepatic lymph follows the gradient of hepatic venous pressures: portal pressure (5–8 mmHg) > hepatic vein/vena cava pressure (~2–3 mmHg).

Portal hypertension of hepatic causes—e.g., sinusoidal fibrosis—can increase lymph flow up to 30-fold. With hepatic synthetic failure, marked hypoalbuminemia aggravates lymph formation secondary to a decline in plasma oncotic pressure. Although increased lymph flow initiates reflex lymphangiogenesis, ultimately enhanced lymph formation distends hepatic lymphatic channels. When lymph drainage insufficiently decompresses hepatic lymphatics, lymph weeps from the subcapsular lymphatic interstitium into the abdominal cavity (ascites). Notably, this process does not occur in animals with PSVA or MVD, as there is no intrahepatic portal hypertension in these disorders.

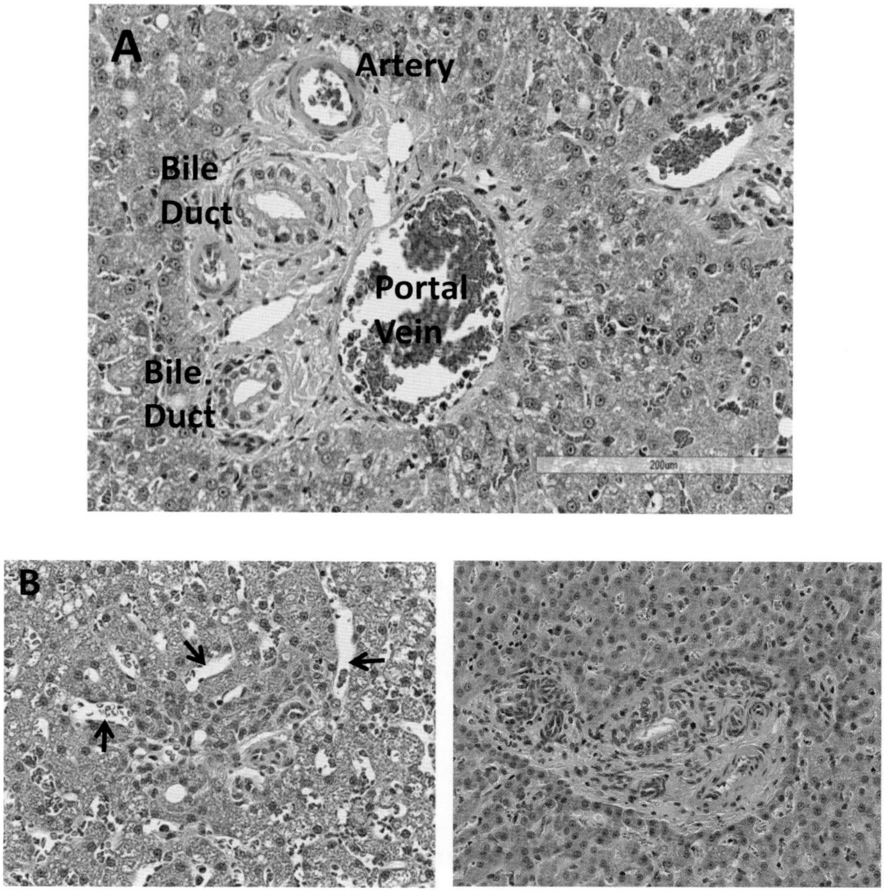

FIGURE 5.1-8 (A) Photomicrograph of hematoxylin and eosin-stained liver section (400× magnification) from a clinically healthy dog displaying a portal tract with 2 bile ducts, 1–2 hepatic arteries, and a single thin-walled portal vein. (B) Photomicrograph of liver sections from two dogs with PSVA stained with hematoxylin and eosin (200× magnification). In each example, note the small hepatocyte size, increased number of thick muscular arteries, and increased biliary profiles (cuboidal epithelial cells) entangled with serpiginous arterials. No portal vein silhouettes are evident. In the panel on the left, note unusual inlet venules intersecting with hepatic sinusoids *(arrows)*. In the right panel, note the densely packed disorganized hepatic cords, tiny hepatocytes, and increased numbers of binucleated hepatocytes. Variation in the severity of these stereotypic changes likely reflects the severity of hepatofugal portal circulation.

PORTOSYSTEMIC VASCULAR ANOMALIES (PSVA)

Although PSVA are congenital, extensive genetic mapping studies have failed to define a single causal gene mutation. Extrahepatic PSVA (E-PSVA) predominantly develop in small breed dogs and cats, whereas intrahepatic PSVA (I-PSVA) predominantly develop in large breed dogs. The most common types of E-PSVA involve shunting through normal portal tributaries rather than through anomalous vasculature. Intrahepatic PSVA are traditionally described according to their position in the liver, as left, central, or right divisional shunts; left divisional I-PSVA would be consistent with a patent ductus venosus.

Any disturbance reducing hepatic portal venous perfusion generates similar histologic features and physiologic effects. Histologic changes are collectively referred to as portal venous hypoperfusion (PVHP)—including small hepatocyte size, increased thick muscular arterial profiles, variable proliferative-like bile duct profiles, lobular atrophy (lobular elements in abnormally close apposition), often inapparent silhouettes of portal veins, and microhepatia. Fig. 5.1-8, illustrates histologic portal tract features in a clinically healthy dog (A) and in two dogs (B) with PVHP due to PSVA.

Embryology of hepatic vascular structure

Embryologic development of hepatic sinusoids, vena cava, and portal vein derives from the vitelline veins (omphalomesenteric veins) interconnecting the yolk sac and sinus venosus (a caudal portion of the cardiac tube). Central segments of the vitelline veins evolve into hepatic sinusoids, the left cranial vitelline segment atrophies, and the right cranial vitelline segment contributes to the hepatic vena cava. Anastomotic and involutional remodeling of the caudal right and left vitelline veins evolve into the portal vein. In the embryo, paired umbilical veins connect the allantois (which develop into the umbilicus) and sinus venosus (which develops into the heart). The umbilical veins subdivide into cranial, middle, and caudal segments. The cranial and right caudal segments undergo involution. The left umbilical vein transports placental blood to the developing liver; one portion perfusing the developing hepatic sinusoids and the other forming the ductus venosus, the principal conduit transporting placental blood (from the left umbilical vein and cranial right vitelline vein) to the developing heart.

Hepatic portions of the caudal vena cava evolve from segments of cardinal veins and the cranial segment of the right vitelline vein. The left and right supracardinal veins evolve into the hemiazygos and azygos veins, respectively. Developmental processes creating hepatic sinusoids, the hepatic portion of the vena cava, and the portal vein are complex, involving a myriad of signaling molecules and crosstalk between different gene signaling pathways. The ductus venosus begins its process of closure shortly after birth (starting within 3 hours) and typically closes by day 6 (except for slower closure in the Irish Wolfhound, a breed predisposed to development of patent ductus venosus).

Selected references

Anatomy of liver and abdominal vasculature

[1] Evans HE, de Lahunta A. The veins. In: Miller's anatomy of the dog. 4th ed. St. Louis: Saunders/Elsevier; 2013. p. 516–20.
[2] Evans HE, de Lahunta A. The digestive apparatus and abdomen. 4th ed. St. Louis: Saunders/Elsevier; 2013. p. 328–33.

Anatomy and diagnosis of portosystemic shunts—dogs and cats

[3] White RN, Parry AT, Shales C. Implications of shunt morphology for the surgical management of extrahepatic portosystemic shunts. Aust Vet J 2018;96(11):433–41.
[4] Parry AT, White RN. Post-temporary ligation intraoperative mesenteric portovenography: comparison with CT angiography for investigation of portosystemic shunts. J Small Anim Pract 2018;59:106–11.
[5] White RN, Shales C, Parry AT. New perspectives on the development of extrahepatic portosystemic shunts. J Small Anim Pract 2017;58(12):669–77.

CHAPTER 5

CASE 5.2

Extrahepatic Bile Duct Obstruction Secondary to Acute Pancreatitis

Shannon M. Palermo[a] **and Mark P. Rondeau**[b]
[a]Department of Internal Medicine, Veterinary Specialists and Emergency Services, Rochester, New York, US
[b]Department of Clinical Sciences and Advanced Medicine, University of Pennsylvania School of Veterinary Medicine, Philadelphia, Pennsylvania, US

Clinical case

History

A 5-year-old male castrated mixed-breed dog presented for evaluation of vomiting and anorexia of 5 days' duration. He was originally treated with subcutaneous fluids and maropitant (Cerenia™; antiemetic—alleiviating nausea and vomiting) by the primary care veterinarian, but his signs did not improve. He had a history of dietary indiscretion 2 days before the onset of vomiting and anorexia but had been otherwise healthy.

Physical examination findings

The dog had diffuse icterus, most notably in the sclerae, pinnae, and mucous membranes. He was tachycardic at 160 bpm and had tacky mucous membranes. The rectal temperature was mildly elevated at 102.9 °F (39.4 °C). He was painful on palpation of his cranial abdomen. The remainder of his physical examination was normal.

> The term icterus refers to a yellow discoloration of the skin, sclera, and mucous membranes secondary to increased serum bilirubin. Icterus may be caused by an increased breakdown of red blood cells (e.g., with hemolysis—called prehepatic icterus) or with diseases resulting in impaired metabolism or excretion of bilirubin—i.e., hepatic or posthepatic icterus, respectively.

Differential diagnoses

Hepatic icterus caused by primary liver disease, such as infection, inflammation, toxicity, drug reaction, sepsis, or neoplasia; posthepatic icterus caused by an obstruction of bile flow such as occurs secondary to pancreatitis, neoplasia (bile duct, pancreas, and duodenum), gallbladder mucocele, severe bile duct inflammation (cholangitis), or duodenal foreign body at the level of major duodenal papilla; prehepatic icterus caused by hemolysis

Diagnostics

Clinicopathologic testing (CBC, chemistry screen, and urinalysis) revealed neutrophilia with a left shift, lymphopenia, and monocytosis; hyperbilirubinemia, hypercholesterolemia, and increased activities of alkaline phosphatase and alanine aminotransferase. Serum canine pancreatic lipase immunoreactivity (cPLI) was increased at 799 μg/L (normal, < 200 μg/L).

Comparative Veterinary Anatomy: A Clinical Approach. https://doi.org/10.1016/B978-0-323-91015-6.00032-7

Abdominal radiographs revealed no evidence of an obstructive small intestinal pattern. Widening of the gastroduodenal angle was appreciated, with an overall loss of detail in the right cranial abdomen. Transabdominal ultrasound revealed a diffusely enlarged and hypoechoic pancreas, with surrounding hyperechoic peripancreatic fat (Fig. 5.2-1A). The gall bladder was distended with a mild amount of intraluminal debris. The common bile duct was dilated at 6 mm (0.02 in.) (normal <3 mm or <0.01 in.) (Fig. 5.2-1B). A scant amount of anechoic abdominal effusion was present in the cranial abdomen.

Diagnosis

Extrahepatic bile duct obstruction secondary to acute pancreatitis

Treatment

To date, there is limited evidence regarding the optimal treatment strategy for extrahepatic bile duct obstruction secondary to pancreatitis, with both medical and surgical treatment options available. As pancreatitis is a medically managed condition, treatment with IV fluids, antiemetics, analgesics, and appropriate nutritional support is

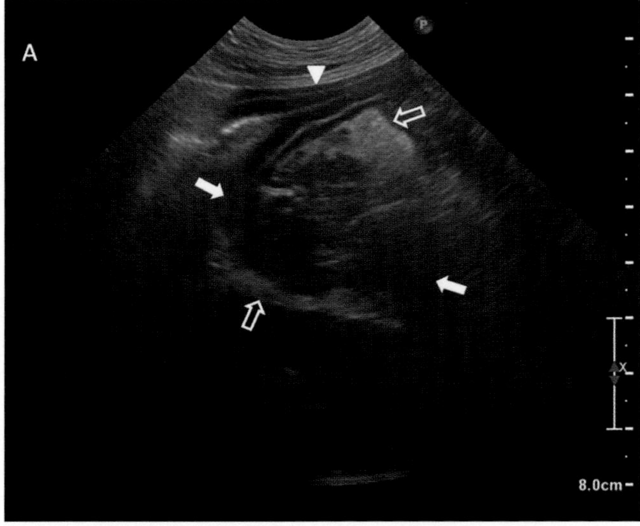

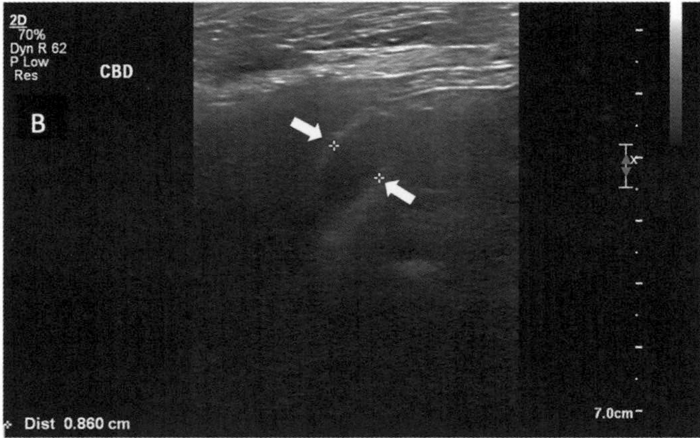

FIGURE 5.2-1 (A) Ultrasonographic appearance of acute pancreatitis in the dog. Note the irregular, hypoechoic right lobe of the pancreas *(solid arrows)* with surrounding hyperechoic mesentery *(open arrows)* adjacent to the descending duodenum *(arrowhead)*. (B) Ultrasonographic appearance of the distended common bile duct *(solid arrows)* proximal to its insertion at the major duodenal papilla as a result of pancreatitis. (Images courtesy of Dr. Jenn Reetz.)

paramount. The use of antibiotics is somewhat controversial and typically reserved for feline patients with suspected triaditis (L. *trias* group of three). Resolution of pancreatitis results in decreased peri-pancreatic inflammation and subsequent normalization of bile flow. In some cases of biliary obstruction, surgery is elected to relieve the common bile duct obstruction. Typically, the common bile duct is flushed, and a temporary stent is placed within the lumen to maintain patency.

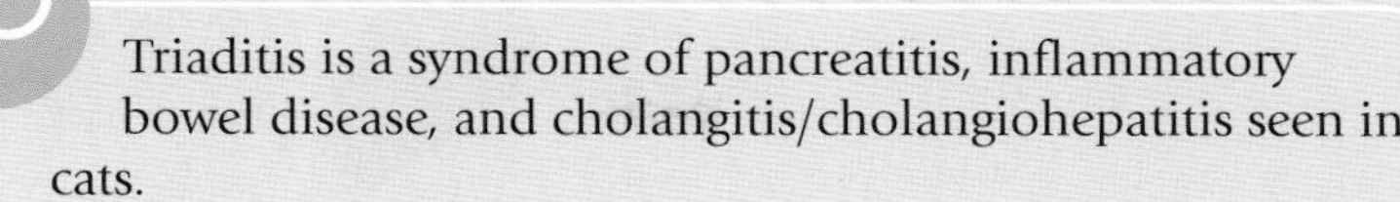
Triaditis is a syndrome of pancreatitis, inflammatory bowel disease, and cholangitis/cholangiohepatitis seen in cats.

Anatomical features in canids and felids

Introduction

The pancreas functions as both an endocrine and exocrine organ and plays an important role in glucose regulation (via the production of insulin and glucagon) as well as digestion (via the production of digestive enzymes). It is made up primarily of acinar and ductal epithelial cells (exocrine), with far fewer islet cells (endocrine) to carry out these various functions. This discussion covers the pancreas and its multiple functions along with its blood supply, lymphatics, and innervation. The gall bladder is discussed briefly in relation to pancreatic function.

Endocrine function of the pancreas

The islet cells (islets of Langerhans) consist of alpha (α), beta (β), and delta (δ) cells. Each cell type produces a different hormone. The most numerous, β cells, secrete insulin in response to hyperglycemia. Insulin affects virtually every organ system in the body, playing a central role in the anabolic metabolism of carbohydrates, proteins, and fats. The principal target sites for insulin are the liver, adipose tissue, and skeletal muscle. Insulin increases the uptake of glucose into cells to be used as energy and prevents lipolysis and proteolysis. The α cells secrete glucagon in response to decreased blood glucose levels, which promotes mobilization of energy in the form of lipolysis, glycogenolysis, and gluconeogenesis. Lastly, the δ cells secrete somatostatin, which inhibits the secretion of other hormones including insulin.

Exocrine function of the pancreas

The exocrine pancreas secretes many enzymes important in the digestion of proteins, fats, and carbohydrates. Protein digestion is catalyzed by trypsin, chymotrypsin, and carboxypeptidase. Several enzymes such as pancreatic lipase and phospholipase are also activated by trypsin to aid in fat digestion. Carbohydrate digestion is facilitated by pancreatic amylase, which hydrolyzes carbohydrates into disaccharides and trisaccharides. These digestive enzymes are released by the pancreas as inactive zymogens that become activated once they reach the small intestine. In the presence of chyme, enteropeptidase secreted by enterocytes activates trypsinogen to form trypsin. Trypsin then activates the remaining zymogens to aid in digestion.

Extrahepatic bililary system

The canine extrahepatic biliary system comprises the gallbladder, cystic duct, hepatic ducts, common bile duct, and major duodenal papilla (Fig. 5.2-2). Bile is produced by the hepatocytes and collected in the hepatic canaliculi. Bile is composed largely of bile acids, which are synthesized from cholesterol. They function to emulsify fat in the intestine and aid in the absorption of nutrients. Over 90% of bile is recirculated via enterohepatic recirculation.

The gallbladder fills continuously through hepatic secretion as passive gallbladder distention occurs. The sphincter of Oddi, located at the terminal portion of the common bile duct, regulates duodenal bile flow. It acts as a one-way valve to regulate the flow of bile into the duodenum while also preventing the reflux of duodenal contents into the common bile duct. Cholecystokinin (CCK) is a hormone produced and secreted by duodenal enterocytes that stimulates gall bladder contraction as well as release of pancreatic digestive enzymes.

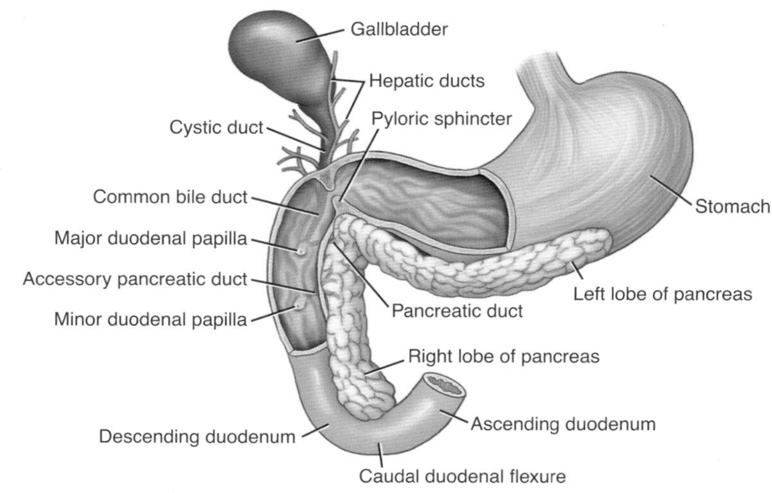

FIGURE 5.2-2 Normal canine anatomy of the extrahepatic biliary system and pancreas.

The anatomic proximity of the pancreas (specifically the body and right lobe) to the insertion of the common bile duct at the major duodenal papilla results in a potential obstruction of the common bile duct in dogs with pancreatitis. Regional inflammation associated with pancreatitis, pancreatic edema, and pancreatic fibrosis (in cases of chronic pancreatitis) may cause mechanical obstruction of the common bile duct resulting in obstruction.

This anatomic difference in the pancreatic ducts between dogs and cats is thought to play a role in the pathophysiology of "triaditis" in cats. It has been postulated that cholangitis/cholangiohepatitis may occur in cats with inflammatory bowel disease secondary to reflux of enteric bacteria into the common bile duct. Pancreatitis may result from bacterial reflux into the pancreatic duct or from pancreatic duct obstruction secondary to cholangitis. It is also possible that an immune-mediated pathogenesis underlies triaditis and that the close anatomic proximity and shared enteric bacterial populations may contribute to the pathogenesis.

Gross pancreatic anatomy

The **pancreas** is comprised of right and left lobes, joined at a central body (Fig. 5.2-2). The **right lobe** lies in the sigmoid flexure of the proximal duodenum, while the **left lobe** is closely associated with the greater curvature of the stomach and sits cranial to the transverse colon.

The canine **pancreatic duct** enters the duodenum at the **major duodenal papilla**, while the **accessory pancreatic duct** enters at the **minor duodenal papilla**. The common bile duct enters separately from the pancreatic duct at the major duodenal papilla. In cats, the pancreatic duct fuses with the **common bile duct** before entering the duodenum at the major duodenal papilla (Fig. 5.2-3).

Regional blood supply, lymphatics, and innervation (Fig. 5.2-4)

The majority of the pancreatic blood supply originates from the **celiac artery**, supplied by the **splenic** and **hepatic arteries**. The right lobe receives blood from the **cranial and caudal pancreaticoduodenal arteries**, while the left lobe receives blood predominantly from the splenic artery.

Lymphatic drainage from the pancreas flows into the **pancreaticoduodenal**, **hepatic**, **splenic**, and **mesenteric lymph nodes**.

The pancreas is innervated by the enteric nervous system and branches of the **vagus nerve**. Pancreatic secretion is stimulated by the parasympathetic nervous system and inhibited by the sympathetic nervous system.

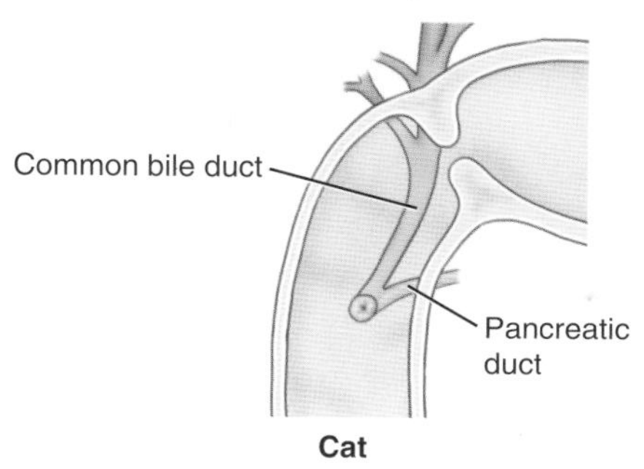

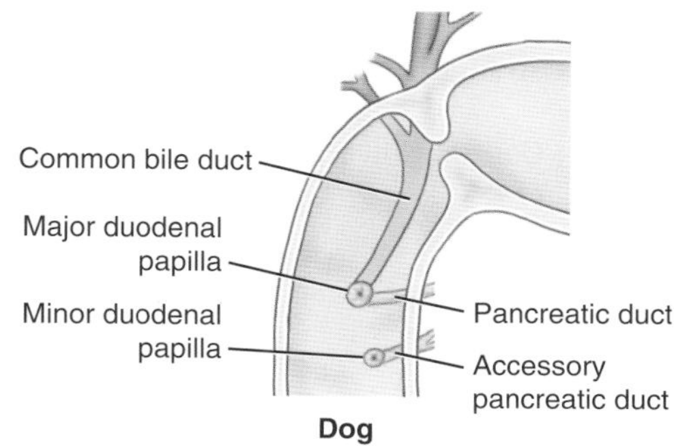

FIGURE 5.2-3 Pancreatic and bile ducts of the dog and cat.

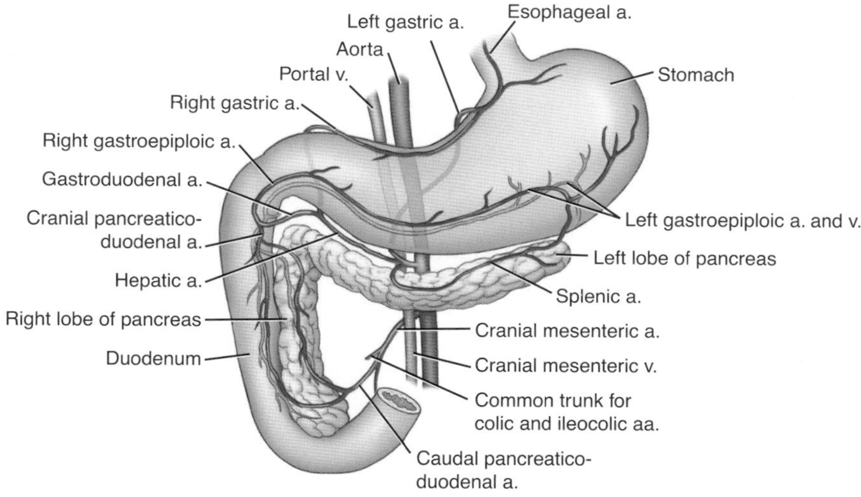

FIGURE 5.2-4 Blood supply of the pancreas.

Selected references

[1] Mansfield C, Beths T. Management of acute pancreatitis in dogs: a critical appraisal with focus on feeding and analgesia. J Small Anim Pract 2015;56:27–39.

[2] Mayhew PD, Richardson RW, Mehler SJ, et al. Choledochal tube stenting for decompression of the extrahepatic portion of the biliary tract in dogs: 13 cases (2002–2005). J Am Vet Med Assoc 2006;228:1209–14.

[3] Radlinsky MA, Fossum TW. Surgery of the extrahepatic biliary system. In: Fossum TW, Cho J, Dewey CW, et al., editors. Small animal surgery. 5th ed. Philadelphia: Elsevier; 2019. p. 571–85.

[4] Washabau RJ, Xenoulis PG, Steiner JM, et al. Pancreas. In: Washabau RJ, Day MJ, editors. Canine and feline gastroenterology. 1st ed. St. Louis: Elsevier; 2013. p. 799–848.

CHAPTER 5

CASE 5.3

Hyperadrenocorticism

Nora S. Grenager[a], Rebecka S. Hess[b], and James A. Orsini[c]
[a]Steinbeck Peninsula Equine Clinics, Menlo Park, California, US
[b]Department of Clinical Sciences and Advanced Medicine, University of Pennsylvania, Philadelphia, Pennsylvania, US
[c]Department of Clinical Studies - New Bolton Center, University of Pennsylvania School of Veterinary Medicine, Kennett Square, Pennsylvania, US

Clinical case

History

An 11-year-old female spayed mixed breed dog was presented for polydipsia, polyuria, and polyphagia; weight gain; and truncal hair loss. The dog had been having difficulty with stairs and jumping up into the car. The owner noticed that the dog had recently developed a "pot-bellied" appearance and was panting more than normal.

> Polydipsia (PD) and polyuria (PU) are copious drinking and urinating, respectively. They typically happen together and present in a wide variety of medical conditions.

Physical examination findings

Relevant findings included generalized abdominal distension; truncal nonpruritic, symmetrical alopecia (hair loss); hypotonia (diminished tone) of the skin; hyperpigmentation of the inguinal skin; and weakness. The vital signs—rectal temperature, pulse, and respiration—were normal.

> Hyperadrenocorticism is high on the list of possible diagnoses in a middle-aged dog presenting for lethargy, PU/PD, abdominal distension, and alopecia.

Differential diagnoses

Systemic disease such as diabetes mellitus, hyperadrenocorticism (Cushing's syndrome), neoplasia, liver disease, renal disease, hypothyroidism, or diabetes insipidus

CUSHING'S SYNDROME

The eponym "Cushing's syndrome" is an "umbrella term" referring to a group of clinical and laboratory abnormalities that result from excessive glucocorticoids. It can be iatrogenic, adrenal-dependent, or pituitary-dependent. Chronic excesses in blood glucocorticoid concentration is the common link in hyperadrenocorticism (HAC). "Cushing's disease" is more specific and relates to dogs and cats in which hypercortisolism is secondary to excessive secretion of adrenocorticotropic hormone (ACTH) by the pituitary gland and is termed pituitary-dependent hyperadrenocorticism (PDH).

Physician and neurosurgeon Harvey Cushing first described this syndrome with clinical manifestations of basophil adenomas of the pituitary gland in 1932. These tumors produced excess ACTH, causing adrenal hyperplasia.

Comparative Veterinary Anatomy: A Clinical Approach. https://doi.org/10.1016/B978-0-323-91015-6.00033-9

Diagnostics

Clinicopathologic testing (complete blood count, serum chemistry profile, and complete urine analysis with culture) revealed lymphopenia, eosinopenia, neutrophilia, monocytosis (collectively referred to as a "stress leukogram"), and thrombocytosis; increased alkaline phosphatase, alanine aminotransferase (ALT), and cholesterol; and hyperglycemia. Urinalysis revealed isosthenuria and mild proteinuria.

Transabdominal ultrasonography revealed unilateral adrenal enlargement—a right adrenal mass measured 15×20 mm (1.5×2.0 cm or 0.6×0.8 in.) with no apparent vascular invasion into the caudal vena cava. The left adrenal was small (thickness was 3 mm or 0.01 in.). The liver was homogeneously hyperechoic with biliary debris. Abdominal radiographs revealed hepatomegaly and mineralization of the right adrenal gland; thoracic radiographs were performed to look for metastases and revealed no abnormal findings. A low-dose dexamethasone suppression test confirmed hyperadrenocorticism. A low endogenous ACTH concentration further supported that this was a glucocorticoid-secreting adrenal tumor.

Diagnosis

Adrenal tumor hyperadrenocorticism (HAC)—suspect unilateral adrenal adenoma

Treatment

The decision in how to treat HAC once confirmed as the working diagnosis depends on the severity of the disease, malignancy, other concurrent diseases, available treatment options, potential adverse events (side effects) of treatment, and clinician preferences. Medical treatment in this dog was instituted using trilostane, a synthetic steroid analog that blocks an enzyme important in the end-product cortisol. There was a good clinical response to medical management based on follow-up examination and testing. Once stabilized, CT of the abdomen and chest was performed to exclude the presence of metastatic disease and determine if the adrenal mass was invading the caudal vena cava. After excluding visible metastatic disease, routine unilateral adrenalectomy (right-sided) was performed under general anesthesia with a course of tapering dexamethasone then prednisolone following.

HYPOTHALAMIC-PITUITARY-ADRENAL (HPA) AXIS

The HPA is the neuroendocrine unit composed of the hypothalamus (brain), the pituitary gland (at the base of the brain), and the adrenal glands (adjacent to the kidneys) and is key in basal homeostasis and the body's stress response. Nerve input to the hypothalamus follows a circadian rhythm with input from stress, secreting corticotropin-releasing hormone (CRH). The CRH secreted in turn stimulates the pituitary pars distalis to secrete ACTH, which circulates to the adrenal cortex stimulating the release of cortisol into the systemic circulation (Fig. 5.3-1).

Hypothalamic-pituitary-adrenal is an example of a negative reaction loop—cortisol reduces its own secretion by feedback to the hypothalamus decreasing CRH, which in turn decreases ACTH secretion from the pituitary pars distalis as well as direct negative feedback of cortisol on the pituitary gland itself.

Normal cortisol secretion in the absence of stress plays vital roles in the body. In the face of stress, cortisol enhances blood vessel tonicity, decreases immune responses, reduces inflammation, and stimulates gluconeogenesis (glucose production from amino acids, lactate, and fats).

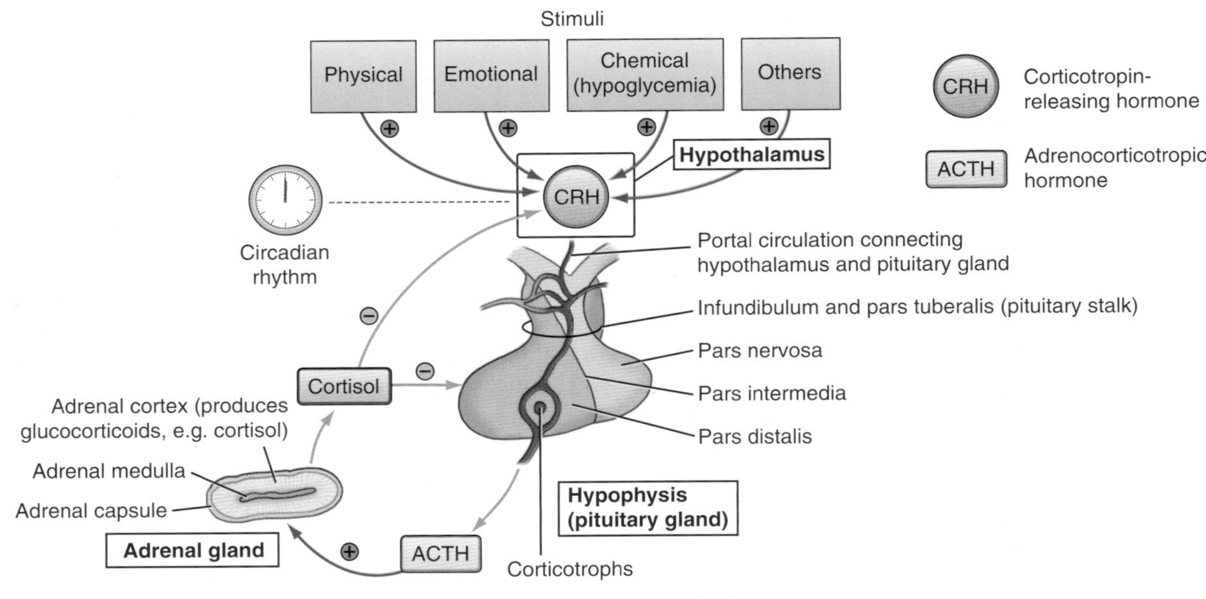

FIGURE 5.3-1 Hypothalamic-pituitary-adrenal axis. The pituitary gland consists of the neurohypophysis (pars nervosa and infundibulum) surrounded by the adenohypophysis (pars tuberalis, pars distalis, and pars intermedia). Alternatively, the pituitary gland can be described as having an anterior lobe (pars tuberalis and pars distalis) and posterior lobe (pars nervosa and pars intermedia).

HYPOADRENOCORTICISM (ADDISON'S DISEASE)

This disease in uncommon in dogs and rare in cats and results from failure of the adrenal glands to secrete adequate quantities of corticosteroids to support normal clinical function. Greater than 85% of the adrenal gland reserve must be depleted before clinical signs are seen. The most common cause of hypoadrenocorticism is primary adrenocortical failure resulting in deficiencies of glucocorticoids (primarily cortisol) and mineralocorticoids (primarily aldosterone).

Electrolyte changes are commonly present in dogs and cats and are the classic hallmark in the addisonian. Na:K ratio < 24 is considered highly suspicious for a diagnosis of hypoadrenocorticism, however this is only useful in those patients with electrolyte changes and can be completely normal in both primary adrenocortical failure and secondary hypoadrenocorticism due to insufficient ACTH production by the pituitary gland.

In certain breeds of dog, the disease is heritable (e.g., Standard Poodle, Nova Scotia Duck Tolling Retriever, and Portuguese Water dog) while other breeds are considered at decreased risk (e.g., Boxer, Dalmatian, and Golden Retriever).

Anatomical features in canids and felids

Introduction

The paired adrenal glands ("suprarenal glands") are components of the endocrine system which includes the hypophysis (pituitary gland), epiphysis (pineal gland), thyroid, parathyroid glands, pancreas (islets), ovaries, testes, and other endocrine components of organs (e.g., kidney, enteroendocrine cells distributed in the gastric and intestinal epithelia).

Function

The adrenal glands produce several important hormones for normal physiological homeostasis (i.e., regulate vascular tone, blood pressure, and volume) and in times of stress (fight or flight). These include catecholamines (epinephrine, norepinephrine) and three steroid hormones—mineralocorticoids (aldosterone), glucocorticoids (cortisol), and androgens (testosterone). The adrenal cortex is divided into three distinct layers, each producing different hormones: (1) zona glomerulosa, the outer layer of the adrenal cortex, produces the mineralocorticoid aldosterone—important in electrolyte balance, especially the retention of sodium (Na^+) and bicarbonate (HCO_3^-), and the excretion of potassium (K^+) and hydrogen (H^+); (2) zona fasciculata, the largest (approximately 70%) and middle layer of the adrenal cortex, synthesizes cortisol and corticosterone—key in metabolic regulation (tend to spare/increase glucose and decrease lipogenesis/increase lipolysis) and immunosuppression; and (3) zona reticularis, the inner-most layer of the adrenal cortex, produces androgens—these are converted to sex hormones in the testes and ovaries. The adrenal medulla, the innermost part of the adrenal gland, is the origin of the production of catecholamines—epinephrine and norepinephrine—which are released throughout the body at times of stress (Fig. 5.3-2).

Gross adrenal anatomy

The paired **adrenal glands** (left and right) are positioned dorsally along the top of the abdomen close to the junction of the thoracolumbar vertebrae. They are located retroperitoneally and lie craniomedial to each corresponding **kidney**. Each is closely associated with a great vessel—the left adrenal with the **aorta** and the right adrenal with the **caudal vena cava**. The phrenicoabdominal vein crosses and indents the ventral surface of each gland. The left lobe of the pancreas lies in close proximity to the left adrenal gland. The right adrenal lies immediately adjacent to the caudate process of the liver.

> The left adrenal gland lies adjacent to the left lobe of the pancreas and must be meticulously separated to minimize the potential for injury to the pancreas (causing traumatic pancreatitis). When performing right adrenalectomy, the hepatorenal ligament must be severed to be able to retract the kidney to access the gland.

The glands are generally elongated, asymmetrical, and their shape molded around their adjacent vessels (Fig. 5.3-3). Clinical experience supports that any short-axis diameter—width or height > 10 mm (1.0 cm or 0.4 in.) or length > 24 mm (2.4 cm or 0.9 in.)—in a medium-sized dog is considered enlarged. Some internal medicine specialists believe the diameter of the adrenal gland is the most reliable indicator of the size and that 7.5 mm (0.75 cm or 0.3 in.) is considered the upper limit of normal for the dog.

> There is much variability in the shape of the adrenals between different breeds of dogs and cats. The glands can be shaped like a bean, "lawn chair," oval, or dumbbell.

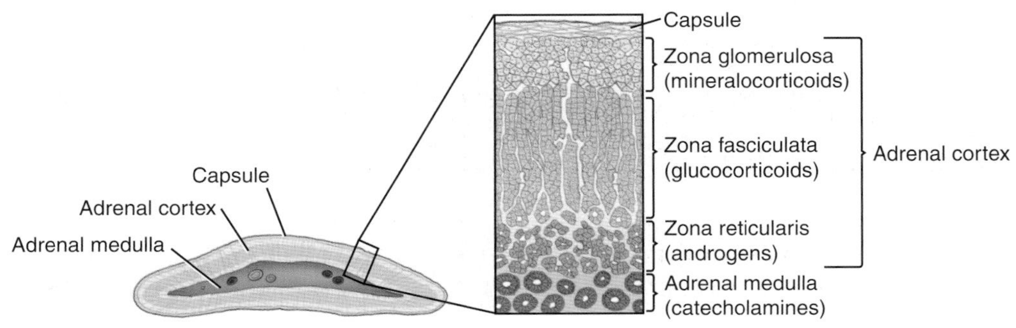

FIGURE 5.3-2 Cross-section of the adrenal gland, its zones, and the hormones they secrete.

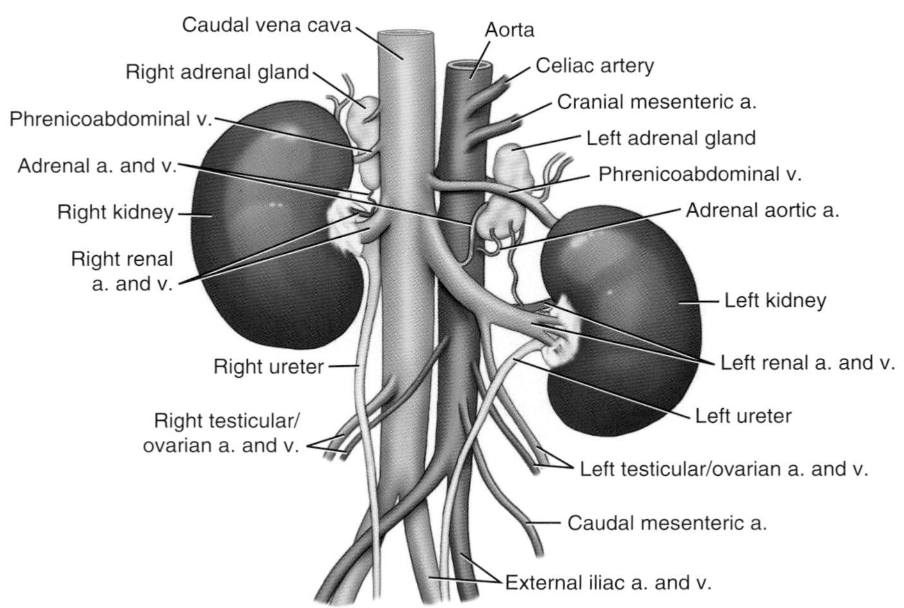

FIGURE 5.3-3 The relationship of the adrenal glands to the regional anatomy.

Adrenal glands are firm with a fibrous capsule. The **cortex** is radially striated and yellow while the **medulla** is darker in color and more consistent in appearance. There is typically no gross distinction between the zones of the cortex, which are only discernible under microscopic evaluation.

296 Innervation, blood supply, and lymphatics associated with the adrenals

> Special care must be taken when retracting the caudal vena cava during adrenalectomy to avoid possibly fatal hemorrhage.

The adrenal glands receive copious blood supply from small branches of many vessels, including the **aorta** and **renal**, **lumbar**, **phrenicoabdominal**, and **cranial mesenteric arteries**. Venous drainage occurs through the **central vein** to the **hilus** of each adrenal gland, then emptying into the caudal vena cava or one of its local tributaries.

Dorsal to each adrenal gland lies the **adrenal plexus** where the **splanchnic nerves** give off branches to the **adrenal ganglia**. The cortex of each adrenal is under hypothalamic control (see side box entitled "HPA axis") and contains fine nerves closely associated with celiac and mesenteric plexuses. The medulla has more visible innervation—comprised of bundles of sympathetic preganglionic fibers (comparable to postganglionic fibers in other tissues).

> The splanchnic nerves must be preserved during adrenal surgery because this network of nerves innervates other abdominal organs.

The **lumbar center** is a series of lymph nodes dorsally located in the abdomen that spread along the abdominal aorta and likely between lumbar transverse processes. The **renal lymph nodes** associated with the kidneys and adrenal glands are generally larger than others in the local lymph node series and flow to the **cisterna chyli**.

PHEOCHROMOCYTOMAS

Pheochromocytomas are the most common, though still rare, tumor of the adrenal medulla. Most often seen in dogs and cows, and occasionally in horses, pheochromocytomas are most commonly tumors of chromaffin cells of the adrenal medulla (or less commonly paragangliomas) that secrete excess catecholamines—epinephrine and/or norepinephrine.

Clinical signs are usually variable and nonspecific, including PU/PD, increased panting, lethargy, weakness/collapse, anorexia, vomiting, and diarrhea. Hypertension is common. Usually solitary, slow-growing masses, they locally invade the caudal vena cava and can metastasize to the liver and regional lymph nodes.

If metastasis is not identified during workup, patients are stabilized with an alpha$_1$-antagonist before careful surgical excision. Due to the catecholamines produced by these masses, these patients may experience marked (occasionally fatal) fluctuations in blood pressure and arrhythmias during the peri-operative period.

In human beings $\geq$90% of pheochromocytomas arise from the adrenal medulla and are linked with a clinical syndrome termed multiple endocrine neoplasia (MEN).

Selected references

[1] Côté E. Clinical veterinary advisor—Dogs and cats. 3rd ed. Elsevier; 2015.
[2] Singh B. Dyce, Sack, and Wensing—Textbook of veterinary anatomy. 5th ed. Elsevier; 2018.
[3] Perez-Alenza MD, Melian C. Hyperadrenocorticism in dogs. In: Ettinger SJ, Feldman EC, editors. Textbook of veterinary internal medicine: Diseases of the dog and the cat. Elsevier; 2017. p. 1795–811.
[4] Hess RS. Hypoadrenocorticism. In: Ettinger SJ, Feldman EC, editors. Textbook of veterinary internal medicine: Diseases of the dog and the cat. Elsevier; 2017. p. 1825–33.

CHAPTER 5

CASE 5.4

Splenic Torsion

David Holt[a], Nora S. Grenager[b], and James A. Orsini[c]
[a]Department of Clinical Sciences and Advanced Medicine, University of Pennsylvania School of Veterinary Medicine, Philadelphia, Pennsylvania, US
[b]Steinbeck Peninsula Equine Clinics, Menlo Park, California, US
[c]Department of Clinical Studies - New Bolton Center, University of Pennsylvania School of Veterinary Medicine, Kennett Square, Pennsylvania, US

Clinical case

History

A 6-year-old male neutered German shepherd presented with a 2-day history of lethargy, anorexia, and one episode of vomiting. The owners believed the dog's abdomen was somewhat distended.

Physical examination findings

The dog appeared quiet and was reluctant to stand or move. The mucous membranes were pale with a CRT of 2–3 s. The dog was tachycardic (180 bpm) with weak and thready femoral pulses (very fine and barely perceptible on palpation). The dog's abdomen was distended and painful on palpation. Abdominal palpation also suggested a "mass effect" (the presence of a mass) in the mid-abdomen.

Differential diagnoses

Abdominal mass originating from the spleen (benign, such as hematoma, or malignant, such as hemangiosarcoma), liver, or bowel (less likely); splenic torsion; GI obstruction (less likely to result in the "mass effect" on abdominal palpation); gastric dilatation and volvulus (less likely because this usually causes a distended, tympanic abdomen)

Diagnostics

Two IV catheters were placed in the cephalic veins, blood was obtained for an expanded database, and IV crystalloid fluid administration was started at 20 mL/kg/h. A rapid focused assessment with sonography in trauma (FAST) ultrasound (Fig. 5.4-1A and B) showed free fluid in the peritoneal cavity and a diffusely enlarged spleen with a lacy, hypoechoic appearance to the parenchyma. The splenic vein was distended. Color flow Doppler ultrasound showed slow blood flow in the hilar splenic vessels. Abdominal radiographs (Figs. 5.4-2A, B shows normal abdominal radiographs; compare to Fig. 5.4-3A, B) revealed a mass effect in the mid-abdomen and a loss of normal peritoneal detail, consistent with the presence of fluid in the peritoneal cavity.

Diagnosis

Splenic torsion

Comparative Veterinary Anatomy: A Clinical Approach. https://doi.org/10.1016/B978-0-323-91015-6.00034-0

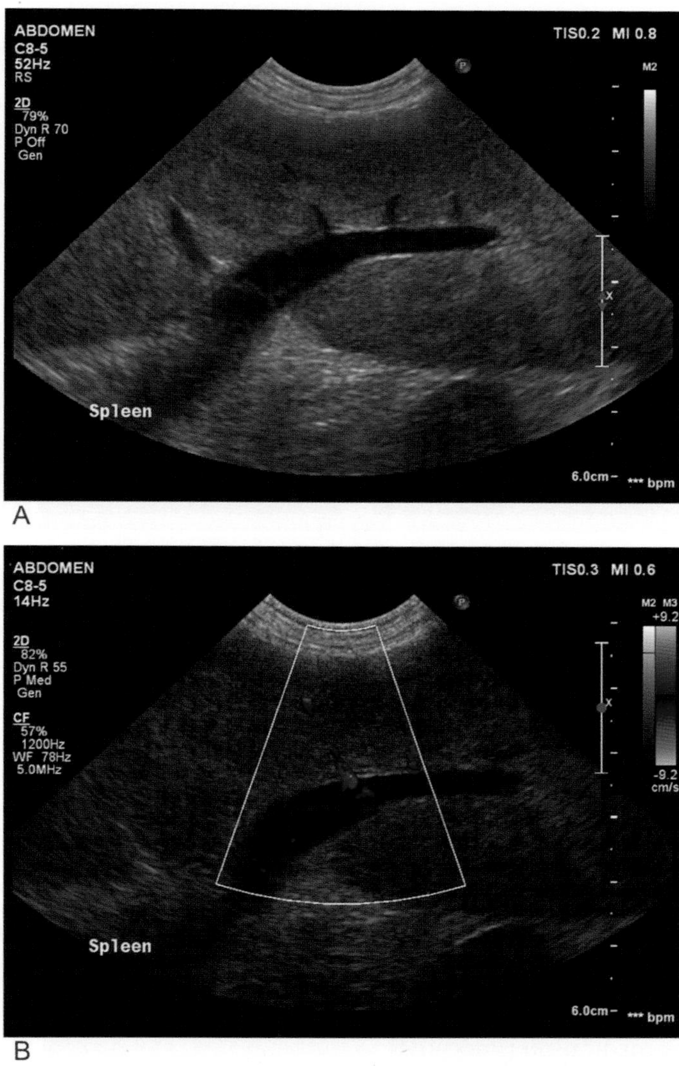

FIGURE 5.4-1 Ultrasound image of the spleen: (A) Sonogram demonstrating a diffusely enlarged spleen with a very distended splenic vein. (B) Color Doppler shows sluggish blood flow in the hilar splenic vessels.

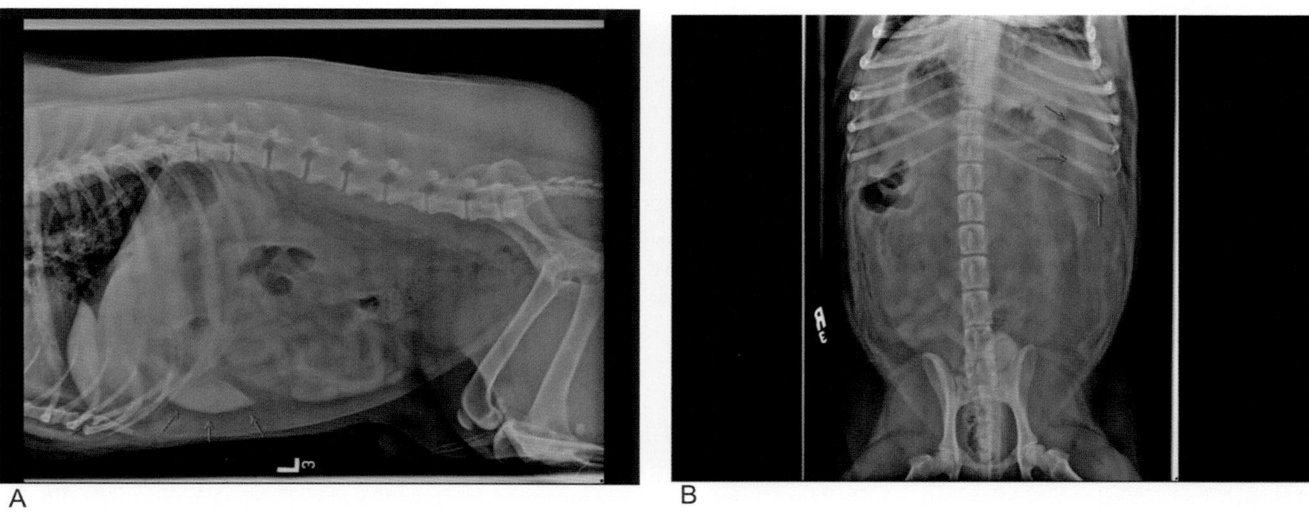

FIGURE 5.4-2 Normal abdominal radiographs of a dog. (A) Lateral radiograph, ventral part of a normal spleen highlighted *(red arrows)*. (B) VD radiograph, dorsal part of a normal spleen highlighted *(red arrows)*.

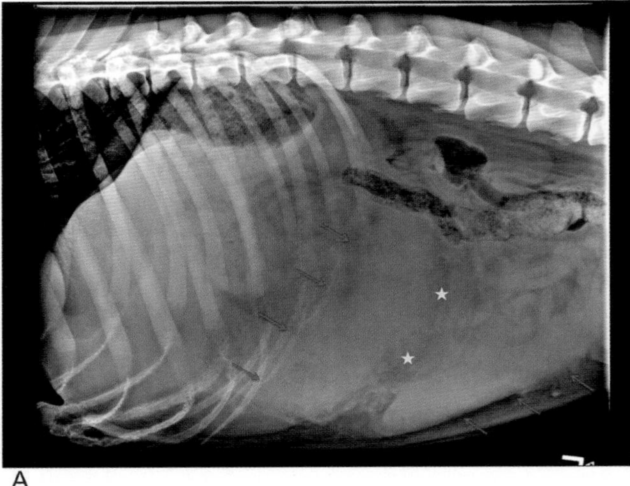

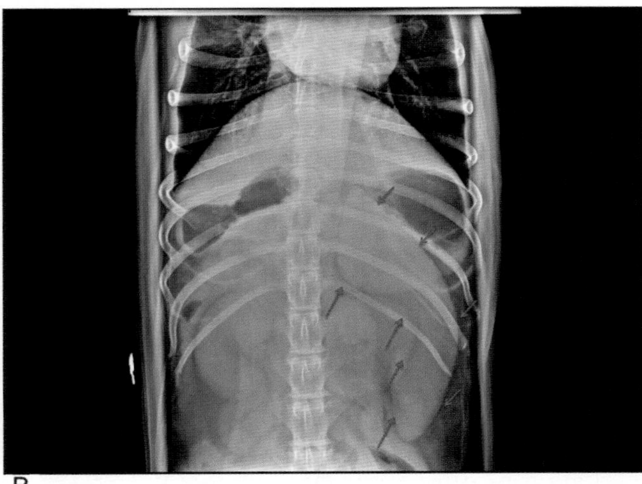

FIGURE 5.4-3 Abdominal radiographs of the dog in this case. (A) Lateral radiograph, twisted (torsed) spleen visible creating a "mass effect" in the mid-abdomen *(red arrows)*; (B) VD radiograph, abnormal splenic position on the VD view *(red arrows)*. *, twisted splenic mesentery.

Treatment

The dog was resuscitated using a rapid infusion of a balanced electrolyte solution to stabilize in preparation for general anesthesia and an emergency exploratory celiotomy (Gr. *celio* belly + Gr. *tomē* cutting). In surgery, a large quantity of serosanguinous fluid was removed from the peritoneal cavity using suction. The splenic torsion was exteriorized from the abdomen and the splenic vessels in the hilar pedicle were carefully separated and ligated. The splenic torsion was not corrected—i.e., untwisted—during surgery because of the potential for release of thrombi and vasoactive compounds into the general circulation. Splenectomy was performed and splenic samples were submitted for histopathology; pathology results were consistent with the surgical diagnosis of splenic torsion.

While still recovering in the ICU, the dog developed ventricular arrhythmias 12 h after surgery, which were treated for 36 h with an IV antiarrhythmic medication (lidocaine).

Anatomical features in canids and felids

Introduction

Accessory splenic nodules ("splenosis") are frequently identified in the omentum of dogs and cats. These are likely revascularized splenic fragments secondary to traumatic events. In exploratory surgery, surgeons may also find suspicious-looking lesions along the borders of the spleen in the dog. These are likely nodular hyperplasia (Gr. *hyper* above, + Gr. *plasis* formation) or siderophilous (Gr. *sideros* iron, + *philein* to love) plaques with no clinical importance.

The spleen (Gr. *splēn* spleen) is part of the reticuloendothelial and lymphoid systems. Its size, shape, and even color can vary with age, breed, and health—i.e., depending on medications or pathologic conditions.

Function

The spleen serves many functions, including RBC management and immune functions such as phagocytosis. The splenic parenchyma consists of red and white pulp; the canine spleen is considered to be sinusoidal while the feline spleen is less well organized—i.e., "nonsinusoidal." The red pulp comprises a series of venous sinuses containing concentrated cellular elements of the blood—i.e., RBCs and monocytes—and cords of reticular cells. Here, hemoglobin is removed from senescent (L. *senescere* to grow old)

RBCs and is reduced to its essential amino acids; the heme portion is metabolized to bilirubin and removed by the liver. The spleen serves as a storage site for RBCs, sequestering 10–20% of a dog's RBCs, and up to 30% of the platelets, at any given time. The spleen can be a site of extramedullary hematopoeisis when necessary and acts as a primary site of hematopoeisis in the fetus. Finally, reticulocytes undergo final maturation as they pass through the splenic tissue (Fig. 5.4-4).

> Splenic trauma (with subsequent rupture of the capsule/tearing of the parenchyma) can lead to substantial intraabdominal hemorrhage (and even death) due to the spleen's highly vascular nature.

The white pulp consists of lymph nodules within the reticuloendothelial supporting structure. This tissue produces lymphocytes, plasma cells, and immunoglobulins. The white pulp has phagocytic properties, removing antibody-coated bacteria and cells.

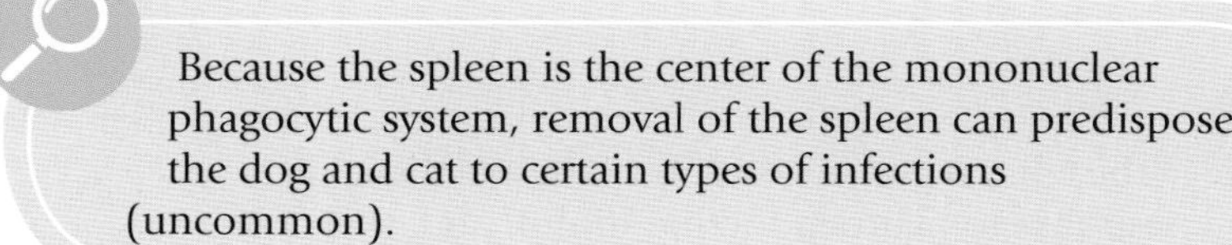

> Because the spleen is the center of the mononuclear phagocytic system, removal of the spleen can predispose the dog and cat to certain types of infections (uncommon).

Spleen

In the dog and cat, the **spleen** lies in a vertical position along the left cranial abdominal wall. It is comfortably positioned between the left crus of the diaphragm, the cardia and fundus of the stomach, and the left kidney. It

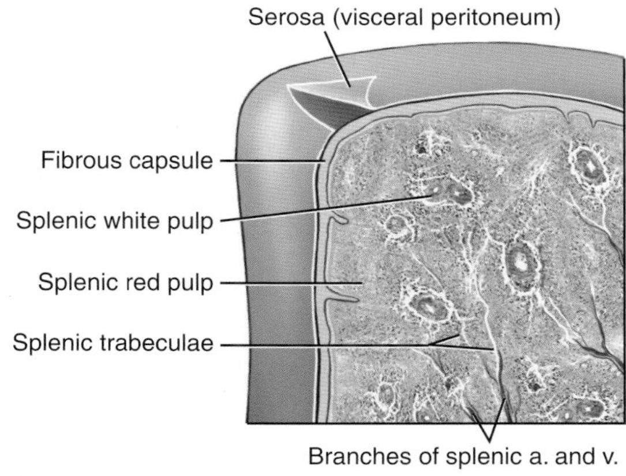

FIGURE 5.4-4 Splenic anatomy in cross-section demonstrating the red and white pulp.

SPLENECTOMY

Splenectomy (L. *splen* spleen + Gr. *ektomē* excision) is complete surgical removal of the spleen. It is a common procedure for splenic disorders including tumor removal (e.g., hemangioma, hemangiosarcoma); and management of traumatic splenic rupture (e.g., hit by car), torsion of the splenic pedicle, splenic abscess, and splenic hematoma; and rarely to treat immune-mediated hemolytic anemia or thrombocytopenia that is nonresponsive to medical therapies.

Interestingly, even though the spleen provides many critical functions, its removal is well tolerated. The liver, lymph nodes, and bone marrow seem able to take over most of its physiological work.

The spleen is somewhat mobile within the abdomen and is impacted by gastric distention as well as its own blood volume. Splenic torsion can occur by itself or be associated with gastric dilatation and volvulus (uncommon), generally in large breed, deep-chested dogs. The etiology of isolated splenic torsion is poorly understood and is likely associated with the mobile nature of the spleen in the abdomen.

The contraction/relaxation of these trabeculae in response to neuroendocrine stimuli and pharmaceuticals facilitate the variable size of the spleen. This allows the spleen to serve as a physiological reserve for additional circulating cells in time of need—e.g., hemorrhagic shock.

is joined to the greater curvature of the stomach by the **gastrosplenic ligament** as part of the **superficial leaf** of the **greater omentum**. It has a convex parietal surface and a concave visceral surface, with a longitudinal/hilar ridge where the vessels and nerves are found, and to which the greater omentum attaches (Fig. 5.4-5).

The spleen is covered by a fibrous capsule rich in smooth muscle and elastic fibers. The spleen varies in size and can increase several-fold in volume when relaxed. It is generally a dumbbell-shaped organ in the dog and cat, with two spherical portions connected by a narrow isthmus. It is grossly red to purple in color, flexible in consistency, and composed of two types of parenchymal tissue—the red and white pulp—contained within a network of fibromuscular trabeculae.

Splenic vascularization, innervation, and lymphatics

Blood enters the spleen via the **splenic artery**, a branch of the **celiac artery**. Before reaching the spleen, a branch is given off to the left portion of the pancreas. The main splenic artery provides several branches to the hilus (L. *hila* a small thing) of the spleen and sends several branches through the gastrosplenic ligament to the greater curvature of the stomach, called the **short gastric arteries**. The **gastroepiploic vessels** branch from the middle of the hilus

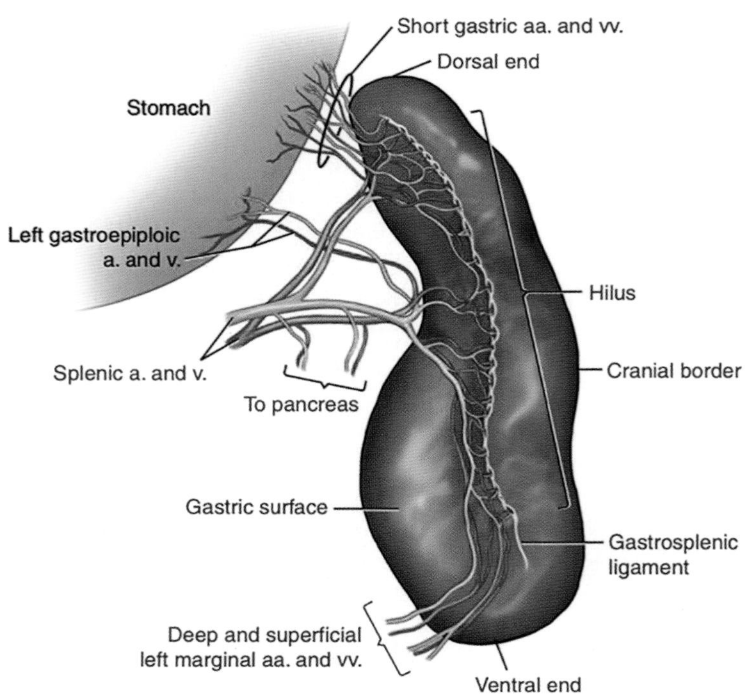

FIGURE 5.4-5 Splenic anatomy—parietal and visceral surfaces with major vessels of the dog and cat.

and insert on the greater curvature of the stomach encased in the gastrosplenic ligament. The **left gastroepiploic vein** drains to the **splenic vein**, which ultimately enters the **portal vein**. In the dog and cat, the splenic artery and vein branch and vascularize independent (though communicating) compartments of the spleen.

During splenectomy, the individual branches of the aforementioned vessels must be ligated close to the spleen to preserve the blood supply to the stomach and pancreas. A surgeon cannot establish appropriate hemostasis by just ligating the main splenic artery and vein because retrograde bleeding will occur through the left gastroepiploic artery and the short gastric arteries. To prevent this from happening, the branches of the splenic vessels entering the hilus are ligated, the gastrosplenic ligament is separated near the spleen, and the short gastric arteries are ligated. The integrity of the short gastric vessels, however, is not critical for circulation to the stomach.

Sympathetic and parasympathetic nerves innervate the spleen and track along with the artery and vein.

Lymph vessels are present in the capsule and trabeculae, but not the splenic pulp. The **splenic lymph nodes** are situated near the splenic vessels and only a few centimeters from the spleen. The lymphatic vessels are efferent (L. *ex*, out + *fere*, to bear) with no afferent (L. *ad*, to + *ferre*, to carry) vessels.

Selected references

[1] DeGroot W, Guiffrida MA, Rubin J, et al. Primary splenic torsion in dogs: 102 cases (1992–2014). J Am Vet Med Assoc 2016;248:661–8.
[2] Dyce KM, Sack WO, Wensing CJG. Textbook of veterinary anatomy 5th ed. Elsevier.
[3] Autran de morais H, Argyle D, O'Brien R. Diseases of the spleen. In: Ettinger SJ, Feldman EC, editors. Textbook of veterinary internal medicine: Diseases of the dog and the cat. Elsevier; 2010. p. 1816–40.

CHAPTER 5

CASE 5.5

Gastric Dilatation and Volvulus

Carol A. Carberry
Department of Surgery, Oradell Animal Hospital, Paramus, New Jersey, US

Clinical case

History

A 9-year-old male intact Doberman Pinscher was presented for acute onset of abdominal distention, nonproductive vomiting, retching, and excessive salivation.

Physical examination findings

On admission, the patient was alert, restless, and appeared uncomfortable. Physical examination revealed tachycardia (140 bpm), tachypnea (36 brpm), bounding femoral pulses, pink mucous membranes with normal capillary refill time, and normal rectal temperature of 99 °F (37.2 °C). The patient resisted abdominal palpation; the abdomen was distended and tympanic. Initial systolic blood pressure was 120 mmHg. Routine bloodwork revealed increased blood lactate (4.2 mmol/L).

Differential diagnoses

Ascites, gastrointestinal obstruction, ileus, small intestinal volvulus, cecocolic volvulus, splenic torsion, abdominal mass, simple gastric dilatation, and gastric dilatation and volvulus (GDV)

Diagnostics

Ventrodorsal abdominal radiographs are often not taken in patients with suspected GDV to avoid respiratory compromise. The respiratory system can be compromised by the direct distention of the stomach cranially, which limits movement of the diaphragm and reduces tidal volume and lung compliance.

Screening right lateral chest and abdominal radiographs were obtained (Fig. 5.5-1). The radiograph showed severe gas distention of the stomach lumen and displacement of the pylorus cranially and dorsally with a characteristic "double bubble" appearance (Fig. 5.5-2). Mildly gas-distended small intestinal segments overlying the central dorsal portion of the gastric silhouette were consistent with the duodenum and additional small intestinal segments displaced caudally by the distended stomach. The spleen was also moved ventrocaudally as a result of the gastric dilatation and malposition. Gas- and fluid-distended small intestinal segments were displaced caudally to the severely distended stomach.

Comparative Veterinary Anatomy: A Clinical Approach. https://doi.org/10.1016/B978-0-323-91015-6.00035-2

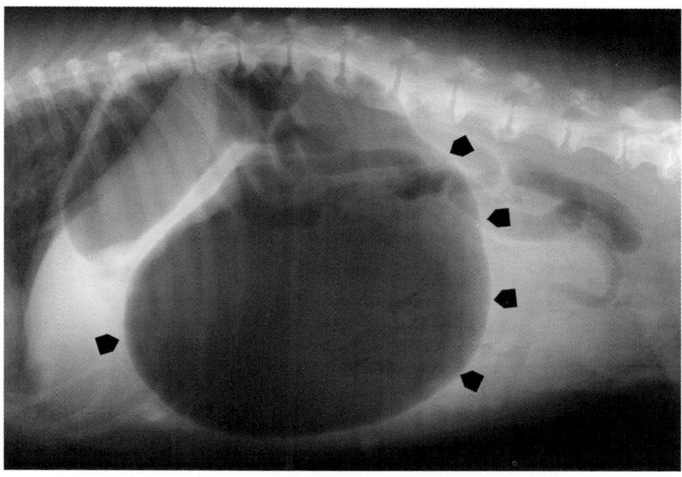

FIGURE 5.5-1 Right lateral radiograph of a 9-year-old male intact Doberman Pinscher with a gastric dilatation and volvulus. Multiple *black arrow heads* outline the gastric dilatation.

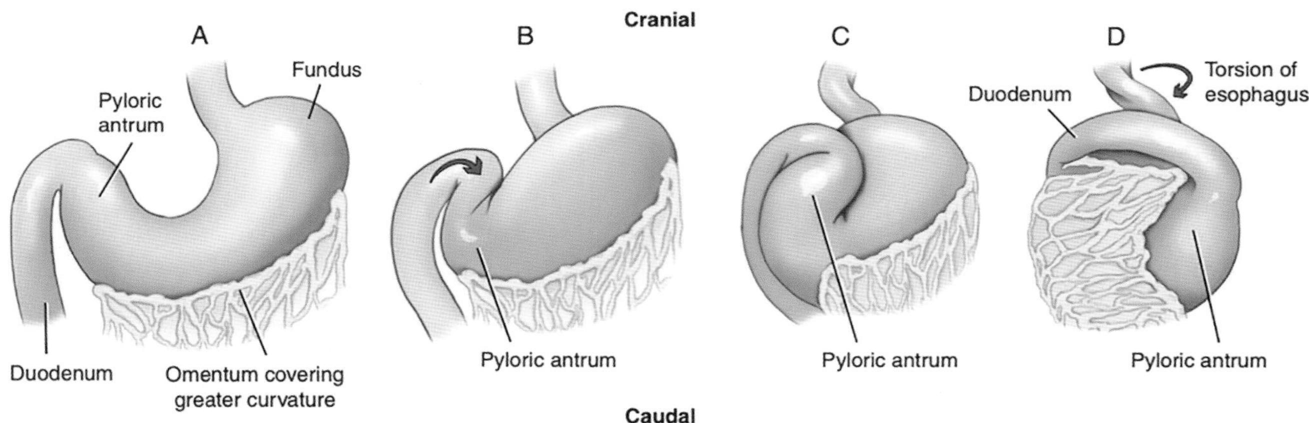

FIGURE 5.5-2 Direction of a "clockwise" stomach rotation in most dogs with gastric dilatation-volvulus as viewed ventrally. (A) The normal position of the esophagus, stomach, pylorus, and duodenum. (B) The pylorus rotates ventrally and laterally to the body of the stomach. (C) The pylorus moves upward, crossing the distended stomach toward the left abdominal wall and stretching the duodenum across the cardia as it crosses the midline. (D) The fundic part of the stomach moves dorsally, the greater curvature is displaced ventrally, and the ventral leaf of the greater omentum covers the ventral aspect of the stomach. This results in a clockwise torsion of the esophagus with the duodenum displaced to the left.

GASTRIC DILATATION AND VOLVULUS (GDV)

Gastric dilitation and volvulus is an acute, life-threatening condition. It most commonly occurs in large, deep-chested dogs including the German Shepherd, Standard Poodle, Great Dane, Boxer, Doberman Pinscher, Saint Bernard, and Irish Setter. It is most common in middle-age to older dogs.

Etiologic factors associated with GDV include anatomic predisposition, hereditary factors, food-filled stomach with postprandial exercise, large amounts of water consumption, and weakened gastric supportive ligaments.

Risk factors associated with gastric dilatation are general anesthesia, aerophagia, high anxiety level, overeating, and gastric and duodenal obstruction.

With GDV, the stomach moves in a clockwise direction as viewed ventrodorsally and caudally. The degree of rotation usually ranges from 90 to 360 degrees, and occasionally rotates up to 540 degrees.

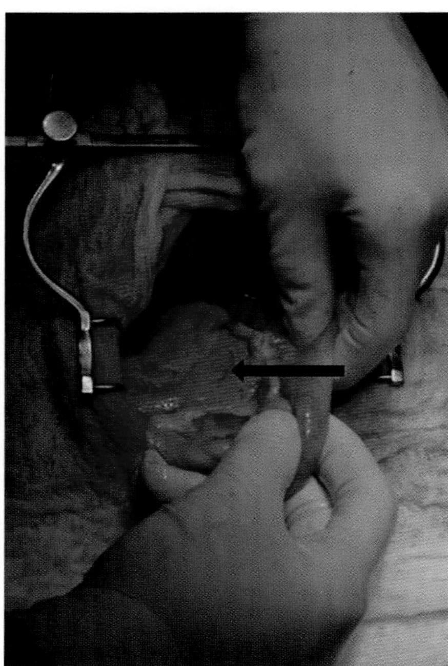

FIGURE 5.5-3 Duodenum is being held. Note the location of the right lobe of the pancreas (*arrow*).

Diagnosis

Gastric dilatation and volvulus (GDV)

Treatment

The patient received circulatory resuscitation, which consisted of a crystalloid fluid bolus at 22 mL/kg. Lactated Ringer's solution (LRS) was administered at 90 mL/kg/h, followed by rapid surgical intervention to reposition the displaced organs and prevent recurrence.

The dog's stomach was decompressed pre-operatively percutaneously with an 18 g IV catheter. The dog was placed under general anesthesia and a ventral midline exploratory laparotomy was performed. The distended stomach was identified and decompressed with an 18 g needle and suction to facilitate repositioning. The stomach and spleen were repositioned by identifying the pyloric portion on the left side of the abdomen and dorsal to the body of the stomach, caudal to the esophagus. The pyloric part also forms a "handle" and is grasped and elevated as the greater curvature is pushed dorsally (toward the spine) to rotate counter-clockwise. The spleen is often engorged and must be rotated with the stomach. The repositioning is complete when the duodenum is on the right side of the abdomen. The duodenum can be further identified by the right lobe of the pancreas, which lies adjacent to it (Fig. 5.5-3). An incisional gastropexy was performed to create a permanent bond between the stomach at the pyloric antrum and the body wall (Fig. 5.5-4). The abdomen was closed in anatomical layers using an absorable suture material; recovery from surgery was uneventful.

Anatomical features in canids and felids

Introduction

This section discusses the stomach and its blood supply, innervation, and lymphatic drainage.

Function

The stomach consists of a single compartment receiving boluses of food and liquid from the esophagus. Gastric juices initiate digestion and muscular contractions progressively move the retained ingesta into the duodenum.

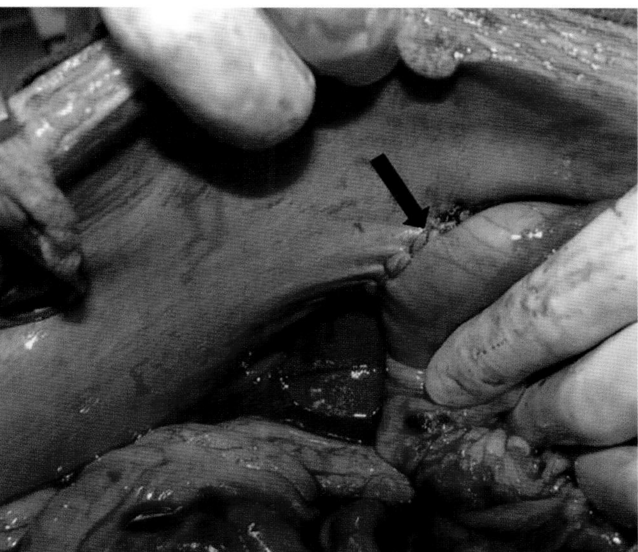

FIGURE 5.5-4 The seromuscular layer of the pyloric antrum is sutured to an incision in the peritoneum and the transversus abdominis muscle of the cranial right ventral abdomen *(arrow)*.

Gross anatomy of the stomach

The shape, size, and position of the stomach depends on the degree of fullness, the amount of ingesta and gas it contains, and the presence of muscular contractions. It is commonly divided into 4 sections: cardia, fundus, body, and pylorus (Fig. 5.5-5). The **cardiac region** of the stomach is located near the esophagus. The **fundus** lies to the left and extends dorsal to the cardia. The **body** of the stomach is the largest portion of the stomach, which reaches from the fundus on the left to the pyloric part on the right. The **pylorus** consists of the pyloric antrum, pyloric canal, and pyloric sphincter.

The **greater curvature** of the stomach is convex and is directed toward the left and ventrally, and is much longer than the lesser curvature. It extends from the cardia to the pylorus and is connected to the head of the spleen by the **gastrosplenic ligament**. The **lesser curvature** is concave, close to the liver, and is directed to the right and dorsally. It also extends from the cardia to the pylorus. The wall of the stomach consists of a mucous membrane, along with muscular and serous layers. The mucous membrane is subdivided into the epithelium, lamina propria, lamina muscularis, and submucosa.

When assessing a GDV from a ventral midline incision approach, the pyloric section of the stomach is displaced to the left, and the proximal section of the duodenum is stretched across the ventral aspect of the cardia. The duodenum extends across the rotated gastroesophageal region. This rotation obstructs stomach outflow, resulting in distention of the stomach. Eructation and vomiting cannot occur in dogs with GDV because the gastroesophageal sphincter is obstructed. Initially with GDV, the veins are compressed as the stomach rotates, causing further congestion of the stomach wall, especially along the greater curvature of the body and fundus. The greater omentum is pulled across the ventral aspect/body of the distended stomach.

Blood supply and lymphatics of the stomach

The main blood supply to the stomach is comprised of the **left** and **right gastric arteries**, which track along the lesser curvature, and the **left** and **right gastroepiploic arteries**, which course along the greater curvature. All these arteries—i.e., the gastric and

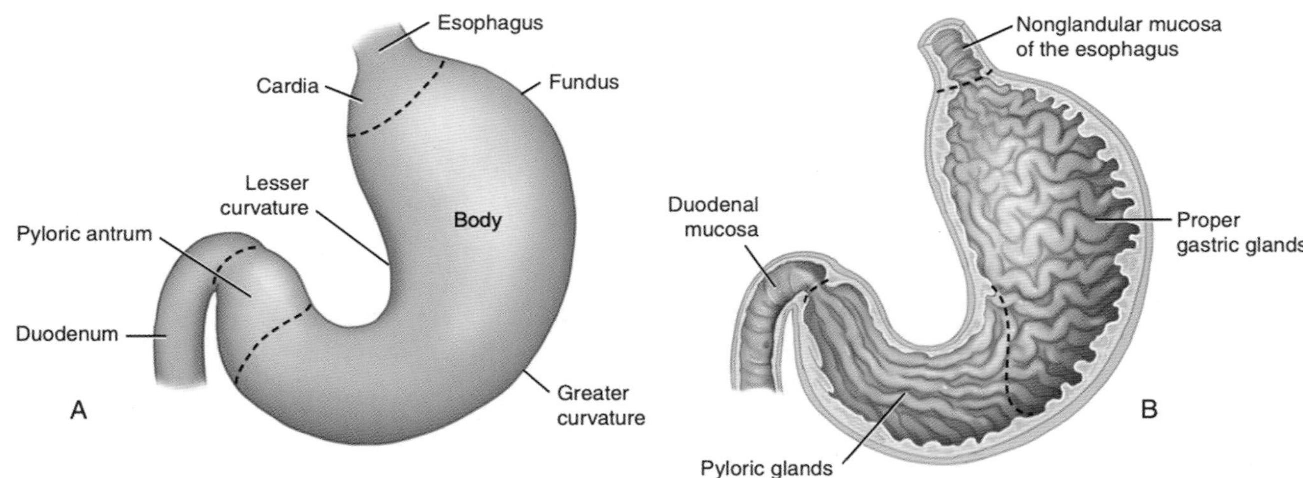

FIGURE 5.5-5 Regions of the canine stomach. (A) Topographical (exterior) anatomy; (B) mucosal (interior) regions.

The gastrosplenic ligament often tears with rupture of the short gastric arteries, as the spleen and stomach rotate in patients with GDV. It is not unusual to find blood in the abdominal cavity and bleeding along the greater curvature of the stomach due to rupture of the short gastric vessels. This may contribute to gastric wall necrosis, most often occurring along the fundus and greater curvature.

gastroepiploic—are initially derived from the **celiac artery**, which has 3 branches—the hepatic, left gastric, and splenic arteries. The **short gastric arteries**, which arise from the **splenic artery**, track through the **gastrosplenic ligament** to supply the greater curvature and fundus of the stomach (Fig. 5.5-6).

The **hepatic artery** terminates as the right gastric artery and the **gastroduodenal artery**. The right gastric artery is the blood supply to the lesser curvature of the stomach and anastomoses with the left gastric artery. The gastroduodenal artery supplies the pylorus and terminates as the right gastroepiploic and **cranial pancreaticoduodenal arteries**. The

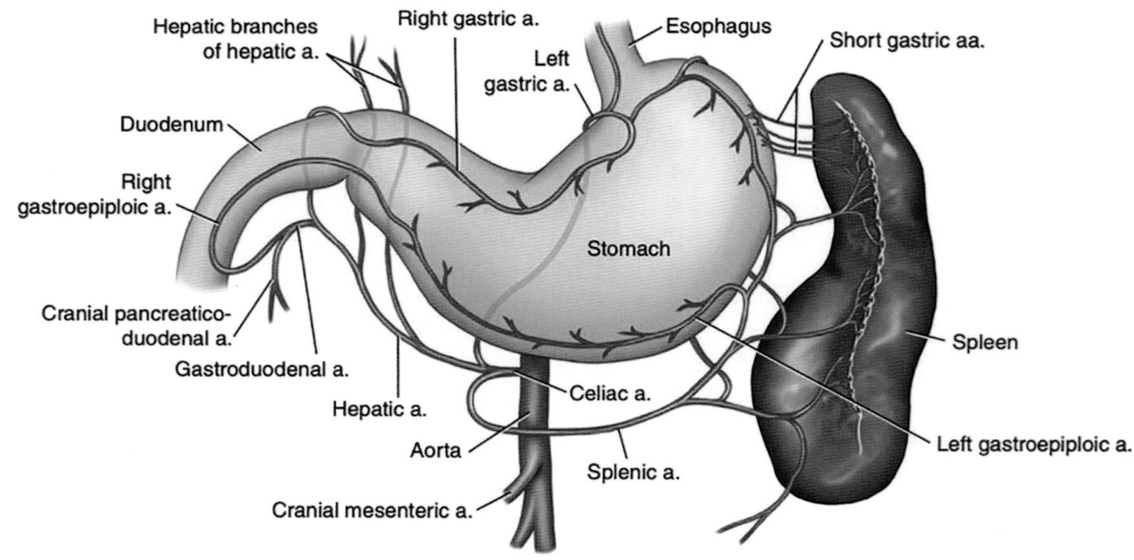

FIGURE 5.5-6 Blood supply of the stomach and the spleen. The celiac artery and its branches constitute the major blood supply.

right gastroepiploic artery tracks in the greater omentum along the greater curvature of the stomach and supplies the stomach and gastric omentum. The right gastroepiploic artery anastomoses with the left gastroepiploic artery, a branch of the splenic artery.

The left gastric artery courses to the lesser curvature of the left side of the stomach near the cardia and supplies blood to the fundus, cardia, and lesser curvature of the stomach; the left gastric artery extends toward the pylorus and anastomoses with the right gastric artery. The splenic artery is the blood supply to the spleen and continues as the left gastroepiploic artery to the greater curvature of the stomach.

The right and left gastric arteries are accompanied by satellite veins. The **left gastric vein** and **left gastroepiploic vein** are tributaries of the **splenic vein**. The **right gastric vein** and **right gastroepiploic vein** are tributaries of the **gastroduodenal vein**. The splenic vein and gastroduodenal vein enter the **portal vein**. The portal vein carries venous blood to the liver.

The venous return is often compromised with GDV. As the stomach dilates in GDV, it can obstruct blood flow from the caudal vena cava and portal vein. This leads to sequestration of blood in the portal system, splanchnic circulation, and caudal vena cava—and a cascade of events resulting in a decrease in venous return, cardiac output, and arterial blood pressure. Gastric distention is also associated with an increase in portal venous pressure, resulting in decreased venous blood flow from the stomach, spleen, pancreas, and intestines, further contributing to the hypotension, hypovolemia, and shock seen in these patients.

Lymphatic drainage is via multiple **gastric nodes** that remove lymph from specific parts of the stomach to the **celiac nodes**, and ultimately draining into the **thoracic duct**.

Innervation of the stomach

The stomach is innervated by parasympathetic fibers from the **vagus nerve**, which carry preganglionic parasympathetic efferent fibers. Sympathetic innervation comes from the **celiac plexus**. The **dorsal** and **ventral vagal trunks** lie on the dorsal and ventral surfaces of the esophagus. Each passes through the diaphragm and tracks along the lesser curvature of the stomach. The ventral vagal trunk supplies the liver, parietal surface of the stomach, and pylorus. The dorsal vagal trunk continues along the lesser curvature to supply the visceral surface of the stomach and the pylorus.

Gastric ligaments and omentum

The **greater omentum** consists of **superficial** and **deep leaves**. It connects the greater curvature of the stomach to the dorsal body wall. From the greater curvature of the stomach, it extends caudally as the ventral/superficial leaf, tracks dorsally at the level of the urinary bladder, and proceeds dorsal to the stomach as the deeper dorsal layer, including the left lobe of the pancreas. The spleen is attached to the greater curvature of the stomach by the **gastrosplenic ligament**, a part of the greater omentum.

Selected references

[1] Miller M, Evans H, DeLahunta A. Miller's guide to the dissection of the dog. Philadelphia: Saunders; 2016.
[2] Miller M, Evans H, Christensen G. Miller's anatomy of the dog. 5th ed. Philadelphia: W.B. Saunders; 2020.
[3] Dyce K, Sack W, Wensing C. Textbook of veterinary anatomy. 5th ed. Elsevier; 2018.
[4] Nickel R, Schummer A, Seiferle E. The viscera of the domestic mammals. Berlin: Parey; 1995.

CHAPTER 5

310

CASE 5.6

Small Intestine Obstruction

David Holt[a], Nora S. Grenager[b], and James A. Orsini[c]
[a]Department of Clinical Sciences and Advanced Medicine, University of Pennsylvania School of Veterinary Medicine, Philadelphia, Pennsylvania, US
[b]Steinbeck Peninsula Equine Clinics, Menlo Park, California, US
[c]Department of Clinical Studies - New Bolton Center, University of Pennsylvania School of Veterinary Medicine, Kennett Square, Pennsylvania, US

Clinical case

History

A 4-year-old male intact Malamute presented with a 3-day history of intermittent vomiting. The dog had a history of eating animal carcasses found on the owners' property.

Physical examination findings

The dog was difficult to handle and restrain, precluding a complete physical examination. The dog was febrile (102.7°F [39.7°C]) and tachycardic (140 bpm), and the femoral pulses were weak upon palpation. The dog's abdomen was tense and palpated to be "doughy."

Differential diagnoses

Gastrointestinal obstruction; intestinal foreign body; pancreatitis; colonic torsion; gastric dilatation and volvulus, mesenteric volvulus (both considered less likely because the dog was clinically stable and lacked the hallmark sign of severe abdominal distension)

Initial treatment and diagnostics

Two IV catheters were placed in the cephalic veins and blood was submitted for an extended database, including electrolytes, pH, bicarbonate, and venous CO_2. The dog was started on IV administration of crystalloids. An initial rapid ultrasound of the abdomen referred to as focused assessment with sonography in trauma (FAST) scan showed no evidence of free peritoneal fluid. Radiographs of the abdomen revealed gas-dilated loops of intestine that appeared to be clustering or lining up, suggesting a linear foreign body (Fig. 5.6-1).

ABDOMINAL RADIOGRAPHS

In properly exposed radiographs without contrast (barium), the small intestine often contains gas and an obstruction is likely present if the small intestinal diameter is >2 times the height of the body of lumbar vertebrae 5 (L5).

Comparative Veterinary Anatomy: A Clinical Approach. https://doi.org/10.1016/B978-0-323-91015-6.00038-8

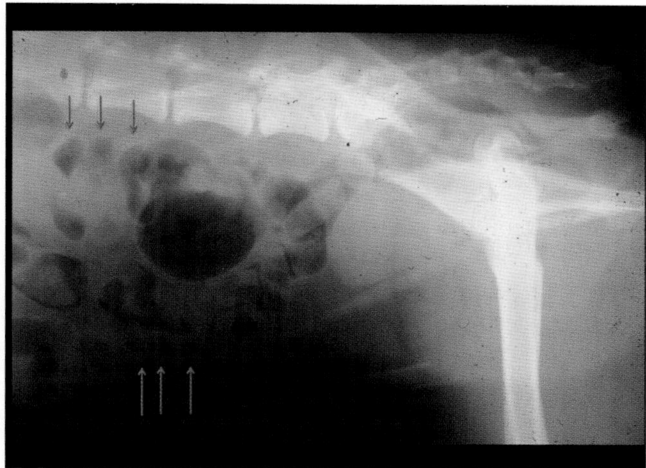

FIGURE 5.6-1 Lateral abdominal radiograph showing plicated (folded) loops of small intestine *(arrows)*.

Diagnosis

Linear foreign body

Treatment

A balanced electrolyte solution was administered IV to correct dehydration and stabilize the patient in preparation for general anesthesia and an exploratory celiotomy (Gr. *koilia* belly + Gr. *ektomē* excision). Soft foreign material was palpated in the stomach and the small intestines were plicated (*L. plicatus* folded). The duodenocolic ligament was divided to facilitate a complete evaluation of the caudal duodenal flexure. The stomach was isolated from the peritoneal cavity and a gastrotomy (Gr. *gastēr* stomach, + Gr. *stomoun* to provide an opening) was performed. A piece of cloth extended from the stomach into the small intestines. A portion of the cloth was pulled from the stomach until tension increased. The cloth was then cut at the pylorus and the gastrotomy was closed using a two-layer inverting suture pattern. The remainder of the cloth was removed from the small intestine via three enterotomies (Gr. *enteron* intestine, + Gr. *stomoun* to provide an opening). The peritoneal cavity was thoroughly lavaged to remove any contamination, and the abdominal incision was closed. The dog was maintained on IV fluids and antinausea medication until it began to eat 2 days after surgery.

Enterotomies are most commonly performed in the dog and cat for the removal of foreign bodies, while intestinal resection and anastomosis (Gr. *anastomōsis* opening, outlet) are used in surgery when the intestine is devitalized due to intussusception (L. *intus* within + *suscipere* to receive), bowel necrosis (Gr. *nekrōsis* deadness), or torsion (Gr. *torsio* to twist) and volvulus.

Anatomical features in canids and felids

Introduction

The small intestine (L. *intestinum tenue* small intestine, internal) or small bowel is part of the gastrointestinal tract and the site where the majority of nutrients and minerals are absorbed. Unlike other domestic animals, the dog and cat have a similar arrangement as that of humans with little difference in the diameter of the small and large

The similarities in the size of the small and large intestine can make abdominal radiographs difficult to interpret in differentiating between small and large intestinal diseases.

intestines. It is positioned between the stomach and large intestine with digestive contributions from the liver (in the form of bile) and the pancreas (pancreatic juices containing enzymes essential for protein digestion).

There are three distinct parts to the small intestine: duodenum (fixed part); jejunum and ileum (mesenteric parts). The duodenum is where intestinal villi (L. *villus* tuft of hair) are first located for the absorption of fluids and nutrients. The jejunum is the longest segment of the small intestine and is important in the absorption of small nutrient molecules. The ileum is the shortest in length and is the terminal part of the small intestine that acts as the site for the absorption of vitamins and bile salts. The average length of the small intestine in the dog measures approximately 1.8–4.8 m and in the cat measures approximately 0.8–1.3 m.

Function

Chewing, crushing, agitating, and mixing take place in the mouth and stomach. Digestion starts in the mouth with enzymes secreted by the salivary glands that break down carbohydrates into disaccharides. Pepsin and acid in the stomach initiate protein digestion. Pancreatic enzymes secreted into the duodenum further digest proteins. Bile from the liver emulsifies fats and allows them to be broken down into fatty acids by pancreatic lipase. The contents of the small intestine change as they move by peristalsis (Gr. *peri* around, + Gr. *stalsis* contraction), starting as semisolid in the duodenum and becoming more liquid by the time they reach the ileum.

Absorption of simple sugars, amino acids, and fatty acids occurs by active or facilitated carrier-mediated transport, or by passive diffusion. Endocytosis (Gr. *endon* within, + Gr. *kytos* hollow vessel) of small antigenic peptides is involved in the neonatal absorption of colostral (maternal) antibodies and remains important to the mucosal immune response in later life.

A net loss of water and electrolytes occurs in diarrhea and leads to a rapid dehydration (L. *de* away, + Gr. *hydōr* water).

The small intestine is also actively involved in the absorption and secretion of water and electrolytes, with a goal of net absorption in a healthy dog and cat.

Duodenum

The cranial or oral (L. *oralis* pertaining to the mouth) part of the small intestine is the **duodenum** (L. *duode'ni* 12 at a time)—so-called because it is about 12 finger-widths in length in humans. It starts at the **pylorus** of the stomach

MICROBIOME AND THE IMMUNE SYSTEM

Normal bacterial flora in the small intestine is an integral part of the health of the bowel with >200 species of aerobic, anaerobic, and facultative anaerobic bacteria identified. The microbiome promotes intestinal motility, digestion, absorption of nutrients, brush-border enzymes, and much more. Equally important is the stimulation of the immune system, with concurrent exclusion of other potential pathogens. Indiscriminate use of antibiotics can adversely affect this delicate balance. The number of bacterial flora (L. *flora* the goddess of flowers) increases from the duodenum to the colon.

The small intestine also serves a general barrier function against pathogens. The mucosa-associated lymphoid tissue (MALT) is found at internal body surfaces. In the small intestine, it is termed gut-associated lymphoid tissue (GALT). The sites comprise Peyer's patches, lymphoid follicles, mesenteric lymph nodes, intestinal lamina propria, and epithelium.

to the right of the midline or median plane. The cranial portion is closely related to the liver passing dorsally and to the right at the level of the ninth intercostal space. The duodenum is attached to the liver by the **hepatoduodenal ligament** and also to the pancreas, encircled by the mesoduodenum. The **cranial flexure** is the point at which the duodenum runs caudally and becomes the **descending duodenum** with a long **mesoduodenum** (Gr. *mesos* middle, intermediate, or moderate) that surrounds the right lobe of the pancreas. The descending duodenum is free of omental attachments and passes near the right dorsolateral abdominal wall and the caudal pole of the right kidney. At the level of the fourth to sixth lumbar vertebrae, the **ascending duodenum** begins as the **caudal flexure** and tracks from right to left around the cecum, ascending colon, and root of the mesentery. This part of the duodenum has a short mesoduodenum and lies between the cecum, ascending colon, and mesenteric root on the right and the descending colon and left kidney on the left. The **duodenocolic fold** is a mesenteric attachment that connects the descending colon and rectum to the ascending duodenum. This fold shapes the **duodenojejunal flexure**, located to the left of the mesenteric root, which turns ventrally and—with a longer mesentery—continues as the jejunum (Figs. 5.6-2 and 5.6-3).

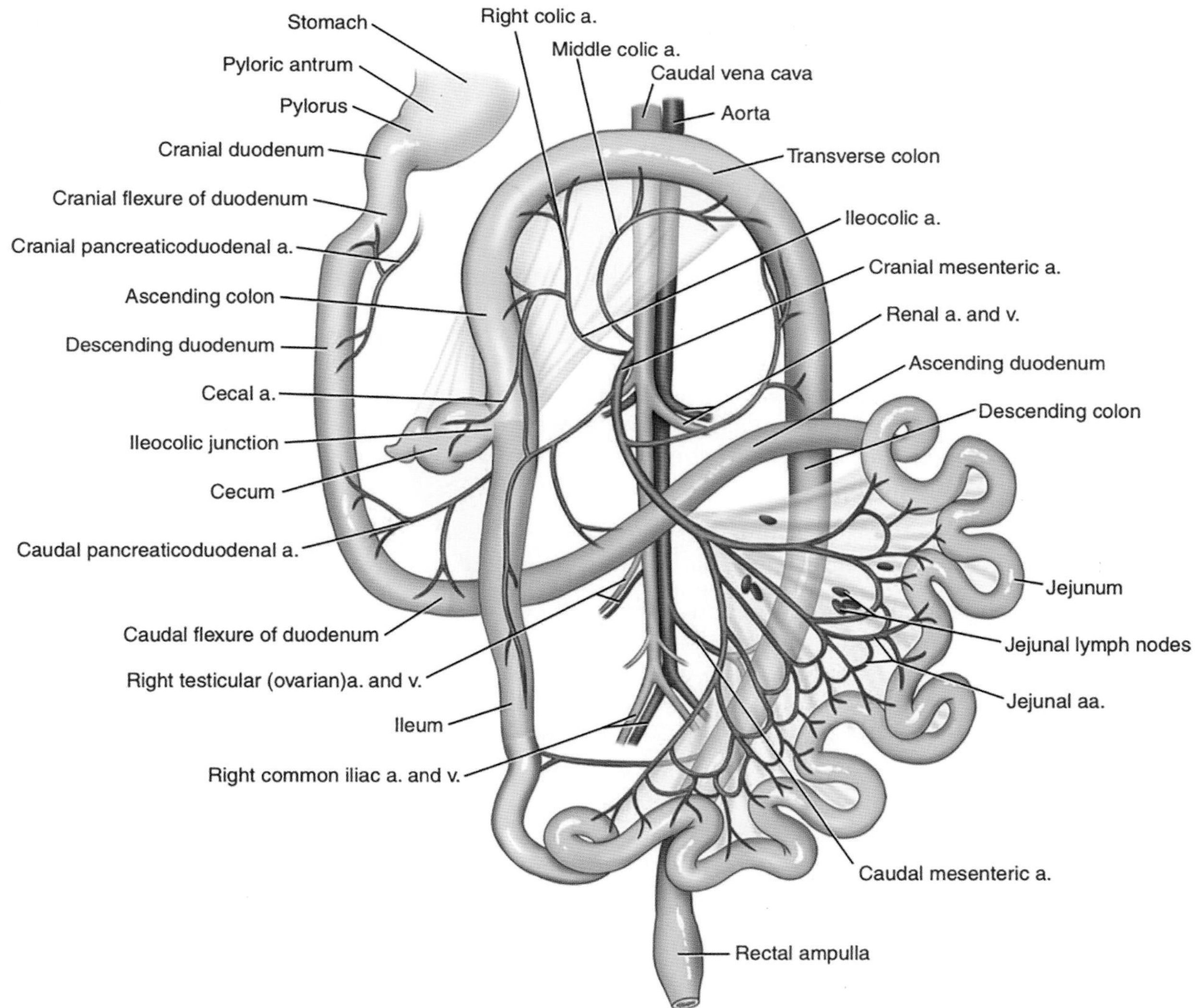

FIGURE 5.6-2 Dog intestinal tract demonstrating arterial supply, formation of the caudal vena cava, and the various segments comprising the gastrointestinal tract (each a different color to highlight the various sections).

Jejunum

When the stomach distends and moves caudally, the jejunum, with its long mesenteric attachment, easily displaces dorsally and to the right. This is a compensatory mechanism of the jejunum and one of the principles for the long mesenteric attachment. This long mesentery allows a surgeon to exteriorize most of the jejunum during exploratory surgery for improved exposure of the organs more dorsally positioned in the abdominal cavity.

Jejunum is the most extensive part of the small intestine with substantial mesentery that meets at the top of the abdominal cavity as the **root of the mesentery**. The jejunum consists of multiple loops (averaging 6–8 large loops) commonly referred to as the "jejunal mass" lying between the stomach and the pelvic inlet. The **greater omentum** (L. *omen'ta* fat skin) blankets the jejunum ventrally and laterally.

Ileum

The **ileum** is the shortest segment of the small intestine and is easily distinguished from the jejunum by the **ileocecal fold**, an antimesenteric fold of the peritoneum which joins it to the colon and an antimesenteric vessel on the serosal surface. The ileum arises caudally from the jejunum, passes cranially, and opens into the ascending colon at the **ileal orifice** at the level of the first or second lumbar vertebra (Fig. 5.6-3).

The ileal orifice or ostium empties into the ascending colon in all domestic animals except the horse, in which the ileal ostium empties into the cecum.

Small intestinal blood supply, innervation, and lymphatics

The blood supply for the small intestine comes mainly from the **cranial mesenteric artery** (Figs. 5.6-2). A portion of the duodenum receives the **cranial pancreaticoduodenal branch**, originating from the **gastroduodenal artery**, a branch of the **celiac artery**. Venous drainage generally follows the arterial blood supply comprising the **gastroduodenal** and **cranial mesenteric veins**. These two veins together with the **splenic**, **caudal mesenteric**, **ileocolic**, **middle colic**, **left gastric**, **right gastroepiploic**, and **cranial pancreaticoduodenal veins** form the **portal vein**, which drains into the liver.

There is sympathetic and parasympathetic innervation to the small intestine: sympathetic fibers communicate through the **celiac** and **cranial mesenteric ganglia**, and the parasympathetic fibers communicate via the **vagal nerve**.

Large volumes of lymphatic drainage are common in the small intestine because many of the breakdown products of digestion are absorbed through the lymphatic system. The major lymph-collecting network is the **thoracic duct system**. The **cisterna** (L. a cistern, reservoir) **chyli**, an expanded part of the vessel before the beginning of the thoracic duct system, receives lymph from the **cranial mesenteric lymph nodes** near the root of the mesentery, the **jejunal lymph nodes** found scattered in the jejunal mesentery, and from the pelvis and pelvic limbs (Fig. 5.6-2).

ABDOMINAL PALPATION

If the abdominal muscles are relaxed in a dog, it is possible to palpate the right descending duodenum and ileo-colic junction. This is accomplished using gentle and firm digital pressure, compressing the duodenum and ileum dorsally against the sublumbar muscles and rolling them in a medial to lateral direction. The descending colon can likewise be palpated on the left, especially if constipation is present.

In the cat, the ileocolic junction can be mistaken for a foreign body.

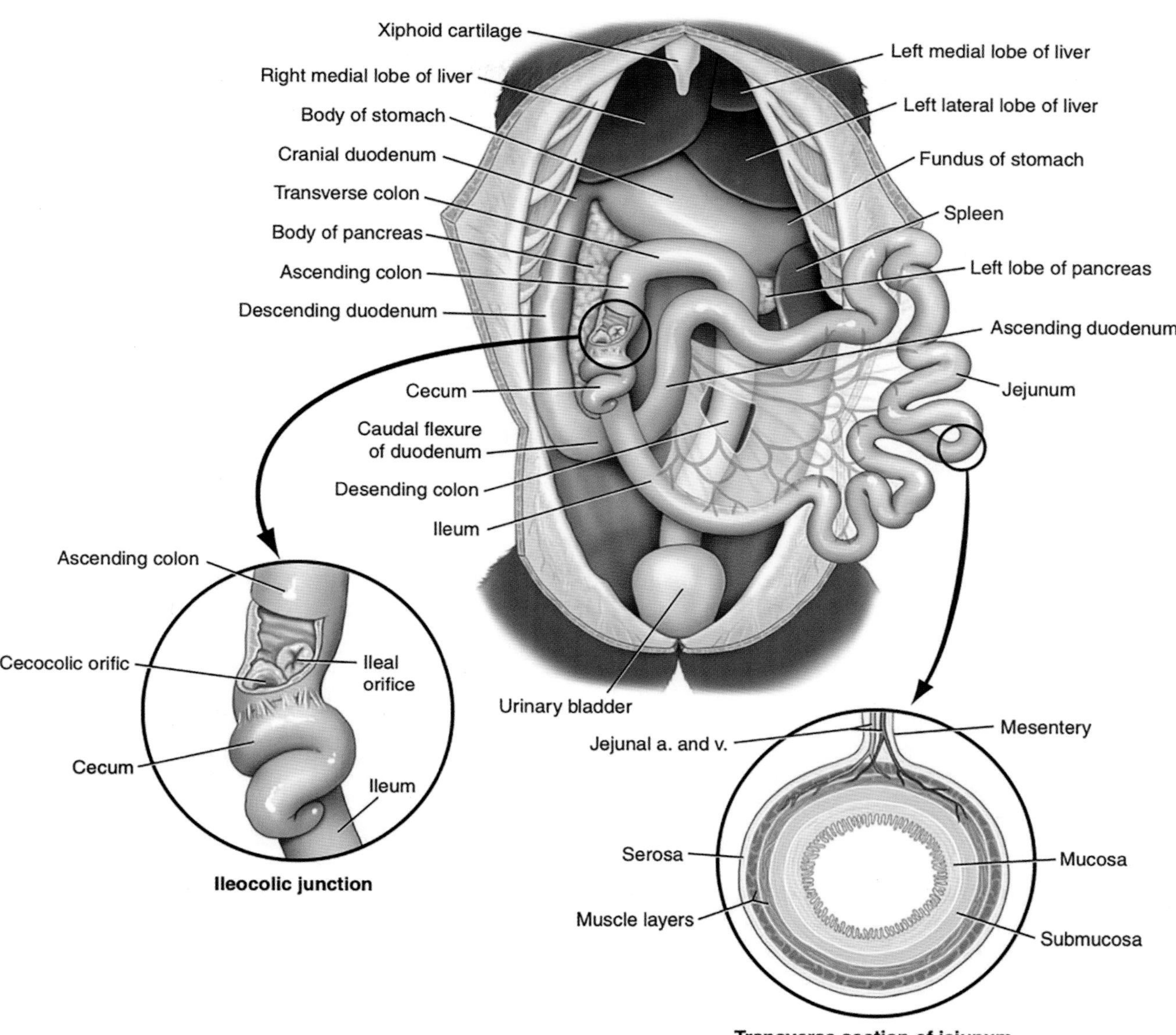

FIGURE 5.6-3 Dog small intestine in situ and its relationship to surrounding abdominal organs. Cutaways: ileocolic junction and a cross section of a segment of jejunum with the mesenteric vessels.

Selected references

[1] Hobday MM, Pachtinger GE, Drobatz KJ, Syring RS. Linear *versus* non-linear gastrointestinal foreign bodies in 499 dogs: clinical presentation, management and short-term outcome. J Small Anim Pract 2014;55:560–5.
[2] Hall EJ, German AJ. Diseases of the small intestine. In: Ettinger SJ, Feldman EC, editors. Textbook of veterinary internal medicine: Diseases of the dog and the cat. Elsevier; 2010. p. 1526–72.
[3] Singh B. Dyce, Sack and Wensing - Textbook of Veterinary Anatomy, 5th ed. St. Louis: Elsevier; 2018.
[4] Getty R. Sisson and Grossman's the anatomy of the domestic animals. 5th ed. W.B. Saunders; 1975.

CASE 5.7

Megacolon

Takanori Sugiyama[a] and Helen M.S. Davies[b]
[a]Animalius Vet, Bayswater, WA, AU
[b]Veterinary BioSciences, The University of Melbourne, Melbourne, Victoria, AU

Clinical case

History

A 15-year-old male neutered domestic mixed-breed cat was presented for chronic, persistent/recurrent constipation of 7 years' duration. The cat had been managed medically with oral lactulose, dioctyl sodium sulfosuccinate (DSS), and cisapride, as well as periodic warm water enema administration. Recognizing that the medications were not helping, the owner discontinued treatment several weeks before presentation. The cat was presented for further evaluation when he once again developed tenesmus.

> Lactulose and DSS are laxatives given to soften the feces, and cisapride is a prokinetic (L. and Gr. pro- "before" + Gr. *kinētikos* producing motion) agent given to stimulate peristalsis. These drugs are used to aid defecation in patients with chronic constipation.

Physical examination findings

The cat's vital signs (heart rate, respiratory rate, rectal temperature) were within the normal limits, and his hydration status and body condition score were clinically normal. Abdominal palpation revealed the presence of a large, hard, tubular mass, measuring approximately 5 × 20 cm (2 × 8 in.). The urinary bladder, which was moderately filled, was palpable ventral to the mass. Palpation of the mass elicited a pain response. No other abnormalities were found on physical examination.

Differential diagnoses

Megacolon, space-occupying mass (e.g., neoplasia)

MEGACOLON

Megacolon (gross distension of the colon with digesta/feces) may be primary or secondary. Primary, or idiopathic, megacolon is the most common form in older cats; 60% of senior cats with megacolon have this form.

Secondary megacolon develops as a result of obstruction aboral to the distension, such as pelvic canal narrowing (fracture or deformity), mechanical obstruction (e.g., prostatomegaly, intraabdominal mass), or a functional obstruction (neurologic or metabolic).

Comparative Veterinary Anatomy: A Clinical Approach. https://doi.org/10.1016/B978-0-323-91015-6.00042-X

Diagnostics

Abdominal radiography revealed massive colonic distension with digesta/feces, extending from the ascending colon to the rectum (Figs. 5.7-1 and 5.7-2). The urinary bladder appeared normal. Routine blood work revealed mild anemia and mild increases in creatinine and glucose; these findings were attributed to chronic kidney disease (anemia and increased creatinine) and stress (mild hyperglycemia), commonly encountered in senior cats.

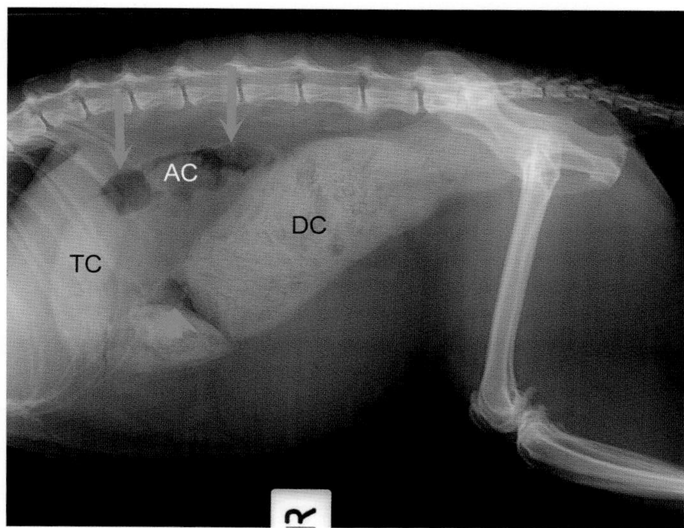

FIGURE 5.7-1 Right lateral radiograph of the cat in this case. The colon is markedly distended with digesta/feces from transverse colon (TC) to rectum (superimposed by the pelvis). Note the accumulation of gas *(arrows)* in the ascending colon (AC) oral, yet caudal, to the fecal impaction. There is also a gas pocket *(arrowhead)* in the proximal third of the descending colon (DC).

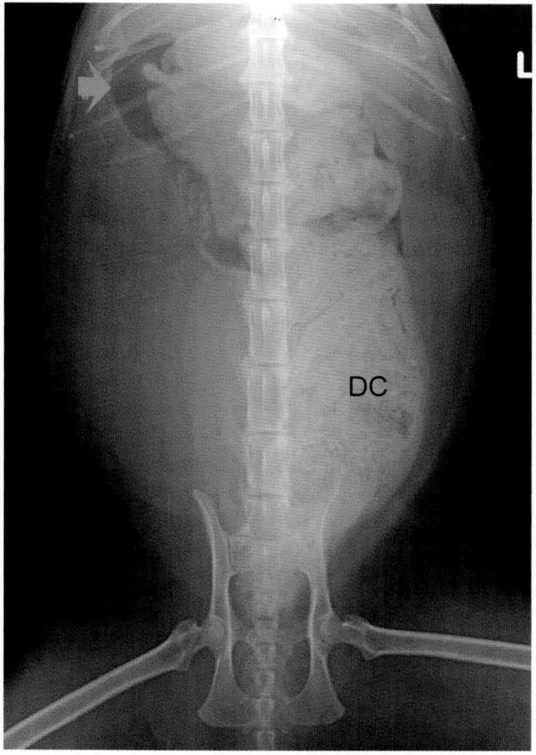

FIGURE 5.7-2 Ventrodorsal radiograph of the cat in this case. The markedly distended descending colon (DC) occupies most of the left caudal abdominal cavity. Note the gas accumulation oral to the obstruction *(arrowhead)*.

Diagnosis

Feline idiopathic megacolon

Treatment

Surgical treatment was recommended because long-term medical management had failed. A ventral midline celiotomy was performed under general anesthesia, with the incision extending from the xiphoid process of the sternum to the pubis. The fecal impaction extended almost the entire length of the colon, although the cecum and ileocolic junction were not affected. The other abdominal organs appeared grossly normal.

A subtotal colectomy was performed between the ascending colon (1 cm or 0.4 in. from the cecum) and the descending colon (1 cm or 0.4 in. from the rectum), and an end-to-end anastomosis was performed to connect the proximal and distal ends of the remaining colon. The cat recovered uneventfully from general anesthesia and surgery; however, he developed diarrhea for 6 weeks postoperatively, which is common following subtotal colectomy, due to temporary disruption of normal colonic function. The diarrhea resolved without treatment, and the cat had no further episodes of constipation.

A primary function of the colon is water resorption from the digesta/feces, so colectomy patients typically pass liquid feces/diarrhea until the distal small intestine and rectum adapt to their new role in water resorption.

Anatomical features in canids and felids

Introduction

The caudal part of the gastrointestinal tract consists of the colon, cecum, and rectum. The colon and cecum are contained within the abdomen. The rectum, which is continuous with the descending colon, begins at the pelvic inlet and is entirely contained within the pelvic canal, where it lies dorsal to the bladder and urethra. In the intact female dog and cat, the uterine body, cervix, and vagina lie between the bladder and the rectum. (The female reproductive tract is described in Cases 6.2 and 6.3.)

Function

In the dog and cat, the large intestine functions to reabsorb water and electrolytes, and concentrate the excretion products of the digestive processes that have mostly been completed within the small intestine. Peristalsis helps to mix the contents and encourage the maximum reabsorption of water and electrolytes. When digesta distends the cecum, it provokes antiperistaltic waves which help to drive the cecal contents back into the colon. There is a proliferation of bacteria and other microorganisms feeding on these remnants of digestion throughout the large intestine, and they and their detritus (L. *deterere* to rub away) form a considerable part of the feces.

Colon

The colon in dogs and cats is much shorter and simpler than that in horses, and the cecum is merely a small diverticulum of the proximal colon (Figs. 5.7-3 and 5.7-4). Also, in contrast to horses, the colon in dogs and cats generally is only a little larger in diameter than the small intestine, and it has neither teniae (longitudinal bands of smooth muscle) nor haustra (sacculations). Furthermore, the stomach (when full) and small intestine lie ventral to the colon in dogs and cats.

The diameter of the colon varies somewhat with the animal's diet, it being larger in dogs that may consume bulky or fermentable fiber such as fruit and vegetables.

During ventral midline celiotomy, the first structure normally encountered in dogs and cats is the small intestine. Surgical access to the colon, as required in this case, involves retraction of the small intestine.

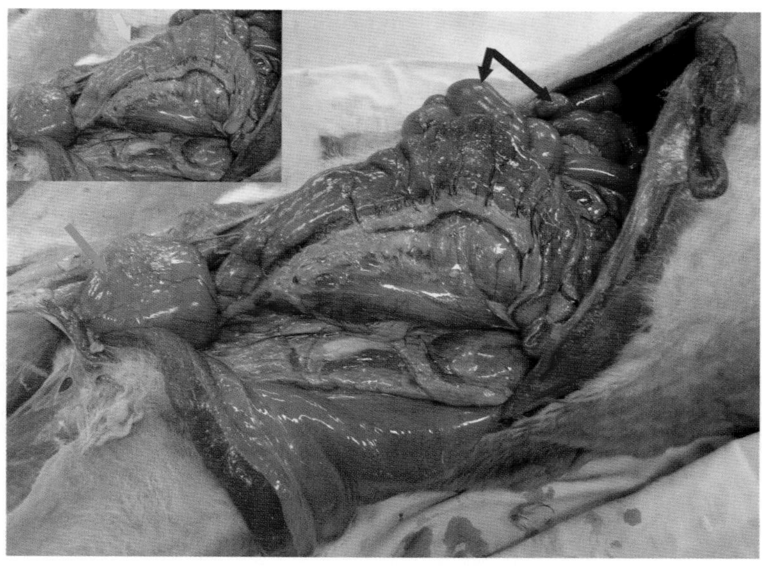

FIGURE 5.7-3 Abdominal organs of the dog, left lateral view. The lateral body wall was incised along the caudal border of the last rib and then reflected caudodorsally. The small intestine *(blue arrows)* is visible behind the descending colon (inset, highlighted in *yellow*), which has been reflected ventrally to show the mesocolon and associated structures (shown in greater detail in Fig. 5.7-6). The descending colon approaches the pelvic inlet dorsal to the urinary bladder *(green arrow)*.

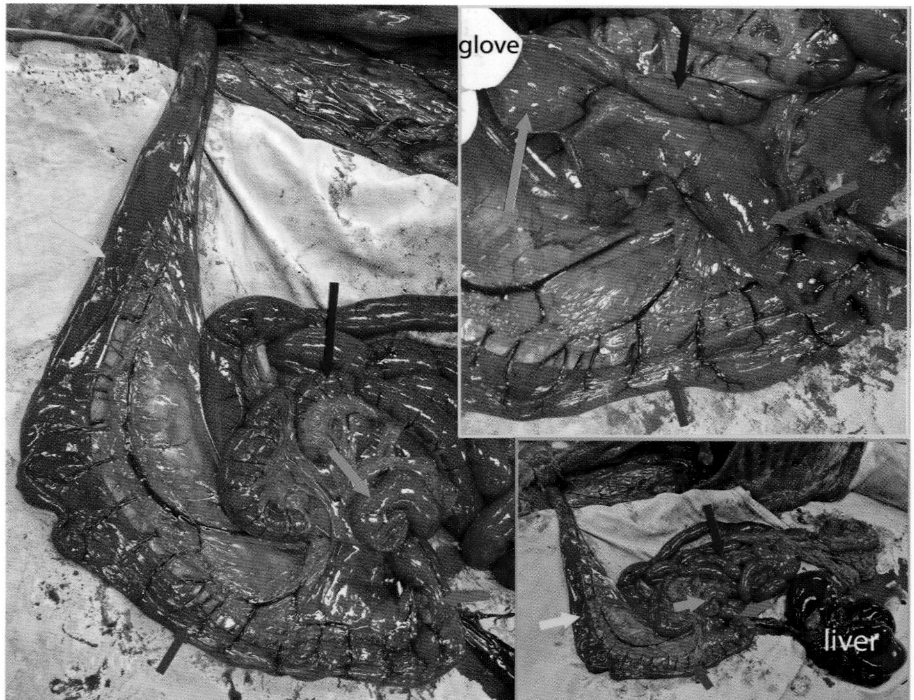

FIGURE 5.7-4 Colon of a dog, ex situ, showing the small intestine *(blue arrow)*, cecum *(green arrow)*, ascending colon *(pink arrow)*, transverse colon *(purple arrow)*, and descending colon *(yellow arrow)*. In the upper-right image, the cecum is being drawn away from the ascending colon to illustrate the relationship between the ileum and the ascending colon. (The ileocolic junction is difficult to appreciate externally.)

The ileocolic valve should be preserved whenever possible during surgery because it is a barrier to the migration of colonic bacteria into the small intestine, which can lead to bacterial overgrowth and thus intestinal inflammation and maldigestion. That said, resection of the ileocolic valve is sometimes required to increase mobility of the proximal intestine and thus reduce tension on the anastomosis following colonic resection.

As the cecum in this case was grossly normal, and owing to its proximity to the ileocolic valve, the proximal colectomy incision was made in the ascending colon 1 cm (0.4 in.) from the cecum in this cat—preserving the cecum and ileocolic valve.

The **colon** begins at the ileocolic junction. Here, a muscular **ileocolic valve** controls the flow of digesta into the colon and limits the backflow of material from the colon to the ileum. This valve consists of a short, broadly conical protrusion, the **ileal papilla**, which faces into the ascending colon. The ileocolic junction is located in the sublumbar region just to the right of midline.

The **cecum** in dogs is a relatively short, corkscrew-shaped diverticulum of the proximal colon. Although variable in size, average cecal length in situ is 5–8 cm (2–3 in.). In cats, the cecum is very small and comma-shaped. The cecum lies immediately aboral and lateral to the ileocolic junction (Fig. 5.7-4). It is usually found between the right flank and the midline, ventral to the duodenum.

Although it is a simple tubular structure, the canine/feline colon is described as comprising 3 parts, which correspond to the terms used in human anatomy: ascending, transverse, and descending (Fig. 5.7-4). The colon is shaped a bit like a question mark—"?"—when viewed from below, such as during ventral midline celiotomy or on a ventrodorsal radiograph of the abdomen. Beginning in the right side of the patient's abdominal cavity, the **ascending colon** passes cranially, running along the medial surface of the duodenum toward the pyloric portion of the stomach, where it takes a sharp left turn (the **right colic flexure**) and crosses midline as the **transverse colon**. It then takes the slighter left turn (the **left colic flexure**) on the left side of the root of the mesenteries. The colon continues in the left sublumbar region as the **descending colon**, coursing caudally along the medial border

PERINEAL HERNIA IN DOGS

Perineal hernia occurs when the muscles of the pelvic diaphragm are weakened and allow subcutaneous herniation of pelvic or abdominal organs adjacent to the anus. It primarily occurs in intact male dogs. The exact pathogenesis is unknown, but an androgenic effect is suspected either primarily or secondarily (e.g., tenesmus caused by prostatomegaly). The defect usually occurs between the external anal sphincter and the rectococcygeus/levator ani complex, which in these cases is severely atrophied.

In most cases, the hernia contains small intestine, colon, omentum, and peritoneal fat. Herniation of the urinary bladder can be a life-threatening emergency, as retroflexion "kinks" the urethra, causing a complete urinary obstruction.

Surgical repair is challenging, as the muscles of the pelvic diaphragm are atrophied and fragile in these cases. Dorsal hernias are relatively easy to close, but ventral hernias are almost always impossible to close without creating a muscle flap from the internal obturator muscle, which is elevated from the ischial table. Sometimes the internal obturator muscle is also atrophied, so it cannot be used to create a sufficient flap to close the perineal defect. Artificial mesh closure and semitendinosus muscle flap have also been described. In addition, cystopexy and colopexy (surgical anchoring of the bladder and colon, respectively) may be performed to prevent reherniation of these organs.

or ventral surface of the left kidney. As it approaches the pelvic inlet, the colon shifts toward midline and continues caudally as the **rectum** (Fig. 5.7-5).

> The ascending and transverse portions of the colon are relatively short in dogs and cats; in fact, the descending portion comprises about two-thirds the length of the colon in these species. In addition, the descending colon is generally more prominent on abdominal radiographs because of the greater density of its contents; in comparison, the transverse colon is often gas-filled.

The colon is suspended from the dorsal body wall throughout the sublumbar region by a ventrally directed fold of parietal peritoneum that forms a short mesentery, the **mesocolon** (Fig. 5.7-6). The blood and nerve supply to the colon track in the mesocolon. These structures are described later, as the distal portion of the descending colon and the proximal portion of the rectum share blood and nerve supplies.

Rectum and anus

The **rectum** is the continuation of the colon through the pelvic canal, terminating at the anus (Fig. 5.7-7). The proximal or cranial portion of the rectum is covered by peritoneum; the distal or caudal portion is retroperitoneal, so it has no serosal (peritoneal) covering.

At the junction of the rectum and the **anus**, the lining is composed of stratified squamous epithelium and contains the **anal glands**—multiple, small, mucus-secreting glands that lubricate the feces for their passage through the anus.

These small glands must not be confused with the **anal sacs**, which are a pair of subcutaneous pouches (left and right) that lie ventrolateral to the anus in dogs and cats, and empty via a narrow duct into the lumen of the anus at the recto-anal junction (Fig. 5.7-7). The anal sac epithelium secretes a malodorous, oily substance that facilitates fecal lubrication and appears to serve a social function (i.e., for communication or recognition). In addition, the anal skin contains large sebaceous glands.

The paired **perianal muscles** form the **pelvic diaphragm**. Together, they depress the tail or, when the tail is raised, they compress the caudal portion of the rectum, thus facilitating defecation. The pelvic diaphragm also helps anchor the anus and prevent herniation of the pelvic or abdominal organs during defecation. The **coccygeus** muscle originates on the ischiatic spine; it courses dorsocaudally, fanning out to insert onto the 2nd to 5th caudal (coccygeal) vertebrae lateral to the anus (Fig. 5.7-7). The **levator ani** muscle originates on the pelvic floor and the shaft of the ilium; it runs caudodorsally and attaches to the fascia and other tissue in the region of the 7th caudal

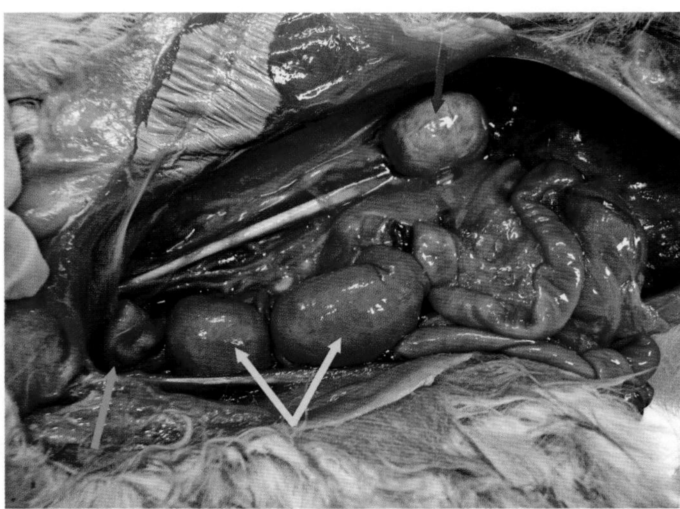

FIGURE 5.7-5 Abdominal organs of a cat, right ventrolateral view. The lateral body wall was incised along the caudal border of the last rib and then reflected caudodorsally. Most of the small intestine has been moved out of the field so that the deeper structures may be seen. Note the gas-filled descending colon *(yellow arrows)*, the empty urinary bladder *(green arrow)*, and the right kidney *(purple arrow)* in the retroperitoneal space.

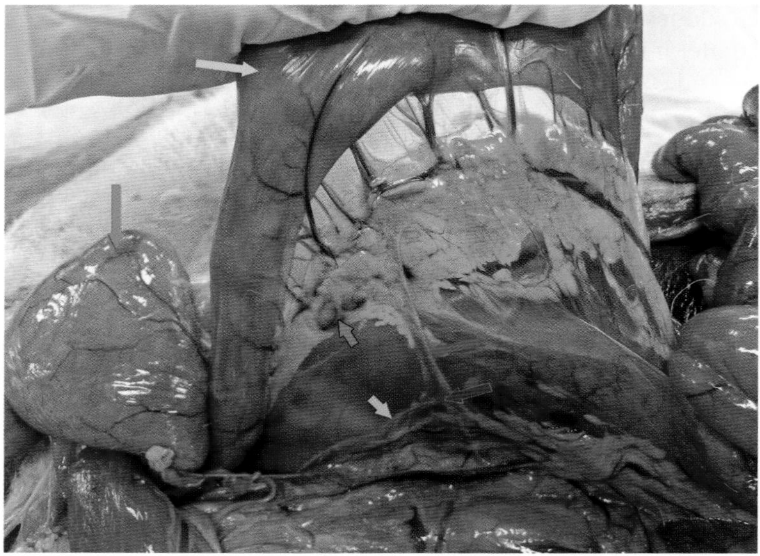

FIGURE 5.7-6 Descending colon of a dog, showing the mesocolon and associated structures. Cranial is to the right and caudal is to the left in this image. The partially filled urinary bladder *(green arrow)* is reflected caudally. Note the descending colon *(yellow arrow)* and, within the transparent mesocolon, the caudal mesenteric artery *(pink arrow)* and ganglion *(dark blue arrow)*, the hypogastric nerve *(light blue arrow)*, and the colonic lymph nodes *(orange arrow)* surrounded by peritoneal fat.

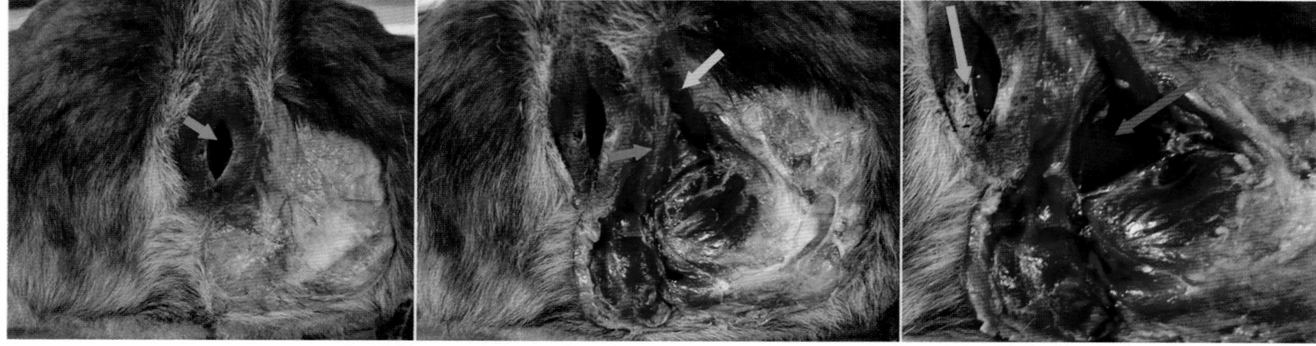

FIGURE 5.7-7 Perineum of a male dog. *Left*: Superficial dissection, showing the anus *(green arrow)* and the subcutaneous tissues to the right of the external anal sphincter. *Center*: Muscles of the pelvic diaphragm, including coccygeus *(yellow arrow)* and levator ani *(blue arrow)*. *Right*: The pelvic diaphragm has been incised to reveal the wall of the rectum *(pink arrow)*. The *turquoise arrow* indicates the external opening of the left anal sac. The bulge of the right anal sac is faintly visible, deep to the levator ani muscle.

ANAL SACCULITIS IN DOGS

The anal sacs often become impacted and inflamed in dogs. “Scooting” or rubbing the anus along the ground is a common behavior in affected dogs, although it is not specific for anal sacculitis. Affected dogs may also have an unpleasant “fishy” odor.

It is common practice for groomers and veterinary technicians to “express” or manually empty the anal sacs whenever the contents are palpable on digital examination of the caudal rectum. However, care is needed, as excessive pressure or frequent expression may, itself, lead to anal sacculitis.

The anal sacs are a site of neoplasia in dogs, particularly anal sac apocrine gland adenocarcinoma. Metastasis to the internal and medial iliac lymph nodes often occurs.

Anal sac disease is far less common in cats than in dogs.

vertebra, merging with the **external anal sphincter** along the way (Fig. 5.7-7). The coccygeus muscles lie lateral to the levator ani muscles either side of the rectum and reinforce their compressive action on the rectum during defecation. The **rectococcygeus** muscle consists of smooth muscle fibers, which continue caudally from the wall of the rectum. The left and right sides meet dorsally and insert on the midline of the 5th and 6th caudal vertebrae, thus acting to shorten the rectum during defecation and aid in evacuation of the fecal column. Lastly, the anal portion of the **retractor clitoridis/penis** muscle consists of smooth muscle fibers that run caudoventrally from the sacrocaudal junction to the external anal sphincter before continuing to the genital insertion.

During surgical resection of anal sac adenocarcinomas, up to 50% of the external anal sphincter muscle may be removed without a significant risk of causing fecal incontinence postoperatively.

Blood supply, lymphatics, and innervation

In general, blood supply to the gastrointestinal tract originates from a sequence of unpaired, median branches of the abdominal aorta which course ventrally within the respective parts of the mesenteries that are suspended from the digestive tract from the dorsal body wall. From cranial to caudal, they are the celiac, cranial mesenteric, and caudal mesenteric arteries. The venous drainage is primarily via the portal vein, which courses cranially to the liver.

The ileocolic junction, cecum, and proximal half of the colon are supplied by branches of the **cranial mesenteric artery**, which arises from the abdominal aorta in the region of the second lumbar vertebra. Of note, the **right** and **middle colic arteries** supply the ascending and transverse portions of the colon, respectively; the proximal part of the descending colon is also supplied by the middle colic artery. The rest of the descending colon and the cranial portion of the rectum are supplied by the **caudal mesenteric artery** (Fig. 5.7-6), which also arises from the abdominal aorta, but in the region of the 5th lumbar vertebra. Its notable branches are the **left colic artery** and the **cranial rectal artery**. The colic branches of the cranial and caudal mesenteric arteries anastomose in the middle of the descending colon.

During subtotal colectomy, the vascular supply to the colon is an important consideration, as ligation or transection of the wrong vessels may result in avascular necrosis of the small intestine or other structures. The cranial mesenteric artery is particularly important, as its branches supply the small intestine and pancreas, in addition to the cecum and the proximal half of the colon. The vessels should be ligated as close as possible to the portion of colon to be resected. It is essential to preserve the mural vascular network (the complex of vessels within the bowel wall) and segmental blood supply. When colotomy (surgical incision into the colon) is required to access the colonic contents, the incision should be made at the antimesenteric border (i.e., furthest from the mesocolon) in order to avoid injury to the blood supply to that part of the colon. This approach also minimizes the risk of blood vessel occlusion when the colotomy incision is closed with sutures or surgical staples.

SURGICAL ACCESS TO THE DISTAL COLON AND RECTUM

There are several surgical approaches to the distal colon and rectum: ventral, dorsal, lateral, rectal mucosal eversion, and rectal pull-through techniques. The ventral approach requires the separation of the pubic symphysis or a pubic bone flap. The dorsal approach requires transection of the rectococcygeus muscle dorsally and possibly the levator ani muscle laterally. The lateral approach is made between the external anal sphincter and the levator ani muscle. Rectal mucosal eversion is simple and less invasive, but exposure is limited. This technique is indicated for incisional biopsy or the removal of small, superficial tumors in the mid- or caudal portion of the rectum. Rectal pull-through may also be used for lesions in the mid- or caudal rectum, but it carries a high risk of causing fecal incontinence postoperatively.

The blood supply to the **rectum** arises from 3 separate sources. As noted earlier, the cranial portion of the rectum is supplied by the cranial rectal artery. The middle portion of the rectum is supplied by the **middle rectal artery**, which is ultimately a branch of the **internal pudendal artery**, itself one of the 2 terminal branches of the **internal iliac artery**. The middle rectal artery anastomoses with the **cranial rectal artery** in the cranial third of the rectum. The caudal portion of the rectum, along with the anus and the perianal muscles, is supplied by the **caudal rectal artery**, which is also a branch of the internal pudendal artery.

The blood supply to the specific area must be considered during surgery of the rectum or involving the distal colon, such as subtotal colectomy. Although anastomosis occurs between the cranial and middle rectal arteries, such "anatomical redundancy" cannot be relied upon when the blood supply to a specific segment is compromised. Care must be taken to preserve the normal blood supply to the colon and rectum during surgery.

The **internal pudendal artery** reaches the perineum on either side of the rectum, along with the **pudendal nerve**. These structures supply the skin around the anus and the more medial regions of the perineum. The more lateral parts of the perineum are supplied with blood via superficial branches of the caudal gluteal artery, which is the second of the 2 terminal branches of the internal iliac artery, both of which traverse the pelvic canal.

Pudendal (L. *pudendum* external genitalia) refers to structures that supply the external genitalia and perineal area.

All of these colonic and cranial rectal vessels track within the mesocolon, along with the veins, lymphatic vessels, and nerves that supply the colon and cranial rectum (Fig. 5.7-6). The cecum and colon are drained by the **caudal mesenteric vein**, which empties into the **portal vein**, and by the **right**, **middle**, and **left colonic lymph nodes**. In contrast, the caudal rectal and perineal veins (tributaries of the internal pudendal and caudal gluteal veins) flow into the caudal vena cava, and the lymph vessels from the anal canal drain into the internal and medial iliac lymph nodes.

The ileocolic junction, cecum, and most of the colon are supplied by branches of the **cranial mesenteric plexus**, which track in the mesentery alongside the cranial mesenteric artery. The middle and caudal portions of the descending colon and the cranial portion of the rectum are supplied by the **caudal mesenteric plexus**, which arises alongside the caudal mesenteric artery (Fig. 5.7-6). These cranial and caudal mesenteric ganglia are part of the sympathetic nervous system. These sympathetic nerves follow the blood vessels, except for the **hypogastric nerves** which supply the pelvic organs; they originate from the caudal mesenteric ganglion and course either side of the mesentery caudally toward the pelvis (Fig. 5.7-6).

The parasympathetic innervation to the proximal portion of the colon is supplied by branches of the **vagus nerve**, which enters the abdominal cavity via the diaphragm on either side of the esophagus. The parasympathetic nerves branch throughout the intestine in 2 layers: one between the muscle layers (myenteric plexus and ganglia) and the other in the submucosa (submucosal ganglia). The parasympathetic supply to the pelvic organs—including the rectum, and to the descending colon as far cranially as the left colic flexure—are the **pelvic nerve** branches from the sacral spinal cord. The skeletal muscle sphincters of the bladder and anus, as well as the muscles of the pelvic diaphragm, are supplied by the **pudendal nerve** and its branches.

Selected references

[1] MacPhail C. Anal sacculectomy. Compend Contin Educ Pract Vet 2008;30(10):530–5.
[2] Yoon HY, Mann FA. Bilateral pubic and ischial osteotomy for surgical management of caudal colonic and rectal masses in six dogs and a cat. J Am Vet Med Assoc 2008;232(7):1016–20.
[3] Danova NA, Robles-Emanuelli JC, Bjorling DE. Surgical excision of primary canine rectal tumors by an anal approach in twenty-three dogs. Vet Surg 2006;35(4):337–40.
[4] Nucci DJ, Liptak JM, Selmic LE, et al. Complications and outcomes following rectal pull-through surgery in dogs with rectal masses: 74 cases (2000-2013). J Am Vet Med Assoc 2014;245(6):684–95.
[5] Shaughnessy M, Monnet E. Internal obturator muscle transposition for treatment of perineal hernia in dogs: 34 cases (1998-2012). J Am Vet Med Assoc 2015;246(3):321–6.

CASE 5.8

Abdominal Wall Hernia

Takanori Sugiyama[a] **and Helen M.S. Davies**[b]
[a]Animalius Vet, Bayswater, WA, AU
[b]Veterinary BioSciences, The University of Melbourne, Melbourne, Victoria, AU

Clinical case

History

A 10-year-old female spayed Russian Blue cat was presented as an emergency, after having been hit by a car.

Physical examination findings

The cat's vital signs (heart rate, respiratory rate, rectal temperature) were within normal limits, with adequate hydration status and body condition score. Abdominal palpation revealed the presence of a large, soft swelling on the right side of the abdominal wall. Gentle palpation of the mass elicited a pain response. A defect in the body wall was easily palpable bordering the mass. Within the mass, gas bubbles were apparent and palpable in tubular structures that were reducible and presumed to be intestine. No other abnormalities were found on physical examination.

"Reducible" in this situation means that the protruding structure could be returned to its normal position ("reduced") with manual pressure. In this case, the contents of the mass could be returned to the abdominal cavity. Reducibility is an important finding with this presentation, as it influences treatment decisions and outcome.

Differential diagnoses

Abdominal hernia, hematoma, gastric dilatation (gaseous distension of the stomach)

Diagnostics

Abdominal radiography revealed discontinuity of the abdominal wall, with small intestine containing digesta and gas herniated through the defect to a subcutaneous position (Figs. 5.8-1 and 5.8-2). The urinary bladder appeared normal, and there were no abnormal findings on thoracic radiographs. Routine bloodwork revealed mild elevations of creatine kinase and glucose, indicating nonspecific muscle damage and stress.

It is important to include the thorax in radiographic examination of dogs and cats with high-impact injuries such as hit-by-car (HBC) because these events often injure ribs and internal thoracic structures. Hemothorax, pneumothorax, and diaphragmatic hernia are common in this situation, and cats may not show overt signs of respiratory compromise on physical examination.

Comparative Veterinary Anatomy: A Clinical Approach. https://doi.org/10.1016/B978-0-323-91015-6.00037-6

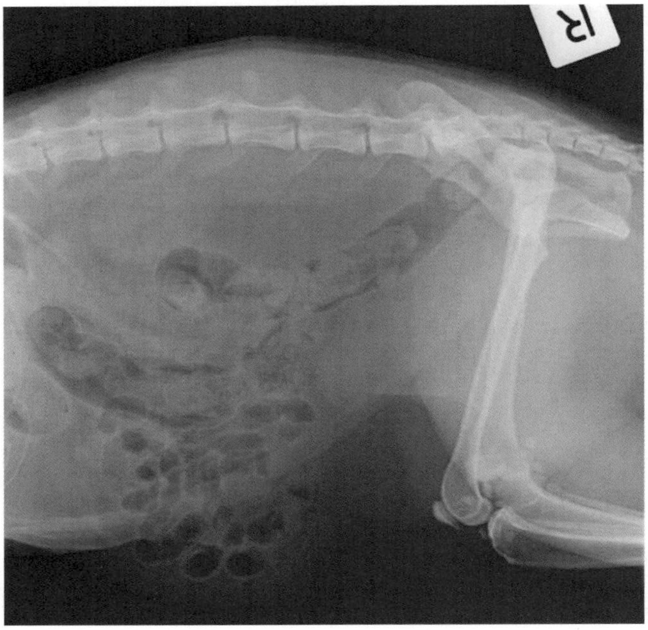

FIGURE 5.8-1 Lateral radiograph of the cat described in this case, depicting herniation of small intestine through the abdominal wall. Both the small intestine and the colon contain digesta/feces and gas, making it easy to see the herniated loops of small intestine under the skin ventral to the body wall on this lateral projection. (The hernia was located on the right lateral abdominal wall, but the herniated bowel assumed a more ventral position in this view because the cat was lying in right lateral recumbency.)

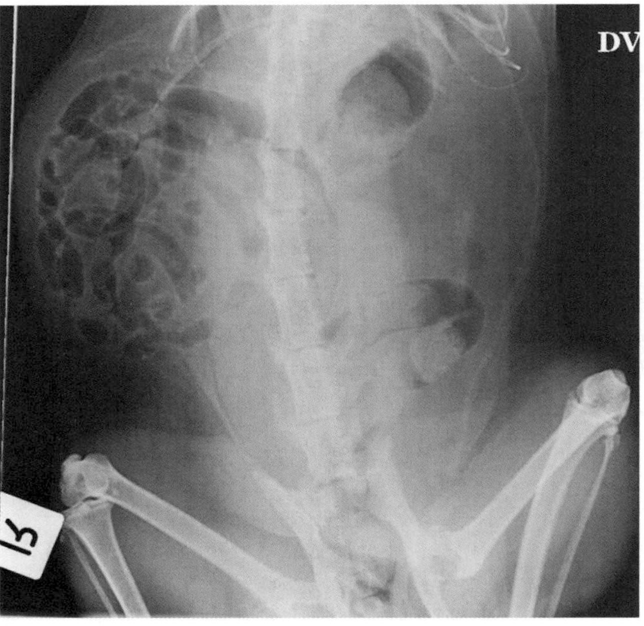

FIGURE 5.8-2 Dorsoventral radiograph of the cat in Fig. 5.8-1, showing the right-sided herniation of small intestine through the abdominal wall.

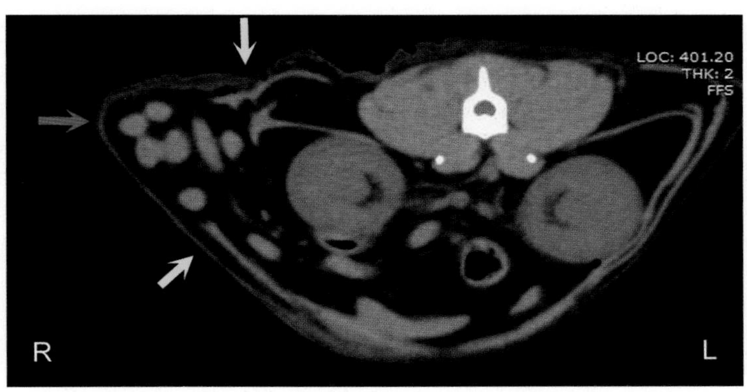

FIGURE 5.8-3 Transverse CT image at the level of the abdominal hernia. The defect in the body wall *(yellow arrows)* and subcutaneous herniation of multiple loops of small intestine *(pink arrow)* on the right side (R) of the body are clear, particularly when compared with the normal left side (L).

The cat was placed under anesthesia for CT. The transverse image showed a defect in the body wall and subcutaneous herniation of multiple loops of small intestine (Fig. 5.8-3).

Diagnosis

Traumatic abdominal hernia

Treatment

A ventral midline celiotomy was performed under general anesthesia with the cat in dorsal recumbency. The herniated small intestine was carefully returned to the peritoneal cavity and examined for signs of traumatic injury or ischemia. The other abdominal organs were also examined for signs of trauma.

> It is important to inspect all the abdominal organs in high-impact injuries such as HBC. Often, the obvious trauma is not the only injury sustained. Ventral midline celiotomy provides the best "picture" of all the abdominal contents.

An approximately 5-cm (2-in.)-long vertical defect was found in the mid-to-dorsal region of the right abdominal wall, 5 cm (2 in.) caudal to the last (13th) rib. The margins of the defect were debrided (Fr. *épluchage* cleaning, picking), and each layer of abdominal muscle was closed separately and anatomically, using a simple continuous suture pattern. The ventral midline incision was closed in 3 separate layers, from deep to superficial: linea alba, subcutaneous tissue, and skin.

> When closing an abdominal wall defect via the peritoneal cavity (e.g., as in the cat of this report during ventral midline celiotomy), the layers are apposed from superficial to deep (i.e., from external abdominal oblique to transversus/rectus abdominis). In contrast, when closing an abdominal incision (such as the ventral midline celiotomy), the layers are apposed in the opposite order, from deep to superficial.

The skin over the body wall defect was severely bruised, and the subcutaneous tissue was hemorrhagic and edematous. As a precaution, an uncertain area of skin and subcutaneous tissue viability was debrided and sutured separately. The cat recovered well from surgery and was discharged after 5 days of supportive care.

Anatomical features in canids and felids

Introduction

The abdominal cavity is bordered cranially by the diaphragm, caudally by the pelvic brim, dorsally by the lumbar spine and hypaxial muscles (and indirectly the caudal portion of the thoracic spine dorsal to the diaphragm), and laterally and ventrally by the caudal portion of the ribcage and sternum cranially and the abdominal musculature and associated fascia caudal to the ribcage.

In adult dogs and cats, there are 5 naturally occurring openings in the abdominal wall: 3 in the diaphragm (esophageal hiatus, aortic hiatus, and vena caval foramen) and the paired inguinal canals (left and right). In the fetus and neonate, a sixth natural opening is found where the umbilical structures enter/exit the ventral body wall. As discussed later, the umbilical opening gradually closes with fibrous tissue after birth. (The umbilical vessels are described in Case 4.6.)

All of these normal openings in the abdominal wall are potential sites for herniation of abdominal contents (see side box entitled "Abdominal hernias.").

Function

The abdominal cavity contains many viscera, including the excretory organs, most of the gastrointestinal tract, and the abdominal parts of the reproductive tract. As such, it needs to be able to accommodate changes in the size of the intestines, bladder, and—in females—a growing fetus. Therefore, the abdominal wall is comprised of differently oriented layers of muscle that can adjust to changing volume, safely support large loads, and effectively contract in multiple directions to reduce the internal compartment when required.

Abdominal contraction is an important function to increase the force on abdominal contents, assisting in defecation, urination, parturition, and the abdominal component of expiration. The intestines, urinary bladder, and female reproductive organs are suspended and supported dorsally by thin double sheets of peritoneum, allowing changing shapes, sizes, and movement independent of other organs. The abdominal wall assists in locomotion by flexing the lumbar spine and lumbosacral joint. Relaxation of the abdominal muscles permits a complete contraction of the diaphragm, assisting in inspiration, and it provides for temporary increases in intestinal load. Sequential stretching of the muscle and fascia accommodates a growing fetus and adapts to an increase in abdominal fat stores.

Peritoneum

The abdominal cavity is lined by a thin mesothelial layer called the **peritoneum**, which gives rise to the term "peritoneal cavity." However, the peritoneal space is mostly a potential space, as the **parietal** and **visceral** layers of the peritoneum are normally separated by only a thin film of fluid, produced by this serosal membrane for lubrication. In normal individuals, the abdominal contents fill the peritoneal cavity, bringing the visceral and parietal surfaces of the peritoneum in direct contact throughout most of the abdomen. In the standing dog and cat, the peritoneal space typically is widest ventrally, as peritoneal fluid naturally gravitates to the lowest part of the peritoneal cavity. In laterally or dorsally recumbent patients, the peritoneal fluid gravitates to the dependent part of the peritoneal cavity.

The parietal peritoneum is closely adhered to the innermost layer of the abdominal wall, which is the deep fascia of transversus abdominis (Figs. 5.8-4 and 5.8-5).

The parietal peritoneum (generally called just the "peritoneum") is incised during celiotomy, and a small amount of peritoneal fluid is normally found within the peritoneal cavity. Care must be taken when incising the abdominal wall because the abdominal contents lie in direct contact with the peritoneum, except when the peritoneal cavity is distended with free fluid or gas. In closing an incision or defect in the linea alba or transversus/rectus abdominis muscle, the incised or torn edge of the peritoneum may be included with the suture needle, thereby closing the peritoneum with the internal layer of the body wall, or the peritoneum may be left to heal separately.

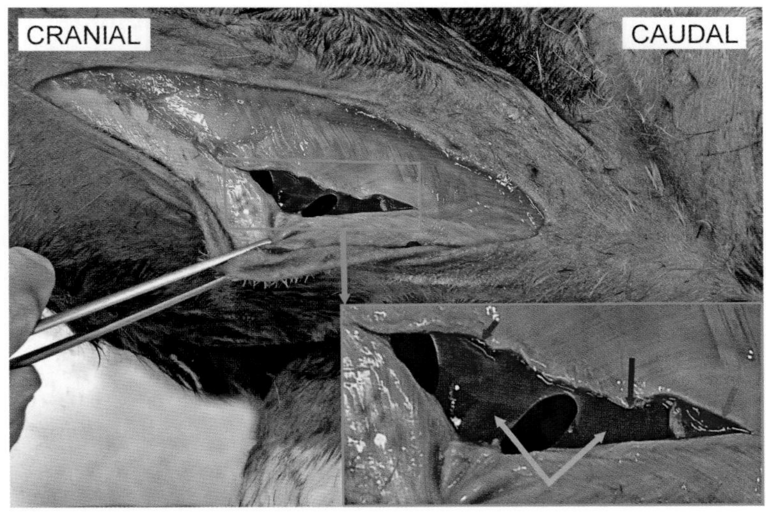

FIGURE 5.8-4 Ventral abdominal wall of a cat. The skin, subcutaneous tissue, and linea alba have been incised to reveal the thin, translucent parietal peritoneum, which lines the abdominal wall *(yellow arrows)*. The peritoneum was inadvertently punctured in two places in this prosection, which illustrates how readily the peritoneal cavity is entered on incising the linea alba *(blue arrows)*. Note, too, that the belly of the rectus abdominis muscle is exposed *(pink arrow)* in the caudal portion of the incision, which deviated slightly from midline.

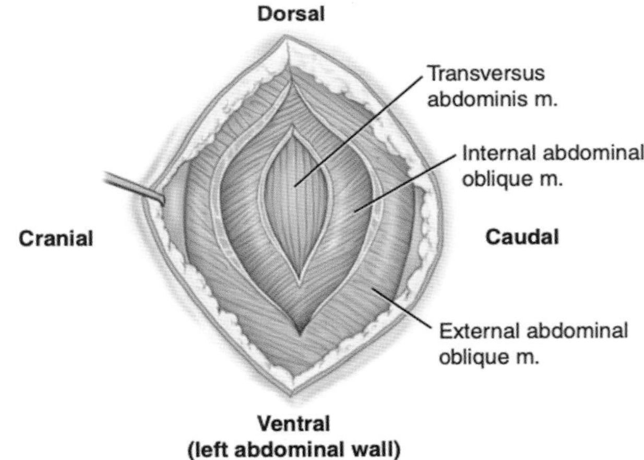

FIGURE 5.8-5 The direction of the muscle fibers of the left lateral abdominal wall from superficial to deep. The rectus abdominis muscle is confined to the ventral region and its fiber direction is cranial to caudal. Key: *m.*, muscle.

Most of the abdominal contents are encased by the visceral layer of the peritoneum, which is a deep invagination of the parietal peritoneum, separated from the parietal surface by folds of peritoneum which suspend these organs within the abdominal cavity. The folds are variously called the **mesentery** (intestine), **mesocolon** (colon), and **mesovarium** (ovary). The **greater** and **lesser omentum** are also peritoneal folds. However, several abdominal organs and vessels are located dorsal or dorsolateral to the parietal peritoneum in the **retroperitoneal space**, which is separated from the peritoneal cavity by the parietal peritoneum. The kidneys and most of the ureters, the adrenal glands, the abdominal aorta, the caudal vena cava, and the sublumbar lymph nodes are all considered retroperitoneal structures.

An aponeurosis (plural, aponeuroses) is the broad, fibrous attachment of a sheet-like muscle such as the external and internal abdominal obliques.

Abdominal tunic

The abdominal muscles and their associated fascial sheaths and aponeuroses form a robust "tunic" for the abdominal contents. From deep to superficial, the paired abdominal muscles are the **transversus abdominis, rectus abdominis, internal abdominal oblique**, and **external abdominal oblique**. The extensive fascial sheaths and shared aponeuroses enable the abdominal muscles to function as a unit. When they contract, the abdominal muscles increase intraabdominal pressure, which facilitates defecation, urination, expiration, and—in pregnant females—parturition. They also flex the thoracolumbar spine and lumbosacral joint.

The linea alba, a.k.a. linea alba abdominis (L. white line), is a readily identifiable landmark during ventral midline celiotomy, lying deep to the subcutaneous tissue. If the linea alba cannot be seen after careful separation of the subcutaneous tissue, then the incision is likely lateral to midline and an effort should be made to find the linea alba. Careful incision along the center of the linea alba allows access to the peritoneal cavity without injuring the adjacent rectus abdominis or transversus abdominis muscle bellies. Incisions through the abdominal muscles results in increased bleeding, greater post-operative swelling and pain, and an increased risk of herniation or wound dehiscence than an incision made through the linea alba.

The aponeuroses of the external and internal abdominal obliques and transversus abdominis muscles on each side of the abdomen meet at the ventral midline to form the **linea alba** ("white line"), a narrow strip of dense fascia that runs craniocaudally from the xiphoid process of the sternum to the prepubic tendon (Figs. 5.8-6 and 5.8-7).

ABDOMINAL HERNIAS

A true hernia comprises an opening in the abdominal wall (typically one of the 5 or 6 normal openings) bounded by a distinct fibrous rim (the hernial ring), through which an outpouching of the parietal peritoneum (the hernial sac) and some type of peritoneal content (fluid, omentum, intestine, etc.) protrudes to lie external to the abdominal cavity. Such hernias usually are congenital, and are named for their location—e.g., inguinal, umbilical. Depending on the type and amount of herniated tissue and the size of the hernial ring—i.e., on the risk and consequences of incarceration and strangulation of the tissue—the hernia may be symptomatic or asymptomatic, and thus may or may not require surgical repair.

Acquired hernias, such as traumatic and incisional hernias, are not "true" hernias because they may occur anywhere in the abdominal wall and do not contain a hernial sac. The herniated abdominal contents lie within the thoracic cavity (diaphragmatic hernia) or subcutis and may even be found external to the body if the skin is also disrupted. Acquired hernias usually are symptomatic and typically require surgical repair, because the inflammatory response in the injured peritoneum and body wall may narrow the defect and result in adhesions to the herniated tissue. This increases the risk of incarceration and strangulation in a defect that normally might be large enough to avoid these complications.

Herniation of abdominal contents through the weakened "pelvic diaphragm" (perineal hernia) is a specific type of acquired hernia seen almost exclusively in adult male dogs. It is discussed in Case 5.7.

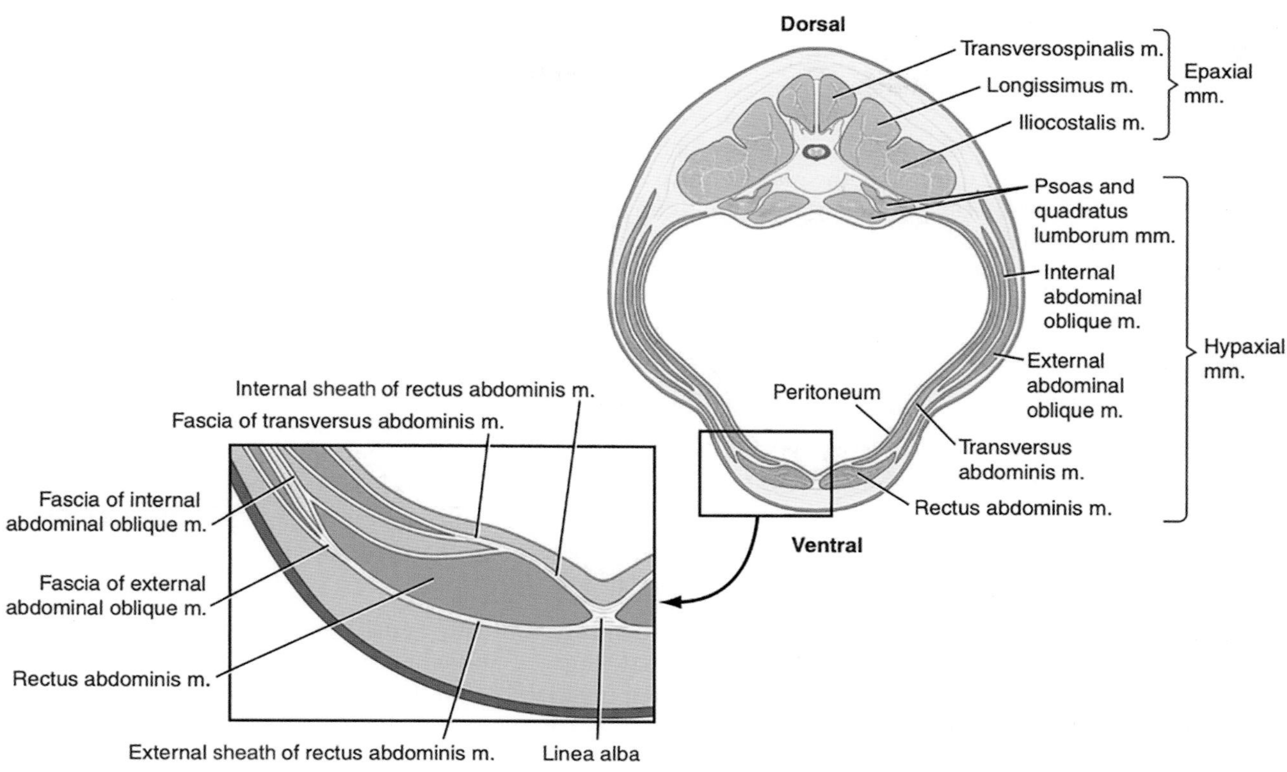

FIGURE 5.8-6 Cross section of the abdominal muscles cranial to the umbilicus with their aponeuroses extending ventrally and medially from the muscle bellies to meet at the linea alba. Caudal to the umbilicus, the whole of the aponeurosis of the internal abdominal oblique and part of the aponeurosis of the transversus abdominis lie external to the rectus abdominis muscle and integrate with its external sheath. Magnified image depicting the fascial and muscle relationships for the rectus abdominis muscle and linea alba. Key: *m.*, muscle; *mm.*, muscles.

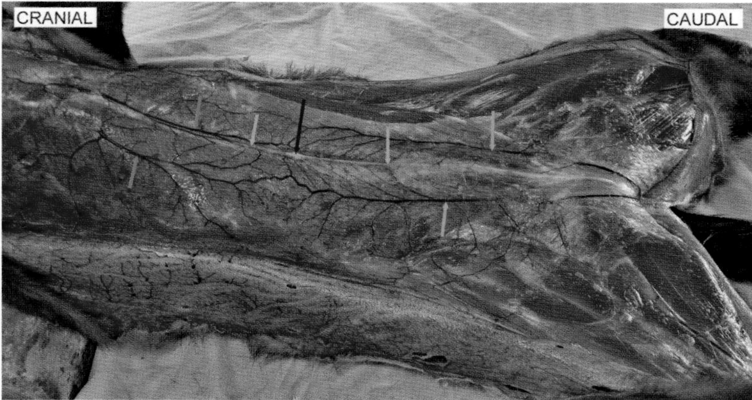

FIGURE 5.8-7 Ventral abdominal wall of a female dog, after removal of the skin and subcutaneous tissues. Note the linea alba *(yellow arrows)*, caudal and cranial superficial epigastric vessels (*aqua and green arrows*, respectively), and umbilical scar *(blue arrow)*.

When suturing the linea alba to close a ventral midline incision, it is important to incorporate the external rectus sheath—while avoiding passing the needle through the rectus abdominis muscle itself or through just the internal rectus sheath. Suturing only the internal rectus sheath does not provide a secure closure, and placing the suture through muscle increases bleeding, post-operative swelling and pain, and the risk of incisional hernia or dehiscence.

This arrangement makes the linea alba weaker in the caudal portion of the abdomen, so incisional hernias most often occur caudal to the umbilicus.

The **external rectus sheath** is the superficial fascial covering of the rectus abdominis muscle. The aponeuroses of the external and internal abdominal oblique muscles merge with the external rectus sheath, adding to its strength (Fig. 5.8-6). The internal rectus sheath is not as strong.

The **external abdominal oblique aponeurosis** lies superficial to the rectus abdominis muscle along its entire attachment, but the relationship of the **internal abdominal oblique aponeurosis** with the rectus abdominis changes from cranial to caudal—in the cranial third of the abdominal wall, the aponeurotic fibers pass both superficial and deep to the rectus abdominis (Fig. 5.8-6), whereas all aponeurotic fibers pass superficial to the rectus abdominis from the umbilicus caudally.

Caudally, the muscle fibers of the internal abdominal oblique curve caudally as they near their aponeurosis, and they form the cranial border of the internal or deep **inguinal ring**. The aponeurosis then sweeps caudally with that of external abdominal oblique. In males, some of the caudal muscle fibers of the internal abdominal oblique extend through the inguinal canal as the **cremaster muscle**, which raises and lowers the testis within the scrotum (described in Case 6.4).

The **flank** is the portion of the lateral abdominal wall that is composed entirely of soft tissue. The **paralumbar fossa** is a triangular depression within the flank dorsally, and it is more obvious in lean animals. It is bordered dorsally by the lateral margins of the lumbar vertebrae and associated lumbar muscles (iliocostalis lumborum), cranially by

INCISIONAL HERNIA AND DEHISCENCE

Incisional hernias are acquired hernias that occur post-operatively somewhere along an incision. Incisional dehiscence is a complete breakdown of the surgical wound closure along part or all of the incision. Either can occur following even routine elective procedures in healthy animals, such as an ovariohysterectomy. These post-operative wound complications usually occur within the first week after surgery, as it normally takes at least 2 weeks for the linea alba to regain even 30% of its original strength. For the first 2 weeks after surgery, the integrity of the wound closure relies primarily on suture strength and placement to resist the tensile forces on the healing incision. As even ordinary occurrences such as slipping, falling, or unrestricted activity may cause the sutures to break or pull through the tissue, pet owners are advised to restrict the patient's activity for at least 2 weeks after surgery to allow the surgical wound to improve in tensile strength.

FLANK SPAY

Flank spay (ovariohysterectomy via the lateral body wall) is still performed at some animal shelters, particularly in immature cats. The incision is made in the right flank, just caudal to the last rib, through the skin, subcutaneous tissue, and muscles of the abdominal wall (external and internal abdominal oblique and transversus abdominis). This approach is not recommended because the abdominal fascia in this location is thin and suturing muscle instead of fascia causes bleeding and greater post-operative swelling and pain. In addition, access to the left ovary and uterine body is poor with this approach, and further traumatizes the incised muscles in the process of identifying and removing the ovaries and uterus.

the caudal border of the last rib, and caudoventrally by the dorsal border of the muscular portion of the internal abdominal oblique muscle.

Blood supply, lymphatics, and innervation

The skin and subcutis of the ventral body wall are mainly supplied by the **cranial** and **caudal superficial epigastric arteries**, and caudodorsally by ventral branches of the **deep circumflex iliac artery**. The caudal superficial epigastric artery is a branch of the external pudendal artery after the latter has exited the inguinal canal. As depicted in Fig. 5.8-7, these superficial arteries run parallel to the linea alba; they may be evident through the skin in thin-skinned individuals with minimal subcutaneous fat. The caudal superficial epigastric arteries lie on either side of the penis in male dogs, where they are relatively minor; but in female dogs and cats, they supply the caudal mammary glands, so they are of substantial size in lactating females.

The ventral and lateral parts of the abdominal wall itself are supplied by branches of the **cranial** and **caudal epigastric arteries**. These arteries lie on the internal surface of the rectus abdominis muscle on either side of the midline. Other arterial supply to the abdominal wall includes the **caudal intercostal**, **musculophrenic**, and **cranial abdominal arteries**. Generally, the venous and lymphatic drainage parallels the arterial supply.

The **superficial inguinal lymph nodes** are located on either side of the prepuce in male dogs and within the fat of the inguinal region in both males and females, just medial to the external inguinal ring.

The abdominal wall is innervated segmentally by sequential **ventral branches** of the **thoracolumbar spinal nerves**. These paired branches (left and right) exit the spine through the intervertebral foramina and course first laterally along the caudal edge of the relevant rib or lumbar vertebral transverse process, and then ventrally in the transversus abdominis muscle or alongside the relevant rib, while regularly sending out branches that extend to the more superficial tissues.

In male dogs, a paramedian incision through the skin and subcutaneous tissue must be made lateral to the prepuce (parapreputial incision) if access to the caudal abdominal cavity is required (Fig. 5.8-8). (If the surgical procedure is completed via a ventral midline celiotomy that ends cranial to the prepuce, then there is no need for a parapreputial incision.) Any branches of the caudal superficial epigastric vessels encountered must be ligated or cauterized to prevent bleeding. The preputial fascia is then elevated to expose the linea alba beneath the ensheathed penis. During surgical wound closure, the preputial fascia is closed separately from, and secondary to, the external rectus sheath; otherwise, abdominal contents may herniate in between the ventral body wall and the prepuce.

Blood supply to the prepuce and mammary glands

Although not part of the abdominal wall, the position and blood supply of the prepuce and mammary glands must be considered when performing a ventral midline or paramedian incision on a male dog or a pregnant/lactating bitch or queen. Of note, the left and right **caudal superficial epigastric** vessels lie just lateral to the ventral midline within the subcutaneous tissue (Fig. 5.8-7). In male dogs, the **preputial artery** also lies in the subcutaneous tissue layer of the prepuce.

In pregnant/lactating bitches and queens, care is required to stay on midline and careful blunt dissection is needed to avoid the superficial epigastric vessels and the engorged left and right mammary glands (Fig. 5.8-9). Particularly in cats, the left and right mammary glands lie in proximity during late pregnancy and early lactation. Leakage of milk into the surgical field from an incised or punctured mammary gland hampers visualization during surgery and causes incisional inflammation and granulation tissue formation resulting in swelling and delayed wound healing.

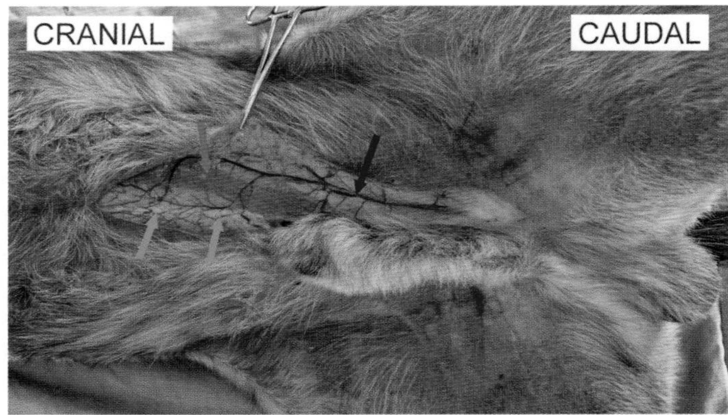

FIGURE 5.8-8 Ventral abdominal wall of a male dog. A ventral midline incision through the skin has been extended caudally as a left parapreputial incision (adjacent to the prepuce). Note the proximity of the caudal superficial epigastric vessels *(blue arrow)* and preputial muscle *(red arrow)* to the linea alba *(green arrows)*. The prepuce must be retracted laterally to access the linea alba in the caudal part of the abdomen.

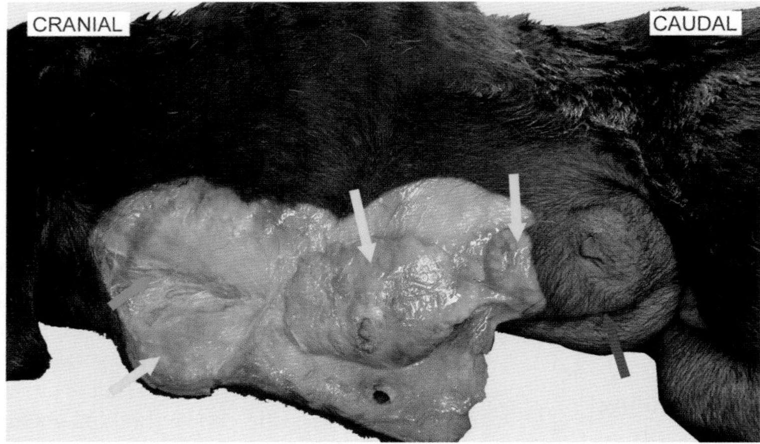

FIGURE 5.8-9 Ventrolateral abdominal wall of a bitch with engorged mammary glands *(yellow arrows)*. The left and right pairs of mammary glands lie in proximity, creating a deep cleft *(pink arrow)* of the ventral midline. In this prosection, the right cranial mammary gland has been reflected ventrally to show the linea alba *(green arrow)* beneath the subcutaneous tissue. (The extensive blood supply of the engorged mammary glands is not appreciable on this postmortem specimen.)

UMBILICAL HERNIA

Umbilical hernia is a herniation of abdominal contents through the umbilical opening in young individuals. Although breeders often attribute it to birth trauma (excessive intraabdominal pressure as the fetus is expelled through the pelvic canal) or excessive traction on the umbilical cord as the bitch/queen bites through the cord, it is generally considered to be a defect of fetal development in which the left and right sides of the ventral abdominal fascia fail to properly fuse by the sixth week of gestation. In dogs, umbilical hernia is believed to be hereditary and may be associated with other congenital or hereditary defects such as cryptorchidism. However, it is, possible that excessive traction on the umbilical cord, ligation of the cord too close to the body wall, or umbilical infection ("navel ill") may lead to acquired umbilical herniation.

Umbilicus

The umbilicus and the umbilical opening in the ventral abdominal wall are unique to fetal and early neonatal life. The umbilical opening consists of a discrete discontinuity in the parietal peritoneum, linea alba, subcutaneous tissue, and skin through which the umbilical arteries and vein exit and enter, respectively, the abdominal cavity of the fetus (umbilical structures are described in Case 4.6).

In the days after birth, the remnant of the umbilical cord shrinks, and any external portion detaches from the body. At the same time, the umbilical opening is closed with fibrous tissue, thereby creating continuity in the ventral midline postnatally. At least in young animals, the umbilical scar is usually visible externally (on the skin of the ventral midline) and internally (within the linea alba; see Fig. 5.8-7).

Selected references

[1] Reina Rodriguez FS, Buckley CT, Milgram J, et al. Biomechanical properties of feline ventral abdominal wall and celiotomy closure techniques. Vet Surg 2018;47:193–203.
[2] Rosin E, Richardson S. Effect of fascial closure technique on strength of healing abdominal incision in the dog. A biomechanical study. Vet Surg 1987;16:269–72.
[3] Selcer BA, Buttrick M, Barstad R, et al. The incidence of thoracic trauma in dogs with skeletal injury. J Small Anim Pract 1987;28:21–7.

CHAPTER 6

PELVIC ORGANS

Helen M.S. Davies, Chapter editor

Female Urogenital System

6.1 Ectopic ureters—*James Flanders* 338
6.2 Pyometra—*Natali Krekeler and Helen M.S. Davies* 346
6.3 Dystocia and the mammary gland—*Natali Krekeler, Helen M.S. Davies, and Cathy Beck* 360

Male Urogenital System

6.4 Benign prostatic hyperplasia—*Magdalena Schrank, Stefano Romagnoli, and Natali Krekeler* 373
6.5 Congenital phimosis—*Magdalena Schrank, Natali Krekeler, Helen M.S. Davies, and Stefano Romagnoli* 385

CHAPTER 6

CASE 6.1

Ectopic Ureters

James Flanders
Department of Clinical Sciences, Cornell University, College of Veterinary Medicine, Ithaca, New York, US

Clinical case

History

A 4-month-old female Siberian husky was presented with chronic urinary incontinence. The owners acquired the dog when she was 7 weeks old and noticed her perineum was constantly wet. The dog dripped urine from her vulva continually; however, she was able to urinate normally.

Physical examination findings

The dog was very bright and active. All physical examination findings were normal, except for the moist perineum. The dog had a small, palpable urinary bladder that was not easily expressed. Urine dribbled on the examination table during the examination.

Differential diagnoses

Cystitis (urge incontinence), idiopathic detrusor instability, overflow incontinence, urethral sphincter mechanism incontinence, urethrovaginal fistula, and ectopic ureter

Diagnostics

A moderately increased number of red and white blood cells along with extracellular coccoid bacteria were seen during microscopic examination of the urine sediment obtained by cystocentesis. The urine pH was 6.8, and urine bacterial culture identified growth of *Staphylococcus aureus*, sensitive to cephalosporin antibiotics. Abdominal ultrasound showed right-sided hydroureter (Fig. 6.1-1) and slight dilation of the right renal pelvis. A ureteral urine jet (Fig. 6.1-V1 in the online version at https://doi.org/10.1016/B978-0-323-91015-6.00039-X) was seen within the urinary bladder. A CT urogram (Fig. 6.1-2) showed the right ureter bypassing the trigone and emptying into the mid-urethra. The left ureter entered the trigone and emptied contrast-containing urine into the urinary bladder lumen (not shown in the figure). Cystoscopy identified a large, ectopic ureteral opening in the mid-pelvic urethra (Fig. 6.1-3) and a normal left ureteral papilla on the dorsal aspect of the urinary bladder (Fig. 6.1-V2 in the online version at https://doi.org/10.1016/B978-0-323-91015-6.00039-X).

Diagnosis

Unilateral right-sided ectopic ureter, right-sided hydroureter, and mild right-sided hydronephrosis

Treatment

The dog was placed under general anesthesia and a cystoscope was introduced through the vaginal vestibule. A holmium-YAG laser fiber was passed through the operating portal of the scope and used to ablate the tissue between

Comparative Veterinary Anatomy: A Clinical Approach. https://doi.org/10.1016/B978-0-323-91015-6.00039-X

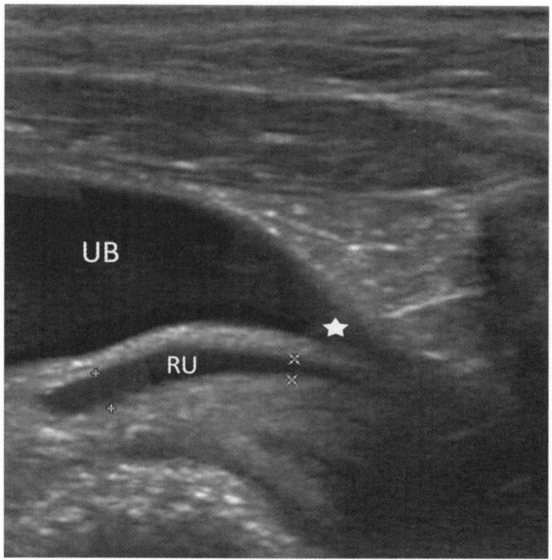

FIGURE 6.1-1 Abdominal ultrasound image of caudal abdomen of female dog showing urinary bladder (UB) and right ureter (RU). The right ureter is dilated to approximately 4 times normal size (marked by *+*) and passes parallel to the bladder neck rather than entering the bladder obliquely (marked by *X*) at the trigone region *(white star)*. This finding supports a diagnosis of right-sided ureteral ectopia.

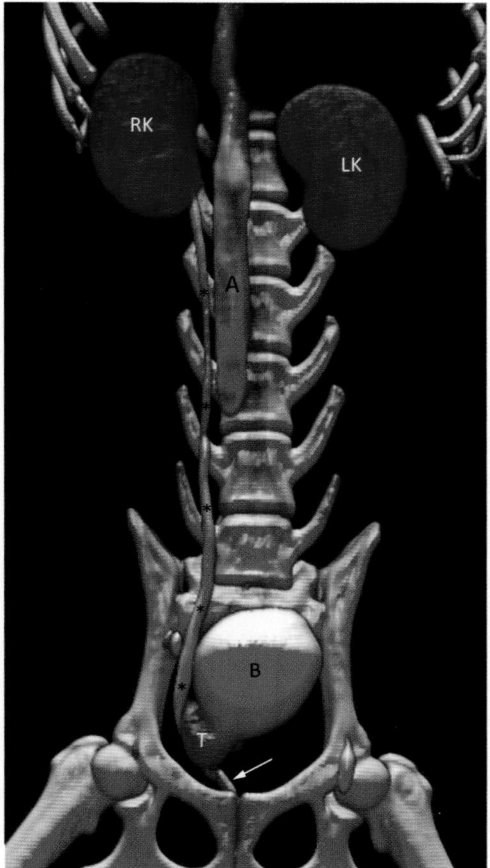

FIGURE 6.1-2 Image of 3D reconstruction of CT contrast urogram of female dog. The dilated right ureter *(black asterisks)* passes from the right kidney (RK) toward the urinary bladder (B). The right ureter bypasses the trigone region of the bladder (T) and continues distally in the pelvic canal *(white arrow)*. The left ureter is not seen in this image.

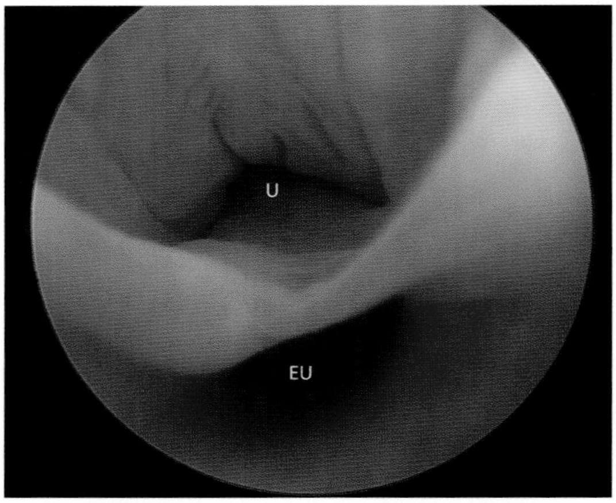

FIGURE 6.1-3 Cystoscopic image of urethra lumen of female dog with the dog in dorsal recumbency. The cystoscope has been introduced through the vaginal vestibule and the urethra has been distended with saline. An ectopic ureter (EU) opens into the urethral lumen separated from the proximal urethra (U) by a tissue membrane.

the ectopic ureter and the urethral lumen (Fig. 6.1-V3 in the online version at https://doi.org/10.1016/B978-0-323-91015-6.00039-X). The division between the urethra and the ectopic ureter was resected along the entire length of the urethra and the proximal urinary bladder neck. As a result, the ureteral opening was moved from the ectopic location to a site within the trigone of the bladder (Fig. 6.1-V4 in the online version at https://doi.org/10.1016/B978-0-323-91015-6.00039-X). The dog was completely continent at recheck examination 1 month later.

Anatomical features in canids and felids

Introduction

The urinary system is divided into upper and lower tracts. The upper tract includes the paired kidneys (which filter blood and produce urine) and the ureters (which transport urine from the kidneys to the bladder). The lower tract includes the urinary bladder (which stores urine); the urethra (a conduit from the bladder to the exterior, which in the male also serves as the site of passage for male reproductive cells [gametes/sperm]); and contributions from the accessory glands (see Case 6.4).

Function

The urinary system and its multiple parts are important in: (1) controlling blood volume and electrolytes—e.g., sodium, potassium, and calcium; (2) regulating blood pressure; (3) controlling pH (acid/base) balance of blood; (4) promoting red blood cell production by the kidney via the production of erythropoietin; (5) contributing to the production of calcitriol—the active form of vitamin D; and (6) producing, storing, and excreting waste products of urine—i.e., urea and uric acid.

Ureters

The **ureters** are paired tubular structures that transmit urine from the kidneys to the urinary bladder. The cranial portion of each ureter drains the renal pelvis and exits the kidney at the renal **hilus**. Each ureter passes caudally in the retroperitoneal space and through the **lateral ligaments** of the bladder to enter the dorsolateral urinary bladder wall

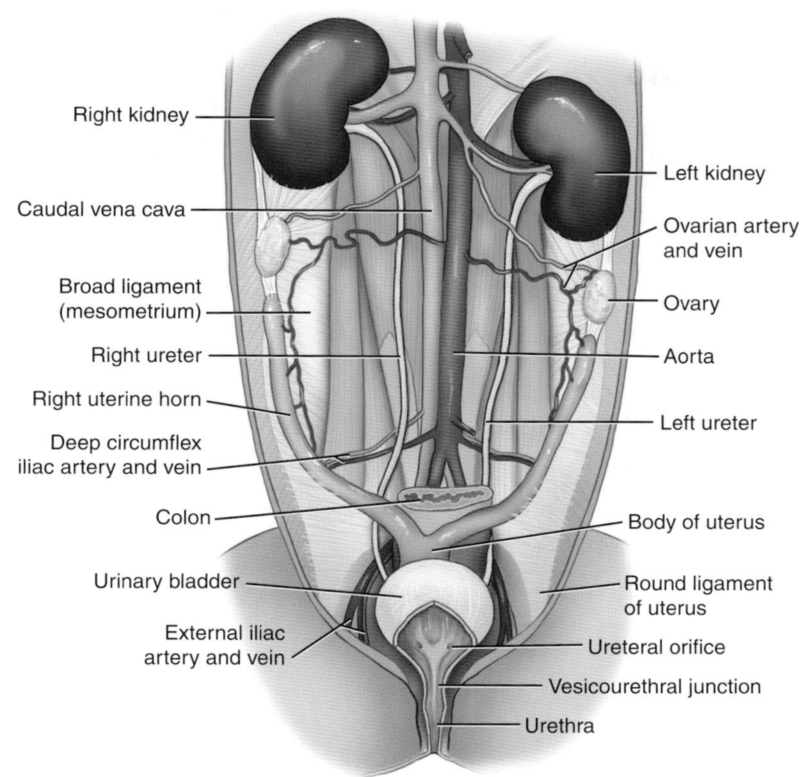

FIGURE 6.1-4 The structures of the urinary tract of the female dog.

near the **vesicourethral junction** (Fig. 6.1-4). In the male dog or cat, the ureter passes dorsal to the **ductus deferens**, and in the female, the ureter is associated with the **broad ligament** of the uterus, before entering the lateral ligament of the bladder. After passing obliquely a short distance through the urinary bladder wall, each ureter terminates at a slit-like opening in the bladder **trigone** or **trigonum** (Gr. *trigonos* triangle)—a triangular region formed by the 2 ureteric orifices and the internal urethral orifice—called the **ureteral papilla** (Fig. 6.1-5). The oblique course of the ureter through the bladder wall creates a one-way valve effect so that as the bladder fills with urine, more pressure is placed on the intramural portion of the ureter, preventing urine reflux.

The traditional means of correcting ectopic ureters is by surgical implantation of the ureters into the urinary bladder through an open abdominal approach. The minimally invasive alternative is termed cystoscopic-guided laser ablation (CLA). Both these techniques are approximately 50% successful in providing continence in dogs with ectopic ureters because coincidental urethral sphincter mechanism incontinence is present in many dogs with ectopic ureters. However, approximately 50% of the dogs that remain incontinent become continent after treatment with an alpha-sympathomimetic drug and/or estrogen.

Blood supply, lymphatics, and innervation of the ureters

The blood supply of the ureter, the **ureteral artery**, is derived from two sources: cranially from the **renal artery** and caudally from the **prostatic** or **vaginal artery**. The venous drainage parallels the arterial supply. The ureter wall is composed of 4 layers—an internal mucosa composed of transitional epithelium and 3 muscular layers: inner longitudinal, middle circular, and outer longitudinal.

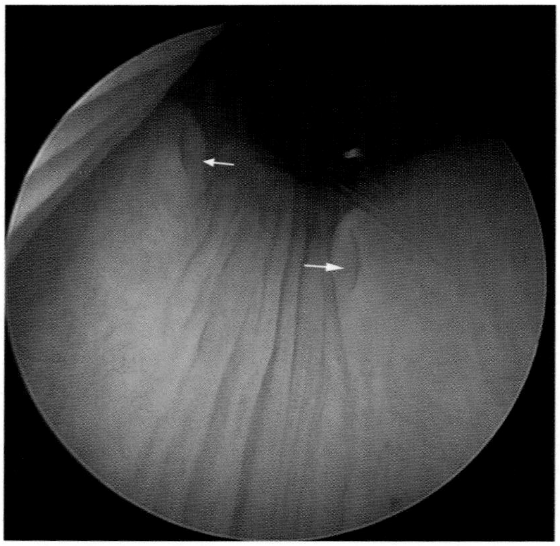

FIGURE 6.1-5 Cystoscopic image of the trigone region of the urinary bladder of a female dog in dorsal recumbency. The cystoscope has been introduced through the vaginal vestibule and the urethra has been distended with saline. The normal ureteral papillae *(white arrows)* are visible at the trigone of the urinary bladder.

IMAGING THE URETERS

Several imaging techniques can be used to diagnose ectopic ureters. In an excretory urogram, intravenously administered iodinated contrast material excreted by the kidneys highlights the path of the ureters in a radiograph. Another radiographic technique, retrograde cystourethrography, has higher diagnostic accuracy than the excretory urogram and consists of retrograde administration of iodinated contrast material through a catheter into the closed vaginal vestibule of female dogs or cats (or the urethra of males). The contrast agent flows up the vagina, urethra, and any ectopic ureters that empty into the distal urogenital tract (Fig. 6.1-6). Computerized tomography can be used to visualize intravenously administered contrast material in the urinary tract, and 3D reconstructions can demonstrate the path of an ectopic ureter with high resolution. Besides radiographic techniques, abdominal ultrasound examination is very useful for evaluating the entire urinary tract and can reveal abnormalities associated with ureteral ectopia in the kidneys, ureters, or urinary bladder. The gold standard for diagnosis of ectopic ureters is cystoscopy, which permits the direct observation of ectopic openings into the distal urogenital tract.

EMBRYOLOGY

During the development of the fetal urogenital tract, the ureters are derived from the metanephric duct, which buds off from the distal mesonephric duct, close to the embryonic cloaca. The metanephric duct normally separates from the mesonephric duct, migrates cranially, and opens into the trigone of the developing urinary bladder. If the metanephric duct does not separate from the mesonephric duct during embryogenesis, the metanephric duct is to be carried caudally with the mesonephric duct and instead opens into the urethra or vestibule. Ectopic ureters are often associated with congenital abnormalities of the distal reproductive tract such as double vagina and paramesonephric remnants (Fig. 6.1-7). Congenital abnormalities of the kidney may also be present.

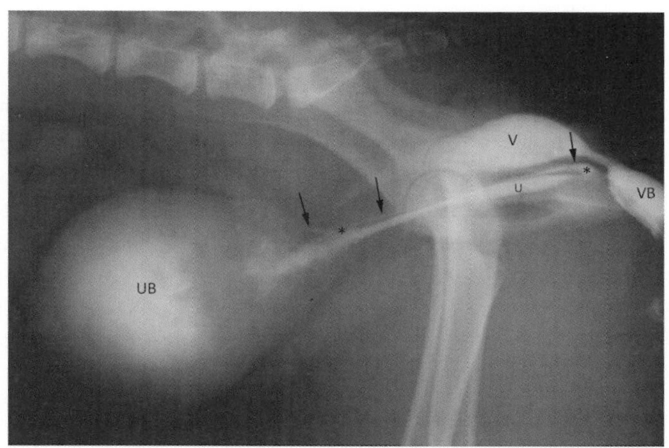

FIGURE 6.1-6 Lateral radiographic image of retrograde cystourethrogram in a female dog. Radiographic contrast material was injected into the vaginal vestibule. The contrast material fills the vestibule (VB) and the vagina (V) dorsally. The contrast material is present within the urethra (U) and the urinary bladder (UB). Contrast has also flowed retrograde from the urethral lumen into an ectopic ureter present within the wall of the urethra *(black arrows)*. There are 2 openings between the ectopic ureter and the urethra *(black asterisks)*, the most distal opening of the ectopic ureter is just proximal to the vestibule and a more proximal opening is present near the vesicourethral junction.

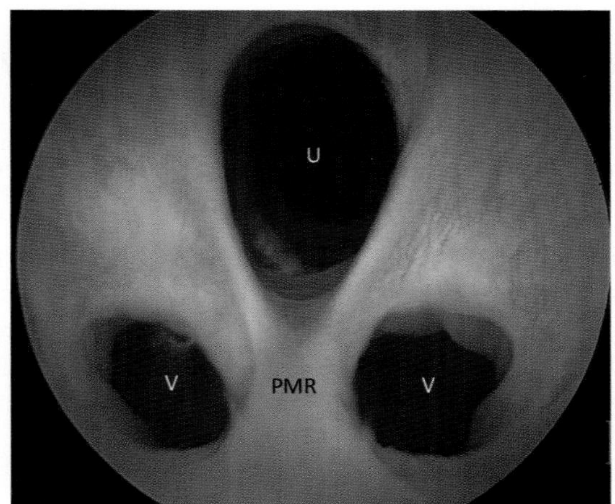

FIGURE 6.1-7 Cystoscopic image of the vaginal vestibule of a female dog in dorsal recumbency. The cystoscope has been introduced into the vaginal vestibule and the vestibule has been distended with saline. The distended urethral orifice (U) is seen above two vaginal openings (V) separated by a paramesonephric remnant (PMR).

URINARY INCONTINENCE

There are multiple etiologies for urinary incontinence in dogs. Differentials should include:

- Neurogenic—reduced expression of alpha-sympathetic receptors at internal urethral sphincter
- Urge incontinence—due to detrusor (L. *detrudere* to push down) muscle hyperactivity, secondary to spinal cord injury
- Overflow incontinence—also secondary to spinal cord injury, affecting bladder detrusor and urethal sphincter
- Ectopic ureters—most common cause, can be unilateral or bilateral and most common in females

Urinary bladder

The **urinary bladder** is a hollow muscular organ positioned in the caudal abdomen. The main functions of the urinary bladder are to store and void urine. The shape of the bladder changes from spherical to tear-shaped as it fills with urine. Urine produced in the kidneys passes through the paired ureters and enters the bladder lumen at the trigone (see Case 6.4), located dorsally in the neck region of the bladder, as discussed earlier. The major, central portion of the bladder is called the **body**, while the cranial, blind end of the bladder is the **apex**. The wall of the urinary bladder consists of 4 layers—inner mucosa or urothelium, submucosa, muscularis, and serosa. The **mucosa** contains specialized transitional epithelial cells interspersed with goblet cells that secrete mucus to help protect the urothelium from potential noxious effects of urine. The **muscular layer** is composed of smooth muscle oriented into inner and outer longitudinal layers and a middle circular layer. The longitudinal muscle fibers become obliquely oriented into the internal **urethral sphincter** at the vesicourethral junction.

The urinary bladder is fixed in the caudal abdomen by the lateral and ventral ligaments of the bladder and is capable of distention as far cranially as the umbilicus. Three peritoneal ligaments extend from the body wall to the bladder: First, the **lateral ligaments** enclose the **round ligaments** (umbilical arteries) in their free borders and are reflections of the dorsal and ventral peritoneum on either side of the bladder. Within the lateral ligaments are found the ureters, nerves, lymphatics, blood vessels, and [in males only] the ductus deferens. The lateral ligaments can be easily identified by the large amount of adipose tissue contained within. Second, the **median ligament of the bladder** is a fold of peritoneum located on the midline of the abdomen between the linea alba and the bladder extending from the umbilicus to the symphysis pubis. The medial ligament attaches the ventral surface of the bladder to the mid-ventral abdominal wall. In the fetus, the urachus and umbilical vasculature extend from the apex of the bladder to the umbilicus within the median ligament.

Blood supply, lymphatics, and innervation of the urinary bladder

Blood supply of the urinary bladder is provided by the **caudal vesical arteries** that are branches of the **prostatic** or **vaginal arteries**. The venous drainage of the urinary bladder flows through the **internal pudendal veins**. Lymphatic drainage from the bladder enters the **lumbar** and **hypogastric lymph nodes**.

The urinary bladder distends as urine accumulates within it, and the storage capacity is augmented by the action of the beta sympathetic nerve fibers within the **hypogastric nerve**. The hypogastric nerve enters the bladder wall through the lateral ligaments of the bladder. The tone of the internal urethral sphincter is augmented by the action of the alpha-sympathetic fibers of the hypogastric nerve. Contraction of the bladder musculature (the detrusor muscle) during micturition is controlled by the parasympathetic **pelvic nerves** that also enter the bladder through the lateral ligaments.

PHYSIOLOGY OF URINARY CONTINENCE

Urinary continence is produced by the activity of 2 muscular sphincters located in the lower urinary tract. The first, the internal urethral sphincter, is a smooth muscle sphincter located at the vesicourethral junction. It is innervated by alpha-sympathetic nerve fibers from the hypogastric nerve and provides involuntary, autonomic control of urinary continence. The second is the external urethral sphincter, composed of skeletal muscle, located at the mid-pelvic urethra in the female dog and just caudal to the prostate gland in the male dog. Innervated by the pudendal nerve, the external urethral sphincter is more important for voluntary control of urinary continence.

Selected references

[1] Berent AC, Weisse C, Mayhew PD, Todd K, Wright M, Bagley D. Evaluation of cystoscopic-guided laser ablation of intramural ectopic ureters in female dogs. J Am Vet Med Assoc 2012;240:716–25.
[2] Cannizzo KL, McLoughlin MA, Mattoon JS, Samii VF, Chew DJ, DiBartola SP. Evaluation of transurethral cystoscopy and excretory urography for diagnosis of ectopic ureters in female dogs: 25 cases (1992-2000). J Am Vet Med Assoc 2003;223:475–81.
[3] Davidson AP, Westropp JL. Diagnosis and management of urinary ectopia. Vet Clin Small Anim 2014;44:343–53.
[4] Ho LK, Troy GC, Waldron DR. Clinical outcomes of surgically managed ectopic ureters in 33 dogs. J Am Anim Hosp Assoc 2011;47:196–202.
[5] Holt PE, Moore AH. Canine ureteral ectopia: an analysis of 175 cases and comparison of surgical treatments. Vet Rec 1995;136:345–9.

CASE 6.2

Pyometra

Natali Krekeler and Helen M.S. Davies
Veterinary Biosciences, The University of Melbourne, Melbourne, Victoria, AU

Clinical case

History

A 9-year-old female Golden Retriever was presented with lethargy, polydipsia (PD), and polyuria (PU). She had been observed to be in estrus ("heat") 6 weeks before presentation.

Pyometra is high on the list of differential diagnoses in an intact bitch presented for lethargy, PD/PU, and abdominal distension 1–3 months after estrus. Fever is an inconsistent finding in bitches and queens with pyometra, as is vaginal discharge. The presence of vaginal discharge depends on the cervical patency: it is observed only when the cervix is open, in which case the discharge is typically hemopurulent. Abdominal palpation is ill-advised in patients suspected of having a closed-cervix pyometra, such as the bitch in this case, because it may result in rupture of the distended and friable (fragile) uterus or drive purulent material via the uterine tubes into the peritoneal cavity.

Physical examination findings

The only significant physical examination findings were abdominal distension and pale mucous membranes. Notably, her rectal temperature was normal, and there was no vaginal discharge. Because pyometra was high on the list of differentials, abdominal palpation was not performed.

Differential diagnoses

Pyometra; pregnancy; mucometra, hemometra, or hydrometra (accumulation of mucus, blood, or serous fluid, respectively, in the uterus); metritis (inflammation of the uterine wall), with or without retained fetal membranes—e.g., with recent fetal loss following unobserved mating; systemic diseases such as diabetes mellitus, diabetes insipidus, hyperadrenocorticism (Cushing's disease), or renal disease

PYOMETRA

Pyometra (Gr. *pyo-* pus + *metra* uterus) is an acute or chronic suppurative inflammation of the uterus. Up to 25% of intact (unneutered) bitches develop pyometra by 10 years of age. Affected bitches most often present 1–3 months after estrus, whereas affected queens (female cats) most often present 2–5 weeks after estrus.

Pyometra is characterized by endometrial hyperplasia (thickening of the uterine mucosa), with cystic dilation of the endometrial glands and accumulation of a neutrophil-rich exudate in the uterine lumen. The patient may or may not be systemically ill (depression, fever, inappetence, weakness, PD/PU, etc.).

Comparative Veterinary Anatomy: A Clinical Approach. https://doi.org/10.1016/B978-0-323-91015-6.00040-6

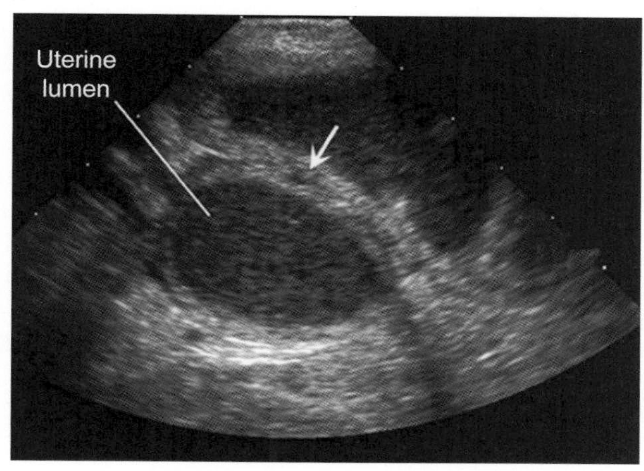

FIGURE 6.2-1 Ultrasonographic image of the caudal abdomen of a bitch with pyometra, transverse view. The uterus is distended with heterogeneous *(speckled)* fluid, and cystic changes are evident in the thickened uterine wall *(arrow)*.

Diagnostics

Transabdominal ultrasonography revealed a convoluted, thick-walled uterus with heterogenic fluid in its lumen, consistent with purulent material (Fig. 6.2-1).

Routine bloodwork showed a markedly elevated white blood cell count which predominantly consisted of marked neutrophilia (increased neutrophil count) with a left shift (increased percentage of immature neutrophils) and toxic changes—a common finding with severe bacterial infection. Serum chemistry revealed elevations in total protein, globulins, alkaline phosphatase, alanine aminotransferase, and urea nitrogen, which together were supportive of a diagnosis of bacterial toxemia.

A urine sample was collected by ultrasound-guided transabdominal cystocentesis. Urinalysis revealed

Ultrasound-guided cystocentesis is strongly advised in these cases to minimize the risk of peritonitis or sample contamination from inadvertent puncture of the distended and infected uterus. The procedure involves inserting a sterile needle into the urinary bladder via the ventral abdominal wall (under ultrasound guidance) and aspirating a urine sample directly from the bladder. Collection of a voided urine sample (i.e., during urination or by catheterization of the bladder via the external urethral orifice) is not recommended, as the urine sample is likely to be contaminated with bacteria from the reproductive tract, even if there is no vaginal discharge.

DIAGNOSTIC IMAGING IN PATIENTS WITH PYOMETRA

Ultrasonography is the method of choice for the diagnosis of pyometra in bitches and queens. Commonly, the uterine wall is thickened and has an irregular endometrial surface, with small hypoechoic areas consistent with cystic changes in the uterine glands (Fig. 6.2-2). However, if the uterus is severely distended with purulent material, the uterine wall can appear abnormally thin.

Radiography shows the enlarged uterus as a distended, tubular structure with a diffuse soft-tissue density, but it does not provide as much clinical information as ultrasound. Furthermore, it is difficult with radiography alone to differentiate pyometra from pregnancy in the first 6 weeks of gestation. Typically, fetal skeletal ossification is not apparent radiographically until approximately 42 days after mating in bitches and queens.

When suitable endoscopic equipment is available, a sample should be obtained directly from the uterine lumen via the cervix, a procedure called hysteroscopy. When the vagina must be sampled instead, a sterile speculum is used to ensure that the swab is collected from the cranial part of the vagina, as it is closest to the site of infection (i.e., the uterus) and is less likely than the caudal vagina or vulva to be contaminated with bacteria from the perineal area. This holds true for open-cervix pyometra as well. Although vaginal discharge in those cases is readily collected from the external surface of the vulva, the sample for microbiological analysis should be collected from the uterus or as close as possible with the equipment available.

proteinuria and bacteriuria, and bacterial culture of the urine yielded a pure growth of *Escherichia coli*.

In addition, a lubricated speculum was carefully inserted into the vagina via the vulva, and the cranial vagina was swabbed for bacterial culture and antimicrobial susceptibility testing.

Diagnosis

Closed cervix pyometra

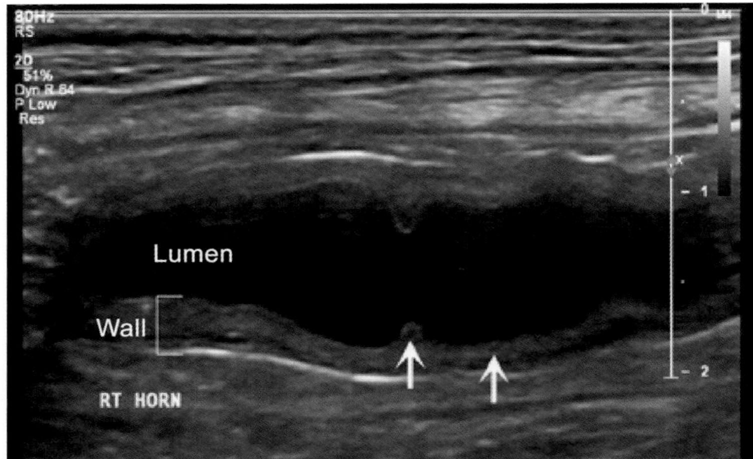

FIGURE 6.2-2 Ultrasonographic image of the right uterine horn of a bitch with pyometra, longitudinal view. The uterine lumen is filled with hypoechoic fluid (which appears mostly black on ultrasound). The uterine wall is thickened and has an irregular endometrial surface, with small hypoechoic areas consistent with cystic changes in the uterine glands *(arrows)*.

ORIGINS OF UTERINE TERMS

Two separate Greek word origins are used in medical terminology describing the uterus. "Metra" is the origin for terms such as pyometra, metritis (inflammation of the uterus), and all those ending in -metrium, such as endometrium (mucosal layer), myometrium (muscular layer), and mesometrium (mesothelial membrane surrounding the uterus). Its specific use may have been in the relationship of the uterus to motherhood (meter, mother) and/or perhaps to a meter or measure of time—i.e., the periodic nature of its processes.

"Hystera" is the origin for terms such as hysterectomy (surgical excision of the uterus) and hysteroscopy (endoscopic examination of the uterus). As the adjective *hysteros* means "coming after" or "following," the specific use of this word may have been in relation to the position and/or function of the uterus—caudal, and in humans inferior, to most of the other viscera, and producing offspring after mating.

Other than convention, there is no reason for the use of 2 different word origins for the same organ in medical terminology.

Treatment

Ovariohysterectomy (surgical removal of the ovaries and uterus; "neuter/spay") was performed under general anesthesia by means of a ventral midline approach. The patient made an uneventful recovery.

Although pyometra may be treated nonsurgically, this approach is generally reserved for young, systemically healthy bitches and queens intended for breeding. In an older, systemically ill patient, as in this case, ovariohysterectomy is recommended. Not only does it immediately remove the source of infection, but it also eliminates the potential for recurrence, which remains a risk when pyometra is treated nonsurgically.

Anatomical features in canids and felids

Introduction

The female reproductive system includes the ovaries, uterine tubes (called oviducts or fallopian tubes in humans), uterus (horns, body, and cervix), vagina, and vulva. The ovaries, uterine tubes, uterine horns, and most or all of the uterine body are contained within the peritoneal cavity. The cervix and vagina are pelvic organs and retroperitoneal in location, except during late pregnancy, when the gravid uterus may pull the cervix cranioventrally over the pelvic brim into the abdomen and remain officially retroperitoneal in relationship to the peritoneal cavity.

All of these organs are described and illustrated below. The clinical emphasis of this section is on the anatomical considerations when performing an ovariohysterectomy in a healthy bitch or queen, with additional notes on patients with pyometra. The anatomical considerations in mating, pregnancy, parturition, and lactation are described in greater detail in Case 6.3.

Function

The structures of the female reproductive system facilitate the production of the future generation: insemination (vulva and vagina); production, maturation, release, transport, and fertilization of the ova (ovaries and uterine tubes); support of the embryos/fetuses during gestation (uterus and cervix); and parturition (uterus, vagina, and vulva). In addition, the vestibule of the vulva provides external protection for the external urethral orifice.

Broad ligaments

Within the peritoneal cavity, the reproductive organs are attached to the dorsolateral body wall by a deep fold of peritoneum called the **broad ligament**. It is analogous to the mesentery which enfolds the intestinal tract, except that the mesentery is a singular structure, whereas the broad ligaments are paired—left and right. In addition to the ovary, uterine tubes, and uterus, the broad ligament encompasses the ovarian and uterine vessels, lymphatics, and nerves.

A common point of confusion is that the name used for the broad ligament changes along its length, depending on what structure it is supporting: **mesovarium** (ovary), **mesosalpinx** (uterine tube), and **mesometrium** (uterine horns and body). The term "broad ligament" is often used for just the mesometrium, but the mesovarium and mesosalpinx are also part of the broad ligament proper.

These peritoneal structures (mesovarium, mesosalpinx, and mesometrium) are readily separated using finger pressure during ovariohysterectomy (OHE). In the process, it is important to remember that the ovarian and uterine vessels, lymphatics, and nerves all track within the broad ligament.

Adding to the confusion, this peritoneal reflection forms the serosal surface of the uterus and continues for a short distance, laterally, to form the **round ligaments of the uterus**. It also forms the **ovarian bursa** which envelops the ovary (described later). Other bands of connective tissue (e.g., the suspensory and proper ligaments of the ovary) are of clinical importance. All of these structures are described in their relevant sections below. The point to emphasize is the commonality or continuity of these structures.

Ovaries and their blood supply, lymphatics, and innervation (Fig. 6.2-3)

During OHE, the best way to locate the ovaries is to identify the uterine horns and trace one horn cranially to its associated ovary, then repeat with the other uterine horn.

In intact bitches and queens, the paired **ovaries** (left and right) typically lie just caudal to their respective kidneys, about halfway between the last rib and the ilial crest. As with the kidneys, the right ovary is usually located slightly further cranially than the left ovary. (Remember: "The left one was left behind.") In multiparous bitches and queens (i.e., those who have had multiple litters), the ovaries may be located further caudally and ventrally than in nulliparous females (those who have never given birth), owing to stretching of the broad ligament during pregnancy.

The size of the ovaries varies according to the stage of the estrous cycle. In the follicular phase (proestrus and estrus), multiple follicles grow as a cohort, each containing one ovum. At ovulation, these ova are released into the uterine tube and the empty ovarian follicles transform into corpora lutea (singular, corpus luteum; literally, "white body"). The estrous cycle in bitches and queens as it relates to clinical anatomy are described further in Case 6.3.

The kidneys are located in the retroperitoneal space, whereas the ovaries are positioned within the peritoneal cavity, attached to the dorsolateral body wall by the particular part of the broad ligament, the **mesovarium**. This peritoneal reflection also forms a pouch, the **ovarian bursa**, which, in cats, partially covers the ovary. In dogs, the ovarian bursa completely envelops the ovary, except for a small ventral opening, in which the infundibulum of the uterine tube is located (described later).

OHE VS. OVE

Ovariohysterectomy (OHE)—surgical removal of the ovaries and uterus—via a ventral midline incision is the procedure commonly known as a "spay" or "neuter." It is the most widely used method for reproductive sterilizing in female dogs and cats.

Spay or spey originates from the Old French *espeer* (cut with a sword), which itself originates from the Latin *spathe* (broad blade). Why spay/spey is used only in relation to the neutering of female animals is lost to time.

Ovariectomy (OVE)—surgical removal of only the ovaries—via laparoscopy is becoming more widely used, especially in Europe. While it requires specialized equipment and training and takes longer to perform, it typically causes less post-operative pain, carries less risk of serious wound breakdown, and results in faster return to normal activities than a conventional spay. An initial concern with OVE was that leaving the uterus in situ would predispose the dog/cat to pyometra at some point, post-operatively. However, pyometra requires the hormonal influence of intact ovaries, so OVE is a viable alternative to OHE for neutering healthy female dogs and cats.

Note: "flank spay" (OHE via the lateral abdominal wall) is not recommended, for reasons discussed in Section 5.8.

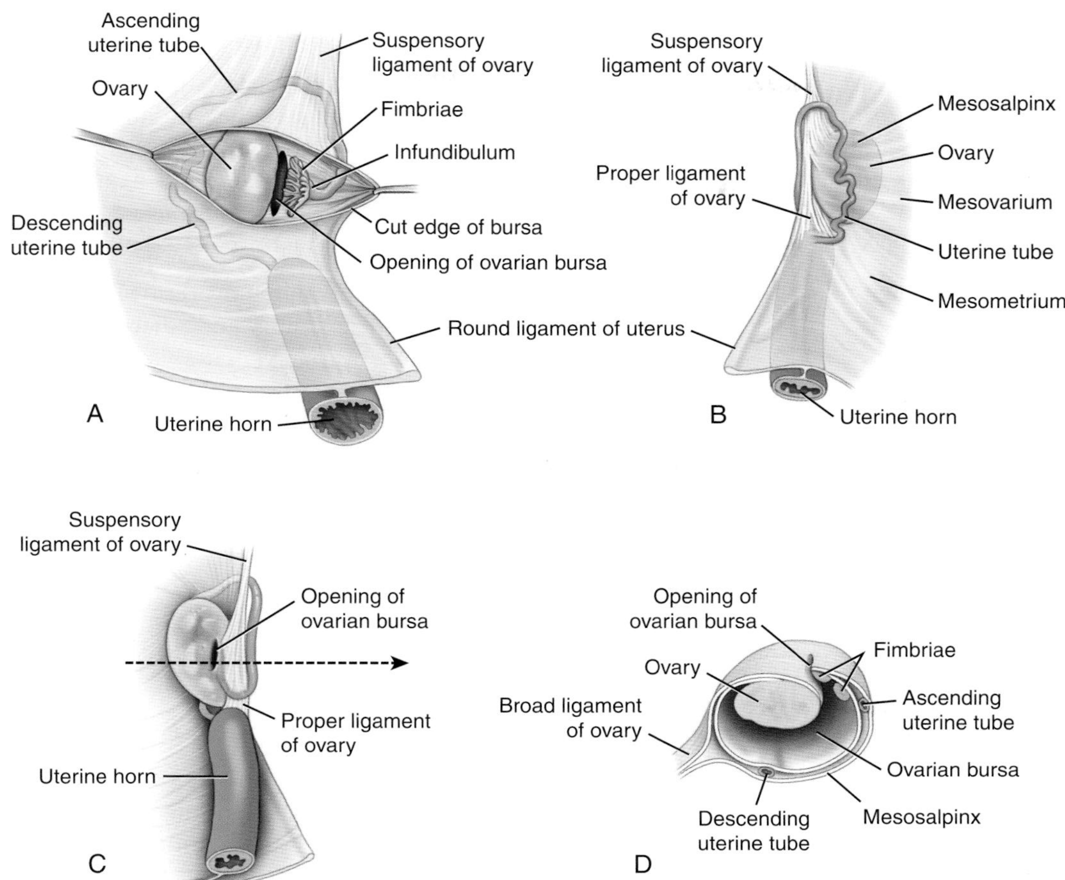

FIGURE 6.2-3 Canine and feline ovaries and their associated supporting structures—mesovarium, ovarian bursa, uterine tube, cranial aspect of uterine horn, suspensory ligament, proper ligament, and "pedicle" (containing ovarian vessels, lymphatics, and nerves). (A) Ovarian bursa opened to reveal ovary within ovarian bursa. (B) Ventrolateral aspect left ovary in situ. (C) Dorsomedial aspect left ovary. (D) Cross section of ovary within the bursa taken at site of the *dashed arrow* in (C).

Cranially, the ovary is attached to the dorsolateral body wall at about the last rib by a band of connective tissue, the **suspensory ligament of the ovary**, which runs between the two peritoneal layers of the mesovarium and thereby forms the cranial free border of the broad ligament.

> During OHE or OVE, the suspensory ligament of the ovary must be manually separated or carefully transected in order to remove the ovary. During open surgery, such as routine OHE, the suspensory ligament can usually be stretched to failure by placing an index finger under it and applying gentle, persistent traction until the ligament gives way. Care must be taken to avoid rupturing the ovarian pedicle (see below) in the process.

Caudally, the ovary is attached to the cranial end of the uterine horn by a caudal continuation of the suspensory ligament, here called the **proper ligament of the ovary**. It is continuous with the round ligament of the uterus (described later).

The ovarian blood vessels, lymphatics, and nerves all run in close proximity through the mesovarium. For this reason, the mesovarium is often termed the **ovarian pedicle** for surgical procedures involving the ovary. The paired **ovarian artery** is a ventral branch of the abdominal aorta that arises midway between the renal artery and the external iliac artery. The size of the ovarian artery varies enormously with the reproductive phase and, particularly,

During OHE or OVE, the ovarian pedicle must be ligated before the ovary is removed. Even in prepubertal patients, bleeding from the ovarian artery may result from inadequate hemostasis of the ovarian pedicle. To minimize the amount of connective tissue included in the ligature, the mesovarium is carefully fenestrated (perforated) immediately caudal to the ovarian vessels before the clamp is applied.

with pregnancy status and stage. In addition to supplying the ovary, small branches also supply the oviduct and uterus, as the ovarian artery continues caudally within the broad ligament and anastomoses with the uterine artery (described later) near the cranial end of the uterine horn.

The paired **ovarian veins** primarily drain each ovary and adjacent oviduct. The right ovarian vein drains directly into the caudal vena cava, and the left ovarian vein drains into the left renal vein. The ovarian **lymphatic vessels** drain into the **lumbar lymph nodes**. The ovaries are innervated by sympathetic nerves from the **renal** and **aortic plexuses**. They accompany the ovarian vessels within the mesovarium.

Applying traction to the suspensory ligament of the ovary can cause a pain response in an anesthetized animal, so a deeper plane of anesthesia or adjunctive analgesics may be needed at this stage of the ovariectomy procedure.

Uterine tubes (Fig. 6.2-3)

The paired uterine tubes (left and right) are known as oviducts or "fallopian tubes" in humans. These narrow tubes transport the ova from the ovary to the uterine horn, following ovulation.

Although the infundibulum of the uterine tube lies near the ovary and is largely contained within the ovarian bursa, the proximal end of the uterine tube is open to the peritoneal cavity. In bitches and queens with pyometra, additional care must be taken to avoid applying pressure to the uterine horns and inadvertently driving purulent material into the oviducts and thus into the peritoneal cavity.

Proximally, the uterine tube widens into a funnel-shaped **infundibulum**, which is fimbriated (fringed). The **ostium**, or proximal opening of the infundibulum, occupies the ventral opening in the **ovarian bursa**, so during ovulation, the infundibulum "catches" the ovum somewhat like a baseball mitt and guides it across the narrow gap between ovary and uterine tube.

After successful mating, fertilization of each ovum occurs in the uterine tube. The zygotes (fertilized ova), and subsequently the early embryos, are then transported through the uterotubal junction into the uterine horn for implantation. The uterine tube remains in close association with the ovary for almost its entire length. It partially wraps around the ovary, beginning on the ventral aspect, then coursing cranially along the medial margin of the ovary to the cranial pole before doubling back along the dorsal aspect of the ovary on its way to meet the uterine horn just caudal to the caudal pole of the ovary. In the process, its portion of the broad ligament, the **mesosalpinx**, contributes to the wall of the **ovarian bursa**. The arterial and nerve supply and the venous and lymphatic drainage for the uterine tube run in the mesosalpinx, and are essentially as described for the ovary.

During routine OHE in a healthy dog or cat, the uterine tube is removed along with the ovary; no separate procedure pertaining to the uterine tube is required. However, during OHE in a patient with pyometra, clamping the proper ligament of the ovary (at the caudal pole of the ovary) together with the uterine tube as soon as the ovary is located, prevents the unintended expression of purulent material into the peritoneal cavity during manipulation of the infected uterine horns.

Uterus and its blood supply, lymphatics, and innervation (Figs. 6.2-4 and 6.2-5)

In bitches and queens, the **uterus** forms an elongated Y: The relatively short **uterine body** (average, 2–3 cm or 0.8–1.2 in. long in dogs) forms the base of the Y, and the pair of long, narrow **uterine horns** (average 12–15 cm or 4.7–5.9 in. long in dogs) forms the left and right arms of the Y. The uterine horns are 5–6 times the length of the uterine body, as the multiple fetuses typical of these litter-producing species develop within the uterine horns.

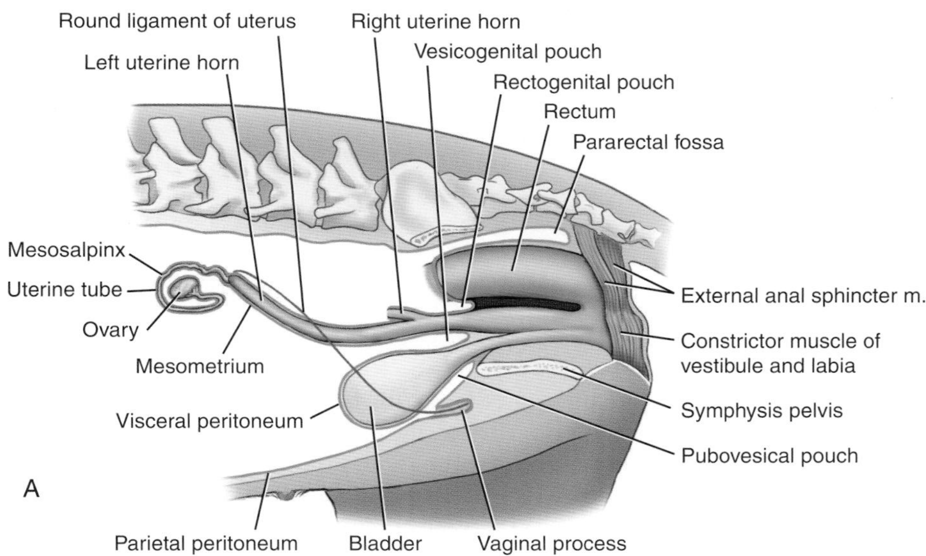

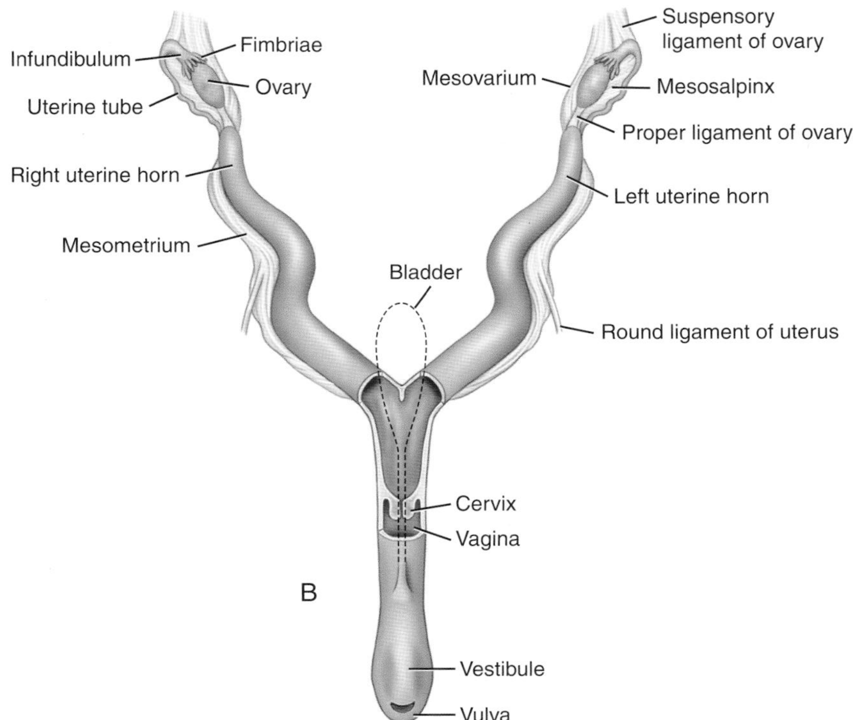

FIGURE 6.2-4 Female reproductive tract lateral (A) and dorsal (B) views, depicting the relative size, length, relationships, and anatomy.

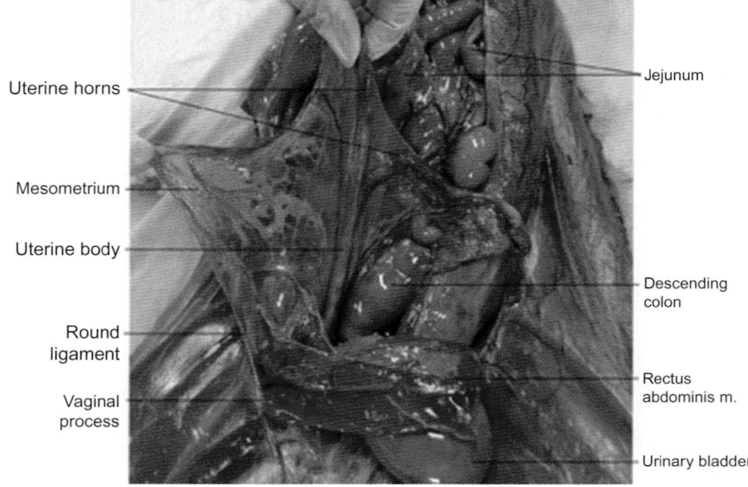

FIGURE 6.2-5 Reproductive tract of a nonpregnant bitch, ventral view. The right uterine horn and broad ligament have been shifted and the urinary bladder reflected caudolaterally for better visualization of the uterus. Key: *m.*, muscle.

The **uterine horns** diverge as they extend cranially from the uterine body and approach the caudal pole of their respective ovary. As mentioned above, the ovary is attached to the cranial end of the uterine horn by the **proper ligament of the ovary**, and the caudal end of the **uterine tube** is continuous with the uterine horn in this region, called the **uterotubal junction**. At their caudal convergence with the uterine body, the left and right uterine horns are connected medially by a triangle of mesometrium (Fig. 6.2-5).

In nonpregnant bitches and queens, the uterine horns lie dorsal to the small intestine and greater omentum and ventral to the colon. During pregnancy, the gravid uterine horns occupy increasingly more of the abdominal cavity, and in late pregnancy, they displace the small intestine to lay side-by-side along the ventral body wall.

In "shelter" practice and in unspayed/intact dogs and cats that have recently been adopted, it is common to find that the bitch or queen is pregnant. When incising the linea alba during OHE, remember the position and relationship of the gravid uterus relative to the ventral abdominal wall (see Case 6.3). This is similarly important when performing on OHE when there is a pyometra with pus-filled friable uterine horns.

Most of the **uterine body** is typically located within the abdominal cavity in nonpregnant bitches and queens, but the caudal part (including the cervix) is situated within the pelvic cavity. During the latter half of pregnancy, the

ASCENDING INFECTION OF THE UTERUS

Escherichia coli is one of the most common causes of pyometra in bitches and queens because it is part of the normal fecal microflora and is commonly found in the perineal area. Ascending infection of the uterus by *E. coli* is most likely to occur during estrus, as the cervix is open currently.

However, clinical signs of pyometra typically do not appear until at least 1 month after estrus in bitches and at least 2 weeks in queens, as the conditions necessary for pyometra to progress include the influence of progesterone on the cervix and endometrium. The patency of the cervix is largely governed by serum progesterone levels, making closed-cervix pyometra more likely in early- to mid-diestrus, whereas open-cervix pyometra is more commonly observed toward the end of diestrus, when serum progesterone levels are declining. Closed-cervix pyometra is a more serious condition than open-cervix pyometra, as purulent material cannot drain from the uterus into the vagina to exit via the vulva if the cervix is closed.

gravid horns draw the entire uterus, including the cervix, cranioventrally over the pelvic brim into the abdominal cavity. In multiparous bitches and queens, the entire uterine body may be located within the abdominal cavity in all stages of the reproductive cycle, owing to the elongation of the broad ligaments.

When performing an OHE, the uterine body is clamped, ligated, and transected as far caudally as possible within the abdominal cavity while ensuring secure ligation of the uterine vessels. If greater access is needed, the laparotomy incision is extended.

The three primary layers of the uterine wall are (from inner to outer) the **mucosa**, **muscularis**, and **serosa**. The muscular layer of the uterus, the **myometrium**, is most significant in relation to conception and pregnancy. Coordinated contractions of the longitudinal and circular smooth muscle layers of the myometrium are necessary for the passage of sperm from the cervix to the oviduct (where fertilization of the ova takes place), as well as for normal parturition.

The mucosal layer, or **endometrium**, needs special mention, as it is the thickest of the three primary layers of the uterine wall. It consists of low columnar epithelium interspersed with the ducts of long, branched, tubular glands within the lamina propria (connective tissue layer deep to the mucosa) and with smaller endometrial glands or crypts. As described in Case 6.3, the endometrium enables embryonic implantation and supports fetal development during pregnancy. The uterine glands produce histiotrophe ("uterine milk"), which provides early nutrition to the embryo until the placenta takes over this role for the remainder of the pregnancy.

The **cervix** is the caudal-most part of the uterine body and it forms the junction between the uterus and the vagina. The lumen of the cervix slopes ventrally from the uterine body to the vagina, at an angle of 45–90 degrees, such that the vaginal opening of the cervix is located near the floor of the cranial vagina.

The cervix is primarily a discrete thickening of the circular smooth muscle layer of the uterine wall, which narrows the uterine lumen for a short distance (average, 1.5–2 cm or 0.6–0.8 in. in dogs) and regulates its diameter according to the animal's reproductive status. Throughout diestrus and during pregnancy, the circular muscle contracts under the influence of progesterone to "close" the cervix; during pregnancy, a mucus plug within the cervical lumen provides an extra barrier against ascending infection of the uterus. During estrus, the circular muscle relaxes under the influence of estrogen to "open" the cervix, thereby allowing sperm to enter the uterus if mating takes place. The circular muscle also relaxes in preparation for parturition, and the entire cervix dilates as the first fetus is expelled by contractions of the uterine and abdominal walls.

The abdominal portion of the uterus, including the horns and body, is contained within the uterine portion of the broad ligament, the **mesometrium**. As mentioned previously, a lateral fold of the mesometrium forms the **round ligament of the uterus**, which extends along the lateral margins of the uterine body and horns (Fig. 6.2-5). Cranially, it merges with the proper ligament of the ovary. Caudally, the round ligament extends toward, and in

During OHE, the mesometrium ("broad ligaments") must be separated in order to remove the uterus. These thin membranes are readily separated by lifting the uterine horns out of the abdominal incision. The more caudal portion at the uterine body, notably the round ligament of the uterus, is manually broken down before applying the clamps at the cervix. At the same time, it is important to avoid injurying the uterine vasculature, which runs in the broad ligament adjacent to the uterus.

UTERINE HORN OR JEJUNUM?

In healthy, nonpregnant bitches and queens, the first structure encountered on entering the peritoneal cavity during routine OHE is normally the jejunum or omentum. Particularly in diestrus (the period following estrus, marked by high progesterone levels), the nongravid uterine horns have a similar appearance to the jejunum, so the two structures can be easily confused at first glance. The uterine horn is easily distinguished from the jejunum by tracing it caudally to the uterine body.

most animals through, the inguinal canal to terminate as the **vaginal process** in the subcutaneous tissue adjacent to the vulva. This component parallels the vaginal tunic of the testis in males (see Case 6.4.)

Because the uterine artery enters the mesometrium near the cervix, applying a transfixation ligature to the base of the uterus is recommended during OHE. By anchoring the suture by incorporating the uterus, the ligature is prevented from slipping off the uterine stump after transection, a complication that can result in severe, potentially fatal, hemorrhage.

The uterus is supplied by the paired **uterine artery**, which is the main branch of the vaginal artery, itself a branch of the internal iliac artery. The uterine artery enters the mesometrium near the cervix and travels cranially along the uterine body and its respective uterine horn. Near the cranial end of the uterine horn, it anastomoses with the ovarian artery, as previously described. The **uterine veins** follow a similar course to the arteries, and both left and right uterine veins drain into the caudal vena cava. The uterine **lymphatic vessels** drain into the **lumbar** and **internal iliac lymph nodes**. The uterus is richly supplied by the sympathetic nerves from the **hypogastric plexus** and by the parasympathetic nerves from the **pelvic plexus**.

Vagina and vulva and their blood supply, lymphatics, and innervation (Figs. 6.2-6 and 6.2-7)

This transverse ridge is a site where abnormalities and anomalies such as strictures and septa, or even just hymenal remnants, may hamper passage of a speculum or endoscope through the vaginal canal. They can also make urethral catheterization difficult, as the external urethral orifice is located just caudal to this site (see below).

The vagina and vulva (including the labia, vestibule, and clitoris) are the pelvic and external components of the female reproductive tract. Depending on breed and age, the **vagina** in dogs is 10–30 cm (3.9–11.8 in.) in length. It extends caudally from the cervix to a transverse mucosal ridge on the floor of the canal, just cranial to the ischiatic arch, which marks the transition from the vagina to the vestibule. This ridge also marks the position of the **hymen**, a protective membrane across the caudal vagina in unmated females.

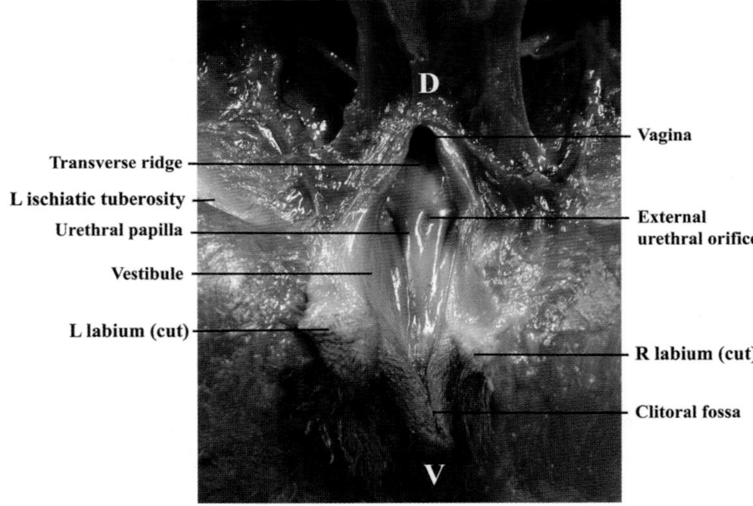

FIGURE 6.2-6 The vestibule of a bitch, caudal view. Most of both labia (left and right) have been removed for better visualization. The external urethral orifice is located in the center of the triangular urethral papilla, which is situated on the midline in the vestibule, just caudal to the transverse ridge of mucosa that marks the junction between the vagina and the vestibule. Key: *L*, left; *R*, right; *D*, dorsal: *V*, ventral.

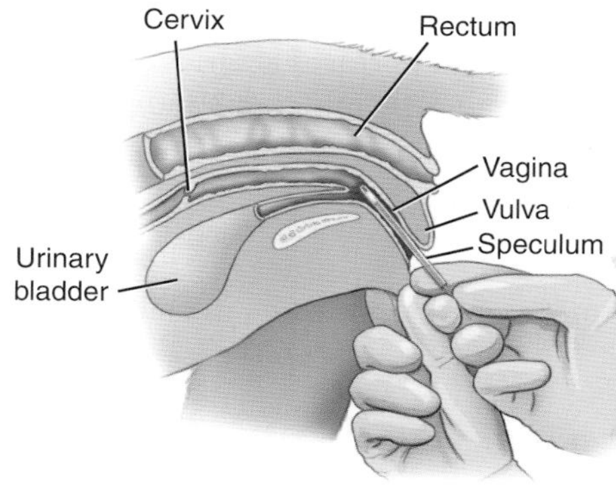

FIGURE 6.2-7 Correct positioning for insertion of a vaginal speculum.

The caudal aspect of the **cervix** protrudes a short distance (0.5–1 cm or 0.2–0.4 in. in dogs) into the cranial vagina. Ventral to the cervix is a small cranial recess in the vagina, the **fornix**, which must not be mistaken for the cervical orifice.

The mucosal surface of the vagina has a series of longitudinal folds that enable the vagina to dilate during copulation and parturition. The outer surface of the vagina is covered by peritoneum, cranially, and by loose connective tissue, caudally. Although the vagina is entirely contained within the pelvic cavity, shallow caudal reflections of peritoneum extend a short distance into the pelvic canal as the **rectouterine pouch** (between rectum dorsally and uterus/vagina ventrally) and the **vesicouterine pouch** (between uterus/vagina dorsally and urinary bladder ventrally). The caudal aspect of the vagina is loosely attached by connective tissue to the rectum dorsally and the urethra ventrally.

As with the more cranial structures, the vasculature, lymphatics, and nerves supplying the vaginal tract within its peritoneal or connective tissue coverings. The **vaginal artery (paired)**, which supplies the vagina, urethra, and vestibule, is a branch of the internal iliac artery. The paired **vaginal vein** drains into the internal pudendal vein. The vaginal **lymphatic vessels** drain into the **internal iliac lymph nodes**. The vagina is innervated by autonomic nerves from the **pelvic plexus** and by sensory branches of the **pudendal nerve**.

The **vestibule** is a short, vertically oriented chamber (2–6 cm or 0.8–2.4 in. long in dogs) between the vagina and the labia ("lips") of the vulva (Fig. 6.2-6). Clinically, its most important feature is the **urethral papilla**, containing the **external urethral orifice**. Immediately caudal to the low transverse ridge that marks the caudal end of the vagina, the urethra opens into the cranial aspect of the vestibule within a small, triangular mound, the urethral papilla. In dogs, the external urethral orifice is located, on average, 4–5 cm (1.6–2.0 in.) craniodorsal to the ventral commissure of the vulva.

> The urethra is easily catheterized in bitches and queens by gently parting the labia to visualize the external urethral orifice within the urethral papilla in the cranial aspect of the vestibule.

The **labia** define the external boundaries of the vulva. The left and right labia meet dorsal and ventral to the external opening of the reproductive tract, forming the **dorsal** and **ventral commissures**. The ventral commissure is located ventral to the ischiatic arch. The **clitoris**, a small mound of erectile tissue, analogous to the penis, is located in the caudoventral aspect of the vestibule, just dorsal and cranial to the ventral commissure. It is readily visualized by gently parting the labia ventrally.

> When a vaginal speculum is used, it is important to consider the vulvar and vaginal anatomy. Owing to the position of the vulva in relation to the ischiatic arch, the speculum must be introduced through the vulva at a steep angle, pointing approximately 80 degrees dorsally from the horizontal plane (Fig. 6.2-7). Once the tip of the speculum reaches the pelvic floor, identified by the transverse ridge just dorsocranial to the urethral papilla, the speculum is reoriented horizontally as it is advanced into the vagina.

The vulva is supplied by the paired **vaginal branches of the internal pudendal arteries** and further paired branches of the internal pudendal arteries including the **ventral perineal arteries** and **arteries of the clitoris**. The **hypogastric** and **pelvic nerves** supply autonomic innervation, and the **pudendal** and **genital nerves** supply sensory innervation to the vulva.

Muscles of the female perineum (Fig. 6.2-8)

The **perineum** is the region of the pelvic outlet and is bounded by the tail dorsally, the ischiatic tuberosities laterally, and the ischiatic arch ventrally. The muscles of the female perineum include those that maintain closure of the external orifices of the female reproductive tract and the digestive tract. They also include muscles of the pelvic diaphragm (**levator ani** and **coccygeus**).

> The pelvic diaphragm helps prevent contents of the pelvic cavity being displaced ("herniated") into the region under the skin of the perineum when intraabdominal pressures are high, such as during straining or parturition.

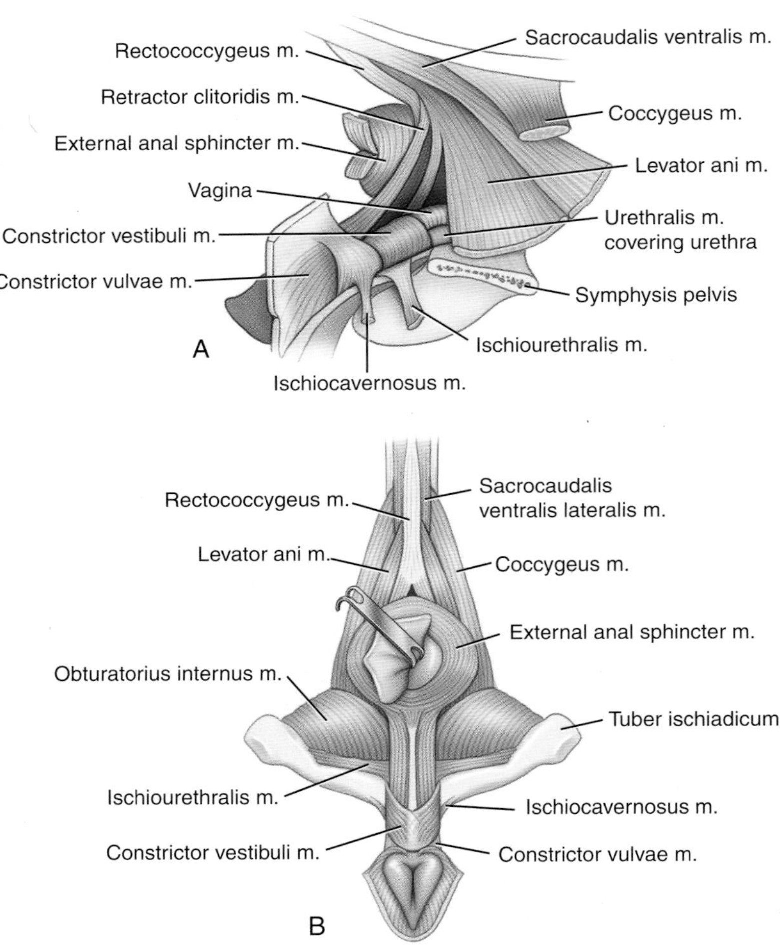

FIGURE 6.2-8 Muscles of the female perineum. (A) Medial view. (B) Caudal view.

Selected references

[1] Hagman R. Canine pyometra: what is new? Reprod Domest Anim 2017;52:288–92.
[2] Hagman R. Pyometra in small animals. Vet Clin N Am Small Anim Pract 2018;48:639–61.
[3] Hollinshead F, Krekeler N. Pyometra in the queen: to spay or not to spay? J Feline Med Surg 2016;18:21–33.
[4] Kutzler MA, Krekeler N, Hollinshead FK. Pyometra. In: Monnet E, editor. Small animal soft tissue surgery. Ames, IA: Wiley-Blackwell; 2013. p. 623–34.

CHAPTER 6

CASE 6.3

Dystocia and the Mammary Gland

Natali Krekeler[a], Helen M.S. Davies[a], and Cathy Beck[b]
[a]Veterinary Biosciences, University of Melbourne, Melbourne, Victoria, AU
[b]Veterinary Hospital, The University of Melbourne, Melbourne, Victoria, AU

Clinical case

History

A 4-year-old female intact Labrador Retriever was presented for dystocia. She had already whelped 6 puppies over the past 9 hours, but the last puppy was born 2 hours ago. Following its birth, the bitch labored weakly for about 20 minutes, but then stopped and had no active labor since. One puppy was still believed to be in utero (L. within the uterus) based on an abdominal radiograph taken a week before her whelping date, which showed the presence of at least 7 fetuses (Fig. 6.3-1).

> If it has been at least 2 hours since the last puppy or kitten is delivered and more are expected based on radiography in late pregnancy, an examination is recommended for the health of the mother and the viability of the remaining fetus(es).

Physical examination findings

The bitch was quiet, but alert. She was panting and tachycardic (heart rate: 164 bpm).

Differential diagnoses

Obstructive dystocia, uterine inertia (primary or secondary)

> Secondary uterine inertia is highly likely in this case, as the bitch seemed to have whelped the first 6 puppies normally but then failed to produce enough abdominal and uterine contractions to expel the remaining fetus. One possible cause was malpresentation or malposition of the fetus (obstructive dystocia), resulting in ineffective contractions and, finally, exhaustion of the uterine muscles. Another possibility was simple exhaustion after whelping the first 6 puppies. In either case, hypocalcemia and/or hypoglycemia are very common and contribute to uterine inertia.

TERMINOLOGY

Whelping is the term used by veterinarians and breeders to describe normal parturition in dogs. It is comparable to foaling in horses, calving in cows, etc. The term used for normal parturition in cats is queening.

Parturition is the medical term for the normal process of giving birth; parturient is the adjective. Periparturient pertains to the hours or days before or after parturition. Pre-partum and post-partum describe the periods before and after parturition, respectively.

Dystocia is the term used to describe abnormal (delayed or exceptionally difficult) parturition. However, its opposite (Gr. *eutocia* normal labor) is seldom used.

Pregnancy and parturition pertain to the mother, gestation and birth (nativity) to the offspring. For example, the fetal puppy or kitten gestates while the bitch or queen is pregnant. The first few days after birth is the neonatal period for the puppy or kitten and the post-partum period for the bitch or queen.

Comparative Veterinary Anatomy: A Clinical Approach. https://doi.org/10.1016/B978-0-323-91015-6.00041-8

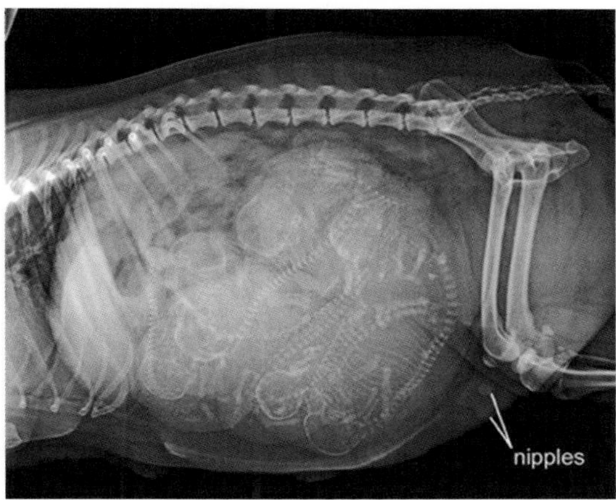

FIGURE 6.3-1 Abdominal radiograph of a bitch during late pregnancy. In addition to the multiple fetal skeletons in the uterus, note the mammary development along the ventral midline (2 of the caudal nipples are labeled). (Image courtesy of Dr. Cathy Beck, University of Melbourne.)

Diagnostics

Transabdominal ultrasonography revealed a fetus within the uterus. The fetus was still alive and had an appropriate heart rate (222 bpm).

An ultrasound examination is strongly advised in order to assess fetal viability. Fetal heart rates in uncompromised, full-term puppies are > 220 bpm. Heart rates < 180 bpm are a sign of fetal distress, and those < 160 bpm are an indication for immediate cesarean section (surgical removal of the fetus(es) via the abdominal wall). The fetus closest to the cervix is often the most compromised during dystocia and, therefore, should always be carefully evaluated. However, using ultrasound to determine the number of remaining fetuses is unreliable at this stage; abdominal radiography is needed if the number of fetuses is in doubt.

Examination of the cranial vagina using a vaginal speculum showed no part of the fetus to be within the cervical canal. (Insertion of a vaginal speculum is described and illustrated in Case 6.2.) In addition, the bitch failed to respond to

DIAGNOSTIC IMAGING IN PREGNANCY

Ultrasonography is the method of choice for pregnancy diagnosis in bitches and queens. In both these litter-bearing species, multiple fetuses per pregnancy is normal. In fact, singletons (single fetus) may increase the likelihood of dystocia because a single fetus can grow larger than multiples (> 1 fetus per litter) typically do.

The first pregnancy exam is commonly performed around 4 weeks after the luteinizing hormone (LH) surge (start of estrus; described later). The gestational sacs normally appear as spherical structures, each with an embryo inside that has definable cardiac activity. Enumeration of gestational sacs, and thus number of fetuses, is most accurate before day 35 post-LH-surge; it becomes unreliable thereafter because of superimposition of the fetuses.

Radiography is useful for pregnancy diagnosis in bitches and queens, starting about 6 weeks after the LH surge, when fetal ossification becomes apparent. Radiography is commonly used in the last week of pregnancy to determine the number of puppies or kittens in the pregnancy.

The Ferguson reflex is a neuroendocrine reflex that is initiated by pressure on, or stretching of, the wall of the cervix or cranial vagina. It results in the release of oxytocin from the pituitary gland, which stimulates uterine contractions in the normal parturient bitch or queen. Speculum examination is of additional help because it may reveal the presence, presentation (position), and size of the fetus in the cervical canal.

If the uterine contractions do not resume within a few minutes after resolving the maternal tachycardia, a small dose of oxytocin can be administered. This is usually unnecessary once the hypocalcemia and hypoglycemia are corrected.

this examination with uterine contractions (termed the "Ferguson reflex"), which was a strong clinical indicator of hypocalcemia.

Diagnosis

Dystocia due to secondary uterine inertia

Treatment

Calcium borogluconate solution was administered IV, slowly "to effect"—i.e., until the maternal heart rate slowed to between 60 and 90 bpm and uterine contractions resumed.

The bitch subsequently gave birth to a final live puppy, followed by the fetal membranes. She recovered uneventfully and nursed her litter without further problems.

Anatomical features in canids and felids

Introduction

The female reproductive system includes the ovaries, uterine tubes, uterus (horns, body, and cervix), vagina, and vulva. These organs are described and illustrated in Case 6.2. The following is a discussion of female reproductive anatomy as it relates specifically to conception, pregnancy, parturition, and lactation. The canine and feline estrous cycles are also discussed as they relate to clinical anatomy. The mammary glands are integumentary structures that are present in both females and males, but they have an exclusively reproductive function specific to post-partum females, so they are described in this section as well.

Function

All the structures of the female reproductive system facilitate the production of a future generation. The vulva and vagina enable insemination and, thus, conception. The ovaries and uterine tubes participate in the production, maturation, release, transport, and fertilization of the ova. The uterus is the site of gestation for the embryos/fetuses, and it supports the fetal membranes, most intimately the chorioallantois. Parturition primarily involves the uterus, vagina, and vulva, with contributions from the muscles of the abdominal wall. Post-partum, the female mammary glands nourish the neonates, ensuring the survival of the newborns until the puppies or kittens are old enough to process solid food.

Canine estrous cycle

The canine estrous cycle is unique in that it lacks the luteolytic mechanism which, in other domestic animals and humans, allows the nonpregnant female to undergo another estrus or ovulation within 1–2 weeks after luteolysis.

Luteolysis is the process by which the **corpus luteum** (CL) is lysed ("declines" or wanes) at the end of diestrus, allowing the development of another ovarian **follicle**. The CL is the ovarian tissue that replaces the follicle after the release of the ovum during ovulation. The CL is responsible for progesterone production that hormonally defines diestrus or the interestrous phase of the estrous cycle.

Canids have an average interestrous interval of between 7 (domestic dogs) and 12 months (wild canids such as wolves and coyotes). They are therefore defined as "monoestrous" animals, whereas cats are polyestrous animals (described later). Because the interestrous interval in domestic dogs is about 7 months, most bitches are normally in estrus approximately twice a year, unless neutered.

The estrous cycle in both monoestrous and polyestrous animals includes 4 distinct phases: proestrus, estrus, diestrus, and anestrus.

> For each of the phases of the estrous cycle, the noun (e.g., estrus) ends with -us, whereas the adjective—e.g., estrous cycle—ends with -ous or with -ual, e.g., estrual phase.

Proestrus in dogs is characterized externally by swelling of the vulva and sanguineous (L. *sanguinolentus* blood-tinged) vulvar discharge. During this phase, **ovarian follicles** enlarge and serum estrogen levels rise (produced by ovaries), reaching their peak in late proestrus. Proestrus lasts an average of 9 days, although it ranges from 3 days to 3 weeks.

The beginning of **estrus** is marked externally by the cessation of the bloody vulvar discharge seen during proestrus and transition to either clear or no discharge and behaviorally by the bitch's receptivity to mating (sometimes referred to as "standing heat"). During estrus, the bitch allows the male to mount, whereas during proestrus she is increasingly attractive to male dogs but usually does not allow mating. The start of estrus is marked hormonally by a rapid increase in serum luteinizing hormone (LH) produced by the pituitary gland; ovulation follows this "LH surge" by a few days. **Oocytes** are ovulated in a relatively immature state and need to undergo further maturation in the **uterine tube** for 48–72 hours before fertilization is possible. Given the hormonal requirements for ovulation, the need for oocyte maturation, and the sequential release of multiple ova in this litter-bearing species, estrus in bitches typically lasts about 8 days.

> The LH surge is defined as the day when serum progesterone levels significantly rise (commonly >2 ng/mL). Serum progesterone levels rise slowly during proestrus and then sharply increase at the time of the LH surge; they continue to increase well into diestrus.

Diestrus follows estrus and lasts for approximately 58 days in pregnant bitches, at which point parturition begins. In nonpregnant bitches, diestrus lasts 60–90 days.

> The precise mating dates are usually known by breeders, but because the point of conception is difficult to determine accurately based solely on mating dates, the duration of pregnancy in bitches varies according to the definition used. Published figures are 56–58 days from the first day of diestrus (end of estrus), 64–66 days from the LH surge (start of estrus), and 58–72 days from the first mating.

Following diestrus, all bitches—post-partum and nonpregnant—enter **anestrus**, a period of relative ovarian quiescence, before the next proestrus phase commences several months later.

PSEUDOPREGNANCY

Pseudopregnancy, also known as "pseudocyesis" (Gr. *pseudēs-* false + Gr. *kyēsis* pregnancy), is a physiologically normal stage for a nonpregnant bitch in diestrus. It is usually "silent," although young bitches in particular may show typical physiological and behavioral signs of the periparturient period, such as mammary development, restlessness, nesting behavior, and caring for her "young" in the form of stuffed toys, 6–8 weeks after estrus. No treatment is necessary.

Pseudopregnancy occurs in queens following ovulation without fertilization. Pseudopregnant queens usually show no signs other than a longer-than-normal interestrous interval.

Feline estrous cycle

Queens are **seasonally polyestrous** females. They are reproductively active in the spring, summer, and autumn, during which the interestrous interval is 2–3 weeks, whereas winter represents a prolonged anestrous period.

Queens are considered **induced ovulators**; mating induces a pituitary LH surge that stimulates ovulation. **Proestrus** in queens is short (average duration 1–2 days), whereas **estrus** lasts about 1 week. If ovulation is triggered during this time and the queen becomes pregnant, **diestrus/pregnancy** lasts 60 days and concludes with parturition. If ovulation occurs without fertilization, multiple CL develop and progesterone is produced for up to 45 days—a state referred to as **pseudopregnancy**.

> Because queens are induced ovulators, they generally do not undergo the repeated nonpregnant diestrous phases seen in domestic animals and humans. However, oriental breeds spontaneously ovulate in up to 30% of cycles, in which case pseudopregnancy may occur—and those queens may be as susceptible to pyometra in a manner similar to bitches (see Case 6.2).

Queens that do not ovulate undergo a period of **postestrus** lasting 8–10 days, during which serum progesterone is not elevated, as no CLs are produced. This phase is followed by another proestrus.

Anatomical considerations in mating

Mating, or copulation, involves most of the female reproductive tract in dogs and the entire tract in cats (generally being induced ovulators). Following intravaginal insemination, transport of sperm through the cervix, uterine body, and uterine horns to the uterine tubes involves rhythmic contractions of the smooth muscle in the walls of the uterus and uterine tubes. The following discussion pertains to changes in the vagina and vulva in relation to the estrous cycle and mating.

The **vaginal mucosa** in bitches and queens is composed of nonglandular, stratified squamous epithelium. There are glands in the submucosa; however, the bulk of the vaginal mucus is of cervical origin. The vaginal epithelium undergoes cyclic changes according to the stage of the estrous cycle. For example, during anestrus in bitches, the squamous epithelium is only 2–3 cell layers thick; during proestrus, it proliferates to 12–20 cell layers and the outer layers cornify (the cells contain keratin and lose their nuclei) in response to rising estrogen levels. In late estrus/early diestrus, the keratinized cells desquamate (slough) into the vaginal lumen as estrogen levels fall. The vaginal epithelium then returns to being just a few cell layers thick during diestrus and anestrus.

The **vulva**—comprising the labia, vestibule, and clitoris—is composed of muscle, connective tissue, cavernous or erectile tissue, and adipose tissue; it is lined by mucosa and covered externally by relatively characteristic skin. The left and right **labia**, or lips, of the vulva meet at the mucocutaneous junction of each and form a complete seal for the vestibule during most of the estrous cycle. The vascularity of the vulva lends itself to congestion during estrus, particularly in bitches. The thin bands of skeletal muscle surrounding the vestibule merge dorsally with the external anal sphincter.

VAGINAL CYTOLOGY

During proestrus and estrus, the squamous cells of the vaginal mucosa increasingly become cornified. Therefore, vaginal cytology may be used to detect or confirm estrus, when necessary.

After parting the lips of the vulva, a saline-moistened sterile swab is gently inserted into the caudal vagina and then rotated over the vaginal mucosa to collect a sample of the surface cells. Any of the stains routinely used for blood smears can be used to check for cornified cells.

The **vestibule** of the vulva is similar in structure to the vagina, except that it has subepithelial lymphatic nodules which are more pronounced caudally. The submucosa of the vestibule has abundant lymphatic vessels and **cavernous tissue**, similar to the cavernous bodies (erectile tissue) of the penis, which undergo congestion during estrus. In bitches, this tissue is well developed and forms the **vestibular bulb**, which comprises discrete areas of erectile tissue homologous to the bulb of the penis in males (see Case 6.5). The **vestibular glands** in the mucosa and submucosa of the vestibule provide lubrication during copulation and also during parturition.

The paired **constrictor vestibuli** muscle wraps the wall of the vestibule, and in dogs it plays an important part in copulation, as this muscle is primarily responsible for the female component of "**tying**." The male dog is prevented from removing his penis from the vagina for a variable period (15–45 minutes) after ejaculation because the constrictor vestibuli muscles of the female contract around the penis, caudal to its bulbus glandis. Tying is not essential for conception, but it is a normal component of the mating process in dogs.

On the floor of the vestibule is the **clitoral fossa** which contains the **clitoris**. The clitoral fossa should not be mistaken for the external urethral opening, which lies several centimeters dorsally and cranially, immediately caudal to the transverse ridge of mucosa that marks the junction of the vestibule and vagina. The urethral opening is discussed in Case 6.2. In queens, the clitoris is a nonerectile, fibroelastic structure, whereas in bitches there is functional cavernous tissue that forms an erectile **corpus cavernosum** and **glans**, similar to the penis in males. There may even be an **os clitoridis**, analogous to the os penis in males, if the bitch has been exposed to endocrine disruptors such as androgens.

Maternal considerations: Pregnancy and parturition (Figs. 6.3-2 and 6.3-3)

Pregnancy and parturition are normal processes, reflected in the anatomy of the female reproductive tract and the accommodations that normally occur during pregnancy and parturition. Below are the most clinically relevant anatomical considerations, discussed in generally the same order in which the specific structures are described in Case 6.2.

The **broad ligament** contains varying amounts of smooth muscle, so it is able to accommodate progressive enlargement of the gravid uterus during pregnancy and allow displacement ventrally and cranially as gestation proceeds. During the latter half of pregnancy, the gravid horns pull the uterus, including the cervix, cranioventrally over the pelvic brim into the abdominal cavity. In late pregnancy, the gravid uterine horns lie side-by-side along the ventral body wall.

> Surgeons must use care when performing a cesarean section because the first abdominal structure encountered on entering the peritoneal cavity is likely one or both gravid uterine horns. Similar care is necessary when performing a hysterotomy (Gr. *hystera* uterus + Gr. *temnein* to cut) because the uterine wall is thin and unintended injury to the puppies and kittens may occur.

The smooth muscle component of the broad ligament allows the uterus to resume its normal position as it involutes (returns to normal size and shape) post-partum. However, in multiparous bitches and queens, the uterus may remain in the abdominal cavity because of the elongation of the broad ligaments from multiple pregnancies.

During the estrous cycle, the **ovaries** change in size and appearance according to the presence and regression of their functional structures (follicles and CL). The ovaries are small before puberty and during anestrus in adult bitches and queens. The ovaries become much more prominent when follicles (proestrus and early estrus) and CL (late estrus and diestrus) are present. In multiparous bitches and queens, the ovaries may be located further caudally and ventrally than in nulliparous and primiparous females, owing to stretching of the broad ligaments, including the mesovarium, and the suspensory ligament of the ovary during pregnancy.

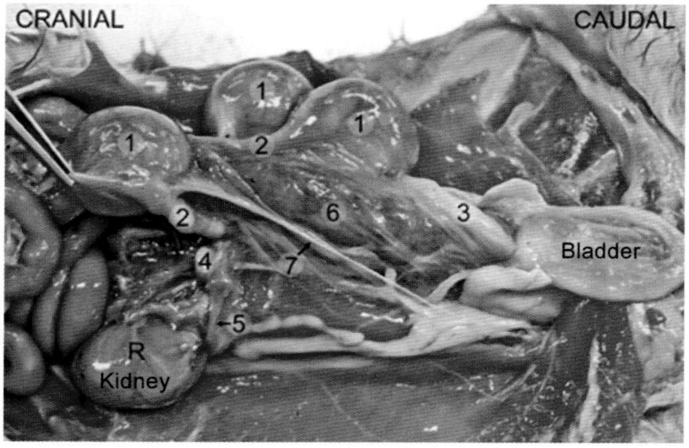

FIGURE 6.3-2 Reproductive tract of a pregnant queen partially in situ, right ventrolateral aspect. The uterus contains 3 fetal kittens (1), each within its own gestational sac. The urinary bladder has been reflected caudally. Key: *1*, fetus within uterine horn; *2*, right uterine horn (the left uterine horn is hidden in this image); *3*, uterine body; *4*, right ovary; *5*, suspensory ligament of the right ovary; *6*, broad ligament (overlying the descending colon); *7*, round ligament of the uterus. (Photo courtesy of Dr. Christina Murray, University of Melbourne.)

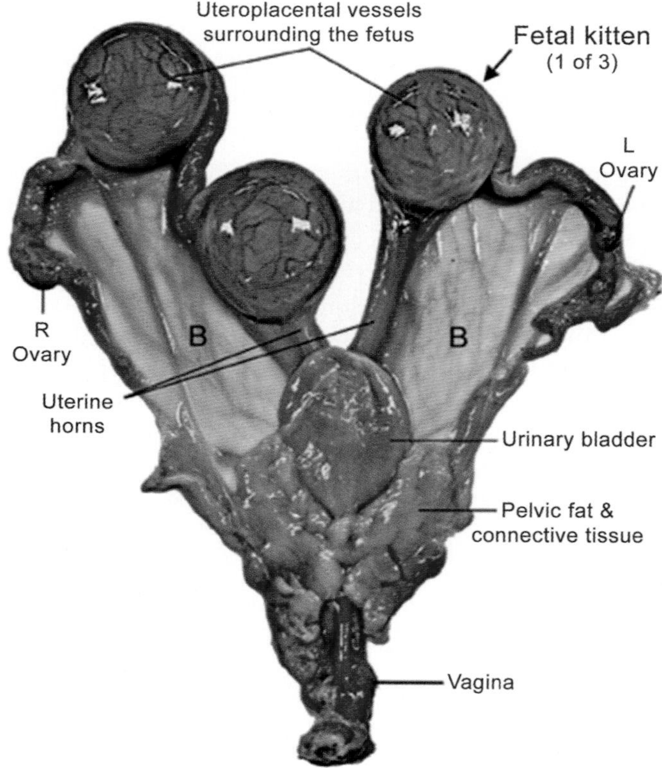

FIGURE 6.3-3 Reproductive tract of the pregnant queen shown in Fig. 6.3-2, ventral aspect. Two of the fetal kittens are in the right uterine horn and the third is in the left uterine horn. *B*, broad ligament; *L*, left; *R*, right. (Photo courtesy of Dr. Christina Murray, University of Melbourne.)

The **cervix** is normally closed during pregnancy. Throughout pregnancy, the circular muscle in the wall of the cervix contracts under the influence of progesterone to close the cervix. A thick mucous plug within the cervical lumen provides an extra barrier against ascending infection of the uterus during pregnancy. This plug is usually expelled as the cervix begins to relax in preparation for parturition. At this time, the circular muscle of the cervix relaxes in response to falling progestogen levels, and the entire cervix dilates as the first fetus is delivered by muscular contractions of the uterine and abdominal walls.

During parturition, the circular and longitudinal layers of smooth muscle in the uterine wall coordinate to move fetuses toward and through the cervix, into the vagina, and out of the dam through the vulva. Contraction of the abdominal wall muscles (called the "abdominal press") contributes by increasing the intraabdominal pressure, helping move the fetuses caudally similar to how the abdominal press facilitates defecation. Intermittent closure of the glottis (the entrance to the larynx) briefly increases the power of the abdominal press during the expulsion of the fetus through the cervix, which is the "bottleneck" or narrowest part of the birth canal.

Parturition is described as occurring in 3 stages: preparation (stage I), active labor (stage II), and expulsion of the fetal membranes (stage III). The **preparatory phase (stage I)** normally lasts 12–24 hours, the variation largely attributable to the lack of obvious signs. During this stage, uterine contractions progressively increase in strength, frequency, and productivity (caudal movement of the fetuses), but outward signs may be limited to restlessness, reclusive or nesting behavior, inappetence, panting, and clear or slightly sanguineous vulvar discharge. **Active labor (stage II)** may take anywhere from 1 to 24 hours to complete, depending on litter size. Each puppy or kitten is delivered separately, normally at intervals of 30–60 minutes. Typically, each set of **fetal membranes** (the allantochorion, described next) is delivered (in **stage III**) separately, either with the amnion or shortly after the puppy or kitten is delivered. However, it is also normal for several fetuses to be delivered before any of their allantochorionic membranes are passed.

Fetal considerations: Placentation and gestation (Figs. 6.3-4–6.3-6)

In female mammals, the developing young are nurtured within the uterus by the **placenta**, a specialized arrangement of semipermeable fetal membranes. Each fetus has its own separate placenta, so litter-bearing species such as dogs and cats typically have multiple placentas per pregnancy.

The placenta is the site of contact between fetal and maternal tissues, where bidirectional exchange (oxygen, fluid, nutrients, and waste) occurs. The maternal component of the interface is the **endometrium**, specifically the endothelium lining the endometrial vessels. The fetal component comprises the chorioallantois, the amnion, and the umbilical cord. The umbilical cord is described separately in the next section.

The outermost layer of the fetal membranes, which fuses with the endometrium, is the **chorion**. Deep to it is the **allantois**. The allantois forms as a diverticulum of the embryonic hindgut, so it gradually fills with fetal waste fluids (**allantoic fluid**) as gestation progresses. The allantoic membrane fuses with the chorion to form the **allantochorion** or **chorioallantois**, depending on whether one is describing this structure from the fetal or maternal side. These fetal membranes form before implantation.

EMBRYO OR FETUS?

There is no universal agreement as to when an embryo becomes a fetus. For our purposes, the embryonic stage may be defined most simply as that which begins shortly after fertilization and ends at implantation in the uterus.

The fetal stage of growth thus begins, and it ends at the time of birth, which begins the neonatal period for the newborn puppy or kitten.

Another definition for the transition from embryonic to fetal life is the end of organogenesis, but clinically that definition is less useful. It is more practical to define the fetal stage as the placental stage, because during birth the fetal part of the placenta is detached from the uterine wall and is, itself, expelled with or shortly after the fetus.

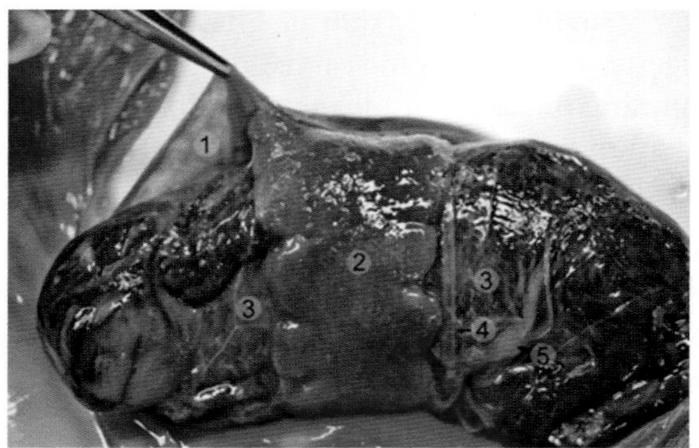

FIGURE 6.3-4 Late-gestational kitten within its chorioallantoic membrane, showing the zonary placentation in this species. Dogs have the same type of placentation. Note the connection between the uterine wall and the placental zone of the chorioallantois, adjacent to the tip of the forceps. Key: *1*, uterine wall, endometrial surface; *2*, placental zone of the chorioallantois; *3*, paraplacental zone of the chorioallantois; *4*, umbilical vessels; *5*, umbilical stalk. (Photo courtesy of Dr. Christina Murray, University of Melbourne.)

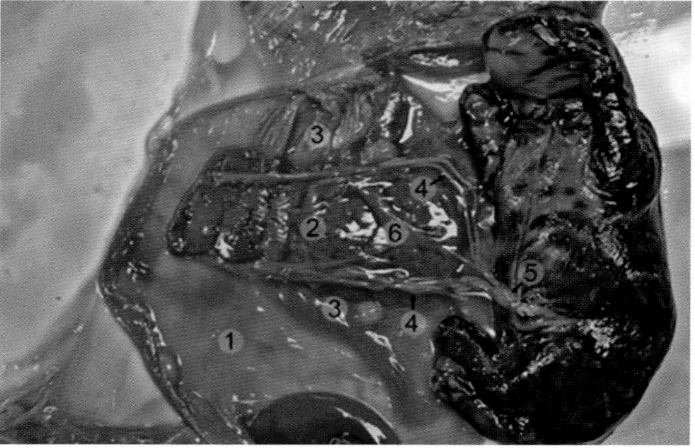

FIGURE 6.3-5 Same specimen as in Fig. 6.3-3. The chorioallantois has been incised and reflected to reveal the fetal kitten still within its amniotic sac, although the amniotic fluid has been removed. Key: *1*, uterine wall, endometrial surface; *2*, placental zone of the chorioallantois; *3*, paraplacental zone of the chorioallantois; *4*, umbilical vessels; *5*, umbilical stalk; *6*, remnant of the yolk sac. (Photo courtesy of Dr. Christina Murray, University of Melbourne.)

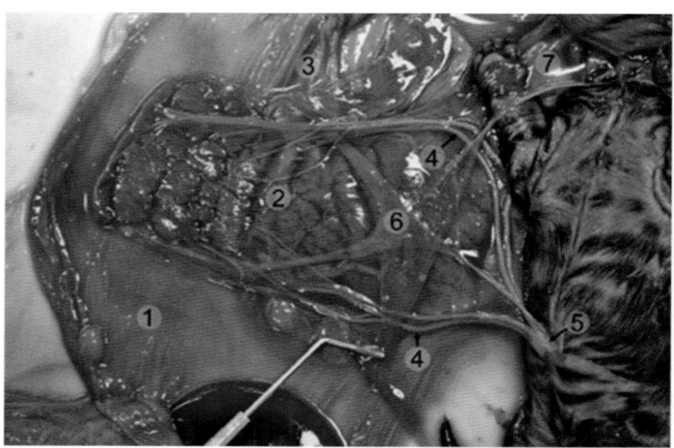

FIGURE 6.3-6 Same specimen as in Figs. 6.3-4 and 6.3-5. The amniotic sac has now been incised and reflected. Key: *1*, uterine wall, endometrial surface; *2*, placental zone of the chorioallantois; *3*, paraplacental zone of the chorioallantois; *4*, umbilical vessels; *5*, umbilical stalk; *6*, remnant of the yolk sac; *7*, amnion. (Photo courtesy of Dr. Christina Murray, University of Melbourne.)

The **blastocyst** is a very early stage of mammalian development. Its inner **cell mass** gives rise to the embryo proper, and its outer or **trophoblast layer** gives rise to the chorionic membrane. Before implantation, the blastocysts move freely within the lumina of the uterine horns. Attachment to the endometrium, or **implantation**, occurs at 14–17 days of gestation in dogs and 11–12 days in cats.

Unlike most other domestic animal species, dogs and cats have a **zonary placenta**: placentation is confined to a **central band** encircling the mid-region of the chorionic vesicle, termed the **placental zone**. The outer layer "invades" the endometrium until it contacts the maternal endothelium. Trophoblast-covered villi then coalesce to form a vascular labyrinth within the endometrium, supporting the growing fetus throughout the remainder of gestation.

Gestation is the period in which the developing young is within the mother's body. The normal length of gestation in dogs is 65 ± 1 day from the LH surge. The normal length of gestation in cats is approximately 62 days postmating—i.e., postovulation. Thus, in both species, gestation normally takes about 9 weeks.

Presumably, the zonary placenta ensures adequate nutrition for each developing fetus in the sometimes "crowded" uterine environment of these litter-bearing species. At the periphery of the placental zone in the endometrium, **marginal hematomas (hemophagous zones)** form where the endothelium has degenerated. These hematomas form visibly green (dogs) or brown (cats) regions and are believed to act as a source of iron for the fetus. The **paraplacental zones** of the fetal membranes are smooth, thin, only loosely juxtaposed to the endometrium, and are believed to be nonfunctional. No fusion occurs between the membranes of adjacent fetuses.

The innermost layer of the fetal membranes is the **amnion**. It is a thin, transparent sac that surrounds the developing fetus in fluid (**amniotic fluid**), thereby providing a buffer/cushion to protect the fetus, while still allowing relatively free movement. Thus, the fetus is enclosed most intimately by the amnion, which contains amniotic fluid; the amniotic sac or vesicle is enclosed by the allantochorion/chorioallantois, which contains the allantoic fluid; and the entire **gestational sac** is contained within the **uterine horn** until parturition.

The caudal or distal part of the allantochorion, and usually the amnion as well, ruptures during the birth process, allowing the puppy or kitten to be delivered from the uterus separate from its placenta (although it is still attached to the placenta via the umbilical cord during parturition). The allantochorion is usually passed shortly afterward. Puppies and kittens are often born still partially or completely enveloped by their amnion. Unless ill or otherwise inhibited in some way, the mother licks or bites the amnion free of the neonate.

The gestational sac ruptures as it is pressed into the cervix by the uterine contractions. The sudden release of allantoic and amniotic fluid is colloquially referred to as the "waters breaking." This step is usually more dramatic in species that typically produce single young, such as horses and humans and less so in litter-bearing species, such as dogs and cats.

PLACENTA TYPES AND DISPOSITION

The placenta is comprised of 2 parts: chorioallantois (fetal part) and endometrium (maternal part). Generally, placenta types are categorized based on "contact" of the chorioallantois and the uterine endometrium: (1) zonary—carnivores, contact is 2–8 cm (0.8–3.1 in.) wide around the circumference of the fetal membranes; (2) discoid—rodents and primates, single focal contact occurs; (3) cotyledonary—ruminants, multiple focal contacts, e.g., 75–120 in cow and 80–90 in the ewe; and (4) diffuse—horse and pig, the total chorioallantois is covered with villi and microvilli connect to (evagination) the endometrium.

Placentae are either deciduate (part of the maternal tissues are shed at parturition—e.g., rodents, carnivores, and humans) or nondeciduate (maternal layers are left intact—e.g., ungulates [except parts of the maternal caruncles in ruminants are removed]).

Full-term puppies and kittens are born fully haired, but with a fine, soft, "downy" neonatal haircoat, whiskers, and eyelashes. The eyelids and external ear canals remain sealed closed until about 2 weeks of age, although the eyelids may open earlier in kittens. In both species, the young start eating solid food at around 2 weeks of age, although they continue to nurse from their mothers until at least 6 weeks of age. Most breeders wean their puppies or kittens between 6 and 8 weeks of age.

Typically, if the umbilical cord is not ruptured during birth, it is ruptured by the mother, or as the neonate begins moving around to find the mammary glands. The short remnant of the umbilical cord normally desiccates (L. *desiccare* to dry up, dehydrate) within hours after birth and generally detaches from the neonate's body after a few days.

Umbilical cord (Figs. 6.3-4–6.3-6)

The **umbilical cord** is a fetal structure that contains 2 umbilical arteries, two umbilical veins, and the urachus, a structure unique to fetal life that connects the fetal urinary bladder to the allantoic cavity.

As described and illustrated in Case 4.6, the paired **umbilical arteries** derive from the fetal abdominal aorta, directly, or from the paired external iliac arteries. Within the fetus, they track alongside the urachus to the umbilicus. After birth, the **urachus** regresses into the **median ligament of the bladder**, while the **remnant umbilical arteries** become the **round ligaments of the bladder**. Outside the fetus, the umbilical arteries divide in the allantochorion to exchange nutrients, waste, and gases (oxygen and carbon dioxide) with the maternal circulation. The blood then returns to the umbilicus in the umbilical veins. Notably, this is the opposite of arteries and veins in animals after birth, in which arteries carry richly oxygenated blood and veins carry relatively deoxygenated blood (high in carbon dioxide) and waste products.

In dogs and cats, the pair of umbilical veins unite into a single **umbilical vein** at the umbilicus. The umbilical vein courses cranially through the ventral abdomen to the liver, where it sends branches to the left and right lobes of the liver before continuing on to the caudal vena cava via the **ductus venosus**. The ductus venosus normally reduces in size as the fetus approaches full term so that the majority of blood then returns to the caudal vena cava via the hepatic veins. In the neonate, the remnant of the umbilical vein may be found tracking along the margin of the **falciform ligament**.

Blood supply, lymphatics, and innervation of the caudal female reproductive tract

The blood supply to the female reproductive tract arises from many branches, including the **uterine branch of the ovarian artery** supplying the cranial part of the uterine horns. The main branch for the uterus is the **uterine artery**, a branch of the **vaginal artery**, which is a branch of the internal pudendal artery. The vaginal artery supplies the caudal part of the uterus, cervix, and a portion of the vagina. The remaining parts of the reproductive tract are supplied

MAMMARY NEOPLASIA

Mammary tumors such as adenocarcinoma and inflammatory carcinoma are relatively common in intact bitches and those spayed after 2 years of age. They are uncommon in bitches spayed shortly after puberty and rare in bitches spayed before puberty. They are also rare in male dogs and in cats.

Mammary tumors reportedly develop in >25% of intact bitches. However, only about 50% are malignant. In contrast, mammary tumors in cats, while uncommon overall, are most likely to be malignant and aggressive.

Diagnosis involves fine-needle aspiration or biopsy of the mass(es). Depending on the tumor type—i.e., invasiveness, risk of recurrence, and the species (more aggressive approach used in cats)—treatment usually includes surgery.

Surgery may involve excision of the mass ("lumpectomy"), a single mammary gland (solitary mastectomy), the entire line of mammary glands on that side of the body (hemimastectomy), or all of the mammary glands (total mastectomy). Depending on the same factors determining the surgical approach, the local lymph node(s) may also be removed. Neutering/spaying the patient after diagnosis appears to have little effect on tumor behavior or recurrence after removal.

by further branches of the **vaginal** and **internal pudendal arteries**. Venous drainage parallels the arterial network draining in the caudal vena cava.

Lymphatic drainage is primarily to the **medial iliac** and **aortic lumbar lymph nodes** and **superficial inguinal lymph nodes**. Innervation is by parasympathetic and sympathetic nerves, through **pelvic nerves** (parasympathetic) and **caudal mesenteric ganglion** and **plexus** (sympathetic), via the **hypogastric** and **pelvic plexus** primarily, with contributions from the **pudendal** and **caudal rectal nerves** from the **sacral plexus**.

Mammary glands (Fig. 6.3-7)

The **mammary glands** are specialized integumentary structures. Dogs and cats of either gender typically have between 4 and 6 pairs of mammary glands (left and right) distributed along the ventrolateral body wall from the pectoral to inguinal region. Most common is 5 pairs of mammary glands: 2 thoracic (cranial, caudal), 2 abdominal (cranial, caudal), and 1 inguinal.

Each mammary gland is accompanied by a simple (hairless) **nipple** on the skin surface through which milk in the lactating female is expressed via several small ducts. Other than in short-haired dogs and cats and multiparous bitches and queens, the nipples and the rest of the mammary glands are inconspicuous except during pregnancy, pseudopregnancy, and lactation. In multiparous bitches and queens, especially those regularly used for breeding, the mammary glands may become pendulous and the nipples and mammary glands more obvious, regardless of pregnancy and lactational status. During pregnancy and lactation, the mammary glands are highly vascular.

When performing a ventral midline laparotomy for any reason, care is needed in pregnant or lactating bitches and queens to avoid the engorged mammary glands. In late pregnancy and early lactation, the left and right mammary glands expand in all directions and may even meet on the ventral midline (see Case 5.8).

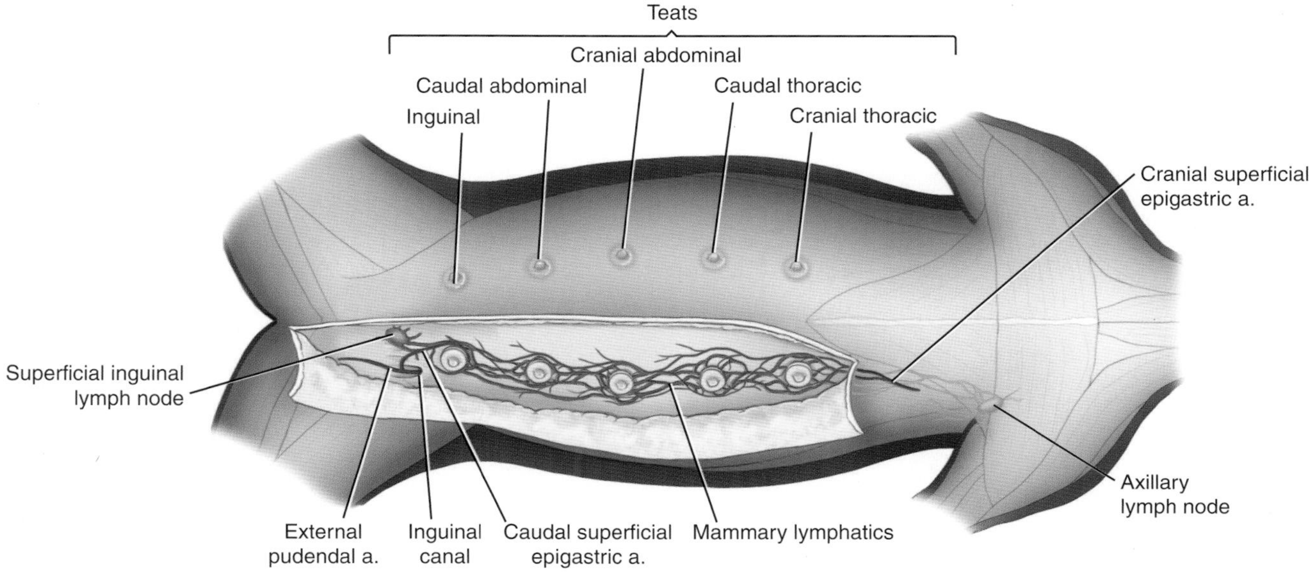

FIGURE 6.3-7 The canine and feline mammary glands, with the skin reflected on the right side to reveal the underlying structures. Cranial is to the right.

Blood supply, lymphatics, and innervation of the mammary gland

The abdominal and inguinal mammary glands are supplied by the paired **caudal superficial epigastric arteries**, which branch from the external pudendal arteries as they exit the inguinal canals and course cranially on either side of the linea alba. These arteries anastomose with the paired **cranial superficial epigastric arteries**, which are continuations of the internal thoracic arteries that penetrate the thoracic wall to supply the cranial abdominal musculature and then supply the skin and the (generally smaller) thoracic mammary glands.

> When performing mastectomy (surgical removal of one or more mammary glands), it may be necessary to ligate both the cranial and caudal arteries supplying the gland(s).

The mammary glands are drained by the **cranial** and **caudal superficial epigastric veins**, which accompany the like-named arteries. Lymphatic drainage of the mammary glands is predominantly toward the **superficial inguinal lymph nodes**, but the thoracic mammary glands generally drain at least in part toward the **axillary lymph nodes** and occasionally elsewhere. Especially during lactation, the lymphatic channels associated with the mammary glands are extensive and interconnected between successive glands in each row (left and right), but mostly the left and right lymphatic channels are separate.

> Because the lymphatic channels within each row of mammary glands (left or right) are extensive and interconnected, hemimastectomy (removal of the entire row of mammary glands on that side) should be considered if a malignant tumor is found in a single gland.

Innervation is via multiple nerves from the lumbar and sacral spinal plexus. The **genitofemoral nerve** has sympathetic efferent and afferent fibers supplying smooth muscle of the teats, blood vessels, and myoepithelial gland cells (efferent) and to carry sensory input to stimulate the secretion of oxytocin for milk "let-down" (afferent).

Selected references

[1] Runcan EE, Coutinho da Silva MA. Whelping and dystocia: maximizing success of medical management. Top Companion Anim Med 2018;33:12–6.
[2] Smith FO. Guide to emergency interception during parturition in the dog and cat. Vet Clin North Am Small Anim Pract 2012;42:489–99.
[3] Sleeckx N, de Rooster H, Veldhuis Kroeze EJ, et al. Canine mammary tumours, an overview. Reprod Domest Anim 2011;46:1112–31.

CASE 6.4

Benign Prostatic Hyperplasia

Magdalena Schrank[a], Stefano Romagnoli[a], and Natali Krekeler[b]
[a]Department of Animal Medicine, Production, and Health, University of Padova, Legnaro, Padova, IT
[b]Veterinary Biosciences, The University of Melbourne, Melbourne, Victoria, AU

Clinical case

History

A 5-year-old male American Staffordshire Terrier was presented when the owner noticed blood dripping from the dog's penis. No other abnormalities were reported.

Physical examination findings

The physical examination was limited, due to the dog's aggressive behavior. No obvious abnormalities were seen. With the dog under sedation, palpation per rectum of the prostate gland and pelvic portion of the urethra revealed an asymmetrically enlarged prostate gland, although it maintained its normal bilobar form. Because the patient was sedated, pain on palpation of the enlarged gland could not be evaluated. The dog's rectal temperature was normal.

Rectal palpation is a valuable, minimally invasive method of evaluating the prostate gland in dogs. It should routinely be part of the general physical examination of an intact male dog. However, a limiting factor is the size of the dog—i.e., distance from the anus to the prostate gland—in relation to the length of the examiner's index finger. The prostate gland can be difficult to reach in large and giant breeds and when an enlarged gland extends over the pelvic brim into the caudal abdomen. In the latter instance, applying pressure with one hand to the caudoventral abdomen (i.e., bimanual examination/palpation), directed caudodorsally, facilitates digital examination of the prostate gland.

BENIGN PROSTATIC HYPERPLASIA (BPH)

In intact or "entire" male dogs the prostate gland enlarges with age, starting as early as 3 years of age. About 95% of these dogs have benign prostatic hyperplasia (nonneoplastic enlargement of the prostate gland) by 9 years of age, although most never show any clinical signs.

In humans, the enlarging prostate gland impinges on the urethra, causing signs of urinary dysfunction early in the course of BPH. However, in dogs, the hyperplastic prostate gland expands outward into the pelvic canal and caudal abdomen; so clinical signs become apparent only when the gland is severely enlarged.

Common presenting complaints in dogs include hematuria, flattened ("ribboned") or tapered feces, and tenesmus. Bloody urethral discharge and hemospermia may also be seen. Stranguria and urinary incontinence are rare in dogs with BPH; they are more often seen with prostatic neoplasia.

Benign prostatis hyperplasia is the most common prostatic disease in male dogs. In affected dogs, the prostatic volume increases twofold to > sixfold compared with unaffected dogs of similar size and weight.

Comparative Veterinary Anatomy: A Clinical Approach. https://doi.org/10.1016/B978-0-323-91015-6.00054-6

Differential diagnoses

Benign prostatic hyperplasia (BPH), prostatitis (inflammation of the prostate gland), trauma to the penis or prepuce, cystitis (inflammation of the urinary bladder)

Diagnostics

Transabdominal ultrasonography revealed an enlarged prostate gland, measuring 5.6 × 5.9 cm (2.2 × 2.3 in.). The normal dimensions in a dog this size are < 3.5 cm (< 1.4 in.) in any plane. Instead of the normal homogenous echodensity, the prostatic parenchyma appeared heterogeneous with multiple anechoic cysts of varying size, but all < 10 mm (< 0.4 in.) diameter. The prostatic capsule was intact, and no intra- or extracapsular masses were visible. Routine bloodwork showed no signs of infection or inflammation. Together, these findings were most consistent with BPH.

On ultrasound, prostatic abscesses appear as hypoechoic areas that may be of mixed echogenicity (heterogeneous), but they are larger than the typical cystic changes of BPH. Although BPH is found in most intact male dogs over 7 years of age, not all dogs show clinical signs. It is worthwhile distinguishing between complicated (clinical) and uncomplicated (subclinical) BPH to determine the appropriate treatment approach. Large intra- or extracapsular prostatic cysts may require surgical treatment.

Diagnosis

Benign prostatic hyperplasia

Treatment

A 5-day course of treatment with osaterone (a steroidal antiandrogen drug) was prescribed. The clinical signs resolved with treatment. Surgical castration of the dog was discussed with the owner to prevent a recurrence. Semen collection before castration, and storage in liquid nitrogen for future use, was also discussed.

Surgical castration (orchidectomy) removes the source of the androgenic hormones that "drive" BPH. A 70% reduction in prostatic size can be expected after castration, although it may take up to 4 months for complete involution to occur. Castration provides a permanent cure for BPH, but obviously the dog's breeding potential is lost in the process unless semen is collected and cryopreserved before surgery. Castration is also beneficial in the treatment of other prostatic diseases, such as prostatitis and prostatic cysts or abscesses, all of which are common sequelae of BPH. However, there is no beneficial effect of castration on prostatic neoplasia.

SINGULAR AND PLURAL TERMS USED IN THERIOGENOLOGY

The following structures are unpaired so are always described using their singular form:

- Scrotum
- Prostate gland
- Bulbourethral gland

These other structures are paired (left and right), so depending on the context, they may be described using their singular or plural form:

- Testis or testes
- Epididymis or epididymides
- Deferent duct or ducts
- Spermatic cord or cords
- Vaginal cavity or cavities
- Prostate lobe or lobes
- Bulbourethral lobe or lobes

Anatomical features in canids and felids

Introduction

The male reproductive system consists of the paired testes, epididymides, deferent ducts, and spermatic cords (comprising the vasculature, lymphatics, and nerves supplying the testis and epididymis), all contained within the scrotum (except for the proximal portion of the deferent duct), and the accessory glands (prostate gland and in cats also the bulbourethral gland), penis, and prepuce. The urethra, while part of the urinary system, also plays an important role in insemination. This section covers the scrotal contents, the proximal portion of the deferent duct, and the accessory glands. The penis and prepuce are described in Case 6.5.

Function

The function of the male reproductive system is the production and delivery of spermatozoa to the female reproductive tract.

Scrotum

The **scrotum** is a sac-like structure that provides external support and protection to the testes, epididymides, distal portion of the deferent ducts, and the spermatic cords. In dogs, the scrotum is located caudoventrally, between and caudal to the thighs (Fig. 6.4-1 and Landscape Fig. 3.0-7). In cats, the scrotum is located caudally, immediately ventral to the anus (Fig. 6.4-2 and Landscape Fig. 3.1-7). In both species, the testes, epididymides, and spermatic cords may all be palpated within the scrotum (Fig. 6.4-3).

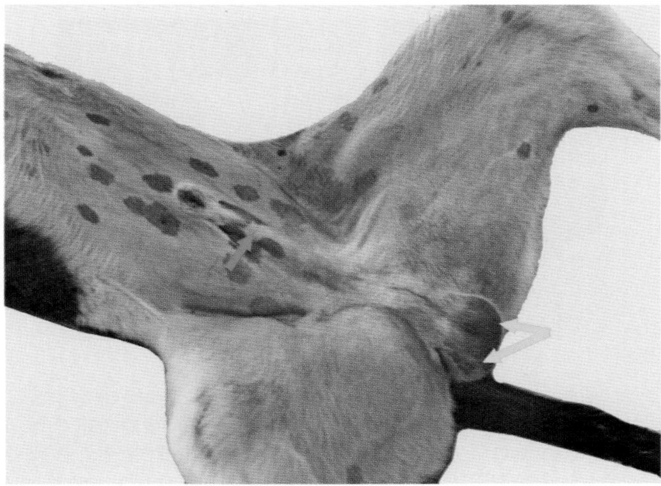

FIGURE 6.4-1 Ventrolateral view of a male dog in dorsal recumbency, showing the prepuce *(green arrow)* and scrotum *(yellow arrows)*. The median raphe of the scrotum is seen as a longitudinal depression between the left and right testes within the scrotum.

INGUINAL HERNIAS

Because the vaginal cavity is continuous with the peritoneal cavity, it is possible for abdominal contents, such as small intestine or omentum, to pass through the inguinal canal into the vaginal cavity or into the subcutaneous tissue, adjacent to the scrotum. These hernias are best described as inguinal, although they are often called "scrotal" hernias. Most inguinal hernias in dogs and cats are congenital. Treatment involves surgical replacement of the herniated organ/tissue into the peritoneal cavity (or resection if devitalized) and closure of the inguinal canal to prevent further herniation. Castration is advised for dogs and cats not intended to be used for breeding.

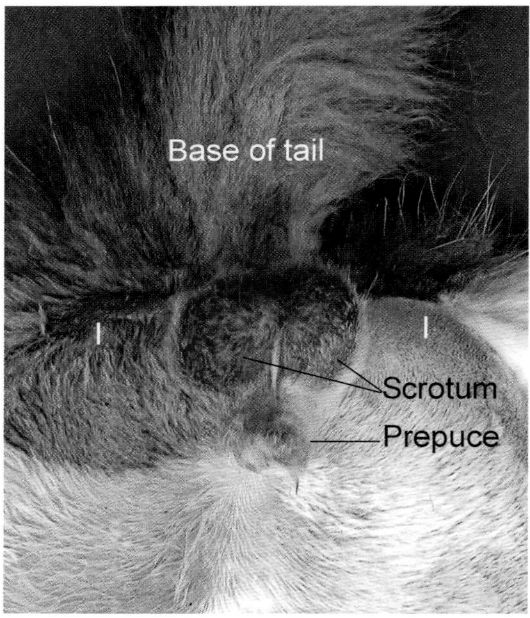

FIGURE 6.4-2 The caudoventral aspect of a male cat. The tail is raised to illustrate the scrotum and prepuce. Because of the camera angle, the anus is concealed in this image by the scrotum and testes. Key: *I*, ischiatic arch.

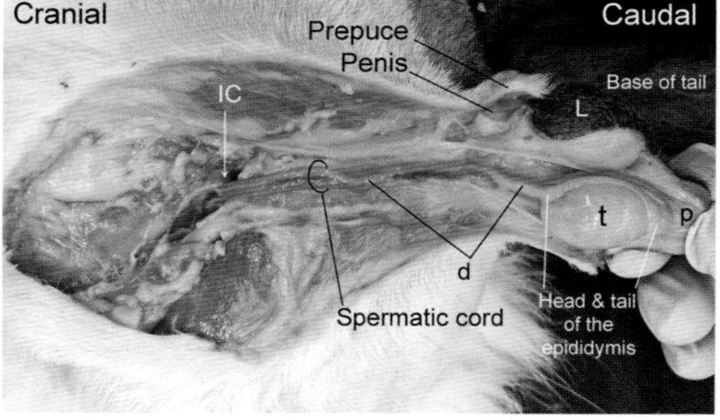

FIGURE 6.4-3 Dissection of a male cat, ventral aspect, highlighting the right testis and spermatic cord. The right scrotal sac is opened and everted to reveal the right testis, epididymis, and spermatic cord. The left testis is still within its scrotal sac, although part of the overlying scrotum has been excised to reveal the scrotal septum (not labeled). Key: *D*, deferent duct; *IC*, inguinal canal (*arrow* points toward canal); *L*, left testis within the left scrotal sac; *P*, parietal layer of the vaginal tunic of the right vaginal cavity; *T*, right testis within the visceral layer of the vaginal tunic.

This concept is of practical importance when performing surgical castration (orchidectomy) and when managing a patient with an inguinal hernia. Castration is described later, after the component anatomy has been discussed. The anatomy of the abdominal wall is described in Case 5.8.

The scrotum and its contents may best be understood as an extension of the abdominal cavity (Fig. 6.4-4). Whereas the female gonads (the ovaries) remain within the abdomen, the male gonads (the testes) normally descend caudoventrally through the inguinal canals into pouches of the ventrolateral body wall (the paired **vaginal cavities** of the scrotum), which are lined with an extension of the peritoneum (the **vaginal tunic**). As with the abdominal cavity and its contents, the vaginal tunic comprises a parietal and a visceral layer, with discrete portions—e.g., the **mesorchium**—surrounding and supporting specific scrotal organs and tissues.

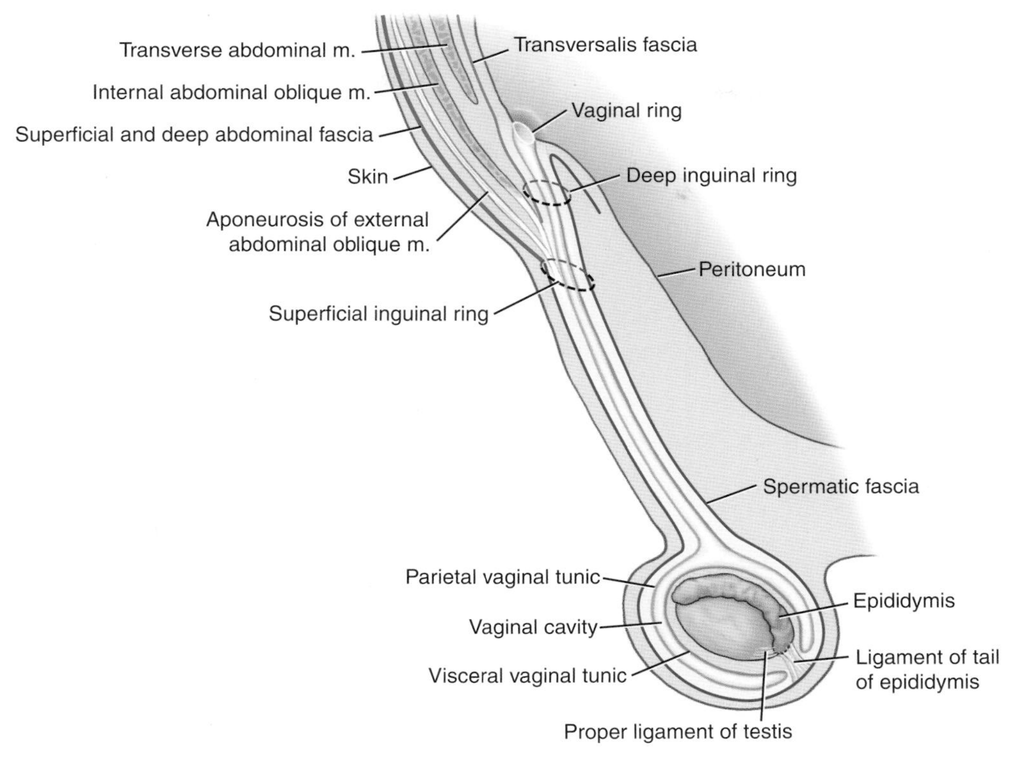

FIGURE 6.4-4 Sagittal view of the scrotum and the caudal abdomen. The image illustrates the relationship between the abdominal muscles and testicular sheaths. Cranial is to the left in this image.

The scrotum is continuous with the abdominal wall and likewise consists of several distinct, although much thinner, tissue layers: skin, subcutis, muscle (highly modified), superficial and deep fascia, and a serous membrane (Figs. 6.4-4 and 6.4-5). The scrotal **skin** has the same structure as the skin of the body, including hair follicles (sparse in dogs), sweat glands, and sebaceous (oil or sebum) glands. However, the subcutis is modified and referred to as the **tunica dartos**. It consists of elastic fibers and smooth muscle, which, by contraction, reduce the dimensions of the scrotum and thus participate in raising the testicular temperature; relaxation of the smooth muscle has the opposite effect.

The cremaster muscle (described later) has a greater role in this process, but the smooth muscle of the tunica dartos is instrumental in protecting the testes from physical and thermal harm.

The tunica dartos and scrotal fascia, together termed the **spermatic fascia**, also form the **scrotal septum**, which divides the scrotum into 2 separate cavities (left and right), 1 for each testis. The location of the septum is visible externally as the median raphe of the scrotum, or the **scrotal raphe**, which is a longitudinal depression between the 2 bulges formed by the testes within the scrotum (Figs. 6.4-1 and 6.4-2).

The paired **cremaster muscles** are important for the maintenance of optimal temperature within the testes (Figs. 6.4-4 and 6.4-5). Each originates as a caudoventral extension of the internal abdominal oblique muscle that continues through the inguinal canal to insert onto the vaginal tunic covering the spermatic cord and testis. When the cremaster muscles contract, they raise the testes toward the body, which subsequently increases the temperature within the testes. Relaxation of the cremaster muscles has the opposite effect.

The serous membrane lining each scrotal cavity is called the **vaginal tunic** or tunica vaginalis. It is continuous with the peritoneum. As with the peritoneum, the vaginal tunic has a **parietal layer**, which lines the scrotal wall and septum, and a **visceral layer**, which covers the scrotal contents. Thus, the scrotal cavity, which contains the testis, is also called the **vaginal cavity**.

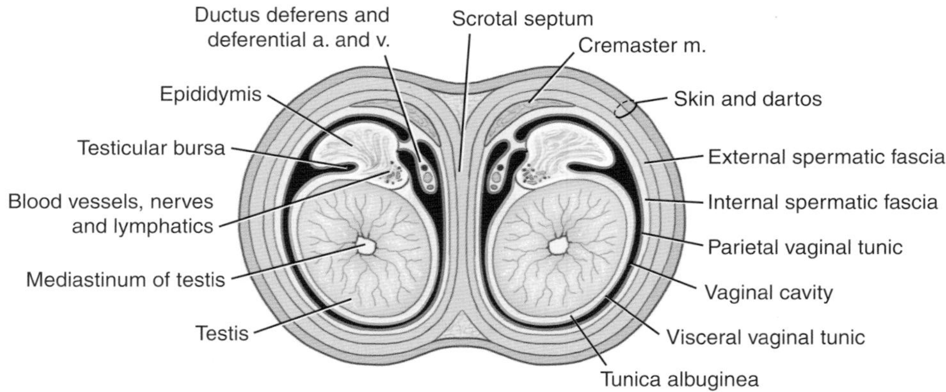

FIGURE 6.4-5 Schematic image of a transversal cut of the testicles and the scrotum.

As discussed later, there are 2 general approaches to castration: "open" and "closed." With open castration, the parietal tunic is incised, and therefore, the vaginal cavity is opened to expose the testis, epididymis, and spermatic cord. With closed castration, the parietal tunic is intact, and the vaginal cavity remains closed.

The **parietal layer** of the vaginal tunic lies deep to the tunica dartos and the spermatic fascia of the scrotal wall and septum. The parietal layer is attached to the spermatic fascia somewhat loosely such that these 2 layers may be identified and easily separated.

The **visceral layer** of the vaginal tunic intimately covers the testis, epididymis, the ligament that anchors the "tail" of the testis to the scrotum (described later), the vessels and nerves of the spermatic cord, and the deferent duct. The visceral layer is a reflection of the parietal layer along the dorsocaudal border of the testis (Figs. 6.4-5 and 6.4-6). Thus, the visceral layer suspends and supports the scrotal contents within the vaginal cavity, just as the visceral layer of the peritoneum suspends and supports the abdominal organs.

Although the visceral layer of the vaginal tunic is a continuous serosal sheet, it is intimately invested around each of the scrotal organs and is described using different names according to the specific structure with which it is associated: the **mesorchium** surrounds the testis; the **mesepididymis** surrounds the epididymis; the **mesofuniculus** surrounds the vessels and nerves of the spermatic cord; and the **mesoductus deferens** is a discrete fold that suspends the deferent duct (ductus deferens) and its vessels and nerves.

Proximally, each testis is suspended within its vaginal cavity by the spermatic cord and the vaginal tunic. Distally, the testis is anchored within the scrotum by the short, strong **ligament of the tail of the epididymis**, which is

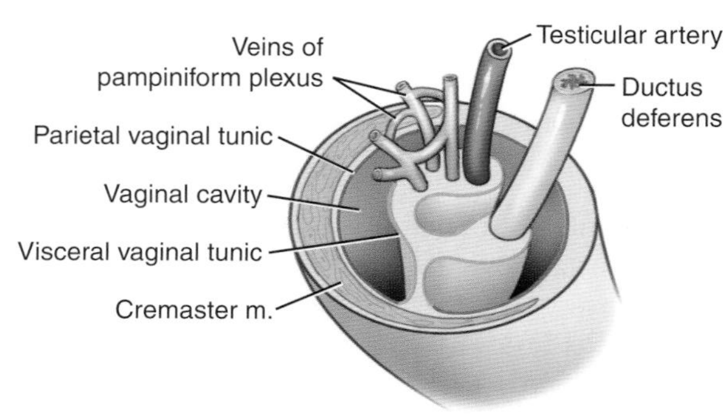

FIGURE 6.4-6 Schematic image of the spermatic cord including vessels and deferent duct within the vaginal tunic.

continuous with the spermatic fascia of the scrotal wall. This ligament is covered by a reflection of the parietal tunic, as illustrated in Fig. 6.4-4. The tail of the epididymis is firmly attached to the caudal pole of the testis by the **proper ligament of the testis**. Thus, the testis is indirectly attached to the scrotal wall.

The ligament of the tail of the epididymis is important during open castration, as it must be transected in order to remove the testis from the vaginal cavity.

The **vaginal cavity** is normally just a potential space, containing only a small amount of serous fluid that lubricates the facing parietal and visceral tunics, allowing them to slide over one another without friction as the individual moves. Thus, the scrotal anatomy ensures that each testis is held in position, yet is still somewhat mobile and protected from friction-induced injury when it moves within its scrotal sac.

Testis

The **testis** is the site of development and differentiation of the spermatozoa, as well as the origin of the male sex hormones. The paired testes are rounded or ovoid organs that are normally located within the scrotum. Slight asymmetry in size and position of the left and right testes is normal. In young dogs and cats, especially before puberty, it may be possible to reposition the testis into the inguinal canal with digital pressure. However, after puberty, the size of the testis prevents its passage back through the inguinal ring.

Cryptorchidism is a condition in which the testis fails to complete its descent into the scrotum during fetal development and instead remains within the abdomen or the inguinal canal. Cryptorchidism may be unilateral or bilateral and is relatively common. In contrast, monorchidism, in which one testis fails to develop, is rare. Because the temperature is higher within the body than in the scrotum, the cryptorchid testis generally does not produce viable spermatozoa; however, it still produces androgens. Thus, orchidectomy (castration) of just the scrotal testis may not prevent undesirable male behavior in unilateral cryptorchids. Furthermore, a cryptorchid testis is at an increased risk of developing a Sertoli cell tumor. For these reasons, localization and removal of the retained testis/testes are advisable in all cryptorchids.

Each testis has a medial and a lateral aspect. Based on the position of the epididymis (discussed later), the testis may also be described as having a head, a tail, a free side, and an epididymal side. The **tail of the testis** may have a rudiment of the Müllerian ducts (embryonic origin of the female reproductive tract), and it is positioned caudodorsally in dogs and cats (Landscape Figs. 3.0-7 and 3.1-7).

TESTICULAR TORSION

Testicular torsion, in which the spermatic cord twists around its long axis (like wringing a towel), is an uncommon but serious condition in dogs. It results in partial or complete occlusion of the vessels supplying the scrotal contents. Scrotal swelling, caused by venous and lymphatic congestion, is typically accompanied by pain and an abnormal gait as the dog attempts to swing the pelvic limb away from the painful testis. Diagnosis is based on palpation (malposition of the epididymis) and ultrasound examination. In many cases, unilateral orchidectomy is required because there is typically irreversible vascular compromise of the affected testis by the time the diagnosis is made.

SERTOLI AND LEYDIG CELLS

In addition to spermatozoa, 2 other types of testicular cells are differentiated. Sertoli cells play an important role in protecting the testis by creating and maintaining the blood-testis-barrier. They are also responsible for the nutrition and transport of spermatozoa, which are produced by the epithelium of the seminiferous tubules. Sertoli cell function is controlled by the pituitary release of follicle stimulating hormone (FSH). Leydig cells produce androgens (male hormones) and oxytocin under the influence of luteinizing hormone (LH) produced by the pituitary gland.

The testicular architecture comprising the parenchyma, septa, and mediastinum can be appreciated on ultrasound examination of the scrotum. Disruption of the normal architecture may occur with various testicular diseases. For example, the mediastinum may be displaced or obliterated with space-occupying processes such as hematoma/seroma formation (pooling of free blood/serum within a tissue), abscessation, or neoplasia.

As mentioned earlier, the testis is covered by the **visceral** layer of the **vaginal tunic**. Deep into this tunic is the testicular **capsule**, or **tunica albuginea**. It is a pale, fibrous membrane that is firmly adhered to the surface of the testis and contributes **septa**, which support the parenchyma of the testis. The septa unite in the center of the testis to create a **mediastinum**.

When the testis is viewed within the vaginal cavity during open castration (discussed later) or dissection (Fig. 6.4-3), it is the visceral layer of the vaginal tunic that is observed as the superficial layer covering the testis. Even so, as both the vaginal tunic and the testicular capsule are translucent, the extensive network of arteries and veins that supply the testis (described later) may be seen coursing over the surface of the testis, beneath the capsule, through the vaginal tunic.

The septa extending from the capsule into the testis divide the parenchyma into numerous lobes, in which the **convoluted seminiferous tubules** may be found. The convoluted seminiferous tubules connect with the **straight seminiferous tubules** and thus empty into the **rete testis**, which is in the **mediastinum** of the testis. Twelve to 16 **efferent ductules** connect the rete testis with the head of the **epididymis**.

Epididymis

The paired **epididymides** are essentially tubular structures that connect the testes to the deferent ducts, and thus to the urethra. In dogs and cats, each epididymis overlies the dorsolateral surface of the testis and is attached to the testis by a short fold of the vaginal tunic. The epididymis may be described in 3 parts: head, body, and tail.

The **head** of the epididymis begins on the craniomedial side of the testis but immediately wraps around the cranial pole of the testis to the lateral side. The **body** of the epididymis thus tracks distally over the dorsolateral surface of the testis (Landscape Figs. 3.0-7 and 3.1-7). The **tail** of the epididymis is attached to the caudal pole of the **testis** by a short ligament, the **proper ligament of the testis**. As described earlier, the tail of the epididymis, and therefore the testis, is attached to the scrotal wall by a short ligament, the **ligament of the tail of the epididymis**.

The extremities or poles of the testis are named according to their relationship with the head and tail of the epididymis. The epididymis is also distinguishable during palpation of the testis. The head and tail of the epididymis each present as prominent, rounded thickenings at either end of the testis.

DISEASES OF THE TESTIS, EPIDIDYMIS, OR SCROTUM

Diseases that may be encountered in clinical practice include:

- Testicular torsion
- Inguinal hernia
- Inflammation of the testis (orchitis) and/or epididymis (epididymitis), caused by viral or bacterial infection
- Cryptorchidism
- Monorchidism (rare)
- Testicular neoplasia: Sertoli-cell tumor, Leydig-cell tumor, other tumors
- Scrotal trauma: whether it includes the scrotal contents, it may result in testicular degeneration by raising the testicular temperature
- Scrotal neoplasia
- Age-related testicular degeneration

The numerous efferent ductules from the rete testis unite in the head of the epididymis to form a single duct, the **epididymal duct**. This extensively coiled duct courses the length of the epididymis and then continues proximally as the deferent duct (described later). In its normal tightly coiled configuration, the epididymal duct is short, extending only to the length of the testis. However, it may be several meters in length if uncoiled.

While the testis is the site of spermatozoan production, the epididymis is the site of maturation and storage of spermatozoa. Storage of mature spermatozoa in the tail of the epididymis is possible for only a limited time before sperm viability begins to decline. The time spermatozoa spend within the epididymis is similar for all domestic species: 10–14 days. During ejaculation, peristaltic (rhythmic, directional) contractions of smooth muscle in the epididymal duct transport the spermatozoa toward and into the deferent duct.

Deferent duct

The paired **deferent ducts**, also called the ductus deferens or vas deferens, course proximally from their origin at the tail of the epididymis, on the medial side of the testes and spermatic cords, in their own thin fold of vaginal tunic, the **mesoductus deferens**. Each duct continues proximally through the inguinal canal to empty into the urethra just caudal to the urinary bladder, adjacent to the prostate gland (described later). The mature spermatozoa stored in the tail of the epididymis are thus delivered into the pelvic portion of the urethra during ejaculation.

After passing through the inguinal canal, the deferent duct lies within a fold of peritoneum along with the ureter on that side, which crosses the duct dorsally on its way to the urinary bladder. The deferent duct opens into the prostatic portion of the urethra, surrounded by the parenchyma of the prostate lobe on that side. The urethral opening of the deferent duct is called the **seminal colliculus**, and it is here that prostatic fluids and spermatozoa are combined to create the ejaculate.

Just distal to the seminal colliculus, the deferent duct dilates into a region called the **ampulla**, which in the dog contains the **ampullary glands**. These tiny accessory glands are absent in cats. However, cats have a **bulbourethral gland**, which is absent in dogs. (The accessory glands are described later.) In some cases, a vestige of the Müllerian ducts may be found between the ampullary glands, a structure referred to as the uterus masculinus, or "male uterus."

Spermatic cord

The paired **spermatic cords** consist of the arteries, veins, lymphatic vessels, and nerves that supply each testis and epididymis. The vessels and nerves all course together within the visceral layer of the vaginal tunic (Fig. 6.4-6), which gives the appearance of a single, cordlike structure (Fig. 6.4-3). The deferent duct passes alongside it in a small fold of the vaginal tunic, the mesoductus. The diameter of the spermatic cord depends on the size of the individual. In addition, it may be increased or decreased in testicular disease. The specific components of the spermatic cord are described next.

Ligation and transection of the spermatic cord during castration inevitably includes the deferent duct. In comparison, vasectomy preserves the vessels and nerves of the spermatic cord and transects only the deferent duct or vas deferens. Vasectomy is seldom used in domestic animals. It is occasionally performed in tomcats to create infertile but otherwise intact males that can be used to identify queens in estrus.

Regional blood supply, lymphatics, and innervation

The main blood supply to the testis is the **testicular artery**, which arises directly from the aorta. Shortly after passing through the inguinal ring, the testicular artery coils within the spermatic cord as it continues distally toward the testis. After arriving at the head of the testis, it continues distally over the epididymal surface of the testis toward the tail of the testis, where it changes direction and returns along the free margin to the head of the testis. Along its course over the testis, it sends numerous small branches into the parenchyma of the testis, toward the mediastinum. These small arteries may be seen beneath the capsule (tunica albuginea) with their paired veins, running over the lateral and medial sides of the testis.

The testicular artery also supplies the epididymis and the deferent duct. The scrotum and vaginal tunic are supplied by the **external pudendal artery** and the **cremasteric artery**, both of which pass through the inguinal ring.

It is believed that this "countercurrent" arrangement of arteries and veins has a role in the regulation of testicular temperature and blood pressure.

The **testicular vein** has a similar course to the testicular artery. It coils within the spermatic cord to create the **pampiniform plexus** in which the veins surround the coiled testicular artery.

Lymphatic vessels from the testis and epididymis drain into the **medial iliac lymph nodes** and **lumbar aortic lymph nodes**. Lymphatic drainage from the scrotum drains into the **superficial inguinal lymph nodes**.

Sympathetic innervation of the testis and epididymis is provided by the **cranial spermatic nerve**, which originates in the lumbar plexus and accompanies the testicular artery. Parasympathetic innervation is provided by the **caudal spermatic nerve**, which originates from the sacral nerves and accompanies the deferent duct. The scrotum is innervated by branches of the **pudendal nerve**. The **genitofemoral nerve** provides sensory innervation to the vaginal tunic and motor innervation to the cremaster muscle.

Anatomical considerations in surgical castration (orchidectomy)

Regardless of the species and choice of surgical approach, it is important to ensure that both testes are in the scrotum before anesthetizing the surgery patient. If there is a nonscrotal testis, it must be located (in the abdomen or inguinal canal) and the surgical plan adjusted accordingly.

Castration of dogs and cats is one of the most common surgical procedures performed in clinical practice. Specific techniques vary in some details, but the procedure can broadly be classified as "**open**" or "**closed**," based on whether the parietal layer of the vaginal tunic is incised, and the vaginal cavity opened. Open castration, in which the parietal tunic is incised, is performed most often. Its main advantage is that the scrotal contents can be directly visualized and the spermatic cord ligated, without having to include the parietal tunic, thereby making the ligature more secure and decreasing the risk of bleeding from the testicular artery and pampiniform plexus in the post-operative period.

The typical approach for uncomplicated castration in dogs includes general anesthesia, with the patient positioned in dorsal recumbency and the pelvic limbs abducted. The prescrotal region—i.e., the base of the prepuce—immediately cranial to the scrotum is clipped of hair and prepared for aseptic surgery. (Cryptorchid castration is beyond the scope of this discussion.)

Before making the skin incision, one testis is moved cranially and medially with the thumb and forefinger from its scrotal position until it lies on midline just cranial to the scrotum. While holding the testis steady, a midline incision is made directly over the testis, through the skin and subcutis of the median raphe and through the spermatic fascia overlying the testis. The length of the incision is relative to the size of the testis in that dog; the incision must be large enough to enable removal of the testis without excessive tension on the margins of the surgical wound. The surgeon must be careful to avoid incising the penile urethra, which lies on the midline beneath the surgical site. By applying increased digital pressure, the testis within the parietal layer of the vaginal tunic is moved through the incision.

When performing an open castration, the parietal layer of the vaginal tunic is incised, exposing the testis, epididymis, and spermatic cord within the vaginal cavity. The testis is still attached to the scrotum by the ligament of the tail of the epididymis, which can be manually separated in most cases to avoid tearing the proper ligament of the testis. After separating the tail of the epididymis from the scrotum, the testis and epididymis are elevated, allowing the spermatic cord (and deferent duct) to be clamped, ligated, and transected. The testis, epididymis, and the portion of the spermatic cord distal to the ligature are then removed. The "stump" of the spermatic cord is checked for bleeding from the transected vessels. The process is repeated with the other testis.

Whether the parietal tunic is sutured depends on the size of the testis and thus the length of the incision. Generally, in large dogs and dogs with an enlarged testis (e.g., testicular neoplasia), 1 or 2 sutures may be placed to close the tunic. Surgical closure of the midline incision classically involves a 2-layer closure (subcutaneous tissue and skin).

Cat castration usually involves making 2 separate vertical incisions over the scrotum; 1 over each testis, or a single incision directly over the scrotal raphe. The castration may be open or closed. When performing a closed **castration**, the testis within its parietal tunic is moved through the scrotal incision and elevated from the body, separating the tunic from the spermatic fascia. The spermatic cord in cats is long enough that, when gently stretched, it is knotted by itself instead of using a suture. The scrotal incisions are typically left open to heal by second intention.

Accessory glands

The number and type of accessory sex glands depend on the species. Dogs have only the prostate gland (and the tiny ampullary glands of the deferent duct), whereas cats have a prostate gland and a bulbourethral gland.

Although cats have prostate and bulbourethral glands, abnormalities involving either are much less common than prostatic disorders in dogs, which include BPH, prostatitis, prostatic and paraprostatic cysts, abscesses, and neoplasia.

In both species, the **prostate gland** is in the cranial portion of the pelvic canal, although it may extend cranially into the caudal abdomen, depending on the size of the urinary bladder and the prostate gland. The prostate gland lies on the ventral midline of the pelvic canal, immediately caudal to the urinary bladder, ventral to the rectum, dorsal to the pubic symphysis, and medial to the abdominal wall. The prostate is divided into 2 symmetrical, oval, or spherical **lobes** (left and right) that are divided by a **median septum**. Both lobes are subdivided into numerous **lobules**, so the surface of the prostate gland can be described as lobular. A fibromuscular capsule encircles the lobes.

The urethra passes between the 2 lobes. The cranial part of the urethra is referred to as the **prostatic urethra**. Fluid produced by the prostate gland empties into the urethra and is expelled during ejaculation, together with the sperm-rich fluid from the deferent ducts that empty into the prostatic urethra, adjacent to the ducts of the prostate gland. In dogs, the prostatic fluid is expelled mainly in the third (final) fraction of the ejaculate.

Blood supply to the prostate gland is by the paired **prostatic artery**, which arises from the paired **internal pudendal artery**. The prostatic arteries enter the gland via numerous branches on its dorsal and dorsolateral surfaces. Before entering the prostate gland, the prostatic arteries send branches to supply the rectum, deferent ducts, the caudal portion of the bladder and ureters, and the pelvic part of the urethra.

Venous drainage is provided by the **prostatic** and **urethral veins**, which drain into the **internal iliac vein**. Numerous lymphatic vessels drain into the **iliac lymph nodes**.

DEVELOPMENT OF THE PROSTATE GLAND

During the embryonic phase of development, the prostate gland develops out of 3 or 4 pairs of buds of the urogenital sinus endoderm and, therefore, shares embryonic origins with the prostatic urethra and the urinary bladder. Androgens promote growth and differentiation of a primitive epithelium into basal (stem) cells, luminal (secretory) cells, and neuroendocrine cells. The basal cells have a key role in the growth and organization of the prostatic ducts and acini (small, sac-like cavities, lined with secretory cells, within a gland). Until puberty, the prostate gland remains a small and inactive organ overlying the proximal urethra. It begins to grow during puberty, under the influence of androgens (testosterone and dihydrotestosterone). The mature prostate gland consists of acini, ductal, and urothelial tissue. The acini are the androgen-responsive part; the ductal and urothelial tissue are not affected by castration.

As the nerves that innervate the prostate gland also innervate the urinary bladder, its detrusor muscle, and the external urethral sphincter, special care must be taken during surgical manipulation of the prostate gland to avoid causing nerve injury that can result in temporary or permanent urinary incontinence.

Innervation of the prostate gland is provided by the paired **hypogastric** (sympathetic) and **pelvic nerves** (parasympathetic). These nerves also innervate the bladder and urethra. Arising from the caudal mesenteric ganglion, the hypogastric nerves enter the pelvic plexus and then run with the arteries of the deferent ducts. The pelvic nerve instead arises from sacral nerves 1–3 and follows the prostatic arteries, joining with branches of the hypogastric nerves to form the pelvic plexus. Sympathetic stimulation of the prostate gland is responsible for the ejection of prostatic fluid during ejaculation, whereas parasympathetic stimulation increases the glandular secretion.

An important clinical aspect of prostatic anatomy is its size, which changes with age as a normal occurrence and may also change with different pathologies. Normal prostatic size in intact male dogs can be described in 3 phases: normal growth in young adults (1–5 years), hyperplasia during middle age (6–10 years), and involution in old age (> 11 years). On ultrasound, the prostate gland in healthy adult dogs has a mean length of 3.4 ± 1.1 cm (1.3 ± 0.4 in.), a mean height of 2.8 ± 0.8 cm (1.1 ± 0.3 in.), and a mean width of 3.3 ± 0.9 cm (1.3 ± 0.4 in.). Given the low incidence of prostatic problems in cats, similar information is not reported for cats.

The bilobed **bulbourethral gland** in cats is found adjacent to the pelvic urethra at the root of the penis (see Case 6.5). The lobes are small and rounded, like tiny peas. Dogs lack a bulbourethral gland.

Selected references

[1] Smith J. Canine prostatic disease: a review of anatomy, pathology, diagnosis, and treatment. Theriogenology 2008;70:375–83.
[2] LeRoy BE, Northrup N. Prostate cancer in dogs: comparative and clinical aspects. Vet J 2009;180:149–62.
[3] Goericke-Pesch S, Hölscher C, Failing K, et al. Functional anatomy and ultrasound examination of the canine penis. Theriogenology 2013;80:24–33.
[4] Schimming BC, Vicentini CA. Ultrastructural features in the epididymis of the dog (*Canis familiaris*, L.). Anat Histol Embryol 2001;30:327–32.
[5] Setchell BP. Testicular blood supply, lymphatic drainage and secretion of fluid. In: Johnson AD, Gomes WR, Vandemark NL, editors. The testis. New York: Academic Press; 1970. p. 101–239.

CASE 6.5

Congenital Phimosis

Magdalena Schrank[a], Natali Krekeler[b], Helen M.S. Davies[b], and Stefano Romagnoli[a]
[a]Department of Animal Medicine, Production and Health, University of Padua, Legnaro (PD), IT
[b]Veterinary Biosciences, The University of Melbourne, Melbourne, Victoria, AU

Clinical case

History

A 5-month-old male mixed-breed dog was presented because the owner had noticed a small, reddish mass protruding from the dog's preputial orifice. The protrusion was present most of the time and was particularly prominent when the dog became excited. In addition, the dog frequently licked the area. No other clinical signs were reported.

> The preputial orifice is the opening at the distal extent of the prepuce, which is the fold of skin that normally covers the relaxed penis. In dogs, the prepuce and penis lie along the caudal aspect of the ventral midline, directed cranially, so the preputial orifice opens cranially (Fig. 6.5-1). In cats, the prepuce and relaxed penis are directed caudally, so the preputial orifice opens caudally (Fig. 6.5-2). When erect, the feline penis is directed cranially to facilitate penetration of the queen (female cat).

Physical examination findings

Physical examination was unremarkable except for the penis and prepuce. The preputial orifice was narrowed to 3 mm (0.12 in.) and the tip of the penis (the glans penis) protruded through the orifice for 2–3 mm (0.07–0.12 in.). The visible portion of the penis was reddened, probably due to persistent licking by the dog. Palpation of the penis within the prepuce was unremarkable. Both testes were palpable within the scrotum and appeared normal.

Differential diagnoses

Phimosis (narrowing of the preputial orifice, preventing normal extrusion of the penis from the prepuce)—congenital or acquired

> Given the signalment (i.e., immature dog) and clinical presentation (i.e., no history or signs of trauma), congenital phimosis was most likely.

Diagnostics

Routine bloodwork (complete blood count and serum biochemistry panel) was normal.

Diagnosis

Congenital phimosis

Comparative Veterinary Anatomy: A Clinical Approach. https://doi.org/10.1016/B978-0-323-91015-6.00043-1

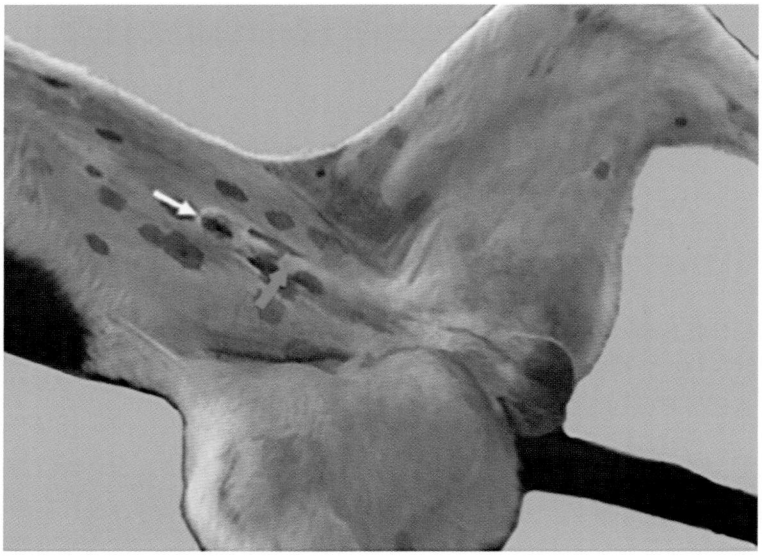

FIGURE 6.5-1 Ventrolateral view of a male dog, highlighting the prepuce *(green arrow)* and preputial orifice *(white arrow)*, which, in dogs, are both directed cranially.

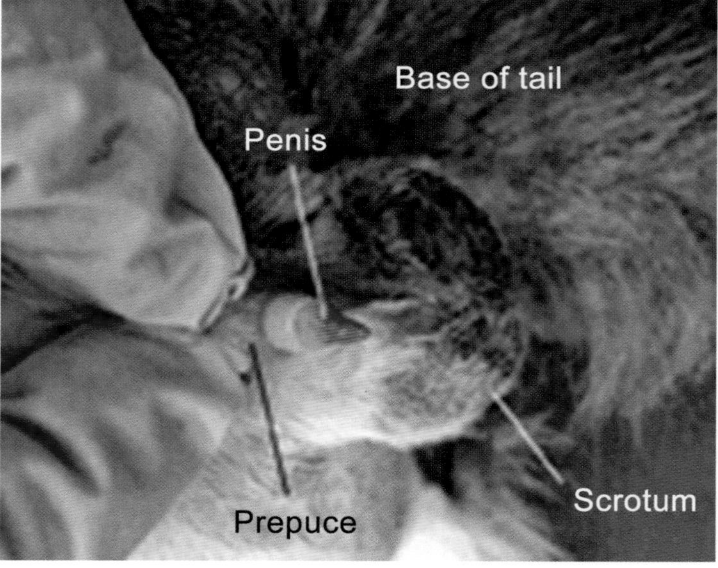

FIGURE 6.5-2 Caudoventral aspect of a male cat, showing the penis being extruded from the prepuce, both of which lie immediately ventral to the scrotum and the base of the tail, directed caudally—except during coitus when the apical penile ligament (not shown) redirects the penis cranially for intromission.

Treatment

Due to the severity of the phimosis, the near-permanent protrusion of the glans penis, and the persistent licking, castration and surgical enlargement of the preputial orifice were performed under general anesthesia. During the post-operative period, the prepuce was pulled back to expose the glans penis every other day to minimize the formation of adhesions between the penis and the surgical site as the prepuce healed. At recheck 3 months after surgery, the preputial orifice was large enough to permit full extrusion of the penis.

Anatomical features in canids and felids

Introduction

This section describes the penis, prepuce, and briefly, the penile portion of the urethra. The scrotum, testes, tubules (epididymis, deferent duct), spermatic cord, and accessory glands are described in Case 6.4. The lower urinary tract is described in Case 6.1.

Function

The penis serves 2 distinct functions—universal (urination) and, specific to intact males, (reproduction). The penis contains the external portion of the urethra, thereby forming the distal or terminal portion of the urinary tract in males. Both the urethra and the erectile tissues of the penis contribute to reproduction, serving as the method for delivery of semen to the female reproductive tract. The prepuce serves the simple function of protecting the relaxed penis.

Penis

Dogs and cats have a musculocavernous **penis** type in contrast to ruminants, which have a fibroelastic penis type. Another notable characteristic of the canine and feline penis is the presence of the **os penis**—a small, elongated bone within the body of the penis. In dogs, the os penis develops within the first few weeks after birth, and in cats, within the first 3 months. In dogs, the os penis is easily palpable and radiographically evident (Fig. 6.5-3), whereas in cats, it is only a few millimeters in length.

The penis may be described as having a **root (crura penis)**, a **body (corpus penis)**, and a **tip (glans penis)**. In dogs, a discrete enlargement at the base of the glans penis, the **bulbus glandis**, plays an important role during coitus (described later). In cats, the glans penis is covered in numerous tiny **spines**. Unlike bitches, queens are induced ovulators (see Case 6.3), and the mechanical stimulation of the penile spines facilitates ovulation during repeated copulation.

The root or base of the penis is formed by the paired **penile crura**, which are attached via tendons to the ischiatic arch and by the proximal portions of the corpus cavernosum and corpus spongiosum, both described below. The penile crura may be considered the origin of the paired corpus cavernosum. The pelvic portion of the urethra continues distally by curving around the caudal aspect of the ischiatic arch and entering the root of the penis.

The urethra continues through the body of the penis, where it is often referred to as the **penile urethra**. In relation to reproductive function, essential components of the corpus penis are the 2 highly vascular, cavitary, or "spongy" bodies, the corpus cavernosum and the corpus spongiosum penis (Fig. 6.5-4). As described later, these are the erectile tissues of the penis. The unpaired **corpus spongiosum penis** encircles the urethra. The larger, paired **corpus cavernosum** is located dorsolaterally in relation to the urethra, which, in dogs, courses through the ventral portion of the corpus penis in the **urethral sulcus**.

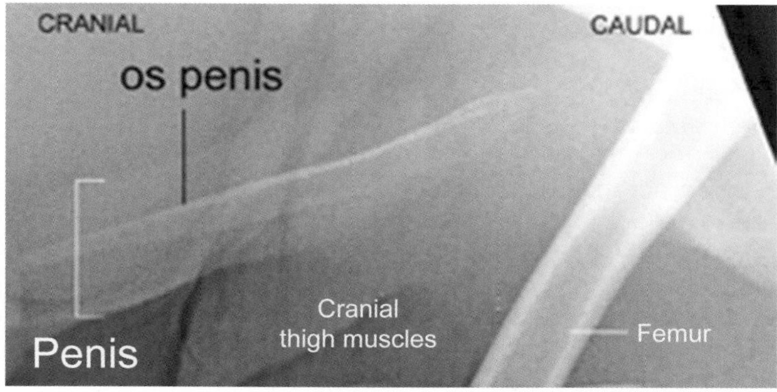

FIGURE 6.5-3 Cropped lateral radiograph of the caudal abdomen of a dog, showing the Os penis within the body of the penis, just cranial to the thigh.

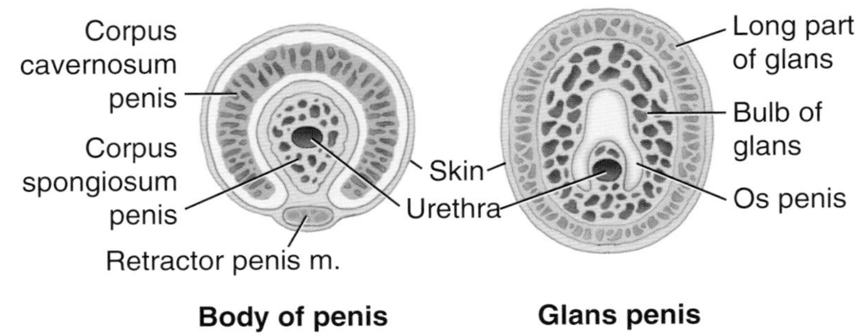

FIGURE 6.5-4 Schematic drawing of the penis in cross section, showing the urethra, corpus cavernosum, and corpus spongiosum (+/– Os penis).

The glans penis is the most distal portion of the penis. In addition to the urethra, it contains its own spongy body, the **corpus spongiosum glandis**, which is a continuation of the corpus spongiosum penis. The glans penis is much longer in dogs than in cats, and in dogs, it is described in 2 parts: the **bulbus glandis**, which marks the transition from corpus penis to the glans penis, and the **pars longa glandis**, which is the distal or cranial "free end" of the glans penis. The bulbus glandis is a discrete enlargement, which becomes particularly engorged during coitus and is responsible for "tying" with the female (described later). The base of the bulbus glandis is also the site at which the epithelium lining the prepuce is reflected onto the skin covering the penis. Thus, the pars longa glandis, cranial to the bulbus glandis, is described as the "free" portion of the penis.

The **urethra** opens externally at the tip of the glans penis as the **external urethral ostium**. In dogs, a small urethral protrusion is visible, the **urethral process**; it is absent in cats.

Prepuce

The **prepuce** is a separate organ whose presence and integrity are important, not only for reproduction but also for good quality of life. Malformation or injury of the prepuce is often difficult to resolve.

The prepuce consists of 2 sheets—the **external lamina** and the **internal lamina**. The outer sheet is composed of normal skin, complete with hair follicles. It has a palpable and faintly visible median line, the **preputial raphe**, which is the continuation of the scrotal raphe (described in Case 6.4).

At the distal end of the prepuce (cranial in dogs, caudal in cats), the **preputial orifice** is formed, where the external lamina folds back on itself to form the internal lamina, which overlies the penis.

DISEASES OF THE PENIS AND/OR PREPUCE

Following is a list of some of the conditions involving the penis and/or prepuce that may be encountered in clinical practice:

- Diphallia—duplication of the penis
  - Fracture of the os penis
- Hypospadias—developmental abnormality in which the urethral opening is located ventral and caudal to its normal position at the tip of the glans penis
- Phimosis—narrowing of the preputial orifice, preventing extrusion of the penis
- Paraphimosis—inability to retract the penis back into the prepuce
- Penile hypoplasia—under-development of the penis
- Penile tumors
- Persistent penile frenulum
- Traumatic or inflammatory lesions (posthitis, balanitis, balanoposthitis)

Dorsally, the internal lamina is attached to the abdominal wall. It, too, is composed of squamous epithelium like the external lamina, with both sweat and sebaceous glands, but the internal lamina lacks hair follicles. In dogs, the internal lamina also contains lymph follicles and special nerve endings called genital bodies.

Smegma is the oily debris that normally forms within the prepuce (even in castrated male dogs). It is composed of desquamated epithelial cells and sebaceous gland secretions of the internal lamina of the prepuce. Cleaning and removing smegma is one reason why male dogs lick the preputial orifice. Another possible reason is discomfort, such as that caused by inflammation of the prepuce (posthitis), glans penis (balanitis), or both (balanoposthitis). In each case, enlarged lymph follicles may be visible as raised, red spots, which cover the internal lamina of the prepuce. Licking may also be considered normal masturbatory behavior in young adult dogs. Masturbation in dogs is of no clinical significance if it does not become so frequent and persistent that it creates posthitis or balanitis.

The **penile frenulum** is a band of connective tissue that connects the penis and prepuce in the fetus and neonate. It normally separates within a few months of birth but may persist in adults, limiting extrusion of the penis from the prepuce. Surgical transection of the frenulum may be required in such cases.

Muscles of the penis and prepuce

Several different muscles insert on the penis or prepuce, thus playing an important role in copulation.

Abnormalities involving any of these copulatory muscles or their innervation may result in the inability to successfully mate.

In dogs, the variable strands of the **cranial preputial muscle** derive from the cutaneous trunci muscle. The cranial preputial muscle originates at the xiphoid region and inserts on the internal lamina of the prepuce. It may appear as a single muscular sheet or as a set of strands that radiate cranially from the preputial area into the surrounding fascia. When this muscle contracts, it draws the prepuce cranially and closes the preputial orifice. This muscle is absent in cats.

In dogs and cats, 3 muscles play an important role in ejaculation: **urethralis**, **bulbospongiosus**, and **ischiocavernosus**. Their rhythmic contractions transport seminal fluids through the urethra. The bulbospongiosus is a continuation of the urethralis muscle. Its contraction empties the distal portion of the penile urethra during ejaculation and urination.

The bulbospongiosus and ischiocavernosus muscles are also important as muscular pumps, which act on the arteries within the corpus cavernosum and corpus spongiosum to cause engorgement and thus erection of the penis. Rhythmic contraction of the ischiocavernosus muscle pumps arterial blood into the spongy tissues and decreases venous return, thus helping to maintain the erection. In addition, compression of a fourth muscle, the **ischiourethralis**, on the dorsal penile vein decreases venous drainage from the spongy bodies of the penis, helping to maintain the erection.

In cats, the **apical penile ligament** is another essential structure, as it cranially flexes the penis during erection—a prerequisite for intromission during coitus in cats, as the penis is directed caudally in its relaxed state.

Blood supply, lymphatics, and innervation of the penis

The main blood supply to the penis and prepuce is the **penile artery**, which originates from the **internal pudendal artery** and has 3 main branches. The **artery of the bulb of the penis** and the **deep penile arteries** primarily supply the corpus cavernosus and corpus spongiosum. The **dorsal penile artery** in dogs supplies the glans penis and the prepuce; in cats, it supplies only the prepuce.

The dorsal penile artery has a **superficial preputial branch**, which anastomoses with **preputial branches** of the **external pudendal artery** that may supplement the dorsal penile artery. Branches of the external pudendal artery also supply the prepuce.

The veins draining the penis and prepuce track with their corresponding arteries and are similarly named. The lymphatic vessels of the penis and prepuce drain into the **superficial inguinal lymph nodes**.

The **dorsal penile nerve**, a branch of the **pudendal nerve**, provides sensory innervation to the penis and prepuce. **Lumbar nerves** also provide sensory fibers to the prepuce, supported by the **superficial perineal nerve**. Fibers from the **pelvic plexus** innervate both the blood vessels and the spongy bodies of the penis. The cranial preputial muscle, present in dogs but not cats, is innervated by the **lateral thoracic nerve**.

The bulbospongiosus muscle acts as a muscular pump on the artery of the bulb of the penis to effect an erection. In dogs, the consequent engorgement of the bulbus glandis results in "tying" during coitus, in which the vestibular muscles in the bitch (see Case 6.3) contract around the glans penis at the base of the bulbus glandis, effectively preventing the male from dismounting for a variable period of time (up to 45 min) after ejaculation.

The preputial branches of the paired caudal superficial epigastric artery, a branch of the external pudendal artery, must be considered when performing a paramedian incision for laparotomy in male dogs (see Case 5.8).

Selected references

[1] Goericke-Pesch S, Hölscher C, Failing K, et al. Functional anatomy and ultrasound examination of the canine penis. Theriogenology 2013;80:24–33.
[2] Olsen D, Salwei R. Surgical correction of a congenital preputial and penile deformity in a dog. J Am Anim Hosp Assoc 2001;37:187–92.
[3] Neihaus SA, Hathcock TL, Boothe DM, et al. Presurgical antiseptic efficacy of chlorhexidine diacetate and providone-iodine in the canine preputial cavity. J Am Anim Hosp Assoc 2011;47:406–12.
[4] Root Kustritz MV. Disorders of the canine penis. Vet Clin North Am Small Anim Pract 2001;31:247–58.

CHAPTER 7

THORACIC LIMB

Helen M.S. Davies, Chapter editor

Proximal Thoracic Limb (Shoulder, Brachium, and Antebrachium)

7.1 Osteochondritis dissecans of the shoulder—*Kimberly A. Agnello* 392

7.2 Incomplete ossification of the humeral condyle—*Takanori Sugiyama, Helen M.S. Davies, and Cathy Beck* 400

Distal Thoracic Limb (Carpus and Manus)

7.3 Carpal valgus deformity—*Lane A. Wallett, Cathy Beck, and Helen M.S. Davies* 416

7.4 Phalangeal fracture—*Ray G. Ferguson and Helen M.S. Davies* 430

Innervation of the Thoracic Limb

7.5 Nerve sheath neoplasm—*Eric N. Glass and Marc Kent* 446

CASE 7.1

Osteochondritis Dissecans of the Shoulder

Kimberly A. Agnello
Department of Clinical Sciences & Advanced Medicine, University of Pennsylvania School of Veterinary Medicine, Philadelphia, Pennsylvania, US

Clinical case

History

A 6-month-old female spayed Labradoodle presented with a 6-week history of intermittent left thoracic limb lameness. The lameness was worse when first rising after rest or after activity. No improvement in clinical signs was observed with the administration of a nonsteroidal anti-inflammatory agent and exercise restriction.

Physical examination findings

At a walk and a trot, the dog displayed a mild decreased weight-bearing lameness of the left thoracic limb. On standing palpation, mild muscle atrophy was palpated over the supraspinous, infraspinatus, and deltoid muscles of the left thoracic limb. The dog had pain during the endpoint of range of motion (ROM) in flexion and extension of the glenohumeral (scapulohumeral) joint, with more pronounced pain in extension.

Differential diagnoses

Osteochondritis dissecans (OCD), chronic fracture of proximal humerus and/or distal scapula, injury of the surrounding soft tissue stabilizers of the shoulder joint, or glenohumeral sepsis

Diagnostics

Bilateral orthogonal radiographs of the glenohumeral joints were performed. The left glenohumeral joint revealed an irregular concavity with a rounded lucent area along the caudal humeral head with associated moderate subchondral sclerosis (Fig. 7.1-1A). The right glenohumeral joint revealed a flattened caudal humeral head with an associated rounded lucent area and mild subchondral sclerosis (Fig. 7.1-1B).

OSTEOCHONDRITIS DISSECANS (OCD)

Osteochondrosis is a developmental disease in which there is a disturbance of the normal process of endochondral ossification at the epiphysis or the physis. When it occurs at the epiphysis, it usually results in a focal area of thickened cartilage that is predisposed to injury. Decreased nutrition and normal stresses of the joint result in vertical clefts in the cartilage and detachment of a cartilage flap. This leads to synovitis and abnormal wear on the opposing cartilage surface, causing pain and lameness. The clinical syndrome is termed osteochondritis dissecans. The caudal humeral head is a common location in the dog and frequently presents with bilateral pathology, generally between the ages of 4–9 months.

Comparative Veterinary Anatomy: A Clinical Approach. https://doi.org/10.1016/B978-0-323-91015-6.00044-3

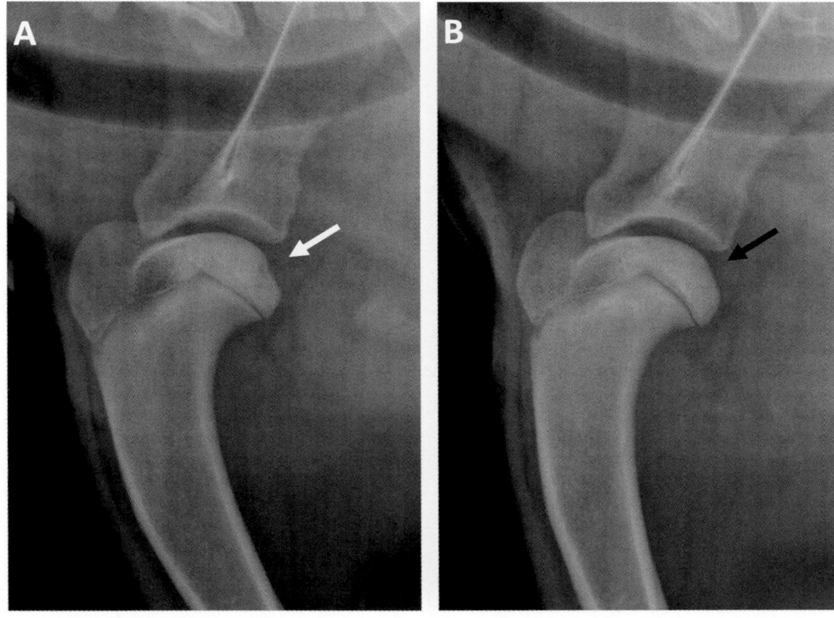

FIGURE 7.1-1 Lateral radiographs from a dog with osteochondrosis. (A) The left lateral projection of the glenohumeral joint shows an irregular concavity with a rounded lucent area along the caudal humeral head with associated moderate subchondral sclerosis *(white arrow)*. (B) The right lateral projection of the glenohumeral joint shows a flattened caudal humeral head with an associated rounded lucent area and mild subchondral sclerosis *(black arrow)*.

A CT scan was also performed of both glenohumeral joints. A large, irregular subchondral bone defect with adjacent subchondral bone sclerosis was seen in the left caudal humeral head on the sagittal and dorsal projections. In addition, a curvilinear structure representing a cartilage flap was identified immediately adjacent to the bone defect (Fig. 7.1-2A and C). The right sagittal and dorsal projections also identified a caudal humeral head defect with adjacent subchondral sclerosis; however, this defect was much smaller than seen on the right and no cartilage flap was visualized (Figs. 7.1-2B, D and 7.1-3).

Diagnosis

Osteochondritis dissecans of the left caudal humeral head, and osteochondrosis of the right caudal humeral head

Caudal humeral head OCD has been reported in cats; however, clinical feline cases are uncommon as compared with the dog.

Treatment

The dog was placed under general anesthesia for bilateral arthroscopic glenohumeral joint exploration via a lateral approach. The camera/scope portal was placed slightly distal and caudal to the acromion process of the scapular spine on the lateral surface of the glenohumeral joint. The instrument portal was also placed on the lateral surface of the joint, just caudal and distal to the camera portal. In the left glenohumeral joint, a large cartilage flap (Fig. 7.1-4A) was identified on the caudal humeral head with generalized joint synovitis; the most severe change was seen in the caudal joint capsule adjacent to the lesion (Fig. 7.1-4D). In addition, there was cartilage injury on the opposing surface of the glenoid, likely due to the abnormal wear of the articular surface from the flap—i.e., a "kissing lesion" (Fig. 7.1-4B). The right glenohumeral joint had only a small superficial indentation in the cartilage identified on the caudal humeral head.

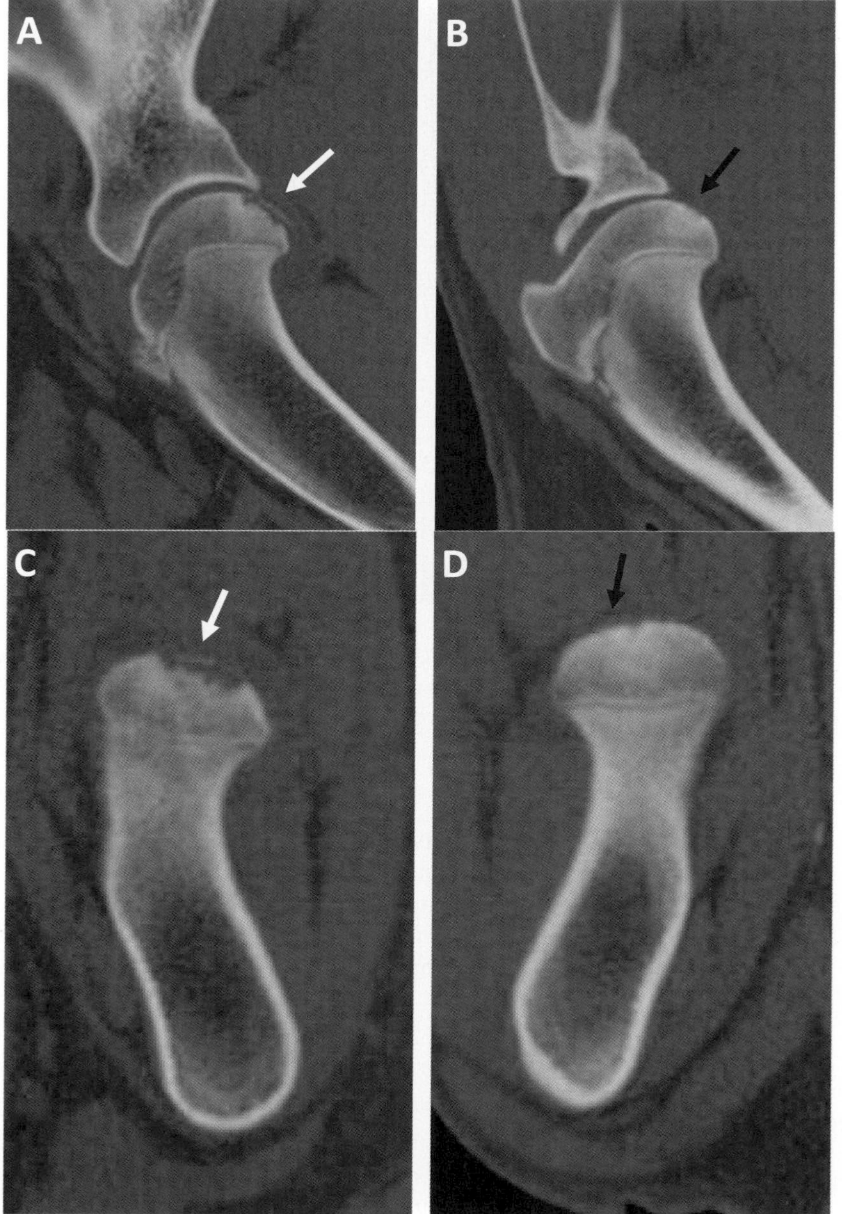

FIGURE 7.1-2 Computed tomography scans of the left and right glenohumeral joints. (A) Sagittal image of the left glenohumeral joint identifying a large subchondral bone defect of the caudal humeral head with associated subchondral bone sclerosis. Immediately adjacent to the bone defect is a mineralized curvilinear structure, which represents the cartilage flap *(white arrow)*. (B) Sagittal image of the right glenohumeral joint showing the smaller subchondral bone defect with associated subchondral sclerosis of the caudal humeral head *(black arrow)*. A cartilage flap was not seen in the right glenohumeral joint. (C) Dorsal image of the left glenohumeral joint identifying the large subchondral bone defect with associated subchondral bone sclerosis of the caudal humeral head and mineralized curvilinear body representing the cartilage flap *(white arrow)*. (D) Dorsal image of the right glenohumeral joint showing a smaller defect and surrounding subchondral bone sclerosis of the caudal humeral head *(black arrow)*.

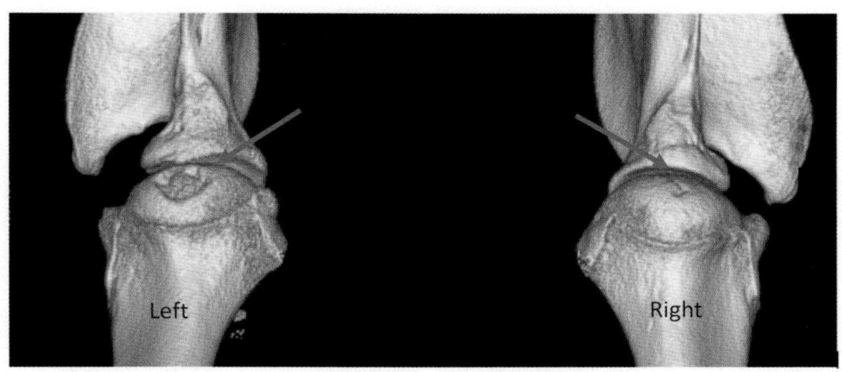

FIGURE 7.1-3 Computed tomography volume reconstructed scan of the left and right glenohumeral joints. The *red arrows* indicate the subchondral bone defect of the caudal humeral head.

The injured cartilage and flap were removed (Fig. 7.1-4C and D), and curettage of the subchondral bone bed within the defect (also referred to as abrasion arthroplasty) (Fig. 7.1-4E and F) was performed.  Some alternatives to curettage of the humeral defect for the treatment of OCD in the dog include resurfacing the defect with an autologous osteochondral graft, an allograph, hemiarthroplasty (for large lesions), or synthetic plugs.

> Removal of the cartilage flap is intended to relieve clinical signs of pain and lameness and decrease abnormal wear on the opposing joint surface with the goal of slowing progression of osteoarthritis. The aim of the subchondral bone debridement is to remove unhealthy cartilage to the level of the subchondral bone plate, ensuring healthy bleeding bone is exposed to stimulate fibrocartilage repair in the defect.

Clinically, the dog did well during her 6-week post-operative recovery, during which exercise was limited to low-impact activities, such as walking. Following surgical recovery, the dog returned to normal activity without evidence of pain or lameness.

Anatomical features in canids and felids

Introduction

The glenohumeral joint is a common source of thoracic limb lameness in the dog; therefore, an understanding the shoulder joint and its associated musculature, tendons, blood supply, and innervation is necessary to diagnose and successfully treat the source of the lameness. The shoulder anatomy of the cat has some anatomical differences noted in the text to follow.

The glenohumeral joint is a diarthrodial "ball and socket" joint between the scapula and the humerus. The distal ventral border of the scapula forms the glenoid cavity and articulates with the proximal head of the humerus. Glenoid coverage of the humeral head is minimal, which allows for a high degree of motion of the shoulder joint. Consequently, stability of the shoulder relies heavily on the surrounding soft tissue support structures.

Function

The anatomy of the shoulder joint allows for a large range of movements, including abduction, adduction, and circumduction; however, flexion and extension remain the main movements. The soft tissue stabilizers of the shoulder joint are passive (static) and active (dynamic) stabilizers. These stabilizers function together to stabilize the shoulder during normal range of motion and movement, and to counteract forces that may destabilize the joint.

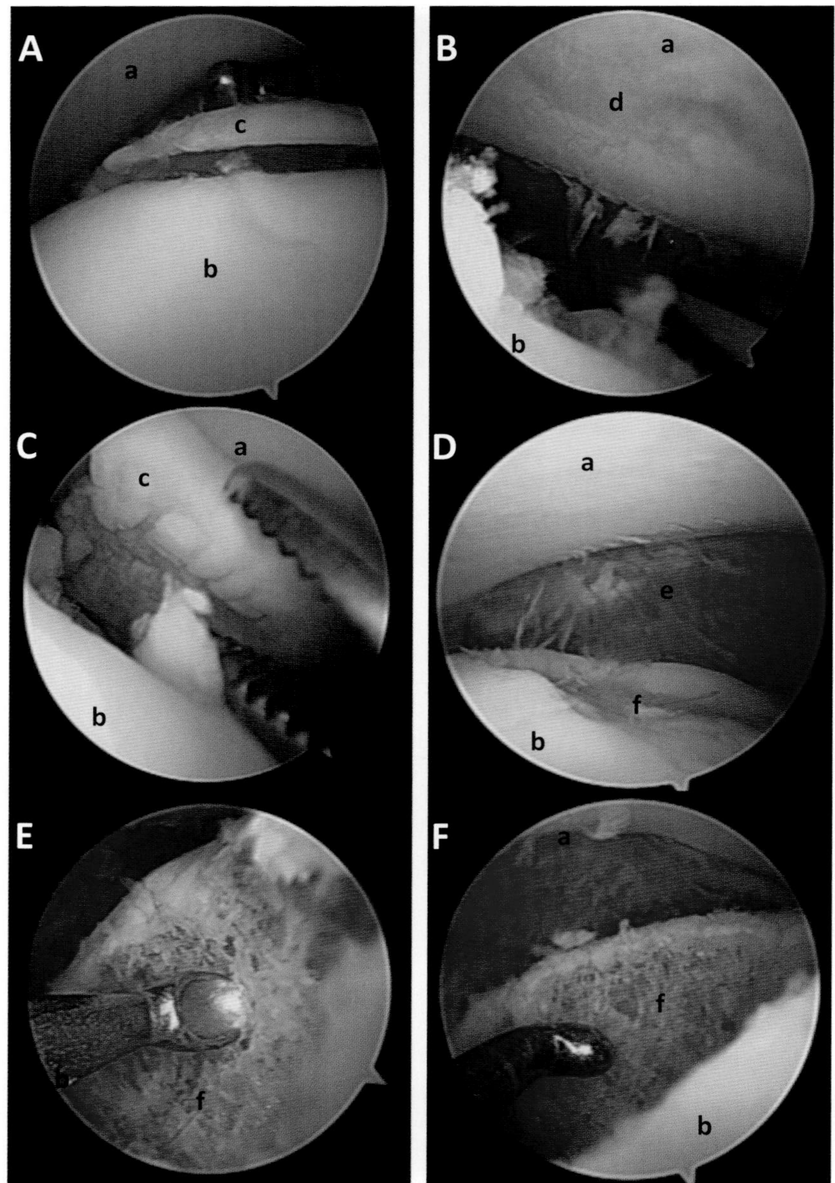

FIGURE 7.1-4 Arthroscopic images of the left glenohumeral joint. (A) The OCD flap (c) can be identified between the glenoid cavity of the scapula and the humeral head (b) arising from the articular surface of the humeral head (b). (B) The articular surface of the glenoid cavity (a) directly adjacent to the cartilage flap has a partial thickness cartilage lesion (d), representing a "kissing lesion." (C) Removal of the cartilage flap (c) with an arthroscopic grasper. (D) Humeral head articular surface after removal of the cartilage flap and associated caudal medial joint synovitis. (E) Debridement of the subchondral bone with a curette. (F) Final defect after curettage with bleeding subchondral bone. Key: *a*=glenoid cavity of the scapula; *b*=humeral head; *c*=OCD cartilage flap; *d*=articular cartilage damage on the glenoid cavity; *e*=synovitis of the caudal medial joint capsule; *f*=articular cartilage defect after the removal of the cartilage flap.

Glenohumeral joint

The passive stabilizers of the shoulder are those in which muscle movement is not necessary for stability, and include limited joint fluid volume with adhesion and cohesion properties, concavity compression (shape/congruity of the glenoid and humeral head), joint capsule, glenohumeral ligaments (medial and lateral), labrum, and the tendon of origin of the biceps brachii muscle. The **medial glenohumeral ligament** is characteristically Y-shaped (Figs. 7.1-5–7.1-9) and the **lateral glenohumeral ligament** is a single band that blends with the joint

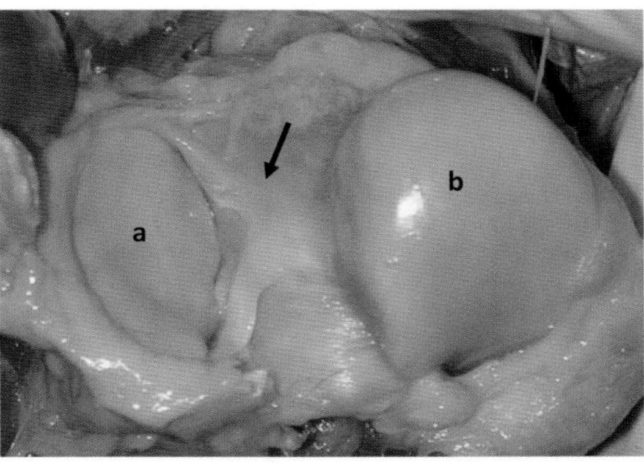

FIGURE 7.1-5 Gross appearance of the Y-shaped medial glenohumeral ligament *(black arrow)*. Key: a = glenoid cavity, b = humeral head.

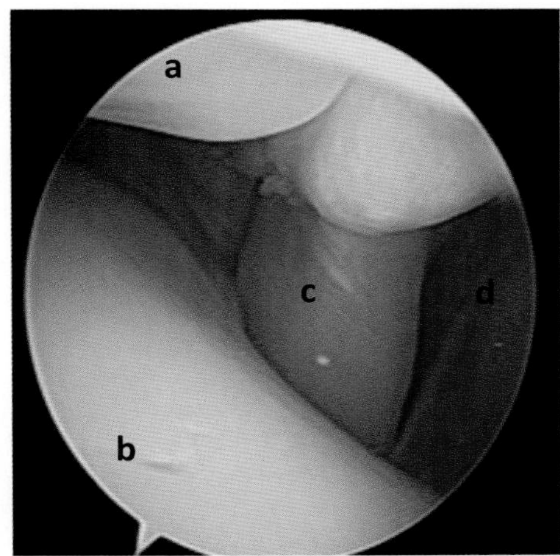

FIGURE 7.1-6 Arthroscopic image of the biceps brachii tendon in the cranial shoulder joint. Key: a = glenoid cavity, b = humeral head, c = biceps brachii tendon, d = joint capsule.

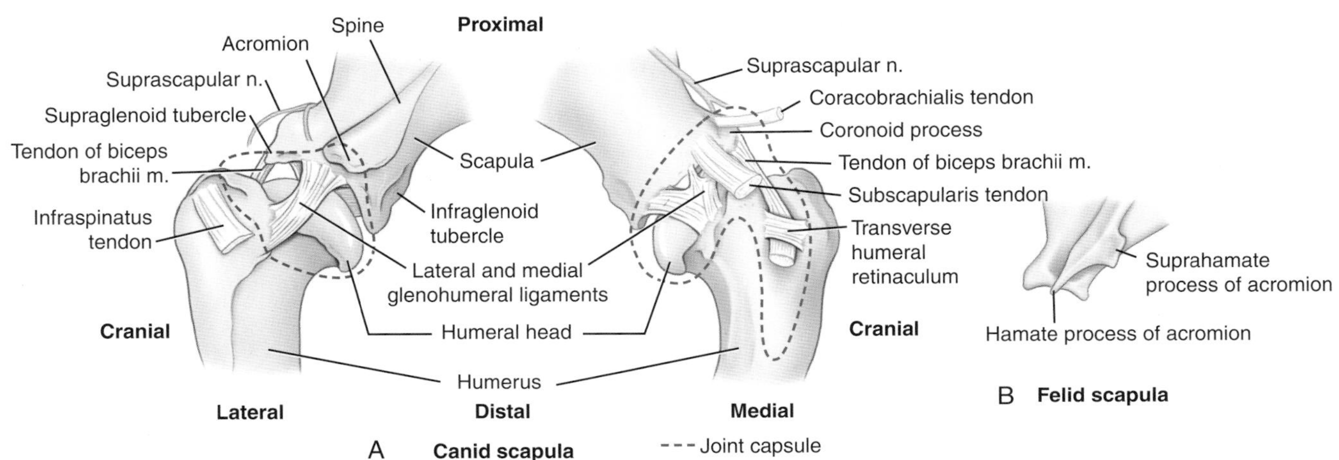

FIGURE 7.1-7 (A) Medial and lateral stabilizers of the canine glenohumeral joint. *Blue dashed line* represents an outline of the joint capsule. (B) The superhamate and hamate process of the feline acromion of the scapular spine. The suprahamate process extends caudally from the distal scapular spine. This structure is lacking in the canine scapula.

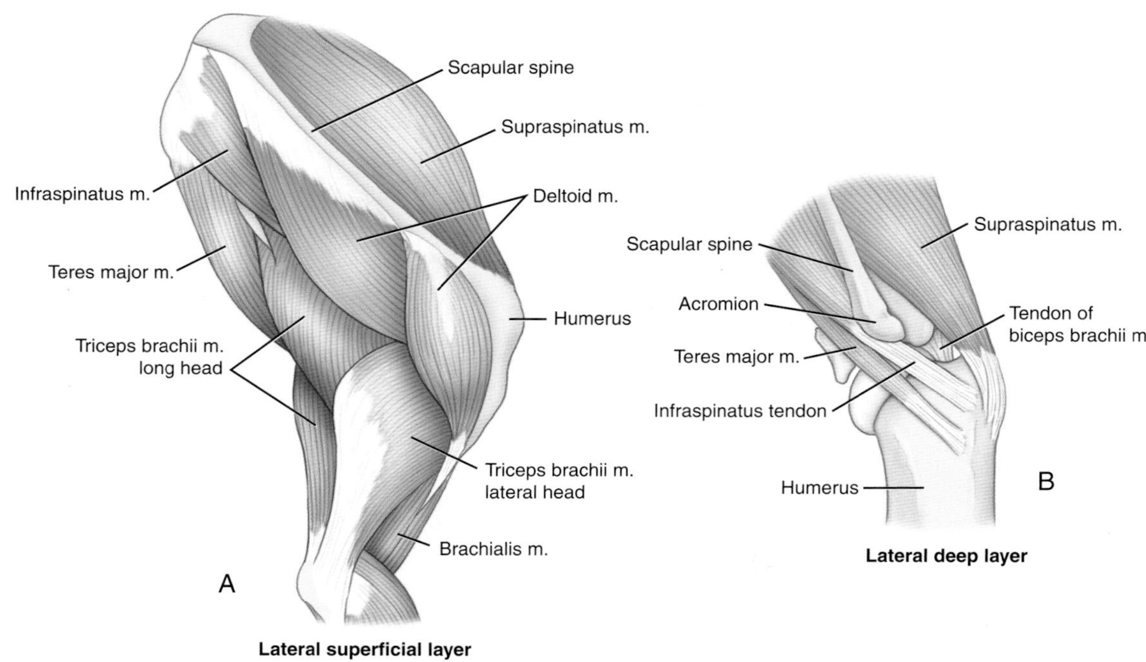

FIGURE 7.1-8 Lateral images of the musculotendinous structures of the canine glenohumeral joint: (A) superficial and (B) deep layers.

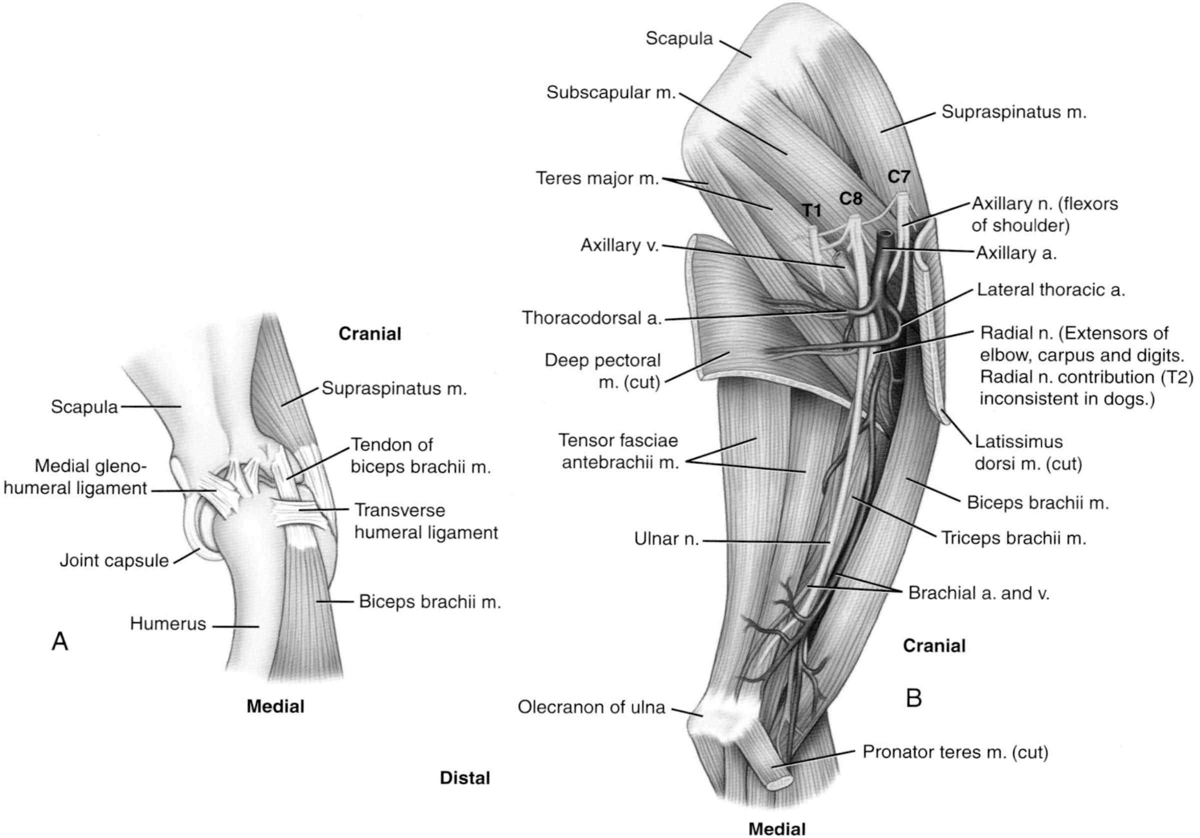

FIGURE 7.1-9 (A) Medial supporting structures of the canine glenohumeral joint. (B) Muscular, vascular, and nervous supply to the medial aspect of the thoracic limb and glenohumeral joint.

capsule (Figs. 7.1-7A and 7.1-9A). The **labrum** is triangular in cross section and is attached to the circumference of the **glenoid**, aiding with joint congruity by deepening the concavity of the glenoid. The **biceps brachii tendon** originates on the **supraglenoid tubercle** of the **scapula** and runs on the cranial surface of the shoulder joint in the **intertubercular groove** (Figs. 7.1-7A and 7.1-9A). The **intertubercular (bicipital) bursa** is the synovial sheath for the biceps brachii tendon and is an extension of the shoulder joint capsule. The biceps tendon can be observed on exploration of the shoulder joint (Fig. 7.1-6).

The active stabilizers of the shoulder include the infraspinatus, supraspinatus, subscapularis, and teres minor muscles, and to a lesser extent the biceps brachii, long head of the triceps brachii, deltoideus, and teres major muscles. Most of these muscles have different functions depending on the position of the joint, and they allow for active movement of the joint along with providing stabilization. The lateral muscles include the supraspinatus, infraspinatus, and the teres minor muscles. The **supraspinatus** and **infraspinatus muscles** provide lateral stability, and the **teres minor**—together with the **infraspinatus muscles**—prevents medial rotation during weight-bearing. The medial muscles include the subscapularis and the teres major muscles. The **subscapularis muscle** stabilizes the joint medially and—with the **teres major muscle**—prevents lateral rotation during weight-bearing. Again, to a lesser degree, the biceps brachii muscle adds stability cranially, the **long head of the triceps brachii** muscle caudally, and the **deltoid muscle** laterally (Figs. 7.1-8 and 7.1-9).

The feline shoulder joint has few anatomic differences, including the presence of a nonfunctional, rudimentary **clavicle**. The feline acromion of the scapular spine has a **suprahamate process** in addition to the hamate process (a laterally flattened prominence in carnivores). The suprahamate process extends caudally from the distal scapular spine (see Fig. 7.1-7B). This structure is lacking in the dog.

Regional blood supply, lymphatics, and innervation

The main blood supply and innervation to the glenohumeral joint is from the **axillary artery** and **nerve**. The **caudal circumflex humeral artery** is a branch of the axillary artery that supplies the joint capsule. The cranial muscles of the glenohumeral joint (biceps brachii and the brachialis muscles) are supplied by the **superficial cervical**, **axillary**, and **brachial arteries**, and are innervated by the **musculocutaneous nerve**. The lateral muscular stabilizers of the shoulder joint (supraspinatus and infraspinatus muscles) are supplied by the superficial cervical artery and the **subscapular nerve**, and the caudal muscles (deltoideus, teres major, and teres minor) are supplied by the subscapular artery and the axillary nerve (Fig. 7.1-9B).

> The axillary artery and nerve run across the caudal and lateral portion of the glenohumeral joint and therefore can be inadvertently injured when using a lateral arthroscopy approach if the portals are placed too far caudal.

The venous drainage is mainly from the **axillary vein** with lymphatic drainage to the **superficial cervical** and **axillary lymph nodes**.

Selected references

[1] Johnston SA. Osteochondritis dissecans of the humeral head. Vet Clin North Am Small Anim Pract 1998;28:33.
[2] Fitzpatrick N, van Terheijden C, Yeadon R, et al. Osteochondral autograft transfer for treatment of osteochondritis dissecans of the caudocentral humeral head in dogs. Vet Surg 2010;39:925.
[3] Sparrow T, Fitzpatrick N, Blunn G. Shoulder joint hemiarthroplasty for treatment of a severe osteochondritis dissecans lesion in a dog. Vet Comp Orthop Traumatol 2014;27:243.
[4] Cook JL, Kuroki K, Bozynski CC, et al. Evaluation of synthetic osteochondral implants. J Knee Surg 2014;27:295.
[5] Gray MJ, Lambrechts NE, Maritz NGJ. A biomechanical investigation of the static stabilizers of the glenohumeral joint in the dog. Vet Comp Orthop Traumatol 2005;18:55.

CHAPTER 7

CASE 7.2

Incomplete Ossification of the Humeral Condyle

Takanori Sugiyama[a], **Helen M.S. Davies**[b], **and Cathy Beck**[b]
[a]Animalius Vet, Bayswater, WA, AU
[b]Veterinary BioSciences, The University of Melbourne, Melbourne, Victoria, AU

Clinical case

History

A 9-month-old female spayed German Shorthaired Pointer was presented for acute, nonweight-bearing lameness in the right thoracic limb. There was no history of trauma or prior lameness.

Physical examination findings

On presentation, the dog was nonweight-bearing on the right thoracic limb. Extension of the right elbow elicited a pain response that was localized to the elbow joint by palpation. No other abnormalities were identified.

Differential diagnoses

Lameness and pain localized to the elbow joint may be caused by several specific abnormalities in this simple "hinge" joint. In a young dog, developmental orthopedic diseases, collectively termed "elbow dysplasia," must be high on the differential list. Specific conditions causing lameness include OCD, FCP, UAP, and IOHC.

Osteochondritis dissecans (OCD) involving the medial humeral condyle; medial coronoid disease, such as fragmented coronoid process (FCP); ununited anconeal process (UAP); incomplete ossification of the humeral condyle (IOHC); articular fracture; soft tissue injury

Diagnostics

Computed tomography was performed on both elbows. In the right elbow, a complete fissure/cleft separated the lateral and medial parts of the humeral condyle, extending from the articular surface of the trochlea to the supratrochlear foramen (Fig. 7.2-1). The left elbow was normal.

Diagnosis

Incomplete ossification of the humeral condyle in the right elbow

Treatment

Under general anesthesia, a transcondylar bone screw was used to "reduce" (compress) the condylar fissure, restore the integrity of the articular surface, and stabilize the humeral condyle. A lateral approach to the right elbow joint

Comparative Veterinary Anatomy: A Clinical Approach. https://doi.org/10.1016/B978-0-323-91015-6.00045-5

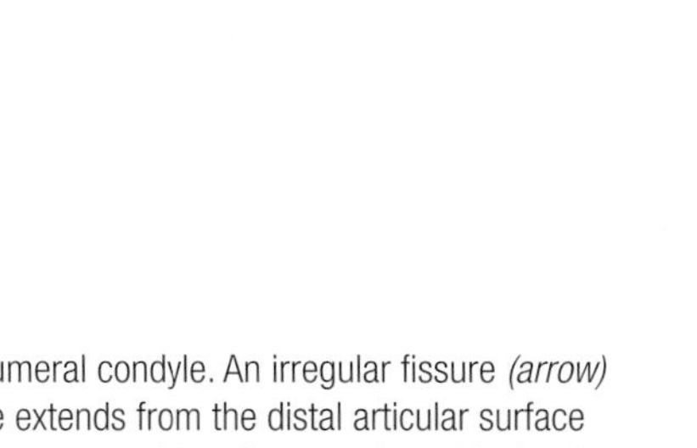

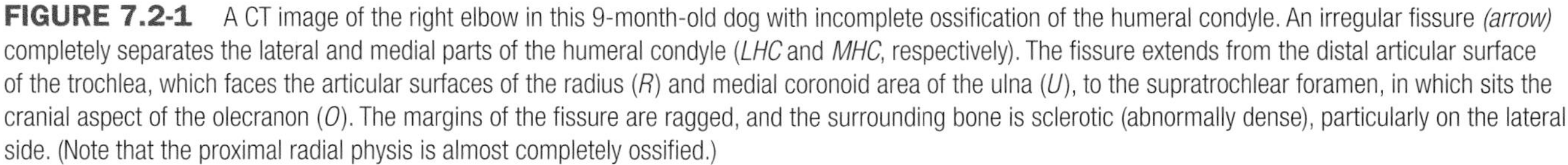

FIGURE 7.2-1 A CT image of the right elbow in this 9-month-old dog with incomplete ossification of the humeral condyle. An irregular fissure *(arrow)* completely separates the lateral and medial parts of the humeral condyle (*LHC* and *MHC*, respectively). The fissure extends from the distal articular surface of the trochlea, which faces the articular surfaces of the radius (*R*) and medial coronoid area of the ulna (*U*), to the supratrochlear foramen, in which sits the cranial aspect of the olecranon (*O*). The margins of the fissure are ragged, and the surrounding bone is sclerotic (abnormally dense), particularly on the lateral side. (Note that the proximal radial physis is almost completely ossified.)

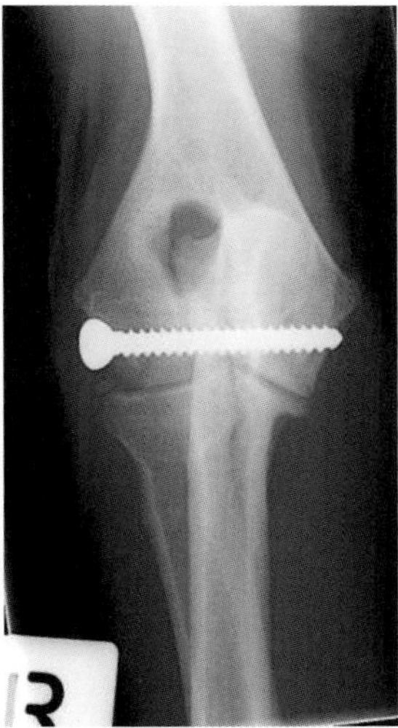

FIGURE 7.2-2 Post-operative radiograph (craniocaudal view) of this patient, showing the transcondylar bone screw inserted to repair the condylar fissure and stabilize the humeral condyle.

was used for the repair. Using fluoroscopic guidance, a Kirschner wire (K-wire) was inserted across the humeral condyle (i.e., transcondylar) as a guide. After confirmation of appropriate K-wire placement, a drill hole was made using cannulated drill bits. A 4.5-mm (0.18-in.) cortical bone screw was then inserted across the fissure in a lag fashion to compress the fissure and restore the integrity of the humeral condyle. Post-operative radiographs confirmed proper positioning of the screw (Fig. 7.2-2). At the follow-up examination, the dog was not lame and had a normal, pain-free range of motion in the right elbow.

Anatomical features in canids and felids

Introduction

The elbow joint is positioned distal to the shoulder joint and proximal to the carpal joints within the thoracic limbs. It is the joint that lies at the junction between the large shoulder muscles distributed in a fan-shaped arrangement towards the dorsal surface of the dog and cat and the more vertically oriented bones and muscles of the rest of the thoracic limb. Thus, in a normal standing posture, the humerus is in an upwardly sloping angle between the elbow and the shoulder, while the radius and ulna are approximately vertical. This case discusses the anatomy of the elbow joint along with regional musculature, blood supply, lymphatics, and innervation.

Function

The elbow acts as a hinge coordinating vertical support from the distal parts of the limb in contact with the ground with the various movements between the shoulder muscles and the chest wall. The shoulder movements include flexion, extension, slight abduction, adduction, and rotation of the shoulder joint, and more extensive proximal, distal, cranial, caudal, medial, and lateral positioning and rotation of the whole limb in contact with the thoracic wall. The elbow joint is the major stabilizer providing support of the trunk from the thoracic limb in whatever position the thoracic limb lands during movement while still functioning as a spring, distributing the weight of the trunk onto the limb in a predictable way aligned with the bony column. Sturdy ligaments and closely connected joint surfaces are needed for effortless function.

Elbow joint

The elbow joint is sometimes called the "cubital joint" (L. *cubitus* the bend of the arm). However, "cubital" is more often used in human anatomy.

The term **elbow** is generally used to describe the region between the distal brachium ("arm") and the proximal antebrachium ("forearm"). The term **elbow joint** specifically describes the articulation between the distal humerus and the proximal radius and ulna.

The elbow joint includes 3 articulations, enclosed by a common joint capsule: the distal humerus articulates with (1) the proximal radius and (2) the proximal ulna, and (3) the proximal radius and ulna articulate immediately distal to their humeral surfaces. (A second radioulnar articulation is also present distal to, and separate from, the elbow joint, described later.)

Distal humerus (Figs. 7.2-3–7.2-5)

The distal end of the **humerus** is called the **condyle**. Although it may appear as 2 condyles (lateral and medial), it is considered a single condyle that is marked centrally by a broad **trochlea**, (L. pulley-shaped part or structure resembling a pulley) which extends from the cranial to caudal margins of the condyle. The trochlea of the elbow is a wide groove in the distal end of the humerus which articulates extensively with the trochlear notch of the ulna. Lateral to the trochlea, on the cranial and distal aspects of the condyle, is the **capitulum** (L. *caput* a little head), which is the humeral surface that articulates with the radial head.

In dogs and cats, unilateral condylar fractures almost always occur on the lateral side of the humeral condyle because of the relationship between the distal humerus and the proximal radius, and the fact that the radius is the main weight-bearing "strut" in the antebrachium of quadrupeds.

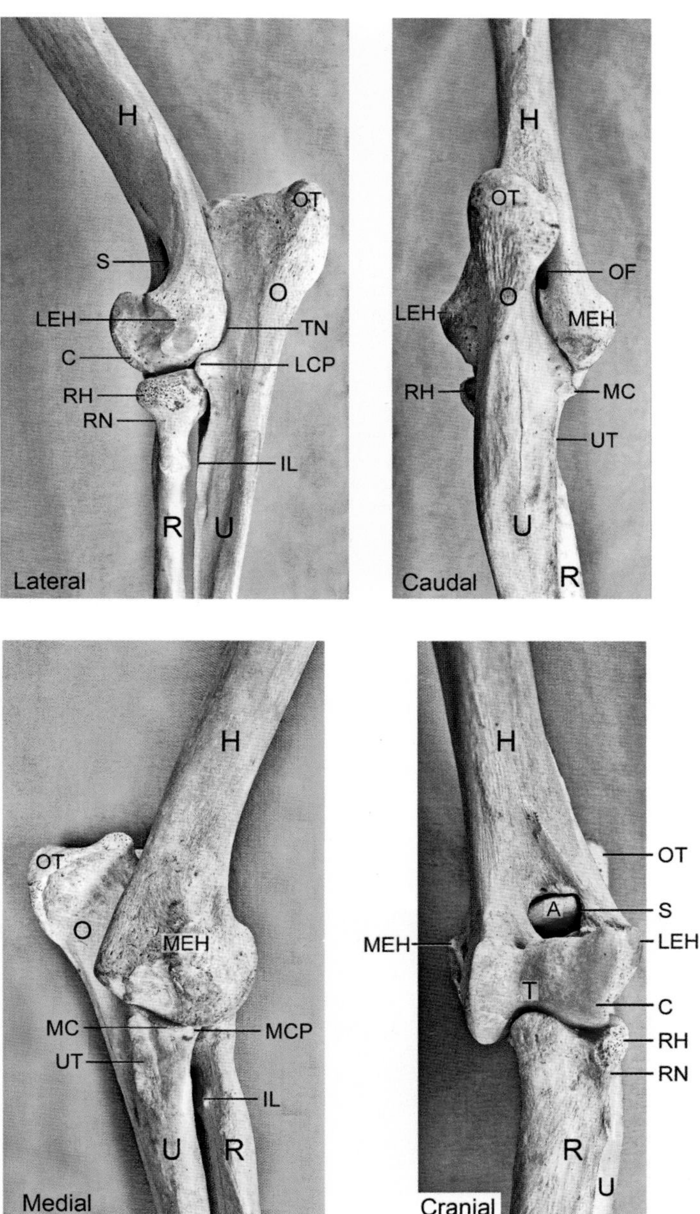

FIGURE 7.2-3 Bones of the canine elbow, left thoracic limb: lateral, caudal, medial, and cranial aspects. Key: *A*, anconeal process of the olecranon; *C*, capitulum of the humerus; *H*, humerus; *IL*, interosseus ligament (attachment site); *LCP*, lateral coronoid process of the ulna; *LEH*, lateral epicondyle of the humerus; *MC*, medial coronoid area; *MCP*, medial coronoid process of the ulna; *MEH*, medial epicondyle of the humerus; *O*, olecranon; *OF*, olecranon fossa of the humerus; *OT*, olecranon tuberosity; *R*, radius; *RH*, radial head; *RN*, radial neck; *S*, supratrochlear foramen of the humerus; *T*, trochlea of the humerus; *TN*, trochlear notch of the olecranon; *U*, ulna; *UT*, ulnar tuberosity (attachment for biceps brachii and brachialis tendons). Note: Not all structures are visible on all views.

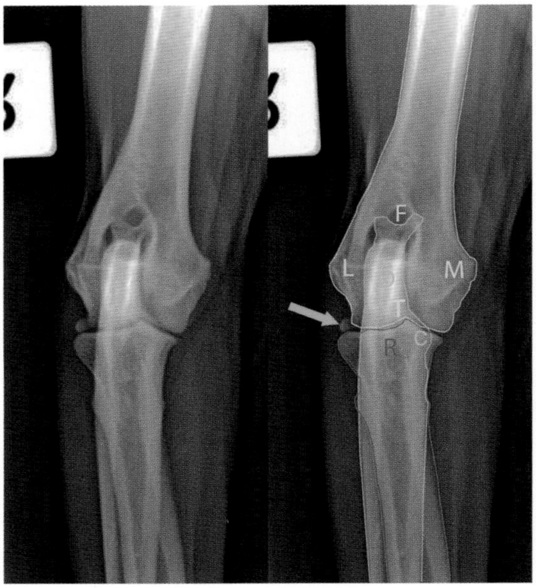

FIGURE 7.2-4 Craniocaudal radiograph of the right elbow of an adult dog. On the right, the humerus is outlined in *yellow*, the radius in *pink*, and the ulna in *blue*. Also marked are the lateral (*L*) and medial (*M*) epicondyles, olecranon fossa (*F*), and trochlea (*T*) of the humerus, the olecranon (*O*) and medial coronoid (*C*) area of the ulna, and the proximal radius (*R*). This dog has a small sesamoid bone *(arrow)* at the origin of the supinator muscle.

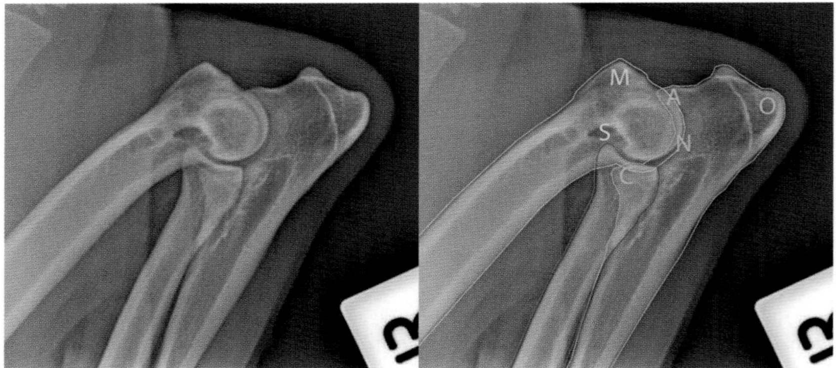

FIGURE 7.2-5 Flexed lateral radiograph of the right elbow of an adult dog. On the right, the humerus is outlined in *yellow*, the radius in *pink*, and the ulna in *blue*. Also marked are the medial epicondyle (*M*) and supratrochlear foramen (*S*) of the humerus and the anconeal process (*A*), olecranon tuberosity (*O*), trochlear notch (*N*), and medial coronoid process (*C*) of the ulna.

Prominent bony eminences on the abaxial surfaces, the **lateral** and **medial epicondyles**, serve as the proximal attachments for the collateral ligaments of the elbow joint and various extensors and flexors of the carpus and digits (described later). The medial epicondyle is significantly larger than the lateral.

The size disparity between the lateral and medial epicondyles is apparent on the craniocaudal radiographic view of the elbow (Fig. 7.2-4), which is helpful if the radiograph is improperly labeled. On the lateral view (a mediolateral projection), the caudal border of the medial epicondyle is more prominent than that of the lateral (Fig. 7.2-5). It does not mean that the projection is obliqued; this disparity is present even when the lateral and medial parts of the humeral condyle are exactly superimposed.

The prominences of the epicondyles are palpable landmarks during surgery for the elbow. When placing a transcondylar screw, a lateral approach is generally recommended because the lateral epicondyle is smaller than the medial epicondyle, providing improved centering of the screw within the epicondyle.

In dogs, a deep **olecranon fossa** lies immediately proximal to the trochlea on the caudal aspect of the humerus. As the name suggests, it accommodates the olecranon of the ulna, particularly during extension of the elbow. On the craniolateral aspect of the humerus is the smaller **radial fossa**, so-named because it lies proximal to the radial head. Both fossae typically communicate through a large **supratrochlear foramen**.

In dogs, the supratrochlear foramen may be noticeable radiographically on lateral and oblique views of the elbow (Fig. 7.2-5). It is largely obscured by the olecranon in craniocaudal views of the elbow.

In cats, the olecranon fossa is much smaller and there is no supratrochlear foramen. Another anatomical difference between the 2 species is that cats have a small, oblique **supracondylar foramen** that perforates the distal humerus on its medial aspect (Fig. 7.2-6). This narrow, linear foramen runs obliquely, parallel to the medial surface of the humerus, and proximal to the medial epicondyle. The median nerve and brachial artery pass through this foramen, while in dogs these structures lie medial to the humerus.

In cats, the supracondylar foramen should not be mistaken for a fissure, fracture, or other bone pathology. The absence of a supratrochlear foramen makes the humeral condyle more resistant to fracture in cats than in dogs. However, supracondylar or distal diaphyseal fractures in cats can result in injury to the median nerve if they involve the supracondylar foramen. This foramen also limits the safe placement of orthopedic pins and screws in the feline elbow.

In puppies and kittens, the **distal humeral physis** ("growth plate") is located just proximal to the condyle, immediately distal to the supratrochlear foramen in dogs (Fig. 7.2-7). Radiographically, this physis is fully ossified ("closed") by 6–8 months of age in dogs and 3–4 months of age in cats. The humeral condyle (the distal humeral epiphysis) itself develops from 3 separate ossification centers: lateral and medial to the trochlear groove, and a small center that forms the **medial epicondyle**. The larger **lateral** and **medial** centers are separated by a cartilaginous plate, which extends from the distal articular surface of the trochlea to the distal humeral physis. Endochondral ossification normally unites these 2 centers between 8 and 12 weeks of age in dogs.

The location of the intracondylar fissure in this case (Fig. 7.2-1) corresponds to the cartilaginous plate that separates these two larger ossification centers before they unite—hence the term "incomplete ossification of the humeral condyle" for this developmental orthopedic disease.

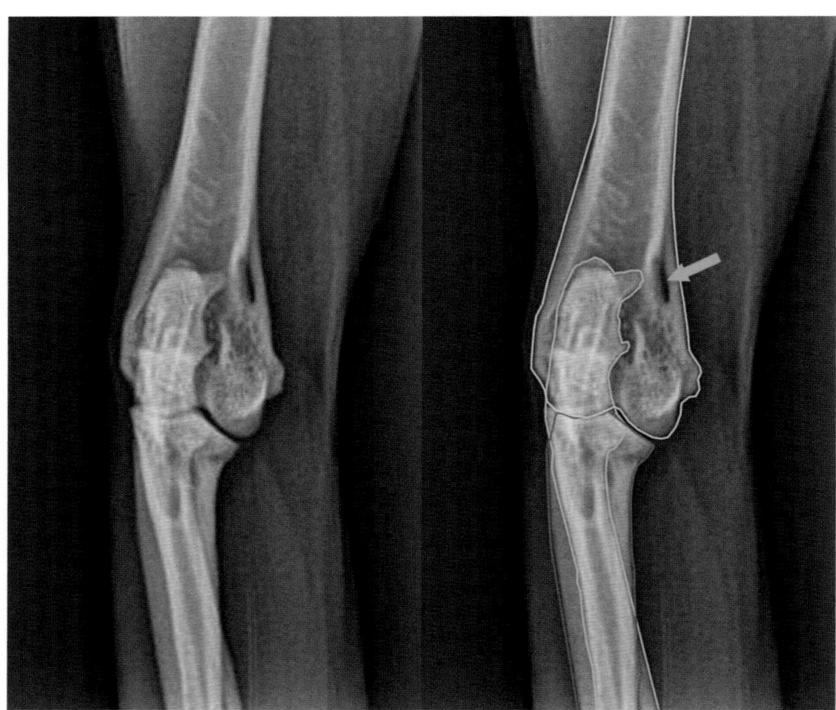

FIGURE 7.2-6 Craniocaudal radiograph of the elbow of an adult cat. On the right, the humerus is outlined in *yellow*, the radius in *pink*, and the ulna in *blue*. Note the supracondylar foramen *(arrow)* proximal to the medial epicondyle of the humerus.

Proximal radius (Figs. 7.2-3–7.2-5)

The proximal end or **head** of the **radius** has a shallow, concave articular surface, the **capital fovea**, that conforms to the shape of the capitulum of the distal humerus with which it articulates. The **articular circumference** is on the caudal margin of the radial head immediately distal to this articular surface. It is a narrow, convex band where the radius articulates with the ulna. The convex radial head fits in the concave radial notch of the ulna, as described later, and both surfaces are covered by joint cartilage. The location and shape of these articulations indicate that the radius has some capacity to rotate in relation to the ulna and to the humerus.

Just lateral to the articular circumference is a bony prominence, the **lateral tuberosity** of the radius, where the joint capsule attaches. A less distinct **medial tuberosity** is located on the craniomedial aspect of the proximal radius. Distal to these tuberosities, the radial metaphysis narrows slightly, yet noticeably, to the radial **neck**. Distal to the neck on the lateral side of the radius is a second bony prominence, to which the lateral collateral ligament attaches (described later).

ELBOW DYSPLASIA IN DOGS

Synchronous growth of the humerus, radius, and ulna is vital for normal development of the elbow joint. The radial and ulnar physes are particularly important for both elbow and carpal joint congruity.

Asynchronous growth leads to incongruity of the articular surfaces, which predisposes to abnormal load distribution and therefore to various manifestations of elbow dysplasia, including OCD of the medial humeral condyle, ununited anconeal process, and medial coronoid disease (including fragmented coronoid process). Medial coronoid disease is described in more detail later.

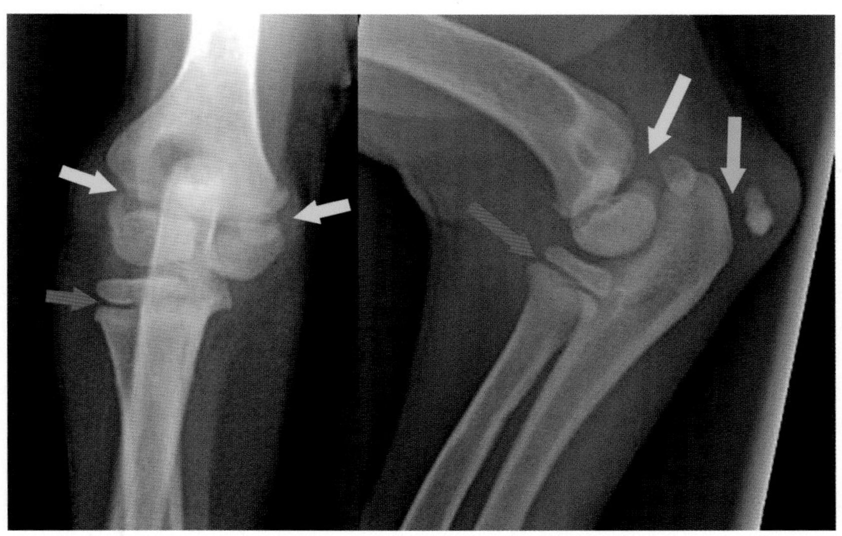

FIGURE 7.2-7 Craniocaudal and lateral radiographs of an 8-weeek-old puppy. The *arrows* point to the distal humeral *(yellow)* and proximal radial *(pink)* physes and to the cartilage separating the ossification center at the olecranon tuberosity *(blue)*.

There is a second, separate **radioulnar articulation** that is located a variable distance distal to (below) the elbow joint. Its radial side consists of a narrow, vertically oriented area of roughened bone on the caudolateral surface of the radius. A short, strong **interosseus ligament** restricts movement across this joint. The **nutrient foramen** of the radius is in its proximal third, obliquely penetrating the caudal cortex of the radius from proximal to distal.

> The nutrient foramen is sometimes evident radiographically, and it must not be mistaken for a fracture line. Likewise, the roughened areas on the caudal radius and cranial ulna where the 2 are joined by the interosseus ligament should not be mistaken for an exostosis (abnormal bone deposition on the outer surface) or synostosis (pathological fusion of 2 bones; osseous union of normally separate bones).

The **proximal radial physis** is located just proximal to the radial neck (Fig. 7.2-7). Longitudinal growth of the radius in young dogs is equally divided between this physis and the distal radial physis, which is located just proximal to

UNUNITED ANCONEAL PROCESS (UAP)

The anconeal process normally ossifies by 5 months of age in most dog breeds, and by 7–8 months of age in Saint Bernards and Basset Hounds. Beyond this age, the presence of a radiolucent line separating the anconeal process from the rest of the olecranon is diagnostic for UAP.

This condition may result from asynchronous growth of the radius and ulna, such that the ulna is shorter than the radius ("short ulna" syndrome), which places abnormal load on the developing anconeal process. However, UAP is occasionally seen in dogs with congruent elbows, suggesting that it may be a form of osteochondrosis (abnormal bone development).

Currently, UAP is treated by proximal ulnar osteotomy (surgical transection of the ulna) just distal to the annular ligament, thus "releasing" the proximal ulna and relieving pressure on the anconeal process. The fragmented anconeal process may be removed or repaired with a lag screw, depending on the clinical case.

the carpus (see Case 7.3). In cats, the proximal radial physis is fully ossified by 6–8 months of age (see Fig. 7.3-1), although in dogs it can range from 4.5 to 11 months of age, depending on the breed (later in large and giant breeds).

Proximal ulna (Figs. 7.2-3–7.2-5)

At the elbow, the **ulna** lies caudal and slightly medial to the radius. This orientation gradually changes as the 2 bones course distally, such that the ulna lies lateral to the radius at the carpus.

The proximal end of the ulna is more extensive than that of the radius and has several clinically important features. Most prominent is the **olecranon**, which is a broad, lateromedially flattened, bony eminence that extends proximally, both dorsal and caudal to the elbow joint. It is the distal attachment for the triceps brachii muscles, which are the primary extensors and stabilizers of the elbow (described later). Its proximal border has a pair of small cranial tuberosities and an extensive caudal eminence—the **tuber olecrani** which is commonly referred to as the "point of the elbow."

The anconeal process is best seen on a flexed lateral radiograph of the elbow (Fig. 7.2-5). Even so, the caudal margins of the humeral condyle are superimposed on it. The anconeal process can be found by following the cranial surface of the proximal olecranon distally to the trochlear notch, or by following the contour of the trochlear notch proximally and cranially. It is important to examine the anconeal process in dogs presented for lameness localized to the elbow, as UAP is a common form of elbow dysplasia in dogs.

The cranial surface of the olecranon is marked distally by the **trochlear notch**, a large, concave articular surface that conforms to—and articulates with—the shape of the trochlea of the humerus. It is also known as the "semilunar notch" because of its half-moon shape when viewed sagittally. The **anconeal process** is an acutely angled (sharp or beak-like) cranial prominence at the proximal margin of the trochlear notch. It lies in the olecranon fossa of the distal humerus, rotating deep into the fossa during elbow extension.

At the base or distal margin of the trochlear notch, the articular surface broadens lateromedially. The lateral aspect narrows craniolaterally to a small prominence, the **lateral coronoid process**. The medial aspect is a broad, craniodistally and medially sloped surface (see Figs. 7.2-1 and 7.2-3), called the **medial coronoid area** whose convex abaxial border is marked by a distinct bony rim and which narrows below the ulna. The cranial margin of the medial coronoid area tapers to a distinct bony prominence, the **medial coronoid process**, which is larger and more cranial than its lateral counterpart.

The medial coronoid area is a common site for medial coronoid disease, including fragmentation of the medial coronoid process. It is an important area to examine radiographically in dogs.

The lateral and medial coronoid processes are identifiable on the lateral and craniocaudal radiographic views of the elbow (the humeral condyle and radial head are superimposed, making the lateral coronoid process difficult to appreciate).

The lateral and medial coronoid processes are separated by a cranially facing concavity, the **radial notch**, where the radius lies and articulates with the ulna. Caudomedial to the radial notch, and distal to the abaxial margin of the medial coronoid area, is the **ulnar tuberosity**, a cranial triangular roughened ridge of bone for the attachment of the tendons of the biceps brachii and brachialis muscles (discussed later).

The **nutrient foramen** of the ulna is located proximally, just distal to the proximal radioulnar articulation, on the cranial surface of the ulna. The proximal ulna does not have a transverse physis and instead an apophysis (Gr. "an offshoot" natural protuberance forming a process, tubercle, or tuberosity) that ossifies by endochondral ossification, except for a separate ossification center for the tuber olecrani (Fig. 7.2-7). The olecranon is completely ossified by 8–9 months (range 5–15 months), depending on the breed (later for large and giant breeds). In cats, the olecranon is completely ossified by 9–12 months of age. The distal ulnar physis is V-shaped and ossifies at 7–15 months of age in dogs and 12–23 months of age in cats.

At the elbow, weight-bearing is equally divided between the radial head and the medial coronoid area of the ulna vs. distal in the antebrachium, where the distal radius assumes nearly 100% of the load.

The distal ulnar physis, proximal to the carpus, is responsible for 100% of the longitudinal growth of the ulna. Physeal injury or pathology (e.g., retained cartilaginous core) involving the distal ulnar physis affects elbow congruity and causes an angular limb deformity (bowing of the limb).

Joint capsule

A single joint capsule encloses the articular surfaces of the distal humerus and the proximal radius and ulna, including the radioulnar articulation, and is thicker cranially than caudally. The joint capsule is reinforced over its cranial aspect by a ligament running obliquely from the craniolateral aspect of the distal humerus, cranial to the lateral epicondyle, to the craniomedial aspect of the proximal ulna, merging with the biceps brachii and brachialis tendons at their insertion (described later). The joint capsule is reinforced over its caudomedial aspect by a short, elastic ligament that runs from the medial surface of the olecranon fossa to the medial surface of the olecranon, just proximal to the anconeal process.

MEDIAL CORONOID DISEASE

Medial coronoid disease is the most common type of elbow dysplasia in dogs and the most common cause of thoracic limb lameness in large- and giant-breed dogs. The cause remains hypothetical but likely involves abnormal loading of the medial coronoid area, with or without abnormal bone development.

Radiographic findings variably include fragmentation or fissuring of the medial coronoid process (FMCP), an indistinct or deformed MCP, irregular or decreased bone density in the MCP, sclerosis (abnormally increased density), or loss of the normal trabecular pattern in the subchondral bone of the medial coronoid area, and osteophyte formation (abnormal bony deposition) at the articular margin of the medial coronoid area.

Computed tomography and arthroscopy of the elbow joint greatly improve the clinician's ability to diagnose and characterize medial coronoid disease. Arthroscopy allows the surgeon to remove MCP fragments and debride unhealthy articular cartilage in the medial coronoid area.

JOINT CAPSULE DISTENSION

Fluid distension of the joint capsule (effusion) is palpable as a soft, fluid-filled swelling immediately dorsocaudal to the lateral and medial epicondyles of the humerus, on either side of the olecranon. In short-haired patients, it may be noticeable as a second "bulge" on the lateral side of the elbow (the first being the lateral epicondyle). The lateral pouch is easier to access for arthrocentesis and for the placement of an egress portal during arthroscopy of the elbow joint. Distension of the joint capsule cranially may also be palpable proximal to the radial head.

Effusion is nonspecific for an intraarticular process that has caused inflammation of the synovium lining the joint. The joint capsule may be distended with pus—i.e., septic arthritis; blood (hemarthrosis); or a soft-tissue mass—e.g., granuloma, neoplasia. These often cannot be distinguished radiographically.

Collateral and annular ligaments (Fig. 7.2-8)

The shape of the humeroulnar joint largely restricts the range of motion in the elbow joint to flexion-extension in the sagittal plane, making the elbow a simple "hinge" joint. In cats, the humeroradial and proximal radioulnar joints allow a significant axial rotation of the radial head, and thus of the antebrachium and manus (paw), providing them greater manual dexterity than dogs. The collateral and annular ligaments of the elbow both allow and limit axial rotation in clinically significant ways.

The medial and lateral collateral ligaments support the abaxial surfaces of the elbow joint; the **lateral collateral ligament** is stronger in dogs. It originates on the lateral epicondyle of the humerus and inserts primarily onto the lateral surface of the radius; some fibers insert onto the adjacent lateral border of the ulna. These 2 insertions are called the **cranial** and **caudal crura**, respectively.

Experimentally, elbow luxation in dogs is impossible without transecting the lateral collateral ligament, emphasizing its importance for joint stability. In cats, elbow luxation is experimentally impossible, unless both collateral ligaments are transected, making them equally important in cats.

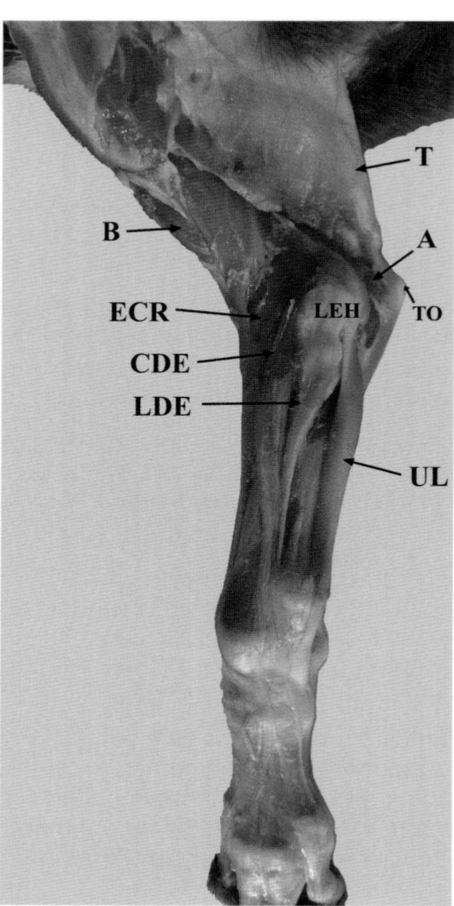

FIGURE 7.2-8 Extensor muscles of the feline elbow, craniolateral aspect. Key: *LEH*, lateral epicondyle of the humerus; *OT*, tuber olecrani; Key: *T*, triceps brachii; *B*, brachialis; *A*, anconeus; *ECR*, extensor carpi radialis muscle and tendon; *CDE*, common digital extensor muscle and tendon; *LDE*, lateral digital extensor; *UL*, extensor carpi ulnaris/ulnaris lateralis. Key: 1, triceps brachii; 2, brachialis; 3, anconeus; 4, extensor carpi radialis muscle and tendon (4′); 5, common digital extensor muscle and tendon (5′); 6, lateral digital extensor; 7, extensor carpi ulnaris/ulnaris lateralis.

The **medial collateral ligament** is the smaller of the 2 in dogs, originating on the medial epicondyle of the humerus. It runs obliquely, distally, and caudally, crossing over the annular ligament and dividing into a weaker **cranial crus** that inserts on the caudal surface of the radius, and a stronger **caudal crus** that inserts mainly onto the adjacent cranial surface of the ulna.

The proximal radioulnar joint is supported cranially by an **annular ligament**, which extends across the cranial surface of the radial head and is covered by the collateral ligaments, blending with the cranial aspect of the joint capsule.

The supporting ligaments are important in defining the normal range of axial rotation (pronation and supination) allowed by the elbow joint. When the elbow is flexed, dogs normally have approximately 30 degrees of pronation (range, 20–40 degrees) and 45 degrees of supination (range, 30–70 degrees) in the carpus and paw. Cats normally have approximately 50 degrees of pronation (range, 30–70 degrees) and 130 degrees of supination (range, 110–150 degrees). Angles of pronation and supination are highly variable among individuals, with a strong correlation between larger body size and smaller angles of rotation.

In humans, pronation of the forearm is the act of placing the hand palm-down, and supination is the act of placing the hand palm-up. In dogs and cats, pronation is the normal neutral position. A cat playing with a toy or a mouse illustrates that the range of rotation is greatest when the elbow is flexed.

Because cats have greater mobility than dogs between the radius and ulna, it is important in cats to avoid permanently surgically joining the radius and ulna—e.g., during fracture repair—to avoid inducing fusion (synostosis) between the radius and ulna.

Extensors and flexors of the elbow and their innervation (Figs. 7.2-9 and 7.2-10)

411

The **triceps brachii** group is the major elbow extensor and stabilizer. The 3 main heads originate on the caudal surface of the scapula (**long head**) and on the lateral crest (**lateral head**) and medial surface (**medial head**) of the proximal humerus; all insert onto the olecranon. In dogs there is a fourth belly of triceps, the **accessory head** which originates from the neck of the humerus and lies medially between the brachialis and the other triceps muscle bellies. Because the long head originates proximal and caudal to the humerus, it also flexes the shoulder joint. A **bursa** lies between the tendon of the long and lateral heads and the tuber olecrani. Additionally, a subcutaneous bursa may be found overlying the tuber olecrani. The triceps group is innervated by the **radial nerve** (described later).

The small anconeal and **tensor fascia antebrachii muscles** may aid in elbow extension. The **anconeal muscle** is a short, wide, sheet-like muscle that covers the proximolateral aspect of the elbow joint.

Surgical access to the anconeal process in dogs with UAP, or to the caudal part of the lateral condyle of the humerus for condylar fracture repair, is facilitated by detaching the humeral origin of the anconeal muscle using a periosteal elevator.

CAMPBELL'S TEST FOR COLLATERAL LIGAMENT INTEGRITY

The integrity of the collateral ligaments of the elbow can be clinically evaluated using Campbell's test. This test is useful for assessing the collateral ligaments after reduction of a traumatic elbow luxation.

For this test, the elbow and carpal joints are maintained at 90 degrees, while the manus is pronated and supinated. Medial collateral ligament integrity is in question if pronation is >40 degrees in dogs and >70 degrees in cats. Lateral collateral ligament integrity is in question if supination is >70 degrees in dogs and >150 degrees in cats.

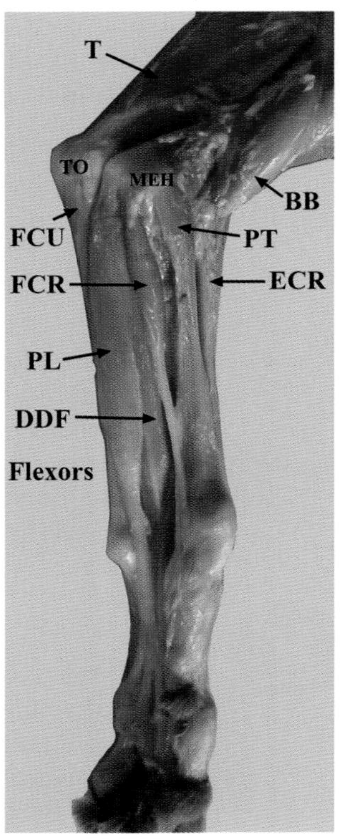

FIGURE 7.2-9 Flexor muscles of the feline elbow, caudomedial aspect. Key: *MEH*, medial epicondyle of the humerus; *OT*, tuber olecrani. Key: *T*, triceps brachii; *BB*, biceps brachii; *PT*, pronator teres; *ECR*, extensor carpi radialis muscle and tendon; *FCU*, flexor carpi ulnaris; *PL*, superficial digital flexor (in cats, palmaris longus) muscle and tendon; *DDF*, deep digital flexor; *FCR*, flexor carpi radialis. Key: 1, triceps brachii; 2, biceps brachii; 3, pronator teres; 4, extensor carpi radialis muscle and tendon (4′); 5, flexor carpi ulnaris; 6, superficial digital flexor (in cats, palmaris longus) muscle and tendon (6′); 7, deep digital flexor; 8, flexor carpi radialis.

The major flexors of the elbow joint are the biceps brachii and brachialis muscles. **Biceps brachii** originates on the supraglenoid tubercle of the scapula, proximal to the shoulder joint, and runs distally over the cranial and medial aspect of the humerus inserting as 2 tendons onto the medial tuberosities of the proximal ulna and radius. The biceps brachii stabilizes and extends the shoulder joint proximally, and stabilizes and flexes the elbow joint distally. Given the axial-to-medial course and craniomedial ulnar insertion, it may also laterally rotate the elbow a small amount, facilitating supination.

Biceps ulnar release procedure (BURP) is sometimes performed in dogs with medial coronoid disease to relieve pressure on the medial aspect of the elbow joint. It involves tenotomy (tendon transection) of the distal insertion of the biceps brachii-brachialis complex onto the proximal ulna, caudodistal to the medial coronoid process.

Brachialis originates on the caudal aspect of the proximal humerus and inserts proximal to biceps brachii. Its actions are restricted to stabilization and flexion of the elbow joint and slight lateral rotation. Both elbow flexors are supplied by the **musculocutaneous nerve**.

The **antebrachial fascia** is a “sleeve” of fascia that binds the muscles of the forearm. Proximally, it is continuous with the fascia associated with the shoulder muscles and continues cranially with a leaf of deep fascia connecting to the humerus. The fascial connection to the humerus significantly restricts elbow extension in the absence of muscle action and stabilizes the extended limb, especially during weight-bearing.

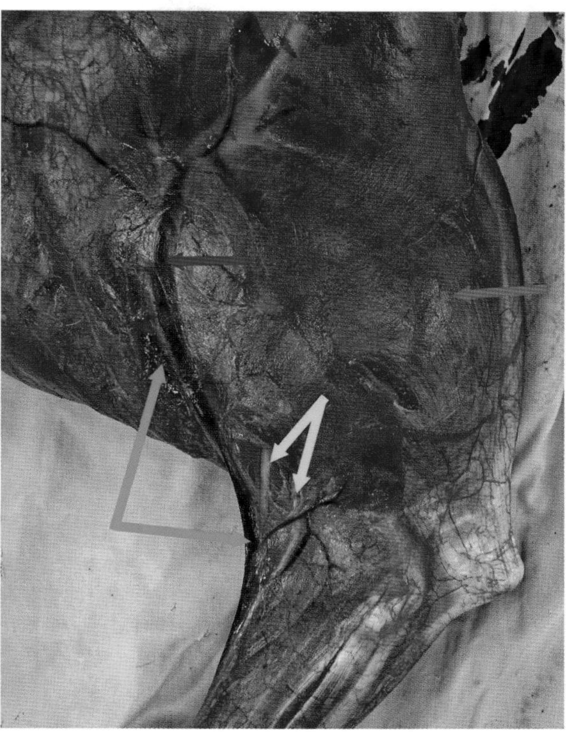

FIGURE 7.2-10 Vessels and nerves of the canine elbow, left thoracic limb, lateral aspect. The skin and subcutaneous tissue have been removed. Branches of the radial nerve *(yellow arrows)* may be seen emerging from beneath the lateral head of the triceps brachii muscle *(pink arrow)* on the craniolateral aspect of the elbow. Note the axillobrachial vein *(blue arrow)* and the cephalic vein *(green arrows)* on the flexor surface of the elbow.

TRAUMATIC ELBOW LUXATION

Traumatic luxation of the elbow is a common injury in dogs and cats. Because the medial part of the humeral condyle is larger than the lateral, luxation most often involves the lateral displacement of the radius and ulna in relation to the humerus. Congenital, atraumatic luxation of the radial head is also reported, particularly in bulldogs.

Closed reduction (nonsurgical correction) is advised unless there are clear indications for open reduction—e.g., concurrent fracture. Closed reduction involves flexing the elbow then rotating the antebrachium medially (pronating) to reseat the anconeal process in the olecranon fossa. While slowly extending the elbow with the antebrachium medially rotated, lateral pressure is applied to the proximal radius and medial pressure to the distal humerus until the trochlear notch is reseated in the trochlear groove. Reduction requires firm force, carefully applied.

After successful reduction, the joint is taken through its full range of motion, and Campbell's test may be used to check the integrity of the collateral ligaments. If the joint is stable, the elbow is immobilized in extension (≥ 135 degrees) using a soft-padded bandage (modified Robert Jones bandage) or splinted bandage for 2–4 weeks. This ensures that the anconeal process remains seated in the olecranon fossa, thereby stabilizing the elbow joint and decreasing the chance of re-luxation.

Muscles originating at the elbow and their innervation (Figs. 7.2-8 and 7.2-9)

The caudal antebrachial muscles are carpal and digital flexors, and include the pronators of the carpus and manus. The muscles arise from the medial epicondyle of the humerus and/or adjacent areas of the medial radius and ulna. From cranial (medial) to caudal, they include **pronator teres**, **flexor carpi radialis**, and the **superficial** and **deep digital flexor muscles**. (The superficial digital flexor in the cat is called the **palmaris longus muscle**.) Further caudally, **flexor carpi ulnaris** originates on the medial epicondyle of the humerus and the medial side of the olecranon. Further distally, **pronator quadratus** runs obliquely (distally and medially) to fill the space between the radius and ulna on their caudal surfaces.

> To access the medial coronoid process, a medial arthrotomy is made by separating the flexor carpi radialis and superficial digital flexor muscles distal to their origins on the medial epicondyle of the humerus. The ulnar nerve (see below) must be identified and carefully retracted caudally to protect it during the procedure.

The flexors that are attached to the ulna are supplied by the **ulnar nerve**, and those attached to the radius, as well as the superficial digital flexor and pronator quadratus, are supplied by the **median nerve**.

The craniolateral antebrachial muscles are carpal and digital extensors. This group also includes supinators of the carpus and manus. These muscles arise from the lateral epicondyle of the humerus and the lateral epicondylar crest proximal to it. From cranial to caudal (lateral), they include **brachioradialis**, **extensor carpi radialis**, **common digital extensor**, **lateral digital extensor**, and **extensor carpi ulnaris** or **ulnaris lateralis**—all of which are innervated by the **radial nerve**.

The **supinator** is a small muscle lying deep to these extensors on the craniolateral aspect of the proximal radius, which aids in supination and is innervated by the radial nerve. Some dogs and cats have a small **sesamoid bone** in its tendon of origin, which may be evident on a craniocaudal or lateral radiograph of the elbow (see Fig. 7.2-4).

> This sesamoid bone should not be mistaken for a chip fracture or avulsion on radiographs or during surgery.

Blood supply, lymphatics, and innervation (Figs. 7.2-10 and 7.2-11)

The main blood supply to the elbow is the **brachial artery**, a major branch of the **axillary artery** crossing the medial aspect at the distal third of humerus in dogs. In cats, the brachial artery runs through the supracondylar foramen of the distal humerus. The brachial artery runs over the craniomedial aspect of the elbow, sending off many small branches proximal and distal to the elbow before continuing into the antebrachium as the **median artery**. The main venous drainage parallels the arterial supply in route and name, except for the **cephalic vein** and the **median cubital veins**.

> Surgical approaches to the elbow joint usually do not risk injury to major vessels, because collateral artery supplies are plentiful and small arterioles can be cauterized for hemostasis. In contrast, surgery involving the humeral diaphysis or antebrachium risks injury to the brachial artery or cephalic vein, respectively.

The major nerves that innervate or run over the elbow derive from the brachial plexus, and include the radial, median, ulnar, and musculocutaneous nerves. The **radial nerve** runs from caudomedial to craniolateral and supplies the triceps muscles, and then rounds the lateral aspect of the humerus in the middle to distal third in dogs and the proximal third in cats. It appears from beneath the cranial edge of the lateral belly of the triceps brachii muscle just proximal to the elbow before traversing the flexor aspect of the elbow and penetrating the carpal and digital extensor muscles on the craniolateral aspect of the forearm.

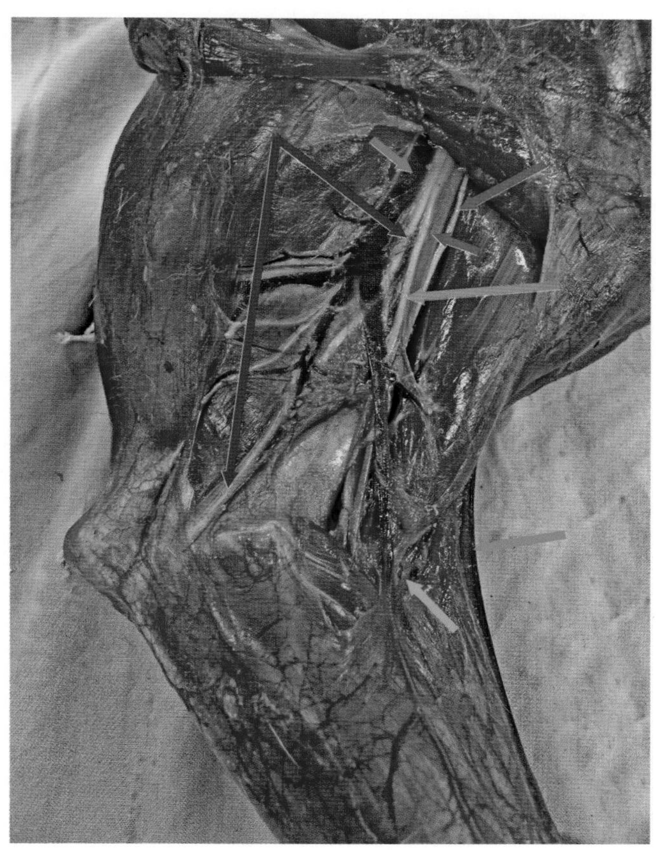

FIGURE 7.2-11 Vessels and nerves of the canine elbow, left thoracic limb, medial aspect. The skin and subcutaneous tissue have been removed. From proximal to distal, the *arrow* key is as follows: brachial vein *(dark green)*, musculocutaneous nerve *(red)*, ulnar nerve *(purple)*, brachial artery *(dark pink)*, median nerve *(orange)*, cephalic vein *(light green)*, and median cubital vein *(light blue)*.

The **median**, **ulnar**, and **musculocutaneous nerves** track caudomedial to the humerus and continue over the medial aspect of the elbow joint. The musculocutaneous nerve supplies the elbow flexors—the biceps brachii and brachialis muscles. The ulnar nerve passes caudally toward the ulna as it crosses the medial surface of the elbow, whereas the median nerve is more closely associated with the brachial artery and follows its more cranial course across the medial aspect of the elbow joint. In cats, the median nerve runs through the supracondylar foramen of the distal humerus.

> Identifying and protecting these nerves during surgery involving the distal humerus and elbow is important. The ulnar nerve (see Fig. 7.2-11) is particularly vulnerable during a medial approach to the elbow.

Selected references

[1] Constantinescu GM, Constantinescu IA. A clinically oriented comprehensive pictorial review of canine elbow anatomy. Vet Surg 2009;38:135–43.

[2] Farrell M, Draffan D, Gemmill T, et al. In vitro validation of a technique for assessment of canine and feline elbow joint collateral ligament integrity and description of a new method for collateral ligament prosthetic replacement. Vet Surg 2007;36:548–56.

[3] Harasen G. Feline orthopedics. Can Vet J 2009;50:669–70.

[4] Moores A. Humeral condylar fractures and incomplete ossification of the humeral condyle in dogs. In Pract 2006;28:391–7.

[5] Chong TE, Davies HMS. Investigating canine elbow joint stabilisation through mechanical constraints of the deep fascia and other soft tissues. J Anat 2018;232:407–21.

CHAPTER 7

CASE 7.3

Carpal Valgus Deformity

Lane A. Wallett[a], Cathy Beck[b], and Helen M.S. Davies[c]
[a]Integrative Physiology and Neuroscience, Washington State University College of Veterinary Medicine, Pullman, Washington, US
[b]Veterinary Hospital, The University of Melbourne, Melbourne, Victoria, AU
[c]Veterinary BioSciences, The University of Melbourne, Melbourne, Victoria, AU

Clinical case

History

A 12-month-old male neutered Curly-Coated Retriever was examined for chronic lameness of the right thoracic limb. Three months earlier, at approximately 9 months of age, the dog jumped from a moving vehicle resulting in an acute lameness. The lameness resolved without treatment within 10 days; however, it recurred 1 month before presentation and persisted.

Valgus is an angular limb deformity in which the portion of the limb distal to the site of injury or malformation deviates laterally in relation to the long axis of the limb. In this case, the forepaw deviated laterally distal to the carpus. Varus is the opposite deformity: the distal portion of the limb deviates medially or toward the median plane of the body.

Physical examination findings

The dog was mildly lame in the right thoracic limb and a carpal valgus deformity was present. Soft tissue swelling was palpable in the carpal region, and extension and flexion of the carpus elicited a pain response. The left carpus appeared normal, and there were no other abnormal findings on physical examination.

Differential diagnoses

Premature physeal closure, carpal laxity syndrome; carpal luxation; fracture of the radius, ulna, or carpal bones; poor conformation

Diagnostics

Radiographs were taken of both thoracic limbs, from distal humerus to proximal metacarpus (Figs. 7.3-1 and 7.3-2). The left thoracic limb was normal. The radiograph of the right thoracic limb showed the carpal valgus deformity to be centered at the distal radius and ulna. The ulnar head was displaced laterally and caudally, and there was

Comparative Veterinary Anatomy: A Clinical Approach. https://doi.org/10.1016/B978-0-323-91015-6.00046-7

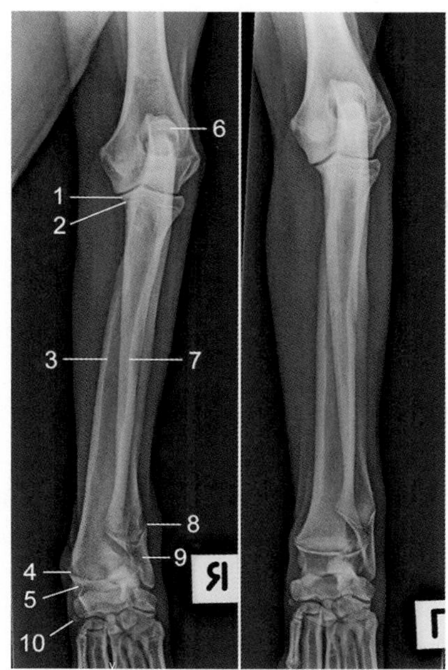

FIGURE 7.3-1 Craniocaudal radiographs of both thoracic limbs of the young dog in this case. In both images, the marker is placed to indicate the lateral aspect of the limb (*R, right or L, left*). The left limb is normal for a 1-year-old dog. The right limb has a carpal valgus deformity (lateral deviation distally), centered on the distal radius and ulna. The ulnar head (9) is displaced laterally, and the distal ulnar physis (8) is asymmetrically narrowed. In addition, there is a loss of muscle mass in the antebrachium. Key: 1, proximal radial epiphysis or radial head; 2, proximal radial physis; 3, radius (diaphysis); 4, distal radial physis; 5, distal radial epiphysis; 6, ossification center of the olecranon; 7, ulna (diaphysis); 8, distal ulnar physis; 9, distal ulnar epiphysis or ulnar head.

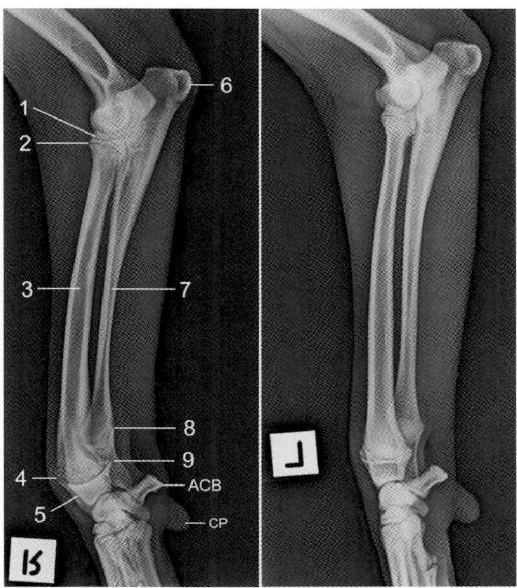

FIGURE 7.3-2 Lateromedial radiographs of both thoracic limbs of the young dog in this case. In both images, the marker is placed to indicate the cranial aspect of the limb (*R, right or L, left*). The left limb is normal for a dog this age. In the right limb, there is cranial bowing of the radius (procurvatum deformity), and its caudal cortex is thickened distally. The ulnar head (9) is displaced caudally, the distal ulnar physis (8) is asymmetrically narrowed, and the metaphysis (the region just proximal to the distal physis) is decreased in opacity and is slightly heterogeneous. Refer to Fig. 7.3-1 for the key. Key: *ACB*, accessory carpal bone; *CP*, carpal pad.

The valgus and procurvatum deformities lead to abnormal alignment and loading of the carpus, leading to chronic lameness and the other physical abnormalities as described for this patient.

asymmetric narrowing of the distal ulnar physis, indicating premature physeal closure. In addition, there was cranial bowing of the radius—a condition termed a procurvatum (L. *curvătus* bent, bowed, curved) deformity of the distal radius.

Diagnosis

Carpal valgus and procurvatum deformity of the distal radius, resulting from premature closure of the distal ulnar physis and disproportionate growth of the radius and ulna

Treatment

Given the patient's age, and thus limited further growth potential in the radial and ulnar physes, surgery was recommended to correct the angular limb deformity and restore normal alignment and loading of the carpus. Under general anesthesia and positioned in dorsal recumbency, the patient underwent a corrective radial osteotomy and ulnar ostectomy. A vertical incision was made over the cranial aspect of the distal antebrachium and extended distally over the dorsomedial aspect of the carpus and proximal metacarpus. The incision made a slight medial deviation over the distal radius and carpus to avoid the low ridge of bone on the distal radius that separates the grooves for the common digital extensor and extensor carpi radialis tendons.

The radius was transected just proximal to its distal physis, the distal portion was realigned with the carpus, and a wedge was removed from the proximal portion to restore the integrity of the radius (wedge osteotomy). The distal ulna was transected just proximal to its distal physis, and a portion of the diaphysis was removed (ostectomy) to create a gap large enough to delay reossification and restoration of ulnar integrity. The radius was stabilized with a bone plate and pin. Post-operative radiographs confirmed good realignment (Fig. 7.3-3). Radiographs taken 10 months after surgery showed resolution of the carpal valgus deformity and reunification of the ulnar diaphysis (Fig. 7.3-4).

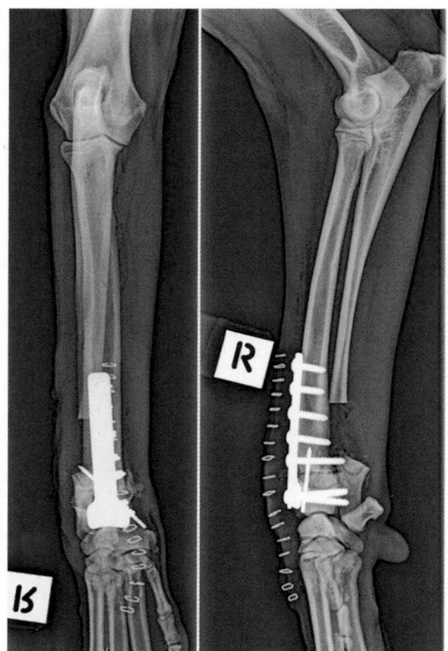

FIGURE 7.3-3 Craniocaudal and lateromedial radiographs of the right thoracic limb immediately after surgery. Note the vertical row of staples used to close the skin incision.

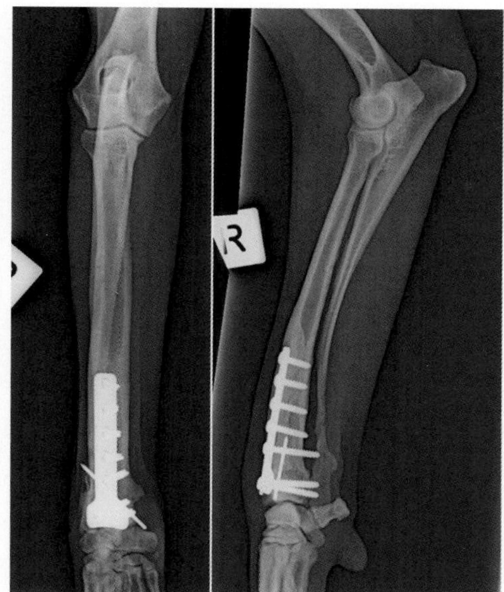

FIGURE 7.3-4 Craniocaudal and lateromedial radiographs of the right thoracic limb 10 months after surgery. The dog was approximately 22 months of age at the time. The physes of the radius and ulna are fully ossified, and the carpal valgus deformity has corrected. In addition, new bone has bridged the gap created by the ulnar ostectomy.

Anatomical features in canids and felids

Introduction

The region of the thoracic limb between the elbow and the carpus is the antebrachium, or forearm, while the region from the carpus to the distal phalanges is the manus, or forepaw (Fig. 7.3-5). The bones of the antebrachium are the radius and ulna, while the manus includes the carpal bones, metacarpal bones, phalanges, and multiple associated sesamoid bones. The metacarpus and digits are described in Case 7.4.

The cranial aspect of the manus, including the carpus, is referred to as the "dorsal aspect." Thus, both terms appear in the following discussion: cranial for the antebrachium and dorsal for the carpus. Similarly, caudal is used for the antebrachium and palmar for the carpus, even though both terms describe the same aspect of the limb. Lateral and medial are used for both the antebrachium and carpus (see Fig. 1-1).

Function

The primary function of the antebrachium and carpus is to facilitate both weight-bearing and locomotion. The ligamentous, tendinous, and fascial connections surrounding the carpus cause the carpal bones to compress together when the limb supports the dog or cat, so that a relatively stable infrastructure is formed between the elbow and the metacarpophalangeal joints. When the limb is unloaded, the 2 rows of carpal bones allow the carpus to flex and the limb to fold, such that the metacarpus is almost parallel to the antebrachium during the full carpal flexion.

A secondary function may perhaps be described as dexterity, although cats and especially dogs have a very limited manual dexterity compared with the human hand, including the carpus (wrist). As described in Case 7.2, the articulations of the proximal radius and ulna with the humerus, and with one another, at the elbow joint allow a moderate amount of axial rotation of the antebrachium and manus, including both pronation (L. *pronatio* facing down, posteriorly, inferiorly facing) and supination (L. *supinatio* facing upward, forward, anteriorly facing, think "s" carrying soup) in addition to the normal range of motion in the carpus (see side box entitled "Planes of movement").

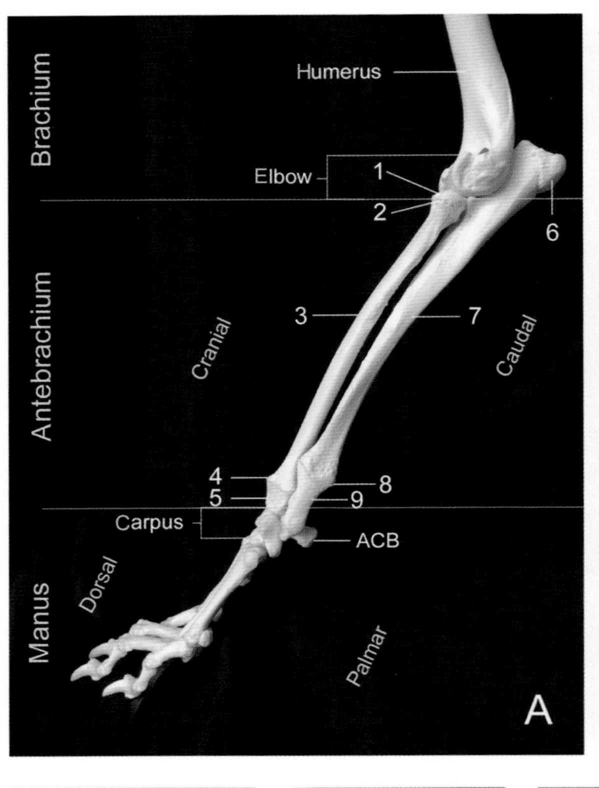

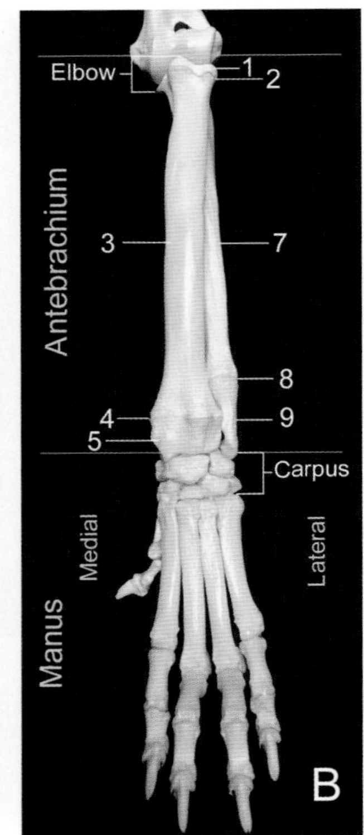

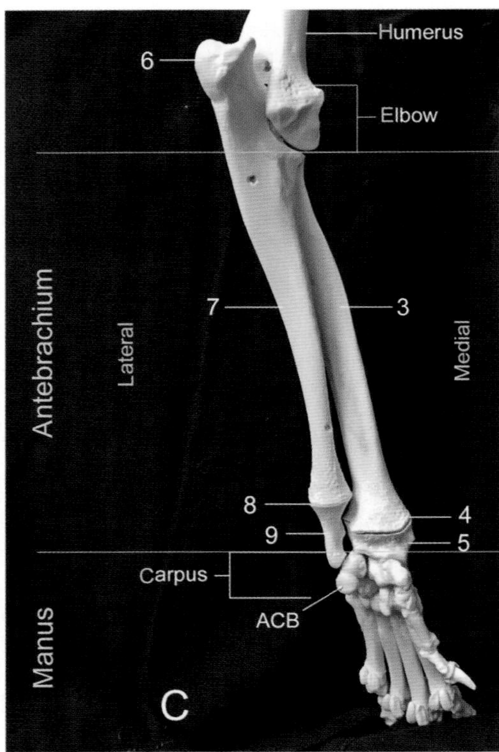

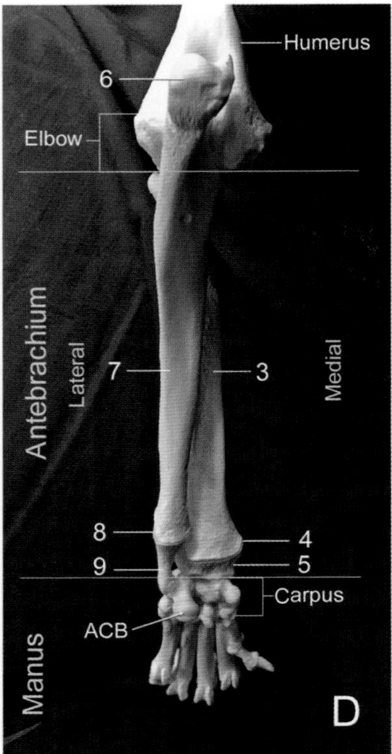

FIGURE 7.3-5 Left antebrachium and manus of a young dog. The physes are not yet fully ossified. (A) Lateral aspect; (B) cranial/dorsal aspect; (C) caudo/palmaromedial aspect; (D) caudal/palmar aspect. Key: 1, proximal radial epiphysis or radial head; 2, proximal radial physis; 3, radius (diaphysis); 4, distal radial physis; 5, distal radial epiphysis; 6, ossification center of the olecranon; 7, ulna (diaphysis); 8, distal ulnar physis; 9, distal ulnar epiphysis or ulnar head. Key: *ACB*, accessory carpal bone. Note: Not all structures are visible on all views.

Ulna (Figs. 7.3-5–7.3-7)

Robust proximally and tapering to its distal extremity, the **ulna** serves to transmit the forces produced by the extensor muscles of the elbow without contributing much to weight-bearing in the limb. With its proximal projection, the **olecranon**, the ulna is longer than the radius and is generally the longest bone in the body of a dog or cat. The proximal ulna is described in Case 7.2.

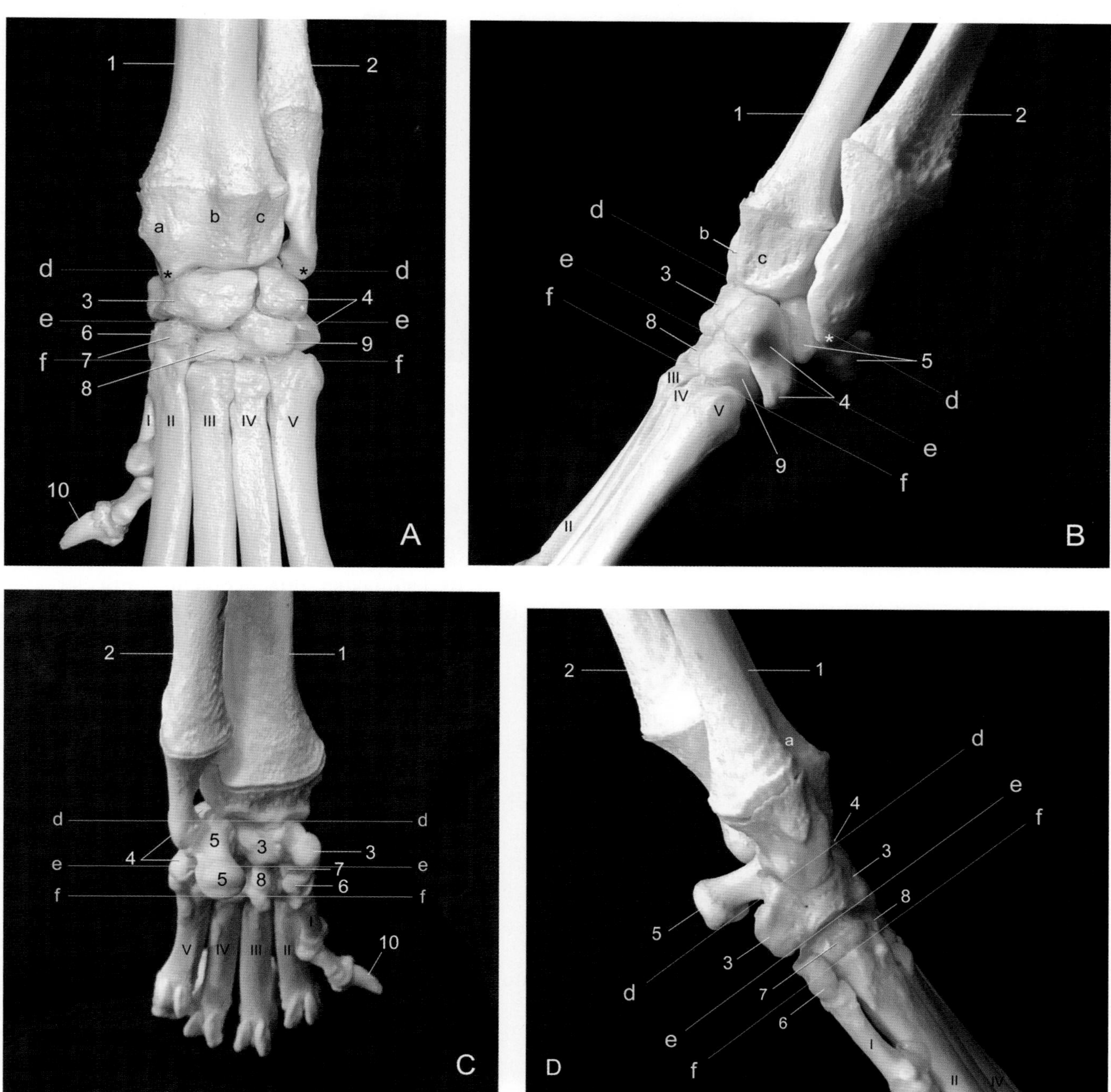

FIGURE 7.3-6 Left carpus of the dog in Fig. 7.3-5. (A) Dorsal aspect; (B) lateral aspect; (C) palmar aspect; (D) medial aspect. Number key: 1, radius; 2, ulna; 3, intermedioradial carpal bone; 4, ulnar carpal bone; 5, accessory carpal bone; 6–9, carpal bones 1–4; 10, distal phalanx of first metacarpal bone; *I–V*, metacarpal bones 1–5. Letter key: *a–c*, grooves for the tendons of abductor digiti I longus (a), extensor carpi radialis (b), and common digital extensor (c); *d*, antebrachiocarpal joint; *e*, middle carpal joint; *f*, carpometacarpal joint. *Asterisk* (*): styloid process of the radius or ulna, as appropriate. Note: Not all structures are visible on all views. The sesamoid bone of the abductor digiti I longus tendon is not included in this skeletal preparation.

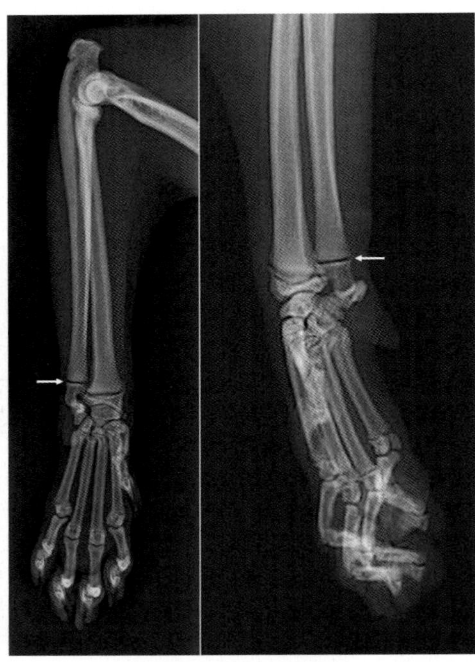

FIGURE 7.3-7 Craniocaudal (*left*) and lateromedial (*right*) radiographs of the antebrachium and manus of a young cat. Note the different shape of the distal ulnar physis *(arrow)* in cats compared with that in dogs (see Fig. 7.3-2).

The **head** of the ulna is its most distal extremity, demarcated from the body of the ulna by a slight notch, which lies at the proximal edge of its distal articulation with the radius. The craniomedial aspect of the head contains the **articular circumference**, a raised surface that articulates with the **ulnar notch** of the radius (described later). The **styloid process** of the ulna is the distal extremity of the ulnar head. Its distomedial surface contains 2 articular surfaces: facing cranially, a concave surface in contact with the **ulnar carpal bone**; and medially, a convex surface in contact with the **accessory carpal bone**.

> Because of its unique conical shape and relatively late ossification, the distal ulnar physis in young dogs is at an increased risk for nondisplaced compression fractures localized to the physeal cartilage (designated as Salter-Harris fracture type V). These fractures are generally undetectable radiographically at the time of injury, and instead may present weeks or months later as gait and/or limb-conformational abnormalities.

Most longitudinal growth (85–100%) in the ulna is produced by the **distal ulnar physis**. In dogs, this physis is somewhat conical, so it appears V-shaped on craniocaudal and lateromedial radiographic projections of the antebrachium or carpus (see Figs. 7.3-1 and 7.3-2). The distal ulnar physis in cats is straight (Fig. 7.3-7). In dogs, this physis is fully ossified ("closed") by 11–12 months of age, although it ranges from 6 to 16 months, depending on breed and more so on height at maturity (later in larger breeds). In cats, this physis normally closes between 15 months and 2 years of age.

Radius (Figs. 7.3-5–7.3-7)

The radius is the primary load-bearing bone in the antebrachium. The proximal portion of the radius is described in Case 7.2. Distally, the cranial aspect is marked by 3 prominent grooves, each corresponding to an extensor tendon. From medial to lateral, these channels contain the tendons of **abductor digiti I (pollicis) longus**, **extensor carpi radialis**, and **common digital extensor**. These muscles and their tendons of insertion are described later.

The distal radius articulates with the ulna via the **ulnar notch**, a concave articular surface located on the lateral aspect of the distal radius. Articulation with the **intermedioradial carpal bone** occurs at 2 locations: the broad, concave **carpal articular surface** of the radius and the lateral aspect of the **styloid process** of the radius, which is a small, beak-like distal projection on the craniomedial aspect of the distal radius (see Fig. 7.3-6A). The radius also articulates, to a lesser degree, with the **ulnar carpal bone** at the carpal articular surface.

The radius and ulna each have a styloid (Gr. *stylos* pillar + *eidos* form) process, which can be a source of confusion. The word "styloid" means resembling a stylus or pen. The styloid process of the ulna is simply its distal tip and is located on the lateral aspect of the distal antebrachium/carpus. The styloid process of the radius is a discrete distal projection of the radius and is located on the craniomedial aspect of the carpus.

The distal radial physis is fully ossified by 11–12 months of age in dogs, although it may range from 5 to 18 months. In cats, this physis normally closes between 15 and 22 months of age.

The periosteum (fibrous sheath) of the radius may be especially thick in young dogs and may need to be resected during corrective osteotomy for growth defects. The periosteum of the distal radius blends with the extensor and flexor retinacula of the carpus (described later), so care should be taken to preserve the retinacula during resection of the periosteum.

Carpus (Figs. 7.3-5–7.3-7)

The **carpus** is comprised of 7 carpal bones arranged in 2 transverse rows (proximal and distal), and the small sesamoid bone, which lies under the tendon of insertion of the abductor digiti I longus muscle. The carpal bones articulate proximally with the radius and ulna, intermediately with one another, and distally with the 5 metacarpal bones. Together, these bones form a compound joint classified as a ginglymus (hinged) joint. The carpus allows flexion and extension in the sagittal plane, lateral and medial bending (abduction and adduction, respectively, of the metacarpus and digits), and axial rotation (pronation and supination) (see side box entitled "Planes of movement").

The small sesamoid bone of the abductor digiti I longus tendon must not be mistaken for a bone fragment on radiographs of the carpus.

PREMATURE CLOSURE OF THE DISTAL ULNAR PHYSIS

Injury to the distal ulnar physis can result in insufficient ulnar growth in relation to radial growth and thus the development of an angular limb deformity (ALD). Trauma to the vulnerable physeal cartilage is the most common cause. For example, injury may occur when a young dog or cat jumps down from too great a height or lands awkwardly, overloading the lateral aspect of the limb. Premature closure of this physis frequently results in relative overgrowth of the radius, leading to cranial bowing of the radius in its mid- to distal diaphysis. This disproportionate growth of the radius and ulna disrupts proper alignment and articulation of the radius and ulna with the carpal bones. While the angular deformity is generally most obvious at the carpus, malformation of the ulna and/or radius may also have consequences at the elbow in the form of joint incongruity (see Case 7.2).

PLANES OF MOVEMENT

Together, the elbow and carpus allow a variety of movements in the manus, including carpal flexion-extension, abduction-adduction (lateral-medial bending at the carpus), and axial rotation of the antebrachium and manus.

Pronation is axial rotation in which the dorsal surface of the forepaw is rotated medially around the long axis of the limb. Supination is the opposite: the dorsal surface of the forepaw is rotated laterally. The normal range of pronation and supination in dogs and cats is discussed in Case 7.2.

The most proximal joint of the carpus is the **antebrachiocarpal joint**, formed at the interface of the **radius** and **ulna** with the proximal row of carpal bones: the **intermedioradial** (formerly radial), **ulnar**, and **accessory carpal bones**. This joint is sometimes referred to as the "radiocarpal joint," as the articulation of the radius and the large intermedioradial carpal bone serves as the main weight-bearing surface of the antebrachiocarpal joint. The accessory carpal bone, articulating with the palmar aspect of the ulnar carpal bone, is the distal attachment site of both the flexor carpi ulnaris and ulnaris lateralis muscles (described later).

The carpal pad is located on the palmar aspect of the carpus, just distal and medial to the accessory carpal bone (see Fig. 7.3-2). It is similar in structure to the metacarpal and digital pads described and illustrated in Case 7.4, although it is more thumb-like in shape than the other pads. Except during extreme carpal hyperextension, the carpal pad is not in contact with the ground.

The **middle carpal joint** is located between the proximal and distal rows of carpal bones, the latter consisting of **carpal bones 1–4**. The first carpal bone is the most medial, and the fourth carpal bone is the most lateral. The distal row of carpal bones articulates distally with the metacarpal bones to form the **carpometacarpal joint**. Within the carpus, the various small joints between individual carpal bones are simply termed the **intercarpal joints**.

Both the carpometacarpal and intercarpal joints are considerably less mobile than the antebrachiocarpal and middle carpal joints. The synovial cavities of the middle carpal and carpometacarpal joints communicate with one another, but the antebrachiocarpal joint does not communicate with any other joint. The carpal joint capsule is thickened both dorsally and palmarly, thus contributing to joint stabilization.

Ligaments of the carpus

Unlike the elbow, there are no long collateral ligaments stabilizing the carpus. Instead, short **medial** and **lateral collateral ligaments** are formed by the combination of deep carpal fascia and discrete thickening of the joint capsule. The medial collateral ligament diverges into straight and oblique sections, both arising proximal to the styloid process of the radius. The lateral collateral ligament begins on the styloid process of the ulna and attaches distally to the ulnar carpal bone. Additional stabilization is provided by the **dorsal radiocarpal**, **palmar radiocarpal**, and **palmar ulnocarpal ligaments**, so-named for their positions, origins, and insertions.

On the dorsal surface of the carpus, the **extensor retinaculum** secures the tendons of the carpal and digital extensor muscles in place as they course distally over the joint. The boundaries of this retinaculum are indistinct. The palmar portion of the carpal fascia is modified to form the **flexor retinaculum**, which splits into deep and superficial layers. The deep layer lies between the tendons of the superficial and deep digital flexor muscles (described later). The superficial layer comprises the palmar boundary of the **carpal canal**. The palmar carpal ligaments and the palmar aspect of the carpal joint capsule unite to form the deep boundary of the canal, and the accessory carpal bone completes the canal laterally (Fig. 7.3-8).

The carpal canal surrounds structures vital to the function of the manus. These structures include the tendons of the superficial and deep digital flexor and the flexor carpi radialis muscles; the median and ulnar nerves; and the median, radial, and caudal interosseous arteries and veins (Fig. 7.3-9).

Antebrachial muscles and their innervation (Figs. 7.3-10–7.3-12)

The muscle bellies of the antebrachium are located primarily in the proximal half of the region, leaving the distal half covered mostly by their tendons of insertion. The distal two-thirds of the radius and ulna are relatively free of muscular attachments and thus are easily accessed without transecting any muscles, with one exception. The **pronator quadratus** muscle occupies the space between the radius and the ulna, and functions to stabilize the bones of the distal limb.

The cranial and craniolateral surfaces of the radius are occupied by the extensors of the carpus and digits. Most of these muscles—including **extensor carpi radialis**, **common digital extensor**, and **lateral digital extensor**—arise

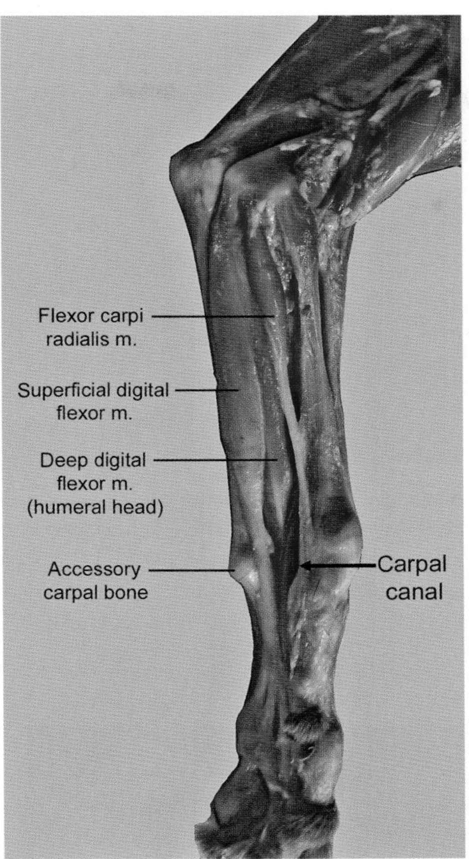

FIGURE 7.3-8 Left thoracic limb of a cat, caudo/palmaromedial aspect. The skin, including the carpal pad, and most of the fascia have been removed to show the structures within the carpal canal. Key: *m.*, muscle.

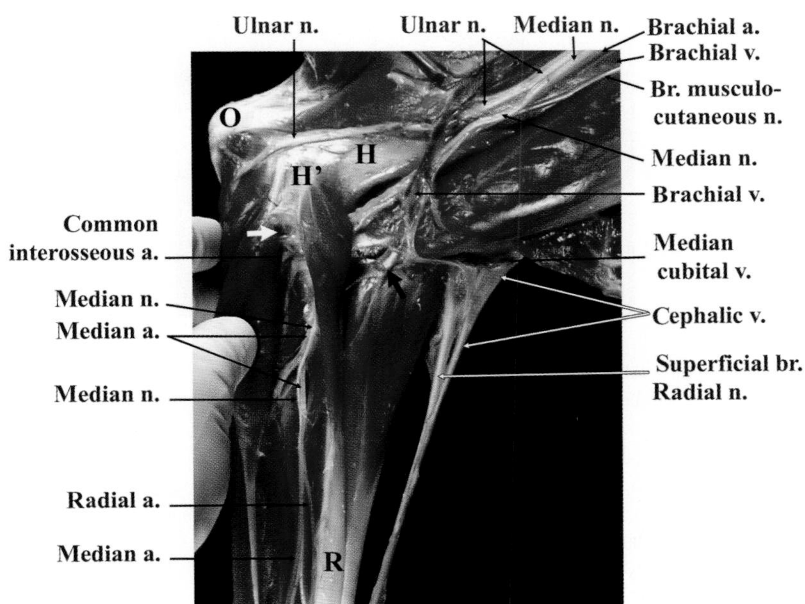

FIGURE 7.3-9 Left elbow and proximal antebrachium of a dog, medial aspect. The skin and most of the fascia have been removed, and the muscle bellies of the superficial digital flexor and flexor carpi radialis (see Fig. 7.3-12) have been reflected caudally to show the major vessels and nerves on the medial aspect of the elbow and antebrachium, including those that course distally to pass through the carpal canal. Key: *H*, humeral diaphysis and medial epicondyle (*H'*); *O*, olecranon; *R*, radius. Key: *a.*, artery; *br.*, branch; *n.*, nerve; *v.*, vein. *White arrow*: location of the elbow joint; *black arrow*: median vein.

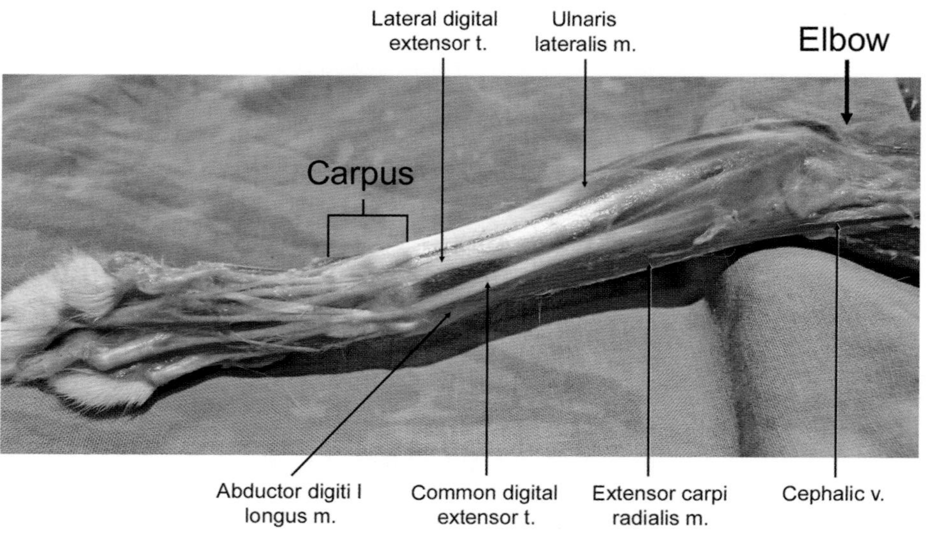

FIGURE 7.3-10 Right thoracic limb of a cat, cranio-/dorsolateral aspect. The skin and superficial fascia have been removed to show the superficial muscles and tendons of the antebrachium and carpus. Key: *m.*, muscle; *t.*, tendon of insertion; *v.*, vein.

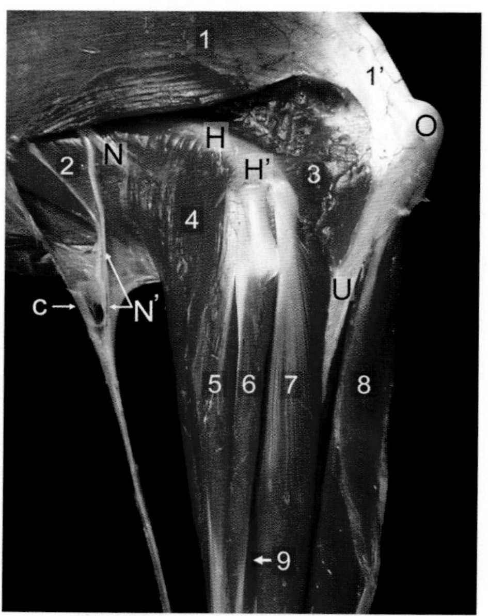

FIGURE 7.3-11 Left elbow and proximal antebrachium of a dog, lateral aspect. The skin and most of the fascia have been removed to reveal the superficial muscles on the lateral aspect of the antebrachium. Key: *m.*, muscle; *n.*, nerve. Number key: 1, lateral head of triceps brachii m. and its tendon of insertion (1′); 2, brachialis m.; 3, anconeus m.; 4, origin of extensor carpi radialis m.; 5–7, muscle bellies of the common digital extensor (5), lateral digital extensor (6), and ulnaris lateralis (7); 8, flexor carpi ulnaris m.; 9, abductor digiti I longus m. Letter key: *C*, cephalic vein; *H*, humeral diaphysis and lateral epicondyle (*H′*); *N*, radial n. and its superficial branch, including the lateral cutaneous antebrachial n. (*N′*); *O*, olecranon; *U*, ulna (body). In life, the superficial branch of the radial n. (*N′*) emerges from beneath the lateral head of the triceps (1) and later disappears under extensor carpi radialis (4). The cephalic vein (*C*) courses proximally in the superficial fascia overlying the superficial muscles.

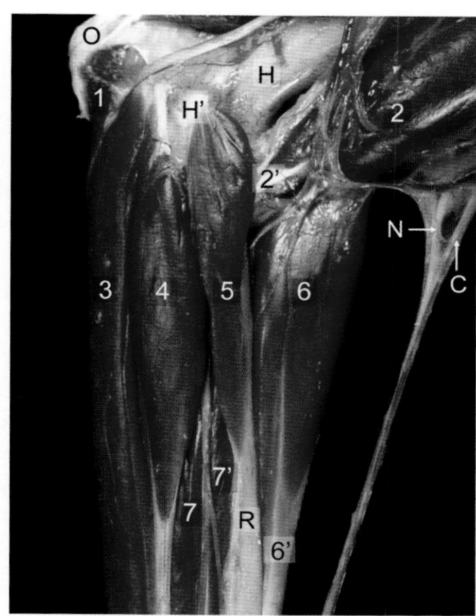

FIGURE 7.3-12 Left elbow and proximal antebrachium of a dog, medial aspect. The skin and most of the fascia have been removed to reveal the origins of the superficial muscles on the medial aspect of the antebrachium. Key: *m.*, muscle; *n.*, nerve. Number key: 1, flexor carpi ulnaris m. (ulnar head); 2, biceps brachii m. and its medial tendon of insertion (2′); 3–5, muscle bellies of superficial digital flexor (3), flexor carpi radialis (4), and pronator teres (5); 6, extensor carpi radialis m. and tendon of insertion (6′); 7, deep digital flexor m., humeral (7) and radial (7') heads. Letter key: *C*, cephalic vein; *H*, humeral diaphysis and medial epicondyle (*H*′); *N*, superficial branch of the radial n. (lateral cutaneous antebrachial n.); *O*, olecranon; *R*, radius. In life, the superficial branch of radial n. (*N*) emerges from beneath the lateral head of triceps brachii m. on the lateral aspect of the distal brachium and later disappears under the extensor carpi radialis m. (6).

from the distal humeral diaphysis (extensor carpi radialis) and lateral epicondyle of the humerus (digital extensors), and all are supplied by branches of the **radial nerve** (Fig. 7.3-11). The extensor tendons pass over the cranial surface of the radius and ulna on their way to their distal attachments (see Case 7.4).

The **ulnaris lateralis** (also called the extensor carpi ulnaris) **muscle** is generally grouped with the carpal extensors, although its actions are more complex. Like the digital extensors, the muscle arises from the lateral epicondyle of the humerus and is innervated by the radial nerve. Unlike the other carpal extensors, however, its distal attachments on the fifth metacarpal bone and the accessory carpal bone position it to influence abduction of the manus and carpal extension with a slight lateral rotation at the carpus. Primarily, this muscle functions to support the carpus when extended and weight-bearing, being active in the last part of the swing phase and the first part of the stance phase at all gaits in dogs. A bursa is generally present beneath the tendon of origin, and another bursa may be present between its tendon of insertion and the ulnar head.

Abductor digiti I (pollicis) longus is a small muscle that arises on the lateral surface of the proximal radius and ulna and inserts on the medial aspect of the first metacarpal bone, accompanied by the small **sesamoid bone** described earlier. For most of its oblique course from lateral to medial, this muscle lies deep to the carpal and digital extensors. Just proximal to the carpus, it emerges from beneath the common digital extensor tendon and passes obliquely over the extensor carpi radialis tendon and the craniomedial aspect of the distal radius, in the most medial of the 3 grooves on the distal radius. There may be a small bursa or synovial sheath between the 2 structures where it crosses the extensor carpi radialis tendon. Abductor digiti I longus is so-named because it abducts the first digit (the "dewclaw") in relation to the 4 weight-bearing metacarpal bones. Depending on which other muscles are contracting, it may also medially bend or adduct the manus, to the limited extent allowed by the canine/feline carpus, and laterally rotate or supinate the manus.

The flexors of the carpus and digits are located on the caudal and caudomedial surfaces of the antebrachium. From caudal to cranial, they include **flexor carpi ulnaris (ulnar and humeral heads)**, **superficial digital flexor**, **deep**

digital flexor (humeral, radial, and ulnar heads), and **flexor carpi radialis muscles**. Another antebrachial muscle of importance is **pronator teres**, the function of which is to pronate the antebrachium and thus the manus. Unlike the carpal and digital flexors, it inserts onto the medial surface of the radius in the proximal third of the antebrachium. Flexor carpi ulnaris inserts on the accessory carpal bone, while flexor carpi radialis inserts on the palmar aspect of the carpus within the carpal canal. The long tendons of the superficial and deep digital flexors continue over the palmar surface of the carpus, within the carpal canal, on their way to inserting on the digits. The **median nerve** supplies the superficial digital flexor muscle and some heads of the deep digital flexor, while the **ulnar nerve** supplies the other muscles of the caudomedial antebrachium (see Fig. 7.3-9).

In addition to the blood vessels and nerves on the caudomedial aspect of the distal forearm, the superficial and deep digital flexor tendons must be carefully avoided during surgical procedures involving the distal radius and/or ulna, such as the radial osteotomy and ulnar ostectomy performed in this case.

Regional blood supply and lymphatics (Figs. 7.3-9–7.3-12)

The innervation of this region is discussed in the text above in the section on antebrachial musculature. The antebrachium and carpus are primarily supplied by 2 branches of the **brachial artery**—the common interosseous and median arteries. The **common interosseous artery** arises from the brachial artery in the proximal antebrachium and almost immediately divides into the cranial interosseous, caudal interosseous, and ulnar arteries. The **cranial interosseous artery** remains in the proximal antebrachium, primarily supplying the common and lateral digital extensors, as well as ulnaris lateralis.

After branching from the common interosseous artery, the **caudal interosseous artery** tracks distally between the radius and the ulna. The caudal interosseous artery supplies the **nutrient arteries of the radius and the ulna**. Muscles supplied by the caudal interosseous artery include the pronator quadratus, deep digital flexor, abductor digiti I longus, common digital extensor, and lateral digital extensor.

The caudal interosseous artery is at an increased risk during osteotomy or ostectomy of the radius or ulna. It must be carefully identified and avoided.

The **ulnar artery** supplies the humeral and ulnar heads of the deep digital flexor, along with flexor carpi ulnaris. The ulnar artery, along with the **ulnar nerve**, travels distally in the limb, first closely associated with the lateral (deep) border of the deep digital flexor, and later with the deep surface of the humeral head of flexor carpi ulnaris. The ulnar artery anastomoses with the deep antebrachial artery in the distal third of the antebrachium, and the resulting vessel then anastomoses with the caudal interosseous artery immediately proximal to the carpal canal.

The branching of the common interosseous from the brachial artery, deep to the pronator teres muscle, marks the beginning of the **median artery**. The median artery is the largest of the antebrachium and is the main source of blood to the distal limb. In the distal antebrachium, the artery crosses obliquely over the humeral head of the deep digital flexor before passing through the carpal canal alongside the median nerve. It then forms the **superficial palmar arch,** which gives rise to the **palmar common digital arteries** (described in Case 7.4).

The **deep antebrachial artery**, a branch of the brachial artery in the proximal antebrachium, is the main supply to the flexor muscles of the antebrachium.

The **radial artery** branches cranially from the median artery in the proximal to mid-antebrachium and follows the caudomedial border of the radius as it tracks distally to the carpus. Its **dorsal carpal branch** supplies the dorsal carpal joint capsule, while its palmar caudal branch travels through the carpal canal to contribute to the **deep palmar arch** (see Case 7.4).

Located superficially on the craniomedial surface of the antebrachium, the **cephalic vein** provides the main path for return of blood from the distal limb. It receives blood from the **radial**, **brachial**, and **accessory cephalic veins**. The more distal path of the cephalic vein is described and illustrated in Case 7.4.

Care should be taken to avoid injuring the cephalic vein and its major tributaries during a craniomedial surgical approach to the radius. Similarly, the radial nerve lies deep to the extensor carpi radialis muscle on the cranial surface of the radius and should be avoided.

The lymphatic vessels of the thoracic limb, including the antebrachium and carpus, drain into the **axillary lymph node** and, in some individuals, the **accessory axillary lymph node**. Together, these nodes are known as the **axillary lymph center**. The axillary lymph node lies medial and caudal to the shoulder joint, resting between the teres major and rectus thoracis muscles. The accessory axillary lymph node is a smaller structure present in approximately 25% of dogs and, even in those individuals, it is often unilateral. It is more caudally located than the axillary lymph node, occupying a space between the latissimus dorsi and deep pectoral muscles.

Owing to its position deep to the shoulder muscles, the axillary lymph center is not palpable on physical examination.

Selected references

[1] Evans HE, de Lahunta A, Miller ME. Miller's anatomy of the dog. 4th ed. Elsevier-Saunders; 2013.
[2] Fossum TE, editor. Small animal surgery. 4th ed. Elsevier-Mosby; 2013.
[3] von Pfeil DJ, DeCamp CE. The epiphyseal plate: physiology, anatomy, and trauma. Compend Contin Educ Pract Vet 2009;31:E1–11.
[4] Quinn M, Ehrhart N, Johnson AL, et al. Realignment of the radius in canine antebrachial growth deformities treated with corrective osteotomy and bilateral (type II) external fixation. Vet Surg 2000;29:558–63.

CHAPTER 7

CASE 7.4

Phalangeal Fracture

Ray G. Ferguson[a] and Helen M.S. Davies[b]
[a]Monash Veterinary Clinic, Oakleigh East, Victoria, AU
[b]Veterinary BioSciences, The University of Melbourne, Melbourne, Victoria, AU

Clinical case

History

Injuries to the 5th digit of a thoracic limb are common in racing Greyhounds, as this lateral digit is at risk for abnormal loading during turns at high speed.

A 2-year-old male intact Greyhound in race training was presented for lameness and loss of racing performance associated with chronic swelling and lateral displacement of the 5th digit on the right thoracic limb ("sprung toe"; Fig. 7.4-1).

Physical examination findings

The swelling was localized to the proximal interphalangeal (PIP) joint of the 5th digit (Fig. 7.4-2). Distal to the swelling, the rest of the digit was displaced laterally. The swelling was firm, and palpation and manipulation of the digit elicited a pain response. No other abnormalities were found on physical examination.

Differential diagnoses

Ruptured collateral ligament(s) with partial luxation (dislocation) of the proximal interphalangeal (PIP) joint; osteoarthritis of the PIP joint; cellulitis; phalangeal fracture; neoplasia

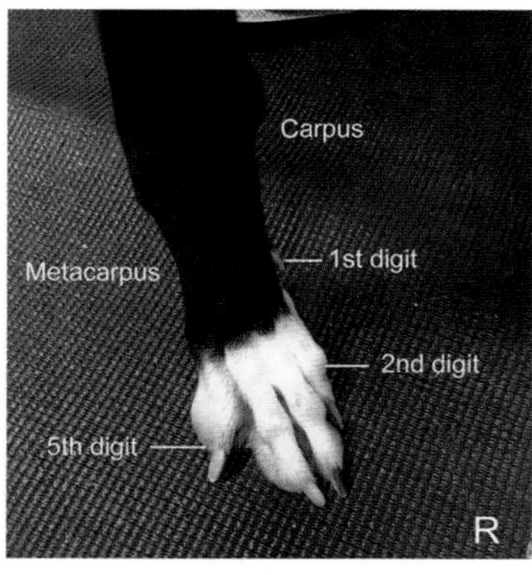

FIGURE 7.4-1 Right distal thoracic limb of the Greyhound in this case at the initial examination. There is swelling of the 5th digit, which is laterally displaced distal to the swelling. The colloquial term for this condition in racing Greyhounds is a "sprung toe."

Comparative Veterinary Anatomy: A Clinical Approach. https://doi.org/10.1016/B978-0-323-91015-6.00047-9

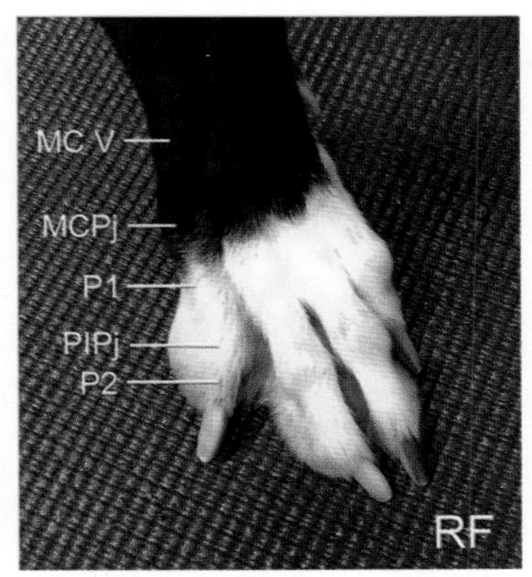

FIGURE 7.4-2 Close-up of the right distal thoracic limb of the Greyhound. The swelling is localized to the PIPj of the 5th digit, which is also the point from which the digit deviates laterally. Key: *PIPj*, proximal interphalangeal joint; *MCPj*, metacarpophalangeal joint (of the 5th digit); *MC V*, 5th metacarpal bone; *P1*, 1st phalanx (5th digit); *P2*, 2nd phalanx (5th digit); *RF*, right fore (thoracic) limb.

Diagnostics

Radiographs of the affected foot revealed a comminuted fracture of the 2nd phalanx (P2) of the 5th digit (Figs. 7.4-3 and 7.4-4). Bony callus formation on the lateral aspect created a fixed valgus deformity of the digit.

Valgus (L. *valgus* turned outward, knock-knee) is a term describing an angular limb deformity of the digit in which the position of the limb/digit distal to the origin of the defect deviates laterally from its normal longitudinal/axial alignment. Varus (L. *varus* turned inward/bow-legged) is the term used to describe an angular limb deformity in which the position of the limb/digit distal to the origin of the defect deviates medially from its normal longitudinal/axial alignment. As a rule, the point of origin of the defect is included in the description of the limb deformity—e.g., carpal valgus. Origin of the deformity originates at the carpus. The same terminology is used in the pelvic limb—e.g., genu (L. *genu* knee) valgus—origin of the deformity originates at the stifle (true "knee").

Diagnosis

Fracture of P2, resulting in ankylosis (bony fusion) of the PIP joint and valgus deformity of the 5th digit of the right thoracic limb

USE OF "META-"

The many anatomical terms used to describe the various structures of the distal thoracic and pelvic limb can be confusing. It is useful to remember that the prefix meta- is a Greek word root, which means "with" or "after." For example, the metacarpal bones are associated with or "come after"—i.e., lie distal to—the carpus; likewise, the metatarsal bones and the tarsus.

Thus, the metatarsophalangeal joint combines a metatarsal bone with a phalangeal bone—specifically, the 1st phalanx of the digit—and as the metatarsus is associated with the tarsus, this joint is in a pelvic limb, distal to (or "after") the tarsus.

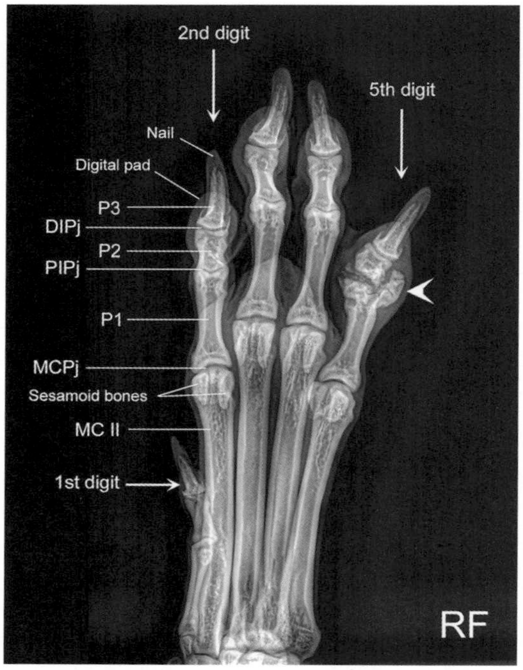

FIGURE 7.4-3 Dorsopalmar radiograph of the Greyhound's right distal thoracic limb, showing a complete, displaced, comminuted fracture of the 2nd phalanx of the 5th digit. A bony callus *(arrowhead)* bridges the fracture and fuses the PIPj laterally. Only the structures of the 2nd digit are labeled, but the same labels apply to digits 3–5 (although the corresponding metacarpal bones are numbered III–V). The outline of the large, single metacarpal pad (not labeled) can be seen beneath the MCPj and proximal P1 of digits 2-5. Key: *DIPj*, distal interphalangeal joint; *MC II*, 2nd metacarpal bone; *MCPj*, metacarpophalangeal joint; *P1*, 1st phalanx; *P2*, 2nd phalanx; *P3*, 3rd phalanx; *PIPj*, proximal interphalangeal joint; *RF*, right fore (thoracic) limb.

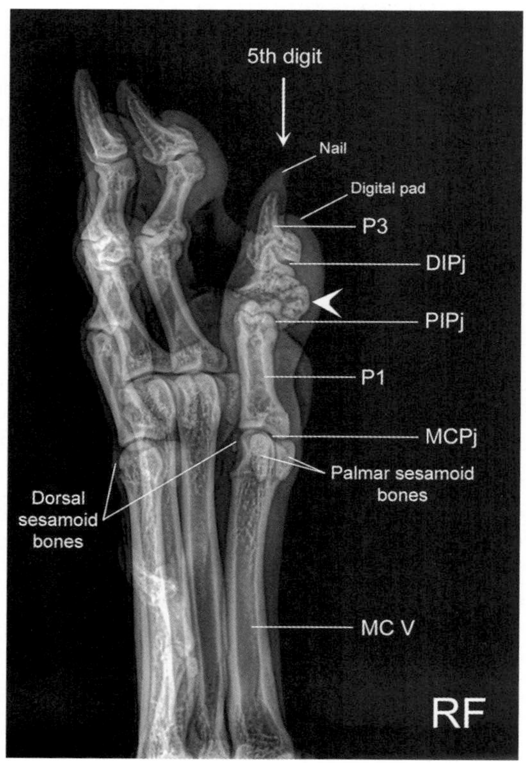

FIGURE 7.4-4 Palmaromedial-dorsolateral oblique radiograph of the Greyhound's right distal thoracic limb, showing the fracture and callus *(arrowhead)* from a more lateral aspect. Key: *DIPj*, distal interphalangeal joint; *MC V*, 5th metacarpal bone; *MCPj*, metacarpophalangeal joint; *P1*, 1st phalanx; *P3*, 3rd phalanx; *PIPj*, proximal interphalangeal joint; *RF*, right fore (thoracic) limb.

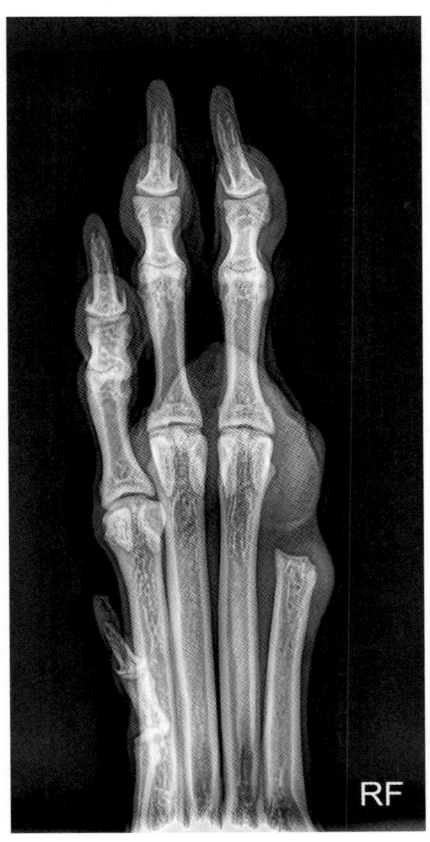

FIGURE 7.4-5 Postoperative radiograph, dorsopalmar view. (Compare this radiograph with that in Fig. 7.4-3.) The 5th digit of the right thoracic limb was amputated at the level of the distal diaphysis of the 5th metacarpal bone, along with the dorsal and palmar sesamoid bones of the 5th metacarpophalangeal joint.

Treatment

Because the prognosis for soundness is poor with conservative management, the owner elected surgical amputation (L. *amputare* to cut away) of the 5th digit. Under general anesthesia, amputation was performed at the level of the distal diaphysis of the 5th metacarpal bone (MC V), removing the entire 5th digit and the dorsal and palmar sesamoid bones of the 5th metacarpophalangeal (MCP) joint (Fig. 7.4-5). The dog made an uneventful recovery and returned to race training 2 months after surgery, winning a race by several lengths 1 month later.

Amputating the digit proximal to the MCP joint eliminates chronic pain originating from weight-bearing on the distal articular surface of MC V if the amputation is performed at the MCP joint, or on the distal articular surface of the 1st phalanx (P1) if the amputation is performed at the PIP joint. It also minimizes the potential for delayed wound healing from persistent synovial fluid production if amputation is performed by disarticulating a joint. Additionally, it is important to resect the digital extensor and flexor tendons at the site of amputation so that the tendon ends do not impede wound healing.

Anatomical features in canids and felids

Introduction

The distal part of the thoracic limb in dogs and cats is referred to as the manus (L. *manus* hand). It is broadly equivalent to the human hand (including the wrist), although structurally and functionally the manus in quadrupeds is quite different from the human hand. The manus in dogs and cats includes the carpus, metacarpus, and digits (4 load-bearing and 1 rudimentary digit). The equivalent structures in the pelvic limb form the pes (L. *pes* foot), consisting of the tarsus, metatarsus, and digits.

In dogs and cats, the term paw is often used interchangeably with "foot" to describe the part of the distal thoracic/pelvic limb that is normally in contact with the ground when the animal is standing. However, because dogs and cats have 4 paws vs. humans having only 2 feet, paw is the preferred term in dogs and cats. The paw contains specialized dermal and epidermal structures that form the protective surfaces of the manus and pes in contact with the ground, notably the pads and nails/claws.

Following is a discussion of the metacarpus/metatarsus and digits. The carpus is described separately in Case 7.3 and the tarsus in Case 8.5. Because there are more similarities in relevant clinical anatomy than differences between the manus and the pes distal to the carpus and tarsus, this section primarily describes the metacarpus and digits of the thoracic limb. Any clinical differences in the anatomy of the pelvic limb are noted.

Function

The metacarpus/metatarsus and digits serve as an interface between the ground surface and the appendicular skeleton when the animal is standing and moving (walking, running, jumping, climbing, pouncing, digging, etc.). The digits are also used in defense, to capture prey, and to dig.

The paws are sensitive to pressure and temperature, and they contain abundant nerve endings to assist in their functions of proprioception and general sensing of the ground surface. The skin of the pads contains a high concentration of sweat glands in both cats and dogs, unlike the rest of the skin surface. Hence, when dogs and cats are overheated, they leave damp footprints. The skin surface of the pads is ridged in a way that helps to prevent slipping on smooth or unstable surfaces, and it thickens further in animals that regularly work or move on hard or abrasive surfaces. The paws of sled dogs may require protection with booties from the abrasive ice crystals, but they do not require protection from the cold. They can control the temperature in their paws through arteriovenous anastomoses and a degree of countercurrent heat exchange in the digital vessels such that they minimize heat loss. In addition, the thick, fat-filled collagenous cushions beneath the pads help to insulate them from the ground surface, whether hot rock or snow and ice.

Bones of the metacarpus/tarsus and digits (Figs. 7.4-6–7.4-8)

From proximal to distal, the principal bones of the **manus** and **pes** distal to the carpus and tarsus include the metacarpal (thoracic limb) or metatarsal (pelvic limb) bones and the phalanges of the digits. In addition, small sesamoid bones are associated with the metacarpophalangeal (thoracic limb) or metatarsophalangeal (pelvic limb) joints and the distal interphalangeal joints.

Each limb has 5 **metacarpal (MC)** or **metatarsal (MT) bones**: 4 long, load-bearing MC or MT bones connecting the carpus or tarsus with the 4 load-bearing digits (see below), and 1 short, rudimentary MC or MT bone connected to the rudimentary 1st digit (when present). The convention is to label the digits 1–5 from medial to lateral, and the MC or MT bones are likewise labeled from I to V. The short MC/MT I is the most medial and MC/MT V is the most lateral. With the notable exception of the small MC/MT I, all of the MC and MT bones are long and rod-like.

Extra or supernumerary digits (polydactyly) may occasionally be found in dogs and cats. Polydactyly is especially common in certain cat breeds. The Briard dog breed has double pelvic limb (hind) dewclaws.

Dogs and cats normally have 5 **digits** on the thoracic limb and 4 digits on the pelvic limb. The difference lies in the inconsistent presence of the rudimentary 1st digit, or "dew claw," in the pelvic limb (commonly absent in the pelvic limb).

The weight-bearing digits are numbered 2–5, from medial to lateral. The 1st digit, when present, is small and nonweight-bearing, lying on the medial aspect of the manus or pes, just distal to the carpus or tarsus.

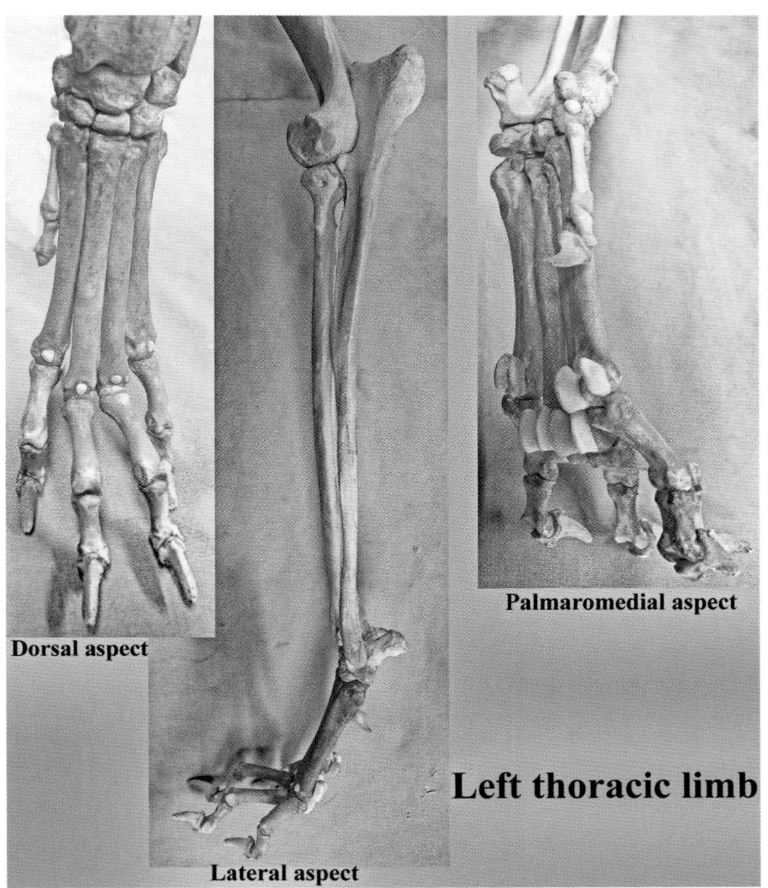

FIGURE 7.4-6 Skeletal components of the left distal thoracic (fore) limb of a Greyhound. The individual perspectives are labeled separately in Figs. 7.4-7 and 7.4-8.

Each weight-bearing digit comprises 3 **phalanges**. The most **proximal phalanx** is the 1st (**P1**), the **middle phalanx** is the 2nd (**P2**), and the **distal phalanx** is the 3rd (**P3**). The 1st digit normally has just 2 phalanges, proximal and distal. Although its distal phalanx is its 2nd and thus may be called P2, it is shaped more like P3 in the weight-bearing digits and likewise supports a nail or claw so may more correctly be called P3. To avoid confusion, it is enough to refer to the phalanges of the 1st digit as simply the proximal and distal phalanx.

In the weight-bearing digits, P1 and P2 are rod-shaped, whereas P3 is modified, with an elongated and slightly curved **ungual process**, which supports the overlying nail or claw. At its base, P3 has a rounded dorsal part, the **extensor process**, onto which its branch of the digital extensor tendon inserts for that digit. A rounded **tubercle** on the palmar or plantar side is where its branch of the deep digital flexor tendon inserts onto that digit.

As described later, the digital extensor and flexor tendons arise from muscle bellies located proximal to the carpus or tarsus. They divide into separate branches (1 for each digit) in the proximal metacarpus or metatarsus and insert onto their respective digits as described earlier. Owing to the digitigrade stance in dogs and cats, the extensor and flexor tendons must change direction over the metacarpo- or metatarsophalangeal joints. To facilitate this process and limit wear on the tendons, each of these joints in the 4 weight-bearing digits has a large pair of **sesamoid bones** on its palmar or plantar surface, within the tendons of the interosseous muscles (described later), which facilitate passage of the deep digital flexor tendon over the joint. On the dorsal surface of the joint, there is a single small, mostly cartilaginous sesamoid beneath the digital extensor tendon. In addition, the DIP joint may have a single, small cartilaginous sesamoid on its dorsal aspect beneath the digital extensor tendon. The metacarpophalangeal joint of the 1st digit has a single palmar sesamoid.

FIGURE 7.4-7 Dorsal aspect of the left distal thoracic limb of a Greyhound, simulating the normal standing position of the skeletal components. This thoracic limb is clearly the left because the small 1st digit is the most medial of the digits. If it were the right thoracic limb, the 1st digit would be located on the opposite side. Key: *CMCj*, carpometacarpal joint; *DIPj*, distal interphalangeal joint; *MC*, metacarpal bone (in this image MC I, II, and V are labeled); *MCPj*, metacarpophalangeal joint; *P*, phalanx (numbered 1–3); *PIPj*, proximal interphalangeal joint.

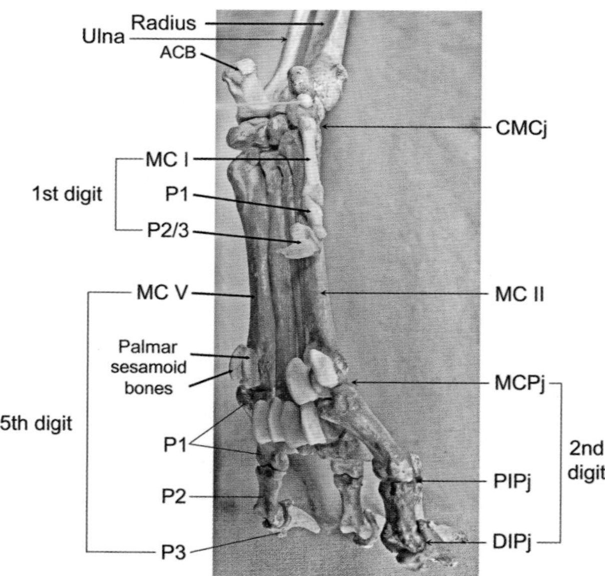

FIGURE 7.4-8 Palmaromedial aspect of the left distal thoracic limb of a Greyhound, simulating the normal standing position of the skeletal components. Note: The 1st digit has only 2 phalanges (proximal and distal); its 2nd phalanx is shaped like P3 in the weight-bearing digits and likewise supports a nail. Key: *ACB*, accessory carpal bone; *CMCj*, carpometacarpal joint; *DIPj*, distal interphalangeal joint; *MC*, metacarpal bone (in this image, MC I, II, and V are labeled); *MCPj*, metacarpophalangeal joint; *P*, phalanx (numbered 1–3); *PIPj*, proximal interphalangeal joint.

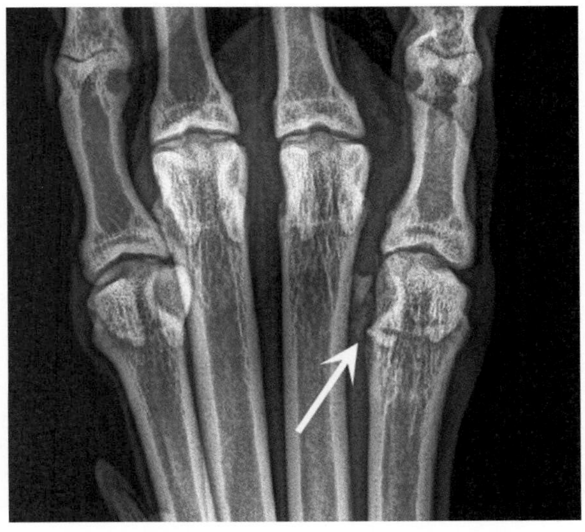

FIGURE 7.4-9 Comminuted fracture of the axial palmar sesamoid bone of the 5th digit in a Greyhound. The *arrow* points to the multiple bone fragments that have separated from the body of the shattered sesamoid bone. This radiograph is not labeled, but the shape of the metacarpal/metatarsal pad suggests that it is a thoracic limb. There is also a fracture line visible on the distal shaft of the proximal phalanx of the fifth digit.

Joints and ligaments of the distal limb (Figs. 7.4-6–7.4-10)

Given the number of metacarpal/metatarsal bones and phalanges, there are many joints within the manus and pes distal to the carpus and tarsus. Each is named according to which bones articulate and their relative position in the limb and to each other. The more proximal of the 2 bones is generally named first. The carpometacarpal and tarsometatarsal joints are described in the sections on the carpus and tarsus. (Cases 7.3 and 8.5, respectively)

SESAMOID BONES

The palmar or plantar sesamoid bones associated with the metacarpo- or metatarsophalangeal joints of the weight-bearing digits are consistently visible on radiographs of the manus or pes, as is the single sesamoid bone of the 1st digit in the manus (see Figs. 7.4-3 and 7.4-4).

The smaller dorsal sesamoid bones of the weight-bearing digits are inconsistently visible radiographically, as they are mostly, and often entirely, cartilaginous. However, their radiographic presence as osseous bodies over the dorsal aspect of these joints, or over a distal interphalangeal joint, is also considered normal.

Fractured palmar or plantar sesamoids are a common cause of lameness in large- and medium-sized dogs. The sesamoid bones of the 2nd and 5th digits are most often fractured (Fig. 7.4-9).

IDENTIFYING THE DIGITS

One way to remember the numbering system for the digits and their associated MC/MT bones is by laying your hand flat, palm down, and numbering your thumb as the 1st digit. Your "pointer" finger then becomes the 2nd digit, your "middle" finger the 3rd, your "ring" finger the 4th, and your "pinky" finger the 5th digit. It makes no difference whether it is your left or right hand; your thumb is the 1st and your little finger the 5th digit.

Laying your hand palm-down in this way is essentially replicating the normal stance of a quadruped. Raising your wrist so that your hand is balancing on your fingertips moves you closer to the normal digitigrade stance of a dog or cat. Lifting your wrist even further, so that your thumb leaves the surface, is closer still, although the normal hyperextension of the metacarpophalangeal and distal interphalangeal joints seen in the dog and cat is not possible with the normal human hand.

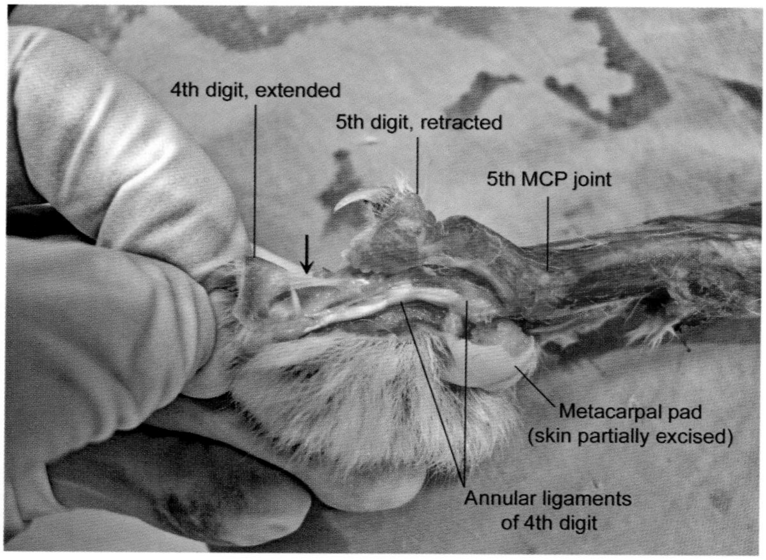

FIGURE 7.4-10 Left manus of a cat, lateral aspect. Most of the skin and superficial fascia have been removed. The 4th digit is being extended to demonstrate the dorsal elastic ligaments *(arrow)* and the annular ligaments of that digit. The 5th digit is in its normal, relaxed/retracted position, in which the distal interphalangeal joint is hyperextended by its dorsal elastic ligaments. Key: *MCP*, metacarpophalangeal.

Each digit has a **metacarpophalangeal (MCP) joint** (thoracic limb) or **metatarsophalangeal (MTP) joint** (pelvic limb) between the MC/MT and P1 of that digit. In each of the 4 weight-bearing digits, the **proximal interphalangeal (PIP) joint** is the articulation between P1 and P2 of that digit. The most distal joint in the weight-bearing digits is the **distal interphalangeal (DIP) joint**, located between P2 and P3 of that digit. The rudimentary 1st digit has a single interphalangeal joint.

All these articulations are separate **synovial joints**, each with its own synovial compartment, joint capsule, and collateral ligaments. Unlike larger and more complex joints such as the carpus and tarsus, which have discrete and layered lateral and medial collateral ligaments, the digital joints are supported on their lateral and medial aspects by modifications of their fibrous joint capsule into **collateral ligaments**. Each ligament is named according to its position (lateral/medial) and the joint it supports. The joints of the metacarpus/metatarsus and digits are thus constrained to flexion and extension in the sagittal plane by the shapes of their articular surfaces and the placement of their collateral ligaments. Normally, minimal lateral bending or axial rotation is possible in these joints.

Retinaculi (singular, **retinaculum** L. a rope, cable, a structure retaining a tissue in place) or **annular ligaments** are present where muscle contraction would cause the tendon to pull away from the bone surface over which it tracks. These rings of fibrous connective tissue encircle the bone(s) and tendons wherever there is a risk of structures separating in this manner. Annular ligaments in the manus and pes pass around the digital flexor and extensor tendons and are named according to their anatomical position and their relationship to other annular ligaments (proximal/distal).

Also important are the **dorsal elastic ligaments** that are a small pair of yellow—i.e., elastin-rich—elastic ligaments that lie on either side of the **digital extensor tendon** branch on each of the weight-bearing digits (Fig. 7.4-10).

DESCRIBING A PARTICULAR JOINT

To properly describe a particular joint in the distal limb, the relevant digit (1–5) and limb (left or right, thoracic or pelvic) needs to be included. It is not necessary to include thoracic or pelvic limb in the description for the metacarpophalangeal or metatarsophalangeal joints, as "carpo-" or "tarso-" identifies the limb as thoracic or pelvic, respectively. However, the proximal and distal interphalangeal joints need to be described in terms of digit, left or right, and thoracic or pelvic.

For example, in the patient in this case, the fracture involved the 2nd phalanx and proximal interphalangeal joint in the 5th digit of the right thoracic limb.

They connect the dorsal surfaces of P2 and P3 in such a way that the DIP joint rests in a position of hyperextension (Figs. 7.4-4 and 7.4-6). Their presence causes the nail or claw to spring up whenever the digital flexor muscles are relaxed. In cats, this claw-retraction mechanism is reinforced by the shape of P2 and P3.

Muscles and tendons of the distal limb (Figs. 7.4-11 and 7.4-12)

The muscles that support and move the MC/MT bones and digits may be divided into extensor and flexor groups. The muscle bellies of the digital extensors lie on the craniolateral surfaces of the antebrachium and crus (i.e., the limb from the carpus/tarsus to the foot). Those of the digital flexors lie on the caudomedial surfaces. The antebrachium is described in Case 7.3 and the crus in Case 8.4.

The digital extensor muscles originate from the distal humerus (thoracic limb) and the distal femur or proximal fibula (pelvic limb), with their main bellies in the antebrachium or crus. The primary digital extensor is the **common digital extensor** (thoracic limb) or **long digital extensor** (pelvic limb) muscle. Just distal to the carpus or tarsus, its long tendon of insertion separates into 4 branches, each inserting onto the **extensor process of P3** of its respective weight-bearing digit (digits 2–5).

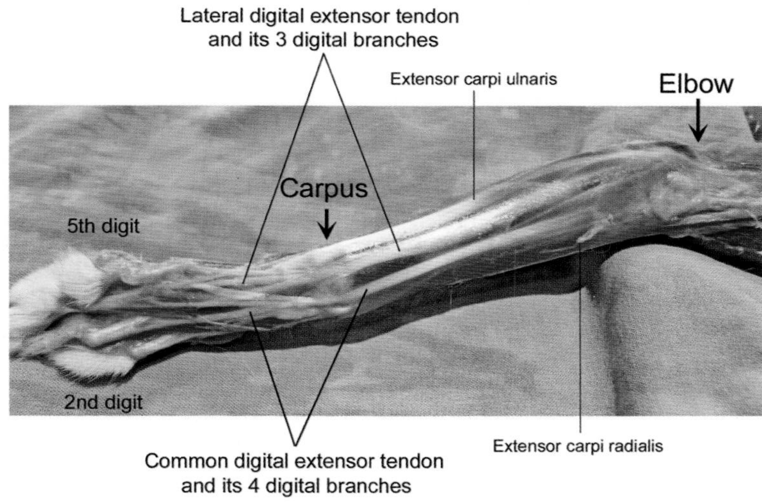

FIGURE 7.4-11 Right thoracic limb of a cat, craniolateral view (dorsolateral view of the manus). Most of the skin and superficial fascia have been removed to show the digital extensor tendons.

DIGITAL ARTHRITIS—AN UNDERDIAGNOSED SOURCE OF LAMENESS

In dogs and cats, the MCP/MTP, PIP, and DIP joints are as predisposed to osteoarthritis and other forms of degenerative joint disease as any other synovial joint. The digits are often overlooked as a cause of chronic lameness unless there is an obvious deformity. It is important to palpate each of the bones and joints in the distal limb during a lameness examination. Each joint is examined for deformity—e.g., swelling, stability, range of movement/motion, and pain on palpation and manipulation. Radiography is important in identifying bony changes. The joint capsule attachments to the bone are usually where the first radiographic changes of osteoarthritis (opaque "fuzziness" at the joint margins) are detected. With more advanced osteoarthritis, there is narrowing of the joint space (due to the loss of cartilage thickness) and bone remodeling (exostoses, "bone spurs") at the joint margins and subchondral sclerosis (Gr. *sclērōsis* hardening).

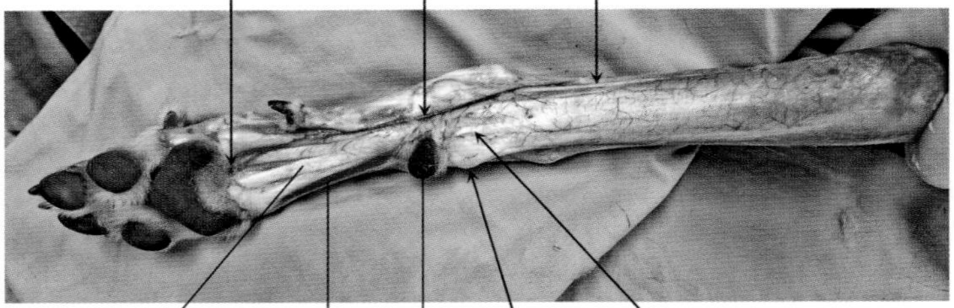

FIGURE 7.4-12 Right thoracic limb of a dog, caudal/palmar view. Most of the skin and superficial fascia have been removed to show the superficial structures on the palmar aspect of the manus. Key: *ACB*, accessory carpal bone; *m.*, muscle; *n.*, nerve; *v.*, vein.

The **lateral digital extensor** muscle is about half the size of the common/long digital extensor. Its tendon splits into 3 branches, which lie deep to those of the common/long digital extensor in the distal part of the manus/pes and insert onto the extensor processes of digits 3–5. Generally, however, they have many more minor connections to the proximal ends of the phalanges, thus reinforcing the joint capsules in the distal manus and pes.

The main function of the digital extensors is to extend the MCP/MTP joints and the interphalangeal joints. In the thoracic limb, they also extend the carpus, and in the pelvic limb, they also flex the tarsus and slightly rotate the distal limb laterally. In dogs and cats, the MCP/MTP and DIP joints are normally in a position of hyperextension when the animal is standing. The digital extensor muscles, via their long tendons of insertion, support the manus and pes in this standing position.

There are 2 primary digital flexors: the **superficial** and **deep digital flexor muscles**. Each has a long tendon of insertion, which begins proximal to the carpus/tarsus, tracks through the **carpal/tarsal canal**, branches distal to these joints, and inserts via its branches onto the palmar/plantar aspect of the digits.

The 4 branches of the **superficial digital flexor tendon** insert on the proximal palmar/plantar surfaces of P2 on the 4 weight-bearing digits. First, each branch forms a small collar or ring around its corresponding deep digital flexor branch, which passes through it over the palmar/plantar surface of the MCP/MTP joint. The **deep digital flexor tendon** divides into 4 or 5 branches, each of which inserts onto the palmar/plantar surface of P3 on its respective digit, including the 1st digit (when present). As its name implies, the deep digital flexor tendon lies deep to the superficial digital flexor tendon along its entire course.

Deep to the branches of the deep digital flexor tendon lie 4 small, fleshy **interosseous muscles**, which cover the palmar/plantar surfaces of the 4 main MC/MT bones. Each muscle arises from the proximal end of its respective MC/MT bone and the carpal/tarsal joint capsule. Distally, each muscle divides into 2 small tendons, each attaching to its respective lateral or medial palmar/plantar **sesamoid,** and then continuing as **sesamoidean ligaments** to the proximal end of P1 in that digit. A smaller extensor branch from each part of the divided interosseous muscle tendon tracks obliquely across the proximal aspect of P1 and joins the common/long digital extensor tendon branch on the dorsal surface of that digit. This arrangement is like the suspensory ligament in the horse and provides additional support to the MCP/MTP joints in their normal weight-bearing angle (hyperextended).

Blood supply and innervation of the distal limb (Figs. 7.4-12 and 7.4-13)

The main blood supply to the manus is the **median artery**, which is the primary continuation of the brachial artery through the antebrachium and into the manus. The **interosseous** and **radial arteries** branch from the median artery in the antebrachium and track distally. They branch further before anastomosing to create the **deep palmar**

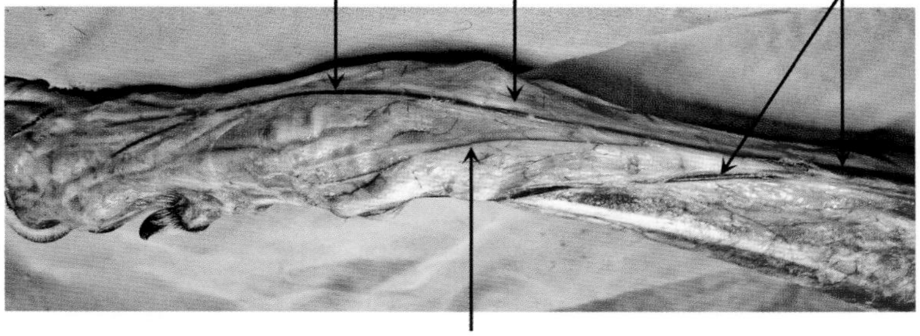

Superficial radial n., lateral branch
Accessory cephalic v.
Cephalic v.
Superficial radial n. & Proximal collateral radial a., medial branches

FIGURE 7.4-13 Right thoracic limb of a dog, cranio/dorsomedial aspect. "In this prosection," the skin and superficial fascia have been removed to show the superficial vessels and nerves on the dorsomedial aspect of the manus. Key: *a.*, artery; *n.*, nerve; *v.*, vein.

arch on the palmar aspect of the metacarpus, beneath the interosseous muscles. The median artery continues into the manus via the **carpal canal** on the palmar aspect of the carpus, medial to the accessory carpal bone, and forms the **superficial palmar arch** beneath the proximal margin of the metacarpal pad, along with a minor anastomotic branch from the interosseous artery (Fig. 7.4-12).

Distally, branches of the **radial artery** cross the medial aspect of the proximal metacarpal region obliquely to supply the metacarpus and digits. Branches from the **superficial** and **deep palmar arches** remain on the palmar aspect of the manus to supply the **palmar** aspect of the metacarpus and digits. Each digit therefore has its own arterial supply on both its dorsal and palmar aspects. However, these arteries continue to anastomose extensively throughout the metacarpus and digits.

A similar arrangement is found in the pelvic limb, with distal branches of the **cranial tibial artery** supplying the dorsum of the metatarsus and digits, and those of the **saphenous artery** supplying the plantar aspect of the metatarsus and digits. As in the thoracic limb, there is extensive anastomosis of dorsal and plantar arteries, both superficial and deep, within the pes.

Each digit has 8 potential arteries (and their satellite veins) distributed around the digit: dorsal and palmar/plantar, superficial and deep, and axial and abaxial. In this context, "axial" refers to the long axis of the limb, not that of the digit. The axis of the limb is considered to bisect the manus/pes between digits 3 and 4. Thus, the axial arteries on digits 2 and 3 mirror in position those of digits 4 and 5. The superficial arteries are named the **dorsal** and **palmar/plantar common digital arteries**. The palmar/plantar common digital arteries are the main supply to the digits and pads. The deep arteries are named the **dorsal** and **palmar metacarpal** or **dorsal** and **plantar metatarsal arteries**, and further distally the **dorsal** and **palmar/plantar proper digital arteries**.

Because the main arterial supply to the digits and pads is superficial and located on the palmar/plantar aspect of the paw, deep lacerations to the pads or the surrounding skin on the underside of the paw may bleed profusely and have the potential to compromise blood supply to one or more of the digits.

The major venous return from the distal thoracic limb is via the **cephalic vein**, which arises from the **superficial palmar venous arch** draining the palmar aspect of the manus. The superficial palmar venous arch lies at the level of the metacarpophalangeal joints on the palmar surface of the manus, deep to the metacarpal pad. The **accessory cephalic vein** drains the dorsum of the manus. The cephalic vein passes proximally on the medial aspect of the carpus to unite with the accessory cephalic vein on the cranial aspect of the distal antebrachium.

The cephalic vein is commonly used for venipuncture in dogs. Generally, the cephalic vein is best accessed proximal to its union with the accessory cephalic vein in the distal antebrachium.

The main venous drainage from the distal pelvic limb is into the **medial** and **lateral saphenous veins** in the dog, and mostly into the **medial saphenous vein** in the cat. The lateral saphenous vein is generally a much smaller vessel in the cat.

If a pelvic limb vessel is needed for venipuncture instead of the cephalic or jugular vein, the medial saphenous vein is used in cats, while the lateral saphenous vein is more commonly used in dogs.

Innervation to the manus is supplied by the radial, median, and ulnar nerves, all of which arise from the brachial plexus. Distally, the superficial branch of the **radial nerve** bifurcates to track on either side of the cephalic and accessory cephalic veins into the digits. The radial nerve is the motor nerve to all the extensor muscles of the elbow, carpal, metacarpophalangeal, and phalangeal joints. It also provides sensory innervation to the lateral aspect of the antebrachium and the dorsum of the manus.

The relevant branches of the **median nerve** innervate the pronator muscles of the manus (those that rotate the manus medially and resist lateral rotation) and the carpal, metacarpal, and digital flexor muscles. The median nerve passes through the carpal canal to supply sensory innervation to the palmar aspect of the manus.

The **ulnar nerve** gives off muscular branches in the antebrachium before forming dorsal and palmar branches. The **dorsal branch** arises in the mid-antebrachium and becomes superficial, supplying sensory innervation to the lateral surface of the manus, mostly the 5th digit. The **palmar branch** passes distally on the lateral side of the carpal canal to provide motor innervation to the interosseous muscles and sensory innervation to the palmar surface of the manus.

A similar arrangement in the pelvic limb results in branches of the **superficial** and **deep peroneal nerves** innervating the dorsum of the pes and those of the **tibial nerve** innervating the plantar aspect of the pes.

Like the digital vessels, the **digital nerves** lie on each side of each digit, arranged (with one exception) as superficial and deep, and are named as for their accompanying vessels. Hence, they are the **dorsal** and **palmar/plantar common digital nerves** if superficial, and the **palmar metacarpal** or **plantar metatarsal nerves** and **palmar/plantar proper digital nerves** if deep. There are no deep digital nerves on the dorsal surface of the digits.

Pads and nails (Fig. 7.4-14)

The distal part of the limb is protected by the "foot pads" and the nails or claws. The **foot pads** comprise a thick subcutaneous cushion of fibrous and elastic connective tissue and fat, termed the **digital cushion**, with a covering of very thick skin, the epidermal layer being the most modified.

The largest pad in each paw is the single **metacarpal** or **metatarsal pad**. In the thoracic limb, it is half-moon- or heart-shaped; in the pelvic limb, it is generally more elongated. This pad extends across the distal metacarpus/metatarsus, beneath the MCP/MTP joints. Distal to this pad are 4 **digital pads**, which are positioned beneath the

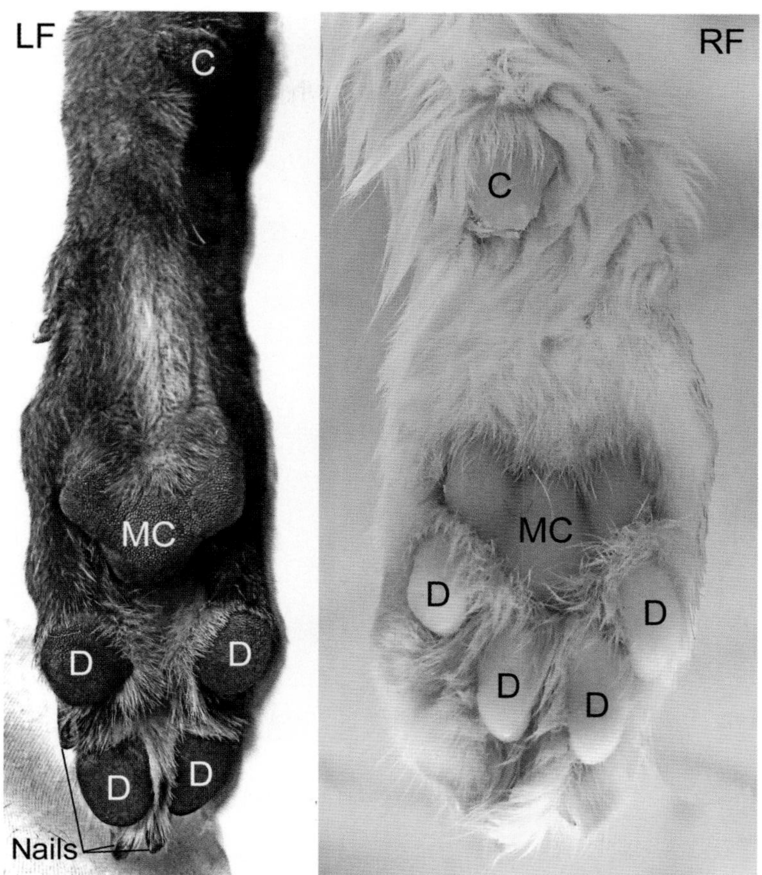

FIGURE 7.4-14 Palmar aspect of the thoracic paw of a dog *(left)* and, enlarged for illustration, a cat *(right)*, showing the carpal (C), metacarpal (MC), and 4 digital (D) pads. Key: *LF*, left fore (thoracic) limb; *RF*, right fore (thoracic) limb.

DIP joints of the 4 weight-bearing digits. An additional single **carpal pad** lies just distal to the accessory carpal bone in the thoracic limb. There is no equivalent to the carpal pad in the pelvic limb.

The metacarpal/metatarsal pads and the digital pads are normally in contact with the ground during locomotion. The carpal pad is not, except at racing speeds, when the carpus may be hyperextended (Fig. 7.4-16), or when running in deep mud or sand. The thick, rugged epidermis of the pads provides a relatively hard-wearing surface for contact with the ground, and the digital cushion serves a shock absorption role during locomotion.

CORNS

A corn is an abnormal, focal thickening of the epidermis of a foot pad in dogs (Fig. 7.4-15). In most cases, a digital pad is affected, usually the pad of digit 3 or 4, occasionally the pad of digit 2 or 5. Rarely, corns are found in one of the metacarpal or metatarsal pads.

Corns predominantly affect dog breeds with soft, thin pads such as Greyhounds and Whippets. They are painful lesions, so lameness is usually the presenting clinical sign.

Medical options generally are unrewarding, and surgical excision of the corn often results in recurrence. Better long-term results may be obtained with a digital flexor tenotomy of the affected digit at the level of P1, midway between the MCP/MTP joint and the PIP joint. The tenotomy alters the way the digital pad is loaded, allowing the corn to resolve with less likelihood of recurrence.

FIGURE 7.4-15 A corn *(arrow)* on the digital pad of a dog. The other 2 weight-bearing digits have been removed from this prosection.

FIGURE 7.4-16 At racing speeds in Greyhounds, the carpus may hyperextend such that the carpal pad is in contact with the ground *(arrow)*.

NAIL/CLAW ABNORMALITIES

In dogs that are exercising excessively on hard, abrasive surfaces, the nails may become over worn such that the underlying dermal vessels and nerves—and even the bone—are exposed. Any distortion of the digit or interference with normal loading of the digit can cause problems with normal nail/claw growth. An overgrown or distorted nail/claw may cause pressure on the claw bed, interfering with normal nail/claw growth and may be painful enough to discourage the dog from exercising—resulting in further interference with normal wear. Such pressure may distort an otherwise normal nail/claw, and the nail/claw may even grow around in a circle or spiral and pierce the tissues of the paw. Thus, abnormalities of nail/growth are best prevented or addressed early with routine nail trimming.

The **nails or claws** are shells of hard keratin, produced by the epidermis that covers the terminal portion of P3 on each digit, including the rudimentary 1st digit. The underlying dense dermis merges with the periosteum of P3. Like the equine/bovine hoof, the canine/feline claw consists of a **claw plate** (analogous to the hoof wall) and a **sole**. In addition, the **claw fold** is a fold of epidermis that covers the proximal portion of the claw plate, like the periople in the hoof and the cuticle of the human fingernail. The nails/claws grow continuously from a **claw bed** at the base of the ungual process of P3.

In dogs, the nails need to be exposed to normal wear—i.e., regular exercise on an abrasive surface—or they will require regular trimming. Cats normally keep their claws sharp and shaped by clawing at preferred surfaces such as tree trunks or scratching posts. This behavior can create management problems in household cats. Because the ungual process and associated vessels and nerves underlie and support the proximal part of the nail/claw, these structures must be avoided when trimming the animal's nails/claws to prevent bleeding.

The cat's claws are normally retracted ("ensheathed") unless purposely extended ("ex-sheathed"). Retraction occurs through recoil of the **dorsal elastic ligaments**, while protraction or extension requires coordinated contraction of both the digital flexor and extensor muscles. The larger digital flexor muscles create the main flexion movement in the PIP joint which can dig the claws into the surface or prey, while coordinated contraction of the smaller and more widely attached digital extensor muscles straighten the DIP joint, protracting and stiffening the digit.

Selected references

[1] Ober CP, Freeman LE. Computed tomographic, magnetic resonance imaging, and cross-sectional anatomic features of the manus in cadavers of dogs without forelimb disease. Am J Vet Res 2009;70:1450–8.

[2] Guilliard MJ, Segboer I, Shearer DH. Corns in dogs; signalment, possible aetiology and response to surgical treatment. J Small Anim Pract 2010;51:162–8.

[3] Hamelin A, Begon D, Conchou F, et al. Clinical characterisation of polydactyly in Maine Coon cats. J Feline Med Surg 2017;19:382–93.

CHAPTER 7

CASE 7.5

Nerve Sheath Neoplasm

Eric N. Glass[a] and Marc Kent[b]
[a]Neurology and Neurosurgery, Red Bank Veterinary Hospital, Tinton Falls, New Jersey, US
[b]Department of Small Animal Medicine and Surgery, University of Georgia College of Veterinary Medicine, Athens, Georgia, US

Clinical case

History

An 11-year-old female spayed Labrador Retriever was presented for a 3-month progressive history of right thoracic limb lameness. Initially, the lameness was subtle; however, over time, it progressed to the right thoracic limb collapsing when the patient placed weight on the limb. In the last 1 month, the owner noticed that the muscles of the limb were atrophied. Additionally, the patient frequently knuckled over and stood on the dorsum of the manus (forepaw).

Physical examination findings

Severe, generalized atrophy of the right thoracic limb was present, but no pain was present on manipulation of the limb. No other abnormalities were identified on routine physical or orthopedic examination.

Neurological examination findings

Mentation

- Normal

Gait/posture

- Right thoracic limb monoparesis; during the stance (weight-bearing) phase of the gait, unable to support weight at the elbow and shoulder, frequently knuckled over onto the dorsum of the manus (forepaw); right shoulder joint was lower than the left

Postural reactions

- Absent hopping and placing of the right thoracic limb; normal left thoracic limb and pelvic limbs, bilaterally

Spinal reflexes/muscle tone/muscle mass/nociception

- Thoracic limbs—right: absent withdrawal reflex, decreased tone (flaccid), severe generalized atrophy of all the muscles of the limb; left: normal withdrawal reflex, muscle tone, and muscle mass
- Pelvic limbs—normal patellar and withdrawal reflexes, normal muscle tone, and normal muscle mass, bilaterally
- Cutaneous trunci reflex—absent cutaneous trunci reflex on the right side. Stimulation on both sides of the vertebral column only elicited movement of the left cutaneous trunci muscle

Cranial nerves

- All cranial nerves were normal

Comparative Veterinary Anatomy: A Clinical Approach. https://doi.org/10.1016/B978-0-323-91015-6.00048-0

Sensory

- Right thoracic limb—reduced sensation to the skin over the dorsum of the foot, the skin just proximal to the metacarpal pad, and to the skin over the lateral manus (forepaw)
- Painful to deep palpation in the axilla

Neuroanatomic diagnosis

Right side; C6 to T2 spinal nerve roots, spinal nerves, or named nerves of the right thoracic limb; less likely a focal or diffuse lesion involving the C6 through T2 spinal cord segments

Differential diagnoses

Neoplasia (e.g., nerve sheath neoplasm [NSN], lymphoma, or metastatic neoplasia); neuritis involving the spinal nerve roots, spinal nerves, brachial plexus, or named nerves of the thoracic limb—e.g., noninfectious/immune-mediated or infectious neuritis such as bacterial (abscess), fungal, or protozoal such as *Neospora caninum, Toxoplasma gondii*

Diagnostics

A complete blood count, serum chemistry profile, and urinalysis were assessed for any concurrent disease that may impact or preclude general anesthesia or identify evidence of a neoplasm that may have metastasized. Three-view thoracic radiographs were performed to screen for evidence of metastatic neoplasia. These diagnostic procedures were within normal limits. The dog was placed under general anesthesia for MRI of the caudal cervical vertebral column and right brachial plexus.

At the level of the C7-T1 intervertebral foramen and extending distally, the right C8 spinal nerve was markedly enlarged compared with the left side (Fig. 7.5-1). The enlarged C8 spinal nerve could be traced to a mass in the area of the brachial plexus. The lateral scapular muscles (infra- and supraspinatus muscles) and medial scapular muscles (subscapularis and serratus ventralis muscles) were atrophied compared with the same muscles on the left side. The epaxial muscles appeared normal in size and signal intensity, bilaterally, which suggested that the lesion did not affect the dorsal branch of the spinal nerve. The enlarged C8 spinal nerve did not extend into the vertebral canal. The epidural space, meninges, and caudal cervical through cranial thoracic spinal cord were normal. Following IV

NERVE SHEATH NEOPLASMS—TREATMENT OPTIONS

In most dogs with neoplasms arising from a nerve, the cell of origin is unknown. Therefore, the broad term "NSN" is used. The majority of NSN are malignant; however, benign types have been reported. Treatment for dogs with malignant NSN consists of definitive or palliative therapies. Definitive intent therapy includes surgery, radiation therapy, or both. Palliative intent therapy consists of analgesics.

Surgery consists of resection of the affected spinal nerves and limb amputation. Occasionally resection of the affected spinal nerve alone is performed.

The "disease-free interval"—i.e., the time between surgery and return of clinical signs, and the "median survival time" with limb amputation both depend on the proximal extent of the lesion with a better prognosis if the disease is confined to the brachial plexus in dogs. Dogs with NSN located within the intervertebral foramen, epidural space, or compressing/invading the spinal cord have a shorter survival time.

Radiation therapy requires specialized protocols—i.e., 3-D plans to include stereotactic or volumetric modulated arc radiotherapy. Radiation without surgery improves clinical signs with a longer survival time than no therapy.

The dog in the present case underwent a limb amputation. At surgery the C8 spinal nerve was identified and dissected as proximal as the intervertebral foramen, at which point it grossly appeared normal in size. The C8 spinal nerve was resected at the level of the intervertebral foramen. The dog recovered well from surgery. The dog remained clinically normal for 9 months at which time it developed cervical pain and right pelvic limb paresis. Despite corticosteroids and analgesics, the right pelvic limb paresis progressed, and the dog was humanely euthanized.

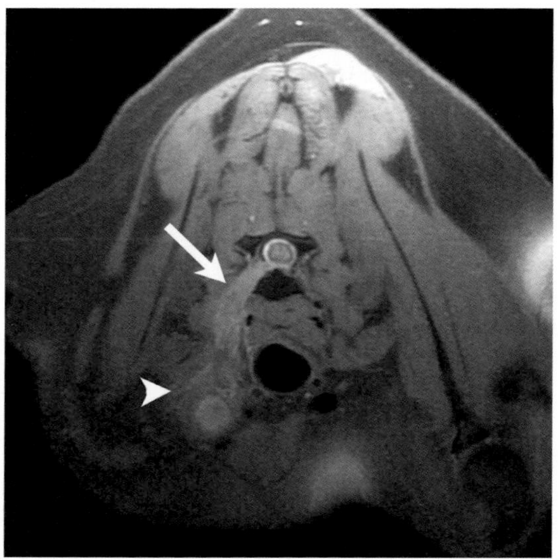

FIGURE 7.5-1 Postcontrast, transverse T1-weighted MRI with fat saturation from the same dog as in Fig. 7.5-1. The enlarged C8 spinal nerve *(arrow)* that connects to a mass (*arrowhead*) in the brachial plexus displays a strong, homogenous contrast enhancement.

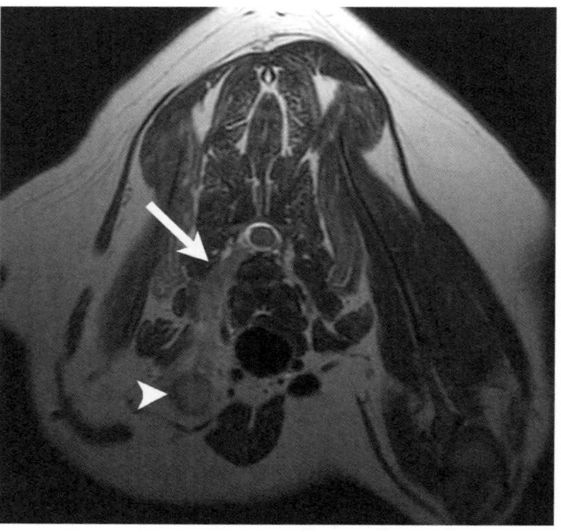

FIGURE 7.5-2 Transverse T2-weighted MRI at the C7-T1 intervertebral disk space in an 11-year-old Labrador Retriever with LMN paresis of the right thoracic limb. Note the enlarged C8 spinal nerve *(arrow)* that extends into a mass *(arrowhead)* in the right brachial plexus. These findings are consistent with a nerve sheath neoplasm.

Nerve sheath neoplasm was the most likely presumptive diagnosis based on the signalment, protracted history, and MRI findings. Less likely was lymphoma involving the C8 spinal nerve and brachial plexus. Lymphoma tends to be a more acute and rapidly progressive disease process and would be more likely in a cat. Neuritis of the C8 spinal nerve or brachial plexus was unlikely given the mass effect within the brachial plexus. Definitive diagnosis may be obtained through cytology or histology of the mass in the brachial plexus; however, in many cases, treatment is pursued based on a presumptive diagnosis.

administration of contrast medium, the enlarged C8 spinal nerve from the C7-T1 intervertebral foramen to the mass in the brachial plexus displayed a marked contrast enhancement (Fig. 7.5-2). No abnormal enhancement was seen in the epidural space, meninges, or spinal cord from C5 through T3 vertebrae.

Diagnosis

Nerve sheath neoplasm (NSN) (presumptive)

Treatment

See side box entitled: "Nerve sheath neoplasms—treatment options."

Anatomical features in canids and felids

Introduction

The correct neuroanatomic diagnosis requires the recognition of signs related to dysfunction of the general somatic effect (GSE) and general somatic afferent (GSA) neurons involving the thoracic limb. Knowledge of the 6 clinically important nerves and their respective origins in the cervical intumescence of the spinal cord is essential to correctly form an accurate neuroanatomic diagnosis. See Cases 3.12 and 14.1 for a description of GSE, GSA, and upper vs. lower motor neurons.

Function

Lower motor neuron (LMN) quality paresis is characterized as an inability to support weight. During the stance (weight-bearing) phase of the gait, the patient collapses on the limb, resulting in overflexion of the joints. When standing, the patient may have a crouched posture. The limb may knuckle over causing the patient to stand on the dorsum of the manus as the affected limb flexes under the patient's weight. When this occurs while walking, the contralateral limb is placed back on the ground earlier than normal, which results in a shortening of the length of the patient's stride. The resultant gait can be described as a short-strided, "choppy" gait. Importantly, a short-strided gait looks very similar to the gait of a patient with an orthopedic lameness where the stride is shortened as the patient off-loads its weight due to pain.

Lower motor neuron weakness results in abnormal postural reactions, such as "hopping" and paw placement tests. Recall that postural reactions are a means of assessing LMN function as well as the ascending general proprioceptive (GP) pathways that convey the position sense of the limb to the brain (GP information projects to the contralateral prosencephalon [somesthetic cerebral cortex] and the ipsilateral cerebellum) and the descending upper motor neurons (UMN). In the case of LMN paresis, the patient lacks the strength to perform postural reactions correctly. With the postural reaction test termed "hopping," the patient may make rapid shorter hops, akin to the short-strided gait observed as the patient walks. In those patients that are too weak to stand, the examiner, providing support for the patient to stand, can compensate for the deficits in the postural reactions.

The key to recognizing LMN paresis is through the evaluation of the patient's ability to bear weight when standing, spinal reflexes, muscle tone of the limbs, and the presence of atrophy. The ability to bear weight in a standing position is provided by the muscles that extend the elbow (primarily the triceps muscle) and carpus (cranial antebrachial muscles). These muscles are innervated by the radial nerve. Therefore, the inability to support weight implies dysfunction of either the extensor muscles of the carpus and elbow, the radial nerves, or the C7, C8, or T1 spinal cord segments. The radial nerve provides the most important contribution to the ability to bear weight and walk with a normal gait. Transection of any one nerve, other than the radial nerve, does not result in a clinically appreciable effect on the gait or the ability to stand.

WITHDRAWAL REFLEX

In the thoracic limb, the only reliable spinal reflex that can be tested is the withdrawal reflex. The withdrawal reflex is performed by applying pressure to a digit and assessing the patient's ability to flex all the joints of the thoracic limb in an effort to pull the limb toward the body. Normal flexion of the carpus, elbow, and shoulder joint requires the normal function of nearly all the named nerves of the thoracic limb—e.g., radial, ulnar, musculocutaneous, axillary, and less so, the suprascapular nerve. With LMN paresis, there is a decreased or absent withdrawal reflex. Other presumed reflexes have been described including the extensor carpi radialis and triceps reflexes. These reflexes are not reliably elicited in normal patients and more likely represent contraction of the muscle in response to being percussed by a reflex hammer.

Cervical intumescence

In general, the **cervical intumescence** is C6 through T2 spinal segments, however, exceptions exist. The GSE neurons that give rise to the nerves of the thoracic limb are located in the ventral gray column of the cervical intumescence, the C6 through T2 spinal cord segments. The GSE axons exit the spinal cord ventrolaterally giving rise to the **ventral spinal roots**. A ventral spinal root joins with a **dorsal root** to form a **spinal nerve**. The spinal nerve exits the vertebral canal via an **intervertebral foramen**. Remember, the 1st cervical spinal nerve exits the vertebral column via the lateral vertebral foramen of the atlas (C1); thereafter, the cervical spinal nerves exit via an intervertebral foramen. In the cervical vertebral column, the spinal nerves exit the vertebral foramen cranial to the vertebra of the same number—i.e., C2 spinal nerve exits between the C1 [atlas] and C2 [axis] vertebrae.

Consequently, there are 8 cervical spinal cord segments and cervical spinal nerves and 7 cervical vertebrae. In the remainder of the spinal cord and vertebral column, there are equal numbers of spinal cord segments, spinal nerves, and vertebrae with the spinal nerve exiting caudal to the vertebra of the same number—i.e., T1 spinal nerve exits between the T1 and T2 vertebrae.

These GSE neurons are also referred to as LMN. The axons of the LMN form the named nerves of the thoracic limb, which are in synaptic contact with the muscles they innervate via a neuromuscular junction. Together, the LMN, its axon, the neuromuscular junction, and the innervated muscle form the **LMN unit.**

> Dysfunction of any part of the LMN unit (GSE neuron, its axon, the neuromuscular junction, or muscle) results in LMN paresis (weakness) or paralysis (loss of voluntary movement). The term "neuromuscular" can be used synonymously with LMN. Clinically, it is challenging to differentiate a lesion affecting any one portion of the LMN unit. Signs related to LMN weakness from a lesion affecting the spinal cord gray matter, nerves, neuromuscular junction, or muscle look similar. In most cases, the most specific neuroanatomic diagnosis may be LMN dysfunction. Dysfunction of the GSA neurons or their processes results in varying degrees of hypalgesia or analgesia.

The GSA neurons provide sensory innervation. The GSA neurons provide for general proprioception, as well as nociception.

After exiting their respective foraminae, the caudal cervical (C6, C7, and C8 spinal nerves) and first 2 thoracic spinal nerves project ventrolaterally from between the longus capitis muscle medially and the superficial portion of the scalenus muscle laterally to coalesce in a network of spinal nerves called the **brachial plexus**. From the brachial plexus form the named nerves of the thoracic limb. As a rule, 2 or more spinal cord segments and spinal nerves contribute to each named nerve of the thoracic limbs (Fig. 7.5-3).

Spinal reflexes as relating to clinical anatomy

Spinal reflexes test the functional integrity of sensory nerves (afferent), specific spinal cord segments, and motor nerves (efferent). A spinal reflex is composed of a sensory nerve that projects to the spinal cord and specific spinal cord segments, where the incoming sensory axons project directly or indirectly through interneurons to synapse on GSE neurons in the ventral gray matter of the spinal cord, and the GSE axons that give rise to the motor nerve, which are in synaptic contact with a muscle (Fig. 7.5-4). Knowledge of the named nerves of the thoracic limbs, the muscles

HORNER SYNDROME WITH BRACHIAL PLEXUS CONDITIONS

In some patients with LMN paresis in a thoracic limb, Horner syndrome (miosis, ptosis, protrusion of the 3rd eyelid, and enophthalmos) also may be observed. Remember that the preganglionic GVE neurons for sympathetic innervation to the eye are in the intermediate gray matter of the T1 through T3 spinal cord segments. Preganglionic GVE axons exit the spinal cord along the T1 through T3 spinal nerves. Instead of synapsing in the ramus communicans, these preganglionic GVE axons destined for the head project cranially via the vagosympathetic trunk to ultimately provide sympathetic innervation to structures of the head. Disease processes that affect the T1 through T3 spinal cord segments, spinal nerves, or the brachial plexus can result in Horner syndrome in the eye ipsilateral to the lesion.

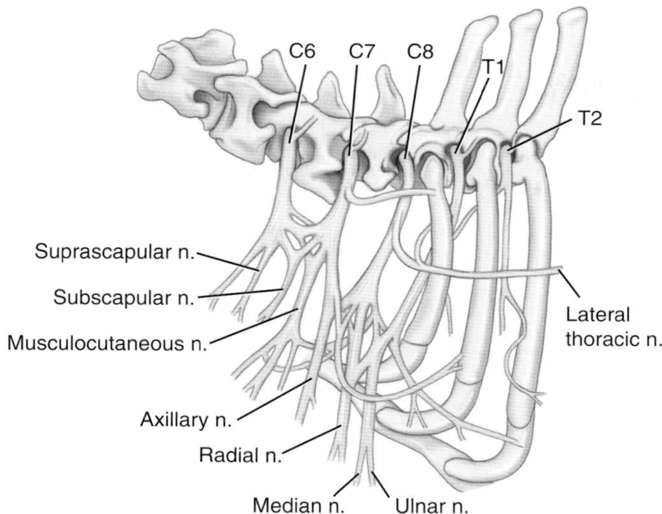

FIGURE 7.5-3 Lateral view of the C6 through T2 spinal nerves, brachial plexus, and nerves of the thoracic limb.

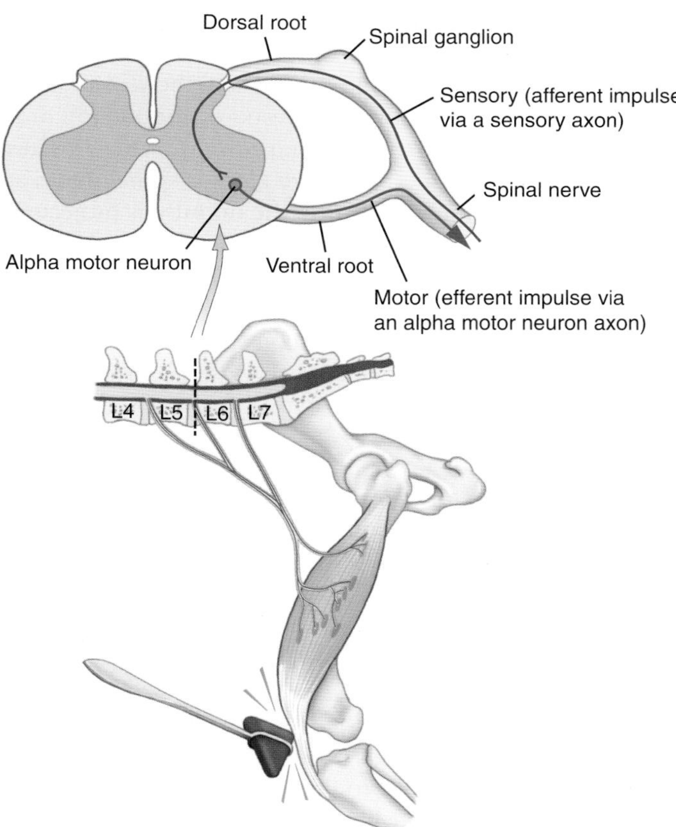

FIGURE 7.5-4 Monosynaptic reflex arc involved in the patellar reflex.

TABLE 7.5-1 Innervation of the thoracic limb.

Spinal cord segments[a]	*Nerve*	*Action/muscles innervated*
C6, C7	Suprascapular	Shoulder extension (supraspinatus, infraspinatus)
(C6), C7, C8	Axillary	Shoulder flexion (teres major, teres minor, deltoideus)
C6, C7, C8	Musculocutaneous	Elbow flexion (biceps brachii, brachialis, coracobrachialis)
C7, C8, T1, (T2)	Radial	All the extensors of the elbow, carpus, and digits (triceps brachii and craniolateral muscles of the antebrachium)
C8, T1, (T2)	Median	Flexors of the carpus and digits (caudomedial muscles)
C8, T1, (T2)	Ulnar	Flexors of the carpus (caudomedial muscles) and digits (flexor digitorum profundus)

[a] Segments in parentheses signify that the contribution is not seen in all dogs.

each nerve innervates, and the spinal cord segments that give rise to each nerve is imperative in forming a correct neuroanatomic localization (Table 7.5-1). Importantly, spinal reflexes do not require the normal function of the spinal cord cranial to the specific spinal cord segments that contain the afferent projecting GSA axons and the GSE neurons and efferent axons. With a disease process that affects any part of the spinal reflex (the afferent axons, spinal cord, or efferent axons), the resultant reflex is decreased or absent. Likewise, a disease process affecting either the GSE neurons in the spinal cord gray matter or its axons results in a decrease in muscular tone (flaccid paresis). The decreased muscle tone can be appreciated by putting the limb through passive range of motion. Within 1–2 weeks, the loss of GSE innervation to the muscle results in a pronounced muscle atrophy (denervation atrophy).

In patients with LMN paresis in a thoracic limb, the cutaneous trunci reflex also may be reduced or absent. The cutaneous trunci reflex is elicited by pinching the skin on both sides of the vertebral column and observing movement of the skin over the lateral aspect of the thorax as a result of contraction of the cutaneous trunci muscle. The reflex begins with stimulation of the skin along the vertebral column, which excites a cutaneous nerve. Impulses travel proximally in the cutaneous nerve to enter the spinal cord along a spinal nerve, approximately 1–2 vertebrae cranial to where the skin is stimulated. Within the spinal cord, the impulse is projected cranially to activate the GSE neurons in the ventral gray column at the C8 and T1 spinal cord segments, bilaterally, that contribute to the lateral thoracic nerve, which innervates the cutaneous trunci muscle. Therefore, stimulation on one side of the vertebral column results in bilateral contraction of the cutaneous trunci muscle. A unilateral lesion affecting the C8 or T1 spinal cord segments, the lateral thoracic nerve, or cutaneous trunci muscle results in a decreased to absent cutaneous trunci reflex on the side ipsilateral to the lesion. However, the contralateral reaction is normal.

Nociception as relating to clinical anatomy

Most of the named nerves in the thoracic limb carry both sensory and motor functions. Therefore, lesions affecting spinal cord segments C6 through T2, the dorsal roots, spinal nerves, brachial plexus, and named nerves may result in decreased to absent nociception (sensation). As a rule, lesions involving the brachial plexus or the nerves are more likely to result in more severe sensory deficits compared with a lesion involving a single spinal cord segment, dorsal nerve root, or spinal nerve. Sensory testing is accomplished by pinching the skin and observing that the patient has a conscious reaction to the stimulus—i.e., seeing the patient turn and look at the skin being pinched or vocalizing.

Observing the limb withdraw (pull away) is a reflex that does not signify that the patient has a normal nociception. A conscious reaction signifies that the stimulus resulted in an impulse that projected along a sensory nerve to enter the spinal cord and ultimately project to the prosencephalon where the stimulus was consciously recognized as being noxious.

The skin innervated by a single nerve is called the **cutaneous area (dermatome)**. The innervation of some areas of skin is provided for by one or more nerves. These areas are called **overlap zones**. An area of skin innervated by only one nerve is called an **autonomous zone** (Fig. 7.5-5). Knowledge of autonomous zones is critical to assessing sensory innervation (Fig. 7.5-6 and Table 7.5-2). Reduced or absent nociception to the skin in an autonomous zone implicates dysfunction of a specific nerve.

For the skin of the manus, the autonomous zone for the radial nerve is the skin over the dorsum of the manus, whereas the autonomous zone for the ulnar nerve is the skin over the lateral aspect of digit V.

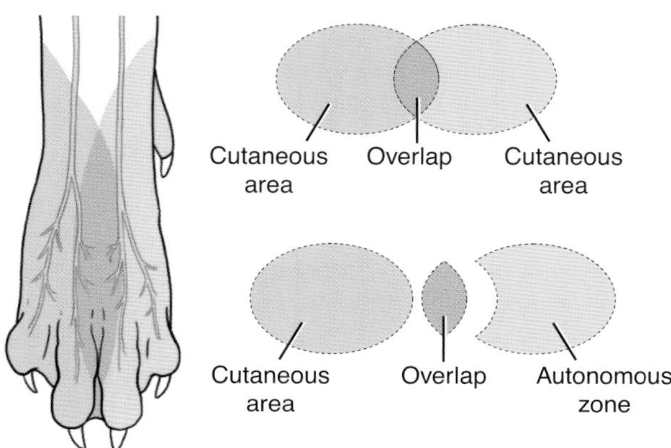

FIGURE 7.5-5 A hypothetical depiction of two cutaneous nerves, a medially and a laterally nerve, innervating the skin of the dorsum of the foot to illustrate the importance of knowing the location of the autonomous zones. Consider a lesion affecting the medially located nerve. If the cutaneous sensation is tested in the overlap zone, the patient will respond given the overlap zone has dual innervation. Only by testing the skin of the autonomous zone of the medial nerve, the examiner can recognize the loss of sensation due to dysfunction of the medially located nerve.

PELVIC LIMB INVOLVEMENT?

An important clinical distinction to make in cases of LMN paresis in a thoracic limb is determining whether the neuroanatomic diagnosis is consistent with a lesion affecting the C6 through T2 spinal cord segments or the axons outside of the CNS—i.e., the spinal roots, spinal nerves, brachial plexus, or the named nerves of the limb). With a disease process affecting the C6 through T2 spinal cord segments, the ascending GP pathways and descending UMN pathways of the ipsilateral pelvic limb are affected. As a result, there are varying degrees of GP ataxia and UMN paresis in the pelvic limb ipsilateral to the lesion affecting the C6 through T2 spinal cord segments, which may be recognized as abnormal postural reactions in the pelvic limb ipsilateral to the thoracic limbs with LMN paresis. If the lesion affects the spinal roots, spinal nerves, brachial plexus, or the named nerves of the limb, the postural reaction of the ipsilateral pelvic limb is normal.

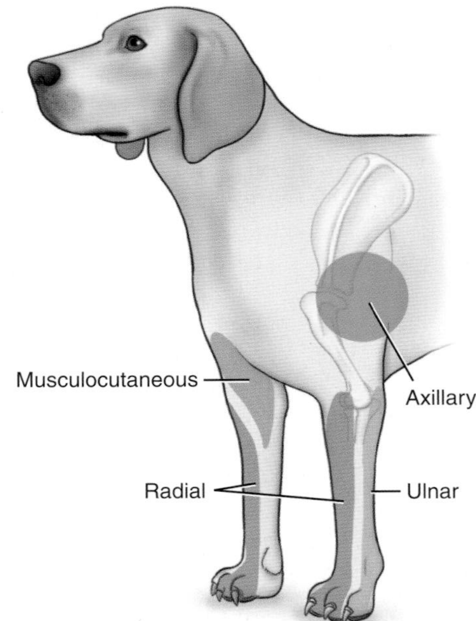

FIGURE 7.5-6 Cutaneous areas (dermatomes) of the radial, ulnar, musculocutaneous, and axillary nerves.

TABLE 7.5-2 Autonomous zones of the thoracic limb.

Nerve	*Area of skin to test autonomous zone*
Axillary	2 cm (0.8 in.) caudal to the shoulder joint
Musculocutaneous	2 cm (0.8 in.) distal to the medial epicondyle of the humerus
Radial	Dorsal surface of digits III, IV
Ulnar	Lateral surface of the digit V

Selected references

[1] Bailey CS, Kitchell RL. Cutaneous sensory testing in the dog. J Vet Intern Med 1987;1:128–35.
[2] Brehm DM, Vite CH, Steinberg HS, et al. A retrospective evaluation of 51 cases of peripheral nerve sheath tumors in the dog. J Am Anim Hosp Assoc 1995;31:349–59.
[3] Evans HE, de Lahunta A. Spinal nerves. In: Miller's anatomy of the dog. Elsevier; 2013. p. 611–57.
[4] Worthman R. Demonstration of specific nerve paralyses in the dog. J Am Vet Med Assoc 1957;131:174.

CHAPTER 8

PELVIC LIMB

Helen M.S. Davies, Chapter editor

Proximal Pelvic Limb (hip, stifle, crus)

8.1 Hip dysplasia—*Christina Murray and Cathy Beck* 456
8.2 Femoral fracture—*Christina Murray and Cathy Beck* 469
8.3 Cranial cruciate ligament tear—*Kimberly A. Agnello* 486
8.4 Tibial fracture—*Lane A. Wallett, Cathy Beck, and Helen M.S. Davies* 494

Distal Pelvic Limb (tarsus and pes)

8.5 Fracture of the tarsal bones—*Chris Boemo, Oday Al-Juhaishi, Zeeshan Akbar, and Helen M.S. Davies* 506

Innervation of the Pelvic Limb

8.6 Degenerative lumbosacral stenosis—*Marc Kent and Eric N. Glass* 520
8.7 Calcaneal tendon injury—*Marc Kent, Eric N. Glass, and Alexander de Lahunta* 531

CHAPTER 8

CASE 8.1

Hip Dysplasia

Christina Murray[a] and Cathy Beck[b]
[a]Veterinary Biosciences, The University of Melbourne, Melbourne, Victoria, AU
[b]Veterinary Hospital, The University of Melbourne, Melbourne, Victoria, AU

Clinical case

History

A 4-year-old female spayed Staffordshire Bull Terrier was presented for bilateral pelvic limb lameness that had a limited response to treatment with a prescribed oral nonsteroidal anti-inflammatory drug (NSAID).

Physical examination findings

The patient was in good health but overweight. She exhibited a mild pain response on palpation of the coxofemoral (hip) joints and resented pelvic limb extension bilaterally. The Ortolani test was positive in both hips.

Even when the Ortolani test is positive, radiography is required for evaluating hip conformation, joint laxity, and the presence and severity of osteoarthritis. Not all puppies with a positive Ortolani test go on to develop osteoarthritis. In adult dogs suspected of having hip dysplasia, a positive Ortolani test is indicative of the disease, but a negative test does not rule it out, because in older dogs the signs may become less distinct with increasingly severe joint changes.

Differential diagnoses

Hip dysplasia, other developmental or acquired musculoskeletal abnormality

Diagnostics

Pelvic radiography confirmed bilateral hip dysplasia, more advanced in the right hip (Fig. 8.1-1A and B; compare to the normal pelvic radiographs shown in Fig. 8.1-2). Both coxofemoral joints were subluxated

ORTOLANI TEST

The Ortolani test is used to identify hip joint laxity, a hallmark of hip dysplasia (Gr. *dys-* abnormal + Gr. *plassein* to form = abnormal development of the hip joint) in dogs. It is best performed with the patient sedated and in lateral or dorsal recumbency. The stifle is flexed to 90°, and gentle pressure is applied along the long axis of the femur—i.e., the femur is pushed in a dorsocaudal direction. In a dog with hip laxity, this maneuver causes dorsolateral subluxation (partial dislocation) of the hip. While maintaining pressure on the femur, the limb is abducted until the femoral head is palpated, which may also be heard as slipping or "clunking" upon returning to the acetabulum. The ability to subluxate the hip, followed by a palpable or audible slip or clunk on reduction of the femoral head, is considered a positive result or "Ortolani sign" of abnormal hip laxity.

The Ortolani maneuver is a sensitive screening test for joint laxity and the potential for progression to osteoarthritis in puppies at 6 months of age. A negative test is a reliable indicator that the dog does not have hip dysplasia. The test is less sensitive in puppies under 4 months of age.

Comparative Veterinary Anatomy: A Clinical Approach. https://doi.org/10.1016/B978-0-323-91015-6.00049-2

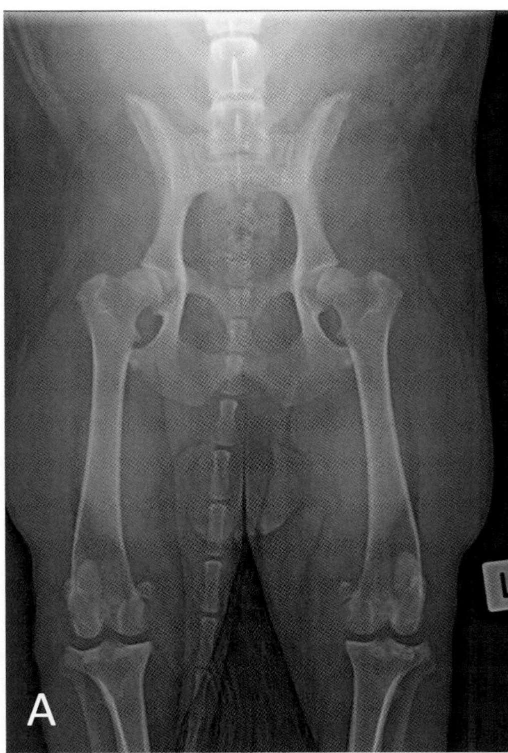

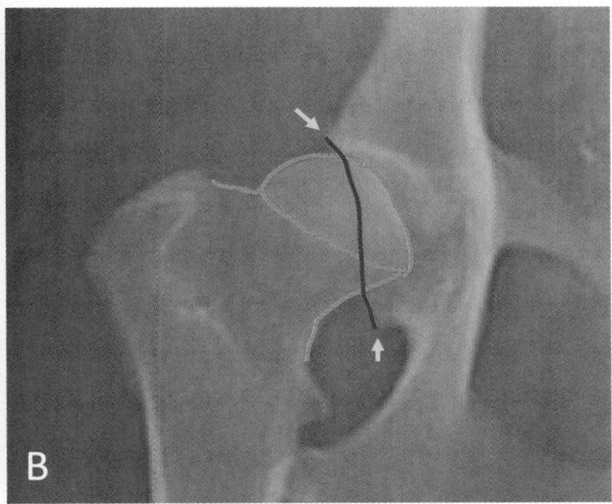

FIGURE 8.1-1 (A) Ventrodorsal (VD) pelvic radiograph of the dog in this case, taken with the pelvic limbs extended and medially rotated (i.e., a "VD extended-hip" view). (B) A coned view of the right hip, radiographically the most severely affected hip. Compare these radiographs with that of a normal dog in Fig. 8.1-2. In this patient, both coxofemoral joints are subluxated; each acetabulum is shallow, covering less than half of the femoral head (1B, *pink line*), and there is osteophyte formation at the acetabular rim (1B, *arrows*). In addition, bony remodeling of the femoral head and neck has created a "mushroom" shape (1B, *blue lines*).

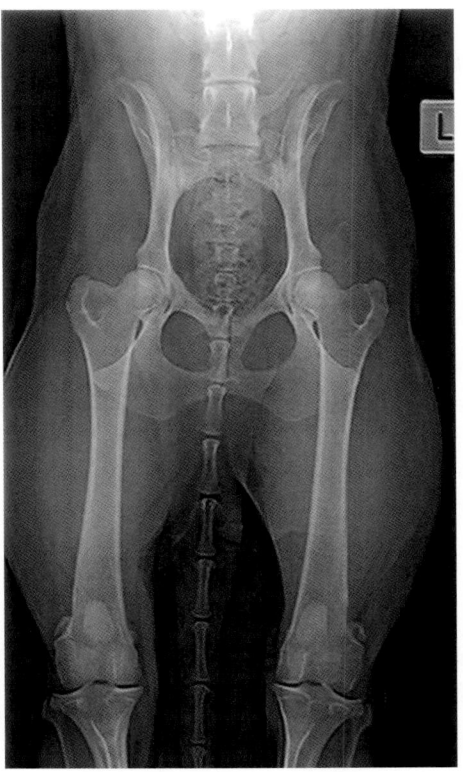

FIGURE 8.1-2 Ventrodorsal pelvic radiograph of a normal dog, a "VD extended-hip" view. Note how the femoral head is seated deep within the acetabulum of each hip and has a rounded shape.

on the ventrodorsal extended-hip view. In each hip, the acetabulum was shallow, covering less than half of the femoral head, and there was osteophyte formation at the acetabular rim. Bony remodeling of the femoral head and neck (right > left hip) created the "mushroom" shape often seen in advanced cases of hip dysplasia.

Diagnosis

Bilateral hip dysplasia

Treatment

The treatment plan included changing the type of oral NSAIDs used for pain relief, the addition of monthly pentosan polysulfate sodium (Cartrophen Vet™) injections, and adjustments to diet and activity. Dietary changes emphasized calorie restriction while providing adequate protein and minerals. Daily low-impact activity was advised for weight management and muscle tone, limited to walks on-leash initially, with gradual increases in both duration and intensity. Had this patient not already been spayed; the owners would also have been recommended not to use her for breeding.

CANINE HIP DYSPLASIA

Hip dysplasia is one of the most common developmental orthopedic diseases in dogs. It is most often seen in large-breed dogs, but it also occurs in small dog breeds and in cats. Abnormal formation of the acetabulum and femoral head results in incongruity and laxity of the coxofemoral joint, which leads to early-onset osteoarthritis. Later, periarticular remodeling and fibrosis may result in joint stiffness.

In most cases, hip dysplasia is bilateral, although it is common for one hip to be more severely affected than the other. Its inheritance in dogs is complex, so it is often described as "polygenic" with environmental contributors such as diet and the type and intensity of activity. Nevertheless, reputable dog breeders screen prospective stud (for breeding) dogs and bitches for hip dysplasia using such programs as the Orthopedic Foundation for Animals (OFA) certification or the Penn Hip Improvement Program (PennHIP). These programs are described later.

Anatomical features in canids and felids

Introduction

The following section describes the coxofemoral (hip) joint. The femur and associated structures of the thigh are described in greater detail in Case 8.2. Because hip dysplasia and osteoarthritis of the hip are common clinical presentations, this section concludes with a detailed discussion of canine hip dysplasia as an anatomical study.

Function

The coxofemoral joint serves as the dynamic junction between the pelvic limb and the axial skeleton, with the pelvis functioning as a rigid frame that is fused to the vertebral column at the sacrum. The coxofemoral joint thus facilitates weight-bearing and locomotion. As described later, this ball-and-socket joint acts somewhat like a "universal" joint, as it allows a broad range of motion in various planes. Its structure and function may make it uniquely vulnerable to developmental defects and degenerative joint disease.

Pelvis

With pelvic osteotomy, one surgical treatment for hip dysplasia, the part of the pelvis containing the acetabulum is incised (osteotomy [Gr. *osteon* bone + Gr. *tomē* a cut]), rotated along its long axis, and re-attached so that the dorsal acetabular rim is positioned more laterally to increase coverage of the dorsal aspect of the femoral head.

The **pelvis** is composed of a pair (left and right) of hip bones, or **os coxae**, that are united ventrally at the **pelvic symphysis** and dorsolaterally via the **sacrum**. In dogs and cats, each hip bone is composed of 4 fused bones: **ilium**, **ischium**, **pubis**, and **acetabular**. All 4 bones contribute to the structure of the **acetabulum**, which is the cup-shaped cavity on the lateral aspect of the os coxae that articulates with the **head of the femur**, forming the **coxofemoral joint**.

Of the bones that form the acetabulum, the ossification centers of the ilium, ischium, and pubis are normally present at birth, while the acetabular bone appears at about 7 weeks of age. The growth plates for these centers close at approximately 4–6 months of age. The other ossification centers of the pelvis are the iliac crest, tuber ischiadicum and the caudal border of the ischium. These centers appear several months after birth and their growth plate closure times start at about 8 months of age, with fusion of the pelvic symphysis starting at about 15 months and reaching completion by 6 years of age. The ossification center of the femoral head appears at about 2 weeks of age, and growth plate closure occurs between 6 and 9 months of age.

Juvenile pubic symphysiodesis (Gr. *symphysis* a growing together + Gr. *desis* binding), another surgical treatment for hip dysplasia, involves surgical closure of the pelvic symphysis earlier than it would naturally close. The goal is to induce lateral acetabular rotation as the puppy grows, thereby increasing femoral head coverage by the dorsal acetabular rim. For best results, this procedure must be done at 3–4 months of age.

Coxofemoral joint (Fig. 8.1-3)

The **acetabular lip** or **labrum** (L. a lip) is a narrow band of fibrocartilage, attached along the rim of the acetabulum, which deepens the acetabulum and thus increases its coverage of the femoral head (discussed later). Within the acetabulum is the **lunate surface** (*L. luna* moon), which is the articular surface located along the inner circumference of the acetabulum. The lunate surface is incomplete ventrally and medially; this gap is called the **acetabular notch**. The **transverse acetabular ligament** extends across the acetabular notch.

The depressed area within the arc of the lunate surface is the **acetabular fossa**. It forms the deepest part of the acetabulum. The **ligament of the head of the femur** is a thick, flattened, intra-articular ligament that attaches to the acetabular fossa and the transverse acetabular ligament, and is

Total hip replacement involves replacing the acetabular cartilage and the femoral head and neck with a prosthetic hip. It is performed only after closure of the growth plates of the acetabulum and femur, which normally occurs by about 12 months of age. The prosthesis comprises an acetabular cup that is fitted into the acetabulum after it has been reamed out (smoothed using a rotary cutting tool), and implanting a femoral head on a stem in the medullary canal of the femoral shaft. Prostheses are available for dogs of all sizes and for cats.

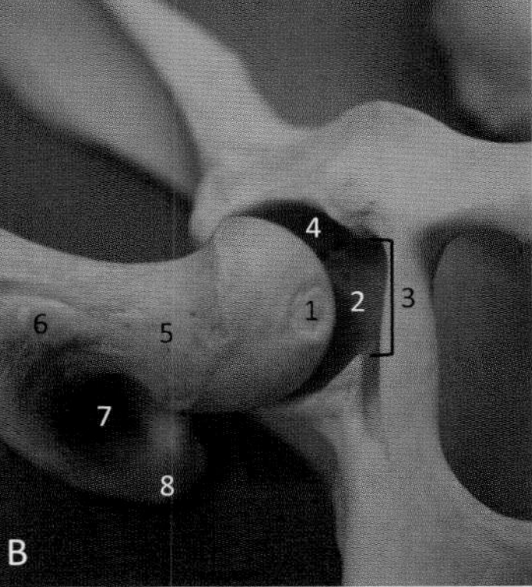

FIGURE 8.1-3 Right coxofemoral joint of the dog, ventral aspect. (A) The hip of a puppy 10–12 weeks of age, showing the proximal and distal attachments of the ligament of the head of the femur. (B) The skeletal elements of an adult dog. Key: 1, fovea; 2, acetabular fossa; 3, acetabular notch; 4, lunate surface; 5, femoral neck; 6, lesser trochanter; 7, trochanteric fossa; 8, greater trochanter. *Arrows:* transverse acetabular ligament.

covered by synovial membrane. The joint capsule (described later) is lined by synovium, which produces the synovial fluid that lubricates the joint surfaces and nourishes the articular cartilage of the acetabulum and femoral head.

Femur (Fig. 8.1-3)

As described and illustrated in Section 8.2, the proximal end of the femur has a head, a neck, and 3 trochanters. The **head** is located on the medial aspect of the proximal femur as a smooth, almost hemispherical expansion on the dorsal and medial aspects of the femoral neck. The **fovea** is a small, circular indentation on the convex articular surface of the head. It forms the site for attachment of the **ligament of the head of the femur**.

The femoral **neck** unites the head with the body of the femur. A ridge of bone extends from the femoral neck along the proximal margin of the femur to meet the greater trochanter. The **trochanteric fossa** is located on the caudal aspect of this ridge of bone and the medial aspect of the greater trochanter. The **greater trochanter** is a large process on the caudolateral aspect of the proximal femur. It has a prominent apex that projects proximally from the femur.

> Femoral neck lengthening is infrequently performed as a surgical option for hip dysplasia, but when used, it is done to position the greater trochanter further from the pelvis and thus increase the lever arm of the abducting and rotating muscles that insert on the greater trochanter. This procedure increases the medially directed forces these muscles can apply to the proximal femur. Intertrochanteric varus osteotomy of the femur is another uncommonly performed procedure in which the femoral neck is repositioned from a relatively upright position to be more perpendicular to the femoral shaft so that the femoral head can be placed more deeply into the acetabulum. This technique is also used to relieve hip pain in dogs with osteoarthritis. Femoral head-and-neck excision is a salvage procedure performed in dogs and cats to relieve pain and improve quality of life when total hip replacement is not an option. The femoral head and neck are removed and a fibrous "false joint" subsequently forms between the proximal femur and the pelvis.

> The apex of the greater trochanter is readily palpable. It is about level with the dorsal surface of the femoral head and can therefore indicate the approximate location of the hip joint. The greater trochanter often becomes more prominent in dogs and cats with chronic hip pain, owing to atrophy of the hip muscles.

The **lesser trochanter** is located on the caudomedial aspect of the femur, near the junction between the neck and body of the femur. A low ridge of bone, the **intertrochanteric crest**, connects the greater and lesser trochanters. The **3rd trochanter** is a small projection on the lateral aspect of the femur, distal to the greater trochanter, at approximately the same level as the lesser trochanter.

Joint capsule, blood supply, and innervation

The coxofemoral **joint capsule** attaches to the os coxae medial to the acetabular rim and to the neck of the femur. There are no defined ligaments in the outer fibrous layer of the capsule, but regions of reinforcement have been identified.

The **blood supply** to the joint capsule and to the femoral neck and head arises from 3 sources: the **lateral circumflex artery** (a branch of the femoral artery), the **medial circumflex artery** (a branch of the deep femoral artery), and the **caudal gluteal artery**. These vessels form a ring around the joint capsule, and branches from this ring ascend the neck to supply the epiphyseal arteries of the femoral head. The joint capsule receives branches from the **obturator nerve** and from a muscular branch that supplies the gemelli, internal obturator, and quadratus femoris muscles, which arises from the **lumbosacral trunk** before it becomes the sciatic nerve.

Associated musculature (Table 8.1-1)

The various directions and range of motion permitted by the hip joint are described next. There are many muscles involved in these actions, and many of these muscles also have a role in stabilizing the hip joint. These muscles may be grouped according to their location, or otherwise by their action on the hip. Given their number and diversity, they are summarized in Table 8.1-1.

TABLE 8.1-1 Muscles of the hip in the dog and cat.

Muscle	*Proximal attachment*	*Distal attachment*	*Action on the hip joint*	*Innervation*
Lumbar hypaxial muscles				
Iliopsoas • Psoas major • Iliacus	• Lumbar vertebrae 2–7 • Ventral surface of ilium	Psoas major and iliacus join to form the iliopsoas muscle which attaches to the lesser trochanter of the femur	Flexion of hip to advance limb	Ventral branches of lumbar nn.
Lateral pelvic muscles				
Tensor fascia latae	Ventral ilium from tuber coxae to alar spine, middle gluteal fascia	Lateral femoral fascia that blends with fascial insertion of biceps and vastus lateralis over the stifle	Flexion of hip, abduction of limb	Cranial gluteal n.
Superficial gluteal	Gluteal and caudal fascia, sacrum, 1st caudal vertebra, proximal sacrotuberous ligament (*dogs only*)	3rd trochanter of femur, tendon also fuses with aponeurosis of tensor fascia latae	Extension of hip joint, abduction of hip	Caudal gluteal n.
Middle gluteal	Gluteal surface of ilium, tuber sacrale and iliac crest, dorsal sacroiliac ligament, deep gluteal fascia	Free end of greater trochanter of femur	Extension of hip, medial rotation of hip, prevention of lateral rotation when weight-bearing	Cranial gluteal n.
Piriformis	Lateral surface of 3rd sacral and 1st caudal vertebrae	Joins tendon of middle gluteal on greater trochanter	Extension of hip, retraction and abduction of limb	Caudal gluteal n.
Deep gluteal	Lateral surface of body of ilium	On greater trochanter distal to the insertion of middle gluteal	Extension of hip, abduction, medial rotation, prevention of lateral rotation when weight-bearing	Cranial gluteal n.
Gluteofemoral—*cats only*	2nd to 4th caudal vertebrae, caudal to superficial gluteal	Extends deep to biceps to attach to fascia lata and lateral aspect of patella	Extension of hip, retraction and abduction of limb	Caudal gluteal n.
Medial pelvic muscles				
Internal obturator	Internal (dorsal) surface of rami of pubis and ischium, ischiatic table, and ischiatic arch	Tendon joins tendons of gemelli to attach to the trochanteric fossa as single tendon	Lateral rotation of hip, prevention of medial rotation when weight-bearing	Sciatic n. (lumbosacral trunk)
Gemelli	Lateral surface of body of ischium	Trochanteric fossa	Lateral rotation of hip, prevention of medial rotation when weight-bearing	Sciatic n. (lumbosacral trunk)
Quadratus femoris	Ventral ischium medial to lateral angle of tuber ischiadicum	Intertrochanteric crest	Extension of hip, lateral rotation of hip, prevention of medial rotation when weight-bearing	Sciatic n. (lumbosacral trunk)
External obturator	External (ventral) surface of pubis and ischium adjacent to symphysis	Joins tendon of internal obturator and gemelli to insert on trochanteric fossa	Lateral rotation of hip, prevention of medial rotation when weight-bearing	Obturator n.

Continued

TABLE 8.1-1 Muscles of the hip in the dog and cat—cont'd

Muscle	*Proximal attachment*	*Distal attachment*	*Action on the hip joint*	*Innervation*
Caudal thigh muscles				
Biceps femoris • Cranial head • Caudal head	• Ventrocaudal end of sacrotuberous ligament (*dogs only*) and the lateral angle of ischiatic tuberosity • Ventral side of lateral angle of ischiatic tuberosity	To patella and patellar ligament and then to tibial tuberosity Via fascia lata and crural fascia to cranial border of tibia and via common calcaneal tendon to tuber calcanei	Hip extensor during stance—antigravity role Abduction of pelvic limb	Sciatic n.
Caudal crural abductor	*Dogs:* sacrotuberous ligament *Cats:* caudal vertebra 2	Crural fascia	Abduction of pelvic limb (with caudal head of biceps)	Sciatic n.
Semitendinosus	Caudal and ventrolateral parts of lateral angle of ischiatic tuberosity	Via crural fascia to medial tibial body and via common calcaneal tendon to tuber calcanei	Extension of hip	Sciatic n.
Semimembranosus	Ventral surface of the rough portion of the ischiatic tuberosity then divides into 2 heads—cranial (femoral) and caudal (tibial)	Cranial head to medial lip of femur and tendon of origin of gastrocnemius, caudal head to medial collateral ligament of stifle and medial tibial condyle	Cranial head extends hip during stance phase, caudal head contributes to hip extension	Sciatic n.
Cranial thigh muscles				
Quadriceps femoris • Rectus femoris	The only head of quadriceps arising from the pelvis, rectus femoris is attached to the body of the ilium, cranial to the acetabulum	Via the patellar ligament to the tibial tuberosity	Flexion of the hip	Femoral n.
Articularis coxae	On ilium adjacent to rectus femoris	Femoral neck	A very small muscle with minimal contribution to hip flexion, may adjust hip joint capsule position	Femoral n.
Medial thigh muscles				
Sartorius • Cranial part • Caudal part	• Iliac crest, cranial ventral iliac spine thoracolumbar fascia • Tuber coxae of ilium	Medial femoral fascia and tendon of rectus femoris and vastus medialis Via crural fascia to cranial border of tibia	Flexion of hip during pelvic limb advancement, adduction	Saphenous n.
Gracilis	Symphyseal tendon of pelvic symphysis from pecten of pubis to ischiatic arch	Cranial border of tibia, crural fascia, and via the common calcaneal tendon to the tuber calcanei	Extension of hip, adduction of limb	Obturator n.
Pectineus	Iliopubic eminence and cartilage, prepubic tendon	Medial lip of caudal rough surface of femur	Adduction of the limb	Obturator n.
Adductors • Longus • Magnus et brevis	• Pubic tubercle • Symphyseal tendon and adjacent parts of ischiatic arch	Proximal part of lateral lip of caudal rough surface of femur near 3rd trochanter Lateral lip of the rough surface of the femur distal to 3rd trochanter, caudal surface, and popliteal surface of femur	Adduction and extension of hip joint	Obturator n.

Normal range of motion (Figs. 8.1-4–8.1-7)

In the normal standing dog and cat, the left and right femurs are in parallel sagittal planes. The **flexor angle** is the angle formed between the long axis of both the pelvis and femur at the cranial aspect of the hip joint. The mean standing flexor angle in the dog is 95°–115°, depending largely on breed, but with considerable individual variation within breeds.

The ball-and-socket construction of the hip joint permits a wide range of movement in the normal individual, including:

- Flexion—decrease in the flexor angle
- Extension—increase in the flexor angle
- Abduction—movement of the pelvic limb away from its perpendicular position in relation to the body
- Adduction—movement of the pelvic limb from its perpendicular position toward the median plane of the body
- Circumduction—movement of the distal limb through a circle such that the limb outlines the surface of a cone
- Internal (medial) rotation of the pelvic limb
- External (lateral) rotation of the pelvic limb

Range of motion is affected by species, breed, body condition, muscle mass, type and level of activity, and joint health. When interpreting the results of different studies of range of motion of the hip, variations in measurement

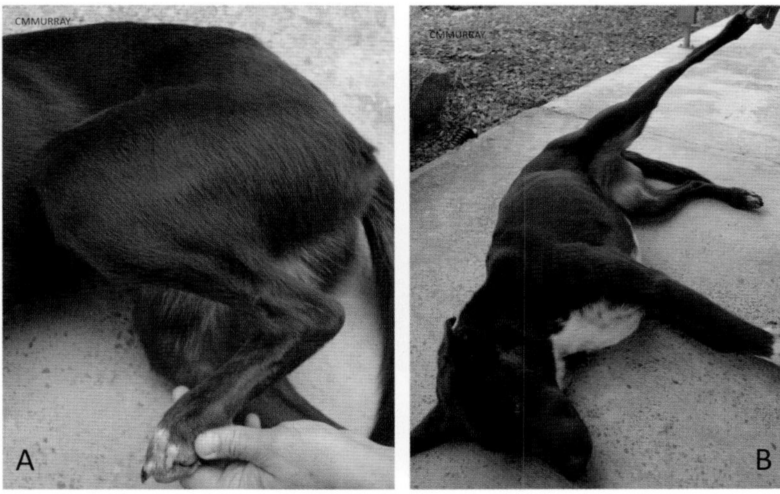

FIGURE 8.1-4 (A) Normal hip flexion in a dog, in which the flexor angle (between the pelvis and femur at the cranial aspect of the hip joint) is decreased. (B) Abduction of the pelvic limb with the hip as the pivot/point of rotation.

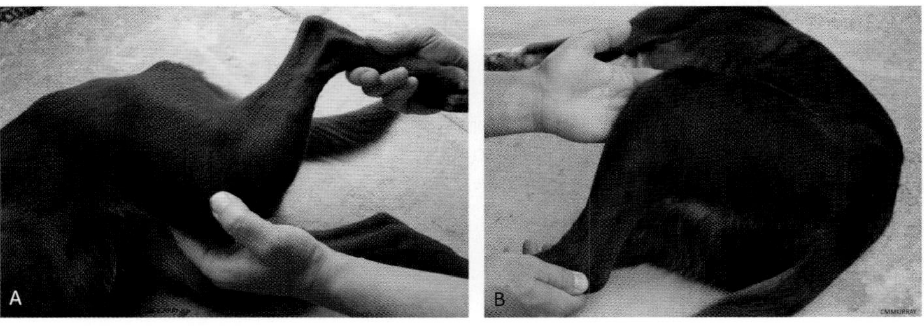

FIGURE 8.1-5 (A) Internal (medial) rotation of the hip in a dog. With the stifle supported in flexion, the distal limb is rotated away from the body. (B) External (lateral) rotation of the hip. With the stifle supported in flexion, the distal limb is rotated toward the body.

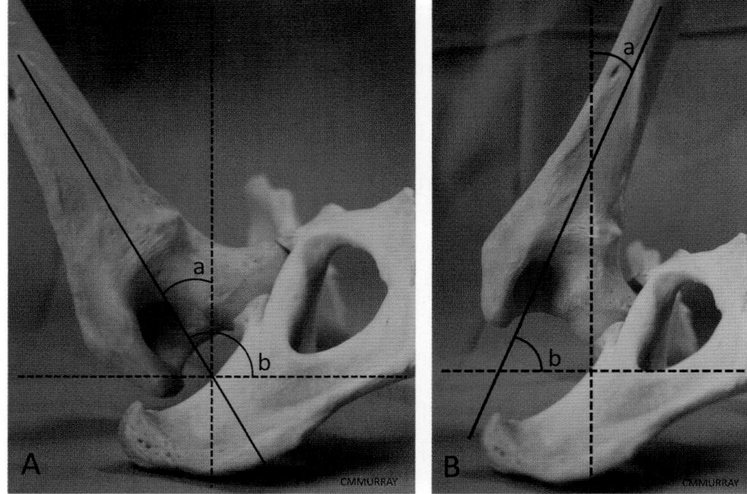

FIGURE 8.1-6 Angles of abduction (A) and adduction (B) of the right coxofemoral joint in the dog, caudal aspect. The pelvis and femur are positioned as for a dog in dorsal recumbency. These angles have been measured and reported using 2 methods. The first measures the angle between the limb and a line perpendicular to the dorsal plane of the body, parallel to the normal standing position of the femur (shown as angle a). The second measures the angle between the limb and a line parallel with the dorsal plane of the body (shown as angle b).

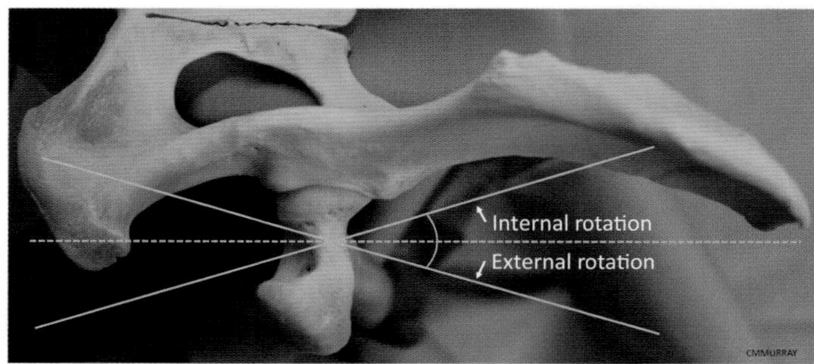

FIGURE 8.1-7 Right hip of a dog, dorsal aspect. The angles of internal and external rotation of the hip are measured in relation to a line parallel to the long axis (median plane) of the dog.

techniques and definitions of angles must be considered. There is reasonable agreement on mean **flexion and extension angles** in the dog—with mean hip flexion in dogs of approximately 50° and mean hip extension of approximately 160°. In cats, the mean hip flexion is approximately 35° and the mean hip extension is approximately 165°.

Abduction, adduction, and rotation angles are less clearly established because these are difficult angles to measure, and results vary depending on the methodology. Fig. 8.1-6 shows the **abduction and adduction angles**, and Fig. 8.1-7 shows the **rotation angles** as measured in different studies. A mean abduction angle of 75° and adduction angle of 35°, shown as angle "a" in Fig. 8.1-6A and B, respectively, have been reported in dogs.

Biomechanics of hip dysplasia

The stabilizers of the hip include the **ligament of the head of the femur**, the **dorsal acetabular rim**, the **joint capsule**, and the **pelvic muscles** acting to retain the femoral head in the acetabulum (Figs. 8.1-8 and 8.1-9). The thin film of synovial fluid within the normal joint contributes to joint stability by acting in concert with the joint capsule to create a vacuum effect within the joint.

Several factors have been suggested as contributing to joint laxity, including a weak joint capsule and lack of muscle development. Increased joint fluid volume has also been suggested, either as an initiating cause of the laxity, or a subsequent contributing factor. Delayed ossification of the bones of the pelvis and femoral head and abnormalities in the development of the acetabulum have also been implicated. A steeper dorsal acetabular slope has been

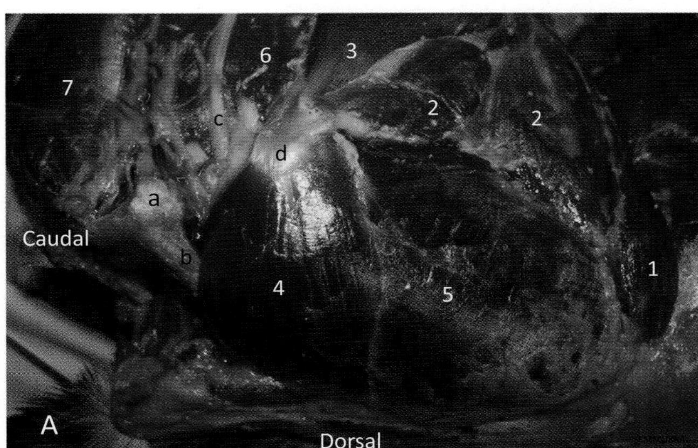

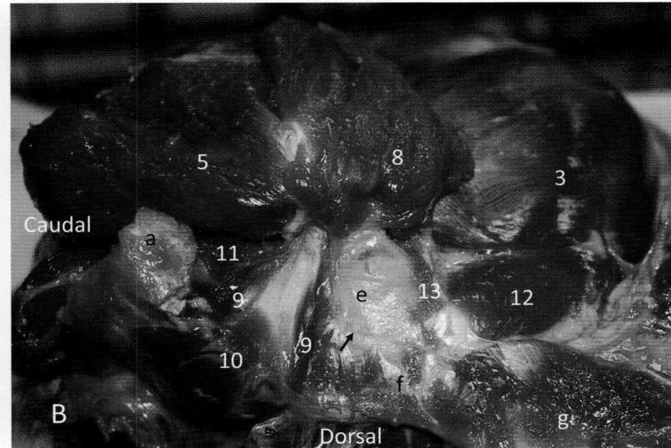

FIGURE 8.1-8 Left hip region of a dog, dorsolateral aspect. The cranial margin of the pelvis and proximal thigh are to the right, and the dorsum of the dog lies along the lower boundary of these images. (A) Superficial structures visible after the removal of the biceps femoris muscle. (B) Deeper prosection; the gluteal muscles are reflected at their insertions onto the femur, and the sacrotuberous ligament, sciatic nerve, and caudal gluteal vessels are removed to expose the dorsal aspect of the hip joint. The sartorius and tensor fascia latae muscles cranial to the joint are also removed. Key for muscles: 1, sartorius (cranial belly); 2, tensor fascia latae; 3, vastus lateralis; 4, superficial gluteal; 5, middle gluteal; 6, adductor; 7, semitendinosus; 8, deep gluteal; 9, gemelli; 10, internal obturator; 11, quadratus femoris; 12, rectus femoris; 13, articularis coxae. Key for additional structures: a, tuber ischiadicum; b, sacrotuberous ligament; c, sciatic nerve; d, greater trochanter of the femur; e, coxofemoral joint capsule; f, body of the ilium; g, gluteal surface of the ilium. The *black arrow* in (B) indicates the dorsal acetabular rim.

CANINE HIP DYSPLASIA—ANATOMICAL CONSIDERATIONS

Canine hip dysplasia is an inherited disease in which there is a predisposition for hip joint laxity that leads to joint subluxation and eventually osteoarthritis of the joint. Multiple genes are involved in its inheritance, and a range of environmental factors influence whether a predisposed individual develops the condition as well as affecting disease severity. Canine hip dysplasia arises when skeletal growth in a genetically susceptible young dog outstrips the pace of muscle growth. In dogs, the period of maximal growth is 3–8 months of age. Hip joint laxity may be detected as early as 2 months of age, but not all affected puppies go on to develop osteoarthritis.

Clinical signs of osteoarthritis, pain, or lameness, are seen in affected dogs beginning at about 4–6 months of age. Risk factors for disease progression include increased food consumption and rapid weight gain during early life. Limiting food intake from 8 weeks of age and continuing into adult life have been shown to reduce the prevalence and severity of hip osteoarthritis in older dogs. Other risk factors include early neutering and excess dietary calcium.

A correlation between pelvic muscle size and hip dysplasia has been reported, with dysplastic dogs having less muscle mass compared with their normal counterparts. Dogs allowed appropriate exercise in early life have been shown to have a decreased risk of developing hip dysplasia, suggesting a protective effect of increased muscle development in the hip area.

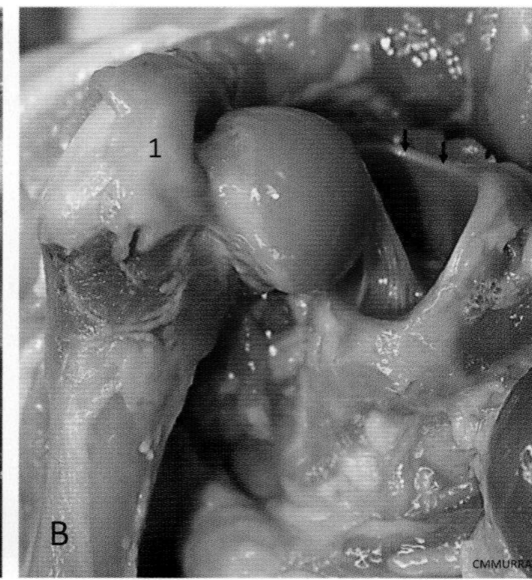

FIGURE 8.1-9 Left coxofemoral joint of the dog. The joint capsule has been opened to demonstrate the stabilizing action of the ligament of the head of the femur in the normal hip. (A) Adult dog, dorsolateral aspect. Note that the ligament does not prevent lateral luxation of the femoral head, indicating the important role of the other hip stabilizers. (B) Puppy, 10–12 weeks old, caudal aspect. Dorsally directed force applied to the femur demonstrates how the ligament limits dorsal subluxation. Key: 1, joint capsule; 2, vastus lateralis muscle (m.); 3, rectus femoris m.; 4, internal obturator m. The *black arrows* in both images indicate the labrum of the dorsal acetabular rim.

associated with a positive Ortolani test and progressive osteoarthritis in young dogs. Increased steepness of this slope results in reduced coverage of the femoral head by the dorsal acetabular rim and permits a stronger dorsolateral thrust of the femoral head against the acetabulum.

Subluxation of the femoral head due to joint laxity, in which the femoral head moves in a lateral direction from the acetabulum, was believed to occur during weight-bearing in dysplastic dogs. However, it is now thought that subluxation is initiated during the swing phase of the gait when the forces exerted by the muscles that advance the limb—the iliopsoas, rectus femoris, and sartorius muscles—predispose the femoral head to lateral subluxation. Upon weight bearing the femoral head is further subluxated, but in a dorsal direction across the dorsal acetabular rim. Stretching of the hip joint capsule triggers a response in which the muscles oriented perpendicular to the joint—the abductors and adductors—are engaged to rapidly reduce the femoral head back into the acetabulum.

The repeated subluxation and reduction of the femoral head injures the cartilage and subchondral bone of the dorsal acetabular rim and femoral head. The ongoing inflammation (osteoarthritis) and abnormal weight-bearing forces cause progressive loss of normal bone conformation, with the acetabulum becoming flatter and wider and the femoral head also flattening. Joint inflammation and abnormal loading also lead to tearing of the joint capsule at its insertions, resulting in formation of periarticular osteophytes ("bone spurs") at joint margins, seen on radiographs as thickening of the femoral neck and bone proliferation at the acetabular rim (see Fig. 8.1-1B).

TREATMENT OPTIONS FOR HIP DYSPLASIA

Many dogs and cats with hip dysplasia can be treated conservatively. Therapy may include exercise modification, physical therapy, weight reduction in overweight animals, NSAIDs for chronic pain management, nutraceuticals such as glucosamine and chondroitin, and stem cell therapy.

Surgical treatment options include:

- Techniques for prevention of disease progression—realign either the acetabulum or the femoral head so that the femoral head is more completely enclosed within the acetabulum. Techniques include pelvic osteotomy or juvenile pubic symphysiodesis; and, less commonly, femoral neck lengthening and intertrochanteric varus osteotomy.
- Total hip replacement—replace the acetabular cartilage and the femoral head and neck with a prosthetic hip.
- Femoral head and neck excision—a salvage procedure in dogs and cats for improved quality of life when total hip replacement is not an option.

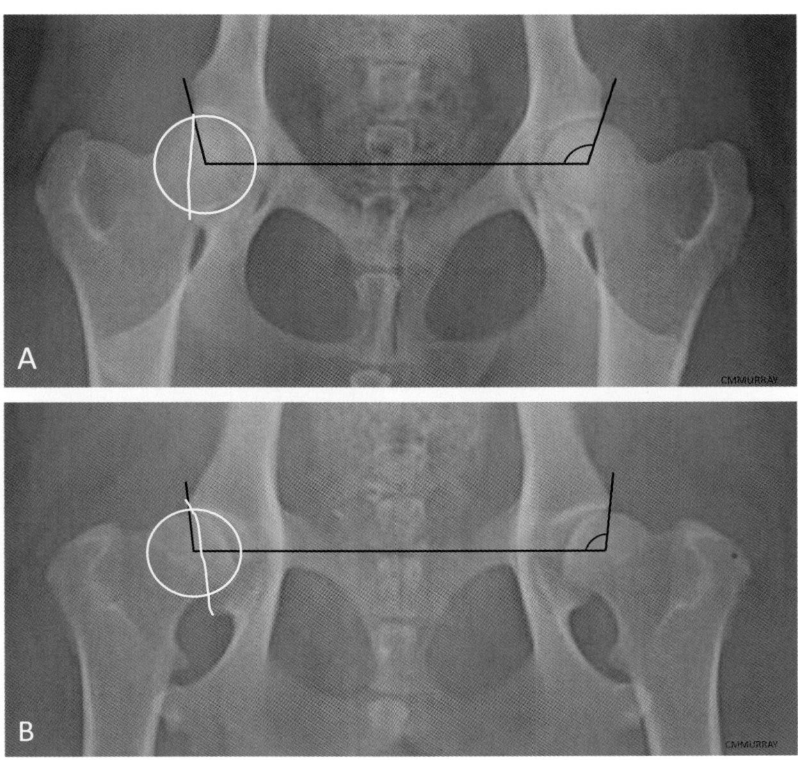

FIGURE 8.1-10 Annotated ventrodorsal radiographs of canine hip joints illustrating the Norberg angle and % femoral head coverage. (A) Normal dog. (B) The patient in this case. The *black lines* illustrate how the Norberg angle (NA) is established and measured. The line transecting the circle on the right hip is the location of the dorsal acetabular rim on the radiograph; the portion of the circle medial to this line represents the percent of femoral head coverage. With hip dysplasia, the NA and femoral head coverage are reduced. An NA of > 105° and femoral head coverage of at least 50% are considered indicative of a normal hip joint. Consistent positioning of the patient is critical for accurate and repeatable measurements.

RADIOGRAPHIC EVALUATION OF THE HIP JOINT

Radiography is the main diagnostic tool used for identification and assessment of hip dysplasia. It is used for assessment of hip conformation, joint laxity, and presence and severity of osteoarthritis. Both CT and MRI are emerging imaging modalities, but not yet routinely used for these cases. Radiographic assessment of hip conformation is a major component of hip dysplasia screening programs used to reduce the incidence of the disease in genetically susceptible populations. Scoring systems used by these programs generally involve subjective assessment of conformation and osteoarthritis as well as objectively measurable parameters. Current radiographic techniques include:

- Hip-extended radiographs—used for subjective assessment of hip conformation and osteoarthritis (Figs. 8.1-1 and 8.1-2). Parameters measured on this projection for further objective assessment include the Norberg Angle (NA) and the Orthopedic Foundation for Animals (OFA) Hip Score—both are methods of measuring the coverage of the femoral head by the acetabulum and thus reflect the relative congruence of the hip joint (Fig. 8.1-10).
- Pennsylvania Hip Improvement Program (PennHIP)—distraction-stress radiographs used to assess the degree of joint laxity, termed the distraction index (DI); the dog or cat is radiographed in dorsal recumbency with its femurs upright in stance position, before and during distraction (using mediolaterally directed pressure) of the proximal femur, in order to objectively measure how far the femoral head may be luxated laterally.
- Dorsolateral subluxation (DLS) score—an alternative means of measurement with the patient in ventral recumbency and the radiograph taken while dorsally directed pressure is applied to the distal femur to induce subluxation.
- Dorsal acetabular rim (DAR) view—a craniocaudal projection of the pelvis to assess the degree of injury to the dorsal acetabular rim and to measure the steepness of the dorsal acetabular slope for surgical planning.

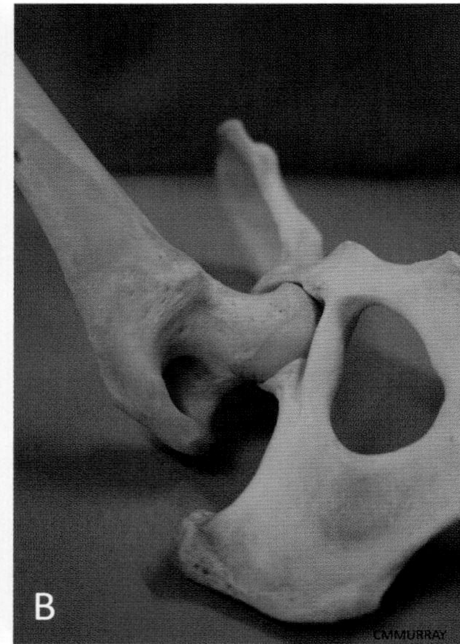

FIGURE 8.1-11 Ortolani maneuver on the right hip of a dog, caudal aspect. The pelvis and femur are positioned as for a dog in dorsal recumbency. (A) The femoral head is luxated dorsolaterally. (B) Abduction of the femur causes the femoral head to return to the acetabulum. Key: 1, greater trochanter of the femur; 2, lesser trochanter; 3, 3rd trochanter; 4, trochanteric fossa; 5, femoral head; 6, femoral neck; 7, lunate surface of acetabulum. The *white arrow* in (A) indicates the dorsal acetabular rim.

CLINICAL EXAMINATION OF THE HIP JOINT

Dogs are presented for the assessment of hip laxity either for screening, as part of a breeding program, or because of clinical signs—e.g., pelvic limb lameness, other gait abnormalities, exercise intolerance, hip pain, or an audible “click” when walking.

On examination, the hips are assessed for pain or crepitus (a crunching or grinding sound or sensation) on manipulation and for a range of motion, which may be reduced in dogs with hip dysplasia. Hip laxity is assessed using the following:

- Ortolani maneuver (Fig. 8.1-11)—described previously.
- Barlow test—the hip is adducted (pelvic limb moved toward the median plane of the body) while the femur is moved in a dorsal direction; when positive, a sudden, dorsally directed subluxation is palpated.
- Bardens test—involves palpation of the greater trochanter for the presence and amount of lateral movement (distraction) of the femoral head when mediolateral pressure is applied to the proximal femur (mainly a preliminary screening test for hip laxity in young puppies 6–8 weeks of age).

Selected references

[1] Butler JR, Gambino J. Canine hip dysplasia: diagnostic imaging. Vet Clin N Am Small Anim Pract 2017;47:777–93.

[2] DeCamp CE, Johnston SA, Déjardin LM, et al. The hip joint. In: Brinker, Piermattei, and Flo’s handbook of small animal orthopaedics and fracture repair. 5th ed. St Louis: Saunders Inc.; 2016. p. 468–517.

[3] Ginja MM, Silvestre AM, Gonzalo-Orden JM, et al. Diagnosis, genetic control and preventative management of canine hip dysplasia: a review. Vet J 2010;184:269–76.

[4] King MD. Etiopathogenesis of canine hip dysplasia, prevalence and genetics. Vet Clin N Am Small Anim Pract 2017;47:753–67.

[5] Perry K. Feline hip dysplasia: a challenge to recognise and treat. J Feline Med Surg 2016;18:203–18.

CHAPTER 8

CASE 8.2

Femoral Fracture

Christina Murray[a] **and Cathy Beck**[b]
[a]Veterinary Biosciences, The University of Melbourne, Melbourne, Victoria, AU
[b]Veterinary Hospital, The University of Melbourne, Melbourne, Victoria, AU

Clinical case

History

A 7-month-old male Kelpie was presented for an acute nonweight-bearing lameness of the left pelvic limb that he developed after jumping out of a car window.

Physical examination findings

The dog was in good general health and condition, other than the nonweight-bearing lameness in the left pelvic limb with noticeable swelling of the thigh. Owing to the severity of the lameness and the peracute nature of the injury, plans were made for radiographs of the left pelvic limb.

Differential diagnoses

Fracture of the femur; severe muscle trauma; hematoma

Diagnostics

Radiography revealed a comminuted fracture of the distal diaphysis of the left femur (Fig. 8.2-1A and B).

A comminuted fracture is defined as one that has more than 2 pieces of bone, indicating greater biomechanical forces sustained by the bone, and is generally caused by a vehicular accident, gunshot, or bite wound. As can be seen in this diaphyseal fracture, there are the 2 bigger pieces of the affected bone—proximal and distal to the fracture—plus 1 or more smaller splintered or crushed fragments at the fracture site.

There was a small skin defect that was not noticed on the initial physical examination; however, the craniocaudal radiograph (Fig. 8.2-1B) revealed a gas opacity surrounding the distal end of the proximal fracture segment, supporting a diagnosis of an open fracture.

Open fractures, in which the fracture communicates with an external wound (previously termed a "compound" fracture), are at a greater risk of post-operative infection than are closed fractures (i.e., skin intact). Fractures involving considerable soft tissue swelling also carry a greater risk of infection, whether they are "open" or "closed," because of the trauma to the soft tissue and the likelihood of compromised blood supply to the traumatized bone.

FEMORAL FRACTURES

In dogs and cats, fractures are more common in the appendicular than in the axial skeleton, and are more likely in the pelvic limb than in the thoracic limb. Femoral fractures account for 15–40% of all fractures in dogs and 25–35% of all fractures in cats. Vehicular trauma and falls are the most common etiologies.

Comparative Veterinary Anatomy: A Clinical Approach. https://doi.org/10.1016/B978-0-323-91015-6.00050-9

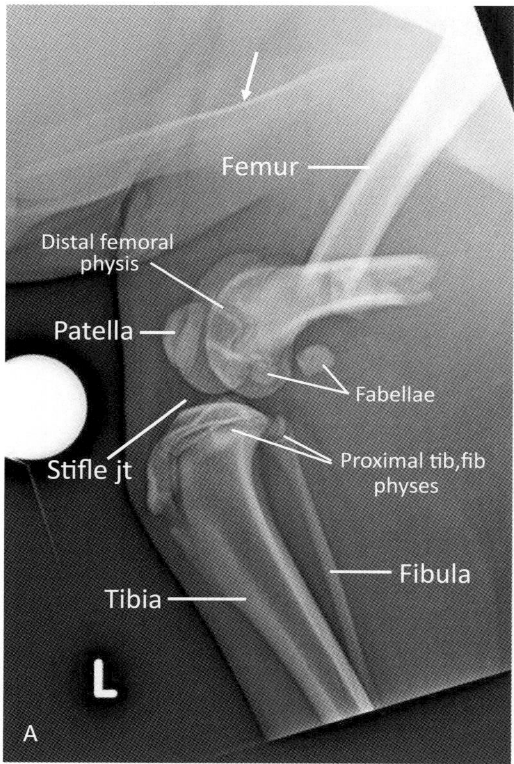

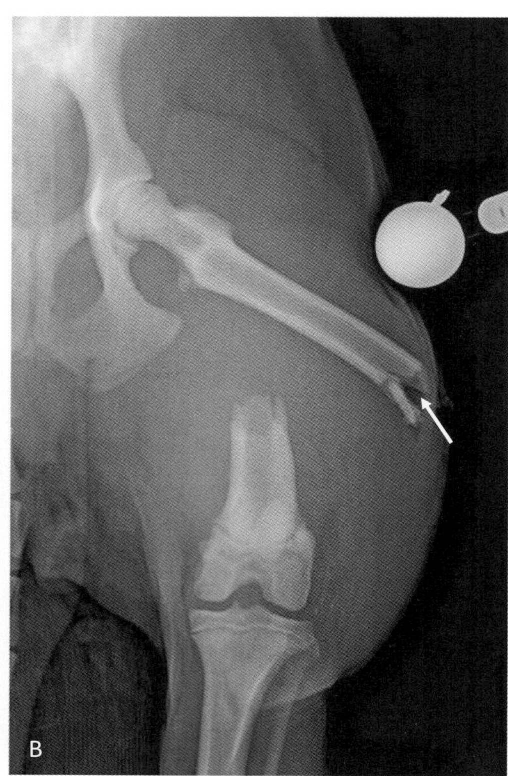

FIGURE 8.2-1 (A) Lateral radiograph of the dog's left pelvic limb, showing a comminuted femoral fracture of the distal diaphysis. Note the normal appearance of the physes of the distal femur and the proximal tibia and fibula in this 7-month-old dog. (The os penis can also be seen on this radiograph [*white arrow*].) Key: *jt*, joint. (B) Craniocaudal radiograph of the dog in this case. There is a marked soft tissue swelling over the lateral aspect of the femur. A gas opacity is seen surrounding the distal end of the proximal fracture fragment (*white arrow*), consistent with an open fracture. The radiopaque circle is a calibration marker used for surgical planning, such as determining appropriate implant sizes for fracture repair.

Diagnosis

Comminuted femoral fracture

Treatment

After the dog was stabilized, he was placed under general anesthesia for fracture fixation. Using a lateral approach, the fracture was reduced, anatomically realigned, and stabilized using a combination of an intramedullary pin and

TREATMENT OF FEMORAL FRACTURES

Femoral fractures carry a relatively higher risk of delayed union or nonunion (failure to heal) and osteomyelitis (which further delays healing) if not treated appropriately for the fracture type and configuration. Conservative management, using splints or bandages, is not recommended because it does not provide enough immobilization of the bone fragments and can even result in further fracture displacement, tissue trauma, and pain.

Internal fixation techniques, using intramedullary pins and bone plates and screws, are generally required for fracture stabilization, normal anatomical alignment, and healing to ensure return to normal function as soon as possible. Presurgery planning requires a review of the relevant anatomy of the region and an understanding of the surgical options for repair of the fracture, to minimize further tissue injury.

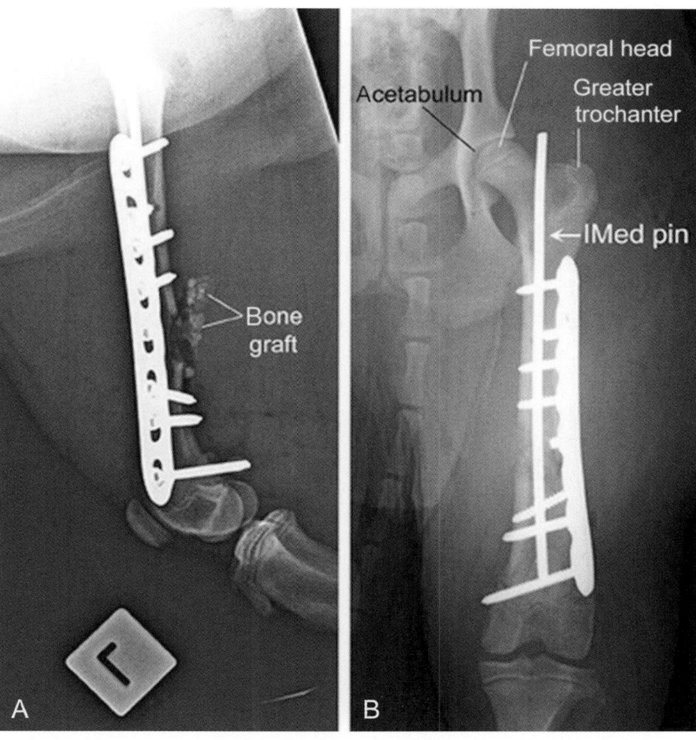

FIGURE 8.2-2 Post-operative radiographs, showing good fracture reduction (realignment) and stabilization using an intramedullary (IMed) pin and a bone plate applied to the lateral surface of the left femoral diaphysis. (A) Lateral projection. Some of the fragments of the cancellous bone graft are seen as radiopacities caudal to the fracture site. (B) Craniocaudal projection. The position of the intramedullary pin is best seen on this view. Also notice that ossification of the head and greater trochanter of the femur is incomplete in this young dog; a *thin line* of physeal cartilage is still present at each site.

bone plate and screws with a cancellous bone graft (Fig. 8.2-2). Related surgical considerations and the approach used in this case follow at the end of the anatomy section. The dog made an uneventful recovery. At suture removal 2 weeks after surgery, only mild left pelvic limb lameness was apparent.

Fracture repair requires the restoration of normal bone alignment and length; however, the principle of biological fixation (osteosynthesis) must also be considered. This principle requires minimal disturbance of smaller bone fragments to preserve the surrounding soft tissues and therefore the extraosseous blood supply to the healing bone.

Anatomical features in canids and felids

Introduction

This section describes the femur and other anatomical structures of the thigh. As discussed below, the femur articulates proximally with the os coxae to form the coxofemoral joint (hip) and articulates distally with the tibia to form the femorotibial joint (stifle or "true" knee). The hip is described in Case 8.1 and the stifle in Case 8.3. The following is a discussion of the femur and the femoral components of these joints.

Function

The femur is one of the largest and longest bones in the body. Its size and shape reflect its role in providing a strong attachment point for major locomotory muscles in the body.

Femur (Figs. 8.2-3–8.2-5)

When viewed from the lateral aspect in the standing position, the **femur** inclines cranially from proximal to distal. Because the stifle joint is situated craniodistal to the hip joint, the flexor angle of the hip is at the cranial aspect of the proximal femur, and the flexor angle of the stifle is at the caudal aspect of the distal femur.

The almost hemispherical **head of the femur** sits in the **acetabulum** of the pelvis, forming the coxofemoral joint (Fig. 8.2-3). A small, circular indentation on the medial surface of the femoral head, termed the **fovea** (L. pit or depression), is the site of attachment for the **ligament of the head of the femur** (formerly, the "round ligament of the femur"). This short, strong ligament attaches the femoral head to the acetabular fossa and helps limit distraction of the femoral head from the acetabulum (i.e., luxation or dislocation of the hip).

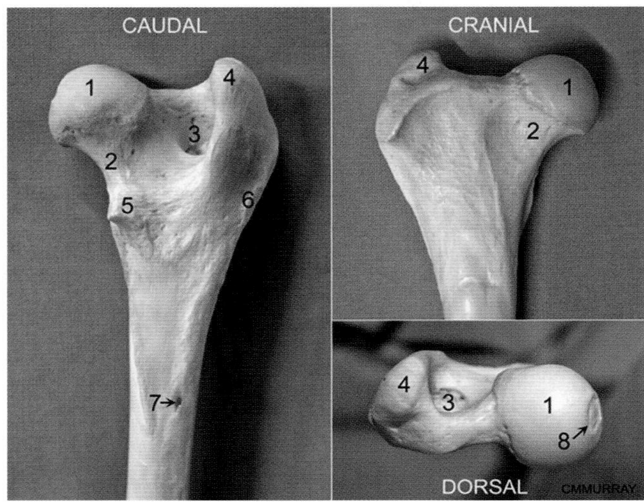

FIGURE 8.2-3 Right femur of a dog, proximal portion, showing the caudal, cranial, and dorsal aspects. Key: 1, femoral head; 2, femoral neck; 3, trochanteric fossa; 4, greater trochanter; 5, lesser trochanter; 6, 3rd trochanter; 7, nutrient foramen; 8, fovea.

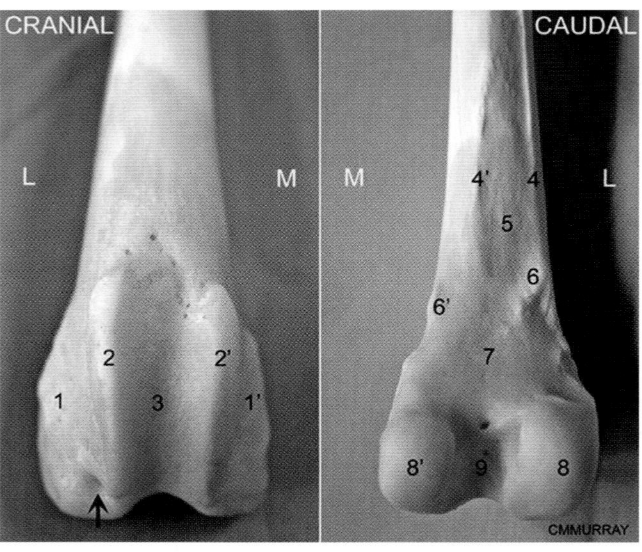

FIGURE 8.2-4 Right femur of a dog, distal portion, showing the cranial and caudal aspects. Key: *L*, lateral; *M*, medial. Key: 1, 1′, lateral and medial epicondyle; 2, 2′, lateral and medial trochlear ridge; 3, trochlea; 4, 4′, lateral and medial lip; 5, rough surface; 6, 6′, lateral and medial supracondylar tuberosity; 7, popliteal surface; 8, 8′, lateral and medial condyle; 9, intercondylar fossa. *Black arrow:* extensor fossa.

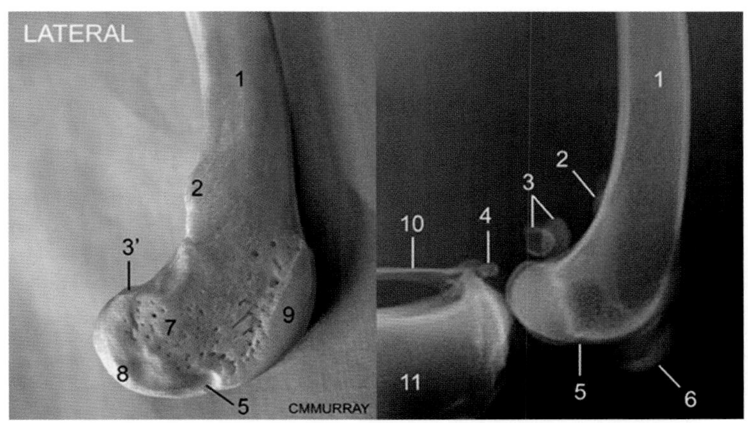

FIGURE 8.2-5 Right femur of a dog, lateral aspect of the distal portion, alongside a lateral radiograph of the stifle. Key: 1, femoral diaphysis; 2, supracondylar tuberosity; 3, medial and lateral fabellae; 3′, articular surface for the lateral fabella; 4, popliteal sesamoid bone; 5, extensor fossa; 6, patella; 7, lateral epicondyle; 8, lateral condyle; 9, lateral lip of the trochlea; 10, fibula; 11, tibia.

Lateral to the femoral head is a short **neck** that connects the head to the body of the femur. The femoral neck is flattened craniodorsally. Lateral to the neck is the **greater trochanter**, a prominent tuberosity with a free apex that projects caudally and dorsally from the proximal femur. The middle and deep gluteal muscles, which are primarily hip extensors, attach to the greater trochanter.

> The greater trochanter is easily palpable in dogs and cats. It is an important clinical landmark because it indicates the position of the coxofemoral joint (located medial to it). The greater trochanter becomes more prominent in dogs and cats with gluteal muscle atrophy, such as those with chronic pelvic limb lameness caused by hip dysplasia or osteoarthritis. The greater trochanter is also an important surgical landmark.

Caudal to the ridge of bone connecting the femoral neck to the greater trochanter is a deep hollow called the **trochanteric fossa**. The **lesser trochanter** is located on the caudomedial aspect of the proximal femur and is connected to the greater trochanter by the intertrochanteric crest. The **3rd trochanter** is found on the lateral aspect of the body of the femur, distal to the apex of the greater trochanter, at about the same level as the lesser trochanter. Muscular attachments to these smaller tuberosities are described later.

The body or **diaphysis** of the femur is long and cylindrical. It is marked along its caudal aspect by a **rough surface** that is bounded by the **medial** and **lateral lips**, which extend distally toward the epicondyles. The semimembranosus and pectineus muscles insert on the medial lip, and the adductor muscles insert on the lateral lip. These and the other muscles of the thigh are described later (Fig. 8.2-4).

The caudal flattened surface of the femur widens distally to form the **popliteal surface**. A pair of small tubercles, the **medial** and **lateral supracondylar tuberosities**, are found at the proximal edge of the popliteal surface. The medial and lateral heads of the gastrocnemius muscle originate on their respective tuberosities, and the superficial digital flexor also originates on the lateral tuberosity.

The distal end of the femur has 3 articular surfaces, including 2 condyles and a trochlea (Figs. 8.2-4 and 8.2-5). The caudally directed **lateral** and **medial condyles** articulate with the lateral and medial condyles, respectively, of the tibia. The femoral condyles are separated by a deep longitudinal indentation, the **intercondylar fossa**.

> A trochlea (L. a "small wheel" or "sheave of a pulley") is a trough or broad, shallow groove that functions much like the sheave or grooved wheel of a pulley.

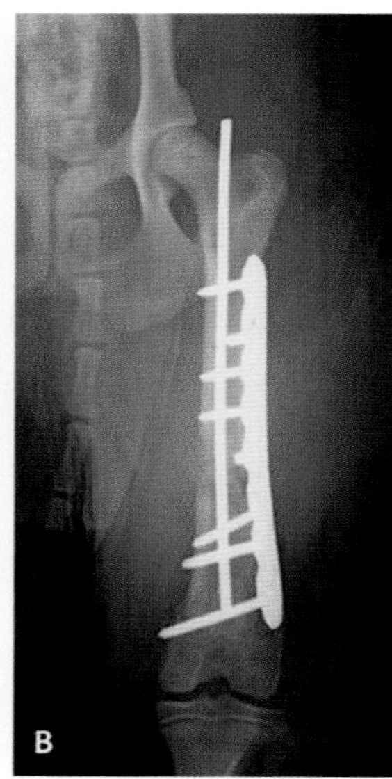

FIGURE 8.2-6 Proximal right femur of a dog (A), caudal aspect, alongside the post-operative radiograph (B) of this case (craniocaudal projection of the left femur). The intramedullary pin exits slightly more medially than desired (it should exit the proximal femur at the trochanteric fossa), but it is clear of the femoral head and neck. Key: 1, femoral head; 2, femoral neck; 3, trochanteric fossa; 4, greater trochanter; 5, lesser trochanter; 6, 3rd trochanter; 7, nutrient foramen.

The femoral **trochlea** is a groove on the cranial surface of the distal femur that is bounded by **medial** and **lateral trochlear ridges**. It articulates with the **patella** (described later). Proximal to the articular surfaces of the condyles are the **medial** and **lateral epicondyles**. The **medial** and **lateral collateral ligaments** of the stifle arise from their respective epicondyles. The **extensor fossa**, the site of origin for the long digital extensor muscle, is a small hollow at the junction between the epicondylar border of the lateral condyle and the lateral ridge of the trochlea.

The extensor fossa can be seen radiographically (Fig. 8.2-5) and should not be mistaken for a lesion in the bone.

During bone development, the **diaphysis** (shaft) and **metaphysis** (wider part at both ends of a long bone next to the epiphysis) of the femur are separated from the **epiphysis** (end of the bone; plural, epiphyses) by a region of growing cartilage, the **physis** (plural, physes), also known as the "epiphyseal plate" or "growth plate." At maturity, the physeal cartilage stops growing and the cartilage is replaced by bone—a time point referred to as "closure of the growth plate." The locations of the growth plates in the femur represent the several **ossification centers** in this bone: the femoral head, greater and lesser trochanters, diaphysis, distal epiphysis, and trochlea. In dogs and cats, femoral growth plate closure times vary, depending on the location, from 3 to 19 months of age, but most are closed by 12 months of age.

The radiographs in this case show the normal radiographic appearance of the proximal and distal femoral physes and the proximal physes of the tibia and fibula in a 7-month-old dog (Fig. 8.2-1).

Sesamoid bones (Fig. 8.2-5)

There are 4 sesamoid bones associated with the distal femur—the patella, lateral and medial fabellae, and popliteal sesamoid bone. The **patella** is located over the cranial aspect of the distal femur, where it articulates with the **trochlea**. It is the sesamoid bone within the tendon of insertion of the **quadriceps femoris muscle** (described later).

The **patellar ligament** is the name given to the distal portion of this tendon, between the patella and the tibial tuberosity. In addition to its protective and mechanical roles, the patella directs the alignment of the quadriceps femoris tendon over the distal femur via its articulation within the trochlear groove. The **parapatellar fibrocartilages** are attached to either side of the patella; they articulate with the trochlear ridges and act to increase the surface area of the patella.

The **lateral** and **medial fabellae** (singular, **fabella**) are located on the caudal aspect of the distal femur. They are the sesamoid bones within the tendons of origin of the lateral and medial heads, respectively, of the gastrocnemius muscle. They articulate with the femur at the caudoproximal margin of their respective **femoral condyles**.

The small **popliteal sesamoid bone** is found at the caudal aspect of the proximal tibia, within the tendon of origin of the popliteus muscle. It articulates with the lateral condyle of the **tibia**.

At birth, the sesamoid bones are entirely cartilaginous. Their ossification centers usually appear by 3 months of age in dogs and 4–5 months in cats. However, the popliteal sesamoid bone is not always ossified in the mature animal.

Muscles of the thigh (Figs. 8.2-7–8.2-11)

The muscles overlying the femur are described below in groups according to their location: cranial, lateral, caudal, and medial. The muscles that insert onto the proximal femur from their origins on the lumbar spine or pelvis, and others that originate on the distal femur and insert distal to the stifle, are described separately.

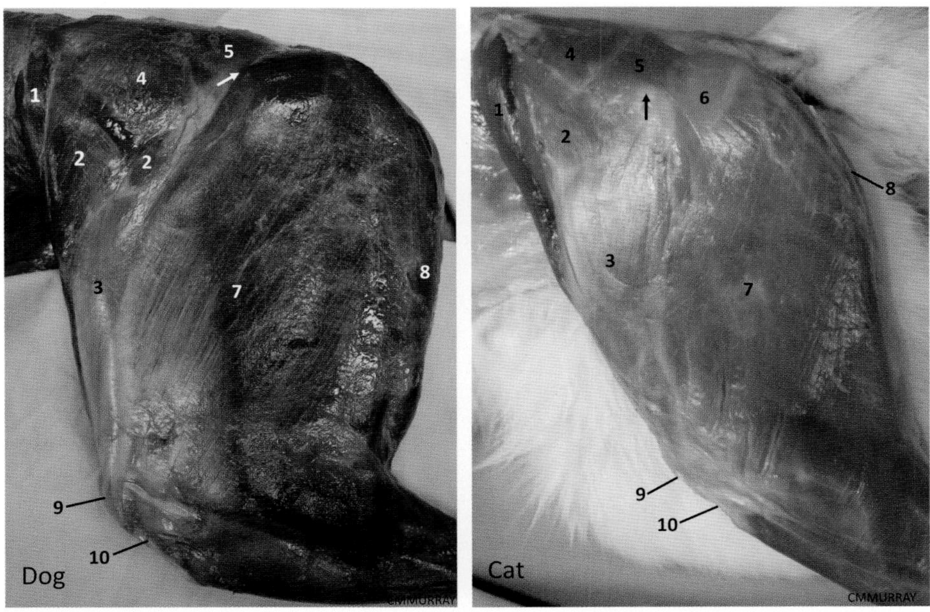

FIGURE 8.2-7 Superficial muscles of the left thigh, lateral aspect, in the dog and cat. Key: *m.*, muscle. Key: 1, sartorius m.; 2, tensor fascia latae m.; 3, fascia lata over the lateral aspect of vastus lateralis m.; 4, middle gluteal m.; 5, superficial gluteal m.; 6, gluteofemoral m. (cat only); 7, biceps femoris m.; 8, semitendinosus m.; 9, patella; 10, patellar ligament. *Arrow* (*white* in dog, *black* in cat): location of the greater trochanter of the femur.

SESAMOID BONES

A sesamoid bone is a small bone that is found within a tendon (or ligament) where the tendon changes direction when passing over a bony prominence. The sesamoid bone acts to protect the tendon from excessive wear and to help direct its alignment. By placing the tendon further away from the axis of the joint, the sesamoid bone also functions to increase the mechanical advantage of the muscle from which the tendon originates.

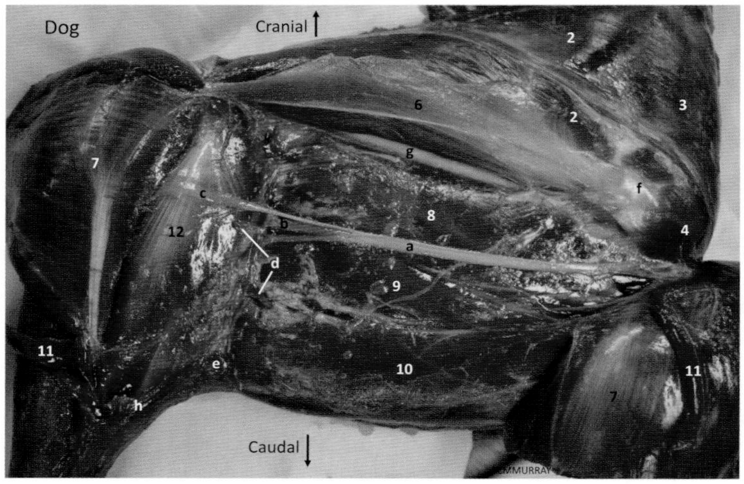

FIGURE 8.2-8 Deep prosection of the left thigh, lateral aspect, in the dog. (The cat is depicted in Fig. 8.2-9.) This image is of the same prosection shown in Fig. 8.2-7 but rotated 90° to the right. The biceps femoris and caudal crural abductor muscles have been transected and reflected to reveal the deeper structures. Key: The following muscles are labeled with numbers: 1, cranial belly of sartorius; 2, tensor fascia latae; 3, middle gluteal; 4, superficial gluteal; 5, gluteofemoral (cat only); 6, fascia lata over the lateral aspect of vastus lateralis; 7, biceps femoris (deep surface); 8, adductor; 9, semimembranosus; 10, semitendinosus; 11, caudal crural abductor; 12, lateral head of gastrocnemius; 13, soleus (cat only). The following structures are labeled with letters: a, sciatic nerve; b, tibial nerve; c, common peroneal nerve; d, distal caudal femoral vessels; e, popliteal lymph node; f, greater trochanter of the femur; g, femoral shaft; h, lateral saphenous vein; i, common calcaneal tendon.

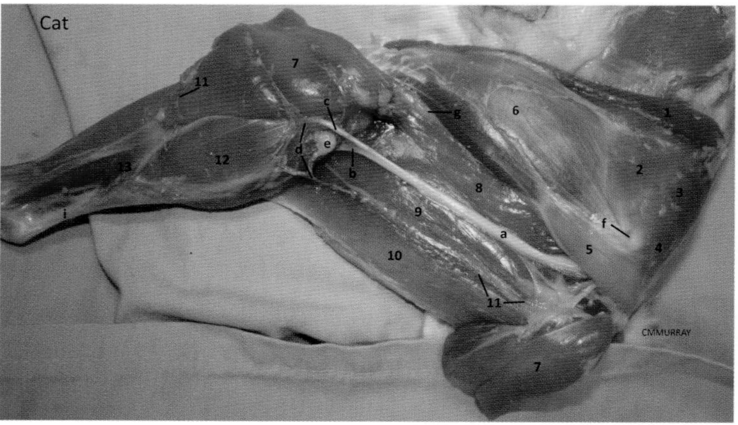

FIGURE 8.2-9 Deep prosection of the left thigh, lateral aspect, in the cat. See Fig. 8.2-8 for further description and key.

The cranial thigh muscles include the sartorius, tensor fascia latae, and quadriceps femoris. In general, they act to stabilize and extend the stifle, and those that cross the coxofemoral joint also stabilize and flex the hip (Figs. 8.2-7 and 8.2-10).

The **sartorius** is the most superficial of the cranial thigh muscles. It is a long, flat muscle that extends from the ilium to the cranial border of the tibia. The sartorius has a **cranial belly** that forms the cranial margin of the thigh, and a **caudal belly** on the craniomedial aspect of the thigh. It acts to flex the hip and stifle during protraction (forward step) and extend the stifle during stance.

The **tensor fascia latae** lies caudal to the cranial belly of the sartorius on the proximolateral aspect of the thigh. It arises on the ilium and gluteal fascia. This muscle has **cranial** and **caudal** parts, each of which attaches via a broad sheet of lateral femoral fascia, the **fascia lata**, to the distal fascia of the biceps femoris and lateral vastus muscles (each described below).

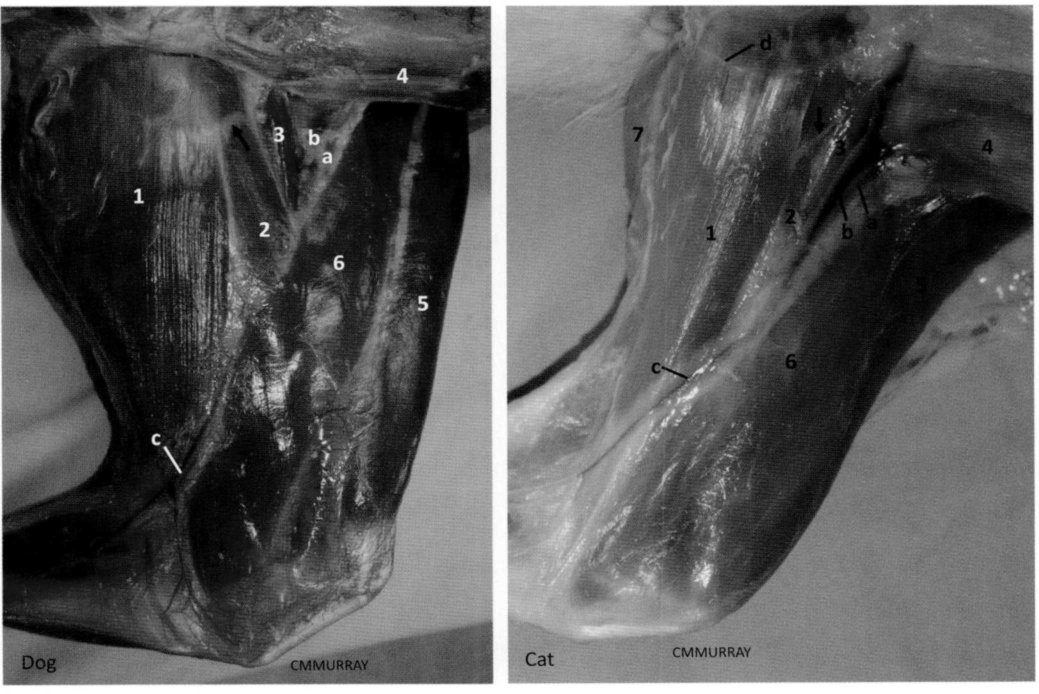

FIGURE 8.2-10 Superficial muscles of the left thigh, medial aspect, in the dog and cat. Key: The following muscles are labeled with numbers: 1, gracilis; 2, adductor; 3, pectineus; 4, aponeurosis of external abdominal oblique; 5, cranial belly of sartorius; 6, caudal belly of sartorius; 7, semitendinosus. The following structures are labeled with letters: a, femoral artery; b, femoral vein; c, saphenous artery, nerve, and medial saphenous vein; d, symphyseal tendon. *Black arrow:* branch of the obturator nerve.

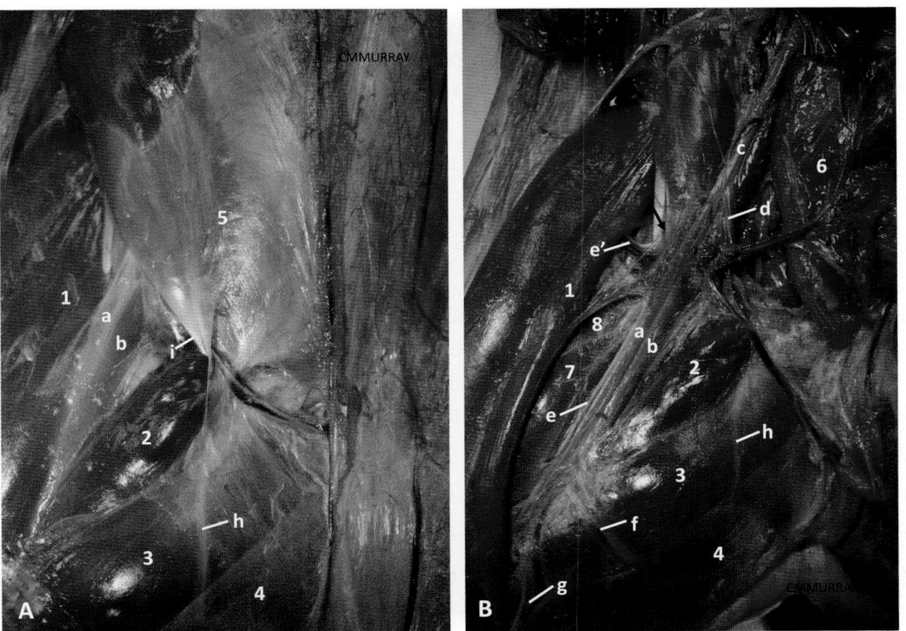

FIGURE 8.2-11 Femoral triangle and vascular lacuna of the right medial thigh in the dog. (A) Superficial structures of the proximal thigh, medial aspect. (B) Abdominal muscles removed to expose the vessels entering the medial thigh from the abdominal cavity. Key: The following muscles are labeled with numbers: 1, caudal belly of sartorius; 2, pectineus; 3, adductor; 4, gracilis; 5, aponeurosis of external abdominal oblique; 6, deep surface of rectus abdominis (reflected medially); 7, vastus medialis; 8, rectus femoris. The following structures are labeled with letters: a, femoral artery; b, femoral vein; c, external iliac artery; d, deep femoral artery; e, saphenous nerve—cutaneous branch (e) and the muscular branch to sartorius, together with the superficial circumflex iliac vessels (e′); f, proximal caudal femoral vessels; g, saphenous artery and medial saphenous vein; h, branch of obturator nerve; i, inguinal ligament. *Black arrow:* emergence of the femoral nerve from within the iliopsoas muscle.

The **quadriceps femoris** lies deep to the sartorius and tensor fascia latae, and forms the bulk of the cranial thigh musculature. It comprises 4 muscles or "heads" (hence, the name *quadriceps*): **rectus femoris** arises on the ilium, whereas **vastus lateralis**, **vastus intermedius**, and **vastus medialis** all arise on the proximal femur. Thus, only the rectus femoris crosses the coxofemoral joint. As their names indicate, the vastus lateralis is the most lateral of the 4 bellies, and vastus medialis is the most medial; these muscles enclose the rectus femoris for much of its length along the femur. The vastus intermedius lies between the other 2 vastus muscles, on the cranial surface of the femur, deep to the rectus femoris. Distally, the quadriceps group forms a single tendon, the **patellar tendon**, that inserts on the tibial tuberosity. As described above, the part of this tendon between the patella and the tibial tuberosity is also referred to as the **patellar ligament**.

The lateral thigh muscles include the biceps femoris, caudal crural abductor, and—in cats—the gluteofemoral muscle.

The **biceps femoris** lies caudal to the tensor fascia latae and forms the bulk of the musculature on the lateral aspect of the thigh. It arises on the tuber ischiadicum, and in dogs also on the **sacrotuberous ligament** (which is absent in cats). The muscle attaches via the crural fascia and patellar tendon to the tibial tuberosity and the cranial border of the tibia (formerly called the tibial crest). It sends a further narrow tendon to the tuber calcanei via the **common calcaneal tendon**. The cranial part of this muscle acts to extend the hip and stifle during standing, while the caudal part flexes the stifle during the swing phase of the stride. This muscle also contributes to extension of the tarsus.

In cats, an additional muscle is interposed between tensor fascia latae and biceps femoris: the **gluteofemoral** muscle. It arises caudal to the superficial gluteal muscle (described later), on caudal vertebrae 2–4, and its tendon of insertion extends deep to the biceps femoris to attach to the fascia lata and the lateral aspect of the patella. This muscle contributes to hip extension and limb abduction.

The **caudal crural abductor** is a narrow, flat muscle that arises on the sacrotuberous ligament in dogs and the 2nd caudal vertebra in cats. It courses deep to the biceps femoris for most of its length before becoming superficial distally. It inserts on the crural fascia on the lateral aspect of the proximal crus, just caudodistal to the biceps femoris. This muscle contributes to pelvic limb abduction and stifle flexion.

The caudal thigh muscles comprise the semitendinosus and semimembranosus. These muscles act to extend the hip during standing and flex the stifle during the swing phase of the stride; the semitendinosus also contributes to extension of the tarsus. Together with the biceps femoris, these muscles make a major contribution to the forward thrust of locomotion.

The **semitendinosus** forms much of the caudal contour of the thigh. It arises on the ischiatic tuberosity and extends between the biceps femoris and semimembranosus to insert as a broad, flat tendon on the medial surface of the proximal tibia, deep to the insertion of the gracilis muscle (described below). An added tendinous band joins with a similar band from the gracilis tendon to attach to the tuber calcanei via the common calcaneal tendon.

The **semimembranosus** arises on the ischiatic tuberosity and, after a short distance, it divides into 2 bellies. Both muscle bellies lie medial to the biceps femoris and semitendinosus, and on the medial aspect of the thigh lie caudal and superficial to the adductor muscles and deep to the gracilis. The **cranial belly** inserts on the distal end of the medial lip of the femur and on the medial tendon of origin of the gastrocnemius muscle (described later). The **caudal belly** passes across the medial aspect of gastrocnemius to insert on the margin of the medial condyle of the tibia.

The medial thigh muscles include the gracilis, pectineus, and the adductors (longus and magnus et brevis). Along with other individual actions, they all contribute to adduction of the pelvic limb (Fig. 8.2-10).

The **gracilis** lies superficial to the semimembranosus and the adductor muscles on the caudal part of the medial thigh. It is a broad muscle, arising from the pelvic symphysis via the symphyseal tendon and inserting along the length of the cranial margin of the proximal tibia. It also contributes a tendinous band to the common calcaneal tendon, so in addition to limb adduction and hip extension, it contributes to extension of the tarsus.

The **pectineus** is the superficial muscle cranial to the gracilis on the medial aspect of the thigh. It arises from the iliopubic eminence and cartilage and the prepubic tendon. Proximally, it has a short, spindle-shaped muscle belly; its tendon then passes between the sartorius, adductor, and vastus medialis muscles to insert on the medial lip of the femur. Its muscle belly forms the caudal margin of the femoral triangle.

The **femoral triangle** is a shallow, V-shaped depression on the medial aspect of the proximal thigh where the femoral artery and vein and the saphenous nerve are superficial, covered only by fascia and skin (Fig. 8.2-11). It is bounded cranially by the caudal belly of the sartorius, caudally by the pectineus muscle belly, and dorsomedially by the **inguinal ligament**.

In dogs and cats, the femoral artery is readily palpable in the femoral triangle. To palpate the arterial pulse, gently press the femoral artery against the shaft of the proximal femur.

The adductors are the deep muscles on the medial aspect of the thigh that, in addition to pelvic limb adduction, act to extend the hip. The **adductor longus** arises on the pubic tubercle and inserts on the proximal part of the lateral lip of the femur, near the 3rd trochanter. Thus, this muscle (and the next) crosses the caudal surface of the femur as it extends distally from its medial origin to its lateral insertion.

The much larger **adductor magnus et brevis** forms the bulk of the caudomedial thigh. It arises on the pelvic symphysis deep to the origin of gracilis, and it inserts on the length of the lateral lip of the femur, from the 3rd trochanter to near the origin of the lateral head of the gastrocnemius.

Muscles inserting on the proximal femur (Fig. 8.2-12A–D)

The following is a brief description of the muscles that arise on the vertebrae and pelvis and insert on the proximal femur.

The **iliopsoas** comprises the **psoas major**, which arises from the ventral aspect of lumbar vertebrae 2–7, and the **iliacus**, which arises from the ventral aspect of the ilium. These muscles join to form the iliopsoas that inserts on the lesser trochanter of the femur. This muscle acts to flex the hip to advance the limb during motion and to stabilize or ventrally flex the lumbar spine.

The **articularis coxae** is a small muscle that crosses the cranial aspect of the coxofemoral joint, extending from the ilium to the cranial aspect of the femoral neck. It makes a minor contribution to hip flexion and, through tensing the hip joint capsule, acts to protect the capsule from injury.

The lateral pelvic muscles include the gluteals (superficial, middle, and deep) and piriformis [and in cats, the gluteofemoral muscle (described above)]. In general, these muscles extend the hip, and each also contributes to one or more of the following actions—abduction, retraction, and medial rotation of the limb.

The **superficial gluteal** has origins that include the gluteal and caudal fascia, the sacrum and 1st caudal vertebra, and, in dogs, the sacrotuberous ligament. It inserts on the 3rd trochanter of the femur. The next 3 muscles all insert on the greater trochanter of the femur, but their origins differ. The **middle gluteal** arises on the ilium, dorsal sacroiliac ligament, and gluteal fascia. Its caudal part lies deep to the superficial gluteal muscle. The **piriformis** arises from the 3rd sacral to the 1st caudal vertebrae, and its tendon of insertion joins that of the middle gluteal. The **deep gluteal** arises on the ilium, deep to the middle gluteal and piriformis muscles.

The **gluteofemoral muscle** in cats lies caudal to the superficial gluteal. The **tensor fascia latae** in dogs and cats may also be included in the lateral pelvic muscle group, although it acts to flex rather than extend the hip. Both muscles are described above, with the muscles of the thigh.

The **medial pelvic muscles** include the internal and external obturator, gemelli, and quadratus femoris. These muscles all contribute to lateral rotation of the hip and prevent medial rotation during weight-bearing. The quadratus femoris also contributes to hip extension.

The **internal obturator** arises on the internal or dorsal surface of the pubis and ischium and covers the dorsal aspect of the obturator foramen. Its narrow tendon passes over the lesser ischiatic notch and between the 2 bellies of the gemelli muscle. Both **gemelli muscle** bellies arise from the lateral surface of the ischiatic body just ventral to the lesser ischiatic notch, and their tendons join that of the internal obturator to insert in the trochanteric fossa of the femur.

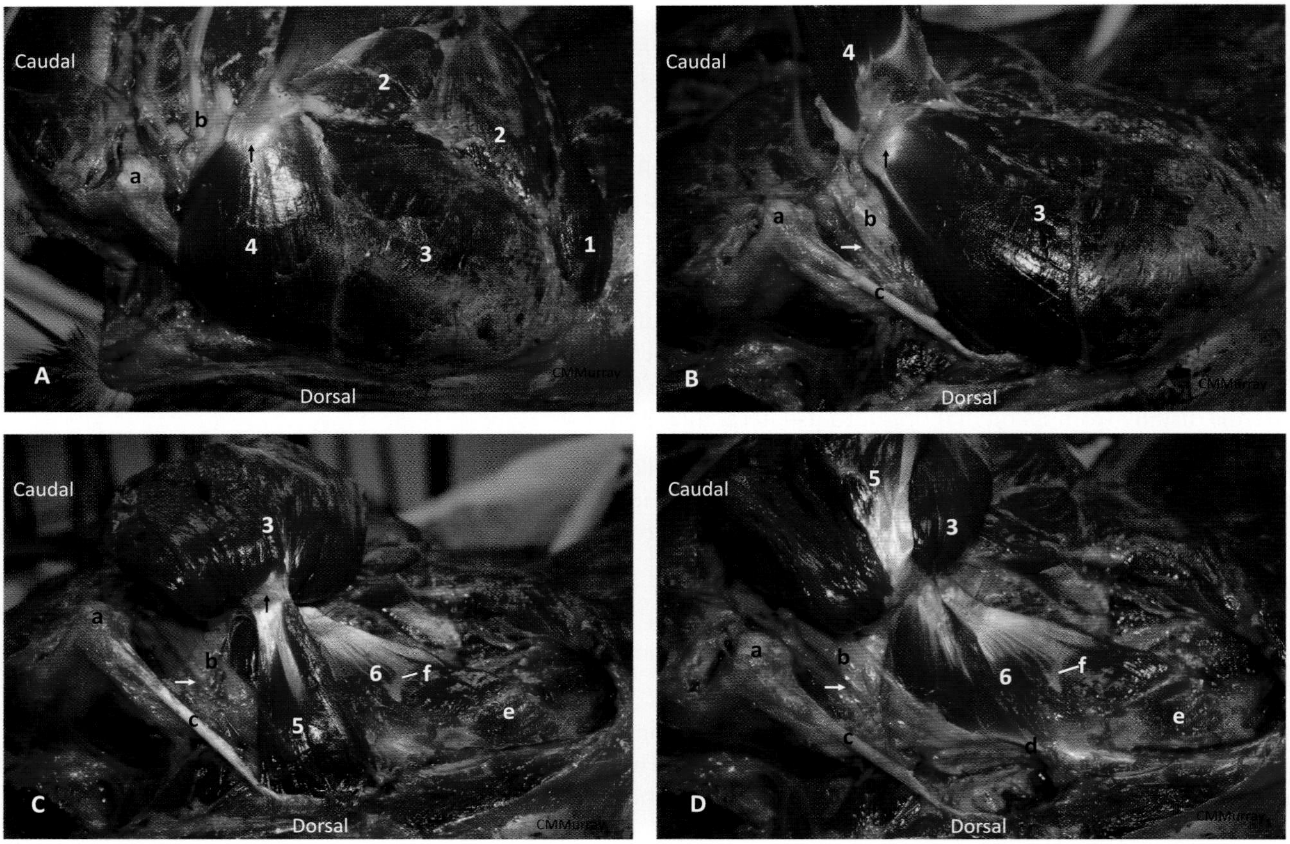

FIGURE 8.2-12 Left hip, dorsolateral aspect, in the dog. Images A–D show sequential prosections from superficial to deep. (A) The biceps femoris has been removed. (B) The superficial gluteal muscle has been reflected at its insertion on the 3rd trochanter of the femur. (C) The middle gluteal muscle has been reflected at its insertion on the greater trochanter of the femur. (D) The piriformis muscle has been reflected at its insertion on the greater trochanter of the femur. Key: The following muscles are labeled with numbers: 1, cranial belly of sartorius; 2, tensor fascia latae; 3, middle gluteal; 4, superficial gluteal; 5, piriformis; 6, deep gluteal. The following structures are labeled with letters: *A*, ischiatic tuberosity; *B*, sciatic nerve; *C*, sacrotuberous ligament; *D*, greater ischiatic notch; *E*, gluteal surface of ilium; *F*, cranial gluteal nerve. *Black arrow:* greater trochanter. *White arrow:* caudal gluteal artery.

The **external obturator** arises on the external or ventral surface of the pubis and ischium and covers the ventral aspect of the obturator foramen. Its tendon joins that of the internal obturator and gemelli to insert in the trochanteric fossa. The **quadratus femoris** (not to be confused with quadriceps femoris) arises from the ventral ischium caudal to the gemelli and lateral to the external obturator. It inserts on the intertrochanteric crest on the caudal aspect of the femur.

Muscles arising on the distal femur

The following is a brief description of the muscles that arise on the distal femur and insert further distally on the pelvic limb.

The **gastrocnemius** is divided into **lateral** and **medial heads**, which arise, respectively, from the lateral and medial supracondylar tuberosities of the distal femur (Figs. 8.2-8 and 8.2-9). The **lateral** and **medial fabellae** are the sesamoid bones found within each tendon of origin. The 2 heads fuse distally to form a single tendon that inserts on the tuber calcanei via the common calcaneal tendon. An extensor of the tarsus, it also contributes to stifle flexion.

The **superficial digital flexor** arises with the lateral head of the gastrocnemius on the lateral supracondylar tuberosity of the femur, and its belly lies mainly within/between the 2 bellies of the gastrocnemius. Its tendon of insertion initially forms part of the common calcaneal tendon and then continues distally from the tuber calcanei to the digits. It acts as a digital flexor, tarsal extensor and stabilizer, and flexor of the stifle.

The **popliteus** lies deep to the gastrocnemius and superficial digital flexor muscles and covers the caudal aspect of the stifle joint. It arises on the lateral aspect of the lateral femoral condyle, beneath the lateral collateral ligament of the stifle, and extends obliquely to insert on the medial border of the proximal tibia. It acts to rotate the tibia medially with respect to the femur during the swing phase of the stride, and it may also have a proprioceptive role regarding stifle position. The **popliteal sesamoid bone** is within its tendon of origin.

The **long digital extensor** originates in the **extensor fossa** of the lateral femoral condyle and inserts on the dorsal aspect of the digits. A digital extensor and flexor of the tarsus, it does not contribute to stifle movement. Lastly, the **articularis genus** is a small muscle that arises on the cranial surface of the distal femur and inserts within the **femoropatellar joint capsule**. It is thought to tense or retract the joint capsule during stifle extension and may also have a proprioceptive role in monitoring stifle movement.

Blood supply and lymphatics of the thigh (Figs. 8.2-8–8.2-12)

The major blood supply to the pelvic limb arises from the paired **external iliac arteries**, which arise from the abdominal **aorta**. Each gives rise to the paired femoral and deep femoral arteries. The **deep femoral artery** branches from the external iliac artery while still within the abdomen. The external iliac artery then continues as the **femoral artery** through the vascular lacuna to the medial aspect of the thigh.

The **vascular lacuna** is the opening in the abdominal wall between the inguinal ligament and the pelvis that allows passage of the femoral and deep femoral arteries and veins to the pelvic limb. It is bounded cranially by the caudal border of the internal abdominal oblique muscle and the inguinal ligament, and medially by the rectus abdominis muscle. (The muscles of the abdominal wall are described in Case 5.8.)

After passing through the vascular lacuna, the **femoral artery** lies superficially in the **femoral triangle** (described earlier) between the femoral vein and the saphenous nerve. It then passes distally, running deep to the gracilis and the caudal belly of the sartorius, along the medial border of the insertion of the adductor muscle, and then deep to the semimembranosus. Its final branch is the **distal caudal femoral artery**. From this point, the femoral artery continues as the **popliteal artery** that passes with the popliteal vein across the popliteal surface of the distal femur, between the heads of the gastrocnemius.

The following describes the major branches of the **femoral artery** in order from proximal to distal. The **superficial circumflex iliac artery** passes dorsocranially to supply the cranial aspect of the proximal thigh. The **lateral circumflex femoral artery** passes cranially, supplying proximal thigh, pelvic muscles, and the joint capsule of the hip. The **proximal caudal femoral artery** supplies the medial aspect of the proximal thigh.

The **saphenous artery** arises from the femoral artery just before it passes deep to the semimembranosus muscle. It runs distally, in company with the **medial saphenous vein** and the saphenous nerve, on the medial surface of the thigh between the gracilis and sartorius, to the crus. It sends a **genicular branch** to the medial aspect of the stifle. The **descending genicular artery** passes between the vastus medialis and semimembranosus muscles, also supplying the stifle joint.

The **middle caudal femoral artery** runs caudodistally to supply the adductor and semimembranosus muscles. The **distal caudal femoral artery** is the last branch of the femoral artery that arises just proximal to the origins of the gastrocnemius muscle. Its branches supply the gastrocnemius, vastus lateralis, adductor, biceps femoris, semimembranosus, and semitendinosus muscles. The large ascending branches form a major supply to the biceps femoris and caudal thigh muscles.

The **deep femoral artery** gives off 3 or 4 intra-abdominal branches before passing through the caudal part of the vascular lacuna to become the **medial circumflex femoral artery**. This artery passes caudally, medial to the femur in the proximal thigh, and supplies the proximal medial thigh and pelvic muscles, the joint capsule of the hip, and the femoral neck. In most dogs, it also supplies the **nutrient artery to the femur** that enters the **nutrient foramen** on the caudal aspect of the femoral diaphysis (Fig. 8.2-6A).

The nutrient foramen can be misinterpreted as a fracture line or cortical defect on radiographs. If in doubt, evaluate a different view of the affected limb or the contralateral limb.

Vascular contributions to the structures of the thigh are also made by branches of the paired deep circumflex iliac and internal iliac arteries, which arise from the aorta proximal and distal to the external iliac arteries, respectively. The **deep circumflex iliac artery** supplies superficial tissues over the lumbar and pelvic regions and the craniolateral aspect of the thigh. The **internal iliac artery** terminates as the caudal gluteal artery and the internal pudendal artery.

The **caudal gluteal artery** initially lies ventral to the sacrum and caudomedial to the iliopsoas muscle. Its branches supply the iliopsoas and pelvic muscles along with the joint capsule of the hip. It continues toward the ischiatic tuberosity and passes over the greater ischiatic notch with its vein and the sciatic nerve; in dogs, it lies deep to the sacrotuberous ligament. Its terminal branches enter the proximal parts of the biceps femoris, semitendinosus, and semimembranosus muscles, and—together with the branches of the distal caudal femoral artery—it comprises the major blood supply to these muscles.

The arteries of the thigh almost all have like-named satellite veins accompanying them. One exception is the **medial saphenous vein** that accompanies the saphenous artery on the medial aspect of the pelvic limb. This vein is so named because in addition to both the medial saphenous and popliteal veins, there is a further major vein entering the distal thigh region from the crus—the **lateral saphenous vein**. This vein, which does not have an accompanying artery, begins on the lateral aspect of the mid crus where its cranial and caudal branches join, ascends on the caudal

BLOOD SUPPLY AFTER FRACTURE

The blood supply to a long bone, such as the femur, consists of 3 groups of vessels: (1) the nutrient artery, which enters the diaphysis via a nutrient foramen and then branches proximally and distally within the medulla; (2) epiphyseal and metaphyseal arteries, which enter at the epiphyseal and metaphyseal regions, respectively, at each end of the bone; and (3) the periosteal arterioles, which enter only in the regions of fascial or ligamentous attachment to the bone.

The branches of the nutrient artery and the metaphyseal vessels anastomose; however in young dogs and cats, the epiphyseal and metaphyseal vessels are separated by the physis until its closure at skeletal maturity, when they too unite. The branches of the nutrient artery also anastomose with the periosteal vessels where present. While both arteries supply the diaphyseal cortex, the periosteal vessels make only a small contribution to the normal bone. However, all these anastomoses are important, as they help maintain blood supply to the bone after injury.

Following injury to the bone, such as a fracture, the blood vessels undergo hypertrophy. In nondisplaced long bone fractures, the intact branches of the nutrient artery remain the major blood supply to the bone for callus formation and fracture healing. Immediately following a fracture, a further external supply of blood forms, called the extraosseous blood supply. It is especially important for healing with displaced fractures, in which the branches of the nutrient artery have been damaged. This extraosseous supply first arises from injured vessels in torn muscles close to the fracture site. It participates in the organization of the external hematoma (pool of free blood within tissues) and subsequent formation of the periosteal callus. The extraosseous blood supply continues to develop in the surrounding soft tissues and supplies detached bone fragments, devitalized bone cortex, and the developing bone callus.

During the period in which the branches of the injured nutrient artery are regenerating, the metaphyseal arteries sustain the medullary blood supply to the diaphyseal cortex. The extraosseous blood supply regresses once the continuity of the blood supply throughout the length of the medulla has been re-established.

surface of the gastrocnemius muscle, and then passes between the biceps femoris and semitendinosus muscles to enter the distal caudal femoral vein in the popliteal fossa.

Lymph from the pelvic limb ultimately drains to the **medial iliac lymph nodes**, found between the origins of the deep circumflex iliac and external iliac arteries. It reaches these nodes either directly or via other lymph nodes that drain regions of the pelvic limb, including the **popliteal**, **superficial inguinal**, and **internal iliac lymph nodes** and—when present—the **sacral**, **external iliac**, and **distal femoral lymph nodes**. The popliteal lymph node is superficially located in the distal thigh and is readily palpable. It is found caudal to the stifle joint and lies embedded in fat between the caudal margins of the biceps femoris and semitendinosus muscles (Figs. 8.2-8 and 8.2-9).

Innervation of the thigh (Figs. 8.2-8–8.2-12)

The nerves that innervate the pelvic limb arise from the **lumbosacral plexus**, which is formed by the ventral branches of lumbar nerves 3–7 and all 3 sacral nerves. The 3 major nerves that innervate the muscles of the thigh that arise from this plexus are, from cranial to caudal, the femoral, obturator, and sciatic nerves.

The **femoral nerve** passes through the iliopsoas muscle, which it supplies, and then enters the quadriceps femoris muscle between the rectus femoris and vastus medialis to supply all 4 heads of the quadriceps group. It also supplies the articularis coxae muscle in the hip. The **saphenous nerve** branches from the femoral nerve before its exit from the iliopsoas. It has a **muscular branch**, supplying the sartorius muscle, and a **cutaneous branch** that runs initially along the cranial aspect of the femoral artery and then continues with the saphenous artery and medial saphenous vein to the medial aspect of the crus.

The **obturator nerve** arises within the iliopsoas muscle and passes caudoventrally along the medial aspect of the shaft of the ilium and then through the cranial part of the **obturator foramen** to supply the muscles on the medial aspect of the thigh. These muscles—the pectineus, gracilis, and the adductors—act primarily to adduct the pelvic limb. The obturator nerve also innervates the external obturator muscle.

The **lumbosacral trunk** is the largest part of the lumbosacral plexus. The cranial and caudal gluteal nerves (described later) arise from this trunk, and the sciatic nerve is the extra-pelvic continuation of it. The **sciatic nerve** is a large nerve that consists of the **tibial** and **common peroneal** (or common fibular) **nerves**, which proximally are closely bound together and thus appear as a single nerve (Figs. 8.2-8–8.2-13). On leaving the pelvis, the sciatic nerve passes medial to the greater trochanter and thus lies dorsal to the trochanteric fossa. It passes deep to the superficial gluteal muscle, across the surface of the gemelli muscle, the tendon of internal obturator, and the lateral aspect of quadratus femoris muscle. A muscular branch innervates the gemelli, internal obturator, and quadratus femoris muscles.

The sciatic nerve then continues along the lateral aspect of the adductor and semimembranosus muscles, deep to the **biceps femoris**. It supplies the biceps femoris, semitendinosus, semimembranosus, and the caudal crural abductor muscles. The sciatic nerve ends in the distal thigh by branching into the common peroneal and tibial nerves. These 2 nerves go on to supply the muscles distal to the stifle.

It is critical to take care when inserting IM pins into the femur to minimize the risk of sciatic nerve injury.

Normograde insertion—the pin is introduced at the proximal end of the proximal fragment and passed through the fracture site into the distal fragment. The pin is passed along the medial surface of the greater trochanter in order to avoid the sciatic nerve, and enters the bone via the trochanteric fossa.

Retrograde insertion—the pin is first introduced into the proximal fragment via the fracture site and then passed distally across the fracture into the distal fragment. The proximal fragment is held in an adducted and extended position, with the pin directed proximally along the craniolateral surface of the medullary cavity, to reduce the risk of injuring the sciatic nerve and femoral head when the pin appears at the proximal end of the femur Fig. 8.2-6).

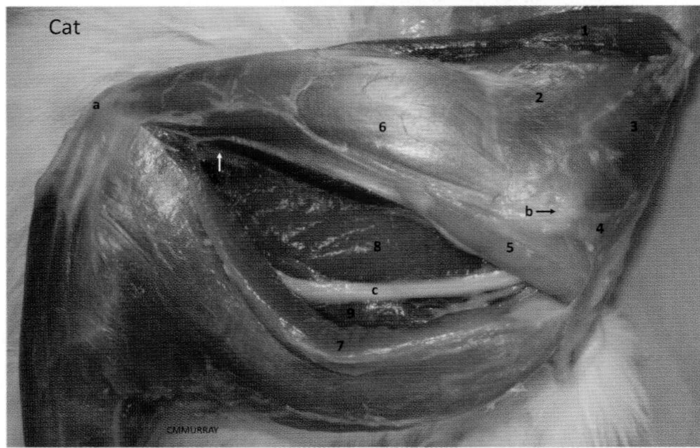

FIGURE 8.2-13 Surgical approach to the femoral shaft in the dog (*upper figure*) and cat (*lower figure*). The skin has been incised along the lateral thigh from the greater trochanter to the patella and retracted (skin removed to show the deeper anatomy). The fascia lata is incised along the cranial border of biceps femoris. Biceps femoris retracted caudally exposing the shaft of the femur. Key: The following muscles are labeled with numbers: 1, cranial belly of sartorius; 2, tensor fascia latae; 3, middle gluteal; 4, superficial gluteal; 5, gluteofemoral (cat only); 6, superficial and deep (6′) leaves of fascia lata over the lateral aspect of vastus lateralis; 7, biceps femoris; 8, adductor; 9, semimembranosus. The following structures are labeled with letters: a, patella; b, greater trochanter; c, sciatic nerve. *White arrow:* femoral shaft.

BIOMECHANICS OF FRACTURE REPAIR

The femur is subject to asymmetric loading because the femoral head projects from its medial aspect proximally to articulate with the acetabulum. Weight-bearing forces are thus greater along the medial side of the bony column of the femur. As a result, the medial side of the femur is subject to greater compressive forces and the lateral side to greater tensile or distracting forces, with the net effect being a tendency for the femur to bend outward/laterally. This is demonstrated in the case described here, in which the proximal fracture segment was displaced laterally within the soft tissues of the thigh.

A bone plate must be applied to the side of the bone that is most often subject to tensile forces, so in the case of the femur, it must be applied to the lateral surface of the diaphysis. If applied to the medial surface, the plate is subject to excessive bending forces, resulting in plate and screw fatigue and failure. Furthermore, because this was a comminuted fracture, with many small fragments in the mid-shaft region that could not be incorporated into the repair, the plate was applied to “bridge” the gap in the bone column. In this situation, because the bone column is incomplete and cannot bear a load, a plate on its own would be incapable of completely neutralizing the weight-bearing forces, and at increased risk of failing. For this reason and in this case, an intramedullary pin was added to the construct. The pin and plate act as parallel beams that share the load, reducing the risk of either failing.

The **cranial gluteal nerve** branches from the lumbosacral trunk and passes cranially across the lateral aspect of the shaft of the ilium, between the middle and deep gluteal muscles, to end in the tensor fascia latae. It supplies each of these muscles. The **caudal gluteal nerve** supplies the superficial gluteal and piriformis muscles, and in cats the gluteofemoral muscle.

SURGICAL APPROACH TO THE FEMUR (FIG. 8.2-13)

The surgical approach must consider the location of major vessels and nerves so that these can be avoided. It is also desirable where possible to follow the anatomical planes between muscles. For a mid-shaft femoral fracture, a lateral surgical approach is used that avoids the major vessels of the thigh located on the medial aspect of the femur. The skin is incised along the lateral aspect of the thigh from the greater trochanter to the level of the patella. The skin is retracted and the superficial leaf of the fascia lata is incised along its margin with the cranial border of biceps femoris. Care must be taken during caudal retraction of the biceps to avoid injury to the sciatic nerve beneath it. The vastus lateralis and vastus intermedius muscles are retracted from the femoral shaft by freeing the fascia that connects these muscles to the bone. Elevation of the adductor muscle is limited to what is needed for access to the fracture, because this muscle contains important sources of blood supply.

Selected references

[1] Beale B. Orthopaedic clinical techniques femur fracture repair. Clin Tech Small Anim Pract 2004;19(3):134–50.
[2] DeCamp CE, Johnston SA, Déjardin LM, et al. Brinker, Piermattei and Flo's handbook of small animal orthopedics and fracture repair. 5th ed. St Louis: Elsevier; 2016.
[3] Piermattei DL, Johnson KA. An Atlas of surgical approaches to the bones and joints of the dog and cat. 4th ed. Philadelphia: Saunders; 2004.
[4] Rhinelander FW. Tibial blood supply in relation to fracture healing. Clin Orthop Rel Res 1974;105:34–81.

CHAPTER 8

CASE 8.3

Cranial Cruciate Ligament Tear

Kimberly A. Agnello
Department of Clinical Sciences & Advanced Medicine, University of Pennsylvania School of Veterinary Medicine, Philadelphia, Pennsylvania, US

Clinical case

History

A 6-year-old female spayed Labrador Retriever was presented with a 4-week history of a left pelvic limb lameness. The dog first became acutely lame after playing in the backyard. Since this incident, the owner noted that the lameness worsened after activity and when the dog was rising after rest. The owner reported no improvement in lameness following treatment with a nonsteroidal anti-inflammatory drug and exercise restriction.

Physical examination findings

The dog had a lameness characterized by decreased weight-bearing at a walk and a trot with mild muscle atrophy of the left pelvic limb. There was moderate left stifle effusion with thickening of the medial joint capsule (medial

CRANIAL DRAW TEST

Also called the "cranial drawer test," this is a passive test for stifle joint instability, which, when positive, means that the cranial cruciate ligament is not intact or is nonfunctional. The examiner is positioned caudal or caudolateral to the dog while the dog is either in lateral recumbency or standing. The index finger of one hand is placed on the patella and the thumb is placed behind the lateral fabella. The index finger of the other hand is placed on the tibial tuberosity and the thumb behind the fibular head. The hand on the femur is kept steady while the tibia is manipulated in a caudal (allowing reduction to its normal position) and then cranial direction. Cranial translation of the tibia is abnormal and indicates a positive test for incompetence of the cranial cruciate ligament. It is important that the fingers are accurately positioned on these boney landmarks as motion of the soft tissues could be misinterpreted as joint instability.

TIBIAL THRUST OR TIBIAL COMPRESSION TEST

This is a dynamic test for stifle joint instability. The examiner is located caudal or caudolateral to the dog while performing the test in lateral recumbency or while the dog is standing. The index finger of one hand is placed on the tibial tuberosity, while the thumb and rest of the fingers are used to grasp the femoral condyles and maintain the stifle in an extended position. The other hand grasps the metatarsals and the tarsocrural joint is flexed and extended, simulating contraction of the gastrocnemius muscle and the tibial compression mechanism—i.e., it is a dynamic test that simulates weight-bearing. Tibial tuberosity movement is monitored by the index finger and motion is considered a positive test, indicating an incompetent cranial cruciate ligament.

Comparative Veterinary Anatomy: A Clinical Approach. https://doi.org/10.1016/B978-0-323-91015-6.00051-0

buttress). Left stifle extension elicited a pain response, and the stifle was unstable with positive cranial draw and tibia thrust tests. A clicking noise was noted when moving the stifle through a range of motion.

Differential diagnoses

Cranial cruciate ligament rupture; neoplasia of the distal femur, proximal tibia, or periarticular soft tissues; septic arthritis; immune-mediated arthropathy; tick-borne infectious diseases

Diagnostics

The dog was sedated for bilateral stifle radiographs (Fig. 8.3-1). The left and right stifle had moderate and mild intracapsular soft tissue opacities, respectively, causing cranial displacement of the intrapatellar fat pad and caudal displacement of the gastrocnemius fascial planes/muscles (left, Fig. 8.3-1A and right, Fig. 8.3-1C). The left tibia was displaced cranially relative to the femur (Fig. 8.3-1A). A small enthesiophyte (Gr. *enthesis* a putting in; insertion

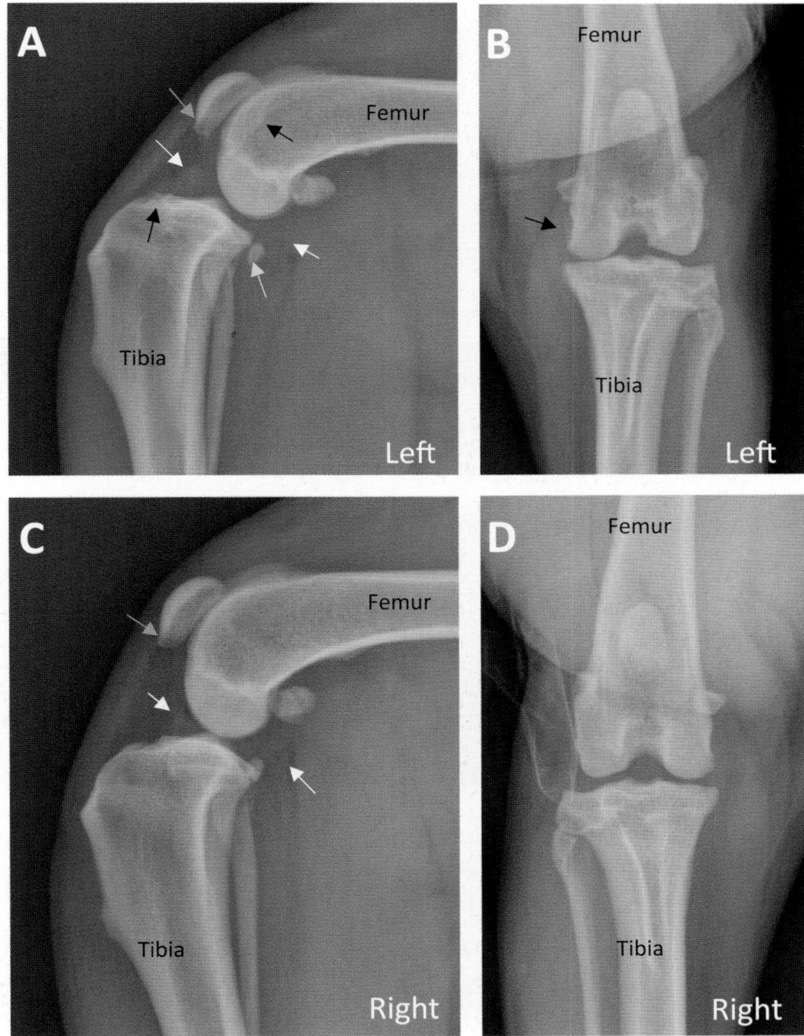

FIGURE 8.3-1 Bilateral stifle radiographs. (A) Lateral radiographic projection of the left stifle. There is moderate intracapsular soft tissue opacity causing cranial displacement of the infrapatellar fat pad, caudal displacement of the gastrocnemius facial planes/muscles (*white arrows*), and caudal distal displacement of the popliteal sesamoid (*yellow arrow*). There is a small enthesiophyte of the distal patella (*blue arrow*). There is mild osteophytosis of the trochlea and tibia (*black arrows*). (B) Craniocaudal projection of the left stifle. There is mild osteophytosis of the medial femoral condyle (*black arrow*). (C) Lateral radiographic projection of the right stifle. There is moderate intracapsular soft tissue opacity causing cranial displacement of the infrapatellar fat pad and caudal displacement of the gastrocnemius fascial planes/muscles (*white arrows*). (D) Craniocaudal projection of the right stifle.

of muscle or ligament to bone) of the distal patella was noted bilaterally (left, Fig. 8.3-1A and right, Fig. 8.3-1C). There was mild osteophytosis (Gr. *osteon* bone + Gr. *phyton* plant; osseous outgrowth; formation of osteophytes) of the trochlea, tibia, and femoral condyles (Fig. 8.3-1A and B). The popliteal sesamoid was moderately displaced caudodistally (Fig. 8.3-1A).

In summary, radiographs showed bilateral stifle effusion or synovitis (moderate left stifle, mild right stifle) and left stifle periarticular osteophytosis and cranial subluxation of the left tibia. While the dog was sedated for stifle radiographs, palpation of the stifles was repeated, confirming a positive cranial draw and tibial thrust of the left stifle with an occasional "clicking" sound; no physical abnormalities were palpated in the right stifle.

The results of the radiographs, together with the patient's signalment and history, convincingly suggested bilateral cranial cruciate disease with a complete tear on the left and a possible associated meniscal tear. Regardless, radiographs alone do not provide a definitive diagnosis, because the ligament cannot be visualized. Definitive diagnosis relies on exploration of the joint (open or arthroscopic approach: Fig. 8.3-2) or visualization of the ligament using imaging techniques—e.g., ultrasound, CT scan, or MRI (Fig. 8.3-3). In this case, arthroscopic examination of the left stifle revealed a complete tear of the cranial cruciate ligament (Fig. 8.3-2A) and a "bucket-handle" tear of the caudal horn of the medial meniscus (Fig. 8.3-2B) with moderate synovitis (Fig. 8.3-2C) throughout the joint.

Diagnosis

Left cranial cruciate ligament rupture and "bucket-handle" tear of the caudal horn of the medial meniscus

Treatment

After left arthroscopic stifle joint exploration, the remnants of the cranial cruciate ligament were arthroscopically debrided along with a partial medial meniscectomy removing only the torn portion of the ligament. An incision on the medial tibia was made, and a tibial plateau leveling osteotomy was performed to neutralize cranial tibial thrust. In this case, the surgeon choose the tibial plateau leveling osteotomy procedure; however, there are a variety of other well-described techniques in the dog for the management of stifle instability.

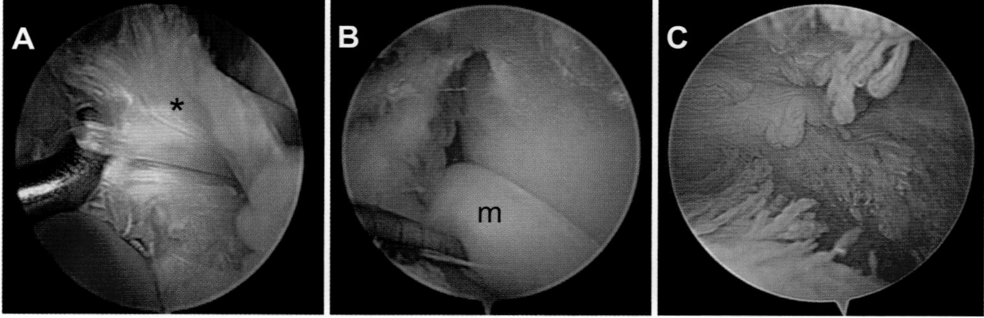

FIGURE 8.3-2 Arthroscopic images of the left stifle. (A) Complete tear of the cranial cruciate ligament is identified with the arthroscopy probe (*). (B) Bucket handle tear of the caudal horn of the medial meniscus. The probe is pulling the torn portion cranially in the joint between the femur and tibia (m). (C) Moderate synovitis of the left stifle joint capsule (entire arthroscopic view). Key: *, complete tear of the cranial cruciate ligament; *m*, bucket handle tear of the caudal horn of the medial meniscus.

TIBIAL PLATEAU LEVELING OSTEOTOMY (TPLO)

The TPLO is a surgical procedure intended to neutralize cranial tibial thrust in dogs with a cranial cruciate ligament-deficient stifle. The procedure involves making a radial osteotomy in the proximal tibia and rotating the bone segment to decrease the tibial plateau slope or tibial plateau angle. This decrease in tibial plateau angle counterbalances tibiofemoral shear forces during weight-bearing in the cranial cruciate-deficient stifle. This results in a dynamic or functionally stable stifle joint when the dog is weight-bearing. There are many techniques reported for stifle stabilization following cranial cruciate ligament injuries; TPLO was the technique chosen for this clinical case.

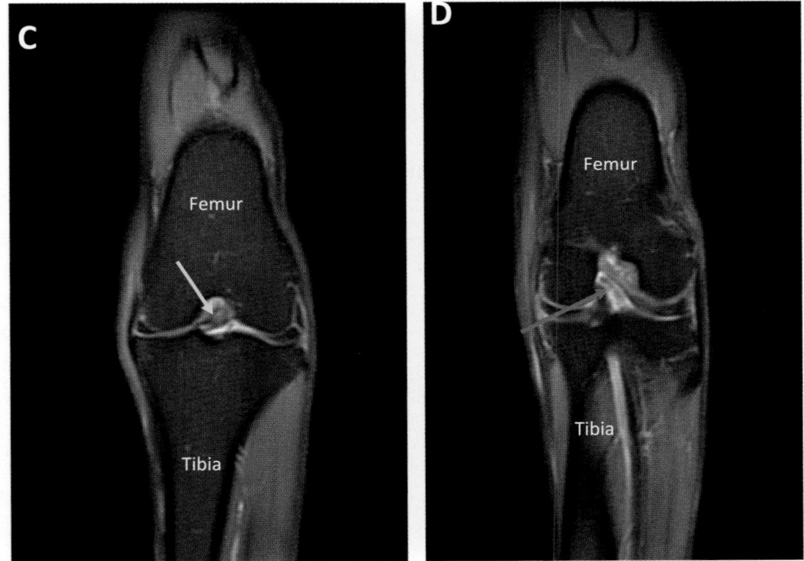

FIGURE 8.3-3 Magnetic resonance imaging of a normal canine stifle joint. (A) Sagittal image of the stifle showing the cranial cruciate ligament (*yellow arrow*). (B) Sagittal image of the stifle with the normal shape of the meniscus (*blue arrow*) highlighting the improved congruity between the femur and tibia. (C) Intact cranial cruciate ligament identified on the coronal or dorsal image (*yellow arrow*). (D) Intact caudal cruciate ligament on the coronal or dorsal image (*red arrow*).

Anatomical features in canids and felids

Introduction

The stifle is categorized as a complex, condylar, synovial joint with interposition of the main femorotibial articular surfaces by fibrocartilage structures called menisci, along with numerous intra- and extra-articular ligaments. The joint forms 3 compartments that all communicate in companion animals: the medial and lateral femorotibial joints and the femoropatellar joint. This case reviews the bones and soft tissues of the stifle joint, with regional blood supply, lymphatics, and innervation. The feline stifle has similar anatomical features.

Function

The main motion of the stifle joint is flexion and extension; however, cranial and caudal displacement, compression and distraction, internal and external rotation, varus and valgus angulation, and medial and lateral translation all can occur in the joint.

Stifle joint

The main articulation of the stifle joint is the **femorotibial joint**. This joint is comprised of convex, rounded **femoral condyles** and the slightly convex, flat surface of the proximal **tibia** (the tibial plateau). The **proximal tibia** is divided into the **medial** and **lateral condyles** where they articulate with associated medial and lateral femoral condyles. The **intercondylar eminence** of the tibia separates these 2 articular surfaces of the tibia, and the **intercondylar tubercles** project from the eminence to articulate with the femur on its abaxial surfaces. The congruity of the femoral condyles and the proximal tibia is improved by the presence of the medial and lateral menisci (discussed below).

The **femoropatellar joint** is comprised of the femoral trochlea bounded by medial and lateral ridges and the concave surface of the patella. The **patella** is the largest sesamoid bone in the body and is found within the tendon of insertion of the **quadriceps muscle** group (rectus femoris, vastus medialis, intermedius, and lateralis muscles). The patella functions to improve the extension mechanism of the quadriceps muscle group by lengthening its moment arm.

Ligaments of the stifle

The 4 femorotibial ligaments include 2 collateral ligaments (medial and lateral) and 2 cruciate ligaments (cranial and caudal) that provide the main support to the stifle joint. The **medial collateral ligament** is attached to the **medial femoral epicondyle** and—as it tracks distally—it blends with the joint capsule, including a strong attachment to the medial meniscus. The **lateral collateral ligament** attaches to the **lateral epicondyle** of the femur and then tracks distally to attach to the head of the **fibula** and the **lateral tibial condyle**. The 2 cruciate ligaments are intra-articular, but extra-synovial. The **cranial** and **caudal cruciate ligaments** are named for their point of insertion on the tibia. The cranial cruciate ligament attaches on the caudomedial aspect of the lateral femoral condyle and runs cranially, medially, and distally to its cranial attachment in the **intercondylar fossa** of the tibia. The caudal cruciate ligament attaches to the lateral surface of the medial femoral condyle, tracks caudodistally, and attaches to the medial edge of the **popliteal notch** of the tibia (caudal to the attachment of the cranial cruciate ligament, Fig. 8.3-4).

> Since the cruciate ligaments are extra-synovial structures, tearing results in extra-synovial tissues entering the joint, which likely contributes to the synovitis seen in cranial cruciate disease. This was the rationale for debriding the remaining torn portions of the cranial cruciate ligament in this case (Fig. 8.3-4).

Menisci and their ligaments

The **menisci** are 2 "C"-shaped fibrocartilaginous structures interposed between the femur and the tibia. They are crescent-shaped and also triangular in cross section. The peripheral portion is thick and attached to the joint capsule, which then tapers axially to a thin free edge. The proximal portion of each meniscus is concave and provides congruity with the rounded femoral condyles, whereas the distal portion of each meniscus is flat and attached to the tibia.

The **cranial** and **caudal meniscotibial ligaments** attach each meniscus to the tibial plateau, and the **intermeniscal ligament** attaches the cranial horns of the medial and lateral menisci to each other (Figs. 8.3-4–8.3-6). The

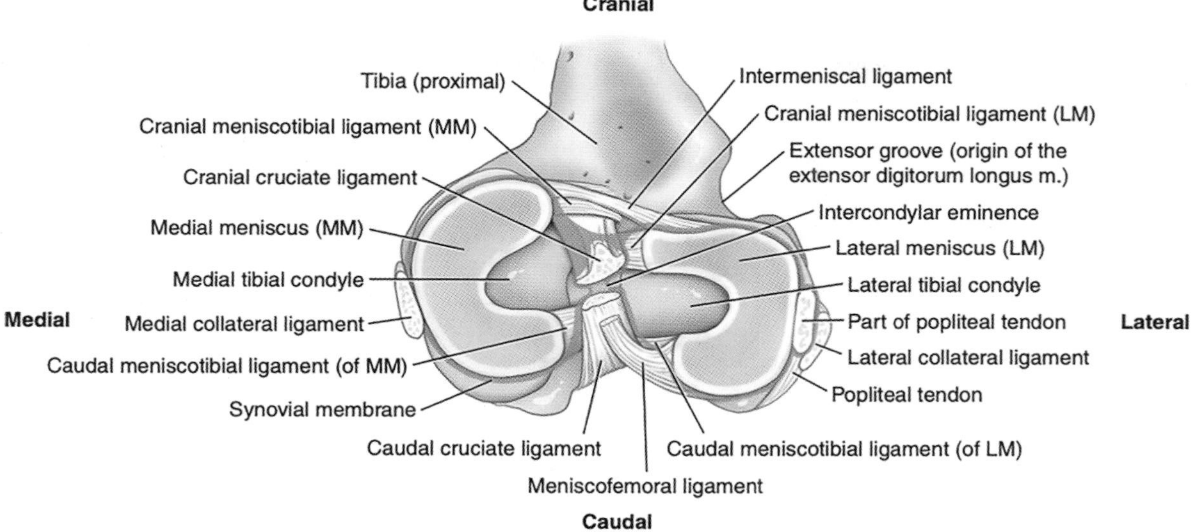

FIGURE 8.3-4 Tibial plateau of the canine stifle with the identification of joint ligaments and synovial membrane coverage.

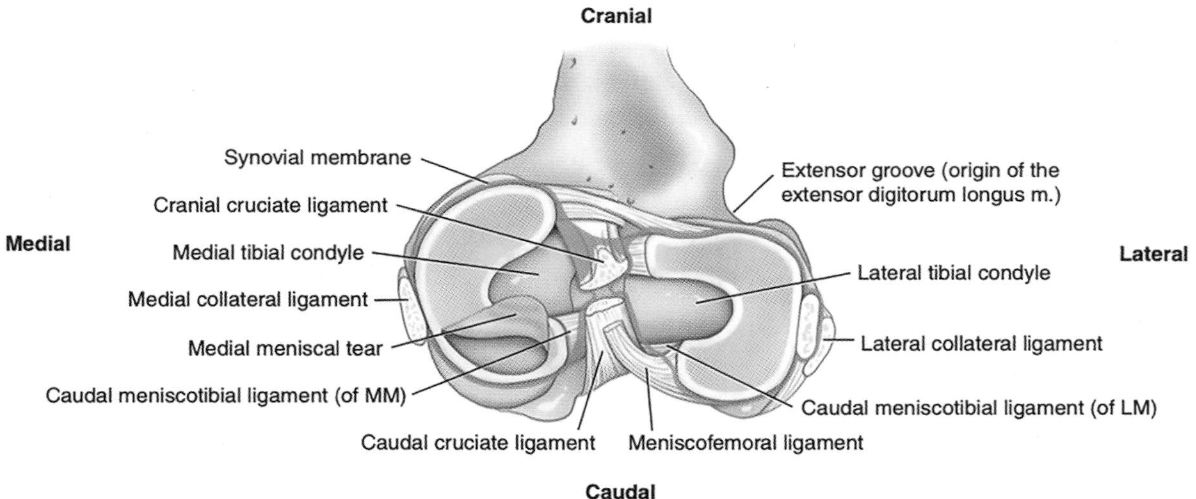

FIGURE 8.3-5 Tibial plateau of the canine stifle joint with a common site of meniscal injury to the caudal horn of the medial meniscus. This type of longitudinal tear in the caudal horn of the medial meniscus with the detachment of the central portion and flipping cranially is called a bucket handle meniscal tear. Key: *LM*, lateral meniscus; *MM*, medial meniscus.

menisci have numerous functions, the most significant of which is the ability to distribute load over a larger area of the joint. The collagen fibers of each meniscus are circumferentially arranged in a crescent-shaped structure.

The radial forces created by weight-bearing—i.e., when the joint is loaded—are resisted by the tensile forces of this circumferential collagen arrangement—called "hoop stress." This is a key function of the meniscus, allowing for redistribution of load over a larger area of the joint, therefore reducing the contact stresses in the joint and preserving the articular cartilage.

The meniscal body attachment varies between the medial and lateral menisci. The **medial meniscus** is firmly attached to

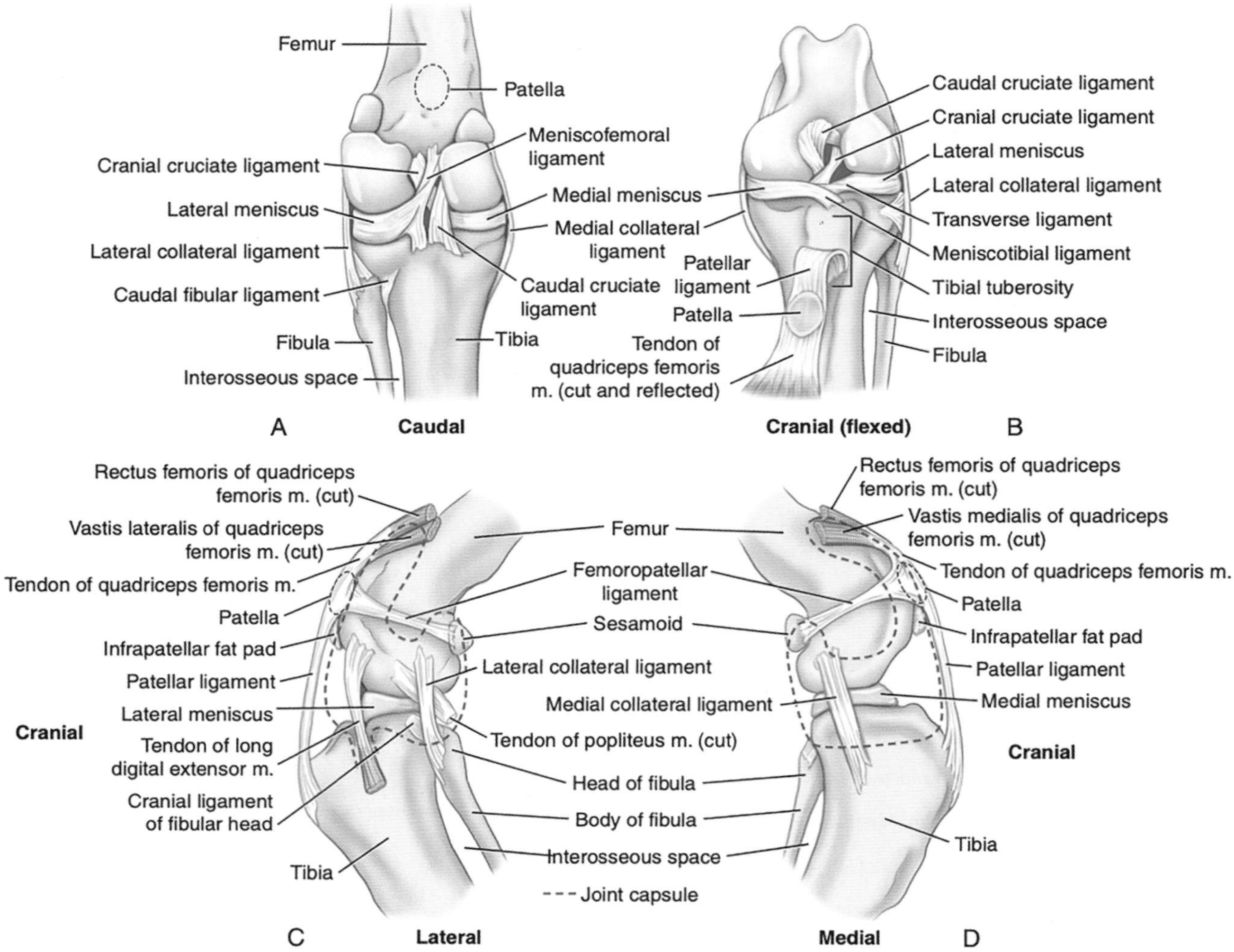

FIGURE 8.3-6 (A) Caudal, (B) cranial, (C) lateral, and (D) medial views of the canine left stifle joint.

CRUCIATE TEARS—PREDISPOSING CONDITIONS

Other diseases of the canine stifle joint can predispose the cranial cruciate ligament to injury or trauma. A complete evaluation of the stifle joint is necessary to determine an appropriate treatment plan to address all clinical disease and improve the prognosis. Patella luxation is a common developmental orthopedic condition in dogs, and when the patella is luxated, there is increased stress placed on the cranial cruciate ligament, likely contributing to its injury over time. In addition, any condition furthering increased inflammation in the joint can result in cranial cruciate ligament pathology, such as OCD, immune-mediated diseases, sepsis, or other infectious/inflammatory diseases. Neoplasia of joint capsule or the surrounding soft tissues can result in increased inflammatory mediators in the joint, ending in ligament injury. Primary bone tumors of the distal femur or proximal tibia can lead to injury at either the origin or insertion of the cranial cruciate ligament.

the medial collateral ligament and joint capsule via the **coronary ligament**, whereas the lateral meniscus lacks these attachments. The **lateral meniscus** has fewer attachments to the joint capsule because the popliteus muscle tendon is interposed between the joint capsule and the lateral meniscus. In addition, the lateral meniscus has an attachment to the femur called the **meniscofemoral ligament** (Figs. 8.3-4 and 8.3-6A). The lateral meniscus is more mobile due to its looser attachment to the tibia, its relationship with the popliteal muscle tendon, and the meniscofemoral ligament.

Because the lateral meniscus moves with the femur, it is less likely to be severely injured requiring surgery when there is instability in the stifle joint as compared with the firmly attached medial meniscus. This was demonstrated in this clinical case, in which the medial meniscus had a characteristic longitudinal tear in the caudal horn that was caused by repetitive trauma of the femoral condyle to the firmly attached caudal medial meniscus. This longitudinal tear can then flip back and forth in the joint (resembling a bucket handle) causing a clicking or clunking sound during the range of motion.

Regional blood supply, lymphatics, and innervation

The blood supply to the cruciate ligaments comes mainly from the soft tissues with insignificant contributions from their osseous attachments. The main blood supply to both cruciate ligaments is from branches of the **middle genicular artery** (arising from the **popliteal artery**), which penetrates the caudal joint capsule to enter the joint centrally and run cranially between the cruciate ligaments.

Lymphatic drainage may be to small local **femoral lymph nodes** and thence to the popliteal lymph nodes; directly to the **popliteal lymph nodes** and thence to the **superficial inguinal lymph nodes**; or directly to the superficial inguinal lymph nodes.

The **medial articular nerve**, a branch from the **saphenous nerve**, contributes the majority of innervation to the stifle joint. The nerves are in the vascularized synovial tissue covering the cruciate ligaments and have various types of nerve endings. The largest number of mechanoreceptors are found in the proximal third of the cranial cruciate ligament.

Blood supply to the menisci is limited and most of the axial portion of the meniscus is considered avascular. The **medial** and **lateral genicular arteries**, which supply the blood to the synovial and capsular tissues, form **perimeniscal plexi** at the peripheral border of the meniscus and supply only 10%–25% of the meniscus. Therefore, most of the axial portion of the menisci rely on nutrition from the synovial fluid through diffusion and mechanical pumping.

Tearing of the meniscus mainly occurs in the avascular portion of the menisci and, subsequently, its healing capabilities are limited. Therefore, treatment of bucket handle medial meniscal tears generally involves removing the torn portion of the meniscus (Fig. 8.3-5).

Innervation of the menisci is not well-delineated; however, nerve supply from the perimeniscal tissue radiates into the periphery of the meniscus. This innervation appears to play mostly proprioceptive and mechanoreceptive roles.

Selected references

[1] Evans HE. The skeleton, arthrology, the muscular system. In: Evans HE, editor. Miller's anatomy of the dog. 3rd ed. Philadelphia: Saunders; 1993. p. 122.
[2] Muir P. Advances in the canine cranial cruciate ligament. American College of Veterinary Surgeons Foundation and Wiley-Blackwell; 2007.
[3] Arnoczky SP, Marshall JL. The cruciate ligaments of the canine stifle: an anatomical and functional analysis. Am J Vet Res 1977;38:1807.
[4] Luther JK, Cook C, Constantinescu I, et al. Clinical and anatomical correlations of the canine meniscus. J Exp Med Surg Res 2007;14:5.
[5] de Rooster H, De Bruin T, van Bree H. Morphologic and functional features of the canine cruciate ligaments. Vet Surg 2006;35:769.

CHAPTER 8

CASE 8.4

Tibial Fracture

Lane A. Wallett[a], Cathy Beck[b], and Helen M.S. Davies[c]
[a]Integrative Physiology and Neuroscience, Washington State University College of Veterinary Medicine, Pullman, Washington, US
[b]Veterinary Hospital, The University of Melbourne, Melbourne, Victoria, AU
[c]Veterinary BioSciences, The University of Melbourne, Melbourne, Victoria, AU

Clinical case

History

A 3-month-old male Cavoodle (Cavalier King Charles Spaniel–Poodle cross) was examined for acute non-weight-bearing lameness of the right pelvic limb. The lameness suddenly appeared earlier the same day, following a fall while playing. The owners reported that the patient became tangled with a playmate mid-fall and immediately after began yelping and limping.

Physical examination findings

> The crus is the portion of the pelvic limb between the stifle (knee) and the tarsus (hock). Its primary weight-bearing bone is the tibia. The much narrower fibula accompanies the tibia on the lateral aspect of the crus.

On presentation, the patient was unable to bear weight on the right pelvic limb, and the right crus (L. stifle to tarsus) was moderately swollen. No wounds were seen on examination of the skin. On gentle palpation, the swelling appeared to be caused by a large, diffuse hematoma. In addition, palpation elicited a pain response and appreciable crepitus (L. a grinding, crunching, or crackling sensation/sound). Further manipulation was discontinued because diagnostic radiography was believed a logical next step to rule out a fracture.

Differential diagnoses

Fracture of the tibia and/or fibula

> **RADIOGRAPHIC DESCRIPTION OF FRACTURE ORIENTATION**
>
> In describing fracture fragments and their radiographic position relative to one another, the arrangement of the pieces of bone is about the location of the distal fragment compared with the proximal fragment: (1) medial vs. lateral; (2) cranial vs. caudal; (3) dorsal vs. palmar; and (4) dorsal vs. plantar. Using this uniform descriptive ordering, veterinarians have consistency in the use of anatomical language.

Comparative Veterinary Anatomy: A Clinical Approach. https://doi.org/10.1016/B978-0-323-91015-6.00052-2

Diagnostics

Radiographs of the right crus revealed a closed, complete, noncomminuted, minimally displaced, spiral fracture of the tibia (Figs. 8.4-1 and 8.4-2).  At the fracture site, the distal fragment was displaced cranially and laterally, relative to the proximal fragment. However, both fragments remained axially positioned and did not disrupt the surrounding soft tissues or the skin—i.e., this was a "closed" fracture. The fibula was intact, and the growth centers of the distal femur, tibia, fibula, and tarsus were normal for the dog's age.

A comminuted fracture is one in which there are multiple pieces of bone. A noncomminuted fracture has just 2 fragments—in this case, proximal and distal to the fracture site.

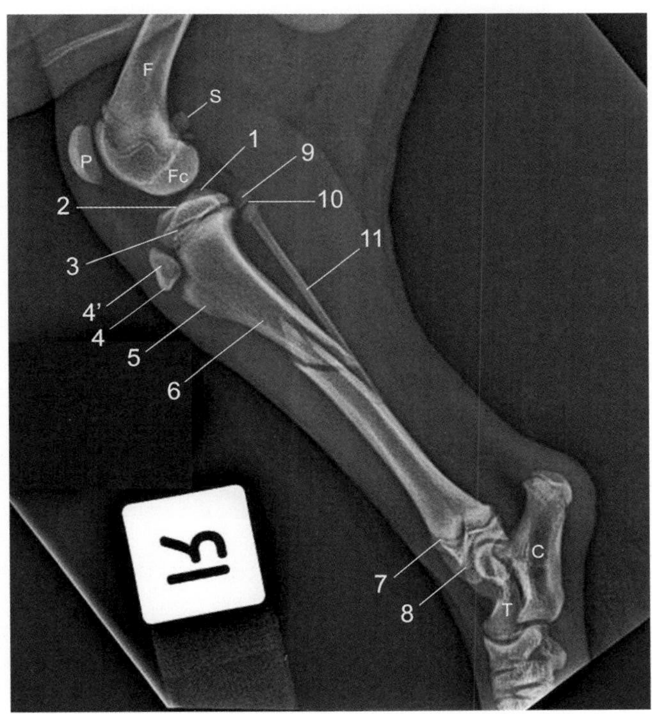

FIGURE 8.4-1 Lateromedial radiograph of the right pelvic limb of the dog in this case, depicting a closed, complete, noncomminuted, minimally displaced, spiral fracture of the tibial diaphysis (6). The craniocaudal projection, presenting the orthogonal view of the fracture, is shown in Fig. 8.4-2. Number key for both radiographs: 1, intercondylar eminence; 1′, intercondylar tubercle (lateral or medial, as applicable) forming the intercondylar eminence; 2, margin of the tibial condyle (lateral/medial); 3, proximal tibial physis; 4, tibial tuberosity; 4′, ossification center of the tibial tuberosity; 5, cranial border (formerly tibial crest); 6, tibia (diaphysis); 7, distal tibial physis; 8, distal tibial epiphysis; 8′, medial malleolus; 9, head of the fibula; 10, proximal fibular physis; 11, fibula (diaphysis); 12, distal fibular physis; 13, lateral malleolus. Letter key for both radiographs: *C*, calcaneus; *F*, femur (diaphysis); *Fc*, femoral condyle (lateral/medial); *L*, lateral; *M*, medial; *P*, patella; *S*, sesamoid bone associated with the origin of the gastrocnemius muscle, lateral/medial head; *T*, talus. Note: Not all structures are apparent on both projections.

CLASSIFICATION OF OPEN FRACTURES

Classifying open fractures is a guide for surgical planning and is helpful in determining a prognosis for the injury. The 3 types of open fractures are classified as follows: (1) a Type I open fracture penetrates the skin from within; (2) a Type II open fracture has a skin wound < 1 cm (< 0.4 in.) in size and the wound communicates with the fracture; and (3) a Type III open fracture is a high-energy injury with major soft tissue trauma that may or may not include bone loss. Type III injuries are further classified as: (a) a IIIa open fracture requiring no major reconstructive soft tissue procedure to cover the bone; (b) a IIIb open fracture, where the unaffected soft tissues are inadequate for closure and require a reconstructive procedure to cover the exposed bone; and (c) a IIIc open fracture, which has arterial blood vessel injury needing repair; generally, these patients are candidates for limb amputation because of the guarded to poor prognosis for limb preservation.

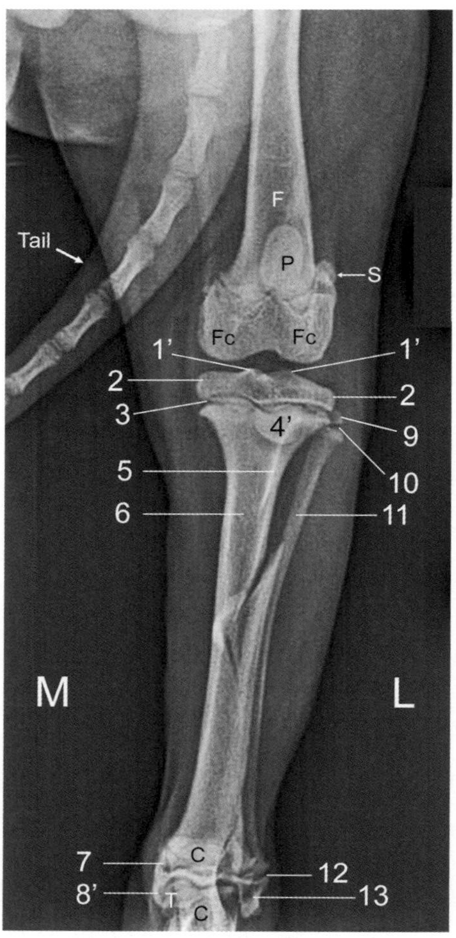

FIGURE 8.4-2 Craniocaudal radiograph of the right pelvic limb of the dog in this case. See Fig. 8.4-1 for number and letter keys.

Diagnosis

Fracture of the tibia

Treatment

Under general anesthesia, and with the dog in right lateral recumbency, the fracture was stabilized using threaded bone pins connected to an external skeletal fixator. Stab incisions were made through the skin, subcutis, and periosteum on the caudomedial aspect of the crus, and 2 pairs of bone pins were placed transversely through both cortices of the tibia. One pair of pins was placed proximal to the fracture and the other pair distal to the fracture, taking care to avoid the fracture site by at least 1 cm (0.4 in.) proximally and distally. Externally, a vertical bar connected the pins to provide stability to the tibia during bone healing. Post-operative radiographs showed adequate fracture alignment (Fig. 8.4-3).

The fixation device was applied to the caudomedial aspect of the crus because the tibia is largely free of muscular attachments on its medial surface. In addition, a slight caudal-to-cranial trajectory allows the surgeon to avoid placing the pins into or through the fibula. However, it is important to avoid the cranial and caudal branches of the saphenous artery and medial saphenous vein on the medial aspect of the proximal tibia.

Radiographs repeated 4 weeks after surgery showed fracture healing to be nearly complete; only a thin radiolucent line remained within the substantial bony callus (Fig. 8.4-4). The external fixator and bone pins were removed under general anesthesia. The dog made a complete recovery and continued to develop and grow normally.

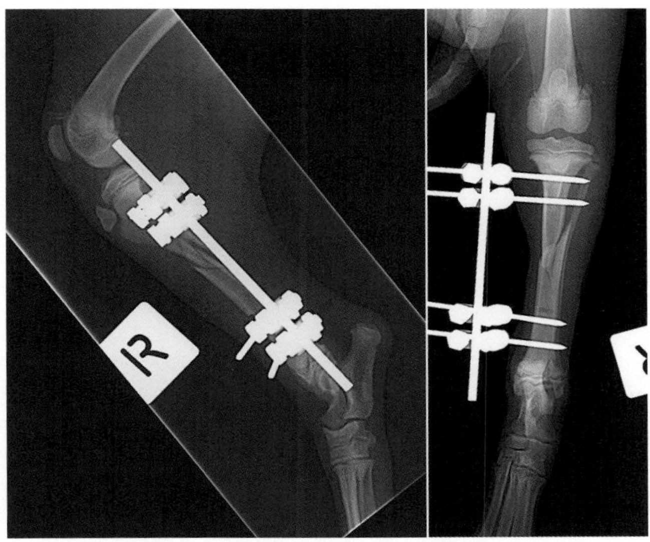

FIGURE 8.4-3 Lateromedial and craniocaudal radiographs of the patient immediately after surgery. Although the fracture fragments remain slightly displaced, good bone repair can be expected with the stability provided by the external skeletal fixator. Any bodyweight borne by this limb is transmitted from the proximal pins to the distal pins via the external device, completely bypassing the fractured diaphysis.

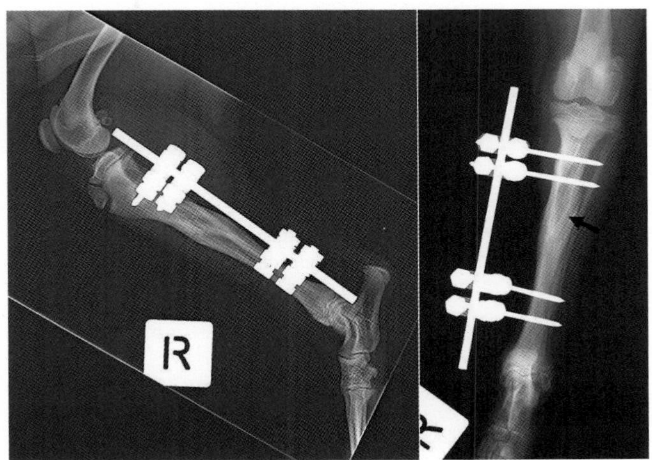

FIGURE 8.4-4 Lateromedial and craniocaudal radiographs of the patient 4 weeks after surgery. The tibial fracture is almost completely healed, and the integrity of the tibia has been restored. Only a faint radiolucent line marking the position and orientation of the fracture is visible within the bony callus on the craniocaudal projection (*arrow*). Laterally, the bony callus extends to the medial aspect of the fibula but does not impinge on the fibula. The growth centers of the femur, tibia, fibula, and tarsus are normal for the dog's age.

Anatomical features in canids and felids

Introduction

The term crus refers to the region of the pelvic limb between the femorotibial joint (stifle or knee) proximally and the tarsus (hock) distally (Fig. 8.4-5). These articulations are described separately in Cases 8.3 and 8.5, respectively. The crus comprises 2 bones: the tibia and fibula. The fibula lies lateral to the tibia for the entire length of the crus.

The joint commonly called the stifle or knee also includes the femoropatellar joint, but because the patella does not articulate with the tibia or fibula, only the femorotibial component of the joint is described in this discussion.

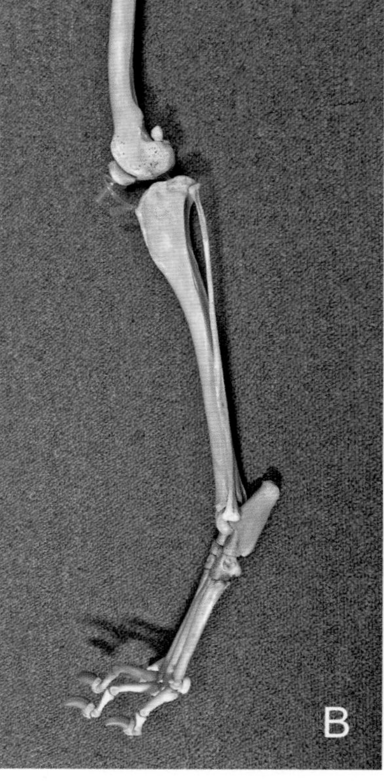

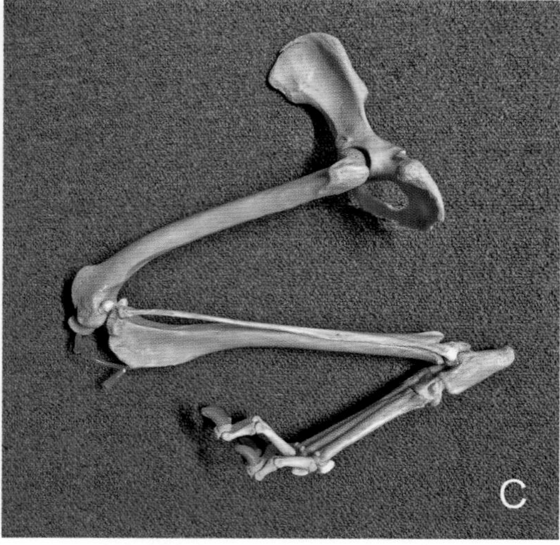

FIGURE 8.4-5 Left pelvic limb of a mature dog, lateral aspect. The bones are positioned to simulate normal stance (A), full extension (B), and flexion (C) of the stifle and tarsus. The pair of hypodermic needles at the cranial aspect of the stifle are securing in place a short length of plastic tubing that represents the distal portion of the patellar tendon, also called the patellar ligament, in dogs and cats. Key: *jt*, joint. Note: the stifle also includes the femoropatellar joint (not labeled) at the cranial aspect of the stifle.

Function

The tibia facilitates weight-bearing and locomotion by providing a rigid strut between the stifle and the tarsus. Both its proximal and distal articulations limit the planes of movement through this part of the pelvic limb essentially to flexion-extension in the sagittal plane (Fig. 8.4-5), thus contributing to the lever-like action of the pelvic limb during locomotion. The fibula is long and thin, and while unsuited for bearing much weight, it serves as an attachment site for several muscles.

Tibia (Figs. 8.4-5 and 8.4-6)

The largest bone in the crus, the **tibia** is the primary weight-bearing bone in this part of the limb. It articulates proximally with the lateral and medial condyles of the **femur** and distally with the talus, which is the more medial and cranial of the two proximal tarsal bones. Laterally, the tibia articulates with the **fibula** at both proximal and distal extremities.

The proximal surface of the tibia is roughly triangular in cross section, with marginal eminences formed by the **tibial tuberosity** cranially (described later) and the margins of the **medial** and **lateral condyles** abaxially. The tibial condyles each have a broad, flattened proximal surface that is slightly convex sagittally and concave laterally.

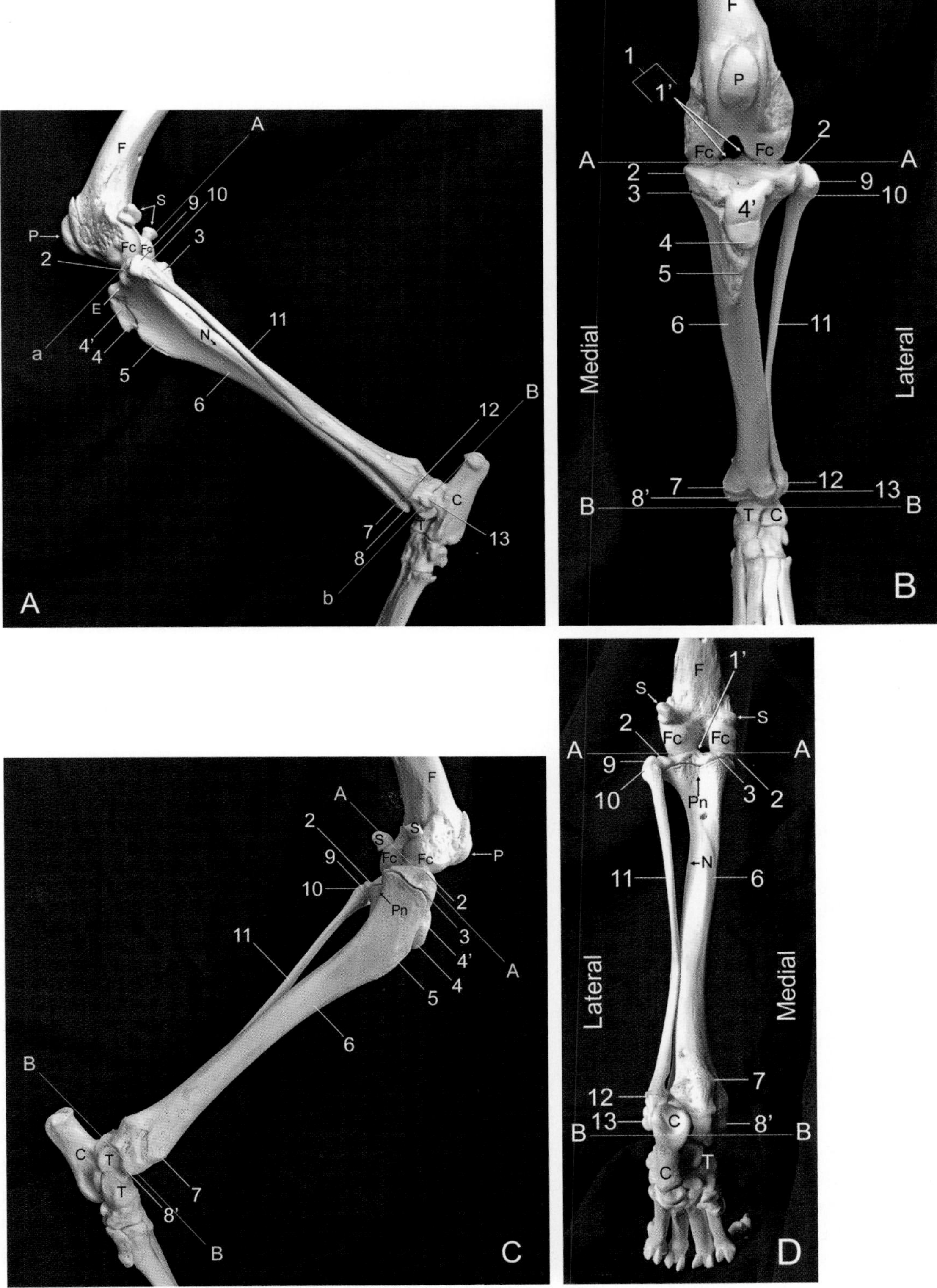

FIGURE 8.4-6 Left pelvic limb of an immature dog; the growth centers are not completely ossified. (A) Lateral aspect; (B) cranial aspect; (C) medial aspect; (D) caudal aspect. As much as possible, the same number and letter keys used in Fig. 8.4-1 are used in this figure. Number key: 1, Intercondylar eminence; 1′, intercondylar tubercle (lateral or medial, as applicable) forming the intercondylar eminence; 2, margin of the tibial condyle (lateral/medial); 3, proximal tibial physis; 4, tibial tuberosity; 4′, ossification center; 5, cranial border (formerly tibial crest); 6, tibia (diaphysis); 7, distal tibial physis; 8, distal tibial epiphysis; 8′, medial malleolus; 9, head of the fibula; 10, proximal fibular physis; 11, fibula (diaphysis); 12, distal fibular physis; 13, lateral malleolus. Letter key: *A*, femorotibial joint; *B*, tarsocrural joint; *C*, calcaneus; *E*, extensor groove; *F*, femur (diaphysis); *Fc*, femoral condyle (lateral/medial); *N*, nutrient foramen (tibia); *P*, patella; *Pn*, popliteal notch; *S*, sesamoid bone associated with the origin of the gastrocnemius muscle, lateral/medial head; *T*, talus. Notes: not all structures are visible on all views. The 2 circular drill holes in the caudal cortex of the tibia (image D) were made during preparation and mounting of the skeleton; they are not to be confused with *N*, which is located more laterally and is elliptical in shape.

The menisci (singular, meniscus) of the femorotibial joint are somewhat restricted, or held in position, by the shape of the proximal articular surface of the tibia, as well as by the strong meniscotibial ligaments that secure them to the proximal tibia.

The intercondylar eminence and tubercles are not readily evident in Fig. 8.4-6 because they are obscured by the axial margins of the femoral condyles. However, they are clearly seen in standard radiographic projections of the stifle as in Figs. 8.4-1 and 8.4-2.

Together, the proximal surface of the tibial condyles form the **proximal articular surface** of the tibia. This surface articulates with the medial and lateral condyles of the femur via the pair of fibrocartilaginous menisci (medial and lateral).

The **intercondylar eminence** is a central peak between the tibial condyles. It includes the offset **medial** and **lateral intercondylar tubercles**, which articulate abaxially with their respective femoral condyles. A pair of depressions immediately cranial and caudal to the intercondylar eminence, the **cranial** and **caudal intercondylar areas**, provide attachment sites for meniscal ligaments. In addition, the cranial cruciate ligament inserts distally on the cranial intercondylar area. The medial and lateral menisci and the cranial and caudal cruciate ligaments are described in Case 8.3.

The proximal aspect of the tibia also contains 2 conspicuous features associated with the musculature of the crus. First, the tendon of origin of the **long digital extensor** muscle crosses distally over the lateral condyle in the **extensor groove** (Fig. 8.4-6A). Second, the caudal aspect of the proximal tibia is marked by the **popliteal notch**, which allows the tendon of origin of the **popliteus** muscle to wrap caudally around the tibia (Fig. 8.4-6D). The belly of the popliteus muscle covers the proximal third of the tibia on its caudal aspect, having originated on the lateral femoral condyle by a long, thin tendon that courses over the stifle joint capsule, beneath the lateral collateral ligament of the stifle. The popliteal tendon contains a small **sesamoid bone** where it crosses the distal edge of the lateral meniscus.

The distinct ridge of bone comprising the cranial border of the proximal tibia was formerly called the "tibial crest." Although cranial border is the correct term, tibial crest is still widely used in clinical practice.

The most prominent features on the cranial aspect of the proximal tibia are the **tibial tuberosity** craniodorsally and the **cranial border** immediately distal to it. Both serve as attachment sites for the extensor muscles of the stifle and digits and the flexors of the tarsus. Of note, the patellar ligament (described later) inserts onto the tibial tuberosity. In contrast, the tarsal extensors and digital flexors originate on the caudal aspect of the proximal tibia and the distal femur.

Distal to these muscular attachments, the **body** or **diaphysis**of the tibia transitions from the triangular shape of its proximal surface to a more cylindrical shape. Caudolaterally, at the junction between the proximal and middle thirds, a **nutrient foramen** provides a route for blood vessels to reach the medullary cavity. The caudodistal surface of the diaphysis may also contain a **vascular groove**, which extends to the distal extremity of the tibia.

Medially, the tibial diaphysis is largely free of muscle and tendons, making it readily accessible for surgical approaches to the tibia.

Proximally, the caudolateral aspect of the **lateral tibial condyle** articulates with the **head of the fibula** (described later). Apart from this articulation, the tibia and fibula remain separate in their proximal halves. They are closely associated in their distal halves, so the lateral surface of the tibia in this region may be referred to as the **interosseous border**. Both the fibular diaphysis and the caudolateral surface of the tibia widen and flatten where they are united by an **interosseous membrane**.

Distally, the tibia broadens and forms a pair of concave articular surfaces, or **cochlea** (singular) (L. spiral, snail shell), separated by an **intermediate ridge**. Each cochlea, lateral and medial, articulates with its respective trochlear ridge on the talus to form the **tarsocrural joint** (see Case 8.5.)

Medially, the distal extremity of the **tibia** is supported by a process cranially and a semilunar notch caudally, the combination of which is called the **medial malleolus**. The lateral malleolus is the distal part of the fibula (described next); it articulates with the lateral surface of the distal tibia and the lateral surface of the trochlea of the talus at the tarsus.

The tibia has 4 growth centers: the proximal and distal physes, the tibial tuberosity, and the medial malleolus (see Figs. 8.4-1 and 8.4-2). In dogs, the medial malleolus usually is fully ossified (closed) by 5 months of age. The proximal and distal physes typically close by 10–11 months of age and the tibial tuberosity a few months earlier, usually by 8 months, although it ranges from 5 to 18 months for all 3 sites, depending on breed (later in larger breeds). In cats, these growth centers usually close between 12 and 18 months of age, although the distal tibial physis may close as early as 9 months.

Fibula (Figs. 8.4-5 and 8.4-6)

The **head of the fibula** is its largest and most proximal part. The head articulates on its medial surface with the **lateral condyle of the tibia**. The fibular **neck** is short and indistinct, blending unnoticeably into the **body or diaphysis** of the fibula.

> The head of the fibula is located on the lateral aspect of the proximal crus, immediately distal to the femorotibial joint. It serves as a palpable landmark for multiple procedures, including the cranial drawer test performed at the stifle (see Case 8.3).

The fibular diaphysis is long and slender, running independently along the lateral aspect of the crus before rejoining the tibia in its distal half. Accordingly, the proximal half of the fibula is roughly cylindrical in cross section, while its distal half is flatter and thinner to facilitate its close association with the tibia. The craniomedial aspect of the fibula, facing the tibia, is called the **interosseous border**. The **interosseous membrane** fills the **interosseous space** and attaches the tibia and fibula along this border.

The fibula is an attachment site for several of the smaller muscles of the crus, and the lateral collateral ligament of the stifle (described later) inserts on the head of the fibula. Craniolaterally, the fibular head and body are the origins for 2 of the digital extensors (lateral digital extensor and extensor digiti I longus), and for the long (longus) and short (brevis) fibularis (peroneus) muscles, which contribute to tarsal flexion and lateral stability. Caudally,

INTRAMEDULLARY PIN PLACEMENT FOR TIBIAL FRACTURE REPAIR

Intramedullary (IM) pins are often used to stabilize tibial fractures. Normograde (proximal-to-distal) placement of the pin is preferable to retrograde placement, because it avoids injury to the articular surfaces of the stifle in dogs and to the patellar ligament in cats.

With normograde placement, the IM pin is carefully introduced into the medullary cavity of the tibia near its proximal extremity and then advanced distally across the reduced (realigned) fracture into the distal diaphysis. The preferred insertion site is on the medial aspect of the proximal tibia, midway between the medial tibial condyle and the tibial tuberosity.

With retrograde placement, the pin is introduced into the proximal fragment at the fracture site and blindly advanced until it exits through the proximal aspect of the tibia; the distal end of the pin is then maneuvered into the distal fragment and directed into the distal diaphysis.

With either approach to IM pin placement, care must be taken to avoid penetrating the distal articular surface of the tibia when advancing the pin into the distal fragment.

The muscles named fibularis were formerly called peroneus and both terms may be found interchangeably in anatomy and surgical text books. Both peroneus and fibularis derive from Greek and Latin words, respectively, for the needle-like pin or clasp of a brooch or buckle. Digiti I was formerly called hallucis—e.g., flexor hallucis longus—as the hallux is the first digit or "big toe" in humans. Given that the first digit or "dewclaw" in dogs and cats is rudimentary and is absent in the pelvic limbs in some individuals, digiti I is the more appropriate term.

the fibula is an attachment site for 2 small tarsal extensors, the caudal tibial muscle and one of the muscles that contributes to the deep digital flexor tendon (flexor digiti I longus), although both originate primarily on the caudal tibia.

The distal extremity of the fibula forms the **lateral malleolus**. It articulates medially with the tibia, the trochlea of the talus, and the craniolateral aspect of the calcaneus. The lateral malleolus is marked by 2 grooves, craniolateral and caudal; the tendon of the fibularis longus muscle rests in the craniolateral groove, while the caudal groove contains tendons of the lateral digital extensor and fibularis brevis muscles.

The fibula has 2 growth centers: the proximal and distal physes. In dogs, both are fully ossified by about 10 months of age, although it ranges from 5 to 12 months for the proximal physis and up to 17 months for the distal physis. In cats, the proximal physis closes between 12 and 17 months of age, and the distal physis a few months earlier, between 9 and 13 months.

Ligaments of the crus (Figs. 8.4-7 and 8.4-8)

The abaxial portions of the tibia and fibula are insertions proximally and origins distally for the collateral ligaments of the stifle and tarsus. The medial and lateral collateral ligaments of the stifle run from their origins on the distal femur to their distal attachments on the proximal tibia and fibula, respectively. The **medial collateral ligament**

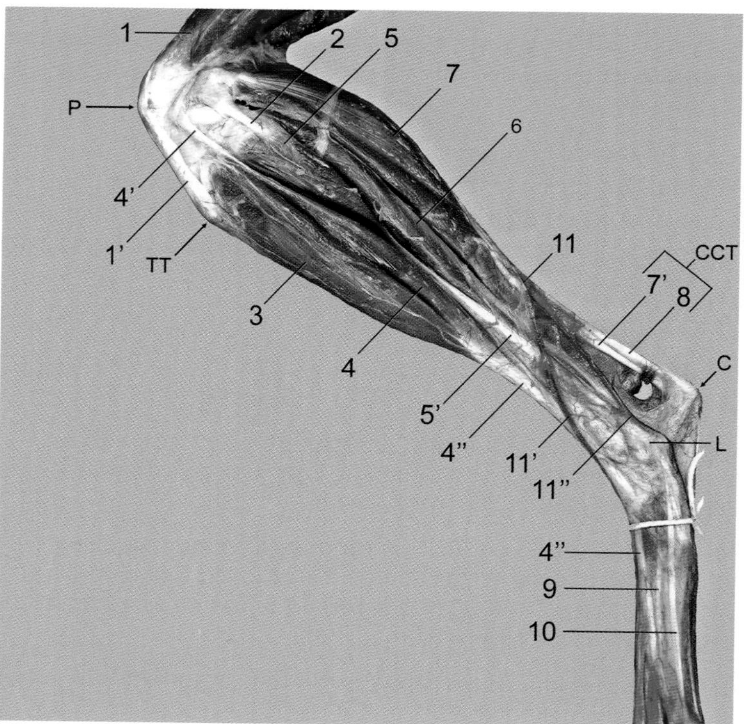

FIGURE 8.4-7 Left crus of a Greyhound, lateral aspect. The skin and most of the superficial fascia have been removed to show the superficial muscles and vessels of the lateral crus. Key: *m.*, muscle; *t.*, tendon; *v.*, vein. **Number key**: 1, Rectus femoris m.; 1′, patellar ligament (patellar t.); 2, lateral collateral ligament of the stifle; 3, cranial tibial m.; 4, long digital extensor m.; 4′, tendon of origin; 4″, tendon of insertion (continues to digits); 5, fibularis longus m.; 5′ tendon of insertion; 6, deep digital flexor m., lateral head; 7, gastrocnemius m., lateral head; 7′, gastrocnemius t.; 8, superficial digital flexor t.; 9, short digital extensor t. (extensor digitorum brevis), lateral head; 10, lateral digital extensor t.; 11, lateral saphenous v.; 11′, cranial branch; 11″, caudal branch. **Letter key**: *C*, tuber calcanei; *CCT*, common calcaneal (Achilles) tendon; *L*, lateral malleolus; *P*, patella; *TT*, tibial tuberosity.

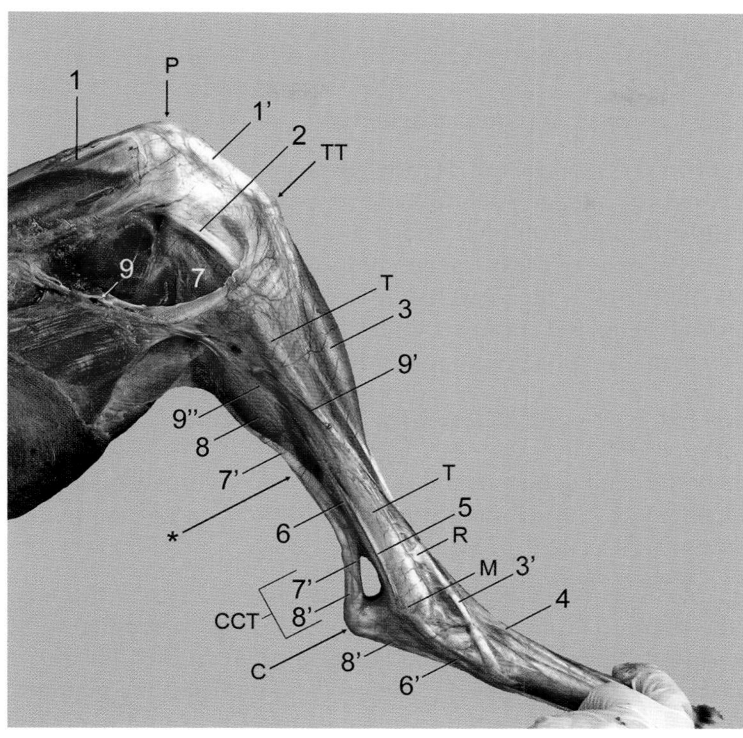

FIGURE 8.4-8 Left crus of a Greyhound, medial aspect. The skin and most of the superficial fascia have been removed to show the superficial muscles and vessels of the medial crus. Key: *a.*, artery; *m.*, muscle; *n.*, nerve; *t.*, tendon; *v.*, vein. As much as possible, the same number and letter keys as Fig. 8.4-7 are used in this figure. **Number key**: 1, Rectus femoris m.; 1′, patellar ligament (patellar t.); 2, medial collateral ligament of the stifle; 3, cranial tibial m.; 3′ tendon of insertion; 4, long digital extensor t.; 5, caudal tibial t.; 6, tendons of long digital flexor m. and flexor digiti I longus m.; 6′, united to form the deep digital flexor t.; 7, gastrocnemius m., medial head; 7′, gastrocnemius t.; 8, superficial digital flexor m.; 8′ tendon of insertion; 9, saphenous a., medial saphenous v., and saphenous n.; 9′ cranial branch of a. and v.; 9″, caudal branch of a. and v. **Letter key**: *C*, tuber calcanei; *CCT*, common calcaneal (Achilles) tendon; *M*, medial malleolus; *P*, patella; *R*, extensor retinaculum; *T*, tibia (diaphysis); *TT*, tibial tuberosity. *Asterisk*: where the gastrocnemius and superficial digital flexor tendons switch position.

inserts just distal to the medial condyle of the tibia, while the **lateral collateral ligament** inserts on the head of the fibula. As it passes over the femorotibial joint, the medial collateral ligament fuses with the medial meniscus. The lateral collateral ligament remains independent of the lateral meniscus—the tendon of the popliteus muscle passing between them as it wraps caudally around the limb. Both menisci are anchored to the proximal articular surface of the tibia via multiple meniscal ligaments described in Case 8.3.

> Because the medial collateral ligament and meniscus are fused, the medial meniscus is significantly more vulnerable than its lateral counterpart to tearing injury associated with excessive twisting and sliding of the femorotibial joint surfaces when the cranial cruciate ligament is ruptured.

Distally, the **medial** and **lateral collateral ligaments** of the tarsus originate on the medial and lateral malleolus, respectively, as described in Case 8.5.

On the cranial surface of the distal tibia lies the obliquely oriented **crural extensor retinaculum**. This retinaculum binds the tendons of insertion of the cranial tibial, long digital extensor, and extensor digiti I longus muscles, preventing the tendons from slipping laterally or pulling away from the surface of the tarsus when their muscles contract.

Proximal muscles involving the crus (Figs. 8.4-7 and 8.4-8)

Many muscles that act on the stifle have their distal attachments in the crus or track distally over the crus to insert on the tarsus. These muscles are described and illustrated in Case 8.2. The primary extensors of the stifle are the

quadriceps femoris muscles, comprising rectus femoris and lateral, intermediate, and medial vastus. This muscle group inserts via a common tendon—the patellar tendon on the tibial tuberosity. The patella is the sesamoid bone of this muscle group and its tendon of insertion. The portion of the patellar tendon that extends between the patella and the tibial tuberosity is also known as the **patellar ligament**.

Flexors of the stifle—the **superficial digital flexor**, **gastrocnemius**, **biceps femoris**, **semitendinosus**, and **gracilis** muscles—contribute to the **common calcaneal tendon** and insert on the tuber calcanei. These muscles are the main extensors of the tarsus. The superficial digital flexor tendon attaches securely to the calcaneal tuberosity and then continues distally on the plantar surface of the calcaneus and the metatarsus. In the metatarsal region, it divides into 4 tendons, each of which inserts onto the proximal interphalangeal joint of the associated weight-bearing digit (digits 2–5). The superficial digital flexor can act to flex the digits, but mostly it stabilizes the stifle, tarsal, metatarsophalangeal, and proximal interphalangeal joints during weight-bearing.

Muscles arising on the crus (Figs. 8.4-7 and 8.4-8)

The muscles of the crus occupy the cranial, lateral, and caudal surfaces of the tibia and fibula; the medial surface of the tibia is free of muscle attachments. The actions of the crural muscles may be complex, involving multiple joints, and do not group functionally as well as they do by location.

The flexors of the tarsus and extensors of the digits lie on the cranial and lateral surfaces of the crus and are innervated by branches of the **fibular** (formerly fibular) **nerve**. The most prominent of these muscles is the **cranial tibial**, which arises from the cranial border and lateral condyle of the tibia and dominates the cranial surface of the bone. Its tendon crosses obliquely over the tarsus to insert on the plantaromedial aspects of the 1st tarsal bone and the 1st and 2nd metatarsal bones. Progressing laterally around the crus, the **long digital extensor**, **extensor digiti I longus**, **fibularis longus**, **lateral digital extensor**, and **fibularis brevis** all originate in the proximal half of the crus.

The extensors of the tarsus and flexors of the digits lie caudally on the crus and are innervated by the **tibial nerve**. The **gastrocnemius**, both its medial and lateral heads, and the **superficial digital flexor** originate on the caudal aspect of the femur. While both muscles insert on the tuber calcanei, the superficial digital flexor continues distally to its distal attachments on the digits, as described earlier. The **deep digital flexor**, comprising medial and lateral heads, arises from the caudal surface of the tibia. Its combined tendon passes distally over the sustentaculum tali of the calcaneus, in the flexor canal, and inserts on the distal phalanges of the weight-bearing digits.

The **popliteus** and **caudal tibial muscles** also occupy the caudal aspect of the crus. Neither is critical to limb function in the dog, but their muscle bellies should be considered in any procedure that requires access to the caudomedial aspect of the proximal crus.

Blood supply, lymphatics, and innervation (Figs. 8.4-7 and 8.4-8)

The **femoral artery** wraps caudally around the stifle, continuing into the crus as the **popliteal artery**. The popliteal then divides into the cranial and caudal tibial arteries, the caudal being much smaller than the cranial. The **cranial tibial artery** runs distally in the **interosseous space** between the tibia and fibula to the craniolateral aspect of the tibia, lying between the tibia and the deep surface of the cranial tibial and long digital extensor muscles. The **caudal tibial artery** enters the **nutrient foramen** of the tibia, at the junction of the proximal and middle thirds of the tibia.

The cranial tibial artery is the main blood supply to the muscles of the crus, via several large branches in the proximal crus. Anastomoses occur between the cranial tibial and both the distal part of the caudal femoral artery and the cranial branch of the saphenous artery, thereby providing alternative routes of blood supply to the pes.

> Because of its proximity to the bone, the cranial tibial artery must be identified and preserved during surgical approaches to the tibia. Surgical approaches to the crus should also consider the superficial vessels, including the cranial and caudal branches of the saphenous artery.

The **saphenous artery** supplies the skin on the medial aspect of the crus and is closely associated with the **medial saphenous vein**. Both artery and vein divide into **cranial** and **caudal branches** in the proximal crus. The **caudal branch** of the saphenous artery lies medial to the tibia and tracks distally, along with a small **satellite vein** and the prominent **tibial nerve**.

Two branches of the **sciatic nerve** supply the muscles of the crus: the **common fibular** and **tibial nerves**. The common fibular divides into **deep** and **superficial fibular nerves** lateral to the stifle, and both run distally to supply the muscles on the craniolateral aspect of the crus. The **tibial nerve** passes between the heads of the gastrocnemius muscle and travels the length of the crus in the space between the common calcaneal tendon and the deep digital flexor muscle, where it may be palpated beneath the skin caudal to the tibia.

Lymphatic drainage of the pes and crus passes through 3 independent afferent systems to reach the regional lymph nodes. The vessels of the **superficial lateral lymph system** run with the dorsal metatarsal veins from the dorsal surface of the pes, follow the lateral saphenous vein across the caudolateral surface of the crus to reach the calcaneal tendon, and finally cross the lateral head of the gastrocnemius to reach the **popliteal lymph node**. The **superficial medial lymph system** runs subcutaneously from the distal crus proximally to the **superficial inguinal lymph node**. The **deep medial afferent lymph system** begins between the calcaneal tendon and the distal tibia, runs proximally to cross the medial head of the gastrocnemius, and follows the gracilis muscle to reach the **iliac lymph nodes**—mainly the medial iliac lymph nodes—via the deep lymphatic vessels of the thigh. Generally, there are communications between these 3 routes of lymphatic drainage in the paw and in the mid-crus and inguinal regions.

> Routinely palpated during physical examination in dogs and cats, the popliteal lymph node is a large, ovoid structure located in the superficial fat immediately caudal to the stifle. In contrast, the superficial inguinal lymph nodes typically are identifiable only when enlarged. They may be found where the caudal superficial epigastric artery branches from the external pudendal artery, medial to the external inguinal ring; in males, they lie adjacent to the root of the penis.

Selected references

[1] Hermanson JW, de Lahunta A, Evans HE. Miller and Evans' anatomy of the dog. 5th ed.. Elsevier Saunders; 2019.
[2] Fossum TW. Small animal surgery. 5th ed.. Elsevier; 2018.
[3] Kealy JK, McAllister H, Graham JP. Diagnostic radiology and ultrasonography of the dog and cat. 5th ed.. Elsevier Saunders; 2011.

CHAPTER 8

CASE 8.5

Fracture of the Tarsal Bones

Chris Boemo[a], Oday Al-Juhaishi[b], Zeeshan Akbar[b], and Helen M.S. Davies[c]
[a]Keysborough Veterinary Practice, Keysborough, Victoria, AU
[b]Veterinary and Agricultural Sciences, The University of Melbourne, Melbourne, Victoria, AU
[c]Veterinary BioSciences, The University of Melbourne, Melbourne, Victoria, AU

Clinical case

History

A 2½-year-old male intact Greyhound with a good racing performance history was presented after sustaining a fracture of the right tarsus (hock) during a race.

Varus is an abnormality of limb conformation in the median or sagittal plane in which the limb deviates medially distal to the point of deformity (in this case, the tarsus). The opposite abnormality is valgus, in which the distal limb deviates laterally. The terms "valgus" and "varus" are adjectives and are used only when joined with the noun they describe—e.g., tarsal varus in this case. These angular limb deformities may be congenital and caused by a developmental abnormality such as asynchronous growth at a physis ("growth plate"), or—as in this case—acquired as the result of injury.

Physical examination findings

The dog was non-weight-bearing on the right pelvic limb. There was moderate swelling of the tarsal region and a mild varus deformity centered at the tarsus. On palpation, there was crepitus (evident displacement of bone fragments in the region of the central tarsal bone), increased joint laxity, and pain on digital pressure over the dorsal aspect of the tarsometatarsal joint.

Differential diagnoses

Tarsal bone fracture(s), tarsal luxation, or possibly both

TARSAL INJURIES IN GREYHOUNDS

The forces acting on the tarsus in quadrupeds are compression, tension, torsion, shearing, and bending. The greater the intensity of exercise or impact, the greater the forces acting on the tarsus (Fig. 8.5-1) and the greater the potential for injury.

Tarsal injuries are, therefore, common in racing Greyhounds. They may be classified as sprains, fractures, luxations (dislocation), or a combination. Fractures and luxations are readily diagnosed, as they typically cause severe lameness and obvious swelling, instability, and deformity of the entire tarsus.

Sprains involve injury to one or more of the tarsal ligaments: dorsal, plantar, lateral, or medial. Desmitis involving one or more of the tarsal ligaments is common in racing Greyhounds. These lesions are graded from 1 (mild) to 3 (severe). Grade 3 sprains and those involving avulsion (tearing away) of a bone fragment at the site of attachment are most readily diagnosed.

Comparative Veterinary Anatomy: A Clinical Approach. https://doi.org/10.1016/B978-0-323-91015-6.00053-4

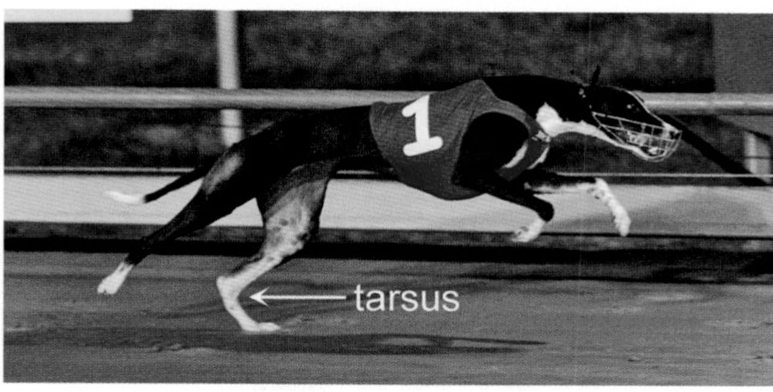

FIGURE 8.5-1 A racing Greyhound, showing the left tarsus under peak load.

Diagnostics

Radiographs of the tarsus revealed a complete, comminuted (multiple-fragment), displaced fracture of the central tarsal bone (Figs. 8.5-2 and 8.5-3). Collapse of the central tarsal bone resulted in a mild tarsal varus deformity, centered on the fractured central tarsal bone. In addition, a hinge-type dorsal slab fracture of the 4th tarsal bone was found, displacing dorsally at the distal articular margin (Fig. 8.5-4). A brief description of the biomechanical process

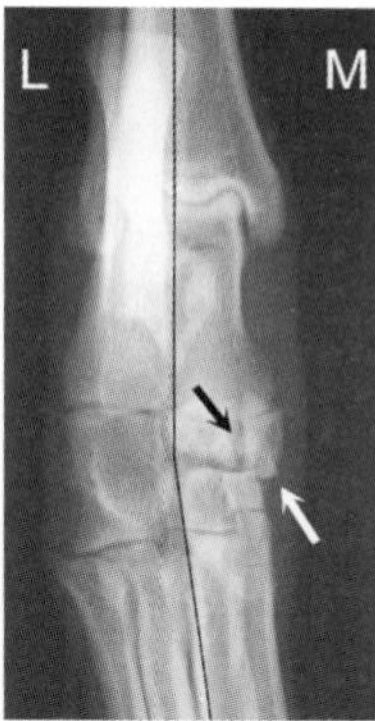

FIGURE 8.5-2 Radiograph of the Greyhound's right tarsus, plantarodorsal view, showing a complete fracture of the central tarsal bone (*black arrow*) with medial displacement of a large fragment (*white arrow*). Collapse of the central tarsal bone has caused a mild tarsal varus deformity (*thin black lines*). Key: *L*, lateral; *M*, medial. (Note: this case pre-dates the widespread use of digital radiography in veterinary practice, so scratches and other artifacts can be seen on these films.)

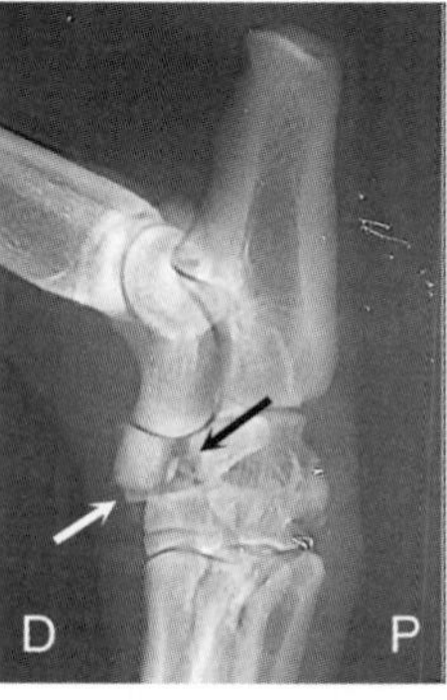

FIGURE 8.5-3 Radiograph of the Greyhound's right tarsus, slightly oblique lateral view, showing the central tarsal bone fracture (*black arrow*) from a lateral perspective. Not only is there dorsal displacement of a large fragment (*white arrow*), there are multiple small fragments within the main fissure and dorsal to the distal intertarsal joint. Key: *D*, dorsal; *P*, plantar.

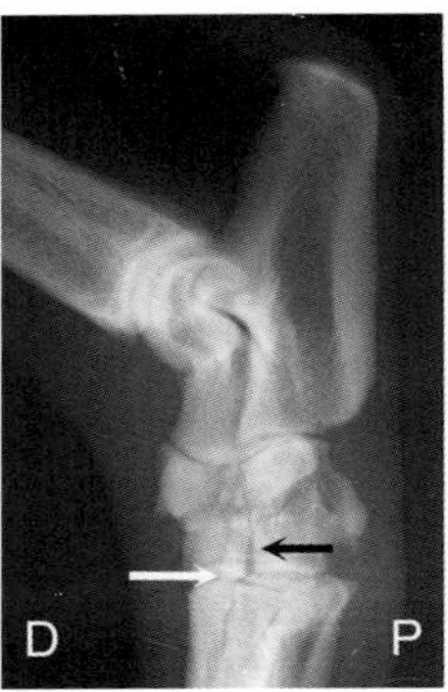

FIGURE 8.5-4 Radiograph of the Greyhound's right tarsus, lateral view. The central tarsal bone fracture is not as clear as on the other views, but this projection reveals a hinge-type dorsal slab fracture of the 4th tarsal bone. The vertical fracture line (*black arrow*) separates a thin dorsal "slab" of bone from the body of the 4th tarsal bone. Distally, the slab is displaced dorsally (*white arrow*), creating the appearance of a fragment that is "hinged" proximally. Key: *D*, dorsal; *P*, plantar.

by which this fracture configuration occurs in racing Greyhounds is discussed in the side boxes entitled "Tarsal injuries in Greyhounds" and "Biomechanics of tarsal fractures."

Diagnosis

Fracture of the central tarsal bone and concurrent dorsal slab fracture of the 4th tarsal bone in the right pelvic limb

Treatment

Under general anesthesia, a cortical bone screw was used to repair the dorsal slab fracture of the 4th tarsal bone (Fig. 8.5-5). The central tarsal bone fracture was repaired with 2 cancellous bone screws—1 extending laterally into the body of the 4th tarsal bone for additional stability; the smaller bone fragments were removed to minimize sequestrum formation.

A sequestrum (plural, sequestra) is a piece of bone that has lost its blood supply, becomes separated from the parent/sound bone, and becomes devitalized. Sequestra delay bone healing and increase the risk for post-operative infection, so small bone fragments are generally removed at the time of fracture repair because they often become potential sequestra.

Post-operative radiographs revealed good fracture reduction and joint realignment. The dog made an uneventful recovery and was almost fully weight-bearing on the right pelvic limb within a week. Repeat radiographs taken 6 weeks after surgery showed good fracture healing with some periosteal callus formation. The dog was retired from racing but was able to be used for breeding starting 7 weeks after surgery.

TARSUS VS. ANKLE

The tarsus in quadrupeds is broadly equivalent to the human ankle, but structurally and functionally, it is very different. Humans have a plantigrade stance, in which the plantar surfaces of the digits and the metatarsal and tarsal bones are either directly or indirectly in contact with the ground. In contrast, dogs and cats have a digitigrade stance, in which only the digits are in contact with the ground.

When viewed from the lateral aspect, the "standing" angle at the ankle in humans—the angle formed by the tibia/fibula and the metatarsal bones—is approximately 90°. In dogs and cats, the tarsus is suspended and supported at a standing angle of approximately 130° (range, 115–145°, depending on breed/size).

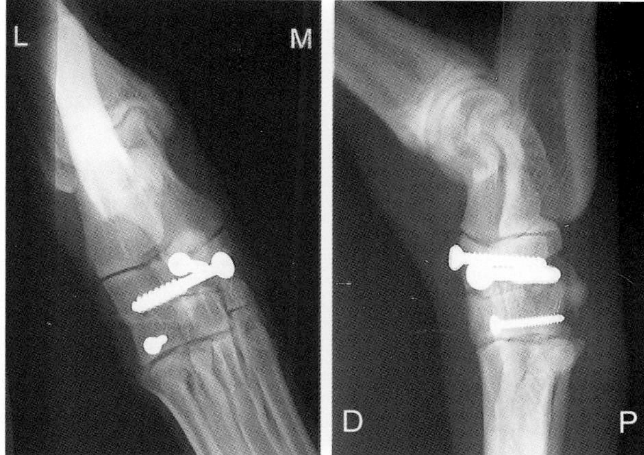

FIGURE 8.5-5 Post-operative radiographs of the Greyhound's right tarsus. Left: plantarodorsal view. Right: lateral view. The proximal 2 screws were placed to repair the central tarsal bone fracture, one of which was extended laterally into the 4th tarsal bone for additional stability. The smaller distal screw was placed to repair the dorsal slab fracture of the 4th tarsal bone. Key: *D*, dorsal; *L*, lateral; *M*, medial; *P*, plantar.

Anatomical features in canids and felids

Introduction

The tarsus, or hock, is a compound and complex joint that is a common site of acute and chronic disease in dogs and cats. Located between the distal aspects of the tibia and fibula and the proximal aspects of the metatarsal bones, the tarsus (Gr. *tarsos* flat of the foot) consists of a group of irregularly shaped tarsal bones that form several small, low-motion intertarsal and tarsometatarsal joints and one high-motion, hinge-type tarsocrural joint (L. *crus* leg). In addition to the series of small ligaments that support the tarsal joints, the long tendons of the digital flexor and extensor muscles track over the plantar and dorsal aspects of the tarsus, respectively, on their way to on the digits.

Traumatic and, in dogs, sports-related injuries commonly cause acute damage, but overall the most common condition affecting the tarsus in dogs and cats is osteoarthritis, a chronic degenerative joint disease that is often multifactorial. The tarsus may also be affected acutely or chronically by infectious (bacterial, mycoplasmal, Lyme disease/borreliosis) or immune-mediated (rheumatoid or poly-) arthritis.

Function

The primary movements through the tarsus are flexion and extension, so in effect the tarsus acts as a hinge joint; almost all of this movement occurs at the tarsocrural joint. The tarsal joints other than the tarsocrural joint are low-motion joints, normally permitting only small amounts of axial rotation and/or lateral bending through the tarsus. These joints primarily serve to accommodate the forces of ground impact, braking, propulsion, turning, sliding, and movement of the body over the foot during the stance phase of the stride (Fig. 8.5-1).

The digital flexors and extensors support and move (extend/flex) the tarsus and influence how it is loaded.

Tarsal bones (Figs. 8.5-6–8.5-8)

In dogs and cats, there are 7 tarsal bones, arranged in 3 irregular rows. The largest 2 tarsal bones—the talus and the calcaneus—are also the most proximal.

The **talus** lies on the dorsomedial aspect of proximal tarsus. It has a relatively compact body, a large **trochlea** (broad groove or trough) on its dorsoproximal surface, and a distal **head** that forms its base. When viewed from the dorsal aspect, the trochlea is oriented vertically, angling by about 25° toward the lateral side from proximal to distal. The talus articulates proximally with the distal tibia and fibula to form the **tarsocrural joint**. The head of the talus

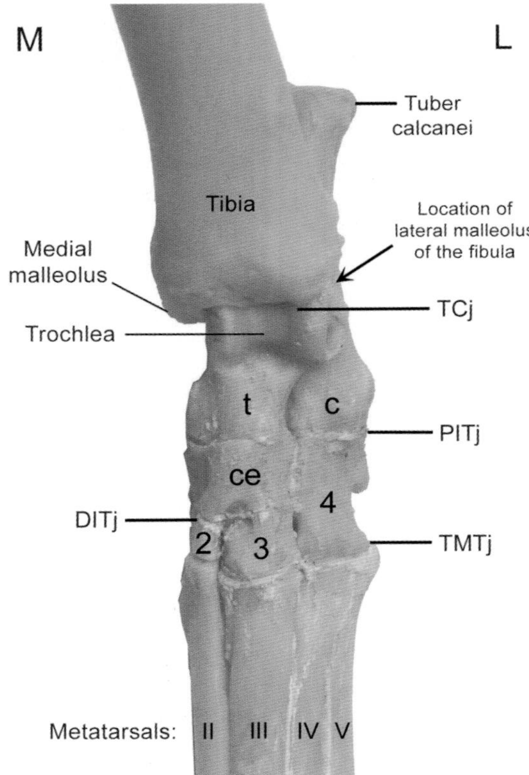

FIGURE 8.5-6 Left tarsus of a dog, dorsal view, absent the fibula. The *arrow* points to the location of the lateral malleolus if the fibula were present. The tarsal bones shown are the talus (t); calcaneus (c); central (ce) tarsal; and the 2nd (2), 3rd (3), and 4th (4) tarsal bones. (The 1st tarsal bone is not visible in this slightly lateral view.) The metatarsal bones seen in this image are numbered with the Roman numerals II–V. From proximal to distal, the labeled joints are the tarsocrural (TCj), proximal intertarsal (PITj), distal intertarsal (DITj), and tarsometatarsal (TMTj). Key: *L*, lateral; *M*, medial.

The slight obliquity of the trochlea of the talus causes the distal limb to be passively rotated or abducted laterally a few degrees when the tarsus is flexed. This action allows the pelvic limbs to be extended cranially beyond and lateral to the thoracic limbs during the gallop. This phase of the stride is best seen in racing Greyhounds at top speed (Fig. 8.5-9). This action is also evident in a sitting cat who is self-grooming the medial thigh or ventral abdomen. Most axial rotation and abduction of the pelvic limb occur at the coxofemoral joint, but the trochlea of the talus ensures an additional degree of passive rotation/abduction of the distal limb.

has a slightly convex distal articular surface, which rests in a shallow fossa in the proximal articular surface of the central tarsal bone, forming part of the **proximal intertarsal joint**.

The **calcaneus** is the largest of the tarsal bones. It occupies the plantarolateral aspect of the proximal tarsus and is intimately associated with the talus which lies dorsal and medial to it. The most notable feature of the calcaneus is the **tuber calcanei**, which extends proximally, caudal to the distal tibia/fibula, and forms the externally visible

DIRECTIONAL TERMS (SEE CHAPTER 1)

When describing the tuber calcanei, there may be some ambiguity in the directional terminology because the calcaneus is situated in the plantar aspect of the tarsus, but its tuberosity extends proximal to the tarsocrural joint, so it is located in the caudal aspect of the distal crus—the region of the pelvic limb between stifle and hock.

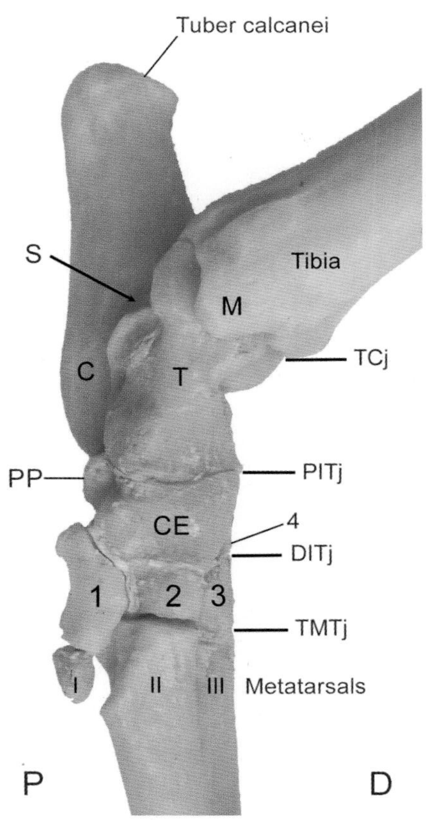

FIGURE 8.5-7 Left tarsus of a dog, medial view. The tarsal bones shown are the talus (T); calcaneus (C); central (CE) tarsal; and the 1st (1), 2nd (2), 3rd (3), and 4th (4) tarsal bones. (The 4th tarsal bone is barely visible in this view.) The position of the sustentaculum tali (S) of the calcaneus is indicated by the *arrow*. The plantar process (PP) of the central tarsal bone is visible just distal to the base of the talus. The medial malleolus (M) of the tibia is also shown on this view. The metatarsal bones visible in this image are numbered with the Roman numerals I–III. From proximal to distal, the labeled joints are the tarsocrural (TCj), proximal intertarsal (PITj), distal intertarsal (DITj), and tarsometatarsal (TMTj). Key: *D*, dorsal; *P*, plantar.

"point of the hock" (Figs. 8.5-10 and 8.5-11). The common calcaneal tendon ("Achilles" tendon in humans) attaches to the tuber calcanei, as described later. Also, of note is the **sustentaculum tali**, a broad plantaromedial extension of the calcaneus (Fig. 8.5-9). It has a shallow groove on its plantar surface in which the deep digital flexor tendon lies as it courses over the plantar aspect of the tarsus (described later).

The middle row of tarsal bones consists of the central tarsal bone and the proximal half of the 4th tarsal bone, whose distal half lies in the distal row of tarsal bones (see below). The compact, cylindrical **central tarsal bone** is located on the medial aspect of the tarsus, directly distal to the talus, with which it articulates. The longer, irregularly cylindrical **4th tarsal bone** lies lateral to it, directly distal to the calcaneus, with which it articulates. The central and 4th tarsal bones each have a prominent **plantar process** for the attachment of the strong plantar ligaments and plantar portion of the joint capsule (described later). The plantar process of the central tarsal bone is located more proximally, just distal to the base of the talus (Figs. 8.5-7 and 8.5-8). The plantar process of the 4th tarsal bone is located more distally, closer to the metatarsal bones than to the calcaneus.

Both plantar processes (central and 4th tarsal bones) are visible on lateral radiographs of the tarsus. The central tarsal bone is smaller, and its body occupies a more dorsal position, so its plantar process is usually superimposed on the body of the 4th tarsal bone and—on some views—the proximal intertarsal joint as well. Usually, the plantar process that is visible on the plantar aspect of the tarsus, such as in Fig. 8.5-2, belongs to the 4th tarsal bone.

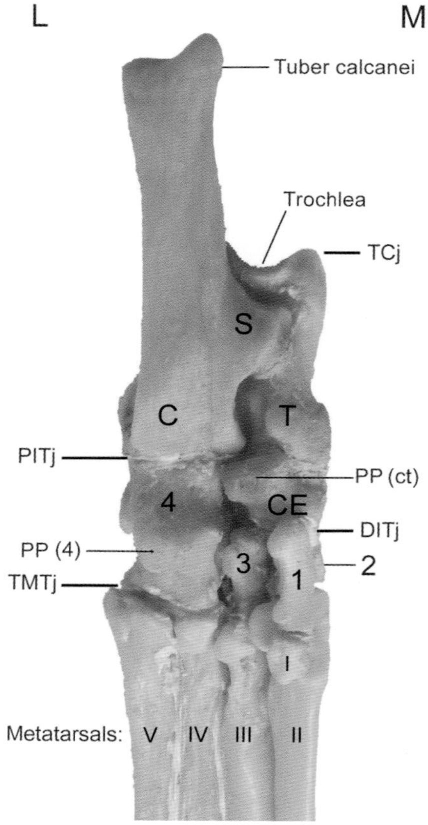

L
M
Tuber calcanei
Trochlea
TCj
S
C
T
PITj
PP (ct)
4
CE
DITj
PP (4)
3
1
2
TMTj
I
Metatarsals: V IV III II

FIGURE 8.5-8 Left tarsus of a dog, plantar view, absent the tibia and fibula. The tarsal bones shown are the talus (T); calcaneus (C); central (CE) tarsal; and the 1st (1), 2nd (2), 3rd (3), and 4th (4) tarsal bones. The sustentaculum tali (S) on the plantaromedial aspect of the calcaneus is best seen on this view. The plantar process (PP) of the central tarsal bone and that of the 4th tarsal bone is also shown on this image. The metatarsal bones are numbered with the Roman numerals I-V. From proximal to distal, the labeled joints are the tarsocrural (TCj), proximal intertarsal (PITj), distal intertarsal (DITj), and tarsometatarsal (TMTj). Key: *L*, lateral; *M*, medial.

FIGURE 8.5-9 Owing in part to the slight lateral orientation of the trochlea of the talus, the distal pelvic limbs are extended cranial and lateral to the thoracic limbs during the gallop. (This image also shows the tarsal flexors in action in the right pelvic limb: the tendons of the cranial tibial, long digital extensor, and lateral digital extensor muscles can each be identified, as they are under tension.)

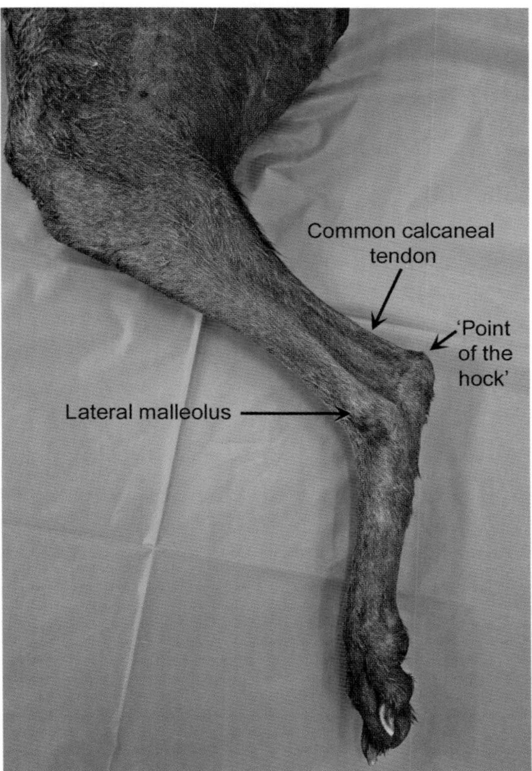

FIGURE 8.5-10 Lateral aspect of the left pelvic limb of a dog. The proximal extent of the tuber calcanei is commonly referred to as the "point of the hock."

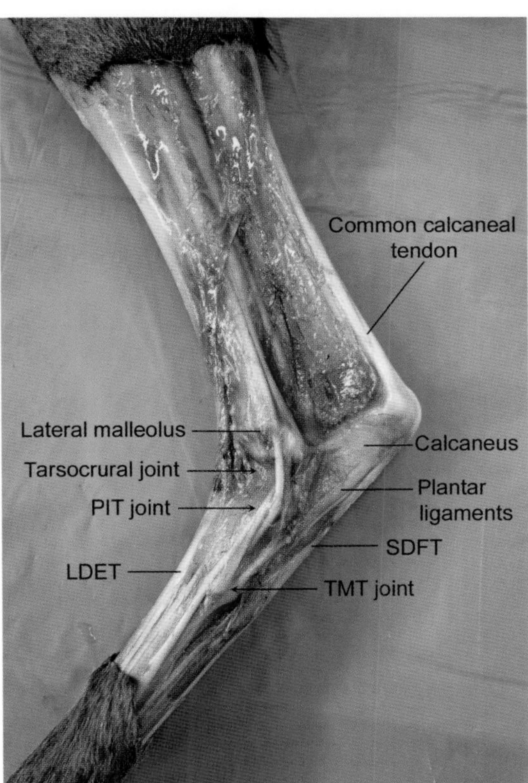

FIGURE 8.5-11 Lateral aspect of the left pelvic limb of a dog, with skin and subcutis removed to demonstrate the structures of the tarsal region beneath the translucent layer of superficial fascia. Key: *LDET*, long digital extensor tendon; *PIT*, proximal intertarsal (joint); *SDFT*, superficial digital flexor tendon; *TMT*, tarsometatarsal (joint).

The distal row of tarsal bones comprises 3 small tarsal bones, numbered 1–3, plus the base of the 4th tarsal bone. The **1st** and **2nd tarsal bones** are the most medial and the 4th tarsal bone is the most lateral; and the **3rd tarsal bone** lies between the 2nd and 4th. The 1st tarsal bone is located plantaromedial to the 2nd tarsal bone (Figs. 8.5-7 and 8.5-8), and it is distinctly different from the others. Whereas tarsal bones 2–4 articulate with load-bearing metatarsal bones (II–V), the 1st tarsal bone forms the base for the rudimentary 1st digit, the "dewclaw," when present (see below). Proximally, tarsal bones 1–3 articulate with the central tarsal bone, whereas the 4th tarsal bone articulates with the calcaneus.

Distal to the tarsus, the structures of the pelvic limb are like those in the thoracic limb (see case 7.4), with 2 notable exceptions. First, the **metatarsal bones** are a little longer than the metacarpal bones. Because the tarsus is also longer than the carpus, the pes (entire region from the tarsus distally) is therefore longer than the manus (entire region from the carpus distally). Second, the **1st digit** of the pelvic limb is highly variable. Some dogs and cats have a fully formed 1st digit like that in the thoracic limb, whereas in others the 1st digit may be absent or be little more than a rudimentary claw with a small osseous component attached only by skin and fibrous tissue.

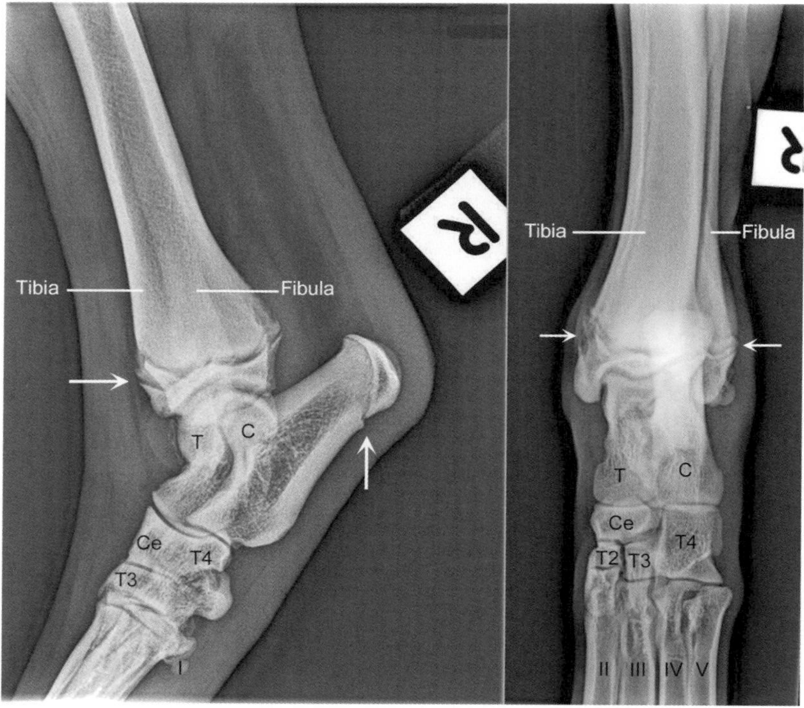

FIGURE 8.5-12 Radiographs of the physes (*arrows*) in the tarsal region of a 5-month-old dog: distal tibia and fibula, and proximal tuber calcanei. Key: *C*, calcaneus; *Ce*, central tarsal bone; *T*, talus; *T2*, 2nd tarsal bone; *T3*, 3rd tarsal bone; *T4*, 4th tarsal bone; *I-V*, 1st to 5th metatarsal bones. [Radiographs courtesy of Dr. Cathy Beck, The University of Melbourne.]

OSTEOCHONDRITIS DISSECANS IN DOGS

Osteochondritis dissecans (OCD) is a form of osteochondrosis, a developmental orthopedic disease that affects weight-bearing joint surfaces in young animals and people. In the tarsus, OCD lesions in dogs most often involve the trochlear ridges of the talus. Although OCD lesions on the medial trochlear ridge are more common, they are typically more superficial than those on the lateral trochlear ridge, which often comprise large osteochondral fragments. The difference lies in the shape of, and thus the biomechanical forces on, the medial and lateral trochlear ridges (Fig. 8.5-12).

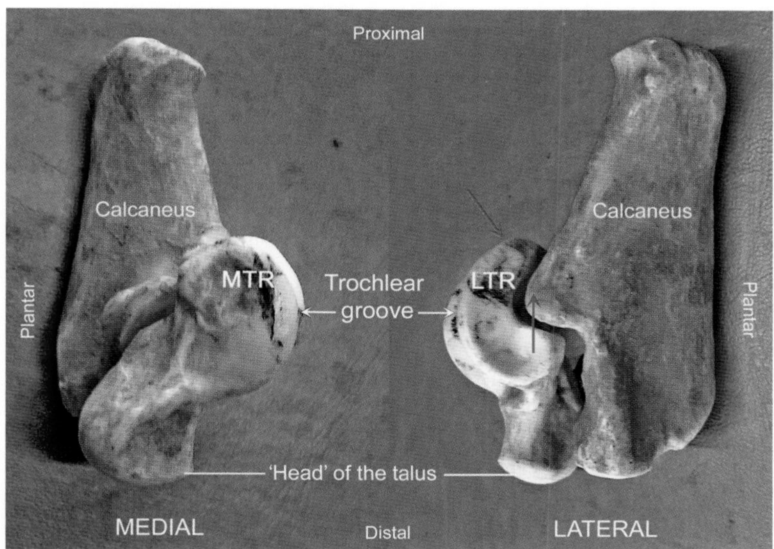

FIGURE 8.5-13 Left talus and calcaneus of a Greyhound, viewed from medial and lateral aspects. Because the medial trochlear ridge (MTR) of the talus is relatively robust, osteochondritis dissecans (OCD) lesions on the MTR usually comprise flattening of the subchondral bone or shallow osteochondral flaps or fragments off the surface of the MTR. In contrast, OCD lesions on the lateral trochlear ridge (LTR) of the talus often consist of large osteochondral fragments that involve the entire proximoplantar aspect of the LTR. As shown by the *red arrows*, this portion of the LTR is compressed between the tibia proximally and the calcaneus distally during weight-bearing.

The distal tibia and fibula are described in Case 8.4.

There are 3 normal **ossification centers** of clinical importance in the tarsal region of young dogs and cats: the distal **tibia** and **fibula** and the proximal aspect of the **tuber calcanei** (Fig. 8.5-13). In most dogs and cats, these sites are all fully ossified by 10–12 months of age, although it ranges from 5 to 18 months in dogs, depending on breed, and from 8 to 14 months in cats. All other bones in the tarsal region typically are fully ossified by 3–4 months of age.

Tarsal joints (Figs. 8.5-6–8.5-8)

The most proximal of the tarsal joints is the **tarsocrural joint**, comprising the distal tibia and fibula proximally and the talus distally. The principal components of this articulation are the sagittal grooves and intermediate ridge of the distal tibia, and the trochlea and the medial and lateral trochlear ridges of the talus. The tarsus also contains several vertical intertarsal joints, the **proximal** and **distal intertarsal joints**(separated by the central tarsal bone), and the **tarsometatarsal joints**.

The fibrous layer of the **joint capsule** covers the entire tarsal joint from the periosteum of the distal tibia and fibula to the proximal aspect of the metatarsal bones. However, separate synovial membranes, extending only to the margins of the facing articular cartilage surfaces, create a series of **synovial sacs/compartments**, the most voluminous of which is the tarsocrural joint. This joint space usually communicates with that of the proximal intertarsal joint. The distal intertarsal and tarsometatarsal joint spaces also communicate with one another, but the 2 proximal synovial cavities do not normally communicate with the 2 distal synovial cavities.

Tarsal ligaments (Figs. 8.5-14 and 8.5-15)

The tarsus is supported by several sets of ligaments, layered from deep to superficial. They run horizontally, vertically, and obliquely over the tarsus to support this complex joint and resist the various forces it experiences.

Of importance are the **medial** and **lateral collateral ligaments** (CL). Each has a long, superficial portion (**long CL**) that spans the entire tarsus and a short, deep portion (**short CL**) that predominantly supports the tarsocrural joint. The long CL originate on the lateral malleolus of the fibula (lateral CL) and the medial malleolus of the tibia (medial CL). They insert onto the proximolateral aspect of metatarsal V (lateral CL) and the 1st tarsal bone and metatarsals I and II (medial CL), each attaching to the lateral/medial aspect of the tarsal bones along their course. The short CL also originate on the malleoli, but they insert primarily onto the calcaneus (lateral CL) and talus (medial CL).

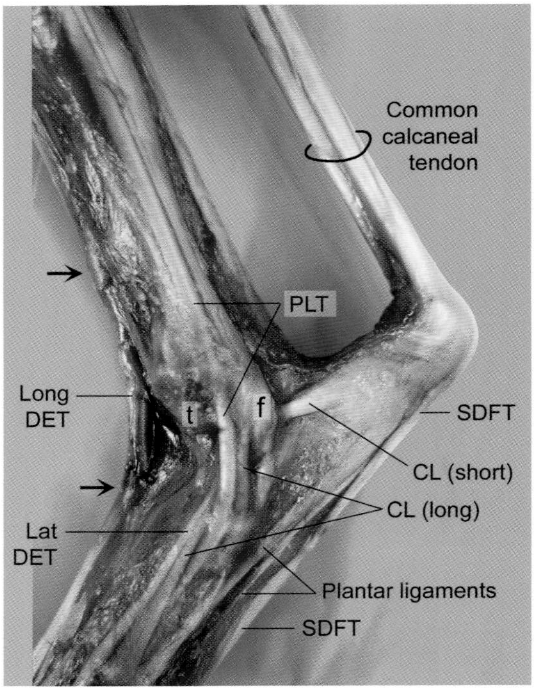

FIGURE 8.5-14 Lateral aspect of the left tarsus of a dog, superficial fascia removed. The *black arrows* point to the pair of transverse ligaments (proximal and distal) which hold the long DET in place as it courses over the tarsocrural joint. The proximal ligament also overlies the cranial tibial tendon (see Fig. 8.5-16). Key: *CL*, lateral collateral ligament (long and short parts); *DET*, digital extensor tendon (long and lateral); *F*, fibula (distal extent); *Lat*, lateral; *PLT*, peroneus longus tendon; *SDFT*, superficial digital flexor tendon; *T*, tibia (distal extent). Note how the SDFT, which is part of the common calcaneal tendon, continues over the point of the hock and tracks along the plantar aspect of the tarsus, superficial to the plantar ligaments of the tarsus, on its way to its insertion on the digits.

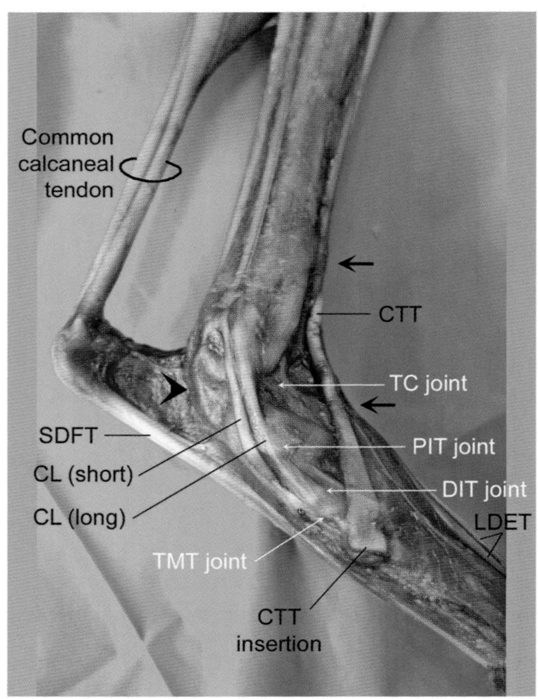

FIGURE 8.5-15 Medial aspect of the left tarsus of a dog, superficial fascia removed. The *black arrows* point to the transverse ligaments shown in Fig. 8.5-15. The proximal ligament covers both the CTT and the LDET. The *black arrowhead* indicates where the deep digital flexor tendon tracks over the sustentaculum tali of the calcaneus. Key: *CL*, medial collateral ligament (long and short parts); *CTT*, cranial tibial tendon; *DIT*, distal intertarsal (joint); *LDET*, long digital extensor tendon (already divided into its digital components); *PIT*, proximal intertarsal (joint); *SDFT*, superficial digital flexor tendon; *TC*, tarsocrural (joint); *TMT*, tarsometatarsal (joint).

The fibrous layer of the **joint capsule** is thickened over the plantarodistal aspect of the tarsus, further supported by a series of short, strong **plantar ligaments.** These structures form the deep surface of the **tarsal canal**, through which pass the deep digital flexor tendon in its tendon sheath and the plantar blood vessels and nerves that supply the metatarsus and digits (discussed later). The superficial aspect of the tarsal canal is composed of the plantar fascia.

> Partial avulsion of these plantar ligaments is common in racing Greyhounds, and Fig. 8.5-5 shows why. This injury causes lameness, poor racing performance, and pain on palpation of the plantar aspect of the tarsus. Complete avulsion results in luxation of the proximal intertarsal joint, which is usually a career-limiting injury.

The **superficial fascia** of the tarsus connects/continues the crural fascia and the fascia of the distal limb (Fig. 8.5-11). It provides additional mechanical strength to the tarsus and supports the blood and nerve supply of this region.

Tendons of the tarsal region (Figs. 8.5-14 and 8.5-15)

Several tendons track over the tarsus from their muscular origins in the proximal limb to their insertions distal to the tarsus. Those on the dorsal and lateral aspects of the tarsus include the **cranial tibial, long digital extensor, lateral digital extensor,** and **peroneus longus.** Generally, these muscles flex the tarsus and provide dorsal/lateral support in addition to their other specific functions, such as extending the metatarsophalangeal joints and digits or laterally rotating/abducting the distal limb.

> These tarsal flexors are seen in action in the right pelvic limb of the Greyhound in Fig. 8.5-9.

The **cranial tibial** muscle is important in relation to the tarsus, as the muscle originates on the craniolateral aspect of the proximal tibia; however, its tendon of insertion crosses the dorsal aspect of the tarsus obliquely before inserting on the medial aspect of the proximal metatarsus (Fig. 8.5-15). Over the cranial aspect of the distal tibia, the cranial tibial tendon and that of the long digital extensor run under a short, obliquely oriented band of connective tissue, the **proximal transverse ligament** or **proximal extensor retinaculum**. A second such ligament, the horizontally oriented **distal transverse ligament** or **distal extensor retinaculum**, overlies the long digital extensor tendon alone as it courses distally over the proximal intertarsal joint, lateral to the cranial tibial tendon. These transverse ligaments serve to hold the tendons in place as they change direction over the tarsocrural joint.

The tendons at the plantar aspect of the tarsus include the common calcaneal, deep digital flexor, and caudal tibial. In dogs and cats, the main components of the **common calcaneal tendon** are the tendons of the **gastrocnemius** muscle (both lateral and medial heads) and the **superficial digital flexor** muscle. There are also minor contributions (accessory tendons) from the semitendinosus, biceps femoris, and gracilis muscles. In addition to their individual actions on the stifle and/or metatarsophalangeal joints and digits, these muscles collectively extend the tarsus and maintain the normal digitigrade stance via their common insertion onto the tuber calcanei. The **superficial digital flexor tendon** also supports the plantar aspect of the tarsus as it tracks distally to its insertions on the digits. The tendon is affixed to the tuber calcanei as it travels over the point of the hock by a pair of short horizontal ligaments (lateral and medial).

> Rupture of the common calcaneal tendon is a serious injury, as it allows the muscles that flex the tarsus free to function unopposed. As a result, the tarsus "drops"—as the limb is hyperflexed—toward a more plantigrade stance.

The **deep digital flexor tendon** courses distally over the plantar aspect of the tarsus, facilitated proximally by the tendon sheath as it crosses the sustentaculum tali of the calcaneus. In dogs and cats, the deep digital flexor has 2 heads: a large **lateral head (flexor hallucis longus)** and a small **medial head (flexor digitorum longus)**. The tendon of the former, commonly referred to simply as the deep digital flexor tendon, courses within the **tarsal canal**, further encased within a tendon sheath, whereas the tendon of the much smaller medial head courses medial to the tarsal canal. Distal to the canal, at the level of the distal row of tarsal bones, both tendons unite and the (common) deep digital flexor tendon continues distally, lying deep to the superficial digital flexor tendon, on its way to its insertions on the digits.

The tendon of the **caudal tibial** muscle is small and lies medial to the deep digital flexors over the plantaromedial aspect of the tarsus, merging with the medial collateral ligament of the tarsus. These tendons, particularly the deep digital flexor, provide additional support to the plantar aspect of the tarsus.

Blood supply, lymphatics, and innervation

The main arterial supply to the dorsal aspect of the tarsus is the **cranial tibial artery**. It is the continuation of the **popliteal artery**, itself the main continuation of the **femoral artery**. The cranial tibial artery courses distally along the cranial aspect of the tibia, deep to the cranial tibial and long digital extensor muscles. Over the distal tibia, the artery continues superficially beneath the proximal extensor retinaculum, between the cranial tibial and long digital extensor tendons. It runs alongside and medial to the long digital extensor tendon over the dorsum of the tarsus. As it continues distally, it becomes the **dorsal pedal artery**. Proximal to the tarsus, the cranial tibial artery sends off a superficial branch, which often anastomoses with the cranial branch of the **saphenous artery** (see below) at the level of the tarsus and supplies the dorsal and lateral aspects of the tarsus.

> The dorsal pedal artery just distal to the tarsus may be used for indirect or direct measurement of blood pressure during anesthesia in dogs and cats. The arterial pulse may also be palpated in the dorsal pedal artery or more proximally in the cranial tibial artery, although the femoral artery is most often used for this purpose.

The main arterial supply to the plantar aspect of the tarsus is the **caudal branch of the saphenous artery**. The saphenous artery arises from the femoral artery in the distal thigh and divides into cranial and caudal branches in the crus (L. a leg-like part). As it approaches the tarsus, the caudal branch sends off several small tarsal arteries and the **medial** and **lateral plantar arteries**, which course within the tarsal canal on either side of the deep digital flexor tendon, over the plantaromedial aspect of the tarsus. Distal to the tarsal canal, the plantar arteries anastomose with each other and with the continuation of the dorsal pedal artery to form the **deep plantar arch** from which the plantar metatarsal arteries originate.

The venous drainage of the tarsus customarily parallels the arterial supply. Distal to the tarsus, the **medial** and **lateral plantar metatarsal veins** derive from the **superficial plantar venous arch**, which lies on the proximal aspect of the **metatarsal pad** and drains the blood from the **plantar digital veins**. The **medial** and **lateral plantar veins** continue proximally over the plantar aspect of the tarsus within the tarsal canal, alongside their corresponding arteries. They then drain into the **medial** and **lateral saphenous veins**. The cranial branch of the medial saphenous vein is a continuation of the medial plantar vein, while the caudal branch of the lateral saphenous vein is a continuation of the lateral plantar vein.

> In dogs and cats, the superficial popliteal lymph nodes are usually palpable in the popliteal fossa (immediately caudal to the stifle joint). Any proximal obstruction to lymphatic drainage may result in lymphedema, which is diffuse swelling distal to the site of obstruction. Lymphedema typically presents as "pitting edema" in which gentle pressure with a finger or thumb leaves a depression for several seconds before the accumulated lymph refills the area.

The lymphatic vessels of the digits, metatarsus, and tarsus drain into the **popliteal lymph center** caudal to the stifle, comprising the **superficial** and **deep popliteal lymph nodes**.

CONTENTS OF THE TARSAL CANAL IN DOGS AND CATS

- Deep digital flexor tendon lateral head, within its tendon sheath
- Medial and lateral plantar arteries
- Medial and lateral plantar veins
- Medial and lateral plantar nerves (branches of the tibial nerve)

The tarsus is innervated by branches of the **tibial** and **peroneal nerves** as they track distally to supply the metatarsus and digits. Both are branches of the **ischiatic nerve**, which originates from the **lumbosacral trunk**. The tibial and peroneal nerves both lie on the medial aspect of the crus—the tibial nerve caudal and the peroneal nerve more cranial.

The **tibial nerve** continues distally over the plantaromedial aspect of the tarsus. As it passes over the proximal tarsus, it divides into the **lateral** and **medial plantar nerves** supplying the plantar aspect of the metatarsus and digits. Both plantar nerves course within the tarsal canal, alongside the deep digital flexor tendon.

The **peroneal nerve** divides into **superficial** and **deep branches** just distal to the stifle, although both take a similar path along the crus and over the tarsus. The nerves become more superficial over the distal tibia and curve slightly to assume a dorsal path over the tarsus. The deep peroneal nerve lies axially over the tarsus, whereas the superficial peroneal nerve lies lateral to it, overlying the medial aspect of the distal fibula. Both peroneal nerves branch to supply the dorsal aspects of the metatarsus and digits.

BIOMECHANICS OF TARSAL FRACTURES

Racing in a counterclockwise direction (curving to the left), the right tarsus experiences the greatest compressive load on its medial aspect and under peak load. At this time, the convex "head" (distal aspect) of the talus is "driven" into the concave proximal articular margin of the central tarsal bone. If loading exceeds the central tarsal bone's capacity, fracture occurs with 3 main pieces: medial, dorsolateral, and plantarolateral with comminution in some cases.

Without the intact central tarsal bone, the 4th tarsal bone, calcaneus, and/or talus may fracture.

Selected references

[1] Boemo CM. Specific tests for tarsal injury. In: Bojrab MJ, Ellison GW, Slocum B, editors. Current techniques in small animal surgery. 4th edn. Baltimore: Williams & Wilkins; 1998. p. 1254.
[2] Dee JF. Tarsal injuries. In: Bloomberg MS, Dee JF, Taylor RA, editors. Canine sports medicine and surgery. Philadelphia: WB Saunders Co.; 1998. p. 120–37.
[3] Guilliard MJ. Fractures of the central tarsal bone in eight racing greyhounds. Vet Rec 2000;147:512–5.
[4] Boemo CM. Greyhound surgical case studies. In: Proceedings, Australian Greyhound Veterinary Association Conference; 2007.
[5] Boemo CM. Hock surgery in the greyhound—techniques and tips. In: Proceedings, Australian Greyhound Veterinary Association Conference; 2012.

CHAPTER 8

CASE 8.6

Degenerative Lumbosacral Stenosis

Marc Kent[a] and Eric N. Glass[b]
[a]Department of Small Animal Medicine and Surgery, University of Georgia College of Veterinary Medicine, Athens, Georgia, US
[b]Neurology and Neurosurgery, Red Bank Veterinary Hospital, Tinton Falls, New Jersey, US

Clinical case

History

A 9-year-old male neutered, German Shepherd dog was presented for a 4-month progressive history of pelvic limb weakness. Initially, the dog displayed trouble standing up with its pelvic limbs from a lying position. Over approximately 4 months, signs progressed to walking slower than normal and having a "crouched" posture in the pelvic limbs while standing. In the previous 3 weeks, the owner noticed that the dog was dribbling urine while walking and leaked urine when lying down. The dog had started squatting to urinate rather than lifting a leg. Also, the owner had noted that the force of the dog's urine stream was reduced. The owner also noticed that occasionally the dog dropped feces as he walked. The owner had not seen the dog wag his tail in the last week before examination.

Physical examination findings

Severe, generalized atrophy of the pelvic limb musculature was noted. No other abnormalities were identified with orthopedic or complete physical examination.

Neurological examination findings

Mentation

- Normal

Gait/posture

- Paraparesis; the dog walked with a shortened stride length in the pelvic limbs; during the weight-bearing (stance) phase of the gait, the dog was plantigrade; posture was plantigrade when standing, giving the dog a "crouched" posture in the pelvic limbs; the gait in the thoracic limbs was normal

Postural reactions

- Thoracic limbs—normal hopping and paw placement bilaterally
- Pelvic limbs—abnormal; delayed hopping with short, choppy-looking hops, slow paw placing bilaterally

Spinal reflexes/muscle tone/muscle mass/nociception

- Thoracic limbs—normal withdrawal reflexes, normal muscle tone, and normal muscle mass bilaterally
- Pelvic limbs—normal to slightly increased patellar reflex, markedly reduced withdrawal reflex, no flexion of the tarsus, only flexed at the coxofemoral joint, reduced muscle tone (flaccid), and atrophy of the caudal thigh musculature and of the cranial and caudal muscles of the crus, bilaterally

Comparative Veterinary Anatomy: A Clinical Approach. https://doi.org/10.1016/B978-0-323-91015-6.00055-8

- Cutaneous trunci reflex—normal bilaterally
- Absent perineal reflex bilaterally; reduced tone of the external anal sphincter with dilated (open) anus; easy to express urine with minimal abdominal pressure
- Reduced tail tone; no voluntary tail movement

Cranial nerves

- Normal

Sensory

- Painful with palpation and pressure applied to the dorsal aspect of the lumbosacral articulation
- Painful with extension of the pelvic limbs and extension of the tail

Neuroanatomic diagnosis

L6 through caudal spinal cord segments, spinal roots, spinal nerves/cauda equina; or nerves of the lumbosacral plexus, including the sciatic, pudendal, pelvic, and caudal nerves

Differential diagnoses

Degenerative lumbosacral stenosis; primary neoplasia (vertebral neoplasm such as osteosarcoma, fibrosarcoma, meningioma, nerve sheath neoplasm); metastatic or multicentric neoplasia—e.g., lymphoma; neuritis involving the spinal nerve roots, spinal nerves/cauda equina, or named nerves—e.g., sciatic, pudendal, pelvic, and caudal nerves; discospondylitis with secondary vertebral canal empyema; infectious neuritis—e.g., bacterial, fungal, protozoal such as *Neospora caninum*, *Toxoplasma gondii*

Diagnostics

A CBC, serum chemistry profile, and urinalysis were assessed for a concurrent disease that may impact or preclude general anesthesia or identify evidence of a neoplasm that may have metastasized. Three-view thoracic radiographs were performed to screen for evidence of metastatic neoplasia. The results of these tests were within normal limits. The dog was placed under general anesthesia and an MRI of the lumbar vertebral column and sacrum revealed a large, herniated intervertebral disc that compromised $> 50\%$ of the vertebral canal at the lumbosacral disc space causing compression of the nerves of the cauda equina. Degeneration (loss of the normal T2-hyperintensity) of the lumbosacral intervertebral disc was present. There also was ventral spondylosis at the lumbosacral intervertebral disc space (Fig. 8.6-1).

Diagnosis

Degenerative lumbosacral stenosis

Treatment

Treatment of degenerative lumbosacral stenosis includes conservative and surgical treatments. Conservative therapy includes exercise restriction along with anti-inflammatory drugs (nonsteroidal anti-inflammatory drugs or corticosteroids). Analgesic medications may also be used in dogs that remain painful despite anti-inflammatory drugs. Alternatively, surgical treatment of degenerative lumbosacral stenosis involves performing a dorsal laminectomy and discectomy. This is achieved by removing the caudal portion of the lamina of L7 and cranial portion of the lamina of the sacrum. The spinal nerves that make up the cauda equina (dura-arachnoid sac and intradural and extradural spinal nerve roots) are retracted to one side of the vertebral canal enabling the resection of the dorsal annulus and nucleus pulposus on the ipsilateral side. This is repeated on the contralateral side. In some instances, stabilization of the lumbosacral articulation is performed in addition to discectomy. A variety of surgical stabilization techniques have been described.

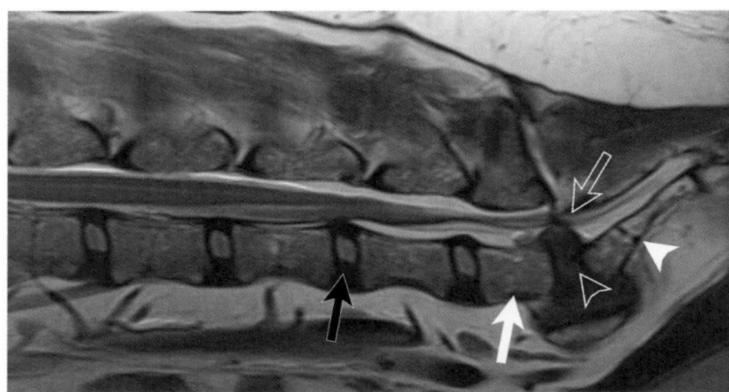

FIGURE 8.6-1 Sagittal T2-weighted MRI of the lumbar vertebral column and sacrum. The lumbosacral intervertebral disc is herniated (*open arrow*) resulting in compression of the cauda equina. In comparison with a normal intervertebral disc (*black arrow*), there is degenerative change to the nucleus pulposus of the lumbosacral intervertebral disc (*open arrowhead*) as evidenced by a loss of normal T2-hyperintensity. Ventrally, there is spondylosis deformans that spans from the L7 vertebra (*white arrow*) to the sacrum (*white arrowhead*).

Given the severity of the neurological dysfunction, the owners elected surgical treatment. Dorsal laminectomy and discectomy was performed. Post-operatively, the dog gradually regained strength in his pelvic limbs such that the gait and pelvic limb strength returned to normal over the course of 1 month. Likewise, the pain over the lumbosacral articulation resolved. While urinary and fecal incontinence improved, the dog still occasionally dribbled urine. The tail function partially returned to the point where the dog could weakly wag his tail.

Anatomical features in canids and felids

Introduction

The correct neuroanatomic diagnosis requires the recognition of signs related to dysfunction of the general somatic efferent (GSE), general somatic afferent (GSA) neurons involving the pelvic limb innervation, general visceral efferent (GVE) neurons, the innervation of sphincter muscles of the bladder and anus, and the innervation of the tail. Knowledge of the 4 clinically important nerves (femoral, sciatic, pudendal, and pelvic nerves) and their respective origins in the lumbar intumescence and sacral spinal cord segments (the L4 through S1 and the sacral spinal cord segments) is imperative to correctly form an accurate neuroanatomic diagnosis.

Function

Lesions that affect either the L4 through S1 spinal cord segment, the respective spinal roots, spinal nerves, lumbosacral plexus, or the named nerves of the pelvic limb result in lower motor neuron (LMN)-quality paresis. A further discussion of this follows after the relevant anatomy is covered.

Lumbar intumescence and sacral spinal cord segments

The **lumbar intumescence** is a grossly visible enlargement of the spinal cord that is composed of the L5, L6, L7, and 1st sacral spinal cord segments. The enlargement is due to the large number of GSE neurons located in the ventral gray column of these segments that ultimately give rise to the nerves of the pelvic limb. Caudal to the lumbar intumescence, the spinal cord tapers toward its termination. This tapering termination is called the **conus medullaris** and is comprised of the sacral and caudal spinal cord segments.

The GSE neurons for the pelvic limb are in the ventral gray column of the L4 through S1 spinal cord segments. The GSE axons exit the spinal cord ventrolaterally to give rise to the **ventral spinal roots**. Each ventral spinal root joins with its respective **dorsal root** forming a **spinal nerve** at each spinal cord segment. The spinal nerve exits the vertebral canal via an **intervertebral foramen**. Together, the L7 spinal nerves along with the sacral and caudal spinal nerves are referred to as the **cauda equina** (based on a resemblance of a horse's tail when viewed dorsally) (Fig. 8.6-2).

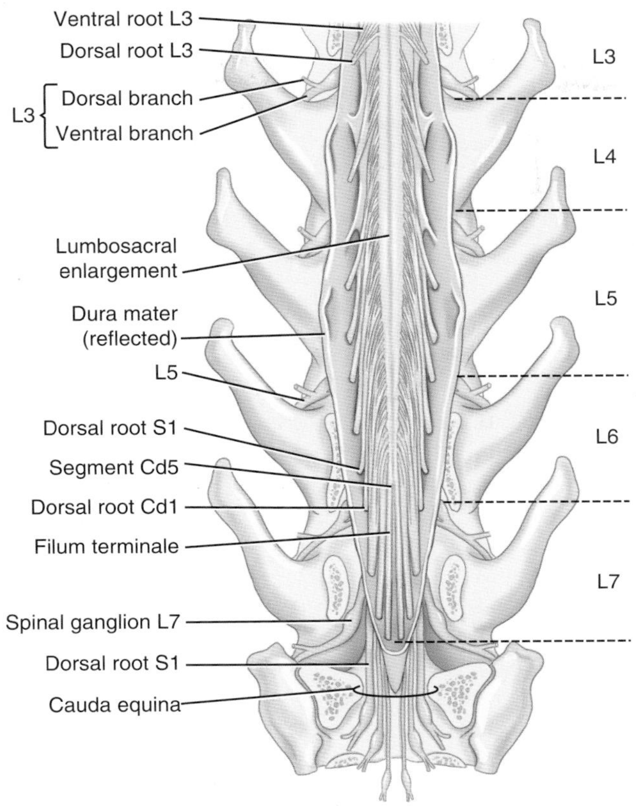

FIGURE 8.6-2 The cauda equina. (Modified from Miller and Evan's Anatomy of the Dog. 5th ed.)

The GSE neurons also are referred to as LMN. The axons of the GSE neurons of the lumbar intumescence and conus medullaris form the named nerves of the pelvic limb, provide the innervation of the external urethral sphincter and external anal sphincter, and provide innervation to the tail musculature. These GSE neurons give rise to axons that are in synaptic contact with the muscles they innervate via a neuromuscular junction. Together, the LMN, its axon, the neuromuscular junction, and the innervated muscle form the **LMN unit**.

The **GSA neurons** are in the spinal ganglia and provide sensory innervation. These GSA neurons provide general proprioception as well as nociception. For cutaneous innervation, GSA neurons have receptors in the skin that send impulses to the spinal cord which project along spinal nerves and their dorsal root to ultimately enter the spinal cord via the dorsal root entry zone in the dorsolateral sulcus to synapse in the dorsal gray column. Proprioception takes a similar pathway from receptors in muscle spindles, tendons, and joints to the dorsal gray column of the spinal cord.

In addition, **GVE neurons** are contained in the intermediate gray matter of the sacral spinal cord segments. The GVE neurons in the sacral spinal cord segments provide parasympathetic innervation to the bladder via the pelvic nerve. These GVE neurons give rise to axons that ultimately provide innervation to the smooth muscles of the bladder that form the detrusor muscle.

Spinal nerves

Analogous to the thoracic limb, the spinal nerves divide into 3 or 4 main branches (dorsal, ventral, communicating, and meningeal [variably present]) after exiting their respective intervertebral foraminae. **Dorsal branches** provide innervation to the epaxial muscles and skin along the dorsal aspect of the body. **Ventral branches** intermingle into a collection of spinal nerves called the **lumbosacral plexus**. They also supply innervation to the skin over the lateral and ventral aspects of the body (Fig. 8.6-3). Emerging from the lumbosacral plexus are the named nerves that provide motor and sensory innervation to the pelvic limb and the innervation of the bladder, distal colon, rectum, and anus, as well as the sensory innervation to the skin of the inguinal region and perineum (Fig. 8.6-4).

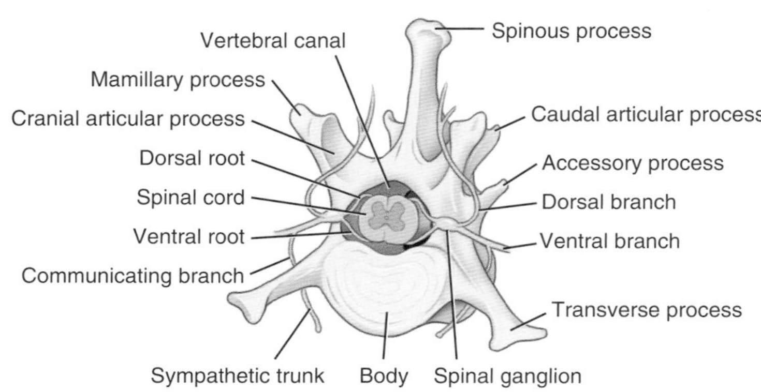

FIGURE 8.6-3 The spinal nerves and their main branches.

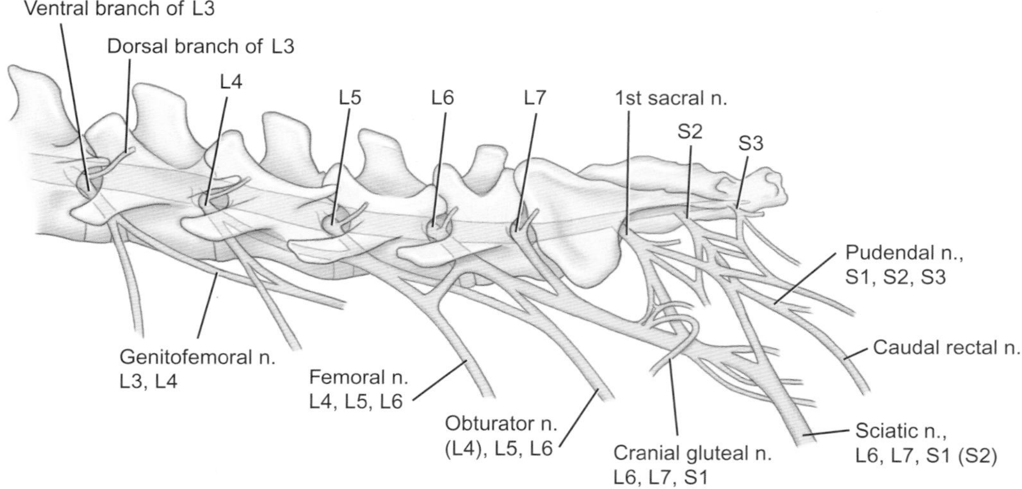

FIGURE 8.6-4 The lumbar, sacral, and caudal spinal cord segments and their spinal nerves.

The 2 clinically important nerves of the pelvic limb are the **femoral** and **sciatic** nerve. The sciatic nerve divides into 2 main (terminal) branches, each of which provides major contributions to normal function of the pelvic limb. Additionally, the 2 main nerves that provide for normal bowel and bladder function are the **pelvic** and **pudendal nerves**, originating from the lumbosacral plexus.

Function as it relates to neuroanatomy

Dysfunction of any part of the LMN unit (GSE neuron, its axon, the neuromuscular junction, or muscle) results in LMN paresis (weakness) or paralysis (loss of voluntary movement). The term "neuromuscular" can be used synonymously with LMN. Dysfunction of the GSA neurons or their processes results in varying degrees of hypalgesia or analgesia. Clinically, it is challenging to differentiate a lesion affecting any one portion of the LMN unit. Signs related to LMN paresis from a lesion affecting the spinal cord gray matter, nerves, neuromuscular junction, or muscle look similar. In most cases, the most specific neuroanatomic diagnosis may be "LMN dysfunction."

Lower motor neuron paresis is characterized as an inability to support weight. During the stance (weight-bearing) phase of the gait, the patient collapses on the limb resulting in overflexion of the joints. When standing, the patient may have a crouched posture. Overflexion of the tarsus results in a plantigrade stance (the tuber calcanei is closer to the ground surface than normal). The foot of the pelvic limb may knuckle over causing the patient to stand on the dorsum of the foot. Evaluation of the gait reveals a shortened stride length, which has a "choppy" appearance. Importantly, a short-strided gait looks very similar to the gait of a patient with orthopedic lameness in which the stride is shortened as the patient off loads its weight due to pain.

Given the prevalence of orthopedic diseases (please see Case 8.7), such as hip dysplasia and disruption of the cranial cruciate ligament, it is imperative to perform an orthopedic examination in order to exclude orthopedic conditions in all patients with gait abnormalities of the pelvic limb.

Lower motor neuron dysfunction also results in abnormal postural reactions. Performance of normal postural reactions necessitates not only the normal function of the LMN, but also the normal function of the ascending general proprioceptive (GP) pathways that convey the position sense of the limb to the brain (GP information projects to the contralateral prosencephalon [somesthetic cerebral cortex] and the ipsilateral cerebellum) and the normal function of the descending upper motor neurons. In the case of LMN paresis, the patient lacks the strength to perform postural reactions correctly. As observed in the gait, patients with LMN dysfunction may make rapid short hops when performing the postural reaction called "hopping." In those patients that are too weak to stand, the deficits in the postural reactions can be compensated for by the examiner providing support for the patient to stand or perform the postural reactions.

Spinal reflexes

The recognition of LMN paresis is through the evaluation of the patient's gait, ability to bear weight when standing, spinal reflexes, muscle tone of the limbs, and the presence of muscle atrophy. A spinal reflex is composed of a sensory nerve that projects to the spinal cord, the specific spinal cord segments where the incoming sensory axons project directly or indirectly through interneurons to synapse on GSE neurons in the ventral gray matter of the spinal cord, and the GSE axons that give rise to the motor nerve which are in synaptic contact with a muscle (see the reflex arc in Fig. 7.5-4). Therefore, spinal reflexes are a means to test the functional integrity of sensory nerves (afferent), specific spinal cord segments, and motor nerves (efferent). Knowledge of the named nerves of the pelvic limb, the muscles each nerve innervates, and the spinal cord segments that give rise to each nerve is imperative in forming a correct neuroanatomic diagnosis (Fig. 8.6-5, Table 8.6-1). Dysfunction of the LMN results in decreased-to-absent spinal reflexes and a loss of muscular tone and strength (flaccid paresis) (Table 8.6-2). Loss of LMN innervation of the muscle also results in a rapid and severe muscle atrophy.

The ability to bear weight in the pelvic limb is mainly provided by the extensors of the stifle and tarsus. Extension of the stifle is accomplished through the action of the quadriceps muscle, which is innervated by the femoral nerve. The GSE neurons that give rise to the femoral nerve reside in the ventral gray column of the L4, L5, and L6 spinal cord segments. Dysfunction of the L4 through the L6 spinal cord segments or the femoral nerve results in an inability to extend the stifle. Consequently, during the weight-bearing (stance) phase of the gait, the stifle flexes under the weight of the patient. When standing, the stifle is in a flexed position, resulting in a crouched posture.

In the pelvic limb, there are 2 reliable spinal reflexes that can be tested—the patellar reflex and the withdrawal reflex. The patellar reflex assesses the functional integrity of the femoral nerve and the L4, L5, and L6 spinal cord segments. The patient is placed in lateral recumbency with the limb relaxed and the stifle flexed. The patellar reflex is elicited by striking the patellar ligament with a reflex hammer and observing extension of the stifle by the action of

PATELLAR REFLEX

Lesions that affect the L4, L5, L6 spinal cord segments, their respective roots, spinal nerves, or femoral nerve result in a diminished-to-absent patellar reflex. Therefore, the finding of a diminished-to-absent patellar reflex signifies the lesion must involve these anatomic structures. Importantly, patients with a lesion that affects the CNS cranial to the L4, L5, L6 spinal cord segments have a normal-to-exaggerated (hyper-reflexive) patellar reflex. The reflex movement (extension of the stifle) may appear exaggerated due to the loss of the descending UMN pathways that typically provide inhibition of the reflex arc. Furthermore, a lesion cranial to the L4, L5, and L6 spinal cord segments also disrupts the ascending GP pathways and causes a postural reaction deficit and general proprioceptive ataxia.

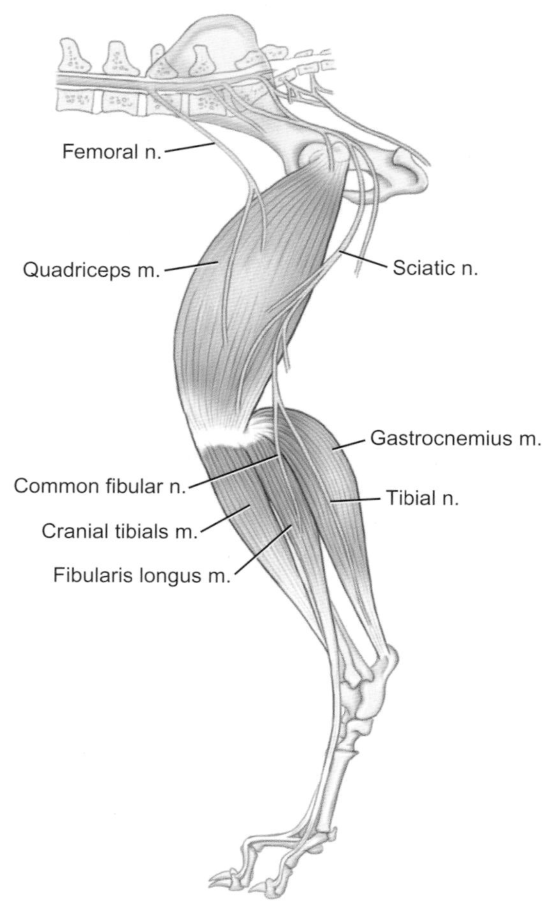

FIGURE 8.6-5 The femoral and sciatic nerves along with a few muscles innervated by these nerves.

TABLE 8.6-1 Pelvic limb innervations.

Spinal cord segments	*Nerve*	*Action/muscles innervated*
L4, L5, L6	Femoral	Extension of the stifle, flexion of the hip
L6, L7, S1	Sciatic nerve	
	• Common fibular	Flexion of the tarsus, extension of the digits
	• Tibial	Flexion of the stifle, extension of the tarsus, and flexion of the digits
S1, S2, S3	Pudendal	External urethral and anal sphincters
S1, S2, S3	Pelvic nerve	Parasympathetic innervation to the detrusor muscle of the bladder

TABLE 8.6-2 Pelvic limb reflexes and posture.

Spinal cord segments	*Nerve*	*Action/muscles innervated*
Patellar	Femoral	Extension of the stifle
Withdrawal	Sciatic (common fibular)	Flexion (flexion at tarsus)
Perineal	Pudendal ([superficial] perineal nerve)	Contraction of the external anal sphincter
Plantigrade stances	Sciatic (tibial)	Paresis of the extensors of the tarsus

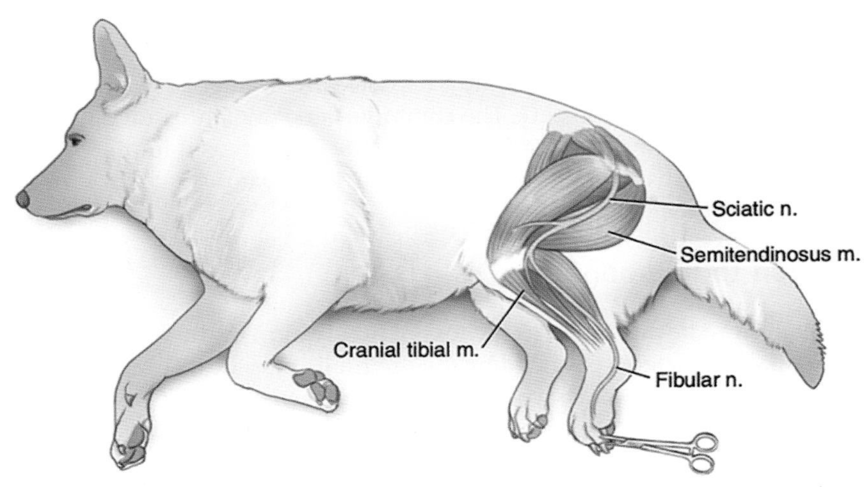

FIGURE 8.6-6 The withdrawal reflex.

the quadriceps muscle. A decreased-to-absent patellar reflex implies the loss of function of the L4, L5, and L6 spinal cord segments, their respective spinal roots, spinal nerves, lumbar plexus femoral nerve, or the quadriceps muscle (see side box entitled "Patellar reflex").

The withdrawal reflex is performed by applying pressure to a digit of the pelvic limb and assessing the patient's ability to flex all the joints of the pelvic limb in an effort to pull the limb toward the body (Fig. 8.6-6). Flexion of the coxofemoral joint is accomplished by the action of several muscles, including the iliopsoas, cranial muscles of the thigh (including the sartorius and rectus femoris muscles), and the tensor fasciae latae muscle, which are innervated by the ventral branches of the lumbar spinal nerves, the femoral nerve, and sciatic nerve, respectively. Hence, flexion of the coxofemoral joint is accomplished by multiple muscles with innervation by several nerves. Therefore, the ability to flex the coxofemoral joint may be preserved even if several spinal cord segments, spinal roots, spinal nerves, or individual nerves are affected (i.e., flexion of the coxofemoral joint is preserved with sciatic nerve lesions). Consequently, the ability to flex the coxofemoral joint contributes less than other findings toward forming an accurate neuroanatomic diagnosis (see side box entitled "Upper vs. lower motor neuron?").

In contrast, flexion of the stifle and tarsus is mainly provided by innervation by the L6, L7, and S1 spinal cord segments, respective spinal roots, spinal nerves, sciatic nerve, or its 2 terminal branches—the common fibular and tibial nerves. Flexion of the stifle is accomplished by the action of the caudal and medial thigh muscles, which are mainly innervated by the sciatic nerve. Flexion of the tarsus is accomplished by the action of the craniolateral crural muscles, which are innervated by the common fibular nerve. The GSE neurons that give rise to the common fibular nerve are in the ventral gray column of the L6, L7, and S1 spinal cord segments. Therefore, careful attention should be focused on the ability of the patient to flex the stifle and tarsus as this is specifically provided for by the sciatic and its terminal branch, the common fibular and tibial nerves. An inability to flex these joints implies dysfunction of the L6, L7, and S1 spinal cord segments, sciatic nerve, or the common fibular nerve.

UPPER VS. LOWER MOTOR NEURON?

There are only 2 potential neuroanatomic diagnoses to account for dogs and cats with one or more paralyzed pelvic limbs (mono- or paraplegia) in which there is a complete inability to flex all the joints of the pelvic limb, including the coxofemoral joint. One neuroanatomic diagnosis would be that the lesion must involve all the spinal cord segments from L4 through S1, their respective spinal roots, spinal nerves, lumbosacral plexus, or the sciatic and femoral nerves—said another way, LMN paralysis of the limb. Alternatively, the lesion may affect the spinal cord cranial to the lumbar intumescence and, therefore, be a lesion that disrupts the GP and UMN pathways; this is often called an "UMN lesion."

Conversely, extension of the tarsus is provided by the thigh and caudal crural muscles. The caudal crural muscles are innervated by the tibial nerve. Like the common fibular nerve, the GSE neurons that give rise to the tibial nerve are located in the ventral gray column of the L6, L7, and S1 spinal cord segments. As mentioned above, an inability to extend the tarsus results in a plantigrade posture. The combination of a plantigrade stance—i.e., an inability to extend the tarsus—and an inability to flex the tarsus (as seen in the present case) implies dysfunction of the L6, L7, S1 spinal cord segments, their respective spinal roots, spinal nerves, lumbar plexus, sciatic nerve, common fibular and tibial nerves, or the muscles of the pelvic limb.

Incontinence as related to neuroanatomy

Urinary and fecal incontinence can be attributed to the loss of innervation of the external sphincters of the urethra and anus. The external urethral sphincter is provided by the urethralis muscle. Its innervation is provided by the GSE neurons in the ventral gray column of the sacral spinal cord segments. These GSE axons pass via the ventral sacral roots into spinal nerves, which coalesce in the lumbosacral plexus and ultimately form the **pudendal nerve**. Likewise, the pudendal nerve gives off the **caudal rectal nerve**, which innervates the external anal sphincter muscle. Function of the external anal sphincter is assessed by evaluating sphincter tone during a digital rectal examination and the perineal reflex. The perineal reflex is performed by pinching the skin lateral to the anus and observing contraction of the external anal sphincter. Additionally, many dogs also flex their tail, which provides information about tail function. The GSE neurons for the tail reside in the ventral gray column of the caudal spinal cord segments.

The lateral horns of the sacral spinal cord segments also contain the preganglionic parasympathetic GVE neurons that primarily function in urine voiding. The axons of the preganglionic parasympathetic GVE neurons course in the pelvic nerve to synapse in pelvic plexus ganglia or ganglia in the bladder wall. Given the close anatomic relationship of the GSE and GVE neurons in the sacral spinal cord segments, roots, and spinal nerves, as well as the pelvic and pudendal nerves, parasympathetic dysfunction also often results in an inability to void urine normally. Urinary and fecal incontinence as a consequence of LMN dysfunction of the external sphincters of the urethra and anus along with the absent perineal reflex implies dysfunction of the sacral spinal cord segments, their respective spinal roots, spinal nerves, sacral or pelvic plexuses, pelvic and pudendal nerves, or the sphincter muscles.

Regional nociception

The femoral nerve, terminal branches of the sciatic nerve (common fibular and tibial nerves), pudendal nerve, and caudal nerves also provide GSA innervation to the skin of the pelvic limb, inguinal region, perineum, and tail, respectively (Table 8.6-3). Therefore, lesions affecting the spinal cord segments L4 through caudal spinal cord segments, their respective dorsal roots, spinal nerves, lumbosacral plexus, and named nerves may result in decreased-to-absent nociception (sensation). Sensory testing is accomplished by pinching the skin and observing that the patient has a conscious reaction of the stimulus—i.e., seeing the patient turn to look at the skin being pinched or vocalizing. Remember, merely observing the limb withdraw (pull away) is a reflex that does not signify that the patient has normal nociception. A conscious reaction signifies that the stimulus resulted in an impulse that projected along a sensory nerve to enter the spinal cord and ultimately projected to the prosencephalon where the stimulus was consciously recognized as being noxious.

The skin innervated by a single nerve is called a **dermatome** or **cutaneous area**. The innervation of some areas of skin is provided for by one or more nerves. These areas are called **overlap zones**. An area of skin innervated by only one nerve is called an **autonomous zone** (see Fig. 7.5-5). Knowledge of autonomous zones is critical to assessing

TABLE 8.6-3 Pelvic limb autonomous zones.

Nerve	*Area of skin to test autonomous zone*
Femoral nerve	Skin on the medial surface of the stifle (level of the condyle)
Common fibular	Dorsal surface of the metatarsus and digits
Tibial	Plantar surface of the metatarsus and digits
Pudendal	Skin lateral to the anus

sensory innervation. Reduced or absent nociception to the skin in an autonomous zone implicates dysfunction of a specific nerve.

Both the **common fibular** and **tibial nerves** provide cutaneous innervation to the crus, tarsus, metatarsus, and skin on the dorsal and plantar aspects of the paw, respectively. The common fibular nerve gives rise to **superficial** and **deep fibular nerves,** both of which supply cutaneous innervation to the skin on the dorsum of the foot.

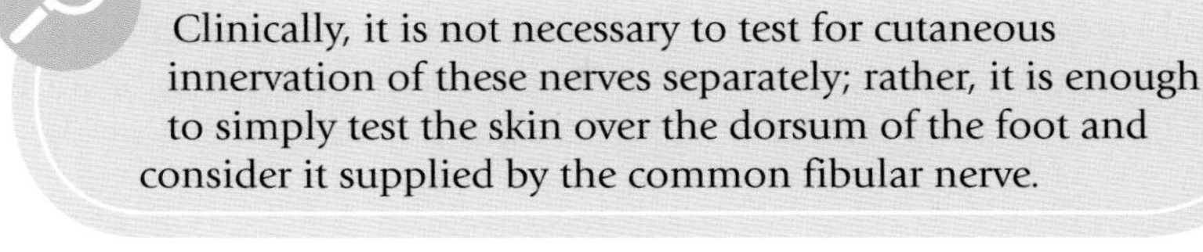

Clinically, it is not necessary to test for cutaneous innervation of these nerves separately; rather, it is enough to simply test the skin over the dorsum of the foot and consider it supplied by the common fibular nerve.

The tibial nerve provides cutaneous innervation to the plantar aspect of the metatarsus and foot. Just proximal to the tarsus, the tibial nerve bifurcates into medial and lateral branches, which provide cutaneous innervation to the plantar aspect of the metatarsus and digits.

Clinically, it is not necessary to test for cutaneous innervation of these nerves separately; rather, it is sufficient to simply test the skin over the plantar aspect of the metatarsus or digits and consider it supplied by the tibial nerve.

The femoral nerve also gives rise to the **saphenous nerve**, which provides cutaneous innervation to the medial aspect of the thigh, crus, tarsus, metatarsus, and 1st digit (when present). There is considerable overlap between the cutaneous innervation of the saphenous nerve and the superficial fibular nerve. Therefore, the saphenous nerve does not have an autonomous zone distal to the tarsus (Fig. 8.6-7).

This is important in cases of dysfunction of the L4, L5, and L6 spinal cord segments, roots, spinal nerves, or femoral nerve. In such cases, cutaneous innervation of the 1st digit would be provided by the superficial fibular nerve. Therefore, testing of cutaneous innervation provided by the saphenous nerve should be done by pinching the skin over the medial aspect of the stifle.

The cutaneous innervation of the skin of the perineum is provided by perineal nerves, which are branches of the pudendal nerve. To test the cutaneous innervation provided for by the perineal nerves, the skin lateral to the anus is stimulated.

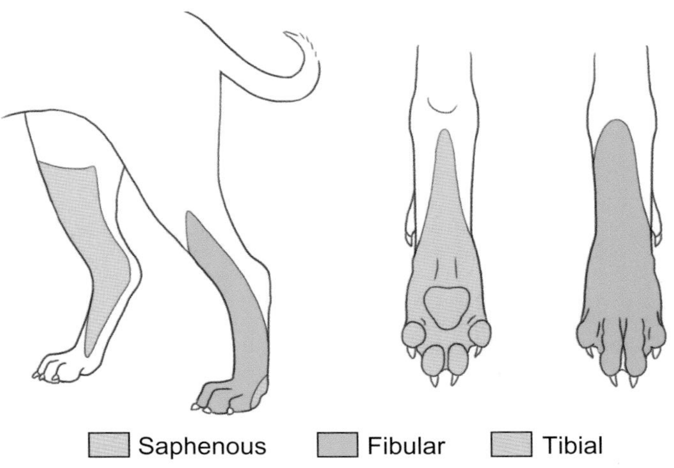

FIGURE 8.6-7 Autonomous zone of the foot of the pelvic limb.

Innervation of the tail

The innervation of the muscles of the tail is provided by the **caudal nerves**. The GSE neurons that give rise to the caudal nerves are located in the caudal spinal cord segments that—together with S2 and S3—make up the conus medullaris. The termination of the spinal cord (end of the conus medullaris) in relation to the vertebral canal varies between species. In the dog, the termination of the spinal cord is located between the L6 and L7 vertebrae. The termination of the spinal cord in large breed dogs tends to be more cranial, whereas in small breed dogs, the termination of the spinal cord tends to be more caudal. In the cat, the spinal cord termination is located between the L7 vertebra and the sacrum. Given the cranial location of the spinal cord termination in relation to the caudal vertebrae, the caudal spinal roots and nerves track a long distance within the vertebral canal before exiting their respective intervertebral foraminae. Once external to the vertebral canal, the caudal nerves take a short course to innervate the tail musculature.

As a result of their long course within the vertebral canal combined with a short course external to the vertebral canal, a lesion that would affect the innervation to the pelvic limbs and cause LMN paresis of the tail must be located within the vertebral canal to affect the L6 through caudal segments or their respective spinal roots or spinal nerves.

Selected references

[1] Meij BP, Bergknut N. Degenerative lumbosacral stenosis in dogs. Vet Clin Small Anim Prac 2010;40:983–1009.
[2] Evans HE, de Lahunta A. Spinal nerves. Miller's anatomy of the dog. Elsevier; 2013. p. 611–57.
[3] Worthman R. Demonstration of specific nerve paralyses in the dog. J Am Vet Med Assoc.
[4] Bailey CS, Kitchell RL. Cutaneous sensory testing in the dog. J Vet Int Med 1987;1:128–35.

CHAPTER 8

CASE 8.7

Calcaneal Tendon Injury

Marc Kent[a], Eric N. Glass[b], and Alexander de Lahunta[c]
[a]Department of Small Animal Medicine and Surgery, University of Georgia College of Veterinary Medicine, Athens, Georgia, US
[b]Neurology and Neurosurgery, Red Bank Veterinary Hospital, Tinton Falls, New Jersey, US
[c]Professor Emeritus, Biomedical Sciences, Cornell University College of Veterinary Medicine, Ithaca, New York, US

Clinical case

History

An 8-year-old male intact German Shorthaired Pointer was presented with a 6-week history of a persistent left pelvic limb gait abnormality. The lameness started when the owners heard a "pop," and the dog was suddenly lame. Initially, the primary care veterinarian suspected a "back problem" and prescribed exercise restriction and nonsteroidal anti-inflammatory medication. Except for the lameness, the owner reported that the dog was otherwise normal and was still active. The dog was used for bird hunting. Apart from occasional lacerations during hunting trips, there was no history of trauma.

Physical examination findings

The physical examination was normal except for the left pelvic limb. The dog was lame in the left pelvic limb (see Fig. 8.7-V1 in the online version at https://doi.org/10.1016/B978-0-323-91015-6.00144-8). The dog was able to walk on the affected pelvic limb. In the stance (weight-bearing) phase of the gait, the dog was plantigrade (over-flexed at the tarsocrural joint) in the left pelvic limb compared with the right (Fig. 8.7-1). When standing, the dog did not bear full weight on the left pelvic limb. Additionally, when standing with a plantigrade stance in the left pelvic limb, the dog's left pelvic limb digits were slightly flexed. There was a swelling, which was mildly painful upon palpation, at the site of the attachment of the common calcanean tendon on the tuber calcanei on the left pelvic limb. When the left stifle was extended, a greater degree of flexion could be achieved in the left tarsocrural joint than in the right tarsocrural joint when the right stifle was extended. Neurologically, the left pelvic limb had normal postural reactions (when the dog's weight was supported), spinal reflexes (patellar and withdrawal reflexes), and sensation in the autonomous zone of the tibial and fibular nerves.

MRI CONFOUNDING ARTIFACTS

Occasionally, a unique MRI artifact may be observed in tendons. In highly organized tissues such as tendons, the water molecules within the tendon align in one direction. Normally, water molecules in tendons interact in such a way that the tendon lacks signal intensity and appears hypointense. However, if the direction of the alignment of water molecules happens to be at a specific angle, called the "magic angle" (approximately 55° to the main axis of the MRI), the interactions between water molecules become altered and signal intensity within the tendon may become evident. On T2-weighted sequences, this results in an increase in signal—i.e., making the tendon appear white rather than black. This artifact may be misinterpreted as a pathologic change.

Comparative Veterinary Anatomy: A Clinical Approach. https://doi.org/10.1016/B978-0-323-91015-6.00144-8

FIGURE 8.7-1 The 8-year-old male intact German Shorthaired Pointer in this case at the time of presentation. The dog was plantigrade (overflexion of the tarsus) in the left pelvic limb secondary to injury of the left common calcanean tendon. Note the flexion of the digits of the left pelvic limb when weight-bearing, which suggests that the superficial digital flexor tendon is intact. (Image captured from video.)

Differential diagnoses

Partial avulsion of the attachment of the common calcanean tendon on the tuber calcanei (associated with suspected traumatic injury to the common calcanean tendon); dysfunction of the lower motor neuron units that contribute to the tibial nerve (i.e., ventral gray matter in the L6-S1 spinal cord segments, their spinal roots; spinal nerves; lumbosacral plexus; or tibial nerve itself)

Diagnostics

This patient was a highly trained hunting dog in whose training the owners had made a substantial financial investment. As a result, an MRI was pursued given its exceptional contrast resolution which would be able to best identify and define pathology in the various soft tissues at the site of attachment of the common calcanean tendon on the tuber calcanei. Moreover, pathologic changes that might involve the distal tibia and fibula as well as the tarsal bones and the tarsocrural joints could be seen with MRI. The MRI revealed a primary lesion involving the attachment of the common calcanean tendon (Fig. 8.7-2). As the tendons of the gastrocnemius and superficial digital flexor muscles and the combined (united) tendons of the biceps femoris, gracilis, and semitendinosus muscles were traced distally, the normally hypointense (signal void—black) combined (united) tendons of the biceps femoris, gracilis, and semitendinosus muscles were replaced by T2-hyperintense tissue at approximately 1–2 cm (0.4–0.8 in.) proximal to the site of attachment on the tuber calcanei. Likewise, the tissues surrounding the tendons of the gastrocnemius and superficial digital flexor muscles were swollen and T2-hyperintense. The site of attachment of the combined (united) tendons of the biceps femoris, gracilis, and semitendinosus muscles and the surrounding tissues displayed an abnormal contrast enhancement. Just proximal to the dorsal surface of the tuber calcanei, only the tendons of the gastrocnemius and superficial digital flexor muscles appeared normal, while the combined (united) tendons of the biceps femoris, gracilis, and semitendinosus muscles were not visualized. Instead, the combined (united) tendons of the biceps femoris, gracilis, and semitendinosus muscles were replaced by abnormal contrast-enhancing tissue on the transverse T1W post-contrast images.

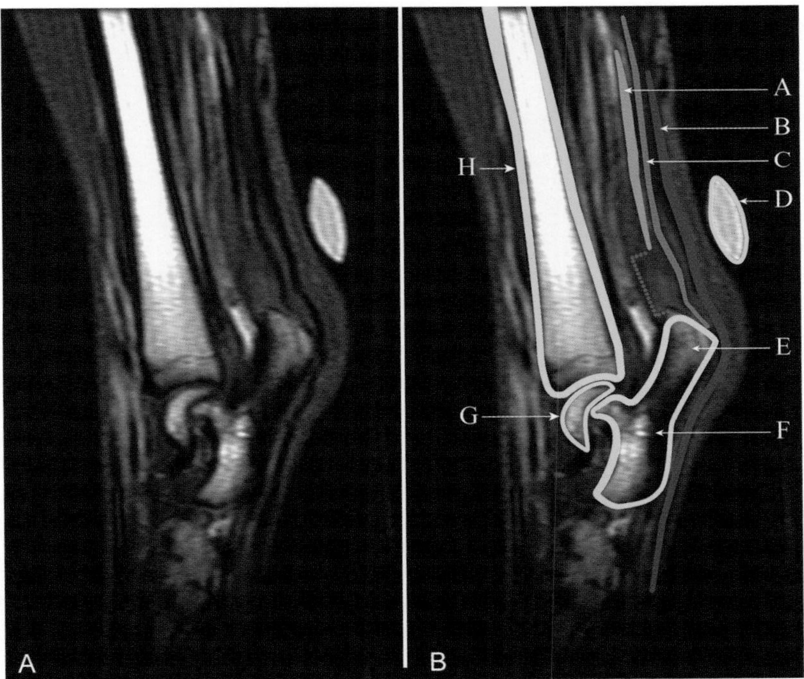

FIGURE 8.7-2 (A) Sagittal T2-weighted MRI of the left distal tibia and tarsus. The distal 2 cm (0.8 in.) of the combined tendons of the biceps femoris, gracilis, and semitendinosus muscles is abruptly disrupted from its attachment to the tuber calcanei (*red bracket*). In place of the normal hypointense (*black*) tendon, there is tissue with an increased T2-weighted signal intensity. The tendons of the gastrocnemius and superficial digital flexor muscles are normal. (B) Annotated image. Key: A—combined tendons of the biceps femoris, gracilis, and semitendinosus muscles; B—tendon of the superficial digital flexor muscle; C—tendon of the gastrocnemius muscle; D—MRI marker; E—tuber calcanei; F—calcaneus; G—trochlea of the talus; H—tibia.

Diagnosis

Partial avulsion of the combined (united) tendons of the biceps femoris, gracilis, and semitendinosus muscles from the dorsal surface of the tuber calcanei with an approximately 2-cm (0.8-in.) long fibrous thickening at the site of partial avulsion of the tendon

Treatment

The dog was placed under general anesthesia and a lateral incision was made centered over the site of attachment of the common calcaneal tendon on the tuber calcanei. The tendon sheath of the common calcanean tendon was incised allowing the visualization of the components of the common calcanean tendon (Fig. 8.7-3). The tendon sheath of the superficial digital flexor muscle was retracted medially. The combined (united) tendons of the biceps femoris, gracilis, and semitendinosus muscles were identified proximally and traced distally to where they were avulsed from the tuber calcanei. The avulsed end of the combined (united) tendons of the biceps femoris, gracilis, and semitendinosus muscles was debrided and anchored to the proximal aspect of the calcaneus using a modified 3-loop pulley suture pattern. Laxity in the tendon of the gastrocnemius muscle was identified and tightened using a locking loop suture passed through a drill hole in the calcaneus. The superficial digital flexor tendon was sutured with a horizontal mattress suture to imbricate (L. *imbricatus* overlapping opposing surfaces) the tendon sheath. Post-operatively, the dog was placed in a full-limb cast (bivalved and taped) from just distal to the stifle to the digits. This allowed for a periodic inspection of the surgical site. External coaptation of the limb was maintained for approximately 2 months, followed by physical rehabilitation. At 3 months post-operative, the dog had regained normal limb function.

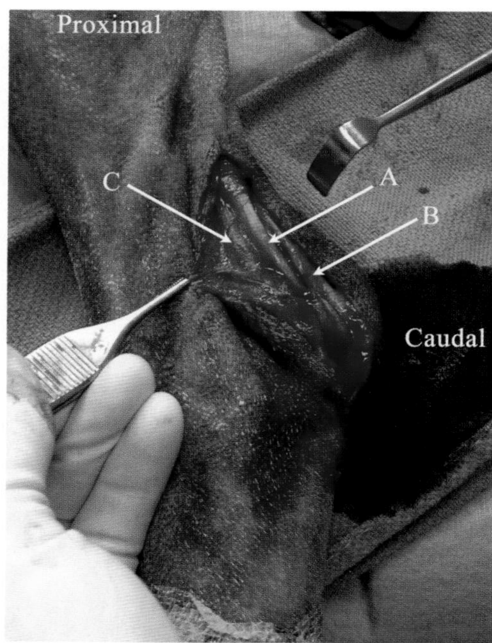

FIGURE 8.7-3 Intra-operative image of the components of the common calcanean tendon. Through a lateral skin incision, the tendon sheath was incised allowing visualization of the tendon of the gastrocnemius (A) and superficial digital flexor (B) muscles. There is abnormal fibrous tissue at the distal aspect of the combined (united) tendons of the biceps femoris, gracilis, and semitendinosus muscles (C).

Anatomical features in canids and felids

Introduction

The normal function of a thoracic or pelvic limb needs the functional integrity of the central nervous system, peripheral nervous system, and musculoskeletal structures. The central nervous system provides for a coordinated integration of the musculature of the limb. Upper motor neurons descend from the brain to initiate the lower motor neurons (LMN) that finally give rise to the nerves of the limb. Lower motor neurons in the cervical and lumbar intumescences send axons via the ventral roots, spinal nerves and plexi (brachial and lumbosacral, respectively) to form the named nerves of the limbs. In turn, these motor nerves are in a synaptic connection with the limb musculature, which results in limb movement. Contraction of a muscle results in either flexion or extension of a joint; the action of the muscle is translated into movement at a joint by the pull exerted by the muscle through a tendinous attachment to a bone. Intuitively, abnormal limb function as evidenced by an abnormal gait may be the consequence of either nervous system dysfunction or an abnormality in the musculoskeletal structures of the limb. Therefore, examination of a dog or cat with an abnormal gait involves an assessment of the nervous system as well as a careful assessment of the musculoskeletal structures, including muscles, tendons, ligaments, and joints.

This section covers the muscles and innervation of the crus and tarsus with special reference to the neuromuscular function of this region. Bones, ligaments, tendons, and other soft tissues, regional blood supply, and lymphatics are covered in Cases 7.3 and 8.5.

Function

Normal posture while standing and maintenance of a normal gait are predicated on the normal musculoskeletal and nervous system function. The tarsal joints are a composite joint that primarily permits flexion and extension. The largest range of motion of the tarsal joints is provided by the tarsocrural joint (mainly created by the articulation of the tibia with the trochlea of the talus). Tension applied to the common calcanean tendon results in extension of the tarsocrural joint. The common calcanean tendon is comprised of the tendons of the gastrocnemius and superficial digital flexor muscles and the combined (united) tendons of the biceps femoris, gracilis, and semitendinosus muscles. Extension of the tarsocrural joint is achieved through the action of the caudal crural muscles, principally the gastrocnemius muscle along with the superficial digital flexor muscles. These muscles maintain the extension of the tarsocrural joint in standing posture and during the stance phase of the gait. Additionally, the actions of the gracilis, semitendinosus, and biceps femoris muscles through their combined (united) tendons also result in extension of the tarsal joint.

As the tarsocrural joint flexes with weight-bearing, the calcaneus rotates ventrally placing tension on the tendon of the superficial digital flexor muscle resulting in flexion of the digits (Fig. 8.7-4). In the present case, seeing abnormal flexion of the digits while standing on the pelvic limb supports the normal integrity of the superficial digital flexor muscle and its tendon (Fig. 8.7-5). Finally, the actions of the biceps femoris, gracilis, and semitendinosus muscles are primarily to extend the stifle as well as the tarsal joints.

Extension of the stifle should also result in extension of the tarsocrural joint. A greater-than-normal amount of flexion of the tarsocrural joint with the stifle in extension suggests the disruption of one or more components of the common calcanean tendon. A similar finding would be expected with LMN paresis (weakness) involving the sciatic nerve or the tibial nerve which innervates the caudal crural muscles. With LMN paresis, there is flaccid muscle tone which allows for a greater degree of stretch of the muscle.

While not pathognomonic, the ability to flex the tarsocrural joint while the stifle is held in extension supports the diagnosis of musculotendinous pathology of one of the components of the common calcanean tendon. In the present case, the diagnosis of an injury to one of the components of the common calcanean tendon was based on pain and swelling at the level of the tuber calcanei. Key to the diagnosis was the observation of the flexed digits when weight-bearing.

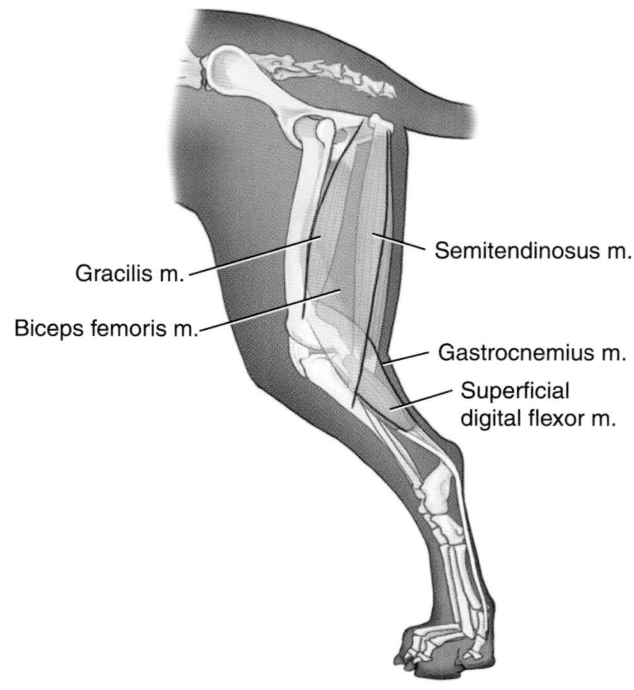

FIGURE 8.7-4 Lateral view of the normal anatomy of the muscles contributing to the common calcanean tendon.

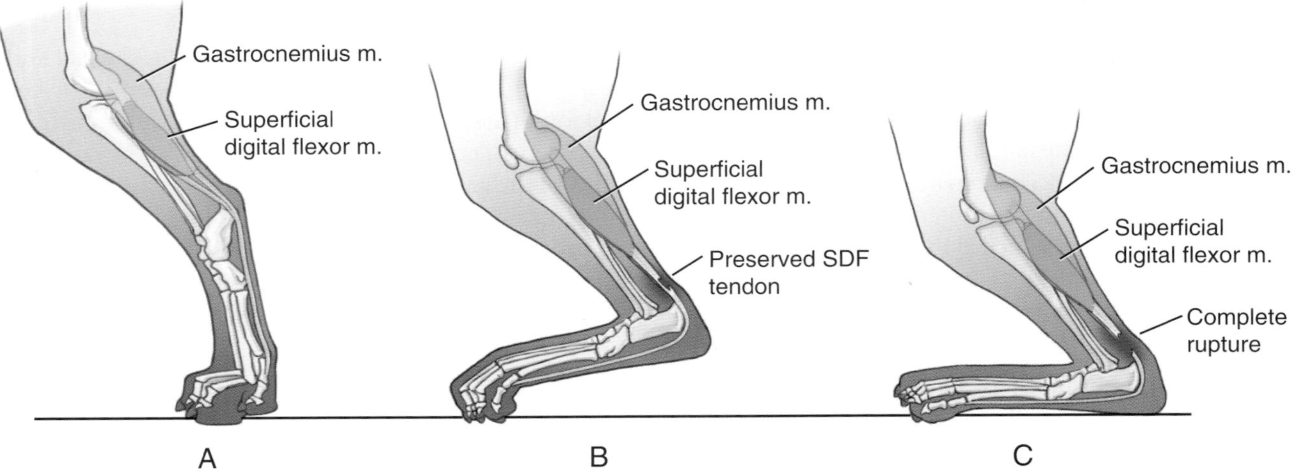

FIGURE 8.7-5 Normal anatomy (A). Consequence of a partial (B) and complete (C) disruption of components of the common calcanean tendon. In both cases, there is a plantigrade stance.

These muscles are innervated by the tibial branch of the sciatic nerve. Consequently, the observation of a plantigrade stance implies a musculo(tendinous)skeletal disorder affecting one or more of the components of the common calcanean tendon (gastrocnemius, superficial digital flexor muscles, biceps femoris, gracilis, and/or semitendinosus muscles), or their attachment on the tuber calcanei. Likewise, pathological alterations (fracture or luxation) of the tarsal joints may result in a plantigrade stance. Alternatively, a plantigrade stance may be the consequence of a lesion affecting the L6 through S1 spinal cord segments, the sciatic nerve, or tibial nerve.

Regional muscles of the crus

The **gastrocnemius** and **superficial digital flexor muscles** are the extensors of the tarsal joints. Their action exerts their effect by tension applied to the **common calcanean tendon**.

The **biceps femoris** muscle has a dual action. The cranial portion of the muscle functions in an antigravity role helping support weight (extension of the hip) when standing. It also helps extend the tarsocrural joint via its contribution to the common calcanean tendon. During the protraction phase (swing phase), the caudal aspect of the biceps femoris muscle causes stifle flexion.

The **semitendinosus muscle** is located between the biceps femoris muscle (lateral) and the semimembranosus muscle (medially). Its action is to extend the hip, flex the stifle, and—through its contribution to the common calcanean tendon—extend the tarsocrural joint.

The **gracilis muscle** is located medially on the thigh. At the cranial aspect of its insertion, it attaches to the cranial border of the tibia, whereas the caudal aspect of the distal end of the terminal aponeurosis contributes to the common calcanean tendon. The gracilis muscle is active during weight-bearing via its action to extend hip, stifle, and tarsocrural joints.

The **common calcanean tendon** encompasses the tendons of the gastrocnemius, superficial digital flexor, biceps femoris, gracilis, and semitendinosus muscles. The gastrocnemius and superficial digital flexor tendons are the main contributors. The common calcanean tendon attaches on the dorsal aspect of the tuber calcanei.

Lumbosacral plexus

This information is covered in Case 8.6.

Regional innervation

Innervation of the muscles responsible for the extension of the tarsal joints is by the **tibial nerve,** which is a branch of the **sciatic nerve** (see Case 8.6 for nerves of the pelvic limb). The GSE cell bodies that give rise to the tibial nerve are located in the ventral gray column in the L6 through S1 spinal cord segments.

Selected references

[1] Cervi M, Brebner N, Liptak J. Short- and long-term outcomes of primary Achilles tendon repair in cats: 21 cases. Vet Comp Orthop Traumatol 2010;23:348–53.
[2] Corr SA, Draffan D, Kulendra E, et al. Retrospective study of Achilles mechanism disruption in 45 dogs. Vet Rec 2010;167:407–11.
[3] Evans HE, de Lahunta A. The muscular system. Miller's anatomy of the dog. Elsevier; 2013. p. 185–280.
[4] Kramer M, Gerwing M, Michele U, et al. Ultrasonographic examination of injuries to the Achilles tendon in dogs and cats. J Small Anim Pract 2001;42:531–5.
[5] Lamb CR, Duvernois A. Ultrasonographic anatomy of the normal canine calcaneal tendon. Vet Radiol Ultrasound 2005;46:326–30.
[6] Rewerts JM, Grooters AM, Payne JT, et al. Atraumatic rupture of the gastrocnemius muscle after corticosteroid administration in a dog. J Am Vet Med Assoc 1997;210:655–7.

CHAPTER 9

INTEGUMENT AND MAMMARY GLAND

Helen M.S. Davies, Chapter editor

9.1 Sebaceous adenitis—*Karen Trainor and Brian Palmeiro* 540

CHAPTER 9

CASE 9.1

Sebaceous Adenitis

Karen Trainor[a] **and Brian Palmeiro**[b]
[a]Dermatopathology Service, Innovative Vet Path, Leawood, Kansas, US
[b]Lehigh Valley Veterinary Dermatology & Fish Hospital, Allentown, Pennsylvania, US

Clinical case

History

A 2-year-old male castrated Vizsla presented with signs of alopecia that started on the head and progressed to affect the trunk. The owners reported that the skin was not pruritic and that he was otherwise in good health with no systemic clinical signs.

Physical examination findings

There were patchy, circular/annular, often coalescing areas of alopecia affecting the head/pinnae and trunk (Fig. 9.1-1). The skin in the affected areas had fine adherent white scale. The hairs epilated (L. *e* out + *pilus* hair) easily, and there was dried keratinaceous debris fixed to the hair shafts (follicular casts). The remainder of the physical examination was within normal limits.

Differential diagnoses

Demodicosis, dermatophytosis (ringworm), bacterial folliculitis, sebaceous adenitis

Diagnostics

Deep skin scrapings were performed to rule out demodicosis. A fungal culture was performed to rule out dermatophytosis. Skin cytology was performed to look for evidence of bacterial pyoderma/folliculitis.

The patient was sedated for skin biopsies. Histologic examination of the skin biopsies (Fig. 9.1-2) demonstrated that multifocal follicular ostia were dilated with orthokeratotic (Gr. *orthos* straight + Gr. *keratos* horn) hyperkeratosis and none of the sections contained sebaceous glands. The epidermis was covered by fronds of orthokeratotic hyperkeratosis extending from the follicular ostia. Moderate to large numbers of foamy macrophages, neutrophils, lymphocytes, and plasma cells were scattered around hair follicles throughout the dermis at the level of the isthmus. No etiologic/infectious agents were observed using routine stains.

Diagnosis

Sebaceous adenitis (Gr. *adenos* gland)

Treatment

Topical therapy is the mainstay of treatment for sebaceous adenitis. This dog was treated with a typical combination of topical therapies, including frequent bathing with moisturizing shampoo, leave-on conditioners, moisturizing sprays, mousses, and topical oil soaks. Nutraceuticals, including oral supplementation with omega-3/omega-6 fatty acids and vitamin A, can also be helpful, typically in combination with other treatments. The dog was also started

Comparative Veterinary Anatomy: A Clinical Approach. https://doi.org/10.1016/B978-0-323-91015-6.00136-9

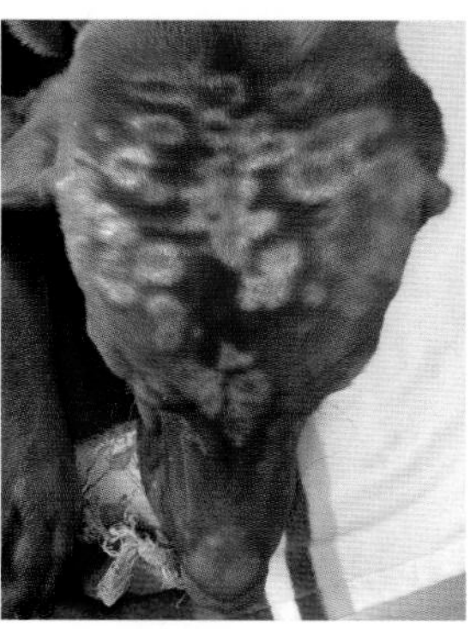

FIGURE 9.1-1 Dog from this case with patchy, circular/annular, often coalescing areas of alopecia, seen here on the head.

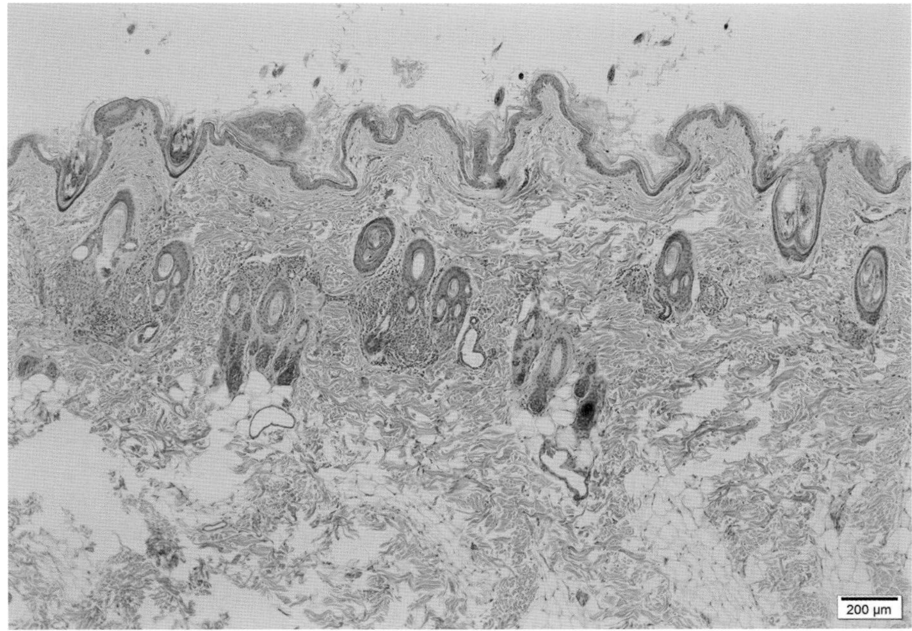

FIGURE 9.1-2 Lymphohistiocytic periadnexal dermatitis at the level of the isthmus with loss of sebaceous glands.

on modified cyclosporine (Atopica®, Novartis) at 5 mg/kg/day. Secondary bacterial and yeast infections must also be managed in cases of sebaceous adenitis, though at the time this dog did not have evidence of infection. Sebaceous adenitis is a condition that typically requires lifelong management to control. At the 45-day follow-up examination, the dog's condition had improved dramatically with almost complete hair regrowth.

Anatomical features in domestic animals

Introduction

The common integument is comprised of skin, hair, pads, claws, and skin glands. The skin functions to protect against mechanical, thermal, and chemical injuries. It also functions to prevent dehydration and infection. It is the largest sensory organ with the ability to detect temperature, pressure, pain, touch, and pruritus. Constriction and dilation of blood vessels aid in thermoregulation, as do variations in the subcutis and erection of the hair coat. Adipose tissue in the subcutis provides a source of energy, and the epidermis helps to synthesize vitamin D. The major layers of the skin are the epidermis, dermis, and subcutis (Fig. 9.1-3).

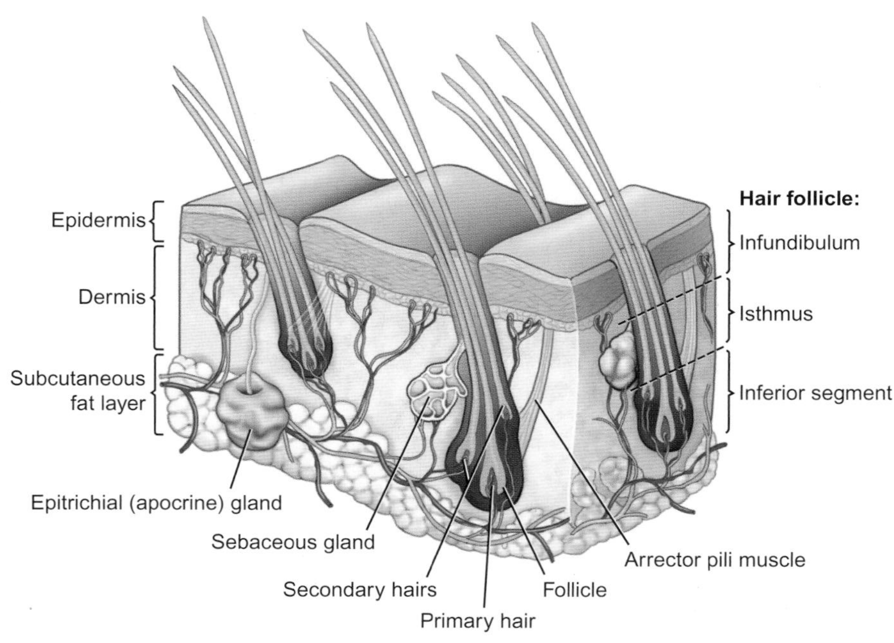

FIGURE 9.1-3 Layers of skin and its components.

Epidermis (Fig. 9.1-4)

The **epidermis** is comprised of 4 layers: stratum basale (basal layer), stratum spinosum, stratum granulosum, and the outermost stratum corneum. There are 4 distinct cell types in the epidermis: keratinocytes (approximately 85%), melanocytes (approximately 5%), Langerhans cells (approximately 3–8%), and Merkel cells (approximately 2%). Mitotic activity of **keratinocytes** in the **basal layer** provides a continuous supply of keratinocytes that are lost due to normal wear and tear; epidermal cells in all other layers are derived from the basal layer. The basal layer is supported by a **basement membrane** along the dermal-epidermal junction. Cells from this layer progress upward into the **stratum spinosum** and remain connected to surrounding keratinocytes by intracellular bridges (**desmosomes**). Desmosomes are the major cell adhesion junctions in the skin and contain various molecular components, including desmogleins and desmocollins.

Desmogleins and desmocollins are targeted by the body in autoimmune conditions like pemphigus foliaceus (see 9.1-5A and B) and pemphigus vulgaris.

Keratinocytes in the spinous layer begin the process of keratinization or cornification. Keratinocytes within the **stratum granulosum** contain numerous keratohyalin granules within their cytoplasm. Keratohyalin granules contain profilaggrin (a precursor of filaggrin), keratin filaments, and loricrin. Filaggrin functions to aggregate, pack, and align keratin filaments during cornification. Cell death occurs along the outer portion of this layer causing the release of lysosomal enzymes during the process of keratinization. The granular layer keratinocytes also have lamellar bodies (initially synthesized in the spinous layer) that fuse with the plasma membrane granulosum/corneum interface, extruding their contents to form intercellular lipid bilayers.

The **stratum corneum** is the most superficial layer, and it is composed of terminally differentiated, dead and dying **corneocytes** filled with mature keratin suspended in an extracellular lipid matrix. This layer is thickest over the paw pads and nose. Desquamation is normal sloughing or shedding of cells from the stratum corneum.

Under normal circumstances, desquamation is unnoticeable. However, if there is an increased turnover of epithelial cells, then scales become apparent clinically. Defects in terminal keratinization and desquamation may result in a group of conditions called ichthyosis.

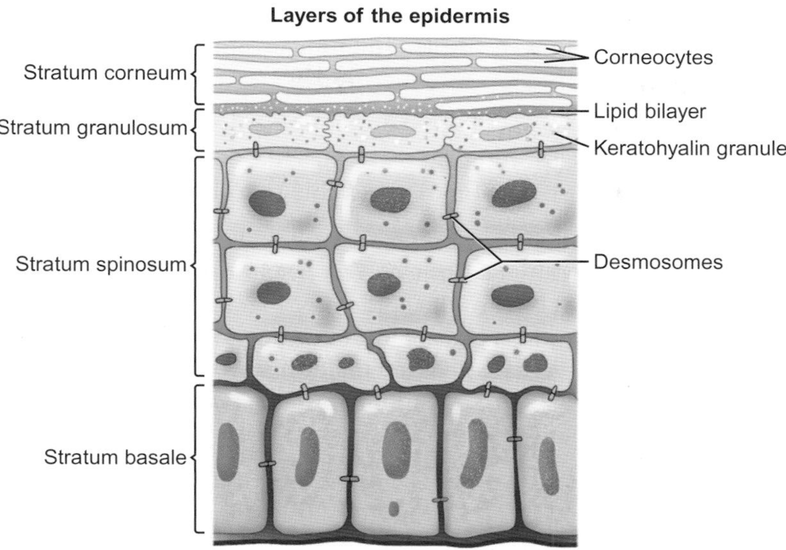

FIGURE 9.1-4 Epidermis.

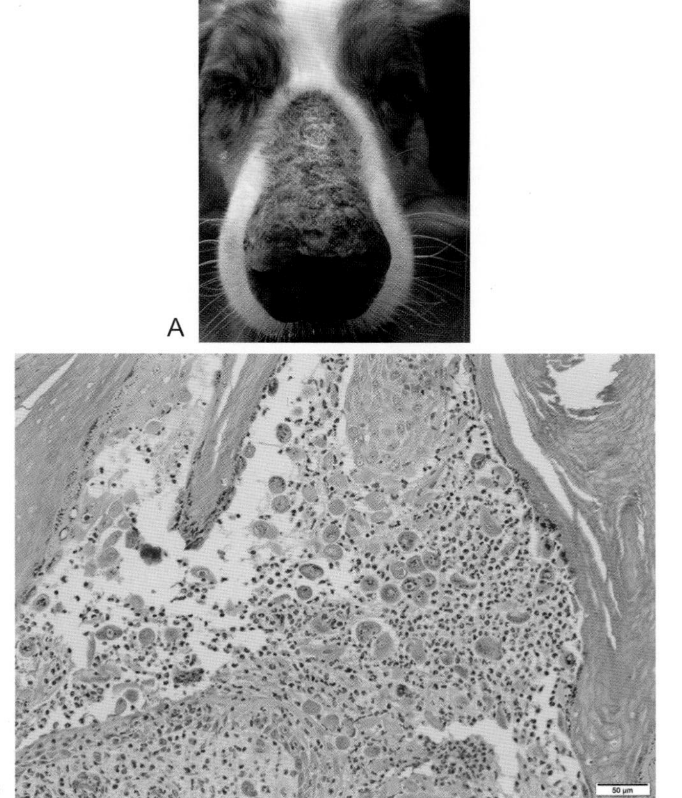

FIGURE 9.1-5 (A) Dog with crusting due to pemphigus foliaceus. (B) Neutrophilic pustular epidermitis with numerous acantholytic keratinocytes and nondegenerate neutrophils within the pustule.

Keratins can be divided into soft keratins (skin) or hard keratins (hair, claw), as well as into alpha keratins (skin, hair) or beta keratins (scale, feather). Keratin is the protein that imparts hardness and strength to hair, hooves, and horns.

Footpads are specialized areas of integument present in dogs and cats that contain thick epidermis to protect against mechanical trauma and act as large, shock-absorbing fat pads. **Atrichial** (Gr. *a-* absence + Gr. *thrix* hair) **sweat glands** are present in the footpads that may improve traction or play a role in scent marking.

In addition to keratinocytes, there are several other non-keratinocyte cell types in the epidermis—these include melanocytes, Langerhans cells, and Merkel cells. **Melanocytes**, derived from neural crest cells, synthesize melanin pigment and are typically found in the basal layer of the epidermis. Pigment in the skin and hair shafts serves to protect against ultraviolet radiation. **Langerhans cells** are found throughout all layers of the epidermis and may also be found around dermal blood vessels. Langerhans cells are major histocompatibility complex (MHC) II antigen-presenting cells and play an important surveillance role in the immune system. **Merkel cells** function as touch receptors in the basal layer. There are no blood or lymphatic vessels in the epidermis.

Dermis

The **dermis** provides vascular support for—and to—the overlying avascular epidermis, and it is composed of fibers (collagen, reticulin, and elastin), ground substance, various cells, epidermal appendages (including hair follicles and associated glands), arrector pili muscles, blood and lymph vessels, and nerves. **Fibroblasts** produce collagen and elastin, which lend tensile strength and elasticity to the skin. **Ground substance** is a mix of proteoglycans and glycoproteins, and it occupies the space between dermal structures.

Normally, mucinous ground substance is not readily noticeable in histologic sections of skin; however, it is prominent in the dermis of Chinese Shar-Pei dogs. It may also be prominent in the skin of some dogs with advanced hypothyroidism (myxedema).

Vasodilation and vasoconstriction of dermal blood vessels are important contributing factors to thermoregulation. **Mast cells** normally reside in the dermis around blood vessels and produce inflammatory mediators such as histamine. Release of intracellular granules containing histamine can be triggered by exposure to allergens in patients with allergic skin disease. Mast cells are critical to the initiation of Type 1 hypersensitivity (allergic) reactions, and they protect against parasitic infections.

Dermal blood vessels and lymphatic vessels are usually accompanied by peripheral somatic sensory and autonomic motor nerve fibers. The region of skin innervated by the extension of a spinal nerve is called a **dermatome**. Peripheral autonomic nerve fibers play a role in the activation of skin glands, contraction of arrector pili muscles, and vasoconstriction or vasodilation. The brindle coat pattern in dogs is distributed along Blaschko lines. These lines represent a linear pattern of embryonic ectodermal proliferation and migration that may be associated with a variety of skin abnormalities such as inflammatory, pigmentary, dysplastic, or dyskeratotic conditions.

Pain or rash affecting a dermatome may be used to localize neurological lesions such as radiculopathy or compression of the nerve root.

These Blaschko lines are distinct from skin tension lines, which are an important feature to consider when closing surgical incisions.

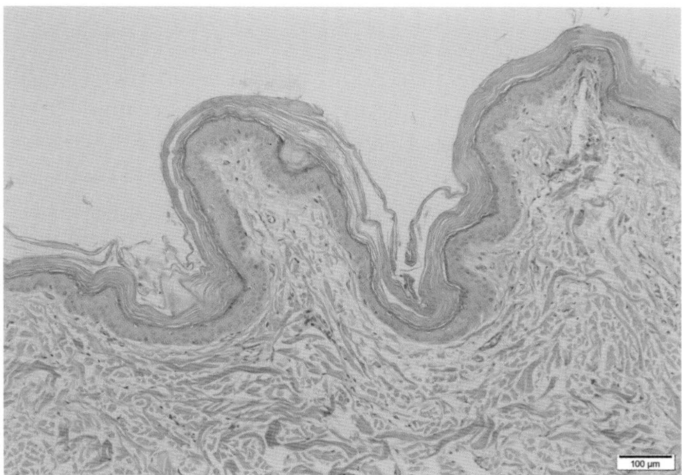

FIGURE 9.1-6 Skin section from a Golden Retriever with nonepidermolytic ichthyosis. Note the brightly eosinophilic, densely compacted, and lamellar orthokeratotic hyperkeratosis on the surface.

ICHTHYOSIS

Ichthyosis is a rare, congenital, hereditary cornification disorder that has been reported to affect dogs, cats, cattle, chickens, pigs, mice, and llamas. Ichthyosis is divided into epidermolytic and nonepidermolytic forms. Epidermolytic ichthyosis has been identified in the Norfolk Terrier and involves a defect in keratin formation (mutation in K10 gene). Nonepidermolytic ichthyosis is more common and involves defects in lipids and structural proteins associated with the keratinization process; it has been reported in several breeds, including the Golden Retriever, Jack Russell Terrier, German Shepherd, Cavalier King Charles Spaniel, West Highland White Terrier, and American Bulldog. For nonepidermolytic forms of ichthyosis, an autosomal recessive mode of inheritance is suspected or confirmed.

Overall, the defects identified occur in the terminal differentiation of keratinocytes and during desquamation, and result in generalized, often adherent scaling that is often worse in more sparsely haired areas of the body such as the inguinal region. The diagnosis is confirmed via skin biopsy, in which the predominant finding is lamellar to compact orthokeratotic hyperkeratosis (Fig. 9.1-6). Other potential findings include hypergranulosis, vacuolated keratinocytes in the granular layer (Golden Retrievers), and epidermolysis (Norfolk Terrier). Genetic tests are available to identify affected animals and carriers in many breeds.

The condition is not curable, and most patients are affected for their entire life. Hydrating shampoos, leave-on conditioners/creams, sprays, and mousses are the mainstay of therapy. Oral omega-3/6 supplementation may also be beneficial. There are also reports of using vitamin A or synthetic derivatives (retinoids).

OTHER KERATINIZED STRUCTURES OF NOTE

Supracarpal and tarsal chestnuts are regions of epidermal thickening without glands found in horses. Ergots are similar areas found on the underside of the fetlock in horses. Ruminant horns, bird beaks, hooves of ungulates, and the claws on paws of carnivores are all highly specialized formations composed of dense keratin. The nasal skin of horses is thin and contains sebaceous and sweat glands in addition to fine hairs and scattered sinus hairs. The nasal planum in carnivores is very thick and lacks hairs and sebaceous or apocrine glands. However, the nasal planum is moistened by the lateral nasal glands which are innervated by parasympathetic branches of the facial nerve.

Hair follicles and adnexa (Figs. 9.1-3 and 9.1-7)

Hair is important in thermal insulation and sensory perception, and acts as a barrier against chemical, physical, and microbial injuries. **Hair follicles** may be primary or secondary, and they may be simple or compound. **Primary hairs** are larger and extend deeper into the dermis and subcutis. They are also associated with sebaceous glands, apocrine glands, and arrector pili muscles (Fig. 9.1-3). **Secondary hairs** are, by comparison, smaller and rooted more superficially in the dermis. These are associated with sebaceous glands, but not with arrector pili muscles or apocrine sweat glands. **Simple follicles** are characterized by a single hair emerging from the follicular opening, whereas **compound follicles** are characterized by multiple hairs emerging from one follicular opening. Horses and ruminants have simple (single) hair follicles evenly distributed throughout the skin. In pigs, these are found in groups of 3. Most of the hair follicles in carnivores are compound follicles, which are formed by a single primary follicle and multiple secondary follicles. These converge at the level of the sebaceous gland duct to form a common follicle that extends towards the surface via a common follicular opening. Both primary (outercoat or guard hairs) and secondary (undercoat) hairs are medullated (myelinated – having a myelin sheath [Gr. *myelos* marrow]) in dogs and cats. In cats, secondary hairs are more numerous than primary hairs.

The hair follicle has a **lower (inferior) portion** at the base, the **isthmus,** which is the entrance of the sebaceous gland and attachment of the arrector pili muscle (if present), and an **infundibulum**, which is the portion extending from the entrance of the sebaceous gland superficially towards the follicular opening. Hair grows in a cyclical pattern characterized by the **anagen** growth phase, **catagen** phase of transition, resting **telogen** phase, and shedding **exogen** phase. The duration of these phases may vary with age, sex, breed, and region of the body and can be modified by various external (photoperiod, temperature, nutrition), physiologic, and pathologic factors.

The hair follicle is divided into 5 components: the **outer root sheath**, **inner root sheath**, the hair itself, **hair matrix**, and **dermal hair papilla**. The base of the hair follicle is formed by the hair root and the dermal papilla, which invaginates into the root; together, the hair root and dermal papilla form the hair bulb. A small plexus of blood vessels in the dermal papilla nourishes the root.

The hair matrix is the most central portion of the root next to the papilla, and it is formed by densely packed cells that proliferate and lead to hair growth. Hair matrix cells are a specialized type of basal cell and form keratin differently from the basal layer of the epidermis and lack keratohyalin, but instead contain more sulfur, which provides more mechanical strength. When hair growth is active, the epithelial cells surrounding the dermal papilla contribute to the 4 innermost layers of the follicle, while the outermost layers are formed by a downward extension from the stratum basale from the surface epithelium.

The **hair shaft** has 3 components: the medulla, the cortex, and the cuticle. The middle layer of the hair shaft is the **cortex**, and it is comprised of cornified cells that are parallel to the length of the hair shaft. The cells forming the cortex contain pigment that imparts color to the hair shaft. This layer is responsible for most of the mechanical functions of hair. The **cuticle** is the external layer, and it is formed by flat, anuclear cornified cells. Primary hairs

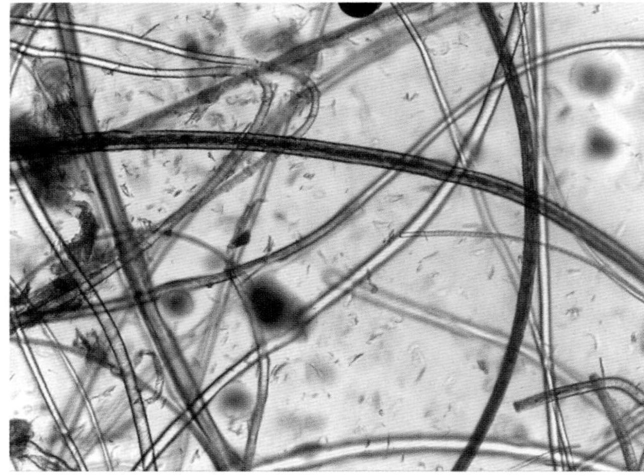

FIGURE 9.1-7 Hairshaft/trichogram (hair plucking) from a dog. Note the inner medulla and outer cortex (100x magnification).

have a thicker **medulla** compared with the cuticle layer, whereas secondary hairs have a more pronounced cuticle compared with the medulla. The cuticle itself is lined by an amorphous layer of cuticular cells forming the epicuticle.

Sebaceous glands typically secrete sebum into hair follicles; however, they may also deposit directly on the surface of the skin; they are present in all mammals except for porpoises and whales. These glands are composed of small basal stem cells around the perimeter that differentiate into mature sebocytes with vacuolated cytoplasm. The innermost, mature cells break down and release an oily substance (sebum) via holocrine secretion. One or more sebaceous glands may be associated with each follicle. Sebum is composed of cholesterol, triglycerides, and phospholipids. Sebum serves as a waterproofing and antimicrobial agent (linoleic acid) and moisturizes the skin, providing a glossy coating to each hair shaft. Each hair follicle along the dorsum and its associated arrector pili muscle and sebaceous glands is referred to as a **pilosebaceous unit**.

The pattern of cytoplasmic vacuolation in mature sebocytes is different between dogs and cats; the vacuoles are smaller and more uniform in cats. Sebaceous glands are more numerous along the dorsum in dogs and sparse along the ventrum—an important factor to consider when sampling for sebaceous adenitis. Sebaceous glands are well developed and densely concentrated on the ventral aspect of the chin in cats (called the **submental organ**). They are also concentrated on the dorsal surface of the tail in dogs (called the **tail gland**). Sebum from sheep is collected and processed as lanolin. Sebaceous glands are absent from some glabrous (L. *glaber* smooth) regions such as claws, hooves, horns, and footpads.

Epitrichial (Gr. *epi-* on + Gr. *trichion* hair) **(apocrine) sweat glands** are distributed throughout all haired skin, but are absent from the nasal planum and footpads. They are located below the sebaceous glands and usually open into the infundibulum of the hair follicle above the sebaceous gland opening. Epitrichial sweat likely has pheromonal and antimicrobial properties. These glands are composed of cuboidal-to-columnar epithelium surrounded by a layer of myoepithelial cells.

Atrichial sweat glands are present in footpads of dogs and cats, and empty directly onto the skin surface.

Arrector pili muscles are smooth muscles in haired skin and are responsible for pulling hair follicles vertically along the dorsum (piloerection). Contraction of these muscles plays a role in communication by raising the hairs on the back, or they may serve to create air pockets in the hair coat to provide insulation. Skeletal muscle is found in dermis from the face, muzzle, and perianal regions.

Hepatoid glands (also called "perianal glands" or "circumanal glands") in dogs are a modified type of sebaceous gland composed of basal cells and mature cells that resemble hepatocytes. These tend to be more prominent in male dogs due to the presence of androgen receptors. Perianal glands are concentrated around the perineal area; however, they can be found circumferentially around the proximal third of the tail, in the dorsal lumbosacral area, lateral to the prepuce, and along the ventral midline as far cranial as the neck. The tail gland of the dog is located on the dorsal tail over the 5th–7th coccygeal vertebrae, whereas in cats the glands run the entire length of the dorsal tail; excess accumulation of glandular secretion in the region of the tail gland is referred to as "stud tail." Carnivores have paired **anal sacs** lateral to the anus. These structures are lined by keratinizing squamous epithelium. Specialized apocrine glands of the anal sac secrete into these sacs.

A variety of conditions are seen in practice which affect the anal sac glands of dogs and present with clinical signs such as scooting and/or frequent licking of the perianal region. These glands can become inflamed (anal sacculitis), impacted, abscessed, or develop neoplasia.

APOCRINE SWEAT GLANDS

Apocrine sweat glands serve to thermoregulate in horses and cattle. Horses produce an albuminous sweat that creates a froth associated with movement. The volume of apocrine (epitrichial) sweat glands increases substantially during the summer compared to winter in Thoroughbreds as opposed to non-Thoroughbred horses. Horses differ from other species with regard to physiologic control of sweating. Specifically, the adrenal medulla secretes adrenergic agonists (humoral mechanism) and autonomic adrenergic nerves (nervous mechanism).

Mechanoreceptors

There are 3 classes of cutaneous rapid-adapting receptor units: 1) Meissner corpuscles, 2) Pacinian corpuscles, and 3) tylotrich hairs. **Meissner corpuscles** are derived from neural crest cells and located in the superficial dermis—they respond to vibration. **Pacinian corpuscles** are relatively large, lamellar structures that sense changes in pressure and vibration. **Tylotrich hairs** are specialized large, primary hair follicles encircled by neurovascular tissue—these are found in association with **tylotrich pads**.

There are 4 types of cutaneous slow-adapting mechanoreceptors: (1) Merkel cell nerve endings, (2) Ruffini corpuscles, (3) tylotrich pads (Fig. 9.1-8), and (4) sinus hairs ("vibrissae" or "tactile hairs"; Fig. 9.1-9). **Merkel cell**

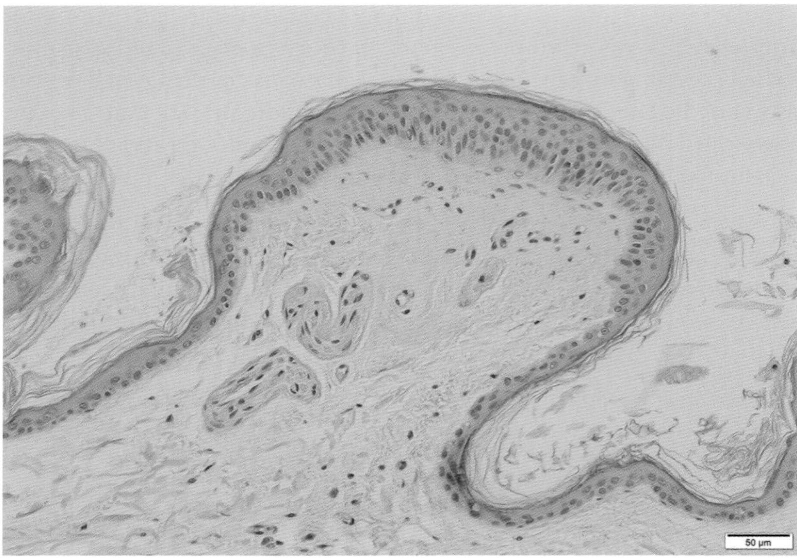

FIGURE 9.1-8 Canine mechanoreceptor tylotrich pad from dorsal skin with focal, plaque-like thickening of the epidermis and associated nerve endings and blood vessels in the subjacent dermis.

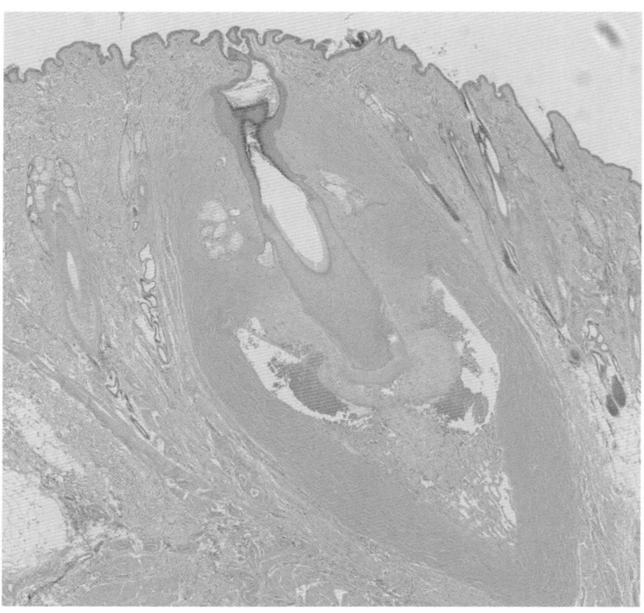

FIGURE 9.1-9 A focal cluster sinus hair follicle from a dog surrounded by normal primary and secondary hair follicles. Note the thickened fibrous capsule and elaborate blood-filled vascular sinus at the base of the vibrissa.

nerve endings are found in the basal layer of the epidermis, and they respond to pressure, edges, and curves. **Ruffini corpuscles** are in the dermis, and they are receptive to skin stretch. **Tylotrich pads** are characterized by a localized area of epidermal thickening with very vascular, well-innervated connective tissue below. **Sinus hairs** are larger than normal hairs and have a large vascular sinus at the base of the follicle. These vibrissae can be found on the face, muzzle, eyelids, lips, and throat. Cats also have them on the palmar aspect of the carpus.

Subcutis

The subcutis is composed of loose connective tissue and fat. The subcutis is not evenly distributed and may be thin or absent from regions, such as the face and ears. This is the tissue that connects the skin to the underlying muscle and bone. It functions in thermogenesis and insulation, as an energy reserve, and as protection/support.

ANHIDROSIS

Anhidrosis is a condition in horses—most commonly occurring in hot, humid climates—characterized by the inability to sweat in response to an adequate stimulus. It is believed to result from a conditioned insensitivity of sweat glands to epinephrine, eventually resulting in secretory cell degeneration. The severity of clinical signs is related to the severity of anhidrosis and chronicity. Affected horses exhibit patchy or inadequate sweating response and are slow to cool after exercise. Acutely, anhidrosis is characterized by labored breathing, flared nostrils, elevated temperature, and tachypnea. Collapse and death may occur in severe cases. Chronic cases tend to demonstrate partial alopecia of the face and neck, excessive scaling, dry hair coat, and a loss of normal skin elasticity. Residual sweating may persist under the mane and saddle and in the axillae/groin.

Diagnosis is based on history, examination, and response to intradermal injections of terbutaline. Anhidrotic horses respond only to 1:1000 dilution of terbutaline and only after 5 h or more, whereas normal horses also respond to weaker dilutions (1:10,000, 1:100,000, and 1:1 million).

No form of medical treatment has consistent benefit, and controlled studies are lacking. The most effective mode of therapy is moving the horse to a cooler, drier climate. Other management changes include air-conditioned, low-humidity stalls; exercising only during cool periods of the day; and cooling with cold water baths.

Selected references

[1] Bacha WJ, Bacha LM. Color atlas of veterinary histology. 2nd ed. Blackwell Publishing; 2006.
[2] Mauldin EA. Canine ichthyosis and related disorders of cornification. Vet Clin North Am Small Anim Pract 2013;43(1):89–97.
[3] Maxie MG. Jubb, Kennedy, and Palmer's pathology of domestic animals. 6th ed. St. Louis: Saunders Elsevier; 2016.
[4] Miller WH, Griffin CE, Campbell KL. Muller and Kirk's small animal dermatology. 7th ed. St. Louis: Saunders Elsevier; 2013.
[5] Simpson A, McKay L. Applied dermatology—sebaceous adenitis in dogs. Compendium 2012;34:10.

SECTION IV

EQUINE CLINICAL CASES

4.0 EQUINE LANDSCAPE FIGURES (1–11) 553
N.S. Grenager, M. Gerard, J.A. Orsini, and A. de Lahunta

CHAPTER 10: AXIAL SKELETON: HEAD, NECK, AND VERTEBRAL COLUMN 565
Mathew Gerard and Amy L. Johnson, Chapter editors

CHAPTER 11: THORAX 693
Sarah Reuss, Chapter editor

CHAPTER 12: ABDOMEN 723
Jarred Williams and Kira Epstein, Chapter editors

CHAPTER 13: PELVIC ORGANS 771
Dirk Vanderwall, Chapter editor

CHAPTER 14: THORACIC LIMB 861
Nick Carlson, Chapter editor

CHAPTER 15: PELVIC LIMB 933
Nick Carlson, Chapter editor

CHAPTER 16: INTEGUMENT AND MAMMARY GLAND 977
Sarah Reuss, Chapter editor

SECTION IV

4.0 Equine Landscape Figures (1–11)

N.S. Grenager[a], M. Gerard[b], J.A. Orsini[c], and A. de Lahunta[d]
[a]Steinbeck Peninsula Equine Clinics, Menlo Park, California, US
[b]Department of Molecular Biomedical Sciences, North Carolina State University College of Veterinary Medicine, Raleigh, North Carolina, US
[c]Department of Clinical Studies - New Bolton Center, University of Pennsylvania School of Veterinary Medicine, Kennett Square, Pennsylvania, US
[d]Professor Emeritus, Biomedical Science, Cornell University College of Veterinary Medicine, Ithaca, New York, NY

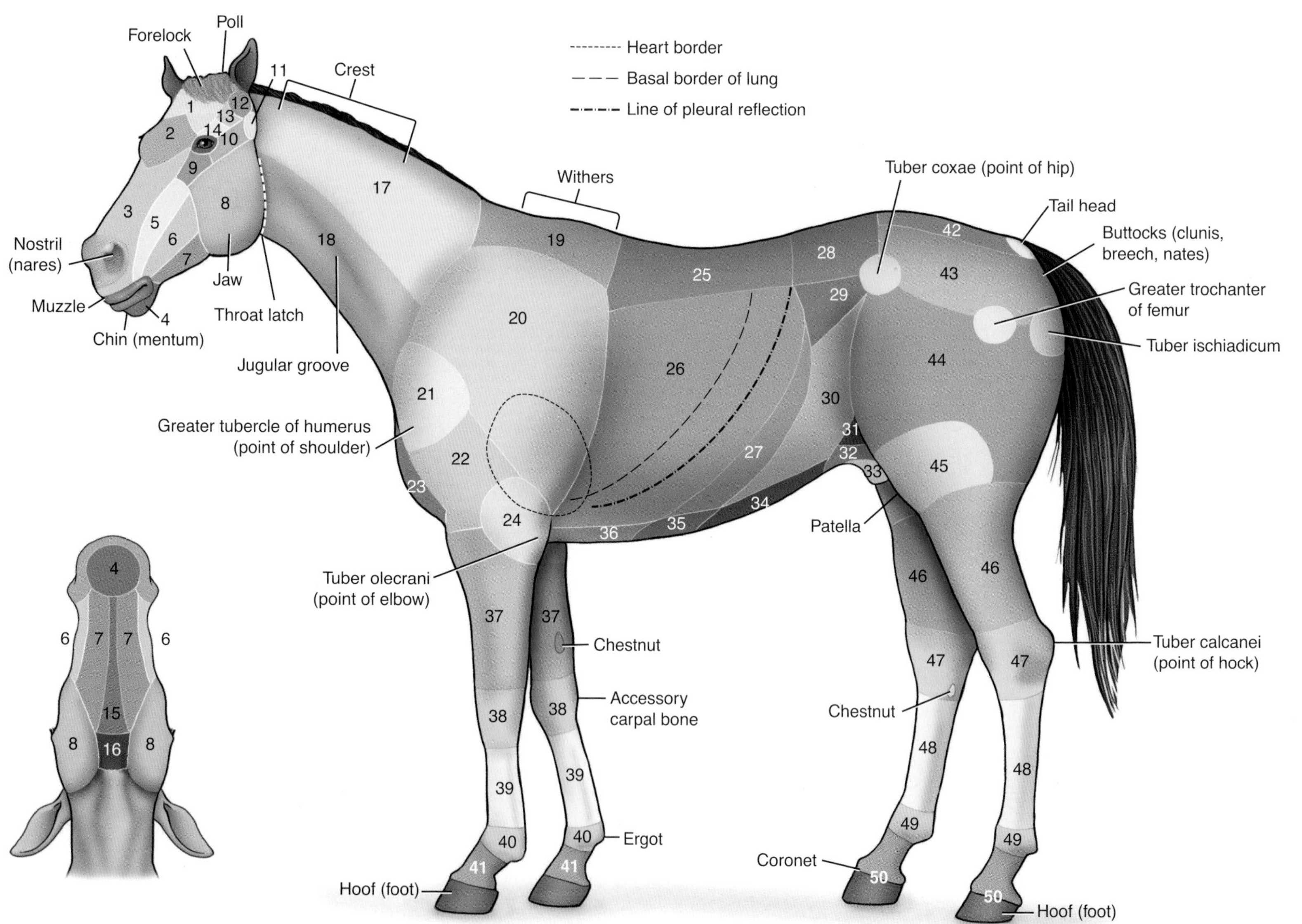

FIGURE 4.0-1 Regional equine anatomy.

Equine anatomical regions: 1 parietal, 2 frontal, 3 nasal, 4 chin, 5 maxillary, 6 buccal, 7 mandibular, 8 masseteric, 9 infraorbital, 10 zygomatic, 11 region of temporomandibular joint, 12 auricular, 13 temporal, 14 supraorbital fossa, 15 intermandibular, 16 subhyoid, 17 dorsal cervical, 18 ventral cervical, 19 interscapular, 20 scapular, 21 shoulder, 22 brachial, 23 pectoral, 24 elbow, 25 dorsum (back), 26 costal, 27 hypochondriac, 28 lumbar (loin), 29 paralumbar fossa, 30 lateral abdominal, 31 inguinal, 32 prepubic, 33 udder/preputial, 34 umbilical, 35 xiphoid, 36 sternal, 37 antebrachial (forearm), 38 carpal (knee), 39 metacarpal (thoracic limb cannon), 40 metacarpo/metaphalangeal (fetlock, ankle), 41 pastern, 42 sacral (croup), 43 gluteal, 44 femoral/thigh, 45 stifle (true knee), 46 crural (leg/gaskin), 47 tarsal (hock), 48 metatarsal (pelvic limb cannon), 49 metatarsophalangeal (fetlock, ankle), and 50 digital.

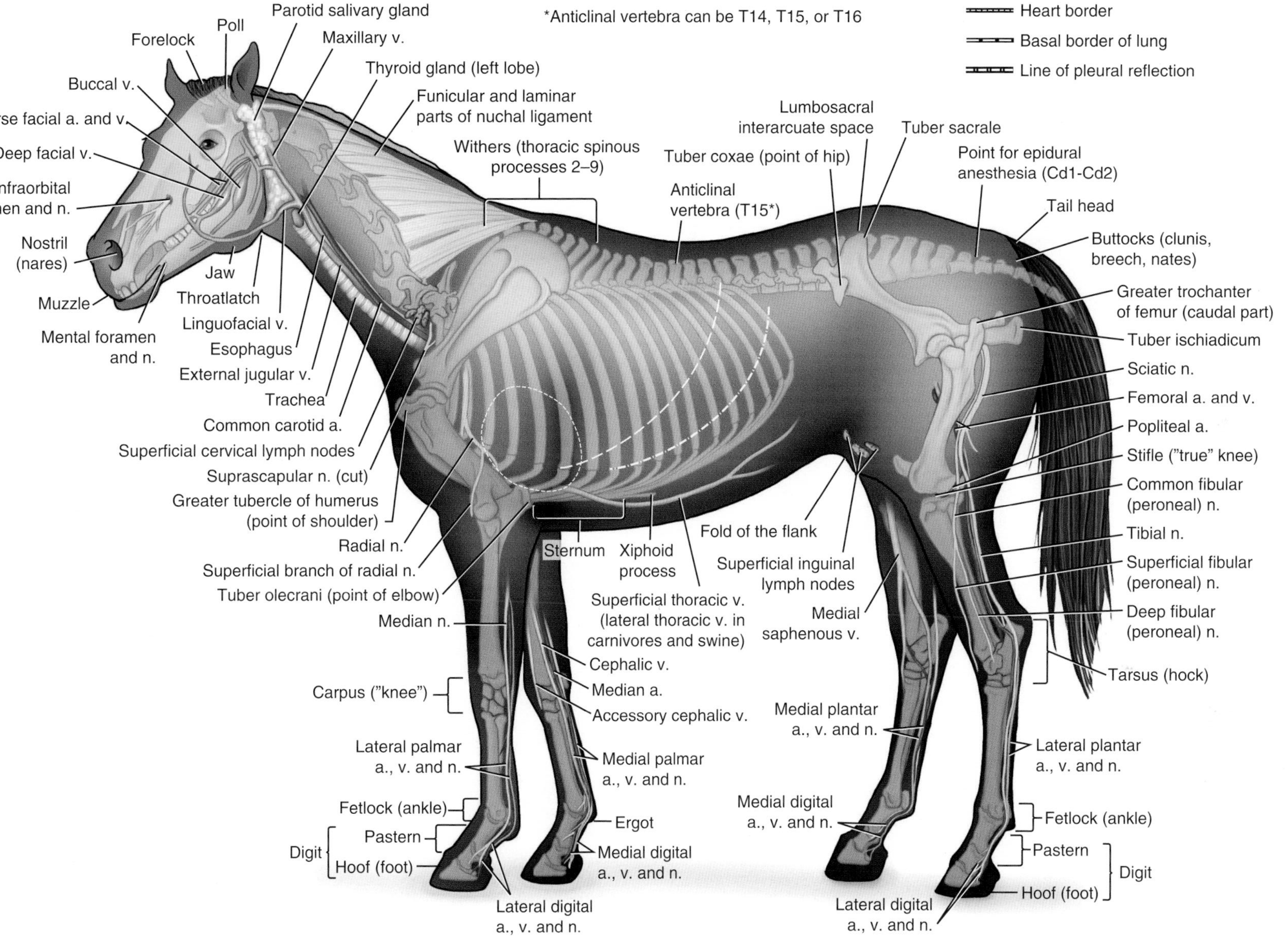

FIGURE 4.0-2 Topographical equine anatomy.

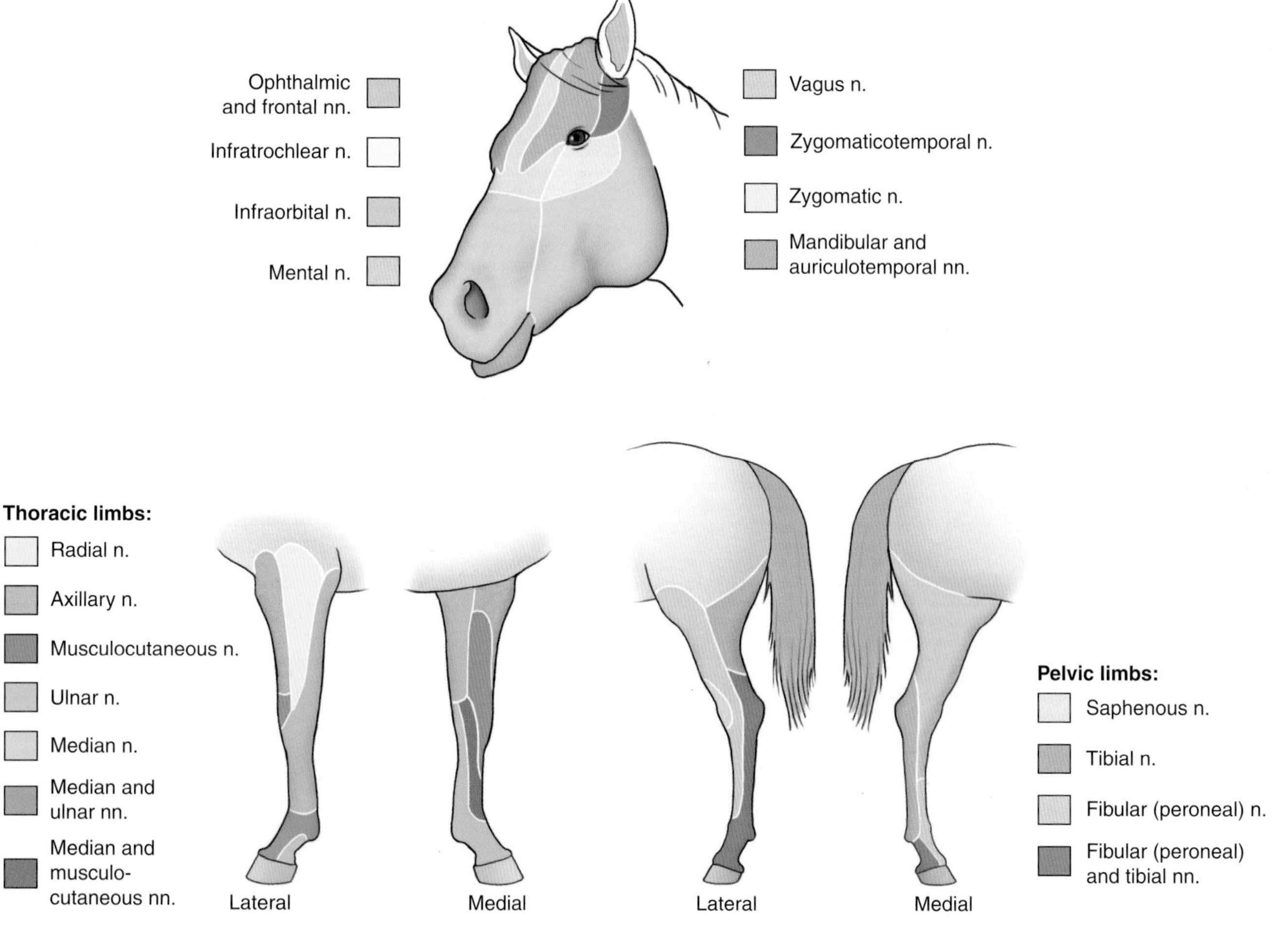

FIGURE 4.0-3 Dermatomes of the equine head and thoracic and pelvic limbs.

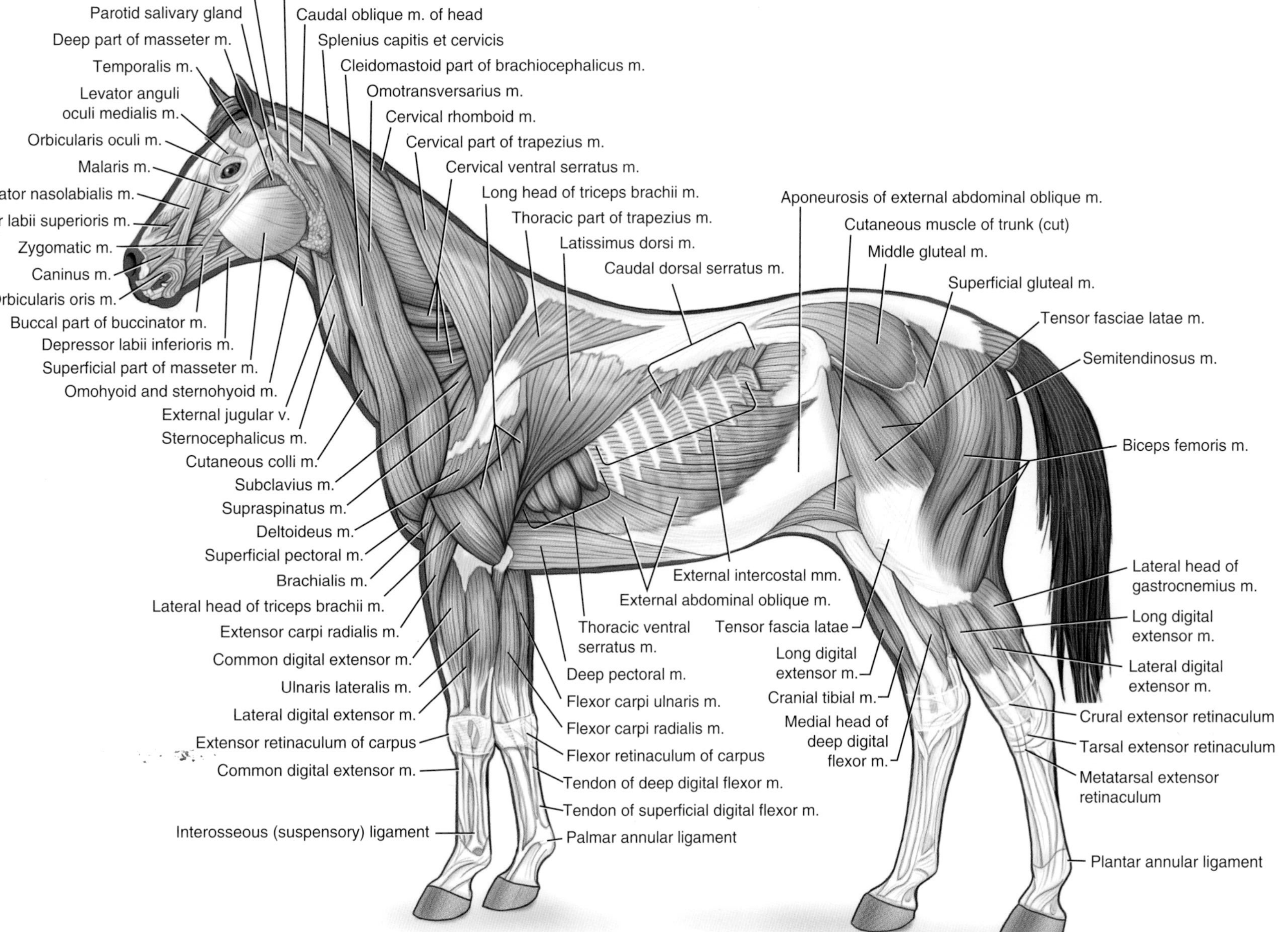

FIGURE 4.0-4 Superficial equine muscle layer.

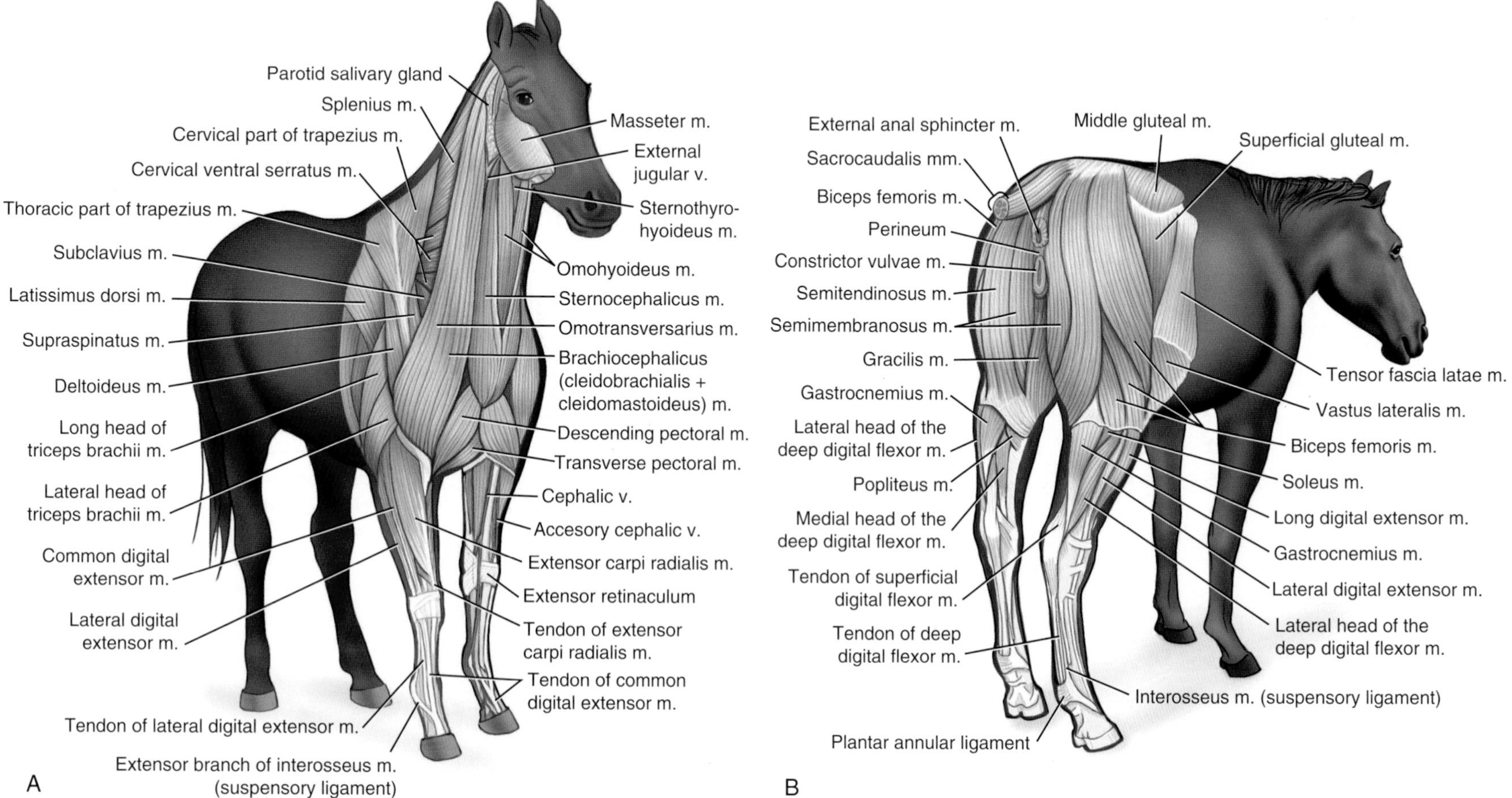

FIGURE 4.0-5 Cranial (A) and caudal (B) views of the equine superficial muscle layers.

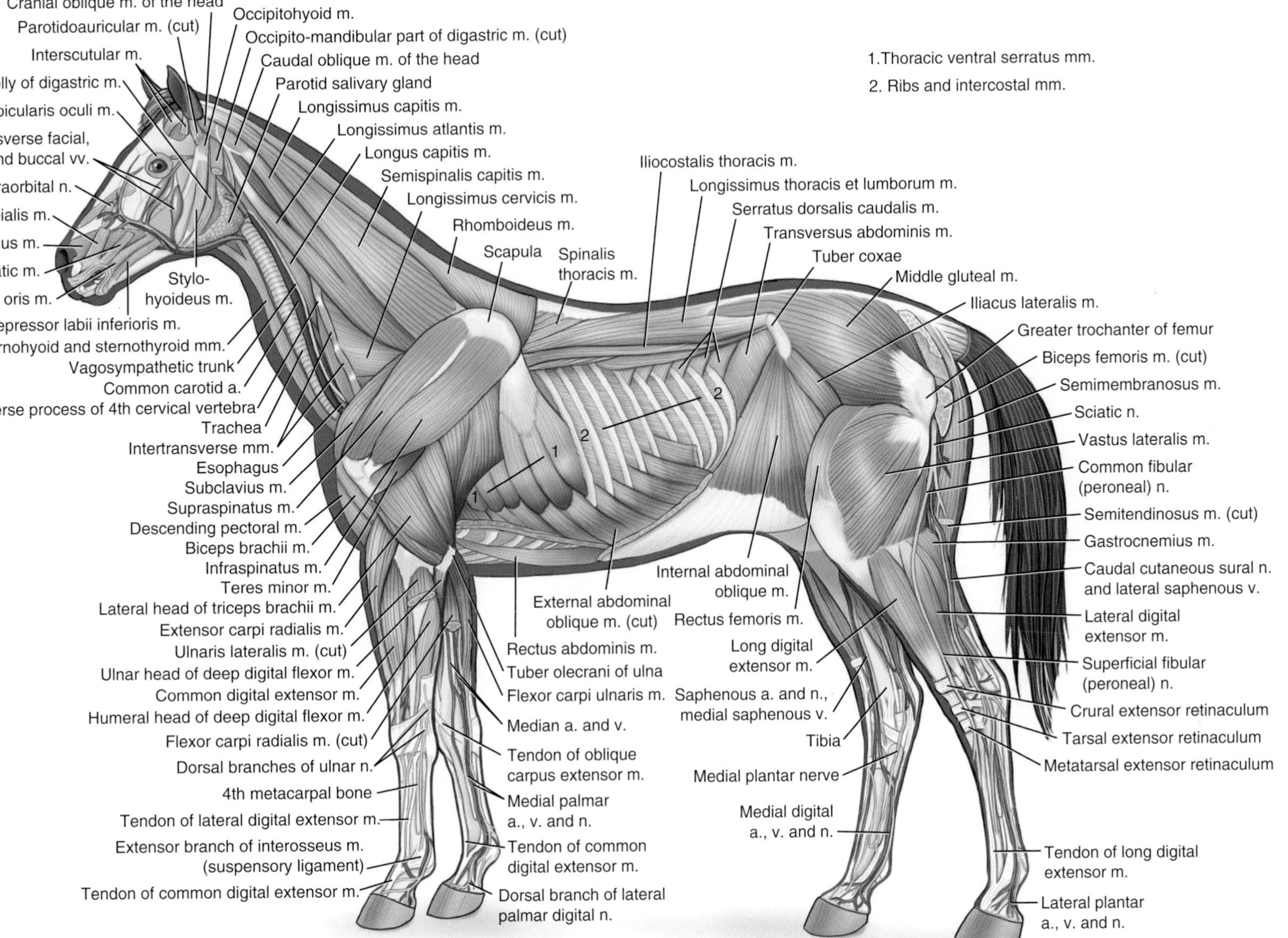

FIGURE 4.0-6 Deep equine muscle layer.

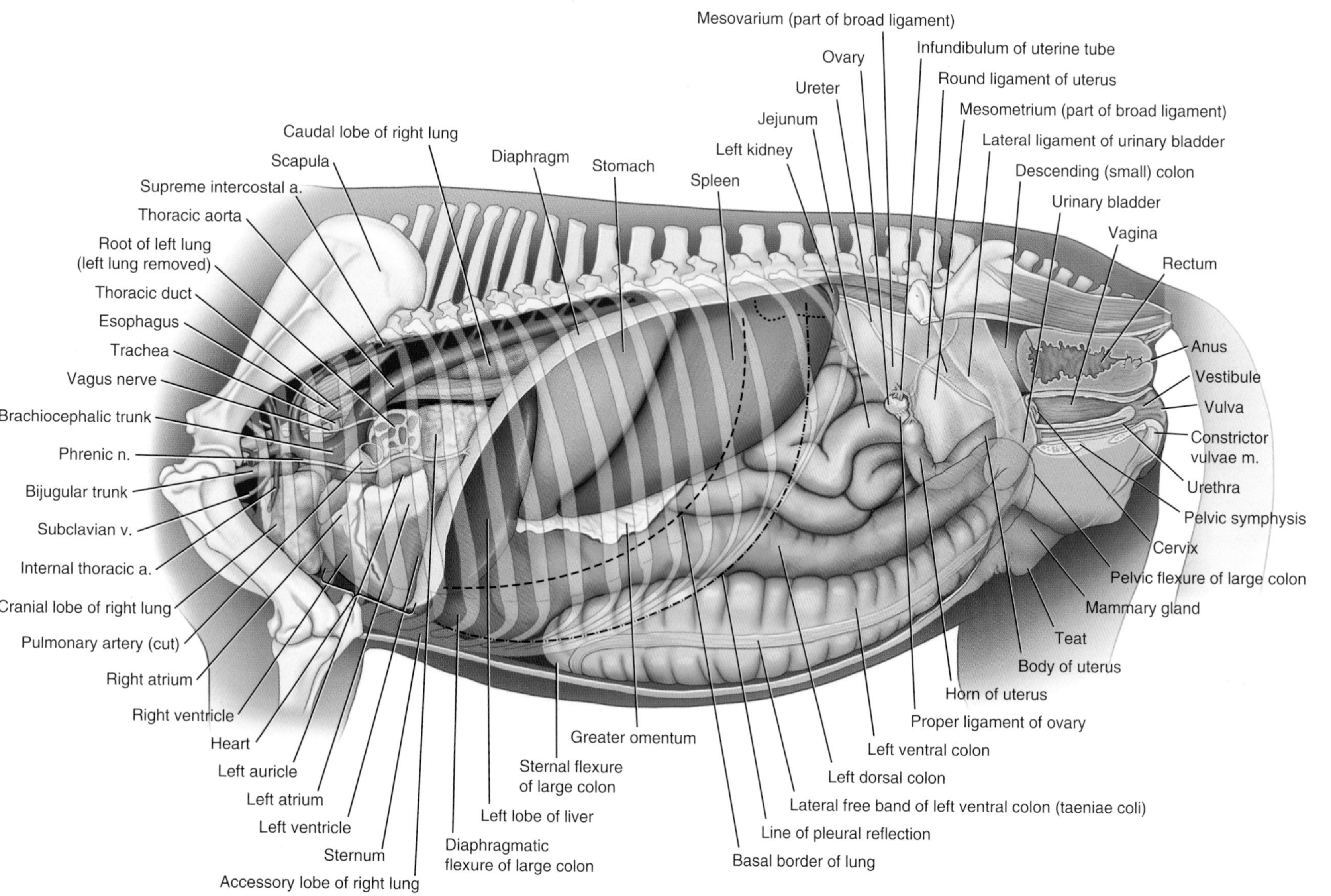

FIGURE 4.0-7 Left view of the equine thoracic and abdominal cavities and pelvis (female).

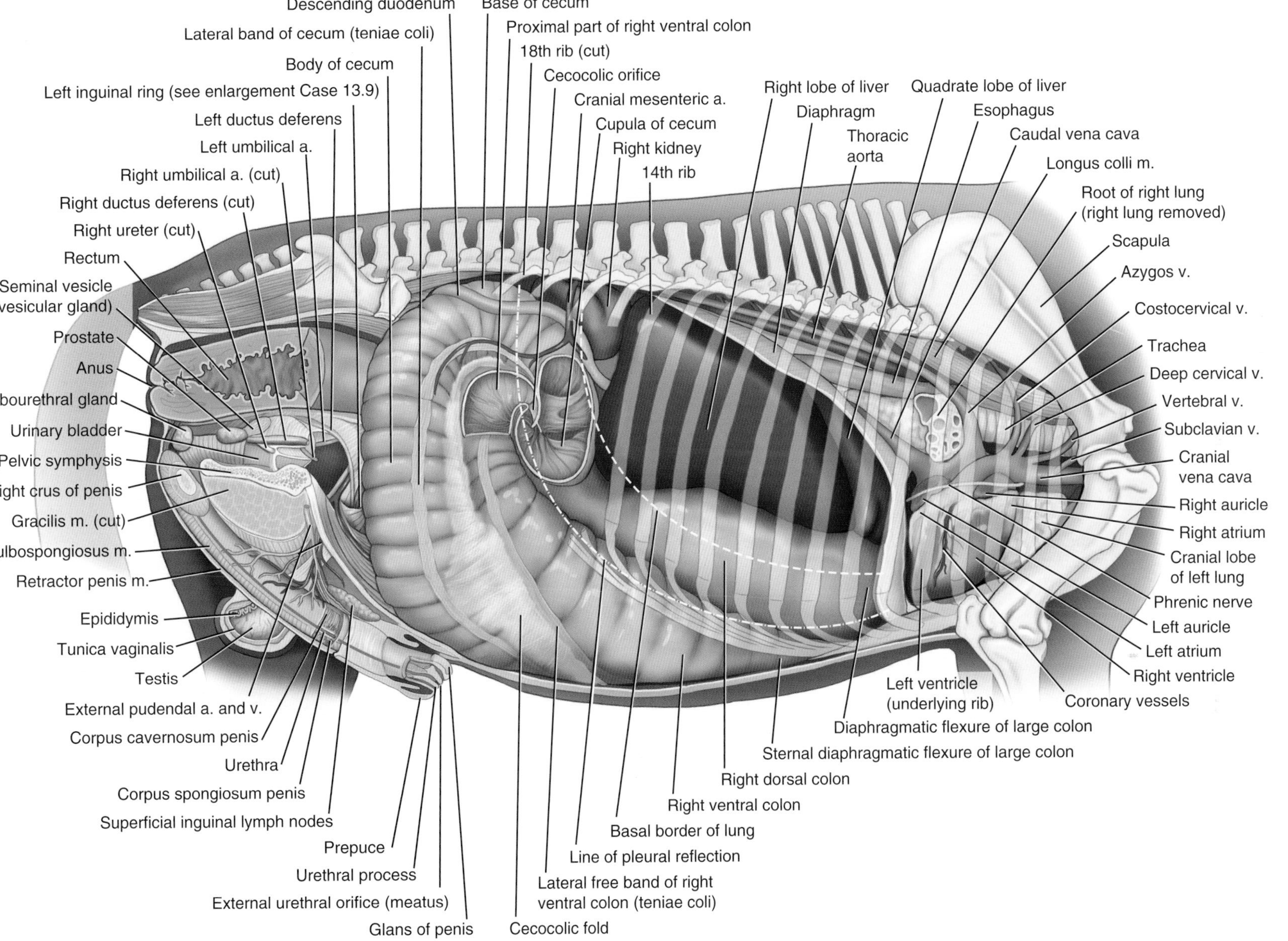

FIGURE 4.0-8 Right view of the equine thoracic and abdominal cavities and pelvis (male).

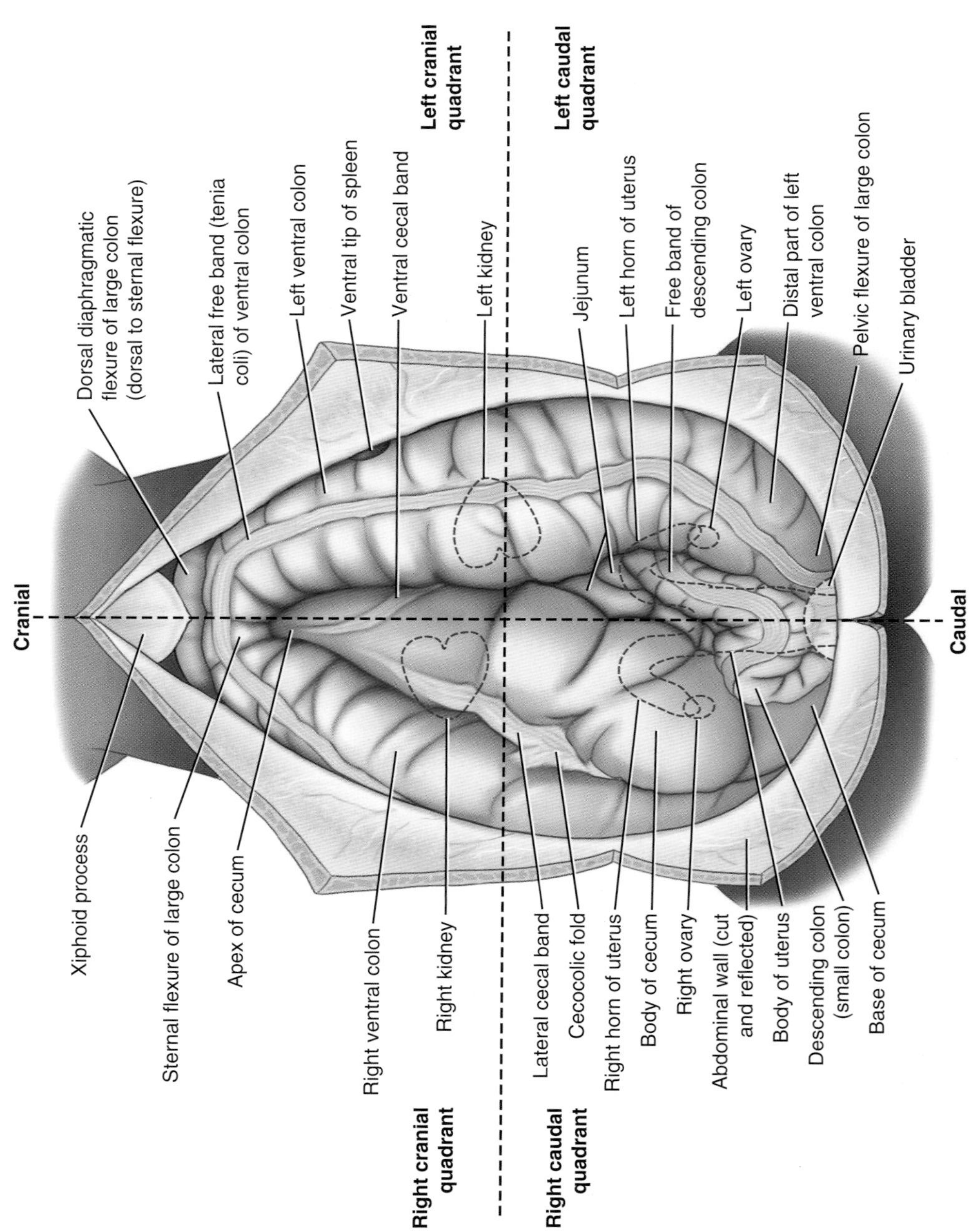

FIGURE 4.0-9 Ventral view of the equine abdominal cavity.

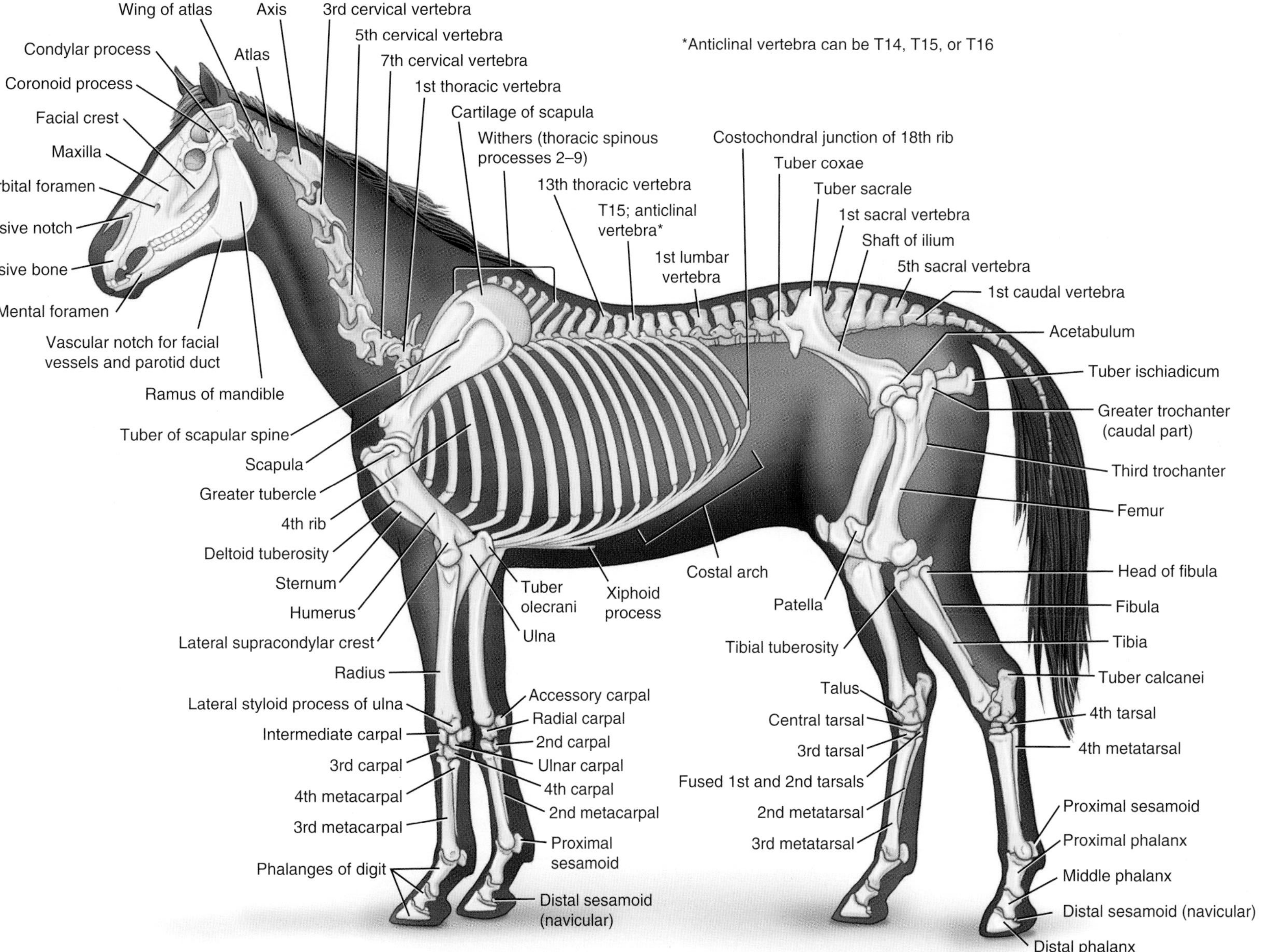

FIGURE 4.0-10 Lateral view of the equine skeleton. *Vertebral Column Formula*: $C_7T_{18}L_6S_5Cd_{15\text{-}21}$.

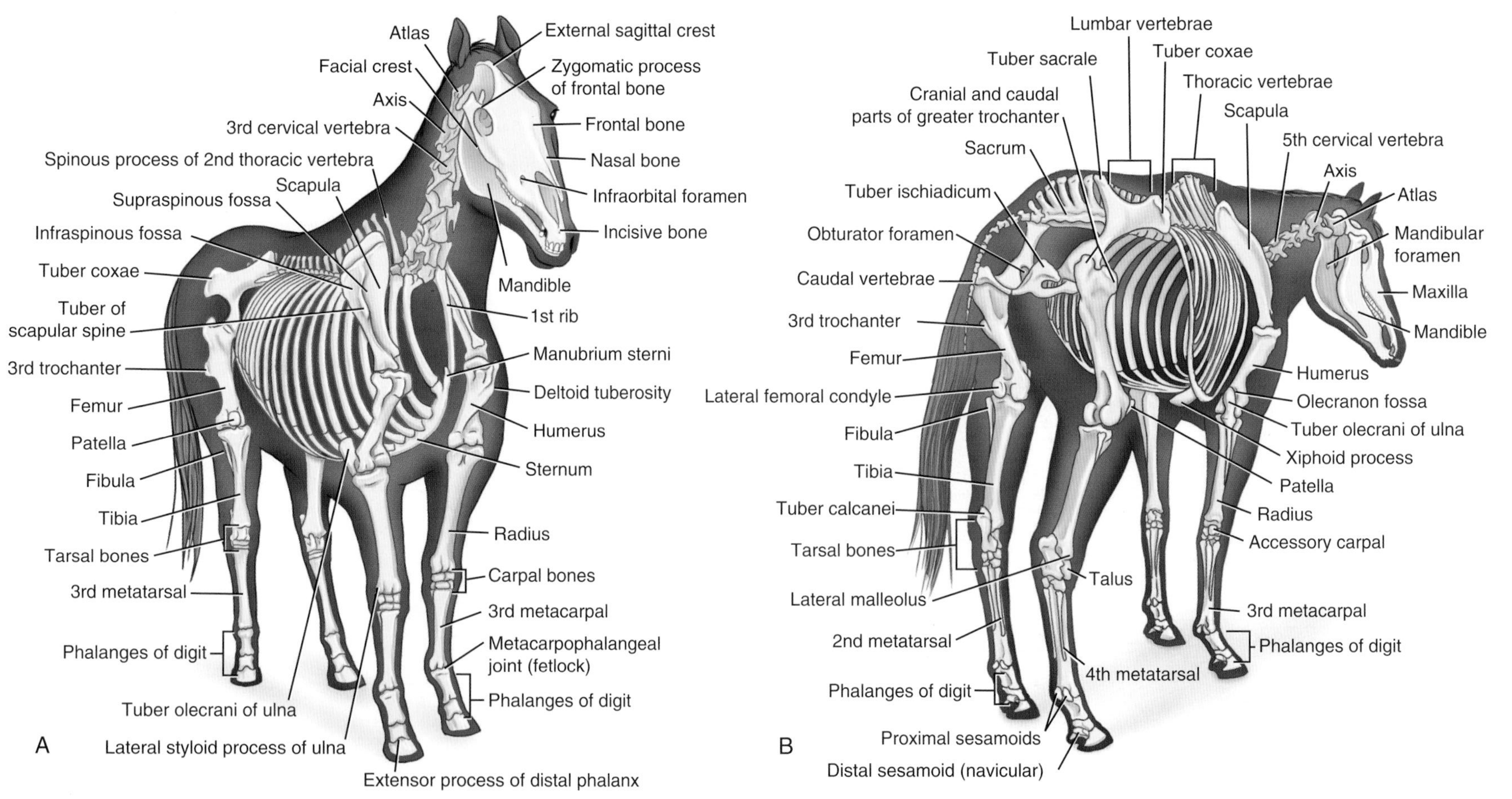

FIGURE 4.0-11 Cranial (A) and caudal (B) views of the equine skeleton.

CHAPTER 10

Axial Skeleton: Head, Neck, and Vertebral Column

Mathew Gerard and Amy L. Johnson, Chapter editors

Eye

10.1 Squamous cell carcinoma—*Amanda Davis and Brian Gilger* ... 566

Mouth

10.2 Septic sialoadenitis—*Timo Prange and Mathew Gerard* ... 576

Paranasal Sinuses

10.3 Paranasal sinus cyst—*Ferenc Toth and James Schumacher* ... 586

10.4 Dental disease and sinusitis—*Callie Fogle and Mathew Gerard* ... 599

Pharynx

10.5 Guttural pouch disease—*Olivier M. Lepage* ... 611

10.6 Dorsal displacement of the soft palate—*Tara R. Shearer and Susan J. Holcombe* ... 624

10.7 Laryngeal hemiplegia—*Eric Parente* ... 636

Cranial Nerves

10.8 Vestibular disease—*William Gilsenan* ... 646

Ear

10.9 Ear sarcoid—*Annette M. McCoy* ... 655

Poll

10.10 Nuchal bursitis—*José García-López* ... 663

Neck

10.11 Esophageal obstruction—*Judith Koenig and Shune Kimura* ... 670

Central Nervous System

10.12 Cervical vertebral osteoarthritis—*Amy L. Johnson* ... 679

10.13 Congenital cerebellar disorder—*Monica Aleman* ... 688

CHAPTER 10

CASE 10.1

Squamous Cell Carcinoma

Amanda Davis[a] and Brian Gilger[b]
[a]Red Bank Veterinary Hospital Ophthalmology, Mount Laurel, New Jersey, US
[b]Department of Clinical Sciences, North Carolina State University College of Veterinary Medicine, Raleigh, North Carolina, US

Clinical case

History

A 15-year-old Tennessee Walking Horse gelding presented with a recurrent mass involving the left eye. A mass of the 3rd eyelid had been surgically removed 3 years before and was confirmed as squamous cell carcinoma. Regrowth of the mass was suspected over the past few months based on an increase in mucoid discharge and medial canthal swelling.

Physical examination findings

There was mild blepharospasm (squinting) of the left eye with mild mucoid discharge at the medial canthus. The palpebral and bulbar conjunctiva in the ventronasal quadrant of the eye were raised and thickened. The 3rd eyelid was absent due to the previous excision surgery. There was neovascularization extending across the nasal 25% of the cornea with a raised, roughened appearance and opacity to the corneal surface (Fig. 10.1-1). The right eye was normal. No other physical examination abnormalities were noted.

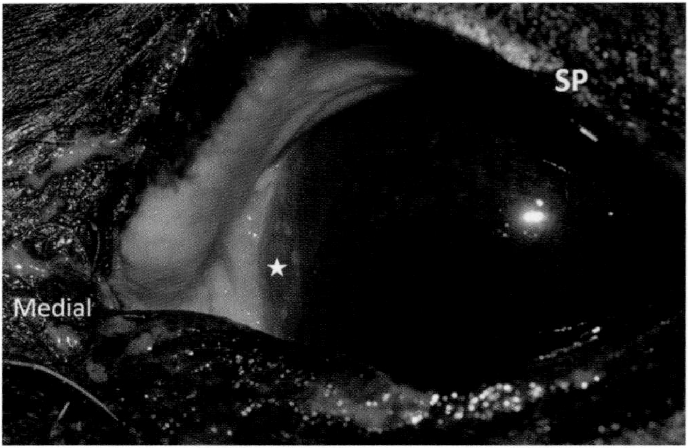

FIGURE 10.1-1 Left eye of a 15-year-old Tennessee Walking horse gelding showing mild mucoid discharge at the medial canthus. There is raised and thickened palpebral and bulbar conjunctiva. Neovascularization *(thin wavy red lines)* extends into the cornea and the corneal surface is raised, roughened, and opaque in the ventronasal quadrant *(star)*. *SP*, superior palpebral.

Comparative Veterinary Anatomy: A Clinical Approach. https://doi.org/10.1016/B978-0-323-91015-6.00137-0

Differential diagnoses

Recurrent and locally invasive squamous cell carcinoma (highly likely), lymphosarcoma, amelanotic melanoma, mastocytoma, immune-mediated keratitis (IMMK), eosinophilic keratitis, granulation tissue, stromal abscess

Diagnostics

The gelding was sedated and the zygomatic branch of the auriculopalpebral nerve (commonly referred to as the "palpebral branch") and the supraorbital nerve were locally anesthetized with 2% lidocaine to allow for a complete ophthalmic examination. Additionally, the surrounding orbital bones and the eyelids were palpated for irregularities and the globe was retropulsed to evaluate for any extension of disease into the retrobulbar space. Palpation and retropulsion were normal.

To further evaluate for local or metastatic spread of disease, the horse was placed under general anesthesia for CT of its head. Disease did not appear to spread beyond the globe and surrounding conjunctival tissue. A 3-mm (0.12-in.) sample of the irregular conjunctival tissue was obtained using a Westcott tenotomy scissor and submitted for histopathology while the horse was under anesthesia. A clinical pathologist confirmed squamous cell carcinoma (Fig. 10.1-2).

Diagnosis

Locally invasive, recurrent, conjunctival and corneolimbal squamous cell carcinoma

Common locations for ocular squamous cell carcinoma include the eyelids, nictitating membrane (3rd eyelid), and lateral limbus (corneoscleral junction). Disease may spread from these sites into the surrounding corneoconjunctival regions. Squamous cell carcinoma of the eyelids may require surgical removal or debulking of the affected region combined with adjunctive therapy to address microscopic disease.

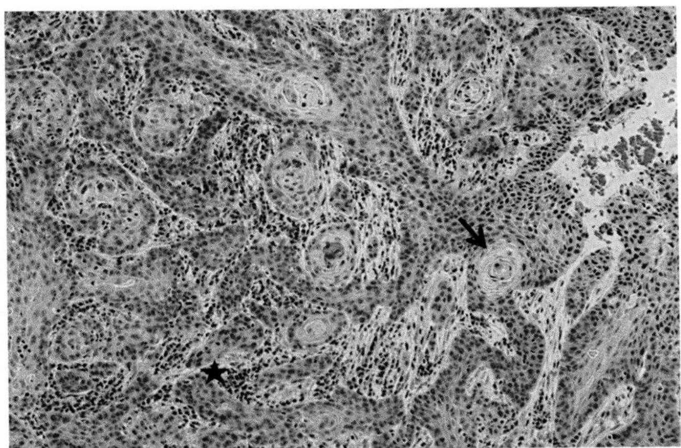

FIGURE 10.1-2 Histopathologic image of squamous cell carcinoma. A population of pleomorphic epithelial neoplastic cells is arranged in cords, nests, and islands. Keratin pearls can be seen *(arrow)* as a result of abnormal squamous cell layering. Lymphoplasmacytic inflammation *(star)* is seen interdispersed among tumor cells.

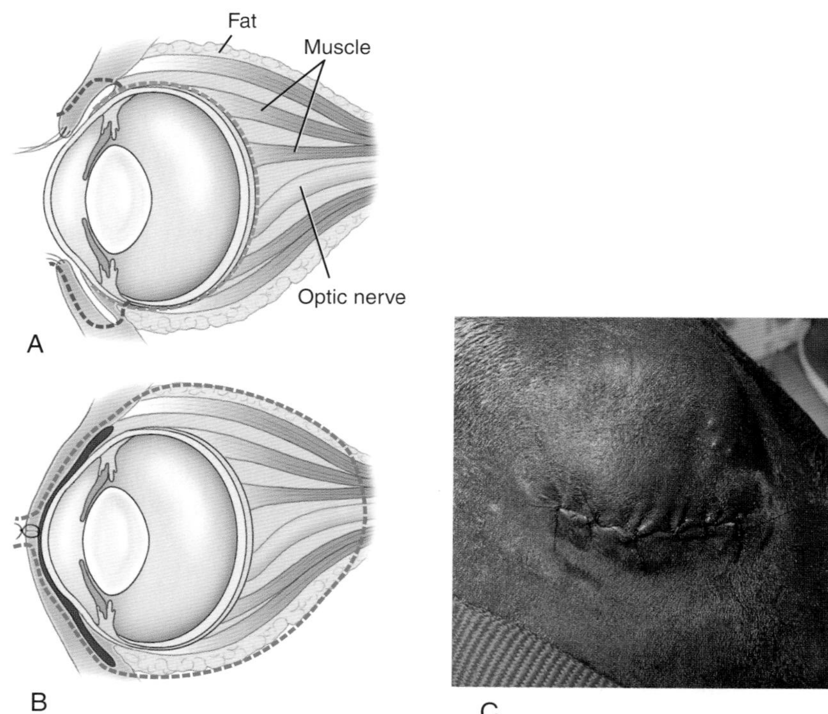

FIGURE 10.1-3 (A) Enucleation via a subconjunctival approach *(orange dotted line)*. The dissection proceeds close to the globe. Once the globe is removed, the eyelids, nictitating membrane, and remaining conjunctiva are also excised *(blue dotted line)* before closing the skin margins. (B) Exenteration via a transpalpebral approach *(green dotted line)*. After the eyelids are sutured closed, dissection proceeds posteriorly adjacent to the orbital walls. (C) Completed enucleation procedure with the final layer of the closure using cruciate skin sutures in this case. Note that the eyelid margins have been resected.

SURGICAL APPROACHES TO ENUCLEATION AND EXENTERATION

Transconjunctival: This approach is primarily used for enucleation (Fig. 10.1-3A). A 360-degree conjunctival incision is performed, and subconjunctival dissection continues posteriorly. The extraocular muscles are identified and transected close to their scleral attachments. The optic nerve is transected close to the scleral surface and without clamping to avoid undue tension on this structure. The globe is removed. The remaining nictitating membrane, conjunctiva, and eyelid margins are then removed. The wound is closed in 3 layers.

Transpalpebral: This approach is used for exenteration and enucleation (Fig. 10.1-3B). Eyelid margins are apposed with suture in a continuous pattern. The skin is incised 5–7 mm (0.20–0.28 in.) from the eyelid margin. The subcutaneous tissues are dissected posteriorly, avoiding puncturing of the conjunctival sac. Dense medial and lateral palpebral ligaments are transected. The extraocular muscles and the optic nerve are transected. The globe, conjunctiva, eyelid margins, and nictitating membrane are then removed as a unit. The wound is closed in 3 layers (Fig. 10.1-3C).

For enucleation, the posterior dissection follows the surface of the globe with muscles and the optic nerve transected close to the scleral surface. For an exenteration, the posterior dissection continues adjacent to the orbital walls to maximize the removal of all orbital tissues. Periosteal elevators, rongeurs, and bone curettes may be indicated. Increased bleeding is expected and addressed by various hemostatic means.

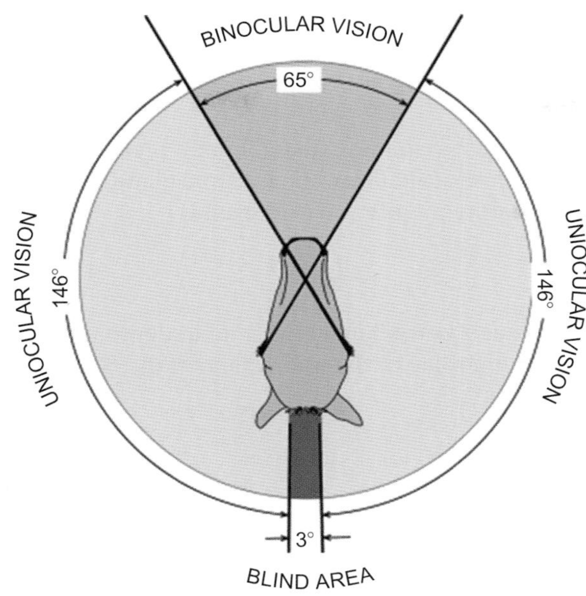

FIGURE 10.1-4 Representative schematic of equine vision. Horses have a wider range of uniocular vision and a small blind spot as a result of lateral globe placement. (From, Slatter's fundamentals of veterinary ophthalmology. 4th ed. Elsevier; 2008)

Treatment

The gelding was placed under general anesthesia for transpalpebral exenteration (L. *ex* out of, away from + Gr. *enteron* bowel)—the removal of the contents of the orbit.

Given the extent of disease in this patient, exenteration was performed as opposed to enucleation. Exenteration is the removal of the bulbus oculi—i.e., the globe—and the remaining orbital contents, including muscles, fat, glands, vessels, nerves, and periorbita (Fig. 10.1-3B). Enucleation is removal of the bulbus oculi only (Fig. 10.1-3A). In both procedures, the eyelid margins, nictitating membrane, and conjunctival tissues are also removed. The equine globe is placed laterally allowing for a wide field of view (approximately 350 degrees) with few small blind spots (Fig. 10.1-4). Enucleation for any reason eliminates a significant portion of the visual field; however, most horses adapt well to this change. Akinesia and appropriate local anesthesia can significantly increase ease of enucleation whether performed with the horse under general anesthesia or standing sedation. Important motor and sensory nerve blocks used in these procedures are covered in Table 10.1-1.

TABLE 10.1-1 Important nerve blocks for akinesia and anesthesia of the equine eye.

Name	***Nerve(s) blocked; function***
Auriculopalpebral	Blocks the zygomatic branch ("palpebral branch") of this nerve arising from cranial nerve VII (facial nerve); motor to the orbicularis oculi muscle which causes eyelid closure
Sensory	Block portions of cranial nerve V; sensory to eyelids (Fig. 10.1-11)
Retrobulbar	Blocks CN II (vision), III, IV, maxillary and ophthalmic branches of V and VI; motor and sensory (see side box entitled "Retrobulbar nerve block" and Fig. 10.1-10)

Anatomical features in equids

Introduction

The major components of the equine ocular system discussed herein include the globe (bulbus oculi) and its surrounding bony orbit with a complete orbital rim, the lacrimal gland and nasolacrimal system, eyelids (palpebrae), conjunctiva, and the nictitating membrane (3rd eyelid).

Function

Eyes function as the body's camera through vision and contribute to depth perception and balance. These multiple functions are only possible by an exact interplay of the many parts of the eye—e.g., cornea, lens, photoreceptors of the retina, and their supporting networks.

Globe (Fig. 10.1-5)

The **globe (bulbus oculi)** consists of 3 layers: The outermost layer is the **fibrous tunic**, which consists of the **cornea** and **sclera**. The middle layer is the **uvea**, which consists of the **iris** and **ciliary body** anteriorly and the **choroid** posteriorly. The innermost, central layer is the **nervous layer,** which consists of the **retina** and most anterior portion of the optic nerve.

The inner globe is clinically divided into anterior and posterior segments by the lens. Within the anterior segment from anterior to posterior are the cornea, the **anterior chamber** containing the **aqueous humor**, and the **iris**. The space between the iris and the lens is referred to as the **posterior chamber**. Posterior to the **lens** is the posterior segment of the eye. The posterior segment consists of the **vitreous humor**, the **retina,** and intraocular segment of the **optic nerve**.

Orbit (Fig. 10.1-6)

The bony **orbit** is formed by portions of the frontal, lacrimal, zygomatic, temporal, sphenoid, and palatine bones. The horse has a complete bony orbital rim. Viewed from the clinical anterior to posterior perspective, the orbital foramina consist of the ethmoidal foramen, optic canal, orbital fissure, and rostral alar foramen.

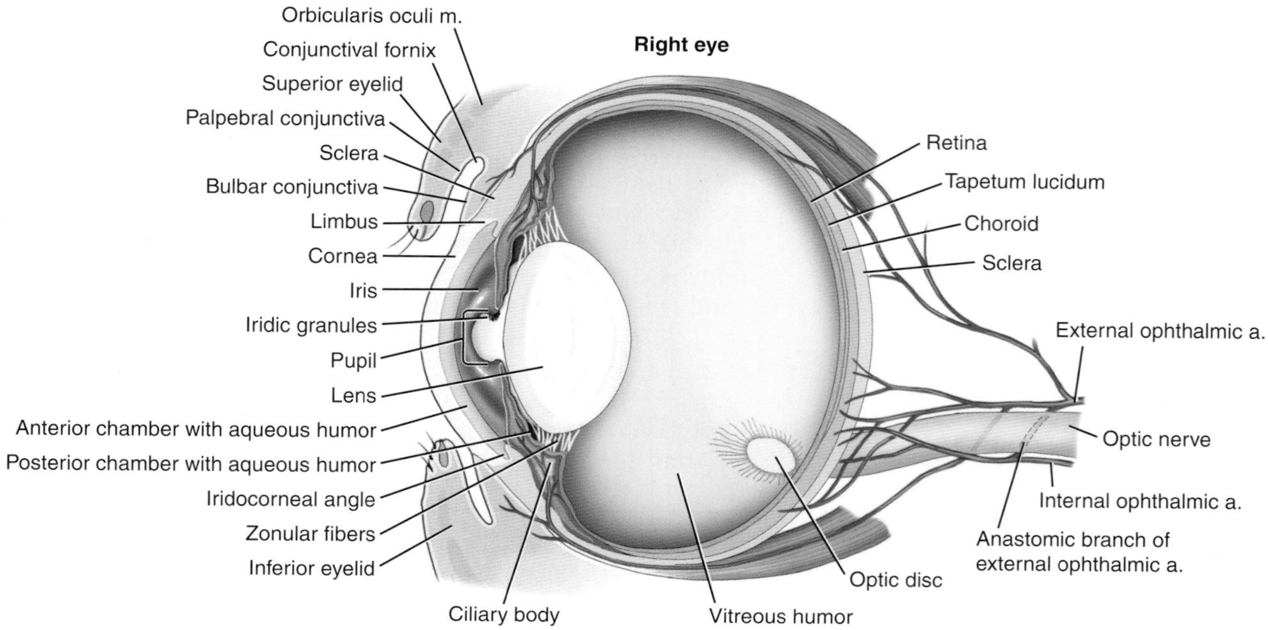

FIGURE 10.1-5 The globe and the structures and spaces within.

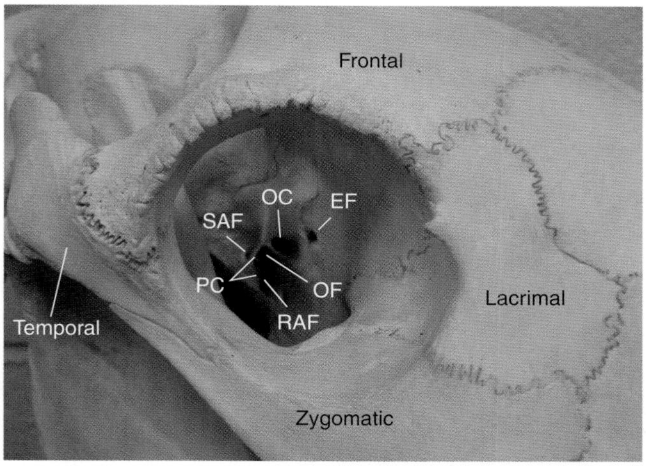

FIGURE 10.1-6 The bony orbital rim is defined by the frontal, lacrimal, zygomatic, and temporal (zygomatic process of) bones. Orbital foramina represent the portals for cranial nerves and vessels to pass between the cranial cavity and other regions of the head. *EF*, ethmoidal foramen; *OC*, optic canal; *OF*, orbital fissure; *RAF*, rostral alar foramen; *SAF*, small alar foramen; *PC*, pterygoid crest.

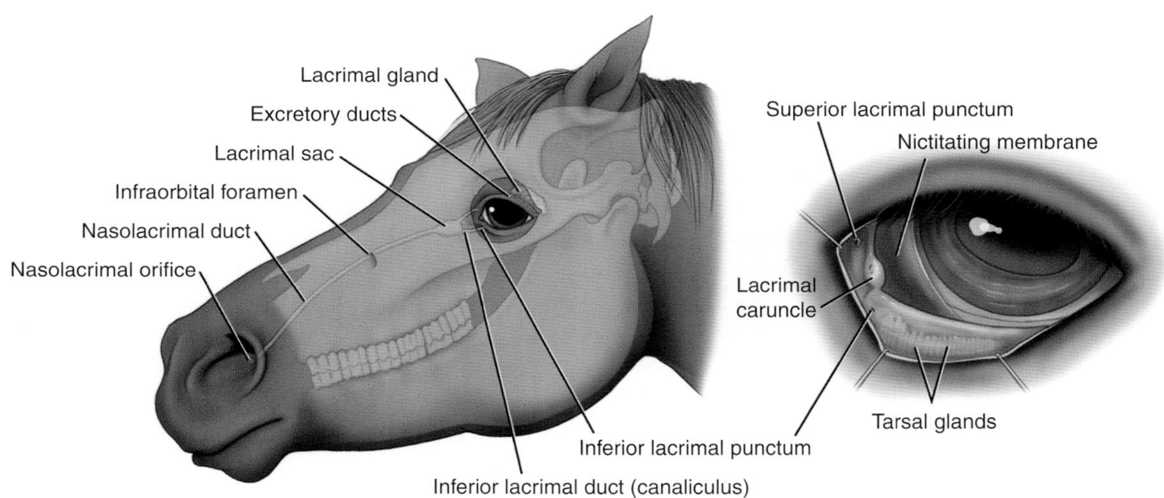

FIGURE 10.1-7 Nasolacrimal apparatus. Tears produced from the lacrimal gland and the superficial gland of the nictitating membrane (not shown) drain from the surface of the eye through lacrimal puncta and into a pathway of ducts, and then drain into the nasal vestibule at the nasolacrimal orifice.

Lacrimal gland and nasolacrimal system (Fig. 10.1-7)

The **lacrimal gland** is a tubuloacinar gland located dorsolateral to the globe. The gland connects to the **conjunctival fornix** of the superior eyelid and releases fluid through ductules. Horses also have a **superficial gland of the nictitating membrane** as discussed below.

Tears drain through a **superior** (upper) and **inferior** (lower) **lacrimal punctum**, which are located 8–9 mm (0.31–0.35 in.) lateral to the medial canthus (medial angle of the eye) on the margin of the respective eyelid. The puncta connect to the **lacrimal ducts** (or canaliculi) and join at a dilation referred to as the **lacrimal sac**. The single **nasolacrimal duct** then extends through the lacrimal and maxillary bones and adjacent nasal cavity soft tissues to exit at the ventral floor of the nasal vestibule as the **nasolacrimal orifice**.

Eyelids (Fig. 10.1-5)

The upper and lower **eyelids** (**superior** and **inferior palpebrae**) consist of 3 layers. These layers include the **external skin** with sebaceous glands; a middle layer containing muscle, connective tissue, and the **tarsal plate**; and the inner **palpebral conjunctiva**, which is in contact with the globe.

Innervation to the eyelids is important when considering analgesic and akinesic blocks for diagnostic and treatment purposes (see Table 10.1-1).

Motor innervation to the eyelid is provided by CN III (to the levator palpebrae superioris m.) and VII (to the orbicularis oculi m.). Sensory innervation to the eyelid is provided by CN V (see Table 10.1-1).

Nictitating membrane

This fat pad is important to consider when surgically excising the nictitating membrane for any reason, but most commonly for the removal of squamous cell carcinoma, as failure to suture the remaining conjunctival tissue can lead to fat prolapse.

The **nictitating membrane** or 3rd eyelid is located in the ventromedial aspect of the orbit, and its movement, dorsolaterally to cover the globe, is a passive process. Anatomically, this structure consists of **conjunctiva**. Enveloped within the conjunctival tissue is a **T-shaped cartilage** with a tubuloacinar **lacrimal gland** (**superficial gland of the 3rd eyelid**) at its base. It is unknown how much this gland contributes to equine tear production. Ventral to the gland is a large **fat pad**.

Corneoconjunctival region

Squamous cell carcinoma in this region generally begins at the limbus at the lateral aspect of the eye. Focal or smaller lesions can be removed through a superficial keratectomy and conjunctivectomy. Following this procedure, it is common to insert a subpalpebral lavage system to administer medications or follow-up therapy such as topical chemotherapy (see side box entitled "SPL placement"). Disease affecting more of the orbit and surrounding structures may require enucleation or exenteration was the case in our clinical patient example.

The **cornea** consists of 4 layers. From anterior to posterior, these are the superficial stratified squamous **epithelium**, the central **stroma**, the basement membrane to the endothelium or **Descemet's membrane**, and the single-layered **endothelium**. The adjacent **bulbar conjunctiva** is composed of non-keratinized stratified columnar or squamous cell epithelium, which has a rich vascular supply and lymphatic system, and is continuous with the corneal epithelium. The **conjunctival ring** is the anatomical junction between the epithelium of the cornea and bulbar conjunctiva, and it is directly adjacent to the limbus.

SUBPALPEBRAL LAVAGE (SPL) PLACEMENT

Treatment of ocular disease may require the use of a special catheter system that allows medications to be delivered to the ocular surface without having to manipulate the eye. Medication is injected into a closed port typically secured to the mane near the caudal end of the neck. The medicine injection is followed with air to push the medication towards the globe through the length of tubing. The medication then exits the tube through the device footplate to contact the ocular surface. The footplate is located under the base of the upper or lower eyelid at the conjunctival fornix.

The horse is sedated and the periorbital region prepared with 1% povidone iodine. The mane should be braided to secure the SPL tubing after placement. Auriculopalpebral and supraorbital nerve blocks are performed as described in Table 10.1-1. The trocar of the SPL system is inserted in the conjunctival fornix of the upper or lower eyelid using an insertion tube from the kit or a gloved finger to position (Fig. 10.1-8A and B). The eyelid is then punctured as far dorsal as possible using the trocar and the tubing passed through and secured through the braids of the mane. The trocar is removed, and a catheter with an injection port is placed on the end of the tubing. This is secured to one braid of the mane using tape and a tongue depressor. The tubing exiting the eyelid is held in place with two pieces of tape that are sutured to the skin (Fig. 10.1-9).

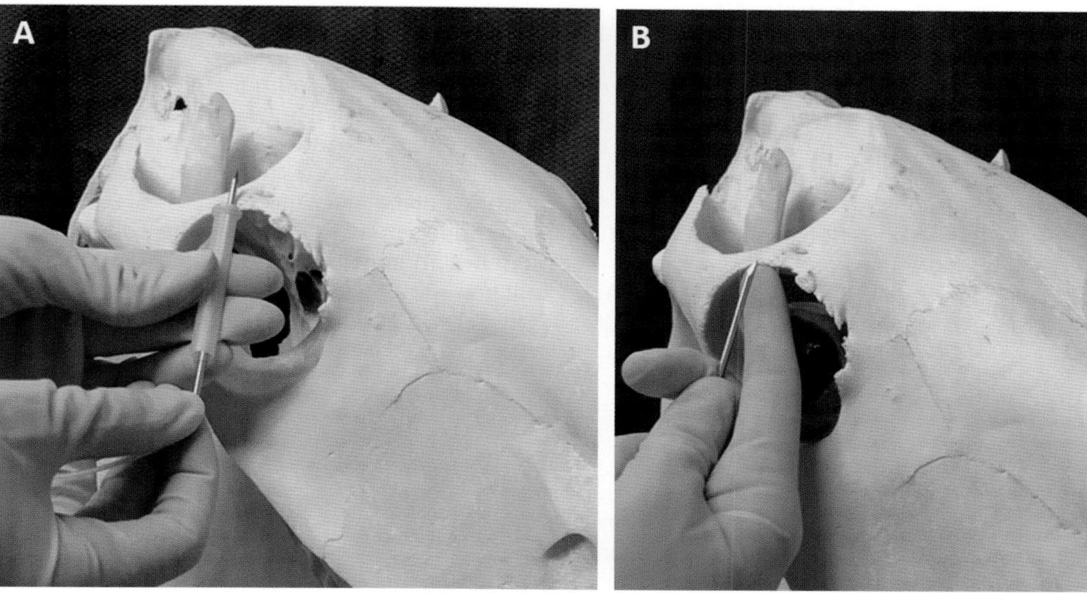

FIGURE 10.1-8 (A) Equine skull model demonstrating a proper positioning of a subpalpebral lavage trocar using an insertion tube. (B) Equine skull model demonstrating a proper positioning of a subpalpebral lavage trocar using a gloved finger.

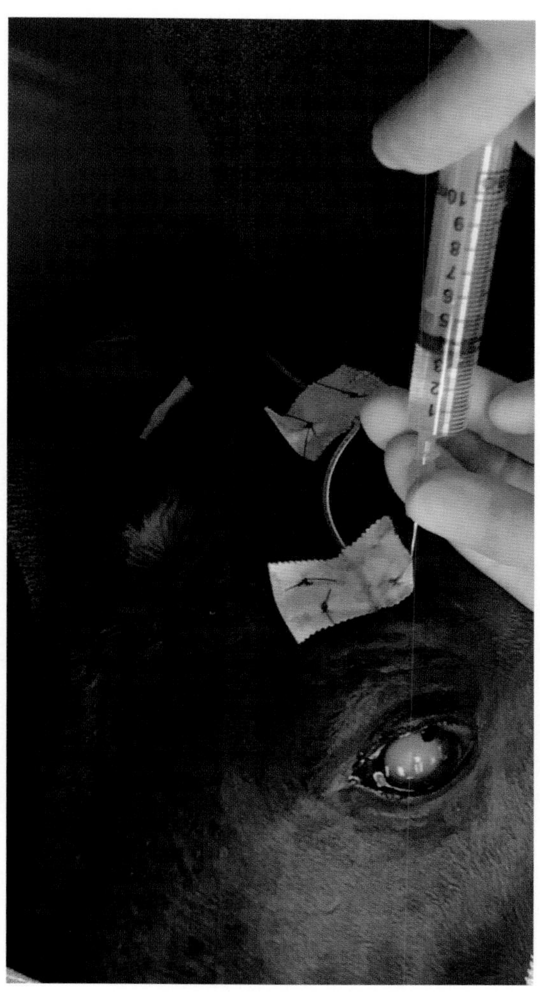

FIGURE 10.1-9 Injection of local anesthetic via a retrobulbar block technique to desensitize the globe and tissues within the retrobulbar orbital cone formed by the periorbita. This foal has had a subpalpebral lavage catheter previously inserted.

SUTURING THE EYELID

Due to its ulcerative nature, squamous cell carcinoma can often present as trauma. Full-thickness eyelid wounds are another common ocular condition in equine patients. The eyelids are well vascularized, which allows for excellent healing of surgical wounds. Excision of traumatized eyelid tissue to ensure an achievement of proper anatomical alignment before closure is rarely indicated. Two-layer closure is ideal for closing eyelid lacerations. Sutures are placed in the subconjunctival layer using 4–0 or 5–0 synthetic absorbable suture. It is important to ensure sutures are not exposed on the surface of the conjunctiva to avoid rubbing on the globe and causing ulceration. The palpebral margin is closed with a figure-of-8 pattern suture that apposes the margin without any gaps. Any misalignment can lead to corneal irritation and ulceration. The remaining skin can be closed with simple interrupted sutures (Fig. 10.1-10).

RETROBULBAR NERVE BLOCK

A retrobulbar nerve block is strongly recommended to assist in procedures such as a keratectomy or enucleation. The preferred site for injection is the orbital fossa, immediately caudal to the palpable posterior margin of the dorsal orbital rim. This area is aseptically prepared. A 22-gauge, 2.5-in. spinal needle is passed through the skin in this location perpendicular to the skull (Fig. 10.1-9). The needle is advanced through the periorbita of the retrobulbar orbital cone (noted by a dorsal deviation to the eye) and into the cone (the eye resumes its normal position and a "popping" or loss of resistance sensation is often appreciated). At this location, before injection, it is important to aspirate to ensure the needle is not within a blood vessel. Then, approximately 8 mL of 2% lidocaine can be injected to desensitize the globe, and adnexal tissues (extraocular muscles, eyelids, lacrimal gland). In addition, vision and the blink reflex are temporarily compromised for the duration of the anesthetic agent.

Blood supply and lymphatics

Innervation to the structures of the eye is discussed in each relevant section above. The eye is primarily supplied by the external and internal ophthalmic arteries, which anastomose about 3 cm (1.2 in.) behind the posterior wall of the globe, adjacent to the optic nerve (Fig. 10.1-5). The **external ophthalmic artery** arises from the maxillary artery and provides branches to the layers of the globe, muscles, eyelids, and lacrimal gland. The smaller **internal**

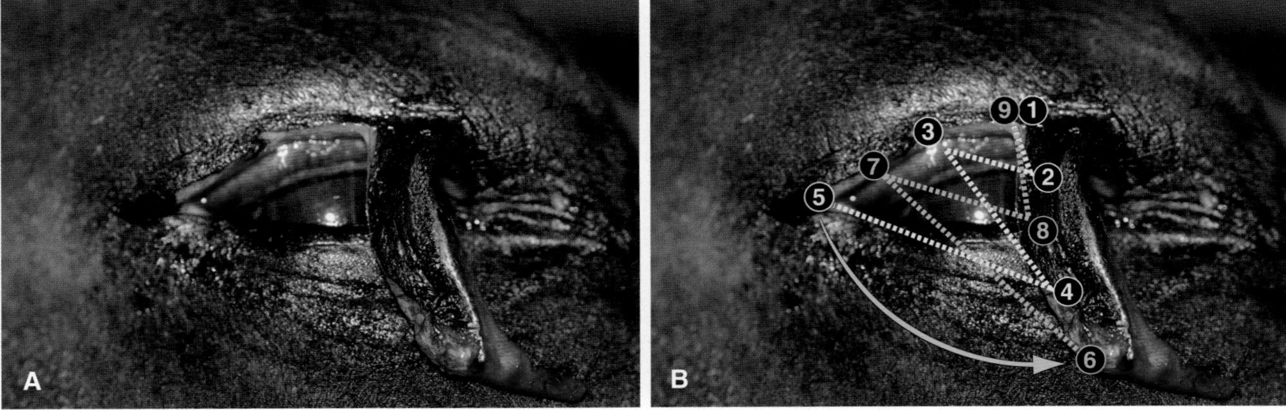

FIGURE 10.1-10 (A) Right superior eyelid laceration with the medial palpebral commissure remaining intact. (B) Two-layer conjunctival closure. A surgical knot using 4–0 or 5–0 Vicryl™ is placed at the apex of the laceration within the conjunctival tissue (1). A continuous suture pattern is placed through the deeper layers of the conjunctival tissue and continued medial to lateral (*light blue*, 1–5). A superficial continuous pattern is performed lateral to medial (*yellow*, 6–9) and ends at the original apex suture. After conjunctival apposition, the palpebral margin is closed with a figure-of-8 pattern (at its lateral margin in this case, and not shown) and the remaining skin is closed with simple interrupted sutures.

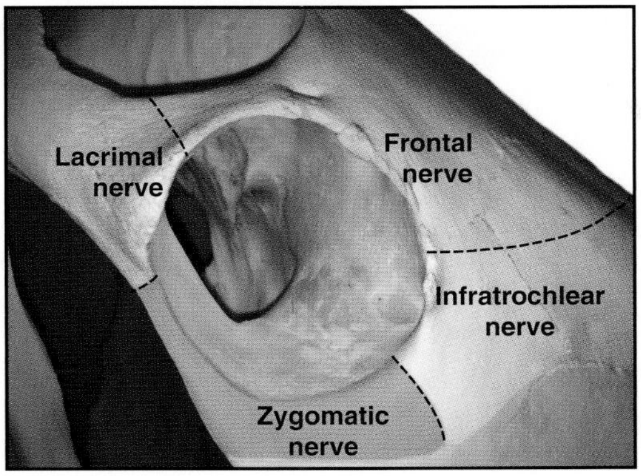

FIGURE 10.1-11 Shading of the skull around the orbit to illustrate the distribution of sensory nerves.

ophthalmic artery arises from the cerebral arterial circle and supplies the optic nerve, which it accompanies to the globe. The **malar artery** branches from the infraorbital artery and passes over the ventral surface of the orbit to reach the medial canthus of the eye. Here, it provides branches to the medial portions of the superior and inferior eyelids.

Extensive venous networks drain the eye and extraocular tissues with much of this blood entering the **deep facial vein** in the horse. Lymphatic drainage of the eye region is principally to the **parotid lymphocenter**, with a portion of eyelid drainage passing to the **mandibular lymph nodes**.

Selected references

[1] de Linde Henriksen M, Brooks DE. Standing ophthalmic surgeries in horses. Vet Clin North Am Equine Pract 2014;30(1):91–110.
[2] Utter ME, Wotman KL, Covert KR. Return to work following unilateral enucleation in 34 horses (2000–2008). Equine Vet J 2010;42(2):156–60.
[3] Clode AB, Miller C, McMullen Jr RJ, Gilger BC. A retrospective comparison of surgical removal and subsequent CO_2 laser ablation versus topical administration of mitomycin C as therapy for equine corneolimbal squamous cell carcinoma. Vet Ophthalmol 2012;15(4):254–62.
[4] Gilger BC. Equine ophthalmology. 3rd edition. Ames: John Wiley & Sons, Inc.; 2017.

CHAPTER 10

CASE 10.2

Septic Sialoadenitis

Timo Prange[a] and Mathew Gerard[b]
[a]Department of Clinical Sciences, North Carolina State University College of Veterinary Medicine, Raleigh, North Carolina, US
[b]Department of Molecular Biomedical Sciences, North Carolina State University College of Veterinary Medicine, Raleigh, North Carolina, US

Clinical case

History

A 21-year-old American Quarter Horse gelding presented with a 4-day history of increasing swelling caudal to the left ramus of the mandible, inappetence, halitosis (malodorous breath), and ptyalism (hypersalivation).

Physical examination findings

The gelding was quiet, alert, and responsive with a normal temperature (100.2 °F [37.9 °C]), slight tachypnea (24 brpm), and mild tachycardia (44 bpm). His oral mucous membranes appeared slightly hyperemic with a mildly prolonged CRT of 3 s. A firm, painful swelling was present caudal to the left ramus of the mandible extending ventrally, involving the parotideal, pharyngeal, laryngeal, and intermandibular regions. The swelling tapered rostrally, allowing palpation of moderately enlarged mandibular lymph nodes. Ptyalism and halitosis were noted during the examination.

Differentials

Lymphadenopathy/lymphadenitis of the retropharyngeal lymph nodes (e.g., *Streptococcus equi* subsp. *equi* (strangles) infection), septic sialoadenitis, trauma, foreign body

Diagnostics

The horse was mildly painful upon palpation of the swelling in the retromandibular fossa. The skin overlying the swollen areas was clipped and thoroughly inspected. No signs of trauma were present and no draining tract was found. The gelding was sedated, and following positioning of a McPherson full-mouth speculum, an oral examination revealed a 2 × 2 cm (0.8 × 0.8 in.) ulcerated area underneath the apex of the tongue in the left lateral sublingual recess (Fig. 10.2-1). An 8-French Foley catheter was inserted into the defect and advanced caudally for 4–5 cm (1.6–2.0 in.). Lavage of the tract produced small amounts of food material, purulent material, and necrotic tissue. Additional superficial ulcerations were found along the lateral surfaces of the tongue (Fig. 10.2-2).

Ultrasonography of the swelling in the retromandibular fossa showed a complex structure in the left mandibular salivary gland, deep to the parotid salivary gland. The contents of the structure were of mixed echogenicity and included numerous hyperechoic foci, likely representing small gas bubbles within an abscess. A smaller, similar structure was identified in the rostral part of the intermandibular swelling. This rostral mass appeared to be associated with the polystomatic sublingual salivary gland. Ultrasound-guided fine-needle aspiration of the mass in the mandibular salivary gland produced foul-smelling, purulent fluid that was submitted for anaerobic and aerobic culture. A CBC showed a mild neutrophilic leukocytosis. Endoscopic evaluation of the nasal passages, pharynx, guttural pouches, larynx, and cranial trachea was unremarkable.

Comparative Veterinary Anatomy: A Clinical Approach. https://doi.org/10.1016/B978-0-323-91015-6.00138-2

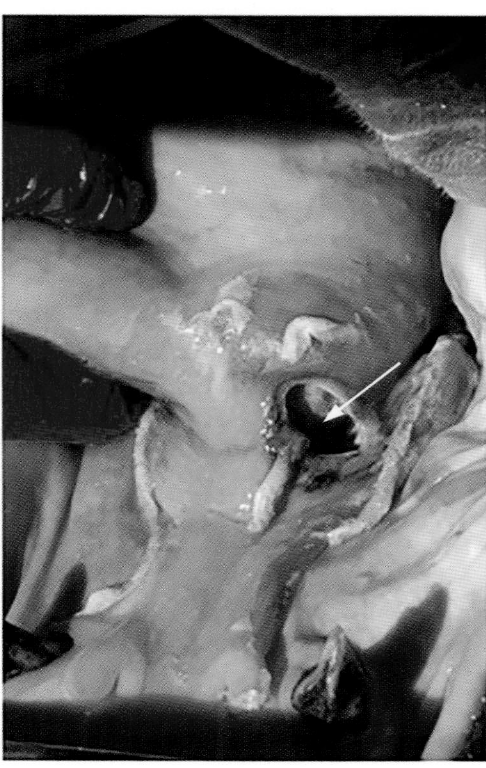

FIGURE 10.2-1 Photograph of rostral oral cavity showing the ulcerated defect ventral to the apex of the tongue *(white arrow)* in the left lateral sublingual recess.

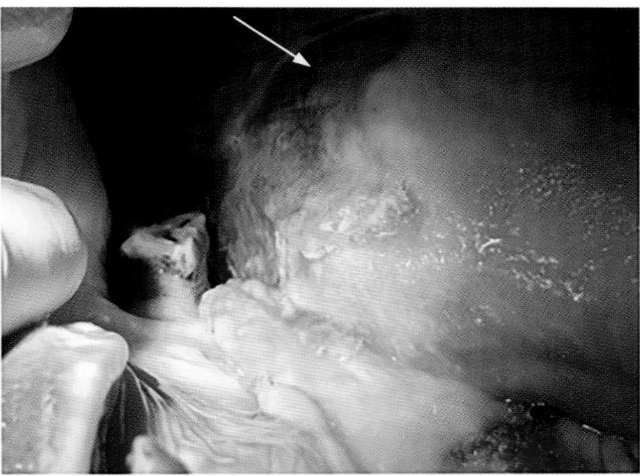

FIGURE 10.2-2 Large superficial ulceration of the right lateral surface of the tongue, adjacent to the first cheek tooth *(white arrow)*.

Diagnosis

Septic sialoadenitis of the left mandibular and polystomatic sublingual salivary glands

Treatment

Under ultrasonographic guidance, the ventral-most aspect of each abscess was identified, and after the site was anesthetized with 2% mepivacaine, the abscesses were open and drained with a #10 scalpel blade. During subsequent lavage with 0.2% povidone iodine solution, communication between the abscess in the mandibular salivary gland and the ulcerated area in the oral cavity via the mandibular duct was confirmed. There was no communication between the sublingual abscess and the oral cavity. Following this initial treatment, the horse was administered procaine penicillin, twice daily, IM, and gentamicin once daily, IV, to provide a broad-spectrum antimicrobial coverage. Flunixin meglumine was administered twice daily, IV, for pain management and anti-inflammatory effects. The microbiological culture grew a gram-negative anaerobic microorganism, identified as a *Fusobacterium* spp. Consequently, antimicrobial management was changed to oral metronidazole, 3 times a day, an antimicrobial effective against anaerobic bacteria. Daily lavages of the abscesses with 0.2% povidone iodine solution and of the oral cavity with a dilute chlorhexidine gluconate solution (5 mL of 2% chlorhexidine gluconate added to 1 L of water) were completed to address the lesions topically. The horse responded well to the treatment, and the daily lavages of the abscesses were discontinued after 10 days. After 14 days, systemic antimicrobials were discontinued and the ulcerations in the oral cavity had healed by day 21. By then, the clinical signs observed upon presentation had resolved, and the horse was discharged from the hospital.

Anatomical features in equids

Introduction

The relatively long and narrow oral cavity (mouth) of the horse is the first part of the alimentary tract and the site of prehension, mastication, and insalivation of food. The oral cavity is bordered by cheeks (lateral), palate (dorsal), sublingual mucosa (ventral), lips (rostral), and palatoglossal arches (caudal). The teeth divide the oral cavity into the centrally located oral cavity proper and the peripherally located vestibule of the oral cavity—i.e., the space between the outer surfaces of the teeth and the inner surface of the cheeks and lips (Fig. 10.2-3).

Function

The most elementary responsibility of the oral cavity is the preparation of food for swallowing. This includes not only the already-mentioned prehension, mastication, and insalivation, but also the intraoral "assessment" of ingested plant material. This is accomplished with the help of the gustatory organ (Organ of taste), i.e., all the taste buds, which are primarily located in the papillae of the tongue.

The relatively small rima oris in horses can make examination of the caudal aspect of the narrow and long oral cavity difficult. Consequently, evaluation and treatment of disorders pertaining to the last cheek teeth (modified Triadan system 10s and 11s) and the body and root of the tongue are challenging.

Muscles and soft tissues of the lips and cheeks

The mouth opening (**rima oris**) is formed by the upper and lower **lips** that meet at the **commissure** of the lips. In the horse, the rounded commissure is located approximately at the level of the second premolar (modified Triadan system 06s), creating a relatively small mouth opening. For comparison, the commissure of the lips in the dog reaches as far caudal as the last premolar (modified Triadan system 08s), allowing the mouth to be opened wide.

The lips of the horse are very mobile and sensitive in order to allow the selection and prehension of food. Their skin is directly connected to the underlying **orbicularis oris muscle**. This circular muscle is the sphincter of the rima oris, and its fibers are not attached to the skeleton. The **levator nasolabialis muscle** arises rostral to the orbita and inserts at the lateral aspect of the upper lip and the lateral nostril. Contraction opens the nostril and raises the upper lip. Partially covered by the preceding muscle, the **levator labii superioris muscle** originates from the lacrimal bone and passes across the infraorbital foramen before forming an aponeurosis with its counterpart and inserting on the upper lip. It is also responsible for raising the upper lip. Originating from the ramus of the mandible, the **depressor labii inferioris muscle** is largely fused with the **buccinator muscle** (Fig. 10.2-3). It inserts into the orbicularis oris muscle, and depresses and retracts the lower lip.

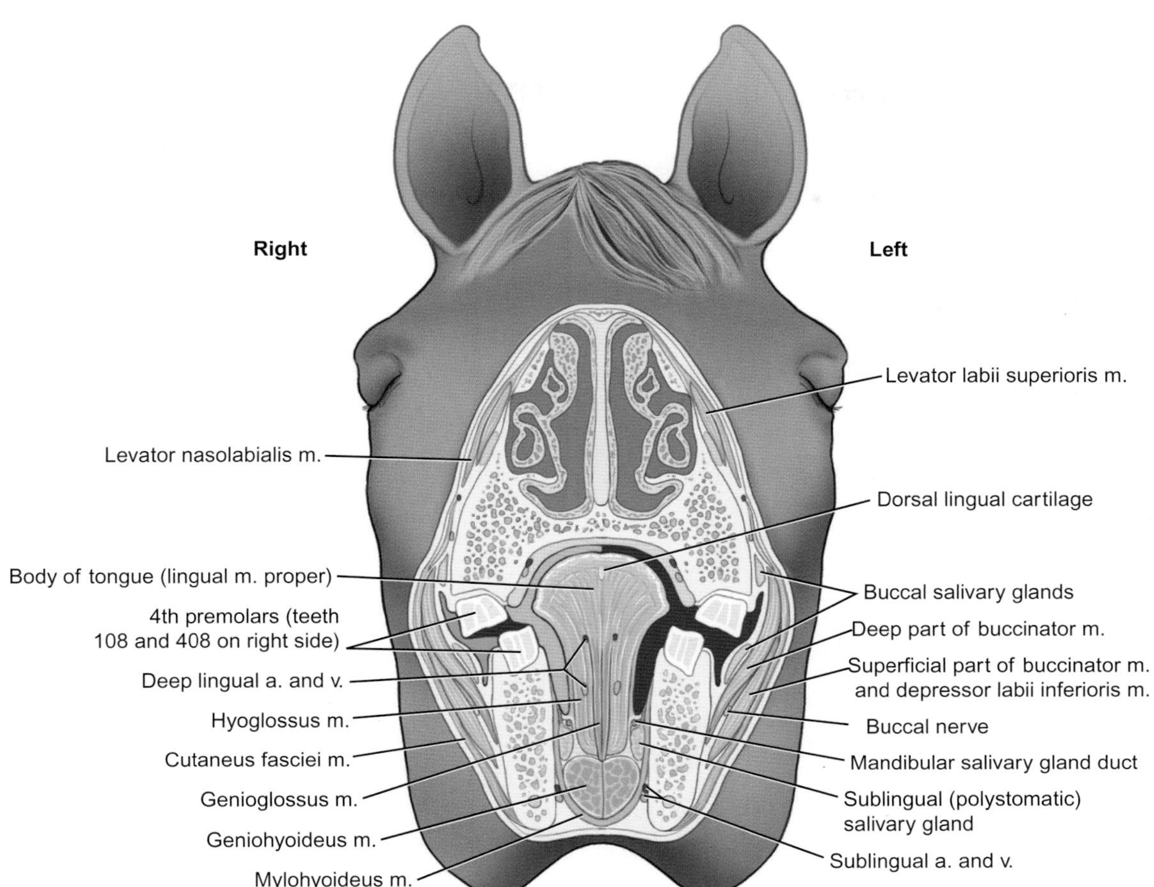

FIGURE 10.2-3 Cross-sectional view at level of 4th premolar tooth (3rd cheek tooth) identifying oral cavity, cheek anatomy, and structure relationships. For one-half of the image, the vestibule of the oral cavity is *shaded green*, and the oral cavity proper is *shaded blue*. View is from the cranial to caudal perspective.

The basic structure of the cheeks (**buccae**) consists, similar to the lips, of a hair-covered skin (without a subcutis), a muscular layer, and the oral mucosa. The **buccinator muscle** is the principle muscle of the cheeks, which form the lateral walls of the oral cavity. Its superficial part attaches along the molar areas of the maxilla and mandible, while the deep part consists of longitudinal fibers that extend caudally to the ramus of the mandible, where it is covered by the **masseter muscle**. Function of the buccinator ensures that food does not accumulate in the vestibule of the oral cavity, but is returned into the oral cavity proper.

> The high mobility of the lips and the absence of a subcutis need to be considered when repairing a lip laceration. The intimate connection between the muscle and skin, as well as the mucosa, can lead to excessive motion and tension on the suture lines, often resulting in suture dehiscence. To minimize the risk of this complication, skin and mucosa are sharply separated from the underlying muscle along the edges of the wound. Tension relieving sutures are placed through the muscle before skin and mucosa are closed in separate layers (Fig. 10.2-4).

Muscles and soft tissues of the tongue

The **tongue** (L. *lingua*; Gr. *glossa*) occupies the greater part of the oral cavity proper and extends caudally into the oropharynx. Primarily composed of the intrinsic striated **lingual muscle proper**, the tongue is divided into 3 anatomical regions: root, body, and apex. The tip of the tongue, the **apex**, is freely moveable and only caudally attached to the floor of the oral cavity by the **lingual frenulum** (Figs. 10.2-3 and 10.2-5).

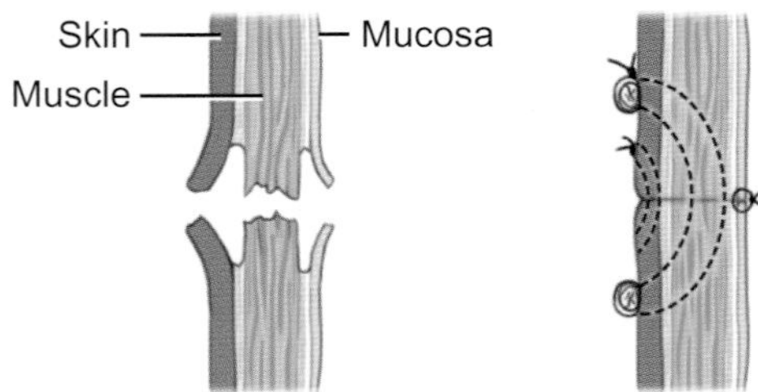

FIGURE 10.2-4 Cross-section of lip structure showing a lip wound *(left)* and the intimate connection between the muscle and skin, as well as the mucosa. To avoid excessive motion and tension on the suture lines, skin and mucosa are sharply separated from the underlying muscle along the edges of the wound. Tension-relieving sutures are placed through the muscle before skin and mucosa are closed in separate layers *(right)*.

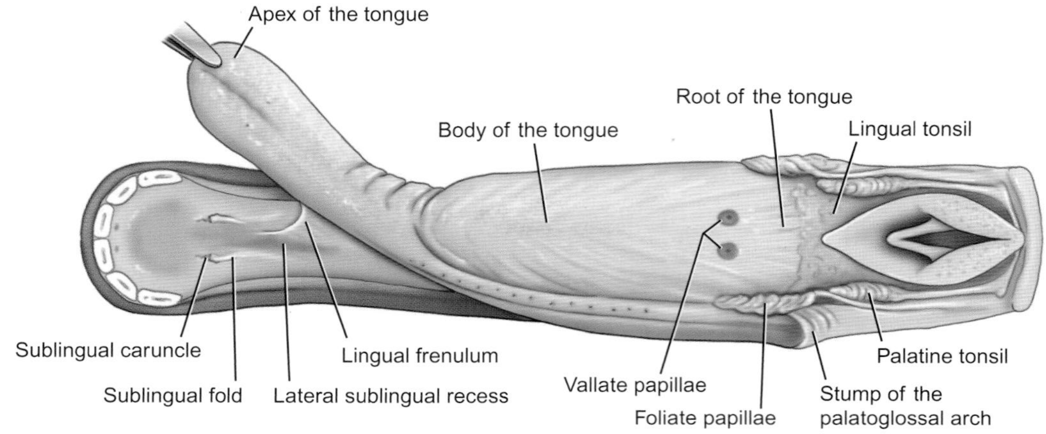

FIGURE 10.2-5 Dorsal view of the tongue and additional structures in the oral cavity proper. The caudal extent of the oral cavity proper is defined by the palatoglossal arches. The oropharynx continues the oral cavity proper, and therefore, the root of the tongue is primarily located in the oropharynx.

The apex has 4 mucosa-covered surfaces: the dorsal surface, 2 lateral borders, and the ventral surface. The middle part of the tongue, the **body**, is ventrally anchored to the mandible by the extrinsic lingual muscles (see later) and has 3 free surfaces. The **root** is defined as the part of the tongue caudal to the **vallate papillae** (Fig. 10.2-5). It is attached to—and supported by—the **lingual process** of the **basihyoid**, and only its dorsal surface is free and covered by mucosa.

The lingual mucosa is thick on the dorsum of the tongue and, via a dense submucosa, closely attached to the underlying musculature. The mucosa is thinner and more easily separated from the lingual muscle proper on the lateral borders and ventral aspect of the apex. The dorsal mucosa of the apex and body presents an abundance of the delicate mechanical **filiform papillae**, interspersed with gustatory (sensory) **fungiform papillae** on the dorsum of the apex and along the lateral edges of the tongue. The 2 remaining types of gustatory papillae in the equine tongue, the vallate papillae (see earlier) and the **foliate papillae**, are found on the border between the body and the root of the tongue, and directly rostral to the palatoglossal arches, respectively. In the submucosa of the root, numerous lymph follicles form a diffuse **lingual tonsil** (Fig. 10.2-5). Adjacent the lingual tonsil, in the ventrolateral walls of the oropharynx, lie the prominent palatine tonsils. Located in the body just below the mucosa and palpable as a median dorsal thickening, the thin **dorsal lingual cartilage** provides an additional support for the tongue (Fig. 10.2-3).

The lingual muscle proper (see earlier) consists of transverse, perpendicular, and longitudinal fibers that do not have a bony attachment. However, they are indirectly connected to the skeleton by extrinsic muscles that enter the tongue and interweave with the intrinsic musculature. The 2 most important extrinsic muscles are the hyoglossus and genioglossus muscles. The **genioglossus** originates from the incisive part of the mandible and enters the tongue ventrally, with its most rostral fibers passing through the lingual frenulum. This muscle is responsible for protrusion of the tongue. The **hyoglossus** counteracts the genioglossus muscle by retracting and depressing the tongue. It arises from the lingual process of the basihyoid bone, along with the thyrohyoid and stylohyoid bones in the horse, and courses toward the dorsal aspect of the median plane of the tongue.

Ventral to these 2 extrinsic muscles of the tongue, the **geniohyoid muscle** connects the rostromedial body of the mandible to the lingual process of the basihyoid bone and functions to move the tongue rostrally. Lying superficial to the geniohyoid, the **mylohyoid muscle** spans the intermandibular space with transverse fibers that are attached to the medial cortex of the mandible on each side. The mylohyoid muscle suspends the tongue and raises the floor of the oral cavity (Fig. 10.2-3). A third extrinsic tongue muscle, the **styloglossus muscle**, attaches between the stylohyoid bone and the apex of the tongue, functioning to retract the tongue caudodorsally and draw the apex to either side when acting singularly.

The **lingual frenulum** is a mucosal fold that arises from the ventral surface of the caudal apex of the tongue and attaches to the floor of the oral cavity. On either side of its attachment is the **lateral sublingual recess**. The lateral border of this recess is formed by the **sublingual fold (plica sublingualis)**, where the openings of the **sublingual polystomatic salivary gland** (see later) are located (Fig. 10.2-5).

Innervation of the mouth, lips, and tongue

The **trigeminal nerve** (CN V) provides sensory innervation to the lips and cheeks via its major **maxillary** (CN-V_2) and **mandibular** (CN-V_3) branches. The **maxillary nerve** supplies the **infraorbital nerve** to the upper lip. The lower lip receives sensory innervation via the **mental nerve** which continues as the **inferior alveolar nerve**. supplied by the mandibular branch. The **buccal nerve**, also a branch of the mandibular nerve, supplies sensory fibers to the cheek mucosa. Motor innervation of the lips and cheeks is derived from the **buccal branches** of the **facial nerve** (CN VII).

Traumatic injury to the buccal branches of the facial nerve can be the result of compression during anesthesia and recovery or by a tight-fitting halter. Iatrogenic damage is a possible complication of oral and maxillofacial surgeries, especially buccotomies. Knowledge of the course of these nerves is therefore critical, especially when planning a surgical approach to the oral cavity (Case 10.4). Clinical signs of injury pertaining to the lips and cheeks include food packing in the ipsilateral vestibule of the oral cavity and asymmetry of the upper and/or lower lips.

Motor innervation of the intrinsic and extrinsic muscles of the tongue is provided by the **hypoglossal nerve** (CN XII). Regular sensory innervation of the lingual mucosa is divided between the **lingual nerve** of the **mandibular nerve** (rostral two-thirds) and the **lingual branch** of the **glossopharyngeal nerve** (CN IX; caudal third). Special sensory innervation for taste is supplied through nerve fibers that join the regular sensory nerves for distribution.

Because the hypoglossal nerve (CN XII) is the only source of motor innervation for the in- and extrinsic tongue muscles, injury to the nerve typically results in abnormal function and tone, asymmetry, and deviation of the tongue. Dysfunction of CN XII can occur following head trauma or be a complication of guttural pouch disease (e.g., guttural pouch mycosis) because the nerve travels along the caudal wall of the guttural pouch. It is therefore recommended to endoscopically examine the guttural pouches in cases of suspected hypoglossal nerve injury (see Case 10.5).

Blood supply to the mouth, lips, and tongue

Branches of the **facial artery** and the **masseter branch of the external carotid a.** provide the blood supply to the cheeks. The **inferior** and **superior labial arteries** from the **facial artery** supply the lips. The **linguofacial trunk** gives rise to the **facial** and **lingual arteries**. The latter enters the tongue medial to the hyoglossus muscle and becomes the

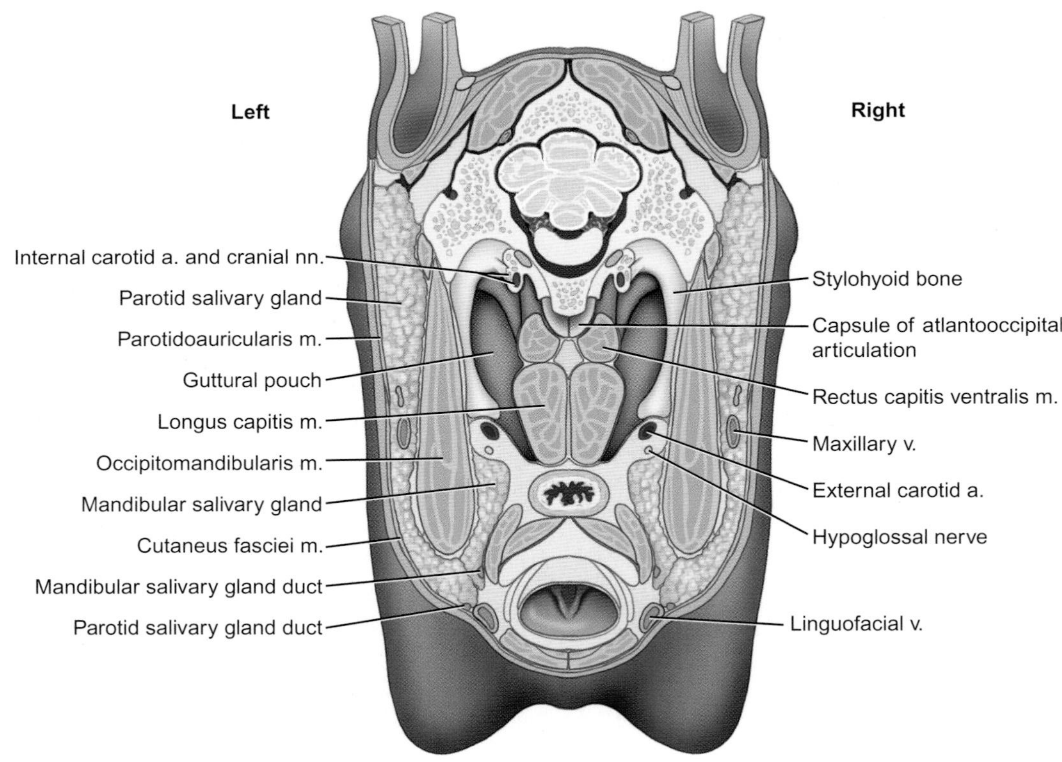

FIGURE 10.2-6 Cross-sectional image of the head at the level of the retromandibular fossa, showing the relationships of the major salivary glands and their ducts to surrounding structures. View is from the caudal to cranial perspective.

deep lingual artery (Fig. 10.2-3). This vessel continues rostrally, releasing numerous **dorsal lingual branches** that course toward the upper surface of the tongue. The facial artery supplies the **sublingual artery** which travels rostrally in the intermandibular space, ending in the sublingual floor of the oral cavity where it supplies the lingual frenulum and surrounding muscles (Fig. 10.2-3). Venous drainage converges into the large **linguofacial** and **maxillary veins** (Fig. 10.2-6), which unite to form the **external jugular vein**.

Salivary glands

The minor and major salivary glands in the horse produce approximately 40 L of saliva per day. Saliva keeps the oral cavity moist, facilitates mastication and deglutition, and contains the enzyme amylase that initiates carbohydrate digestion. The **minor salivary glands** are comprised of microscopic glandular tissue in the lips, tongue, palate, and cheeks. Only the glandular tissue in the cheeks forms the macroscopically noticeable **dorsal** and **ventral buccal salivary glands** (Figs. 10.2-3 and 10.2-7A and B) that are located above and below the buccinator muscle, respectively. Mucous secretions of the minor salivary glands are directly released into the oral cavity. Although these glands are of limited clinical importance, disease of the buccal salivary glands can lead to clinical signs.

The parotid, mandibular (Figs. 10.2-6 and 10.2-7A and B), and polystomatic sublingual (Figs. 10.2-3 and 10.2-7B) salivary glands constitute the **major salivary glands** in the horse. Compared with their minor counterparts, they produce more serous secretions that drain through ducts into the oral cavity.

Expanding cranio-caudally from the ramus of the mandible to the wing of the atlas and dorso-ventrally from the base of the ear to the **linguofacial vein**, the **parotid salivary gland** is the largest salivary gland in the horse. It is enclosed by the parotid fascia, which sends trabeculations into the glandular tissue, dividing the gland into distinct lobules. The major collecting ducts for the glandular secretions follow these trabeculations to merge in the cranio-ventral aspect of the parotid salivary gland to form a single **parotid duct**. This duct crosses the tendon of the

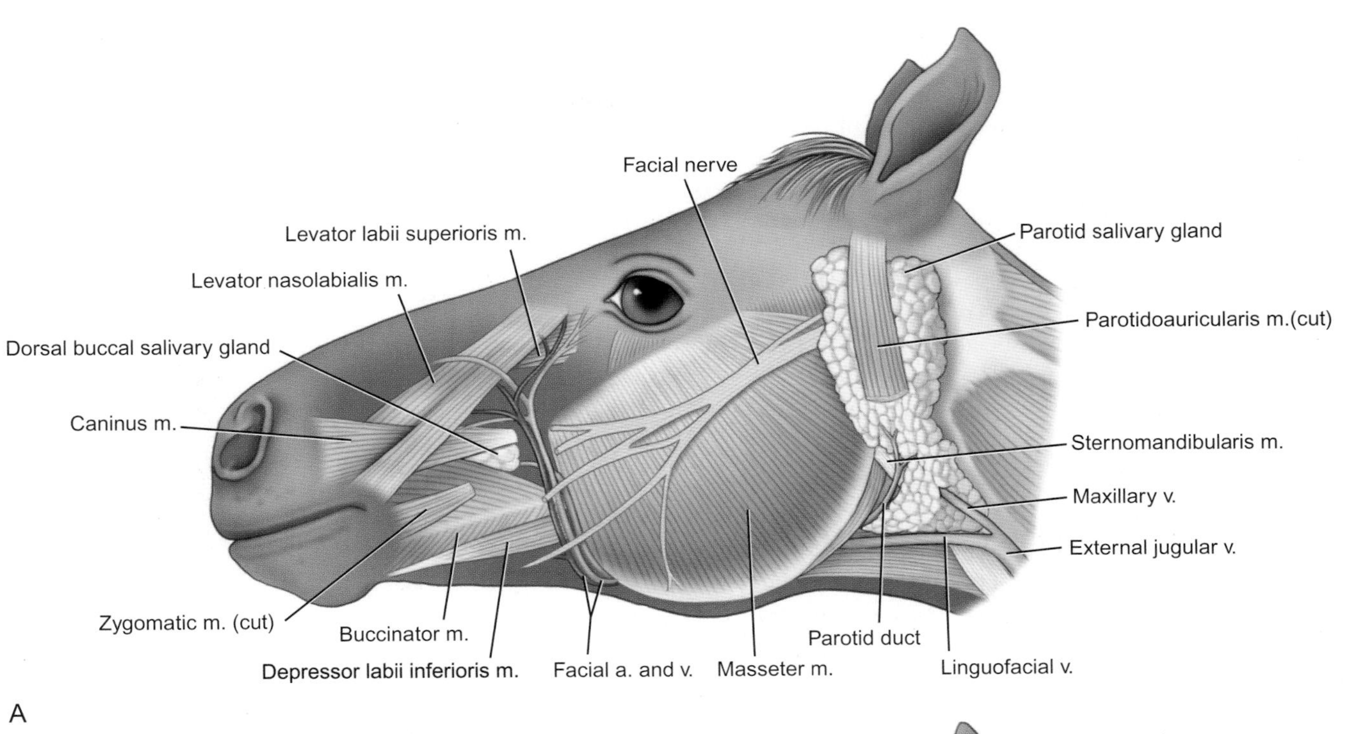

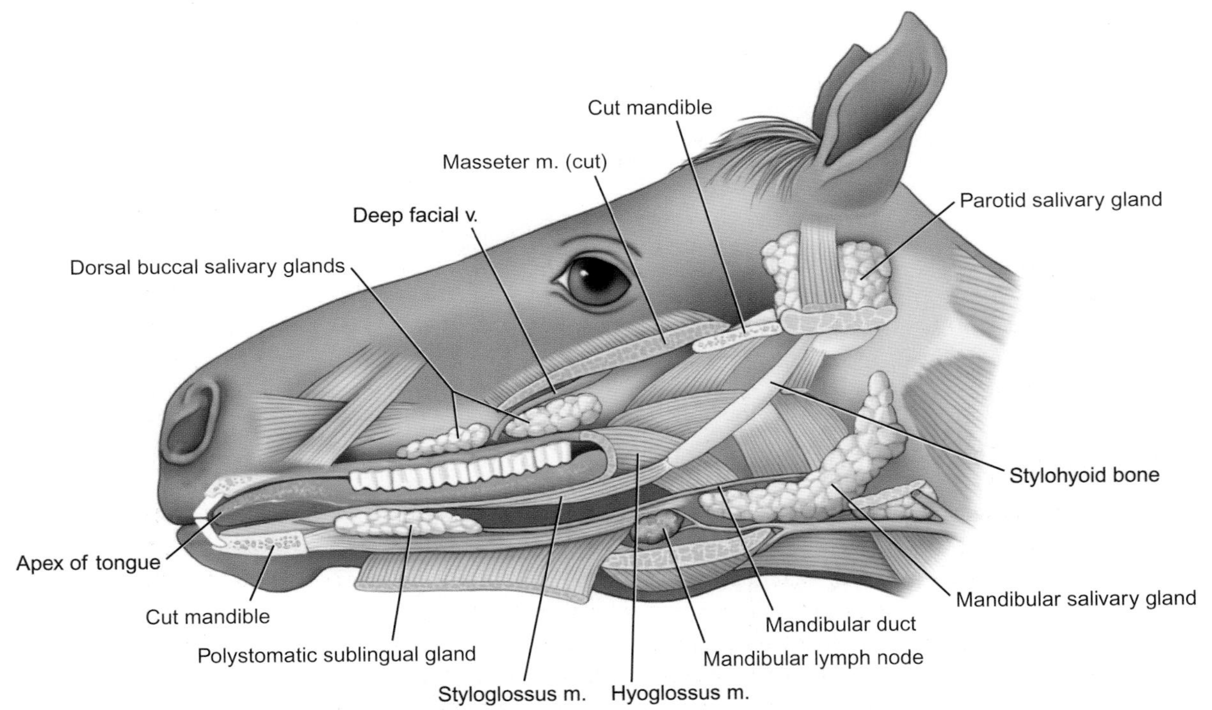

FIGURE 10.2-7 Superficial (A) and deep (B) views of the lateral head of the horse, focusing on the salivary glands and related structures.

sternomandibularis muscle (Fig. 10.2-7A) before continuing rostrally on the medial side of the mandible. Together with—and immediately caudal to—the facial artery and vein, the parotid duct passes medially to laterally over the **facial notch** in the free ventral border of the mandible and enters the face on the rostral edge of the masseter muscle. For approximately 5 cm (2.0 in.), the duct follows the border of the masseter muscle before turning rostrally, crossing under the facial vessels to enter the vestibule of the oral cavity via the **parotid papilla** at the level of the upper 3rd or 4th premolars (modified Triadan 07/08 teeth).

Sialoliths are concretions that may develop in any salivary gland and its duct; however, the parotid salivary duct is most commonly affected in the horse (see side box entitled "Sialoliths"). Melanomas are tumors that can be found in most aging gray horses, where they typically develop on the ventral tail, the perineum, and the external genitalia. However, another predilection site is the parotid salivary gland, where tumors can reach considerable size (Fig. 10.2-8). Surgical removal, especially of large masses, is complicated because of the critical anatomical structures associated with the parotid salivary gland.

Knowledge of the surrounding anatomical structures and their association with the parotid gland is important when contemplating a surgical procedure in the parotideal region. Branches of the 2nd cervical nerve (C2) cross the gland's lateral (superficial) surface, which is covered by the **parotidoauricularis muscle** (Figs. 10.2-6 and 10.2-7A) and the cutaneous muscle of the face. The **maxillary vein** passes through the central aspect of the gland, while the **linguofacial vein** courses along its ventral border (Fig. 10.2-6). The segment of the parotid salivary gland located between these 2 large vessels abuts the **mandibular salivary gland** medially. Other clinically important anatomical structures associated with the irregular medial (deep) surface of the parotid salivary gland include: (1) branches of the internal and external carotid arteries and of the maxillary vein; (2) the vagosympathetic trunk, facial (CN VII), glossopharyngeal (CN IX), and hypoglossal nerves (CN XII); (3) the parotid and retropharyngeal lymph nodes; and (4) the guttural pouch and stylohyoid bone.

The substantially smaller, crescent-shaped **mandibular salivary gland** (Fig. 10.2-7B) is located between the wing of the atlas and the basihyoid bone. Its lateral (superficial) surface is in direct contact with the maxillary vein, the parotid salivary gland, and the tendon of the sternomandibularis muscle, which passes between the 2 salivary glands. The flexor muscles of the head, along with the guttural pouch, larynx, common carotid artery, and vagosympathetic trunk, are located deep (medial) to the mandibular salivary gland. The lingual vein separates the rostral end of the gland from the adjoining mandibular lymph nodes. A single **mandibular duct** emerges from the central part of the gland and travels rostrally along its dorsal edge. The duct continues submucosally in the floor of

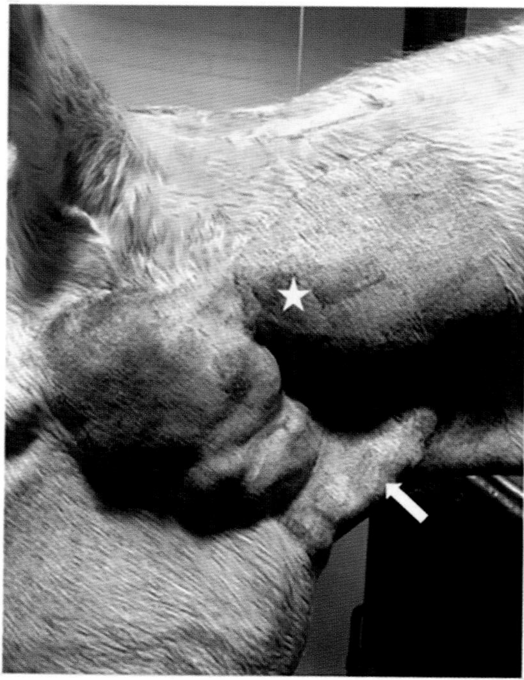

FIGURE 10.2-8 Melanomas in the left parotid salivary gland in a 25-year-old Warmblood gelding. The tumors are causing gross irregular external swelling of the parotid region. The hair has been clipped for ultrasonographic evaluation. The palpable wing of the atlas *(star)* and distended linguofacial vein *(arrow)* are identified. Rostral is to the left in this image.

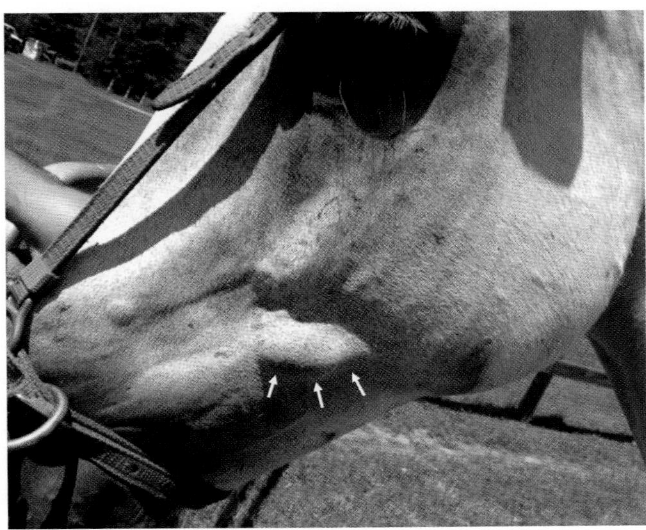

FIGURE 10.2-9 The external swelling created by a parotid duct sialolith is visible at the rostral margin of the masseter muscle and ventral to the rostral extent of the facial crest. *White arrows* define the ventral margin of the swelling. (Photo courtesy of Dr. Cindi Prestage.)

the oral cavity, adjacent to the inner surface of the polystomatic sublingual salivary gland. The opening of the duct is located on a flattened papilla, the **sublingual caruncle**, that marks the end of the **sublingual fold** and can be identified easily rostrolateral to the **lingual frenulum** (Fig. 10.2-5).

The third major salivary gland, the **polystomatic sublingual gland** (Fig. 10.2-7B), is located submucosally in the lateral sublingual recess, stretching from the mandibular symphysis caudally to the last mandibular premolar or first molar (modified Triadan 08/09 teeth). Numerous small ducts release secretions of the sublingual glands through openings along the **sublingual fold**.

SIALOLITHS

Salivary stones (sialoliths) are concretions of calcium carbonate and organic matter that may appear smooth or spiculated, and be gray, light tan, or white. They are often ovoid-shaped, developing within the confines of a salivary duct, and usually occur singularly. Septic sialoadenitis may precede or be a consequence of sialolith formation. A large stone can be visualized and palpated externally as a firm, non-painful swelling, often at the rostral margin of the masseter muscle and facial crest (Fig. 10.2-9). A swelling may be seen orally, bulging through the cheek mucosa in the case of a parotid duct stone. A stone can cause acute or chronic duct obstruction and subsequent swelling and pain of the associated salivary gland and preceding portion of the duct. Removal of the stone is the definitive treatment. Surgical access to the salivary duct externally or via the oral cavity is often required.

Selected references

[1] Kilcoyne I, Watson JL, Spier SJ, Whitcomb MB, Vaughan B. Sialoadenitis in equids. Equine Vet J 2015;47:54–9.

[2] Budras K-D, Sack WO, Röck S, Horowitz A, Berg R. The head. In: Budras K-D, Röck S, editors. Anatomy of the horse. 5th ed. Hannover, Germany: Schlütersche Verlagsgesellschaft mbH & Co; 2009. p. 32–51.

[3] Pusterla N, Latson KM, Wilson WD, Whitcomb MB. Metallic foreign bodies in the tongues of 16 horses. Vet Rec 2006;159(15):485.

[4] Lang HM, Panizzi L, Smyth TT, Plaxton AE, Lohmann KL, Barber SM. Management and long-term outcome of partial glossectomy in 2 horses. Can Vet J 2014;55(3):263–7.

CHAPTER 10

CASE 10.3

Paranasal Sinus Cyst

Ferenc Toth[a] and James Schumacher[b]
[a]Veterinary Clinical Sciences, University of Minnesota College of Veterinary Medicine, St. Paul, Minnesota, US
[b]Department of Large Animal Clinical Sciences, University of Tennessee College of Veterinary Medicine, Knoxville, Tennessee, US

Clinical case

History

A 12-year-old American Quarter Horse mare was presented because of an abnormal respiratory noise heard when the horse was in work or at rest. Airflow from the right nostril was decreased. These signs were apparent to the owner for about 6 weeks before presentation.

A soft whistling sound heard during inhalation and exhalation suggests narrowing of one or both nasal passage(s). Normal respiratory effort combined with asymmetrical airflow from the nostrils indicates that the problem is unilateral or likely worse on one side. The symmetry of airflow is easily assessed in the horse by cupping both hands and holding them in front of the horse's nostrils. The findings in this case likely support obstruction of the right nasal passage. If a horse is presented in respiratory distress and has minimal airflow from both nostrils, suggesting obstruction of the upper respiratory tract, a temporary tracheostomy is recommended before continuing the clinical examination.

Physical examination findings

The respiratory rate and effort were normal at rest, but a soft whistling sound was heard during both inhalation and exhalation and was noticeably referable to the right side of the horse's head. In addition, airflow from the right nostril was appreciably reduced compared with that of the left. No nasal discharge was reported by the owner or observed during physical examination. No other abnormalities were found on physical examination.

Differential diagnoses

A space occupying lesion originating in the right nasal passage or right paranasal sinuses due to infection (bacterial or fungal sinusitis, or fungal rhinitis), neoplasia, or a paranasal sinus cyst

General anesthesia—or more specifically, recovery to the standing position after general anesthesia—is more difficult and carries more risk in horses than in small animals, so diagnostic procedures in horses are typically performed while the horse is standing, with sedation as needed. Furthermore, it is easier to appreciate the presence of free fluid in normally air-filled cavities, such as the paranasal sinuses and guttural pouches, in the standing horse. Free fluid in these structures creates a horizontal line at the gas-fluid interface on a lateral radiograph in the standing horse.

Diagnostics

The horse was sedated and both nasal passages and the nasopharynx were examined endoscopically using a flexible videoendoscope. The left nasal passage was normal (Fig. 10.3-1), as was the nasopharynx, but a complete evaluation of the right nasal passage was prevented by medial deviation of the dorsal and ventral nasal conchae.

Comparative Veterinary Anatomy: A Clinical Approach. https://doi.org/10.1016/B978-0-323-91015-6.00139-4

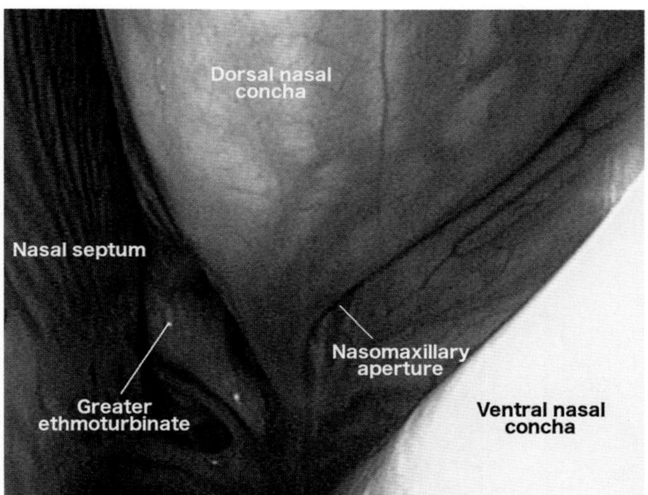

FIGURE 10.3-1 Endoscopic view of the caudal aspect of the normal left nasal passage of a horse. Important structures observed are the nasal septum, dorsal nasal concha, middle nasal concha (greater ethmoturbinate), and nasomaxillary aperture, which is the communication between the nasal cavity and the maxillary sinuses. Each nasal concha surrounds its respective conchal sinus.

Radiographs of the head were obtained with the horse sedated. Laterolateral ("lateral," Fig. 10.3-2A) and dorsoventral (Fig. 10.3-2B) projections showed the presence of a large, moderately radiodense mass occupying the right conchofrontal sinus. The left paranasal sinuses appeared radiographically normal. The mass appeared to fill the right conchofrontal sinus and extended rostrally from the region of the right ethmoidal labyrinth to the apex of the right maxillary 3rd premolar (tooth 107 in the modified Triadan system—see Case 10.4). It distorted the right nasal conchae, deviating them medially and thus compressing the right common nasal meatus. Although the mass had a radiodensity consistent with fluid, no "fluid line" was seen on the lateral projection. Sinocentesis of the right conchofrontal sinus yielded abundant, watery, honey-colored fluid (Fig. 10.3-3).

Sinocentesis (centesis; Gr. *kentēsis* perforation or tapping), or aspiration of a sinus (L. *sino*—pertaining to a sinus), is a procedure in which a needle or catheter is inserted into the paranasal sinus of interest for sampling its contents or for direct administration of treatments. The needle is inserted through a small skin incision after a hole is made in the overlying bone with a surgical drill, pin, or trephine (a toothed instrument used for cutting a circular hole in bone). By enlarging the opening in the sinus, the sinuses can be examined via sinoscopy. Sinocentesis can usually be performed in the sedated horse using local anesthesia. Normally, the paranasal sinuses contain no accumulated fluid, so the presence of fluid in this case, along with its volume and character, and the radiographic appearance of the mass supported a probable diagnosis.

Diagnosis

Presumptive paranasal sinus cyst involving the right conchofrontal sinus

Treatment

The horse was sedated and restrained in equine stocks. The right paranasal sinuses were desensitized by anesthetizing the right maxillary and ethmoidal nerves, and local anesthetic solution was injected subcutaneously at the proposed site of incision over the right conchofrontal sinus. After preparing the horse for surgery, a frontonasal osteoplastic flap ("bone flap") was created to access the conchofrontal sinus. The sinus space contained a large, grayish-white, soft, fluid-filled cystic structure that filled the sinus, distorting its normal anatomical landmarks. The cyst was removed, the sinus lavaged (irrigated) with a sterile isotonic solution, and the conchofrontal and maxillary sinuses

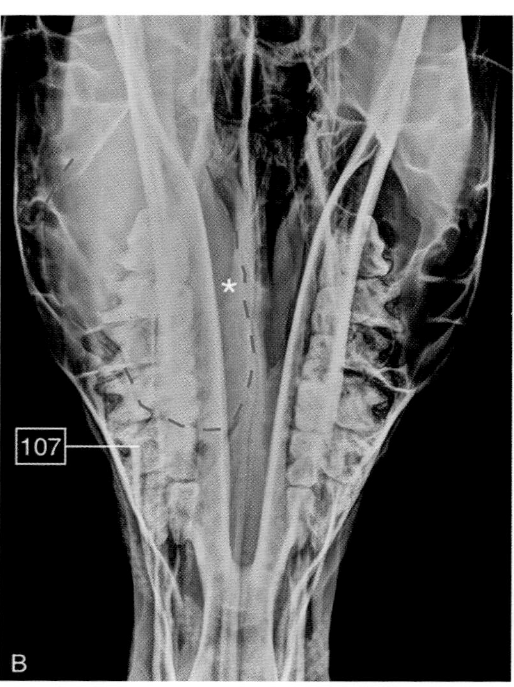

FIGURE 10.3-2 (A) Laterolateral radiograph of an equine head. A mass of moderate radiographic density within the paranasal sinuses is partially *outlined in red.* The mass extends from the ethmoidal labyrinth (not seen) to the apex of the maxillary 3rd premolar tooth (Triadan 107) and fills the conchofrontal sinus. The *arrow* points to a portion of a woven halter encircling the nose of the horse. The side of the head affected with disease cannot be determined on a laterolateral radiograph. The dorsoventral radiographic projection (Fig. 10.3-2B) shows that the affected side of the head is the right side. (B) Dorsoventral radiograph of an equine head. A *dashed red line* traces the outline of a mass of moderate radiographic density within the right paranasal sinuses. On this view, it is observed that the mass has expanded the paranasal sinuses rostrally, to the level of the 3rd premolar tooth (Triadan 107). The mass has deviated the right nasal conchae medially toward the nasal septum, compressing the right common nasal meatus *(asterisk).* The left paranasal sinuses are filled only with air.

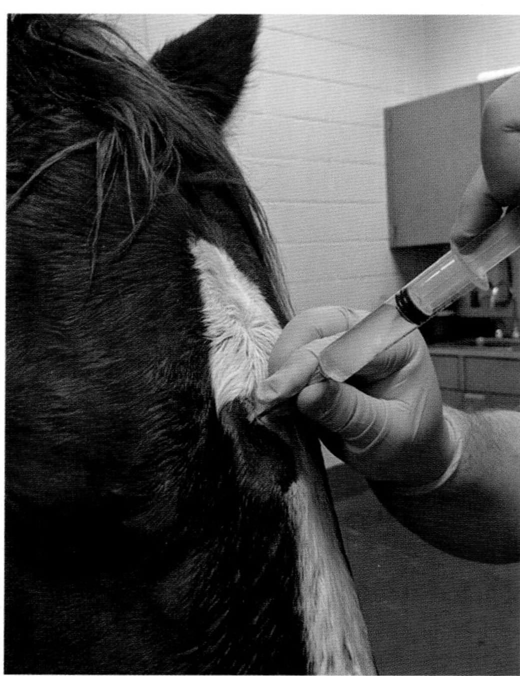

FIGURE 10.3-3 Sinocentesis of the right paranasal sinuses. A hypodermic needle has been inserted into the right conchofrontal sinus through a small hole created in the right frontal bone with a Steinmann pin. The syringe is filled with a honey-colored fluid, typical of that found within a paranasal sinus cyst.

were packed with sterile gauze to control bleeding. The osteoplastic flap was repositioned and secured with sutures, and the overlying skin was apposed. The packing was removed 3 days after surgery through a separate opening in the sinuses created at the time of surgery; the horse recovered completely and no longer displayed signs of paranasal sinus disease.

Anatomical features in equids

Introduction

The nasal cavity and paranasal sinuses form the first part of the upper respiratory tract and are closely related. As the name indicates, the paranasal sinuses are situated adjacent to (Gr. *para-*) the nasal cavity. The paranasal sinuses are part of the upper respiratory tract because they communicate with the nasal cavity in health and disease.

Function

The nasal cavity provides a conduit for air entering through the nostrils to reach the nasopharynx and eventually the lungs. The nasal cavity is also important for filtering and conditioning inspired air. The paranasal sinuses may be considered nasal diverticula. Their function remains hypothetical, but 2 probable roles include the following: (1) decrease the weight of the skull without requiring a decrease in its size and (2) provide additional reverberation during phonation. Because the sinuses are lined with respiratory epithelium and communicate with the nasal cavity, they also likely play a role in the immunological surveillance and defense of the upper airways. This section covers the anatomy of the nasal cavity and paranasal sinuses.

Nasal cavity

The nasal cavity begins at the **nares (nostrils)** and ends at the **choanae** (Gr. *choanē* funnel). The choanae are the openings between the caudal extent of the nasal cavity and the nasopharynx. The nasal cavity is divided into right and left sides or compartments by the **nasal septum** (clinically, right and left nasal cavities are commonly referred to; however anatomically, there is one nasal cavity subdivided by the nasal septum). Diseases restricted to the right or left side of the nasal cavity, paranasal sinuses, or lacrimal system typically result in ipsilateral nasal discharge.

> The horse described in this case had no nasal discharge for 2 reasons: The contents of the cyst were contained within the cystic membrane, and the enlarging cyst distorted the sinus structures blocking the nasomaxillary aperture (from which drainage would have occurred).

The nares are supported by the **alar cartilages**. Opening (dilation) of the nares is mediated by the levator nasolabialis, caninus, lateralis nasi, and dilatator naris apicalis muscles, the latter 2 attaching to the alar cartilage. The **nasal diverticulum** ("false nostril") is a cutaneous blind sac identified at the dorsolateral aspect of the nares occupying the **nasoincisive notch**. It is separated from the nasal cavity by the **alar fold**, which originates on the lamina of the alar cartilage and attaches to the rostral portion of the cartilaginous ventral nasal concha. The **nasolacrimal duct** terminates at the **nasolacrimal opening**, located in the ventral aspect of the **nasal vestibule**, caudal to the nares, at the junction of the nasal mucosa and skin.

Entry into the left or right side of the nasal cavity is gained through the ventromedial aspect of the ipsilateral naris, between the nasal septum and the ventral aspect of the alar fold. The nasal septum is covered by highly vascularized mucosa and is primarily composed of cartilage. The caudodorsal aspect (fundus) of each side of the nasal cavity contains the lateral masses of the ethmoid bone, which form the ethmoidal labyrinth (see later). The left and right sides of the nasal cavity are further divided into 3 **meatuses** (L. pathway or passage) by the dorsal and ventral nasal conchae (Fig. 10.3-4). The **ventral nasal meatus** has the largest diameter, providing a direct pathway for airflow to the nasopharynx and a preferred pathway for passage of a nasogastric tube (see Case 10.11). The **dorsal nasal meatus** extends to the ethmoidal labyrinth and thus transmits air to the olfactory region. The **middle nasal meatus** is located between the dorsal and ventral nasal conchae and extends caudally to the nasal opening of the

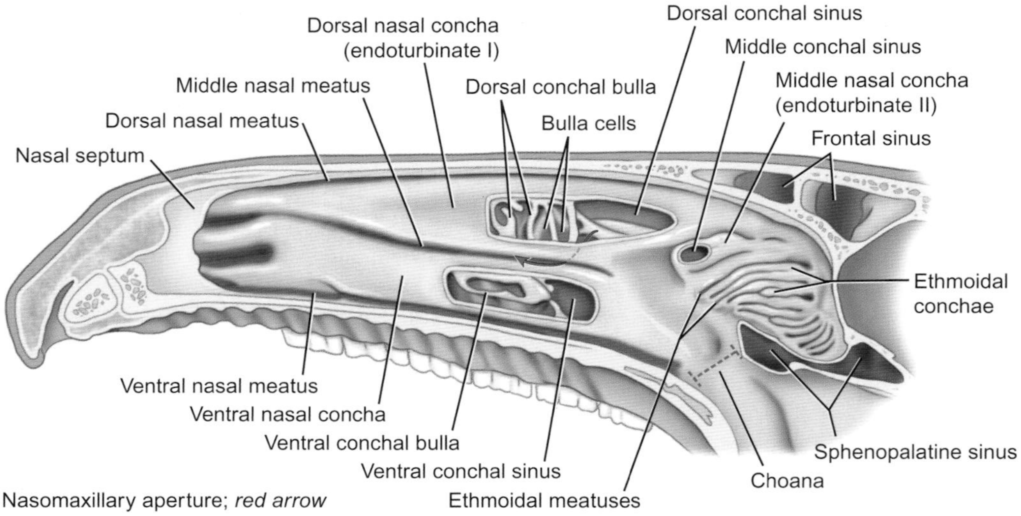

FIGURE 10.3-4 The right side of the equine nasal cavity, with the nasal septum removed. The *dashed red arrow* indicates the location of the nasomaxillary aperture at the caudal end of the middle nasal meatus. The choana *(dashed red line)* is the opening between the nasal cavity and the nasopharynx. The ethmoidal labyrinth consists of the ethmoidal conchae, their inner spaces (cells or sinuses), and the ethmoidal meatuses. Therefore, both the caudal extent of the dorsal nasal concha and the middle nasal concha are part of the ethmoidal labyrinth.

nasomaxillary aperture, the slit-like communication between the nasal cavity and the maxillary sinuses. A fourth meatus, the **common nasal meatus** is the vertical space located between the nasal septum and the medial aspect of the conchae and it is continuous laterally with the other 3 meatuses.

The **dorsal** and **ventral nasal conchae** are comprised of single scrolls of mucosa-covered, thin bone that attach to the nasal, maxillary, and ethmoid bones at the lateral and caudal aspects of the nasal cavity. Developmentally, the caudal extent of the dorsal nasal concha is a continuation of endoturbinate I of the ethmoidal labyrinth. The dorsal concha is scrolled ventrally, and the ventral concha is scrolled dorsally. The dorsal and ventral conchae are separated into rostral and caudal portions by a transverse septum. The rostral portion of each concha contains an air-filled bulla, namely the **dorsal conchal bulla (bulla conchalis dorsalis)** (*L. bulla* a large vesicle) and the **ventral conchal bulla (bulla conchalis ventralis)** (Fig. 10.3-4). The latter should not be confused with the thin bullous dorsal aspect of the maxillary sinus septum, termed the **maxillary septal bulla**, but commonly referred to incorrectly as the

"ANATOMY" OF NASAL DISCHARGE

Unilateral nasal discharge generally indicates a disease process that is confined to the ipsilateral nasal passage or paranasal sinuses—i.e., rostral to the nasopharynx.

Bilateral nasal discharge indicates either bilateral nasal or paranasal sinus disease; disease of one or both guttural pouches; or a disease process involving the respiratory or alimentary tract caudal to the nasal cavity—e.g., 1) a horse with *Streptococcus equi equi* infection ("strangles") and drainage of a ruptured retropharyngeal lymph node abscess into the nasopharynx (often via a guttural pouch) resulting in bilateral mucopurulent nasal discharge; 2) esophageal obstruction ("choke") or dysphagia caused by dysfunction of pharyngeal structures may result in discharge of saliva, food, and water from both nostrils.

Guttural pouch disease, such as empyema (accumulation of purulent material) and mycosis (fungal infection causing vascular erosion and spontaneous bleeding), can present variably when these diseases affect one guttural pouch. While bilateral nasal discharge may indeed occur, it is common for the horse to present with unilateral nasal discharge or discharge that is obviously more profuse or consistent on one side than the other, particularly when the horse lowers its head to eat or drink. Although the guttural pouches open caudal to the choanae, their opening is on the lateral wall of the nasopharynx and thus discharge from a diseased guttural pouch exits more easily through the ipsilateral nasal passage. (The guttural pouches are described in detail in Case 10.5.)

"ventral conchal bulla." The dorsal conchal bulla is further subdivided by transverse septa into small compartments called **cells**. Contained within the caudal portion of the dorsal and ventral nasal conchae are their respective conchal sinuses (described in the paranasal sinus section below).

The **ethmoidal labyrinth** is the complex of **ethmoidal conchae** (also called ethmoturbinates) plus the small cellular spaces within the conchae and the meatuses between the conchae. This labyrinth projects from the ethmoid bone in the caudodorsal extent of the nasal cavity on each side. The ethmoturbinates are subdivided into endoturbinates and ectoturbinates. Endoturbinates are identified within the nasal cavity, and ectoturbinates extend into the frontal sinus. Endoturbinate I forms the caudal part of the dorsal nasal concha. The **middle nasal concha** (also known as the greater ethmoturbinate) represents endoturbinate II. This relatively short concha is "sandwiched" between the dorsal nasal concha and the remaining ethmoidal conchae ventral to it. The middle nasal concha contains a sinus (described in the paranasal sinus section below).

Blood supply, lymphatics, and innervation of the nasal cavity

The nasal cavity receives its blood supply through branches of the external ethmoidal, sphenopalatine, major palatine, infraorbital, and facial arteries. Blood supply to the nares is provided primarily by the **superior labial artery** (branch of the facial artery) and the **dorsal** and **lateral nasal arteries** (branches of the infraorbital or facial arteries). There is also an anastomotic branch from the major palatine arteries that passes through the **interincisive canal** to join branches of the superior labial artery. Superficial venous drainage of the nasal cavity occurs via the dorsal and lateral nasal veins. Deeper drainage is via the sphenopalatine and major palatine veins, which connect to the deep facial vein. These vessels eventually unite with the facial vein.

Lymph drains to the **medial** and **lateral retropharyngeal** and **mandibular lymph node centers**.

Sensory innervation is provided by the **trigeminal (maxillary branch)** and **olfactory** (special sensory for smell) nerves. Motor innervation is derived from branches of the **facial nerve**.

Injury to the dorsal buccal branch of the facial nerve results in dysfunction of the ipsilateral (L. *ipse* self + *latus* side; same side) external naris, which in turn results in respiratory impairment.

Paranasal sinuses

The **paranasal sinuses** of the horse consist of a complex of interconnected, air-filled chambers that are divided into equal and mirror-image left and right halves. The sinuses are completely encased by the bones of the skull and extend from the forehead to the rostral limit of the facial crest (Fig. 10.3-5).

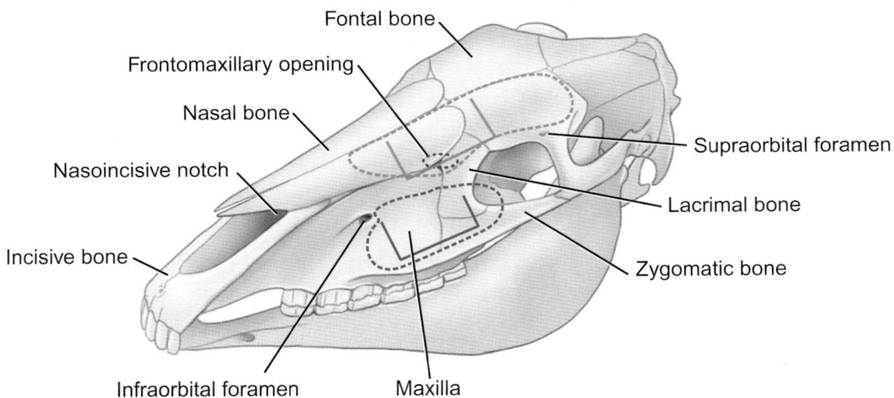

FIGURE 10.3-5 An equine skull showing the boundaries of the paranasal sinuses (conchofrontal in *orange dotted line*, maxillary in *blue dotted line*). The sites of incision to create osteoplastic (plastic surgery of the bones) frontonasal and maxillary flaps are depicted with *orange and blue lines*, respectively. The frontomaxillary opening is approximated by the *purple dotted oval*.

There are 6 pairs of sinuses (left and right), although they are broadly divided into 4 categories:

- Frontal (also called conchofrontal when combined with the dorsal conchal sinus)
- Conchal—dorsal, middle, and ventral
- Maxillary (rostral and caudal)
- Sphenopalatine

Unless the septum between the frontal sinuses has been injured by trauma or disease, the left and right sides do not communicate. The exception to this is the small sphenopalatine sinus, which has considerable individual variation and sometimes communicates with its counterpart. On each side of the head, however, all 6 sinuses communicate, either directly or indirectly, with the nasal cavity via the **nasomaxillary aperture/opening**, which is in the caudal part of the middle nasal meatus (see the nasal cavity section above).

Frontal or conchofrontal sinus

The **frontal sinus** is the largest of the paranasal sinuses in surface area. It is roughly triangular in shape and primarily occupies the forehead region and the area between the eyes (Fig. 10.3-6). It extends laterally into the zygomatic process of the frontal bone and caudally almost as far as a transverse plane that connects both temporomandibular

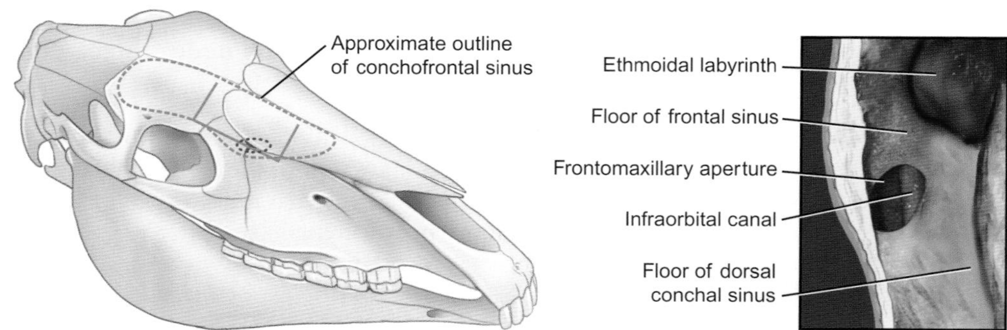

FIGURE 10.3-6 Left: the *orange line* on this equine skull shows the site of incision for creating an osteoplastic frontonasal flap to expose the right conchofrontal sinus *(dashed orange line)*. Right: structures observed through such a flap include the ethmoidal labyrinth, the floor of the frontal sinus and dorsal conchal sinus, and the frontomaxillary aperture. The infraorbital canal, lying within the caudal maxillary sinus, can be seen through the frontomaxillary aperture. The opening to the sphenopalatine sinus (not seen in this picture) lies medial to the infraorbital canal at the caudal aspect of the caudal maxillary sinus.

PROGRESSIVE ETHMOID HEMATOMA (PEH)

Progressive ethmoid hematoma is an expansile ("progressive") vascular lesion of horses that usually originates in the ethmoidal labyrinth. The cause is unknown, and even its classification—e.g., neoplastic, dysplastic, or traumatic—is uncertain. In most cases, the lesion is unilateral and may be confined to the conchofrontal sinus or protrude ventrally into the caudodorsal region of the nasal cavity on the same side; it may enter the maxillary sinuses and the sphenopalatine sinus.

The slowly expanding mass is likely to bleed spontaneously, so typically the first sign of PEH is ipsilateral epistaxis (Gr. nosebleed). This soft tissue mass is observable on routine radiographs of the skull, but it is easily missed when small, because it is hidden by the ethmoidal labyrinth or nasal conchae. Lesions that project into the caudodorsal recess of the nasal cavity may also be seen during rhinoscopy. While often unilateral, PEH may be bilateral, so the contralateral side should always be evaluated by endoscopy.

Treatment involves surgical resection or chemical ablation of the mass via a sinusotomy or a transendoscopic approach through the nasal cavity. Progressive ethmoid hematoma often recurs, so the site is periodically monitored by rhinoscopy; regrowth is eliminated using a laser or by chemical ablation.

joints. Its rostral limit is about level with the 5th cheek tooth/2nd molar (Triadan 110/210). On the outside of the skull, its rostral limit is about where the dorsum or bridge of the nose begins to widen as it approaches the eyes.

At its rostromedial aspect, the frontal sinus communicates widely with the **dorsal conchal sinus**, and the combined sinuses are often referred to as the **conchofrontal sinus.** The conchofrontal sinus is covered by the frontal, lacrimal, and nasal bones.

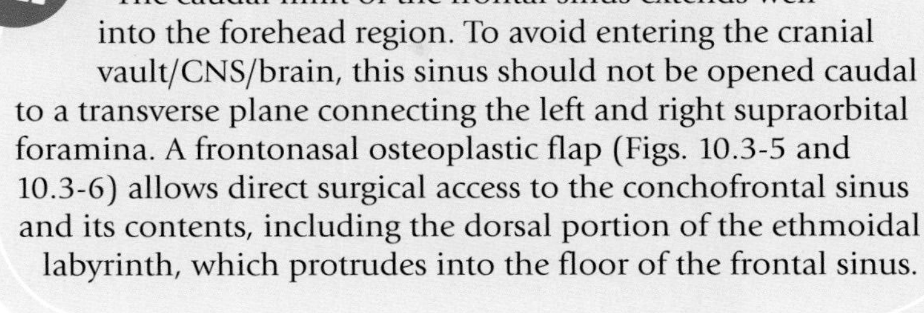
The caudal limit of the frontal sinus extends well into the forehead region. To avoid entering the cranial vault/CNS/brain, this sinus should not be opened caudal to a transverse plane connecting the left and right supraorbital foramina. A frontonasal osteoplastic flap (Figs. 10.3-5 and 10.3-6) allows direct surgical access to the conchofrontal sinus and its contents, including the dorsal portion of the ethmoidal labyrinth, which protrudes into the floor of the frontal sinus.

On the floor of the conchofrontal sinus, rostrolateral to the ethmoidal labyrinth, is the large, ovoid **frontomaxillary aperture/opening** through which the conchofrontal sinus freely communicates with the **caudal maxillary sinus** (Fig. 10.3-6). This opening is typically 4–5 cm (1.6–2.0 in.) long and 2–3 cm (0.8–1.2 in.) wide.

The **nasolacrimal duct** it tracks rostrally and slightly ventrally in the external wall of the sinuses on its path from the eye to the nostril. Through the sinus region, it is enclosed within the osseous **lacrimal canal**, which is situated laterally between the frontal sinus dorsally and the caudal maxillary sinus ventrally.

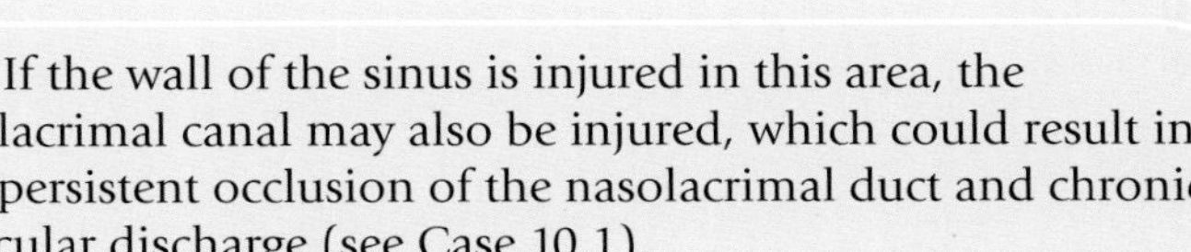
If the wall of the sinus is injured in this area, the lacrimal canal may also be injured, which could result in persistent occlusion of the nasolacrimal duct and chronic ocular discharge (see Case 10.1).

Conchal sinuses

The **dorsal conchal sinus** is little more than a rostromedial extension of the frontal sinus (Figs. 10.3-4, 10.3-6, and 10.3-7). As its name implies, it is the most dorsal or superficial of the conchal sinuses, lying immediately beneath the frontal and nasal bones where the bridge of the nose begins to widen. The nasal wall of the dorsal conchal sinus is formed by an extension of endoturbinate I, a subpart of the ethmoidal conchae.

The **middle conchal sinus** is a small chamber encased by the middle nasal concha, also called the **greater ethmoturbinate.** This concha represents endoturbinate II of the ethmoidal conchae, and it projects rostral to much of the ethmoidal labyrinth (Figs. 10.3-4 and 10.3-7). It is located ventral to the conchofrontal sinus, and medial to the caudal maxillary sinus, with which it communicates.

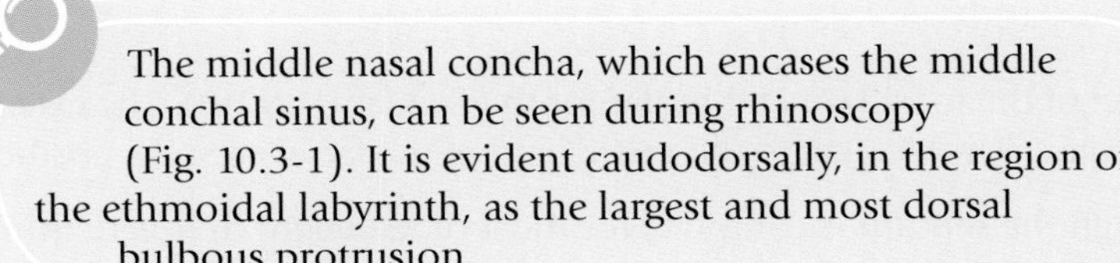
The middle nasal concha, which encases the middle conchal sinus, can be seen during rhinoscopy (Fig. 10.3-1). It is evident caudodorsally, in the region of the ethmoidal labyrinth, as the largest and most dorsal bulbous protrusion.

ABSENCE OF NASAL DISCHARGE IN SINUS DISEASE

The nasomaxillary aperture is normally the only means by which all 6 ipsilateral paranasal sinuses communicate with the nasal cavity. Because this aperture is small, it may be occluded by a disease process that results in mucosal swelling or accumulation of a discharge, such as mucus, pus, blood clots, or fibrin.

NOTE: Absence of nasal discharge does not mean absence of sinus disease. Signs of pain on palpation or change in percussion over the sinus, protrusion of the overlying facial bones, and/or radiographic changes involving the sinuses or maxillary cheek teeth often indicate sinus disease, even if the horse does not have nasal discharge.

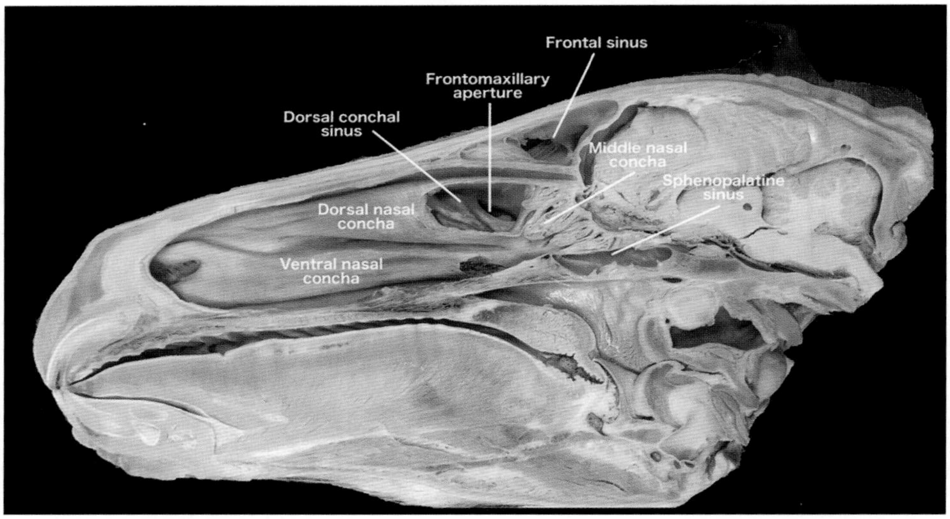

FIGURE 10.3-7 Right half of a sagittally sectioned equine skull showing the structures of the right nasal passage and their association with the paranasal sinuses.

The **ventral conchal sinus** (Figs. 10.3-4 and 10.3-8) is located more rostrally than the other 2 conchal sinuses. It is contained within the caudal part of the **ventral nasal concha**, and it communicates dorsally with the rostral maxillary sinus through the long, narrow **conchomaxillary aperture**.

Maxillary sinuses

The **maxillary sinus** is the largest, in volume, of the paranasal sinuses. It is separated by a thin bony septum into 2 chambers—the larger **caudal** and the smaller **rostral** compartments (Fig. 10.3-8). The septum runs obliquely in a medial, caudal, and dorsal direction from its lateral margin, which is usually located at about the level of the maxillary 2nd molar tooth (Triadan 110/210). The external landmark for the septum in most horses is about 5 cm (2 in.) caudal to the rostral end of the facial crest. Its position, however, is quite variable.

The bony septum suggests that the rostral and caudal maxillary sinuses are 1 structure, not 2. While they can be described separately, both communicate with the nasal cavity through a single opening. Both sinuses also contain dental/tooth apices and reserve crowns.

Although the septum is complete for most of its height, it is very thin and rounded and often fenestrated (has openings) dorsally, so it is easily injured. This thin, dorsal expansion of the maxillary septum is termed the **maxillary septal bulla**. Importantly, the rostral and caudal maxillary sinus cavities each communicate with the middle nasal meatus through the shared, slit-like **nasomaxillary aperture**.

DENTAL DISEASE AND SINUSITIS (SEE ALSO CASE 10.4)

Maxillary sinusitis is common in horses secondary to apical/root infections of the last 3 or 4 maxillary cheek teeth. The alveolar bone that covers the dental apices is thin and easily injured by the buildup of fluid around the apex of the infected tooth.

Unilateral mucopurulent nasal discharge is a common presenting complaint in horses with tooth root infection of the maxillary molars.

The maxillary sinus occupies most of the lateral margin of the face (Fig. 10.3-5), running adjacent to the facial crest from the floor of the bony orbit to an imaginary line connecting the rostral limit of the facial crest and the infraorbital foramen (about halfway between the eye and the nostril). Externally, the bony orbit marks the caudal-most extent of the caudal maxillary sinus cavity, and the infraorbital foramen marks the rostral-most extent of the rostral maxillary sinus cavity.

Medially, the maxillary sinus is bordered by the maxilla, ventral nasal concha, and infraorbital canal, and in its caudal extent by a small part of the ethmoidal labyrinth. Dorsolaterally, it is bordered by the maxilla and the lacrimal and zygomatic bones. When the skull is viewed from the lateral aspect, an imaginary line running caudally from the infraorbital foramen to the medial canthus of the eye, and parallel with the facial crest, marks the dorsal extent of the maxillary sinus (Fig. 10.3-5).

The caudal maxillary sinus communicates dorsally with the conchofrontal sinus via the **frontomaxillary aperture/opening**. Caudomedially, the caudal maxillary sinus also communicates with the sphenopalatine sinus, and medially it communicates with the middle conchal sinus. The conchomaxillary aperture/opening provides communication between the rostral maxillary sinus and the ventral conchal sinus.

The ventral margin or floor of the maxillary sinus comprises the molar portion of the **maxilla**, which forms the thin plates of **alveolar bone** that cover the apices and reserve crowns of the last 3 or 4 maxillary cheek teeth—the 4th premolar (Triadan 108/208) and all 3 molars (Triadan 109–111 and 209–211).

There is individual variation not only in the location of the septum that separates the rostral and caudal maxillary sinus cavities, but also in how many tooth roots reside within each chamber of the maxillary sinus (see Case 10.4).

In foals and young horses, the maxillary sinus cavities are almost filled by the reserve crowns. With age and dental wear, the maxillary sinuses progressively vacate, so that by the horse's late teens or twenties, all that remain within the maxillary sinuses are the small, short protuberances of the tooth roots. Presented another way, the maxillary sinuses expand ventrally with age, owing to the continuous eruption of the cheek teeth.

The osseous **infraorbital canal** (Fig. 10.3-8) is important anatomically because it encloses the infraorbital nerve and vessels along their rostrally directed path through the maxillary sinuses. In the caudal maxillary sinus, the infraorbital canal partially divides the chamber into medial and lateral compartments, whereas the infraorbital canal and its supporting bone plate bound the rostral maxillary sinus medially. The infraorbital nerve exits the skull at the infraorbital foramen.

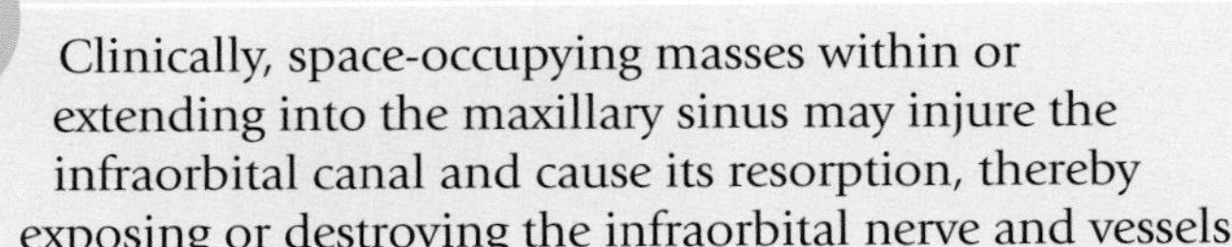

Clinically, space-occupying masses within or extending into the maxillary sinus may injure the infraorbital canal and cause its resorption, thereby exposing or destroying the infraorbital nerve and vessels.

SPHENOPALATINE DISEASE AND CRANIAL NERVES

The sphenopalatine sinus is next to the cranial vault and close to cranial nerves (CN) II–VI (Fig. 10.3-9); clinical signs caused by disease of the sphenopalatine sinus not normally seen with paranasal sinus disease may include the following: blindness, strabismus (deviation of the eye), and facial sensitivity. Dorsal distortion of the sinus may compress the optic nerves and chiasm, resulting in blindness. Inflammation or compression of CN III, IV, VI, and the ophthalmic branch of CN V, which pass through the orbital fissure adjacent to the sphenopalatine sinus, may cause facial hypersensitivity or strabismus.

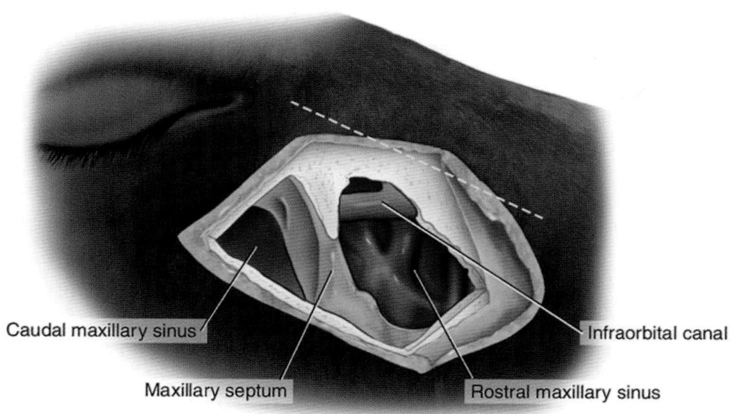

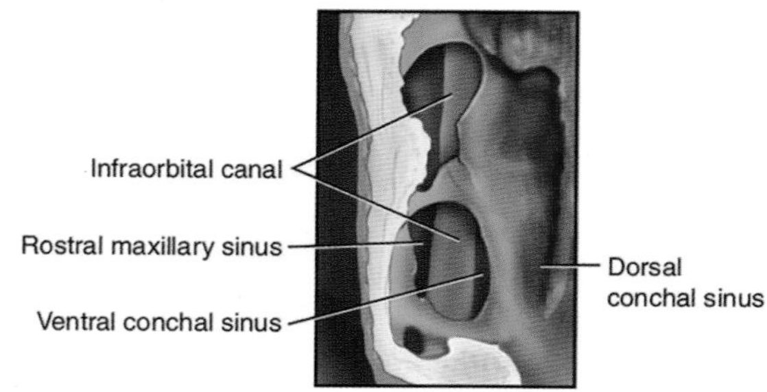

FIGURE 10.3-8 The *top illustration* shows the relationship between the caudal and rostral equine maxillary sinuses, which are separated by the obliquely oriented maxillary septum. The *bottom illustration* shows the relationship of the ventral conchal sinus, which is contained within the caudal part of the ventral nasal concha, to the rostral maxillary sinus, as viewed from a dorsal perspective (with the skull sectioned at the level of the *dashed yellow line*). These 2 sinuses communicate with each other dorsally through the long, narrow conchomaxillary aperture.

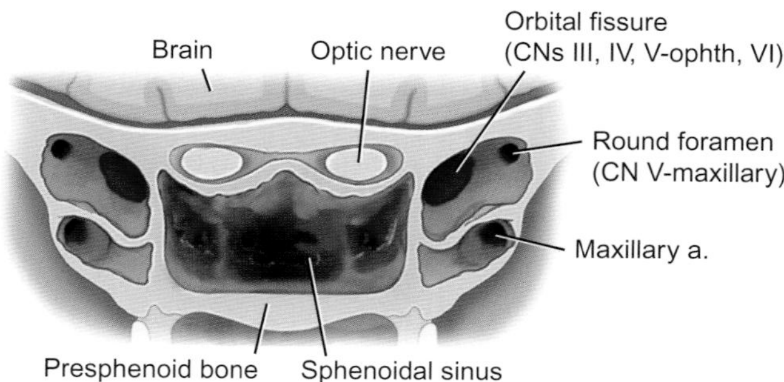

FIGURE 10.3-9 The equine skull sectioned transversely through the sphenopalatine sinuses.

Sphenopalatine sinus

The small sphenoid and palatine sinuses frequently communicate with each other, and therefore, they are referred to as the **sphenopalatine sinus**. This sinus is located deep within the skull, caudoventral to the ethmoidal labyrinth and ventral to the rostral portion of the **cranial vault**

> The sphenopalatine sinus is surgically accessed dorsolaterally through a frontonasal or maxillary osteoplastic flap, or ventrally through a hole made in the presphenoid bone via a pharyngotomy or laryngotomy, using rhinoscopic guidance.

(Fig. 10.3-7). It is surrounded by parts of the palatine, sphenoid, ethmoid, and vomer bones. A median bony septum is not a consistent finding, particularly in the sphenoid part of the sinus. The sphenopalatine sinus communicates with the caudal maxillary sinus medial to the infraorbital canal.

Blood supply, lymphatics, and innervation of the paranasal sinuses

The blood supply to the paranasal sinuses includes branches of the **sphenopalatine artery** and the **external ethmoidal artery**. The **maxillary artery** supplies these vessels. Venous drainage is via the **sphenopalatine vein**, which merges into the **deep facial vein**. Additionally, venous drainage occurs via the **external ethmoidal vein**, into the **ophthalmic plexus** and subsequently through anastomotic branches to the **deep facial** and **maxillary veins**.

Lymphatic drainage of the paranasal sinuses is to the **lateral** and **medial retropharyngeal lymph node centers**.

Sensory innervation to the paranasal sinuses is provided primarily by branches of the trigeminal nerve (CN V).

The **maxillary nerve** is a major sensory branch of the trigeminal nerve, and it innervates the ipsilateral paranasal sinuses, nasal passage, and maxillary dental arcade, including the alveoli and gingivae. The **ethmoidal nerve** innervates the caudodorsal region of the ipsilateral nasal passage, the caudal aspect of the nasal septum, and the caudal aspect of the frontal sinus, including the ethmoidal labyrinth.

These 2 nerves are of clinical importance, because they can be anesthetized ("blocked") to facilitate pain-free diagnostic or therapeutic procedures of the nasal cavity or paranasal sinuses (see side box entitled "Paranasal sinus and nasal cavity nerve blocks").

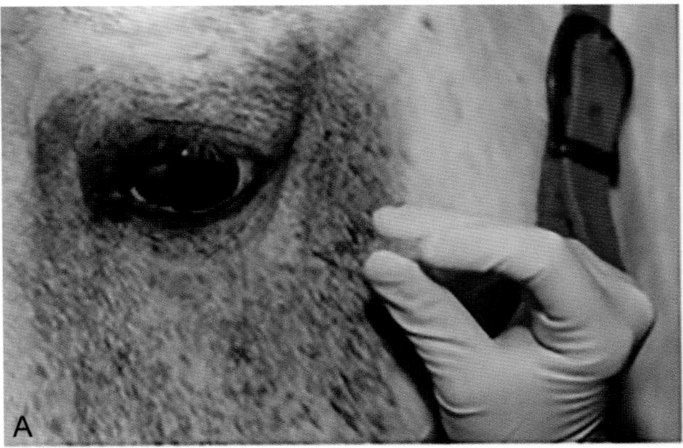

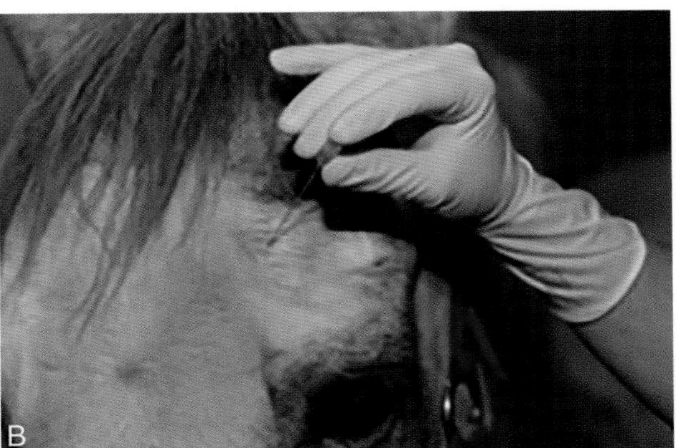

FIGURE 10.3-10 (A) Site of insertion of the needle for a maxillary nerve block. (B) Site of insertion of the needle for an ethmoidal nerve block.

PARANASAL SINUS AND NASAL CAVITY NERVE BLOCKS

Maxillary n. anesthesia: insert a 20-ga, 9-cm (3.5-in.) long spinal needle ventral to the zygomatic arch at the level of the lateral canthus of the eye; pass the needle perpendicular to the sagittal plane of the skull, through the masseter m. ~5 cm (~2 in.) to reach the extra-periorbital fat pad; inject 15–20 mL of anesthetic and wait 10–15 min for effect (Fig. 10.3-10A).

Ethmoidal n. anesthesia: insert a 20-ga., 6-cm (2.4-in.) long spinal needle at the rostromedial part of the supraorbital fossa at the notch where the caudal aspect of the zygomatic process of the frontal bone meets the zygomatic process of the temporal bone. The needle is inserted to its hub (6 cm or 2.4 in.) in a rostroventral and medial direction; inject 5 mL of local anesthetic and wait 5 min for effect (Fig. 10.3-10B).

Selected references

[1] Tremaine H, Freeman DE. Disorders of the paranasal sinuses. In: McGorum B, Dixon P, Robinson E, Schumacher J, editors. Equine respiratory medicine and surgery. Oxford: Elsevier Science; 2006. p. 393–407.
[2] Robinson NE, Furlow PW. Anatomy of the respiratory system. In: McGorum B, Dixon P, Robinson E, Schumacher J, editors. Equine respiratory medicine and surgery. Oxford: Elsevier Science; 2006. p. 317–407.
[3] Freeman DE. Paranasal sinuses. In: Beech J, editor. Equine respiratory disorders. Philadelphia, PA: Lea and Febiger; 1991. p. 275–305.
[4] Freeman DE. Sinus disease. Vet Clin N Am 2003;19(1):209–43.

CASE 10.4

Dental Disease and Sinusitis

Callie Fogle[a] and Mathew Gerard[b]
[a]Department of Clinical Sciences, North Carolina State University College of Veterinary Medicine, Raleigh, North Carolina, US
[b]Department of Molecular Biomedical Sciences, North Carolina State University College of Veterinary Medicine, Raleigh, North Carolina, US

Clinical case

History

A 12-year-old Arabian-Thoroughbred cross gelding presented with unilateral, right-sided mucopurulent nasal discharge of 2 months' duration. The gelding had been evaluated 1 month before and was prescribed 2 weeks of oral trimethoprim sulfamethoxazole for a possible respiratory infection. The discharge resolved while the horse was on antimicrobials, but recurred within 48 h once the medication was discontinued. The gelding had no history of other major medical or surgical illness. He was vaccinated twice a year against Eastern and Western encephalitis, West Nile, tetanus, herpes, and influenza, and once yearly against rabies. There were no other sick horses on the farm, and the gelding was used for trail riding on the owner's property.

Physical examination findings

At presentation, the gelding was alert and responsive with a heart rate of 36 bpm. His mucous membranes were pale pink, with a capillary refill time of less than 2 s. His respiratory rate was 24 brpm, and his heart, lungs, and gastrointestinal system on auscultation were normal. Rectal temperature was 99.9 °F (37.7 °C).

Examination of his head and cervical region identified unilateral right-sided copious, malodorous mucopurulent nasal discharge (Fig. 10.4-1). Airflow from the right, affected nostril was normal and equal to the unaffected left side. External bony structures of the head were symmetrical. Mild atrophy of the right temporalis muscle was appreciated. Percussion of the right maxillary paranasal sinus yielded a relatively dull sound when compared to percussion over the left maxillary paranasal sinus (Fig. 10.4-2). The right mandibular lymph nodes were moderately enlarged, firm, and nonpainful to palpation. Palpation of the throat latch (caudal intermandibular region) and larynx was unremarkable and did not elicit a cough. Palpation of the cervical trachea was also unremarkable.

Differential diagnoses

Dental disease with secondary sinusitis, paranasal sinus cyst, neoplasia, primary bacterial sinusitis, guttural pouch empyema, retropharyngeal lymph node abscessation (e.g., secondary to *Streptococcus equi equi* ("strangles") infection)

Diagnostics

The gelding was sedated with a combination of detomidine and butorphanol to facilitate placement of a full mouth speculum and thorough oral examination. The Triadan 109 and 209 teeth (maxillary right and left first molar teeth) had occlusal defects (infundibular caries) packed with feed material. Evaluation of the 109 tooth following removal of the impacted feed material revealed the carious (L. *cariosus* rottenness) lesion to involve most of the occlusal surface of the tooth, with a depth of approximately 1.5 cm (0.6 in.) (Fig. 10.4-3). Secondary fractures of

Comparative Veterinary Anatomy: A Clinical Approach. https://doi.org/10.1016/B978-0-323-91015-6.00140-0

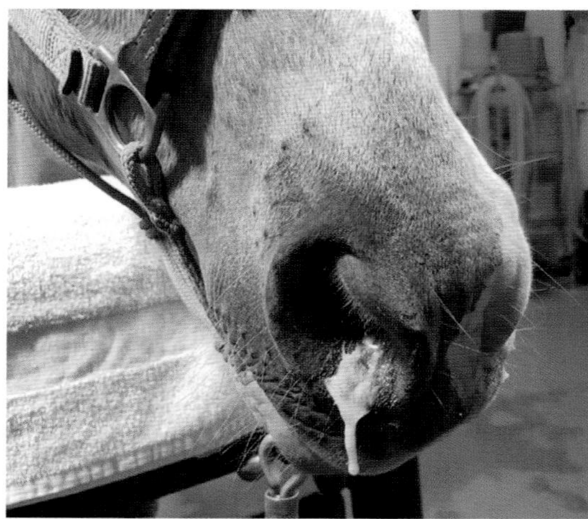

FIGURE 10.4-1 Copious unilateral, mucopurulent nasal discharge at the right nares.

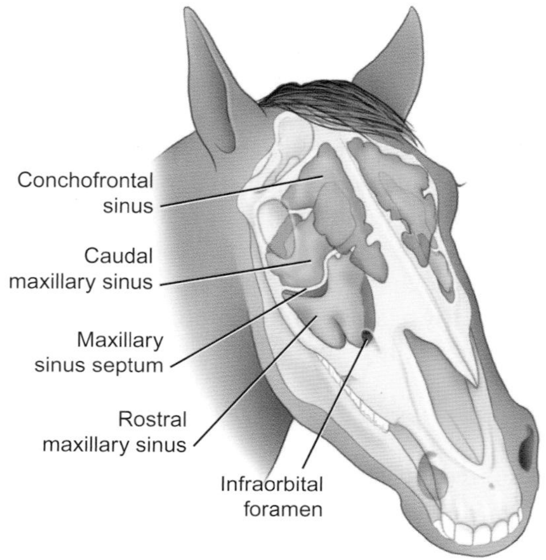

FIGURE 10.4-2 The region of the conchofrontal and caudal and rostral maxillary sinuses, which may be percussed to detect any changes in sound.

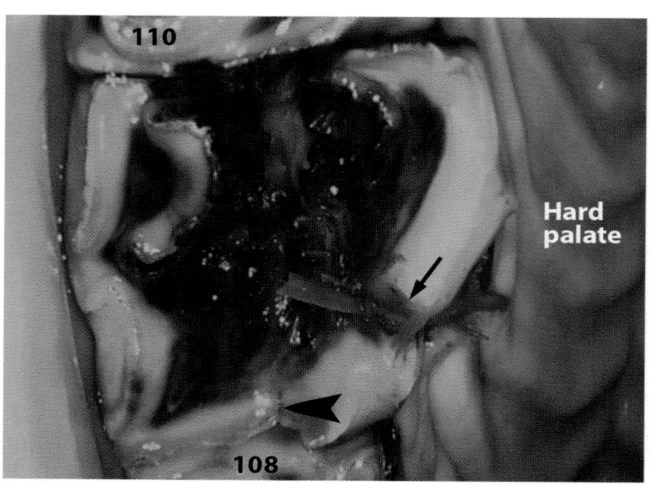

FIGURE 10.4-3 Endoscopic intraoral examination of tooth 109 occlusal surface revealed a severe central carious lesion, seen here, with secondary palatal *(black arrow)* and mesial *(arrow head)* fractures. Grass fragments are lodged in the palatal fracture plane.

the palatal and mesial edge of the tooth were noted, with a blade of grass lodged between 2 of the fragments. The occlusal defect of the 209 tooth was shallower, and the tooth was not fractured. The opposing 409 tooth was slightly overgrown (Fig. 10.4-4C).

Skull radiographs revealed fluid within the right rostral and ventral conchal maxillary sinuses, consistent with sinusitis of those spaces (Figs. 10.4-4A–D). Discrete fluid lines were seen on the lateral projection, and an opacity consistent with soft tissue or fluid was noted within the right rostral maxillary sinus on the dorsoventral (DV) projection. The fracture of tooth 109 was visible on the right maxillary oblique projection, along with moderate periapical sclerosis. Overgrowth of tooth 409 was also visible, most easily seen on the right maxillary oblique projection. On the DV projection, the infundibulum of the 209 tooth was widened with a central gas opacity, consistent with an infundibular caries.

Diagnosis

Dental (infundibular) caries of tooth 109 with secondary paranasal sinusitis

Treatment

Standing extraction of tooth 109 and trephination of the rostral maxillary sinus were elected (see side box entitled "Additional surgical approaches to extracting cheek teeth"). Heavy sedation of the horse and a right maxillary nerve block (see Case 10.3) provided restraint and analgesia. The 109 tooth was extracted orally. However, the diseased tooth root was fractured during the extraction and could not be removed via oral manipulation in its entirety. Consequently, the right rostral maxillary sinus was trephined, and the remaining 109 root fragments were repulsed into the oral cavity. The rostral maxillary sinus and ventral conchal sinus (see Case 10.3) were filled with purulent

CLINICAL REASONING FOR THIS CASE PRESENTATION

Dental disease with secondary sinusitis was considered most likely due to the nature of the discharge (malodorous, mucopurulent, and unilateral) together with right mandibular lymphadenopathy, atrophy of the right temporalis muscle, and dull percussion sounds over the right paranasal sinuses. A horse with a paranasal sinus cyst might present similarly, but is less likely to exhibit malodorous nasal discharge and unilateral temporalis muscle atrophy and may be more likely to have asymmetry of the frontal or maxillary bones overlying the sinus cyst. Neoplasia, too, might have a similar presentation, but is less common in horses of this age and might also exhibit deformity of the bony structures of the head. Primary bacterial sinusitis is more common in younger horses and often has a nonodorous discharge. A horse with guttural pouch empyema or lymph node abscessation due to strangles frequently has an elevated body temperature, nonodorous bilateral nasal discharge, and would be less likely to have atrophy of the right temporalis muscle.

COMPLICATIONS OF CHEEK TOOTH EXTRACTION—SURGICAL-RELATED TRAUMA

Soft tissue trauma—Trauma to the infraorbital nerve can occur when a maxillary cheek tooth is being repulsed, as it passes through the infraorbital canal above the palatal roots of the cheek teeth, and it is very closely related to the apices of those teeth in young horses. With external approaches to the maxillary cheek teeth (06–08s especially), trauma to the dorsal buccal branches of the facial nerve can occur. Significant intra-operative bleeding can result from injury to the major palatine artery during extraction of a maxillary cheek tooth. Trauma to the ventral buccal branches of the facial nerve and/or parotid duct is a possible complication during mandibular cheek tooth extraction. Short-term neuropraxis or long-term nerve damage can result from intra-operative nerve trauma. Parotid duct trauma results in leakage of saliva from the surgical incision, which may resolve with time or need additional surgical treatment.

Bony or dental trauma—Trauma to adjacent teeth or surrounding supporting bone can result from misplacement of surgical extraction instruments during an oral extraction. Adjacent teeth or supporting bone can also be injured during repulsion of cheek teeth if the dental punch is not properly aligned with the apex of the diseased tooth. Mandibular fracture can occur during or after tooth extraction in cases with significant dental disease and related weakened mandibular bone.

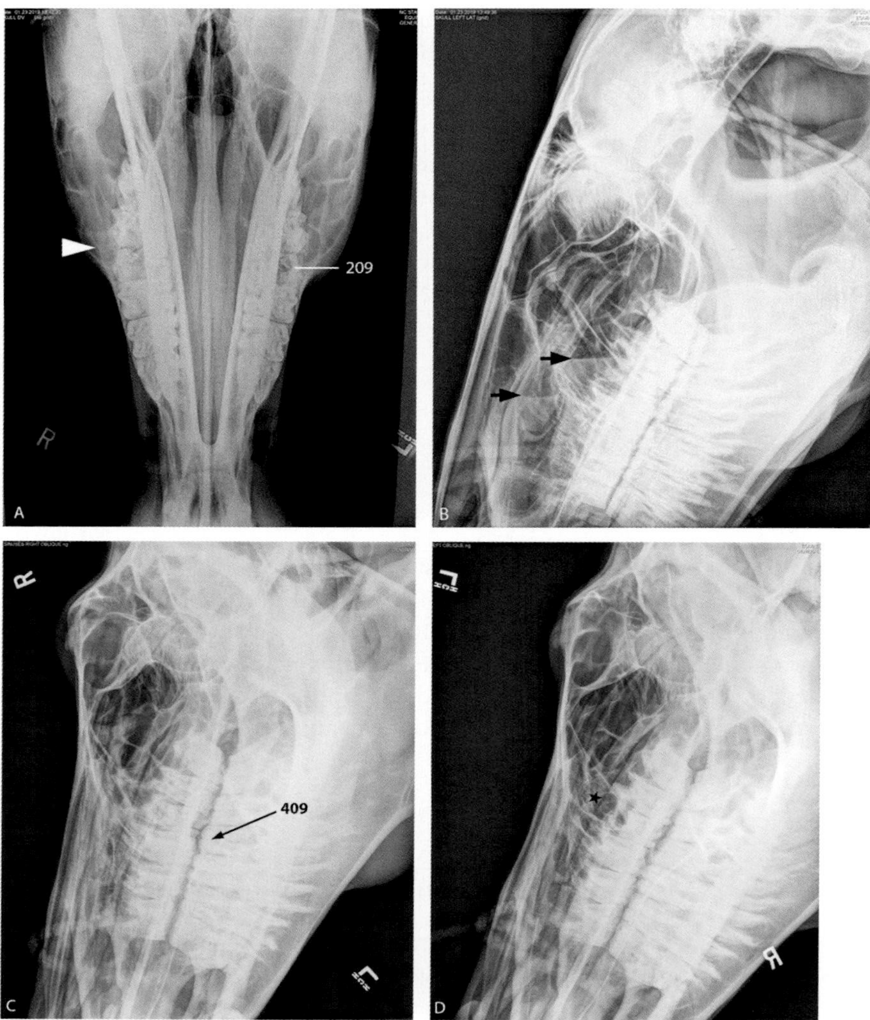

FIGURE 10.4-4 (A) The dorsoventral projection reveals a fluid or soft tissue opacity superimposed over the right maxillary sinus compartment *(arrowhead)*. The infundibulum of the 209 tooth was wide with a central gas opacity, consistent with infundibular caries. (B) The lateral projection reveals horizontal fluid lines *(arrows)* within the rostral maxillary sinus and ventral conchal sinus. The "fluid line" is the interface between gas *(black)* and fluid *(light gray)* opacities on the radiograph. (C) The right maxillary oblique projection reveals moderate sclerosis surrounding the 109 tooth roots, with a fissure evident extending to the occlusal surface of the 109 tooth. The right rostral maxillary sinus appears generally opaquer than the left rostral maxillary sinus (see D). Tooth 409 is mildly overgrown compared to the rest of the mandibular arcade. (D) The left maxillary oblique projection reveals gas opacity (more normal in appearance) within the left rostral maxillary sinus *(black star)*.

material, and a sample was obtained for culture and sensitivity. The sinus was lavaged with sterile saline and the tooth socket was sealed on its oral side with a plug made of dental impression material to prevent food entering the socket and contaminating the sinus. A Foley catheter was passed via the trephination portal and secured within the sinus to allow for daily sinus lavage, and the trephination portal skin flap was sutured closed. The horse was discharged into the owner's care the following day. The horse was administered antimicrobials orally for 2 weeks and nonsteroidal anti-inflammatory medication for 4 days for pain management. The owners were instructed to lavage the sinus via the Foley catheter with 1 L of saline once daily for 2 weeks. The sutures and Foley catheter were removed at the 2-week recheck. The dental plug within the socket was removed to evaluate healing and the plug was then replaced. At the 4-week recheck, granulation tissue had formed within the socket and sealed the communication between the sinus and oral cavity, so the dental plug was permanently removed. An oral examination was recommended every 6 months for routine dental care and reduction of overgrowth of the opposing mandibular 409 tooth to prevent possible malocclusion.

Anatomical features in equids

Introduction

Equine dental disease and paranasal sinusitis secondary to dental disease are common in horses. Familiarity with normal dental anatomy and the spatial relationships of the teeth with surrounding structures is important for careful evaluation and comprehensive treatment of horses with dental disease and secondary complications. This section will cover equine dentition, dental anatomy, sinuses as they relate to dental anatomy, and the periodontium. Sinus anatomy is covered in Case 10.3.

Function

Equine teeth have cutting, crushing, and grinding functions, serving to aid in ingestion and mastication of feed materials before swallowing. Incisor teeth provide a cutting action, while premolar and molar teeth serve crushing and grinding functions. Canine teeth have little evident function and are typically not in occlusal contact with their opposite number. The first premolar, when present, is usually vestigial and not in occlusal contact with any other tooth.

Dentition formulas, eruption times, and Triadan tooth terminology (Fig. 10.4-5)

The **deciduous** (temporary) dental formula for the horse is represented by the following formula: [I 3/3; C 0/0; PM 3/3; M 0/0] × 2 = 24 teeth. There are no deciduous canines, first premolars, or molars.

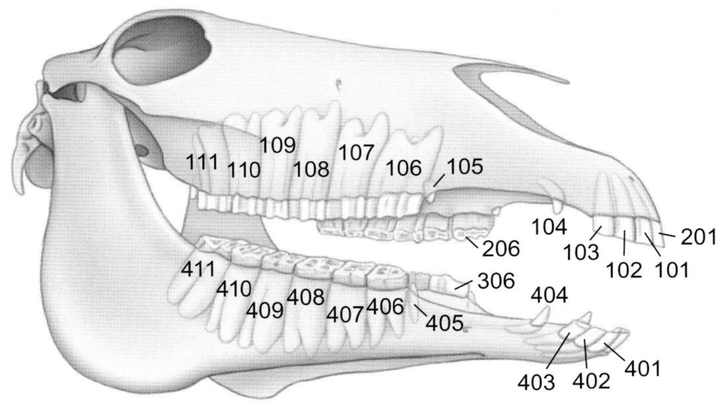

FIGURE 10.4-5 The modified Triadan numbering of teeth.

> **COMPLICATIONS OF CHEEK TOOTH EXTRACTION—HEALING AND INFECTION**
>
> Delayed healing or nonhealing alveolus—A delay in, or complete lack of, the healing of the alveolus can occur when a portion of the tooth, usually the root, remains after extraction. Post-extraction radiographs are important for confirmation that the entire tooth has been extracted or repulsed. Presence of a sequestrum, or a dead fragment of alveolar or other supporting bony structure, can also result in delayed healing. This is more commonly seen after repulsions. A loose or missing alveolar dental plug can result in an infection, and delayed healing, of the alveolus.
>
> Oro-sinus or oronasal fistula—In some cases, a communication between the oral cavity and sinus, or between the oral cavity and nasal passage, persists after the remainder of the alveolus has healed. This communication can result in food material entering and accumulating in the sinus or nasal passage, and secondary bacterial infection.
>
> Persistent sinusitis in the absence of fistulization, alveolar sequestration, or retained dental fragments—Sinusitis that persists despite a lack of unhealthy tissue or communication with the oral cavity is often due to inspissated material within one or more paranasal sinus compartments. Advanced three-dimensional imaging, such as a CT scan, may be necessary to identify the persistently infected compartment. Treatment frequently requires additional trephination and exposure of the infected compartment and debridement and lavage to resolve.

The **permanent** dentition of the mature horse is represented by the following formula: [I 3/3; C (1)/(1); PM 3(4)/3(4); M 3/3] × 2 = 36–44 teeth. Therefore, in a normal horse, up to 11 teeth may be present in each dental arcade. Uncommonly, supernumerary teeth may be present. Canines are typically absent, unerupted, or vestigial in female horses, and the first premolar teeth (referred to as the **wolf** teeth in the horse) are commonly absent. When present, the first premolar teeth are usually found in the maxilla and rarely in the mandible. **Cheek teeth** is a collective term for the 6 closely approximated premolars 2–4 and molars 1–3 in each arcade.

Approximate eruption times for deciduous and permanent teeth provide the most reliable aging cues in the horse (Table 10.4-1). The normal horse has permanent dental arcades by 5 years of age. Up to this age, there is considerable dental and periodontal activity present as the deciduous teeth develop, erupt, wear, and are shed, and permanent teeth develop, erupt, and come into wear. There are slight breed variations in eruption times.

For ease of medical record-keeping and communicating accurate identification, a number has been ascribed to each tooth, based on a modification of the **Triadan tooth numbering system**. From the viewpoint of standing in front of the horse and looking at it, the head is divided into quadrants, with a dental arcade in each quadrant. The 100 quadrant is the horse's maxillary right, the 200 quadrant the maxillary left, the 300 quadrant the mandibular left, and the 400 quadrant the horse's mandibular right. Teeth are numbered sequentially from the first incisor (number 01) to the third molar (number 11) within each arcade, and then the tooth number is preceded by its quadrant identifier. This 3-digit number allows specific labeling of each tooth present in the normal horse (Fig. 10.4-5). For example, the 109 tooth is the first molar of the right maxillary dental arcade. Absent teeth still retain their number. For deciduous teeth, the individual tooth number remains the same, and the quadrants are numbered 500, 600, 700, and 800 in the same clockwise sequence (as looking at the horse), starting in the horse's maxillary right quadrant. Therefore, for example, the 703 tooth is the third deciduous incisor of the left mandibular arcade.

Dental anatomy and physiology

Structurally, teeth consist of dentin, enamel, cement, and pulp (Fig. 10.4-6D). **Dentin** is a cream to pale yellow calcified tissue, consisting of 70% mineral and 30% organic compounds and water. Dentin makes up the bulk of the tooth and is divided into primary, secondary, and tertiary types. Secondary dentin is laid down over the dental (pulp) cavity subocclusally to protect the pulp horns and vital tissues of the cavity from being exposed secondary to tooth wear. **Enamel** is the most dense and brittle substance in the body, with a crystalline mineral matrix. Peripheral enamel wraps around the dentin in a simple layer (on the incisors, canines, first premolar) or forms deep folds in a tortuous path around the dentin of the premolars 2–4 and molars. In addition, during development enamel forms an invagination on the occlusal surface of incisors and 2 invaginations on the occlusal surface of maxillary cheek

TABLE 10.4-1 Average eruption schedule for equine teeth.

Tooth (Triadan #)[a]	*Deciduous*	*Permanent*
1st incisor (01)	Birth or first week	2½ years
2nd incisor (02)	4–6 weeks	3½ years
3rd incisor (03)	6–9 months	4½ years
Canine (04)	N/A	4–5 years
1st premolar (05)	N/A	5–6 months
2nd premolar (06)	Birth or first 2 weeks	2½ years
3rd premolar (07)	Birth or first 2 weeks	3 years
4th premolar (08)	Birth or first 2 weeks	4 years
1st molar (09)	N/A	9–12 months
2nd molar (10)	N/A	2 years
3rd molar (11)	N/A	3½–4 years

[a]Juvenile or deciduous teeth are identified by replacing the first digit with 5, 6, 7, or 8; e.g., 203 for the permanent tooth (left upper 3rd incisor) would be identified by the number 603 for the deciduous tooth.
Adapted and modified from the American Association of Equine Practitioners, Guide for Determining the Age of the Horse, 2007.

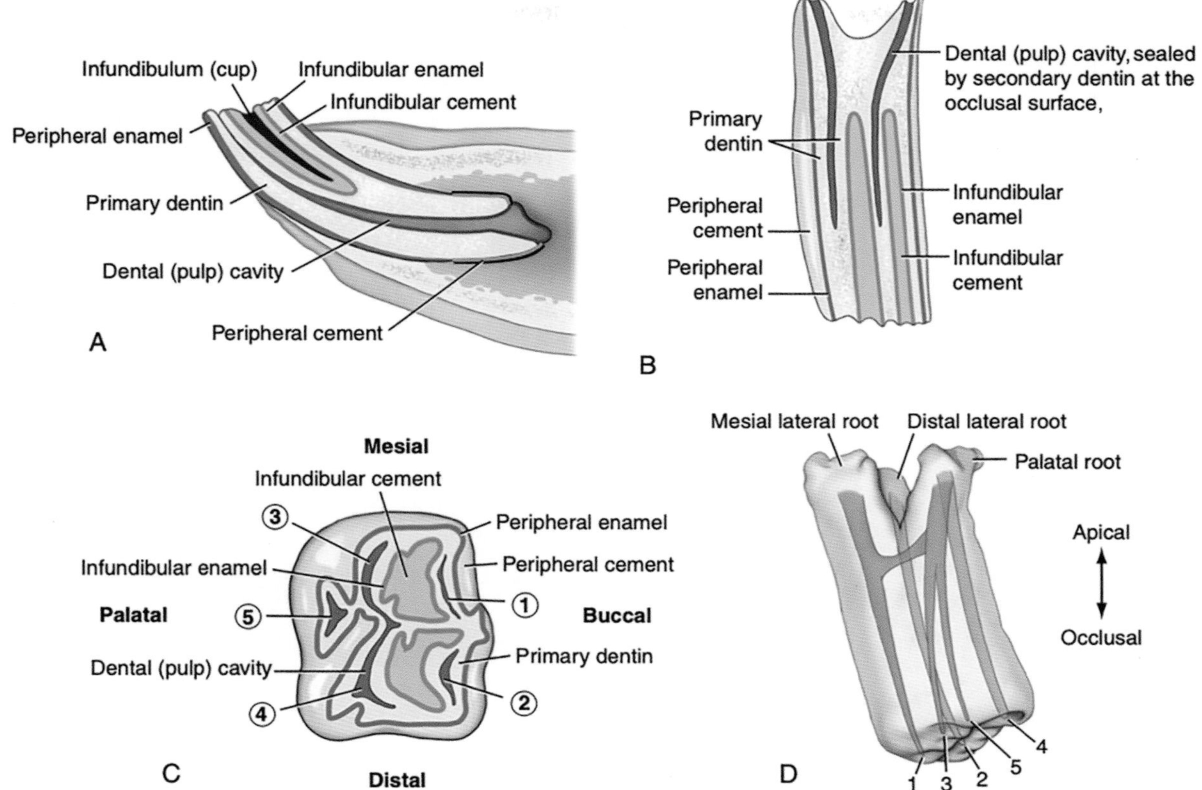

FIGURE 10.4-6 (A) Anatomy of a sectioned incisor. (B) Anatomy of a molariform tooth longitudinally sectioned and (C) of its occlusal surface. (D) A representative configuration of pulp horns of an equine cheek tooth, with the pulp horn number matching the numbers in image C.

teeth. These enamel invaginations are referred to as **infundibula** and appear as separate rings of infundibular enamel contained within the peripheral enamel boundary.

Cement, with a matrix similar to bone, covers the root and body of teeth (**peripheral cement**) and partially fills the infundibula of incisor teeth and maxillary cheek teeth (**infundibular cement**). Infundibular caries is the destruction and dissolution of the cement found in this location. The **pulp** is the vital supply of the tooth—that is, connective tissues, vessels, nerves, and lymphatics. It is contained in the **dental (pulp) cavity** of the tooth. The pulp structures enter and exit the tooth at its apical extent (and subsequently the developed tooth root(s)). Equine incisors have 1 or 2 pulp horns (2 is more common), canine and wolf teeth have a single pulp horn, and the cheek teeth have 5–8, depending on the tooth (Fig. 10.4-6C–D).

Teeth are defined as having 5 surfaces (Fig. 10.4-7B and C). The **occlusal** (masticatory, grinding, cutting) surface faces the ipsilateral upper or lower arcade. The inner surface facing the oral cavity is the **palatal** (maxillary teeth) or **lingual** (mandibular teeth) surface. The external surface facing the vestibule of the oral cavity (and lips and cheeks) is the **vestibular** (or **buccal** for cheek teeth and **labial** for incisors and canines) surface. The **rostral or mesial** surface faces rostrally and—in the case of incisors—is the surface toward midline. The **caudal or distal** surface faces caudally, or away from midline in the case of incisors. In lieu of mesial and distal, the term **contact surface** is used for the closely abutting surface of cheek teeth in the horse. Use of surface terminology allows for the specific description of where on a tooth a lesion is present.

For example, in this Fig. 10.4-3 the occlusal surface of tooth 109 has a non-displaced fissure fracture in its mesial margin, running mesial to distal, 5 mm (0.2 in.) from the mesiopalatal corner.

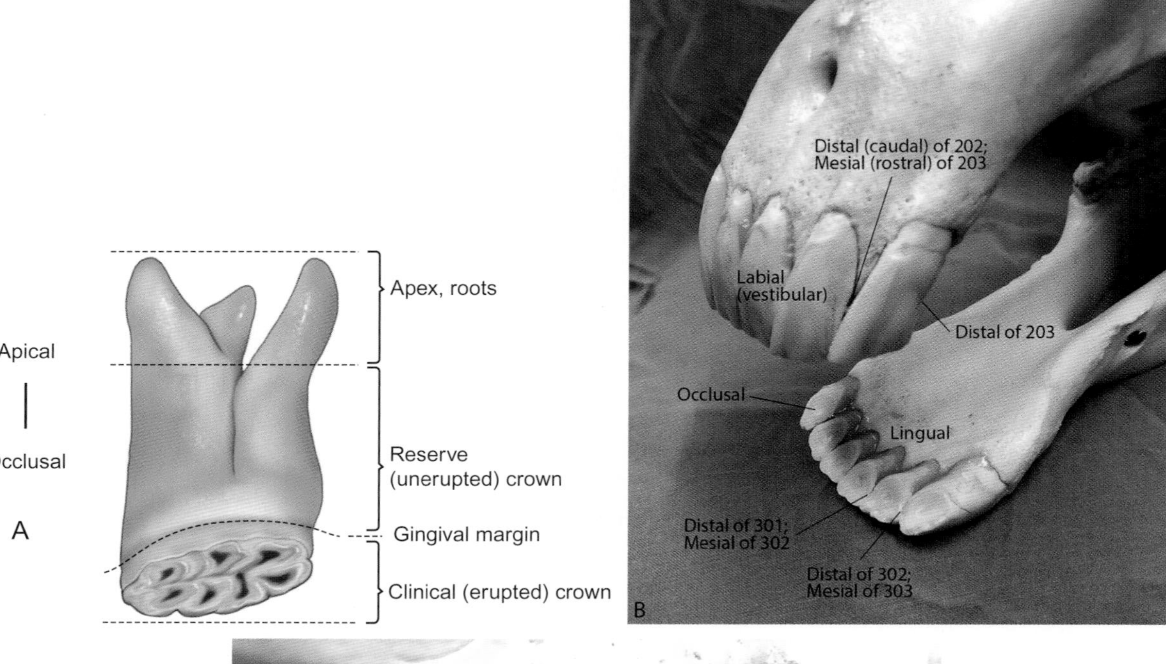

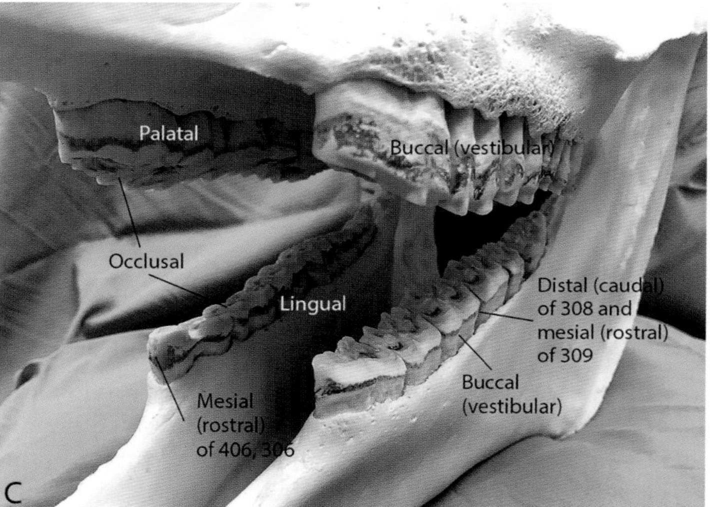

FIGURE 10.4-7 (A) Parts and directional terms for a hypsodont maxillary cheek tooth. (B) Anatomical surfaces of incisor teeth. (C) Anatomical surfaces of cheek teeth.

Before extracting a diseased tooth, it is important to know the number and location of roots in the tooth.

Equine teeth have a **crown** and **apex** (Fig. 10.4-7A). The apex (**apical** aspect) of a tooth is the site of root development, and **occlusal** refers to the masticatory end of the tooth. The crown is subdivided into 2 parts: **clinical** (that portion of the tooth visualized on oral exam) and **reserve** (the portion of the crown below the gingival margin). Equine cheek teeth develop characteristic root structures as the tooth matures and erupts. Incisors, canines, and wolf teeth are single-rooted teeth, mandibular cheek teeth develop 2 roots in a mesial-distal alignment, and maxillary cheek teeth develop 3 roots, 2 located laterally in mesial and distal positions, and 1 larger root located medially (palatal) (Fig. 10.4-6D).

Equine incisors and cheek teeth are classified as hypsodont—that is, having long crowns. Hypsodont teeth develop and grow to a certain length and slowly erupt throughout much of the life of the horse. Eruption rates are about 2–3 mm (0.07–0.11 in.)/year, which is similar to the rate of dietary abrasive wear at the occlusal surface. The wearing of the occlusal surface of hypsodont teeth leads to architectural changes in progressively exposed dental tissues (enamel, dentin, and cement). Canines have long crowns (mostly reserve), but do not continuously erupt like the cheek teeth. The first premolar is brachydont, with a short crown, an intervening neck, and a root of variable length.

Changes in appearance of the occlusal surface of the mandibular incisors provide many cues to assist in aging a horse.

Periodontium

The tissues surrounding the tooth include the **gingiva**, the **periodontal ligament**, and the **alveolus**. In some texts, the cement is also considered part of the **periodontium**. The tightly attached gingiva transitions to more loosely attached oral mucosa at the **gingivomucosal junction**.

Teeth and anatomical relationships

Erupting and recently erupted permanent mandibular premolars 2–4 and the first molar have apices very closely related to the ventral mandibular cortex. Due to the natural curve of the caudal body of the mandible as it transitions to the **ramus** (known as the **Curve of Spee**), the apices of molars 2 and 3 are relatively distant from the ventral cortex. The apical portion of developing and erupting teeth is highly metabolically active and forms an **eruption cyst**. The eruption cyst at the apex of a permanent, erupting premolar may manifest as a focal swelling of the ventral mandible due to its proximity to the cortex.

These swellings are known as "eruption bumps" and are typically self-limiting. Eruption bumps are anticipated to be present in both mandibles at the age when permanent premolars are replacing temporary premolars (2–5 years of age) and then regress within 1–2 years. These normal swellings should not be misdiagnosed as apical dental abscesses or trauma to the mandible, although eruption cysts can rarely become infected. Determining the horse's age and obtaining radiographs allows accurate assessment if questions arise.

The apices of the maxillary molars and premolar 4 are directly related to the **maxillary sinus**. The Triadan 11 teeth project into the **caudal maxillary sinus**, the 10 teeth are associated with the caudal and **rostral maxillary sinus** and dividing **maxillary sinus septum**, and the 09 and 08 teeth are associated with the rostral maxillary sinus. The infraorbital canal passes in direct contact with the dorsomedial margin of the maxillary alveoli in young horses. With increasing age, the infraorbital canal increases its distance from the apical region of the alveoli, supported by a vertical bony plate that forms the lateral wall of the **ventral conchal sinus**. The amount of space in the maxillary sinus becomes larger with age, primarily due to tooth eruption and a decrease of the apices of the 08–11 maxillary teeth. The dental arcades migrate rostrally in relation to the sinuses over the life of the teeth; however, the 08 tooth usually remains associated with the rostral maxillary sinus.

The relationship of the infraorbital canal to the apices of the teeth must be considered during dental and sinus procedures and when apical infections are present (see side box entitled "Complications of cheek tooth extraction—surgical-related trauma").

In young equids, trephination of the rostral maxillary sinus is difficult at best and should be avoided if possible (unless specifically being done for a primary dental procedure). The risk of trauma to the underlying apices of the cheek teeth is high due to minimal sinus space (see side box entitled "Trephination").

Innervation and nerve blocks

Blocking/anesthesia of the infraorbital nerve as it leaves the infraorbital canal is only effective for soft tissue anesthesia of the rostral face. Local anesthetic must pass caudally in the canal to reach the alveolar branches that supply incisors, canines, and the rostral cheek teeth. The infraorbital/maxillary nerve may also be blocked before it enters the maxillary foramen using a lateral approach, ventral to the zygomatic arch (see Case 10.3).

The inferior alveolar nerve is routinely blocked at this location via an external or intraoral approach to affect analgesia of the entire ipsilateral arcade. The lingual nerve, also a branch of the mandibular nerve, may be inadvertently blocked if needle position is inaccurate or a large volume of local anesthetic is used, and it diffuses to reach the lingual nerve. Blocking the lingual nerve desensitizes the tongue on the affected side, and this may result in self-mutilation of the tongue due to loss of sensation.

Blocking/anesthesia of the mental nerve does not desensitize the teeth. Injecting local anesthetic via the mental foramen into the mandibular canal may reach rostral alveolar branches supplying incisors, canines, and rostral cheek teeth.

The maxillary dental arcade is innervated by **superior alveolar branches (caudal, middle, and rostral)** of the **infraorbital nerve**, which continues the **maxillary branch** of the **trigeminal nerve** (CN V). The caudal superior alveolar nerve exits the infraorbital nerve immediately before it enters the **maxillary foramen** and infraorbital canal, and the middle and rostral superior alveolar nerves branch off before the infraorbital nerve exits rostrally at the **infraorbital foramen**.

The mandibular dental arcade is innervated by **alveolar branches** of the **inferior alveolar nerve**, a branch of the **mandibular nerve** (which is from CN V). The inferior alveolar nerve enters the **mandibular canal** at the **mandibular foramen** on the medial ramus of the mandible.

Rostrally, the inferior alveolar nerve exits the mandibular canal at the **mental foramen** on the lateral surface of the body of the mandible. At this point, the inferior alveolar nerve becomes the **mental nerve**, providing sensory innervation to soft tissues of the lower rostral face.

TREPHINATION

Minimally invasive approaches to dental repulsion use imaging (radiographs), external facial landmarks, and oral triangulation to identify the appropriate location for creating a small hole to pass a Steinmann pin or similar-style dental punch. This instrument is seated directly onto the apex of the diseased tooth (confirmed by radiographs), and then bone mallet is used to repel the tooth normograde into the oral cavity. Routine trephination of the maxillary or conchofrontal sinus is frequently indicated to obtain a sample of purulent material for culture and sensitivity testing and to aid in lavage of an infected sinus. Knowledge of sinus anatomy and communication between compartments is critical in selection of the specific sinus cavity for culture and lavage and to minimize complications. The locations described for standard trephination sites are shown in Fig. 10.4-8.

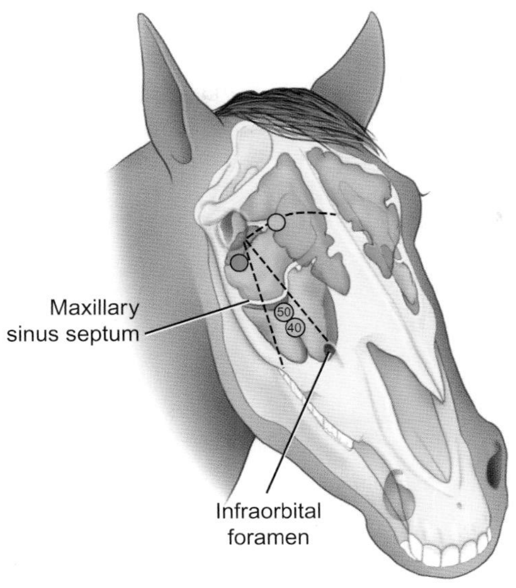

FIGURE 10.4-8 Diagram depicting locations for trephination of the conchofrontal sinus *(orange)*, the caudal maxillary sinus *(green)*, and the rostral maxillary sinus *(blue)*. To minimize injury to the underlying dental apices, the caudal entry (50) into the rostral maxillary sinus is recommended in horses < 6 years of age and the rostral site (40) can be used in horses ≥ 6 years of age. 50 and 40 refer to the percent distance from the rostral end of the facial crest to the medial canthus of the eye.

Blood supply and lymphatics of the teeth

Maxillary dental arcades receive their blood supply from branches of the **infraorbital artery**, a continuation of the **maxillary artery** as it enters the infraorbital canal. Mandibular teeth are supplied by dental branches from the **inferior alveolar artery**, also a primary branch of the maxillary artery. Dental lymphatic drainage is to the **mandibular** and **retropharyngeal lymph node centers**.

ADDITIONAL SURGICAL APPROACHES TO EXTRACTING CHEEK TEETH

Standing oral extraction is generally the preferred method for extraction of teeth because it is associated with fewer complications. However, when the clinical crown is fractured or missing, oral extraction may not be successful. These cases require a second attempt with an alternative extraction method. Repulsion of the tooth into the oral cavity, through a sinusotomy or via resection of bone and alveolus, is a traditional alternative method. In this approach, a dental punch is positioned directly over the apex of the tooth, and a bone mallet is used to tap the tooth out of the alveolus and into the oral cavity for removal. Another option is a surgical approach to the lateral aspect of the tooth—for sectioning or for loosening for oral extraction—via a buccotomy or via a lateral alveolar resection. A newer alternative method is the minimally invasive transbuccal approach with intradental screw placement. The tooth is first loosened orally, if possible. Then, a buccal stab incision is made for introduction of a cannula and screw. With radiographic guidance, the screw is inserted into the diseased tooth, and a slotted mallet is used to tap the tooth out into the oral cavity.

Any approach through the cheek has important anatomic considerations (Fig. 10.4-9). The dorsal and ventral buccal branches of the facial nerve traverse the lateral aspect of the masseter muscle in a dorsocaudal to rostroventral direction, and they frequently lie in the path of the proposed incision. The nerve branches are externally palpable and often visible and can be avoided through adjustment in incision location and careful retraction if encountered in the surgical approach. The facial artery and vein, along with the parotid duct, cross the ventral border of the mandible at the rostral margin of the masseter muscle and extend rostrodorsally on the lateral face, continuing along the margin of the masseter muscle. Vascular branches pass rostrally from the vessels. The parotid duct turns rostrally at the level of the maxillary cheek teeth and opens into the vestibule of the oral cavity in line with the 07 or 08 maxillary tooth. Pre-operative identification of the facial nerves, vessels, and parotid duct through visualization and palpation, together with careful intra-operative dissection, is critical to avoiding iatrogenic trauma during extraction procedures. Retrograde catheterization of the parotid duct to aid its identification may be performed.

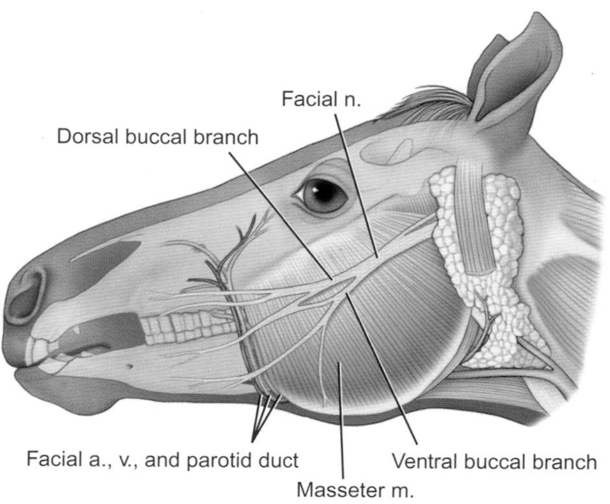

FIGURE 10.4-9 Superficial structures relevant to the lateral buccotomy approach for accessing and removing cheek teeth.

Selected references

[1] Dixon PM, Hawkes C, Townsend N. Complications of equine oral surgery. Vet Clin Equine 2009;24:499–514.
[2] Englisch LM, Rott P, Lüpke M, Seifert H, Staszyk C. Anatomy of equine incisors: pulp horns and subocclusal dentine thickness. Equine Vet J 2018;50(6):854–60.
[3] Liuti T, Reardon R, Dixon PM. Computed tomographic assessment of equine maxillary cheek teeth anatomical relationships, and paranasal sinus volumes. Vet Rec 2017;181:452.
[4] Hawkes CS, Easley J, Barakzai SZ, Dixon PM. Treatment of oromaxillary fistulae in nine standing horses (2002–2006). Equine Vet J 2008;40(6):546–51.
[5] Langeneckert F, Witte T, Schellenberger F, Czech C, Aebischer D, Vidondo B, Koch C. Cheek tooth extraction via a minimally invasive transbuccal approach and intradental screw placement in 54 equids. Vet Surg 2015;44:1012–20.

CHAPTER 10

CASE 10.5

Guttural Pouch Disease

Olivier M. Lepage
Equine Health Center, Veterinary School of Lyon, VetAgro Sup, Marcy l'Etoile, FR

Clinical case

History

A 6-year-old Belgian Warmblood mare was admitted for bilateral epistaxis. Clinical signs appeared 2 days before, when the rider noticed traces of blood from both nostrils (Fig. 10.5-1). Bilateral epistaxis after training was observed the following day and persisted for 1 h before spontaneously resolving. The primary care veterinarian referred the mare for further evaluation.

Physical examination findings

On admission the day after the second epistaxis event, the mare had mild bilateral serous nasal discharge. The remainder of the physical examination was normal.

Initial differential diagnoses

Guttural pouch disease, paranasal sinus disease, nasal cavity disease, and lower airway disease—e.g., exercise-induced pulmonary hemorrhage (EIPH)

Diagnostics

Bilateral epistaxis suggested the source of the bleeding was located caudal to the nasal septum—i.e., caudal to the nasal cavity. The mare was lightly sedated for upper respiratory tract endoscopy, which revealed the presence of a blood clot at the pharyngeal orifice of the right guttural pouch (Fig. 10.5-2). The remainder of the upper respiratory tract, including the trachea, was normal.

Differential diagnoses

Causes of epistaxis originating from the guttural pouch include bacterial and/or mycotic infection, ruptured aneurysm (see side box entitled "Aneurysm"), foreign body, fracture of the stylohyoid bone, and rupture of the longus capitis and rectus capitis ventralis muscles

Additional diagnostics

Endoscopic examination (see side box entitled "Performing guttural pouch endoscopy") of the left guttural pouch was within normal limits. The right guttural pouch was then endoscopically examined, dislodging the visible blood clot on entry into the pouch. Additional blood clots were noted on the floor of the pouch, and a suspected mycotic lesion was noted on the roof of the medial compartment, covering a portion of the internal carotid artery (ICA) (Fig. 10.5-3). Although not considered essential for diagnosis and treatment planning, a guttural pouch wash

Comparative Veterinary Anatomy: A Clinical Approach. https://doi.org/10.1016/B978-0-323-91015-6.00141-2

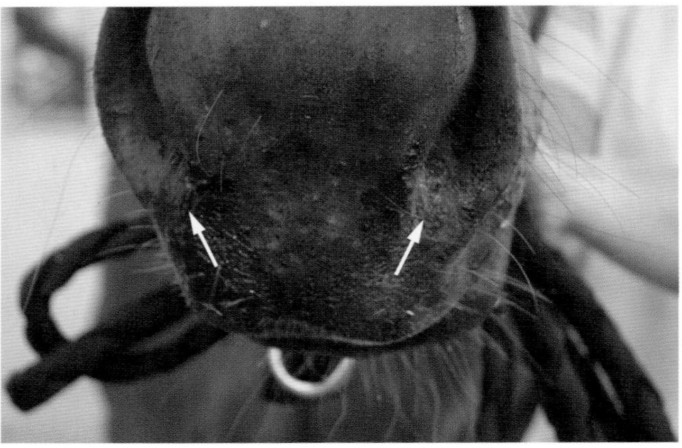

FIGURE 10.5-1 Horse presented with traces of bilateral epistaxis *(arrows)*.

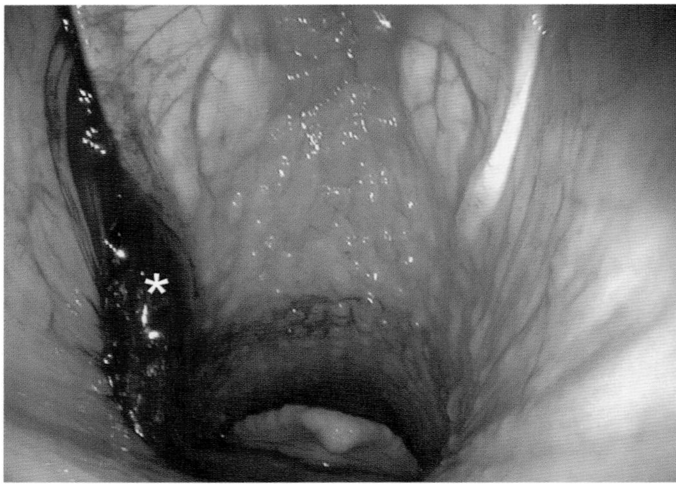

FIGURE 10.5-2 Endoscopic examination of the upper respiratory tract showing the presence of a blood clot at the pharyngeal orifice of the right guttural pouch *(white star)*.

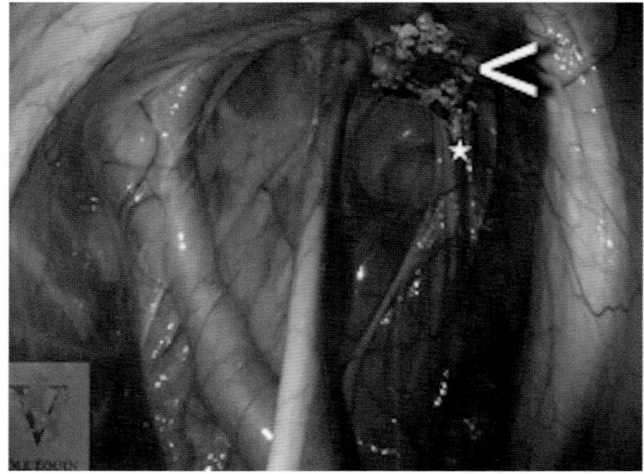

FIGURE 10.5-3 Endoscopic examination of the right guttural pouch reveals the presence of a suspected mycotic lesion located on the roof of the medial compartment *(white arrowhead)*, obscuring the internal carotid artery *(white star)*.

(lavage) was performed for epidemiological purposes. Cytology and culture of the recovered fluid revealed filaments and fungal spores, and growth of *Aspergillus* spp., respectively. A transendoscopic biopsy of the lesion should be considered if the lesion is not associated with a blood vessel; however, similar to a wash, this procedure is not routinely performed for case management.

Treatment

The recommended treatment for guttural pouch mycosis is vessel occlusion of the affected blood vessels to prevent bleeding from occurring or recurring. Surgical treatment should be initiated immediately once the disease is confirmed because of the ever-present risk of life-threatening hemorrhage. A transarterial coil embolization (TACE) technique was used (see side box entitled "TACE") to occlude the ICA under general anesthesia. As an adjunct treatment for the fungal infection and as a therapy to possibly reduce the severity or risk of neurological dysfunction, topical oxygen therapy (TOT) was performed during the TACE procedure and continued for a several days. The mare also received penicillin and flunixin meglumine peri-operatively. Endoscopic examination of the affected guttural pouch at 7 weeks post-operatively revealed the resolution of the fungal lesion with residual local scarring of the guttural pouch wall. There were no reported complications at the 12-month follow-up examination.

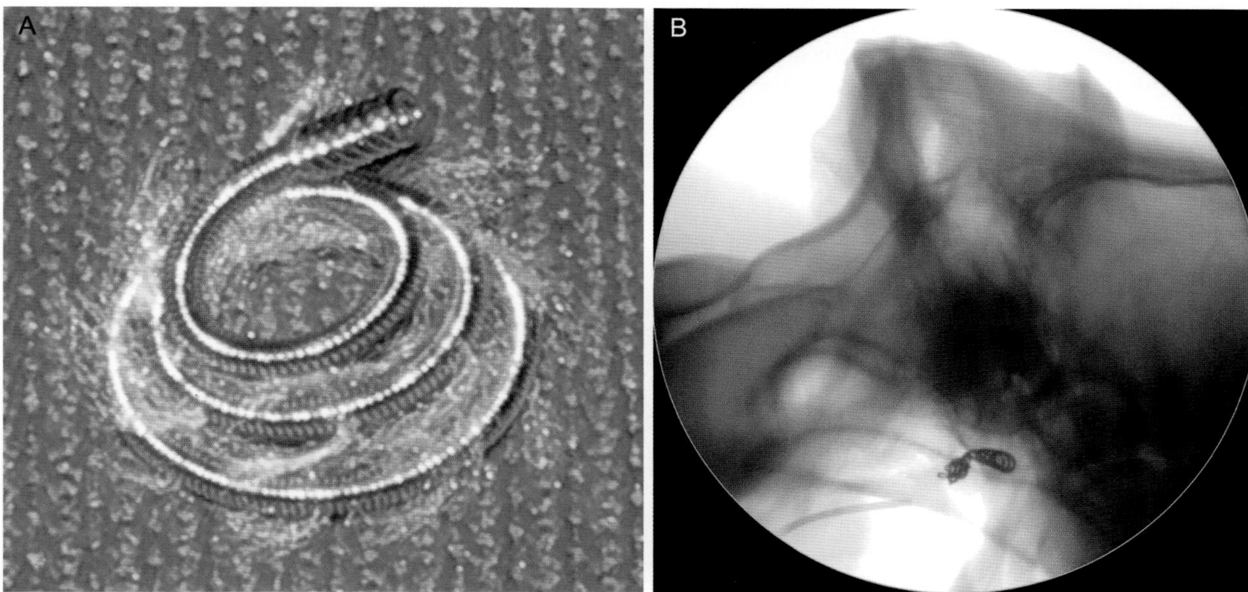

FIGURE 10.5-4 (A) Photograph showing a coil coated transversely with synthetic fibers. This type of coil is used for vascular occlusion of the affected artery in the transarterial coil embolization procedure. (B) Fluoroscopic image showing a coil deployed in the sigmoid flexure of the internal carotid artery with coil size selected based on the internal diameter of the artery.

TRANSARTERIAL COIL EMBOLIZATION (TACE) AND ANGIOGRAPHIC ANATOMY

Methods of vascular occlusion include TACE, use of a transarterial nitinol vascular plug, use of detachable balloons, and insertion of a balloon-tipped venous thrombectomy catheter. Transarterial coil embolization is the preferred method because it is minimally invasive and allows the occlusion of the affected vessel on the cranial (brain) and cardiac sides of the damaged artery, preventing retrograde and normograde blood flow, respectively, from reaching the compromised section of artery. Coils are permanent mechanical occluding devices of different lengths and diameters (Fig. 10.5-4A and B) and are selected based on the internal diameter of the artery. The coils are delivered via a catheter system introduced into the surgically exposed common carotid artery and are deployed in the appropriate locations using fluoroscopic imaging and angiography.

Anatomical features in equids

Introduction

The guttural pouch is a unique structure of the horse among domesticated animals. It is an air-filled outpouching or diverticulum of the auditory tube (Eustachian tube), with an approximate volume of 300–500 mL. The auditory tube passes from the tympanic cavity of the middle ear to the nasopharynx and allows for atmospheric pressure equalization across the tympanic membrane (eardrum). There are left and right guttural pouches, and their walls are comprised of a transparent, glistening respiratory mucosa.

Function

Guttural pouches may contribute to the regulation of arterial blood temperature and cooling circulation to the brain to below body temperature. Regulation of intracranial blood pressure is an additional proposed function.

Guttural pouch location and structure (Figs. 10.5-5–10.5-7)

With guttural pouch mycosis, infection and destruction of tissues may result in a pharyngeal fistula (Fig. 10.5-8) or loss of the median septum between the pouches.

The guttural pouches occupy the region of the head caudal to the rami of the mandibles, dorsocaudal to the nasopharynx and larynx, and ventral to the base of the skull and wing of the atlas. They lie deep to the parotid salivary glands. Left and right guttural pouches are separated on midline by the **median septum**, the thin membrane where their medial walls are adjacent to each other.

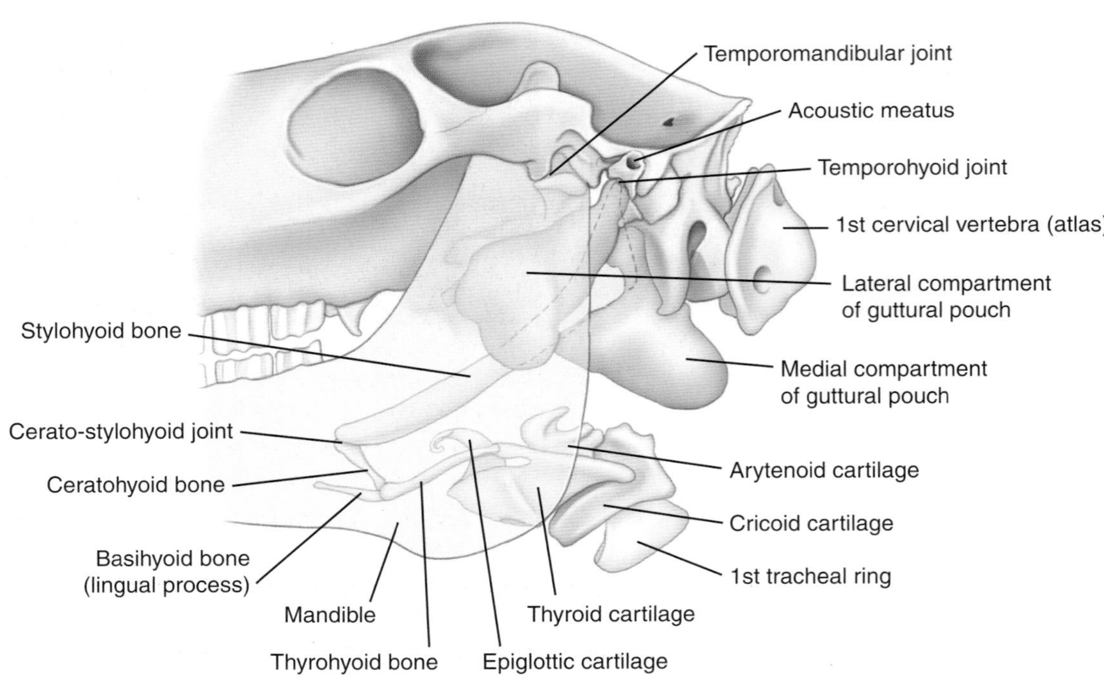

FIGURE 10.5-5 Anatomical drawing of the equine guttural pouches in relation to the head and neck.

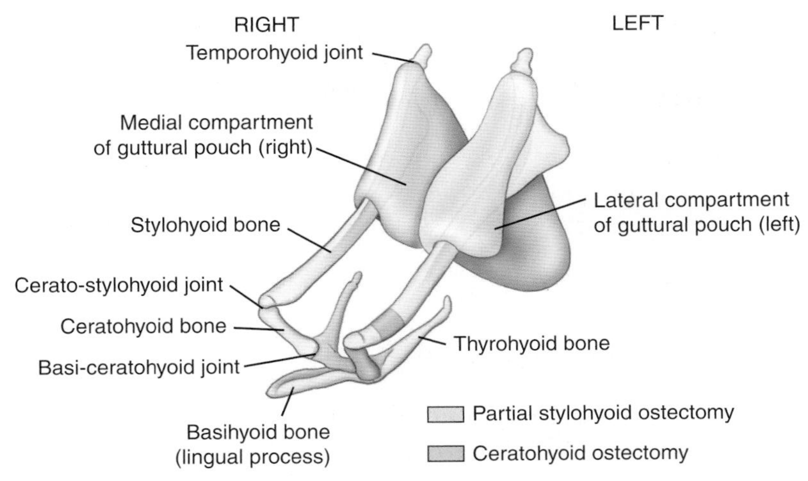

FIGURE 10.5-6 Equine hyoid apparatus, marking the site of partial stylohyoid and ceratohyoid ostectomies.

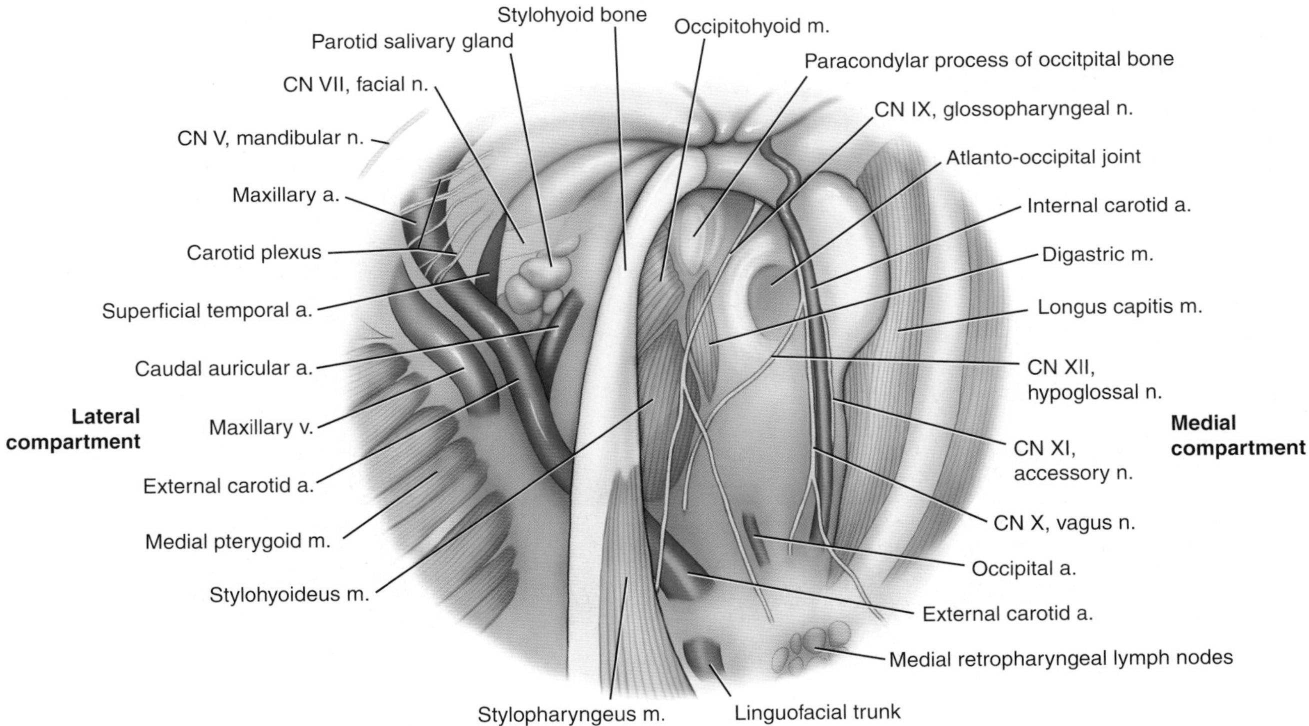

FIGURE 10.5-7 Anatomy of the internal wall of the equine guttural pouch and anatomical relationships with adjacent structures.

Each guttural pouch communicates with the nasopharynx by its ipsilateral pharyngeal orifice of the auditory tube (Fig. 10.5-9A). The pharyngeal orifice is funnel-shaped, and the medial lamina of the orifice is composed of fibrocartilage. The pharyngeal orifice dilate during swallowing and are otherwise normally passively closed.

Caudodorsally, the median septum is expanded and filled by the presence of the ventral muscles of the head—the **longus capitis** and **rectus capitis ventralis muscles**—as they insert on the **basioccipital bone**. Each guttural pouch is divided into a larger medial and smaller lateral compartment by the **stylohyoid bone** (a bone of the hyoid apparatus—see side box entitled "Hyoid apparatus") (Figs. 10.5-5, 10.5-6, and 10.5-9B) and the **stylopharyngeus muscle** (Figs. 10.5-7 and 10.5-9C).

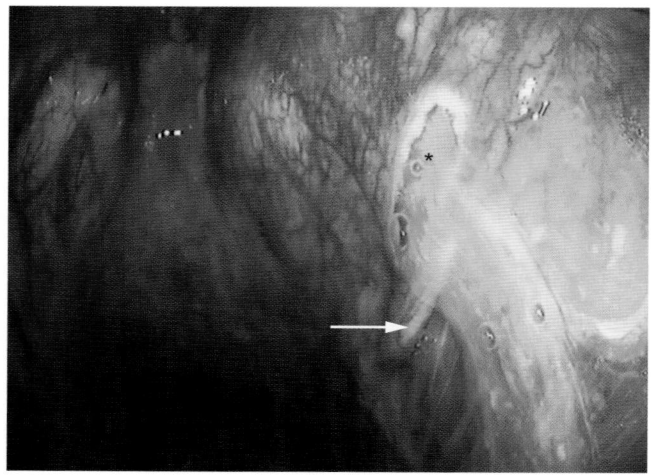

FIGURE 10.5-8 Erosion of the pharyngeal wall *(black star)* adjacent to the left guttural pouch orifice (ventral margin indicated by the *black arrow*), secondary to fungal infection. Mucopurulent exudate is exiting from the fistula and flowing over the cartilage flap of the orifice of the guttural pouch, and down the wall of the nasopharynx.

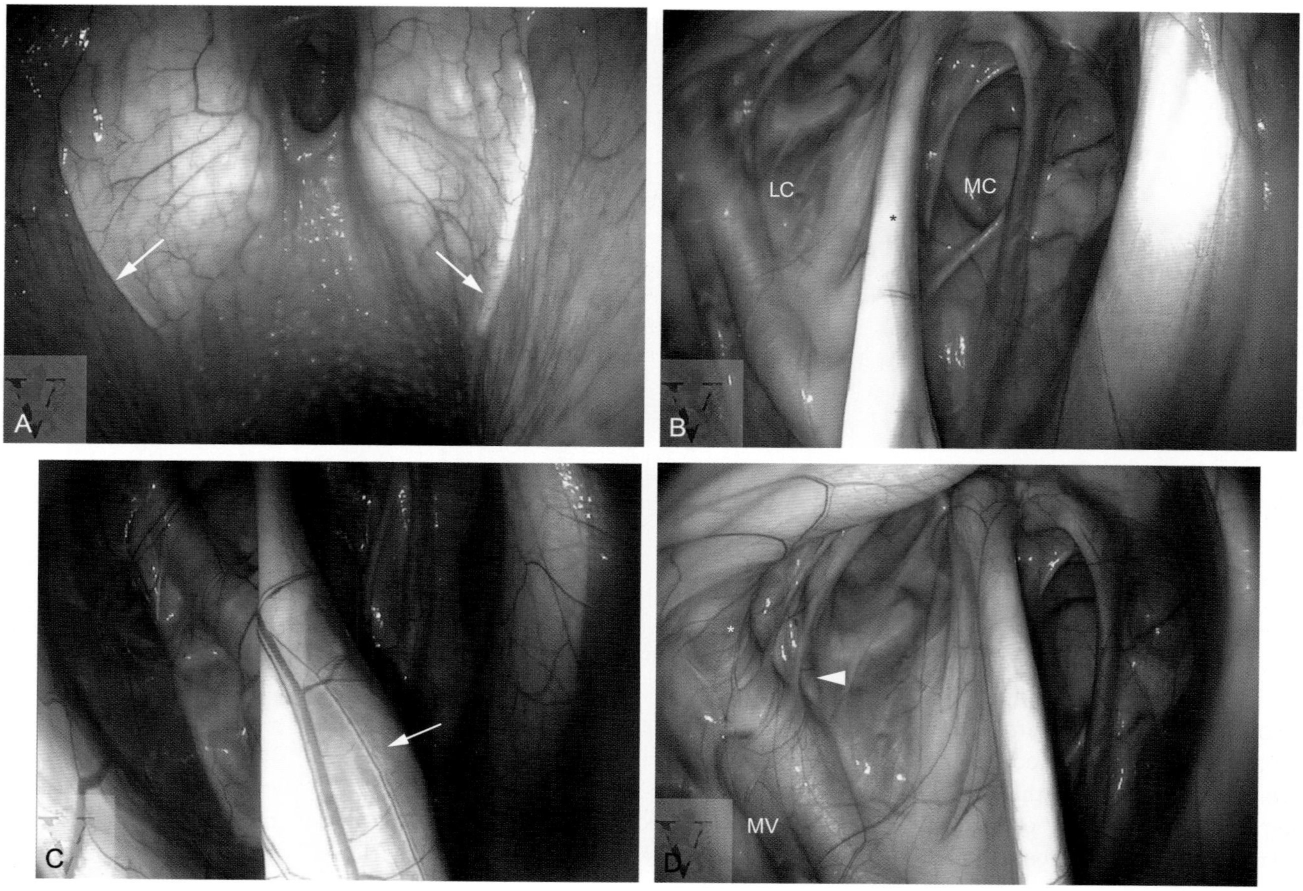

FIGURE 10.5-9 Endoscopic images of the nasopharynx and right guttural pouch. (A) Nasopharynx showing the fibrocartilage flaps *(arrows)* at the pharyngeal orifice of the auditory tubes. (B) Stylohyoid bone *(star)* dividing the lateral (LC) and medial (MC) compartments. (C) Insertion of the stylopharyngeus muscle *(arrow)* on the distal part of the stylohyoid bone. (D) The external carotid artery continues as the maxillary artery *(star)* after the superficial temporal a. branches off *(arrowhead)*. Before this, the caudal auricular artery is noted branching from the external carotid a. and tracking dorsally. Ventral and parallel to the external carotid and maxillary arteries is the maxillary vein (MV).

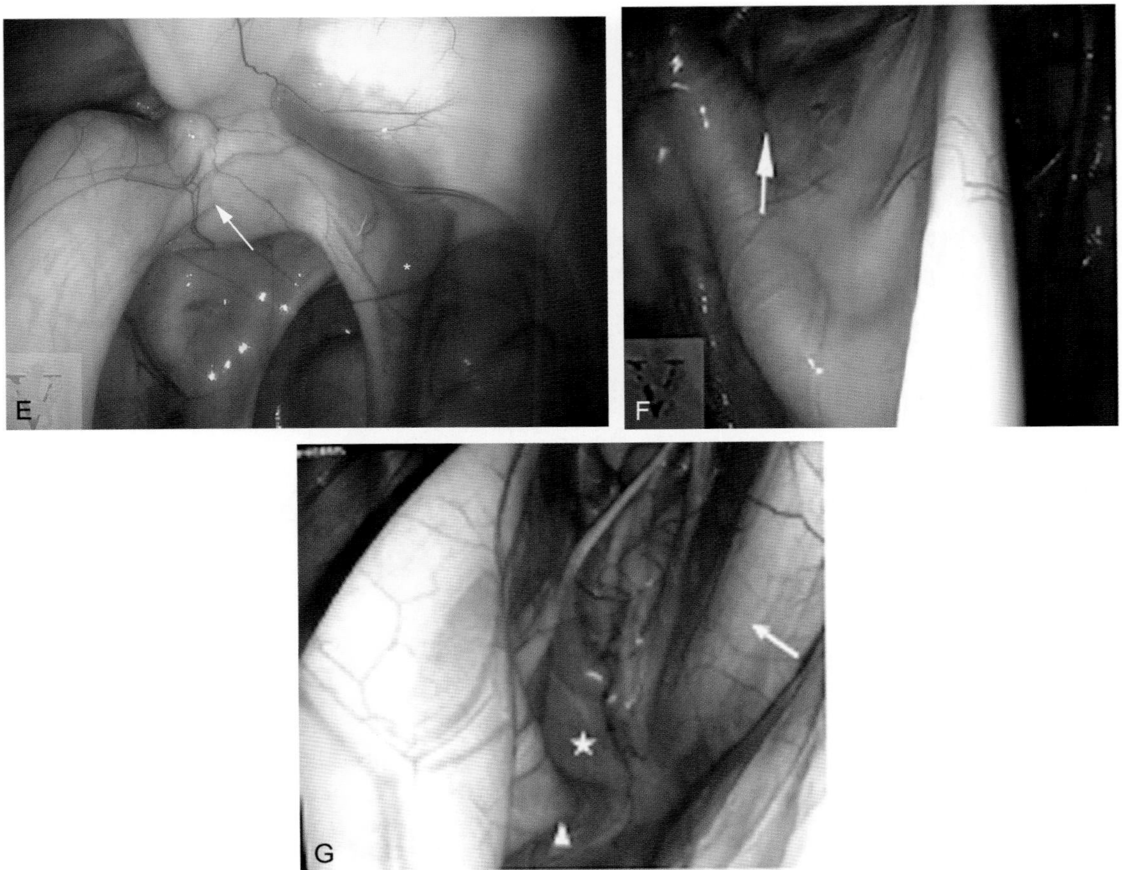

FIGURE 10.5-9, CONT'D (E) The temporohyoid joint *(arrow)* lies caudodorsally, and approximately 1 cm (0.4 in.) rostral and medial to this structure is the internal carotid artery *(star)*. The sigmoidal bending of the artery is seen through the mucosa dorsally and is not always seen on endoscopy. (F) The parotid salivary gland *(arrow)* is in contact with the caudolateral wall below the facial nerve, in the small space between the external carotid a. and the caudal auricular a. The gland is rarely observed from inside the pouch, except when not covered by adipose tissue, and then it can be seen through the transparent mucosa. (G) Ventrally, deep to the pouch mucosa of the medial compartment, are the medial retropharyngeal lymph nodes *(star)*, slightly transparent nodules sometimes visible in the adipose tissue. On the right side of the image, the median septum of the pouch is limited by the longus capitis muscle, which is identified by a red fleshy part divided by a vertically oriented white tendon *(arrow)*. On the left side of the image, the stylopharyngeus muscle and the external carotid a. *(arrowhead)* are visible.

Structures adjacent to the guttural pouches

The guttural pouches have important anatomical relationships with adjacent structures (Figs. 10.5-5, 10.5-7, and 10.5-10), particularly the ventromedially located **medial retropharyngeal lymph nodes** and vital neurovascular structures throughout the region. The pouches are very closely related dorsomedially to the **atlanto-occipital joint**. More dorsolaterally, each pouch is adjacent to the **occipitomandibularis muscle**, the caudal belly of the **digastricus muscle**, and the **paracondylar process** of the occipital bone. The lateral compartment is associated with the deep surface of the **parotid salivary gland** and the **pterygoid muscles**. Multiple vessels and nerves track in the walls of both guttural pouch compartments (Figs. 10.5-7 and 10.5-9). The internal carotid artery and cranial nerves IX, X, XI, and XII course in the caudal wall of the medial compartment. Ventrally, the external

Injury to the vessels and nerves that track through the guttural pouch accounts for many of the clinical signs observed in horses with guttural pouch disease. For example, dysphagia, with its sequela of nasal discharge containing food and saliva, occurs secondary to injury to CN's IX and X (Fig. 10.5-11). Laryngeal hemiplegia, and its sequela of exercise intolerance and abnormal airway noise, may occur secondary to CN X injury because the recurrent laryngeal nerve fibers track in CN X.

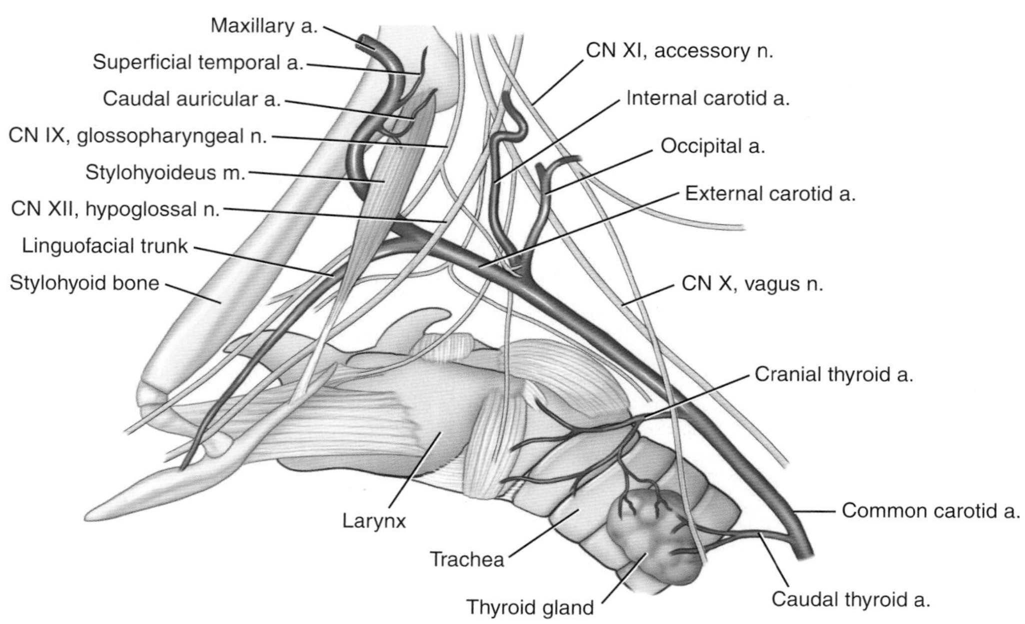

FIGURE 10.5-10 Left lateral view of arteries and nerves associated with the equine guttural pouch.

Surgeons performing vascular occlusion (Fig. 10.5-12) must be familiar with the vascular anatomy of the guttural pouch region and be aware of the many normal variations in the anatomy that may be encountered. Fluoroscopy and angiography are used to plot the arterial anatomy in real time during TACE (Fig. 10.5-13).

carotid artery is associated with the floor of the medial compartment before it passes deep to the stylohyoid bone and then is associated with the wall of the lateral compartment. In the lateral compartment, the external carotid artery gives off the caudal auricular artery close to the stylohyoid bone, and then continues as the maxillary artery in the dorsal wall of the lateral compartment, after the superficial temporal artery branches off. The maxillary vein lies adjacent to the external carotid artery in the lateral compartment. Cranial nerves VII and VIII are adjacent to the temporohyoid joint, and CN VII is also related to the dorsocaudal wall of the lateral compartment.

PERFORMING GUTTURAL POUCH ENDOSCOPY

To enter the guttural pouch, the fibrocartilage flap covering the pharyngeal orifice of the auditory tube must be elevated to advance the endoscope (Fig.10.5-9A). This can be accomplished with a flexible endoscopic biopsy instrument. The endoscope is passed ventral and medial in the nasal cavity and then is directed toward the dorsolateral wall of the nasopharynx to reach the ipsilateral pharyngeal orifice. The biopsy instrument is advanced beyond the tip of the endoscope and passed under the dorsal aspect of the flap to enter the guttural pouch until 10–15 cm (3.9–5.9 in.) of the biopsy instrument is “seated” in the guttural pouch. The endoscope is then rotated (counterclockwise for the right guttural pouch and clockwise for the left one), such that the biopsy instrument opens the flap allowing the endoscope to be advanced into the guttural pouch using the biopsy instrument as a “guide wire.” Once the endoscope is inside the guttural pouch, the biopsy instrument is retracted back into the biopsy channel. The pouches are then systematically examined and digital pictures taken for the medical record. If guttural pouch disease is suspected or identified, a resting endoscopic examination of the nasopharynx and larynx (see Cases 10.6 and 10.7) is also performed to determine if there are any functional or anatomic abnormalities due to injury to cranial nerves associated with the guttural pouch wall.

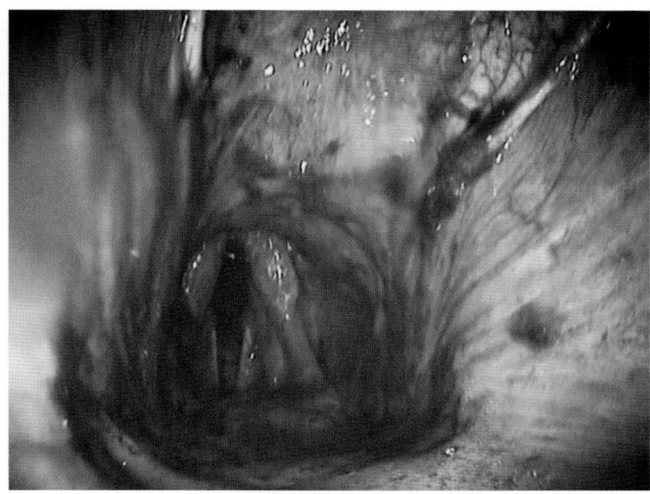

FIGURE 10.5-11 Horse diagnosed with guttural pouch mycosis. Endoscopic examination shows an accumulation of food and saliva in the nasopharynx, consistent with dysphagia. This is also seen externally as a bilateral nasal discharge. Injury to CN's IX and X can result in this clinical finding.

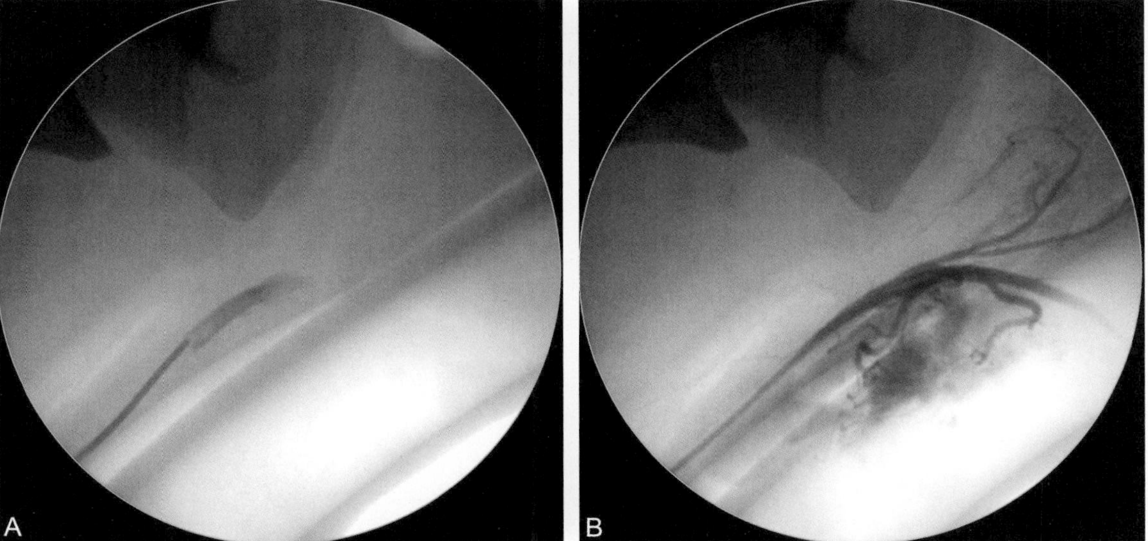

FIGURE 10.5-12 Fluoroscopic angiography: (A) The tip of the angiographic catheter is in the common carotid artery, and contrast (radiopaque) medium is injected to depict its branching. (B) Misdirection of the catheter into the cranial thyroid artery, as evidenced by the typical branching of this vessel when contrast medium is injected.

(Continued)

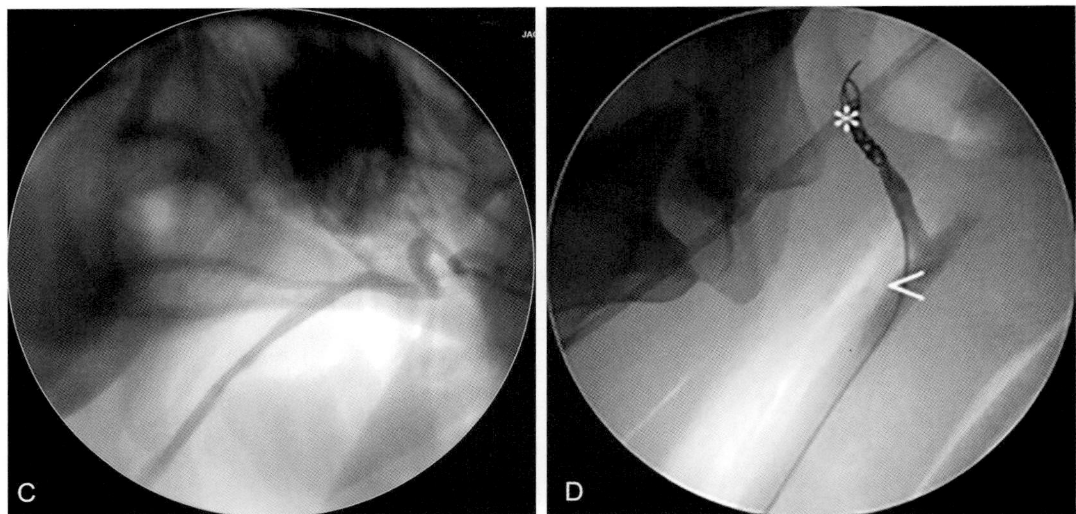

FIGURE 10.5-12, CONT'D (C) Contrast medium is injected in the cranial internal carotid artery to depict the sigmoid flexure, the location for coil occlusion to prevent retrograde flow from the arterial circle of Willis. (D) Coil occlusion (*) of the origin (cardiac side of the lesion) of the internal carotid a. just after branching from the common carotid artery *(white arrow head)*.

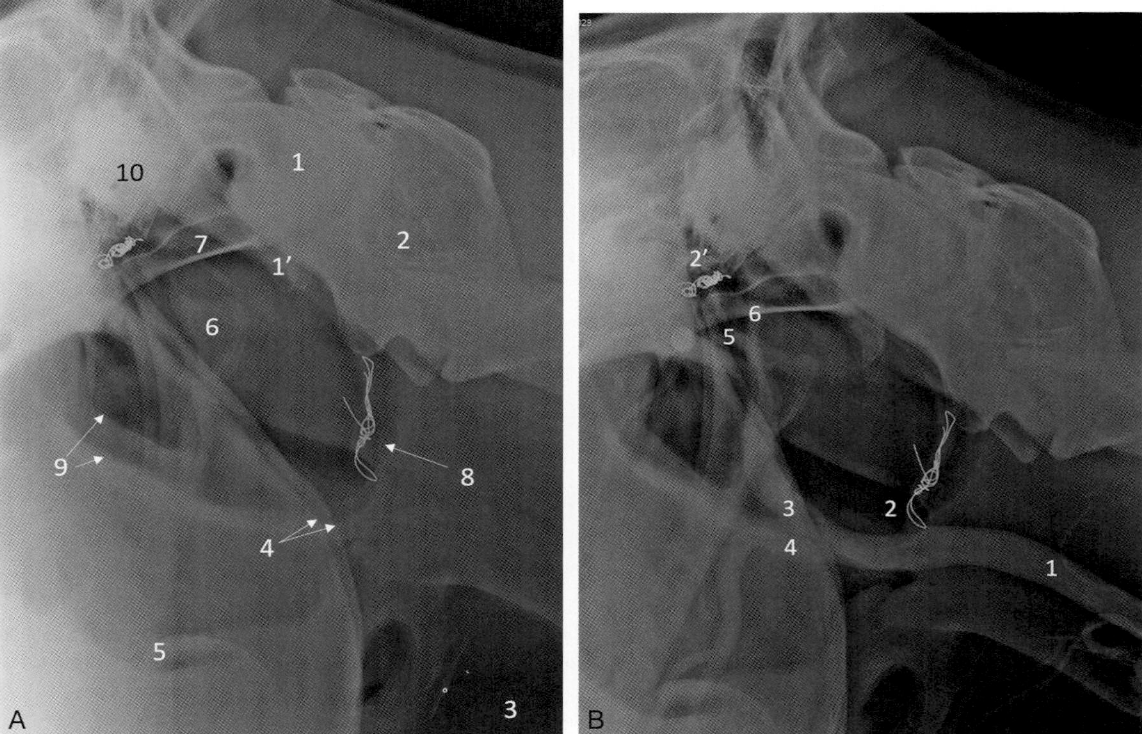

FIGURE 10.5-13 Diagnostic imaging *(lateral view)* of the guttural pouch region of an adult horse after a standing coil embolization procedure of the internal carotid a. (A) Radiography: 1-occipital condyle and 1′-occipital paracondylar process, 2-atlas (C1), 3-rostral trachea, 4-mandibular rami, 5-epiglottis, 6-stylohyoid bone, 7-basisphenoid, 8-caudal border of the guttural pouch, 9-rostral border of both guttural pouches, 10-tympanic bulla. (B) Angiography: 1-common carotid a., 2-coil in the origin ("cardiac side of lesion") of the internal carotid a., with one coil ending engaged in a small-diameter branching vessel, and 2′-coil in the sigmoid flexure ("cranial or brain side of lesion") of the internal carotid a.; 3-external carotid a., 4-linguofacial trunk., 5-maxillary a., 6-caudal auricular a. Note: occipital a. and superficial temporal a. are not visualized in this angiogram.

Hyoid apparatus

The **hyoid apparatus** (Figs. 10.5-5 and 10.5-6) is a series of articulating bones suspended from the ventral skull between the rami of the mandibles. This apparatus functions to support the larynx, pharynx, and tongue. There are paired **stylohyoid, ceratohyoid,** and **thyrohyoid** bones, with the left and right sides connected on the midline by the unpaired, transversely located **basihyoid** bone. The basihyoid bone in the horse, and to a lesser extent in the ox, has a prominent blade-shaped lingual process protruding rostrally into the root of the tongue. The horse lacks a separate epihyoid bone that is found in carnivores and the ox. In the horse, the epihyoid fuses to the stylohyoid bone. The thyrohyoid bone attaches to the rostral cornu of the thyroid cartilage of the larynx and the stylohyoid bone articulates—via a short, partly cartilaginous **tympanohyoid**—with the petrous part of the temporal bone (at its mastoid process), forming the **temporohyoid joint** (THJ). Numerous muscles attach to the hyoid apparatus, including extrinsic muscles of the tongue—hyoglossus, styloglossus—and the geniohyoideus, mylohyoideus, sternohyoideus, omohyoideus, and thyrohyoideus muscles.

Removal of a segment of the hyoid apparatus (Figs. 10.5-6 and 10.5-7) is the recommended surgical procedure to eliminate painful and mechanical forces on the THJ in horses with temporohyoid osteoarthropathy (see side box entitled "Temporohyoid osteoarthropathy"). Ceratohyoid ostectomy is the preferred procedure (Fig. 10.5-6).

ANEURYSM

In the differential diagnosis list for guttural pouch epistaxis, ruptured aneurysm (Gr. *aneurysma* a widening), although uncommon, is an important consideration, and second only to guttural pouch mycosis as a cause of epistaxis. An aneurysm develops when part of an artery wall weakens, allowing it to enlarge abnormally. In humans, aneurysms may be congenital or acquired, and in horses, the cause is unknown. The site of an aneurysm can be diagnosed by endoscopy only if the horse is not actively bleeding or if the abnormal artery wall is not obscured by blood clots. Treatment to prevent further bleeding requires intra-operative fluoroscopy and angiography to fully evaluate the aneurysm and then TACE.

TEMPOROHYOID OSTEOARTHROPATHY

Temporohyoid osteoarthropathy (THO) is a chronic disease characterized by inflammation, thickening, and possibly fracture of the bones of the THJ and damage to adjacent cranial nerves (VII, VIII). Proposed etiologies include inner or middle ear infection from a hematogenous origin, non-septic degenerative joint disease, and extension of otitis externa or guttural pouch infection. Diagnosis of THO is based on (1) clinical signs related to dysfunction of CN VII, e.g., ear droop, muzzle deviation, lack of palpebral reflex and consequent exposure keratitis, and CN VIII, e.g., head tilt, circling; (2) other behavioral changes related to pain such as head shaking; (3) endoscopy of the guttural pouches showing stylohyoid bone and THJ dystrophy; and (4) radiography and/or CT of the head. The preferred imaging technique is CT, and it is the most specific and sensitive diagnostic test allowing for a comprehensive assessment of the hyoid apparatus and related structures. Temporohyoid osteoarthropathy can occur bilaterally, and it is therefore important to examine both guttural pouches. The endpoint of the disease is ankylosis of the THJ. In these cases, mechanical forces induced by normal movements of the tongue and larynx transmitted through the hyoid apparatus can cause fracture of the petrous temporal bone and, less commonly, the stylohyoid bone.

Medical treatment includes broad-spectrum antibiotics and nonsteroidal anti-inflammatory drugs, and is often combined with surgical treatment (e.g., ceratohyoidectomy, Fig. 10.5-6) with the goal of reducing the risk of fracture of the stylohyoid or temporal bone.

Guttural pouch endoscopic anatomy (see side box entitled "Performing guttural pouch endoscopy")

Understanding and recognizing normal endoscopic anatomy (Figs. 10.5-7 and 10.5-9A–G) is important in assessing the guttural pouch and assists in identifying disease and changes that occur during and after the treatment of the disease (Fig. 10.5-14A–C).

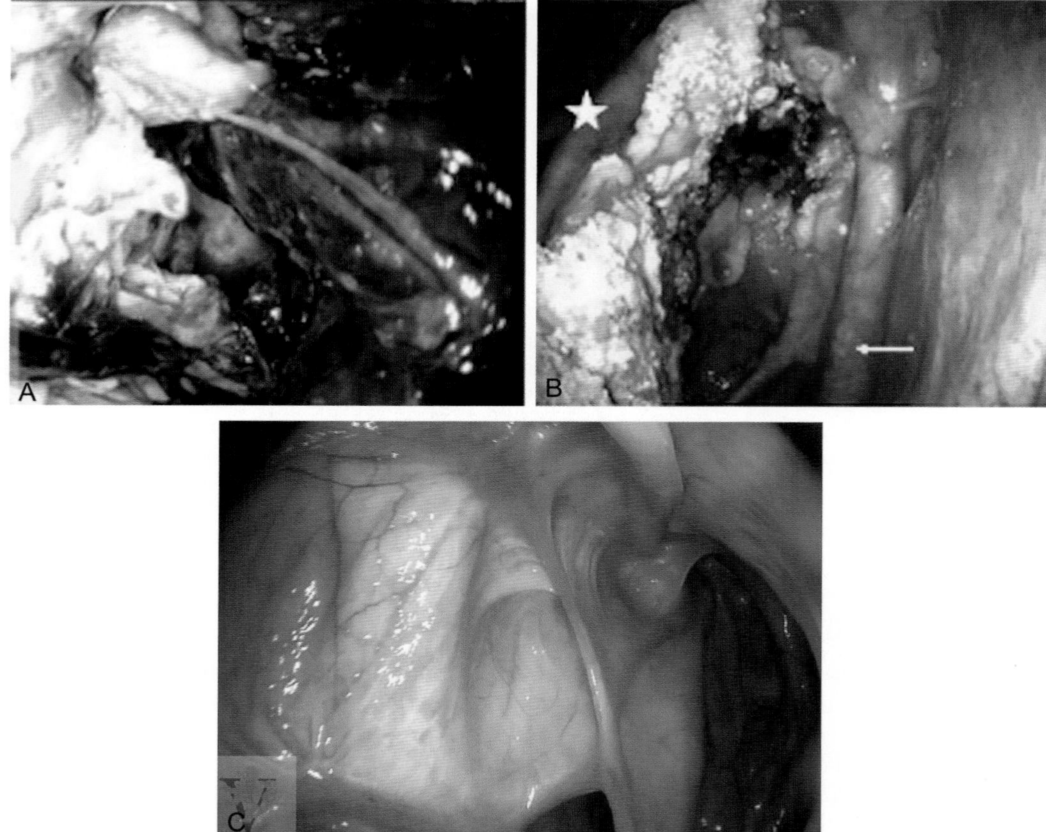

FIGURE 10.5-14 Endoscopic images of guttural pouch mycosis lesions during various phases of healing: (A) Inflammatory reaction and blood clots observed on admission before any treatment. (B) Resolving medial compartment fungal infection with inflammatory tissue proliferation, and a contraction and resorption of blood clots. The internal carotid artery *(arrow)* and stylohyoid bone *(star)* are identified. (C) Residual scar tissue lines the guttural pouch following a complete resolution of a fungal infection.

Selected references

[1] Baptiste KE, Naylor JM, Bailey J, Barbers EM, Post K, Thornhill J. A function for guttural pouches in the horse. Nature 2000;403:382–3.
[2] Khairuddin NH, Sullivan M, Pollock PJ. Angiographic variation of the internal carotid artery and its branches in horses. Vet Surg 2015;44:784–9.
[3] Lepage OM, Cadoré J-L, Perron-Lepage MF. The mystery of fungal infection in the guttural pouches. Vet J 2004;168:60–4.
[4] Lepage OM. Challenges associated with the diagnosis and management of guttural pouch epistaxis in equids. Equine Vet Educ 2015;28:372–8.
[5] Oliver ST, Hardy J. Ceratohyoidectomy for treatment of equine temporohyoid osteoarthropathy (15 cases). Can Vet J 2015;56:382–6.
[6] Piat P. Proposal of a standardized method to recognize the anatomic structures visible during guttural pouch endoscopy in the horse. Implications in the management of guttural pouch diseases. Thesis no. 27, ENV Lyon, 2008. http://www2.vetagro-sup.fr/bib/fondoc/th_sout/dl.php?file=2008lyon027.pdf.

CHAPTER 10

CASE 10.6

Dorsal displacement of the soft palate

Tara R. Shearer and Susan J. Holcombe
Department of Large Animal Clinical Sciences, Michigan State University College of Veterinary Medicine, East Lansing, Michigan, US

Clinical case

History

A 3-year-old Thoroughbred filly in race training presented with a 4- to 6-week history of poor performance and an abnormal respiratory sound during intense exercise. The trainer described a snoring or gurgling noise that was heard only when the filly was galloped. Associated with the noise, the filly would suddenly "quit" (slow down substantially). She was otherwise normal.

Physical examination findings

Physical examination was unremarkable except for a 2nd-degree atrioventricular (AV) block on cardiac auscultation. No upper respiratory sound was audible at rest; normal airflow was present from both nostrils, and auscultation of the trachea was normal. Palpation of the intermandibular region revealed normal mandibular lymph nodes (often incorrectly referred to as the "submandibular lymph nodes"). The larynx and trachea were palpably normal, and no cough could be elicited on manipulation of the larynx or trachea. Because lameness is a common cause of poor performance, the filly's gait was assessed at the trot; no lameness was observed.

Second-degree AV block is a common arrhythmia in healthy, fit horses at rest. High parasympathetic (vagal) tone results in bradycardia and a regularly dropped beat (e.g., once every 4 beats). Notably the 4th heart sound (S4, associated with atrial contraction) can be heard in the silence of the dropped beat. This filly's resting heart rate was 25 bpm, whereas the average resting heart rate for an adult horse is 30–40 bpm. With benign 2nd-degree AV blocks such as this one, raising the horse's heart rate by taking the horse for a brisk walk resolves the arrhythmia. A 2nd-degree AV block that does not disappear with activity is uncommon in horses and warrants further investigation.

Asymmetry of the larynx may be palpable in horses with recurrent laryngeal neuropathy (formerly called "idiopathic laryngeal hemiplegia"). This distal axonopathy of the recurrent laryngeal nerve(s) causes atrophy of the dorsal and lateral cricoarytenoid muscles, most commonly on the left side, which is discernible as palpable asymmetry of the larynx when the disease is advanced. Recurrent laryngeal neuropathy is common in Thoroughbreds, so palpation of the larynx is an important component of the workup in any case involving abnormal respiratory sounds heard primarily or exclusively during exercise. The structure and function of the equine larynx are discussed in depth in Case 10.7.

Differential diagnoses

Nasopharyngeal collapse, recurrent laryngeal neuropathy, dorsal displacement of the soft palate, palatal instability, epiglottic entrapment or retroversion, and subepiglottic cyst

Comparative Veterinary Anatomy: A Clinical Approach. https://doi.org/10.1016/B978-0-323-91015-6.00142-4

Diagnostics

Endoscopic examination of the nasal passages, nasopharynx, larynx, guttural pouches, and rostral trachea was performed at rest. All observed structures were normal and functioned well at rest. Swallowing was induced by spraying water through the endoscope to stimulate pharyngeal and laryngeal sensory receptors. During swallowing, normal contraction of the nasopharynx was noted. Brief nasal occlusion was then performed to observe the responses of the nasopharynx and larynx to increased inspiratory pressure. The responses were normal: both arytenoid cartilages of the larynx abducted fully and the nasopharynx dilated, with mild dorsal pharyngeal collapse (Fig. 10.6-1). Lastly, the endoscope was withdrawn to the junction of the hard and soft palate so that the rostral soft palate could be observed; it too was normal (Fig. 10.6-2).

Resting endoscopic examination did not identify the problem in this case, but it ruled out many of the possible causes of poor performance and associated airway noise, including subepiglottic cyst, epiglottic entrapment, grade 3 or 4 recurrent laryngeal neuropathy, and static obstructions, such as nasopharyngeal cicatrix (see side box entitled "Nasopharyngeal cicatrix"). Exercising endoscopy is required to identify dynamic causes of abnormal airway noise and poor performance, such as dorsal displacement of the soft palate, palatal instability, rostral soft palate collapse, epiglottic retroversion, and nasopharyngeal collapse.

Because no apparent cause for the respiratory noise was found during the resting endoscopic examination, a remote dynamic overground endoscopic examination was performed the next day during training at the racetrack. This examination revealed intermittent billowing of the caudal free margin of the soft palate and loss of visualization of the epiglottis, a characteristic of intermittent dorsal displacement of the soft palate (Fig. 10.6-3).

Dorsal displacement of the soft palate causes expiratory obstruction that is performance-limiting in some horses because it decreases minute ventilation (the volume of air entering or exiting the lungs per minute). The billowing caudal free margin of the soft palate creates the "gurgling, snoring" noise often described in horses with this condition.

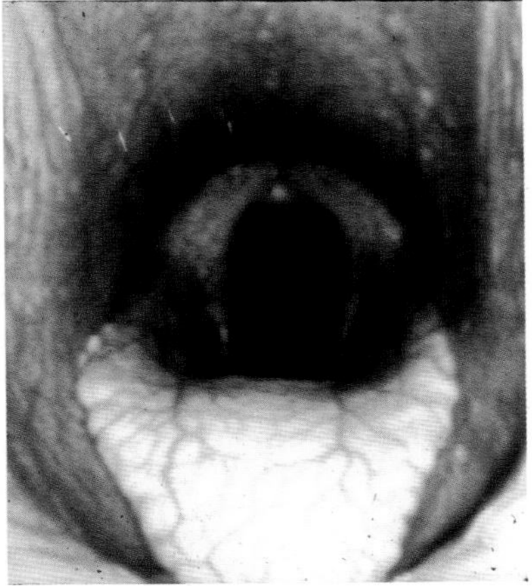

FIGURE 10.6-1 Endoscopic view of the caudal nasopharynx and entrance to the larynx during brief nasal occlusion, showing symmetrical abduction of the arytenoid cartilages bilaterally and the epiglottis positioned dorsal to the soft palate. The roof of the pharynx dips mildly (i.e., dorsal pharyngeal collapse).

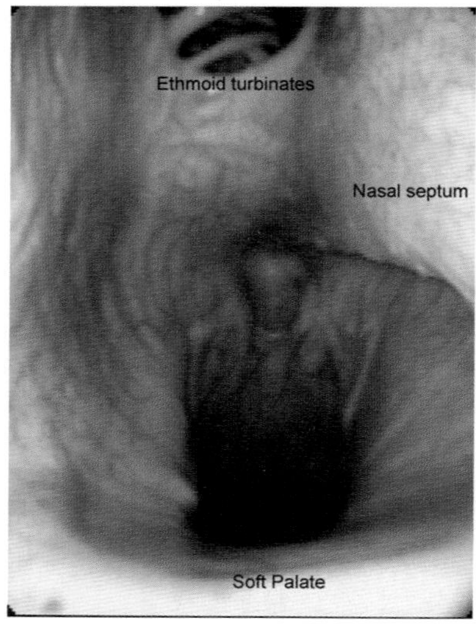

FIGURE 10.6-2 Endoscopic view of the rostral nasopharynx showing the appearance of the normal soft palate.

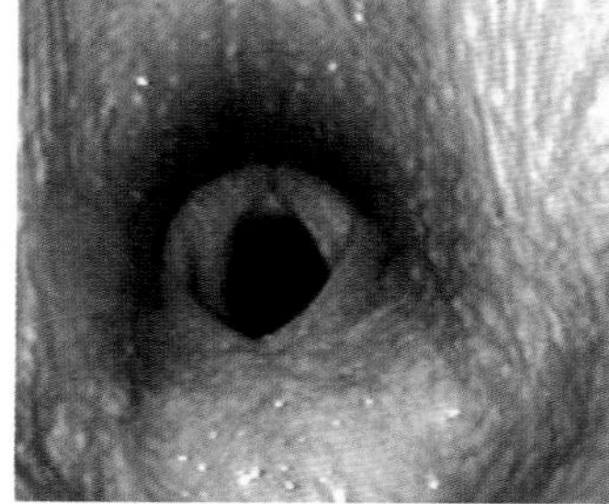

FIGURE 10.6-3 Endoscopic view of dorsal displacement of the soft palate. The epiglottis is not seen (compare with Fig. 10.6-1) because it is ventral to the soft palate.

Diagnosis

Intermittent dorsal displacement of the soft palate (DDSP)

Treatment

After discussion of the treatment options (see side box entitled "Management options for DDSP in racehorses"), the owner elected the recommended laryngeal tie-forward and sternothyroid tenotomy procedures. The filly was anesthetized and placed in dorsal recumbency. A bilateral sternothyroid tenotomy was performed by transecting the

NASOPHARYNGEAL CICATRIX

Nasopharyngeal cicatrix is an uncommon condition in which severe inflammation of the nasopharynx results in scarring and a partial obstruction of the caudal nasopharynx by a web of fibrous strands (Fig. 10.6-4). Almost exclusively seen in Texas, an environmental allergen is the suspected cause. If scarring is extensive and airway obstruction is severe, a permanent tracheostomy may be needed (Fig. 10.6-5). Permanent tracheostomy may be indicated for any permanent impairment of the nasopharynx or larynx, such as bilateral laryngeal hemiplegia or bilateral arytenoid chondrosis.

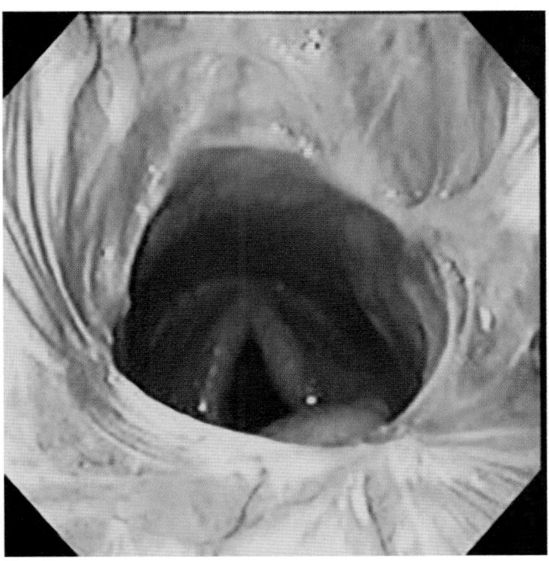

FIGURE 10.6-4 Endoscopic view of extensive scarring of the nasopharynx causing a constricting ring of tissue. Structures of the larynx are visible caudal to the scarred area. (Image courtesy M. Gerard.)

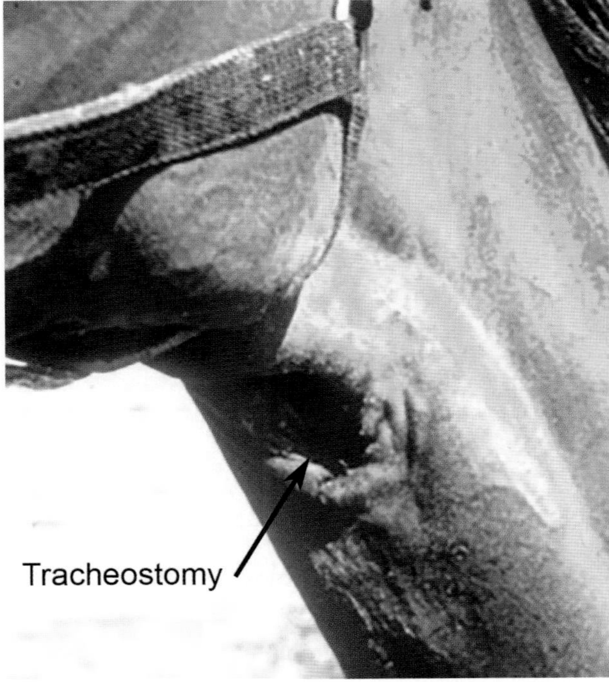

FIGURE 10.6-5 Horse with a permanent tracheostomy (*arrow*).

sternothyroideus tendon at the level of the thyroid cartilage of the larynx (Fig. 10.6-6). Performing the tenotomy first permits unimpeded advancement of the thyroid cartilage during the tie-forward procedure.

For the tie-forward, 2 bites of no. 5™ (ultra-high-molecular-weight polyethylene) suture were placed through the right lamina of the thyroid cartilage, and repeated in the left lamina. The right and left sutures were then passed

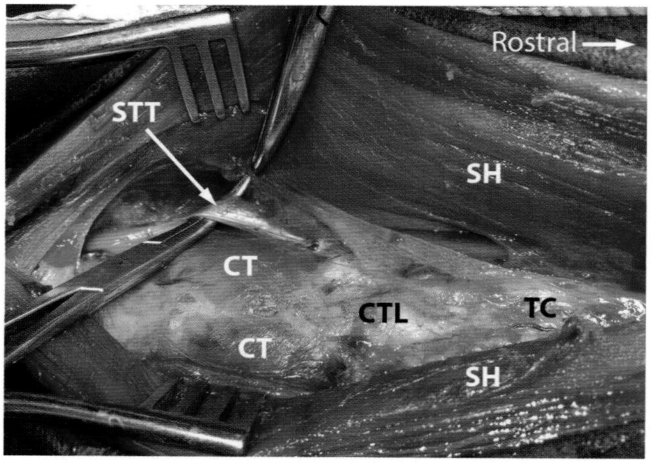

FIGURE 10.6-6 Surgical approach for a tie-forward procedure. The sternothyroideus tendon of attachment is isolated and elevated by the hemostat and is ready for transection. Key: *SH*, sternohyoideus m.; *CT*, cricothyroideus m.; *CTL*, cricothyroid ligament; *TC*, thyroid cartilage; *STT*, sternothyroideus tendon.

around the basihyoid bone in a specific pattern and sequence and the suture ends subsequently tied across the lingual process of the basihyoid bone. While the sutures were being tied, an assistant raised the horse's nose to produce a head-neck angle of approximately 90 degree of flexion. The rostral edge of the thyroid cartilage was advanced approximately 1 cm (0.4 in.), rostral to the caudal edge of the basihyoid bone in this tie-forward procedure.

The filly had an uncomplicated post-operative period and successfully returned to racing 60 days after surgery, winning her next 3 races.

MANAGEMENT OPTIONS FOR DDSP IN RACEHORSES

The laryngeal advancement procedure or "tie-forward" is currently recommended to secure the position of the epiglottis rostrodorsal to the caudal margin of the soft palate, thus preventing abnormal displacement of the soft palate. The tie-forward procedure effectively ties the larynx forward (rostrally) and dorsal to the caudal margin of the soft palate by creating a thyrohyoideus muscle prosthesis.

Another option is myectomy (removing a segment of the muscle) or transection of the sternothyroideus and sternohyoideus muscles near their insertions (Figs. 10.6-6 and 10.6-7). This procedure reduces caudal retraction of the larynx; however, data supporting its efficacy disagree.

Staphylectomy, or trimming of the caudal free margin of the soft palate, is a third option. This procedure involves the resection of the caudal margin of the soft palate using a transendoscopic, laser-assisted technique, or by sharp excision via a laryngotomy (see Case 10.7). Laser-assisted thermoplasty of the caudal soft palate may be selected over removing the caudal margin.

Staphylectomy and sternothyroid tenotomy/myectomy may also be combined; this approach is known as the Llewellyn procedure. The sternothyroideus tendons are transected where they insert on the thyroid cartilage (Fig. 10.6-6), and the sternohyoideus myectomy is performed at the level of the laryngotomy. This procedure has fallen out of favor, as data supporting its efficacy are lacking.

Lastly, conservative treatment includes 30 days of rest with turn out, anti-inflammatory medication (dexamethasone at a tapering dose over 30 days), and retraining over the next 30–60 days.

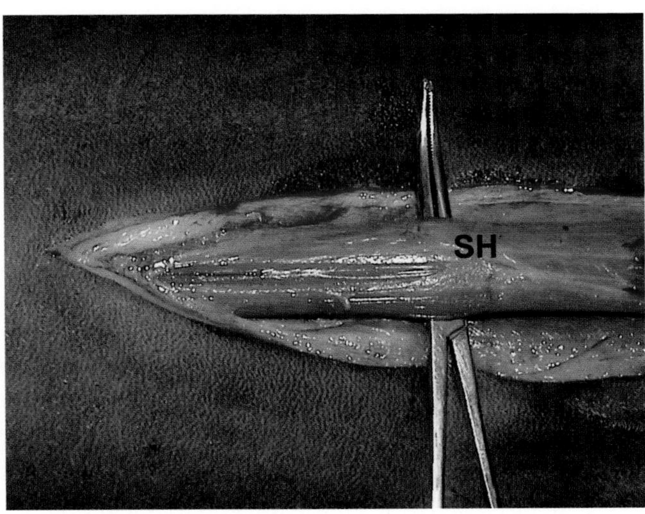

FIGURE 10.6-7 A photo showing the cranioventral aspect of the equine neck, with an incision through the skin and subcutaneous tissues. The sternohyoid (*SH*) muscle has been isolated for transection.

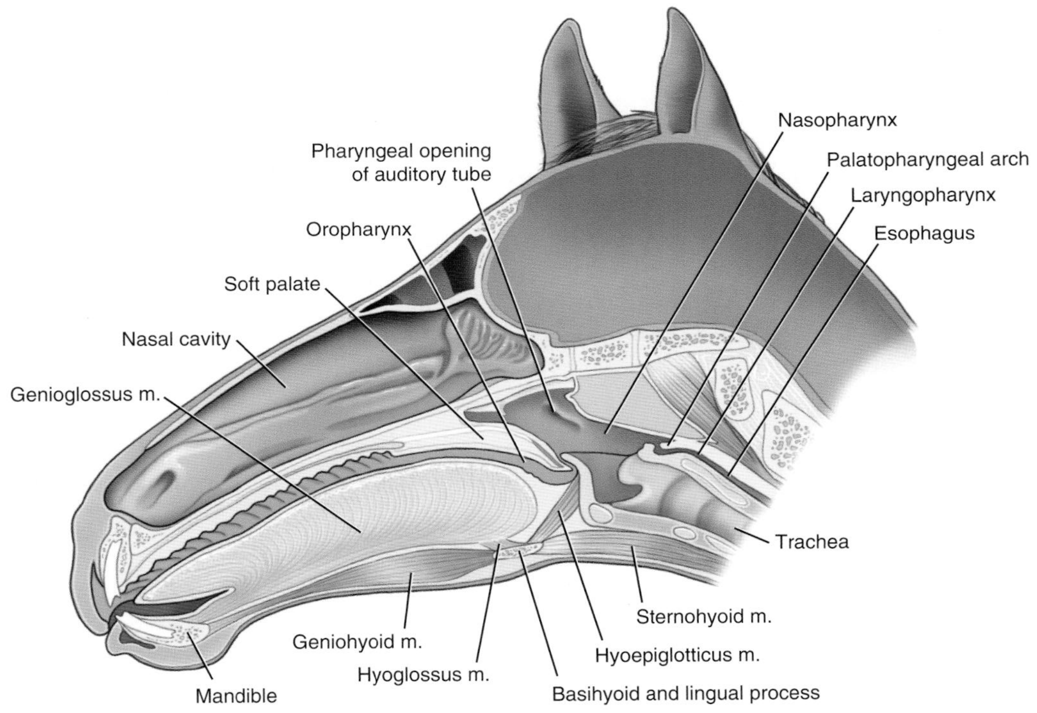

FIGURE 10.6-8 The 3 parts of the equine pharynx: nasopharynx, orophraynx, and laryngopharynx.

Anatomical features in equids

Introduction

Horses and other equidae have an elongated soft palate. The soft palate divides the rostral pharynx into 2 compartments: the nasopharynx dorsally and the oropharynx ventrally. The caudal portion of the pharynx, the laryngopharynx, continues the rostral parts and lies lateral and dorsal to the larynx (Fig. 10.6-8). This section will discuss pharyngeal anatomy and the muscles of the nasopharynx. Laryngeal anatomy is covered in Case 10.7.

Function

The pharyngeal region is where the respiratory and digestive tracts cross over on their respective paths to the larynx and esophagus. The pharynx functions to maintain airflow and facilitate swallowing of food and liquids. Deglutition (swallowing) is a complex physiological and anatomical process requiring fine neuromuscular coordination of the respiratory and digestive tracts and their intersection at the pharynx (see side box entitled "Anatomy and function of deglutition").

Pharyngeal anatomy

The equine **soft palate** extends to the base of the epiglottis, so except during swallowing, the epiglottis rests on top of the soft palate (Fig. 10.6-8). The caudal free margin of the soft palate is continuous laterally with the **palatopharyngeal arches**, folds of mucosa that pass dorsolaterally and caudally, meeting on the dorsal midline of the caudal extent of the nasopharynx. These arches lie caudal to the **corniculate processes** of the **arytenoid cartilages** of the larynx.

The soft palate and palatopharyngeal arches together form the **intrapharyngeal opening**, the caudal opening of the nasopharynx (Fig. 10.6-9). The rostral structures of the larynx—that is, the **epiglottis** and the corniculate processes of the arytenoid cartilages—project through this opening (somewhat akin to a button through a button hole), and thus, an airtight seal is formed between the nasopharynx and the entrance to the larynx (*aditus laryngis*). Therefore, air flows directly through the nasal passages, nasopharynx, and laryngeal cavity and does not normally pass into the oropharynx or laryngopharynx. This arrangement is primarily what makes horses obligate nasal breathers.

> Unless the caudal free margin of the soft palate is displaced to lie above the epiglottis, horses are constrained to breathing only through the nostrils. Even when the soft palate is displaced, oral respiration is inefficient (and noisy) because it is impeded by the free margin of the soft palate in close proximity to the entrance of the larynx.

From dorsal to ventral, the soft palate consists of 4 layers: (1) nasopharyngeal mucosa, which is continuous with—and resembles—the respiratory mucosa of the nasal cavity; (2) aponeurosis and palatine muscles; (3) palatine glands, a layer approximately 1 cm (0.4 in.) thick in the rostral portion of the soft palate, with mucus-secreting glands that open into the oropharynx; and (4) oropharyngeal mucosa, which is continuous with, and resembles, the mucosa of the hard palate (Fig. 10.6-10).

The equine **nasopharynx** is a membranous and muscular tube. It is suspended rostrally from the **pterygoid** and **palatine** bones and is anchored caudally on portions of the **hyoid apparatus** and **laryngeal cartilages**. However, the

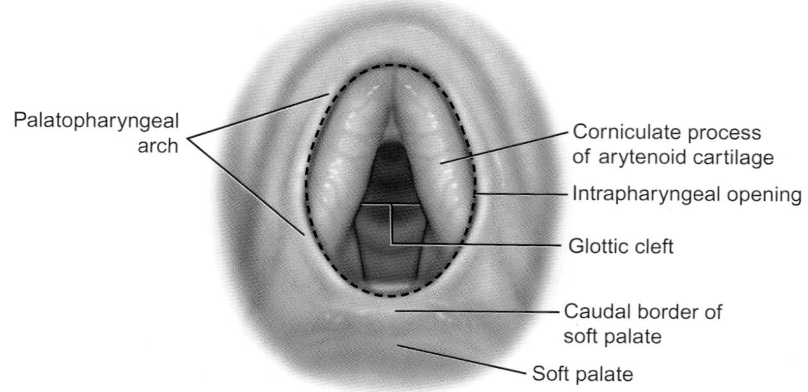

FIGURE 10.6-9 Schematic of dorsal displacement of the soft palate, showing the intrapharyngeal opening (outlined by *dashed line*) formed by the caudal border of the soft palate and the palatopharyngeal arches, which unite dorsally completing the ring of tissue. The epiglottis and corniculate processes of the arytenoid cartilages normally project through this opening, forming an airtight seal between the larynx and nasopharynx.

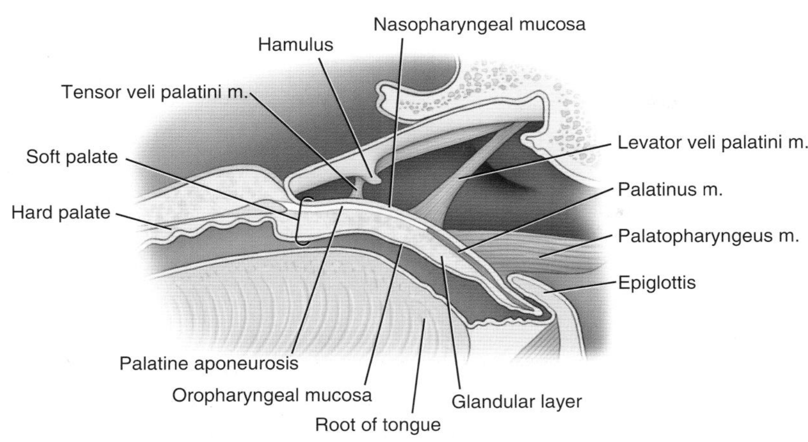

FIGURE 10.6-10 Diagram showing the layers of the soft palate and intrinsic muscles of the nasopharynx.

nasopharynx has no direct bone or cartilage support, so its musculature is particularly important for its proper function. The muscular component of the equine nasopharynx may be divided into intrinsic and extrinsic muscles.

> During strenuous exercise, airflow velocities and airway pressures fluctuate tremendously, so appropriate sensory and motor activity must occur to maintain airway patency and the normal relationship between the nasopharynx and larynx. Failure of one or more of these anatomic structures or neuromuscular activities results in a constellation of clinical disorders, ranging from a relatively benign airway noise to exercise intolerance.

Intrinsic muscles of the nasopharynx and their innervation

The intrinsic muscles insert onto the soft palate (which forms the floor of the nasopharynx) or onto the walls and roof of the nasopharynx (Figs. 10.6-10 and 10.6-11). Most contribute to the stability of the nasopharynx by tensing and expanding the floor and/or walls of the nasopharynx. All of the following muscles are paired (left and right).

The **palatinus** muscle is a fusiform (spindle-shaped) muscle that originates on the caudal aspect of the palatine aponeurosis (described below) and runs through the middle of the soft palate to insert near its caudal free edge. Some lateral fibers continue caudodorsally into the mucosal folds of the palatopharyngeal arches (sometimes referred to as the "pillars" of the soft palate). The palatinus is innervated by the **pharyngeal branch** of the **vagus nerve**. It functions to shorten the soft palate and depress it toward the tongue, thereby expanding the nasopharynx.

The **palatopharyngeus** muscle arises from the palatine aponeurosis (lateral to palatinus) and from the palatine and pterygoid bones. Its fibers run caudally in the lateral wall of the pharynx and insert onto the dorsal edge of the thyroid cartilage of the larynx, with some fibers continuing to insert onto the fibrous median raphe in the roof of the caudal pharynx. It, too, is innervated by the pharyngeal branch of the vagus nerve and functions to shorten the soft palate and depress it toward the tongue. The muscle also helps stabilize the walls of the nasopharynx.

The **tensor veli palatini** muscle is a fusiform muscle that originates on the muscular process of the tympanic part of the temporal bone, the pterygoid bone, and the lateral lamina of the auditory tube; it travels rostroventrally in the lateral wall of the nasopharynx. Its tendon courses around the hamulus (hook-like process) of the pterygoid bone and then expands into the **palatine aponeurosis**, which attaches to the caudal margin of the hard palate. The aponeurosis is thicker rostrally and thins caudally. The tensor veli palatini is innervated by the **mandibular branch** of the **trigeminal nerve**. It tenses the rostral aspect of the soft palate by using the hamulus as a pulley; it thus expands the nasopharynx by depressing the rostral soft palate during inspiration.

The **levator veli palatini** muscle arises from the muscular process of the tympanic part of the temporal bone and the lateral lamina of the auditory tube; it inserts within the soft palate, dorsal to the glandular layer. It is innervated by the pharyngeal branch of the vagus nerve. Unlike the preceding 3 muscles, it elevates the soft palate during swallowing, thereby collapsing the nasopharynx and, in particular, closing its caudal opening.

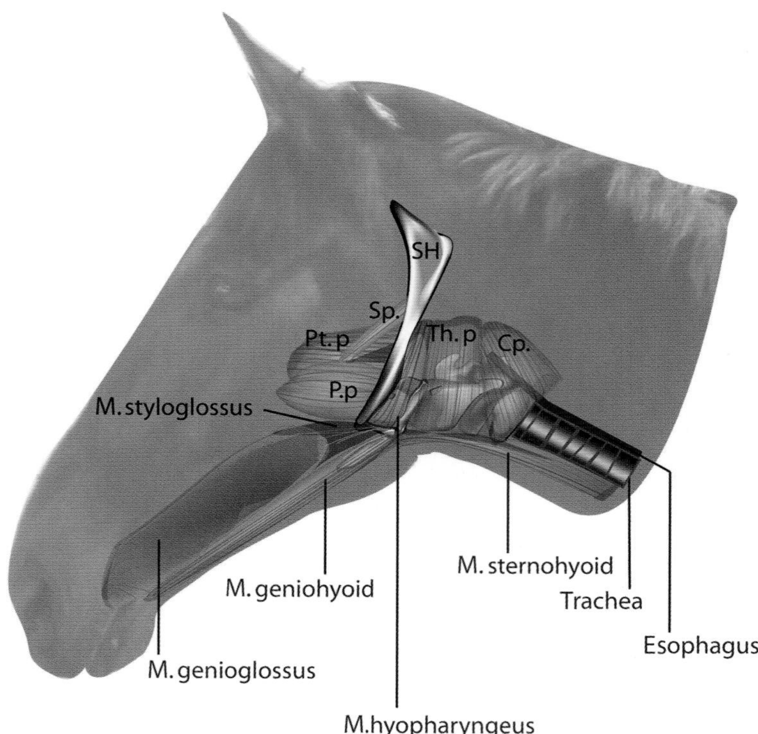

FIGURE 10.6-11 Diagram showing extrinsic muscles of the equine nasopharynx and those aiding swallowing. Key: *P.p*, Palatopharyngeus m.; *Pt.p*, Pterygopharyngeus m.; *Sp.*, Stylopharyngeus m.; *Th.p*, Thyropharyngeus m.; *Cp.*, Cricopharyngeus m.; *SH*, Stylohyoid bone.

The **stylopharyngeus** muscle has rostral and caudal components (Figs. 10.6-11 and 10.6-14A). The rostral component is a pharyngeal constrictor; it originates on the fibrous median raphe in the roof of the pharynx and inserts on the medial rostral (ventral) end of the stylohyoid bone (see Cases 10.3 and 10.7). It is innervated by the **glossopharyngeal nerve**. The caudal component is a pharyngeal dilator, providing tension on the roof of the nasopharynx to prevent its collapse. The caudal part originates on the medial aspect of the caudal (dorsal) third of the stylohyoid bone and courses ventrally and rostrally to attach on the dorsolateral wall of the pharynx.

Extrinsic muscles of the nasopharynx and their innervation

The extrinsic muscles (Figs. 10.6-8 and 10.6-11) are those that affect the function of the nasopharynx by increasing its diameter or changing the position of the larynx to facilitate its proper function. All of the following muscles are paired (left and right).

The **geniohyoideus muscle** originates with the genioglossus (see below) on the medial rostral aspect of the mandible and inserts on the basihyoid bone and its lingual process. It is innervated by the **hypoglossal nerve**. It draws the basihyoid rostrally and helps protrude the tongue.

The **genioglossus** muscle is a large, fan-shaped muscle that originates on the medial surface of the mandible, just caudal to the mandibular symphysis, and radiates throughout the tongue. Some of its fibers continue caudally to insert onto the basihyoid bone. It is innervated by the **hypoglossal nerve**. Its contraction depresses the tongue, particularly in its middle portion. The caudal fibers protrude the tongue, the middle fibers depress the tongue, and the rostral fibers retract the tip of the tongue. The fibers that extend to the basihyoid bone draw the basihyoid rostrally.

> The action of the genioglossus is somewhat reproduced by a "tongue tie." This simple device is a soft strap that is applied over the tongue and secured below the lower jaw to prevent movement of the tongue during exercise. It is sometimes used in the conservative management of intermittent DDSP, usually with a limited effect.

The **styloglossus** muscle lies on the lateral aspect of the tongue, originating on the lateral aspect of the stylohyoid bone and attaching at the apex of the tongue. It retracts the tongue. The **hyoglossus** muscle lies on the ventrolateral aspect of the tongue, and it inserts between the styloglossus laterally and the genioglossus medially; it originates from the basihyoid and thyrohyoid bones, and then fibers pass into root and body of the tongue, with some fibers extending to the apex. It retracts and depresses the tongue. Both the styloglossus and hyoglossus muscles are innervated by the **hypoglossal nerve**.

The **sternohyoideus** and **sternothyroideus** muscles are a pair of long, thin muscles that are located along the ventral aspect of the trachea, originating on the sternum and inserting on the hyoid apparatus and the larynx, respectively. They apply caudal traction to the hyoid apparatus and larynx. Specifically, both originate on the cartilage of the manubrium. The sternohyoideus inserts on the basihyoid bone and its lingual process, while the sternothyroideus inserts on the caudal abaxial aspect of the lamina of the thyroid cartilage of the larynx. Both muscles receive their motor innervation from the **ventral branches of the 1st and 2nd cervical nerves**. These muscles are colloquially known as the "strap" muscles. The **omohyoideus** is a third "strap" muscle. Unlike the previous 2 it originates from the subscapular fascia adjacent to the shoulder joint. It inserts onto the basihyoid bone and its lingual

The modified Forssell's procedure for the management of cribbing in horses involves the removal of large sections of the sternothyroideus, sternohyoideus, and omohyoideus muscles from the cranial ventral neck region. Rather than also removing a large portion of the sternocephalicus muscle, as was directed in the initial technique, the modified procedure describes doing a neurectomy of the nerve supplying this muscle. Therefore, an approximately 10 cm (3.9 in.) section of the ventral branch of the accessory nerve (CN XI) is isolated and excised, just proximal to where the nerve enters the muscle. The modified procedure is cosmetically more acceptable and easier to perform. Cribbing is well-recognized as a behavioral vice in which a horse bites down on a structure with its upper incisors (often a fence rail or stall door) then ventroflexes its head and neck, retracts its larynx, and pulls backwards. Chronic cribbing is associated with an increased risk of colic (due to aerophagia [Gr. *aēr-* air + Gr. *phagein* to eat]; excessive swallowing of air resulting in abdominal distention) and excessive rates of incisor wear. If conservative treatments are unsuccessful (e.g., environmental management, use of a cribbing strap), surgery is recommended. Surgery removes or denervates the primary muscles involved in the act of cribbing, and prognosis for resolution of the vice following surgery is considered good.

ANATOMY AND FUNCTION OF DEGLUTITION

Oral, pharyngeal, and esophageal phases of swallowing occur in sequence, with the oral cavity, pharynx, larynx, and esophagus functioning to move food and liquid from the oral cavity to the esophagus, while protecting the airway.

Sensory and motor innervation of the structures involved is extensive. The intrinsic and extrinsic muscles of the nasopharynx play key roles in the process. The mucosa of the soft palate, lateral naso pharyngeal walls, and dorsal naso pharynx is innervated by sensory branches of the glossopharyngeal nerve. The mucosa covering the epiglottis, arytenoids, and piriform recesses of the laryngopharynx receives sensory innervation from the internal branch of the cranial laryngeal nerve, a branch of the vagus nerve (Fig. 10.6-12).

Progressive and rapid pharyngeal contraction during swallowing occurs when the pharyngeal constrictors—including the palatopharyngeus and pterygopharyngeus muscles (cranial group), hyopharyngeus muscle (middle), and thyropharyngeus and cricopharyngeus muscles (caudal group)—actively contract (Fig. 10.6-11).

The soft palate moves dorsally during swallowing (primarily due to the action of the levator veli palatini) to seal the caudal opening to the nasopharynx and therefore prevent food and liquid passing retrograde into that part of the airway. The tensor veli palatini contracts to stiffen the rostral soft palate, and this muscle also opens the pharyngeal openings of the auditory tubes (guttural pouches) in the horse (Fig. 10.6-13).

The larynx is protected by retroversion of the epiglottis, sealing the entrance to the larynx, and by adduction of the arytenoid cartilages, sealing the glottic cleft (*rima glottidis*). At the end of the swallowing sequence, contraction of the hyoepiglotticus muscle aids return of the epiglottis to its normal position dorsal to the soft palate. Contraction of the right and left stylopharyngeus muscles lifts the pharyngeal roof dorsally (Fig. 10.6-14A), and this can be seen endoscopically as 2 concavities in the dorsal naso pharyngeal wall (Fig. 10.6-14B).

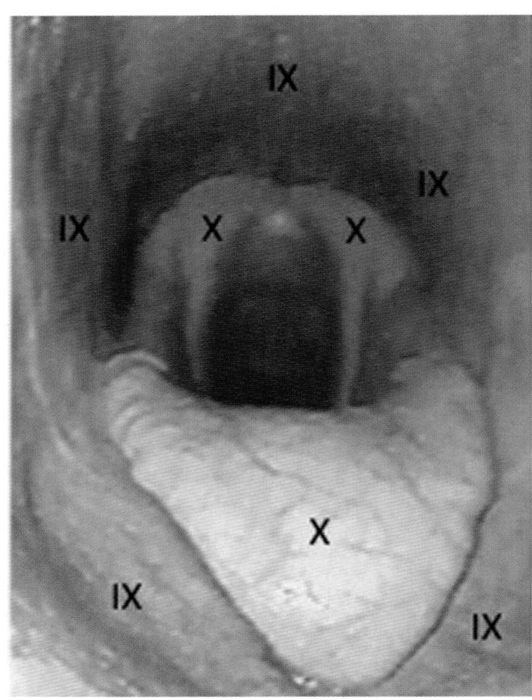

FIGURE 10.6-12 Endoscopic view of the equine nasopharynx and larynx demonstrating sensory innervation of the soft palate, naso pharyngeal walls, and dorsal naso pharynx by CN IX (glossopharyngeal nerve). The sensory innervation of the epiglottis and arytenoids is from CN X (vagus, via the internal branch of the cranial laryngeal nerve).

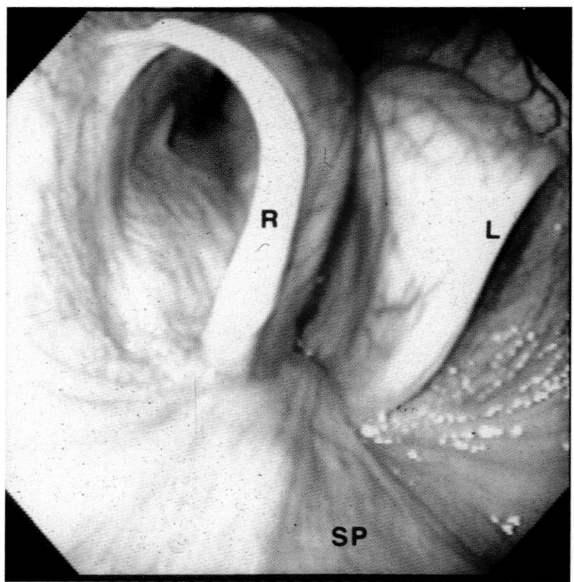

FIGURE 10.6-13 Contraction of the tensor veli palatini muscle causes stiffening of the rostral soft palate and dilation of the pharyngeal openings of the auditory tubes. Key: *R*, right; *L*, left; *SP*, soft palate.

process along with sternohyoideus, and it too retracts the hyoid bone; it is supplied by the **ventral** (motor) **branch** of the **1st cervical nerve**.

The **hyoepiglotticus** muscle is a small muscle that, together with its elastic sheath, connects the basihyoid bone and the ventral aspect of the epiglottic cartilage near its base. It is innervated by the **hypoglossal nerve**. When it contracts, the hyoepiglotticus m. draws the basihyoid bone and epiglottis closer together and pulls the epiglottis

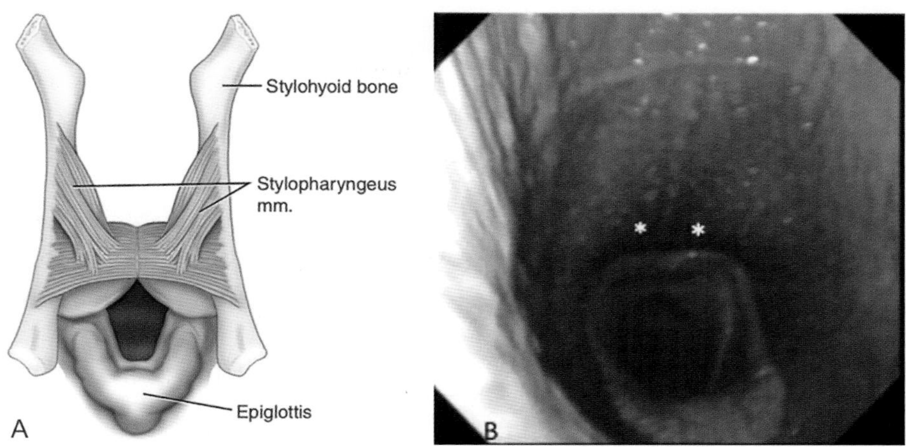

FIGURE 10.6-14 (A) Schematic diagram of the location and attachment of the stylopharyngeus muscles to the dorsal aspect of the naso pharynx. (B) Contraction of the stylopharyngeus is evident during endoscopic examination as two concavities (**) in the dorsal pharyngeal wall.

ventrally, toward the base of the tongue, thereby increasing the ventral dimension of the entrance to the larynx (*aditus laryngis*).

The **thyrohyoideus** muscle extends from the lateral lamina of the thyroid cartilage of the larynx to the caudal aspect of the thyrohyoid bone. It is considered to be innervated by the **hypoglossal nerve** and/or the **1st cervical nerve**. Depending on the actions of the other muscles that attach to the hyoid apparatus or the larynx, it either draws the larynx rostrally or the hyoid apparatus caudally. It enhances soft palate stability during exercise when it moves the larynx rostrally.

Blood supply and lymphatics

The main blood supply to the muscles of the pharynx and soft palate derive from the **common carotid**, **external carotid**, **linguofacial trunk**, and **maxillary arteries** with their corresponding veins draining the region. The **cranial deep cervical, retropharyngeal, and mandibular lymph nodes** serve as the collecting system for the lymphatic vessels. Innervation is discussed along with the musculature earlier in this case.

Selected references

[1] Barakzai SZ, Johnson VS, Baird DH, Bladon B, Lane JG. Assessment of the efficacy of composite surgery for the treatment of dorsal displacement of the soft palate in a group of 53 racing thoroughbreds (1990–1996). Equine Vet J 2004;36(4):175–9.

[2] Smith JJ, Embertson RM. Sternothyroideus myotomy, staphylectomy and oral caudal soft palate photothermoplasty for treatment of dorsal displacement of the soft palate in 102 thoroughbred racehorses. Vet Surg 2005;34(1):5–10.

[3] Cheetham J, Pigott JH, Hermanson JW, Campoy L, Soderholm LV, Thorson LM, Ducharme NG. Role of the hypoglossal nerve in equine nasopharyngeal stability. J Appl Physiol 2009;107:471–7.

[4] Allen KJ, Lane JG, Woodford NS, Franklin SH. Severe collapse of the rostral soft palate as a source of abnormal respiratory noise in six ponies and horses. Equine Vet J 2007;39:562–6.

[5] Holcombe SJ, Derksen FJ, Stick JA, Robinson NE. Effect of bilateral tenectomy of the tensor veli palatini muscle on soft palate function in horses. AJVR 1997;58(3):317–21.

CASE 10.7

Laryngeal hemiplegia

Eric Parente
Department of Clinical Studies - New Bolton Center, University of Pennsylvania School of Veterinary Medicine, Kennett Square, Pennsylvania, US

Clinical case

History

A 3-year-old male Thoroughbred racehorse was presented with a history of decreased performance and increasingly loud respiratory noise during exercise. He reportedly started the race well, but "had no finish." A veterinarian performed an endoscopic examination and thought he had a "weak flap."

Physical examination findings

General physical examination findings were normal, including rebreathing evaluation of the lungs. Airflow through both nostrils was symmetrical. Laryngeal palpation revealed asymmetry in which the left muscular process of the arytenoid cartilage was more easily palpable than the right.

Differential diagnoses

Recurrent laryngeal neuropathy (RLN), arytenoid chondropathy, laryngeal dysplasia, intermittent dorsal displacement of the soft palate, and medial deviation of the aryepiglottic folds

Diagnostics

Resting endoscopic examination of the upper airway was performed without sedation, using a nose twitch for restraint. The examination revealed a normal, symmetrical structural appearance of the arytenoid cartilages. There was movement of both arytenoids, but when swallowing was stimulated by flushing water through the operating portal of the endoscope, the colt was unable to abduct the left arytenoid fully as compared with the right arytenoid (Fig. 10.7-1A and B).

An ultrasound examination of the larynx was performed using an 8- to 10-MHz linear transducer, comparing the right and left sides of the larynx. The lateral margin of the arytenoid was bilaterally smooth and similar, and the only difference between both sides was that the left cricoarytenoideus lateralis muscle was more echogenic than the right (Fig. 10.7-2).

An over-ground exercise endoscopy (a.k.a. "dynamic endoscopy"; see Section 2.1/Endoscopy) was performed with the horse breezing 5 furlongs. During the entire test period, the right arytenoid was fully abducted. In the early part of the test, the left arytenoid was 70–80% abducted and stable, but toward the end of the exercise test, this arytenoid became increasingly unstable, dynamically collapsing on inspiration (Fig. 10.7-3) (see side box entitled "Assessment of laryngeal function"). In addition, an increase in inspiratory noise was heard in the latter stages of the test. There were no other abnormalities noted.

Comparative Veterinary Anatomy: A Clinical Approach. https://doi.org/10.1016/B978-0-323-91015-6.00143-6

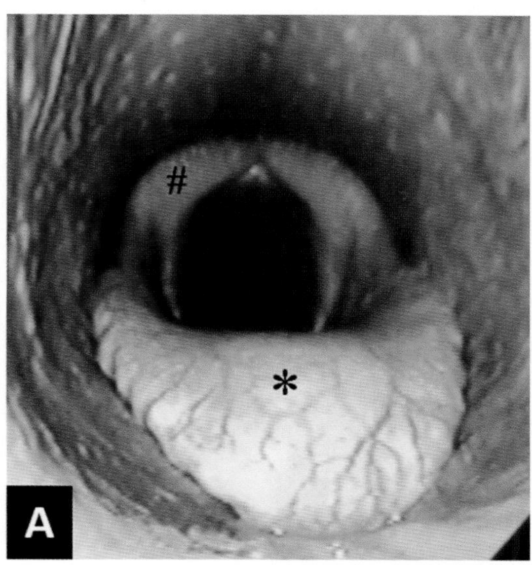

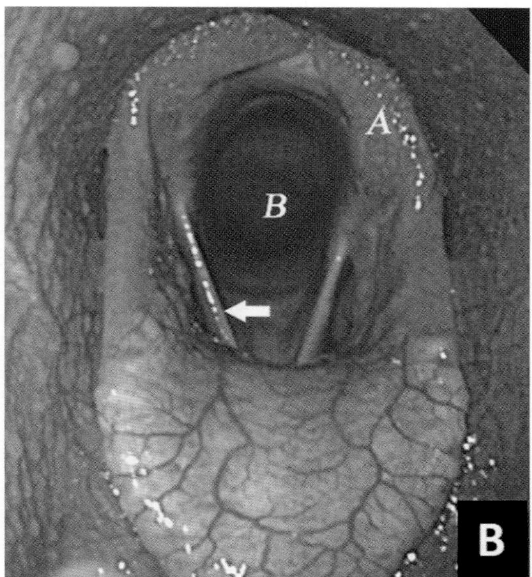

FIGURE 10.7-1 (A) Endoscopic view of a normal larynx with full symmetric abduction of both arytenoids. Key: *#*—right arytenoid, corniculate process; ***—epiglottis. (B) Endoscopic view of the larynx of a horse with an incomplete abduction of the left arytenoid. Key: *A*—left arytenoid, corniculate process; *B*—location of the glottis, which includes the glottic cleft (space) bound by the arytenoid cartilages and vocal cords; *arrow*—right vocal cord.

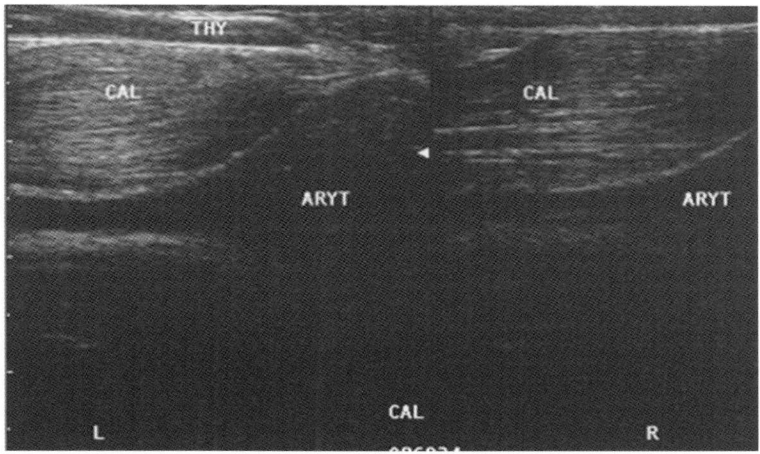

FIGURE 10.7-2 Ultrasonographic image of right and left cricoarytenoideus lateralis (*CAL*) muscles showing an increased echogenicity of the left CAL muscle. Key: *THY*—thyroid cartilage; *ARYT*—arytenoid cartilage.

Diagnosis

Recurrent laryngeal neuropathy (a.k.a. left laryngeal hemiplegia)

> Recurrent laryngeal neuropathy is an idiopathic neuropathy that almost uniformly and clinically affects the left side and rarely the right. This results in a weak cricoarytenoideus dorsalis muscle that is unable to attain full abduction of the arytenoid at rest, and unable to maintain even partial abduction under moderate to intense respiratory stress or exercise.

Treatment

Laryngoplasty and left vocal cordectomy were recommended. The colt was placed under general anesthesia with right nasotracheal intubation and placed in right lateral recumbency. Transendoscopic laser (laser is an acronym for light amplification by stimulated emission of radiation) resection of the left vocal cord was first performed. Then, a lateral approach was made to the larynx, ventral to the linguofacial vein (Fig. 10.7-5A).

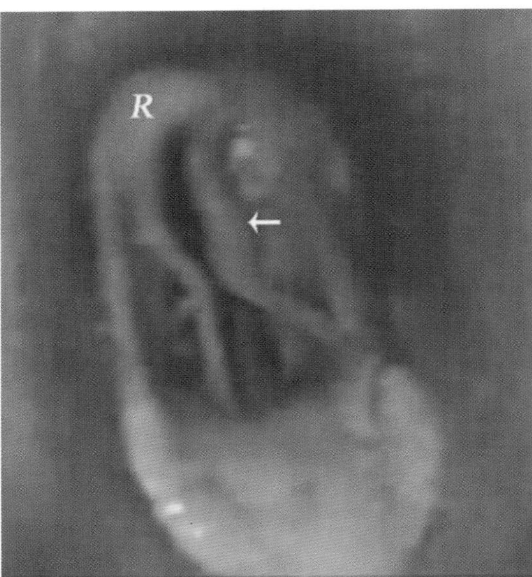

FIGURE 10.7-3 Endoscopic image of collapsed left arytenoid during over-ground endoscopic examination. Key: *R*—right arytenoid maintaining full abduction; *arrow*—collapsed left arytenoid obstructing the glottic cleft.

The cricopharyngeus muscle (see side box entitled "Laryngoplasty") was retracted rostrally to expose the muscular process of the left arytenoid cartilage, lying dorsal to the dorsal margin of the thyroid cartilage and rostral to the cricoarytenoideus dorsalis (CAD) muscle. The left CAD muscle was markedly atrophied. The attachment of the CAD muscle to the muscular process was transected, exposing the lateral capsule of the cricoarytenoid joint, and the joint was then opened (Fig. 10.7-5B and C). The cartilage of the articular surfaces was debrided to facilitate ankylosis (Gr. *ankylōsis* joint consolidation or fusion) of the joint for increased stability. Two separate suture loops were placed through the dorsal caudal edge of the cricoid cartilage and then passed through the muscular process of the arytenoid to mimic the position of the CAD muscle and act as prosthetic replacements. With the endoscope positioned within the nasopharynx to view the larynx, the suture loops were individually tightened and knotted to pull the arytenoid to approximate a 90% abduction. The soft tissues were closed anatomically, and the horse recovered uneventfully from general anesthesia.

Post-operative endoscopic examination the following day revealed approximately 80% fixed abduction of the left arytenoid, minimal upper airway inflammation, and no evidence of dysphagia (Fig. 10.7-4). After 4 weeks of stall rest and hand-walking, recheck endoscopic examination confirmed good abduction of the left arytenoid; the horse returned to race successfully.

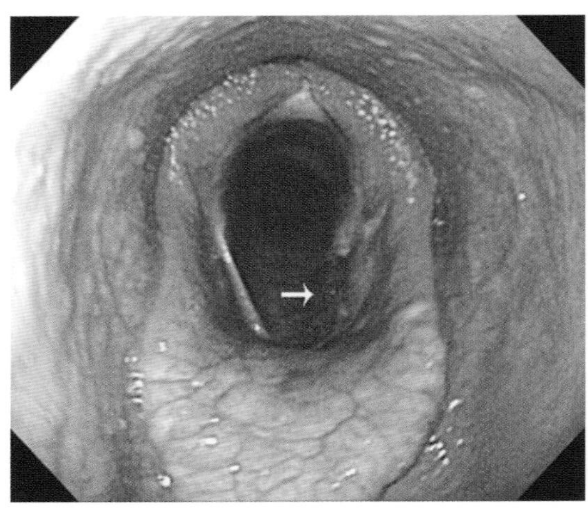

FIGURE 10.7-4 Post-operative endoscopic view after laryngoplasty. *Arrow*—site of resected left vocal cord.

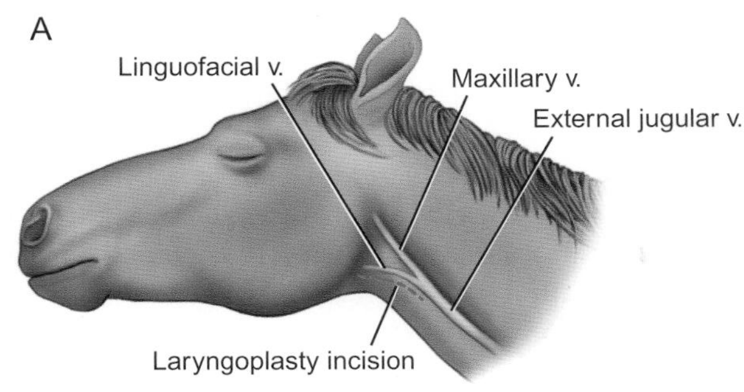

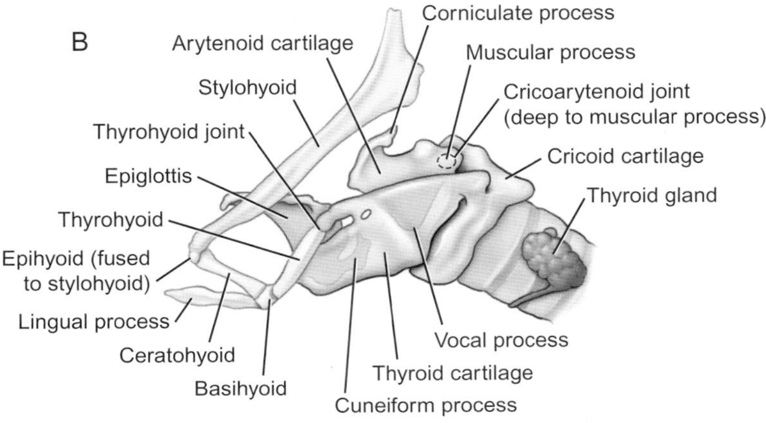

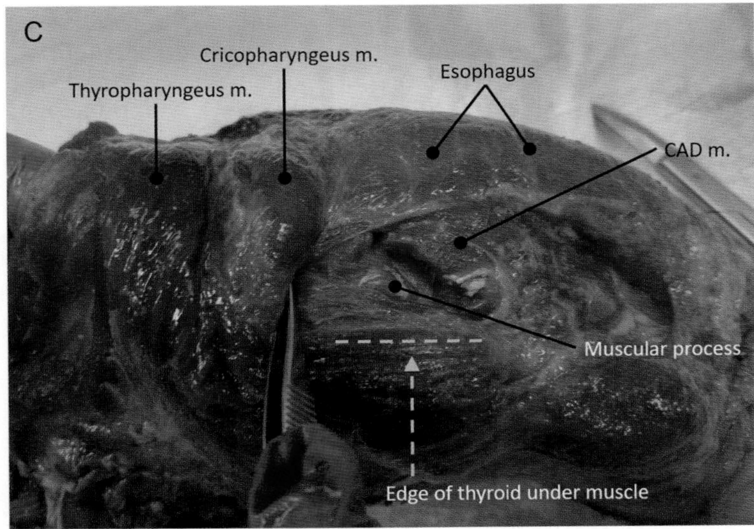

FIGURE 10.7-5 (A) Left lateral aspect of head demonstrating the location of the laryngoplasty incision ventral to the linguofacial vein. (B) Left schematic of laryngeal cartilages and hyoid apparatus with soft tissues removed. (C) Left dorsolateral view of larynx with the cricopharyngeal muscle retracted rostrally by the tissue forceps. The cricoarytenoideus dorsalis (CAD) muscle has been transected at its insertion on the muscular process of the arytenoid cartilage. Deep to this process is the cricoarytenoid joint.

Anatomical features in equids

Introduction

The larynx consists of 4 cartilages: the epiglottis, paired left and right arytenoids, thyroid, and cricoid. These cartilages are connected and stabilized through muscles, joints, and ligaments, with respiratory mucosa lining the lumen of the larynx. The narrowest portion of the larynx is the glottis, which consists of the rima glottidis (or glottic cleft) bound on each side by the vocal fold (clinically referred to as the vocal cord) and arytenoid cartilage. All the intrinsic muscles of the larynx, except the cricothyroideus muscle, are innervated by branches of the caudal laryngeal nerve, which is the terminal nerve of the recurrent laryngeal nerve. The primary abductor of the arytenoid cartilage is the CAD muscle. This case covers anatomy of the larynx along with its blood supply, lymphatics, and innervation. Nasopharyngeal anatomy is covered in Case 10.6.

Function

The larynx is the anatomical structure between the pharynx and the trachea. The arytenoids of the larynx should open symmetrically and completely during intense exercise—i.e., maximal opening of the glottis (Fig. 10.7-1A)—and close completely during swallowing to prevent food and liquids from entering the airway as they are moved from the oropharynx through the laryngopharynx and into the esophagus. The larynx also functions to produce vocal sounds.

LARYNGOPLASTY

Traditionally, the muscular process of the arytenoid is approached by separating the thyro- and cricopharyngeus muscles. The cranial laryngeal nerve travels between these muscles and can theoretically be injured when dissecting in this region. Thus, approaching the muscular process from caudal to the cricopharyngeus muscle avoids this potential complication. Furthermore, when the suture is passed from the cricoid to the muscular process, an approach between the muscle bellies requires that the suture is passed blindly under the cricopharyngeus muscle and the suture could twist. By approaching the muscular process from caudal to the cricopharyngeus muscle, the sutures can be seen passing in their entirety from the cricoid to the muscular process and confirmed to be positioned correctly and straight.

Suture placement risks

The suture passing through the cricoid cartilage can enter the lumen of the larynx and result in post-operative complications. The esophagus is directly dorsal to the midline of the cricoid cartilage and must be moved away from the suture needle so as not to be inadvertently penetrated. Suture passing into the esophagus may cause problems with swallowing, esophageal dysfunction, and/or infection.

ASSESSMENT OF LARYNGEAL FUNCTION

Laryngeal function is best assessed by upper airway endoscopy. At rest, the movement of the arytenoids should be comparatively synchronous. Most importantly, full symmetric abduction and adduction should be achieved after a stimulated swallow or during nasal occlusion. Inability of (typically) the left side to abduct an arytenoid symmetrically and completely can be consistent with RLN, dysplasia, or arytenoid chondropathy (Fig. 10.7-6A and B). Differentiating among the 3 is based on whether a structural or functional abnormality is causing the decreased abduction, and it is critical to obtain an accurate diagnosis so the correct treatment can be performed.

The degree of laryngeal motion has been graded on multiple established scales. The most common is a 4-level system in which the 4th grade indicates a complete inability of an arytenoid to move and a 3rd level means compromised mobility. A horse with a grade 3 larynx may be able to maintain abduction under low-level exercise and not have a clinical problem. Yet, the horse with a grade 3 larynx under more strenuous exercise, such as racing, would likely be unable to maintain abduction and therefore have some respiratory compromise. An exercising endoscopic examination should be performed if there is a question of what is happening during exercising conditions.

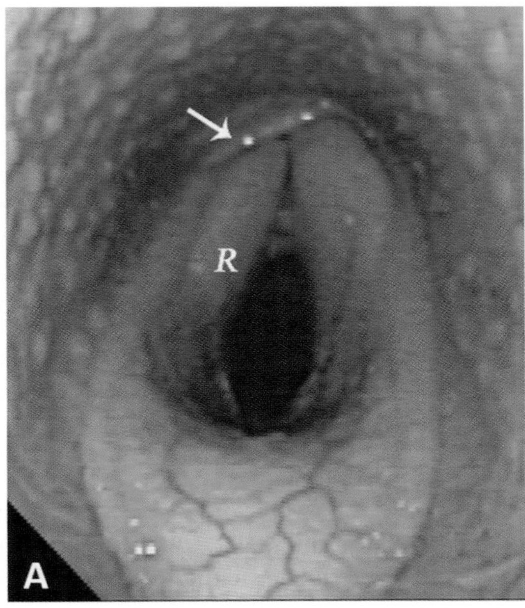

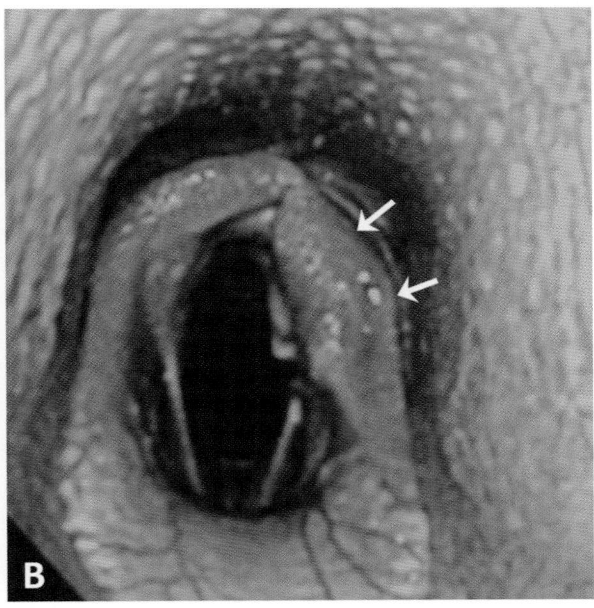

FIGURE 10.7-6 (A) Laryngeal dysplasia. *Arrow*—palatopharyngeal arch visible over the non-abducted (*R*) right arytenoid. (B) Arytenoid chondropathy. *Arrows* are pointing to abnormally thickened and non-abducted left arytenoid, causing narrowing of the glottis.

Laryngeal cartilages

The **epiglottis** is a single, flat, leaf-shaped cartilage. Its base attaches to the **thyroid cartilage** via the **thyroepiglottic ligament** at the ventral glottis, and the tip of the epiglottis extends rostrally. The paired **cuneiform processes**, which are found on the **arytenoid cartilage** in the canine, in the horse project dorsally from the caudolateral corners of the base of the epiglottis (Fig. 10.7-5B). The paired **vestibular folds** envelop the cuneiform processes. Another paired mucosal fold, the **aryepiglottic fold**, connects the epiglottis to the arytenoid cartilage on each side. Horses are obligate nasal breathers and should always have the epiglottis positioned above the free edge of the caudal border of the soft palate on endoscopic examination.

The aryepiglottic fold may collapse into the opening of the larynx during high-intensity exercise, causing respiratory obstruction and exercise intolerance. This condition is known as "medial deviation of the aryepiglottic folds" and is diagnosed during exercising endoscopy.

In some horses, the soft palate moves above the epiglottis during intense exercise, a condition known as "intermittent dorsal displacement of the soft palate" (see Case 10.6).

The arytenoid cartilages are paired with a large body and 3 major processes in the horse: muscular, corniculate, and vocal (Fig. 10.7-5B). The **corniculate process** is the large, rostrally located process covered by stippled respiratory mucosa and is plainly evident on endoscopic examination (Fig. 10.7-1A). The **vocal process** at the ventral edge of the body is the dorsal point of attachment for the **vocal cord** (see next section for more information on the vocal cord). The vocal cord attaches ventrally to the thyroid cartilage on the midline of the glottis. The **muscular process**, not seen on endoscopy, is just dorsal to the articulation of the arytenoid cartilage with the cricoid. It is

Arytenoid chondropathy (Fig. 10.7-6B) is pathology of one or both arytenoids—typically a sequela of infection. This disease can mimic RLN if the chondropathy is mild. Laryngeal dysplasia (Fig. 10.7-6A) is a congenital abnormality of the laryngeal cartilages that can also mimic RLN because of the inability to fully abduct the arytenoid; however, this is a congenital malformation of multiple laryngeal cartilages (see side boxes "Arytenoid chondropathy or chondritis" and "Laryngeal dysplasia").

A laryngotomy, to enter the lumen of the larynx, is performed by incising the cricothyroid ligament (see side box "Laryngotomy").

the attachment site for the CAD muscle and **cricoarytenoideus lateralis** (CAL) muscle. The left and right arytenoids form individual saddle joints with the dorsal rostral aspect of the cricoid cartilage (Fig. 10.7-5B). This acts as the "hinge" for the arytenoid to abduct or adduct around. Contraction of the CAD muscle results in abduction, and contraction of the CAL muscle results in adduction, of the arytenoid cartilages.

The **thyroid** cartilage is boat hull-shaped with left and right **lamina** (plates) that project dorsally from their ventrorostral midline fusion along the lateral side of the arytenoid cartilages (Fig. 10.7-5B). Caudodorsally, the thyroid lamina articulates with the cricoid cartilage. Ventrally, there is a deep, "V-shaped" caudal thyroid notch between the lamina, spanned by the **cricothyroid ligament**. This ligament attaches caudally to the ventral rostral margin of the cricoid cartilage. Near the rostral dorsal margin of the thyroid, the **thyrohyoid bone** of the hyoid apparatus articulates with the thyroid lamina (Fig. 10.7-5B).

ARYTENOID CHONDROPATHY OR CHONDRITIS

This can imitate RLN since it usually causes a decreased abduction of the affected arytenoid. The important difference is that the abduction is inhibited by a structural change in the arytenoid cartilage that prevents abduction vs. with RLN the structure is normal and the abduction is limited by an inability of the muscle to abduct the cartilage or maintain abduction.

The structural changes with arytenoid chondropathy can usually be appreciated on upper airway endoscopy but the severity is quite variable and infrequently can be misinterpreted as RLN. Beyond decreased abduction, abnormalities to look for with chondropathy are any differences in the shape or size of the corniculate process, mucosal disruptions and/or granulation tissue on the medial (luminal) surface of the arytenoid, and displacement of the palatopharyngeal arch lateral to the corniculate process (Fig. 10.7-6B). If there is any question on upper airway endoscopy, ultrasound examination of the larynx can be helpful in making the distinction. Ultrasound typically detects an irregular margin in the lateral body of the arytenoid and normal echogenicity of the CAL muscle associated with chondropathy vs. a regular lateral margin of the arytenoid body and increased echogenicity of the CAL muscle with RLN.

LARYNGEAL DYSPLASIA

This can have an endoscopic appearance similar to RLN in that a single arytenoid may not be able to achieve full abduction. However, most often the right side is affected, which is an important sign that the horse does not have RLN, and often the structure of the arytenoids is not symmetrical. Occasionally the palatopharyngeal arch is draped over the apex of the affected arytenoid (Fig. 10.7-6A). Palpation of the larynx does not reveal a prominent muscular process because of the deformation of the cartilage structures and often because the wing of the thyroid is displaced dorsally, covering part of the muscular process. Ultrasound or MRI can be used to recognize the structural abnormalities associated with dysplasia, giving a definitive diagnosis.

The **cricoid** (Gr. *krikos* ring + *eidos* form) cartilage forms a complete ring at the caudal aspect of the larynx and is attached to the first tracheal cartilage by the **cricotracheal ligament**. The cricoid provides lumen size and stability, and articulates with the arytenoid and thyroid cartilages. This cartilage is much wider dorsally (called the "lamina") and narrower laterally and ventrally (the "arch") (Fig. 10.7-5B). The lamina is divided by a median crest, which may be palpated at surgery as a useful landmark.

Vocal cord and laryngeal ventricle

The clinical term, **vocal cord**—i.e., the vocal fold or *plica vocalis*—is a soft tissue structure that originates on the vocal process of each arytenoid, and attaches to the thyroid cartilage on the ventral midline of the glottis (Figs. 10.7-1B and 10.7-8D). The cord consists of a **vocal ligament** and **vocalis muscle** covered by respiratory mucosa. The cord forms the medial wall of the **laryngeal ventricle** (clinically referred to as the "saccule"). The laryngeal ventricle is a space lined with mucosa extending laterally to the vocal cord and dorsally between the lateral side of the arytenoid body and the medial side of the thyroid lamina. Endoscopically, the entrance to the laryngeal ventricle is identified as the opening between the vestibular fold and the vocal cord (Fig. 10.7-1B).

When treating RLN, the ipsilateral vocal cord is commonly removed (a.k.a. "vocal cordectomy," Fig. 10.7-4) in addition to the laryngoplasty procedure, since only partial fixed abduction of the arytenoid cartilage is created with laryngoplasty. With laryngoplasty alone, the vocal cord remains prone to being dynamically pulled into the airway during strenuous exercise. Some clinicians advocate removing the lining of the saccule ("sacculectomy" or "ventriculectomy") along with the cord (the combined surgery is a "ventriculocordectomy") in order to create additional fibrous stability of the "tied-back" arytenoid.

During vocalization, the arytenoids are adducted, moving the vocal cords into the airway to create sound. During normal, full inspiration, the arytenoids should be completely abducted; thus, the vocal cords are pulled flat against the lateral side of the glottis, and the ventricle is collapsed.

Pharyngeal musculature attached to the larynx

The **cricopharyngeus** and **thyropharyngeus muscles** are paired muscles arising from the cricoid and thyroid cartilages, respectively. The left and right muscles meet on midline dorsal to the larynx and laryngopharynx. These muscles form the caudal muscular roof of the laryngopharynx, acting as constrictors of the pharyngeal lumen, and therefore functioning as a cranial esophageal sphincter. The esophagus begins at the caudal margin of the laryngopharynx—directly dorsal to the cricoid cartilage (Fig. 10.7-5C).

Laryngeal innervation

The **left** and **right recurrent laryngeal nerves** provide the clinically important innervation to the larynx. Each nerve originates from the ipsilateral **vagus nerve** (CN X). The left and right vagus nerves travel caudally to the thorax from their origin in the myelencephalon division of the brain stem. On each side of the neck, the vagus nerve and the cranially passing sympathetic trunk travel together as the vagosympathetic trunk, and are enclosed in the carotid sheath with the common carotid artery. The left recurrent laryngeal nerve branches from the vagus nerve at the heart base and wraps caudally around the aortic arch, whereas the right recurrent laryngeal nerve leaves the vagus nerve more cranially, passing around the right subclavian artery. On its respective side of the neck each nerve returns cranially to the larynx (passing within the carotid sheath).

Because of the different pathways of the recurrent laryngeal nerves in the thorax, the left recurrent laryngeal nerve is longer than the right. This is one explanation as to why the left nerve is clinically affected more often than the right. In left-sided, clinically affected horses, histologic examination of the right CAD muscle also shows evidence of denervation, but not as significant as is seen on the left.

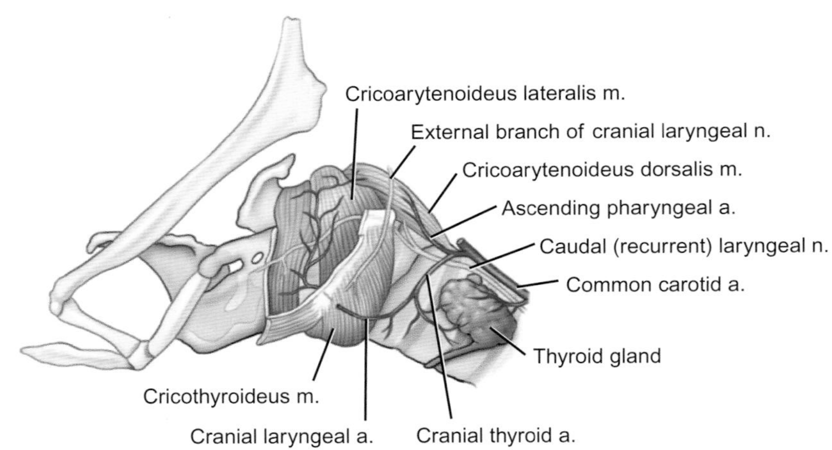

FIGURE 10.7-7 Schematic of laryngeal vasculature and innervation as viewed from the left side.

Therefore, RLN is reflected in changes in muscles beyond the CAD muscle. This is clinically important because the CAD muscle cannot be imaged ultrasonographically, but the CAL muscle can (as in this case). Evidence of increased echogenicity of the CAL muscle is an indication of RLN (Fig. 10.7-2).

The recurrent laryngeal nerve terminates as the **caudal laryngeal nerve** at the caudal border of the cricoid cartilage, and it innervates all the intrinsic muscles of the larynx except the cricothyroideus (which is supplied by the **cranial laryngeal nerve**, a branch of the vagus) (Fig. 10.7-7).

Laryngeal blood supply and lymphatics

The **common carotid artery** passes immediately dorsolateral to the larynx, bilaterally, and is palpable within the surgical site for laryngoplasty. Associated with the common carotid artery are the **vagosympathetic trunk** and the terminal portion of the recurrent laryngeal nerve. The **cranial thyroid artery** branches off the common carotid artery 1–3 cm (0.4–1.2 in.) caudal to the cricoid cartilage, and is the primary supply to the larynx and thyroid gland (Fig. 10.7-7).

The **ascending pharyngeal artery** branches early from the cranial thyroid artery and passes over the dorsal aspect of the CAD muscle. After supplying branches to the thyroid gland, the cranial thyroid artery continues ventrally and rostrally as the **cranial laryngeal artery** supplying branches to the larynx. The bilobed **thyroid gland** is often palpable adjacent to the trachea, as bilateral, freely moveable soft tissue masses, caudal to the larynx.

The **lateral** and **medial retropharyngeal lymph nodes** serve as the principal lymphatic collecting centers for the larynx.

LARYNGOTOMY

A laryngotomy is a surgical approach between the thyroid and cricoid cartilages that allows access to the lumen of the larynx and surrounding structures. After a linear incision on ventral midline through the skin, the paired sternohyoideus muscles are separated to expose the cricothyroid ligament (Fig. 10.7-8A and B). The ligament and underlying mucosa are incised in a caudal-to-rostral direction to enter the lumen of the larynx (Fig. 10.7-8C). The laryngotomy approach can be used to remove the vocal cord, the laryngeal ventricle, subepiglottic tissues associated with epiglottic entrapments, or diseased arytenoid cartilage, as in the case of arytenoid chondropathy (Fig. 10.7-8D). Because the lumen of the larynx is not sterile, the skin is generally left open to heal by second intention at the conclusion of the procedure.

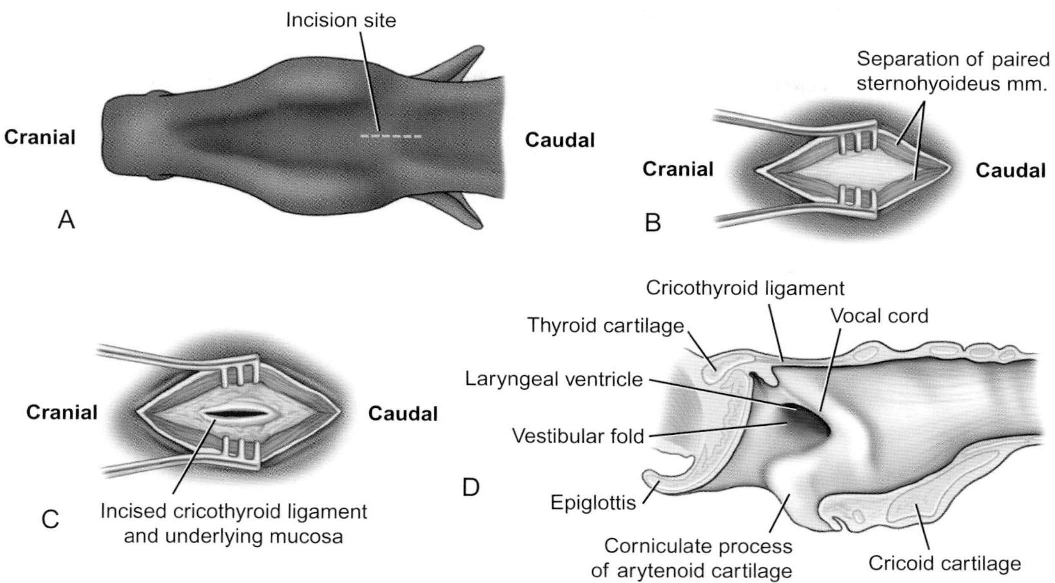

FIGURE 10.7-8 (A) Schematic of ventral head and neck of horse, indicating site of laryngotomy incision. Cranial is to the left. (B) The paired sternohyoideus muscles are separated on midline to expose the underlying cricothyroid ligament, coved by adipose tissue in this image. (C) Following exposure of the cricothyroid ligament, it is incised from caudal to cranial, along with the underlying mucosa to enter the laryngeal lumen. (D) Sagittal view of the lumen of the larynx indicating structures viewed via a laryngotomy through the cricothyroid ligament. Cranial is to the left and the dorsal aspect of the larynx is ventral, simulating the view with a horse in dorsal recumbency.

Selected references

[1] Parente EJ, Birks EK, Habecker P. A modified laryngoplasty approach promoting ankylosis of the cricoarytenoid joint. Vet Surg 2011;40(2):204–10.

[2] Brandenberger O, Martens A, Robert C, et al. Anatomy of the vestibulum esophagi and surgical implications during prosthetic laryngoplasty in horses. Vet Surg 2018;47(7):942–50.

[3] Derksen F, Parente EJ, Tessier C, Franklin SH, Chalmers HJ. Diagnostic techniques in equine upper respiratory tract disease. In: Auer J, Stick J, editors. Equine surgery. 4th ed. Philadelphia: Elsevier; 2012. p. 536–56.

CHAPTER 10

CASE 10.8

Vestibular disease

William Gilsenan
Internal Medicine, Rood and Riddle Equine Hospital, Lexington, Kentucky, US

Clinical case

History

A 7-month-old Thoroughbred weanling colt was presented following an acute onset of neurologic signs. Earlier in the day, the colt was apparently normal when being led outside to a paddock. While being led, the colt became startled and reared up. He fell backward and landed on his poll. Following this incident, he was able to stand, but was unsteady on his feet and a head tilt was observed. Two hours after the incident, the colt was noted to have blood coming from his nose and mouth, which prompted referral.

Physical examination findings

On an initial evaluation, the colt was alert but agitated. The colt was mildly tachycardic (52 bpm) with normal respiratory rate and effort and normal rectal temperature. Mucous membranes were pink and moist with a normal CRT. Dried blood was present at both nasal openings, but airflow was symmetric and subjectively normal in volume through both nasal passages. No upper respiratory noises were observed. A marked symmetric, firm swelling was present in the intermandibular region and on the ventral aspect of the cranial third of the neck.

A marked right-sided head tilt was present (Fig. 10.8-1). Pupillary light reflexes and menace responses were normal in both eyes; bilateral ventral strabismus was present, and the colt was blinking frequently. No abnormal nystagmus was observed. Facial sensation was present bilaterally. Tongue tone was normal. When offered hay, the colt was able to prehend, chew, and swallow without difficulty and with good appetite. The colt had a tendency to lean and circle to the right. At a walk, the colt would drift to the right. No proprioceptive deficits were observed. Cutaneous trunci reflexes were normal. Tail tone and anal tone were normal. Based on neurologic examination, the colt's clinical signs were considered consistent with vestibular disease and localized either to the brainstem (central vestibular system) or to the vestibular components of the inner ear or CN VIII (peripheral vestibular system).

Differential diagnoses (see Table 10.8-1)

Central vestibular disease (basilar skull fracture, equine protozoal myeloencephalitis, traumatic brain injury) or peripheral vestibular disease (otitis media-interna, petrous temporal bone fracture, temporohyoid osteoarthropathy)

Diagnostics

The colt was sedated with xylazine (0.3 mg/kg, IV) and butorphanol (0.01 mg/kg, IV) for upper airway endoscopy. On endoscopic examination, dried blood was present on the pharyngeal wall at the openings of both guttural pouches (Fig. 10.8-2). Raised nodules on the dorsal pharyngeal wall were consistent with lymphoid hyperplasia, a common finding in young horses. The medial compartment of the right guttural pouch was obliterated by an extramural mass (Fig. 10.8-3). The epithelial lining of the medial compartment of the right guttural pouch was dark red-purple. Both guttural pouches, including the stylohyoid bones, were otherwise grossly normal.

Radiographs of the skull were obtained while the colt was sedated (Fig. 10.8-4). On the lateral view of the caudal skull, a fracture was present along the junction of the caudal aspect of the basisphenoid bone and the rostral aspect

Comparative Veterinary Anatomy: A Clinical Approach. https://doi.org/10.1016/B978-0-323-91015-6.00146-1

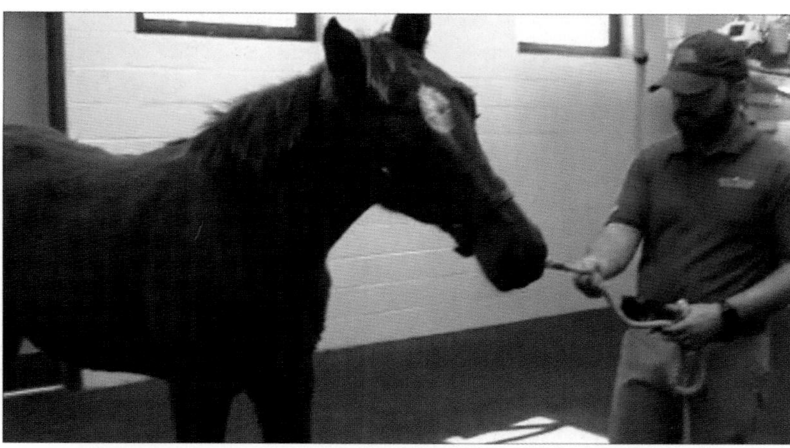

FIGURE 10.8-1 A right-sided head tilt was noted on an initial evaluation.

TABLE 10.8-1 Selected differential diagnoses for central vs. peripheral vestibular disease.

Central	*Peripheral*
Basilar skull fracture	Otitis interna/media
Equine protozoal myeloencephalitis	Petrous temporal bone fracture
Intracranial mass—e.g., abscess, neoplasm	Temporohyoid osteoarthropathy
Traumatic brain injury	Polyneuritis equi
Viral encephalitis (EEE, WEE, WNV, EHV)	

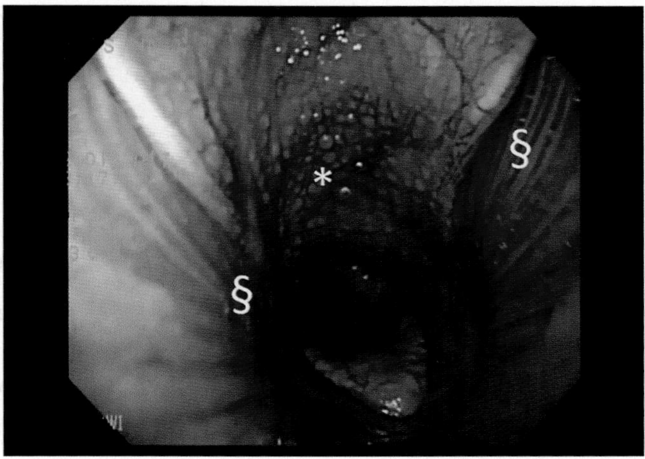

FIGURE 10.8-2 Endoscopic view of the pharynx from the right nasal passage. Lymphoid hyperplasia (*) is present on the dorsal wall of the pharynx. Dried blood is present at the openings of the guttural pouches (§).

of the basioccipital bone. The caudal aspect of the basisphenoid bone was mildly displaced ventrally. A soft tissue or fluid opacity with small gas lucencies was observed in the region of the guttural pouches, consistent with hematoma formation due to the rupture of the cranial neck musculature.

Diagnosis

Basilar skull fracture, longus capitis/rectus capitis ventralis muscular rupture, and subsequent vestibular disease (peripheral more likely than central) (see Table 10.8-2)

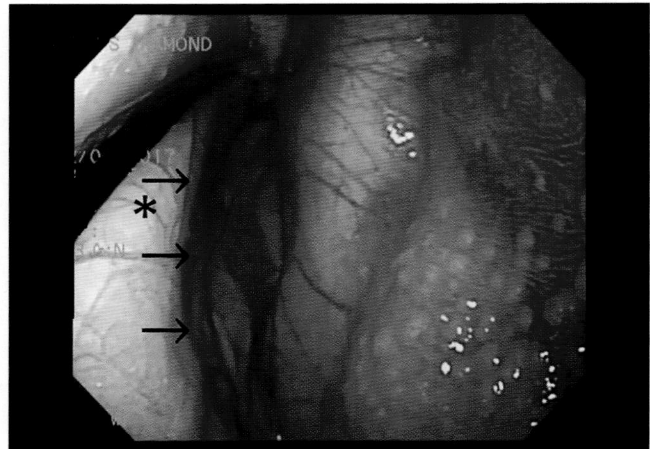

FIGURE 10.8-3 Endoscopic view of the right guttural pouch. *Arrows* (→) indicate the medial compartment of the guttural pouch. The stylohyoid bone (*) marks the delineation between the medial and lateral compartments.

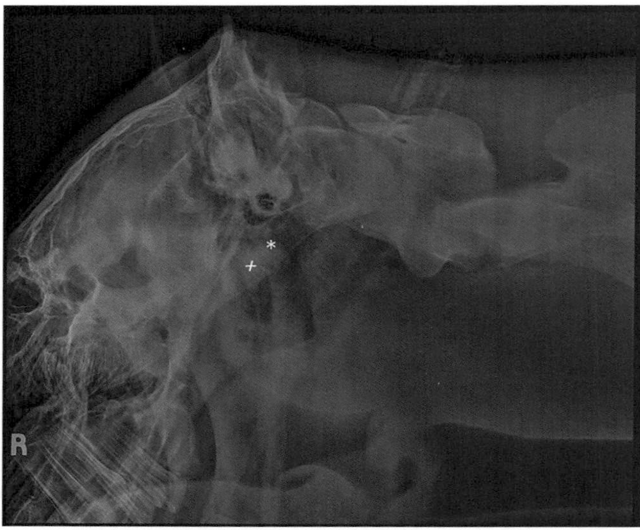

FIGURE 10.8-4 Right lateral radiographic view of the skull. A basilar skull fracture is present between the basisphenoid bone (+) and the basioccipital bone (*).

TABLE 10.8-2 Clinical signs observed in central vs. peripheral vestibular disease.

	Central	***Peripheral***
Mentation	Dull to obtunded	Normal
Head tilt	Present with poll tilted toward affected side (usually; exception = paradoxical vestibular disease)	Present with poll tilted toward affected side
Cranial nerve deficits	Multiple cranial nerves may be affected (CNVIII + others)	CNVIII; CNVII may be affected
Nystagmus	May be present	May be present
Orientation of nystagmus	Varies with head position; direction can vary	Does not vary with head position; often horizontal, with fast phase away from affected side
Proprioceptive deficits	+/– General proprioceptive ataxia +/– Upper motor neuron paresis	None

Treatment

The colt was started on IV antimicrobial therapy with potassium penicillin (22,000 IU/kg, IV, q6h) and gentamicin (6.6 mg/kg, IV, q24h), and anti-inflammatory therapy with flunixin meglumine (1.1 mg/kg, IV, q12h). Omeprazole therapy (1 mg/kg, PO, q24h) was administered as a preventive for gastric ulcers. Continuous isotonic crystalloid fluid therapy (2 mL/kg/hr) was instituted following the administration of 7.2% sodium chloride solution (4 mL/kg, IV).

The colt's condition remained stable following an initial evaluation. The colt remained alert, maintained a good appetite with normal water intake, and continued to have normal vital parameters. After 48 h of hospitalization, the colt's head tilt had neither improved nor worsened. At a walk, the colt tended to drift and circle to the right less compared with at admission. The colt was discharged from the hospital for further care at the farm. It was recommended to turn the colt out in a small paddock alone to maintain the muscle strength. Continued therapy with antimicrobials and flunixin was recommended for 7 days of treatment and omeprazole for 21 days. On the follow-up examination 14 days after admission, the colt's head tilt had improved slightly but had not resolved. Gait abnormalities had also lessened in severity but were still apparent at a walk. The colt was evaluated 4 months later. At this time, neither a head tilt nor gait abnormalities were perceptible on a neurologic evaluation.

Anatomical features in equids

Introduction

When evaluated in isolation from the mandible, the equine skull has been described as a 4-sided pyramid with the base of the pyramid being its caudal surface. These 4 surfaces are the dorsal or frontal surface, the 2 lateral surfaces, and the ventral or basal surface. The orbits are set on the lateral aspect of the skull. As a prey species, this feature renders the scope of the horse's peripheral vision to be far superior to that of humans or predator species. The orbits are also set relatively caudal on the skull, which is a typical feature of grazing species. The dorsal surface of the skull is broad. Rostrally, the skull encases the nasal passages, which are long and slender. The equine brain is encased in the caudal half of the skull.

This section describes the skull, brainstem, cranial nerves, cervical musculature, and guttural pouches. Additional information on the guttural pouch and its components can be found in Case 10.5.

Function

The vestibular system (central and peripheral) in health, in conjunction with the visual system, affords the sense of balance and the specifics about body position, enabling rapid compensatory movements in response to self-produced and externally generated forces. The vestibular system receives sensory information on head movements and body position with reference to gravity and sends signals to the central nervous system (CNS) for processing and, combined with other sensory information, approximates head and body orientation. The major vestibular system components are located in the inner ear and include the utricle, saccule, and lateral, superior, and posterior semicircular canals.

VESTIBULAR DISEASE

Signs of vestibular disease can result when there is dysfunction centrally (in the CNS, such as the brainstem or cerebellum) or peripherally (in the inner ear or in CN VIII). The vestibular system is a crucial component of the nervous system. It gathers information that allows the body to maintain balance and equilibrium. When one side of the vestibular system becomes dysfunctional, an imbalance in sensory input results. As a result, the brain receives less sensory input from the affected side and the animal begins to lean, fall, or circle toward that side to achieve a new "balance." It can be difficult to discern central vs. peripheral vestibular disease on the physical examination and neurologic examination. Table 10.8-2 summarizes the clinical signs of central vs. peripheral vestibular disease. Suspicion of central vestibular disease should heighten if the clinician observes multiple cranial nerve reflex deficits, the patient has a decreased level of consciousness/awareness—due to interference with the ascending reticular activating system—or general proprioceptive ataxia, and upper motor neuron paresis is present. Additionally, nystagmus that changes direction with head position and truly vertical nystagmus are associated with central rather than peripheral vestibular disease.

Ventral surface of the skull

The ventral aspect of the skull consists of 3 regions; from caudal to rostral, these include the **cranial region**, **choanal region**, and **palatine region** (Fig. 10.8-5).

The cranial region comprises the base of the skull. Caudally, the cranial region begins at the **foramen magnum** as the base of the **occipital bone**. The occipital bone forms the caudal part of the skull and part of the ventral surface of the skull. The **base of the occipital bone** extends laterally to the **jugular processes** of the occipital bone. It also extends rostrally toward bony tubercles that form the junction of the occipital bone and the body of the **basisphenoid bone**. These tubercles serve as the attachment site for the **longus capitis** and **rectus capitis ventralis** muscles (Fig. 10.8-6). The rostral extent of the basisphenoid bone articulates with the **vomer** (vomeronasal organ), which marks the boundary between the cranial and choanal regions of the ventral aspect of the skull.

The choanal region comprises the pharyngeal opening to the nasal cavity. The vomer is a thin bone that extends rostrally along the medial axis of the choanal region, dividing the region into 2 choanae. The vomer forms the ventral portion of the nasal septum at its attachment to the palatine bone rostrally. The choanal and palatine regions of the skull meet at caudal border of the **horizontal plate of the palatine bone**. This marks the caudal extent of the hard palate and the palatine region. The palatine region comprises the rostral half of the base of the skull, including the hard palate and the floor of the nasal passages.

Floor of the cranium, brainstem, and cranial nerves

The dorsal surfaces of the basilar part of the occipital bone and the basisphenoid bone form the **floor of the cranium** (Fig. 10.8-7). The floor of the cranium consists of 3 fossae: the **rostral cranial fossa**, **middle cranial fossa**, and **caudal cranial fossa**. The rostral cranial fossa is formed by the **presphenoid bone** and serves to support the

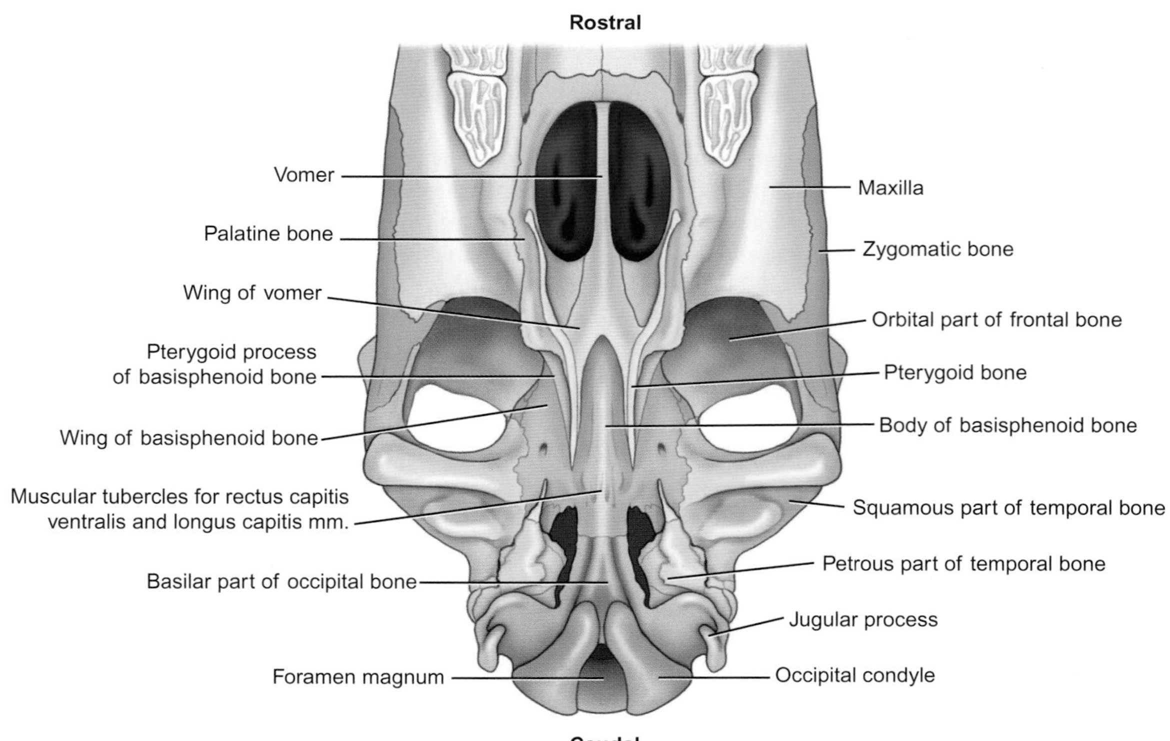

FIGURE 10.8-5 Ventral surface of the equine skull without mandible.

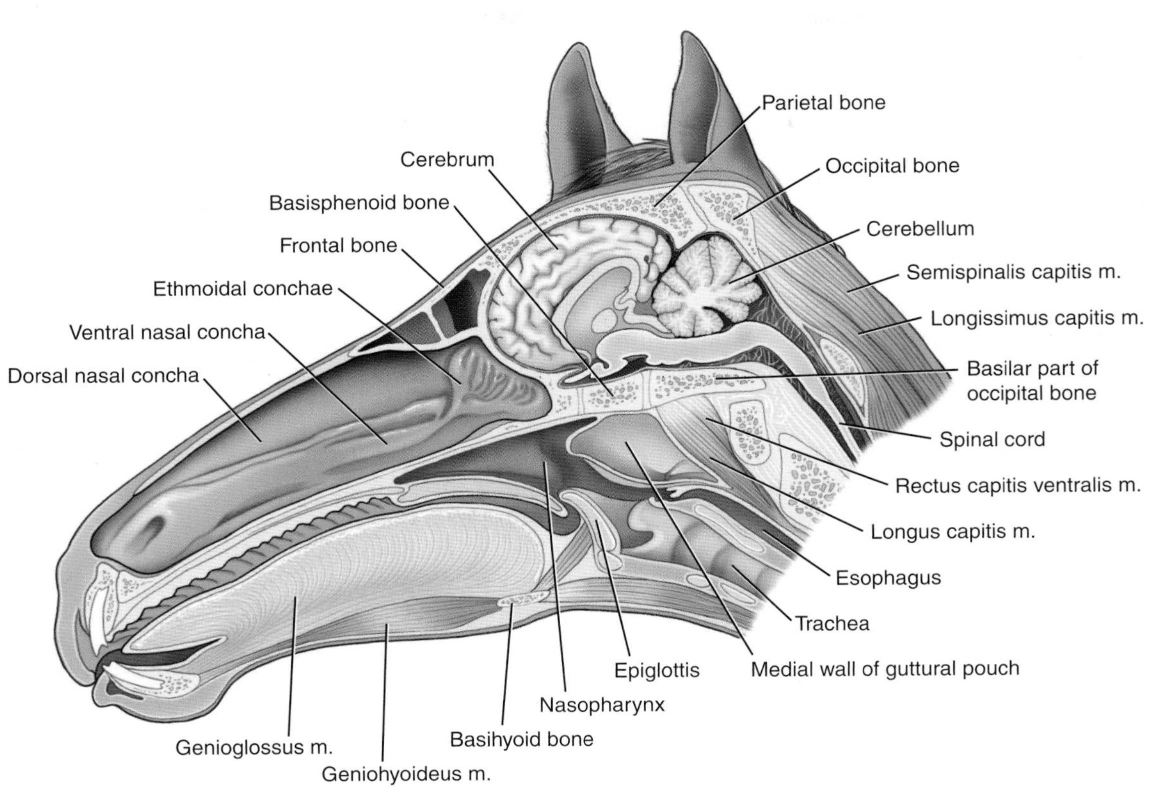

FIGURE 10.8-6 Sagittal cross section of the equine skull.

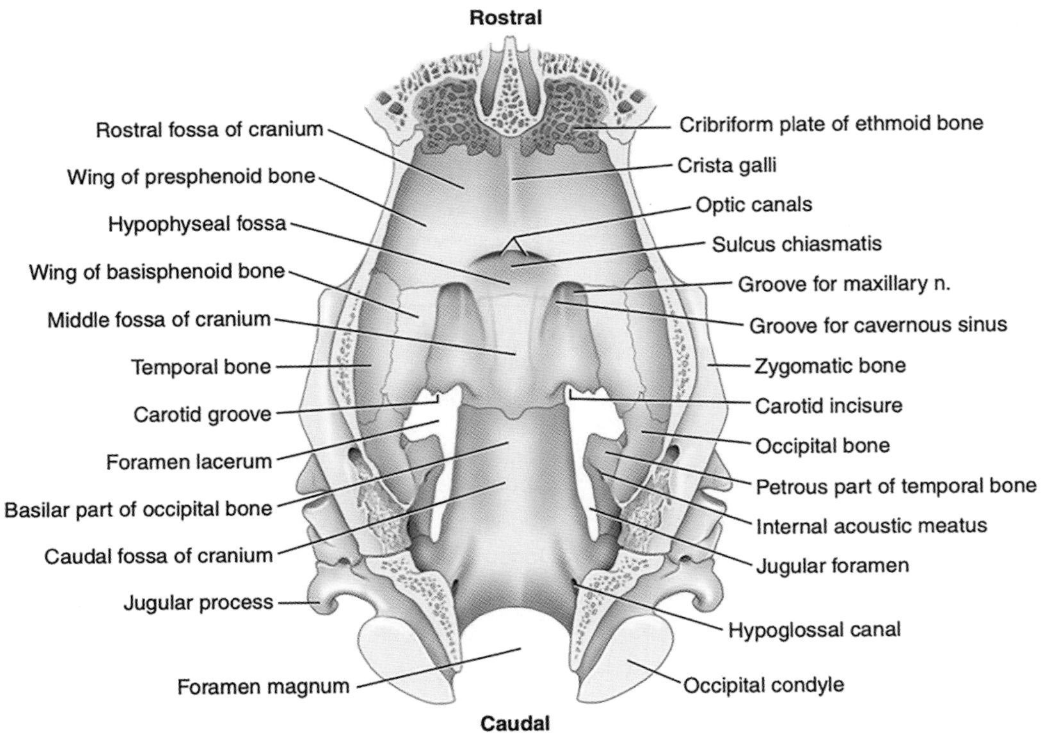

FIGURE 10.8-7 Floor of the cranium.

frontal and **olfactory** parts of the cerebrum. The **olfactory nerve** (CN I) is composed of nerve bundles that project rostrally from the olfactory bulb through the **cribriform plate**. The **optic groove** is a transverse depression in the caudal section of the rostral cranial fossa. The optic groove extends rostrally to give rise to the **optic canal**, from which the **optic nerve** (CN II) exits the skull.

The middle cranial fossa corresponds to the basisphenoid bone. The **hypophyseal fossa** is present axially in the middle cranial fossa. Lateral to the hypophyseal fossa are 2 grooves. The more medial groove accommodates the **oculomotor nerve** (CN III), the **trochlear nerve** (CN IV), the **ophthalmic branch of trigeminal nerve** (CN V), and the **abducent nerve** (CN VI) as they exit the skull at the **orbital fissure**. The lateral groove accommodates the **maxillary branch of the trigeminal nerve** (CN V) to its exit at the **round foramen**. The **mandibular branch of the trigeminal nerve** (CN V) passes through the **oval notch of the foramen lacerum (oval foramen)** as it exits the cranium. As it exits, it courses between the wing of the basisphenoid bone and the muscular process of the petrous part of the temporal bone. See Table 10.8-3 for more information.

The caudal cranial fossa corresponds to the basilar part of the occipital bone. The caudal cranial fossa supports the **medulla oblongata**, the **pons**, and the **cerebellum**. The **jugular foramina** and **hypoglossal canals** are present at the lateral aspects of the caudal cranial fossa. The **medulla oblongata** lies on the base of the occipital bone (Fig. 10.8-8). It is adjacent to the **pons** at its rostral aspect and is continuous with the **spinal cord** at its most caudal extent. The ventral aspect of the medulla oblongata serves as the origin for several of the cranial nerve rootlets—glossopharyngeal, vagus, accessory, and hypoglossal nerves.

The **abducent** nerve (CN VI) arises caudal to the pons and lateral to the **pyramids** of the medulla oblongata. The pyramids are longitudinal bands containing motor fibers (**pyramidal tracts**) which track in a rostro-caudal orientation along the midline of the ventral surface of the medulla oblongata. The **facial nerve** (CN VII) originates from the lateral aspect of the medulla oblongata, just caudal to the pons. The **vestibulocochlear nerve** (CN VIII) also originates from the lateral aspect of the medulla oblongata.

The facial and vestibulocochlear nerves track into the **internal acoustic meatus** together. The facial nerve emerges from the skull at the **stylomastoid foramen** and tracks along the dorsolateral aspect of the **guttural pouch** before it branches extensively across the face. The vestibulocochlear nerve courses away from the facial nerve once it enters the internal acoustic meatus. It splits into the **vestibular part** and the **cochlear part**. The vestibular part innervates the structures of the **internal ear**. The cochlear part innervates the structures of the **cochlea**. The vestibulocochlear nerve does not exit the skull.

The **glossopharyngeal nerve** (CN IX), the **vagus nerve** (CN X), and the **accessory nerve** (CN XI) track through the jugular foramen as they exit the skull. The **hypoglossal nerve** (CN XII) exits the skull via the **hypoglossal canal**. The

TABLE 10.8-3 Sites from which cranial nerves exit the skull.

Foramen/canal	*Cranial nerve*
Cribriform plate	Olfactory
Optic canal	Optic
Orbital fissure	Oculomotor Trochlear Trigeminal (ophthalmic branch) Abducent
Round foramen	Trigeminal (maxillary branch)
Oval foramen	Trigeminal (mandibular branch)
Stylomastoid foramen	Facial
Jugular foramen	Glossopharyngeal Vagal Accessory
Hypoglossal canal	Hypoglossal

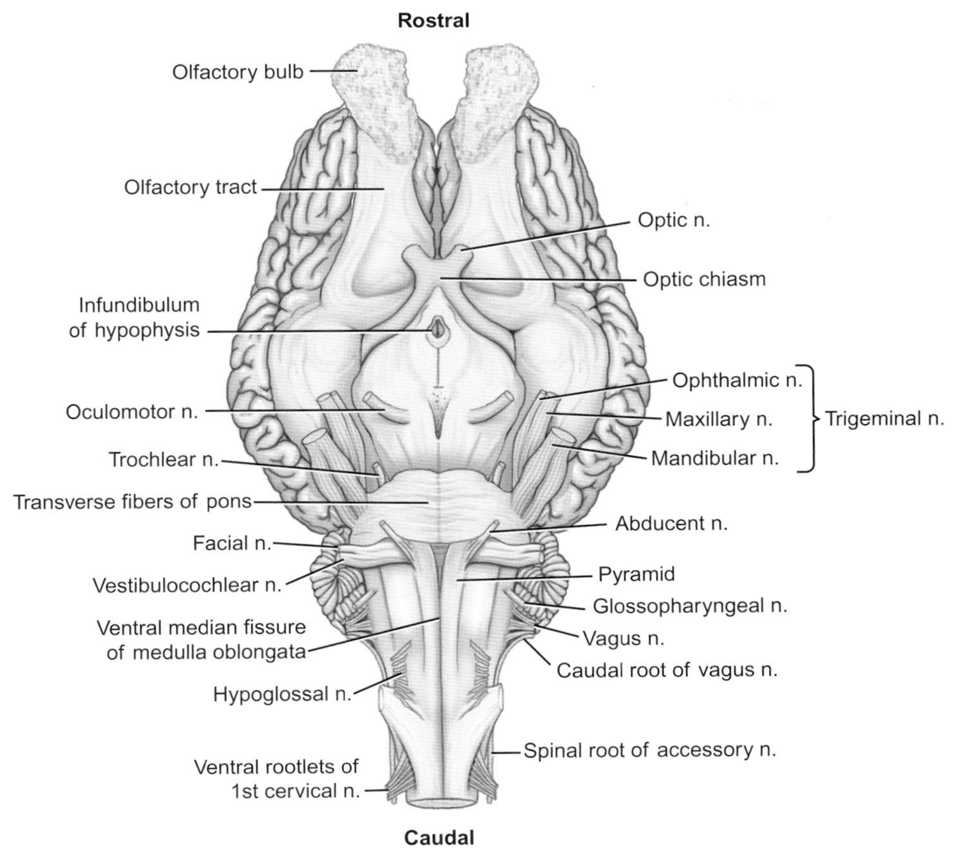

FIGURE 10.8-8 Ventral surface of the brainstem.

glossopharyngeal nerve, the vagus nerve, accessory nerve, and hypoglossal nerve track together caudally and ventrally with the **internal carotid artery** along the medial aspect of the guttural pouch. This bundle of structures creates a mucosal fold that indents into the **medial compartment of the guttural pouch.**

Cervical musculature

There are 12 pairs of muscles that lie ventral and lateral to the **cervical vertebrae**. Of these, 3 pairs originate from the cervical vertebrae and have insertions on the ventral surface of the skull.

The **longus capitis** is the largest of these muscles. It originates from the **transverse processes** of the **3rd**, **4th**, and **5th cervical vertebrae**. It inserts on the bony tubercles that serve as the junction of the basilar part of the occipital bone and the body of the basisphenoid bone. The pair of longus capitis muscles function to flex the head

If a horse hits its poll with enough force, rapid extension of the head can cause the base of the skull to fracture from traction forces exerted by these muscles during reflex contraction. The base of the skull supports and lies near the brainstem. The brainstem is the site at which the cranial nerve roots arise and contains the ascending reticular activating system, which regulates wakefulness. Upper motor neuron and general proprioceptive tracts also course through the brainstem. Consequently, horses with basilar skull fractures might have multiple cranial nerve reflex deficits, a dull or obtunded mentation, proprioceptive ataxia, and weakness. Bleeding into—or rupture of—the longus capitis and rectus capitis ventralis muscles can result in peripheral nerve dysfunction of cranial nerves VII, IX–XII.

and can also function independently to incline the head to one side or the other. These muscles are deep in the neck and are bounded dorsally by the cervical vertebrae. At the level of their insertion points, the longus capitis muscles lie dorsal to the **pharynx** and between the **guttural pouches**.

The **rectus capitis ventralis** is a smaller muscle that lies dorsal to the longus capitis. The rectus capitis ventralis originates on the ventral arch of the **atlas (1st cervical vertebra)** and inserts on the basilar part of the occipital bone to flex the **atlantooccipital articulation**. At their insertion, the rectus capitis ventralis muscles lie between the guttural pouches.

The **rectus capitis lateralis** is the smallest of this muscle group. This pair of muscles originates on the atlas, lateral to the origin of the rectus capitis ventralis muscles. The muscles insert on the jugular processes of the occipital bones and also serve to flex the atlantooccipital articulation. This pair of muscles is not closely bounded by the guttural pouches.

Guttural pouches (also see Case 10.5)

The horse has 2 symmetric guttural pouches, each a diverticulum of the **auditory tube**. Each pouch is bounded dorsally by the base of the skull and the atlas. The longus capitis muscle and the rectus capitis ventralis muscle pass along the dorsomedial aspect of the guttural pouches. The ventral aspect of the guttural pouch is adjacent to the dorsal **pharyngeal wall** and the entry to the **esophagus**. Laterally, it is covered by the **pterygoid muscles**, the **parotid glands**, and the **mandibular glands**. The 2 guttural pouches border each other medially and are separated by a median septum. Each guttural pouch is divided into **medial** (larger) and **lateral** (smaller) **compartments** by the **stylohyoid bone**.

TRAUMATIC BRAIN INJURY

Traumatic brain injury (TBI) can result from basilar skull fractures due to concussive forces on the brain, subsequent brain swelling, or intracranial hemorrhage. Any of these events can disrupt cerebral blood flow and increase intracranial pressure, resulting in ischemia of the brain. Many horses with TBI present with multiple neurologic signs. These can include abnormal mentation, nystagmus, ataxia, and decreased-to-absent pupillary light response (PLR). An important aspect of treating TBI is the maintenance of cerebral blood flow, which can be accomplished using isotonic crystalloid fluid therapy. Overhydration should be avoided due to its potential to increase intracranial pressure. Administration of hypertonic saline is a helpful component of therapy for TBI because it increases intravascular volume while simultaneously decreasing intracranial pressure. Mannitol may accomplish similar goals, but it is not as effective an osmotic agent as hypertonic saline. Corticosteroid therapy does not appear to improve outcome in cases of TBI. Antimicrobial therapy is recommended following head trauma in horses due to the potential for open skull fractures and secondary septic meningitis.

Selected references

[1] Furr M, Reed S. Equine neurology. 2nd ed. Philadelphia: Wiley-Blackwell; 2015.
[2] Getty R. Sisson and Grossman's the anatomy of the domestic animals. 5th ed. vol. 1. Philadelphia: Saunders; 1975.
[3] Reed S, Bayly W, Sellon D. Equine internal medicine. 4th ed. St. Louis: Elsevier; 2018.
[4] Singh B. Dyce, Sack and Wensing's textbook of veterinary anatomy. 5th ed. St. Louis: Elsevier; 2018.

CASE 10.9

Ear sarcoid

Annette M. McCoy
Equine Surgery, Veterinary Clinical Medicine, University of Illinois Urbana-Champaign College of Veterinary Medicine, Urbana, Illinois, US

Clinical case

History

A 6-year-old Shire mare presented with multiple masses affecting both ears. The masses had been growing slowly for several months and had recently become ulcerated. Therapy had not been undertaken by the owner before presentation.

Physical examination findings

The mare was bright, alert, and responsive. Vital parameters were within normal limits. Multiple ulcerated masses, roughly spherical in shape and ranging in size from 2 to 4 cm (0.8 to 1.6 in.) in diameter, were located on the pinnae of the left and right ears (Fig. 10.9-1). The remainder of the physical examination was normal.

Differential diagnoses

Sarcoid, fibrosarcoma, melanoma, habronemiasis

Diagnostics

The mare was sedated to facilitate closer inspection of the masses. The internal auricular branch and great auricular nerve of each ear were blocked with 2% mepivacaine to achieve regional anesthesia of the pinnae. The masses were debulked surgically and at that time a biopsy was submitted for histopathological examination to confirm the presumptive diagnosis.

Diagnosis

Sarcoid, ulcerative fibroblastic form

Treatment

Excision of grossly visible sarcoid tissue via CO_2 laser, followed by intra- and perilesional placement of cisplatin beads, was performed under standing sedation and regional anesthesia of each ear. Three treatments of cisplatin bead placement, scheduled at 2–3-week intervals, were required to resolve the sarcoids.

Comparative Veterinary Anatomy: A Clinical Approach. https://doi.org/10.1016/B978-0-323-91015-6.00155-2

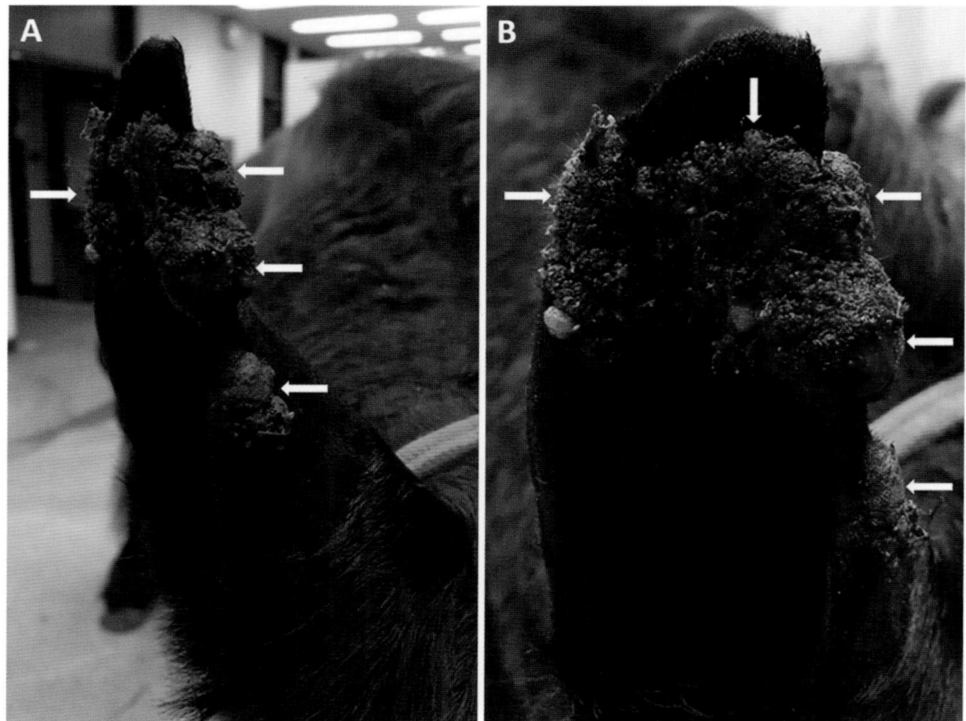

FIGURE 10.9-1 Masses on the pinna of a 6-year-old Shire mare *(white arrows)*. (A) Cranial view. (B) Dorsolateral view. (Images courtesy of S.D. Gutierrez-Nibyerbo.)

Anatomical features in equids

Introduction

This section discusses the external, middle, and internal parts of the equine ear, along with its blood supply, lymphatics, and innervation.

Function

Horses rely heavily on hearing to assess their environment, and the ears are highly mobile, each with approximately 270-degree range of motion. This mobility also allows for expression of a range of emotions (Fig. 10.9-2A and B).

External ear (auris externa)—gross anatomy and musculature

The external ear consists of the visible part of the ear—that is, the auricle (auricula or pinna) and the external acoustic meatus. The **auricle** is comprised primarily of cartilage (**auricular cartilage**), with scarce connective tissue deep to the skin. The **external acoustic meatus** is the canal extending from the base of the auricular cartilage to the **tympanic membrane**. The funnel-shaped base (**concha auriculae**) of the auricular cartilage is supported by an **annular (anular) cartilage** at its attachment to the opening of the osseous external acoustic meatus in the tympanic part of the temporal bone of the skull.

There are multiple muscles that originate from the head and neck and insert on the external ear, providing its mobility (Fig. 10.9-3). These muscle groups insert on the rostral, dorsal, caudal, and lateral aspects of the pinna, and there are deep rotator muscles located at the base of the auricular cartilage. Only the most clinically relevant muscles are described further. The **parotidoauricularis muscle** passes over the lateral aspect of the parotid salivary gland and traverses dorsally to insert at the base of the external opening of the auricular cartilage, near the junction of the rostral and caudal borders of the cartilage. This muscle retracts the ear ventrally and caudally. The **zygomaticoauricularis**

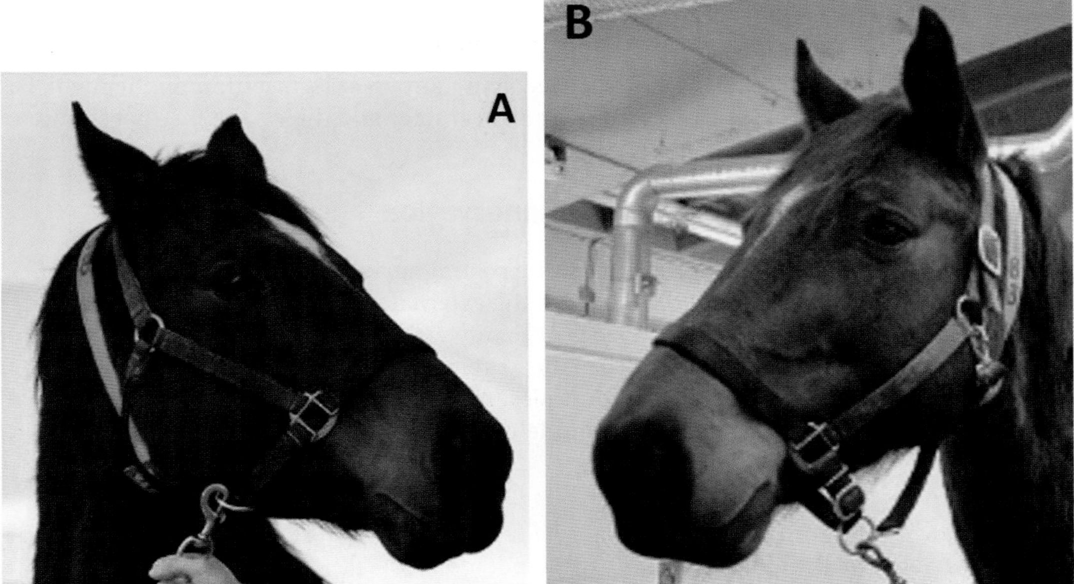

FIGURE 10.9-2 The horse uses its ears to convey emotions. (A) Ears back suggest irritation, anger, or nervousness. (B) Ears forward suggest alertness.

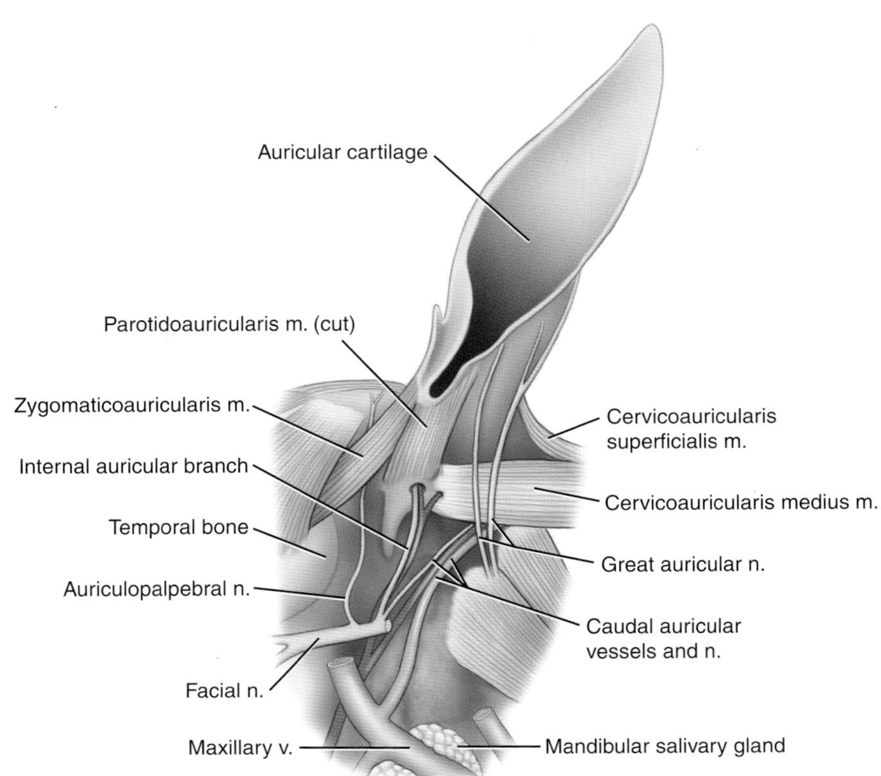

FIGURE 10.9-3 Relevant muscular and neurovascular anatomy of the equine ear *(lateral view)*. Note that the internal auricular branch is often referred to as the internal auricular nerve.

muscle originates from the caudal aspect of the zygomatic process of the temporal bone and inserts on the dorsolateral aspect of the auricular cartilage, immediately rostral to the parotidoauricularis muscle. This muscle, and other rostral auricular muscles, holds the ear erect and facing forward. The **cervicoauricularis superficialis muscle** originates from the nuchal crest of the skull as a broad sheet and narrows at its insertion on the medial aspect of the base of the pinna. This muscle adducts and raises the auricular cartilage and directs the internal surface of the ear laterally.

External ear—blood supply, lymphatics, and innervation

Blood supply to the external ear is provided primarily by the **rostral auricular artery** (from the **superficial temporal artery**) and the **caudal auricular artery** (Fig. 10.9-3) (from the **external carotid artery**). A rostral branch from the **occipital artery** also contributes. Venous drainage is primarily via the **rostral auricular vein** to the **superficial temporal vein**, and via the **caudal auricular vein**. The latter 2 veins drain to the **maxillary vein**. Lymph from the external ear drains primarily to the **parotid lymph nodes**. Afferents also flow to the **lateral retropharyngeal**, **superficial cervical**, and **cranial deep cervical lymph node centers**.

The major muscles of the ear are innervated by motor branches of the **facial nerve**—the **caudal auricular nerve** and **rostral auricular branches** from the **auriculopalpebral nerve**. An important exception is the cervicoauricularis superficialis muscle, which is innervated by the **great auricular nerve** (sensory and motor functions), a branch of the **2nd cervical nerve** (Figs. 10.9-3 and 10.9-4A). Branches of the **1st cervical nerve** also contribute to the innervation of the small ear muscles. The inner surface of the pinna is largely innervated by the sensory **internal auricular branch** (identified as the internal auricular nerve in other texts) (Figs. 10.9-3 and 10.9-4B), which arises from the facial nerve as it exits the stylomastoid foramen. The internal auricular branch enters the auditory meatus through a palpable foramen located ventral to the visible ventral "notch" at the base of the external ear. This foramen is deep to the parotidoauricularis muscle and its location varies slightly between individuals.

The fact that the cervicoauricularis superficialis muscle is innervated by the great auricular nerve—not a branch of the facial nerve—has clinical implications. This means the ear may still exhibit some movement in cases of facial nerve paralysis.

The inner and outer surfaces of the pinna may be desensitized by blocking the internal auricular branch and the great auricular nerve (Fig. 10.9-4). This facilitates minor procedures such as biopsy, mass removal, and laceration repair, particularly in horses that are already sensitive to having their ears handled. The internal auricular branch is blocked by placing a 25-g, 5/8″ needle into the palpable foramen described earlier, and depositing 2 mL of local anesthetic (Fig. 10.9-5B). The great auricular nerve is palpated on the caudal aspect of the base of the pinna, and 3 mL of local anesthetic is deposited subcutaneously, directly adjacent to the nerve, using a 22-g 1″ needle (Fig. 10.9-5A). The ear is typically droopy after the onset of anesthesia, as the motor function of the cervicoauricularis superficialis muscle is also blocked, but normal ear position is restored after the anesthesia wears off.

Middle ear (auris media)—gross anatomy and musculature

The middle ear is comprised of the tympanic cavity, tympanic membrane, auditory ossicles, and auditory (Eustachian) tube (tuba auditiva) (Fig. 10.9-6). The **auditory tube** connects the tympanic cavity with the **nasopharynx**. The **guttural pouches** of the horse are caudoventral enlargements of the auditory tubes and are therefore part of the middle ear anatomy in this species (see Case 10.5). The **tympanic cavity** is an air-filled space, lined by mucosa and defined by the tympanic membrane, the inner ear, and the tympanic and petrous parts of the temporal bone. The **auditory ossicles** (Fig. 10.9-7)—comprised primarily of the **malleus** (hammer), **incus** (anvil), and **stapes** (stirrup)—lie within the tympanic cavity and transmit sound vibrations to the inner ear. A fourth, tiny ossicle, the **os lenticulare**, is located between the incus and stapes. These delicate bones articulate with each other and are attached to the walls of the tympanic cavity via a network of small ligaments.

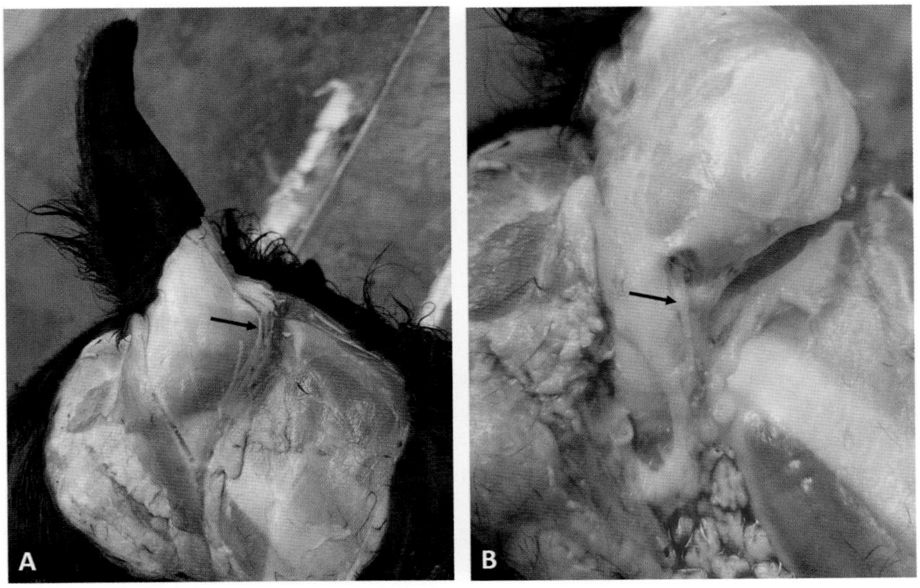

FIGURE 10.9-4 Gross prosection of a cadaveric specimen showing the great auricular nerve (A, *arrow*) and the internal auricular branch (B, *arrow*).

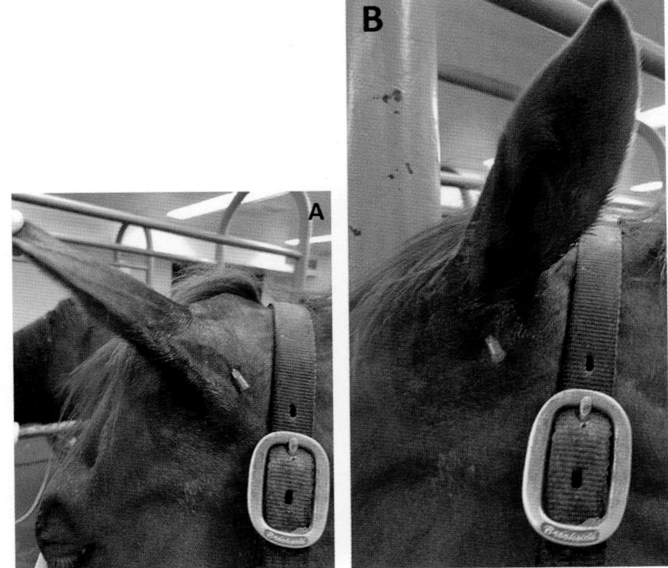

FIGURE 10.9-5 Placement of needles for the great auricular nerve block (A) and internal auricular branch block (B) in a standing, unsedated horse.

There are two muscles associated with the auditory ossicles (Fig. 10.9-7). The **tensor tympani muscle** arises from the osseous portion of the auditory tube and inserts on the malleus; as its name implies, its primary action is to tense the tympanic membrane. The **stapedius muscle** arises from the posterior wall of the tympanic cavity and inserts on the stapes; this muscle tenses the annular ligament and connects the stapes to the inner ear.

The middle ear is separated from the inner ear by the **labyrinthine wall**. There are 2 fenestra in the labyrinthine wall: 1) the **fenestra vestibule** (vestibular window), which is closed by the footplate of the stapes (attached via its annular ligament); and 2) the **fenestra cochleae** (cochlear window), which is covered by a thin membrane (**secondary tympanic membrane**).

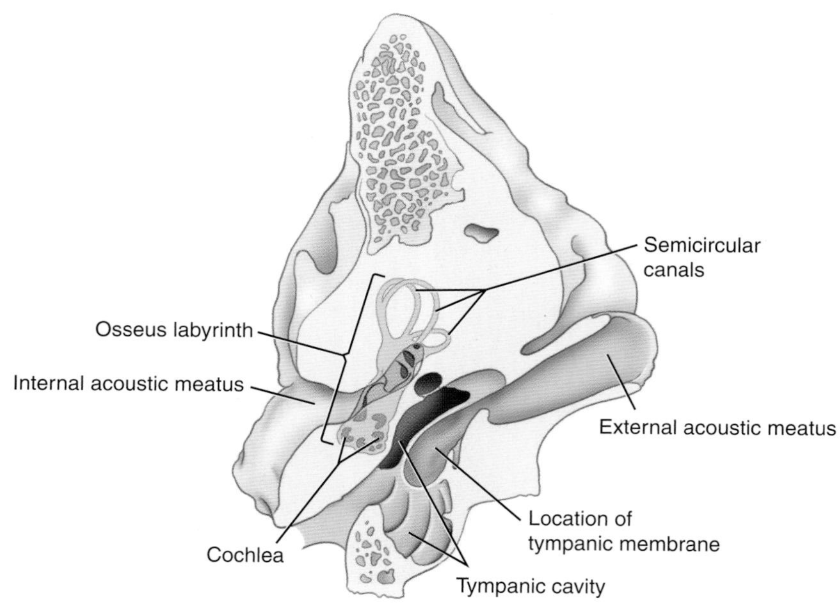

FIGURE 10.9-6 A transverse section through the petrous and tympanic parts of the left temporal bone of a horse, viewed from the rostral aspect. Structures of the middle and inner ear are identified. Sound enters via the external acoustic meatus. Cranial nerves VII and VIII pass through the internal acoustic meatus.

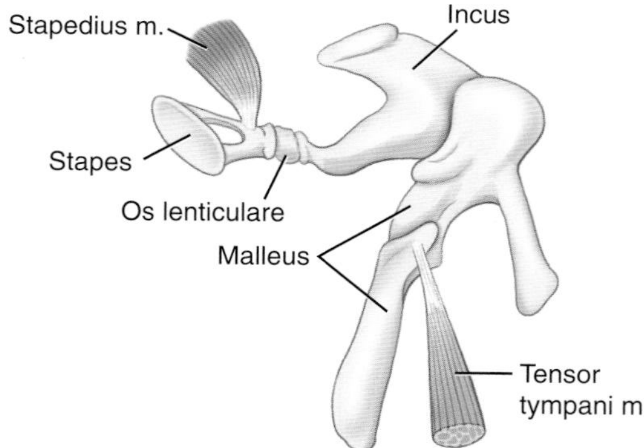

FIGURE 10.9-7 Auditory ossicles showing their relationships with each other and the attachment of the 2 muscles that act on the malleus and the stapes.

Middle ear—blood supply, lymphatics, and innervation

The **deep auricular artery**, arising from the **caudal auricular artery**, provides branches that enter the tympanic cavity via the stylomastoid foramen. The **maxillary artery** provides a **rostral tympanic artery** that enters the middle ear via the auditory tube. Venous drainage returns to the regional large veins, including the **superficial temporal**, **caudal auricular**, and **occipital veins**.

Lymphatic drainage of the middle ear is similar to the external ear.

The **mandibular branch of the trigeminal nerve** (CN V) supplies the tensor tympani muscle, and the **facial nerve** supplies the stapedius muscle. The mucous membrane lining the middle ear contains small lymph nodules and mucous glands and is innervated by the **tympanic plexus**. This plexus receives branches from the **tympanic nerve** (branch of the glossopharyngeal nerve) and from the **internal carotid plexus**.

Inner ear (auris interna)—gross anatomy

The inner ear (**labyrinth**) is housed within the petrous part of the temporal bone and consists of the membranous labyrinth located within the osseous labyrinth (Fig. 10.9-6). The **membranous labyrinth** contains endolymph and the sensory fibers of the **vestibular** and **cochlear nerves**. The **perilymphatic space** between the membranous and osseous labyrinth is filled with a fluid known as perilymph.

The **osseous labyrinth** is divided into 3 portions: (1) the **vestibule**, centrally located and communicates with the middle ear via the fenestra vestibule; (2) the **cochlea**, located anteriorly; and (3) the **semicircular canals** (dorsal, posterior, and lateral), located posteriorly. The medial wall of the vestibule is the fundus of the **internal acoustic meatus**, through which nerve fibers of the **vestibulocochlear n. (CN VIII)** pass en route from the membranous

FIGURE 10.9-8 Mucoid fluid draining from a sinus tract at the rostrolateral base of a horse's right ear, characteristic of discharge from a dentigerous cyst. There is crusty dried discharge coating the skin and hair around the tract opening.

DENTIGEROUS CYSTS (A.K.A. CONCHAL CYSTS OR "EAR TEETH")

Dentigerous cysts are congenital defects caused by a failure of closure of the 1st branchial cleft during embryonic development. Dental tissues, including enamel, dentin, and cementum, are found in the cyst and may be arranged in the form of a complete or partial tooth. Dentigerous cysts may be recognized by a palpable swelling in the region of the base of the ear, and/or by the drainage of mucoid cystic fluid from a sinus tract, often associated with the external ear margins (Fig. 10.9-8). Radiography and CT are useful diagnostic aids and for surgical planning (Fig. 10.9-9). Dentigerous cysts rarely cause clinical signs beyond intermittent drainage of cystic fluid and are usually removed for cosmetic reasons, concerns about discomfort, drainage, or distortion of surrounding anatomy. Differential diagnoses include foreign body and abscess; however, these are typically more painful and of an acute onset. If surgical excision is elected, the cyst and its draining tract are removed in their entirety to prevent recurrence. Meticulous excision of the mass and draining track is required to preserve the auricular muscles, rostral auricular artery, and auriculopalpebral nerve. If the tooth material is firmly attached to the skull, careful elevation with a chisel or osteotome is performed, minimizing trauma to the underlying bones of the cranium. With focused debridement (surgical dissection), risk of injury to deeper structures in the area, including the parotid salivary gland, the base of the auricular cartilage, and associated vascular and nerve structures, is avoidable.

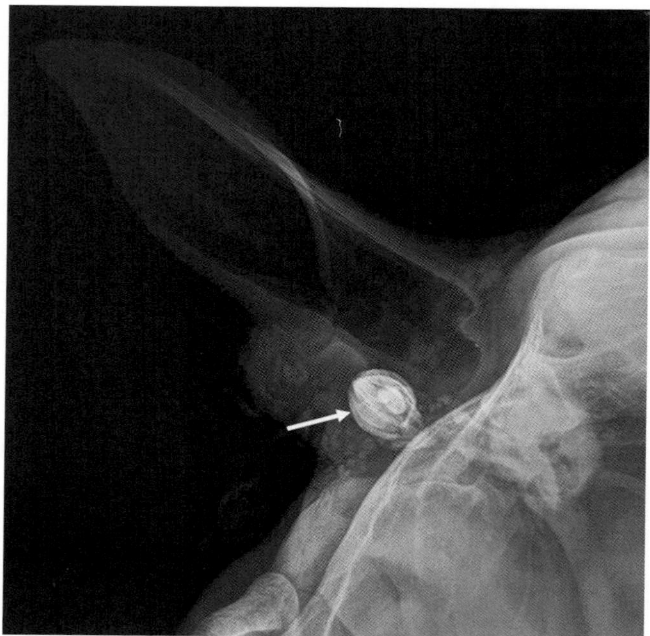

FIGURE 10.9-9 Oblique radiograph of the right ear base and associated skull region of a horse showing a discrete circular radio-opaque structure *(white arrow)* consistent with the appearance of a dental body within a dentigerous cyst.

labyrinth. The membranous labyrinth is divided into 4 parts: (1) **utricle** and (2) **saccule**, which lie within the ventricle; (3) **semicircular ducts**, which lie within the semicircular canals; and (4) **cochlear duct**, which lies within the cochlea.

Inner ear—blood supply, lymphatics, and innervation

The **vestibulocochlear nerve (CN VIII)** provides innervation to the inner ear. **Hair cells** are specialized epithelial cells that are found within the **maculae acousticae**, which are thickenings on the inner walls of the utricle and saccule. Fibrils of the **vestibular nerve** (**utricular** and **saccular branches**) arborize around the base of these hair cells and are a vital part of the vestibular system that mediates balance and equilibrium. Similarly, hair cells (and associated supporting cells) make up the **spiral organ (organ of Corti)**, located within the ampullae of the semicircular ducts. The nerve fibers surrounding the base of these hair cells originate from the **cochlear nerve** and are therefore important in hearing.

The **labyrinthine artery** (which has a variable origin off the **basilar** or **caudal cerebellar artery**) provides **vestibular** and **cochlear branches** to the inner ear, entering via the internal acoustic meatus. **Labyrinthine veins** return blood to the **basilar sinus**. Lymphatic drainage of the inner ear is not well described for the horse.

Selected references

[1] Cerasoli I, Cornillie P, Gasthuys F, Gielen I, Schauvliege S. A novel approach for regional anesthesia of the auricular region in horses: an anatomic and imaging study. Vet Anes Analg 2017;44:656–64.

[2] McCoy A, Schaefer E, Malone E. How to perform effective blocks of the equine ear. Proc Am Assoc Equine Pract 2007;53:397–8.

[3] Sommerauer S, Snyder A, Breuer J, Schusser GF. A technique for examining the external ear canal in standing sedated horses. J Equine Vet Sci 2013;33:1124–30.

CASE 10.10

Nuchal Bursitis

José García-López
Clinical Sciences, Tufts University Cummings School of Veterinary Medicine, North Grafton, Massachusetts, US

Clinical case

History

A 12-year-old Hanoverian gelding used as a Grand Prix dressage horse presented with a month-long history of reluctance and unwillingness to flex his neck and head and maintain a proper frame during exercise. Changing the bridle and mouth bit did not influence clinical signs. The gelding underwent a 10-day course of oral phenylbutazone, which resulted in partial improvement of his clinical signs. There was no history of trauma to the head or neck region and there had been no change in the horse's exercise routine.

Physical examination findings

Upon presentation, the gelding primarily maintained his head and neck in an extended position. The poll region appeared grossly enlarged, particularly on the left (Fig. 10.10-1A and B). Palpation of the poll region elicited a strong and repeatable pain response. There was no evidence of a break in the skin or drainage from the area. Pulse and respiration were normal (36 bpm and 16 brpm, respectively) and rectal temperature was normal at 100.0 °F (37.8 °C).

Differential diagnoses

Nuchal bursitis (nonseptic or septic, see side box entitled "Septic nuchal and supraspinous bursitis"), vertebral fracture, occipital bone fracture, nuchal ligament tear

Diagnostics

The gelding was sedated to facilitate additional examination. Even with sedation, palpation and manipulation of the poll region elicited a painful response, causing the horse to throw his head around. Palpation of the area revealed

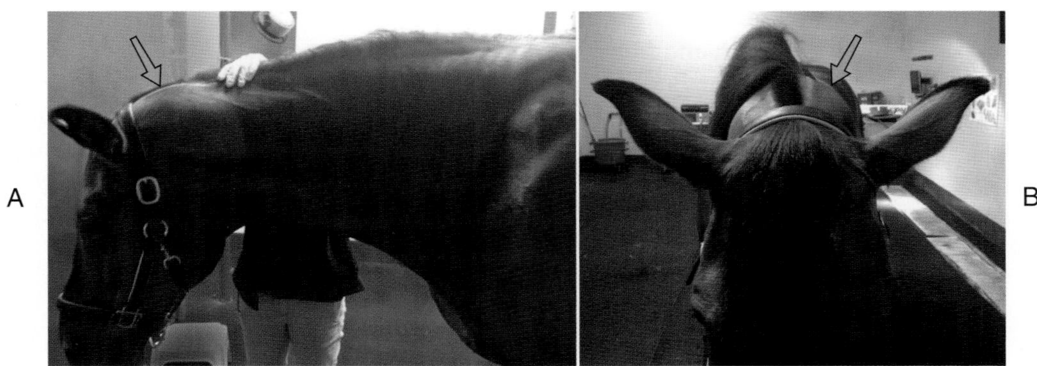

FIGURE 10.10-1 Asymmetrical swelling in the poll region is visible, with the left side larger than the right *(orange arrows)*. A. Lateral view. B. Cranial view.

Comparative Veterinary Anatomy: A Clinical Approach. https://doi.org/10.1016/B978-0-323-91015-6.00145-X

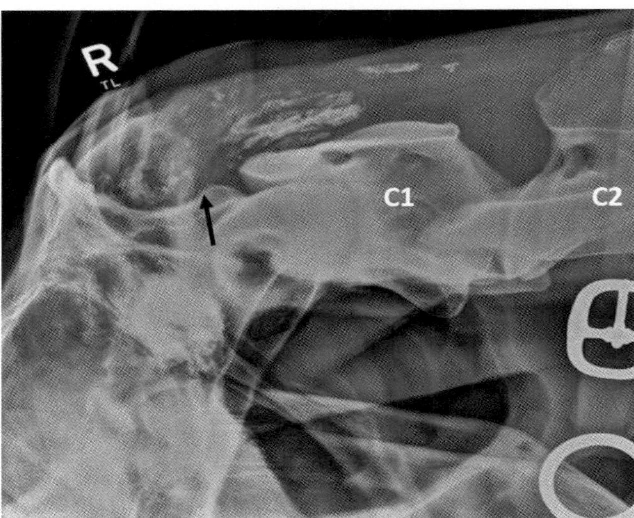

FIGURE 10.10-2 Lateral radiograph of the poll region of a horse. Extensive mineralization dorsal and cranial to C1 vertebra and a mild bony irregularity *(black arrow)* of the caudal aspect of the occipital bone is visible.

a focal, approximately 14 × 6 cm (5.5 × 2.4 in.), hard and nonmovable swelling more prominent on the left and dorsolateral to the 1st cervical vertebra (C1, atlas).

A lateral radiograph of the poll region was obtained and revealed extensive mineralization dorsal and cranial to C1, as well as a mild bony irregularity of the caudal aspect of the occipital bone (Fig. 10.10-2). Ultrasound examination of the poll region revealed marked mineralization of the nuchal ligament and periligamentous soft tissues. A large, partially mineralized soft tissue density was seen deep to the nuchal ligament, overlying C1. This soft tissue density was consistent with an abnormal cranial nuchal bursa containing an extensive amount of mineralized tissue within a thickened bursal capsule. The bursa had a mostly homogeneous hypoechoic to echogenic appearance with a mild to moderate amount of homogeneously echogenic fluid seen on the left (Figs. 10.10-3 and 10.10-4). An ultrasound-guided sample of fluid was obtained and submitted for fluid analysis and culture and sensitivity. The analysis revealed a total nucleated cell count of 3200 cells/μl (primarily degenerate neutrophils and a few macrophages) and a total protein of 3.2 g/dL. No bacteria were seen. Aerobic and anaerobic cultures were negative.

Diagnosis

Cranial nuchal bursitis (nonseptic)

Treatment

Exploratory nuchal bursoscopy was performed with the gelding under general anesthesia and positioned in right lateral recumbency. For improved support and to access scope portals for bursoscopy, a 3-L fluid bag was placed under the neck at the level of C1. The cranial nuchal bursa was identified sonographically and distended with 60 mL of a polyionic solution. Endoscopic examination of the cranial nuchal bursa revealed a large quantity of debris consisting of proteinaceous cellular material with mineralization (Fig. 10.10-5). Using manual and motorized synovial resectors, the proteinaceous material was removed (Fig. 10.10-6) and samples were submitted for histopathology. After lavage and exploration of the cranial nuchal bursa, the surgical incisions were closed and 5 mL of Polyglycan[1] was injected into the bursa. The gelding had an uneventful anesthetic recovery.

[1]Polyglycan (10 mL): Hyaluronic acid sodium (50 mg), sodium chondroitin sulfate (1000 mg), *N*-acetyl-D-glucosamine (1000 mg), Bimeda® USA, Oakbrook Terrace, IL 60181.

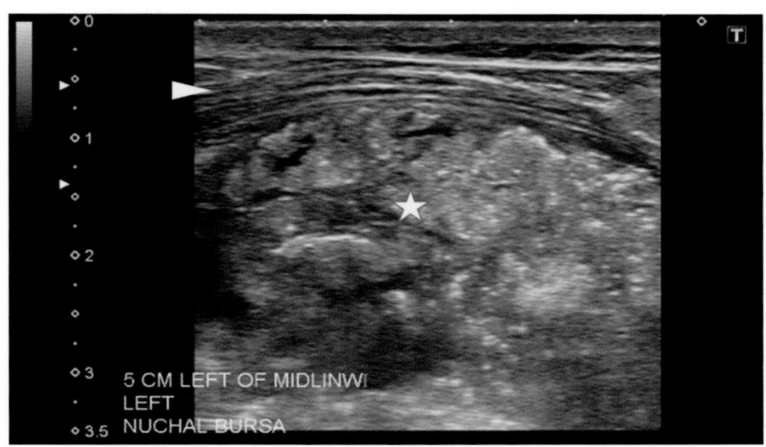

FIGURE 10.10-3 Ultrasonographic longitudinal view of the cranial poll, dorsal to C1. A large, partially mineralized soft tissue density (*white star* is in the center of the mass) was seen deep to the nuchal ligament *(white arrowhead)* overlying C1.

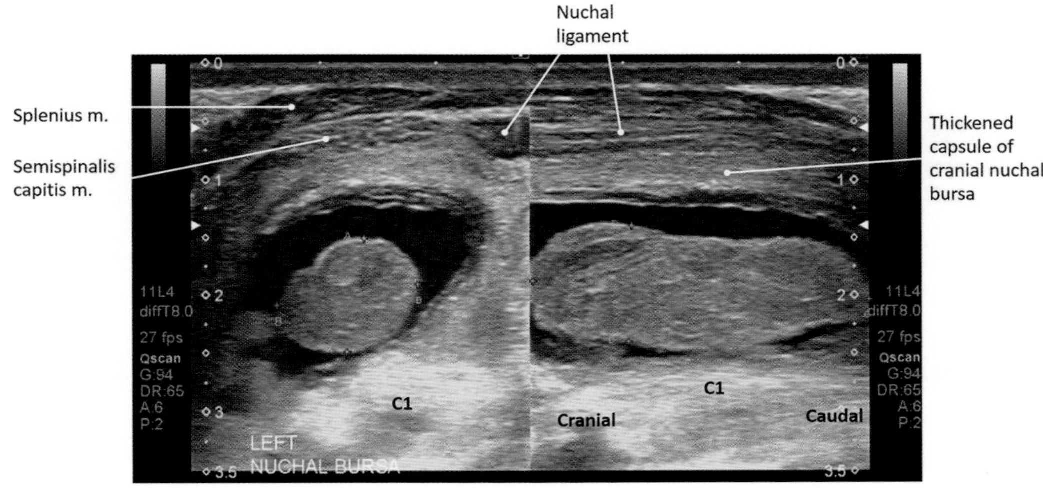

FIGURE 10.10-4 Cross-sectional *(left frame)* and longitudinal ultrasonographic images of the left side of the cranial nuchal bursa, showing homogeneously echogenic fluid *(dark regions)* surrounding the soft tissue mass "floating" in the fluid.

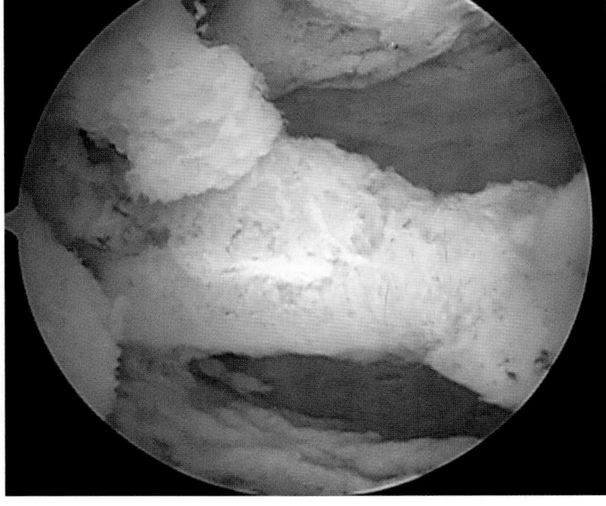

FIGURE 10.10-5 Endoscopic view of the cranial nuchal bursa. A large amount of white debris consisting of proteinaceous cellular material with the areas of mineralization is visible.

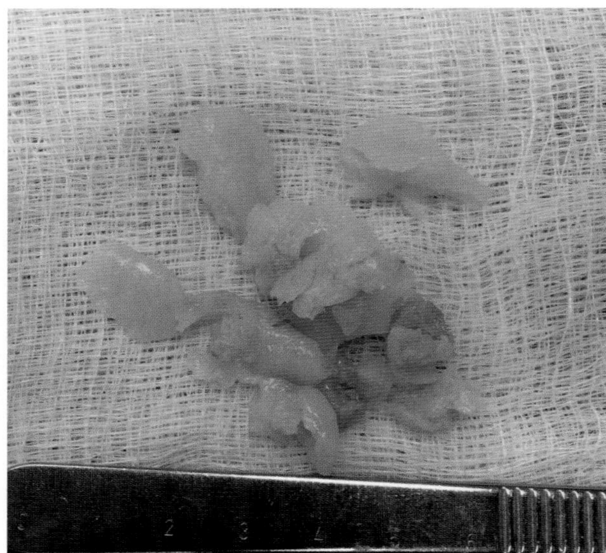

FIGURE 10.10-6 A portion of the debris removed from the cranial nuchal bursa using manual and motorized synovial resectors. Scalpel handle is included as a reference point.

Post-operatively, the gelding was maintained on IV antibiotics and anti-inflammatories for 3 and 10 days, respectively. The swelling over the poll area resolved with residual thickening of the poll area. Results of the biopsy reported abundant amorphous acellular proteinaceous material with a few foci of degenerative neutrophils. After 6 months of rest and an incremental increase in controlled exercise, the horse was able to return to competition and clinical signs did not recur.

Anatomical features in equids

Introduction

The nuchal ligament is a paired elastic structure that passively supports the weight of the horse's head and neck and assists dorsal neck muscles in lifting and extending the head and neck. There are 3 associated bursae that will also be discussed in this section.

Function

The nuchal ligament functions to divide the dorsal cervical muscles into right and left groups. This independently developed structure is adapted for running, supporting the weight of the head, and allowing sufficient flexibility so that the horse can easily raise and lower its head when grazing and running.

Nuchal ligament

The **nuchal ligament** (*ligamentum nuchae*) consists of 2 clearly defined parts: funicular (funiculus nuchae) and laminar (lamina nuchae) (Fig. 10.10-7A and B). The **funicular (dorsal)** part is a paired, thick cord that extends from the highest spinous processes of the withers—i.e., T3–T4—to the external occipital protuberance of the skull. The funicular part is continuous with the supraspinous ligament, which passes along the thoracolumbar vertebral column. The paired **laminar part** forms broad sheets extending from the funicular part and the spinous processes of T2 and T3 to attach to the spinous processes of C2 to C7. The laminar part is fenestrated between the insertions of the ligament to each cervical vertebra.

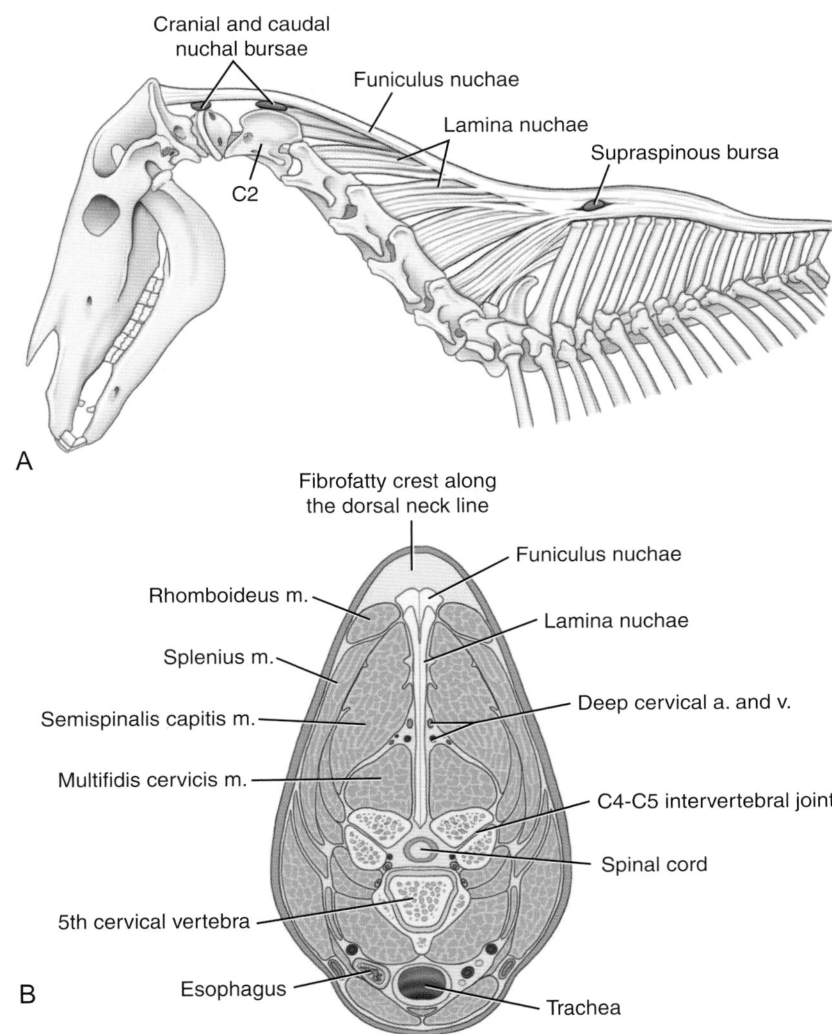

FIGURE 10.10-7 (A) The equine nuchal ligament (funicular and laminar parts) and associated bursae. The lamina nuchae is fenestrated (shown as gaps in the ligament) between its insertions on the cervical vertebrae. (B) Cross-section through the C4–C5 vertebral junction demonstrating the paired parts of the nuchal ligament and closely associated muscles.

COMPARATIVE FEATURES OF THE NUCHAL LIGAMENT

The cat lacks a nuchal ligament while the dog has a paired, cordlike ligament where the funicular (L. a cord) portion inserts on the spinous process of the axis (C2). The nuchal ligament in the cow is similar to that of the horse, except the funicular portion extends further caudally in the thoracic region attaching to each side of the supraspinous ligament. Furthermore, the laminar part is divided into cranial (paired and extending to spinous processes of cranial cervical vertebrae) and caudal (unpaired and extending from the spinous process of T1 to the spinous processes of caudal cervical vertebrae) parts. Cattle also lack nuchal ligament bursae.

Nuchal ligament bursae

In horses, there are 2 bursae associated with the nuchal ligament in the cervical region, and there is 1 bursa located caudally over the thoracic vertebrae (Fig. 10.10-7A). The **cervical bursae** are the **atlantal** or **cranial nuchal bursa** (bursa subligamentosa nuchalis cranialis), which is consistently present between the dorsal arch of the atlas (C1) and the funicular part of the nuchal ligament, and the **caudal nuchal bursa** (bursa subligamentosa nuchalis caudalis), which is inconsistently present in horses and is located between the spinous process of the axis (C2) and the funicular part of the nuchal ligament. The **thoracic nuchal bursa** is the **supraspinous bursa** (bursa subligamentosa supraspinalis) located between the nuchal ligament and spinous processes of T2–T3.

Cranial nuchal bursa

The **cranial nuchal bursa** is a bilobed structure located ventrolaterally to the funicular part of the nuchal ligament (Fig. 10.10-8). The bursa is divided by an incomplete septum composed of **synovium**, a variable amount of adipose tissue, and bundles of fibers extending from the laminar part of the nuchal ligament. The nondistended bursa lies on the middle and caudal thirds of C1, over its dorsal arch, and extends caudally over the articulation between C1 and C2. The contours of the cranial nuchal bursa follow the shape of the nuchal ligament dorsally and the atlas ventrally. The bursa is delineated laterally and ventrolaterally by the **splenius muscles** and tendons of insertion of the **semispinalis capitis muscles**. The **rectus capitis dorsalis muscles** are found deep to the bursa, passing between the semispinalis capitis muscle and the dorsal surface of the atlas.

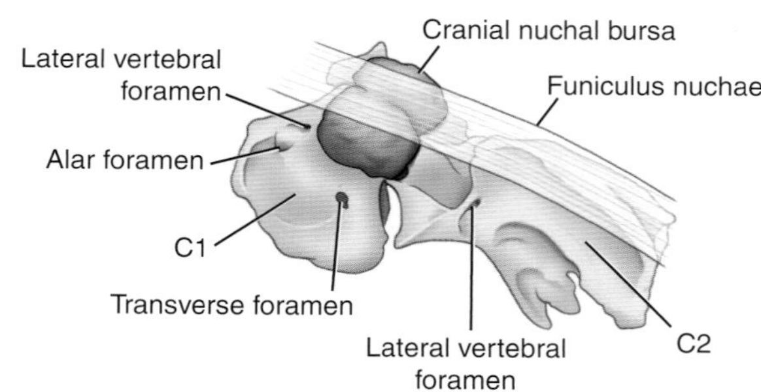

FIGURE 10.10-8 The bilobed cranial nuchal bursa lying over the middle and caudal thirds of the dorsal arch of C1 and extending caudally over the articulation between C1 and C2. The bursa is divided by an incomplete midline septum.

SEPTIC NUCHAL AND SUPRASPINOUS BURSITIS

Although currently relatively uncommon, septic nuchal bursitis (known as "poll evil") and septic supraspinous bursitis ("fistulous withers," Fig. 10.10-9) are 2 conditions that should be considered when evaluating horses with swelling and pain in the poll or wither regions. Due to an effective eradication program in North America, *Brucella abortus/suis* is no longer the etiologic agent of main concern; however, testing is recommended in cases where septic nuchal or supraspinous bursitis is suspected due to its public health implications. Although there are reports of managing septic nuchal bursitis endoscopically, most cases of both poll evil and fistulous withers require thorough debridement of the affected area due to the amount of tissue affected, to facilitate drainage, and to permit healing by second intention. The use of a dye, such as methylene blue, injected into the affected structure before debridement can facilitate identification of infected and necrotic tissues to be removed. These cases typically require multiple debridements and long-term antibiotic therapy. Not uncommonly, horses with fistulous withers develop septic osteomyelitis of the spinous processes immediately ventral to the supraspinous bursa. In these cases, a partial ostectomy of the spinous processes may be required to resolve the infection. Prognosis for both locations for bursal infection is guarded; horses seropositive for *B. abortus* should be humanely euthanized due to public health concerns.

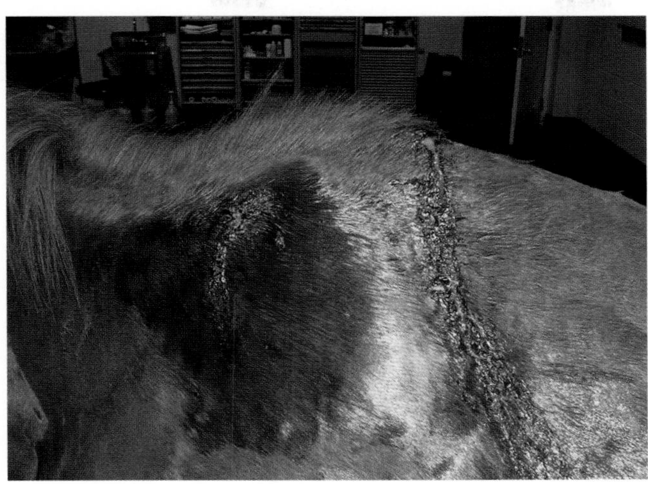

FIGURE 10.10-9 Draining tracts at the withers of a horse with infection of the supraspinous bursa ("fistulous withers").(Photograph courtesy of M. Gerard.)

When not diseased, the cranial nuchal bursa contains only a small amount of synovial fluid, making visualization on ultrasound almost impossible.

When fully distended (a volume of 60 mL), the cranial nuchal bursa extends over half of the dorsal surface of the atlas in a craniocaudal direction and extends laterally beyond the margins of the laminar part of the nuchal ligament. Fibers of the laminar part of the nuchal ligament extend cranially to attach to the funicular part of the nuchal ligament at the level of the atlas. At this location, there is no bony attachment of the nuchal ligament to the atlas. More caudally, the laminar fibers contact and attach to the dorsal aspect of C2 and on each remaining cervical vertebra—i.e., to C7.

Regional blood supply, lymphatics, and innervation

Blood vessels emerging from the **alar foramen** of C1 (Fig. 10.10-8)—including the **ramus descendens** of the **vertebral artery**, the **ramus occipitalis** of the **occipital artery**, and their accompanying veins—are located at the most craniolateral aspect of the bursa. These vessels track laterally to the cranial nuchal bursa in a caudal direction. Ventral to the bursa, branches of the **deep cervical artery** are located between the laminar part of the nuchal ligament and the deep surface of the semispinalis capitis muscle (Fig. 10.10-7B).

Lymphatic drainage of the head and ventral parts of the neck are via a group of lymph nodes located in the parotid, pharyngeal, and deep cervical regions referred to as the **parotid**, **retropharyngeal** (lateral and medial groups), and **deep cervical lymph nodes**, respectively. **Superficial cervical lymph nodes** also drain neck structures and their efferents pass to the **deep cervical lymph nodes**. Drainage from the upper part of the head is primarily via the **medial retropharyngeal lymph nodes**.

The dorsal branch of the **C1 spinal nerve** also exits through the alar foramen, providing branches that track cranially and caudally, and which remain lateral to the cranial nuchal bursa.

Selected references

[1] García-López JM, Jenei T, Chope K, Bubeck KA. Diagnosis and management of cranial and caudal nuchal bursitis in four horses. J Am Vet Med Assoc 2010;237:823–9.

[2] Abuja GA, García-López JM, Manso-Díaz G, Spoormakers TJ, Taeymans O. The cranial nuchal bursa: anatomy, ultrasonography, magnetic resonance imaging and endoscopic approach. Equine Vet J 2014;46:745–50.

[3] Bergen AL, Abuja GA, Bubeck KA, et al. Diagnosis, treatment and outcome of cranial nuchal bursitis in 30 horses. Equine Vet J 2018;50:465–9.

[4] Schramme M, Schumacher J. Management of bursitis. In: Auer JA, Stick JA, Kummerle JM, Prange T, editors. Equine surgery. 5th ed. St. Louis, MO: Elsevier Inc.; 2019. p. 1403.

CHAPTER 10

CASE 10.11

Esophageal Obstruction

Judith Koenig[a] and Shune Kimura[b]
[a]University of Guelph Ontario Veterinary College, Guelph, Ontario, CA
[b]Department of Clinical Sciences, Auburn University College of Veterinary Medicine, Auburn, Alabama, US

Clinical case

History

A 14-year-old Dutch Warmblood gelding was presented with a 1-day history of sudden onset of anorexia and a bilateral green nasal discharge. He was observed to repeatedly cough make unsuccessful efforts to swallow.

Physical examination findings

On presentation, bilateral nasal discharge containing green ingesta was observed (Fig. 10.11-1), along with ptyalism (Gr. *ptyalismos* excessive flow), agitation, and sweating. His mucous membranes were pink, yet slightly tacky with a delayed CRT of 3 s. Skin tenting was slightly delayed. He was mildly tachycardic (48 bpm) and auscultation over the cranioventral lung field on both sides revealed crackles. Gastrointestinal sounds were normal in all 4 quadrants.

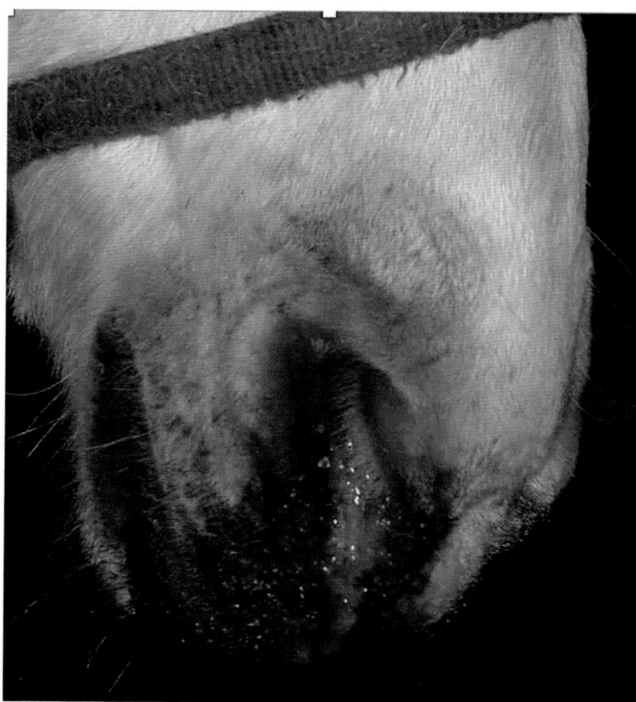

FIGURE 10.11-1 Right nasal discharge, comprising food material and saliva, was observed at both nostrils.

Comparative Veterinary Anatomy: A Clinical Approach. https://doi.org/10.1016/B978-0-323-91015-6.00147-3

Rectal temperature was 100 °F (37.8 °C). Palpation of the left jugular groove revealed mild swelling in the mid-cervical region. A nasogastric tube was passed through the left nostril into the esophagus and could not be advanced beyond the palpable swelling in the mid-cervical region.

Differential diagnoses

Intraluminal esophageal obstruction ("choke") due to food impaction or a foreign body; external trauma to the esophagus—e.g., a kick from another horse or colliding with a fixed obstacle, resulting in severe swelling; dysphagia, due to equine protozoal myeloencephalitis, grass sickness, or other neurological disease; and space-occupying neck lesion—e.g., abscess, neoplasia

Diagnostics

A neurological examination revealed no abnormalities. The gelding was sedated for an oral examination, which was normal. Endoscopic examination of the nasopharynx and both guttural pouches appeared normal, except for traces of food material and saliva present in the nasopharynx. Small amounts of food material, increased mucus, and hyperemia of the mucosa were seen in the cranial trachea. The endoscope was advanced into the esophagus, and the esophageal lumen was completely obstructed by food material appearing to contain apple pieces and mucus approximately 100 cm (39.4 in.) caudal to the right nostril (Fig. 10.11-2). Due to the food material seen in

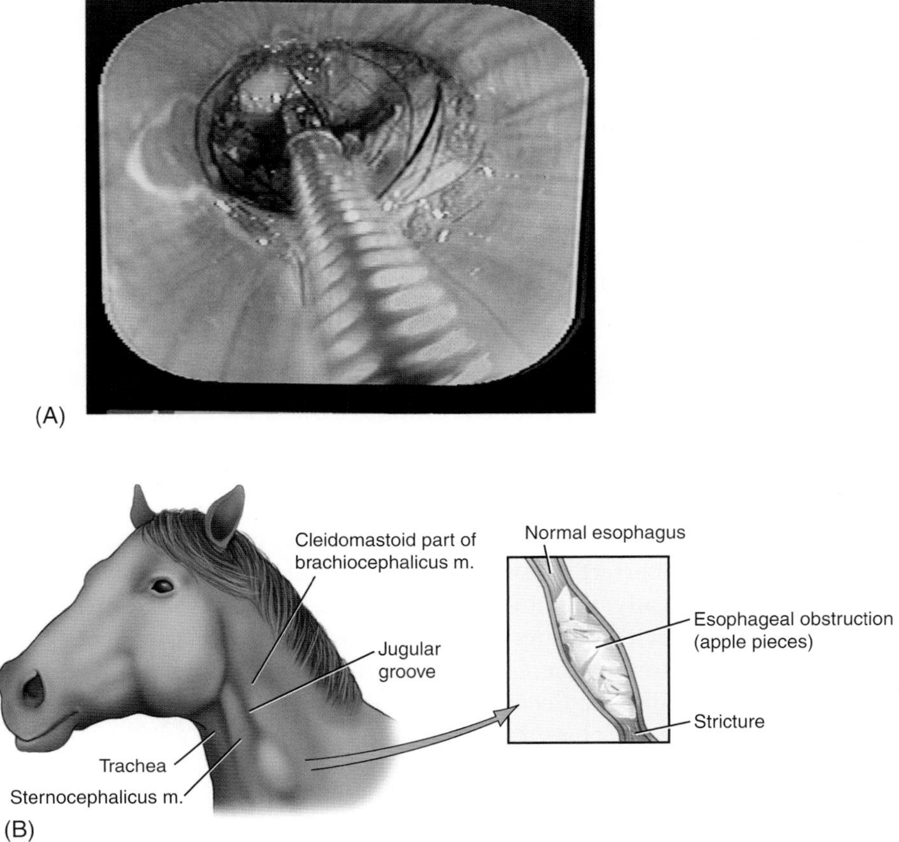

FIGURE 10.11-2 (A) Endoscopic view of a "choke" (an obstruction comprised of hair, food, and shavings) lodged inside the esophagus. Note that an endoscopic biopsy instrument is being used to try to dislodge the obstruction. (B) A midcervical esophageal obstruction. Notice the "bulge" in the jugular groove representing the site of the esophageal obstruction.

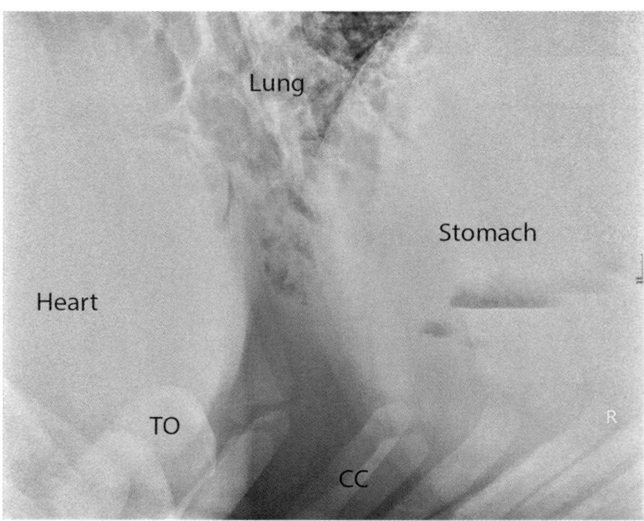

FIGURE 10.11-3 Lateral radiograph of the equine thorax showing consolidation (opacification) of the lung in the caudoventral lung field (cranioventral thorax), consistent with aspiration pneumonia. Key: TO, tuber olecrani; CC, costal cartilages (ossified).

the trachea and the crackles heard on auscultation, thoracic radiographs were obtained and confirmed evidence of aspiration pneumonia (Fig. 10.11-3).

Diagnosis

Esophageal obstruction/impaction (frequently referred to as "choke," even though the trachea/airway is unaffected) with secondary dehydration and aspiration pneumonia

Treatment

The horse was started on IV NSAIDs and broad-spectrum antimicrobial therapy. Following sedation, a gentle attempt was made to dislodge the obstruction and push it into the stomach with a nasogastric tube; this was not successful. Therefore, an endotracheal tube was placed into the esophagus, and the cuff was inflated to prevent liquid and food material from entering the nasopharynx and the trachea. A nasogastric tube was then advanced through the endotracheal tube into the esophagus. With the endotracheal tube cuff inflated, warm water was gently pumped through the nasogastric tube into the esophagus to fragment the obstructing material and flush it retrograde through the inserted tubes or move it further aborally. Care was taken not to excessively distend the esophagus to avoid injury and possible rupture. After repeated unsuccessful lavage and breakdown of the esophageal obstruction, the horse was placed in an unbedded stall, held off food and water, and administered IV fluid therapy to correct serum electrolyte imbalances and dehydration. Twelve hours later, the horse was reevaluated endoscopically; the obstruction was unchanged. A second lavage procedure was performed under sedation as before, with no success in resolving the obstruction. The mucosa cranial to the obstruction was edematous and hyperemic. Increasingly, concerns of possible esophageal rupture (see side box entitled "Esophageal rupture") guided a decision to perform an esophagotomy (Gr. *eosin* to carry + *phagos* to eat + Gr. *tome* a cutting) to remove the obstructing material (see side box entitled "Surgical approaches to the cervical esophagus"). General anesthesia and surgery proceeded routinely without complications. A feeding tube was placed through the esophagotomy and subsequently removed after 14 days (Fig. 10.11-4). The esophagotomy was managed by second intention healing after tube removal. The horse was discharged and administered broad-spectrum antimicrobial therapy for a further 6 weeks to treat the pneumonia. There were no long-term complications.

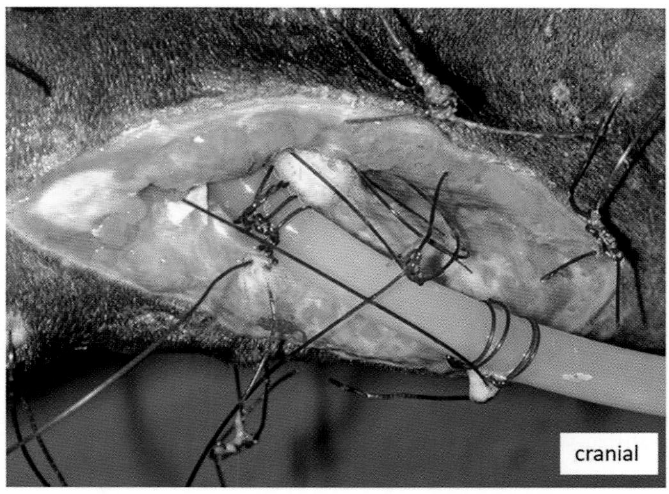

FIGURE 10.11-4 A feeding tube is seen passing into the esophagus that was placed after an esophagotomy was performed to remove an obstruction. The tube is usually removed 10–14 days after surgery, and the esophagus and layers of the overlying tissues are left to heal by second intention.

ESOPHAGEAL RUPTURE

Rupture of the esophagus can occur if an obstruction results in necrosis of the esophageal wall, secondary to blunt trauma to the neck (such as from a kick), or it may be iatrogenically induced through an incorrectly passed nasogastric tube. A ruptured esophagus allows the leakage of saliva, water, and food material into the surrounding tissues. This results in severe swelling of the tissues secondary to cellulitis and bacterial contamination, quickly leading to tissue necrosis. Left untreated, cervical swelling can compress the trachea, resulting in respiratory distress, septicemia, and possible neurological deficits due to inflammation of closely associated nerves (vagosympathetic trunk and recurrent laryngeal nerve). Saliva and other contaminated matter can migrate caudally along fascial planes adjacent to the esophagus. Ultimately, contamination can pass through the thoracic inlet into the mediastinum and the pleural cavities, resulting in septic mediastinitis and pleuritis, respectively. A key to successful treatment is timely surgical drainage, debridment, and lavage of the diseased tissues.

SURGICAL APPROACHES TO THE CERVICAL ESOPHAGUS

A ventral approach is used for surgeries involving the cranial one-third of the cervical esophagus. The paired sternohyoideus and sternothyroideus muscles are separated along the midline, exposing the trachea; the trachea is retracted to the left or right; the loose adventitia is separated until the ventral part of the esophagus is located on the midline.

A lateral or ventrolateral approach (Fig. 10.11-5) is used for surgeries involving the middle or caudal one-third of the cervical esophagus, or for the placement of a feeding tube. The approach is typically on the left side. The skin and subcutaneous tissues are incised immediately ventral to the external jugular vein; the vein is retracted dorsally, and dissection continued between the sternocephalicus (sternomandibularis) and brachiocephalicus muscles to expose the deeper omohyoideus muscle, which is then incised (Fig. 10.11-6A). The carotid sheath and its contents are retracted dorsally (Fig. 10.11-6B); the esophagus is identified with the aid of a preplaced nasogastric tube.

The esophagus is exposed in each approach with carefully placed retractors and then, using a scalpel, is longitudinally incised through adventitial, muscular, submucosal, and mucosal layers (Fig. 10.11-6C), preserving the blood supply to the esophagus. A healthy esophagus is usually closed in 2 layers (Fig. 10.11-7A and B): the innermost layer, the mucosa and submucosa combined, is the primary holding layer, and the esophageal musculature is the outer layer. The adventitia is included in the muscle layer. The fascia surrounding the neck muscles and soft tissues are of clinical importance because the deep cervical fascia attaches caudally to the 1st rib, sternum, and the endothoracic fascia of the thorax. Any post-operative complication—e.g., infection—may gravitate along and deep to the fascial layers in to the thoracic cavity, resulting in mediastinitis or pleuritis (see side box entitled "Esophageal rupture").

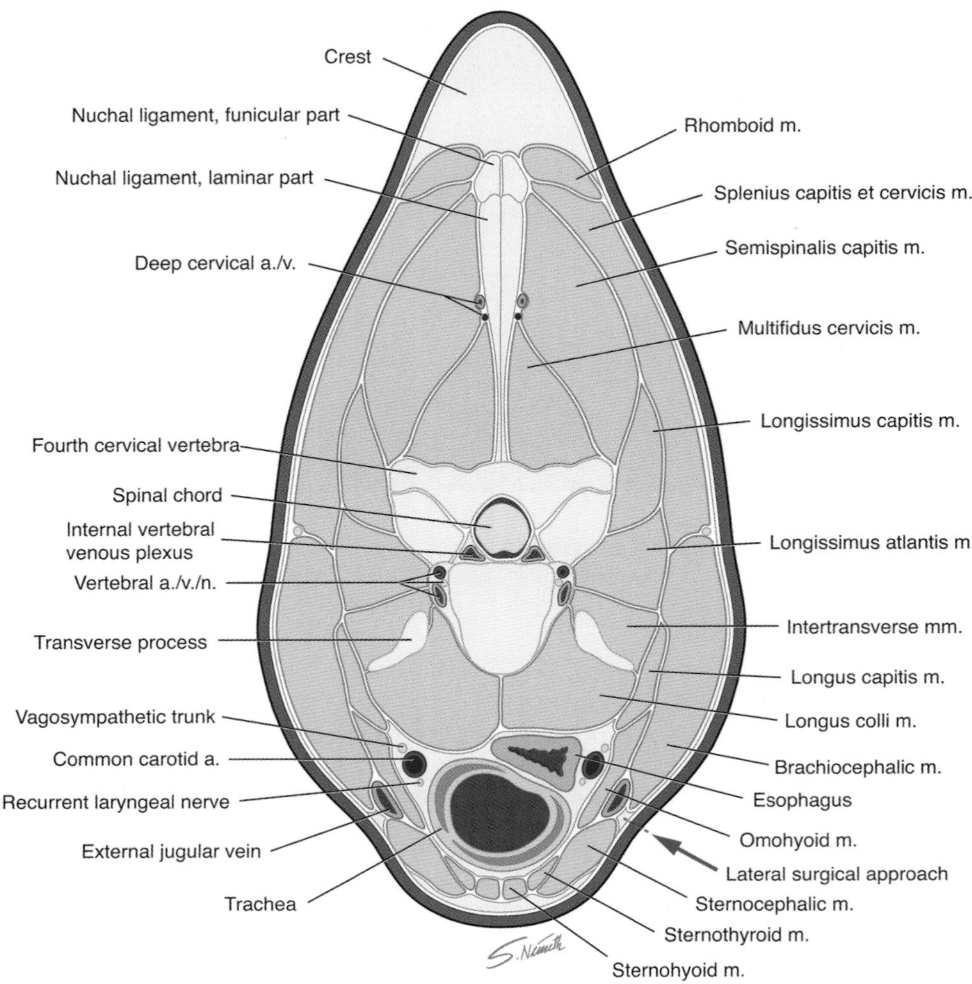

FIGURE 10.11-5 Cross-sectional anatomy of the neck of the horse at the level of the 5th cervical vertebra. The *red arrow* and *dashed red line* indicate the location for lateral/ventrolateral surgical access to the middle and caudal cervical esophagus.

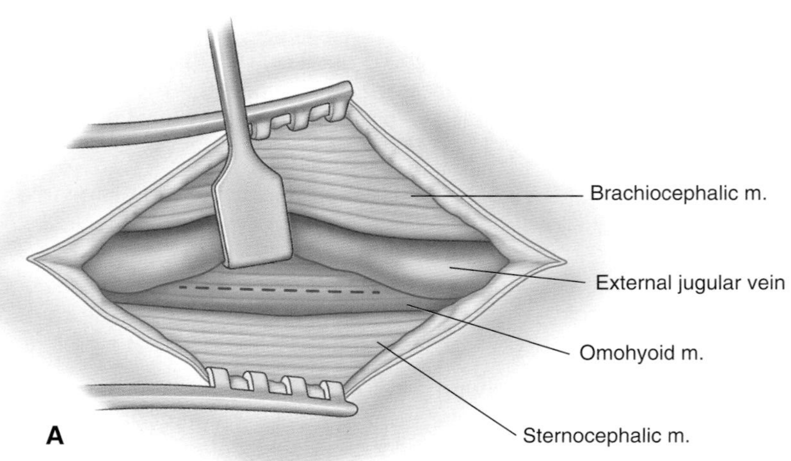

FIGURE 10.11-6 The surgical anatomy encountered during the ventrolateral approach to the esophagus in a horse approximating the level of the 5th cervical vertebra. (A) After the sternocephalicus and brachiocephalicus muscles are separated, continued deep dissection in the jugular groove ventral to the dorsally retracted external jugular vein exposes the omohyoideus muscle, which is then sharply incised.

(Continued)

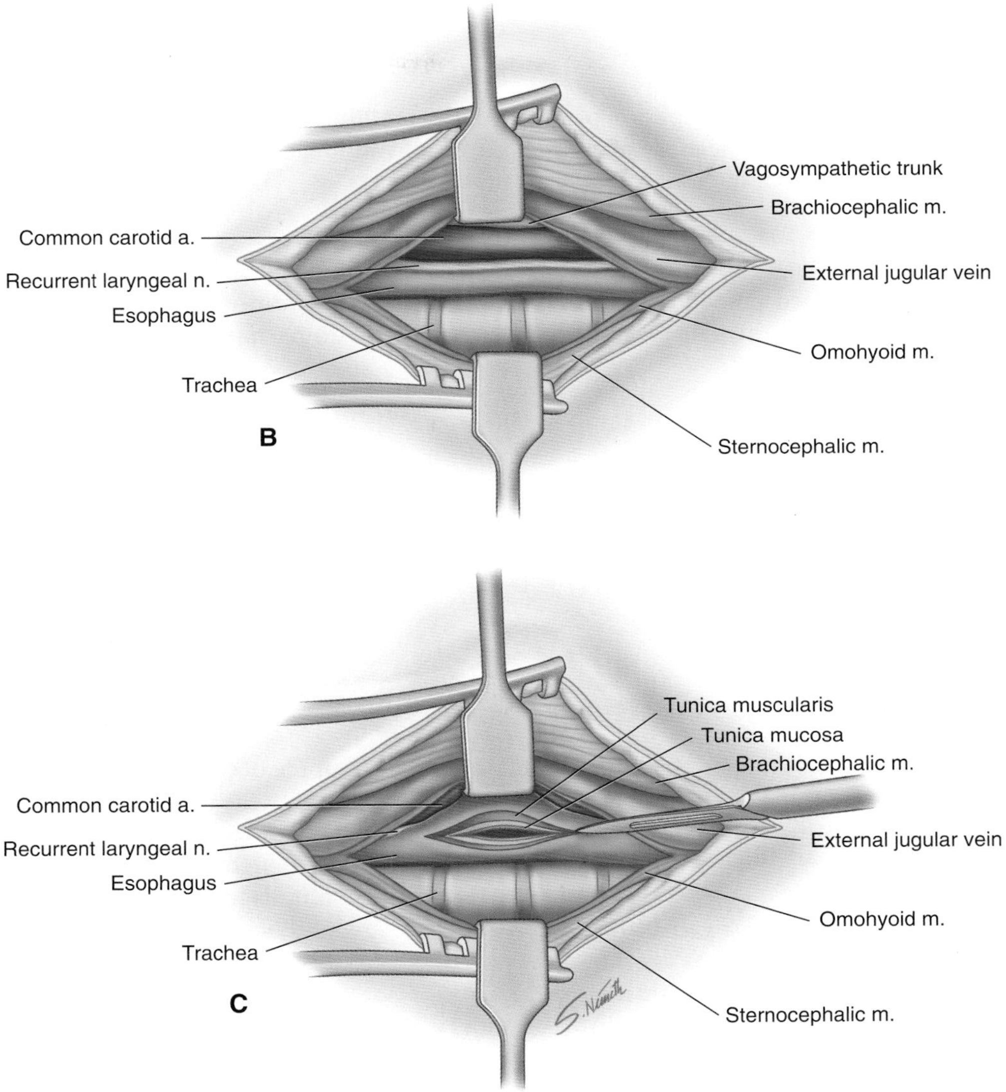

Vagosympathetic trunk
Brachiocephalic m.
Common carotid a.
External jugular vein
Recurrent laryngeal n.
Esophagus
Omohyoid m.
Trachea
Sternocephalic m.
B
Tunica muscularis
Tunica mucosa
Brachiocephalic m.
Common carotid a.
External jugular vein
Recurrent laryngeal n.
Esophagus
Omohyoid m.
Trachea
Sternocephalic m.
C

FIGURE 10.11-6, CONT'D (B) After incising the omohyoideus muscle, a careful dissection is performed to identify the esophagus, trachea, and the left carotid sheath containing the common carotid artery, vagosympathetic trunk, and recurrent laryngeal nerve. (C) After retracting the carotid sheath structures dorsally, the esophagus is exposed and then is sharply incised longitudinally through adventitial, muscular, submucosal, and mucosal layers.

Tunica mucosa
Tela submucosa
Tunica adventitia
Tunica muscularis
Esophagus
A
B

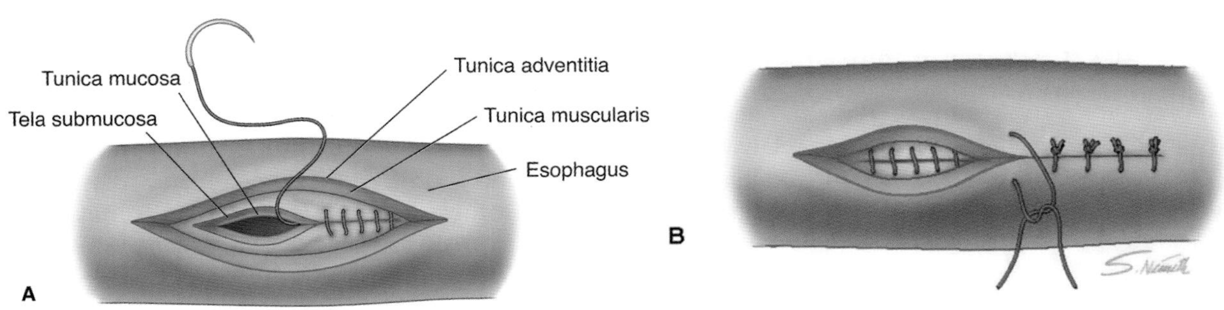

FIGURE 10.11-7 (A) The first layer closed in an esophagotomy is the mucosa and submucosal layers. (B). The second layer of the closure is the muscular coat and any available adventitia.

Anatomical features in equids

Introduction

The digestive system of the horse, like all species, begins with the mouth and moves aborally. Once food is ingested and chewed, the food (bolus) enters the esophagus. The process of swallowing (see Case 10.6) ensures food does not normally enter the trachea. The esophagus is reviewed in this section.

Function

The esophagus transports ingested/prehended matter (food, water, oral medications) from the mouth to the stomach. Once swallowed, food is carried by peristalsis into the equine stomach.

Esophageal impactions typically occur in locations where the lumen of the esophagus narrows: most common is caudal to the larynx in the cervical portion of the esophagus at the level of the 2nd cervical vertebra; the thoracic inlet; the base of the heart; or the terminal portion of the esophagus.

Esophagus

The **esophagus** is a muscular, tubular structure up to 2 m (78.7 in.) in length in mature horses. Transit time for a food bolus through the esophagus is ≤20 s. The esophagus is commonly divided into 3 general parts based on location: cervical, thoracic, and abdominal.

Cervical part of the esophagus (see Landscape Fig. 4.0-2)

The esophagus originates at the caudal border of the short esophageal vestibule of the **laryngopharynx**, immediately dorsal to the **cricoid cartilage** of the larynx. Its path initially continues dorsal to the trachea and ventral to the **longus colli muscle**. At approximately the 4th cervical vertebra, which is the junction of the cranial and middle thirds of the length of the cervical esophagus, the esophagus passes to the left of the trachea and lies closer to the skin. In some horses, the esophagus passes to the right of the trachea—however, this is uncommon. The cervical esophagus encompasses 50% of the total length of the esophagus.

In this region, the tip of a nasogastric tube or a bolus of food may be seen passing by as the skin moves over the tube or food. Of the 3 parts of the esophagus, the cervical part is the most accessible surgically.

Adjacent to the cervical esophagus (bilaterally in the cranial part, and then on the left as the esophagus moves to the left side) lies the **carotid sheath** surrounding the common carotid artery, vagosympathetic trunk, recurrent laryngeal nerve, and lymphatic tracheal trunk. The **external jugular vein** lies more superficially within the neck and is separated from the esophagus by the **omohyoideus muscle** in the cranial region (Fig. 10.11-5) (see side box entitled "Surgical approaches to the cervical esophagus").

Thoracic part of the esophagus (see Landscape Fig. 4.0-7)

At the thoracic inlet, the beginning of the thoracic esophagus lies ventrolateral to the trachea, and between it and the left 1st rib. The esophagus returns to a relatively dorsal position within the cranial **mediastinum**, passing dorsal to the base of the heart and tracheal bifurcation, and crossing the aorta on the right. In the caudal mediastinum, continuing towards the diaphragm, the esophagus is accompanied by the **dorsal** and **ventral vagal trunks**. The **esophageal hiatus** (an opening in the right **crus of the diaphragm** to allow the esophagus and vagal trunks to pass through) is located on a transverse plane, level with the 14th thoracic vertebra and approximately 12 cm (4.7 in.) below the ventral surface of the vertebral column.

Abdominal part of the esophagus

The abdominal part of the esophagus is the shortest span, running from the esophageal hiatus of the diaphragm to the cardia of the stomach (**pars cardiaca**). The end of the esophagus is marked by its position oral to the **cardiac sphincter**, which in the horse is thicker compared with other species.

At times, it may be difficult to pass a nasogastric tube through the cardiac sphincter, especially in horses with a distended stomach. Patience and the introduction of a small amount of water may facilitate passage through this sphincter. Nasogastric intubation is a valuable tool for the diagnosis and management of esophageal disorders and colic, as well as the means by which veterinarians can provide enteral (Gr. *enteron* intestine; by way of) fluid therapy. Therefore, an understanding of, and comfort with, equine esophageal anatomy is essential in successfully performing and interpreting this diagnostic and therapeutic procedure (see side box entitled "Nasogastric intubation").

Microscopic esophageal anatomy

Microscopically, the esophagus consists of several layers (from internal to external): tunica mucosa, tela submucosa, tunica muscularis, and tunica adventitia (in the cervical portion) or tunica serosa (in the thoracic and abdominal portions).

The **tunica mucosa** is the inner elastic layer which provides a barrier function, allows movement, and gives structure to the esophagus. These functions are achieved by the 3 layers of the mucosa: stratified squamous epithelium, lamina propria, and lamina muscularis. The **stratified squamous epithelium** provides an inner barrier for the esophagus. The **lamina propria** consists of collagen and elastic fibers that provide structural support. The **lamina muscularis** consists of smooth muscle bundles that take part in peristaltic activity. The mucosa, together with its submucosal tissue, is freely movable relative to the muscular layer. The mucosal layer lies as deep longitudinal folds in the non-distended esophagus. Deep and immediately adjacent to the mucosa lies the **tela submucosa**. This layer contains the blood supply (arteries and veins), lymphatic vessels, and nerves. In some species, glands are present in the submucosa throughout the esophagus to aid in lubricating food; however, in the horse these are only present at the junction of the pharynx and esophagus.

NASOGASTRIC INTUBATION

Nasogastric intubation is a common and a relatively benign procedure when done correctly. The most injurious complication (though uncommon) is to inadvertently intubate the trachea rather than the esophagus, and not recognize this error before instilling a foreign substance. Remember that the trachea is a cartilaginous and rigid structure in comparison with the collapsed, soft tissue structure of the esophagus. Because of this difference, passing a nasogastric tube into the trachea results in the loss of resistance compared to if it were being directed down the esophageal lumen. The clinical impression is like passing a tube into an open pipe that would pass with little effort, as compared to the esophagus, a more elastic structure, in which greater effort is needed to overcome the resistance created by the collapsed mucosa encircling the tube.

Additionally, when properly placed, the tube can be seen subcutaneously passing through the esophagus on the left side of the neck as previously mentioned.

Lastly, upon entering the stomach, the smell and presence of feed material should be confirmed before administering anything via the tube.

The **tunica muscularis** receives the paired **lateral longitudinal esophageal muscles** (which form a functional [cranial] esophageal sphincter) at its origin, dorsal to the cricoid cartilage, and then continues as 2 muscular layers. These muscular layers are initially arranged elliptically, then spirally as they pass to the stomach in a distinct inner circular layer and an outer longitudinal muscle layer. The muscular layer is striated for most of its length, transitioning to smooth muscle around the level of the base of the heart. At the origin of the esophagus, the tunica muscularis is about 0.5 cm (0.2 in.) thick, and this thickness progressively increases towards the stomach to approximately 1.2–1.5 cm (0.5–0.6 in.) at the cardia.

The tunica muscularis is the target for most drugs used in the medical management of esophageal obstruction. Sedatives such as xylazine or detomidine and anticholinergics such as atropine or N-butylscopolammonium bromide (Buscopan™) help to relax different muscle types to increase the diameter of the esophagus and aid movement of the food bolus. Oxytocin, normally used to increase (uterine) or decrease (cervical) muscle tone in the female reproductive tract by binding to oxytocin receptors in smooth muscle, also acts on muscle cells in the esophagus, allowing for easier passage of the impaction.

The **tunica adventitia** is a loose, fibrous connective tissue layer surrounding the tunica muscularis that allows the esophagus to move independently of adjacent tissues as it expands over variably sized food boluses and as the horse lowers and raises its neck. This layer also includes vessels (blood and lymph) and nerve fibers. In the caudal one-third of the equine esophagus—i.e., starting at the thoracic inlet—this layer is the **tunica serosa**, which is a fluid-secreting membrane that lines and borders body cavities. These differences in structures stem from their location—serosa covers structures within the pleural, pericardial, peritoneal, and vaginal cavities, while adventitia exists outside of these cavities.

Esophageal blood supply, lymphatics, and innervation

The blood supply of the esophagus originates from branches of the **common carotid artery**, the **bronchoesophageal artery**, and the **gastric arteries**. Lymphatic drainage for the cervical part of the esophagus is via the **cranial, middle**, and **caudal deep cervical lymph nodes**, and that of the thoracic and short part of the abdominal esophagus is via the **cranial** and **caudal mediastinal lymph nodes**. Nerves regulating the function of the esophagus are provided by the **glossopharyngeal** and **vagus nerves** (CN IX and X), **sympathetic trunk**, and the **myenteric ganglion cells** in the muscle layers. The striated muscle is supplied by **pharyngeal** and **esophageal branches** of the vagus nerve, and the smooth muscle by both the intrinsic and autonomic nervous systems.

Selected references

[1] Meyer GA, et al. The effect of oxytocin on contractility of the equine oesophagus: a potential treatment for oesophageal obstruction. Equine Vet J 2000;32(2):151–5.

[2] Orsini JA, Schaer BD. Gastrointestinal system. In: Orsini JA, Divers TJ, editors. Equine emergencies: Treatment and procedures. Saunders; 2013. p. 177–81.

[3] Koenig JB, Silveira A, Cribb NC, Piat P, Laverty S, Sorge US. Clinical indications, complications, and long-term outcome of esophageal surgeries in 27 horses. Can Vet J 2016;57(12):1257–62.

[4] Feige K, Schwarzwald C, Fürst A, Kaser-Hotz B. Esophageal obstruction in horses: a retrospective study of 34 cases. Can Vet J 2000;41(3):207–10.

CASE 10.12

Cervical Vertebral Osteoarthritis

Amy L. Johnson
Large Animal Medicine and Neurology, Department of Clinical Studies - New Bolton Center, University of Pennsylvania School of Veterinary Medicine, Kennett Square, Pennsylvania, US

Clinical case

History

A 9-year-old Irish Sport Horse gelding was presented for lameness evaluation. The gelding was competing successfully as a preliminary level eventer until 3 weeks before presentation, at which time he appeared painful. The gelding displayed a right thoracic limb lameness characterized by a shortened cranial phase. The gelding was also sensitive to palpation of the right cervical musculature and thoracic epaxial musculature. The gelding was reluctant to bend to the right and tended to raise his head in the air when cantering to the left. Mild symmetrical muscle atrophy across his top line was observed at the time of presentation. The referring veterinarian had attempted to localize the lameness using diagnostic analgesia, but no improvement was reported up to, and including, the level of median and ulnar nerve blocks.

Regional anesthesia of the median and ulnar nerves desensitizes the entire thoracic limb below the distal radius.

Physical examination findings

The gelding was quiet and alert with normal vital parameters and cardiothoracic auscultation. The horse had mild symmetrical muscle atrophy along the top line and gluteal region, with easily palpable vertebral dorsal spinous processes in the thoracolumbar region. Palpation revealed moderate sensitivity over the epaxial musculature, more on the right than on the left. The gelding was reluctant to bend his neck to the right when enticed with a food reward.

Gait evaluation in hand revealed a mild (grade 1/5) left thoracic limb lameness, exacerbated slightly by circling to the left, and a mild (grade 1/5) right thoracic limb lameness most obvious when circling to the right. The thoracic limb lamenesses were improved when lunged on soft ground, and therefore, they were considered most likely related to foot pain. During ridden (a.k.a. under saddle) evaluation, a much more severe lameness became apparent and the gelding showed a markedly shortened stride and toe drag with the right thoracic limb at the walk. This lameness was subtle when walking with a loose rein and low head carriage, but significantly exacerbated by increasing rein contact and asking the horse to raise and flex his head and neck into a frame. The gait abnormality was significantly improved at the trot and canter. The right thoracic limb lameness apparent under saddle was considered most likely due to a caudal cervical abnormality based on historical difficulty in localization using diagnostic analgesia and the strong influence of head and neck position.

Neurologic examination revealed equivocal proprioceptive deficits in all 4 limbs. These deficits included infrequent scuffing of the toes, mild circumduction of the outside limbs during tight circles, and over-reaching with the thoracic limbs when walking down a hill, particularly with the head raised. Attitude, behavior, and cranial nerve examination were within normal limits. These findings were considered of minimal clinical importance compared with the prominent right thoracic limb lameness during the ridden exercise.

Comparative Veterinary Anatomy: A Clinical Approach. https://doi.org/10.1016/B978-0-323-91015-6.00148-5

Differential diagnoses

Caudal cervical osteoarthritis and foraminal stenosis (most likely); less likely on the differential list—discospondylitis, discospondylosis, meningitis/meningomyelitis, neoplasia, and previous trauma

Diagnostics

Survey laterolateral cervical vertebral radiographs were obtained (Fig. 10.12-1A–D). The radiographs revealed a mild enlargement and osteoarthritis of the caudal cervical articulations, including C5–6 and C6–7. No obvious narrowing of the vertebral canal was present. Computed tomography was recommended to further investigate the caudal cervical region.

Standing robotic cone-beam CT (see Section 2.4) of the caudal cervical region showed caudal cervical osteoarthritis with enlargement of the articular processes and marked foraminal stenosis at C5–6 on the right side with moderate foraminal stenosis at C5–6 on the left side (Fig. 10.12-2A–C).

Diagnosis

Caudal cervical osteoarthritis causing foraminal stenosis at C5–6 (with right side more severe than the left side)

Treatment

A period of rest from forced exercise was recommended. A tapering course of a nonsteroidal anti-inflammatory drug, phenylbutazone, was prescribed. Ultrasound-guided intra-articular steroid injections were performed at C5–6 and C6–7. Finally, stretching and core-strengthening ground exercises were recommended. However, due to the marked bony changes and inability to address the foraminal stenosis directly, a guarded prognosis was given for return to consistent work under saddle, even with reduced expectations.

Relevant anatomical features in equids

Introduction

Like all mammals, the horse has 7 cervical vertebrae in the neck that surround the cervical spinal cord. The cervical vertebral column can be affected by several disease processes; cervical osteoarthritis and cervical vertebral stenotic myelopathy are commonly diagnosed causes of neck pain and neurologic signs, respectively. The nuchal ligament is intimately associated with the dorsal aspect of the cervical vertebral column, and the cervical spinal cord is confined within the cervical vertebral column. This case describes the anatomy of the cervical vertebrae, cervical spinal cord and spinal nerves, and the nuchal ligament. Additional information on the atlanto-occipital joint, nuchal ligament, and regional bursae can be found in Case 10.10. Regional blood supply and lymphatics are well covered in Case 10.10.

COLLECTION OF CSF IN THE CERVICAL REGION

Cerebrospinal fluid (CSF) collection in the horse can be performed at two locations in the cervical region. The atlanto-occipital approach is performed similarly in the horse as in other species. The horse is placed under general anesthesia in lateral recumbency with the head held in a flexed position (chin towards chest) such that the long axis of the head forms an approximately 90-degree angle with the long axis of the neck. The mane is clipped from the occipital protuberance (poll) caudally for approximately 6 in. (15.2 cm) (to the caudal aspect of the atlas), and this area is aseptically prepared. A 3.5-in. (9-cm) spinal needle is inserted on midline at the level of the palpable cranial edge of the wing of the atlas. The needle is advanced perpendicular to the crest in a line pointing to the bottom jaw until the tip is within the subarachnoid space and CSF is collected. The atlanto-axial (C1–C2) approach can be performed in the horse with ultrasound guidance using standing sedation. For this procedure, a small area on the dorsocranial neck is aseptically prepared, centered approximately 1.2 in. (3 cm) ventral from dorsal midline and immediately caudal to the caudal edge of the wing of the atlas, approximately one hand's breadth caudal to the ear. Ultrasound is used to visualize the spinal cord surrounded by CSF, and a 3.5-in. (9-cm) spinal needle is placed using ultrasound guidance to direct the needle dorsomedially into the subarachnoid space.

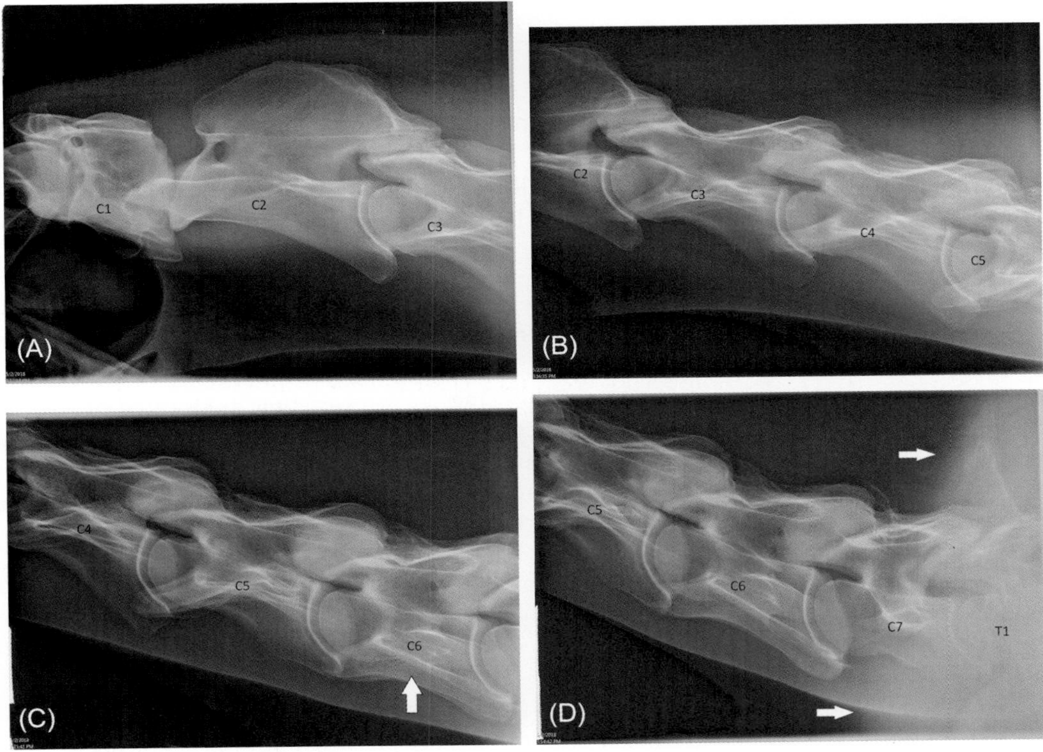

FIGURE 10.12-1 Survey laterolateral radiographs from the 9-year-old Irish Sport Horse gelding described in this case study. (A) Cranial cervical vertebrae (C1, C2, cranial C3). (B) Cranial to mid-cervical vertebrae (caudal C2, C3, C4, cranial C5). (C) Mid- to caudal cervical vertebrae (caudal C4, C5, C6). Note the "sled runner" appearance of the ventral transverse processes of C6 *(arrow)*. (D) Caudal cervical vertebrae (caudal C5, C6, C7, cranial T1). Note that C7 is shorter than the other cervical vertebrae. C7–T1 is well visualized in this image but is often challenging to see on radiographs because of the overlying shoulder musculature (seen here as the hazy opacity overlying the right side of the radiograph; *white arrows*).

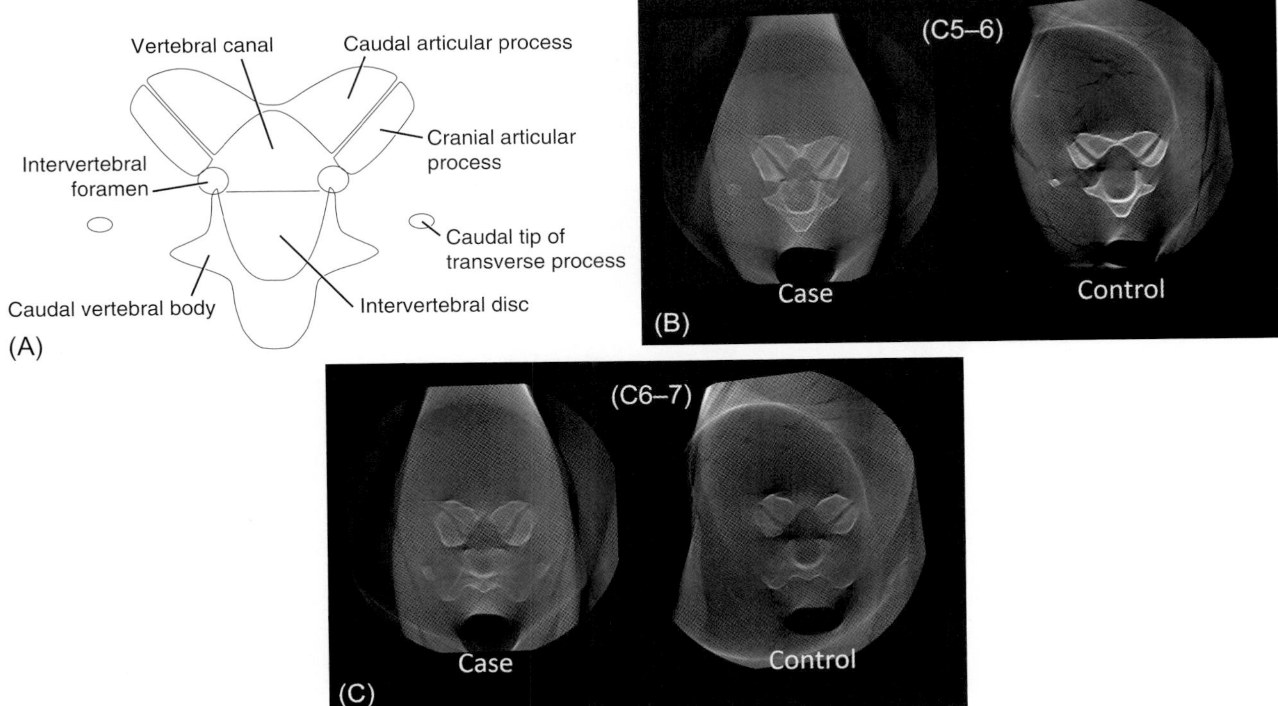

FIGURE 10.12-2 Transverse images of the caudal cervical vertebral column at the level of C5–6 and C6–7 articulations. (A) Labeled diagram of structures visible on transverse CT images. (B) Transverse standing robotic cone-beam CT image at the level of C5–6 of the 9-year-old Irish Sport Horse gelding ("Case") compared with a normal horse ("Control"). See Fig. 10.12-2A for anatomical features. (C) Transverse standing robotic cone-beam CT image at the level of C6–7 of the 9-year-old Irish Sport Horse gelding ("Case") compared with a normal horse ("Control").

Function

Neuronal pathways within the cervical spinal cord transmit signals from the brain to effector lower motor neurons, and sensory information from the body back to the brain. The cervical vertebral column, nuchal ligament, and cervical musculature support the weight of the head and allow excellent mobility so the horse can lower its head to the ground, raise it straight into the air, and bend its head laterally to reach the top of its pelvis with its nose.

> The relatively ventral location of the caudal cervical vertebrae is important to remember in choosing the appropriate location for IM injections in the cervical region (Fig. 10.12-3). Injections that are too close to the vertebral column can cause increased pain or even neurologic signs. The ideal window for cervical intramuscular injection is approximately one hand's breadth below the crest, one hand's breadth cranial to the scapula, and one hand's breadth above the vertebral column.

Cervical vertebrae

The first 2 vertebrae are located dorsally within the neck, but more caudal cervical vertebrae are located progressively more ventral in the neck, and the lowest point of the vertebral column is located at the **cervicothoracic junction**, dorsal to the shoulder joints.

The **1st cervical vertebra** (C1 or **atlas**) articulates with the caudal aspect of the skull at the **atlanto-occipital joint**, formed between the condyles of the skull and the corresponding concavities of the atlas. This joint generally has a single synovial cavity due to ventral convergence, and its movement is restricted to flexion and extension in the sagittal plane (nodding). The **wing** of the atlas is a prominent visible and palpable landmark. If one hand is placed over the wing of the atlas, the position of subsequent vertebrae can be estimated, approximately one hand's breadth for each vertebra.

The **2nd cervical vertebra** (C2 or **axis**) has a prominent **spine** dorsally and **dens** cranially. The **atlanto-axial** (C1–C2) **joint** is formed by the articular surfaces of the **ventral arch** of the atlas and the **body** and dens of the axis. There is a single synovial cavity and ligaments that secure the dens to the adjacent ventral arch of the atlas. The atlanto-axial joint allows rotation around a longitudinal axis (side-to-side movement).

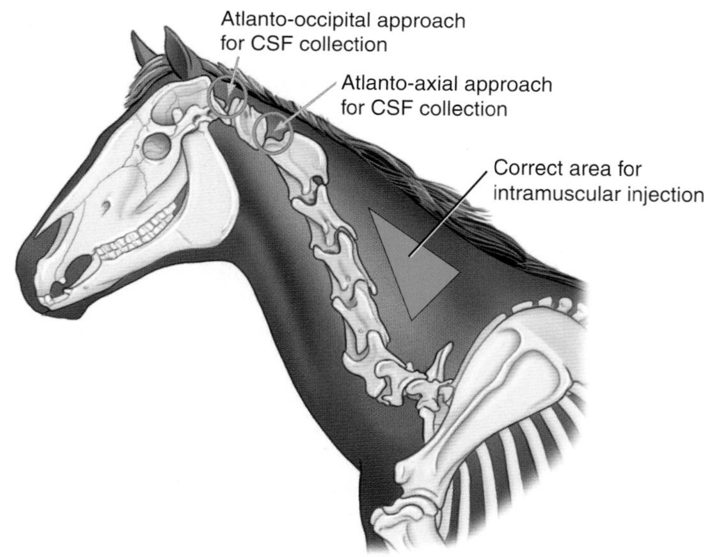

FIGURE 10.12-3 Diagram of the cervical region of the horse depicting the location of the cervical vertebrae, the "safe" area for intramuscular injection, and the location of 2 possible approaches for cerebrospinal fluid (CSF) collection. (Modified from Dyce KM, Sack WO, Wensing CJG. Textbook of veterinary anatomy, 2nd ed., Fig. 19-1, p. 514.)

The remaining cervical vertebrae (C3–C7) have rudimentary **spinous processes**, large divided **transverse processes**, and **cranial** and **caudal articular processes** ending in broad, roughly oval articular surfaces. From C2–C3 through C7–T1, the vertebrae articulate with each other by both an **intercentral articulation** and paired zygapophyseal (Gr. *zygon* yoke, + Gr. *faisis* nature, to grow = the 4 paired processes of a vertebra articulating with the vertebra above and below) synovial joints termed **articular process joints**. The articular surfaces are commonly referred to as **facets**, and the articular process joints as **facet joints**. Within each joint, the caudal articular processes of the cranial vertebra are axial to the cranial articular processes of the caudal vertebra (Figs. 10.12-2 and 10.12-4).

Plain radiographs are commonly used to assess for pathological changes in the cervical vertebral column and especially in the articular process joints. Normal joints have smooth margins, with easily identifiable intervertebral foramina (Fig. 10.12-5). The morphology of the vertebrae should be critically assessed for evidence of congenital, developmental, or degenerative disease. Many horses, particularly Warmblood breeds, have anatomical variation in the 6th cervical vertebra, with transposition of the ventral lamina of the transverse process onto C7. The clinical significance remains uncertain, with many normal horses having a morphologic variation. Some horses can have marked degenerative changes in the caudal cervical region, with fragmentation and approaching ankyloses (Gr. *ankylōsis* fusion of a joint due to progressive disease) (Fig. 10.12-6).

Radiographically, C1 and C2 are easily identified due to their unique shapes. The next 3 cervical vertebrae (C3, C4, and C5) are similarly shaped, and it can be difficult to distinguish between the C3–4 and C4–5 articulations unless C2 or C6 are present in the image as reference points. The transverse processes of C6 have an extra ventral lamina that projects caudally and ventrally, giving the ventral silhouette of this vertebra the appearance of a "sled runner" on lateral radiographs. Compared with C3–C6, C7 has a shorter vertebral body and sometimes a small dorsal spinous process protruding over the C6–7 articular process joint. This spinous process should not be mistaken for an osteophyte. Depending on radiographic equipment and technique, the C7–T1 articulation might be visible on radiographs; the articulation of the 1st rib with T1 is apparent as well as the more pronounced dorsal spinous process of T1.

The **ligamentum flavum** spans the dorsal half of the vertebral canal as thin elastic sheets contiguous with the articular process joint capsules, connecting adjacent vertebral arches. The **intervertebral foramina** are the spaces between the articular processes of the cranial vertebra and the vertebral body of the caudal vertebra at each articulation. **Intervertebral disks** are present from the C2–C3 articulation caudally; these disks are relatively thin. Although each consists of a peripheral **annulus fibrosus** and central **nucleus pulposus**, the boundary between these parts is less distinct than in most other species.

Intervertebral disc disease is a less important syndrome in horses than in dogs due to the difference in structure of the disc, namely, the indistinct nucleus pulposus. Aging changes can occur, including dehydration and fragmentation of the outer fibrous part, but calcification of the nucleus pulposus rarely occurs. High-velocity extrusion of the nucleus pulposus is not a clinical problem, and disc protrusion is considered relatively rare.

Conversely, articular process joint disease is very common in horses. This disease can be considered developmental (osteochondrosis) or degenerative (osteoarthritis). Proliferative changes in the articular process joints and ligamentum flavum can cause cervical pain, dorsolateral spinal cord compression, or foraminal stenosis depending on exactly where the bony or soft tissue hypertrophy occurs.

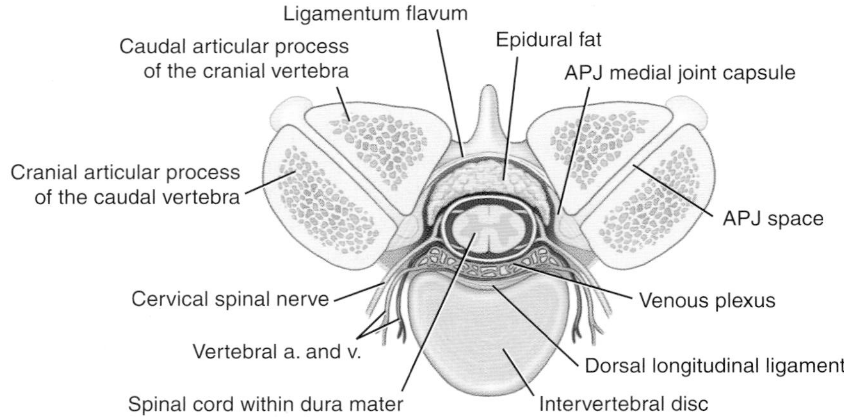

FIGURE 10.12-4 Cross-sectional drawing demonstrating the relationship between the articular process joint (APJ) and contents of the vertebral canal at the level of the intervertebral foramina: (Modified from Hepburn R. 2015.)

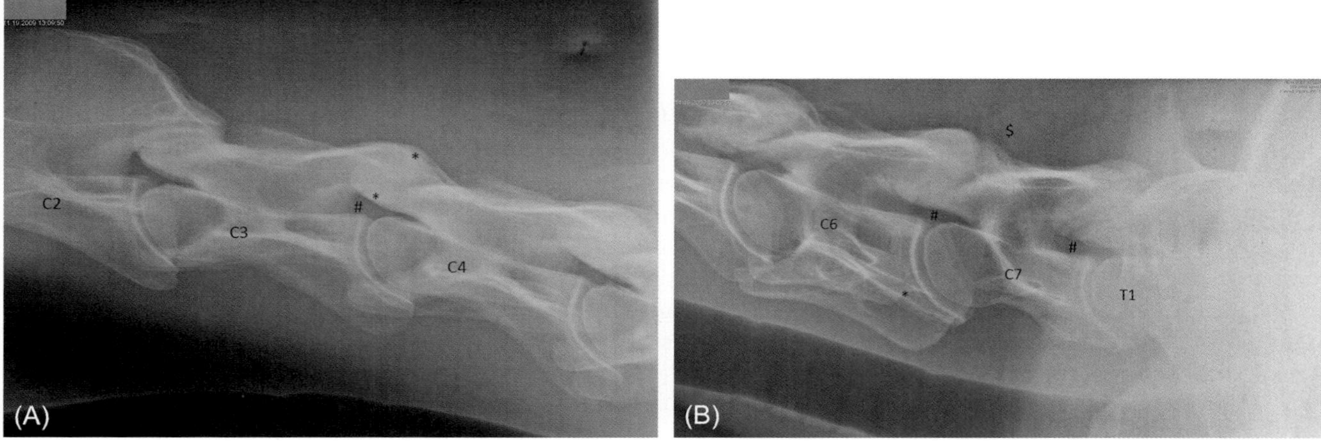

FIGURE 10.12-5 Normal cervical radiographs from an 8-year-old Warmblood horse. (A) Cranial cervical region. Note the smooth ventral and dorsal margins of the superimposed articular process joints (*) and the smoothly margined, triangular-shaped lucency representing the superimposed intervertebral foramina (#). (B) Caudal cervical region. Note the smooth ventral and dorsal margins of the superimposed articular process joints at each articulation, and the prominent, unobscured intervertebral foramina (#). This horse had an anomalous C6 vertebra, with asymmetric partial absence of the ventral lamina of the C6 transverse process (*). Also, note the normal C7 spinous process ($), which should not be mistaken for an osteophyte.

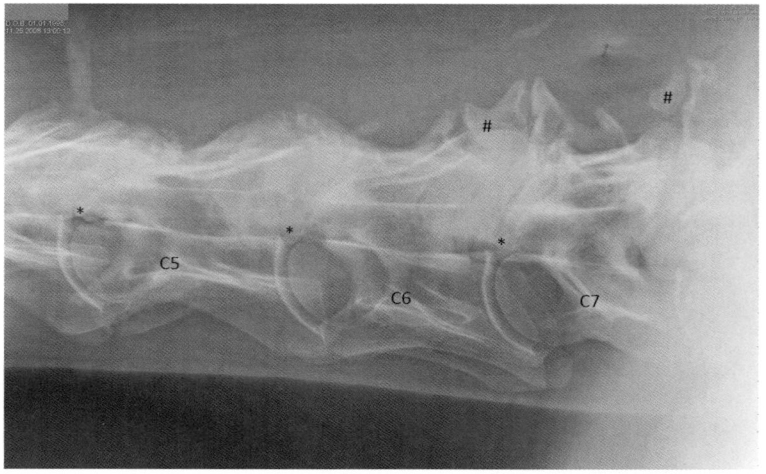

FIGURE 10.12-6 Caudal cervical radiograph from a 13-year-old Thoroughbred. Severe osteoarthritic changes are present at all projected articulations, with multiple osteophytes and bony fragmentation (#). All intervertebral foramina are obscured by the bony proliferation (*).

Cervical spinal cord and spinal nerves

The **spinal cord** is an elongated, relatively cylindrical structure that lies within the vertebral canal, surrounded by fluid. The spinal cord thickens at the **cervical** and **lumbar intumescences** (L. *intumescentia* a swelling). The spinal cord can be divided into segments corresponding to the **somites** (Gr. *sōma* body). The cervical spinal cord has 8 segments and, hence, 8 paired spinal nerves formed by the union of dorsal and ventral roots. These spinal nerves leave the vertebral canal through the appropriate intervertebral foramen. The 1st paired nerves emerge through the lateral foramina of the atlas, and the 2nd through 7th paired spinal nerves emerge cranial to the vertebra of the same number. The 8th spinal nerves emerge between the last cervical and 1st thoracic vertebrae. The last 3 segments (C6, C7, and C8) are part of the cervicothoracic intumescence and contribute to the brachial plexus.

"Somite" is an embryological term meaning the blocks into which the paraxial mesoderm is segregated to each side of the neural tube and notochord. Vertebrae are formed by contributions from the medial portions (sclerotomes) of 2 somites of each side, and the muscles of the vertebral column are derived from the lateral portions (myotomes) of the somites. Each myotome attracts a single nerve that grows from the adjacent neural tube.

Caudal cervical articular process joint arthritis that causes foraminal stenosis can injure C6, C7, and C8 spinal nerves, leading to pain and thoracic limb lameness as seen in the horse in this case study.

CERVICAL VERTEBRAL STENOTIC MYELOPATHY (CVSM)

Cervical vertebral stenotic myelopathy is a common cause of ataxia and weakness in horses, caused by narrowing of the cervical vertebral canal and compression of the spinal cord. Malalignment, malformation, and malarticulation of the cervical vertebrae are frequently present. Compression of the spinal cord can occur constantly or only in certain head and neck positions, such as flexed or extended. Cervical vertebral stenotic myelopathy has been divided into 2 types, affecting younger and older horses. Type I CVSM affects young horses and involves developmental abnormalities of the cervical vertebral column, including malformation of the vertebral canal, enlargement of the physes, extension of the dorsal aspect of the vertebral arch, angulation and possibly fixation between adjacent vertebrae, and malformation of the articular processes due to osteochondrosis. Type II CVSM affects older horses and involves osteoarthritic changes of the articular processes; changes include malformation with degenerative joint disease of the articular processes, wedging of the vertebral canal, periarticular proliferation, synovial or epidural cysts, and fractures of the articular processes. There is a great deal of overlap between these types; young horses might have substantial degenerative changes, while older horses might have malformations that remains subclinical until osteoarthritic changes develop at an older age.

Radiographic diagnosis of CVSM is usually achieved by looking for morphologic indicators, as well as performing more objective minimum sagittal diameter ratio measurements. Radiographic indicators include subluxation (dorsal angulation of the more caudal vertebra), physeal enlargement with dorsal projection of the caudal physis, osteoarthritis and bony proliferation of the articular processes, osteochondrosis changes at the articular processes, and caudal extension of the dorsocaudal vertebral arch over the cranial physis of the adjacent vertebra (Fig. 10.12-7). Both intra- and intervertebral minimum sagittal diameter ratios have been described. Published cutoffs for intravertebral ratios are 0.52 for C3–4, C4–5, and C5–6 and 0.56 for C6–7. Ratios below the cutoffs indicate an increased risk of having CVSM; however, they do not confirm the diagnosis, nor do they accurately indicate the site of compression (Fig. 10.12-8). Intervertebral measurements $\leq$0.485 identified all 8 cases of CVSM in a more recent study of 26 horses.

Myelography—i.e., injection of iodinated contrast material into the subarachnoid space to opacify the CSF and outline the position of the spinal cord within the vertebral canal—is considered a more definitive test for spinal cord compression. Radiographs are obtained with the head and neck in neutral, flexed, and extended positions. Reduction of the dorsal and ventral contrast columns at a specific site indicates likely spinal cord compression (Fig. 10.12-9).

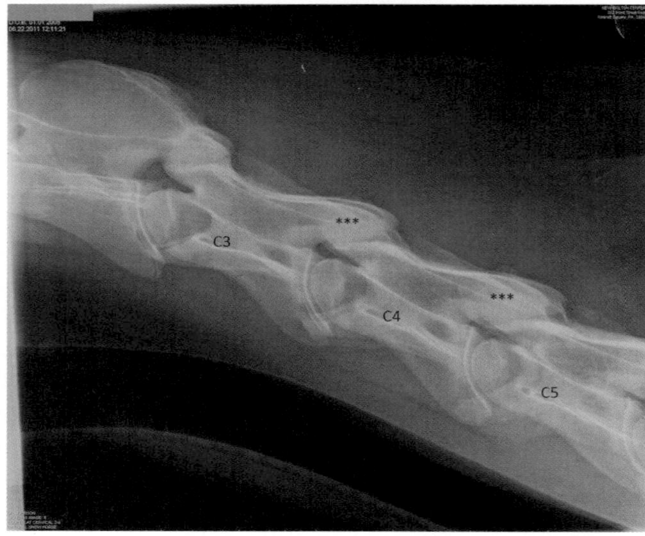

FIGURE 10.12-7 Radiograph from a 3-year-old National Show Horse gelding. Note the caudal extension of the dorsocaudal vertebral arch over the cranial aspect of the adjacent vertebra at C3–4 and C4–5 (***).

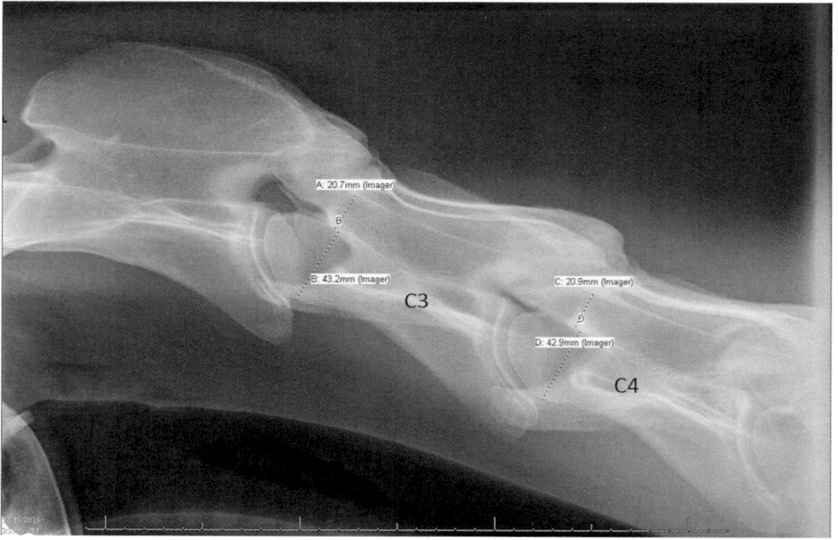

FIGURE 10.12-8 Intravertebral sagittal diameter ratio measurements from a 5-year-old Warmblood horse. Both ratios are abnormal and less than 50% (C3 = 48%, C4 = 49%).

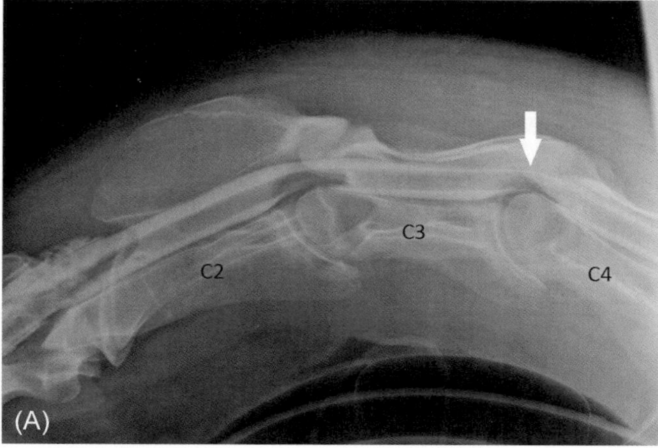

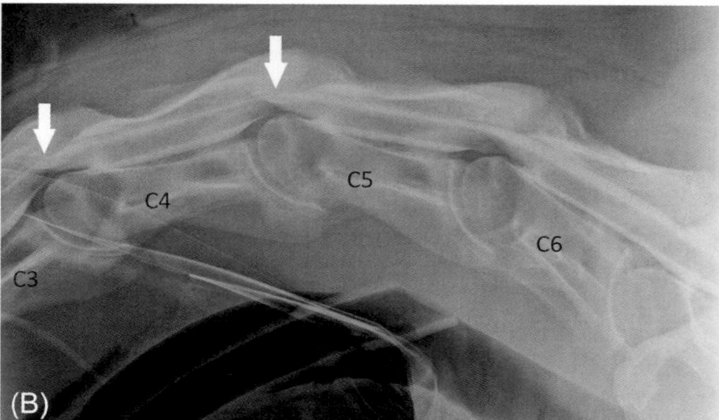

FIGURE 10.12-9 (A) and (B) Myelographic images from the 3-year-old National Show Horse gelding (survey radiograph shown in Fig. 10.12-7). Marked reduction of the dorsal and ventral contrast columns indicates likely spinal cord compression present at C3–C4 and C4–C5 *(white arrows)*.

Nuchal ligament and crest (also see Case 10.10)

The **nuchal ligament** is massively developed in the horse and connects the skull to the withers. It has 2 distinct, paired parts (see Fig. 10.10-7A and B). Its **dorsal (funicular)** part is structurally like a thick cord that extends from the external occipital protuberance of the skull to the highest spines of the withers (thoracic vertebrae). The **ventral (laminar)** part forms a fenestrated sheet closely attached to the funicular part and extending ventrally to fill the space from the funicular part to the cervical vertebrae, with bundles of elastic fibers that run cranioventrally from the funicular part and spines of T2 and T3 to attach to C2–C7. Fatty fibrous tissue located dorsal to the funicular part of the nuchal ligament forms the "crest" of the neck.

Stallions have a particularly heavy or "cresty" neck due to strong development of this tissue. Horses with endocrine disease, such as equine metabolic syndrome (EMS) or pituitary pars intermedia dysfunction (PPID), frequently have excess fatty deposition in this region, leading to a thick, overdeveloped crest.

Synovial bursae are located between the funicular part and the underlying bony structures to minimize friction. One, the **nuchal bursa**, is consistently present above the dorsal arch of the atlas. A second is inconsistently present over the spine of the axis. A third, the **supraspinous bursa**, is invariably found over the most prominent processes of the withers.

Historically, infections of the nuchal bursa and supraspinous bursa were common. The former was referred to as "poll-evil," and the latter was referred to as "fistulous withers." *Brucella abortus* was a commonly isolated and problematic organism because of the zoonotic potential. Today, aseptic bursitis (see Case 10.10) is likely more common than septic bursitis, although attempts to treat the bursitis can result in secondary infections that are difficult to eliminate and cause chronic intermittent swelling and drainage from these sites.

Regional blood supply, lymphatics, and innervation

This information is well covered in Case 10.10.

Selected references

[1] Dyce KM, Sack WO, Wensing CJG. Chapter 19: The neck, back, and vertebral column of the horse. In: Dyce KM, Sack WO, CJG W, editors. Textbook of veterinary anatomy. 2nd ed. Philadelphia, PA: W.B. Saunders Company; 1996.
[2] Hepburn R. Chapter 30: Cervical articular process disease, fractures, and other axial skeletal disorders. In: Furr M, Reed S, editors. Equine neurology. 2nd ed. Ames, Iowa: John Wiley & Sons, Inc; 2015.
[3] Mayhew IG. Collection of cerebrospinal fluid from the horse. Cornell Vet 1975;65(4):500–11.
[4] Hahn CN, Handel I, Green SL, et al. Assessment of the utility of using intra- and intervertebral minimum sagittal diameter ratios in the diagnosis of cervical vertebral malformation in horses. Vet Radiol Ultrasound 2008;49:1–6.
[5] Johnson AL, Reed S. Chapter 28: Cervical vertebral stenotic myelopathy. In: Furr M, Reed S, editors. Equine Neurology. 2nd ed. Ames, Iowa: Wiley-Blackwell; 2015.

CHAPTER 10

CASE 10.13

Congenital Cerebellar Disorder

Monica Aleman
Medicine and Epidemiology, University of California-Davis School of Veterinary Medicine, Davis, California, US

Clinical case

History

A 1-month-old Arabian filly was presented for progressive gait abnormalities and hyper-reactive behavior to physical stimuli. The filly was born apparently healthy from a healthy mare with a normal gestational length. Birth was observed and considered normal. Physical examination by a veterinarian did not reveal abnormalities at that time. At 2 weeks of age, the filly displayed an increasingly "bouncy" gait, as described by the owner. At the time, it was also noted that upon nursing, the filly had trouble controlling the movement of her head. These abnormalities progressed and prompted referral to a veterinary hospital.

Physical examination findings

The filly had an appropriate body condition, and her physical examination parameters were within normal limits. She appeared anxious and avoided being approached.

Neurological examination findings

The filly had normal behavior other than appearing anxious, and her mentation was bright, alert, and responsive. There were no cranial nerve abnormalities except for a lack of menace response with intact vision bilaterally and normal ability to blink. The filly showed intention tremors of the head. Segmental reflexes were all within normal limits. Postural reactions and proprioception were difficult to assess due to her constant movement and avoidance of handling, however appeared abnormal in all limbs (Fig. 10.13-1). The filly had severe ataxia (grade 4 out of 5 on a modified scoring system) and a dysmetric gait in all limbs that was more pronounced in the thoracic limbs. Urination and defecation were observed and appeared normal other than a wide-based posture. The neuroanatomical localization was cerebellum.

FIGURE 10.13-1 Arabian filly with wide-based stance. (Figure captured from video.)

Differential diagnoses

Cerebellar abiotrophy, cerebellar hypoplasia, other congenital anomalies affecting the cerebellum

Diagnostics

A CBC and biochemistry panel were within reference values. Radiographs of the cervical vertebral column and caudal skull were normal. Cerebrospinal fluid from the atlanto-occipital area had normal cytology and biochemical profile.

Comparative Veterinary Anatomy: A Clinical Approach. https://doi.org/10.1016/B978-0-323-91015-6.00149-7

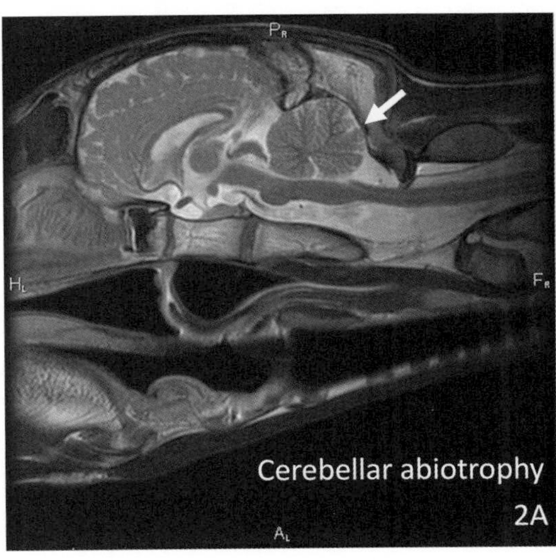

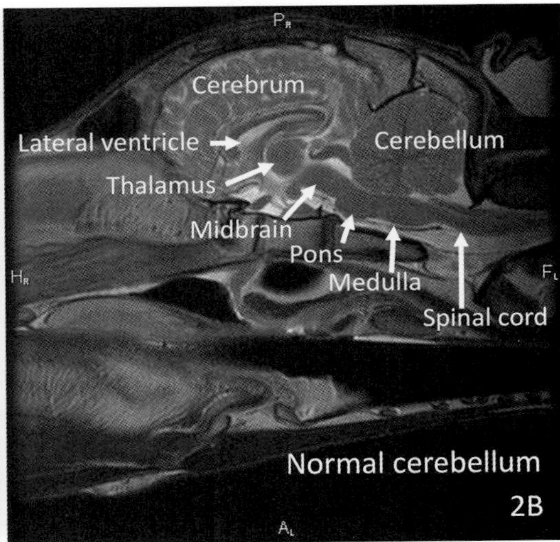

FIGURE 10.13-2 Magnetic resonance image (T2 sequence). (A) Filly with cerebellar abiotrophy. Note the smaller cerebellum showing prominent sulci and increased fluid signal (CSF) around the cerebellum *(arrow)*. (B) Healthy colt with normal cerebellum. Note the difference in both the size of the cerebellum and the fluid signal around the cerebellum between the affected *(arrow)* and unaffected foals.

On MRI, the cerebellum was mildly and uniformly decreased in size (Fig. 10.13-2). No relevant findings were identified upon contrast administration. The small size of the cerebellum was suspected to represent cerebellar atrophy or abiotrophy.

Diagnosis

Cerebellar disease—most likely cerebellar abiotrophy

Treatment

There is no specific or curative treatment for this condition.

Anatomical features in equids

Introduction

The cerebellum is one of the three major functional areas of the brain. The other 2 areas are the cerebrothalamus (forebrain or prosencephalon) and brain stem.

Function

The cerebellum is responsible for the coordination of all somatic motor activity but does not initiate movement. The cerebellum receives information for intended movement, movement in progress, and control of posture. The coordination of movement involves input from sensory projections from muscles of the body to the cerebellum and feedback circuits between the cerebellum and motor centers of the pyramidal and extrapyramidal systems.

Gross anatomy of the cerebellum

The **cerebellum** lies in the caudal fossa, above the pons and medulla, and is separated from the cerebral hemispheres by the membranous **tentorium cerebelli**. It is connected to the brain stem by three **peduncles (rostral**, **middle**, and **caudal)** on each side. The cerebellum itself consists of two large **lateral hemispheres** and a median ridge called the **vermis**. Transverse fissures divide the small **caudal flocculonodular lobe** from the larger mass, which is divided into **caudal** and **rostral lobes**.

Therefore, diseases that affect the Purkinje cells, such as cerebellar abiotrophy, result in substantial loss of inhibition and marked loss of coordination of rate, range, and force of motion. This can be noted as a slight bouncy gait in the neonatal foal that changes to a more coordinated movement as the foal matures. This contrasts with predator species in which cerebellar maturation takes longer to occur (weeks to months post-birth). For example, neonatal puppies and kittens are unable to walk in the days after birth and do not gain the ability to ambulate more normally until several weeks after birth, maintaining a bouncy or jerky gait for a substantially longer period than foals.

Mentation and behavior are normal with cerebellar disease, provided other regions of the brain are unaffected. Because the cerebellum coordinates functions associated with somatic muscles of the same side, unilateral lesions of the cerebellum cause ipsilateral signs.

Histological anatomy of the cerebellum

As a prey animal, the horse is born with a more developed cerebellum compared to predator species. The three histological layers of the **cerebellar cortex** (**molecular**, **Purkinje**, and **granular**; Fig. 10.13-3) appear more distinct in the neonatal foal than in predator species. It is important to point out that the only outgoing fibers of the cerebellar cortex are the axons of the Purkinje cells, which have an inhibitory input to the cerebellar nuclei.

Cerebellar tracts and their functions

Functionally, the cerebellum can be divided into three regions: vestibular, proprioceptive, and feedback. The main abnormalities with cerebellar disease are those associated with the coordination of movement.

The *afferent pathways* to the cerebellum include the following:

- **Corticopontocerebellar** (pyramidal system)
  - From primary motor area to contralateral cerebellar cortex
- **Extrapyramidal**
  - Direct
    - From ipsilateral vestibular nuclei (balance)
    - From contralateral tectum (vision and sound)
  - Indirect
- **Spinocerebellar**
  - Ipsilateral pathways that terminate in the cerebellar cortex and transmit proprioceptive information from muscle spindles and Golgi tendon organs

The *efferent pathways* from the cerebellum include the following:

- **Neuron 1 in cerebellar cortex: Purkinje cell**
- **Neuron 2 in cerebellar nucleus**
  - Projects to thalamus (neuron 3), which projects to:
    - Cerebral cortex (pyramidal)
    - Cerebral cortex to globus pallidus (extrapyramidal)

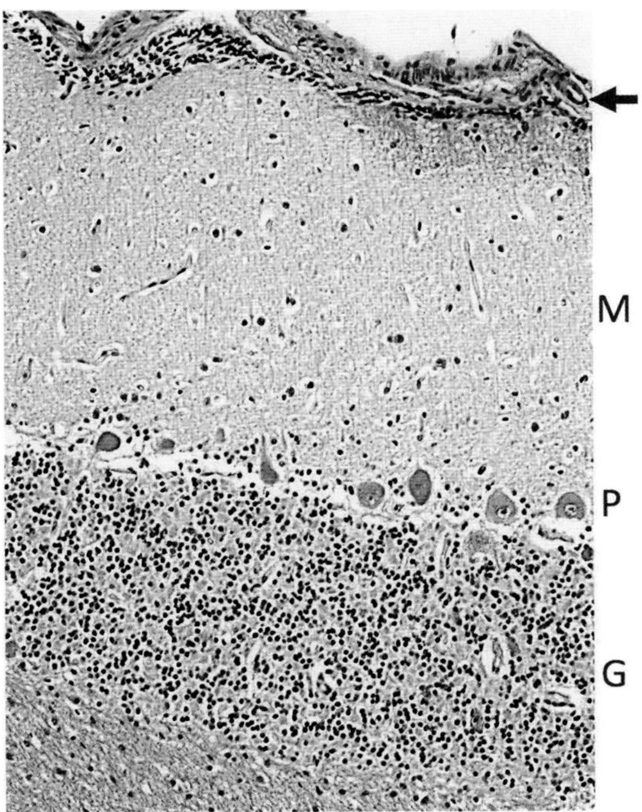

FIGURE 10.13-3 Histological layers of the cerebellum: *Top*, molecular (M); *middle*, Purkinje cells (P); *bottom*, granular (G). Cell migration is still occurring from the surface of the cerebellar cortex to one of the three layers *(arrow)*.

SIGNS OF CEREBELLAR DISEASE

Depending on lesion location within the three functional areas of the cerebellum, three main clinical syndromes can occur; however, most often a combination of signs are observed.

Vestibulocerebellar
- Swaying
- Wide-based stance
- Unsteadiness or staggering
- Positional nystagmus

Spinocerebellar
- Hypertonia
- Exaggeration of reflexes

Pontocerebellar
- Asynergia (lack of synchrony, movement out of proportion)
- Dysmetria (erratic length and height of the stride, jabbing movements of the head)
- Intention tremors

Other signs include lack of menace response and anisocoria. For these signs, it is important to rule out ocular disease or injury to other parts of the nervous system.

CEREBELLAR ABIOTROPHY

Cerebellar abiotrophy is the most common disorder of the cerebellum in horses with a genetic etiology in the Arabian breed. Sporadically, other breeds such as Gotland pony, Oldenburg, Thoroughbred, and Paso Fino have also been reported. Cerebellar abiotrophy refers to the premature degeneration of the Purkinje cells that occur after the cerebellum is formed. Foals can be born clinically normal or develop signs (especially ataxia, dysmetria, and intention tremors) shortly after birth or within the first 6 months of life.

Cerebellar abiotrophy is inherited as an autosomal recessive trait in Arabians and is associated with a single nucleotide polymorphism resulting in loss of Purkinje neurons with secondary loss of the granular cell layer of the cerebellum and proliferation of Bergmann glia.

Other cerebellar disorders seen include cerebellar hypoplasia, resulting from arrested development of the cerebellum, and Dandy-Walker syndrome, caused by a midline defect of the cerebellum and cystic dilation of the 4th ventricle. Infectious agents such as *Streptococcus equi* and other bacteria that cause meningitis might also affect the cerebellum. Aberrant parasitic migration and *Halicephalobus gingivalis* have been reported to cause cerebellar disease.

Selected references

[1] King SA. Cerebellum. In: Physiological and clinical anatomy of the domestic mammals, 1. Blackwell Publishing; 2008. p. 171–82.
[2] Holliday TA. Clinical signs of acute and chronic experimental lesions of the cerebellum. Vet Sci Commun 1979;3:259.
[3] DeBowes RM, Leipold HW, Turner-Beatty M. Cerebellar abiotrophy. Vet Clin North Am Equine Pract 1987;3:345.
[4] Brault LS, Cooper CA, Famula TR, et al. Mapping of equine cerebellar abiotrophy to ECA2 and identification of a potential causative mutation affecting expression of MUTYH. Genomics 2011;97:121–9.
[5] Blanco A, Moyano R, Vivo J. Purkinje cell apoptosis in Arabian horses with cerebellar abiotrophy. J Vet Med A Physiol Pathol Clin Med 2006;53:286–7.

CHAPTER 11

THORAX

Sarah Reuss, Chapter editor

Heart

11.1 Tetralogy of Fallot—*Cristobal Navas de Solis* 694
11.2 Mitral regurgitation—*Kari Bevevino and Cristobal Navas de Solis* 704

Pleura, Mediastinum, and Lungs

11.3 Pleuropneumonia—*Michelle Coleman* 714

CHAPTER 11

CASE 11.1

Tetralogy of Fallot

Cristobal Navas de Solis
Cardiology and Internal Medicine, Department of Clinical Studies - New Bolton Center, University of Pennsylvania School of Veterinary Medicine, Philadelphia, Pennsylvania, US

Clinical case

History

A 4-month-old Thoroughbred colt presented for the correction of angular limb deformities.

Physical examination findings

The colt was quiet and alert. The mucous membranes were pink and moist, and the CRT was less than 2 s. The heart rate was 72 bpm with a regular rhythm and audible S1, S2, and S3. Normal respiratory sounds and rate (24 brpm) were auscultated over the trachea and thorax. Peripheral pulses and jugular veins were normal. The foal had a grade 4/6 pansystolic, band-shaped, coarse murmur with the point of maximal intensity over the tricuspid valve area and a 6/6 pansystolic, crescendo-decrescendo, coarse murmur with the point of maximal intensity over the pulmonic valve area. Due to the presence of bilateral murmurs with the above-mentioned description, complex congenital heart disease was suspected and an echocardiogram was recommended.

A systolic murmur that is only right-sided in a horse is tricuspid regurgitation until proven otherwise. However, if the right-sided murmur is accompanied by a left-sided murmur over the pulmonic valve area, congenital heart disease becomes the most likely differential diagnosis.

GRADING OF CARDIAC MURMURS

Grade I: Very soft, focal murmur that is only detected in a quiet environment after extended auscultation
Grade II: Soft, focal murmur that is readily audible (softer than S1 and S2)
Grade III: Moderately loud murmur with some radiation (similar intensity to S1 and S2)
Grade IV: Very loud murmur that radiates widely (louder than S1 and S2)
Grade V: Very loud murmur with a palpable thrill
Grade VI: Very loud murmur with a thrill that is audible with the stethoscope not in contact with the chest wall

Comparative Veterinary Anatomy: A Clinical Approach. https://doi.org/10.1016/B978-0-323-91015-6.00150-3

Differential diagnoses

Ventricular septal defect (VSD) or tetralogy of Fallot (TOF)

Diagnostics

Echocardiogram showed a right ventricular outflow obstruction caused by pulmonic and subpulmonic stenosis, an overriding aorta, a large VSD, and right ventricular hypertrophy (Figs. 11.1-1–11.1-3). Additionally, a defect in the atrial septum was found, making a diagnosis of pentalogy of Fallot.

The VSD was large and measured 4.5 cm on a long axis (Fig. 11.1-4) and 2.2 cm on a short axis (Fig. 11.1-5). The maximal velocity of the left-to-right shunt (Fig. 11.1-6) was 4.06 m/s, the maximal velocity of the pulmonary outflow was 3.78 m/s (Fig. 11.1-7), and the velocity of the left-to-right ASD was 2.6 m/s. The velocity of the left-to-right shunt and outflow suggested a somewhat restrictive VSD and considerable right ventricular outflow obstruction.

Ventricular septal defects are, by far, the most common congenital heart defects of foals and the most common location is perimembranous—in the left ventricular outflow tract beneath the aortic valve. In this case, a combination of murmurs made the clinicians suspect complex congenital heart disease.

Tetralogy of Fallot is the most common complex congenital heart disease of horses. The left-sided murmur is louder in these cases because pulmonic stenosis (or right ventricular outflow obstruction) is part of TOF. Horses with subpulmonic VSDs, located beneath the pulmonic valve and communicating with the right ventricular outflow tract above the supraventricular crest, can also display a left-sided murmur that is louder than the right-sided murmur. In cases of TOF—and also pentalogy of Fallot—the severity of the right ventricular outflow tract obstruction dictates the severity of the clinical progression.

Diagnosis

Pentalogy of Fallot

Treatment

The owner elected humane euthanasia because of the poor prognosis for the foal to become a performance athlete. Necropsy examination confirmed the diagnosis (Figs. 11.1-8–11.1-10).

MOST COMMON COMBINATIONS OF MURMURS IN CONGENITAL HEART DISEASE

(1) A 4 to 6/6 pansystolic, band-shaped, and coarse murmur with the point of maximal intensity over the tricuspid valve area on the right side of the thorax. This is the “shunt murmur”—blood is shunted left to right due to the higher left ventricular pressures.
(2) A murmur on the left, softer than the one on the right, that is holosystolic or pansystolic, crescendo-decrescendo, and blowing or coarse with the point of maximal intensity over the pulmonic valve area.

This is the “relative pulmonic stenosis” murmur. The pulmonic valve or artery is not actually stenotic but is “relatively” stenotic in relation to the amount of blood ejected through it after the left-to-right shunt.

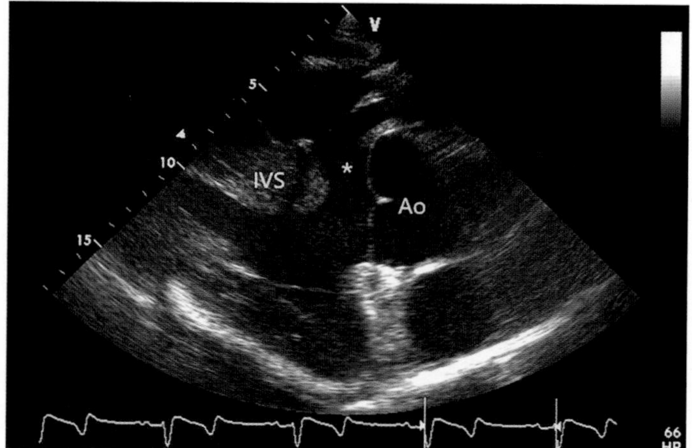

FIGURE 11.1-1 Sonogram showing a left ventricular outflow tract view from the right parasternal window. This image shows the aortic root (*Ao*) overriding (displaced to the right, dextroposed) the interventricular septum (*IVS*) and a large VSD (*).

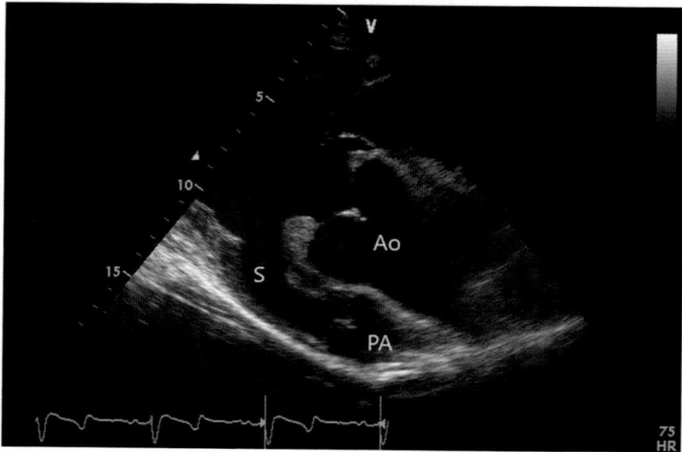

FIGURE 11.1-2 Sonogram showing a right ventricular outflow tract view from the right parasternal window. In this image, the stenotic pulmonary artery (*PA*) and valve can be observed accompanying the subpulmonic infundibular stenosis (*S*). The aorta can be observed opening to the right ventricle and the difference in size between the aortic root (*Ao*) and stenotic pulmonary artery can be clearly observed.

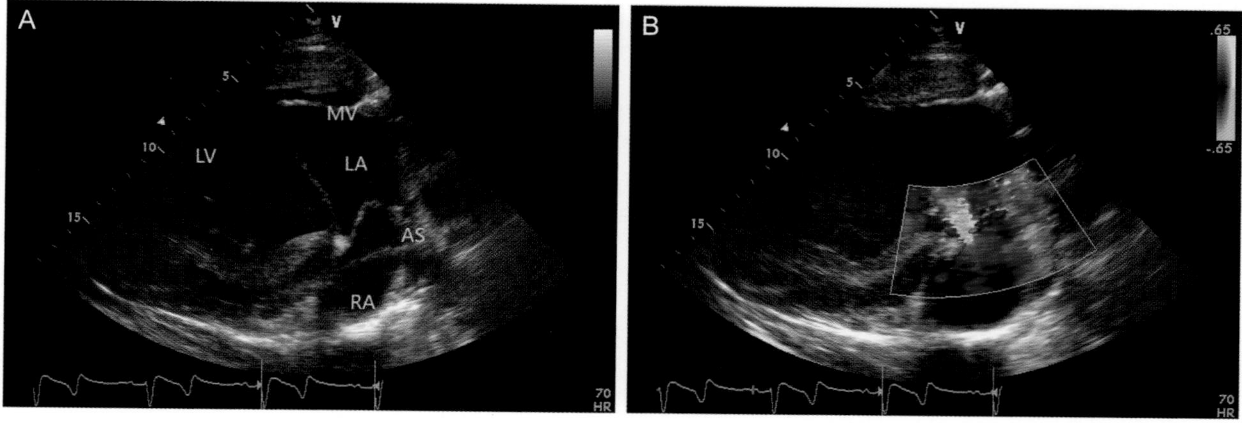

FIGURE 11.1-3 Sonogram showing a 2-chamber view from the left parasternal window. (A) The interatrial septum (*AS*) appeared as a thin membrane. (B) Color flow Doppler demonstrates a bidirectional shunt through an atrial septal defect at the level of the foramen ovale. Key: *LA*, left atrium; *LV*, left ventricle; *MV*, mitral valve; *RA*, right atrium.

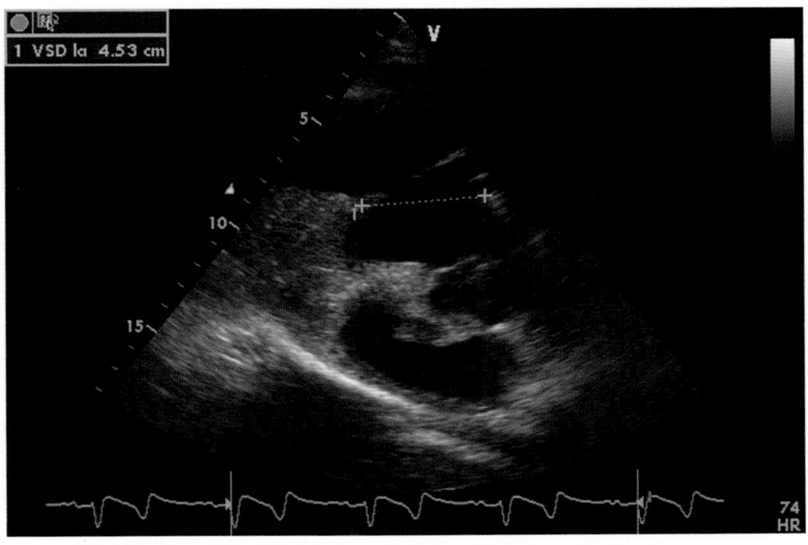

FIGURE 11.1-4 Sonogram showing the measurement of the large VSD (4.5 cm or 1.8 in.) on a long axis.

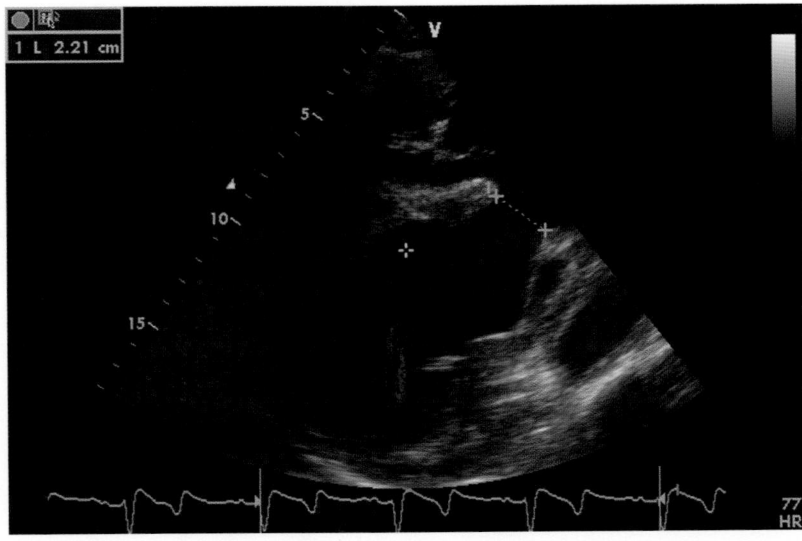

FIGURE 11.1-5 Sonogram showing the measurement of the large VSD (2.2 cm or 0.9 in.) on a short axis.

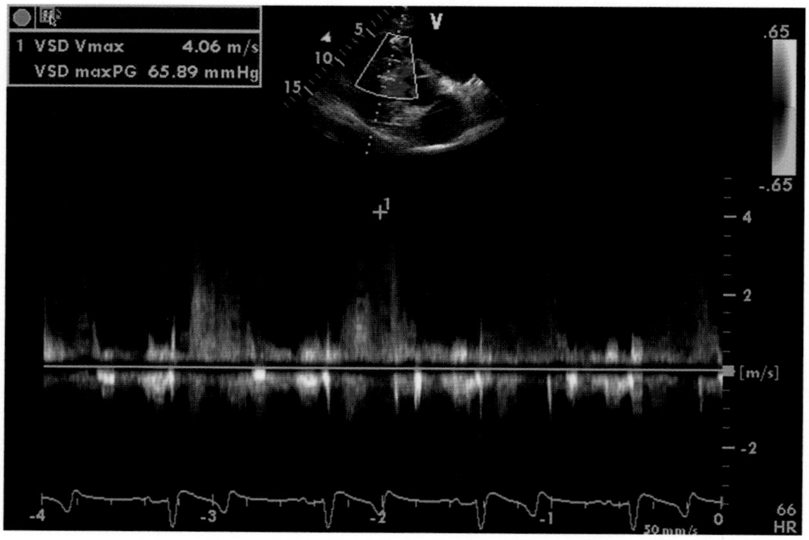

FIGURE 11.1-6 Sonogram showing the maximal velocity of the left-to-right shunt was 4.06 m/s.

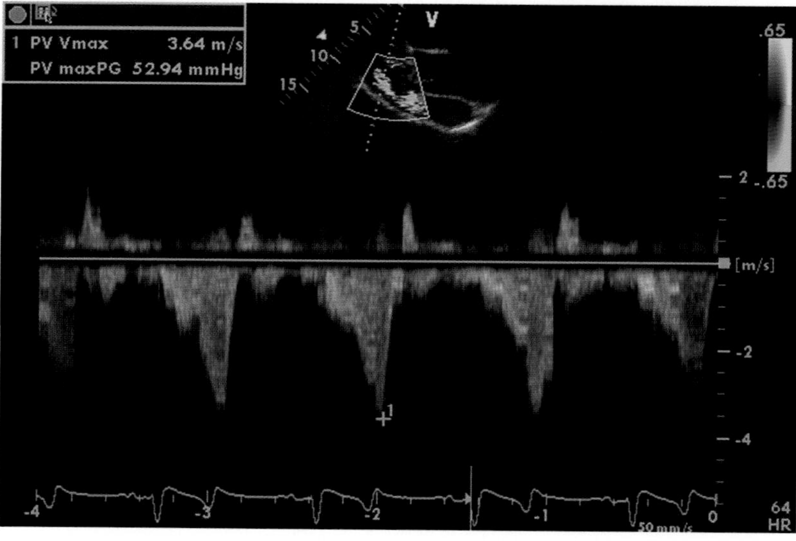

FIGURE 11.1-7 Sonogram showing the maximal velocity of the pulmonary outflow was 3.78 m/s and the velocity of the left-to-right ASD was 2.6 m/s.

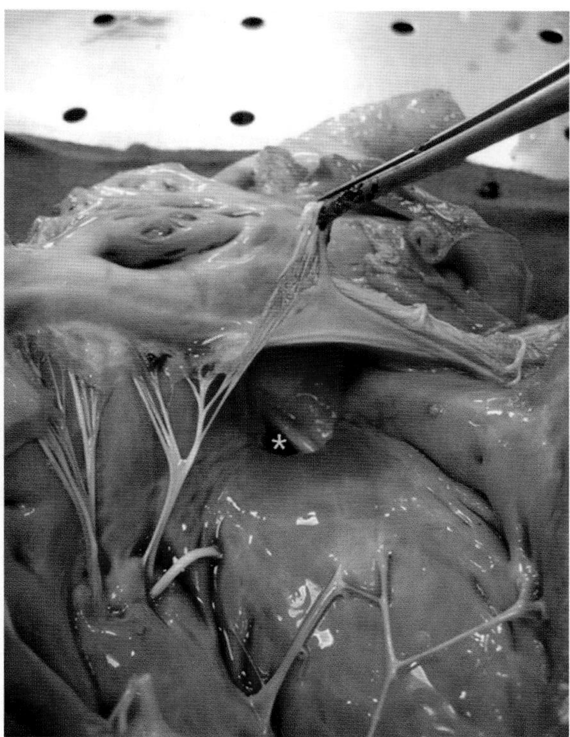

FIGURE 11.1-8 Necropsy photograph demonstrating the large VSD (*) from the left ventricle under the mitral valve that is held by the forceps.

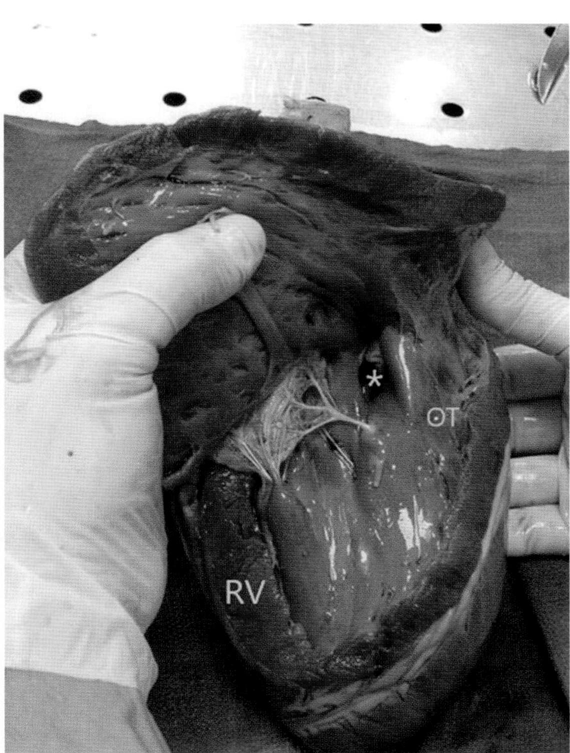

FIGURE 11.1-9 Necropsy photograph demonstrating the VSD (*), the right ventricular outflow tract (*OT*) obstruction, and the thickened right ventricular (*RV*) wall when the right ventricle was dissected.

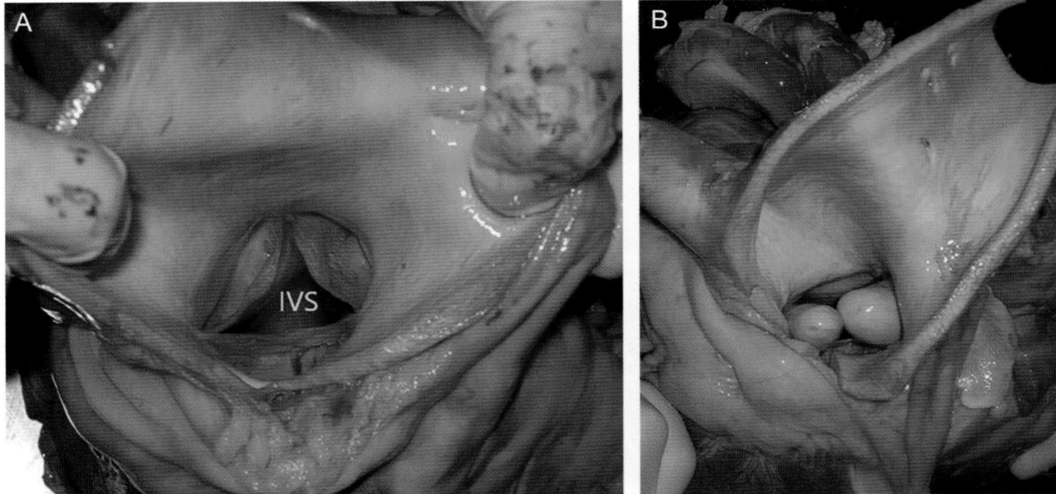

FIGURE 11.1-10 (A) Necropsy photograph showing the aortic valve from the aortic root and the interventricular septum (*IVS*) highlighting the overriding aorta. (B) One gloved finger is in the right ventricle and the other in the left ventricle.

Anatomical features in equids

Introduction

This section covers the embryology and developmental anatomy of the heart as it relates to the most commonly seen congenital cardiac defects in horses. Additional cardiac anatomy is covered in Case 11.2.

Developmental anatomy

Tetralogy of Fallot (TOF) is a combination of 4 defects with a common embryological origin— i.e., there is only one developmental abnormality. It is easy to understand this abnormality when it is described as an unequal division of the conus. The **conus** is the single major fetal vessel that divides and forms the pulmonary artery and aorta. In the most basic description of TOF, the pulmonary artery does not develop as it normally should, so the aorta occupies more space to make up for the difference. This is more properly explained as underdevelopment of the **subpulmonary infundibulum** that results from the failure of the conal septum to expand posteriorly (caudally

VENTRICULAR SEPTAL DEFECT (VSD) (FIGS. 11.1-1, -4, -5, -8, -9, AND -10)

Ventricular septal defect is the most common congenital heart disease of horses. Prognosis of horses with VSDs is generally determined by the size of the VSD and the pressure gradient between the left and right ventricles. Using both echocardiography and following the modified Bernoulli equation—i.e., difference in pressure = 4 × velocity^2—the pressure gradient is estimated by measuring the velocity of the shunt. The prognosis can be simplified by the rule:

- Ventricular septal defect < 2.5 cm (< 1.0 in.) (or < 1/3 diameter of the aorta) + shunt velocity > 4 m/s is associated with a good prognosis for life and possible athletic career
- Ventricular septal defect > 3.5 cm (> 1.4 in.) + shunt velocity < 3 m/s is associated with a shortened life expectancy

Other factors—such as the presence of other abnormalities or lesions in the aortic valve due to its adjacent location to the VSD (perimembranous in most instances)—can be relevant. Horses with small, restrictive VSDs (Fig. 11.1-11) can have a normal life expectancy and normal athletic careers although rarely in sports that involve high-intensity exercise such as racing or eventing.

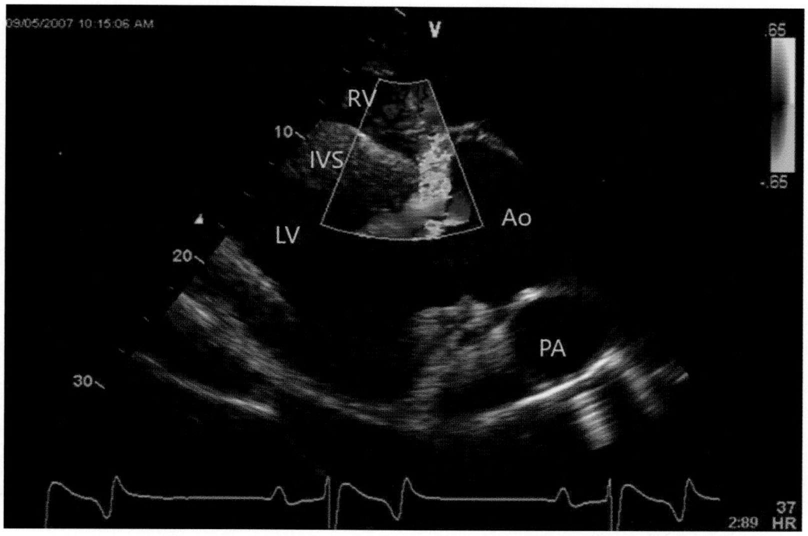

FIGURE 11.1-11 Sonogram showing a typical restrictive perimembranous VSD with color flow Doppler showing the turbulence created by the left-to-right shunt in a different patient than that described in this case. In this image, the right coronary cusp of the aorta is partially "plugging the hole" making the VSD effectively smaller. This position of the aortic valve in relationship to the VSD makes the aortic valve more prone to degenerative changes over time. Key: *Ao*, aortic root; *PA*, pulmonic artery; *LV*, left ventricle; *IVS*, interventricular septum; *RV*, right ventricle.

in the horse), also inferiorly and to the right, causing anterior (or cranial in the horse) displacement of the right ventricular outflow tract.

The **ventricular septal defect** is a consequence of a too-small subpulmonary infundibulum that does not fill the space above the interventricular septum. The overriding (dextroposed) aorta is a consequence of the same process, as it extends too far anteriorly (cranially) (see side box entitled "Ventricular septal defect (VSD)").

The **right ventricular hypertrophy** is a consequence of persistent high pressure in the right ventricle mainly caused by the right ventricular outflow tract obstruction.

Growth of the interventricular septum starts at the apex and ends at the central portion (at the endocardial cushions). Defects in the interventricular septum can be anatomically described as **atrioventricular** (at the right ventricular inlet under the septal leaflet of the tricuspid valve), **perimembranous** (more cranial than the previous, resulting in fibrous continuity of the aortic and tricuspid valves), **doubly committed** (below the pulmonary valve resulting in fibrous continuity of the major vessels), and **muscular** (surrounded by muscle) (Fig. 11.1-11).

The division of the heart into its 4 chambers occurs during fetal life. The separation between the atria includes the valve of the **oval foramen**, which remains open during fetal life and often for several days after birth. Atrial septal defects can occur at different levels including the ventral interatrial septum above the tricuspid valve (primum ASD), the oval fossa (secundum ASD), high in the dorsal atrial septum near the caval inflows (sinus venosus ASD), or can be associated with anomalous pulmonary venous drainage.

Embryology of tetralogy of Fallot (Fig. 11.1-12)

The understanding that TOF is one abnormality makes it easier to understand the condition, and experts in congenital heart disease playfully use the term "monology of Stensen" to illustrate this concept. Clinicians that work on species that can benefit from surgical management of this problem consider the central role of the small size of the subpulmonary infundibulum as key to the management. Étienne-Louis Arthur Fallot used the term "blue malady" in a case series that greatly contributed to the understanding of this disease. Fallot also understood the presence of 4 nonrandom-associated defects in children with this condition and the term "Tetralogy of Fallot" was coined in 1924 after Fallot's death.

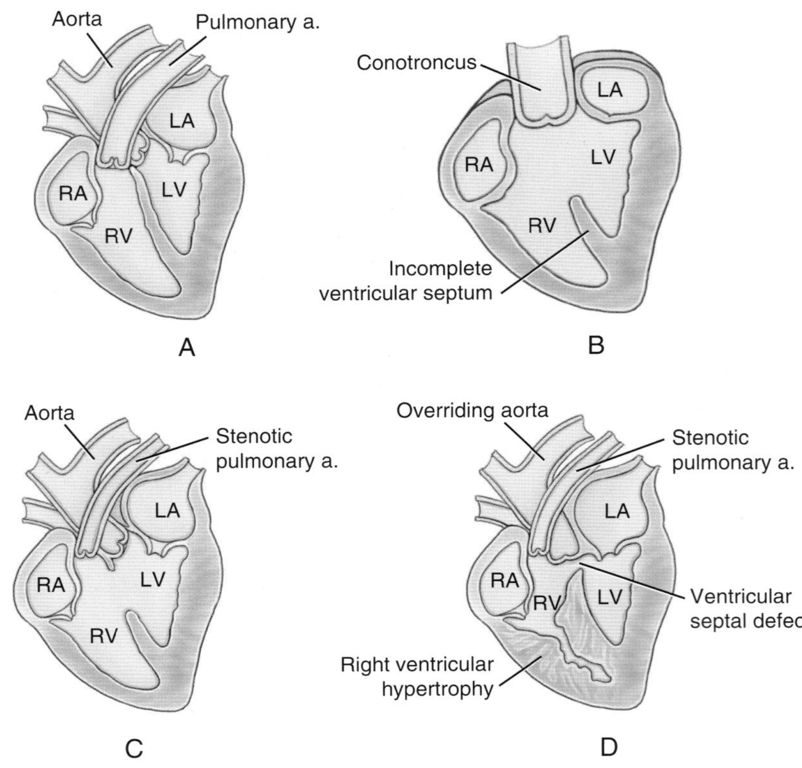

FIGURE 11.1-12 These are oversimplified diagrams to explain the development of TOF. (A) Normal heart. (B) Fetal heart before conus division and growth of the interventricular septum. (C) The normal progression is a nearly equal division of the conus, rotation of the pulmonary artery and aorta, and growth of a complete interventricular septum. If the conus "maldivides" (subpulmonary infundibulum underdevelops) (D), the pulmonary artery becomes stenotic, the aorta overrides, and a VSD develops (D). The right ventricular outflow tract obstruction eventually leads to RV hypertrophy (D). The four components of the TOF are (1) pulmonic stenosis, (2) overriding aorta, (3) VSD, and (4) right ventricular hypertrophy. Key: *LA*, left atrium; *LV*, left ventricle; *RA*, right atrium; *RV*, right ventricle.

DEVELOPMENTAL ANOMALIES OF THE HEART AS CLINICAL COMPLAINTS (FIG. 11.1-13)

- Atrial septal abnormalities: (1) foramen ovale—generally not a clinical problem; (2) interatrial septal defects—ostium primum (ventral part of the septum) and ostium secundum (upper part of the septum) defects; and (3) common atrium—interatrial septum does not develop
- Ventricular septal defect: interventricular septum fails to close
- Tetralogy of Fallot: see the section on TOF
- Transposition of the great vessels: aorta comes from the right ventricle (instead of left), and the pulmonary trunk comes from the left ventricle (instead of right)
- Persistent truncus arteriosus: pulmonary trunk and aorta are combined
- Valvular stenosis: strictures of valves
- Valvular atresia: valve(s) fail to develop
- Dextrocardia: heart deviates to the right
- Ectopia cordis: heart on the thoracic surface at birth

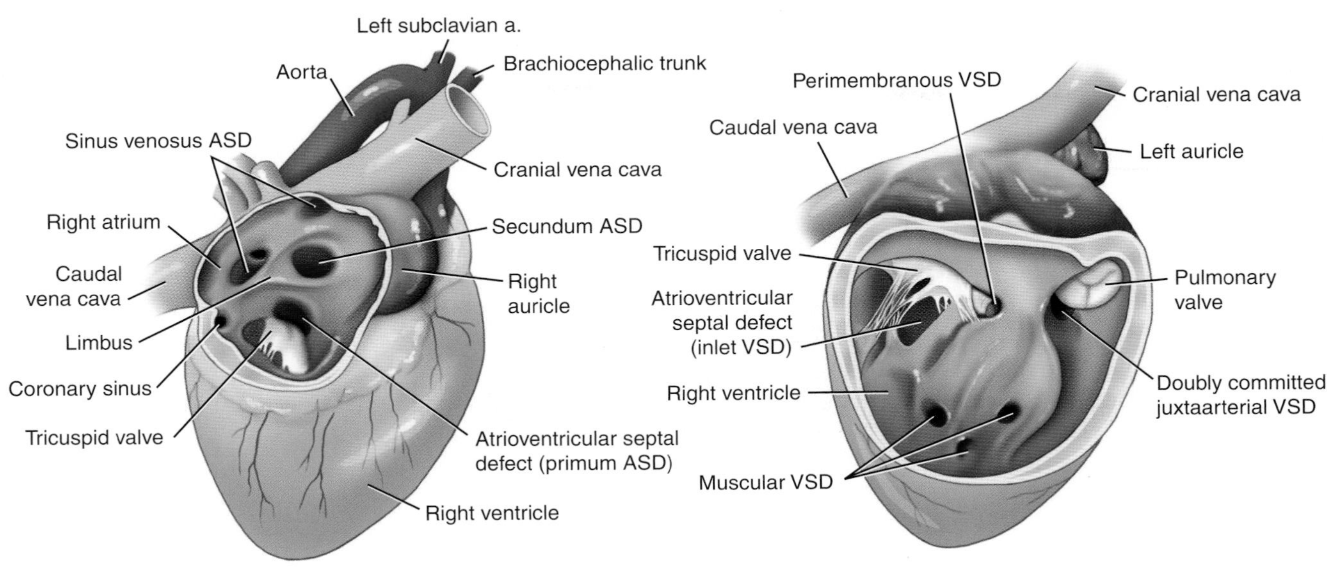

FIGURE 11.1-13 The anatomic locations of congenital heart defects involving the atrial septum (*left image*) and ventricular septum (*right image*) of animals. (Courtesy of Brian A. Scansen, Colorado State University.)

Selected references

[1] Scansen BA. Equine congenital heart disease. Vet Clin North Am Equine Pract 2019;35:103–17.
[2] Marr CM. The equine neonatal cardiovascular system in health and disease. Vet Clin North Am Equine Pract 2015;31:545–65.
[3] Dufourni A, Decloedt A, De Clercq D, Saey V, Chiers K, van Loon G. Reversed patent ductus arteriosus and multiple congenital malformations in an 8-day-old Arabo-Friesian foal. Equine Vet Educ 2018;30:315–21.
[4] Hall TL, Magdesian KG, Kittleson MD. Congenital cardiac defects in neonatal foals: 18 cases (1992-2007). J Vet Intern Med 2010;24:206–12.

CASE 11.2

Mitral Regurgitation

Kari Bevevino[a] and Cristobal Navas de Solis[b]
[a]Roaring Fork Equine Medical Center, Glenwood Springs, Colorado, US
[b]Cardiology and Internal Medicine, Department of Clinical Studies - New Bolton Center, University of Pennsylvania School of Veterinary Medicine, Philadelphia, Pennsylvania, US

Clinical case

History

A 3-year-old Standardbred gelding presented with a history of poor performance, increased heart and respiratory rates, cardiac arrhythmia, and ventral edema.

Physical examination findings

The horse was depressed with dark pink mucous membranes and a normal CRT. The jugular veins were distended, and there were prominent jugular pulses (Fig. 11.2-1). Moderate ventral and preputial edema was present. The heart rate was 90 bpm with an irregular rhythm. The peripheral pulses were weak and varied in intensity with an irregular rhythm. There was a grade 5/6 holosystolic, band-shaped, coarse murmur with the point of maximal intensity over the mitral valve area and a 4/6 holosystolic, band-shaped, coarse murmur with the point of maximal intensity over the tricuspid valve area. Rectal temperature was slightly increased at 101.7°F (38.7°C), likely due to environmental temperatures because the horse was just transported on a hot and humid day. Diffuse crackles were auscultated over both hemithoraces, and the respiratory rate was increased (52 brpm).

> In cases of mitral valve insufficiency, the murmur is caused by audible turbulent flow created by the abnormal mitral valve allowing regurgitation from the left ventricle to the left atrium during systole. A loud, left-sided systolic murmur in a horse should be considered mitral regurgitation until proven otherwise.

Differential diagnoses

Valvular regurgitation, bacterial endocarditis, pericarditis, heart failure, pleuropneumonia, pulmonary edema, cardiac arrhythmia

Diagnostics

An echocardiogram revealed enlargement of the right ventricle, pulmonary artery, left atrium, and left ventricle (Fig. 11.2-2). Left ventricular function was decreased considering the amount of volume overload. The tricuspid valve was mildly thickened, and the mitral valve was moderately thickened. There was a linear echo of the septal leaflet of the mitral valve and it moved chaotically in the left atrium and left ventricle during diastole, consistent with a ruptured chorda tendinea. Color flow Doppler showed large jets of mitral (Fig. 11.2-3) and tricuspid

Comparative Veterinary Anatomy: A Clinical Approach. https://doi.org/10.1016/B978-0-323-91015-6.00151-5

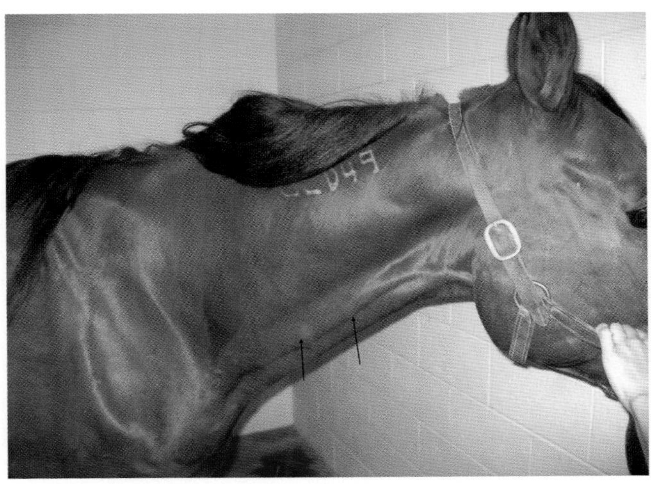

FIGURE 11.2-1 On presentation, the Standardbred gelding in this case had apparent distention of the right jugular vein (*arrows*) that could be seen with the horse's head held in a neutral position. This distention was one of the clinical signs associated with congestive heart failure and was due to the increased volume in the right heart that prevented normal emptying of the jugular veins.

regurgitation and clinically insignificant jets of aortic and pulmonic regurgitation. An electrocardiogram confirmed atrial fibrillation (Fig. 11.2-4). Sonographic examination of the thorax revealed mild pleural effusion and diffuse, coalescing comet tail artifacts consistent with pulmonary edema (Fig. 11.2-5).

Diagnosis

Heart failure caused by severe mitral valve disease and a ruptured chorda tendinea, together with tricuspid valvular disease and pulmonary hypertension

Treatment

Humane euthanasia was recommended due to the severity of the heart disease. Necropsy examination confirmed the clinical and echocardiographic diagnosis (Fig. 11.2-6A–C).

Tricuspid valve disease in combination with pulmonary hypertension was causing signs of right-sided heart failure in this horse. Pulmonary edema, pleural effusion, pulmonary hypertension, peripheral edema, jugular distention, atrial fibrillation, and jugular pulses are all consequences of biventricular heart failure. Thickening of the atrioventricular valves suggested that valvulitis could have been the underlying cause of the cardiac disease and progressed slowly over time. Atrial fibrillation and/or rupture of the chorda tendinea may have caused the recent acute decompensation of cardiac disease in this case.

MITRAL VALVE INSUFFICIENCY

The lesions that lead to mitral valve insufficiency can be degenerative, inflammatory, congenital, infectious, or idiopathic. Echocardiography can be used to evaluate the degree of regurgitation as well as—sometimes—the specific valvular lesion. In many cases, the cause of the mitral valve disease can only be surmised. As with this case, pre- and—more specifically—postmortem findings supported an inflammatory condition (valvulitis). In early stages of mild mitral regurgitation, cardiac output can be maintained and no apparent clinical signs are recognized. However, as mitral regurgitation progresses, cardiac output becomes compromised and clinical signs—such as exercise intolerance, tachycardia, tachypnea, increased respiratory effort, and/or lethargy—may become apparent. An increasing volume of regurgitated blood at the mitral valve leads to volume overload of the left atrium and eventually atrial dilatation and an increase in left atrial pressures. This can progress to cause pulmonary hypertension and right-sided heart failure.

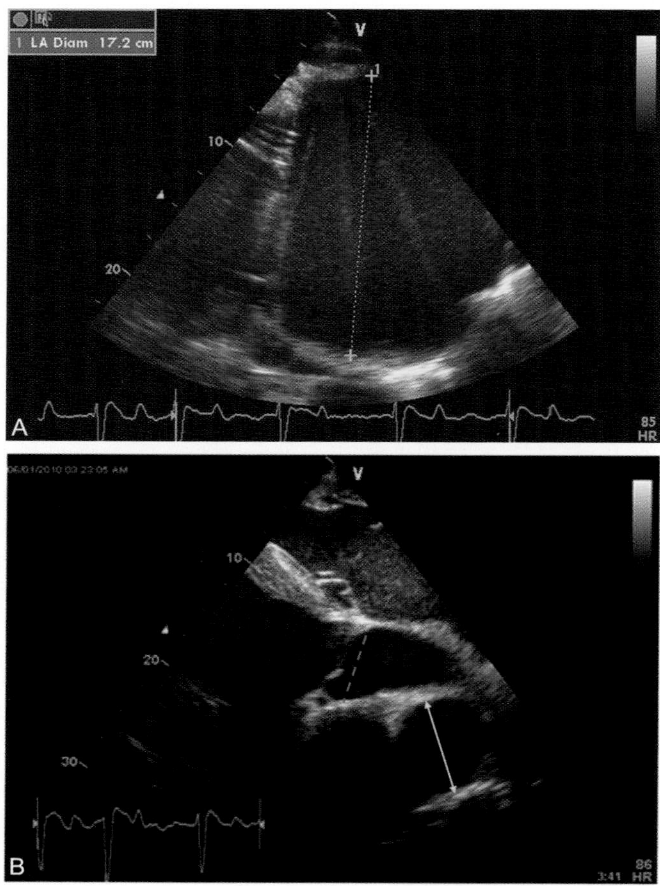

FIGURE 11.2-2 (A) A left parasternal long-axis view of the LV, MV, and LA. The LA is greatly enlarged with a diameter of 17.2 cm (6.7 in.). The continuous ECG can be used to identify the phase of the cardiac cycle at which the measurement is obtained. This imaged was obtained at the end of the systole. Although the atrial filling was occurring concurrently, the diameter was still markedly greater than normal. The normal mean 2D echocardiography measurement of LA diameter at the end of systole is 12.87 cm (5.1 in.). (B) A right parasternal left ventricular outflow tract view of a horse with pulmonary hypertension. Note that the diameter of the pulmonary artery *(arrowed line)* is noticeably greater than the diameter of the aortic root *(dashed line)*. Key: *LV*, left ventricle; *MV*, mitral valve; *LA*, left atrium.

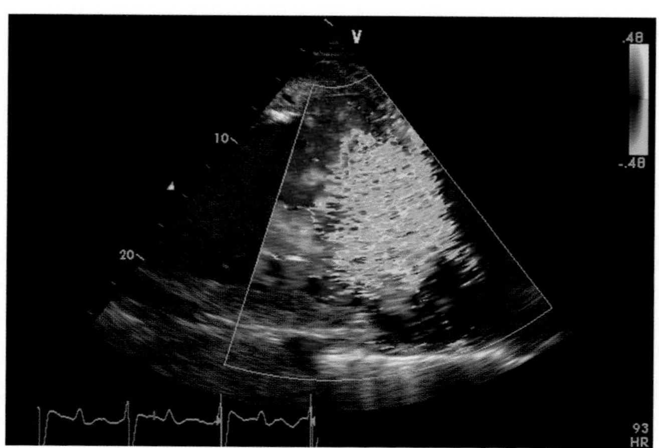

FIGURE 11.2-3 A left parasternal long-axis view of the LV, MV, and LA from the Standardbred gelding in this case with color Doppler placed over the mitral valve revealing a large volume of regurgitation recognized as the mixed color pattern in this still image. Continuous ECG aids in identifying the timing of the regurgitation during the cardiac cycle. This image was obtained during systole, which supports a diagnosis of mitral regurgitation. Key: *LV*, left ventricle; *MV*, mitral valve; *LA*, left atrium.

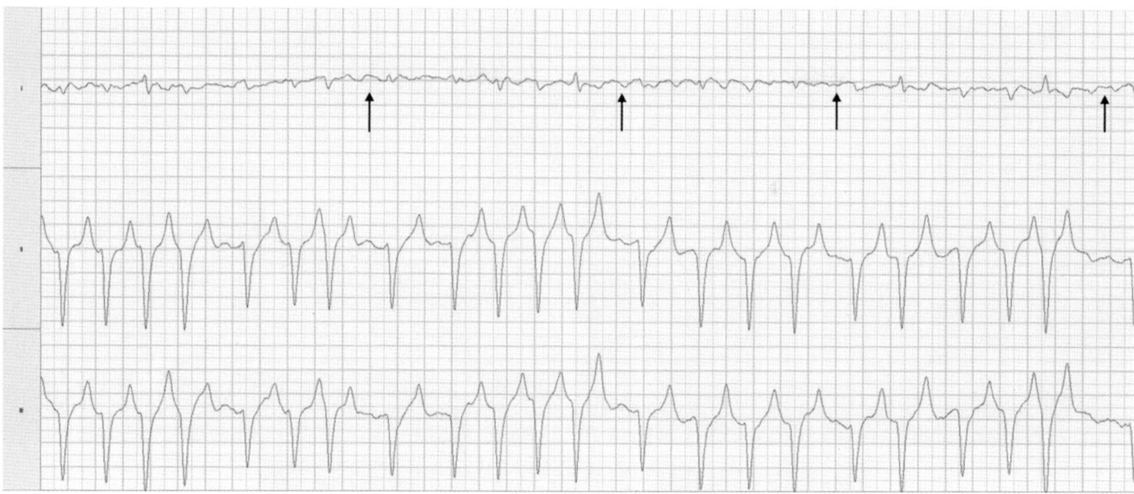

FIGURE 11.2-4 A 3-lead ECG of a horse in atrial fibrillation with significant tachycardia. P waves are absent throughout all leads and instead "f," or fibrillation, waves are present (*arrows*) in the baseline. The QRS complexes that are present are normal in morphology and duration; however, the rate is irregularly irregular, hence the auscultated arrhythmia.

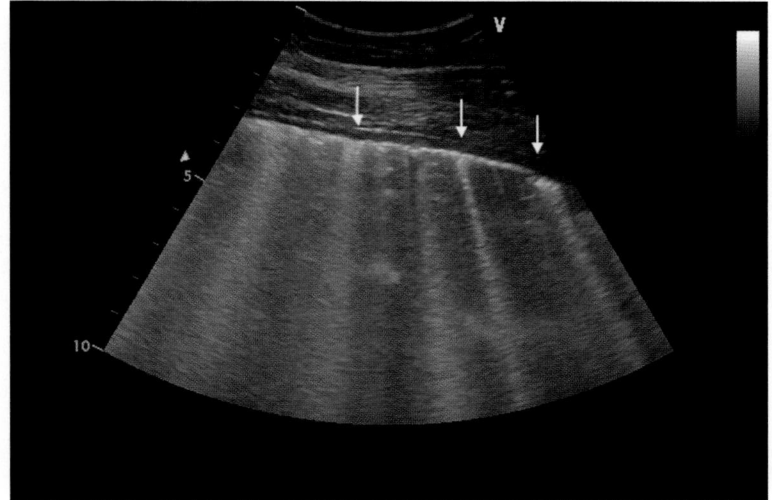

FIGURE 11.2-5 Ultrasonographic image of the thorax from the Standardbred gelding in this case. The hyperechoic line of the pleural surface has multiple comet tails (*arrows*). Comet tails are the result of roughening of the pleural surface, which leads to reverberation artifact.

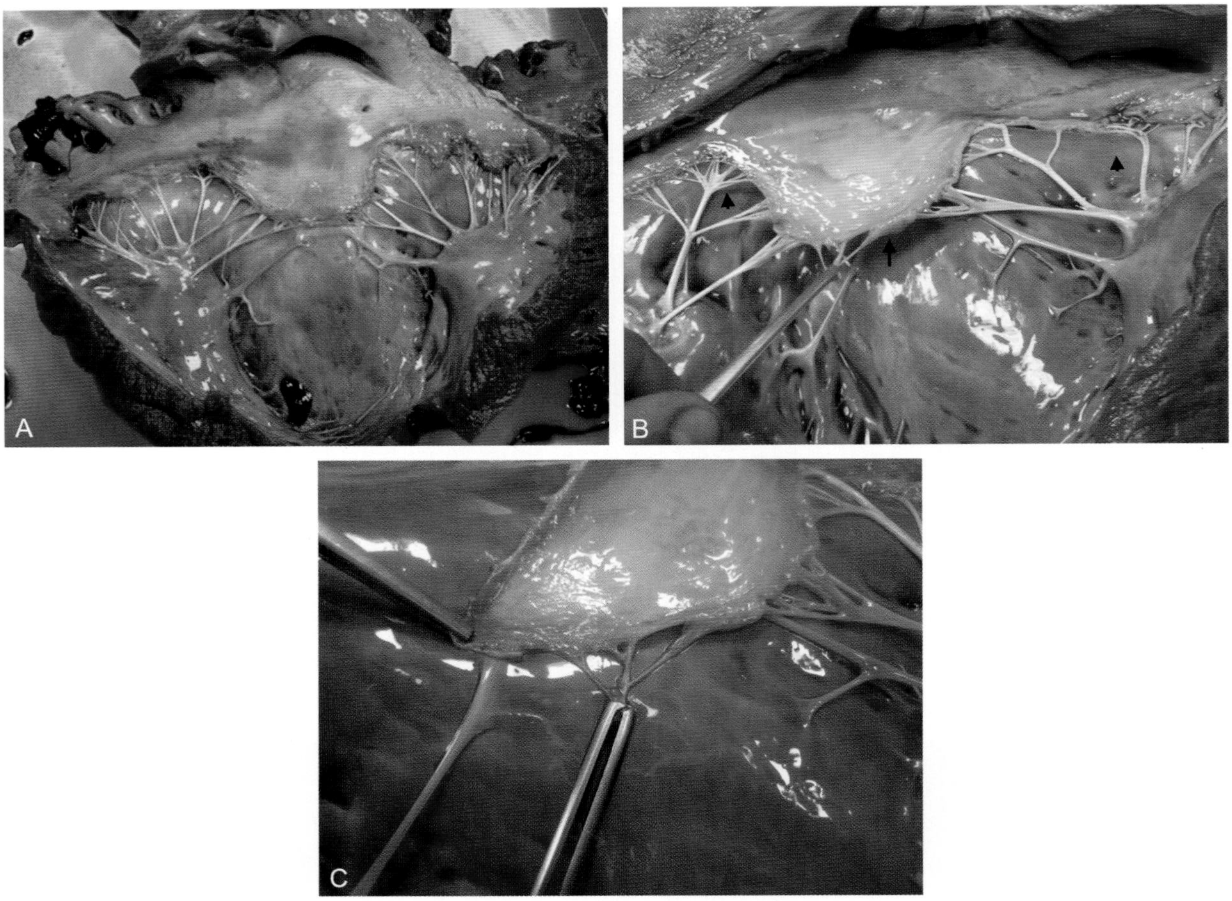

FIGURE 11.2-6 (A) Necropsy evaluation of the mitral valve. (B) The septal leaflet *(black arrow)* can be seen with its associated, intact, chordae tendineae. The accessory leaflets *(arrowheads)* reside on either side of the septal valve. This image illustrates the differing importance between chordae tendineae of the heart valves. (C) A leaflet of the mitral valve with disruption of its chordal attachments (held by the instrument). One can see that a large portion of the leaflet lacks the anchoring effects of the chordae, which likely resulted in mitral valve prolapse and acute, life-threatening disease.

Anatomical features in equids

Introduction

This case illustrates a clinically severe presentation of mitral regurgitation (MR), the most common valvular disease that causes clinical problems in horses, such as exercise intolerance or congestive heart failure. The physical examination, echocardiographic, and necropsy findings help illustrate the correlation between anatomical structures involved, the pathophysiological processes that alter these structures, and the pathophysiological consequences.

Function

The equine heart functions as a rhythmic pump that continually contracts in a coordinated fashion as the central organ in the circulatory system, working to receive, oxygenate, and distribute blood throughout the body (Fig. 11.2-7). The circulatory system consists of systemic circulation (moves blood to and from the body) and pulmonary circulation (moves blood to and from the lungs). Blood in the pulmonary circulation exchanges carbon dioxide for oxygen in the lungs via respiration, while the systemic circulation transports oxygen to the body and returns carbon dioxide and low-oxygenated blood to the heart for exchange in the lungs. Besides supplying oxygen and removing carbon dioxide, the systemic circulation transports nutrients to the body's tissues and removes metabolic by-products produced by other tissues and organs (e.g., to the kidneys and liver for breakdown and excretion).

> Because of this continuity, a disease affecting one side or structure of the heart can have deleterious effects on the remainder of the heart if the disease becomes severe enough, or when the inherent physiologic reserve factors of the heart become exhausted.

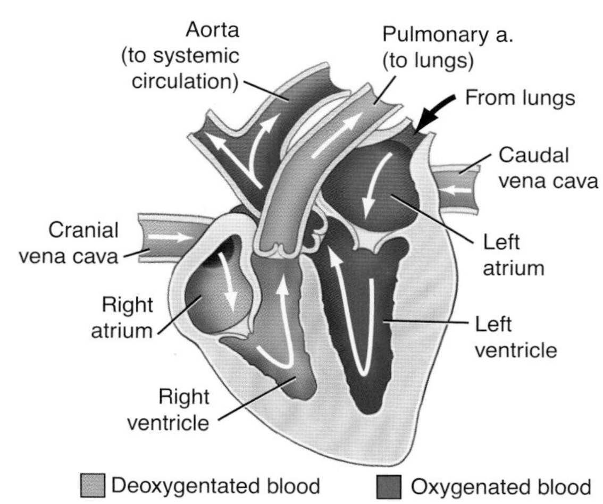

FIGURE 11.2-7 Blood flow through the heart. Deoxygenated blood is received from systemic circulation to the right side of the heart where it is pumped into the pulmonary vessels via the pulmonary artery to the lungs. The left side of the heart receives oxygen-rich blood from the lungs via the pulmonary vein where it is pumped into systemic circulation via the aorta.

PHYSIOLOGY OF NORMAL HEART SOUNDS

Ventricular systole ensues as the myocytes of the ventricle begin to contract. As contraction progresses, the pressures within the ventricle exceed atrial pressures and the atrioventricular (AV) valves (mitral and tricuspid) close. The closure of the AV valves is associated with the first heart sound ("lub," S1) heard during auscultation. Ventricular pressures continue to climb and once they exceed pressure in the major vessels (aorta and pulmonary artery), the semilunar valves (aortic and pulmonary) open, allowing ejection of blood into systemic and pulmonary circulations. Thus, in the normal, healthy heart, ventricular contraction (systole) leads to the ejection of blood into the major vessels while the AV valves are closed to avoid backflow into the atria. As diastole begins, the aortic and pulmonary valves close causing the second heart sound ("dub," S2). During early diastole, there is rapid, passive, filling of the ventricles, which can occasionally be auscultated as the S3 sound in normal horses. At the end of diastole, the atria contract (atrial systole) leading to an additional push of blood into the ventricles. This causes a presystolic heart sound (S4).

Electrical activity of the heart

Coordinated electrical activity of the heart is essential for appropriate cardiac function. In the normal heart, conduction is spontaneously initiated at the **sinoatrial (SA) node**, which is located subepicardially at the junction of the right auricle and the cranial vena cava. The SA node is the pacemaker of the heart and arrhythmias develop if its inherent automaticity is lost or altered. The generated impulse from the SA node spreads throughout the atria and also through intranodal, specialized fiber tracts to the **atrioventricular (AV) node**, located in the ventral atrial septum. The AV node acts as a conduit to electrical activity between the atria and ventricles which are electrically insulated and separated by a band of tissue known as the **annulus fibrosis**. Although the conduction from the SA to the AV node is relatively rapid, the impulse is slowed at the AV node, which is greatly influenced by vagal tone. Once the impulse passes through the AV node, it is rapidly conducted through the **bundle of His** and specialized **Purkinje fibers** within the interventricular septum and free walls of the right and left ventricles. Due to their relatively large cardiac mass, horses have **moderator bands** crossing from the interventricular septum to the free wall of the ventricles, which contain conductive tissue (more on this below).

Disruption of these normal electrical impulses at any stage of the cardiac cycle can lead to arrhythmias.

Left side of the heart (Fig. 11.2-8)

The **left atrium** (LA) occupies the caudal heart base and consists of a large, cavernous chamber with a small, muscular appendage known as the **left auricle**, which extends laterally and cranially. The walls of the LA consist of specialized **myocytes** that communicate via tight junctions to facilitate coordinated impulse conduction. The LA receives oxygenated blood via 7–8 **pulmonary veins** that empty into the caudal and right aspects of the atrium. Along the ventral aspect of the chamber lies the mitral valve, which separates the LA from the left ventricle (LV).

It is important to remember that not all chordae tendineae are equal in size or relevance. For example, a disrupted chorda tendinea of a small part of an accessory leaflet may lead to minor valvular regurgitation, sparing the patient from severe, acute disease. Alternatively, rupture of a large chorda tendinea attached to the large septal leaflet may cause significant, acute, life-threatening disease (Fig. 11.2-6C).

The **mitral valve**, or left atrioventricular valve, consists of 2 large leaflets: the **septal** and the **parietal** (or, free wall), hence its alternative name, the "bicuspid valve" (Fig. 11.2-9). In addition, there are small leaflets that are often called accessory leaflets. In horses, these are commonly referred to as the **cranial accessory** and **caudal accessory leaflets**. Leaflets are composed primarily of collagenous tissue and are covered by the endocardium. They are anchored to the **papillary muscles** of the left ventricle via a series of **chordae tendineae** (Fig. 11.2-6A and B).

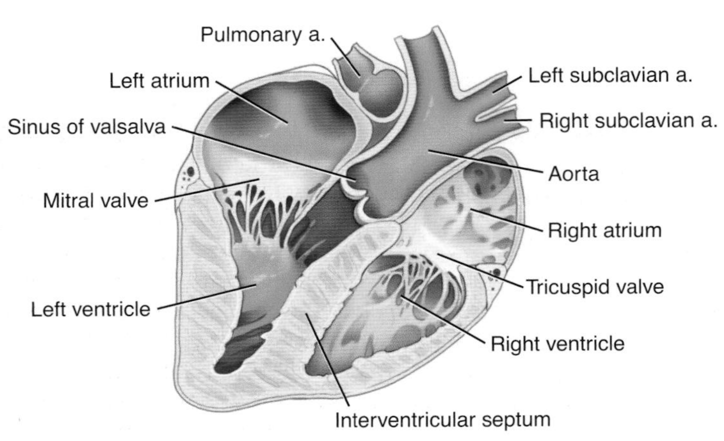

FIGURE 11.2-8 The heart from a right-sided view.

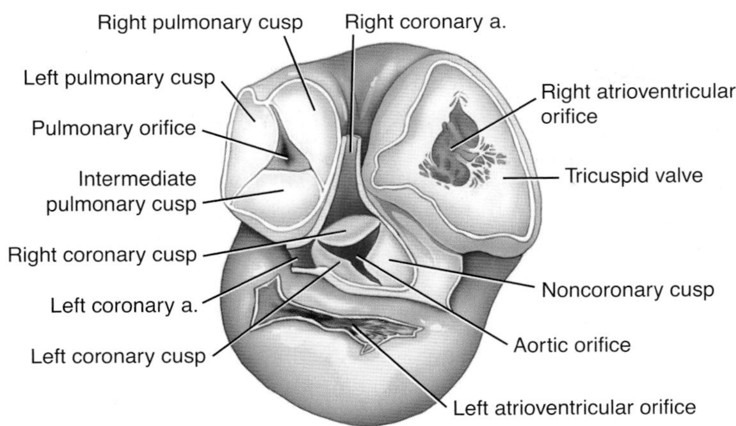

FIGURE 11.2-9 A cross-section through the heart base illustrating the valve leaflets. (Adapted from Marr C., Bowen M., editors. Cardiology of the horse. 2nd ed.)

The LV is a conical-shaped chamber that receives blood from the LA via the mitral valve. The portion of the chamber that divides the LV from the right ventricle (RV) is known as the **interventricular septum** (IVS), and the remainder of the wall is referred to as the **free wall**, which includes the **apex** of the heart. Except for the dorsal extent of the IVS, the walls of the LV consist of a thick **myocardium** composed of specialized myocytes that facilitate rapid electrical conduction. The robust myocardium allows powerful forces to be generated to eject blood into systemic circulation via the aorta. **Moderator bands**, also known as septomarginal trabecula, are bands of fibromuscular tissue that extend across both the left and right ventricles, from the septum to free wall. In addition to the fibromuscular tissue, the bands have also been recognized to have conductive cells and act as an extension of the bundle of His. The size of the bands varies greatly between individuals.

The aorta and its associated valve contribute to the left ventricular outflow tract. The **aortic valve** consists of 3 **cusps**: **right coronary**, **left coronary**, and **noncoronary** (Fig. 11.2-9). The cusps are half-moon shaped and are anchored to the fibrocartilaginous aortic annulus. The aortic valve is in close proximity to the septal leaflet of the mitral valve, and the fibrous tissue that comprises the valves is continuous from mitral to aortic. The origin of the aorta is bulbous in shape and known as the **sinus of Valsalva**. Branching from the sinus of Valsalva are the **right** and **left coronary arteries**, which supply blood to the myocardium. Beyond the sinus, the aorta becomes tubular and is termed the **ascending aorta**.

PULMONARY HYPERTENSION

In cases of significant, progressive mitral regurgitation, the left atrium continues to experience volume overload leading to an increase in pulmonary venous pressures due to lack of appropriate forward flow. This causes increased pulmonary capillary pressures and eventually, pulmonary edema. Crackles can be auscultated on physical examination in patients with severe pulmonary edema, and this edema can be recognized on thoracic ultrasound as diffuse and coalescing comet tail artifacts of the pleural surface. Pulmonary hypertension develops as a result of an increase in pressure caused by left-sided heart disease being transferred to the right side of the heart, as well as the hypoxia-induced vasoconstriction. Severe pulmonary hypertension can lead to enlargement/dilation of the pulmonary artery. This enlargement can secondarily cause blood flow regurgitation at the level of the pulmonary valves during diastole. Severe enlargement of the pulmonary artery can result in the rupture of the pulmonary artery and sudden death. Fig. 11.2-2B illustrates the increase in diameter of the pulmonary artery relative to the aorta in a horse with severe, acute pulmonary hypertension.

Right side of the heart (Fig. 11.2-10)

The **right atrium** (RA) occupies the cranial heart base and, like the LA, consists of a large, cavernous chamber and a small, muscular appendage known as the **right auricle,** which projects cranially and to the left. The walls of the RA consist of specialized myocytes that serve the same function as those in the walls of the LA. The RA receives deoxygenated blood from systemic circulation; the **cranial vena cava** enters dorsally, the **caudal vena cava** enters caudally, and the **azygous vein** enters between the cavae. In addition, the **coronary sinus**, which drains the coronary circulation, enters the ventral to the caudal vena cava. Along the ventral floor of the RA is the tricuspid valve, which separates the RA from the right ventricle (RV).

The **tricuspid valve**, or the right atrioventricular valve, is so-named because it consists of 3 collagenous **leaflets**: **septal**, **parietal**, and **angular** (Fig. 11.2-9). As its name suggests, the septal leaflet lies adjacent to the interventricular septum. The parietal leaflet lies along the right margin, which offers an alternative name—the right leaflet; in humans, the leaflet is also termed the posterior leaflet. The angular leaflet lies adjacent to the right ventricular outflow tract and is also termed the left or anterior leaflet. In a similar manner as the mitral valve, the leaflets are anchored to the **papillary muscles** of the RV via **chordae tendineae**. Thus, the leaflets of the tricuspid valve can undergo disruption of chordal attachments, as mentioned with the mitral valve.

The RV is a chamber that is smaller in size than the LV, with a thinner myocardium. It may have moderator bands, similar to the LV. Within the thorax, the RV sits cranially to the LV, with its apex oriented ventrally. In the cross-sectional view, the RV has a crescent shape. It has a U-shaped projection dorsally and to the left in which deoxygenated blood is ejected across the pulmonary valve into the pulmonary artery.

The **pulmonary artery** and associated valve sit dorsal and superficial to the aorta. The **pulmonary valve** consists of 3 half-moon-shaped **cusps**: **right**, **left**, and **intermediate** (Fig. 11.2-9). The pulmonary artery eventually splits into the **right** and **left pulmonary arteries**, which further branch into smaller vessels that become closely associated with gas-exchanging airways, allowing for oxygenation of the blood before it is drained into the left atrium via the pulmonary veins.

Cardiac blood supply, lymphatics, and innervation

The heart has a rich blood supply receiving upward of 15% of the left ventricular output via the **coronary arteries** that originate from several sinuses above the semilunar cusps at the origin of the aorta. The **left coronary artery**, originating adjacent to the left coronary cusp, is the largest of these, with the trunk tracking toward the apex of the heart as the **circumflex branch**. The **right coronary artery**, originating adjacent to the

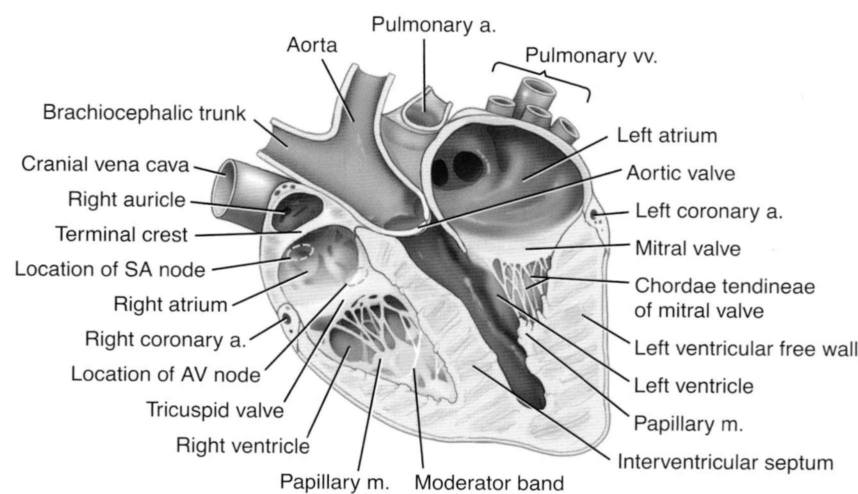

FIGURE 11.2-10 The heart from a left-sided view.

right coronary cusp, tracks between the right auricle and pulmonary trunk. The coronary arteries have multiple branches that supply the atrial and ventricular walls. The coronary artery system lacks anastomotic branches.

A sudden blockage of one of the coronary artery branches results in a myocardial infarction (*L. infarcire* to stuff in), commonly referred to as a "heart attack." This causes blood flow to be interrupted or blocked to a part of the heart, leading to ischemia and necrosis of the heart muscle. While common in people, this is uncommon in horses and other animal species.

The **great cardiac vein** is the avenue for the return of blood to the heart through a dedicated opening into the right atrium via the coronary sinus. The smallest cardiac veins, called **Thebesian veins**, are tiny, valveless veins that open directly in the walls of all 4 chambers of the heart.

The lymph nodes of the thorax comprise 4 **lymph centers**: **dorsal thoracic**, **ventral thoracic**, **mediastinal**, and **bronchial**. The mediastinal center has **cranial**, **middle** (base of the heart), and **caudal** mediastinal groups. These drain to large lymph collecting vessels at the cranial opening of the thoracic cavity.

The heart is innervated by a combination of sympathetic and parasympathetic nerve fibers. The sympathetic nerve fibers are contributions from the caudal cervical and the first few thoracic ganglia of the **sympathetic trunk**, and the parasympathetic nerve fibers are branches of the **vagus nerve**. The nerve fibers from the vagus nerve may be direct branches, or nerve fibers from the recurrent laryngeal nerve ending on nerve cells in the heart wall, primarily in the sinoatrial and atrioventricular nodes.

Selected references

[1] Bonagura JD. Overview of equine cardiac disease. Vet Clin North Am Equine Pract 2019;35(1):1–22. https://doi.org/10.1016/j.cveq.2019.01.001.
[2] Marr CM, Bowen IM. Introduction to cardiac anatomy and physiology. In: Cardiology of the horse. 2nd ed. Saunders/Elsevier; 2010. p. 3–19.
[3] Marr CM, Bowen IM. Electrophysiology and arrythmogenesis. In: Cardiology of the horse. 2nd ed. Saunders/Elsevier; 2010. p. 59–73.
[4] Pasquini C, Spurgeon T, Pasquini S. Circulatory system. In: Anatomy of domestic animals. 10th ed. Sudz Publishing; 2003. p. 381–414.

CHAPTER 11

CASE 11.3

Pleuropneumonia

Michelle Coleman
Large Animal Clinical Sciences, Texas A&M University College of Veterinary Medicine, College Station, Texas, US

Clinical case

History

A 7-year-old Thoroughbred gelding used for barrel racing presented for evaluation of ventral edema and fever of 48-h duration. Approximately 2 weeks before, the gelding was transported ~ 1200 miles (750 km) for a competition, where he competed successfully. Upon returning home, the gelding displayed mild, intermittent signs of colic and was treated symptomatically by the owner with flunixin meglumine for several days. Signs of colic were improved; however, ventral edema developed and fever was noted, prompting referral for further evaluation.

Physical examination findings

Upon presentation, the gelding was quiet and in fair body condition (BCS 4/9) with a bodyweight of 420 kg (926 lbs). The gelding had a rectal temperature of 102.8 °F (39.3 °C), heart rate of 52 bpm, and respiratory rate of 40 brpm with increased effort. Mucous membranes were pink and slightly tacky with a CRT of 2 s. Cardiac auscultation revealed a normal sinus rhythm; however, thoracic auscultation disclosed attenuation of the lung sounds in the cranioventral thorax and distinct cardiac auscultation, bilaterally. There was bilateral mucopurulent nasal discharge. The gelding had marked pectoral and ventral edema (Fig. 11.3-1). Gastrointestinal sounds were present bilaterally, with normal feces passed during the examination. Digital pulses were normal in all 4 feet.

Differential diagnoses

Pulmonary disease was suspected—infectious pleuropneumonia, neoplasia

Diagnostics

Given the increased respiratory rate and effort, a rebreathing examination was not believed necessary to improve auscultation of the thorax. Thoracic ultrasound revealed bilateral fibrinous pleuropneumonia. In the left hemithorax, the hyperechoic echo of the lung was smooth and regular and glided with respiratory motions from the 16th to the 11th intercostal spaces (ICS) (Fig. 11.3-2) except for a few comet tail artifacts (fading vertical echogenic artifacts; see Chapter 2, Section 2.3). At the ventral aspect of the 11th ICS there were multiple loculated pockets of anechoic fluid (Fig. 11.3-3). There was echoic material consistent with fibrin seen within the fluid and ventrally consolidated lung. Similar findings were identified in the right hemithorax. Additionally, fluid and fibrin were noted in the cranial mediastinum (Fig. 11.3-4).

Comparative Veterinary Anatomy: A Clinical Approach. https://doi.org/10.1016/B978-0-323-91015-6.00152-7

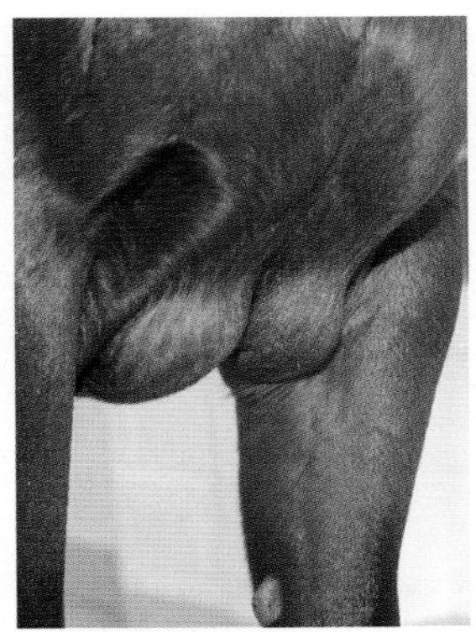

FIGURE 11.3-1 Pectoral edema of a 7-year-old Thoroughbred gelding that was diagnosed with bacterial pleuropneumonia.

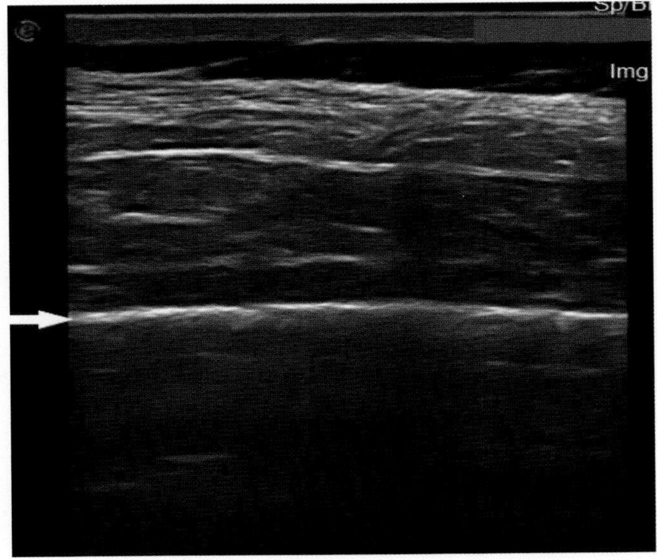

FIGURE 11.3-2 Sonogram of the normal lung at the left 14th intercostal space. The hyperechoic line *(white arrow)* represents the normal pleural surface. This image was obtained at the left 14th intercostal space using a linear probe operating at 5.0 MHz at a depth of 8 cm (3.1 in.).

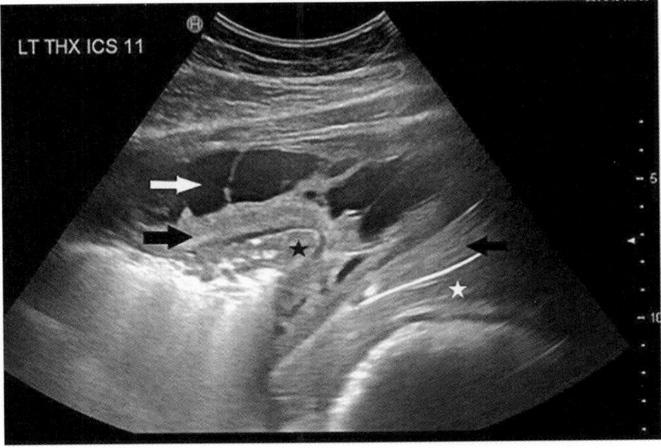

FIGURE 11.3-3 Sonogram of the left thorax at the 11th intercostal space. There is severe fibrin deposition (*black arrows*) within the pleural space and along with the diaphragm (*white star*), resulting in loculation of fluid (*white arrow*). The ventral tip of the lung is consolidated (*black star*). This image was obtained using a curvilinear probe operating at 4.0 MHz at a depth of 15 cm (5.9 in.).

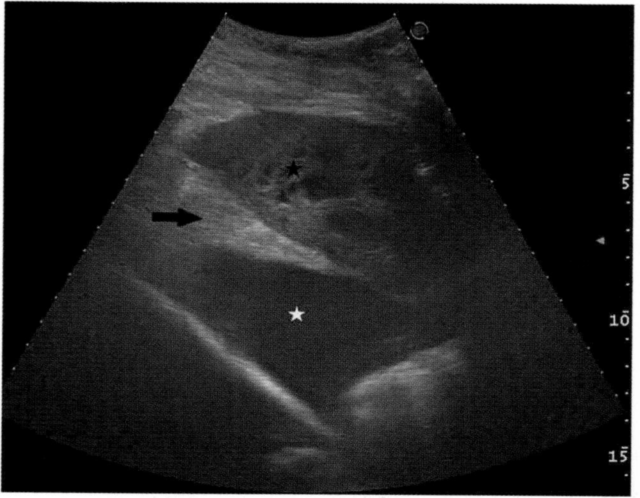

FIGURE 11.3-4 Sonogram taken from the left 4th intercostal space of mediastinum (*black arrow*) that divides the thorax into 2 cavities. In this horse with bacterial pleuropneumonia, fibrin accumulation is seen in the left (*black star*) cranial mediastinum, and fluid accumulation is seen in the right (*white star*) cranial mediastinum. This image was obtained using a curvilinear probe operating at 4.0 MHz at a depth of 15 cm (5.9 in.).

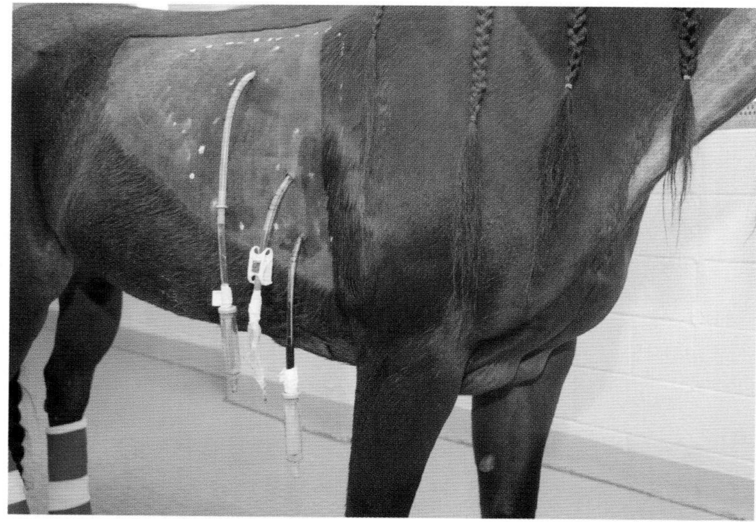

FIGURE 11.3-5 Multiple 28-French chest tubes with one-way Heimlich valves placed in the pleural space to allow for drainage of pleural fluid.

Chest tubes were placed bilaterally at the 11th ICS on the left and 10th, 8th, and 7th ICS on the right (Fig. 11.3-5). A total of 4 and 6 L of pleural fluid was obtained from the left and right hemithoraces, respectively. The cytological evaluation of fluid from each side of the thorax and a tracheal aspiration was consistent with septic, suppurative inflammation with intracellular bacterial cocci. No neoplastic or fungal organisms were identified cytologically. Cultures from each site were submitted for the identification of the causative bacterial organism(s) and antimicrobial susceptibility.

Following the drainage of the pleural fluid, lateral radiographs of the thorax revealed a diffuse bronchial pattern. Scant fluid was present within the ventral pleural space ventral to the caudal vena cava. The chest tube terminated just dorsal to the pleural fluid. Gas opacities dorsal to the pleural fluid created a linear horizontal gas-fluid interface, likely secondary to a pneumothorax from chest tube placement (Fig. 11.3-6A and B).

Pneumothorax, or the presence of gas within the pleural space, may be classified as open or closed and spontaneous, traumatic, or iatrogenic. The most common cause of pneumothorax in the horse is pleuropneumonia, secondary to placement of chest tubes or bullous emphysema.

Diagnosis

Bilateral, septic, fibrinous pleuropneumonia

Treatment

Broad-spectrum antimicrobial therapy consisted of potassium penicillin (22,000 U/kg, IV, q 6 h), gentamicin (6.6 mg/kg, PO, q 24 h), and metronidazole (10 mg/kg, PO, q 8 h). Fluid therapy at a maintenance rate (50 mL/kg/day) was initiated with lactated Ringer's solution (LRS). For pain and inflammation, 1.1 mg/kg of flunixin

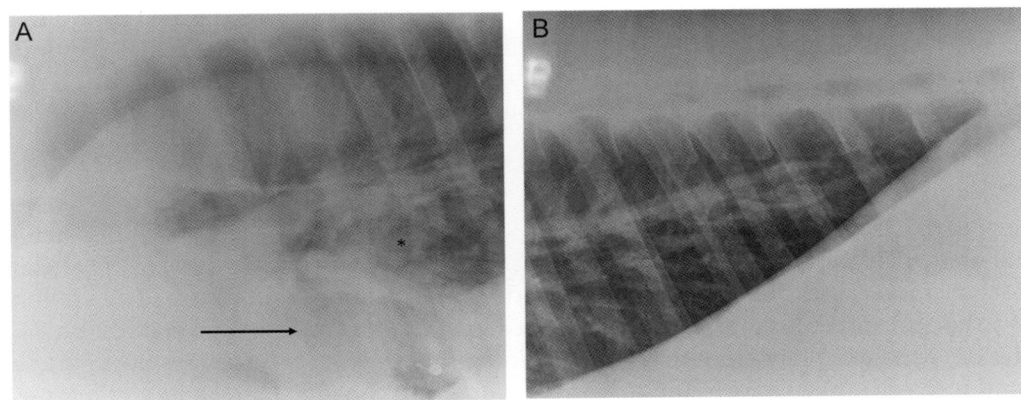

FIGURE 11.3-6 Lateral thoracic radiographs of the thorax of a 7-year-old Thoroughbred gelding with bacterial pleuropneumonia. (A) Craniodrosal view. Increased pleural fluid was present in the ventral pleural space, ventral to the caudal vena cava (*black arrow*). Gas opacity was dorsal to the pleural fluid (*asterisk*). (B) Caudodorsal view. Findings were consistent with a diffuse bronchial pattern.

TRACHEAL ASPIRATION

A tracheal aspiration (TA) may be performed either percutaneously via a through-the-needle catheter inserted between 2 tracheal rings (i.e., transtracheal aspiration [TTA]) or through the biopsy channel of an endoscope, using a commercially available guarded, sterile catheter. The latter method is sometimes preferred, as it enables visual assessment of the airway. With either method, fluid is aspirated directly from the trachea, and thus, the sample is representative of cells originating throughout the lung and traveling to the trachea via mucociliary clearance. The cellular morphology is less well preserved compared to the BAL; however, focal pulmonary inflammation is represented from a TA sample, which could be missed by a BAL. Because this technique is performed aseptically, culture results may be useful if the etiology is suspected to be infectious. Finally, both of these tracheal aspirate techniques are considered safe and easy to perform.

meglumine was administered IV q 12 h. Due to the substantial amount of fibrin accumulation within the pleural space, drainage of fluid from the chest tubes was blocked, necessitating the placement of additional chest tubes (Fig. 11.3-5). The pleural space was lavaged with 10 L of LRS, q 12 h, via the indwelling thoracic tubes. Following lavage, the tubes were maintained with a one-way Heimlich valve to permit continuous drainage of the fluid. Sonographic assessment of the pulmonary disease was monitored over the next 10 days. The chest tubes were removed once production of excess pleural fluid. Culture results revealed a *Streptococcus zooepidemicus* subspecies *equi* in both the pleural and tracheal aspirate samples. The injectable antimicrobial therapy was discontinued, and the gelding was started on an oral antimicrobial therapy (chloramphenicol, 50 mg/kg, PO, q 6 h), which was considered an appropriate choice based on microbial sensitivity.

The gelding returned home and was maintained on antimicrobials; the thorax was regularly monitored sonographically. No further fluid accumulation was appreciated, and the gelding remained bright with normal vital parameters. Based on clinical and sonographic improvement, the antimicrobials were discontinued after 8 weeks of therapy. At 6 months after the onset of the pleuropneumonia, the gelding was gradually returned to work and was in full work at 1 year.

Anatomical features in equids

Introduction

The thoracic cavity of horses is the second largest of the body cavities, bounded dorsally by the thoracic vertebrae and the ligaments and muscles connecting them, laterally by the ribs and intercostal muscles, and ventrally by the sternum, cartilage of the sternal ribs, and associated muscles. The posterior wall, formed by the diaphragm, is very oblique and convex. The ventral floor is about half as long as the dorsal wall. The cranial aperture is narrow, bounded dorsally by the 1st thoracic vertebrae and laterally by the first pair of ribs, and contains the longus colli muscles, trachea, esophagus, vessels, nerves, and lymph nodes.

Function

The primary function of the respiratory system is the transport of oxygen and carbon dioxide between the environment and tissues. Other less obvious but important functions of the lungs include (1) pH balance—increase in carbon dioxide retention causes the body to become acidic; (2) filtration—removes small air embolisms; (3) protection—a shock absorber for the heart in high-speed injuries; (4) protection from infection—lungs secrete immunoglobulin A; (5) mucociliary clearance—mucus lining the respiratory tract traps dust particles and bacteria; (6) blood reservoir—lungs vary in the amount of blood they contain at any point in time, e.g., exercise; (7) whinny—horses, like all species, need air moving to communicate.

Airways

The **trachea** extends from the larynx to the **hilus** of the lungs. It is formed of dorsally incomplete hyaline cartilaginous rings connected by **annular ligaments** that are covered by adventitia and lined by a mucous membrane. The dorsal ends of the tracheal cartilages are connected by the smooth muscle of the **trachealis muscle**.

The trachea occupies a median position, except at its termination, where it is pushed right by the arch of the aorta. The cervical trachea is adjacent to the **longus colli muscles** and is briefly ventral to the esophagus. It is related laterally to the lobes of the thyroid gland; the carotid arteries; the jugular veins; the vagus, sympathetic, and recurrent laryngeal nerves; and the tracheal lymph ducts and cervical lymph glands.

The thoracic part of the trachea passes between the pleural sacs and terminates by bifurcating into the right and left mainstem bronchi at the level of the 5th or 6th ICS, then entering the **hilum** of each lung. The trachea is related proximally to the longus colli muscle, and then more distally to the esophagus. On the left of the thoracic trachea lie the aortic arch, left brachial artery, and thoracic duct. The right of the thoracic trachea is crossed by the azygos vein, the dorsocervical and vertebral vessels, and the right vagus nerve. Ventrally, the thoracic trachea is related to the anterior vena cava, the brachiocephalic and common carotid trunks, and the cardiac and left recurrent laryngeal nerves.

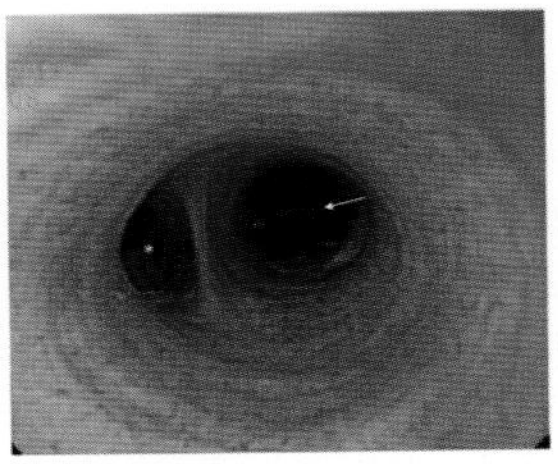

FIGURE 11.3-7 Endoscopic image of a normal trachea branching at the carina of the trachea into the right (*asterisk*) and left (*arrow*) primary bronchi.

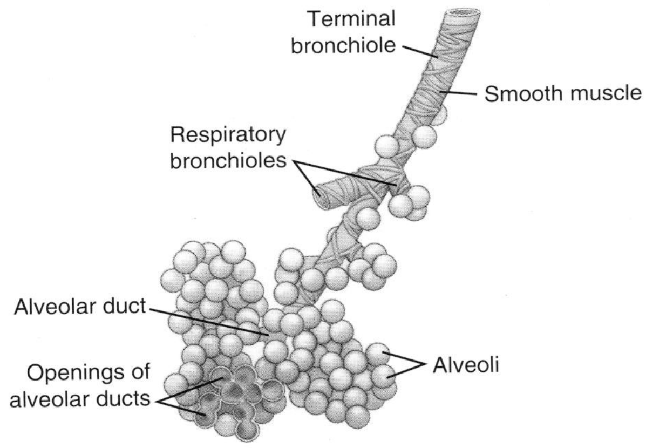

FIGURE 11.3-8 Terminal airways of the horse. Fresh air flows in from the segmental bronchi through the terminal bronchioles, ultimately being delivered to the alveoli for gas exchange.

The right and left primary **bronchi** (singular, bronchus; Fig. 11.3-7) then enter the hilum of each lung. Each bronchus further divides into **lobar** (or secondary bronchi), **segmental**, and **subsegmental bronchi** (or tertiary bronchi), which ventilate each bronchopulmonary segment within a lung lobe. The segmental bronchi divide into **bronchioles** (Fig. 11.3-8). The structure of the bronchi is similar to that of the trachea; however, the cartilaginous framework consists of plates rather than rings.

Bronchioles are small tubular structures which lack cartilaginous support and arise from the segmental bronchi. These are the last strictly conductive branches of the respiratory tract. Terminal bronchioles give rise to respiratory bronchioles or open directly into **alveolar ducts**. The respiratory bronchioles function in the hematogenous exchange of gas. Alveolar ducts are surrounded by **alveoli** and terminate at the **alveolar sacs**, which primarily function in gas exchange.

BRONCHOALVEOLAR LAVAGE

When pulmonary disease is suspected, sampling of the lower airways is indicated to characterize the type, degree, and origin of disease. The most common methods of sampling include tracheal aspiration and bronchoalveolar lavage (BAL) and the method selected depends on the suspected diagnosis, based on historical information and physical examination findings. Bronchoalveolar lavage is a method of recovery of respiratory secretions lining the peripheral airways and alveoli in a discrete part of the lung; therefore, it is more applicable to conditions causing diffuse lower respiratory tract inflammation, such as equine asthma and exercise-induced pulmonary hemorrhage. A BAL can be performed either endoscopically, to permit direct visualization of the lung segment to be lavaged, or blindly using a commercially available, cuffed nasotracheal catheter. The catheter or endoscope is passed through the nasal passages through the trachea until it is wedged in the distal airway. Sterile isotonic fluid is instilled into the terminal bronchus and then immediately aspirated for cytologic evaluation. Because the technique requires the endoscope or catheter to be passed through the nasal passages, the sample obtained is contaminated, precluding use of the sample for bacterial culture and antimicrobial susceptibility testing.

Pleural effusion, or accumulation of fluid in the pleural space, most often results from imbalances of Starling's law of fluid flux. Bacterial pneumonia and lung abscessation are among the most common causes of pleural effusion in horses; however, pleural effusion can develop as a result of a vast number of other conditions, including neoplasia, thoracic trauma, diaphragmatic herniation, hypoproteinemia, pericarditis, peritonitis, viral and fungal infections, equine infectious anemia, congestive heart failure, and hepatic disease. Cytological and microbiological examination of both pleural and tracheal aspirate fluid are important in helping determine the etiology of an effusion.

Fenestrations (L. *fenestra* window) in the mediastinum are normal in the equid. These fenestrations are clinically important as they may allow a unilateral pneumothorax to become bilateral. Importantly, in cases of pleuropneumonia, accumulation of fibrin often results in obstructing or blocking the fenestrations. While this may help limit a bilateral pneumothorax, it also limits simultaneous lavage of both sides of the thorax.

Pleural cavity

The serosa lining the thoracic cavity and the thoracic organs is known as the **pleura** (Gr. rib, side). The pleura forms 2 pleural sacs on either side of the mediastinum and is comprised of **visceral pleura** on the surface of the lung and **parietal pleura** lining the walls of the thorax, including the mediastinum and diaphragm (see Case 12.7). The pleural cavities lie within the pleural sacs, between the visceral and parietal layers of the pleura.

The heart, esophagus, trachea, aorta, and thymus contribute in forming the **mediastinum**, which divides the thorax into 2 cavities. The pleura that lines the mediastinum is called the parietal pleura and, in places, the mediastinum is formed only by contact between the two mediastinal pleura. The mediastinum in these locations is often incomplete or fenestrated.

The **lungs** are a paired organ, occupying much of the thoracic cavity, and function primarily to exchange oxygen for carbon dioxide in the blood. Equine lungs differ from those of other domestic species in that they lack deep interlobar fissures and distinct lung lobes. Horses have five **lobes,** including the **left cranial** and **caudal** lobes, and the **right cranial**, **caudal**, and **accessory** lobes. Other species also have a right middle lobe, though this is not present in the equids (Fig. 11.3-9A–C).

The **costal surface** of the lungs is convex and lies against the lateral thoracic wall. The mediastinal surface is much less extensive and has a large cavity adapted to the pericardium and heart.

Blood supply and lymphatics of the lungs

The major source of blood supply to the lungs is the **pulmonary circulation**, which delivers blood to the alveoli for gas exchange and alveolar nutrient supply. The distribution of pulmonary arterial blood flow to the lungs depends

BACTERIAL LOWER RESPIRATORY TRACT INFECTIONS

Bacterial infections of the lower respiratory tract are common in adult horses. Bronchopneumonia refers to infection of the bronchi and parenchyma of the lung. Infection that extends to the pleural space is known as pleuropneumonia. Diagnostic evaluation using thoracic ultrasound and radiographs helps determine the location of infection and guides therapy.

LUNG COMPARISONS AND NOMENCLATURE

The left lung has two lobes, cranial and caudal in all species *except* the horse. The right lung has four lobes—cranial, middle, caudal, and accessory. Horses, the anomaly, do not have a middle right lobe. The cranial lobe of the left lung is additionally divided into two parts—cranial and caudal—excluding horses. Outdated terms used to describe the four parts of the right lung are apical/cranial, cardiac/middle, and diaphragmatic/caudal lobes. The accessory lobe has been referred to as the intermediate, or azygos lobe.

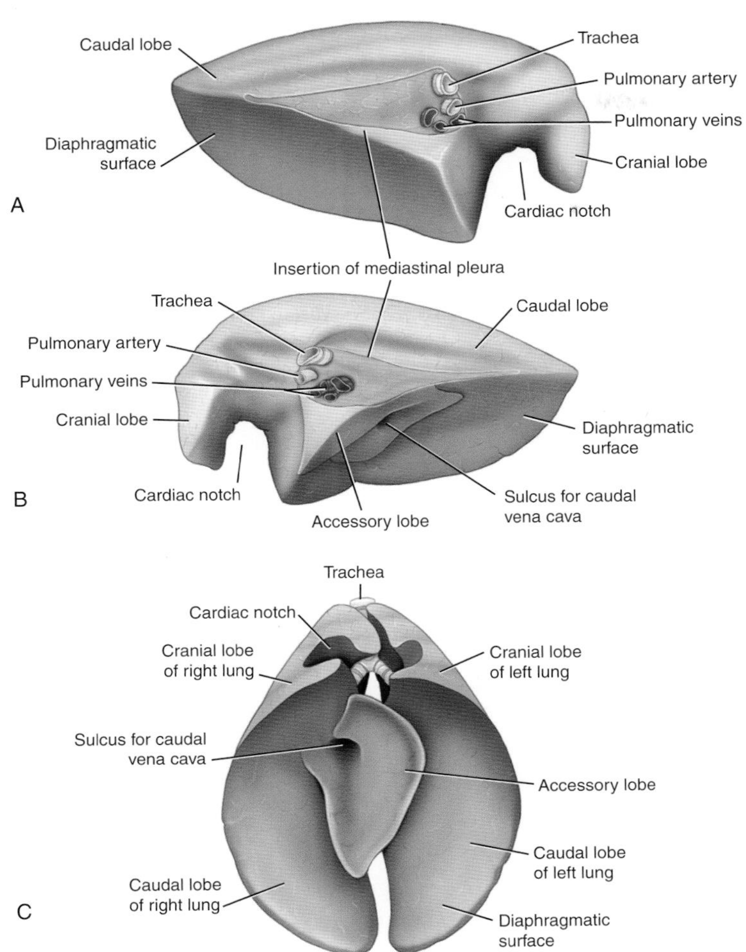

FIGURE 11.3-9 Equine lung lobes. (A) Medial surface of the left lung. (B) Medial surface of the right lung. (C) Ventral/diaphragmatic surfaces of lungs as viewed if horse were in dorsal recumbency.

upon mechanical forces (i.e., gravity, pulmonary arterial pressure, pulmonary venous pressure, and alveolar pressure) and vasoactive agents (i.e., catecholamines, histamines, and eicosanoids). In a standing horse, the most dorsal lung is less well-perfused than the dependent lung. Changes in vascular resistance help match ventilation and perfusion to optimize gas exchange. **Bronchial circulation** also provides blood flow to the lungs and nutritional support to the lymphatics. Bronchial circulation provides arterial blood to the pleural surface and anastomotic connections with the alveolar-capillary bed from the pulmonary circulation. Again, pressure in the bronchial and pulmonary microvasculature affects the magnitude of anastomotic pulmonary microvasculature.

The lymph nodes of the thorax consist of aggregates of small nodes. The **right, left,** and **middle tracheobronchial lymph nodes** surround the tracheal bifurcation. **Pulmonary lymph nodes** are covered by lung tissue and located inconsistently throughout the thorax. The **cranial mediastinal lymph nodes** are congruent with the caudal deep cervical lymph nodes. The **middle mediastinal nodes** are dorsal to the hilus, and the **caudal mediastinal nodes** lie between the aorta and the esophagus within the caudal mediastinum. The **cranial sternal lymph nodes** are situated cranial to the internal thoracic vessels and are difficult to distinguish from the cranial mediastinal nodes. Adjacent to where the diaphragm attaches to the sternum lie the **caudal sternal lymph nodes**. The dorsal aspect of ICS 3 through 16 contain intercostal lymph nodes bilaterally. **Thoracic aortic lymph nodes** are present dorsal to the aorta opposite the 3rd and 17th thoracic vertebrae. Often these lymph nodes are absent from the 9th to 14th vertebrae on the right.

The efferent lymphatics originate from the lymph nodes of the thoracic cavity and communicate or join the veins at the level of the thoracic inlet or **cranial thoracic duct**. The thoracic duct joins the aorta, crossing the trachea before typically terminating at the cranial vena cava.

Selected references

[1] Arroyo MG, Slovis NM, Moore GE, et al. Factors associated with survival in 97 horses with septic pleuropneumonia. J Vet Intern Med 2017;31:894–900.
[2] Boy MG, Sweeney CR. Pneumothorax in horses: 40 cases (1980-1997). J Am Vet Med Assoc 2000;216:1955–9.
[3] Hare WCD. Equine respiratory system. In: Sisson S, Grossman JD, editors. The anatomy of domestic animals, vol. 1. WB Saunders Co; 1975. p. 498–523.
[4] Hewson J, Arroyo LG. Respiratory disease: diagnostic approaches in the horse. Vet Clin North Am Equine Pract 2015;31:307–36.
[5] Hoffman AM. Bronchoalveolar lavage: sampling technique and guidelines for cytologic preparation and interpretation. Vet Clin North Am Equine Pract 2008;24:423–35. vii–viii.

CHAPTER 12

ABDOMEN

Jarred Williams and Kira Epstein, Chapter editors

Stomach, Spleen, and Small Intestine

12.1 Gastric ulcer disease—*Cynthia Xue, Katie Withowski, Alexandra St. Pierre, and Kira Epstein* 724

12.2 Duodenitis-proximal jejunitis—*Katherine Christie and Jarred Williams* 732

12.3 Epiploic foramen entrapment—*Jesse Tyma, José Goni, and Jarred Williams* 737

Cecum and Colon

12.4 Cecal impaction—*Euan Murray and Kira Epstein* 744

12.5 Large colon volvulus—*Jessica Bramski and Kira Epstein* 750

12.6 Small colon enterolith—*Phillip Kieffer and Kira Epstein* 756

Organs

12.7 Cholangiocarcinoma—*Brina Lopez and Kira Epstein* 763

CASE 12.1

Gastric Ulcer Disease

Cynthia Xue[a], Katie Withowski[b], Alexandra St. Pierre[c], and Kira Epstein[d]
[a]Department of Clinical Sciences, Ross University School of Veterinary Medicine, St. Kitts and Nevis, WI
[b]North Carolina State University School of Veterinary Medicine, Raleigh, North Carolina, US
[c]Hamilton Wenham Veterinary Clinic, South Hamilton, Massachusetts, US
[d]Department of Large Animal Medicine, University of Georgia College of Veterinary Medicine, Athens, Georgia, US

Clinical case

History

A 16-year-old Appendix Quarter Horse mare was presented for signs of intermittent mild colic of a few weeks' duration. In her previous episodes, she was treated with flunixin meglumine by the owners, which seemed to resolve her clinical signs. On the morning of the presentation, however, the clinical signs persisted and did not respond to symptomatic treatment as they had before. The owners also expressed concern that, over this same period, she had become more "girthy" and more irritable, both on the ground and under the saddle.

Physical examination findings

The mare presented with normal vital parameters: 100.2 °F (37.8 °C), 40 bpm, 20 brpm. Her mucous membranes were pink and moist with a CRT of < 2 seconds. She pawed during her initial assessment but stood quietly afterward. Abdominal auscultation revealed that GI borborygmi were decreased in all quadrants. She passed normal feces on the trailer.

Differential diagnoses

Equine gastric ulcer syndrome, enterolithiasis, gas colic, intermittent colonic displacement, neoplasia, liver disease, and ovarian pain

Diagnostics

The mare was sedated for abdominal palpation per rectum, which revealed no palpable abnormalities. A nasogastric tube was passed, and no net reflux was obtained. Routine bloodwork (CBC, fibrinogen, and serum biochemistry) was normal. The mare was held off feed for the rest of the day and that night. She showed no more signs of colic. Gastroscopy was performed the following morning under standing sedation with a 3-m endoscope. Two squamous ulcers were identified at the margo plicatus of the greater curvature, and there were diffuse areas of hyperkeratosis on the squamous mucosa (Fig. 12.1-1). No ulcers were seen on the lesser curvature. The glandular mucosa and esophagus were normal.

Comparative Veterinary Anatomy: A Clinical Approach. https://doi.org/10.1016/B978-0-323-91015-6.00153-9

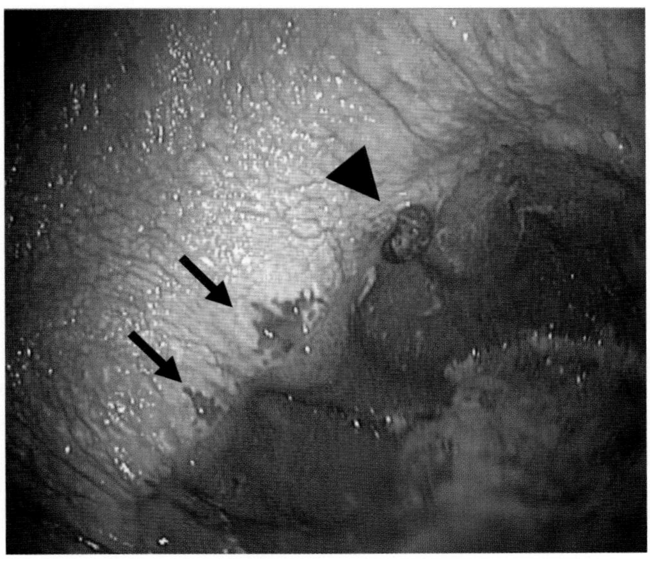

FIGURE 12.1-1 Gastroscopy showing squamous ulcers *(arrows)* along the margo plicatus of the greater curvature surrounded by areas of hyperkeratosis *(yellow, thickened mucosa)*. The *Gasterophilus* bot larva *(black arrowhead)* adjacent to the ulcers is a common incidental finding.

Diagnosis

Equine squamous gastric disease (ESGD), which is part of the equine gastric ulcer syndrome (EGUS) complex

Treatment

The mare was started on oral sucralfate and omeprazole, administered at least 1 h apart. She was scheduled for a repeat gastroscopy in 4 weeks at which time results would dictate modification of her current therapeutic plan. Additional recommendations were to add alfalfa hay to her diet and allow as much turnout in pasture as possible.

Anatomical features in equids

Introduction

This section covers the stomach, omentum, ligamentous attachments of the stomach to surrounding structures, spleen, regional blood supply and innervation, and lymphatic drainage.

EQUINE GASTRIC ULCER SYNDROME (EGUS)

This is an umbrella term that encompasses all erosive and ulcerative diseases of the stomach, including ulceration of the squamous mucosa (equine squamous gastric disease or ESGD) and glandular mucosa (equine glandular gastric disease or EGGD). Equine squamous gastric disease occurs most commonly adjacent to the margo plicatus and has a high prevalence (up to 93%) in horses under high-stress and intensive management conditions such as racehorses. Risk factors for ESGD include intensive exercise training, signs of stress, and dietary factors. Equine glandular gastric disease is relatively poorly characterized. It appears to occur most commonly at the pylorus and occurs in up to 65% of racehorses. Risk factors for EGGD are relatively unknown, though it is also suspected to be a multifactorial disease.

Function

The process of digestion begins in the stomach where acid and pepsin are secreted. The acidic environment of the stomach may also play a role in protection from ingested pathogens. The mucosa of the stomach is susceptible to digestion by acid and pepsin and is injured if the protective mechanisms of the normal stomach are disrupted. Digesta remains in the stomach for 2–6 h. However, horses have evolved to eat for 18–20 h a day, and initiation of peristalsis to move food into the small intestine is caused by the introduction of new food into the stomach. Thus, it is uncommon for the equine stomach to be completely empty at any time.

The spleen is made up of both red pulp and white pulp; the red pulp is the venous and arterial supply supported by the trabeculae sponge, and the white pulp is comprised of reticuloendothelial cells (lymphocytes and macrophages that provide its immune functions). The spleen serves as a storage site for red blood cells (RBCs) and up to 50% of the RBC volume in a horse can be stored in the spleen at any time. These cells are released into the circulation when needed and those that are degenerate are filtered and removed from circulation. The equine spleen can relax and contract associated with signaling from the sympathetic nervous system. This means that an excited or nervous horse can release large numbers of stored RBCs into circulation. In addition, the spleen acts as a filter and phagocytizes senescent RBCs, platelets, parasites, and even bacteria via its macrophage activity. The spleen is also the site of IgM and iron production.

Stomach (Fig. 12.1-2)

The equine **stomach** is divided into the **cardia**, **fundus**, **body**, **antrum**, and **pylorus**. The stomach has a **greater** (convex, longer) and **lesser** (concave, shorter) **curvature** between the cardia and the pylorus when viewed in cross-section (Fig. 12.1-3). The stomach lies to the left of the median plane except for the pylorus, which sits on the right. Due to the curvature of the stomach, in situ the cardia and pylorus are spatially adjacent to one another, separated by the lesser curvature (Fig. 12.1-3 and Landscape Fig. 4.0-7).

The nonglandular or proventricular part of the stomach is a common site for the larvae of the bot fly (*Gastrophilus intestinalis*). Small crater-like mucosal perforations caused by these larvae are commonly seen on gastroscopy (see Chapter 2.1) and found near the cardia.

The normal equine stomach has a capacity of 5–15 L. Based on the mucosa, the stomach can be divided into **nonglandular** **(squamous or proventricular part)** and **glandular** parts separated by the **margo plicatus**, an anatomically distinct region and the junction where the squamous mucosa and the glandular mucosa meet, creating a raised irregular

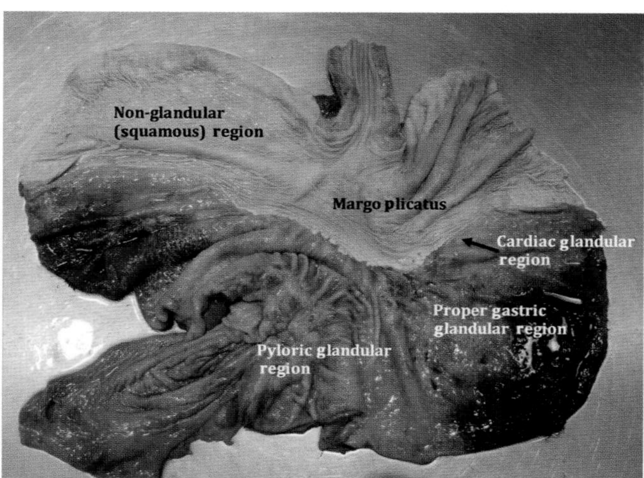

FIGURE 12.1-2 Stomach, ex situ and sectioned sagittally along the greater curvature. Note the margo plicatus dividing the squamous mucosa (*above*) and the glandular mucosa, with cardiac *(black arrow)*, proper gastric, and pyloric regions (*below*).

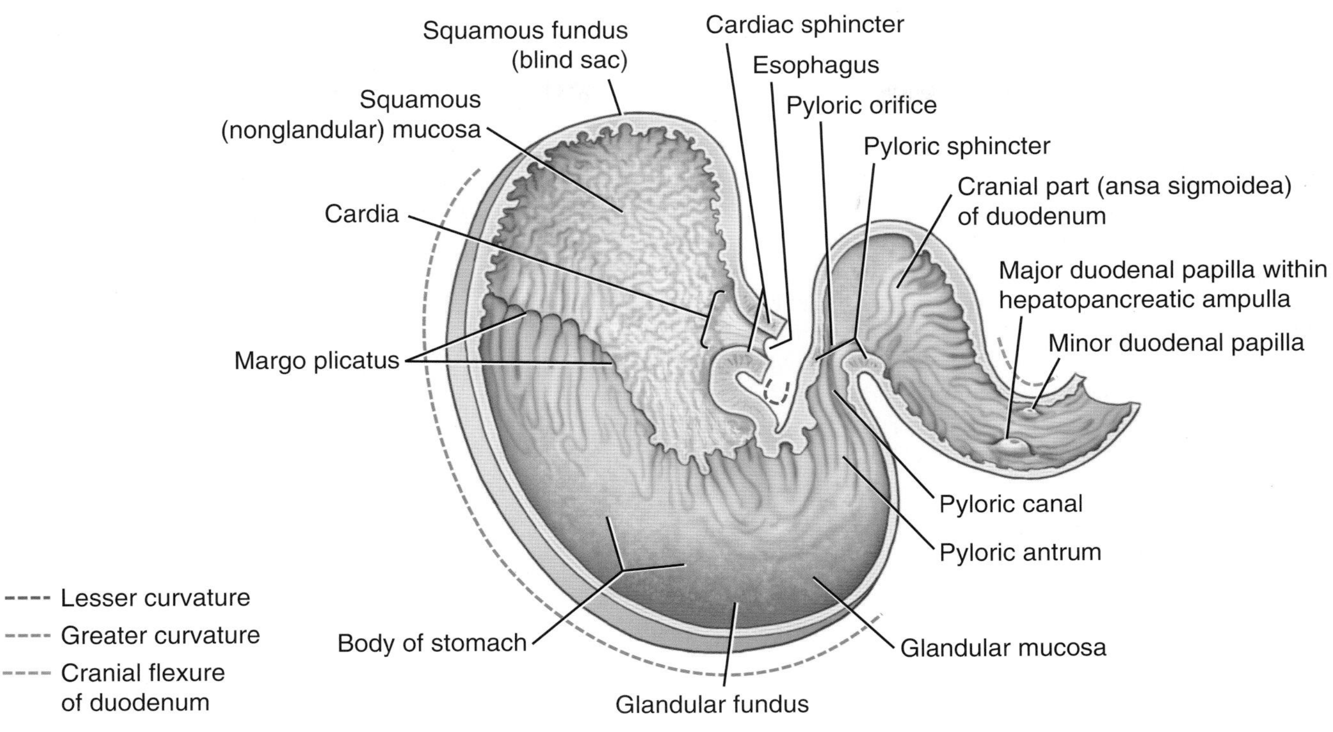

FIGURE 12.1-3 Cross-sectional anatomy of the stomach with labeled cardia, fundus, body, antrum, pylorus, greater and lesser curvature, glandular and nonglandular region, and margo plicatus.

ridge circumferentially around the gastric mucosa. The nonglandular stomach is typified by a nonkeratinized stratified squamous mucosa; its function is not clearly understood. The region is rarely in contact with gastric contents and does not have a secretory function. The glandular stomach has a simple columnar epithelium.

The distal esophagus joins the **cardia**, the entrance of the stomach, which is surrounded by a muscular **cardiac sphincter**. This region is so-named due to its proximity to the heart and is fixed to the diaphragm at the **esophageal hiatus**. The cardia opens to the fundus, which is bordered dorsally by the **saccus secus** (or blind sac). The anatomic region of the cardia should not be confused with the functional cardiac glandular region.

The squamous mucosa of the horse's stomach is akin to the distal esophagus in humans. Most other animals do not have this portion of the stomach. Squamous gastric disorder is therefore most like esophageal reflux disorder in humans (see side box entitled "Gastroesophageal reflux disease (GERD) in humans").

The muscular cardiac sphincter, coupled with the oblique angle at which the esophagus enters the stomach, is believed to contribute to horses' inability to vomit. The muscular cardiac sphincter closes tightly when the stomach is full, making it difficult for ingesta to escape in a retrograde fashion. For this reason, a nasogastric tube should be passed in virtually all acute abdomen (colic) horses to prevent gastric rupture with subsequent septic peritonitis and death.

Pyloric stenosis or stricture in foals can be acquired or congenital. Acquired is more common and is due to fibrosis and contraction during the healing of pyloric gastric ulcers. Clinical signs most often occur as the foal transitions from milk to solid feed. Clinical signs are variable and include colic, ptyalism/sialorrhea (Gr. *ptyalismos* excessive flow of saliva), bruxism (Gr. *brychein* to gnash the teeth), depression, anorexia, difficulty swallowing, unthriftiness, weight loss, and diarrhea. Diagnosis is based on history, gastroscopy, and contrast radiographic gastric emptying studies. Medical management consists of gastric decompression via NGT, treatment of mucosal ulceration, and supportive care—i.e., parenteral nutrition if indicated. Surgical bypass, gastroduodenostomy or gastrojejunostomy, may be required. The prognosis for long-term, post-operative survival (>2 years) is fair to good for pyloric strictures (69% survival). Prognosis decreases when the duodenum is involved in the stricture or there are adhesions.

The **fundus** is the dilation of the stomach near the cardia with the **body** defined as the area bordered by the convex greater curvature. The body comprises the largest portion of the stomach. The **antrum** is the transition zone between the glandular mucosa of the body and the pylorus. The mucosa in this region has no secretory function. The **pylorus** is the egress from the stomach, defined anatomically as the narrowed, ventral portion of the body leading into another muscular sphincter that closes off the stomach from the duodenum.

Soft tissue attachments of the stomach

Externally, the stomach has attachments to the omentum, spleen (gastrosplenic ligament), diaphragm (gastrophrenic ligament), and liver (gastrohepatic ligament). The **greater omentum** arises from the greater curvature and the **lesser omentum** from the lesser curvature. The parietal surface of the stomach sits adjacent to the diaphragm and the left lobe of the liver, while the visceral surface, facing caudoventrally, sits adjacent to the jejunum, pancreas, and both ascending and descending colons. The two omenta and the visceral surface of the stomach encompass the **omental bursa**, a sac-like potential space that communicates with the peritoneal cavity via a narrow opening known as the **epiploic foramen** (see Case 12.3).

The stomach is attached to surrounding organs by peritoneal folds, termed ligaments comprising a double fold of peritoneum that forms connections between 2 independent viscera, or between a viscus and the abdominal wall. The greater curvature is attached to the diaphragm by the **gastrophrenic ligament**, which is contiguous with the **phrenicosplenic** and **gastrosplenic** (Fig. 12.1-4) **ligaments**. The gastrophrenic ligament is a continuation of the greater omentum, attaching the greater curvature of the stomach in the region of the cardia to the **crura of the diaphragm**.

The gastrosplenic ligament is contiguous with the greater omentum and attaches the stomach to the hilus of the spleen. Its blood supply originates from the **gastroepiploic arteries**. The gastrosplenic ligament is proposed to

GASTROESOPHAGEAL REFLUX DISEASE (GERD) IN HUMANS (COMPARATIVE INFORMATION)

In humans, stomach ingesta refluxes into the esophagus causing the symptoms termed "heartburn" because of irritation of the esophagus (adjacent to the heart, mimicing "heart pain") by stomach acid.

Chronic gastroesophageal reflux disease can result in scarring and stricture of the esophagus, requiring stretching (dilating) of the esophagus (termed *bougienage*, Fr. increases the diameter of a tubular organ). Medical treatment for GERD is like the treatment for EGUS, targeting the parietal cell (acid-producing cell in the stomach) to increase the pH of stomach (reduce acid secretion). Two groups of drugs—histamine-receptor antagonists—e.g., cimetidine, ranitidine, and famotidine—and proton pump inhibitors—e.g., omeprazole—are commonly prescribed.

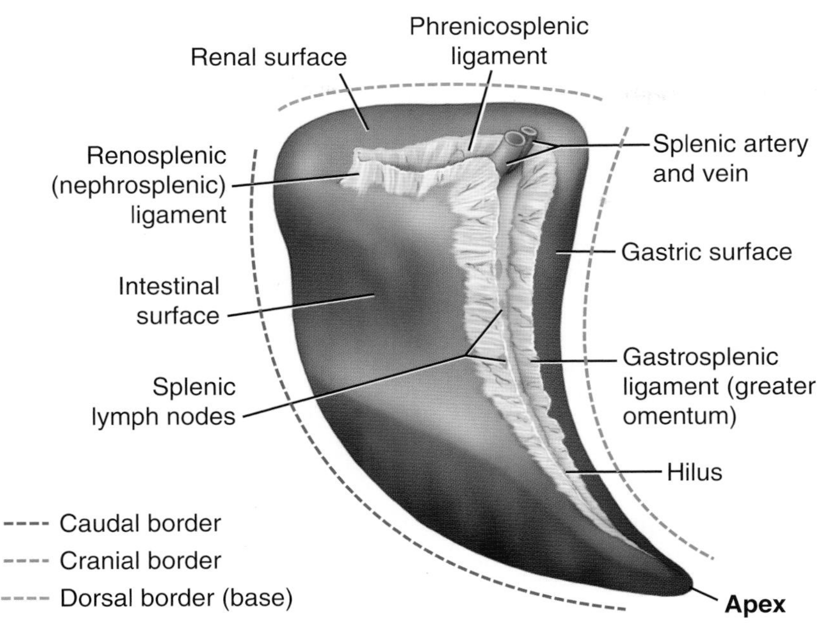

FIGURE 12.1-4 Equine spleen (visceral surface) and associated ligaments.

function as a suspensory apparatus and may serve to maintain both the spleen and stomach in anatomically appropriate positions within the abdomen. The **gastrohepatic** and **hepatoduodenal ligaments**, originating in the lesser omentum, attach the lesser curvature of the stomach and a portion of the duodenum to the liver.

This purported function of the gastrosplenic ligament is supported by a case report in which gastric torsion was identified concurrently with a complete rupture of the dorsal aspect of the gastrosplenic ligament. Entrapment of the bowel within the gastrosplenic ligament is one cause of strangulating small intestinal colic. Midjejunum, distal jejunum, and proximal ileum are most commonly found entrapped in the gastrosplenic ligament. Incarceration of the proximal jejunum does not appear to occur, likely due to its shorter mesentery not allowing for entrapment through a cranially located rent. Isolated case reports and recent retrospective data have documented the incarceration of the large and small colons in the gastrosplenic ligament.

Spleen

Within the abdomen, the **spleen** is a left-sided, roughly sickle-shaped viscus with a broad dorsal base tapering toward a cranioventrally positioned apex. The **greater omentum** attaches to the visceral aspect of the spleen and provides an anatomical division between the narrow gastric and considerably larger intestinal surfaces of this organ. A third, much smaller caudodorsal division of the spleen lies adjacent to the left kidney and left crus of the diaphragm, known as the renal surface. The spleen is connected to the diaphragm, left kidney, and greater curvature of the stomach through the **phrenicosplenic**, **renosplenic**, and **gastrosplenic ligaments**, respectively (Fig. 12.1-4).

On abdominal palpation per rectum, the caudal border of the spleen can be evaluated at, or extending just caudal to, the last rib. Rounding of this caudal border can be indicative of splenic engorgement and is detectable even with limited palpation experience. Certain gastrointestinal disorders can alter the accessibility of the spleen on palpation per rectum. Gastric distension and dilation can result in a greater portion of the spleen being readily palpable, although the contour and overall size should remain unaffected. Nephrosplenic entrapment of the large colon can preclude palpation of the spleen due to ventral displacement of the spleen by a fluid- or gas-filled large colon.

Renosplenic (nephrosplenic) entrapment of the large colon is a common displacement of the large colon and cause of colic. Diagnosis can be made on abdominal palpation per rectum. Transabdominal ultrasound revealing large colon obscuring visualization of the left kidney and obscuring visualization of the spleen has little predictive value. However, the ability to visualize the kidney adjacent to the spleen is a good method to rule out nephrosplenic entrapment. Phenylephrine administered to shrink the spleen can be used as an adjunctive therapy in both medical (jogging or rolling) and surgical management of left dorsal displacement of the large colon. The renosplenic (nephrosplenic) space can be surgically closed to prevent recurrence.

The **renosplenic (nephrosplenic) space** (Fig 12.1-5) is defined as the space formed by the renosplenic (nephrosplenic) ligament, the dorsal edge of the spleen, the dorsal body wall, and the left kidney. The left kidney lies medial to the spleen. The splenic artery, vein, and nerves enter the spleen at the **splenic hilus**, a long groove present on the visceral surface of the spleen, which protects these vital structures. This area is just caudal to the area of the greater curvature of the stomach and can be identified readily on ultrasound.

Gastric and splenic blood supply

It is reasonable to assume that disruption of the short gastric arteries may occur simultaneous to—and because of—gastrosplenic ligament tearing. The clinical significance of this vascular insult is likely minimal and mitigated by the extensive collateral circulation.

Blood supply to the stomach is primarily by way of the **celiac artery** and its tributaries, the **left gastric artery**, **hepatic artery** (branches to the **right gastric** and **gastroduodenal arteries** that further branches to the **right gastroepiploic artery**), and **splenic artery** (branches to the **short gastric** and **left gastroepiploic arteries**). The left gastric artery supplies the cardia, and the left and right gastric arteries supply the lesser curvature of the stomach. The gastroduodenal artery supplies a portion of the pylorus in addition to the proximal portion of the small intestine. The short gastric arteries supply the fundus. The left gastroepiploic artery supplies and passes along the greater curvature before joining back with the right gastroepiploic artery. Venous drainage parallels the

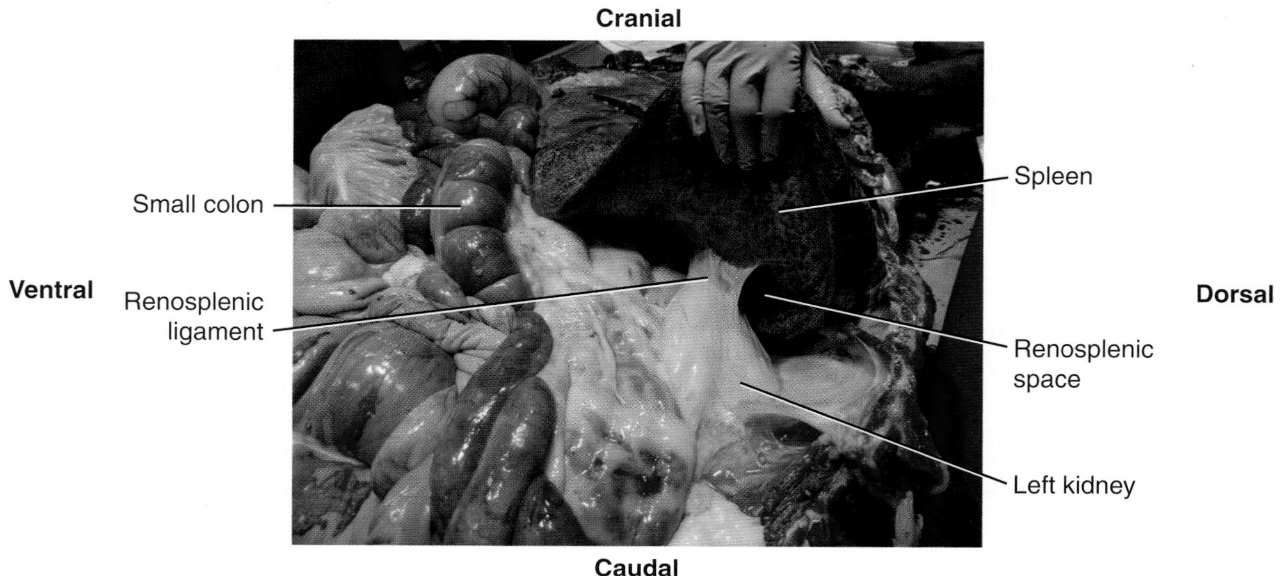

FIGURE 12.1-5 Normal equine renosplenic (nephrosplenic) space identified during routine necropsy.

arterial supply; the **portal vein** is formed by the joining of the **caudal mesenteric**, **cranial mesenteric**, and **splenic veins**.

Within the hilus of the spleen, the **celiac artery** branches to the **splenic artery,** which then supplies both the spleen and the **short gastric arteries** within the greater curvature of the stomach. The **splenic vein** drains to the **portal vein**.

Gastric and splenic lymphatics

Lymphatic drainage of the stomach is closely associated with its vasculature. Clusters of lymph nodes are found around the 3 main branches of the celiac artery—left gastric artery, hepatic artery, and splenic artery—and along the vessels' routes. One group is found in the lesser curvature, another group is associated with the caudal portion of the left gastric artery, a few are found at the saccus secus (blind sac) and ventral pylorus, and a chain follows the path of the gastroepiploic arteries. These lymph nodes are generally small and diffuse. The spleen is drained via the **splenic lymph nodes**, which drain to the **celiac lymph nodes**. The celiac trunk drains to the **cisterna chyli**, a dilated portion of the thoracic duct.

Gastric and splenic nerve supply

The stomach receives both sympathetic (fibers travel with the arteries) and parasympathetic (vagal trunks) innervation. Parasympathetic plexuses within the submucosa and between muscle layers are involved in local reflexes that allow the stomach to react to direct stimulation of its walls. Parasympathetic fibers also terminate on intragastric endocrine cells for the stimulation of digestion.

Innervation of the splenic plexus is via the **major splanchnic nerve** (sympathetic) and the **vagus nerve** (parasympathetic). The mesenteric ganglia supply nerves through an associated network of plexuses for the abdominal viscera.

Selected references

[1] Blikslager AT, Wilson DA. Stomach and spleen. In: Auer JA, Stick JA, editors. Equine surgery. 4th ed. Philadelphia: Saunders; 2012. p. 389–402.
[2] Budras K, Sack W, Rook S. Anatomy of the horse. 6th ed. Hannover, Germany: Schlütersche; 2011.
[3] Dyce KM, Sack WO, Wensing CJG. Textbook of veterinary anatomy. Saunders Elsevier: St. Louis, MO; 2010.
[4] Anon. European college of equine internal medicine consensus statement—equine gastric ulcer syndrome in adult horses. JVIM 2015;29(5):1288–99.
[5] Sprayberry KA, Robinson NE. Equine gastric ulcer syndrome. In: Current therapies in equine medicine; 2015. p. 7E.

CASE 12.2

Duodenitis-proximal Jejunitis

Katherine Christie[a] and Jarred Williams[b]
[a]Internal Medicine, Rood and Riddle Equine Hospital, Lexington, Kentucky, US
[b]Large Animal Medicine, University of Georgia College of Veterinary Medicine, Athens, Georgia, US

Clinical case

History

A 7-year-old, 450-kg (992-lb) American Quarter horse mare was presented with signs of colic and progressive depression. The mare was last seen to be acting normally approximately 2 hours before being found down in the field. Initial management included the administration of flunixin meglumine by the owners, which did not resolve the mare's discomfort. There was no previous history of colic. No significant alterations had been made to the mare's routine except for a new round bale of fescue hay. The owners noted that the outside of the round bale showed some evidence of mold, but that this layer had been removed before feeding.

Physical examination findings

On presentation, the mare was quiet and depressed with a low head carriage. Clinical signs of hypovolemia and systemic inflammation (i.e., the systemic inflammatory response syndrome or SIRS) were present. Oral mucous membranes were tacky and injected (bright pink) with a moderate toxic line and a prolonged capillary refill time of 3 seconds. The mare was tachycardic (80 bpm) with no auscultated cardiac murmurs or arrhythmias. Rectal temperature was 102.1 °F (38.9 °C). Gastrointestinal sounds were decreased to absent in all quadrants. Distal extremities were cool to the touch and digital pulses were normal. The mare was sedated to allow for further diagnostics. The tachycardia was improved but persisted (68 bpm) following sedation.

Differential diagnoses

Duodenitis-proximal jejunitis (anterior enteritis, proximal enteritis), nonstrangulating obstruction of the small intestine (ileal impaction), or early strangulating obstruction of the small intestine (lipoma, epiploic entrapment, small intestinal volvulus)

Diagnostics

Nasogastric intubation yielded 12 L of net gastric reflux that was orange-brown in color and malodorous. The mare became more comfortable following gastric decompression but remained dull and depressed for the remainder of her evaluation. Abdominal palpation per rectum revealed mild distention of multiple loops of the small intestine. Firm but indentable ingesta was present in the large colon.

Transabdominal ultrasound identified multiple loops of mildly fluid-distended (4–5 cm or 1.6–2.0 in. in diameter) small intestine with decreased motility and increased wall thickness (7 mm or 0.3 in.) in both inguinal regions (Fig. 12.2-1). The duodenum was identified along the right side of the abdomen at the cranial pole of the right kidney and demonstrated decreased motility, as well as a corrugated appearance with a wall thickness of greater than 6 mm (0.2 in.). A subjectively increased amount of anechoic, free abdominal fluid was appreciated.

Comparative Veterinary Anatomy: A Clinical Approach. https://doi.org/10.1016/B978-0-323-91015-6.00154-0

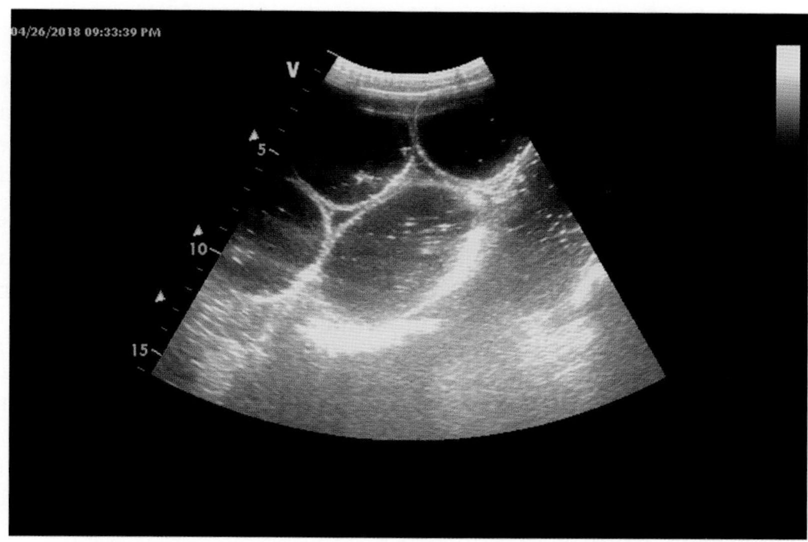

FIGURE 12.2-1 Abdominal ultrasound image of distended small intestine.

Additional diagnostics

Abdominocentesis yielded grossly turbid and orange-yellow peritoneal fluid. Further evaluation of the fluid revealed a total nucleated cell count that was within normal limits (1500 cells/μL), increased total solids (TS) of 3.6 g/dL, and a lactate of 3.0 mmol/L (compared to peripheral lactate of 3.5 mmol/L). Bloodwork was consistent with dehydration, acute infection, and SIRS. Nasogastric reflux was submitted for infectious disease testing (PCR for *Clostridium* and *Salmonella* spp.), which was negative.

Diagnosis

Duodenitis-proximal jejunitis (DPJ), also known as anterior enteritis (AE)

Treatment

The mare was administered IV fluids and the nasogastric tube was left indwelling to allow decompression of the stomach every other hour. A total of 52 L of net gastric reflux was obtained in the initial 18 h of hospitalization. Flunixin meglumine was administered at an anti-inflammatory dose, and ice boots were placed on all 4 ft. for the prevention of laminitis. Additional therapy included polymyxin B (anti-endotoxic) and a lidocaine CRI (both analgesic and anti-inflammatory).

The nasogastric tube was removed after approximately 72 h and oral feeding was gradually reintroduced. The mare tolerated refeeding well and did not show any additional signs of colic. At the time of discharge, 6 days following the initial presentation, the mare was bright and alert with a normal physical examination. Repeat chemistry and complete blood count were within normal limits and no additional complications were identified.

Anatomical features in equids

Introduction

The relevant anatomical structures involved in the overall disease process of DPJ are examined in this section, including the blood supply, lymphatics, and innervation of the small intestine. The discussion also covers the duodenum, exocrine pancreas, and common bile duct.

Function

Functionally, the pancreas is both an exocrine and endocrine organ with the secretory exocrine activities implicated in DPJ. Pancreatic secretions contain digestive proenzymes responsible for the breakdown of dietary carbohydrates, proteins, and fats as well as bicarbonate ions, which help neutralize the pH of the gastric contents entering the duodenum. Normal duodenogastric reflux is composed of water, sodium, bicarbonate, and bile salts.

The duodenum is short and consistent in position with the sigmoid flexure being the widest portion. Here, many digestive enzymes secreted by the pancreas and liver enter via the major pancreatic and bile ducts through a single papilla; the minor pancreatic duct opens on a small papilla on the opposite side. Pancreatic enzymes come into the small intestine in response to the hormone cholecystokinin, while secretin, another hormone produced in the small intestine, promotes the release of bicarbonate into the duodenum to neutralize the acid from the stomach.

The paracrine hormones of the pancreatic islet cells are important in synthesizing and releasing hormones from their respective endocrine cells and then binding to specific receptors in different cells and, by doing this, affect their function. The following are some examples of these hormones and their respective functions: (1) glucose/insulin—activate beta cells and inhibit alpha cells; (2) glycogen/glucagon—activate alpha cells that in turn stimulate beta cells and delta cells; and (3) somatostatin—role in inhibiting alpha cells and beta cells.

There are a large number of G protein-coupled receptors that regulate the secretion of insulin, glucagon, and somatostatin from pancreatic islet cells. In humans, these receptors are targets for the drugs used to treat type-2 diabetes and may have a place in veterinary medicine in treating insulin dysregulation.

Duodenum and its attachments

The **duodenum** is the first segment of the small intestine and is approximately 1 m (39.3 in.) in length. Originating from the stomach at the **pylorus**, the cranial duodenum forms a **sigmoid flexure** adjacent to the liver. The **major** and **minor duodenal papillae** lie within the second bend of the sigmoid flexure, which progresses into the **descending duodenum** (Fig. 12.2-2). This aspect of the duodenum continues to the right kidney at which point the **caudal flexure** courses around the caudal aspect of the root of the mesentery. Attachments to both the base of the cecum and transverse colon (i.e., the **duodenocolic fold**) exist at the level of the caudal flexure and these attachments help tether the duodenum in a constant position. The **ascending duodenum** continues from the caudal flexure and transitions into the beginning of the jejunum.

Unlike other sections of the equine small intestine, the attachments of the duodenum to adjacent organs, as well as to the body wall, are short, allowing for little movement and providing the duodenum a fixed position in the abdomen. The **hepatoduodenal ligament** anchors the cranial portion of the duodenum to the liver. The bile duct and pancreatic ducts run within this structure on their journey to the proximal duodenum. The **mesoduodenum**, a continuation of the hepatoduodenal ligament, forms short attachments between the duodenum and adjacent structures, including the liver, pancreas, right kidney, base of the cecum, and caudal side of the root of the mesentery.

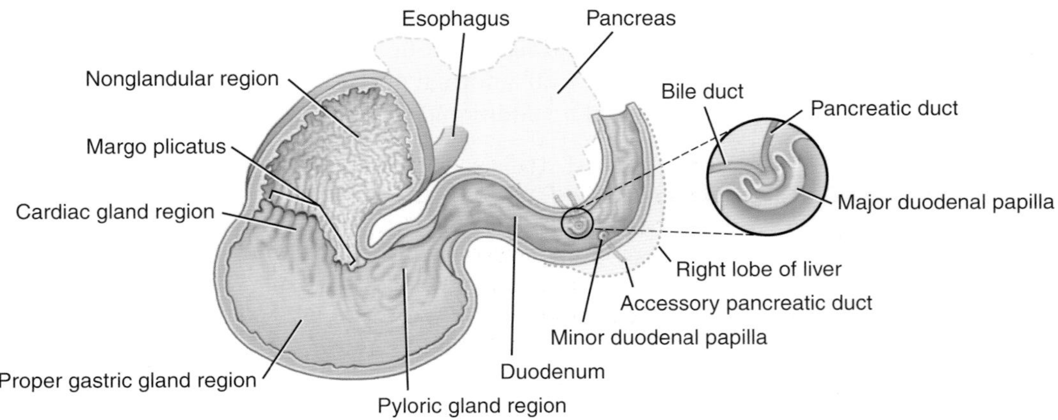

FIGURE 12.2-2 The equine duodenum and its papillae.

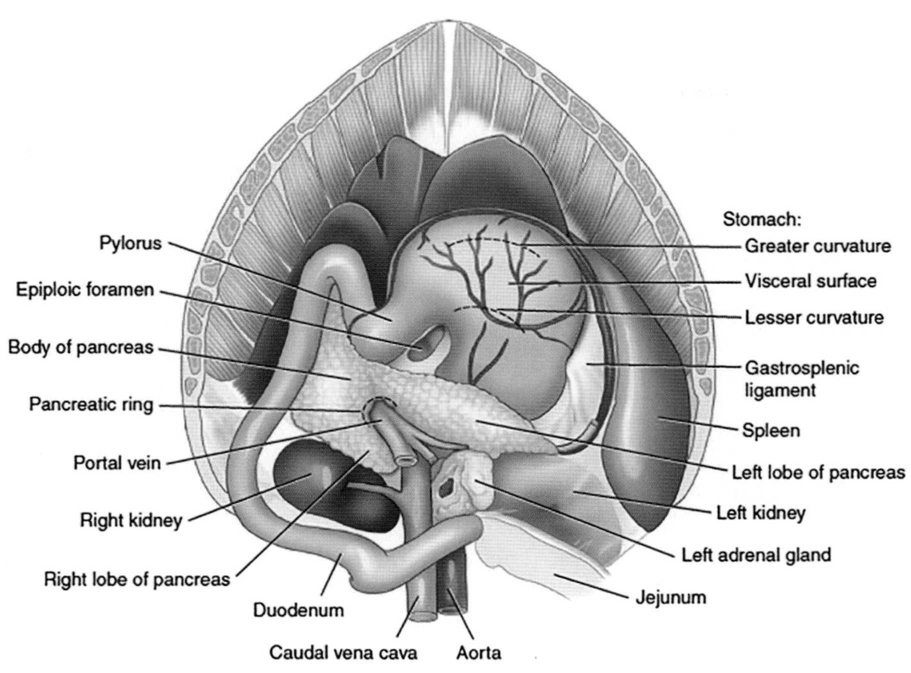

FIGURE 12.2-3 Cross-section of the equine abdomen (view is caudal to cranial) illustrating the relationship of the duodenum, pancreas, and surrounding organs.

Following the caudal flexure, the mesoduodenum splits into 2 parts—the **duodenocolic fold** and a continuation as the mesentery of the jejunum.

Exocrine pancreas

The **pancreas** in the horse is a triangular organ comprised of 3 lobes that are less anatomically distinct than they are in other species (Fig. 12.2-3). The **body** of the pancreas is its most cranial aspect and is associated with the cranial duodenum and the portal vein. The **left lobe** of the pancreas is located medially and has associations with the stomach and the dorsal wall of the abdomen. The **right lobe** is adjacent to the descending duodenum and approaches both the right kidney and the base of the cecum.

Pancreatic enzymes are produced by cells that are arranged in acini (exocrine cell clusters) and that empty into a shared **intralobular duct**. In the horse, these ducts ultimately empty into the **main** and **accessory pancreatic ducts**, which in turn open into the **major duodenal papilla** (with the common bile duct) and the **minor duodenal papilla**, respectively.

GALLBLADDER

In most species, except for equids, the gallbladder is a small hollow organ where bile is stored and concentrated before it is released into the small intestine. The structure and position of the gallbladder can vary significantly among animal species. It receives bile produced by the liver via the common hepatic duct, stores it, and then releases it via the common bile duct into the duodenum, where the bile helps in the digestion of fats.

The gallbladder can be affected by gallstones, formed by material that cannot be dissolved—cholesterol or bilirubin. These “stones” may cause abdominal pain localized to the upper-right quadrant of the abdomen and are often treated with surgery via a cholecystectomy (L. *cholecyst* gallbladder + Gr. *ektomē* excision). Cholelithiasis is the formation of biliary calculi in horses; hepatoliths are calculi within the intrahepatic ducts and choledocholiths are calculi in the common bile duct. Both choledocholithotripsy (crushing choleliths in the common bile duct) and choledochotomy (surgical removal via the common bile duct) have been successfully performed in horses.

This arrangement makes hepatic injury secondary to ascending infection or inflammation from the proximal small intestine likely in cases of DPJ.

Common bile duct

The **common bile duct** of the horse arises as a convergence of hepatic ducts, which are larger in diameter when compared to other species. The distal opening of the common bile duct is located at the **major duodenal papilla** (Fig. 12.2-2). Because horses do not have a gallbladder, a continuous stream of bile is secreted from the common bile duct, which contributes to normal duodenogastric reflux.

Blood supply, lymphatics, and innervation

The major blood supply to the duodenum and pancreas is via the **gastroduodenal** and **cranial pancreaticoduodenal arteries**, which are branches of the **celiac artery**. The venous drainage parallels the arterial supply with the **cranial pancreaticoduodenal** and **gastroduodenal veins** draining into the **portal vein**.

Lymph from regional nodes drains through a **celiac trunk** near the celiac artery to the **cisterna chyli**.

The innervation to the abdominal viscera is supplied via **major** and **minor splanchnic nerve plexuses** associated with the **mesenteric ganglia**.

Selected references

[1] Arroyo LG, et al. Duodenitis-proximal jejunitis in horses after experimental administration of *Clostridium difficile* toxins. J Vet Intern Med 2017;31:158–63.
[2] Freeman DE. Duodenitis-proximal jejunitis. Equine Vet Educ 2000;12(6):322–32.
[3] Davis JL, Pusterla N. Duodenitis-proximal jejunitis. In: Smith BP, editor. Large animal internal medicine. 5th ed. St. Louis: Elsevier (Mosby); 2015. p. 696–9.
[4] Davis JL, Jones SL. Suppurative cholangiohepatitis and enteritis in adult horses. J Vet Intern Med 2003;17:583.

CASE 12.3

Epiploic Foramen Entrapment

Jesse Tyma[a], José Goni[b], and Jarred Williams[c]
[a]Mid-Atlantic Equine Medical Center, Ringoes, New Jersey, US
[b]Department of Veterinary Clinical Sciences, Large Animal Medicine, Purdue University College of Veterinary Medicine, West Lafayette, Indiana, US
[c]Large Animal Medicine, University of Georgia College of Veterinary Medicine, Athens, Georgia, US

Clinical case

History

A 13-year-old Thoroughbred gelding was presented with a history of acute colic. The gelding was first noted to be demonstrating the signs of colic 3h before presentation after being brought inside from pasture turnout. He was sweaty and depressed, inappetent, and began to flank-watch in sternal recumbency. The owners had not observed the passage of manure for at least 12 hours. Despite treatment with IV flunixin meglumine and 30 minutes of hand-walking, the signs of colic progressed; the gelding was repeatedly pawing and trying to roll. The gelding had no known history of colic. There were no recent changes in exercise, turnout, or feeding management. The gelding was reported to crib.

"Cribbing" (or "wind-sucking") is a behavioral stereotypy in which the horse compulsively puts the upper incisors firmly on a hard object—e.g., stall door, water bucket, fence—and pulls back, arching the neck and gulping air into the upper esophagus with a distinct grunting sound. Besides the wear and tear to property and the horse's incisors, it has been linked to gas colic, epiploic foramen entrapment, and temporohyoid osteoarthropathy.

Physical examination findings

On presentation, the gelding was anxious but responsive and was intermittently pawing. Mild bilateral facial abrasions were observed dorsal to the eyes. The gelding was tachycardic at 66 bpm, with a normal respiratory rate and rectal temperature. His mucous membranes were pale pink and slightly tacky with a prolonged CRT of 3 seconds. Jugular fill was mildly prolonged bilaterally. Borborygmi were decreased to absent in all abdominal quadrants.

Differential diagnoses

Colic of unknown etiology pending further workup

Diagnostics

The gelding was sedated and a nasogastric tube was passed, yielding 2 L of nonodorous net nasogastric reflux containing some feed material. Peripheral bloodwork was consistent with moderate dehydration. Abdominal palpation per rectum revealed a few dry fecal balls in the rectum and additional firm chains of fecal balls in the small colon. A few loops of moderately distended segments of the small bowel were palpable at the cranial extent of the rectal examination (mid-abdomen), to the right of midline.

Comparative Veterinary Anatomy: A Clinical Approach. https://doi.org/10.1016/B978-0-323-91015-6.00057-1

Fast localized abdominal sonography of horses (FLASH) relies on the use of 7 transcutaneous abdominal windows, allowing for rapid ultrasonographic evaluation of the acute abdomen. This technique can quickly add diagnostic information to the colic examination, often identifying small intestinal lesions that need surgical correction.

Transabdominal ultrasound revealed several loops of mildly thickened (wall thickness of 4–5 mm or 0.16–0.20 in. in cross-section), amotile loops of small bowel in the right latero-cranioventral abdomen. Along the ventral midline and in both the left and right inguinal regions, several loops of moderately distended small bowel (5–6 cm or 2.0–2.4 in. in diameter) were observed in cross-section (Fig. 12.3-1). The wall thickness of these loops was within normal limits, and there was no evidence of luminal sedimentation of ingesta. There was a subjective mild increase in free peritoneal fluid, which appeared anechoic in nature.

Abdominocentesis produced a stream of red peritoneal fluid. The peritoneal fluid had a moderately increased lactate and mildly increased total protein with a normal total nucleated cell count. These values were thought to be consistent with a strangulating disease process, indicating compromise of bowel viability secondary to ischemia.

Differential diagnoses

Strangulating obstruction of the small intestine—e.g., pedunculated lipoma, epiploic entrapment, small intestinal volvulus; duodenitis-proximal jejunitis (a.k.a. anterior enteritis, proximal enteritis)

Treatment

Exploratory laparotomy under general anesthesia revealed entrapment of the distal jejunum within the epiploic foramen in a left-to-right direction. An approximately 2-m (78.7-in.) segment of entrapped distal jejunum was reduced and exteriorized for evaluation. The affected jejunal segment was diffusely thickened with dark purple serosa and edematous, hemorrhagic mesentery with poor vascular pulse quality. An approximately 3-m (118.1-in.) segment of distal jejunum was identified for resection, and—following ligation of feeder jejunal and arcuate vessels and resection of the devitalized bowel—an end-to-end jejunoileal anastomosis was performed. Recovery from anesthesia was uneventful. Post-operatively, the gelding received supportive care including IV fluids and anti-inflammatory

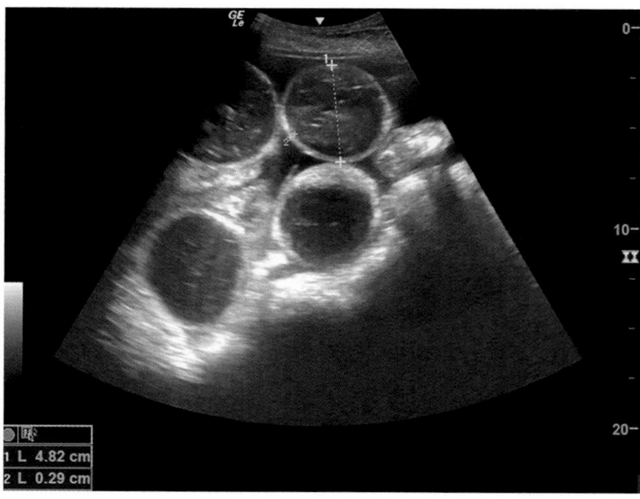

FIGURE 12.3-1 Transabdominal ultrasound image taken along the ventral midline of a horse showing multiple loops of moderately distended small bowel (5–6 cm or 2.0–2.4 in. in diameter) in cross-section. The wall thickness of these loops was within normal limits, and there was minimal to no evidence of luminal sedimentation of ingesta. (Figure courtesy of Nora Grenager.)

medications. Food was slowly reintroduced beginning 36 h post-operatively, and the gelding was discharged 5 days after the surgery with no reported problems.

Diagnosis

Epiploic foramen entrapment of the distal jejunum

Anatomical features in equids

Introduction

The following relevant anatomy is covered in this section: the jejunum and ileum and their blood supply; the epiploic foramen; and mesodiverticular bands and their origins. The duodenum is discussed in Case 12.2.

Function

The small intestine has digestive, absorptive, and immunological functions. Most of the chemical digestion is of proteins, lipids, and carbohydrates. Many of the digestive enzymes that act in the small intestine are secreted by the pancreas and liver and enter the small intestine via the major, minor, and accessory pancreatic and common bile ducts (see Case 12.2). Pancreatic enzymes and bile from the liver respond to the hormones cholecystokinin and secretin in response to ingested food.

Digested food passes into the blood vessels in the wall of the intestine through either diffusion or active transport. The small intestinal mucosa is covered with villi whose function is to increase the quantity of surface area available for the absorption of nutrients. The majority of the absorption of nutrients takes place in the jejunum.

The small intestine supports the body's immune system with a mucosa rich in nodules of lymphoid tissue, singularly or in clumps called Peyer's patches. Peyer's patches are in greater numbers toward the juncture with the large intestine and are part of the lymphatic system. The noticeable Peyer's patches described at the ileocolic junction in other species are absent in the horse. Peyer's patches provide a site for antigen-processing for the immune system.

Small intestinal cross-sectional anatomy and physiology

The small intestine is composed of 4 distinct histological layers or tunics: mucosa, submucosa, muscularis, and serosa. The **tunica mucosa** is lined by **villi**, which—at their tips—contain the **microvilli brush border** that serves to maximize surface area for the absorption of nutrients (Fig. 12.3-2). A thin layer of smooth muscle, the **lamina muscularis**, lines the mucosa. The **tunica submucosa**, a loose connective tissue deep to the **mucosa**, carries blood vessels and nerves to the bowel. The **tunica muscularis** consists of 2 distinct muscle layers: an **inner circular layer** and an **outer longitudinal layer** that propel boluses of ingesta progressively through the gastrointestinal tract in a

RISK FACTORS FOR SMALL INTESTINAL EPIPLOIC FORAMEN ENTRAPMENT

- Signalment:
  Taller horses
  Thoroughbreds and their crosses
- Behavior:
  Crib-biting/wind-sucking
- Environment:
  Winter months
  Increased periods of stall confinement
- Medical history:
  Recent episode of colic

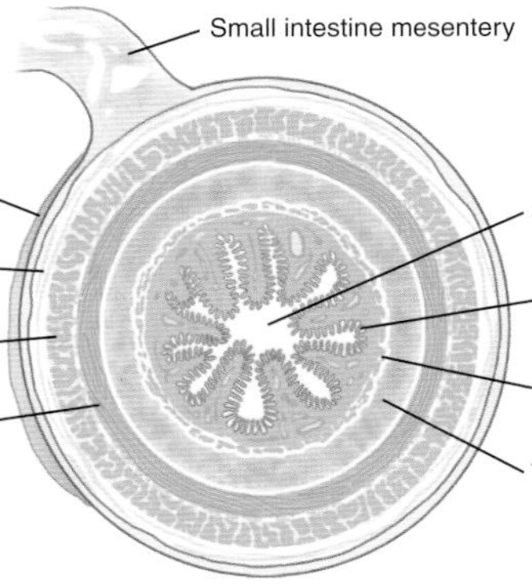

Small intestine mesentery
Visceral peritoneum
Tunica serosa
Longitudinal tunica muscularis
Circular tunica muscularis
Small intestine lumen
Tunica mucosa
Lamina muscularis
Tunica submucosa

FIGURE 12.3-2 Cross-section of equine jejunum showing the layers.

coordinated wave-like contraction. The **tunica serosa** is a layer of loose connective tissue surrounding the muscularis that is continuous with the **visceral peritoneum**.

Neural control of the gastrointestinal tract is based on the **enteric nervous system**, which contains approximately 100 million neurons that independently and meticulously regulate gastrointestinal movement and secretions via local and autonomic signals. The **myenteric plexus (Auerbach's plexus)** of neurons is found between the 2 layers of muscle in the **tunica muscularis** and is the primary controller of gastrointestinal movement, while the **submucosal plexus (Meissner's plexus)** regulates local blood flow and gastrointestinal secretions from the submucosa. Extrinsic sympathetic and parasympathetic fibers supply both plexuses, serving to augment and inhibit constitutive gastrointestinal functions.

Jejunum and its blood supply

The **jejunum** is the middle segment of the small intestine (between the **duodenum** and the **ileum**) and is the longest in length, measuring approximately 17–28 m (55.8–91.9 ft.) in adult horses. The jejunum has a long mesentery allowing for greater mobility of these bowel segments.

The jejunal arterial blood supply is from the **cranial mesenteric artery**, which divides into smaller vascular arcades within the mesentery; these arcades are composed of the **major jejunal arteries**, which reach each neighboring jejunal artery via an **arcuate vessel**. The terminal branches of the arcuate vessels penetrate the bowel wall. The venous return runs parallel to the arterial supply and collects in the **portal vein**.

PORTAL VEIN AND CAUDAL VENA CAVA—ASSOCIATION WITH CATASTROPHIC COMPLICATIONS?

The portal vein lies immediately ventral to the epiploic foramen, passing within the hepatoduodenal ligament and mesoduodenum. Laceration of the portal vein with uncontrollable bleeding and hemoabdomen has been associated with epiploic foramen entrapment. Laceration of the portal vein can also occur spontaneously, secondary to tension caused by a strangulated portion of bowel, or iatrogenically during the manipulation of incarcerated bowel or attempted manual dilation of the foramen. Because the epiploic foramen and portal vein are surgically inaccessible during a ventral midline celiotomy, this complication is invariably fatal. The caudal vena cava lies dorsal to the right lobes of the liver (and the caudate lobe, which serves as the immediate dorsal boundary of the foramen). While catastrophic rupture of this vessel has been reported in association with epiploic foramen entrapment, it is less commonly involved than the portal vein.

Ileum and its blood supply

The **ileum** is the last segment of the small intestine (continuing from the jejunum) and is the shortest in length at approximately 0.7 m (27.6 in.) in adult horses. The ileum is grossly distinct from the jejunum by means of 2 primary features. The first is a prominent antimesenteric band, the **ileocecal fold**, which continues as the **dorsal band of the cecum**. The second is wall thickness. The ileal walls are palpably thicker than those of the jejunum, associated with an increase in tunica muscularis thickness as the small intestine approaches the large intestine. At the terminal end of the ileum is the **ileocecal orifice**, which sits inside of a papilla that projects into the base of the cecum (see Case 12.4). There are 3 muscular layers within this papilla contributing to a sphincter-like mechanism controlling the emptying of ileal contents into the **cecum**.

The arterial blood supply to the ileum is derived from the **ileocecal artery**, a branch of the **cranial mesenteric artery**. The ileocecal artery anastomoses with the aforementioned **jejunal arteries**. Venous return runs parallel to the arterial supply and drains into the **portal vein**.

Lymphatics of the jejunum and ileum

Many lymph nodes are located at the root of the mesentery and receive lymph from the jejunum and ileum. Lymph collected in these **mesenteric lymph nodes** is connected to an intestinal trunk joining the **cisterna chyli**.

Epiploic foramen

The **epiploic foramen** is a small, natural, slit-like opening that act as a potential space of communication between the **peritoneal cavity** and the **omental bursa** in the right cranial abdomen (Fig. 12.3-3, see also Case 12.7). The epiploic foramen initially opens into a small area called the **omental vestibule**, then expands into the **caudal recess of the omental bursa**, a potential space enclosed by the **greater omentum**. This omental bursa exists between the stomach and intestines. The epiploic foramen is approximately 4-cm (1.6-in.) long (2–3 fingers' width on palpation at surgery) in the average 450-kg (992-lb) horse. A positive correlation has been established between the size of the opening and body weight.

The anatomy of the epiploic foramen is 3-dimensionally complex and historically has been described inconsistently, with a lack of detail of surrounding structures and significant discrepancies between textbook sources. The boundaries of the epiploic foramen have been described across sources as including the following structures: **caudate lobe** of the **liver**, **caudal vena cava**, **hepatoduodenal ligament**, **portal vein**, **hepatogastric ligament**, **pancreas** (right lobe or entirety), and a fold continuous with the **lesser omentum** (Fig. 12.3-4A and B). Recent work has focused on elucidating the 3-D association of the surrounding structures. Ultimately the foramen is described as bound dorsally, caudally, and ventrally

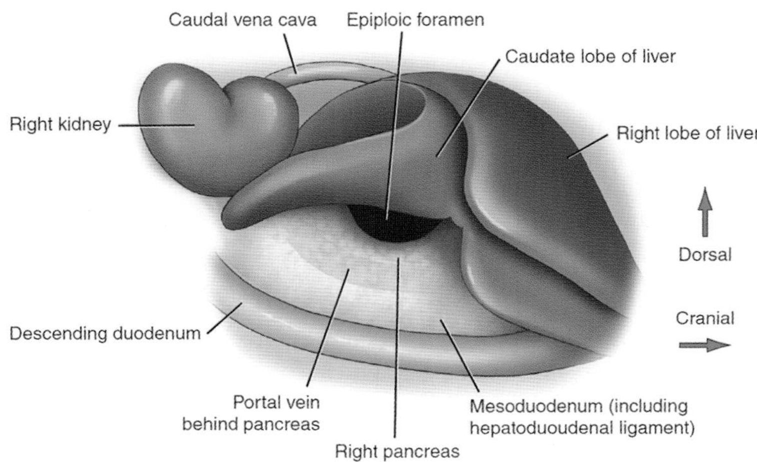

FIGURE 12.3-3 Location of the epiploic foramen in the horse.

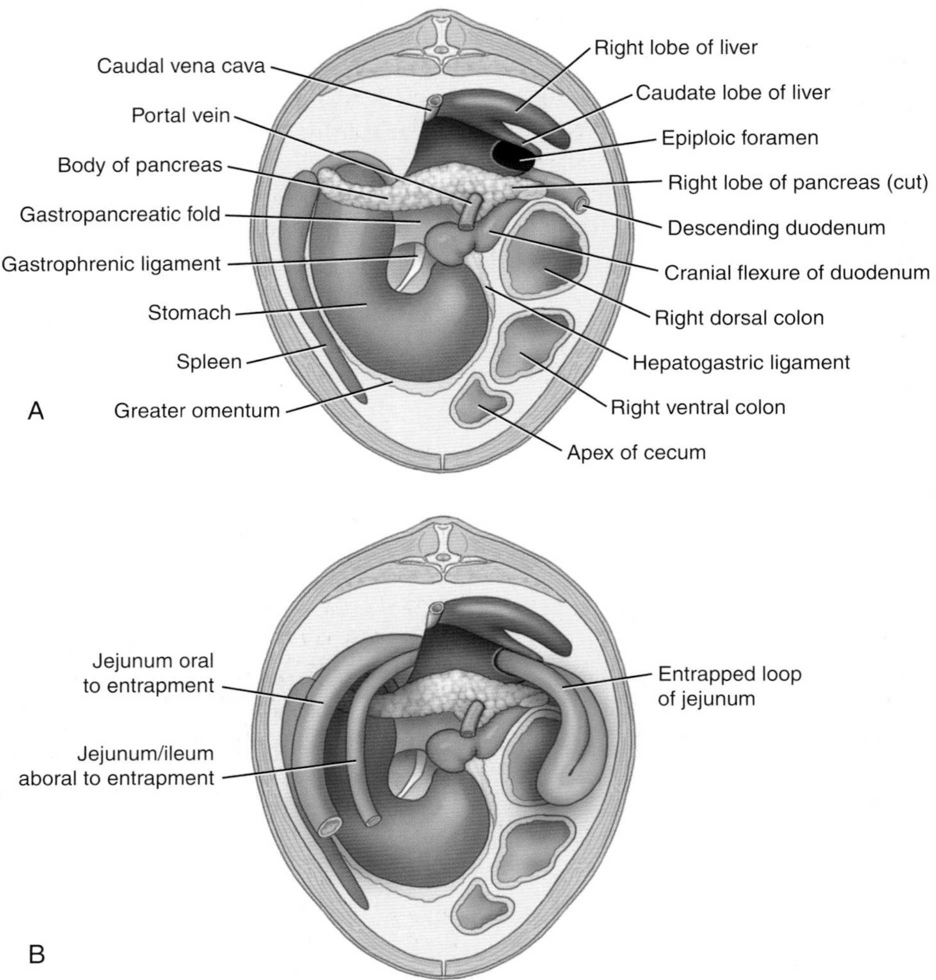

FIGURE 12.3-4 Transverse view of the equine abdomen illustrating the boundaries of the epiploic foramen (A) and an entrapped loop of jejunum (B).

Reported length of bowel involved in epiploic foramen incarceration ranges from 8 cm to 18 m (3.1 to 708.7 in.). Left-to-right (axial to abaxial) entrapment is far more common (97%–100% of reported cases) (Fig. 12.3-4B) than right-to-left (abaxial to axial) entrapment. The ileum is the segment of the bowel most commonly associated with epiploic foramen entrapment and is reportedly involved in 81% of cases. The jejunum is also commonly entrapped. Although rare, the cecum can become incarcerated within the epiploic foramen.

by the **gastropancreatic fold**, cranially and ventrally by the **portal vein**, and craniodorsally by the **caudate lobe of the liver**. Identification of the large role of the **gastropancreatic fold** had been previously overlooked, presumptively having been missed as a distinct anatomical structure due to its intimate relationship with the pancreas and/or its close association with both the hepatogastric and hepatoduodenal ligaments.

Mesodiverticular band

The **mesodiverticular band** is a congenital remnant of the vitelline artery or structures of the omphalomesentery that extends from the mesentery to the antimesenteric border of the jejunum or ileum (Fig. 12.3-4). Vitelline arteries (branches of the aorta) play an important role in the vitelline circulation of blood to and from the yolk sac of a fetus. Usually, these anatomic structures completely disappear between the 7th and 8th weeks of fetal development, respectively. When regression fails, the result is a persistent fibrous band connecting the mesenteric and antimesenteric borders. More specifically, failure of regression of either the distal segment of the right vitelline artery or the whole left vitelline artery creates a mesodiverticular band, whereas incomplete regression of the vitelline duct may create an intestinal fistula, a vitelloumbilical band, or a Meckel's diverticulum.

These structures are typically benign; though they can behave like a fulcrum around which a segmental volvulus can occur, or they can have a small tear, resulting in incarceration and strangulation of the small intestine (in particular, jejunum).

MESODIVERTICULAR BANDS AND COLIC

A retrospective study of 224 horses with small intestinal strangulating obstruction reported that mesodiverticular bands accounted for only 1.33% (3/224) of the cases, whereas strangulation caused by other types of bands termed "fibrous bands" accounted for 3.57% (8/224) of the cases. Mesodiverticular bands are oftentimes found incidentally. In a case series of 17 horses in which mesodiverticular bands where encountered during colic surgery, only 5 horses (29.4%) had lesions associated with the bands. Despite being a congenital defect, horses may show clinical signs associated with the defect over a wide age range: 40 days to 9 years old.

Selected references

[1] Archer DC, Pinchbeck GL, French NP, et al. Risk factors for epiploic foramen entrapment colic in a UK horse population: a prospective case-control study. Equine Vet J 2008;40:405–10.
[2] Freeman DE. Small intestine. In: Auer JA, Stick JA, editors. Equine surgery. 4th ed. St. Louis, MO: Elsevier; 2012.
[3] Freeman DE. Surgery of the Small Intestine. Vet Clin North Am Equine Pract 1997;13:261–301.
[4] Freeman DE, Pearn AR. Anatomy of the vestibule of the omental bursa and epiploic foramen in the horse. Equine Vet J 2015;47:83–90.

CASE 12.4

Cecal Impaction

Euan Murray[a] and Kira Epstein[b]
[a]BluePearl Veterinary Partners, Louisville, Kentucky, US
[b]Department of Large Animal Medicine, College of Veterinary Medicine University of Georgia, Athens, Georgia, US

Clinical case

History

A 15-year-old American Quarter Horse mare was presented for the evaluation of colic. At 36 hours before presentation, she had progressive signs of loss of interest in food and colic. On examination before referral, palpation per rectum revealed a large distended structure in the right caudal abdomen. No gastric reflux was reported on nasogastric intubation. She remained uncomfortable despite the administration of the sedative/analgesic xylazine. She had undergone an elective arthroscopy of her stifle 5 days before and had been discharged after 3 days of hospitalization. She was receiving oral phenylbutazone at the dose of 2.2 mg/kg q12h.

Physical examination findings

On initial examination, the mare was showing signs of moderate colic, including flank watching, pawing, and kicking at her abdomen. She was mildly tachycardic and tachypneic at 56 bpm and 36 brpm, respectively. She had a prolonged skin tent, pink and tacky mucous membranes, and a CRT of 2 seconds. Abdominal auscultation revealed decreased GI borborygmi in all abdominal quadrants.

> Risk factors for cecal impactions include decreased exercise, stall confinement, hospitalization (with or without general anesthesia), nonsteroidal anti-inflammatory administration, poor dentition, poor quality hay/roughage, decreased water consumption, and tapeworm (*Anoplocephala perfoliata*) infestation. Importantly, this case had multiple risk factors for cecal impaction in the history, placing cecal impaction high on the list of differential diagnoses.

Differential diagnoses

Right-sided large intestinal distension associated with diseases of the cecum—nonstrangulating (cecal impaction) or strangulating—cecocecal or cecocolic intussusception; large colon disorders—nonstrangulating (right dorsal displacement, impaction)

Diagnostics

After sedation, a nasogastric tube was passed and no net reflux was obtained. Abdominal palpation per rectum confirmed marked distension of the cecum—it was not possible to palpate beyond the dorsal aspect of the structure, and there were cecal teniae (bands) passing caudodorsal to cranioventral on the ventral/medial aspect—with fluid and ingesta. The ventral/medial band was painful on palpation, immovable, and thickened. Results of routine bloodwork (CBC and serum biochemistry screen) were consistent with dehydration. The mare became painful again after the effects of sedation had worn off.

Comparative Veterinary Anatomy: A Clinical Approach. https://doi.org/10.1016/B978-0-323-91015-6.00056-X

Diagnosis

Cecal impaction

Treatment

Based on the diagnosis and clinical signs, surgery was recommended and performed. Following the induction of general anesthesia, a routine ventral midline celiotomy (Gr. *koilia* belly + Gr. *tome* a cutting) was made. Upon entering the abdomen, a fluid- and ingesta-filled cecum was immediately identified. The cecum appeared severely inflamed and thickened. It was exteriorized and a typhlotomy (Gr. *typhlon* cecum + Gr. *tome* a cutting) was performed to evacuate the contents. A side-to-side jejunocolostomy was performed to bypass the severely compromised cecum. The mare recovered uneventfully from anesthesia. Post-operatively, she received prescribed antimicrobials and anti-inflammatories, and a slow reintroduction of food with careful monitoring.

The signs of cecal impaction can be subtle. Progression can occur rapidly, leading to cecal rupture, septic peritonitis, and death in up to 43% of all cecal impactions. Because of these facts, recognition of cecal impaction must be rapid and cases must be monitored closely. Medical management can be successful; however, the decision to go to surgery should not be delayed if the history, physical examination, and ancillary diagnostics dictate otherwise.

Cecal impactions are often divided into 2 categories although significant overlap makes clear differentiation difficult. Primary cecal impaction—excessive accumulation of dry ingesta—is not unlike a large colon impaction. Secondary cecal impaction—most commonly associated with recent hospitalization, change in exercise, etc.—is believed to be caused by dysmotility and results in an accumulation of fluid and ingesta. In cases consistent with secondary impaction, like this case, some surgeons elect to bypass the cecum with a jejunocolostomy to avoid recurrence. The efficacy of this procedure is not clear, and surgeon discretion and experience is the determining factor as to whether the procedure is performed.

Anatomical features in equids

Introduction

The cecum is discussed in this chapter, along with its blood supply, lymphatics, and nerve supply. The cecal attachments to the colon and ileum are also covered.

Function

The 2 main functions of the cecum are the absorption of large volumes of water and electrolytes, and the microbial digestion of soluble and insoluble carbohydrates before introduction to the ascending colon. The capacity of the cecum is approximately 30–35 L, and it absorbs the largest volume of water from the gastrointestinal tract compared to the remainder of the digestive tract. The cecum in horses is significantly larger than that of most other domestic species so that it serves—along with the large colon—as an important site of microbial digestion. The cecum also has the highest concentration of microbes compared to the rest of the large intestine, so it may break down and absorb amino acids that remain in the ingesta after passage through the small intestine. The microbial fauna of the cecum is like that of the large colon.

There are 4 different peristaltic patterns identified in the equine cecum, 3 of which mix the cecal contents, and the last of which empties the cecum. Pattern I is initiated at the apex and is conducted to the cupula (cranial cecal base). Pattern II is conducted from the caudal cecal base to the apex, and Pattern III is conducted from the cupula to the apex. Pattern IV is initiated by the pacemaker in the cecal apex and occurs every 3 min in the fed horse. When ingesta has undergone adequate "fermentation initiation," contractions beginning at the apex push it dorsally back into the base, where it spills over the edge created by the ventral fold into the cupula. The wave of contraction continues through the cecocolic orifice, sending ingesta into the right ventral colon. Gravity opposes the movement of the ingesta dorsally, and the pacemaker at the apex is key to the cecal evacuation process (Pattern IV).

Cecum

The **cecum** is a comma-shaped organ approximately 1 m/3.3 ft in length and is considered part of the **ascending colon**. It has **haustra** (sacculations) and 4 **teniae** (bands). Ingesta enters the cecum from the **ileum** through the **ileocecal orifice** and exits through the **cecocolic orifice** into the **right ventral colon** after fermentation and absorption of water and nutrients. There are several anatomic parts to the cecum—**base** (caudal base and **cupula**), **body**, and **apex** (Fig. 12.4-1).

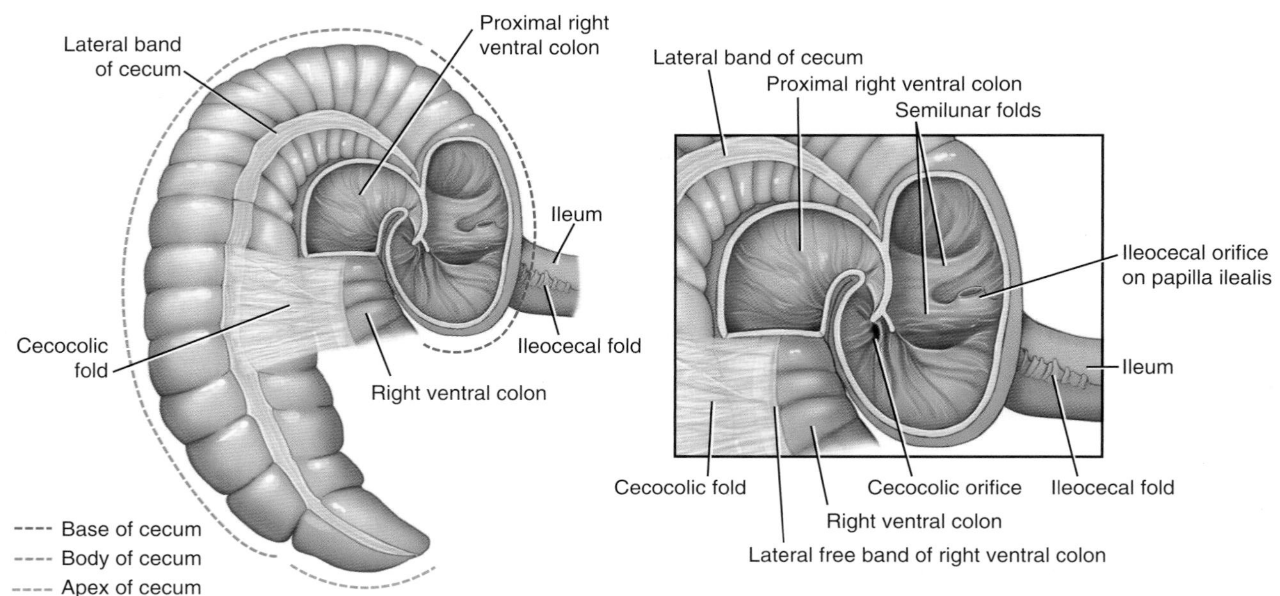

FIGURE 12.4-1 The equine cecum showing the base, apex, and body, as well as ileocecal and cecocolic orifice.

COMPARATIVE ANATOMICAL CECAL FEATURES

The cat has proportionately the shortest cecum, which becomes comparatively longer in the dog, pig, ruminant, and horse. The horse is unique in that the cecum has 2 openings: (1) the ileal ostium (L. a door or opening); and (2) the cecocolic ostium. In all other domestic animals, the ileal ostium joins the colon directly.

CECOCOLIC AND CECOCECAL INTUSSUSCEPTION

These occur relatively rarely, and the etiology remains unknown. Suspected causes include drug-induced alterations in cecal or colonic motility, verminous arteritis, and mucosal inflammation from tapeworm infestation. Rapid surgical intervention carries a good prognosis, whereas delaying treatment can lead to peritonitis and cecal rupture, or even a chronic condition involving weight loss and soft feces. Reducing the intussusception via colotomy and partial typhlectomy has a good prognosis. Bypassing the lesion via ileocolostomy is reserved until other surgical options have been attempted.

The **base of the cecum** extends from about the 15th rib to the **tuber coxae** on the right side, occupying much of the dorsal abdominal wall on the right side of the abdomen. The cecum has sturdy attachments, which anchor it securely in position within the equine abdomen. Dorsally, it has a retroperitoneal attachment to the right kidney and right lobe of the pancreas, which extends caudally to the level of the 2nd lumbar vertebra. Cranially, it has attachments to the right dorsal colon. Medially, the base fuses with the transverse colon and the root of the mesentery.

In cases of cecal rupture secondary to impaction, it is the medial or caudal surface of the base, which tends to overstretch and rupture first. The distended base is also what is typically palpated on abdominal examination per rectum in cases of cecal impaction.

The entrance (**ileocecal orifice**) and exit (**cecocolic orifice**) of the cecum are found in the base, separated by a ventral/transverse fold. The ileum enters the base of the cecum at the **ileal papilla**, located in the caudal part of the fold. This conical structure protrudes into the lumen of the cecum and contains the ileocecal orifice. It is found approximately 7 cm (2.8 in.) in to the right of the median plane, in the **lesser curvature** of the base of the cecum, and is level with the 1st or 2nd lumbar vertebra. A network of veins at the tip of this papilla can become engorged, leading to the closure of the orifice. Ingesta is moved from the ileum into the cecum via this ileocecal orifice, and this functional sphincter prevents the reflux of cecal contents back into the small intestine (ileum).

The cranial pocket (**cupula**) of the cecal base contains the cecocolic orifice in its caudal wall, and some consider this anatomical segment to be a part of the right ventral colon. As mentioned, the cecocolic orifice is the exit from the cecum into the ascending colon (right ventral colon). It is not actually a sphincter, but rather a transverse slit that acts as a valve when the ascending colon contracts. It is bordered by twofolds of tissue and is located just caudal and lateral to the ileal papilla due to the extreme curvature of the ventral portion of the base of the cecum.

The base merges with the **body** of the cecum, which initially extends ventrally along the right flank. The body then curves cranioventrally and follows the caudal surface of the right ventral colon. As it reaches the abdominal floor, it is forced medially, and the lumen begins to narrow. It terminates at the **apex**, which is found near the ventral midline approximately 20 cm (7.9 in.) caudal to the **xiphoid cartilage**, lying between the right and left parts of the ventral colon.

The apex of the cecum should be the first portion of the gastrointestinal tract encountered using a ventral midline celiotomy approach during exploratory surgery. It is often referred to as the "road map of the abdomen" because it is the presenting organ and provides orientation for exploring the small intestine and large colon via the ileocecal and cecocolic bands, respectively.

As previously noted, the cecum has haustra and 4 teniae. The bands are named simply for their anatomic location—**dorsal**, **ventral**, **medial**, and **lateral** (Fig. 12.4-2). The bands arise from the base, and the dorsal and medial bands (and sometimes lateral) extend to the apex. The ventral band joins the medial band near the apex. The ileocecal ligament is continuous with the **antimesenteric band**/artery of the ileum and the dorsal band of the cecum. The **cecocolic ligament** is continuous with the lateral cecal band and the lateral band of the right ventral colon.

The lateral band is the cecocolic fold and the dorsal band is the ileocecal fold. To remember this, think "L's go together" (lateral and large intestine) and "D's go together" (dorsal and duodenum—once you trace the ileum orally). If the cecum cannot be identified in its normal position, it is likely that the cause of colic is a right dorsal displacement of the large colon or a large colon volvulus.

The teniae are a combination of external longitudinal muscle and elastic fibers, the shortening of which results in haustra formation. Peristalsis of the muscular layers of the cecum and contraction of the teniae cause the haustra to continuously change size and position, helping to move ingesta against gravity from the apex to the base.

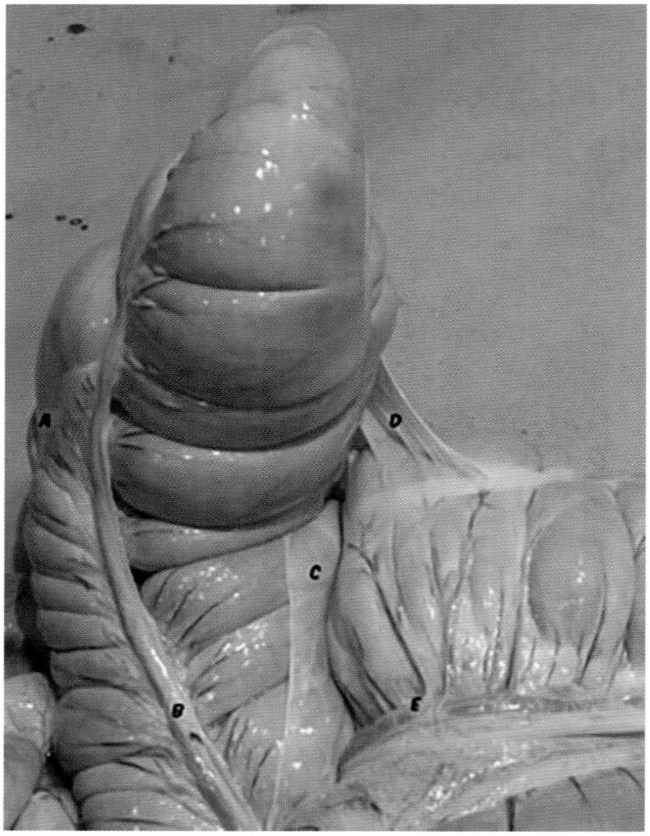

FIGURE 12.4-2 The relationship between the cecum and the large colon is viewed from the left ventral abdomen (horse in dorsal recumbency). The ventral (*A*) and/or medial (*B*) bands are the cecal bands that are palpable on rectal examination. The medial band (*B*) also contains the medial cecal artery and vein. The ileocecal fold (*C*) runs from the dorsal band of the cecum to the antimesenteric ileum. The sturdy cecocolic ligament (*D*) passes from the lateral band of the cecum to the right ventral colon.

Cecal blood supply, lymphatics, and innervation

The medial and lateral cecal bands contain the **medial** and **lateral vascular bundles** (Fig. 12.4-2). The cecum receives its blood supply from the **cecal artery**, which arises from the **ileocolic artery** via the root of the mesentery. It branches into **medial** and **lateral cecal arteries**, and these follow the **medial** and **lateral cecal bands,** respectively, accompanied by their venous counterparts. There is minimal mixing of their 2 circulations, and the medial cecal artery is the main blood supply for the apex. Each artery has a complex **rete** that forms a mesh-like network around the respective veins. Vessels from these retia supply the cecal tissue and lymph nodes and form an intricate network in the submucosa. Cecal venous drainage parallels the arterial supply as the **cranial** and **caudal mesenteric veins** that empty in the **portal vein**.

There are numerous **lymph nodes** found along the cecal artery and root of the mesentery, which receive lymph drainage from the cecum. The lymph then travels to the **cisterna chyli** via an intestinal trunk. Cecal innervation comes from nerves originating from the **mesenteric ganglia**.

The cecal lymph nodes are one of the few sets of intraabdominal lymph nodes visible during a transabdominal ultrasound. They are imaged in the right flank, adjacent to the cecal mesenteric vessels.

Selected references

[1] Clayton HM, Flood PF, Rosenstein DS. Clinical anatomy of the horse. Mosby Elsevier; 2015.
[2] Dart AJ, Snyder JR, Julian D, Hinds DM. Microvascular circulation of the cecum in horses. Am J Vet Res 1991;52(9):1545–50.
[3] Hubert JD, Hardy J, Holcombe SJ, Moore RM. Cecal amputation within the right ventral colon for surgical treatment of nonreducible cecocolic intussusception in 8 horses. Vet Surg 2000;29(4):317–25.
[4] Rakestraw PC, Hardy J. Large intestine. In: Auer JA, Stick JA, editors. Equine surgery. 4th ed. Missouri: Elsevier Saunders; 2012. p. 454.
[5] Sherlock CE, Eggleston RB. Clinical signs, treatment, and prognosis for horses with impaction of the cranial aspect of the base of the cecum: 7 cases (2000-2010). J Am Vet Med Assoc 2013;243(11):1596–601.

CHAPTER 12

CASE 12.5

Large Colon Volvulus

Jessica Bramski[a] **and Kira Epstein**[b]
[a]Department of Large Animal Medicine, University of Georgia College of Veterinary Medicine, Athens, Georgia, US
[b]Department of Large Animal Medicine, University of Georgia College of Veterinary Medicine, Athens, Georgia, US

Clinical case

History

A 9-year-old Thoroughbred broodmare was presented with the signs of severe colic of 1 hour's duration. She was found in her stall rolling during routine overnight checks. The farm manager administered 600 mg of flunixin meglumine orally and called the primary care veterinarian, who promptly referred her to a surgical facility. The broodmare had a 3-day-old foal by her side. The foaling was observed and reported as uneventful.

Physical examination findings

The broodmare presented covered in shavings with multiple abrasions on her face; she was pawing and tachycardic with a heart rate of 60 bpm. Her mucous membranes were pale pink and slightly tacky with a CRT of 2 seconds. Borborygmi were absent in all quadrants and her abdomen appeared moderately distended. The mare required sedation with 5 mg of detomidine and 5 mg of butorphanol IV to assist in the examination due to repeated pacing, pawing, and knee-buckling (wanting to lie down). Blood was collected at the time of sedation for a PCV/TP and a venous blood gas and electrolyte analysis.

Broodmares in the immediate post-partum period or the first 4 months of gestation are predisposed to large colon volvulus or displacement. If large colon volvulus is on the differential list, prompt referral to a surgical facility can improve the survival rate following surgical correction of the lesion. Within 3–4 hours of the initial insult, irreversible ischemic damage of the mucosa can occur, and the prognosis deteriorates.

Pre-operative factors that have been associated with decreased prognosis in horses with large colon volvulus include increased heart rate, PCV, and peripheral lactate.

Differential diagnoses

Large colon volvulus, uterine artery rupture, strangulating obstruction of the small intestine

Diagnostics

Nasogastric intubation yielded no net reflux. Abdominal palpation per rectum revealed a large, severely gas-distended viscus within the pelvic canal, preventing deeper abdominal palpation per rectum and examination. Results of the bloodwork revealed moderate dehydration and hyperlactatemia.

The mare required additional sedation with 5 mg detomidine IV for continued examination. A brief abdominal ultrasound revealed large colon sacculations along the ventral midline with a severely thickened large colon wall (1.18 cm or 0.46 in., normal <0.3–0.4 cm or <0.12–0.16 in.) (Fig. 12.5-1). No other physical abnormalities were identified.

Comparative Veterinary Anatomy: A Clinical Approach. https://doi.org/10.1016/B978-0-323-91015-6.00058-3

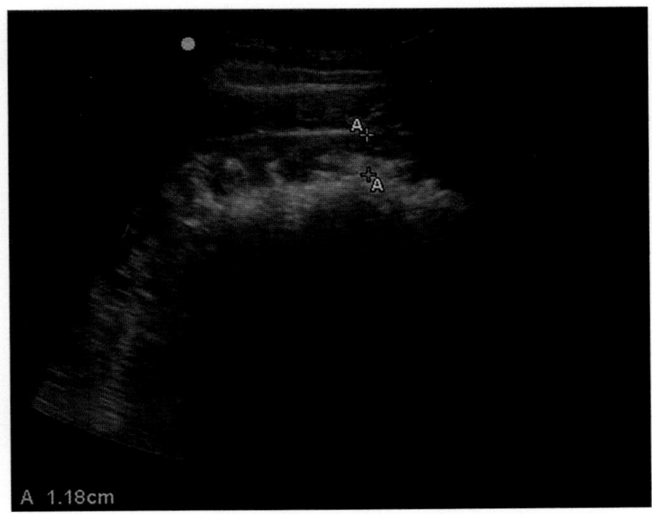

FIGURE 12.5-1 Transabdominal ultrasound of a horse showing increased colon wall thickness in the cranioventral abdomen (wall is demarcated by + signs). This has been documented in horses with large colon volvulus.

The mare continued to paw and repeatedly wanted to lie down despite the sedation/analgesic drug combinations. Exploratory celiotomy was recommended and elected due to persistent unrelenting pain in the face of sedation.

Diagnosis

Large colon volvulus (≥270° rotation)

Treatment

A ventral midline celiotomy was performed and a 360° volvulus beginning at the cecal base in an anticlockwise/dorsomedial direction was identified (Fig. 12.5-2). The serosa was purple-red in color in situ, indicating compromised bowel. The line of demarcation of compromised bowel was oral to the cecocolic ligament. A pelvic flexure enterotomy was performed to empty the contents of the large colon during which the darkened mucosa bled and the serosa began to normalize in color (Fig. 12.5-3). Arterial pulses were detected in the colic branch of the ileocolic artery and the right colic artery through a markedly edematous mesentery. Peristalsis was very weak but present. Due to the location of the volvulus and these findings, large colon resection and anastomosis was not performed. The mare recovered uneventfully from anesthesia and was maintained on intensive supportive care in the immediate post-operative period. She recovered without complications and was discharged from the hospital 10 days after admission.

VOLVULUS OR TORSION?

"Volvulus" (L. *volvere* to twist around) and torsion (L. *torsio* from *tourquēre*, to twist) can be confusing terms when it comes to describing types of bowel lesions that are caused by twisting, like twisting of a wet towel to wring the water out. The difference is in the structure that serves as the long axis of the twist: the bowel itself—*torsion*, or its mesentery and associated vasculature—*volvulus*. Think capital T: torsion involves the transverse or horizontal part—the bowel, whereas volvulus involves the vertical part—the mesentery. This terminology is more confusing when used regarding the large colon because the large colon lumen and mesentery are parallel/the same except at the pelvic flexure. For this reason, when the large colon twists, the result is generally both a large colon torsion and a large colon volvulus. Because the twist most often occurs near the base of the colon at the cecocolic ligament, the majority of the colonic blood supply is affected and results in the compromise of the entire large colon.

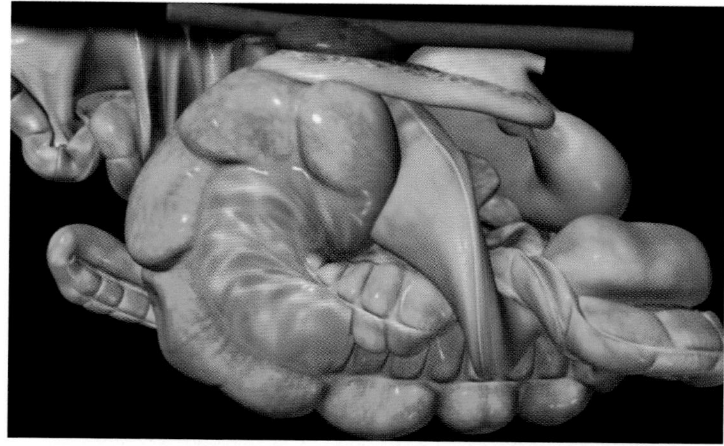

FIGURE 12.5-2 A 360° volvulus of the large colon as viewed from the right body wall in the standing horse. Venous and lymphatic occlusion begins at 180° and progresses to complete arterial occlusion at 270°. The colon may continue to twist further. Both 360° and 720° twists still imaging of the haustra ventrally.

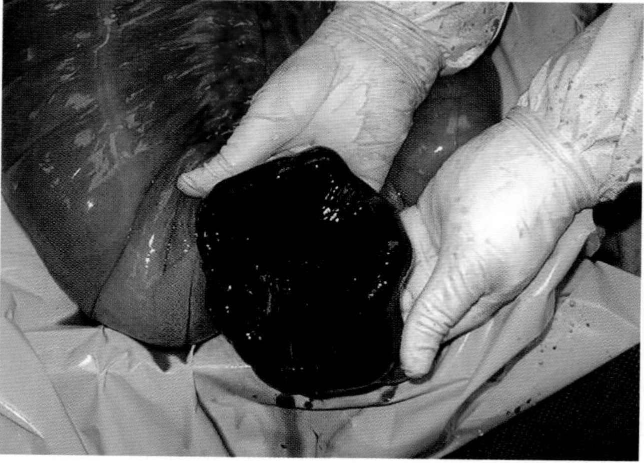

FIGURE 12.5-3 A pelvic flexure enterotomy performed during exploratory laparotomy. The colonic wall has been everted to show the dark mucosa.

Anatomical features in equids

Introduction

Discussed in this chapter are the large—or ascending—colon, including the right ventral colon, sternal flexure, left ventral colon, pelvic flexure, left dorsal colon, diaphragmatic flexure, and right dorsal colon. Also included are the mesocolon, colonic vasculature, and colonic attachments to both the cecum (cecocolic fold) and body wall. The cecum is discussed in Case 12.4.

Function

As ingesta passes through the gastrointestinal tract, it passes from the ileum to the cecum and from the cecum to the ascending colon. The details of the course of ingesta through the large colon are outlined in the ascending colon section that follows.

The function of the ascending colon is mainly storage of ingesta, fermentation, and absorption of electrolytes and large amounts of water. The ascending colon, in combination with the cecum, absorbs approximately 96% of the sodium and chloride, 75% of potassium and phosphate, and 95% of water leaving the ileum. The volume of water absorbed each day is equivalent to the horse's entire extracellular volume.

Fauna in the ventral colon is like the cecum but differs from the fauna of the dorsal colon. Overall, the fauna of the large colon and cecum are relatively similar between individuals when compared to other parts of the gastrointestinal tract. Four different classes of protozoa (Fig. 12.5-4) and many phyla of bacteria live in the cecum and colon. Ciliates and Firmicutes are the predominant classes of protozoa and phyla of bacteria, respectively. Together with fungi, bacteriophages, and archaea, they function to break down plant material and produce short-chain fatty acids such as acetate, propionate, and butyrate, which are an important source of energy for the gastrointestinal tract and the entire horse, and are transported into the circulation preferentially in that order. The microbiome of the large colon and cecum also produce vitamin K and the B vitamins (other than niacin).

The ascending colon can store up to 60–80 L of ingesta. Contractions of the large colon allow for the separation of small, digested feed material from the larger matter that requires additional digestion. With retropulsive contractions, the smaller particles are moved aborally, whereas the larger feed material is pushed orally. These contractions are regulated by a colon pacemaker in the left dorsal colon that is located approximately 30 cm (11.8 in.) aboral to the site in the left ventral colon where the medial and lateral ventral teniae terminate.

Ascending (large) colon

The **ascending colon** is the portion of the large intestine that begins at the cupula of the cecum and ends at the junction of the **right dorsal colon** and **transverse colon.** From the cecum, ingesta passes into the **right ventral colon**, which courses cranially along the right costal arch to become the **left ventral colon** at the **sternal flexure**,

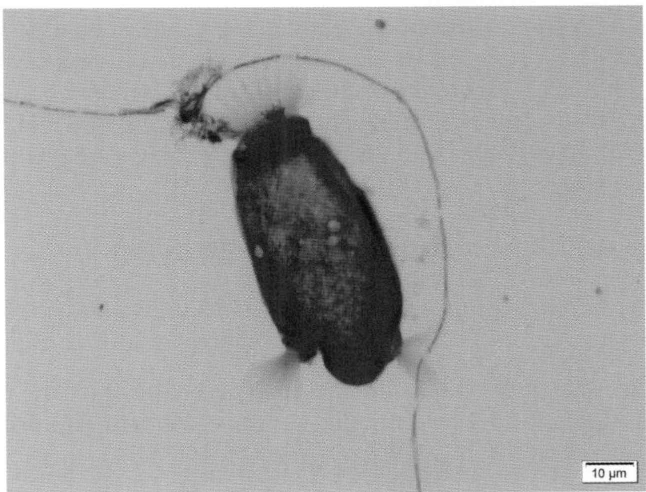

FIGURE 12.5-4 Microscopic photograph of cecal protozoa identified on the cytologic evaluation of peritoneal fluid obtained by abdominocentesis. This abdominocentesis confirmed contamination secondary to rupture; cecal rupture was confirmed on necropsy after humane euthanasia. Protozoa in the proximal large colon would be like those in the cecum.

LARGE COLON IMPACTIONS

Large colon impactions can be caused by ingesta, hair, sand, and/or other foreign materials. Accumulation and obstruction typically occur orad to where there is a narrowing of the lumen at either the pelvic flexure (left ventral colon) or right colic flexure (right dorsal colon). Impactions of the pelvic flexure can be diagnosed on examination per rectum, whereas impactions in the right dorsal colon cannot. Sand can be diagnosed on abdominal radiographs. Medical management with enteral fluids and laxatives is frequently successful. However, severe cases that become gas distended and painful require surgical evacuation.

During an exploratory celiotomy, the pelvic flexure is the most common site used for performing an enterotomy to empty the colon. It is easily exteriorized and can be placed on an operating tray to minimize contamination of the abdomen during surgery.

The large colon is attached only on the right side of the abdomen, by the right ventral colon's attachment to the cecum and the right dorsal colon's attachment to the dorsal body wall. This anatomic characteristic is what allows a large colon displacement or volvulus to occur.

which then passes caudally along the left ventral abdominal wall. At the **pelvic flexure,** near the pelvic inlet, the left ventral colon takes a sharp turn dorsally to become the narrower **left dorsal colon,** which then tracks cranially to the **diaphragmatic flexure.** The diaphragmatic flexure is where the left dorsal colon becomes the wider **right dorsal colon (ampulla coli)** and then becomes the short **transverse colon**. The right dorsal colon shares a common attachment to the dorsal body wall with the dorsal cecum through a short, thick mesentery.

The anatomical parts of the colon can be differentiated by their anatomic features, including the presence or absence of **haustra** (sacculations) and the number and character of **teniae** (bands) (Fig. 12.5-5). The right and left ventral colons have haustra, whereas the left and right dorsal colons do not. The teniae coli in the colon are made up of smooth muscle and collagen and aid in physical support and contractions of the colon. The bands located in the right and left ventral colon are high in elastin and have minimal smooth muscle to provide for better support of the colon with the distension that naturally occurs during fermentation. The 2 ventral bands of the ventral colon are free, whereas the 2 dorsal bands are within the mesenteric attachment. The lateral band of the right ventral colon is continuous with the lateral cecal band to make up the **cecocolic band/fold**. Toward the end of the left ventral colon, the haustra and 3 of the bands disappear leaving the smooth pelvic flexure supported by only 1 band in the mesenteric attachment. The 1 mesenteric band continues in the left dorsal colon. Two more free bands are added at the diaphragmatic flexure resulting in 3 teniae on the right dorsal colon. The teniae of the right dorsal colon have higher smooth muscle content consistent with their role in the propulsive movement of ingesta into the narrower transverse colon.

Blood supply, lymphatics, and innervation of the large colon

The mesentery of the large colon, or **mesocolon**, contains the blood supply and lymph nodes of the colon. It is short, running between the ventral and dorsal colons, maintaining their anatomic alignment.

The medial dorsal band of the ventral colons within the mesentery contains the **colic branch** of the **ileocolic artery,** which is the main blood supply for the right and left ventral colon. The **right colic artery** is adjacent to the

FIGURE 12.5-5 The ascending colon as examined at necropsy. The right ventral colon with its sacculations (*A*). The medial free teniae of the left ventral colon (*B*). The pelvic flexure (*C*). The narrow left dorsal colon (*D*). The wide right dorsal colon (*E*). Note the apex of the cecum at the upper left corner of the photo.

mesenteric band of the dorsal colons and supplies the right and left dorsal colon. These arteries meet and join at the pelvic flexure. Along their course, they branch every 2 cm (0.8 in.) and join with vessels that are oral and aboral (**colonic rete**). These branches pass into the colonic tissue and enter the submucosa where arterioles continue into the mucosa, which contains an extensive capillary network. The mucosal venules are sacculated due to smooth muscle contractions and—as a result—store a large volume of blood.

Colonic venous drainage parallels the arterial supply as the **cranial** and **caudal mesenteric veins** that empty in the **portal vein**.

There are numerous **lymph nodes** found along the colic arteries and root of the mesentery, which receive lymph drainage from the colon. The lymph then travels to the **cisterna chyli** via an intestinal trunk. Colonic innervation comes from nerves originating from the **mesenteric ganglia**.

Selected references

[1] Hackett ES, Embertson RM, Hopper SA, Woodie JB, Ruggles AJ. Duration of disease influences survival to discharge of Thoroughbred mares with surgically treated large colon volvulus. Equine Vet J 2015;47:650–4. https://doi.org/10.1111/evj.12358.
[2] Johnston K, Holcombe SJ, Hauptman JG. Plasma lactate as a predictor of colonic viability and survival after 360° volvulus of the ascending colon in horses. Vet Surg 2007;36:563–7. https://doi.org/10.1111/j.1532-950X.2007.00305.x.
[3] Julliand V, Grimm P. The microbiome of the horse hindgut: history and current knowledge. J Anim Sci 2016;94:2262–74.
[4] Pease AP, Scrivani PV, Erb HN, Cook VL. Accuracy of increased large-intestine wall thickness during ultrasonography for diagnosing large-colon torsion in 42 horses. Vet Radiol Ultrasound 2004;45:220–4. https://doi.org/10.1111/j.1740-8261.2004.04038.x.
[5] Rakestraw PC, Hardy J. In: Auer JA, Stick JA, editors. Equine surgery. 4th ed. St. Louis, MO: Elsevier; 2012. p. 486–513.

CHAPTER 12

CASE 12.6

Small Colon Enterolith

Phillip Kieffer[a] **and Kira Epstein**[b]
[a]Evidensia Specialisthastsjukhus Helsingborg, Helsingborg, SE
[b]Department of Large Animal Medicine, University of Georgia College of Veterinary Medicine, Athens, Georgia, US

Clinical case

History

A 12-year-old American Quarter Horse gelding was presented for an acute moderate to severe colic of approximately 3 hours' duration that failed to respond to symptomatic treatment with flunixin meglumine and a combination of detomidine and butorphanol. Over the past several months, he had 4 episodes of mild to moderate colic characterized by an acute onset and substantial gas distension of the large intestine on abdominal palpation per rectum. The patient had lived in California for most of his life and was fed a commercial sweet feed and alfalfa hay.

Physical examination findings

On presentation, the patient was quiet, alert, and responsive and had noticeable abdominal distension. The heart and respiratory rates were mild to moderately increased at 54 bpm and 30 brpm, respectively. Mucous membranes were pale pink and tacky with a CRT of 2.5 seconds. Borborygmi were decreased in all 4 abdominal quadrants.

Differential diagnoses

Nonstrangulating large intestinal causes of colic including spasmodic colic, intraluminal obstruction (impaction, enterolith, fecalith), or displacement

ENTEROLITHS—BETWEEN A ROCK AND A HARD PLACE

Enteroliths (Fig. 12.6-1A) (Gr. *enteron* intestine + Gr. *lithos* stone) are concretions of mineral formed within the viscera of the gastrointestinal tract. They account for up to 15% of colics and 27.5% of surgical colics in endemic areas (rates are much lower in nonendemic areas). Enterolith formation is commonly associated with a nidus (Fig. 12.6-1B), or a small piece of foreign material, that becomes encased by the mineral layers. There may be one "stone" (round in shape) or multiple stones (irregular in shape with one or more flat surfaces). Common sites of obstruction include the pelvic flexure, transverse colon, and small colon.

Risk factors for enterolith formation include diets high in alfalfa and being stalled. Arabian and Miniature Horses have historically been thought to be at increased risk; however, the role of diet and management is now thought to be of higher importance. Horses with enteroliths usually have a higher pH, higher mineral content, and lower moisture content within their ingesta and fecal matter. Areas endemic for enteroliths are those that traditionally feed high proportions of alfalfa with limited pasture and include parts of California, Texas, Arizona, and Florida. Cases of enterolithiasis that survive surgery have an excellent prognosis.

Comparative Veterinary Anatomy: A Clinical Approach. https://doi.org/10.1016/B978-0-323-91015-6.00059-5

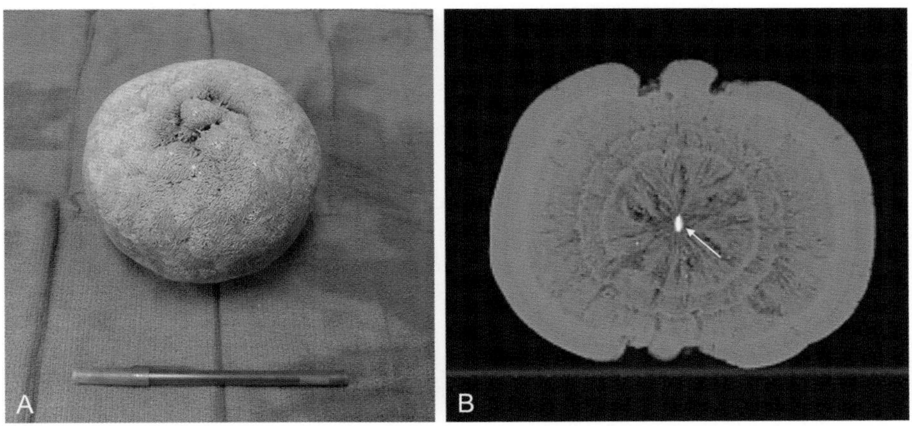

FIGURE 12.6-1 (A) Enterolith following surgical removal. (B) Computed tomography image of an enterolith showing the radiodense nidus *(arrow)*.

Diagnostics

The patient was sedated for additional diagnostics following the initial physical examination. A nasogastric tube was passed, and no reflux was obtained. Abdominal palpation per rectum revealed marked gas distension of the small and large colon. No feces were present in the rectum.

Routine bloodwork was consistent with mild dehydration. Abdominocentesis yielded slightly turbid peritoneal fluid with a mildly increased total white cell count of 7500 cells/dL and total protein of 2.8 g/dL. Radiographs of the abdomen revealed gas distension of the large intestine and an enterolith (Fig. 12.6-2).

Abdominal radiographs are part of a routine colic workup in geographic locations where enteroliths and/or sand are common causes of colic. Radiographs have been shown to be 76.9–85% sensitive and 93–96% specific for identifying enteroliths. Radiographs are more sensitive to enteroliths in the large colon than in the small colon.

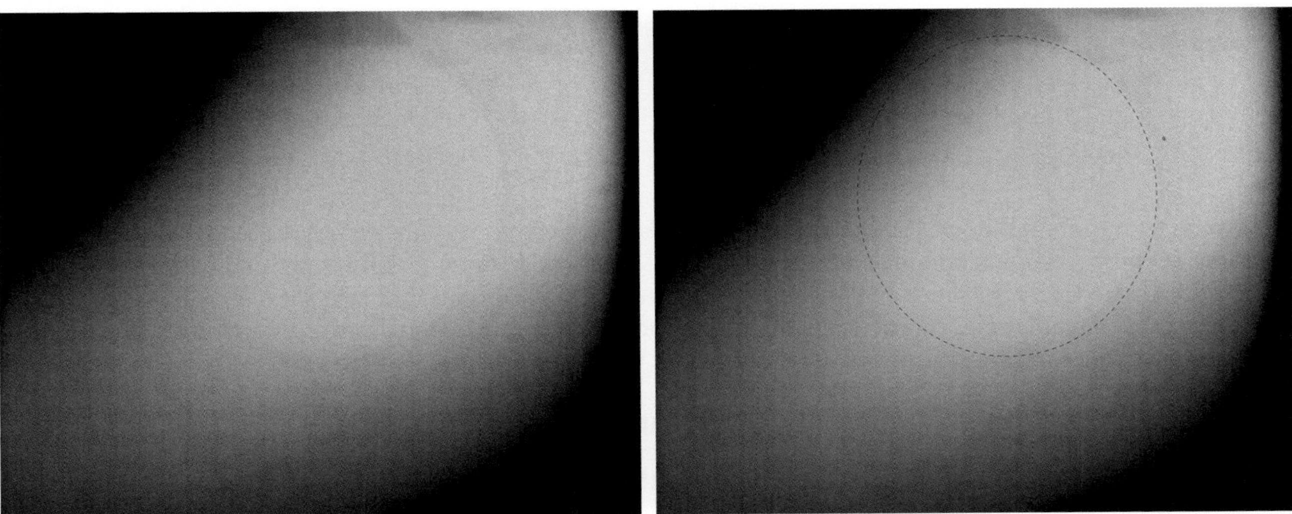

FIGURE 12.6-2 Lateral abdominal radiograph of a horse showing an enterolith. Left: unmarked, right: dashed line highlights the enterolith.

Diagnosis

Enterolithiasis

Treatment

An emergency laparotomy under general anesthesia was recommended and performed. Following gas decompression of the cecum and large colon, an enterolith in the small colon was identified. The enterolith was removed via a small colon enterotomy with minimal contamination. The gelding recovered without problems post-operatively and was discharged from the hospital 7 days later.

Anatomical features in equids

Introduction

Discussed in this chapter are the transverse colon, small—or descending—colon, and rectum. Also included are the mesocolon, blood supply, and microanatomy.

Function

The transverse colon serves as the channel between the right dorsal colon (see Case 12.5) and the small (descending) colon. The primary function of the small colon is the development of well-formed fecal balls from the ingesta, and their transportation to the rectum for removal. Some water resorption occurs in the small colon; however, under normal conditions, this is quite small compared to the amount of resorption that occurs in the large intestine and cecum. The rectum stores fecal balls until defecation occurs through the anal canal.

Transverse colon—anatomy, blood supply, and innervation

The **transverse colon** is closely attached to the dorsal body wall by a short **mesentery** just cranial to the **cranial mesenteric artery**. The transverse colon is the connection between the right dorsal colon and the small colon. There are also attachments to the **pancreas** and directly to the dorsal body wall. The transverse colon tracks from the right to the left side of the abdomen, making its length approximately the width of the dorsal abdominal wall. The diameter of the colon lumen decreases considerably at the junction of the right dorsal colon and the transverse colon and continues to decrease until it unites with the small colon.

> This abrupt decrease in luminal diameter makes the transverse colon 1 of the 3 most common sites of enterolith obstruction. The other 2 are the right dorsal colon and the small colon, as in this case.

The transverse colon shares the 2-band arrangement with the small colon though the bands are less prominent. No **haustra** (sacculations) are present in the transverse colon, reappearing only after the transition to the small colon. Blood supply to the transverse colon is by the **middle colic artery**, a branch of the **cranial mesenteric artery**. Venous drainage parallels the arterial supply, and the **cranial** and **caudal mesenteric veins** drain into the **portal vein**. The transverse colon receives afferent innervation from the **lumbosacral dorsal root ganglia**, as well as **splanchnic nerves** from T5-L2.

Small (descending) colon—anatomy, blood supply, and innervation

The **small colon** is approximately 3.5 m/11.5 ft in length. The transition from the termination of the transverse colon to the small colon occurs on the left side of the dorsal midline, cranial to the root of the mesentery. This area is firmly attached to the dorsal body wall by a short mesentery. This is also the site of colonic attachment of the

duodenocolic ligament. The small colon then passes along the left side of the root of the mesentery, with the bulk of the small colon residing in the left caudodorsal abdomen, before entering the pelvic canal and terminating at the rectum. Although this is the most common location for the small colon to reside, it has a long mesentery and can freely move into several locations throughout the abdomen without clinical consequences.

> The duodenocolic ligament connects the aboral duodenum to the oral part of the small colon. The duodenocolic ligament serves as an important landmark in exploratory laparotomy helping surgeons identify the most oral (beginning) part of the jejunum. It is palpable as an antimesenteric triangular band of tissue at the junction of the jejunum and duodenum deep within the abdomen during a ventral midline celiotomy. Because of this contiguity of the duodenum and jejunum to the large intestine, nasogastric reflux may occur even with large colon disease.

The small colon has distinctive haustra, often containing fecal balls. The small colon has 2 **teniae coli** (bands)—1 muscular, broad **antimesenteric band** and a **mesenteric band** (Fig. 12.6-3). The small colon has an average diameter of 6–8 cm (2.4–3.1 in.), making it larger than most of the small intestine but considerably smaller than the rest of the colon. The microanatomy of the small colon is like the other tubular organs and consists of 4 layers—**mucosa**, **submucosa**, **muscularis**, and **serosa** from the lumen outward. Unlike the more orad (L. *os, oris* mouth) gastrointestinal viscera, villi are absent from the mucosal surface. Instead, many **goblet cells** line simple tubular glands, supplying the large amounts of mucus that is secreted to coat the forming fecal balls. The teniae of the small colon are formed by cells from the muscularis.

The long small colon mesentery contains the blood vessels, nerves, lymphatic vessels, and **caudal mesenteric lymph nodes**. Unlike the usually thin and transparent mesentery of the small intestine, the mesentery of the small colon almost invariably contains a thick layer of adipose tissue, creating an opaque mesentery (**mesocolon**) (Fig. 12.6-3). The more orad small colon mesentery is suspended from the root of the mesentery, and attachment continues along the dorsal body wall to the pelvic canal where it is continuous with the **mesorectum**.

> Surgeons have reported a higher incidence of post-operative complications with small colon surgery compared to some other segments of the GI tract. This is believed to be due to a high concentration of aerobic and anaerobic bacteria in this part of the intestinal tract, along with a high concentration of collagenase, mechanical stress placed on enterotomy sites because of the size and shape of fecal balls, and varied blood supply. Attention to flawless surgical technique and timely reintroduction of food after surgery improves the prognosis.

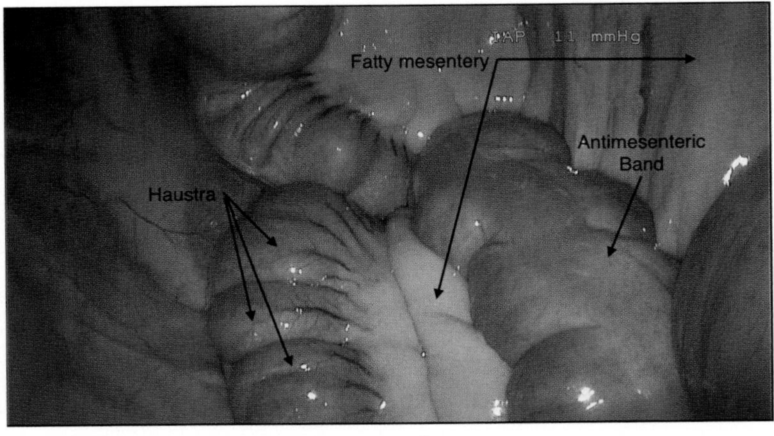

FIGURE 12.6-3 Laparoscopic picture of the small colon depicting haustra, wide antimesenteric band, and typical fatty mesentery.

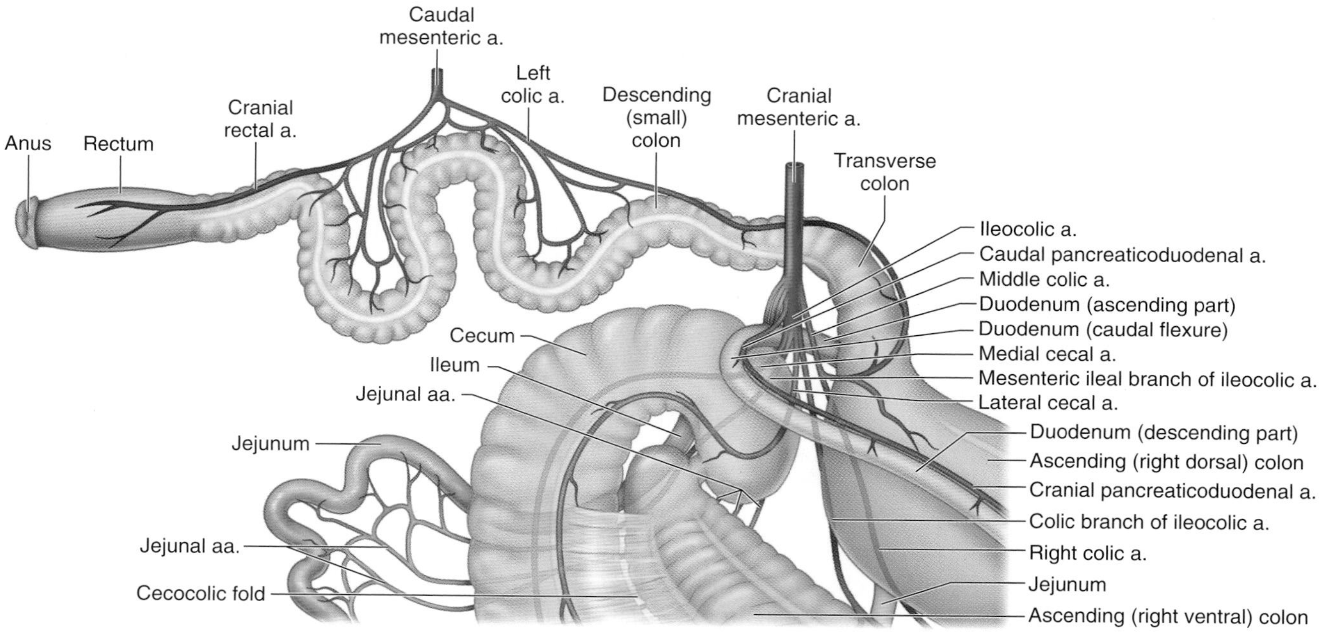

FIGURE 12.6-4 The blood supply to the equine small colon—note the arcades and vascular plexuses. Note the different segments of bowel each have a different color.

The blood supply to the small colon originates with the **caudal mesenteric artery** (Fig. 12.6-4). The caudal mesenteric artery divides into the **left colic** and **cranial rectal arteries**. These arteries give rise to large **arcuate vessels**, which in turn supply the **marginal arteries**. From the marginal arteries, the blood is supplied to **small arteries** in the mesocolon and the **secondary branching arcade** of the small colon. Blood moves from the secondary arcade to the **short arteries** (supplying the mesenteric aspect of the small colon) and **long arteries**. The long arteries anastomose on the antimesenteric part of the small colon after traveling hemicircumferentially in both directions from the secondary arcade.

The larger blood vessels of the small colon supply 2 vascular plexuses within the wall of the small colon. The first is the **intramuscular plexus**, lying between the longitudinal and circular muscle layers of the tunica muscularis, and the second is the large **submucosal plexus**. Venous drainage parallels the arterial supply and the **cranial** and **caudal mesenteric veins** then drain into the **portal vein**.

The small colon receives afferent innervation from the **lumbosacral dorsal root ganglia** as well as **splanchnic nerves** from T5-L2.

Rectum—anatomy, blood supply, and innervation

The **rectum** and **anus** represent the termination of the gastrointestinal tract. The transition from the small colon to the rectum occurs at the level of the **pelvic brim**, resulting in a total rectal length of around 30 cm (11.8 in.). The rectum has both intra- and extraperitoneal segments. The oral portion of the rectum is intraperitoneal and becomes extraperitoneal as it travels through the pelvic canal, terminating as the anus. The number of goblet cells increases as one moves aborally through the rectum. Goblet cells are absent from the anus proper. The rectum ends at the

RECTAL AMPULLA

The rectal ampulla (L. a jug) is a flask-like dilation of a tubular structure (the rectum) and comprises the terminal part of the intestine. This term is used in the horse, dog, and cow. The most important use of the rectum in large animal veterinary medicine is its use in abdominal palpation. Rectal prolapse occurs in the horse and other species and may require surgical resection and anastomosis if the prolapsed tissue is nonviable.

rectal ampulla, where the rectum dilates to store feces before passage through the anal canal during defecation. Here, the **rectococcygeus muscles** transit dorsally and caudally to anchor the rectum to the spine.

Blood to the rectum is supplied by the **middle rectal artery** (a branch of the **vaginal/prostatic artery**) and the **caudal rectal artery** (a branch of the **internal pudendal artery**), which also supplies blood to the anus. Venous drainage parallels arterial supply, as described for the small colon. Rectal innervation is provided by the S2–S4 nerves via the **pudendal nerve, caudal rectal nerve**, and **pelvic nerves** (providing parasympathetic nerve fibers).

The rectum is comprised of the same 4 layers as the small colon. The transition from the rectum to anus is marked by the end of the laminae muscularis. The anus is primarily comprised of **external** and **internal anal sphincters** and supporting small musculature. The external sphincter is made up of skeletal muscle, and the internal anal sphincter is comprised of muscle fibers like those found in the circular layer of the muscularis.

In mares, the anus is separated from the female urogenital tract by the **perineal body**, comprised of fibrous and muscular tissue. In the stallion and gelding, the bulk of this tissue is absent, resulting in a thin separation between the urogenital tract and the rectum/anus. The anus ends at the **anorectal line**, a mucocutaneous junction where the rectal mucosa (columnar epithelium) becomes skin (stratified squamous epithelium).

Rectal tears are divided into 4 grades based on which rectal layers are compromised: grade I—mucosa and submucosa are disrupted; grade II—muscularis layer is separated, causing the mucosa and submucosa to prolapse; grade III—involves all layers except the serosa (IIIa) or mesorectum and retroperitoneal tissue (IIIb); and grade IV—all layers are included and this is the most clinically devastating grade because the peritoneal cavity becomes contaminated with feces, resulting in septic peritonitis with a grave prognosis.

Perineal lacerations occur during parturition if the foal's limb(s), and sometimes head, are forced caudally and dorsally while coming through the birth canal. This is more commonly seen in the primipara (L. *prima* first + *parere* to bring forth, produce) mare. These lacerations are graded in degrees. First-degree—the mucosa of vagina and vulva affected; second-degree—submucosa and muscularis of the vulva, anal sphincter, and the perineal body are affected; and third-degree—a complete laceration through the rectovaginal septum, muscularis of the rectum and vagina, and the perineal body (see side box entitled "Rectovaginal fistula").

Regional lymphatics

There are numerous **lymph nodes** found along the regional arteries and root of the mesentery, which receive lymph drainage from the transverse and small colon. The lymph then travels to the **cisterna chyli** via an intestinal trunk. Numerous lymph nodes that drain the rectum are also located near the termination and parietal branches of the aorta. Additionally, the **anorectal nodes** are positioned dorsal to the caudal rectum. Lymph from the pelvic walls and viscera also drains to the **medial iliac nodes**, which additionally collect lymph from the **superficial** and **deep**

RECTOVAGINAL FISTULA

A rectovaginal fistula (L. pipe) is an abnormal passage between the lumens of the rectum and vagina/vestibule. Rectovaginal fistulas are formed primarily by the wayward penetration of a foal's foot (most common) or head through the dorsal wall of the vagina, the perineal body, and ventral rectum during unattended parturition. For true fistulas, the penetrating body part is repositioned before completion of the 2nd stage of parturition (expulsion of the foal), resulting in an intact perineal body and anal sphincter. If the foal's foot and leg breach the anal sphincter, perineal body, and vulva, this is classified as a third-degree, complete, rectovaginal tear. Fistulas also occur after failed surgical repair of other rectovaginal injuries.

Surgical repair using good surgical principles generally have a good prognosis with the mare capable of conceiving and delivering a subsequent term pregnancy.

inguinal nodes, which in turn drain lymph from the superficial pelvic structures and pelvic limb, respectively. Finally, these lymph nodes drain into the **aortic lumbar nodes** or, sometimes, a **lumbar trunk** may develop.

Selected references

[1] Kelleher ME, Puchalski SM, Drake C, le Jeune SS. Use of digital abdominal radiography for the diagnosis of enterolithiasis in equids: 238 cases (2008–2011). J Am Vet Med Assoc 2014;245:1126–9.
[2] McMaster M, Caldwell F, Schumacher J, McMaster J, Hanson R. A review of equine rectal tears and current methods of treatment. Equine Vet Educ 2015;27:208–9. https://doi.org/10.1111/eve.12266.
[3] Pierce RL, Fischer AT, Rohrbach BW, Klohnen A. Postoperative complications and survival after enterolith removal from the ascending or descending colon in horses. Vet Surg 2010;39:609–15. https://doi.org/10.1111/j.1532-950X.2010.00647.x.
[4] Rakestraw PC, Hardy J. Large intestine. In: Equine surgery. 4th ed. Saunders W.B.; 2012. p. 454–94.

CASE 12.7

Cholangiocarcinoma

Brina Lopez[a] **and Kira Epstein**[b]
[a]Midwestern University College of Veterinary Medicine, Glendale, Arizona, US
[b]Department of Large Animal Medicine, University of Georgia College of Veterinary Medicine, Athens, Georgia, US

Clinical case

History

A 16-year-old Percheron gelding was presented for a fever of 1 day's duration. The gelding's appetite had decreased appreciably over the past 2 days, and he was noted to be increasingly lethargic.

Physical examination findings

The gelding was dull but responsive with tachycardia (64 bpm) and an increased rectal temperature of 103.0 °F (39.4 °C). Mucous membranes were slightly icteric and moist with a normal CRT. Thoracic auscultation at rest was normal, and borborygmi were decreased in all abdominal quadrants. No gait abnormalities were noted. Initial blood work revealed evidence of mild dehydration along with marked hypoglycemia and a moderate increase in the concentrations of lactate, creatinine, bilirubin, and globulins. Liver enzymes were markedly increased.

Differential diagnoses

Acute hepatitis, chronic active hepatitis, hepatotoxicity, cholangiohepatitis, cholelithiasis, hepatic abscess, and neoplasia

Diagnostics

Additional bloodwork was obtained to evaluate liver function and revealed markedly increased concentrations of triglycerides, ammonia, and bile acids. All coagulation parameters were normal. Abdominal palpation per rectum was normal. Transabdominal ultrasonography revealed a diffuse heterogeneous appearance of the liver with rounding of the liver margins (Fig. 12.7-1). An ultrasound-guided transcutaneous liver biopsy was performed and submitted for histopathology and culture.

Liver biopsy is the most definitive means of diagnosing liver disease and providing the most accurate prognosis. Complications secondary to liver biopsies are rare and include colic, bleeding, peritonitis, pneumothorax, pleuritis, and tumor seeding. Biopsies for the suspected hepatic disease are most frequently collected transcutaneously via the right liver lobe, although left liver lobe biopsies can be performed. Ultrasound guidance is strongly recommended over blind sampling at the recommended anatomic landmarks (just below a line drawn between the dorsal aspect of the tuber coxae and the point of the elbow joint in the right 11th–14th intercostal spaces). Using ultrasound guidance, the clinician minimizes the likelihood of complications (sampling lung or large colon, for example) and maximizes the chance of obtaining a diagnostic sample. The location selected generally includes an area with abnormal hepatic architecture (if present) that is free of blood vessels and nearby organs (e.g., gastrointestinal tract and lung). The left lobes of the liver have a much smaller window of imaging ultrasonographically compared to the right.

Diagnosis

Cholangiocarcinoma, based on histopathology

Comparative Veterinary Anatomy: A Clinical Approach. https://doi.org/10.1016/B978-0-323-91015-6.00070-4

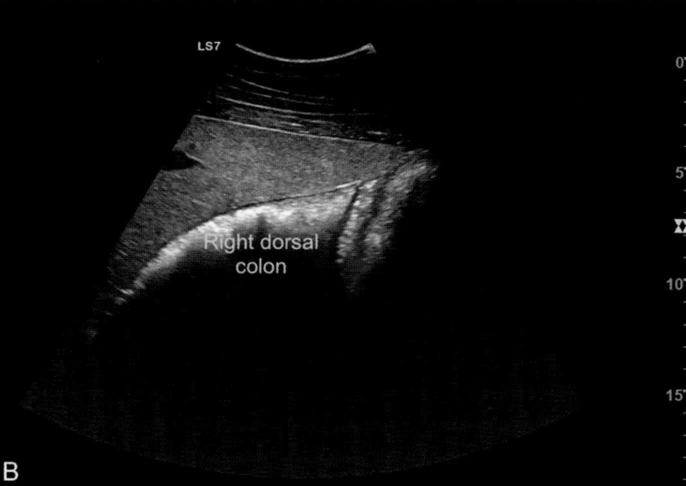

FIGURE 12.7-1 (A) Ultrasound image obtained from a horse's right side, mid-abdomen at the 14th intercostal space. Note the mixed echogenicity of the liver. (B) Ultrasound image of a normal liver adjacent to the large colon viewed from the right side. Compared to image (A) note the difference in the homogeneity of the echogenicity and the sharp, triangular margin. (Images courtesy of Nora Grenager, VMD, DACVIM.)

HEMOPERITONEUM

Hemoperitoneum, also called "hematoperitoneum" or "hemoabdomen," is defined as blood in the peritoneal cavity. The blood accumulates in the space between the inner lining of the abdominal wall (parietal layer of peritoneum) and the internal abdominal organs (visceral layer of peritoneum). Unlike in other species, hemoperitoneum is rarely treated surgically; in most cases, horses are treated medically with supportive care, including blood transfusion, antifibrinolytics, and possibly procoagulant therapies. Medical management is more common in horses due to the difficulty in identifying and treating the more common causes of hemoperitoneum surgically in addition to severe hemodynamic instability making these cases poor anesthetic candidates. The abdominal cavity is greatly distensible and may easily hold more than the entire circulating blood volume of an average-sized horse. Therefore, rapid blood loss into the abdomen likely progresses to hypovolemic shock and, if not recognized and treated rapidly, may lead to death.

Treatment

Medical management for hepatic insufficiency and broad-spectrum antibiotics were initiated while awaiting the histopathology. Despite treatment, the gelding's tachycardia persisted, and he began displaying mild signs of abdominal discomfort. Reassessment of bloodwork was consistent with potential blood loss—i.e., decreased PCV and total solids concentration and increased lactate concentration. Bleeding from the biopsy site was considered likely and an abdominocentesis was performed. Results from the peritoneal fluid analysis were consistent with a hemoabdomen. Liver biopsy results were consistent with hepatic neoplasia and humane euthanasia was elected. A complete necropsy examination was performed and confirmed primary hepatic neoplasia with metastasis to the lungs. Regional lymph nodes exhibited reactive hyperplasia.

Anatomical features in equids

Introduction

The liver, including gross and some relevant histologic anatomy, is discussed in combination with the peritoneal cavity in this section. The abdominal musculature of the horse is anatomically and clinically similar to that in the canine; see Case 5.8 for a description along with Landscape Figs. 4.0-4, 4.0-6, and 4.0-8.

Function

The liver has many functions including nutrient, protein, carbohydrate, and lipid metabolism and energy production, bile formation and excretion, detoxification, hematopoiesis, and the storage of vitamins and minerals.

The peritoneal lining of the abdominal cavity supports many of the abdominal organs and functions as a conduit for their blood vessels, lymphatic vessels, and nerves. The abdominal cavity—the space bounded by the vertebrae, abdominal muscles, diaphragm, and pelvic floor—is different from the intraperitoneal space (also called the peritoneal cavity), which is located within the abdominal cavity but covered in the peritoneum. The structures within the intraperitoneal space are "intraperitoneal"—e.g., the stomach and intestines—and the structures in the abdominal cavity behind the intraperitoneal space are called "retroperitoneal"—e.g., the kidneys—and finally the structures caudal to the intraperitoneal space are called "subperitoneal"—e.g., the urinary bladder and rectum.

Liver

The equine **liver** makes up 1.5% of the total body weight of an adult horse. Located within—and protected by—the ribcage, it does not contact the ventral abdominal floor and is situated primarily to the right of the midline. The most cranial part of the liver occupies the ventral third of the 6th intercostal space on the left and extends caudodorsally on the right to the 15th intercostal space (Fig. 12.7-2). Importantly, while most of the liver lies to the right of midline, right liver lobe atrophy in adult horses has been reported. It is unclear if the atrophy is a normal anatomic variation or a result of chronic compression secondary to large colon distention. The equine liver can be divided into 4 lobes: **caudate**, **left**, **quadrate**, and **right lobes** (Fig. 12.7-2). The visceral surface of the liver contains a **hilus**, a portal through which blood vessels, nerves, and the hepatic duct pass.

Hepatic ligaments

The liver is held in place within the abdominal cavity by 6 ligaments. Two **triangular ligaments** secure the right and left liver lobes to the diaphragm, laterally. The **coronary ligament** secures the liver to the diaphragm more centrally. The **round ligament**, formerly the umbilical vein in utero, is situated between the quadrate and left liver lobes and passes caudally within the free border of the falciform ligament to the umbilicus (see Case 13.10). The **hepatorenal ligament** secures the caudate lobe of the liver to the right kidney and base of the cecum (Fig. 12.7-3).

Hepatic blood supply, lymphatics, and innervation

Blood flow to and through the liver is unique and related to its many functions. Receiving approximately one-third of the cardiac output, normal cellular metabolism and function are supported by oxygen-rich blood supplied by the **hepatic artery**. In contrast, the **portal vein** carries deoxygenated blood and nutrients absorbed from the GI tract to the liver. Here, the nutrients are stored, metabolized, or packaged for transport to other tissues. Blood is then filtered in the hepatic sinusoids before entering the **central vein** followed by the **caudal vena cava**. The **hepatic lymph nodes** are identified surrounding the portal vein at the hilus of the liver. Sympathetic and parasympathetic innervations to the liver are provided by periarterial plexuses and the vagal trunks, respectively.

CHOLELITHIASIS

Cholelithiasis is a condition in which stones are formed within the biliary ducts and can be symptomatic or an incidental finding. In horses, choleliths are frequently composed of brown pigments of bilirubin. Bacterial cholangiohepatitis appears to be an important component of cholelith formation, although the exact pathophysiology is unknown. Common clinical signs include fever, icterus, and abdominal pain. While serum SDH, GGT, and bile acids are frequently increased in horses with cholelithiasis, elevations in these parameters are indistinguishable from other causes of liver disease. The classical histologic sign of bile duct obstruction on liver biopsy is concentric fibrosis around the bile ducts. *Escherichia coli* and *Enterococcus* spp. are commonly isolated from liver biopsy cultures. Treatment consists of surgical removal of obstructive choledocholiths along with anti-inflammatory and antimicrobial therapy for septic cholangiohepatitis. The reported prognosis for long-term survival ranges from 30% to 60%.

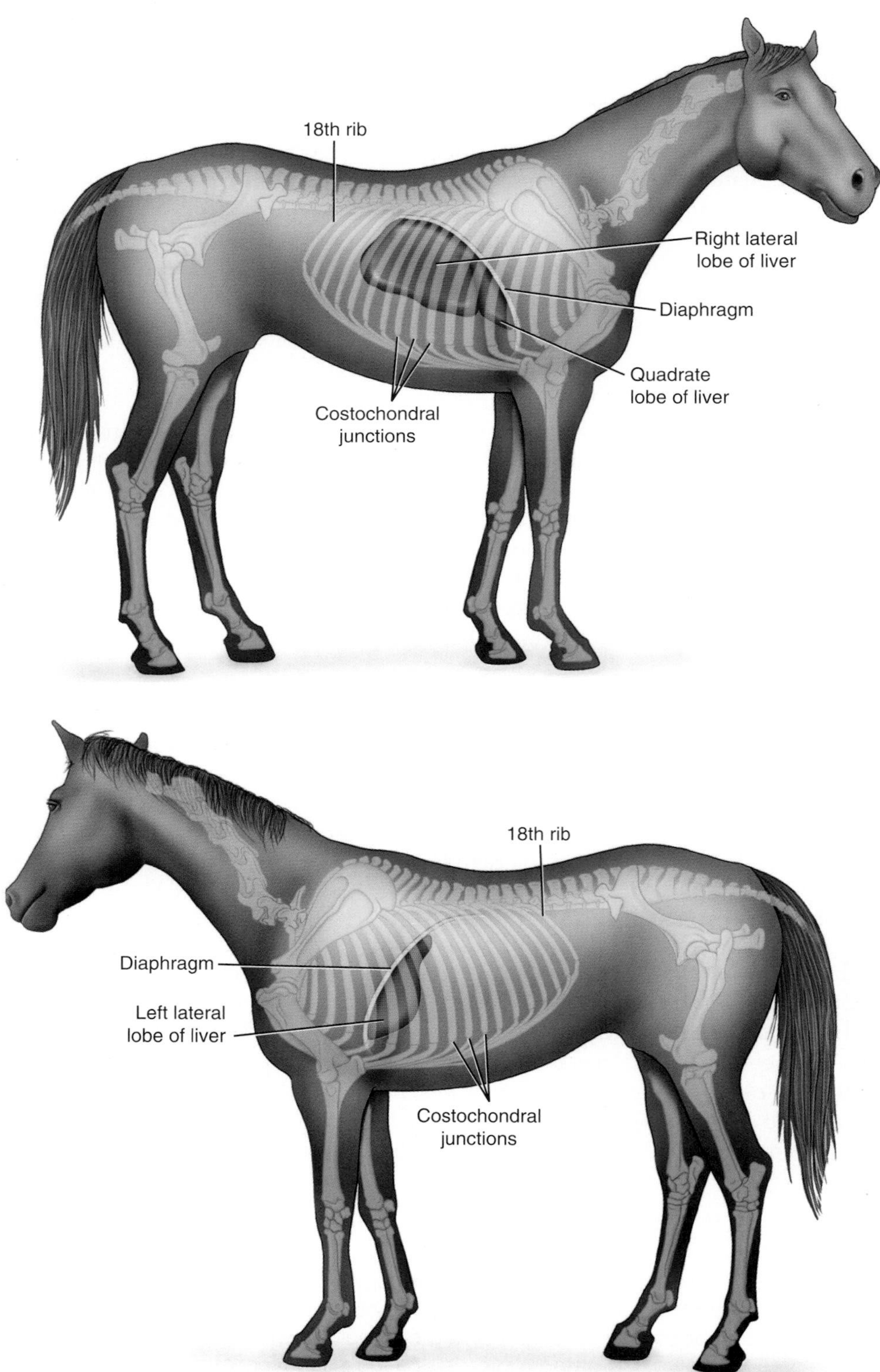

FIGURE 12.7-2 Illustration depicting the location of the equine liver.

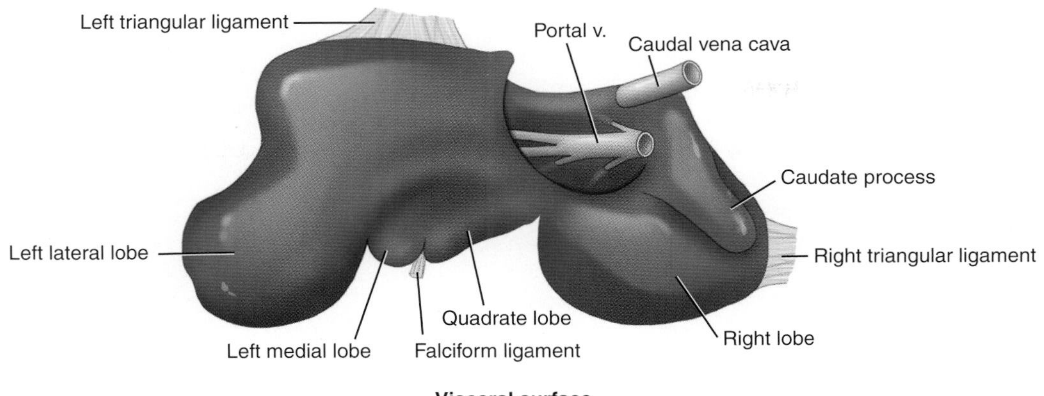

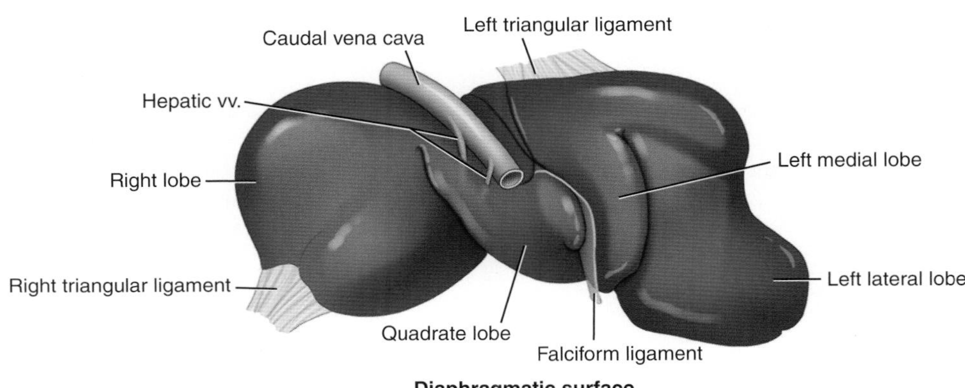

FIGURE 12.7-3 Equine liver anatomy including liver lobes, hilus, and ligamentous attachments.

Hepatic histology (Fig. 12.7-4)

The liver can be divided into lobules to facilitate histologic descriptions of lesions. Lobules can be defined by anatomical, functional, and metabolic characteristics. The classic **hepatic lobule** divides the liver anatomically and is created by invaginations of the connective tissue capsule into the liver parenchyma. The hepatic lobule appears roughly hexagonal on a cross-section with the central vein located in the center of the lobule and portal tracts at the corners. **Portal tracts** or **triads** are made up of a bile duct, hepatic artery, hepatic portal vein, and a lymphatic vessel (although difficult to see histologically).

A **portal lobule** instead divides the liver into functional units centered on the portal triad with the central veins at each of the 3 corners. The **acinus lobule** is a metabolic unit that is further divided into 3 zones based on oxygenation of the hepatic parenchyma in that region. The acinus is centered on a line connecting portal triads and extends to the central veins on either side. Zone I is located adjacent to the portal triad and thus receives the most oxygenated blood. Hepatocytes in this location are specialized in performing the oxidative functions of the liver. Importantly, this zone is the least susceptible to ischemic injury. Conversely, zone III is the furthest from the oxygenated blood, being situated near the central veins and, therefore, the most susceptible to ischemia and toxic injury.

Peritoneum

The **peritoneum** (Fig. 12.7-5) is a serous membrane, which lines the **intraperitoneal/peritoneal cavity**. Thus, the peritoneal cavity is the space within the abdominal cavity between the parietal and visceral peritoneum. There is a small volume of fluid in the peritoneal cavity normally but nothing else. Based on location, there are 2 major

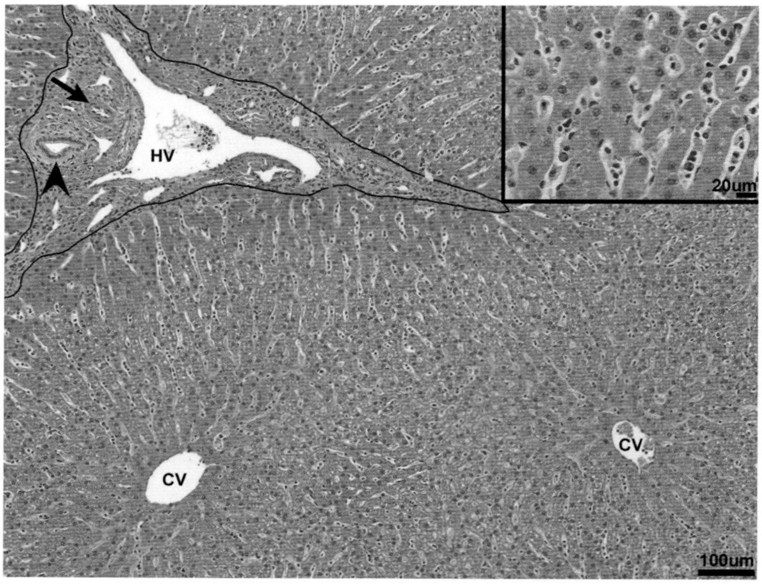

FIGURE 12.7-4 Histopathologic photomicrograph of the liver (10 × mag) highlighting the portal triad (*outlined*) with hepatic artery (*arrow*), vein (*HV*) and bile duct (*arrowhead*) and the central veins (*CV*). *Inset* (40 × mag) depicting hepatic cords interspersed with sinusoids containing erythrocytes admixed with neutrophils, rare eosinophils, and Kupffer cells. Key: *CV*, central vein; *HV*, hepatic vein; *arrowhead*, bile duct; *arrow*, hepatic artery. (Figure courtesy of Dr. Julie Engiles, University of Pennsylvania.)

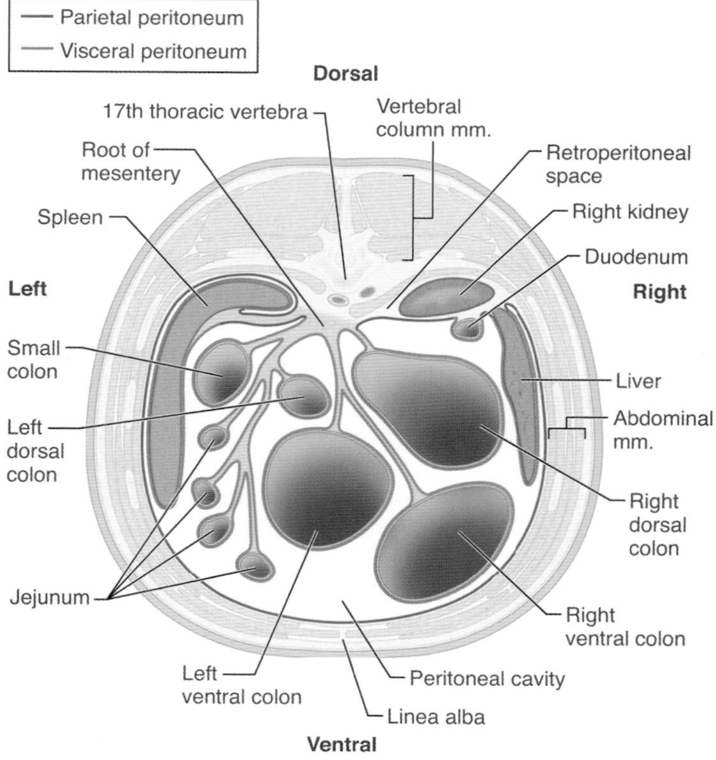

FIGURE 12.7-5 Transverse section schematic defining the peritoneum and peritoneal cavity.

divisions of the peritoneum: (1) **visceral peritoneum**—covering the digestive and urogenital systems (tube-shaped, serous membrane covering (complete and incomplete), and reflected at various places over abdominal and pelvic viscera); and (2) **parietal peritoneum**—lining the wall of the abdominal cavity and includes all peritoneum not directly attached to the abdominal viscera. The parietal peritoneum is connected to the visceral peritoneum through several components, all consisting of 2 distinct layers united. These peritoneal connections include:

- Greater omentum—connects the greater curvature of the stomach to the dorsal body wall
- Lesser omentum—connects the lesser curvature of the stomach and duodenum to the liver
- Mesoduoenum—supports the duodenum
- Mesentery of the small intestine—supports the jejunum and ileum
- Mesocolon and mesorectum—support the colon and rectum
- Medial and lateral ligaments of the bladder—support the urinary bladder; an important distinction is that the medial ligament is the caudal vestige of the ventral mesentery between the umbilicus and the urinary bladder vs. the lateral ligaments, which are folds of the peritoneum and not the round ligaments of the bladder, and are the vestiges of the umbilical arteries
- Broad ligament of the uterus, suspensory ligaments of the ovaries, mesorchium, and mesoductus deferens—support and suspend their respective reproductive structures
- Hepatorenal ligament—supports the liver and kidneys
- Triangular (paired right and left) ligaments and coronary ligament of the liver—these ligaments are peritoneum from the diaphragm to the liver; specifically, the triangular ligaments attach to the right and left lobes of the liver and the coronary ligament connects the iver to the caudal vena cava
- Falciform ligament—a vestige of the cranioventral mesentery between the umbilicus and liver; this should not be confused with the round ligament of the liver, which is enveloped by the falciform ligament and represents the rudimentary umbilical vein

The peritoneal cavity space also has several parts and pouches (diverticula), including:

- Omental bursa—a segment between the greater omentum, which communicates with the peritoneal cavity through the **epiploic foramen** (the area surrounded by the caudate lobe of the liver, gastropancreatic fold, and portal vein, and a site for small intestine herniation and incarceration)
- Ovarian bursa—opens to the peritoneal cavity via the ovarian fossa (see Fig. 13.4-6A and B in Case 13.4)
- Testicular bursa—the analogous or homologous to the ovarian bursa
- Rectouterine, vesicouterine, and pubovesical pouches/spaces—diverticula (L. *divertere* to turn) between the rectum and uterus, uterus and bladder, and bladder and bottom of the pelvic cavity, respectively

The epiploic foramen (L. *foramen epiploicum*) is also called the "foramen of Winslow" (after the anatomist, Jacob B. Winslow) and is uncommonly referred to as the "aditus" (L. entrance or approach), and finally as the "foramen omentale." It is the passageway into the omental bursa from the peritoneal cavity (see Case 12.3). Entrapment and strangulation of various parts of the intestinal tract have been reported and include the ileum, jejunum, cecum, and duodenum. The entrapment is generally in a left-to-right direction (>95%) with the rare case of right-to-left entrapment. The ileum alone, or in combination with the jejunum, is the most common piece of intestine involved. Surgery is needed to correct the herniation with resection and anastomosis of any nonviable intestine.

Selected references

[1] Barton M. Cholelithiasis in horses. In: Proceedings from the seventeenth annual veterinary medical forum of the American College of Veterinary Internal Medicine; 1999.
[2] Barton MH. Disorders of the liver. In: Reed S, Baylay W, editors. Equine internal medicine. 3th ed. St. Louis, MO: Saunders; 2010. p. 939–75.
[3] Beeler-Marfisi J, Arroyo L, Caswell JL, Delay J, Bienzle D. Equine primary liver tumors: a case series and review of the literature. J Vet Diagn Invest 2010;22(2):174–83.
[4] Johns IC, Sweeney RW. Coagulation abnormalities and complications after percutaneous liver biopsy in horses. J Vet Intern Med 2008;22(1):185–9.
[5] Sammons SC, Norman TE, Chaffin MK, Cohen ND. Ultrasonographic visualization of the liver in sites recommended for blind percutaneous liver biopsy in horses. J Am Vet Med Assoc 2014;245(8):939–43.

CHAPTER 13

PELVIC ORGANS

Dirk Vanderwall, Chapter editor

Female Urogenital System

13.1 Urovagina—*Jennifer Linton* 772
13.2 Second-degree perineal laceration—*Candace Lyman and G. Reed Holyoak* 779
13.3 Uterine artery rupture—*Maria Ferrer* 786
13.4 Granulosa cell tumor—*Dirk Vanderwall* 793
13.5 Endometrial cysts—*Carlos Pinto and Luis Henrique de Aguiar* 801
13.6 Oviduct/uterine tube obstruction—*Candace Lyman and Patricia Sertich* 808

Male Urogenital System

13.7 Squamous cell carcinoma of the penis—*David Levine and Carrie Jacobs* 816
13.8 Seminal vesiculitis—*Malgorzata Pozor* 825
13.9 Inguinal hernia—*Nora S. Grenager, Mathew P. Gerard, and James A. Orsini* 834

Urinary Bladder and Urachal Remnant

13.10 Omphalitis and bladder rupture—*Singen Elliott and Jarred Williams* 845
13.11 Cystic calculus—*Tamara Dobbie* 851

CASE 13.1

Urovagina

Jennifer Linton
B.W. Furlong & Associates, Oldwick, New Jersey, US

Clinical case

History

A 4-year-old Thoroughbred mare was presented 2 days after foaling as a healthy companion to her sick foal. The foaling was reportedly without problems and the mare was healthy at the time of referral.

Physical examination findings

The mare was bright, alert, and responsive. Vital parameters were within normal limits. There was a malodorous discharge on the tail and pelvic limbs. When the mare moved in the stall, a similar malodorous discharge was noted at the vulva. Perineal conformation was poor, with a sunken anus and vulvar lips that were not in contact.

Differential diagnoses

Urine pooling, endometritis, metritis

Diagnostics

Palpation per rectum was performed and revealed a large, fluid-filled uterus and an open cervix. The ovaries were not palpated at that time. Transrectal ultrasonography revealed echogenic fluid within the vagina, uterine body, and right uterine horn (Fig. 13.1-1). The cervix was imaged within the echogenic fluid (Fig. 13.1-2). The echogenicity of the fluid within the tubular reproductive tract was like that of the fluid imaged in the bladder. A speculum examination was performed and dark yellow to brown fluid was observed in the cranial vagina with the external os of the cervix immersed in the fluid. The vaginal mucosa was red.

Diagnosis

Urovagina

Treatment

In the immediate post-partum period, the urovagina was treated conservatively using uterine lavage with 7 L of sterile saline serving to dilute the urine in the uterus and also creating a siphon to remove the urine from the uterus. After the removal of the intrauterine urine, oxytocin was administered to stimulate uterine contractions. Repeated doses

Comparative Veterinary Anatomy: A Clinical Approach. https://doi.org/10.1016/B978-0-323-91015-6.00060-1

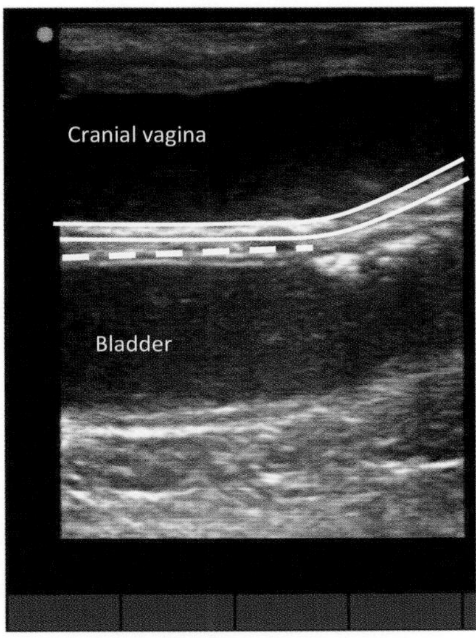

FIGURE 13.1-1 Transrectal ultrasound image of the vagina and urinary bladder of the mare, in this case with fluid (urine) present in both cavities. The fluid is separated by the ventral vaginal wall *(outlined in solid white)* and the dorsal bladder wall *(white dashed line)*.

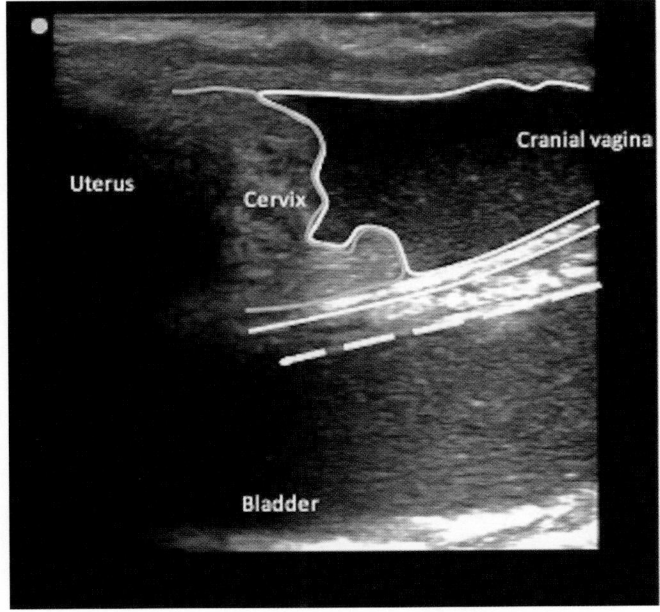

FIGURE 13.1-2 Transrectal ultrasound image of fluid (urine) in the cranial vagina *(outlined in white)*, effacing the cervix *(outlined in blue)*. The urine in the cranial vagina is separated from the urine in the urinary bladder by the ventral vaginal wall *(solid white lines)* and the dorsal bladder wall *(white dashed line)*.

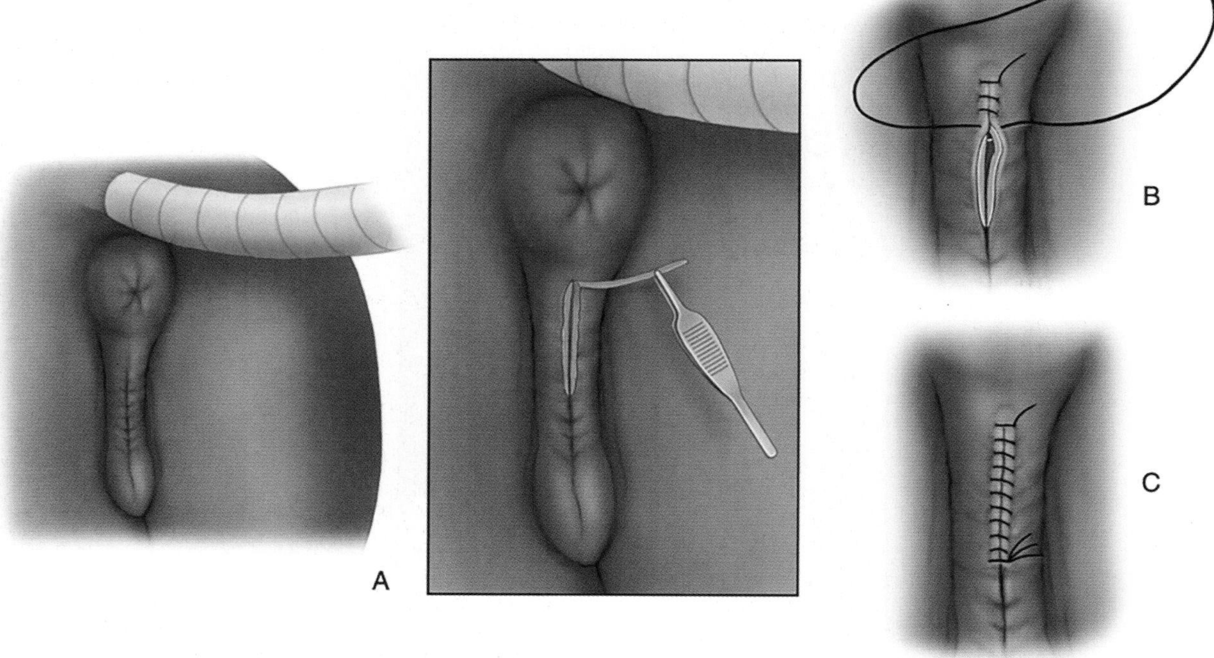

FIGURE 13.1-3 Caslick procedure: (A, left figure) Patient preparation of the perineal area and local anesthesia of the planned procedure. (A, right figure) A thin layer of vestibular mucosa is resected. (B) Wound edges are apposed using a simple continuous suture pattern (depicted here) or a Ford interlocking suture pattern extending from the dorsal vulvar commissure to the level of the pelvic brim. (C) Suturing ventral to the pelvic brim may prevent urine from emptying from the vestibule normally, resulting in urine pooling in the cranial vagina. Sutures are removed in 10–14 days.

A temporary Caslick procedure is used when access to the vagina or uterus is needed within the first 10–14 days but the perineal conformation needs to be improved. For example, a temporary Caslick could be placed in a post-partum mare with poor perineal conformation that is rebred within the 2 weeks or a mare with fluid accumulation within the uterus (urometra, pyometra) that requires repeated intrauterine therapy daily but has poor perineal conformation.

of oxytocin were administered to aid in continued uterine contractions to expel urine within the uterus and encourage uterine involution. A temporary Caslick procedure was used by suturing the incised borders of dorsal vulvar lips to the level of the pelvic brim leaving the ventral aspect of the vulva open for excretion of urine (Fig. 13.1-3). Oxytocin administration was continued until the mare had completed her first ovulation and entered diestrus (approximately 14 days after foaling).

Anatomical features in equids

Introduction

The vagina, vestibule, and vulva serve as the anatomical convergence of the urinary and reproductive tracts. The vagina and vestibule are lined by stratified squamous epithelium that, unlike in other species, does not cornify appreciably during estrus, while the vulva is covered with highly glandular skin. The cervix, uterus, uterine tubes (oviducts), ovaries, and urinary bladder are discussed in the remaining cases in Chapter 13.

Function

The caudal reproductive tract (vulva, vestibule, and vagina) plays an important role in breeding, pregnancy, parturition, and urination. The vestibule and vagina serve as an expandable space, allowing for penile intromission during breeding. The vulva and vestibulovaginal ring are physical barriers (along with the cervix) that prevent uterine and placental contamination and infection in the nonpregnant and pregnant mare, respectively. During parturition, the vulva, vestibule, and vagina stretch to allow for delivery of the foal; urine exits the urethra at the ventral aspect of the vestibulovaginal ring and is directed caudally through the vestibule and vulvar lips during urination (Fig. 13.1-4).

Vulva—anatomy, blood supply, and innervation

The **vulva** and **clitoris** develop from the genital tubercle, initially residing in an intermediate location until day 55 in the equine embryo. After day 55, the tubercle migrates, either to the caudal umbilicus in the male or ventral to the tail head in the female. The vulva is comprised of 2 **lips** of tissue that unite at a sharp **dorsal commissure** and ventrally at a more rounded **ventral commissure**. The vulva is covered by highly glandular, usually pigmented skin and acts as the

> Transrectal ultrasonographic fetal sexing is best performed at days 60–75 after this migration and before the fetus is too large to be readily visualized per rectum.

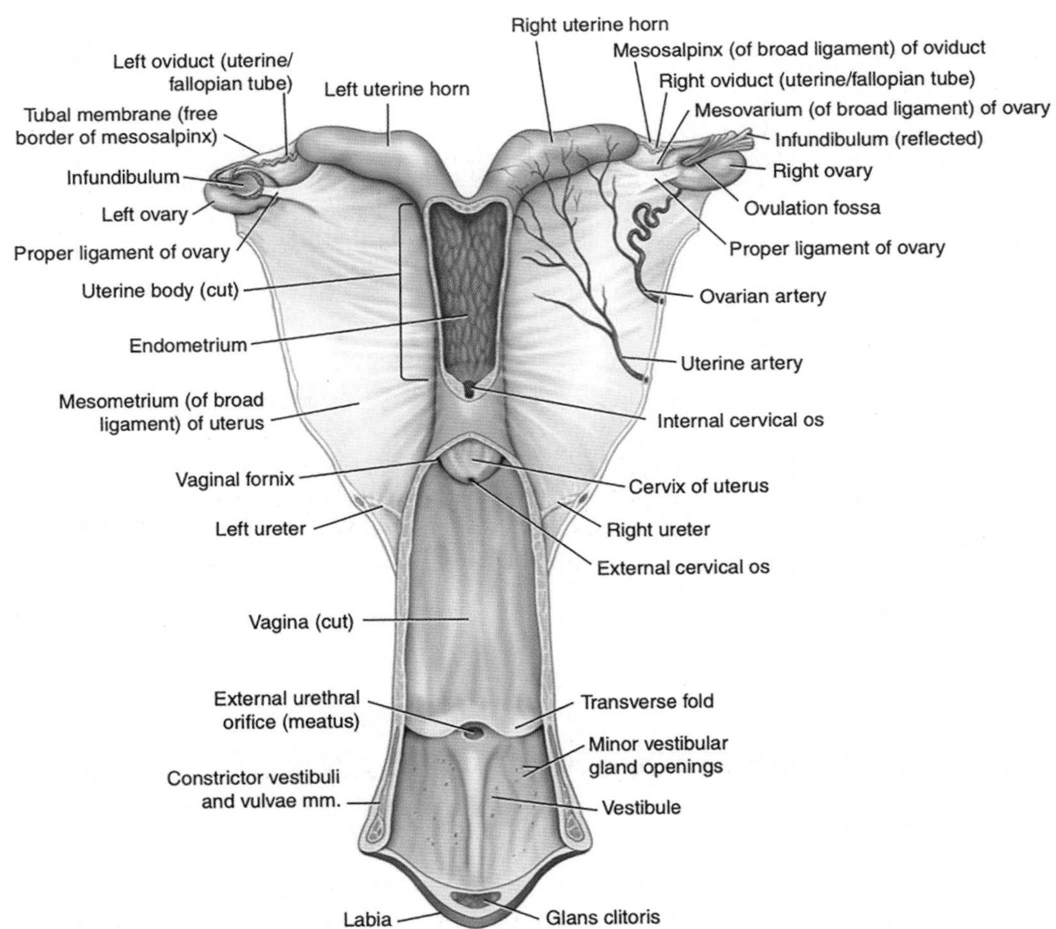

FIGURE 13.1-4 In this dorsoventral view of the equine female reproductive tract, the vulva, vestibule, and vagina are displayed on cut section. The ventrally located clitoris is located just inside the vulvar lips, which provide a physical barrier to protect the tubular reproductive tract. The vestibulovaginal ring (i.e., transverse fold) can be seen just cranial to the entrance of the external urethral orifice (meatus) into the vestibule, and it demarcates the transition from vestibule to vagina. The vagina, a potential space that accommodates the penis during copulation and the foal during parturition, connects the vestibule to the cervix, which can be seen protruding into the vaginal cavity.

Good perineal conformation is important to maintain fertility of the mare. In mares with less than 75% of the vulva below the pelvic bone, with an angle of >20% forward from the vertical, or with poor vulvar closure, a Caslick procedure (Fig. 13.1-3) can be performed to augment the perineal conformation. This procedure apposes the dorsal aspect of the vulvar lips and decreases the amount of foreign material that can be carried into the vestibule.

The clitoris is generally not a site of clinical disease, but the clitoral sinus may house *Taylorella equigenitalis*, the causative agent of contagious equine metritis. *T. equigenitalis* has been implicated in early pregnancy loss and transient endometritis. *T. equigenitalis* is generally transmitted by asymptomatic stallions, but has been eradicated in many countries and is commonly considered a reportable disease.

outer-most (first) physical barrier to protect the reproductive tract from ascending infection or inflammatory mediators. The vulvar lips should be vertical and well-apposed, with 75% of the vulva below the level of the pelvic bone. The vulvar angle is typically upright or at less than 10% from vertical.

The **clitoris,** within the **clitoral sinus,** is located at the ventral commissure of the vulvar lips. The clitoris, a bundle of nerves and muscle that is the homolog to the glans penis (balanus) in the male, is a small mound of tissue. The **constrictor vulvae muscle** everts the vulvar lips to expose the clitoral glans during periods of estrus (sexual receptivity) and is innervated by the **deep perineal nerve** and the **caudal rectal nerve**. The **superficial perineal nerve**, a branch of the pudendal nerve, provides sensory innervation to the vulva. The **dorsal labial branches** of the abdominal aorta and caudal vena cava supply and drain blood to and from the dorsal aspect of the vulva, respectively. The **ventral labial branch** of the **caudal mammary vessels** supplies blood to the ventral aspect of the vulva. Clitoral blood flow is supplied and drained by the **dorsal** and **deep vessels of the clitoris** (branches of the caudal vena cava/abdominal aorta and caudal mammary vessels, respectively) and the **middle artery of the clitoris**, a branch of the **obturator artery**. Lymph from the vulva flows to the **medial iliac lymph nodes** via the **anorectal lymph node** and to the **deep inguinal lymph nodes** via the **superficial inguinal lymph node**.

Vestibule—anatomy, blood supply, and innervation

The **vestibule** is the caudal-most part of the female tubular reproductive tract that is formed from an invagination of the urogenital sinus in the developing embryo. The vestibule contains the **Bartholin's glands**, which are responsible for the production of mucus in the vestibule. The vestibule functions as a potential space that receives the stallion's penis during intromission. The cranial border of the vestibule is demarcated by the **vestibulovaginal ring**. The external urethral orifice enters the vestibule at the base of the vestibulovaginal ring with urine passing caudally through the vestibule during micturition (urination).

REPRODUCTIVE ANOMALIES IN MARES

In rare cases, the paramesonephric ducts do not fully fuse in the vaginal canal and a remnant of the ducts can remain, causing pain when the stallion intromits (intromission L. *intro*- within + L. *mittere* to send = insertion of the penis into the vagina). In most cases, the remnant persists as a small band of tissue tracking from the dorsal vaginal wall to the ventral wall and is easily transected, under sedation, eliminating the remnant. However, incomplete fusion of the paramesonephric ducts can occur anywhere along the tubular reproductive tract, resulting in a double uterus, double cervix, or double vagina. These are rarely reported in the mare.

In cattle, freemartinism (associated with a male and female co-twin) may also result in abnormalities of the caudal reproductive tract such as vaginal aplasia. However, this disorder is not seen in the horse, which is thought to be due to the fact that placental fusion and comixing of fetal blood occurs later in gestation in horses compared to cattle. The external genitalia may also be ambiguous, such as an enlarged clitoris, a ventrally located vulva, or fused vulvar lips. In these cases, careful characterization of the internal gonads by palpation, ultrasonography, and hormone testing is recommended.

The **constrictor vestibuli**, innervated by the **deep perineal** and **caudal rectal nerves**, is present at the ventral aspect of the vestibule and is responsible for constriction of the vestibule. This muscle is covered with collagenous connective tissue, adipose tissue, and a venous plexus, which is formed by the **vestibular branch of the internal pudendal vessels** along with the **vessels of the vestibular bulb** (a branch of the **ventral perineal vessels**). Lymph from the vestibule flows to the **medial iliac lymph nodes** via the **anorectal lymph node**.

Vestibulovaginal ring—anatomy, blood supply, and innervation

The **vestibulovaginal ring** (or vestibulovaginal fold) is the second physical barrier that prevents ascending infection of the uterus and is formed where the invagination of the urogenital sinus (which forms the vestibule) meets the caudal aspect of the **paramesonephric ducts** (which form the vagina, cervix, uterus, and oviducts). The **hymen**, a thin membrane that exists at the level of the vestibulovaginal ring, is a remnant of this fusion and is generally easily ruptured during sexual intercourse or when performing reproductive procedures manually such as artificial insemination.

A persistent hymen (a thickening of this tissue) is easily treated by transecting the persistent membrane with scissors when it is present. Signs of a persistent hymen may include an egg-like structure protruding through the vulvar lips, anechoic uterine fluid, or difficulty performing a vaginal procedure, such as a uterine culture.

The competency of the vestibulovaginal ring is tested by performing the "windsucker test"—the vulvar lips are parted, and the clinician listens for air drawn in through the vestibulovaginal ring toward the cranial vagina. This procedure is not reliable when performed after the mare has been sedated. There are currently no surgical procedures available to augment the vestibulovaginal ring.

Vagina—anatomy, blood supply, and innervation

The **vagina** is a potential space that receives the penis during intromission. The vagina is formed as the caudal-most aspect of the paramesonephric ducts, which also form the oviducts, uterus, and cervix. The

Trauma during parturition can result in the formation of adhesions in the vagina, which need to be lysed (released) with blunt or sharp dissection before rebreeding.

CAUSES OF UROVAGINA AND UROMETRA

There is currently no surgical procedure to improve the function of the vestibulovaginal ring after injury during parturition, a common occurrence. During vaginal speculum examination, the clinician should carefully evaluate the competency of the vestibulovaginal ring when the speculum is removed from the vaginal vault. The conformation of the caudal reproductive tract, particularly the location of the vestibulovaginal ring, may change with injury to the vestibulovaginal ring, a heavy (fluid-filled or post-partum) uterus, muscle wasting, or musculoskeletal hind end issues. As the uterus falls forward or the mare is unable to posture to urinate, the reproductive tract is pulled cranially (Gr. *splanchnoptosis* sagging viscera), and the location of the urethral orifice is also altered, moving in a cranial direction. Therefore, urine flows cranially, collecting in the vagina (vestibulovaginal reflux) rather than being excreted caudally through the vestibule, causing urine pooling (urovagina), which can lead to urometra (urine present in the uterine lumen) if the cervix is impaired or unhealthy.

Evaluation of the vagina can be performed by manual/digital examination and visual inspection with a speculum. Speculum examination allows for observation of the vaginal walls and assessment for inflammation, adhesions, or the presence of fluid in the vaginal vault. Yellow fluid within the vaginal vault is consistent with urine pooling, particularly in conjunction with vaginal wall inflammation and minimal intrauterine fluid.

Although vaginal bleeding is not a common clinical presentation, when observed, it is most frequently associated with bleeding from vascular varicosities (L. varix an enlarged, twisting vessel; varicose veins) in the vagina and/or vestibulovaginal ring in older mares during late gestation. Treatment generally consists of benign neglect (observation), although chemical cautery or ligation can be performed if needed.

external os of the cervix resides in the cranial vagina and, when under the influence of progesterone, protrudes into the cranial vagina. The external os is suspended within the vagina by a **dorsal frenulum** (L. *frenum* bridle, small fold of mucous membrane that limits the movement of an organ or part). The vagina is an elastic and accommodating organ that changes shape when, during ejaculation, the bell of the stallion's penis distends the vagina, opens the cervix, and deposits semen directly into the uterus.

Blood supply to the vagina is provided by the **vaginal artery**, a branch of the **internal pudendal artery**. Deoxygenated blood and metabolic by-products are returned to the central circulation via the **internal pudendal vein** by way of the **vaginal vein**. Lymph from the vagina flows to the **medial iliac lymph nodes** via the **anorectal lymph node**.

Vaginal nerve supply comes from the **renal**, **aortic**, **uterine**, and **pelvic plexuses**.

Regional blood supply, lymphatics, and innervation

Specific blood supply and innervations are listed in each of the sections herein.

Selected references

[1] Budras K, Sack WW, Röck S. Anatomy of the horse: an illustrated text. 2nd ed; 1994.
[2] Dascanio JJ. External reproductive anatomy. In: McKinnon AO, Squires EL, Vaala WE, Varner DD, editors. Equine reproduction; 2011. p. 1577–81 [chapter 164].
[3] Kainer RA. Internal reproductive anatomy. In: McKinnon AO, Squires EL, Vaala WE, Varner DD, editors. Equine reproduction; 2011. p. 1582–97 [chapter 165].

CASE 13.2

Second-Degree Perineal Laceration

Candace Lyman[a] and G. Reed Holyoak[b]
[a]Auburn University College of Veterinary Medicine, Auburn, Alabama, US
[b]Oklahoma State University College of Veterinary Medicine, Stillwater, Oklahoma, US

Clinical case

History

A 4-year-old, primiparous American Quarter Horse mare foaled in the pasture at home unattended. The owner, not anticipating that the mare would foal for about another 2 weeks, was surprised to find the mare with a foal at her side during early morning feeding and estimated that the foal was less than 12 hours old. The owner immediately noticed that the mare had sustained an injury to her vulva and that the fetal membranes had not passed; therefore, the owner transported the mare and foal to a university hospital for examination and treatment.

Physical examination findings

During the initial external reproductive examination, it was noted that a suture from a previously placed Caslick procedure (vulvuloplasty) had not been removed. Subsequently, because the portion of the dorsal vulvar commissure that had been surgically treated was not opened before foaling, the vulva was torn during the 2nd stage of parturition. The laceration started approximately 5 cm (2.0 in.) distal to the dorsal commissure of the vulva, extending approximately 6 cm (2.4 in.) lateral to the right labium (12 cm or 4.7 in. in length) and was swollen and edematous (causing it to droop ventrally), with minimum gross contamination (Fig. 13.2-1). The left labium was excoriated (L. *excoriare* abrasion of the skin) along the mucocutaneous junction but was intact. About 20 cm (7.9 in.)

PERINEAL LACERATIONS IN THE MARE

Injury to the outermost (caudal-most) part of the mare's reproductive and gastrointestinal tract—i.e., vulva, perineum, and rectum—occurs most frequently as a result of breeding or foaling events. Significant injury that occurs at foaling is likely due to the rapid force at which the foal, with an inappropriately oriented nose or foot, is expelled during stage II of labor or when foaling takes place in the presence of an intact Caslick procedure—i.e., suture closure of the most dorsal portion of the vulva, as in this case.

There are 3 types of perineal lacerations—1st-, 2nd-, and 3rd-degree. Many 1st-degree lacerations heal without problems and involve the mucosa of the vestibule and dorsal commissure of the vulva, although some cases may require a Caslick procedure to achieve the best outcome. Second-degree lacerations involve the mucosa and submucosa of the vestibule, constrictor vulvae muscle, and perineal body. If not accompanied by significant edema, inflammation, and infection, 2nd-degree perineal lacerations may be repaired immediately (as in this case) using a Caslick procedure; otherwise, repair is delayed for 2–4 weeks. Third-degree lacerations are defined as a complete disruption of the rectovestibular septum, rectum, perineal body, and anal sphincter resulting in a common communication between the rectum and vestibule. Third-degree perineal lacerations require repair of the rectovestibular septum, anal sphincter, and perineal body using one of several surgery techniques following a minimum delay of 3–4 weeks post-partum to allow the inflammation, edema, and tissues to heal before attempting a reconstruction procedure.

Comparative Veterinary Anatomy: A Clinical Approach. https://doi.org/10.1016/B978-0-323-91015-6.00137-0

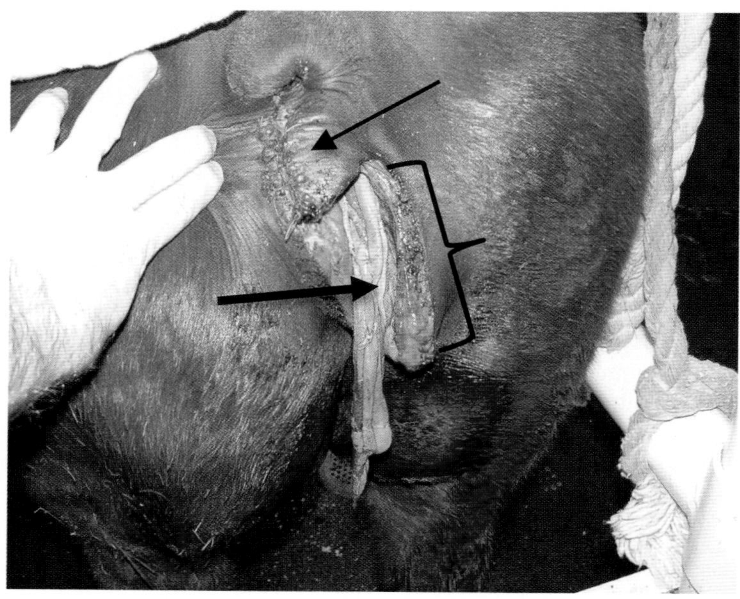

FIGURE 13.2-1 Image of the injury at presentation. A Caslick suture is still apparent in the dorsal part of the vulva *(narrow arrow)*; about 5 cm (2.0 in.) below the dorsal commissure of the vulva, the right labium has a laceration that resulted in an edematous tissue flap *(bracket)*. Fetal membranes that had yet to be passed *(black arrow)* were seen protruding from the mare's vulva.

of umbilicus and chorioallantoic membranes were evident, protruding from the vestibule. All other physical examination findings were normal, except for mild tachycardia that was attributed to pain.

Retained fetal membranes occur when there has been a failure of delivery of the allantochorionic membrane (plus or minus the amniotic membrane) within a specified time frame; this time frame is generally 30 minutes to 12 hours (although classically considered retained if not passed within 3 hours in mares). Retained fetal membranes increase the risk of metritis, laminitis, and septicemia and are potentially life-threatening.

Differential diagnoses

Retained fetal membranes with a superficial vulvar laceration, 1st-degree perineal laceration, or 2nd-degree perineal laceration

Diagnostics

A more complete physical examination of the vulva, vestibule, and vagina revealed that the laceration extended into the vestibule but did not involve the perineum or perineal body. There was no evidence of trauma to the rectovestibular septum, however the injury to the right vestibular wall extended deeper and involved the constrictor vestibuli and vulvae muscle. Digital examination of the uterus and associated fetal membranes revealed continued attachment of the fetal membranes to the endometrial lining of the uterus.

Diagnosis

Second-degree perineal laceration with retained fetal membranes

Treatment

The retained fetal membranes were successfully treated using the Burn's technique (see side box entitled "Burn's technique"). The 2nd-degree perineal laceration was repaired in 2 separate layers using a horizontal mattress suture pattern (a tension-relieving suture pattern) for the deeper layer and a simple interrupted suture pattern in the skin, with planned removal in 14 days, before breeding on the 2nd (post-foaling) estrous cycle. The mare was bred via

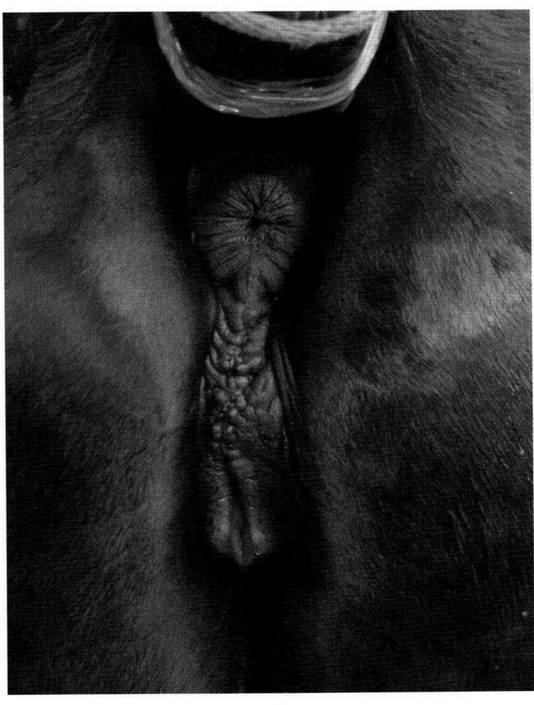

FIGURE 13.2-2 The mare in this case at day 55 postfoaling; the vulva is healed.

artificial insemination and an embryonic vesicle was confirmed by ultrasound 25 days after ovulation. Healing of the 2nd-degree perineal laceration was complete 55 days post-injury (Fig. 13.2-2).

Anatomical features in equids

Introduction

The caudal reproductive tract of the mare is comprised of the perineum, the clitoris, the vestibule and its associated musculature, the rectovestibular septum, and the vagina. Unlike much of the intraabdominally located members of the female reproductive tract, these structures are positioned within the pelvic girdle and the pelvic outlet. This case presents the perineal anatomy of the mare. Additional detail on the caudal female reproductive tract, including blood supply, lymphatics, and innervation is in Case 13.1.

Function

The perineum, vagina, and everything in between as described below are anatomical structures responsible for preventing fecal and environmental contamination to a mare's reproductive tract. Additionally, this portion of mare

BURN'S TECHNIQUE

The Burn's technique is performed by filling the allantoic cavity with saline, or another uterine-appropriate crystalloid fluid, and tying off the caudal portion of the membranes with umbilical tape or large zip ties. Once the fluid has been infused into the allantoic cavity, a low dose of oxytocin is administered; a mare within a few hours of parturition requires as little as 5–10 IU of oxytocin IM to promote physiologic uterine contractions without inducing significant abdominal discomfort and straining.

This technique assists in the removal of the fetal membranes by encouraging uterine stretch, subsequent uterine contraction, and separation of the microvilli from their attachments. Many mares are reported to pass the entire chorioallantoic sac within 30 minutes.

anatomy functions to facilitate the act of natural breeding in the equid not only by providing increased mucus secretions during estrus, but also by actively stabilizing the stallion's penis within the caudal reproductive tract of the mare. Injury to these structures ultimately results in a meaningful decrease in mare fertility, but the effects of injury can be mitigated or eliminated with timely and appropriate treatment and repair.

Clitoris (Fig. 13.2-3)

Just beyond the ventral portion of the vulvar commissure, the highly innervated **glans clitoris** is positioned within a cavity termed the **fossa clitoris**; the glans clitoris is also partially covered on the dorsal part by a frenulum (L. *frenum* "bridle," restraining structure or part), or a fold of tissue, referred to as the **prepuce** of the clitoris. The body of the clitoris—i.e., the **corpus clitoris**—is the homolog to the corpus cavernosum penis in the stallion. It is about 5 cm (2.0 in.) in length and attached to the pelvic ischial arch by 2 crura (L. leg-like part). The clitoris also contains multiple ventral sinuses.

The clitoral sinuses are a haven for pathogenic bacteria and therefore removal of sinus debris is important when treating a mare for endometritis.

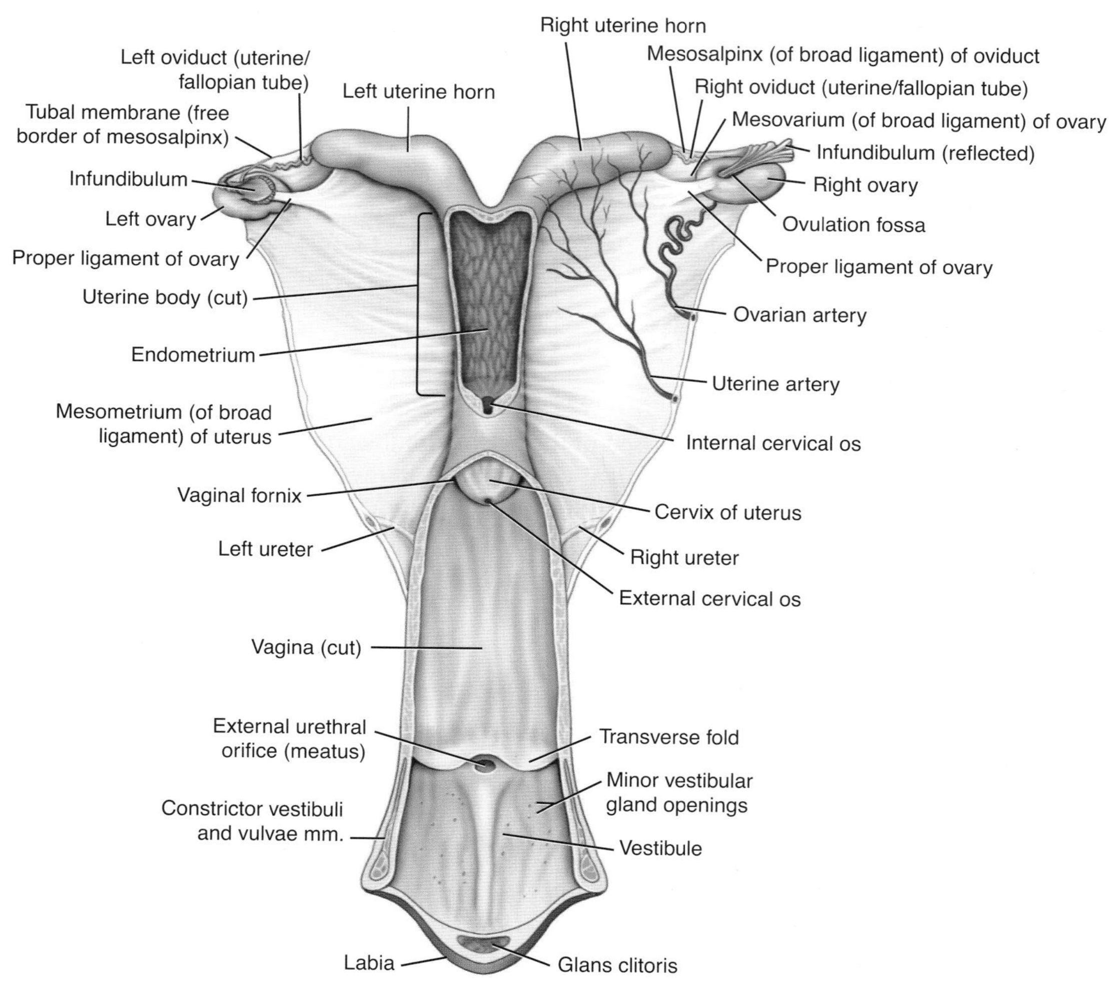

FIGURE 13.2-3 Dorsoventral view of the mare's tubular reproductive tract.

Constrictor vulvae muscles and perineal body (Figs. 13.2-3 and 13.2-4)

Muscles called the **constrictor vulvae** are located within the individual labia and serve to provide vulvar closure. These vulvae muscles likely work in synchrony with the **constrictor vestibuli musculature** during copulation to provide support to the penis. The **perineal body** is a palpable, pyramid-shaped mass of tissue located just beneath the surface of the perineum. It provides a point of insertion for the various muscles and fascia associated with both the anus and the walls of the perineum. Without the support of an intact perineal body, the integrity of the perineum and vestibule is compromised in such a way that the perineum becomes "sunken" and recesses cranially.

During estrus, another function of the constrictor vulvae muscles is to actively evert the labia—i.e., winking—thereby providing a visual cue of estrus and breeding receptivity to a stallion.

When the integrity of the perineal body is compromised, the vulva's ability to provide anatomic closure is affected. This may lead to not only pneumovagina, but also to fecal and urinary contamination of the caudal reproductive tract.

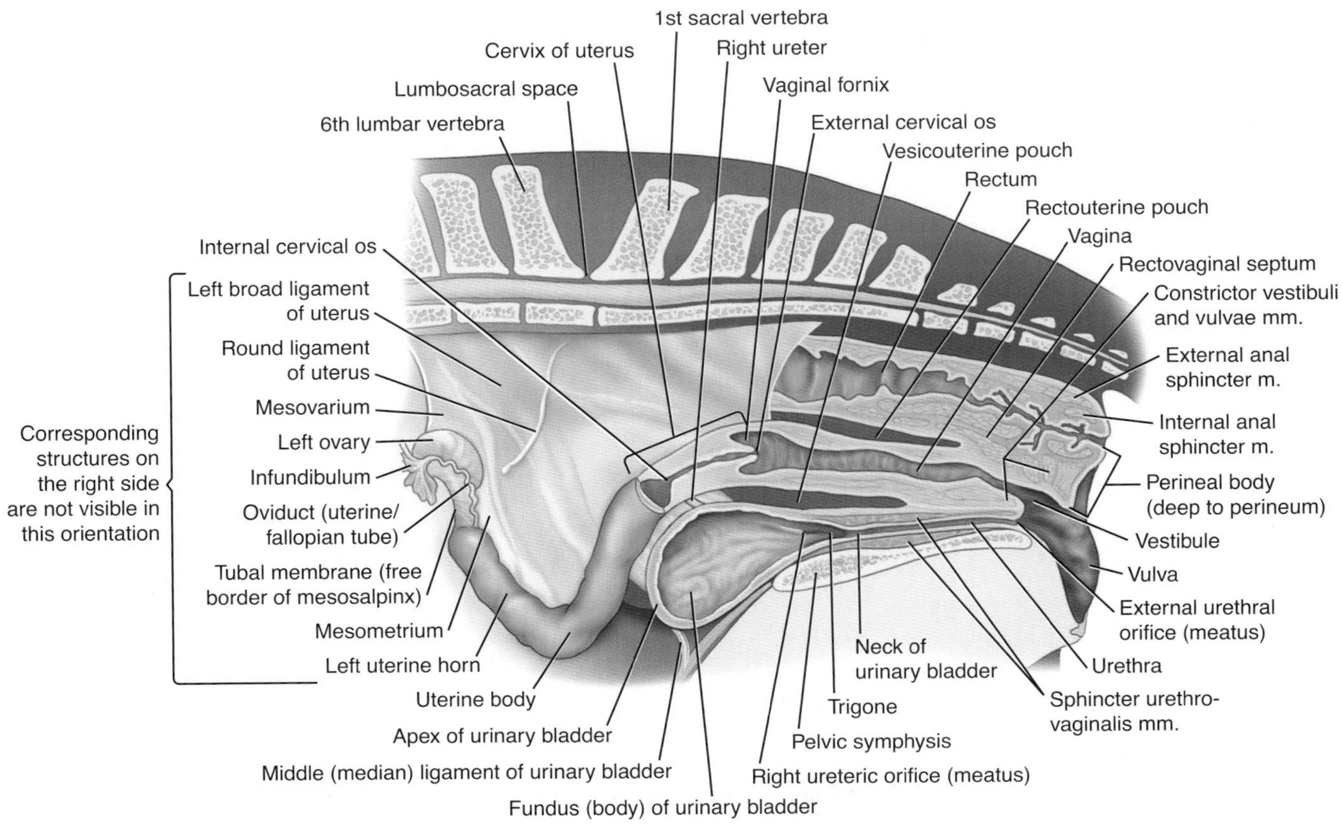

FIGURE 13.2-4 Lateral view of the mare reproductive tract.

Repair of this septum is the 1st step during reconstruction of a 3rd-degree tear and includes cranial dissection of the septum 3–5 cm (1.2–2.0 in.) beyond the tear, creating separate rectal and vestibular shelves. After sufficient cranial dissection is complete, the goal is to appose 2 lateral margins to create the rectal and vestibular shelves without tension.

Rectovestibular septum (Fig. 13.2-5)

The rectovestibular septum is a fascial separation or division that lies on the dorsal surface of the vestibule, with the fascial plane diving deep into the constrictor vulva muscle and extending dorsocranially to the rectum.

Vestibule (Fig. 13.2-3)

The **vestibular bulbs**, located dorsolateral to the clitoris, are oval bodies of erectile tissue with a purple color. This embedded tissue and the deeper **constrictor vestibuli muscles,** located within the walls of the vestibule, contribute to the act of breeding by providing stabilizing support for the body of the penis. Slightly cranial and ventrolateral to the clitoris are 2 lines of papillae that function as openings to the minor **vestibular glands**. Believed to be similar in function to the male accessory sex glands, these glands contribute mucous secretions to the caudal tract of the mare.

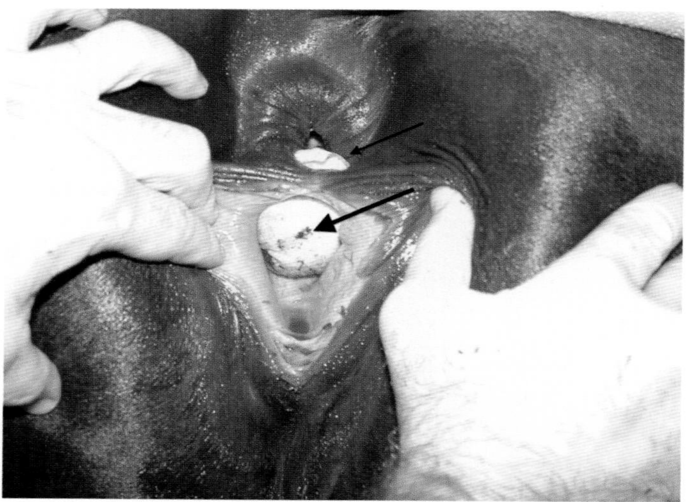

FIGURE 13.2-5 A rectovestibular fistula demonstrated by a single piece of gauze placed between the vestibule *(large arrow)* and rectum *(small arrow)*.

RECTOVAGINAL FISTULA

It is important to recognize that 1st-, 2nd-, or 3rd-degree perineal lacerations differ from the commonly referred to rectovestibular or rectovaginal fistula—i.e., an "RV fistula." A fistula is the opening that exists between 2 adjoined tubular organs—i.e., the vestibule or the vagina and the rectum, Fig. 13.2-5). This condition can arise not only from a foaling injury, but also from a failed surgical repair of a 3rd-degree perineal laceration. Externally, it may not be obvious that a fistula lies deep into the perineum, but manual exploration reveals an open communication between the dorsally located, caudal rectum near the anal sphincter and the ventrally located vagina or vestibule. Manure contamination of the caudal reproductive tract that occurs due to the presence of a fistula has a detrimental impact on the mare's fertility because of the chronic contamination and infection of the reproductive tract.

The **external urethral orifice** is at the cranial border of the vestibule, on the ventral midline. Overlying the urethral orifice is the remnant of the hymen called the **transverse fold** and an area referred to as the **vestibulovaginal junction**. This junction is the site where the union between the mesodermal (the future vagina) and the ectodermal (the future vestibule) tissues occurs during fetal development.

Vagina (Figs. 13.2-3 and 13.2-4)

Located just cranial to the vestibulovaginal junction is the caudal part of the **vagina**. The vaginal cavity extends cranially from this anatomic junction, ending blindly where the cervix projects caudally from a discoid depression referred to as the vaginal **fornix** (L. arch-like structure or vault-like space).

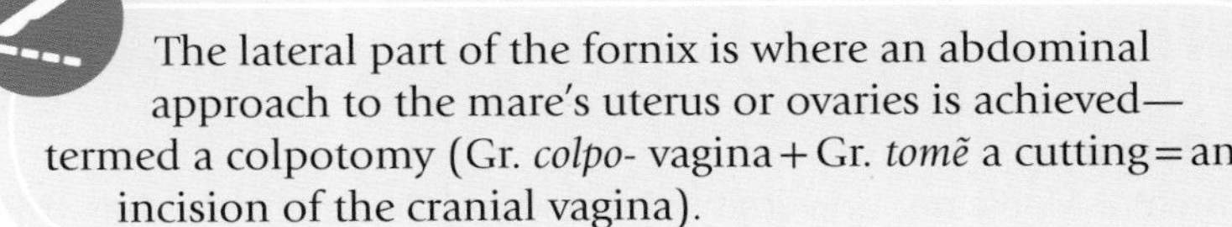

The lateral part of the fornix is where an abdominal approach to the mare's uterus or ovaries is achieved—termed a colpotomy (Gr. *colpo-* vagina + Gr. *tomē* a cutting = an incision of the cranial vagina).

Selected references

[1] Ginther OJ. Reproductive anatomy. In: Reproductive biology of the mare: basic and applied aspects. 2nd ed. Cross Plains, WI: Equiservices; 1992. p. 1–40 [chapter 1].
[2] Woodie JB. Vulva, vestibule, vagina, and cervix. In: Auer JA, Stick JA, Kümmerle JM, Prange T, editors. Equine surgery. 5th ed. St. Louis, MO: Elsevier; 2019. p. 1065–82.
[3] Threlfall WR. Retained fetal membranes. In: McKinnon AO, Squires EL, Vaala WE, Varner DD, editors. Equine reproduction. 2nd ed. West Sussex, UK: Wiley-Blackwell; 2011. p. 2520–9.

CASE 13.3

Uterine Artery Rupture

Maria Ferrer
Large Animal Medicine, University of Georgia College of Veterinary Medicine, Athens, Georgia, US

Clinical case

History

A 16-year-old multiparous Thoroughbred mare was presented to the hospital with her foal 6 hours after delivery with signs of acute abdominal pain/colic. Pregnancy and parturition were reportedly uncomplicated.

Physical examination findings

On presentation, the mare showed severe signs of pain and hypovolemic shock—trembling, pawing, and trying to lie down and roll in the examination room. There was marked tachycardia (L. *tachy* swift + Gr. *kardia* heart) and tachypnea (L. *tachy* swift + Gr. *pnoē* breathing) with a capillary refill time of > 3 seconds. The mucous membranes were pale and tacky, and the distal extremities were cold to the touch.

Differential diagnoses

Gastrointestinal colic, peri-partum hemorrhage, uterine rupture, gastrointestinal rupture

Diagnostics

The mare received a combination of xylazine and butorphanol for sedation and analgesia. Because of her unrelenting pain, she was not placed in stocks for restraint during the examination. Transabdominal ultrasound revealed a large volume of free fluid within the peritoneal cavity (Fig. 13.3-1A). The fluid was heterogeneous, with swirling of free hyperechoic elements. The appearance was consistent with active bleeding resulting in hemoabdomen (Gr. *haima* blood + L. *abhere* to hide, belly, venter). Transrectal palpation and ultrasound revealed a heterogeneous mass within the right broad ligament with an appearance consistent with a hematoma (Fig. 13.3-1B). Abdominocentesis confirmed a diagnosis of hemoabdomen. No other abnormalities were identified on palpation or ultrasound of the gastrointestinal tract. While nasogastric intubation was indicated for completeness of acute abdomen workup, it was deferred to minimize potentially stressful interventions, preventing further exacerbation of bleeding.

Diagnosis

Post-partum hemorrhage due to rupture of the uterine artery

Treatment

Resuscitation efforts were immediately initiated with placement of a jugular IV catheter and administration of crystalloid replacement IV fluids, followed by maintenance fluid therapy. While fluid replacement may elevate blood pressure and exacerbate bleeding, IV fluids were administered at a measured rate to address the hypovolemia.

Comparative Veterinary Anatomy: A Clinical Approach. https://doi.org/10.1016/B978-0-323-91015-6.00062-5

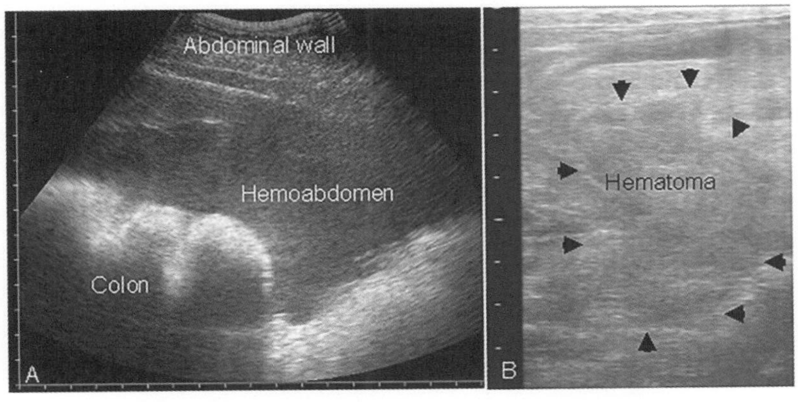

FIGURE 13.3-1 (A) Transabdominal ultrasonographic image from a mare presenting with a ruptured uterine artery and hemoabdomen. (B) Transrectal ultrasonographic image of a recent hematoma present in the broad ligament. The margins of the hematoma are marked with *arrowheads*.

The mare also received a blood transfusion to restore blood volume, essential clotting factors, platelets, and red blood cell components. Aminocaproic acid was administered IV to promote stabilization of the blood clot and Yunnan Baiyao was given PO as a hemostatic agent. The mare was placed in a quiet, darkened stall and sedated with detomidine and butorphanol to control pain, minimize stress, and decrease excitement with the clinical goal of minimizing additional bleeding, allowing the clot to form and stabilize. Acepromazine was avoided because of its hypotensive effects. The foal was allowed to remain with the mare to reduce stress, but physically separated by a stall divider to prevent injury during the mare's painful episodes. The bleeding was controlled, and the mare clinically stabilized based on a return of normal heart and respiratory rates and improved comfort. Treatment with aminocaproic acid, flunixin meglumine, and antibiotics was continued for 3 days until discharge from the hospital without further problems.

Anatomical features in equids

Introduction

The broad ligament is a double-layered membrane of peritoneum suspending the female genital tract from the abdominal and pelvic walls. The embryologic origin of the broad ligament is the urogenital mesentery, which is the peritoneum covering the mesonephros and genital blastema (Gr. *blastēma* shoot: group of cells that produce an offspring). The broad ligament is formed after the fusion of the Müllerian ducts, when the 2 sheets of peritoneum join together enveloping the reproductive organs.

Function

The broad ligaments provide support to the genital tract and help maintain its position within the caudal abdomen and pelvic cavity. Blood vessels, lymphatic vessels, and nerves that supply and innervate the genital tract run between the 2 layers of the broad ligament. The broad ligaments extend into the serosal surface of the genital organs, providing additional protection to the organs. They also contain a large quantity of smooth muscle that is continuous with the outer longitudinal layer of the myometrium.

The broad ligament

The paired left and right **broad ligaments** attach the internal genitalia to the abdominal and pelvic walls, extending from the 3rd or 4th lumbar vertebra to the 4th sacral vertebra. Each broad ligament consists of 2 serous membranes that are continuous with the serous lining of the peritoneal cavity and reproductive organs. Arteries, veins, lymphatics,

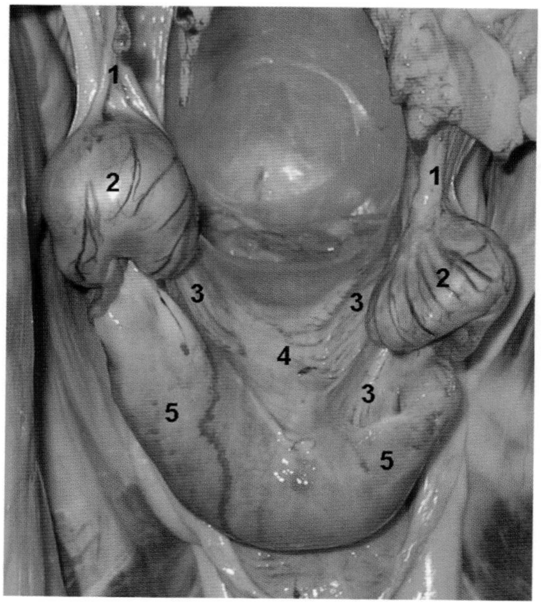

FIGURE 13.3-2 The broad ligament suspends the genital tract from the abdominal and pelvic wall. Key: 1, mesovarium; 2, ovaries; 3, mesometrium; 4, uterine body; 5, uterine horns.

The dorsal attachment of the broad ligament has clinical implications in examination of the genital tract per rectum. The genital tract is suspended from the abdomen in a V-shape and cannot be retracted. Palpation of the uterine body is performed by following this structure cranially with a flat hand swiping the floor of the pelvis. The uterine horns can be elevated using a scooping motion to allow palpation of the ventral or free surface of the uterus.

and nerves supplying and innervating the genital tract and smooth muscle are positioned between these 2 layers. The right and left ligaments converge over the uterine body and cervix. The visceral attachment of the ligaments and main point of entry of vessels and nerves are on the dorsal aspect of the organs, leaving the ventral aspect of each organ as the free surface (Fig. 13.3-2).

The parts of the broad ligaments that suspend the ovaries, uterine tubes (oviducts), and uterus are called the mesovarium, mesosalpinx, and mesometrium, respectively (Figs. 13.3-2 and 13.3-3A–D). The **mesovarium** suspends the ovaries between the 3rd and 5th lumbar vertebrae. While they are usually suspended just cranial to the shaft of the ilium at the 2 and 10 o'clock positions (right and left, respectively; from a caudal-cranial direction), their positions vary with stretching of the broad ligaments in multiparous, pregnant, or post-partum mares, and with distention of intestinal viscera. Most of the ovarian surface, except for the ventral border, is covered by mesovarium. An incomplete **ovarian bursa** is formed by the **ovulation fossa** (cranial), the tip of the uterine horn (caudal), mesosalpinx and oviduct (lateral), and the proper ligament of the ovary (medial) (Fig. 13.3-3B).

During ovariectomy in standing mares, local anesthetic is administered to the mesovarium to block the nerve receptors supplying the ovary, thereby providing local analgesia. The mesovarium is also ligated or crushed for hemostasis to prevent bleeding from the ovarian pedicle post-operatively.

The **proper ligament of the ovary** is a fold of the broad ligament that connects the tip of the uterine horn with the caudal pole of the ovary. The nerves and vessels supplying and draining the ovary lie within the mesovarium and enter the ovary on the dorsal surface (**hilus**). The length and stretching of the broad ligaments permit exteriorization of the ovaries through a flank or ventral midline laparotomy.

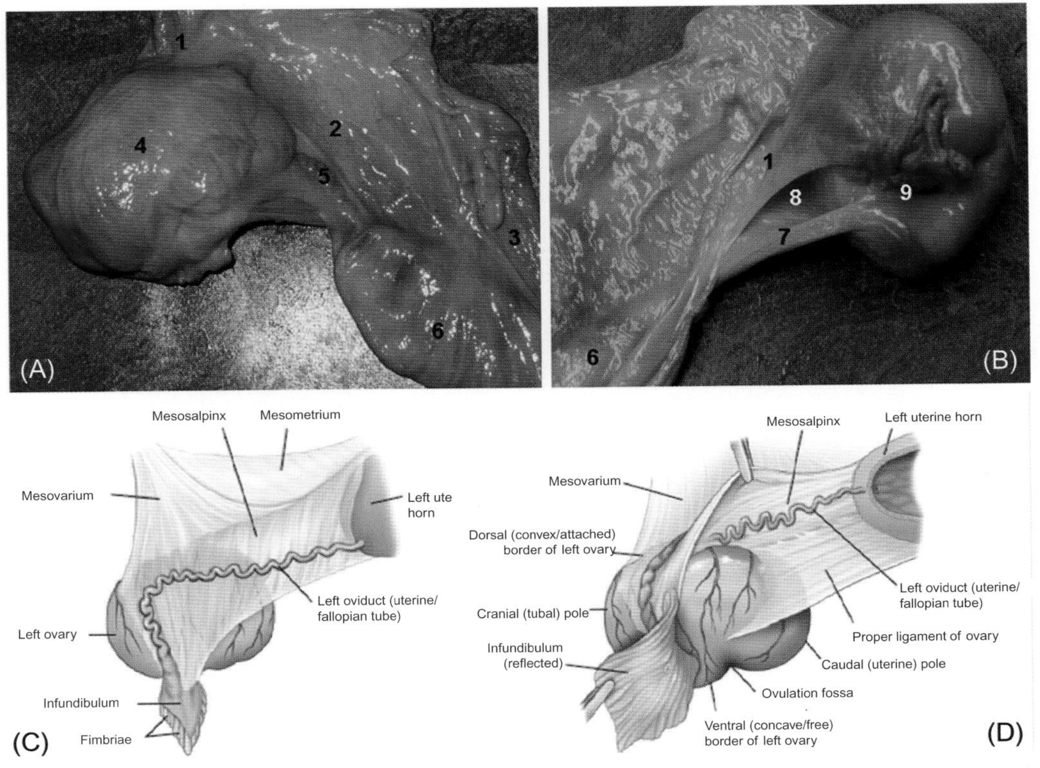

FIGURE 13.3-3 (A) The broad ligament is divided into the (*1*) mesovarium, (*2*) mesosalpinx, and (*3*) mesometrium, which suspend the (*4*) ovary, (*5*) uterine tube (oviduct), and (*6*) uterus, respectively. (B) The (*1*) mesosalpinx, (*6*) tip of the uterine horn, (*7*) proper ligament of the ovary, and (*9*) ovulation fossa form the limits of the (*8*) ovarian bursa. Lateral (C) and medial view (D) of the structures in figures (A) and (B), respectively.

The **mesosalpinx** supports the uterine tubes (oviducts) along their dorsal border. The free border of the mesosalpinx is often referred to as the **tubal membrane**. Contained within the mesosalpinx may be found 2 vestigial structures that are remnants of the Wolffian ducts (a.k.a. "the duct of Wolff")—the **epoophoron** and **paroophoron**.

The epoophoron and paroophoron can sometimes be seen as parovarian cystic structures during ultrasonographic examination. Although their clinical significance is unknown, from a practical standpoint, parovarian cysts can initially be confused with an ovarian follicle when examining the ovary with transrectal ultrasonography (Fig. 13.3-4).

UTERINE TORSION—HORSE VS. COW

Uterine torsion in the mare generally occurs between the 9th and 10th month of pregnancy (average gestation length 345 days) and the twist occurs cranial to the cervix. In the cow, uterine torsion occurs at the end of gestation (with an average gestational length 283 days), and the twist is usually caudal to the cervix. This is an important clinical distinction because a mare with a uterine torsion usually presents with signs of acute abdomen/colic at 9–10 months of gestation, and the reproductive cause of colic is not evident on vaginal examination, but is evident only on rectal examination. This is in contrast to the cow, in which vaginal examination may help with diagnosis of uterine torsion. In both cases, surgery is required to correct the uterine torsion if not resolved by a conservative approach—i.e., rolling the cow or horse to untwist the uterus.

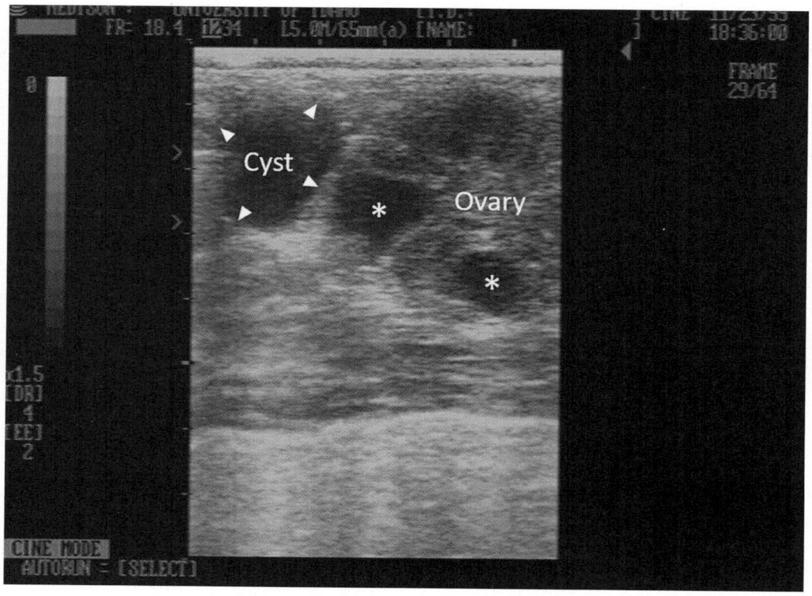

FIGURE 13.3-4 Transrectal ultrasonographic image of an ovary and adjacent paraovarian cyst in a mare. Each gradation along the top and left sides of the image equals 10 mm (0.4 in.). Paraovarian cysts appear as anechoic round structures adjacent to the ovary. The paraovarian cyst in this image (*arrowheads*) had a diameter of approximately 20 mm (0.8 in.). The ovary contained two small (<20 mm [0.8 in.]) follicles (*asterisks*). (Image courtesy of Dirk Vanderwall.)

The length of the mesometrium in some mares allows for a rotation of the gravid uterus along its longitudinal axis causing a uterine torsion. An untreated uterine torsion of more than 180° compromises the blood supply to the uterus, resulting in reduced placental perfusion and fetal death, and/or uterine rupture and maternal death if not recognized and corrected with surgery early in the clinical process. Palpation per rectum of the broad ligaments is essential for both diagnosis of uterine torsion and determination of the direction of the torsion. For example, when the torsion occurs in a clockwise direction (from a caudal-to-cranial viewpoint), the left broad ligament is palpated as a tight band running over the dorsal aspect of the uterus toward the right side of the abdomen. The right ligament is palpated as a tight band running ventrally.

The most caudal and broadest part of the broad ligament is the **mesometrium.** The cranial part is defined by the **round ligament of the uterus**, which is a prominent structure tracking along the lateral surface of each broad ligament. The round ligament blends with the peritoneum near the inguinal ring and is thought to be homologous to the gubernaculum in the male.

Blood supply, lymphatics, and innervation

The blood supply of the reproductive tract supports the metabolic requirements of the reproductive organs and transports hormones to and from the reproductive system, adrenal gland, and pituitary gland.

The uterus is supplied by 3 sets of paired (right and left) arteries: the uterine branch of the vaginal artery, also called **caudal uterine artery**; the uterine artery, also called **middle uterine artery** (a branch of the external iliac artery); and the uterine branch of the ovarian artery, also called **cranial uterine artery** (Fig. 13.3.-5). The main blood supply is provided by the (middle) uterine artery, which forms a cranial branch that supplies the proximal uterine horn, and a caudal branch that supports the distal uterine horn and body. The uterine artery enters the dorsal surface of the uterus with branches throughout the uterine wall, between the 2 layers of the myometrium.

The uterine branch of the **vaginal artery** (caudal uterine artery) passes along the lateral side of the cervix and uterine body to anastomose with the caudal branch of the middle uterine artery. The **ovarian artery** is in the cranial portion of the broad ligament. It bifurcates into an ovarian branch and a uterine branch. The uterine branch of the ovarian artery

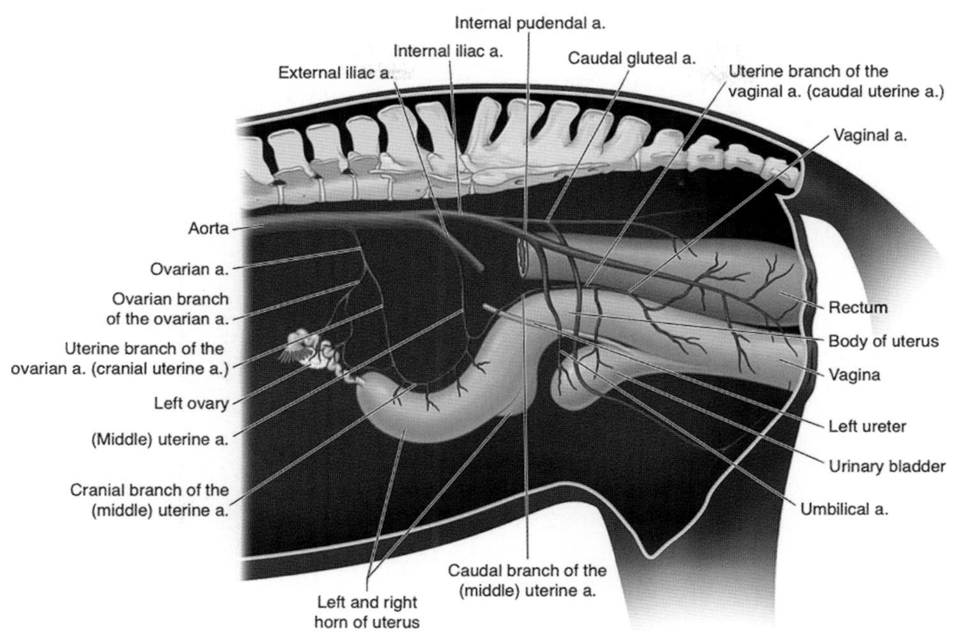

FIGURE 13.3.-5 Blood supply to the reproductive tract of the mare.

(the **cranial uterine artery**) anastomoses with the cranial branch of the middle uterine artery and supplies the uterine tubes (oviducts), mesosalpinx, and tip of the uterine horns. The corresponding veins drain the reproductive tract. Lymphatic vessels in the broad ligament are quite prominent and drain interstitial fluid from the tissue to the **uterine**, **medial iliac**, and **aortic lymph nodes**.

Nerves supplying the genital tract are also located within the broad ligaments. Parasympathetic fibers originate from the **pelvic nerves** in the sacral area. Sympathetic fibers from the **hypogastric nerves** and pelvic plexus originate from the caudal mesenteric ganglion, located by the origin of the caudal mesenteric artery. Two pairs of nerves proceed caudally from the ganglion. The **internal spermatic nerve** runs along the ovarian artery and supplies the ovaries, oviducts, and tip of the uterine horns. The other nerves follow the aorta and enter the pelvic cavity, where they anastomose with each other and with branches of the 3rd and 4th sacral nerves. The **pudendal** and **caudal rectal nerves** from the sacral plexus carry motor fibers. The sensory fibers only supply the vulva.

Any of the previously mentioned blood vessels can rupture during parturition or in the post-partum period; however, the right uterine artery has the highest incidence of rupture. The predisposition of this vessel to rupture is believed to be due to displacement of the gravid uterus to the left by the cecum, resulting in increased tension in the right broad ligament. There are several clinical scenarios associated with peri-partum hemorrhage. Bleeding may dissect the broad ligament, resulting in rupture of the ligament and bleeding directly into the abdomen resulting in peracute (L. *peracutus* excessively acute or sharp) hypovolemic shock, rapid demise, and death. In other cases, bleeding may be contained within the 2 layers of the broad ligament or the serosal layer of the uterus, resulting in a hematoma. These hematomas are palpated and visualized per rectum with ultrasonography (Fig. 13.3-6). Bleeding confined to the broad ligament or serosal layer of the uterus carries a more favorable prognosis than bleeding directly into the peritoneal cavity. Aggressive therapy improves the chances of survival when bleeding occurs into the peritoneal cavity. However, rupture of the hematoma and further bleeding into the abdomen can occur, resulting in death several days or weeks after the initial bleed. Finally, small uterine mural vessels can rupture, leading to bleeding into the uterine lumen. While the rate of recurrence for peri-partum hemorrhage is not known, 49% of surviving mares in one study produced at least one foal in subsequent years.

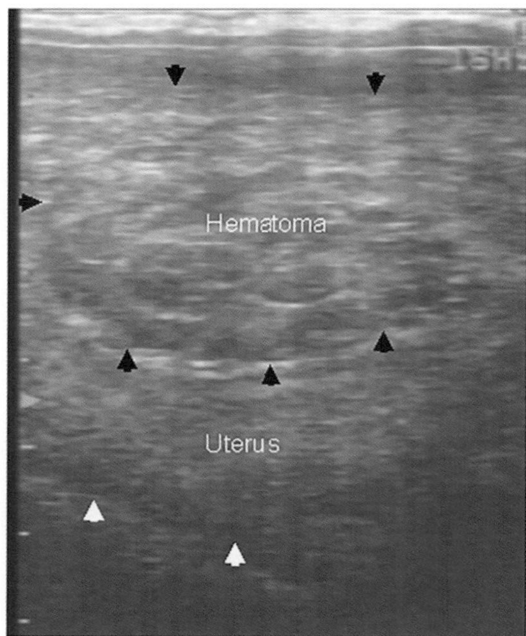

FIGURE 13.3-6 Transrectal ultrasonographic image of a hematoma (margins demarcated with *black arrowheads*) contained within the serosal layer of the uterus (ventral edge of the uterus marked with *white arrowheads*).

Selected references

[1] Arnold CE, Payne M, Thompson JA, Slovis NM, Bain FT. Periparturient hemorrhage in mares: 73 cases (1998–2005). J Am Vet Med Assoc 2008;232:1345–51.
[2] Ginther OJ. Reproductive anatomy. In: Ginther OJ, editor. Reproductive biology of the mare. Basic and applied aspects. 2nd ed. Cross Plains, WI: Equiservices; 1992.
[3] LeBlanc MM, Johnson RD, Calderwood Mays MB, Valderrama C. Lymphatic clearance of India ink in reproductively normal mares and mares susceptible to endometritis. Biol Reprod 1995;52:501–6.

CHAPTER 13

CASE 13.4

Granulosa Cell Tumor

Dirk Vanderwall
Animal, Dairy, and Veterinary Sciences, Utah State University School of Veterinary Medicine, Logan, Utah, US

Clinical case

History

A 3-year-old American Quarter Horse filly was retired from race training in the fall in preparation for breeding the following spring. To hasten the onset of seasonal ovulatory activity, the filly was placed under an artificial long-day photo-period (16 hours total light/day) beginning December 1.

Physical examination findings

In late January, transrectal palpation and ultrasonography were performed to assess the status of the filly's reproductive tract. Examination revealed that the left ovary was small (31 mm [1.2 in.] long × 12 mm [0.5 in.] wide × 23 mm [0.9 in.] high), firm, and devoid of follicular activity. In contrast, the right ovary was much larger (approximately 70 mm or 2.8 in. in diameter), lacked a palpable ovulation fossa, and was characterized ultrasonographically by a large, thick-walled cavity containing echogenic fluid. The uterus and cervix were flaccid, consistent with a seasonally anovulatory state, and suggestive that the right ovary did not contain a functional corpus luteum (CL) producing progesterone. No stallion was available to assess the filly's reproductive behavior, but when she was turned out with other mares, she behaved normally. The filly was reevaluated in mid-February (17 days after the initial examination), at which time there was no appreciable change in the status of either ovary (Fig. 13.4-1), tubular genitalia, or behavior.

Differential diagnoses

Anovulatory hemorrhagic follicle, ovarian hematoma, granulosa cell tumor (GCT), or other type of ovarian tumor of the right ovary

Diagnostics

A blood sample was collected from the jugular vein and submitted for endocrine testing for progesterone, inhibin, testosterone, and anti-Müllerian hormone (AMH; Table 13.4-1). The inhibin and testosterone results were borderline and marginally elevated, respectively, making them inconclusive. In contrast, the AMH level was nearly 5 times higher than normal.

Diagnosis

Presumptive diagnosis of a GCT of the right ovary

Treatment

The right ovary was surgically removed via colpotomy (incision in the vagina providing entry into the cul-de-sac of the peritoneal cavity) using a chain ecraseur (Fr. crusher). After surgery, the mare was confined to a box stall for

Comparative Veterinary Anatomy: A Clinical Approach. https://doi.org/10.1016/B978-0-323-91015-6.00063-7

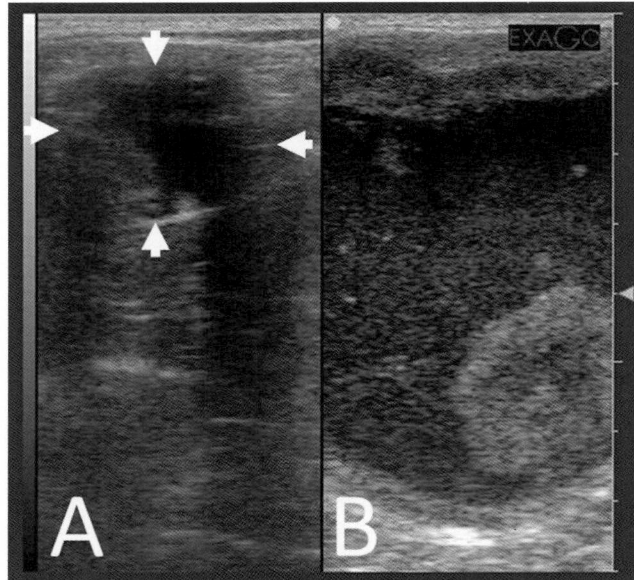

FIGURE 13.4-1 Transrectal ultrasonographic images of the left (A) and right (B) ovaries of a 3-year-old American Quarter Horse filly. The left ovary was small, firm, and devoid of follicular activity (*arrows* demarcate outer margins of the ovary). The right ovary was approximately 70 mm (2.8 in.) in diameter, lacked a palpable ovulation fossa, and consisted of a large, thick-walled cavity containing echogenic fluid. Each gradation along the right side of the image equals 10 mm (0.4 in.). (Reproduced from J Am Vet Med Assoc 2013;243:791–793, with permission.)

TABLE 13.4-1 Endocrine testing results for a 3-year-old American Quarter Horse filly with a presumptive GCT of the right ovary.

Hormone	*Test result*	*Reference range*[a]	*Laboratory designation*
Inhibin	0.78 ng/mL	0.1–0.7 ng/mL	Borderline elevated
Testosterone	56.5 pg/mL	20–45 pg/mL	Marginally elevated
AMH	19.0 ng/mL	≤4.2 ng/mL	Elevated
Progesterone	0.3 ng/mL	<0.5 ng/mL[b]	Absence of luteal tissue

[a] Normal values for a nonpregnant mare.
[b] Indicative of the absence of luteal tissue.

GRANULOSA CELL TUMORS IN MARES

The most common ovarian neoplasm in mares, GCTs account for >85% of all tumors of the reproductive tract in mares. Other less common ovarian tumors are teratoma, dysgerminoma, cystadenoma, and carcinoma. Histopathologically, GCTs are sex cord-stromal tumors comprised primarily of granulosa cells or a mixture of granulosa and thecal cells. There is no breed or age predilection for GCTs in mares, and although most GCTs are diagnosed in sexually mature mares, they have been diagnosed in prepubertal fillies. In adult horses, GCTs have been diagnosed in maiden, barren, pregnant, and post-partum mares. Although there are exceptions, most GCTs are unilateral, benign tumors associated with cessation of cyclical reproductive activity that is typified by the presence of a small, inactive contralateral ovary. Affected mares generally exhibit 1 of the 3 behavioral patterns: (1) anestrus; (2) continuous or intermittent estrus—i.e., nymphomania; or (3) stallion-like behavior. The type of behavior reflects the specific tumor cell type(s) involved and the resultant steroid hormone production (if any). Although the "classic" ultrasonographic appearance of a GCT is typified by a multicystic or honeycomb-like architecture, they can be more uniformly echo-dense, or comprised of one or more large fluid-filled cavities like the mare in this report. Removal of the affected ovary is curative and is generally followed by the resumption of normal activity of the remaining ovary.

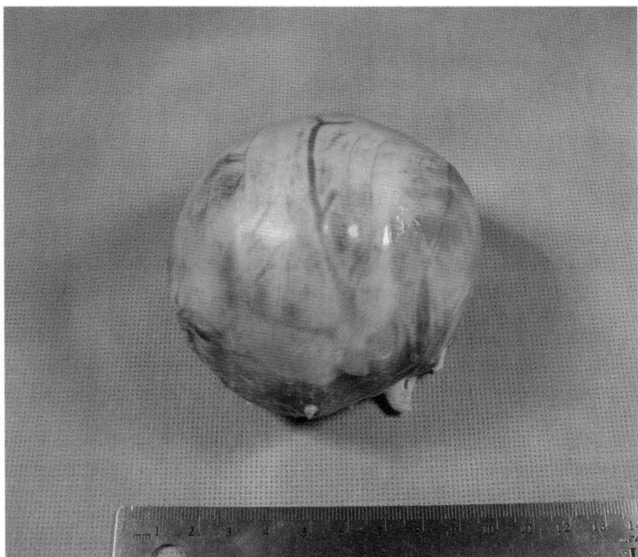

FIGURE 13.4-2 Photograph of the excised right ovary from a 3-year-old American Quarter Horse filly that was confirmed to be a GCT. Scale in mm is shown in the image.

1 week during which time nonsteroidal anti-inflammatory and prophylactic antibiotics were administered. On gross examination, the excised ovary was spherical with a diameter of 7.5 cm (3.0 in.) (Fig. 13.4-2). When bisected, the ovary contained a single central cyst filled with viscous yellow fluid. Histopathologically, the cyst wall consisted of vascularized fibrous tissue bordering both granulosa and theca cell components, confirming the diagnosis of a GCT.

Anatomical features in equids

Introduction

The major parts of the female reproductive system include the ovaries—i.e., female gonads; the tubular genitalia, comprised of the uterine tubes (oviducts), uterus, cervix, vagina, vestibule, and vulva; and the mammary gland (see Landscape Fig. 4.0-7 and Figs. 13.4-3 and 13.4-4). More than half of the cranial portion of the reproductive tract lies within the abdominal cavity, while the caudal components are located in the pelvic cavity. The abdominal components of the reproductive tract are suspended from the abdominal wall by 2 large ligamentous sheets of tissue called the broad ligaments, which arise from the sublumbar region. Each broad ligament is subdivided into 3 continuous sections: (1) the mesometrium that attaches to the uterus, (2) the mesovarium that attaches to the ovary, and (3) the mesosalpinx that attaches to the uterine tube (oviduct) via a projection from the lateral surface of the mesovarium. The components of the tubular genitalia and mammary gland are discussed in other clinical cases (Cases 13.3/13.5 and 16.2, respectively).

Function

After puberty, generally occurring at 12–18 months of age, the ovaries are responsible for: (1) cyclical maturation and release (at ovulation) of a mature oocyte and (2) synthesis and secretion of reproductive hormones, e.g., estrogen and progesterone, that—in addition to other reproductive hormones of the hypothalamic-pituitary-gonadal/ovarian axis, and hormones from other tissues/organs, such as prostaglandin F2α [PGF2α] from the endometrium—orchestrate the entire range of postpubertal female reproductive phenomena (behavioral and physiological).

Ovaries

At birth, equine **ovaries** are oval in shape with a centrally located medullary zone and a peripheral cortical zone that is completely encapsulated by a layer of tissue called the **germinal epithelium**. The **medullary zone** is comprised

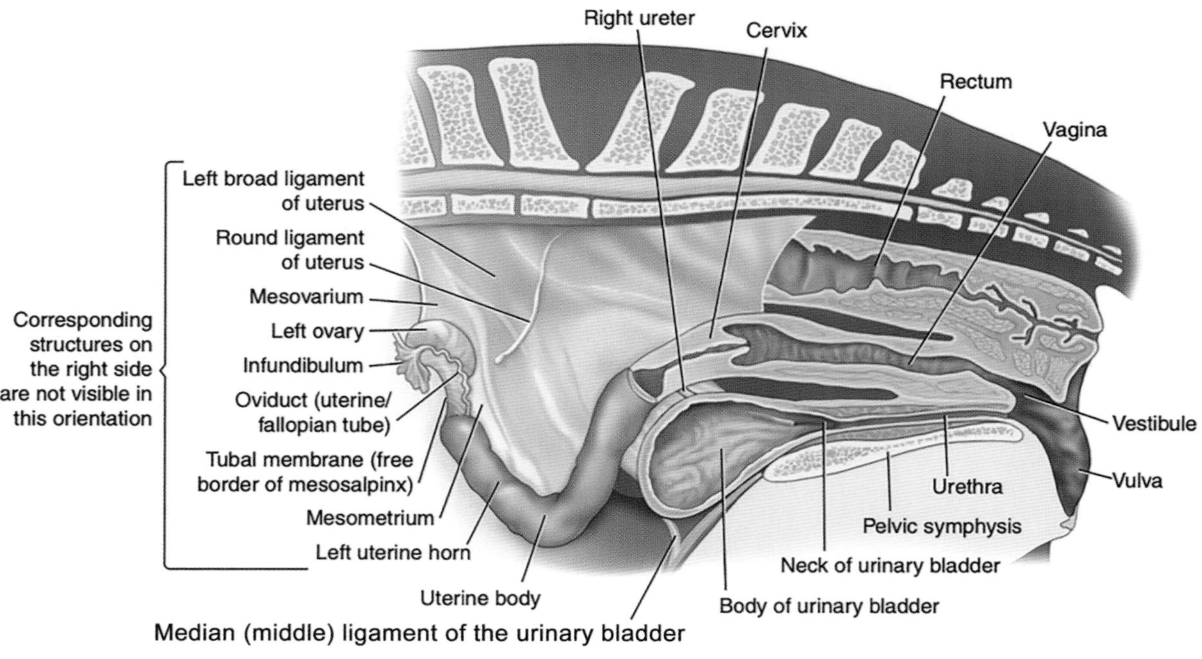

FIGURE 13.4-3 Lateral view of the female equine reproductive tract and adjacent structures (mammary gland not shown).

predominantly of connective tissue, blood vessels, lymphatic vessels, and nerves, while the **cortical zone** contains the female gametes—i.e., oocytes—and supporting cellular components. However, at as early as 5–7 months of age, the organization of the ovary undergoes a dramatic structural transformation that involves invagination of the peripheral ovarian cortex into the centrally located medulla, essentially transposing the location of the 2 zones. One result of this transformation is an adult ovary that appears kidney-bean shaped due to the presence of a notch-like depression called the **ovulation fossa**, which is the only portion of the adult ovary that remains covered by germinal epithelium (Fig. 13.4-5A and B). The remainder (majority) of the ovary, with the medulla now superficial to the cortex, is covered by visceral peritoneum—i.e., **mesothelium**.

This "inside-out" anatomical relationship of the ovarian cortex and medulla is unique to the horse, and it has important clinical implications. For example, the event of ovulation—i.e., follicular rupture and evacuation—only occurs at the ovulation fossa, whereas in other species rupture of a preovulatory follicle can occur at any location on the exterior surface of the ovary (dictated by the specific location of the preovulatory follicle).

The location of the ovaries in the mare makes them readily accessible for evaluation via transrectal palpation and ultrasonography, such that these complementary diagnostic modalities are important techniques used by veterinarians for the routine reproductive management of mares, as well as the diagnosis of ovarian abnormalities such as the GCT in this case report.

In adult mares, the ovaries are located dorsally in the abdomen, cranioventral to the iliac wings of the pelvis in the same plane as the 3rd to 5th lumbar vertebrae (the left ovary is generally located more caudal than the right ovary).

For descriptive purposes, the equine ovary has been assigned 2 surfaces (medial and lateral), 2 borders (dorsal/convex and ventral/concave), and 2 poles (cranial and caudal (Fig. 13.4-6A and B). Additionally, the **dorsal/convex border** is referred to as the "attached border" because of its direct connection to the mesovarium of the broad ligament. The **ventral/concave border** (containing the ovulation fossa) is referred to as the "free border" because it does not have a similar attachment to

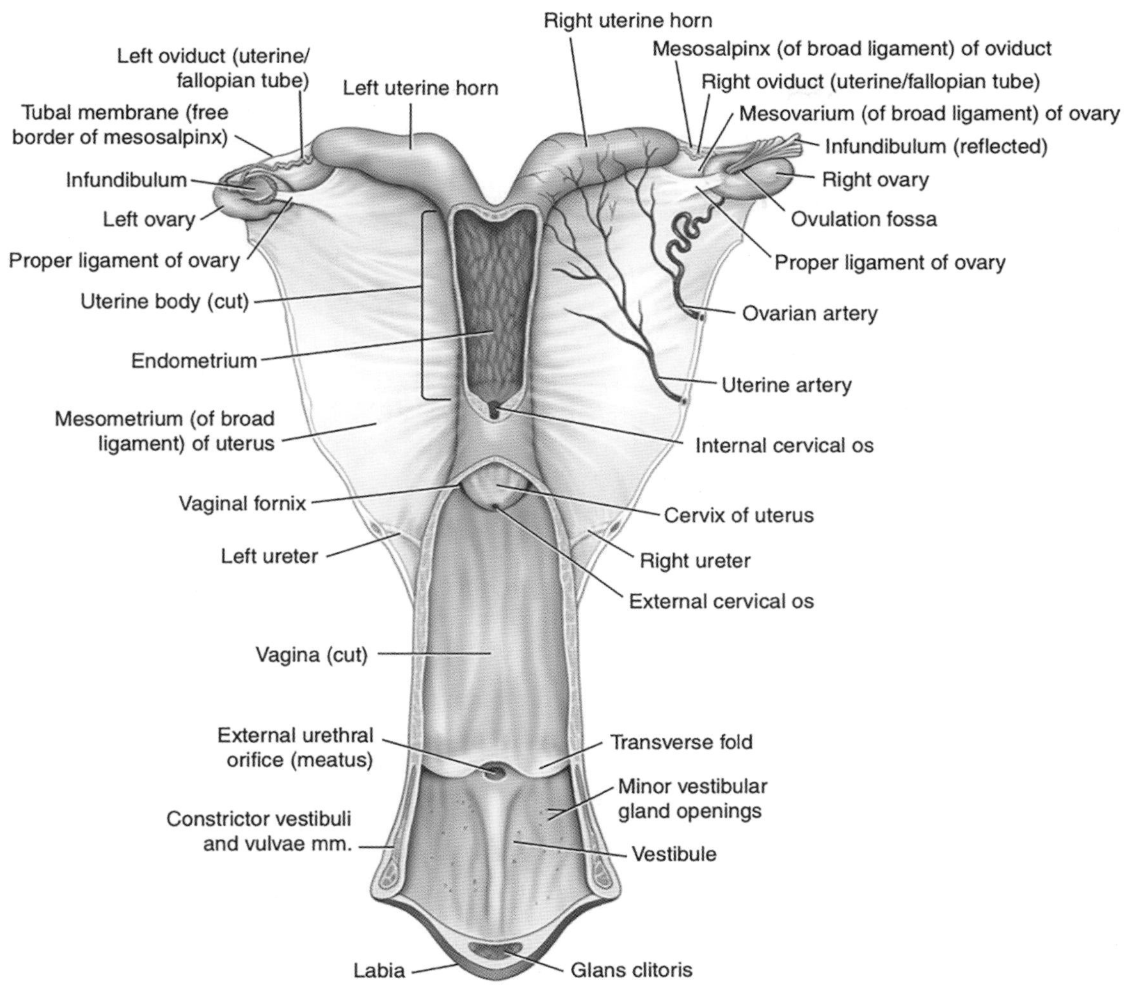

FIGURE 13.4-4 Dorsal view of the female equine reproductive tract.

adjacent structures. In addition, the **cranial pole** is referred to as the "tubal pole," and the **caudal pole** is referred to as the "uterine pole," due to their close proximity to those structures, respectively. The ovaries of an adult mare are generally 4–8 cm (1.6–3.1 in.) in length, 3–6 cm (1.2–2.4 in.) in width, and 3–5 cm (1.2–2.0 in.) in height.

The overall size of the ovaries changes dramatically in concert with both seasonal and cyclical changes in reproductive/ovarian activity that is discussed in the side box entitled "Physiological basis for seasonal and cyclical changes in ovarian activity."

Ovarian supporting structures

The ovaries are suspended from the abdominal wall by a portion of the broad ligament called the **mesovarium**. In addition to its role in tethering the ovary to the abdominal wall, the **mesovarium** provides the tissue framework through which blood vessels—i.e., **ovarian artery** and **vein**—lymphatic vessels, and nerves reach the ovary. Importantly, the vessels and nerves reach the ovary along the dorsal/convex border where they spread over the medial and lateral surfaces. Therefore, it is the convex surface of the equine ovary that serves as the functional "hilus," while in other organs—e.g., the kidneys—the hilus is located in a depression on the concave surface. Another aspect of this unique (to the mare) anatomical relationship between the mesovarium and ovary is that

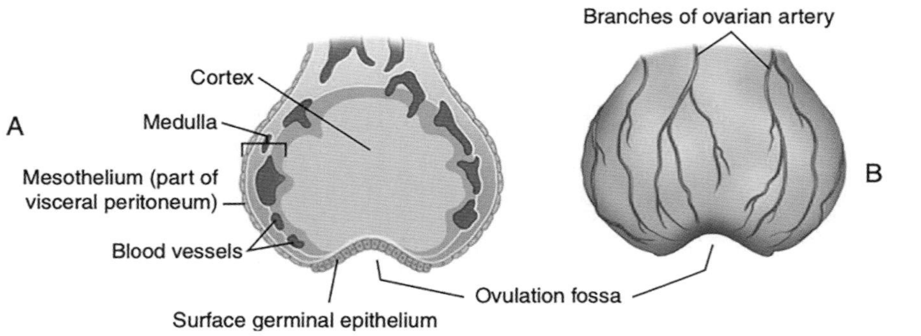

FIGURE 13.4-5 (A) Unique "inside-out" orientation of the equine ovarian medulla and cortex. See text for a complete description. (B) The relationship between the mesovarium/mesothelium and ovary. The attachment of the mesovarium/mesothelium to the ovary extends a considerable distance over the medial and lateral surfaces. This results in comparatively large, and grossly visible, blood vessels tracking in a relatively parallel fashion from the dorsal/convex to ventral/concave borders, giving the ovaries a distinctive striped appearance.

the attachment to the ovary extends a considerable distance over the medial and lateral surfaces. This results in comparatively large, and grossly visible, blood vessels tracking in a relatively parallel fashion from the dorsal/convex to ventral/concave borders, giving the ovaries a distinctive striped appearance (Fig. 13.4-5A and B).

The ovary and several adjacent structures contribute to the formation of a pouch-like structure called the **ovarian bursa** that opens ventrally into the abdominal cavity (Fig. 13.4-6A and B). The entirety of the ovarian bursa is

MESONEPHRIC REMNANTS

Shortly after fertilization, the mesonephros and the mesonephric duct begin to differentiate. The mesonephric duct system begins to degenerate by 2–3 months of gestation. Degeneration is never complete, however, and several structures may persist in the normal mare and include: (1) hydatid of Morgagni (likely of paramesonephric or Müllerian origin), (2) vesicular appendage, (3) epoophoron, (4) paroophoron, and (5) Gartner's duct or canal (ductus epoophori longitudinalis).

Vestiges of the mesonephric duct system are usually present to varying degrees in the broad ligament. The regressing mesonephric or Wolffian duct system begins near the tubal ostium and runs parallel to the tube in the broad ligament, entering the wall of the uterus near the level of the internal os. Vestiges are variable with respect to place and size. Occasionally, a small structure called the "appendix vesiculosa" (sessile hydatid) may be found below the tubal ostium at the edge of the broad ligament at the distal end of the mesonephric duct and is believed to be its blind end.

In the lateral third of the mesovarium lies the epoophoron, consisting of 8–13 tubules running from the mesonephric duct toward the ovary. They are of little clinical significance, although benign cysts are believed to occasionally arise in them.

Farther caudad along the regressing mesonephric duct are a small group of mesonephric tubules called the paroophoron. They are usually found only in neonates and are clinically unimportant.

Following along the course of the vestiges of the mesonephric duct can be found remnants of the duct, called Gartner's duct (ductus epoophori longitudinalis). Coiled tubes often occur in the lower part of the supravaginal cervical wall, where they are called the ampulla. In some cases, cystic remnants may be found in the vaginal walls, where their presence may cause diagnostic confusion. In rare cases, Gartner's duct may form an ectopic ureter because of having kept its ureteral connection.

Although believed to be paramesonephric rather than mesonephric in origin, clear pedunculated hydatid or cystic structures arising at the ostium at the end of the tube are commonly found. These are called the hydatids of Morgagni or appendix vesiculosa and are generally harmless.

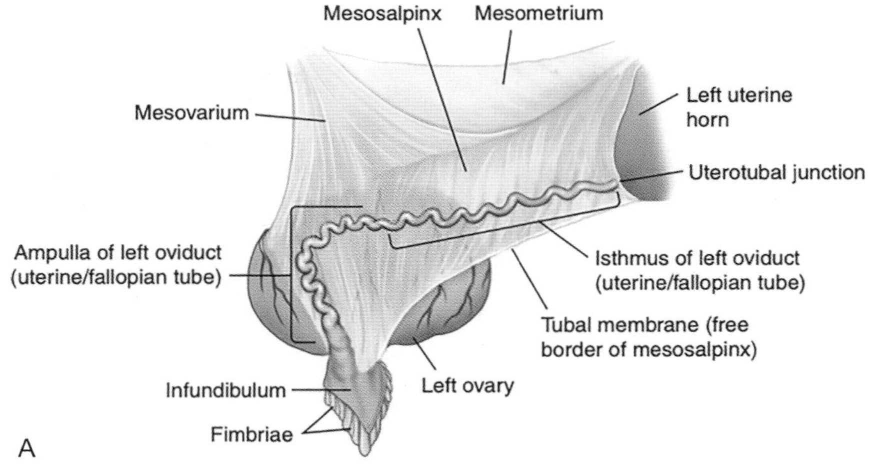

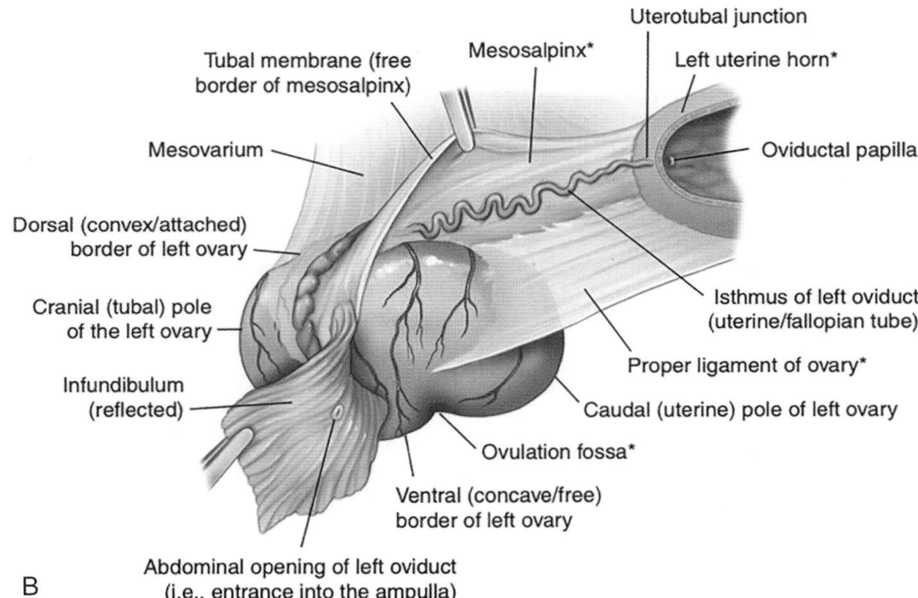

FIGURE 13.4-6 Anatomical interrelationships among the ovary, mesovarium, mesosalpinx, mesometrium, uterine tube (oviduct), and uterine horn. Structures marked with an *asterisk* contribute to the formation of the ovarian bursa. See text for complete description. (A) Appearance in situ. (B) The infundibulum and tubal membrane have been reflected to show the abdominal opening of the left uterine tube (oviduct).

formed/demarcated by the ovulation fossa (cranially), mesosalpinx (laterally), proper (or round) ligament of the ovary (medially), and tip of the uterine horn (caudally). The **proper (or round) ligament** of the ovary is a strong band of fibromuscular tissue that connects the tip of the uterine horn to the caudal pole of the ovary. 

No specific physiological function or role has been established for the ovarian bursa in the mare.

Blood supply, lymphatics, and innervation of the ovaries

Arterial blood supply to the ovary (from the aorta) is provided by the **ovarian branch** of the **ovarian artery** (a **uterine branch** of the ovarian artery serves the tip of the uterine horn). Similarly, venous blood leaves the ovary through several small veins that coalesce to form the **ovarian branch** of the **ovarian vein** (that is joined by a uterine branch of the ovarian vein that drains the tip of the uterine horn), which connects to the caudal vena cava.

Lymphatic drainage from the ovary occurs via lymphatic vessels that pass through the mesovarium to the **lumbar aortic lymph nodes**.

The ovary is innervated by sympathetic and parasympathetic neurons of the autonomic nervous system that are closely associated with the ovarian blood vessels.

PHYSIOLOGICAL BASIS FOR SEASONAL AND CYCLICAL CHANGES IN OVARIAN ACTIVITY

The mare is defined as a long-day, seasonally polyestrous breeder. As a result of changing day length over the course of the calendar year, a nonpregnant mare generally falls into 1 of 4 reproductive states: (1) a noncycling, anovulatory state during the short days of winter, (2) an ovulatory state—i.e., "cycling"—during the long days of the late spring and summer, (3) a spring "transitional" phase between the winter anovulatory and summer ovulatory states, or (4) a fall "transitional" phase between the summer ovulatory and winter anovulatory states.

During the shortest days of the year (December and January in the Northern Hemisphere), 85–90% of nonpregnant mares are in the winter anovulatory phase associated with quiescence of the hypothalamic-anterior pituitary-ovarian axis. Clinically, these mares have small, firm ovaries with follicles <20 mm (<0.76 in.). The winter anovulatory phase is followed by a spring transitional phase that is characterized by an increasing plane of ovarian follicular activity—development of multiple follicles >30 mm (>1.18 in.), which ultimately leads to the first ovulation of the year. Following the first ovulation of the year, nonpregnant mares should have repeating estrous cycles, each encompassing one interovulatory interval of approximately 21–22 days. Each estrous cycle is comprised of a period of estrus, i.e., "heat," lasting approximately 1 week and a period of diestrus lasting approximately 2 weeks during which time the mare is out of "heat." Estrus is characterized by growth of a dominant follicle at a rate of 3–5 mm (0.12–0.20 in.)/day until it reaches preovulatory size (generally >40 mm [1.6 in.]) and ovulates. The event of ovulation is followed by luteinization of the follicular cells resulting in formation of a CL that secretes progesterone. The CL and its attendant production of progesterone keep the mare in diestrus for approximately 14–16 days. In nonpregnant mares, CL function ceases after approximately 2 weeks as a result of the luteolytic effect of PGF2α secretion from the endometrium, which brings the mare back into estrus of the subsequent estrous cycle. When day length decreases during the late summer and early fall, the hypothalamic-pituitary-ovarian axis again becomes quiescent causing most mares to enter the winter anovulatory phase.

Selected references

[1] Ball BA, Almeida J, Conley AJ. Determination of serum anti-Müllerian hormone concentrations for the diagnosis of granulosa-cell tumours in mares. Equine Vet J 2013;45:199–203.
[2] Bergfelt DR. Anatomy and physiology of the mare. In: Samper JC, editor. Equine breeding management and artificial insemination. 2nd ed. St. Louis, MO: Saunders Elsevier; 2009. p. 113–31 [chapter 11].
[3] Ginther OJ. Reproductive anatomy. In: Reproductive biology of the mare: basic and applied aspects. 2nd ed. Cross Plains, WI: Equiservices; 1992. p. 1–40 [chapter 1].
[4] McCue PM, Roser JF, Munro CJ, et al. Granulosa cell tumors of the equine ovary. Vet Clin North Am Equine Pract 2006;22:799–817.
[5] Nickel R, Schummer A, Seiferle E. The viscera of domestic animals. 2nd revised ed. New York, NY: Springer-Verlag; 1979. p. 385–9.

CHAPTER 13

CASE 13.5

Endometrial Cysts

Carlos Pinto[a] and Luis Henrique de Aguiar[b]
[a]Theriogenology, Veterinary Clinical Sciences, Louisiana State University School of Veterinary Medicine, Baton Rouge, Louisiana, US
[b]Department of Veterinary Clinical Sciences, Louisiana State University School of Veterinary Medicine, Baton Rouge, Louisiana, US

Clinical case

History

A pregnant 19-year-old Thoroughbred mare was purchased as a broodmare at a new farm. After foaling in early February, the mare was bred twice by natural cover and ovulated during the 1st week in March. At several reproductive examinations beginning 14 days postovulation and biweekly examinations thereafter, the farm's resident veterinarian remained uncertain whether the mare was or was not pregnant due to the presence of several uterine cysts. The mare had not shown any signs of behavioral estrus during the month, she was being examined ultrasonographically, and the corpus luteum (CL) appeared to continue to be functional; the mare also did not show behavioral estrus during the month she was being examined ultrasonographically.

The 2 main types of breeding in horses include natural cover and artificial insemination. In natural cover, the stallion physically mates with the mare. With artificial insemination, a semen sample is manually deposited in the uterus of a mare via an insemination pipette inserted through the cervix per vaginum.

FUNCTIONAL CORPUS LUTEUM

In a normal estrous cycle, a corpus luteum (CL) is formed after ovulation occurs. It secretes progesterone for approximately 14–15 days after ovulation. In the absence of pregnancy, the CL regresses, allowing the mare to return to estrus. A CL may appear functional on transrectal ultrasonography based on its size (3–4 cm or 1.2–1.6 in. on average) and degree of vascularization as assessed by color Doppler ultrasonography. Blood concentrations of progesterone > 1–2 ng/mL indicate an active (functional) CL.

EARLY PREGNANCY DIAGNOSIS

Pregnancy diagnosis in horses is typically done using transrectal ultrasonography. On day 12 of pregnancy, the spherical embryonic vesicle measuring 10–12 mm (0.4–0.5 in.) is a hypoechoic (black) round structure. At approximately day 21, a hyperechoic (white) embryo proper is imaged within the embryonic vesicle, commonly at the bottom of the vesicle (between 4 and 8 o'clock positions). At day 32, the embryo proper is seen in the equatorial region of the vesicle.

Comparative Veterinary Anatomy: A Clinical Approach. https://doi.org/10.1016/B978-0-323-91015-6.00064-9

Physical examination findings

The mare was referred to a theriogenology specialty clinic in April. Transrectal palpation and ultrasonography were performed to assess the status of the mare's reproductive tract. The uterus and cervix had a typical diestrus tone (in diestrus, under the influence of progesterone, the cervix should have a firm tone and feel tubular when palpated per rectum). Color Doppler ultrasonography revealed that the left ovary contained a well-developed, vascularized CL; the right ovary had 2 follicles each measuring 35 mm (1.4 in.) in diameter. No uterine edema or intrauterine fluid was noted during the ultrasound examination. Several cystic structures, single or clustered, measuring approximately 5–35 mm (0.2–1.4 in.) in diameter were imaged throughout the uterine body with a few isolated cystic structures also present at the base of both uterine horns. A major cluster of cystic structures had more significantly irregular margins and appeared to occlude the uterine lumen. A typical embryonic vesicle compatible with a presumptive, approximately 32-day, pregnancy was not seen. Therefore, the mare was diagnosed as not being pregnant.

Differential diagnoses

Uterine cysts, uterine neoplasia

Diagnostics

Sonohysterography (the infusion of a sterile isotonic fluid into the uterus to enhance detailed imaging of the uterine lumen and its lining) was performed to more accurately map the uterine cysts and to rule out the presence of a cystic neoplastic mass. The cervix was catheterized with a 34-Fr silicone catheter and 2 L of lactated Ringer's solution was infused into the uterus, followed by immediate ultrasound examination of the uterus per rectum. Most of the cystic structures presented a narrow pedicle and were pedunculated, floating in the fluid-filled uterus (Fig. 13.5-1). One large cluster of cysts in the uterine body appeared to be sessile.

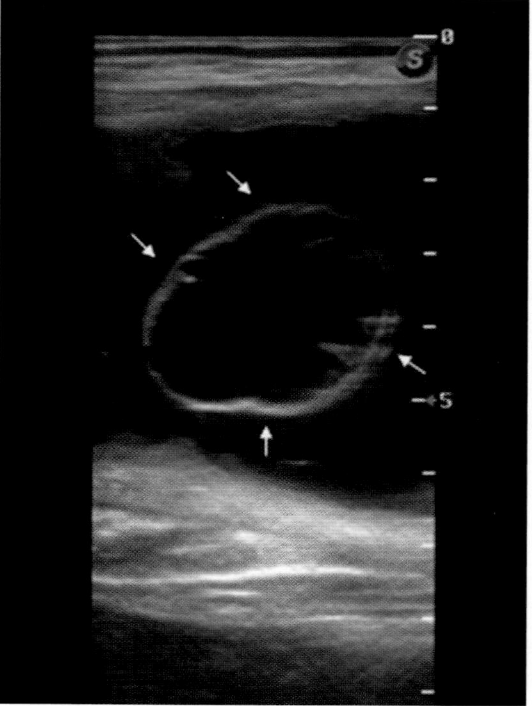

FIGURE 13.5-1 Transrectal ultrasonographic image of an endometrial cyst after intrauterine infusion of 2 L of lactated Ringer's solution. Note how the cyst *(white arrows)* appears floating within the hypoechoic *(dark)* intraluminal fluid. Tick marks = 1 cm (0.4 in.).

Diagnosis

Uterine cysts—a.k.a. endometrial cysts

Treatment

Cystic ablation (L. *ablatus* removed) via hysteroscopy was chosen to reduce the number of uterine cysts that could potentially interfere with the establishment of pregnancy. The mare was sedated and the cervix was manually dilated. Lidocaine was topically instilled on the endometrial surface. Large, pedunculated, narrow-based cysts were physically excised using nonabsorbable suture placed in a snare loop inserted into 2 plastic 53.3-cm (21-in). long infusion pipettes taped together (Fig. 13.5-2). After removal of the large cysts, laser ablation of the smaller cysts was performed, as well as cauterization of the beds of the larger cysts, thereby destroying the cystic lining in the endometrial surface. Cystic fluid and blood were aspirated from the uterus via the endoscope as needed to improve visualization.

Transrectal ultrasonography was performed immediately after the hysteroscopy. Mild pneumouterus (Gr. *pneuma* air + Gr. *hystera* uterus = air in the uterus) and intraluminal fluid were noted, as well as few small cysts (< 10 mm or < 0.4 in.) remaining in the mid-right uterine horn. Post-operative management consisted of a single treatment of a luteolytic dose of conventional PGF2α (2.5 mg IM) and flunixin meglumine (1.1 mg/kg IV), and a course of sulfamethoxazole and trimethoprim (30 mg/kg, PO q12h for 7 days). The mare was discharged to return to the farm and conceived on the 2nd estrus after ablation of the uterine cysts.

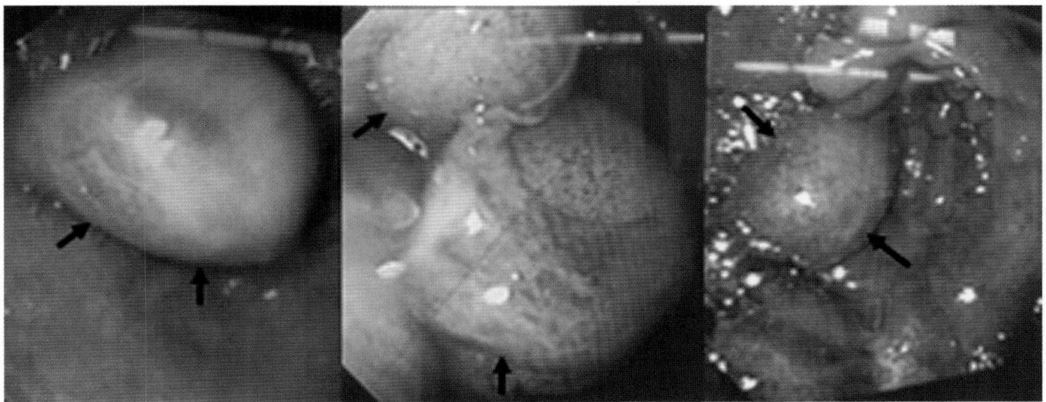

FIGURE 13.5-2 Large endometrial cysts as seen during hysteroscopy *(arrows)*. These large cysts were manually excised using an improvised snare.

UTERINE CYSTS

Uterine cysts can be glandular or lymphatic in origin. It appears that degenerative changes in the uterus leading to glandular fibrosis may predispose the uterus to develop cystic glandular dilations; these lesions are typically microscopic and not commonly detected by reproductive ultrasonography. In contrast, lymphatic lacunae may lead to relatively large accumulations of fluid, forming sacculations that protrude from the endometrial surface. These can measured from as small as a few millimeters to as large as several centimeters and are pedunculated or sessile (L. *sessilis* not pedunculated, attached by a base). Most uterine cysts detected by ultrasonography that protrude into the uterine lumen are lymphatic.

Uterine cysts can be found in the uterine horns or in the uterine body. Although it has been shown that mares with uterine cysts may become pregnant and carry a foal to term, establishment of pregnancy may be prevented in mares with multiple large (> 3 cm or > 1.2 in.) cystic areas (single cyst or cluster of multiloculated cysts). It is suspected that pregnancies that develop in close association with large cystic areas are at a higher risk of early embryonic loss.

Anatomical features in equids

Introduction

This section discusses the uterus, which is a tubular organ that communicates cranially with the uterine tubes and caudally with the vagina via the cervix (neck) of the uterus (cervix uteri). Embryonic and fetal development (except for the first 5–6 days) occurs in the uterine lumen, ending with parturition (delivery, birth of the fetus). Tissues and membranes of the developing embryo increasingly communicate with the maternal endometrium (lining of the uterus) to form a chorioallantoic placenta, an epitheliochorial type of diffuse placentation. The remainder of the female reproductive tract is covered in Cases 13.3, 13.4, and 13.6.

Function

The main functions of the uterus are to receive semen at the time of breeding (natural or artificial) and then—if fertilization occurs—to receive the early embryo and serve as the environment that provides nutrition for the development of an embryo into a fetus. The conceptus stays in the uterus until term, enclosed by extra-embryonic membranes that connect the fetus to the uterus by forming a functional placenta. When at term, the equine fetus is mature and ready to survive following parturition.

The cervix functions as the most cranial physical barrier in the reproductive tract, protecting the body of the uterus from the external environment. It serves to facilitate coitus, transport sperm, and clear the uterus during estrus; it closes during diestrus and pregnancy to isolate the future conceptus from potential contamination or other disturbances occurring in the caudal reproductive tract.

Uterus

The **uterus** is located in the sublumbar region of the abdomen, near the pelvic inlet, where the caudal portion of the body of the uterus and cervix are situated (see Case 13.4). The uterine **body** has a cylindrical shape, flattened in a dorsal-ventral direction. Cranially, the uterus bifurcates into left and right uterine **horns** (cornua) that extend cranially and slightly dorsally toward their respective ovaries. Dorsally, the uterus lies near the rectum and ventrally sits just above the urinary bladder and pelvic flexure of the colon. Cranially, the uterine horns are in contact with intestine. The ventral portions of both uterine horns are free while the dorsal regions of the uterine horns are suspended by ligaments derived from the paired left and right **broad ligaments** of the uterus (see Case 13.3).

The mare's uterus has 5–10 longitudinal **endometrial folds** that extend from the uterine body through the whole of each uterine horn. The core of the endometrial folds is composed of connective tissue devoid of secretory glands, but outlined by the glandular lamina propria of the luminal epithelium.

The endometrial folds become edematous during estrus and assume a characteristic echotexture (see Section 2.3) that aid in staging the estrous cycle.

PLACENTA TYPES

There are 4 types of placentation in pregnancy: (1) diffuse placenta: in horses and pigs, the chorionic villi are uniformly distributed over the surface of the chorion (Gr. membrane); (2) cotyledon placenta: in ruminants, chorionic villi are in multiple groups—cotyledons—punctuated by smooth chorion between cotyledons; (3) zonary placenta: in carnivores, villi are in bands around the center of the chorionic sac; and (4) discoid placenta: in primates and rodents, villi are restricted to a disc-shaped area on the chorion.

Deciduate (L. *deciduus* falling off, shedding) placenta (carnivores) and nondeciduate placenta (horse, ruminants, and pig) are differentiated based on whether there is a loss of endometrium at birth (deciduate) vs. minimal maternal endometrial tissue loss (nondeciduate) because of the simple appositional arrangement.

The uterus is composed of 3 distinct layers of tissue: perimetrium (tunica serosa), myometrium (tunica muscularis), and endometrium (tunica mucosa). The **perimetrium** is mainly composed of connective tissue and vessels (small arteries and veins; lymphatics). The **myometrium** has a composition like that of intestines—i.e., the fibers of the smooth muscle are arranged in an inner circular layer and an outer longitudinal layer. The **endometrium** consists of 2 epithelial types—luminal epithelium and glandular epithelium. The luminal epithelium changes according to the physiologic status of the mare, varying from low cuboidal to tall columnar. For example, it can be low cuboidal under the influence of progesterone during diestrus or pregnancy, and tall columnar—and even pseudostratified—during estrus.

Below the luminal epithelium lies the **stratum compactum** followed by the **stratum spongiosum**. The **stratum basalis** separates the deep layers of the endometrium from the myometrium. The stratum compactum is a densely organized area composed of stromal cells and few capillaries, whereas the stratum spongiosum has a relatively low cellular density. Blood vessels and immune cells are found throughout these 2 stratified zones. The endometrial glands have ducts that communicate with the coiled, tubular endometrial glands, which are scattered throughout the lamina propria (stratum compactum and stratum spongiosum), depositing their secretions into the uterine lumen.

Glandular secretions ("uterine milk"), also known as histotroph, provide early nutrition to the embryo as soon as it arrives to the body of the uterus. Horses are known for having a protracted preimplantation period, and thus histotroph is important in this species for early support of the embryo.

Broad ligaments of the uterus (also see Case 13.3)

The **broad ligaments of the uterus** are membranous bands derived from the peritoneum and, owing to their attachment to the uterus, are responsible for keeping the uterus suspended in the abdomen, surrounded by the intestines. The **mesometrium, mesosalpinx**, and **mesovarium** are the main parts of the broad ligament and they support the abdominal position of uterus, uterine tubes, and ovaries, respectively. Derivatives of mesometrium, mesosalpinx, and mesovarium are the **round ligament of the uterus**, **proper ligament of the uterus**, and the **suspensory ligament of the ovary**, respectively. The cervix does not have any membrane derived from the peritoneum.

Cervix

The **cervix uteri (cervix)** allows communication between the body of the uterus and the vagina. The cervix can be divided into 2 major portions in relation to their anatomical position. First, the **portio prevaginalis** refers to the cervical portion that is cranial to the **fornix vaginae** and contains the **ostium uteri internum** (internal cervical ostium) that opens into the caudal part of the body of the uterus. Second, the **portio vaginalis** is the part of the cervix that is caudal to the vaginal fornix and contains the **ostium uteri externum** (external cervical ostium, see Case 13.4). The portio vaginalis protrudes into the cranial vaginal lumen and may or may not touch the vaginal

ENDOMETRIAL EDEMA

In a reproductively, healthy, cycling mare, the appearance of endometrial edema typically follows declining or low levels of blood progesterone, indicating that the mare is "coming into estrus" or is in estrus. This feature of the mare uterine anatomy and physiology enables clinicians to use sequential ultrasonography examinations to follow mares during their estrous cycles and to accurately identify mares in estrus without the aid of a teasing stallion. When endometrial edema is present, the cross-section of a uterine horn resembles a "wagon wheel" or a cut "grapefruit" because the edematous folds are hypoechoic, alternating with more echoic areas of the endometrium between the folds (Fig. 13.5-3A). Under progesterone influence, the endometrial folds are not readily apparent; cross-sections of uterine horns become smaller in diameter and rounder than when in estrus (Fig. 13.5-3B).

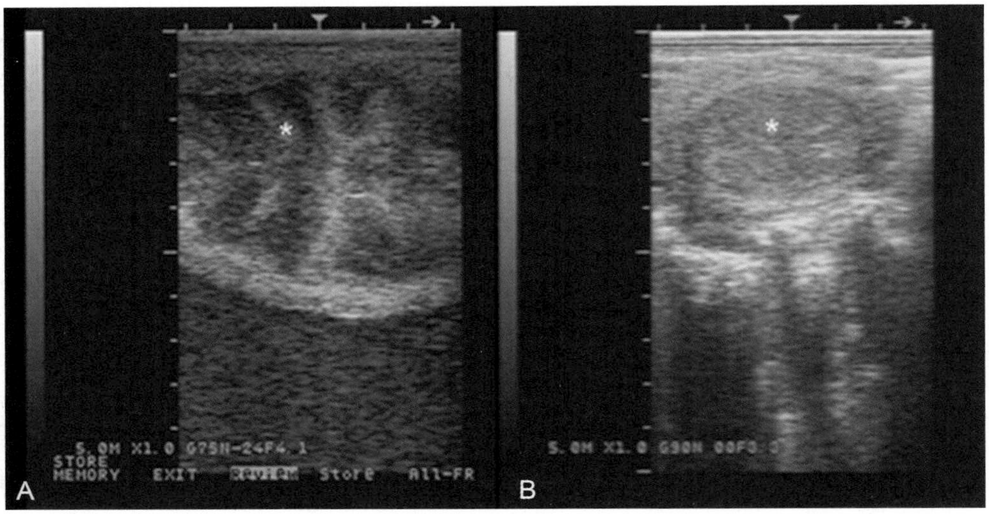

FIGURE 13.5-3 Transrectal ultrasonographic images of cross-sections of the mid-portions of uterine horns of mare when in estrus (A, *asterisk* denoting edematous folds) and diestrus (B, *asterisk* denoting nonedematous wall). Note the marked decrease in uterine diameter during diestrus. Tick marks = 1 cm (0.4 in.).

floor, depending on the stage of the estrous cycle. Under progesterone influence, the cervix portio vaginalis appears "high and tight," suspended away from the vaginal floor by the **dorsal cervical frenulum**. The smooth muscle, like the arrangement seen in the uterus and uterine tubes, is composed of both an inner circular and an outer longitudinal layer that are continuous with the smooth muscle body of the vagina and uterus. The cervix also has longitudinal mucosal folds that are continuous with the longitudinal endometrial folds. The cervical mucosa is composed of simple columnar epithelium, with ciliated and nonciliated epithelial cells arranged in an alternating pattern. Mucin-producing goblet-like cells also follow an alternate pattern between the epithelial cells.

Blood supply, lymphatics, and innervation

In the horse, 3 main arteries supply blood to the reproductive tract: the ovarian, uterine, and vaginal arteries. The **ovarian artery** branches off directly from the caudal abdominal aorta and then further branches: the 1st branch (the **ovarian artery proper**) tracks through the mesovarium and further successive branches course over the ovarian surface; the 2nd branch supplies blood to the uterine tubes and cranial portion of the uterine horns. The **external iliac artery** branches to the main **uterine artery**; a small uterine artery also arises from the **vaginal branch** of the **internal pudendal artery**. The uterine artery further extends cranially and caudally. The **vaginal artery**, also known as the **caudal uterine artery**, originates from the internal pudendal artery and supplies blood to the caudal portion of the uterine body, cervix, and vagina.

The names of the veins are homonyms of the arteries; veins are invariably adjacent to their corresponding arteries. However, the arteries supplying the ovaries are not in close juxtaposition to the uterine veins in mares.

This arrangement explains the difference in how the uterine PGF2 alpha elicits luteolysis in horses vs. ruminants. Ruminants have marked utero-ovarian arterial-venous contact, forming a countercurrent exchange mechanism that allows luteolytic amounts of PGF2 alpha from the uterus to directly reach the CL. In contrast, PGF2 alpha in mares only travels to the CL via the systemic circulation to cause luteolysis.

The lymph vessels are numerous in the mare reproductive tract, contributing to substantial drainage from the uterus, passing to the **internal iliac** and **lumbar aortic lymph nodes**.

Parasympathetic innervation of the uterus comes from pelvic nerves in the sacral area. Specifically, the **pudendal nerve** and sympathetic innervation come primarily from the **hypogastric nerves** and **pelvic plexus**, running alongside the arteries and their branches. In the caudal portion of the reproductive tract—especially the vulva and perineum—there are many sensory nerve fibers. The outer myometrial smooth muscle layers are rich in nerve bundles, in contrast to the fine nerve fibers found in the inner myometrial smooth muscle layers, the intramyometrial vascular zone, and the endometrium. Overall, uterine innervation is denser in the myometrial region than in the endometrium. The cervix has a more distinct innervation (bundles and fine nerves fibers) than the uterine horns and body.

This explains why a surgical incision in the vulvar tissues elicits more marked signs of pain than an incision in the uterine wall.

Selected references

[1] Bae SE, Corcoran BM, Watson ED. Organisation of uterine innervation in the mare: distribution of immunoreactivities for the general neuronal markers protein gene product 9.5 and PAN-N. Equine Vet J 2001;33(3):323–5.
[2] Ginther OJ, Pierson RA. Ultrasonic anatomy and pathology of the equine uterus. Theriogenology 1984;21(3):505–16.
[3] Katila T. The equine cervix. Pferdeheilkunde 2012;28(1):35–8.
[4] Kenney RM. Cyclic and pathologic changes of the mare endometrium as detected by biopsy, with a note on early embryonic death. J Am Vet Med Assoc 1978;172(3):241–62.
[5] Stanton MB, Steiner JV, Pugh DG. Endometrial cysts in the mare. J Equine Vet 2004;24(1):14–9.

CHAPTER 13

CASE 13.6

Oviduct/Uterine Tube Obstruction

Candace Lyman[a] **and Patricia Sertich**[b]
[a]Equine and Small Animal, Auburn University College of Veterinary Medicine, Auburn, Alabama, US
[b]Department of Clinical Studies - New Bolton Center, University of Pennsylvania School of Veterinary Medicine, Kennett Square, Pennsylvania, US

Clinical case

History

A 5-year-old, nulliparous (L. *nullus* none + L. *parere* to bring forth, produce), barren Standardbred mare was presented with a history of infertility extending over a period of 2 breeding seasons. Breeding attempts involved artificial insemination with semen from 3 different stallions that were known to be fertile. In the first year of breeding, the mare was inseminated on 8 estrous cycles, while the following year the mare was inseminated on 4 estrous cycles, and none of the breeding attempts resulted in a pregnancy. As a result, the mare was referred to a university teaching hospital for further evaluation. The mare's previous breeding records indicated interovulatory periods were regular and estrous cycles had been properly managed by veterinarians with respect to the breeding-to-ovulation interval. Therefore, the cause of infertility was not evident from review of the breeding records.

Physical examination findings

The mare was bright, alert, and responsive on presentation, and no abnormalities were found on physical examination. To evaluate her current reproductive behavior, the mare was exposed to a pen-housed stallion across a tease rail; the mare's behavior was considered indifferent to the stallion's advances, as it was neither strongly receptive nor rejective. The mare had excellent perineal conformation with well-apposed, vertical labia. No vulvar discharge was present. The mare's mammary gland was inactive and small, appropriate for a maiden mare.

Breeding soundness examination was performed. Speculum and digital vaginal examination revealed a competent vestibulovaginal sphincter ring. No vaginal exudate or fluid was present. The cervix was competent, and its serosal and muscular layers were intact. Palpation and ultrasonography per rectum following IV sedation revealed a 7-cm-long (3.0-in.), closed cervix. The uterus was well-toned with palpable endometrial folds. No uterine fluid was imaged, and the endometrial folds were not edematous on ultrasonography. The left ovary was 4.7 cm (1.9 in.) in width × 6.5 cm (2.6 in.) in length × 4.5 cm (1.8 in.) in height and contained a 46 × 22 mm (1.8 × 0.9 in.) "lacey"—i.e., centrally nonechogenic—corpus luteum (CL); and a 27 × 35 mm (1.1 × 1.4 in.) follicle with multiple small follicles and a palpable ovulation fossa. The right ovary was 4 cm (1.6 in.) in width × 5.3 cm (2.1 in.) in length × 4 cm (1.6 in.) in height, and it contained a 27 × 16 mm (1.1 × 0.6 in.) follicle with multiple small follicles and a palpable ovulation fossa. When palpating the left and right mesosalpinx (the portion of the broad ligament that supports the uterine tube [oviduct]), a firm, cord-like structure was identified on each side.

> The normal uterine tube typically has a discrete, malleable, and muscular characteristic that is not detectable by palpation per rectum.

Comparative Veterinary Anatomy: A Clinical Approach. https://doi.org/10.1016/B978-0-323-91015-6.00065-0

Differential diagnoses

Primary infertility as a result of one, or a combination, of the following conditions: endometritis, congenital uterine tubal abnormality, acquired uterine tubal pathology, or abnormalities of gonadal sex determination

Diagnostics

Aerobic culture and susceptibility of an endometrial swab yielded a light growth of *Actinomyces* spp. and *Bacillus* spp.; both organisms had a broad antibiotic susceptibility and were not interpreted as a significant cause of endometritis. Histological evaluation of an endometrial biopsy sample classified the sample as a Kenney-Doig Category IIA due to mild periglandular fibrosis and mild, diffuse, lymphocytic endometritis (see side box entitled "Grading of uterine biopsies"). The degree of uterine inflammation was mild, and therapy was not warranted. Hysteroscopy performed under standing IV sedation revealed no abnormalities; healthy endometrium was seen with normal luminal anatomical orientation. Karyotype analysis revealed 64XX (normal mare).

A karyotype analysis is a laboratory procedure in which an individual's full set of chromosomes is evaluated for aberrations (abnormalities in number, size, shape, etc.). This evaluation can be performed using a blood sample or virtually any tissue from the body.

Diagnosis

Presumptive diagnosis of bilateral uterine tube obstruction

Treatment

In order to visually assess the uterine tubes and administer treatment (if warranted based upon findings at the time of the examination), a standing bilateral abdominal laparoscopic surgery was performed using IV sedation and regional anesthesia. The left uterine tube was of normal character except for a perceived slight bulge near the junction of the ampulla and the isthmus (Fig. 13.6-1).

Both the ampulla and isthmus of the right uterine tube appeared grossly distended; the mesosalpinx and mesometrium also had a pronounced increase in vascularity (Fig. 13.6-2). Because of the altered appearance of both uterine tubes, laparoscopic administration of a prostaglandin E_2 (PGE_2) analogue, dinoprostone gel (Prepidil® Gel, Pfizer, NY, NY), was performed; 0.25 mg of PGE_2 was applied to the serosal surface of the ampulla and isthmus of both uterine tubes (Fig. 13.6-3).

This treatment was used because of PGE_2's ability to induce uterine tubal dilation and muscular contractions, properties that are useful in evacuating the uterine tube of inspissated (L. *inspissatus* from in intensive + *spissare* to thicken) debris.

GRADING OF UTERINE BIOPSIES

Efforts at predicting a mare's ability to get pregnant and carry the foal to term may be guided by obtaining an endometrial biopsy. In this process, a histological specimen is microscopically evaluated, and endometrial characteristics such as inflammation, fibrosis, and glandular degeneration are noted. Additionally, clinical patient information is combined with the microscopic findings so that the sample may be categorized into one of 4 categories: I, IIA, IIB, and III. While several grading schemes have been proposed, the 4-category system established by Drs. Kenney and Doig in 1986 is considered the international standard. Per this category system, a grade I sample is "normal" endometrium with only slight and sparsely scattered inflammation or fibrosis, giving a > 80% chance of conceiving and maintaining a pregnancy to term. This is in obvious contrast to a grade III sample where "severe, irreversible changes including fibrosis and inflammation" indicate a < 10% chance of conceiving and maintaining a pregnancy until term. It should not be overlooked that these percentage values are merely estimates and do not take into consideration important factors such as stallion fertility and appropriate breeding management or timing of breeding.

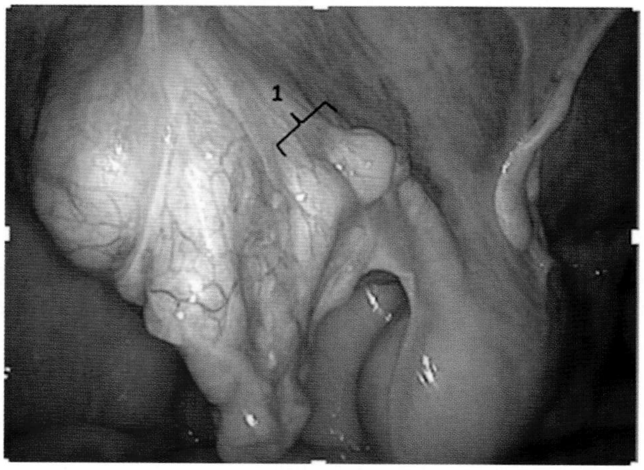

FIGURE 13.6-1 Left uterine tube of a mare with an apparent enlargement (1) near the junction of the ampulla and the isthmus—i.e., the ampullary-isthmic junction.

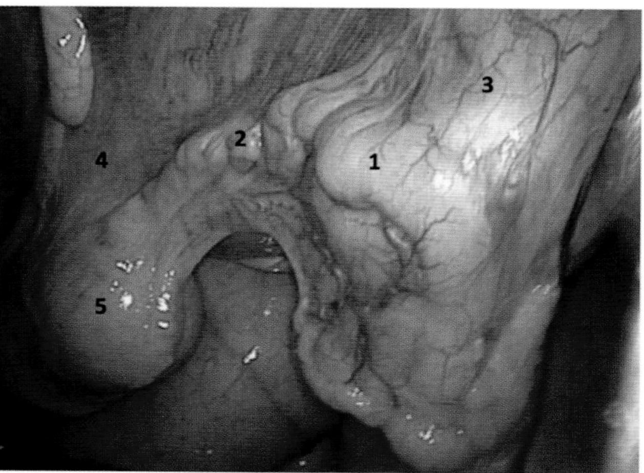

FIGURE 13.6-2 Right uterine tube with gross distention of both the ampulla (1) and isthmus (2); the mesosalpinx (3) and mesometrium (4) have a marked increase in vascularity. Also pictured is the tip of the right uterine horn (5).

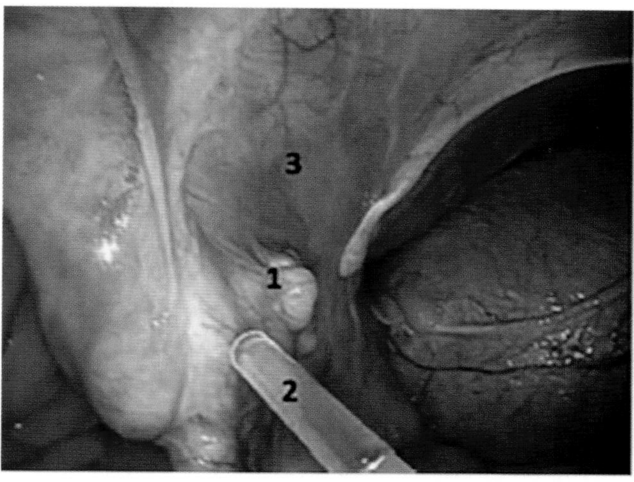

FIGURE 13.6-3 Capture image of the topical PGE_2 gel dispensed on the serosal surface of the uterine tube (1) via a pipette (2) during laparoscopy. Also pictured is the mesosalpinx (3).

Two days—i.e., day 2—after laparoscopic application of the PGE_2 gel, the mare spontaneously ovulated; on day 6, the mare's diestrus was shortened by administering 250 µg of cloprostenol sodium (Estrumate®, Merck Animal Health, Madison, NJ) IM. Soon after, the mare progressed into estrus and a dominant follicle (36 × 40 mm or 1.4 × 1.6 in.) was induced to ovulate with 2500 IU of IV hCG (Chorulon®, Merck Animal Health, Madison, NJ) on day 9 after the application of the PGE_2 gel was performed. Additionally, on day 9, the mare was artificially inseminated with 420 million progressively motile, morphologically normal sperm (420×10^6 PMMNS). On day 10, ovulation was confirmed, and no further treatment was required. The mare was subsequently confirmed pregnant with transrectal ultrasonography.

Anatomical features in equids

Introduction

Flanked by the ovary on one side and the distal end of the uterine horn on the other, the uterine tube (oviduct) is an integral component of the tubular genitalia that is essential for the interaction of female and male gametes, and necessary for the establishment of pregnancy. Each of the two uterine tubes is surrounded by a portion of the sheet-like, ligamentous structure referred to as the broad ligament; the broad ligament is continuously attached and stretches dorsally from the region of the 3rd to 4th lumbar vertebra to the level of the 4th sacral vertebra (Figs. 13.6-4 and 13.6-5). This case discusses the anatomy of the uterine tube and mesosalpinx; the remainder of the equine female reproductive tract is covered in Cases 13.3–13.5.

Function

The uterine tube is responsible for capturing and transporting the oocyte from the ovary and facilitating the transport of spermatozoa from the uterus to the site of fertilization. After fertilization in the ampulla and early embryonic development, the embryo secretes PGE_2, which initiates embryo transport through the isthmus into the uterus between the 5th and 6th day after ovulation. In contrast, unfertilized oocytes are retained within the uterine tubal lumen to degenerate and are nonviable. Not a passive organ, studies have demonstrated that the uterine tube produces PGE_2 and PGF_2 in vitro in response to endogenous or exogenous stimuli. Ovarian steroids, adrenergic nerves, nitric oxide, oxytocin, prostaglandins, and cytokines—e.g., tumor necrosis factor—alter muscular and ciliary activities in the uterine tube, which can ultimately impact individual gamete or embryo transport therein.

PROSTAGLANDINS: A STANDARD IN EQUINE REPRODUCTION

Prostaglandins (PG) are potent lipid compounds that stimulate multiple diverse physiological responses when released from cells within the body or administered exogenously. The effects of body tissues exposed to PG include (but are not limited to) vasodilation, inflammation, pain stimulus, blood clot formation, bronchial constriction, promotion of parturition, and luteolysis.

In equine reproduction, $PGF_{2}\alpha$, PGE_1, and PGE_2 are 3 PGs commonly used for their effects on the reproductive tract. By way of luteolysis, $PGF_{2}\alpha$ (cloprostenol, dinoprost) is used for its potential to shorten diestrus—i.e., "short cycle" a mare, hasten estrus, and terminate pregnancy; it also stimulates continuous uterine contractions for fluid evacuation. A synthetic PGE_1 (misoprostol) is effective both in the treatment of gastric ulcers and may be effective in causing cervical dilation or uterine tubal muscular motility and dilation when used topically. Even if administered orally, it is still effective in initiating uterine contractions. PGE_2 (dinoprostone gel), similar to PGE_1, is best known for its use in equine reproduction to re-establish the patency of blocked mare uterine tubes. Topical application by laparotomy or laparoscopy results in uterine tubal dilation and muscular contractions to move thickened uterine tubal contents into the uterus for clearance.

These PG compounds have become an important part of reproduction management in horses since their usefulness was discovered in the early 1970s.

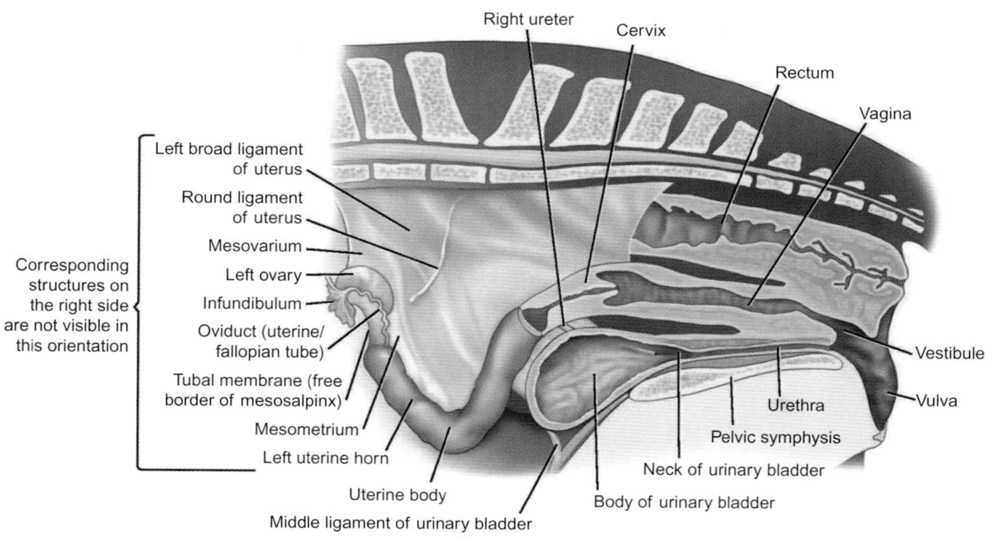

FIGURE 13.6-4 The broad ligament attaches at the 3rd to 4th lumbar vertebra and extends to the level of the 4th sacral vertebra.

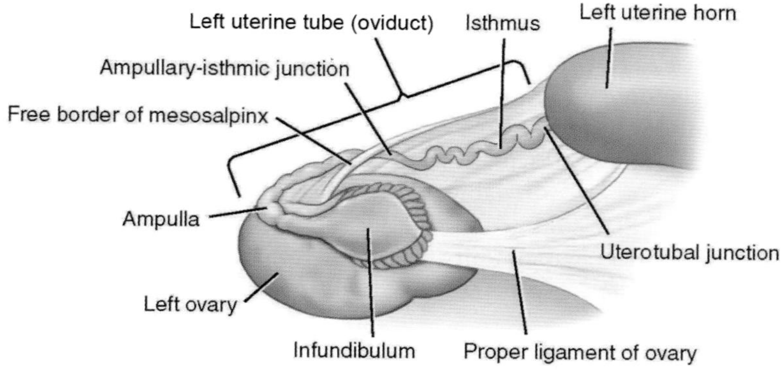

FIGURE 13.6-5 Five distinct regions of the equine uterine tube: infundibulum, ampulla, ampullary-isthmic junction, isthmus, and uterotubal junction.

Uterine tube

The **uterine tube**, or **oviduct**, has 5 distinct regions, each with a different function and appearance (Fig. 13.6-5)—infundibulum, ampulla, ampullary-isthmic junction, isthmus, and uterotubal junction. The **infundibulum** is a funnel-shaped sheet of tissue that has irregular **fimbriae** (L. border or fringe). Tumescence of the infundibulum occurs near the time of ovulation and positions the infundibulum near the ovulation fossa of the ovary to facilitate capture of the oocyte upon its release from the ovulation fossa. Portions of the fimbriae are attached to the cranial pole of the ovary, which allows the rest of the infundibulum to expand over the ovulation fossa. The abdominal **ostium** from the infundibulum into the lumen of the ampulla is about 6 mm (0.2 in.) in diameter.

The **ampulla** (Fig. 13.6-6) is the most active region of fluid secretion from the mucosal epithelial cells, enabling it to be classically known as the site of fertilization and early embryo development in mammals. The ampullary-isthmic junction is the region of uterine tube that functionally retains unfertilized oocytes; alternatively, it is also the region that, in response to embryonic secretion of PGE_2, allows the passage of an embryo through the isthmus into the uterus by day 6 after ovulation. The **isthmus** facilitates spermatozoa transport and, substantiated in other species, there is evidence that the isthmus is an anatomical site for spermatozoa storage in the mare. The **uterotubal junction** (UTJ) connects the uterine tube to the tip of the uterine horn. It is at this location where preferential

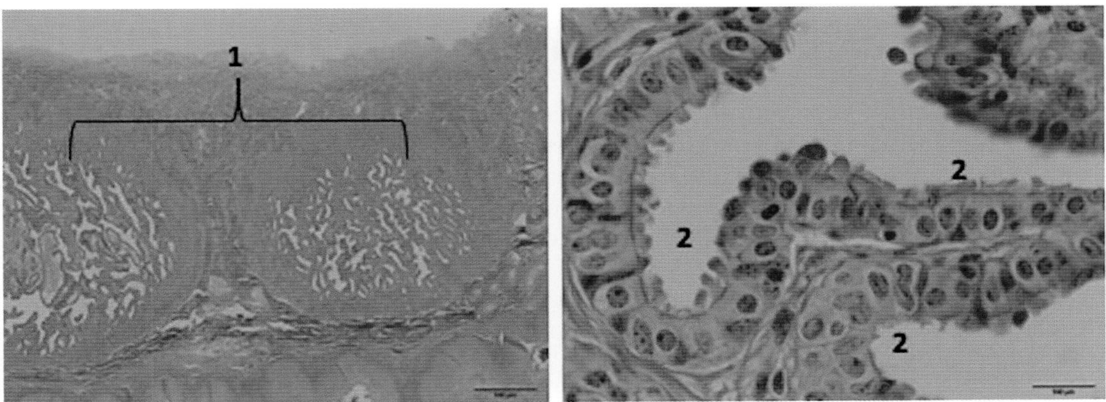

FIGURE 13.6-6 Histopathology cross-sectional image of the equine uterine tube at the level of the ampulla. Left: note how multiple cross sections were obtained with one tissue cut (1), demonstrating the undulating nature of the ampulla. Right: note the highly tortuous mucosal epithelial folds (2) in the lumen of the ampulla.

selection of motile and morphologically normal spermatozoa occurs. The UTJ includes many folds and terminates with a 2–3 mm (0.08–0.12 in.) uterine **ostium** located on a small papilla. There has also been suggestion that, like the isthmus, the UTJ may act as a secondary reservoir for spermatozoa.

The outer serosal surface of the uterine tube is continuous with the serosa of the other abdominal reproductive organs; additionally, a fibrous supportive tissue (continuous with the fibrous layers of the broad ligaments) overlays the muscularis layers. A functional muscular layer exists and is divided into 2 layers: The outer muscular layer fibers are longitudinally arranged, while the inner fibers are circumferentially oriented. The infundibulum contains few muscle fibers, while the middle portion of the ampulla contains a thin muscular layer. The isthmus possesses the thickest layer of muscle (Fig. 13.6-7).

The inner, luminal mucosa provides a fluid medium that is necessary for both oocyte and spermatozoa transport and nourishment; its degree of plication varies depending on location. The ampulla with its tortuous external appearance has appropriately the highest degree of mucosal folding, which functionally provides an appropriate site for oocyte nourishment and fertilization (Fig. 13.6-6). The isthmus is lined with ciliated and nonciliated epithelial

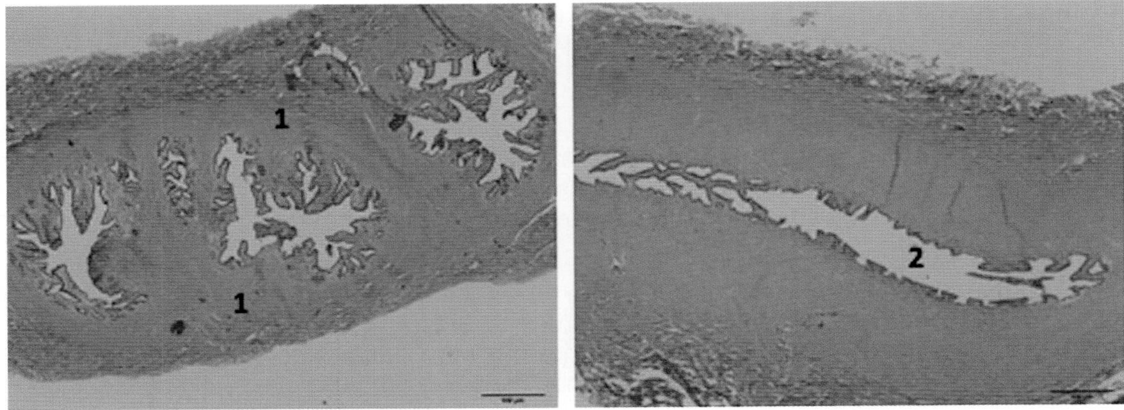

FIGURE 13.6-7 Histopathology cross-sectional image of the equine uterine tube at the level of the isthmus. Left: note the increased thickness of the circumferential and longitudinal muscular layers (1) that is unique to the isthmus. Right: the lumen of the isthmus has a lower surface area due to decreased mucosal epithelial folding (2).

Since significant changes in spermatozoa integrity can be expected to occur in varying degrees after cryopreservation—i.e., structural acrosome changes, increased intracellular calcium fluctuations, and alterations to membrane fluidity—it is to be expected that uterine tubal binding and therefore spermatozoa longevity are abbreviated when using cryopreserved compared with fresh semen. While not nearly to the same degree of detriment, processing and cooling procedures performed with fresh ejaculates can also result in spermatozoa capacitation of acrosome-intact spermatozoa, also altering their ability to bind to the oviduct.

cells dispersed among longitudinal folds (Fig. 13.6-7). These uterine tubal epithelial cells are thought to prevent spermatozoa capacitation by way of maintaining basal levels of calcium once spermatozoa are attached to the uterine tube. This ultimately results in the prolongation of spermatozoa viability until the appropriate time of spermatozoa release for fertilization.

Blood supply, lymphatics, and innervation

The blood supply to the uterine tube is derived from the abdominal aorta via the ovarian artery. The **ovarian artery** provides 2 branches: the **ovarian branch** and the uterine branch. It is the **uterine branch**—a.k.a. the **cranial uterine artery**—that provides blood to the uterine tube and to the anterior part of the uterine horn. Venous drainage from the uterine tube is by way of the **uterine branch of the ovarian vein**. The uterine branch of the ovarian vein is an important source of venous drainage for the entire uterus.

Lymphatic drainage for the ovary is inconsistent and involves a small lymph node in the mesovarium. Additional ovarian lymph drainage can also occur via the **subperitoneal nodes** (lumbales aortici [L.] or **lumbar aortic lymph nodes**) that are located along the abdominal aorta and caudal vena cava and are firmly adhered to the peritoneum. Lymphatics in the reproductive tract, particularly the broad ligament, are difficult to see except in late pregnancy during which the lymphatics are prominent. These lymphatics drain to the **iliac lymph nodes** and finally to the lumbar aortic lymph nodes.

Sympathetic and parasympathetic nerve fibers from the **renal**, **aortic**, **uterine**, and **pelvic plexuses** innervate the ovaries, tubular reproductive tract, and vagina.

HYDROSALPINX AND SALPINGITIS: MARES ARE NOT ELEGANT COWS

Contrary to the mare, salpingitis and subsequent hydrosalpinx has historically been a cause of infertility in bovids. Both were thought to occur, in part, due to the previous practice of manual enucleation of the CL per rectum. Hemorrhage descending from the ovary would result in hematoma formation in the ovarian bursa, fibrous adhesion formation, and oviductal occlusion. With the development of injectable PG, manual enucleation of the CL is no longer recommended. An additional precursor to salpingitis can be inflammation or pathological fluid within the uterus of the bovid; these are more likely to result in uterine tubal pathology since no uterine tubal papillae exist to prevent reflux; rather, the tip of the uterine horn simply attenuates to form a uterine tube—i.e., oviduct. Not surprisingly, large animal clinicians may be tempted to erroneously extrapolate information from one species to the other in cases of infertility in the mare.

With rare exceptions, salpingitis and hydrosalpinx have not been associated with infertility in the mare. Perhaps this is due to the dorsal position of the uterine tubes at the tips of the uterine horns and the prominent musculature of the uterine tubal papillae, as both characteristics could potentially be adept in preventing uterine tubal reflux. While studies have shown contradicting evidence, it appears that histopathological evidence of salpingitis is more common than previously believed—in many of these studies, uterine tubal inflammation was classified as "slight" and limited to lymphocytes. So, while inflammation may have been identified more frequently than originally expected based on clinical impression, unlike cattle, these findings may not have a significant effect on fertility. Rarely, hydrosalpinx has been diagnosed in mares related to ascending infection or traumatic injury. Congenital segmental aplasia is also a viable explanation, but with a limited frequency, of hydrosalpinx in the mare.

Mesosalpinx and associated structures

The connective tissue directly surrounding and supporting the uterine tube (Fig. 13.6-5) originates from the broad ligament and is termed the **mesosalpinx**. (Fig. 13.6-5). The mesosalpinx arises from the lateral surface of the **mesovarium**, which is the portion of the broad ligament that surrounds and supports the ovary. (The reader is referred to Case 13.3 for more information on the broad ligament.)

Selected references

[1] Ginther OJ. Reproductive anatomy. In: Reproductive biology of the mare: basic and applied aspects. 2nd ed. Cross Plains, WI: Equiservices; 1992. p. 1–40 [chapter 1].
[2] Kainer RA. Internal reproductive anatomy. In: McKinnon AO, Squires EL, Vaala WE, Varner DD, editors. Equine reproduction. 2nd ed. West Sussex, United Kingdom: Wiley-Blackwell; 2011. p. 1582–97 [chapter 165].
[3] Mouguelar H, Diaz T, Borghi D, et al. Morphometric study of the mare oviductal mucosa at different reproductive stages. Anat Rec 2015;298:1950–9.
[4] Sertich PL. Clinical management of the equine oviduct. Clin Theriogenol 2013;5:523–8.
[5] Kenney RM, Doig PA. Equine endometrial biopsy. In: Morrow DA, editor. Current therapy in theriogenology. vol 2. Philadelphia, PA: WB Saunders Company; 1986. p. 723–9.

CHAPTER 13

CASE 13.7

Squamous Cell Carcinoma of the Penis

David Levine[a] and Carrie Jacobs[b]
[a]Department of Clinical Studies - New Bolton Center, University of Pennsylvania School of Veterinary Medicine, Kennett Square, Pennsylvania, US
[b]Equine Orthopedic Surgery, North Carolina State University School of Veterinary Medicine, Raleigh, North Carolina, US

Clinical case

History

A 15-year-old Paint gelding was presented with a several-month history of odor originating from the prepuce and more recent development of a bloody discharge from the prepuce.

Physical examination findings

The gelding was bright, alert, and responsive with normal vital parameters. Sedation was administered to facilitate examination of the penis and prepuce. A large (6 cm or 2.4 in. long × 6 cm or 2.4 in. wide × 4 cm or 1.6 in. high) friable, cauliflower-like mass was present on the glans penis which distorted the normal anatomy of the glans and obstructed assessment of the urethral process (Fig. 13.7-1A). A moderate amount of malodorous, hemorrhagic discharge was noted during manipulation and examination of the region. Following examination, the gelding was observed to urinate normally.

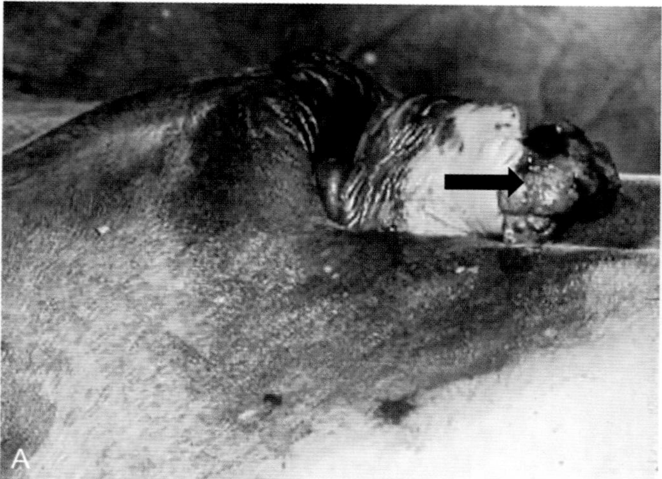

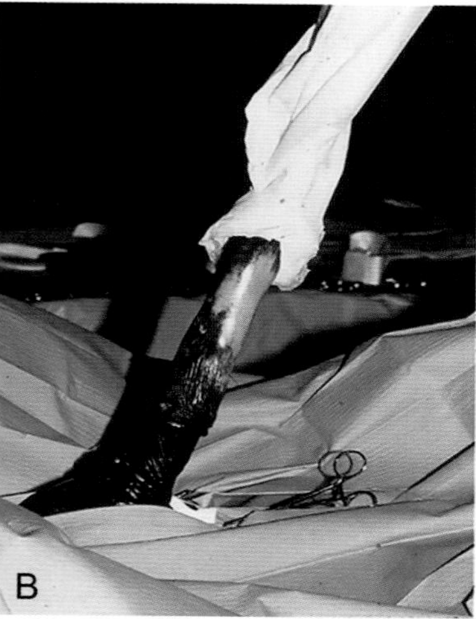

FIGURE 13.7-1 (A) Squamous cell carcinoma of the glans penis *(arrow)*. (B) Pre-operative image of surgery site. (Photo courtesy of William J. Donawick.)

Comparative Veterinary Anatomy: A Clinical Approach. https://doi.org/10.1016/B978-0-323-91015-6.00066-2

Differential diagnoses

Neoplasia (SCC), sarcoid, habronemiasis, trauma (hematoma)

Diagnostics

Based on the signalment (an older Paint horse with nonpigmented genitalia) and presenting clinical signs, a presumptive diagnosis of SCC was made. Definitive diagnosis requires a biopsy and histopathologic evaluation. Identification of potential metastasis to regional inguinal lymph nodes can be performed by identifying enlarged lymph nodes on abdominal palpation per rectum and/or with ultrasound examination.

Diagnosis

Squamous cell carcinoma

Treatment

The gelding was prepared for surgery and placed under general anesthesia (Fig. 13.7-1B). A partial penile amputation (phallectomy) including resection of the prepuce was performed (Figs. 13.7-2A–F and 13.7-3). Histopathology confirmed a well-differentiated SCC with nests of squamous epithelial cells arising from the epidermis and extending into the dermis. Multiple mitotic figures (Fig. 13.7-4A and B) were identified with eosinophilic cytoplasm and large, vesicular nuclei.

> Surgical options for penile tumors in horses include partial phallectomy, partial phallectomy with sheath ablation, or en bloc penile resection. The choice of surgical procedure is dependent on the extent and type of tumor, owner finances, and surgeon preference. Frequently, surgical removal is also combined with chemotherapy.

SQUAMOUS CELL CARCINOMA OF THE PENIS IN HORSES

Penile neoplasia is common in horses and squamous cell carcinoma (SCC) is the most common tumor to affect the penis and prepuce. These tumors are usually diagnosed in older horses and in horses lacking penile skin pigment, such as Appaloosas and Paint horses. Squamous cell carcinoma of the penis and prepuce generally originates on the glans penis and within the internal borders of the prepuce. These tumors are generally slow growing but can metastasize to the regional lymph nodes—i.e., the inguinal lymph nodes.

Diagnosis is frequently based on clinical signs that include malodorous, purulent, and/or hemorrhagic preputial discharge, blood associated with urination, difficulty urinating, inability to extend the penis, preputial edema, and/or a mass or ulceration observed on the end of the penis. Definitive diagnosis of any abnormal mass requires a biopsy and histopathologic examination.

Treatment options vary depending on the extent and invasive nature of the tumor and include cryotherapy, chemotherapy, and surgical resection in the form of partial phallectomy, partial phallectomy with sheath ablation, or en bloc penile resection. Frequently, a combination of the treatment options is recommended. Prognosis and rate of recurrence are highly variable (55–81%) and depend on the severity and extent of the tumor, in addition to the type of treatment. Recurrence rates are higher with partial phallectomy (25.6%) compared with en bloc penile resection (17.9%), along with a higher rate of recurrence (25%) when metastasis to the inguinal lymph nodes is identified.

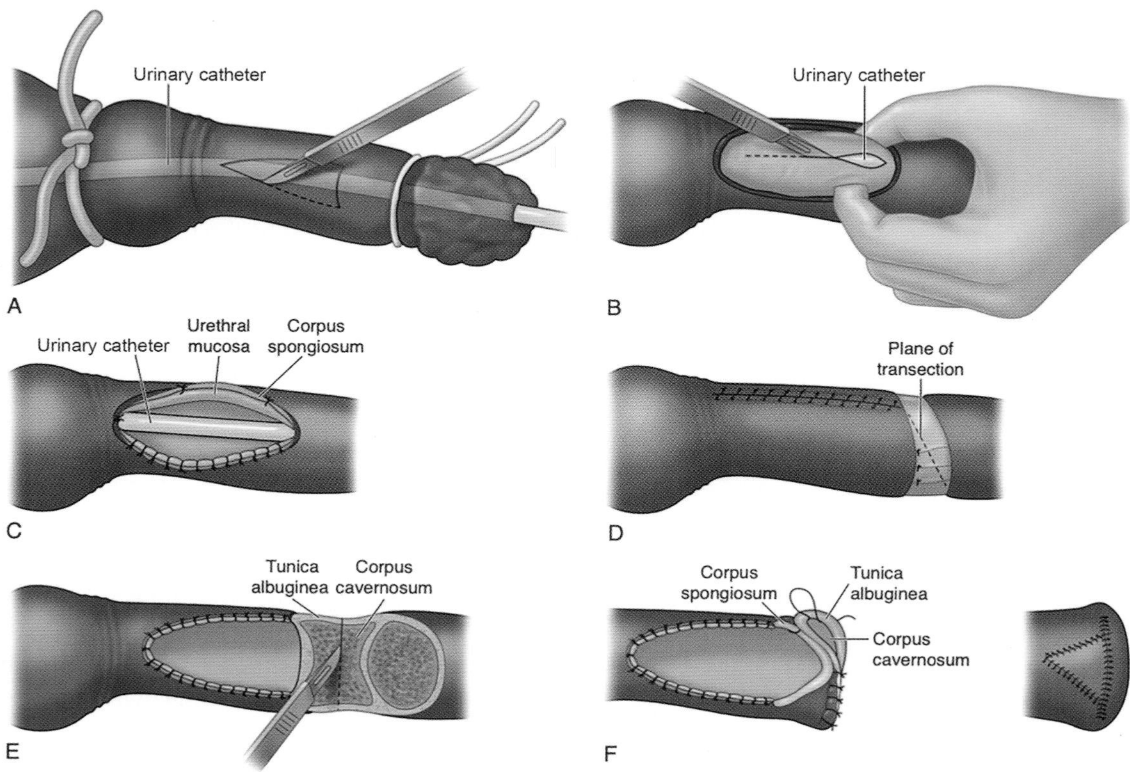

FIGURE 13.7-2 Partial penile amputation, a.k.a. partial phallectomy. The figures depict a horse with a SCC: (A) pre-operative appearance; (B–E) intra-operative procedure; and (F) the post-operative result.

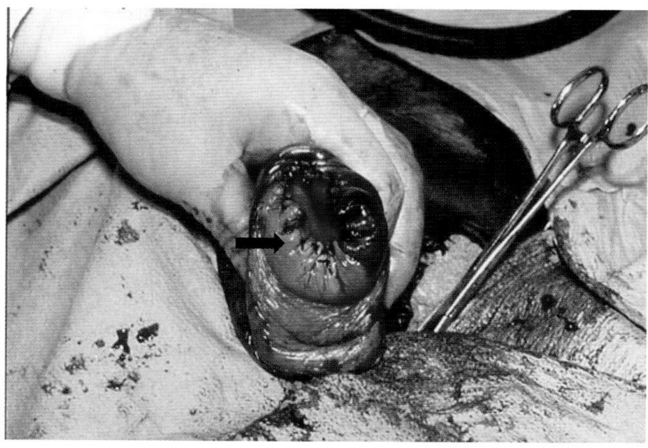

FIGURE 13.7-3 Post-operative image after completion of partial phallectomy *(arrow)*. (Photo courtesy of William J. Donawick.)

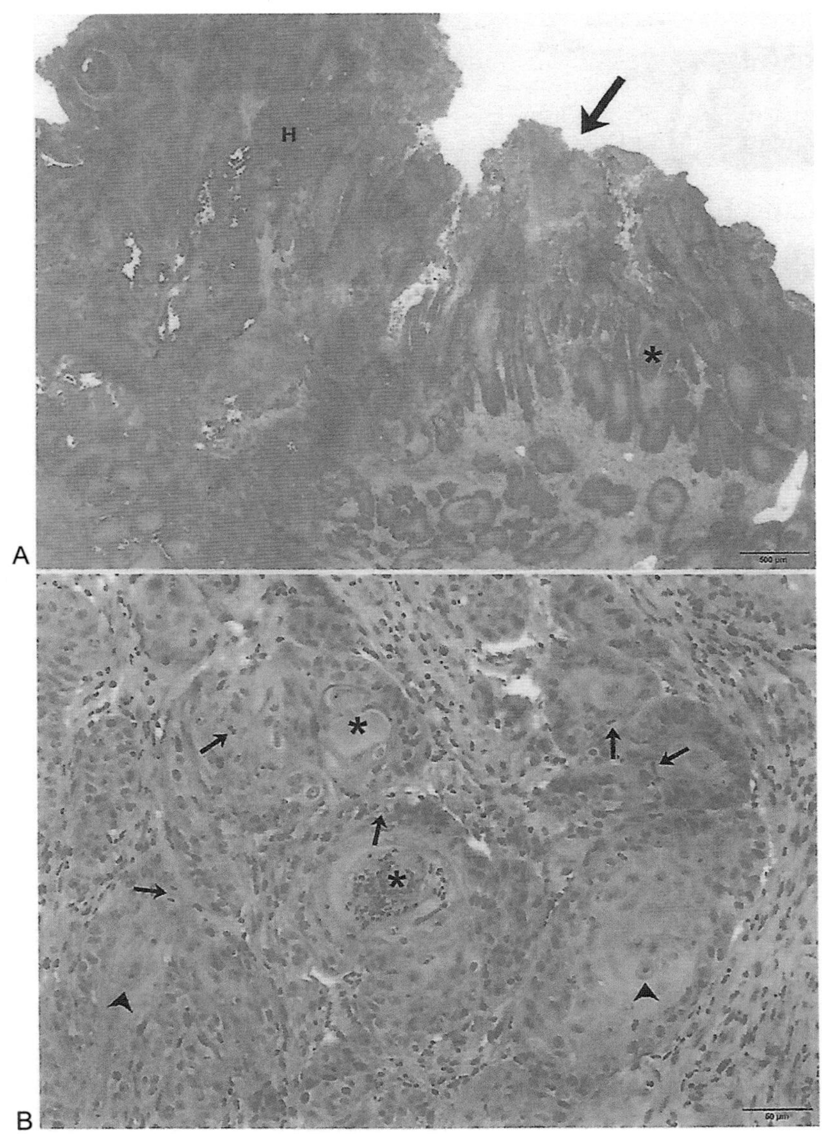

FIGURE 13.7-4 (A) Low (2×) magnification photomicrograph of an exophytic (Gr. *exō-* outside + Gr. *phyein* to grow = growing outward) SCC excised from the glans penis depicting superficial necrosis with ulceration *(arrow)* and hemorrhage (*H*). Papillary projections with invasive cords and nests of malignant stratified squamous epithelium efface the glans, forming outer basal and inner spinous layers with progressive keratinization and central keratin pearls *(asterisk)*. H scale bar = 500 μm. (B) High (20×) magnification photomicrograph of invasive neoplastic squamous islands depicting multiple mitotic figures *(arrows)* and anisokaryosis (Gr. *anisos* unequal, uneven + Gr. *karyon* nucleus + Gr. *osis* process = variation in the size of the cell nuclei) *(arrowheads)* with central accumulations of necrotic sloughed keratinocytes *(asterisks)* and neutrophilic inflammation. H scale bar = 50 μm. (Images courtesy of Dr. Julie Engiles, Department of Pathobiology, University of Pennsylvania.)

Anatomical features in equids

Introduction

The major parts of the male reproductive organs include the testicles (testes, male gonads) and spermatic cords; the accessory sex glands (seminal vesicles, prostate, bulbourethral glands, and ampullae of the ductus deferens); the prepuce; and the penis. This case discusses the anatomy of the penis and prepuce, along with associated blood supply, lymphatics, and innervation. The testis, spermatic cord, and inguinal ring are discussed in Case 13.9, and the accessory sex glands in Case 13.8.

Function

In intact males, the penis functions for urination and copulation. In castrated males—referred to as geldings—the penis functions for urination only. The equine penis is classified as musculocavernosus and is composed of cavernous erectile tissue that fills with blood for enlargement during erection.

Penis (Fig. 13.7-5A and B)

The equine **penis** is divided into 3 parts—root, body, and glans penis. The **root** serves as the origin of the penis where it attaches to the pelvis at the **ischial arch** by 2 **crura**. The **body**, also known as the "shaft of the penis," is the main part of the penis and begins at the convergence of the crural attachments of the root. The **glans penis** is the expansion present at the end of the penile body with multiple components. The **corona glandis** (Fig. 13.7-6) is the circular edge of the glans penis and the **collum glandis** is the constriction located proximal to the corona glandis.

Urethra (Fig. 13.7-5B)

The proximal or **pelvic urethra** lies on the ventral aspect of the pelvic canal and connects the trigone of the bladder to the **penile portion of the urethra**. The accessory reproductive glands of the stallion (the seminal vesicles, ampullae, prostate, and bulbourethral glands) are closely associated with the pelvic urethra. The **ampullae** and **seminal vesicles** drain into the **colliculus seminalis**, which is in the caudal part of the proximal urethra. The openings of the **prostate** are located lateral to the colliculus seminalis. The **bulbourethral gland** drains in 2 rows into the urethra as the latter exits the pelvis around the ischial arch. The **distal urethra** terminates as a free end, termed the "urethral process," and extends from the **fossa glandis**, a depression within the end of the glans penis. The fossa glandis has dorsal diverticula known as the **urethral diverticulum.**

The urethral diverticulum is a reservoir for venereal disease. This diverticulum readily fills with smegma—a mixture of sebaceous and epithelial debris. When this smegma hardens, it is colloquially referred to as "a bean" and, in geldings, it typically needs to be routinely cleaned to minimize discomfort to the horse when urinating.

Prepuce

The **prepuce** serves to encompass the retracted, inactive, free end of the penis and is composed of internal and external folds. The part of the penis distal to the preputial attachment is referred to as the "free end of the penis" and is approximately 15–20 cm (5.9–7.9 in.) in length. The telescoping nature of the penis during retraction results in inner and external folds of the prepuce. When the penis is retracted, the inner fold forms 2 distinct folds—an **inner** and **outer lamina**, separated by the **preputial ring** (Fig. 13.7-6).

Muscles of the penis

Important muscles of the penis include the ishiocavernosus, urethralis, bulbospongiosus, and the retractor penile muscles. The **ischiocavernosus muscle** arises from the tuber ischiadicum and sacrotuberous ligament and inserts

COMPARATIVE PENILE ANATOMY

The anatomy of the penis is entirely variable between species. Variation seen in the corpus cavernosus produces either a fibroelastic type penis, as seen in the boar and ruminants, or a musculocavernosus penis, seen in the stallion and the dog. In the fibroelastic penis, the corpus cavernosus is divided into small blood spaces which are divided by tough fibroelastic tissue. In the musculocavernosus penis, the spaces for blood are larger and are divided by more muscular septa. A larger volume of blood is required to achieve erection in the musculocavernosus penis vs. the fibroelastic penis.

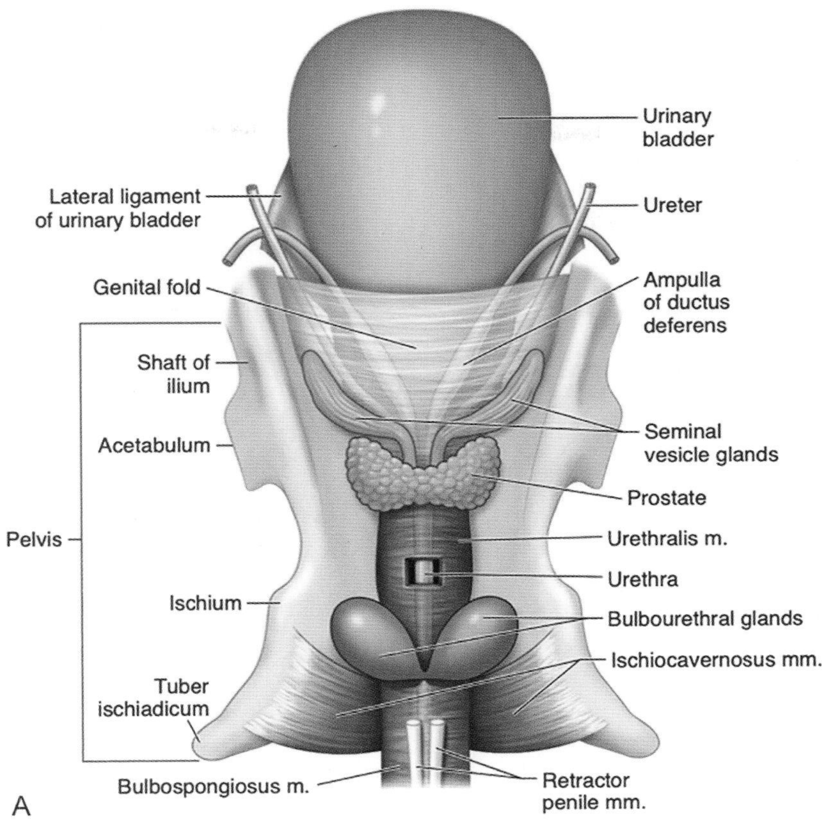

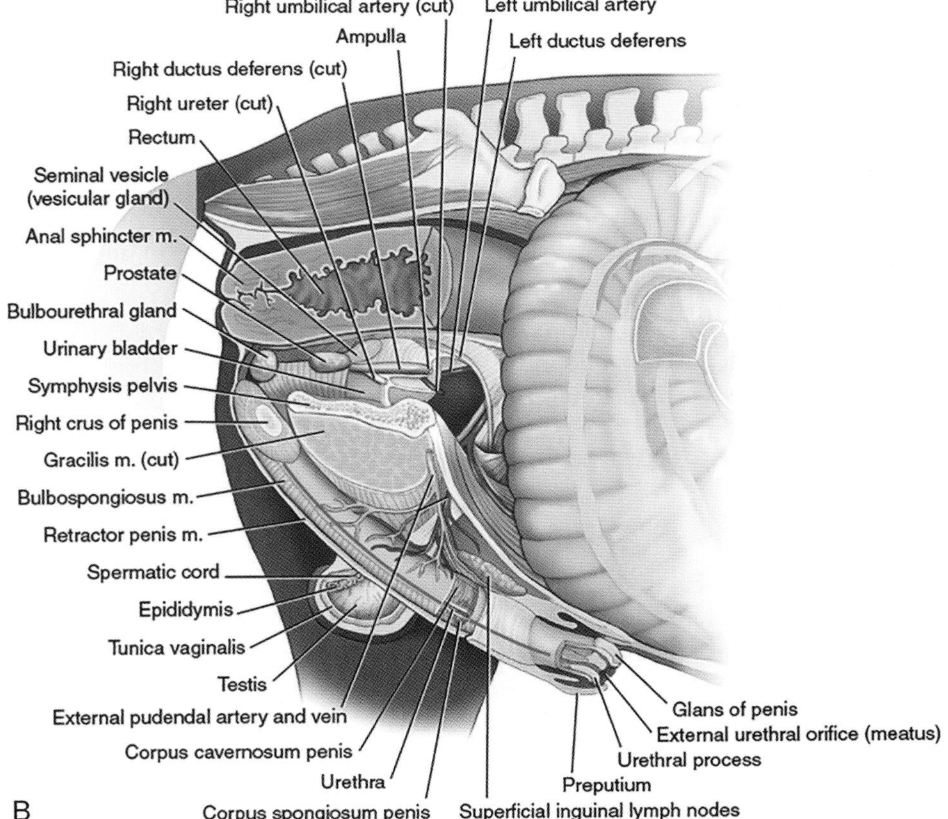

FIGURE 13.7-5 (A) Image representing the genitourinary tract of the stallion (dorsal view), emphasizing the relationship of the accessory sex glands to the urinary bladder and pelvic canal. (B) Image depicting the genitourinary tract of the stallion (lateral view) with relevant clinical anatomy. (Adapted from Dyce KM, Sack WO, Wensing CJG. Textbook of veterinary anatomy. 3rd ed. Philadelphia, PA: Saunders; 2002. p. 561–3.)

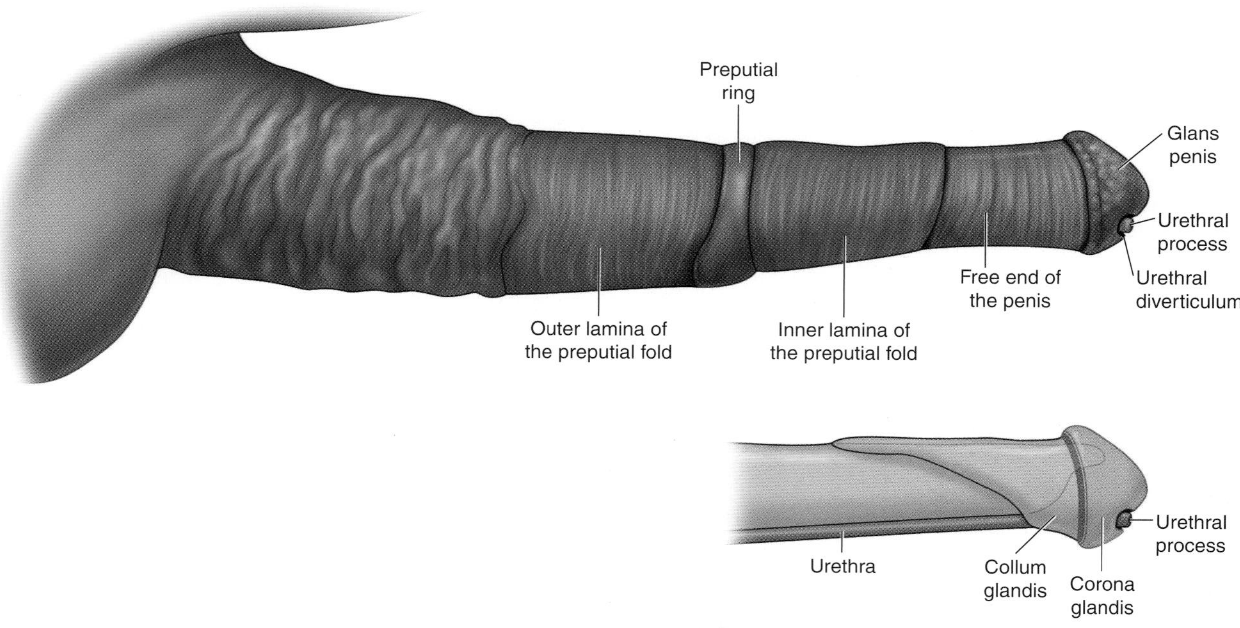

FIGURE 13.7-6 Extended penis of the stallion and its relationship to the prepuce. *Inset*: free end of the penis with the skin removed.

on the crura and body of the penis. This muscle supports maintaining an erection during sexual arousal because it compresses the penis against the ischium, impeding venous return. It also assists in elevating the penis for copulation.

The **urethralis muscle** surrounds the pelvic urethra and bulbourethral glands. Contraction of this muscle ejects seminal fluid during ejaculation. At the end of urination, contraction of this muscle empties urine from the urethra.

The **bulbospongiosus muscle** covers the ventral aspect of the corpus spongiosum for the entire length of the penis. Contraction of this muscle occurs in waves during ejaculation and leads to increases in pressure in the corpus spongiosum, discharging semen from the urethra.

The **retractor penile muscles** are paired, longitudinal muscles that originate from the 1st coccygeal vertebrae, loop around the rectum, then track along the ventral aspect of the penis—ventral to the bulbospongiosus muscles—and insert on the glans penis. These muscles retract the penis within the prepuce.

Regional blood supply, lymphatics, and innervation

Because of the musculocavernous nature of the penis in the stallion, the use of a tourniquet for improved hemostasis is important during partial or full phallectomy surgical procedures. This can be accomplished using a Penrose drain secured around the base of the penis.

The major blood supply to the penis originates from the internal and external pudendal arteries and the obturator artery. The **internal pudendal artery** delivers blood to the pelvic urethra and continues as the **corpus spongiosum**. The **dorsal penile artery** supplies blood to the erectile tissue and arises from the **external pudendal artery**. The **caudal superficial epigastric artery** also arises from the external pudendal artery and supplies branches to the prepuce. The **obturator artery** distributes blood to the **corpus cavernosum** tissue.

Blood is removed from the body and glans penis via the **venous plexus of the penis** to the **external** and **obturator veins**. The **internal pudendal vein** returns blood from the root of the penis. Lymphatics from the penis drain to **the inguinal lymph nodes** and are important in the spread and metastasis of neoplasia from the penis.

Nerve supply to the penis is via the **pudendal, deep perineal,** and **caudal rectal nerves**. The pudendal and caudal rectal nerves originate from spinal nerves S2–S4. The pudendal nerve gives rise to the deep perineal nerve at the level of the lesser sciatic foramen. The dorsal nerves of the pelvis originate from the pudendal nerves. The bulbospongiosus, ischiocavernosus, and rectractor penile muscles are innervated by the deep perineal and caudal rectal nerves.

Standing examination and surgery of the penis is facilitated by anesthetizing/blocking the pudendal nerves at the level of the ischial arch. The pudendal nerve block desensitizes/blocks both the sensory nerves of the penis and the retractor penis muscle allowing examination without sensation and retraction.

Functional penile anatomy

The equine penis is largely formed of cavernous space, which fills with blood during erection (Fig. 13.7-7). The **corpus cavernosum penis** forms most of the cavernous tissue of the penile body and is located on the dorsal aspect of the penis. At its distal end, the corpus cavernosum organizes into a central projection with 2 blunt projections located ventrolaterally. The smaller and more ventrally located cavernous tissue, the **corpus spongiosum penis**, surrounds the urethra as it passes through the penile body. At its distal end, the corpus spongiosum enlarges into the glans penis. The **tunica albuginea** is a thick fibroelastic tissue that surrounds the cavernous spaces. This tissue has thinner fibrous trabeculae which form the framework of the cavernous spaces. Both endothelial cells and smooth muscle bundles line the cavernous spaces. In conjunction with the retractor penile muscles, the tone within the muscle bundles serves to maintain the penis within the prepuce (Fig. 13.7-7).

Increased arterial blood flow to the cavernous spaces during sexual arousal leads to an erection. Deep coiled branches of the **deep artery of the penis** relax, dilate, and straighten while the smooth muscle bundles relax to allow blood to fill the cavernous spaces. The rigid **tunica albuginea** limits continued filling, resulting in lengthening and stiffening of the penis. In addition to increased arterial flow to both the corpus cavernosum and corpus spongiosum, decreased venous return from the corpus cavernosum plays an important role in erection. Increased parasympathetic outflow is responsible for erection while increased sympathetic outflow results in loss of erection through increased venous outflow and return of the branches of the penile artery to their normal, coiled state.

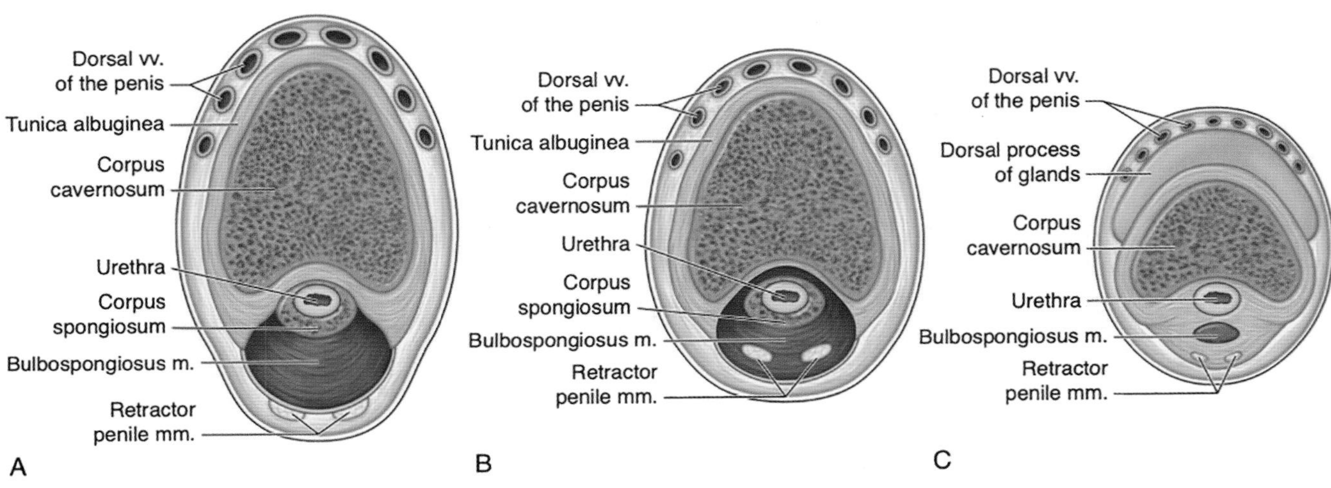

FIGURE 13.7-7 Cross-sectional anatomy of the penis: (A) distal to the root; (B) midshaft; and (C) free end.

PRIAPISM VS. PARAPHIMOSIS VS. PHIMOSIS

Priapism (L. *priapismus*; Gr. *priapismos*) is referred to as a "persistent erection" and can result from disturbance in arterial supply to the penis and/or venous outflow from the corpus cavernosum.

Paraphimosis (Gr. *para-* to, at, from the side of + Gr. *phimoun* to muzzle + Gr. *-osis* a process) is the inability to retract the penis into the prepuce and can occur as a result of penile edema from a variety of causes, such as preputial or penile trauma. Edema within the tissues increases the weight of the penis, the retractor penile muscles become fatigued, and the penis protrudes from the prepuce. This can also be seen in debilitated and underweight horses as a result of decreased retractor penile muscle tone.

Phimosis (Gr. *phimosis* a muzzling or closure) is the inability to protrude the penis from the prepuce. This is most often a result of adhesions within the prepuce or a stricture at the opening of the prepuce.

Selected references

[1] Van den Top JG, de Heer N, Klein WR, et al. Penile and preputial tumours in the horse: A retrospective study of 114 affected horses. Equine Vet J 2008;40:528–32.

[2] Van den Top JG, de Heer N, Klein WR, et al. Penile and preputial squamous cell carcinoma in the horse: a retrospective study of 77 affected horses. Equine Vet J 2008;40:533–7.

[3] Mair TS, Walmsley JP, Philips TJ. Surgical treatment of 45 horses affected by squamous cell carcinoma of the penis and prepuce. Equine Vet J 2000;5:406–10.

[4] Schumacher J. Penis and prepuce. In: Auer JA, Stick JA, Kummerle JM, Prange T, editors. Equine surgery. 5th ed. St. Louis, MO: Elsevier; 2019. p. 1034–64 [chapter 61].

[5] Dyce KM, Sack WO, Wensing CJG. Textbook of veterinary anatomy. 3rd ed. Philadelphia, PA: Saunders; 2002. p. 561–3.

CHAPTER 13

CASE 13.8

Seminal Vesiculitis

Malgorzata Pozor
Large Animal Clinical Sciences, University of Florida College of Veterinary Medicine, Gainesville, Florida, US

Clinical case

History

A 29-year-old Arabian stallion was presented with a history of inability to exteriorize the penis to urinate, abnormal posturing (wide-base stance), and chronic weight loss. The gait of the horse was abnormally widened and slow in the hind end, suggesting mild discomfort. The results of the CBC, urinalysis, and blood chemistry were normal. During initial transrectal palpation and ultrasound evaluation performed by the referring veterinarian, a large fluid-filled mass was found on the right side of the rectum, approximately 25 cm (9.8 in.) cranial to the anus. A presumptive diagnosis of a pelvic abscess was made, and the horse was placed on 9 days of treatment with trimethoprim-sulfamethoxazole (30 mg/kg, q12h, PO). Because of a lack of improvement after treatment, the horse was referred to a tertiary care hospital for further workup and treatment.

Physical examination findings

Upon presentation, all vital signs were within normal limits. There was evidence of chronic balanoposthitis (Gr. *balano-* acorn, relationship of the glans penis or glans clitoridis + Gr. *posthē* prepuce + Gr. *itis* disease = inflammation of the glans penis and prepuce) caused by frequent urination into the preputial cavity. Initial blood work revealed signs of active chronic inflammation (mild lymphopenia: 0.96 K/μL; mild neutrophilia: 8.07 K/μL; toxic changes: 1 +; hyperfibrinogenemia: 800 mg/dL; elevated serum amyloid A [SAA]: 772 μg/mL). Transrectal palpation and ultrasonography revealed a normal ultrasonographic appearance of the bulbourethral glands and prostate (Fig. 13.8-1A and B). Both ampullae of the ducti deferentes appeared normal, with the right ampulla slightly larger than the left (Fig. 13.8-1C and D). The colliculus seminalis was prominent and contained a cystic structure with slightly echogenic contents (the midline cyst of the colliculus seminalis) (Fig. 13.8-1E). The excretory ducts of the vesicular glands contained numerous small, grit-like deposits (Fig. 13.8-1F). The left vesicular gland was normal in size with slightly echogenic contents containing numerous hyperechoic speckles (Fig. 13.8-1G). The right vesicular gland was abnormally large and easily palpable per rectum, as it extended beyond the pelvic brim, and it had a thickened wall and distended lumen on ultrasonographic evaluation (Fig. 13.8-1H). The excretory duct of the right vesicular gland was also abnormally enlarged. This gland contained nonuniformly echogenic material with hyperechoic speckles throughout the lumen. Manual and ultrasonographic examination of the scrotum revealed 2 testes in normal orientation that were small and flaccid. Both testes had nonuniform echogenicity consistent with fibrotic changes likely due to the stallion's advanced age. The epididymal duct and the ductus deferens on the right side were enlarged, compared to the left side where they were prominent but normal.

Differential diagnoses

Seminal vesiculitis, pelvic abscess, neoplastic mass of the right vesicular gland

Comparative Veterinary Anatomy: A Clinical Approach. https://doi.org/10.1016/B978-0-323-91015-6.00067-4

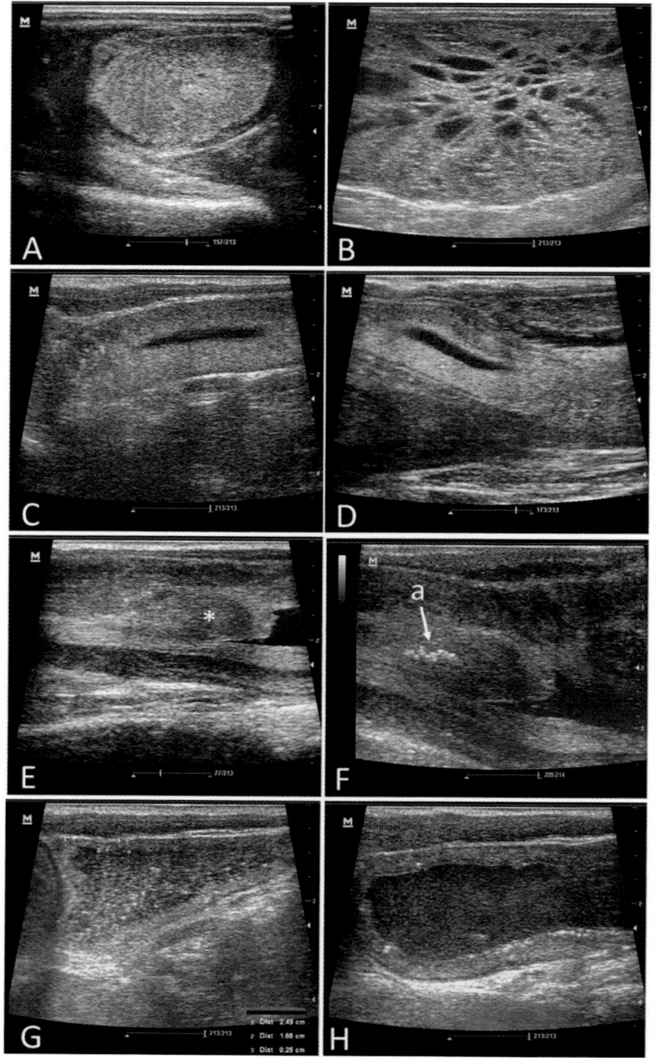

FIGURE 13.8-1 Initial ultrasonographic images of the accessory sex glands and the colliculus seminalis of the stallion with seminal vesiculitis. (A) Bulbourethral gland; (B) prostatic lobe; (C) ampulla of the ductus deferens; (D) terminal portion of the ampulla diving under the isthmus of the prostate; (E) midline cyst of the colliculus seminalis (*); (F) grit-like concretions in the excretory duct of the vesicular gland *(a)*; (G) left vesicular gland; (H) right vesicular gland.

Diagnostics

Several unsuccessful attempts to collect semen were made to evaluate the nonspermatozoal cell content in the semen. The stallion mounted a phantom (dummy) numerous times, inserted the penis into the artificial vagina, but did not thrust normally, which was believed to be associated with the chronic balanoposthitis. Further attempts to collect semen were postponed. The stallion was tranquilized with xylazine to allow massage of the vesicular glands per rectum, and a moderate amount of fluid was expressed into the urethral lumen. A sterile urinary catheter was introduced into the urethra to the level of the colliculus seminalis, and the fluid that was expressed from the vesicular glands was submitted for bacterial culture and cytology. Bacterial culture was negative, and no cellular content was seen on cytological evaluation. The owner declined more advanced diagnostic and therapeutic procedures such as urethroscopy and cannulation of the vesicular glands.

Diagnosis

Presumed bilateral seminal vesiculitis, balanoposthitis preventing ejaculation, and partial or complete occlusion of the excretory duct of the right vesicular gland at the colliculus seminalis

Treatment

The stallion was treated with a 9-day course of oral enrofloxacin. The owner reported a significant improvement after 3 days of treatment; however, all signs returned several days after treatment was discontinued. One month of antimicrobial therapy with enrofloxacin was administered yielding similar results. The owner agreed to bring the horse back to the hospital for urethroscopy and local treatment of the vesicular glands. At that time, all vital signs were normal and laboratory tests revealed signs of active chronic inflammation (white blood cells = 16.4 K/μL; neutrophils 13 K/μL; band neutrophils = 1.3 K/μL; monocytes = 1.3 K/μL; fibrinogen = 900 g/dL; SAA = 934 μg/mL). Upon transrectal palpation and ultrasonography, both vesicular glands were equally enlarged with thickened walls and were distended with moderately echogenic contents. The excretory ducts of these glands were dilated extending to the colliculus seminalis and had numerous hyperechoic deposits.

The horse was confined in stocks, sedated, and prepared for urethroscopy and bilateral catheterization of the vesicular glands. A 1-m (39.4-in.) long, 10-mm (0.4-in.) diameter sterilized flexible endoscope (see Section 2.1) was inserted into the stallion's urethra and the colliculus seminalis area was clearly visualized. The entire area of the colliculus seminalis was edematous and the dorsal aspect of the urethra was bulging abnormally into the urethral lumen on both sides of the colliculus. The cranial aspect of the colliculus was edematous, preventing direct insertion of the endoscope into the vesicular glands. However, both vesicular glands were catheterized using a thin sterile catheter introduced into the operating channel of the endoscope (Fig. 13.8-2). Thick, cheese-like, and gelatinous contents were aspirated from each gland (Fig. 13.8-2A) and submitted for bacterial culture and cytology. Each vesicular gland was flushed with sterile saline and infused with antimicrobials (Fig. 13.8-2B). Flunixin meglumine was administered IV and treatment with trimethoprim-sulfamethoxazole was initiated. The cytology of the aspirate of the left vesicular gland revealed a moderate number of well-preserved neutrophils (Fig. 13.8-3A) consistent with subacute, mild seminal vesiculitis. The aspirate from the right vesicular gland had a high number of degenerate neutrophils (Fig. 13.8-3B) consistent with chronic, severe seminal vesiculitis. Bacterial culture yielded growth of β-hemolytic *Streptococcus equi* subspecies *zooepidemicus*. Transrectal ultrasound evaluation of the internal genitalia performed 3 days later revealed that both vesicular glands had decreased in size and the glandular wall was thinner than on initial evaluation; the stallion was discharged from the hospital. The owner did not observe any clinical signs during or after completing the prescribed treatment.

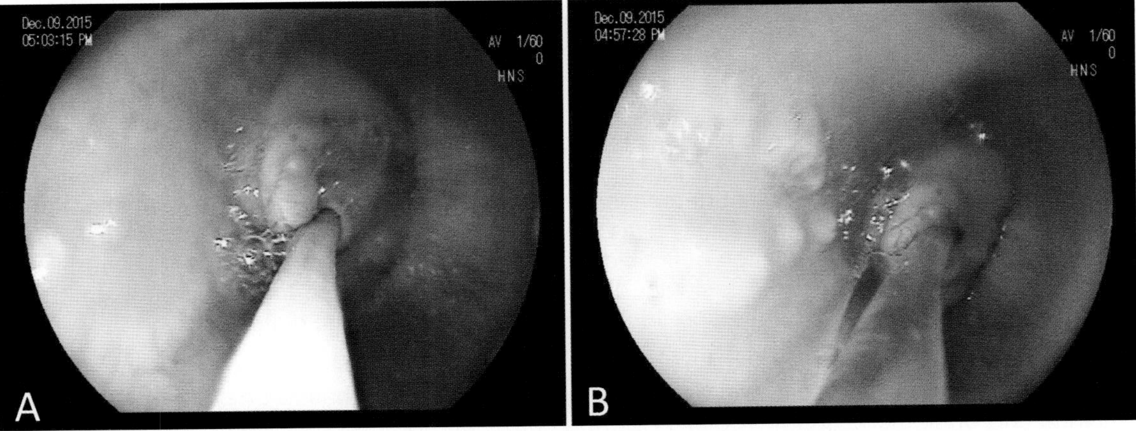

FIGURE 13.8-2 Catheterization of the vesicular glands of the stallion with seminal vesiculitis using a flexible endoscope. (A) Aspiration of purulent material from the infected vesicular gland. (B) Infusion of antimicrobials into the vesicular gland through a thin sterile catheter.

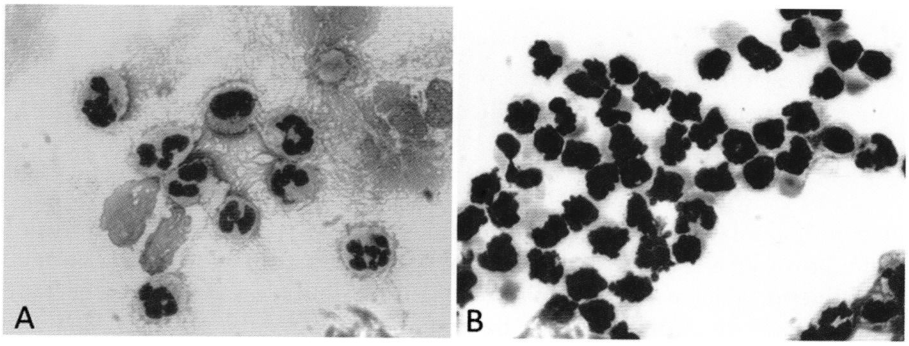

A B

FIGURE 13.8-3 Microscopic images of smears of purulent material aspirated from the vesicular glands of the stallion with seminal vesiculitis. (A) Well-preserved polymorphonuclear neutrophils (PMNs) in the aspirate collected from the left vesicular gland. (B) Degenerated PMNs in the aspirate collected from the right vesicular gland (Romanowsky stain; magnification ×1000).

Anatomical features in equids

Introduction

The stallion reproductive tract consists of the external and internal genitalia. This section describes the internal organs of the male reproductive tract, including the pelvic urethra, ampullae of the paired ducti deferentes, vesicular glands, prostate, and bulbourethral glands (please see Case 13.7, Figs. 13.7-4, 13.8-4, and 13.8-5). Information on the remainder of the male equine reproductive tract is in Cases 13.7 and 13.9.

Function

The ducti deferentes (singular, ductus deferens) carry sperm with secretions from the testes, epididymides, and the ampullae to the pelvic urethra through the ejaculatory ducts that open on the colliculus seminalis. The accessory sex glands produce variable amounts of secretions that are delivered directly to the urethra during emission and/or ejaculation through their own excretory ducts. These secretions form seminal plasma (the fluid portion of the ejaculate), serve as a vehicle for sperm, and also contain a mixture of enzymes, amino acids, hormones, and proteins that play a variety of roles. For example, specific proteins in seminal plasma modulate binding of neutrophils to spermatozoa within the female reproductive tract to either protect viable sperm from phagocytosis or to promote phagocytosis of nonviable sperm. Other components of seminal plasma affect the expression of pro-inflammatory cytokines in the mare's endometrium.

SEMINAL VESICULITIS

Bacterial infection of vesicular glands in stallions is uncommon. The pathogenesis is currently believed to be an ascending infection from the urethra or reflux of urine or semen with no apparent clinical signs, except that fertility is usually affected because of decreased semen quality, and mares bred by the affected stallion may develop a uterine infection. Seminal vesiculitis may become a systemic disease, causing signs of colic, change in gait, dysuria, and/or ejaculation difficulties. Diagnosis is based mainly on the following findings: semen evaluation, enlarged and painful vesicular glands, and variable ultrasonographic findings. Samples for cytologic evaluation and bacterial culture should be collected to confirm the diagnosis.

Treatment of seminal vesiculitis in stallions is achieved by urethroscopy with lavage of the lumen of the affected gland with sterile saline, followed by 3–5 infusions of antibiotics based on culture and sensitivity. Enrofloxacin, a fluoroquinolone antimicrobial, is a good candidate for systemic treatment of infections of the reproductive tract.

Surgical removal of the vesicular glands, termed vesiculectomy, is a treatment option that may result in ejaculatory dysfunction.

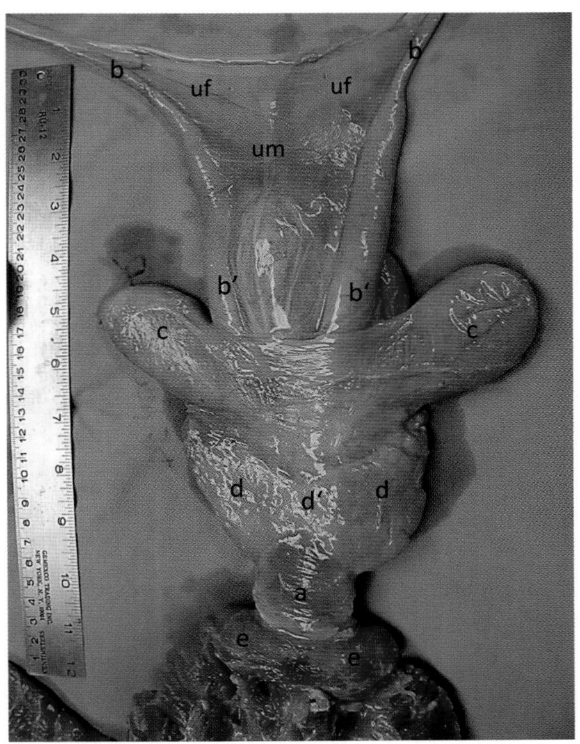

FIGURE 13.8-4 Anatomical prosection of the internal reproductive tract of a stallion, dorsal view. Key: pelvic urethra with the urethralis muscle (a); ducti deferentes (b) ampullae of ducti deferentes (b′); vesicular glands (c); prostate lobes (d) and prostatic isthmus (d′); bulbourethral glands (e); urogenital fold (uf); uterus masculinus (um).

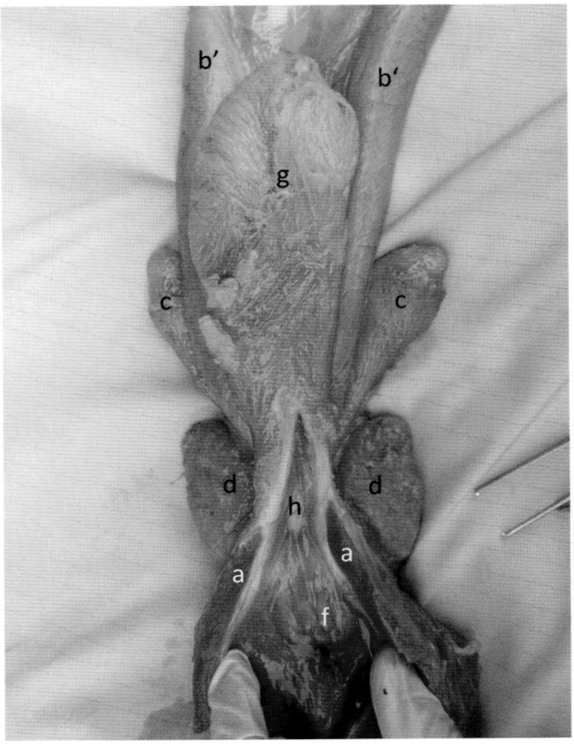

FIGURE 13.8-5 Anatomical prosection of the internal reproductive tract of a stallion, ventral view, with pelvic urethra sectioned longitudinally to expose the colliculus seminalis. Key: urethralis muscle (a); ampullae of ducti deferentes (b′); vesicular glands (c); prostate lobes (d); colliculus seminalis (f); urinary bladder (g); urethral crest (h).

Urethra

The male urethra is divided into pelvic and spongy parts. The **pelvic urethra** is associated with the accessory sex glands, and all the excretory ducts of these glands open into its lumen. The mucosa of the pelvic urethra is covered by transitional epithelium and it has numerous alveolar glands with small openings on its lateral aspect. Beneath the mucosa, there is a vascular layer and a thin smooth muscle layer. The thick, circular, striated **urethralis muscle** surrounds the pelvic urethra (Figs. 13.8-4a and 13.8-5a), starting from the most distal portion of the prostate and continuing until the transition between the pelvic and spongy urethra at the pelvic outlet, before the **ischial arch**. The **spongy urethra** begins here and turns ventrally, tracks around the ischial arch, and joins the penile body (**penile urethra**). The penile urethra is surrounded by the **corpus spongiosum**, and it runs along the body of the penis in the urethral groove of the **corpus cavernosum**. The **bulbospongiosus muscle** runs along the ventral aspect of the penile body and encloses the corpus spongiosum of the urethra.

Contractions of the bulbospongiosus muscle contribute significantly to the process of ejaculation.

Colliculus seminalis

Sperm is deposited into the urethral lumen during the emission process, which directly proceeds ejaculation. Semen is deposited through the **ejaculatory ducts** (Fig. 13.8-7) that open on a dorsal prominence of the urethral mucosa called the **colliculus seminalis** (Figs. 13.8-5f, 13.8-6, and 13.8-7f). This structure lies between the bulbourethral glands and the caudal edge of the prostate. Each ejaculatory duct is a result of the fusion of the excretory duct of one vesicular gland and the most terminal and narrow portion of the ipsilateral ductus deferens (ampullar excretory duct). The ejaculatory ducts are short and wide with no sphincter. Approximately 15% of stallions do not have ejaculatory ducts. In these individuals, the excretory ducts of the vesicular glands and ampullar excretory ducts do not fuse but instead open separately on the colliculus seminalis.

The anatomy of the ejaculatory ducts allows for relatively easy insertion of a flexible endoscope or thin catheter into the lumen of each vesicular gland, through the ejaculatory orifice (Fig. 13.8-6eo).

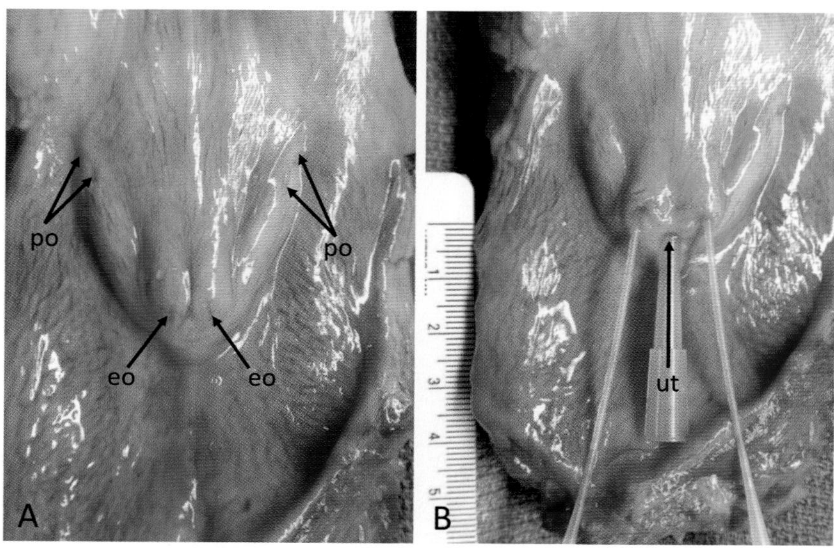

FIGURE 13.8-6 Anatomical prosection of the colliculus seminalis of a stallion, pelvic urethra sectioned longitudinally. *Key*: orifice of the prostatic duct (po); ejaculatory orifice (eo); utriculus seminalis orifice (ut).

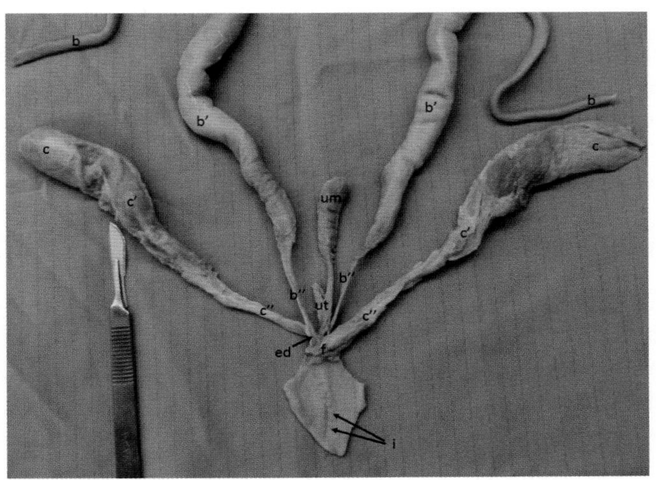

FIGURE 13.8-7 Anatomical prosection of the ducti deferentes, vesicular glands, and embryological remnants in the stallion reproductive tract, fixed in 10% formalin and dissected. Key: ducti deferentes (b), ampullae of ducti deferentes (b′), ampullar excretory ducts (b″); vesicular glands—fundus (c), body (c′), excretory duct (c″); ejaculatory duct (ed); colliculus seminalis (f); utriculus masculinus (ut); cystic uterus masculinus (um); openings of the excretory ducts of the bulbourethral glands (i).

Ampullae of the ducti deferentes

Each **ductus deferens** (vas deferens, deferent duct) tracks from the tail of the epididymis through the inguinal canal, abdominal and pelvic cavities, and opens into the ejaculatory duct or directly into the urethra. Initially, its diameter is narrow (Figs. 13.8-4b and 13.8-7b); however, at the level of the urinary bladder, its wall becomes glandular and thick, forming a distinct structure called the **ampullae** (Figs. 13.8-4b′, 13.8-5b′, and 13.8-7b′). Both ampullae run in a caudal-medial direction over the dorsal surface of the urinary bladder toward the urethra. They descend underneath the isthmus of the prostate and gradually lose their glandular component to become narrow terminal portions of the ducti deferentes, which are also called **ampullar excretory ducts** (Fig. 13.8-7b″).

The ampullae are suspended between layers of the **urogenital fold** (Fig. 13.8-4uf) together with the vesicular glands and ureters. A rudimentary **uterus masculinus** is often present within the urogenital fold, between the ampullae, on the dorsal side of the urinary bladder, or underneath the prostatic isthmus (see side box entitled "Uterus and utriculus masculinus"). The uterus masculinus may consist of a body and 2 processes (or horns); it may be a short tube or exist as a solid band of tissue (Fig. 13.8-4um). The glandular epithelium of the uterus masculinus is capable of producing secretions which contribute to the formation of small or large cysts.

While these cysts, if large, may contribute to ejaculatory problems in stallions, they are often found in individuals with normal ejaculatory patterns as well.

UTERUS AND UTRICULUS MASCULINUS

Another anatomical variant of the colliculus seminalis that has clinical implications is the presence of rudimentary remnants of the female ductal system such as uterus masculinus (Fig. 13.8-7um) or utriculus masculinus (Fig. 13.8-7ut), often called "vagina masculinus." The latter usually has a single opening between the ejaculatory orifices (Fig. 13.8-6ut). If there is no communication between the lumen of the utriculus masculinus and that of the pelvic urethra, fluid-filled cysts can form, which can interfere with emission and ejaculation. These cysts are called "midline cysts of the colliculus seminalis."

The mucosa of the ductus deferens is lined with a pseudostratified columnar epithelium, which often becomes a simple columnar epithelium closer to the end of the duct. The ampullae contain simple branched tubule-alveolar glands lined with tall columnar and cuboidal cells, with occasional basal cells. The mucosa is folded, which allows for significant expansion of the lumen of the ampulla during ejaculation. There are numerous depressions and diverticula in the mucosa into which the secretions of these glands are deposited. In addition, these depressions often contain spermatozoa as well as crystals and calcified concretions.

Vesicular glands (seminal vesicles)

The **vesicular glands** in the horse are elongated hollow sacs which lie dorsolateral to the ampullae of the ducti deferentes (Figs. 13.8-4c and 13.8-5c). Each gland consists of a fundus, body, and excretory duct. The **fundus** of each vesicular gland may extend over the pelvic brim if its lumen is filled with large amounts of secretions. Only the anterior portions of the vesicular glands are covered by peritoneum, and these protrude from the cranial edge of the urogenital fold into the abdominal cavity. The remaining portions of the glands lie extraperitoneal.

The glands have a muscular wall consisting of 2 planes of longitudinal fibers and a circular layer between them. Their lumens are lined with a thick mucous membrane with short, branched, tubule-alveolar glands separated by thin trabeculae. The **excretory duct** of each gland descends under the prostate, similar to the ampulla of the ductus deferens, and joins the ipsilateral ampullar excretory duct to form a short **ejaculatory duct**, which opens at the colliculus seminalis (Fig. 13.8-5f).

Although seminal vesiculitis rarely occurs in stallions, it has serious consequences and is difficult to treat (see side box entitled "Seminal vesiculitis").

Prostate pathology is rare in horses. Prostate masses are occasionally seen in geldings with dysuria. There is one reported case of a poorly differentiated carcinoma of primary prostate origin.

Vesicular glands in stallions produce variable amounts of gel, which is expelled as distinct boluses into the pelvic urethra during the second half of ejaculation, following the cessation of ampullar and prostatic activities. Therefore, the gel fraction of the equine ejaculate contains a low number of sperm.

Prostate

The equine **prostate gland** consists of 2 large **lobes** and the **isthmus**, which lies on the neck of the bladder and connects both lobes (Figs. 13.8-4d, d′ and 13.8-5d). The dorsal surface of the prostate resembles the shape of a butterfly, with the lobes as wings (Fig. 13.8-4d) and the isthmus as the body (Fig. 13.8-4d′). The lobes are thick and have somewhat prismatic shapes. They encroach

AMPULLARY OCCLUSION

Ampullar content starts moving through its excurrent ducts approximately 3 s before ejaculation—i.e., the emission process. In this way, sperm-rich content is delivered into the urethral lumen. In most cases, this process ends before urethral contractions and ejaculation begin. Occasionally, the lumen of one or both distal ampullae or their excurrent ducts become occluded with inspissated sperm or thick acellular material. The clinical signs associated with the condition include oligospermia, azoospermia, a low percentage of morphologically normal sperm, a high percentage of tailless sperm heads, and palpable enlargement of the distal ampullae. Ultrasound evaluation reveals a distended lumen of the ampulla with either hypoechoic or hyperechoic contents in the ampullar lumen or ampullar excurrent duct. Manual massage of the ampullae and/or administration of oxytocin (20 I.U., IV) 5–10 min before ejaculation usually helps solve this problem. Several attempts may be necessary to remove the blockage. The diameter and contents of the ampullar lumen should appreciably decrease when treatment is successful. Infections of ampullae are extremely rare. One case of severe ampullitis associated with *Pseudomonas aeruginosa* infection and seminal vesiculitis has been reported in a stallion.

the neck of the bladder and proximal urethra from both sides, but do not surround these structures as seen in other species. The isthmus is thin and located on the dorsal surface of the neck of the bladder. It covers the excretory ducts of the vesicular glands and ampullae as well as the uterus masculinus. The most distal part of the prostate is covered by the **urethralis muscle**.

The prostate is surrounded by a fibromuscular capsule and is divided into spheroid and ovoid lobules by thick trabeculae. Each lobule has a central space called a **diverticulum** in which the secretions are collected. The **secretory tubules** (alveoli) are lined by cuboidal epithelium, which changes into a striated columnar or transitional epithelium at the terminal portions of the ducts. Prostatic secretions are delivered to the urethral lumen via **prostatic ducts** that open lateral to the colliculus seminalis, with 15–30 slit-like openings on each side (Fig. 13.8-6po). There is no disseminate portion (i.e., lacking a glandular layer in the wall of the urethra) of a prostate in a horse. Prostatic secretions are delivered into the pelvic urethra during emission and the first half of ejaculation. This process contributes to the continuous and progressive dilution of sperm during ejaculation in consecutive ejaculatory jets.

Bulbourethral glands

The paired **bulbourethral glands** are located on the dorsolateral aspect of the pelvic urethra at the pelvic outlet, close to the ischial arch. The glands have an ovoid shape, are approximately walnut-sized, and their long axes are directed obliquely cranially and laterally (Fig. 13.8-4e). Each gland is encapsulated by the **bulbourethralis muscle** (also called the "bulboglandularis muscle"), which is a continuation of the urethralis muscle. They deliver their secretions to the pelvic urethra through 12–16 ducts (6–8 ducts per gland), which open as 2 parallel rows of small papilla-like openings on the dorsal aspect of the urethra, close to the midline plane (Fig. 13.8-7i).

The bulbourethral glands in stallions have a fibro-elastic capsule that is thinner and has fewer muscle fibers than that of the prostate. The glands are tubule-alveolar, with alveoli lined by cuboidal epithelium, collecting ducts lined by cuboidal or columnar epithelium, and large intraglandular ducts lined by pseudostratified columnar epithelium. The excretory ducts are lined by transitional epithelium. The bulbourethral glands do not contribute to equine ejaculate. They produce watery presperm fluid, which is expelled before emission and ejaculation to wash away any residual urine.

Blood supply, lymphatics, and innervation

The **internal iliac arteries** (terminal branches of the aorta) are short and separate into **internal pudendal** and **caudal gluteal arteries**. The internal pudendal artery is primarily visceral (organs) in its distribution while the **prostatic artery**, the most important branch of the internal pudendal artery, supplies blood to the bladder, urethra, and accessory genital glands. Similar to other part of the body, the pathway of the veins parallels that of the arteries. Lymph nodes of the pelvis comprise many individual nodes joining as the lateral and medial (major collecting system for the pelvic walls) **iliac nodes** and drain into the **aortic lumbar nodes** or **lumbar trunk**.

The lumbosacral plexus begins with the 4th lumbar nerve and ends with the 2nd sacral nerve (L4–S2), traverses the pelvis, and gives off important branches to the pelvic viscera including the **pudendal, deep perineal, caudal rectal,** and **pelvic nerves**. The pelvic nerves (S2–S4) are composed of parasympathetic fibers.

Several nerves—obturator, cranial, and caudal gluteals, and sciatic—arise from the lumbosacral plexus, track through the bony pelvis, and pass through or near the obturator foramen and lesser sciatic foramen. These nerves are subject to injury when fractures of the pelvis occur, or they may be compressed during foaling.

Selected references

[1] Dellman HD, Eurell J. Textbook of veterinary histology. Baltimore: Williams & Wilkins; 1998. p. 226–46.
[2] Love CC, Riera FL, Oristaglio RM, et al. Sperm occluded (plugged) ampullae in the stallion. Proc Soc Theriogenol 1992;117–25.
[3] Mancill CC. Clinical commentary: clinical and sub-clinical seminal vesiculitis in the stallion. Equine Vet Educ 2010;22:220–2.
[4] Nickel R, Schummer A, Seiferle E. The viscera of domestic animals. 2nd revised ed. New York: Springer-Verlag; 1979. p. 340–8.
[5] Pozor M, Macpherson ML, Troedsson MH, et al. Midline cysts of colliculus seminalis causing ejaculatory problems in stallions. J Equine Vet Sci 2011;31:722–31.

CHAPTER 13

CASE 13.9

Inguinal Hernia

Nora S. Grenager[a], Mathew P. Gerard[b], and James A. Orsini[c]
[a]Internal Medicine, Steinbeck Peninsula Equine Clinics, Menlo Park, California, US
[b]Department of Molecular Biomedical Sciences, North Carolina State University College of Veterinary Medicine, Raleigh, North Carolina, US
[c]Department of Clinical Studies - New Bolton Center, School of Veterinary Medicine University of Pennsylvania, Kennett Square, Pennsylvania, US

Clinical case

History

A 3-year-old Standardbred stallion presented with signs of moderate colic of 2 hours' duration with increasing intensity. He was initially treated with flunixin meglumine by the trainer, but the signs did not improve. He had no history of earlier colic and had raced successfully the day before presentation.

Physical examination findings

The stallion was moderately tachycardic (heart rate of 56 bpm), slightly sweaty (i.e., diaphoresis/sudoresis [Gr. *diaphorēsis* profuse perspiration]), exhibiting flehmen (Gr. to bare the upper teeth), and was pawing. The mucous membranes were pale pink and slightly tacky with a CRT of 1.5 seconds. Rectal temperature was 99.7 °F (37.6 °C). Gastrointestinal borborygmi were decreased in all quadrants, and no pings were ausculted (L. *auscultare* to listen to). The left half of the scrotum was approximately twice the size of the right (Fig. 13.9-1).

Differential diagnoses

Inguinal hernia, testicular torsion, orchitis, testicular neoplasia

Diagnostics

The stallion was sedated for palpation of the scrotum, which revealed generalized thickening of the left scrotal sac, correct anatomical position of the testis (head of epididymis was cranial), and indiscrete turgid soft tissue palpable within the sac dorsal to—and seemingly separate from—the testis. It was not possible to definitively palpate the left spermatic cord. The right scrotum and testicle palpated to be normal.

Abdominal palpation per rectum revealed mild gas distention of the large colon, no tight colonic bands, normal nephrosplenic space, and several "bicycle tire-sized" loops of distended small intestine in the left caudal abdomen. Small intestine was palpated entering the left inguinal canal. The entrance to the right inguinal canal was normal.

Nasogastric intubation yielded 5 L net of mildly malodorous, feed-colored nasogastric reflux.

Comparative Veterinary Anatomy: A Clinical Approach. https://doi.org/10.1016/B978-0-323-91015-6.00068-6

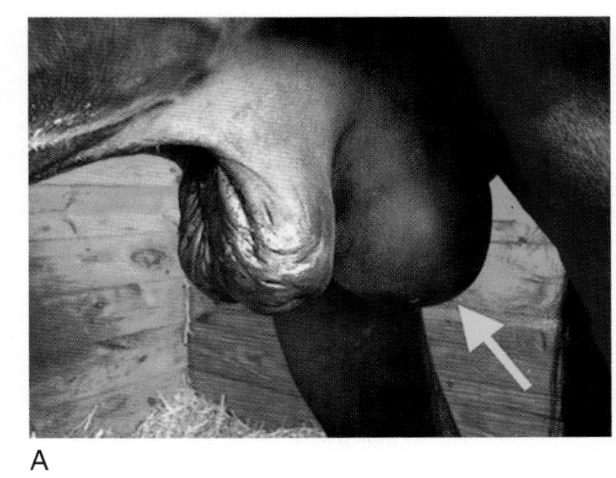

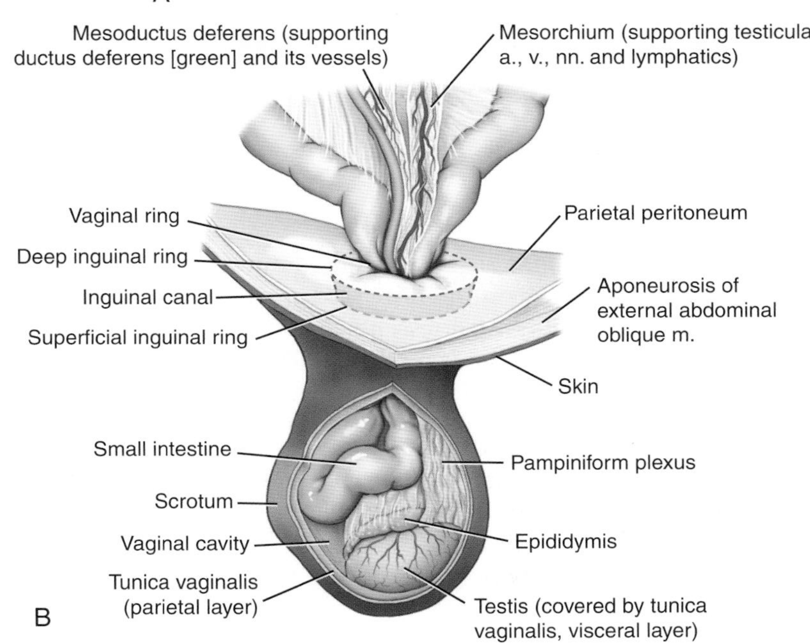

FIGURE 13.9-1 (A) The scrotum of the stallion in this case is grossly enlarged, particularly on the left side dorsal to the testis *(arrow)*. (B) Illustration depicting an indirect inguinal hernia involving the small intestine. (*Photo courtesy of Dr. Regina Turner, University of Pennsylvania.*)

Abdominal ultrasound revealed a mild increase in anechoic peritoneal fluid, a few loops of amotile, distended small intestine with wall thickness of approximately 3–4 mm (0.12–0.16 in.; normal is ≤3 mm [0.12 in.]) in the left flank, and decreased duodenal motility. The scrotal ultrasound (Fig. 13.9-2) revealed at least 2 loops of amotile, distended small intestine with wall thickness of approximately 3–4 mm (0.12–0.16 in.) within the left vaginal cavity, adjacent to a normal-appearing testis. The right testis appeared normal with a slight increase in anechoic fluid in the vaginal cavity.

Diagnosis

Unilateral inguinal hernia

Treatment

Corrective surgery with hemi-castration was recommended. The stallion was placed under general anesthesia for exploratory celiotomy via a ventral midline incision. A second incision via an inguinal approach over the left superficial inguinal ring exposed the intestine entrapped in the vaginal cavity. An indirect inguinal hernia was

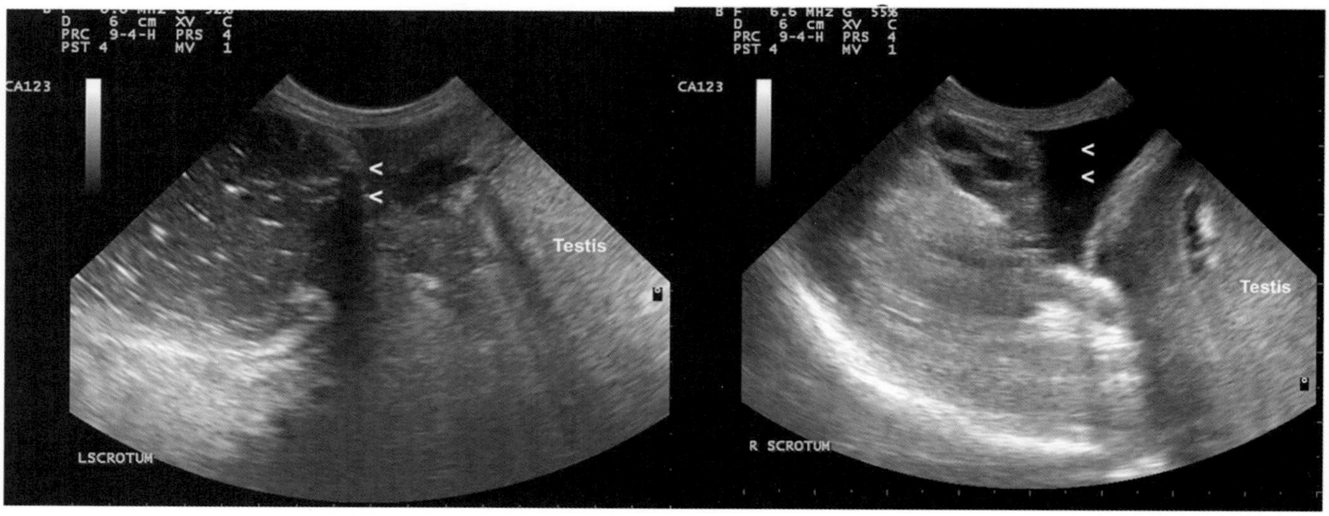

FIGURE 13.9-2 Ultrasonographic images of the scrotum. (A) Left scrotal sac, showing a loop of distended small intestine *(arrowheads)* next to a normal left testis. (B) Right scrotal sac for comparison. There is a small amount of anechoic fluid *(arrowheads)* in the vaginal cavity and a normal testis.

diagnosed (Fig. 13.9-1B). The herniated small intestine was reduced through the inguinal canal into the abdomen. The intestine directly oral to the entrapped segment was distended and mildly thickened. The entrapped portion of intestine was purple and cool with poor viability. After resection and anastomosis (jejunojejunostomy—i.e., between 2 parts of the jejunum), open castration was performed on the left testis and the spermatic cord and vaginal tunic were ligated as proximally as possible. The superficial inguinal ring was closed with suture to prevent repeated inguinal herniation. The stallion had an uneventful anesthetic and clinical recovery.

The stallion remained at increased risk of an inguinal hernia via the right inguinal canal; however, the owner wanted to preserve the option of using the stallion for breeding and therefore only consented to a unilateral castration.

Anatomical features in equids

Introduction

The major parts of the male reproductive system include the testes; epididymis and ductus deferens associated with each testis; the penile urethra, penis, scrotum, and accessory sex glands (ampulla of the ductus deferentes, vesicular glands, prostate, and bulbourethral glands). The penis, prepuce, and penile urethra are discussed in Case 13.7 and information about the accessory sex glands is in Case 13.8.

Function

Each testis produces spermatozoa within the microscopic seminiferous tubules and testosterone within the interstitial cells between the tubules. Sperm mature in the epididymis and, at the time of ejaculation, the accessory sex glands supply added fluids to the semen to nourish and extend the life of the sperm. The ductus deferens is continuous with the epididymis and transports spermatozoa from the tail of the epididymis to the urethra.

Testes and scrotum (Figs. 13.9-3 and 13.9-4)

Male equids have paired **testes** (singular, testis) that lie within the scrotum. The testes are ellipsoid in shape, though slightly compressed from side to side. The **scrotum** lies below the pubic brim and the testes lie horizontally within the scrotum with their long axis almost parallel to the horse's vertebral column. Strong contraction of the **cremaster muscle**—which acts on the cranial pole of the testis via its insertion on the vaginal tunic—can pull the testis to an almost vertical position. The cremaster muscle is a slip of muscle off the caudal margin of the internal abdominal

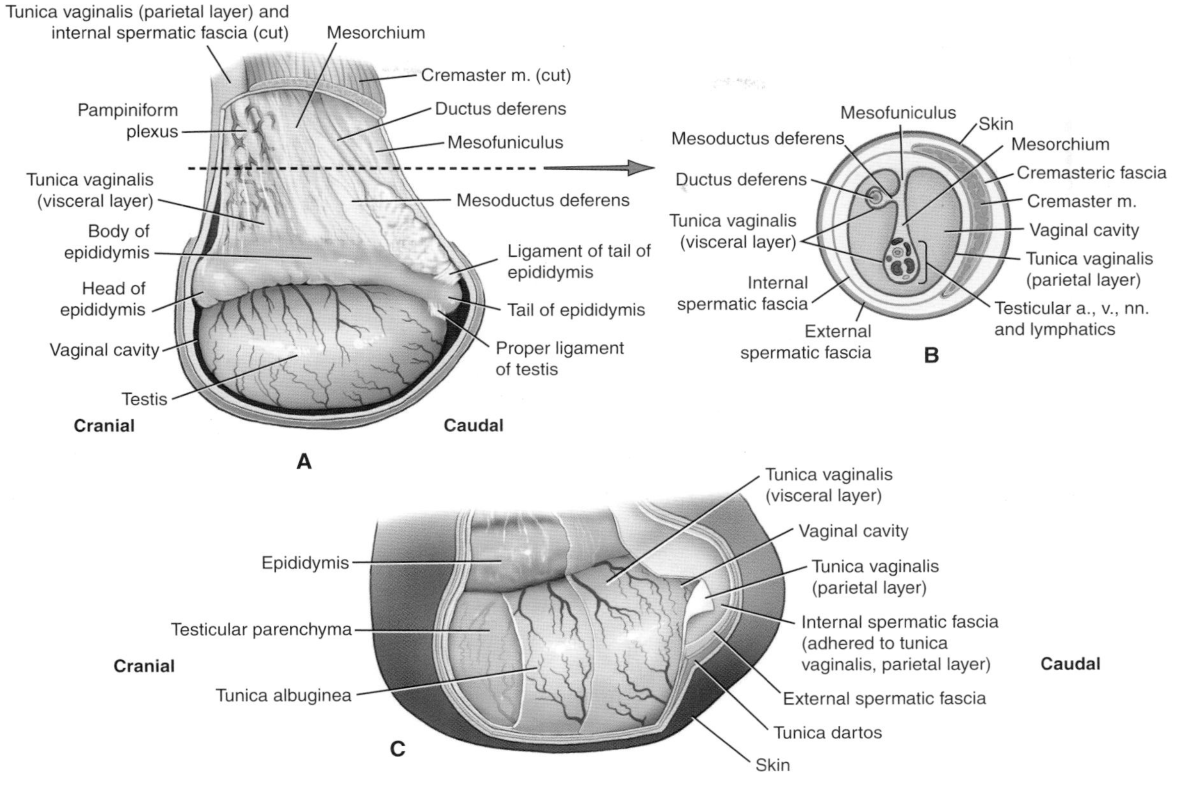

FIGURE 13.9-3 Equine testis, epididymis, and associated tissues. (A) Lateral view—parietal part of tunica vaginalis and internal spermatic fascia cut. (B) Cross-sectional view of spermatic cord, vaginal cavity, and surrounding tissues. (C) Tissue layers covering testis.

oblique muscle. The muscle passes through the inguinal canal, associated with the lateral external surface of the vaginal tunic and contained within its own fascia.

The parenchyma of each testis is encapsulated by the **tunica albuginea** and the **visceral vaginal tunic**.

Trabeculae from the tunica albuginea penetrate the testis, dividing it into lobules and providing support. The **parietal vaginal tunic** is an outpouching of the parietal peritoneum, and passes through the inguinal canal (described later). The name change from "peritoneum" to "vaginal tunic" occurs at the point of evagination (i.e., at the vaginal ring, described later). Visceral peritoneum surrounding testicular neurovascular and lymphatic structures, and surrounding the ductus deferens and its vessels, becomes **visceral vaginal tunic** at the vaginal ring. The **vaginal canal** and **cavity**, which lie between the visceral and parietal vaginal tunics are a diverticulum of the peritoneal cavity.

The continuous nature of the peritoneal cavity into the scrotum's vaginal cavity leads to several possible situations. First, there is a risk of peritonitis when there is infection within the vaginal cavity or following castration. Second, intestine can travel through the inguinal canal and enter the vaginal cavity (i.e., an indirect inguinal hernia).

TESTICULAR NEOPLASIA

This is uncommon in stallions and does not typically cause testicular enlargement. The most common types of neoplasia to affect the testes are benign and arise from germinal cells, including: teratomas, seminomas, teratocarcinomas, and embryocarcinomas. Nongerminal tumors arising from stromal cells include: Sertoli and Leydig cell tumors. Malignant seminomas have been reported.

CASTRATION—SURGICAL OPTIONS

Orchidectomy, orchiectomy, emasculation, gelding, and cutting are all terms that are synonymous with the word "castration"—one of the most common surgical procedures performed in horses as a means of sterilization. Castration is usually used for 1 of 2 reasons—first, the genetic pool of the stallion is not considered a beneficial contributor to future progeny and/or second, to eliminate the potentially aggressive behavior of stallions.

Castration can be done with the horse standing or in recumbency. In a standing castration, the horse is sedated and the scrotum, testes, and spermatic cord are locally anesthetized; in a recumbent castration, the horse is anesthetized (typically with IV anesthetics). The instrument to perform the castration or emasculation (L. *emasculare* to castrate) is termed an "emasculator." Common emasculators used for equine castration include White's, Reimer, and Serra. The Henderson Equine Castrating Instrument is also gaining popularity. Which procedure and which emasculator are used is at the surgeon's discretion based on the stallion's temperament, age, and available help and facilities.

Donkeys and mules are often especially challenging to castrate using a standing procedure because their lively or agile nature creates a potentially unsafe situation for the surgeon.

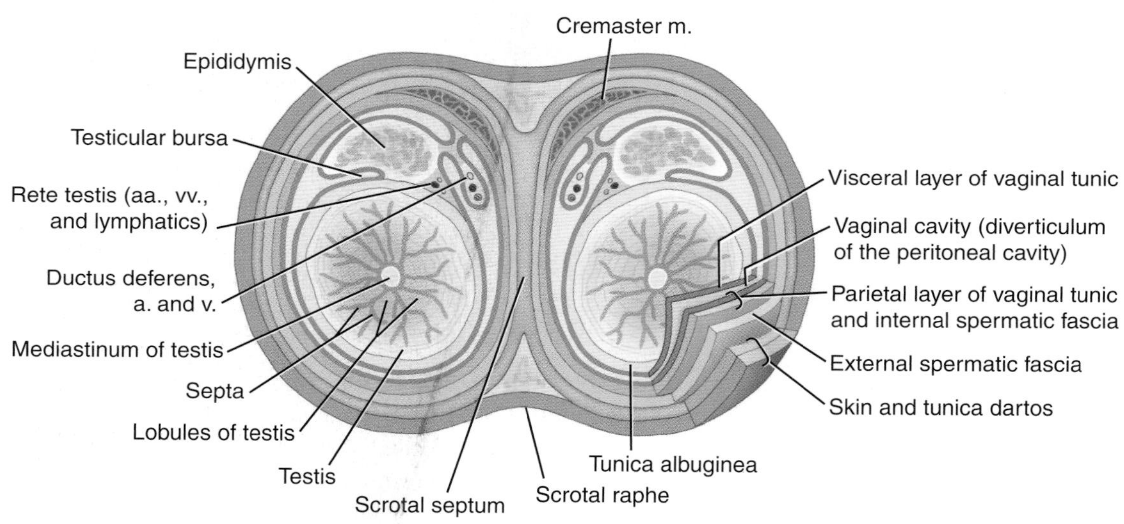

FIGURE 13.9-4 Transverse section through the scrotum, testes, epididymis.

The terms "vaginal cavity" and "vaginal process" are used interchangeably; however, the former refers to a space and the latter to the tissues creating the walls of the space—i.e., the parietal and visceral layers of the vaginal tunic. Frequently, the term "vaginal tunic" is used clinically when referring to the parietal layer of the tunic. Specifically, it should be understood that 2 tissue layers are actually included in the clinical "vaginal tunic." The parietal vaginal tunic is inseparable from the overlying internal spermatic fascia, which contributes to the opaque appearance and toughness of the combined tissues. The internal spermatic fascia continues the transversalis fascia of the abdomen at the vaginal ring. At the superficial inguinal ring the clinical vaginal tunic and attached cremaster muscle are joined by a covering layer of external spermatic fascia (see below). When the term "vaginal tunic" is used in this text, without further qualification, it is referring to the adhered internal spermatic fascia and parietal vaginal tunic.

The scrotal wall is composed of lightly haired skin with a modified subcutaneous layer of closely adherent connective tissue and smooth muscle called the **tunica dartos**. The tunica dartos surrounding left and right scrotal sacs combines on midline between the sacs to form the **scrotal septum**, which divides the scrotal cavity into 2 roughly symmetrical halves. The **scrotal (median) raphe** is the line formed where the skin of the right and left halves of the scrotum joins on the midline. The **external spermatic fascia** (i.e., the **scrotal fascia**), which lies between the tunica dartos and the vaginal tunic, consists of loose connective tissue that allows the testes to move freely within the scrotum. The external spermatic fascia is continuous with the fascia covering the external abdominal oblique muscle. This fascia reflects onto the vaginal tunic as it exits the inguinal canal at the superficial inguinal ring. The **scrotal ligament**, continuing the ligament of the tail of the epididymis, lies between the parietal vaginal tunic and the tunica dartos of the scrotum at the caudal pole of the testis. The scrotal ligament is a remnant of the **gubernaculum** (L. helm, rudder, a structure that guides) **testis**.

The 2 main types of castration are open and closed, defined by whether or not the vaginal tunic and cavity are opened. In closed castration, the vaginal tunic stays intact and the tunic, the spermatic cord within, and the cremaster muscle on the outside of the tunic are emasculated together. In open castration, the parietal layer of the vaginal tunic is incised but the visceral layer of the vaginal tunic remains intact; the vascular spermatic cord and ductus deferens are separately emasculated. The visceral layer of the vaginal tunic and tunica albuginea are breached if the parenchyma of the testis is incised due to the intimate relationship of the visceral layer and the tunica albuginea. Open castration is preferred in mature stallions with larger testes as it decreases the risk of post-operative bleeding because the vascular spermatic cord is isolated from other tissues and less tissue is incorporated during emasculation and/or ligation.

During routine castration, 2 scrotal skin incisions are made parallel and a few centimeters abaxial to the scrotal (median) raphe, along the long axis of the testes. Alternatively, a section of scrotal skin, including the scrotal raphe, is completely removed, to expose both testes through one incision and allowing for improved drainage post-operatively.

The scrotal ligament is used to help find a retained testis during cryptorchid surgery, because of its attachment to the parietal vaginal tunic. It can often be seen at the medial or lateral cranial aspect of the superficial inguinal ring. Gentle traction on this ligament can evert the vaginal process (vaginal cavity). Opening of the vaginal process leads to identifying the tail of the epididymis because it is attached to the former via the ligament of the tail of the epididymis.

CRYPTORCHIDISM

A cryptorchid (Gr. *kryptos* hidden + Gr. *orchis* testis) testis is one that does not properly descend to its final location in the scrotum. The gubernaculum testis extends from the caudal pole of each embryonic testis to the inguinal canal. The testes follow the gubernacula testes from their position near the kidneys into the inguinal canal and, finally, the scrotum. The driving force behind testicular descent is not fully understood.

Cryptorchidism is more common in Quarter Horses, Percherons, Saddlebreds, and ponies; it is considered heritable. It can be unilateral or bilateral, and the location can be abdominal or inguinal. Right and left retained testes are equally common, but left retained testes are more likely to be abdominal while right retained testes are more likely to be inguinal. Laparoscopic (Gr. *lapara* flank + Gr. *skopein* to examine) surgery can be used to find and remove a retained abdominal testis. Laparoscopic surgery is considered a "minimally invasive" surgery procedure.

In humans, the lower part of the peritoneal diverticulum stays patent as the tunica vaginalis testis while the upper part becomes obliterated. This obliteration normally happens during the first post-natal year or even later; the right side is obliterated later than the left. This factor may be one explanation for the higher incidence of right-sided inguinal hernias in people. A similar phenomenon may occur in the horse.

Testes and scrotum—blood supply, lymphatics, and innervation (Figs. 13.9-3 and 13.9-5)

Testicular vessels include the testicular artery and vein traveling within the spermatic cord, along with the lymphatics and nerves. The arteries are direct branches of the abdominal aorta. The testicular veins return to the caudal vena cava. The **testicular artery** courses through the mesorchium caudally along the epididymal border of the testis, passes around the caudal pole of the testis, and returns cranially along the free border of the testis. Branches to the medial and lateral surfaces of the testis exit the testicular artery along the free border. The **testicular vein** forms the **pampiniform plexus** around the testicular artery. The lymphatics of the testis drain into the **medial iliac** and **lumbar aortic lymph nodes**. The **testicular plexus** is formed of autonomic and visceral sensory nerves traveling with the artery and vein. Scrotal blood supply comes from the **external pudendal vessels**; the **cremasteric artery** supplies the cremaster muscle. Scrotal nerve supply comes from the **iliohypogastric nerve** (L1), **ilioinguinal nerve** (L2), **genitofemoral nerve** (L3 > L2, L4), and **perineal nerve** (S2–S5). The vessels and nerves enter and exit the scrotum outside the vaginal tunic. The scrotal lymphatics drain to the **superficial inguinal lymph nodes** ("scrotal" lymph nodes).

The superficial inguinal lymph nodes may be enlarged (lymphadenopathy [L. *lymph-* water + Gr. *adēn* gland + Gr. -pathy *patheia*, from *pathos* disease = disease of the lymph nodes]) with scrotal conditions such as infections secondary to castration or neoplasia that extend beyond the vaginal tunic. Conditions that are limited to the testes—e.g., orchitis, testicular neoplasia, and some post-castration complications—would alternatively drain to the medial iliac lymph nodes, which are not palpable transcutaneously.

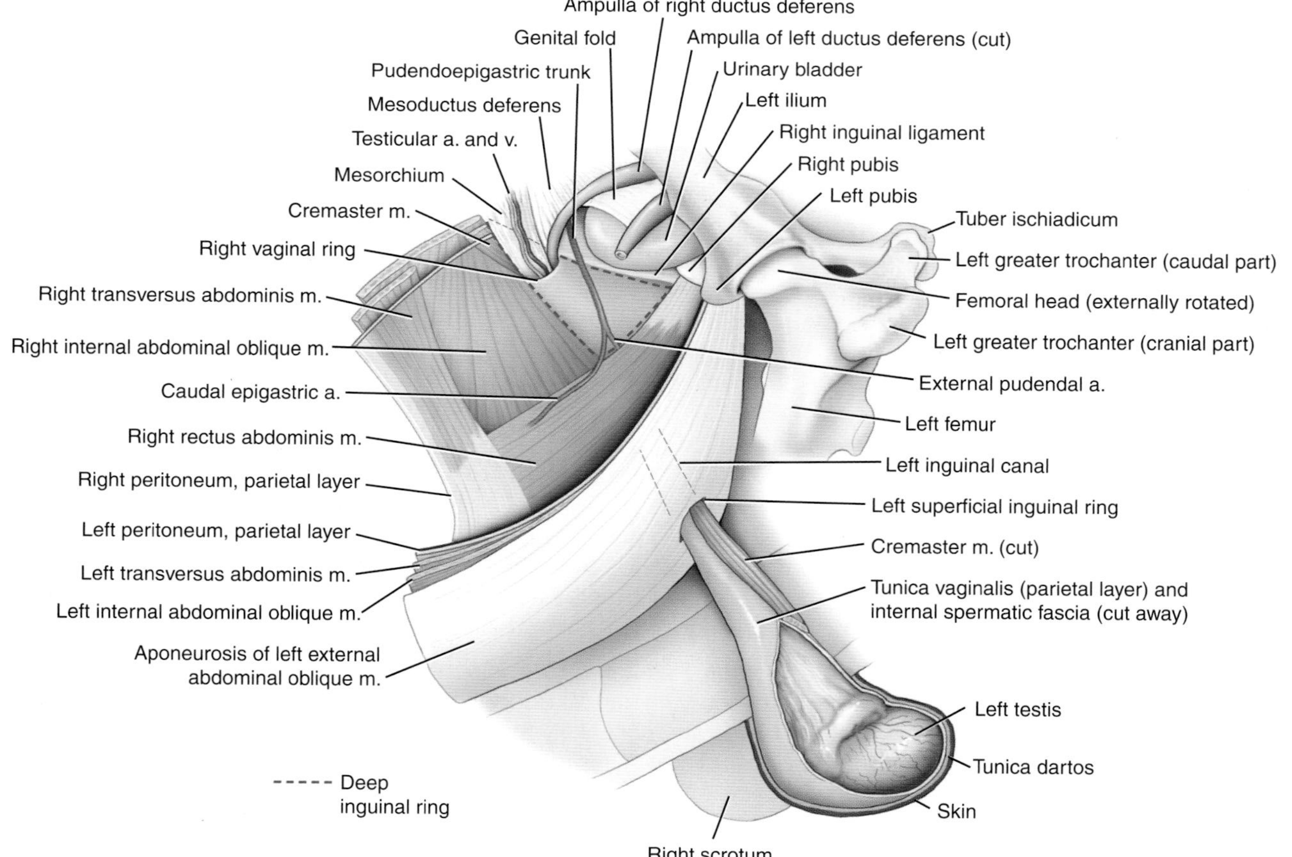

FIGURE 13.9-5 Deep and superficial equine inguinal rings, vaginal ring, blood vessels, and associated tissues.

Epididymis (Fig. 13.9-3**)**

The **epididymis** is attached to—and lies along the dorsal (and slightly lateral) border of—the testis. The laterally **located testicular bursa** is the potential space between the epididymis and testis where a slight lateral overlap occurs. The enlarged head of the epididymis is at the cranial aspect of the testis, while the enlarged tail is at the caudal aspect of the testis. The epididymis is also covered by a serous layer, the visceral layer of the vaginal tunic. The **proper ligament of the testis** is the fold between the testis and the tail of the epididymis. The **ligament of the tail of the epididymis**, which connects the tail of the epididymis to the parietal vaginal tunic, is the thickened distal free border of the **mesorchium** (homologue of the **mesovarium** in mares, Gr. *meso*-in the middle or intermediate). The mesorchium is quite thick in equids and is a connecting serous membrane continuous with the visceral vaginal tunic containing the testicular neurovascular structures. Similarly, the visceral vaginal tunic surrounding the ductus deferens and its vessels is continuous with its connecting **mesoductus deferens**. The mesoductus deferens joins to the mesorchium and at this junction a single connecting membrane, the **mesofuniculus**, continues to the parietal vaginal tunic. The proper ligament of the testis and the ligament of the tail of the epididymis are remnants of the gubernaculum testis.

In the area of the head of the epididymis and on the surface of the testis, you commonly find small outgrowths of tissue referred to as **appendix** (L. *appendere* to hang upon) **testis**. These are remnants of the paramesonephric ducts and are of no clinical significance.

> The proper location of the testis can be confirmed by palpation of the epididymis (i.e., in stallions with testicular enlargement, swelling, or pain) and this technique can be used to rule out testicular torsion.

> The ligament of the tail of the epididymis is relatively thick and must be transected during open castration. In standing castrations, traction on the testes and vaginal tunic is done by perforating the mesofuniculus. A finger is hooked through the perforation and anchored over the ligament of the tail of the epididymis when downward tension is applied.

> Paramesonephric (or Müllerian) ducts are paired ducts of the embryo on the lateral sides of the urogenital ridge which terminate at the sinus tubercle in the primitive urogenital sinus. In the female, they form the fallopian tubes, uterus, cervix, and the upper one-third of the vagina. In the male, these ducts degenerate or become a vestigial (L. *vestigium* remnant, footprint) structure or appendix testis; the contiguous mesonephric ducts develop into male reproductive organs.

Ductus deferentes and accessory sex glands

This information is covered in Case 13.8.

Spermatic cord (Fig. 13.9-6C**)**

The anatomic **spermatic cord** for each testis is comprised of the mesorchium and visceral tunic containing the testicular artery, vein, lymphatics, and plexus of nerves (located at the cranial aspect of the cord), and the mesoductus deferens and visceral tunic containing the ductus deferens and its vessels (at the caudal aspect of the

TESTICULAR DESCENT

Testicular descent is slightly different in equids compared to other domestic species. The testes of the fetal colt greatly enlarge between the 100th and 250th days of gestation, delaying the testes at the vaginal ring on approximately the 120th day. The testes do not continue migration until their size has reduced, arriving in the scrotum within 2 weeks of birth (before or after). The vaginal ring contracts soon after birth, making it difficult for a retained testis to finish the descent.

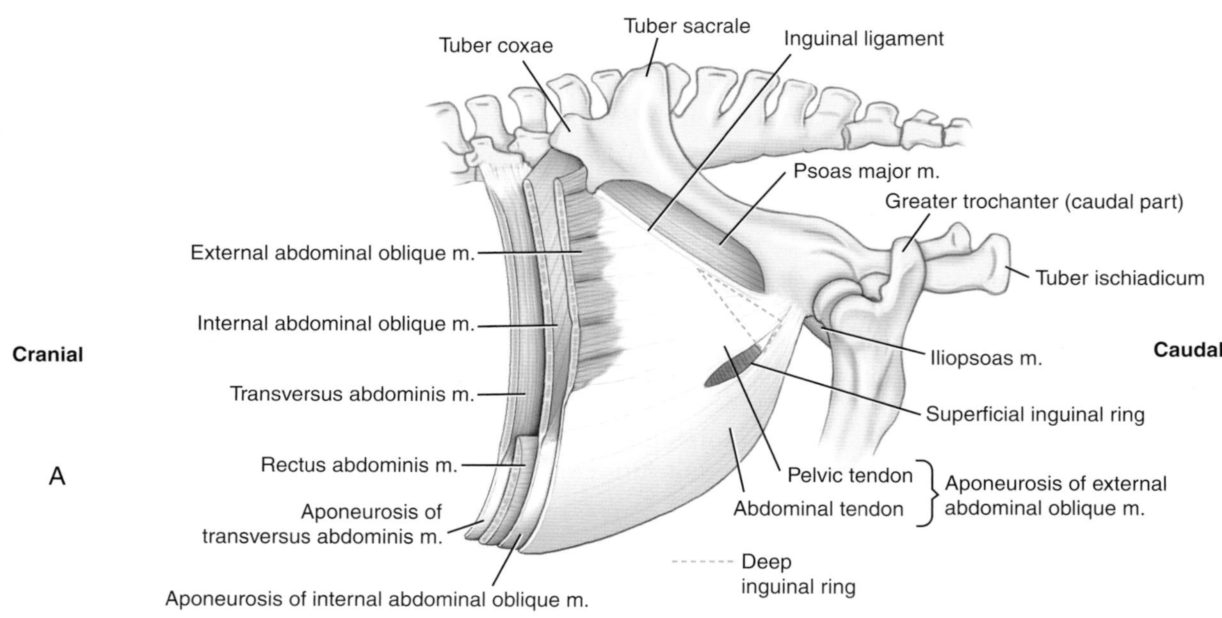

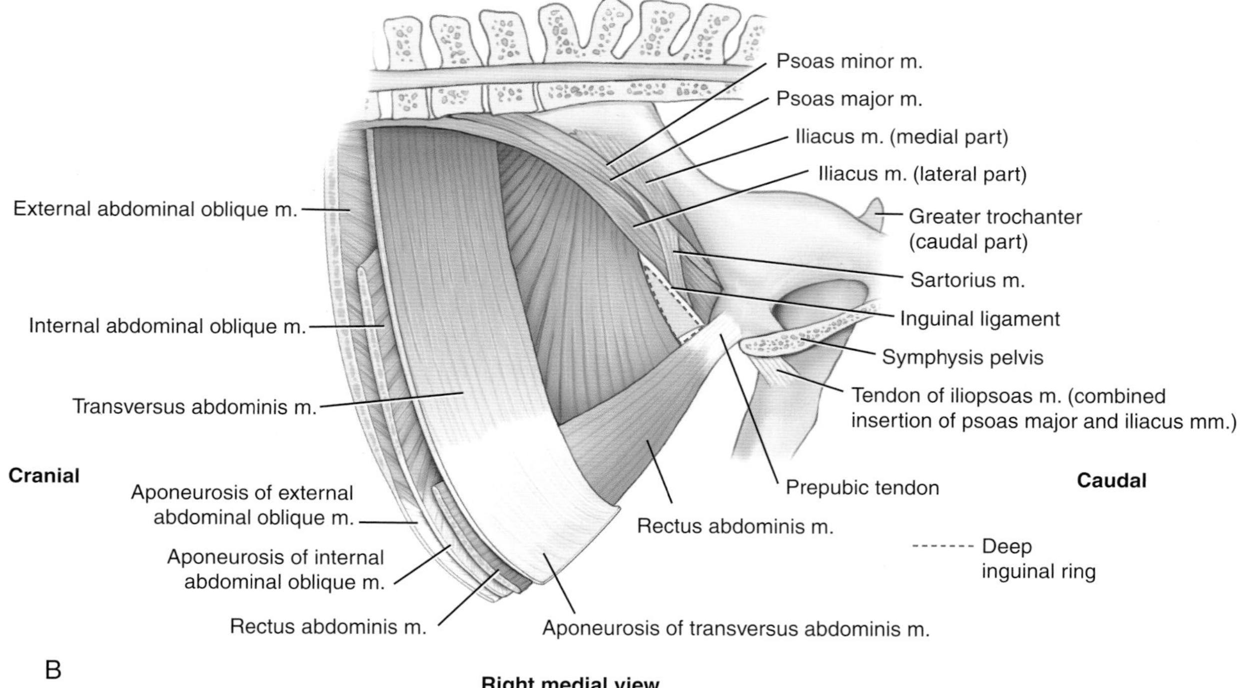

FIGURE 13.9-6 Equine inguinal canal and rings, illustrating the muscles forming the borders of these structures. (A) Lateral view of the left superficial inguinal ring. The triangular outline of the left deep inguinal ring shows the relative relationship of the superficial and deep rings. The inguinal canal is the fissure-like pathway between the superficial and deep inguinal rings. (B) Medial view of the triangular-shaped right deep inguinal ring.

(Continued)

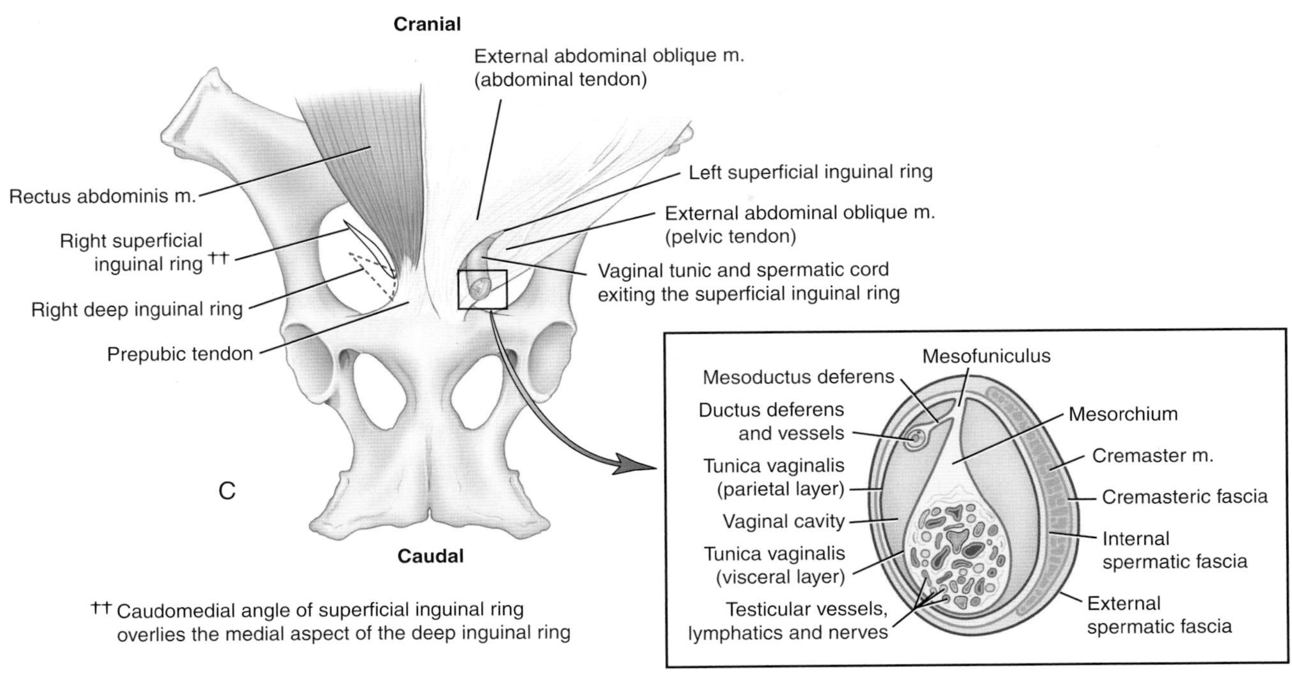

FIGURE 13.9-6, CONT'D (C) Ventral view of the abdominal wall and relative locations of the superficial and deep inguinal rings and intervening inguinal canal. *Insert*: cross-section of the spermatic cord, vaginal tunic and related structures after exiting the inguinal canal at the superficial inguinal ring.

cord). The mesofuniculus connects the spermatic cord to the parietal layer of the vaginal tunic. The spermatic cord begins at the vaginal ring and ends at the testis. In clinical usage of the term "spermatic cord," the anatomic structure is expanded to include the parietal vaginal tunic and its adherent internal spermatic fascia, along with the cremaster muscle with its surrounding cremasteric fascia.

> The external spermatic (i.e., scrotal) fascia is the loose connective tissue "stripped" off the "clinical spermatic cord" in preparation for a closed castration.

Inguinal canal

The **inguinal canal** (Figs. 13.9-5 and 13.9-6) is a compressed, fissure-like pathway between abdominal wall muscles, providing a passage through the caudoventral abdominal wall for structures to enter and leave the abdominal cavity. The canal is entered/exited at the **deep inguinal ring** internally and the **superficial inguinal ring** externally. The vaginal process evaginates through the inguinal canal, developing into the extensive vaginal tunic in the male horse, and remaining a vestigial or, more commonly, an absent structure in the female horse. In the male, normal testicular descent traverses the

> The deep inguinal ring can be palpated per rectum. The inguinal canal is wide in horses, making inguinal hernias more common in this species. There are 2 main types of inguinal hernias—direct and indirect, with the former more common in foals and usually due to a defect in the fascia of the caudal wall of the inguinal canal or other abdominal wall failure, and the latter more common in adult horses. In direct congenital inguinal hernias, the intestine (typically small intestine) does not penetrate the vaginal tunic and instead travels between the vaginal tunic and scrotal or preputial tissue. The direct inguinal hernia in the adult horse occurs when there is a breach in the vaginal tunic and scrotal fascia with small intestine sitting subcutaneously in the inguinal or scrotal area. The correct term for this clinical finding in the adult is a ruptured inguinal hernia and not a direct inguinal hernia. In indirect acquired inguinal hernias, the intestine passes through the vaginal and deep inguinal rings into the vaginal cavity (Fig. 13.9-1B).

inguinal canal, with the testis drawing through the attached spermatic cord structures. In addition, the cremaster muscle passes through the inguinal canal in the male, attached to the external surface of the vaginal tunic. In the female horse, the round ligament of the uterus passes to the deep inguinal ring, adhering to fascia at this location, and does not typically enter the canal. In both male and female, the genitofemoral nerve, external pudendal vasculature, and efferent lymphatic vessels from the superficial inguinal lymph nodes pass through the canal. In the male, it is important to recognize these structures are traveling through the canal external to the vaginal tunic. More detail about the inguinal canal follows to help clarify the anatomy of this complex region.

The deep (or internal) inguinal ring, not really a ring, is more triangular shaped and is formed by the combination of 3 overlapping abdominal muscles; it is located immediately craniolateral to the cranial margin of the pubis. The base of the triangle (i.e., the medial border of the ring, closest to ventral midline) is formed by the lateral margin of the rectus abdominis muscle at its insertion on the pubic bone (referred to as the **prepubic tendon** at this location). The cranial border of the triangle is the fleshy, slightly oblique caudal margin of the internal abdominal oblique muscle. The caudal border of the triangle is formed by the curved, thickened, caudal margin of the aponeurosis of the external abdominal oblique muscle (referred to as the "**inguinal ligament**"). The cranial and caudal borders converge dorsally on the abdominal wall to form the apex of the triangular space, approximately 10–15 cm (3.9–5.9 in.) from the base. Viewed from within the abdominal cavity (i.e., laparoscopically; see Chapter 2, Section 2.1), the lateral surface of the deep inguinal ring is limited by the aponeurosis of the external abdominal oblique muscle, extending craniodorsally from its inguinal ligament. This same aponeurosis forms the superficial inguinal ring, in the area located ventral to where it contributes to the deep inguinal ring (see below).

The internal (peritoneal) surface of the deep inguinal ring is covered by parietal peritoneum (see Fig. 12.7-5) and a deeper layer of transversalis fascia. These 2 layers turn acutely, evaginate through the deep inguinal ring, and continue as the vaginal tunic in the male. The distinct oval boundary formed where the tissue layers change direction is termed the **vaginal ring** (anulus vaginalis). The vaginal ring defines the opening into the vaginal cavity and is the connection between the parietal peritoneum and the beginning of the vaginal tunic. The vaginal ring lies within the borders of the deep inguinal ring. The approximate diameter of the vaginal ring is 3–5 cm (1.2–2.0 in.).

The superficial (or external) inguinal ring is an elongate oval formed by a division in the aponeurosis of the external abdominal oblique muscle. The split is created by the separation of the aponeurosis into 2 broad tendons of insertion, the abdominal and pelvic tendons. The abdominal tendon inserts on the linea alba and the prepubic tendon and its edge provides the medial crus (border) of the superficial inguinal ring. The edge of the pelvic tendon forms the lateral crus of the superficial inguinal ring and this part of the aponeurosis of the external abdominal oblique muscle inserts on the tuber coxae and prepubic tendon and terminates as the inguinal ligament. The pelvic tendon is the portion of the external abdominal oblique aponeurosis described previously that also limits the lateral surface of the deep inguinal ring, and in doing so, it is forming the lateral wall of the inguinal canal. The long axis of the superficial inguinal ring runs caudomedial to craniolateral. The caudomedial angle is closely related to the underlying deep inguinal ring, resulting in the inguinal canal being short in this region and longer (see Fig. 13.9-6) at the craniolateral angle (up to 15 cm [5.9 in.]).

Selected references

[1] Auer JA, Stick JA, Kümmerle JM, Prage T. Equine surgery. 5th ed. St. Louis: Elsevier; 2018.
[2] Orsini JA, Divers TJ, editors. Equine emergencies: treatment and procedures. 4th ed. St. Louis: Elsevier; 2014.
[3] van der Velden MA. Surgical treatment of acquired inguinal hernia in the horse: a review of 51 cases. Equine Vet J 1988;20(3):173–7.
[4] Weaver AD. Acquired incarcerated inguinal hernia: a review of 13 horses. Can Vet J 1987;28(4):195–9.
[5] Wilderjans H, Meulyzer M, Simon O. Standing laparoscopic peritoneal flap hernioplasty technique for preventing recurrence of acquired strangulating inguinal herniation in stallions. Vet Surg 2012;41(2):292–9. https://doi.org/10.1111/j.1532-950X.2011.

CHAPTER 13

CASE 13.10

Omphalitis and Bladder Rupture

Singen Elliott[a] and Jarred Williams[b]
[a]Mid-Atlantic Equine Medical Center, Ringoes, New Jersey, US
[b]Large Animal Medicine, University of Georgia College of Veterinary Medicine, Athens, Georgia, US

Clinical case

History

A 1-week-old Thoroughbred colt presented to the hospital with a 2-day history of lethargy. He was originally observed to have a fever and wet umbilicus. He was started on antibiotics and flunixin meglumine and initially responded well. However, he became lethargic again 2 days later and the fever returned. On the morning of presentation, he developed marked ventral edema and was referred to the teaching hospital.

Physical examination findings

On presentation, the colt was lethargic, bradycardic (42 bpm), and febrile (103.1 °F [39.5 °C]). Mucous membranes were pink and moist with a CRT of 2 seconds. Borborygmi were slightly decreased in all abdominal quadrants. There was a moderate amount of ventral edema extending caudally into the prepuce. The umbilicus was enlarged, sensitive to palpation, and warm. He was persistently flank-watching.

Differential diagnoses

Umbilical abscess, umbilical hernia, sepsis, peritonitis

Diagnostics

Blood was aseptically drawn for a culture, CBC, serum chemistry panel, and serum amyloid A (SAA). Abnormalities included a neutrophilic, monocytic leukopenia with a left shift; hyperfibrinogenemia of 800 mg/dL; increased lactate of 5.7 mmol/L; azotemia (BUN 52 mg/dL and creatinine 3.1 mg/dL); hyponatremia (128 mmol/L), hyperkalemia (6.8 mmol/L), and hypochloremia (90 mmol/L); and increased SAA of 1392 ng/mL.

Ultrasound of the abdominal cavity and umbilicus revealed an increased quantity of hypoechoic free abdominal fluid with 'lollipop'-like structures floating within (Fig. 13.10-1). The urinary bladder was not identified. The umbilical vein and arteries all measured < 10 mm (0.4 in.) in diameter (Fig. 13.10-2). An encapsulated area of hyperechoic fluid measuring approximately 4 cm (1.6 in.) in diameter was seen within the urachus (Fig. 13.10-3). An ECG was performed because of the electrolyte abnormalities and bradycardia. There was increased widening of the QRS complex and P wave, as well as increased duration of the P-R interval—consistent with hyperkalemia.

> Infection of the umbilical remnants is not uncommon in neonatal foals. Clinicians treating neonatal and young foals must be proficient in ultrasound of the following structures: umbilical stump, urachus, umbilical vein, and the 2 umbilical arteries.

Comparative Veterinary Anatomy: A Clinical Approach. https://doi.org/10.1016/B978-0-323-91015-6.00069-8

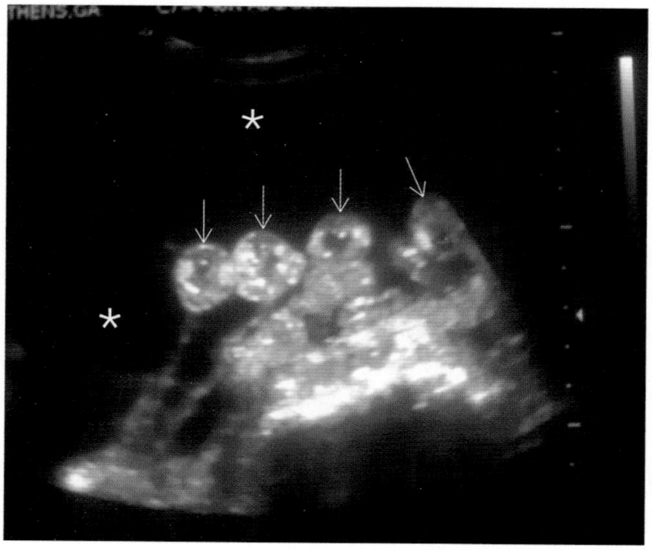

FIGURE 13.10-1 Transabdominal ultrasound image of the abdominal cavity of the foal in this case. Note the increased hypoechoic free abdominal fluid (*asterisk*). Additionally, soft tissue structures are pronounced within the increased fluid, creating a characteristic 'lollipop' appearance (*arrows*).

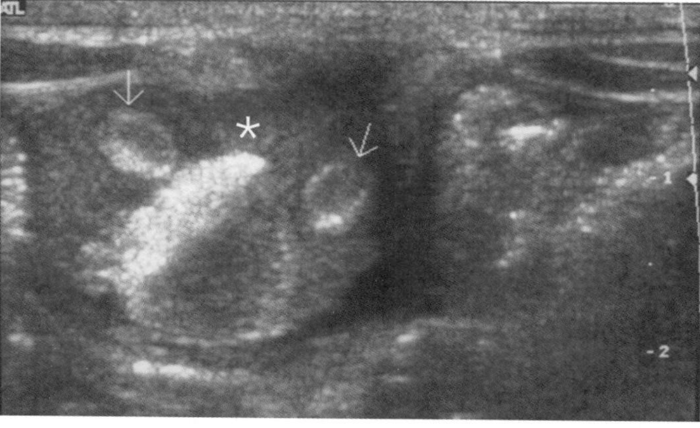

FIGURE 13.10-2 Ultrasound image of the umbilicus. The umbilical vein (not seen here), urachus (*asterisk*), and arteries (*arrows*) measured normal for the foal's age and size (< 10 mm [0.4 in.] in diameter).

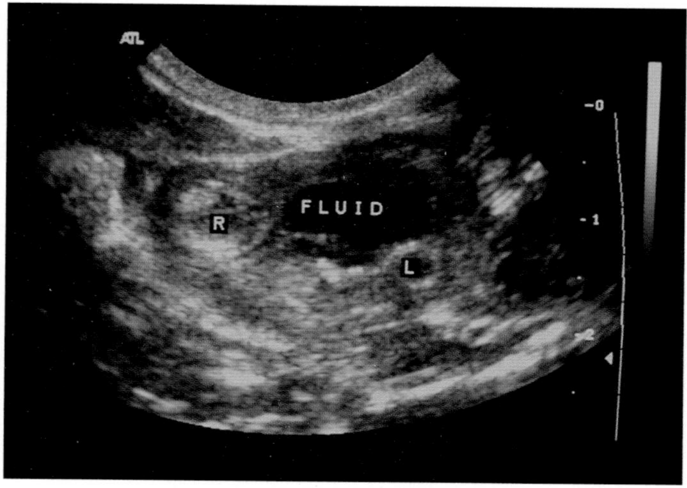

FIGURE 13.10-3 Ultrasound image of the urachus. An encapsulated mass containing a region of hyperechoic fluid is seen within the urachus. Key: *R*, right umbilical artery; *L*, left umbilical artery.

Abdominocentesis yielded a copious amount of clear, straw-colored fluid. Fluid analysis was performed while the fluid was drained from the abdomen and revealed a normal white blood cell count and total protein, increased lactate of 6.1 mmol/L, and increased creatinine of 7.3 mg/dL (twice the peripheral value, consistent with uroperitoneum).

Diagnosis

Uroperitoneum and urachal abscess—likely due to a ruptured urinary bladder

Treatment

Medical stabilization was followed by general anesthesia for umbilical resection and ventral midline exploratory celiotomy. On entering the abdominal cavity, a copious amount of free fluid was removed by suction. The umbilicus was carefully resected from the body wall. The umbilical vessels were unremarkable, ligated, and transected. The urachus contained a large, firm, flocculent mass. Adjacent to the origin of the urachus on the urinary bladder, there was an approximately 3-cm (1.2-in.), full-thickness tear located along the dorsal margin. The urachus was carefully resected, leaving the abscess untouched to allow removal with the umbilicus *en bloc*. The margins of the tear in the bladder were resected and the defect repaired.

The foal recovered from general anesthesia without problems. The urachal abscess was submitted for culture and sensitivity. The foal was discharged 3 days later on antimicrobials and continued to do well at home.

Anatomical features in equids

Introduction

In the fetal mammal, the umbilicus is the entry and exit point for all blood, nutrients, and waste. The umbilicus is intimately associated with the liver and the urinary system. The major parts of the umbilicus include: one urachus, which attaches to the urinary bladder; one umbilical vein, which attaches to the liver; and 2 umbilical arteries, which attach to the urinary bladder. The relevant anatomy of the umbilicus, especially as it relates to the urinary system, is discussed in this section, including the umbilical structures, kidneys, ureters, bladder, and urethra. Additional information about the urinary bladder is also presented in Case 13.11.

Function

In placental mammals, the umbilical cord, also called the navel string, birth cord, or funiculus (*L. a cord*) umbilicalis, is the passage between the developing fetus and the placenta. During prenatal development, the umbilical cord is physiologically and genetically part of the fetus and normally contains 2 arteries—the umbilical arteries, and one vein—the umbilical vein, enveloped in Wharton's jelly (a jelly-like intercellular substance of the umbilical cord). The umbilical vein supplies the fetus with oxygenated, nutrient-rich blood from the placenta, whereas the fetal heart pumps oxygen- and nutrient-depleted blood through the umbilical arteries back to the placenta (see Case 4.6) The urachus drains the fetal urinary bladder via the umbilicus to the allantoic cavity.

Umbilicus and umbilical remnants (Fig. 13.10-4)

In utero, maternal blood supply enters the fetus through the umbilical cord in the single **umbilical vein**. The umbilical vein passes cranially along ventral midline and into the liver before continuing to the heart via the caudal vena cava. Blood then perfuses the rest of the body and ultimately drains back into the umbilical cord via paired **umbilical arteries**. These arteries arise abaxially along either side of the bladder as branches off the **internal pudendal arteries**.

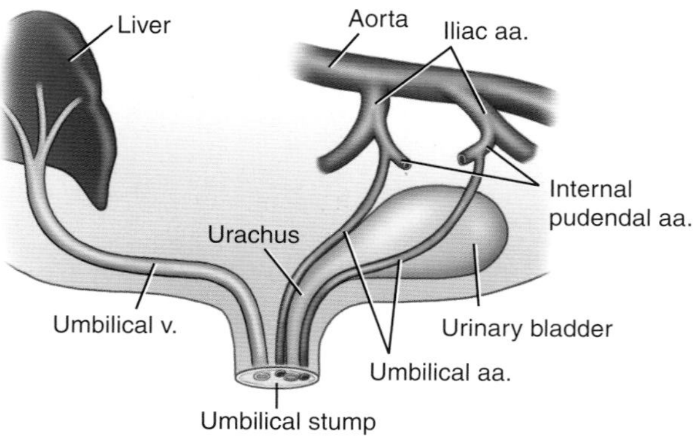

FIGURE 13.10-4 Umbilicus and umbilical remnant structures.

In small animals, fat (yellow) deposition in the falciform ligament can sometimes complicate ventral midline celiotomies, particularly with cranial approaches to the abdomen used to access the spleen, liver, and stomach. The falciform ligament is commonly excised during ventral celiotomies in dogs to prevent including it in the closure for concern that it could delay healing and contribute to incisional problems. However, this is seldom a problem when entering the abdomen in the horse, due to the cranial and dorsal location of the liver, as well as the more typically caudal incision made during equine celiotomies.

Urinary bladder rupture in foals most commonly occurs at the junction of the urachus and apex of the bladder, or on the dorsal surface where the bladder wall is thinnest. Bladder rupture followed by uroperitoneum commonly occurs during parturition, especially in colts.

After birth, the umbilical vein becomes the **round ligament of the liver**, residing in the free border of the **falciform ligament**. This ligament fuses with the **linea alba** for most of its length in the adult horse. As it approaches the liver, the falciform ligament and the round ligament of the liver connect to the liver, splitting the left medial liver lobe and the quadrate lobe at the notch of the round ligament. The umbilical arteries become the paired (left and right) **round ligaments of the bladder**, which run along the cranial margin of the paired (left and right) **lateral ligaments of the bladder**.

As blood circulates through the fetus, the fetal kidneys filter it and an ultra-filtrate of the blood (urine) is excreted into the bladder. The bladder is connected through its apex to the umbilicus via the **urachus**, which is continuous with the **allantois**, allowing urine to flow into the **allantoic sac**. The urachus fibroses and regresses after parturition and becomes the **middle (median) ligament of the bladder**.

UROLITHIASIS (SEE ALSO CASE 13.11)

Bladder stones, composed of calcium carbonate crystals, are the most common type of urolith (Gr. *ouron* urine + Gr. *lithos* stone = urinary calculus) in horses. Presenting clinical complaint and signs include hematuria, stranguria, pollakiuria (Gr. *pollakis* often + *ouron* urine + *-ia* state or condition = frequent passage of urine), pyuria, and incontinence. The classical presenting complaint is that the horse consistently has bloody urine (hematuria) post-exercise. Frequently, male horses demonstrate stranguria (Gr. *stranx* drop + *ouron* urine = painful passage of urine), repeatedly dropping the penis and posturing to urinate but voiding small amounts or no urine. Surgical removal of bladder stones is the treatment of choice using laparocystotomy, perineal urethrotomy and lithotriptors (the use of an instrument for crushing calculi in the urinary bladder), or the use of electrohydraulic or extracorporeal shock wave lithotripsy to break up the stone into pieces small enough to be removed from the bladder by suction or voiding. Successful dietary management, as in dogs and cats, is unlikely to replace surgical treatment of urolithiasis in horses. However, legume hays and supplements with calcium should be avoided and the addition of salt to increase water intake and urine output with pasture turnout is useful in minimizing recurrence of stone formation.

Urinary bladder (Fig. 13.10-5)

The **urinary bladder** consists of a **neck**, **body**, and **apex**. The neck of the bladder lies on the pelvic floor and is continuous with the **urethra** (see Fig. 13.11-3).

The size and location of the body and apex vary depending on the amount of urine in the bladder. When devoid of urine, the urinary bladder in the average adult horse is only about the size of a closed fist and resides in the pelvic cavity. However, when full, the body and apex expand and move cranially to lie on the ventral abdominal wall.

> The urethra in the mare is short and wide. During times of increased intra-abdominal pressure, such as parturition, the bladder can be inverted through the urethra and out the vulva. This is known as a 'prolapsed bladder' and should be distinguished from premature separation of the placenta or 'a red bag' delivery, which requires immediate intervention to save the foal.

> A normal, empty bladder is difficult to distinguish on palpation per rectum. However, a full bladder should be easily identified, typically palpable 'wrist-deep.' A clinician might also be able to palpate uroliths within the bladder or, less commonly, in the pelvic urethra in males.

Expansion and contraction of the bladder are possible due to the **transitional epithelium** of the bladder mucosa, as well as the 3-layered **detrusor muscles**. On the dorsal aspect of the neck of the bladder lies the **trigone**, where the ureters empty into the bladder. The ureters pass through the wall of the bladder obliquely, allowing the pressure of a full bladder to collapse the ureteral openings and prevent retrograde flow of urine. The bladder is attached to the body wall by the paired lateral ligaments of the bladder (discussed previously), including the round ligament of the bladder in the cranial free margin and the singular **middle (median) ligament of the bladder**, attaching the ventral surface to the linea alba.

Blood supply, lymphatics, and innervation (Fig. 13.10-6)

The bladder receives its blood supply and innervation through the lateral ligaments of the bladder. The **vaginal** or **prostatic artery** is the main blood supply and is supplemented by smaller umbilical arteries. The **vesical arteries**, **veins**, and **lymphatics** are the direct supply to and from the bladder. The **cranial** and **caudal vesical arteries**, which supply the bladder, are branches of the **internal pudendal arteries**. Similarly, the **caudal vesical vein** flows into the **internal pudendal vein** to the internal iliac vein eventually draining into the **caudal vena cava.** The **vesical lymph ducts** flow to the **medial iliac lymph nodes**.

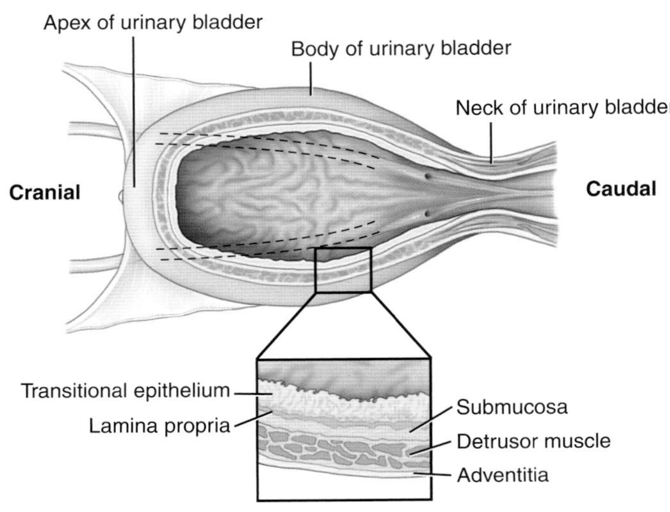

FIGURE 13.10-5 Equine urinary bladder anatomy (ventral-dorsal view), with inset representing the tissue layers of the bladder. *Dashed dotted lines* represent the ureters terminating at the ureteric orifice (meatus).

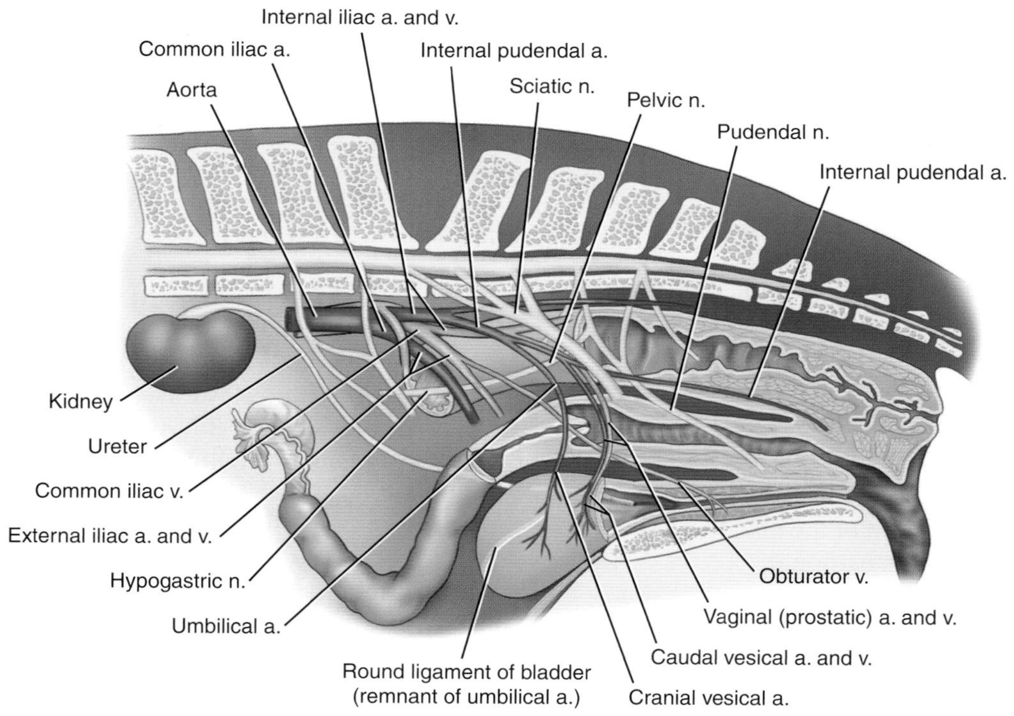

FIGURE 13.10-6 Lateral view of the equine female caudal abdomen and pelvis focused on the urogenital system anatomy.

> Local anesthesia of the pudendal nerve affects only the external pudendal branches and should not affect micturition or lead to temporary incontinence.

Innervation to the bladder consists of the pudendal, pelvic, and hypogastric nerves. The **pudendal nerve** supplies the somatic control of the bladder, allowing for conscious micturition and closing of the external urethral sphincter. Autonomic fibers innervate the bladder through the sympathetic hypogastric and parasympathetic pelvic nerves. The parasympathetic **pelvic nerves** facilitate micturition through contraction of the innervated detrusor muscles of the bladder and contain sensory stretch receptors, which provide feedback to the brain. The **hypogastric nerve** provides sympathetic control, which relaxes the detrusor muscles allowing the bladder to fill and is also responsible for constriction of the internal urethral sphincter.

Selected references

[1] Popesko P. Atlas of topographical anatomy of the domestic animal. 2nd ed. vol. 3; 1977. p. 146–55.
[2] Dyce KM, Sack WO, Wensing CJG. Textbook of veterinary anatomy. 3rd ed; 2002. p. 545–67.
[3] Schott II HC, Woodie JB, Auer J, Stick J. Bladder. In: Equine surgery. 4th ed. St. Louis: Elsevier; 2012. p. 927–39.

CHAPTER 13

CASE 13.11

Cystic Calculus

Tamara Dobbie
Department of Clinical Studies - New Bolton Center, University of Pennsylvania School of Veterinary Medicine, Kennett Square, Pennsylvania, US

Clinical case

History

A 15-year-old Thoroughbred cross mare was presented for pollakiuria (frequent urination). For the past 3 months, the mare was frequently observed to posture as if to urinate, and instead pass small amounts of urine, repeatedly exposing the clitoris. As these signs were persistent, the owner was unsure whether the problem was related to the mare's reproductive or urinary tracts.

"Pollakiuria" refers to the frequent, voluntary passage of small amounts of urine, whereas "incontinence" refers to the involuntary passage of urine, which may be in repeated small amounts or a near-constant trickle. Each has several possible causes, not all of which involve a urinary tract abnormality.

Physical examination findings

There were no visible abnormalities in the perineal area, vaginal vestibule, or clitoral fossa. Transrectal palpation revealed a small urinary bladder. Transrectal ultrasonographic examination confirmed that the bladder contained only a small amount of echogenic urine.

The size of the bladder on rectal palpation in large animals, and on abdominal palpation in small animals, is an important finding in any case involving abnormal urination. A patient with a mechanical or functional (upper motor neuron) obstruction to urine outflow has a large, distended bladder that cannot easily be expressed with manual pressure. A patient with a lower motor neuron lesion that affects the bladder/urethral sphincter has a large, distended, atonic bladder that is easily expressed with manual pressure. In contrast, a patient with lower urinary tract inflammation and no outflow obstruction typically has a small, empty bladder, particularly if pollakiuria is a presenting complaint.

It is common for equine urine to appear moderately echogenic (lots of white speckles) sonographically, as equine urine normally contains abundant calcium carbonate and mucoproteins. The more concentrated the urine, the more echogenic it appears; conversely, the more dilute the urine, the less echogenic it appears.

As the lower urinary tract appeared normal, the reproductive tract was examined thoroughly, using both transrectal palpation and ultrasonography. The uterus was toned with no evidence of intrauterine fluid or uterine edema. These findings were typical of a mare in mid-diestrus. The right ovary was 4 cm (1.6 in.) in length, with one 24-mm (0.9-in.) follicle and multiple smaller follicles. These findings are normal for a mare in diestrus.

Comparative Veterinary Anatomy: A Clinical Approach. https://doi.org/10.1016/B978-0-323-91015-6.00078-9

The left ovary was twice the size of the right ovary; it was 8 cm (3.1 in.) in length, with only a few small follicles and a "meaty" texture. A large disparity in ovarian size in a mare showing signs of persistent estrus may indicate ovarian neoplasia, such as a granulosa-theca cell tumor (GCT), so a blood sample was submitted for a GCT panel in addition to routine hematology (complete blood count), serum chemistry panel, and measurement of plasma fibrinogen concentration.

All bloodwork was within the normal limits, including the GCT panel (for more information on GCTs, see Case 13.4). The serum progesterone concentration was 13.1 ng/mL, a level indicative of a functional corpus luteum (normal ovarian activity). Together, these findings supported the conclusion that the mare's reproductive system was normal for the time of year and unlikely to be the cause of the observed urinary abnormalities.

It is ideal to collect a sample of urine for biochemical and microscopic analysis ("urinalysis"), and for bacterial culture and antibiotic susceptibility testing ("culture and sensitivity"), before starting antibiotic therapy when possible. However, experience-based (empirical) decisions are often made in clinical practice for a variety of reasons—i.e., efficient use of time and medical expenses.

As the mare's signs may also have been caused by lower urogenital tract inflammation, including infectious causes, the mare was treated empirically with an oral, broad-spectrum antibiotic drug that would help achieve therapeutic concentrations in the urine (trimethoprim-sulfamethoxazole). The clinical signs improved with treatment but did not completely resolve.

The mare was re-evaluated 2 months later when clinical signs returned to their pretreatment status. At that time, transrectal palpation and ultrasonography revealed a small, hard, intensely echogenic mass in the bladder, approximately 3 cm (1.2 in.) in diameter (Fig. 13.11-1). The reproductive tract appeared normal, with a corpus luteum on the left ovary and multiple small follicles on the right ovary.

Differential diagnoses

Cystic calculus (bladder stone), calcified neoplasm, or sabulous urolithiasis

Diagnostics

The mare was sedated for endoscopic (see Section 2.1) examination of the urinary bladder (cystoscopy) and possible surgical removal of the mass. With the mare restrained in stocks and the tail bandaged and tied to one side out of the way, the perineal area was cleaned and a sterile equine urinary catheter ("stallion catheter") was inserted to drain the bladder of urine and thus improve endoscopic visualization. Following lubrication with sterile lidocaine gel, a flexible, sterilized endoscope was passed through the urethra into the bladder for direct examination of the bladder lumen. A small, white, spiculated (spikey) stone was identified; in addition, the adjacent mucosa appeared inflamed and in places hemorrhagic (Fig. 13.11-2). The stone appeared solitary, although it was not possible to examine the portion of the bladder cranial to the stone without risking mucosal injury, so the decision was made to proceed with surgical removal of the stone and then re-examine the bladder lumen post-operatively.

SIGNS OF ESTRUS IN MARES

Mares in estrus ("in heat" or "in season") frequently posture as if to urinate ("squat"), pass small amounts of urine ("squirt"), and repeatedly evert the vulvar lips to expose the clitoris ("wink"). This behavior is normal for mares in estrus, but it is not specific to estrus. Although less common, inflammation involving the distal urogenital tract (bladder, urethra, vagina, vulva, clitoris, clitoral fossa) may cause this behavior independent of the mare's estrous cycle.

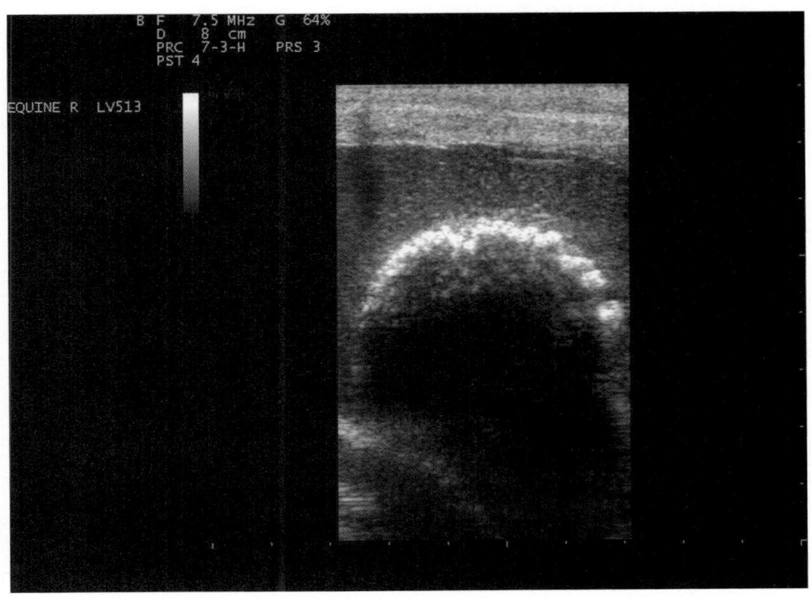

FIGURE 13.11-1 Ultrasonographic image of the mare's bladder, obtained per rectum. There is an intensely echogenic mass within the bladder, adjacent to the ventral bladder wall, which is lying in its normal position on the floor of the pelvic canal. The dorsal surface of the mass (bright-white, curved line) is so dense that it prevents the ultrasound wave from penetrating the mass, leaving only a black shadow beneath. Also, note the small amount of echogenic urine within the bladder surrounding the mass. It is not possible from this image to determine whether the mass is adherent to the bladder wall or free within the bladder lumen.

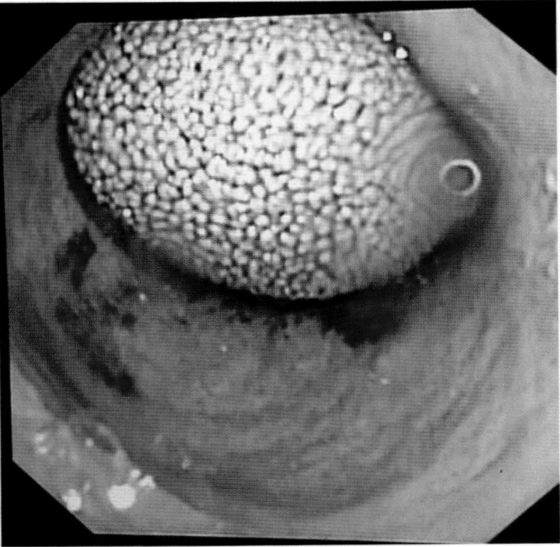

FIGURE 13.11-2 Cystoscopic view of the bladder lumen, depicting a spiculated urinary calculus. The adjacent bladder mucosa appears inflamed and in places hemorrhagic. Although it appears large in this image, the stone measured only 3 cm (1.2 in.) in diameter on ultrasound. By emptying the bladder of urine to facilitate cystoscopy, the bladder wall contracts around the mass.

Diagnosis

Cystic calculus (bladder stone)

Cystic calculi are removed using: (1) surgical forceps, or Foerster sponge forceps—for removing small stones; (2) lithotrite (Gr. *lithos* stone + Gr. *tribein* to rub)—an instrument for crushing a stone; or (3) lithotripsy or litholapaxy (*litho-* + Gr. *lapaxis* evacuation)—fragmenting the calculus and washing the pieces of stone out of the bladder.

Treatment

The stone was small enough that it could be removed through the urethra using specially designed forceps. Before extraction, a small (2-mm or 0.08 in.) incision was made ventral to the external urethral orifice to enlarge it, facilitating the removal of the stone.

Following removal of the stone, the bladder and urethra were re-examined endoscopically for signs of iatrogenic (Gr. *iatros* physician + Gr. *gennan* to produce) injury. No other calculi were found in the bladder, and there was minimal mucosal injury; the bladder and urethra were gently lavaged with warm sterile saline solution to remove residual grit and cellular debris. Post-operative pain was managed with sterile lidocaine gel applied to the external urethral orifice, systemic flunixin meglumine and phenazopyridine (a lower urinary tract analgesic), and prophylactic antimicrobials (trimethoprim-sulfamethoxazole) were administered.

Anatomical features in equids

Introduction

The following material covers the relevant clinical anatomy of the urinary bladder, including ureters and urethra, in the female and male horse, along with its blood supply, lymphatics, and innervation. Additional information on the equine urinary bladder is also presented in Case 13.10.

Function

The urinary bladder is a hollow muscular organ that collects and stores urine produced by the kidneys. Urine is typically voided from the bladder during urination, a complex physiologic process involving sympathetic, parasympathetic, and voluntary nervous control. Horses typically urinate about every 4 h and produce 5–15 L of

SABULOUS UROLITHIASIS

This term describes a syndrome of bladder paralysis and accumulation of urine sediment in the ventral aspect of the bladder, often confused with cystic calculi. Affected horses (more commonly males) typically present with urinary incontinence. Transrectal palpation reveals a large, distended bladder from which urine can easily be expressed with manual pressure. Transrectal ultrasonography reveals hyperechoic material overlying the ventral bladder wall. Once the bladder is empty, the sabulous material can be indented with firm pressure, differentiating it from a bladder stone.

Horse urine typically contains a large amount of calcium carbonate crystals, which can accumulate in the bladder when the bladder is not fully emptied. There are several theories as to the underlying cause(s), including lumbosacral pain precluding a normal stance for urination and therefore emptying, and peripheral nerve or CNS deficits.

Unfortunately, horses with sabulous urolithiasis typically do not present until they become incontinent. Therapy includes routine (ranging from monthly to yearly) lavage of the urinary bladder to remove the sabulous debris, medications to facilitate bladder emptying, and broad-spectrum antimicrobials. The condition carries a poor prognosis unless the bladder paralysis can be resolved. Long-term complications of this condition include urine scalding of the hind legs and recurrent bladder infections.

urine per day. The bladder has a capacity of 3–4 L of urine. The calcium carbonate crystals in horse urine give it a milky yellow appearance. Sometimes, the urine appears reddish-orange in color due to the oxidation of the naturally occurring urocatechin (from the catechu plant and also called "catechol" or "catechuic acid") by light.

People often mistake this normal discoloration of the urine for hematuria, especially during the winter months when the urine is seen in the snow.

Urinary bladder

The equine **urinary bladder** is comprised of 3 parts: the **apex**, **body**, and **neck** (Fig. 13.11-3). The bladder **neck** is fixed within the pelvic cavity, but the **apex** is free to move, and it changes location depending on the volume of urine in the bladder. When empty, the equine bladder is about the size of a fist and resides entirely within the pelvis. When moderately full, the apex of the bladder drapes over the pelvic brim (the cranial margin of the pelvis) and overhangs into the peritoneal cavity. When completely full, the weight of the urine draws the bladder apex toward the floor of the peritoneal cavity, in the direction of the umbilicus.

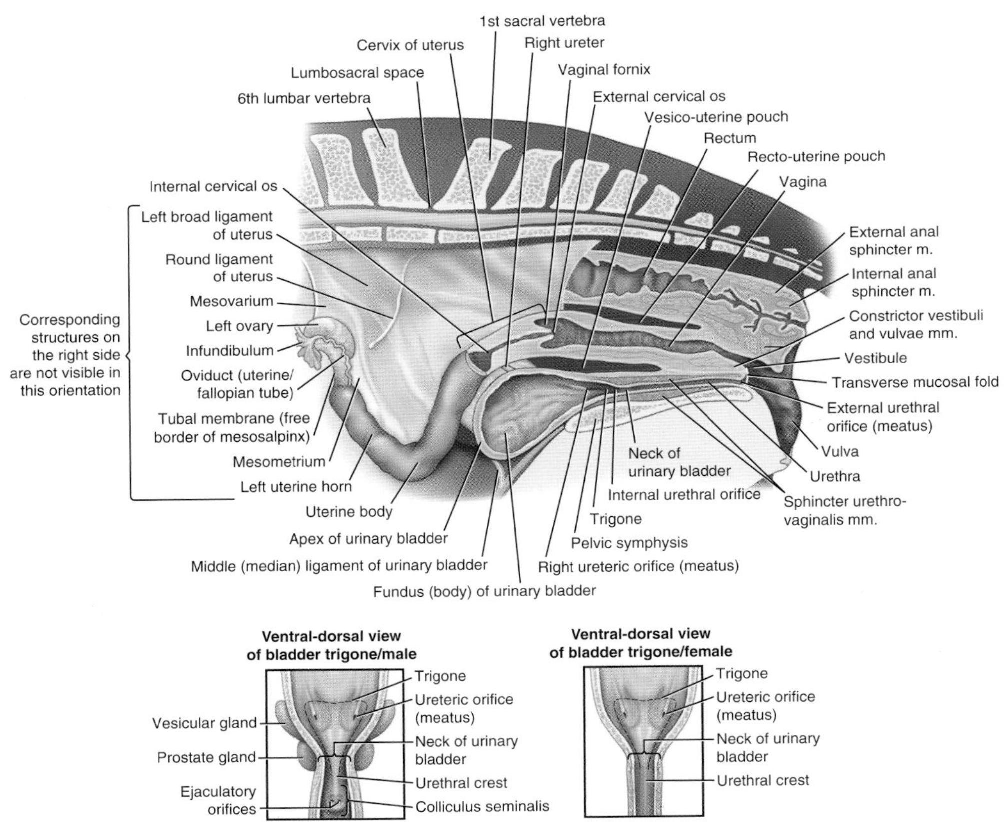

FIGURE 13.11-3 Lateral view of the female genitourinary tract. Figure inlays are the male (left) and female (right) trigone of the urinary bladder.

FETAL UMBILICAL REMNANTS

The round ligaments of the bladder, which are the remnants of the fetal umbilical arteries, are contained within the free edge of the lateral ligaments of the bladder. Fetal circulation is detailed in Case 13.10.

Urinary bladder displacement can occur in mares and is the result of repeated abdominal contractions or straining, often associated with parturition or, less frequently, colic. The bladder can prolapse through the urethra and into the vestibule or, if there is trauma in the ventral vagina or perineum, the bladder can prolapse through the tear. Correction involves catheterizing the bladder to facilitate emptying and then replacing the bladder and repairing the vaginal or perineal laceration. In the case of bladder prolapse, the bladder can be reduced manually, often with the aid of a urethral sphincterotomy. Uncommonly, some cases of bladder prolapse require surgical correction using a ventral midline approach.

The orientation and general position of the bladder are maintained by lateral ligaments and one middle (median) ligament. The **lateral ligaments** (paired left and right) attach the bladder to the walls of the pelvic canal, and the **middle (median) ligament** attaches the apex of the bladder to the floor of the pelvic canal and the caudal-most portion of the linea alba. Together, these ligaments prevent the bladder from displacing or rotating (twisting) when a horse rolls.

The **ureters** (paired left and right) enter the bladder obliquely on its dorsal surface, near the bladder neck (Fig 13.11-4). The oblique angle at which each ureter meets the bladder serves somewhat as a one-way valve, effectively preventing the backflow of urine into the ureters as the bladder fills.

The wall of the urinary bladder is composed of 4 layers: mucosa, submucosa, muscularis, and serosa (Fig 13.11-4). The **mucosa** is pale pink, thin, folded when the bladder is empty, and composed of transitional epithelium. The folded, transitional epithelium affords the mucosa the ability to expand and contract as the bladder fills and empties,

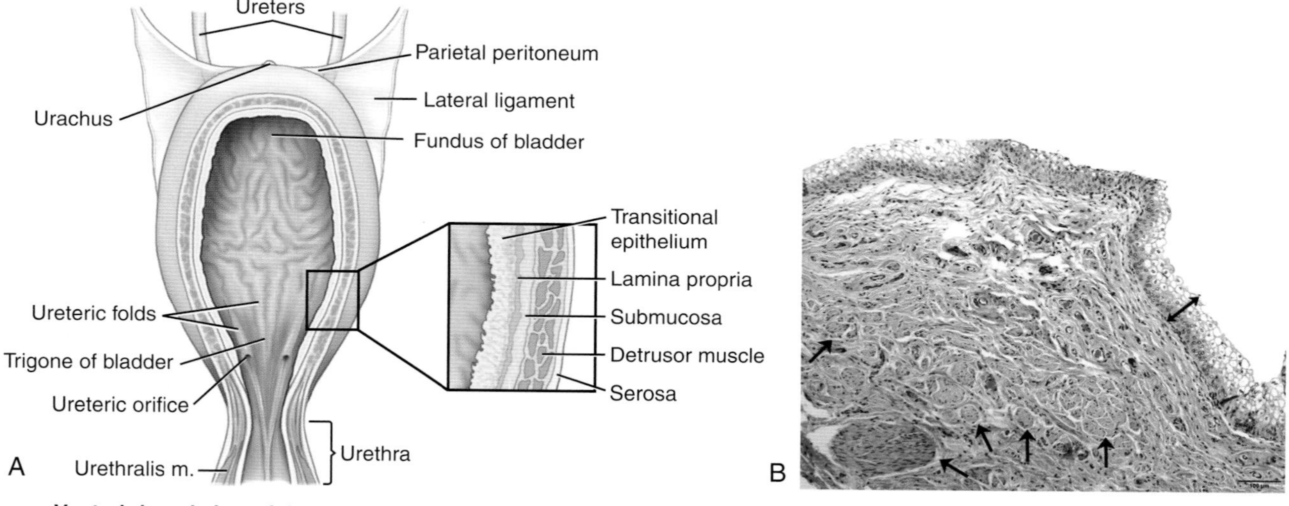

FIGURE 13.11-4 Anatomy and histology of the urinary bladder. (A) Gross anatomy of the bladder. Figure inlays show layers of the urinary bladder wall. (B) H&E photomicrograph of the urinary bladder from a horse. The mucosa is composed of transitional epithelium *(double-headed arrow)*. Within the deeper layers of the lamina propria are scattered smooth muscle bundles *(arrows)* that comprise the inner layer of the detrusor muscle. (Image courtesy of Dr. Julie Engiles.)

VESICOVAGINAL REFLUX—URINE POOLING/UROVAGINA

The fact that the urethra empties into the cranial vestibule in the mare can have clinical and pathological consequences in broodmares. Structural abnormalities involving the vulva, vagina, and/or cervix because of foaling trauma, malnutrition, poor pelvic conformation, or other factors can lead to urine backflow into—and pooling in—the cranial vagina, and even the body of the uterus. A surgical technique, termed "urethroplasty" or "urethral extension" may be helpful in correcting this problem in these mares. This procedure mobilizes a portion of the walls of the vestibule to effectively extend the external urethral orifice caudally and therefore allow urine to be voided normally (see Case 13.1).

respectively. The **submucosa**, which contains blood vessels and nerves, is highly elastic and anchors the mucosa to the muscularis layer. The **muscularis,** or muscular layer, (also termed the **detrusor muscle**) has a complex arrangement of smooth muscle fibers arranged in spiral, longitudinal, and circular bundles. A circular layer of smooth muscle is present at the neck of the bladder, effectively forming a sphincter. The urethral wall contains these same layers, including smooth muscle fibers arranged in both longitudinal and circular configurations. Lastly, the **serosa** is a thin sheet that overlies the muscular layer of the bladder.

Urethra in the mare

The bladder neck opens into the **urethra** at the **internal urethral orifice**. In the mare, the urethra is relatively short, approximately 5–7.5 cm (2–3 in.) in length in the average-size mare and has a diameter that readily admits one finger. With care, the female urethra can easily be dilated, permitting not only the removal of small stones, such as in this case, but also allowing the digital correction of a bladder prolapse.

The external urethral orifice in the mare opens onto the floor of the cranial **vestibule**, at the caudal extent of the vaginal vault and thus marking the transition from vagina to vestibule (Fig. 13.11-3). The entrance to the urethra in the mare is seen by separating the lips of the vulva to expose the vestibule. The external urethral orifice is loosely covered by a transverse fold of mucosa, which crosses the floor of the vestibule at this level.

This soft tissue fold is easily elevated allowing access to the urethra. In fillies, this transverse fold forms part of the hymen, which separates the vaginal vault from the vestibule during development.

Urethra in the stallion and gelding

In the stallion and gelding, the urethra is long, extending from the bladder to the glans penis (Fig. 13.11-5). In the average-size horse, the portion of the urethra that is contained within the pelvic canal is 10–12 cm (4–4.7 in.) in

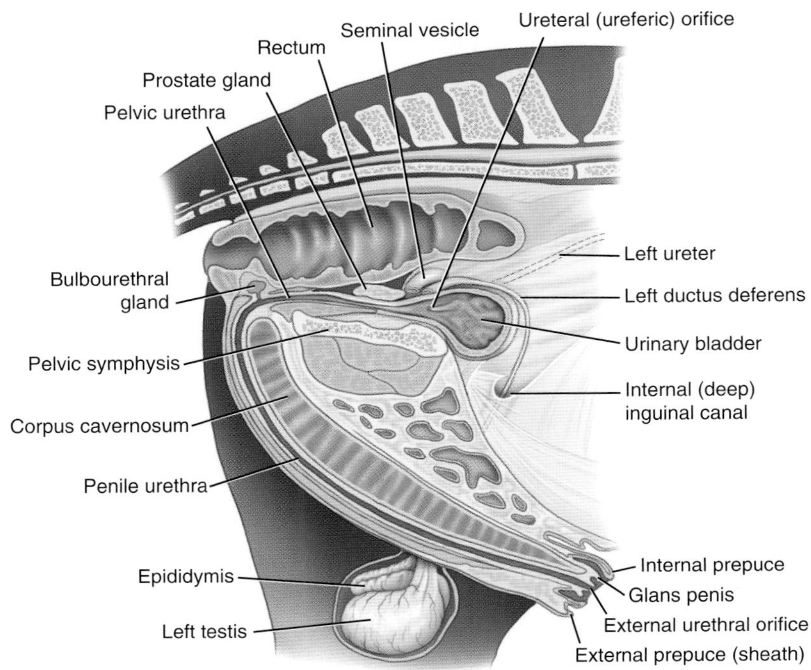

FIGURE 13.11-5 Lateral view of the male equine genitourinary tract.

It is difficult for stallions and geldings to spontaneously pass even small bladder stones. Surgical removal of cystic calculi via the intact urethra, as described in the mare in this case, is impossible in stallions and geldings. Instead, stones small enough to be moved through the pelvic urethra may be surgically removed by perineal urethrotomy, a procedure in which the urethra is accessed via an incision in the perineal region, either just below the ischium (subischial) or adjacent to the rectum (pararectal). Larger stones must be removed by cystotomy, a procedure in which the bladder is accessed via either laparotomy (ventral midline or paramedian approach) or laparoscopy (minimally invasive surgery using a surgical endoscope). In some cases, laser or mechanical lithotripsy is used to break down large stones into smaller fragments for easier removal via perineal urethrotomy. Urethral calculi that are lodged in the penile portion of the urethra can often be removed endoscopically, although extraction may still require a urethrotomy incision.

length, while the extrapelvic or penile portion is at a minimum 50 cm (20 in.) in length when the penis is flaccid (penile length in the erect horse increases by at least 50%).

Urethral diameter is roughly 1.5 cm (0.6 in.), except in the part between the prostate and bulbourethral glands, where the diameter increases to several centimeters. In addition to its length and small diameter, the male urethra bends sharply around the ischial arch.

At its origin, the urethra is surrounded by a circular layer of smooth muscle, but the remainder of the male urethra is surrounded by a layer of erectile tissue. Outside this erectile tissue is smooth and striated muscle: **urethralis muscle** within the pelvic cavity, **ischiourethalis muscles** around the ischial arch, and **bulbospongiosus muscle** surrounding the penile urethra. For more information on the equine penis, see Case 13.7.

Blood supply, lymphatics, and innervation

The blood supply to the bladder is predominantly from the **internal pudendal arteries**, with smaller branches from the **obturator arteries**. Venous drainage is via the **internal pudendal veins**. The lymphatic vessels form plexuses in the bladder wall, submucosa, and muscularis layers, and eventually drain into the **internal iliac** and **lumbar lymph nodes**.

Neurologic control of bladder function is complex and requires the coordination of the autonomic (sympathetic and parasympathetic) and somatic nervous systems. Sympathetic nervous supply is provided by the **hypogastric nerve**, which causes relaxation of the detrusor muscle, thereby promoting urine retention. Parasympathetic nervous supply originates from the sacral spinal cord (S1–S2), forming the **pelvic nerve**, and causes contraction of the detrusor muscle, thereby stimulating urination. Somatic innervation is provided by a branch of the **pudendal nerve**. Somatic innervation controls the external urethral sphincter, thereby providing voluntary control over micturition. Control of the internal urethral sphincter is regulated by the sympathetic and parasympathetic nervous system. Sensory nerves within the bladder travel to the brain and signal when the bladder is full and when there is a need to urinate.

SYNDROMES OF EQUINE BLADDER DYSFUNCTION

Loss of bladder function is an uncommon problem in the horse and is often only recognized once the horse is incontinent. The 3 main syndromes are: upper motor neuron (UMN) or spastic bladder, lower motor neuron (LMN) or paralytic bladder, and nonneurogenic or myogenic bladder.

An UMN bladder is a distended bladder that is difficult to express with the lesion residing in the brainstem or spinal cord cranial to the lumbosacral intumescence (*L. intumescentia* a swelling). There is increased urethral sphincter tone, so the bladder only empties once the pressure within the bladder exceeds the resistance of the urethra.

A LMN bladder is a distended bladder that is easy to express due to the loss of detrusor function and reduced contractility of the urethral sphincters. Lesions affecting the spinal cord (L1–S5) can give rise to a LMN bladder. An UMN bladder can become a LMN bladder after prolonged bladder distension and subsequent stretching of the detrusor muscles.

Nonneurogenic causes of bladder dysfunction may be associated with detrusor muscle inflammation—e.g., presence of a bladder stone—or disorders that prevent complete bladder emptying—e.g., musculoskeletal disease.

Treatment for horses with bladder dysfunction should be directed at identifying and treating the underlying cause and then promoting bladder emptying. Regular catheterization or an indwelling urinary catheter facilitates bladder emptying. Phenoxybenzamine can be used in cases with UMN bladders in an attempt to eliminate urethral resistance and facilitate bladder emptying. Bethanechol chloride can be used in cases of LMN bladders that are not completely atonic to stimulate detrusor muscle activity. Antimicrobial therapy is an important part of treatment, especially in horses with intermittent or indwelling urinary catheters. Grooming and hygiene are needed to prevent urine scalding of the legs and feeding nonleguminous hay reduces the calcium carbonate crystals in the urine. Occasionally, lavage of the bladder is required to remove sabulous debris.

Selected references

[1] Getty R. Sisson and Grossman's: the anatomy of the domestic animals. 5th ed. vol. 1. W.B. Saunders Company; 1975.
[2] Reed SM, Bayly WM, Sellon DC. Equine internal medicine. 2nd ed. Elsevier Saunders; 2004.
[3] Budras KD, Sack WO. In: Rock S, editor. Anatomy of the horse. 6th ed. Hanover, Germany: Schlutersche Verlagsgesellschaft mbH & Co.; 2011.

CHAPTER 14

THORACIC LIMB

Nick Carlson, Chapter editor

Proximal Thoracic Limb (Shoulder, Brachium, and Antebrachium)

14.1 Radial neuropathy—*Laura Johnstone* 862
14.2 Supraglenoid tubercle fracture—*Nick Carlson* 873
14.3 Ulnar fracture—*Nick Carlson* 882
14.4 Radial fracture—*Liberty Getman* 891

Distal Thoracic Limb (Carpus and Manus)

14.5 Superficial digital flexor tendonitis—*Nick Carlson* 898
14.6 Osteochondral fragment of the metacarpophalangeal joint—*Nick Carlson* 905
14.7 Fracture of the 2nd phalanx—*Nick Carlson* 912
14.8 Foreign body penetration of the hoof—*Nick Carlson* 918
14.9 Laminitis—*Nick Carlson* 925

CASE 14.1

Radial Neuropathy

Laura Johnstone
Equine Veterinary Clinic, Massey University, Palmerston North, NZ

Clinical case

History

An 11-year-old Warmblood mare was presented for chronic, progressive lameness of the left thoracic limb of 5 months' duration. An extensive orthopedic examination had been performed with no diagnosis. Palmar digital, low 4-point (anesthetizing the palmar and palmar metacarpal nerves), glenohumeral joint, carpal joints, elbow joint, and bicipital bursa diagnostic nerve blocks produced no improvement in lameness.

Physical examination findings

Routine physical examination parameters were within normal limits. Examination revealed severe muscle atrophy of the left proximal thoracic limb, most prominently of the pectoral and triceps musculature. The mare frequently stood with the left forelimb flexed and the dorsal aspect of the hoof resting on the ground (Fig. 14.1-1). The left tuber olecrani was displaced distally in relation to the trunk (a.k.a., "dropped elbow" sign). However, minimal discomfort was observed when the mare was bearing weight on the limb. The flexor withdrawal reflex was present in all limbs. No other spinal reflexes were assessed. Gait evaluation revealed a 4/5 left thoracic limb lameness, characterized by a shortened stride length, marked head nod, and tremor and knuckling of the fetlock joint that was exaggerated when backing or trotting. The pelvic limb gait showed mild abnormalities—both pelvic limbs showed delayed protraction, occasional circumduction and pivoting when circled, and the right pelvic limb had increased force of placement. Cranial nerve examination revealed no abnormalities except mild ptosis of the left eye.

> Atrophy of the triceps muscle indicated the lesion to the radial general somatic efferent lower motor neurons was proximal, while involvement of the pectoral nerves indicated the involvement of multiple nerves. This narrowed the neuroanatomical diagnosis to the following structures: C7, C8, and T1 spinal cord segments; spinal nerve roots and spinal nerves; brachial plexus; and radial and pectoral nerves. Mild upper motor neuron general proprioceptive deficits of the pelvic limbs would be consistent with disruption of white matter in the cervical or thoracic spinal cord. To refine the neuroanatomic diagnosis to a single lesion, ptosis might be caused by disruption of the sympathetic nerve pathway at the level of the cranial thoracic spinal cord segments, spinal nerve roots and spinal nerves or the thoracic sympathetic trunk, cervicothoracic ganglion, or middle cervical ganglion in proximity to the brachial plexus. Alternatively, the disease might be multifocal, affecting the palpebral branch of the facial nerve or the oculomotor nerve.

Neuroanatomical localization

Extensor weakness associated with radial neuropathy

Differential diagnoses

Trauma (e.g., vertebral, rib, humeral, or elbow fracture; shoulder compression; hyperextension of the limb in a backward direction; and extreme abduction

Comparative Veterinary Anatomy: A Clinical Approach. https://doi.org/10.1016/B978-0-323-91015-6.00071-6

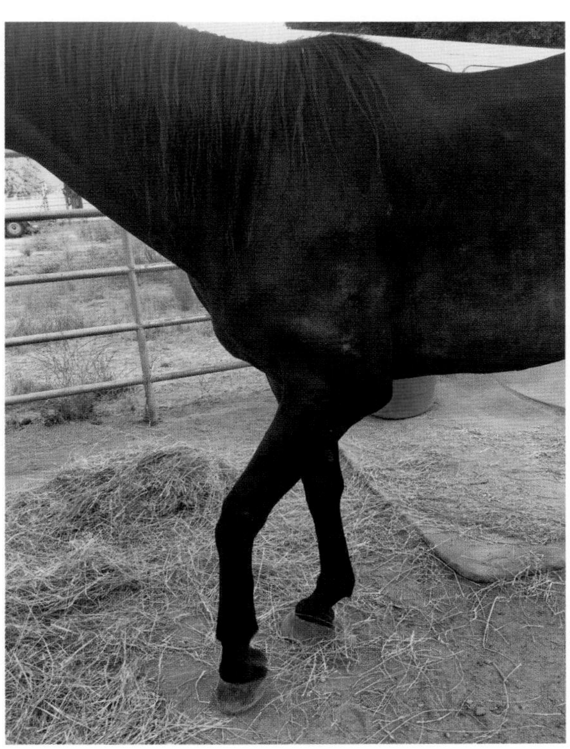

FIGURE 14.1-1 A horse with left radial nerve paralysis. Note the atrophy of the triceps muscle and the "dropped elbow" stance. (Image courtesy of Nora Grenager.)

of the shoulder), equine protozoal myeloencephalitis (EPM), Lyme neuroborreliosis, lymphoma, polyneuritis equi (though typically involves the nerves of the cauda equina)

Diagnostics

Electromyography (EMG) recorded prolonged insertional activity, positive sharp waves, fibrillation potentials, and complex repetitive discharges in the supraspinatus, infraspinatus, triceps brachii, extensor carpi radialis, and pectoralis muscles of both the left and right thoracic limbs. Ultrasound of the left axilla revealed no abnormalities aside from diffuse muscle atrophy of the triceps and pectoral muscles. Ultrasound and radiography of the caudal

ELECTROMYOGRAPHY (EMG)

Electromyography detects the electrical activity of a motor unit by inserting needle electrodes into the muscle in question and recording electrical activity with an amplifier on an oscilloscope. The motor unit includes the neuron, neuromuscular junction, and muscle fibers innervated by that neuron. As the nerve fiber reaches the muscle, it divides into several terminal branches, each branch innervating one muscle fiber. These branches end in irregular, club-shaped synaptic ends that are imbedded in the depressed motor endplate of the muscle cell, forming the neuromuscular junction. Each muscle cell is innervated by the axonal branch of one lower motor neuron (LMN) only. A normal EMG recording shows electrical activity when the needle electrode is inserted into the muscle. However, spontaneous activity in a relaxed muscle that persists after insertion indicates an abnormality of some part of the motor unit. Positive sharp waves, fibrillation potentials, and complex repetitive discharges, as seen in the EMG in this case, strongly indicate denervation and, therefore, a deficit of the LMN. Disuse atrophy occurs rapidly in the horse. Electromyography helps distinguish neurogenic atrophy from disuse atrophy.

cervical and cranial thoracic vertebrae revealed no abnormalities. Analysis of CSF was normal. Antibody tests for the causative organisms of EPM and Lyme were negative in serum and cerebrospinal fluid.

Diagnosis

The bilateral nature of muscle denervation suggested a lesion involving the caudal cervical intumescence; the chronic and progressive nature made lymphoma most likely

Treatment

A course of steroids was administered as an antitumor agent. The mare showed slight improvement but regressed when the steroid dose was tapered. Based on the poor prognosis and progression, the owner elected euthanasia. Necropsy confirmed lymphoma of the nervous system. The lesion primarily involved gross enlargement of the caudal cervical intumescence and the left C8 spinal nerve root and spinal nerve. Histopathology showed neoplastic lymphocytes within the epineurium, infiltrating the nerve fascicle to surround individual neurons. The C8 spinal cord segment had mild pressure-induced changes and secondary, Wallerian-like white matter degeneration present in both the ascending and descending pathways, cranial and caudal to C8, respectively. No lesions were detected in the palpebral nerve.

Anatomical features in equids

Introduction

An understanding of neuroanatomy is extremely useful when diagnosing neurological diseases as it allows a clinician to relate clinical signs and gait deficits to precise neuroanatomic sites. This chapter describes the anatomy of the general somatic efferent, the general somatic afferent, and the sympathetic general visceral efferent systems as they pertain to this peripheral neuropathy case study.

Peripheral nervous system

The **peripheral nervous system** (PNS) provides 2-way communication between the **CNS** (brain and spinal cord) and the rest of the body. It consists of 12 pairs of cranial nerves, pairs of spinal roots and spinal nerves, autonomic nerve trunks with associated ganglia, nerve plexuses, and nerves. In the PNS, neuronal processes and their myelin sheaths are collectively termed **nerve fibers**. Each nerve fiber is surrounded by a layer of connective tissue, termed **endoneurium**. Bundles of nerve fibers form **fascicles**, and each fascicle is enclosed by connective tissue called the **perineurium**. A cable of fascicles is collectively termed a **nerve** and is surrounded by connective tissue called **epineurium** (Fig. 14.1-2).

NEUROPATHOLOGIC CLASSIFICATION OF PERIPHERAL NERVE TRAUMA

Traumatic nerve lesions are classified into 3 categories according to the structures injured within the nerve fiber. (1) Neuropraxia involves compression without loss of axon or sheath integrity, resulting in temporary loss of function but no morphological change. (2) Axonotmesis is caused by crushing, which injures the axon but preserves the myelin sheath. (3) Neurotmesis is severing of the axon and myelin sheath.

Muscle atrophy occurs with axonotmesis and neurotmesis and becomes clinically prominent in 1–3 weeks. Reinnervation occurs by 2 mechanisms. The first, “collateral sprouting,” depends on the incomplete loss of axons to a muscle with surviving axons “sprouting” terminal branches that establish new connections with denervated muscle units. The second is axonal regrowth at a rate of 1 mm (0.04 in.)/day. Most horses with trauma-induced peripheral neuropathy of the thoracic limb return to athletic soundness following an adequate period of rest.

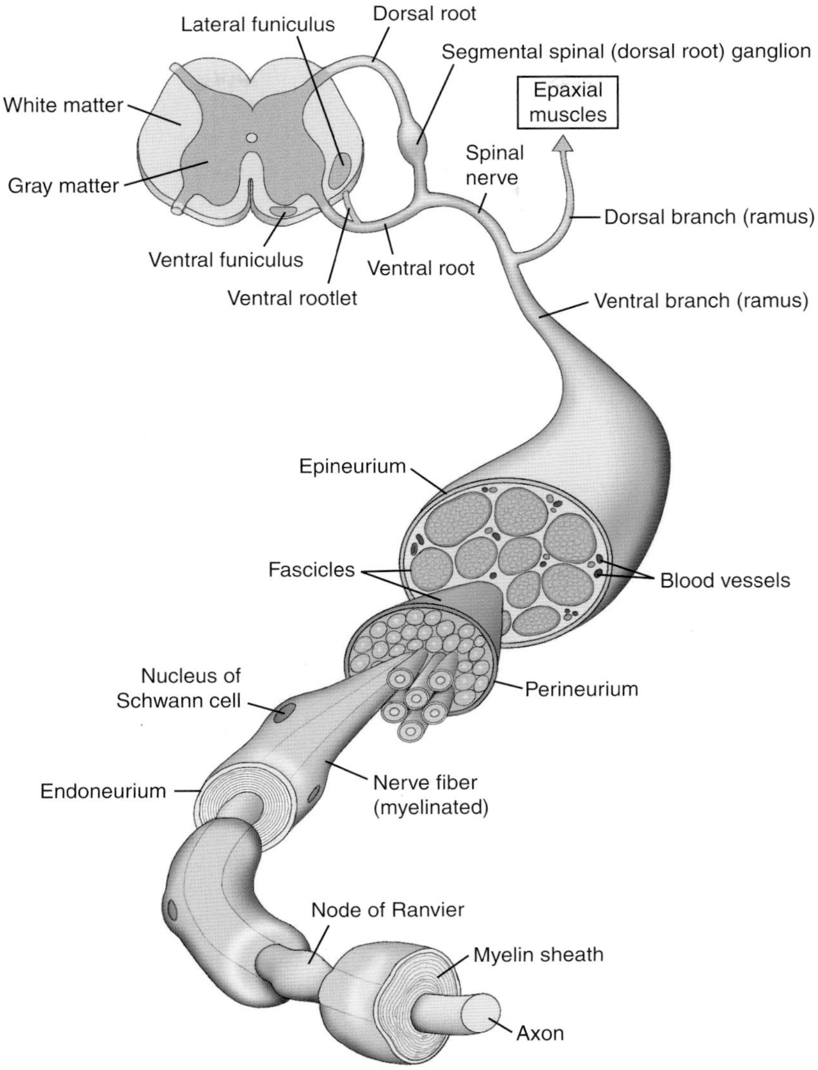

FIGURE 14.1-2 Schematic transection of the equine spinal cord and a spinal nerve.

Nerves contain both **afferent (sensory) fibers**, which convey information from sensory receptors to the CNS, and **efferent (motor) fibers**, which send signals from the CNS to the peripheral effector organs. These functional subdivisions of the nervous system have somatic and visceral components. Efferent neurons in the PNS are also referred to as **lower motor neurons** (LMN). **General somatic efferent** (GSE) LMN innervate striated voluntary skeletal muscle, while **general visceral efferent** (GVE) LMN belong to the autonomic nervous system and innervate the smooth muscle associated with blood vessels and visceral structures, glands, and cardiac muscles.

General somatic efferent lower motor neurons

The GSE LMN cell bodies are located in the **ventral gray columns** of the spinal cord with neurons innervating appendicular muscles being located lateral to those innervating axial muscles. The enlargement of the spinal cord at the level of the thoracic and pelvic limbs is due to the large number of LMN cell bodies in the lateral ventral gray column, and these enlargements are called the **cervicothoracic intumescence** (L. *intumescentia* a swelling) and the **lumbosacral intumescence**, respectively. The GSE-LMN axons exit the spinal cord by segmental ventral rootlets between the lateral and ventral funiculi of the white matter. Rootlets combine to form **ventral spinal roots** that then join the dorsal root to form the mixed afferent and efferent **spinal nerves** (Fig. 14.1-2), which leave the vertebral canal through the **intervertebral foramina**. There are 7 cervical vertebrae but 8 cervical spinal cord segments. Therefore, the first 7 spinal nerves exit the vertebral canal through the intervertebral foramina cranial to the vertebrae of the same number. The spinal nerve of the 8th cervical spinal cord segment exits cranial to the 1st thoracic vertebra. The remaining spinal nerves leave the vertebral canal through the intervertebral foramina that is caudal to the vertebra of the same number.

This relationship between intervertebral foramen and spinal nerves is established during embryonic development. "Ascensus medullae spinalis" refers to the shortening of the spinal cord relative to the vertebral column caused by the unequal growth of these 2 structures. Spinal nerves grow in length to accommodate for the displacement. The clinical relevance is twofold: firstly, spinal cord segments are located cranial to the vertebra of the same number such that a fracture of C6 might traumatize both C7 and C8 spinal cord segments; secondly, the 4th and 5th sacral spinal cord segments align with the lumbosacral space in the horse, and needle insertion can occur through these segments during cerebrospinal fluid centesis, which is usually of no clinical significance but rarely results in clinical signs of injury. For more details on vertebral formulae, please refer to Case 10.12.

Upon exiting the intervertebral foramen, each spinal nerve divides into **dorsal** and **ventral branches (rami)**. The dorsal branch innervates the epaxial muscles of the trunk. The ventral branch innervates the hypaxial muscles of the trunk and the muscles of the limbs. Both dorsal and ventral branches form interconnections with neighboring nerves, forming a continuous, insignificant **plexus**. The interconnections of the ventral spinal nerves originating from the cervicothoracic and lumbosacral intumescences are distinct and form the **brachial** and **lumbosacral plexuses**, respectively. In horses, the **brachial plexus** receives contributions from the last 3 cervical and the first 2 thoracic spinal nerves (C6–T2). It is located in the axilla, emerging between parts of the scalenus muscle. The plexus allows for regrouping and reassociation of nerve fibers so that nerves that emerge distal to the plexus are composed of nerve fibers derived from 2 or more spinal segments (Table 14.1-1). Peripheral nerves extend from the brachial plexus to innervate the muscles of the thoracic limb (Fig. 14.1-3).

The **radial nerve** is one of the larger peripheral nerves emerging from the brachial plexus and is the sole provider of innervation to the extensor muscles of all joints distal to the shoulder. The radial nerve extends from the brachial plexus caudolaterally, diving between the medial and long heads of the triceps muscle and distributing terminal branches to the triceps muscle group. It emerges along the caudolateral surface of the humerus where it distributes branches to the extensor muscles of the carpus and digit.

The marked atrophy of the triceps musculature in this case indicated that radial neuropathy occurred proximal to the origin of the triceps muscle. The radial nerve is prone to trauma at several locations. Proximally, the radial nerve can be compressed between the scapula and the 1st rib, secondary to falling or pressure during recumbency. Distally, the radial nerve lies subcutaneously on the surface of the brachialis muscle and is vulnerable to kicks or pressure during recumbency. It can also be torn or compressed by trauma to the humerus or the elbow joint.

TABLE 14.1-1 Spinal nerve contributions to peripheral nerves of the brachial plexus.

Spinal nerves (those in parentheses are inconsistent between individuals)	*Peripheral nerve (Fig. 14.1-3)*	*Muscle innervated* (Fig. 14.1-3)*
C6, C7	Suprascapular *(pink)*	Supraspinatus (A) Infraspinatus (B)
(C6), C7	Subscapular *(orange)*	Subscapularis (C)
(C6), C7, C8	Musculocutaneous *(yellow)*	Biceps brachii (D) Brachialis (E) Coracobrachialis (F)
(C6), C7, C8	Axillary *(green)*	Deltoideus (G) Teres major (H) Teres minor (I) (Subscapularis)
C7, C8	Cranial pectoral	Pectoralis descendens** Pectoralis transversus**
C8, T1	Caudal pectoral	Pectoralis profundus**
(C7), C8, T1, (T2)	Radial *(light blue)*	Triceps brachii (J) Extensor carpi radialis (K) Common digital extensor (L) Lateral digital extensor (M) Ulnaris lateralis (N)
C8, T1, T2	Median *(dark blue)*	Flexor carpi radialis (O) Superficial digital flexor (P) (Deep digital flexor)
T1, T2	Ulnar *(purple)*	Flexor carpi ulnaris (Q) Deep digital flexor (R)

* Muscles in parentheses indicate that the nerve only supplies a portion of the muscle.
** Pectoral muscles do not have letters because they are not included in the figure as their origin is not on the forelimb.

General somatic afferent neurons

Spinal reflexes can be tested during a neurologic evaluation and used to localize lesions to portions of the PNS or spinal cord. The sensory portion of these reflex arcs involves the **general somatic afferent** (that sense temperature, touch, pressure, and noxious stimuli) and **general proprioceptive** (that sense movement in a muscle, tendon, or joint) systems. The dendritic zone of peripheral afferent neurons has endings that are modified to form a receptor organ that converts a form of energy (the stimulus) into a neuronal impulse. The afferent axons course proximally within nerves, plexuses, and spinal nerves and then diverge from efferent axons into the **dorsal spinal root** where their cell bodies form the **dorsal root ganglion** (Fig. 14.1-2). The axons continue proximally in the dorsal spinal root to enter the spinal cord dorsolaterally.

The sensory axons within a peripheral nerve are divided between multiple spinal cord segments. Therefore, sensory deficits are more obvious when the lesion is localized to a peripheral nerve compared to a lesion involving a single spinal nerve, dorsal root, or spinal cord segment. Thus, loss of sensation to a region indicates either a peripheral nerve lesion or an extensive lesion involving multiple spinal nerves, dorsal roots, or spinal cord segments.

In a simple spinal reflex pathway, the afferent axon synapses with an **internuncial** (L. *internuncius* an intermediary, connection between nerves) **neuron**, which then stimulates the appropriate GSE LMN, resulting in muscle contraction (Fig. 14.1-4). If there is a lesion along any part of the reflex pathway, the reflex will be reduced. As their name suggests, spinal reflexes can occur without any ascending or descending input from the CNS. However, sensory information from the reflex arc is relayed to the higher centers, and motor pathways from the brain do influence the spinal reflex. Cerebral responses associated with perception include changes in facial expression, head movement, and phonation (can be difficult to assess in stoic animals). **Upper motor neurons** (UMN) are confined to the CNS,

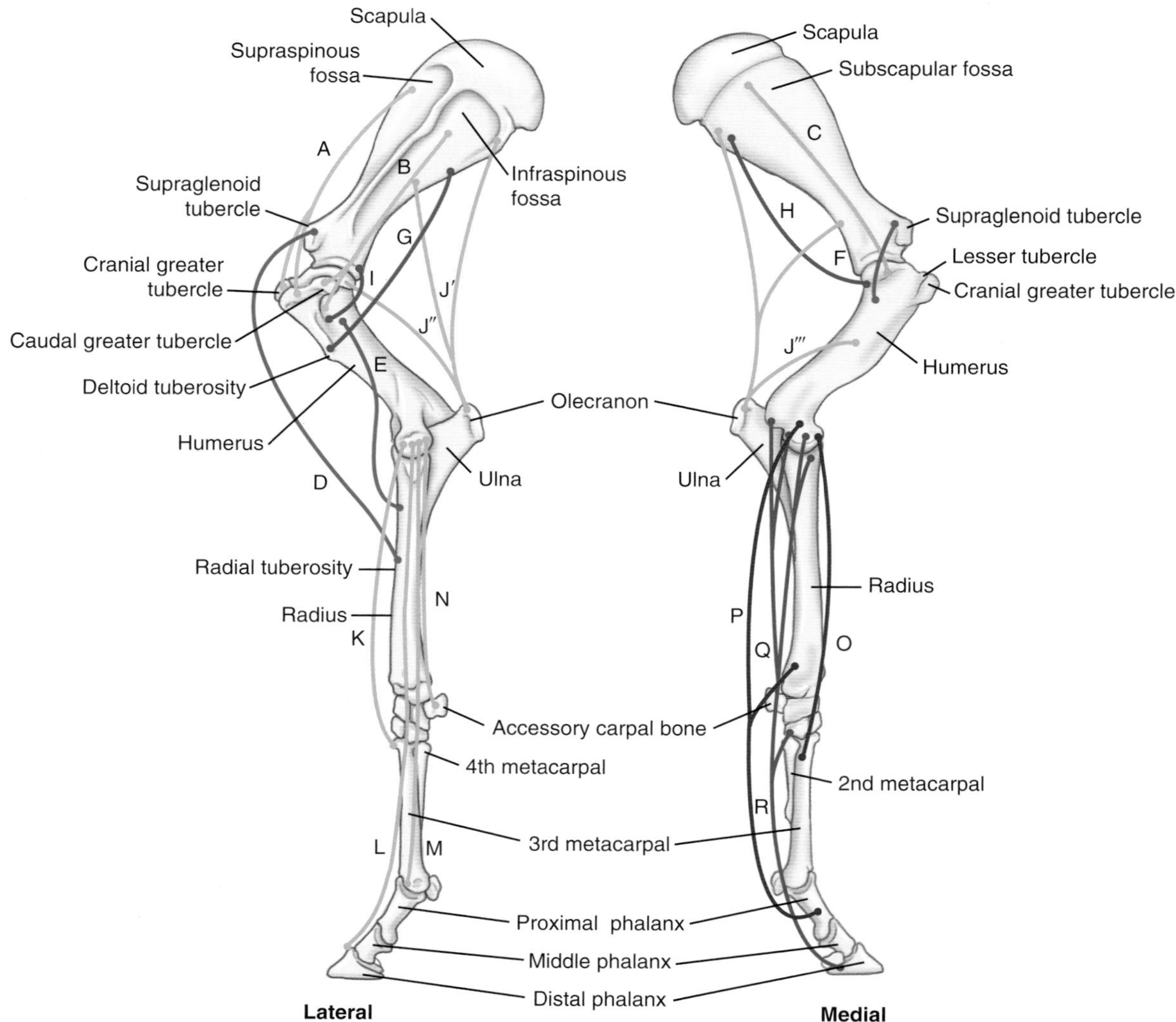

FIGURE 14.1-3 Muscles of the equine thoracic limb and their innervation. Key: See Table 14.1-1.

Upper motor neuron lesions cranial to the reflex arc will cause the reflex to be exaggerated. In horses, assessment of limb reflexes requires the patient to be restrained in lateral recumbency, largely limiting their use to neonates and recumbent horses. However, if a horse can stand and walk, it can be assumed that reflex function in the limbs is intact.

Table 14.1-2 details reflexes that might be useful in horses.

and in the horse, they descend from the UMN centers largely via the **rubrospinal, reticulospinal,** and **vestibulospinal tracts**. Typically, UMN have a calming effect on reflexes.

The **lateral cutaneous antebrachial nerve** is a purely sensory branch of the radial nerve that supplies the skin over the lateral aspect of the thoracic limb. An **autonomous zone** is the mapped portion of the body surface that is innervated by only one peripheral nerve, allowing analgesia of this zone to indicate a lesion of the associated peripheral nerve (Fig.14.1-5). Autonomous zones are relatively small and show

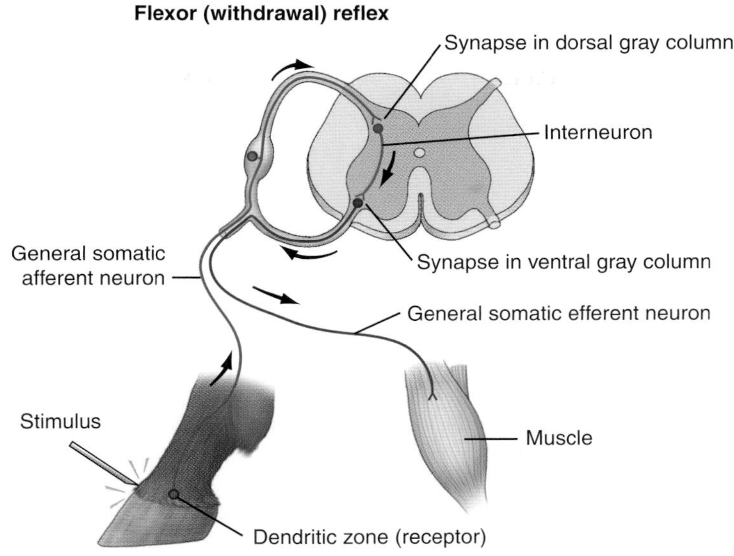

FIGURE 14.1-4 Schematic representation of the polysynaptic flexor reflex response to compression of the coronary band.

TABLE 14.1-2 Spinal reflexes of clinical importance in equine neurology.

Reflex initiation	*Afferent peripheral nerve*	*Spinal cord segments*	*Level in vertebral canal*	*Efferent LMN*	*Reflex reaction*	*Comment*
Triceps—ballotement of the distal portion of the long head of the triceps and its tendon of insertion	Radial	C7–T2	C6–T1	Radial	Contraction of the triceps muscle, extension of the elbow	Can be difficult to demonstrate in adult recumbent horses
Extensor carpi radialis—ballottement of the musculotendinous portion of the muscle	Radial	C7–T2	C6–T1	Radial	Extension of the carpus	Not always present in normal adult horses
Thoracic limb flexor reflex—compression of the coronary band by forceps	Median and ulnar	C6–T2	C5–T1	Axillary, musculocutaneous, median, and ulnar	Flexion of the fetlock, carpus, elbow, and shoulder	Conscious perception of stimulus can also be assessed
Patellar reflex—ballottement of the middle patellar ligament	Femoral	L4–L5	L3–L4	Femoral	Contraction of the quadriceps muscle, extension of the stifle	
Pelvic limb flexor reflex—compression of the coronary band by forceps	Sciatic	L5–S3	L4–L6	Sciatic	Flexion of the fetlock, hock and stifle	Conscious perception of stimulus can also be assessed

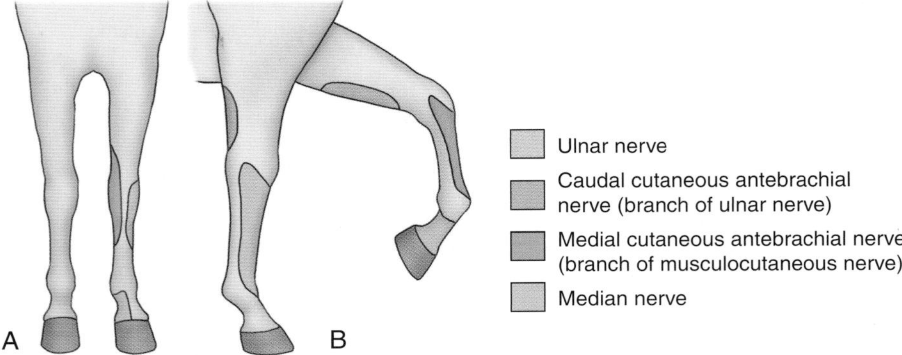

FIGURE 14.1-5 Autonomous zones of cutaneous innervation of the equine thoracic limb. (A) Cranial view. (B) Right lateral view.

interindividual variability. Unlike other domestic species, the lateral cutaneous antebrachial nerve of the horse terminates proximal to the carpus, and its cutaneous distribution overlaps with the cutaneous branches from other peripheral nerves so that there is no autonomous zone for the radial nerve in the horse.

Ptosis: cranial nerves and the sympathetic general visceral efferent neuron

A horse with ocular pain will also have a smaller palpebral opening and lower eyelash angle because of blepharospasm.

Differentials for ptosis (Gr. *ptōsis* fall, drooping of upper eyelid) can be determined by applying anatomic knowledge of the 3 dorsal eyelid muscles and their associated innervation.

Two muscles of the dorsal eyelid are voluntary striated skeletal muscles, which receive innervation from GSE LMN in **cranial nerves.** They will be mentioned here for the sake of completeness, but for further description of cranial nerve anatomy in the horse, refer to Case 10.8. The **levator palpebrae superioris** muscle is the main muscle elevating the eyelid that is under voluntary control and it is innervated by the oculomotor nerve (CN III). The **levator anguli oculi medialis** muscle is innervated by the palpebral branch of the facial nerve (CN VII). The cell bodies of these GSE LMNs are in the **brainstem nuclei** located between the mesencephalon and the caudal medulla. Like peripheral and spinal nerves, cranial nerves also contain afferent and GVE neurons.

The third muscle involved in dorsal eyelid elevation is the **orbitalis (Müller's) muscle**, and it is most relevant to the current discussion of the cervical intumescence. This smooth muscle consists of periorbital circular fibers and 2 sheets of fibers that extend into the lids—the **superior** and **inferior tarsus muscles**. The orbitalis muscle is innervated by GVE LMN of the sympathetic division of the autonomic nervous system.

The **autonomic nervous system** is an involuntary system that maintains internal physiologic homeostasis. As with the somatic system, afferent neurons provide sensory information which is processed by centers in the brain and relayed by brainstem and spinal cord pathways to activate LMN. The autonomic LMN is a 2-neuron system: the preganglionic neuron extends from the CNS to synapse in peripheral ganglia with postganglionic neurons, which then innervate smooth muscle associated with blood vessels, visceral structures, glands, and cardiac muscle. **Sympathetic UMNs** descend in the intermediolateral funiculus of the spinal cord and connect with preganglionic neuronal cell bodies located in the lateral gray column from T1 to L4–L5 spinal cord segments. This contrasts with the **parasympathetic system,** whose preganglionic cell bodies are restricted to the brainstem nuclei of the oculomotor, facial, glossopharyngeal, and vagus nerves, and the sacral spinal cord segments.

Horner syndrome is a collection of clinical signs associated with a lesion of the sympathetic nervous system pathway that provides innervation to the head (Fig. 14.1-6; see also the side box entitled "Unique features of Horner syndrome in the horse".) Preganglionic neurons from the first three or four thoracic spinal cord segments contribute

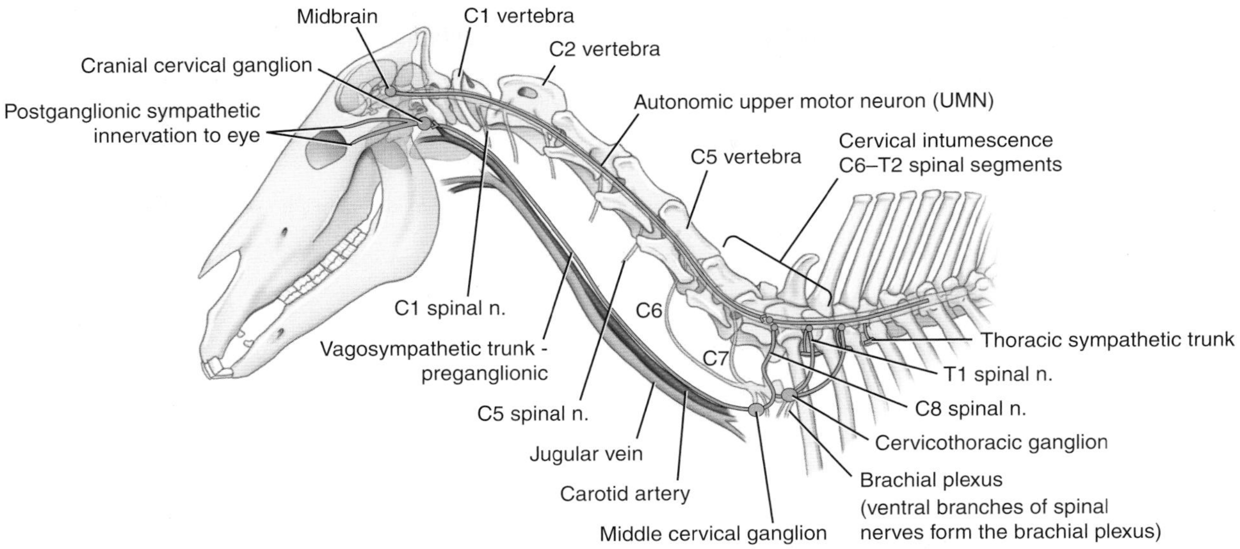

FIGURE 14.1-6 Sympathetic general visceral efferent pathway to the head.

to this pathway. These axons pass through the ventral gray column and travel in the ventral spinal roots and the proximal portion of the segmental spinal nerve. As the spinal nerve emerges from the intervertebral foramen, the preganglionic axons divide to join the **thoracic sympathetic trunk**, passing through the **cervicothoracic ganglion (stellate ganglion)** near the brachial plexus and the middle cervical ganglia to continue cranially in the **cervical sympathetic trunk**. This trunk runs with the vagus nerve, becoming the **vagosympathetic trunk**, and is closely associated with the carotid artery. Preganglionic neurons synapse with postganglionic neurons in the **cranial cervical ganglion**, which in the horse is a fusiform, greyish-pink ganglion located in the caudodorsal wall of the medial compartment of the guttural pouch. Postganglionic axons extend from the cranial cervical ganglion alongside blood vessels and within cranial and spinal nerves to innervate the smooth muscles of blood vessels and the sweat glands of the skin in the head and cranial cervical area. The postganglionic pathway to the eye is poorly defined.

UNIQUE FEATURES OF HORNER SYNDROME IN THE HORSE

Sympathetic denervation in all other domestic animals results in hypohidrosis (Gr. *hypo* under + Gr. *hidrōsis* sweating, reduced perspiration) or anhidrosis (Gr. absence of sweating). Conversely, excessive sweating of skin that is denervated is the most prominent clinical sign of acute sympathetic denervation in the horse. Marked sweating persists for 24–48 h after which patchy sweating might be induced by exciting the animal. In chronic cases, sweating may decrease with time and only be present at the base of the ear. The cause of sweating is debated but likely related to sudden vasodilation, increased blood flow, and subsequent flooding with catecholamines.

Horses, unlike small animals or man, have an eyelid muscle named the arrector ciliorum muscle, also under sympathetic control. This muscle inserts on the eyelashes (similar to the arrectores pilorum of hair). Its role is to increase the angle of the eyelashes of the upper eyelid and, therefore, denervation contributes to ptosis.

Miosis and enophthalmos are rarely observed in the horse associated with Horner syndrome.

Sympathetic neuronal fibers do not pass through the petrosal bone. Therefore, unlike small animals, otitis media in horses and farm animals does not cause Horner syndrome.

Selected references

[1] Emond A-L, et al. Peripheral neuropathy of a forelimb in horses: 27 cases (2000-2013). J Am Vet Med Assoc 2016;249(10):1187–95.
[2] Lopez MJ, Nordberg C, Trostle S. Fracture of the 7th cervical and 1st thoracic vertebrae presenting as radial nerve paralysis in a horse. Can Vet J 1997;38(2):112.
[3] Johnstone LK, et al. Retrospective evaluation of horses diagnosed with neuroborreliosis on postmortem examination: 16 cases (2004-2015). J Vet Intern Med 2016;30(4):1305–12.
[4] Morrison LR. Lymphoproliferative disease with features of lymphoma in the central nervous system of a horse. J Comp Pathol 2008;139(4):256–61.
[5] Durham AC, et al. Two hundred three cases of equine lymphoma classified according to the World Health Organization (WHO) classification criteria. Vet Pathol 2013;50(1):86–93.

CASE 14.2

Supraglenoid Tubercle Fracture

Nick Carlson
Steinbeck Peninsula Equine Clinics, Salinas, California, US

Clinical case

History

A 1-year-old American Quarter Horse filly presented with an acute onset of severe right thoracic limb lameness. The owners noticed a superficial wound over the point of the shoulder associated with moderate swelling in the region. The filly was transported to an equine clinic for further evaluation.

Physical examination findings

The filly was grade 4+/5 lame on the right thoracic limb with a limited cranial phase of the stride. No crepitus was appreciated in the region; however, the horse was resistant to palpation of the swelling over the point of the shoulder. The horse could bear full weight on the limb and resisted any attempt to flex or extend the proximal limb.

Differential diagnoses

Fracture of the supraglenoid tubercle, proximal humerus, greater or lesser tubercles of the humerus, or scapular neck; synovial infection of the bicipital bursa, scapulohumeral joint, or infraspinatus bursa

Diagnostics

Ultrasound of the right shoulder region showed the wound tracking to the bicipital bursa. The synovium was thickened, and there was increased fluid in the bursa. A small osseous (bony) fragment was also identified originating from the humeral tubercle adjacent to the biceps brachii tendon (Fig. 14.2-1). Radiographs of the scapulohumeral joint revealed a complete displaced fracture of the supraglenoid tubercle that extended into the cranial glenoid cavity (Fig. 14.2-2).

Diagnosis

Complete displaced fracture of the supraglenoid tubercle and septic bicipital bursitis secondary to a traumatic wound

Treatment

The fracture was repaired with open reduction and internal fixation using a combination of cortical bone screws and cerclage wire under general anesthesia. Concurrently, wound debridement and septic bursa irrigation (lavage) were performed. The filly was maintained on systemic antimicrobials and repeated injections of intrabursal antibiotics over subsequent days for local treatment of the infection. The filly's lameness gradually improved; however, the filly never regained full soundness for her intended use as a riding horse.

Comparative Veterinary Anatomy: A Clinical Approach. https://doi.org/10.1016/B978-0-323-91015-6.00072-8

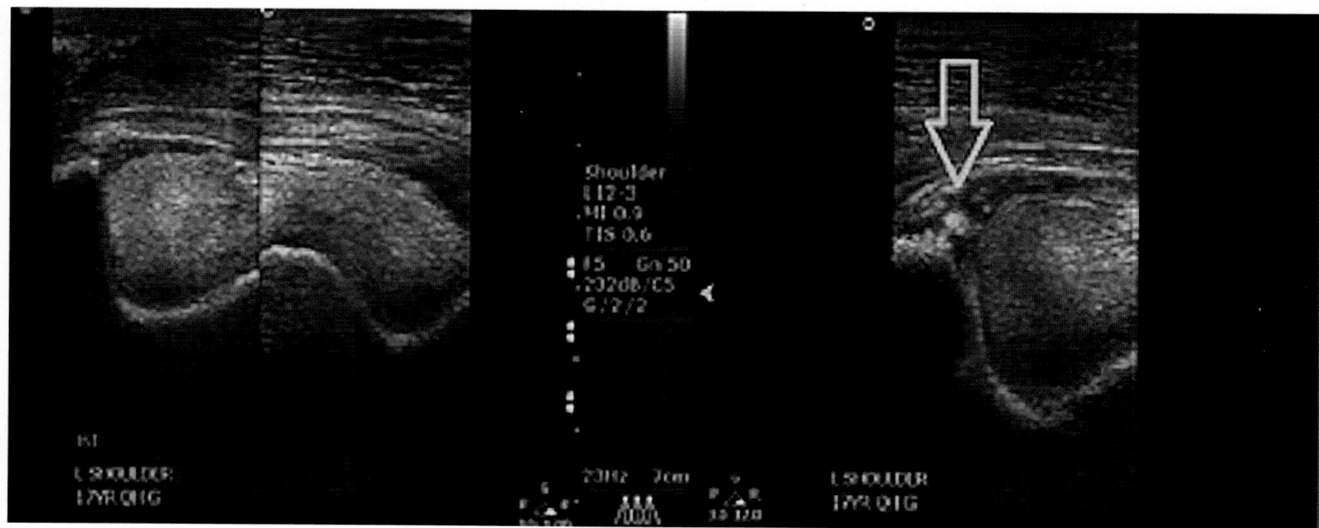

FIGURE 14.2-1 Ultrasound of the left biceps tendon as it travels through the bicipital bursa (*left image*). A small fragment off the greater tubercle of the humerus is observed *(yellow arrow)* secondary to trauma to the region (*right image*).

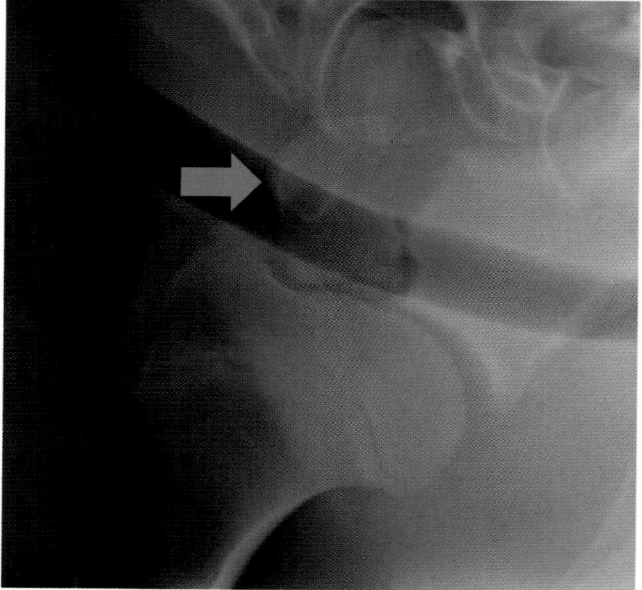

FIGURE 14.2-2 Lateromedial radiograph of a complete comminuted fracture *(arrow)* of the supraglenod tubercle with articular involvement.

Anatomical features in equids

Introduction

The major structures in the scapulohumeral (shoulder) region include the scapula, humerus, scapulohumeral joint, local bursas, brachial plexus, regional blood supply, and surrounding musculature.

Function

The scapulohumeral, glenohumeral (Gr. *glene* eyeball + *oid* form of, L. *humerus* shoulder), or shoulder joint is the most mobile joint in the body but is unusual because it lacks collateral ligaments like other joints. It has the characteristics of a spheroidal (ball and socket) joint, but clinically it behaves like a hinge (Gr. *ginglymus* hinge) joint (synovial joint with movement in one plane—extension and flexion) with extension and flexion in a sagittal plane. In support of this joint's mobility, stability is reliant on surrounding groups of muscles and their respective tendons on the lateral, medial, cranial, and caudal borders of the joint.

The lateral groups of muscles supporting the shoulder are the supraspinatus, infraspinatus, deltoideus, and teres minor. The medial group of muscles includes the subscapularis, teres major, coracobrachialis, and a clinically insignificant muscle, the capsularis. Cranial muscle stabilization includes the biceps brachii and supraspinatus while caudal support is provided by the long head of the triceps brachii. Transverse mobility in other hinge joints is normally restricted by collateral ligaments; however, in the shoulder joint the collateral ligaments are substituted by the tendons of the infraspinatus, primarily, and to a lesser extent the supraspinatus, laterally and the subscapularis medially.

Scapula (Fig. 14.2-3)

The **scapula** is a broad, flat bone bisected on the lateral surface by the scapular spine. Noteworthy anatomic landmarks include the **scapular cartilage**, **scapular spine**, **supraglenoid tubercle**, **coracoid process**, and **glenoid cavity**. Unlike cats and dogs, the clavicle is absent in the horse. Horses also lack an acromion on the distal part of the scapula. The scapula has 4 centers of ossification: the body of the scapula and the scapular cartilage (fused at birth), the cranial portion of the glenoid cavity of the scapula (fused at 5 months), and the supraglenoid tubercle (fused between 12 and 24 months).

The most common fracture of the scapula is through the supraglenoid tubercle, usually seen in younger animals (<2 years old) associated with direct trauma and a concurrent avulsion injury from the biceps. The goal of surgery to repair supraglenoid tubercle fractures is to restore congruity of the glenoid cavity. Depending on the size of the fragment, it can be removed or repaired with cortical bone screws. The supraglenoid tubercle is the origin of the biceps tendon and can place considerable tensile forces on fixations of these fractures. Other fractures seen are stress fractures of the body of the scapula in young racehorses, complete scapular body fractures, and fractures of the scapular spine.

ROTATOR CUFF AND GLENOID LABRUM

The rotator cuff in people is a group of muscles and their respective tendons of the shoulder that produce a high tensile force, helping to pull the head of the humerus into the glenoid cavity. The rotator cuff group of muscles and tendons stabilize the shoulder. The 4 muscles that comprise the rotator cuff are the supraspinatus, infraspinatus, teres minor, and subscapularis muscles.

The glenoid labrum is a fibrocartilaginous rim attached around the margin of the glenoid cavity. In humans, the shoulder joint is considered a ball and socket joint and the glenoid fossa/cavity of the scapula is actually shallow and small, covering only a third of the head of the humerus; the glenoid cavity is deepened by the glenoid labrum.

The labrum (L. *labra* edge, brim, or lip) is triangular-shaped and fixed to the circumference of the glenoid cavity. It is continuous proximally with the tendon of the long head of the biceps brachii and gives off 2 fascicles to blend with the fibrous tissue of the labrum.

Rotator cuff and glenoid labrum tears are common orthopedic problems of the shoulder in humans, many of which are treated arthroscopically.

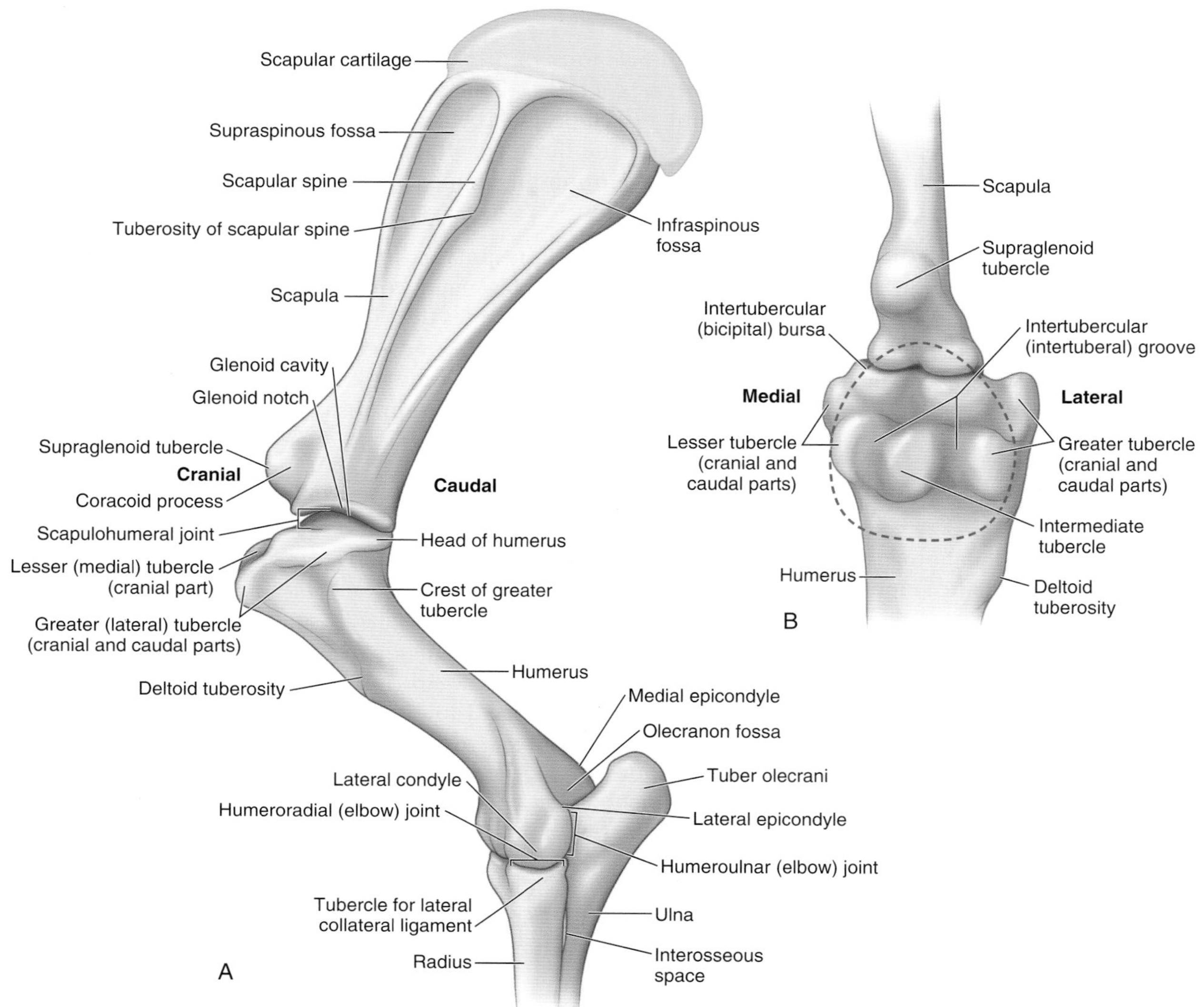

FIGURE 14.2-3 Bones, joints, and bursae of the equine shoulder region: (A) lateral view and (B) cranial view.

Brachial plexus and innervation of the proximal thoracic limb (Fig. 14.2-4)

The **brachial plexus** is comprised of the set of nerves arising from the ventral branches of the C6–T2 spinal nerves. These nerves supply most of the muscles and skin sensation in the thoracic limb (Table 14.2-1). More detail regarding the brachial plexus can be found in Case 14.1.

Loss of function to these muscles (and surrounding muscles of the shoulder) associated with injury to the suprascapular nerve or brachial plexus leads to instability of the shoulder joint with variable degrees of subluxation. This can also occur temporarily if local anesthetic is misdirected outside the joint space during a scapulohumeral joint block.

Scapulohumeral joint (Fig.14.2-5)

The **scapulohumeral (shoulder) joint** is a spheroid-type joint between the **glenoid cavity** of the scapula and the **humeral head**. Unlike other joints, there are no true collateral ligaments supporting this joint. The tendons of the **subscapularis** (medially) and the **infraspinatus** and **supraspinatus** (laterally) muscles act like collateral ligaments.

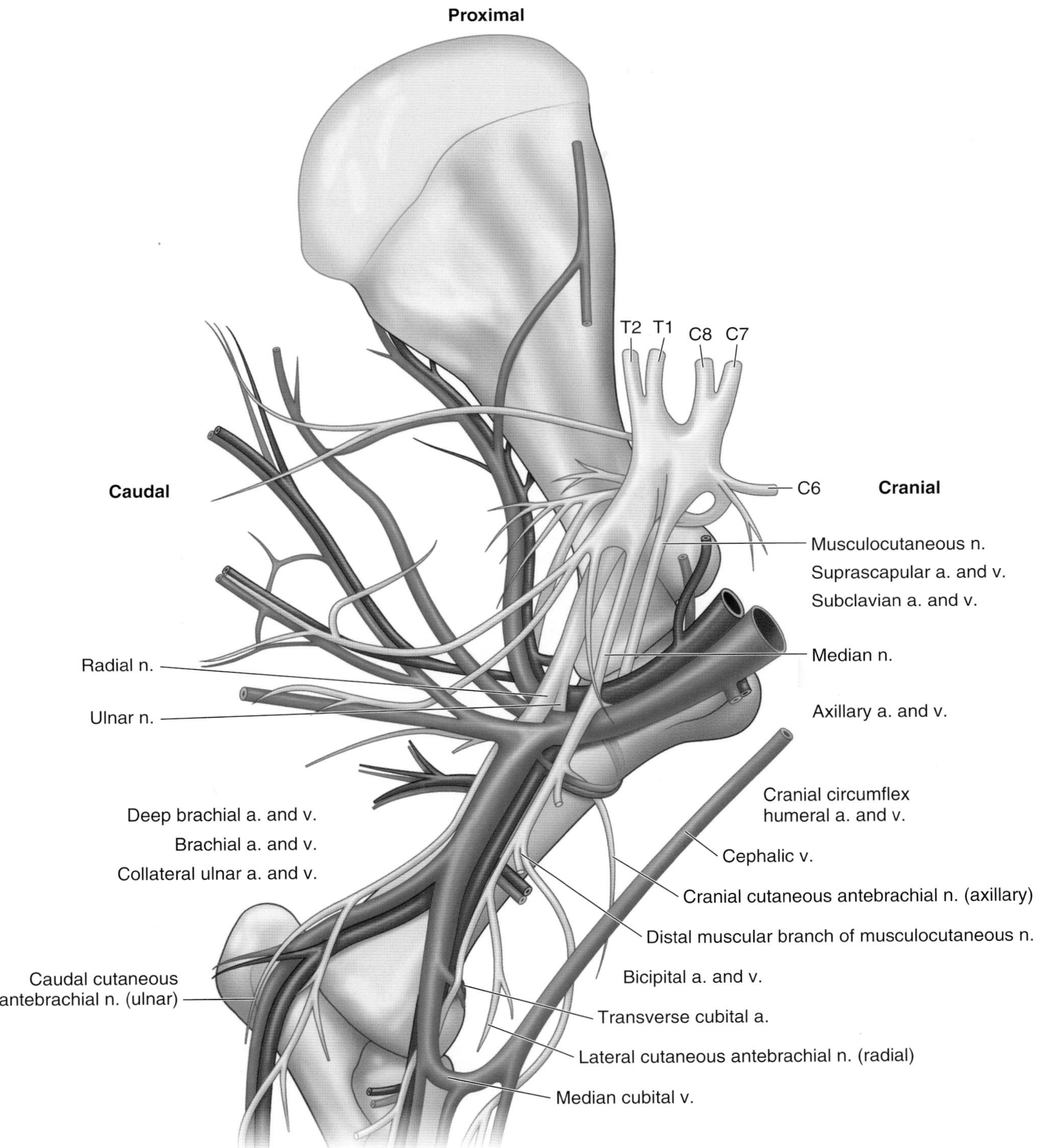

FIGURE 14.2-4 Medial view of the blood vessels and nerves of the equine proximal thoracic limb.

TABLE 14.2-1 The brachial plexus is comprised of the set of nerves arising from the ventral branches of the C6–T2.

*Nerves of the brachial plexus**
• **Suprascapular n. (C6-7)** ○ The nerve travels lateral and ventral between the supraspinatus and subscapularis muscles before curving around the cranial edge of the scapula proximal to the scapular tuberosity and courses into the supraspinous and infraspinous fossae ○ Innervates the infraspinatus and supraspinatus muscles ○ Injury creates atrophy of these muscles in a clinical condition called "sweeney."
• **Radial n. (C8-T1)** ○ Travels with the ulnar nerve between the long and medial heads of the triceps muscle coursing distally around the humerus in the musculospiral groove and then crossing the lateral aspect of the elbow before innervating the extensor muscle of the digits ○ Innervates flexor muscles of the shoulder and extensor muscles of the elbow, carpus, and digits ○ Injury to the proximal nerve can lead to a dropped elbow appearance and an inability to bear weight due to loss of triceps function ○ If injured at the level of the elbow the affected horse develops a toe drag
• **Ulnar n. (T1-T2)** ○ It descends along the medial aspect of the leg with major arteries and terminates in palmar cutaneous branches above the carpus ○ Innervates extensor muscles of the elbow and flexor muscles of the carpus and digits
• **Median n. (C8-T1)** ○ The nerve courses along the medial aspect of the thoracic limb with major arteries and terminates on palmar cutaneous branches proximal to the carpus ○ Innervates extensor muscles of the elbow and flexor muscles of the carpus and digits ○ Damage to median nerve, along with the ulnar nerve, creates a "tin-solider" gait with decreased flexion of the carpus and digit and a dragging of the toe during the protraction phase of the stride
• **Musculocutaneous n. (C7-8)** ○ Descends medially on the coracobrachialis muscles and joins the median nerve beneath the brachial artery before terminating into the biceps, pectoral, and brachialis muscles ○ Innervates flexor muscles of the elbow
• **Axillary n. (C7-8)** ○ Innervates flexors of the shoulder
• **Long thoracic n. (C7)** ○ Innervates the ventral serratus muscles.
• **Thoracodorsal n. (C8)** ○ Innervates the latissimus dorsi muscle.
• **Lateral thoracic n. (T1)** ○ Innervates the cutaneous trunci muscle and provides cutaneous innervation of the ventrolateral abdominal wall

*These nerves supply the majority of the muscles and skin sensation in the thoracic limb.

Regional bursae (Fig. 14.2-5)

The **bicipital (intertubercular) bursa** lies between the proximal **biceps tendon** and the **intertubercular groove** of the humerus. Its synovial membrane is covered dorsally by the **subclavian muscle** and medially by the **pectoral profundus muscle**. The lateral surface around the lateral tubercle is only covered with skin, leaving it at risk for injury. It is a relatively sizable synovial structure that envelops both lobes of the biceps tendon, leaving just the cranial-most surface outside of the bursa. The bursa could more fittingly be described as a tendon sheath because of its embryological development and function.

REGIONAL ARTHROCENTESIS—SCAPULOHUMERAL JOINT AND BICIPITAL BURSA

Arthrocentesis of the scapulohumeral joint is performed using topographical anatomy landmarks by insertion of a needle at the palpable depression between the cranial and caudal parts of the lateral humeral tuberosity. The needle should be parallel to the ground and directed toward the opposite elbow. Ultrasound-guided needle placement reduces the risk of misdirecting the needle periarticularly (outside of the joint).

Centesis of the bicipital bursa is performed via a needle directed laterally just cranial to the humerus approximately 4 cm (1.6 in.) proximal to the distal deltoid tuberosity. There is the occasional communication between the bursa and scapulohumeral joint.

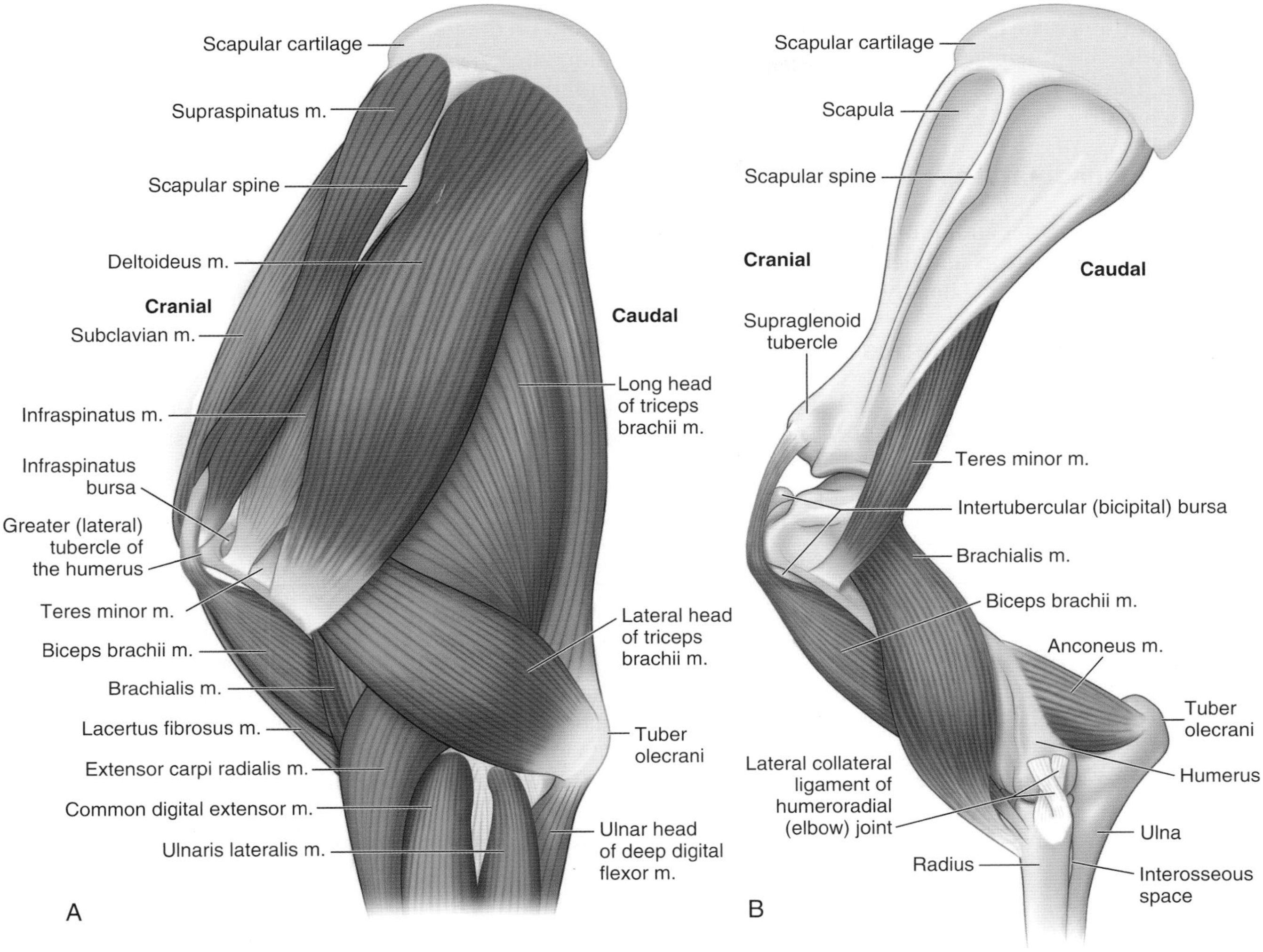

FIGURE 14.2-5 Lateral views of the equine shoulder region: (A) superficial and (B) deep.

The **infraspinatus bursa** is located between the **infraspinatus tendon** and **caudal eminence of the greater tubercle**. It is surrounded by the **omotransversus muscle** laterally, **teres minor** and **deltoideus muscles** caudally, and the greater tubercle medially. Trauma to the greater tubercle can also involve the infraspinatus bursa (as it can the bicipital bursa).

Proximal humeral and greater tubercle stress fractures can be managed conservatively. Displaced fractures carry a poorer prognosis. Internal fixation can be successful in smaller-sized horses, and smaller displaced fragments may be managed with surgical removal. Displaced fractures may require open reduction and internal fixation. Fractures of the deltoid tuberosity are uncommon and associated with impact trauma. Middle and distal diaphyseal humeral fractures are typically catastrophic in nature. Laceration of the radial nerve can occur with these fractures because its path tracks over the cranial diaphysis, further reducing the prognosis for these fractures due to an inability to extend the elbow, leading to loss of the stay apparatus.

Proximal humerus (Fig. 14.2-3)

Notable landmarks on the proximal humerus include the **greater (lateral)** and **lesser (medial) tubercles**. Both tubercles have cranial and caudal parts; however, the cranial aspect of the greater tubercle is the most prominent and palpable feature known as the "point of the shoulder." Between the greater and lesser tubercles lie the **intertubercular groove** and **intermediate tubercle** over which the biceps tendon travels. The **deltoid tuberosity** is another palpable landmark and is found on the lateral body of the humerus. The smaller **teres major tuberosity** is found on the medial body of the humerus and is not palpable. The proximal humerus has 3 centers of ossification: the greater tubercle, the humeral head, and the diaphysis. The separate centers of ossification merge by 3–4 months of age, and the proximal humeral physis closes at 24–36 months.

Regional musculature (see Landscape Figs. 4.0-4–4.0-6)

Muscles of the shoulder and humeral region can be divided into lateral and medial muscle groups (Table 14.2-2). Most function as extensors and flexors of the shoulder and elbow joints, while some play other roles in stabilizing the shoulder medially and laterally.

Blood supply and lymphatics (Fig. 14.2-4)

The **subclavian artery** and **vein** become the **axillary vessels** (a. and v.) that travel on the medial aspect of the limb with the median and ulnar nerves. Their branches supply the muscles of the shoulder and arm. The **suprascapular**

TABLE 14.2-2 Lateral (A) and medial (B) muscles of the shoulder and humeral region.

Muscle (innervation)	*Origin*	*Insertion*	*Function*
(A) Lateral muscles of the scapula and humerus			
Supraspinatus (Suprascapular n.)	Supraspinous fossa, scapular spine	Greater and lesser tubercles of humerus	Extends shoulder, acts as collateral ligament
Infraspinatus (Suprascapular n.)	Infraspinous fossa, scapular spine	Lateral surface of greater tubercle distal to insertion of supraspinatus	Acts as collateral ligament, secondary abductor
Deltoideus (Axillary n.)	Scapular spine, caudal edge of scapula	Deltoid tuberosity of humerus	Flexes shoulder and abdus the limb
Teres minor (Axillary n.)	Distal half of the caudal edge of scapula	Proximal to deltoid tuberosity	Flexes shoulder
Triceps—long head (Radial n.)	Caudal edge of scapula	Tuber olecrani	Extends elbow, flexes shoulder
Triceps—medial head (Radial n.)	Medial middle third of humerus	Tuber olecrani	Extends elbow
Triceps—lateral head (Radial n.)	Deltoid tuberosity	Tuber olecrani	Extends elbow
(B) Medial muscles of the scapula and humerus			
Teres major (Axillary n.)	Caudal edge of scapula and subscapularis	Teres tuberosity of the humerus	Flexes shoulder
Articularis humeri (Axillary n.)	Just proximal to the medial rim of glenoid cavity	Neck of humerus	Tenses shoulder
Subscapularis (Axillary and subscapular n.)	Subscapular fossa of scapula	Lesser tuberosity of humerus	Primarily extends shoulder and secondarily stabilzes the shoulder joint
Coracobrachialis (Musculocutaneous n.)	Coracoid process of the scapula	Proximal medial surface of the humerus	Extends shoulder, adducts the limb
Biceps brachii (Musculocutaneous n.)	Supraglenoid tubercle	Radial tuberosity and blends with medial collateral ligament of the elbow	Extends shoulder, flexes elbow, stabilizes carpus (indirect)
Brachialis (Musculocutaneous n.)	Proximal caudal surface of the humerus	Proximal medial aspect of the radius	Flexes elbow
Tensor fasciae antebrachii (Radial n.)	Caudal edge of scapula and insertion of latissimus dorsi	Deep fascia of forearm and olecranon	Tenses forearm fascia and extends elbow joint

artery branches cranially and tracks dorsally to supply the cranial scapular structures while the **subscapular artery** branches caudally and follows the caudal scapula. The **deep brachial artery** branches from the axillary artery and travels into the head of the triceps muscle. The **transverse cubital artery** and **collateral ulnar artery** branch just proximal to the elbow to supply the muscles of the antebrachium. The axillary artery ends as the **brachial artery** in the mid-humerus where it continues to track medially and crosses the elbow cranial to the medial collateral ligament.

Most veins of the thoracic limb mirror their accompanying arteries and assume the same name as their arterial counterparts. The **cephalic** (larger) and **accessory cephalic** (smaller) **veins** are the exception, located superficially on the craniomedial (cephalic v.) and cranial (accessory cephalic v.) side of the antebrachium; the 2 veins join near the elbow. A large **median vein** branches close to the junction of the cephalic and accessory cephalic and passes on the medial side at the distal aspect of the biceps brachii.

Lymphatic flow collects in the distal limb and drains to the **cubital lymph nodes** proximal to the elbow. It continues caudodistally along the teres major toward the **axillary lymph nodes,** which transfer lymphatic drainage to the **caudal deep cervical nodes** and then into veins at the thoracic inlet.

Selected references

[1] Dyson S. The elbow, brachium, and shoulder. In: Ross MW, Dyson S, editors. Diagnosis and management of lameness in the horse. 2nd ed. Saunders; 2011. p. 456–74.
[2] Fortier LA. The shoulder. In: Auer JA, Stick JA, editors. Equine surgery. 4th ed. Saunders; 2012. p. 1379–88.
[3] Butler JA, Colles CM, Dyson SJ, Kold SE, Poulos PW, editors. Clinical radiology of the horse. 3rd ed. Wiley-Blackwell; 2008. p. 273–320.

CASE 14.3

Ulnar Fracture

Nick Carlson
Steinbeck Peninsula Equine Clinics, Salinas, California, US

Clinical case

History

An 8-month-old Warmblood filly was found nonweight-bearing on the right thoracic limb while out in pasture. The owner called a veterinarian to have the filly examined in the field because the filly was reluctant to move, and the owner was unable to load her onto a trailer.

Physical examination findings

The filly was grade 4+/5 lame on the right thoracic limb. There was a "dropped elbow" appearance due to an inability to fix the carpus in extension. Mild swelling was present over the lateral elbow; the filly resented palpation of the area.

Differential diagnoses

Ulnar fracture, humeral fracture, and radial nerve paralysis

Diagnostics

Radiographs of the elbow revealed a fracture at the caudal third of the apophyseal (Gr. "an offshoot"; a bony process, tubercle, or tuberosity; no direct joint articulation with another bone; anatomical location for insertion of a tendon or ligament) physis, which propagated into the metaphysis of the olecranon and exited at the proximal aspect of the trochlear notch near the anconeal process. Lateral displacement of the fracture fragment was observed on the craniocaudal view (Fig. 14.3-1).

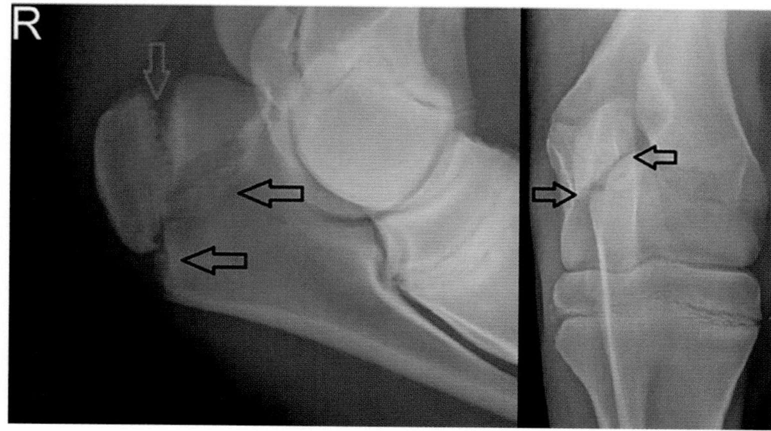

FIGURE 14.3-1 Flexed lateral *(left)* and cranial-caudal *(right)* radiographs of the elbow showing the fracture through the olecranon and apophysis, noted with *black arrows*. The apophysis is shown with the *red arrow*.

Comparative Veterinary Anatomy: A Clinical Approach. https://doi.org/10.1016/B978-0-323-91015-6.00073-X

Diagnosis

Type 1b ulnar fracture

Treatment

The filly was stabilized by placing a caudal splint to the level of the proximal antebrachium to fix the carpus in extension, allowing her to bear weight on the limb. She was clinically more comfortable and calmer after splint placement. The filly was placed under general anesthesia, and using open reduction and internal fixation surgical principles, the fracture was reduced into anatomic alignment and repaired with a bone plate and cortical screws. The filly recovered uneventfully from surgery and was weight-bearing on the limb immediately after surgery. The horse was sound at 1-year follow-up.

Anatomical features in equids

Introduction

The major structures in the cubital joint (elbow) region include the humerus, shaft of the ulna, olecranon, radius, surrounding musculature, nerves, and vessels.

Function

The function of the elbow joint is to extend and flex the upper limb and advance the limb. The range of movement in the elbow is expansive from a vertical extended position (0°) to an almost nearly horizontal flexed position of 150° in a sagittal plane. Muscles contributing to elbow function are all flexors—biceps brachii and brachialis—and extensors—triceps and anconeus. Because of the ability of the elbow joint to move from a stable position (extension) to a more mobile position (flexion), it is often referred to as an example of a "snap joint." The wide angle of flexion at the elbow made possible by this anatomy—almost 180° range—allows the bones of the antebrachium (forearm; i.e., radius and ulna) to be brought almost in parallel to the humerus.

Distal humerus (Fig. 14.3-2)

Distally, both the **lateral** and **medial epicondyles** of the **humerus** are prominent palpable structures. The **condyles** of the humerus are distal to the epicondyles. Cranially there is a shallow **radial fossa**, and caudally there is a deeper **olecranon fossa** that provides space for the anconeal process of the olecranon when the limb is weight-bearing.

There are 3 centers of ossification in the distal humerus: diaphysis, distal epiphysis, and epiphysis of the medial epicondyle. The distal humeral physes close between 11–24 months of age.

FIVE TYPES OF OLECRANON FRACTURES

Type 1: Physeal separation (type 1a) or Salter-Harris type II articular fracture (type 1b) involving the trochlear notch near the anconeal process. Both types are seen in young animals
Type 2: A simple articular fracture through the midtrochlear notch
Type 3: A nonarticular fracture of the proximal metaphyseal region of the olecranon
Type 4: Like a type 2 fracture, but is a comminuted articular fracture through the trochlear notch
Type 5: An oblique fracture traversing the bone in a proximal and cranial direction to enter the distal aspect of the trochlear notch; this is the most common fracture in horses > 1 year of age

Open reduction and fixation of these fractures carries a good prognosis in most cases.

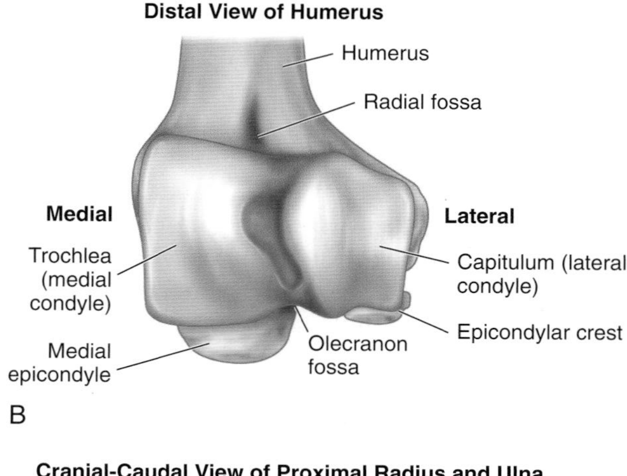

FIGURE 14.3-2 Bones of the equine elbow (cubital) region.

Ulnar fractures are usually a result of a direct blow (e.g., kick or a young horse in training rearing up and falling onto the elbow). The horse typically stands with a dropped elbow due to the inability of the triceps to effectively extend the elbow and a loss of the passive stay apparatus. The loss of the stay apparatus from an olecranon fracture can be relieved by fixing the carpal joints in extension. This is accomplished with a caudally placed splint that extends from the point of the elbow to the distal limb. This allows the horse to place weight on the fractured limb, reduces mental stress, and minimizes fatigue on the other limb while the horse is being transported to a hospital for treatment.

Ulna (Fig. 14.3-2)

The **olecranon** is the proximal part of the ulna and the **tuber olecrani** is known as the "point of the elbow," which serves as an insertion point for the extensors of the elbow. The **trochlear notch** is an articulating depression in the ulna that ends proximally with the **anconeal process**. In horses, the more distal **styloid process** of the ulna fuses with the midradius on the caudolateral aspect of the bone.

The ulna has a single proximal apophysis that may have a separate center of ossification from the anconeal process. The epiphysis grows until approximately 12 months of age and the apophysis closes between 24–36 months of age.

Proximal radius (Fig. 14.3-2)

The **head** is the proximal portion of the radius that articulates with the humerus. There are **medial** and **lateral eminences** where the long superficial and short deep medial and lateral collateral ligaments insert. The **radial**

tuberosity is located on the craniomedial aspect of the bone. The body of the radius has a slight curve when viewed from lateral to medial, leaving the cranial aspect of the bone convex and the caudal portion of the bone concave. There is no muscle coverage over the craniomedial radius. The ulna joins the proximal radial body distal to an interosseous space between the two bones. There is a single proximal epiphysis that is skeletally mature by 11–24 months of age.

Radial fractures are often the result of trauma (e.g., a kick). These fractures often travel in a spiral or oblique configuration with a variable degree of comminution. They are typically open on the medial side of the limb due to the lack of soft tissues covering this area. Affected horses present with a variable degree of lameness and often have a wound from a kick (Fig. 14.3-3). Displaced fractures are easy to diagnose because horses are nonweight-bearing with a flexed carpus and fetlock, and—when they attempt to bear weight—there is a valgus angulation of the radius at the point of the fracture site. The limb should be placed in a full-limb Robert-Jones bandage and 2 splints applied at right angles to each other for maximum support; first, a lateral splint from the ground to midscapula, and second, a caudal splint from the ground to the olecranon. The goal of splinting is to prevent abduction of the limb during weight-bearing and to prevent conversion of a nondisplaced to a displaced fracture, or closed to an open fracture. Nonsurgical management is possible for nondisplaced fractures by combining splinting and use of a tie-stall. Many radial fractures are amendable to repair with open reduction and internal fixation, depending on the configuration.

Regional musculature and innervation (see Landscape Figs. 4.0-4–4.0-6)

Muscles of the distal humerus and proximal radius can be divided into flexor and extensor muscle groups (Table 14.3-1). Most of the extensors arise on the craniolateral aspect of the distal humerus and proximal radius, and—except for the

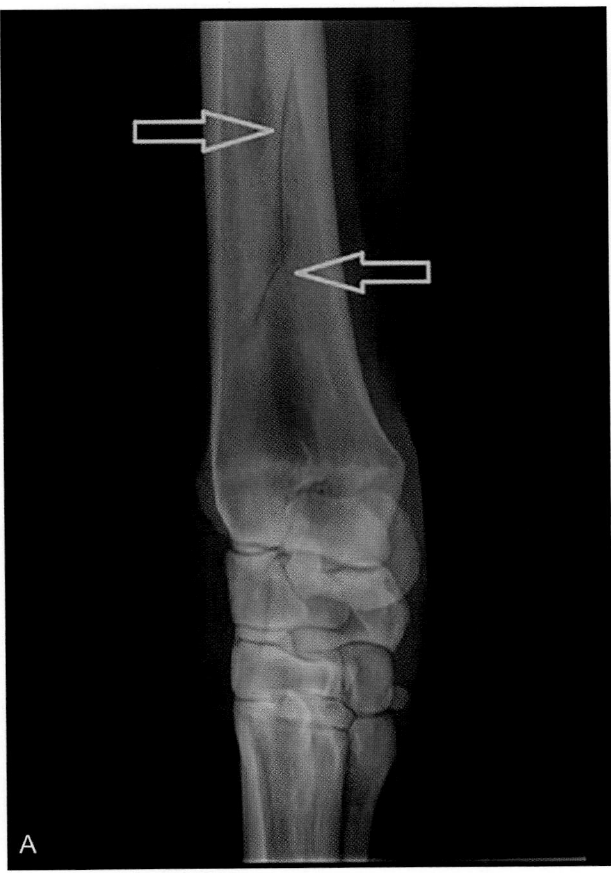

FIGURE 14.3-3 Craniomedial-caudolateral radiograph (left) of the distal radius and carpus showing an incomplete nondisplaced fracture of the radius *(yellow arrows)* following a kick from a pasture mate. This horse was managed successfully with a tie-stall and splinting (right).

TABLE 14.3-1 Muscles of the distal humerus and proximal radius.

Muscle	*Origin*	*Insertion*	*Function*
Anterior muscles (flexion)			
Biceps brachii	Coracoid process; tubercle proximal to the glenoid cavity	Radial tuberosity	Flexes and supinates antebrachium
Brachialis	Cranial aspect of distal humerus	Coronoid process of ulna	Flexes the elbow
Brachioradialis	Lateral supracondylar ridge at the distal end of humerus	Base of the styloid process of the radius	Flexes the antebrachium
Posterior muscles (extension)			
Triceps brachii	Infraglenoid tubercle of the scapula; posterior shaft of the humerus; posterior humeral shaft distal to the radial groove	Olecranon process of the ulna	Extends the antebrachium
Anconeus	Lateral epicondyle of the humerus	Lateral aspect of the olecranon process of the ulna	Extends elbow
Anterior muscles (pronation/medial rotation of the forearm)			
Pronator teres	Medial epicondyle of the humerus; coronoid process of the ulna	Lateral radius	Pronates the antebrachium
Pronator quadratus	Distal portion of the anterior ulnar shaft	Distal surface of the anterior radius	Assists in pronating the antebrachium
Posterior muscles (supination/lateral rotation of the forearm)			
Supinator	Lateral epicondyle of the humerus; proximal ulna	Proximal end of the radius	Supinates the antebrachium

ulnaris lateralis muscle—all are extensors of the carpus, innervated by the **radial nerve**. The flexor group arises from the caudomedial humerus and track over the caudal aspect of the radius. All are innervated by the **median** and **ulnar nerves** and are flexors of the carpus (and some flex the digits).

Cubital (elbow) joint (Fig. 14.3-4)

The **elbow joint** is a hinge-type joint with motion that is limited almost exclusively to flexion and extension. It is supported by prominent **long** and **short collateral ligaments—medial** and **lateral**. There is a superficial and deep portion of the medial collateral ligament and a single, broad lateral collateral ligament. The lateral aspect of the joint has minimal soft tissue coverage and is more susceptible to trauma and puncture wounds. There is a small bursa associated with the **ulnaris lateralis muscle** that may communicate with the elbow joint, which has clinical relevance when a wound involves this bursa.

CUBITAL (ELBOW) JOINT ARTHROCENTESIS

There are 3 sites for arthrocentesis of the cubital (elbow) joint. All are performed in weight-bearing. The first involves placing a needle approximately 3.5 cm (1.4 in.) proximal to the lateral tuberosity of the radius and 2.5 cm (1.0 in.) cranial to the lateral collateral ligament. Deposition of local anesthetic peri-articularly in this location can induce temporary radial nerve dysfunction (paresis or paralysis) due to its proximity to the injection site. Alternatively, a needle can be placed medial to the lateral collateral ligament in a caudolateral approach to the joint. The proximolateral aspect of the caudal pouch can be located at the depression between the cranial olecranon and caudal aspect of the lateral epicondyle of the humerus. This is approximately 3–4 cm (1.2–1.6 in.) caudal to the lateral epicondyle, with the needle advanced 4–8 cm (1.6–3.1 in.) in a distal slightly craniomedial direction.

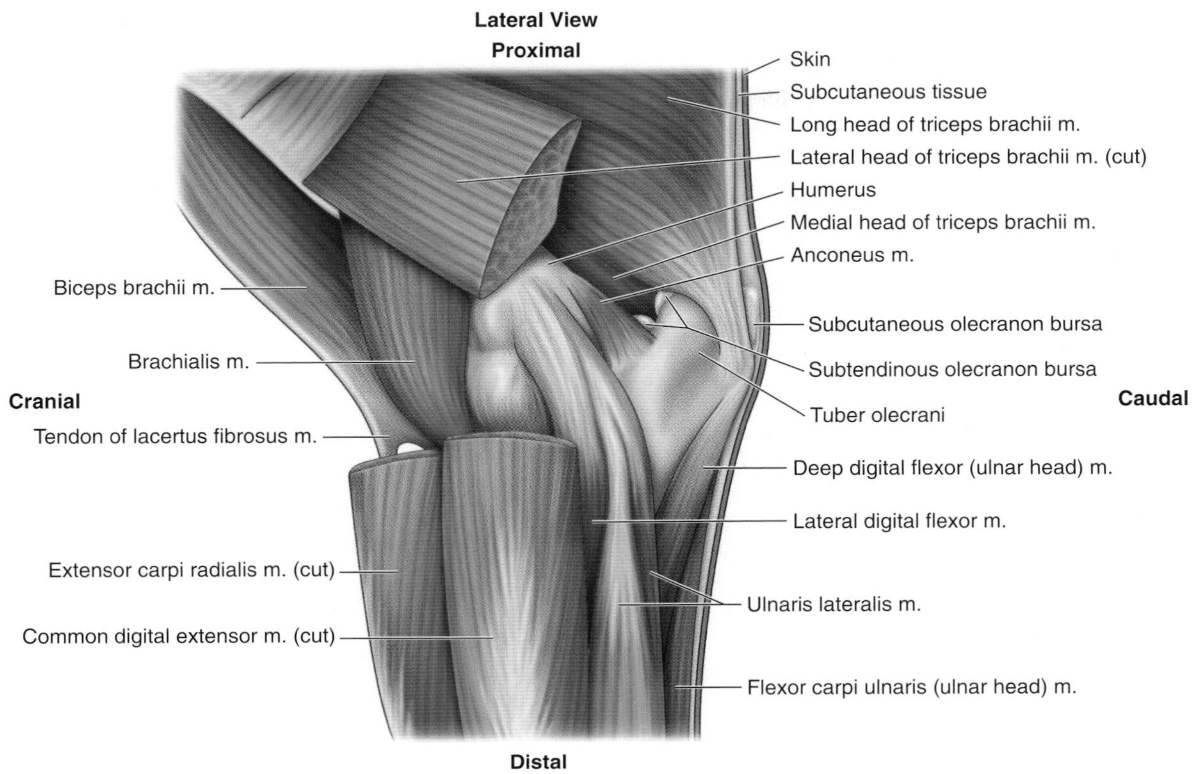

FIGURE 14.3-4 Muscles, joints, and bursae of the equine elbow region.

Regional bursae (Fig. 14.3-4)

The **olecranon tuberosity bursa**, also known as a **shoe boil** or **capped elbow**, is a false or acquired bursa that is a subcutaneous swelling between the skin and the tuber olecrani. It is not to be confused with the **triceps bursa**, which is a true/anatomical bursa that lies deeper between the olecranon and tendon of insertion of the triceps.

This false bursa is often caused by repetitive self-trauma from the shoe of the affected limb, traumatizing the point of the elbow when the horse is lying down. Acutely, the swelling contains fluid, but chronically can become fibrous in nature, and is mostly a cosmetic issue unless the swelling becomes infected. Although anecdotal evidence of successful treatment of "shoe boils" with injection of steroids or sclerosing agents exists, surgical management is the treatment of choice and includes incision and drainage of the bursa, with a gravity drain placed or en bloc resection of the bursa.

Regional nerves of the elbow (Fig. 14.3-5)

Noteworthy nerves tracking from the distal humerus to the radius include the radial, median, and ulnar nerves.

The **radial nerve** passes between the medial and long heads of the triceps muscles and tracks over the caudal humerus to the lateral aspect of the limb to innervate the extensors of the carpus and digits. There is a sensory branch that continues as the **lateral cutaneous antebrachial nerve** and supplies sensation to the lateral aspect of the

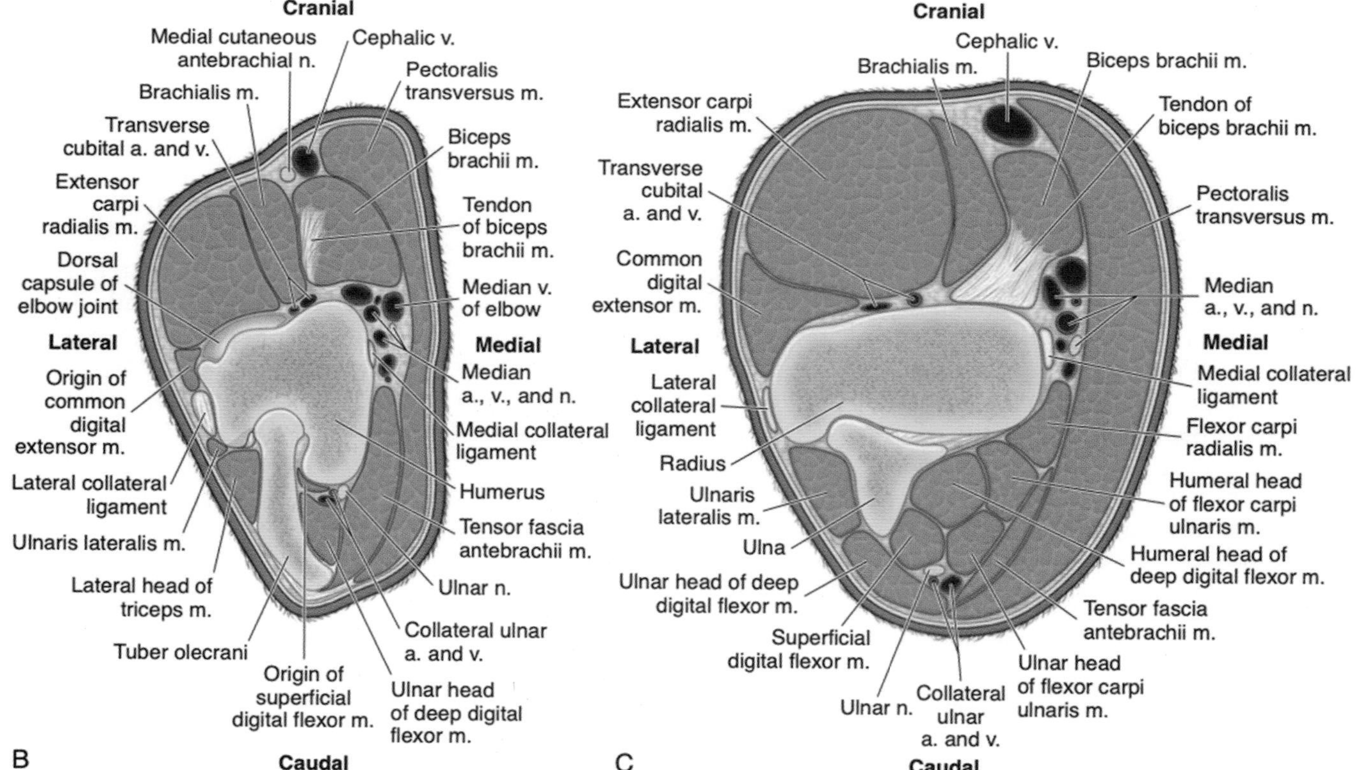

FIGURE 14.3-5 (A) Arteries, veins, and nerves supplying the distal equine humerus, elbow, and proximal radius and ulna. (B) Cross-section of the equine elbow at the level of the tuber olecrani (at level of *dashed line* in part A). (C) Cross-section in the region of the equine elbow joint (at level of *dashed line* in part A).

antebrachium to about the level of the midcarpus. The radial nerve provides innervation to the extensor muscles of all joints distal to the shoulder.

If the proximal part of the radial nerve is injured, the horse cannot fix/extend the limb due to the loss of triceps function, leading to loss of the passive stay apparatus (see Case 14.2). Alternatively, injury to the nerve closer to the elbow results in loss of extension of the distal limb, resulting in the characteristic toe drag. If the limb is manually placed in a neutral standing position, the horse remains weight-bearing because the triceps function is preserved. Distal radial paralysis can occur from ischemia due to prolonged periods of lateral recumbency. This risk can be reduced during general anesthesia by placing the horse on a well-padded mat, extending the down limb, and appropriate pre-operative planning to avoid long anesthesia time.

Beginning medially, the **median nerve** tracks with the brachial artery and moves to the caudal aspect of the limb as it nears the elbow. It passes along the caudomedial radius between the flexor muscles of the radius with the median artery and vein and eventually branches into the **medial** and **lateral palmar nerves**. It innervates some of the flexors of the antebrachium and provides sensation to the palmar metacarpus, metacarpo-/metatarsophalangeal joint (fetlock), and everything distal to this.

The **ulnar nerve** tracks with the brachial artery in the proximal arm and divides caudally where it passes over the medial epicondyle of the humerus, supplying branches to the flexor muscles. It continues as a sensory nerve following the ulnar head of the deep flexor muscles in the caudal limb and divides into the dorsal and palmar branches just above the carpus.

A regional nerve block can be performed on the median and ulnar nerve in the proximal antebrachium for diagnostic analgesia purposes, because it desensitizes everything distally except superficial skin sensation of the medial and dorsal parts of the antebrachium. To block the median nerve, a needle is inserted medially on the caudal aspect of the radius, approximately 10 cm (3.9 in.) distal to the elbow joint and distal to the superficial pectoral muscle. To block the ulnar nerve, a needle is placed approximately 10 cm (3.9 in.) proximal to the accessory carpal bone between the ulnaris lateralis and flexor carpi ulnaris muscles. Additional cutaneous analgesia can be performed by blocking the musculocutaneous nerve with needles placed adjacent to the accessory cephalic and cephalic veins, about halfway between the elbow and the carpus. This blocks the skin sensation on the medial and dorsal parts of the antebrachium.

Blood supply and lymphatics (Fig. 14.3-5)

Just proximal to the elbow, the **brachial artery** on the medial humerus supplies branches to the cranially located **transverse cubital artery** and **caudal collateral radial,** and then the **ulnar arteries**. The brachial artery is located cranial to the medial collateral ligament of the elbow, between the ligament and the **pectoralis transversus muscle,** as it tracks with the median nerve. It sends off a branch to the radius called the **common interosseous artery** before becoming the **median artery**, which tracks along the caudal aspect of the antebrachium before dividing into 3 branches just proximal to the carpus.

Most of the veins of the thoracic limb follow the arteries and may further replicate where they accompany larger arteries—e.g., the **brachial veins**. There are several superficial cutaneous veins that are plainly visible—the **cephalic vein** (larger) is on the craniomedial side of the antebrachium, while the **accessory cephalic vein** (smaller) is more cranially located. The cephalic vein joins the **brachial vein** via the **median cubital vein** at the elbow and rises proximally in a channel between the brachiocephalicus m. and pectoralis descendens m. before joining the **external jugular vein** at the base of the neck.

Because of the location of the brachial vein, it is at increased risk of injury when any traumatic event—for example, when a horse impales itself on a wooden stake or sustains a high-tensile wire accident.

Lymphatic drainage from the distal part of the thoracic limb travels medially to the **cubital lymph nodes** on the proximomedial aspect of the elbow. Lymphatic flow continues caudodistally along the teres major muscle, toward the **axillary lymph nodes,** which sends lymphatic drainage to the **caudal deep cervical nodes** before emptying into the axillary vein at the thoracic inlet.

Selected references

[1] Dyson S. The elbow, brachium, and shoulder. In: Ross MW, Dyson S, editors. Diagnosis and management of lameness in the horse. 2nd ed. Saunders; 2011. p. 456–74.
[2] Watkins JP. The radius and ulna. In: Auer JA, Stick JA, editors. Equine surgery. 4th ed. Saunders; 2012. p. 1363–87.
[3] Butler JA, Colles CM, Dyson SJ, Kold SE, Poulos PW, editors. Clinical radiology of the horse. 3rd ed. Wiley-Blackwell; 2008. p. 273–320.

CASE 14.4

Radial Fracture

Liberty Getman
Tennessee Equine Hospital, Thompson's Station, Tennessee, US

Clinical case

History

An 8-year-old Thoroughbred gelding was admitted for evaluation of multiple wounds to the right thoracic limb sustained after falling over a timber jump during a steeplechase race. The horse was treated with 2 g phenylbutazone IV and tetanus toxoid IM. The wounds were cleaned and bandaged before referral.

Physical examination findings

The gelding was quiet, alert, and lame at the walk, but able to bear full weight on the right thoracic limb. There were multiple lacerations to the limb, the most significant on the dorsal and lateral parts of the carpus (Fig. 14.4-1). Abnormal vital signs included mild tachycardia and tachypnea, both of which were presumed to be pain-related.

Differential diagnoses

Laceration without joint communication, laceration with joint communication (antebrachiocarpal joint, middle carpal joint), fracture (carpus or radius), and laceration with tendon sheath communication +/− tendon trauma (extensor carpi radialis tendon, common digital extensor tendon, lateral digital extensor tendon, ulnaris lateralis tendon)

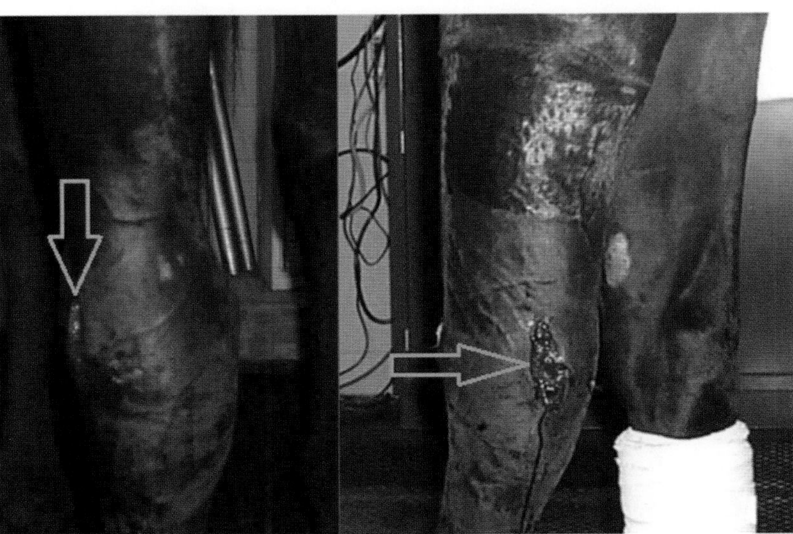

FIGURE 14.4-1 The image on the left shows the dorsal aspect of the left carpus with lacerations on the lateral and dorsolateral aspect of the distal radius *(blue arrow)*. The image on the right depicts the same wound *(orange arrow)* observed from the lateral aspect of the limb.

Comparative Veterinary Anatomy: A Clinical Approach. https://doi.org/10.1016/B978-0-323-91015-6.00074-1

Diagnostics

Radiographs of the carpus revealed a complete, minimally displaced fracture of the distal lateral radius that communicated with the antebrachiocarpal joint (Fig. 14.4-2). The horse was sedated and both wounds were aseptically prepared and then evaluated with a sterile gloved hand and a malleable probe. The fracture fragment was palpable through the lateral wound, confirming communication of the laceration with both the fracture and the antebrachiocarpal joint. The dorsal wound was probed and no communication could be confirmed with the joint. Therefore, using aseptic technique, the middle carpal joint was distended with 35 mL of sterile saline; no saline was observed to egress (exit) from the wound and the joint appeared distended. The syringe was disconnected so the fluid could egress from the needle; the joint was treated with 500 mg of amikacin before removing the needle from the joint.

Diagnosis

Complete, displaced fracture of the distal lateral radius communicating with the antebrachiocarpal joint; skin laceration communicating with the antebrachiocarpal joint (lateral wound); and skin laceration without joint communication (dorsal wound)

Treatment

The horse was placed under general anesthesia and the wounds were debrided. Arthroscopy of the antebrachiocarpal joint was performed to lavage the joint and confirm fracture reduction at the joint surface. The fracture was reduced and repaired with three 5.5-mm (0.22-in.) cortical bone screws placed in a lag fashion (Fig. 14.4-3). Both lacerations were closed anatomically and the antebrachiocarpal joint was treated with 500 mg of amikacin. The horse was recovered from general anesthesia using a "pool recovery system" to minimize the risk of reinjury during standing from general anesthesia.

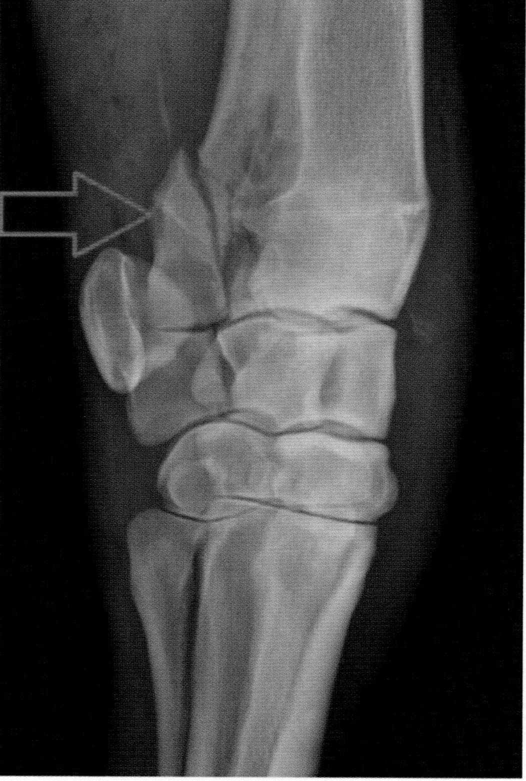

FIGURE 14.4-2 A dorsomedial palmarolateral oblique radiograph showing a large, minimally displaced fracture *(arrow)* through the distal radius with an articular component.

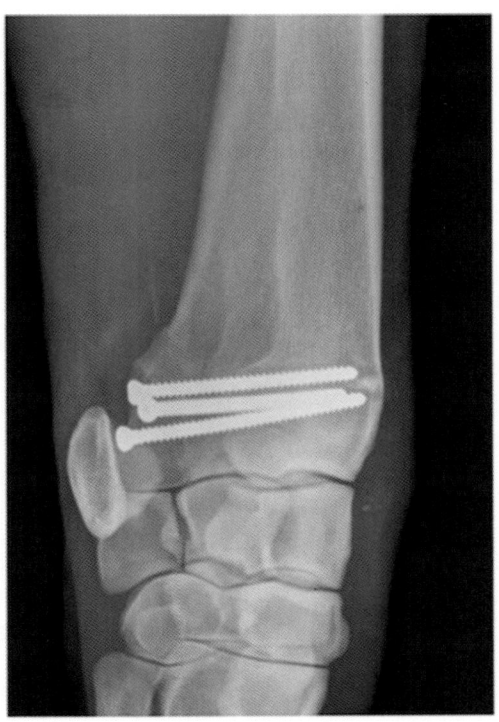

FIGURE 14.4-3 Post-operative dorsomedial palmarolateral oblique radiograph showing reduction of the radial fracture with three 5.5-mm (0.22-in.) cortical bone screws placed in a lag fashion, demonstrating good anatomic reconstruction of the articular surface and compression of the fracture fragment to the parent bone.

Anatomical features in equids

Introduction

The equine carpus is a combination of complex synovial structures involving 3 joints and the carpal canal, which functions as an important synovial structure for the flexor tendons as they track distally over the carpus and multiple bones. In addition, there are several small tendon sheaths of the overlying extensor tendons. This case covers anatomy of the carpus, its blood supply and innervation, and a description of the local soft tissues.

Function

The equine carpus is a high-motion joint composed of 2 rows of carpal bones positioned between the distal radius and the proximal metacarpal bones, creating 3 joint spaces: (1) the antebrachiocarpal joint; (2) the middle carpal joint; and (3) the carpometacarpal joint. The antebrachiocarpal and the middle carpal joints are hinge-type (ginglymus—Gr. *ginglymos* hinge) joints, while the low-motion carpometacarpal joint is arthrodial (Gr. *arthrōdia* particular kind of articulation) in function. The antebrachiocarpal joint range of motion (ROM) is the most extensive of the 3 joints with up to 100° of flexion; the middle carpal joint ROM approaches 45°, and the carpometacarpal joint allows no flexion but undergoes shear stress.

ANGULAR LIMB DEFORMITIES

Angular limb deformities are medial or lateral deviations of one or both thoracic and/or pelvic limbs described by naming the joint of origin and the direction of deviation of the limb distal to the joint. The problem is frequently encountered in growing foals—a common site is the carpus/distal radius, which in these cases has disproportionate growth of the epiphysis and metaphysis. A combination of radiographic and clinical evaluation is important in formulating a treatment plan.

Injection of both the antebrachiocarpal joint and the middle carpal joint together is typically performed over the dorsal joint surface with the carpus held in flexion using aseptic technique. The needle is commonly placed in the depression palpated between the extensor bundles of the common digital extensor and extensor carpi radialis tendons, between either the distal radius and the proximal row of carpal bones, or between the proximal and the distal row of carpal bones for the proximal and middle carpal joints, respectively. Synovial fluid is readily obtained with a 3.75-cm (1.5-in.) needle. Injection into the carpometacarpal joint separately is not performed because it directly communicates with the middle carpal joint.

These palmar parts of the carpometacarpal joint are important to remember because unintended desensitization of the carpometacarpal joint—and by extension of the middle carpal joint +/− antebrachiocarpal joint—can occur when diagnostic analgesia of the proximal metacarpal region is performed. This is important to keep in mind during lameness evaluations when using diagnostic analgesia.

Carpal joints (Fig. 14.4-4)

The **carpus** is composed of 3 joints: the **antebrachiocarpal joint**, the **middle carpal joint**, and the **carpometacarpal joint**. The middle carpal joint and the carpometacarpal joint anatomically communicate. Communication between the antebrachiocarpal and middle carpal joints is not well defined; studies report that diffusion of local anesthetic agents occurs between these joints in 84–96% of the joints evaluated. The carpometacarpal joint has long extensions of the palmar part of the joint pouch that are near the proximal suspensory ligament and the proximal palmar one-third of the metacarpal region.

Carpal bones (Fig. 14.4-5)

The bones comprising the carpus are the distal radius; the proximal row of carpal bones—**radial carpal bone**, **intermediate carpal bone**, **ulnar carpal bone**, and

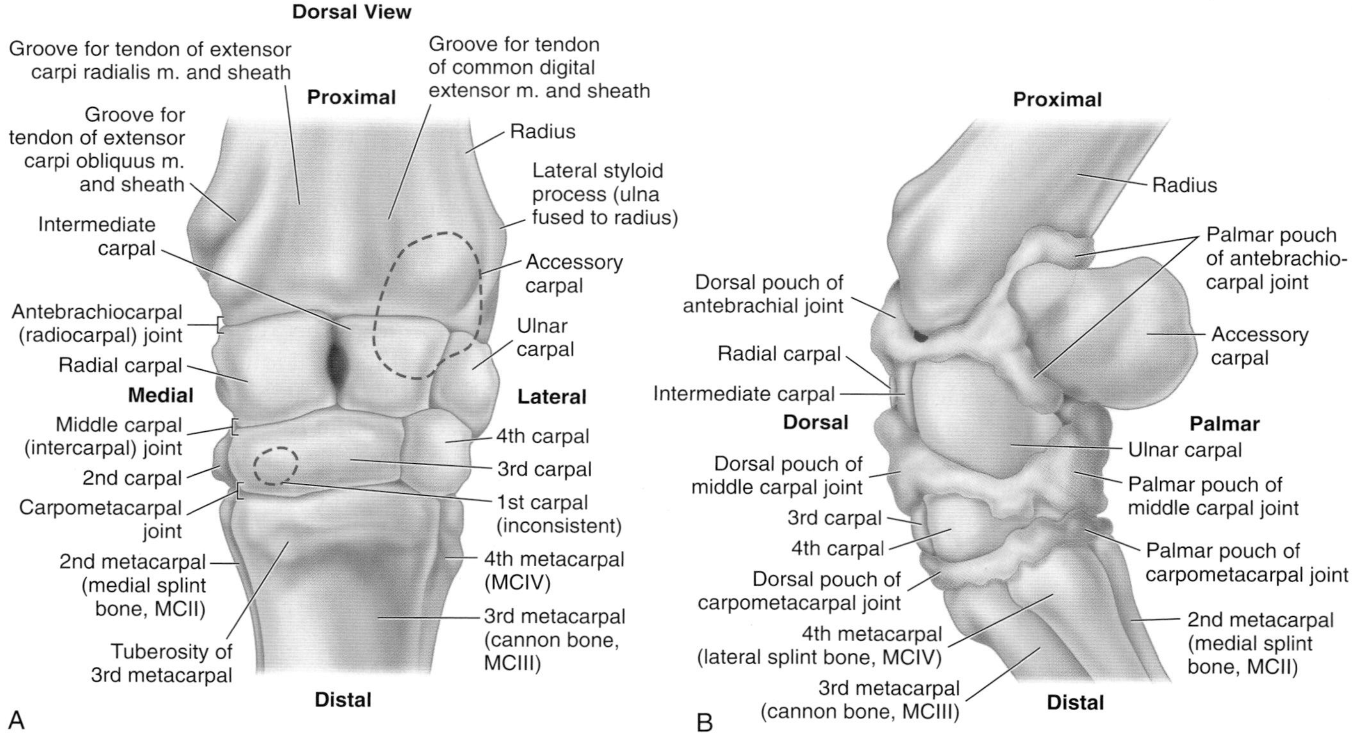

FIGURE 14.4-4 The equine carpal bones and joints. (A) Dorsal view. (B) Flexed lateral view.

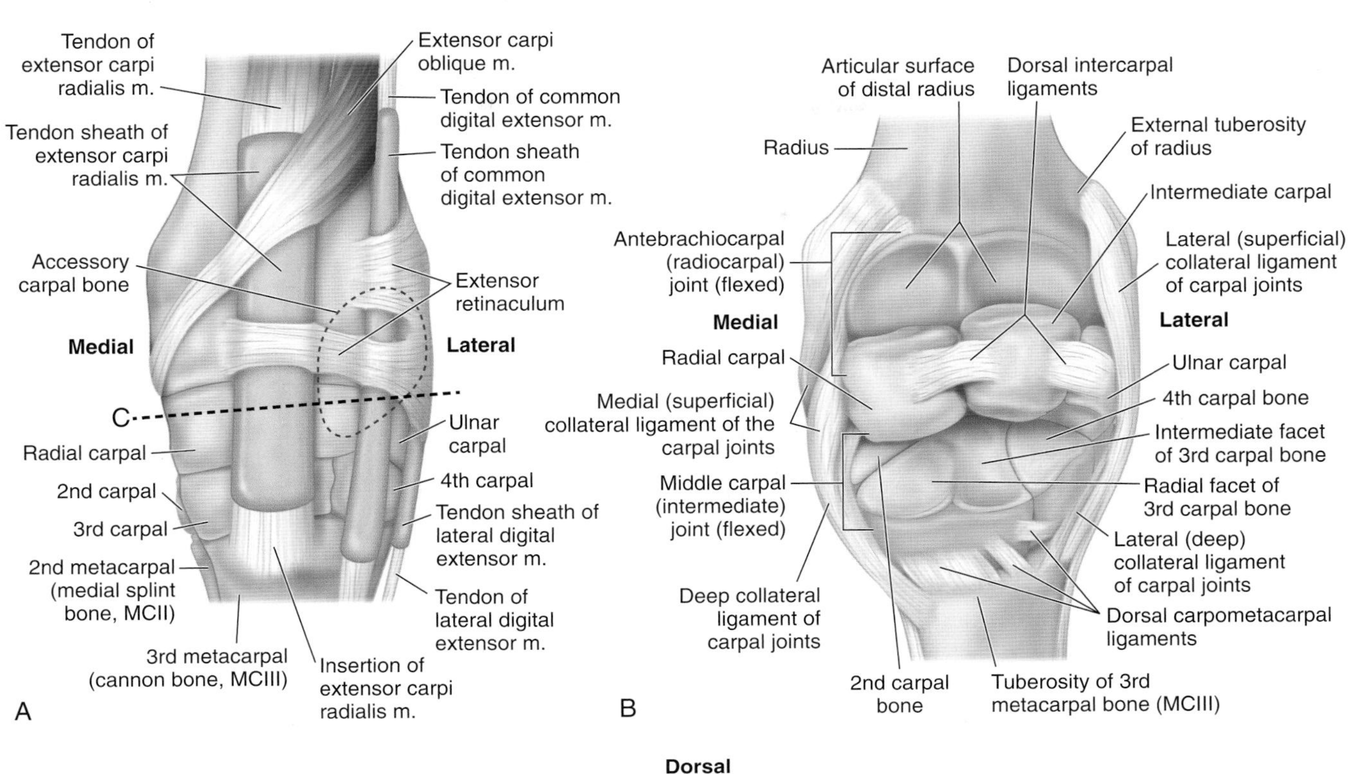

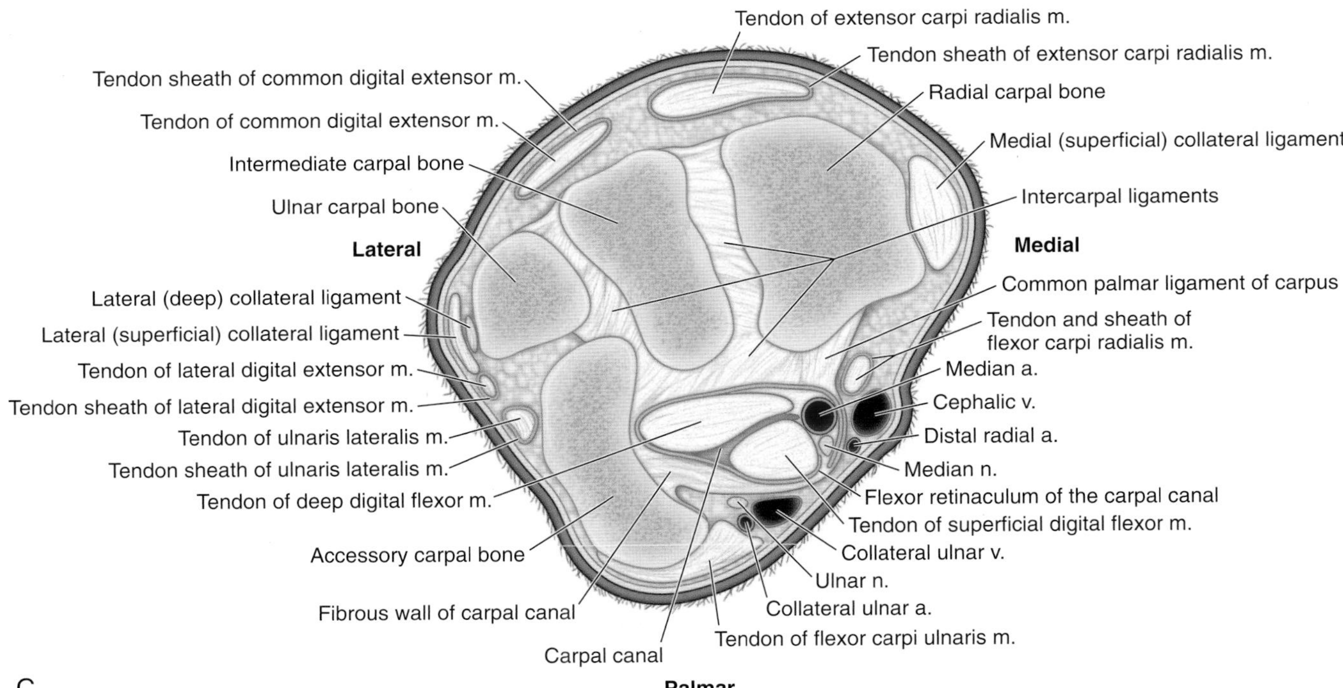

FIGURE 14.4-5 Muscles, ligaments, tendons, and their sheaths of the equine carpal region. (A) Dorsal view. (B) Tangential (skyline) view. (C) Transverse section at the level of *dashed line* marked "C" in part (A).

Osteochondral chip fractures are a common cause of lameness in racehorses. These are caused by repetitive trauma during training and racing. The fragments typically occur on the dorsal aspect of the bone surfaces. Most commonly affected are the distal radial carpal bone, proximal 3rd carpal bone, and the distal intermediate carpal bone in the middle carpal joint. In the antebrachiocarpal joint, the proximal intermediate and the proximal radial carpal bones and distal lateral and medial radius are most frequently affected. These are typically treated arthroscopically with good outcomes depending on the degree of cartilage and bone injury. Slab fractures of the carpal bones are another injury seen more commonly in racehorses because of chronic remodeling of bones due to the stress of training and racing. The 3rd carpal bone is the most common bone affected with frontal slab fractures involving the radial facet (Fr. *facette* hard surface on a bone; L. *fovea* pit or depression). The radial carpal bone is the next most common bone to be affected with slab fractures of the intermediate, 4th, and ulnar carpal bones rarely reported. Depending on the size of the bone fragment, internal fixation is often the treatment of choice using the principle of lag screw fixation to compress the fracture fragment.

These vestigial 1st and 5th carpal bones can be misinterpreted as fracture fragments on radiographs.

accessory carpal bone; the distal row of carpal bones—**2nd carpal bone**, **3rd carpal bone**, and **4th carpal bone**; and the metacarpal bones—**2nd metacarpal bone**, **3rd metacarpal bone**, and **4th metacarpal bone**. In some horses, there is a vestigial **1st carpal bone**, and less commonly a vestigial **5th carpal bone**.

Soft tissues of the carpus (Figs. 14.4-5 and 14.4-6)

There are **intercarpal ligaments** passing between each of the carpal bones holding them in place. **Medial** and **lateral collateral ligaments** support the whole carpus. The lateral collateral ligament originates from the **lateral styloid process** of the radius and has a variety of insertions on the ulnar and 4th carpal bones, and the 3rd and 4th metacarpal bones. The medial collateral ligament originates from the **medial styloid process of the radius** and has a variety of insertions on the radial and 2nd carpal bones and 2nd and 3rd metacarpal bones.

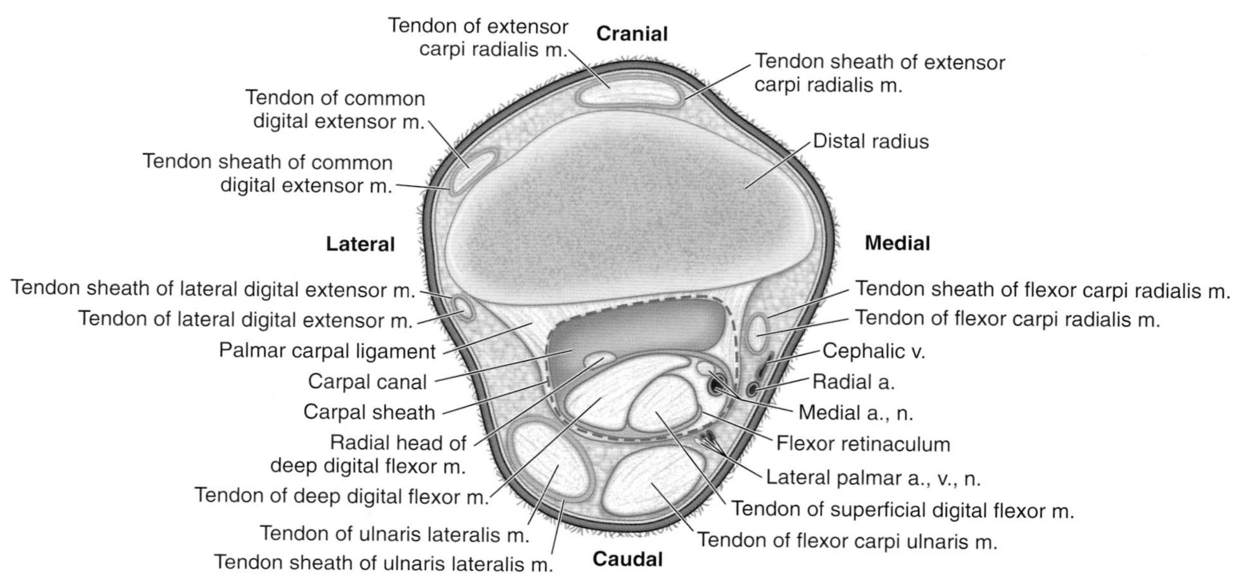

FIGURE 14.4-6 Transverse section at the level of the distal radius of a horse showing the carpal canal, tendons and their sheaths, blood vessels, and nerves.

Other soft tissue structures of the carpus are the **extensor carpi radialis tendon** (most dorsal tendon), **common digital extensor tendon** (dorsolateral), **lateral digital extensor tendon** (lateral), and **ulnaris lateralis** (palmarolateral to the lateral digital extensor tendon). Each of these tendons has an associated tendon sheath. The **palmar carpal ligament** is an important thick broad fibrous layer of the joint capsule that spans the palmar aspect of the carpus and serves as the dorsal surface of the carpal canal. Deep to the palmar carpal ligament is a group of smaller supportive palmar ligaments (**radiocarpal ligament**, **palmar middle carpal ligament**, and **palmar carpometacarpal ligament**). Together these ligaments act to prevent joint hyperextension.

The **carpal canal** lies palmar to the carpus (Fig. 14.4-6). The carpal sheath contains the superficial and deep digital flexor tendons and the passageway for blood vessels and nerves supplying the distal limb. The carpal canal is surrounded dorsally by the palmar carpal ligament, laterally by the accessory carpal bone, and palmaromedially by the flexor retinaculum. Medial to the carpal sheath, but not within the sheath, is the **flexor carpi radialis tendon** and its associated sheath.

Blood supply to the carpus (Fig. 14.4-6)

In the distal radius, the **median artery** divides into 3 vessels—small branches of the palmar branch of median and radial artery contribute to the **deep palmar arch,** which then forms the **lateral** and **medial metacarpal arteries;** the main trunk of the median artery becomes the **medial palmar artery**—primary blood supply to the distal limb. Major venous drainage of the distal limb occurs through the **medial** and **lateral palmar veins** distally, and then the medially located **cephalic and accessory cephalic veins.** The **median cubital vein,** at the level of the elbow, connects the cephalic vein to the **brachial vein** and then to the **jugular vein** at the base of the neck.

Injuries to the pectoral region involving the brachiocephalic and pectoralis descendens muscles frequently include trauma to the brachial vein. Therefore, these wounds have the potential for substantial blood loss and the need for large volume fluid replacement as part of emergency treatment practice.

Lymphatic drainage is primarily via the **cubital lymph nodes** (lymph from the distal parts of the limb) and the **superficial cervical lymph nodes** (lymph from the skin of the upper limb and deeper structures).

Innervation of the carpus (Fig. 14.4-6)

In the caudal distal radius, the **ulnar nerve** gives off a deep branch that receives some branches from the **median nerve** before becoming the **lateral palmar nerve** on the axial side of the accessory carpal bone. This nerve passes distally and—at the level of the proximal metacarpus—gives off the **palmar metacarpal nerves** and continues distally as the **lateral palmar nerve**. The **median nerve** travels on the palmaromedial surface of the carpus where it becomes the **medial palmar nerve.**

A high 4-point nerve block is typically used to anesthetize the proximal metacarpus with placement of needles in the palmar metacarpal and palmar nerves at the level of the distal heads of the splint bones. However, an alternative method is to place a needle into/over the lateral palmar nerve at the level of the accessory carpal bone to anesthetize both the palmar metacarpal nerves and the lateral palmar nerve. This helps isolate a proximal suspensory injury without unintentionally anesthetizing the carpal joints. The medial palmar nerve is then blocked at the same location as the high 4-point block, just dorsal to the flexor tendons at the level of the distal head of the medial splint bones.

Selected references

[1] Martin GS, McIlwraith CW. Arthroscopic anatomy of the intercarpal and radiocarpal joints of the horse. Equine Vet J 1985;17:373–6.
[2] Sisson S, Grossman JD, Getty R. The anatomy of the domestic animals. 5th ed. Philadelphia: W.B. Saunders Co.; 1975. p. 354–7.
[3] Getman LM, McKnight LM, Richardson DW. Comparison of magnetic resonance contrast arthrography and arthroscopic anatomy of the equine palmar lateral outpouching of the middle carpal joint. Vet Radiol Ultrasound 2007;48(6):493–500.

CHAPTER 14

CASE 14.5

Superficial Digital Flexor Tendonitis

Nick Carlson
Steinbeck Peninsula Equine Clinics, Salinas, California, US

Clinical case

History

A 5-year-old Warmblood gelding presented for swelling on the palmar aspect of the mid-metacarpus of the right thoracic limb. The swelling had been present for 24 hours and was first noted when the horse returned from a weekend horseshow and the trainer removed the shipping bandages. They requested an examination to confirm the swelling was just a "bandage bow."

Physical examination findings

The horse was a grade 2-/5 lame on the right thoracic limb, trotting in a circle in both clockwise and counterclockwise directions. The swelling over the flexor tendons in the middle third of the right metacarpus was warm and sensitive to palpation. The horse was not positive to lower limb flexion tests or hoof testers.

Differential diagnoses

Tendonitis/desmitis of the flexor tendons, inferior check, or suspensory ligament; focal cellulitis; and focal trauma to the subcutaneous (peritendinous) tissues surrounding the superficial and deep flexor tendons

Diagnostics

Ultrasound of the flexor tendons in the metacarpus revealed a hypoechoic "core" lesion in the lateral half of the superficial digital flexor tendon (SDFT). The remaining tendons and ligaments were normal (Fig. 14.5-1).

Diagnosis

Superficial digital flexor tendonitis (tendinitis)

Treatment

The horse was initially treated with a combination of NSAIDs, hydrotherapy, and compression wraps. Stem cells—a regenerative medicine treatment—were injected into the lesion after the acute inflammation of the tendon subsided. The horse was started on a graduated rehabilitation program over 10 months, with intermittent reexaminations for soundness and ultrasonographic evaluation for healing of the SDFT. The horse resumed training after the final reevaluation and was back in competition 12 months after the initial injury.

Comparative Veterinary Anatomy: A Clinical Approach. https://doi.org/10.1016/B978-0-323-91015-6.00075-3

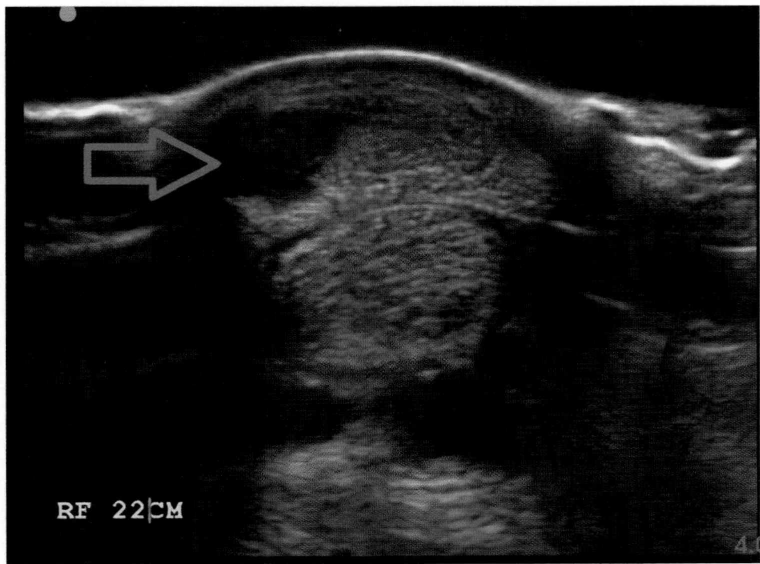

FIGURE 14.5-1 Short-axis ultrasound image from the mid-metacarpus showing a hypoechoic core lesion on the lateral aspect of the superficial digital flexor tendon *(arrow)*.

Anatomical features in equids

Introduction

The major anatomical structures in the metacarpus/-tarsus include the extensor tendons, superficial and deep digital flexor tendons, the accessory (inferior check) ligament of the deep digital flexor tendon, and the suspensory ligament. Clinically relevant differences between the thoracic and pelvic limbs are discussed. The regional blood supply, lymphatics, and nerves are discussed in Case 15.5.

Function

The superficial and deep digital flexor tendons (*flexor digitorum profundus* L. deep) are flexors of the carpus, metacarpophalangeal, and interphalangeal joints, and are important structures in supporting the metacarpo-/metatarsophalangeal (fetlock) joint. The common (thoracic limb) and long (pelvic limb) digital extensors function to extend the digit. The term "proximal sesamoidean ligament" applies to the interosseous muscle (suspensory ligament). The term "distal sesamoidean ligaments" refers collectively to the straight, oblique, and cruciate sesamoidean ligaments. "Suspensory apparatus" collectively comprises the interosseous muscle and the distal sesamoidean ligaments. The interosseous muscle (suspensory ligament) divides into 2 sizable, palpable, and narrow branches that insert on the concave abaxial surfaces of the proximal sesamoid bones, and pass laterally and medially around the 1st phalanx as extensor branches (unsubstantial in size) to join the common/long digital extensor tendon. The tension of the interosseous on the sesamoid bones is transmitted to the palmar/plantar surfaces of the 1st and 2nd phalanges (PI and PII) by the distal sesamoid ligaments. They have 2 functions when the limb is weight-bearing and the fetlock joint is overextended: (1) stiffen the digit, resisting the tendency of the deep digital flexor tendon to flex the distal interphalangeal joint when it is under tension by the descending fetlock joint and (2) form part of the interosseous-sesamoid suspensory apparatus (passive stay apparatus) that supports the fetlock joint.

Extensor tendons (see Landscape Figs. 4.0-4–4.0-6)

Complete rupture of the common/long digital extensor tendon initially presents with the horse standing on the toe because of an inability to extend the distal limb. However, over time, horses compensate by throwing the hoof forward to extend the toe.

In the thoracic limb, the **common digital extensor muscle** originates on the lateral condyle of the humerus and becomes tendinous from the carpus to its insertion on the extensor process of the 3rd phalanx. The **lateral digital extensor muscle** originates on the lateral epicondyle of the humerus and inserts on the lateral aspect of the 1st phalanx.

In the pelvic limb, the **long digital extensor muscle** originates with the **fibularis (peroneus) tertius** in the extensor fossa of the distal femur. Its insertion tendon begins proximal to the hock and inserts on the phalanges dorsally, with its most prominent insertion on the extensor process of the 3rd phalanx. It is encased in a tendon sheath over the tarsus, which continues to the proximal metatarsus. The main **lateral digital extensor muscle** body arises from the lateral collateral ligament of the stifle, with small contributions to the muscle originating on the lateral proximal tibia and fibula. It becomes a tendon above the tarsus and is enclosed in a tendon sheath as it tracks over the tarsus. It joins the long digital extensor tendon at the proximal to mid-cannon bone.

Stringhalt (equine reflex hypertonia) is a gait abnormality, causing an involuntary hyperflexion of the tarsal joint, during which the limb jerks toward the abdomen when moving in the cranial phase of the stride. It can be idiopathic, secondary to trauma of the extensor tendons, or associated with plant toxicosis. Excision of a portion of the lateral digital extensor tendon and muscle at the level of the tarsus improves some cases of stringhalt.

Superficial digital flexor tendon (see Landscape Fig. 4.0-4 and Fig. 14.5-2)

In the thoracic limb, the **superficial digital flexor muscle** originates from the medial epicondyle of the humerus. Its function is to flex the carpus and the metacarpophalangeal joint. The **superficial digital flexor tendon** (SDFT) arises

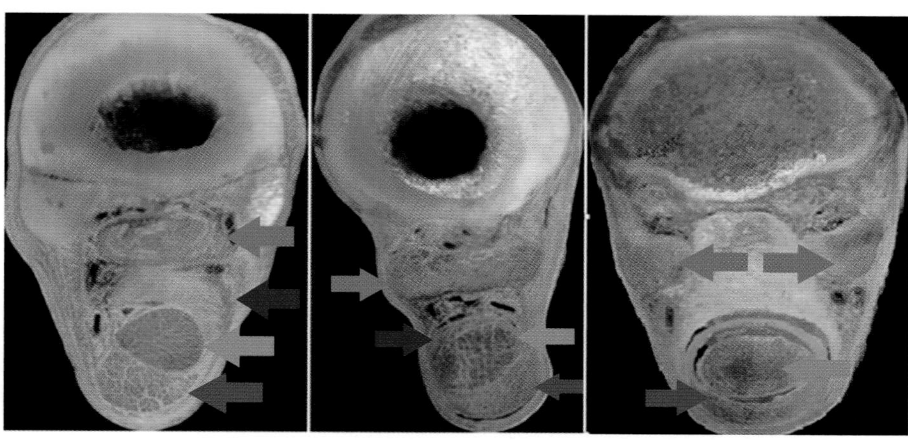

FIGURE 14.5-2 Cross-sectional anatomy of the equine thoracic limb from proximal to distal (left to right) in a postmortem specimen. The suspensory ligament (SL) *(green arrows)*, accessory ligament of the deep digital flexor tendon (ALDDFT) *(red arrows)*, deep digital flexor tendon (DDFT) *(blue arrow)*, and superficial digital flexor tendon (SDFT) *(purple arrow)* are visible. Note the branching of the SL, eventual merging of the ALDDFT to the DDFT, and the formation of manica flexoria (outlined in *pink*) with the SDFT surrounding the DDFT.

from its muscle belly in the distal radius. It is joined by the **accessory (superior check) ligament of the SDFT** in the caudomedial aspect of the radius proximal to the accessory carpal bone. In the pelvic limb, the superficial digital flexor is mainly tendinous and originates from the supracondylar fossa of the femur. It does not have an accessory ligament in the pelvic limb. It is the most palmar/plantar tendinous/ligamentous structure in the metacarpus/-tarsus (Fig. 14.5-3). The tendon changes from circular to oval in cross-section as it enters the metacarpal region. The tendon takes on an asymmetrical crescent shape with the medial border thicker than the lateral margin in the mid-metacarpal/-tarsal region. The SDFT becomes thinner and widens in the distal metacarpus. In the distal metacarpus/-tarsus, it forms a thin ring around the deep digital flexor tendon called **manica flexoria**, as it enters the **digital flexor tendon sheath** above the metacarpo-/metatarsophalangeal joint.

Congenital flexural deformities can be observed in the carpus/tarsus, fetlock, pastern, and coffin joint, and are present at birth. Acquired flexural deformities are typically seen later in development and are observed in the fetlock, proximal interphalangeal (pastern), and distal interphalangeal (coffin) joints. Acquired flexural deformities occur most frequently after a period of rapid long bone growth. However, they can also occur secondary to a painful injury that causes the limb to be unloaded from full weight-bearing for some period of time, with the thoracic limbs more commonly affected. Cases refractory to conservative management—e.g., dietary management, controlled exercise, analgesics, podiatry, cast application, and administration of IV oxytetracycline—may need surgical correction. Deformities involving the fetlock joint may benefit from a desmotomy (Gr. *desmo* band, ligament, Gr. *-tomē* incision) of the accessory ligament of the SDFT or DDFT, and in more severe cases may require transection of both ligaments.

Deep digital flexor tendon (see Landscape Fig. 4.0-4 and Fig. 14.5-2)

The function of the **deep digital flexor tendon** (DDFT) is to flex the carpus and the digits. The **deep digital flexor muscle** has 3 points of origin in the thoracic limb: the **humeral**, **radial**, and **ulnar heads**. In the metacarpal region, the tendon has a round cross-sectional area. It lies between the SDFT and the accessory ligament of the DDFT in the proximal half of the metacarpus. The tendon has an ovoid appearance throughout the metacarpus. The accessory ligament of the digital flexor tendon joins with the DDFT in the mid-metacarpal region. The tendon starts to widen and becomes more elliptical in shape as it enters the fetlock region with the SDFT.

In the pelvic limb, the DDFT is formed by the larger **lateral digital flexor tendon** and the smaller **medial digital flexor tendon** (Fig. 14.5-3). The lateral digital flexor tendon enters the proximal metatarsus after passing over the sustentaculum tali in the **tarsal sheath**. The smaller medial digital flexor tendon passes over the proximal tubercle of the talus on the medial aspect of the talus within its own tendon sheath, before joining the lateral digital flexor tendon.

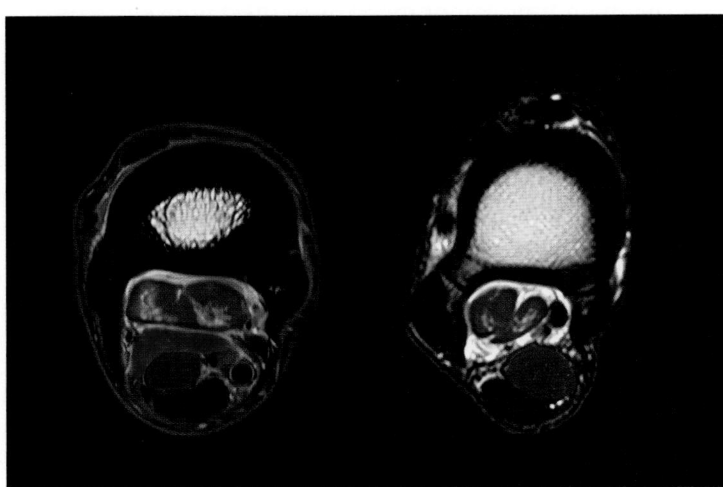

FIGURE 14.5-3 Difference in anatomy of the proximal metacarpus *(left)* and metatarsus *(right)* demonstrated on cross-sectional MRI. The suspensory ligament (SL) *(green)*, accessory ligament of the deep digital flexor tendon *(red)*, deep digital flexor tendon (DDFT) *(blue)*, and superficial digital flexor tendon (SDFT) *(purple)* are visible.

A desmotomy of the ALDDFT may be indicated in the cases of chronic desmitis of the ligament that are unresponsive to conservative management. As the ligament heals after the desmotomy, the defect fills with scar tissue that effectively lengthens the ALDDFT and theoretically reduces its predisposition to reinjury. This procedure is also performed in young horses with a flexural deformity of the thoracic limb, originating within the deep digital flexor muscle-tendon unit.

Accessory ligament of the deep digital flexor tendon (see Landscape Fig. 4.0-4 and Fig. 14.5-2)

In the thoracic limb, the **accessory ligament of the deep digital flexor tendon** (ALDDFT) is a continuation of the **common palmar ligament of the carpus** from the 3rd carpal bone. It tracks distally between the DDFT and the suspensory ligament until it merges with the DDFT at the level of the mid-metacarpus. It has a rectangular cross-sectional appearance in the proximal metacarpus and is more crescent-shaped as it tracks toward its union with the DDFT.

In the pelvic limb the ALDDFT is typically smaller than in the thoracic limb. In some horses, it may only be present as a small sheet, connecting the DDFT to the **short plantar ligament of the tarsus**.

Suspensory ligament (see Landscape Fig. 4.0-4 and Fig. 14.5-2)

The **suspensory ligament** (SL)—also referred to as the "interosseous muscle"—has an important role in the "stay apparatus" and locomotion (Fig. 14.5-4 and see side box entitled "The stay apparatus" in Case 15.3 and Fig. 15.3-4). While it is predominately ligamentous, there are individual variations in the amount of muscle present. In the thoracic limb, the SL originates in the proximal palmar metacarpus, with the parts originating from the **palmar carpal ligament** and **palmar carpal joint capsule**. The SL extends distally and divides above the metacarpophalangeal joint into the medial and lateral branches that insert on their respective proximal sesamoid bones. Below the metacarpophalangeal joint, there are small dorsal extensor branches that join the **common digital extensor tendon**.

Horses have one main interosseous muscle—i.e., the suspensory ligament—and 2 insignificant interosseous muscles. In contrast, cows have 2 fused interosseous muscles and dogs and cats have 4.

The origin of the pelvic limb SL is primarily the proximal plantar 3rd metatarsal bone with accessory attachments from the 4th tarsal bone and the calcaneus. While the thoracic limb origin has 2 lobes and is more rectangular on cross-section, the division is less prominent in the pelvic limb and it is rounder in appearance on cross-section (Fig. 14.5-3). The main function of the SL is to prevent overextension of the metacarpo-/metatarsophalangeal joint.

TRAUMATIC DISRUPTION OF THE SUSPENSORY APPARATUS

Disruption of the suspensory apparatus is usually an acute racing injury from a comminuted fracture of the proximal sesamoid bones—i.e., a "breakdown injury" in racehorses—or a laceration of the suspensory apparatus. It can also occur as a chronic problem in older horses due to degeneration of the SL (Fig. 14.5-5). If a loss of the SL is combined with the loss of the SDFT and DDFT, the horse's toe does not contact the ground and presents with severe loss of the fetlock support. Emergency first aid and treatment options depend on the severity of the injury and the anatomical structures involved. These cases may require emergency splinting, casting, and podiatry to support the limb in an attempt to minimize the loss of the tendon/ligament function. Surgery for arthrodesis (Gr. *arthron* joint + Gr. *desis* binding = surgical fixation of a joint resulting in fusion of the joint) of the fetlock joint is the treatment of choice.

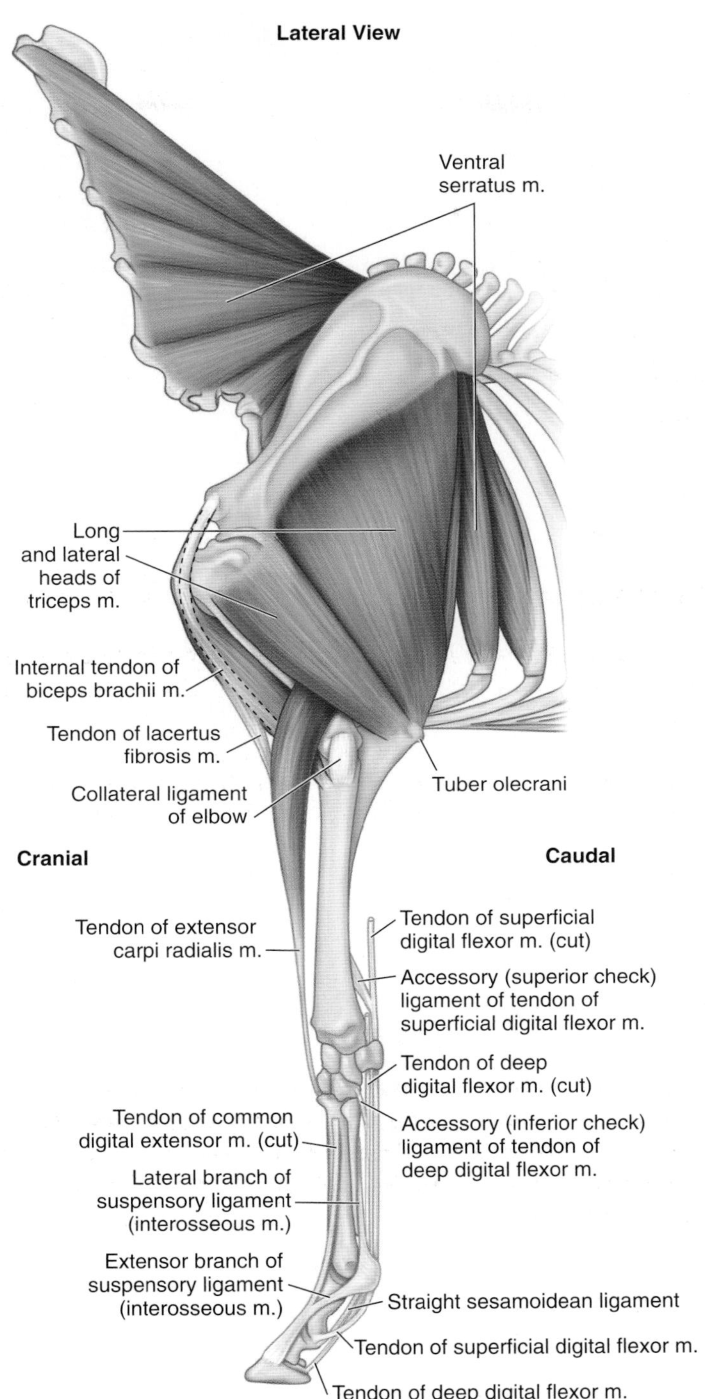

FIGURE 14.5-4 Stay apparatus of the equine thoracic limb.

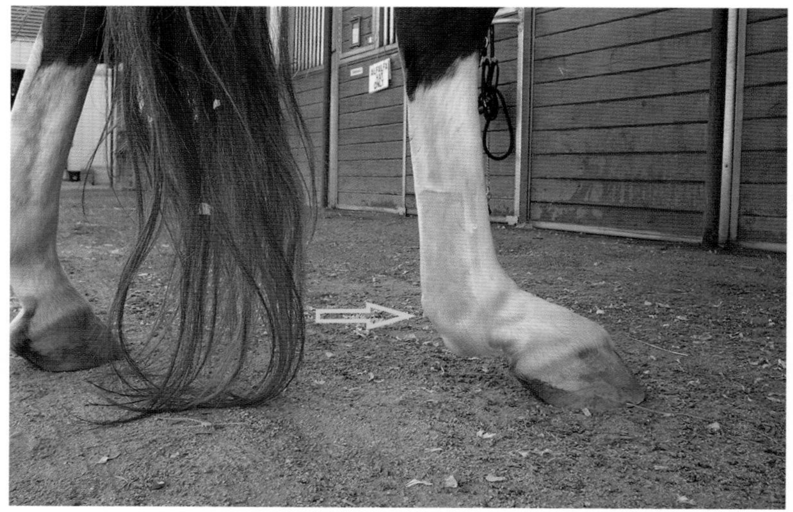

FIGURE 14.5-5 Chronic degenerative desmitis of the pelvic limb suspensory ligament of a horse, leading to loss of fetlock support *(arrow)* (hair was removed for ultrasound examination).

Selected references

[1] Ross MW, Dyson S, Genovese RL, Jorgensen JS. Superficial digital flexor tenond, deep digital flexor tendon, accessory ligament of the deep digital flexor tendon, and suspensory apparatus. In: Ross MW, Dyson S, editors. Diagnosis and management of lameness in the horse. 2nd ed. Saunders; 2011. p. 706–60.

[2] Avella CS, Smith RKW. Diagnosis and management of tendon and ligament disorders. In: Auer JA, Stick JA, editors. Equine surgery. 4th ed. Saunders; 2012. p. 1157–88.

[3] Denoix JM. Essentials of clinical anatomy of the equine locomotor system. 1st ed. Taylor and Francis Group; 2019. p. 77–98.

CASE 14.6

Osteochondral Fragment of the Metacarpophalangeal Joint

Nick Carlson
Steinbeck Peninsula Equine Clinics, Salinas, California, US

Clinical case

History

A 7-year-old Thoroughbred mare transitioning from racing to 3-day eventing was presented for a left thoracic limb lameness of 2–3 weeks' duration. The owner noticed swelling in the metacarpophalangeal (fetlock) joint and was treating the swelling with NSAIDs, hydrotherapy, and compression wraps.

Physical examination findings

The horse was grade 3/5 lame on the left thoracic limb. There was moderate effusion in the left fetlock joint and the horse was resistant to flexion of the left distal thoracic limb compared to the right distal thoracic limb. Flexion of the distal limb resulted in increased lameness. Intraarticular injection of the left metacarpophalangeal joint with a local anesthetic resolved the lameness.

Differential diagnoses

Osteoarthritis or soft tissue injury of the collateral ligaments of the metacarpophalangeal joint; injury of the suspensory branches; osteochondritis dissecans; and traumatic fracture of the distal 3rd metacarpal bone, proximal 1st phalanx, or proximal sesamoid bones

Diagnostics

Radiographs confirmed a traumatic osteochondral (chip) fracture of the dorsomedial aspect of the proximal 1st phalanx (Fig. 14.6-1).

Diagnosis

Chip fracture of the proximal 1st phalanx

Treatment

The fragment was removed arthroscopically under general anesthesia. After 3 months of routine post-operative rehabilitation, the horse successfully resumed training.

Comparative Veterinary Anatomy: A Clinical Approach. https://doi.org/10.1016/B978-0-323-91015-6.00076-5

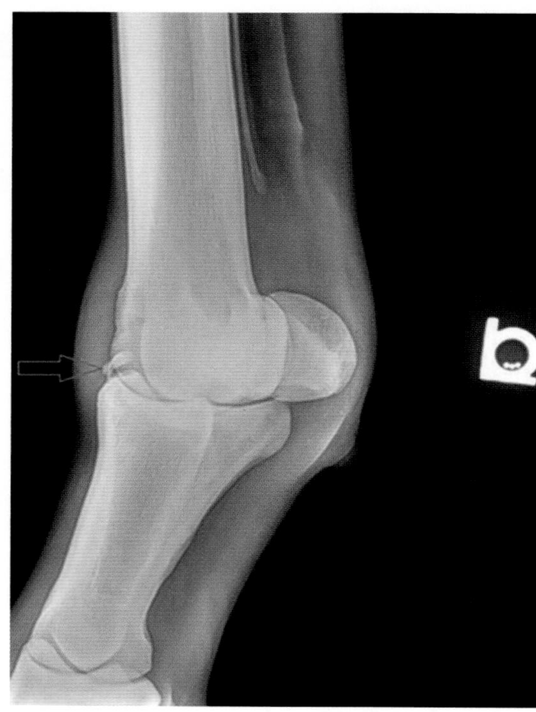

FIGURE 14.6-1 Dorsolateral-palmaromedial oblique radiograph showing an osteochondral fragment *(green arrow)* on the dorsomedial aspect of the proximal 1st phalanx.

Anatomical features in equids

Introduction

The metacarpophalangeal joint (in the thoracic limb) and the metatarsophalangeal joint (in the pelvic limb) are commonly known as the "fetlock joint." The major structures in the fetlock region include the distal 3rd metacarpal/metatarsal bone or MC/MTIII (known as the "cannon bone"), paired proximal sesamoid bones, 1st phalanx, supporting tendons and ligaments, and regional vessels. Any clinically relevant differences between the thoracic and pelvic limbs are discussed.

Function

The metacarpo-/metatarsophalangeal joints are high-motion, high-load joints, especially in equine athletes—e.g., racehorses. This makes these joints, particularly the metacarpophalangeal joint, at increased risk of injury when performing at maximum speeds. Multiple forces of torsion, hyperextension, and compression place the fetlock joints in jeopardy of a variety of injuries. The suspensory apparatus, superficial and deep digital flexor tendons, and collateral ligaments are important soft tissue components that countervail the extreme forces the metacarpo-/metatarsophalangeal joints undergo.

Distal 3rd metacarpal/metatarsal bone (see Landscape Figs. 4.0-10 and 4.0-11)

The articular surface of the distal **3rd metacarpal bone (MCIII/MTIII)** is divided by a sagittal ridge with the convex cartilage surfaces matching the corresponding sagittal groove and the concave joint surfaces of the 1st phalanx. The **medial** (wider) and **lateral** (narrower) **condyles** of the distal bone are the origins of the collateral ligaments of the metacarpo-/metatarsophalangeal joint and the proximal sesamoid bones. The distal physis closes at 3–6 months of age.

First phalanx

The **1st phalanx** is broader proximally where it interfaces with MC/MTIII. Wide **tubercles** are palpable on the medial and lateral proximal aspects of the bone. Distally, palpable **condyles** are noted with a convex distal articular surface, containing a shallow groove corresponding to a small axial ridge of the proximal joint surface of the 2nd phalanx. The proximal physis closes by approximately 1 year of age while the distal physis is closed at birth.

Proximal sesamoid bones

The paired **proximal sesamoid bones** are located at the palmar/plantar aspect of the metacarpo-/metatarsophalangeal joint. The concave abaxial surface of each acts as the insertion for the suspensory ligament branches. At the **axial surfaces**, an **intersesamoidean ligament** with an overlying palmar/plantar **fibrocartilaginous** surface unites the bones. This fibrocartilage forms the **proximal scutum** (L. shield), allowing the flexor tendons to pass over the palmar/plantar surfaces of the fetlock region with reduced friction. The base of the bones is the origin of the **distal sesamoidean ligaments** (Fig. 14.6-2). Together these structures play an important role in the passive stay apparatus (see Case 14.5, Fig. 14.5-5 and Case 15.3, Fig. 15.3-4 for thoracic and pelvic limbs, respectively). The **deep** and **superficial digital flexor tendons** glide over the palmar/plantar aspect of the proximal sesamoid bones as part of the digital flexor tendon sheath (see below).

Metacarpo-/metatarsophalangeal joint (fetlock joint) (Fig. 14.6-3)

The metacarpo-/metatarsophalangeal joint comprises the distal aspect of MC/MTIII, the 1st phalanx, and the dorsal surfaces of the paired proximal sesamoid bones. The lateral and medial joint surfaces are divided by the **sagittal ridge** of MC/MTIII and the corresponding **sagittal groove** of the 1st phalanx. The joint is stabilized by paired long and short **medial** and **lateral collateral ligaments** between the metacarpus/-tarsus and 1st phalanx. The **medial** and **lateral collateral sesamoidean ligaments** stabilize the proximal sesamoid bones to the metacarpal condyle and the

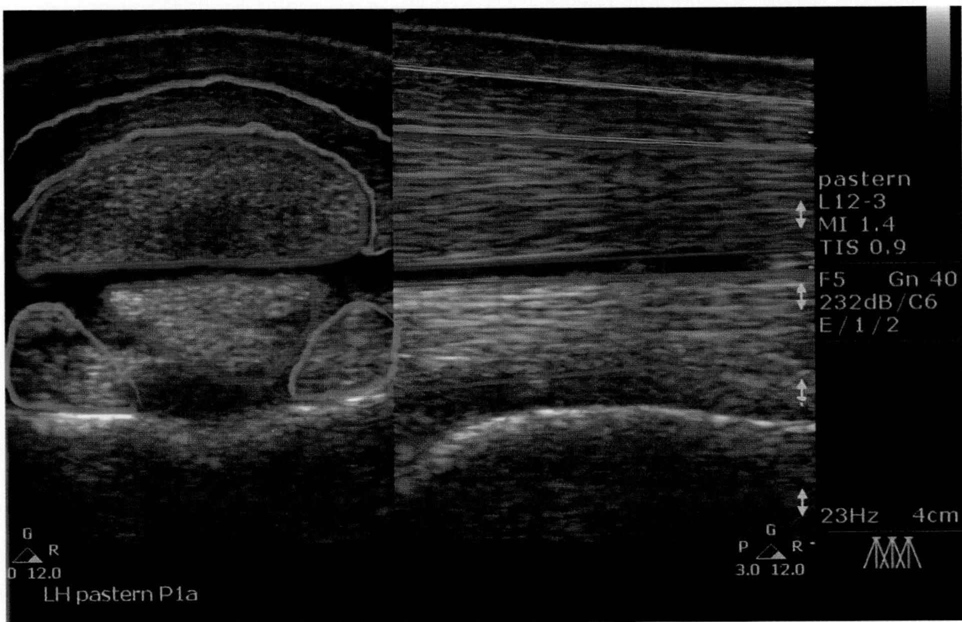

FIGURE 14.6-2 Ultrasound image of the proximal pastern region. The cross-sectional image on the left depicts the superficial digital flexor tendon (SDFT/*green*), deep digital flexor tendon (DDFT/*red*), straight distal sesamoidean ligament (SDSL/*purple*), and oblique distal sesamoidean ligament (ODSL/*blue*); the longitudinal image on the right depicts the SDFT *(green)*, DDFT *(red)*, and SDSL *(purple)*. The ODSLs are not imaged in long axis because they are abaxial to the SDSL.

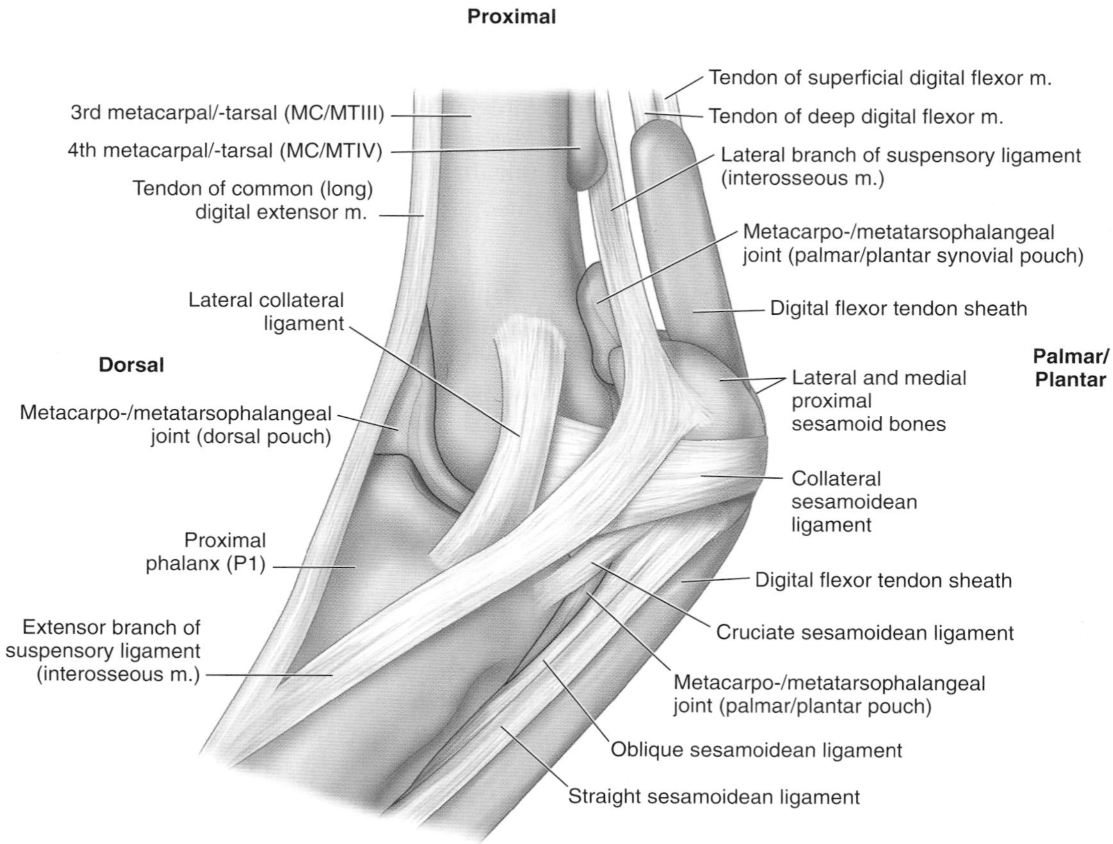

FIGURE 14.6-3 Lateral view of the equine metacarpo-/metatarsophalangeal (fetlock) joint, ligaments, tendons, and their associated tendon sheaths.

There are 3 main approaches to the metacarpo-/metatarsophalangeal joint for arthrocentesis or arthroscopy: (1) the palmar/plantar proximal synovial outpouching between MC/MTIII and the suspensory ligament branch; (2) further distal with the limb held in flexion and placing the needle through the lateral collateral sesamoidean ligament between palmar/plantar MC/MTIII and the sesamoid bone; and (3) the dorsal pouch by placing a needle just proximal to the palpable joint space under the lateral border of the common/long digital extensor tendon.

proximal tubercles of the 1st phalanx. There is a **synovial pad** located on the proximal dorsal articular rim of the distal MC/MTIII to cushion the bone from the impact of the 1st phalanx during hyperextension of the metacarpo-/metatarsophalangeal joint. There are prominent proximal palmar/plantar joint pouches that lie between MC/MTIII and the suspensory ligament branches. The **dorsal synovial pouch** is bisected by the **common (long) digital extensor tendon**.

Palmar/plantar distal sesamoidean ligaments (Fig. 14.6-4)

The **distal sesamoidean ligaments** are a series of ligaments originating on the distal aspect (or base) of the proximal sesamoid bones that act as a continuation of the medial and lateral branches of the suspensory ligament from the fetlock region distally. The paired **short sesamoidean ligaments** are the deepest ligamentous structure and insert on the proximal palmar/plantar aspect of the 1st phalanx. The paired **cruciate sesamoidean ligaments** are the next deepest structure and originate from the intersesamoidean ligament on the basilar surface of the sesamoids and insert while crossing to the opposite proximal palmar/plantar eminence of the 1st phalanx. The more substantial

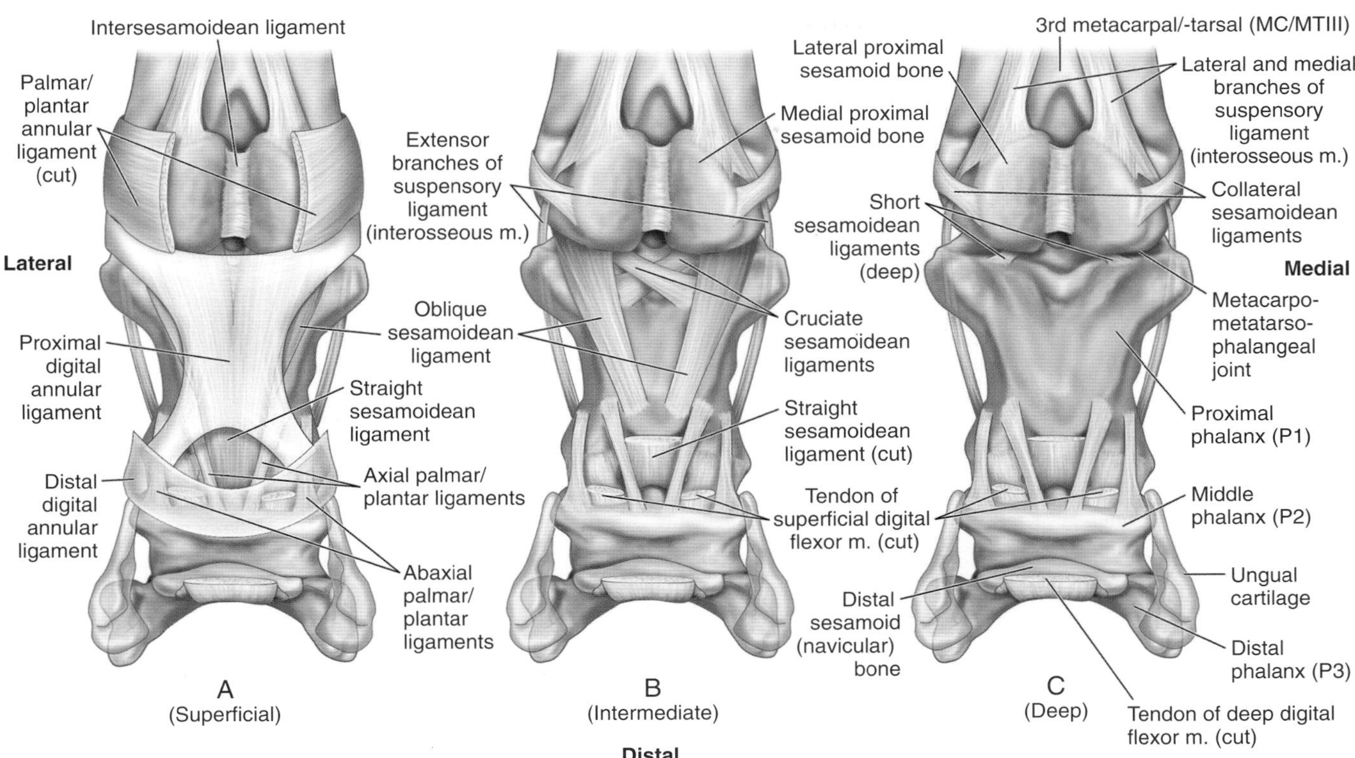

FIGURE 14.6-4 Palmar/plantar view of the distal sesamoidean ligaments of the equine limb. (A) Superficial. (B) Intermediate. (C) Deep.

oblique sesamoidean ligaments are the second-most superficial and insert on the axial triangular area of the distal palmar/plantar aspect of the 1st phalanx. A single **straight sesamoidean ligament** is the most superficial, arising from the base of each proximal sesamoid bone, converging, and inserting as one ligament on the proximal palmar/plantar aspect of the 2nd phalanx, blending in with the substantial fibrocartilage in the area (Fig. 14.6-2).

Digital flexor tendon sheath (DFTS) (Fig. 14.6-5)

The **digital flexor tendon sheath** (DFTS) is located at the palmar/plantar aspect of the distal limb, beginning just above the metacarpo-/metatarsophalangeal joint and extending to the middle of the 2nd phalanx. Proximally, it encircles the superficial and deep digital flexor tendons as they track over the palmar/plantar aspect of the proximal sesamoid bones. The deep digital flexor tendon (DDFT) passes over the fetlock region in a sleeve formed by the superficial flexor tendon (SDFT) called the **manica flexoria.** The deep part of the sleeve splits at the midpoint of the 1st phalanx, allowing the SDFT to attach to the

Several options exist for tenocentesis of the DFTS: (1) a proximal outpouching above the palmar/plantar annular ligament is readily palpated, but can be a difficult location from which to obtain fluid due to multiple synovial folds; (2) alternatively, a needle can be passed through the palmar/plantar annular ligament between the axial surface of the lateral sesamoid bone and the flexor tendons, or a needle can be placed in the synovial outpouching between the palmar/plantar annular ligament and the proximal digital annular ligament; and (3) a needle can be placed in the distal outpouching between the proximal and distal digital annular ligaments on the palmar/plantar midline over the 1st phalanx.

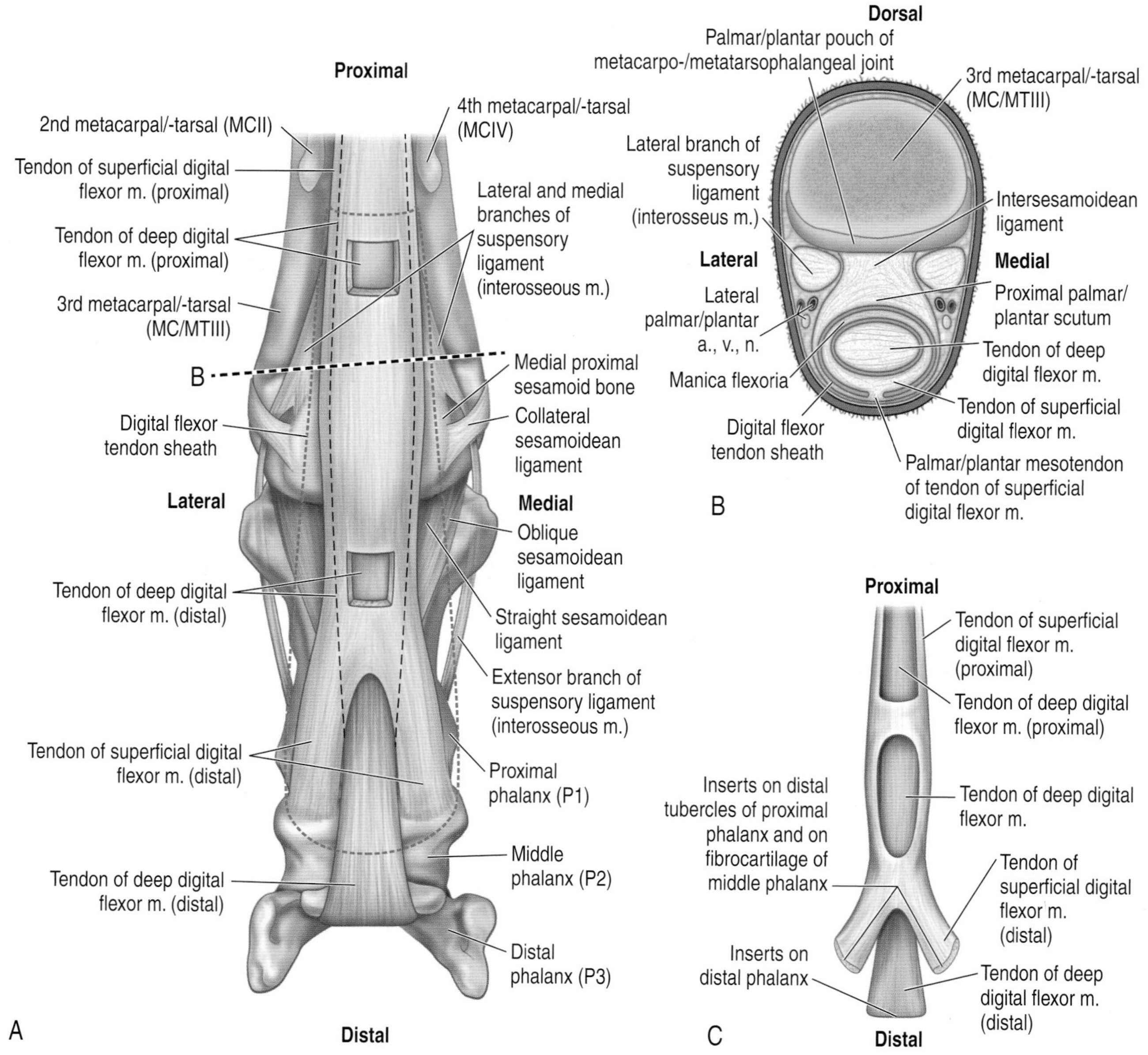

FIGURE 14.6-5 The equine digital flexor tendon sheath. (A) Palmar/plantar view. (B) Cross-section at the distal metacarpus/-tarsus to highlight the manica flexoria and flexor tendons. (C) Dorsal view of the orientation of the superficial and deep digital flexor tendons.

distal tubercles of the 1st phalanx and the fibrocartilage of the 2nd phalanx. This puts the DDFT in a superficial position where it can be palpated in the distal aspect of the DFTS. The distal aspect of the DFTS travels over the straight sesamoidean ligament and the **middle scutum,** and contains the DDFT. At the middle of the 2nd phalanx, the DFTS encounters the **T-ligament,** separating it from the navicular bursa.

The DDFT and SDFT are confined within the DFTS by a series of palmar/plantar annular ligaments. The **palmar/plantar annular ligament** arises from the palmar/plantar border of the proximal sesamoid bones and confines the SDFT and DDFT between the palmar/plantar ligament and the palmar/plantar surface of the proximal sesamoid bones and the intersesamoidean ligament. Two digital annular ligaments restrict the DDFT. The **proximal digital annular ligament** is X-shaped with the upper portions originating on the proximal palmar/plantar aspect of the 1st phalanx and the distal arms merge with the SDFT as it attaches on the 1st phalanx. The **distal digital annular ligament** originates on the medial/lateral aspect of the 1st phalanx and forms a sheet of fascia that runs over the distal aspect of the palmar/plantar pastern region, covering the DDFT as it enters the hoof. Within the hoof, the distal digital annular ligament coalesces into the fibrous hoof cartilage and digital cushion, with a part following the insertion of the DDFT on the flexor surface of the 3rd phalanx.

Blood supply, lymphatics, and innervation (see Landscape Fig. 4.0-2)

The palmar/plantar digital neurovascular bundle is a palpable structure on the abaxial surface of the medial and lateral sesamoid bones of the metacarpo-/metatarsophalangeal joint. The **palmar/plantar digital arteries** continue distally as the major blood supply to the distal limb with their respective veins. The **palmar/plantar digital nerves** give off a dorsal branch at the mid-pastern region while the main nerve continues distally to innervate the digit. The dorsal fetlock is also innervated by the terminal branches of the **palmar/plantar metacarpal nerves.** Lymphatic drainage of the hoof is carried to lymphatics that move medially just proximal to the metacarpo-/metatarsophalangeal joint.

Abaxial (basisesamoid) nerve block is performed by inserting needles on the axial surface of each palpable neurovascular bundle at the level of the distal fetlock. Vein, artery, and nerve (otherwise known as "VAN") in a dorsopalmar/-plantar order represents the regular grouping of the neurovascular bundle, with the artery being deeper than the vein and nerve in the distal one-fourth of the metacarpus/-tarsus and coursing over the sesamoids as described in this nerve block. This block anesthetizes the palmar/plantar pastern region, the lower third of the dorsal pastern region, and all structures in the hoof. A low 4-point nerve block is needed to anesthetize the metacarpo-/metatarsophalangeal joint by placement of needles over the medial and lateral palmar/plantar digital nerves, just proximal to the proximal outpouching of the DFTS. A second set of needles are placed into the skin just distal to the "button/bell" of the 2nd and 4th metacarpophalangeal and metatarsophalangeal (splint) bones and proximal to the palmar outpouching of the metacarpo-/metatarsophalangeal joint to anesthetize the palmar metacarpal/metatarsal nerves.

REGIONAL LIMB PERFUSION

The palmar/plantar digital veins are frequently used for regional limb perfusion of the distal limb—e.g., the foot. This may be performed to infuse a high concentration of antimicrobials for the treatment of a foot infection, or other targeted treatments (stem cell therapy), or for imaging studies using a contrast material to evaluate blood supply of the foot in cases of laminitis (see Case 14.9). In short, a tourniquet is placed around the distal metacarpus/-tarsus just proximal to the fetlock joint, and a palpable and prominent digital vein passing over the abaxial surface of the sesamoid bones (medial or lateral) is used for catheter insertion and infusion of the sterile medium. The tourniquet is left in place for ≤30 min to allow high levels of the chosen medication to diffuse into the surrounding tissues.

Selected references

[1] Richardson DW, Dyson S. The metacarpophalangeal joint. In: Ross MW, Dyson S, editors. Diagnosis and management of lameness in the horse. 2nd ed. Saunders; 2011. p. 394–411.
[2] Butler JA, Colles CM, Dyson SJ, Kold SE, Poulos PW, editors. Clinical radiology of the horse. 3rd ed. Wiley-Blackwell; 2008. p. 189–232.
[3] Nixon AJ. Phalanges and metacarpophalangeal and metatarsophalangeal joints. In: Auer JA, Stick JA, editors. Equine surgery. 4th ed. Saunders; 2012. p. 1300–24.
[4] Denoix JM. The equine distal limb. Manson Publishing; 2000. p. 243–373.

CASE 14.7

Fracture of the 2nd Phalanx

Nick Carlson
Steinbeck Peninsula Equine Clinics, Salinas, California, US

Clinical case

History

A 13-year-old American Quarter Horse mare used for roping presented with an acute right thoracic limb lameness of 3 days' duration. The owner initially treated the lameness as a suspected hoof abscess, but was concerned because of the lack of improvement in the lameness and increased heat and swelling in the distal limb.

Physical examination findings

The horse was 5/5 lame (nonweight-bearing) on the right thoracic limb. There was mild heat and edema of the distal limb, extending from the coronary band to just proximal to the metacarpophalangeal (fetlock) joint. The horse resisted palpation and flexion of the distal limb. There was a moderate increase in the digital pulse on the affected limb, and the horse was negative to hoof tester application.

Differential diagnoses

Fracture of the phalanges or navicular (distal sesamoid) bone, synovial infection of a distal limb structure, septic osteitis of the 3rd phalanx, cellulitis, or hoof abscess

Diagnostics

Digital radiographs of the distal limb confirmed a comminuted fracture of the 2nd phalanx (Fig. 14.7-1).

Diagnosis

A closed, mildly displaced, comminuted fracture of the 2nd phalanx involving the proximal and distal interphalangeal joints

EMERGENCY ORTHOPEDIC CARE

A well-applied splint or other external coaptation is central in the triage of fractures involving the distal limb. Commercial splints are available and allow for rapid application, thereby protecting the limb from additional injury and reducing stress for the patient. If a commercial splint is not available, the heel should be elevated so the distal limb is horizontal. A dorsal splint made of a 2×4-in. plank of wood or sturdy PVC pipe should be secured—using nonflexible tape, such as duct tape—over a well-placed bandage, extending from the toe to the proximal metacarpus/-tarsus (Fig. 14.7-2). In the pelvic limb, it may be difficult to apply a dorsal splint, in which case the splint can be applied on the plantar side of the limb. It is important to ensure that the hoof is incorporated into the splint to reduce further trauma and displacement of the fracture.

Comparative Veterinary Anatomy: A Clinical Approach. https://doi.org/10.1016/B978-0-323-91015-6.00077-7

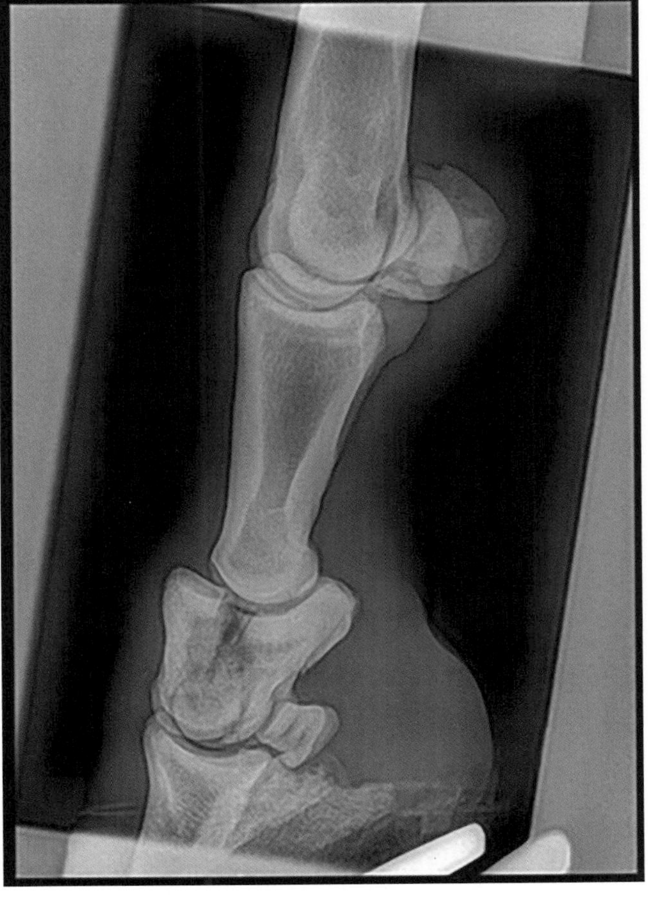

FIGURE 14.7-1 Lateral radiograph showing a comminuted fracture of the 2nd phalanx. Note the horse is not weight-bearing and the quality of the radiograph is consistent with an acute emergency (no evidence of external coaptation on the distal limb).

Treatment

The mare was placed in a dorsal splint and heavy cotton bandage (Fig. 14.7-2) designed to support and align the phalanges and minimize displacement of the fracture fragments in preparation for transport to the referral center for surgery. The fracture was surgically repaired under general anesthesia using open reduction and internal fixation. Special care was taken to ensure proper alignment of the articular surface of the distal interphalangeal (a.k.a. "coffin") joint. The proximal interphalangeal joint was surgically fused (i.e., an arthrodesis) as part of the fracture repair. The horse was maintained on peri-operative antimicrobials and nonsteroidal anti-inflammatory drugs and in a half-leg cast for 12 weeks. The horse was ultimately only pasture-sound due to osteoarthritis in the distal interphalangeal joint that precluded athletic use.

Depending on the breed, conformation, discipline, and traumatic events, some horses are more predisposed to developing arthritis in the proximal interphalangeal joint colloquially referred to as "high ring bone." Conversely, arthritis of the distal interphalangeal joint is colloquially referred to as "low ring bone."

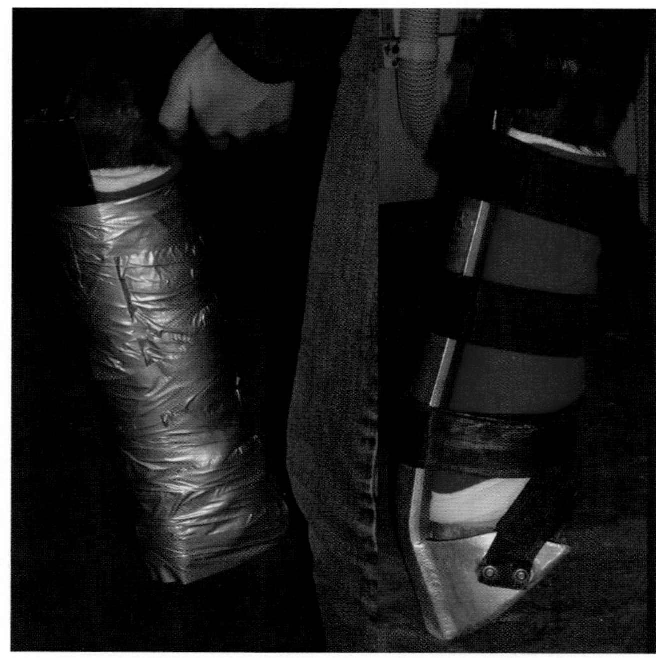

FIGURE 14.7-2 Application of a homemade *(left)* and commercially manufactured (Kimsey™) *(right)* splint for fractures below the fetlock. A similar splint would also be appropriate for horses with loss of the suspensory apparatus due to traumatic disruption of the suspensory apparatus—a.k.a. a "breakdown injury."

Anatomical features in equids

Introduction

The clinically important anatomical elements in the pastern and foot include the distal part of the 1st phalanx (long pastern bone; discussed in Case 14.6), the 2nd phalanx (short pastern bone), and soft tissues—local synovial structures, tendons, ligaments, and neurovascular structures. Additional information on the soft tissue structures of the palmar/plantar pastern region can be found in Case 14.6. There are no clinically significant differences in this region between the thoracic and the pelvic limbs except where noted in the information that follows.

Function

The 2nd phalanx functions as part of the digital skeleton or manus (L. the hand) and is the middle of the 3 phalanges. The phalanges—especially the 3rd phalanx of ungulates (L. *ungula* hoof)—carry and shape the hoof, and are classified as long bones. The middle or intermediate 2nd phalanx is not only in-between in location, but also in size, because it is shorter and more compressed. All 3 phalanges articulate with one another through interphalangeal articulations—the proximal and distal interphalangeal joints. The 2nd phalanx functions as an important insertion site for ligaments—e.g., distal sesamoidean ligaments and tendons (e.g., superficial digital flexor tendon) and it plays a vital role in the passive stay apparatus of the thoracic limb.

Second phalanx (Fig. 14.7-3)

The **2nd phalanx** is commonly called the "short pastern bone." It is similar in shape to the 1st phalanx, but is half the length and more cuboidal in shape. The proximal articular surface is concave and corresponds to the convex surfaces of the distal 1st phalanx. The distal articular surface is convex with a shallow groove that corresponds to the small ridge on the axial articular surface of the concave 3rd phalanx. The distal epiphysis is fused at birth. The proximal physis closes at approximately 8–12 months, but is functionally closed at 8 weeks of age.

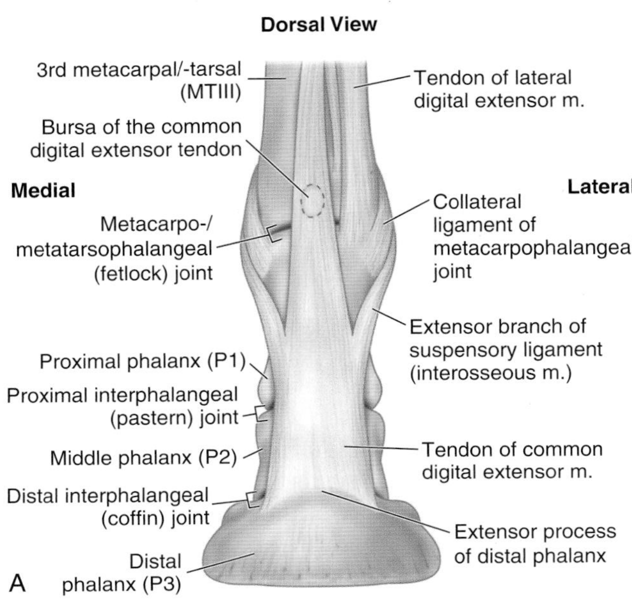

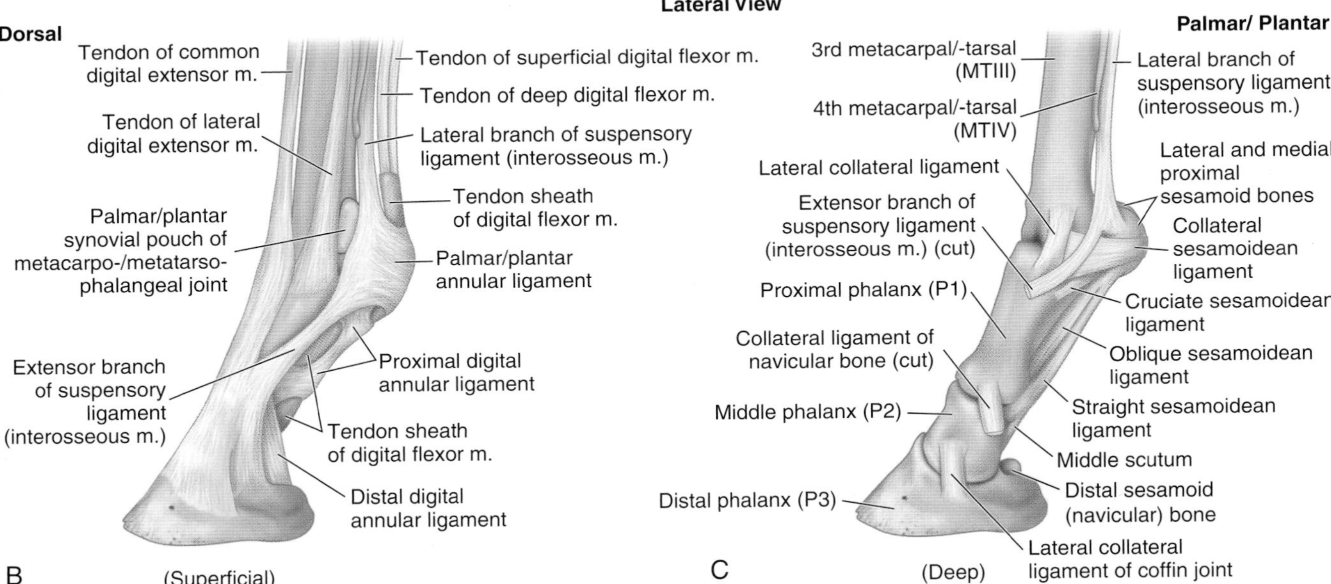

FIGURE 14.7-3 The equine pastern and foot. (A) Dorsal-palmar/plantar view. (B) Superficial lateral-medial view. (C) Deep lateral-medial view (superficial and deep digital flexors are removed).

Injection or arthrocentesis of the proximal interphalangeal joint is accomplished by inserting a needle into either the dorsal or palmar/plantar joint pouches. The dorsal pouch is entered with the horse weight-bearing and the needle placed just lateral to the common/long digital extensor tendon at the level of the distal palmar process of the 1st phalanx, with the needle oriented horizontal to the ground and directed axially into the joint. The palmar/plantar approach is performed with the limb in flexion and the needle inserted in the "V-shaped" notch formed between the distal palmar aspect of the 1st phalanx and the superficial digital flexor tendon before its insertion on the 2nd phalanx. The needle is directed distomedially until synovial fluid is obtained.

Proximal interphalangeal joint (Fig. 14.7-3)

The proximal interphalangeal joint is commonly known as the "pastern joint." There are proximal dorsal and palmar/plantar outpouchings of this joint. The **common digital extensor tendon** (thoracic limb) passes over the joint dorsally. There is a synovial recess extending proximally from the pastern joint beneath the common digital extensor tendon on the dorsal aspect of the 1st phalanx that is often palpable as a small fluctuant bulge. There is a small **extensor bursa** between the tendon and the proximodorsal pouch of the metacarpo-/metatarsophalangeal (fetlock) joint before proceeding distally toward the pastern joint but no bursa overlies the pastern joint dorsally. There are medial and lateral **collateral ligaments** that originate on the distal 1st phalanx and attach on the proximal 2nd phalanx. These collateral ligaments limit abduction/adduction and rotational motion of the joint.

The palmar aspect of the joint is supported by the **straight distal sesamoidean ligament** that originates on the distal aspects of the proximal sesamoid bones of the fetlock and inserts on the proximal palmar/plantar aspect of the 2nd phalanx. **Axial** and **abaxial palmar/plantar ligaments** of the pastern joint further reduce the range of motion of the pastern joint and support the palmar/plantar aspect of the joint. The axial palmar ligaments are found on each side (medial and lateral) of the straight sesamoidean ligament on the palmar/plantar surface of the middle part of the 1st phalanx and insert on the proximal 2nd phalanx. The abaxial palmar ligaments originate on the middle aspect of the 1st phalanx and track distally across the pastern to insert on the abaxial aspects of the 2nd phalanx near the insertions of the superficial digital flexor tendon.

The **middle scutum** is a thick fibrocartilaginous region on the proximal palmar/plantar aspect of the 2nd phalanx that augments the deep digital flexor tendon in gliding over the proximal palmar/plantar surface of the 2nd phalanx. The middle scutum is formed at the insertion of the straight sesamoidean ligament, axial palmar pastern ligaments, and branches of the superficial digital flexor tendon.

Blood supply, lymphatics, and innervation

A palmar digital nerve block is performed over the medial and lateral palmar/plantar digital nerves with needles placed on the axial aspect of each palpable neurovascular bundle in the distal pastern, most often just proximal to the palpable collateral cartilage of the hoof. This nerve block is classically considered to anesthetize structures in the heel and palmar/plantar one-third of the hoof; however, it likely variably anesthetizes more of the foot than previously believed. Lameness associated with the navicular bone, navicular bursa, distal deep digital flexor tendon, portions of the coffin bone, and possibly the distal interphalangeal joint generally are improved with this nerve block.

Clinically, the major blood supply continues as the paired **palmar/plantar digital arteries** and **veins** along the lateral and medial aspects of the deep digital flexor tendon. Dorsal branches are seen in the 1st and 2nd phalanges that join with one another on the dorsal aspect of their respective bones. The paired **palmar/plantar digital nerves** travel axial to these vessels as they track toward the hoof. In both the thoracic and the pelvic limbs, there is a **dorsal branch** of the digital nerves that innervates the dorsal aspect of the pastern and digit. In the pelvic limb, the **medial** and **lateral dorsal metatarsal nerves** also contribute to sensation of the dorsal surface of the fetlock and pastern region. Lymphatic drainage is like that discussed in Case 15.5.

Selected references

[1] Ruggles AJ. The proximal and middle phalanges and proximal interphalangeal joint. In: Ross MW, Dyson S, editors. Diagnosis and management of lameness in the horse. (2nd ed. Saunders; 2011. p. 387–93.
[2] Butler JA, Colles CM, Dyson SJ, Kold SE, Poulos PW, editors. Clinical radiology of the horse. 3rd ed. Wiley-Blackwell; 2008. p. 53–189.
[3] Nixon AJ. Phalanges and the metacarpophalangeal and metatarsophalangeal joints. In: Auer JA, Stick JA, editors. Equine surgery. 4th ed. Saunders; 2012. p. 1300–24.
[4] Denoix JM. The equine distal limb. Manson Publishing; 2000. p. 129–242.

CHAPTER 14

CASE 14.8

Foreign Body Penetration of the Hoof

Nick Carlson
Steinbeck Peninsula Equine Clinics, Salinas, California, US

Clinical case

History

A 15-year-old American Quarter Horse gelding was lame on the right thoracic limb at the walk coming in from pasture. The owner had pulled a nail from his hoof 2 days before and was administering 2 grams of phenylbutazone orally every 12 hours since then (Fig. 14.8-1). The horse was considerably better after the nail was removed, but became progressively more sore to the point of being unwilling to fully weight-bear on the limb. The morning of presentation, the owner also noticed that the distal limb—from the mid-cannon bone region to the coronary band—was swollen and warm to the touch (Fig. 14.8-1).

Physical examination findings

Initial physical examination findings included a rectal temperature of 102.8 °F (39.3 °C), a heart rate of 52 bpm, and a respiratory rate of 12 brpm. The horse was 4+/5 lame on the right thoracic limb. There was moderate, warm pitting edema that was sensitive to palpation from the mid-metacarpus to the coronary band. Marked effusion was present in the digital flexor tendon sheath with bounding digital pulses. The horse was clinically sensitive to distal limb flexion and hoof tester application across the heel of the affected limb. The sole and frog were inspected for evidence of the nail entry point in the sole, but no identifiable track was found.

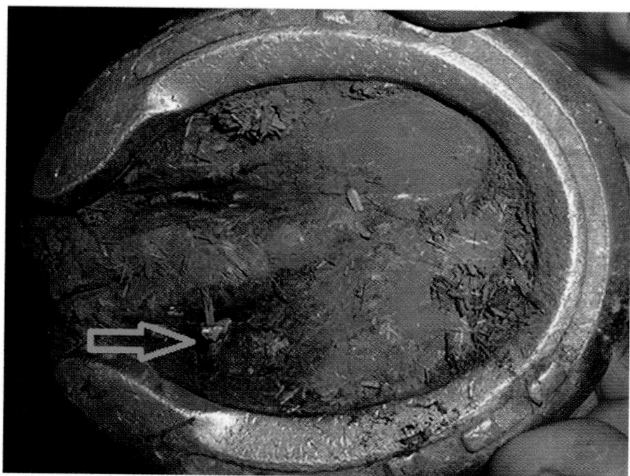

FIGURE 14.8-1 A nail *(green arrow)* observed in the medial sulcus of the frog. Nails located in this region of the foot warrant further evaluation with radiographs and synovial centesis of the distal interphalangeal joint, navicular bursa, and deep digital flexor tendon sheath based on the proximity of the foreign body to these important, sterile structures.

Comparative Veterinary Anatomy: A Clinical Approach. https://doi.org/10.1016/B978-0-323-91015-6.00079-0

Differential diagnoses

Subsolar abscess; septic osteitis of the 3rd phalanx; infection of the distal interphalangeal joint, the navicular bursa, and/or the digital flexor tendon sheath; cellulitis; laminitis; and 3rd phalanx or navicular bone fracture

Diagnostics

Initial survey radiographs of the hoof were normal. Routine laboratory studies, CBC and clinical chemistry, were normal except for a fibrinogen of 551 mg/dL. Synovial samples of the distal interphalangeal joint and the digital flexor tendon sheath were aseptically obtained under standing sedation and regional anesthesia; sampling of the navicular bursa did not yield any fluid (Fig. 14.8-2). The fluid sample from the distal interphalangeal joint had an increased total protein of 5.1 g/dL and an increased total nucleated cell count of 61,000 cells/dL. The fluid sample from the digital flexor tendon sheath had an increased total protein of 5.6 g/dL and an increased total nucleated cell count of 67,000 cells/dL.

Diagnosis

Foreign body penetration (a.k.a. "street nail"), resulting in an infection in the distal interphalangeal joint, navicular bursa (suspect; confirmed at surgery), and digital flexor tendon sheath

Treatment

The horse was started on broad-spectrum systemic antimicrobials; arthroscopic evaluation and lavage of the affected synovial structures was performed under general anesthesia. At surgery, navicular bursa involvement was confirmed tenoscopically (examination of the tendon sheath with a scope) by creating a communication between the T-ligament separating the digital flexor tendon sheath and the bursa. After debridement and lavage of the affected synovial structures, the horse was treated using a combination of regional and systemic antimicrobials. A second lavage of the affected synovial structures was performed 2 days after the first surgery. The horse was continued on regional treatments (regional limb perfusions and injection of antimicrobials directly into the affected synovial structures) for 5 additional days and maintained on systemic antimicrobials for 30 days. The horse progressively improved in clinical comfort during hospitalization, which was supported by declining synovial nucleated cell counts and serum fibrinogen. The horse was sound at the walk when discharged from the hospital, but never returned to his previous athletic level.

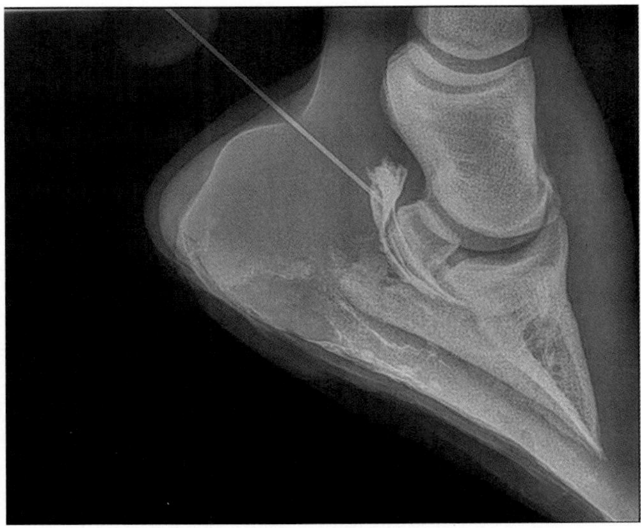

FIGURE 14.8-2 Latero-medial radiograph of the distal limb of a horse showing centesis and injection of radioopaque contrast medium of/into the navicular bursa of a horse.

Anatomical features in equids

Introduction

The major structures in the region of the distal interphalangeal joint include the distal aspect of the 2nd phalanx, the 3rd phalanx, the navicular bone, the distal interphalangeal joint, the navicular bursa, the digital flexor tendon sheath, surrounding tendons and ligaments, and regional nerves and blood vessels. There are no clinically significant differences between the thoracic and pelvic limbs in this region. Laminar structure, function, regional innervation, and blood supply are discussed in Case 14.9.

Third phalanx (Fig. 14.8-3)

The **3rd phalanx** (distal phalanx) is often referred to as the **coffin bone** due to the coffin-like nature of the hoof capsule encasing the bone. It is a semicircular, wedge-shaped bone that is largely "spongy" in appearance due to abundant vascular channels that pass through it. The palmar/plantar, dorsal, medial, and lateral surfaces each have a slope that matches the hoof capsule. The parietal and solar surfaces interconnect with the dermal tissue of the hoof capsule via a complex laminar system (see Case 14.9).

The prominent dorsoproximal **extensor process** is the site for insertion of the common/long digital extensor tendons. There are depressions on each side of the extensor process that serve as the insertion sites of the **collateral ligaments** of the distal interphalangeal joint.

Collateral cartilages arise from the palmar/plantar wings of the 3rd phalanx. The deep base region of the cartilage extends caudally toward the heel and axially from the 3rd phalanx with the components of the cartilage frequently extending into portions of the digital cushion. The proximal region of the collateral cartilage extends proximally and in a palmar/plantar direction from the 3rd phalanx and can be palpated in the bulbs of the heel.

> The collateral cartilages have various stages of calcification termed "side bone." This calcification may be associated with lameness if there is substantial calcification or a fracture occurs in the ossified cartilage. However, this is often an incidental finding and not associated with lameness. Infection of the cartilage, often secondary to trauma, is a condition called "quittor" and requires surgical debridement to resolve chronic draining tracks associated with the infected cartilage.

The solar surface of the 3rd phalanx is crescent-shaped and concave. The **crena** is a variably sized, axially located concavity observed on the distal palmar/plantar surface of this bone. Caudal and proximal to the solar surface is the flexor surface, which is the insertion site for the deep digital flexor tendon. The **impar ligament** attaches palmaro-/plantaroproximally between the flexor surface and the articular margin of the 3rd phalanx (Fig. 14.8-3).

> Radiographic lucencies on the solar margin of the coffin bone observed on a 60° dorsopalmar/-plantar radiograph may be normal anatomic variations of the crena or associated with pathologic conditions, such as keratoma and septic pedal osteitis. Do not confuse these with fractures (Fig. 14.8-4).

ARTHROCENTESIS OF THE DISTAL INTERPHALANGEAL JOINT (FIG. 14.8-5)

There are 2 main sites for centesis of the distal interphalangeal joint. Both are performed with the horse weight-bearing. The dorsal pouch is approached by inserting a needle above the coronary band with the needle perpendicular to the ground passing it through the common digital extensor tendon. Alternatively, the dorsal approach can be attempted by passing a needle on the medial or lateral side of the extensor tendon, aiming toward the middle sagittal plane of the hoof behind (palmar/plantar to) the extensor process of the coffin bone.

The palmar or plantar pouch is approached by inserting a needle at a 45° angle distally at the level of the distal palmar/plantar aspect of the 2nd phalanx axial to the proximal edge of the collateral cartilage of the 3rd phalanx.

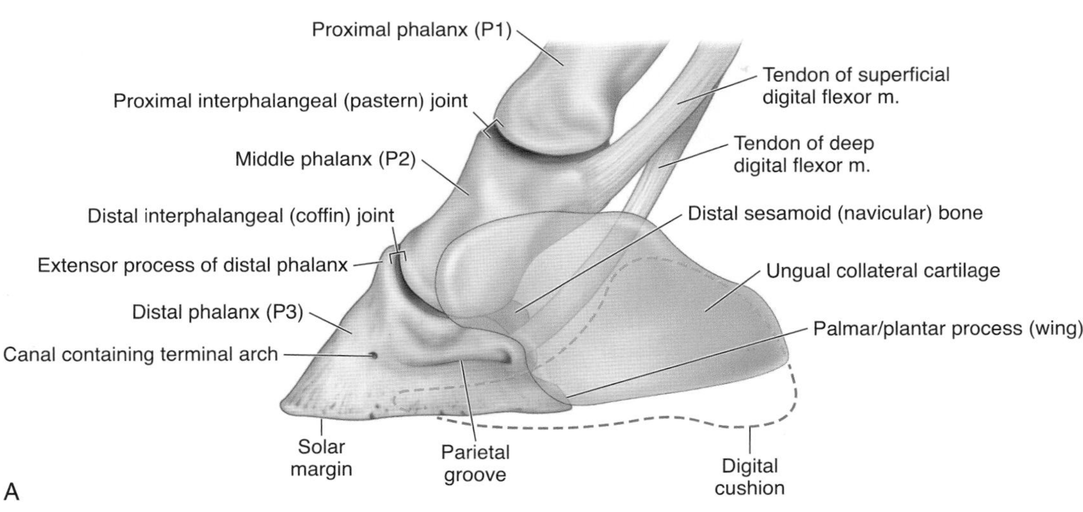

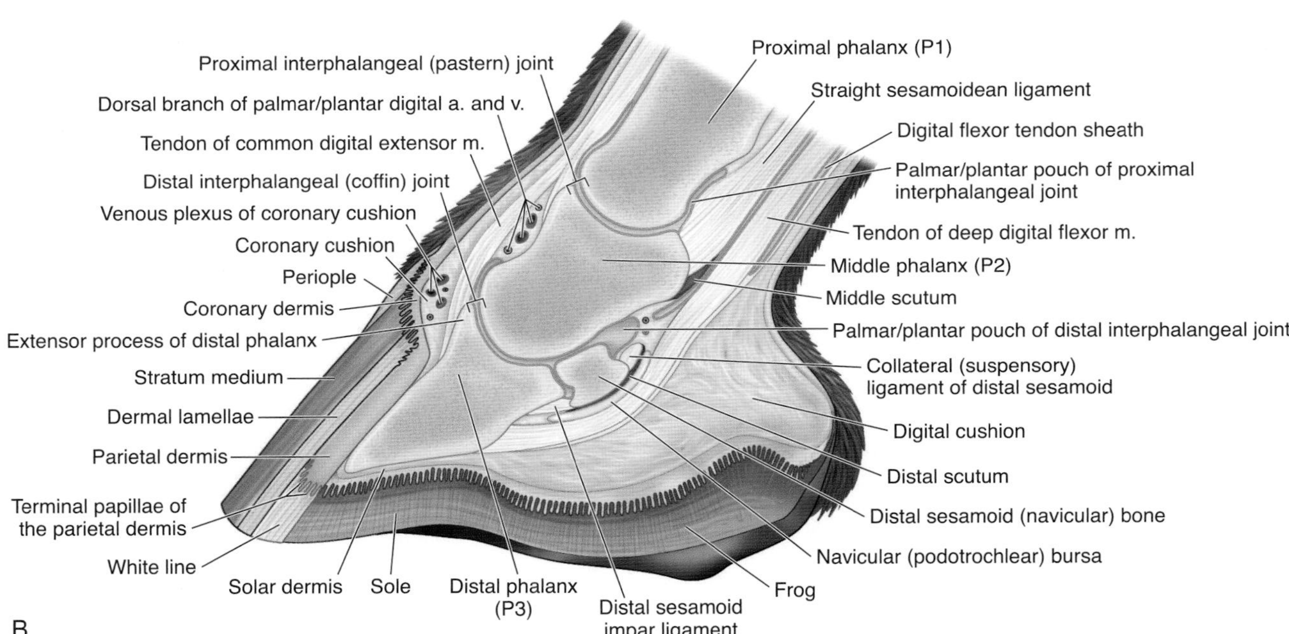

FIGURE 14.8-3 (A) Lateral view of the bones and collateral cartilage of the equine foot. (B) Sagittal view of the soft tissues and synovial structures of the equine distal limb.

The distal interphalangeal joint (Fig. 14.8-5)

The **distal interphalangeal joint** is often called the **coffin joint**. This is a composite saddle-type joint that allows primarily flexion and extension, with limited side-to-side and rotational movement. The bones that comprise the joint are the distal 2nd phalanx, the proximal 3rd phalanx, and the dorsal/articular cortex of the navicular bone.

Soft tissues around the joint include the common (thoracic limb) digital or long (pelvic limb) extensor tendon that insert on the extensor process of the 3rd phalanx and acts to flex the elbow joint and extend the carpus and the digit in the thoracic limb and flex the tarsus and extend the digit in the pelvic limb; the medial and lateral collateral ligaments of the distal interphalangeal joint that stabilize the joint; the impar ligament; the T-ligament; the navicular

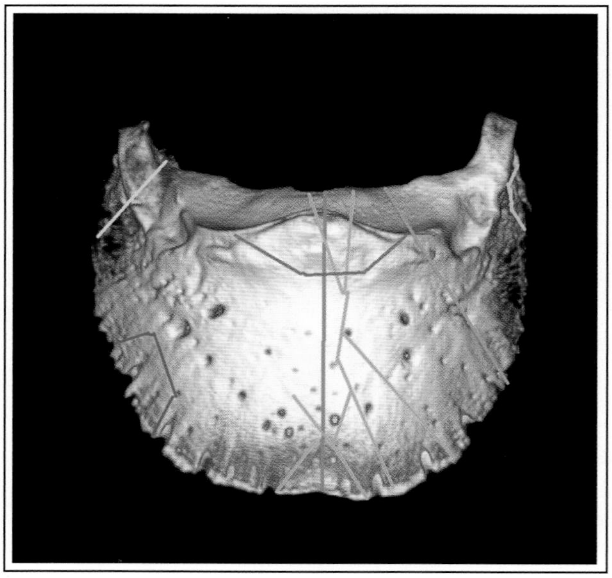

FIGURE 14.8-4 The 7 types of common coffin bone fractures. Abaxial fracture without joint involvement *(yellow)*; abaxial fracture with joint involvement *(blue)*; axial and perisagittal fracture with joint involvement *(dark green)*; fracture(s) of the extensor process *(purple)*; comminuted fracture with joint involvement *(lime green)*; solar margin fracture *(red)*; palmar/plantar process fracture (foals only-*pink*).

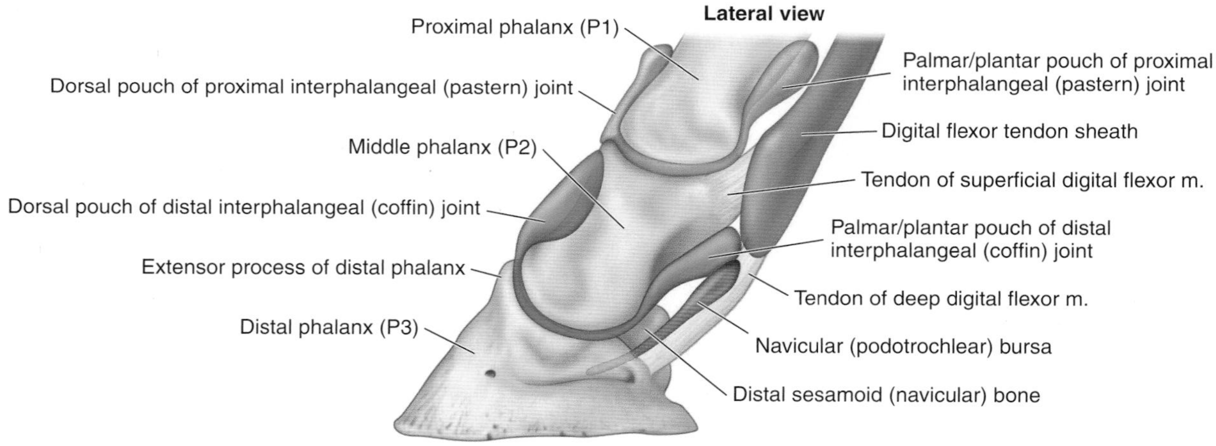

FIGURE 14.8-5 Synovial structures of the distal limb.

bursa; and the digital flexor tendon sheath. The impar ligament tracks between the 3rd phalanx and the navicular bone and separates the palmar/plantar distal interphalangeal joint from the distal navicular bursa. The T-ligament is discussed further below.

Deep digital flexor tendon (Fig. 14.8-3)

The **deep digital flexor tendon** (DDFT) is bilobed in the pastern as it descends into the hoof capsule. The **distal scutum**, a fibrocartilaginous portion of the DDFT, is just proximal to the navicular bone immediately palmar/plantar and distal to the middle scutum (see Case 14.7 for additional detail). The middle and distal scutum allow the DDFT to slide along the palmar/plantar surface of the 2nd phalanx. Distal to the distal scutum, the DDFT becomes thinner and wider as it moves beyond the navicular region to broadly insert on the flexor surface of the 3rd phalanx. Deep to the DDFT on the palmar/plantar aspect of the hoof is the **digital cushion**, a collection of collagenous and fatty tissue proximal to the frog, which functions to dampen the energy incurred during the impact portion of the stride.

Distal sesamoid bone (Fig. 14.8-3)

The **distal sesamoid bone** is commonly referred to as the **navicular bone**. This bone is supported on the palmar/plantar aspect of the digit by a ligamentous sling that attaches to the bone on the proximal medial and lateral surfaces. The origin of these **medial** and **lateral collateral (suspensory) ligaments** of the navicular bone is the dorsomedial/-lateral aspect of the distal 1st phalanx. There are additional ligamentous attachments of the navicular bone in this region that pass between the 2nd phalanx and the medial/lateral aspect of the navicular bone. Additionally, there are short ligamentous attachments between the navicular bone and the 3rd phalanx and its collateral cartilages.

The **T-ligament** (a.k.a. transverse lamina, proximal sesamoid ligament) is a thin, elastic sheet of fibrous tissue forming attachments between the proximal palmar/plantar edge of the navicular bone, the palmar aspect of the 2nd phalanx, and the deep surface of the DDFT (Fig. 14.8-3). This separates the palmar pouch of the distal interphalangeal joint, the distal digital flexor tendon sheath, and the navicular bursa. The distal surface of the navicular bone is the origin of the impar ligament (see above). The dorsal articulating surface of the navicular bone interfaces with the 2nd and 3rd phalanges in the distal interphalangeal joint. The palmar/plantar flexor surface of the navicular bone provides a surface over which the bilobed DDFT tracks.

The **navicular bursa** lies between the navicular bone and DDFT. The proximal and distal limits of the navicular bursa are defined by the T-ligament and the impar ligament, respectively (Fig. 14.8-5). The navicular bursa, especially when distended, can have large medial and lateral proximal outpouchings that can extend to the level of the proximal collateral cartilages of the 3rd phalanx.

The navicular bursa has medial and lateral outpouchings in the heel on either side of the DDFT. It is also in close proximity to the palmar/plantar pouch of the distal interphalangeal joint. Both structures are axial and deep to the palpable collateral cartilage of the coffin bone. Therefore, any significant heel bulb laceration that involves the tissue axial to the collateral cartilage should be investigated for possible synovial sepsis of the navicular bursa, distal interphalangeal joint, and possibly the tendon sheath of the DDFT.

CENTESIS OF THE NAVICULAR BURSA (FIG. 14.8-2)

Centesis of the navicular bursa is typically performed using radiographic guidance, placing a spinal needle above the hair line at the heels, horizontally through the deep digital flexor tendon, until it contacts the navicular bone. This has been done in the radiograph in Fig. 14.8-2, in which the needle is visible and radiopaque contrast has been injected into the navicular bursa. An ultrasound-guided technique also exists, which guides a needle to the medial or lateral proximal outpouching of the bursa by placing the needle just proximal to the collateral cartilage between the digital flexor tendon sheath and the neurovascular bundle. Use of nonguided injection techniques typically leads to missing the bursa or unintentionally entering the distal interphalangeal joint or the digital flexor tendon sheath.

NAVICULAR SYNDROME

"Caudal heel syndrome" or navicular syndrome has been historically evaluated based on the radiographic changes that occur in the navicular bone in horses with lameness that respond to anesthesia/analgesia of the palmar/plantar digital nerves. However, lameness in this area is likely multifactorial, involving the navicular bursa, deep digital flexor tendon, impar ligament, and collateral ligaments of the navicular bone. The use of ultrasound and MRI more accurately defines the clinical lesions causing pain in this area. Management of these cases relies on local and systemic anti-inflammatories, controlled rehabilitation programs, and therapeutic shoeing.

Selected references

[1] Dyson SJ. The distal phalanx and distal interphalangeal joint. In: Ross MW, Dyson S, editors. Diagnosis and management of lameness in the horse. 2nd ed. Saunders; 2011. p. 349–66.
[2] Butler JA, Colles CM, Dyson SJ, Kold SE, Poulos PW, editors. Clinical radiology of the horse. 3rd ed. Wiley-Blackwell; 2008. p. 53–189.
[3] Furst AE, Lischer CJ. Foot. In: Auer JA, Stick JA, editors. Equine surgery. 4th ed. Saunders; 2012. p. 1264–99.
[4] Denoix JM. The equine distal limb. Manson Publishing; 2000. p. 243–373.

CHAPTER 14

CASE 14.9

Laminitis

Nick Carlson
Steinbeck Peninsula Equine Clinics, Salinas, California, US

Clinical case

History

A 10-year-old Warmblood gelding presented with a history of nasal discharge, cough, and fever. In addition to signs consistent with a respiratory infection, the horse had increased digital pulses in the thoracic and pelvic limbs. He was diagnosed with pleuropneumonia and treated intensively with broad-spectrum antimicrobials, anti-inflammatories, thoracic cavity drainage (thoracocentesis/pleurocentesis), and cryotherapy of all 4 distal limbs. While his respiratory condition improved, his digital pulses progressively increased and he soon became reluctant to walk from his stall.

Physical examination findings

The horse had bounding digital pulses in both thoracic limbs and increased digital pulses in both pelvic limbs. All 4 hooves were warm to the touch and a palpable depression was noted along the coronary bands of both hind feet. The horse was reluctant to pick up the front feet and had significant hoof tester sensitivity at the toe of both front hooves, particularly in front of the apex of the frog. Due to his degree of pain, the horse was unwilling to pick up the hind feet for hoof testing.

Differential diagnoses

Laminitis; subsolar abscess; septic osteitis of the 3rd phalanx; and 3rd phalanx or navicular bone fracture

BOVINE LAMINITIS

Equine and bovine laminitis are dissimilar in etiology, treatment, and prognosis. Although acute laminitis occurs in both species and can be caused by grain overload, there are many other etiologies that cause equine laminitis. The distinction between the 2 species is anatomic with the laminae of the equine hoof being more extensive than that of the cow. The bovine foot lacks secondary epidermal lamellae, which are important in increasing the surface area attachments for the suspensory apparatus of the 3rd phalanx in the hoof wall.

There is a subclinical laminitis phenomenon described in cows, which really refers to a disorder caused by claw horn disruption. This clinical entity is of substantial economic importance to the dairy industry, because it predisposes cattle to sole ulcers (Case 21.3), white line disease, and "necrosis syndrome" of the toe. These foot diseases are reported in most developed countries and are of considerable concern in high-producing, intensively managed dairy herds.

Comparative Veterinary Anatomy: A Clinical Approach. https://doi.org/10.1016/B978-0-323-91015-6.00080-7

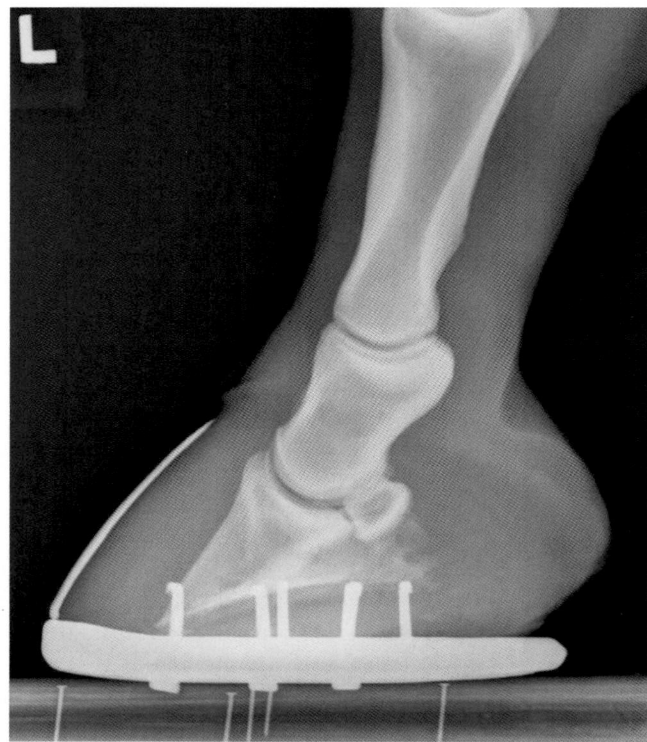

FIGURE 14.9-1 Lateromedial radiograph of the left distal thoracic limb of the horse in this case. Note the barium paste applied to the dorsal hoof wall. The 3rd phalanx is rotated with respect to the hoof wall. See Fig. 14.9-4 for more on how to interpret radiographs in horses with laminitis.

Diagnostics

Lateromedial and dorsopalmar/dorsoplantar radiographs of all four limbs were performed to evaluate for radiographic evidence of laminitis. Rotation of the 3rd phalanx with respect to the hoof wall was seen in both front feet and sinking of the 3rd phalanx in the hoof capsule was observed in both hind feet (Fig. 14.9-1).

Diagnosis

Laminitis in all four feet with moderate rotation in both thoracic limbs and severe sinking of the 3rd phalanx in both pelvic limbs

Treatment

Multimodal medical therapy was initiated in an attempt to manage the horse's pain. Efforts were made to stabilize the feet with therapeutic shoeing. However, despite aggressive therapy the horse remained in persistent pain; repeated radiographic examination showed continued progression of the laminitis in the hind feet. Humane euthanasia was recommended because of the inability to manage the clinical pain and the progressive changes seen on the radiographs of all four feet; the owner agreed because of the grave prognosis.

Anatomical features in equids

Introduction

The equine hoof is a keratinized structure that surrounds the 3rd phalanx in each foot. The major structures include the distal 2nd phalanx, navicular bone, 3rd phalanx (also known as the "coffin bone"), distal interphalangeal joint (also known as the "coffin joint"), navicular bursa, digital flexor tendon sheath, palmar/plantar digital veins/arteries/nerves

(V.A.N. relationship), the suspensory apparatus of the distal phalanx (SADP) best known as "laminar network," the digital cushion, the frog and sole, and the hoof wall. The distal interphalangeal joint, the navicular bursa, and the digital flexor tendon sheath are discussed in Case 14.8.

Function

The equine hoof is an elastic, flexible structure that—when placed under a heavy load/force on contacting the ground—adapts and dispels much of the associated energy. The hoof serves many important functions, including shock absorption, body weight support, traction, and as an auxiliary pump for venous blood return to the heart. The laminar interface serves as an important distributor of energy on ground impact of the foot; 90% percent of the energy of ground contact in the foot is moderated by the time it reaches the 1st phalanx. Thus, the well-known phrase "no hoof, no horse" highlights the clinical importance of the equine hoof.

Hoof capsule

Externally, the hoof is divided into the coronet, toe, quarter, and heel. On the bottom of the hoof (the solar surface), major landmarks include the **toe**, **quarter**, **heel**, **bars**, and a **frog**, surrounded by **collateral grooves** (sulci) and divided by a **central groove** (sulcus) (Fig. 14.9-2). The hoof is divided into epidermal and dermal tissues. The wall consists of 3 layers: **stratum externum** (tectorium), **stratum medium**, and **stratum internum** (stratum lamellum) (Fig. 14.9-3). The dermal tissue of the 3rd phalanx (a.k.a. the "sensitive laminae") and the cartilage of the hoof interact with the hoof wall at the level of the stratum internum where primary horny laminae (a.k.a. "insensitive laminae") connect to the primary dermal laminae and secondary interdigitating laminae of the wall. The sensitive laminae are richly supplied with blood vessels and nerves. (There is more detail regarding the laminae below.)

The dermis can be divided into 5 parts: perioplic, coronary, laminar, solar, and frog (Fig. 14.9-3). All but the **laminar corium** have papillae that extend into the epidermis. The laminar corium is the dermal tissue of the laminae and is a collection of blood vessels and connective tissue. At the site of the dermal papillae, the epidermis produces tubular and nontubular horn. The **perioplic dermis (corium)** of the hoof is continuous with the dermis of the skin. **Coronary dermis (corium)** is thicker and located distal to the perioplic dermis at the coronary groove. The coronary dermal papillae are the origin of the tubular and nontubular horn of the wall. The **laminar dermis (corium)** is

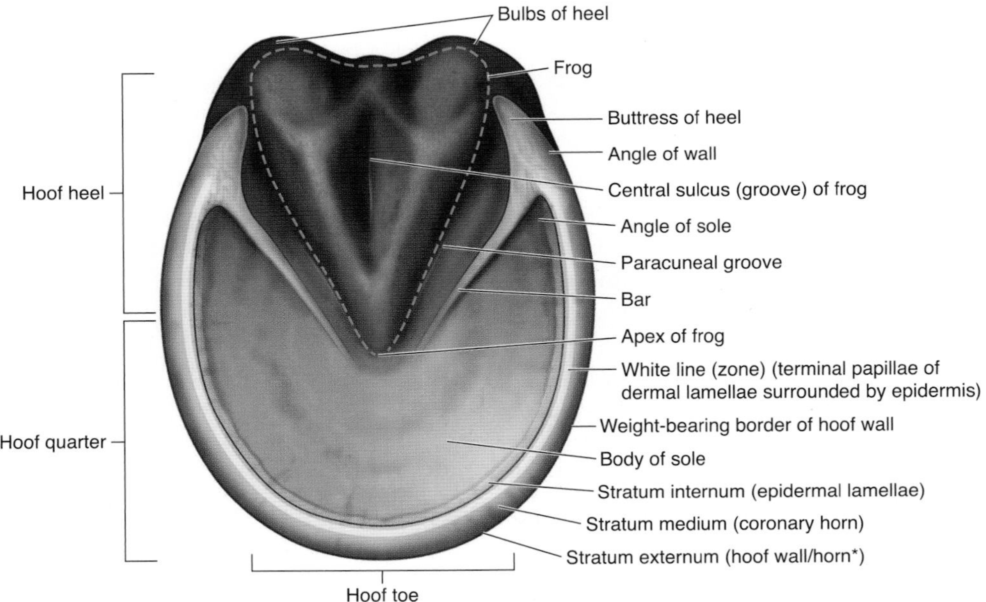

FIGURE 14.9-2 External features of the equine hoof.

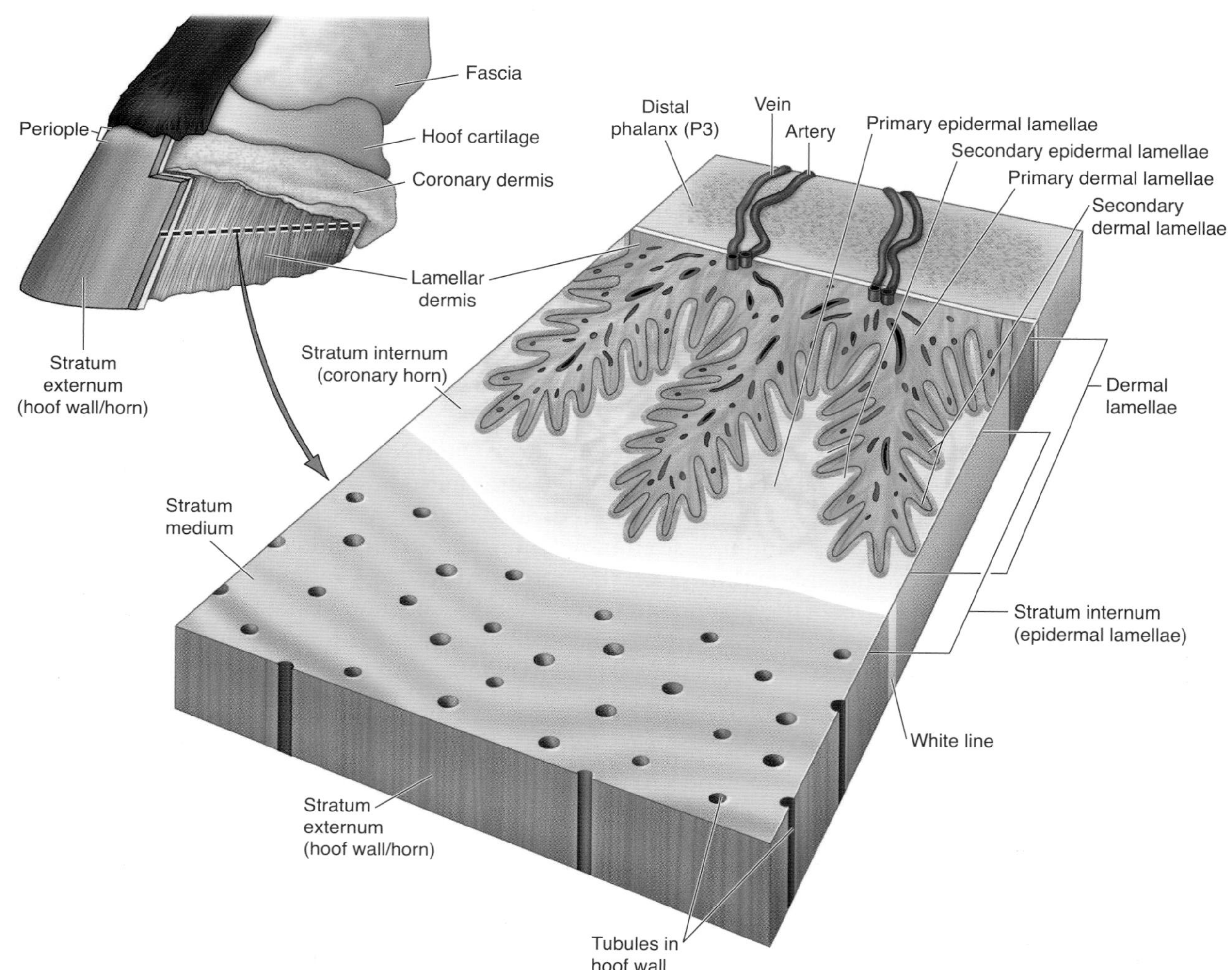

FIGURE 14.9-3 Sectional anatomy of the equine hoof wall.

> The terms "laminae" and "lamellae" are often used interchangeably in various texts. However, histologically "laminae" refer to the live dermal tissue covering the 3rd phalanx. The term "lamellae" refers to the innermost layer of the epidermis (hoof), which interdigitates with the laminae to anchor the 3rd phalanx to the hoof.

composed of 600 primary dermal laminae interdigitating with the epidermal lamellae (insensitive), which connect the 3rd phalanx to the hoof wall on the dorsal and lateral surfaces. The papillae in this area are elongated into the primary laminae that are perpendicular to the parietal surface of the 3rd phalanx. The secondary laminae are oriented at acute angles to the primary laminae. These laminae are tightly bound to the laminar epidermis of the hoof. The **solar dermis** is located on the bottom of the 3rd phalanx. The dermis of the frog blends with the digital cushion in this region.

The **epidermis** forms the outer layer that hardens into the hoof. Cells in a deep germinal layer divide and push away from the dermis and stiffen as they move away. The **perioplic epidermis** is the junction between the hoof and skin at the proximal and dorsal edge of the coronet. It is grossly observed as a thin shiny external layer of the hoof wall. The **coronary epidermis** is the highly keratinized hoof wall that extends distally from the underlying dermis and forms the bulk of the hoof wall. **Laminar epidermis** is the inner layer of the hoof wall that connects the hoof to the 3rd phalanx.

Over the dermis' laminae, the **germinal epidermis** moves cells perpendicularly and away from the interdigitating laminar dermis and epidermis to critically increase the surface area and strength of the bond between the hoof and toe. Nonpigmented horn between the stratum medium and stratum internum continues to the ground surface to form the **white line (zona alba)**, the junction between the wall and sole. The **solar epidermis** is like the coronary epidermis, and the frog epidermis is a similar tissue, but it is more elastic and not fully keratinized.

White line disease is a condition that starts as a separation of the stratum medium and stratum internum of the nonpigmented horn of the hoof. Typically due to poor hoof management or conformation (long toe, underrun heels) the normal tissue is stressed and stretched, weakening the tight link between the tissues to allow moisture, bacteria, and fungi to "wick" along the tissues. These opportunistic microbes further degenerate the tissue, spreading separation along the hoof. Diagnosis is made on external examination of the hoof combined with radiographs to evaluate the extent and severity. Treatment requires removal of the diseased hoof, application of topical disinfectants, and therapeutic shoeing to support the hoof if a substantial amount of dorsal hoof wall requires resection.

Secondary epidermal lamellae

The hoof's 3rd phalangeal attachment apparatus is unimaginably robust, allowing the hoof wall and the 3rd phalanx to move in harmony and detach only when laminitis disrupts the normal lamellar anatomy. The **stratum internum**, the innermost layer of the hoof wall and bars of the equine foot, bears the **stratum lamellae** (L. *lamina* layer of leaves) and includes 550–600 **primary epidermal lamellae** projecting from its surface in uniform rows. The function of the epidermal lamellae is the suspension of the 3rd phalanx—i.e., the SADP—and to increase the surface area for the attachment of the many collagen fibers originating from the parietal surface of the 3rd phalanx. **Secondary epidermal lamellae** serve the function of increasing the surface area further. These secondary epidermal lamellae form on the inner coronary groove with the growth of a basal cell layer, creating folds of secondary lamellae. These folds elongate to form 150–200 secondary lamellae along the length of each of the 550–600 primary lamellae. This complex and interdigitating network of lamellae—primary and secondary epidermal and dermal tissue—are calculated to average 0.8 m^2 (31.2 $in.^2$) (equivalent to the surface area of human skin). This large surface area provides the suspensory network for the 3rd phalanx and allows adaptability of the interdigitating lamellae architecture to decrease stress, and it further assures even energy transfer during peak loading of the equine foot.

Innervation of the distal limb

The **medial** and **lateral palmar digital nerves** continue from the metacarpus/-tarsus on the medial/lateral palmar/plantar metacarpo-/metatarsophalangeal (fetlock) region abaxial to the sesamoid bones. As the nerves track distally to the proximal interphalangeal joint (pastern), they give off the **dorsal branches** of the **digital nerves**. The nerves continue distally in the

Palmar/plantar digital nerve block can be performed in the distal limb by subcutaneous perineural injection of local anesthetic over the medial and lateral neurovascular bundles just proximal to—or at the level of—the collateral cartilages of the 3rd phalanx. This usually desensitizes the palmar/plantar aspect of the anatomical structures within the hoof capsule and the caudal one-third to complete anesthesia of the sole.

LAMINITIS

Laminitis is inflammation of the lamina with injury and loss of lamellar attachments between the hoof wall and 3rd phalanx. There are multiple causes of laminitis including endocrinopathic, sepsis-associated (systemic inflammatory response syndrome—SIRS), excessive unilateral weight-bearing (support or contralateral limb), and trauma. Prevention and management include a multimodal approach of cryotherapy (cold treatment), analgesia (pain relief), anti-inflammatories, stall rest with soft bedding, frog support, and deep digital flexor tenotomy to realign the 3rd phalanx postrotation. See Fig. 14.9-4 for examples of 2 classic radiographic findings seen with laminitis.

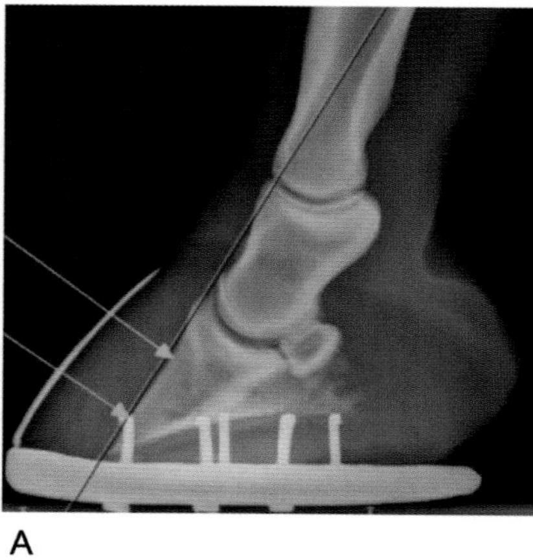

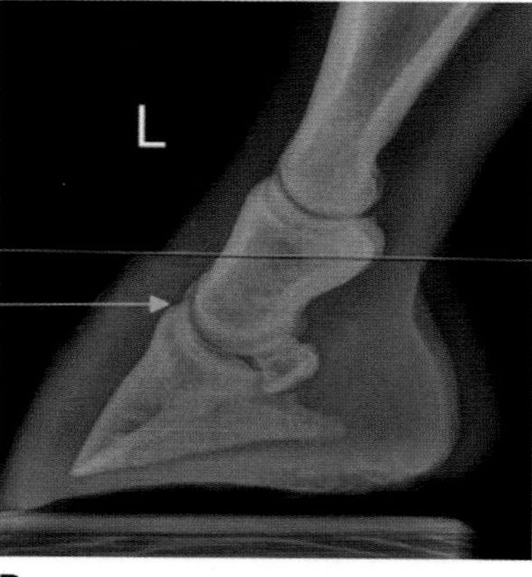

FIGURE 14.9-4 Lateromedial radiographs of a horse's hoof showing different features of radiographically apparent laminitis. (A) Rotation—note the steep angle of the dorsal 3rd phalanx relative to the dorsal hoof wall. (B) Sinking—note the significant drop of the 3rd phalanx below the coronary band (the *blue line* demarcates the position of the dorsal coronary band and the *green arrow* shows the extensor process of the 3rd phalanx noticeably below the level of the coronary band).

Some cases of chronic lameness (such as navicular syndrome) associated with heel pain that are not responsive to medical management can be treated surgically via a palmar/plantar digital neurectomy. Segments of the medial and lateral nerves at the level of the pastern are isolated from the neurovascular bundle and resected to provide pain relief that can last several years until the nerves regenerate.

Venograms are performed by placing a tourniquet around the metacarpo-/metatarsophanalangeal joint and injecting a radiopaque contrast medium into the palmar/plantar digital vein distal to the tourniquet site. Radiographs are taken after injection of the contrast medium to evaluate the regional blood supply (or lack thereof) in the distal extremity. Venograms are used to help determine the degree of vascular compromise secondary to laminitis or better define a space-occupying mass in the hoof capsule, such as a keratoma (Fig. 14.9-6).

Because of the extensive vessel anastomoses for the hoof, a laceration involving either the medial or lateral digital arteries supplying the hoof does not adversely compromise the blood supply to the horse's hoof. Laminitis can also severely affect the blood supply to the laminae secondary to rotation or sinking of the 3rd phalanx, which compresses the blood vessels that generate hoof growth.

pastern region, just lateral/medial to the deep digital flexor tendon. The nerves travel with the major vessels of the region in a neurovascular bundle that enter the hoof capsule axial to the collateral cartilage of the 3rd phalanx.

Regional vessels (Fig. 14.9-5)

The **lateral** and **medial palmar/plantar digital arteries** accompany the palmar/plantar digital nerves on the palmar/plantar aspect of the distal limb. Dorsal branches are seen coming off at the 1st and 2nd phalanx that join with one another on the dorsal aspect of their respective bones. The palmar/plantar digital vessels continue into the hoof capsule axial to the collateral cartilage. At the proximal 3rd phalanx, the digital arteries give off palmar branches, which travel dorsally to become the **dorsal arterial branches of the 3rd phalanx**. These vessels travel on the parietal (L. *parietalis* of or pertaining to the walls of a cavity) surface of the 3rd phalanx in a "notch" called the **parietal groove** and anastomose with one another. On the solar surface of the 3rd phalanx, the digital arteries enter the **solar foramen,** anastomosing in the bone to form the **terminal arch**. Branches from the terminal arch track through bony canals to the parietal surface/margins and provide blood supply to the laminar dermis via the **circumflex marginal artery**.

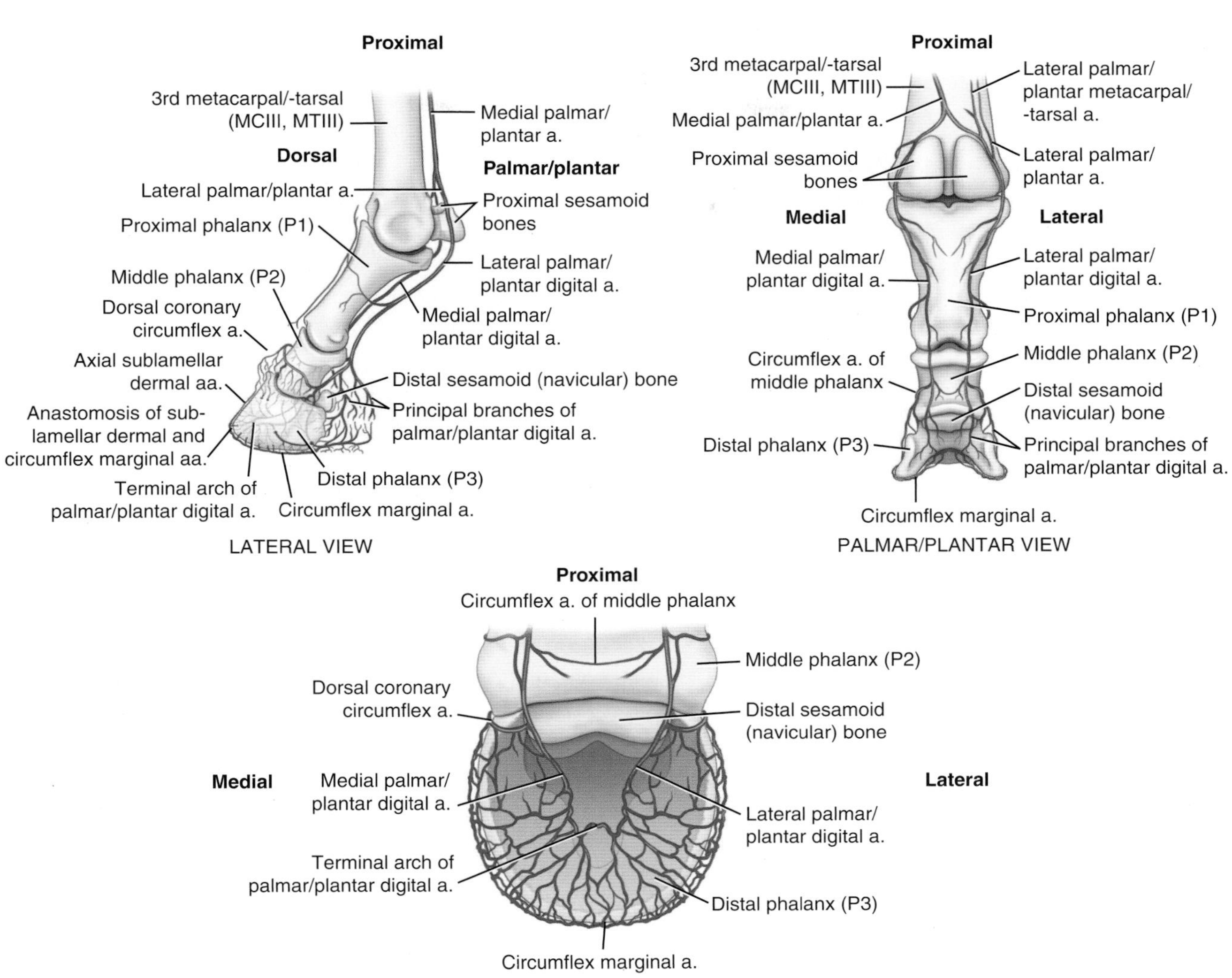

FIGURE 14.9-5 Arteries of the equine digit and hoof region. (A) Lateral-medial view. (B) Dorsopalmar/dorsoplantar view. (C) Dorsoproximal-palmarodistal oblique (dorsoventral/solar) view.

DISEASES THAT CAN ALTER 3RD PHALANX MORPHOLOGY

Several conditions can impact the morphology observed radiographically. Keratomas are space-occupying masses of abnormal horn tissue that lead to pressure necrosis of the 3rd phalanx bone, causing a well-demarcated lucency typically on the solar margin of the bone. It should not be confused with the crena, a normal anatomical feature on the distal aspect of the dorsal 3rd phalanx.

Pedal osteitis is caused by chronic inflammation (due to repetitive concussion on hard surfaces). The characteristic widening of the vascular channels observed radiographically is caused by an increase in blood flow to this region of the bone due to chronic inflammation.

Alternatively, septic pedal osteitis is caused by a septic process of the 3rd phalanx, leading to a focal lysis of the bone.

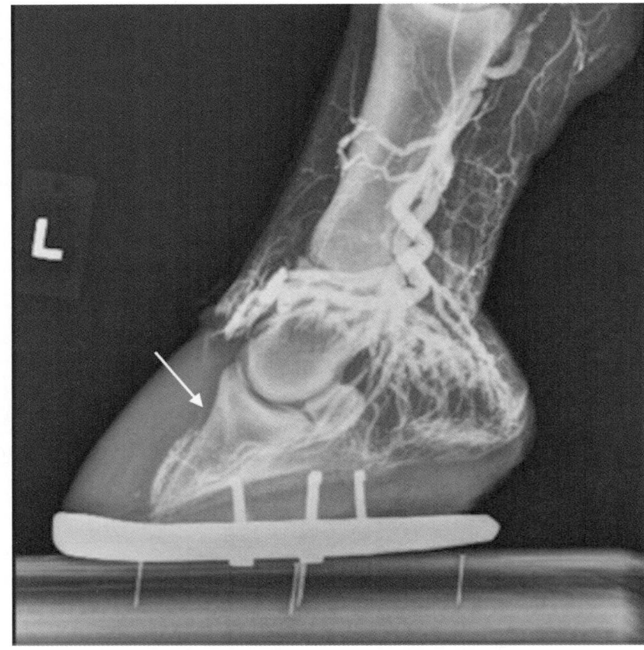

FIGURE 14.9-6 A clinical example of a contrast venogram, lateromedial view, performed on a horse with laminitis and significant rotation of the 3rd phalanx. Notice the lack of contrast dye along the dorsal surface of the 3rd phalanx *(arrow)*.

Selected references

[1] Pollitt CC. Laminitis. In: Ross MW, Dyson S, editors. Diagnosis and management of lameness in the horse. 2nd ed. Saunders; 2011. p. 366–86.
[2] Butler JA, Colles CM, Dyson SJ, Kold SE, Poulos PW, editors. Clinical radiology of the horse. 3rd ed. Wiley-Blackwell; 2008. p. 53–189.
[3] Pollitt CC. Laminitis. In: Floyd AE, Mansmann RA, editors. Equine podiatry. 1st ed. Saunders; 2007. p. 313–76.

CHAPTER 15

PELVIC LIMB

Nick Carlson, Chapter editor

Proximal Pelvic Limb (Hip, Stifle, Crus)

15.1 Coxofemoral joint luxation—*Nick Carlson* 934

15.2 Osteochondritis dissecans—*Sarah DeSante* 942

15.3 Disruption of the fibularis (peroneus) tertius—*Sarah DeSante and Nick Carlson* 952

Distal Pelvic Limb (Tarsus and Pes)

15.4 Gastrocnemius tendonitis—*Sarah DeSante and Nick Carlson* 958

15.5 Fracture of the 4th metatarsal bone—*Nick Carlson* 969

CASE 15.1

Coxofemoral Joint Luxation

Nick Carlson
Steinbeck Peninsula Equine Clinics, Salinas, California, US

Clinical case

History

A 5-year-old miniature horse gelding presented for severe acute right pelvic limb lameness of an unknown cause. The gelding had no prior history of lameness and lived in a small pasture by himself. The referring veterinarian was unable to coax the horse to weight-bear on the limb, and when attempting to examine the limb, the gelding became anxious. The gelding was referred to an equine clinic for further clinical workup and diagnostic imaging.

Physical examination findings

The horse was apprehensive while unloading from the trailer, unable to weight-bear on the limb, and his lameness was graded as 5/5 on the AAEP lameness scale. There was no swelling or increase in digital pulses in the limb, and he resisted palpation of the proximal limb. No crepitus or clicking sounds were auscultated when manipulating the proximal limb, both of which would have supported a fracture diagnosis. There was no resistance to palpation of the pelvis over the sacroiliac and tuber coxae region.

Differential diagnoses

Femoral fracture, pelvic fracture, coxofemoral (hip) joint luxation

Coxofemoral joint luxation is most common in ponies and miniature horses secondary to a fall or attack while struggling to pull away from an entrapped limb. This can also be observed in a patient contending with severe upward fixation of the patella or in foals and adult horses struggling to stand with full-limb casts. Treatment is generally only attempted in miniature horses and small ponies.

Diagnostics

Ultrasound and radiographs of the pelvis found no evidence of fracture. Right-sided coxofemoral joint luxation was identified on radiographs (Fig. 15.1-1).

Diagnosis

Unilateral coxofemoral joint luxation

Treatment

The gelding was placed under general anesthesia for open reduction of the dislocation followed by joint stabilization using an orthopedic cable system. The goal of the cable system is to substitute for the torn soft tissue structures—i.e., the femoral and accessory ligaments. The horse went on to develop moderate coxofemoral joint osteoarthritis; however, he was comfortable and able to ambulate without difficulty.

Comparative Veterinary Anatomy: A Clinical Approach. https://doi.org/10.1016/B978-0-323-91015-6.00081-9

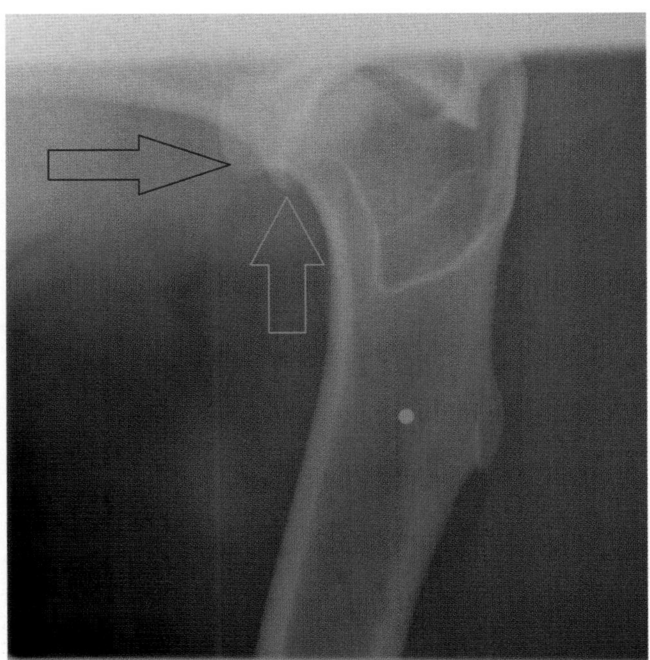

FIGURE 15.1-1 A ventrodorsal radiograph of the gelding in this case obtained under general anesthesia showing coxofemoral (hip) luxation. The head of the femur *(red arrow)* is displaced cranial and dorsal to the acetabulum *(green arrow)*. The radiopaque BB seen overlying the proximal femur was used as a marker for presurgical planning.

Anatomical features in equids

Introduction

The applicable clinical anatomy of the equine hip includes the pelvis, sacroiliac joint, pelvic ligaments, coxofemoral joint, femur, trochanteric bursa, muscles of the pelvis and thigh, and regional blood supply and nerves.

Function

The equine pelvis primarily functions to bear the weight of the upper body when standing and to transfer the weight from the axial skeleton to the appendicular skeleton when standing and moving. The pelvis also provides attachments for, and supports, the mechanical forces of the strong muscles of locomotion. Compared to the shoulder cingulum (L. *cingere* to gird, girdle, and bony support), the pelvic cingulum is strong and inflexible.

The pelvis also serves a secondary function in housing and protecting the pelvic and abdominopelvic viscera—i.e., the lower parts of the urinary tracts and internal reproductive organs—and provides a site of attachment for external reproductive organs and their associated muscles and mucous membranes.

Pelvis (Fig. 15.1-2)

The **os coxae** is composed of 3 parts—the ilium, ischium, and pubis, which meet at the **acetabulum** where the head of the femur articulates, forming the **coxofemoral joint**. The **ilium** is divided into the large ilial wing, dorsal tuber sacrale, ventrolateral tuber coxae, and medial ischiatic spine. The **ilial wing** has gluteal (outer) and pelvic (inner) surfaces. The middle and deep gluteal muscles attach to the gluteal surface, whereas the psoas minor muscle originates from its tubercule on the pelvic surface. The **tuber sacrale** is the highest point of the pelvis and is the site of attachment of the dorsal sacroiliac ligament. The **tuber coxae** are the most lateral part of the pelvis and each one forms the external landmark commonly referred to as the "point of the hip." The tuber coxae are the origin of the tensor fasciae latae and superficial gluteal muscles. The tuber sacrale and tuber coxae physes close at approximately 6–7 years of age.

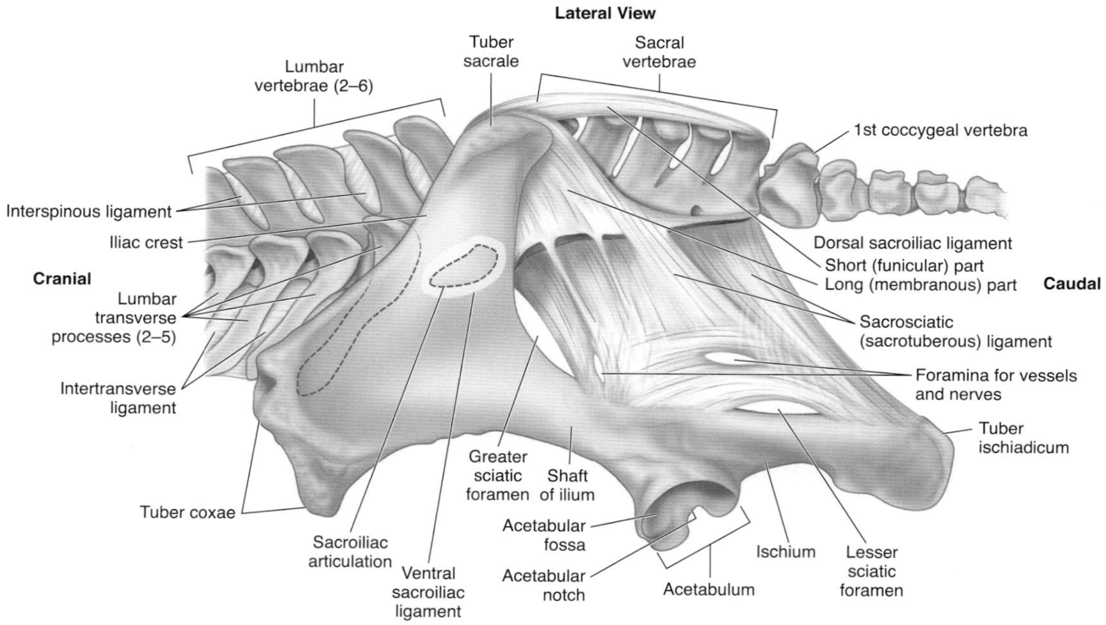

FIGURE 15.1-2 Lumbar, sacral, and coccygeal vertebrae, hemipelvis, and ligaments comprising the equine pelvic cingulum (girdle).

The ventral portion of the pelvis is formed caudally by the **ischium**, which has both pelvic and ventral surfaces. The palpable **tuber ischiae** (singular: **tuber ischiadicum**) are the most caudal parts located on either side of the horse's tail. The **tuber ischiadicum** physes closes around 5–6 years of age.

The **pubis** forms the cranial part of the ventral floor of the pelvis. The pubic symphysis closes at approximately 5.5–7 years of age. There is another symphysis between each pubis and ischium, which is fused at 10–12 months of age. The **prepubic tendon** blends into the pelvic symphysis ventrally and is a collection of the tendons of origin of the pectineus muscles, the pelvic tendons of the rectus abdominis, internal/external abdominal oblique muscles, and the tendon of origin of the cranial portion of the gracilis muscle. The **obturator foramen** is an ovoid hole between the pubis and ischium where a groove exists for the obturator nerve and vessels.

Sacroilliac joint and ligaments of the pelvis (Fig. 15.1-2)

The **sacroiliac joint** is a unique articulation between the sacrum and pelvis. Unlike other joints, where 2 hyaline cartilage surfaces intersect, the hyaline cartilage surface of the sacrum intersects with the fibrocartilaginous surface of the ilium. The joint is enclosed in a strong fibrous capsule and is supported by 3 strong sacroiliac ligaments: the **dorsal** and **ventral sacroiliac ligaments** along with the **interosseous sacral ligament**. The larger dorsal sacroiliac ligament has 2 parts: one that connects the sacral vertebral spines to the tuber sacrale and another connecting the lateral sacrum and medial tuber sacrale to the lateral surface of the sacrosciatic ligament. The ventral sacroiliac ligament connects the sacrum to the ventromedial aspect of the tuber sacrale and iliac wing.

The **sacrosciatic ligament** is prominent in cattle and horses and is sparse to absent in smaller species. It tracks from the lateral crest of the sacrum and transverse processes of the first 2 caudal vertebrae to the ischial spine and tuber ischiadicum. There are 2 openings identified between the sacrosciatic ligament and its insertion on the pelvis: the **greater** and **lesser ischial (sciatic) foramina**. The greater ischial foramen provides a pathway for the sciatic nerve and cranial gluteal artery. No important anatomical structures pass through the lesser ischial foramen in the horse, unlike cattle in which the caudal gluteal artery tracks through it. The **iliolumbar ligament** also provides stability and is a lateral extension of the **intertransverse ligament** that is found between the lumbar vertebrae. The ligament widens caudally and inserts on the wing of the ilium, ventral to the origin of the **longissimus muscle**.

Coxofemoral joint and proximal femur (Fig. 15.1-3)

The **femoral head** interfaces with the acetabulum of the pelvis to form the **coxofemoral joint**—i.e., the "hip" joint. The apex of the fovea of the femoral head is lacking cartilage and is the attachment for the **ligament of the head of the femur** and the **accessory ligament from the prepubic tendon**. These accessory ligaments of the hip are unique to equids

> Arthrocentesis of the coxofemoral joint entails placement of a 15.2-cm (6-in.) spinal needle at the angle formed between the cranial and caudal processes of the greater trochanter of the femur. The needle is inserted in a craniomedial direction and directed distally just above the femoral neck until a loss of resistance (a "pop") is felt when the needle penetrates the joint capsule. Ultrasound is helpful in correctly placing the needle; synovial fluid should be retrieved before injection to avoid extrasynovial injection and its subsequent adverse event of local anesthetic around the sciatic nerve causing paresis (Gr. relaxation) or paralysis of the limb.

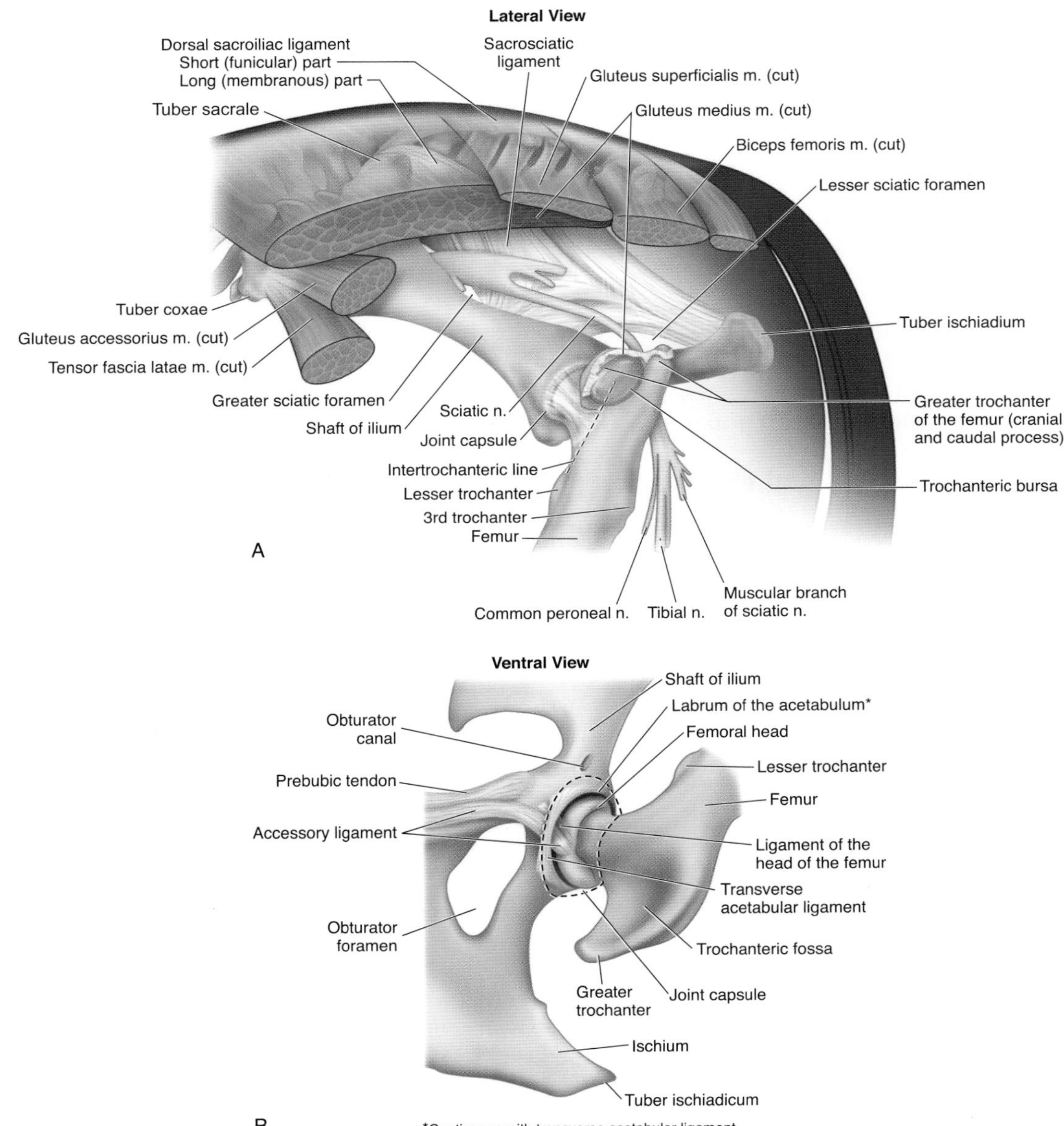

FIGURE 15.1-3 (A) Lateral view of the equine pelvis depicting the superficial muscles, ligaments, and nerves. (B) Ventral view of the equine hemipelvis highlighting the coxofemoral joint and associated ligaments.

Inflammation of the trochanteric bursa is called "whorlbone" in layman's terms. It can be a source of primary lameness but is often considered a secondary condition brought on by abnormal use of the pelvic limb from a stifle or tarsal condition. It is diagnosed by palpation, thermography, and ultrasound and is confirmed by improvement in lameness after injection of local anesthetic into the bursa. Treatment includes addressing issues in the distal limb, injection of the bursa with corticosteroids, shockwave therapy (extracorporeal shockwave therapy/ESWT), and rest.

For centesis of the trochanteric bursa, a 8.9- to 15.2-cm (3.5–6 in.) spinal needle is inserted perpendicular to the skin directly over the cranial aspect of the greater trochanter until the needle contacts the bone. Ultrasound is helpful in guiding the injection to improve accuracy. Synovial fluid is rarely obtained. A local anesthetic can be injected until there is resistance to injection.

among domestic species. The **femoral neck** is less defined in horses as compared to other species. The **greater trochanter** is located on the proximal lateral aspect of the bone with a cranial and caudal pole, which are palpable landmarks on the horse's hip. The **trochanteric fossa** is defined laterally by the ventral caudal part of the greater trochanter and medially by the **lesser trochanter**. A palpable **third trochanter** is on the lateral part of the bone at the junction of the proximal and middle third of the femur.

Trochanteric bursa

The **trochanteric bursa** is located between the tendon of insertion of the **accessory gluteal muscle** and the cranial process of the greater trochanter of the femur.

Regional musculature (see Landscape Figs. 4.0-4–4.0-6)

Pelvic limb muscles, including the hip, can be divided into the gluteal muscle group (Table 15.1-1), caudal thigh muscle group (Table 15.1-2), deep muscles of the coxofemoral joint (Table 15.1-3), and adductor muscles of the thigh (Table 15.1-4). Most function as extensors and flexors of the coxofemoral joint, whereas others play roles in protraction (L. *protrahere* to drag forth), rotation, abduction (laterally), and adduction (medially) of the pelvic limb.

Blood supply, lymphatics, and innervation (see Table 15.1-5)

The major arteries to the pelvic limb branch from the aorta as the **external iliac artery**. This gives off a **deep femoral** arterial branch before becoming the **femoral artery** in the proximal femur. Midfemur, it splits into the **saphenous artery** craniomedially and the **caudal femoral artery**. Another branch of the aorta becomes the **obturator artery** that tracks the obturator nerve passing through the obturator foramen of the caudal ischium. The obturator artery eventually merges with the caudal femoral artery at the level of the stifle in the caudomedial thigh.

TABLE 15.1-1 Gluteal muscle group.

Muscle (innervation)	*Origin*	*Insertion*	*Function*
Superficial gluteal (Cranial and caudal gluteal n.)	Tuber coxae	Third trochanter and fascia lata	Flexes the hip Protract and abduct the limb
Middle gluteal (Cranial gluteal n.)	Longissimus lumborum, gluteal surface of ilium, sacrum, sacroiliac and sacrosciatic ligaments	Greater trochanter	Extends the hip Abducts the limb
Deep gluteal (Cranial gluteal n.)	Ischial spine	Greater trochanter	Abducts the limb
Accessory gluteal (Cranial gluteal n.)	The gluteal surface of the ilium	Distal to the greater trochanter	Extends the hip Abducts the limb
Tensor fasciae latae (Cranial gluteal n.)	Tuber coxae	Patella, lateral patellar ligament, cranial tibia, and third trochanter	Flexes the hip, protracts the limb, extends the stifle

TABLE 15.1-2 Caudal thigh muscle group.

Muscle (innervation)	*Origin*	*Insertion*	*Function*
Semitendinosus (Caudal gluteal and sciatic nn.)	Pelvic head: tuber ischiadicum Vertebral head: last sacral and first 2 caudal vertebrae, sacrosciatic ligament, similar to the origin of the biceps m.	Medial aspect of the tibia and the crural fascia with a tarsal tendon that joins the common calcanean tendon	Extends the hip, stifle, and tarsus while weight-bearing Unweighted, it flexes the stifle; retracts and adducts the limb
Semimembranosus (Caudal gluteal and sciatic nn.)	Pelvic head: ventromedial tuber ischiadicum Vertebral head: first caudal vertebra	Medial condyles of the femur and tibia	Extends the hip and stifle during weight-bearing When unloaded it retracts and adducts the limb inward
Biceps femoris (Caudal gluteal and sciatic nn.)	Pelvic head: tuber ischiadicum Vertebral head: last 3 sacral vertebrae on their spinous and transverse processes	Patella, the lateral and middle patellar ligaments, cranial tibia, and via a tendon on the calcaneus	Extends the hip and stifle, abducts the hindlimb, and extends the tarsus

TABLE 15.1-3 Deep muscles of the coxofemoral joint.

Muscle (innervation)	*Origin*	*Insertion*	*Function*
Gemelli (Sciatic n.)	Dorsal ischium	Trochanteric fossa of the femur	Rotates the thigh outward
Internal obturator (Sciatic n.)	Internal ischium and pubis along the obturator foramen to the pelvic symphysis	Trochanteric fossa of the femur	Rotates the thigh outward
Quadratus femoris (Sciatic n.)	Ventral ischium	Caudal femur near the third trochanter	Extends the hip joint
External obturator (Obturator n.)	Ventral pelvis along the border of obturator foramen	Trochanteric fossa of the femur	Rotates the thigh outward along with limb adduction

TABLE 15.1-4 Adductor muscles of the thigh.

Muscle (innervation)	*Origin*	*Insertion*	*Function*
Gracilis (Obturator n.)	Pelvic symphysis and symphyseal tendon	Medial patellar ligament, cranial tibia	Adducts the limb with some stifle extension
Adductor (Obturator n.)	Ventral pelvis and symphyseal tendon	Caudal femur and medial epicondyle of the femur	Adduction of the limb
Pectineus (Obturator and femoral nn.)	Pubis and iliopubic eminence	Medial surface of the femur	Adducts the limb with flexion of the hip joint

The deep veins of the pelvic limb generally follow the arteries. There are also superficial trunks, which include the **medial** and **lateral saphenous veins**. On either side of the calcanean tendon, the saphenous veins pass between the calcanean tendon and the caudal muscles. The medial saphenous vein unites with the **femoral vein** whereas the lateral saphenous vein unites with the **caudal femoral vein** at the level of the stifle.

Lymphatic flow from the distal limb collects in the **popliteal lymph nodes** in the caudal proximal stifle between the biceps femoris and semitendinosus muscles. It is then carried to the **deep inguinal nodes** in the femoral triangle, which then drain to the **medial iliac nodes**. Additional lymph drainage from the croup and cranial thigh drain to the **subiliac nodes** on the cranial aspect between the tuber coxae and patella and also drain to the medial iliac nodes.

TABLE 15.1-5 Nerves of the lumbosacral plexus innervating the pelvis and pelvic limb.

Lateral cutaneous femoral nerve (L3–4)
- Crosses the dorsal surface of the deep circumflex iliac vessels following its caudal branch through the body via the flank fold
- Innervates the psoas major, skin on the cranial thigh and stifle

Genitofemoral nerve (L2–4)
- This nerve crosses the deep circumflex iliac vessels and passes through the medial portion of the superficial inguinal ring
- Innervates the internal abdominal obliques and cremaster; the skin of prepuce, scrotum, or udder
- The femoral branch innervates the medial skin of the thigh

Femoral nerve (L4–6)
- Supplies sublumbar muscles and quadriceps
- The major branch is the saphenous nerve supplying skin over the medial limb from the thigh to the fetlock
- Injury to the femoral nerve leads to the inability to fix the stifle and support the weight on the affected limb

Obturator nerve (L4–6)
- Innervates adductor muscles of the proximal limb
- Injury (foaling or pelvic fracture) leads to partial to full inability to adduct the limb

Cranial gluteal nerve (L6–S2)
- Travels through the greater sciatic foramen to supply the gluteus medius, piriformis, and accessory (deep) gluteal muscles; and tensor fasciae latae

Sciatic nerve (L5–S2)
- Travels through the greater sciatic foramen of the pelvis passing over the sacrosciatic ligament and tracking deep to the biceps femoris
- Total loss of sciatic function impacts the function of hamstring (tendons of the biceps femoris, semitendinosus, and semimembranosus muscles) and distal limb muscles; however, because the quadriceps enable fixation of the stifle and the reciprocal apparatus between the stifle and tarsus, the horse is able to maintain weight-bearing
- At the level of the hip joint, it divides into the peroneal and tibial nerves, which travel together before dividing at the stifle
- The peroneal nerve provides a cutaneous branch that innervates skin to the lateral limb starting at the level of the biceps femoris
- Peroneal nerve supplies extensor muscle group of the pelvic limb
    - Loss of peroneal n. function causes an inability to extend the digit, causing the hoof to rest on its dorsal surface
- The tibial nerve supplies a cutaneous branch that innervates the plantarolateral tarsus and cannon bone to the fetlock
    - Loss of tibial n. function leads to slight sagging of the tarsus and loss of flexor function of the limb and is of minimal clinical significance

Caudal gluteal nerve (L6–S2)
- Passes through the greater sciatic foramen to the gluteus superficialis, and then to the heads of the biceps femoris and semitendinosus
- Innervates the vertebral heads of biceps, semitendinosus, and semimembranosus muscles

Caudal cutaneous femoral nerve (S1–2)
- Follows the dorsal border of the sciatic n. before turning ventrally into the limb passing over the tuber ischiadicum to end subcutaneously on the caudal surface of the thigh
- Innervates skin on the caudal thigh

Pudendal nerve (S2–4)
- Starts on the medial surface of the sacrosciatic ligament and moves to the lesser sciatic foramen to communicate with the caudal cutaneous femoral nerve
- Innervates cutaneous trunci muscle and cutaneous innervation to the ventrolateral abdominal wall

Caudal rectal nerve (S4–5)
- Passes caudoventrally and supplies sensation to the rectum, anal canal, and perineum as well as motor innervation to the perineal musculature

Innervation of the pelvic limb begins as branches of the lumbosacral plexus and its peripheral nerves (Table 15.1-5). The **cranial** and **caudal gluteal nerves** innervate the lateral group of hindquarter muscles. The **femoral, obturator,** and **sciatic nerves** are the most clinically important: (1) femoral n. supplies the sublumbar muscles with multiple branches innervating the quadriceps muscles before giving off the **saphenous n.** (see Case 15.4, innervation) supplying skin sensation on the medial part of the femoral area distal to the fetlock; it also innervates the sartorius muscle; (2) the obturator n. innervates the adductor muscle group—pectineus, gracilis, and adductor and obturator externus muscles; and (3) the sciatic n. (a.k.a. ischiatic n.) is the largest nerve of the lumbosacral plexus, and it divides at the level of the stifle (L. *genu* knee) into the **tibial** and **common peroneal nerves.** The sciatic n. innervates the caudal hip and thigh muscles with the exception of the obturator externus muscle.

Injury to the sciatic n. results in a partial (paresis) or paralysis of the caudal thigh muscles resulting in an unstable tarsus; however, the pelvic limb's ability to bear weight is unaffected because the femoral nerve innervates the quadriceps femoris muscle.

Selected references

[1] Dyson S. Lumbosacral and pelvic injuries in sports and pleasure horses. In: Diagnosis and management of lameness in the horse. 2nd ed. Saunders; 2011. p. 571–82.
[2] Butler JA, Colles CM, Dyson SJ, Kold SE, Poulos PW, editors. Clinical radiology of the horse. 3rd ed. Wiley-Blackwell; 2008. p. 53–189.
[3] Richardson DW. Femur and pelvis. In: Auer JA, Stick JA, editors. Equine surgery. 4th ed. Saunder; 2012. p. 1442–52.

CASE 15.2

Osteochondritis Dissecans

Sarah DeSante
Steinbeck Peninsula Equine Clinics, Salinas, California, US

Clinical case

History

A 2-year-old Warmblood gelding was evaluated as part of a prepurchase examination. He had been turned out in pasture and received daily ground training.

Physical examination findings

The gelding was noted to have a grade 1/5 left pelvic limb lameness when evaluated on both soft and hard ground. Moderate distention of the left femoropatellar joint could be appreciated when viewed from the side, and the femoropatellar joint capsule was distended and subjectively thickened on palpation. Swelling in the right femoropatellar joint was also appreciated but to a lesser degree. The medial and lateral femorotibial joints palpated normally, and there was no atrophy of the gluteal or quadriceps muscles. The horse was well-behaved for manipulation and flexion of his right pelvic limb but resented a similar examination of his left pelvic limb. The left pelvic limb lameness increased to a 2+/5 during the course of the examination, and the caudal extension of left stifle significantly exaggerated the lameness.

Effusion (L. *effusio* a pouring out) is a common medical term referring to fluid leaking into tissue or an anatomical space—e.g., a joint—as an exudate (high content of protein >3.0 g/dL) or transudate (low content of protein <2.5 g/dL).

Differential diagnoses

Subchondral cystic lesion; osteochondritis dissecans; soft tissue injury to stifle—meniscus, collateral ligament, cruciate ligament, or patellar ligament

Diagnostics

Intraarticular (intrasynovial) anesthesia was not performed in this case due to the palpable distention of the femoropatellar joint and increased lameness with caudal extension of the stifle. Instead, radiographs were performed to evaluate the stifle (Fig. 15.2-1). Irregularity of the subchondral bone along the lateral trochlear ridge could be seen on the caudolateral-craniomedial oblique and lateral views of both the left and right stifles. Ultrasonographic examination was performed to further evaluate the lateral trochlear ridges of each stifle and both were confirmed to have discrete defects in the subchondral bone. Abnormally thick and echogenic ("mottled") cartilage was also seen, separated from mineralized fragments by hypoechoic regions of presumed necrotic tissue (Fig. 15.2-2A).

Comparative Veterinary Anatomy: A Clinical Approach. https://doi.org/10.1016/B978-0-323-91015-6.00082-0

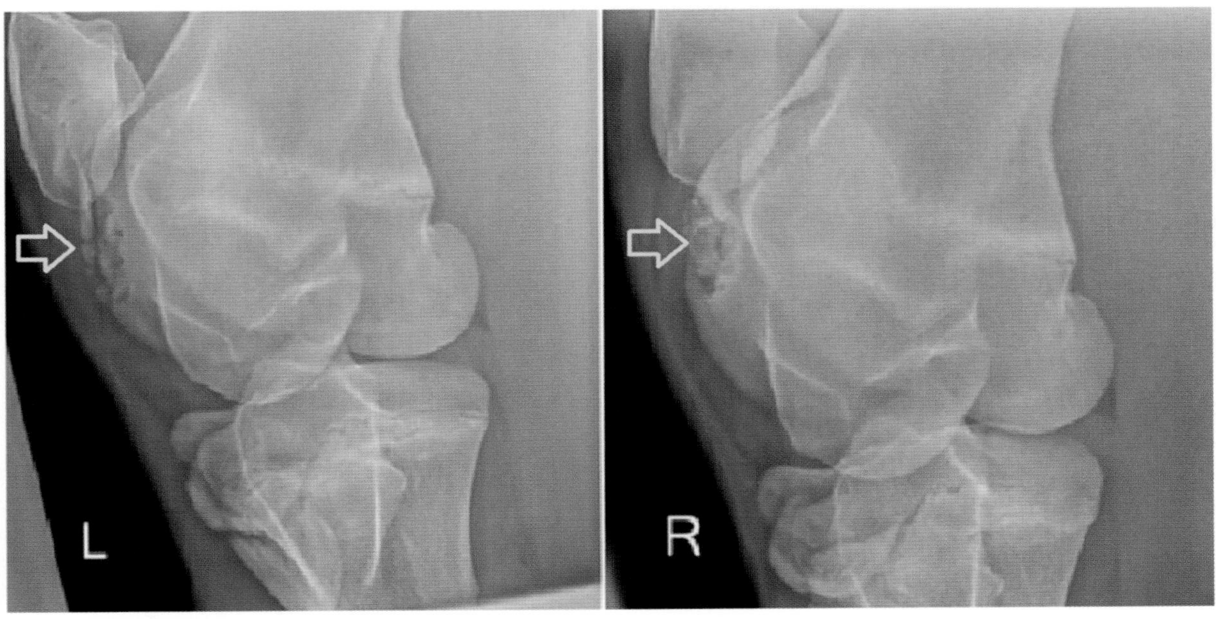

FIGURE 15.2-1 Caudolateral-craniomedial oblique radiographs of both stifles demonstrating multiple OCD lesions on the lateral trochlear ridge of the femur *(yellow arrows)*.

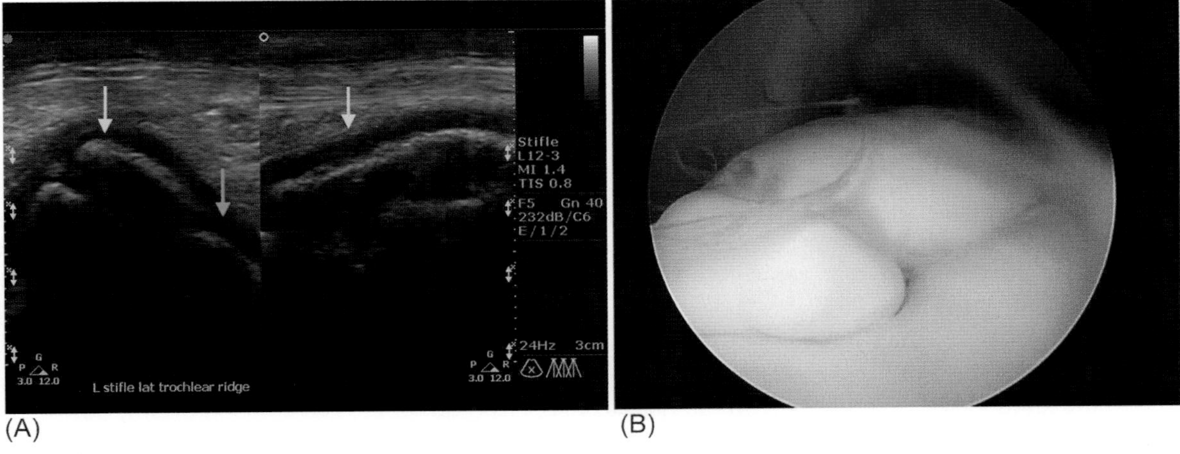

FIGURE 15.2-2 (A) Ultrasound of the lateral trochlear ridge of the stifle in cross-sectional (short-axis) and longitudinal (long-axis) views demonstrating the subchondral bone defect *(yellow arrows)*. The *green arrow* in the cross-sectional (short-axis) view depicts the normal subchondral bone and cartilage adjacent to the lesion. (B) An arthroscopic image of the same stifle before debridement of the lesions.

Diagnosis

Osteochondritis dissecans (OCD) of the lateral trochlear ridge of both femurs

> Osteochondritis dissecans most commonly affects the lateral trochlear ridge of the stifle, but can also be located on the medial trochlear ridge, the trochlear groove, the articular surface of the patella, or in a combination of these locations. Surgical treatment (arthroscopy) to remove these osteochondral fragments is commonly recommended if athletic soundness is the goal.

Treatment

Arthroscopic debridement of the bilateral lesions (Fig. 15.2-2A) was performed under general anesthesia followed by a period of rest and rehabilitation. The horse was treated with a series of intraarticular injections (with interleukin-1 receptor antagonist protein [IRAP]) to help decrease inflammation and improve healing, as well as IV sodium hyaluronate and IM polysulfated glycosaminoglycan therapy.

Anatomical features in equids

Introduction

The stifle anatomy discussed in this section includes the femur, tibia, fibula, patella, stifle joint, ligaments of the stifle joint, and menisci of the stifle joint. Regional musculature, blood supply, innervation, and lymphatics are also covered.

Function

The stifle joint or genual (L. *genu* the knee) joint, commonly referred to as the stifle, is the most complex joint in the body and is part of the pelvic limb of quadrupeds such as the horse. It is the equivalent of the human knee and is the largest synovial joint. Despite its complexity, it functions as a hinge joint undergoing flexion and extension. The stifle joint includes 3 bones—femur, patella, and tibia—and 3 joints—femoropatellar, medial femorotibial, and lateral femorotibial.

The stifle joint consists of the femorotibial articulation between the femoral and tibial condyles, the femoropatellar articulation between the femoral trochlea and the patella, and the proximal tibiofibular articulation between the cranial and caudal ligaments attaching the fibula to the tibia.

The stifle joint has a "locking mechanism" wherein one pelvic limb can support a greater part of the body weight and allow the other pelvic limb to rest while the horse remains standing. The arrangement is an important component in its function as part of the passive stay-apparatus (see Case 15.3).

ARTHROCENTESIS OF THE STIFLE JOINTS

To access the medial femorotibial (MFT) joint a needle is inserted in a depression between the medial patellar ligament and the tendon of the sartorius muscle about 2.5 cm (1 in.) above the medial tibial plateau. Advance the needle about 2.5 cm (1 in.) in a cranial to caudal direction parallel to the ground and a plane that bisects the limb longitudinally.

There are 2 techniques to enter the lateral femorotibial (LFT) joint: (1) the needle is inserted about 2.5 cm (1 in.) in depth, slightly caudal to the lateral patellar ligament, cranial to the long digital extensor tendon, just proximal to the lateral tibial plateau; and (2) the needle is inserted about 2.5 cm (1 in.) in depth, between the lateral collateral ligament and the origin of the long digital extensor tendon, just proximal to the lateral tibial plateau.

There are 3 primary techniques to enter the femoropatellar (FP) joint: (1) a needle is inserted 2.5–3.8 cm (1–1.5 in.) proximal to the tibial tuberosity between the medial and middle patellar ligaments. Direct the needle parallel to the ground if weight-bearing, or proximally (directly under the patella) if the limb is slightly flexed; (2) a needle is inserted perpendicular to the long axis of the limb about 5.1 cm (2 in.) over the lateral tibial plateau, just behind the caudal edge of the lateral patellar ligament/lateral trochlear ridge of the femur, until bone is contacted or synovial fluid is aspirated (not common in normal FP joints). Slightly withdraw the needle before aspirating or injecting; and (3) a needle is inserted just distal to the apex of the patella on either side of the middle patellar ligament.

There is also an approach to enter all 3 joints with one needle insertion. Introduce a 8.9-cm (3.5-in.), 18–20 g. spinal needle 1.9 cm (0.8 in.) proximal to the tibial plateau between the lateral and middle patellar ligaments with the stifle slightly flexed. Direct the needle caudomedially parallel to the tibial crest to enter the MFT joint. Withdraw the needle to the subcutaneous tissue then direct the needle caudolaterally parallel to the tibial crest to enter the LFT joint. Withdraw the needle to the subcutaneous tissue and redirect it proximally and under the patella, entering the FP joint.

Femur (Fig. 15.2-3)

The proximal **femur** articulates with the pelvis (coxofemoral joint; see Case 15.1), whereas the distal femur articulates with the tibia and is part of the stifle joint. The **medial** and **lateral condyles** are located at the caudal, distal end of the femur and are separated by an **intercondylar fossa**, which allows passage of the cruciate ligaments. Both condyles protrude cranially to form **trochlear ridges**. The medial trochlear ridge is wider, extends more proximally, and is more rounded than the lateral trochlear ridge proximally. The trochlea has a wide and deep groove (the **trochlear groove**) between them, which serves as a gliding surface for articulation with the patella.

> Subchondral bone cysts are most commonly observed in the medial femoral condyle (Fig. 15.2-4). They occur most frequently in younger animals, and treatment options include injection of corticosteroids into the cyst, enucleation of the cyst, or placement of a cortical bone screw through the cyst.

> The prominent medial trochlear ridge partly explains why patellar luxation—during which the patella slips out of the trochlear groove to the medial side—is not as common in horses (other than in miniature horses) as it is in other species.

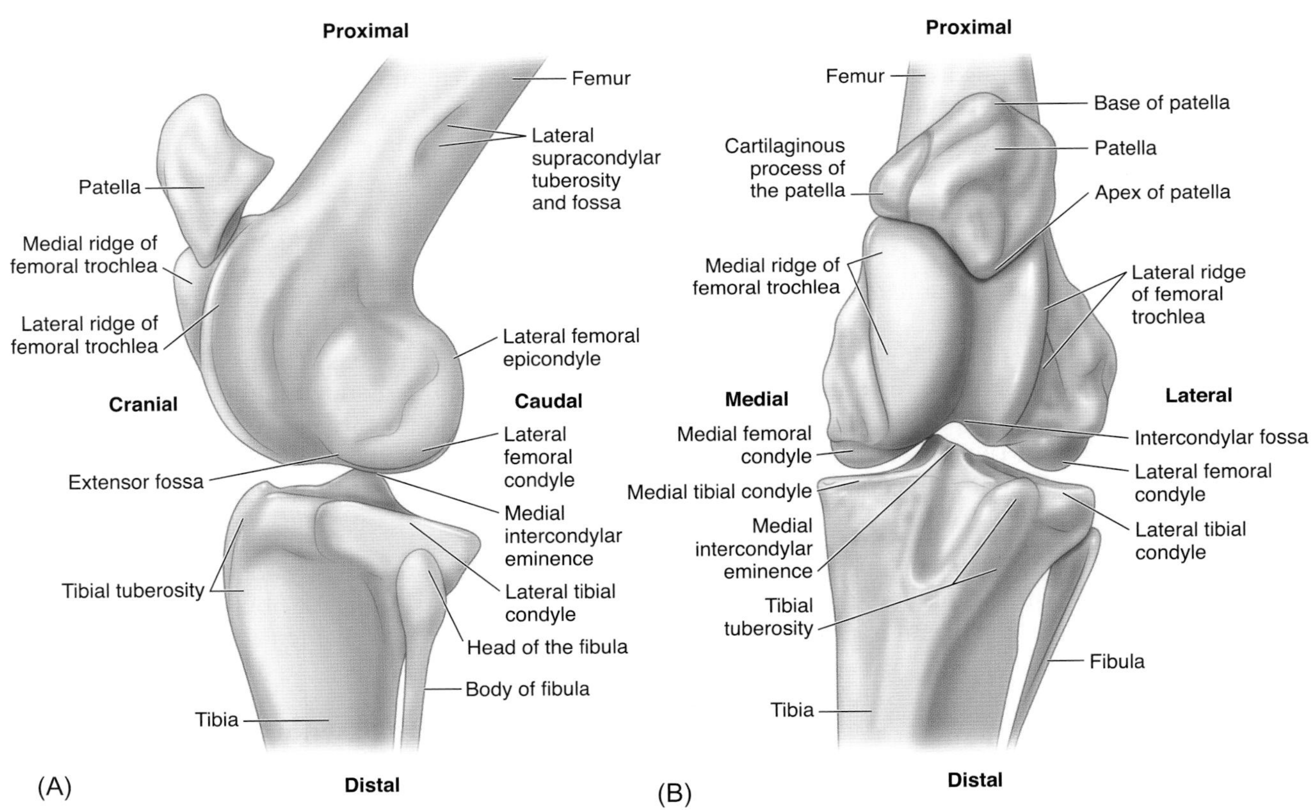

FIGURE 15.2-3 Bones of the equine stifle region. (A) Lateral view and (B) cranial view.

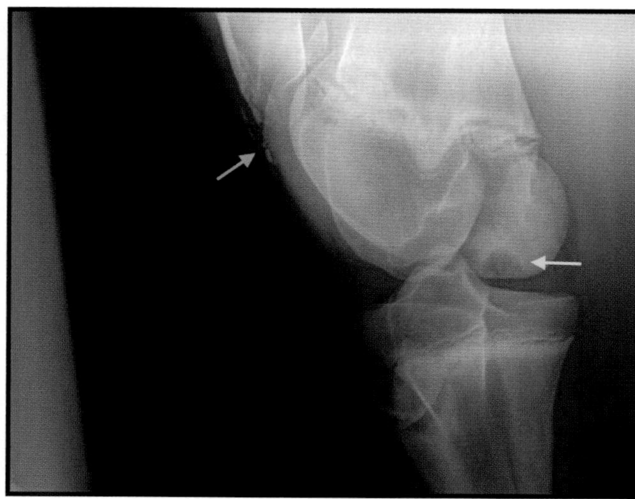

FIGURE 15.2-4 A caudolateral-craniomedial oblique radiograph of a horse's stifle showing a subchondral cyst in the medial femoral condyle *(yellow arrow)* and multiple osteochondral fragments in the femoropatellar joint *(green arrow)*.

Patella (Fig. 15.2-3)

The **patella** is a triangular-shaped sesamoid (Gr. *sēsamon* sesame + *eidos* form) bone that has a proximal **base**, distal **apex**, and a **sagittal ridge** along its articular surface that fits between the trochlear ridges of the femur. It has a plate of fibrocartilage, the **parapatellar fibrocartilage**, along its medial border that curves over the medial ridge of the trochlea. The patella glides proximally on the femoral trochlea when the stifle is in extension and distally when it is in flexion.

Tibia and fibula (Fig. 15.2-3)

The **tibia** has a triangular-shaped articular surface with the femur. The cranial part of the tibia consists of a palpable **cranial tibial tuberosity** where the 3 patellar ligaments insert. There is also a large, palpable **extensor groove** along the craniolateral part of the tibia that is the combined origin of the **fibularis (peroneus) tertius** and **long digital extensor muscles**. A large **medial intercondylar tubercle** and smaller **lateral intercondylar tubercle** sit between the tibial condyles. In foals, the apophysis of the tibial tuberosity is separated from the proximal tibial epiphysis and proximal tibial metaphysis. The physis between the apophysis and epiphysis closes between 9–12 months of age. The physis between the apophysis and metaphysis closes between 30–36 months of age. The proximal tibial physis closes at 24–30 months of age.

The **fibula** is situated along the lateral border of the tibia. It is rudimentary and structurally does not support any weight in the horse. The fibula ends about one-half to two-thirds down the distal tibia; the fibula's its distal end fuses with the tibia to form the **lateral malleolus**. An **interosseous ligament** spans the interface of the fibula and tibia.

Previously, it was believed that lesions within the bones of the stifle were more common than those in its soft tissue counterparts. However, advances in diagnostic imaging—e.g., ultrasound and MRI—and surgery—e.g., arthroscopy—have provided us with the ability to recognize an increasing number of injuries that can occur involving the soft tissue components of the stifle joint.

Regional synovial structures (Fig. 15.2-5)

There are 2 articulations within the stifle: the femoropatellar joint and the femorotibial joint. The stifle consists of 3 synovial compartments: the femoropatellar joint, the medial femorotibial joint, and the lateral femorotibial joint.

The **femoropatellar (FP) joint** is the largest of 3 synovial structures in the stifle and the largest joint in the body. It is formed by the articulation between the patella and the trochleae of the femur. The joint capsule inserts on the abaxial part of the trochlear ridges and extends proximally to form a **suprapatellar recess** under the quadriceps muscle. A large **infrapatellar fat pad** is located cranial to the joint capsule and extends proximal and distal to the patella.

Distention of the FP joint can be prominent and is best seen when standing at the horse's side. Effusion is palpated between the patellar ligaments and distal to the patella, but it may be necessary to compare to the other stifle when the effusion is subtle. One needs to be aware that the infrapatellar fat pad may simulate filling of the femoropatellar joint.

The **medial** and **lateral femorotibial** (MFT and LFT, respectively) **joints** are divided by a **median septum**. The 2 separate compartments do not communicate in a normal joint but may communicate after trauma. The **medial** and **lateral menisci** between the femoral and tibial condyles form a consistent articulation. The joint capsule of the LFT joint extends into the extensor groove of the tibia, beneath the common origin of the long digital extensor and fibularis (peroneus) tertius muscles.

Effusion in the MFT joint can often be palpated when it is present, however, effusion in the LFT joint can be more difficult to appreciate. Effusion in the FT joints can be palpated just cranial or caudal to the collateral ligaments of the stifle. Effusion within the LFT joint may also be palpated underneath the previously described common origin of the long digital extensor tendon and fibularis (peroneus) tertius.

The MFT joint commonly communicates with the FP joint (60–80% of the time) through a slit-like opening at the distal part of the medial trochlear ridge. Communication between the FP joint and the LFT joint is uncommon (occurs about 3% of the time) and occurs at the distal part of the lateral trochlear ridge.

It is important to remember which joints in the stifle communicate because pathology in the MFT joint can manifest as an effusion in the FP joint and vice versa, and because diagnostic anesthesia/analgesia in one joint may affect the other.

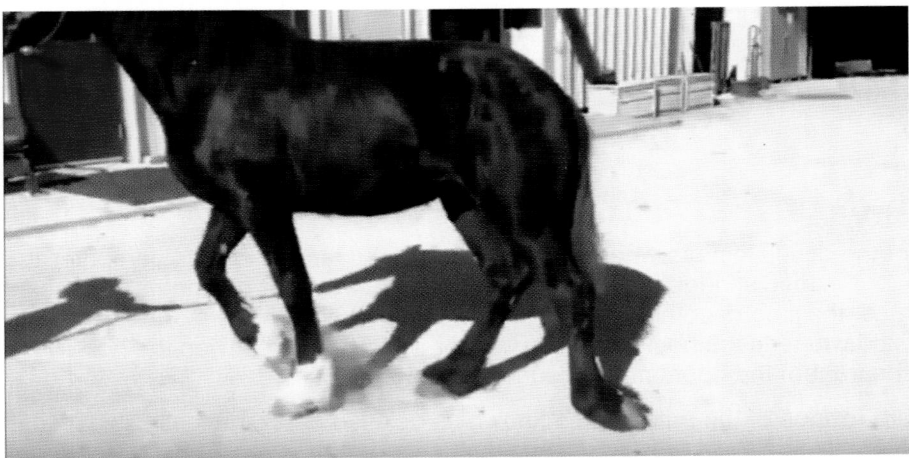

FIGURE 15.2-5 This horse has a complete upward fixation of the patella and is unable to unlock the stifle on its own, causing the horse to drag the toe.

Ligaments of the stifle (Fig. 15.2-6)

The patellar ligaments in horses differ from most other species, including humans, who have just one patellar ligament in the front of the joint connecting the patella to the tibia. Desmitis of the patellar ligaments is not commonly reported and most often occurs in the middle patellar ligament as a result of direct trauma.

Desmotomy of the medial patellar ligament (MPL) is the most common procedure performed to treat intermittent upward fixation of the patella (see side box entitled "Upward fixation of the patella"). However, it can be associated with fragmentation of the patella due to the resultant instability created. An alternative method is MPL splitting, which is believed to enlarge and thicken the MPL, limiting the patella's ability to engage the top of the medial trochlear ridge and "lock" or catch.

The patella has medial, middle, and lateral patellar ligaments that insert on the tibial tuberosity. They are fundamentally tendons of insertion of the **quadriceps femoris** and **biceps femoris muscles** and therefore serve to transmit the action of these muscles to the tibia. The **medial patellar ligament** originates with the **parapatellar fibrocartilage** along the medial and distal border of the patella. It attaches medially to a groove on the cranial aspect of the tibial tuberosity. This ligament is joined by the common **aponeurosis** of the **gracilis** and **sartorius muscles** and serves as an insertion for the **vastus medialis**. The **middle patellar ligament** originates cranially on the patella just proximal to the apex and inserts on the distal part of the groove on the cranial aspect of the tibial tuberosity. The **lateral patellar ligament** originates from the lateral aspect of the patella and inserts on the lateral part of the tibial crest. The biceps femoris muscle has a tendon of insertion that blends with the lateral patellar ligament and together they make up part of the **fascia latae**—the fibrous insertion of the tensor fascia latae muscle on the proximal patella. There are also 2 femoropatellar ligaments that reinforce the joint medially and laterally and help keep the patella in the trochlear groove between the 2 femoral trochleae. These ligaments originate and insert from the patella to the medial and lateral epicondyles of the femur.

Collateral ligament injury is primarily seen in the medial collateral ligament vs. the lateral. This may occur in conjunction with meniscal injury and/or cranial cruciate ligament injury.

The collateral ligaments originate on the medial and lateral epicondyle of the femur. The **medial collateral ligament** inserts on the medial condyle of the tibia with an intermediate attachment to the medial meniscus. The **lateral collateral ligament** lies over the popliteal tendon and inserts on the head of the fibula with no intermediate attachments.

UPWARD FIXATION OF THE PATELLA

Upward fixation of the patella occurs when the medial patellar ligament and its parapatellar fibrocartilage remain hooked over the medial ridge of the femoral trochlea at the start of limb flexion. As a result of the reciprocal apparatus, the limb becomes locked in extension. When complete upward fixation occurs, the fetlock can be flexed whereas the hock and stifle cannot flex. Therefore, when the horse is forced to move forward it drags the front of the hoof on the ground (Fig. 15.2-5).

Intermittent upward fixation of the patella (also known as delayed patellar release) is a milder form of the condition where there is a delay of patellar release during limb protraction. This causes a jerky motion in the limb and is most evident as the horse transitions from a canter to a trot.

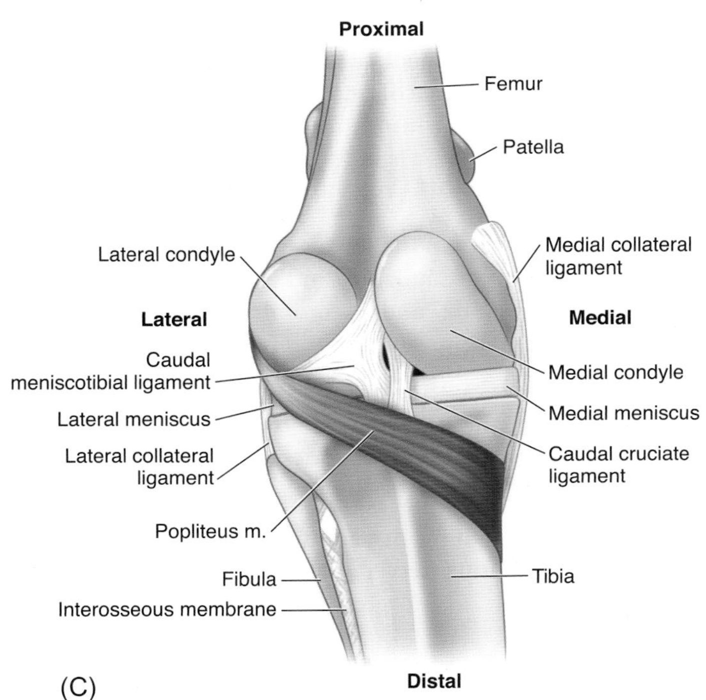

FIGURE 15.2-6 Ligaments and menisci of the equine stifle region. (A) Lateral view, (B) cranial view, and (C) caudal view.

The cruciate ligaments are at the center of the stifle joint and connect the femur to the tibia. These ligaments cross each other to form an "X" and are responsible for the cranio-caudal stability of the joint. They are named by their tibial attachments. The **cranial cruciate ligament** originates craniomedial to the intercondylar eminence of the tibia and inserts caudolaterally in the intercondylar fossa of the femur. The **caudal cruciate ligament** is the more substantial of cruciates. It originates caudomedially on the tibia (from the **popliteal notch**) and inserts craniomedially in the intercondylar fossa of the femur. Although uncommon, injury most often occurs to the cranial cruciate ligament.

Menisci (Fig. 15.2-6)

There are 2 crescent-shaped **menisci** in the stifle joint (medial and lateral) positioned between the femoral and tibial condyles. The menisci are fibrocartilaginous in nature and are thicker peripherally and thinner along their concave edge. The proximal part of each meniscus is concave (to fit the femoral condyles) whereas the distal portion conforms to the tibial surface. During flexion of the stifle, the menisci slide caudally, while they slide cranially during extension.

> Injuries (tears) in the meniscus can be vertical or longitudinal and can occur in both the lateral and medial menisci. However, medial meniscal injuries are more common than lateral meniscal injuries, and the cranial pole is the most common location. This is because the medial meniscus remains in a more fixed position while being subjected to a combination of rotational and sliding forces during motion. Injury often happens in conjunction with injury to the articular cartilage, as well as to the cranial or caudal (less common) meniscal ligaments.

Both the medial and lateral menisci are attached to the tibia via **cranial** and **caudal (meniscotibial) ligaments**. These ligaments attach cranially and caudally to the medial or lateral intercondylar tubercles of the tibia. The medial meniscus gains additional support through an intermediate attachment to the medial collateral ligament. The lateral meniscus gains additional support through the **meniscofemoral ligament**, which passes from the caudal meniscus to the caudal part of the intercondylar fossa of the femur.

Blood supply, lymphatics, and innervation (see Landscape Fig. 4.0-2)

The **femoral artery** becomes the **saphenous artery** in the lower third of the medial femur. As it approaches the stifle joint, it gives off descending vessels into the joint. Caudal to, and at the level of, the femoral condyles, it branches into the **popliteal** and **caudal femoral arteries**. The caudal femoral artery is also fed by the **obturator artery** at this level. The **popliteal**, **caudal femoral**, and **saphenous veins** all drain into the **femoral vein**.

Lymphatic drainage from the distal limb travels on the medial aspect of the tibia and drains into the **popliteal lymph nodes** that are located just proximal to the stifle between the semitendinosus and biceps femoris muscles. Lymphatic drainage continues to the **deep inguinal lymph nodes**.

Innervation of the stifle and the surrounding soft tissues is primarily through the peripheral nerves of the lumbosacral plexus. Three nerves with clinical significance are the **femoral, obturator,** and **sciatic nerves**. The femoral nerve and its major branch, the **saphenous nerve**, innervate the quadriceps muscle. The obturator nerve

KEY CLINICAL FACTS ABOUT REGIONAL PHYSES

- The tibial tuberosity does not fuse until at least 3 years of age—its physis should not be confused with a fracture
- The fibula is not evident on radiographs until about 2 months of age
- Radiolucent lines (from separate centers of ossification) can persist distal to the head of the fibula: 1–3 are possible; do not confuse these with fibular fractures
- In young foals, the margins of the femoral trochleas and the patella are irregular for the first 3 months of life because of incomplete ossification; do not confuse these with an OCD lesion

innervates the adductor muscles—pectineus, gracilis, adductor, and obturator externus muscles. The sciatic nerve supplies branches for 2 important nerves—the **tibial** and **peroneal nerves**. The peroneal nerve supplies skin sensation over the lateral aspect of the pelvic limb through the **lateral cutaneous sural nerve** before dividing at the level of the stifle into the **superficial** and **deep peroneal nerves**. The superficial peroneal nerve innervates the lateral extensors of the limb and the more distal parts of the pelvic limb. The deep peroneal nerve supplies the innervation for the dorsolateral group of muscles of the pelvic limb. The tibial nerve supplies skin sensation over the plantarolateral parts of the hock and metatarsal bones via the **caudal cutaneous sural nerve**, eventually coursing distally to the level of the calcaneus where it divides into the **medial** and **lateral metatarsal nerves**. Before the distal divisions, the tibial nerve supplies branches to the gastrocnemius, popliteus, and caudal muscles in the stifle region.

CLINICAL SEQUELAE TO PERIPHERAL NERVE INJURY

Any injury to the femoral nerve (uncommon) or its branches (e.g., the saphenous n.) results in the inability of the quadriceps muscle to fix the stifle, which results in loss of the affected limb's ability to support the weight of the horse.

Impairment of the obturator n. can occur with pelvic injury or as a sequela to foaling and affects the horse's ability to adduct the limb.

Peroneal nerve injury causes loss of skin sensation on the dorsolateral parts of the distal limb and affects the ability of the horse to extend the digit, causing the hoof to rest on its dorsal surface. Peroneal nerve impairment is usually secondary to intrapelvic injury of the sciatic nerve or occurs at the level of the fibula where the peroneal nerve is superficial and more vulnerable to injury. Tibial nerve injury causes loss of cutaneous and deep sensation distal to the stifle and has a less severe effect on the gait—a mild drop in the tarsus during weight-bearing is seen.

Selected references

[1] Walmsley JP. The stifle. In: Ross MW, Dyson S, editors. Diagnosis and management of lameness in the horse. 2nd ed. Saunders; 2011. p. 533–49.
[2] McIlwaith CW. In: Nixon AJ, Wright IM, editors. Diagnostic and surgical arthroscopy in the horse. 4th ed. Elsevier; 2015. p. 175–242.
[3] Butler JA, Colles CM, Dyson SJ, Kold SE, Poulos PW, editors. Clinical radiology of the horse. 3rd ed. Wiley-Blackwell; 2008. p. 363–412.

CASE 15.3

Disruption of the Fibularis (Peroneus) Tertius

Sarah DeSante and Nick Carlson
Steinbeck Peninsula Equine Clinics, Salinas, California, US

Clinical case

History

A 28-year-old Arabian mare was seen to slip and fall in pasture. The mare was guarding her left pelvic limb following the incident; a veterinarian was called to evaluate the acute lameness.

Physical examination findings

The mare was fully weight-bearing on her left pelvic limb on presentation and showed a grade 4/5 lameness at the walk. She was noted to have little flexion of the tarsus (hock) during the cranial phase of the stride. No heat, pain, or swelling was appreciated in the limb, and the digital pulses were normal; there was no sensitivity to hoof testers. The mare did not resent manipulation of the limb; however, the tarsus could be extended when the stifle was flexed. Indentation of the superficial digital flexor tendon was apparent with this maneuver (Fig. 15.3-1).

Differential diagnoses

Fibularis (peroneus) tertius tear

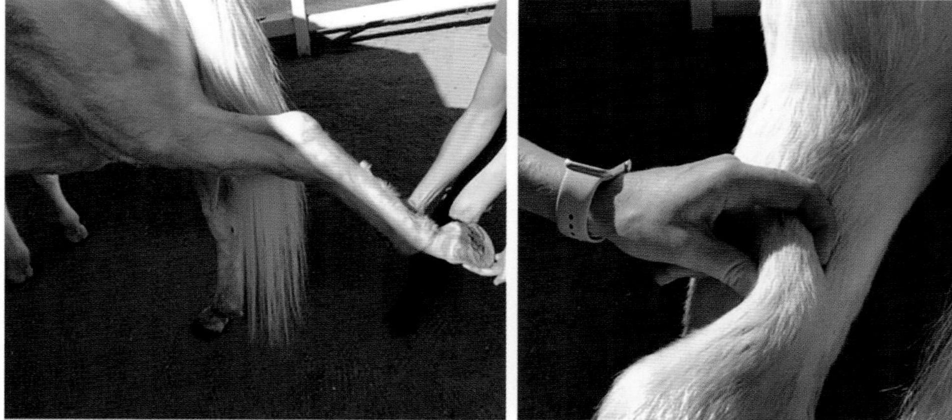

FIGURE 15.3-1 Clinical findings of a torn fibularis (peroneus) tertius showing how the veterinarian is able to flex the stifle and extend the hock (left) with laxity of the combined gastrocnemius-superficial digital flexor tendon proximal to the crus (right) due to loss of the reciprocal apparatus.

Comparative Veterinary Anatomy: A Clinical Approach. https://doi.org/10.1016/B978-0-323-91015-6.00083-2

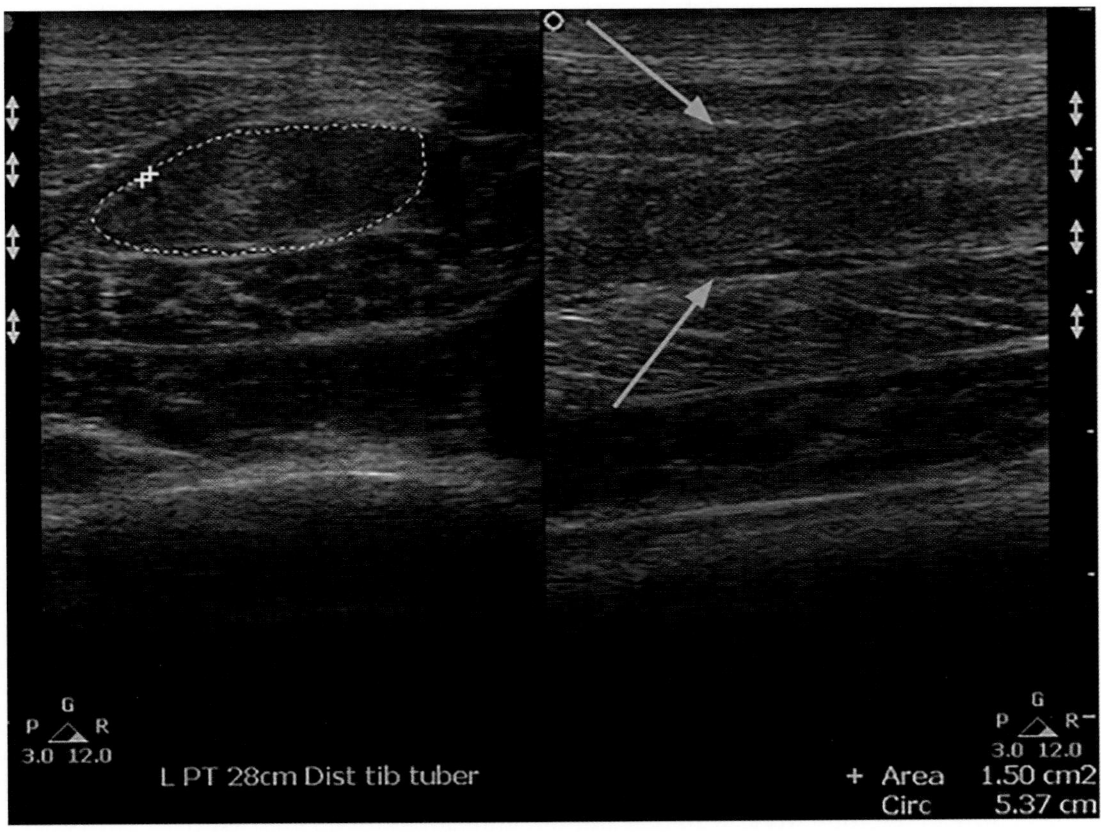

FIGURE 15.3-2 Injury to the fibularis (peroneus) tertius shown ultrasonographically. The short-axis view (left) characterizes the fibularis tertius highlighted by the *white dashed line*. Much of the structure is hypoechoic (darker than normal) due to a loss of normal fiber pattern from the injury. The loss of normal linear fiber structure can be further appreciated on the long-axis view (right) with the superficial and deeper borders of the fibularis tertius highlighted by the *green arrows*.

Diagnostics

Radiographs of the stifle were normal. Anti-inflammatory medication and stall rest were initiated; the stifle and crus were ultrasonographically examined the following day. The fibularis (peroneus) tertius was enlarged at the midbody with a "wavy" and mildly hypoechoic fiber pattern (Fig. 15.3-2).

Diagnosis

Midbody tear of the fibularis (peroneus) tertius

RECIPROCAL APPARATUS (FIG. 15.3-3)

The fibularis (peroneus) tertius and the superficial digital flexor tendon form the reciprocal apparatus. When the stifle is extended, it applies a pull on the superficial digital flexor tendon—a reciprocal or shared movement. Since this structure originates on the femur (see Table 15.3-1 "Muscles of the stifle") and inserts on the tarsus (hock), this pull causes the tarsus to extend concurrently with the stifle. When the stifle flexes, the fibularis tertius mechanically flexes the tarsus because it tracks from the cranial, lateral aspect of the femur to the caudal, lateral aspect of the tarsus and metatarsus.

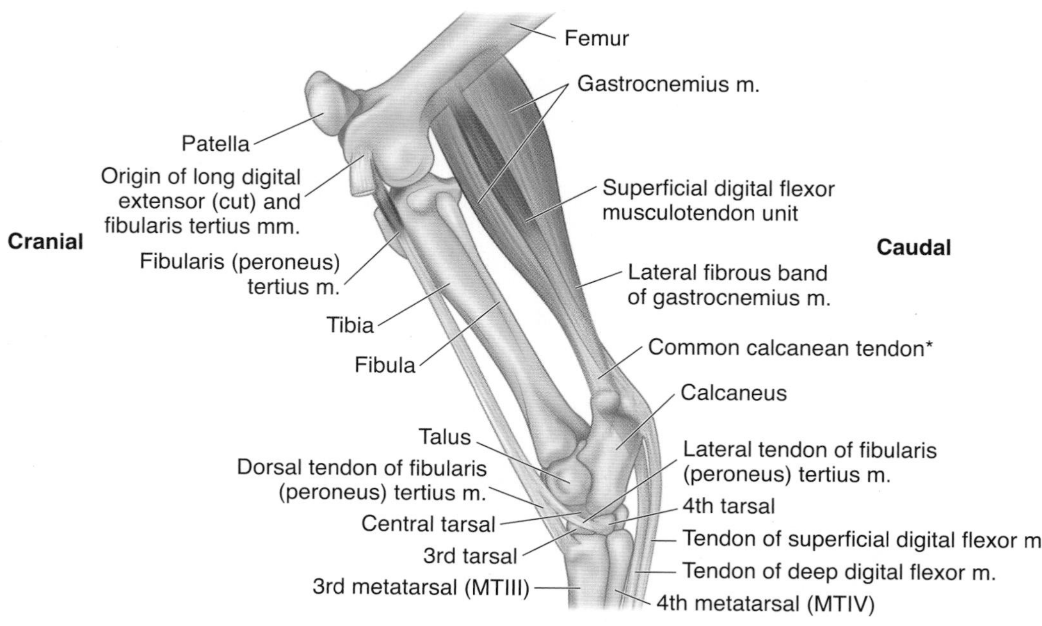

FIGURE 15.3-3 Lateral view of the reciprocal apparatus of the equine pelvic limb.

Treatment

The mare was kept on stall rest for 5 weeks with oral anti-inflammatories for the first few days. The lameness improved within the first week. A follow-up ultrasound examination revealed the tendon remained increased in size, but the fiber pattern had improved. The mare was already retired and sound at the walk, so no further evaluations or treatments were recommended.

Anatomical features in equids

Introduction

This section covers the muscles of the pelvic limb (Tables 15.3-1–15.3-4). The bones, nerves, blood vessels, and joints of the femur and stifle area are discussed in Case 15.2. This case is important because it involves an understanding of the reciprocal apparatus for a clinical diagnosis. While there are only a few differentials that allow flexion of the stifle with extension of the tarsus, it is important to understand the related surrounding musculature to better understand how injury to each structure may affect overall limb function.

The affected muscles in this case are depicted in the Landscape Figs. 4.0-4–4.0-6.

TABLE 15.3-1 Muscles associated with the equine stifle.

Muscle	*Origin*	*Insertion*	*Function*	*Innervation*
Fibularis (peroneus) tertius	Between the lateral trochlear ridge and the lateral condyle of the femur (in the extensor fossa) through a common tendon with the long digital extensor tendon	Tracks cranially through the extensor sulcus of the tibia before inserting on the 3rd tarsal and 3rd metatarsal bones as well as the 4th tarsal bone and calcaneus	Flexes stifle and tarsus	Fibular (peroneal) n.
Superficial digital flexor tendon	Supracondylar fossa of the femur	Tuber calcanei; 1st and 2nd phalanges	Extends stifle and tarsus; flexes digit	Tibial n.

TABLE 15.3-2 Extensors of the equine stifle.

Muscle	Origin	Insertion	Function	Innervation
Tensor fasciae latae	Tuber coxae	Indirectly (through the fascia latae) to the lateral patella, lateral patellar ligament, and cranial border of the tibia	Extends stifle; flexes hip	Cranial gluteal n.
Biceps femoris	Vertebral head: spinous and transverse processes of the last 3 sacral vertebrae; sacrosciatic ligament; tail fascia	Patella; lateral and middle patellar ligaments; cranial border of tibia; crural fascia	Extends stifle	Caudal gluteal n.; sciatic n
Quadriceps femoris				
a. Rectus femoris	Body of ilium cranial and dorsal to acetabulum	Base and cranial surface of the patella	Extends stifle and flexes hip	Femoral n.
b. Vastus lateralis	Proximolateral on femur	Craniolateral patella; tendon of the rectus femoris	Extends stifle	Femoral n.
c. Vastus medialis	Proximomedial on femur	Medial border of the patella and its parapatellar fibrocartilage; proximal medial patellar ligament; tendon of rectus femoris	Extends stifle	Femoral n.
d. Vastus intermedius	Proximodorsal on femur	Base of patella; femoropatellar joint capsule	Extends stifle; raises the femoropatellar joint capsule during extension	Femoral n.

TABLE 15.3-3 Flexors of the equine stifle.

Muscle	Origin	Insertion	Function	Innervation
Popliteus	Lateral condyle of femur	Caudomedial border of tibia	Flexes stifle	Tibial n.
Gastrocnemius	Medial and lateral heads from corresponding supracondylar tuberosities of femur	As part of the common calcanean tendon on the tuber calcanei	Flexes stifle	Tibial n.
Biceps femoris	Pelvic head: tuber ischiadicum	On the calcaneus via its tarsal tendon	Flexes stifle (with this caudal division)	Caudal gluteal n.; sciatic n.

TABLE 15.3-4 Primary hip flexors and extensors with stifle flexion and rotation.

Muscle	Origin	Insertion	Function	Innervation
Semitendinosus	Vertebral head: last sacral and first 2 caudal vertebrae; sacrosciatic ligament; tail fascia Pelvic head: ventral aspect of tuber ischiadicum	Cranial border of tibia; crural fascia; on calcaneus via its tarsal tendon	Limb supporting weight: extends hip, stifle, tarsus Limb not supporting weight: flexes stifle and rotates it medially	Caudal gluteal n.; sciatic n.
Semimembranosus	Vertebral head: first caudal vertebra; sacrosciatic ligament Pelvic head: ventromedial aspect of tuber ischiadicum	Medial epicondyle of femur	Limb supporting weight: extends hip, stifle Limb not supporting weight: retracts, adducts, and rotates the limb inward	Caudal gluteal n.; sciatic n.

The fibularis tertius (formerly called "peroneus tertius") can tear anywhere along its course and is usually due to overextension of the hock. Injury at the origin of the muscle can cause an avulsion fracture in the extensor fossa and associated swelling in the femoropatellar joint. Radiographs are helpful to rule out an avulsion fracture.

Function

The fibularis (peroneus) tertius virtually all tendon in its formation. It is important as the connection between the stifle and tarsus and as a part of the reciprocal apparatus. Injury to the fibularis (peroneus) tertius, resulting in a separation or division of the tendon fibers, leads to the classic uncoupling in the ability of the tarsus and stifle to flex simultaneously. A rupture of the fibularis (peroneus) tertius on physical examination demonstrates the characteristic finding of the ability to extend the tarsus while flexing the stifle.

Other muscles related to the stifle (due to the location of their insertion)

The **sartorius muscle** attaches to the medial patellar ligament and the tibial tuberosity. Its function is primarily to flex the hip and adduct the limb. The **gracilis muscle** attaches to the medial patellar ligament and the medial collateral ligament; its function is to adduct the limb. The **adductor muscle** inserts proximally on the femur, on the medial epicondyle of the femur, and on the medial collateral ligament, and it functions to adduct the limb, extend the hip, and rotate the femur medially.

Blood supply, lymphatics, and innervation

Please see Case 15.2 and Landscape Figs. 4.0-2 and 4.0-6 for this information.

STAY APPARATUS—PELVIC LIMB (FIG. 15.3-4)

The stay apparatus in the pelvic limb prevents flexion of the stifle and tarsal joints, and overextension of the metatarsophalangeal joint, when the horse is at rest. This allows the horse to fully weight-bear on its pelvic limb with minimal muscular effort. The quadriceps femoris muscle is relatively relaxed in this position.

When a horse is standing squarely on both pelvic limbs, the quadriceps femoris and tensor fasciae latae contract to pull the patella, parapatellar fibrocartilage, and medial patellar ligament proximally over the medial trochlear ridge of the femur. The patella rests at the distal end of the trochlea when in this position. When the horse rests one pelvic limb on the toe, the patella in the supporting (other) limb rotates medially and the medial patellar ligament—with its fibrocartilage—slides further caudally on the trochlea to fully "lock" it in place. Because of the reciprocal apparatus, the tarsus also becomes locked in extension. The metatarsophalangeal joint is fixed in extension during rest and is supported by the superficial digital flexor tendon (SDFT), the deep digital flexor tendon (DDFT), the suspensory apparatus, and the distal sesamoidean ligaments. The patella is released by contraction of the quadriceps femoris and the lateral pull from the tensor fasciae latae and biceps femoris muscles.

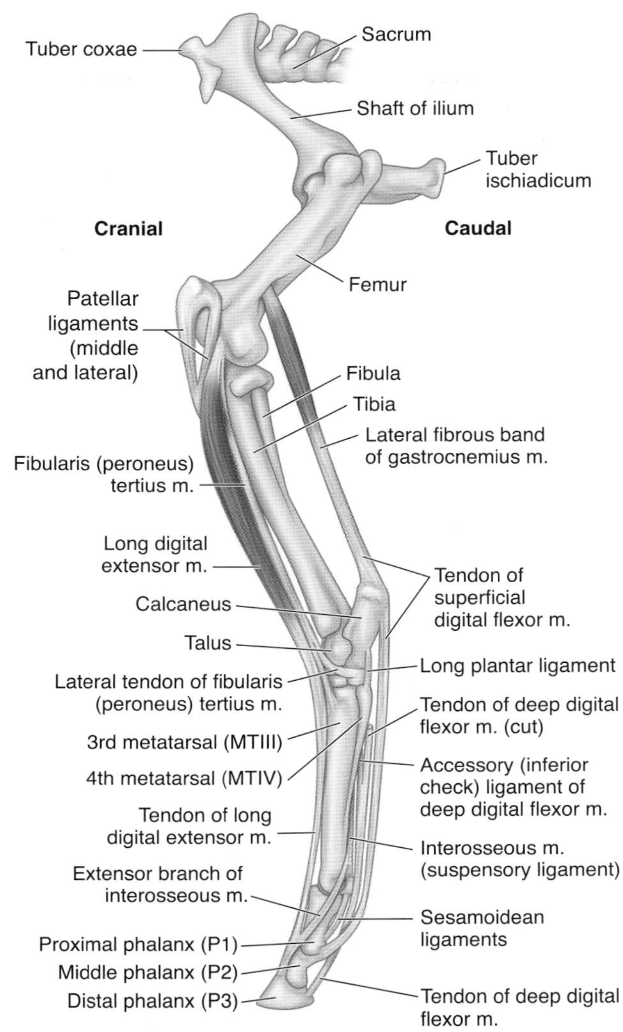

FIGURE 15.3-4 Lateral view of the stay apparatus of the equine pelvic limb.

STAY APPARATUS—THORACIC LIMB (SEE FIG. 14.5-4)

Like the pelvic limb, the stay apparatus in the thoracic limb allows the horse to weight-bear with decreased muscular effort. In the proximal limb, the scapula is suspended from the thorax via the sling of muscle called the ventral serratus muscles. The opposing actions of the muscle-tendon units of the biceps brachii and triceps help fix the shoulder and elbow joints. Further distally, the opposing actions of the extensor carpi radialis and the SDFT and DDFT and their accessory ligaments enable fixation of the carpus. In the distal limb, the metacarpophalangeal joint is fixed in extension during rest and is supported by the SDFT, the DDFT, the suspensory apparatus, and the distal sesamoidean ligaments like in the pelvic limb.

Selected references

[1] Baxter GM. Adams and Stashak's lameness in horses. 6th ed. Wiley-Blackwell; 2011.
[2] Dyce KM, Sack WO, Wensing CJG. Textbook of veterinary anatomy. 4th ed. Elsevier; 2010.
[3] Hinchcliff KW, Kaneps AJ, Geor RJ. Equine sports medicine and surgery. 2nd ed. Elsevier; 2014.
[4] Orsini JA, Divers TJ, editors. Equine emergencies: treatment and procedures. 4th ed. Elsevier; 2014.
[5] Ross MW, Dyson SJ. Diagnosis and management of lameness in the horse. 2nd ed. Elsevier; 2011.

CHAPTER 15

CASE 15.4

Gastrocnemius Tendonitis

Sarah DeSante and Nick Carlson
Steinbeck Peninsula Equine Clinics, Salinas, California, US

Clinical case

History

A 1.5-year-old Gypsy Vanner stallion presented for evaluation of an acute severe left pelvic limb lameness of unknown origin.

Physical examination findings

The stallion was grade 4/5 lame on the left pelvic limb at presentation with significant swelling over the calcaneus (Fig. 15.4-1).

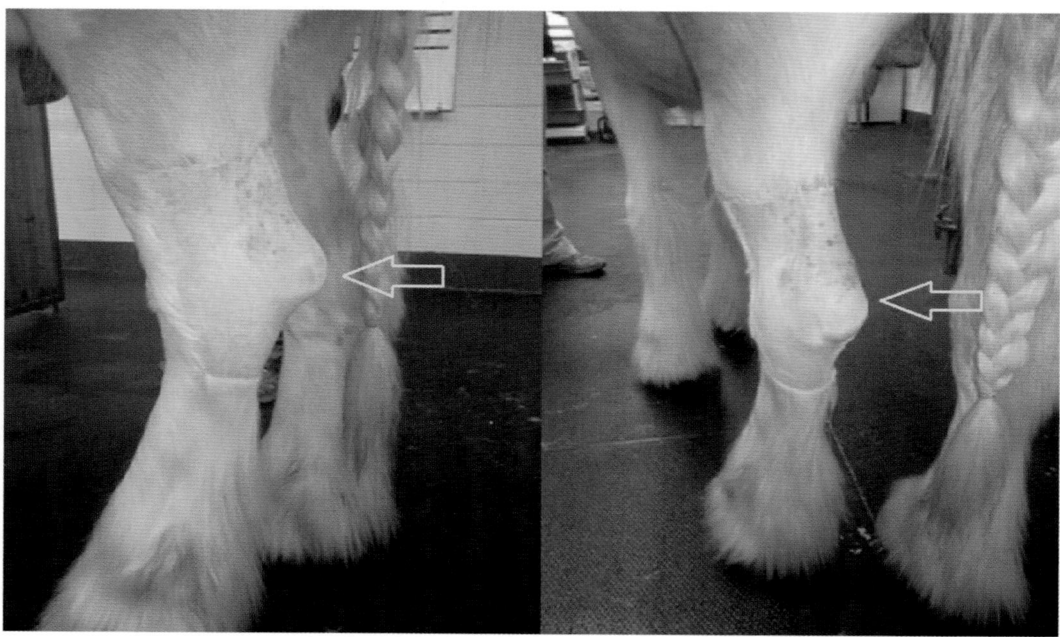

FIGURE 15.4-1 Swelling *(yellow arrow)* proximal to and at the level of the calcaneus, lateral view (left) and caudal view (right). Hair was clipped for ultrasonographic examination.

Comparative Veterinary Anatomy: A Clinical Approach. https://doi.org/10.1016/B978-0-323-91015-6.00084-4

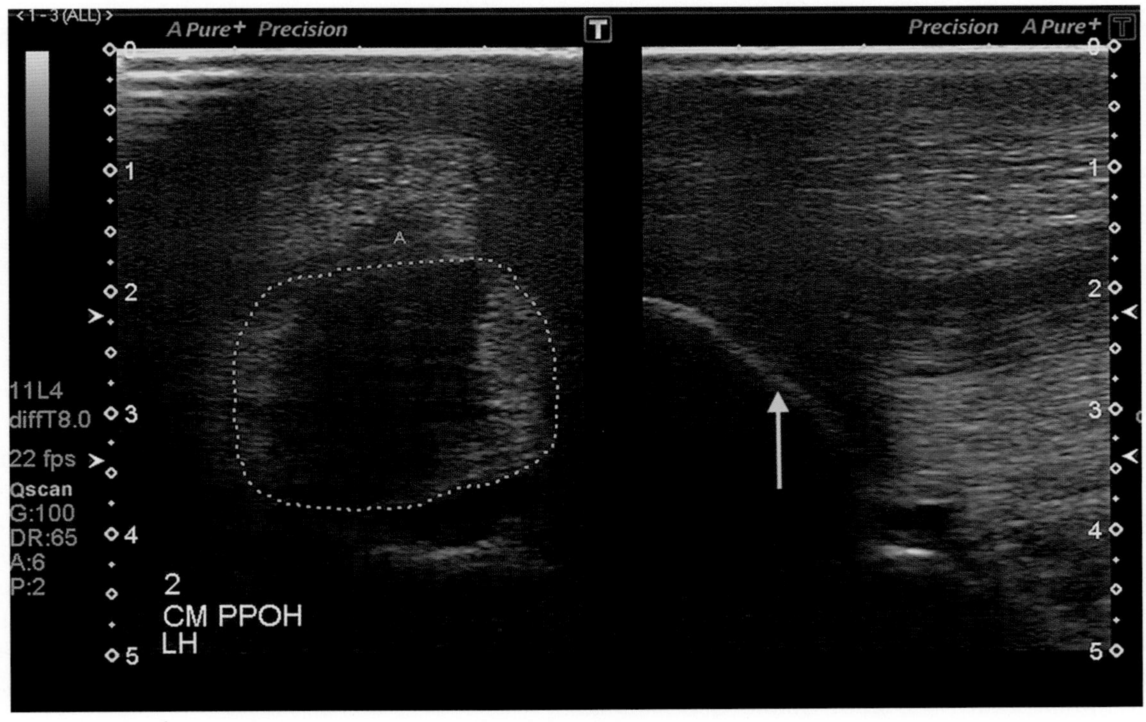

FIGURE 15.4-2 Ultrasonographic images (short-axis on the left and longitudinal axis on the right) showing a hypoechoic core lesion through a large percentage of the gastrocnemius tendon at its insertion on the calcaneus (the tendon is outlined in *blue*). The longitudinal axis view depicts the loss of fibers at the tendon's insertion on the calcaneus *(yellow arrow)*. The superficial digital flexor tendon is superficial to the gastrocnemius at this level.

Differential diagnoses

Superficial digital flexor injury; gastrocnemius injury; bursitis of the intertendinous calcanean bursa, gastrocnemius calcaneal bursa, or subcutaneous calcaneal bursa

Diagnostics

Ultrasound of the region identified significant enlargement of the distal one-third of the gastrocnemius tendon, particularly at its insertion on the tuber calcanei. There was a large hypoechoic region proximal to the point of the hock with a corresponding disrupted fiber pattern. There was an almost complete loss of linear fibers at the tendon's insertion on the tuber calcanei (Fig. 15.4-2).

Diagnosis

Severe gastrocnemius tendonitis (incomplete rupture)

Treatment

The horse was treated with stall rest, nonsteroidal anti-inflammatory drugs (phenylbutazone), and cryotherapy (cold treatment). Hand-walking was gradually increased over several months. He was re-evaluated 5 months later and remained grade 1/5 lame on the left pelvic limb.

Anatomical features in equids

Introduction

Important structures of the equine tarsus (hock) discussed in this case include the distal tibia; the tarsal bones; the tarsal joints (tarsocrural or tibiotarsal, proximal intertarsal, distal intertarsal, and tarsometatarsal); tendons of the gastrocnemius, superficial digital flexor, deep digital flexor, tibialis cranialis, fibularis (peroneus) tertius, and tibialis caudalis; cunean tendon; long plantar ligament; the calcaneal bursa and other regional bursae; tibial and peroneal nerves; and regional blood supply and lymphatics.

Tibial fractures are uncommon in horses and have a guarded prognosis. Depending on the type of tibial fracture, open reduction and internal fixation in conjunction with external coaptation carry the best prognosis. This is similar to radial fractures. It is difficult to use external coaptation alone (e.g., splints, casts) for definitive stabilization/treatment of a complete tibial fracture because of pelvic limb conformation and difficulty in both immobilizing the fracture, and protecting the overlying soft tissues (i.e., blood vessels, nerves, and muscle) from additional injury. The biomechanics of tibial loading reveals that the tension side of the tibia is the craniolateral aspect of the bone. Therefore, if external coaptation is used for emergency treatment, the splint should be placed on the lateral aspect of the limb, essentially dividing the crus into 2 parts between the tarsus (distal part) and stifle (proximal part). The splint should extend the entire length of the pelvic limb. Generally, 2 plates and screws are used for internal fixation (osteosynthesis) in tibial fracture repair.

The distal intermediate ridge of the tibia is the most common site of osteochondrosis dissecans (OCD) in the equine tarsus (Fig. 15.4-3). This is followed by the lateral trochlear ridge, medial malleolus, and medial trochlear ridge in order of frequency.

Function

During standing or rest, the equine tarsus functions as part of the passive stay apparatus (see Case 15.3). During motion, the tarsus helps to provide propulsion (raising and lowering the hoof during the swing phase) or absorb concussion/impact during the weight-bearing phase. Most of the motion in the tarsus occurs at the tarsocrural joint, but there are rotational and sliding movements at the distal tarsal joints.

Tibia

The tibial region is commonly referred to as the crus or "gaskin." The cross-section of the **tibia** in the proximal and midcrus is triangular with muscle coverage on the craniolateral and caudal aspects of the bone. The medial surface of the bone is easily palpable with only skin and subcutaneous coverage. The distal articular surface of the tibia is called the **cochlea**, and it is composed of 2 grooves divided by the **distal intermediate ridge**. Abaxial to the cochlea are the **medial** and **lateral malleoli**. The physes of the medial and lateral malleoli may be incompletely closed at birth. There is a separate center of ossification of the lateral malleolus that is typically united with the tibia by 3 months of age.

Tarsal bones (see Landscape Figs. 4.0-10 and 4.0-11)

The tarsus is comprised of 3 rows of bones—proximal, intermediate, and distal. The proximal row of tarsal bones includes the **talus** and the **calcaneus**. The proximodorsal aspect of the talus has 2 trochlear ridges that articulate with the cochlea of the tibia. The calcaneus shares a synovial joint with the plantar aspect of the talus. The "point of the hock" is the prominent **tuber calcanei**, which serves as an insertion site for the gastrocnemius tendon. The **sustentaculum tali** is on the plantaromedial aspect of the calcaneus and is the surface over which the deep digital flexor tendon tracks. There is a center of ossification at the tuber calcanei that fuses at 16–24 months.

The intermediate row of tarsal bones consists of only the **central tarsal bone,** which is a broad, flat cuboidal bone that articulates between the talus and calcaneus. The distal row of bones is formed plantaromedially by the relatively small **1st and 2nd tarsal bones** (typically fused) followed by the larger **3rd tarsal bone** and the lateral **4th tarsal**

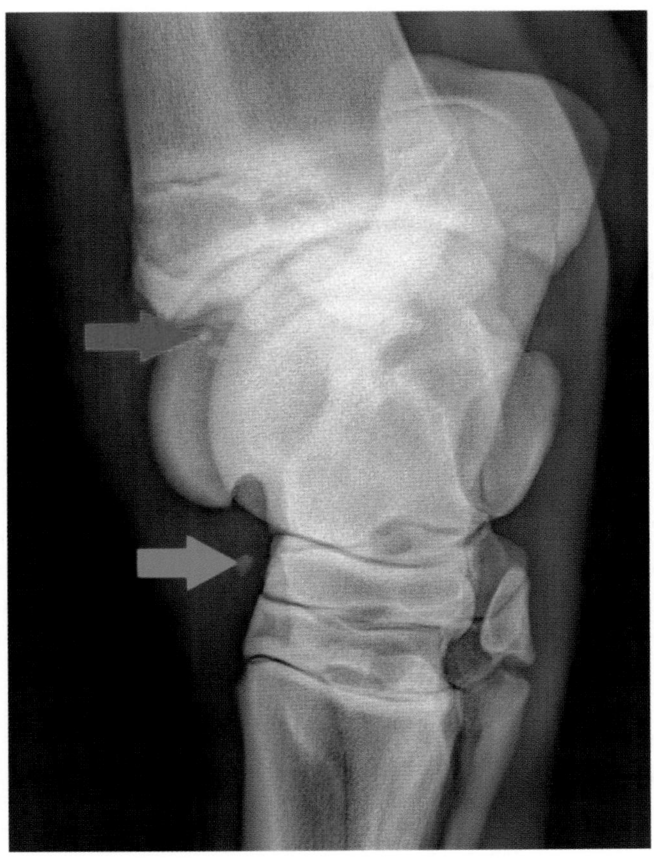

FIGURE 15.4-3 Dorsomedial-plantarolateral oblique radiograph of the tarsus depicting OCD lesions of the distal intermediate ridge of the tibia *(red and green arrows)*. One fragment traveled through the communication between the tarsocrural and proximal intertarsal joints, becoming lodged in the dorsal outpouching of the latter *(green arrow)*.

bone. These bones articulate proximally with the central tarsal bone and distally with the 2nd, 3rd, and 4th metatarsal bones (MTII, III, and IV, respectively). The 4th tarsal bone, physically as large as the combined central and numbered tarsal bones, also articulates with the talus and calcaneus. The intermediate and distal rows of tarsal bones are cuboidal with no physeal growth after birth.

Premature foals with incomplete ossification of the tarsal (or carpal) cuboidal bones may undergo cuboidal bone crushing/collapse if allowed to ambulate before the bones ossify. This results in juvenile-onset osteoarthritis in the joints associated with these bones.

The tarsal bones are held in place by several small **intertarsal ligaments**. Larger **medial** and **lateral collateral ligaments** (MCL and LCL, respectively) of the tarsus originate from the tibial malleoli and have various attachments or insertions on the proximal and distal row of tarsal bones, as well as the 2nd and 4th metatarsal bones. These collateral ligaments each have four components, but the short and long components of each ligament are the most commonly evaluated diagnostically.

The superficial (long) component of the LCL originates from the caudolateral aspect of the distal tibia/lateral malleolus. Its primary insertion is onto the distolateral aspect of the calcaneus. Additional fibers extend to the 4th tarsal and 4th metatarsal (MTIV) bones.

The deep (short) component of the LCL has its primary origin located dorsal to the superficial LCL origin on the distal tibia. It tracks in a nearly transverse direction to its insertion onto the midportion of the calcaneus.

The superficial (long) MCL originates from the midportion of the medial malleolus, and its primary insertion is onto the distal tuberosity of the talus. Additional fibers continue distally toward their insertions onto the central tarsal, 3rd tarsal, and 3rd metatarsal (MTIII) bones.

The deep (short) MCL originates slightly cranial and distal to the superficial MCL origin on the medial malleolus and courses in a slightly transverse orientation toward its primary insertion onto the distomedial surface of the sustentaculum tali of the calcaneus.

Tarsal joints (Fig. 15.4-4)

The joints that comprise the tarsus include the **tarsocrural**, **talocalcaneal**, **proximal intertarsal**, **distal intertarsal**, and **tarsometatarsal**. The tarsocrural joint communicates directly with the proximal intertarsal joint. The proximal and distal intertarsal joints rarely communicate (< 10% of the time). Communication between the distal intertarsal and tarsometatarsal joints occurs more frequently (25–38% of the time).

Regional musculature (see Landscape Figs. 4.0-4–4.0-6 and Tables 15.4-1 and 15.4-2)

The **gastrocnemius muscle** originates from the medial and lateral **supracondylar tuberosities** of the femur. The **soleus muscle** originates on the head of the fibula and joins the lateral head of the gastrocnemius in the caudal midcrus. The tendon of insertion of the gastrocnemius begins superficial to the superficial digital flexor tendon (SDFT), then tracks laterally around the SDFT to a location deep to the SDFT before it inserts on the **tuber calcanei**.

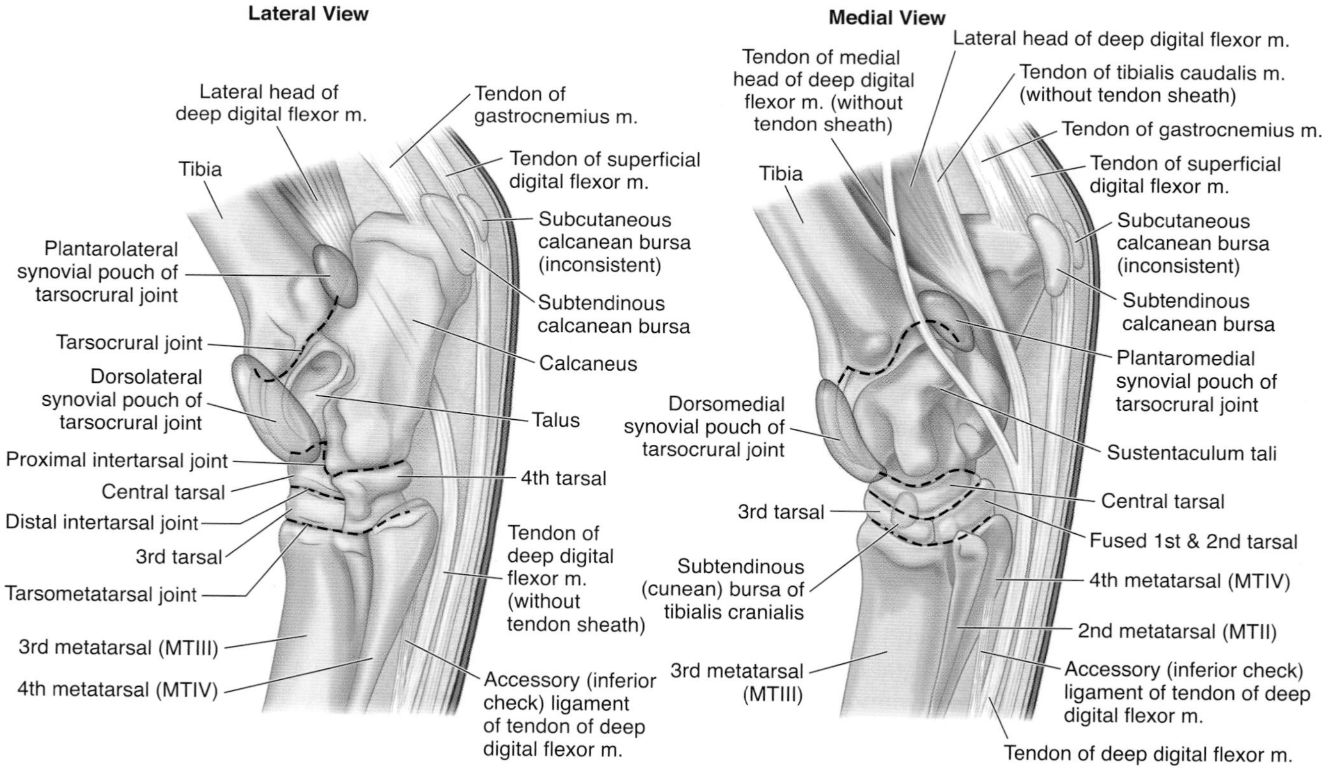

FIGURE 15.4-4 Joints and bursae of the equine tarsus (regional tendon sheaths are shown in Fig. 15.4-5).

TABLE 15.4-1 Flexors of the equine tarsus and extensors of the equine digit.

Muscle	*Origin*	*Insertion*	*Function*
Lateral digital extensor (peroneal n.)	Lateral collateral ligament of the stifle, a portion of the tibia and fibula	Unites with the long digital extensor tendon before inserting on the extensor process of the 3rd phalanx	Extends the digit, flexes the tarsus
Long digital extensor (peroneal n.)	Lateral condyle of the femur	Extensor process of 3rd phalanx with attachments on the dorsal 1st and 2nd phalanges	Extends the digit, flexes the tarsus
Fibularis (peroneus) tertius (peroneal n.)	Lateral condyle of the femur	Branches into 4 parts before inserting on the calcaneus, 3rd tarsal bone, 4th tarsal bone, proximal MTIII, and MTIV	The reciprocal mechanism between the stifle and the tarsus
Tibialis cranialis (peroneal n.)	Lateral condyle and lateral tuberosity of the tibia	The dorsal branch on proximal MTIII; the medial branch continues as the cunean tendon and inserts on the fused 1st and 2nd tarsal bones and MTII	Flexes the tarsus

TABLE 15.4-2 Extensors of the equine tarsus.

Muscle	*Origin*	*Insertion*	*Function*
Deep digital flexor (tibial n.)	Lateral head: caudal surface of tibia Tibialis caudalis: caudal surface of the tibia Medial head: lateral tibial condyle	Plantar aspect of the 3rd phalanx	Extends the tarsus and flexes the digit
Superficial digital flexor (tibial n.)	Supracondylar fossa of the femur	Plantar aspect of the 1st phalanx and collateral tubercles of the 2nd phalanx	The proximal part acts as a member of the reciprocal apparatus, distal part supports fetlock and digits with some digital flexion
Gastrocnemius (tibial n.)	Medial and lateral heads form from medial and lateral supracondylar tuberosities of the femur	Tuber calcanei	Extends the hock and flexes the stifle

The **superficial digital flexor muscle** originates in the **supracondylar fossa** of the caudal femur and is deep to the gastrocnemius tendon in the midcrus. Before it reaches the calcaneus, it spirals around the gastrocnemius tendon to become the more superficial/caudal structure. The SDFT has medial and lateral retinacular attachments as it passes over the point of the hock. It then continues distally in a similar fashion to the thoracic limb (see Case 14.5).

Rupture of the retinacular attachments of the SDFT can result in subluxation or luxation of the tendon from the calcaneus. Luxation more commonly occurs laterally because the medial retinaculum is less fibrous. A medial luxation occurs if the lateral retinaculum fails. Various surgical techniques have been employed to correct this problem with inconsistent results. Conservative management can be successful if complete luxation occurs; however, the horse may have a permanent mechanical lameness with a jerking/yanking motion in the pelvic limb. Partial luxations (in which the SDFT intermittently slides off the calcaneus but can be replaced) that do not stabilize with conservative management may benefit from transection of the remaining retinacular attachments to achieve a permanent luxation.

The **deep digital flexor tendon** (DDFT) is composed of 3 muscle heads—the tibialis caudalis (or superficial head) and the medial and lateral heads. These heads originate on the **lateral condyle** of the fibula and the caudal and medial aspects of the proximal tibia, respectively. The **medial head** of the DDFT tracks medially, and its narrow tendon and sheath pass the tarsus in a groove on the **medial malleolus**. Distal to the hock, its small tendon unites with the singular DDFT. The **lateral head** of the DDFT and **tibialis caudalis muscle** originate from the caudal tibia, and their common tendon passes over the medial tarsus, traveling past the **sustentaculum tali** of the calcaneus. The **tarsal sheath** surrounds this tendon as it tracks over the sustentaculum tali. The medial head of the DDFT has its own small tendon sheath that typically communicates with the larger tarsal sheath. The DDFT continues below the hock in a similar fashion to the thoracic limb, except there is a less well-developed **accessory (inferior) ligament** (see Case 14.5).

The **tibialis cranialis muscle** passes over the cranial tibia deep to the **fibularis (peroneus) tertius muscle** before its tendon of insertion splits into dorsal and medial branches at the level of the talus. At this point, the tendon of insertion tracks over the dorsal branch of the fibularis (peroneus) tertius. The medial branch is known as the **cunean tendon** (see below) (Fig. 15.4-5).

The **fibularis (peroneus) tertius** muscle passes down the cranial tibia before dividing at the tarsus into dorsal and lateral branches (Fig. 15.4-5). It tracks superficial to the **tibialis cranialis** before dividing, after which the dorsal branch tracks deep to the tendon of insertion of the tibialis cranialis. The major function of the fibularis (peroneus) tertius is in the reciprocal mechanism of the pelvic limb, causing the hock to flex when the stifle is in flexion (see side box entitled "Passive stay apparatus" in Case 15.3).

The **lateral digital extensor** muscle originates from the lateral patellar ligament along the proximal lateral tibia and fibula. Its tendon originates just above the tarsus and passes over the dorsolateral aspect of the tarsus before it unites with the long digital extensor tendon in the metacarpus. A tendon sheath envelops this tendon as it passes over the tarsus.

Injury to the lateral digital extensor tendon and its tendon sheath can cause a condition termed "stringhalt" that is characterized by a sudden spasmodic contraction (hyperflexion) of the pelvic limb—most pronounced when the horse is backed or turned tightly on the affected limb. The severity of the condition varies, and proposed etiologies include neurologic, traumatic, or toxin exposure. Surgical treatment consists of transection of the tendon as it tracks through its tendon sheath and eliminates or reduces the clinical signs in some (but not all) cases.

TARSAL ARTHROCENTESIS

Arthrocentesis (aseptic puncture and aspiration of a joint) of the tarsometatarsal joint is approached on the caudolateral aspect above the head of the 4th metatarsal bone (MTIV). The needle is inserted above the head of MTIV (lateral to the SDFT) with the needle directed in a dorsomedial direction and angled slightly distally.

The distal intertarsal joint is typically approached medially (although a lateral approach exists with radiographic guidance). The needle is inserted at the distal border of the cunean tendon into a vaguely palpable concavity between the articulation of the 1st and 2nd tarsal bones and the 3rd and central tarsal bones. The needle is directed laterally and horizontally.

To enter the tarsocrural joint, the needle can be placed into the palpable outpouchings of the dorsolateral, dorsomedial, plantarolateral, or plantaromedial joint space. If using the plantaromedial approach, it is important to differentiate between the tarsocrural joint, tarsal sheath, and calcaneal bursa.

Centesis (Gr. *kentēsis* perforation or tapping) of the tarsal sheath can be performed proximally via a palpable synovial structure (if distended) caudal to the plantaromedial outpouching of the tarsocrural joint. Alternatively, it can be entered distally just below the tarsus on the plantar aspect of the limb. The often-communicating tendon sheath of the medial head of the DDFT can be identified ultrasonographically if centesis of this structure is required.

When distended, there are distinct medial and lateral outpouchings of the calcaneal bursa located just proximal to the tuber calcanei. Small outpouchings can also be identified distal to the tuber calcanei. Needle placement for aspiration of this bursa can occur at any of the palpable outpouchings.

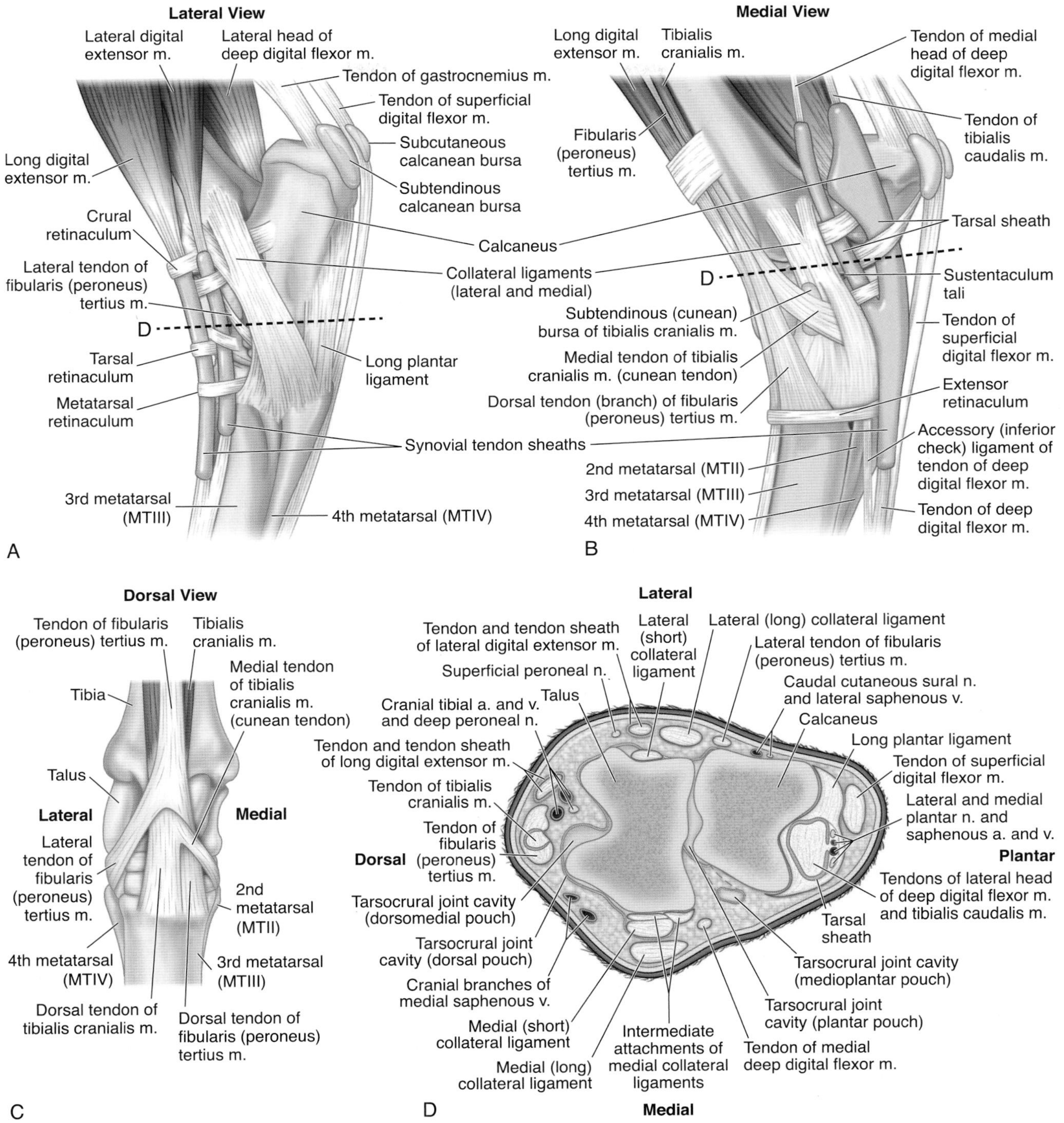

Lateral View
Lateral digital extensor m.
Lateral head of deep digital flexor m.
Tendon of gastrocnemius m.
Tendon of superficial digital flexor m.
Subcutaneous calcanean bursa
Long digital extensor m.
Subtendinous calcanean bursa
Crural retinaculum
Calcaneus
Lateral tendon of fibularis (peroneus) tertius m.
Collateral ligaments (lateral and medial)
D
Tarsal retinaculum
Long plantar ligament
Metatarsal retinaculum
Synovial tendon sheaths
3rd metatarsal (MTIII)
4th metatarsal (MTIV)
A

Medial View
Long digital extensor m.
Tibialis cranialis m.
Tendon of medial head of deep digital flexor m.
Tendon of tibialis caudalis m.
Fibularis (peroneus) tertius m.
Tarsal sheath
D
Sustentaculum tali
Subtendinous (cunean) bursa of tibialis cranialis m.
Tendon of superficial digital flexor m.
Medial tendon of tibialis cranialis m. (cunean tendon)
Extensor retinaculum
Dorsal tendon (branch) of fibularis (peroneus) tertius m.
Accessory (inferior check) ligament of tendon of deep digital flexor m.
2nd metatarsal (MTII)
3rd metatarsal (MTIII)
4th metatarsal (MTIV)
Tendon of deep digital flexor m.
B

Dorsal View
Tendon of fibularis (peroneus) tertius m.
Tibialis cranialis m.
Medial tendon of tibialis cranialis m. (cunean tendon)
Tibia
Talus
Lateral
Medial
Lateral tendon of fibularis (peroneus) tertius m.
2nd metatarsal (MTII)
4th metatarsal (MTIV)
3rd metatarsal (MTIII)
Dorsal tendon of tibialis cranialis m.
Dorsal tendon of fibularis (peroneus) tertius m.
C

Lateral
Tendon and tendon sheath of lateral digital extensor m.
Lateral (short) collateral ligament
Lateral (long) collateral ligament
Lateral tendon of fibularis (peroneus) tertius m.
Superficial peroneal n.
Caudal cutaneous sural n. and lateral saphenous v.
Talus
Cranial tibial a. and v. and deep peroneal n.
Calcaneus
Tendon and tendon sheath of long digital extensor m.
Long plantar ligament
Tendon of superficial digital flexor m.
Tendon of tibialis cranialis m.
Lateral and medial plantar n. and saphenous a. and v.
Tendon of fibularis (peroneus) tertius m.
Dorsal
Plantar
Tendons of lateral head of deep digital flexor m. and tibialis caudalis m.
Tarsocrural joint cavity (dorsomedial pouch)
Tarsal sheath
Tarsocrural joint cavity (dorsal pouch)
Tarsocrural joint cavity (medioplantar pouch)
Cranial branches of medial saphenous v.
Tarsocrural joint cavity (plantar pouch)
Medial (short) collateral ligament
Intermediate attachments of medial collateral ligaments
Tendon of medial deep digital flexor m.
Medial (long) collateral ligament
Medial
D

FIGURE 15.4-5 Equine tarsus. (A) Lateral view of the tarsus depicting the muscles, tendons, ligaments, synovial sheaths, and bursae. (B) Medial view of the same region. (C) Dorsal view of the tarsus depicting the arrangement of the tibialis cranialis and fibularis (peroneus) tersius muscles. (D) Cross-section of the tarsus at the level of the talus and calcaneus including the vessels and nerves. Notice the 3 heads of the deep digital flexor muscle: lateral (largest) and medial heads of the deep digital flexor muscle and tibialis caudalis muscle. (Figure adapted from Dyce.)

Swelling over the point of the hock is commonly referred to as a "capped hock." This swelling may be due to inflammation in the subcutaneous bursa, gastrocnemius bursa, or a false (acquired) bursa created by chronic trauma/rubbing of the area.

Calcaneal bursae (Fig. 15.4-4)

The **calcaneal bursa** is located at the tuber calcanei and lies between the SDFT and gastrocnemius tendons. In most horses, this bursa has a small, direct connection with the **gastrocnemius bursa,** which is located dorsal to the insertion of the gastrocnemius tendon on the tuber calcanei. A **subcutaneous bursa** is located on the tuber calcanei between the SDFT and the skin and shares communication with the calcaneal bursa in 30–50% of horses.

Desmitis of the long plantar ligament is termed "curb." The thickening of the ligament conveys a convex appearance to the plantar aspect of the hock. Ultrasound examination is required to differentiate between lesions of the long plantar ligament vs. the SDFT.

Long plantar ligament (Fig. 15.4-5)

The **long plantar ligament** is a prominent ligament that originates on the plantar aspect of the calcaneus and travels over the 4th tarsal bone before inserting on the 4th metatarsal bone (MTIV).

The cunean bursa is commonly entered during arthrocentesis of the distal intertarsal joint when approached from the medial side. It can be a target of therapeutic injection to alleviate tarsal pain associated with arthritis in the distal joints.

Cunean tendon (Fig. 15.4-5)

The **cunean tendon** originates as part of the tibialis cranialis muscle and inserts on the fused 1st and 2nd tarsal bones. The **cunean bursa** lies between the cunean tendon and the medial collateral ligament at the level of the distal intertarsal joint.

Muscle groups of the tibia and tarsus

See Landscape Figs. 4.0-4–4.0-6 and Tables 15.4-1 and 15.4-2.

Regional innervation (see Landscape Figs. 4.0-2 and 4.0-3)

Horses with tibial neuropathy exhibit hyperflexion of the tarsus without knuckling.

The tibial nerve can be "blocked/anesthetized" for diagnostic or clinical analgesia. Anesthetizing this nerve blocks sensation in the plantar aspect of the distal limb. With the limb held in flexion, the tibial nerve is palpable cranial to the gastrocnemius/superficial digital flexor muscles and caudal to the deep digital flexor muscle. The nerve is anesthetized approximately 1.5 cm (4 in.) proximal to the point of the hock on the medial aspect of the limb.

The **tibial nerve** innervates the extensors of the tarsus and flexors of the digits.
The tibial nerve and its branches provide cutaneous sensation from the caudomedial and caudolateral aspect of the tibia to the distal limb. The dorsomedial tibia and tarsus are supplied by the **saphenous nerve**.

The tibial nerve tracks on the caudomedial aspect of the tibia and—as it passes over the plantar aspect of the calcaneus—divides into the **lateral** and **medial plantar nerves**. The lateral plantar nerve gives off a deep branch in the distal tarsus that becomes the **lateral** and **medial metatarsal nerves**.

The **fibular (peroneal) nerve** innervates the flexors of the tarsus and extensors of the distal limb. The nerve tracks down the dorsolateral aspect of the tibia and tarsus, supplying cutaneous sensation to the dorsolateral aspect of the limb. Below the tarsus, the peroneal nerve separates into the **medial** and **lateral dorsal metatarsal nerves**.

Loss of fibular (peroneal) nerve function causes knuckling of the distal pelvic limb with limited flexion of the tarsus. A peroneal nerve block anesthetizes the dorsal structures of the tarsus and distal limb. Combining a tibial nerve block with the peroneal nerve block eliminates sensation from the tarsus distally. The peroneal nerve is anesthetized approximately 10.2 cm (4 in.) proximal to the tuber calcanei in the groove between the long and lateral digital extensor muscles. The needle is directed into the depression, aiming slightly caudally to contact the caudal edge of the tibia. The superficial branch supplying skin sensation is blocked by continuing to inject anesthetic subcutaneously while removing the needle.

Blood supply and lymphatics (Fig. 15.4-6 and Landscape Fig. 4.0-2)

The **saphenous artery**, a branch of the **femoral artery**, (a continuation of the **external iliac artery**) and **deep** and **caudal femoral arteries** (muscular branches of the femoral artery) join each other and also serve as an alternative blood flow should there be a blockage of the saphenous artery (Fig. 15.4-6). The part of the femoral artery at the level of the caudal stifle, termed the **popliteal artery**, branches into the **cranial** and **caudal tibial arteries**. The cranial tibial artery tracks in a dorsolateral position, becoming the **dorsal pedal artery** at the tarsus and then the **dorsal metatarsal artery** (major blood supply to the foot and a readily available site to take the arterial pulse), eventually becoming the **lateral** and **medial digital arteries** located between the 3rd and 4th metatarsal bones (MTIII and IV), following a similar pattern to that of the thoracic limb. Small branches from the saphenous artery join the dorsal metatarsal artery at the distal 3rd metatarsal bone (MTIII). The caudal tibial artery has a short anastomosing branch with the saphenous at the level of the tarsus, and a longer branch that tracks proximally connecting with the caudal femoral artery. The saphenous artery, supported by the various anastomosing connections, becomes the **lateral** and **medial plantar arteries**, which clinically, along with the deeper plantar metatarsal arteries, either combine with the dorsal metatarsal artery or simply diminish.

Venous drainage is complementary to the arterial supply with important superficial trunks—the **lateral** and **medial saphenous veins**. The medial saphenous vein joins the **femoral vein,** whereas the lateral saphenous vein joins the **caudal femoral vein** at the level of the stifle.

The medial vein is very noticeable on the dorsal medial aspect of the tarsus and should not be confused with an effusion (L. *effusio* a pouring out) of the tarsocrural joint.

Most lymphatic drainage occurs on the medial aspect of the metatarsus between the flexor tendons and continues over the medial tarsus to the **popliteal lymph nodes**. The popliteal lymph nodes are located between the biceps femoris and semitendinosus muscles at a level just proximal to the stifle. These lymph nodes empty into the **inguinal lymph nodes** at the femoral triangle. A second lymphatic system for the croup and cranial thigh empty into the **subiliac lymph nodes**, which are located between the tuber coxae and patella, and flow into the **medial iliac lymph nodes.**

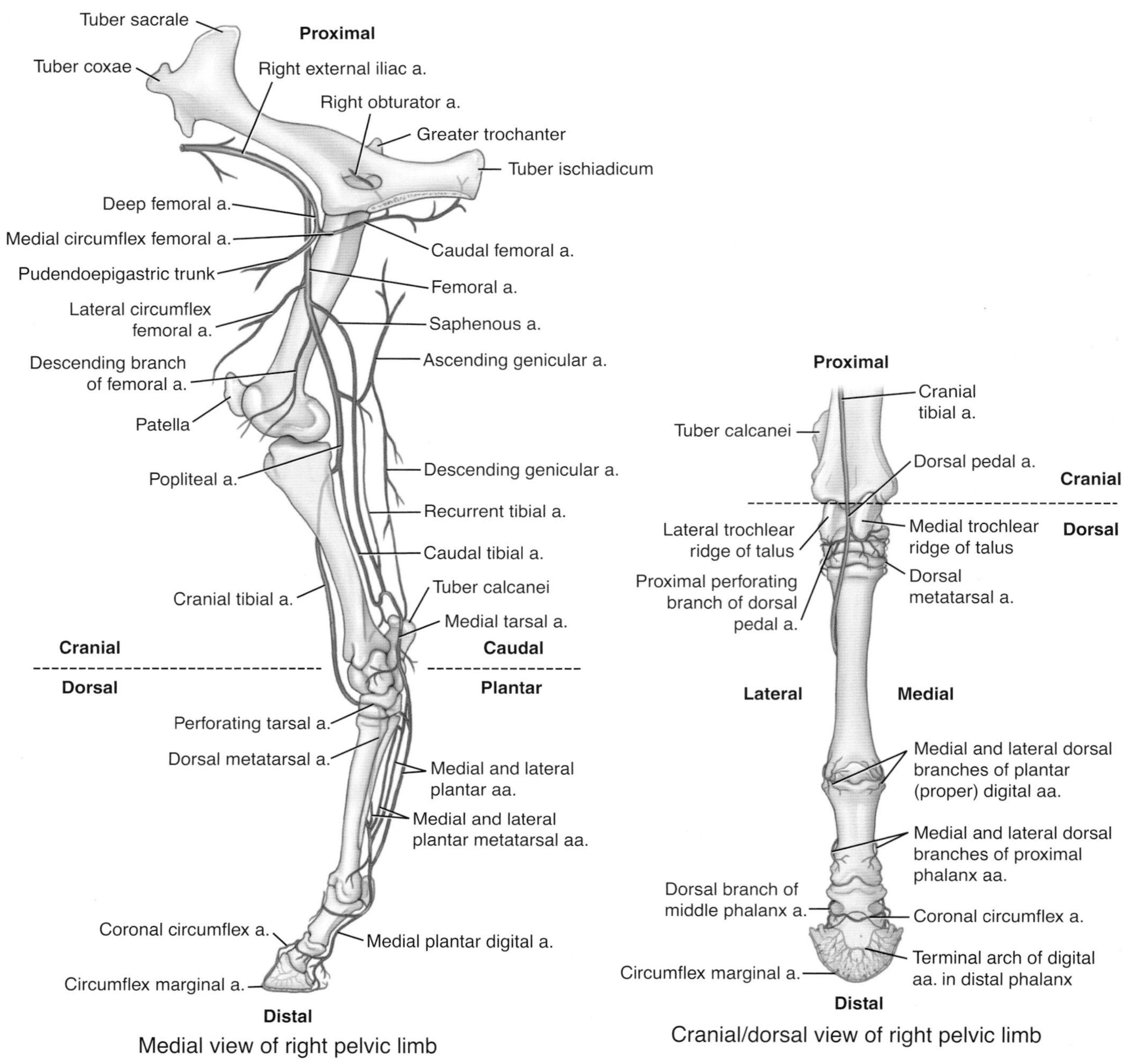

FIGURE 15.4-6 Blood supply of the equine pelvic limb. Left: medial view. Right: cranial/dorsal view.

Selected references

[1] Dyson S, Ross MW. The tarsus. In: Ross MW, Dyson S, editors. Diagnosis and management of lameness in the horse. 2nd ed. Saunders; 2011. p. 508–31.

[2] Butler JA, Colles CM, Dyson SJ, Kold SE, Poulos PW, editors. Clinical radiology of the horse. 3rd ed. Wiley-Blackwell; 2008. p. 321–62.

[3] Auer JA. The tarsus. In: Auer JA, Stick JA, editors. Equine surgery. 4th ed. Saunders; 2012. p. 1388–409.

CASE 15.5

Fracture of the 4th Metatarsal Bone

Nick Carlson
Steinbeck Peninsula Equine Clinics, Salinas, California, US

Clinical case

History

A 7-year-old Warmblood gelding was referred into the clinic for a laceration over the left lateral metatarsus. The owner found the horse that morning and believed the gelding stuck his leg through a fence panel and it got stuck, or a neighboring mare kicked him through the fence.

Physical examination findings

The horse was a grade 4/5 lame on the left pelvic limb but could fully bear weight on the limb. There was a deep laceration on the lateral mid-metatarsus with substantial local swelling. The horse would not tolerate examination of the limb without sedation.

Differential diagnoses

4th metatarsal bone (MTIV) fracture, incomplete fracture of the 3rd metatarsal bone (MTIII), wound sepsis with secondary cellulitis

Diagnostics

Radiographs of the wound revealed an open comminuted fracture of the mid-MTIV (Fig. 15.5-1).

Diagnosis

Open, comminuted, nonarticular fracture of the distal two-thirds of MTIV

Treatment

The patient was placed under general anesthesia, the surgery site aseptically prepared, and the distal two-thirds of MTIV was removed because of the high degree of comminution and contamination. The horse was maintained on systemic antibiotics and anti-inflammatories during the initial recovery period. The horse recovered without complications and returned to full work after 4 months of rehabilitation.

Comparative Veterinary Anatomy: A Clinical Approach. https://doi.org/10.1016/B978-0-323-91015-6.00085-6

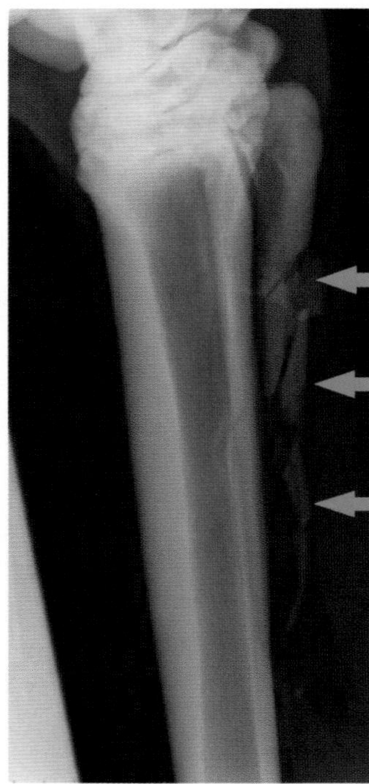

FIGURE 15.5-1 Pre-operative lateral radiograph of a horse with a comminuted fracture of MTIV *(green arrow)*.

Anatomical features in equids

Introduction

The bones of this region include the 3rd metacarpal/-tarsal bone ("cannon bone"; MC/MTIII), the 2nd metacarpal/-tarsal bone (MC/MTII), and the 4th metacarpal/-tarsal bone (MC/MTIV) (the latter 2 being the medial and lateral "splint bones," respectively). Regional blood supply, lymphatics, and innervation are also discussed. The soft tissue structures and the function of the metacarpus/metatarsus are discussed in Case 14.5. The metacarpus and metatarsus—in the thoracic and pelvic limbs, respectively—are anatomically clinically similar except where noted.

Function

Horses only have 3 metacarpal/-tarsal bones, as they are missing both the 1st and 5th bones. The 3rd metacarpal/-tarsal (MC/MTIII) bone is a major weight-bearing structure—heavily loaded and commonly injured—functioning as part of the appendicular skeleton in articulating with the carpus/hock proximally and metacarpo-/metatarsophalangeal joint distally. It also serves as the construct for the important soft tissues of the distal limb (see Case 14.5). The 2nd and 4th metacarpal/-tarsal bones (called "splint bones"; MC/MTII and IV) are vestigial metacarpal/metatarsal bones. Although vestigial, they play important roles in weight-bearing and can be a source of clinical lameness.

The 3rd metacarpal/-tarsal bone (see Landscape Figs. 4.0-10 and 4.0-11)

The third metatarsal bone (MTIII) is slightly longer and rounder in cross-section compared to the more ovoid MCIII. Depending on the horse's discipline, significant remodeling of this bone can occur that can lead to pathologic changes in the bone. The distal aspect of the bone at its articulation with the 1st phalanx has **medial** (wide) and **lateral** (narrow) **condyles** and a **sagittal ridge** that fits into the **sagittal groove** on the proximal **1st phalanx**. The proximal metacarpal/-tarsal physis is closed at birth. The distal metacarpal/-tarsal physis closes radiographically at 3–6 months.

Condylar fractures of MC/MTIII are a common injury in racehorses. The lateral condyle is more frequently involved than the medial condyle, and these fractures are typically repaired with internal fixation—i.e., cortical screw(s). Medial condylar fractures are more complicated, and many of them spiral proximally. They frequently require a combination of plate fixation and cortical screws. Diaphyseal fractures are often traumatic in origin. A successful repair can be accomplished due to relatively easy anatomical (i.e., minimal soft tissues) access to the bone. However, complications are not uncommon with metatarsal fracture repairs because of the absence of soft tissue coverage, meager blood supply, and their frequent comminuted nature. External coaptation in the form of a cast or caudal splint should be applied for emergency care to minimize further injury to the bone and soft tissues. The coaptation should run from the hoof to the elbow or stifle and be augmented by a second splint on the medial or lateral aspect of the limb, depending on the configuration of the fracture. Because of the anatomic limitations of the pelvic limb, it is difficult to extend a caudal splint past the tuber calcanei, so the limb must always be supported with a second splint on the medial or lateral side of the limb, extending proximally to the tarsus.

The 2nd and 4th metacarpal/metatarsal bones (see Landscape Figs. 4.0-10 and 4.0-11)

The lateral bone is MC/MTIV (4th) and the medial bone is MC/MTII (2nd). In addition to MCII, MCIV and MTII both have proximal articular surfaces with the carpus or tarsus. Though MTIV has minimal articulation with the 4th tarsal bone and is not important for weight-bearing, it is an insertion site for the **lateral collateral ligament of the tarsus**, making it important in joint stabilization. These bones are attached to MCIII/MTIII via the **interosseous ligament**. A firm fascial covering over the proximal one-third of the palmar/plantar tendons of MC/MTIII tracks from MC/MTII to MC/MTIV. The bones end distally with a bony bump/swelling called a **button**. There is a thin fibrous band that extends from the buttons of the splints toward the proximal sesamoid bones of the metacarpo-/metatarsophalangeal joint.

Inflammation of the interosseous ligament can lead to proliferative bone formation between MC/MTII or MC/MTIV and MC/MTIII. Mild cases can be treated conservatively, but meaningfully enlarged "splints"—as they are commonly called—can impinge upon neurovascular and soft tissue structures in the palmar/plantar aspect of the limb, requiring surgical removal of that portion of the splint bone using a segmental ostectomy, partial amputation, or debulking of the excess callus and removal of the periosteum of the affected area.

Fractures of MC/MTII and MC/MTIV are not uncommon because of their location and minimal soft tissue coverage. They are often seen with kick-associated trauma. Mid-body and distal fractures can sometimes be treated conservatively if there is minimal callus formation. If surgery is required, the bone distal to the fracture is removed or a segment of the bone involving the callus and fracture may be removed. Proximal fractures may be treated in some cases with segmental ostectomy, but often require internal fixation to provide stability to the joint. The MTIV can be removed in its entirety; however, the horse is at risk of tarsal luxation because of the partial loss of insertion of the lateral collateral ligament.

Innervation of the metacarpus (Fig. 15.5-2)

Medial and **lateral palmar nerves** arise by the division of the median nerve proximal to the carpus. Within the carpal canal palmar to the carpus, the lateral **palmar branch of the**

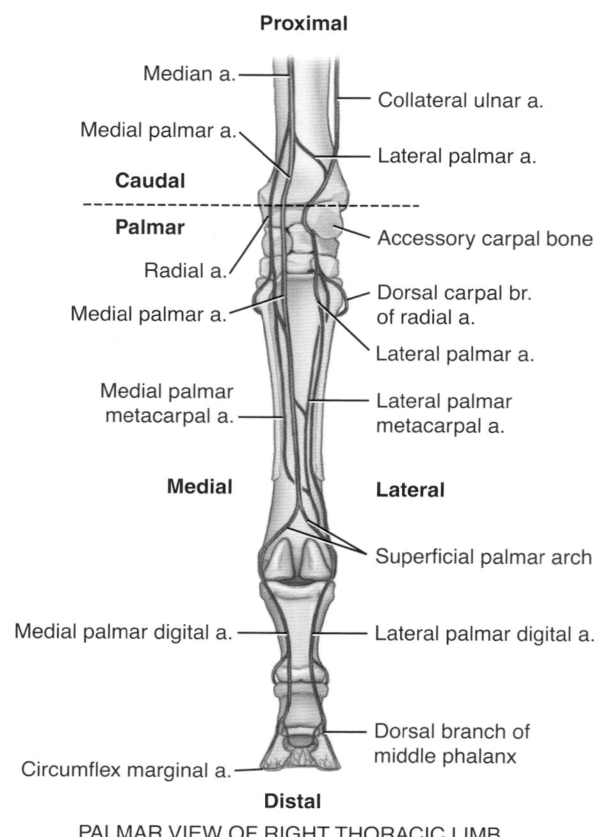

FIGURE 15.5-2 Blood supply of the equine metacarpus—palmar view.

Conventionally a high 4-point anesthetic block is performed by placement of needles axial to MCII and MCIV in the proximal one-third of the metacarpus to anesthetize the palmar metacarpal nerves, and the placement of needles between the deep digital flexor tendon and suspensory ligament to anesthetize the palmar nerves. It is possible to enter the carpometacarpal joint—and middle carpal joint via their communication—when performing the palmar metacarpal nerve blocks, and the distal carpal canal when performing the palmar nerve blocks. The palmar metacarpal nerves and the lateral palmar nerve can alternatively be blocked via needle placement on the axial surface of the accessory carpal bone anesthetizing the lateral palmar nerve before its deep branches. This eliminates unintentional anesthesia of the carpal joints but probably penetrates the carpal canal. Therefore, an aseptic technique is recommended when performing this block.

ulnar nerve joins with the **lateral branch of the median nerve** to form the **lateral palmar nerve**. The **deep palmar branch of the ulnar nerve** supplies sensation to the SL, part of the carpal joint, and the accessory ligament of the deep digital flexor tendon and ends as the **palmar metacarpal nerves**. The dorsal branch of the ulnar nerve tracks on the outside of the metacarpus but does not continue distal to the fetlock. In the distal carpus, the lateral palmar nerve gives off deep branches called the **medial** and **lateral palmar metacarpal nerves** that travel distally between MCIII and MCII or MCIV, medially and laterally, respectively.

At the level of the buttons of the splints, the nerves travel more superficially as they travel over the metacarpophalangeal joint. The **medial** and **lateral palmar nerves** track down the limb deep to the flexor tendons and continue as the **palmar digital nerves** at the level of the metacarpophalangeal joint, giving cutaneous branches to the dorsolateral and dorsomedial sides of the fetlock and proximal pastern. At the level of the midmetacarpus, there is a **communicating branch** between the medial and lateral palmar nerves that travels superficially over the flexor bundle (see Case 14.5).

The low 4-point anesthetic block is performed to anesthetize the medial and lateral palmar metacarpal nerves by inserting the needle just below the button of the splint, avoiding the proximal palmar outpouching of the metacarpophalangeal joint. The medial and lateral palmar nerves are also blocked by placing a needle in the area between the suspensory ligament and deep flexor tendon, just proximal to the digital flexor tendon sheath and distal to the communicating branch between the 2 nerves (see Case 14.5).

Innervation of the metatarsus (Fig. 15.5-3)

At the level of the tarsus, the **tibial nerve** branches into the **medial** and **lateral plantar nerves**. The lateral plantar nerve gives off a deep branch that becomes the **medial** and **lateral plantar metatarsal nerves**. Like the thoracic

A high 4-point anesthetic block of the metatarsus can be performed like that of the metacarpus, but there is a risk of entering the tarsometatarsal joint or distal tarsal sheath. If trying to isolate suspensory origin lameness, the origin of the suspensory ligament can be infiltrated directly or the deep branch of the lateral plantar nerve can be blocked by inserting a needle just plantar to MTIV at the junction where the bone transitions from the obliquely contoured head to the more vertically oriented body of the bone.

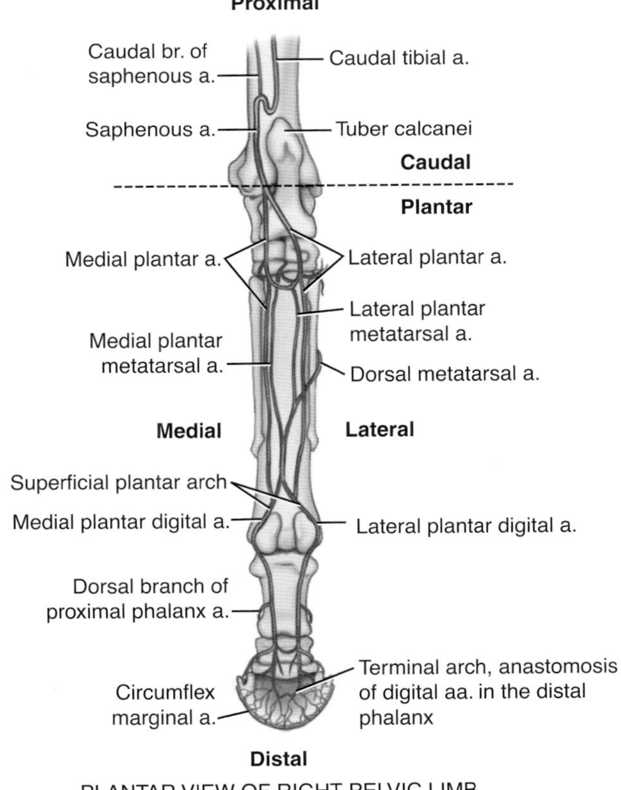

FIGURE 15.5-3 Blood supply of the equine metatarsus—plantar view.

In the distal metacarpus/-tarsus, a dorsal line/ring block is required if the goal is to anesthetize the dorsal metacarpal/-tarsal nerves (i.e., to diagnose pain associated with periostitis of dorsodistal MCIII ["bucked shins"] or villonodular synovitis of the dorsal pad in the metacarpo-/metatarsophalangeal joint). This is performed around the dorsal aspect of the metacarpus/-tarsus starting from where the medial and lateral palmar/plantar metacarpal/-tarsal nerves are anesthetized adjacent to the buttons of the splints when performing a low four-point nerve block.

limb, the plantar nerves track distally and share a **communicating branch** on the plantar aspect of the flexor tendons at the level of the midmetatarsus. The medial and lateral plantar metatarsal nerves pass between MTIII and MTII or MTIV medially and laterally, respectively. In the proximal dorsal metatarsus, there are the **superficial** and **deep peroneal nerves**. The **deep peroneal nerve** branches into the **medial** and **lateral dorsal metatarsal nerves** that join with the **medial** and **deep plantar metatarsal nerves** at the level of the metatarsophalangeal (fetlock) joint. The blood supply beginning at the distal metatarsus is like that of the thoracic limb.

FASCIOTOMY FOR CHRONIC SUSPENSORY DESMITIS

Chronic suspensory desmitis in the pelvic limb refractory to conservative and medical management may benefit from a fasciotomy (L. *fasciae* band+Gr. *temnein* to cut) of the deep fascia surrounding the proximal suspensory ligament (SL) between MTII and MTIV. Theoretically, it is believed to reduce pain caused by a compartmental syndrome—increased tissue pressure in a confined anatomical space causing decreased blood flow with ischemia and pain—from a chronically enlarged SL. This procedure can be combined with neurectomy of the deep branch of the lateral plantar nerve that gives off the plantar metatarsal nerves, supplying innervation to the SL. While this may alleviate the pain, further degeneration and breakdown of the SL may occur without proper case selection.

Metacarpal blood supply and lymphatics (Fig. 15.5-2)

There are 4 sets of blood vessels in the metacarpus: (1) **dorsal common digital**; (2) **dorsal metacarpal**; (3) **palmar common digital**; and (4) **palmar metacarpal**. The **medial palmar artery**, the main artery to the digit and hoof, is on the medial side in the depression between the suspensory ligament and the DDFT. The **lateral palmar artery** is anatomically and clinically insignificant. The **medial vein** is the continuation of the **radial vein** and is joined by the **cephalic vein**.

The **median artery** gives off deeper branches called the **radial artery** and the **palmar branch of the median artery** while also supplying a branch that merges with the **collateral ulnar artery** and continuing as the **medial palmar artery** in the distal radius. The palmar branch of the medial artery and radial artery share a **deep palmar arch** in the proximal metatarsus while continuing distally as the **medial** and **lateral palmar metacarpal arteries**. The median artery continues distally to become the **medial palmar artery**. The medial and lateral palmar artery and palmar metacarpal arteries communicate distally at the **superficial palmar arch** below MCII and MCIV.

The medial palmar artery is the main supply to the digits where it divides into the **digital arteries** proximal to the metacarpophalangeal joint. It does receive a lesser supply from the communicating branch of the medial and lateral palmar arteries.

Venous drainage of the thoracic limb generally parallels the arterial supply and is covered in Case 14.4. Lymphatic drainage typically tracks with the medial palmar vessels and continues its course medially through the carpal canal and forearm to the **cubital lymph nodes** on the proximomedial aspect of the elbow.

Metatarsal blood supply and lymphatics (Fig. 15.5-3)

The blood vessels of the metatarsus are: (1) **dorsal common digital**; (2) **dorsal metatarsal**; (3) **plantar common digital**, and (4) **plantar metatarsal**. Proximal to the tarsus, the **saphenous artery** gives off branches that track on the plantar metatarsus called the **medial** and **lateral plantar arteries**. The **cranial tibial artery** becomes the **dorsal pedal artery** at the proximal metatarsus and branches as the **dorsal 3rd metatarsal artery**—or great metatarsal artery—the main supply of the distal metatarsus. It continues down the dorsolateral metatarsus between MTIII and MTIV. In the distal metatarsus, the dorsal metatarsal artery passes between MTIII and MTIV as the **distal perforating artery**. The dorsal pedal artery also gives off a deeper anastomotic branch that tracks between MTIV and the 4th tarsal bone and forms an anastomosis—the **deep plantar arch**—with the lateral plantar artery. **Plantar metatarsal arteries** arise from the deep plantar arch. These arteries join the dorsal metatarsal artery's distal perforating artery along with the medial and lateral plantar arteries in the distal metatarsus to become the **medial** and **lateral digital arteries** at the level of the metatarsophalangeal joint. Similar to the thoracic limb, the medial and lateral plantar arteries are superficially located whereas the medial and lateral plantar metatarsal arteries are deep.

> Because of its location, the dorsal 3rd metatarsal artery has an increased risk of being traumatized with fractures of the 4th metatarsal bone.

The venous drainage of the distal pelvic generally limb parallels the arterial supply and is covered in Case 15.4. In the pelvic limb, most lymphatic drainage occurs on the medial aspect of the metatarsus between the flexor tendons and continues over the medial tarsus to the **popliteal lymph nodes**. The popliteal nodes are located between the biceps and semitendinosus muscles and then drain to the **deep inguinal** and **subiliac lymph nodes**.

Selected references

[1] Dyson S. The metacarpal region. In: Ross MW, Dyson S, editors. Diagnosis and management of lameness in the horse. 2nd ed. Saunders; 2011. p. 411–26.
[2] Butler JA, Colles CM, Dyson SJ, Kold SE, Poulos PW, editors. Clinical radiology of the horse. 3rd ed. Wiley-Blackwell; 2008. p. 189–232.
[3] Richardson DW. Third metacarpal and metatarsal bones. In: Auer JA, Stick JA, editors. Equine surgery. 4th ed. Saunders; 2012. p. 1325–48.

CHAPTER 16

INTEGUMENT AND MAMMARY GLAND

Sarah Reuss, Chapter editor

16.1 Hereditary equine regional dermal asthenia—*Daniela Luethy* 978
16.2 Mastitis—*Maria Ferrer* 984

978

CASE 16.1

Hereditary Equine Regional Dermal Asthenia

Daniela Luethy
Large Animal Clinical Sciences, University of Florida College of Veterinary Medicine, Gainesville, Florida, US

Clinical case

History

A 3-year-old American Quarter Horse gelding presented with multiple areas of skin sloughing on the thorax and dorsal midline. The horse had recently been started under saddle as a cutting horse. He resented being saddled and often bucked under saddle.

Hyperextensibility in the skin means that the skin has more stretch than normal for the animal's age. Together with increased skin fragility (tendency for skin injury) in this case, hyperextensible skin is a notable finding, as few conditions create this specific clinical picture. In a clinically dehydrated animal, the skin may be slow to return to its normal arrangement after being tented or pinched up, but the skin is not more extensible than normal. Alopecia simply means the abnormal lack or loss of hair.

Physical examination findings

Physical examination was normal, except for large areas of hyperextensible skin on the dorsal midline and multiple regions of ulceration and alopecia bilaterally along the thorax, most apparent in the area covered by the saddle and saddle pad (Fig. 16.1-1).

Differential diagnoses

Dermatophilosis ("rain rot" or "rain scald"), hereditary equine regional dermal asthenia (HERDA), pemphigus foliaceus, lupus, drug reaction, other immune-mediated disease, epitheliogenesis imperfecta, multifocal trauma

Diagnostics

The gelding was sedated to facilitate a skin scraping. The scraping, consisting of crusts and hair, was analyzed microscopically for evidence of ectoparasites (mites) or dermatophilus (*Dermatophilus congolensis*). No abnormalities were seen.

Hereditary equine regional dermal asthenia-affected skin can be more difficult to biopsy than normal skin, because the skin of a horse with HERDA spontaneously splits through the intermediate layer of the dermis into superficial and deep layers. The superficial dermis is not enough tissue sample for identification of the typical histopathologic lesions of HERDA. Care must be taken to ensure that the biopsy sample includes deep dermis. In addition, dermal layers may separate during processing, which may also confound interpretation.

A full-thickness skin biopsy was then performed. Histopathologic abnormalities included thinning and fraying of collagen bundles, with increased space between bundles, within the intermediate dermis. These findings, along with the characteristic clinical signs, were supportive of a presumptive diagnosis of HERDA.

Comparative Veterinary Anatomy: A Clinical Approach. https://doi.org/10.1016/B978-0-323-91015-6.00086-8

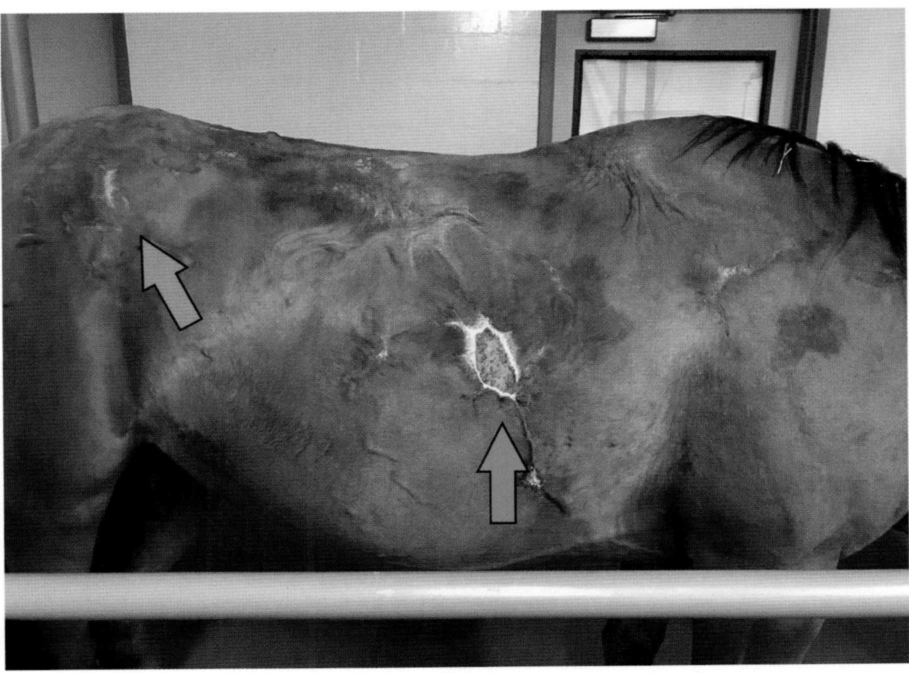

FIGURE 16.1-1 Several areas of ulcerated skin *(green arrow)* in various stages of repair are seen in the saddle area. This horse was in training as a cutting horse, so he was ridden in a Western saddle. The skirt of the saddle or the thick saddle pad used with these saddles may extend caudally to the tuber coxae, where areas of alopecia *(orange arrow)* are seen on this horse.

A hair sample was submitted to a veterinary genetics laboratory to determine the horse's status—normal (N/N), carrier (N/Hrd), or affected (Hrd/Hrd)—for HERDA, an autosomal recessive trait in horses. This test revealed the gelding to be homozygous (Hrd/Hrd) for the HERDA gene.

Diagnosis

Hereditary equine regional dermal asthenia (HERDA)

Treatment

Currently, there is no specific treatment for HERDA. The goal of management is to reduce skin trauma and thus reduce the incidence and severity of skin lesions. Management strategies include reducing exposure to sunlight and excessive heat, using padded halters and other tack, avoiding accidental injury, and using fly control to minimize self-trauma. In some horses, management is not successful, and the horse must be retired or humanely euthanized.

Anatomical features in equids

Introduction

The integumentary system, or skin, is the largest organ in the body. It is comprised of the epidermis, dermis, subcutis, and supporting structures (sebaceous glands, sweat glands, blood and lymph vessels, nerves, and hair) (Fig. 16.1-2), all of which are discussed in this section. The thickness of equine skin varies with anatomical location. Generally, skin thickness decreases dorsally to ventrally on the trunk and proximally to distally on the limbs.

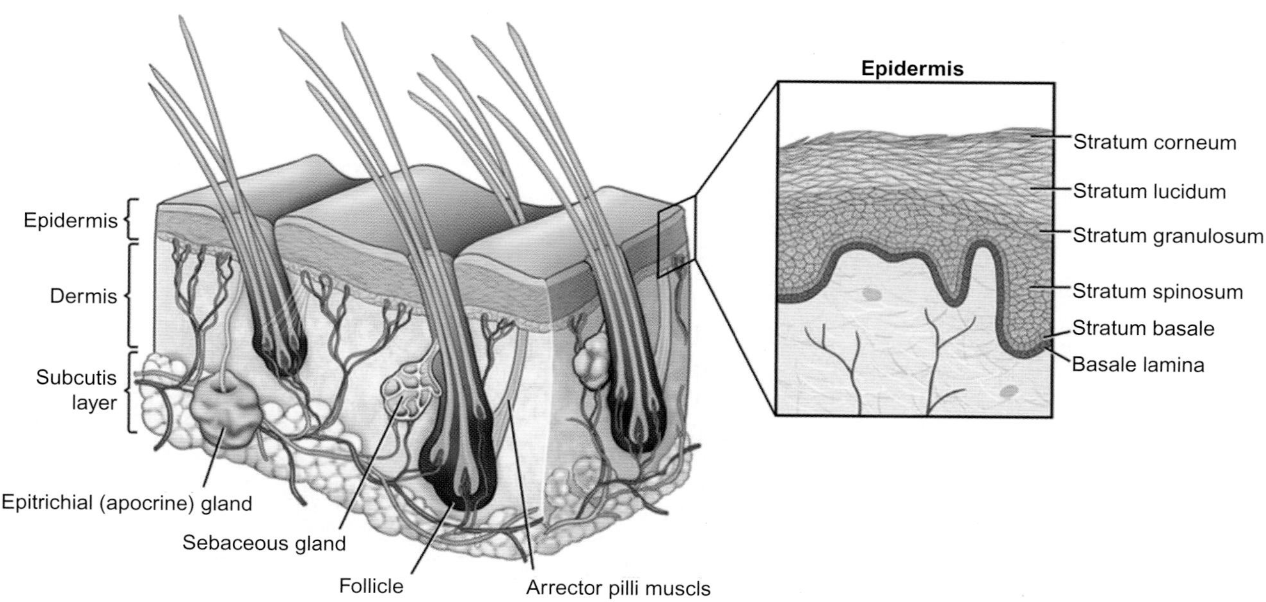

FIGURE 16.1-2 Cross-sectional view of equine skin, illustrating the epidermis, dermis, and subcutis layers, as well as supporting structures. Inset is an enlarged cross-sectional view of the epidermis, illustrating the distinct layers of the epidermis.

Function

The skin serves several important functions: first, skin serves as a protective barrier against potentially harmful environmental factors, including external pathogens, while surrounding and protecting the internal environment to prevent loss of water, electrolytes, and other substances. The skin defines the body shape, and its flexibility allows movement. Skin is also responsible for producing keratinized structures, such as hair and hooves.

Second, the skin is a sensory organ that perceives temperature and senses potentially noxious stimuli and mechanical forces, such as pressure and touch. Temperature regulation involves the skin, with the hair coat, cutaneous blood flow, and sweat gland production all contributing to temperature control. Third, specialized cells in the skin react to foreign antigens and are involved in immunosurveillance and the subsequent immune response. Fourth, skin can act as a reservoir for molecules, including electrolytes, vitamins, water, fats, carbohydrates, and proteins. Lastly, pigment in the skin helps minimize damage from solar radiation and determines the horse's coat color.

EQUINE COAT COLOR DETERMINATION

Equine coat color is determined by the interaction of 2 main genes, named Agouti ("A") and Extension ("E"). The Agouti gene controls the dissemination of black pigment, either concentrating black pigment in the "points" (lower limbs, ears, mane, tail), such as in the bay coat color, or distributing the pigment over the entire body, such as in the chestnut coat color. The Extension gene regulates the production of red and black pigment. These basic colors can then be diluted by 5 additional genes: Champagne, Cream, Dun, Pearl, and Silver. Additional genes (Dominant White, Appaloosa, Tobiano, Overo) are involved in white spotting patterns. Genetic tests are available for several of the genes involved in coat color in horses.

Epidermis

The **epidermis** is the outermost layer of skin. It is generally 5–7 cell layers thick and is classified into 5 main layers, listed from innermost to outermost: **stratum basale**, **stratum spinosum**, **stratum granulosum**, **stratum lucidum**, and **stratum corneum** (Fig 16.1-2). The epidermis is composed of keratinocytes, melanocytes, Merkel cells, and Langerhans cells. **Melanocytes** are derived from neural crest cells and migrate into the epidermis, where they produce melanin pigment. **Merkel cells** are specialized mechanoreceptors found in the stratum basale. **Langerhans cells** are antigen-presenting cells involved in immunosurveillance and the immune response.

Dermis

Deep to the epidermis is the **dermis**. These 2 layers are attached at a basement membrane zone via hemidesmosomes, which are tiny adhesion structures that act as "spot-welds" between the cell membrane and the adjacent basement membrane. The dermis arises from the mesoderm and is the major determinant of overall skin thickness. The dermis is comprised of dermal fibers (collagen, elastin, etc.), ground substance (consisting of glycosaminoglycans and proteoglycans); and cells (fibroblasts and dermal dendrocytes). The arrector pili muscles, blood vessels, lymph vessels, and nerves are all contained within the dermis (Fig. 16.1-2). Mast cells are also found within the dermis.

Subcutis

The **subcutis**, or hypodermis, lies under the dermis in most areas of the body, although certain areas, such as the lips, eyelids, and external ears, contain no subcutis. The superficial aspect of the subcutis penetrates the dermis (Fig. 16.1-2). The subcutis is integral for steroid metabolism and estrogen production, and it also serves as an energy reservoir (it can accommodate large amounts of adipose tissue), helps with thermoregulation, and provides protection for both the skin and the deeper tissues.

Hair

Hair serves several important functions: it provides insulation and sensory input and it acts as a barrier against injury to the underlying skin. Coat color also contributes to thermoregulation.

Blood supply of the skin

The cutaneous microcirculation is composed of 3 main intercommunicating plexuses: the **deep plexus** at the junction of the dermis and subcutis, the **middle plexus** which lies near the sebaceous glands, and the **superficial plexus** below the epidermis. **Arteriovenous anastomoses**—normal anatomical connections directly between arteries and veins—are important in the skin and are particularly common in the coronary band and the dermis of the hoof.

HEREDITARY EQUINE REGIONAL DERMAL ASTHENIA (HERDA)

Hereditary equine regional dermal asthenia is a genetic disorder of horses affecting the mid to deep dermis. It is caused by a missense mutation in the gene that codes for cyclophilin B, an enzyme involved in the rate-limiting step of fibrillary collagen synthesis. This disease has an autosomal recessive mode of inheritance and is like Ehlers-Danlos syndrome in humans. Hereditary equine regional dermal asthenia is most commonly seen in American Quarter horses, but it has been reported in several horse breeds. Skin hyperextensibility is a prominent feature, and skin fragility and sloughing are also common. Clinical signs are often not seen until approximately 2 years of age, which is typically when horses are started under saddle. The severity and extent of lesions may vary between horses. Ocular, musculoskeletal, immunologic, and cardiac manifestations have also been reported. Environmental factors may play a role, with both ambient temperature and ultraviolet light potentially causing more severe clinical signs.

Cutaneous lymphatics

Lymphatic vessels arise within the dermis and are essential for control of the cutaneous microcirculation. Lymph is carried via lymphatics to peripheral lymph nodes, where antigens may interact with immune components. Lymphatic vessels are easily traumatized, so skin injury can easily affect lymphatic flow.

Innervation of the skin

Nerves are contained within the skin in close association with blood vessels, keratinocytes, Langerhans cells, and fibroblasts. Each spinal nerve innervates a region of skin called a **dermatome** (see Landscape Fig. 4.0-3). Cutaneous nerves control vascular tone, regulate secretion of sebaceous and apocrine glands, provide sensation, and release neuropeptides which can activate keratinocytes, mast cells, and endothelial cells. Autonomic sympathetic fibers provide motor innervation to the skin (specifically, to the arrector pili muscles), while sensory fibers are contained within myelinated postganglionic sympathetic fibers. Thermoreceptors sense skin temperature and respond to either decreasing or increasing skin temperature, depending on the category of thermoreceptor. Mechanoreceptors, such as **Pacinian corpuscles** and **Meissner corpuscles**, sense skin movement and pressure. Nociceptors sense potentially harmful ("noxious") stimuli and elicit the sensation of pain or discomfort; they are important for preventing skin injury and protecting injured areas during repair.

The skin is highly innervated by sensory nerves, so local anesthesia is needed to desensitize the skin before any painful procedures, such as skin biopsy and surgical repair of lacerations.

Sebaceous glands

Sebaceous glands are present in all haired skin (Fig. 16.1-2). These glands secrete an oily substance (sebum) that forms a thin surface coating which helps maintain skin moisture. Sebum also acts as a chemical defense to protect the body from environmental pathogens.

When an animal is in negative energy balance, there may be inadequate sebaceous gland function and secretions, so the hair coat can become dry and dull.

SWOLLEN LEG—CELLULITIS OR LYMPHANGITIS?

Cellulitis in horses refers to infection or inflammation of the subcutaneous tissue, most often involving one or more limbs. There is often diffuse swelling of the affected area, with heat and pain on palpation and pitting edema (the depression remains for several seconds after digital pressure is applied). Lameness of the affected limb is also common along with fever. The cause is often unknown, but skin injury is likely a contributing factor.

Lymphangitis refers to infection or inflammation associated with injury to the regional lymphatic system. Horses with lymphangitis present with signs like cellulitis, although the infection is generally located deeper within the tissues of the limb, signs are often more severe and extensive, and the condition is often more chronic. Cellulitis may injure the lymphatic system, which then can precipitate lymphangitis.

ANHIDROSIS

Anhidrosis, or the inability to sweat, is a condition seen in horses in warm or hot climates, particularly when high ambient temperatures are combined with high humidity. This condition is seen most often in athletes and other working horses. Because sweat production is under ß2-adrenergic regulation, anhidrosis may be diagnosed by lack of response to subcutaneous injection of a ß2 agonist, such as terbutaline. For more information, see the side box entitled "Anhidrosis" in Case 9.1.

Sweat glands

Below the sebaceous glands are the **apocrine glands** that produce sweat for thermoregulation (Fig. 16.1-2). Horses differ from other species in that sweat production is controlled both by hormones released from the adrenal medulla into the circulation and by direct autonomic adrenergic neurons.

Sweat production is mainly under ß2-adrenergic control, so epinephrine injection causes sweating.

Selected references

[1] Scott DW, Miller WH. The structure and function of the skin. In: Scott DW, Miller WH, editors. Equine dermatology. 2nd ed. Maryland Heights, MO: Elsevier Saunders; 2011. p. 1–34.

[2] Theoret C. Physiology of wound healing. In: Theoret C, Schumacher J, editors. Equine wound management. 3rd ed. Ames, IA: John Wiley & Sons; 2017. p. 1–13.

[3] Rashmir-Raven A. Heritable equine regional dermal asthenia. Vet Clin North Am Equine Pract 2013;29:689–702.

CHAPTER 16

CASE 16.2

Mastitis

Maria Ferrer
Large Animal Medicine, University of Georgia College of Veterinary Medicine, Athens, Georgia, US

Clinical case

History

A 12-year-old multiparous Warmblood mare presented to the hospital 3 days after foaling with signs of depression and inappetence. Pregnancy, parturition, and passage of the placenta were uncomplicated.

Physical examination findings

On presentation, the mare's heart and respiratory rates were within normal range. The mucous membranes were pink and moist, and CRT was <2 seconds; rectal temperature was 102 °F (38.9 °C). Normal GI sounds were present in all quadrants. The mammary gland was asymmetrical; the left mamma (*L. mammarius* breast, mammary gland) was edematous, hot, and painful on palpation (Fig. 16.2-1), and the skin covering this mamma was reddened.

Differential diagnoses

Mastitis, mammary neoplasia, mammary abscess, mammary hematoma

Diagnostics

Ultrasound of the left mamma revealed an enlarged milk cistern compared with the right mamma (Fig. 16.2-2). Mammary gland secretions were manually expressed from both sides. Gross appearance was white and milky; cytology of the secretion revealed a large number of degenerate neutrophils, some with intracellular bacteria in the

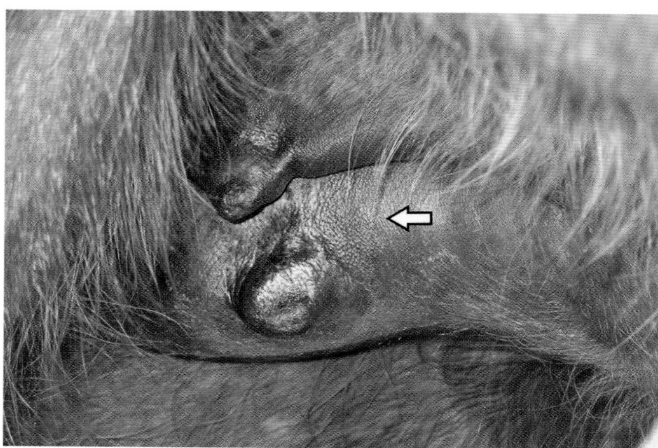

FIGURE 16.2-1 Gross appearance of the mammary gland of a post-partum mare presenting with depression and inappetence. The left mamma *(arrow)* was swollen and painful.

Comparative Veterinary Anatomy: A Clinical Approach. https://doi.org/10.1016/B978-0-323-91015-6.00087-X

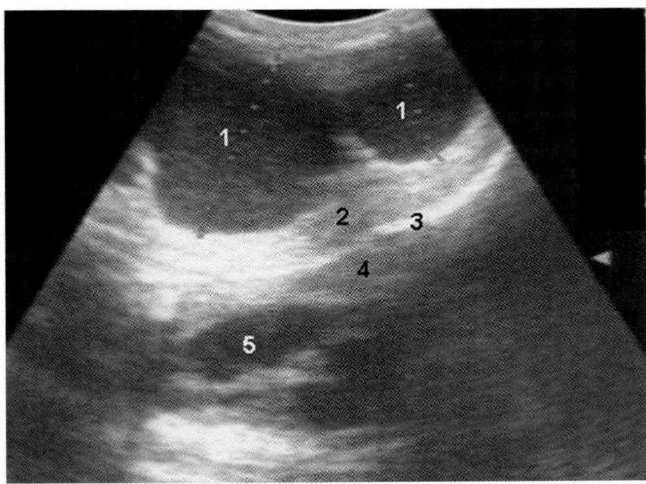

FIGURE 16.2-2 Ultrasonographic image of the mammary gland of a post-partum mare presenting with depression and inappetence. The glandular portion of the milk cistern was enlarged in the left mamma (1), compared with the right mamma (5). The parenchyma of the left (2) and right (4) mammae had a normal homogeneous appearance. The bright hyperechoic line (3) is the septum at the intermammary groove.

secretions from the left mamma. Bacterial culture yielded growth of a moderate amount of *Escherichia coli*, sensitive to gentamicin, amikacin, ceftiofur, and enrofloxacin, among others.

Diagnosis

Mastitis

Treatment

The mare was treated with a systemic antimicrobial (ceftiofur sodium) for 7 days to inhibit bacterial growth. An intramammary antimicrobial formulation (cephapirin sodium) was also infused once on day 1 of treatment into each of the 2 left teat ducts. The foal was allowed nurse during the mare's treatment period. A nonsteroidal anti-inflammatory drug (flunixin meglumine) was administered systemically for 3 days as an anti-inflammatory, antipyretic, and analgesic. The mare's rectal temperature returned to normal within 24 hours of starting treatment. By the end of the treatment regimen, the appearance of the mammary gland had returned to normal, showing no edema, heat, or pain. The mare was discharged from the hospital, and no problems were reported by the owner after discharge.

Anatomical features in equids

Introduction

The mammary gland (mamma) is an exocrine gland of the reproductive system. It is a modified and enlarged sweat gland and is considered part of the common integument. Unlike most mammalian species, external teats are rarely found in male horses. Initial development occurs after puberty in response to reproductive hormones. Further, hyperplasia and hypertrophy, as well as lactation, occur at the end of gestation. This section covers the mammary gland along with its innervation and vasculature.

Function

The mammary gland produces colostrum near the end of gestation. Colostrum is a thick, yellowish secretion, rich in carbohydrates, fat, proteins, electrolytes, and immunoglobulins. Ingestion of colostrum during the first hours of life is essential for the foal to acquire protective antibodies, as well as to maintain a normal glycemic and hydration status. Colostrum also has laxative properties that are important in helping foals eliminate meconium. After foaling, the mammary gland produces milk, which is a watery, white secretion. Milk is the only source of nutrition for the foal during the first months of life.

Mammary gland

The equine mammary gland lies between 2 folds of abdominal skin. The gland is divided into symmetrical **left** and **right mammae** by a septum along the **longitudinal intermammary groove**. Each mamma consists of a glandular portion and a teat and is flattened laterally. The skin is sparsely haired, with numerous sebaceous and sweat glands. A layer of dark-colored sebum covers the skin of the intermammary groove and protects the mammae and reduces friction during locomotion.

The glandular portion of each mamma is organized into **cranial** and **caudal lobes** (Fig. 16.2-3). Within the lobes, secretory **epithelial cells** are organized in **alveoli**, which open into small **ducts**. The alveoli and ducts are grouped into clusters called **lobules** and are surrounded by **myoepithelial cells**, important for milk ejection. The **lactiferous ducts** converge into a **milk cistern**, one for each lobe. The cistern has a glandular portion or **glandular cistern** and a teat portion or **papillary cistern**. The papillary cistern opens into an **orifice** at the teat through the **papillary duct**. There are 2 orifices, one for each lobe, that open into a depression at the apex of the teat.

> This anatomical disposition has implications for local treatment of mastitis. Since the anterior and posterior lobes, and their respective cisterns, do not communicate with each other, intramammary antibiotics should be infused through both orifices of the teat to reach therapeutic drug concentrations within each cistern and lobe.

Support for each mamma is provided by the **lateral** and **medial suspensory ligaments** (see Case 23.2, Fig. 23.2-4). Supportive trabeculae project inward from the ligaments, enclose the lobes, and branch into finer lobular connective tissue that surrounds ducts and alveoli.

Blood supply, lymphatics, and innervation

Hormones inducing mammary development, lactation, and milk ejection arrive to the mammary gland via its blood supply. The blood supply is also important for milk production, since milk is essentially a blood filtrate. The main blood supply and drainage are provided by the **external pudendal artery** and **vein**. The external pudendal artery descends through the inguinal canal and enters the caudal part of the gland where it branches into the **cranial** and **caudal mammary arteries**. A **venous plexus** is present at the base of each mamma, which drains to the contralateral external pudendal veins. Blood also drains from the plexus cranially into the **caudal superficial epigastric vein** and then drains into the **superficial vein** of the thoracic wall. The plexus also drains caudally to the **obturator vein** or branches of the **internal pudendal vein**.

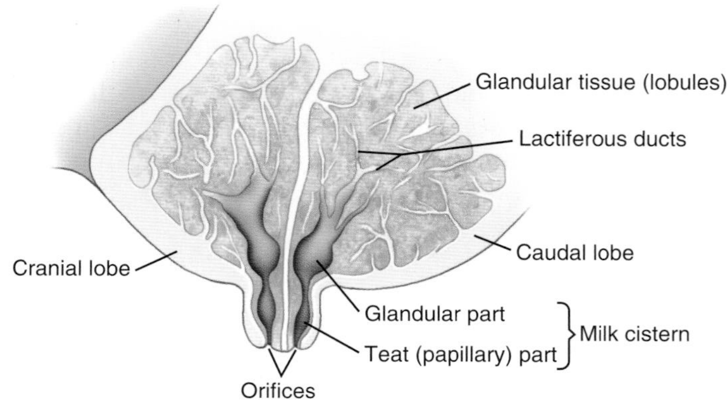

FIGURE 16.2-3 Anatomy of the equine mammary gland. Cranial is to the left.

Lymphatic drainage is provided by lymphatic vessels that converge into **superficial inguinal lymph nodes** at the base of each mamma. The main innervation is by the **genitofemoral nerve** coursing through the inguinal canal. The skin is supplied by nerves of the flank (presumably the iliohypogastric and ilioinguinal nerves, similar to what is reported in the cow) and a descending branch from the **pudendal nerve**.

The superficial inguinal lymph nodes can be enlarged in cases of mastitis.

Mammary gland changes with gestation and lactation

Major changes in mammary gland size and secretions occur at the end of gestation. Initiation of mammary gland development and lactation requires the coordinated action of several hormones. Prolactin is produced in the anterior pituitary and acts synergistically with estrogens and progestogens to promote development of ducts and alveoli, respectively. Prolactin also stimulates milk production and secretion. While the mammary gland is exposed to high concentrations of estrogens and progestogens during gestation, prolactin concentrations increase during the last week of gestation in the mare and remain high after foaling in support of lactation. Mammary gland development can first be seen 2–4 weeks prepartum. As parturition approaches, a small amount of gray watery secretions can be expressed from the mammary glands. These secretions become progressively whiter and thicker to become the thick immunoglobulin-rich colostrum at foaling. Oxytocin also plays an important role in milk ejection, since it stimulates contraction of the myoepithelial cells that surround the alveoli and ducts.

The changes in prepartum mammary gland secretion electrolytes can be monitored and used to predict labor/foaling. Calcium concentration increases in mammary gland secretions in the last 72 h before foaling. This rise is associated with fetal readiness for birth.

PREMATURE OR INAPPROPRIATE MAMMARY GLAND DEVELOPMENT

Some conditions associated with placental insufficiency and chronic fetal stress may induce premature mammary gland development or lactation (<300 d of gestation). The most common cause of premature lactation in the mare is placentitis. Other differential diagnoses are twin pregnancy, uterine body pregnancy, or placental separation, and pituitary pars intermedia dysfunction. On the other hand, agalactia is a failure to produce milk or colostrum, most commonly caused by fescue toxicosis. Tall fescue can be infested with an endophyte that produces ergopeptines. These indole alkaloids can interfere with prolactin secretion and can constrict blood vessels. Other causes of agalactia or hypogalactia (not enough milk production) include advancing age, debilitation, malnutrition, or first parity. Foals from mares with inappropriate lactation (both premature lactation and agalactia) may fail to receive adequate amounts of protective antibodies from colostrum, which increases the risk for neonatal septicemia.

ULTRASOUND OF THE MAMMARY GLAND

As demonstrated in this case, mammary gland can be evaluated with ultrasound (Fig. 16.2-4). The intermammary septum appears as a hyperechoic line between the 2 mammae. Normal parenchyma is homogeneous and hyperechoic. The lactiferous ducts and milk cistern are identified as anechoic or hypoechoic structures and are more prominent in lactating or late-gestation mares. However, the duct system may be visualized in some mares that chronically lactate or accumulate small amounts of secretions in the milk cistern. The blood vessels are also more prominent in lactating mares and can be distinguished from the duct system with color flow Doppler (Fig. 16.2-4C). In cases of mastitis (Fig. 16.2-2), the mammary gland secretions may become more cellular due to the presence of neutrophils, yielding a more echogenic and heterogeneous image of the ducts and cistern. The glandular tissue may also appear more echogenic and heterogeneous due to the presence of fibrosis or inflammatory cell infiltration. Abscesses may be identified as focal cystic areas, while neoplasia usually appears as a focal mass with loss of normal architecture of the parenchyma.

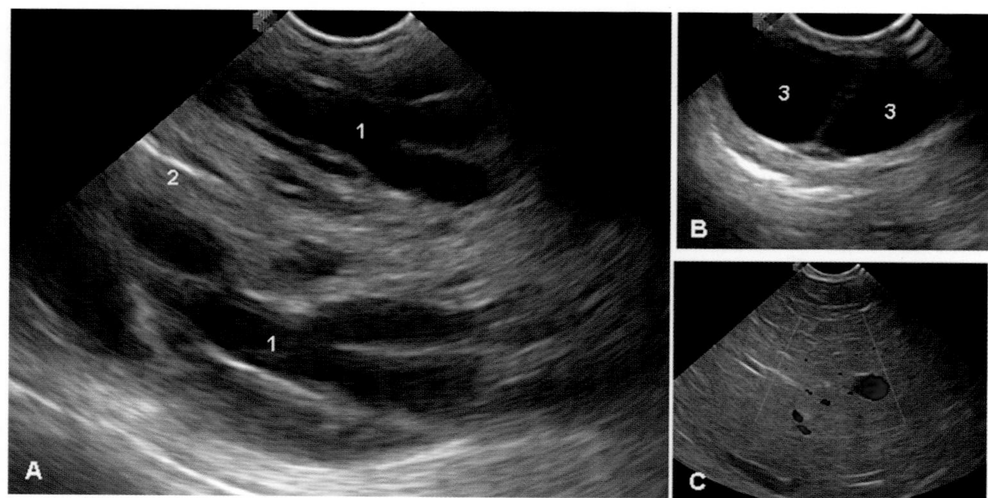

FIGURE 16.2-4 (A) Ultrasonographic image of the glandular portion of the 2 mammae. (B) Ultrasonographic image of one teat showing cranial and caudal papillary cisterns. (C) Color Doppler image of a blood vessel within the mammary gland parenchyma. Key: 1, Glandular portion of the milk cisterns of the left and right mammae; 2, septum at the intermammary groove; 3, cranial and caudal papillary cisterns.

Selected references

[1] Chavette P. Lactation in the mare. Equine Vet Educ 1997;9:62–7.

[2] Kainer RA. Internal reproductive anatomy. In: AO MK, Squires EL, Vaala WE, Varner DD, editors. Equine reproduction. 2nd ed. Ames, IO: Blackwell Publishing Ltd.; 2011. p. 1582–97.

SECTION V

FARM ANIMAL CLINICAL CASES

5.0 BOVINE LANDSCAPE FIGURES (1–9) 991
Nora S. Grenager, James A. Orsini, André Desrochers and Alexander de Lahunta

CHAPTER 17: AXIAL SKELETON: HEAD, NECK, AND VERTEBRAL COLUMN 1001
André Desrochers, Chapter editor

CHAPTER 18: THORAX 1041
André Desrochers, Chapter editor

CHAPTER 19: ABDOMEN 1055
André Desrochers, Chapter editor

CHAPTER 20: PELVIC ORGANS 1105
André Desrochers, Chapter editor

CHAPTER 21: THORACIC LIMB 1145
André Desrochers, Chapter editor

CHAPTER 22: PELVIC LIMB 1193
André Desrochers, Chapter editor

CHAPTER 23: INTEGUMENT AND MAMMARY GLAND/UDDER 1221
André Desrochers, Chapter editor

SECTION V

5.0 BOVINE LANDSCAPE FIGURES (1–9)

Nora S. Grenager[a], James A. Orsini[b], André Desrochers[c], and Alexander de Lahunta[d]

[a]Steinbeck Peninsula Equine Clinics, Menlo Park, California, US

[b]Department of Clinical Studies - New Bolton Center, University of Pennsylvania School of Veterinary Medicine, Kennett Square, Pennsylvania, US

[c]Department of Clinical Sciences, Université de Montréal, Faculty of Veterinary Medicine, St-Hyacinthe, Québec, CA

[d]Professor Emeritus, Biomedical Sciences, College of Veterinary Medicine, Cornell University, Ithaca, New York, US

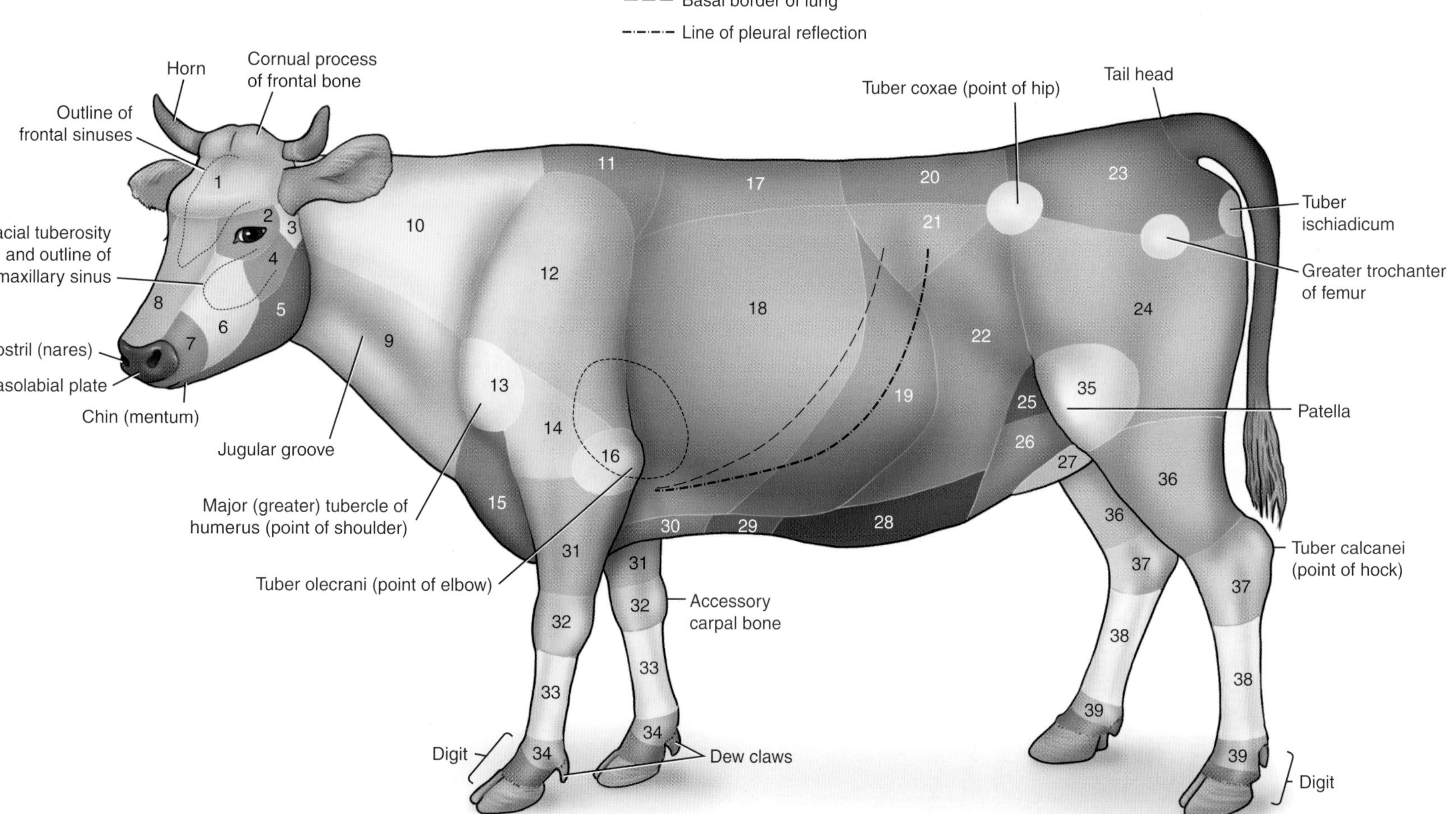

FIGURE 5.0-1 Regional bovine anatomy. 1, frontal; 2, orbital; 3, temporal; 4, zygomatic; 5, mandibular; 6, premaxillary; 7, maxillary; 8, nasal; 9, ventral cervical; 10, dorsal cervical; 11, interscapular; 12, thoracic appendage; 13, shoulder; 14, brachial; 15, presternal; 16, elbow; 17, dorsum (back); 18, costal; 19, hypochondriae; 20, lumbar (loin); 21, paralumbar fossa; 22, lateral abdominal; 23, sacral (croup); 24, femoral; 25, inguinal; 26, prepubic; 27, udder/prepucial, 28, umbilical; 29, xiphoid; 30, sternal; 31, antebrachial (forearm); 32, carpal; 33, metacarpal; 34, digital (ankle); 35, stifle ("true" knee) 36, crural (leg/gaskin); 37, tarsal (hock); 38, metatarsal; 39, digital.

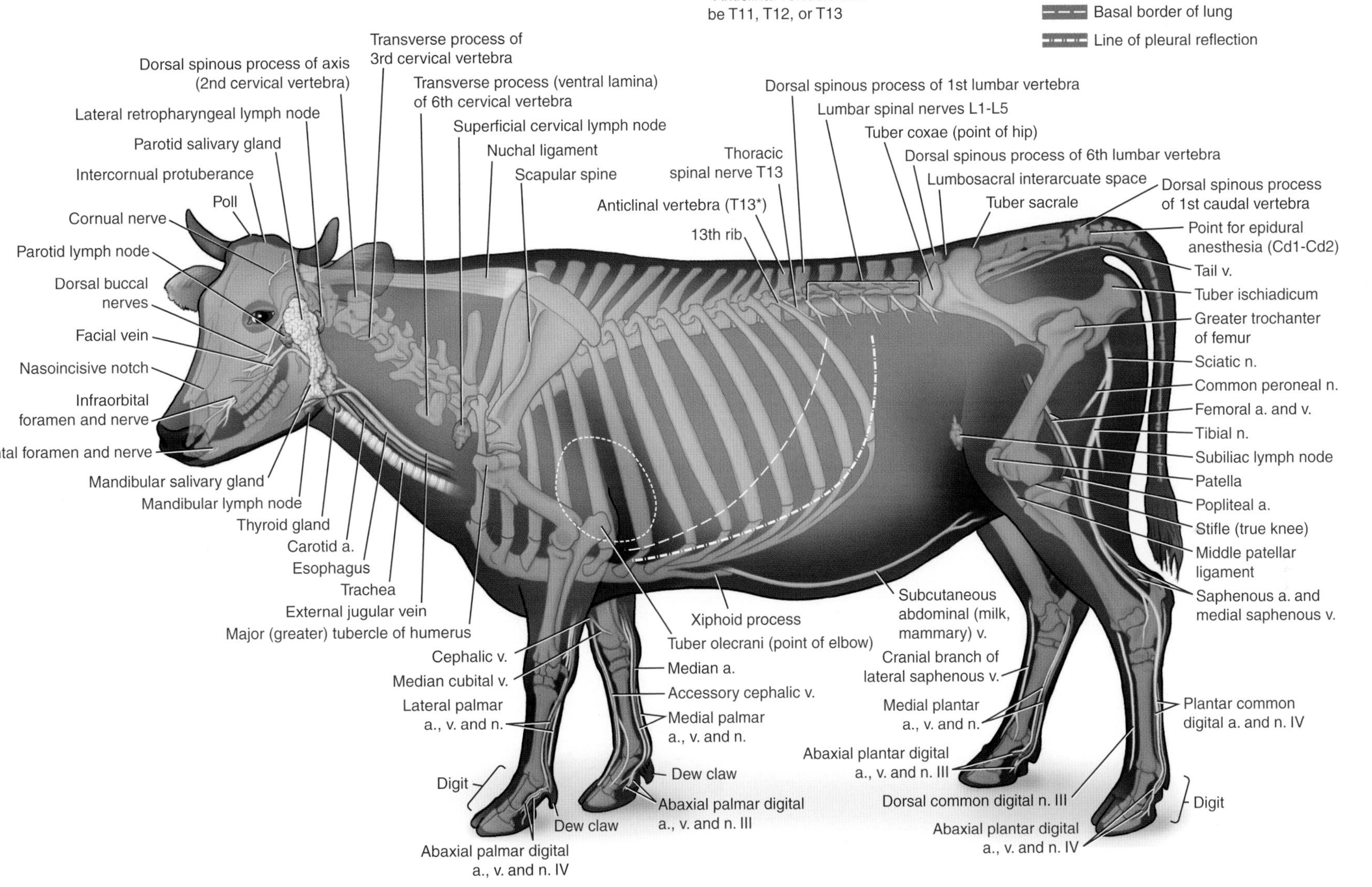

FIGURE 5.0-2 Topographical bovine anatomy.

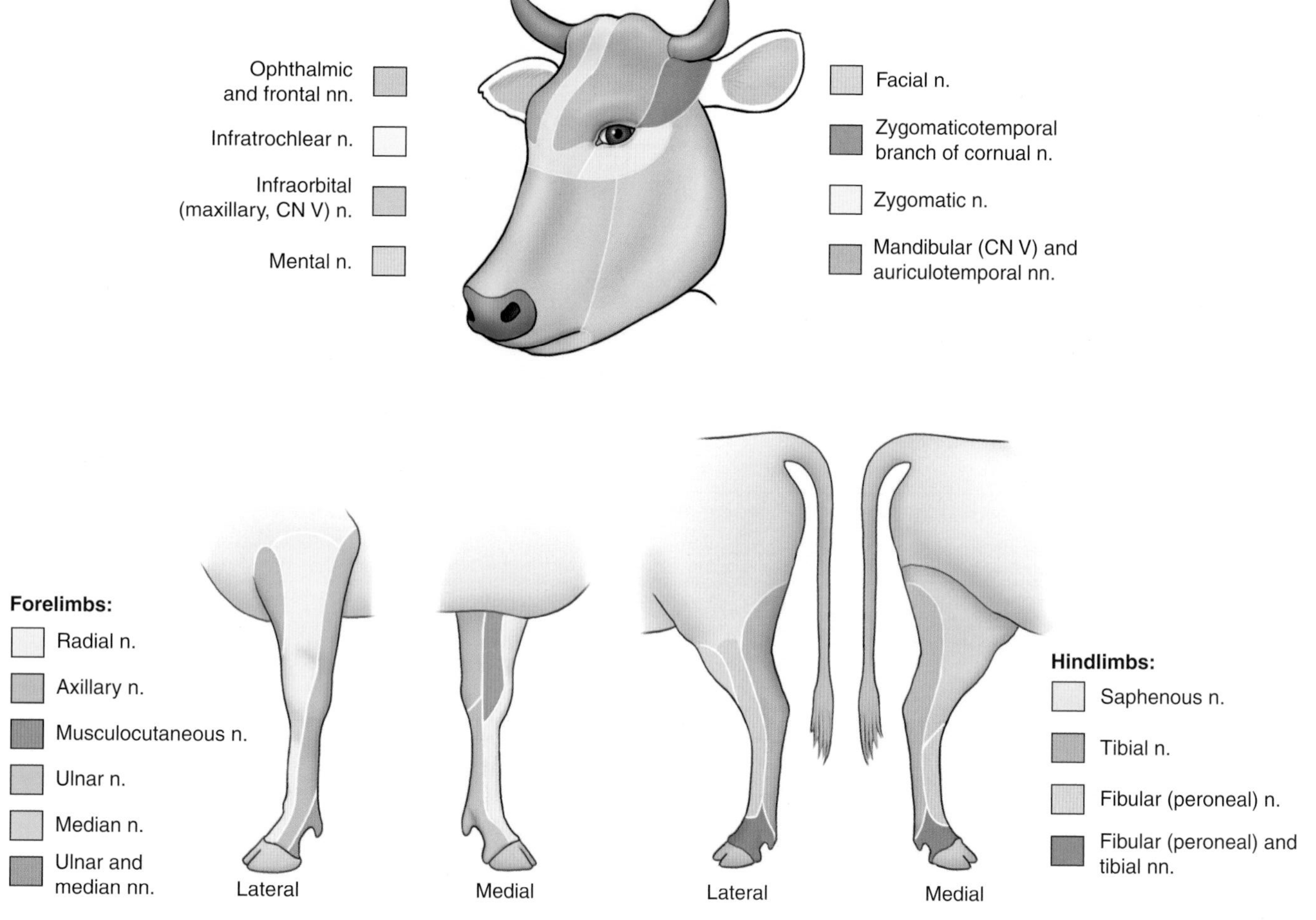

FIGURE 5.0-3 Dermatomes of the bovine head and thoracic and pelvic limbs.

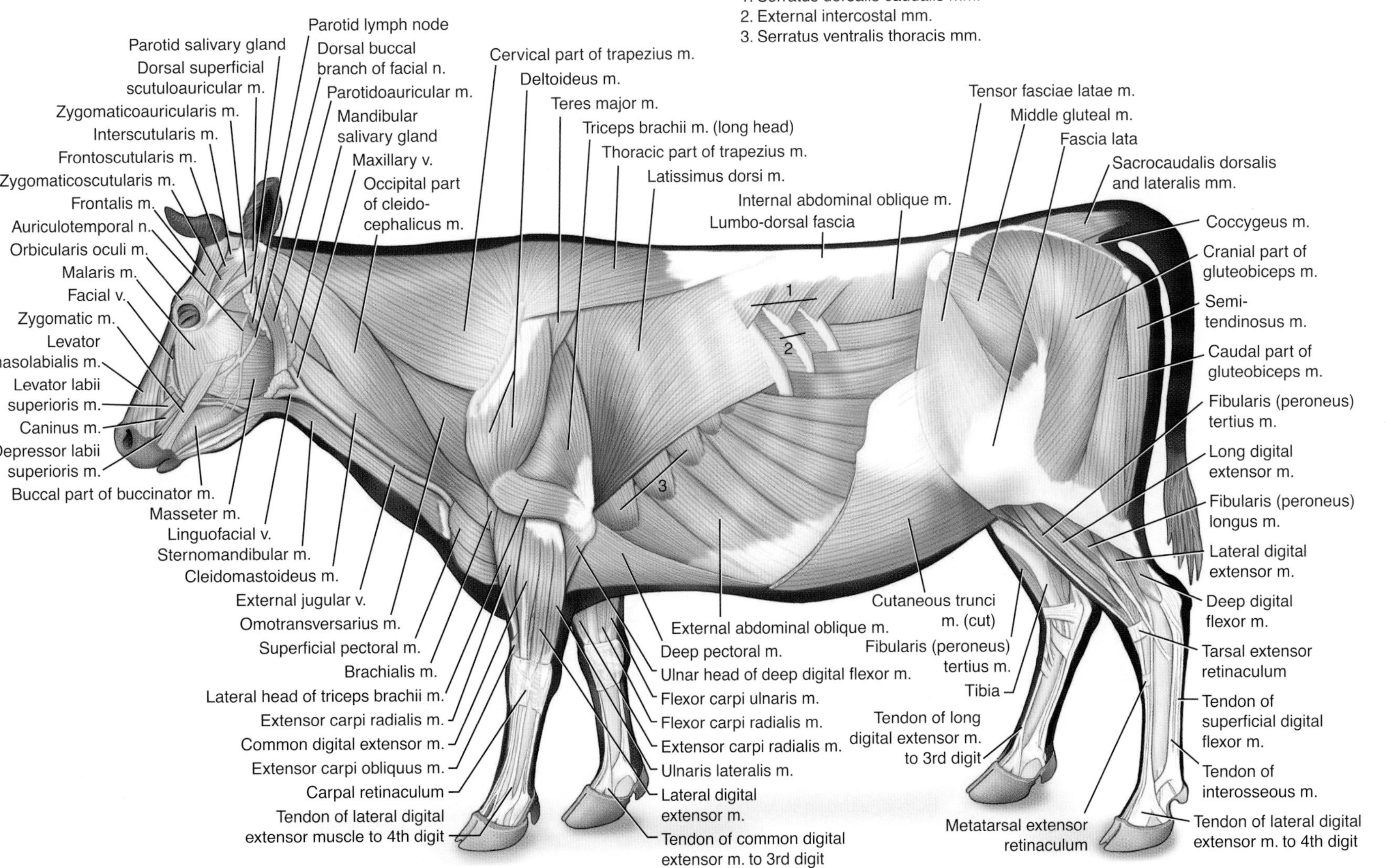

FIGURE 5.0-4 Superficial bovine muscle layer.

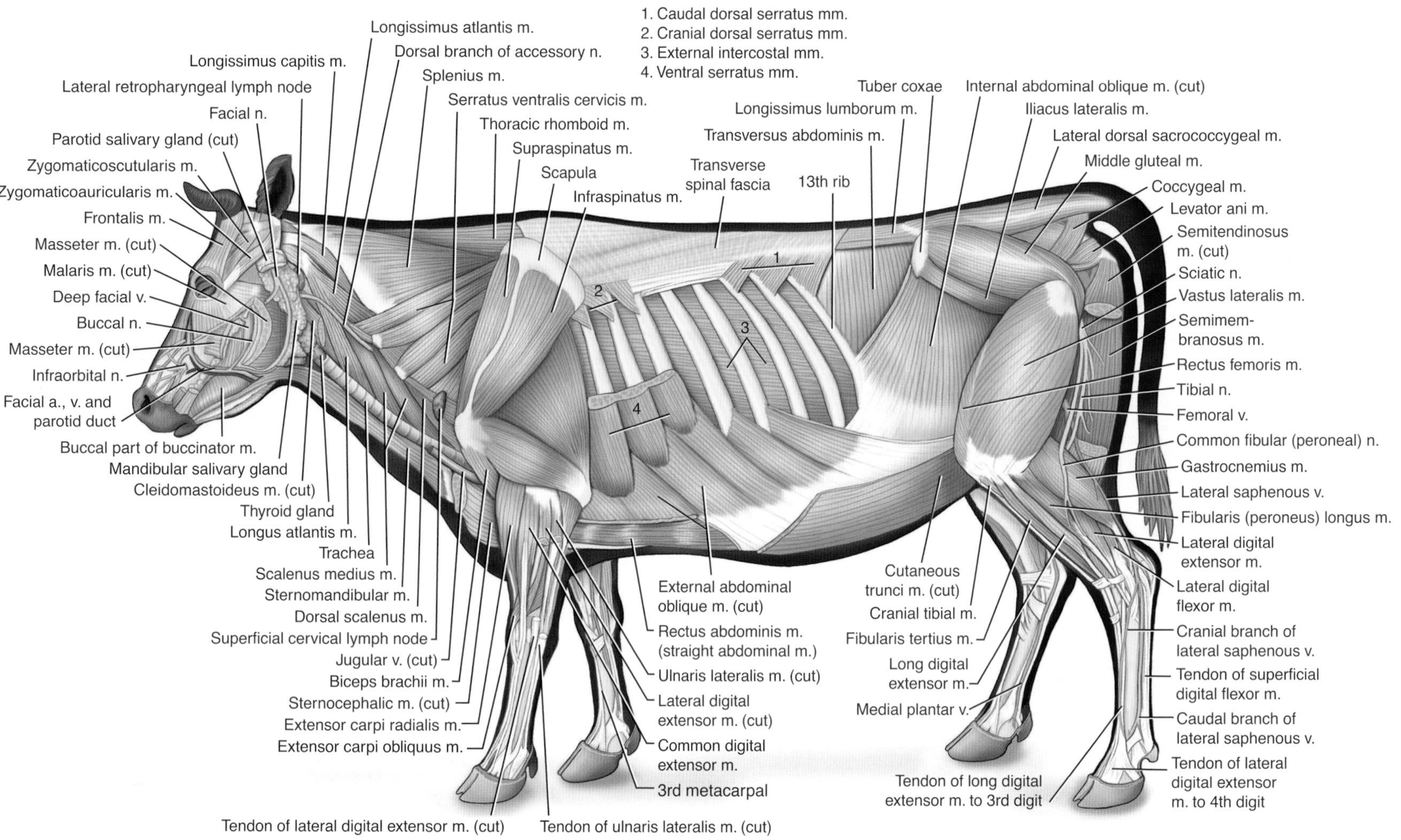

FIGURE 5.0-5 Deep bovine muscle layer.

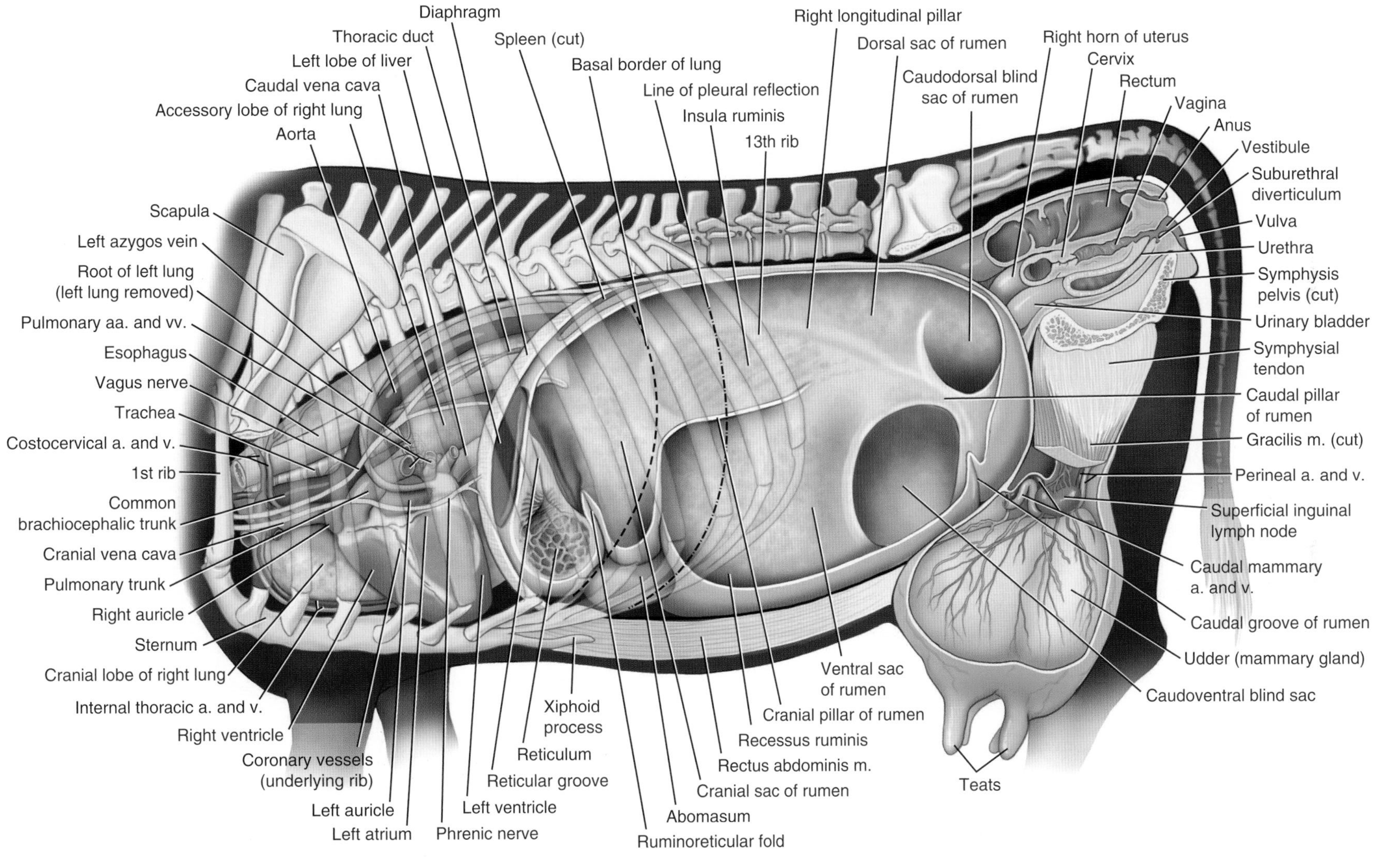

FIGURE 5.0-6 Left view of the bovine thoracic and abdominal cavities and pelvis (female).

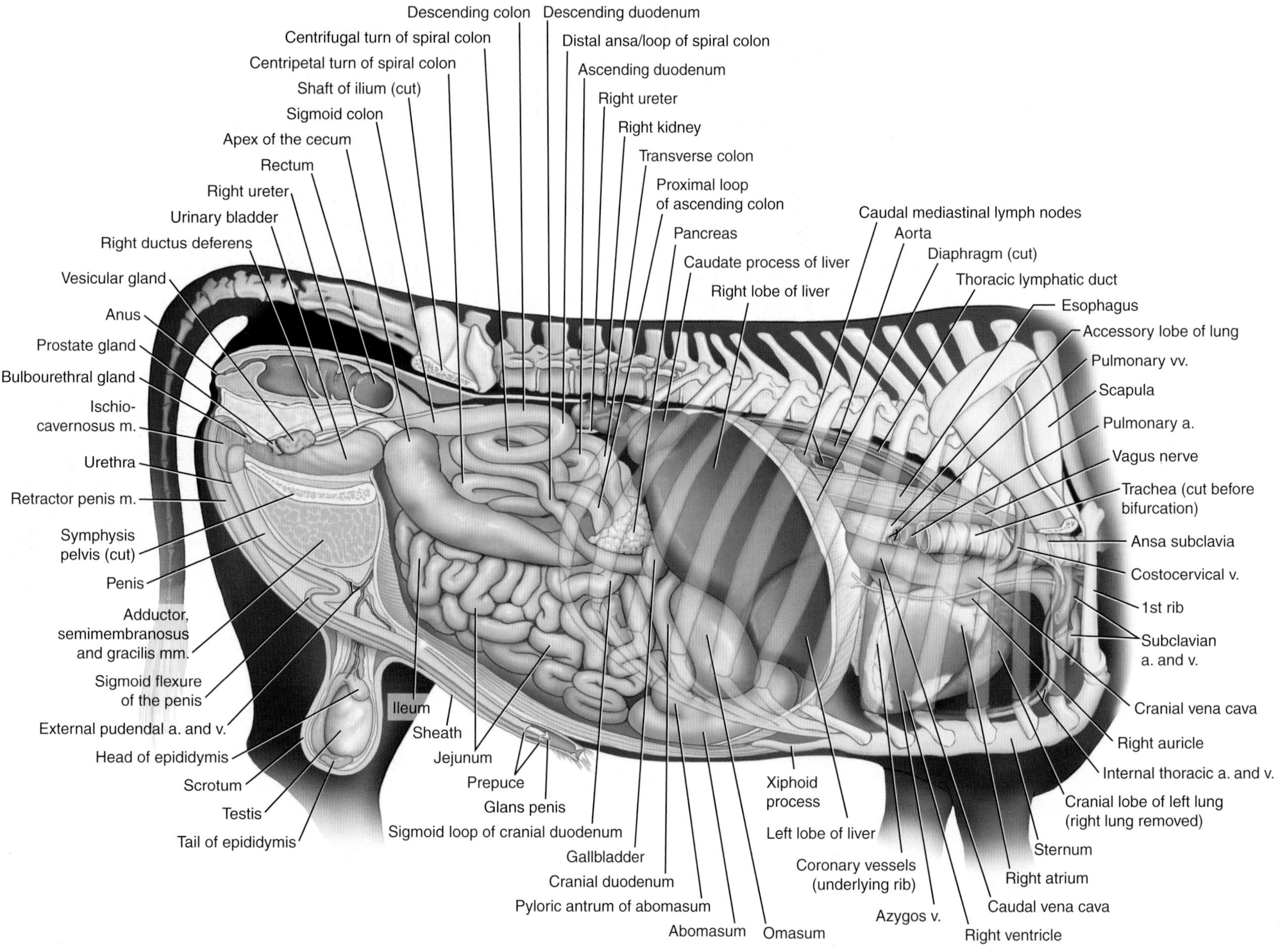

FIGURE 5.0-7 Right view of the bovine thoracic and abdominal cavities and pelvis (male).

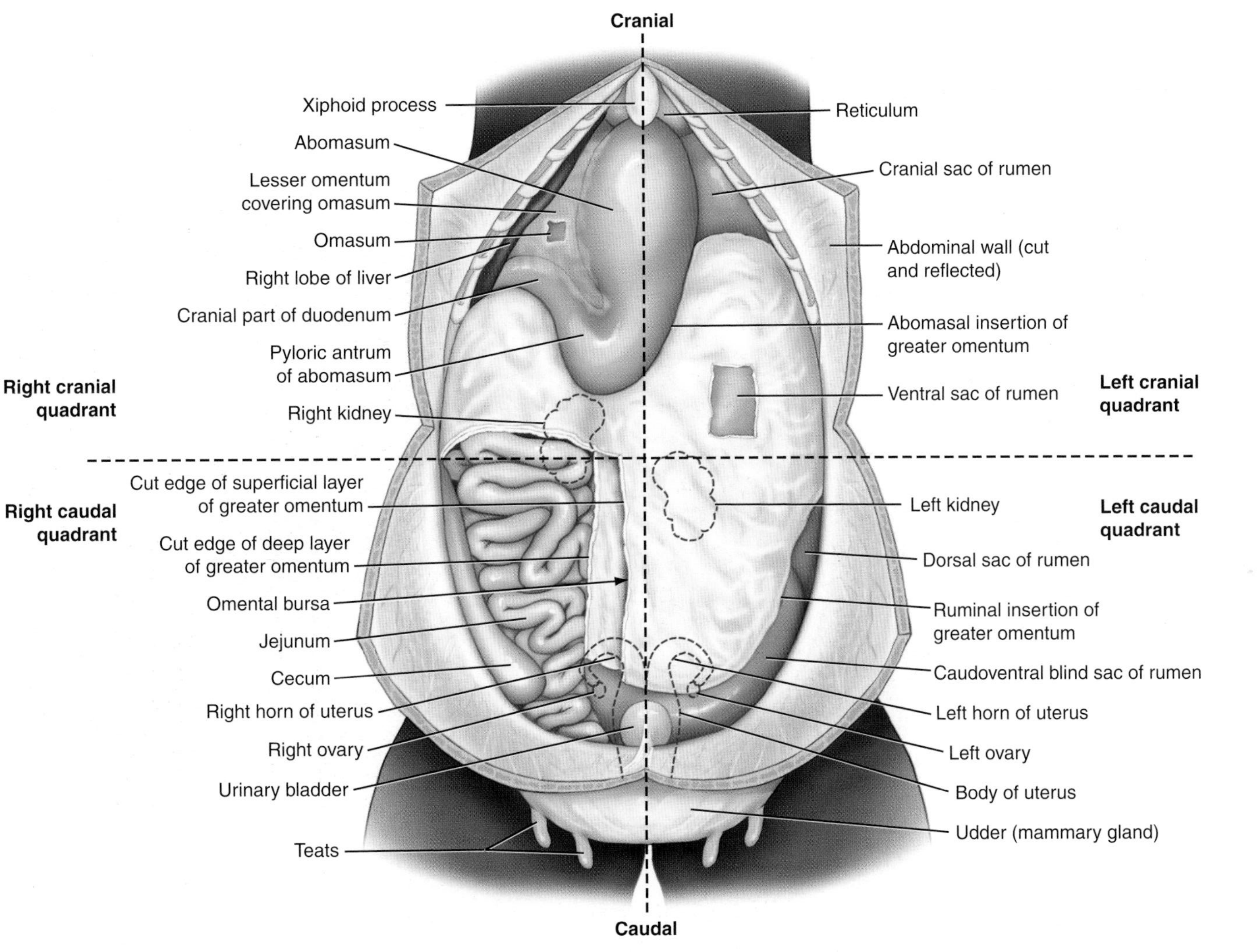

FIGURE 5.0-8 Ventral view of the bovine abdominal cavity.

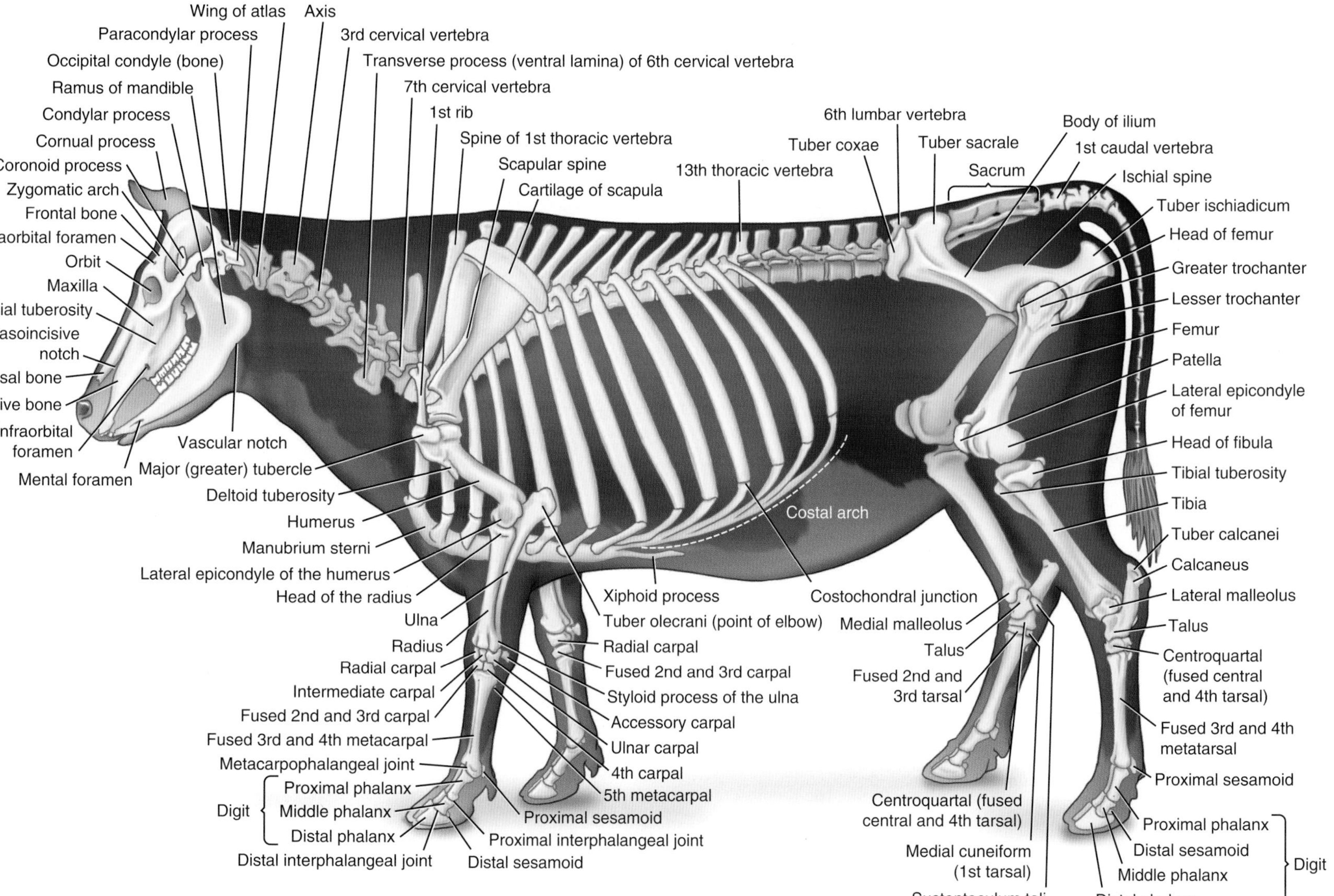

FIGURE 5.0-9 Lateral view of the bovine skeleton. *Vertebral column formula*: $C_7T_{13}\ L_6\ S_5\ Cd_{18\text{-}20}$.

CHAPTER 17

AXIAL SKELETON: HEAD, NECK, AND VERTEBRAL COLUMN

André Desrochers, Chapter editor

Head and Neck

17.1 Maxillary sinusitis—*Caroline Constant* ... 1002
17.2 Tooth root abscess—*Andrew Niehaus* ... 1013
17.3 Dehorning—*Marjolaine Rousseau* ... 1020

Vertebral Column

17.4 Spinal lymphoma—*André Desrochers and Gilles Fecteau* ... 1030

CHAPTER 17

CASE 17.1

Maxillary Sinusitis

Caroline Constant
AO Research Institute, Davos Platz, CH

Clinical case

History

A 4-year-old Holstein lactating cow presented for nasal discharge. The owner first reported unilateral mucopurulent nasal discharge 3 weeks before presentation, which had progressed to bilateral nasal discharge 1 week before presentation. The cow had been treated with IM antibiotics (penicillin G procaine) for 10 days with no resolution of the discharge. Additionally, there was progressive facial swelling noticed during the 3-week period. The owner reported a recent decrease in appetite and milk production.

Percussion of the sinuses is part of a normal physical examination. While it is not reliable to detect sinus disease, a dull percussion suggests a space-occupying substance within the sinuses, such as a mass or exudate. To percuss the sinuses, the fingers of one hand are tapped sharply against the overlying bones of the affected sinuses (see Fig. 17.1-3) immediately followed by percussion of the normal side for comparison.

Physical examination findings

Upon initial clinical examination, the cow was alert and appropriately responsive. An increased inspiratory noise was evident. The cow was moderately tachycardic and tachypneic, and mildly febrile. There was purulent nasal discharge from the left nostril and moderate facial swelling over the left maxillary sinus, creating mild facial distortion (Figs. 17.1-1 and 17.1-2). Percussion of the sinuses were dull compared to the right. Mild epiphora (Gr. *epiphora* sudden burst; tearing of the eye) was noted from the left eye.

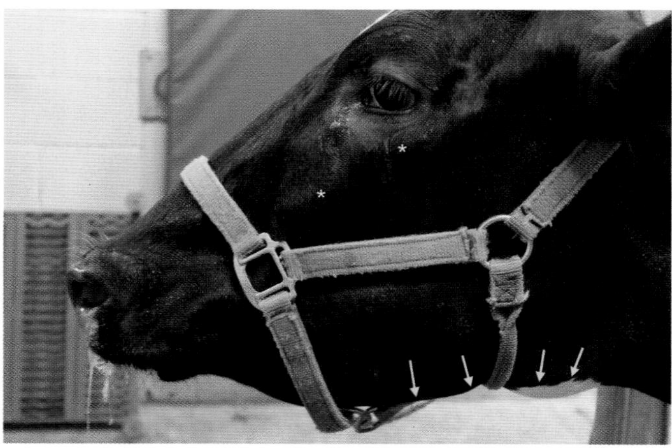

FIGURE 17.1-1 Cow at presentation with unilateral nasal discharge, moderate swelling over the left maxillary sinus *(asterisk)*, and submandibular edema *(arrows)*.

Comparative Veterinary Anatomy: A Clinical Approach. https://doi.org/10.1016/B978-0-323-91015-6.00088-1

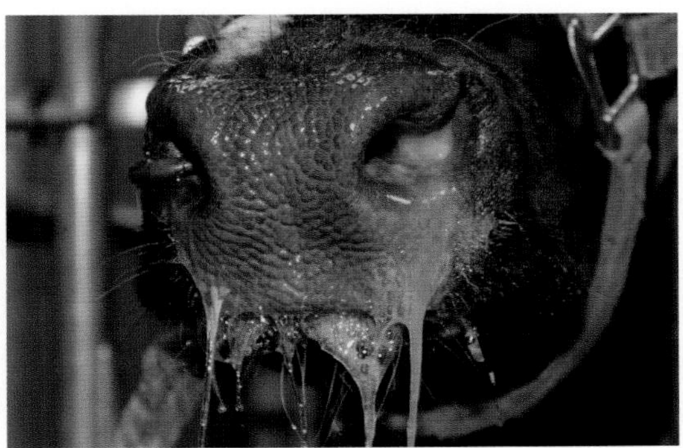

FIGURE 17.1-2 Unilateral purulent discharge apparent from the left nostril on initial clinical examination. Oxygen insufflation was administered through the right nostril.

FIGURE 17.1-3 Topography of the paranasal sinuses, which are the most clinically important in cattle and defined during percussion. Key: 1, maxillary sinus; 2, dorsal conchal sinus; 3, rostral frontal sinus and 4, caudal frontal sinus.

Thoracic auscultation revealed increased lung sounds without abnormal/increased bronchovesicular noises. The remainder of the physical examination parameters was within normal limits. The cow was sedated with 0.05 mg/kg xylazine IV to perform an oral endoscopic examination. Mild periodontitis was noted but no other oral abnormalities or packing of excess feed material in the mouth was observed.

Differential diagnoses

Sinusitis, trauma, and sinus cyst

Diagnostics

Radiographs of the sinuses were taken using a radio-opaque liquid disk marker placed on the x-ray cassette. This marker indicates the lowest point on the radiograph (via gravity) and helps with interpreting fluid lines. Four projections were taken: lateral, dorsoventral, and 30° lateral right and left dorsolateral obliques.

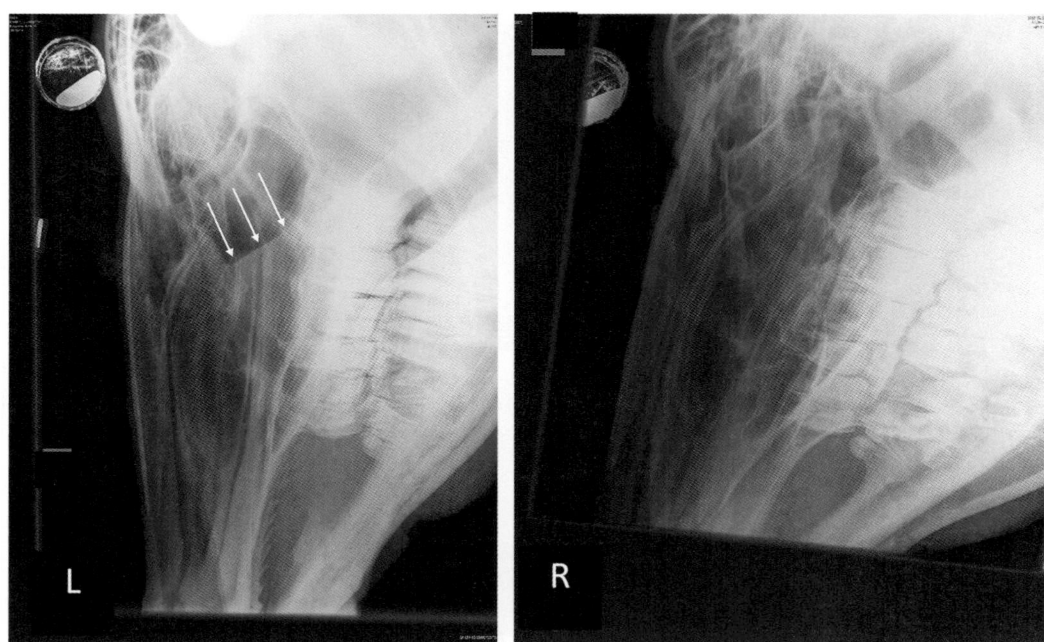

FIGURE 17.1-4 Left (L) and right (R) 30° dorsolateral oblique radiographic views of the maxillary sinuses. The left maxillary sinus radiograph has a fluid line *(arrows)* compatible with fluid in the left maxillary sinus. This fluid line is absent in the right maxillary sinus. Note the disk filled with radio-opaque liquid marker placed in the top left corner of each radiograph, helping to demonstrate that the fluid line parallels the sinus fluid line.

Abnormal fluid components identified by a fluid line were seen within the sinus airspaces. The dorsolateral oblique projections confirmed the suspected location of the fluid components in the left maxillary sinus. The fluid line was noted on the left maxillary sinus projection, but absent on the right projection (Fig. 17.1-4). No other radiographic abnormalities were identified.

Diagnosis

Primary sinusitis of the left maxillary sinus

Treatment

Treatment included sinus irrigation (L. *irrigatio* in, into + *rigare* to carry water = to wash out, lavage, irrigate) with a large volume of warm sterile physiologic saline and systemic antimicrobial therapy pending bacterial culture and sensitivity testing.

A trephination (L. *trepanatio, trephina* removing a circular disk of bone) of the left maxillary sinus was planned for debridement (removal of foreign and devitalized tissue) and irrigation.

The cow was restrained in a standing stock with the head restrained with a halter (Fig. 17.1-5A). Xylazine, butorphanol, and analgesic doses of ketamine (0.05 mg/kg xylazine, 0.025 mg/kg butorphanol, and 0.1 mg/kg ketamine) were administered IM for standing sedation. After aseptic preparation of the surgery site, the area was infiltrated with the local anesthetic, lidocaine. A 3-cm (1.2-in.) diameter full-thickness circular skin incision was made over the left maxillary bone above the facial crest extending to the periosteum. The circular piece of skin was excised. A 19-mm (0.75-in.) Galt trephine was used to remove a section of bone, providing access into the caudal maxillary sinus (Fig. 17.1-5B). A sample of the fluid in the sinus was obtained for bacterial culture, followed by lavage with a large volume of warm, sterile, and physiologic saline (Fig. 17.1-5C).

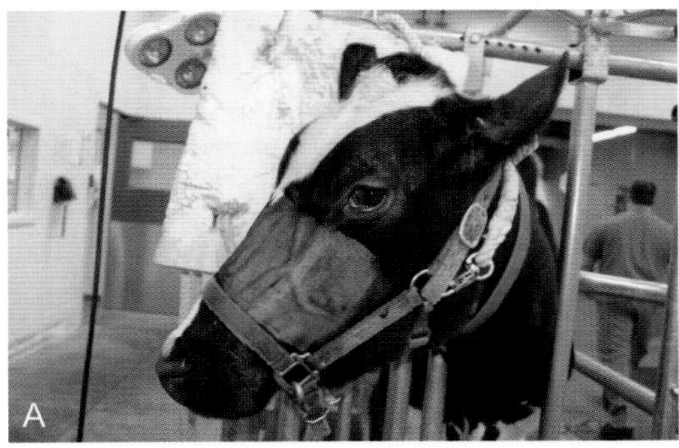
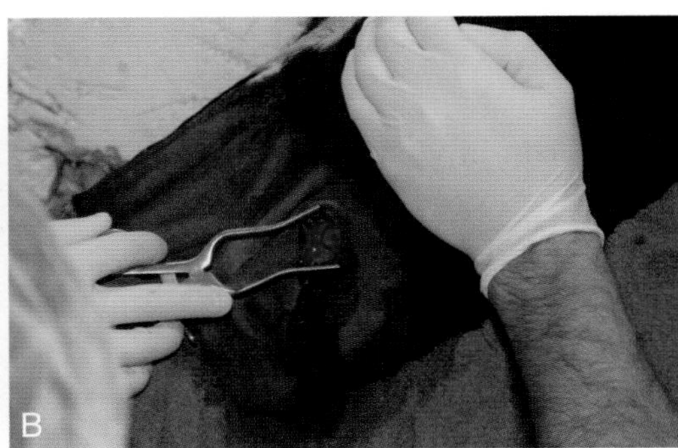
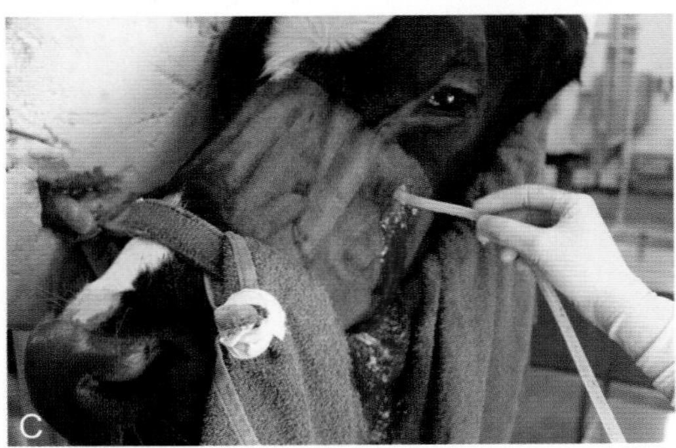

FIGURE 17.1-5 Lavage of the left maxillary sinus through trephination. (A) The cow was restrained in a standing stock for aseptic preparation of the surgery site after local analgesia/anesthesia. (B) The Galt trephine bone cut is apparent through the circular skin incision. (C) The bone disk was removed, and lavage was performed with a large volume of sterile physiologic saline.

The trephine site was left open for daily lavage. A bandage covered the incision in between treatments to minimize environmental contaminants; the incision was allowed to heal by second intention. Antimicrobial therapy, based on culture and sensitivity testing, was continued for 21 days, and nonsteroidal anti-inflammatory drugs were administered according to the cow's clinical comfort and frequency of rumination.

Antimicrobials must be used judiciously in farm animals following the indication and dosage on the label or on the prescription. Additionally, specific antimicrobials (class 1, including fluoroquinolones and ceftiofur) can only be administered based on culture and sensitivity, and if the bacteria are resistant to other classes of antibiotics. These precautions aim to decrease food residue and the development of antibiotic resistance.

Two months after discharge, the owner reported that the cow was doing well with resolution of the nasal discharge. Her appetite and milk production steadily improved post-operatively. The skin covering the trephination healed with no noticeable facial deformation.

Anatomical features in bovids

Introduction

Surgeries for diseases of the upper airway in cattle most commonly involve the frontal or maxillary sinuses. Whatever the cause of sinusitis, trephination and facial bone osteotomy (bone flap) are the most common surgical procedures performed. Although the landmarks for the surgical approaches of sinuses in horses are well described, important anatomical differences exist in ruminants. An understanding of the differences is important for a successful outcome. This section covers the paranasal sinuses of ruminants. The paranasal sinuses of sheep and goats are like cattle, with relevant anatomical differences described where needed.

Function

A sinus (L. a hollow, channel, or fold) is a hollow cavity. There are many sinuses in the body; however, the term "sinus"—or more specifically "paranasal sinuses"—describes the hollow cavities in the head, specifically the cavities around the nose and eyes that drain into the nasal passages. The multiple sinus cavities are all interconnected.

The sinus cavity walls are lined with mucosa and covered with mucus. The mucus supports and maintains moist and healthy tissue; entraps bacteria; and humidifies, warms, and filters air on the way to the lower respiratory tract.

Nasal cavity

The incomplete nasal septum of cattle is of clinical importance during examination of nasal discharge. The origin of bilateral discharge could still derive from a unilateral problem (Fig 17.1-6).

The **nasal cavity** of cattle is small, and much of the internal space is occupied by the conchae (L.; Gr. *konchē* shell). The nasal cavity is divided into equal halves by the **vomer bone** and the **nasal septum**. In contrast to horses, the nasal septum is incomplete at its caudal aspect, resulting in a communication between the right and left nasal passages to the nasopharynx.

The **nasal conchae** are thin scrolls of cartilage and bone present in each nasal cavity. Each nasal passage is divided into 3 meatuses (L., a way, path, course) by the major conchae: **dorsal**, **middle**, and **ventral**. They all branch from the **common meatus** positioned against the nasal septum.

The **ventral meatus** is the most important because of its principal role in air conduction during breathing. The **middle meatus** allows communication between the nasal cavity and certain sinuses (more information in the following section on paranasal sinuses). The rest of the nasal cavity is divided by abundant **ethmoid** (Gr. *ēthmos* sieve, mesh + *eidos* form) **conchae** (Fig 17.1-6).

Traumatic or surgical injury to the nasal mucosa can cause a noticeable bleeding and, when controlled with packing, ordinarily is clinically insignificant.

The wall of the nasal passages is covered by a thick and highly vascularized mucous membrane. Part of the olfactory system, the **vomeronasal organ**, is enclosed ventrally in this mucous membrane.

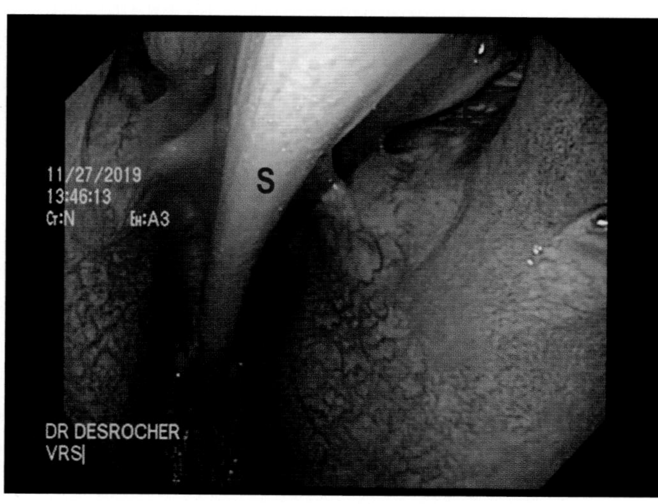

FIGURE 17.1-6 Endoscopic view of the nasal cavity of an adult cow. The nasal septum is incomplete at the caudal aspect of the nasopharynx (*S*). The right and left ethmoid conchae are seen (*E*).

Paranasal sinuses

After birth, the paranasal sinuses progressively expand over several years until they reach full size in the mature cow. The sinuses of the newborn calf are poorly developed. After maturity, the maxillary sinus continues to develop as it adapts to the eruption of the cheek teeth.

The paranasal sinuses are clinically important because they are susceptible to infection extending from the nasal cavity or from the roots of the caudal upper cheek teeth.

The complete paranasal sinus system of cattle consists of 6 pairs of paranasal sinuses: the frontal, maxillary, palatine, lacrimal, sphenoid, and conchal (Figs. 17.1-3 and 17.1-7). All the sinuses communicate with the nasal cavity, directly or indirectly. The maxillary, lacrimal, palatine, and conchal (dorsal and ventral) sinuses open into the **middle nasal meatuses**. The frontal, sphenoid, and conchal (middle) sinuses each open into the respective **ethmoidal meatus** in the caudal aspect of the nasal cavity (Figs. 17.1-8 and 17.1-9).

Any of the sinuses may be come infected, but the sinuses of greatest clinical importance are the maxillary and caudal frontal sinuses. The maxillary sinus is often affected secondary to dental problems. The frontal sinuses can become infected when a mature animal is dehorned and the open sinuses are exposed to the contaminated environment and flies.

The **frontal sinus** occupies the dorsal part of the skull. It extends from a point between the medial canthus of the eye and the caudal margins of the orbit to the horns, or the caudal part of the skull (Fig. 17.1-7). It communicates with the nasal passage via multiple fenestrations (L. *fenestratus* furnished with windows; an opening or open area) into the ethmoid meatus. Unlike horses, the frontal sinus of cattle comprises several compartments. Although there is a rostral compartment, the caudal compartment is the largest, extending primarily within the frontal bone, and further divides in 2 diverticula. The first diverticulum (L. *divertere* to turn aside; a circumscribed pouch or sac), the **cornual diverticulum**,

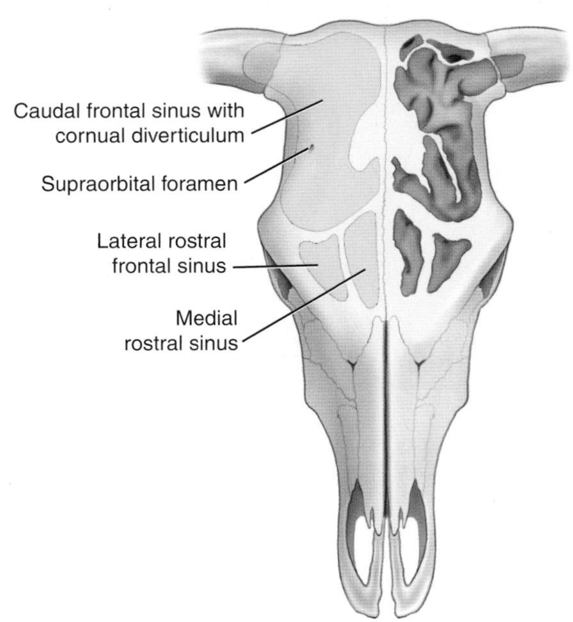

FIGURE 17.1-7 Dorsal projection of the frontal sinuses.

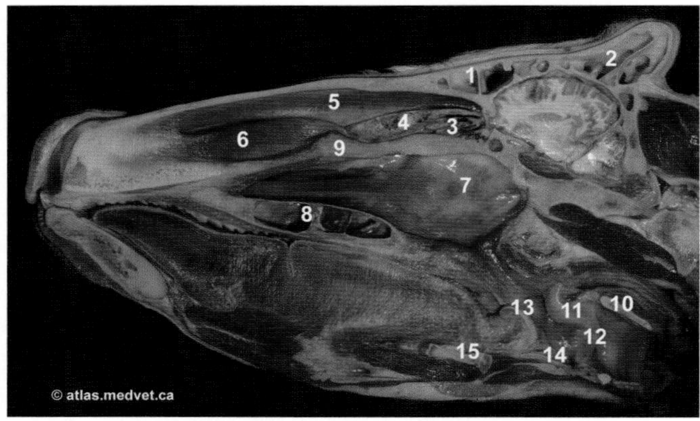

FIGURE 17.1-8 Sagittal section of a prosected head. Key: 1, cranial frontal sinus; 2, caudal frontal sinus; 3, ethmoid conchae; 4, middle concha; 5, dorsal concha; 6, ventral concha; 7, nasopharynx; 8, palatine sinus; 9, vomer bone; 10, cricoid cartilage; 11, arytenoid cartilage; 12, vocal fold; 13, epiglottic cartilage; 14, thyroid cartilage; and 15, basihyoid bone.

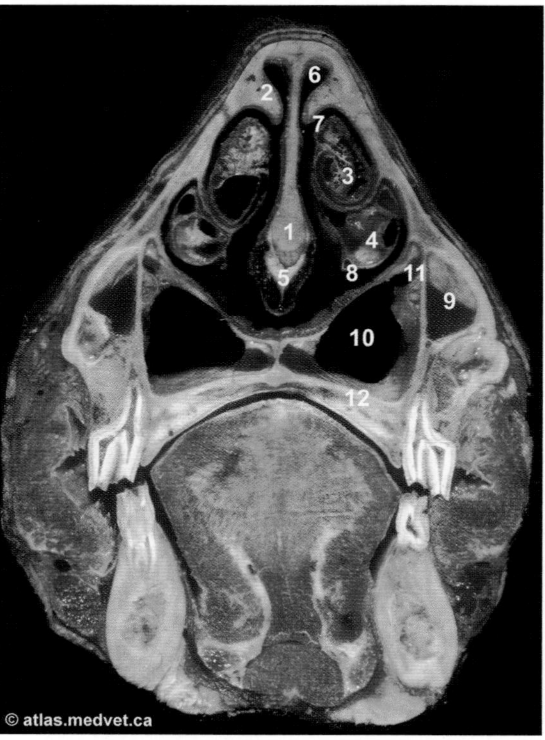

FIGURE 17.1-9 Transverse section of a prosection at the level of the last premolars. Key: 1, nasal septum; 2, dorsal concha; 3, middle concha; 4, ventral concha; 5, vomer bone; 6, dorsal meatus; 7, middle meatus; 8, ventral meatus; 9, maxillary sinus; 10, palatine sinus; 11, infraorbital canal; and 12, hard palate.

extends into the lateral and nuchal (posterior aspect) walls, and into the horn of mature cattle. The second diverticulum, the **postorbital diverticulum**, is located behind the orbit and covers the dorsal part of the brain. The major cavity of the caudal portion of the frontal sinus is subdivided by irregular and perforate septa.

The caudal frontal sinus is divided and named differently in small ruminants; the cornual diverticulum is divided in a small medial subcompartment at its base, while the larger, lateral subcompartment extends into the horn.

The location of the frontal sinus gives it a role in the protection of the cranial cavity. Its extent makes identification of the exact location of the brain difficult.

The **maxillary sinus** is the largest sinus and occupies most of the paranasal cavity covering the caudal upper cheek teeth. The alveoli of the three molar teeth (Triadan 109/209, 110/210, and 111/211) usually enter this sinus. The maxillary sinus extends from the caudal premolars to the last molar. It communicates with

The frontal sinus extends into the horn of the mature bovid (if present) through the cornual diverticulum. Frontal sinusitis is a common complication of surgical dehorning or “horn tipping” in which the frontal sinus is entered. The frontal sinus should also be carefully examined after traumatic fracture of the horn. The septa of the caudal frontal sinus make successful lavage of purulent frontal sinusitis difficult.

The larvae of oestrid flies commonly infest the frontal sinuses in sheep. The preferred area for surgical treatment is rostral to the horn (if present) or medial to the middle of the orbital rim (Fig. 17.1-10A and B). This location is preferred because there is no risk of injury to the frontal vein.

During humane slaughter, the bolt or bullet should pass through the shallowest region of the frontal sinus, which is at the intersection of two diagonals formed by imaginary lines drawn from the horn bases to the contralateral eyes.

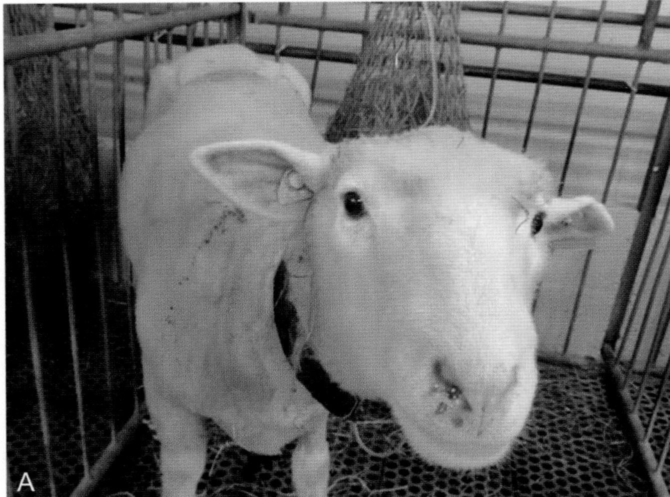

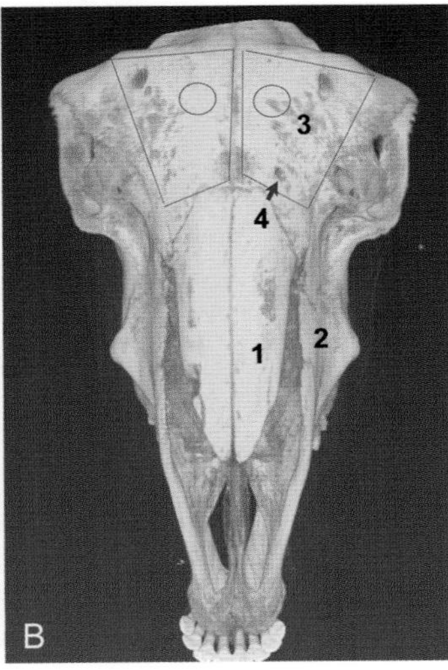

FIGURE 17.1-10 (A) Sheep presented with mild, bilateral, clear mucoid nasal discharge (associated with frontal sinusitis) and a loss of condition with a body score of 2/5. (B) Dorsal view of the 3D reconstruction of the dehorned sheep skull depicting the limits of the frontal sinuses (*orange lines*). The *circles* indicate the site of trephination of the frontal sinuses. Key: 1, nasal bone; 2, maxilla; 3, frontal bone; and 4, supra-orbital canal.

This compressed passage allows drainage of purulent material during sinusitis or the passage may be obliterated by thick secretions, precluding drainage.

the nasal cavity via the **nasomaxillary opening** located high in its medial wall. The maxillary sinus is separated from the **palatine sinus** by the infraorbital canal. There is a large communication between the two over the infraorbital canal.

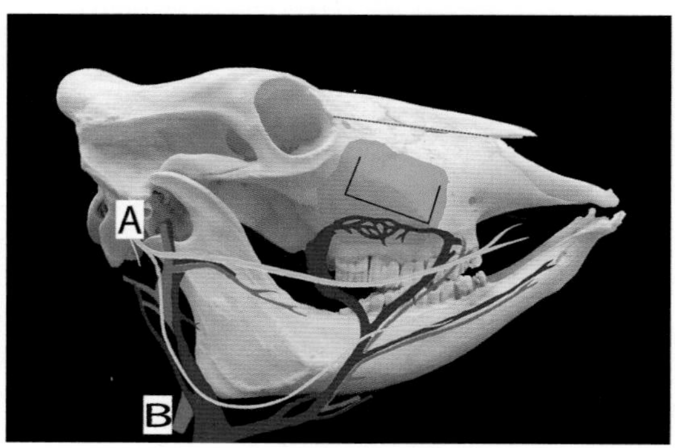

FIGURE 17.1-11 The surgery site for the bone flap technique to expose the maxillary sinus. The boundaries outlined by *black lines* correspond to the skin and periosteal incisions. The bone flap hinges at its dorsal margin. It provides maximal exposure of the maxillary sinus while protecting the nasolacrimal duct. The nasolacrimal duct course is depicted by the *purple bold dashed line* from the medial canthus to the incisive notch. The other relevant structures for surgical extraction of infected cheek teeth through maxillary sinus flap are illustrated. Key: Facial nerve *(A, yellow)* dividing into 2 branches; facial artery *(B, red)* and vein *(B, blue)*; *pink section*, region of the maxillary sinus. (Modified from an original publication in Constant C, Nichols S, Marchionatti E, et al. Cheek teeth apical infection in cattle: diagnosis, surgical extraction, and prognosis. Vet Surg 2019;48:760–9. https://doi.org/10.1111/vsu.13197.)

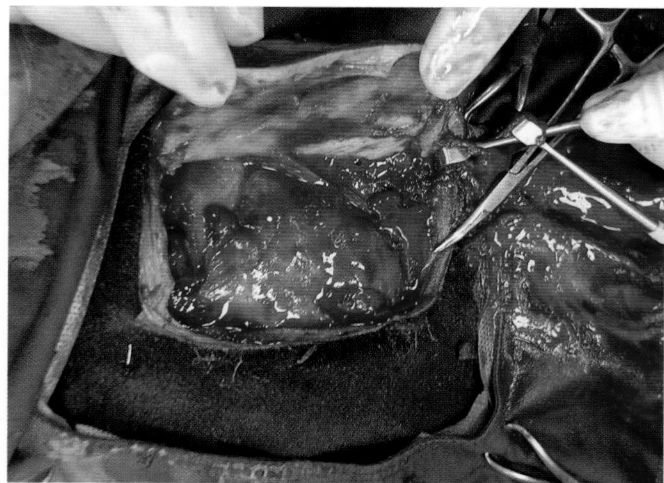

FIGURE 17.1-12 A bone flap exposing the maxillary sinus for a sinonasal cyst in an adult cow. It is a non-neoplastic, fluid-filled mass occupying the nasal conchae along with the maxillary and palatine sinuses. The wall of the cyst is seen filling the cavity.

MAXILLARY SINUSITIS

The location of the maxillary sinus makes it more predisposed to disease compared to other sinuses. Primary sinusitis may result from extension of infection from the nasal cavity. Secondary sinusitis may result from a maxillary cheek tooth periapical abscess. Special attention and care to the facial vein passing over the ventral limit of the sinus is important when surgically entering the sinus (Figs. 17.1-11 and 17.1-12). In addition, the lowest point of the sinus (located above the second molar) is identified/trephined to provide for the best drainage of accumulated material.

The maxillary sinus is shallower in small ruminants than in cattle. It adopts a more pyramidal shape. The limits of the maxillary sinus are: a horizontal line drawn at the level of facial tubercle ventrally, the border of the orbital rim caudally, and a line drawn from the medial eye canthus to the infraorbital foramen dorsally (Fig. 17.1-13). The **infraorbital canal** passes dorsally through the maxillary sinus (Fig. 17.1-14).

The infraorbital nerve is conveyed in the bony infraorbital canal within the sinus and should be preserved during surgery to avoid injury.

Blood supply, lymphatics, and innervation

The principal blood supply to the sinuses is from the **maxillary artery** and drainage is from the **maxillary vein**. The maxillary vessels divide into the **infraorbital** and **descending palatine arteries** and their accompanying **veins**. The nasal cavity's blood supply is provided by the **sphenopalatine, ethmoidal branches**, and **greater palatine arteries,** along with their corresponding veins.

The **mandibular** and **parotid lymph nodes** drain the lymphatics from the head. The mandibular lymph nodes are easily palpated between the rami of the mandible close to the larynx. The parotid lymph nodes are larger, located just caudal to the masseter muscles, and are partially covered by the parotid salivary gland.

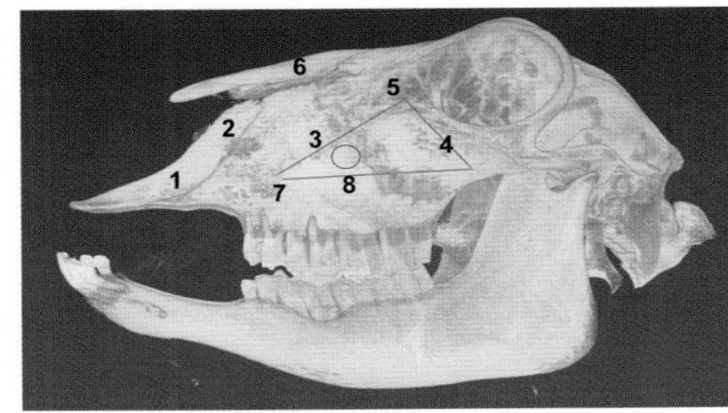

FIGURE 17.1-13 Lateral view of the 3D reconstruction of a dehorned sheep skull depicting the limits of the left maxillary sinuses (*orange lines*). The *circle* indicates the site of trephination of the left maxillary sinuses. Key: 1, incisive bone; 2, nasal process; 3, maxilla; 4, zygomatic bone; 5, lacrimal bone; 6, nasal bone; 7, infraorbital foramen; and 8, facial tubercle.

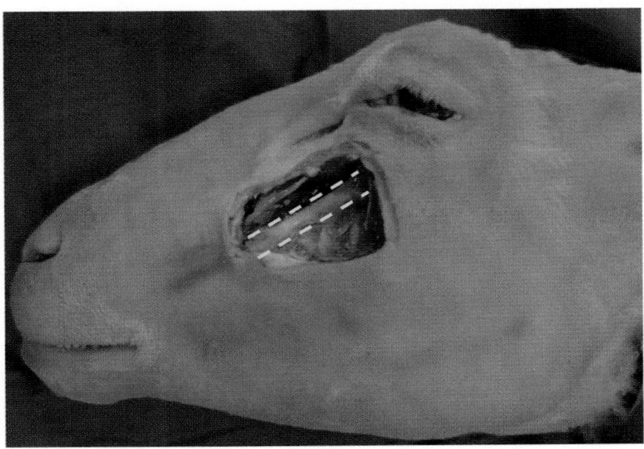

FIGURE 17.1-14 Lateral view of a prosected sheep head depicting the course of the infraorbital nerve within the infraorbital canal, highlighted by the *dashed lines* within the sinuses.

The nasal cavities derive sensory innervation from the **trigeminal nerve**. It is divided into 2 major branches: the **ophthalmic** and **maxillary nerves**.

Selected references

[1] Constant C, Nichols S, Marchionatti E, et al. Cheek teeth apical infection in cattle: diagnosis, surgical extraction, and prognosis. Vet Surg 2019;48:760–9. https://doi.org/10.1111/vsu.13197.

[2] Ducharme NG, Desrochers A, Mulon PY, Nichols S. Surgery of the bovine (adult) respiratory and cardiovascular systems. In: Fubini SL, Ducharme NG, editors. Farm animal surgery. 2nd ed. Missouri: Elsevier; 2017. p. 193–222.

[3] Sisson S. Ruminant osteology. In: Sisson G, Grossman J, editors. Anatomy of domestic animals. Philadelphia: WB Saunders Co.; 1975. p. 741–74.

[4] Awaad AD, Abdel Maksound MKM, Fathy MZ. Surgical anatomy of the nasal and paranasal sinuses in Egyptian native sheep (*Ovis aries*) using computed tomography and cross sectioning. Anat Histol Embryol 2019;48:279–89.

[5] Anderson DE, DeBowes RM, Gaughan EM, Yvorchuk KE, St-Jean G. Endoscopic evaluation of the nasopharynx, pharynx, and larynx of Jersey cows. Am J Vet Res 1994;55:901–4.

CASE 17.2

Tooth Root Abscess

Andrew Niehaus
Veterinary Clinical Sciences, The Ohio State University College of Veterinary Medicine, Columbus, Ohio, US

Clinical case

History

A 7-year-old male intact alpaca presented for evaluation of a swelling on the left ventral mandible first noticed 1 year ago. The owner reported a draining tract and had been treated with antimicrobials (florfenicol and ceftiofur) with some improvement in clinical signs; however, the swelling and pain returned when antimicrobials were stopped (Fig. 17.2-1). Recently, his ingestion of grain had declined; however, he was eating enough hay to maintain body condition with no reported change in body weight since the swelling first appeared. The caretaker had manually expressed purulent material from the site and flushed it with an iodine solution. There were no other health problems reported.

Physical examination findings

On presentation, the patient was alert and appropriately responsive. There was a hard (bony) swelling on the left ventral hemi-mandible. The alpaca resisted digital pressure on the swelling. A draining tract was noted on the ventral aspect of the swelling with dried purulent exudate on the skin around the draining tract. The remainder of the physical examination findings were normal.

The patient was sedated with 0.3 mg/kg of xylazine IV to facilitate a thorough oral exam. No missing teeth, foreign material (including excess feed) in the mouth, or gingiva lesions were observed. Teeth in the affected area were not loose when manipulated.

FIGURE 17.2-1 A llama with mild swelling and a draining tract on the right ventral mandible (note the small scab—*arrow*) similar to that as described in this case. The *arrow* points to the opening of the draining tract.

Comparative Veterinary Anatomy: A Clinical Approach. https://doi.org/10.1016/B978-0-323-91015-6.00089-3

Differential diagnoses

Periapical tooth root abscess, actinomycosis/generalized osteomyelitis of the mandible, mandibular fracture, sequestrum, and neoplasia

Diagnostics

The alpaca was sedated, a radiolucent mouth gag placed in the mouth, and 4 radiographic views were taken as follows: dorsoventral, lateromedial, a 45° right dorsolateral oblique, and a 45° left dorsolateral oblique (Fig. 17.2-2).

Radiographs revealed moderate alveolar lysis with adjacent sclerosis along the rostral aspect of the caudal root of the left mandibular second molar (Triadan 310) and mild alveolar lysis along the rostral aspect of the cranial root of the second molar. There was also moderate alveolar lysis and adjacent sclerosis of the caudal root of the right mandibular second molar. No other bony abnormalities were identified.

A metal teat cannula, used as a probe, was gently inserted into the draining tract, and a 45° left dorsolateral oblique radiograph was taken, highlighting the left hemimandible. The teat cannula was superimposed over the caudal root of the second mandibular molar (Fig. 17.2-3).

Diagnosis

Bilateral tooth root abscess of the second mandibular molar, more severe on the left side

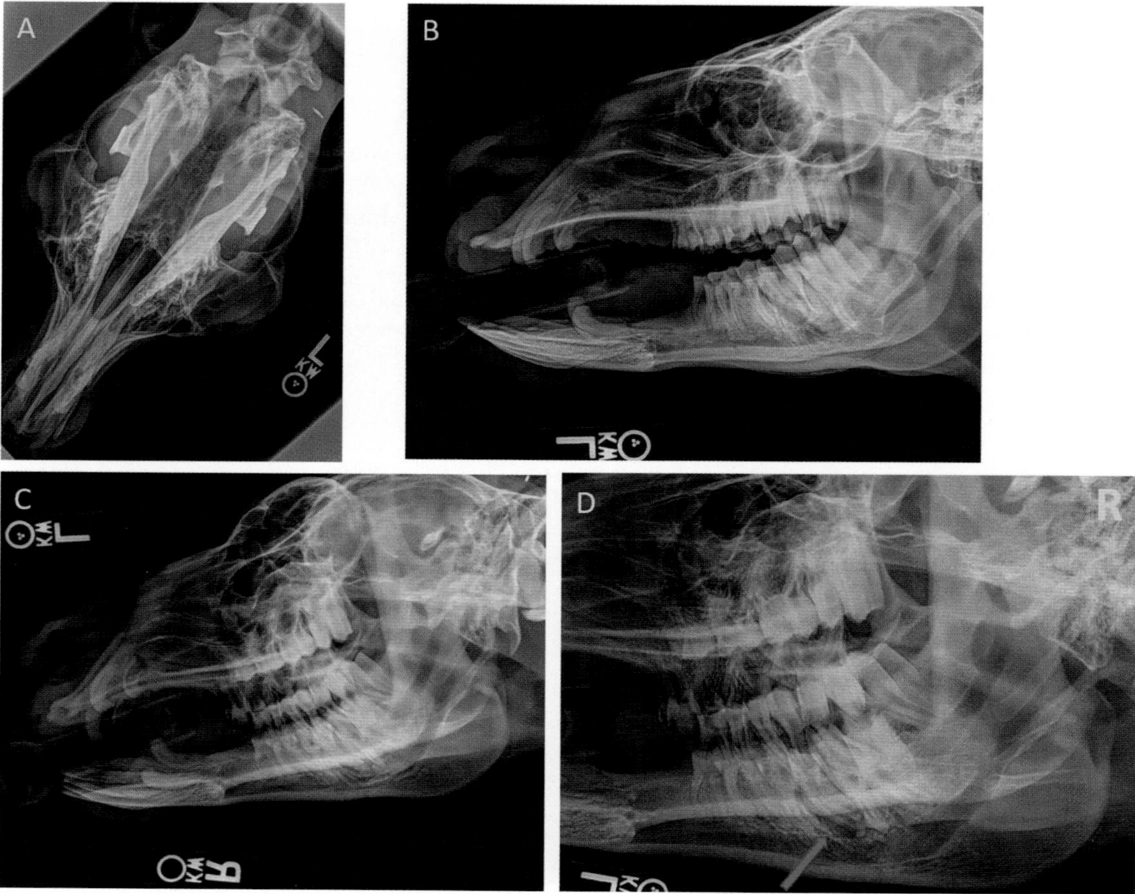

FIGURE 17.2-2 Radiographic views of a llama's jaw with swelling and a draining tract on the left side. (A) Dorsoventral. (B) Lateral. (C) 45° right dorsolateral oblique. (D) 45° left dorsolateral oblique projections. The *blue arrow points* to an area of alveolar lysis along the rostral aspect of the caudal root of the left mandibular second molar.

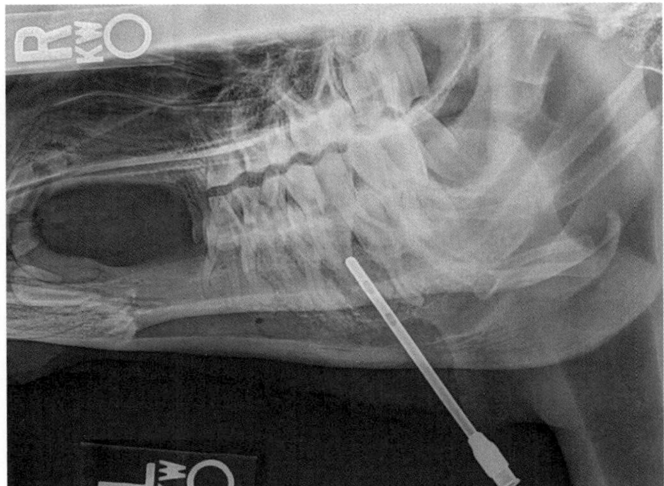

FIGURE 17.2-3 A 45° left dorsolateral ventromedial oblique radiographic view after insertion of a teat cannula into the draining tract.

Treatment

The alpaca was placed under general anesthesia and a pneumatic burr was used to remove the alveolar bone lateral to the left 2nd mandibular molar, exposing the tooth roots. The periodontal ligament surrounding the tooth roots was loosened with dental elevators until the entire tooth was loose, and the 2nd molar was extracted. A small bone sequestrum was found in the area and was removed.

Bleeding from the bone was controlled with bone wax. A sample for culture and sensitivity was taken from both the tooth root and the sequestrum. The area was lavaged with sterile saline to remove any debris associated with the debridement. The area was packed with Betadine®-soaked cotton packing material and the wound was partially closed with the packing material protruding through the ventral aspect of the incision to keep the incision open for ventral drainage. The packing was removed the day after surgery and the opening lavaged; this was repeated for several days while hospitalized. The patient was discharged with instructions to flush the wound daily until closed.

Three months postsurgery, the owner reported that the alpaca was doing well. The incision site was healed with mild residual swelling on both sides of the face, but there was no draining tract and the alpaca was eating normally and gaining weight.

Anatomical features in alpacas and llamas

Introduction

This section covers the dental anatomy, with regional blood supply, innervation, and lymphatics, of camelids.

Function

Camelids have well-developed cheek teeth (premolars and molars) and less-developed incisors and canines; they are thus adapted to grinding forages. Like ruminants, the upper incisors are mostly absent and replaced with a dental pad, which is composed of connective tissue covered with epithelium. The exception is the 3rd upper incisor (see side box entitled "Fighting teeth").

Tooth anatomy

Each tooth has a crown, a neck, and a root. The **crown** is the part exposed in the oral cavity, the **root** is the part attached to the bone of the mandible or maxilla, and the **neck** forms the junction between the crown and the root at the gumline. The tip of the root opposite the crown is referred to as the **apex**. The **apical foramen** communicates with the pulp cavity and serve as a passageway for the nerve and blood vessel (Fig. 17.2-4).

Each tooth has 5 surfaces (see Fig. 1-2 in Clinical anatomy nomenclature): **lingual**—facing the tongue, **labial** or **buccal**—facing the cheek, **occlusal**—facing the opposite arcade, **mesial**—facing the front of the mouth, and **distal**—facing to the back of the mouth. The mesial surface and the distal surface are referred to as "contact surfaces" because they contact the adjacent tooth.

The **pulp cavity** in the center of the tooth contains the sensitive neurovascular structures and is important for tooth viability. The pulp cavity is encased in a hard material termed **dentin**. At the root, dentin is encased in **cementum** and the **periodontal ligament** connects the cementum to the surrounding bone. At the crown, dentin is encased by **enamel** (Fig. 17.2-4). Enamel is the hardest and densest material of the tooth.

Dental anatomy

Llamas and alpacas have 2 premolars (PM3 and PM4) and 3 molars (M1, M2, and M3) present in both the mandibular and maxillary arcades. The 3rd premolar is less well-developed and is frequently missing in adults. Note that there are no 1st and 2nd premolars in camelids (Fig. 17.2-5). The dental formula for llamas and alpacas is summarized in Table 17.2-1, while Table 17.2-2 summarizes the replacement of the deciduous teeth with permanent teeth.

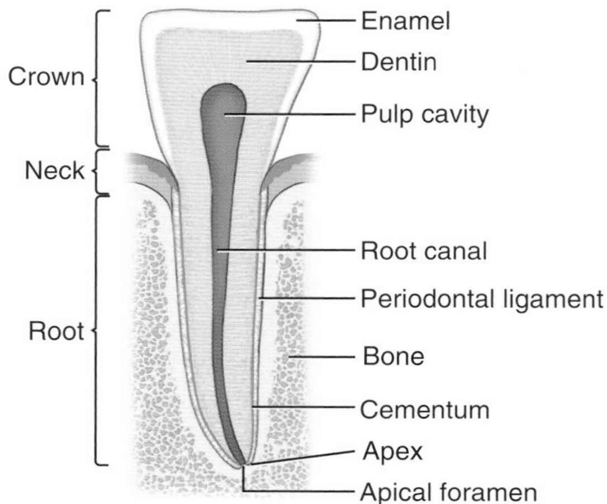

FIGURE 17.2-4 Parts and composition of the llama and alpaca tooth.

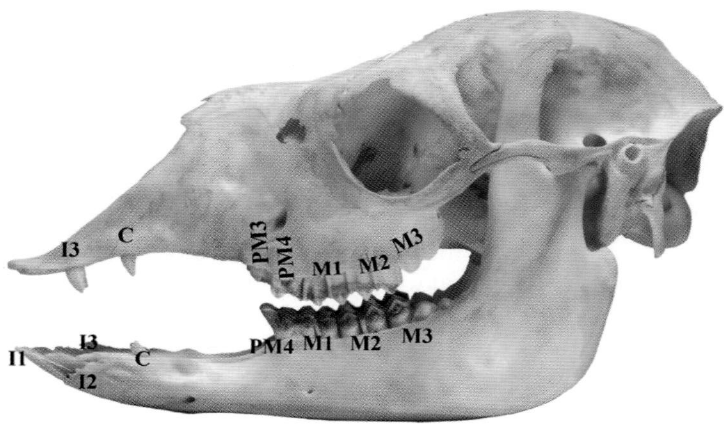

FIGURE 17.2-5 Skull of male llama with teeth of the maxillary and mandibular arcades labeled. *Key*: I = incisors, C = canines, PM = premolars, M = molars.

TABLE 17.2-1 Dental formula for llamas and alpacas.

		Incisors	*Canine*	*Premolars*	*Molars*	*Total*
Deciduous	Maxillary	1	1	2–3	–	2X=8–10
	Mandibular	3	1	1–2	–	2X=10–12
Permanent	Maxillary	1	1	1–2	3	2X=12–14
	Mandibular	3	1	1–2	3	2X=16–18

TABLE 17.2-2 Permanent dental eruption dates for llamas and alpacas.

I1	2–2.5 years	PM3	3.5–5 years
I2	3–3.25 years	PM4	3.5–5 years
I3	3.1–6 years	M1	6–9 months
C	2–7 years (most erupt by 3.5 years)	M2	1.5–2 years
		M3	2.75–3.75 years

Eruption dates are variable. Deciduous premolars are erupted at birth. Deciduous incisors are usually erupted at birth. Unerupted incisors are a classic sign of prematurity.

The shape of individual check teeth varies considerably. In the maxillary arcades, PM3 is small and triangular (1×1×1 cm [0.4×0.4×0.4 in].) with 3 roots and no infundibulum, and PM4 is rectangular (1.5×1 cm [0.6×0.4 in.]) and has 1 medial and 2 lateral roots with 1 infundibulum. All 3 maxillary molars are larger (approximately 1.5×2.5 cm [0.6×1.0 in.]) and have 4 roots and 2 infundibuli. In the mandibular arcade, PM3, if present, is conical (0.8×0.5 cm [0.3×0.2 in.]) and possesses 2 roots and no infundibulum. The mandibular PM4 is triangular (1×0.6 cm [0.4×0.2 in.]) and has 2 divergent roots and no infundibulum. The 3 mandibular molars are rectangular in shape. Moving caudally, the molars progressively become larger and more complex. M1 is approximately 1.7×1 cm (0.7×0.4 in.), while M2 is 2.4×2.0 cm (0.9×0.8 in.) and M3 is 3×1.2 cm (1.2×0.5 in.). M1 and M2 both have 2 roots and 2 infundibuli. M3 has 3 roots (the caudal 2 are fused) with 2 infundibuli. The rostral and caudal roots of the cheek teeth do not communicate with each other in camelids.

Tooth disease can be confined to the pulp cavity of 1 cheek tooth root and not involve the adjacent pulp cavity of the other root. A surgeon also has the option of splitting a tooth in the case of focal disease and removing 1 root (and associated pulp cavity) without disrupting the other (Fig. 17.2-6).

The mandibular and maxillary cheek teeth do not completely appose each other. The maxillary teeth are in a labial position to the mandibular teeth, which lie more lingually.

Wear of mandibular cheek teeth is irregular, creating enamel points on the labial side of the maxillary arcade and the lingual side of the mandibular arcade. Unlike in horses, these usually are not a clinical problem and routine floating of the teeth to a more level confirmation is generally unnecessary.

FIGHTING TEETH

Males have well-developed upper 3rd incisors and upper and lower canines which are referred to as the "fighting teeth." These teeth are adapted as weapons used in fighting other males. Females and castrated males may or may not possess permanent canines. The deciduous canines are small and rarely erupt in females and in only about 5% of males.

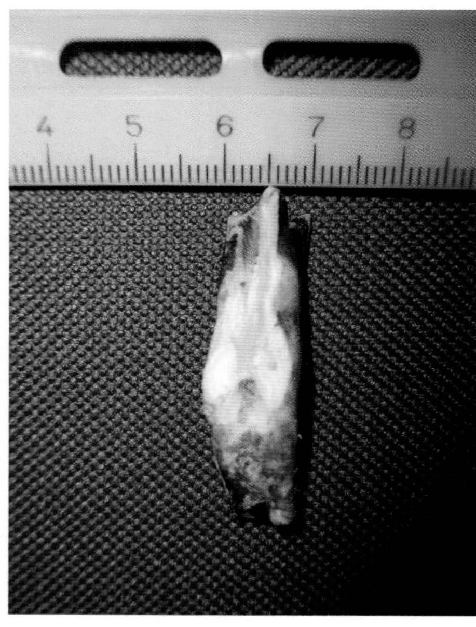

FIGURE 17.2-6 Cut surface of the caudal root of a molar that was split demonstrating the absence of a communication between rostral and caudal roots. Only the caudal root was removed during surgery in this patient. The image depicts the cut surface in the foreground. Intact dentin covers the cut surface; no pulp cavity is exposed. The scale is in cm.

Blood supply, lymphatics, and innervation

A **mandibular canal** runs from the mid-ramus caudally to the angle of the mandible and rostrally to the diastema. The mandibular canal opens as the **mandibular foramen** caudally on the medial aspect of the ramus and rostrally on the lateral aspect of the diastema as the **mental foramen**. The **inferior alveolar artery** and **vein** run through the mandibular canal, and this provides the main blood supply to the mandibular teeth. The blood vessels emerge through the mental foramen as the **mental arteries** and **vein**. The maxillary artery branches to give off the **infraorbital**, **palatine**, and **sphenopalatine arteries**. Collectively, these supply blood to the maxilla and associated structures.

The inferior alveolar nerve can be blocked to provide regional anesthesia to the mandible for dental procedures or it can be used to augment general anesthesia for a lighter plane of general anesthesia during painful dental procedures. It can also be used to provide short-term analgesia post-operatively. The inferior alveolar nerve can be blocked in 2 locations: first, at the mandibular foramen on the medial aspect of the mandible, and second, by inserting a small needle through the mental foramen at the rostral aspect of the mandible and infiltrating local anesthetic through the mandibular canal (Fig. 17.2-7).

The **mandibular nerve** branches to give off the **inferior alveolar nerve**, which runs through the mandibular canal with the inferior alveolar artery and vein and emerges from the mental foramen as the **mental nerve**. This nerve innervates the mandibular teeth. The maxillary teeth are innervated by the **maxillary nerve** and its continuation, the **infraorbital nerve**.

The **parotid**, **mandibular**, and **retropharyngeal lymph nodes** receive lymph drainage from the head region. There is collateral drainage provided by these nodes. Most of the lymph eventually passes through the lateral retropharyngeal node and continues down the neck eventually reaching the thoracic duct.

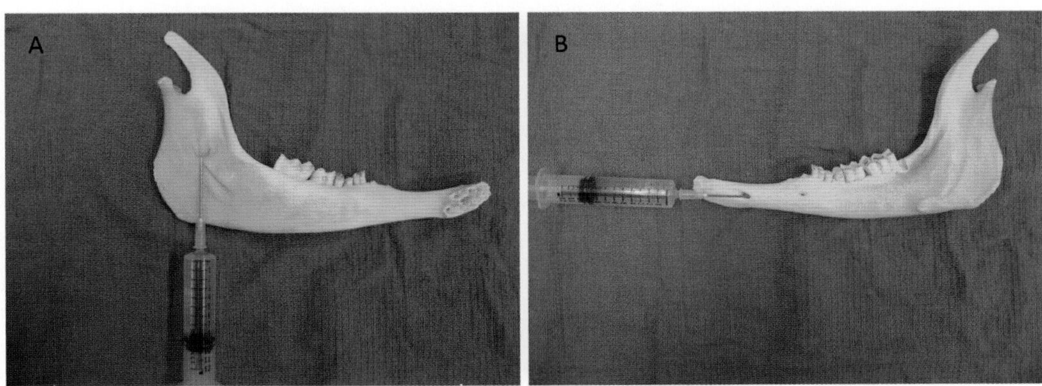

FIGURE 17.2-7 Needle placement demonstrating the location for performing an inferior alveolar nerve block. (A) Medial view of the mandible. A 3.8-cm (1.5-in.) needle is inserted just rostral to the angle of the mandible, perpendicular to the body of the mandible at its depth to reach the mandibular foramen. (B) The same needle is inserted into the mental foramen and local anesthetic is infiltrated into the mandibular canal.

Selected references

[1] Fowler M. Digestive system. In: Fowler M, editor. Medicine and surgery of camelids. John Wiley & Sons; 2011. p. 351–402.
[2] Cebra CK. Disorders of the digestive system. In: Cebra C, Anderson DE, Tibary A, et al., editors. Llama and alpaca care: medicine, surgery, reproduction, nutrition, and herd health. 1st ed. St. Louis, MO: Elsevier; 2014. p. 477–536.
[3] Niehaus A. Dental disease in llamas and alpacas. Vet Clin North Am Food Anim Pract 2009;25:281–93.

CASE 17.3

Dehorning

Marjolaine Rousseau
Department of Clinical Sciences, Université de Montréal, Faculty of Veterinary Medicine, Saint-Hyacinthe, Québec, CA

Clinical case

History

A 9-month-old show Holstein heifer calf presented for an inappropriate poll (intercornual protuberance) silhouette and bilateral partial horn regrowth. This heifer calf had been dehorned at 3 months of age using a Barnes dehorner. Her weight was estimated at 321 kg (706 lbs) by measuring her girth circumference using a commercially available tape. She reportedly had a normal appetite since birth and no history of illness.

Physical examination findings

The heifer was bright and alert. Her vital signs (rectal temperature; heart and respiratory rates) were normal; gingival mucous membranes were pink and moist with a normal CRT. Examinations of her digestive, cardiac, respiratory, mammary, and locomotor systems were normal. Two approximately 2 × 2 cm (0.8 × 0.8 in.) regrown horns were palpated on either side of her poll (Fig. 17.3-1) and were well-attached to the frontal bone. Palpation of the frontal bones, percussion of the frontal sinuses, and subjective evaluation of the nasal air outflow did not reveal any clinical abnormalities.

Diagnosis

Bilateral partial horn regrowth

Treatment

Cosmetic dehorning was elected in this show heifer to minimize noticeable scarring of the poll and to achieve primary healing of the surgical sites during the fly season. Peri-operative antimicrobials and anti-inflammatories

METHODS TO DISBUD AND DEHORN CATTLE

Disbudding is usually performed by thermal cautery using a hot iron or by chemical cautery using a caustic paste of sodium, potassium, and/or calcium hydroxide applied to the horn bud. Disbudding can also incorporate a cutting method when a tube dehorner (or Roberts dehorner) is twisted over the horn bud, making a circular incision around it.

Dehorning is achieved using other types of cutting methods with a various-sized scoop/gouge dehorner (also called Barnes dehorner), Keystone guillotine dehorner, electric guillotine dehorner, or Gigli/obstetrical wire. The defect created during this procedure, no matter how wide, is left to heal by second intention without horn development. However, primary closure of the skin over the defect created by the excision of the horn can be achieved using a cosmetic dehorning technique. Drainage from the dehorning site is impossible when the skin is closed over the defect. Therefore, aseptic technique is necessary.

Comparative Veterinary Anatomy: A Clinical Approach. https://doi.org/10.1016/B978-0-323-91015-6.00090-X

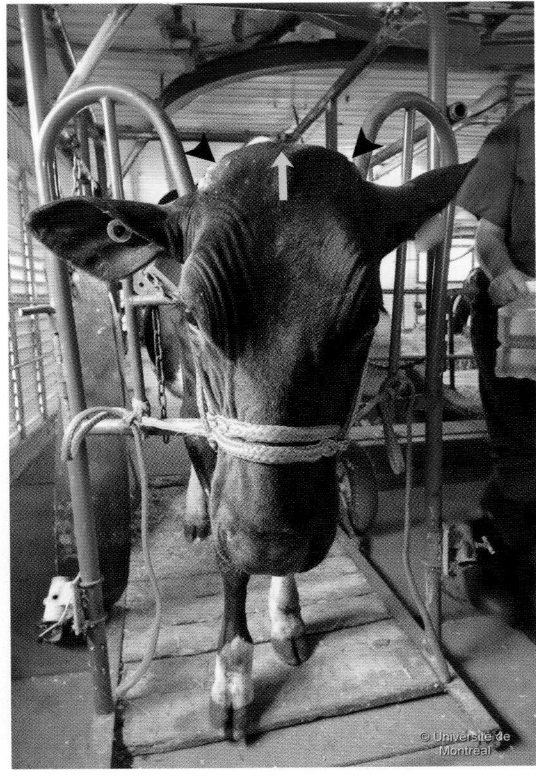

FIGURE 17.3-1 Photograph of the 9-month-old show Holstein heifer calf presented for a "squared" silhouette of her poll (intercornual protuberance, *arrow*) and regrown horns *(arrowheads)* after being dehorned 6 months earlier. The owner could not show the heifer with this blemish. Note that the heifer's head is restrained using two halters to facilitate intra-operative handling of her head without repositioning halters and possibly contaminating the surgery sites.

were administered. The heifer was sedated and restrained in a standing hoof trimming chute. The right and left cornual nerves were anesthetized by local infiltration of lidocaine immediately ventral to the temporal line of the frontal bone approximately half-way between the lateral aspect of the orbit and the base of the regrown horn (Fig. 17.3-2). To ensure absolute desensitization of the horn region in this mature bovine patient, lidocaine was also infiltrated around the caudal part of the base of each regrown horn. Analgesia was confirmed by loss of skin sensation around the base of each horn.

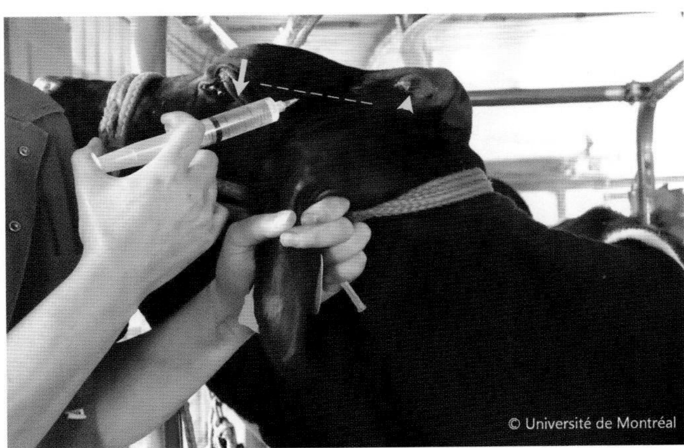

FIGURE 17.3-2 Photograph of a left cornual nerve block being performed using an 18-gauge 0.6-cm (1.5-in.) needle attached to a 25-mL syringe. Lidocaine is injected immediately ventral to the temporal line of the frontal bone *(dashed line)*, approximately half-way between the lateral canthus of the eye *(arrow)* and base of the regrown horn *(arrowhead)*.

Following aseptic preparation and surgical draping of the cornual and intercornual regions, an elliptical skin incision from a point approximately 5 cm (2.0 in.) dorsal to—and ending approximately 5 cm (2.0 in.) ventral to—the base of the left horn was performed and extended through all tissue layers until the underlying bone was reached. The skin surrounding the elliptical incision was sharply undermined and elevated from the underlying bone. The regrown horn was excised using a large, sterile Barnes dehorner (Fig. 17.3-3). While the dehorner was pressed against the exposed bone, the handles were separated, bringing the dehorner blades together to excise the regrown horn and surrounding tissue. This exposed the caudal compartment of the frontal sinus (Fig. 17.3-4). The skin was sutured and the right, regrown horn was surgically removed using the same technique (Fig. 17.3-5). A modified figure-of-8 bandage was applied to the surgical site and removed 2 days post-operatively; suture removal was performed 14 days postsurgery. The heifer had no post-operative complications and achieved 1st intention healing of the surgical incisions.

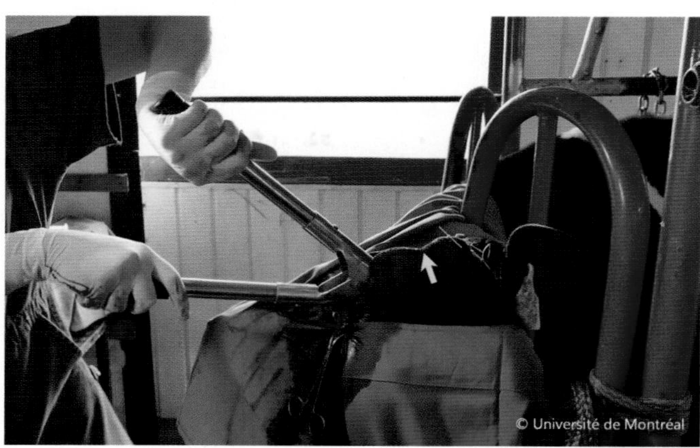

FIGURE 17.3-3 An intra-operative image depicting the jaws of the sterile Barnes dehorner placed at the base of the right regrown horn, positioned directly on the frontal bone, ensuring complete removal of cornual germinal epithelium and preventing horn regrowth. To ensure the dehorning resulted in a sharp, slightly pointed silhouette of the poll (intercornual protuberance, *arrow*), the blades and handles of the Barnes dehorner should be placed over the horn to excise in the same axis as the bony ridge of the lateral aspect of the poll (intercornual protuberance).

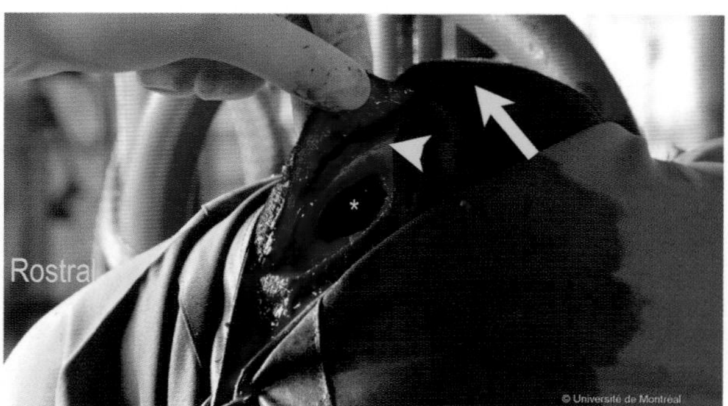

FIGURE 17.3-4 An intra-operative image of the left cornual surgical site. The regrown horn and surrounding tissue were excised, exposing the caudal compartment of the frontal sinus *(star)*. *Arrow*—caudal aspect of the poll (intercornual protuberance). *Arrowhead*—exposed frontal bone.

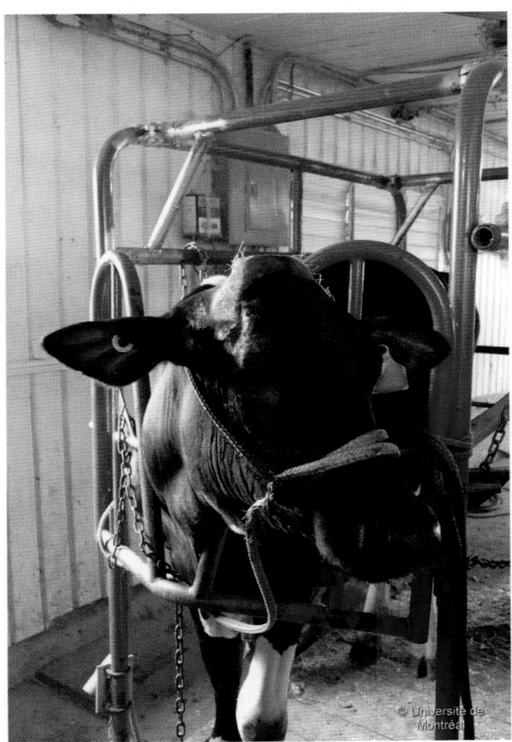

FIGURE 17.3-5 Photograph showing the post-operative appearance of the heifer's poll (intercornual protuberance) after cosmetic dehorning.

Anatomical features in bovids

Introduction

Paired horns (L. *cornuhorn*) are present in both sexes, except in naturally hornless or polled breeds of livestock—e.g., Aberdeen Angus, Polled Hereford, and Polled Shorthorn. Horns are permanent and grow continuously throughout life. Their size and conformation vary according to the species of domestic ruminant, breed, age, and gender. In some breeds of cattle (e.g., Texas Longhorn, Scottish Highland), sheep (e.g., Jacob), and goats (e.g., Boer), breed associations may require the presence of horns in purebred individuals.

For economic and management purposes, and for the safety of humans and other animals, domestic ruminants can be disbudded (removal of the horn bud) at a young age or dehorned (removal of the developed horn) at a later stage of maturity.

Function

Horns are used to establish and maintain of hierarchical levels in wild ruminants. Horns are part of the integumentary system, as are the skin, hair, dewclaws, and claws or hooves. Horns have a bony base (part of the frontal bones) and are supplied by the cornual processes.

Horn anatomy in cattle

In domestic ruminants, the **horn**, also called **cornu**, consists of an epidermal horn sheath protecting a bony center, the **cornual process**. The horn can be subdivided into 3 regions: **base** (basis cornu), **body** (corpus cornus), and **apex** (apex cornus). Horns usually have a conical shape and are located at the caudolateral aspect of the skull in a temporal position in cattle (Fig. 17.3-6). In goats, horns are near the parietal bone, just caudal to the orbits

FIGURE 17.3-6 A left lateral image of an adult bovine skull. The conical-shaped horns are in a temporal position at the caudal and lateral aspects of the skull on either side of the poll (intercornual protuberance). On this skull, the right horn still maintains its horn sheath *(A)* while the cornual process of the frontal bone *(B)* is exposed on the left side. Small vascular canals are seen within the cornual process.

(Fig. 17.3-7). Horns extend in lateral and dorsal directions in cattle and in a caudal direction in goats. (More detailed information on goats is included below.)

Horn development in cattle

In cattle, the **horn bud** starts to develop during fetal life. At approximately 70 days of gestation, the horn bud is visible and formed of several layers of keratinocytes within the epidermis of the frontal region of the head. Nerve bundles are found within the dermis of the horn bud at 3 months of gestation, grow during the rest of fetal life, and are well-developed just before birth. Hair follicles are found deep to the horn bud at 4–5 months of gestation.

At birth, a **trichoglyph** (whorl or swirl of hair) is visible over the future location of each horn in the frontal region. The hair follicles and glands at this specific location atropy (decrease in size) one month after birth. The keratinocytes within the epidermis proliferate to form pointed **horn buds** that develop outward from the skin. During the same period, a bony process, the **cornual process (processus cornualis) of the frontal bone**, develops and constitutes the center of the developing horn.

Paranasal sinuses of young calves are under-developed. The cornual process continues to develop as a solid process until 5 months of age. Starting at approximately 6 months of age, the mucosa of the caudal compartment of the

Disbudding prevents horn development by heat cauterization or chemical burning of the germinal epidermal layer of the horn bud before the cornual process of the frontal bone attaches the horn bud to the skull at approximately 2 months of age (Fig. 17.3-8). During disbudding, the hot iron is applied to the skin at the base of the horn bud until a white ring of bone is seen around the application site (portable or gas dehorner) or until a copper-colored ring is present around the application site (corded or electric dehorner). During dehorning, it is important to remove a haired skin ring at the base of the horn to ensure complete removal of the germinal epithelium responsible for cornual formation. Otherwise, a variably sized horn, often called a “scur,” regrows, usually without attachment to the skull.

Caution is advised in overheating tissues around the horn bud when disbudding a young calf, as heat could be transmitted to the calvarium and meninges, causing thermal meningitis.

FIGURE 17.3-7 A right caudolateral image of an adult male goat (buck) skull. The base of horns in goats is located immediately caudal to the orbits. Their horns usually grow in a caudal direction. Both horns on this skull still maintain their respective horn sheaths on which circumferential grooves and ridges are evident due to intermittent production of the horn sheath.

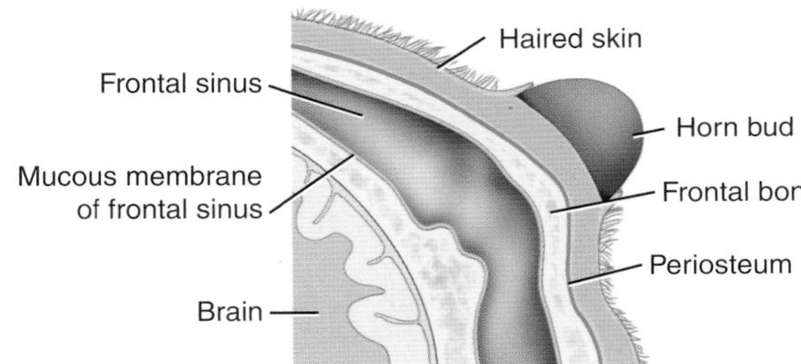

FIGURE 17.3-8 Unattached horn bud. Before 2 months of age, the horn bud is "floating" within the skin and not attached to the underlying frontal bone.

frontal sinus starts to invade the solid cornual process until the entire process is cavitary. The apex of the process remains dense. This phenomenon is called "pneumatization" of the cornual process. This air-filled structure thus formed is called the **cornual diverticulum of the caudal compartment of the frontal sinus** (Fig. 17.3-9). The cornual process and diverticulum of the frontal sinus become progressively larger as the animal ages.

> As depicted in Fig. 17.3-4, dehorning of older calves or adult cattle exposes the frontal sinus due to the large communication of the caudal compartment of the frontal sinus with its cornual diverticulum. Similarly, cutting the tip of a horn through the cornual process of the frontal bone, or the presence of a fracture of the horn at the level of its base or body, exposes the caudal compartment of the frontal sinus. These situations carry the risk of bacterial and parasitic sinusitis. Goats are at increased risk of this complication compared to cattle. For this reason, cosmetic dehorning (aseptic removal of the horn with primary skin closure) is preferred over standard dehorning (defect left to heal by second intention) in adult goats.

The **horn sheath** is made of cornified epidermal tissue produced by the **cornual epidermis**. The **cornual dermis** is well-vascularized and innervated. It is fused to the periosteum of the cornual process, covering it entirely (Fig. 17.3-9). The cornual dermis contains apically directed **papillae** that serve as a template for the formation of horn. Epithelial cells of the **cornual epidermis** form **tubular horn** over the dermal papillae and **intertubular horn** over the interpapillary regions of the cornual dermis. Since dermal papillae run

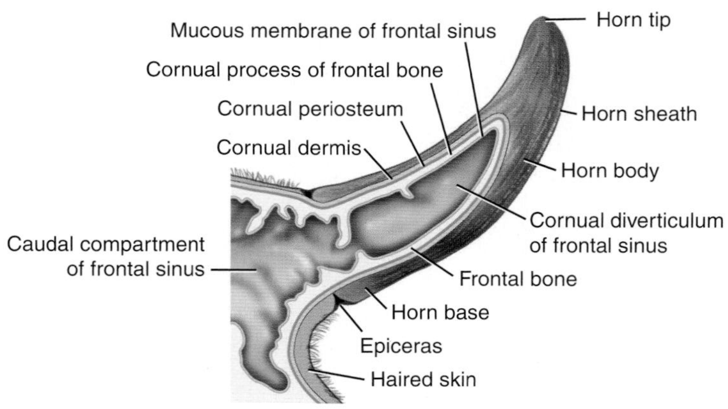

FIGURE 17.3-9 Developed bovine horn.

The rate of horn production decreases during the peripartum period (the period from the last month of pregnancy to several months postdelivery) due to the associated stress and increased metabolic demands, resulting in visible circumferential grooves on the external surface of the horn. A cow's age (number of calvings) may be estimated by counting these yearly grooves.

When dehorning, it is important to excise the horn from its base as close as possible to the frontal bone just before the dorsal and ventral branches of the cornual artery enter the bone of the cornual process. This ensures that hemostatic forceps or the hot iron can provide functional hemostasis. If not, and the horn is excised distal to its base (or in the case of a horn fracture), blood vessels will retract into their bone channels when severed and hemostasis using hemostatic forceps or hot iron is either not possible or difficult to achieve, possibly leading to severe bleeding.

lengthwise, and newly formed horn push older horn toward the apex, horns grow predominantly in length rather than width. A ring of soft horn (**epiceras**) is produced at the base of the horn by a transitional epidermis that blends into haired skin.

Blood supply, lymphatic drainage, and innervation of the horn in cattle

The **cornual artery** and **vein** supply and drain the horn, respectively, and are branches of the **superficial temporal artery** and **vein**, respectively. Before reaching the base of the horn, the cornual artery tracks along the temporal line of the frontal bone as it approaches the horn and then ramifies into dorsal and ventral branches immediately before the cornual base to supply the dermis and bone of the horn (Fig. 17.3-10). These branches provide smaller branches that track in canals within the cornual process (Fig. 17.3-6).

HORN FRACTURE

Traumatic events can fracture horns. This typically occurs when a fully horned animal gets its head trapped and tries to back out or accidentally strikes the horn against a fixed object—e.g., headgate, post, and wall. If the fracture occurs within the apex of the horn, it may not involve the cornual process of the frontal bone nor the cornual diverticulum of the frontal sinus. In this situation, no further treatment is needed. However, if the fracture occurs within the body or the base of the horn, bleeding, an open wound, and exposure of the frontal sinus are likely to occur. An emergency dehorning performed proximal to the base of the horn is needed for successful treatment. Remember—dehorning is usually an elective procedure.

The **parotid lymph node**, located along the rostral border of the parotid salivary gland, drains lymph from the horn.

The horn is mainly innervated by the **cornual nerve**, a branch of the **zygomaticotemporal nerve**, which is a branch of the **maxillary division of the trigeminal nerve (CN V)**. The cornual branch arises within the orbit, leaves the orbit caudal to the zygomatic process of the frontal bone, and courses caudally along the temporal line of the frontal bone before reaching the cornual base where it ramifies into 2–3 branches (Fig. 17.3-10).

In cattle, a cornual nerve block should be used to anesthetize the horn before disbudding and dehorning. The injection site is just ventral to the temporal line of the frontal bone (palpable landmark) and approximately mid-way between the cornual base and the lateral canthus of the eye (Fig. 17.3-2). In older cattle with well-developed horns, cutaneous branches of the greater auricular nerve (2nd cervical nerve) should also be desensitized by local infiltration of a local anesthetic mid-way between the base of the horn and the base of the ear or by a partial ring block at the caudal aspect of the base of the horn.

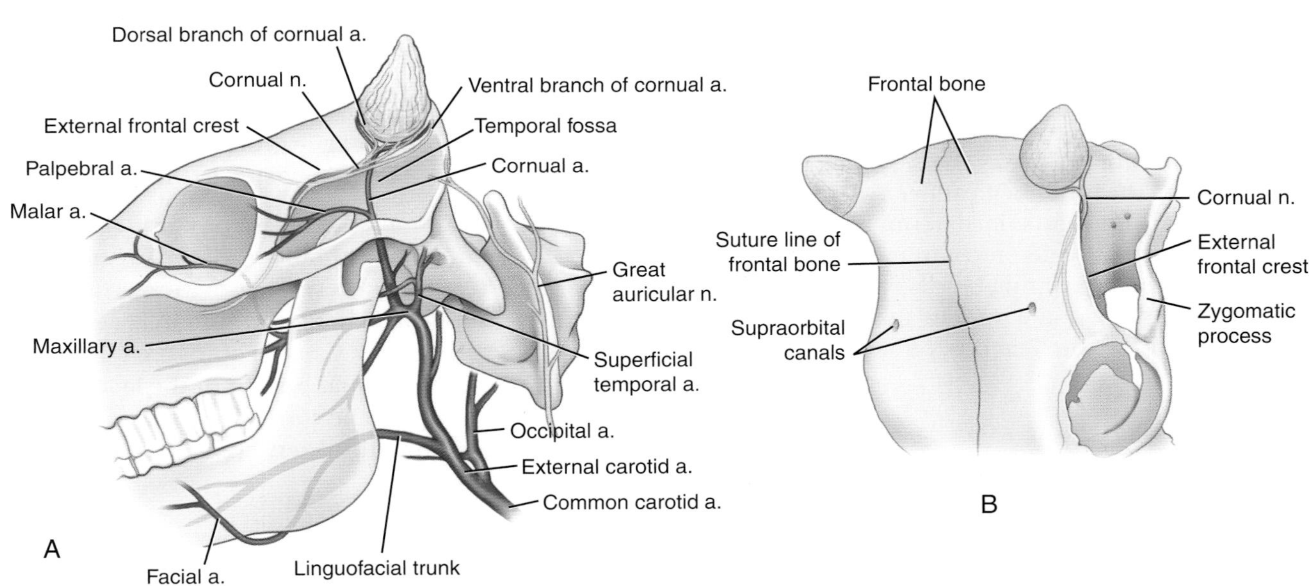

FIGURE 17.3-10 (A) Blood supply to the horn of cattle. (B) Innervation to the horn of cattle.

HORN "TIPPING"

Horn tipping represents the amputation of the tip of the horn (i.e., the nonsensitive end of the horn) without cutting through the cornual process or the cornual diverticulum of the frontal sinus, to make the horn tipless sharp. Horn "tipping" is indicated when the tip of an overgrown horn penetrates, or is likely to penetrate, the skin of the animal (ingrown horns) or when the tip of the horn blocks the sight of the animal. This procedure is also commonly performed on rodeo bulls, stockers, and feedlot cattle for safety reasons.

Use caution not to overheat tissues around the horn bud when disbudding goat kids. Prolonged application of heat and pressure by a hot iron on the horn bud can result in thermal injury to the fine calvarium and meninges, causing thermal meningitis. When dehorning adult goats, additional caution is needed to avoid exposing the brain by inadvertently fracturing the parietal bone or by excising too much bone at the base of the horn—especially caudally.

Descenting (i.e., removal of scent glands of the head) can be performed at the time of disbudding in young bucklings by cauterizing this skin region using a hot iron.

These nerves are anesthetized before disbudding and dehorning (see Fig. 17.3-11). The injection site for the cornual branch of the lacrimal nerve is mid-way between the lateral canthus of the eye and the caudal aspect of the base of the horn. The infratrochlear nerve is anesthetized in a short line block over the dorsomedial rim of the orbit, immediately before the nerve divides into cornual and frontal branches. Additionally, a partial ring block is performed at the caudal aspect of the base of the horn to anesthetize the great auricular nerve (branch of the 2nd cervical nerve) when dehorning adult goats.

Horn anatomy and development of the goat

The basic cornual anatomy in goats is like cattle. The horn also consists of a bony core protected by an **epidermal horn sheath**. While the cornual process of cattle is round in cross-section, the cornual process in goats is oval in cross-section. In goats, the **cornual process** does not grow from the frontal bone as it does in cattle. Instead, it is formed by a separate center of ossification that fuses to the frontal bone. Horn sheath production is intermittent in goats, resulting in several circumferential grooves and ridges (normally 8–14 ridges) seen on the external surface of the horn produced every year (Fig. 17.3-7).

The frontal sinus of goats is relatively shallower than in cattle, especially in younger goats. In addition, the base of the cornual process is near the **parietal bone** of goats. Male goats have scent glands located caudomedially to the horn buds.

Also similar to cattle, the **cornual artery** and **vein** are branches of the **superficial temporal artery** and **vein**, respectively, and supply and drain the horn in goats, respectively. Caprine horns are innervated by the **cornual branch of the lacrimal nerve** and the **cornual branch of the infratrochlear nerve** (Fig. 17.3-11).

FIGURE 17.3-11 A right lateral image of an adult female goat skull (horn sheaths are missing) demonstrating the injection sites for the cornual nerve block in goats. A third injection site is needed for an adult goat (not pictured). Key: *Star,* injection sites for the cornual branch of the lacrimal nerve; *Dashed line*, injection site for the cornual branch of the infratrochlear nerve.

Selected references

[1] Hackett CH, Hackett RP, Nydam CW, Nydam DV, Gilbert RO. Surgery of the bovine (adult) integumentary system. In: Fubini SL, Ducharme NG, editors. Farm animal surgery. 2nd ed. St. Louis, MO: Elsevier; 2017. p. 179–92.
[2] Mansour M, Wilhite R, Rowe J. The head, neck, and vertebral column. In: Guide to ruminant anatomy: dissection and clinical aspects. 1st ed. Hoboken, NJ: John Wiley & Sons; 2018. p. 1–64.
[3] McGeady TA, Quinn PJ, Fitzpatrick ES, Ryan MT, Kilroy D, Lonergan P. Integumentary system. In: Veterinary embryology. 2nd ed. Ames, IA: John Wiley & Sons; 2017. p. 314–30.
[4] Smith MC, Sherman DM. Dehorning and descenting. In: Goat medicine. 2nd ed. Ames, IA: Wiley-Blackwell; 2009. p. 723–31.

CASE 17.4

Spinal Lymphoma

André Desrochers and Gilles Fecteau
Department of Clinical Sciences, Université de Montréal, Faculty of Veterinary Medicine, Saint-Hyacinthe, Québec, CA

Clinical case

History

A 6-year-old Holstein cow was presented for difficulty in getting up and poor milk production. She had calved uneventfully 7 days before presentation. She otherwise appeared healthy, but getting her to stand was difficult, and her milk production at this stage was markedly decreased compared to the year before. Her appetite was reported to be satisfactory.

Physical examination findings

On presentation, the cow was bright, alert, and responsive. She was ambulatory but appeared slightly weak in the pelvic limbs. Her body condition score was 2–3/5. Her mucous membranes were slightly pale, but moist with a normal CRT. Her rectal temperature, heart rate, and respiratory rate were 101.6°F (38°C), 90 bpm, and 28 brpm, respectively. Cardiac auscultation was normal. The left prescapular (superficial cervical) and right prefemoral (subiliac) lymph nodes were mildly enlarged. Rumen auscultation revealed normal motility, and percussion of both sides of the abdomen was normal with no areas of resonance—i.e., pings. The udder palpated normally, and a California Mastitis Test (CMT) was negative in all quarters. A brief neurological examination revealed bilateral weakness without proprioceptive deficits. During a slow walk, it was possible to easily shift the cow to the right or left by gently pulling on the tail.

Differential diagnoses

Spinal lymphoma, traumatic spinal cord injury, lumbar muscle injury including sacroiliac subluxation, and metabolic imbalance

Diagnostics

Serum was submitted for detection of antibodies (using ELISA) against bovine leukosis virus (BLV) and was positive. Fine-needle aspiration of an enlarged prescapular lymph node revealed a population of large immature lymphocytes with abnormal nuclei (large, round, and undifferentiated nuclei). Radiographic examination of the lumbar area was inconclusive because of the size of the cow. A serum biochemistry profile (SBP) revealed a mild increase in muscle enzymes and no electrolyte abnormalities. The CBC showed a lymphocytic leukocytosis. Cytological evaluation of the CSF aspiration was difficult because of blood contamination; however, abnormal lymphocytes were seen and compatible with a diagnosis of spinal lymphoma.

Diagnosis

Spinal lymphoma associated with BLV infection, with secondary pelvic limb paresis

Comparative Veterinary Anatomy: A Clinical Approach. https://doi.org/10.1016/B978-0-323-91015-6.00098-4

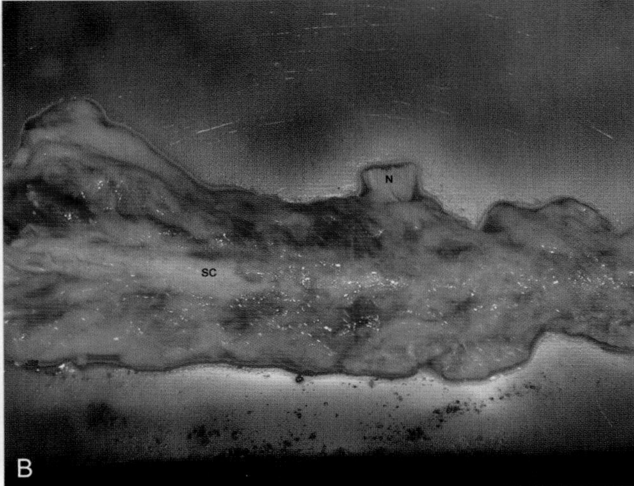

FIGURE 17.4-1 Necropsy specimen of the caudal vertebrae of a cow with lymphoma. (A) Infiltration of lymphoid tissue in the spinal canal. The lymphoid tissue is invading the spinal canal *(blue arrows)*. (B) The spinal cord is isolated. The abundant yellowish tissue corresponds to the lymphoid tissue. Key: VS, vertebral spine; VB, vertebral body; ID, intervertebral discs; SD, spinal cord; N, nerve. SC, spinal cord; N, nerve.

Treatment

The cow was humanely euthanized due to the poor prognosis. Necropsy confirmed the presumptive diagnosis of spinal lymphoma (Fig. 17.4-1A and B).

Anatomical features in ruminants

Introduction

The bovine spinal cord and canal are similar to other species; the clinically relevant differences are highlighted in this section. Adult cattle with vertebral column lesions are not surgical candidates because of their large size and muscle mass, and the challenges of managing them after a hypothetical vertebral surgery (Fig. 17.4-2). Their large size makes post-operative immobilization effectively impossible. Moreover, the type of lesions frequently diagnosed cannot be successfully treated surgically because of the advanced stage of the disease at the time of diagnosis. This section discusses the caudal parts of the vertebral column and the relevant associated structures, with an emphasis on neuroanatomy.

Function

The vertebral column encircles the spinal cord. Because the important nervous structures travel from the head to the trunk, the large, dense vertebrae offer protection to this fragile, critical network.

Osteology of the vertebral column

Cattle have 6 lumbar vertebrae, 5 sacral segments, and 18–20 caudal vertebrae. In sheep, the number of vertebrae in the caudal part of the vertebral column varies from other farm animals—they may have 7 lumbar vertebrae, a sacrum consisting of 4 segments, and 16–20 caudal vertebrae in the tail.

The tail of lambs is often amputated (docked) as a routine husbandry procedure for improved hygiene and animal health. When performed at a young age with appropriate analgesia, this is an accepted routine procedure.

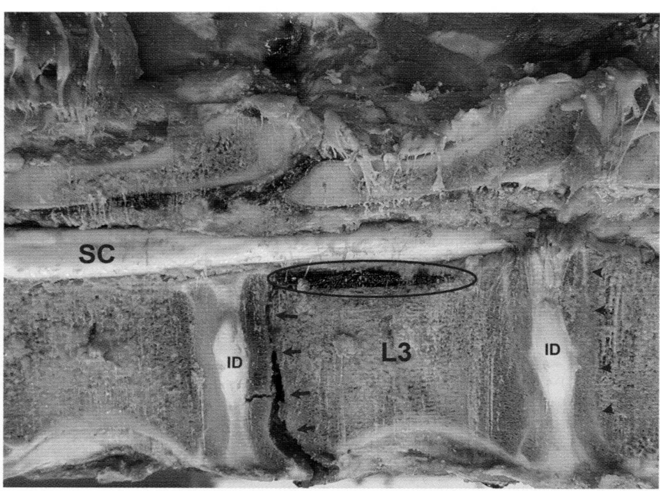

FIGURE 17.4-2 A necropsy photo from a 15-month-old heifer that was presented because she could not stand for 2 days. A physeal fracture *(arrows)* of L3 was diagnosed through longitudinal sectioning of the vertebral column. The normal physis of L4 is identified with *arrowheads.* A subdural hematoma is *circled.* Key: SC, spinal cord; ID, intervertebral disc.

One notable difference between caprine (goats) and ovine (sheep) species is the ability to dorsiflex the last **caudal vertebrae**. Goats are able to dorsiflex the caudal vertebrae, while the fused caudal vertebrae of sheep do not allow this movement (Fig. 17.4-3).

The 6th lumbar vertebra has shortened transverse processes adjacent to the tuber coxae, making it difficult to palpate. Therefore, the transverse processes of the 5th lumbar vertebra can be palpated cranial to the tuber coxae.

Fracture of the sacrum and tail can be stabilized with internal fixation using plates and screws. These fractures typically occur when cows in estrus are mounting each other. There is usually an associated ventral displacement of the distal segment of the column (Fig. 17.4-4). Sacroiliac luxation or subluxation can also occur following trauma. Ventral displacement of sacral vertebrae between the 2 wings of the ilium creates lumbar pain and an unwillingness to stand.

Any part of the pelvis may fracture. Two sites must be affected for the pelvis to become unstable and create a clinically relevant problem. Viewing the pelvis as a "square," if only one side of the square is fractured, it remains relatively stable; however, if 2 sides are fractured, the "square" becomes unstable. Internal fixation is difficult because the pelvis is covered by a large muscle mass (Fig. 17.4-6).

The **lumbar vertebrae** in cattle are longer than in horses, with elongated and thinner transverse processes. The 1st lumbar vertebra has the shortest transverse process, and, progressing caudally, the transverse processes are longer with the distal part projecting cranially. (See side box entitled "Paravertebral blocks" in Case 20.2.)

The **sacrum** is formed by 5 completely fused segments. The 1st segment is wider with wings articulating with the pelvis at the level of the **tuber sacrale**. The dorsal spinous processes are completely fused, forming the **median sacral crest,** while the articular processes are partially fused, forming the **lateral sacral crest**.

Pelvic osteology

The **pelvis** is comprised of the 2 **coxae**, united by the **pubic symphysis** ventrally, and the sacrum articulating with the tuber sacrale. The angle of the pelvis in cattle is more horizontal than in horses. Each coxa is comprised of the ilium, ischium, and pubis (Fig. 17.4-5).

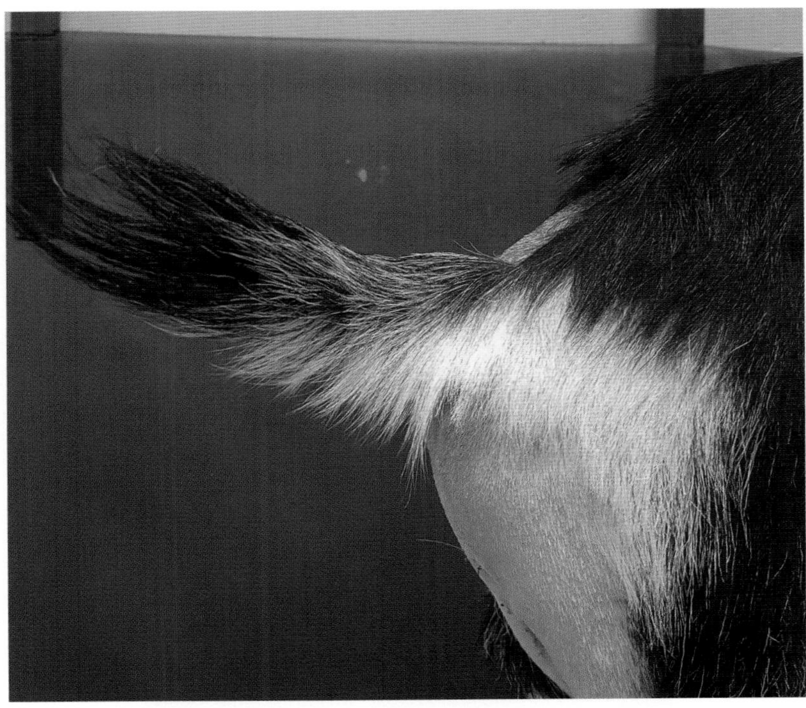

FIGURE 17.4-3 The tail of a goat with normal dorsiflexion.

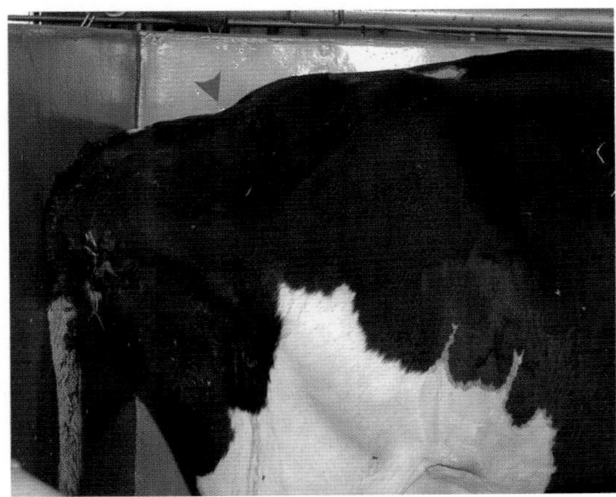

FIGURE 17.4-4 Sacral fracture in an adult cow. The *arrowhead* highlights the depression from the fractured sacrum.

Muscles of the caudal spine

The **longissimus lumborum** is a long muscle that is continuous with the **longissimus thoracis and cervicis**. It originates at the cranioventral aspect of the ilium and inserts on the transverse processes of the thoracic vertebrae and the last rib, and it functions to extend the dorsum (back). The **deep epaxial muscles** are multiple small muscles that lie between each vertebra. The **intertransversarii lumborum** muscles are located between each transverse vertebral process.

This muscle is traversed with the needle when blocking the ventral branch of spinal nerves during a paravertebral block.

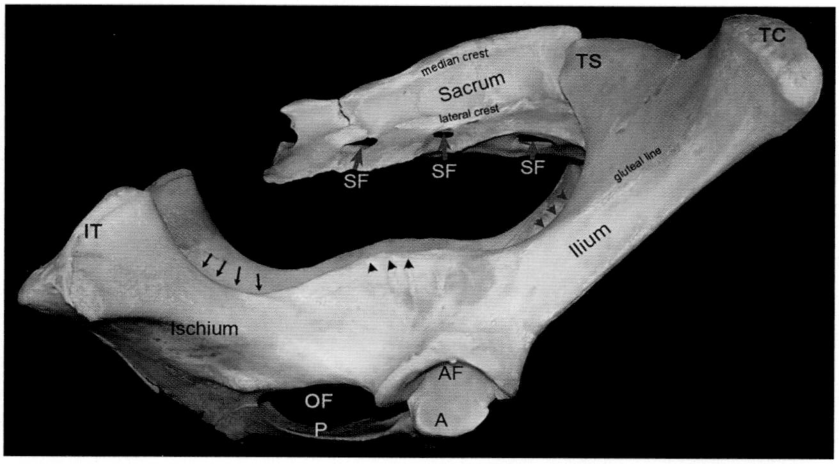

FIGURE 17.4-5 Right lateral view of the bovine pelvis. Key: IT, tuber ischiadicum; OF, obturator foramen; P, pubis; A, acetabulum; AF, acetabular fossa; TS, tuber sacrale; TC, tuber coxae; SF, sacral foramen; black arrows, lesser ischiatic notch; black arrowheads, ischiatic spine; blue arrowheads, greater ischiatic notch.

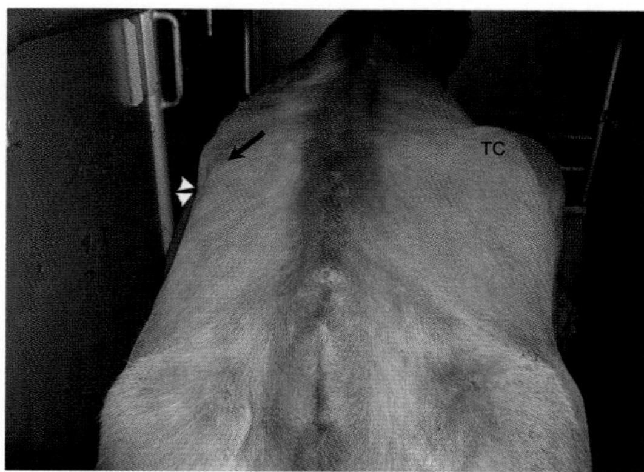

FIGURE 17.4-6 Dorsal view of a standing bovid (the top of the image is cranial). This Jersey cow has a displaced fracture of the left ilium *(arrow)*. Muscle traction displaced the fragment ventrally *(arrow heads)*. Key: TC, tuber coxae.

In cattle, CSF is collected at the level of the lumbosacral cisterna between the 1st segment of the sacrum and the last lumbar vertebra. A 9-cm (3.5-in.) long, 18G spinal needle is used to collect spinal fluid. The puncture site is on midline slightly caudal to a line drawn between the tuber coxae. The puncture site can be easily palpated as the space between the sacrum and the last lumbar spinous process (Fig. 17.4-7).

Spinal cord

There are 2 notable features of bovine spinal cord anatomy that merit discussion: (1) The spinal cord (**conus medullaris**) terminates at the level of S1–S2, and (2) the **dura mater** extends caudally to the 4th caudal vertebra. The CSF is contained in the subarachnoid space (between the dura and arachnoid mater), which terminates at the level of the 3rd or 4th sacral segment.

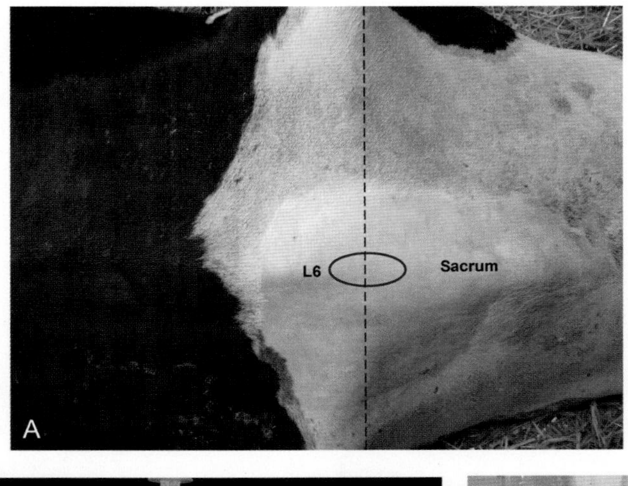

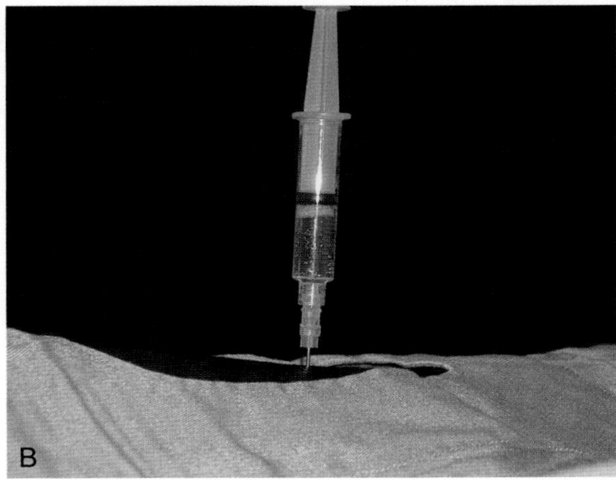

FIGURE 17.4-7 Collection of CSF in a standing cow. (A) Landmark for the puncture site (dorsal view) is caudal to a *line* drawn between the tuber coxae and between L6 and the sacrum *(blue circle)*; (B) the needle is in place for fluid aspiration; (C) normal clear CSF (right) and xanthochromia (yellow discoloration) (left).

Peripheral nervous system

When anesthetizing/blocking L3, cattle may become unstable on their feet because L4 may also be blocked, affecting the quadriceps muscles of the pelvic limb.

The **lumbar nerves** exit at their respective intervertebral foramina (Fig. 17.4-8). The 1st and 2nd lumbar nerves (L1 and L2) have branches to the psoas and quadratus lumborum and primarily innervate the muscles of the flank (see Case 20.2). The 2nd lumbar nerve (L2) may supply a branch to the **genitofemoral nerve**. The 3rd lumbar (L3) nerve provides small branches to the psoas and quadratus lumborum and branches to the genitofemoral nerve. The genitofemoral nerve (passing through the inguinal canal) is therefore a combination of branches of L3 and L4, providing innervation to the mammary gland.

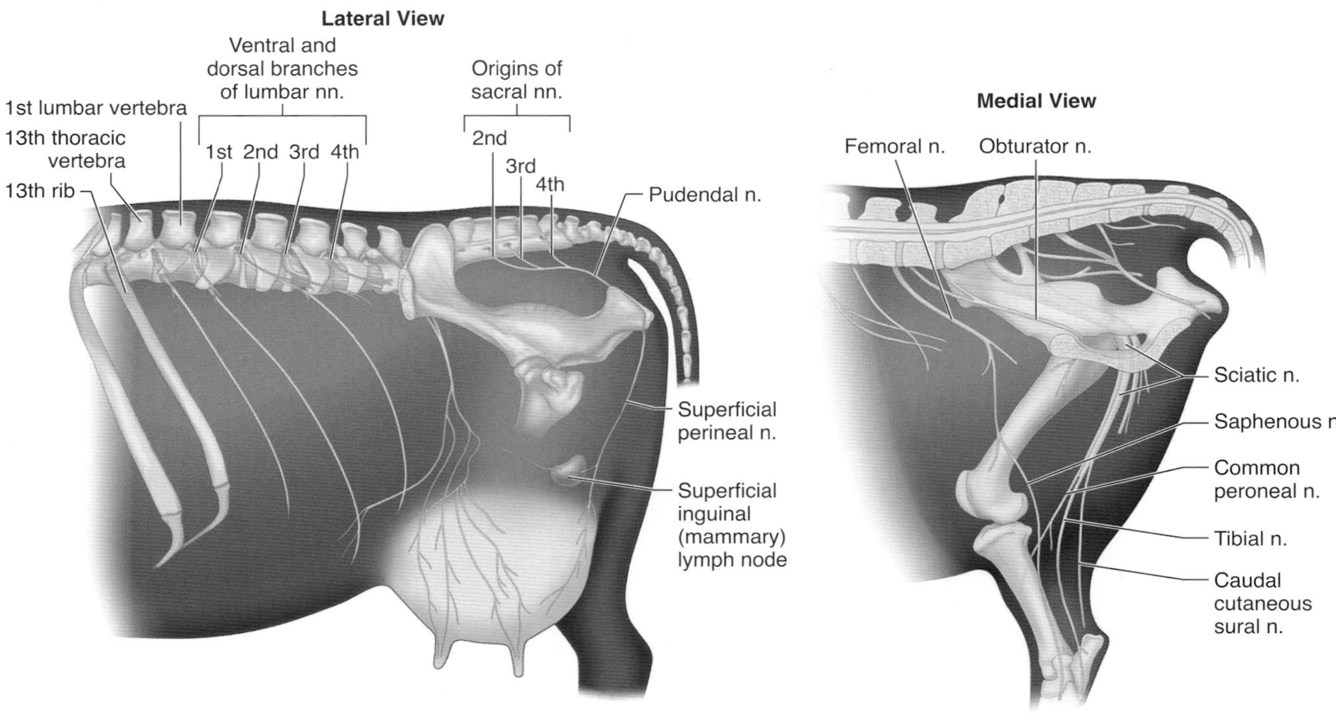

FIGURE 17.4-8 Left: lateral view of a standing cow with emphasis on the lumbar and sacral nerves. Right: medial view with emphasis on the sciatic and obturator nerves.

Femoral nerve paresis or paralysis can occur during parturition, affecting either the cow or the calf. Femoral nerve trauma is more common in calves following a dystocia. The affected calf or cow has marked weakness of the quadriceps muscle when walking; this should be differentiated from the lateral patellar luxation that can occur with femoral nerve injury in calves.

The femoral nerve can be specifically anesthetized with a paravertebral approach, using ultrasound guidance, and has been described for diagnosis of a newly reported variation of spastic paresis affecting the quadriceps muscles of Belgian Blue calves. The most common clinical presentation of spastic paresis affects the gastrocnemius muscles rather than the quadriceps.

The ventral branch of L4 contributes to the formation of the **lateral cutaneous femoral**, **femoral**, **obturator**, and **genitofemoral nerves**. The femoral nerve is formed by branches from L4, L5, and sometimes L6. It is the largest nerve of the lumbar plexus and innervates the quadriceps muscles.

The **saphenous nerve** is a branch of the femoral nerve that tracks along the medial aspect of the thigh on the caudal border of the sartorius muscle. It supplies cutaneous branches to the medial aspect of the stifle and the tarsus. The **obturator nerve** receives branches from L4, but mainly L5 and L6. It courses cranioventrally into the obturator foramen to innervate the adductor muscles: the adductor, pectineus, gracilis, and obturator externus muscles.

The **sacral plexus** is the origin for the following nerves: cranial and caudal gluteal, and **ischiatic**. The **cranial gluteal nerve** innervates the gluteus medius and profundus as well as the tensor fascia latae. The **caudal gluteal nerve** innervates the gluteus medius and the gluteobiceps muscles. The **sciatic nerve** is formed primarily by L6, S1, and S2. In sheep, because they have 7 lumbar vertebrae, L7 primarily forms the sciatic nerve. Nearing the greater trochanter, the sciatic nerve sends branches into the semimembranosus, semitendinosus, and gluteobiceps muscles. Distally, at the middle of the thigh, it divides into 2 major branches: the **fibular** and **tibial nerves**.

The obturator nerve can be injured during calving due to feto-pelvic disproportion. The affected cow may be unable to stand with both pelvic limbs abducted. If the cow is able to stand, straps/hobbles should be attached to the pelvic limbs to prevent abduction (Fig. 17.4-9). Additionally, the affected cow can be "floated" in a tank specially designed for buoyancy to assist her in remaining standing and decrease long periods of recumbency (Fig. 17.4-10).

The sciatic nerve can also be injured during calving or secondary to prolonged recumbency (milk fever) on a hard surface. Clinical signs are variable depending on which branches are affected (Fig. 17.4-11).

The **sacral nerves** originate as 5 pairs from the **conus medullaris**, or **conus terminalis** (lower end of the spinal cord). Like the lumbar nerves, there are dorsal and ventral branches. The dorsal branches exit through the **dorsal sacral foramina,** providing medial (muscular) and lateral (cutaneous) branches to the gluteal muscles and tail. The ventral branches exit through the **pelvic sacral foramina**. The first 2 sacral nerves are part of the **lumbosacral trunk**. The last sacral nerves form the pudendal, perineal, and caudal rectal nerves.

Blood supply and lymphatics

Blood supply to the lumbar spine is by way of the paired **lumbar arteries** and drainage is via the **lumbar veins** entering the body of each vertebra. The paired vertebral arteries originate from the dorsal part of the abdominal aorta. The lumbar veins accompanying the arteries empty into the caudal vena cava.

The **lumbo-aortic lymph nodes** are adjacent to the abdominal aorta and caudal vena cava concealed in fat. The lymph nodes drain the lumbar vertebrae and the last thoracic vertebra, surrounding muscles, ilium, and regional peritoneum.

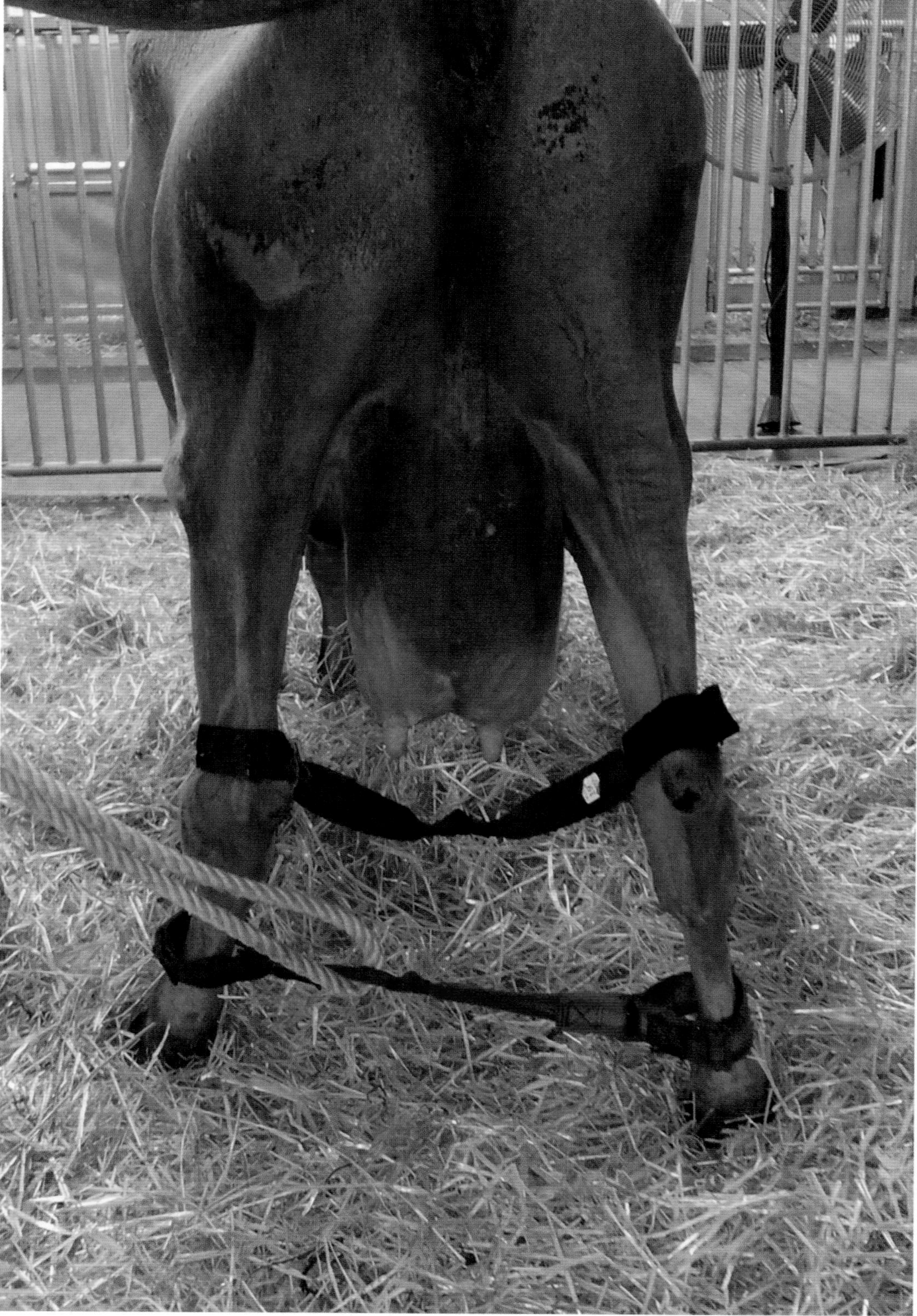

FIGURE 17.4-9 A cow with postcalving obturator nerve paralysis has hobbles placed to prevent her from abducting (moving away from the median plane) the pelvic limbs.

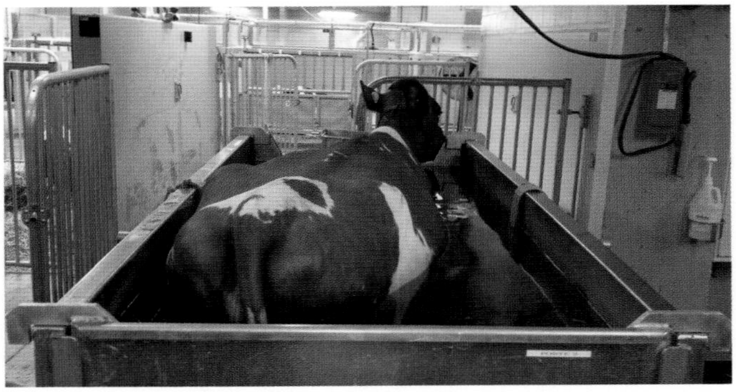

FIGURE 17.4-10 A cow standing and "floating" in a specially designed tank. This cow had an injury to the obturator nerves during calving, making it difficult to rise and stand.

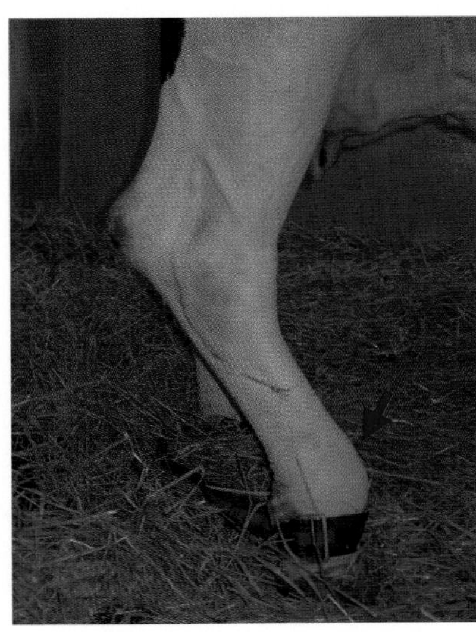

FIGURE 17.4-11 Adult cow affected with a peroneal paralysis of the right pelvic limb (*arrow* is highlighting the fetlock flexion).

Selected references

[1] Ghoshal NG. Spinal nerves. In: Sisson G, Grossman J, editors. Anatomy of domestic animals. Philadelphia: WB Saunders Co.; 1975. p. 1124–50.
[2] De Vlanynck CA, Pille F, Hauspie S, et al. Evaluation of three approaches for performing ultrasonography-guided anesthetic blockade of the femoral nerve in calves. Am J Vet Res 2013;74:750–6.
[3] Desrochers A, Fecteau G. Downer cows: case management and appropriate use of a floating tank. In: Fubini SL, Ducharme NG, editors. Farm animal surgery. 2nd ed. Elsevier; 2017. p. 39–49.
[4] Roe JM. Bovine paravertebral analgesia: radiographic analysis and suggested method for improvement. Vet Rec 1986;119:236–8.

CHAPTER 18

THORAX

André Desrochers, Chapter editor

Heart and Lungs

18.1 Pericarditis—*Marie-Eve Fecteau* 1042
18.2 Endocarditis and atrial lymphoma—*Marie-Eve Fecteau and Gilles Fecteau* 1049

CHAPTER 18

CASE 18.1

Pericarditis

Marie-Eve Fecteau
Food Animal Medicine and Surgery, Department of Clinical Studies - New Bolton Center, University of Pennsylvania School of Veterinary Medicine, Kennett Square, Pennsylvania, US

Clinical case

History

A 5-year-old Holstein cow presented for decreased milk production, anorexia, and intermittent fevers of 10 days' duration. The cow had been treated on the farm with 2 different antimicrobials, but did not improve. She was 4 weeks post-partum at the time of presentation.

Physical examination findings

On initial physical examination, the cow was dull but responsive. She stood with her elbows abducted and appeared reluctant to move. She had a body condition score of 3/5 (1 = emaciated, 3 = obese). She was febrile (103.9 °F [39.9 °C]), tachycardic (120 bpm), and tachypneic (60 brpm). Her mucous membranes were hyperemic with a prolonged CRT. On cardiothoracic auscultation, the heart sounds were muffled and had a distinct splashing sound. Her lungs sounds were absent ventrally but present dorsally. The cow occasionally had a moist cough during examination. There was marked submandibular and brisket edema and distended and pulsating jugular veins (Fig. 18.1-1).

Differential diagnoses

Traumatic reticulopericarditis, septic pericarditis (from hematogenous spread or direct extension from the pleura), idiopathic pericarditis, and neoplasia

Diagnostics

Clinical pathology revealed a mild nonregenerative anemia (hematocrit 20%), increased total protein (8.8 g/dL), hyperfibrinogenemia (1500 mg/dL), leukocytosis ($25 \times 10^3/\mu L$) associated with neutrophilia ($17 \times 10^3/\mu L$), mild

CARDIAC NEOPLASIA

Lymphosarcoma with pericardial involvement and fluid accumulation within the pericardial sac can lead to signs of congestive heart failure similar to those observed in cows with traumatic pericarditis. With lymphosarcoma associated with bovine leukosis virus, the heart is rarely the only organ affected and signs of other organs' involvement may be detectable on physical examination. Masses can be detected over the pericardium or within the heart chambers (typically right atrium) on echocardiography. Pericardiocentesis with fluid analysis helps differentiate cardiac neoplasia from other etiologies of pericarditis. In these cases, pericardial fluid is typically serosanguineous, with variable protein concentration and red and white blood cell counts. The predominant white blood cells are lymphocytes, which can have a neoplastic appearance. Prognosis is grave for these cows because most have multicentric involvement.

Comparative Veterinary Anatomy: A Clinical Approach. https://doi.org/10.1016/B978-0-323-91015-6.00091-1

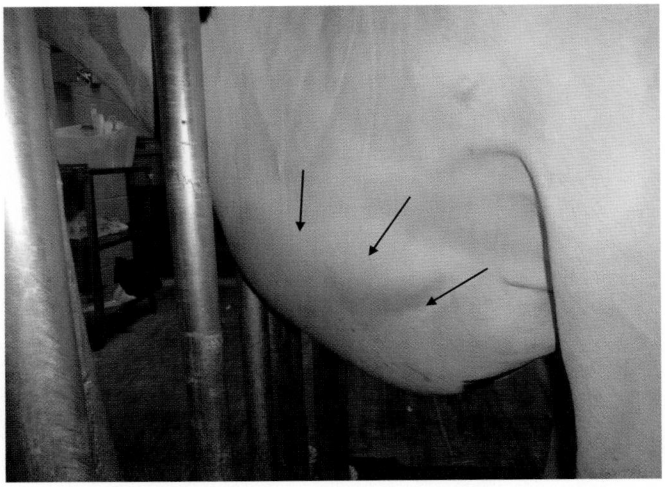

FIGURE 18.1-1 Cow on hospital admission with brisket edema *(arrows)*. (Image courtesy of Dr. André Desrochers, Université de Montréal.)

azotemia (creatinine, 2.5 mg/dL), and mild increases in liver enzymes (GGT, 100 IU/L; SDH, 52 IU/L). Radiographic evaluation of the cow's cranioventral abdomen and thorax revealed fluid and gas accumulation in the pericardium, as well as a metallic linear foreign body visible at the junction of the caudoventral thorax and cranioventral abdomen (Fig. 18.1-2).

An ECG revealed sinus tachycardia and decreased amplitude of the QRS complexes.

An echocardiogram confirmed a large volume of hyperechoic pericardial effusion with numerous echoic strands of fibrin within the effusion and covering the epicardium, as well as hyperechoic pinpoint echoes consistent with free gas within the pericardium (Fig. 18.1-3). A mild to moderate amount of hypoechoic pleural effusion was also noted bilaterally.

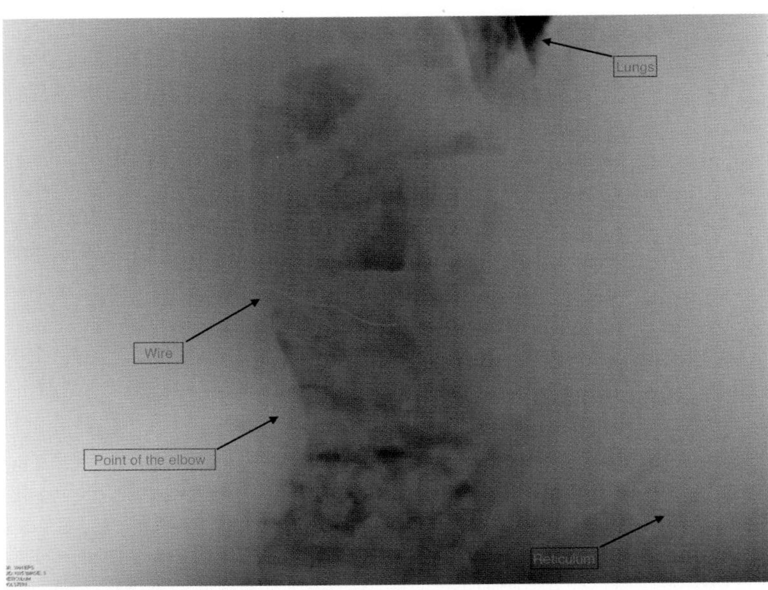

FIGURE 18.1-2 Lateral radiograph of the cranioventral abdomen and caudoventral thorax depicting a linear, metallic, foreign body penetrating the diaphragm and entering the thoracic cavity.

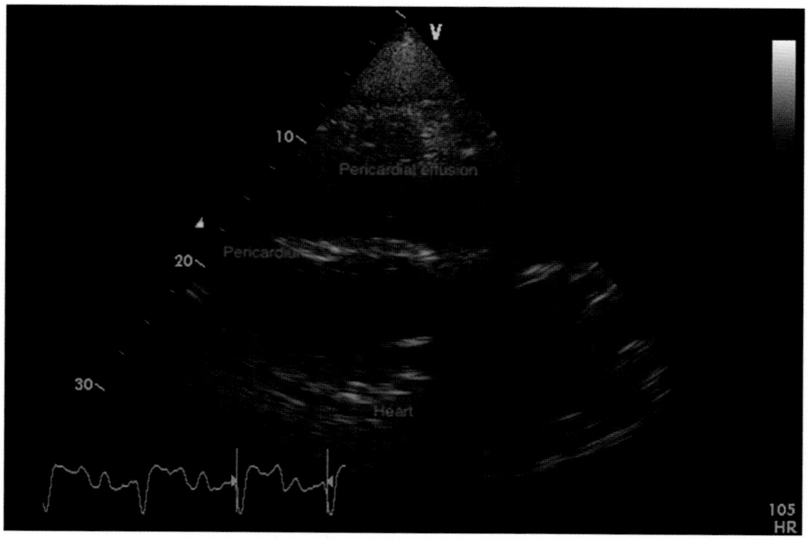

FIGURE 18.1-3 Ultrasonographic image depicting hyperechoic pericardial effusion with a thickened pericardium in a cow.

Pericardiocentesis can also be performed with a spinal needle (Fig. 18.1-4) or small cannula if the intent is to obtain fluid to differentiate between septic (often traumatic), idiopathic, and neoplastic origins. The site should be selected ultrasonographically, but is typically at the left 5th intercostal space, just dorsal to the point of the elbow. Fluid should be submitted for cytology and bacterial culture (or Gram stain), when appropriate. With septic pericarditis, the fluid appears purulent and may have a foul odor. It typically has a high protein concentration (normal < 2.5 g/dL) and high white blood cell count (normal < 5000/μL), with neutrophils being the predominant cells. Bacteria are usually seen on gram-stained smears.

Pericardiocentesis (Gr. *peri* around + Gr. *kardio* heart + Gr. *kentēsis* perforation or tapping) was performed under ultrasonographic guidance. The left 5th intercostal space, approximately 5 cm (2.0 in.) dorsal to the olecranon, was selected and prepared aseptically and locally anesthetized. A large-bore thoracic catheter (Argyle tube) was inserted through the intercostal space and into the pericardial space to allow for pericardial fluid collection and effective drainage of the effusion (Fig. 18.1-5). A total of 3.5 L of tan, malodorous, foamy fluid containing strands of fibrin was obtained from the pericardium. Fluid analysis revealed an increased protein concentration with an increased white blood cell count consisting mainly of neutrophils. Microbial cultures revealed a mixed growth of *Trueperella pyogenes* (an anaerobic gram-positive bacteria) and 2 gram-negative coliform bacteria.

IDIOPATHIC PERICARDITIS

Occasionally, a cow presents with signs of congestive heart failure resembling those seen with traumatic pericarditis, with the exception that the pericardial fluid retrieved is characterized as an aseptic inflammatory exudate and is hemorrhagic in appearance. Fluid analysis usually reveals a high red blood cell count, moderate-to-high protein concentration, and moderate-to-high white blood cell count, consisting primarily of lymphocytes or macrophages. Bacterial growth is usually negative because these cases of pericarditis are sterile/aseptic. These cows typically respond favorably to pericardial aspiration/drainage with lavage, along with systemic antimicrobial therapy and anti-inflammatories.

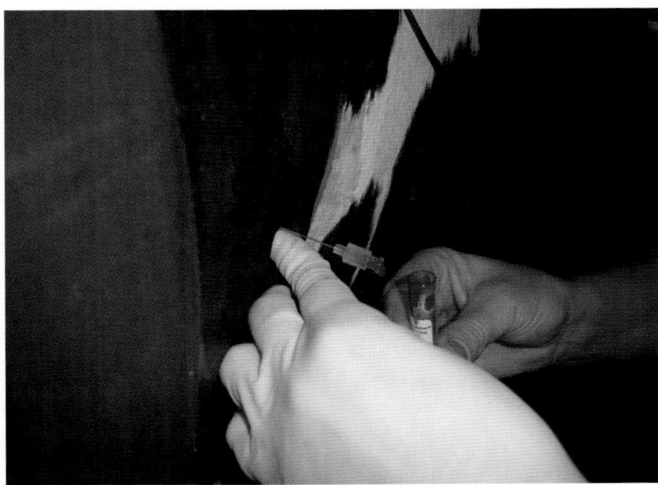

FIGURE 18.1-4 Pericardiocentesis performed with a spinal needle in a cow. The needle is inserted under ultrasonographic guidance at the level of the 5th intercostal space on the left side, just caudal to the point of the elbow.

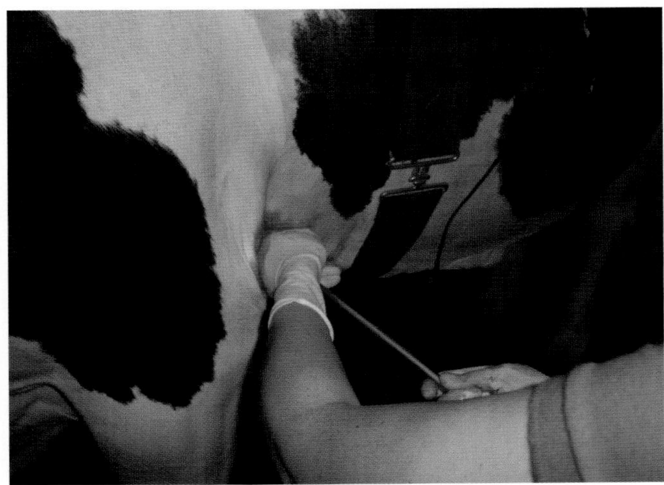

FIGURE 18.1-5 Inserting a large-bore thoracic drain into the pericardium of a cow to drain the pericardial effusion. The drain is inserted at the level of the 5th intercostal space on the left side, just caudal to the point of the elbow.

Diagnosis

Traumatic reticulopericarditis

Treatment

Following initial drainage, the thoracic drain was sutured in place and the pericardium was lavaged with sterile fluids. Broad-spectrum antimicrobial therapy was initiated while awaiting culture results. Forty-eight hours following initial drainage, when the cow's comfort and respiratory and heart rates were improved, a lateral thoracotomy (Gr. *thōrax* chest + Gr. *tomē* a cutting = surgical incision of the chest wall) was performed with the cow standing. A portion of the left 5th rib was resected to allow better access to the pleural cavity. The fibrous tract caused by the migration of the foreign body was identified and explored, and the linear foreign body was removed. The pericardium was then incised and drained. Thick caseous material mixed with fibrin strands was manually removed from the pericardium, and the pericardial sac was lavaged with warm sterile fluids. The pericardium was then sutured (marsupialized [L. *marsupium* pouch]) to the skin to allow for further drainage and lavage. A drain was placed in the dorsal aspect of the left hemithorax, and the pneumothorax (Gr. *pneumo-* breath + Gr. *thōrax* thorax/chest = accumulation of air or gas in the pleural space) was evacuated. Post-operatively, the cow was maintained on broad-spectrum systemic antimicrobials, nonsteroidal anti-inflammatories, and wound care for the thoracotomy site. She was also administered a magnet orally to prevent future episodes of traumatic reticuloperitonitis. Her long-term prognosis was considered guarded to poor due to the need for long-term antimicrobial therapy and the possibility for development of constrictive pericarditis.

Anatomical features in ruminants

Introduction

The pericardium has a close relationship with the diaphragm, which in turn is closely associated with the reticulum on the abdominal side. This makes the pericardium prone to extension of inflammatory processes originating from the reticulum. This section describes the pericardium and mediastinum. The ruminant heart is discussed in Case 18.2.

Function

The pericardium is a double-layer—fibrous outermost layer and serous innermost layer—sac of fibroelastic/fibroserous tissue encasing the heart, anchoring it to the mediastinum (a passageway for many structures that pass through the thorax), and forming a cavity filled with a serous fluid. It protects the heart from external shock (shock absorption) and provides lubrication of the heart during each cardiac contraction. The position of the heart in the mediastinum thus limits its motion. The pericardium also isolates or shields the heart from infections associated with other mediastinal and thoracic cavity organs—i.e., the lungs—and limits the heart from excessive distention associated with an acute volume overload.

Pericardium (Fig. 18.1-6)

The **pericardium** is the fibroserous sac that encloses the heart and comprises a strong, inelastic fibrous layer and a serous layer. The **fibrous layer** is attached dorsally to the large vessels at the base of the heart, and firmly attached ventrally to the middle of the caudal half of the thoracic surface of the sternum by the paired **sternopericardiac ligaments (right** and **left)**. The **serous layer** is a closed sac (the **pericardial sac**) surrounded by the fibrous pericardium and invaginated by the heart; it normally contains a small amount of clear serous fluid. The serous layer consists of 2 parts—**parietal** (which lines the fibrous layer) and **visceral** (which covers the heart and great vessels). The visceral part is also referred to as the **epicardium**. On the left side, the greater part of the pericardium is in contact with the lateral thoracic wall. On the right side, the pericardium may be covered by the lung and have no contact with the thoracic wall.

Like in most veterinary species, the pulmonic valve is best auscultated in the 3rd intercostal space, while the aortic and mitral valves are best auscultated in the 4th and 5th intercostal spaces, respectively, on the left. The tricuspid valve is best auscultated in the 3rd intercostal space on the right side.

In cattle, approximately 70% of the heart is on the left side of the median plane of the thorax. The **base** of the heart is **adjacent to** the thoracic wall from the 2nd to the 5th intercostal spaces, and the **apex** is **adjacent to** the 6th costochondral junction. On the right, the heart stretches between the 3rd and 4th intercostal spaces, directly deep to the elbow.

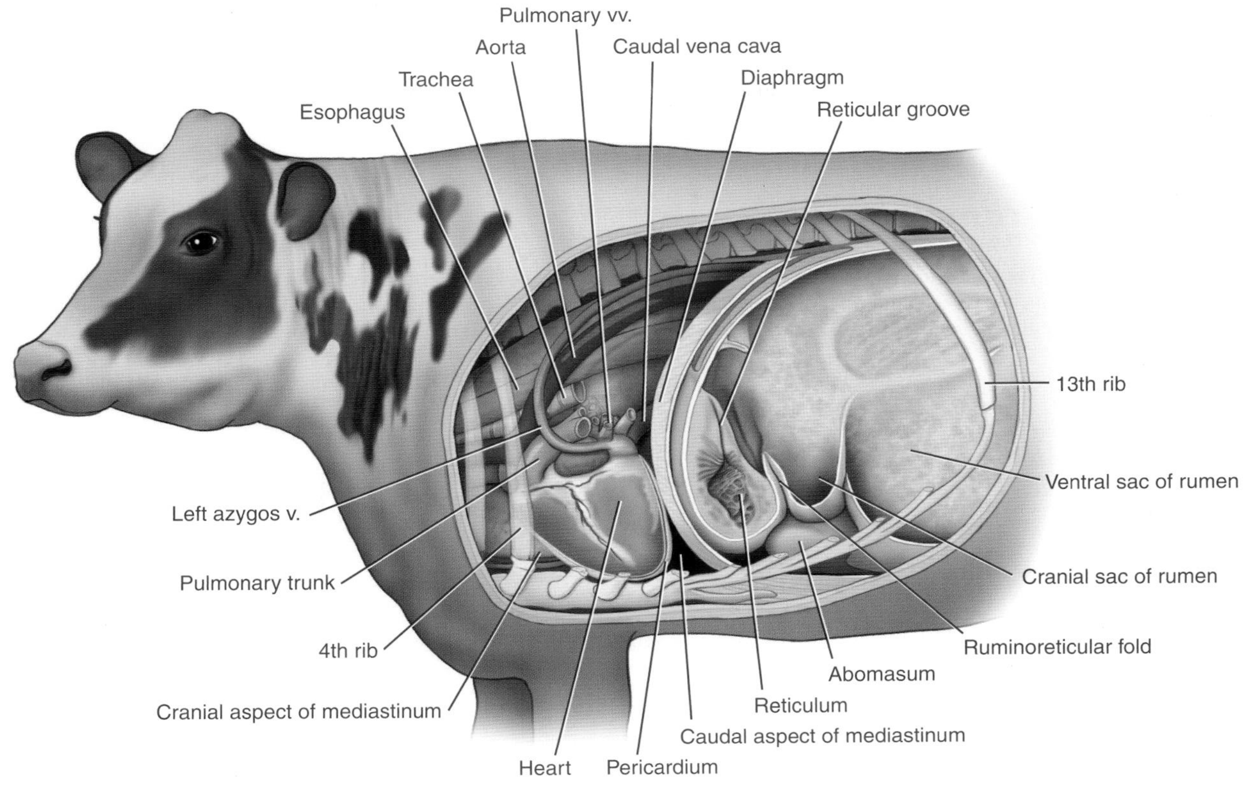

FIGURE 18.1-6 Anatomical relationship between reticulum, diaphragm, and pericardium in a bovid.

In cattle, the apex of the heart contacts the dome of the **diaphragm**, and the **reticulum** in the abdominal cavity lies on the caudal side of the diaphragm.

Sharp metallic objects—e.g., nails, wires, or other foreign material—that are accidentally swallowed by the cow can accumulate in the reticulum and later penetrate the wall of the reticulum, the diaphragm, and the adjacent pericardial sac resulting in infection of the pericardium (septic pericarditis). The tissues of the pericardium thicken and fluid builds up within the pericardial sac, leading to heart failure. Fluid accumulation within the pericardium results in decreased distensibility of the heart, which in turn impairs the ability of the heart to fill during diastole. Impaired ventricular filling in turn increases atrial pressure, negatively affecting not only cardiac output, but also venous return to the heart. The visible consequences on physical examination include jugular vein distention, jugular pulses, and dependent edema (brisket, submandibular, ventral), as well as signs of respiratory distress in advanced cases.

Mediastinum

The heart is situated in the mediastinal space (or **mediastinum**), which is created by the junction of the two pleural cavities near the midline of the thorax. In addition to containing the heart, the mediastinum also contains the esophagus and trachea, the vessels passing to and from the neck and thoracic limbs, the great vessels of the heart, an assembly of lymph nodes, the thoracic duct, the remnants of the thymus (in adults), and various nerves including the vagal trunks. The mediastinum of cattle is robust and forms an impervious barrier between the right and left pleural cavities.

Inflammation (e.g., severe pleuropneumonia) or a mass in the mediastinum (e.g., an abscess or thymoma) can lead to dysfunction of the esophagus or vagus nerve in affected cattle. The complete mediastinum in cattle means that infection or pneumothorax of one hemithorax typically remains unilateral.

Blood supply, lymphatics, and innervation

Blood supply to the mediastinum is from the **internal thoracic arteries** and like-named veins are responsible for drainage. Innervation is from the **vagal trunk**. Lymphatic drainage from the heart goes to the **mediastinal lymph nodes**.

Selected references

[1] Singh B, editor. The thorax of the ruminant. In: Dyce, Sack and Wensing's textbook of veterinary anatomy. 5th ed. Elsevier; 2017. p. 658–63.
[2] Singh B, editor. The rumen and reticulum. In: Dyce, Sack and Wensing's textbook of veterinary anatomy. 5th ed. Elsevier; 2017. p. 670–5.
[3] Dee Fails A, Magee C, editors. Anatomy of the cardiovascular system. In: Anatomy and physiology of farm animals. 8th ed. Wiley Blackwell; 2018. p. 316–60.
[4] Fubini SL, Ducharme NG, editors. Surgery of the ruminant forestomach compartments. In: Farm animal surgery. 2nd ed. Elsevier; 2004. p. 249–58.

CASE 18.2

Endocarditis and Atrial Lymphoma

Marie-Eve Fecteau[a] and Gilles Fecteau[b]
[a]Food Animal Medicine and Surgery, Department of Clinical Studies - New Bolton Center, University of Pennsylvania School of Veterinary Medicine, Kennett Square, Pennsylvania, US
[b]Department of Clinical Sciences, Université de Montréal, Faculty of Veterinary Medicine, Saint Hyacinthe, Québec, CA

Clinical case

History

A 5-year-old Holstein cow presented for poor milk production and weight loss after she calved uneventfully 30 days previously. Her appetite was reported to be fair. A magnet was administered (see Case 19.1) when the cow was diagnosed pregnant as a first-calf heifer.

Physical examination findings

On presentation, the cow was bright, alert, and responsive. She was ambulatory but appeared slightly weak in the hind end. Her body condition score was 2/5 (1 = emaciated, 5 = obese). Her mucous membranes were slightly pale, but moist with a normal CRT. Her rectal temperature, heart rate, and respiratory rate were 101.6 °F (38.7 °C), 110 bpm, and 40 brpm, respectively. Cardiac auscultation did not reveal murmurs or arrhythmias, but the heart sounds were muffled. Bilateral jugular vein distention and brisket edema were present (Figs. 18.2-1 and 18.2-2). The prescapular (superficial cervical) and prefemoral (superficial

> The pericardial sac normally contains a small amount of pericardial fluid. Fluid accumulation within the pericardial cavity can lead to muffling of the heart sounds or splashing heart sounds (also referred to as "washing machine" sounds). In cattle, the 2 most common clinical entities involving the pericardium and pericardial cavity are traumatic reticulopericarditis and idiopathic hemorrhagic pericarditis (discussed in Case 18.1).

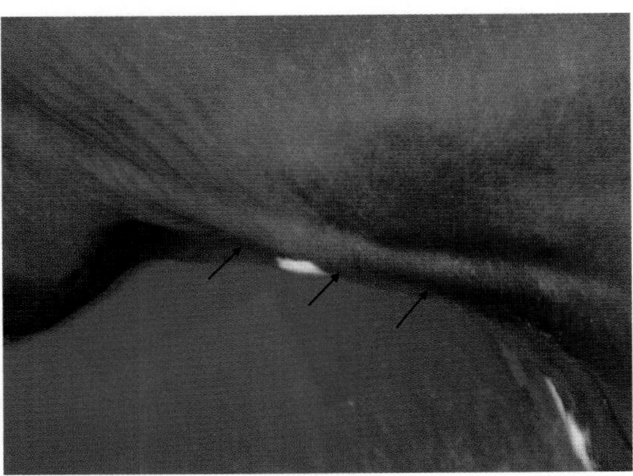

FIGURE 18.2-1 Jugular vein *(arrows)* distention in a cow with cardiac lymphoma. (Image courtesy of Dr. Virginia Reef, University of Pennsylvania.)

Comparative Veterinary Anatomy: A Clinical Approach. https://doi.org/10.1016/B978-0-323-91015-6.00092-3

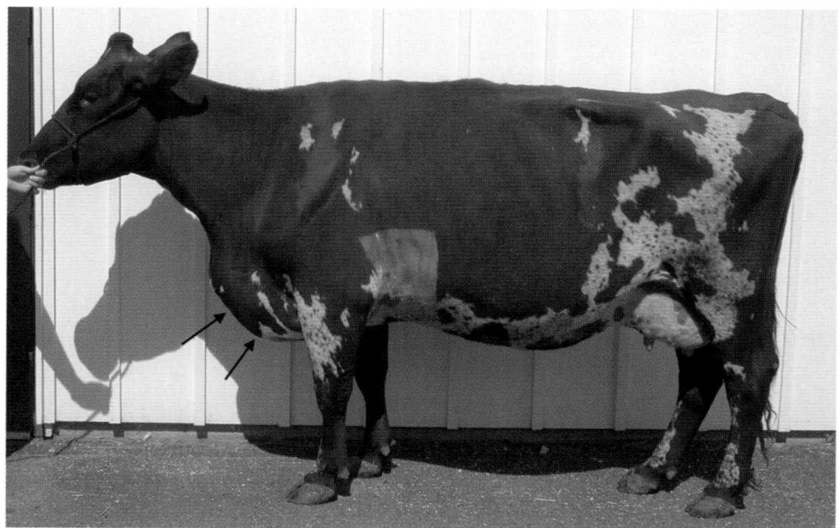

FIGURE 18.2-2 Brisket edema *(arrows)* in a cow with cardiac lymphoma. (Image courtesy of Dr. Virginia Reef, University of Pennsylvania.)

inguinal) lymph nodes were enlarged. The rumen was hypomotile, and percussion of both sides of the abdomen revealed no areas of resonance (a.k.a. "pings"). Abdominal palpation per rectum identified enlarged subiliac lymph nodes and normal manure. Her udder palpated normally, and a California Mastitis Test was negative in all quadrants.

Differential diagnoses

Traumatic reticulopericarditis, bacterial endocarditis, cardiac lymphoma, idiopathic hemorrhagic pericarditis, septic myocarditis, monensin toxicity, white muscle disease, high altitude disease, and red Holstein myocarditis

Diagnostics

A compass, used to detect the direction of a magnetic field, confirmed that a magnet was present in the cranial abdomen. Echocardiography revealed moderate hypoechoic pericardial effusion (Fig. 18.2-3) but no signs of

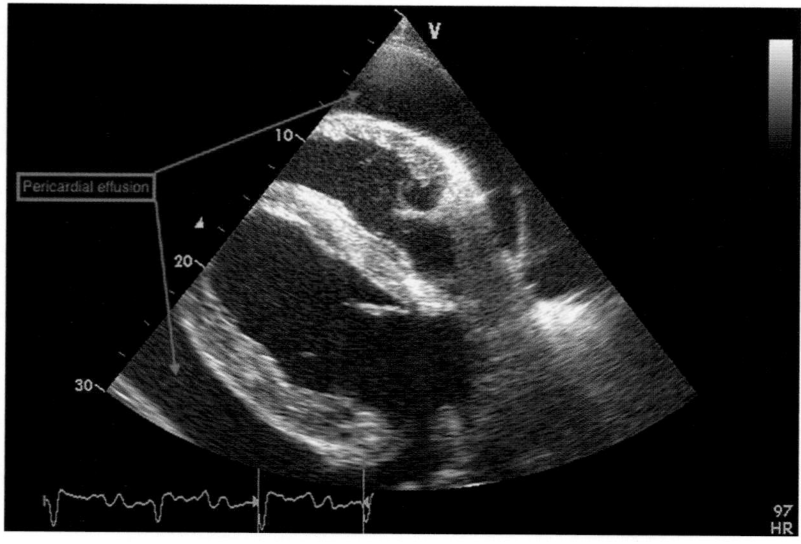

FIGURE 18.2-3 Ultrasound demonstrating moderate hypoechoic pericardial effusion in a cow.

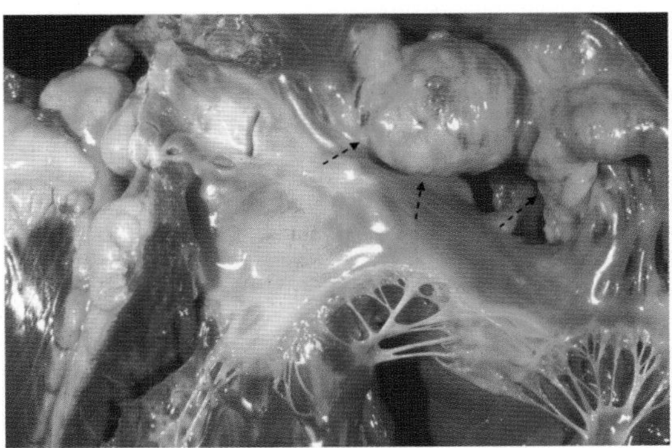

FIGURE 18.2-4 Necropsy image of a cow with lymphoma depicting right atrial myocardial infiltration by a tumor and extension of the tumor into the atrial lumen *(arrows)*. (Image courtesy of Dr. Perry Habecker, University of Pennsylvania.)

valvular lesions. The right atrial chamber was enlarged and had a thickened wall. Serum submitted for detection of antibodies (ELISA) against bovine leukosis virus (BLV) was positive. Fine-needle aspiration of an enlarged prescapular lymph node revealed a population of large immature lymphocytes with abnormal nuclei. Ultrasound-guided pericardiocentesis yielded serosanguineous fluid. Cytologic evaluation of the pericardial fluid revealed an absence of bacterial organisms and identified a mononuclear leukocytosis consisting mainly of neoplastic lymphocytes.

Diagnosis

Right atrial lymphoma and congestive heart failure

Treatment

No treatment was recommended because of the poor prognosis, and the cow was humanely euthanized. Necropsy examination confirmed the presumptive diagnosis of lymphoma (Fig. 18.2-4).

Anatomical features in ruminants

Introduction

The bovine heart is similar to other species, but relevant anatomical features that are specific to cattle can sometimes complicate cardiac auscultation and/or prevent a complete and satisfactory echocardiographic examination. This section covers the clinical anatomy of the bovine heart. The clinical anatomy of the pericardium and mediastinum is covered in Case 18.1.

Cattle have narrow intercostal spaces and an unusually cranial (compared to other species) location of the heart within the thoracic cavity, requiring the stethoscope (or the ultrasound probe) to be placed more cranially in the axillary area around the third intercostal space. Cattle that are exceptionally large—e.g., bulls or overweight cows—may be even more difficult to auscultate because of increased chest wall thickness. Heart sounds are easier to auscultate on the left side. The pulmonic, aortic, and mitral valves are best auscultated in the left 3rd, 4th, and 5th intercostal spaces, respectively. The tricuspid valve is best heard in the right 3rd intercostal space.

Function

The heart is the central organ that pumps blood continuously through the blood vessels by rhythmic contractions, supplying oxygen and life-sustaining nutrients to the tissues and removing carbon dioxide and other wastes.

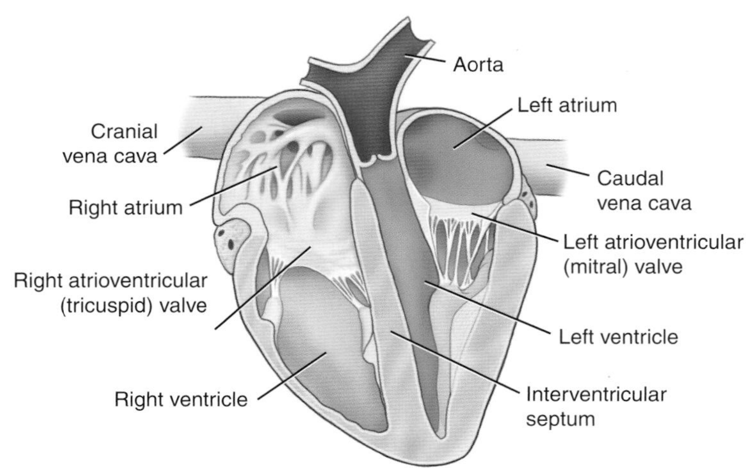

FIGURE 18.2-5 Anatomy of the bovine heart in sagittal section. The pulmonary trunk is not visible in this view (removed with cross-sectioning).

Heart and great vessels (Fig. 18.2-5)

The adult heart has 4 chambers: **right atrium**, **right ventricle**, **left atrium**, and **left ventricle**. The 2 atria and 2 ventricles are separated by an internal septum, but the atrium and ventricle of each side communicate through a large opening—i.e., valve. The heart essentially consists of 2 pumps (right and left).

Right heart: Blood returning to the heart from the systemic circulation is delivered to the right atrium by the **cranial** and **caudal venae cavae**. Venous blood returning from the myocardium is returned directly to the right atrium by way of the **coronary veins**. From the right atrium, deoxygenated blood passes through the **right atrioventricular (AV) valve** (or **tricuspid valve**) into the right ventricle. From the right ventricle, the deoxygenated blood passes through the **pulmonic valve** and enters the **pulmonary trunk**, which divides into **right** and **left pulmonary arteries**, carrying the blood to the respective lungs.

Left heart: From the lungs, newly oxygenated blood returns to the left atrium via the **pulmonary veins**, then through the **left AV valve** (or **mitral valve**) into the left ventricle. The left ventricle then pumps the oxygenated blood through the **aortic valve** into the **aorta** to all parts of the body, including the myocardium via the **left** and **right coronary arteries**.

BACTERIAL ENDOCARDITIS

Another clinical entity that is more commonly seen in older cows is bacterial endocarditis. The condition frequently affects the tricuspid valve—although any valve can be affected—and is thought to be secondary to a primary septic focus causing recurrent bacteremia. The primary infection site can still be active at presentation or may have already resolved. While the clinical signs may or may not include evidence of heart failure as described in this case, most affected individuals are chronically ill with recurrent low-grade fevers. A systolic heart murmur is not uncommon with lesions affecting the tricuspid or mitral valves, but endocarditis may occur in the absence of an auscultable murmur. Laboratory tests typically reveal a chronic active inflammatory process (mature neutrophilia, hyperfibrinogenemia, increased globulin concentration, mild nonregenerative anemia). The valves can be imaged ultrasonographically, and while severe, large lesions are obvious, the more subtle changes seen in acute cases can be difficult to identify antemortem (Fig. 18.2-6). *Trueperella pyogenes* is the most common isolate from blood and endocardial lesions of affected cattle, but other bacterial agents such as *Staphylococcus* spp., *Streptococcus* spp., or gram-negative organisms are also isolated. Treatment relies on long-term antimicrobial therapy adjusted after the antibiogram (culture and susceptibility), if available. Prognosis is guarded, especially for patients showing signs of congestive heart failure.

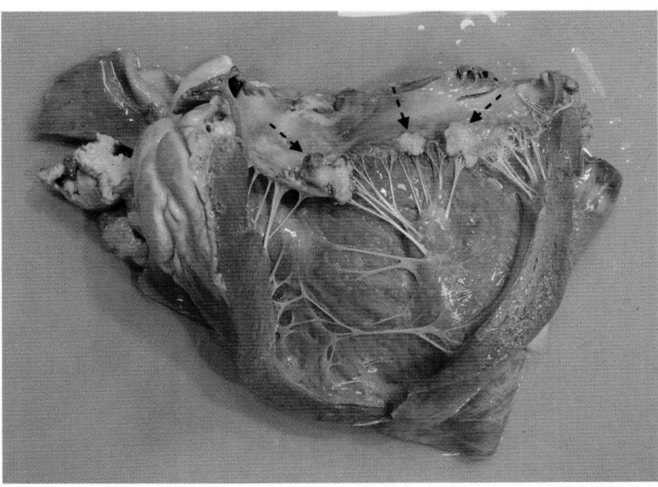

FIGURE 18.2-6 Necropsy image of a cow with bacterial endocarditis *(arrows)*. (Image courtesy of Dr. Perry Habecker, University of Pennsylvania.)

Blood supply, lymphatics, and innervation

The heart muscle (myocardium) receives its own blood supply from a network of arteries called the **coronary arteries**. Two major coronary arteries (**right** and **left coronary arteries**) branch from the aorta near the junction of the aorta and the left ventricle, and lie on the surface of the heart, each giving way to smaller arteries that penetrate from the surface into the cardiac muscle mass. Because the left side of the heart is larger and more muscular, the left coronary artery is larger than the right coronary artery.

The left coronary artery (LCA) originates from the aorta above the aortic valve and supplies blood to the left side of the heart. The main stem/trunk of the LCA branches as the **paraconal** (cone-shaped) or **interventricular branch**. The LCA continues as a **circumflex** (curved) **branch** in the **coronary groove** coursing toward the caudal part of the heart and ending as the **right subsinusoidal** (curved) or **interventricular branch**, following the **right interventricular groove** in ruminants (and carnivores).

Other terms used to describe the left coronary artery include the left main coronary artery (LMCA) and the left main stem coronary artery (LMS).

Most of the venous blood from the myocardium is returned to the right atrium by way of the **coronary veins**, which empty directly into the right atrium via the **coronary sinus** (terminal part of the **great cardiac vein**), adjacent to the opening of the caudal vena cava. Lymphatic drainage from the heart goes to the **mediastinal lymph nodes**, located in the cranial mediastinum.

The heart is innervated by sympathetic and parasympathetic fibers from the autonomic part of the peripheral nervous system. The network of nerves supplying the heart is called the **cardiac plexus**. It receives contributions from the **right** and **left vagus nerves**, and contributions from the sympathetic trunk. The heart must alternatively contract and relax in order to pump and receive blood via an intrinsic myogenic phenomenon. The conduction system consists of the **sinoatrial node** (S-A node), the **atrioventricular node** (AV node), the **atrioventricular bundle**, and the terminal conducting or **Purkinje fibers**. The S-A node (a.k.a. the pacemaker) is a small mass of modified heart muscle located at the junction of the cranial vena cava and the right atrium. The AV node is smaller and lies near the opening of the coronary sinus.

Selected references

[1] Singh B, editor. The cardiovascular system. In: Dyce, Sack and Wensing's textbook of veterinary anatomy. 5th ed. Elsevier; 2017. p. 210–51.
[2] Dee Fails A, Magee C, editors. Anatomy of the cardiovascular system. In: Anatomy and physiology of farm animals. 8th ed. Wiley Blackwell; 2018. p. 315–30.
[3] Dee Fails A, Magee C, editors. Physiology of the heart and circulation. In: Anatomy and physiology of farm animals. 8th ed. Wiley Blackwell; 2018. p. 331–47.
[4] Smith BP, Van Metre DC, Pusterla N, editors. Diseases of the cardiovascular system. In: Large animal internal medicine. 6th ed. Elsevier; 2014. p. 478–514.

CHAPTER 19

ABDOMEN

André Desrochers, Chapter editor

Forestomachs (Rumen, Reticulum, Omasum, and Abomasum)
19.1 Traumatic reticuloperitonitis—*Emma Marchionatti* 1056
19.2 Left displacement of the abomasum—*Brent C. Credille and Susan Fubini* 1063

Small and Large Intestine
19.3 Intestinal volvulus—*David E. Anderson* 1069

Cecum
19.4 Cecal dilatation/volvulus—*Emma Marchionatti* 1077

Liver
19.5 Caudal vena cava syndrome—*Julie Berman* 1082
19.6 Hepatic lipidosis—*Julie Berman* 1090

Kidney
19.7 Hydronephrosis—*André Desrochers* 1097

CHAPTER 19

CASE 19.1

Traumatic Reticuloperitonitis

Emma Marchionatti
Clinic for Ruminants, Vetsuisse Faculty, University of Bern, Bern, CH

Clinical case

History

A 7-year-old Jersey cow presented for sudden anorexia and decreased milk production of 4 days' duration. The referring veterinarian initially treated the cow with procaine penicillin G and administered a magnet, but without a clinical improvement.

Physical examination findings

The cow was slightly depressed but responsive with normal heart and respiratory rates (HR 72 bpm, RR 24 brpm). Rectal temperature was mildly increased 102.7 °F (39.3 °C). Rumen sounds were decreased in amplitude and frequency (1 rumen contraction in 2 minutes). The cow presented with an arched back and tense abdomen. The withers pinch test was abnormal—while the cow did not grunt, no ventroflexion was seen.

Differential diagnoses

Traumatic reticuloperitonitis (a.k.a. "hardware disease"), abomasal ulcers, and hepatic abscess

Diagnostics

Hematological and serum biochemical analyses showed neutrophilia (17,000/μL) with a left shift and hyperfibrinogenemia (9 g/L). Transabdominal ultrasound (Fig. 19.1-1) revealed the presence of mixed echogenicity free fluid and fibrin deposits between the reticulum, the abdominal wall, the ventral sac of the rumen, and the abomasum. Reticular contractions were apparent during the ultrasound examination.

Cranial abdominal radiography (Fig. 19.1-2) revealed a linear metallic foreign body penetrating the cranial abdomen through the caudoventral reticular wall. Multiple small gas bubbles were visible just ventral to the foreign body and reticulum. The magnet was observed in the atrium of the rumen.

Diagnosis

Traumatic reticuloperitonitis (TRP)

Treatment

A left flank rumenotomy was performed to retrieve the foreign body. A proximal paravertebral block (T13, L1, L2) was used for flank anesthesia. A 25-cm (9.8-in.) long surgical incision was made in the left paralumbar fossa starting 10 cm (3.9 in.) ventral to the transverse processes and 5 cm (2.0 in.) caudal to the last rib. Following the skin incision, the external and internal abdominal oblique muscles, transversus abdominis muscle, and peritoneum were incised,

Comparative Veterinary Anatomy: A Clinical Approach. https://doi.org/10.1016/B978-0-323-91015-6.00093-5

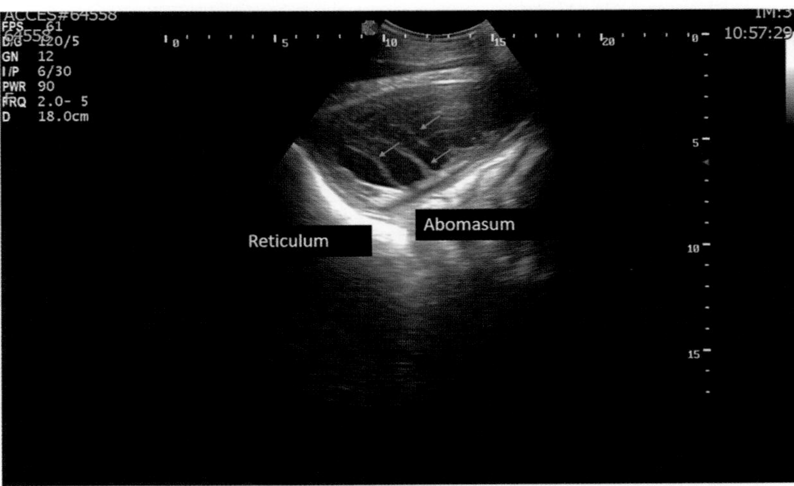

FIGURE 19.1-1 Ultrasonographic image of the cranial ventral abdomen (using a convex 2.5 MHz probe) of a cow showing mixed echogenicity fluid and fibrin deposits (*arrows*) between the reticulum and the abomasum.

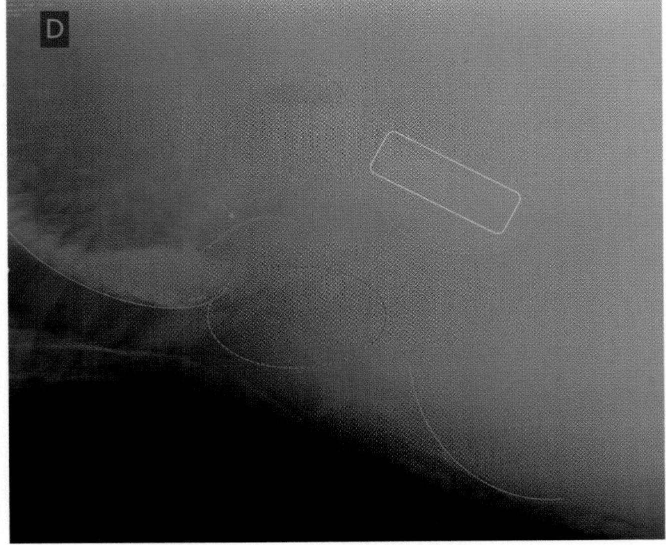

FIGURE 19.1-2 Lateral radiographic view of the cranial ventral abdomen illustrating a linear metallic foreign body (*green arrow*) penetrating the cranial abdomen through the caudoventral reticular wall; gas formation is visible caudoventrally and dorsally to the reticulum (*dashed red lines*); *blue lines* highlight the reticulum, atrium, and ventral sac of the rumen profiles; and the *yellow line* highlights the magnet in the atrium of the rumen.

from superficial to deep. Exploration of the cranial abdomen was not performed to minimize disruption of adhesions and dissemination of the local infection. The dorsocaudal sac of the rumen was fixed to a Weingart ring and incised, taking care to avoid contamination of the abdomen. The reticulum, atrium, cardia, and reticulo-omasal orifice were thoroughly palpated, and the foreign body was retrieved. The magnet was replaced in the reticulum. A regimen of anti-inflammatories and broad-spectrum (including anaerobic coverage) antimicrobials were administered peri-operatively. Prolonged antimicrobial treatment was considered necessary to treat the septic peritonitis.

Anatomical features in ruminants

Introduction

Ruminants' forestomachs consist of 3 nonglandular compartments (reticulum, rumen, and omasum) lined with keratinized stratified squamous epithelium—the rumen, reticulum, and omasum. These precede a fourth compartment lined with glandular mucosa—the abomasum or "true stomach". The abomasum is discussed in Case 19.2.

Function

The forestomach contains numerous microorganisms (protozoa and bacteria) responsible for extensive anaerobic degradation of nutrients. Fermentation end products, such as volatile fatty acids, are absorbed from the digestive tract primarily by the forestomach.

Reticulum (Fig. 19.1-3)

The **reticulum** is situated cranial to the rumen and lies against the diaphragm. The left reticular part contacts the spleen while the right is adjacent to the left lobe of the liver, the omasum, and the abomasum. The most ventral aspect contacts the xiphoid. The mucosa of the inner surface of the reticulum forms a network of crests covered with small papillae, which give it a honeycomb appearance. The wall is strongly muscled and mobile; during forestomach contractions, ingesta are passed back and forth to and from the rumen. In cattle, the reticulum has a capacity of 10–20 L. The **reticular groove** runs from the cardia to the **reticulo-omasal orifice**, beyond which it continues via the **omasal groove**.

Mechanical extraluminal obstruction of the reticulo-omasal groove by a mass (abscess) and adhesions, such as seen in TRP cases, results in type 2 vagal indigestion (also called omasal transport failure). Motility of the rumen may continue normally, but the ability of the reticulum and omasum to move ingesta into the abomasum is impaired.

Rumen (Fig. 19.1-3)

The **rumen** extends from the diaphragm to the pelvis occupying almost all of the left side of the abdominal cavity. The parietal surface lies against the left and ventral abdominal wall and contacts the spleen. The visceral surface contacts the left kidney, intestines, liver, omasum, and abomasum. The **right** and **left longitudinal grooves**, connected by **cranial** and **caudal grooves**, divide the rumen into dorsal and ventral sacs. The right and left longitudinal grooves supply the dorsal **right** and **left accessory grooves**. The longitudinal grooves are the attachment sites for the **superficial** and **deep leaves of the greater omentum**. At the caudal end of the rumen, transverse **dorsal** and **ventral coronary grooves** define the **caudodorsal** and **caudoventral blind sacs** of the rumen. Internally, the ridges formed by the exterior grooves are referred to as **pillars**. The **atrium** or craniodorsal blind sac, which lies between the cranial pillar and the ruminoreticular fold, opens into the reticulum. The pillars and their contractions contribute to an efficient mixing of the rumen contents, ensuring adequate fermentation. The inner surface of the rumen is covered by papillae. The rumen is the largest of the forestomach compartments, and its capacity may reach 150 L in cattle.

Omasum (Fig. 19.1-3)

The **omasum** lies on the right side of the cranial abdominal cavity, between the reticulum, rumen, and abomasum. Most of its surface is covered by the lesser omentum. The **reticulo-omasal orifice** regulates the transport of ingesta to the omasum and the **omaso-abomasal orifice** regulates it to the abomasum. The omasum contains multiple rows of different height muscular laminae covered with small papillae. It has a capacity of 7–18 L in cattle.

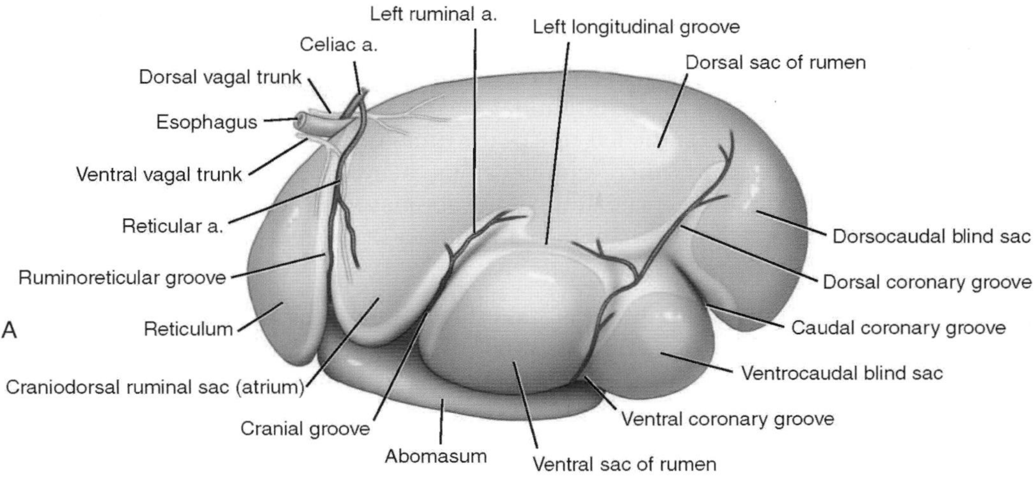

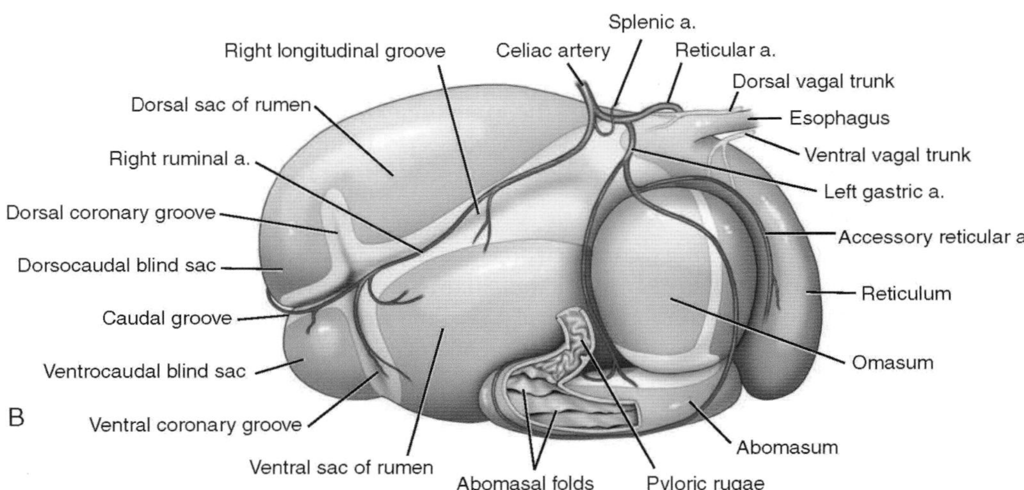

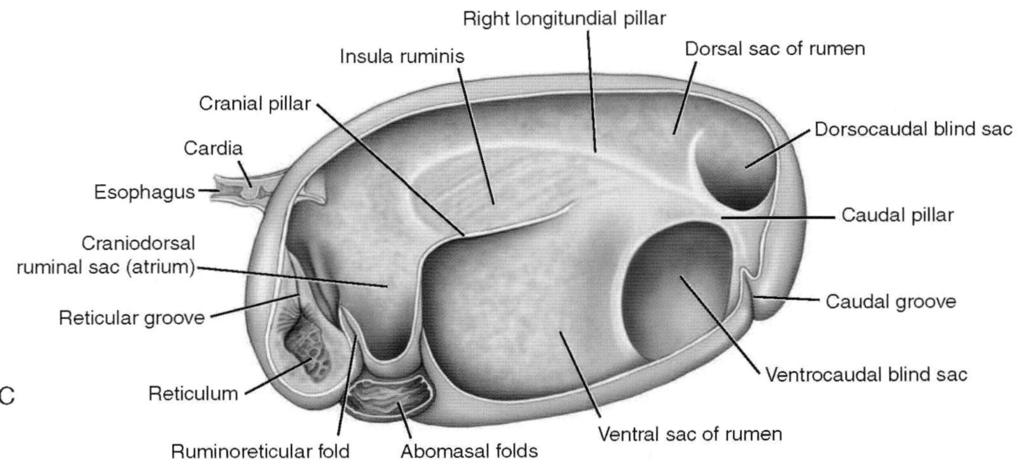

FIGURE 19.1-3 Anatomy of the ruminant's forestomach: (A) parietal surface; (B) visceral surface; (C) inner surface.

Blood supply, lymphatics, and innervation (Fig. 19.1-3)

Arterial supply to the forestomach derives primarily from the **celiac artery** and 2 of its branches, the **splenic** and **left gastric arteries**. The splenic artery gives off the right ruminal artery that flows along the right accessory groove and anastomoses with the **left ruminal artery** on the left side of the rumen. The celiac artery also gives off the **reticular artery**, which passes dorsally over the rumen, then ventrally along the ruminoreticular groove. The left gastric artery supplies the omasum and extends to the lesser curvature of the abomasum. The **accessory reticular artery** arises from the left gastric artery and tracks along the diaphragmatic surface of the reticulum. The veins have corresponding courses and branch and drain into the portal vein.

Lymph drains to multiple small lymph nodes spread around the forestomachs and especially in those lymph nodes in the ruminal grooves and around the omasal and abomasal curves. These small peripheral lymph nodes drain to larger atrial nodes located between the cardia and omasum and finally to the cisterna chyli.

The rumen is innervated mainly by the **dorsal vagal trunk**, while the atrium of the rumen, the reticulum, and the omasum are innervated by both vagal trunks. In addition to the parasympathetic innervation, sympathetic branches derive from the **celiac ganglia**.

Forestomach contractions

Reticulorumen (forestomach) contractions serve to mix the forestomach content, to move it forward to the omasum or backward for rumination, and to aid in removing gases from the rumen. Motility can be divided into 3 types—primary contractions associated with ruminal mixing, contractions related to rumination, and secondary contractions related to eructation of gas.

The most frequent contractions are the **primary contractions**, usually 2–3 per 2 minutes and are more frequent during feeding. Mixing contractions start with a biphasic reticular contraction. The first contraction forces rough, fibrous material from the top of the reticulum to the dorsal sac of the rumen. After a momentary relaxation, the second reticular contraction empties the finely dispersed content from its lower portion to the atrium of the rumen. The reticulo-omasal orifice opens during the second contraction, and a small volume of well-fermented material flows to the omasum. Once the reticulo-omasal orifice closes, an omasal contraction forces the feed material into the abomasum through the omasal-abomasal orifice. A rumen contraction cycle begins in the atrium of the rumen and the cranial pillar, forcing the ingesta over the ruminoreticular fold back into the reticulum. The dorsal rumen sac contraction then forces the content of the upper part of the rumen toward the caudodorsal blind sac. The contraction of this plus that of the caudal pillar forces the contents dorsally and cranially. Following the contraction in the dorsal rumen sac, the ventral rumen sac contracts moving ingesta backward to the caudoventral blind sac. During these circular movements, the well-digested content gradually sinks and—when the caudoventral sac contracts—it is moved forward over the cranial pillar and into the atrium of the rumen.

During **rumination,** an extra reticular contraction precedes the usual biphasic one. This contraction removes the freshly swallowed feed from the cardia and replaces it with ingesta that have already undergone some fermentation. Simultaneous with the extra reticular contraction, the cardia relaxes and negative pressure is created within the thorax, favoring the movement of feed into the esophagus. This process allows the food to reenter the mouth to be chewed again (i.e., "chewing cud"). Rumination reduces the size of the feed particles, exposing new surfaces for bacterial fermentation.

Secondary contractions start in the caudal blind sacs, move cranially, and push the gas cap toward the atrium of the rumen. Collection of gas around the cardia stimulates eructation. The frequency of secondary contractions is usually one every 2–3 primary contractions (Fig. 19.1-4).

Regulation of forestomach contractions

Forestomach motility is mainly regulated via long reflexes. Sensory cells in the forestomach wall send information to the **dorsal vagal nuclei** in the brain stem via the vagal nerves. Two types of sensory cells are present in the forestomach wall: stretch-sensitive sensory cells (**tension receptors**) and sensory cells with both mechano- and chemoreceptors (**epithelial receptors**). The tension receptors react to wall stretch, are activated during rumen wall contractions, and stimulate movements of the forestomach. The epithelial receptors, in addition to reacting to

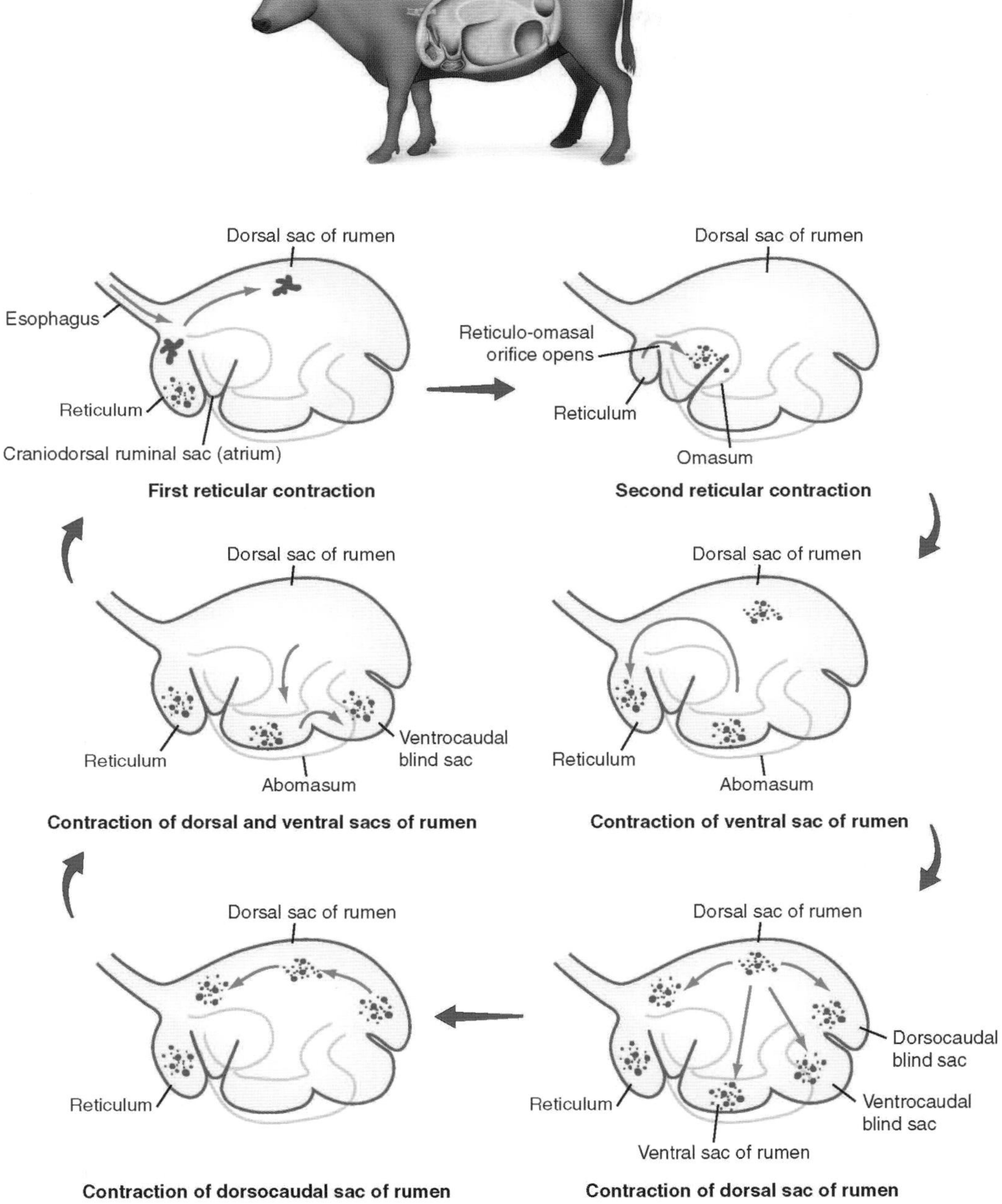

FIGURE 19.1-4 Pattern of ruminant forestomach contractions.

BLOAT (FIG. 19.1-5)

Accumulation of gas in the rumen (bloat) occurs due to impaired eructation (L. *eructatio* belching). Rumen bloating can be caused by blockage of the passage of gas through the esophagus as a result of an intraesophageal obstruction (i.e., “choke”) or by external pressure on the esophagus by a mass. Consequently, gas collects in the gas cap of the rumen. Gas may also collect in the rumen content in the form of small bubbles (foam) in disorders of fermentation. In both cases, gas cannot be eructated and rumen pressure rises quickly, with consequent respiratory and circulatory distress due to compression on the diaphragm. Forestomach contractions are reduced or stopped because of strong stimulation of the epithelial receptors.

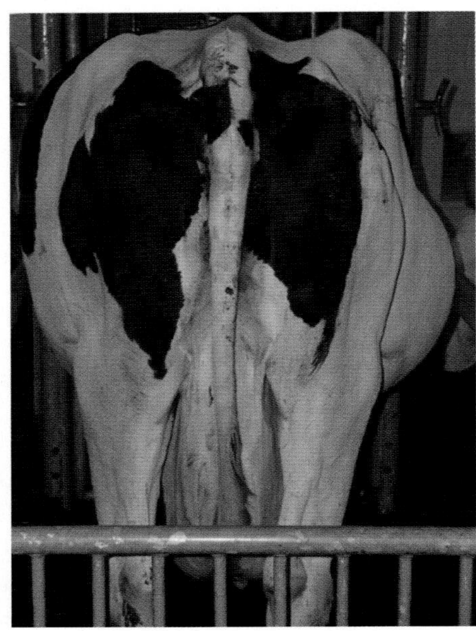

FIGURE 19.1-5 Holstein cow presenting with rumen bloat (*arrow*).

stretch stimuli, also react to changes in pH, osmolarity, and volatile fatty acid concentrations, and have an inhibitory effect on forestomach movements. The dorsal vagal nuclei are responsible for the central regulation of forestomach motility. Efferent nerve fibers travel to the forestomach through the vagal nerves.

Selected references

[1] Budras KD, Habel RE. Bovine anatomy. 2nd ed; 2011. p. 70–3.
[2] Klein BG. Cunningham's textbook of veterinary physiology; 2003. p. 328–36.
[3] Fubini SL, Ducharme NG. Farm animal surgery. 2nd ed; 2017. p. 249–60.
[4] Lozier JW, Niehaus AJ. Surgery of the forestomach. Vet Clin North Am Food Anim Pract 2016;32:617–8.
[5] Braun U, et al. Clinical and laboratory findings in 503 cattle with traumatic reticuloperitonitis. BMC Vet Res 2018;15:66.

CASE 19.2

Left Displacement of the Abomasum

Brenton C. Credille[a] **and Susan Fubini**[b]
[a]Food Animal Health and Management Program, Department of Population Health, University of Georgia College of Veterinary Medicine, Athens, Georgia, US
[b]Large Animal Surgery, Cornell University College of Veterinary Medicine, Ithaca, New York, US

Clinical case

History

A valuable 4-year-old Holstein cow presented for vague complaints of "not doing well" since freshening—i.e., calving—2 weeks ago. The cow cleaned—i.e., passed the placenta—24 h after calving and had no history of mastitis or other infectious or metabolic disorders. The cow was passing scant, loose manure, was inappetent, and was not cleaning her nose. There was mild distention in the left paralumbar fossa and ventral abdomen.

> Cows keep their noses clean (grooming) by frequent licking. This normal pattern of behavior prevents them from inhaling the many microbes that are forced out during eructation and is a clinical sign that the cow feels well.

Physical examination findings

On physical examination, the rectal temperature was 103 °F (39.4 °C), the heart rate was 80 bpm, and the respiratory rate was 24 brpm. The vulvar mucous membranes were light pink and moist with a CRT of 2 s. The eyes were slightly sunken and the skin turgor (L. *turgidus* fullness, congested) somewhat prolonged. The cow had a left-sided ping on a

COMMON POST-PARTUM CONDITIONS IN COWS

- Mastitis
- Metritis
- Injury or lameness
- Milk fever
- Ketosis
- Displaced abomasum
- Pneumonia
- Enteritis

Comparative Veterinary Anatomy: A Clinical Approach. https://doi.org/10.1016/B978-0-323-91015-6.00094-7

Succussion is when the clinician ballots (jostles) the ventral abdomen with a closed hand (fist) and the fluid within a viscus makes a "splashy" sound. Positive succussion indicates fluid accumulation in a gastrointestinal viscus and can be indicative of distention proximal to an obstruction or ileus (i.e., intestinal stasis).

line from the tuber coxae to elbow at the level of the 9th –13th rib (Fig. 19.2-1). There was some succussion (L. *succussio* a shaking from beneath, a splashing sound of fluid in a body cavity) of fluid in the lower left paralumbar fossa, and the cow resented external palpation of the cranial ventral abdomen.

Differential diagnoses

Left displaced abomasum (LDA), LDA with perforated abomasal ulcer, ruminal distention—e.g., due to indigestion dietary indiscretion, or ileus, pneumoperitoneum (air within abdominal cavity)

Diagnostics

Abdominal palpation per rectum revealed a distended gas cap on the dorsal sac of the rumen, scant manure, and the sensation of free air in the abdomen. Transabdominal ultrasound revealed displacement of the rumen from the left body wall and hypoechoic abomasal contents in the ventral left abdominal region (Fig. 19.2-2). The CBC showed a neutrophilic leukocytosis and hyperfibrinogenemia. Serum biochemical profile was indicative of a hypochloremic, hypokalemic, and metabolic alkalosis.

Diagnosis

Left displaced abomasum with perforating abomasal ulcer

Treatment

A right paramedian celiotomy (Gr. *koilia* belly + Gr. *tomē* a cutting = surgical incision of the abdominal cavity) was planned in order to access the greater curvature of the abomasum. The cow was sedated with an alpha-2 agonist (xylazine) and, using a casting rope and a V-trough, stabilized in dorsal recumbency. The right paramedian area was

FIGURE 19.2-1 Outline of a left-sided abdominal ping located between the 9th and 13th rib in a cow with a left displaced abomasum.

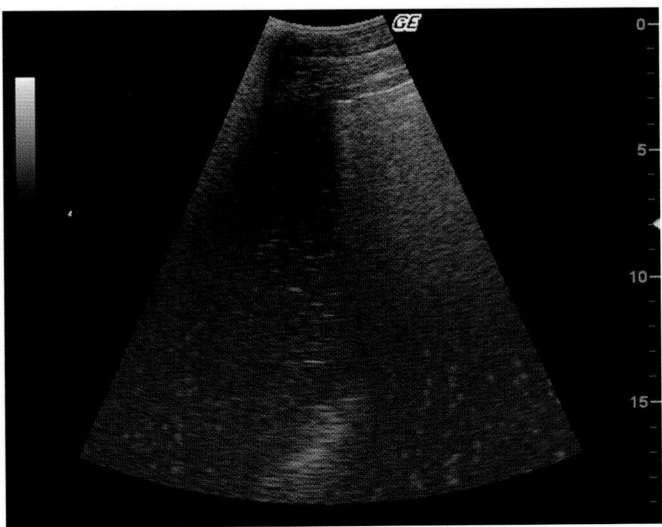

FIGURE 19.2-2 Transabdominal ultrasonographic image of a left displaced abomasum. Dorsal is to the right, ventral is to the left, and the lumen is to the bottom of the picture. The hyperechoic flecks within the lumen are ingesta suspended in sequestered fluid. (Image courtesy of Sebastien Buczinski, DMV, Université de Montréal.)

prepared using standard aseptic technique, and the area was desensitized with L-shaped regional anesthesia. Abdominal exploration revealed that there were adhesions of the abomasal body to the left body wall, presumably from a perforated abomasal ulcer. Using a combination of blunt and sharp dissection, these adhesions were carefully broken down, and the surgeon was careful to stay lateral and as close to the body wall as possible to minimize unintentional entry into the abomasum. Once all adhesions were handled, a right paramedian abomasopexy (Gr. *-pẽxis* fixation) was performed. The cow recovered uneventfully from the procedure and was back in full lactation in 7 days.

There are various ways to anesthetize the surgical site for this procedure, including infiltration of local anesthetic along the intended line of the skin incision (a "line" block) and an "inverted L" block, which provides anesthetic to the nerves supplying the right paramedian area without distorting the tissues. If a flank incision is planned a paravertebral local anesthetic could be used. The paravertebral technique desensitizes the innervation to the flank as the nerves cross the dorsal spines or transverse processes of the last thoracic and first few lumbar vertebrae.

SURGICAL APPROACHES TO PREVENT DISPLACED ABOMASUM

Right paramedian abomasopexy (as described in this case) is a common approach. This approach gives excellent access to the greater curvature of the abomasum and provides direct fixation of the abomasum to the body wall. Disadvantages: The cow must be in dorsal recumbency; it can be difficult to position the cow; and it can cause breathing difficulty and risk of aspiration pneumonia.

Right paralumbar fossa omentopexy is a preferred approach by many surgeons. The procedure is performed in the standing cow, providing good access to the gastro intestinal tract and the pyloric portion of the abomasum. Incorporation of a fold of omentum in the surgical closure is an indirect method of fixation and may not be as secure as other reported techniques. Nevertheless, it is possible to put several sutures in the pyloric part of the abomasum antropyloropexy, a technique that potentially provides a more secure attachment. Disadvantages include limited access to the greater curvature of the abomasum.

Toggle pin fixation is a less precise method than surgery, but can be highly effective. The cow is placed in dorsal recumbency and rolled from side to side so the LDA rises to the ventral abdomen (with the cow on her back). The area is "pinged," and the clinician inserts 2 T-shaped toggles and ties them together. Disadvantages are that the toggle placement is performed blindly, and it is possible to toggle the wrong (or no) viscus.

Anatomical features in ruminants

Introduction

The ruminant forestomach is arranged in 3 distinct compartments: rumen, reticulum, and omasum. The abomasum is the "true" stomach. Collectively, these organs occupy nearly 75% of the abdominal cavity, filling the entirety of the left hemiabdomen and a large portion of the right. This case discusses the abomasum, while the rumen, reticulum, and omasum are discussed in Case 19.1.

Function

The first 3 compartments are broadly referred to as the "forestomach" or "proventriculus" and together function to process and digest the feedstuffs ingested by ruminants. The abomasum is comparable to the simple stomach of monogastric species and secretes hydrochloric acid and other digestive enzymes to further digest feed.

Abomasum

The **abomasum** lies primarily on the abdominal floor as an expanded cavity with the **fundus** near the xiphoid, the **body** between the ventral compartment of the rumen, and the **pyloric part** inclining to the right to join the duodenum at the pylorus. Generally, the abomasum fills the entire right cranioventral abdomen, extending caudally to the right of midline to a point beyond the 13th rib (Fig. 19.2-3). The abomasum is divided into 2 parts—the larger part includes the fundus and body, and the smaller part comprises the pyloric portion. The fundus and body form a pear-shaped sac, the cranial-most extent of which begins 15 cm (5.9 in.) caudal to the **xiphoid process** in adult cattle and crosses over the midline to the left. This portion of the abomasum contacts the body wall between the reticulum and ventral ruminal sac, structures to which the abomasum is connected by muscle bundles. The pyloric region of the abomasum is narrower and more uniform than the fundic region. The pyloric region passes transversely toward the right body wall and ascends dorsally to terminate at the pylorus—the region immediately caudal to the omasum. The positions and relationship of the abomasum to other abdominal structures vary depending on the fill of different portions of the forestomach, intrinsic abomasal activity, and contractions of the rumen and reticulum. Age and pregnancy also influence the position of the abomasum.

> Abomasal length decreases and abomasal width increases during the last 3 months of gestation in response to the progressively developing fetus in the uterus. In addition, the abomasum assumes a position more on the left side of the abdomen during this same time period. In the first 14 days after calving, the uterus returns to its more caudal and right sagittal [non-gravid] position, thus allowing the abomasum to return to its normal position.

Omentum

The **greater omentum** is firmly attached on both sides in the caudal part of the abdomen. The greater omentum attaches to the greater curvature of the abomasum and then continues to the pylorus and medial surface of the ascending duodenum, forming an "omental sling" ventral to the intestines. The space created by this sling is termed the **supraomental recess**. The **lesser omentum** arises from the visceral surface of the liver and attaches to first the lesser curvature of the abomasum and then to the first part of the duodenum.

Blood supply, lymphatics, and innervation

Abomasal blood supply is comprised of the **left gastric** and **left gastroepiploic** branches of the **celiac artery**. These vessels follow the lesser and greater curvatures of the organ, respectively. Veins draining the abomasum generally parallel the aforementioned arteries. These veins eventually join branches of veins that arise from the rumen and spleen (**splenic vein**) and empty into the **portal vein**.

Multiple small lymph node chains spread around the forestomach and abomasum. Many of these nodes lie near the greater and lesser curvatures of the organ, and these nodes direct their efferent flows to the **hepatic lymph nodes**.

FIGURE 19.2-3 Computer-generated illustration of the normal anatomic location of the abomasum. Key: 1, left lobe of the liver; 2, reticulum; 3, abomasum; 4, pyloric antrum; 5, omasum; 6, gall bladder; 7, right lobe of the liver; 8, caudate lobe of the liber; 9, right kidney; 10, descending duodenum; 11, lesser omentum; 12, greater omentum; 13, mesoduodenum. (Image courtesy of the Université de Montréal.)

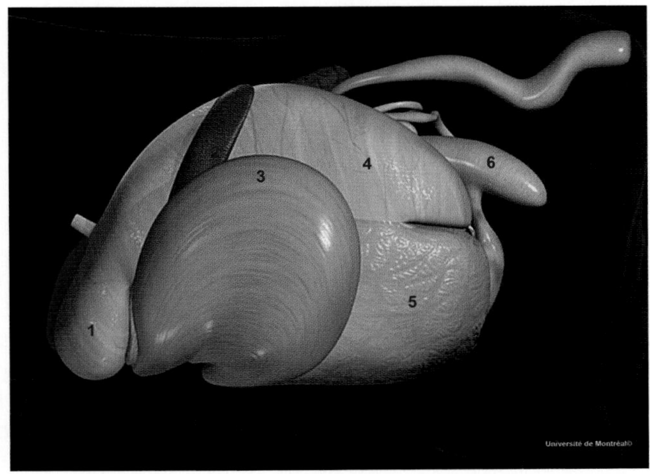

FIGURE 19.2-4 The normal anatomic relationship of the abomasum and other forestomach compartments. Key: 1, reticulum; 2, spleen; 3, greater curvature of the left displaced abomasum; 4, dorsal sac of the rumen; 5, ventral sac of the rumen covered by the greater omentum; 6, apex of the cecum. (Image courtesy of the Université de Montréal.)

ABOMASAL DISPLACEMENT

Displacement of the abomasum is an abnormal position of the abomasum in the abdominal cavity. The abomasum may be displaced to either the left or right side of the abdomen, though left-sided displacements account for approximately 90% of all displacements. With a left displaced abomasum (LDA), the abomasum moves from its normal anatomic position on the ventral abdomen to a location between the rumen and body wall on the left side of the cow (Fig. 19.2-4). In contrast, in a right displaced abomasum (RDA) the abomasum moves from the ventral abdomen to a position lateral to the omental sling on the cow's right side. Abomasal displacements occur most commonly during the first 6 weeks post-partum (L. after delivery) with more than 80% of all cases occurring within the first 30 days after parturition. Additionally, these conditions most often occur in older cattle and a majority of affected individuals are 4–6 years of age—i.e., in their 2nd to 4th lactation. Risk factors for the development of displaced abomasum include hypocalcemia, hyperketonemia, acute puerperal (L. *puerperalis* the period from the end of the 3rd stage of labor until completion of uterine involution) metritis, and retained fetal membranes. Comorbidities (L. *co-* with + L. *morbidus* disease condition or state) are often present in cattle with a displaced abomasum and include acute puerperal metritis, mastitis, enteritis, and retained fetal membranes.

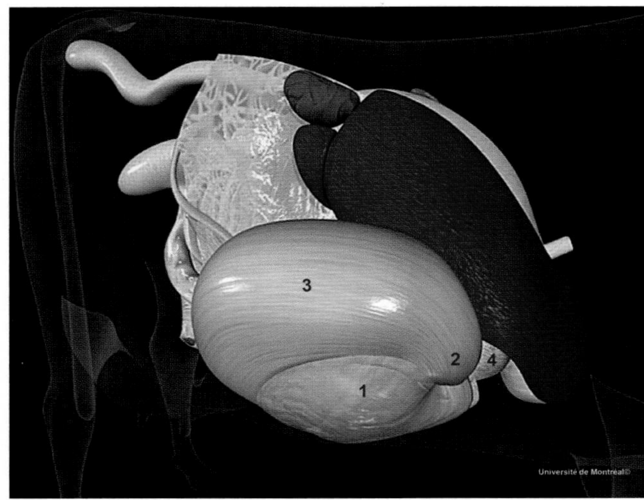

FIGURE 19.2-5 Computer-generated illustration of an abomasal volvulus. Key: 1, lesser omentum; 2, pyloric antrum; 3, greater curvature of the abomasum; 4, omasum. (Image courtesy of the Université de Montréal.)

The abomasum is innervated by branches of the **vagus nerve**. The **dorsal vagal trunk** innervates the majority of the abomasum, while the **ventral vagal trunk** innervates the right face of the abomasum. A long branch of the ventral vagal trunk travels through the lesser omentum to innervate the pylorus.

ABOMASAL VOLVULUS

Abomasal volvulus, like abomasal displacements, is believed to be multifactorial in etiology. It is also believed that RDA precedes the development of abomasal volvulus in most cases, making rapid diagnosis and surgical correction of RDA important to prevent its progression. In naturally occurring cases of abomasal volvulus the reticulum, omasum, and abomasum are included in the rotation. A large ping can be heard over the right hemiabdomen, extending from the 8th intercostal space to the middle of the paralumbar fossa. Succussible fluid is typical with ballottement (Fr. a tossing about; a palpation maneuver to test for a floating object) of the right flank. Rotation of the organs usually begins at the reticulo-omasal junction with the duodenum looping around the omasum, leading to counterclockwise rotation of the abomasum and omasum as viewed from the right and rear of the cow. The duodenum then moves medial to the omasal body and finally around the neck of the omasum (Fig 19.2-5). Continued secretion of hydrochloric acid into the obstructed abomasum leads to increased intraluminal pressure and, eventually, mucosal ischemia. Because the ventral vagal trunk and gastric vessels are involved in the twist, ischemia of the entire organ may occur. Injury to the vagal nerve (either a stretch injury, pressure, or ischemia as a result of the abomasal volvulus) can lead to chronic rumen distention.

Selected references

[1] Dyce KM, et al. The abdomen of the ruminants. Textbook of veterinary anatomy. Philadelphia, PA: Saunders; 2002. p. 666–90.
[2] Frandson RD, et al. Anatomy of the digestive system. Anatomy and physiology of farm animals. Baltimore, MD: Lippincott, Williams, and Wilkins; 2003. p. 306–28.

CASE 19.3

Intestinal Volvulus

David E. Anderson
Department of Large Animal Clinical Sciences, University of Tennessee College of Veterinary Medicine, Knoxville, Tennessee, US

Clinical case

History

A 3-year-old Holstein cow presented for acute onset of severe abdominal pain, generalized abdominal bloating, absence of manure production, and anorexia. The cow was 30 days in milk and had normal milk production up until approximately 12 hours before the onset of clinical signs. On the morning of presentation, the cow failed to come up with the herd to be milked. When the owner found the cow, the cow was breathing rapidly and was restless, as demonstrated by lying down and getting up repeatedly, and seemed to be distressed in general.

Physical examination findings

Physical examination revealed considerable bilateral abdominal distention, more severe on the right side. The cow was restless, tachycardic with a heart rate of 120 bpm, tachypneic with a respiratory rate of 60 brpm, and hypothermic with a rectal temperature of 96° F (36 °C). The cow was approximately 10% clinically dehydrated as judged by sunken eyes within the orbit and evidence of prolonged skin tenting of the upper eyelid and along the lateral aspect of the neck. No rumen contractions were auscultated, no manure output was noted, and no urine was produced during examination. Abdominal palpation per rectum identified multiple distended loops of small intestine in the middle and right quadrants of the abdomen. Transabdominal ultrasound examination confirmed multiple distended loops of small intestine (Fig. 19.3-1) with a mild increase in peritoneal fluid volume. Intestinal loops were active, but no productive motility observed.

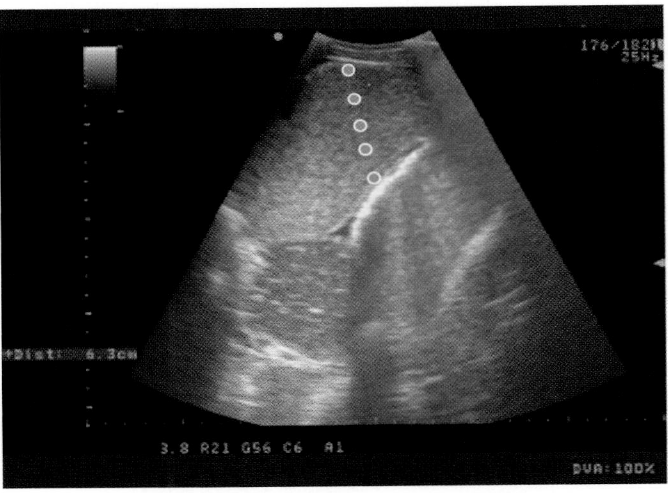

FIGURE 19.3-1 Transabdominal ultrasonographic image of the right ventral abdomen of a cow imaged using a 3.5-MHz curvilinear probe. Distended jejunal intestinal loops are enlarged with echogenic material. The diameter of one loop (*green dots*) is 6.3 cm (2.5 in.) (*N*=2–4 cm or 0.8–1.6 in.).

Comparative Veterinary Anatomy: A Clinical Approach. https://doi.org/10.1016/B978-0-323-91015-6.00095-9

Differential diagnoses

Small intestinal, spiral colon, abomasal, or jejunoileal flange volvulus; obstruction of abomasal outflow; intestinal distention due to ileus; intraluminal obstruction of the intestine; acute rumen tympany; viral, bacterial, or hemorrhagic enteritis caused by *Salmonella* spp. or *Clostridium* spp.; and hemorrhagic bowel syndrome (HBS)

Diagnostics

In cases of acute severe abdominal pain with distended small intestine, the physical examination is often shortened to the minimum database required to make a decision for exploratory surgery. In this cow, results of physical, rectal, and ultrasonographic examinations were consistent with a small intestinal obstruction. Stall-side serum biochemistry analyses revealed the presence of a hypochloremic ($Cl^- = 85$ mEq/L), hypokalemic ($K^+ = 2.5$ mEq/L), and metabolic alkalosis ($HCO_3 = 45$ mEq/L).

Diagnosis

Small intestinal volvulus

Treatment

The cow was prepared for exploratory laparotomy. Ampicillin trihydrate (6 mg/kg, IM) and flunixin meglumine (1 mg/kg, IV) were administered, and 4 L of 5% hypertonic saline were administered IV over 20 min pre-operatively, followed by infusion of a balanced ionic salt solution including NaCl and KCl. Pain management included the administration of butorphanol tartrate (0.025 mg/kg, IV). Over the course of 30 min, the cow's clinical condition improved, and the cow was able to remain standing. A right flank laparotomy incision was made approximately 10 cm (3.9 in.) ventral to the transverse process of the 3rd lumbar vertebra and was continued ventrally for 20 cm (7.9 in.) to a point approximately 10 cm (3.9 in.) caudal to the costochondral arch of the 13th rib. The incision was continued through the skin; subcutaneous tissues; external abdominal oblique, internal abdominal oblique, and transverse abdominis muscles; and peritoneum. On entering the peritoneal cavity, multiple loops of distended small intestine were encountered (Fig. 19.3-2). Abdominal exploratory was performed through manual palpation; a segment of the intestine and mesentery were found to be twisted in a clockwise direction within the intestinal mesentery (Fig. 19.3-3). Through manual manipulation of the intestine, the intestinal mass was untwisted in a counterclockwise direction until it was sufficiently mobile to exteriorize from the incision. Intestinal volvulus of the jejunoileal flange was diagnosed. The twist in the jejunoileal flange was corrected and replaced into the abdomen. The abdomen was closed routinely in 3 layers. The cow recovered without complications and completed her lactation.

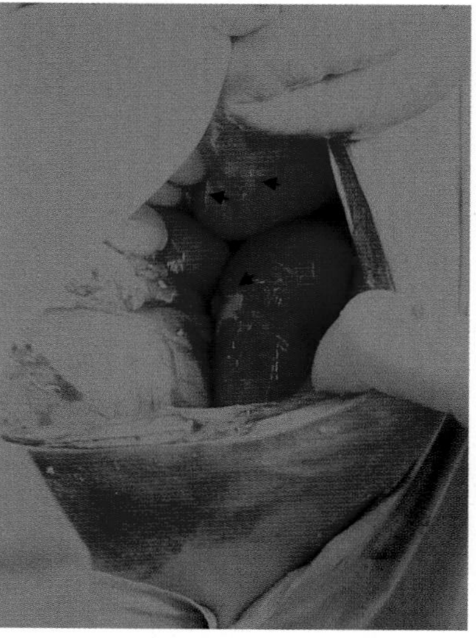

FIGURE 19.3-2 Intra-operative image of a right flank laparotomy of a cow. Distended jejunum is present at the surgical incision with fibrin covering the jejunum (*black arrows*).

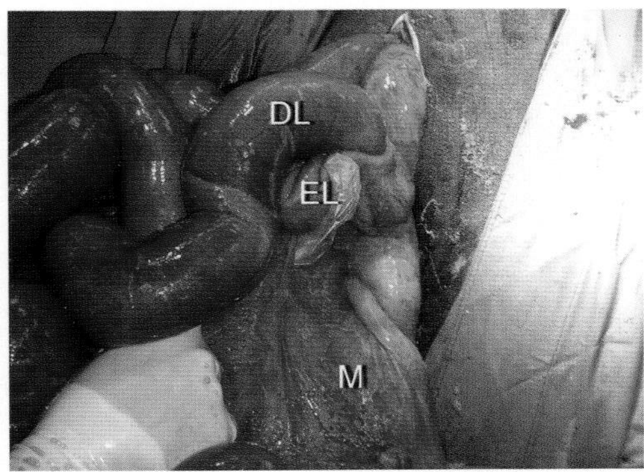

FIGURE 19.3-3 Distended jejunum is partially exteriorized to locate the intestinal obstruction in this cow. The presence of markedly distended intestinal loops (*DL*) with empty intestinal loops (*EL*) confirmed the diagnosis of an obstruction. Further exploration was needed to find the specific area of obstruction. Exteriorization of the jejunum was limited in this case because tension on the short mesentery (*M*) caused a pain response with the patient wanting to lay down. Therefore, the volvulus had to be corrected by correcting the jejunal volvulus within the abdomen.

Anatomical features in bovids

Introduction

The bovine gastrointestinal tract is one of the most anatomically complex and interesting gastrointestinal systems in the animal kingdom. It is a complex system including multiple compartments. It is comprised of the oral cavity, esophagus, rumen, reticulum, omasum, abomasum, duodenum with sigmoid flexure, jejunum, ileum, cecum, spiral colon, descending colon, rectum, and anus. This section covers the small intestine, along with its blood supply, lymphatic drainage, and innervation.

Function

The small intestine is responsible for digestion and absorption of nutrients and the progressive movement of ingesta toward the cecum and large intestine. The small intestine is the segment of bowel located between the abomasum and the cecum and is divided into 3 parts: duodenum, jejunum, and ileum. The duodenum also receives secretions from the pancreas through the accessory pancreatic duct and the gall bladder via the bile duct at the level of the descending duodenum.

Small intestine

The forestomach compartments of cattle occupy more than 50% of the abdominal cavity in non-gravid cattle, so the **small intestine** is confined to the right dorsal and ventral aspects of the abdomen, depending on the gravid (L. *gravida* heavy, pregnant) state of the cow. The small intestine of cattle can be up to 60 m (196.9 ft.) in length. The **duodenum** leaves the abomasum in the right ventral aspect of the abdomen and courses cranially to a position deep to the right lobe of the liver, where it forms the **sigmoid loop**. The cranial part of the **descending duodenum** leaves the sigmoid loop and continues caudally to the caudal margin of the **supraomental recess (omental sling)** (Fig. 19.3-6), where it turns immediately and cranially to become the **ascending duodenum**. The supraomental recess is the

PREVENTING INTESTINAL VOLVULUS

Prevention of intestinal volvulus in cattle is difficult because of a lack of evidence-based information regarding risk factors. Factors proposed to increase the risk of intestinal volvulus are the consumption of rapidly fermentable feeds such as lush spring pastures, grain, and diets deficient in fiber content. Intestinal volvulus may be seen more often in dairy breeds because of the relatively larger abdominal space and exposure to diets that promote a high volume of gas production. Legume-based feeds may also suppress intestinal motility due to the generation of by-products of digestion such as volatile fatty acids. Recommendations for prevention of intestinal volvulus include slow adaptation to new diets by transitioning cattle over several days or weeks as needed.

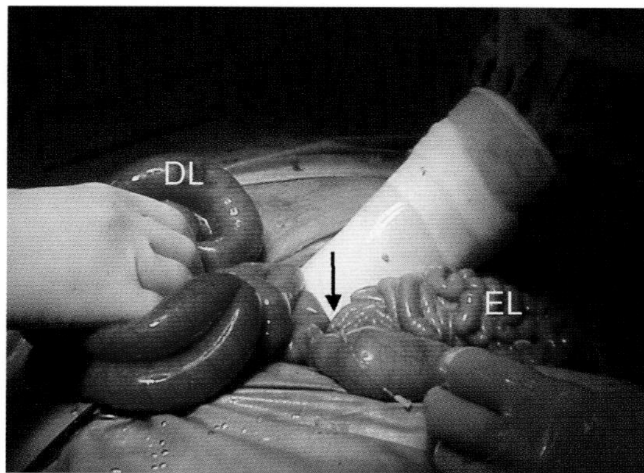

FIGURE 19.3-4 Intra-operative image of a calf in lateral recumbency for a right flank laparotomy. The jejunum is partially exteriorized to identify the volvulus of the ileal flange (*arrow*). The obstructed orad (L. *oris* mouth + *ad* toward = toward the mouth) jejunum is distended (*DL*), while the jejunum aboral (away from the mouth) to the obstruction is empty (*EL*).

JEJUNOILEAL FLANGE VOLVULUS

Volvulus refers to rotation of viscera about the mesenteric attachment; torsion refers to the rotation of viscera around its own axis. The most severe form of intestinal volvulus originates from the root of the mesentery and involves the entire small intestine. Volvulus of the root of the mesentery causes obstruction of both venous outflow and arterial blood supply to the intestines. Ischemic necrosis of the intestine rapidly occurs, causing severe metabolic acidosis, shock, and death.

Volvulus of the jejunoileal flange (Fig. 19.3-4) refers to volvulus of the distal jejunum and proximal ileum where the mesentery is longest. The long mesentery and associated bowel is referred to as the "flange" (collar or rib for strength or attachment), and it may rotate about its mesentery without involving the remaining small intestine. Arterial occlusion may not occur with volvulus of the jejunoileal flange, possibly because extensive fat deposits within the mesentery can prevent compression of the muscular wall of the arteries until the volvulus becomes severe. However, obstruction of outflow of venous blood may be equally detrimental because of transmural edema, shunting of blood away from the mucosa, and progressive ischemia. Cattle of any breed, age, or gender may be affected by intestinal volvulus without seasonal effect.

HEMORRHAGIC BOWEL SYNDROME OR JEJUNAL HEMORRHAGE SYNDROME (FIG. 19.3-5)

Hemorrhagic bowel syndrome is an emerging, often fatal, disease of adult dairy cattle, most often occurring during the first 3–4 months of lactation. It frequently presents as an acute abdomen (colic) needing immediate exploratory surgery; other references in the vernacular are "bloody gut" and "dead gut." The subserosa of the jejunum is hemorrhagic, and the lumen of the bowel is filled with blood clots. The prognosis for survival is poor with a fatality rate > 85%; however, survival is possible with early recognition and intervention. A combination of surgery, to resect unhealthy bowel and anastomose healthy segments of bowel, in conjunction with clot removal and supportive medical care—i.e., blood transfusions, crystalloid fluids, anti-inflammatory medications, analgesics, and antimicrobials—has improved the outcome. Hemorrhagic bowel syndrome is a multifactorial disease with many risk factors, including feeding silage, total mixed ration, finely ground corn; high-producing cows in early lactation; and free-choice feeding. Two organisms are implicated in the disease—*Clostridium perfringens* Type A beta-toxin positive and *Aspergillus fumigatus*—and both are opportunists due to the combination of factors or can act as primary pathogens. The syndrome is reported more frequently in large milking herds and those fed a total mixed ration.

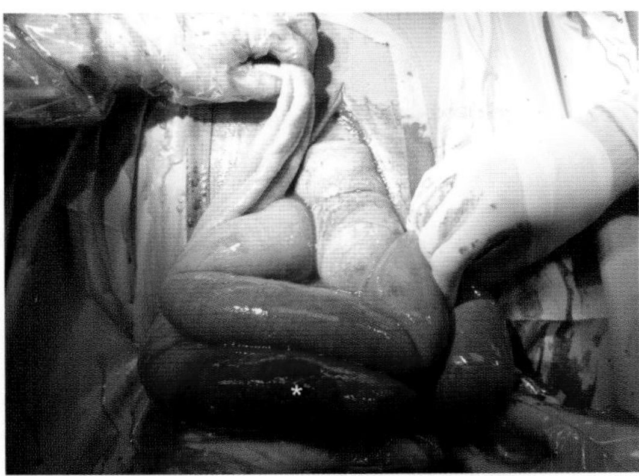

FIGURE 19.3-5 Intra-operative image of a standing cow with jejunal hemorrhage syndrome; the jejunum is partially exteriorized. The discolored, reddish segment (*) of the jejunum contains a large organized hematoma obstructing the normal flow of ingesta.

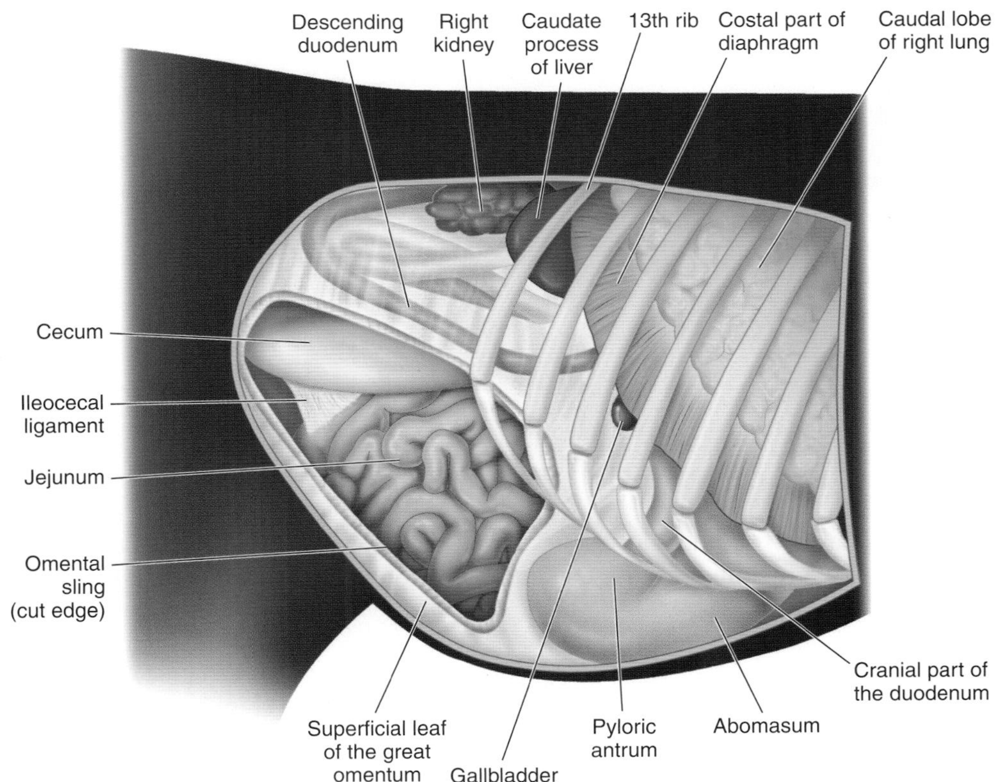

FIGURE 19.3-6 Topographical anatomy of the right abdomen of a standing adult cow.

space containing the small and large intestine surrounded by the deep layer of the greater omentum. The ascending duodenum tracks cranially and dorsally and then tracks cranial and medial to the cranial root of the mesentery. The small intestine continues as the **jejunum**, which is highly convoluted and connected to the intestinal mesentery.

SMALL INTESTINAL DISEASE IN CALVES

Congenital abnormalities are more common in calves in addition to urachal remnants, umbilical infections with adhesions, and intussusceptions. Interestingly, intussusceptions in calves occur throughout the intestinal tract and are proposed to develop because calves have less mesenteric fat compared to adults. General anesthesia is preferred for exploratory surgery using a right flank approach. Volvulus, entrapment of intestine in adhesions, umbilical remnants, and intussusception are the most common findings at surgery.

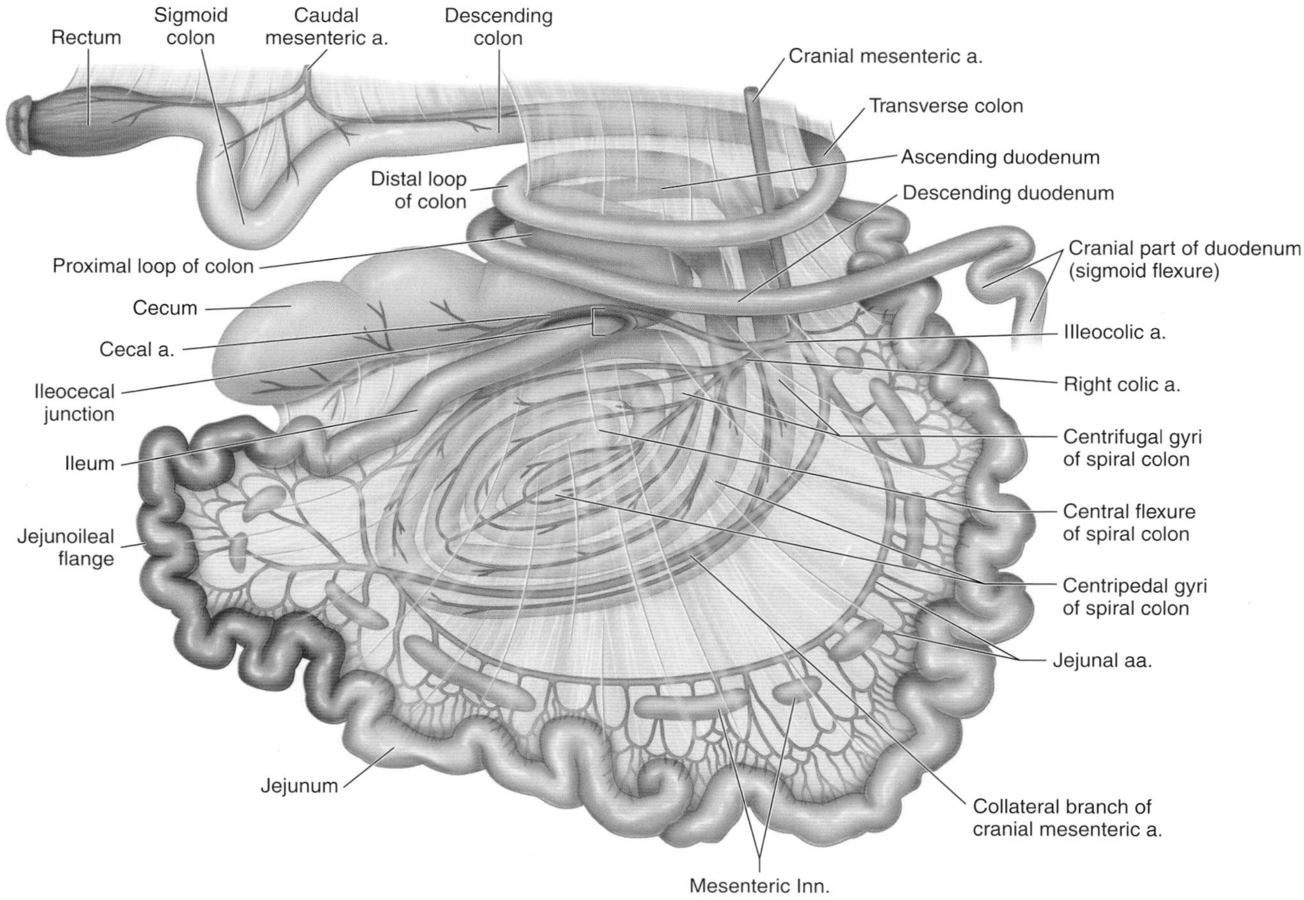

FIGURE 19.3-7 Right view of the bovine gastrointestinal tract and its blood supply. Note the different colors used to highlight different segments of bowel.

The distal portion of the jejunum continues along an extension of the mesentery and joins the **ileum** (Fig. 19.3-7). The section of long mesentery containing the distal jejunum that connects to the proximal ileum is referred to as the **jejunoileal flange** (anatomically it should be called the "jejunal flange" but that is not the accepted terminology). The ileum then courses cranially until it joins the base of the cecum and the proximal ascending colon. The junction of the ileum, cecum, and colon allows free movement of ingesta without a distinct muscular orifice. The **ileocecal junction** then enters the ascending colon through its proximal loop, which feeds into a spiral loop.

The small intestinal mesentery of cattle is diffusely infiltrated with adipose tissue (Fig. 19.3-8).

> Excessive accumulations of fat in the mesentery makes intestinal surgery more difficult. The fat may have a protective effect in some cases of volvulus, as it may limit the obstruction of arterial inflow and venous outflow and minimize the severity of strangulation. Intestinal resection is made more complicated by fatty infiltrates because the fat obscures identification of blood vessels and extrudes from cut edges, exacerbating the ill effects of handling the bowel. Finally, the presence of excessive fat may interfere with blood clotting and may increase the risk of post-operative hemorrhage associated with cut or ligated vessels.

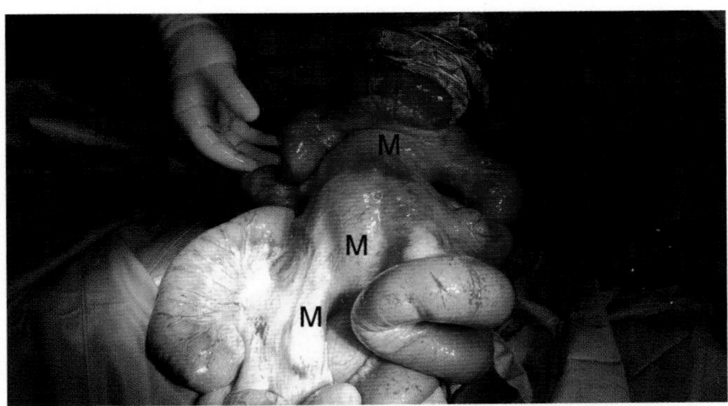

FIGURE 19.3-8 Intra-operative image of partially exteriorized jejunum through a right flank laparotomy incision. The jejunoileal flange is cradled by the surgeon with the long mesentery (*M*). The fatty mesentery in cattle makes recognizing blood vessels difficult.

Blood supply, lymphatics, and innervation

The abdominal aorta gives rise to the celiac artery and the cranial mesenteric artery Fig. 19.3-7. The **celiac artery** provides the major blood supply to the rumen, reticulum, omasum, abomasum, and proximal duodenum. Blood supply to the large and small intestine originates from the **cranial mesenteric artery**. The cranial root of the mesentery arises from the abdominal aorta and forms arcades supplying the abomasum, small intestine, and large intestine. Arterial branches to the small intestine of cattle originate from extensions of the cranial mesenteric artery in the midportion of the mesentery, where they course to their individual segments. The various branches of the cranial mesenteric artery supply multiple segments of the small intestine via various arcades and branches. The cranial mesenteric artery also gives rise to the **caudal pancreaticoduodenal artery**, supplying the pancreas and the ascending duodenum, and **ileocolic artery**.

The cranial mesenteric artery further contributes to the **middle colic artery** and then the **ileocolic** and **right colic arteries**. The ileocolic artery delivers blood to the distal ileum, cecum, and descending and ascending colons. The cranial mesenteric artery comprises a long continuous loop through the intestinal mesenteries with successively smaller branching arcades, ultimately feeding individual segments of the small intestine. The main branch of the cranial mesenteric artery furnishes arcuate blood supply to individual intestinal segments. The collateral branch of the cranial mesenteric artery provides redundant blood supply to the mesenteries and lymph nodes.

The **jejunal** and **ileal veins** accompany the corresponding arteries. They drain into the **cranial mesenteric vein**, terminating into the **portal vein**. The **gastroduodenal vein** also joins the portal vein.

STANDING EXPLORATORY LAPAROTOMY IN CATTLE (FIG. 19.3-9)

This is possible in cattle because of their demeanor and the ability to perform effective regional anesthesia and analgesia in the paralumbar region. Spinal innervation to the skin, paralumbar musculature, and peritoneum arises from the 13th thoracic and 1st and 2nd lumbar nerves. These nerves exit their respective foramina from the spinal column and traverse in caudal, lateral, and ventral directions. Shortly after exiting the spinal column, the nerves separate into dorsal and ventral branches. The dorsal branch innervates the skin and external abdominal oblique muscle. The ventral branch primarily innervates the peritoneum, transversus, and internal abdominal oblique muscles. Perineural anesthetic blockade, such as proximal paravertebral and distal paravertebral nerve blocks, is effective at providing surgical analgesia and anesthesia to the structures such that incision and manipulation are possible without significant adverse perception by the patient. Some clinicians prefer regional blocks—as opposed to perineural blocks—as a way of improving the success rate for surgical procedures. The most common regional blocks used include the "line" block and the "7" block (a.k.a. "inverted L block"). Through volume administration, these blocks are able to anesthetize multiple peripheral small nerve branches that arise from the principal spinal nerves.

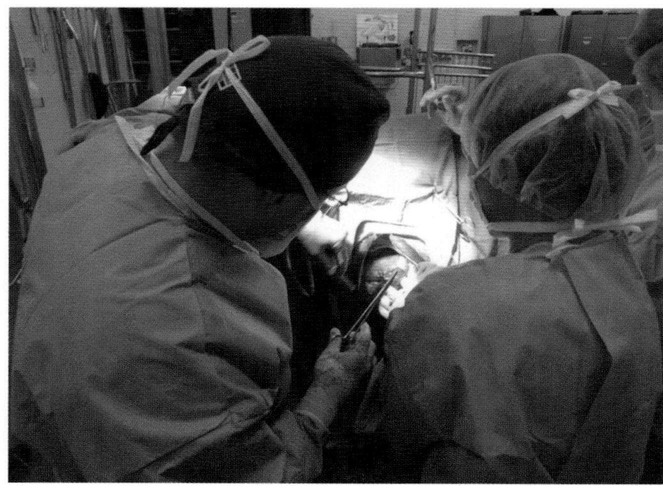

FIGURE 19.3-9 Intestinal surgery performed via a right flank incision on a standing cow under sedation and regional paravertebral anesthesia.

Intestinal lymphatics for the duodenum drain to the **hepatic**, **pancreaticoduodenal**, **cranial mesenteric**, and **cecal lymph nodes**, while the jejunal lymphatics drain to the **jejunal lymph nodes**, and finally the ileum drains to the **jejunal**, **cecal**, and **colic lymph nodes**. The jejunal lymph nodes are clustered close to the mesenteric border of the intestine and form a band. **Peyer's patches** (folliculi lymphatici aggregati) are an extensive accumulation of large, solitary lymphoid tissue nodules in the small intestine and increase in number as the small intestine nears the large intestine.

The principal nerve supply to the intestinal tract is provided by the **vagus nerve**, which gives parasympathetic input. The vagus nerve exits the cranial cavity through the jugular foramen and courses as the vagosympathetic trunk and through the cervical region before entering the thoracic cavity. The vagus nerve divides into dorsal and ventral branches and continues through the mediastinum until emerging through the hiatus of the diaphragm as **dorsal** and **ventral vagal trunks**. These trunks provide parasympathetic innervation to the forestomachs, small intestine, cecum, ascending colon, and transverse colon. Sympathetic nerve supply is provided by the lumbar spinal segment nerves arising from L1 to L3 spinal cord segments. The **superior mesenteric ganglion** gives rise to the major innervation of the intestinal tract. The inferior mesentery supplies sympathetic innervation to the colon.

Selected references

[1] Anderson DE, Constable PD, St-Jean G, Hull BL. Small-intestinal volvulus in cattle: 35 cases (1967-1992). J Am Vet Med Assoc 1993;203:1178–83.

[2] Braun U. Ultrasonography in gastrointestinal disease in cattle. Vet J 2003;166:112–24.

[3] Fubini SL, Smith DF, Tithof PK, et al. Volvulus of the distal part of the jejunoileum in four cows. Vet Surg 1986;15:150–2.

[4] Tulleners EP. Surgical correction of volvulus of the root of the mesentery in calves. J Am Vet Med Assoc 1981;179:998–9.

[5] Vogel SR, Nichols S, Buczinski S, Desrochers A, Babkine M, Veillette M, Francoz D, Dore E, Fecteau G, Belanger AM, Badillo M. Duodenal obstruction caused by duodenal sigmoid flexure volvulus in dairy cattle: 29 cases (2006–2010). J Am Vet Med Assoc 2012;241(5):621–5.

CASE 19.4

Cecal Dilatation/Volvulus

Emma Marchionatti
Clinic for Ruminants, Vetsuisse Faculty, University of Bern, Bern, CH

Clinical case

History

A 5-year-old Holstein cow presented for reduced appetite, decreased milk production, and markedly decreased fecal output, with signs of moderate colic of 6 hours' duration. The referring veterinarian treated the cow with IV flunixin meglumine and calcium borogluconate without improvement of clinical signs.

Physical examination findings

Upon presentation, the cow was slightly depressed with moderate tachycardia (HR 96 bpm). Respiratory rate (RR 32 brpm) and rectal temperature (101.8 °F [38.8 °C]) were within normal limits. Signs of colic included shifting weight, restlessness, and intermittent kicking at the right flank. The right paralumbar fossa was moderately distended. Ruminal motility and intestinal peristalsis were reduced. Right abdominal percussion and succussion (L. *succussio* a shaking from beneath) with auscultation were positive dorsally from the tuber coxae to the 13th rib. Abdominal palpation per rectum revealed no feces and the presence of mucus. A distended tubular hollow organ with a smooth surface was palpated in the right upper quadrant, coursing transversely cranial to the pelvic inlet. The cecal apex could not be identified upon palpation.

Differential diagnoses

Cecal dilatation, cecal volvulus (dorsal or ventral retroflexion), and abomasal volvulus

Diagnostics

The PCV (36%) and total protein (85 g/L or 8.5 mg/dL) were increased, consistent with dehydration. Total serum calcium (1.8 mmoL/L or 7.2 mg/dL) and potassium (3.4 mmoL/L or 3.4 mEq/L) concentrations were decreased. Abdominal ultrasound was performed and revealed the presence of a gas- and fluid-distended viscus adjacent to the abdominal wall in the right paralumbar fossa. A mild increase in anechoic peritoneal fluid was visible in the right ventral abdomen. The pylorus could be identified in the right ventral abdomen at the level of the entrance of the mammary vein (subcutaneous abdominal vein).

Diagnosis

Cecal volvulus (dorsal or ventral retroflexion)

Treatment

A standing right flank exploratory laparotomy was performed, and cecal ventral retroflexion was confirmed. The cecum and the proximal loop of the ascending colon were exteriorized (Fig. 19.4-1), decompressed by typhlotomy (Gr. *typhlon* cecum + Gr. *tomē* a cutting) at the cecal apex, and the displacement was corrected.

Comparative Veterinary Anatomy: A Clinical Approach. https://doi.org/10.1016/B978-0-323-91015-6.00096-0

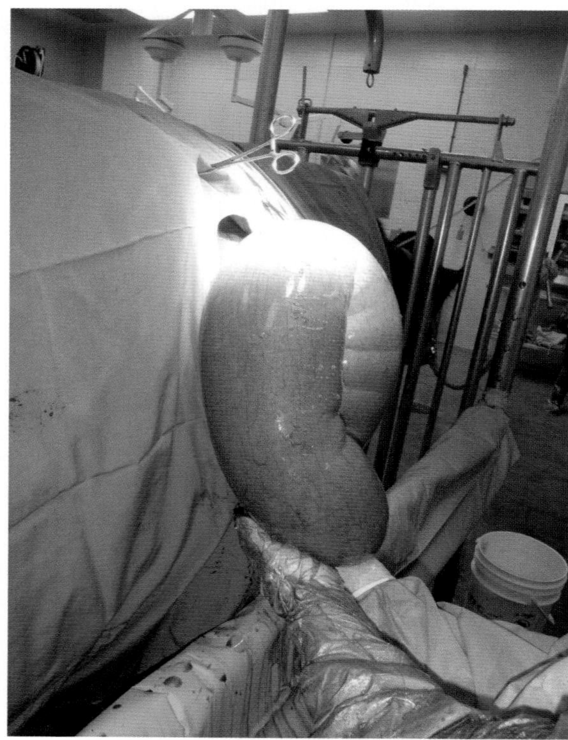

FIGURE 19.4-1 Intra-operative image of the distended cecum exteriorized from the right flank laparotomy of the 5-year-old Holstein cow presented in this case. Cranial is to the right.

Peri-operative IV fluids were supplemented with calcium borogluconate and potassium chloride, and broad-spectrum antimicrobials were administered.

The cow's attitude was brighter with no signs of abdominal pain (colic) and a return to a normal heart rate. Her appetite steadily improved along with an increase in milk production. The cow was discharged from the hospital 72 hours post-operatively.

Anatomical features in ruminants

Introduction

The large intestine (LI) is aboral to the small intestine and consists of the cecum, colon, and rectum. The entire intestinal tract is confined to the right abdominal cavity due to the presence of the large rumen. The LI is moderate in length (8–11 m [26.2–36.1 ft]), but its lumen—and thus its capacity (30–40 L)—is smaller compared to the horse. Unlike the horse, it has neither bands nor sacculations; thus, its mucosal surface is comparatively less extensive.

This case focuses on the cecum, colon, and rectum, along with their blood supply, innervation, and lymphatic drainage.

Function

The most important functions of the LI are recovering fluid and electrolytes from ingesta entering from the ileum, microbial fermentative digestion, and storage of feces. The importance of these functions varies between species and translates into differences in colon size and shape. The major determinant for colon size across species is the degree of importance of colonic fermentation to the animal's energy needs. In ruminants, most of the fermentation occurs in the rumen, so microbial degradation of nutrients in the LI is of less importance compared to other simple-stomached herbivores like horses. Cecocolic fermentation in cattle accounts for 11–17% of total dietary cellulose,

2–11% of total starch, and 20% of soluble carbohydrates. Unlike the small intestine, the LI does not have mucosal villi. However, its absorption of nutrients and fermentation products (volatile fatty acids) is of great importance.

Cecum

The **cecum** is the first compartment of the LI. It extends caudally in the dorsal part of the right abdominal cavity, from the **ileocecocolic junction** to the pelvic inlet. It has a cylindrical shape and is 50–70 cm (19.7–27.6 in.) in length. The cecum is composed of a body and a free, rounded blind **apex**, located at the pelvic inlet. The **ileocecal fold** connects the ileum and the body of the cecum. The **ileal papilla** and **ileal sphincter** (lying therein) mark the boundary between the cecum and the colon. However, the **body** of the cecum is continuous with the colon with no change in luminal diameter at the **cecocolic orifice** (Fig. 19.4-2). The ileal sphincter, composed of a ring of circular muscle, prevents retrograde movement of colonic contents into the ileum. During ileal peristaltic activity, the sphincter relaxes, allowing passage of ingesta into the cecum and colon.

Cecal and spiral colon motility disturbances might be implicated in the development of cecal dilation/dislocation. "Cecal dilatation" refers to distention of the cecum without displacement or a twist. "Cecal torsion" refers to rotation of the cecal apex along its long axis. Rotation in the region of the ileocecocolic region in the sagittal plane is termed "cecal volvulus." The cecum may rotate cranially and ventrally (ventral retroflexion) or dorsally (dorsal retroflexion). Cecocolic and cecocecal intussusception occur more frequently in calves than adults, probably related to the minimal mesenteric fat present in young animals.

Colon

The **colon** is the longest part of the LI (5–10 m [16.4–32.8 ft]) and consists of the ascending colon, transverse colon, and descending colon (see Case 19.3, Fig. 19.3-7). The **ascending colon** is the longest portion and consists of 3 parts: the proximal loop, the spiral loop, and the distal loop. The **proximal loop** is continuous with the cecum and has an S-shape. It runs cranially toward the right kidney and doubles back dorsal to the first part of the cecum, ventral to the descending duodenum. It then turns mediodorsally around the mesentery and runs cranially along its left side. At the level of the kidney, it turns ventrally into the narrower spiral loop. The **spiral loop** forms a flat disk inside the mesentery and is usually

Irregular motility patterns—e.g., subsequent to diarrhea—may lead to spiral colon intussusception. Even if uncommon, this condition is more frequently seen in calves than adults. In calves, intraluminal obstruction of the spiral colon—caused by trichophytobezoars (Gr. *tricho-* hair + Gr. *phyton* plant + *bezoar* concretion of various character found in stomach or intestine of man or animal) or fibrin plugs following diarrhea episodes—is not infrequent. The ascending colon is one of the segments most commonly affected by intestinal atresia. Atresia (Gr. *a* absence + Gr. *trēsis* a hole + *-ia* a condition = congenital absence of a normal orifice or tubular organ) coli is most frequently located in the mid-spiral loop.

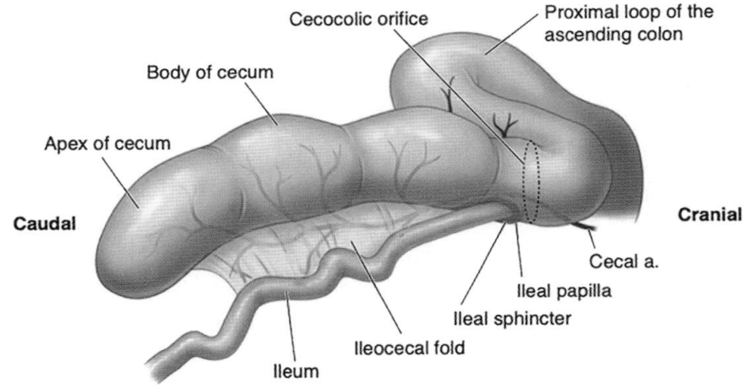

FIGURE 19.4-2 Anatomy of the bovine cecum and proximal loop of the ascending colon.

formed by 2 **centripetal** (L. *centrum, center* middle point + L. *petere* to seek = moving toward a center) **gyri**, the **central flexure**, and 2 **centrifugal** (L. *centrum center* middle point + *fugere* to flee = moving away from a center) **gyri**. The outer centrifugal gyrus continues into the narrower **distal loop**, which runs dorsocaudally on the left side of the mesentery, ventral to the **ascending duodenum** and medial to the proximal loop. It then turns around the mesentery and runs cranially on the right side of the mesentery. The short **transverse colon** follows. It turns around the cranial mesenteric artery from right to left and continues into the descending colon.

The **descending colon** runs caudally and is continuous with the rectum without any structural transition. Its cranial part is adherent to the ascending duodenum through the **duodenocolic fold**. The **mesocolon** lengthens in the distal part of the descending colon. The **sigmoid colon** portion of the LI anatomically begins at the pelvic inlet and is the start of the rectum.

Rectal prolapse occurs most commonly in beef cattle and small ruminants. Predisposing factors include colitis, diarrhea, and tenesmus following dystocia. Rectal prolapse is classified as one of 4 types. Type I involves only the rectal mucosa protruding from the anus. In type II, all layers of the rectum are completely prolapsed. Type III occurs when descending colon intussuscepts into the rectum, in addition to a type II lesion. Type IV occurs when the peritoneal rectum or descending colon forms an intussusception through the anus. Treatment options include replacement and purse-string suture application, submucosal resection, or prolapse amputation. Prognosis is good to fair depending on the condition of the prolapsed tissues and the elimination of predisposing factors.

Rectum

The **rectum** is the terminal portion of the LI. It begins at the pelvic inlet and has a short **mesorectum**. Its retroperitoneal portion is continuous with the **anal canal** and **anal sphincter**.

Blood supply, lymphatics, and innervation

Blood supply to the large intestine originates from the **cranial** and **caudal mesenteric arteries** (see Case 19.3, Fig. 19.3-7). The **ileocolic artery** arises directly from the cranial mesenteric artery and gives off the **right colic arteries** to the centrifugal gyri and the distal loop of the colon, and the **colic branches** to the centripetal gyri and the proximal loop of the colon. The **cecal artery** arises from the ileocolic artery and runs in the ileocecal fold, where it gives origin to the **antimesenteric ileal branch**. In cattle, the **collateral branch** arises from the cranial mesenteric artery, with which it anastomoses after supplying the last centrifugal gyrus. The **middle colic artery** arises from the cranial mesenteric artery and supplies blood to the transverse colon. The caudal mesenteric artery provides the **left colic artery** to the descending colon and the **cranial rectal artery** to the rectum. The caudal portion of the rectum is vascularized by the **middle** and **caudal rectal arteries,** derived from the **internal iliac artery**.

The veins have a corresponding pathway and empty into the **portal vein** and its main branches. Lymph drainage is supplied by the **cranial mesenteric**, **celiac**, **colic**, and **caudal mesenteric lymph nodes**.

Large intestinal innervation is provided by parasympathetic (**vagal nerve** and **pelvic nerve**) and sympathetic fibers. Both nerve fiber types originate from the **cranial** and **caudal mesenteric** and **hypogastric ganglia**. Nerve fibers leave the ganglia and follow the arteries until the mesenteric border where they expand on the intestinal surface and enter the intestinal wall to form the **myenteric** and **submucosal plexuses**.

Large intestine motility

In ruminants, ingesta take approximately 24 hours to pass through the LI. Mixing is the principal activity in the LI, allowing contact of feed particles to microbial fermentation and to intestinal mucosa for absorption. Mixing is achieved primarily by segmental contractions combined with peristaltic and antiperistaltic contractions. Antiperistaltic movements are most pronounced in the first part of the colon, to fill the cecum and facilitate fermentation in this segment. Motility of the spiral colon is associated with that of the ileum and proximal colon, and shows the typical phases and organization of migrating myoelectric complexes.

Selected references

[1] Budras KD, Habel RE. Bovine anatomy. 2nd ed; 2011. p. 76–7.
[2] Fubini SL, Ducharme NG. Farm animal surgery. 2nd ed; 2017. p. 317–28. 512–518.
[3] Braun U, et al. Clinical findings and treatment in cattle with cecal dilation. BMC Vet Res 2012;8:75.
[4] Kunz-Kirchhofer C, et al. Myoelectric activity of the ileum, cecum, proximal loop of the ascending colon, and spiral colon in cows with naturally occurring cecal dilation-dislocation. Am J Vet Res 2010;71:304–13.

CHAPTER 19

CASE 19.5

Caudal Vena Cava Syndrome

Julie Berman
Centre Hospitalier Universitaire Vetérinaire (CHUV), Université de Montréal, Québec, CA

Clinical case

History

A 6-year-old Holstein dairy cow, 120 days in milk, presented for epistaxis, anorexia, and decreased milk production. Three days before presentation, the cow had a severe episode of bilateral epistaxis. The bleeding stopped, but the cow remained anorexic and developed melena. No treatment was administered at the farm. No other cows in the herd (70 lactation cows) were sick.

Physical examination findings

At admission, the cow had a body condition score of 2/5 (1= emaciated; 5= obese). She was tachycardic (150 bpm) and tachypneic (48 brpm) with a normal rectal temperature. Thoracic auscultation revealed bilaterally increased lung sounds without abnormal crackles or wheezes. No nasal discharge was present. No epistaxis was present on admission. The rumen was hypomotile (1 contraction in 2 minutes), and no "ping" was identified. The feces were dark and a fecal occult blood test was positive, confirming the suspicion of melena.

Differential diagnoses

Nasal/pharyngeal or retropharyngeal trauma, abscess, or foreign body; infection of the paranasal sinuses (unlikely given the absence of purulent nasal discharge); lung embolus from caudal vena cava thrombosis (more likely because of mild pulmonary signs)

Diagnostics

A CBC revealed a moderate normocytic, normochromic, non-regenerative anemia (hematocrit = 17%; hemoglobin = 63 g/L; erythrocytes = 4.06 M/μL). Bloodwork was consistent with an active chronic inflammatory process with hyperfibrinogenemia (fibrinogen = 7 g/L) and hyperglobulinemia (47.2 g/L). Serum biochemistry revealed increased urea (14.94 mmoL/L) suggestive of digestion of blood—i.e., from that swallowed during the episode of epistaxis—a finding compatible with the presence of melena. There was a mild increase in liver enzymes—AST = 171 U/L; GGT = 69 U/L; and GLDH = 54 U/L.

Abdominal ultrasonography revealed readily apparent liver vascularization with an enlarged and rounded vena cava (diameter, 4 cm [1.6 in.]) (Fig. 19.5-1B, compare to normal in Fig. 19.5-1A), compatible with the presence of a thrombus. Thoracic ultrasound revealed bilateral consolidation of the caudal lung lobes (Fig. 19.5-1C). Airway endoscopy revealed no abnormalities of the nasal, pharyngeal, or laryngeal regions. Tracheal wash was performed via a double-guarded sampling catheter passed through the endoscope's biopsy channel. Tracheal wash cytology showed suppurative septic inflammation, and the bacterial culture isolated *Fusobacterium necrophorum*.

Comparative Veterinary Anatomy: A Clinical Approach. https://doi.org/10.1016/B978-0-323-91015-6.00097-2

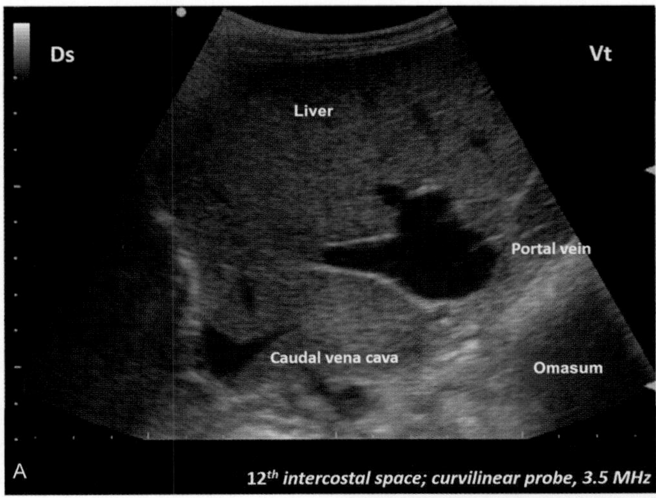

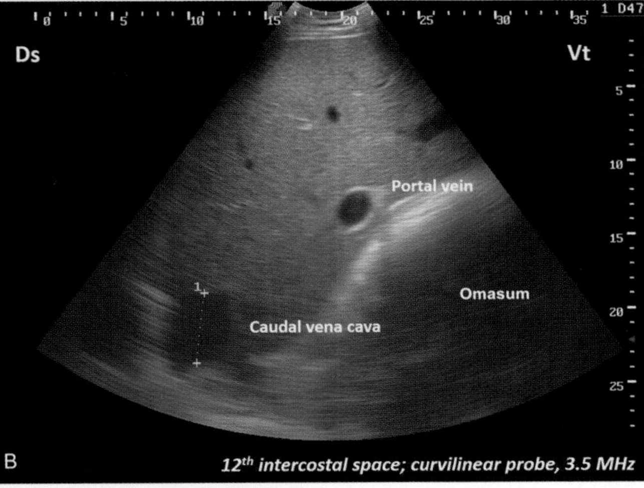

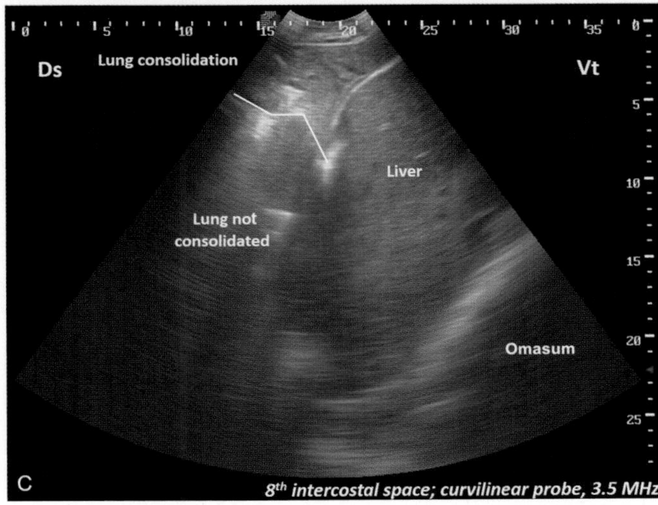

FIGURE 19.5-1 Ultrasound image of (A) a normal liver (12th intercostal space); (B) a liver of a cow with caudal vena cava syndrome (12th intercostal space); (C) the same cow with caudal vena cava syndrome in a more proximal view assessing the right caudal lobe of the lung (8th intercostal space). Normally, the caudal vena cava is triangular (A). In caudal vena cava syndrome, the caudal vena cava is enlarged (diameter, 4 cm [1.6 in.]) and rounded (B). The caudal lobe of the right lung is consolidated (C). Key: *Ds*, dorsal; *Vt*, ventral.

Diagnosis

Lung embolus from caudal vena cava thrombosis

Treatment

Given the poor prognosis, no treatment was considered and humane euthanasia was recommended.

Anatomical features in ruminants

Introduction

This section covers liver function, its vascularization, and its fetal circulation. Gross liver anatomy is covered in Case 19.6. A review of lung anatomy is included because the lungs are ultimately affected with caudal vena cava thrombosis, and hemoptysis and epistaxis are commonly observed.

Function

Although the liver is a discrete organ, it performs many different functions that interrelate with one another. In all species, the liver's different functions include the following:

- Storage of vitamins and iron
- Formation of coagulation factors
- Metabolism of carbohydrates, proteins, fats, hormones, and foreign chemicals
- Formation of bile
- Filtration and storage of blood

Most of these functions—metabolic, formation of bile, formation of coagulation factors, and storage of vitamins and iron—are performed by the hepatic parenchyma, especially the hepatocytes. The last function—filtration and storage of blood—is attributable to the unique blood supply of the liver and its connections with other organs.

Metabolic functions of the liver

The liver is a large, chemically reactant pool of cells (**hepatocytes**) with a high metabolic rate that processes and synthesizes multiple substances that are transported to other areas of the body. In contrast to other species, the bovine liver plays a major role in lipid metabolism (and to a lesser extent the metabolism of carbohydrates and proteins). Indeed, the bovine liver has unique functions because much of the dietary energy is absorbed as volatile fatty acids and not glucose. Thus, the major substances that must be absorbed are chylomicrons (which transport lipids), volatile fatty acids, additional fatty acids (e.g., non-esterified fatty acids), and much of the glycerol obtained by mobilization of fat from adipose tissue (Fig. 19.5-2). Glucose is still needed (in large amounts in lactating cows) but must be produced by gluconeogenesis, of which 85% takes place in the liver.

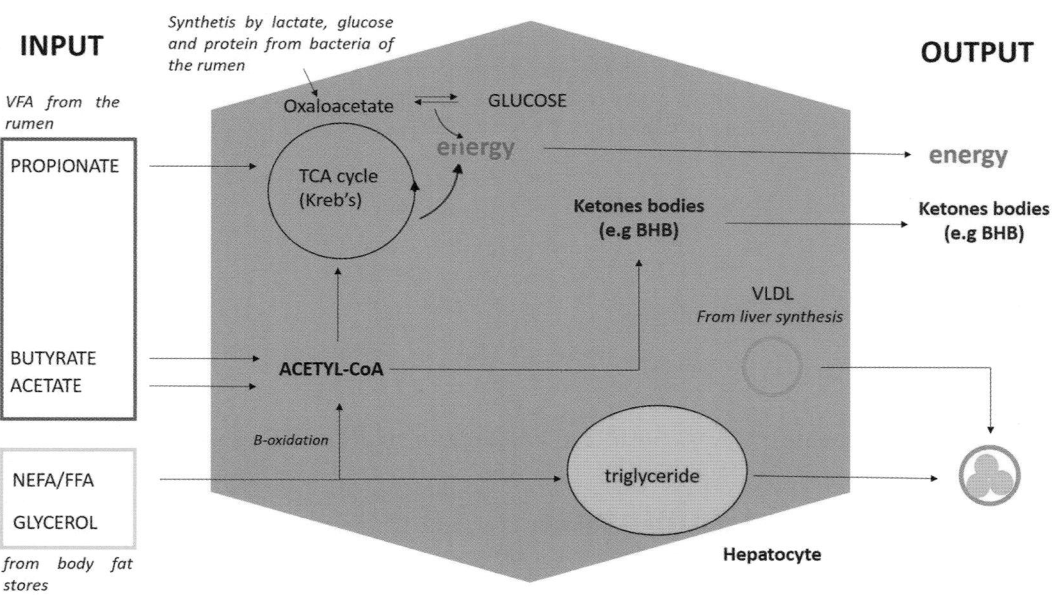

FIGURE 19.5-2 The metabolic functions of the liver. The hepatocytes of the liver are responsible for the transformation of the lipids mobilized from body fat stores (non-esterified fatty acids [NEFA], free fatty acids [FFA]) and volatile fatty acids (VFA) from the rumen (propionate, butyrate, and acetate) into energy and ketone bodies.

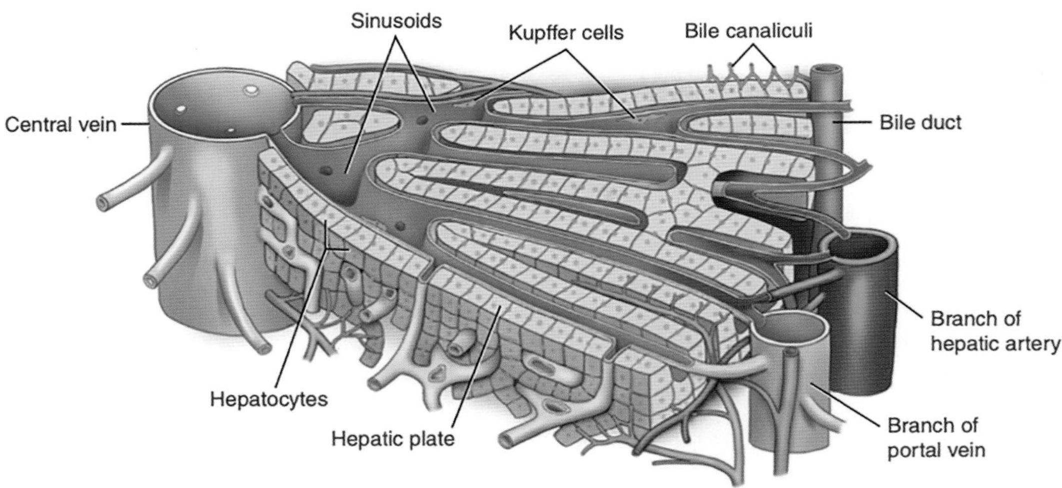

FIGURE 19.5-3 The hepatic parenchymal blood supply. The hepatic artery and portal vein send their blood into a common network of anastomosing small vascular channels called the sinusoids. The peri-sinusoidal space is the main site where material is transferred between blood-filled sinusoids and hepatocytes.

Filtration function of the liver

Blood flowing through intestinal capillaries picks up many bacteria from the intestines (Fig. 19.5-4). Phagocytic **Kupffer cells** (i.e., liver-specific macrophages) line hepatic sinusoids and remove these bacteria from the blood before it passes into the systemic circulation.

Storage of blood

Because the liver is an expandable venous organ, it can store large quantities of blood. Its normal blood volume, including both that in the **hepatic veins** and in the **hepatic sinuses**, makes up almost 10% of the body's total blood volume. Thus, the liver can act as a valuable blood reservoir in times of excess or diminished blood volume.

Blood supply of the liver parenchyma

Inside the parenchyma, the liver receives blood from 2 vessels, the **hepatic artery**, which perfuses the liver with oxygenated blood from the aorta, and the **portal vein**, which carries blood from the digestive tract and spleen to the liver (Fig. 19.5-4). Blood in the portal vein draining from the digestive tract is rich in amino acids, lipids, and carbohydrates absorbed from the small intestine. Blood in the portal vein draining from the spleen is rich in hemoglobin breakdown products. The portal vein divides into 2 branches: the right one vascularizes the right lobe and the caudal process, and the left branches toward the left lobe. The hepatic artery and portal vein send their blood into a common network of anastomosing small vascular channels called the **sinusoids** (Fig. 19.5-3).

MECHANISMS OF ASCITES

In cardiac failure with peripheral congestion, elevated pressures in the right atrium cause secondary increased pressures in the liver, termed "hepatic congestion." When pressures in the hepatic veins rise above a certain threshold, excessive amounts of fluid begin a transudation (L. *trans* through + L. *sudare* to sweat = fluid passing through a membrane as a result of hydrodynamic forces) into the lymph and escape through the outer surface of the liver capsule directly into the abdominal cavity, causing ascites (L. Gr. *askitēs* from *askos* bag = accumulation of serous fluid in the abdominal cavity).

Obstruction of portal flow through the liver also causes elevated capillary pressures in the entire portal vascular system of the gastrointestinal tract. This results in edema of the gut wall and transudation of fluid through the serosa of the gut into the abdominal cavity, causing ascites.

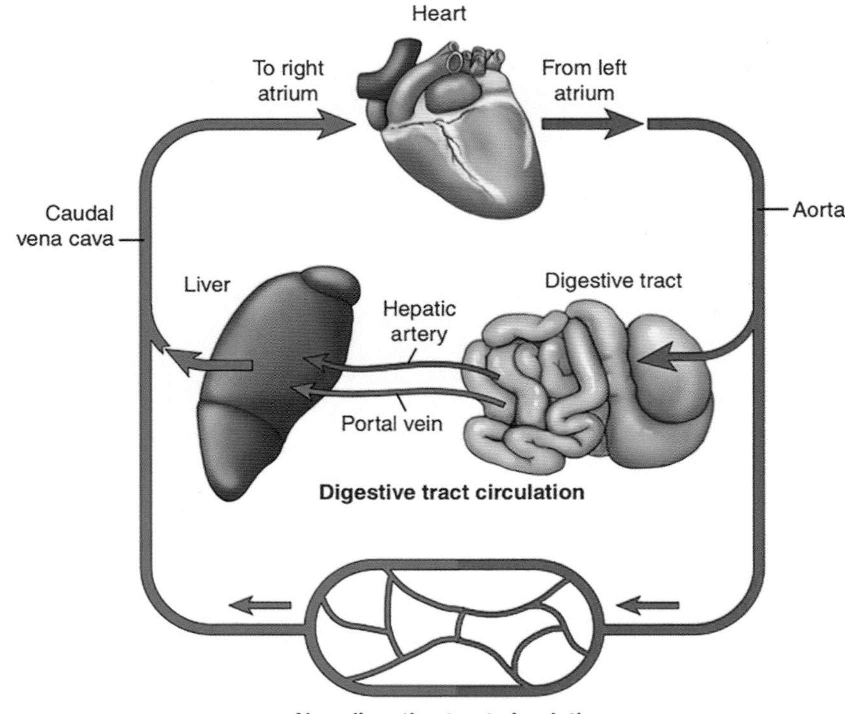

FIGURE 19.5-4 The vascularization of the liver and its vascular connection with the digestive tract and non-digestive tract circulation. The blood from the digestive tract is filtered by the liver via the portal vein. The caudal vena cava drains blood from abdominal contents (non-digestive) and regions located caudal to the diaphragm—i.e., pelvic limbs, pelvis, and abdomen. The caudal vena cava is separated into 3 parts—lumbar, hepatic, and thoracic.

The **peri-sinusoidal space** (or "space of Disse") is the main site where material is transferred between blood-filled sinusoids and hepatocytes. Blood which has passed through functioning liver parenchyma enters terminal **hepatic venules**—central veins of the lobules—which then fuse to form larger **hepatic vein** branches (Fig. 19.5-3). Hepatic veins are devoid of valves and open separately into the **caudal vena cava** as it passes through the liver on its way to the right atrium.

The **caudal vena cava** drains blood from all the parts of the body located caudal to the diaphragm—pelvic limbs, pelvis, and abdomen (Fig. 19.5-4). The caudal vena cava has 3 main parts: lumbar, hepatic, and thoracic. It originates from the **common iliac veins** and runs adjacent to the abdominal aorta ("lumbar part"). It then courses

HEPATIC ABSCESSES AND VENA CAVA SYNDROME

Hepatic abscesses result from entry and establishment of *Fusobacterium necrophorum,* either alone or with other bacteria (especially *Trueperella pyogenes*). The most common route of entry of bacteria into the liver is the portal vein.

Hepatic abscesses adjacent to the caudal vena cava are the most common cause of vena cava syndrome. Septic emboli detach from the thrombus and reach the lung through the pulmonary artery system. Smaller emboli lodge in arterioles, where they cause arterial thromboembolism, arteritis, endarteritis, and pulmonary abscesses. In some cases, a perivascular abscess not only erodes an arterial wall to produce an aneurysm but also simultaneously erodes a bronchial wall. When the aneurysm ruptures, the abscess cavity channels the blood into the bronchus, resulting in massive hemoptysis and epistaxis. Liver abscesses can be secondary to a primary ruminal wall infection, which in turn is secondary to acid-induced ruminitis. Ruminitis makes the wall susceptible to invasion and colonization by *F. necrophorum*. Once colonization has occurred, *F. necrophorum* can enter the blood or cause ruminal wall abscesses and subsequently shed bacterial emboli to the portal circulation, resulting in abscessation (Fig. 19.5-5).

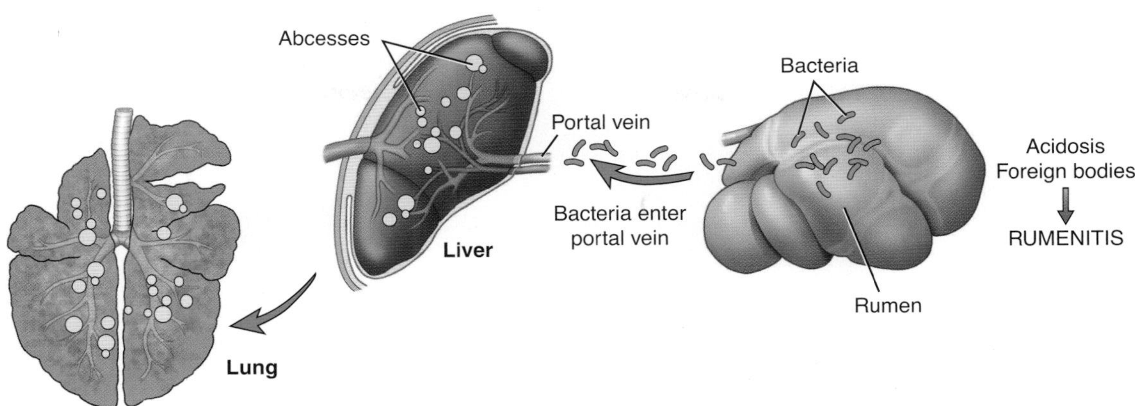

FIGURE 19.5-5 The hepatic parenchymal blood supply. The hepatic artery and portal vein send their blood into a common network of anastomosing small vascular channels called the sinusoids. The peri-sinusoidal space is the main site where material is transferred between blood-filled sinusoids and hepatocytes.

through the liver, anastamosing with hepatic veins, and then passes through the diaphragm ("hepatic part"). Finally, it enters the thoracic cavity and terminates into the right atrium ("thoracic part").

Liver and umbilical (fetal) circulations

During fetal circulation, the umbilical veins directly connect with the portal vein and caudal vena cava via the ductus venosus. The **ductus venosus** and the **right umbilical vein** persist until birth, while the **left umbilical vein** becomes the **falciform ligament** during gestation. By several days after birth, the ductus venosus has closed and the umbilical vein becomes the **round ligament of the liver** (see Case 4.6, Fig. 4.6-4).

> Persistence of fetal circulation after birth can be pathologic. The ductus venosus can persist, causing a connection between the caudal vena cava and the portal vein—termed an "intrahepatic portosystemic shunt." The umbilical vein can persist because of infection—i.e., a patent urachus or septic omphalitis—which can spread easily to the liver, to the caudal vena cava via the ductus venous, or—finally—into systemic circulation.

Anatomy of the lungs

> The umbilical vein can become infected at birth from ascending infection, called omphalophlebitis (Gr. *omphalos* navel + Gr. *phleps* vein + Gr. *nosos* disease, inflammation of a part). If infection reaches the liver it may form abscesses. It has grave consequences if not treated because the infection can spread hematogenously (Gr. *haimatos* blood + Gr. *genēs* born) from the very vascularized liver, causing sepsis, septic arthritis, osteomyelitis, meningitis, and peritonitis. Therefore, an infected umbilical vein must be surgically resected (Fig. 19.5-6).

The left and right **lungs** occupy most of the thoracic cavity. The lungs are covered by the **pulmonary pleura** from their invagination in the **pleural sac**. They are attached by the **pulmonary ligament** formed by the **mediastinal pleura** and the **hilus pulmonis**. The right lung is larger than the left because it has an accessory lobe. The right lung is divided by fissures into 4 lobes: cranial (apical), middle (cardiac), caudal (diaphragmatic), and accessory (intermediate) (Fig. 19.5-7). The **cranial lobe** is subdivided into cranial and caudal lobes as well. The cranial lobe extends to the left, cranial to the heart, which forms the **cardiac impression**.

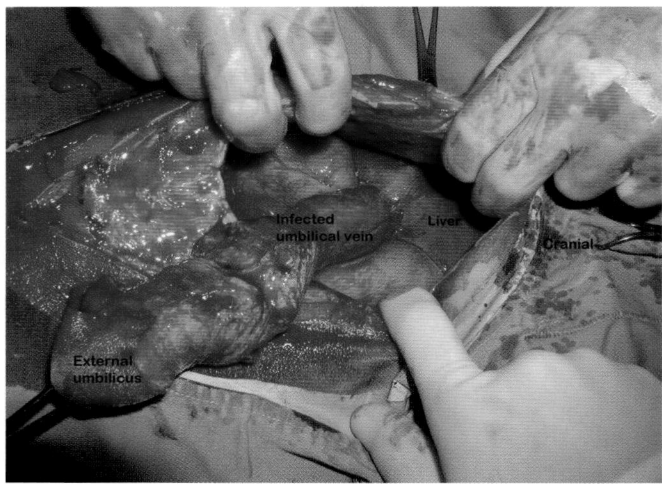

FIGURE 19.5-6 Intra-operative view of a young calf with an infected umbilicus. The calf is in dorsal recumbency, and the abdomen was entered through a cranial midline incision, close to the sternum. The umbilical vein was carefully dissected from the umbilicus, leaving only the skin and the external umbilicus. The vein clearly enters the liver in this photo.

The right cranial lobe has a distinctive bronchus directly from the trachea, cranial to the bifurcation. The **middle lobe** is elongated and it extends ventrally; its medial surface is the caudal aspect of the cardiac impression. The **caudal lobe** is the largest with a large diaphragmatic surface. The **accessory lobe** is situated medial to the caudal lobe. The left lung is divided like the right lung but without the accessory lobe **(**Fig. 19.5-8**)**.

The right cranial lobe is often affected in cases of descending infection as this bronchus is the first one encountered by inhaled bacteria and viruses colonizing the lung. Infection of the caudal lobes is usually only from hematogenous origin, as in the case of septic embolization of the vena cava.

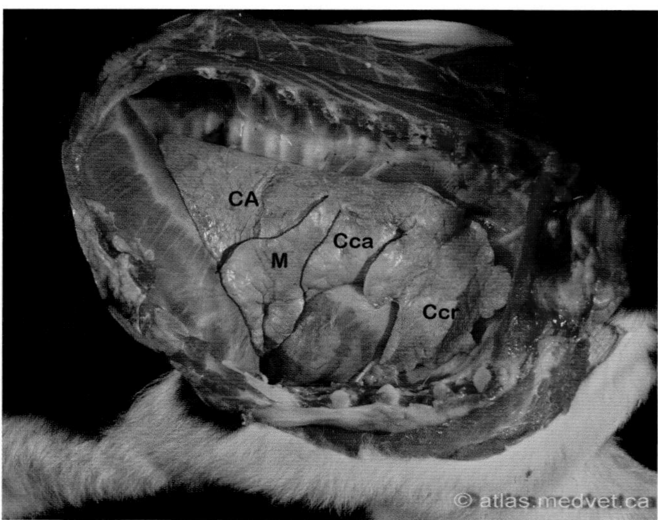

FIGURE 19.5-7 Prosection of a calf. The thoracic wall has been removed to show the right lung. The *dark lines* highlight the interlobar fissures. Key: *Ccr*, cranial part of the cranial lobe; *Cca*, caudal part of the cranial lobe; *M*, middle lobe; *CA*, caudal lobe. The accessory lobe is medial and cannot be seen in this view.

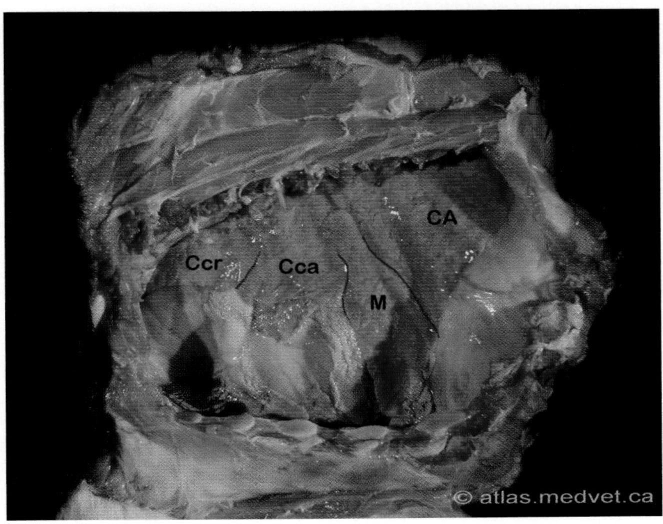

FIGURE 19.5-8 Prosection of a calf. The thoracic wall has been removed to show the left lung. The *dark lines* highlight the interlobar fissures. Key: *Ccr*, cranial part of the cranial lobe; *Cca*, caudal part of the cranial lobe; *M*, middle lobe; *CA*, caudal lobe.

Blood supply, lymphatics, and innervation of the lungs

The **pulmonary arteries** carry venous blood to the lungs from the right ventricle. The **pulmonary veins** bring oxygenated blood to the left atrium from the lungs. The lymphatic circulation of the thoracic cavity is complex. It has multiple lymphocenters but the lungs' lymphatic vessels are derived from the **mediastinal** and **bronchial lymphocenters** with afferent and efferent vessels. The **pulmonary plexus** innervates the lungs, pleura, and bronchial tree. It is an autonomic plexus from pulmonary branches of the parasympathetic nerves—mainly the vagal nerve and the sympathetic trunk.

A sympathetic stimulus—i.e., "fight or flight"—incites bronchodilation, vasoconstriction, and inhibition of glandular secretion. A parasympathetic stimulus creates the opposite—bronchoconstriction, vasodilation, and glandular secretion.

Selected references

[1] Barone R. Topographie abdominale de la vache. In: Splanchnologie II: Anatomie comparée des mammifères domestiques (Tome 4). 3ème édition. Paris: Vigog; 2001. p. 743–5.
[2] Budras KD, Habel RE, Mülling CK, Greenough PR, Jahrmärker G, Richter R, Starke D. Thoracic cavity. Chapter 6, In: Bovine anatomy: an illustrated text. 5th ed. Hannovre: Schlütersche; 2011. p. 62–4.
[3] Budras KD, Habel RE, Mülling CK, Greenough PR, Jahrmärker G, Richter R, Starke D. Abdominal wall and abdominal cavity. Chapter 7, In: Bovine anatomy: an illustrated text. 5th ed. Hannovre: Schlütersche; 2011. p. 66–77.
[4] Kumar V, Abbas AK, Aster JC. Disease of the liver. Chapter 16, In: Robbins basic pathology. Elsevier Health Sciences; 2017. p. 637–76.
[5] Smith BP. Disease of the hepatobiliary system. Chapter 33, In: Large animal internal medicine. 5th ed. Mosby, MO: Elsevier; 2015. p. 843–72.

CASE 19.6

Hepatic Lipidosis

Julie Berman
Centre Hospitalier Universitaire Vetérinaire (CHUV), Université de Montréal, Québec, CA

Clinical case

History

A 2-year-old, primiparous Holstein dairy cow presented with a history of anorexia, drop in milk production, and weight loss. She had calved 16 days earlier with a body condition score (BCS) of 4/5 (1= emaciated; 5= obese). After calving, she developed ketosis (β-hydroxybutyrate [BHB] > 2.6 mmoL/L) and was treated by the owner with IV boluses of calcium borogluconate and dextrose, but appetite and milk production did not improve. No other cows in the herd (50 dairy cows) were sick.

Physical examination findings

On admission, the cow had a BCS of 3/5. Rectal temperature and respiratory rate were normal, but she was mildly tachycardic at 88 bpm. All mucous membranes—ocular, vulvar, and buccal—were icteric (Fig. 19.6-1). The cow was approximately 5% dehydrated. The rumen was amotile, and no "ping" was auscultated on percussion and auscultation. Abdominal palpation per rectum revealed scant, dry feces, a smooth left kidney of normal size but without palpable lobulation, and a small rumen. Palpation of the left kidney did not elicit a pain response.

Differential diagnoses

Hepatic lipidosis, cholangiohepatitis, hepatic abscessation, leptospirosis, pyrrolizidine alkaloid toxicity, aflatoxicosis, and liver flukes

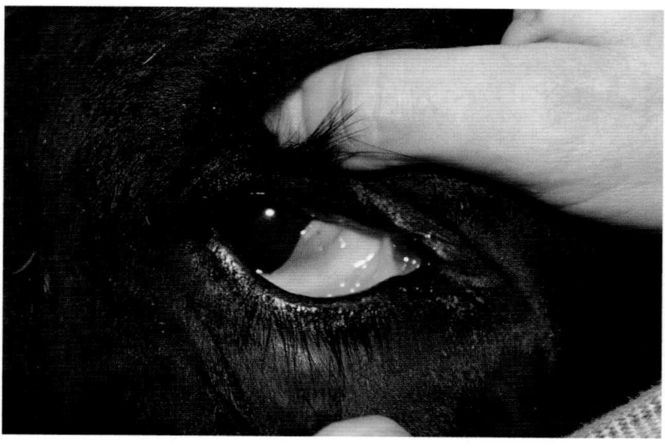

FIGURE 19.6-1 Ocular/scleral icterus in a Holstein cow.

Comparative Veterinary Anatomy: A Clinical Approach. https://doi.org/10.1016/B978-0-323-91015-6.00099-6

Diagnostics

A routine CBC was within normal limits. Serum biochemistry revealed a marked increase of liver laboratory values (total bilirubin = 223.8 µmoL/L; AST = 672 U/L; GLDH = 735 U/L; and GGT = 387 U/L). β-Hydroxybutyrate levels were normal. Transabdominal ultrasound revealed a decrease in vascularization of the liver (Fig. 19.6-2). The right kidney was surrounded by heterogeneous hyperechoic tissue consistent with the appearance of adipose tissue. Transcutaneous liver biopsy was performed (see side box entitled "Liver biopsy in cattle"), and histopathology showed severe and diffuse vacuolation of hepatocytes compatible with severe hepatic lipidosis (Fig. 19.6-3).

Diagnosis

Hepatic lipidosis

Treatment

The prognosis was guarded due to the severity of the liver disease. In general, the principle of therapy for hepatic lipidosis is to correct the negative energy balance by providing energy while improving fat mobilization from the liver. Continuous rate infusion of dextrose was initiated, and appetite was stimulated by transfaunating with rumen fluid from a normal cow once a day. The liver enzymes decreased over the first 5 days, and the cow was discharged to the client.

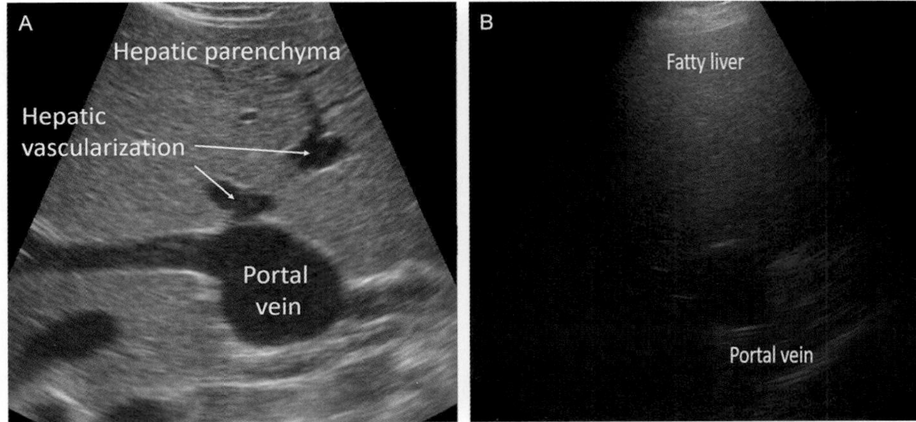

FIGURE 19.6-2 Transabdominal ultrasonographic imaging in a cow comparing a normal liver (A) with a fatty liver secondary to a hepatic lipidosis (B). The hepatic parenchyma vascularization is hardly visible in the liver with hepatic lipidosis.

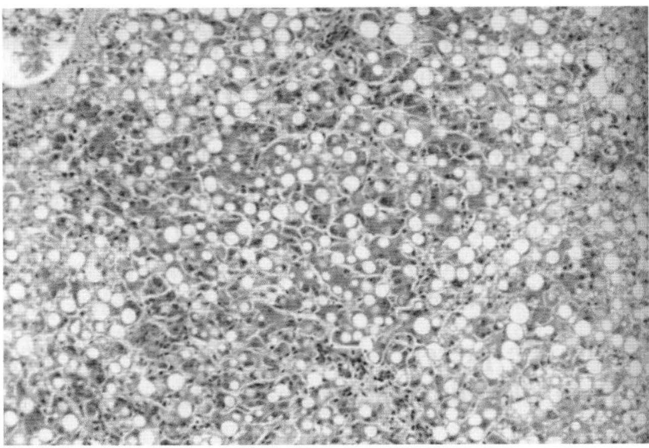

FIGURE 19.6-3 Histological appearance of hepatic lipidosis. The vacuoles (optically empty) within the hepatocytes represent lipid accumulation.

Anatomical features in ruminantss

Introduction

This section covers the gross anatomy of the liver and hepatic ligaments. Liver function, vascularization, and fetal circulation are discussed in case 19.5.

The liver may be easily compressed due to its intra-abdominal location adjacent to the rib cage (it has nowhere to go). Liver enzymes can therefore be mildly elevated during the following: right abomasal dilation/volvulus, severe bloat, small and large intestinal disease, cecal dilation/torsion or retroflexion, or advanced gestation.

Macroscopic morphology and location of the liver

The liver lies in the abdominal cavity, attached to the diaphragm, and extends from the ventral aspect of the 6th to the 12th intercostal spaces (Fig. 19.6-4). Macroscopically, the normal liver appears reddish-brown in the adult (Fig. 19.6-5).

The liver is almost entirely found on the right side of the abdomen in ruminants—except for a small cranial portion (left lobe)—due to the large rumen on the left side of the abdomen. Because of its location, almost the entire liver can be easily palpated during right flank laparotomy (only the left lobe remains difficult to reach because of its depth).

In contrast to the dog and the horse, lobulation of the bovine liver is not obvious. The **caudate process**, in contact medially with the right kidney, continues into the **right lobe**—in contact medially with the pancreas. The right lobe is separated medially from the **quadrate lobe** by the **gallbladder**. These two lobes, right and quadrate, are in contact with the sigmoid flexure of the duodenum. Finally, the liver extends ventrally into the **left lobe** (Fig. 19.6-4).

The left lobe is in contact with the cranial abdomen and is thus a predilection site for abscess formation by migration of a reticular foreign body.

Hepatic ligaments

Hepatic ligaments affix the liver's position within the abdominal cavity. Dorsally, the **right triangular ligament** attaches the diaphragmatic surface of the caudate process to the dorsal abdominal wall and, dorsomedial to it, the **hepatorenal ligament** connects the visceral surface of the caudate process to the right kidney (Fig. 19.6-5).

Ventrally, the **left triangular ligament** extends from the medial visceral surface of the left lobe to the diaphragm near the esophageal hiatus. The **round ligament—**the free border of the falciform ligament—is the remnant of the umbilical vein and courses from the umbilicus to the liver (Fig. 19.6-4). However, in many adults the round ligament has disappeared.

On the diaphragmatic surface, the **coronary ligament** attaches the right lobe of the liver to the diaphragm. Its line of attachment to the liver passes from the left triangular ligament, ventral to the caudal vena cava, along the right side of the caudal vena cava. On the right lobe, it divides into two laminae that surround the **bare area** (area nuda). The bare area of the liver is a large triangular area on the diaphragmatic surface of the liver. It is attached directly to the diaphragm by loose connective tissue.

The **falciform ligament** is attached to the diaphragmatic surface of the liver on a line from the coronary ligament at the **foramen venae cava** (hole in the diaphragm where the cranial vena cava passes through) to the fissure for the round ligament. It separates the liver into the right and the left lobes. It is attached to the diaphragm on a horizontal line from the foramen venae cava to the costochondral junction. However, this diaphragmatic attachment is essentially a secondary adhesion resulting from the displacement of the liver to the right in ruminants. Like the round ligament, the falciform ligament has regressed in many adults.

A
Hepato-renal ligament
Aorta
Right triangular ligament
Caudal vena cava
Caudate process of liver
Pancreas
Splenic a. and v.
Portal v.
Diaphragm
Right lobe of liver
Sigmoid flexure of duodenum
Spleen
Quadrate lobe of liver
Gallbladder
Esophagus
Descending duodenum
Falciform and round ligaments
Left triangular ligament
Lesser omentum
Left lobe of liver

B

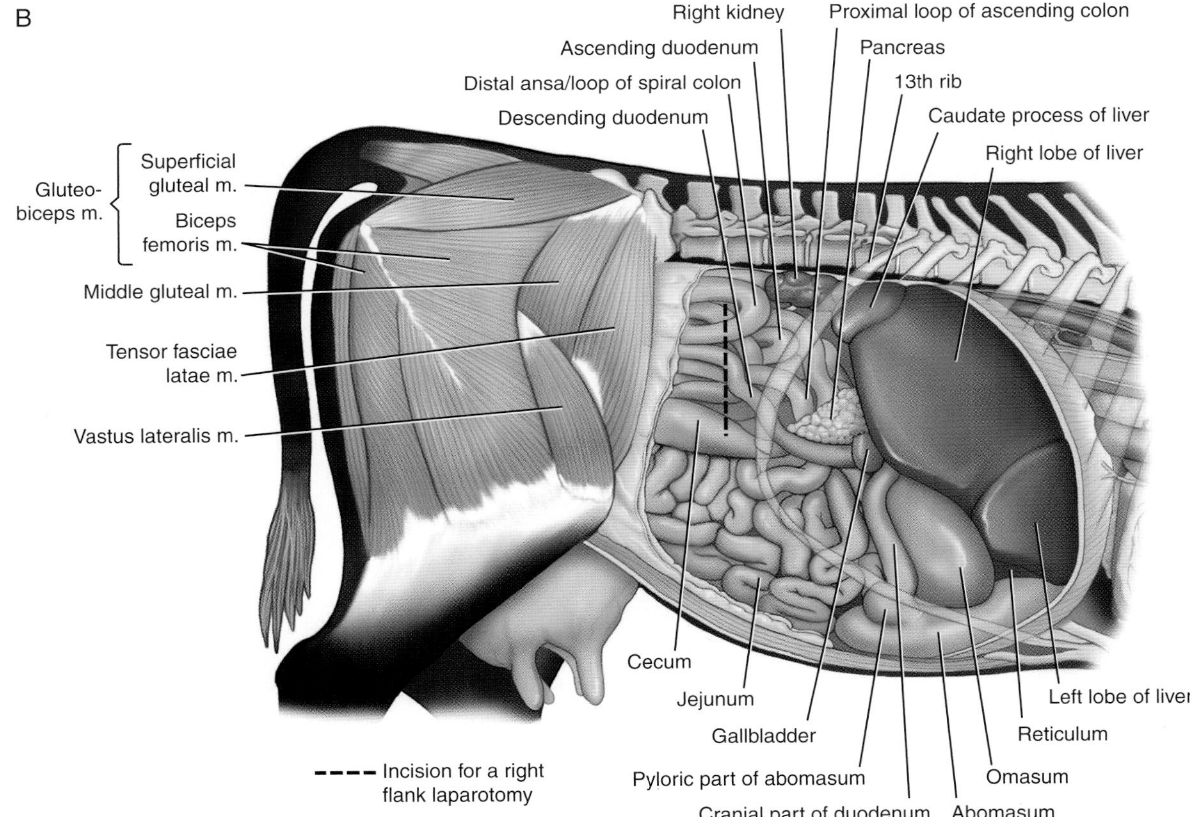

FIGURE 19.6-4 The location in situ (A) and anatomy (B) of the liver. (A) Note the hepatic ligaments. (B) The incision for a right flank laparotomy is illustrated (*dashed line*) to highlight its proximity to the liver.

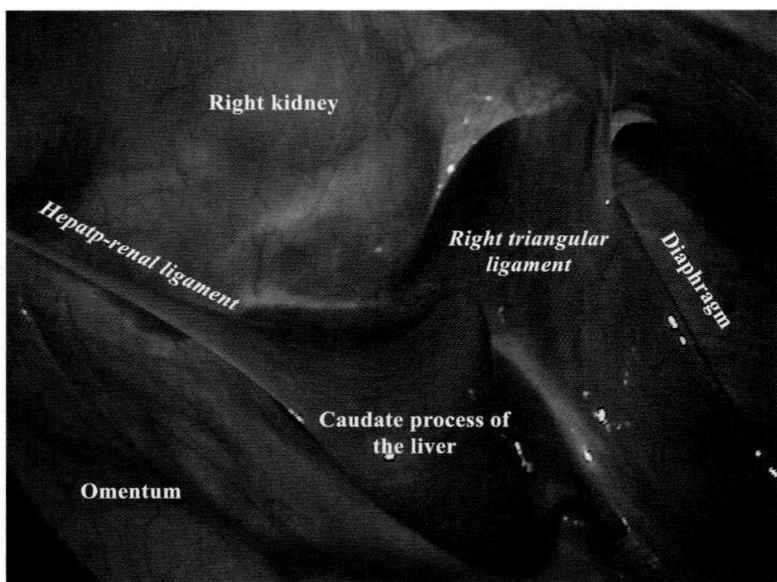

FIGURE 19.6-5 Internal laparoscopic view of the dorsal part of the liver in an adult cow. The scope is inserted dorsally, behind the 13rd rib, and oriented ventro-cranially. The caudate process of the liver is visible as well as his relationship with the right kidney, the diaphragm, and the omentum. The liver is attached dorsally to the right kidney via the hepatorenal ligament and to the abdominal cavity via the right triangular ligament.

Blood supply, lymphatics, and innervation

The liver is supplied by 2 main blood vessels entering on the visceral surface (porta hepatis): the hepatic artery and the portal vein. The **portal vein** brings venous blood to the liver from the spleen, pancreas, and small intestine so that the liver can process the nutrients and byproducts of digestion. It supplies approximately 75% of the blood the liver receives. The **hepatic arteries** supply arterial blood to the liver and account for the remainder of its blood flow.

Lymphatic drainage goes to the **hepatic (portal) lymph nodes**, grouped around the porta hepatis, and the **accessory hepatic lymph nodes**, found on the dorsal border of the liver near the caudal vena cava. Outflow of lymph occurs through the **hepatic trunk**.

Innervation of the liver is governed by the **hepatic nervous plexus**, which runs adjacent with the hepatic artery and portal vein. It receives sympathetic fibers from the celiac plexus and parasympathetic fibers from the dorsal and ventral vagal trunks.

LIVER BIOPSY IN CATTLE

Liver biopsy is a relatively safe and simple procedure in cattle. Cattle need to be suitably retrained, but usually sedation is not necessary. Many biopsy instruments are available, and the Tru-Cut biopsy needle (Fig. 19.6-6) works well. The skin over the biopsy site should be clipped and prepared for aseptic insertion of the needle. Local infiltration of 2% lidocaine into the skin and intercostal muscles reduces the response on initial insertion of the biopsy needle, but there may still be a response when the pleura and peritoneum are penetrated by the biopsy needle. A small stab incision in the skin at the site of insertion is made with a scalpel blade to facilitate introduction of the biopsy needle.

The biopsy site in cattle is located either by ultrasound or by extending a horizontal line cranially from the middle of the paralumbar fossa to the 11th intercostal space on the right side. It can also be located at the intersection of 2 imaginary lines: one from the tube coxae to the elbow and the second, a horizontal line at the level of the greater trochanter (Fig. 19.6-7). At the 11th intercostal space on this line, the needle is directed slightly cranial and ventral (toward the reticulum). Identification of the liver with ultrasound is preferable to a “blind” stick, especially when a hepatic abscess is suspected.

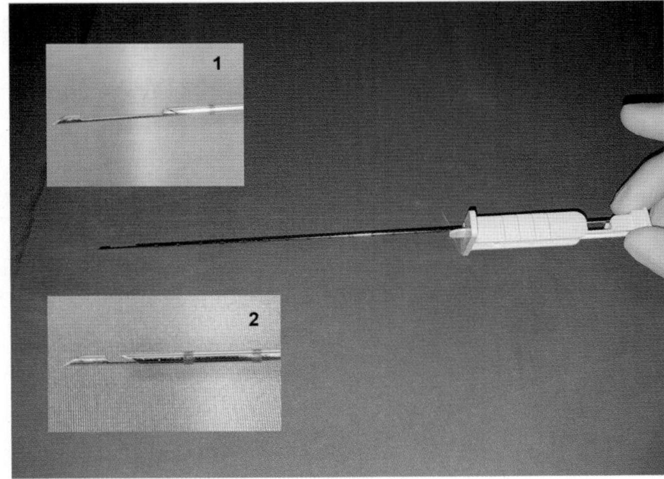

FIGURE 19.6-6 The Tru-Cut biopsy instrument, used to diagnose hepatic lipidosis in a cow. The sample is obtained by sliding the sleeve of the instrument over its tip. 1—the biopsy chamber is open; 2—the cutting sleeve is almost closed over the chamber.

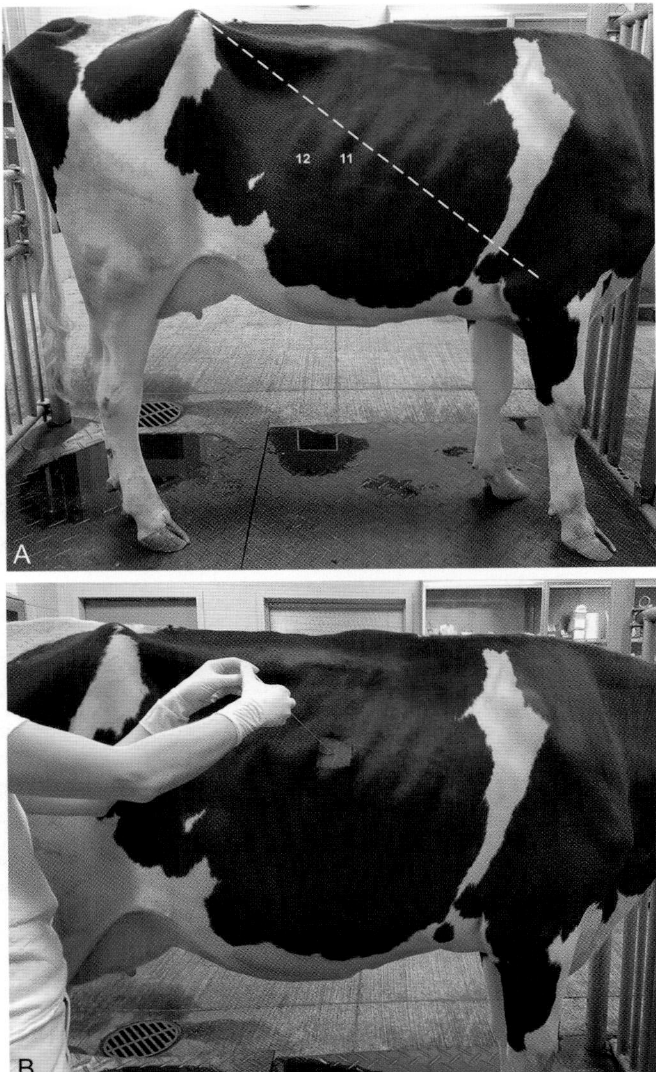

FIGURE 19.6-7 Liver biopsy in a cow. (A) The liver biopsy site is located by the intersection of a line between the tuber coxae and the elbow and a horizontal line at the level of the greater trochanter. The intercostal spaces 11 and 12 are identified. (B) The biopsy needle is directed toward the reticulum or the left elbow.

Selected references

[1] Barone R. Topographie abdominale de la vache. In: Splanchnologie II: Anatomie comparée des mammifères domestiques (Tome 4). 3ème édition. Paris: Vigog; 2001. p. 743–5.
[2] Guyton AC, Hall JE. The liver as an organ. Chapter 70, In: Textbook of medical physiology. 12th ed. Philadelphia: Elsevier Saunders; 2011. p. 837–42.
[3] Kumar V, Abbas AK, Aster JC. Disease of the liver. Chapter 16, In: Robbins basic pathology. Elsevier Health Sciences; 2017. p. 637–76.
[4] Smith BP. Disease of the hepatobiliary system. Chapter 33, In: Large animal internal medicine. 5th ed. Missouri: Elsevier Mosby; 2015. p. 843–72.
[5] Budras KD, et al. Spleen, liver, pancreas, and lymph nodes. Chapter 5, In: Bovine anatomy: an illustrated text. Hannover: Schlütersche; 2011. p. 74.

CHAPTER 19

CASE 19.7

Hydronephrosis

André Desrochers
Department of Clinical Sciences, Université de Montréal, Faculty of Veterinary Medicine, St-Hyacinthe, Québec, CA

Clinical case

History

A 4-year-old Holstein cow presented for mild chronic colic, decreased milk production, and intermittent anorexia. She had been treated without a clinical response using antibiotics and NSAIDs for the last 2 months. She calved without problems 3 months ago and was routinely treated for a retained placenta.

Physical examination findings

The cow was slightly depressed on admission with adequate hydration status. Her body condition score was 2.5/5. Her abdominal profile revealed a small rumen with rumen contractions decreased in frequency and strength. Rectal temperature was 101.8°F (38.8°C), and she had a heart rate of 100 bpm and respiratory rate of 28 brpm. The "withers test" (a test for cranioventral abdominal pain like reticulopericarditis, see case 19.1) was negative—i.e., no pain was elicited when pinching and pushing down on her withers to make her dorsiflex. Percussion and auscultation (pings) were negative on the left and right sides of the abdomen. However, the right paralumbar fossa, just caudal to the last rib, was slightly elevated/distended compared to normal. Abdominal palpation per rectum revealed an enlarged organ in the right dorsocranial quadrant. This organ seemed attached to the dorsal wall of the abdomen and was barely palpable with the extremities of the hand. The rest of the abdominal palpation was normal.

Differential diagnoses

Pyelonephritis, hydronephrosis (congenital or acquired due to urolithiasis), liver abscess, and retroperitoneal abscess

Diagnostics

Transabdominal ultrasound of the right dorsal paralumbar fossa revealed that the enlarged viscus palpated rectally and pushing away the last rib was the right kidney. Although the cranial pole had a normal architecture, the caudal pole showed a large thin-walled mass full of hypoechoic fluid compatible with hydronephrosis. The calices were enlarged with fluid (10 cm or 3.9 in. diameter) with thin cortices (Fig. 19.7-1). The architecture of the left kidney was confirmed to be normal based on palpation and ultrasound examination per rectum.

Serum chemistry and hematologic profiles showed a mild stress leukogram with normal BUN and creatinine indicating normal renal function. Cystoscopy using a flexible endoscope (see Section 2.1) showed normal urine in the bladder. Visual evaluation of the trigone revealed urine flowing normally from the left ureter but not from the right ureter. A urine sample was analyzed and was normal.

Diagnosis

Right-sided hydronephrosis

Comparative Veterinary Anatomy: A Clinical Approach. https://doi.org/10.1016/B978-0-323-91015-6.00100-X

FIGURE 19.7-1 Transabdominal ultrasound image of the right kidney of a cow. The calices are severely distended (*C*), causing thinning of the cortex and medulla (*CM*) due to the progressive distention.

Treatment

A unilateral nephrectomy was performed under standing sedation and regional anesthesia via the right paralumbar fossa. A proximal paravertebral block of T13, L1, and L2 provided appropriate anesthesia/analgesia to perform the surgery. The right paralumbar fossa was surgically prepared. Through a 20-cm (7.9-in.) flank incision, the affected kidney was approached by dividing the peritoneum and entering the retroperitoneal space. Topical anesthesia with 2% lidocaine provided further analgesia during kidney dissection; the hilus was isolated and the renal artery was triple-ligated first with nonabsorbable suture followed by the renal vein. The ureter was separated, ligated, and transected caudally and the kidney was removed (Fig. 19.7-2A and B). The retroperitoneal space was left unsutured. The right paralumbar fossa incision was closed anatomically.

The cow recovered without complications in spite of a transient increase in BUN and creatinine post-operatively that resolved with supportive care.

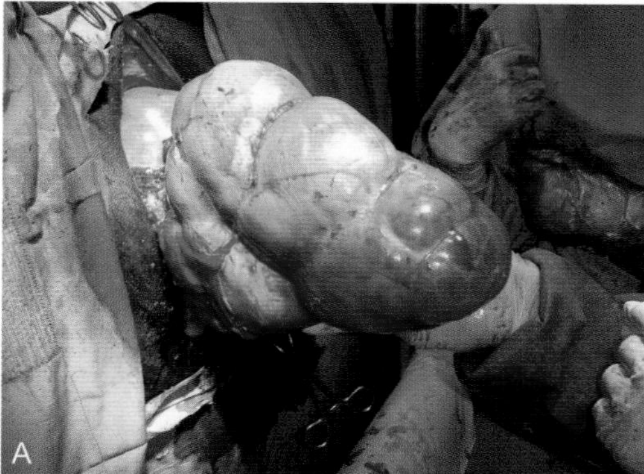

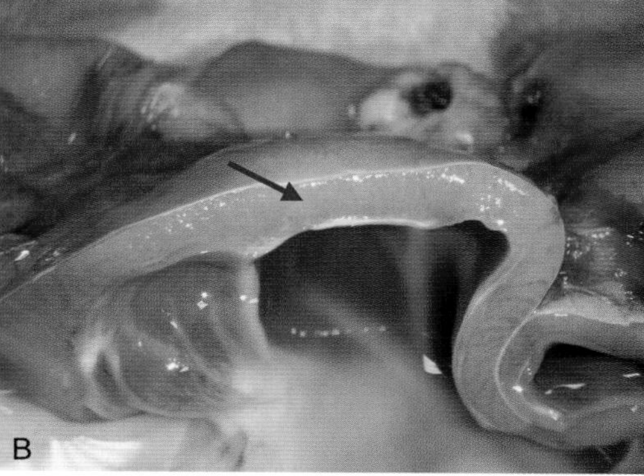

FIGURE 19.7-2 (A) The right kidney exteriorized after ligation of the vessels and ureter. (B) The translucent caudal pole was transected where the thin cortex and medulla (*arrow*) are observed.

Anatomical features in ruminants

Introduction

The urinary system in bovids includes the kidneys and their ureters, the adrenal glands (due to location), the bladder, the urethral recess (males only), the penis (males only), and the urethra, along with the urachus in the fetus. The accessory sex glands and the penis of ruminants are covered in Case 20.4.

Function

The (genito-)urinary system has 2 major functions: body fluid homeostasis and reproduction (the latter is not covered in this section). The kidney is responsible primarily for maintaining body fluid volume—e.g., balancing electrolytes by retaining sodium and excreting potassium, as well as maintaining blood pH homeostasis (Gr. *homoios* like, resembling + Gr. *stasis* standing). It also removes waste products of metabolism—i.e., urea and creatinine. Its capacity to balance electrolytes makes it an essential component of the complex multipart acid–base balance system in concert with blood and lungs. It has a role in regulating blood pressure through the body fluid system and the renin-angiotensin system. Renin is synthesized and stored in the juxtaglomerular cells of the kidneys. It increases blood pressure indirectly by acting on the plasma peptide angiotensin. Erythropoietin is secreted by the renal cortex in response to constant low oxygen in renal arteries and stimulates erythropoiesis. Vitamin D is transformed into calcitriol, its active form, at the level of the tubules, which then is part of the system that regulates calcium and phosphorous levels in the body.

The paired (left and right) ureters drain urine from their respective ipsilateral kidneys and enter the urinary bladder at the trigone.

Kidneys

Kidneys of other ruminants (sheep, goats) are not lobulated and grossly resemble the kidneys of dogs and cats.

Kidneys of cattle are unique because they are lobulated and lack a pelvis. They are brownish-red in color with approximately 20 **lobules** of different sizes. Bovine kidneys are surrounded by perirenal adipose/fat tissue that infiltrates the fissures between the lobes. The right kidney is approximately 20 cm (7.9 in.) long in the adult bovid and flattened dorsoventrally, lies ventral to the last ribs and the first 3 lumbar transverse processes, and in close contact with the subvertebral muscles. The ventral surface is in contact with the liver, pancreas, duodenum, and colon (Fig. 19.7-3). The **hilus** (indentation where vessels enter and

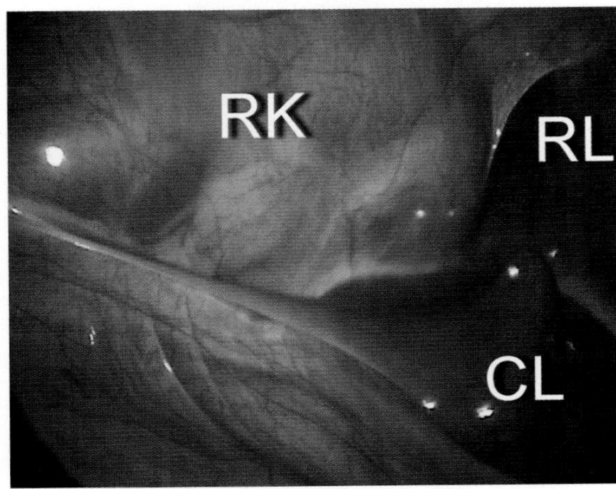

FIGURE 19.7-3 Laparoscopic view of the right dorsal quadrant of the abdomen. Key: *RK*, right kidney; *CL*, caudate lobe of the liver; *RL*, right lobe of the liver.

exit) is on the cranial part of the ventral surface near the medial border. The artery is dorsal, the vein is in the middle, and the ureter is ventral within the hilus. The cranial extremity is well-seated in the renal impression of the liver. The left kidney is shorter than the right, pendulous in the median plane, slightly pushed to the right of midline by a filled rumen, and ventral between the 3rd and 5th lumbar vertebrae with the descending colon to its right. Its hilus is craniolateral to its convex dorsal surface.

According to the 2017 Nomina Anatomica Veterinaria, "The calices pelvis are included in the Pelvis renalis. The latter is defined as the space which collects the urine and is situated within the Sinus renalis at the beginning of the ureter." Calix and calyx are synonymous terms.

The structure of the collecting system of bovine kidneys is also unique because there is no pelvis, which is defined as an enlargement of the ureter inside the kidney. Each lobule contains a **cortex**, **medulla**, **pyramid**, and **papilla** that goes into **minor calices** (L. Gr. *kalyx* cup-shaped organ or cavity) and two **major calices** (cranial and caudal) (Figs. 19.7-4 and 19.7-5).

Renal blood supply, lymphatics, and innervation

The paired (right and left) **renal arteries** originate from the descending aorta at the level of the 2nd lumbar vertebra. The right renal artery is short and slightly cranial to the left renal artery, which is longer. The **renal veins** drain into the caudal vena cava, and their course is parallel to their respective arteries.

The **renal lymph nodes** are positioned along the renal vessels in various numbers. They drain the kidneys and adrenal glands and drain into the **cisterna chyli**. The kidneys have both parasympathetic and sympathetic innervation. The **parasympathetic trunk** provides vasomotor supply to the kidneys while the **renal plexus** supplies sympathetic innervation (including pain receptors) from the **sympathetic trunk**.

Ureters

The **left ureter** emerges from the hilus facing the right side, curves, and lies on the dorsal and caudal surface of the kidney, then tracking along the left side before entering the urinary bladder. The **right ureter** emerges ventrally

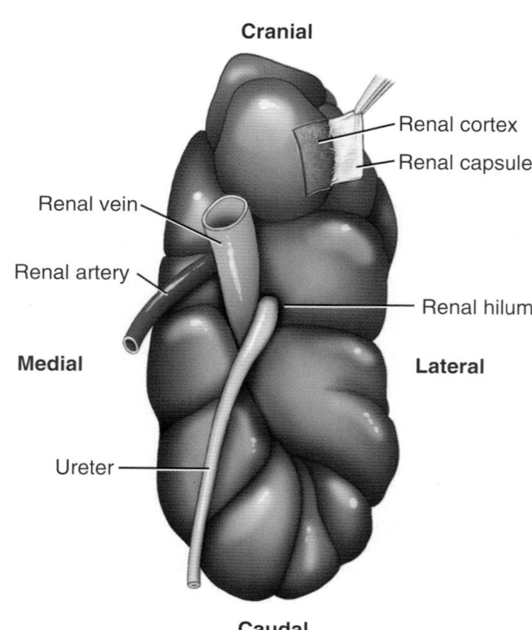

FIGURE 19.7-4 The renal artery, vein, and the ureter are in the hilus of this polylobulated bovine kidney.

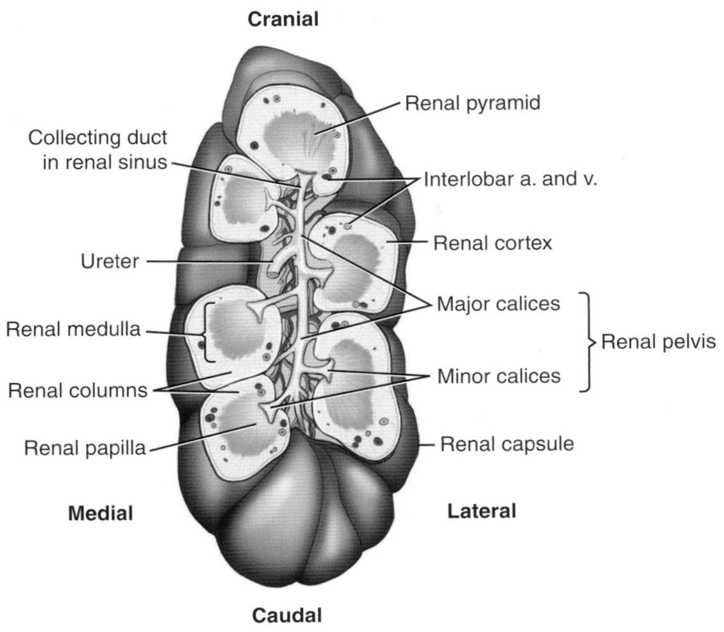

Cranial

Renal pyramid

Collecting duct in renal sinus

Interlobar a. and v.

Renal cortex

Ureter

Major calices

Renal medulla

Renal pelvis

Minor calices

Renal columns

Renal papilla

Renal capsule

Medial

Lateral

Caudal

FIGURE 19.7-5 Bovine kidney with the surface removed to show the internal features.

and medially tracking along the lateral surface of the vena cava on the right side, crossing the external iliac arteries before entering the urinary bladder. Both ureters open dorsally in the neck of the bladder at the **trigone** (Fig. 19.7-6). The ureters are innervated by the **ureteric plexus**, comprised of sympathetic and parasympathetic fibers.

> Ureters can be catheterized by inserting a small catheter through the instrument portal of the flexible endoscope. Each kidney can be sampled separately for individual analysis. In addition, urethral, ureteral, and renal calculi can be fragmented by shock wave lithotripsy (and ureteroscopy for ureteroliths) with equipment introduced via the urethra.

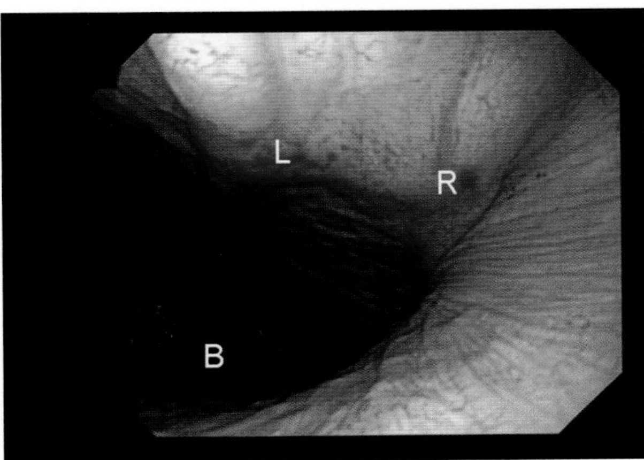

FIGURE 19.7-6 A flexible endoscope was inserted through the urethra to evaluate the flow from the ureter into the bladder at the trigone. Key: *R* right ureter opening; *L*: left ureter opening; *B*: distended bladder from insufflated air.

Urinary bladder

The **urinary bladder** is large compared to the horse and extends further cranial in the abdomen. The **apex** and **body** are in the peritoneal cavity and stabilized by the 2 **lateral ligaments**, remnants of the umbilical arteries. The **median (middle) ligament** (remnant of the urachus) attaches the bladder to the pelvic symphysis. The **neck** of the bladder is extraperitoneal and pelvic in its location as it continues as the pelvic urethra.

The innervation of the bladder is complex, combining the autonomic and somatic nervous systems. The bladder itself is innervated by sympathetic and parasympathetic nerves providing, among other things, information regarding sensation to the CNS. The internal sphincter is under the control of the autonomic nervous system via the **hypogastric nerve** while the voluntarily controlled external sphincter is innervated by the **pudendal nerve**.

> The specific function of the urethral recess is unknown. The fold of mucous membrane containing the opening of the ducts of the bulbourethral glands may act as a one-way valve preventing urine or semen from flowing back into the pelvic urethra. Clinically, the urethral recess prevents retrograde passage of a urethral catheter, complicating the management of urolithiasis in these species. Urethral catheterization can be achieved in spite of this by making a proximal incision close to the anus (perineal urethrostomy) to insert the catheter into the pelvic urethra almost horizontally or after performing a perineal penectomy by cutting the crus of the penis and elevating it horizontally.

> The urethral process must be excised close to the glans penis when obstructed in urolithiasis and is often black and necrotic secondary to the obstruction. The affected individual is relieved immediately after removal. Additional surgery is needed if normal urination is not resumed after removing the urethral process.

Male urethra

The **pelvic urethra** is short and gently curves around the ischial arch. A **urethral recess** is present before becoming the penile part of the urethra. This recess is also present in swine, small ruminants, and camelids. This recess extends caudodorsally and is surrounded by the bulbospongiosus muscles (Fig. 19.7-7). The ducts of the **bulbourethral glands** open into the recess. The ram and the buck have an additional urethral process, an extension of 3–4 cm (1.2–1.6 in.) of the glans penis. Blood supply and innervation of the urethra are covered in case 20.4.

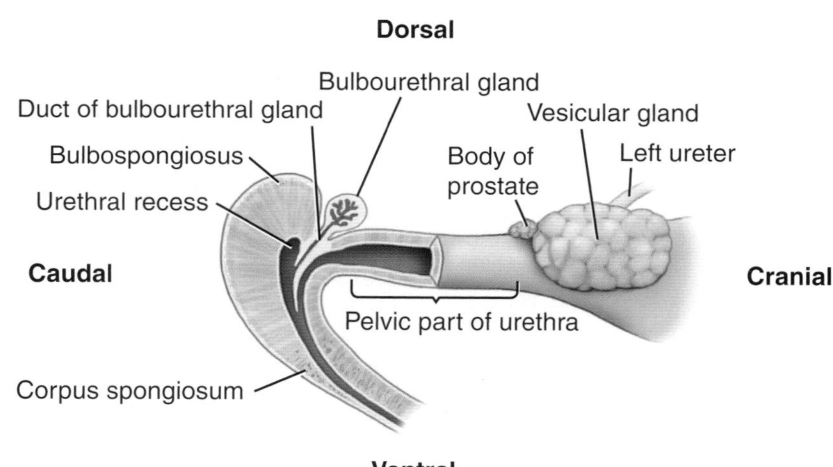

FIGURE 19.7-7 The urethral recess prevents retrograde passage of a catheter in cattle, small ruminants, and pigs.

Female urethra

The female urethra is about the same length as the male pelvic urethra and is located on the floor of the vagina. It opens on the ventral floor of the **vaginal vestibule** as the **urethral orifice**. This orifice has a **suburethral diverticulum** ventral and caudal to the urethral opening. This orifice is also described in goats and is shallow, allowing easy access for catheterization (Fig. 19.7-8).

Urovagina, or urine pooling or vestibulovaginal reflux, is defined as the collection of urine in the cranial portion of the vagina due to poor reproductive conformation, previous trauma secondary to a dystocia, or a sinking of the abdominal and pelvic viscera (uterus and cervix) below their natural anatomical position termed splanchnoptosis (Gr. *splanchnos* viscus + Gr. *ptosis* fall). Urine can also enter the uterus and adversely affect fertility. Several surgical procedures are designed to extend the urethra caudally, including urethroplasty (Gr. *ourēthra* urethra + Gr. *plassein* to form), diverting urine flow caudally.

The suburethral diverticulum is a blind end that must be avoided during catheterization.

Urachus

The **urachus** or allantoic duct is a fetal structure that allows urine from the bladder to drain into the allantois through the umbilicus. After parturition and rupture of the umbilicus, the urachus retracts inside the abdomen and closes. It stays as a scar at the apex of the bladder (Fig. 19.7-9).

The urachus is the umbilical remnant most commonly infected in calves. Unlike in the foal, it is rarely patent after birth. Although it may become infected and enlarged, it never communicates directly with the urinary bladder after birth. However, affected calves may have cystitis from incomplete bladder emptying because of excessive traction from the infected urachus.

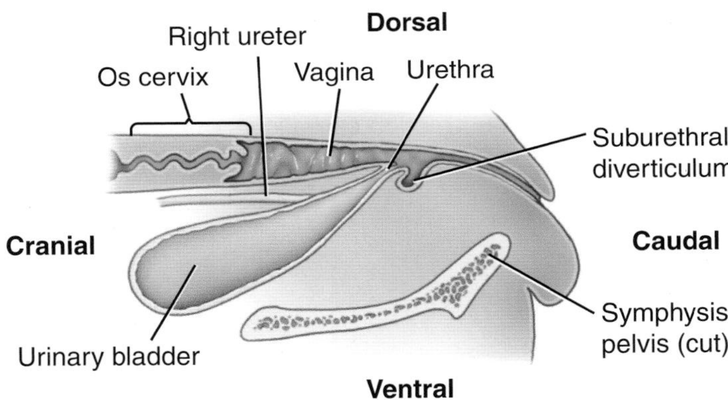

FIGURE 19.7-8 The suburethral diverticulum is ventral to the urethra in female bovids and must be avoided when catheterizing the bladder.

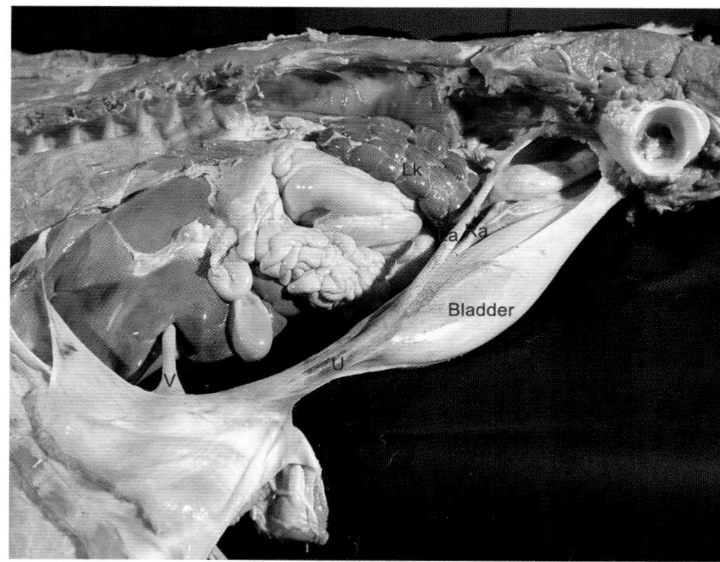

FIGURE 19.7-9 Left view of a newborn calf prosection with an emphasis on the umbilical structures. Key: *Ra*, right umbilical artery; *LA*, left umbilical artery; *V*, umbilical vein; *u*, urachus; *LK*, left kidney.

UROLITHIASIS

A common problem in feedlot steers, male small ruminants, and Vietnamese Pot-bellied boar, urolithiasis is diet-related and more common in castrated males. Uroliths generally are comprised of calcium apatite, phosphatic-based (e.g., struvite), or calcium carbonate. The obstructing calculus is often located at the sigmoid flexure or urethral process in small ruminants with multiple additional calculi of different sizes found in the bladder. Chronic obstruction can result in necrosis and rupture of the urethra or bladder termed "water belly"; surgery is needed to save the affected individual. Uroliths in females also occur; however, the larger urethral diameter usually precludes obstruction.

Surgery for relief and treatment of the obstruction includes a perineal urethrostomy, bypassing urine flow from the obstructed/ruptured area and performed in the perineal area or caudal to the scrotum as a salvage procedure. In breeders or companion small ruminants, the penis is preserved if viable vs. transected and sutured to the skin as a salvage procedure in steers. As part of the penile preservation procedure, a tube cystotomy provides immediate relief of the obstruction and allows the urethra to heal. A Foley catheter is used in small ruminants and is placed in the bladder and secured with a purse string suture, exiting ventrally at the skin.

Selected references

[1] Garret PD. Urethral recess in male goats, sheep, cattle and swine. J Am Vet Med Assoc 1987;191(6):689–91.

[2] Sisson S. Ruminant urogenital system. In: Sisson, Grossman, editors. The anatomy of domestic animals. Philadelphia: W.B. Saunders Co; 1975. p. 937–54.

[3] Budras KD, Wünshe A. Pelvic cavity and inguinal region, including urinary and genital organs. In: Budras KD, Habel RE, editors. Bovine anatomy. Hannover: Schlütersche GmbH & Co; 2003. p. 84–6.

[4] Vogel SR, Desrochers A, Babkine M, Mulon PY, Nichols S. Unilateral nephrectomy in 10 cattle. Vet Surg 2011;20:233–9.

CHAPTER 20

PELVIC ORGANS

André Desrochers, Chapter editor

Female Urogenital System

20.1 Perineal laceration—*Michael Pesato and Billy I. Smith* 1106
20.2 Dystocia with cesarean section—*Andrew Niehaus* 1112

Male Urogenital System

20.3 Urolithiasis—*Marie-Eve Fecteau* 1127
20.4 Penile hematoma—*David E. Anderson* 1135

CHAPTER 20

CASE 20.1

Perineal Laceration

Michael Pesato[a] **and Billy I. Smith**[b]
[a]Department of Pathobiology and Population Medicine, Mississippi State University College of Veterinary Medicine, Mississippi State, Mississippi, US
[b]Department of Clinical Studies - New Bolton Center, University of Pennsylvania School of Veterinary Medicine, Kennett Square, Pennsylvania, US

Clinical case

There are three stages of labor:

- Stage 1 of labor lasts 4–24 hours and is characterized by cervical relaxation and active contractions of both the longitudinal and circular muscle fibers of the uterine wall. The calf begins to move into the appropriate presentation, position, and posture (anterior longitudinal, dorsosacral with thoracic limbs and neck extended), and the cervix continues to dilate. Changes in behavior include increased restlessness and raising the tail when standing.
- Stage 2 of labor is characterized by uterine contractions, entrance of the fetus into the dilated birth canal; rupture of the allantoic sac; abdominal contractions or labor; and fetal expulsion through the vulva. Stage 2 continue with the amniotic sac seen at the vulvar labia and ruptures, releasing amniotic fluid. This stage lasts 1–2 hours and is complete when the calf is delivered.
- Stage 3 of labor involves placental expulsion. The placenta is generally discharged from the uterus within 8 hours and is considered retained after 12 hours.

History

A 2-year-old first calf Holstein heifer presented for dystocia (Gr. *dys-* difficult, painful + Gr. *tokos* birth) to the farm personnel responsible for tending to the cattle in labor. A farm employee reported that the heifer had started in stage 2 labor with strong abdominal contractions approximately 2 hours earlier, but had failed to progress in normal parturition timing (see side box entitled "Normal parturition in bovids (Eutocia)"). They examined the heifer to determine if there was a reason for the lack of progression to stage 3 of labor, and to decide if it was necessary to assist with the labor.

Vaginal examination revealed that the fetus was in normal anterior longitudinal presentation, normal dorsosacral position, normal posture with front legs and head extended, and the head laying between the thoracic limbs (Fig. 20.1-1). The caudal vagina and vulva were poorly dilated. The farm employee spent some time manually dilating and lubricating the vagina and vulva after which obstetric (OB) chains were placed on both the calf's thoracic limbs and manual assistance was used to help deliver the calf. After 15 minutes of pulling manually, it was decided to use a "calf puller" to deliver the calf. The calf was successfully delivered vaginally, but the large size of the calf led to trauma to the dorsal part of the vulva, and the veterinarian was called to evaluate the cow.

Physical examination findings

On presentation, the cow was systemically healthy with normal temperature, pulse, and respiration. There was soft tissue trauma (including a tear) of the dorsal commissure of the vulva and perineal body. The wound surfaces were irregular, jagged, and readily bled on manipulation, and the cow was painful when this area was evaluated.

Comparative Veterinary Anatomy: A Clinical Approach. https://doi.org/10.1016/B978-0-323-91015-6.00101-1

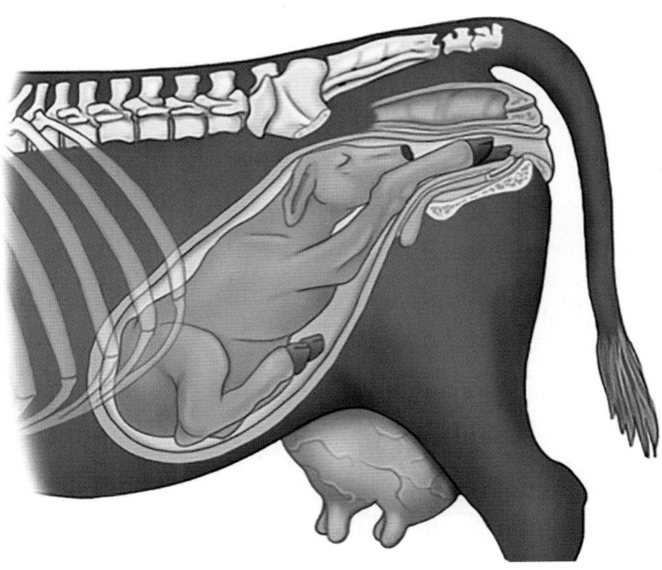

FIGURE 20.1-1 The normal presentation, position, and posture of an unborn calf entering the birth canal of a cow.

NORMAL PARTURITION IN BOVIDS (EUTOCIA, FIG. 20.1-1)

Once a cow has entered into stage 2 labor and is exhibiting strong abdominal contractions followed by moments of rest, the birth of the calf is usually complete within minutes to a few hours. The normal birth of a calf has a regular progression with the presence of the "placenta/calf unit" first noticed as it separates the vulvar lips. Once the amnion (Gr. "bowl" membrane enveloping the fetus, water bag/sac) is seen at the vulvar lips, the progression of the delivery should be assessed every 30 minutes. The following protocol is used to safeguard the delivery of a healthy calf with minimal injury to the cow:

- Visible amnion (amniotic sac, water bag)
  - Wait 30 minutes and reassess.
  - If things are progressing along normally, the amnion ruptures and the calf's feet and nose should be seen moving through the vulvar labia with each strong abdominal contraction.
- Visible feet and nose
  - Wait 30 minutes and reassess.
  - If things are progressing along normally, one should next notice the calf's feet, legs, nose, and face (with eyes) moving through the vulvar labia with each strong abdominal contraction.
- Visible legs and face
  - Wait 30 minutes and reassess.
  - If parturition is progressing normally, one should next notice the calf's feet, legs, nose, face, and whole head moving through the vulvar labia after a strong abdominal contraction.
- Legs and head
  - Pull!
  - If parturition is progressing normally, the cow should continue to move the calf completely through the birth canal and conclude the delivery. However, many times cows need assistance (manual or mechanical) in completing this stage of parturition.

Differential diagnoses

Perineal laceration (first-, second-, or third-degree), rectovestibular fistula, and rectovaginal fistula

Diagnostics

Wound exploration revealed involvement of skin, vulvar mucous membranes, and the perineal body. The laceration extended into the anal sphincter and created a cloaca (L. "drain"—a common passage for gastrointestinal, urinary, and reproductive systems) connecting the vulva and rectum and disrupting the rectal mucosa.

In orthopedics, a cloaca is an opening in the involucrum (L. *involutio; in* into + *volvere* to roll) encircling the sequestrum (L. a piece of separated dead necrotic bone).

Diagnosis

Third-degree perineal laceration

Treatment

A one-step Goetze procedure was performed (see side box entitled "Surgical repair of third-degree perineal lacerations") using standing sedation and regional anesthesia. The cow recovered uneventfully and continued to be a productive member of the herd.

Anatomical features in ruminants

Introduction

This section describes the relevant anatomical features of the perineal region, specifically focusing on the reproductive tract components involved in third-degree perineal lacerations. The specific tissues covered include the vulva, vestibule, and vagina.

Development and function

The bovine female tubular genital organs consist of the uterine tubes, uterus, vagina, vestibule, and vulva. Except for the vestibule, they originate from the paired paramesonephric ducts, which extend from the gonads to the urogenital sinus at the caudal end of the embryo. Fusion of these genital ducts results in a common cervix and uterus, from which the genital ducts extend like "horns" to a position in close vicinity to the ovaries.

The main function of the vagina, vestibule, and vulva is as part of the birth canal. The vagina primarily serves as a copulatory organ, but also serves to expel urine during micturition and as a passive birth canal. The vestibule is part of both the urinary and genital systems, as the external urethral opening is located here. The vulvar lips are the external portion of the reproductive tract and act as an additional barrier against infection.

CLASSIFICATION OF PERINEAL LACERATION

First-degree laceration—involves mucosa of the vestibule, vagina, and vulva and the skin of the dorsal commissure of the vulva.

Second-degree laceration—the vestibular mucosa and submucosa, the muscularis of the vulva and perineal body including the constrictor vulvae muscle are affected.

Third-degree laceration—complete disruption of the rectovaginal septum, muscularis of the rectum and vagina, the perineal body, and anal sphincter creating a cloaca (Fig. 20.1-2).

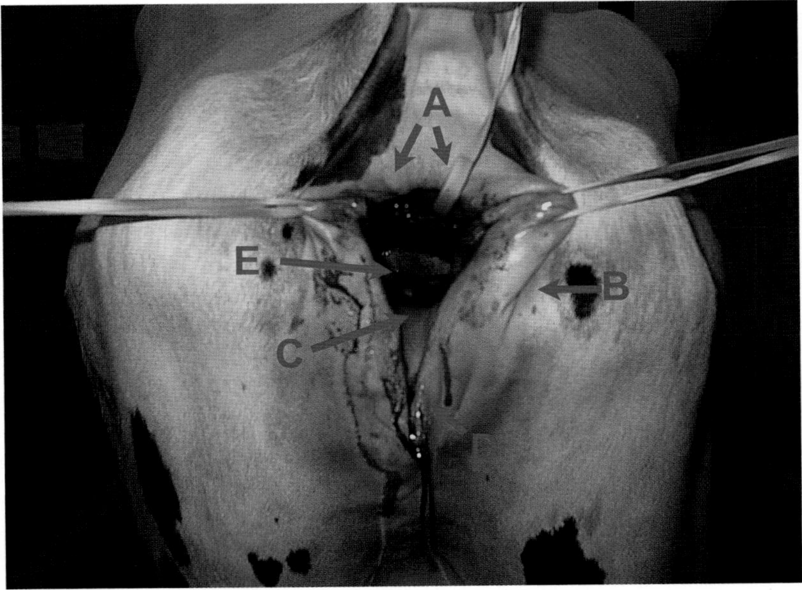

FIGURE 20.1-2 Third-degree perineal laceration. Key: A, dorsal anal sphincter; B, dorsal vulvar commissure; C, vaginal mucosa; D, ventral vulvar commissure; E, rectovaginal septum.

Vulva

The **vulva** is the terminal—and the only external—part of the bovine female genital tract. The vulva consists of paired left and right **labium**, and the 2 labia join to form the **dorsal** and **ventral commissures**. The skin of the labia resembles the common integument and has many sebaceous and sweat glands and hair follicles. The underlying tissue is mostly composed of adipose tissue with small bundles of smooth muscle. These muscles bundles are known as the **constrictor vulvae muscle**, and they help keep the labia apposed.

SURGICAL REPAIR OF THIRD-DEGREE PERINEAL LACERATIONS

While it is possible to correct and surgically repair first- and second-degree perineal lacerations at the time of the occurrence, it is generally recommended to wait to repair third-degree perineal lacerations. A delay of 4–6 weeks allows for a more secure repair because it gives time for the rectal and vestibular mucosa to heal, inflammation to subside, and fibrous tissue to develop. There are 2 surgical techniques that are classically used to repair third-degree perineal lacerations—the Goetze (1938) technique (most commonly used) and the Aanes (1964) technique. The Goetze technique is a one-step procedure while the Aanes technique is a 2-step procedure. Before performing surgery, caudal epidural analgesia is placed to anesthetize the perineal region (see Case 20.2). Stay sutures or tissue forceps are placed at the dorsal commissure of the vulva for tissue retraction to improve exposure of the affected tissues. The vaginal mucosa is then separated from the rectovestibular septum with the dissection starting cranially and extending caudally and laterally along the junction of the rectal and vaginal mucosa. The vaginal mucosa is separated from the underlying tissues at least 4 cm (1.6 in.) cranially, allowing for the subcutaneous tissue beneath to be exposed and subsequently sutured to create a shelf between the rectum and the vagina. The vaginal mucosa is then sutured to the newly created shelf using a purse string-like pattern with absorbable suture material. Lastly, the perineal skin is sutured. First- and second-degree perineal lacerations are generally left to heal by secondary intention (Fig. 20.1-3).

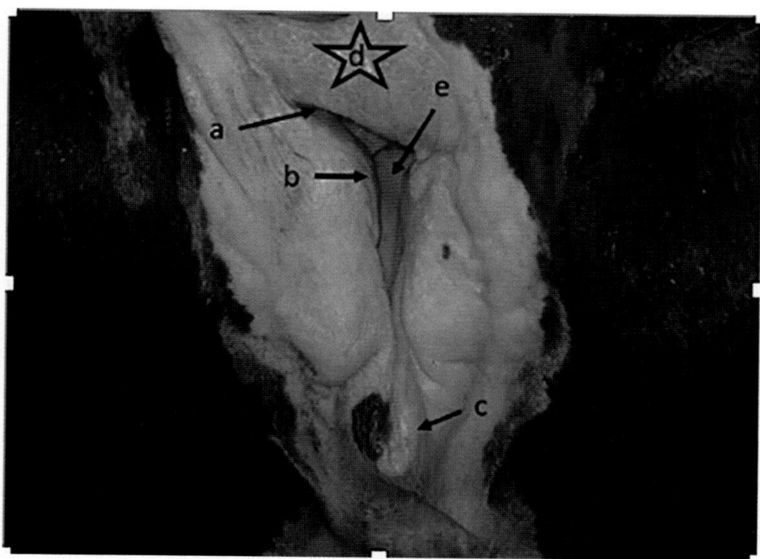

FIGURE 20.1-3 A second-degree perineal laceration of the dorsal vulvar commissure extending cranially and involving the vestibule and vagina that has healed by second intention. Key: a, perineal laceration end point; b, dorsal vulvar commissure; c, ventral vulvar commissure; d in a star, perineal body; e, vaginal mucosa.

Vestibule

The **vestibule** is the anatomical area of the caudal vagina between the external urethral orifice cranially and the labia of the vulva caudally. On the floor of the vestibule lies a blind pouch known as the **suburethral diverticulum** (see Fig. 19.7-8).

The function of suburethral diverticulum is unknown but it serves as a landmark for insertion of a urinary catheter in the cow.

Vagina

The **vagina** is a tubular structure located in the caudal portion of the pelvic canal. It constitutes the copulatory organ, receives the penis, and acts as a receptacle for seminal fluid. The vagina is composed of a poorly organized muscle layer and a well-developed mucosal epithelium. The mucosa of the cranial vagina near the cervix is comprised of columnar epithelium and has a secretory function. In this area, the **cervix** projects into the vagina, creating a crypt or vault known as the **fornix vagina**. The fornix vagina is also lined with columnar epithelium and is highly secretory, especially during estrus.

The bull deposits semen at the fornix vagina during natural breeding.

The caudal vaginal mucosa changes from columnar to stratified squamous epithelium with the stratified squamous epithelium changing in thickness during the estrous cycle.

The squamous epithelium increases in thickness during estrus to mechanically protect the vagina during copulation and to protect the blood vessels in the submucosa from bacteria and other microorganisms introduced during copulation.

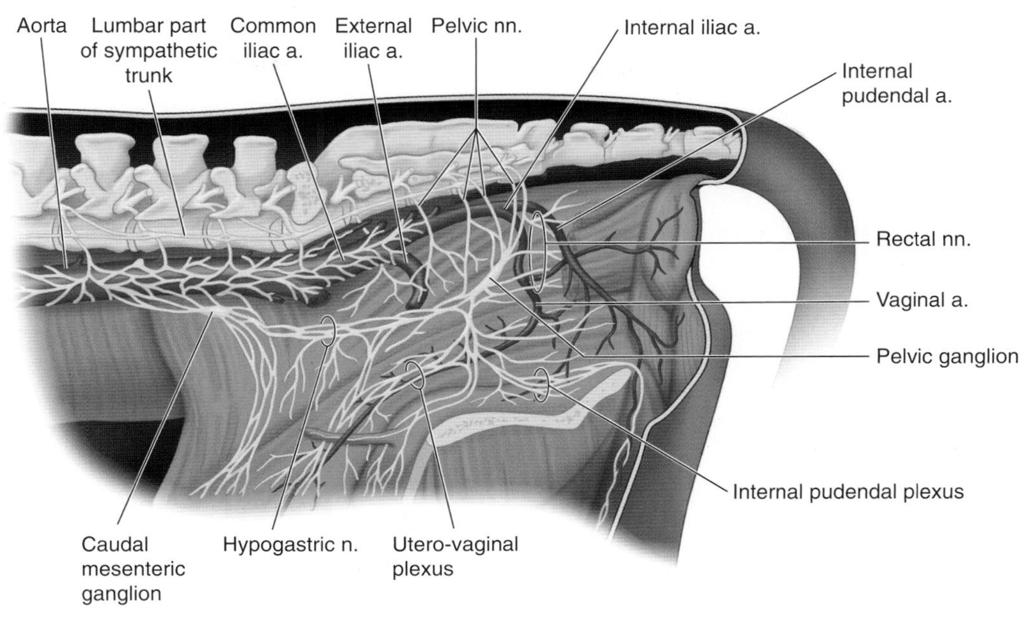

FIGURE 20.1-4 Blood supply and innervation to the bovine perineal area.

Blood supply, lymphatics, and innervation (Fig. 20.1-4)

The **uterine branch** of the **urogenital artery** supplies the caudal part of the uterus, cervix, and parts of the vagina. The urogenital artery is a branch of the **internal iliac artery**. The remaining parts of the genital tract are supplied by branches of the urogenital and **internal pudendal arteries**. Venous return is by various satellite vessels that empty into the **caudal vena cava**.

Lymph from the vagina and vulva is drained to the **medial iliac** and **external sacral lymph nodes**. Lymph from the ventral commissure of the vulva may also pass to the **superficial inguinal lymph nodes**.

Innervation of the female genital tract is both autonomic and voluntary. Parasympathetic innervation comes from the lumbosacral plexus and reaches the genital tract via the **pelvic nerves**. Sympathetic innervation comes from the caudal mesenteric ganglion and plexus and reaches the organs via the **hypogastric nerves** and **pelvic plexus**. The **pudendal**, **caudal rectal**, and **pelvic nerves** are sacral in origin and supply sensory and motor fibers to the pelvic viscera and perineum. The pudendal nerve divides into **deep perineal** and **distal cutaneous** branches; the deep perineal branch supplies both visceral and somatic structures of the caudal pelvic region. The distal cutaneous branch supplies structures of the ventral perineum before becoming superficial and emerging from the ischiorectal fossa to innervate the vulva and perineal skin.

Selected references

[1] Parkinson TJ, Noakes DE, Veterinary reproduction and obstetrics, 10th ed., Elsevier; 2018.
[2] Proudfoot KL, Jensen MB, Heegaard PMH, von Keyserlingk MAG. Effect of moving dairy cows at different stages of labor on behavior during parturition. J Dairy Sci 2013;96:1638–46.
[3] Roberts SJ. Veterinary obstetrics and genital diseases (theriogenology). Parturiton. Woodstock, VT: Published by Author; 1986. p. 245–59 [Chapter VI].
[4] Schummer A, Nickel R, Wolfgang OS. The viscera of the domestic mammals. Female genital organs. New York, NY: Published by Springer-Verlag; 1979. p. 351–89.
[5] Senger PL. Pathways to pregnancy and parturition, 2nd ed. Concurrent Conceptions Inc.; 2005.

CHAPTER 20

CASE 20.2

Dystocia With Cesarean Section

Andrew Niehaus
Veterinary Clinical Sciences, The Ohio State University College of Veterinary Medicine, Columbus, Ohio, US

Clinical case

History

A 21-month-old Holstein cross heifer was presented for dystocia (Gr. *dys-* difficult + Gr. *tokos* birth). The owner reported that she was in labor when he came home from work approximately 3 hours earlier. At that time, the amniotic sac was ruptured, and he saw 2 hooves and the nose protruding from the vulva but nothing else was seen (Fig. 20.2-1). After observing for 15 minutes, no progress was noted so he put calving chains on the legs and pulled for another 30 min without progression to delivery.

Physical examination findings

 Upon the first examination, the heifer was bright and alert while periodically straining. The heifer was mildly tachycardic, moderately tachypneic, and mildly febrile. The remainder of the physical examination parameters were within normal limits.

Two feet were protruding from the vulva, and the muzzle of the calf could be seen. The vulva was aseptically prepared for a vaginal examination. Vaginal palpation determined that the cervix was fully

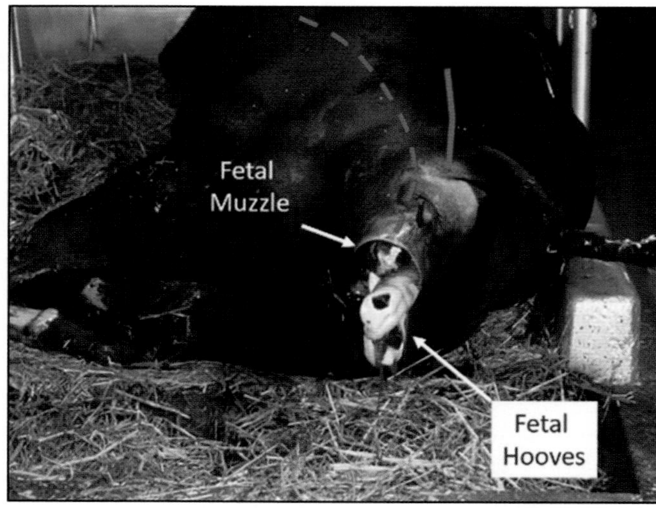

FIGURE 20.2-1 This heifer is in stage 2 labor and is actively trying to deliver a calf. The calf is in anterior presentation (head and thoracic limbs first). The hooves and muzzle are apparent on examination. The calf's dorsum, indicated by the *dashed red line*, is against the dam's sacrum (marked by the *solid red line*), signifying a "dorsosacral" position. The posture of this calf is normal because both the thoracic limbs and neck are extended.

Comparative Veterinary Anatomy: A Clinical Approach. https://doi.org/10.1016/B978-0-323-91015-6.00102-3

dilated. The calf was in anterior presentation, dorsosacral position, with both thoracic limbs and neck fully extended. The head was palpable, and a weak suckle response was elicited from the calf by placing fingers in its mouth.

Calving chains were placed on the thoracic limbs and traction was applied to deliver the calf vaginally. When traction was applied on the chains, the head and neck would retroflex. The calf was rotated at an oblique angle to maximize the width of the dam's pelvis at a diagonal, followed by another attempt at delivering the calf vaginally. With no appreciable progress after 20 minutes of traction, a Cesarean section was recommended to improve the chances of a viable calf.

"The calf was rotated at an oblique angle to maximize the width of the dam's pelvis at a diagonal, followed by another attempt at delivering the calf vaginally." The dam's pelvis should be thought of as a rectangle, not a circle. Because of this, the diagonal has a greater length than the width or height (Fig. 20.2-2).

Diagnosis

Fetomaternal mismatch leading to dystocia

Treatment

A caudal epidural was performed (Fig. 20.2-3) to improve the heifer's comfort and minimize straining. This was followed by a paravertebral nerve block to anesthetize the left paralumbar fossa by blocking the last thoracic, and 1st and 2nd lumbar spinal nerves (T13, L1, and L2). A vertical 40-cm (15.7-in.) incision was made in the left paralumbar fossa to enter the abdomen. The rumen was manually moved cranially and the uterus was identified. The hock and fetlock of the fetus's right pelvic limb were grasped and used to pull the calf and uterus to the abdominal incision before making the hysterotomy (Gr. *hystera uterus* + Gr. *temnein* to cut) incision. The uterus was partially exteriorized from the abdomen, and the fetus's right metatarsus was used to hold the uterus at the incision (Fig. 20.2-4). An approximately 40-cm (15.7-in.) long incision was made through the uterine wall, the overlying amniotic sac was manually ruptured exposing the calf's pelvic limbs, and sterile calving chains were placed on the legs. Traction was applied to the calving chains in a dorsocaudal direction, delivering a 49-kg (108-lb), live bull calf. The uterus was closed using a single-layer inverting suture pattern followed by anatomical closure of the body wall in three layers.

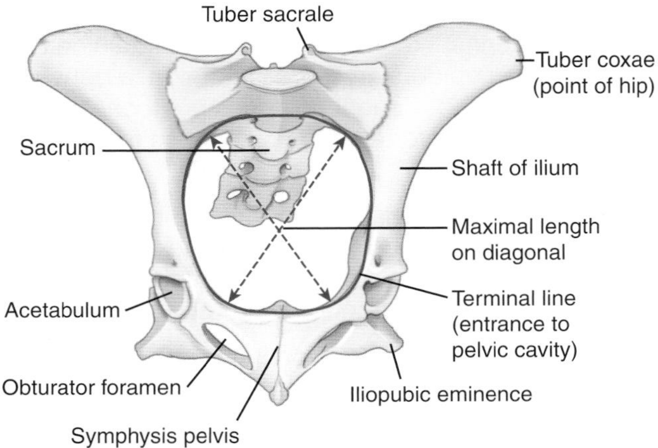

FIGURE 20.2-2 The diagonal of the pelvic opening (*dashed blue arrows*) has a greater length than the width or height. Therefore, lining up the fetus with the diagonal will maximize the room for the fetus to be delivered.

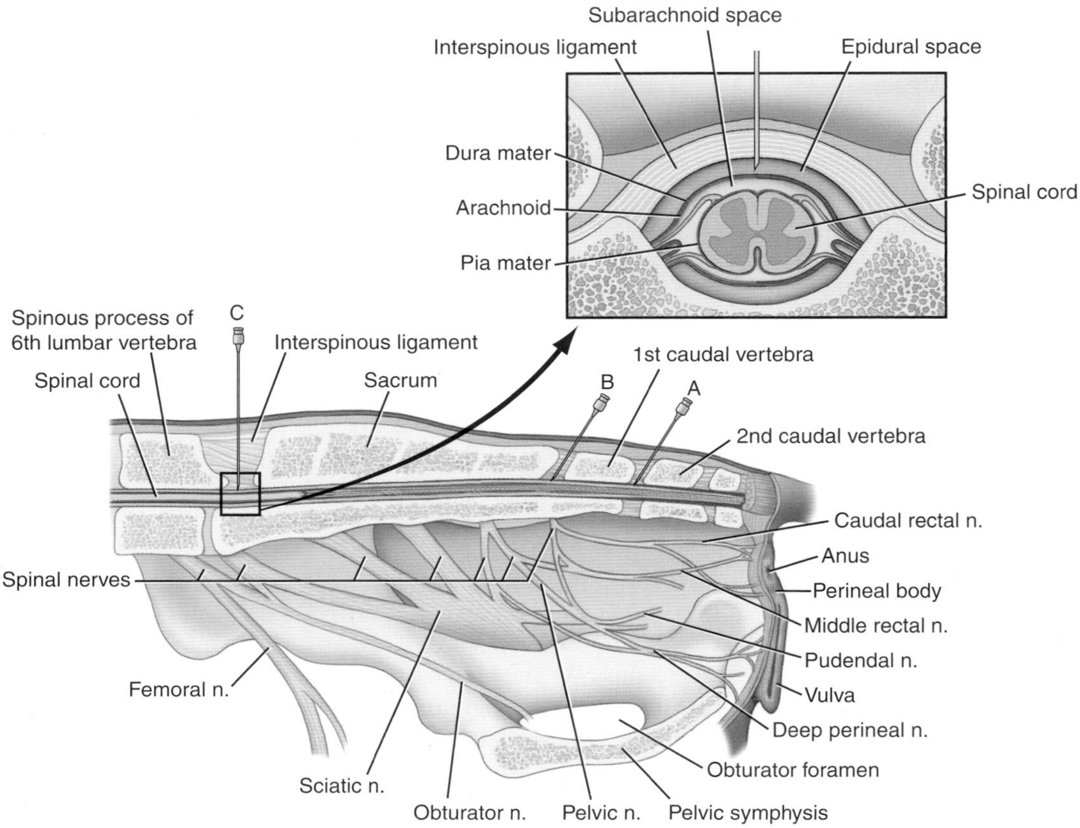

FIGURE 20.2-3 Longitudinal section of caudal vertebral bodies showing locations for performing lumbosacral epidural (*C*), or caudal epidural (*A, B*). The inset is a cross-sectional diagram at the lumbosacral level. Note that the needle is inserted to the epidural space, but at this location, it is possible to penetrate the subarachnoid space, which lies below the epidural space. Also shown are the nerves of the pelvis demonstrating their spinal root origins. (Adapted from Dyce, Sack, Wensing Textbook of Veterinary Anatomy, 2010.)

EPIDURAL ANESTHESIA

Low-volume—e.g., 1–2 mL of 2% lidocaine per 100 kg (220 lbs) or 8–15 mL total for average cow—epidural anesthesia is useful to desensitize the perineal region and decrease straining during obstetrical manipulation. The most common location to perform an epidural in cattle is the space between the sacral and coccygeal vertebrae—i.e., a caudal epidural (Fig. 20.2-3). Caudal epidurals are technically easy to perform with a low risk for complications. The alternative location for epidural placement is the lumbosacral location.

The goal is to anesthetize the nerve roots that innervate the somatic structures of the caudal pelvic region, including the ventral perineum, the vulvar and perineal skin, the vagina, the clitoris, and some of the skin to the udder. Low-volume epidural anesthesia does not provide anesthesia to the body wall appropriate for flank surgery. The pudendal nerve, the caudal rectal nerve, and the pelvic nerves originate entirely from sacral spinal nerves (S_3–S_5); these nerves are anesthetized with a caudal epidural.

Remember that the obturator nerve has nerve roots originating from spinal nerve L_4–L_6 and innervates the adductors of the hind legs. The femoral nerve originating from L_4–L_6 innervates the quadriceps unit. The sciatic nerve originates from L_6-S_2 and gives rise to the peroneal and tibial nerves, which innervate extensors of the digit and hock, respectively (Fig. 20.2-3). If these spinal nerves are anesthetized during an epidural, there is a high likelihood that the patient will experience nerve deficits to its pelvic limbs, increasing the chances of recumbency. Factors that increase how far cranial the local anesthetic migrates following an epidural include the location of the epidural (lumbosacral vs. sacrococcygeal) and the volume of local anesthetic injected.

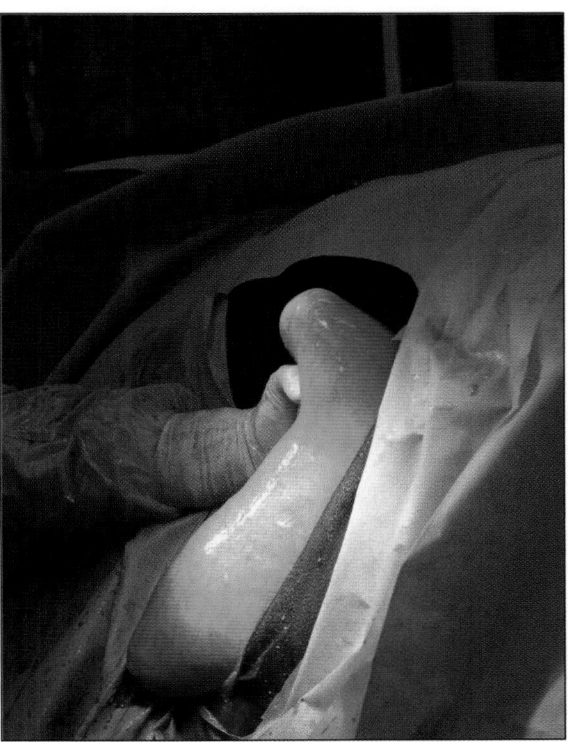

FIGURE 20.2-4 The tarsus and the metatarsophalangeal joint of the calf were palpated through the uterus and used to draw the fetus to the incision. The rigid metatarsus with the tarsus and metatarsophalangeal joint become a "handle" used to exteriorize the fetus and uterus to the flank incision. If the metatarsus can be maneuvered through the incision, it becomes self-retaining in the incision.

Anatomical features in ruminants

Introduction

Cesarean sections (C-section) are one of the most common surgical procedures performed in cattle. Successful C-sections require a basic understanding of surgical anatomy. The procedure is most frequently performed in the standing cow using a local or regional anesthetic block. Therefore, an understanding of the neuroanatomy and a familiarity with the abdominal viscera and reproductive structures are important. The major relevant structures are covered here as are the muscles of the body wall; the supporting framework, nerves, and vessels of the uterus; and the omentum.

PROXIMAL AND DISTAL PARAVERTEBRAL NERVE BLOCKS

The proximal and distal paravertebral nerve blocks are regional blocks used to anesthetize the flank in the area of the paralumbar fossa. The nerves that innervate this area are the last thoracic and the first 2 lumbar nerves (T_{13}, L_1, and L_2). “Proximal” and “distal” refer to the proximity to the spinal cord at which the nerves are blocked. The proximal paravertebral block anesthetizes the nerves immediately after they emerge from the vertebral column. The distal paravertebral nerve block anesthetizes the nerves at the lateral edge of the transverse processes of the lumbar vertebrae. Regardless of which block is chosen, the same area of the flank should be anesthetized (Fig. 20.2-5).

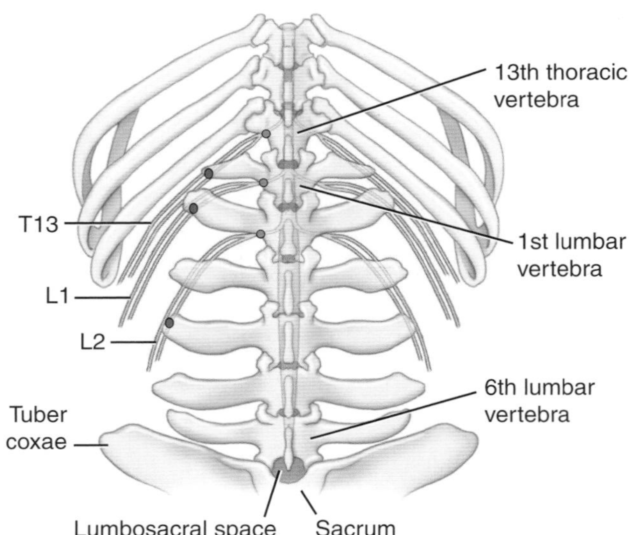

FIGURE 20.2-5 Neuroanatomy for performing a paravertebral nerve block. The paralumbar fossa is innervated by the nerves T13, L1, and L2. Blocking these nerves, either immediately as they emerge from the intervertebral foramen (a proximal paravertebral nerve block, *blue circles*) or more laterally at the lateral edge of the transverse processes (a distal paravertebral nerve block, *red ovals*), anesthetizes the paralumbar fossa.

Function

The muscles and fascia of the abdominal wall provide structural support to the bovine abdomen. It is important that these structures are anatomically apposed to restore the structural integrity to the abdomen during abdominal surgical procedures.

The physiological functions of the uterus are to provide oxygenation, nutrition, and a hospitable environment for the fetus during gestation. The uterus is elastic during the development of the fetus and is mainly responsible for expulsion of the fetus during normal parturition. The ovaries are the primary reproductive organs in the female and produce oocytes and sex hormones. The oviduct, also referred to as the salpinx, uterine tube, or fallopian tube, connects the ovaries and the uterus serving as a channel for transporting the oocytes from the ovaries to the uterus. The oviducts also serve as the site for fertilization.

OTHER BLOCKS FOR FLANK SURGERY

Line blocks and inverted "L" blocks are commonly employed in the field to achieve local anesthesia for surgery. These blocks are technically easy to perform; however, a larger dose of local anesthetic is required. With the line block, edema from local anesthetic at the surgical site can compromise tissue (leading to friability) and delay healing of surgical wounds. This block affords less versatility during surgery with regard to changing the surgical location or size of the incision. An inverted "L" block is performed by injecting local anesthetic along the ventral aspects of the lumbar transverse processes and ventrally along the caudal border of the last rib. The goal is that all nerves innervating the paralumbar fossa are desensitized by infiltrating local anesthetic along these lines (Fig. 20.2-6).

FLANK CELIOTOMIES

Most flank celiotomies in cattle are done in the paralumbar fossa area. This is the triangle that is bounded by the tuber coxae caudally, the transverse processes of the lumbar vertebrae dorsally, and the 13th rib cranially (Fig. 20.2-6). This approach provides good access to much of the abdominal viscera, including the reproductive structures. The typical approach for C-sections in ruminants is via the left paralumbar fossa. The left flank offers an advantage because the rumen lies on the left side of the abdomen and aids in retaining the abdominal GI viscera as the large, gravid uterus is exteriorized to the abdominal incision (Fig. 20.2-7).

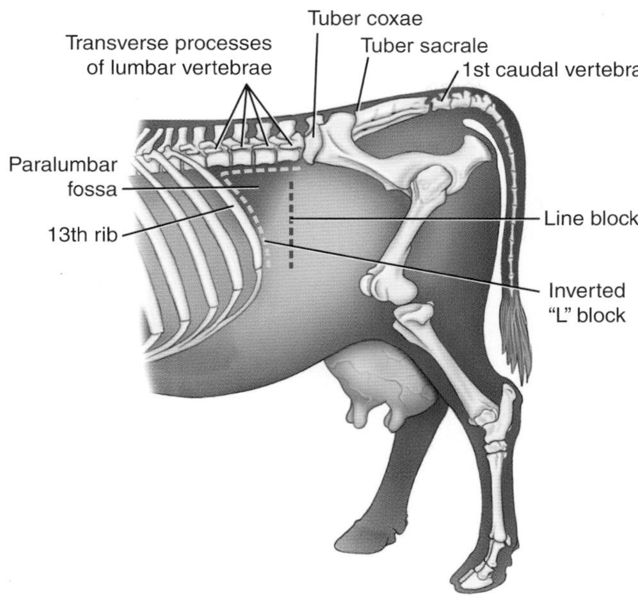

FIGURE 20.2-6 The landmarks for the paralumbar fossa (caudally the tuber coxae, dorsally the transverse processes of the lumbar vertebrae, and cranioventrally the last rib). Also shown are locations for the injection of local anesthetic for performing the inverted "L" block (*dashed blue line*) and the line block (*dashed yellow line*).

Body wall in the flank

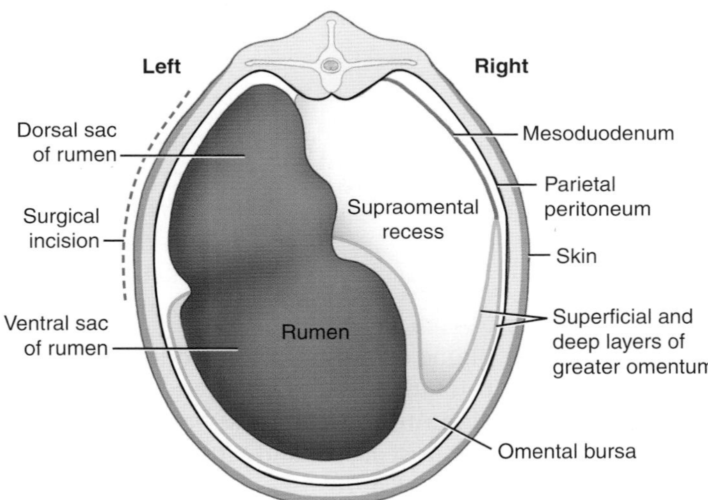

FIGURE 20.2-7 Cross-sectional image of the bovine abdomen illustrating omental anatomy. The omental sling is comprised of deep and superficial leaves. The rumen, predominantly on the left side of the abdomen, acts as a "visceral retainer," keeping small intestine within the supraomental recess.

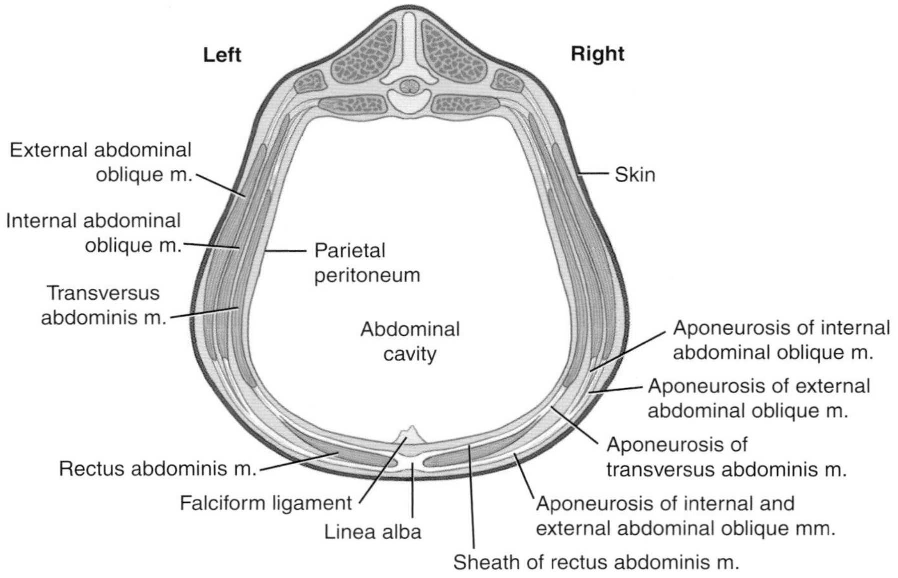

FIGURE 20.2-8 Cross-sectional figure of the bovine abdomen depicting the body wall layers. An incision in the paralumbar fossa penetrates through the skin, external abdominal oblique, internal abdominal oblique, and the transverse abdominis muscles. (Adapted from Dyce, Sack, Wensing Textbook of Veterinary Anatomy, 2010.)

The body wall in the flank area consists of 3 major muscles that provide a support to the abdomen. Superficially, the **external abdominal oblique muscle** is located just beneath the skin and subcutaneous tissues, and is the first large muscle encountered during a flank procedure. The external abdominal oblique muscle attaches dorsally to the thoracolumbar fascia and courses ventrally where it forms an aponeurosis with the fascia of the rectus abdominis m. and the linea alba. The external abdominal oblique m. is identified by its fiber orientation. Cranially, the fibers run in a craniodorsal-to-caudoventral direction except in the flank area where the fibers course almost horizontally.

> The fascia of the external abdominal oblique m. is substantial and is considered the "holding layer" when suturing a flank incision. It is important that this fascia is securely closed to maintain the integrity of the body wall and prevent evisceration.

Immediately deep to the external abdominal oblique m. is the **internal abdominal oblique muscle**. The fibers of the internal abdominal oblique m. course in a caudodorsal-to-cranioventral direction. Deep to the internal abdominal oblique muscle is the **transverse abdominis muscle**. These fibers have a dorsoventral orientation. This muscle layer is thinner than the other superficial muscle layers, and care should be taken when incising this muscle because the parietal peritoneum lies directly beneath (Fig. 20.2-8).

The dorsal portion of the abdominal cavity is under negative pressure in the standing cow compared with the atmosphere, and an audible negative pressure (sucking) sound is heard as air enters the abdominal cavity when the peritoneum is incised. The peritoneum provides an air-tight seal to the abdominal cavity.

> Air entering the abdominal cavity during a celiotomy escapes through small defects in the peritoneum after closure and commonly results in a benign subcutaneous emphysema (Gr. an inflation, accumulation of air in tissues or organs) post-operatively.

Omental sling and supraomental recess

The **greater omentum** is a complex and robust structure within the bovine abdomen. It is comprised of 2 sheets ("leaves")—superficial and deep—that are continuous with each other, creating a potential space in the middle called the **omental bursa**. Together, the superficial and deep leaves attach to the mesoduodenum on the right side and to the rumen on the left. This effectively cradles the abdominal GI viscera and is therefore often referred to as the **omental sling**. The omental sling extends from the cranial flexure of the duodenum along the descending duodenum to its caudal flexure and is open on the caudal aspect. The **supraomental recess** is the space contained within the omental sling where most abdominal viscera reside (Fig. 20.2-7).

With a left flank incision, the rumen confines viscera within the supraomental recess from the incision to help prevent evisceration. Therefore, this is the preferred surgery site for standing C-section.

Uterus

The ruminant **uterus** is bicornuate, meaning it is comprised of 2 **horns**. A gravid uterus can contain a fetus in either horn. The nongravid bovine uterine horns curl ventrally, caudally, and laterally, and connect to the **ovaries** at the caudoventral aspect of the pelvic inlet. The tips of the horns are attached to the ovaries through small convoluted tubular structures called **oviducts**. The uterine horns merge together and form the **uterine body** caudally. The relative size of the uterine body to the horns is deceptively small because a large part of the apparent uterine body is the 2 horns lying side-by-side and contained within a common serosal and muscular covering. The actual length of the uterine body is approximately 3 cm (1.2 in.) and is only apparent when the uterus is opened. The nongravid uterus may be contained almost entirely within the pelvic inlet. As a fetus develops within the uterus, the horns stretch cranially and ventrally into the dependent portion of the abdomen. The gravid uterus may or may not be confined within the omental sling.

If the uterus is within the omental sling, the greater omentum will obstruct the surgeon's access to the uterus if performing a ventral approach. In this case, the uterus must be moved cranially to remove it from the omental sling before performing a hysterotomy.

Cervix

The bovine **cervix** is an approximately 8–10-cm (3.1–3.9-in.)-long fibromuscular structure positioned at the caudal-most aspect of the uterus. The cervix forms the connection between the uterine body and the vagina. It consists of several interlocking folds that form irregular surface projections on its interior aspect.

These interlocking folds make it difficult to pass instruments through the cervix during procedures such as artificial insemination and embryo transfer.

Uterine and ovarian ligaments

The uterine and ovarian ligaments provide a support for the female reproductive organs and maintain their spatial relationships with each other. The uterus, ovaries, and oviducts are supported by the **mesometrium**, **mesovarium**,

FLANK INCISION ANGLE

Accepted techniques for flank incisions include vertically directed incisions and incisions that are angled from a caudodorsal-to-cranioventral direction. An oblique incision is oriented in the direction of the gravid uterus and therefore makes exteriorization of the uterus easier, especially with large calves (Fig. 20.2-9). The disadvantage of the oblique incision is that the weight of the abdominal viscera (mainly the rumen) tends to pull ventrally on the ventral aspect of the incision and can make closure more challenging.

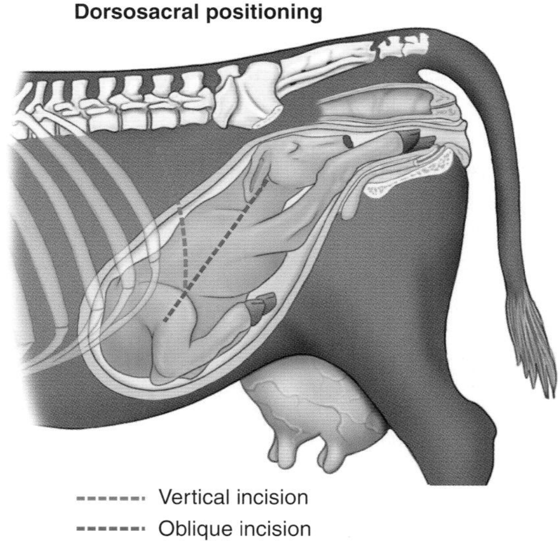

FIGURE 20.2-9 Left flank approaches to access the uterus for a C-section. The *dashed red line* represents the angle of the incision for an oblique approach. The incision is angled in a caudodorsal-to-cranioventral direction. Notice how the oblique incision lines up more closely with the direction of the uterus, facilitating exteriorizing the uterus. The *dashed blue line* represents the location for a vertical abdominal approach.

and **mesosalpinx**, respectively, and are collectively referred to as the **broad ligament** (Fig. 20.2-10). The cranial border of the mesovarium is known as the **suspensory ligament of the ovary**. The broad ligament attaches dorsally to the caudodorsal abdomen.

The dorsal and ventral **intercornual ligaments** attach the left and right uterine horns near the bifurcation. The **proper ligament of the ovary** attaches the ovary to the mesometrium of the uterus.

Uterine blood supply increases significantly during gestation. In the third trimester, blood flow to the gravid and nongravid horns can exceed 14 and 6.5 L/min, respectively.

Blood supply and innervation of the female reproductive tract

The blood supply to the uterus is plentiful with considerable collateral circulation. The blood flow to the uterus is supplied primarily by the **uterine artery**, a branch of

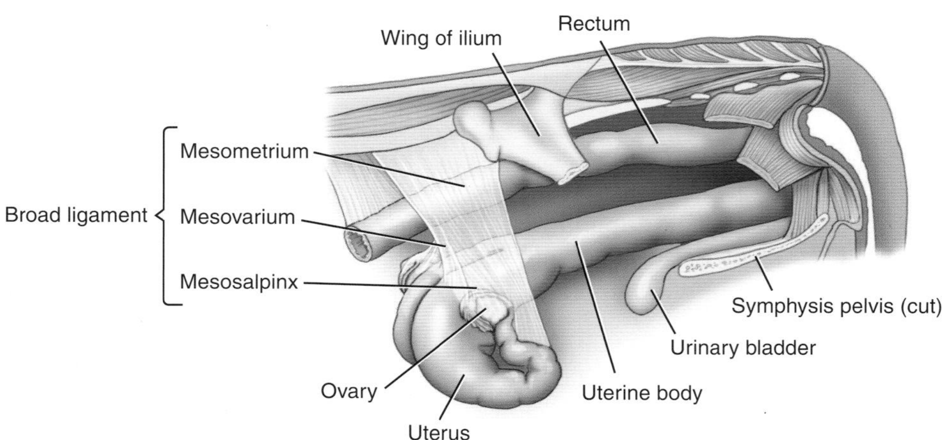

FIGURE 20.2-10 Broad ligament of the cow. The broad ligament is formed from the confluence of the mesometrium, the mesovarium, and the mesosalpinx.

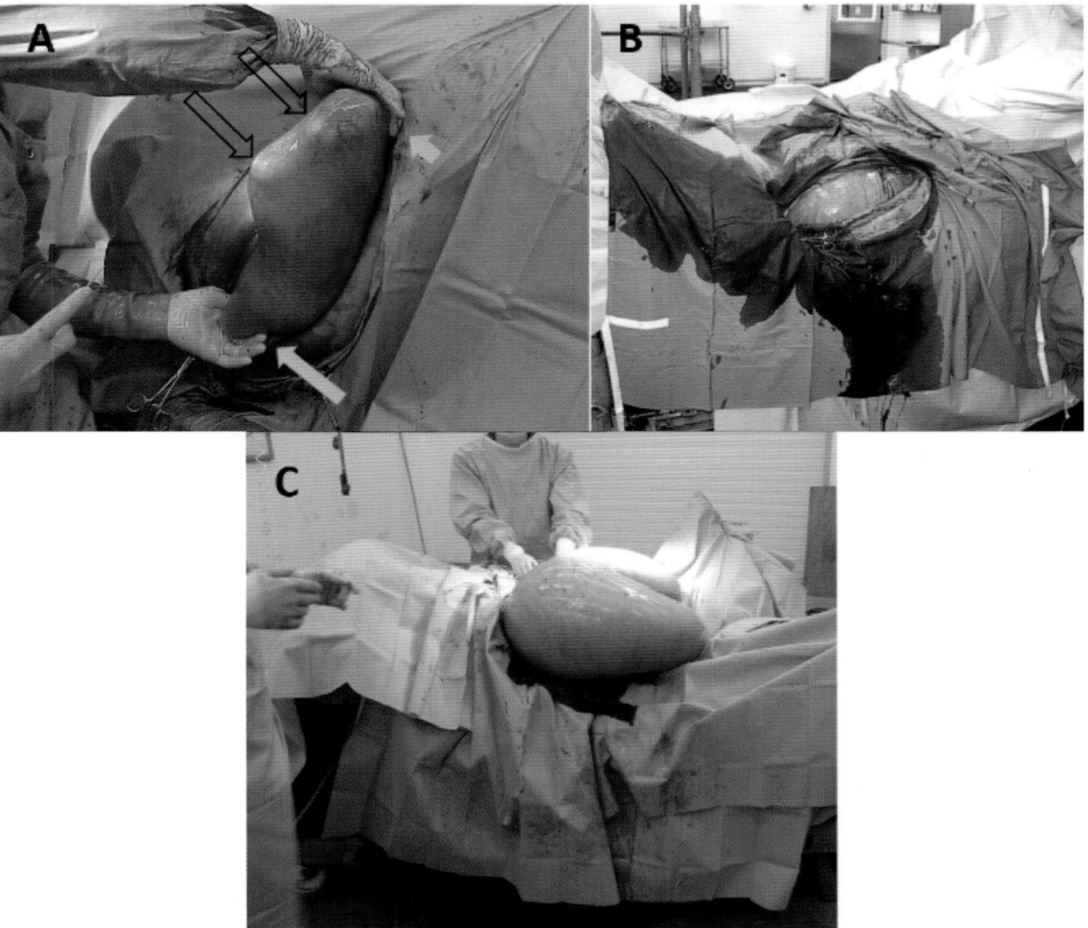

FIGURE 20.2-11 (A) Standing left flank C-section with the metatarsus of a calf exteriorized within the uterus before making a hysterotomy incision. The surgeon exteriorizes the uterus by holding onto the fetus's calcaneus with the left hand (*long white arrow*) and the metatarsophalaneal joint with the right hand (*short white arrow*). The tips of both pelvic limb hooves are compressed against the uterine wall creating 2 small, pale white areas (*open arrows*). (B) Intra-operative picture of a cow in dorsal recumbency undergoing a C-section via a ventrolateral approach. (C) Ventral midline C-section with both uterine horns exteriorized before the hysterotomy incision. The main advantage of a ventral approach for C-section is the ability to exteriorize more uterus, thus minimizing abdominal contamination. (Courtesy of www.medvet.umontreal.ca/mediatheque.)

EXTERIORIZING THE UTERUS DURING C-SECTION

In cases of elective C-sections, C-sections with a closed cervix, or C-sections in farm animals that have had a minimal fetal manipulation before the procedure, the uterine environment is likely clean enough that contamination of the abdomen from fetal fluid is innocuous. However, significant contamination of the abdominal cavity with uterine fluid can result in peritonitis and it is therefore important to exteriorize as much of the uterus as possible before making the hysterotomy incision. The ventral approach allows substantially more uterus to be exposed compared with the standing flank approach (Fig. 20.2-11A–C) and is the recommended approach when the uterine environment is severely contaminated.

Fetal presentation also affects the surgeon's ability to bring the fetus to the surgical incision. The pelvic limbs of a calf are easier to grasp than the thoracic limbs because the calcaneus provides a convenient "handle" to grasp the uterus (Fig. 20.2-11A–C). In anterior presentation—i.e., thoracic limbs engaged in the birth canal—the pelvic limbs are accessible to pull the uterus to the incision vs. in posterior presentation—i.e., backward—the thoracic limbs and head are used to pull the fetus to the incision.

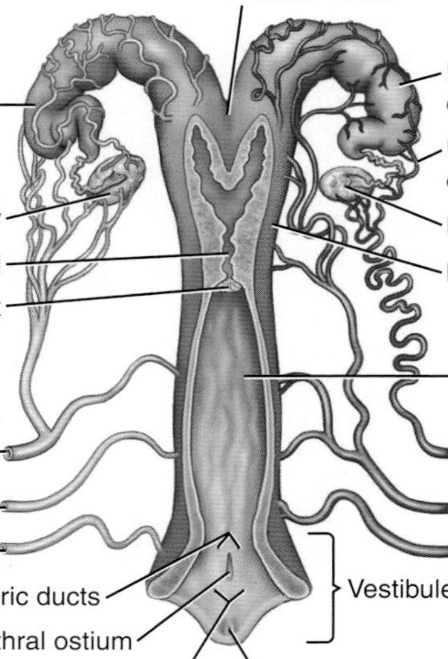

FIGURE 20.2-12 Semi-schematic dorsal view of the blood supply to the bovine reproductive tract (cow). The arteries are depicted on the right side, and the veins on the left side.

the internal iliac artery, and travels within the broad ligament. Several smaller vessels branch from the uterine artery at the surface of the uterus. The uterine a. anastomoses with the **ovarian artery** at the tip of each uterine horn and the **uterine branch of the vaginal artery** caudally.

The ovarian a. is considerably smaller than the uterine. It provides blood flow to the ovary and has a **uterine branch** that travels into the uterine horn and forms an anastomosis with the uterine artery. The ovarian a. is torturous and has extensive contact with the **ovarian vein**, enabling transfer of sex hormones, produced in the ovary, from venous to arterial circulation.

> The antimesometrial border of the uterus is less vascularized than the mesometrial border, and therefore less prone to bleeding when incised during hysterotomy.

Venous drainage of the uterus is provided mostly by the **uterine branch of the ovarian vein** with minor contributions by the **accessory vaginal vein** and the **uterine vein** (Fig. 20.2-12).

Lymph from the reproductive organs generally goes directly to the **medial iliac** and **sacral lymph nodes** located near the bifurcation of the aorta. Lymph from the medial iliac lymph nodes enters the **lumbar lymphatic trunk** at the point of its origin. The **deep inguinal/iliofemoral lymph nodes** may receive lymph from the reproductive organs.

CAESAREAN SECTION IN EWES

Dystocias requiring C-section in sheep also occur. A flank C-section can be performed standing as described for cattle or under general anesthesia. A left flank incision is preferred to the right flank for the same reasons given for cattle. A ventral approach, under general anesthesia, is preferred to minimize the restraint challenges and provides the surgeon access to both uterine horns compared with the flank approach. Like cattle, the ventral approach also allows improved exteriorization of the uterus, minimizing abdominal contamination. When performing a ventral approach, it is important to recognize and avoid the midline branch of the subcutaneous abdominal v. because of the robust blood flow through this vein in an advanced pregnant ewe and the potential for consequential blood loss (Fig. 20.2-13).

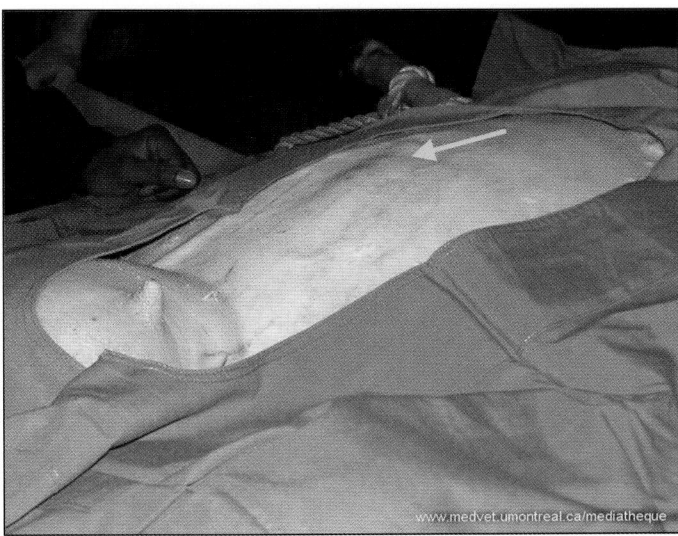

FIGURE 20.2-13 Ewe prepped for ventral midline C-section. Note the midline branch of the subcutaneous abdominal (cranial superficial epigastric) vein (*yellow arrow*). This vein has a large volume blood flow in the advanced pregnant ewe. Cranial is to the right in this image.

The smooth muscle of the uterus is innervated by sympathetic nerves. The post-ganglionic sympathetic nerves originate from the last thoracic and 1st lumbar segments of the spinal cord and travel through the **hypogastric nerve** to the pelvic plexus. At the **pelvic plexus**, the pre-ganglionic and post-ganglionic neurons synapse and continue to supply the uterine muscles.

Vagina and vestibule

The **vagina** is a tubular structure positioned caudal to the cervix, which serves as a connection between the uterus and the vestibule. Although the **vestibule** is frequently, and incorrectly, referred to as the caudal part of the vagina, it is a separate structure that begins at the urethral opening (the **external urethral orifice)** and extends caudally to the vulva. In the young, virgin heifer, a thin mucosal fold, the **hymen**, separates the vagina from the vestibule which ruptures spontaneously or following coitus. The length of the vagina is approximately 3 times that of the vestibule. The **clitoris** is present on the ventral aspect of the caudal vestibule. Both the vagina and vestibule serve in copulation and parturition. The most cranial part of the vagina is termed the **fornix** and lies within the abdominal cavity, specifically the **rectogenital pouch**. The fornix is an area where the vagina retroflexes before its cervical attachment. Most of the vagina is retroperitoneal. The vagina and vestibule are bounded dorsally by the rectum; ventrally by the bladder, urethra, and pubis; and laterally by the right and left shafts of the ilium (Fig. 20.2-14).

> The fornix serves as a site for entry to the abdomen in surgical approach termed "colpotomy" (Gr. *kolpos* vagina + Gr. *tomē* a cutting).

The vestibule is a component of both the reproductive and urinary tracts. It is less distensible than the vagina. It contains glands that secrete mucous that aids in coitus and parturition. The **major vestibular glands lie** in a depression caudolateral to the urethral opening (Fig. 20.2-12). The vestibular mucosa is typically darker in this region covering the glands.

The vagina is a muscular organ and is similar in structure to the uterus. It is lined by stratified squamous epithelium. Glands are present in the cranial vagina. The vagina is highly distensible.

> The vagina can be greatly stretched during parturition. This distensibility contributes to its susceptibility to prolapse through the vulva.

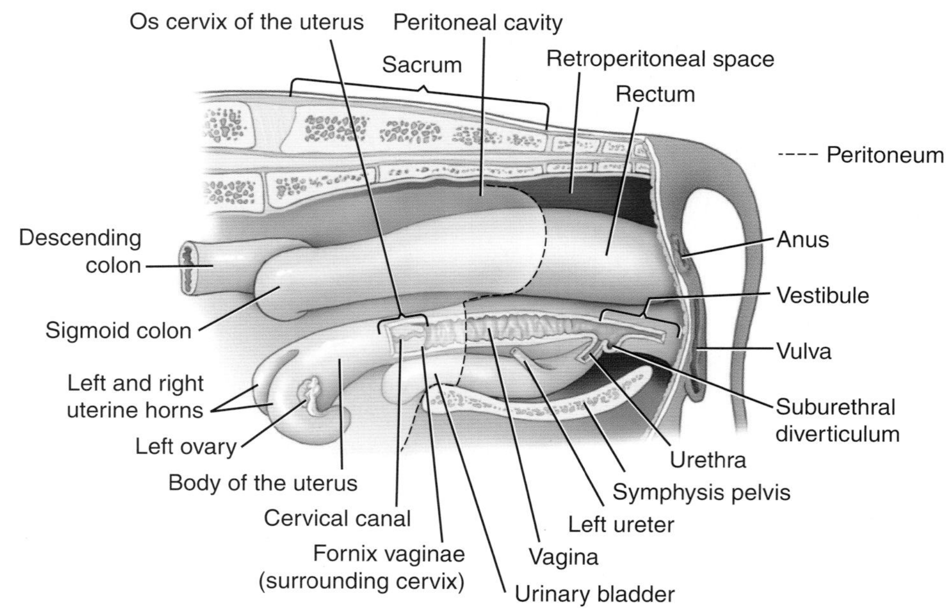

FIGURE 20.2-14 Bovine vaginal and vestibular anatomy. Note the cranial part of the vagina is in the peritoneal cavity, while most remains in a retroperitoneal position (distinction shown by the *dashed line*). The suburethral diverticulum is a blind pouch and immediately caudal and ventral to the urethral opening at the intersection of the vagina and vestibule.

The suburethral diverticulum can make catheterization difficult when attempting to pass a urinary catheter.

The vaginal artery can rupture during normal parturition and dystocia, causing life-threatening bleeding if not promptly controlled.

The peritoneal-covered dorsal fornix region is relatively avascular, making it a preferred location for performing a colpotomy.

The **urethra** is a short structure in the female and opens on the floor at the junction between the vagina and the vestibule. It opens with a blind pouch termed the **suburethral diverticulum**. The suburethral diverticulum opens immediately caudal to the true urethral opening (external urethral orifice). Both the diverticulum and the urethra are enclosed within the **urethralis muscle**. The urethralis muscle is innervated via the pudendal nerve and separate autonomic innervation.

Blood supply to the vagina and vestibule are by the **uterine artery** and the smaller **vaginal artery**. Venous drainage is by the **ovarian**, **vaginal**, and the **accessory vaginal veins** (Fig. 20.2-12). Most of the blood supply tracks on the ventral and lateral sides of the vagina.

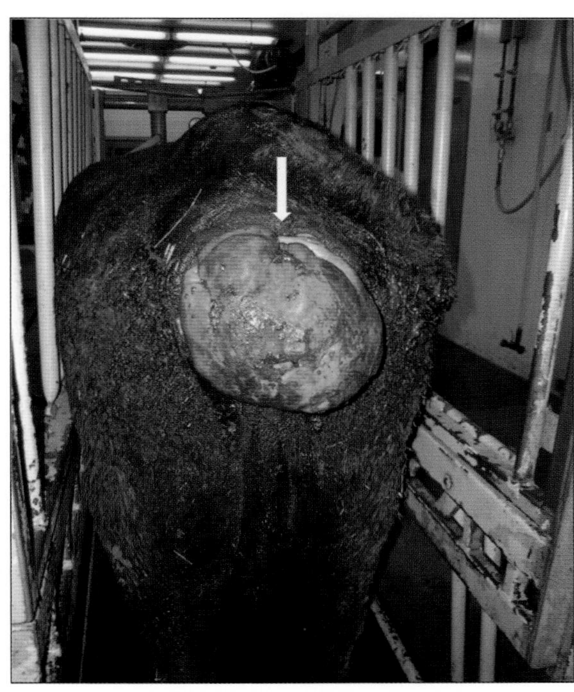

FIGURE 20.2-15 Vestibular and vaginal prolapse in a beef cow. Most of the vestibule and vagina are protruding through the vulva. The cervix is apparent dorsally (*arrow*).

VAGINAL PROLAPSE

Vaginal prolapse is a condition that occurs when the vaginal wall projects caudally through the vulva (Fig. 20.2-15), It usually includes a part of the vestibule and can vary in severity.

Risk factors that increase intra-abdominal or intrapelvic pressure include late pregnancy, ruminal distension, intrapelvic fat, prolonged recumbency, decreased resistance of the perineum due to hormonal-induced relaxation before parturition, and multiparous (> 2 pregnancies with a live fetus) cows.

As a vaginal prolapse becomes chronic, the vaginal tissue (highly vascular) becomes edematous, with further compromise of venous return, resulting in a worsening of clinical signs. Rectal prolapse may accompany a vaginal prolapse, and the bladder may become entrapped in a severe vaginal prolapse causing dysuria (Gr. *dys-* difficult, abnormal + Gr. *ouron* urine + Gr. *-ia* state or condition).

Treatment involves reducing the prolapse and then retaining the vaginal tissue within the pelvic cavity using epidural anesthesia to both decrease straining and provide analgesia.

Hypertonic saline or granulated sugar is helpful in reducing the edema of the vaginal tissue in conjunction with gentle force to reduce the prolapse, which does two things: 1) improves venous return and 2) compresses the tissue and, with this, decreases the fluid in the tissue. Vaginal and vestibular retention methods include a Caslick procedure or a Buhner suture to reestablish the constrictor vestibular muscle function. Two other techniques are (1) surgical "buttons" placed through the dorsal vaginal wall and sacrosciatic ligament exiting lateral to the sacrum causing an inflammatory reaction and adhesions of the vagina to the paravaginal tissues and (2) suturing the cervix cranioventrally to the prepubic tendon.

Selected references

[1] Dyce KM, Sack WO, Wensing CJG. Textbook of veterinary anatomy. 4th ed. St. Louis, MO: Saunders/Elsevier; 2010.
[2] Budras K, Habel RE, editors. Bovine anatomy: an illustrated text. 1st ed. Hannover, Germany: Schlütersche; 2003.
[3] Nishida T, Hosoda K, Matsuyama H, et al. Collateral uterine blood flow in Holstein cows during the third trimester of pregnancy. J Reprod Dev 2006;52(5):663–8.
[4] Russe M. Neural control of the reproductive tract in the cow as it relates to parturition. In: Calving problems and early viability of the calf; 1979. p. 293–6.
[5] Winkler JK. Repair of bovine vaginal prolapse by cervical fixation. J Am Vet Med Assoc 1966;149:768–71.

CASE 20.3

Urolithiasis

Marie-Eve Fecteau
Department of Clinical Studies - New Bolton Center, University of Pennsylvania School of Veterinary Medicine, Kennett Square, Pennsylvania, US

Clinical case

History

A 6-year-old crossbred, castrated male goat was presented for decreased appetite and signs of abdominal discomfort of 12 hours' duration that included repeated stretching and posturing to urinate. The goat was kept as a pet and was part of a small herd of 6 castrated male goats. His diet consisted mainly of pasture grazing, with supplemental alfalfa hay and concentrates.

Physical examination findings

On physical examination, the goat was bright, alert, and responsive. His rectal temperature and respiratory rate were normal, but was tachycardic (120 bpm). The goat was in good body condition, had pink mucous membranes, moderate scleral injection, and appeared well-hydrated. He had 2 rumen contractions per minute and was passing normal manure. During physical examination, the goat stood in a sawhorse position (Fig. 20.3-1) and was occasionally tail-flagging and vocalizing. A few drops of urine were observed at his prepuce.

Differential diagnoses

Obstructive urolithiasis, urinary bladder neoplasm (i.e., enzootic hematuria), urinary tract infection, ulcerative posthitis (Gr. *posthē* foreskin, prepuce + *-itis* inflammation = inflammation of the prepuce), and intestinal obstruction

UROABDOMEN

Occasionally, a small ruminant with chronic urethral obstruction presents with more severe signs of illness and is recumbent with dull mentation, severe dehydration, scleral injection, cardiac arrhythmia, and abdominal distention. A ruptured bladder with secondary uroabdomen (Gr. *ouon* urine + L. *abdere* to hide, belly) should be suspected. Small ruminants with uroabdomen typically have more severe electrolyte and acid-base disturbances, such as marked hyperkalemia, hyponatremia, hypochloremia, and acidemia. Hyperkalemia can lead to life-threatening bradyarrhythmia.

Abdominal ultrasound reveals abdominal effusion (Fig. 20.3-2). It is important to remember that the urinary bladder may still appear distended on ultrasound, even in cases of uroabdomen, as most cases develop pinpoint perforations in the bladder wall secondary to chronic distention, rather than a specific identifiable tear. Confirmation of a uroabdomen relies on obtaining a sample of peritoneal effusion and comparing its creatinine concentration to the patient's serum or plasma creatinine value. The ratio of abdominal fluid to peripheral creatinine should be at least 2:1. These individuals require pre-operative stabilization of their electrolyte and acid-base disturbances.

Comparative Veterinary Anatomy: A Clinical Approach. https://doi.org/10.1016/B978-0-323-91015-6.00107-2

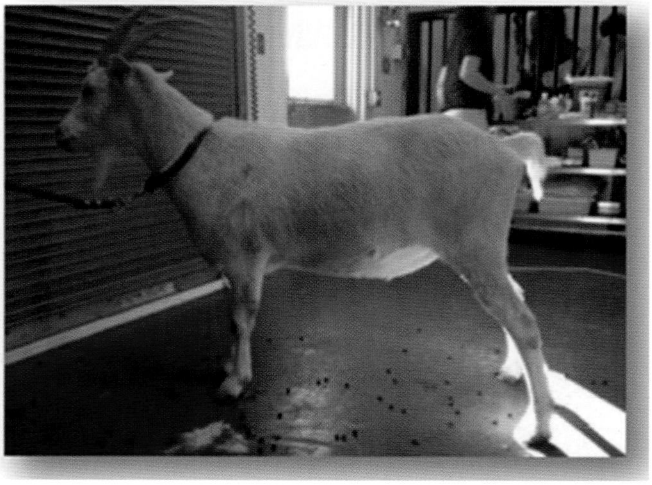

FIGURE 20.3-1 Goat at presentation demonstrating a "sawhorse" stance.

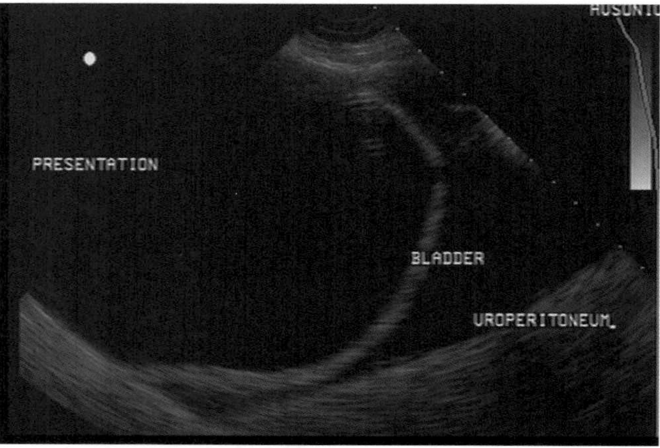

FIGURE 20.3-2 Transabdominal ultrasound image from a small ruminant presented with signs of obstructive urolithiasis. The image shows a large, distended urinary bladder and a large amount of anechoic abdominal fluid, most likely uroperitoneum from bladder leakage due to chronic distention.

Diagnostics

Clinical pathology revealed moderate azotemia and mild hyperkalemia.

Transabdominal ultrasound revealed a large, distended urinary bladder, measuring 8.7 cm (3.4 in.) (*N* = <3–4 cm or <1.2–1.6 in.) in diameter (Fig. 20.3-3). Abdominal radiographs revealed several radiopaque uroliths in the ventral-most aspect of the distended urinary bladder, as well as multiple round radiopaque uroliths in the distal penile urethra (Fig. 20.3-4).

Diagnosis

Obstructive urolithiasis with acute urethral obstruction

Treatment

An IV catheter was placed in the jugular vein and the goat was stabilized with IV crystalloids to address any electrolyte abnormalities and acid-base disturbances. An ECG was performed to rule out a tachyarrhythmia and

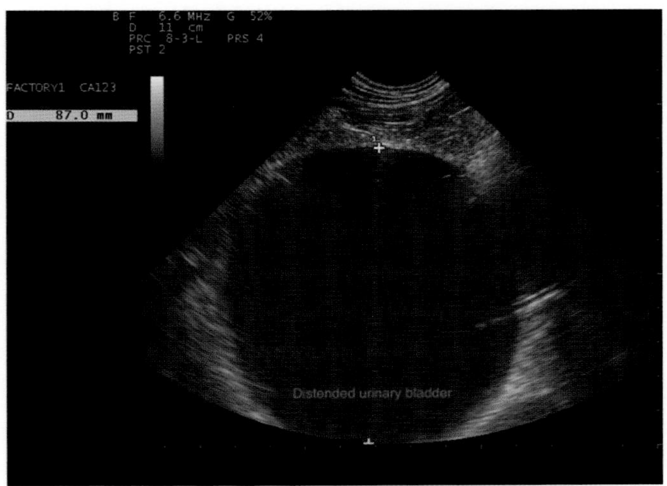

FIGURE 20.3-3 Transabdominal ultrasound image obtained on presentation. The urinary bladder measures 8.7 cm (3.4 in.) in diameter and is filled with anechoic urine.

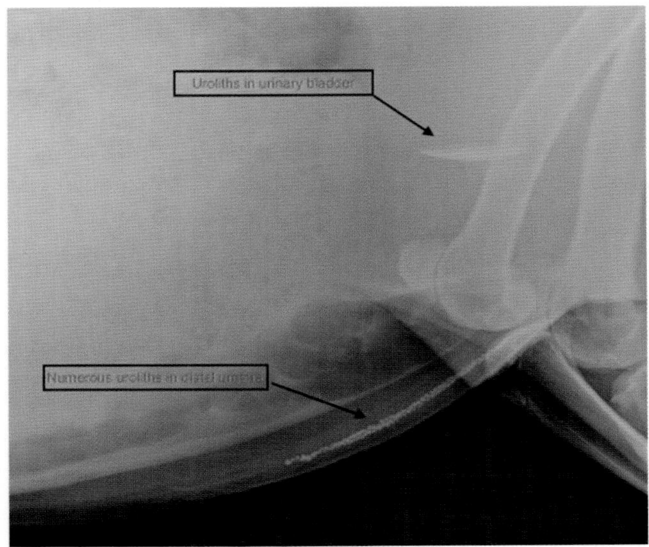

FIGURE 20.3-4 Abdominal radiographs showing several radiopaque uroliths in the ventral aspect of a distended urinary bladder, as well as multiple, round radiopaque uroliths in the distal penile urethra.

revealed sinus tachycardia with a regular rhythm. The goat was then anesthetized, placed in dorsal recumbency, and the penis was exteriorized to examine the urethral process. Several small round uroliths were identified within the urethral process and the process was amputated (Fig. 20.3-5). A paramedian approach to the abdomen was made and the urinary bladder was isolated and decompressed using a needle and sterile suction. A cystotomy (Gr. *kystis* sac or bladder + Gr. *stoma* opening) was made on the ventral aspect of the bladder and numerous uroliths were retrieved using a combination of a bladder spoon and suction. The uroliths were submitted for analysis and later revealed to be composed of calcium carbonate.

A catheter was advanced into the trigone of the bladder and the urethra was gently lavaged/irrigated until normal flow was obtained. A balloon catheter was placed in the bladder via separate stab incisions through the body wall and the urinary bladder to divert urine flow away from the urethra postoperatively, enabling the urethral inflammation to subside. The body wall was closed anatomically in 2 layers, and the goat recovered uneventfully from anesthesia.

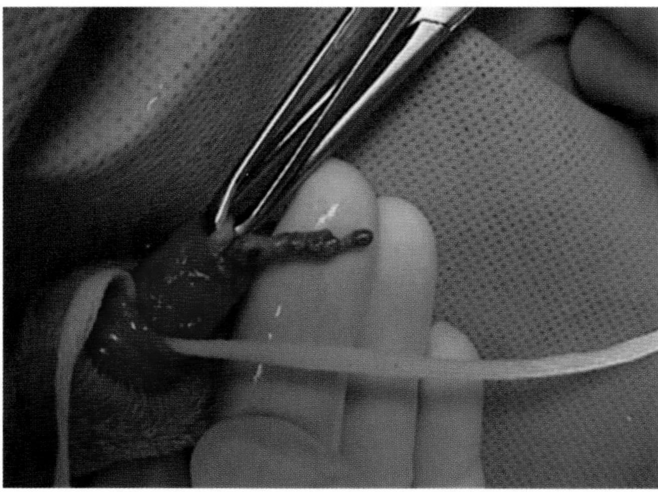

FIGURE 20.3-5 Urethral process (resting on the middle digit) distended with urine and obstructed with several round, copper-colored uroliths—most likely calcium carbonate.

Treatment of obstructive urolithiasis relies mostly on surgery because medical therapy alone—i.e., urine acidification and analgesics—is rarely successful. This condition is considered a surgical emergency. Surgical options include approaches to the urinary bladder (cystotomy, tube cystostomy, and permanent cystostomy) and approaches to the urethra (urethral process amputation, perineal urethrostomy, and urethrotomy). Each surgical option has its advantages and disadvantages and the procedure should be selected based on the patient's intended use, clinical presentation, and surgeon's experience and preference.

Post-operatively, the balloon catheter was occluded for increasing periods of time and urination monitored. The balloon catheter was removed 14 days after surgery and the goat was discharged from the hospital. Dietary recommendations were made to avoid feeds that are rich in calcium, such as grains, concentrates, and alfalfa hay.

Anatomical features in ruminants

Introduction

This section covers the anatomy related to the urinary bladder, urethra, and urethral process of male ruminants. While this case focuses on small ruminants, the anatomy of the urogenital system of the bull is similar to that of small ruminants with a few exceptions, covered in the side box entitled "Cattle urogenital system differences."

Function

Urine is produced and excreted by the kidneys, passes into each ureter originating in the renal pelvis, travels through the ureters, and empties into the urinary bladder near its neck at the trigone before it is excreted from the body by the process of urination (micturition; L. *micturire* to urinate) via the urethra. The urinary bladder can hold approximately 250–500 mL of urine in small ruminants and 1–2 L in adult cattle. As urine accumulates, the rugae (L. *ruga* a ridge, wrinkle, or fold) of the urinary bladder mucosa flatten out and the bladder wall thins and distends, enabling the bladder to store a large volume of urine without significantly increasing the internal pressure of the bladder. Urination includes a coordinated process between the central, autonomic, and somatic nervous systems, while micturition centers are located in the pons and the cerebral cortex.

Urinary bladder and its ligaments

The **urinary bladder** is a hollow muscular organ that varies in size and position with the amount of urine it contains. The empty contracted bladder is thick-walled and sits on the floor of the pelvic cavity, the neck being the only part of the bladder without a peritoneal covering. As the bladder fills with urine, its wall thins, and it enlarges cranially into the abdominal cavity. The bladder is supported by lateral and medial ligaments. The single **middle** (median) **ligament** is a median triangular fold that is formed by reflection of the peritoneum from the ventral surface of the bladder onto the ventral wall of the pelvis and abdomen. The paired **lateral ligaments** of the bladder stretch from the lateral aspects of the bladder to the lateral pelvic walls. Each lateral ligament contains in its free edge a round, firm band, the **round ligament**, which is a remnant of the fetal umbilical artery. The cranial rounded blind end of the bladder is termed the **apex**. The middle part, the **body,** presents 2 surfaces–dorsal and ventral. The caudal part, or **neck**, is narrow and joins the urethra. The smooth muscles of the bladder form a smooth muscle sphincter that controls passage of urine into the urethra.

Urethra in male ruminants (Fig. 20.3-6)

The male small ruminant **urethra** is long (30–50 cm or 11.8–19.7 in.) and narrow and is comprised (from proximal to distal) of the **pelvic urethra**, the **perineal urethra**, and the **penile urethra**. The pelvic urethra is continuous with the neck of the urinary bladder and travels within the pelvis. The pelvic urethra receives the ductus deferens and ducts from accessory sex glands. The pelvic urethra is surrounded by the striated skeletal **urethralis muscle**, over which the male exercises voluntary control by way of the **pudendal nerve**. At the level of the ischiatic arch, the

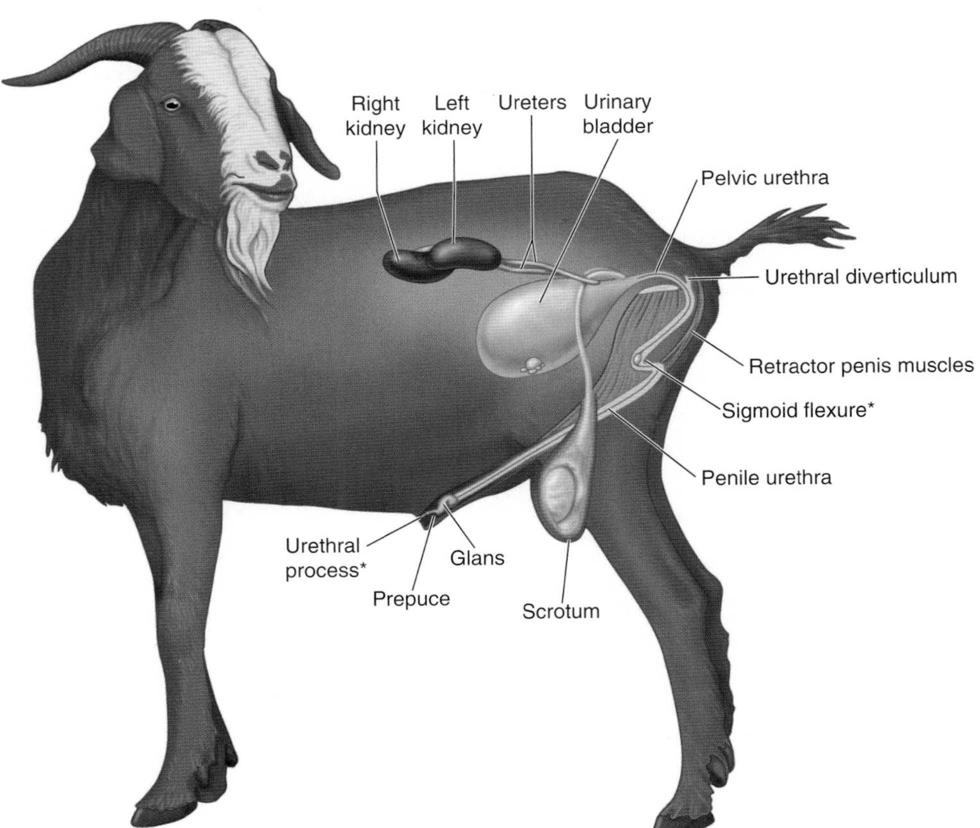

FIGURE 20.3-6 Anatomy of the male caprine urethra (the most common obstruction sites are marked with an *asterisk*). Note the uroliths illustrated in the urinary bladder and at the sigmoid flexure.

The urethral diverticulum is clinically relevant when dealing with obstructive urolithiasis as its location and orientation prevent the passage of a urethral catheter in a retrograde fashion (the tip of the catheter invariably ends up in the diverticulum instead of passing into the pelvic urethra).

The sigmoid flexure, especially the distal curve, is a common site of obstruction in small ruminants as urine flow is thought to be slower in that area.

urethra becomes narrower and curves ventrally toward the perineum.

The **urethral diverticulum** is located at the junction between the pelvic urethra and the perineal urethra. The **perineal urethra**, located between the ventral aspect of the anus and the caudal aspect of the scrotum, forms an S-shaped curve called the **sigmoid flexure**, which is comprised of **proximal** and **distal curves**. This flexure is accordioned when fully retracted and is effaced during erection. The penis of the ruminants is **fibroelastic** and contains a dense system of trabeculae and cavernous blood spaces (**corpus cavernosum**), extending the length of the penis. Engorgement of the corpus cavernosum with blood gives rise to an erection, which manifests itself by straightening of the sigmoid flexure and protrusion of the penis from the prepuce. The **retractor penis muscles** pass on either side of the distal curve of the sigmoid flexure.

The urethral process is another common site of obstruction because of its narrow and slightly twisted lumen. In castrated male small ruminants, especially those castrated at a young age, the urethral process is often adherent to the glans and/or the free part of the penis and has to be carefully dissected free before amputation.

The most distal aspect of the urethra is called the **urethral process** or vermiform appendage and is specific to small ruminants. It projects beyond the **glans** by about 2.5 cm (1.0 in.) in goats and 4 cm (1.6 in.) in rams (not present in the bull). At the end of this tapered process is the narrow **external urethral orifice**. The urethral process of the small ruminant is composed of erectile tissue and has a function during breeding.

Blood supply, lymphatics, and innervation

The urinary bladder principally obtains its blood supply from the **internal pudendal arteries** with drainage via the **internal pudendal vein)**, but branches also arise from the **obturator** and **umbilical arteries**. The penis is supplied by 3 arteries: the internal pudendal artery, the **obturator artery**, and the **external pudendal artery**. The internal pudendal artery divides into 3 branches, including the **artery of the bulb** (supplying the bulb and corpus

CATTLE UROGENITAL SYSTEM DIFFERENCES

The anatomy of a bull's penis and urethra is similar to that of small ruminants, with 2 important differences: (1) the bull's urethral process does not project beyond the glans as it does in small ruminants and (2) the distal/free end of the penis is mildly twisted counterclockwise when viewed from behind.

RUPTURED URETHRA

Small ruminants can present with various degrees of swelling at the perineum and/or ventral abdomen, likely indicating a rupture of the urethra (Fig. 20.3-7). The urethra typically ruptures secondary to pressure necrosis at the site of obstruction, often at or around the sigmoid flexure. A suspected ruptured urethra can be confirmed with contrast radiography (Fig. 20.3-8).

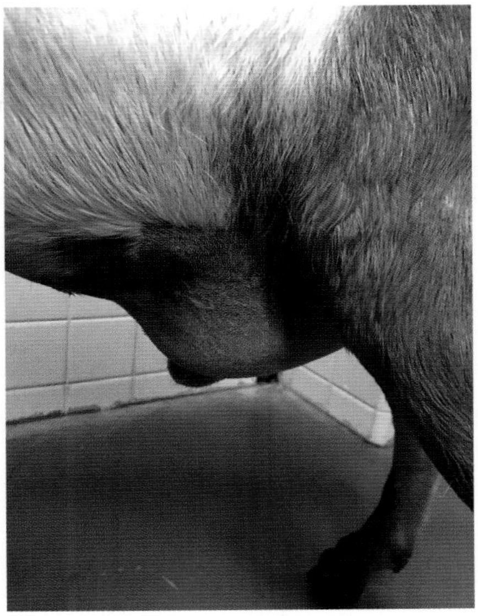

FIGURE 20.3-7 Goat presented with signs of obstructive urolithiasis. Note the ventral swelling around the prepuce suggesting urethral rupture. (Photo credit to Dr. Shannon Hinton.)

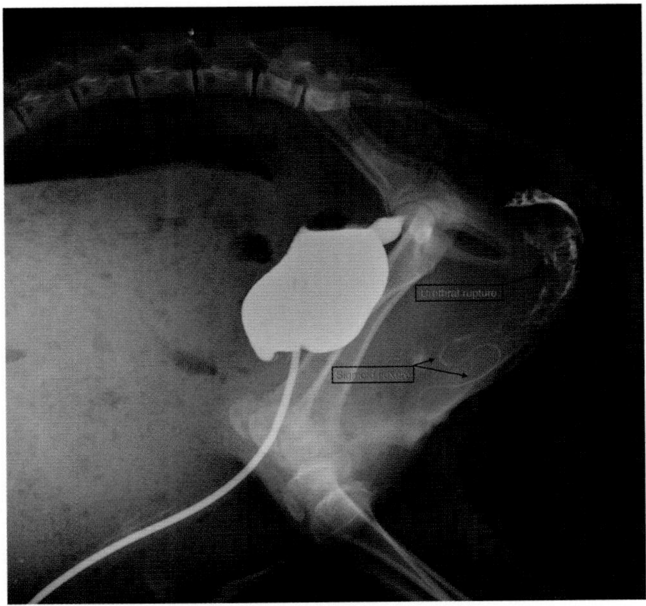

FIGURE 20.3-8 Contrast urethrography confirming a ruptured urethra. Note the contrast medium extravasating (escape of blood or other substance into the tissues) in the perineal region, proximal to the sigmoid flexure.

spongiosum); the **deep artery** of the penis (supplying the crus); and the **dorsal artery** (supplying the glans and prepuce). All 3 are accompanied by satellite veins that drain both the tissues and blood spaces within the spongy and cavernous bodies (responsible for erection) and further drain to the **external pudendal veins**.

The lymphatics from the urinary bladder go to the **internal iliac** and **lumbar lymph nodes**. The lymph vessels of the penis go to the **scrotal lymph nodes**.

Innervation of the bladder wall and sphincters is complex and involves both the sympathetic and parasympathetic nervous systems. The sympathetic portion relies on the **hypogastric nerve**, originating from spinal cord segments

T10–L2, and is responsible for the inhibition of detrusor muscle contraction during bladder filling, as well as contraction of the internal urethral sphincter. Parasympathetic innervation relies on the **pelvic nerve**, which originates from the spinal nerve segments S2–S4, and is responsible for contraction of the detrusor muscle during urination, as well as relaxation of the internal urethral sphincter. Somatic control of the bladder is also derived from the spinal nerve segments S2–S4 via the **pudendal nerve**, which acts on the external urethral sphincter. The pudendal nerve is also responsible for the innervation of the penis and prepuce. The paired **dorsal nerves**, which run with the dorsal arteries, are responsible for the sensation to the apex of the penis and are essential for attainment of full erection.

Selected references

[1] Singh, B., The male reproductive organs. Dyce, Sack and Wensing's textbook of veterinary anatomy, 5th ed., Elsevier, 702-707.
[2] Fails, A.D., Magee, C., The urinary system. Anatomy and physiology of farm animals, 8th ed., Wiley Blackwell, 387-388.
[3] Fubini S.L., Ducharme N.G., Urolithiasis. Farm animal surgery, 2nd ed., Elsevier, 584-595.

CASE 20.4

Penile Hematoma

David E. Anderson
Department of Large Animal Clinical Sciences, University of Tennessee College of Veterinary Medicine, Knoxville, Tennessee, US

Clinical case

History

A 2-year-old polled Hereford bull was presented for an evaluation of caudal ventral abdominal swelling. The bull had been placed into a pasture with 20 cows for natural service breeding approximately 2 weeks before presentation. The owner checked on the cattle approximately once a week. Ten days before presentation, the bull and all cows were visually normal. Three days before presentation, the owner noted a substantial swelling of approximately 10 cm (3.9 in.) in diameter along the bull's caudal ventral abdomen. The owner was unable to remove the bull from the breeding pasture at that time. Approximately 2 days later, the owner noted that the swelling had increased to approximately 30 cm (11.8 in.) in diameter, located caudal to the preputial orifice along the ventral aspect of the abdomen. The owner arranged to remove the bull from the pasture and bring the bull to the veterinary hospital for evaluation.

Physical examination findings

Upon presentation to the hospital, the bull was noted to have a large swelling along the ventral abdomen cranial to the neck of the scrotum (Fig. 20.4-1). The bull was placed into a cattle-restraint chute to facilitate examination. Physical examination findings revealed a normal temperature, pulse, respiration rate, rumen contractions, body condition score, and palpable lymph nodes. Abdominal palpation per rectum revealed a normal rumen, no distended intestinal viscus, and normal internal inguinal rings. Ventral abdominal palpation revealed an ovoid, soft, easily indented swelling, approximately 30 cm (11.8 in.) in diameter (frontal plane; lateral side-to-side axis) by 15 cm (5.9 in.) (frontal plane; dorsal-ventral axis) by 40 cm (15.7 in.) long, centered on the ventral midline along the penis and ending cranial to the neck of the scrotum. The bull did not react meaningfully during palpation of the mass unless firm pressure was applied and this was interpreted as minimally painful swelling. The bull was observed to urinate and defecate normally during the examination.

Differential diagnoses

Urethral rupture secondary to obstructive urolithiasis or trauma, preputial laceration, parapreputial/penile abscess, penile hematoma, abdominal hernia, ectopic testis, subcutaneous abscess or hematoma, penile or preputial neoplasia (e.g., penile papilloma), ventral edema, sepsis of the testicle or epididymis, and scrotal neoplasia

Diagnostics

Ultrasonographic examination of the swelling revealed a hypoechoic confluent mass measuring approximately 25 cm (9.8 in.) in diameter in the axis of the frontal plane, 30 cm (11.8 in.) in diameter in the axis of the sagittal plane, and 15 cm (5.9 in.) in diameter in the dorsal-ventral axis. Fine septa were noted within the hypoechoic mass. The mass did not appear to be encapsulated but did have a well-defined area.

Comparative Veterinary Anatomy: A Clinical Approach. https://doi.org/10.1016/B978-0-323-91015-6.00111-4

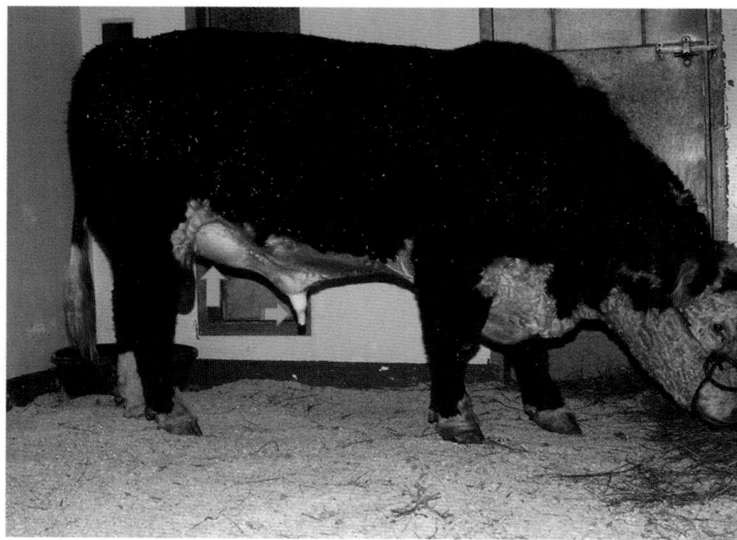

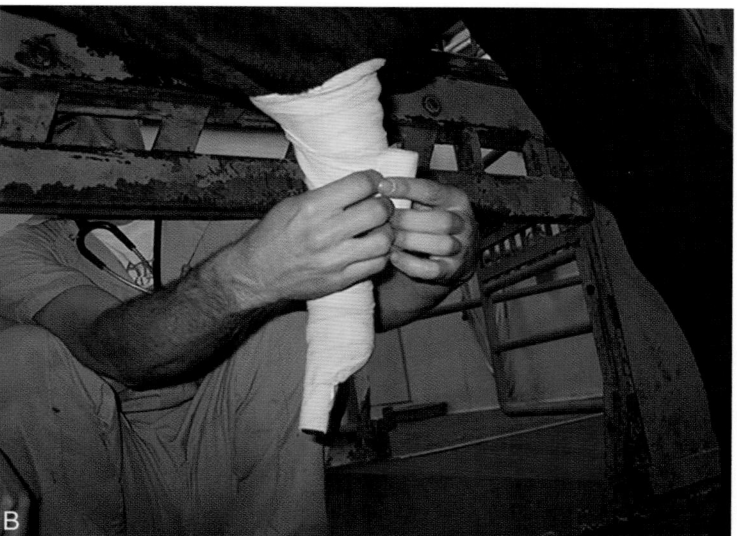

FIGURE 20.4-1 (A) Adult polled Hereford bull with common location of swelling (*vertical arrow*) associated with penile hematoma. A preputial retainer is attached (*horizontal arrow*) to the end of the sheath to prevent secondary preputial prolapse. (B) A temporary preputial retainer is placed using part of a nasogastric tube (inserted within the prepuce), roll of gauze (if needed to protect the prepuce), and adhesive tape (applied to the end of the tube before insertion and including the sheath as much as possible to minimize dislodgement).

The skin overlying the ventral and lateral aspects of the mass was clipped and aseptically prepared for an ultrasound-guided fine needle aspiration. The fine needle aspiration yielded dark red serosanguineous fluid containing what appeared to be fragments of clotted blood. Cytologic examination revealed the presence of abundant numbers of red blood cells, occasional white blood cells, proteinaceous material, and no evidence of bacteria. Microbial culture results did not yield any bacterial growth.

Diagnosis

Simple penile hematoma with associated soft tissue edema

Treatment

Penile hematoma in bulls can be treated medically or surgically, and many breeders choose to cull bulls with penile hematomas because of the risk of complications (see side box entitled "Penile hematoma"). Medical therapy is generally recommended for small penile hematomas and in bulls that have limited economic value. Medical therapy includes administration of nonsteroidal anti-inflammatory and antimicrobial drugs, hydrotherapy of the swelling, and sexual rest by isolation for a minimum of 60 days. Small hematomas have a fair prognosis for returning to breeding soundness.

The owner of this bull elected surgical treatment, which is recommended for large penile hematomas (> 20-cm or 7.9-in. diameter), persistent penile hematomas, and in bulls of high economic value. The bull was routinely held off feed and water (for 24 hours) before surgery to minimize the risk of bloating and regurgitation of rumen fluid during surgery, which severely compromises ventilation during general anesthesia. The bull was placed under general anesthesia in right lateral recumbency, allowing for distention of the rumen during surgery and to minimize compromising diaphragmatic excursion (L. *excurrere* to run out from = movements), which would interfere with normal ventilation. After routine surgical preparation, a 10-cm (3.9-in.) ventral lateral incision was made over the point of maximum swelling on the lateral aspect of the penis immediately cranial to the distal sigmoid flexure (Fig. 20.4-2). The incision was continued into the hematoma, the clotted blood evacuated, and a sample obtained for microbial culture. The hematoma cavity was then lavaged with 1 L of sterile saline and the

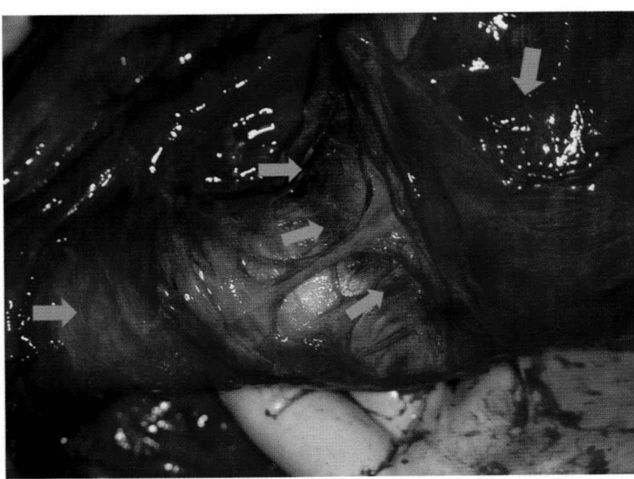

FIGURE 20.4-2 Intra-operative image of the cranial margin of a penile hematoma in a 2-year-old Angus bull, originating from the penis. Key: *gray arrow* depicts the swelling caused by the penile hematoma; *orange arrows* (3) depict the cranial margin of the penile hematoma; *blue arrow* depicts the shaft of the penis.

PENILE HEMATOMA

Penile hematoma is caused by rupture of the tunica albuginea (broken/fractured penis) on the dorsal aspect of the penis at the location of the distal sigmoid flexure opposite to the ventral attachment of the paired retractor penis muscles. This area of the tunica albuginea is relatively thinner and the vascular spaces of the corpus cavernosum penis (CCP) are relatively larger, resulting in a remarkable increase in CCP pressure when the penis is abruptly bent while erect. Penile hematoma in bulls is caused by breeding accidents or failed intromission (L. *intro* within + L. *mittere* to send) of the bull's penis entering the vagina of the cow and is more common in young, inexperienced bulls. The fully engorged penis becomes acutely bent when the bull contacts the perineal area of the cow, resulting in a sharp bend at the location of the dorsal aspect of the tunica albuginea at the distal sigmoid flexure opposite to the attachment of the paired retractor penis muscles. The acute bending of the penis, and resistance by the retractor penis muscles, causes the sudden bending of the penis and compression of the CCP, inordinate increase in CCP pressure (> 75,000 mmHg), rupture of the tunica albuginea, and escape of the blood contained in the CCP. Initial blood accumulation is no more than the volume of the CCP, approximately 200 mL. However, repeated attempts by the bull to breed result in additional bleeding and enlargement of the swelling. The size of the swelling is relatively proportional to the number of times after initial rupture of the tunica albuginea that the bull continues to try to breed cows. Complications of penile hematoma include abscessation, preputial adhesions preventing penile extension, vascular shunting resulting in impotence and damage to the dorsal nerve of the penis (causing an inability to direct and enter the vagina), and pain—all culminating in decreased libido.

residual fluids suctioned from the surgical field. The sigmoid flexure of the penis was exteriorized from the incision and inspected (Fig. 20.4-3). There was a tear of the tunica albuginea of the corpus cavernosum penis (CCP) on the dorsal aspect of the penis immediately distal to the sigmoid flexure and opposite to the point of attachment of the retractor penis muscles, which is common with penile hematomas. The margins of the tear in the tunica albuginea were debrided and the tear was closed with a synthetic absorbable suture material using a cruciate pattern for secure closure of the tunica albuginea. No attempt was made to close the dead space created by the hematoma around the penis because of the potential for this to promote restrictive scar tissue, thereby preventing extension of the penis. Subcutaneous tissues and skin were closed anatomically using standard surgical technique. A penile and preputial retaining tube was applied after surgery and maintained for 5 days to ensure that the preputial prolapse did not occur (Fig. 20.4-4).

Bulls with penile hematoma often have secondary preputial prolapse from the sheath edema and altered blood supply. The retaining tube is a flexible plastic tube secured inside the preputial orifice to prevent prolapse. It is retained by taping the plastic tube to the end of the sheath.

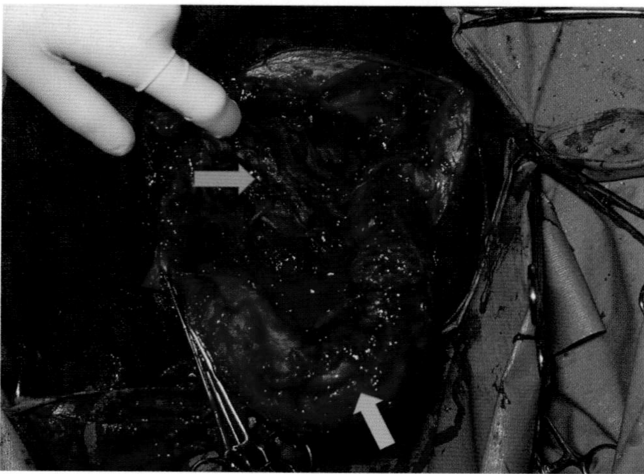

FIGURE 20.4-3 Intra-operative image of penile hematoma in an Angus bull after surgical evacuation and debridement. Key: *blue arrow* depicts the distal sigmoid flexure of the penis; *orange arrows* depict the interior capsule of the hematoma after debridement.

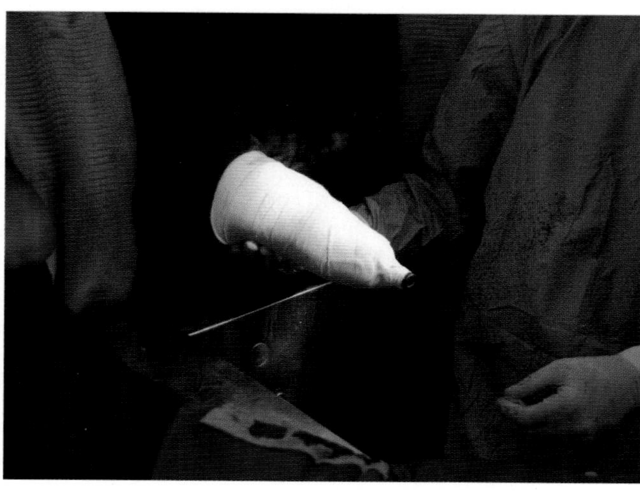

FIGURE 20.4-4 Temporary preputial retainer has been placed to prevent post-operative preputial prolapse. Adhesive bandaging tape was applied to the end of the tube and wrapped over the sheath to a point proximal to the preputial orifice.

After surgery, the bull was maintained in stall confinement for 2 weeks until skin suture removal. He was isolated from cows until the surgery site was completely healed at 60 days to minimize sexual stimulation and recurrence of the hematoma. Antimicrobial drug therapy was continued for 7 days and nonsteroidal anti-inflammatory drug therapy for 2 weeks after surgery. Beginning 5 days postoperatively, low-pressure hydrotherapy of the surgical area was instituted and continued throughout the period of stall confinement as an additional treatment to reduce swelling along the ventral abdomen.

When appropriate periods of sexual rest and separation for 60 days are followed, approximately 1 in 4 bulls experience a recurrence of the penile hematoma in subsequent breeding seasons. Approximately 1 of 2 bulls experience recurrence of the penile hematoma if they are returned to breeding service in less than 60 days after treatment.

Anatomical features in ruminants

Introduction

The clinically important features of this anatomical region in the bull include the skin, subcutaneous tissues, sheath, prepuce and fornix, penis and distal sigmoid flexure, urethra, retractor penis muscles, penile and preputial vascular supply, and dorsal nerve of the penis.

Function

The penis is used for urine excretion from the urinary bladder and in sexual reproduction as a conduit to deposit sperm into the female vagina.

Penis

Bulls have a **fibrovascular penis** that has a relatively fixed diameter and length. The penis is composed of an external fibrous tissue layer, the **tunica albuginea**. The tunica albuginea encases the CCP and creates a vascular compartment that is pressurized during erection. The CCP is composed of 2 **crura** which extend throughout the body of the penis.

DIFFERENTIAL DIAGNOSES FOR CAUDAL VENTRAL SWELLING IN A BULL

- Urethral rupture secondary to obstructive urolithiasis or trauma
- Preputial laceration with infection
- Para-preputial/penile abscess
- Penile hematoma
- Abdominal hernia
- Inguinal hernia
- Ectopic testis
- Subcutaneous abscess
- Subcutaneous hematoma
- Penile or preputial neoplasia
- Ventral edema
- Orchitis or epididymitis
- Scrotal neoplasia

Spiral deviations of the penis are considered a congenital defect in bulls, affecting the ability of the bull to direct the penis into the vagina of the cow. The most common deviations are spiral, ventral, and S-shaped abnormalities, and they require surgery to correct the defect. Fascia lata autogenous grafting is one surgery technique used to correct the deviation with fibrous tissue, thereby preventing rotation of the penis (Fig 20.4-5).

Each of these fibrovascular compartments is pressurized independently during erection. The urethra is surrounded by the **corpus spongiosum penis** (CSP), which provides rigidity around the urethra during erection and ejaculation. The **urethra** is located on the ventral aspect of the penis and the course of the urethra can be palpated as an indentation or groove extending into the main body of the ventral penis. The **dorsal apical ligament** tracks dorsally along the penis, originating at the dorsal aspect of the **distal sigmoid flexure** and terminating at the dorsal and bilateral parts of the **glans penis**. The purpose of the dorsal apical ligament is to maintain the erect penis in a straight line during breeding.

The **prepuce** attaches to the penis, creating the **fornix**. The portion of the penis extending distal to the fornix is referred to as the **free portion** of the penis and terminates as the **glans penis** with the emergence of the urethra.

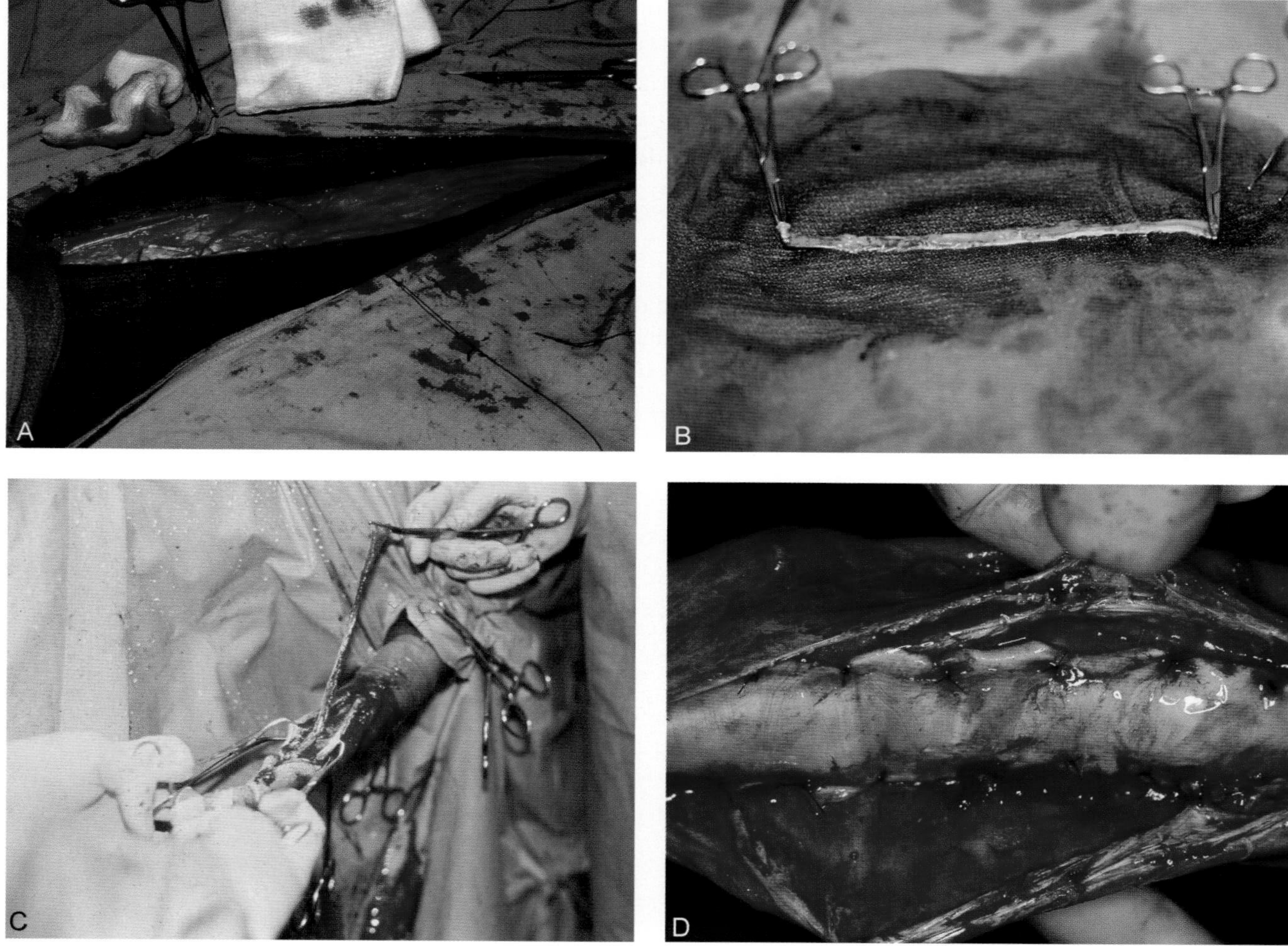

FIGURE 20.4-5 (A) Surgical harvesting of fascia latae graft from the lateral aspect of the thigh. (B) Preparation of the fascia latae autograft before implantation on the penis. (C) Attachment of the autograft to the distal end of the penis. The dorsal apical ligament has been divided and the tunica albuginea exposed; the graft is sutured to the tunica. (D) The dorsal apical ligament is reconstructed using simple interrupted sutures and the autograft has been incorporated into the closure.

The penis of adult bulls approaches 100 cm (39.4 in.) in length of which approximately 10–12 cm (3.9–4.7 in.) of the distal portion of the penis extends beyond the preputial attachment to the penis (fornix). The prepuce of most bulls is 30–40 cm (11.8–15.7 in.) long, although the prepuce of *Bos indicus* bulls is considerably larger. The preputial orifice is generally located 5–10 cm (2.0–3.9 in.) caudal to the umbilicus and may be 4–10 cm (1.6–3.9 in.) in diameter.

Functional anatomy of the penis

The penis is retracted into the protective sheath by the **retractor penis muscles**. In the nonengorged penis, the retractor penis muscles can flex the penis in such a way that a **sigmoid flexure** (creating the letter "S") is created. The proximal sigmoid flexure resides at the level of the neck of the scrotum and passes between the spermatic cords in an extrascrotal location. The distal sigmoid flexure resides within or immediately cranial to the area of the neck of the scrotum. The retracted position of the nonengorged penis protects the distal segment of the penis from exposure to the environment and prevents these tissues from being traumatized.

In the engorged penis, the pressures ($N \sim 15{,}000$ mmHg) within the CCP and CSP result in extension of the penile shaft with enough rigidity to permit breeding and the retractor penis muscles are relaxed.

Prepuce and sheath

The **sheath** of the bull is the outer skin covering, while the inward hairless epithelial tissue, the **prepuce**, connects the sheath to the penis at the fornix. The junction of the sheath and the prepuce is referred to as the **preputial orifice**. The preputial orifice does not have a sphincter muscle. However, the **protractor prepuce** and **retractor prepuce muscles** provide defined anatomical support for the cranial prepuce.

These muscles diminish the likelihood of preputial prolapse or exposure of the prepuce to either the external environment or trauma. Polled breeds of cattle often have an underdeveloped protractor prepuce muscle, resulting in relaxation of the preputial orifice and increased risk of exposure or trauma to the preputial tissues.

Blood supply, lymphatics, and innervation

The arterial supply to the penis comes from the abdominal aorta, which gives rise to the **internal iliac artery**, from which the **internal pudendal artery** originates and ultimately leads to the **artery of the penis**. The artery of the penis terminates as the **dorsal artery of the penis** and the **deep artery of the penis**, the latter of which supplies the CCP. The bull CCP can hold up to 200–250 mL of blood. Arterial branches arise from the artery of the penis and provide blood supply to the CSP (Fig 20.4-6).

Venous drainage of the CCP occurs exclusively in the region of the ischium at the crus penis via the **deep veins of the penis**, which then drain into the **penile veins**. The corpus of the penis does not normally have vessels connecting the CCP with the paired dorsal penile veins (Fig. 20.4-7). Venous return from the remainder of the penis and prepuce occurs in the **dorsal veins of the penis,** which track along the dorsal aspect of the penis.

After ejaculation, the crura of the penis relax and blood is drained, allowing the penis to return to the sigmoid flexure shape and retract into the sheath. During erection, the CCP functions as a closed system, meaning that venous drainage does not return to normal until the relaxation of the crura.

*Common point of insertion for abdominal mm.

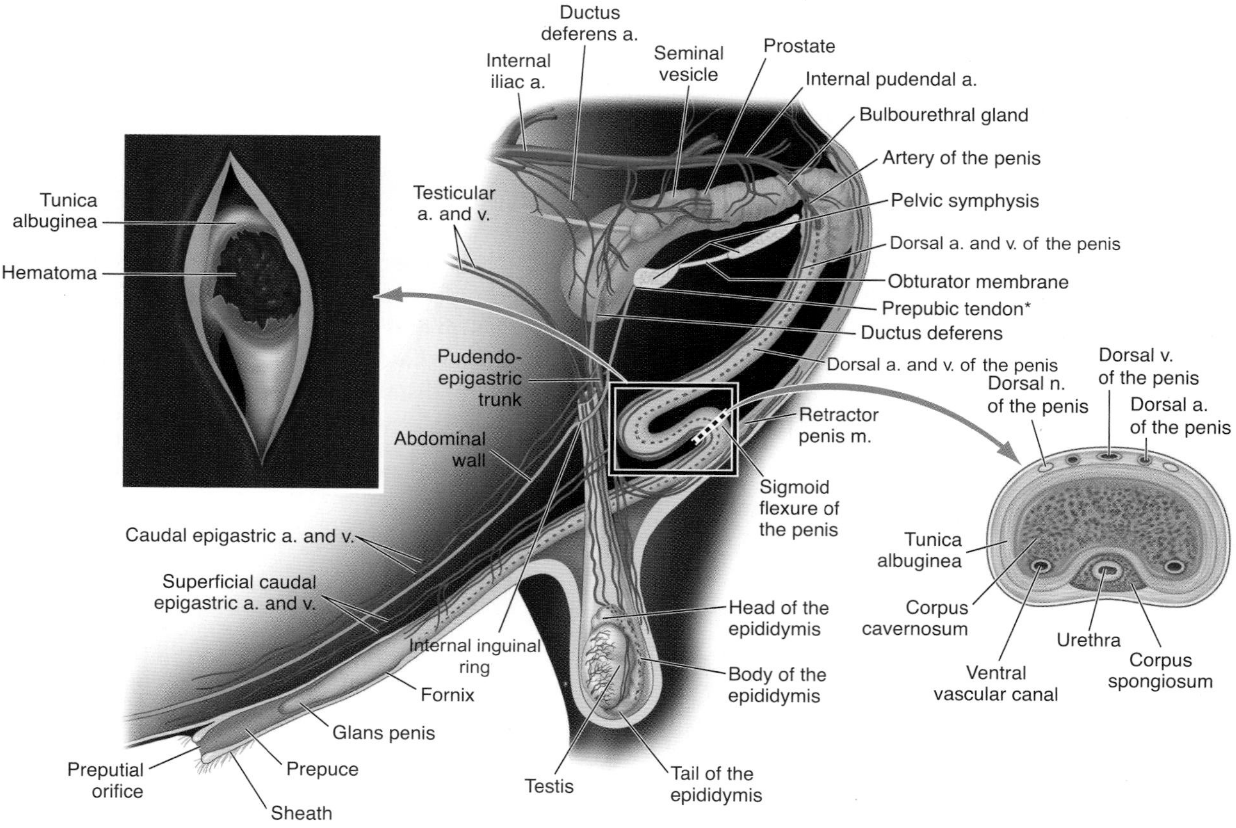

FIGURE 20.4-6 The blood supply of the bovine penis. Inset: Illustration showing a penile hematoma.

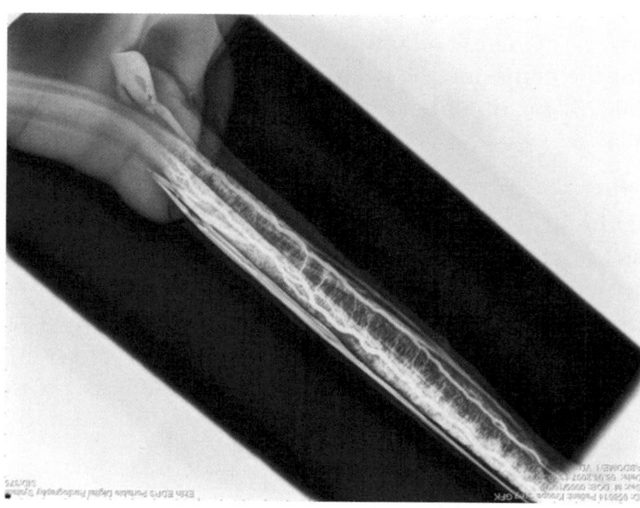

FIGURE 20.4-7 Positive contrast corpus cavernosography demonstrating filling of the CCP with a radiopaque contrast agent.

Lymphatic drainage from the penis and the preputial area is to the **superficial** and **deep inguinal lymph nodes** primarily, and from here to the **lumbar aortic lymph nodes** at the aortic bifurcation.

These nerve endings have receptors which help the bull locate the vagina and vestibule of the female during breeding. The dorsal nerve of the penis functions to detect heat as a means of identifying the vagina during breeding. The pudendal nerves can be blocked with local anesthesia for penile surgery.

Innervation of the penis arises from the **pudendal nerve**, predominantly composed of sensory nerve fibers. The pudendal nerve arises from the 2nd through the 4th sacral spinal cord segments and the paired nerves track along the medial and dorsal aspects of the penis medial to the penile arteries. The nerves are closely associated with the tunica albuginea. As the nerves course distally, they are located more laterally in the area of the sigmoid flexure. Distal to the distal sigmoid flexure, the nerves again are dorsally positioned. The dorsal nerves branch extensively along the dorsolateral and ventral aspects of the glans penis.

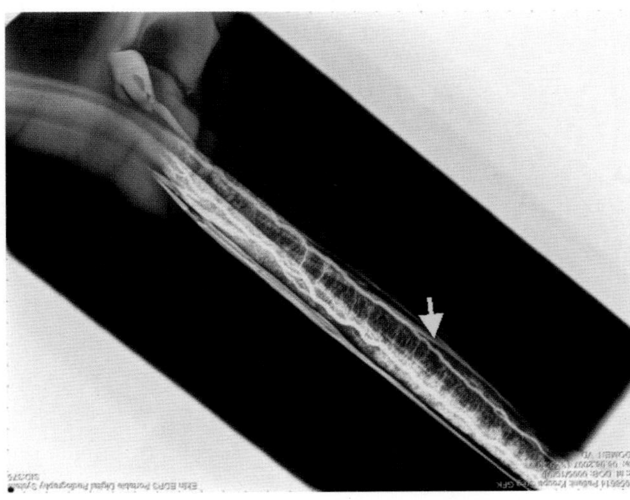

FIGURE 20.4-8 Positive contrast corpus cavernosography demonstrating abnormal venous drainage (*arrow*) of the CCP consistent with a vascular shunt.

VASCULAR SHUNTS

Under abnormal circumstances, such as congenital defects or trauma, venous drainage from the penis occurs via vascular shunts (i.e., a bypass), which impede erection because of the inability to pressurize the CCP (Fig. 20.4-8). Vascular shunts are small vessels, usually from the CCP and outside the tunica albuginea, which prevent erection.

Congenital shunts are multiple and generally cannot be surgically treated. In these cases, the penis cannot achieve rigid erection and deviates ventrally as a flaccid or partially engorged penis. The bull cannot achieve vaginal intromission during mounting and therefore is unable to breed.

An acquired shunt is unique and is identified by positive contrast cavernosography. This imaging technique is achieved by injecting contrast material into the CCP and obtaining radiographic images until the shunt is identified. If the shunt is located using cavernosography, it may be surgically corrected by isolating and ligating the shunting vessel or closing the surrounding tunica albuginea.

Selected references

[1] Musser JM, St-Jean G, Vestweber JG, Pejsa TG. Penile hematoma in bulls: 60 cases (1979–1990). J Am Vet Med Assoc 1992;201(9):1416–8.
[2] Anderson DE. Surgery of the prepuce and penis. Vet Clin North Am Food Anim Pract 2008;24(2):245–51. v-vi https://doi.org/10.1016/j.cvfa.2008.02.002.
[3] Hopper RM. Management of male reproductive tract injuries and disease. Vet Clin North Am Food Anim Pract 2016;32(2):497–510. https://doi.org/10.1016/j.cvfa.2016.01.015.
[4] Anderson DE, St-Jean G, Desrochers A, Hoskinson JJ. Use of Doppler ultrasonography and positive-contrast corpus cavernosography to evaluate a persistent penile hematoma in a bull. J Am Vet Med Assoc 1996;209(9):1611–4.
[5] Wolfe DF, Moll HD, editors. Large Animal Urogenital Surgery. Williams and Wilkens Pub; 1999.

CHAPTER 21

THORACIC LIMB

André Desrochers, Chapter editor

Thoracic Limb

21.1 Shoulder luxation—*Marjolaine Rousseau* ... 1146

21.2 Septic arthritis of the distal interphalangeal joint—*Karl Nuss* ... 1161

21.3 Sole ulcer—*Karl Nuss* ... 1169

21.4 Metacarpal fracture—*André Desrochers* ... 1181

CASE 21.1

Shoulder Luxation

Marjolaine Rousseau
Department of Clinical Sciences, Université de Montréal, Faculty of Veterinary Medicine, Saint-Hyacinthe, Québec, CA

Clinical case

History

A 4.5-year-old male intact Suri alpaca was presented with a severe right thoracic limb lameness that appeared suddenly the night before. He had been a healthy herd sire, housed in a pasture with 8 other intact male alpacas, and had no history of previous lameness. Before presentation, he had been treated with flunixin meglumine by his referring veterinarian without improvement.

Physical examination findings

Upon physical examination, the alpaca was bright and alert. His rectal temperature was 98.6°F (37°C) and he was tachycardic (100 bpm) and tachypneic (60 rpm). He was nonweight-bearing lame on the right thoracic limb. Palpation of the affected limb revealed the presence of swelling and pain localized to the shoulder. The greater tubercle of the humerus was more prominent on palpation, while the acromion was difficult to palpate. The rest of the thoracic limb examination was normal.

Differential diagnoses

Shoulder luxation or subluxation, fracture of the distal aspect of the scapula and/or proximal aspect of the humerus, septic arthritis of the shoulder, infraspinatus bursitis, intertubercular bursitis, and brachial plexus injury

Diagnostics

The alpaca was sedated for a radiographic examination of the shoulder (scapulohumeral) joint. A mediolateral projection of the right shoulder was taken while the alpaca was standing and revealed widening of the joint space compatible with a luxation of the shoulder (Fig. 21.1-1A). The alpaca had to be "cast"—i.e., placed in lateral recumbency—for the second orthogonal view (craniocaudal view) due to difficulties in restraining him for the radiograph. The widening seen on the mediolateral view was not apparent on the craniocaudal view, although a soft tissue swelling lateral to the shoulder joint could be seen. The bony structures and joints were aligned (Fig. 21.1-1B). It was determined that during repositioning of the alpaca for the orthogonal radiographic view, the joint luxation was reduced. The marked improvement in the lameness after radiographs confirmed the clinical belief of a shoulder luxation on presentation and its subsequent correction.

Diagnosis

Right scapulohumeral luxation

Comparative Veterinary Anatomy: A Clinical Approach. https://doi.org/10.1016/B978-0-323-91015-6.00103-5

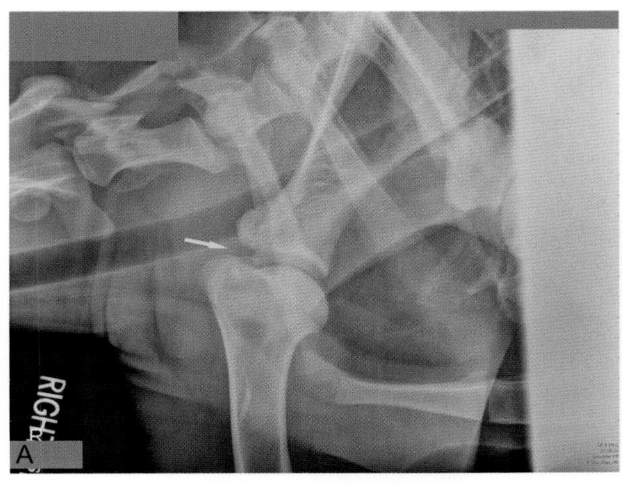

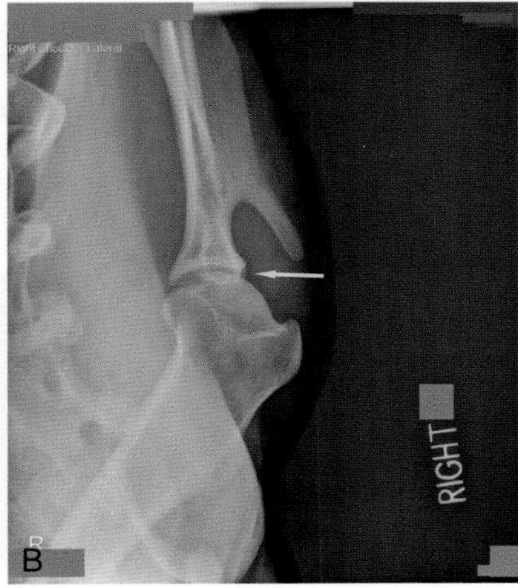

FIGURE 21.1-1 (A) Mediolateral radiographic view of the right shoulder of the alpaca in this case. Joint space is increased (*arrow*). (B) Craniocaudal radiographic view of the right shoulder. There is moderate soft tissue swelling on the lateral aspect of the shoulder. The joint is well-aligned (*arrow*).

Treatment

Closed reduction and external coaptation were recommended. A modified Velpeau sling was placed on the flexed upper thoracic limb to secure the closed reduction of the shoulder (Fig. 21.1-2). Before discharging the alpaca 3 days later, mediolateral and craniocaudal radiographs of the right shoulder were repeated to confirm normal alignment of the scapulohumeral joint. The owner was instructed to restrict the alpaca to stall confinement for 6–8 weeks and maintain the Velpeau sling for an additional 2 weeks.

> The Velpeau sling/bandage is used to immobilize scapular fractures, dislocations of the shoulder joint after reduction, and for any medical reason to suspend weight-bearing of the thoracic limb. It is applied using a long piece of combine roll or similar material, long enough to go around the animal's thorax approximately 3.5 times. The sling incorporates the radius, shoulder, and thorax. It is secured to the body wall to prevent the elbow from moving caudally and the distal limb from moving cranially and is left in place for 10–14 days before changing or removing.

After the reintroduction of this alpaca to the pasture with other herd sires, he was found acutely nonweight-bearing lame on his right thoracic limb approximately 40 days after discharge from the hospital. He was referred to the hospital for surgical treatment (lateral extracapsular tension sutures). Radiographs were taken in a similar manner and revealed scapulohumeral luxation like at the first visit. A closed reduction of the shoulder joint was successfully achieved and maintained in place again, using a modified Velpeau sling until a scheduled surgical stabilization of the joint.

The alpaca was placed in left lateral recumbency under general anesthesia. A craniolateral approach was used, which included a fasciotomy and a deltoid myotomy to access the scapulohumeral joint. A small tear of the joint capsule was identified. Two 4.0-mm (0.16-in.) suture anchors were placed in the scapular neck, one 4.5-mm (0.18-in.) cortical screw placed in the greater tubercle, and 2 tension band sutures were placed laterally to prevent luxation of the humeral head. The surgical site was closed in 3 layers. The alpaca recovered from anesthesia without problems. He was discharged from the hospital 3 days post-operatively with detailed aftercare instructions: a modified Velpeau sling for 10 days, strict stall confinement for 2–3 months, and separation from other intact males. Approximately 80 days post-operatively, the right shoulder luxation recurred after the alpaca attempted to jump out of his stall. Radiographs revealed a lateral luxation of the right shoulder (Fig. 21.1-3). Due to financial limitations, the alpaca was treated with closed reduction and dry lot confinement. At the 12-month follow-up, the alpaca had not had a recurrence of the lameness.

FIGURE 21.1-2 A Velpeau bandage stabilizing the shoulder after scapulohumeral luxation and reduction.

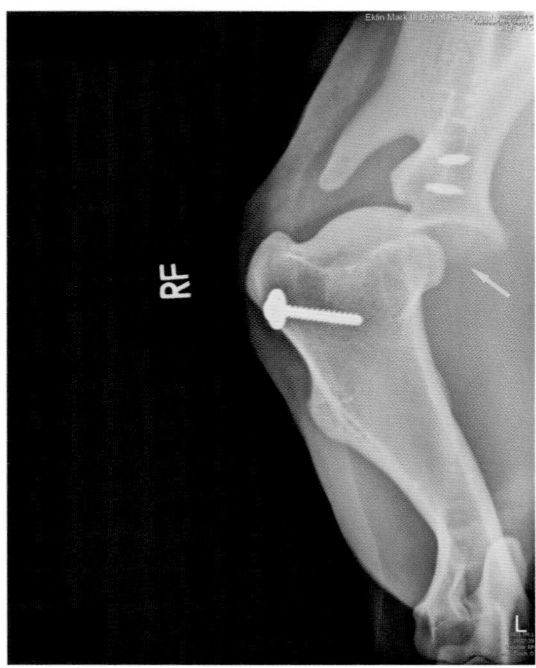

FIGURE 21.1-3 Post-operative craniocaudal radiographic view of the right shoulder. There is luxation of the shoulder joint (*arrow*). The orthopedic implants can be clearly seen on this view: 2 bone anchors on the scapula and 1 cortical screw and washer in the humerus.

Anatomical features in ruminants

Introduction

Bones of the thoracic limb and their associated eminences and cavities in alpacas are similar to cattle; however, there is insufficient information currently available for the alpaca regarding the muscles, tendons, blood vessels, and nerves. The focus of this section is on the proximal part of the thoracic limb and its clinical anatomy, described primarily in cattle because they are the most common ruminant species seen in clinical practice.

Function

The complete myo-arthro-skeletal system of the thoracic limb is important not only for locomotion, but also to allow the animal to express certain behaviors, such as postural adjustments during rest, feeding, stretching, and grooming. The integrity of this system is also important for standing up and lying down. Injuries or abnormalities of the bones, joints, ligaments, and muscles are likely to cause lameness. In cattle, as in most quadruped mammals, the bones of the thoracic limb have several important functions: they support the muscular system of the limb, support the trunk, produce blood cells, and store and release minerals and fat. They allow movement of the limb by serving as attachment/insertion points for the tendons and muscles. The bones of the proximal thoracic limb transmit the forces produced by muscle contraction. Bones act as levers while the joints serve as support points (fulcrums) between 2 or more bones. The muscles of the thoracic limb dynamically control the change in the angulation of the joints during locomotion, particularly during the swing phase of the stride, and stiffen the limb and its joints during the support phase.

Bony ends are covered with articular cartilage where the bones meet to form joints. The joints of the proximal thoracic limb allow a physiologic range of extension and flexion in movement. The joint surfaces of adjacent bones are enclosed in a space filled with joint fluid that allows the bones to move effortlessly between each other. High-motion joints are stabilized by the collateral ligaments and surrounding muscles.

Synovial fluid, a.k.a. synovia, is a viscous fluid in the synovial joints. The major role of synovial fluid is to minimize friction between the articular cartilages of the joints. Synovial fluid is also a small portion of the transcellular fluid of extracellular fluid.

Bones of the pectoral girdle, brachium, and antebrachium

The **scapula**, a.k.a. the shoulder blade, represents the most proximal bone of the thoracic limb and major bone of the pectoral girdle (Fig. 21.1-4). It is attached to the thorax by extrinsic muscles. It lays flat against the cranial part of the thoracic wall. Its dorsal extremity is cartilaginous to help with shock absorption and provides a larger surface of attachment to the trunk. With age, the cartilage ossifies. The **scapular spine** extends from the proximal border to the distal angle of the scapula and divides the lateral surface of the scapula into 2 fossae, in which the supraspinatus muscle resides cranially and the infraspinatus muscle caudally. The spine terminates distally with the prominent **acromion** located near the distal border of the scapula. The scapula articulates with the humerus via its **glenoid cavity**, forming the **shoulder joint**, also called the scapulohumeral or glenohumeral joint. The **supraglenoid tubercle** is an important protuberance located at the cranial aspect of the glenoid cavity and is the site of attachment for the biceps muscle.

Winged shoulders are a common flaw in which cattle, mainly dairy breeds—e.g., Jersey—stand with both shoulders and elbows abducted. This abnormal position of the thoracic limbs usually appears to be insignificant for the life and locomotion of the cow. In conformation judging of show cattle, however, it can cause lower scores as relates to the ideal body type.

The **humerus** is the only bone of the brachium (Fig. 21.1-5). The proximal part of the humerus bears the head, along with the greater and lesser tubercles of the humerus. The convex **head of the humerus** articulates with the concave

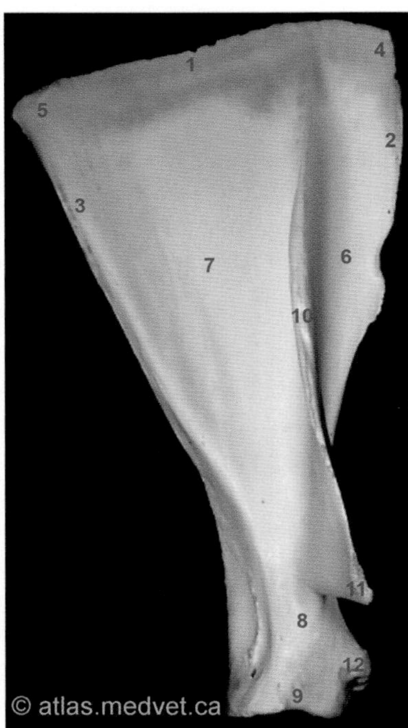

FIGURE 21.1-4 Lateral aspect of a right bovine scapula. Key: 1, dorsal border; 2, cranial border; 3, caudal border; 4, cranial angle; 5, caudal angle; 6, supraspinous fossa; 7, infraspinous fossa; 8, neck; 9, glenoid cavity; 10, scapular spine; 11, acromion; 12, supraglenoid tubercle.

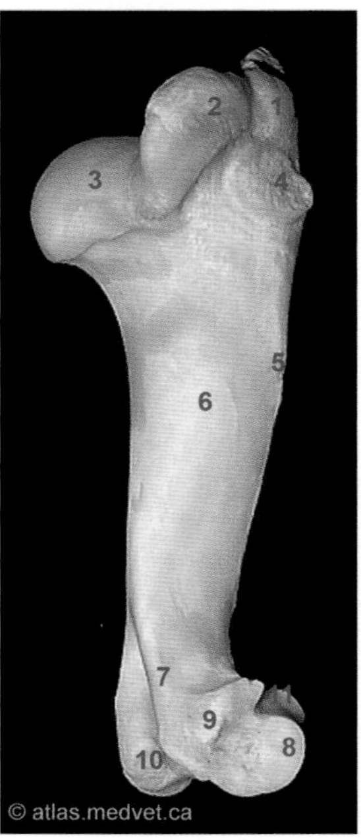

FIGURE 21.1-5 Lateral aspect of a right bovine humerus. Key: 1, major tubercle (cranial part); 2, major tubercle (caudal part); 3, head; 4, infraspinatus surface; 5, deltoid tuberosity; 6, brachialis groove; 7, lateral supracondylar crest; 8, lateral condyles; 9, lateral epicondyle; 10, medial epicondyle; 11, radial fossa.

glenoid cavity of the scapula to form the shoulder joint. It is separated from the shaft of the humerus by the **neck of the humerus**. The **greater and lesser tubercles of the humerus** are located craniolateral and craniomedial to the head of the humerus, respectively. Both tubercles are divided into a cranial and a caudal part. The **bicipital groove** is located between those tubercles and contains the tendon of origin of the biceps muscle.

Just proximal to the mid-level of the humeral shaft are 2 bony landmarks—the **deltoid tuberosity** laterally and the **teres major tuberosity** medially. The distal end of the humerus is comprised of many bony landmarks—the larger **medial condyle** (trochlea), the **lateral condyle** (capitulum), the **medial epicondyle**, the **lateral epicondyle**, the **olecranon fossa**, and the **radial fossa**. The humeral condyle is angled cranially as compared to the long axis of the humerus and articulates with the radius and ulna (i.e., the **elbow joint**). The lateral and medial epicondyles consist of important protuberances located on each side of the humeral condyle, pointing caudolaterally and caudomedially, respectively. The olecranon fossa is a large depression located at the distal and caudal aspects of the humerus. It separates the lateral and medial epicondyles and articulates with the olecranon of the ulna. The radial fossa is a slight depression located on the cranial part of the humeral condyle.

The **radius** and **ulna** comprise the 2 bones of the antebrachium. The ulna is thin and complete in the bovid (Fig. 21.1-6A and B) and continues distally to the end of the radius. The proximal part of the ulna is called olecranon and is formed by the **olecranon** and the **anconeal process**. The olecranon represents the most proximal point of the elbow. The anconeal process points cranially at the most proximal part of the **trochlear notch**. The trochlear notch of the ulna and the radial head articulate with the humeral condyle (elbow joint). The body of the ulna is fused to the body of the radius, except proximally and distally at the **proximal** and **distal interosseous spaces**, respectively. The most distal extremity of the ulna terminates as the **styloid process of the ulna**.

During an orthopedic examination of a thoracic limb in a domestic ruminant, the cranial and caudal scapular angles, scapular spine, greater tubercle, and deltoid tuberosity of the humerus are the palpable landmarks of the brachium. The tuber olecrani and lateral styloid process are the palpable landmarks of the ulna.

The **radius** is broad, shorter than the ulna, and it bears most of the weight (Fig. 21.1-6A and B) in the thoracic limb. The **radial head** is the most proximal part of the radius and includes a **radial tuberosity** on its medial aspect. The medial portion of the **body of the radius** is subcutaneous and is bordered by the flexor muscle group (also called the caudomedial group of muscles of the antebrachium) caudally and the extensor muscle group (also called the craniolateral group of muscles of the antebrachium) cranially (Fig. 21.1-7). The transverse section of the body of the radius is "crescent-shaped," with its convex surface oriented cranially and its concave surface oriented caudally. The distal part of the radius ends in the **styloid process of the radius** on its medial side. The radial trochlea and distal ulna articulate with the proximal row of carpal bones (i.e., the radiocarpal or antebrachiocarpal joint).

One type of external fixation technique for the treatment of radial or metacarpal fractures involves transfixation pins and external coaptation (cast placement). During insertion of transfixation pins into the radius, it is important to know that the transverse section of the radius is "crescent-shaped," as described here. Transfixation pins need to be inserted in a lateromedial plane without engaging the caudal cortex of the radius.

Major muscles and tendons of the pectoral girdle, brachium, and antebrachium

The extrinsic muscles of the thoracic limb that attach to the axial skeleton (Fig. 21.1-8) consist of the trapezius (cervical and thoracic parts), rhomboideus (cervical and thoracic parts), brachiocephalicus (cleidocephalicus and cleidobrachialis muscles), omotransversarius, latissimus dorsi, superficial pectoral (transverse and ascending parts), deep pectoral, subclavius, and serratus ventralis muscles (cervical and thoracic parts).

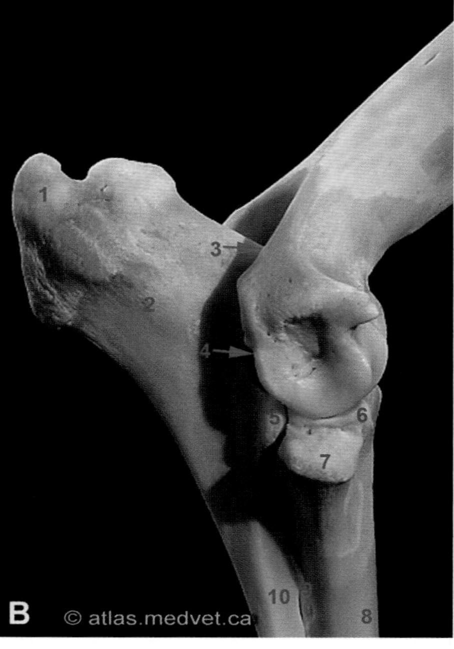

FIGURE 21.1-6 (A) Lateral aspect of a right bovine radius and ulna. (B). Lateral aspect of a right bovine elbow. Key: 1, tuber olecrani; 2, olecranon; 3, anconeal process; 4, trochlear notch; 5, lateral coronoid process; 6, radial head; 7, lateral tuberosity of the radius; 8, body of the radius; 9, proximal interosseous space; 10, body of the ulna; 11, distal interosseous space; 12, head of the ulna; 13, trochlea of the radius; 14, lateral styloid process.

Bos taurus indicus cattle—e.g., Zebu, Brahman, Santa Gertrudis—can be recognized by a hump on their backs. This hump is formed by widening of the rhomboideus muscle in some breeds, whereas, in others the hump is simply comprised of fat. Therefore, the hump can vary in position and content depending on the breed of cattle.

The **trapezius muscle** elevates the scapula when it contracts. It originates on the funicular part of the nuchal ligament and inserts on the spine of the scapula. Contraction of the **rhomboideus muscle** pulls the scapula proximally and originates on the funicular part of the nuchal ligament and thoracic spines and inserts on the scapular cartilage.

During locomotion, the **brachiocephalicus muscle** extends the shoulder joint and advances the thoracic limb. During weight-bearing, unilateral contraction of the brachiocephalicus muscle results in lateral displacement of the head and neck whereas bilateral contraction of the brachiocephalicus muscles results in ventral displacement of the head and neck. This muscle originates on the clavicular intersection and inserts on the humerus, nuchal crest, funicular nuchae, and temporal bone. When the **omotransversarius muscle** contracts, it pulls the neck laterally. It originates on the scapular acromion and inserts on the axis and atlas.

The omotransversarius muscle covers the superficial cervical lymph node, which is typically palpable on physical examination.

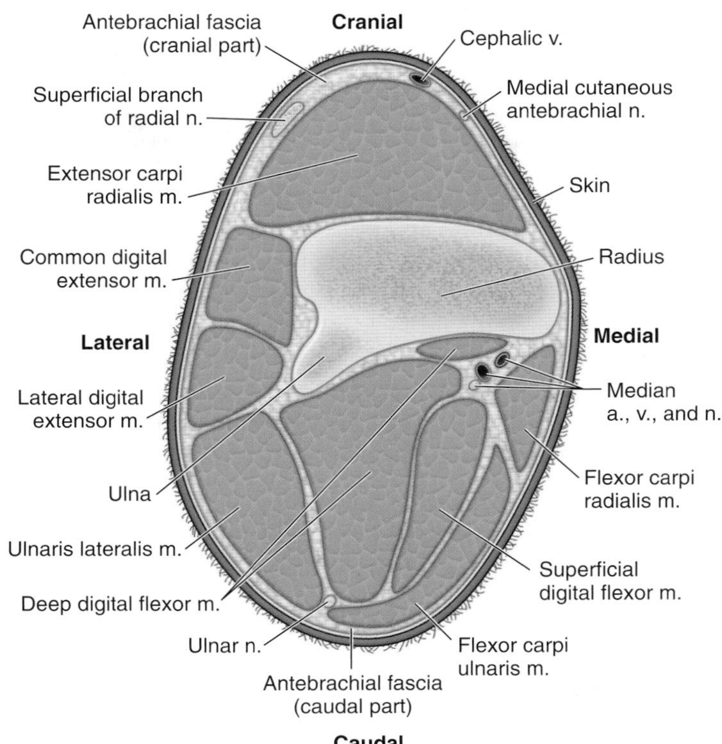

FIGURE 21.1-7 Transverse section of the distal radius and ulna illustrating the muscles of the antebrachium (forearm) of the ruminant.

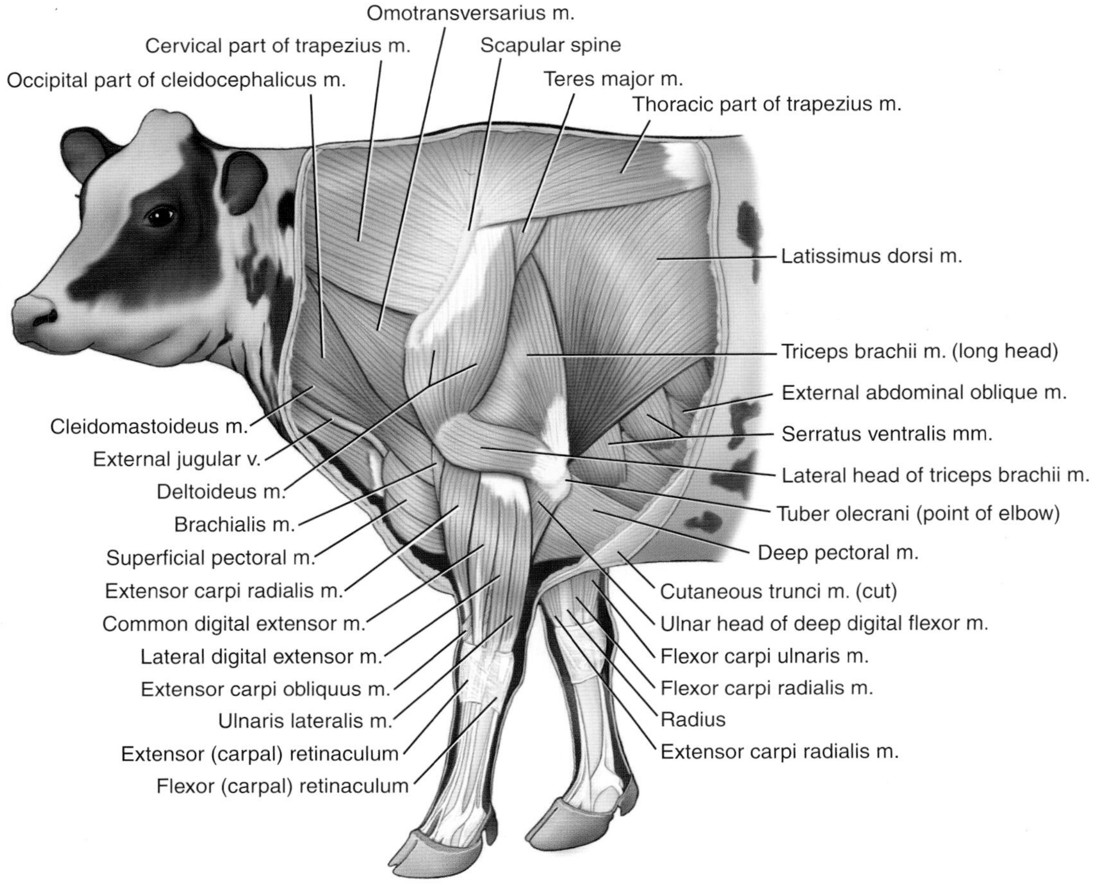

FIGURE 21.1-8 Muscles and tendons of the pectoral girdle, brachium, and antebrachium of the bovid.

Contraction of the **latissimus dorsi muscle** pulls the thoracic limb caudally when in a nonweight-bearing position and pulls the trunk cranially when the thoracic limb is weight-bearing. It originates on the thoracolumbar fascia (dorsal midline of the thoracolumbar region) and inserts on the teres major tuberosity of the humerus. The **superficial pectoral muscle** adducts the nonweight-bearing thoracic limb while preventing abduction during weight-bearing when it contracts. It originates on the sternum and inserts on the humeral crest. Contraction of the **deep pectoral muscle** adducts the thoracic limb and pulls it caudally when nonweight-bearing. It originates on the sternum and inserts on the lesser and greater tubercles of the humerus. The cervical part of the **serratus ventralis muscle** pulls the scapula cranially when the thoracic limb swings caudally. Conversely, the thoracic part of this muscle pulls the scapula caudally when the thoracic limb swings cranially. The serratus ventralis muscle originates on cervical vertebrae 2–7 (C2–C7) and first 7–8 ribs and inserts on the costal surface of the scapula. More importantly, the right and left serratus ventralis muscles, along with the superficial pectoral muscles, act as a sling to support the trunk and its weight between both thoracic limbs. The **subclavius muscle** is underdeveloped in cattle; it originates from the 1st rib cartilage and inserts on the clavicular intersection.

The intrinsic muscles of the proximal part of the thoracic limb (Fig. 21.1-8) with origins and insertions on the appendicular skeleton consist of the supraspinatus, infraspinatus, deltoideus, teres minor, teres major, triceps brachii, subscapularis, anconeus, tensor fasciae antebrachii, coracobrachialis, biceps brachii, and brachialis muscles.

The **supraspinatus** and **infraspinatus muscles** are mainly located lateral to the shoulder, bracing and supporting this joint by functioning as collateral ligaments. The supraspinatus muscle originates in the supraspinous fossa of the scapula and inserts on the lesser and greater tubercles of the humerus, functioning to extend the shoulder joint. The **infraspinatus muscle** originates in the supraspinous fossa of the scapula and inserts on the greater tubercle of the humerus. The **deltoideus muscle** functions in flexing the shoulder joint, covering most of the infraspinatus muscle. It originates on the scapular spine and acromion and inserts on the deltoid tuberosity of the humerus. The **teres minor muscle** is located deep to the deltoideus muscle and lies against the shoulder joint capsule laterally. It originates on the distal part of the caudal border of the scapula and inserts on the teres minor tuberosity of the humerus. The **teres major muscle** flexes the shoulder joint when it contracts and originates on the caudal angle of the scapula and inserts on the teres major tuberosity of the humerus (proximomedially) along with the latissimus dorsi muscle tendon of insertion.

The **triceps brachii muscle** is a broad muscle located in the triangular space bordered by the caudal boundary of the scapula, that of the humerus, and an imaginary line from the caudal edge of the scapula to the tuber olecrani. It consists of 4 heads: long, lateral, accessory, and medial. The **accessory head** is small or absent in ruminants; the **long head** originates on the caudal aspect of the scapula, the **lateral** and **medial heads** originate on the proximal humerus, and all combine to insert on the tuber olecrani. The triceps brachii muscle extends the elbow joint when it contracts, while contraction of the long head of the triceps brachii muscle also flexes the shoulder joint. The **subscapularis muscle** also stabilizes the shoulder joint on its medial side; it originates on the subscapular fossa and inserts on the lesser tubercle of the humerus.

"FLYING" SCAPULA

A bilateral rupture of the main muscle supporting the thorax between the thoracic limbs (primarily the serratus ventralis muscle) is an infrequent condition of traumatic origin found in excited cattle newly turned out in a new environment, such as a pasture. Periods of excessive running cause the muscle tissue to tear. This causes the thorax to drop between the thoracic limbs, resulting in the dorsal border of the scapula (cartilage) protruding above the level of the thoracic spinal column, with the impression of a humpback appearance of the affected individual. Deficiency in vitamin E and/or selenium has been implicated in the pathophysiology of this disorder. Along with the history and physical examination of the affected individual, clinical suspicion of this disorder is supported by elevated muscular enzymes (CK and AST) on serum biochemistry analysis.

The **anconeus** and **tensor fasciae antebrachii muscles** are minor extensors of the elbow joint. Contraction of the **coracobrachialis muscle** adducts and flexes the shoulder joint while concurrently stabilizing the joint. It originates from the coracoid process of the scapula and inserts on the distomedial aspect of the distal humerus. The **biceps brachii muscle** flexes the elbow and extends the shoulder; it originates on the supraglenoid tubercle and inserts on the proximomedial aspect of the radius. The tendon of origin of the biceps brachii muscle is covered by the **transverse (humeral) retinaculum,** which holds the tendon in place within the intertubercular groove. The **brachialis muscle** flexes the elbow joint; it originates on the proximocaudal part of the humerus and inserts on the proximomedial part of the radius.

A large number of muscles cover the antebrachium and are divided into craniolateral and caudomedial groups. The craniolateral muscles are located on the cranial and lateral part of the antebrachium and consist of extensor muscles of the carpus and digits. The caudomedial muscles are located on the caudal and medial part of the antebrachium and are flexors of the carpus and digits.

The craniolateral muscles of the antebrachium (Fig. 21.1-7) consist of the extensor carpi radialis, common digital extensor, lateral digital extensor, ulnaris lateralis, and extensor carpi obliquus muscles, from a cranial to caudolateral direction and all extend the carpus. The **common** and **lateral digital extensor muscles** also extend the digits. The **extensor carpi radialis muscle** originates on the lateral supracondylar crest of the humerus and radial fossa and inserts on the metacarpal tuberosity of the fused metacarpal 3 and 4. The **common digital extensor muscle** has a lateral and a medial head and 3 tendons of insertion. Both heads originate on the lateral epicondyle of the humerus. The tendon of the lateral head divides into 2 smaller tendons and they insert on the extensor processes of the 3rd phalanx of the medial (digit III) and lateral digit (digit IV); it is also called the **common extensor tendon of digits III/IV**. The tendon of insertion of the medial head inserts on the 2nd phalanx of the medial digit (digit III) and is called the **medial digital extensor tendon**. The **lateral digital extensor muscle** originates on the lateral epicondyle of the humerus and lateral collateral ligament of the elbow and inserts on the 2nd and 3rd phalanges of the lateral digit (digit IV). The **ulnaris lateralis muscle** originates on the lateral epicondyle of the humerus and inserts on the accessory carpal bone and the lateroproximal aspect of the fused 3rd and 4th metacarpal bones and can flex or extend the carpus depending on the concurrent action of other muscles. The **extensor carpi obliquus muscle** originates on the radius (body) and inserts on the medioproximal aspect of the fused 3rd and 4th metacarpal bones.

> The surgical approach to the radius for internal fixation of a radial fracture depends on the fracture configuration and the type of internal fixation used. Briefly, the patient is placed in lateral recumbency with the affected limb uppermost. A cranially directed curvilinear skin incision is performed from the lateral epicondyle of the humerus to the distal lateral tuberosity of the radius. A plane of dissection is created between the extensor carpi radialis and common digital extensor muscles. The radial nerve is identified, examined, and protected. Bone plates and screws are used to repair the radial fracture.

The caudomedial muscles of the antebrachium consist of flexor carpi ulnaris, flexor carpi radialis, superficial digital flexor, and deep digital flexor muscles. The **flexor carpi ulnaris muscle** originates on the medial epicondyle of the humerus (humeral head) and olecranon (ulnar head) and inserts on the accessory carpal bone and functions to flex the carpus and extend the elbow. The **flexor carpi radialis muscle** originates on the medial epicondyle of the humerus and inserts on the proximomedial aspect of the fused 3rd and 4th metacarpal bones; it flexes the carpus. The **superficial digital flexor muscle** originates on the medial epicondyle of the humerus and inserts on the palmar aspect of the medial and lateral 2nd phalanx (digits III and IV); it flexes the carpus, metacarpophalangeal (fetlock), and proximal interphalangeal (pastern) joints and extends the elbow. The **deep digital flexor muscle** has 3 heads: the **ulnar**, **radial,** and **humeral heads**. They originate on the olecranon, proximomedial aspect of the radius, and medial epicondyle of the humerus, respectively, and insert on the flexor tuberosity of the 3rd phalanx of both digits (digits III and IV) by a split tendon. Contraction of its muscle belly results in flexion of the carpus, metacarpophalangeal (fetlock), proximal interphalangeal (pastern), and distal interphalangeal (coffin) joints, along with extension of the elbow.

Major synovial structures of the brachium and antebrachium

Shoulder luxation is an uncommon clinical occurrence in ruminants. The supraspinatus and infraspinatus muscles support the shoulder laterally and the subscapularis muscle supports it medially, all functioning as collateral ligaments. The biceps brachii and triceps brachii muscles also support the cranial and caudal parts of the joint, respectively.

The glenoid cavity of the scapula and the humeral head are the 2 bony structures comprising the **shoulder** (scapulohumeral or glenohumeral) **joint**. Movement of the shoulder joint is restricted to flexion and extension in ruminants. The shoulder joint lacks collateral ligaments in cattle, instead relying on muscles and tendons surrounding the joint for stability.

The **elbow joint** consists of 3 articulations: the humeroulnar, humeroradial, and proximal radioulnar joints. The humeral condyle articulates distally with the radial head (**humeroradial joint**) and the trochlear notch of the ulna (**humeroulnar joint**). Furthermore, the articular circumference of the proximal radius articulates with the radial notch of the ulna (**proximal radioulnar joint**). The elbow joint is located at the level of the ventral aspect of the 4th and 5th ribs and stabilized by the **lateral** and **medial collateral ligaments,** originating from the lateral and medial humeral epicondyles and inserting on the proximal lateral and medial margins of the radius.

The **infraspinatus bursa** is a subtendinous synovial bursa located deep to the superficial part of the **infraspinatus tendon**. This large bursa protects the tendon as it passes over the lateral aspect of the greater tubercle of the humerus.

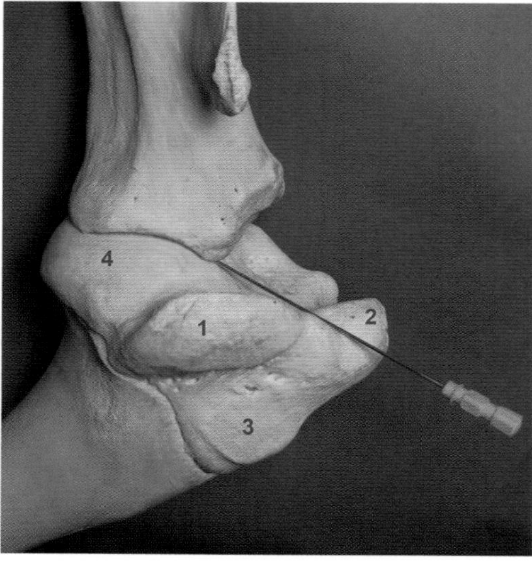

FIGURE 21.1-9 Lateral view of a bovine right shoulder, indicating the direction of a spinal needle for arthrocentesis of the scapulohumeral joint. Key: 1, caudal part of the greater tubercle; 2, cranial part of the greater tubercle; 3, infraspinatus surface; 4, humeral head.

JOINT INJECTION OR ARTHROCENTESIS IN THE THORACIC LIMB

To access the shoulder joint, an 18-gauge × 8.9-cm (3.5-in.) spinal needle is inserted cranial to the tendon of insertion of the infraspinatus muscle proximal to its insertion on the greater tubercle (Fig. 21.1-9).
To access the elbow joint, a hypodermic needle is inserted laterally in the olecranon fossa between the lateral epicondyle of the humerus and the olecranon (Fig. 21.1-10).

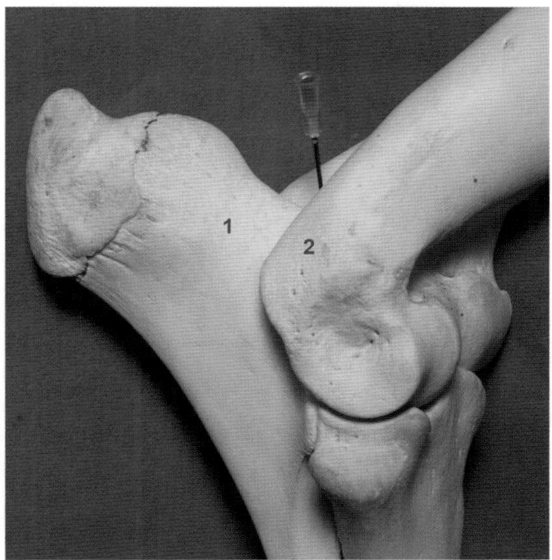

FIGURE 21.1-10 Lateral view of a bovine right elbow depicting one site for arthrocentesis. The needle is inserted between the olecranon *(1)* and the lateral supracondylar crest of the humerus *(2)* in the olecranon fossa.

This bursa may be lacking in sheep and goats. The **intertubercular bursa** is a large synovial bursa located deep and on both sides of the tendon of origin of the biceps brachii as it passes in the intertubercular groove. The **bursa of triceps brachii** is located at the insertion of the combined triceps brachii tendon on the tuber olecrani. The **subcutaneous olecranon bursa** is located on the caudal aspect of the olecranon.

In cattle and horses, the intertubercular bursa is an independent synovial structure, while in dogs, sheep, and goats the deeper part of the tendon of origin of biceps brachii is protected by a pouch of the shoulder joint, instead of having an independent synovial bursa.

Innervation of the brachium and antebrachium

The nerves of the brachium and antebrachium are supplied by the **brachial plexus,** which is formed by the ventral branches of C6–T2 spinal nerves (Fig. 21.1-11).

A brachial plexus nerve block is described in cattle. It is used to anesthetize the thoracic limb for surgery—e.g., arthrotomy (Fig. 21.1-12).

INJURY TO THE BRACHIAL PLEXUS IN NEONATAL CALVES

Forceful and/or prolonged traction applied to the thoracic limb of a calf during delivery can cause nerve injury to the brachial plexus. This infrequent condition is encountered in neonatal calves born in longitudinal anterior presentation. This condition seems to occur more frequently when the standing cow in labor suddenly decides to lay down with the calf halfway delivered. The clinical signs vary according to the severity of the injury. Clinical signs—including an inability to bear weight and an inability to extend the elbow, carpus, and fetlock—look like those of radial nerve paresis or paralysis; however, other nerves—e.g., the musculocutaneous, median, and ulnar nerves—can be affected.

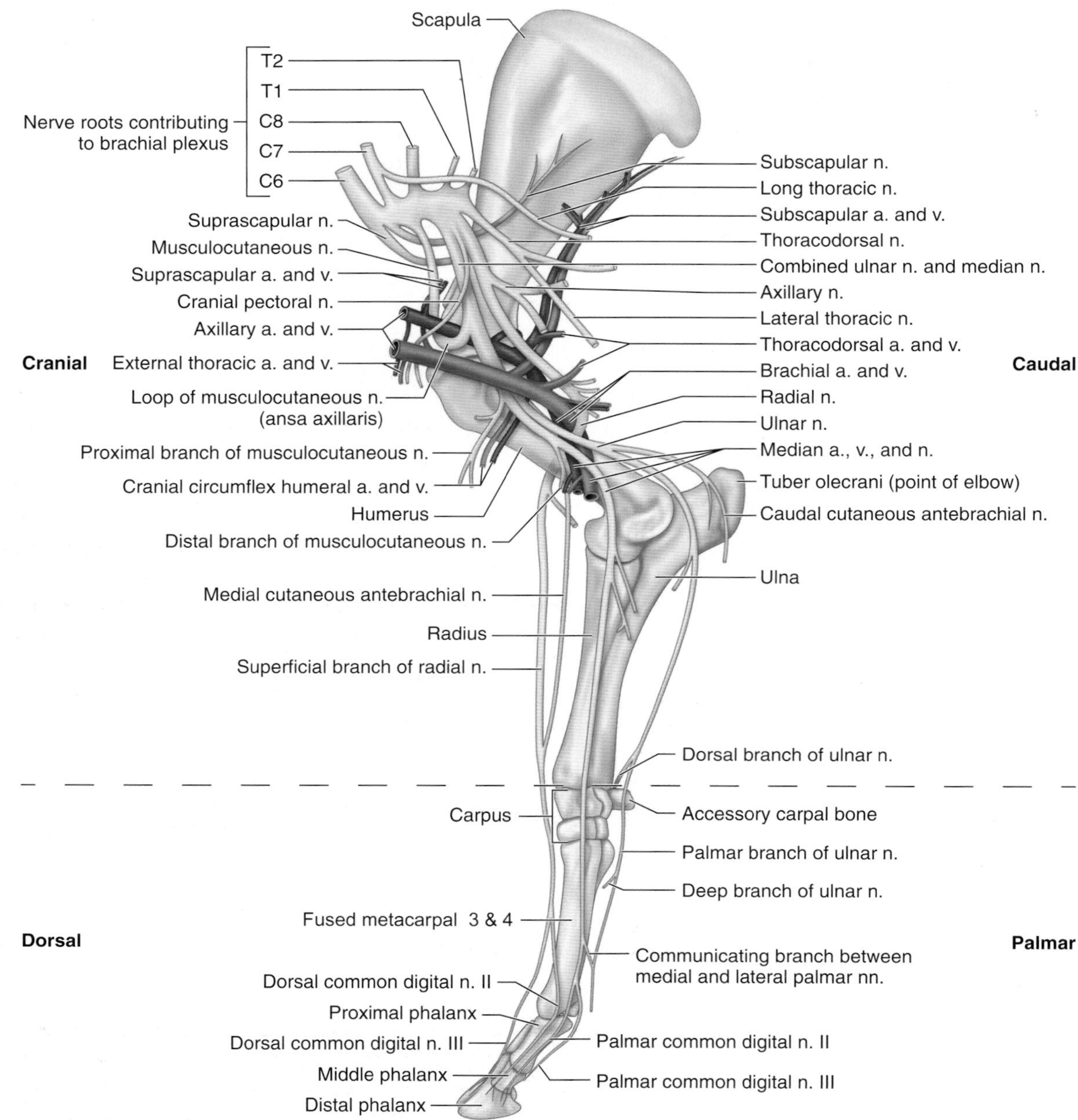

FIGURE 21.1-11 Medial view of the right thoracic limb of a bovid illustrating the nerves of the brachium and antebrachium bovine.

The **suprascapular nerve** (from C6–C7; motor) passes around the cranial border of the scapula to innervate the supraspinatus and infraspinatus muscles. The **subscapular nerve** (from C7–C8; motor) innervates the subscapularis muscle. The **axillary nerve** (from C7–C8; mixed) innervates the shoulder joint, subscapularis, teres major, teres minor, and deltoideus muscles. Its cutaneous branch innervates the skin over the shoulder to the craniolateral surface of the mid-antebrachium. The **musculocutaneous nerve** (from C6–C8; mixed) innervates the coracobrachialis, biceps brachii, and brachialis muscles. Its cutaneous branch innervates the skin over the medial aspect of the forearm. The **median nerve** (from C8–T2; mixed) provides muscular branches to the pronator teres, flexor carpi radialis, flexor carpi ulnaris, flexor digitorum superficialis, and radial and humeral heads of the flexor

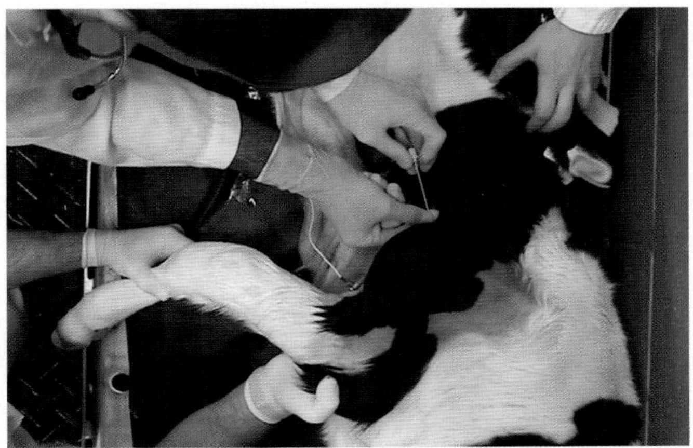

FIGURE 21.1-12 A left brachial plexus block is performed on a calf. An 18-gauge, 15-cm (6-in.) needle is inserted between the shoulder and thorax by abducting the limb. A nerve stimulator is attached to the needle to precisely locate the nerves to be anesthetized.

digitorum profundus muscles. The **ulnar nerve** (from C8–T2; mixed) provides muscular branches principally to the flexor carpi ulnaris and flexor digitorum superficialis and profundus muscles. The **radial nerve** (from C7–T1; mixed) tracks between the medial and long heads of the triceps brachii muscle in a lateral direction and over the supracondylar crest of the humerus.
It mainly innervates the triceps brachii and extensor muscles of the carpus.

Injury to the radial nerve results in an inability to extend the elbow, resulting in a dropped elbow appearance (Fig. 21.1-13). Other conditions can mimic a dropped elbow, e.g., olecranon fracture, radial fracture, and myopathy of the triceps muscle.

When placing a cow in lateral recumbency, it is important to prevent radial nerve paralysis by positioning the bottom thoracic limb in extension and pulled cranially. The supracondylar crest of the humerus, where the radial nerve tracks, should be well-padded to prevent nerve compression.

Major blood supply and lymphatics of the brachium and antebrachium

The principal blood supply to the thoracic limb is the **axillary artery**, located deep and yet palpable as it turns around the 1st rib. The axillary artery supplies the shoulder region by giving off the external thoracic, suprascapular, and **subscapular arteries** before it becomes the **brachial artery**. The latter supplies the brachium through multiple branches. It continues as the **median artery** supplying the antebrachium. After giving off the **deep antebrachial** and **radial arteries** in the middle of the antebrachium, the median artery passes through the carpal canal along with the median nerve.

The **axillary, brachial**, and **median veins**parallel the arteries of similar names and represent the major veins draining the antebrachium and brachium in cattle. A few superficial veins are clinically relevant: the **cephalic** and **accessory cephalic veins**. The accessory cephalic vein drains blood from the **dorsal common digital vein III** to the cephalic vein. The **median cubital vein** connects the brachial vein to the cephalic vein.

The **axillary lymph nodes** are found within the axilla caudal to the shoulder joint near the 2nd intercostal space, deep to the teres major muscle, and on the lateral surface of the 1st rib. The axillary lymph nodes drain the lymph

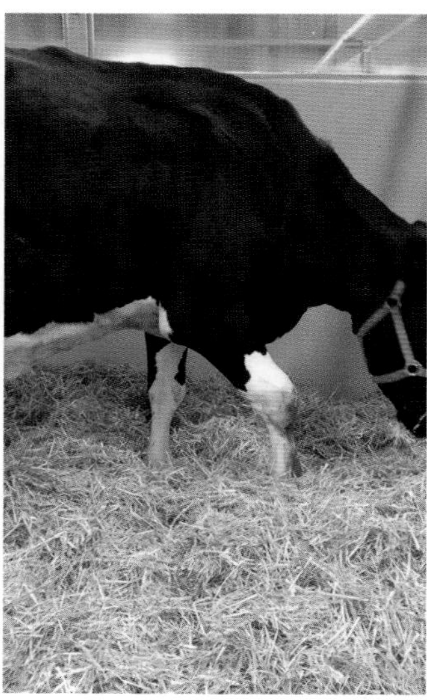

FIGURE 21.1-13 An adult Holstein cow with the classic stance of right radial nerve paresis. The elbow is dropped with a flexed carpus.

The superficial cervical lymph node(s) is palpable on physical examination and may be enlarged with cellulitis or phlegmon (Gr. *phlegmonē* suppurative or gangrenous inflammatory reaction) of the neck and/or proximal thoracic limb secondary to an injection site infection—e.g., subcutaneous injection of calcium gluconate—or lymphoma.

from the shoulder, brachium, and antebrachium. Their efferent lymphatics drain to the **caudal deep cervical lymph nodes**, then to the thoracic duct (left side) or right tracheal duct (right side), and finally to major veins located near the thoracic inlet. The **superficial cervical lymph nodes** drain the skin of the neck and proximal limb.

Selected references

[1] Dellmann HD, McClure RC. Ruminant nervous system. In: Sisson G, Grossman J, editors. Anatomy of domestic animals. Philadelphia: WB Saunders Co; 1975. p. 1065–179.

[2] Singh B, editor. The forelimb of the ruminant. In: Dyce, Sack and Wensing's textbook of veterinary anatomy. 5th ed. St. Louis, MO: Elsevier; 2018. p. 715–28.

[3] Liebich HG, Maierl J, König HE. Forelimb or thoracic limb (membra thoracica). In: König HE, Liebich HG, editors. Veterinary anatomy of domestic mammals. 4th ed. Stuttgart, Germany: Schattauer GmbH; 2009. p. 145–214.

[4] Mansour M, Wilhite R, Rowe J. The forelimb. In: Guide to ruminant anatomy: dissection and clinical aspects. 1st ed. Hoboken, NJ: John Wiley & Sons; 2018. p. 173–215.

[5] Atlas d'anatomie et d'imagerie médicale [Internet]. Montreal (QC): Université de Montréal; 2019. Membre thoracique [cited 2019 December 2]; [about 12 screens]. Available from: http://atlas.medvet.ca.

CASE 21.2

Septic Arthritis of the Distal Interphalangeal Joint

Karl Nuss
Farm Animal Surgery Section, Farm Animals, Vetsuisse-Faculty University of Zürich, Zürich, CH

Clinical case

History

A 4-year-old dairy cow with chronic white line disease of the lateral claw of the left pelvic limb was presented for treatment. The cow had been repeatedly trimmed over a 2-week period by the hoof trimmer. Loose horn was removed, the foot was cleaned, the lateral claw was wrapped, and a wooden block was applied to the medial claw. After initial improvement, the cow's comfort decreased as well as weight-bearing on her left pelvic limb, noticed several days before referral to the clinic.

Physical examination findings

The cow was recumbent for long periods of time during the day, had difficulty rising, and was in poor body condition (BSC 2/5). Her rumen volume was decreased and her abdomen was tucked up. Her left pelvic limb had generalized muscle atrophy. The cow was tachycardic with a heart rate of 96 bpm, tachypneic with a respiratory rate of 52 brpm, and had an elevated rectal temperature of 102.7°F (39.3°C). The cow was nonweight-bearing on the affected leg when standing and she positioned it cranially and medially under the abdomen. There was a grade 4/5 supporting limb lameness at a walk despite the wooden block that was glued to the medial claw.

Differential diagnoses

Heel abscess, interdigital phlegmon (Gr. *phlegmonē* general inflammatory/infection reaction), hematogenous septic arthritis of a phalangeal joint, or septic lesion further proximal in the limb

Diagnostics

The cow was sedated, positioned, and restrained on a tilt table in right lateral recumbency. The left pelvic limb was supported by a brace proximal to the tarsus, the hair on the lower limb was clipped to the midtarsal level, and the foot was cleaned. There was marked swelling of the coronet and heel region of the lateral claw abaxially, and a granulating wound with an evident sinus tract at the midsection of the lateral claw wall was identified (Fig. 21.2-1).

Passive movements of the claw (flexion, extension, and rotation) elicited a pain response. A metal probe introduced into the sinus tract contacted bone at 1.5-cm (0.6-in.) of depth from the insertion point.

Two orthogonal-view radiographs were taken of the affected limb. The dorsoplantar radiograph (Fig. 21.2-2A) showed soft tissue swelling of the affected toe, the claw wall defect, osteolysis in the region of the flexor tubercle, and radiopacities abaxial to the bone outline of the 3rd phalanx, distal sesamoid, and 2nd phalanx. The abaxio-axial ("interdigital") radiographic view showed several lesions in the region of the flexor tubercle and at the distal outline of the distal sesamoid bone (Fig. 21.2-2B).

Ultrasonographic images obtained from the dorsal and plantar aspects of the lateral claw revealed distension of the distal interphalangeal joint (DIPJ) with moderately hyperechoic-mixed echogenic joint fluid (Fig. 21.2-3), indicative

Comparative Veterinary Anatomy: A Clinical Approach. https://doi.org/10.1016/B978-0-323-91015-6.00104-7

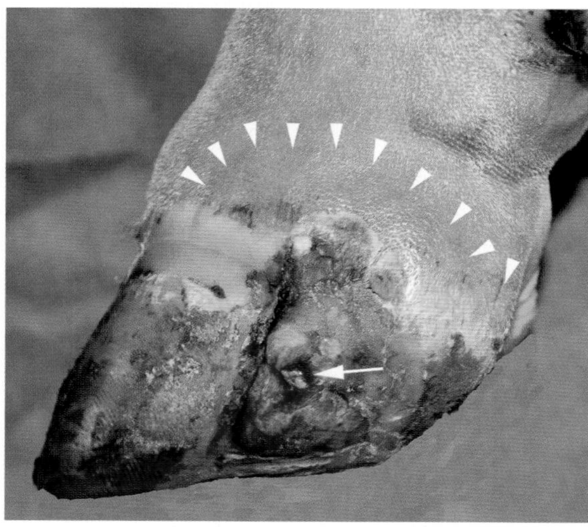

FIGURE 21.2-1 Deep sepsis attributable to white line disease of the lateral hind claw of the left pelvic limb. The *arrowheads* delineate a semicircular area of erythema and swelling of the skin adjacent to the coronet and heel region. The plantar half of the abaxial claw wall has a large defect with purulent exudate draining from a sinus tract at the center of the defect halfway between the sole and the coronet (*arrow*). The tract leads to the distal sesamoid bone and the flexor tubercle of the 3rd phalanx.

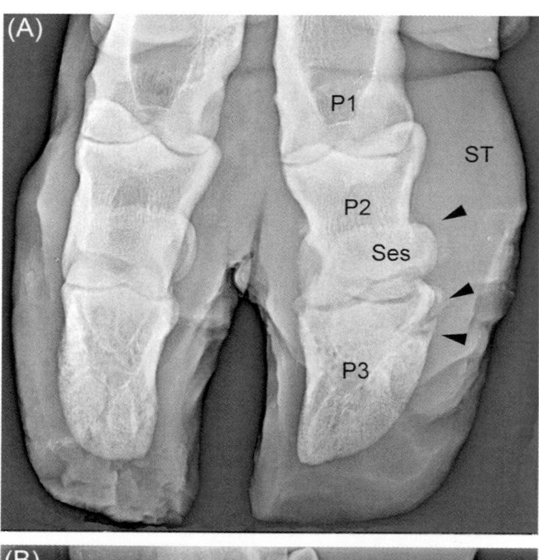

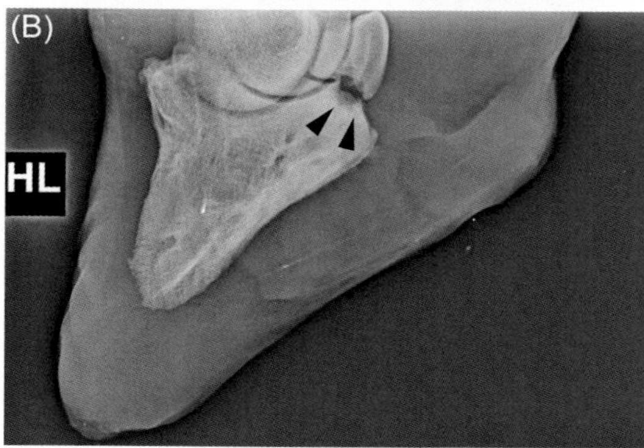

FIGURE 21.2-2 (A) Dorsoplantar radiographic view of the digits of the left pelvic limb. The lateral claw has radiographic changes in the region of the flexor tubercle of the 3rd phalanx and radiopacities (*arrowheads*) along the abaxial surface of the bones, mild widening of the distal interphalangeal joint space, soft tissue swelling (*ST*), and an irregular abaxial contour of the lateral claw horn. Key: *P1, P2, P3,* 1st, 2nd, and 3rd phalanges; *Ses,* distal sesamoid bone; *ST,* soft tissue. (B) Abaxio-axial radiographic view of the lateral claw of the left pelvic limb. The cassette has been placed in the interdigital cleft. There are irregularities of the flexor tubercle (*arrowheads*).

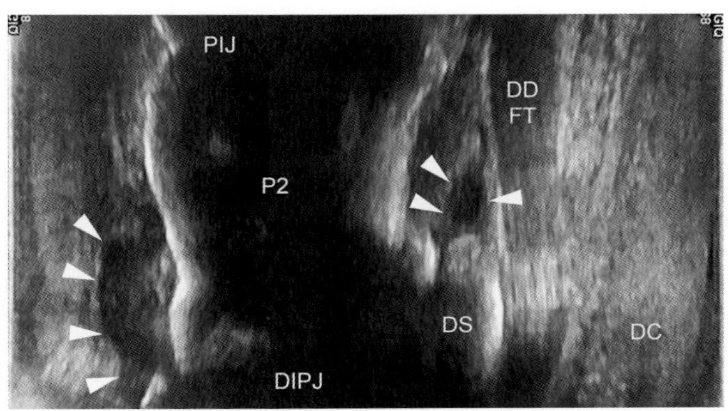

FIGURE 21.2-3 Ultrasonogram of a septic distal interphalangeal joint, comprised of the images obtained in a dorsal and plantar scanning plane. *White arrowheads* indicate the outline of the distal interphalangeal joint space, which is distended by moderately echogenic fluid. Key: *PIJ,* proximal interphalangeal joint; *P2,* 2nd phalanx; *DIPJ,* distal interphalangeal joint; *DS,* distal sesamoid; *DDFT,* deep digital flexor tendon; *DC,* digital cushion.

of septic arthritis. Evaluation of the fistula in a plantar-to-dorsal direction, starting at the soft heel region, showed irregular contours of the 3rd phalanx and the distal sesamoid bone, along with an abscess involving the latter. There were no ultrasonographic signs of digital phlebitis (Gr. *phlebos* vein + *-itis* inflammation = inflammation of a vein).

Diagnosis

Deep sepsis of the foot caused by a white line infection, involving the abaxial part of the flexor tubercle and the distal sesamoid bone, and with subsequent septic arthritis of the DIPJ

ULTRASONOGRAPHY FOR LAMENESS DIAGNOSIS

Ultrasonography provides a quick and objective overview of the soft tissue structures of the bovine digit; however, detailed knowledge of the relevant anatomy and ultrasound scanning planes are a prerequisite. Standard sagittal scanning planes are commonly used when scanning the digits of a cow. The ultrasound probe is applied on both the dorsal and palmar/plantar aspects and evaluation includes the coronet. Two ultrasonographic views of a septic DIPJ that have been reconfigured with a software program are shown in Fig. 21.2-3.

JOINT RESECTION

Resection is the surgical removal of part of an organ or a structure. Resection of the DIPJ implies removal of the relevant joint surfaces of the pedal bone and the 2nd phalanx, complete removal (extirpation) of the distal sesamoid, and removal of the distal end of the deep digital flexor tendon. In bovids, several joints can be successfully treated using arthrotomy, debridement, and resection.

Chronic infections, characterized by fibrin and purulent debris in the joint (Fig. 21.2-4A), are commonly seen in dairy cows. These cases require aggressive surgical treatment because simple joint debridement is replete with complications.

The goal of DIPJ resection is to promote a rapid, less painful, and stable union of the 2nd and 3rd phalanges (facilitated ankylosis), and allow preservation of a weight-bearing limb. It can take up to 2 years for the joint to ankylose. Finally, the surgical fixation of a (generally nonseptic) joint is termed arthrodesis (Gr. *arthron* joint + Gr. *desis* binding).

Fibrinopurulent infection of the DIPJ accompanied by bone necrosis (Fig. 21.2-4B) is usually not responsive to joint resection, and amputation of the affected claw is indicated.

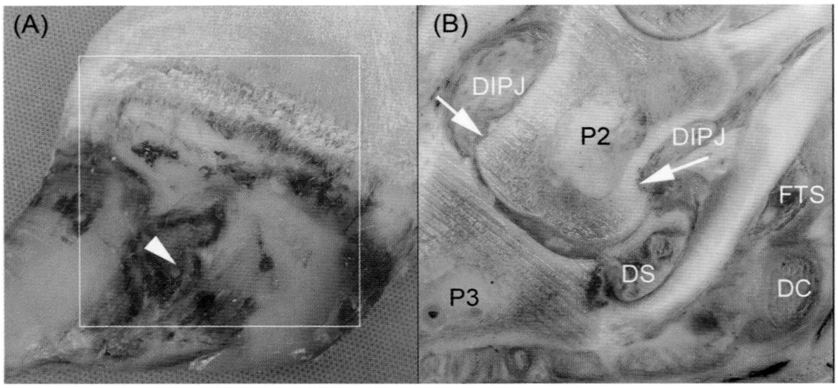

FIGURE 21.2-4 (A) Complicated white line disease of the lateral claw of the left pelvic limb in a dairy cow, treated unsuccessfully with NSAIDs and antimicrobials for 2 weeks. The claw was removed after humane euthanasia. Swelling of the coronet indicates deep sepsis. A fistulous tract (*arrowhead*) at the abaxial claw wall leads to the pedal joint. (B) Sagittal view of a the foot as denoted by the white box in (A). The distal sesamoid bone (*DS*) is almost completely destroyed by necrosis, and there is purulent podotrochlear bursitis and fibrinopurulent infection of the distal interphalangeal joint (*DIPJ*); the flexor tubercle of the 3rd phalanx (*P3*) is also necrotic. Signs of osteitis are apparent at the surfaces of P2 (*white arrows*). The deep digital flexor tendon is only minimally affected, but the podotrochlear bursa is filled with fibrinopurulent material. The common digital flexor tendon sheath (*FTS*) and the digital cushion (*DC*) are interspersed with apostematous (Gr. *apostēma* an abscess) lesions. At this stage of septic arthritis, resection of the DIPJ is no longer a treatment option: amputation is the only viable choice. Key: *P2,* 2nd phalanx with bone marrow; *P3,* 3rd phalanx with bone marrow.

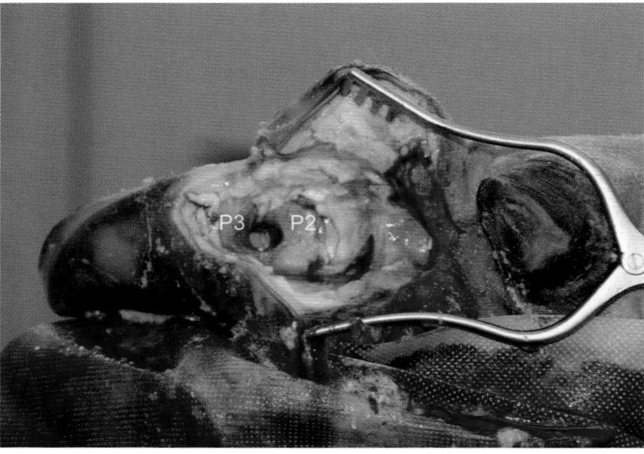

FIGURE 21.2-5 Completed resection of the distal interphalangeal joint. The insertional end of the deep digital flexor tendon, the distal sesamoid, and the cartilage of the 3rd phalanx have been removed with a burr. Key: *P3*, 3rd phalanx; *P2*, 2nd phalanx.

Treatment

After discussing treatment options and prognosis with the owner, a technique to preserve the affected claw was chosen. The cow was placed under general anesthesia in right lateral recumbency. After aseptic preparation, IV regional anesthesia was performed via the abaxial digital vein. This was followed by a step-wise resection of the deep flexor tendon, extirpation (L. *extirpare* to root out + *stirps* root = complete removal of tissue) of the distal sesamoid bone, and ablation (L. *ablatus* removal) of the DIPJ. To achieve this, a skin incision from the base of the dewclaws to the soft heel was made at the plantar aspect of the claw. The sinus tract and the infected end of the deep digital flexor tendon were removed through the digital cushion, followed by excision of the distal sesamoid bone and the flexor tubercle.

A drill hole was made through the DIPJ to create semicircular gaps in the middle of the 2nd and 3rd phalanges without opening the medullary cavities (Fig. 21.2-5).

Anatomical features in ruminants

Introduction

In cattle, weight-bearing forces are transferred to the 3rd and 4th digits, with the limb axis running between them. While the metacarpo-/metatarsophalangeal joints communicate via a palmar-/plantar-located opening, there is no communication between the 2 proximal interphalangeal and the 2 DIPJs. This section covers the DIPJ and the suspensory mechanism of the fetlock (i.e., the metacarpo-/metatarsophalangeal joint), along with the blood supply, lymphatics, and innervation of the digits. The claw itself is discussed in Case 21.3. The distal aspects of the thoracic and pelvic limbs have some anatomic differences, but in general the anatomy is similar except where noted.

Direct trauma to the DIPJ is rarely seen because the joint capsules of the DIPJ are located almost entirely within the claw horn capsule. The most frequent cause of septic inflammation of the DIPJ is an ascending infection from a claw lesion–e.g., white line separation.

Function

The DIPJ of the lateral claw contacts the ground first in the thoracic and pelvic limbs when walking, which stresses the lateral digit more than the medial digit, resulting in an increased number of lateral claw lesions in the pelvic limb. The structure of the horn capsule, digital cushion, and suspensory apparatus of the pedal bone together are important in distributing the forces of static and dynamic weight-bearing (see Case 21.3). The strong axial distal cruciate ligaments prevent spreading of the claws when weight-bearing, while the ligaments of the distal sesamoid bone prevent overextension of the DIPJ.

Structures forming the distal interphalangeal joint

The **claw capsule** encases the **3rd phalanx** and its medullary cavity, the **distal sesamoid bone**, and the distal part of the **2nd phalanx** along with parts of its medullary cavity (Fig. 21.2-6). The articular surface of the 3rd phalanx has a **synovial groove**. Synovial structures contained in the claw capsule include the DIPJ, the podotrochlear bursa, and the distal end of the common digital flexor tendon sheath. Only synovial linings and elastic collagenous tissue divide these 3 synovial structures.

The synovial groove is regularly seen on radiographs and should not be confused with osteolysis (Gr. *osteon* bone + Gr. *lysis* dissolution = loss of bone).

With deep sepsis, bacteria can easily cross these soft tissue boundaries.

WHITE LINE DISEASE IN BOVIDS

White line disease is characterized by an ascending infection from the sole along the laminar horn leaflets. At the proximal end of the leaflets, the tubular horn of the coronary segment forms a barrier to infection; however, this barrier may be overwhelmed, resulting in proximal migration of infection and eventual formation of a fistula (L. "pipe," an abnormal communication between the surface of the body and the internal structure or between two organs) at the coronet. This outcome is an advantage because it allows drainage of purulent material. When this does not occur, an abscess forms within the claw capsule partway between the sole and the coronet (Fig. 21.2-1). A progressively enlarging abscess leads to involvement of the adjoining structures in the area of the distal sesamoid bone, causing an infection of the sesamoid ligaments, the sesamoid bone itself (Fig. 21.2-5), and at the insertion of the deep digital flexor tendon in an abaxial-to-axial direction.

White line disease is typically located abaxial to the lateral claw of a pelvic limb.

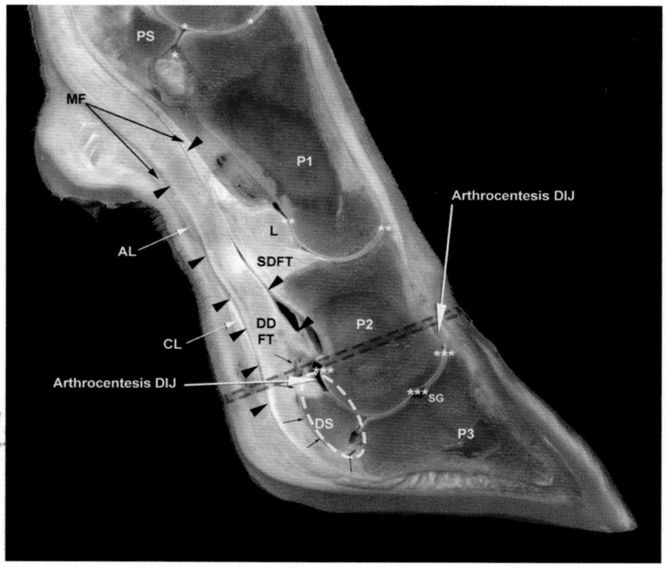

FIGURE 21.2-6 Sagittal section of a pelvic limb digit. Key: *Small black arrows*, podotrochlear bursa palmar/plantar to the distal sesamoid bone; *dotted oval*, area of abscessation and necrosis in a septic white line; *PS*, proximal sesamoid bone; *P1*, 1st phalanx 1; *P2*, 2nd phalanx; *P3*, 3rd phalanx; *, metatarsophalangeal joint; **, proximal interphalangeal joint; ***, distal interphalangeal joint; *SG*, synovial groove of the pedal bone; *DS*, distal sesamoid bone; *long black arrows*, manica flexoria (flexor sleeve, arising from the confluence of the superficial flexor tendon and the check ligament of the interosseous muscle); *SDFT*, superficial digital flexor tendon close to its insertion at the flexor tubercle of P2; *L*, caudal ligament of P2; *arrowheads*, common digital flexor tendon sheath; *AL*, annular ligament of P1; *CL*, distal cruciate (interdigital) ligament; *long white arrow on the dorsal aspect labeled "DIJ"*, site for dorsal arthrocentesis of the DIPJ; *long white arrow on the plantar aspect labeled "DIJ"*, site for lateral arthrocentesis of the DIPJ; *semitransparent dotted purple line*, perioplic border of the claw. *Small black arrows* palmar/plantar to the distal sesamoid outline the podotrochlear bursa. The *dotted oval* indicates the area of abscessation and necrosis in a septic white line.

Arthrocentesis of the DIPJ is performed by inserting a needle 1 cm (0.4 in.) proximal to the coronet and abaxial to the insertion of the long digital extensor tendon at the extensor process, and by directing it distally and slightly palmarly/plantarly. With sepsis and joint distension, horizontal insertion of the needle directed toward the dorsal pouch and the 2nd phalanx is usually successful. In addition, the palmar/plantar pouch is accessible from the lateral approach. Palpation of the skin just above the coronet reveals an indentation between the flexor tuberculum of the 2nd phalanx and the deep digital flexor tendon. The needle is directed from immediately proximal to the coronet, distally and axially toward the distal sesamoid (Fig. 21.2-6). Joint lavage can also be performed using these arthrocentesis sites.

The dorsal pouch of the DIPJ extends slightly above the level of the coronet. The distal sesamoid bone acts as a palmar/plantar reinforcement of the capsule of the DIPJ and as a pulley for the deep digital flexor tendon. It has a small distal joint facet (i.e., a small plane surface in a joint that articulates with another structure) to the pedal bone and a large dorsal contact surface with the 2nd phalanx. It is connected to the 3rd phalanx via a **distal ligament** and to the 2nd phalanx by several sturdy **collateral sesamoidean ligaments**. Between the distal sesamoid and the deep digital flexor tendon lies the **podotrochlear bursa** (Fig. 21.2-6).

Collateral ligaments (axial and abaxial) connect the 2nd and 3rd phalanges and can extend from the 1st phalanx to the 3rd phalanx. The **axial ligaments** are generally more substantial than the **abaxial ligaments** and prevent the claws from separating during weight-bearing. At the palmar/plantar side of the digit, the fascia is reinforced to form the **proximal** and **distal annular ligaments** (Fig. 21.2-6).

Proximal suspensory mechanism of the digits

The suspensory ligament (or **interosseous medius muscle**) and the **sesamoidean ligaments** of the fetlock are important components of the proximal suspensory mechanism that unites the digits. At rest, the fetlock joint is hyperextended; the metacarpo-/metatarsophalangeal and the proximal interphalangeal joints act together to stabilize the fetlock during load and prevent hyperextension. During flexion and extension, abduction of the metacarpo-/metatarsophalangeal joint is restricted by collateral ligaments and the sagittal crests of the metacarpal/-tarsal condyles.

Functionally, the metacarpo-/metatarsophalangeal joint and the 2nd interphalangeal joint are closely connected. The 2nd phalanx is interposed between the metacarpo-/metatarsophalangeal suspensory apparatus proximally and the 3rd phalanx suspensory apparatus distally. Therefore, the 3 phalanges of the bovine digit form a sturdy and flexible column.

> This anatomical arrangement requires that all digital joints must be incorporated in a cast when a metacarpal/-tarsal bone fracture is treated with external coaptation.

The range of motion of the DIPJ is second only to the metacarpo-/metatarso phalangeal joint (when comparing the joints of the bovine distal limb), with the proximal interphalangeal joint having the least relative range of motion because of its interposed position between the proximal and distal suspensory mechanisms. Mobility of the DIPJ is not as important in cows as it is in horses because of their primary gait (walking) and the additional claw that shares weight-bearing (compared to one hoof per limb); therefore, bovids function well with a fused DIPJ.

> Joint fusion of the metacarpo-/metatarso phalangeal joints can be performed in cattle because of this unique relationship of locomotion and anatomy.

Blood supply, lymphatics, and innervation to the claw

The main blood supply to the claws is by the **palmar/plantar arteries**; the **dorsal artery** plays a minor role. The palmar/plantar arteries branch at the digital cushion and the larger **axial arteries** track along the interdigital space to form the **arcus terminalis** after entering the 3rd phalanx. Cross connections and anastomoses among large and small arteries are common.

Blood drainage occurs via 3 veins—a major vein runs dorsally, and 2 smaller veins run abaxially and axially. Arteries and veins form extensive networks at different levels of the claw for horn production and maintenance of suspensory apparatus function. Arteriovenous anastomoses occur only occasionally in cattle and appear to play a minor role in disease processes such as laminitis.

> Digital veins are used for regional IV anesthesia and to administer antimicrobials using a tourniquet. A tourniquet is generally placed at the level of the proximal metacarpus/-tarsus. Distension of the digital veins readily allows venipuncture and infusion of drugs, achieving a brief high regional concentration of the drug and reduced need for systemic administration. The tourniquet is left in place for approximately 30 minutes in case of IV antibimicrobials and approximately 60 minutes in case of anesthesia.

In the thoracic limb, lymphatic vessels drain to the palpable **superficial cervical lymph nodes**. These lymph nodes are found cranial to the supraspinatus muscles of the scapula and are 7–9 cm (2.8–3.5 in.) long and 1–2 cm (0.4–0.8 in.) wide. They are covered by the brachiocephalicus and omotransversarius muscles. In the pelvic limb, lymph fluid from the foot drains to the **deep popliteal lymph nodes**. The deep popliteal lymph node is a 3- to 4.5-cm (1.2- to 1.8-in.) round structure located caudal and distal to the stifle, embedded in the adipose tissue at the proximal aspect of the gastrocnemius muscle.

Innervation of the digits of the thoracic limb is supplied by three nerves that originate from the brachial plexus—radial, ulnar, and median. The **radial nerve** branches supply the medial digit and the axial part of the lateral digit. The dorsal branch of the **ulnar nerve** (located dorsally) innervates the abaxial aspect of the lateral digit. The nerve supply of the palmar part of the metacarpus consists mainly of the palmar branches of the **median nerve**, innervating the palmar parts of the lateral and medial digits, and the palmar branch of the **ulnar nerve**, innervating the palmarolateral part of the metacarpus.

In the pelvic limb, the **fibular nerve** divides into superficial and deep branches in the proximal one-third of the leg. The **superficial fibular nerve** tracks along the lateral digital extensor muscle and the dorsal metatarsal vein over the dorsal aspect of the tarsus. After innervating the dorsal muscles of the leg, the **deep fibular nerve** accompanies the long digital extensor tendon under the extensor retinaculum to innervate the tarsus and short digital extensor muscle. The deep fibular nerve continues along the dorsal aspect of the metatarsal bones as the **dorsal metatarsal nerve**. The dorsal metatarsal nerve anastomoses with the superficial fibular nerve fibers and the plantar digital nerves at the level of the interdigital space.

Local anesthesia is best performed at the proximal aspect of the metatarsus/-carpus where the superficial fibular nerve tracks adjacent to the dorsal metatarsal vein before the nerve divides into branches for the medial and lateral digits more distally.

The plantar region is innervated by the medial and lateral branches of the **tibial nerve,** which divides distally into the **lateral** and **medial branches** and again into **axial** and **abaxial branches** for the respective regions of the digits. The **plantar** and **dorsal nerve branches** communicate in the interdigital space and are used for regional anesthesia (see Case 21.3).

Selected references

[1] Anderson DE, Desrochers A, van Amstel SR. Surgical procedures of the distal limb for treatment of sepsis in cattle. Vet Clin North Am: Food An Pract 2017;33:329–50.
[2] Budras K, Habel RE. Bovine anatomy. 2nd ed. Hannover: Schlueitersche Verlagsanstalt, Thieme Medical and Scientific Publishers; 2011.
[3] Dyce KM, Sack WO, Wensing CJG. Textbook of veterinary anatomy. 4th ed. St. Louis, Missouri: Saunders Elsevier; 2010.
[4] Nuss K. Surgery of the distal limb. Vet Clin North Am Food Anim Pract 2016;32(3):753–75.
[5] Vollmerhaus B. Lymphatic system. In: Nickel R, Schummer A, Seiferle E, editors. Textbook of the anatomy of domestic animals, vol. III. Berlin und Hamburg: Paul Parey; 1976. p. 276–450.

CHAPTER 21

CASE 21.3

Sole Ulcer

Karl Nuss
Farm Animal Surgery Section, Farm Animal Department, Vetsuisse-Faculty University of Zürich, Zürich, CH

Clinical case

History

A 4-year-old Brown Swiss cow from a free-stall barn with concrete flooring and a manure scraper system was identified as having an abnormal gait during routine locomotion scoring. The cow walked with an arched back and abducted the right pelvic limb when standing. The cow was 50 days in her current lactation and had a daily production of 40 kg (88 lbs) of milk, a good appetite, and a body condition score (BCS) of 2/5 (1= emaciated; 5= obese). No abnormalities were observed at the time of foot trimming 5 months before.

Physical examination findings

The cardiovascular and respiratory systems were normal, and the rectal temperature was 101.8 °F (38.8 °C). The cow had a typical cow-hocked conformation and shifted her weight between the hind feet. The gait was irregular with short, hesitant steps, and the hind feet pointed laterally. The cow was diagnosed with a grade 2/5 supporting limb lameness of the right pelvic limb.

Diagnostics

The cow was placed in a foot trimming chute for examination of the digits. There was marked asymmetry between the paired digits, and the lateral digits were larger in both hind feet. The heels were overgrown and abraded (Fig. 21.3-1). Hoof testers were applied to the digits and pressure to the caudal axial sole region of the lateral digit of the right hind foot elicited a pain reaction. Close examination of the foot revealed poorly attached and structured horn, eliminating the axial groove, with underlying inflamed and partially necrotic tissue (Fig. 21.3-1).

Routine trimming of solar horn revealed inflammation of the corium in the axial region of the sole at the transition from the hard, distal bulb to soft, proximal bulb in both hind digits. The lateral digit was trimmed to a solar horn thickness of 3–5 mm (0.1–0.2 in.); however, the corium was spared. Trimming of the medial digit was minimized to maintain a sole profile that was 5 mm (0.2 in.) higher than that of the lateral digit, while establishing a functional, weight-bearing solar surface.

Diagnosis

Sole ulcer of the lateral digit of the right pelvic limb

Treatment

The right pelvic foot was thoroughly cleaned and then aseptically prepared before administration of local anesthetic for a 3-point nerve block (Fig. 21.3-2, see side box entitled "The 3-point nerve block"). Abnormal and undermined solar horn was removed with a hoof knife, and the solar horn surrounding the ulcer was thinned in a funnel-shaped

Comparative Veterinary Anatomy: A Clinical Approach. https://doi.org/10.1016/B978-0-323-91015-6.00105-9

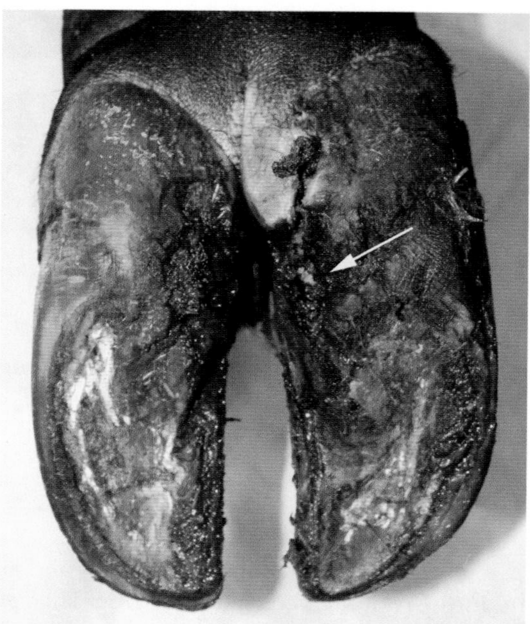

FIGURE 21.3-1 Right pelvic limb foot before trimming. The lateral digit (right) is larger than the medial digit. The axial slope of the lateral digit is covered by the poorly attached, soft heel horn. A defect *(arrow)* is apparent. Pressure applied to this area with hoof testers elicited a pain reaction.

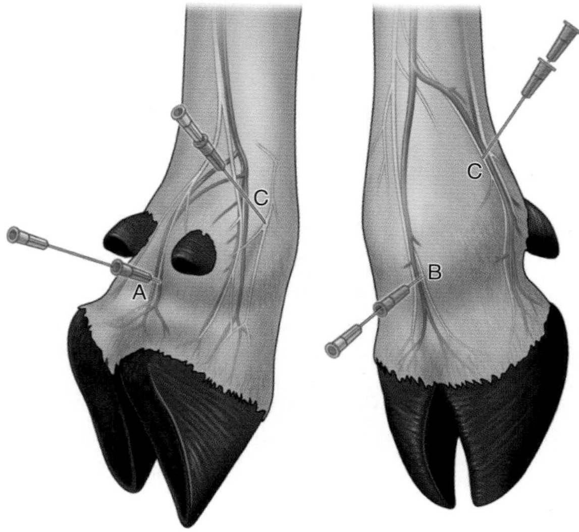

FIGURE 21.3-2 Anatomical sites for local anesthesia of the digital nerves. Key: A = Palmar/plantar injection site, B = dorsal injection site; both injection sites together block the interdigital nerve branches. C = Abaxial injection site for the palmar/plantar digital nerves.

fashion. The foot was cleaned again, and a scalpel was used to remove all horn covering the edge of the ulcer along with a small amount of necrotic tissue surrounding and covering the ulcer. Healthy corium and the adipose pad of the heel were spared to encourage rapid regeneration and horn production (Fig. 21.3-3). The ulcer was then covered with a povidone-iodine-soaked gauze pad and the digit was bandaged. A hoof block was applied to the medial digit to elevate the affected digit (Fig. 21.3-4).

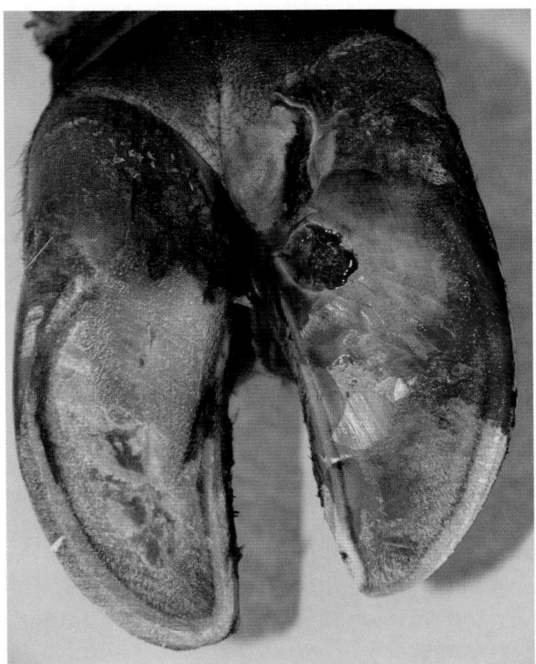

FIGURE 21.3-3 Lateral sole ulcer area completely exposed (under local anesthesia); all the poorly attached horn has been removed and the horn surrounding the ulcer has been tapered. The lateral digit sole has been trimmed to a level lower than that of the medial digit.

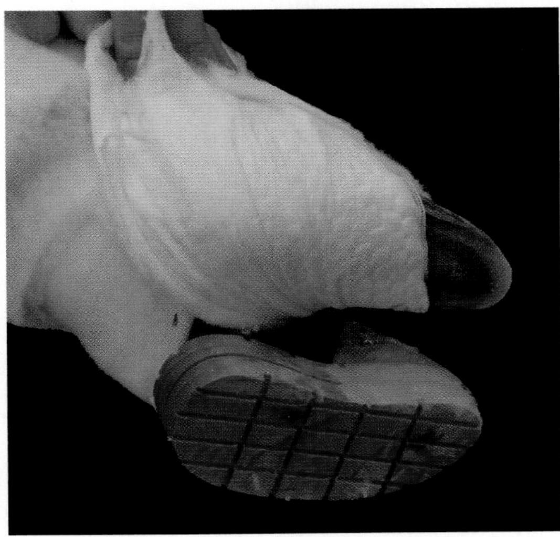

FIGURE 21.3-4 The ulcer is covered with a bandage and a wooden block has been glued to the medial claw to protect the lateral digit.

The cow was confined in a hospital pen with soft rubber mats and ample straw bedding, which was changed several times a day to keep the bandage dry. The bandage was changed every 3–5 days, and superficial cornification of the corium of the previous ulcer site was apparent after 2 weeks. Three weeks after the start of the treatment, the cow was discharged from the clinic. The wooden block was removed at 5 weeks when mature horn covering the ulcer site allowed weight-bearing.

Anatomical features in ruminants

Introduction

Lameness is one of the most common health problems in the dairy industry today and is considered among the greatest welfare issues for dairy cows. Because digit diseases are the primary cause of lameness, comprehensive knowledge of the digit is important for the prevention of lameness. The digit consists of all the structures enclosed in the horn capsule including the 3rd phalanx, the distal sesamoid bone, parts of the 2nd phalanx, the flexor and extensor tendons, collateral ligaments, and the highly differentiated subcutis and corium. This discussion focuses on the anatomical structures that serve as padding or are involved in horn production; blood and nerve supply are discussed in Cases 21.2 and 21.4.

Function

The horn capsule functions as an interface between the limbs of the bovid and the ground with multiple purposes—most importantly, protection of the inner structures of the digit. The digital horn is a poor heat conductor and can modify its hardness relatively quickly through a process of absorption or evaporation of water, facilitating adjustments to environmental conditions. For example, in a wet environment, the horn absorbs water and becomes softer, whereas in dry conditions, it releases water and hardens. This feature is the key in the pathogenesis of digital horn lesions associated with softening, and thus weakening of the digit in humid-free stall housing conditions with hard flooring. Finally, the digit also serves as a sensitive tactile organ by virtue of its abundant nerve supply.

> The slightly concave shape of the sole also creates a palmar/plantar shift in weight-bearing and thus chronic overloading of the heel region in cows on hard flooring. NOTE: Digits of permanently pastured cows are almost symmetrical, whereas digits of stabled cows change shape; the lateral hind digits are larger because they are usually subjected to higher chronic stress overload than the medial digits.

The solar surface is slightly concave, facilitating full contact of the digit on soft ground. This enables sequential loading of different parts of the sole and aids in reducing excessive horn. The lateral and medial hind digits are typically asymmetrical, likely because in cattle the lateral digit is usually slightly longer than the medial digit, leading to increased weight-bearing of the lateral digit. Therefore, the lateral digit is typically larger in the hind feet. The difference in size between the paired digits is less pronounced in the front feet, but in contrast to the hind feet, the medial digit carries more weight and is slightly larger than the contralateral digit. The reason for this is currently unknown.

THE 3-POINT NERVE BLOCK (FIG. 21.3-2)

Local anesthesia of the digit is required for pain-free treatment of a sole ulcer and other associated lesions. Injection of local anesthetic solution into the interdigital space blocks the axial digital nerves of both digits (Fig. 21.3-2A, B). This nerve block allows simple surgical procedures involving the interdigital space and is performed from the dorsal and palmar/plantar aspects. The palmar/plantar injection site is located at the level of a prominent horizontal skin fold (Fig. 21.3-2A), and the dorsal injection site is 1–2 cm (0.4–0.8 in.) proximal to a less distinct skin fold (Fig. 21.3-2B). From each of these injection sites, 10–12 mL of 2% local anesthetic solution is evenly distributed in the interdigital space.

Alternatively, the total local anesthetic dose can be introduced at a single injection site using a 9–11-cm (3.5–4.3-in.) needle; this is best achieved using a palmar/plantar approach, especially in cows placed in a foot-trimming chute.

For pain-free treatment of a sole ulcer, the anesthetic solution must also be injected at a third site located on the abaxial aspect of the lateral or medial digit (Fig. 21.3-2C), depending on the location of the ulcer, to block the lateral or medial plantar/palmar digital nerve. This nerve branch tracks in the indentation that can be palpated slightly proximal and dorsal to the dewclaw between the medial interosseous muscle and the digital flexor tendons. Before the needle is inserted, the skin is tented and the needle is placed subcutaneously to avoid unintended puncture of the common digital flexor tendon sheath. The needle is then advanced toward the indentation and about 10 mL of 2% anesthetic solution is injected. Anesthesia takes effect within 10 minutes.

Anatomy of the claw

Each of the limbs in cattle has 2 digits and 2 dewclaws, which correspond to the 3rd and 4th, and 2nd and 5th digits, respectively (Fig. 21.3-5). The **digits** are responsible for weight-bearing, whereas the dewclaws provide support only on deep or steep ground. The digits are separated by a hairless **interdigital space**, which widens appreciably dorsally, but not palmarly/plantarly, because the heel bulbs are fixed close together to provide cushioning during

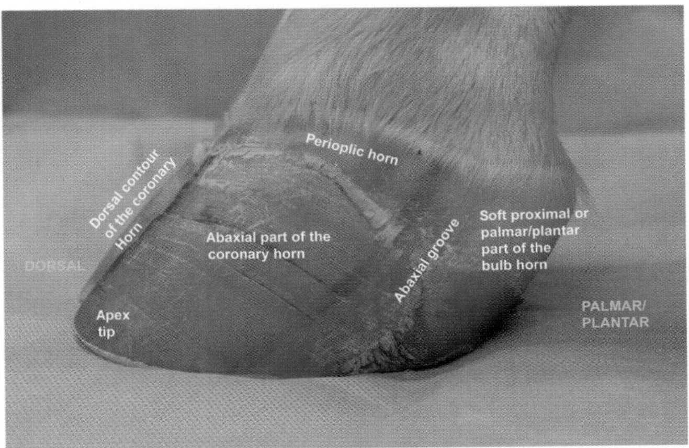

(A)

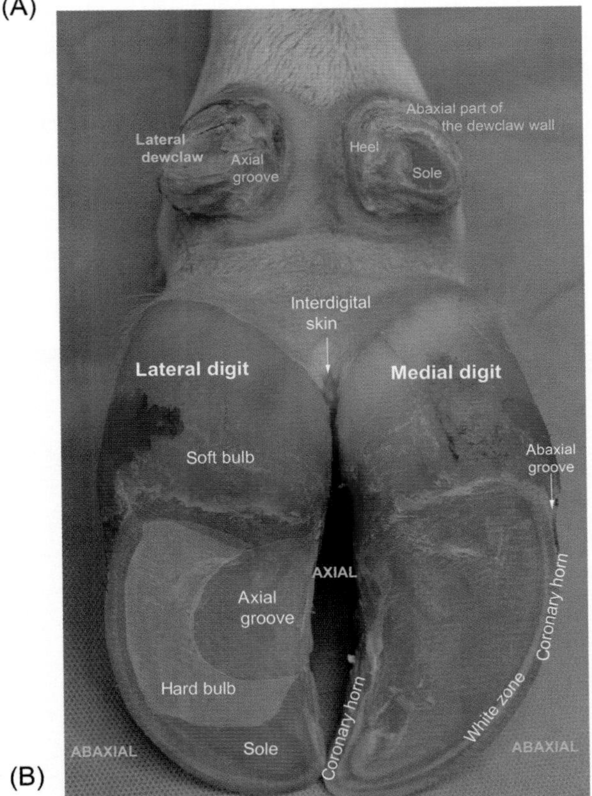

(B)

FIGURE 21.3-5 Anatomical terminology for location and structures of the bovine digit. The digit has dorsal and palmar/plantar, and axial and abaxial aspects. Axial means toward the interdigital space or limb axis and abaxial means away from the interdigital space or limb axis. Thus, the lateral and medial digits both have axial and abaxial surfaces. The abaxial groove (A) marks the transition from the digital wall to the soft (proximal) bulb. The axial groove (B) is situated at the soft bulbar horn and starts at the axial end of the coronary horn. It marks the transition from the hard, distal bulb to the soft proximal bulb. Locations toward the toe are referred to as distal or apical and caudal regions are referred to as proximal or palmar (thoracic limbs), or proximal or plantar (pelvic limbs).

ground contact. Excessive spreading of the digits is prevented by strong **interdigital ligaments**, particularly the **distal cruciate interdigital ligament** and the **interdigital collateral ligaments** of the fetlock joint, which provide a close functional connection between the digits.

It is important to avoid injuring these phalanges during digit trimming.

The **dewclaws** (vestigial digits) are much smaller and have the same basic structure. They are connected to the fasciae of the fetlock region by proximal, transverse, and distal ligaments. The strong distal ligament, also called the "tendon of the dewclaw," passes distally and fuses with the collagen tissue of the digital cushion abaxially. The dewclaws contain small 3rd phalanges and sometimes also 2nd phalanges.

Segments of the digit

The structure of the **horn capsule** corresponds to the structure of the skin; however, the **epidermis** is much thicker and is composed of 5 different segments: **perioplic**, **coronary**, **lamellar**, **bulbar**, and **solar**. These segments are closely integrated even though some borders are distinct; e.g., at the axial and abaxial grooves (Fig. 21.3-5). The **corium** consists of a papillary component that nourishes the basal epidermal cells that divide and differentiate into keratinized cells. Conversely, the reticular component of the corium provides the firm connection with the underlying tissues. The corium has an identical structure to the epidermis, comparable to a matrix and patrix (i.e., the template for matrix development). **Perioplic**, **coronary**, **lamellar**, **bulbar**, and **solar corium** can be differentiated (Fig. 21.3-6). The **laminae** of the corium interdigitate tightly with the **epidermic lamellae** of the digital wall, forming a strong bond between the 3rd phalanx and the horn capsule. The lamellar structure serves to greatly increase the surface area, which ensures a strong connection and exchange of oxygen and metabolites.

The **subcutis** is well-developed in some regions of the digit and absent in others. There is no subcutis in the region of the interdigitations between the laminar corium and the epidermic lamellae; however, the subcutis forms the **coronary** and **digital cushions**, which consist of collagen fibers and adipose tissue (Figs. 21.3-7 and 21.3-8A). The coronary cushion lies below the coronary corium and is the largest on the dorsal side. It is compressed by the movements of the pedal bone at the end of the standing phase and with advancement of the limb. The digital

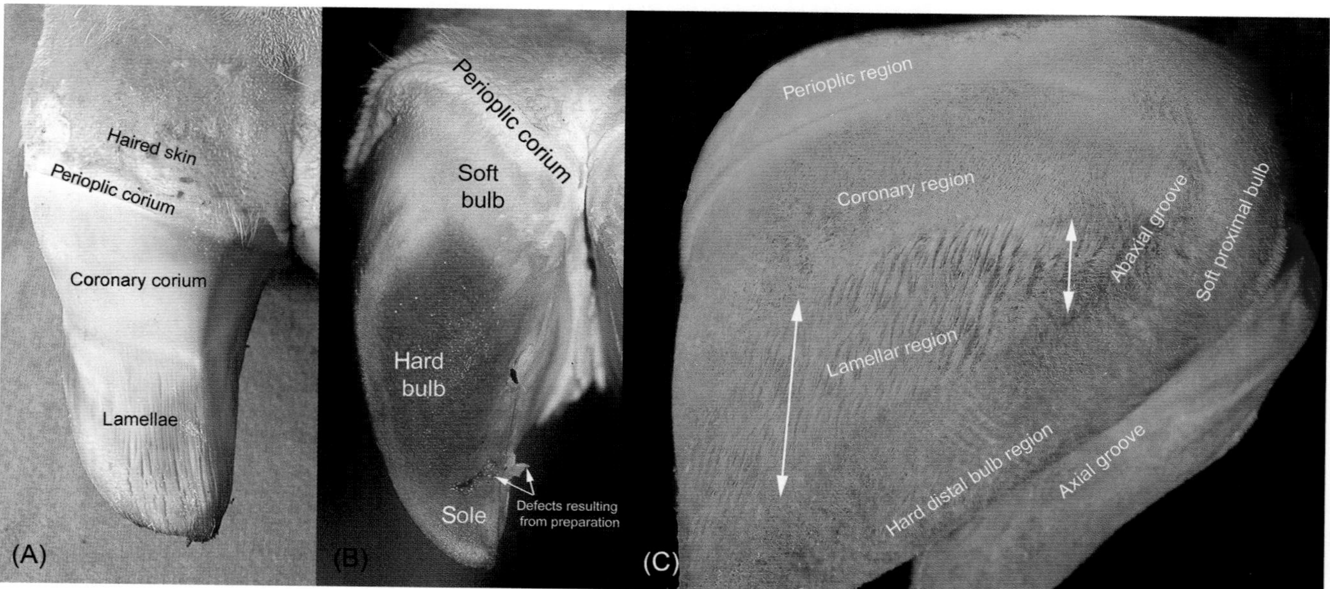

FIGURE 21.3-6 Corium surface: (A) dorsal view and (B) solar view, with relevant corium segments of the digits. (C) Corrosion cast of the corium vessels depicts the same segments found in the hoof and the corium. (Image C is courtesy of Dr. Pete Ossent.)

cushion begins as a 5-mm (0.2-in.) thick layer from the palmar/plantar end of the sole, supports the pedal bone, and tracks to the soft bulb horn, where it is ≥20 mm (0.8 in.) thick. It distributes the mechanical forces during landing of the foot and the first half of the stance phase. The mechanical forces are transmitted to the surrounding structures via the distal cruciate interdigital ligament and the soft horn bulb.

After removal of the digital horn capsules, the previously discussed segments of the digit are readily apparent on the inside of the horn capsules and on the corium. The **perioplic segment** (Figs. 21.3-6A and 21.3-7B) is approximately 10 mm (0.4 in.) long and consists of soft tubular horn with corium papillae underneath. The horn here forms a delicate glaze-like layer that only covers the upper third of the digital wall (Fig. 21.3-7B). It serves as a connection between horn capsule and haired skin and protects the soft coronary horn underneath.

Moving in a palmar/plantar direction, the perioplic segment transitions gradually into the heel bulb and interdigital skin (Figs. 21.3-5 and 21.3-6B). The **coronary segment** consists of a hard, weight-bearing tubular horn, which constitutes the bulk of the digital wall. Even though the base of this segment is about 2–3 cm (0.8–1.2 in.) wide, the coronary horn is only 5–8 mm (0.2–0.3 in.) in thickness, which illustrates the dense spatial arrangement of the horn tubules. Three layers of coronary horn can often be differentiated grossly (Fig. 21.3-7B and C): (1) the outermost layer is dark with relatively thin horn tubules; (2) the middle layer is usually free of pigment and has numerous thick horn tubules; and (3) the inner layer has thin tubules.

The adjacent **wall segment** consists of about 1300–1800 **epidermic (horn) lamellae** that interdigitate with the lamellae of the corium. These lamellae are approximately 2–3 cm (0.8–1.2 in.) long in the dorsal region, becoming shorter toward the palmar/plantar region (Fig. 21.3-6B, arrows), and they are absent in the interdigital space at the level of the axial groove (more to follow). This arrangement of the dermal and epidermal lamellae results in both an enormous increase in the surface area for nutritional supply and in a tight bond between the dermis and epidermis. The tensile strength of this connection varies with the region of the wall, and in healthy cattle is particularly strong abaxially.

The so-called **white zone** consists of lamellar horn and connects the wall of the digit to the sole and bulb. This zone can be divided into 3 different structures: (1) a line at the inner side of the coronary horn sometimes visible grossly (thin white line shown in Fig. 21.3-7C and D); (2) approaching the sole following the lamellar horn which is the middle part of the white zone and is easily recognized; and (3) the zone at the terminal horn formed by the lamellae at the point where they are curved toward the sole (Fig 21.3-7C and D). Abaxially at the digit, the white line extends to the end of the soft bulb. Axially, it ends in the apical third together with the coronary digit wall; this point marks the **axial groove**, representing a nonweight-bearing part of the sole. The axial groove begins at the axial end of the coronary horn and marks the transition from the hard, distal bulb to the soft, proximal bulb (Fig. 21.3-5B). The **abaxial groove** is an indentation in the digital wall and marks the transition to the soft (proximal) bulb (Fig. 21.3-5B). Likewise, the border between the hard and soft bulb corresponds approximately to the line connecting the axial and abaxial ends of the white zone. This line also separates: (1) the cranial part of the digit, where the 3rd phalanx is firmly attached to the digital wall by the lamellae; (2) the caudal part of the digit where there are no lamellae; and (3) the digital cushion as part of the pedal bone support system.

One of the most important features of practical digit trimming is to maintain the axial groove. The concave shape of the axial groove in the trimmed digit near the flexor tubercle of the 3rd phalanx aids in preventing bruising of the corium.

DOES SURFACE AREA MATTER?

Horses have primary and secondary epidermal and dermal lamellae, which greatly increases the surface area of contact and bonding of the hoof wall to the 3rd phalanx, averaging 0.8–1.5 m^2 (2.6–4.9 $ft.^2$). The bovine digit's inner surface area is much smaller because it lacks the secondary lamellae in the epidermal and dermal lamellae. This likely does not matter for the bovine digit because the mechanical forces on the digit are shared by 2 digits (claws/toes) for each foot and there is a big difference in the bovine digit's function—i.e., nonathletic, low-speed walking; and different intensity of loading—and a bovid's lifestyle.

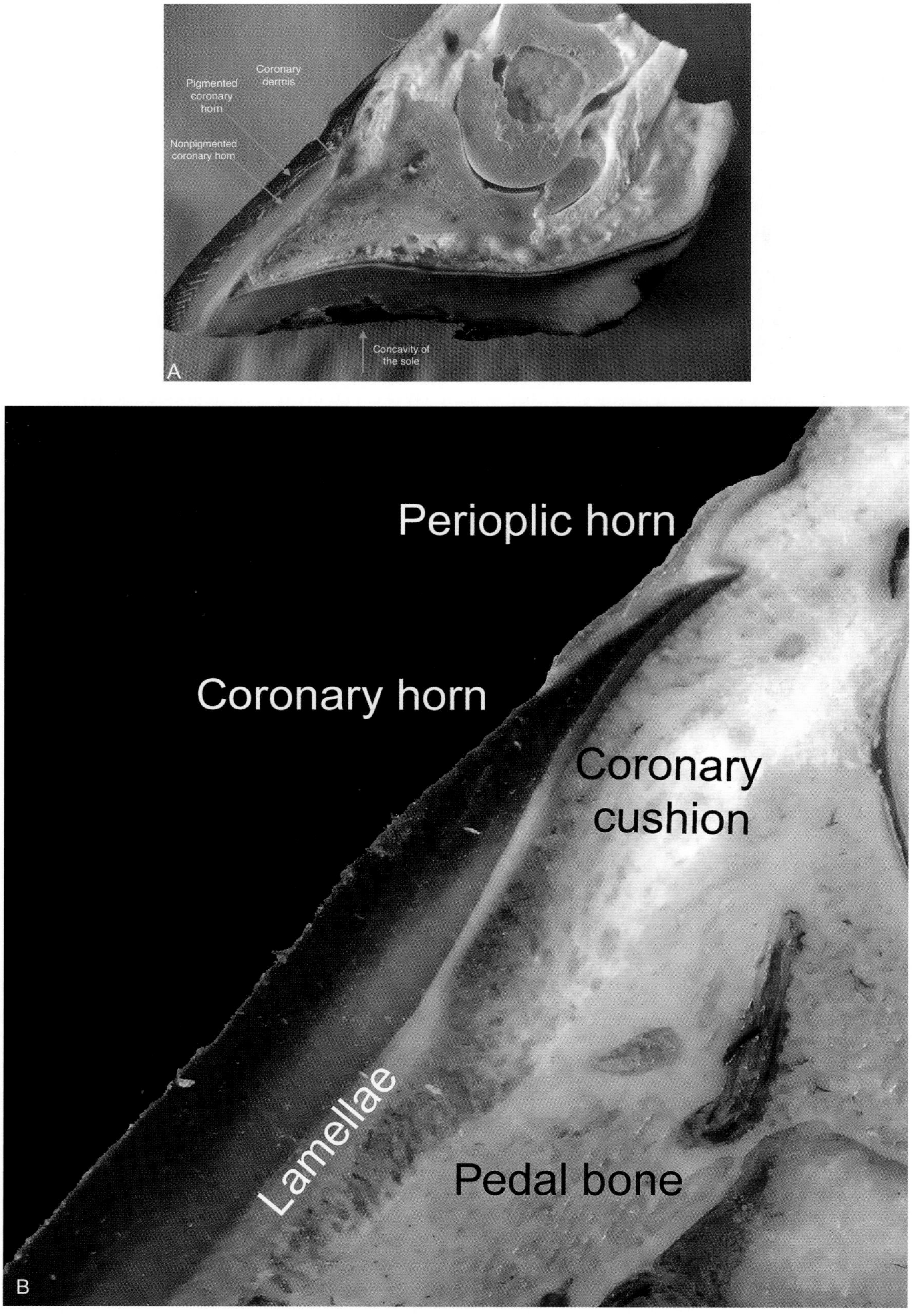

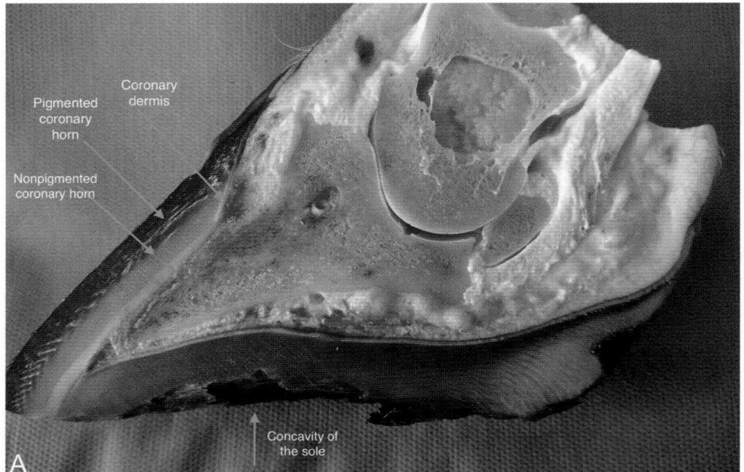

FIGURE 21.3-7 Special features of digit segments shown in sagittal sections. (A) Note the concave surface of the sole horn in the overview (*arrow*). (B) Dorsal coronet region. Perioplic horn covers the soft coronary horn. Pigmented and unpigmented hard coronary horn can be seen more distally. The coronary cushion is between the coronary corium and the pedal bone. The interdigitating dermal and epidermal lamellae begin immediately below the distal end of the coronary corium.

(Continued)

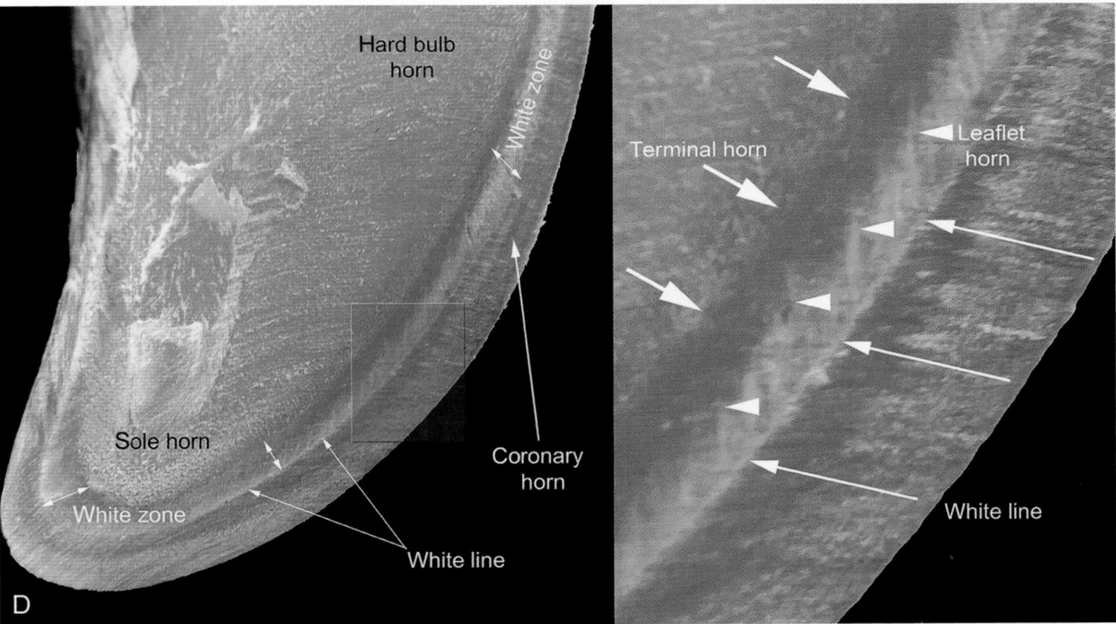

FIGURE 21.3-7, CONT'D (C) Apical (distal) end of a digit depicting pigmented and unpigmented coronary horn (*CH*). Adjacent to the CH is the white zone (*WZ*, *double arrow*), consisting of the white line (*WL*), the lamellar horn (*LH*), and the terminal horn of the lamellae (*TH*). The border between the white line and the coronary horn is outlined by *arrows*, and the border between the white line and the sole horn (*SH*) is outlined by *arrowheads*. (D) Structures of the white zone (*double arrows*) viewed from the sole surface of the digit (left: overview; right: magnified view). *White line*, lamellar horn, and the terminal horn of the lamellae can be seen.

The lack of subcutis at the sole segment means injuries in this region of the digit can have devastating effects, including infection and pedal bone osteomyelitis.

The **sole segment** appears as a narrow crescent-shaped structure on the inside of the white line at the toe; there is no subcutis at this region. The **bulb segment** starts seamlessly after the sole segment and supplements the area of the sole caudally in the form of the "hard" heel (Fig. 21.3-5B). The bulb segment is supported by the digital cushion, a multicompartment structure composed of collagen and adipose tissue that extends from the 3rd phalanx to the heel (Fig. 21.3-8A).

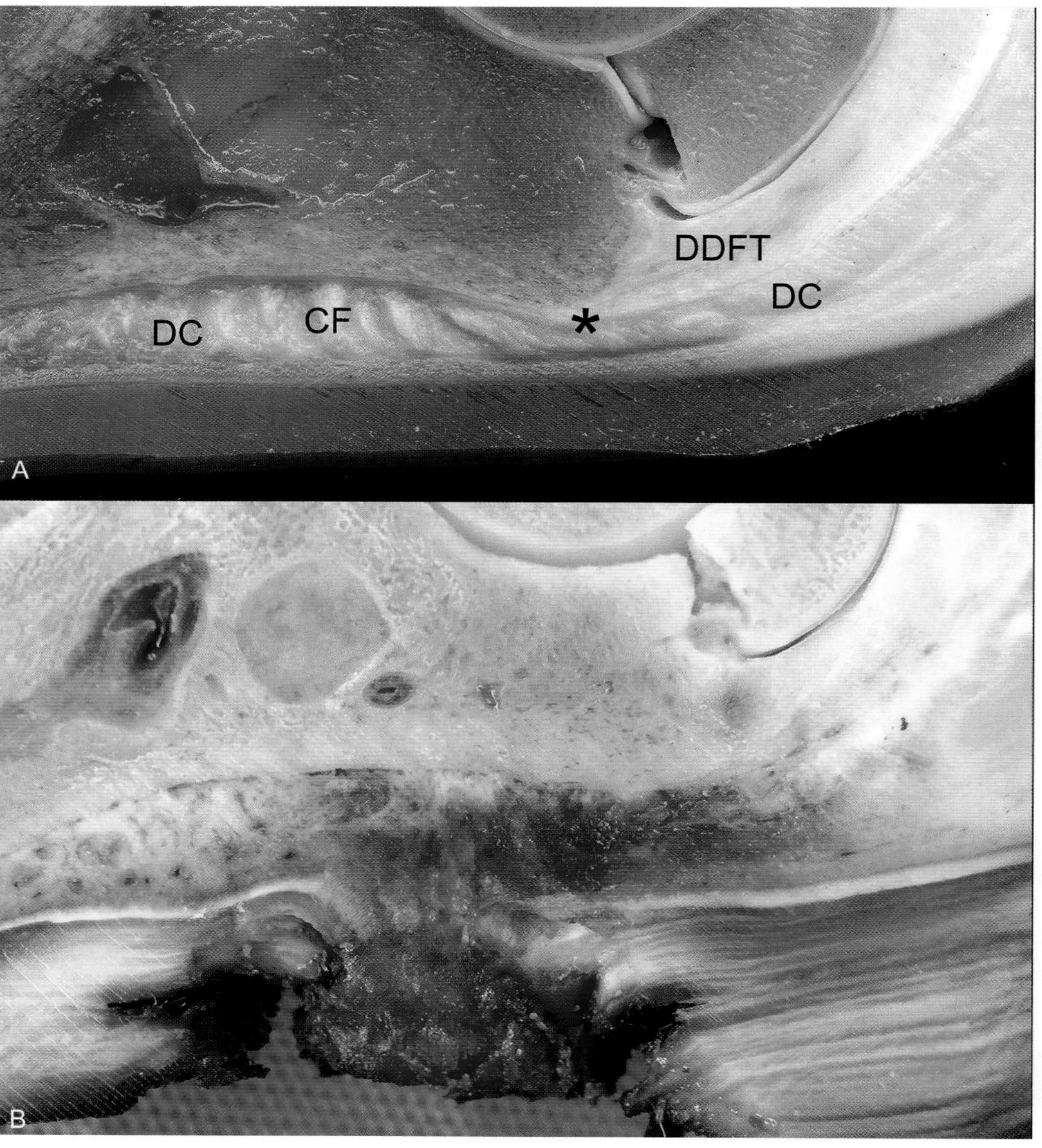

FIGURE 21.3-8 Sagittal sections of a healthy digit (top) and a digit with sole ulcer (bottom) in the plane of the flexor tubercle. (A) The digital cushion (*DC*) consists of collagen and adipose tissue and is located directly undersurface and palmar/plantar to the 3rd phalanx. Three large sagittal adipose pads are arranged undersurface the 3rd phalanx and others are randomly distributed in the bulb of the heel. The surrounding collagen fibers track from the corium to the 3rd phalanx. The digital cushion is normally fused with the distal end of the deep flexor tendon (*). (B) The collagen fibers that run from the periosteum of the 3rd phalanx to the epidermis are inflamed, hemorrhagic, and covered with granulation tissue in cattle with an ulcer. The horn capsule and the corium are perforated. In this case, the ulcer is not directly underneath the tuberculum flexorium, but the inflammation has spread through the digital cushion proximally. (Image A is courtesy of Dr. Susanne Kretschmar.)

The suspensory apparatus of the 3rd phalanx

The suspensory apparatus of the 3rd phalanx includes all the structures involved in anchoring the bone to the immediate and adjacent surrounding tissues. The term "suspensory apparatus" derives from the fact that the 3rd phalanx is "suspended" from the horn capsule by virtue of these structures. The suspensory apparatus absorbs the mechanical forces exerted by the weight of the bovid and transforms them into tensile forces. It is essentially formed by strong collagen fibers, which run from the 3rd phalanx across the subcutis and corium to the basement membrane of the epidermis. These fibers are stretched when the digit is loaded and effectively carry the body weight. The suspensory apparatus is especially strong in the region of the lamellae. The flexor tendons and the distal sesamoid bones, which are anchored by collateral ligaments of the interphalangeal joints, aid in absorption of pressure forces in the soft bulb region. Contraction of the digital flexor muscles may also be involved in shock absorption by raising the flexor tubercle and possibly relieving pressure on the digital cushion by virtue of integration of the flexor tendon in the digital cushion.

In the bulbar part of the digit, pressure forces are disseminated by a combination of complex structures that include the adipose pads, collagen fibers (Fig. 21.3-8A), collateral ligaments, the distal cruciate interdigital ligament, and the deep digital flexor tendon.

Blood supply, lymphatics and innervation

Please see Case 21.4 for a detailed description.

THE DIGIT MECHANISM

The "digit mechanism" is another adaptation of the softer digit components to weight-bearing forces—the perioplic horn and soft coronary horn are pushed inward against the coronary cushion, absorbing the pressure forces. The caudal part of the 3rd phalanx sinks during ground contact and the pressure on the underlying tissues is minimized by the axial groove on the weight-bearing surface of the soft bulb horn. At the same time, the digit capsule widens in the heel bulb region to accommodate the expansion of the digital cushion. On hard flooring, however, this increases the stress to the abaxial groove and the abaxial suspensory apparatus, and is a factor in the pathogenesis of white line disease.

SOLE ULCER (RUSTERHOLZ SOLE ULCER)

In principle, a sole ulcer can occur in any digit; however, the lateral digits of the pelvic limbs are most commonly affected. Multiple etiologies have been identified for these ulcers—including laminitis, poor trimming/management, and endocrine disorders. However, biomechanical pathogenesis is considered most likely because the lateral digits of the pelvic limbs are subjected to overload when walking and standing.

The typical anatomical location is the area of the axial groove where the corium and subcutis are displaced axially when the flexor tubercle of the 3rd phalanx sinks during weight-bearing. In addition to this characteristic location, a sole ulcer at the heel (i.e., a heel ulcer) was recently reported and explained as being caused by osteophyte or enthesophyte formation on the tuberculum flexorium in a palmar/plantar plane.

An ulcer is characterized in general by loss of the protective horn layer and the presence of inflammation and granulation tissue. It is generally believed that a sole ulcer starts in the digit because of bruising of the corium and the basal cell layers of the epidermis (Fig. 21.3-8B). Chronic repetitive bruising of the corium results in ischemia, leading to necrosis of the corium and—in turn—interruption of keratinization. Eventually, this causes loss of the horn layer (protective layer), contamination, and bacterial invasion, all causing painful inflammation. The compromised corium and digital cushion react by producing granulation tissue to repair the injured tissues.

Selected references

[1] Toussaint Raven E. Cattle footcare and claw trimming. In: Reprint of 3rd impression 1989. Ipswich, UK: Farming Press; 1992.
[2] Yavari S, Khraim N, Szura G, et al. Evaluation of intravenous regional anaesthesia and four-point nerve block efficacy in the distal hind limb of dairy cows. BMC Vet Res 2017;13.
[3] Thomas HJ, Miguel-Pacheco GG, Bollard NJ, et al. Evaluation of treatments for claw horn lesions in dairy cows in a randomized controlled trial. J Dairy Sci 2015;98:4477–86.
[4] Muggli E, Weidmann E, Kircher P, et al. Radiographic measurement of hindlimb digit length in standing heifers. Anat Histol Embryol 2016;45:463–8.
[5] Budras K, Habel RE. Bovine Anatomy. 2nd ed. Hannover: Schluetersche Verlagsanstalt, Thieme Medical and Scientific Publishers; 2011.

CHAPTER 21

CASE 21.4

Metacarpal Fracture

André Desrochers
Department of Clinical Sciences, Université de Montréal, Faculty of Veterinary Medicine, St-Hyacinthe, Québec, CA

Clinical case

History

An 11-month-old female Holstein was presented for nonweight-bearing lameness of the left thoracic limb after she was found with her leg caught under a gate. The referring veterinarian was concerned about a closed fracture of the distal limb. After bandaging and splinting the limb, the heifer was referred to the veterinary hospital for further workup and treatment.

Physical examination findings

Upon arrival at the hospital, the heifer was bright and alert. The splint and bandage were in place and well-positioned. The physical examination was within normal limits. The heifer was sedated with IV xylazine (0.05 mg/kg) and restrained on a table in right lateral recumbency for further examination. After splint and bandage removal, instability of the distal limb was confirmed with no evidence of skin wounds or an open fracture.

Differential diagnoses

Fracture of the metacarpus, tendon and/or soft tissue injury, and septic arthritis of the metacarpophalangeal or carpal joints

Diagnostics

After mild sedation, radiographic views (dorsopalmar, lateromedial, and 2 obliques) of the distal limb were obtained, revealing a nonarticular complete spiral diaphyseal fracture of the metacarpal (fused 3rd and 4th metacarpal bones) bones.

Diagnosis

Complete, closed, spiral diaphyseal fracture of the left metacarpal bones

Treatment

The heifer was placed under general anesthesia and the left thoracic limb was aseptically prepared for surgical reduction and immobilization of the fracture. Transfixation pins were inserted proximal and distal to the fracture for best alignment and reduction, and the construct was incorporated in a full-limb cast for immobilization (Figs. 21.4-1 and 21.4-2).

Comparative Veterinary Anatomy: A Clinical Approach. https://doi.org/10.1016/B978-0-323-91015-6.00106-0

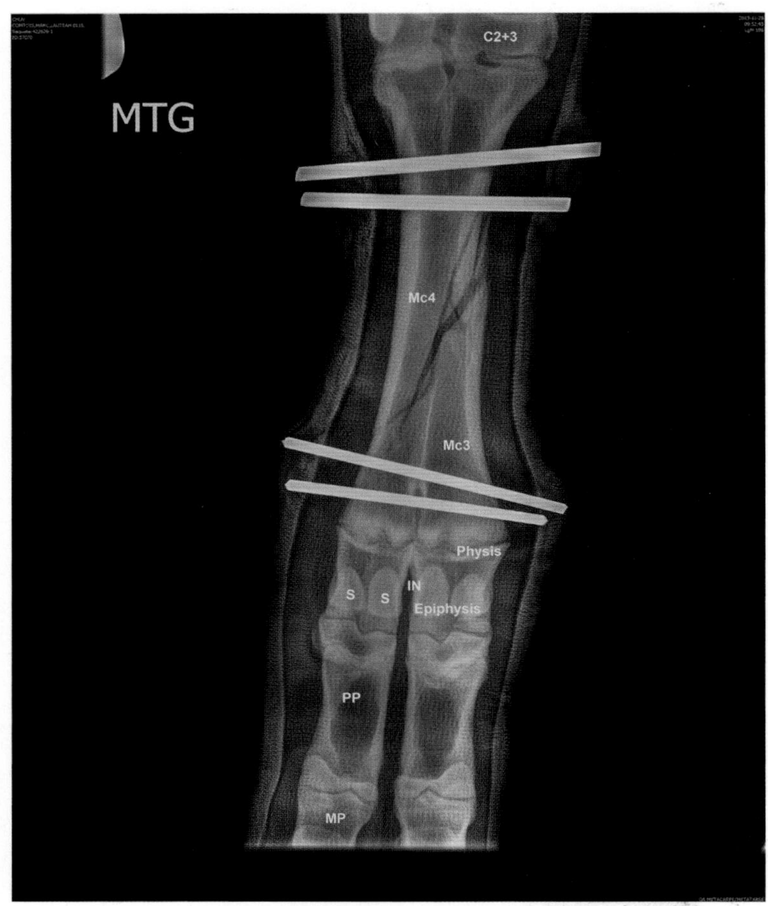

FIGURE 21.4-1 Dorsopalmar radiographic view of the left thoracic limb. A complete, closed, spiral metacarpal fracture was reduced and immobilized with a transfixation cast. Four transcortical/transfixation pins, 2 pins each proximal and distal to the fracture, stabilize the fracture to support bone healing. Key: *C2+ 3*, fused 2nd & 3rd carpal bones; *Mc3 & Mc4*, fused metacarpals 3 & 4, proximal sesamoid bone; *IN*, intertrochlear notch; *PP*, proximal phalanx; *MP*, middle phalanx.

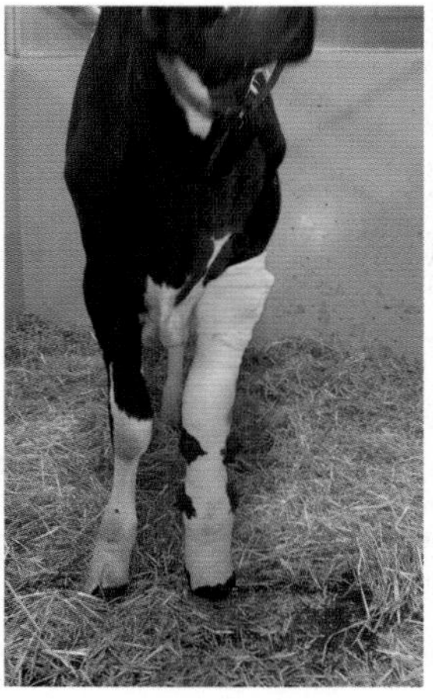

FIGURE 21.4-2 Post-operatively, the heifer is fully weight-bearing on the transfixation cast.

Anatomical features in ruminants

Introduction

Most ruminants have 2 separate digits and therefore have 2 fused metacarpi (metacarpals 3 & 4) or metatarsi (metatarsal 3 & 4) as a single bone. This section describes the anatomy of the distal limb including the carpus and metacarpus, regional synovial structures, blood supply, lymphatics, and innervation. Understanding distal neurovascular structures can be challenging because of the many branches and anastomoses, along with names that change along their course to the distal extremity. Emphasis is placed on structures specifically involved in pathological processes, surgery, or anesthetic procedures. The phalangeal anatomy is described in detail in Cases 21.2 and 21.3.

Function

The distal limb functions to support the weight of the ruminant providing for efficient and effective movement in advancing the limb. The distal limb absorbs energy associated with the impact and the loading of the limb, with the vibrations of impact largely diminished by the time they are transmitted to the proximal phalanx. The weight of the ruminant is not uniformly distributed across the ground surface of the digits and is static only if the ruminant is standing vs. moving, when it is dynamic. The extensor movement is opposed by an equal and opposite flexor movement, facilitated by the metacarpo-/metatarsophalangeal, carpal, and tarsal joints.

Carpus

The **carpus** is a compound joint composed of the radius and ulna proximally, 6 carpal bones, and the fused 3rd and 4th metacarpal bones distally. The carpal bones are arranged in 2 rows. The proximal row of the carpal bones consists of the **accessory**, **ulnar**, **intermediate**, and **radial** carpal bones. The fused 2nd and 3rd carpal (C2 + 3) bones and the 4th carpal (C4) bone represent the distal row (Fig. 21.4-3).

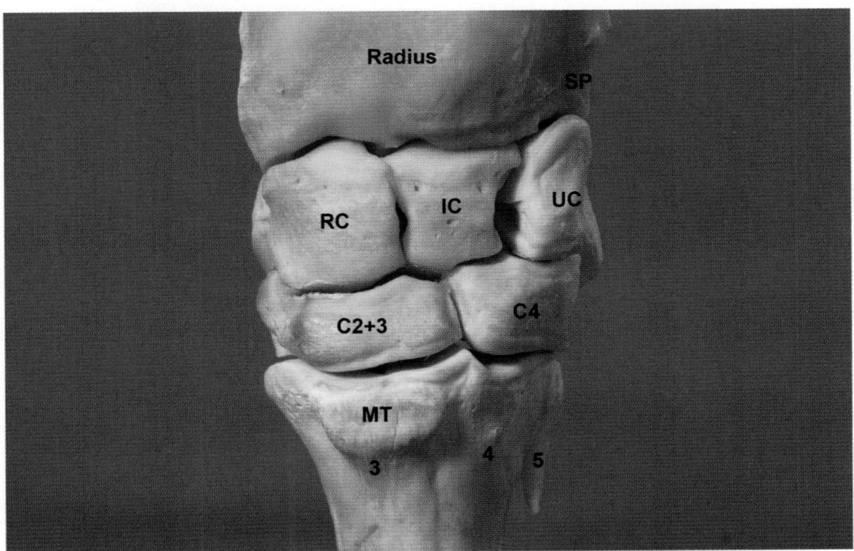

FIGURE 21.4-3 Bones of the carpal joints: Key: *SP*, styloid process (ulna); *RC*, radiocarpal bone; *IC*, intermediate carpal bone; *UC*, ulnar carpal bone; *C2 + 3*, fused 2nd and 3rd carpal bones; *C4*, 4th carpal bone; *MT*, metacarpal tuberosity; 3, 4, 5, metacarpal bones 3, 4, and 5 (rudimentary).

The carpus is the most commonly infected joint in cattle. Septic carpal joints most often occur secondary to a systemic or remote infection, such as an umbilical infection or pneumonia. Since the RC joint does not always communicate with the MC and CM joints, it must be lavaged/irrigated separately if infected (Fig. 21.4-4A and B).

The carpus is divided into 3 distinct joints. The **antebrachiocarpal** (AC) **joint** is formed by the fused radius and ulna and the proximal row of carpal bones. The **middle** (MC) **carpal joint** is formed by a joint space between the 2 rows of carpal

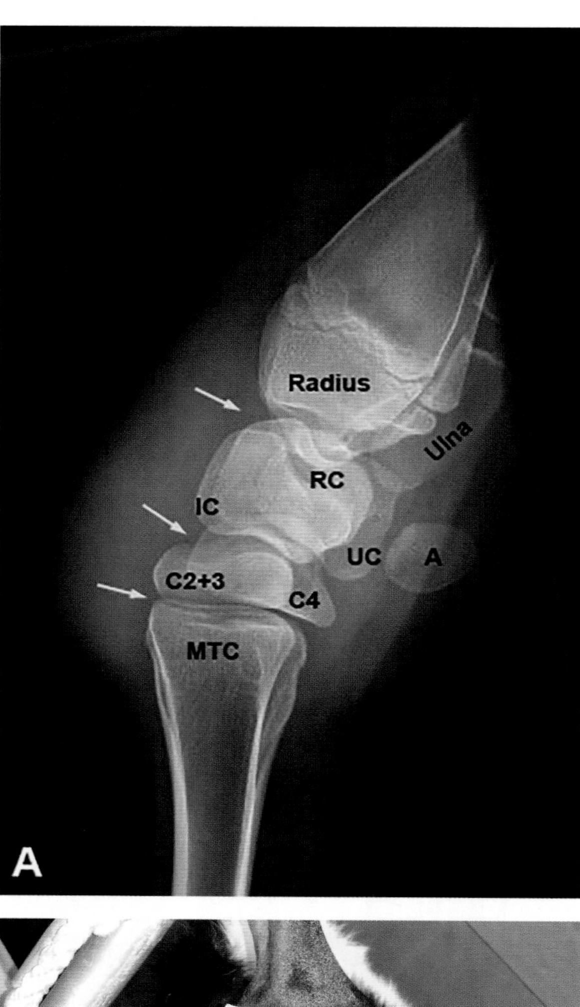

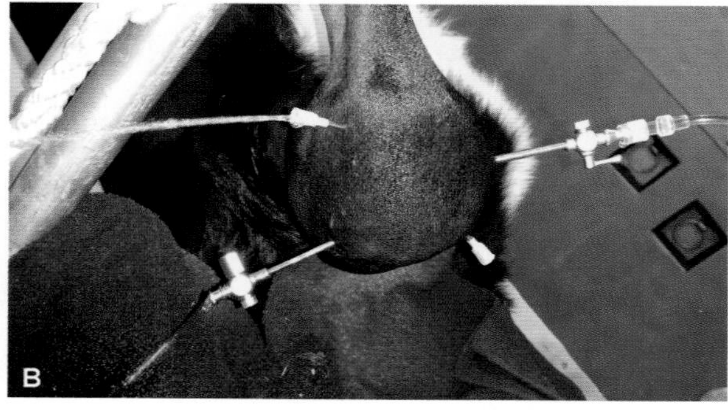

FIGURE 21.4-4 (A) Lateromedial radiographic view of the right carpus in a calf with septic arthritis. The joint space is increased (*arrows*; demonstrating proper needle placement for joint lavage). Key: *RC*, radiocarpal bone; *IC*, intermediate carpal bone; *UC*, ulnar carpal bone; *A*, accessory carpal bone; *C2+ 3*, fused 2nd and 3rd carpal bones; *C4*, 4th carpal bone; *MTC*, metacarpal bones 3 & 4. (B) Calf with septic arthritis of the carpal joints. Needles and cannulae are inserted in each of the 3 joints.

bones. The **carpometacarpal** (CM) **joint** is formed by the distal row of carpal bones and the proximal part of the fused 3rd and 4th metacarpal bones.

The carpal joint functions as a ginglymus (L.; Gr. *ginglymos* hinge) joint, with the AC joint being the largest of the 3 joints forming the carpus and structurally being an ellipsoidal joint (Fig. 21.4-3). The MC and CM joints add a limited amount of range of motion to the ginglymus action of the carpus but are more gliding in their movement and may be called a plane joint. The MC and CM joints always communicate with each other while the AC communicates with the MC in approximately 13% of cattle.

The ligaments of the carpus are like the horse except the **collateral ligaments** are weaker. The number of **intercarpal ligaments** is different from the horse because ruminants have one less carpal bone. There are numerous intercarpal ligaments between the accessory carpal bone and the adjacent carpal bones as well as the long bones.

Carpal ligament injuries are uncommon in ruminants.

Metacarpal/-tarsal skeletal anatomy and joints

The distal limb in cattle is composed of 4 digits that are numbered from medial to lateral as 2nd, 3rd, 4th, and 5th. It parallels the thoracic and pelvic limbs distal to the carpus and tarsus, respectively. The open space between the 2 digits is termed the **interungular area** (a.k.a. interdigital space). The accessory 2nd and 5th digits (**dewclaws**) are positioned on the palmar or plantar aspect of the fetlock. These vestigial digits consist of only rudimentary phalanges and do not articulate with the rest of the skeleton. The 3rd and 4th are functional and their distal part is enclosed in the **claws**. Their skeleton consists of the fused 3rd and 4th metacarpal or metatarsal bones that divide at their distal end into 2 parts by the **intertrochlear notch** (Fig. 21.4-1). The distal epiphysis fuses at 24–30 months; however, maximum bone growth occurs during the first year.

Sheep have a sinus interdigitalis—blind pouch—dorsal and proximal to the coronary band of the claw. This sinus or pouch opens to the skin and contains apocrine glands believed to play a role in marking their territory. The pouch is considered part of the common integumentary system.

The metacarpal or metatarsal bones articulatea with the proximal phalanges and the paired proximal sesamoid bones located at the palmar or plantar aspect to form the **metacarpophalangeal** or **metatarsophalangeal** (fetlock) **joint**. The metacarpophalangeal or metatarsophalangeal joints are classified as ginglymus joints. They are divided into 2 synovial sacs/compartments, which always communicate at the palmar or plantar aspect. Communication is at the level of the proximal sesamoid bones between the interdigital band of the interosseous muscle and the metacarpal or metatarsal bones (Fig. 21.4-5). The synovial sacs form dorsal and palmar or plantar pouches. The dorsal pouch lies between the distal aspect of the distal metacarpal or metatarsal bones and the extensor tendons. The palmar or plantar pouch extends more proximally than the dorsal pouch, beneath the abaxial and axial branches of the interosseous muscles and the deep digital flexor tendons.

While not part of the metacarpus/-tarsus, it is important to describe the phalanges here. The 1st (proximal) phalanx is longer than the 2nd (middle) phalanx. Its proximal articular surface is concave in a dorsopalmar/dorsoplantar plane and is divided by a sagittal groove into 2 areas, of which the abaxial component is the largest and highest. Its distal articular surface is smaller than the proximal and is convex in shape. The distal epiphysis fuses at 18–24 months. The 1st phalanx articulates with the 2nd phalanx at the **proximal interphalangeal joint** (PIJ) or the pastern joint; its synovial compartment is small. The dorsal pouch lies between the fibrous joint capsule and the extensor tendons along the distal aspect of the 1st phalanx. The palmar or plantar pouch is larger than the dorsal pouch and located underneath the terminal portion of the superficial digital flexor tendon. Collateral ligaments and a palmar ligament support the PIJ joint. The 3rd (distal) phalanx is completely enclosed within the claw. The anatomy of the foot is described in detail in Case 21.3.

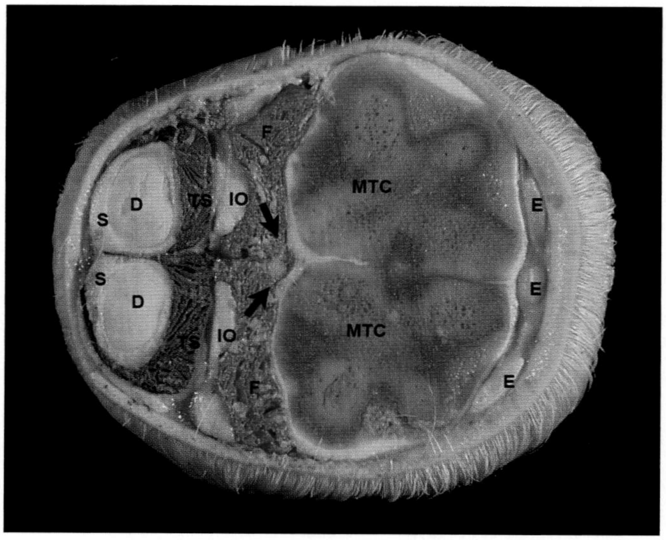

FIGURE 21.4-5 Transverse section of a prosected limb at the level of the metacarpophalangeal joint just proximal to the proximal sesamoid bones. The joint was infused with red latex and the synovial sheath of the flexor tendons was infused with blue latex. The communication between the medial and lateral metacarpophalangeal joints is depicted at the palmar aspect of the metacarpal bones (MTC) (*arrows*). Key: *D*: deep digital flexor tendon; *S*, superficial digital flexor tendon; *E*, extensor tendons; *TS*, tendon sheath: *IO*, interosseous muscle.

Regional tendon sheaths

The fetlock joint has dorsal and palmar or plantar tendon sheaths. The **dorsal (extensor) tendon sheath** allows the common digital extensor (thoracic limb) and long digital extensor (pelvic limb) to glide over the joint. At the proximal aspect of the metacarpus/-tarsus, the common digital extensor tendon in both the thoracic and pelvic limbs plus the long digital extensor tendon in only the pelvic limb divide distally and contribute two tendons to digits 3 and 4. The medial muscle belly of the extensor tendons in the thoracic and pelvic limbs is larger and inserts on the 2nd phalanx, and it is referred to as the "proper extensor of the digit." The lateral muscle belly divides into 2 extensor tendons that insert on the distal phalanges of digits 3 and 4.

The proximity of the medial and lateral tendon sheaths to the claw makes them prone to deep digital flexor tendon sepsis. Therefore, close evaluation is needed during clinical examination of suspected distal interphalangeal joint sepsis (Fig. 21.4-6). Additionally, clinical experience necessitates consideration of the lateral and medial tendon sheaths as noncommunicating or independent structures. Fibrin clots can block the communication between the lateral and medial compartments and therefore they need to be lavaged/irrigated separately if infected.

The **tendon sheaths of the superficial and deep digital flexor muscles** extend from the distal palmar or plantar aspect of the metacarpal or metatarsal bones, divide in distinct portions, and terminate at the midsection of the 2nd phalanx. The medial and lateral tendon sheaths may communicate at their proximal section. Palmar or plantar and digital ligaments increase the support of the flexor tendons and provide demarcation of their sheaths' cul-de-sacs.

Regional muscle-tendon units

Flexion and extension of the joints of the limbs are performed by contraction and relaxation of flexor or extensor muscle-tendon units with this description centered on the muscle-tendon structures of the lower limb. Importantly, the triceps muscle and its innervation (radial nerve) are essential in being able to weight-bear on the thoracic limb because of the need to extend the elbow. Similarly, the quadriceps muscle unit and its innervation (femoral nerve) are essential to pelvic limb weight-bearing because of the need to extend the stifle. The gastrocnemius muscle and its innervation (tibial nerve) are also important for pelvic limb weight-bearing because it extends the tarsus (hock).

FIGURE 21.4-6 The tendon sheaths of the digital flexor tendons were injected with blue latex in this prosection. Right side of prosection; annular ligaments are removed, and the superficial digital flexor tendon (*SF*) is opened to depict the deep digital flexor tendon (*DF*). Key: *PAL*, proximal annular ligament; *PALP*, proximal annular ligament of the proximal phalanx; *DALP*, distal annular ligament of the proximal phalanx: *DIL*, distal interdigital ligament.

Flexor muscle-tendon units and their innervations

Flexor units of the distal thoracic limb include the flexor carpi radialis muscle, flexor carpi ulnaris muscle (ulnar and humeral heads), superficial digital flexor muscle, deep digital flexor muscle, and interosseous muscle. The **flexor carpi radialis muscle** originates on the medial epicondyle of the humerus, attaches to the palmar aspect of the proximal metacarpus, and is innervated by the median nerve. The **ulnar head** of the **flexor carpi ulnaris muscle** originates from the olecranon, inserts on the accessory carpal bone, and is innervated by the ulnar nerve. The **humeral head** of the flexor carpi ulnaris muscle originates from the medial epicondyle of the humerus, inserts on the accessory carpal bone, and is innervated by the ulnar nerve. These 3 muscle groups act to flex the carpus.

Dysfunction of these muscles impairs flexion of the carpus during the nonweight-bearing phase of the stride.

The **superficial digital flexor muscle** originates from the medial epicondyle of the humerus, inserts on the palmar surface of the 2nd phalanx, and is innervated by the median nerve. The **deep digital flexor muscle** originates from the humerus, radius, and ulna, inserts on the flexor tuberosity of the 3rd phalanx, and is innervated by the median and ulnar nerves. These muscles flex the carpus and digits.

Disruption of the superficial digital flexor tendon results in mild hyperextension of the fetlock and digit. Disruption of the deep digital flexor tendon results in mild hyperextension of the fetlock and severe hyperextension of the digit.

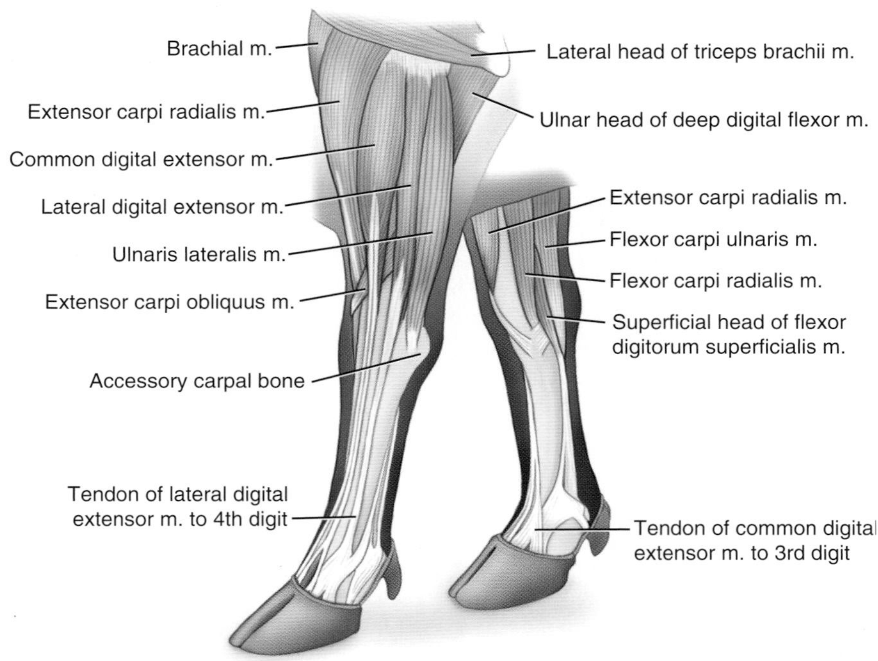

FIGURE 21.4-7 Lateral and medial projections of the clinically important muscles and tendons of the bovine thoracic limbs.

The tendons of the superficial and deep digital flexor muscles are paired, with one branch of each supporting each digit (Fig. 21.4-7).

Disruption of the interosseous muscle results in hyperextension of the fetlock.

The **interosseous muscle** is more ligamentous than muscular histologically. This structure originates from the proximal and palmar aspects of metacarpals 2/3/4/5, inserts on the sesamoid bones of the fetlock, and supplies extensor branches, joining the extensor tendons dorsally and distally. It is innervated by the deep branch of the ulnar nerve. This muscle acts to support the foot.

The flexor units of the digits of the pelvic limb include the superficial digital flexor muscle and deep digital flexor muscle. The **superficial digital flexor muscle** originates on the caudal distal femur, inserts on first the tuber calcanei and then the plantar aspect of the 2nd phalanx, and is innervated by the tibial nerve. The superficial digital flexor tendon tracks along the lateral aspect of the gastrocnemius tendon to a position directly overlying this tendon and then tracks over and plantar to the calcaneus. The **deep digital flexor muscle** originates from the caudal proximal aspect of the tibia and fibula, inserts on the flexor tuberosity of the plantar surface of the distal phalanx, and is

FLEXURAL DEFORMITIES

Flexural deformities are not common in ruminants compared to horses. The superficial and deep digital flexors of the thoracic limb are more commonly affected. The surgical technique often used in calves to correct this condition is tenotomy of both flexor tendons. Infrequently, contraction of the carpus is diagnosed. It should be differentiated from arthrogryposis (Gr. *arthron* joint + *Gr. grypōsis* a bending, persistent flexure or contracture of a joint), which is a congenital disease frequently associated with other important and potentially life-threatening congenital abnormalities, like cleft palate.

innervated by the tibial nerve. The deep digital flexor tendon tracks along the medial aspect of the calcaneus. Tarsal flexion occurs due to contraction of the cranial tibial, long digital extensor, fibularis (peroneus) longus, fibularis (peroneus) tertius, and lateral digital extensor muscles.

Extensor muscle-tendon units and their innervations

The extensor units of the distal thoracic limb include the extensor carpi radialis muscle, common digital extensor muscle, lateral digital extensor muscle, and extensor carpi ulnaris. The **extensor carpi radialis muscle** originates from the lateral epicondyle of the humerus, inserts on the dorsal proximal aspect of the metacarpus, and is innervated by the radial nerve. This muscle flexes the elbow and extends the carpus. The **common digital extensor muscle** originates from the lateral epicondyle of the humerus, inserts on the dorsal aspect of the 3rd phalanx, and is innervated by the radial nerve. The **lateral digital extensor muscle** originates from the lateral epicondyle of the humerus, inserts on the 2nd phalanx, and is innervated by the radial nerve. The common and lateral digital extensor muscles extend the carpus and digits.

The extensor units of the digits of the pelvic limb include the long digital extensor muscle and the lateral digital extensor muscle. The **long digital extensor muscle** originates in the extensor fossa of the femur, inserts on the extensor process of the 3rd phalanx, and is innervated by the peroneal nerve. The **lateral digital extensor muscle** originates from the fibula, inserts on the lateral aspect of the digits, and is innervated by the peroneal nerve. These muscles extend the digits and flex the tarsus. Tarsal extension occurs due to contraction of the gastrocnemius, superficial digital flexor, soleus, semitendinosus, and biceps femoris muscles.

Blood supply, lymphatics, and innervation (Fig. 21.4-8)

The thoracic digits receive their blood supply from the **median artery** with collateral circulation from the **ulnar** and **radial arteries**. The median artery tracks along the caudal and medial surfaces of the antebrachium through the carpal canal and along the palmar axial aspect of the metacarpal bones. The **distal deep palmar branch** arises from the median artery and divides into the **palmar common digital artery 3 and 4**. The **dorsal** and palmar common digital arteries communicate via an **interdigital artery** positioned between the 1st phalanges.

Distally, the blood supply to the digits is achieved primarily by the **axial** and **abaxial palmar proper digital arteries 3 and 4** and the **dorsal common digital arteries 3 and 4**. The proper axial digital artery enters the 3rd phalanx in a foramen on the axial surface of the extensor process and forms a terminal arch within the 3rd phalanx with the smaller abaxial proper digital artery. Branches from the **terminal arch** exit the 3rd phalanx via openings on the surface of the bone and supply the corium. The blood supply of the pelvic digit is provided by the **saphenous** and **cranial tibial arteries**. Digital artery architecture of the pelvic digit is similar to that of the thoracic digit.

The veins are positioned more superficially than the arteries. The **dorsal proper digital veins** and **axial** and **abaxial proper plantar digital veins** drain the venous blood from the claw. They follow the path of the proper digital arteries to empty proximally into the **saphenous vein** in the pelvic limb and the **cephalic vein** in the thoracic limb. The cephalic vein is located on the medial and dorsal aspect of the radius.

Intravenous anesthesia distal to tourniquet placement is commonly used in combination with other analgesic modalities for surgery of the distal limb—e.g., joint resection of interphalangeal joints, digital amputation, and fetlock lavage. It is also used to reduce distal metacarpal fractures. Proximal to the carpus, the cephalic vein is easy to access if IV catheter placement or an injection is needed—e.g., for anesthesia or antibiotics distal to a tourniquet. If distal analgesia of the digit is required, the tourniquet is placed below the carpus and the injection is performed in the dorsal metacarpal vein (Fig. 21.4-9A and B). An alternate site for IV injection is the dorsal common digital vein 3 located between the proximal phalanges. Distal pelvic limb anesthesia is achieved by injecting the saphenous vein at the lateral aspect of the tarsus or the abaxial proper plantar digital vein just proximal to the lateral aspect of the fetlock. The dorsal common digital vein is also accessible for IV injection distal to a tourniquet.

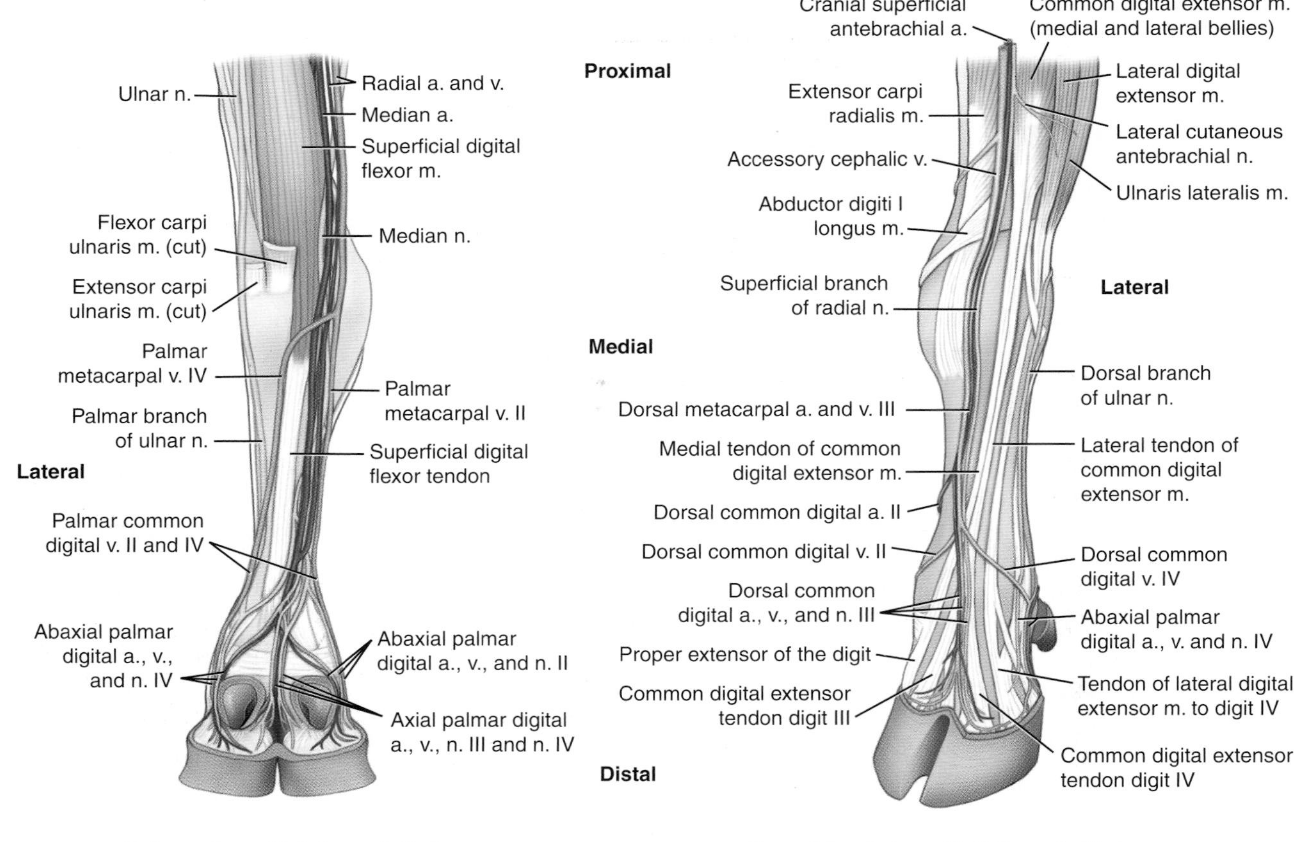

FIGURE 21.4-8 Dorsolateral and palmar views of the muscles, tendons, arteries, veins, and nerves of the bovine distal thoracic limb.

The **proper axillary node** and **axillary lymph nodes of the 1st rib** provide afferent and efferent lymphatic vessels to the thoracic limb. The proper axillary lymph node is the largest. It is situated medial and caudal to the shoulder joint. The axillary lymph nodes are medial to the shoulder joint on the lateral surface of the 1st rib. Lymphatics of the pelvic limb are covered in Chapter 22.

> This feature must be considered when performing a nerve block for diagnostic purposes or when providing presurgical analgesia.

Innervation of the distal thoracic limb is provided palmarly by the **median nerve** and the **palmar branch of the ulnar nerve** and dorsally by the **superficial branch of the radial nerve** with a **cutaneous branch of the musculocutaneous nerve**. Distally, the nerves parallel the vessels and each digit is supplied with dorsal, palmar, abaxial, and axial branches. The dorsal aspect of the pelvic limb is innervated by the **superficial** and **deep branches of the fibular nerve**. Its plantar aspect is innervated by the **tibial nerve** that divides at the caudal aspect of the tarsus into the **lateral** and **medial plantar nerves**. The distal innervation in the pelvic limb is similar to that of the thoracic limb digits.

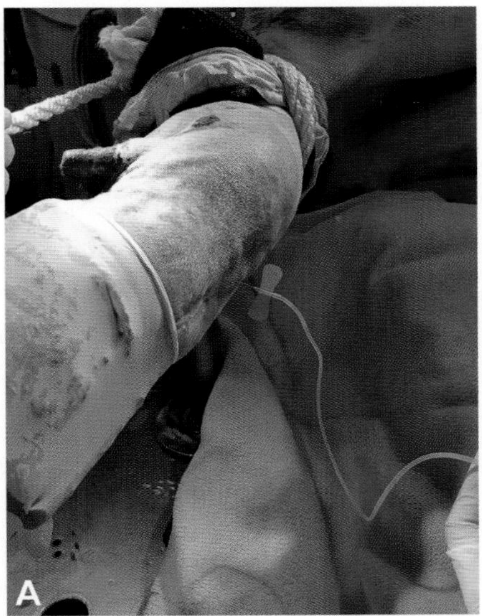

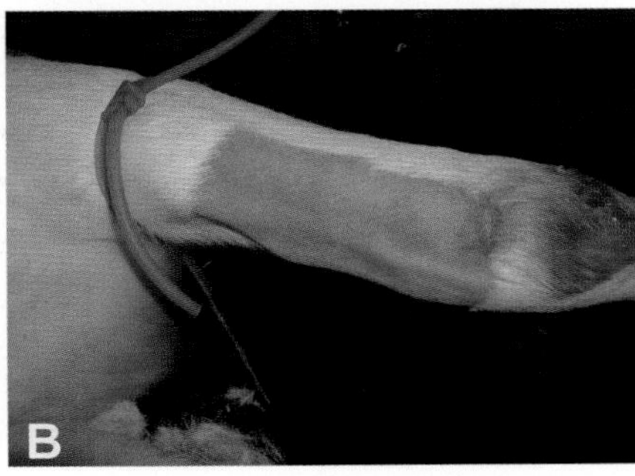

FIGURE 21.4-9 Intravenous injection distal to a tourniquet in a cow. (A) The needle is inserted in the common dorsal digital vein. (B) The cephalic vein is distended and ready for injection.

Selected references

[1] Sisson S. Ruminant syndesmology. In: Sisson G, Grossman J, editors. Anatomy of domestic animals. Philadelphia: WB Saunders Co; 1975. p. 787–90.
[2] Nickel R, Schummer A, Seiferle E, et al. The locomotor system of the domestic mammals. Berlin: Verlag Paul Parey; 1986. p. 184–7.
[3] Barone R. Anatomie comparé des mammifères domestiques: arthrologie et myologie. Paris: Vigot; 1980. p. 162–72.
[4] Dyce KM, Sack WO, Wensing CJG. Textbook of veterinary anatomy. Philadelphia: WB Saunders Co; 1987. p. 697–716.

CHAPTER 22

PELVIC LIMB

André Desrochers, Chapter editor

22.1 Coxofemoral luxation—*Emma Marchionatti* 1194
22.2 Patellar luxation—*David E. Anderson* 1198
22.3 Cranial cruciate ligament tear—*Emma Marchionatti and Caroline Constant* 1205
22.4 Gastrocnemius rupture—*André Desrochers* 1211

CHAPTER 22

CASE 22.1

Coxofemoral Luxation

Emma Marchionatti
Clinic for Ruminants, Vetsuisse Faculty, University of Bern, Bern, CH

Clinical case

History

A 4-year-old Holstein cow presented for recumbency of 36 hours' duration. The owner saw the cow fall with both pelvic limbs abducted. The cow had calved 12 hours before the accident, was treated for milk fever, and responded with an improvement in the signs of hypersensitivity, head-bobbing, ear-twitching, and fine tremors but remained in sternal recumbency. The referring veterinarian examined the cow, placed hobbles on the rear legs, and administered a dose of flunixin meglumine; however, the cow was still unable to stand.

Physical examination findings

The cow was bright and alert with a heart rate of 88 bpm and a rectal temperature of 100.2 °F (37.9 °C). Moderate swelling was present over the right thigh and coxofemoral joint. The relative position of the right greater trochanter to that of the tuber coxae and the tuber ischiadicum was more ventral and caudal than normal. Coxofemoral joint manipulation revealed crepitus (L. *crepitare* to crackle) and greater range of motion of the proximal femur during abduction.

Differential diagnoses

Caudoventral coxofemoral luxation and femoral head/neck fracture

Diagnostics

The femoral head was palpated in the obturator foramen during palpation per rectum. Pelvic radiography (Fig. 22.1-1) revealed a caudal displacement of the femoral head outside the acetabulum. No fracture was identified involving the femur or the pelvis.

Diagnosis

Caudoventral coxofemoral luxation

Treatment

The prognosis was poor based on the duration of the condition and the size of the cow. The owner elected no further treatment and the cow was humanely euthanized.

Anatomical features in ruminants

Introduction

The coxofemoral (acetabulofemoral or hip) joint is a spheroid-type joint that connects the femur to the pelvis and—more specifically—is the joint between the femoral head and acetabulum. Joint unity relies on the health of the joint

Comparative Veterinary Anatomy: A Clinical Approach. https://doi.org/10.1016/B978-0-323-91015-6.00108-4

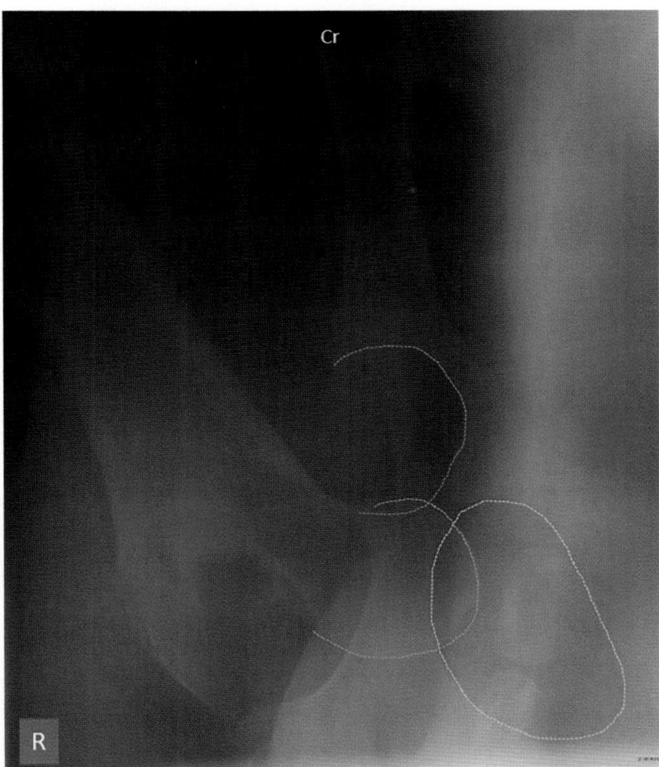

FIGURE 22.1-1 Radiographs of the right caudoventral coxofemoral luxation from the Holstein presented in this case. Note the femoral head (*dashed green line*) outside the acetabular cavity (*dashed blue line*) and caudally displaced toward the obturator foramen (*dashed yellow line*). Cranial is to the top of the image.

capsule, joining of the femoral head and acetabulum, and the gluteal muscles. This case discusses the coxofemoral joint and associated musculature, along with regional blood supply, lymphatics, and innervation.

Function

In cattle, the coxofemoral joint primarily flexes and extends. Abduction and adduction movements are limited. During flexion, the femoral head rolls in the acetabulum so that the distal femur moves craniodorsally. The femoral head tends to subluxate from the acetabulum caudally, but it is retained by the joint capsule, the gluteal muscles, and the femoral head ligament. During extension, the opposite mechanism takes place. Cranial luxation of the femoral head is prevented by the femoral head ligament, the joint capsule, and the long head of the quadriceps muscle. An additional function of the coxofemoral joint is to support the weight of the body in both static—e.g., standing—and dynamic—e.g., walking—positions. The coxofemoral joints are important in balance and the pelvic inclination angle.

COXOFEMORAL LUXATION

Coxofemoral luxation can occur in cattle of any age associated with a traumatic injury. Predisposing factors include milk fever, obturator paresis, and estrus mounting behavior. Coxofemoral luxation may follow initial trauma when a cow or bull falls with both pelvic limbs abducted or may be a result of struggling to stand. In calves, coxofemoral luxation may also follow dystocia. Craniodorsal luxation is more common (especially in young calves), while caudoventral luxation—i.e., caudal and ventral to the pelvis—is also seen in adult cattle, while luxation into the obturator foramen is the least common. Cattle with craniodorsal luxation are usually ambulatory and have moderate lameness compared to other luxations. Diagnosis is based on the altered conformation of the coxofemoral region along with abnormal movements, especially hyperabduction, during limb manipulations. Closed reduction by manipulation can produce good results if performed within 12–24 hours, but recurrence is common. Surgical reduction and stabilization can achieve good results if bone and muscles are minimally compromised.

Joint surfaces

The **os coxae** (bony pelvis) receives the femoral head in the **acetabulum,** a large cavity at the junction of its three components (ilium, ischium, and pubis). A prominent border, interrupted ventrally by the **acetabular notch**, surrounds its semilunar joint surface. The **acetabular fossa**, the deep depression in the center of the acetabular cavity, provides the attachment for the femoral head ligament. A corresponding ligament insertion notch on the femoral head—the **fovea capitis**—creates an indentation in the normal spheroid shape. The **femoral head** elongates in a transverse plane and the **femoral neck** is poorly defined in cattle. The **greater trochanter** is located laterally and is the site of attachment of the gluteal muscles. Two physes are in the proximal femur: the femoral capital physis and the greater trochanter physis. These 2 physes close at approximately 3.5 years of age. A fibrocartilaginous rim—the **labrum acetabulare** (this fibrocartilaginous acetabular rim surrounding the joint is also found in the shoulder joint in some species)—deepens the acetabular joint surface. It is continuous with the bony margin and the **transverse acetabular ligament**, which in turn crosses the acetabular notch (Fig. 22.1-2).

Femoral capital physeal fracture is the most common coxofemoral injury in newborn calves and young cattle. It is frequently caused by forced extraction during dystocia or as a result of a fall on a hard surface. Acute moderate-to-severe lameness attributable to the coxofemoral area is usually seen without an alteration of the regional conformation. Surgical reduction and fracture fixation are recommended.

Joint capsule and ligaments

The joint capsule is attached to the margin of the acetabulum, covering the labrum acetabulare and the transverse acetabular ligament, to insert on the femoral neck (Fig. 22.1-2). It is thinner at its medial aspect.

The short and circular **femoral head ligament** joins the femoral fovea capitis to the acetabular fossa (Fig. 22.1-2). Its principal function is to limit flexion and abduction movements. Unlike horses, cattle do not have a femoral head accessory ligament, which connects the femoral fovea capitis to the prepubic tendon. Its absence allows increased laterality in movement compared to horses.

Coxofemoral muscles

Powerful muscle masses cover and add strength to the coxofemoral joint. Among them are the thigh and the pelvic muscles. The former includes the **tensor fasciae latae**, **biceps femoris**, and **quadriceps femoris muscles**. The pelvic muscles are subdivided in gluteal and deep hip muscles. The **gluteal muscles (middle, accessory, and deep)** originate at the bony pelvis and insert at the greater trochanter of the femur, primarily allowing extension of the coxofemoral joint. The deep hip muscles (**piriformis**, **lateral rotator**, **external obturator**, and **quadratus femoris**) limit adduction and rotation of the limb.

COXOFEMORAL LUXATION—TREATMENT OPTIONS

Correction of hip luxation involves either a closed or open reduction requiring considerable traction and general anesthesia. Traction on the distal limb combined with external rotation is most helpful for a craniodorsal luxation because it rotates the femoral head cranially and aids passage of the femoral head over the dorsal acetabular rim. Closed reduction for a craniodorsal luxation may have a low success rate if the fibrocartilaginous rim (labrum acetabulare), joint capsule strands, or hematoma are trapped between the acetabulum and femoral head during reduction. Open reduction has 2 important advantages: (1) the femoral head is reseated in the acetabulum without interposed tissue and (2) the dorsal joint capsule is strengthened using a combination of orthopedic screws, washers, and a nonabsorbable suture material placed between the dorsal acetabular rim and the greater trochanter. Post-operatively, affected individuals are confined to a stall for several months for healing and to minimize repeat luxation. The prognosis is improved if the hip luxation is recognized early and treated within 12 hours of injury.

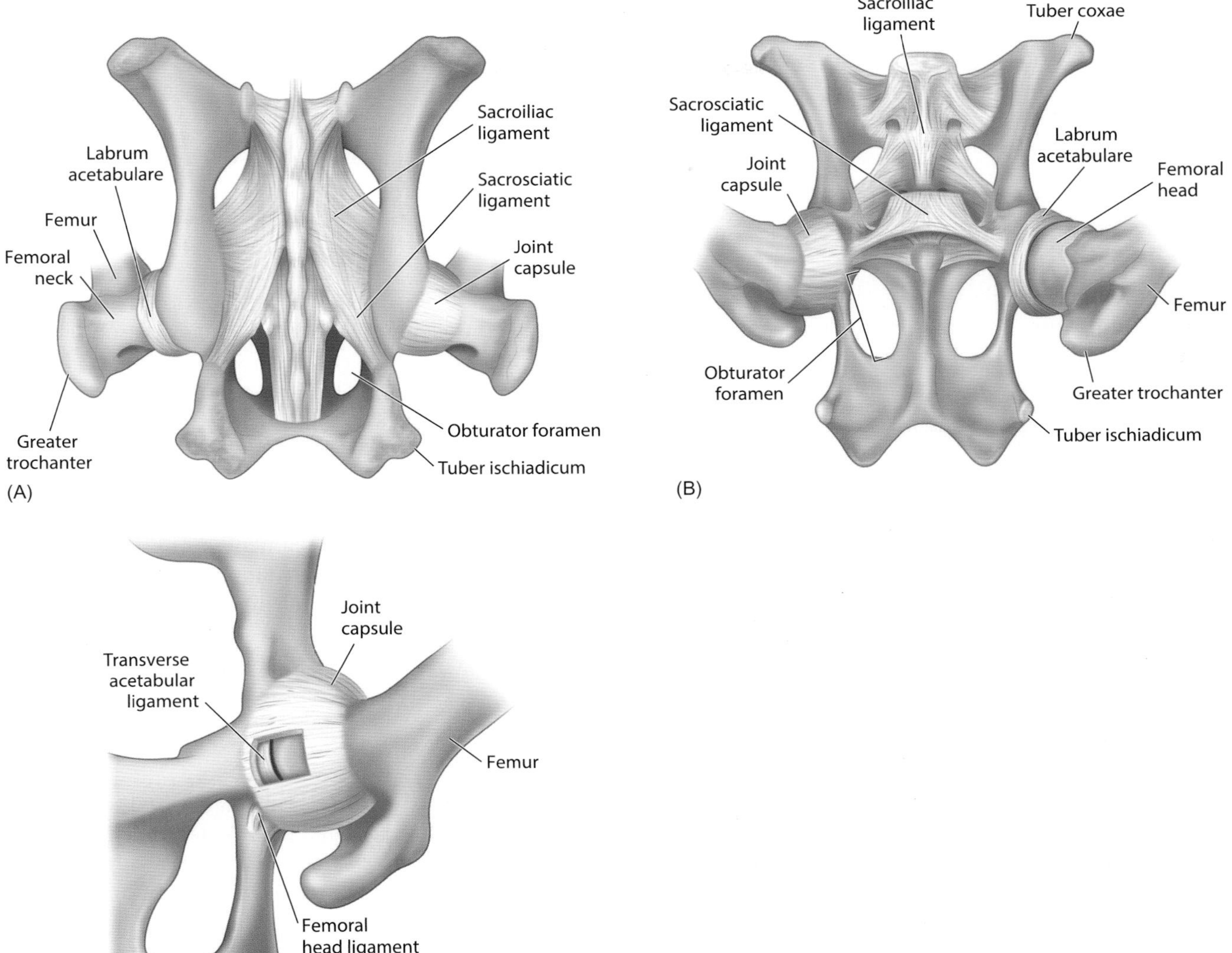

FIGURE 22.1-2 The bovine pelvis and coxofemoral joint. (A) Dorsal view of pelvis; (B) ventral view of the pelvis; and (C) ventral view of the coxofemoral joint.

Blood supply, lymphatics, and innervation

Blood supply to the coxofemoral joint originates from the **circumflex femoral arteries** and branches of the **deep thigh artery** (a. profunda femoris). Small contributions also arise from the **foveal artery**, a small vessel in the femoral head ligament—a branch of the **obturator artery**. Veins draining the pelvis are satellites of the arteries and are comprised primarily of the **deep femoral** and **internal iliac veins** emptying into the **external iliac veins**.

Lymphatics drain into the **iliosacral**, **iliofemoral**, **inguinofemoral**, and **ischiatic lymph centers**. Articular branches from the **obturator**, **femoral**, **sciatic**, and **gluteal nerves** innervate the joint capsule.

Selected references

[1] Budras KD, Habel RE. Bovine anatomy. 2nd ed; 2011. p. 28–9.
[2] Fubini SL, Ducharme NG. Farm animal surgery. 2nd ed; 2017. p. 380–2.
[3] Marchionatti E, et al. Traumatic conditions of the coxofemoral joint: luxation, femoral head-neck fracture, acetabular fracture. Vet Clin North Am Food Anim Pract 2014;30:247–64.

CHAPTER 22

CASE 22.2

Patellar Luxation

David E. Anderson
Department of Large Animal Clinical Sciences, University of Tennessee College of Veterinary Medicine, Knoxville, Tennessee, US

Clinical case

History

A 5-year-old male castrated alpaca was presented for acute onset of a nonweight-bearing lameness. The owners reported that the alpaca had an abnormal gait for several months but had become severely lame following an incident with a herd mate, in which the alpaca was knocked down. The alpaca was able to rise immediately after the incident but had great difficulty in using the left pelvic limb. The owners noted that he seemed to collapse in the hind quarters on that side when attempting to bear weight and had to quickly position the right pelvic limb to enable rising.

Physical examination findings

Physical examination on the farm revealed that the alpaca was in good body condition (body condition score 6/10) and had normal physical examination parameters. Palpation of the pelvic limbs revealed that the left pelvic limb had noticeable muscular atrophy, most obviously affecting the quadriceps and biceps femoris muscles. Palpation of the patella revealed a lateral position relative to the trochlear groove, resting midway between the lateral trochlear ridge and the lateral femoral epicondyle. The patella was movable and somewhat painful on palpation; no other abnormalities were noted.

The alpaca was then sedated and placed in right lateral recumbency to allow palpation of the left stifle. The laterally dislocated (luxated) patella could be repositioned between the lateral and medial trochlea manually; however, the patella would immediately dislocate on flexion and extension of the limb. The alpaca was repositioned in left lateral recumbency for right stifle palpation. The right patella was in its normal position between the lateral and medial femoral trochlea and could not be manually dislocated, even when the limb was hyperextended. The right patella was slightly mobile and would pivot medial and lateral but remained in its normal anatomical location throughout the range of motion and stress tests.

Differential diagnoses

Lateral left patellar dislocation, quadriceps muscle tear, patellar fracture, patellar ligament disruption, septic arthritis of the stifle, distal femoral or proximal tibial fracture, femoral nerve paralysis, and femoral-patellar ligament disruption

Diagnostics

Ultrasonographic examination of the left stifle in this alpaca revealed mild joint effusion and mild synovial proliferation throughout the femorotibial and femoropatellar joint cavities; lateral deviation of the patella and patellar tendon was observed. No other abnormalities were identified. Radiographic examination of the left stifle revealed a grade 3 lateral patellar dislocation affecting the left stifle joint (Fig. 22.2-1).

Comparative Veterinary Anatomy: A Clinical Approach. https://doi.org/10.1016/B978-0-323-91015-6.00109-6

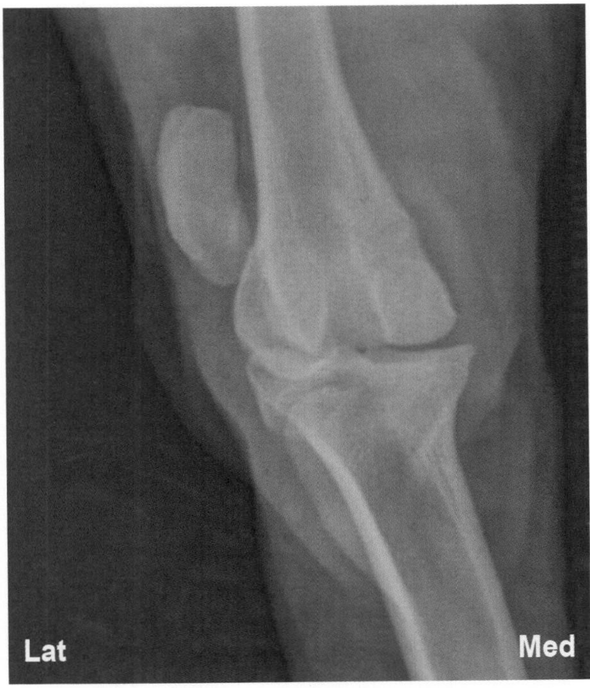

FIGURE 22.2-1 Cranial-caudal radiographic view of the left stifle of the alpaca in this case. The patella is located lateral to the lateral trochlear ridge of the femur (grade 3 lateral patellar dislocation).

Diagnosis

Grade 3 lateral patellar dislocation of the left stifle

Grade 1 patellar dislocation of the patella is characterized by a normal position of the patella during ambulation and the potential to manually dislocate the patella followed by the patella immediately returning to its normal position. Grade 2 patellar dislocation is characterized by mild anatomic changes in the trochlear groove and the patella resulting from abnormal wear. In these cases, the patella remains dislocated until manually replaced in the trochlear groove. The patella remains in the joint for variable periods of time and periodically dislocates spontaneously. Grade 3 patellar dislocation is characterized by more significant anatomical changes including atrophy of the associated trochlear ridge and degenerative changes on the articular surface of the patella. There is spontaneous dislocation that remains dislocated unless manually reduced. Grade 4 patellar dislocation is characterized by complete patellar dislocation that cannot be manually reduced and requires surgical correction to maintain its position in the trochlear groove.

Treatment

Surgical correction of lateral patellar dislocation was recommended to the owners based on the severity of the luxation. The alpaca was placed under general anesthesia and the surgical site was aseptically prepared. A craniolateral skin incision was made, beginning at the distal aspect of the midportion of the femur extending adjacent to the lateral aspect of the patella and ending at the lateral aspect of the tibial crest. The skin incision was continued through the fascial layers surrounding the stifle joint and an arthrotomy of the femoropatellar and femorotibial joints was created. Release of the lateral femoropatellar fascia and lateral femoropatellar ligament was performed in the process of creating the arthrotomy to mobilize the patella, prevent a lateral pull on the patella, and facilitate replacement in the trochlear groove. The trochlear groove and lateral trochlear ridge were examined for

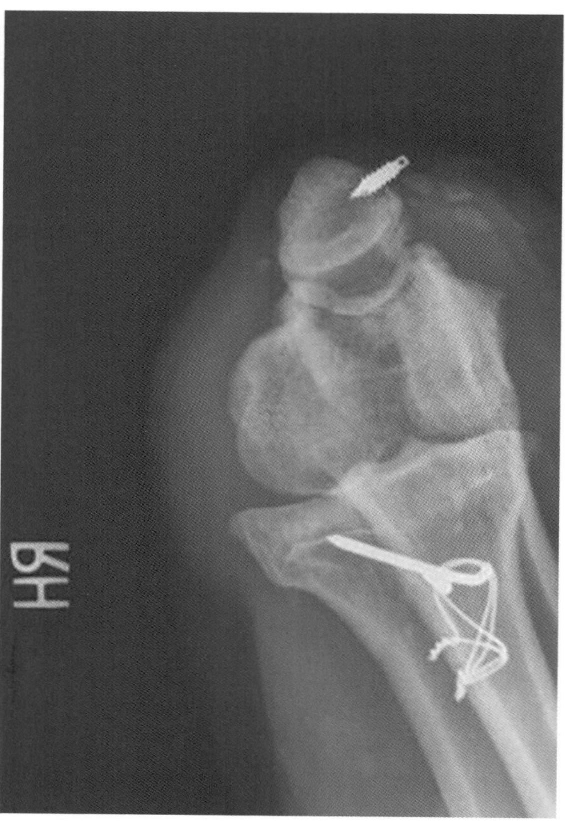

FIGURE 22.2-2 Proximal-to-distal ("skyline") radiographic view of the trochlear groove and right stifle of an alpaca with lateral patellar dislocation after surgical treatment by trochlear recession, tibial crest transposition, and femoropatellar ligament enhancement.

> Trochlear recession involves creating a V-shaped wedge of bone and cartilage and removing a small and similar-shaped wedge in the subchondral bone beneath the first trochlear wedge, resulting in a deepening or recession of the bone and cartilage (Fig. 22.2-3).

cartilage pathology and the depth of the trochlear groove was evaluated. Patellar movement throughout the normal range of motion of the stifle was normal with the patella remaining in the trochlear groove. A grade 3 patellar dislocation usually requires trochlear recession to deepen the groove and insure normal patellar positioning. This may be done in combination with medial joint imbrication and, in some cases, augmentation of the medial femoropatellar ligament or tibial crest transposition may be required to adequately stabilize the patella (Fig. 22.2-2).

After creating the trochlear recession and imbricating (L. *imbricatus*: *imbrex* tile = overlapping of the apposing tissues like tiles or shingles) the medial patellar tissues (Fig. 22.2-4), the stifle was put through full range of motion to ensure that the patella glided normally in the trochlear groove without dislocating. The arthrotomy was closed in anatomical tissue layers and a bandage was placed covering the surgical site and maintained for 7 days after surgery. Antimicrobial and anti-inflammatory medications were administered pre-operatively and for 7 days after surgery. Two weeks of stall confinement was recommended before gradually initiating physical therapy and a controlled exercise program. The alpaca resumed his normal routine and physical activities 3 months after surgery.

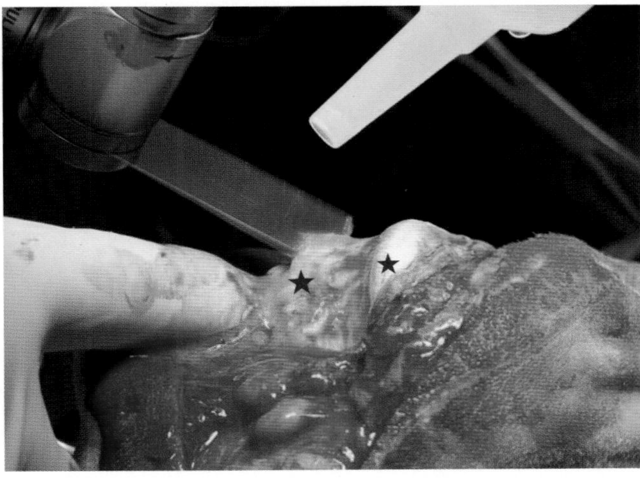

FIGURE 22.2-3 Intra-operative image of "V" trochlear groove recession (between *stars*) using a pneumatic osteotomy saw.

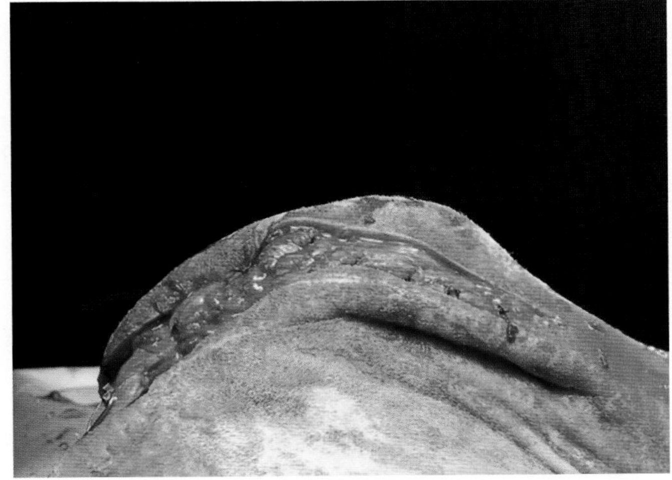

FIGURE 22.2-4 Intra-operative image following completion of a medial joint imbrication after tibial crest recession.

Anatomical features in camelids

Introduction

The stifle joint of llamas and alpacas (camelids) is similar in composition to ovine and canine stifles. This section covers the stifle along with its blood supply, lymphatics, and innervation. The stifle joint of bovids is covered in Case 22.3.

Function

The function of the stifle joint is to flex and extend the pelvic limb during ambulation, recumbency, and standing. The extensor muscles of the stifle attach to the patella and join in the fascial planes, merging with the patellar tendon and attaching to the tibial crest. As the muscles contract, they exert a force across the joint and through the patella to push the femur caudally—and the tibia cranially—to extend the joint. As the extensor muscles relax, the angle of the joint returns to its normal position.

Llamas and alpacas have relatively oblique angles to their stifle and tarsus with an average angle of approximately 140° for the stifle and 145° for the tarsus. When recumbent, the stifle undergoes extreme flexion, resulting in the femur and the tibia nearly reaching parallel. This also involves some rotation of the tibia relative to the femur and proximal displacement of the patella when camelids assume sternal recumbency, referred to as a "cushed" position. Unlike dogs, llamas and alpacas do not have a fabella.

Bones of the stifle

The bones of the stifle joint include the femur, patella, and tibia. The **femur** runs between the pelvis, where it articulates through its **head** with the acetabulum, and the tibia, where the **femoral condyles** rest on the menisci which serve as a cartilage buffer, separating the femur from resting directly on the condyles of the proximal tibia. The **patella** serves to transition forces across the stifle joint, provides a point of attachment for the extensor muscles of the stifle, and acts as a connection to the **tibial crest** via the primary patellar ligament.

Soft tissues of the stifle

The stifle of llamas and alpacas is a single, common joint cavity like that of sheep, camels, dogs, and humans; it does not have separate compartments as in horses and cattle. Therefore, the **femoropatellar** and **femorotibial compartments** freely communicate and constitute the anatomic features of the joint. There is a single **patellar ligament** that originates from the combined fascia surrounding the patella and inserts on the **tibial crest**. In some llamas and alpacas, the parapatellar fascia is oriented such that they are positioned similarly to the medial and lateral patellar tendons of cattle. However, the fascia does not form a distinct ligament and therefore is not considered part of the patellar tendon complex. The fascia attaching to the proximal and cranial portion of the patella is comprised of the common tendinous insertions of the **quadriceps muscles** (4 heads—vastus lateralis, vastus medialis, vastus intermedius, and rectus femoris muscles) with contributions from the **tensor fasciae latae**. The **trochlear groove** of camelids, unlike dogs, is relatively shallow (Fig. 22.2-5). The

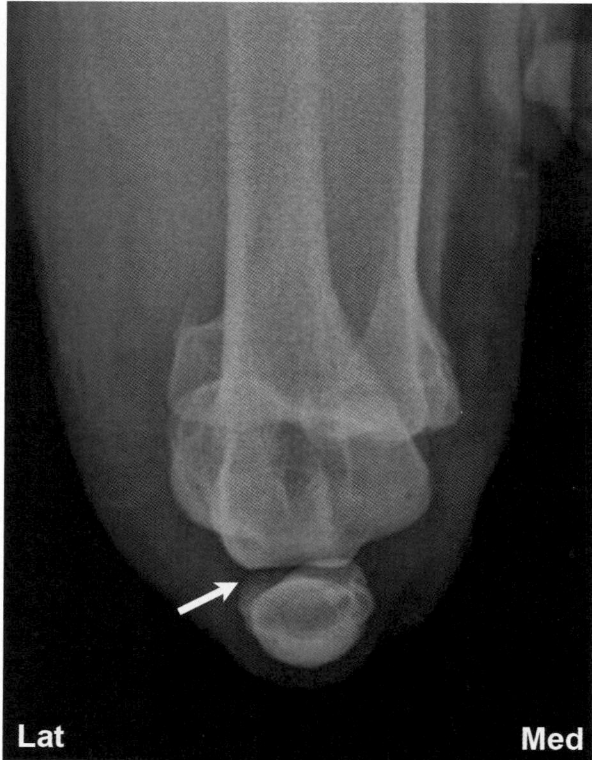

FIGURE 22.2-5 Proximal-to-distal tangential ("skyline") radiographic view of the patella and trochlear groove in an immature alpaca. Note the shallow trochlear groove (*arrow*) limiting the position of the patella relative to the femur.

patella normally is positioned superficial to the trochlear groove, presumably as an adaptation for the extreme hyperflexion typical of the "cushed" sitting posture of camelids.

> These anatomical features allow for surgical procedures that have been developed for patellar dislocation in the dog to be applied to camelids. This includes joint capsule release, joint imbrication, femoral patellar ligament prosthesis or enhancement, trochlear groove "V" recession, trochlear groove block recession, and tibial crest transposition. Among these techniques, tibial crest transposition has the highest complication and failure rates and therefore is not recommended when compared with other techniques. The principal indication for performing a tibial crest transposition is if the patella cannot be maintained in the trochlear groove after the trochlear recession procedure (Fig. 22.2-6). This may be seen in cases where growth distortion of the proximal tibia has occurred as a result of prolonged patellar luxation.

The bones of the stifle joint are stabilized through a complex system of ligaments and musculotendinous units. The major ligaments of the stifle are the medial and lateral collateral ligaments, caudal and cranial cruciate ligaments, femoral patellar ligaments, and meniscal ligaments. The principal tendinous unit is the patellar tendon, which helps to stabilize the stifle.

There are medial and lateral retinacula, referred to as **femoral patellar ligaments**, that contribute to the stability of the patella within the trochlear groove. Within the joint, there are **lateral** and **medial menisci**, **cranial** and **caudal cruciate ligaments**, and **medial** and **lateral collateral ligaments**—all similar in anatomy to that of dogs (see Case 8.3). The lateral collateral ligament is more difficult to identify and is composed of the fused tendons of the **fibularis (peroneus) tertius** and **long digital extensor muscles**.

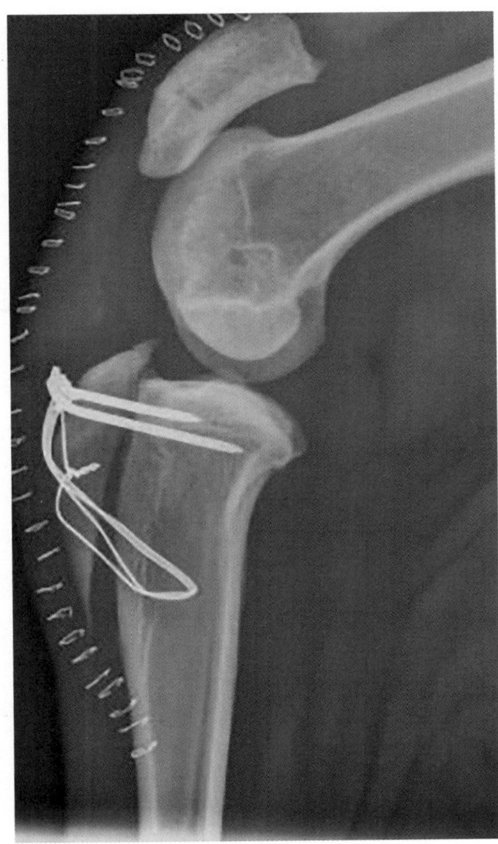

FIGURE 22.2-6 Lateral radiographic view of the left stifle in an alpaca after tibial crest transposition. The tibial crest osteotomy has been reattached to the tibia using orthopedic pins and wire for both stabilization and to resist tension from the extensor muscles of the stifle.

Blood supply, lymphatics, and innervation

The main arterial supply to the pelvic limb of camelids originates from the abdominal aorta, giving rise to the **external iliac artery**, which branches into the **femoral artery,** supplying the many parts of the pelvic limb as the **popliteal artery** and further dividing into the **cranial** and **caudal tibial arteries** distally. The **saphenous artery**, another branch of the femoral artery, is superficially positioned on the gracilis muscle (medially). The arterial system tracks along the medial side of the limb, initially entering the limb cranial to the femur and then coursing across the distal diaphysis of the femur to the caudal aspect of the stifle. The arteries remain near the femur and tibia to the level of the midtibial diaphysis.

The saphenous artery is easily accessible for taking a pulse.

Venous drainage is comprised of deep and superficial systems of vessels. The **medial** and **lateral** (larger) **saphenous veins** and their branches make up the superficial venous system while the deep venous system is comprised of satellites of the respective arteries.

The **popliteal lymph nodes** reside in the popliteal fossa (caudal stifle) and collect lymph from the distal limb with branches to the **ischial** and **deep inguinal nodes**. The **subiliac lymph nodes** drain lymph from the skin over the stifle with branches to the deep inguinal nodes.

Nerve supply to the pelvic limb begins at the **lumbosacral intumescence**, which gives rise to the **sciatic nerve**. The sciatic nerve splits into the major functional nerves of the pelvic limb including the femoral, peroneal, and tibial nerves at the level of the proximal femur. The **femoral nerve** innervates the quadriceps muscles and is responsible for the ability to extend the stifle. The **tibial nerve** tracks along the caudal aspect of the femur and innervates the gastrocnemius muscles and flexor muscles that are responsible for stifle flexion, tarsal extension, and flexion of the digits. The **peroneal nerve** tracks cranially and innervates the muscle groups responsible for flexion of the tarsus and extension of the digits.

Injury to the femoral nerve compromises the ability of the quadriceps muscles to stabilize the patella. Nerve injury, or extensive muscle atrophy for other reasons, results in increased patellar mobility and may predispose to further injury, resulting in traumatic luxation of the patella.

Selected references

[1] Van Hoogmoed L, Snyder JR, Vasseur P. Surgical repair of patellar luxation in llamas: 7 cases (1980–1996). J Am Vet Med Assoc 1998;212(6):860–5.

[2] Abuja GA, Kowaleski MP, García-López JM. Management of bilateral patellar luxation in an alpaca. Vet Surg 2014;43(4):459–64. https://doi.org/10.1111/j.1532-950X.2014.12118.x.

[3] Furman SM, Fortier LA, Schnabel LV, Krotscheck U. Trochlear block recession in an alpaca with traumatic lateral patellar luxation. Vet Surg 2009;38(3):421–5. https://doi.org/10.1111/j.1532-950X.2008.00444.x.

[4] Reed SK, Semevolos SA, Newman KD, Anderson DE. In: Cebra C, Anderson DE, Tibary A, Van Saun RJ, Johnson KW, editors. Musculoskeletal surgery. ST Louis, MO: Elsevier; 2014. p. 669–90.

CHAPTER 22

CASE 22.3

Cranial Cruciate Ligament Tear

Emma Marchionatti[a] and Caroline Constant[b]
[a]Clinic for Ruminants, Vetsuisse Faculty, University of Bern, Bern, CH
[b]AO Research Institute, Davos Platz, CH

Clinical case

History

An 8-year-old, 6-month pregnant Holstein cow presented for left pelvic limb lameness of 2 days' duration. She was out on pasture with no known history of injury. No treatment was administered at the farm.

Physical examination findings

On presentation, the cow was tachycardic (104 bpm) and tachypneic (48 brpm). The rectal temperature was in the normal range (100.9 °F [38.3 °C]). There was edema in the left stifle region with noticeable femorotibial joint effusion. The cow was 2/5 lame on the left pelvic limb with femorotibial cranio-caudal joint instability demonstrated by a clicking sound heard when the cow shifted weight or walked. The cranial draw (see Case 8.3) test/sign was positive for the left stifle.

Differential diagnoses

Cranial cruciate ligament (CrCL) tear, meniscal tear, septic arthritis of the stifle, hemarthrosis

SEPTIC ARTHRITIS IN CALVES

Any calf with omphalophlebitis (Gr. *omphalos* navel + Gr. *phleps* vein + Gr. *itis* inflammation), commonly termed navel ill, is at increased risk of septicemia and septic arthritis. If failure of passive immunity—i.e., insufficient maternal antibodies from colostrum—is also part of the history, this further increases the risk of multiple infected joints. The onset is generally acute and severe with one or more swollen joints. The stifle joint is a common joint affected and calves are usually markedly lame; other joints affected include the carpus, tarsus, and fetlock.

Blood cultures enhance the chances of identifying the causative agent, especially in a febrile calf. *Mycoplasma bovis* is one organism isolated, particularly in clinical cases where the umbilicus is normal on physical and ultrasonographic examination.

Treatment entails a combination of antibiotics (systemic and intraarticular), anti-inflammatory medications, and joint lavage. Other treatments in chronically affected calves include regional limb perfusion arthroscopy, and antibiotic-impregnated collagen implants, establishing a slow release of higher concentration of antibiotics than would normally be reached by systemic administration. The prognosis is poor with multiple joints involved, increased duration of infection, and an inability to isolate the etiologic agent and thus specifically target antibiotic therapy.

Comparative Veterinary Anatomy: A Clinical Approach. https://doi.org/10.1016/B978-0-323-91015-6.00110-2

Diagnostics

Arthrocentesis of the left lateral femorotibial and femoropatellar joints yielded blood-tinged synovial fluid. Cytological analysis was compatible with hemarthrosis without evidence of microorganisms, consistent with synovial inflammation due to acute trauma.

The lateromedial radiographic view (Fig. 22.3-1) of the left stifle showed increased opacity in the femorotibial joints due to an increase in intraarticular fluid volume and cranial displacement of the tibial plateau, compatible with CrCL rupture.

Stifle ultrasound revealed anechoic effusion in all stifle compartments, with a more significant accumulation in the lateral and medial femorotibial joints. The medial meniscus (Fig. 22.3-2) was moderately heterogeneous with a small focal defect and an ill-defined margin. The lateral meniscus had a normal ultrasonographic appearance.

Diagnosis

Cranial cruciate ligament tear with hemarthrosis and possible medial meniscal tear

Treatment

Nonsteroidal anti-inflammatory drugs were used for analgesic and anti-inflammatory purposes. Once the edema and swelling resolved, elective intraarticular ligament replacement using synthetic implants was performed. Complete CrCL rupture and medial meniscal tear (cranial horn) were confirmed during surgery. Post-operative rehabilitation started 3 weeks postsurgery with increasing levels of exercise intensity. The cow maintained a long-term, intermittent grade 2/5 lameness referable to the left pelvic limb.

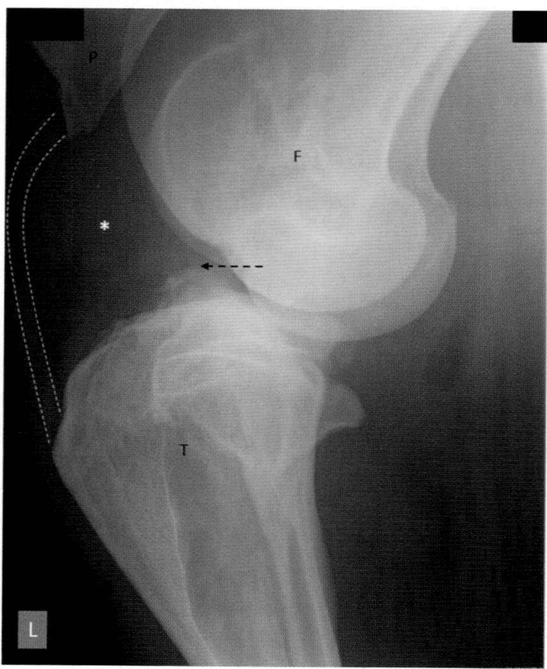

FIGURE 22.3-1 Lateromedial radiographic view of the left stifle showing cranial displacement of the tibial plateau (*arrow*) and increased opacity in the femorotibial joint (*) suggestive of increased intraarticular fluid volume and a CrCL tear. Note the cranial displacement of the median patellar ligament (*dashed yellow lines*) due to the joint distention and cranial displacement of the tibia. Key: *F:* femur, *T:* tibia, *P:* patella.

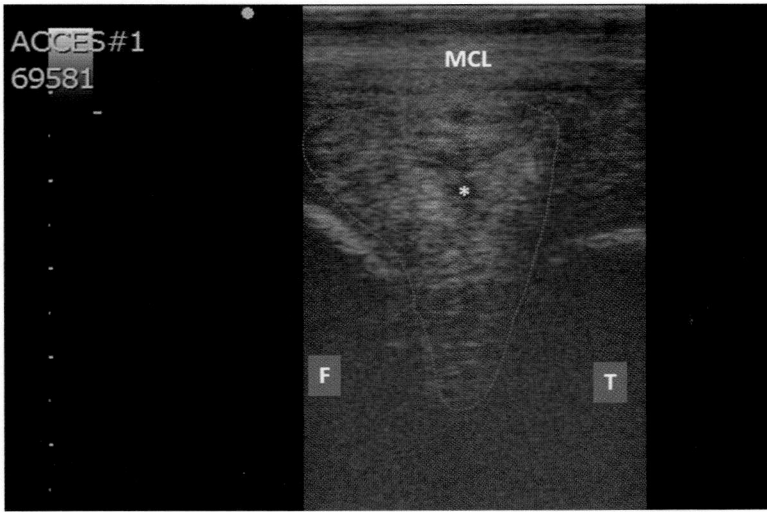

FIGURE 22.3-2 Stifle ultrasound showing the medial meniscus (*dotted line*) with a moderately heterogeneous appearance, a small focal defect (*), and an ill-defined margin. Key: *F:* femur; *T:* tibia; *MCL:* medial collateral ligament.

Anatomical features in ruminants

Introduction

The stifle joint or genual (L. *genu* the knee) joint is comprised of the femoropatellar joint, located between the patella and the femoral trochlea, and the two femorotibial joints between each set of femoral and tibial condyles. The stifle is classified as a diarthrodial, condylar joint and is the most complex joint in the body. This case focuses on the bones and soft tissues of the stifle (primarily bovine), along with its blood supply, lymphatics, and innervation. The stifle of the sheep is more similar to that of the camelid and is discussed in Case 22.2.

Function

The stifle joint plays an important role in the maintenance of normal posture. Primarily, it allows extension and flexion movements with minimal rotational function. Stifle extension mainly involves the tensor fasciae latae, the biceps femoris, and the quadriceps femoris muscles. Stifle flexion primarily involves the semitendinosus and gastrocnemius muscles, the superficial digital flexor, and the flexors of the tarsus—the digital extensor and tibialis cranialis muscles.

Stifle joint (Fig. 22.3-3)

The **femoropatellar joint** qualifies as a sledge joint—i.e., a joint with a gliding element and distinct "runners," the femoral trochlea. The femorotibial joints are divided based on location (lateral and medial) and do not communicate directly. The medial femorotibial joint always communicates with the femoropatellar joint, whereas the lateral femorotibial joint communicates with the femoropatellar joint in approximately 60% of animals. The stifle joint is comprised of the patella, the wide and deep **trochlear groove**, and the trochlear ridges of the distal femur. The **medial trochlear ridge** in cattle, as well as in horses, is more prominent and plays an important role in passively holding the femorotibial angle at rest. The **patella** is a sesamoid bone located in the terminal tendon of the quadriceps femoris muscle. Its articular surface is enlarged by two parapatellar cartilaginous processes, of which the lateral one is less developed.

Intermittent or permanent upward fixation of the patella is due to the hooking/locking of the medial patellar ligament over the prominent upper extremity of the medial trochlear ridge. Even though it is less frequently seen in cattle than in horses, it is reported to affect young working cattle. The stifle and hock are periodically fixed in extension, which gives the limb an unusual rigidity. Subsequent hyperflexion of the fetlock joint causes the toes to be dragged along the ground during walking.

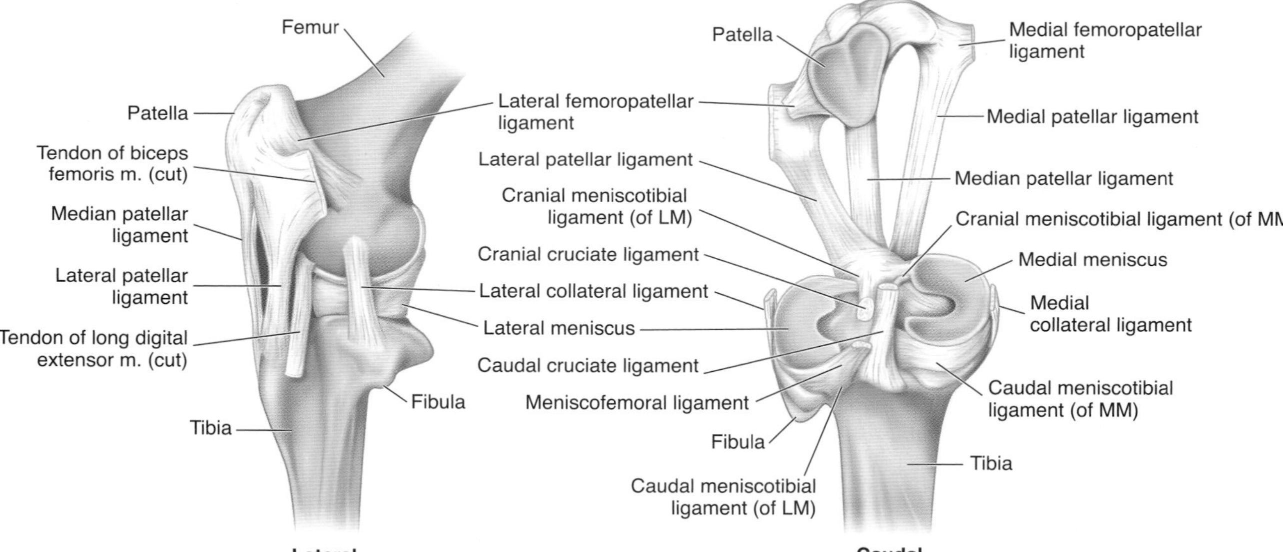

FIGURE 22.3-3 Anatomy of the bovine stifle joint. (A) Caudal view. (B) Cranial view. (C) Lateral view. (D) Caudal view with femur removed. Keys: *LM*, lateral meniscus; *MM*, medial meniscus.

The **femoropatellar joint** has a large suprapatellar pouch and both lateral and medial cul-de-sacs on each side of the trochlear ridges.

The **lateral** and **medial femorotibial joints** are comprised of the femoral condyles, the tibial condyles, the tibial eminences, and the menisci. The **lateral** and **medial femoral condyles** are parallel to each other and are separated by a deep **intercondylar fossa**. The lateral and medial tibial condyles form a flattened articular surface called the **tibial plateau**, between which the **intercondylar eminences** (intercondylar tubercles) rise.

The **lateral** and **medial menisci** each have a crescent shape and provide a congruent surface between the femoral and tibial condyles. Moreover, they allow an even distribution of load, act as shock absorbers for the articular cartilage, and contribute to friction reduction within the joint. The cranial and caudal extremities of each meniscus extend into the **cranial** and **caudal meniscotibial ligaments**. The lateral meniscus has an additional **caudal meniscofemoral ligament**. The convex border of the medial meniscus is firmly attached to the medial collateral ligament, while the lateral meniscus is in contact with the lateral collateral ligament. The femorotibial joints are divided into cranial and caudal pouches by the femoral condyles. The lateral femorotibial joint has two larger pouches. The **subpopliteal pouch** surrounds the tendon of origin of the popliteal muscle, and the **extensor pouch** ensheathes the tendon of the long digital extensor muscle at its origin from the extensor fossa.

The medial meniscus is more prone to concurrent injury than the lateral meniscus. When the medial collateral ligament is injured, the medial meniscus is secondarily injured because of its fixed attachment to the medial collateral ligament. Instead, the lateral meniscus is separated from the lateral collateral ligament by the popliteal tendon. Individuals affected by stretching or rupture of the medial collateral ligament and meniscal detachment typically show lameness, pain, and swelling over the medial collateral ligament. Laxity of the collateral ligament can be palpated when varus (L. bent inward) stress is applied to the stifle, whereas applying valgus (L. bent outward) stress increases the width of the medial joint space (mediolateral instability). Moreover, a separation between the collateral ligament and the meniscus can usually be palpated. There is no rotational instability, and the cranial draw sign is negative unless the CrCL is injured as well.

The **lateral** and **medial femoropatellar ligaments** connect the parapatellar cartilaginous processes to their corresponding femoral epicondyles. In cattle, there are 3 **patellar ligaments**: lateral, median, and medial (medial may be regarded as the continuation of the tendon of the quadriceps femoris muscle). These patellar ligaments join the apex of the patella to the tibial tuberosity. The lateral patellar ligament receives the insertion of the biceps femoris muscle. The femoropatellar and patellar ligaments prevent patellar luxation.

Lateral luxation of the patella may be of congenital (secondary to femoral trochlear groove hypoplasia), neurogenic (femoral nerve paresis/paralysis and quadriceps femoris muscle atrophy), or traumatic origin. Affected individuals are unable to maintain the stifle in extension and adopt a characteristic crouching position. Medial patellar luxation is rarely encountered in cattle.

The **lateral** and **medial collateral ligaments** run from the femoral epicondyles to the proximal tibia. They provide lateromedial support to the stifle. The cruciate ligaments are intraarticular and are considered extrasynovial. They provide rotational and craniocaudal support to the stifle. The **CrCL** originates from the medial side of the lateral

SURGICAL OPTIONS FOR CrCL TEARS

Conservative treatment of CrCL disruption generally has a poor outcome because of the secondary changes that result from joint instability leading to osteoarthritis.

Two surgical procedures to improve femorotibial joint stability are as follows: (1) extraarticular and (2) intraarticular stabilization techniques.

Extraarticular stabilization involves imbrication (L. *imbricatus* overlapping of opposing surfaces) of tissues at the level of the lateral femoropatellar ligament and extending to the tibial crest. The same procedure is performed on the medial side of the stifle joint centered over the medial femoropatellar ligament. Multiple rows of overlapping imbricating sutures are placed on both sides to strengthen the periarticular tissues, predominately comprised of fascia (L. "band").

The intraarticular procedures include a gluteobiceps tendon replacement for the CrCL and a synthetic CrCL using a synthetic material.

Cranial cruciate ligament replacement procedures have the best prognosis for return to normal ambulation. Stifle imbrication is more successful in lighter weight individuals—i.e., <400 kg (882 lbs).

Cranial cruciate ligament rupture is a common cause of lameness originating from the stifle in adult cattle. In breeding bulls, the pathology is usually secondary to meniscal injury and degenerative osteoarthritis associated with straight tarsocrural joint conformation. However, CrCL rupture in dairy cattle is most likely due to trauma, leading to secondary degenerative osteoarthritic changes and meniscal injury. In both cases, lameness is severe in acute cases and decreases in chronic injuries. A clicking sound may be heard at the walk usually due to an associated meniscal tear and joint instability. Moderate-to-severe joint effusion, pain, cranial draw, and increased internal rotation may be elicited upon manipulation of the stifle joint. In cranio-caudal-deficient stifles, the lateral radiographic view shows the femoral condyles displaced caudally in relation to the tibial eminences. Avulsion fragments of the tibial eminences may be present in acute and traumatic lesions. Osteoarthritic changes can also be seen in chronic cases. The caudal cruciate ligament is rarely affected.

femoral condyle near the intercondylar fossa and tracks distally, cranially, and medially before its insertion at the base of the intercondylar eminences of the tibia. Its fibers are grouped into two fascicles whose tibial insertions are separated by the **craniomeniscotibial ligament** of the lateral meniscus. The CrCL is stretched during flexion of the stifle and provides rotational stability to the stifle, with its more important function in preventing cranial displacement of the tibial plateau in relation to the femur. The **caudal cruciate ligament** (CdCL) lies caudal and medial to the CrCL. It originates from the intercondylar fossa of the femur and runs in a distocaudal direction to insert on the popliteal notch on the caudal aspect of the tibia. The CdCL has an antagonistic action to the CrCL.

Blood supply, lymphatics, and innervation

The blood supply to the stifle is provided by the **popliteal** and **saphenous arteries**, both of which are branches of the **femoral artery**. The veins parallel in a corresponding course.

Lymphatic drainage is via the **popliteal lymph node** in the popliteal fossa and the large **subiliac lymph node**. Lymph nodes of lesser significance include a small **coxal lymph node**, a group of **gluteal lymph nodes**, **ischial lymph nodes,** and a **tuberal lymph node** that all drain lymph from various parts of the pelvic limb.

The stifle joint is innervated by articular branches of the **femoral**, **tibial**, and **peroneal nerves**.

Selected references

[1] Budras KD, Habel RE. Bovine anatomy. 2nd ed; 2011. p. 28–9.
[2] Fubini SL, Ducharme NG. Farm animal surgery. 2nd ed; 2017. p. 367–75.
[3] Pentecost R, Niehaus A. Stifle disorders: cranial cruciate ligament, meniscus, upward fixation of the patella. Vet Clin North Am Food Anim Pract 2014;30:265–81.

CASE 22.4

Gastrocnemius Rupture

André Desrochers
Department of Clinical Sciences, Université de Montréal, Faculty of Veterinary Medicine, St-Hyacinthe, Québec, CA

Clinical case

History

A 5-year-old Holstein cow was down for 2 days at the farm after routinely delivering a 40-kg (88-lb) healthy heifer. Following calving, the cow had difficulty getting up and was treated for hypocalcemia. The following day she made no attempt to stand despite repeated treatments for hypocalcemia. Her appetite and manure were considered normal.

Physical examination findings

The cow was sternally recumbent but bright and alert. Rectal temperature was 102.0 °F (38.9 °C), with a heart rate of 92 bpm and a respiratory rate of 44 brpm. There were 2 strong and complete rumen contractions per 3 minutes of auscultation (normal is 2 contractions/3 minutes). A complete musculoskeletal examination was performed with the cow in lateral recumbency. Swelling was noticed at the caudal aspect of the right tibia in the region of the gastrocnemius muscle. The tendinous portion of the gastrocnemius was considered intact based on the physical examination (Fig. 22.4-1).

Differential diagnoses

Tibial fracture, gastrocnemius muscle rupture, calcaneal tendon rupture, trauma to the tibial branch of the sciatic nerve, caudal limb abscess

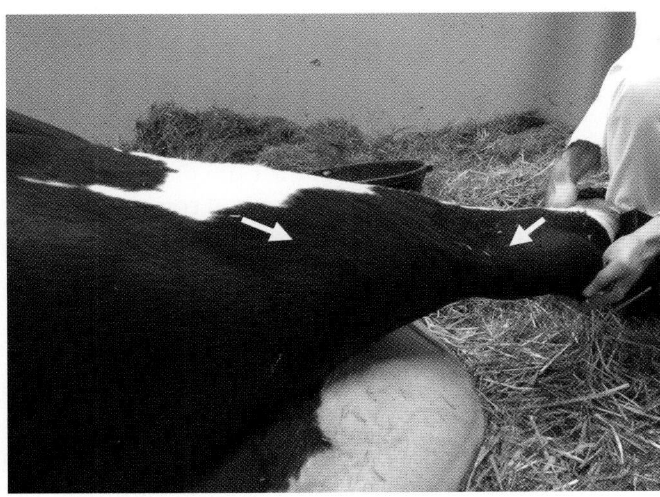

FIGURE 22.4-1 Adult Holstein cow in lateral recumbency. The distal and caudal aspect of the limb is swollen (*arrows*).

Comparative Veterinary Anatomy: A Clinical Approach. https://doi.org/10.1016/B978-0-323-91015-6.00167-9

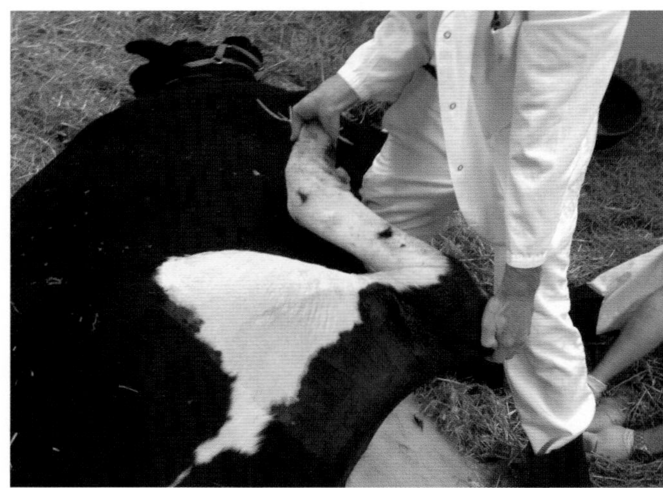

FIGURE 22.4-2 With the stifle extended, the tarsus of this cow can be fully flexed. This is consistent with gastrocnemius rupture.

Diagnostics

The pelvic limb was manipulated and palpated with the cow in lateral recumbency to rule out a fracture. With the stifle partially extended, the tarsus was flexed past the normal range of motion, confirming rupture of the gastrocnemius muscle (Fig. 22.4-2). Ultrasonography of the caudal limb revealed that the muscles were hyperechoic and heterogenous with loss of the architecture, compatible with a hematoma and muscle contusion.

Diagnosis

Gastrocnemius rupture

Treatment

Considering the poor prognosis for gastrocnemius muscle rupture in adult cattle, the cow was humanely euthanized. Thomas-splint casting can be attempted on younger and lighter cattle if they can stand. A Robert Jones bandage can also be used on adult cattle with partial rupture, if the cow can stand. With either of these treatments, the cow should be confined in a small, well-bedded box stall for a minimum of 1 month.

Clinical anatomy in ruminants

Introduction

The reciprocal apparatus is well-described in equids as part of the passive-stay apparatus (see Case 15.3). Cattle have a similar arrangement with the fibularis (peroneus) tertius at the cranial aspect of the limb and the gastrocnemius caudally. However, the fibularis tertius muscle is fleshy and superficial compared to that in equids. This section describes the relevant musculature of the ruminant reciprocal apparatus; bones, bursae, and synovial structures of the tarsus; and regional blood supply, lymphatics, and innervation.

Peripartum dairy cattle are prone to metabolic diseases, such as hypocalcemia, nerve injury secondary to dystocia, or prolonged recumbency. Rising and standing require a considerable effort, making cattle affected with these conditions susceptible to injury (especially if the floor is hard and slippery).

Function

The muscles of the pelvic limb allow the stifle and tarsus to flex or extend in synergy. The reciprocal apparatus is principally responsible for the coupled action between the stifle and tarsus. Any excessive action forcing the limb to extend while the other joint is flexed may injure the muscles (Fig. 22.4-3A and B).

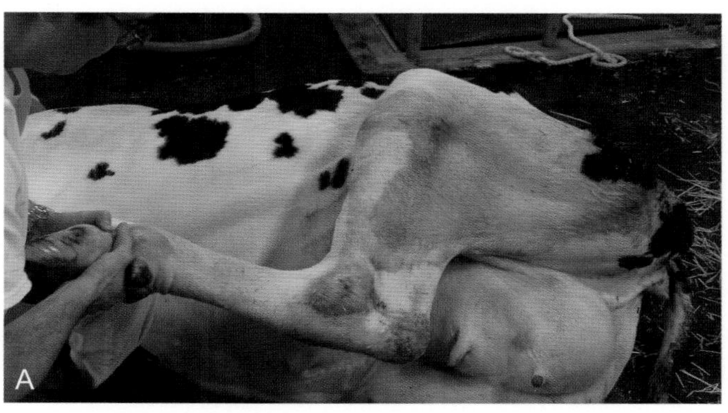

FIGURE 22.4-3 These 2 images from different cows demonstrate a strong reciprocal apparatus with a normal range for the gastrocnemius (A) and fibularis tertius (B) muscles.

Muscles of the reciprocal apparatus

There are 2 groups of muscles involved in the reciprocal apparatus—the dorsolateral and caudal groups. The dorsolateral group includes the following muscles from medial to lateral: cranial tibial, fibularis (peroneus) tertius, long digital extensor, fibularis (peroneus) longus, and the lateral digital extensor.

The **cranial tibial muscle** lies deep on the tibia originating at its cranial border and the lateroproximal portion of its body. Its insertion is the distal aspect of the tarsus and proximal metatarsal bones. It flexes the tarsus.

The **fibularis (peroneus) tertius** muscle is fleshy when compared to that in horses. It originates from the extensor fossa of the distal femur along with the long digital extensor muscle. It inserts on the distal row of tarsal bones and the proximal aspect of the metatarsal bones, where it is perforated by the tendon of the cranial tibial muscle. It flexes the tarsus.

FIBULARIS TERTIUS INJURY AND TREATMENT

The fibularis tertius can rupture if the tarsus is hyperextended while the stifle remains flexed (Fig. 22.4-4). This situation is often encountered when a cow slips, and the limb slides caudally. Its rupture renders tarsus/tarsal flexion more difficult; the cow may stumble because of difficulty in advancing the foot without contacting the ground. Based on ultrasound findings, the “rupture” is usually associated with muscle tearing rather than tendon rupture, precluding the surgical option of a suture reconstruction (tenorrhaphy). It should not prevent the cow from standing or walking. There is no specific treatment for this condition other than stall rest for many weeks; the prognosis is good.

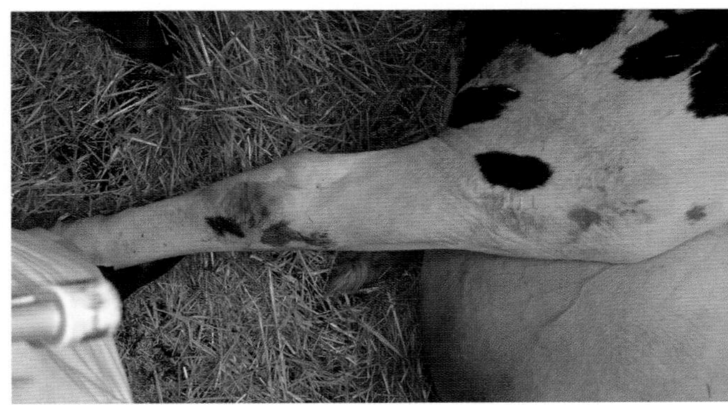

FIGURE 22.4-4 A cow with a ruptured fibularis (peroneus) tertius. The stifle is flexed while the tarsus is extended.

The **long digital extensor** (extensor digitorum longus) muscle originates from the extensor fossa of the femur. For most of its course, it is fused with the fibularis tertius. Its medial belly is the **proper extensor muscle of the 3rd digit** that passes through the extensor groove of the tibia and eventually attaches to the 3rd digit. Its lateral belly travels through the **tarsal retinaculum** and bifurcates proximal to the fetlock, attaching to the extensor process of the 3rd and 4th digits' 3rd (distal) phalanges.

The **fibularis (peroneus) longus** muscle originates from the lateral condyle of the tibia and its collateral ligament. It is superficial and the most lateral muscle of its group. Its insertion is on the 1st tarsal bone and the proximal aspect of the metatarsal bones. It flexes the tarsus and rotates it medially.

The **lateral digital extensor** muscle lies between the extensor and flexor muscle groups. Its origin is the base of the collateral ligament at the lateral condyle of the tibia and the vestigial head of the fibula. Its insertion is on the 4th digit, and it extends the digit (Fig. 22.4-5).

The caudal muscle group includes the following muscles from caudal to cranial—gastrocnemius, superficial digital flexor, deep digital flexor, lateral digital flexor, and soleus.

The **gastrocnemius** muscle is large with **medial** and **lateral** heads. Each head attaches to its respective supracondylar tuberosity and epicondyle of the femur (Fig. 22.4-6). Its strong tendinous portion inserts on the tuber calcanei of the tarsus. It extends the tarsus and flexes the stifle.

The **common calcaneal tendon** is composed of the gastrocnemius tendon, superficial flexor tendon, and tarsal tendons of the biceps femoris and semitendinosus muscles.

The **soleus** muscle lies on the cranial edge of the long digital flexor and originates at the head of the fibula and inserts on the tuber calcanei via the lateral head of the gastrocnemius and its tendon. It, with the gastrocnemius, functions in extending the tarsus.

The **superficial digital flexor muscle** originates from the supracondylar fossa of the femur and is situated between the heads of the gastrocnemius muscle. Its tendinous portion is cranial to the calcaneus and spirals around it medially and caudally over the tuber calcanei, where it attaches (Fig. 22.4-7). Distally, its tendinous portion inserts on the plantar surfaces of the 2nd (middle) phalanges of the 3rd and 4th digits. It extends the tarsus/tarsal flexion and flexes the digits.

The **deep digital flexor muscle** originates from the caudal aspect of the tibia, including its condyle, and the caudal aspect of the head of the fibula. It has 3 parts: the superficial (caudal tibial), medial, and lateral (deep head). The **caudal tibial muscle** is small and joins the larger **lateral head**. Their fused tendons pass over the **sustentaculum tali** and continue distally along the plantar aspect of the metatarsi. The **medial head** courses obliquely across the tibia from its lateral origin. Its tendon tracks on the medial surface of the tarsus adjacent to the collateral ligament. It eventually joins the common tendon of the caudal tibial muscle and its own lateral head at the plantar and proximal aspect of the metatarsi. The insertion of the common deep digital flexor tendon is the 3rd phalanx of the 3rd and 4th digits on the flexor tuberosity (see also Case 21.2). It flexes the digits and extends the tarsus.

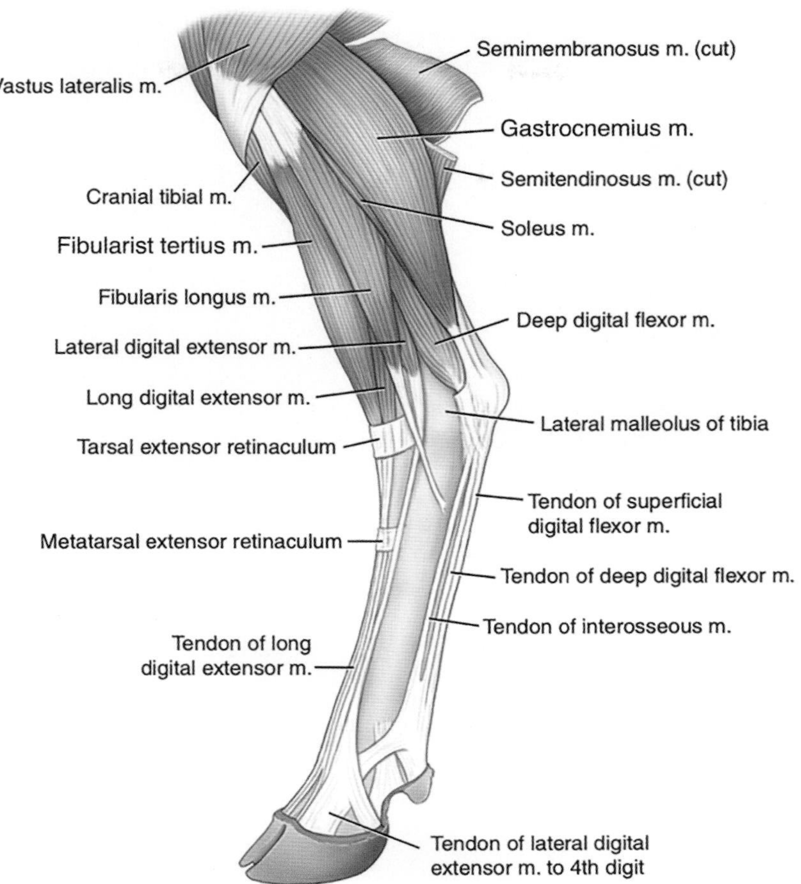

FIGURE 22.4-5 Anatomy of the pelvic limb of an adult cow (lateral view).

FIGURE 22.4-6 Necropsy image showing a ruptured gastrocnemius muscle of a cow. Note that the tendinous portion is intact (*arrows*).

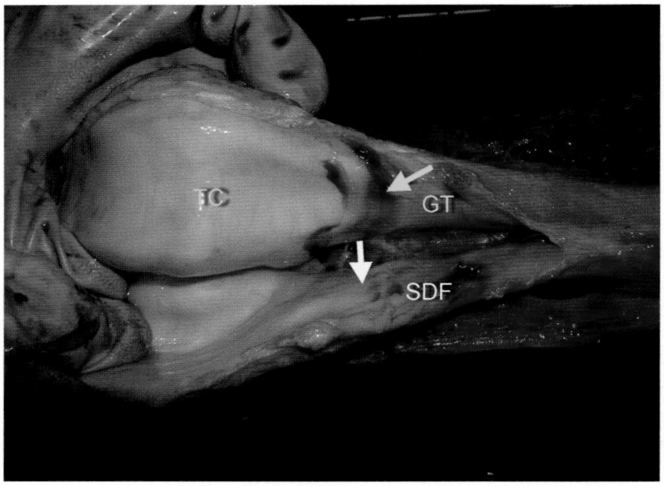

FIGURE 22.4-7 Necropsy specimen showing the region of the calcaneus. The superficial digital flexor tendon (*SDF*) is medially displaced (*white arrow*). The gastrocnemius tendon (*GT*) attaches on the tuber calcanei (*TC*) (*yellow arrow*).

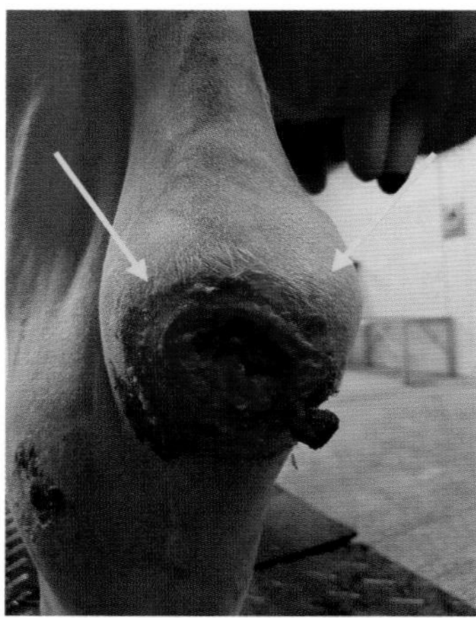

FIGURE 22.4-8 An adult cow with a tarsal bursitis secondary to chronic recumbency (*arrows*).

Bursae and synovial sheaths of the tarsal region

There are many bursae and synovial sheaths around the tarsal joints. The **gastrocnemius bursa** is cranial to the gastrocnemius tendon close to the tuber calcanei. The larger **bursa of the superficial digital flexor tendon** rests on the gastrocnemius and the tuber calcanei.

TUBER CALCANEI BURSITIS

This is rarely seen as a clinical problem in cattle and occurs secondary to trauma or a laceration with subsequent true or false bursitis development, depending on the location of the injury. It is difficult to treat because of the poor soft tissue coverage, limited blood supply from the tendons and surrounding soft tissues, and difficulty in immobilizing a limb with constant motion (Fig. 22.4-8).

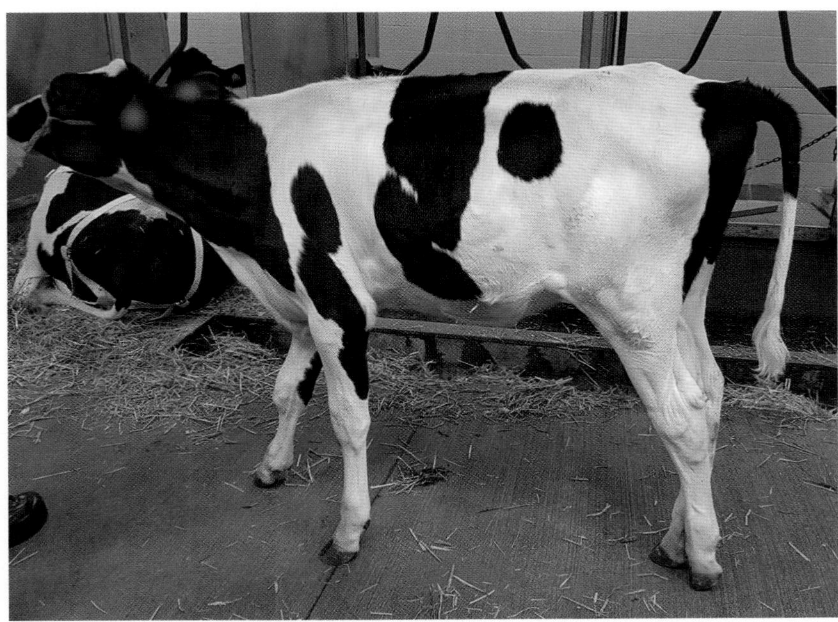

FIGURE 22.4-9 An 8-month-old Holstein heifer with spastic paresis of the left pelvic limb. The tarsus is straight and the leg is held slightly caudal. Note the elevated tail, which is a common clinical finding with this condition.

Bones and joints of the tarsus (Figs. 22.4-10A and B and 22.4-11)

The tarsus is comprised of 5 tarsal bones: the **talus** with a proximal and distal **trochlea**, the **calcaneus**, the **fused central and 4th tarsal (TC + 4 or centroquartal) bones**, the **fused 2nd and 3rd tarsal (T2 + 3 or intermediolateral cuneiform) bones**, and the **1st tarsal bone (T1 or medial cuneiform)**. The tarsus is a compound joint comprised of the **tarsocrural joint**, several **intertarsal joints**, and the **tarsometatarsal joint**. The tarsocrural joint is a ginglymus joint (movement in only 1 plane) formed proximally by the distal end of the tibia and fibula (lateral malleolus), which articulate with the proximal trochlea of the talus, and the articulation formed by the lateral malleolus and lateral aspect of the calcaneus (Figs. 22.4-10 and 22.4-11).

The **talocalcaneocentral** joint, which is a ginglymus joint in cattle, is formed by the distal trochlea of the talus, the proximal aspect of centroquartal bone, and the articulation between the plantar aspect of the talus and the calcaneus. The calcaneus articulates distally at the lateral aspect of centroquartal bone to form the **calcaneoquartal joint**, which together with the talocalcaneocentral joint forms the **proximal intertarsal joint**. In cattle, the tarsocrural and proximal intertarsal joints communicate on the plantar aspect. The **centrodistal joint** (commonly

SPASTIC PARESIS

Spastic paresis is a genetic disease characterized by spastic contracture of the muscles of the pelvic limbs, leading to extension of the stifle and the tarsus (Fig. 22.4-9). It is usually unilateral with clinical signs beginning at 4–6 months of age, depending on the breed. It is hereditary disease, and affected cattle should not be bred. Two surgical procedures are described to manage the spastic contractures and raise the affected individual to a market weight: (1) partial tibial neurectomy and (2) partial or complete tenectomy of the tendinous portion of the gastrocnemius muscle.

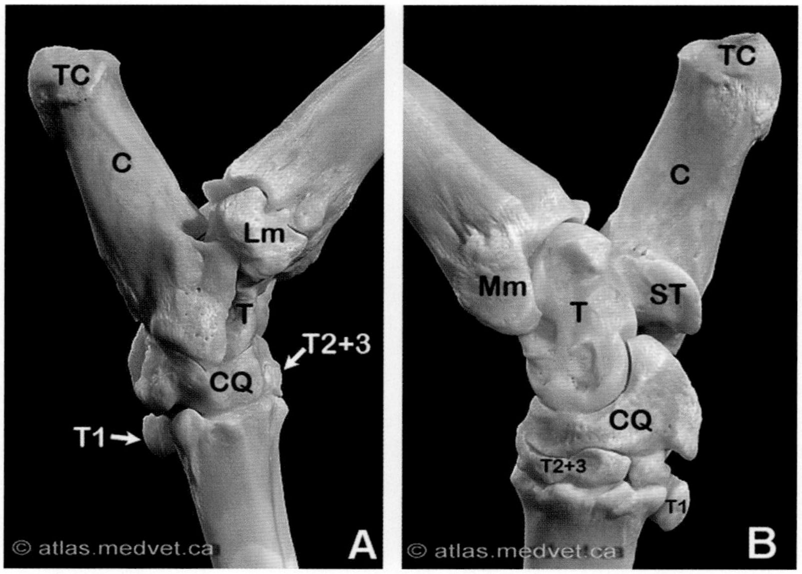

FIGURE 22.4-10 Bones of the bovine tarsal joint. (A) Lateral view. (B) Medial view. Key: *TC:* tuber calcanei; *C:* calcaneus; *CQ:* centroquartal bone; *T:* talus; *ST:* sustentaculum tali; *Lm:* lateral malleolus; *Mm:* medial malleolus; *T1:* first tarsal bone; *T2+ 3:* fused 2nd and 3rd tarsal bones.

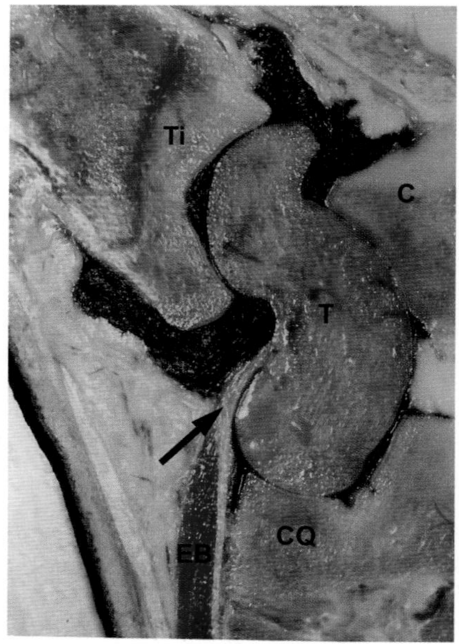

FIGURE 22.4-11 Sagittal view of a bovine prosection with blue latex injected into the tarsocrural joint. The *arrow* depicts the insertion of the EB and the joint capsule separating the tarsocrural and the proximal intertarsal joints. Key: *Ti:* tibia; *T:* talus; *C:* calcaneus; *CQ:* centroquartal bone; *EB:* digital extensor brevis.

called the **distal intertarsal joint)** is formed by the centroquartal and intermediolateral cuneiform bones and the separate medial cuneiform bone. The intermediolateral cuneiform bone, medial cuneiform bone, and the fused 3rd and 4th metatarsal bones articulate together to form the **tarsometatarsal joint**. The tarsometatarsal joint and distal intertarsal joint permit only limited gliding movement and therefore are classified as plane joints.

The distal intertarsal and tarsometatarsal joints communicate about 30% of the time while the tarsocrural joint and the talocalcaneocentral (proximal intertarsal) joints always communicate. There are no communications between the proximal intertarsal and distal intertarsal joints.

The tarsus is a common site for septic arthritis in cattle. The tarsocrural and proximal intertarsal joints should be copiously lavaged using multiple needle sites in the various pouches of the joint. The distal intertarsal and tarsometatarsal joints can also be infected, and this generally occurs secondary to repetitive trauma from poor housing—i.e., lying on hard flooring. Osteochondrosis lesions also affect the tarsus and common sites include the lateral malleoli and trochlear ridge. The needles or portal for the arthroscope should be placed dorsolateral to the digital extensors for sampling and treatment. The lateral saphenous vein courses across this area and should be identified and avoided during treatment. Dorsomedially, the needle is inserted just caudal and medial to the extensor tendons. The plantarolateral synovial pouch is spacious and can be safely accessed between the caudolateral aspect of the tibia and the tarsal tendon sheath. The plantaromedial synovial pouch is smaller but can be accessed between the medial head of the deep digital flexor tendon and the tarsal tendon sheath.

Blood supply, lymphatics, and innervation

The **femoral artery** supplies blood to the distal limb. It branches at the medial and proximal aspect of the stifle to form the **descending genicular artery** cranially; the **popliteal artery** caudally; the **saphenous artery** caudally, distally, and medially; and the **cranial tibial artery** distally and laterally.

Generally, the veins have the same names as their arterially counterparts and track together. On the lateral aspect of the tarsus are the **superficial cranial** and **caudal branches** of the **saphenous vein.** They unite at the distal and caudal aspect of the tibia with the **lateral saphenous vein**, traveling in the popliteal space and more proximally between the semimembranosus and biceps femoris muscles. The lateral saphenous vein empties into the **medial circumflex** and **deep femoral veins**.

These veins should be identified and avoided when performing arthrocentesis, arthroscopy, or arthrotomy procedures.

Distal limb lymph flows to the **popliteal lymph nodes** in the popliteal space and, in cattle, there is only a **deep popliteal lymph node**. The popliteal lymph node is located on the gastrocnemius muscle between the semimembranosus and the biceps femoris muscles and measures approximately 3 × 3 cm (1.2 × 1.2 in.).

The **sciatic nerve** divides at mid-thigh into the **common peroneal** and **tibial nerves**. The common peroneal nerve innervates the cranial aspect of the limb (flexion of the tarsus and extension of the digits) while the tibial nerve innervates the caudal aspect (extension of the tarsus and flexion of the digits).

Injury to the common peroneal nerve results in loss of skin sensation on the dorsal aspect of the distal limb and knuckling of the fetlock. Tibial nerve injury causes loss of skin sensation on the plantar aspect and the also causes calcaneus to fall or sink.

Selected references

[1] Barone R, Lombard M. Le jarret du boeuf et son fonctionement. Réc Méd Vét 1968;119:1141–66.
[2] Getty R. Fascia and muscles of the appendages. In: Sisson G, Grossman J, editors. Anatomy of domestic animals. Philadelphia: WB Saunders Co; 1975. p. 831–60.
[3] Sisson S. Ruminant osteology. In: Sisson G, Grossman J, editors. Anatomy of domestic animals. Philadelphia: WB Saunders Co; 1975. p. 741–74.
[4] Ashdown RR, Done SH. Color Atlas of veterinary anatomy. Vol 1: the ruminants. 6. New York: Gower Medical Publishing; 1984. p. 1–7.22.
[5] Desrochers A. Characterization of the anatomic communications of the carpus, fetlock, stifle and tarsus in cattle using intraarticular latex and positive constrast arthrography [Master's thesis]. Manhattan, KS: Kansas State University; 1995.

CHAPTER 23

INTEGUMENT AND MAMMARY GLAND/UDDER

André Desrochers, Chapter editor

23.1 Contagious ecthyma—*Cynthia M. Faux and Luise King* 1222
23.2 Chronic udder abscess—*Sylvain Nichols* 1236
23.3 Teat obstruction—*Sylvain Nichols* 1242

CHAPTER 23

CASE 23.1

Contagious Ecthyma

Cynthia M. Faux and Luise King
University of Arizona College of Veterinary Medicine, Oro Valley, Arizona, US

Clinical case

History

A small purebred Nubian goat breeding and milking herd consisting of 36 milking does, 5 adult breeding bucks, and 16 new kids (5 males, 11 females) is evaluated.

> Nubian goats, also known as Anglo-Nubian goats, are a domestic dairy breed known for their long, elegant drooping ears, Roman nose, and lively personalities. Developed originally in the United Kingdom by crossing local milking goats with bucks imported from the Middle East, Nubian goats have become a popular breed. They have short hair and come in a variety of coat colors and patterns.

The rancher also buys kids from other breeders to increase genetic diversity. The kids are sold across the county, and one is expected to be sold to another breeder out of state in 2 months. The 5 male kids are housed in a separate pen. The rancher has called his local veterinary clinic because the 5 male kids have multiple raised "bumps" around their muzzle (lips, nose, and mouth). Three of the kids also have similar bumps around the genital area and, on presentation, 2 of the kids were anorectic (Gr. *anorektos* without appetite for) and did not greet the rancher in their normal fashion of standing on their back feet and nuzzling him.

Physical examination findings

Vital signs (temperature, pulse, respiratory rate/effort) were normal in all affected kids. Cardiorespiratory auscultation was normal. Cursory musculoskeletal and neurological examinations were normal. All 5 male kids had similar muzzle lesions and 2 appeared to be severely affected. The muzzle lesions were best described as crusty, proliferative lesions with macules (L. *macula* spot), papules (L. *papula* pimple), pustules (L. *pustula* containing pus), and scabs (Fig. 23.1-1). Lesions were present on the lips, nose, gums, and a few on the genitalia.

> In human and veterinary medicine, specific terms are used to describe integumentary lesions. The important terms from this physical exam include macule—discolored spot on the skin; papule—small, circumscribed, superficial elevation of the skin; and pustule—collection of pus beneath the epidermis affecting a hair follicle or sweat pore.

Differential diagnoses

Contagious ecthyma (sore mouth, orf), lice and/or mites infestation, dermatophytosis, fly strike, pox viruses, foot and mouth disease, and dermatophilosis (rain scald/rot)

Comparative Veterinary Anatomy: A Clinical Approach. https://doi.org/10.1016/B978-0-323-91015-6.00112-6

FIGURE 23.1-1 Crusty lesions on the lips of a Nubian goat, typical of contagious ecthyma (orf).

Diagnostics

Diagnosis of contagious ecthyma (caused by the orf virus in the genus *Parapoxvirus)* is made based on history of new goats introduced into the herd, presence of crusting at the mucocutaneous junction of the mouth, and ruling out other differential diagnoses. Because the kids lacked oral ulcerations and erosions, foot and mouth disease (aphthous [L. Gr. *aphtha* thrush] fever) and bluetongue were low on the differential list.

Electron microscopy can be used to confirm that a parapoxvirus is present but is not specific to the orf virus. Both RT-PCR and histopathology can be used for a definitive diagnosis but are rarely needed.

Diagnosis

Contagious ecthyma (presumptive)

Treatment

Supportive care was instituted, including soft foods, fly protection, and antimicrobials administration to treat any secondary bacterial infections. Preventing flies from laying eggs and causing fly strike (infestation of the skin of susceptible animals with maggots) is important for recovery. In this case, supportive care, soft food, and fly protection were all that were needed with full recovery in 6 weeks.

Contagious ecthyma is a zoonotic (Gr. *zōon* animal + Gr. *nosos* disease) disease, typically causing skin lesions in humans. Therefore, proper precautions are recommended to minimize the risk to handlers (wear gloves and wash hands).

Prognosis for contagious ecthyma is good; morbidity is high in a naïve herd/flock, but mortality is generally low. Nursing young are susceptible to starvation because of oral lesions and/or lesions on the dam's udder leading to reluctance to nurse.

Anatomical features in ruminants

Introduction

The integument includes the skin and its accessory structures—horns, hooves, antlers, nails, and mammary glands. Because of the importance of these integumentary accessory structures, hooves, horns, and mammary glands are described in more detail in other sections (see Cases 17.3, 21.3, and 23.2, respectively).

Skin in livestock is structurally like other mammalian species with a superficial epidermal layer overlying the dermis. Skin includes a rich neurovascular part, lymphatic supply, and various glands—including sweat, sebaceous, and scent glands. Thickness of the skin varies between species, with cattle typically having thicker skin than small ruminants and swine (Fig. 23.1-2). Much of the thickness of bovine skin can be attributed to a thick collagen layer within the deep dermis. The skin entirely covers the body of mammalian species until it meets with the mucous membranes.

Function

Skin is a critical organ that provides physical protection against the environment (sunlight, microbes, physical insults, water loss, etc.), proper thermoregulation (heating and cooling), and sensory input from the surroundings. Thermoregulatory efficiency in a cold environment is augmented with piloerection, which improves insulation, and growing a longer hair coat during winter months at higher latitudes.

Hair coat

Body hair is a defining characteristic of mammals. The dense, relatively short **pelage** (Fr. hairy coats of mammals) of most livestock is usually referred to as a **hair coat** vs. fur, which is generally considered to have a softer, denser pile.

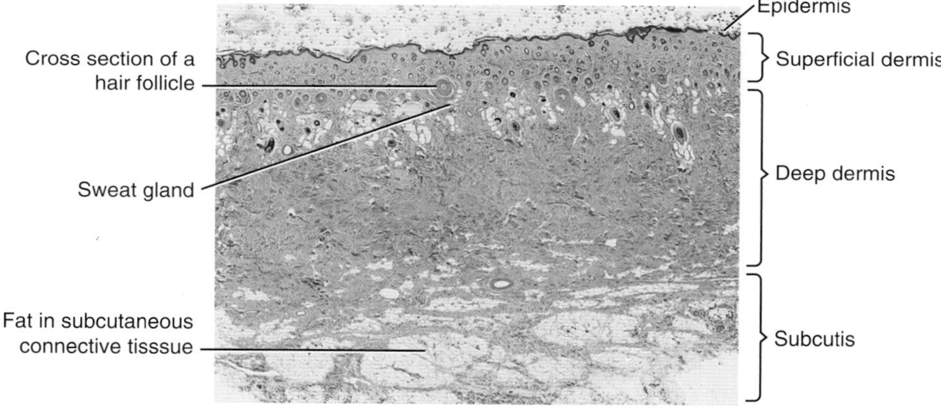

FIGURE 23.1-2 Photomicrograph of bovine skin. (Photomicrograph courtesy of Patrick D. Wilson, MS, DVM, Texas A&M University.)

HAIR COAT VS. FUR

There is not an absolute distinction between hair coat and fur, and colloquial variation in terminology is common. Because hair is a protection against the elements, longer haired breeds are typically associated with the northern latitudes, such as the Scottish Highland breed of cattle, or higher elevations, such as the Cashmere goat.

The morphology of the hair fiber varies between species in characteristics such as the fiber diameter, length, texture, and color. The density and distribution of hair follicles across the body varies, even on an individual. Hair fibers are composed of alpha-keratin with other components, such as pigment and lipids. A hair shaft bears an outer layer of cells called the cuticle which surrounds the cortex and acts as a protective barrier. The cuticle can be thin, just one cell thick in fine wool, or thick, especially in coarser hairs. The cortex makes up the bulk of the hair fiber and determines many of the hair's characteristics. The central core of the hair, when present, is called the medulla.

Hair develops from a hair follicle embedded within the dermal layer (see Fig. 9.1-3). The hair growth cycle in livestock follows the same rhythm as small animals—growth occurs in the **anagen** phase and the resting stage is the **telogen** phase. Between anagen and telogen phases is a transitional stage called the **catagen** phase (i.e., a brief period in which hair growth stops and resting phase starts). New hair fibers are produced during anagen, the hair stays attached during most of telogen, and the new hair pushes the old hair out of the follicle if it has not already been shed (see Case 9.1 for more information).

The hair coat also varies with life stage between neonate, juvenile, and adult, and even varies with season—winter vs. summer, where the periodic shedding, or molting, allows for seasonal thermoregulatory adaptations of the hair coat. Although less typical of domestic livestock, the hair coat of newborns can differ from the parents' in coloration or patterns, providing additional camouflage benefits, such as in fawns or javelina piglets (Fig. 23.1-4).

Cattle groom themselves and others, creating circular areas referred to as "cowlicks." This grooming leads to ingestion of hair, producing hairballs or trichobezoars (Gr. *thrix* hair + *bezoar* Farsi *pādzahr* antidote to poison = a hairball), especially in calves (Fig. 23.1-3).

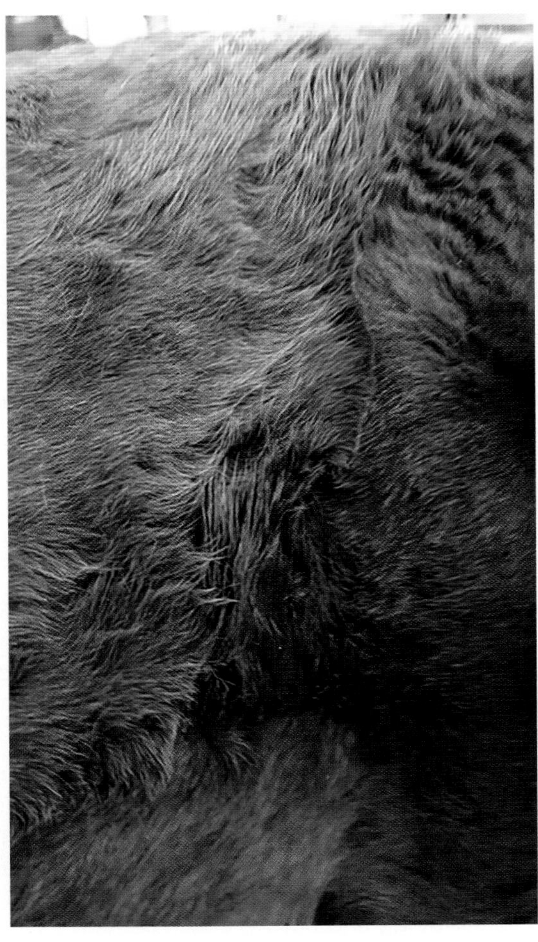

FIGURE 23.1-3 A Jersey cow in winter coat showing the length of hair as well as grooming evidence noted by the swirls of hair (cowlick).

FIGURE 23.1-4 The camouflaging dappled coat of this triplet set of White-tailed deer fawns contrasts with the solid color of the adult doe.

FIGURE 23.1-5 (A) Barbados Blackbelly rams, a breed of hair sheep. (Photo courtesy of Judy Jarnberg.) (B) Newborn Barbados Blackbelly lamb, depicting the long and fine texture of the neonate hair coat. A longer hair coat provides superior insulation for the newborn. (Photo courtesy of Pam Hand, DVM.)

COATS OF SHEEP

Ancestral sheep (and modern "hair" sheep) shed their coat. The coat consists of outer guard hairs and the inner, softer under coat. The modern fleece sheep primarily have the under coat on the body with few guard hairs. The fleece can grow continuously, sometimes to great lengths. The wool coat of sheep, hair goats, and South American camelids is referred to as a fleece and is usually mechanically shorn from these animals annually.

The hair sheep, such as the Barbados Blackbelly, are bred primarily as a protein source (Fig. 23.1-5). These breeds typically shed, not to be confused with a dermatological disorder (Fig. 23.1-6). Fig. 23.1-7 depicts a wool sheep with patches of fleece rubbed off, requiring further examination. Because wool grows continuously, adverse systemic physiological events, such as high fevers or pregnancy toxemia, can lead to an interruption in the growth of the fiber, resulting in a defect in the fiber shaft. The wool can then "break." Wool break can lead to the entire fleece being shed in 1 piece. The fleece can also be easily pulled from the body (Fig. 23.1-8).

FIGURE 23.1-6 Barbados Blackbelly sheep shedding. This is a normal finding in hair sheep. (Photo courtesy of Pam Hand, DVM.)

FIGURE 23.1-7 Suffolk ram showing areas of wool that have been rubbed off (*arrow*).

Cattle have a **simple** follicular arrangement—each hair follicle has a **sebaceous** and a **sweat gland** with an **arrector pili muscle**. The arrector pili muscle pulls on the hair follicle to raise the hair and "fluff up" the coat, supplying an insulating layer of air. In contrast, sheep and goat hair follicles are **compound**, with several **primary hairs** accompanied by **secondary**, smaller **hairs**. The primary hairs have sebaceous and sweat glands. The secondary hairs have only sebaceous glands and lack the arrector pili muscle. Wooled sheep have a greater number of secondary hairs, which are finer in texture, than the primary hairs.

The generally short coat of cattle and goats is formed of hairs with **determinate growth**, meaning the hair growth ceases after a certain length/time. The contrasting long hairs of manes, tails, or beards of goats are achieved by extended growth phases (anagen) and may be **indeterminate** (Fig. 23.1-9).

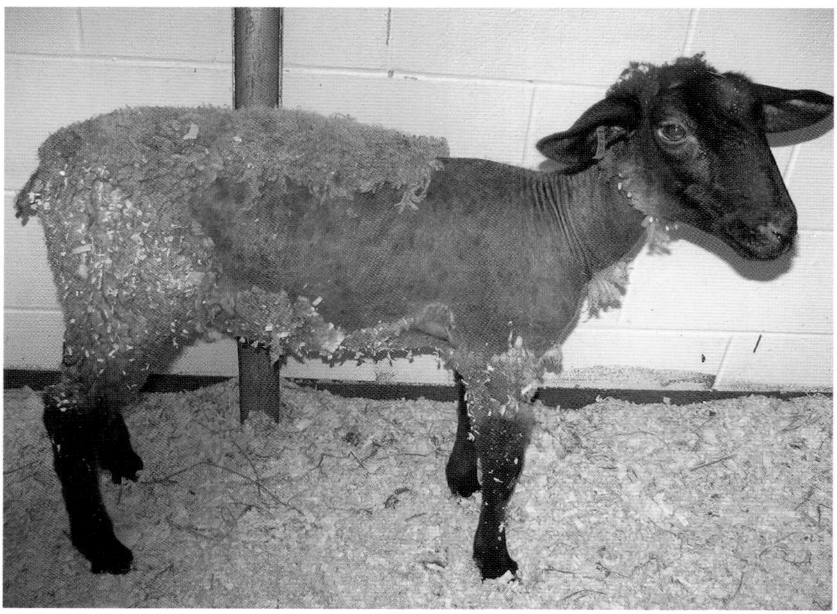

FIGURE 23.1-8 A sheep with a "wool break". The wool can be peeled from the skin in sheets. (Photo courtesy of George Barrington, DVM, PhD, Washington State University.)

FIGURE 23.1-9 Long beard of an adult buck goat.

There are various types of hair. **Guard hairs** are generally stiffer and form the outer hair coat of animals and grow from primary follicles, while the **undercoat**, derived from the secondary follicles, is a softer, fluffier texture and supplies insulation. The outer hair coat also affords a smoother, water-repellant surface. **Tactile hairs**, or **vibrissae**, such as whiskers offer sensory input, while **eyelashes** protect the globe.

Fiber-producing farm animals are raised, in whole or in part, for their hair coat. These include many breeds of sheep, South American camelids, and fiber-producing goats, such as Angora and Cashmere goats.

Wool is simply a special hair fiber of sheep while the term “fleece” usually refers to the “coat” of wool, especially once it has been shorn from the animal. Wool is also a term used to describe the dense soft pelage of camelids, goats, yaks, musk ox, and even some rabbits. The soft wool of sheep is produced by secondary hair follicles, while stiffer, courser wool fibers grow from primary follicles. Multiple secondary follicles can share a common orifice with a primary follicle, with multiple fibers arising from the single opening (Fig. 23.1-10).

Lanolin is the product of sheep sebaceous glands (Fig. 23.1-10). Also called “wool grease” and “wool wax,” lanolin helps protect and waterproof sheep’s skin. Lanolin is extracted from wool during processing and is used in many commercial applications, including cosmetics and lubricants (Fig. 23.1-11).

Wool differs from hair because it typically lacks the medulla/central hollow region. Wool breeds—e.g., Merino, Rambouillet, Corriedale—also have hair, particularly on the face and distal limbs.

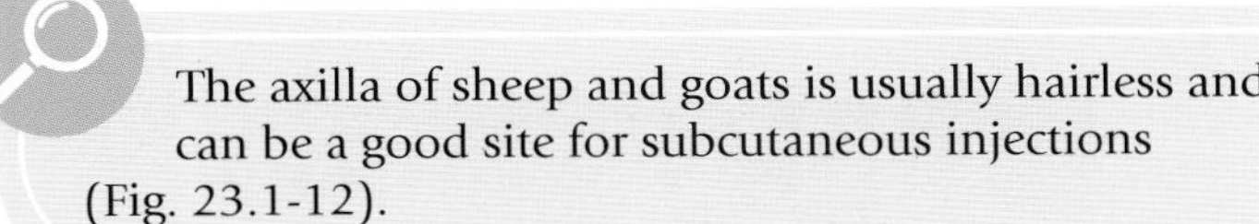
The axilla of sheep and goats is usually hairless and can be a good site for subcutaneous injections (Fig. 23.1-12).

Accessory structures of the integument

The veterinarian needs to be familiar with the many variations of livestock skin and its appendages.

Many breeds of goats, and occasionally sheep and swine, have **wattles**, which are finger-like projections on the cranial ventral neck region (Fig. 23.1-13). Wattles are covered in skin and often have a cartilaginous core. Goats can have 1, 2, or no wattles; wattles can also occur ectopically—i.e., outside of their usual location. The function

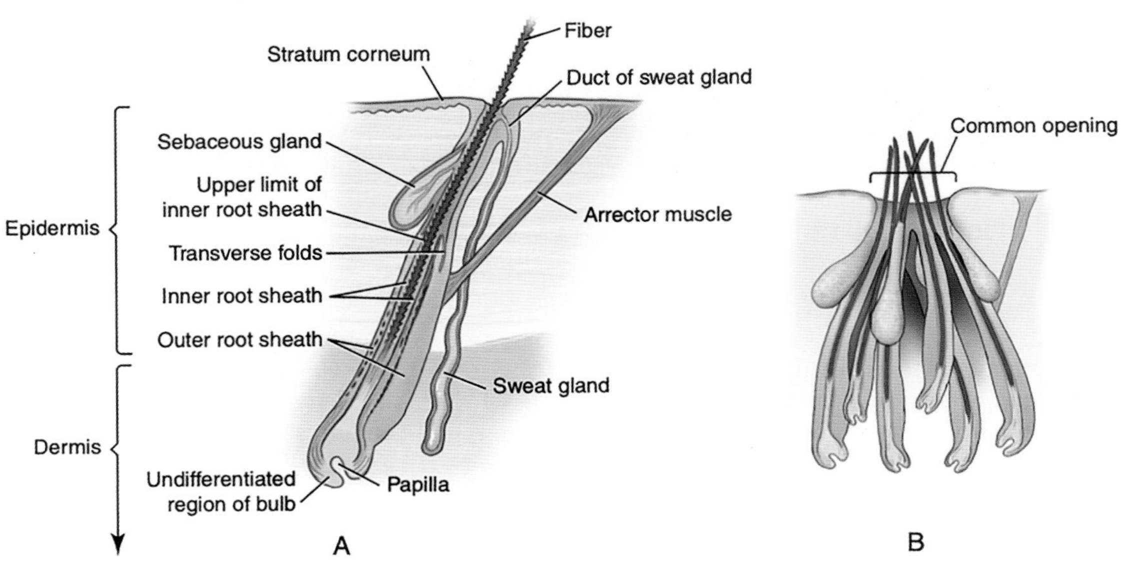

FIGURE 23.1-10 Hair follicles of ruminants. (A) Anatomy of a primary hair follicle. (B) Numerous secondary follicles, with their fibers shown exiting a single orifice.

FIGURE 23.1-11 The wool on this sheep has been spread to show the fiber length and skin. The dark speckles shown by *arrows* are lanolin.

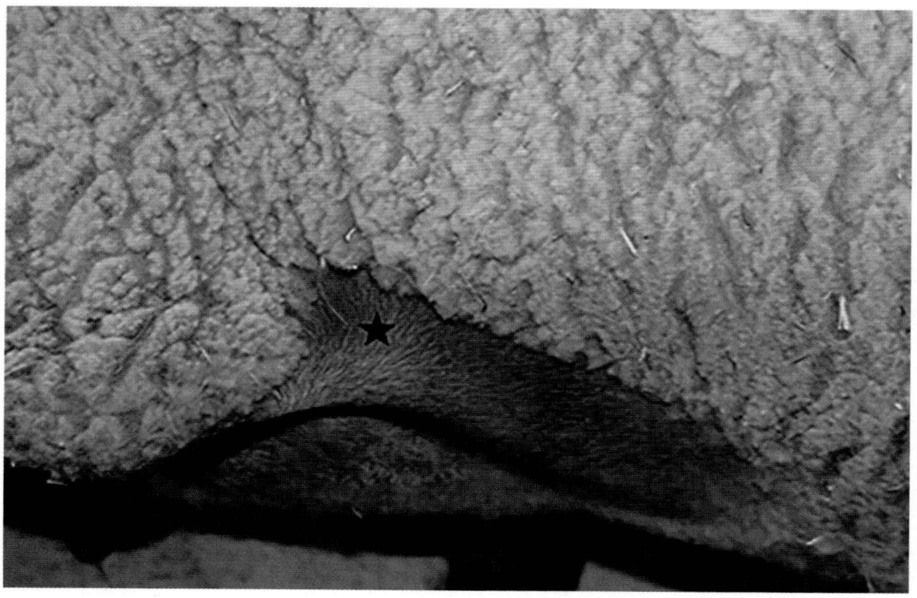

FIGURE 23.1-12 The axillary region of a sheep, showing the sparse hair covering the skin (*star*).

Bucks urinate on their beard, face, and legs during mating season which may cause urine scald dermatitis of the skin, treated with baths and topical therapies. The reason for this behavior is unknown. During the mating season, male goats produce a characteristic strong and aromatic odor secreted from sebaceous scent glands located caudomedially to the base of the horns. These glands are often destroyed by dehorning.

of wattles is unknown; however, they are heritable and commonly found in many of the dairy breeds. Wattles are occasionally seen in sheep and some breeds of swine—e.g., the Red Wattle hog and the kunekune.

Goats of both sexes may have a tuft of hair on the chin called a **beard**, which may be particularly well-developed in the buck (Fig. 23.1-9).

FIGURE 23.1-13 Wattle on the cranial ventral neck region of a goat (*arrow*). Note the goat's beard to the left of the photo.

Sheep have 3 glandular pouches: (1) **infraorbital pouches** located rostral to the medial canthus of each eye; (2) **inguinal pouches** lie bilateral to the mammary gland region in males and females; and (3) **interdigital pouches** found on the dorsal surface of the interdigital space between digits III and IV (Fig. 23.1-14).

The sternum in camelids—and occasionally sheep and goats—can have a noticeable **callus** (Fig. 23.1-15).

The duct of the interdigital pouch can become blocked, causing swelling, inflammation, and infection. This results in pain and lameness referable to the interdigital area.

Goats and sheep often rest on the dorsal carpus when lying down, causing a callus and alopecia. This is a normal finding and should not be misinterpreted as a clinical problem (Fig. 23.1-16).

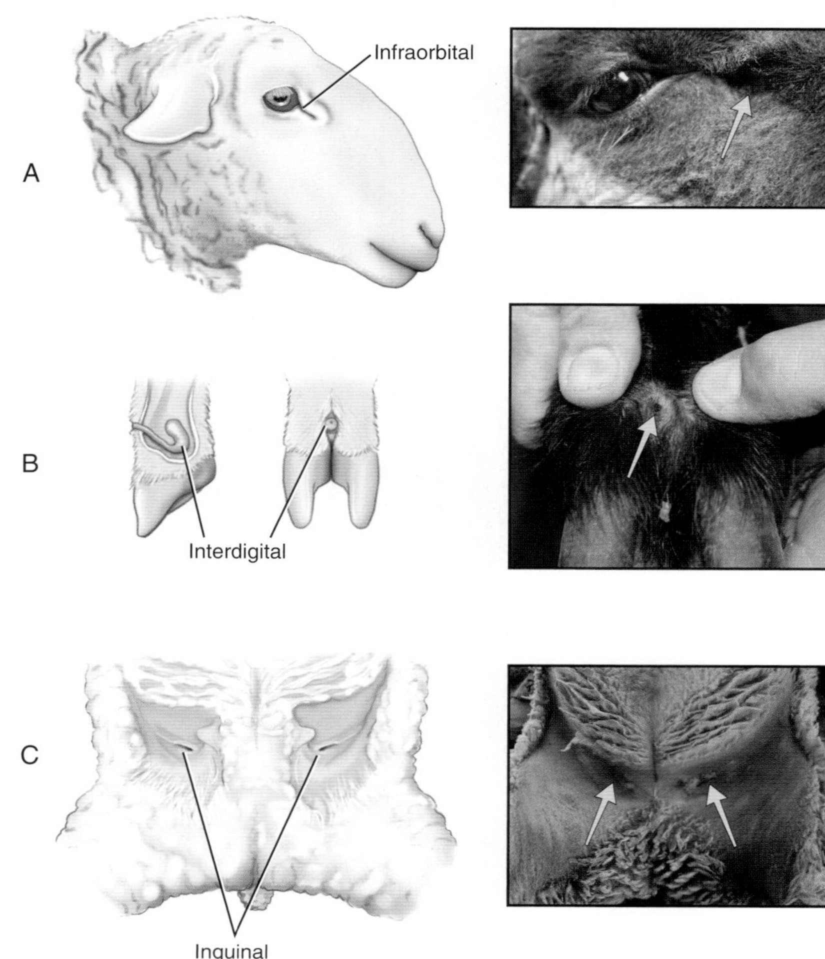

FIGURE 23.1-14 The various cutaneous pouches found on sheep. (A) Infraorbital (*arrow*). (B) Interdigital—the extent of the interdigital pouch (*arrow*). (C) Inguinal (*arrows*).

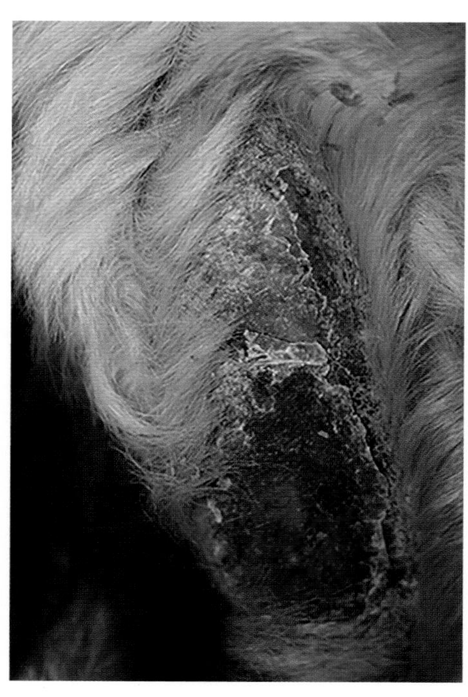

FIGURE 23.1-15 Sternal callus of an alpaca.

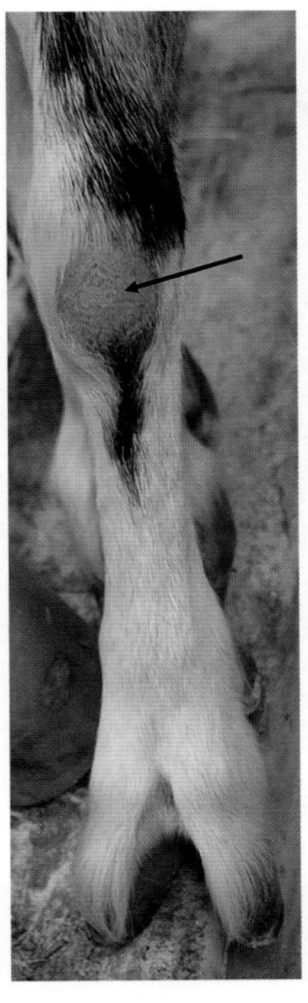

FIGURE 23.1-16 Normal callus on the dorsal aspect of the caprine carpus (*arrow*).

Added functions of skin

Thermoregulation in cattle is aided by a **dewlap**—particularly in cattle of "zebu" or *Bos indicus* heritage—and this increased surface area aids with heat evaporation. The prepuce can also be enlarged to help in heat exchange.

The **planum nasale** is typically sparsely haired and flat in both pigs and cattle. Pigs use their snout for rooting, so the skin is thickened and has short vibrissae. Tactile nerve endings allow for the snout to be principally sensitive (Fig. 23.1-17). The pig hair coat is typically sparse and consists of primarily stiff bristles. The flat planum nasale of cattle (Fig. 23.1-18) is kept clean and moist by the tongue. The ductal openings of the nasolabial glands are plainly noticeable to the eye.

Livestock (and horses) can instinctively "twitch" their skin to deflect insects or other irritants. The cutaneous trunci muscles underlie and attach to the skin over much of the dorsal and lateral thorax. The cutaneous trunci are innervated by the lateral thoracic nerve. 

Pinching of the skin over the cutaneous trunci should elicit a skin twitch reflex in a normal animal.

FIGURE 23.1-17 Pig snout with obvious sensitive vibrissae (hairs).

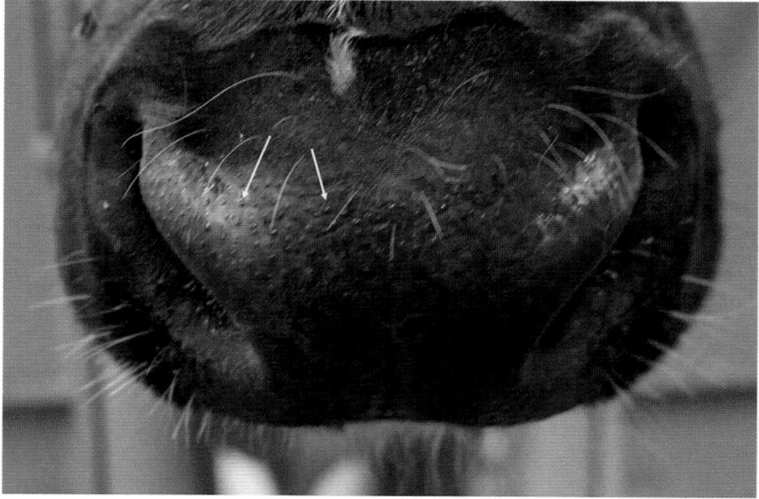

FIGURE 23.1-18 Planum nasale of a bovid highlighting nasolabial gland openings (*arrows*).

Blood supply, lymphatics, and innervation of the skin

The blood supply, lymphatics, and innervation of the skin, as well as the follicles and glands, are located primarily within the dermis.

The blood supply to the integumentary system is provided by the cutaneous circulation and is important in thermoregulation. There are 3 primary types of cutaneous circulation: (1) direct—blood vessels derived directly from the main arterial trunks that drain into the main venous vessels; (2) musculocutaneous—cutaneous vessels arise from intramuscular vessels after penetrating muscles and distributing in the subcutaneous tissue; and (3) fasciocutaneous—blood vessels consist of perforating branches from vessels found beneath the deep fascia. The cutaneous circulation consists of many capillary and arteriovenous anastomoses, especially in the extremities, enabling thermoregulation.

The dermal lymphatics parallel that of the venous system in 2 main areas in the skin: (1) superficial lymphatic network associated with the superficial venous system of capillaries and venules and (2) deep lymphatic network at the dermal/subcutaneous junction. The network of lymphatics empties into collecting lymphatics, then into larger lymphatic trunks, and finally to regional lymph nodes. The lymphatic vessels follow similar paths of the blood vessels.

Innervation for much of the integumentary system supplies sensory capabilities and includes **Pacinian corpuscles** (corpusculum lamellosum)—nerve endings sensitive to pressure and vibration and **Meissner's corpuscles** (corpusculum tactus)—tactile, medium-sized nerve endings in the skin and extremities. The integumentary system receives its innervation via spinal and cranial nerves, primarily associated with the autonomic nervous system.

Selected references

[1] Pugh DG, Baird AN. Sheep and goat medicine. 2nd ed. Maryland Heights, MO: Elsevier; 2012.
[2] Singh B. Dyce, Sack and Wensing—textbook of veterinary anatomy. 5th ed. St. Louis: Elsevier; 2018.
[3] Sisson S, Grossman JD, Getty R. Sisson and Grossman's The anatomy of the domestic animals. 5th ed. vol. II. Philadelphia: Saunders; 1975.
[4] Spickler AR. Contagious ecthyma. Available at: http://www.cfsph.iastate.edu/DiseaseInfo/factsheets.php; 2015.

CHAPTER 23

CASE 23.2

Chronic Udder Abscess

Sylvain Nichols
Department of Clinical Sciences, Université de Montréal, St-Hyacinthe, Québec, CA

Clinical case

History

A 10-year-old Angus cow was evaluated for an enlarged udder associated with a draining tract (Fig. 23.2-1). The cow had not calved for the past several years and was kept as a companion animal. The attending veterinarian had prescribed several treatments for what he believed was chronic mastitis, without success.

Physical examination findings

On presentation, the cow was bright and alert. Her vital parameters were within normal limits; she had a body condition score of 4 out of 5 (1= emaciated; 5= obese). Her udder was firm and measured 30 cm (11.8 in.) in diameter, which was enlarged for a dry cow. Numerous fistulous tracts containing necrotic material and purulent discharge were present on the skin of the udder. The right front quarter was the most severely affected quarter.

FIGURE 23.2-1 A 10-year-old Angus cow with multiple abscesses within its right mammary gland. Bloody purulent material is draining from a fistula (L. "pipe"; abnormal passage or communication) at the caudal aspect of the gland (*red arrow*).

Comparative Veterinary Anatomy: A Clinical Approach. https://doi.org/10.1016/B978-0-323-91015-6.00113-8

Differential diagnoses

Chronic udder abscess, neoplasia

Diagnostics

An ultrasonographic evaluation of the affected quarter was performed using a curvilinear 3.5-MHz probe. Multiple cavities containing heterogeneous material compatible with purulent material were identified.

Diagnosis

Chronic udder abscess

Treatment

The cow was placed under general anesthesia and positioned in dorsal recumbency for a total mastectomy. Before beginning surgery, the draining tracts were flushed and packed with iodine-soaked gauze, covered with a sterile hand towel, and secured in place to reduce intra-operative contamination of the surgical site. A 90-cm (35.4-in.) long fusiform skin incision was created around the udder. Using a combination of blunt and sharp dissection, the lateral suspensory ligaments, the mammary vein, the ventral labial vein, and the mammary branch of the perineal artery were isolated, ligated, and transected. The lateral suspensory ligament was sharply incised around the circumference of the udder to expose the pudendal arteries and veins (Fig. 23.2-2), which were triple-ligated and transected. Finally, the udder was separated from the body wall by transecting (L. through, across + L. *sectio* a cut) the median suspensory ligament close to the body wall.

The subcutaneous tissue was apposed at the cranial and caudal aspects of the incision, and the skin was approximated using an absorbable suture material in a simple interrupted pattern. To reduce tension on the skin closure, the middle of the incision was closed perpendicular to the original incision, creating a cross-pattern (Fig. 23.2-3).

The cow recovered from general anesthesia without problems. The rear legs were initially hobbled together to minimize tension on the skin closure. The sutures were incrementally removed beginning on day 14 post-operatively; the cow was discharged from the hospital 3 weeks after surgery.

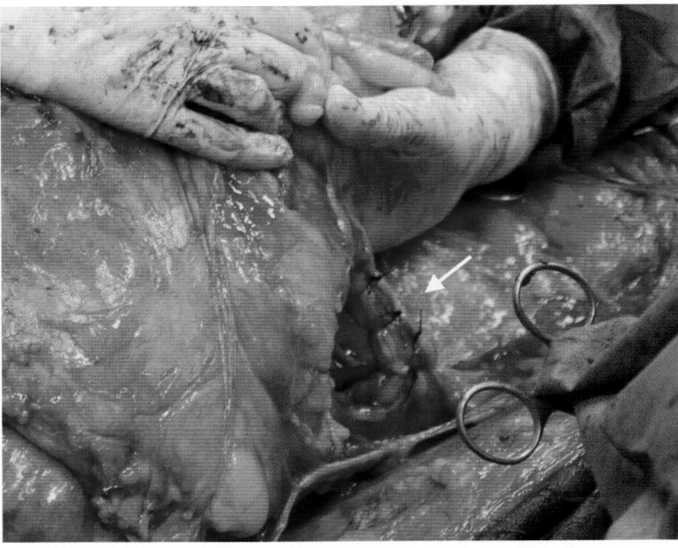

FIGURE 23.2-2 Intra-operative image of the pudendal artery and vein isolated at the level of the inguinal canal. The vessels are triple-ligated (*white arrow*) to prevent post-operative bleeding.

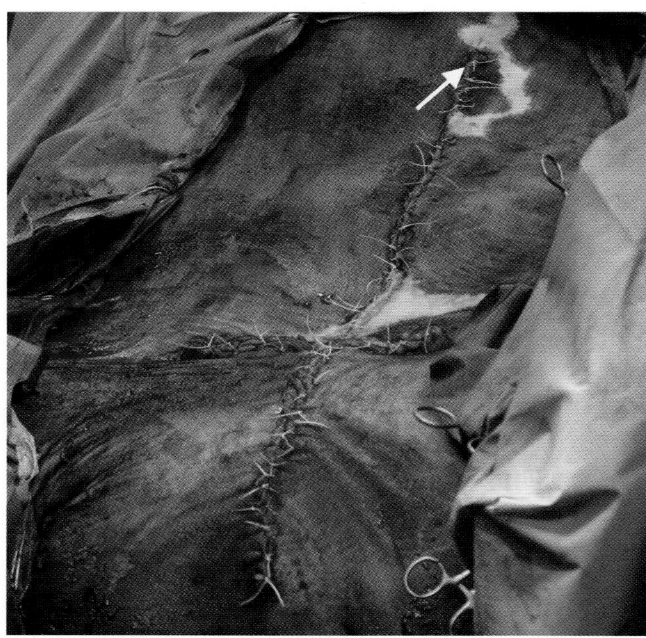

FIGURE 23.2-3 To avoid tension on the skin incision, the center of the incision is closed transversally, creating a cruciate pattern. In this image, the longest suture line corresponds to the long axis of the cow. The *white arrow* is pointing cranially.

Anatomical features in ruminants

Introduction

The udder of cattle is composed of 4 "quarters" or mammary glands. In lactating cows, its size is exceptionally large. The most common pathology involving the udder is mastitis. Depending on the etiologic agent, mastitis can affect the long-term health and productivity of the cow. Teat anatomy is covered in Case 23.3.

Function

The mammary gland is an exocrine gland in mammals—e.g., cows—that produce milk to nourish their young and for milk production (e.g., dairy herds). A mamma (L. the breast, mammary gland, lactiferous gland, udder) is called the breast in primates and udder in ruminants—e.g., cows, goats, and sheep. The mammary gland is considered part of the common integument system and is a modified cutaneous glandular structure with each gland producing milk in the alveoli. Lactation is the production of milk for nursing that occurs following a gestation/pregnancy of 9 months in cows. Lactation occurs under the hormonal influence of estrogen, progesterone, prolactin, and oxytocin.

> In a few mammalian species, male lactation can also occur. Lactorrhea or galactorrhea (Gr. *galaktos* milk + Gr. *rhoia* flow) is the spontaneous flow of milk by the glands that can occur in any mammal.

Mammary gland

Each **mammary gland** (also called **quarters**, of which cattle have 4) is formed by multiple **lobules**. Lobules are formed by a collection of **alveoli**. Milk, secreted by the alveoli, is carried by **alveolar ducts** into the **lactiferous ducts**. Those ducts carry the milk to the **collecting ducts**. A dozen of those ducts drain milk into the **gland cistern** (sinus)

(Fig. 23.2-4A and B). From there, the milk goes into the **teat cistern** and the **papillary canal** before being expressed by a suckling calf, by hand, or by a milking machine. The junction between the gland and teat cisterns is called the **annular folds**.

Suspensory ligaments

The **suspensory ligaments** surround the udder (Fig. 23.2-4A). They are separated into medial and lateral ligaments, and each of these ligaments has a right and left component. A distinctive layer of connective tissue is present between the right and left medial suspensory ligaments. The **lateral suspensory ligament** arises from the lateral border of the inguinal ring and the symphyseal tendon. It eventually splits into outer (**medial femoral fascia**) and inner (**mammary fascia**) sheathes. It is denser proximal to the body wall. As it travels distally on the udder, it becomes smaller as numerous lamellae detach and penetrate the parenchyma to hold the quarter firmly against the body wall.

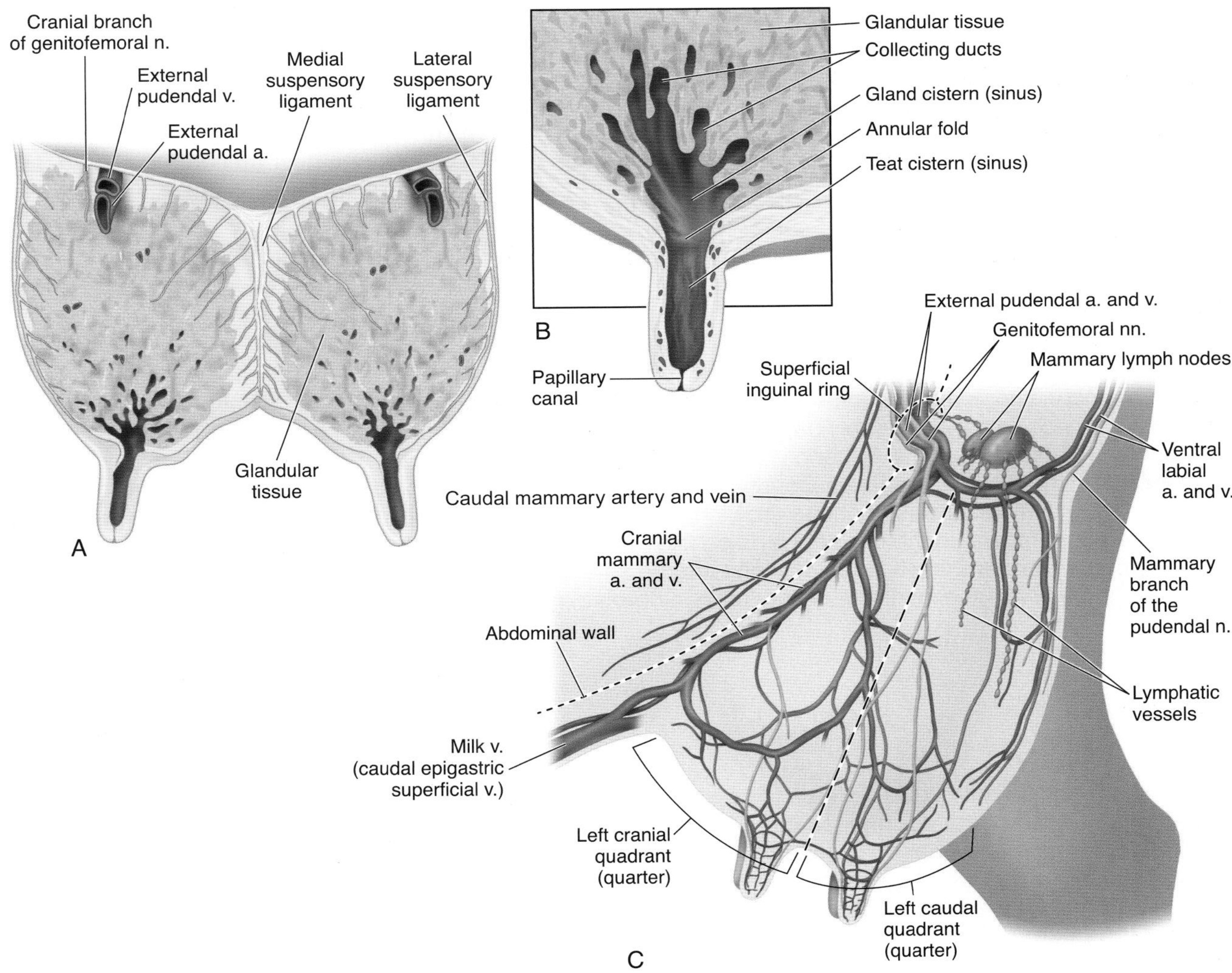

FIGURE 23.2-4 (A) A transverse section of the hind quarters of an adult dairy cow depicting the suspensory apparatus of the udder. (B) A longitudinal section of the teat and multiple glands depicting the milk collecting and entering system of a cow. (C) Blood supply lymphatics, and innervation of the udder of a lactating cow.

The **medial suspensory ligament** arises from the tunica flavia and borders the linea alba and the symphyseal tendon. It is thicker than its lateral counterpart. Like the lateral ligament, the medial ligament becomes smaller as it travels distally on the quarters.

Blood supply, lymphatics, and innervation of the mammary gland

Six hundred liters of blood needs to flow through the udder to produce 1 liter of milk.

A remarkable network of blood vessels matures within and around the udder, from birth to the production of the first drop of milk (Fig. 23.2-4C). Arterial blood supply comes from the **external pudendal arteries** and the **ventral labial artery**, the latter of which comes from the **perineal artery**. The external pudendal arteries divide and form the **cranial** and **caudal mammary arteries**. The caudal mammary artery joins with the ventral labial artery.

Even though it is often large and readily accessible, the milk vein should not be used for venipuncture in cows because of the potential complications associated with thrombophlebitis of such a large vessel. This is particularly true in dairy cows whose livelihood depends on a healthy udder and the continuous production of milk.

Venous drainage is complex and consists of three pathways: the **milk vein (superficial epigastric vein)**, which is the largest vessel visible coursing subcutaneously along the ventral abdomen, the **external pudendal vein**, and the **ventral labial vein**. The cranial and caudal mammary veins are connected cranially to the milk vein, centrally, at their junction, to the external pudendal vein, and caudally to the ventral labial vein. The flow within the mammary (cranial and caudal) veins is bidirectional, creating an impressive vascular ring at the base of the udder.

UDDER DEVELOPMENT

Small mammary glands and teats are present at birth in both males and females. At this time, the collecting system is composed only of a small papillary canal and teat cistern. The udder remains in this stage until puberty when—under the influence of estrogen in heifers—connective tissue and fat invade the small primitive quarters. The epithelial buds proliferate and small collecting ducts appear. During the first pregnancy, the epithelial buds continue growing and dividing, creating the lactiferous ducts. In the second half of gestation, fat within the udder is replaced by glandular tissue. As parturition (calving) approaches, a lumen appears—under the influence of progesterone and estrogen—within the alveoli. The lactocytes begin milk production under the influence of the hormone prolactin.

UDDER EDEMA

Post-partum physiologic udder edema is common in first-calf heifers. It is located over the udder and sometimes on the cranial abdominal and perineal area. It is generally not painful but may predispose the cow to mastitis and impede mechanical milking. When severe, it may cause ulceration of the inner thigh when the udder rubs against the pelvic limbs at the walk. The etiology of physiologic udder edema is multifactorial: (1) the age at first calving; (2) the diet before calving; (3) the immature venous circle draining the udder of heifers; and (4) genetics. Anastomoses occur between the epigastric superficial vein (milk vein) and the cranial mammary vein. Before calving, competent valves within the venous circle prevent multidirectional flow of blood, possibly increasing the risk of udder edema. Time resolves post-partum physiologic udder edema. However, in severe cases, administration of a diuretic may hasten the resolution of udder edema.

Lymphatic drainage is achieved via the **superficial inguinal lymph nodes** ("mammary lymph nodes"). Those nodes are located below the thick suspensory ligament at the proximal aspect of the rear quarters. From there, lymph is carried by the efferent lymphatic vessels through the inguinal canal to the **iliofemoral lymph nodes**.

> The superficial inguinal lymph nodes are palpated by firmly grasping the proximal part of the rear quarters.

The skin and teats of the forequarters are innervated by the **iliohypogastric** and **ilioinguinal nerves** and the **cranial branch of the genitofemoral nerve**. The skin and teats of the rear quarters are innervated by the **caudal branch of the genitofemoral nerve** and the **mammary branch of the pudendal nerve**. The body of the udder is innervated by the **cranial** and **caudal branches of the genitofemoral nerve**, both tracking through the inguinal canal (Fig. 23.2-4C).

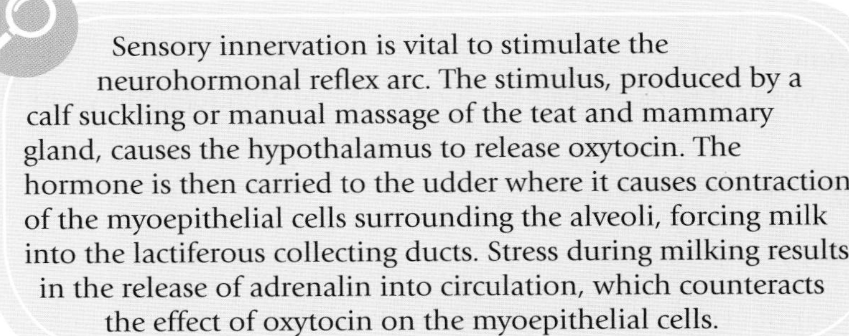

Selected references

[1] Cable C, Fubini SL. Surgery of the mammary gland. In: Fubini, Ducharme, editors. Farm animal surgery. 2nd ed. St-Louis: Elsevier; 2017. p. 482–5.

[2] Budras KD. Habel RE. In: Budras, Habel, editors. Bovine anatomy. 1st ed. Hannover: Schlütersche; 2003. p. 88–91.

[3] Akers RM. A 100-year review: mammary development and lactation. J Dairy Sci 2017;100:10332–52.

CHAPTER 23

CASE 23.3

Teat Obstruction

Sylvain Nichols
Department of Clinical Sciences, Université de Montréal, St-Hyacinthe, Québec, CA

Clinical case

History

A 5-year-old Holstein cow was evaluated for the complaint that it was difficult to milk her left rear quarter. The problem started 10 days before presentation when blood was noticed at the external sphincter and mechanical milking time had noticeably increased. Since that time, the quarter had been drained twice daily using a plastic teat cannula. However, the day before presentation, the teat cannula was difficult to insert, and milk fragments were observed in the drainage. The cow had calved 8 weeks before and had not been bred back. She was producing 50 kg (110 lbs) of milk/day before the incident.

Physical examination findings

The cow's physical examination parameters were within normal limits. On palpation, the distal end of the left rear teat was thickened and painful. The quarter was firm because 24 hours had passed since her last milking; little milk could be expressed by hand from this quarter.

Differential diagnoses

Complete or partial rupture of the papillary canal, thelitis (Gr. *thēlē* nipple + Gr. *-itis* inflammation), and clinical mastitis

Diagnostics

A small plastic cannula was carefully introduced into the papillary canal. Some resistance was encountered 5 mm (0.2 in.) into the canal. Milk was obtained for California mastitis test (CMT) evaluation and bacteriological culture. The CMT yielded a score of 3 out of 3.

An ultrasonographic evaluation of the affected teat was performed using a linear 7.5-MHz probe beginning at the annular fold at the base of the teat. A mass attached to the proximal aspect of the papillary canal was identified (Figs. 23.3-1 and 23.3-2). The distal part of the papillary canal appeared intact. Milk fragments were seen in the teat cistern.

Diagnosis

Rupture and inversion of the proximal part of the papillary canal

Comparative Veterinary Anatomy: A Clinical Approach. https://doi.org/10.1016/B978-0-323-91015-6.00114-X

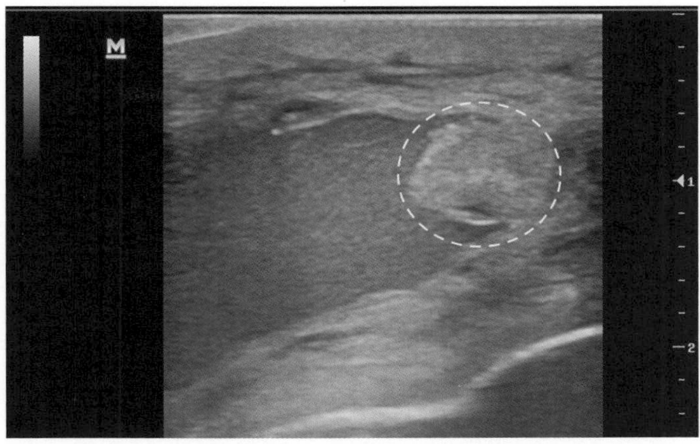

FIGURE 23.3-1 A longitudinal ultrasonographic image of a teat obtained using a linear 7.5-mHz probe. A mass is identified at the distal end of the teat cistern (*dotted white circle*).

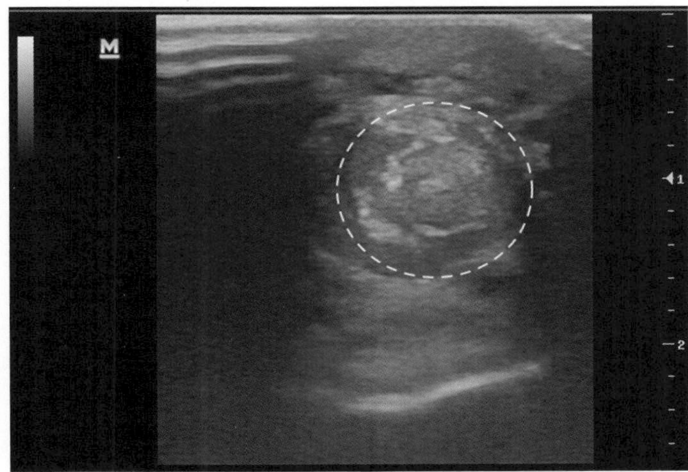

FIGURE 23.3-2 A transverse ultrasonographic image of the distal end of a teat obtained using a linear 7.5-mHz probe. A mass is identified in the normal location of the rosette of Fürstenberg (*dotted white circle*).

Treatment

Excision of the inverted papillary canal via theloscopy (teat endoscopy) under standing sedation was recommended. Local anesthesia was performed with IV regional teat perfusion and an elastrator (elastic) rubber band used as a tourniquet. Using a 22-gauge butterfly needle, 1 ml of 2% lidocaine was infused into one of the multiple longitudinal veins of the teat (see Fig. 23.2-4B).

After aseptic preparation of the surgery site, the theloscope was introduced in the teat cistern using a lateral approach. Briefly, a sharp obturator was introduced in the papillary canal and driven through the lateral wall at the junction of the proximal and middle third of the teat cistern. The theloscopic sheath was slid over the obturator (from outside in), the obturator removed, and the theloscope inserted in the sheath. The teat was distended with sterile saline, and the distal aspect of the teat was examined (Fig. 23.3-3). A mass corresponding to the inverted papillary duct was evident at the distal aspect of the teat cistern (Fig. 23.3-4) and removed by inserting an instrument

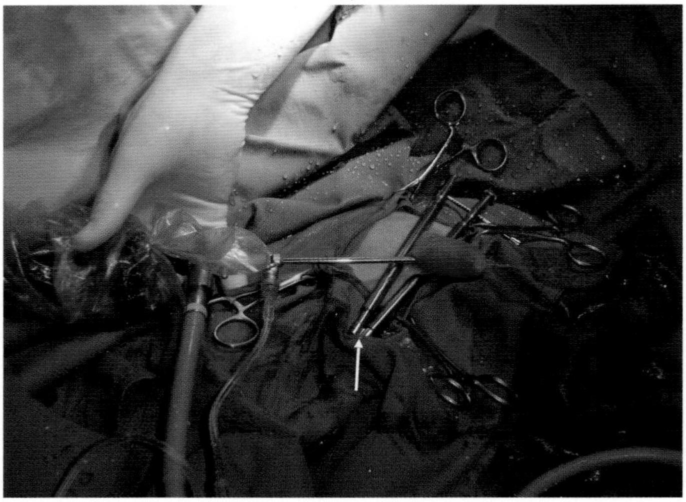

FIGURE 23.3-3 Lateral theloscopy in a cow. The scope and its sheath are inserted through the lateral wall of the teat. In this image, the procedure is done with the cow in dorsal recumbency. A teat clamp is used rather than a rubber band (*white arrow*).

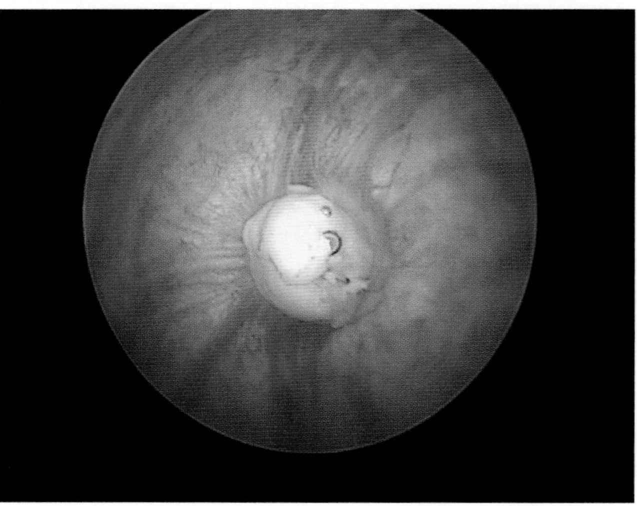

FIGURE 23.3-4 A mass, corresponding to the inverted papillary canal, is identified at the distal aspect of the teat cistern. A complete obstruction of the remaining papillary canal is caused by this mass.

through the papillary canal (Fig. 23.3-5). The teat cistern was flushed, and a silicone plug was introduced in the papillary canal. The incision created in the teat wall was closed with a simple interrupted suture. The tourniquet was removed, and intramammary antibiotics were infused in the quarter.

The teat was passively milked (by inserting a teat cannula into the teat cistern) for 10 days post-operatively before resuming mechanical milking. In between milking, a silicone plug was used to maintain dilatation of the papillary canal during healing. The milk culture yielded *Streptococcus dysgalactiae.* Intramammary antibiotics were used following the recommended guidelines for antimicrobial use in lactating cows; the CMT improved with treatment.

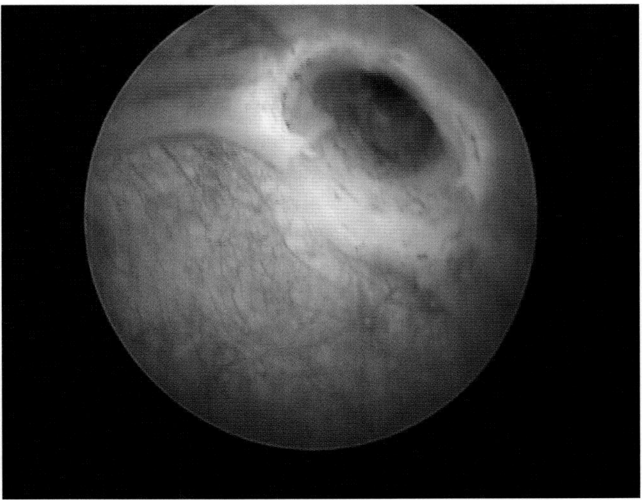

FIGURE 23.3-5 The mass, seen in Fig. 23.3-4, was removed using a stenosis cutter (Dr. Fritz Tuttlingen/Germany). The papillary canal is now open, draining the quarter.

Anatomical features in ruminants

Introduction

Dairy cows produce a large volume of milk. Teats are essential to drain the milk produced by each quarter. Any pathology affecting the teat also affects the normal flow of milk and predisposes the quarter to mastitis, which is a serious infection if caused by coliform bacteria. The bovine udder is discussed in Case 23.2.

Function

Milk is carried from the alveoli to the teat cistern (sinus) by the lactiferous ducts. It is then expressed by a nursing calf, a milking machine, or by hand.

Teat anatomy

In cattle, the mammary gland has 4 independent **quarters** and a single **teat** drains each quarter. The milk is produced by the **alveoli** before being carried to the **gland cistern** (sinus) by the **lactiferous and collecting ducts**. Thereafter, it goes in the **teat cistern** (sinus) before exiting through the **papillary canal**. A fold (**annular fold**), which contains a large circular blood vessel (**Fürstenberg ring**), is located at the junction between the gland and the teat cistern. The mucosa within the teat cistern joins toward the papillary canal to form the **rosette** (Fr. a little rose) **of Fürstenberg.** Finally, the **external sphincter** is located at the end of the papillary canal (see Fig. 23.2-4B).

The rosette of Fürstenberg is a valve-like structure between the teat cistern and teat canal. It is anatomically part of the internal streak canal of the teat, radiates proximally into the teat cistern, and is considered a physical barrier for bacteria without resisting milk flow from the teat.

The distal end of the teat is important in protecting the quarter from an ascending infection. Between milkings, smooth muscle surrounding the papillary canal contracts to close the external sphincter. The papillary canal also fills with keratin to further protect the quarter against mastitis.

The surgical procedure of sharply incising and opening the teat cistern is termed "thelotomy" (Gr. *thēlē* nipple/teat of the mammary gland + Gr. *tomē* a cutting or incision). A teat cannula is inserted through the papillary canal in the teat cistern and serves as a guide. The skin incision is parallel to the length of the teat and is centered to avoid incising the Fürstenberg ring or papillary canal. Bleeding vessels are meticulously clamped using hemostatic forceps. A small opening is created with the scalpel blade to fit one jaw of small Metzenbaum scissors in the teat cistern to extend the incision. Through a thelotomy (Fig. 23.3-6) incision, a pedunculated (papilloma) or free-floating mass can be removed. Any localized fibrous tissue of the teat cistern is excised. The incision is closed in 3 layers. The mucosa and the muscular layer are closed anatomically with a simple continuous pattern, and the skin is closed using an interrupted pattern. Small-gauge, absorbable suture material for the inner layers and similar size, nonabsorbable suture material for the skin are recommended to achieve primary healing of the thelotomy incision.

The skin overlying the teat in ruminants is thin and has no hair. The muscular layer composed of smooth muscle and connective tissues is thick and includes blood vessels that travel parallel to the teat wall. The innermost layer of the teat is the mucosa, which is pink and shiny compared to the muscular layer.

Blood supply, lymphatics, and innervation

Beginning at the Fürstenberg ring, the arteries and veins travel longitudinally toward the distal end of the teat. Their origins and that of the veins, lymphatics, and innervation are described in Case 23.2. The sensory innervation of the teat is essential to stimulate the production of oxytocin by the neurohypophysis. The stimulus of a suckling calf or a massaging hand is conducted by the afferent nerves to the CNS where, in the nuclei of the hypothalamus, the oxytocin is produced.

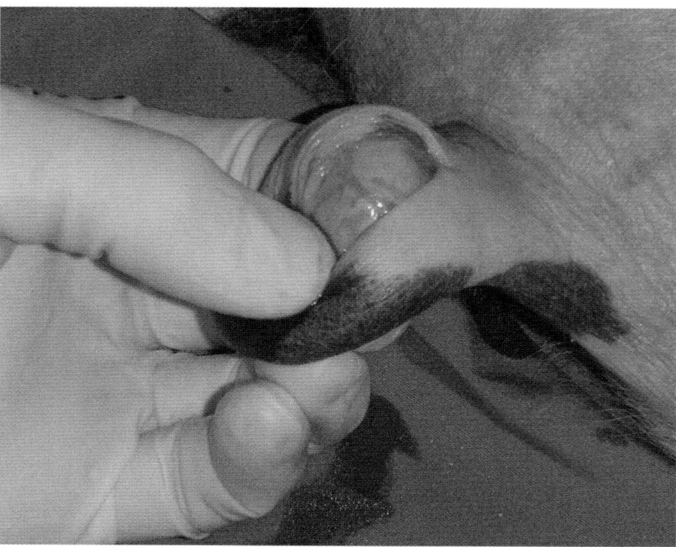

FIGURE 23.3-6 Thelotomy incision. The incision is longitudinal and is centered to avoid both the distal (rosette of Fürstenberg and papillary canal) and proximal structures (Fürstenberg ring).

PATHOLOGIES OF THE DISTAL TEAT

Pathologies affecting the distal end of the teat include sphincter anomalies secondary to overmilking, partial or complete rupture with or without inversion/eversion of the papillary canal (Fig. 23.2-4), and fibrosis of the rosette of Fürstenberg. These pathologies are best treated by theloscopy when the papillary canal or the rosette is involved or by distal end amputation when the sphincter is affected.

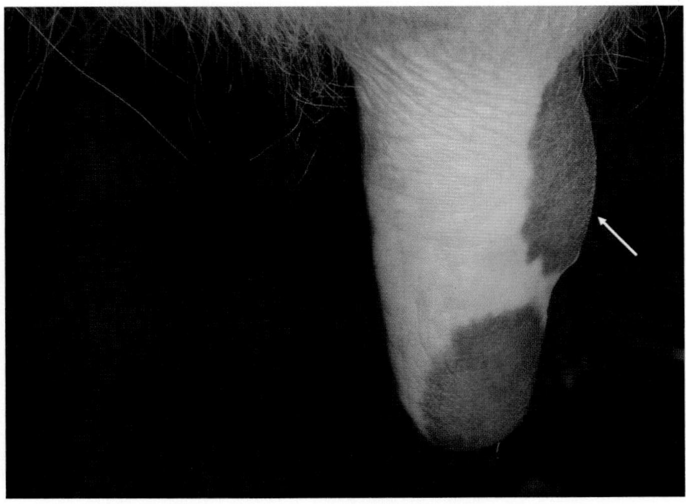

FIGURE 23.3-7 Image of a lactating dairy heifer with an accessory teat connected to the primary teat. The swelling, identified on the right side of this figure, represents the accessory teat (*white arrow*).

SUPERNUMERARY TEATS

Supernumerary teats with or without a gland cistern are a common finding in cattle. They can be located before, between, or behind the primary teats and are usually removed at a young age. Sometimes, they can be located directly on a primary teat and are then termed "accessory teats" (Fig. 23.3-7) and may go unrecognized until calving. Some accessory teats compromise milking by obstructing the lumen of the main teat cistern and therefore need to be surgically removed or anastomosed to the main teat cistern to reestablish normal milk flow from the gland to the teat cistern. The surgical procedure used depends on the surgeon's experience and the volume of milk produced by the accessory gland.

Selected references

[1] Nichols S. Teat laceration repair in cattle. Vet Clin North Am Food Anim Pract 2008;24:295–305.
[2] Couture Y, Mulon PY. Procedures and surgeries of the teat. Vet Clin North Am Food Anim Pract 2005;21:173–204.
[3] Geishauser T, Querengässer K, Querengässer J. Teat endoscopy (theloscopy for diagnosis and therapy of milk flow disorders in dairy cows). Vet Clin North Am Food Anim Pract 2005;21:205–25.
[4] Sisson S. Ruminant urogenital system. In: Sisson, Grossman, editors. The anatomy of domestic animals. Philadelphia: W.B. Saunders Co; 1975. p. 937–54.

SECTION VI

AVIAN CLINICAL CASES

6.0 AVIAN LANDSCAPE FIGURES (1–7) 1251
Cynthia M. Faux, Marcie L. Logsdon, Nora S. Grenager, James A. Orsini, Alexander de Lahunta

CHAPTER 24: ADAPTATIONS TO FLIGHT 1259
Cynthia M. Faux and Marcie L. Logsdon, Chapter editors

CHAPTER 25: HEAD AND NECK 1263
Cynthia M. Faux and Marcie L. Logsdon, Chapter editors

CHAPTER 26: THORACO-ABDOMINAL CAVITY 1315
Cynthia M. Faux and Marcie L. Logsdon, Chapter editors

CHAPTER 27: THORACIC AND PELVIC LIMB 1365
Cynthia M. Faux and Marcie L. Logsdon, Chapter editors

CHAPTER 28: INTEGUMENT/FEATHERS 1399
Cynthia M. Faux and Marcie L. Logsdon, Chapter editors

SECTION VI

6.0 AVIAN LANDSCAPE FIGURES (1–7)

Cynthia M. Faux[a], Marcie L. Logsdon[b], Nora S. Grenager[c],
James A. Orsini[d] Alexander de Lahunta[e]

[a]College of Veterinary Medicine, University of Arizona, Oro Valley, Arizona, US

[b]Exotics & Wildlife Department, College of Veterinary Medicine, Washington State University, Pullman, Washington, US

[c]Steinbeck Peninsula Equine Clinics, Menlo Park, California, US

[d]Department of Clinical Studies - New Bolton Center, University of Pennsylvania School of Veterinary Medicine, Kennett Square, Pennsylvania, US

[e]Professor Emeritus, Biomedical Sciences, College of Veterinary Medicine, Cornell University, Ithaca, New York, US

Avian anatomical regions: 1, primary feathers; 2, primary coverts; 3, secondary feathers; 4, secondary coverts; 5, patagium (propatagium); 6, carpus; 7, alula; 8, mantle; 9, nape; 10, crown; 11, supralores; 12, lores; 13, supercilium; 14, malars; 15, breast; 16, belly; 17, flank; 18, side; 19, rump; 20, vent; 21, rectrices.

FIGURE 6.0-1 Regional avian anatomy.

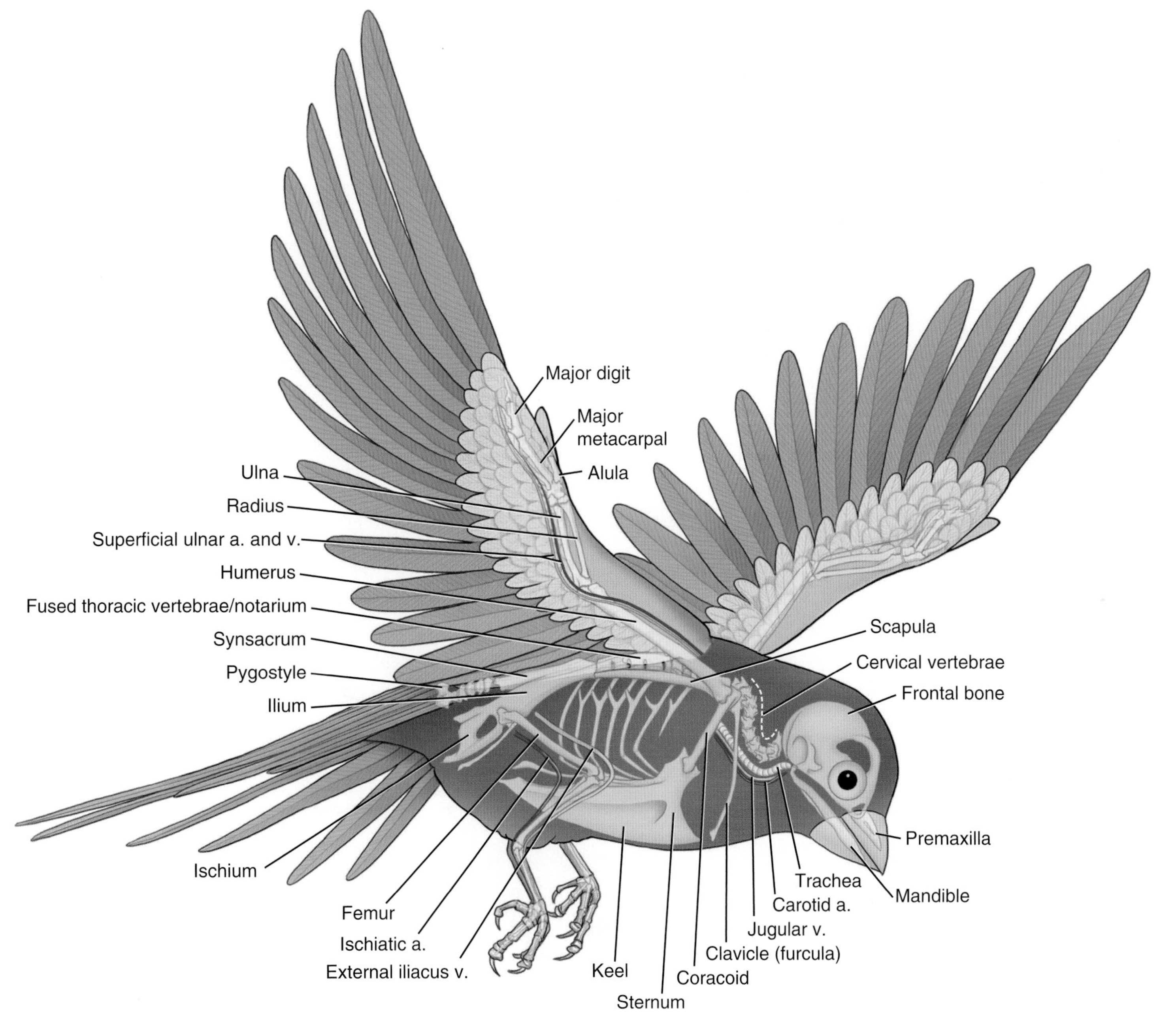

FIGURE 6.0-2 Topographical avian anatomy.

FIGURE 6.0-3 Avian plumage.

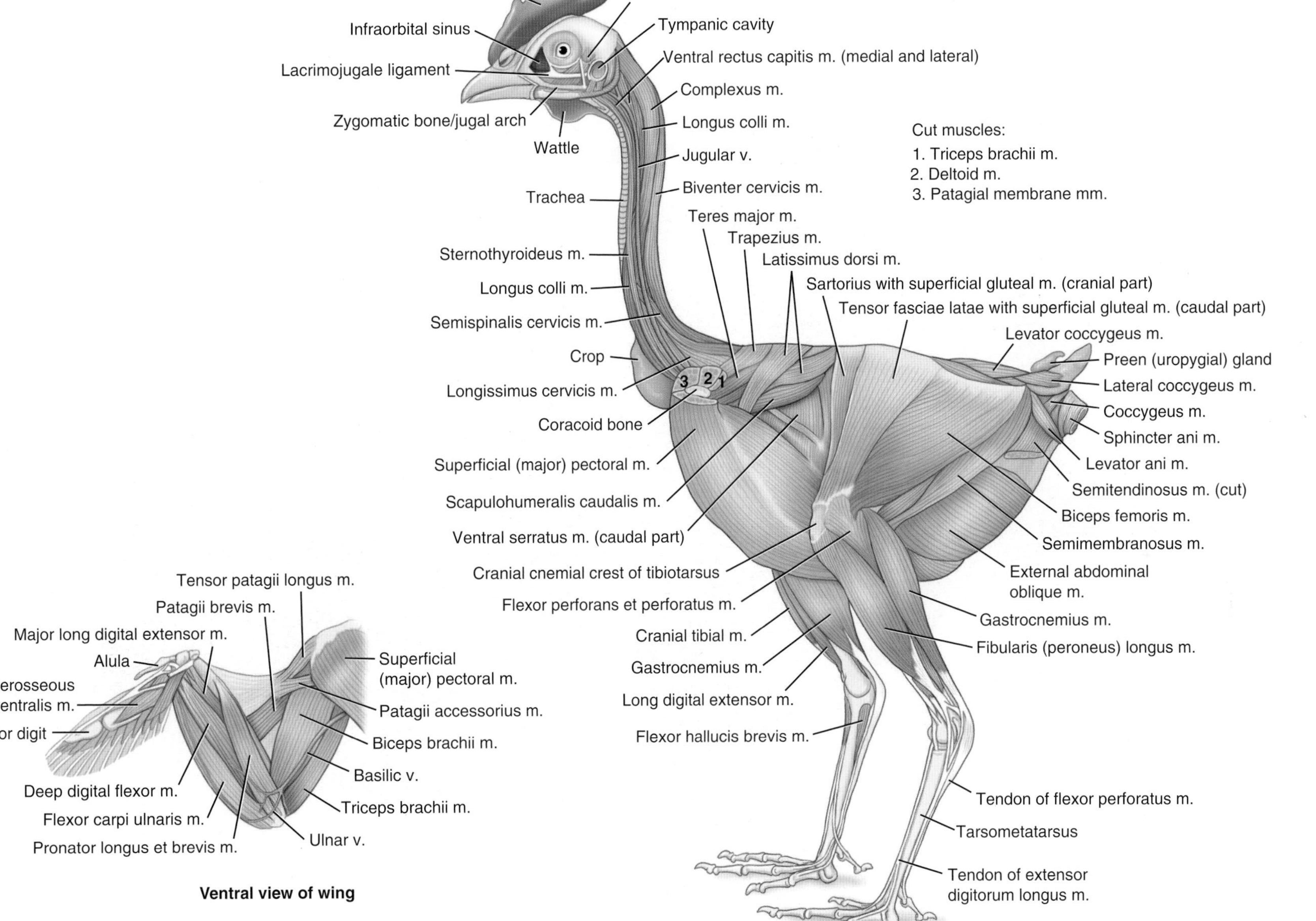

FIGURE 6.0-4 Superficial avian muscle layer.

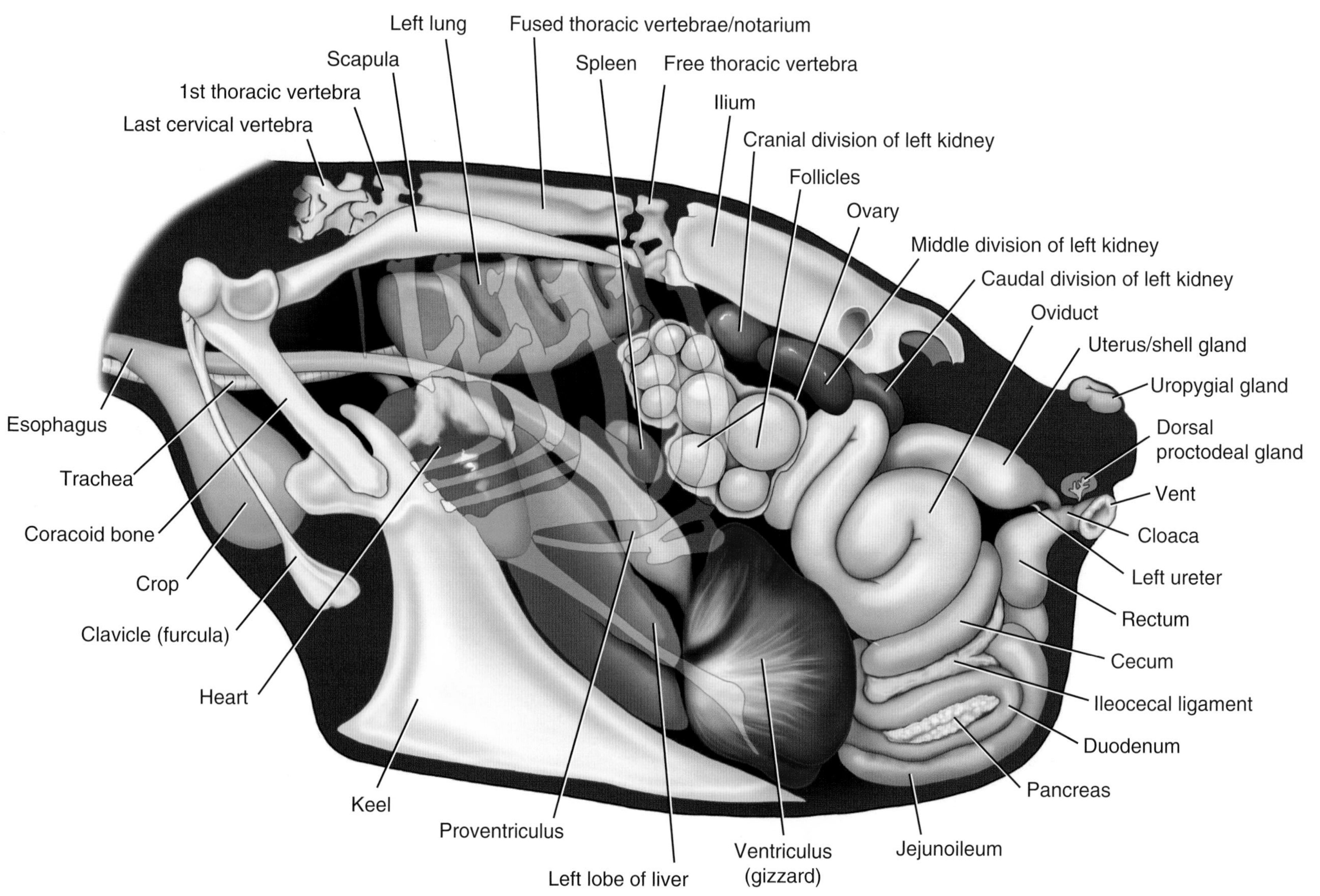

FIGURE 6.0-5 Left view of the avian thoracic and abdominal cavities and pelvis (female).

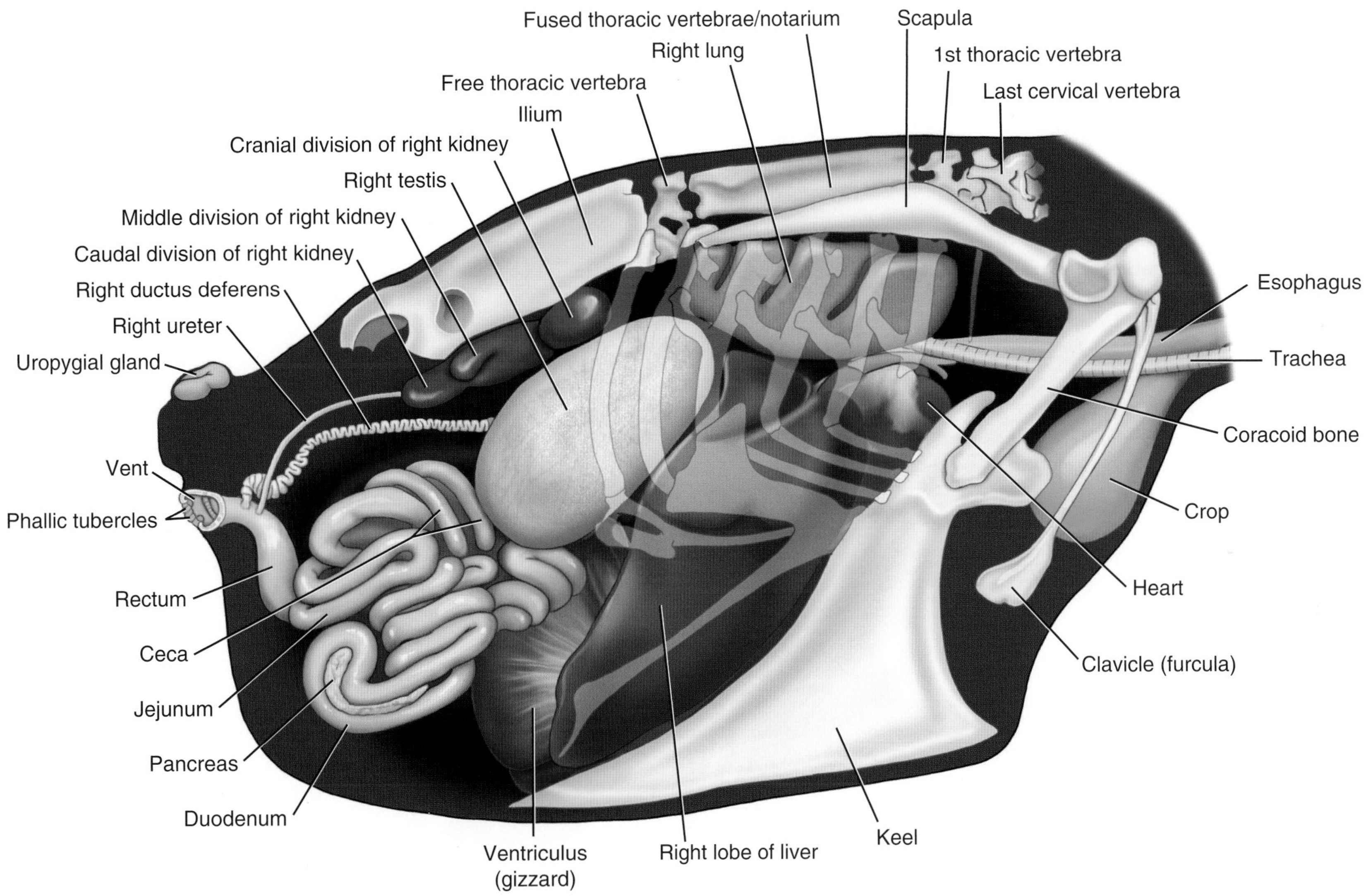

FIGURE 6.0-6 Right view of the avian thoracic and abdominal cavities and pelvis (male).

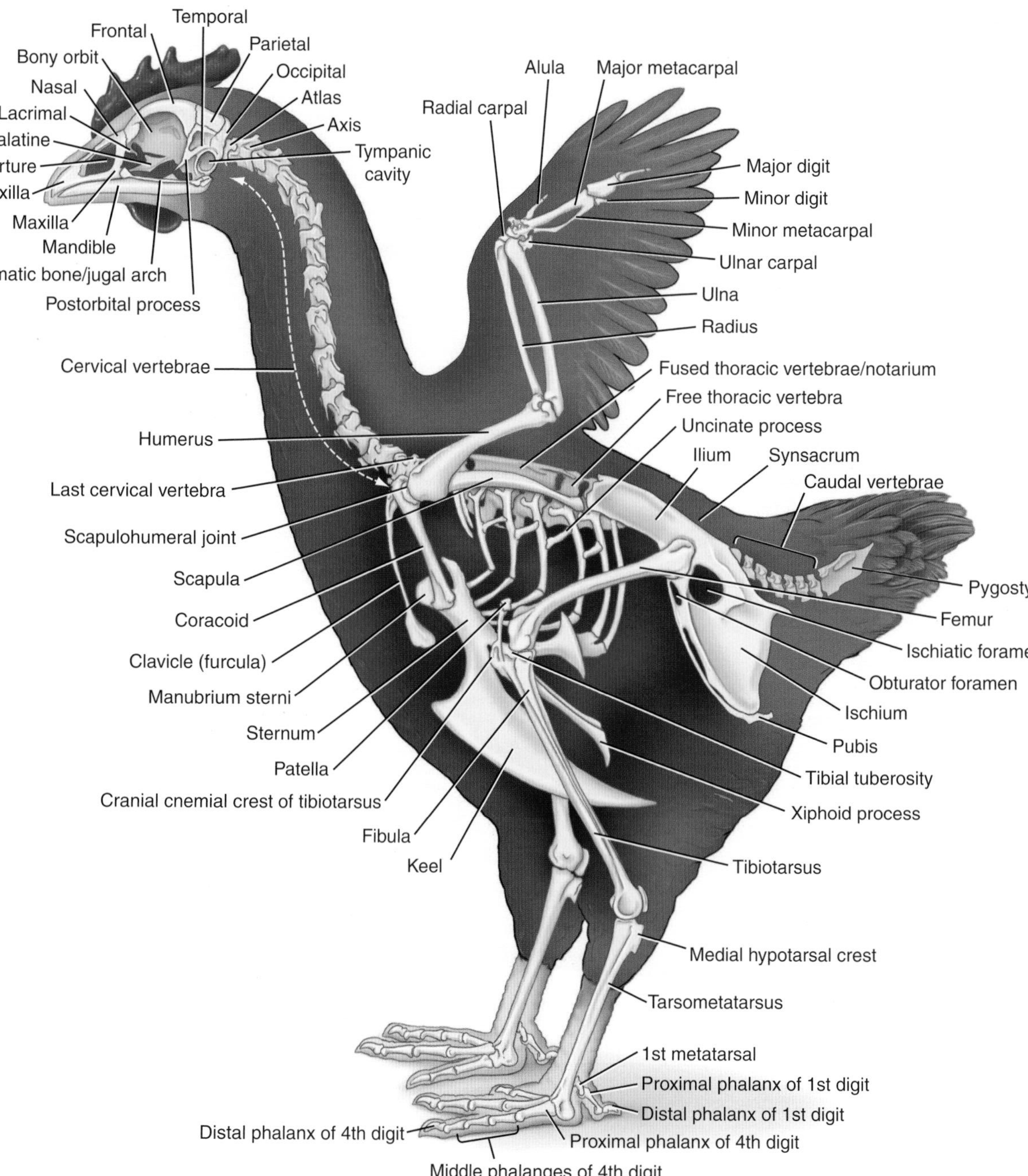

FIGURE 6.0-7 Lateral view of the avian skeleton. Vertebral column formula: $C_{8\text{-}25}T_7L/S_1Cd_{4\text{-}9}P_1$.

CHAPTER 24

ADAPTATIONS TO FLIGHT

Cynthia M. Faux and Marcie L. Logsdon

CHAPTER 24

CASE 24

Adaptations to Flight

Cynthia M. Faux[a] **and Marcie L. Logsdon**[b]
[a]University of Arizona College of Veterinary Medicine, Oro Valley, Arizona, US
[b]Exotics and Wildlife Department, Washington State University College of Veterinary Medicine, Pullman, Washington, US

Birds fill our skies and world with color, song, and companionship. We admire their abilities, antics, and adaptability. As veterinarians, we interact with birds as patients and we must recognize the wide species variability in this diverse group of vertebrates. A canary is not simply a small condor.

In the following chapters, the reader is introduced to, if not already familiar with, the comparative anatomy of birds. It is hoped that we successfully portray birds as, well, birds, and not merely describe their exceptional features as "unlike mammals." Defining them as what they *are not*, vs. what they *are*, does both vertebrate lineages disservice.

The avian story begins deep within the Mesozoic with a lineage of small bipedal carnivorous dinosaurs. Feathers, once considered the very definition of "bird," have been found adorning fossils of clearly nonflying dinosaurs. Thus, the feather predates flight. As more fossils are discovered, the line between what is bird and what is dinosaur continues to blur and we are left to conclude that birds are "dinosaurs" in the same sense that humans are mammals.

Their royal pedigree notwithstanding, all modern birds—the Neornithes—are descended from a lineage of flying, toothless birds, which survived the massive extinction event at the end of the Cretaceous.

As all modern birds are descended from a flying ancestor—even those that no longer fly—their anatomy reflects both the history and constraints of this foundation, as well as the anatomical necessities required for flight.

The avian skeleton is lightweight and sturdy. Invaginations of the respiratory system create pneumatized bones. "Hollow" bones retain the strength of a thin-walled cylinder with strut-like reinforcements. Yet, in birds where excess buoyancy might be a disadvantage, such as penguins, most bones have lost their thin walls and pneumatic features.

A rigid airframe has evolved that minimizes the muscular effort for maintaining trunk posture while airborne. The synsacrum and pygostyle are notable fusions of the vertebral column in all birds, and multiple lineages have independently reinforced their skeletons further through the acquisition of a notarium—additional fusion of thoracic vertebrae. Bones of the manus have been fused and/or reduced, which provides stability to the wing. Unlike mammals, the number of cervical vertebrae varies widely, with long-necked species unsurprisingly possessing more vertebrae than their short-necked cousins.

The lightweight avian skull possesses hinges and joints, allowing movement of the upper jaw. Beak and cranial adaptations allow birds to fill a massive range of environmental niches—from aquatic to arid, tropical to arctic—and exploit diets ranging from carnivory to herbivory and a wide spectrum in between. The avian eye differs in significant ways, both anatomically and physiologically, from that of mammals. Birds distinguish wavelengths in the ultraviolet range, providing them a perception of the world and of each other that is unavailable to us—i.e., their world is not our world.

Feathers provide the airfoil required for flight, as well as a streamlined body contour. As we will see in Chapter 28, feathers not only allow for flight but also provide protection, display, and even sensory functions.

Comparative Veterinary Anatomy: A Clinical Approach. https://doi.org/10.1016/B978-0-323-91015-6.00115-1

Flight puts constraints on avian anatomy in the sense that excess weight is eschewed. During the breeding season, the intra-abdominal testicles of males of many species enlarge, only to regress once the season is over. Likewise, females retain only 1 functional oviduct and lay eggs sequentially (vs. in a batch, as do alligators and turtles) and embryonic development takes place outside the body of the bird.

The avian kidney excretes uric acid, which requires less water to eliminate than does excretion of ammonia or urea, and birds do not fill a vesicular bladder with nitrogenous waste.

The high energy requirements for powered flight are accommodated by efficient cardiovascular and respiratory systems. The avian heart is 4-chambered; however, the proportions of the chambers and anatomy of the valves differ from those of mammals. The respiratory system of birds is markedly different from that of mammals, employing an air sac system and a "4-stroke" respiratory pattern to deliver a continuous flow of fresh air across the lung tissue.

These following chapters will hopefully prove useful to students in their quest for understanding the basics of avian anatomy.

CHAPTER 25

HEAD AND NECK

Cynthia M. Faux and Marcie L. Logsdon, Chapter editors

25.1 Infraorbital sinusitis—*Cynthia M. Faux and Marcie L. Logsdon* 1264
25.2 Crop impaction—*Cynthia M. Faux and Marcie L. Logsdon* 1271
25.3 Syringeal obstruction—*Cynthia M. Faux and Marcie L. Logsdon* 1276
25.4 Beak fracture—*C.M. Faux, M.L. Logsdon, and L. Lossi* 1285
25.5 Obstruction of external ear canal—*Cynthia M. Faux and Marcie L. Logsdon* 1299
25.6 Ocular trauma—*Cynthia M. Faux and Marcie L. Logsdon* 1305

CHAPTER 25

CASE 25.1

Infraorbital Sinusitis

Cynthia M. Faux[a] and Marcie L. Logsdon[b]
[a]University of Arizona College of Veterinary Medicine, Oro Valley, Arizona, US
[b]Exotics & Wildlife Department, Washington State University College of Veterinary Medicine, Pullman, Washington, US

Clinical case

Many parrot species have long life spans. When properly fed and housed, African gray parrots typically live for 40–60 years in captivity. Thus, 34 years of age is not considered "old" for an African gray parrot.

History

A 34-year-old female African gray parrot (*Psittacus erithacus*; Fig. 25.1-1) was presented for a "bulging" left eye and associated soft tissue swelling. The swelling was first noticed 2 months before presentation and was slowly worsening. The bird's attitude and appetite were unaffected, although she had a long history of mild feather-destructive behavior ("feather plucking"). She was fed an inadequate diet, consisting mostly of seeds, nuts, and human food.

FIGURE 25.1-1 A healthy African gray parrot.

Comparative Veterinary Anatomy: A Clinical Approach. https://doi.org/10.1016/B978-0-323-91015-6.00116-3

Physical examination findings

The bird was bright, alert, and responsive but nervous. She had poor feather quality overall, with many covert feathers absent, and broken tail feathers that showed signs of "barbering." In addition, she was slightly over-conditioned; her body condition score was 3.5 (1 = emaciated, 5 = obese). There was periorbital swelling and moderate exophthalmos of the left eye, although the direct pupillary light reflex (PLR) was normal and the anterior chamber of the eye was clear. (The avian eye is described in Case 25.6.)

Covert feathers are feathers that cover others. For example, the covert feathers on the dorsal surface of the wings cover the base of the primary and secondary flight feathers, smoothing air flow over the wings during flight. Barbering in birds is the act of biting or removing feathers—the term coming from cutting hair (barbering) in people. It can be a sign of anxiety, boredom, or an underlying medical condition. (Feathers are described in detail in Case 28.1.)

Differential diagnoses

Infraorbital sinusitis (with the accumulation of inflammatory debris); retrobulbar mass, such as a tumor or abscess; vitamin A deficiency

Infraorbital sinusitis is relatively common in pet parrots and may be caused by bacterial or fungal (e.g., *Aspergillus* spp.) infection. In mammals with sinusitis, nasal discharge may be the primary presenting complaint. In birds, however, periorbital swelling is typically the primary or sole clinical sign because of the unique anatomical features of the avian infraorbital sinus.

Diagnostics

Computed tomography of the skull revealed a mass consisting of mixed soft tissue and fluid densities, located medial and rostral to the left globe (Fig. 25.1-2). The mass extended to the frontoparietal bone, and ventrally it appeared to involve the infraorbital sinus. On other CT images, secondary rhinitis was also apparent with fluid seen the normally air-filled nasal conchae.

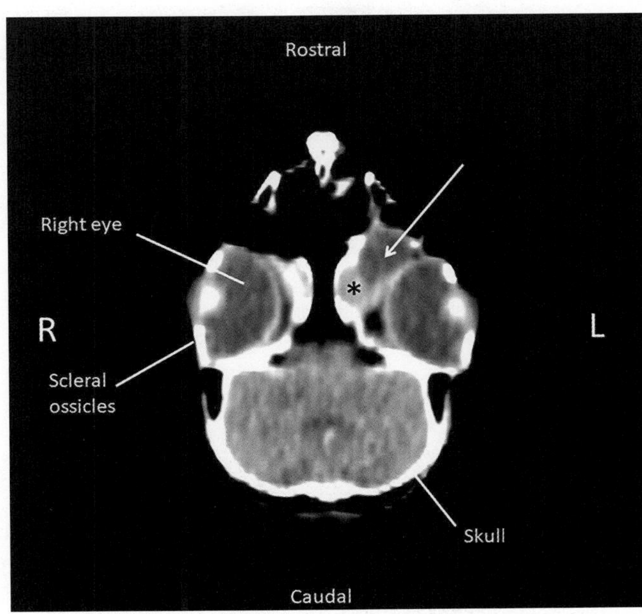

FIGURE 25.1-2 Computed tomographic image of the patient's head. A soft tissue mass (*asterisk*) medial to the left eye is displacing the eye laterally. (Compare the position of the left and right eyes relative to the skull.) In addition, there is a fluid in the infraorbital sinus rostromedial to the left eye (*arrow*).

When sample collection or lavage is required, the infraorbital sinus in birds is usually approached either ventral to the globe or by placing the needle at the commissure of the mouth and directing it dorsally into the sinus (Figs. 25.1-3–25.1-5).

The mass was sampled by performing a fine-needle aspirate under general anesthesia; the needle placement, direction, and depth were guided by the CT images. Aspiration yielded approximately 0.25 mL of clear fluid, which was sent for fluid analysis and cytology. The fluid had low cellularity and low protein concentration, and no bacteria were seen on microscopic examination. A few uniform clusters of epithelial cells were found, but these cells had no cytologic features of malignancy.

The low number of cells obtained from the fine-needle aspirate precluded a definitive diagnosis. However, epithelial cell hyperplasia is a common finding with vitamin A deficiency in birds—a diagnosis that may be assumed in this case based on the bird's poor diet.

Diagnosis

Presumptive diagnosis of epithelial cell hyperplasia and secondary infraorbital sinusitis because of chronic vitamin A deficiency

FIGURE 25.1-3 Needle placement in the infraorbital sinus of a macaw, option 1: Insert the needle through the skin ventral to the eye, dorsal to the jugal bar (see Fig. 25.1-5), and direct it medially and slightly rostrally. (Note: This procedure requires proper restraint to avoid injuring the eye.)

VITAMIN A DEFICIENCY IN BIRDS

Hypovitaminosis A is common in pet birds fed an inappropriate diet, such as one consisting predominantly of seeds. Adequate vitamin A is required for healthy epithelial tissues, including the integument (skin and feathers) and the respiratory epithelium, so chronic vitamin A deficiency often leads to epithelial cell metaplasia or hyperplasia. Common clinical signs include thickening and blunting of the choanal papillae in the roof of the mouth (see Case 25.3), hyperkeratosis and/or loss of normal texture on the foot pads (see Case 27.3), and accumulation of debris in the nostrils, often with secondary infection of the nasal cavity and/or infraorbital sinus.

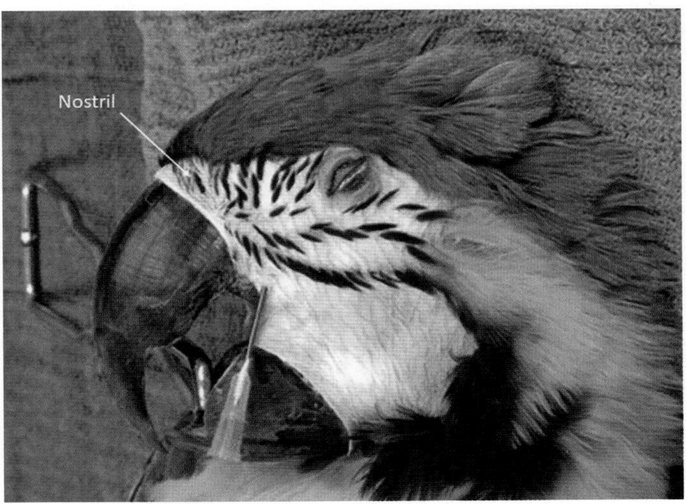

FIGURE 25.1-4 Needle placement in the infraorbital sinus of a macaw, option 2: Insert the needle through the skin at the commissure of the mouth and direct it toward a point midway between the eye and the nostril. Keep the needle as parallel to the side of the head as possible as it is passes under the jugal bar (see Fig. 25.1-5) into the sinus.

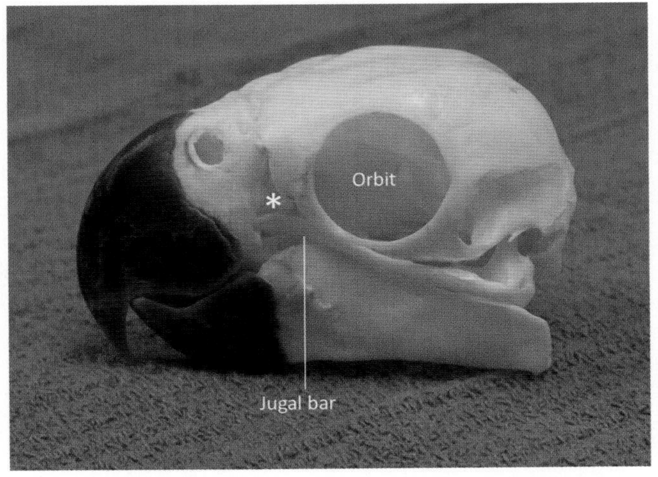

FIGURE 25.1-5 Skull of a Lesser sulfur-crested cockatoo, showing the main portion of the infraorbital sinus (*asterisk*) and the bony landmarks for needle placement.

Treatment

The owners declined further diagnostics and elected conservative management based on the presumptive diagnosis. They were resistant to changing the bird's diet, so the patient was sent home on an indefinite course of anti-inflammatory therapy, a short course of antimicrobial therapy, and oral vitamin A supplementation.

If neoplasia had been confirmed, then nonsurgical treatment modalities such as radiation therapy would have been recommended. As described in Case 25.6, enucleation carries a high risk of life-threatening hemorrhage in birds, as would surgical excision of the retrobulbar mass in this case.

Anatomical features in avian species

Infraorbital sinus (Figs. 25.1-5 and 25.1-6)

A paranasal sinus (Gr. *para-* beside, alongside) is a cavity within the skull that lies adjacent to, and communicates with, the nasal cavity. The infraorbital sinus (L. *infra-* below + *orbita* the eye in this context) lies ventral and rostral to the eye. It is broadly equivalent to the maxillary sinus in mammals. In zoological literature it is sometimes called the antorbital sinus (L. *ante-* before).

The paired **infraorbital sinus** (left and right) is the only paranasal sinus in birds. It is best characterized in psittacines (e.g., parrots) and galliformes (e.g., chickens), as infection of this sinus is common in these groups.

The infraorbital sinus is a large, roughly pyramidal-shaped cavity on the lateral aspect of the skull, rostroventral to the globe, lying within the **infraorbital fossa**. The bony limits of the fossa, while incomplete and quite variable among species, generally include the **palatine bone** ventromedially, the **jugal** and **lacrimal bones** laterally (ventro- and dorsolateral, respectively), and portions of the **nasal bone** (which supports the upper beak) rostrally.

The rostral aspect of the sinus extends toward, and in some species even into, the upper beak. For example, in Amazon parrots, the sinus extends rostrally into the beak. The caudal extent of the sinus, where the sinus wraps around the eye, also varies by species. In parrots, the sinus extends ventral and caudal to the eye, and a transverse canal connects to the contralateral sinus (on the other side of the head).

As skin forms its lateral boundary, distension of the infraorbital sinus with fluid or cellular material is readily discernible clinically as a swelling on the side of the face rostroventral to the eye. Such swellings typically are unilateral, although bilateral facial swelling may be seen with bilateral sinusitis and in species in which the left and right infraorbital sinuses normally communicate (e.g., parrots).

However, a distinctive feature of this sinus in birds is that its dimensions are largely defined by soft tissues rather than bone. It is bounded caudodorsally by the globe, ventrally by the oral cavity (specifically, the roof of the mouth), medially by the wall of the nasal cavity, and laterally by the skin of the head.

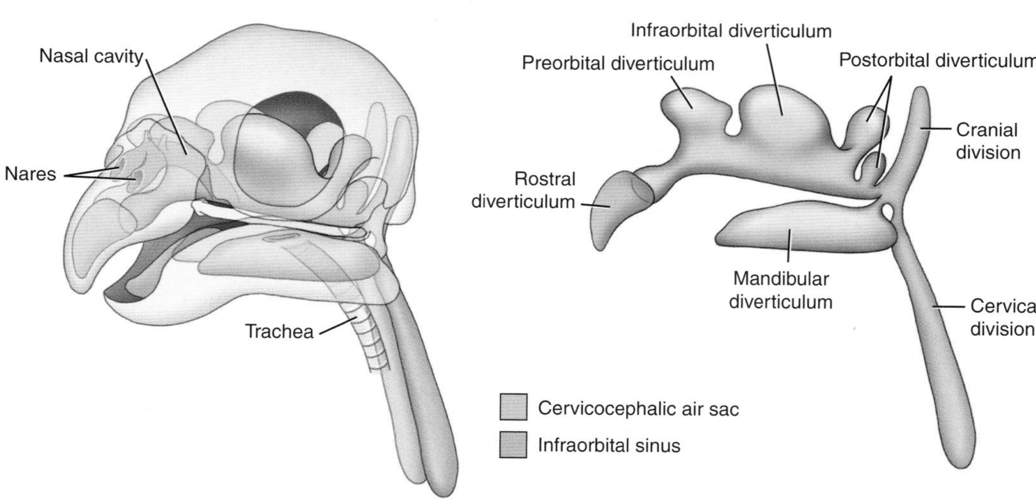

FIGURE 25.1-6 Outline of the infraorbital sinus and its extensive diverticula (rostral, pre- and postorbital, axial infraorbital, and mandibular).

Each infraorbital sinus communicates with the nasal cavity via 2 bony openings, both of which are situated in the dorsal aspect of the sinus cavity. One opens into the dorsal aspect of the nasal cavity dorsal to the choanal opening, and the other opens into the ventral aspect of the **caudal nasal conchae**, which are in the caudodorsal aspect of the nasal cavity. Thus, the infraorbital sinus in birds lacks a direct ventral avenue for drainage. The sinus is lined by stratified squamous epithelium rostrally, transitioning to the typical respiratory mucosa (ciliated columnar epithelium) caudally.

As in mammals, sinusitis in birds may result in excessive mucus production and fluid accumulation within the sinus cavity. But unlike mammals, this mucus material cannot drain from the sinus via a ventral aperture into the nasal cavity.

Depending on the species, various **diverticula** of the infraorbital sinus "pneumatize" the skull.

These diverticula create recesses which can be extensive and communicate with the cervicothoracic air sacs (see below), potentially complicating the diagnosis and treatment of sinusitis, particularly in psittacines such as parrots. Infraorbital sinusitis can result in occlusion and abnormal hyperinflation of these diverticula (see Fig. 25.1-7).

FIGURE 25.1-7 A pair of cockatiels (*Nymphicus hollandicus*). The bird on the left of the image has a chronic hyperinflation of the cervicocephalic air sac, resulting in chronic, regional subcutaneous emphysema. The bird on the right is normal.

NASAL (SALT) GLAND

Some avian species, including budgerigars and chickens, have a paired nasal gland, left and right, that lies within the orbit, dorsonasal (dorsomedial) to the globe. Its duct penetrates the frontal bone and enters the nasal cavity. The gland is bilobed in many birds, although chickens have only a single medial lobe.

The nasal gland actively secretes sodium chloride—therefore, it is also referred to as the salt gland—which assists in osmoregulation (regulation of extracellular sodium concentration) by providing an extra-renal mechanism for salt excretion. This renal "assist" is particularly important in marine and some desert birds. For example, in pelagic ("open sea") marine birds with only sea water to drink, the nasal gland allows them to remain at sea without returning to land for fresh water.

Associated structures

The **infraorbital nerve** travels rostrally within the medial wall of the infraorbital sinus. It is variably noticeable during sinusoscopy or sinusotomy, depending on the species. This **maxillary branch** of the **trigeminal nerve** (CNV) innervates the rostral portion of the lower eyelid, the conjunctiva, and the base of the nictitating membrane.

The **nasolacrimal duct**, the confluence of the **dorsal** and **ventral nasolacrimal canaliculi** that drain the **conjunctival sac**, also passes within the medial wall of the infraorbital sinus. It drains into the middle compartment of the nasal cavity, dorsal to the choanal opening. (The avian nasal cavity is discussed in Case 25.3.)

Cervicocephalic diverticula

The cervicocephalic diverticula are not known to be connected to the pulmonary air sac system (see Case 26.3). However, the extent of these diverticula in some species can complicate treatment of conditions involving the head or neck.

In addition to diverticula of the infraorbital sinus, the avian skull may be pneumatized by diverticula of the **tympanic cavity** (ear canal). Collectively, these air-filled structures may be extensive, in some species continuing subcutaneously down the neck to the shoulder region; hence, the term **cervicocephalic diverticula** is used for these structures. In parrots, these diverticula can extend down the neck, around the crop, and across the shoulders.

Selected references

[1] Jankowski G. Hypovitaminosis. In: Tully T, editor. Clinical veterinary advisor: birds and exotic pets. St Louis: Elsevier; 2013. p. 199–201.
[2] Artmann A, Henninger W. Psittacine paranasal sinus—a new definition of compartments. J Zoo Wildl Med 2001;32:447–58.
[3] Heard DJ. Avian respiratory anatomy and physiology. Seminars in Avian and Exotic Pet Medicine 1997;6:172–9.
[4] Baumel JJ, King AS, Breazile JE, Evans HE, Vanden Berge JC, editors. Handbook of Avian Anatomy: Nomina Anatomica Avium. 2nd ed. Cambridge, MA: Nuttall Ornithological Club; 1993.

CHAPTER 25

CASE 25.2

Crop Impaction

Cynthia M. Faux[a] and Marcie L. Logsdon[b]
[a]University of Arizona College of Veterinary Medicine, Oro Valley, Arizona, US
[b]Exotics & Wildlife Department, Washington State University College of Veterinary Medicine, Pullman, Washington, US

Clinical case

History

A 10-day-old Goffin's cockatoo (*Cacatua goffini*) was presented for a suspected crop impaction. The chick was being hand-raised by the owner and was housed on wood (aspen) shavings. The chick was seen performing a characteristic feeding response directed against the aspen bedding material. The chick was still being successfully fed at frequent intervals, but was consuming smaller than usual amounts of its hand-feeding formula.

> Young parrots display a characteristic feeding response designed to stimulate the regurgitation of crop contents by the parents. In response to the parent (or a surrogate feeding instrument such as a spoon or syringe), the young bird extends its neck, raises its head, and performs a pulsing motion with its mouth.

Physical examination findings

The chick was mostly featherless, appropriate for the age. It was bright, alert, and responsive, and displayed a strong feeding response. The crop was visibly and palpably distended with foreign material suspected to be wood shavings (Fig. 25.2-1). The rest of the physical examination was normal.

Differential diagnoses

Crop impaction (foreign body or ingluviolith), *Candida* or *Trichomonas* plaque, abnormal thickening of the crop wall

Diagnostics

The diagnosis was made from the history and physical examination. In young birds, the skin and crop wall are so thin that they are translucent, so it is easy to see and feel foreign material in a distended crop. In an older bird,

> **CROP IMPACTION**
>
> Impaction of the crop (the ingluvies) is most common in young parrots that have not developed the ability to differentiate between food and nonfood components. Often, it is nest material that causes an impaction. Occasionally, older parrots present with foreign body impactions of the crop, consisting of ingested perch material (usually from older, frayed sisal or rope perches). Chickens occasionally ingest large quantities of mowed grass, which may cause a crop impaction. In contrast, ingluvial stones (ingluvioliths) are very rare.

Comparative Veterinary Anatomy: A Clinical Approach. https://doi.org/10.1016/B978-0-323-91015-6.00117-5

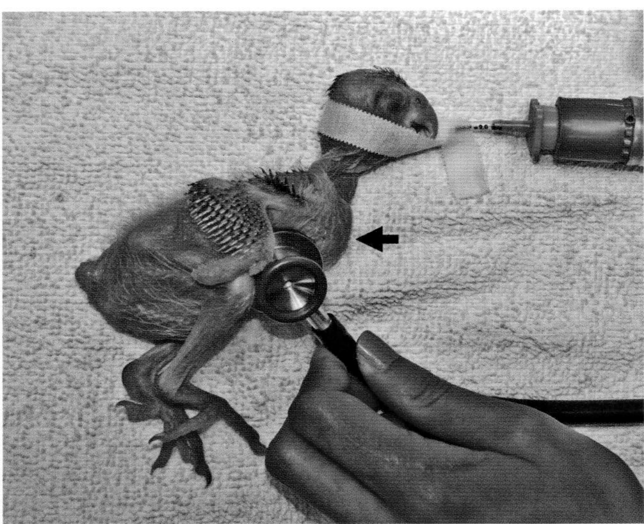

FIGURE 25.2-1 The Goffin's cockatoo chick, anesthetized in preparation for surgery. The crop (*arrow*) is grossly distended with material that was subsequently confirmed to be the wood shavings used as bedding.

positive-contrast radiography or endoscopic examination of the esophagus can be helpful in confirming and characterizing an impaction.

Diagnosis

Crop impaction

Treatment

Under general anesthesia, the wood shavings were surgically removed via ingluviotomy (surgical incision into the crop) (Figs. 25.2-2 and 25.2-3). Following recovery from anesthesia, the chick was monitored closely and administered oral dextrose until it was strong enough to eat the hand-feeding formula. The chick made a full recovery.

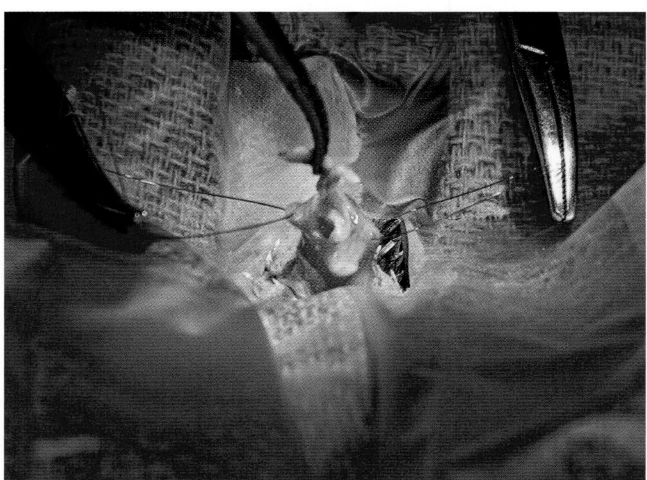

FIGURE 25.2-2 Intra-operative photo of the wood shavings being removed from the chick's crop via an ingluviotomy. A stay suture was placed through the crop wall on each side of the incision to stabilize and exteriorize the crop during the surgery and minimize the risk of contamination of the surrounding tissues with crop contents. Note the translucent wall of the crop is very thin.

FIGURE 25.2-3 The wood shavings that were removed from the chick's distended crop.

Anatomical features in avian species

Introduction

The crop, or ingluvies, is a diverticulum in the cervical portion of the esophagus in birds. The crop is neither universal nor uniform among avian species. Seagulls, penguins, ostriches, owls, and geese are examples of birds that lack a distinct crop. Waterfowl generally have only a rudimentary crop, which is little more than a long, shallow dilation of the cervical esophagus. The Griffon vulture has a simple, open diverticulum off the esophagus. Galliformes (chickens, turkeys, quail, pheasants, etc.), pigeons, and parrots are examples of birds that have a distinct crop with a well-developed, sac-like diverticulum off the esophagus.

The size and shape of the crop is species-specific and reflects each species' adaptation to diet, environment, and feeding behavior. Generally, species with a crop are adapted for rapid feed ingestion under unsafe conditions, allowing for later digestion in a safer place. Thus, omnivores and herbivores, including granivores (grain-eaters), have a larger crop than carnivores.

Function

An important function of the crop is the temporary storage of recently ingested food. In addition to the behavioral considerations and feeding patterns, the presence of food in the stomach (proventriculus and/or ventriculus) determines how soon after ingestion the crop empties into the distal esophagus and thus into the proventriculus. Some microbial fermentation begins in the crop, but as food may reside in the crop for as little as a few seconds—and a few hours at most—its contribution to digestion and gastrointestinal health is relatively small. The crop is moderately vascular (Fig. 25.2-2), so nutrient absorption (glucose, some amino acids, and vitamins) occurs across its mucosa, although this too is limited by the normally brief transit time of food in the crop.

"Sour crop" is a microbial overgrowth and fermentation of ingesta within the crop. The condition can occur in both granivorous or carnivorous species due to delayed crop emptying, primarily caused by spoiled or rotten food, or secondarily due to other systemic disease.

Many avian species also use the crop to feed their nestlings. In these birds, ingested food is regurgitated to the young or obtained directly from the parents' mouth by the chick. Some species, notably pigeons, doves, and flamingos, secrete "crop milk," a thick liquid derived from ingested food mixed with desquamated cells of the crop lining. In these birds, both parents (male and female) produce crop milk, under the influence of prolactin, which induces hyperplasia of the crop mucosa. Thickening of the crop wall is normal in these species when they are raising chicks.

Lastly, the crop may be used as an acoustic resonance chamber to amplify the voice and/or may be filled with air as a display, such as during courtship, as is observed in some species of pigeons, grouse, and cranes.

Crop (ingluvies) (see Landscape Figs. 6.0-5 and 6.0-6)

The position of the **crop (ingluvies)** varies somewhat with species. For most of its course from pharynx to thoracic inlet, the cervical part of the avian **esophagus** lies to the right of midline, ventral to the jugular vein, vagus nerve, and thymus, and dorsal to the trachea. As it approaches the thoracic inlet (immediately cranial to the clavicle and pectoral muscles), the esophagus shifts to midline, where it expands ventrolaterally to form the crop. In most species, such as galliformes and falcons, the crop mainly lies to the right of the trachea. Pigeons and doves have a pair of large, lateral diverticula (one on each side of the trachea; Fig. 25.2-4) and an additional small, median sac. Parrots, too, have a pair of lateral diverticula, although the right is generally larger than the left.

A cranial, left-lateral approach is recommended when performing an ingluviotomy. Incising the crop to the left of midline and as far cranially as possible minimizes the amount of tension on the sutured incision when the crop is filled with food. It also ensures that post-operative feeding, using an oroesophageal feeding tube, will not strain the incision.

The wall of the crop has the same general structure as the esophagus in birds and mammals: mucosa, lamina propria, muscularis, and adventitia. It is best characterized in the chicken, as follows: (1) The mucosa consists of incompletely keratinized, stratified squamous epithelium; (2) the thin lamina propria mostly comprises connective tissue, although the portion nearest the esophageal wall contains mucous glands which empty onto the mucosal surface of the crop; (3) the muscular layer is composed of smooth muscle fibers arranged as an inner circular and an outer longitudinal layer; (4) the adventitia is attached to the overlying skin and the clavicle by connective tissue, and to the sternum by small, superficial muscles that help maintain its position.

On its inner surface, the empty crop has multiple longitudinal folds that allow it to expand with food when necessary. Although the crop is distensible, its wall is thin and is easily traumatized.

In parrots and pigeons, the muscular layer of the crop may function similarly to esophageal sphincters in mammals, thereby controlling the flow of food into the stomach.

When removing foreign material from the crop, atraumatic forceps and tissue handling techniques must be used to avoid rupture of the crop wall. Because the wall is so thin, attention must also be taken to ensure that the crop wall is sutured independently of the overlying skin.

Blood supply and innervation

In the pigeon, the crop and adjacent esophagus are supplied by the left and right **ingluviales arteries**, which are branches of the **vagus artery** off the **common carotid**. Numerous small veins drain the cervical esophagus and crop, emptying into the **jugular vein**. In the chicken, the cervical esophagus is innervated by an **esophageal plexus**, formed by branches of the **glossopharyngeal nerve** (CN IX) and the **hypoglossal nerve** (CN XII). Near the crop,

FIGURE 25.2-4 A Eurasian collared dove (*Streptopelia decaocto*) nestling, following gavage (forced feeding via a "stomach" tube) feeding. The enormous crop occupies the entire cranial aspect of this young bird's body, from just beneath her beak to her feet, expanding on both sides of midline to create a "cleavage." This nestling weighed 110 g and her crop easily accommodated 28 mL of liquid food by gavage. (The normal gavage rate in young birds is 3 mL per 100 g bodyweight.)

branches of CN IX anastomose with recurrent branches of the **vagus nerve** (CN X) and innervate the crop and adjacent esophagus. This complex innervation is consistent with a well-coordinated and integrated motility pattern in the esophagus and crop.

Selected references

[1] Kierończyk B, Rawski M, Długosz J, Świątkiewicz S, Józefiak D. Avian crop function—a review. Ann Anim Sci 2016;16:1–26.
[2] McLelland J. Aves digestive system. In Sisson and Grossman's The anatomy of the domestic animals 5th ed. Getty R (ed). W.B. Saunders Co., Philadelphia. pp. 1866–1868.

CHAPTER 25

CASE 25.3

Syringeal Obstruction

Cynthia M. Faux[a] **and Marcie L. Logsdon**[b]
[a]University of Arizona College of Veterinary Medicine, Oro Valley, Arizona, US
[b]Exotics & Wildlife Department, Washington State University College of Veterinary Medicine, Pullman, Washington, US

Clinical case

History

A 9-year-old male Harris's hawk (*Parabuteo unicinctus;* Fig. 25.3-1) used for falconry was presented for acute onset of dyspnea and dysphonia after a successful hunt in which he caught a large rabbit. In capturing the rabbit, the hawk was observed to engage in a struggle with his prey on the dusty ground, in which he was knocked into a sagebrush bush. His appetite had been normal, although his owner noticed that it took the bird longer than normal to eat his food, and he frequently paused while eating to catch his breath. He was up to date on his West Nile virus vaccination.

> Harris's hawks, also called Bay-winged hawks, are commonly kept for use in falconry, which is the art of hunting wild game with the use of a trained raptor.

Physical examination findings

The bird was bright, alert, and responsive, and his body condition score was 3 (1 = emaciated, 5 = obese). His oral mucous membranes were pink and moist, and his laryngeal inlet appeared normal. However, a faint wheeze was audible during inspiration, and he began open-mouth breathing with even slight exertion or stress. On auscultation, the wheeze was most prominent over the cervical trachea at the thoracic inlet.

> Thorough auscultation of the respiratory tract in birds involves ausculting the cervical trachea as well as the lungs and the extensive system of coelomic air sacs (described in Case 26.3). If upper airway obstruction is suspected, it is useful to auscult the thoracic inlet with the bird's beak both open and closed, if possible. If the wheeze disappears when the bird's mouth is open, the obstruction is likely confined to the nasal passages or choana (caudal opening of the nasal cavity). In this case, the wheeze persisted when the patient was open-mouth breathing, indicating that the obstruction was located caudal to the choana.

ASPERGILLOSIS

Aspergillus fumigatus is an opportunistic fungus that is ubiquitous in the environment. Aspergillosis (the disease caused by this fungus) may be systemic or confined to the respiratory tract, depending on the manner of exposure and the immune status of the patient. Raptors, penguins, waterfowl, and certain species of parrots are especially prone to aspergillosis. The condition is often precipitated by a stressful change in husbandry or being housed in a warmer environment than is natural for the species.

Comparative Veterinary Anatomy: A Clinical Approach. https://doi.org/10.1016/B978-0-323-91015-6.00118-7

FIGURE 25.3-1 The Harris's hawk in this case. (Photo courtesy of Henry Moore Jr., Biomedical Communications Unit, Washington State University.)

Differential diagnoses

Traumatic injury to the larynx or trachea, aspergillosis (specifically, fungal granuloma in the trachea or syrinx), air sacculitis (fungal or bacterial), pneumonia, respiratory parasitism, inhalation of a foreign body

Diagnostics

The bird was anesthetized for diagnostic evaluation. Whole-body radiographs were considered within normal limits. Routine blood test results were consistent with acute inflammation and muscle injury (i.e., trauma sustained on impact). His serum *Aspergillus*-specific antibody titer was interpreted as moderately positive but nondiagnostic. His serum *Aspergillus* galactomannan assay (test for a polysaccharide produced by this fungus) was weak-positive, supportive of normal environmental exposure.

> This test is useful in the diagnosis of aspergillosis, but it must be interpreted in combination with the signalment. In some species, such as parrots, any positive titer is significant and likely reflects true infection. In other species, including raptors, positive titers are common and result from normal environmental exposure.

Coelomic endoscopy revealed generalized inflammation of the caudal thoracic and abdominal air sacs, but no fungal plaques or parasites were evident.

Screening tracheal endoscopy revealed mucosal inflammation and mucus accumulation in the syrinx, worse on the left side (opening to the left primary bronchus; Fig. 25.3-4). A mucosal swab was subsequently positive on bacterial culture for *Ornithobacterium* spp., an opportunistic or secondary bacterial pathogen in birds; however, fungal culture was negative.

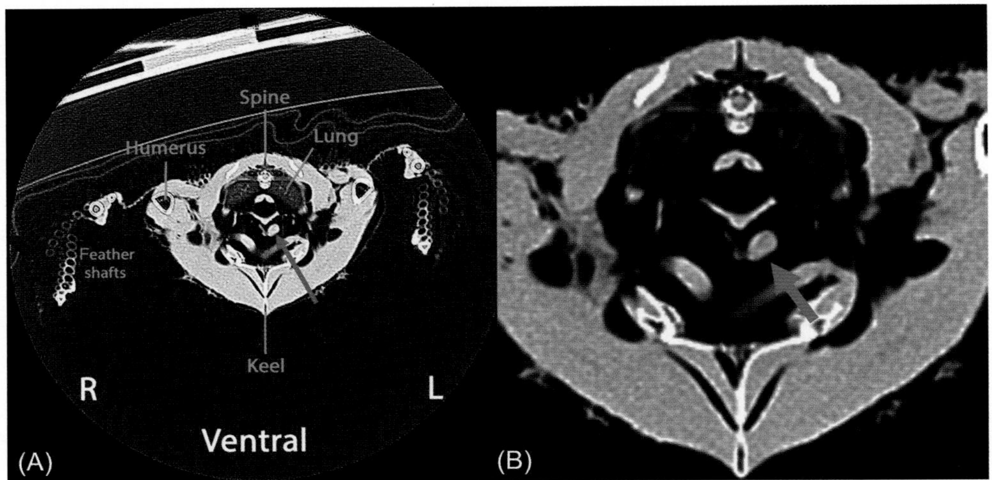

FIGURE 25.3-2 Transverse (CT) images of the hawk described in this case. (A) Whole body CT, at the level of the syrinx. The syringeal lumen on the bird's left side, which is the origin of the left primary bronchus, is completely occluded (*pink arrow*). (B) Enlarged image. In addition to occlusion of the syrinx on the left side (*pink arrow*), the adjacent medial tympanic membrane on the right side is visible dorsomedially. The rest of the right syringeal wall is not seen on this image.

Computed tomography revealed blockage of the left primary bronchus, starting at the syrinx and extending approximately 7 mm (0.3 in.) caudally into the bronchus (Figs. 25.3-2 and 25.3-3). On the enhanced CT image (Fig. 25.3-2B), the obstruction appeared to consist of a dense core surrounded by less dense material (presumed to be mucus and inflammatory debris).

Diagnosis

Syringeal foreign body

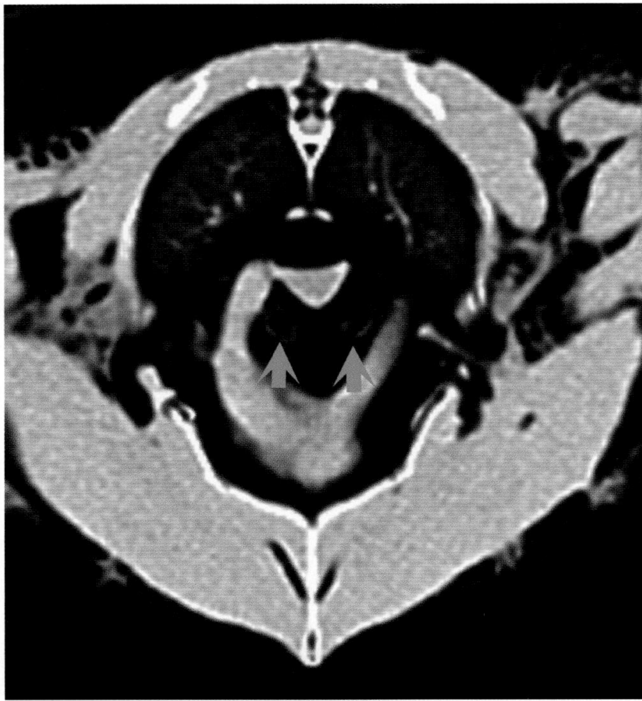

FIGURE 25.3-3 Transverse image just caudal to that shown in Fig. 25.3-2. Both left and right primary bronchi are normal (*green arrows*), indicating that the obstruction was confined to the syrinx and first few millimeters of the left primary bronchus.

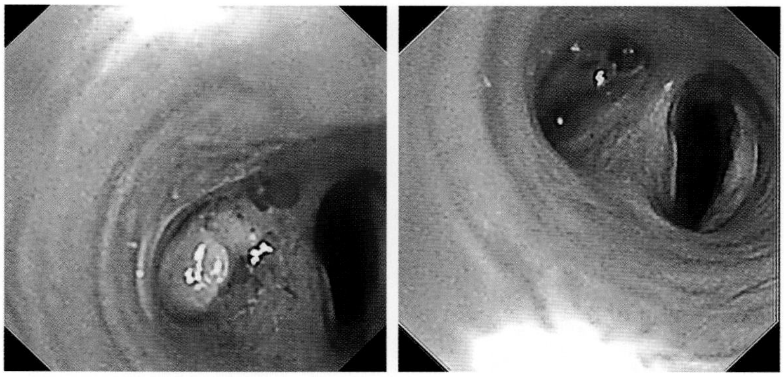

FIGURE 25.3-4 Still images of the syrinx, taken during tracheoscopy. Left: Before the removal of the obstruction. The entire syrinx is inflamed, but only the left side is occluded. (The bird was in dorsal recumbency for this procedure, so the bird's left side is on the left in this view, which looks down at the syrinx from the trachea.) Right: During endoscopic removal of the obstruction; multiple passes with biopsy forceps were required to clear the obstruction.

Treatment

With the bird still under general anesthesia, the obstruction was removed via tracheal endoscopy (Fig. 25.3-4). It consisted of a plug of mucus containing a small amount of inorganic material, presumed to be ground substrate (coarse dirt or gravel) that was inhaled during the hunt.

The patient showed an immediate improvement on recovery from anesthesia. The prophylactic course of antimicrobial and antifungal drugs started at admission were completed, and the hawk returned to his normal hunting capacity, although a subtle voice change persisted.

Respiratory infections can progress rapidly in birds, so any treatment delay while awaiting diagnostic test results can be severely detrimental. Antimicrobial and antifungal drugs were administered both systemically and via nebulization to ensure that therapeutic concentrations were achieved in the tracheal lumen, syrinx, bronchi, and air sacs.

Anatomical features in avian species

Introduction

This section covers the upper respiratory tract, from the nares (nostrils) to the syrinx. The bronchi, lungs, air sacs, and respiratory muscles are described in Case 26.3.

Function

The main function of the syrinx is to vocalize and support the pulmonary airway during respiration. The syrinx minimizes the collapse and compression of the principal bronchi primarily during expiration.

Nares and nasal cavity

In most species of birds, the paired **nares** (singular, **naris**) consist of round, ovoid, or slit-like openings in the upper beak (Fig. 25.3-5). Commonly, the nares are located near the base of the beak. However, in kiwis, they are located at the tip of the beak, and some diving birds, such as gannets, have no external nares at all. Numerous species, including galliformes (chickens and other landfowl), have a stiff, horny **operculum** within the dorsal part of each

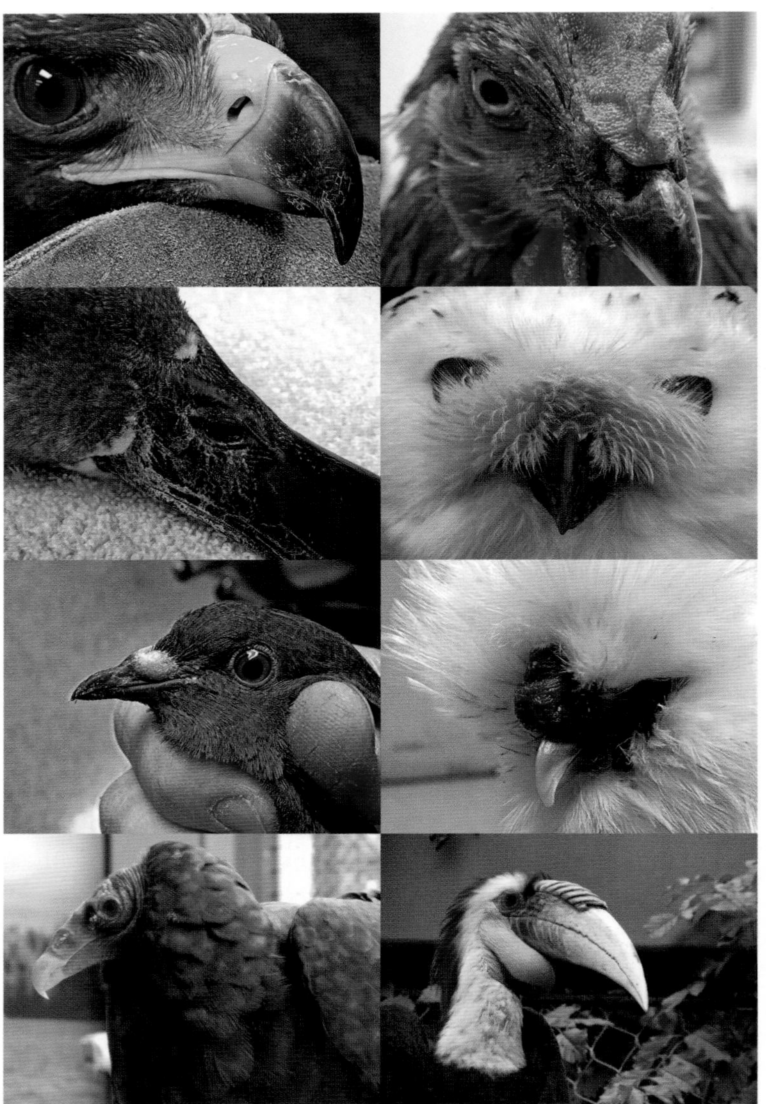

FIGURE 25.3-5 Variations in the nares among avian species. Top left: Golden eagle. Top right: Domestic chicken. Note the horny operculum covering the dorsal aspect of the nares. Row 2, left: Domestic duck. Row 2, right: Snowy owl. An extensive spray of bristle-like feathers protects the exposed nares and provides sensory input. Row 3, left: Juvenile common pigeon. The fleshy, pink cere will decrease in relative size as this young bird matures. Row 3, right: Silkie rooster. This chicken breed has been intensively selected for its elaborate plumage and comb, which in this bird almost completely covers the nares. Bottom left: Turkey vulture. There is no septum between the left and right nares in this species, so you can see straight through the nares to the other side and beyond. Bottom right: Wreathed hornbill. The elaborate covering over the nares is called the casque. It is thought to aid in vocalizations; it also provides a degree of sexual dimorphism in some hornbill species.

naris, which helps prevent foreign material from entering the nasal cavity. In diving birds, the operculum seals the naris closed when the bird dives to prevent water from entering the nasal passages.

The nares may be surrounded by feathers, bristles, fleshy skin, or simply the horny tissue of the beak. Some species, such as budgerigars ("budgies") and pigeons, have a fleshy **cere** around the nostrils. In budgies, the color of the cere is sexually dimorphic (see Case 26.5) while in pigeons, it is not.

The **nasal cavity** is divided by bony septa into three compartments: **rostral** (vestibule), **middle** (respiratory), and **caudal** (olfactory). Rostral, middle, and olfactory **conchae** fill these respective compartments, all of which communicate via small ostia (bony portals). Olfactory epithelium lines the caudal compartment. A complete **bony**

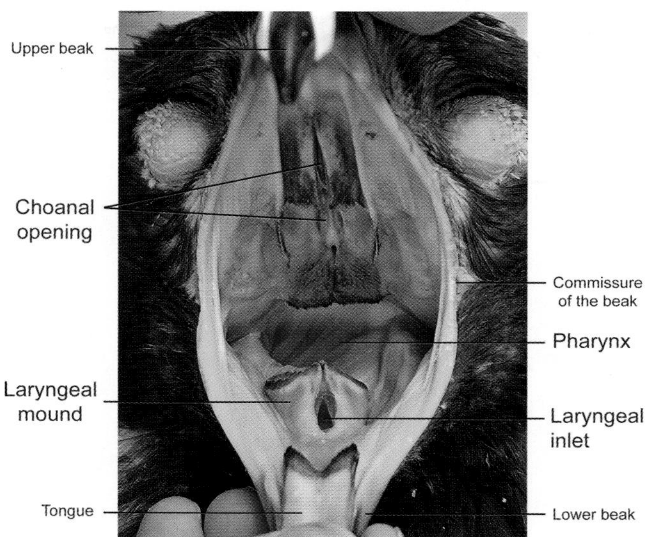

FIGURE 25.3-6 Oropharynx of an anesthetized golden eagle. The beak is held open to show the slit-like choanal opening in the roof of the mouth and the glottis, just caudal to the base of the tongue. The glottis aligns with the choanal opening when the beak is closed, leading to the trachea.

septum divides the nasal cavity into left and right sides in many species. The middle compartment is the largest of the 3, and both left and right sides communicate with the oral cavity and pharynx via a single median, slit-like **choanal opening** in the roof of the mouth (Fig. 25.3-6).

In avian patients, the choanal slit is often swabbed to collect upper respiratory secretions for laboratory analysis (e.g., cytology, microbiology). It is accessed by opening the beak and visualizing the roof of the mouth.

Cervicocephalic diverticula or air sacs "pneumatize" the skull. These diverticula, which are aerated via the nasal cavity, are extensive in some species. They are described further in Case 25.1.

Pharynx

Birds lack a soft palate, so the **pharynx** is not divided into oropharynx and nasopharynx as it is in mammals. Food is prevented from entering the choanal opening when the bird swallows by the combined actions of the tongue and the pharyngeal muscles.

Larynx

In birds, the **larynx** lies on the floor of the pharynx, and the slit-like **laryngeal inlet** (glottis) faces rostrodorsally or dorsally, toward the choanal opening (see Fig. 25.3-6). The glottis is in the center of a broad, triangular **laryngeal mound**, which is situated just caudal to, and in some species is partially covered by, a transverse mucosal fold at the base of the tongue. Deep to the glottis, the lumen of the larynx opens into a broad, dorsoventrally flattened space, the **laryngeal cavity**, which continues caudally as the trachea.

Birds have paired **arytenoid cartilages** and unpaired **cricoid** and **procricoid cartilages**. Portions of the laryngeal cartilages may ossify, especially the procricoid and body of the arytenoids. Birds do not have an epiglottis nor a thyroid cartilage. Birds also lack the laryngeal ("vocal") folds and lateral ventricles used for vocalization in mammals.

Remember that birds lack an epiglottis, so when examining the oral cavity and pharynx, the glottis is readily visible at the base of the tongue.

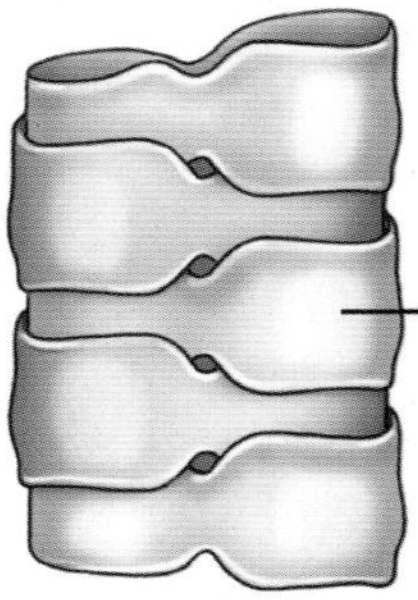

Cartilaginous ring

FIGURE 25.3-7 Overlapping and interlocking avian tracheal rings.

Trachea

The **trachea** in birds is longer than that of a similarly sized mammal. The increase in resistance to airflow caused by the increase in length is offset by a larger relative tracheal diameter. The trachea narrows slightly cranial to the **syrinx**, but up to that point the avian trachea is proportionately wide, and the shape of the syrinx facilitates airflow. In addition, the cartilaginous tracheal rings in birds are complete, which prevents tracheal narrowing under inspiratory or external pressure.

Because the tracheal rings in birds are complete, it is inadvisable to inflate the cuff when using a cuffed endotracheal tube during anesthesia, because it may lead to pressure necrosis of the tracheal mucosa and potentially the cartilage rings also. Noncuffed endotracheal tubes are available.

Only a few of the **tracheal rings** in birds are "simple" (plain rings of cartilage); the majority have alternating and overlapping narrow and wide portions, interlocking with adjacent rings as shown in Fig. 25.3-7. This arrangement provides both strength and flexibility to the trachea, resisting collapse while allowing the impressive range of motion in the neck that is characteristic of birds.

A **tracheal loop**, or **ansa trachealis**, is found in swans, cranes, and various other species. The extra loop(s) of trachea is accommodated within or adjacent to the sternum in most species, external to the coelomic cavity. The trachea then proceeds through the thoracic inlet to the syrinx just as in birds that lack tracheal loops (Fig. 25.3-8).

TRACHEOSCOPY IN BIRDS

Tracheal endoscopy (tracheoscopy) in birds can sometimes be performed under inhalation anesthesia delivered via facemask, which is removed briefly to allow the passage of the endoscope. However, longer procedures require placement of a cannula into a caudal thoracic or abdominal air sac. Anesthetic gas is administered via the cannula using positive-pressure ventilation (oxygenated air and inhalant anesthetic are pumped into the respiratory system under controlled pressure).

Air sac cannulation may also be used in birds with upper airway stricture or blockage. Therapeutically, it accomplishes the same goal as tracheostomy (surgically created opening, or stoma, in the trachea) in mammals.

In birds, the neck is comparatively long and flexible, and the trachea narrows just cranial to the syrinx. It is thus necessary to extend the bird's neck, and therein straighten the trachea, to avoid tracheal trauma if using a rigid endoscope for tracheoscopy.

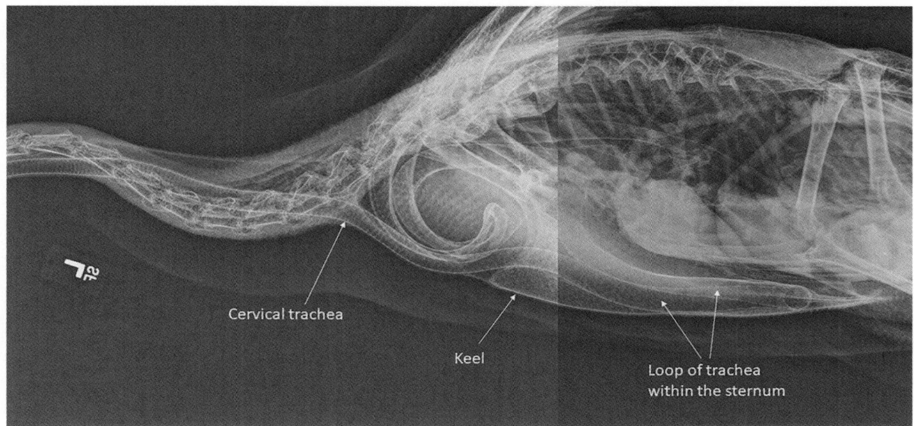

FIGURE 25.3-8 Lateral radiograph of a Trumpeter swan (*Cygnus buccinator*) showing the extremely elongated tracheal loop within the sternum. (Courtesy Washington State University, College of Veterinary Medicine Radiology Department.)

A couple of other species variations are of clinical note. Emus have a large diverticulum in the ventral aspect of the cervical trachea, just cranial to the thoracic inlet, which the bird inflates to generate a "booming" sound for display during the breeding season. New World vultures have a similar diverticulum. In penguins, the caudal trachea is divided by a septum, which effectively means that the tracheal bifurcation occurs further cranially (in the cervical trachea) than in other birds.

In emus and vultures, the neck must be wrapped for general anesthesia to prevent the tracheal diverticulum from inflating. In penguins, the endotracheal tube must not be advanced as far caudally as in other birds, because it is possible for the tube to pass into a single primary bronchus and thus ventilate only one lung.

Syrinx

In most avian species, the trachea terminates as the **syrinx**, which is the vocal organ of birds. There is great variability of the syrinx among species, and even between males and females of the same species. The syrinx may be tracheal, bronchial, or tracheobronchial (the predominant form).

Because the airway narrows at the bifurcation, the syrinx is a relatively common site of airway obstruction.

The chicken is often used as the prototypical bird for the description of the syrinx, which in this species is tracheobronchial: i.e., the syrinx is situated at the tracheobronchial bifurcation. It is formed of 4 cartilaginous components: the **cranial**, **intermediate**, and **caudal cartilages**, and the **pessulus** (Fig. 25.3-9). The **cranial cartilages** are a continuation of the tracheal rings, although here the diameter of the airway widens, so this region is referred to as the **tympanum**. The **intermediate cartilages** follow and form a slight narrowing of the syrinx as it bifurcates into the left and right primary **bronchi**. The intermediate cartilages are C-shaped and attach to the **pessulus**, a triangular-shaped cartilage that lies at the base of the bifurcation and supports the medial tympanic membranes (described below). The **caudal cartilages** are paired (left and right) and form the origins of the primary bronchi. The caudal cartilages are also incomplete rings. Variably, the cartilages of the syrinx may ossify.

Vocalization is produced by the vibration of 2 pairs of thin membranes which form part of the wall of the syrinx. The **lateral tympanic membrane** extends from the caudal edge of the last intermediate cartilage to the cranial edges of the first caudal cartilages, attaching dorsally and ventrally to the pessulus as well. The pair of **medial tympanic membranes** form the medial walls of the primary bronchi at their origins. These medial membranes attach to the

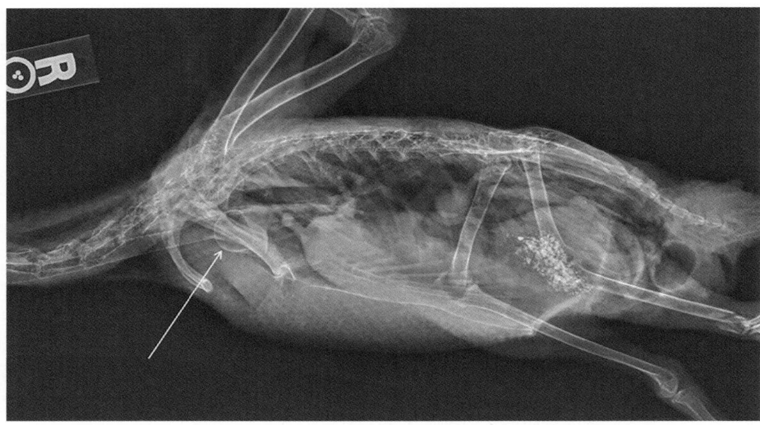

FIGURE 25.3-9 Lateral radiograph of a domestic duck showing the characteristic bullous syrinx of this species (*arrow*). Compare the path of the trachea to that of the swan in Fig. 25.3-8.

pessulus cranially and to the caudal cartilages caudally. The medial tympanic membranes are the primary means of vocalization.

Dysphonia (abnormal vocal sounds) is a hallmark of syringeal disease. In addition to foreign body obstruction, syringeal disease may be caused by inflammation or granuloma formation within the syrinx itself or by Aspergillus infection of the interclavicular air sac.

Muscular control of the syrinx, and thus of vocal sounds, is complex and variable. The syrinx lies within the **interclavicular air sac**, which modulates the pressure against the tympanic membranes. Intrinsic and extrinsic muscles of the syrinx provide vocal control.

There is great variation in syringeal structure and function among avian species, making even broad generalizations difficult. For example, psittacines such as parrots lack a pessulus. In some species of ducks, males possess a **tracheal bulla** proximal to the syrinx. Other species possess a [usually] left-sided **syringeal bulla**, which may be ossified (Fig. 25.3-9).

Selected references

[1] Kirchgessner M. Aspergillosis. In: Mayer J, Donnelly TM, editors. Clinical veterinary advisor: birds and exotic pets. St. Louis, MO: Saunders Inc; 2013. p. 155–7.
[2] Powers LV. Air sac tube placement. In: Mayer J, Donnelly TM, editors. Clinical veterinary advisor: birds and exotic pets. St. Louis, MO: Saunders Inc; 2013. p. 537.
[3] King AS. Aves respiratory system. In: Getty R, editor. Sisson and Grossman's The anatomy of the domestic animals. 5th ed, vol II. Philadelphia: WB Saunders Co; 1975. p. 1883–902.
[4] Dyce KM, Sack WO, Wensing CJG. Textbook of veterinary anatomy. 4th ed. Philadelphia: Saunders; 2010.

CASE 25.4

Beak Fracture

Cynthia M. Faux[a], Marcie L. Logsdon[b], and Laura Lossi[c]
[a]University of Arizona College of Veterinary Medicine, Oro Valley, Arizona, US
[b]Exotics & Wildlife Department, Washington State University College of Veterinary Medicine, Pullman, Washington, US
[c]Department of Veterinary Science, University of Turin, Turin, IT

Clinical case

History

An 8-year-old female Eclectus parrot (*Eclectus roratus;* Fig. 25.4-1) presented as an emergency for bleeding from the beak. She was reported to have been startled, falling off her owner's shoulder and onto a hard floor.

Physical examination findings

The bird was very nervous during the examination. The tip of her rhinotheca (a sheath of keratinized tissue covering the upper beak) was missing, presumably broken off, with a small amount of active bleeding from the exposed vascular portion of the dermis.

In addition, she showed signs of chronic feather-destructive behavior: most of the covert feathers over her chest and back were missing, and her flight feathers were broken and frayed. (Feathers are described in Case 28.1). She was markedly over-conditioned, with a body condition score of 5 (1 = emaciated, 5 = obese). The remainder of her physical examination was unremarkable.

Differential diagnoses

Traumatic fracture of the tip of the rhinotheca

Diagnostics

None were considered needed in this case.

With more extensive beak trauma, radiographs are advisable to rule out an underlying fracture of the premaxilla or, more caudally, the nasal bone, maxilla, and/or jugal bar (Fig. 25.4-2). Radiography can also be helpful in planning corrective beak trimming in birds with overgrowth or deformity of the beak.

RELATED TERMINOLOGY

- cere – area of thin, soft keratin around the nostrils in some birds
- culmen – dorsal midline of the rhinotheca
- gape – orifice of the open beak
- gnathotheca – keratinous sheath of the lower beak
- gonys – ventral midline of the gnathotheca
- rhamphotheca – keratinous component of the beak; divided into rhinotheca (upper) and gnathotheca (lower)
- rhinotheca – keratinous sheath of the upper beak
- rictus – angle of the mouth; the fleshy region from the corners of the mouth to the lateral margins of the rhino- and gnathotheca
- rostrum – beak, including the bony component and the overlying rhamphotheca
- tomium – lateral margin ("cutting edge") of the rhinotheca and gnathotheca; plural, tomia

Comparative Veterinary Anatomy: A Clinical Approach. https://doi.org/10.1016/B978-0-323-91015-6.00119-9

FIGURE 25.4-1 The female Eclectus parrot in this case, before her beak injury (left). Eclectus parrots are one of the few parrot species that exhibit visible or marked feather color sexual dimorphism (Gr. *dis* twice + Gr. *morphē* form = characteristics typically linked with gender): Females are red and purple, while males are blue and green. A male Eclectus is shown on the right. Both parrots above show signs of poor feather quality and feather-destructive behavior, which is quite common in this species and is a likely explanation of why the patient fell to the floor when attempting to fly.

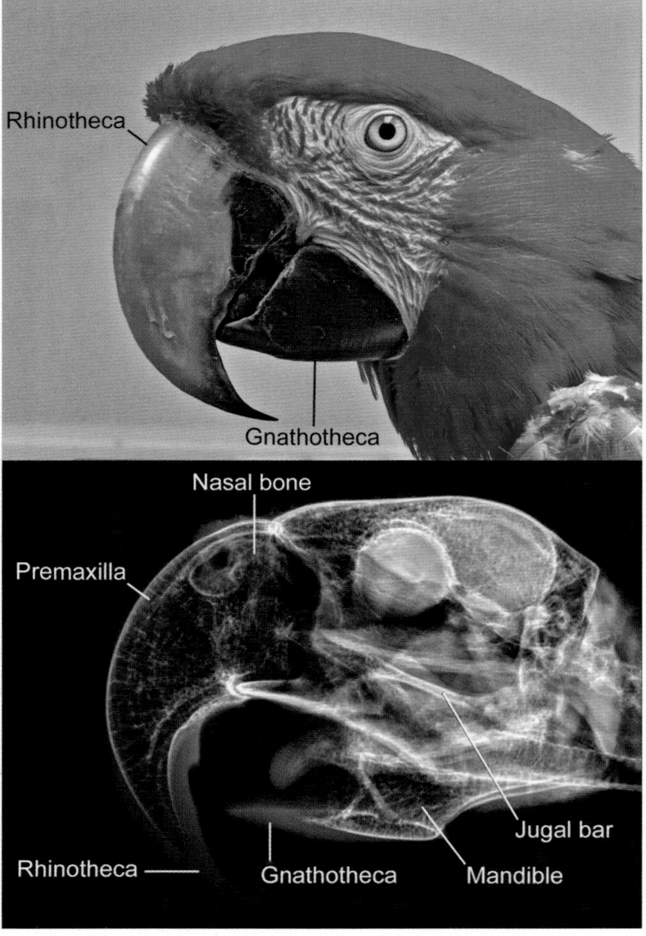

FIGURE 25.4-2 A Scarlet macaw (*Ara macao*) and a lateral radiograph of the same bird under general anesthesia. Radiographically, the rhinotheca and gnathotheca can be seen as uniform "gray" tissue overlying the premaxilla and mandible, respectively. The bony features of the avian skull are further illustrated in Figs. 25.4-10–25.4-12.

Diagnosis

Traumatic fracture of the tip of the rhinotheca

Treatment

A silver-nitrate stick was applied to the tip of the rhinotheca to cauterize the dermal vessels to control the bleeding. Anti-inflammatory therapy was administered for pain relief. No other treatment was considered necessary for patient care, because only the keratinous tissue of the beak tip was lost and would regenerate. Furthermore, the loss of the tip of the rhinotheca would not prevent the bird from eating normally in the meantime.

Anatomical features in avian species

Introduction

The beak varies widely in appearance and structure in order to meet its many functions in the diverse avian realm. This case discusses the anatomy of the beak and oral cavity of the bird.

Function

The beak serves in the primary capacity of prehension and, in some species, mechanical breakdown of food. Depending on the species, the beak is also used for killing prey, defense, preening, manipulating objects, nest construction, searching for food, feeding nestlings, heat regulation, communication, and courtship. As described in Case 28.1, the beak is important in preening (feather-maintenance) behavior. Birds use the beak to ruffle and sort their feathers, pick off parasites, "re-zip" the vanes as needed, and—in some species—to groom their partners. The beak may also be used to ward off danger by making threatening noises. For example, when disturbed, owl nestlings stand tall, stare at the threat, and emit loud clicks by snapping their beaks repeatedly.

Beak—shape and external features

The beak or bill is called the **rostrum** (L. *rostrum* beak) and comprises an underlying skeletal component and an overlying keratinous sheath, the **rhamphotheca** (G. *rhamphos* beak + *theca* case or sheath). The rhamphotheca is divided into the **rhinotheca** (G. *rhinos* nose) of the upper beak and the **gnathotheca** (G. *gnathos* jaw) of the lower beak.

Beak shape varies widely among species and reflects the bird's diet and feeding habits (Fig. 25.4-3A–I). In most species, the **nares** or nostrils are located near the base (caudal extent) of the rhinotheca, although there is considerable species variation in nostril placement, shape, and accompaniments—e.g., operculum, feathers, cere. The nares and their relationship to the nasal cavity are described and illustrated in Case 25.3.

> Beak overgrowths and deformities such as "scissor beak" are relatively common in pet birds and can significantly impact the bird's health, so it is important to evaluate the beak shape during physical examination.

SCISSOR BEAK

Scissor beak is a deformity that consists of malalignment of the rhinotheca in relation to the gnathotheca such that the upper and lower halves of the beak are offset like scissor blades (Fig. 25.4-4). The underlying skeletal structures (premaxillae, nasal bones, maxillae, jugal bar, and/or mandibles) may also be deformed because of uneven growth.

Typically, the deformity worsens over time with continuous growth and use of the beak. Without necessary corrective beak trimming and bracing, the deformity may lead to difficulty prehending food. At that point corrective surgery may be required. Nonsurgical interventions are most successful in young birds, as they retain some growth potential in the skeletal components of the beak.

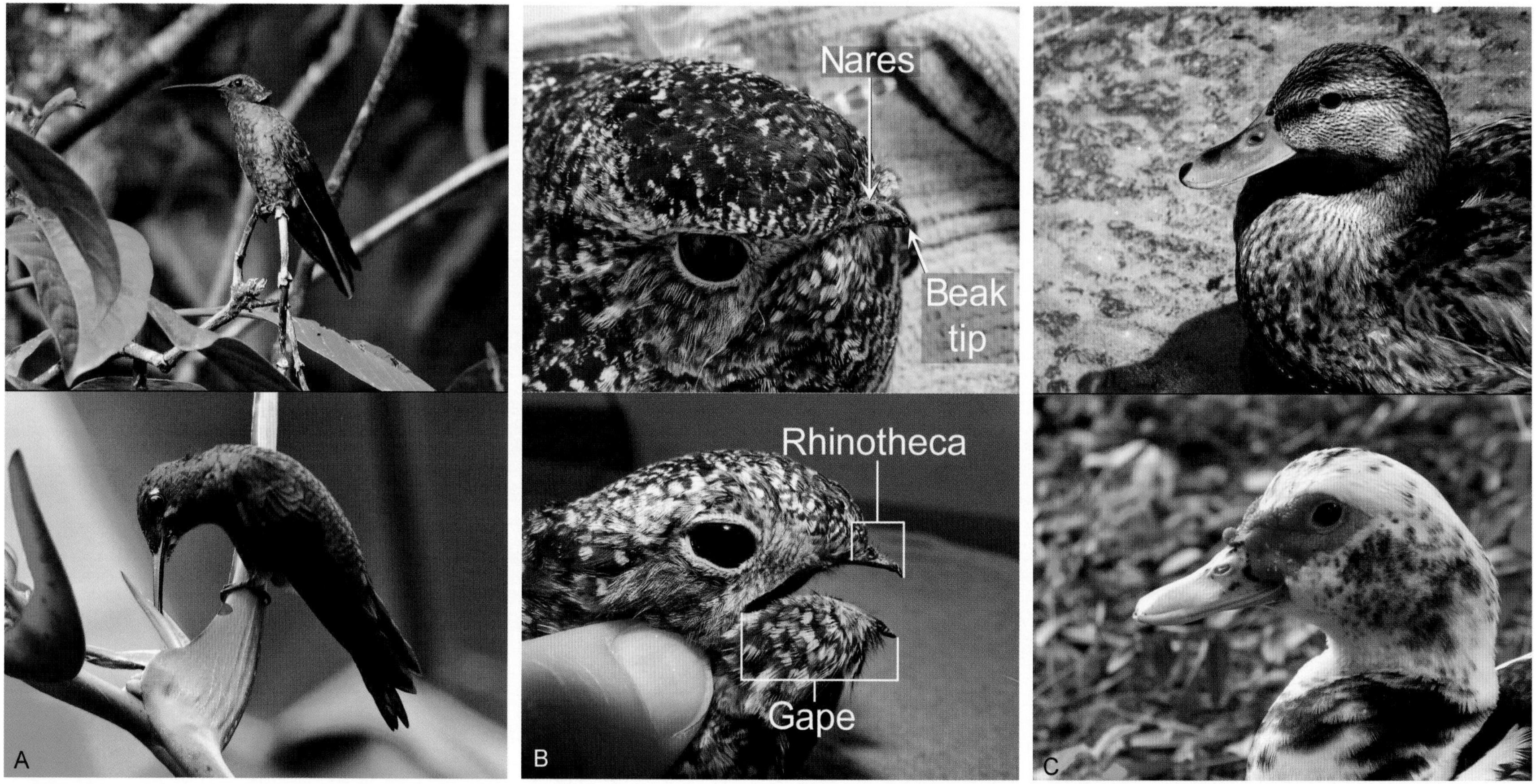

FIGURE 25.4-3 Beak shape varies according to diet and feeding habits. (A) Hummingbirds eat nectar and insects; their long, narrow beaks are adapted to feeding from flowers. Top: Sparkling violet-ear hummingbird (*Colibri coruscans*). Bottom: Green-crowned brilliant hummingbird (*Heliodoxa jacula*). (B) The Common nighthawk (*Chordeiles minor*) is neither a hawk nor strictly a night hunter. It is a typically crepuscular (L. *crepusculum* twight, glimmering) insectivore with a small rhamphotheca, but a large gape, well-suited to catching insects in flight. (C) Ducks have a broad, flat beak or bill that is used as both shovel and sieve to dredge shallow water and mud for food. The fine, comb- or brush-like projections that ducks, geese, and swans have along the tomium (lateral margin of the beak) are shown in Fig. 25.4-7. Top: a female Mallard (*Anas platyrhynchos*). Its species name, *platyrhynchos,* is derived from Latin and Greek word roots meaning "wide, flat, or broad beak." Bottom: Muscovy duck (*Cairina moschata*).

FIGURE 25.4-3, CONT'D (D) The Black-capped chickadee (top; *Poecile atricapillus*) and the American crow (bottom; *Corvus brachyrhynchos*) are both omnivorous passerine species, yet their beaks are quite different, reflecting their particular niches. The species name for this crow, *brachyrhynchos,* is derived from Greek word roots meaning "short beak," perhaps to distinguish this bird from the Common raven (*Corvus corax*), a closely related species with a larger and more robust beak. (E) Two small Ecuadorean birds of similar size but different diets and beak shape. Top: The Bananaquit (*Coereba flaveola*) specializes in eating fruits and nectar; unlike hummingbirds, it punches a hole in the side of the flower to access the nectar, so it does not facilitate pollination as do hummingbirds. Bottom: The Black-backed grosbeak (*Pheucticus aureoventris*) is an omnivore. (F) The Hooded merganser (top; *Lophodytes cucullatus*) and the American white pelican (bottom; *Pelicanus erythrorhynchos*) are both waterbirds that feed on fish and smaller aquatic creatures, but their specific beak shapes reflect the size of their prey and their particular hunting and eating behaviors. The species name for this pelican, *erythrorhynchos,* is derived from Greek word roots meaning "red beak." The pelican's characteristic throat pouch is better illustrated in Fig. 25.4-9.

(Continued)

FIGURE 25.4-3, CONT'D (G) The Peregrine falcon (top; *Falco peregrinus*) and the Blue-and-gold macaw (bottom; *Ara ararauna*) both have a rhinotheca that normally overhangs the gnathotheca at its tip, yet the first is a raptorial carnivore and the second is an omnivore (eating seeds, nuts, fruit, other plant parts, and grubs). The falcon uses the hook-like rostral projection to efficiently kill its prey, whereas the parrot's beak functions as a dexterous multitool. Note the small ventral projection on the lateral margin of the falcon's upper beak (inset), called the "tomial tooth" or "falcon's tooth." The notch formed between beak tip and tomial tooth is used to snap the neck of prey, predominantly other birds. (H) Hornbills are subtropical/tropical omnivores that eat fruit, insects, and small animals. Top: Southern ground hornbill (*Bucorvus leadbeateri*) is primarily a carnivore. Bottom: In contrast, the Wreathed hornbill (*Rhyticeros undulatus*) is primarily frugivorous. (I) Two more examples of the diversity in head and beak shape with diet and habitat among birds native to the same continent (North America). Top: The Cedar waxwing (*Bombycilla cedrorum*) eats insects and fruit. Bottom: The Wood duck (*Aix sponsa*) eats plants and insects.

In some birds—including raptors, parrots, and turkeys—the keratin at the base of the rhinotheca is softer around the nares, and constitutes a discrete area called the **cere** (L. *cera* wax), so-named because it may appear pale or waxy in comparison with the rest of the rhinotheca (Fig. 25.4-3G). In many of these birds, the cere surrounds the nostrils (Fig. 25.4-4). In pigeons, a fleshy or bulbous cere lies caudal and dorsal to the nares. In budgerigars, the color of the cere demonstrates sexual dimorphism: The cere is blue in adult males and brown or pink in adult females (Fig. 25.4-5); it is pink in juvenile budgies of either gender.

In male budgerigars, a change of cere color from blue to brown is clinically meaningful, as it indicates the influence of female hormones, such as occurs with Sertoli cell tumors (see Case 26.5).

FIGURE 25.4-4 Top: A Green-cheeked conure (*Pyrrhura molinae*) with an acquired malocclusion of the beak ("scissor beak"), 4 weeks after sustaining an injury to its gnathotheca in a dog attack. Bottom: The same bird after a corrective beak trim. Note the permanent full thickness defect in the mandible. This bird will require maintenance beak trims for the remainder of its life, but can successfully manipulate food.

FIGURE 25.4-5 A pair of Budgerigars (*Melopsittacus undulatus*), male (left) and female (right). "Budgies" display a sexual dimorphism in the normal color of the cere: blue in adult males and brown or pink in adult females.

The **culmen** (L. *culmen* top or summit) is the dorsal midline of the rhinotheca. Culmen length, measured from beak tip to a defined landmark at the base of the rhinotheca, is data commonly collected in field studies. The specific caudal landmark used—e.g., caudal margin of the nares, feathered skin—varies by species and by the researcher collecting the data.

The **gonys** is the ventral midline of the gnathotheca. The point at which the left and right halves of the gnathotheca diverge caudally as they follow their respective mandibular ramus is called the **gonydeal angle**. Whereas the rhinotheca is a single sheath that covers the outer surface of the entire upper beak, the gnathotheca divides caudally for a variable distance into left and right rami (Fig. 25.4-6). Skin usually fills the area between these rami, but in some birds, notably pelicans, the gonys is short and the inter-ramal region bears a distinctive unfeathered pouch.

In puffins, which have a long gnathotheca, the outer layer of the rhinotheca and gnathotheca is shed, thus changing to the characteristic multicolored beak of the breeding season. In many gull species, the gnathotheca has a red or orange spot near its distal end. The gull chicks peck at the spot, thus triggering the parent to feed the chick.

Although "scissor beak" is a serious beak deformity in most species, crossbills (*Loxia* spp.) are a genus of seed-eaters in which this unusual relationship of the upper and lower beak is normal and allows the bird to efficiently extract conifer seeds from cones (Fig. 25.4-6). In birds such as finches and other passerine species, the tips of the rhinotheca and gnathotheca may meet precisely (Fig. 25.4-3D). However, other species such as parrots and raptors have a "hooked" rhinotheca, which overhangs the gnathotheca (Fig. 25.4-3G). In skimmers (*Rynchops* spp., literally "beak face"), which are birds that hunt by skimming the beak through the surface of the water, the gnathotheca is longer than the rhinotheca.

> It is important to recognize the normal appearance of the avian species being examined before making a diagnosis of beak malformation.

FIGURE 25.4-6 This juvenile Red crossbill (*Loxia curvirostra*) shows the normal scissored beak occlusion for which these birds are named. (Photo courtesy of Gail Collins.)

BEAK TRIMMING

Corrective beak trimming may be necessary with beak overgrowths or malformations (Fig. 25.4-4). In less severe cases, it is possible to simply trim excess keratin with an instrument appropriate for the bird's size. The need for general anesthesia is determined by the severity of the malocclusion and the individual bird's temperament.

More extensive shaping and removal of excess keratin can be facilitated with a rotary tool that has a file tip. However, care must be taken to prevent thermal injury to the underlying tissues.

It is also important to remember that improper trimming and/or overtrimming may lead to malocclusion, pain, and potentially considerable bleeding.

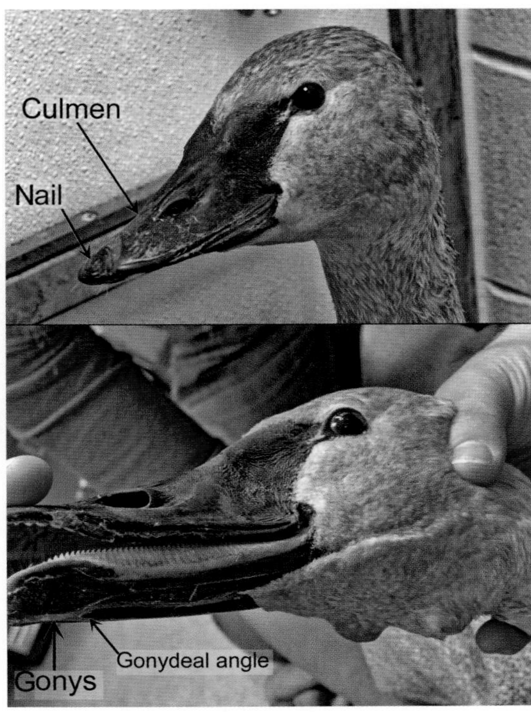

FIGURE 25.4-7 A juvenile Tundra swan (*Cygnus columbianus*), demonstrating the "nail" at the dorsal tip of the rhinotheca (see text) and the gonydeal angle on the ventral midline of the gnathotheca. In this species, the skin between the left and right rami of the gnathotheca is black and featherless, making it seem as though the gonydeal angle is located much further caudally. Also note the comb- or brush-like projections along the tomium (lateral margin) of the gnathotheca and the serrations along the tomium of the rhinotheca in this waterbird.

Ducks, geese, and swans have a thick plate or hook of hard keratin called the **nail** at the tip of the otherwise relatively soft rhinotheca (Fig. 25.4-7). Special sensory nerve endings that act as mechanoreceptors are located beneath the nail, so presumably this part of the beak tip is adapted for searching within the water (or its silty bottom) for food.

In many species, the full-term embryo and newly hatched chick has an **egg tooth.** This is a discrete ridge or point of calcified keratin at the tip of the rhinotheca, gnathotheca, or both, which is used by the chick to break the eggshell during hatching. The egg tooth is lost within a few days or weeks of hatching.

Oral cavity

Beak shape influences the shape of the oral cavity, although general features of the oral cavity and pharynx are similar among avian species. When the beak is fully open, the orifice is referred to as the **gape**. In some altricial species (those whose chicks are entirely dependent on their parents), the chick's gape is brightly colored (Fig. 25.4-8) and may even be reflective to ultraviolet light. This is presumably to positively affect feeding behavior by the parent, particularly when there is competition among siblings for food.

The lateral margins or **tomia** (G. *tomos* to cut; singular, **tomium**) of the rhinotheca and gnathotheca are the "cutting edges" of the beak in many species. The facing or opposing surfaces of the upper and lower tomia range from blunt to sharp and even serrated, depending on the bird's diet and feeding habits. In various waterbirds, the tomia are lined with short comb- or brush-like projections (Fig. 25.4-7), which help the bird grasp and keep hold of slippery prey such as fish or to filter mud for small crustaceans or plant matter. The sharp hook at the tip of the rhinotheca in raptors (Fig. 25.4-3G, top) is also used like a tooth to tear food. In seedeaters (Fig. 25.4-3G, bottom), the rostral portion of the beak can crush the husk of hard seeds and nuts.

The **rictus** (L. *rictus* open mouth), or angle of the mouth, refers to the fleshy region from the corners of the mouth to where it blends into the tomia of the upper and lower beaks. In some species, the rictus is very short, while in others—such as the Common nighthawk shown in Fig. 25.4-3B—it is extensive. The rictus is

FIGURE 25.4-8 European starling (*Sturnus vulgaris*) chicks have brightly colored—and large—gapes. The small spot of red on the head of the chick on the lower left is a nontoxic marking used by the facility to distinguish individual animals.

particularly impressive in European starling chicks (Fig. 25.4-8), in which it is much longer and broader than the rhamphotheca, extending laterally from the skull as a fleshy fold when the beak is closed and creating a gape that is much larger than the chick's head when the beak is opened. This fleshy tissue gradually reduces and the rhamphotheca and underlying skeletal structures grow as the chick matures, such that the rictus of the adult starling is quite unremarkable.

Dermis and epidermis of the beak

The rhamphotheca is heavily keratinized integument with modified epidermal and dermal components. The highly vascular **dermis** is firmly attached to the periosteum of the underlying bones (described later). The **epidermis** includes a thick and hard **stratum corneum**, the cells of which contain free calcium phosphate and orientated crystals of hydroxyapatite, which increase its strength and resistance to wear.

Like nails or talons, the rhamphotheca grows continuously to replace the keratinous tissue lost through wear. The germinal cells are located at the base (caudal extent) of the rhino/gnathotheca. New keratinized cells then migrate along the rhamphotheca in a rostro-lateral direction.

The keratinous tissue is not uniform in thickness over the entire surface of the rhamphotheca. It is thicker at the lateral margins and tip of the rhinotheca (e.g., see the radiograph in Fig. 25.4-2). These are the surfaces that cut/tear/pierce/crush or otherwise experience the greatest wear. The degree of mechanical abrasion and, thus, fray or wear varies greatly among individuals and within populations due to different foraging styles and seasonal fluctuations.

As with feather color (see Case 28.1), beak color is determined by pigments within the epidermis of the rhamphotheca. In some species, beak color and shape vary seasonally and may change in advance of the breeding season. Some birds shed part of the beak at the end of the breeding season. For example, male and female American white pelicans develop a discrete "bill horn" (Fig. 25.4-9), which shed at the end of the breeding season. Beak shape and color may differ even within a species due to sexual dimorphism; examples include hornbills and many shorebirds.

Bones and joints of the beak (Figs. 25.4-10–25.4-12)

Cranial kinesis refers to the movement between the bones of the skull. In the broadest sense, it includes lowering and raising of the mandible to open and close the mouth, and to other movements of the mandible such as rostral-caudal and lateral movements. However, in birds, this term is generally applied to movements of the upper jaw in relation to the neurocranium (the part of the skull housing the brain and its peripheral structures, including the eyes and ears) and within the upper jaw itself, as described later.

The paired **premaxilla** and **nasal bones** form the main support for the rhinotheca. In adult birds, these bones are fused to form a rigid, triangular block-like structure. Caudally, the nasal bone articulates with the frontal bone rostromedial to the orbit, forming the **nasofrontal joint** or **craniofacial hinge**. This functional joint is an important component of cranial kinesis in birds. The upper beak is "hinged" at its articulation with the frontal bone and thus with the neurocranium. In parrots, the nasofrontal joint is a synovial joint, complete with a synovial membrane and synovial fluid.

FIGURE 25.4-9 An adult American white pelican (*Pelicanus erythrorhynchos*), showing the prominent "bill horn" (*arrow*) and more brightly colored beak and throat pouch that these birds display during the breeding season.

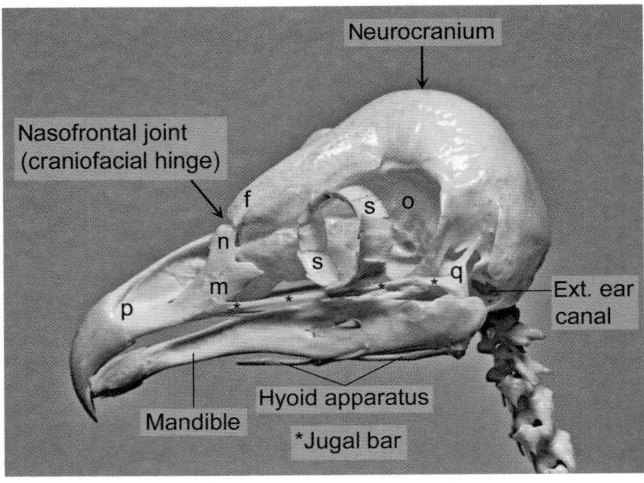

FIGURE 25.4-10 Skull of a Barn owl, lateral view. Key: *ext.*, external; *f*, frontal bone; *m*, maxilla; *n*, nasal bone; *o*, bony orbit; *p*, premaxilla; *q*, quadrate bone; *s*, scleral ossicles. The jugal bar is identified by *asterisks*. Parts of the pterygoid and palatine bones are visible deep to the dorsal margin of the mandible and ventromedial to the jugal bar. These structures are better illustrated in Fig. 25.4-12B.

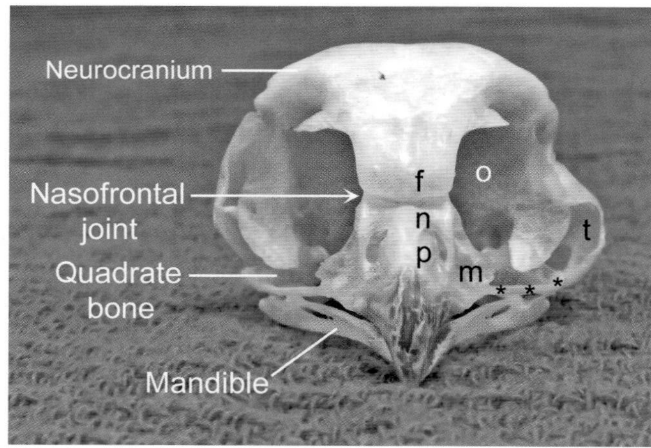

FIGURE 25.4-11 Skull of a Northern saw-whet owl, rostrocaudal view, without the scleral ossicles. Most of the rhinotheca and gnathotheca have been lost in the skeletal preparation. Key: *f*, frontal bone; *m*, maxilla; *n*, nasal bone; *o*, bony orbit; *p*, premaxilla; *t*, the opening of the acoustic meatus. Jugal bar is indicated with asterisks. (All of these bones are paired, but only the right side of the figure is labeled.)

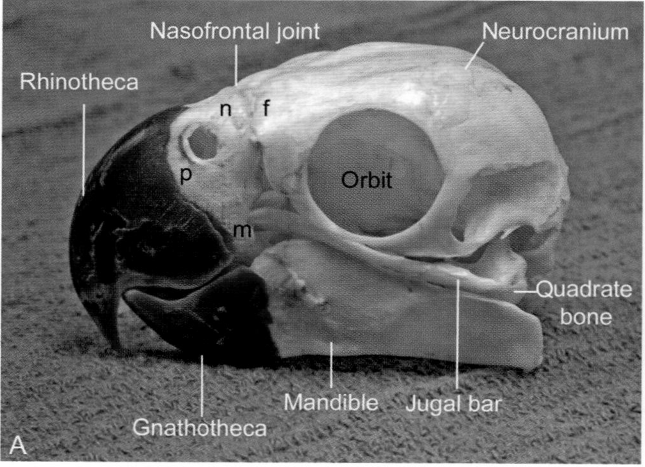

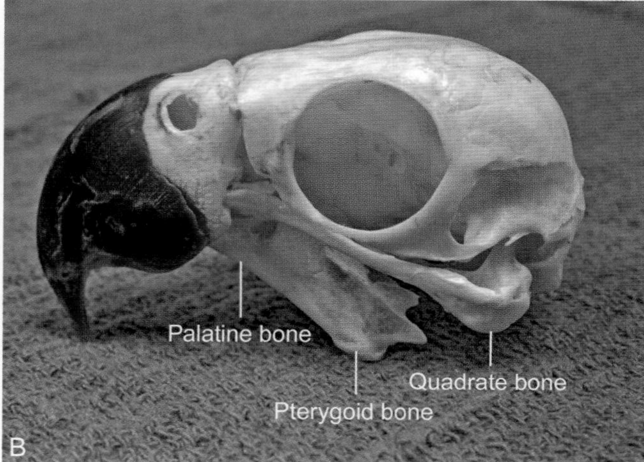

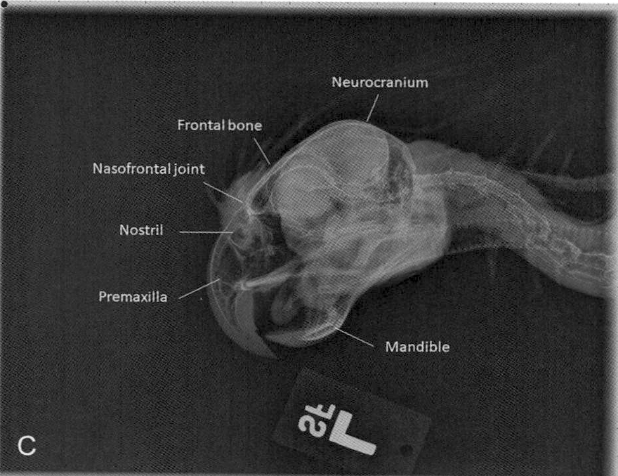

FIGURE 25.4-12 Skull of a Sulfur-crested cockatoo. (A) Lateral view. Key: *f*, frontal bone; *m*, maxilla; *n*, nasal bone; *p*, premaxilla. (B) The mandible was removed to show the pterygoid and palatine bones that lie deep to it. The quadrate bone is also better seen in this image. In this preparation, the pterygoid and quadrate bones have been separated at their articulation for better depiction. (C) Lateral radiograph of a Sulfur-crested cockatoo.

The avian **maxilla** is reduced to a small bone that is fused to the caudolateral aspect of the nasal bone. It forms the rostral portion of the jugal arch or **jugal bar**, which is the union of 3 thin bones that extend, rod-like, from the nasal bone to the quadrate bone (described below), which lies caudoventral to the orbit. From rostral to caudal, the jugal bar comprises the **jugal process** of the maxilla, the **jugal bone**, and the **quadratojugal bone**. These bones are fused into the jugal bar, which forms the ventrolateral bony margin of the infraorbital sinus (see Case 25.1).

The paired **mandible** supports the gnathotheca. There are 2 mandibular rami (L. *ramus* branch), left and right, which meet at the rostral midline to form the mandibular symphysis. Thus, the paired mandible functions as a single structure. Depending on beak/bill shape, the mandibular symphysis may form either a "V" or "U" shape.

The paired **quadrate bone** is a small yet pivotal bone in the jaw mechanism of birds. It is irregularly shaped and located caudoventral to the orbit and rostroventral to the external acoustic meatus (external ear canal) on the ventrolateral aspect of the neurocranium. The quadrate bone has 4 separate articulations. At its caudodorsal aspect, it articulates with the **neurocranium** in a small joint. At its ventrolateral aspect, it articulates with the mandible and separately with the jugal bar. Ventromedially, it articulates with the **pterygoid bone** and thus indirectly with the palatine and vomer bones, which together form the bony base of the upper beak.

When the bird's mandible is lowered or raised by the muscles acting upon it, this action is transmitted to the upper beak via the quadrate bone and its rostral connections (the jugal bar and pterygoid bone). As a result, the upper beak is reciprocally raised or lowered, thereby opening or closing the gape, respectively. This concerted action, which also involves the rotation of the upper beak at the craniofacial hinge, is the basic type of cranial kinesis in birds, called "prokinesis." More elaborate forms of cranial kinesis are found in some species, in which either the rostral portion of the upper beak ("rhynchokinesis") or the entire upper beak ("amphikinesis") can be flexed dorsoventrally, owing to flexibility of nasal bones and jugal bar.

Rostro-caudal and lateral movements of the mandible in relation to the upper beak also occur via the quadrate bone. Such refined movements are needed, for example, when a parrot is manipulating an object or cracking open a hard seed or nut with its beak.

Muscles acting on the beak

Due to cranial kinesis, movement of the beak is much more complex than is movement of the mandible in mammals. Most of the muscles involved have multiple functions.

Generally, there are 7 pairs of muscles involved in beak movement. Some act synergistically to lower the mandible and raise the upper beak, thus opening the gape. Others raise the mandible to close the gape. Muscles that do the former are generally weaker than those that do the latter.

Blood supply and innervation of the rostrum

The beak and nasal and oral cavities are quite vascular and are supplied by branches of the **external carotid artery**. An outline of the arterial supply to the bird's head forms a general outline of its rostrum.

BEAK AS SENSORY ORGAN

Despite its hard, keratinous covering, the beak is a highly sensitive structure that provides birds with a wide range of sensory information.

For example, the rhamphotheca in some birds contains Herbst receptors—mechanoreceptors that are sensitive to vibration and pressure. In wading birds, Herbst receptors in the gnathotheca are believed to enable the bird to sense prey hidden within wet sand or mud. The area in which these receptors are concentrated is sometimes called the "bill tip organ." It is found in birds that forage by probing and parrots.

The beak is a sensory organ, so beak trimming must be done with great care and precision.

The sensory nerves of the rostrum are branches of the **trigeminal nerve** (cranial nerve V). The rhinotheca is innervated by the **maxillary** and **ophthalmic nerves** and the gnathotheca by the **mandibular nerve**. The maxillary and ophthalmic nerves are purely sensory, whereas the mandibular nerve contains both sensory and motor components. It thus provides primary motor innervation to the masticatory muscles, except for the mandibular depressor muscle, which is innervated by the **facial nerve**.

Acknowledgments

The authors thank the Washington State University College of Veterinary Medicine Department of Veterinary Anatomy for preparation of the avian skulls.

Selected references

[1] Getty R. Sisson and Grossman's The anatomy of the domestic animals. 5th ed. Volume II. Philadelphia: WB Saunders; 1975.
[2] Baumel JJ. Handbook of avian anatomy: Nomina Anatomica Avium. 2nd ed. Nuttall Ornithological Club (USA), World Association of Veterinary Anatomists (USA), International Committee on Avian Nomenclature; 1993.
[3] King AS, McLelland J. Birds: Their structure and function. 2nd ed. Philadelphia: Baillière Tindall; 1984.

CHAPTER 25

CASE 25.5

Obstruction of External Ear Canal

Cynthia M. Faux[a] and Marcie L. Logsdon[b]
[a]University of Arizona College of Veterinary Medicine, Oro Valley, Arizona, US
[b]Exotics & Wildlife Department, Washington State University College of Veterinary Medicine, Pullman, Washington, US

Clinical case

History

An adult short-eared owl (*Asio flammeus;* Fig. 25.5-1) was presented for suspected head trauma. The bird had been found on the roadside, assumed to have been hit by a car. Before referral, the primary-care veterinarian reported finding blood on the owl's head and mouth, and had sutured what appeared to be a large laceration on the side of the head.

The opening to the external acoustic meatus in birds can appear very different to that in mammals. In many birds, it is easily mistaken for a tear or laceration (Fig. 25.5-2).

FIGURE 25.5-1 An adult short-eared owl.

Comparative Veterinary Anatomy: A Clinical Approach. https://doi.org/10.1016/B978-0-323-91015-6.00120-5

FIGURE 25.5-2 Opening of the right external acoustic meatus of a short-eared owl.

Physical examination findings

Initial examination revealed a likely female owl, weighing 410 g, in good physical condition and with normal wing and leg posture, but with a small amount of dried blood on the feathers around her mouth. She was alert but appeared unsteady when perching. Ophthalmologic examination showed no signs of ocular trauma. Closer examination of the sutured laceration on the side of her head revealed that the opening to the external acoustic meatus had been sutured closed. Once the sutures were removed, dried blood was observed within the ear canal, but no source of active bleeding was identified.

Differential diagnoses

Head trauma, musculoskeletal injury

Diagnostics

Routine radiography, consisting of the 2 standard orthogonal views (lateral and ventrodorsal), was performed to check for signs of additional trauma. No abnormalities were found.

SHORT-EARED OWLS

These owls are widely distributed across North America and are often observed during daylight. Small tufts of feathers on the dorsolateral aspect of the head resemble mammalian ears, but they are so small as to be barely visible—hence the common name, "short-eared" owl. (In contrast, long-eared owls have much more obvious "ear" tufts.) These tufts of feathers are unrelated to the ear.

Because these owls prefer open habitats for hunting, they are sometimes hit by cars when swooping across a roadway in pursuit of prey.

Diagnosis

Head trauma, with iatrogenic (man-made) obstruction of the external acoustic meatus

Treatment

Following the removal of the sutures occluding the external acoustic meatus, the owl was kept for observation. Her mental state improved over the next 48 hours, and her ability to perch normalized within a day. She was returned to the location where she was found and released 1 week after presentation.

Anatomical features in avian species

Introduction

This section describes the avian ear. As its basic structure and functions (hearing and balance) are the same in birds and mammals, only the features or variations specific to birds are described here.

Function

Sound is critically important to birds, as it plays key roles in social interactions, territoriality, reproduction (in particular, courtship), hunting, and—in some species—echolocation (navigation by reflection of sound waves).

External ear canal

The paired (left and right) **external acoustic meatus** is located on the lateral aspect of the head. In some species, the external orifice is visible as a circular or oval opening on the side of the head (Fig. 25.5-3). Birds do not possess external auricles, or pinnae, at the opening of the ear canal. However, in most species, the opening is protected by specialized, barbless contour feathers called **ear coverts** (Fig. 25.5-4). They arise rostral to the opening and lie flat, oriented caudally, over the opening, thus decreasing air turbulence, which would otherwise impede hearing when the bird is in flight. In aquatic diving birds, these feathers help keep water out of the external ear canal.

These specialized feathers at the caudal aspect of the opening may serve a similar function in helping to channel sound into the acoustic meatus (Fig. 25.5-5).

Various species have an **operculum auris**, which is a small fold of skin and muscle that lies either rostral or caudal to the external orifice and can be pulled over the opening as needed. For example, budgerigars have a pre-aural

FIGURE 25.5-3 In some species, such as the rhea (left) and the helmeted guineafowl (right), the entrance to the external acoustic meatus is visible as a circular or ovoid opening on the side of the head (*arrow*). (Photos courtesy of Daniel Field.)

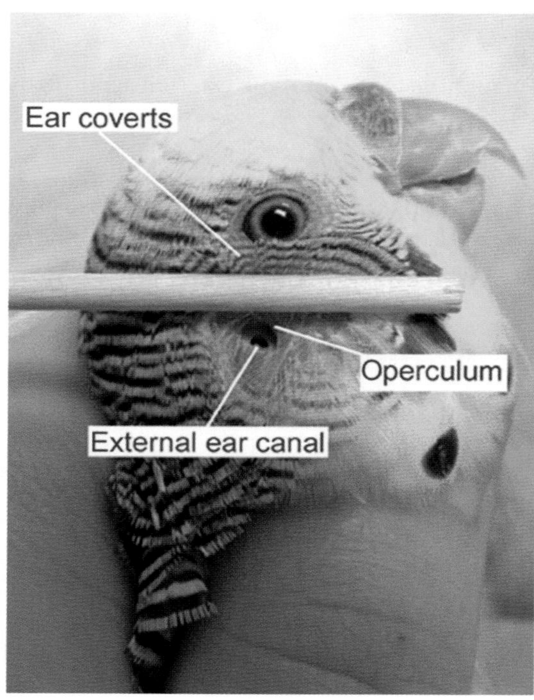

FIGURE 25.5-4 External ear of a budgerigar. In this photo, the ear coverts (specialized feathers that cover the ear canal) are being lifted to allow visualization of the external ear canal. The operculum is a small skin fold that lies rostral to the ear canal.

FIGURE 25.5-5 An anesthetized golden eagle. The ear coverts have been pulled aside from the external ear canal to show the opening, which normally is not visible in this species, and to show the specialized feathers caudal to the orifice which help channel sound into the ear.

ACOUSTIC ABILITIES IN BIRDS

Many birds have a narrower range of pitch discrimination than humans. However, birds have a temporal resolution 10 times that of humans. In other words, a complex avian song would have to be slowed to 1/10th speed for a human to distinguish every detail.

Cave-dwelling and night-flying species may use echo location to navigate. Birds produce a lower frequency of sound than bats for echo location, so they can use this technique for avoiding objects but not for locating and targeting prey such as insects.

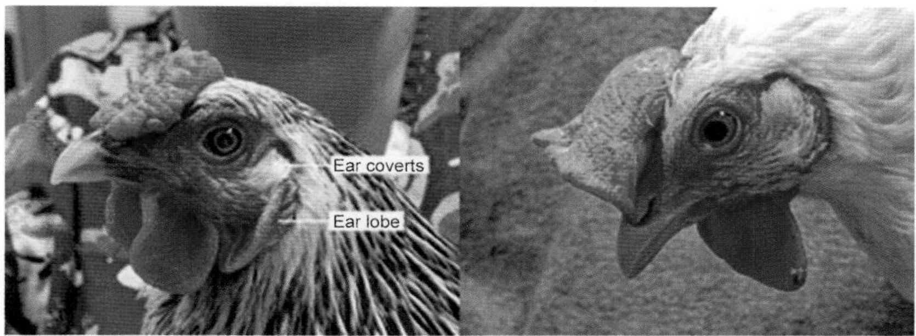

FIGURE 25.5-6 Ear lobe color is associated with egg shell color in many domestic chicken breeds. Left: This hen produces speckled-brown egg shells. Although her ear coverts are white, her ear lobes are red. Right: This hen has white ear coverts and white ear lobes, and produces white-shelled eggs.

operculum (L. a lid or covering structure, Fig. 25.5-4). Some species have both rostral and caudal folds. Pre- and post-aural opercula are believed to assist in sound localization, and—in diving birds—to occlude the opening of the external ear canal.

Domestic chickens have **ear lobes**, which are fleshy, featherless skin folds ventral to the opening of the external ear canal. Ear lobe color is associated with egg shell color in many breeds (Fig. 25.5-6). Generally, breeds with white ear lobes produce white egg shells, and breeds with red or brown ear lobes produce pigmented egg shells that are some variation of brown, blue, or green, depending on the breed.

Middle ear

As in mammals, the external ear canal is separated from the **tympanic cavity** of the middle ear by the **tympanic membrane** or "ear drum." An **auditory tube** connects the tympanic cavity to the pharynx. Unlike in mammals, vibrations of the tympanic membrane are transmitted to the inner ear via a single bony ossicle—the **columella**—and extracolumellar cartilage. The shape of the columella varies among avian species. Generally, it consists of a bony rod with a foot-plate that is in contact with the tympanic membrane. The **columellar muscle** can alter tension in the tympanic membrane.

> The columella is homologous with the mammalian stapes ("stirrup"). The mammalian malleus and incus ("hammer" and "anvil") are homologous with the quadrate and articular bones of the avian and reptile skull.

Inner ear

As in mammals, the inner ear consists of the **cochlea** (auditory apparatus) and the **vestibular organ** (balance apparatus), although the cochlea is relatively shorter in birds. Bony semicircular canals of the vestibular organ contain the semicircular ducts, which function in maintaining balance. The rostral and caudal semicircular canals are vertically oriented, while the lateral canal is horizontally oriented. Neuroepithelial mechanoreceptor cells within the semicircular ducts act to orient the bird to the position and movement of the head in space.

> It is likely that the owl in this case was initially unsteady when perching because head trauma caused a simple concussion or minor vestibular damage.

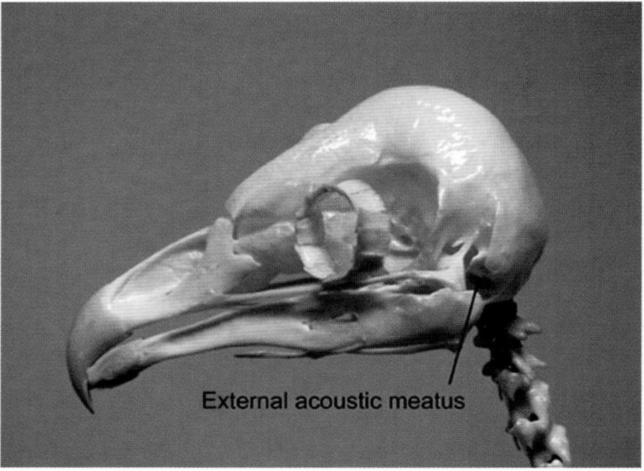

FIGURE 25.5-7 Left: An adult barn owl. The *arrows* point to the location of the external acoustic meatus on each side of the head, showing the asymmetrical placement in this species. Note the parabolic arrangement of feathers around the face creating the facial disc. Right: Compare the apparently flat shape of the face in the figure on the left with the actual shape of the skull.

FIGURE 25.5-8 Image of an adult Northern Saw-whet owl with overlying illustration of the skull depicting the relationship of the underlying bony structures in relation to the superficial feathered facial disc. The asymmetry of the openings of the right and left external acoustic meatuses (*arrows*) enhances sound localization in this fierce little hunter.

SOUND LOCALIZATION IN OWLS

Sound localization is notably achieved in owls via the facial disc—the parabolic arrangement of feathers around the face that acts like a "satellite dish" (Fig. 25.5-7). Stiff feathers, called the facial ruff, support the smooth surface feathers of the facial disc which direct sound waves to the opening of the external acoustic meatus. In many owl species, the bony acoustic meatuses are asymmetrically placed in the skull (Fig. 25.5-8); they are situated more dorsally on one side compared to the other, so sound waves are received "in stereo" and at slightly different angles, which enhances sound localization.

Selected references

[1] Sisson S, Grossman JD, Getty R. Sisson and Grossman's the anatomy of the domestic animals. 5th ed. Volume II. Philadelphia: Saunders; 1975.
[2] Baumel JJ, editor. Handbook of avian anatomy: Nomina Anatomica Avium. 2nd ed. Nuttall Ornithological Club (USA), World Association of Veterinary Anatomists (USA), International Committee on Avian Nomenclature; 1993.
[3] King AS, McLelland J. Birds: Their structure and function. 2nd ed. London; Philadelphia: Baillière Tindall; 1984.

CHAPTER 25

CASE 25.6

Ocular Trauma

Cynthia M. Faux[a] and Marcie L. Logsdon[b]
[a]University of Arizona College of Veterinary Medicine, Oro Valley, Arizona, US
[b]Exotics and Wildlife Department, Washington State University College of Veterinary Medicine, Pullman, Washington, US

Clinical case

History

An adult male great horned owl (*Bubo virginianus*; Fig. 25.6-1) was referred by a wildlife rehabilitation center for flight evaluation and possible pre-release conditioning. The bird had arrived at the center 3 months earlier after sustaining head trauma from being hit by a car. The owl's left eye had obvious signs of trauma and was presumed to be nonvisual. The staff had released the owl after a period of rehabilitation, but shortly afterward he was found down again, in poor body condition, and was unwilling or unable to fly. After supportive care and feeding, the bird was referred to the veterinary clinic for evaluation.

Policies vary among wildlife rehabilitation centers, but one-eyed owls are often considered releasable, owing in part to owls' reliance on their auditory abilities for hunting, and partly based on their eye position. The forward-facing eyes of owls create much more overlap in the field of vision such that blindness in one eye results in less loss in the total field of vision than it would in diurnal species such as hawks and falcons, whose eyes are positioned more laterally. Most rehabilitation centers that release one-eyed owls require that the bird first pass a live-prey test, proving the ability to successfully hunt. This owl had not been evaluated in this manner.

FIGURE 25.6-1 An adult great horned owl.

Comparative Veterinary Anatomy: A Clinical Approach. https://doi.org/10.1016/B978-0-323-91015-6.00121-7

One method of evaluating body condition in birds is by palpating the keel, as many avian species deposit subcutaneous fat in this location. Palpating the keel is also a means of evaluating pectoral muscle mass, an important muscle group used in flight. In this case, the bird had been in captivity long enough to build up fat reserves, but the marked loss of pectoral muscle mass made it evident that he had not been flying for some time, likely not since the first injury. His worn and unkempt feathers were another sign of chronic grounding and debilitation.

The pupillary light reflex (PLR) is the constriction of the pupil in response to bright light. A normal PLR means that there is an intact retina and cranial nerves II (optic) and III (oculomotor), as well as functionally normal anterior and posterior segments of the eye. The PLR in birds is discussed in more detail later. The dazzle reflex is an involuntary aversion response to bright light. Common responses in birds are blinking, protrusion of the nictitating membrane, and/or evasive movement of the head. A normal dazzle reflex means that there is a functional retina as well as cranial nerves II and VII (facial). In this case, the right eye was partially responsive to bright light.

Physical examination findings

The owl was alert and responsive, but hyper-reactive to physical stimulation, even for a wild bird. He was moderately overweight, having a body condition score of 4 out of 5 (1 = emaciated, 5 = obese). Even so, he had marked muscle wasting over the keel and his feathers were worn and unkempt.

On ocular examination, phthisis bulbi (shrunken, nonfunctional eye) was present on the left (Fig. 25.6-2). The right eye appeared normal in size, but the pupil was abnormally dilated for the ambient light, and the pupillary light reflex was incomplete; in addition, the pecten oculus (black body in the fundic region) appeared abnormal (Fig. 25.6-3). The dazzle reflex was intact.

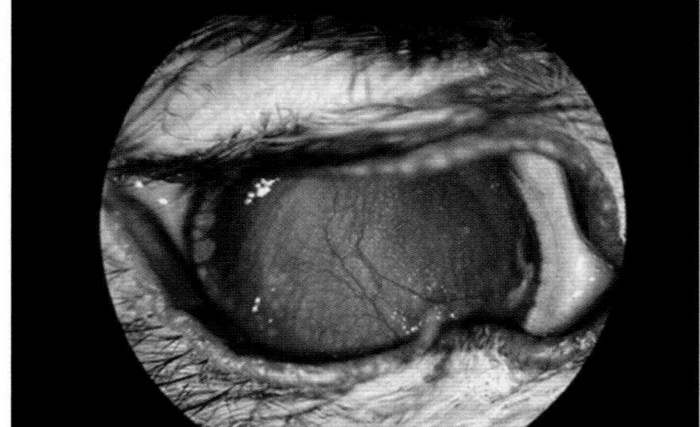

FIGURE 25.6-2 Left eye of the owl in this case, 3 months after head trauma, showing phthisis bulbi: The eyelids are wrinkled, causing the palpebral fissure to be misshapen, because the injured globe (eyeball) has shrunk. Evidence of past trauma also includes the superficial blood vessels tracking across the cornea, the surface of which appears irregular and cloudy; and opacity of the lens (cataract formation), which is preventing the visualization of the fundus. In addition, the pupil is widely dilated and unresponsive to light (absent pupillary light reflex). (Photo courtesy of Terri Alessio, DVM, DACVO.)

GREAT HORNED OWLS

Great horned owls are a large species of owl found throughout North America. They are cosmopolitan in their habits, being present in areas as diverse as deserts, forests, mountains, and plains. The females are typically larger than the males, so presumptive gender identification can be based on body size.

Great horned owls are typically nocturnal hunters, so they may be hit by cars as they chase their prey across a roadway between dusk and dawn. Vehicle headlights may contribute by “dazzling” these nocturnal raptors.

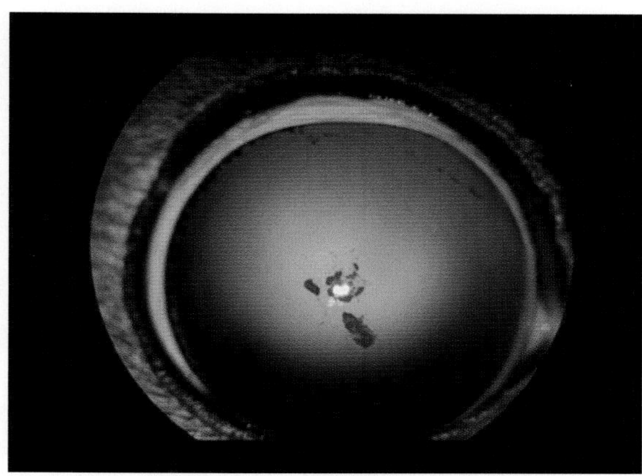

FIGURE 25.6-3 Right eye of the owl in this case. The pupil is widely dilated (mydriatic), and although the lens and the aqueous and vitreous fluids are clear, disruption of the pecten oculus is apparent as scattered dark patches in the fundic region. Fundoscopic examination is necessary to evaluate the fundus in greater detail (see Fig. 25.6-4). (Photo courtesy of Terri Alessio, DVM, DACVO.)

Differential diagnoses

Bilateral ocular trauma resulting in marked vision loss; musculoskeletal injury preventing flight

Diagnostics

Whole-body radiography showed no abnormalities. Additional diagnostics were not possible or necessary in the left, phithic, eye. Fundoscopic examination of the right eye revealed extensive retinal trauma, with areas of retinal detachment and disruption of the pecten oculus (Fig. 25.6-4). Intraocular pressure of the right eye was measured with a tonometer and was increased at 20 mmHg (normal for large owls: 10.6 ± 3.6 mmHg).

Fundoscopy is examination of the fundus (L. *fundus* bottom), or the "deep" structures of the eye, using an ophthalmoscope. Fundoscopy is necessary to examine the retina—and in birds the pecten oculus—in detail.

The increase in intraocular pressure (reported as millimeters of mercury, mmHg) in the right eye in this case was a sign of secondary glaucoma. The head trauma this bird sustained had injured the eye such that normal intraocular pressure could no longer be maintained. The resulting glaucoma caused further impairment to the ocular structures, particularly the retinal cells, and vision loss eventually progressed to total loss of sight.

OCULAR TRAUMA IN BIRDS

Ocular trauma is common with traumatic injuries in wild birds. This patient's eyes were clearly abnormal on physical examination, but that is not always the case with injured birds. For example, retinal detachment is not apparent externally. Therefore, it is important to always perform a fundoscopic examination in wildlife patients that present for traumatic injuries, as the identification and classification of ocular trauma may guide treatment decisions. Ocular trauma can result in secondary glaucoma, residual uveitis, or cataracts in the affected eye(s). Even partial blindness can be devastating in wild birds, particularly in raptors that rely primarily on their vision when hunting, such as eagles, hawks, and falcons. Euthanasia is sometimes the most humane option.

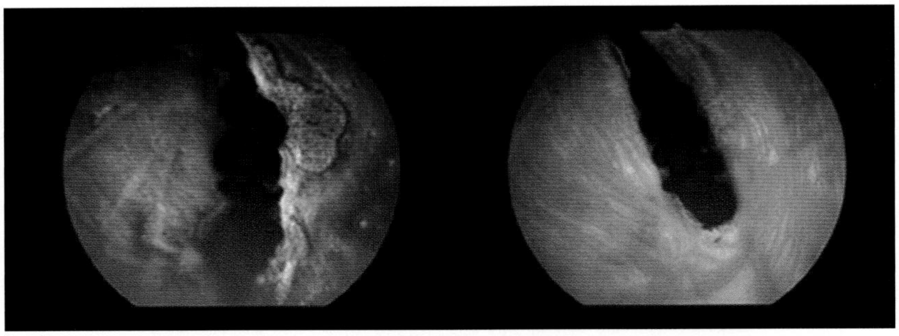

FIGURE 25.6-4 Fundoscopic images of the injured owl (left) and a normal great horned owl (right). As seen in the normal eye, the retina is unpigmented in this nocturnal species, allowing the visualization of the underlying choroidal blood vessels and the white sclera beneath them. The pecten oculus is the prominent "pleated" (multifolded), dark-brown structure in the center of these images. It appears slightly out of focus even in the normal eye, as it is protruding from the posterior wall of the globe toward the viewer, and the camera is focused on the retina. In the injured eye, separation of the translucent retinal membrane from the underlying choroid is apparent in several places, indicating multiple areas of retinal detachment. (Photos courtesy of Terri Alessio, DVM, DACVO.)

In ophthalmology, OS is the abbreviation used for the left eye (oculus sinister), and OD is used for the right eye (oculus dexter) (see chapter 1—Clinical Anatomy Nomenclature).

Birds can adapt to losing small areas of functional retina, especially if the fovea (retinal area of greatest visual acuity) is not involved, but in this case the retinal injury was too extensive. Federal guidelines in the United States prohibit rehoming of blind raptors except with special exemptions. As this bird was hyper-reactive to handling, he would not have thrived in captivity.

Diagnosis

Bilateral ocular trauma with phthisis bulbi OS and retinal detachment and secondary glaucoma OD, resulting in near-total and likely progressive blindness

Treatment

As this bird could not fend for himself and was not a good candidate for permanent placement as an educational bird, he was humanely euthanized.

Anatomical features in avian species

Introduction

In many respects, avian and mammalian eyes are essentially the same in terms of their basic anatomy and physiology. However, the avian eye has some unique features that are of clinical importance and will be discussed here.

Function

Eyes provide for vision and contribute to depth perception and balance. The avian eye is large in relation to body and skull size, emphasizing the importance of vision in birds.

VISUAL FIELD IN BIRDS

Depending on the species, and particularly on eye placement, the static horizontal field of vision in birds ranges from about 200° to 360°, and stereopsis (overlap of left and right visual fields, allowing binocular vision) ranges from 0° to 70°. As a point of comparison, humans have a static horizontal visual field and stereopsis of about 210° and 114°, respectively.

Eye position

The avian eye is large in relation to body and skull size, indicating the importance of vision in birds. The position of the eyes on the head varies with species. For example, in owls, the eyes are set in the front of the skull; in chickens, the eyes are situated laterally; and those of woodcocks are set more dorsally and caudally (Fig. 25.6-5).

Orbit and extraocular muscles

The bony **orbit** in birds is typically more shallow than that in mammals, and it is incomplete, particularly in the dorsal and temporal (lateral) regions (Fig. 25.6-6).

Together with large ocular size and lack of surrounding soft tissue, these orbital features make the avian eye entirely vulnerable to injury in cases of head trauma.

FIGURE 25.6-5 Eye position and orientation vary among avian species. Top row: Lateral eye placement in the domestic chicken (left) and red-lored Amazon parrot (*Amazona autumnalis;* right) means that these birds must cock the head to one side and use monocular vision when focusing on an object. Middle row: In the Sunbittern (*Eurypyga helias;* left and right), the eyes are placed low and fairly forward on the head, facilitating the standing/stealth-hunting strategy of these wetland birds. They use binocular vision when hunting prey, such as the fly on this rock (right), which the bird succeeded in catching. Bottom row: The forward eye placement and orientation in the northern saw-whet owl (*Aegolius acadicus;* left) enhances binocular vision but limits the static horizontal field of vision laterally and caudally. In contrast, the dorsocaudal eye placement in American woodcock (*Scolopax minor;* right), here viewed from above, prohibits binocular vision but allows a horizontal field of almost 360°. (Our thanks to the Charles R. Conner Museum of Natural History, Washington State University, for access to the woodcock specimen.)

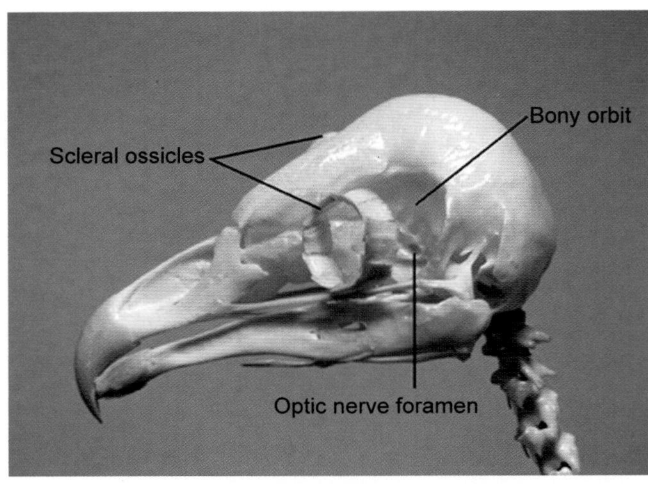

FIGURE 25.6-6 Skull of a barn owl (*Tyto alba*), showing the shallow and incomplete bony orbit, the position of the optic nerve foramen within the orbit, and the ring of scleral ossicles. Note that the scleral ossicles in this species are oriented for the forward eye placement, as seen in the northern saw-whet owl in Fig. 25.6-5 (bottom, left).

In birds, the **globe** (eyeball) fits snugly in the shallow orbit, and its movement is limited. Birds do have **extraocular muscles**, but they are poorly developed in comparison with mammalian eyes, so birds move their eyes, and thus extend their range of vision, primarily by moving their head.

For this reason, nystagmus (repetitive, involuntary movement of the eye) typically manifests as the movement of the entire head in birds.

Birds do not have a **retractor bulbi** muscle, which in mammals retracts the globe more deeply into the orbit. The snug fit within the orbit precludes the posterior movement of the globe in birds; instead, a relatively more extensive and mobile nictitating membrane (discussed later) helps protect the surface of the eye from injury.

Another important difference between birds and mammals is that the bony **septum** separating the left and right eyes is very thin in birds, and in some species, it is composed almost entirely of connective tissue.

Birds do not have an indirect or consensual PLR (where light shone in one eye causes the pupil in the other eye to constrict), as their optic nerves completely decussate (cross) at the optic chiasm. However, the bony septum separating the 2 eyes is thin enough in some avian species to allow bright light to pass from one eye to the other, which may result in a bilateral PLR that mimics the consensual reflex in mammals.

TERMINOLOGY

Seldom are the terms “anterior” and “posterior” used in veterinary anatomy, but the head is one location where the use of these terms occurs (see Fig. 1-3). When describing the eye, anterior refers to a position at or toward the front of the eye (e.g., the cornea) and posterior to a position at or toward the back of the eye (e.g., the retina).

Along the same lines, temporal refers to the lateral aspect (near or toward the temple) and nasal to the medial aspect of the eye.

Globe

The shape of the **globe** also varies by species (Fig. 25.6-7). Most avian species have a "flat" eye shape, in which the anterior–posterior length is shorter than the dorsal-ventral width. The curvature of the cornea is less (i.e., "flatter") in such eyes than in more classically shaped (spherical) eyes, although the posterior part of the globe stays more-or-less spherical.

Another common eye shape is the roughly spherica "globose" eye; it is common in hawks and falcons. Yet another, the "tubular" eye shape, characteristic of owls, is greatest in anterior–posterior length. These eyes are not tubular in the sense that they are tube-like or cylindrical; they are simply longer anteroposteriorly than they are wide dorsoventrally, at least in the anterior segment. The anterior segment is not as wide dorsoventrally as that of the fundic region, so in cross section these "tubular"-shaped eyes have a greatly expanded fundus in which the radius of the fundic curvature is much greater than that of the corneal curvature.

In birds, the shape of the globe is maintained by both cartilage and bony support within the sclera—in particular, by a circular series (ring) of overlapping thin bony plates called the **scleral ossicles**. There are 10–18 scleral ossicles, and they are located just caudal to the **limbus** (the junction of the cornea and the sclera) (Fig. 25.6-8). The scleral ossicles also support the ciliary muscles which alter the curvature of the lens (discussed later). Cartilage reinforces the posterior part of the sclera, with pockets of ossification occurring as **posterior scleral ossicles**.

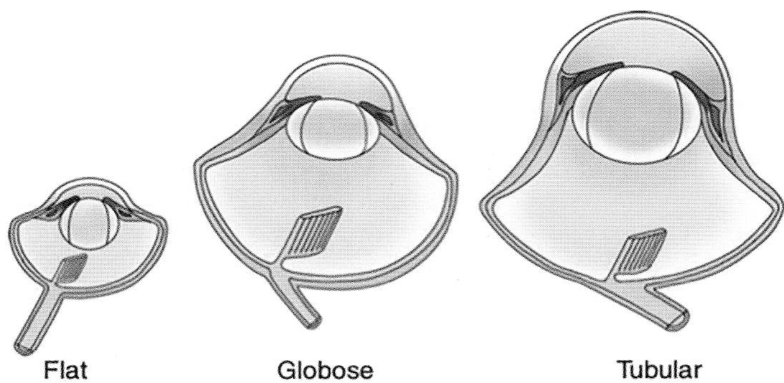

FIGURE 25.6-7 Avian eye shapes: "flat," "globose," and "tubular" examples.

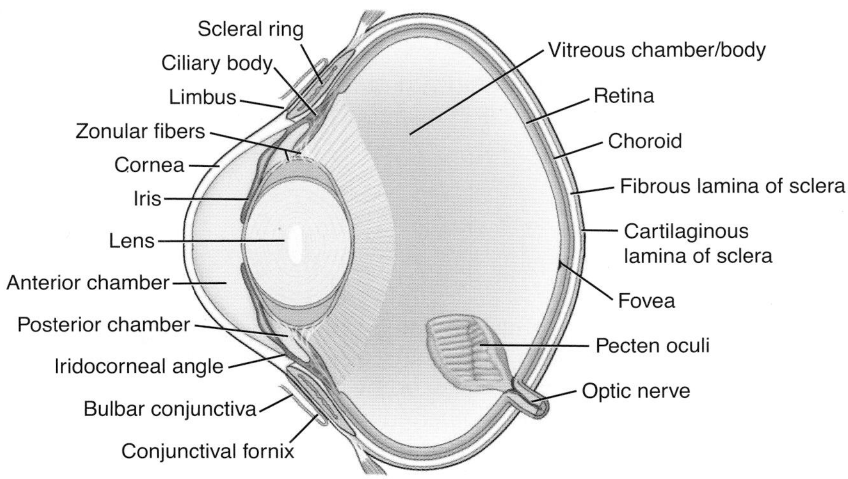

FIGURE 25.6-8 Illustration of the avian eye.

Anterior segment

In birds, the **lens** is soft, which allows a significant change in focal length, what is known as visual accommodation. In nocturnal species, the lens is almost spherical, while in diurnal species, it is flattened anteriorly. Visual accommodation in birds is also carried out by changing the curvature of the cornea and the tone of the iris muscles.

Conscious control of pupil diameter complicates the evaluation of the PLR in birds. Furthermore, the pharmacologic agents used as mydriatics (to dilate the pupil) in mammals are ineffective in birds, as these agents are smooth muscle relaxants.

Of note, the **ciliary muscles** control lens shape and the **sphincter muscles** of the **iris** control pupil diameter. Of particular note, these muscles are composed mostly of striated muscle, unlike the smooth muscle composition in mammals. Some of the ciliary muscle fibers also attach to the periphery of the cornea, so the contraction of the ciliary muscles also reduces the corneal curvature. These muscles—and therefore visual accommodation and pupil diameter—are all under conscious or voluntary control in birds.

In some avian species such as parrots, eye "flashing" or "pinning,"—in which the pupils are rapidly dilated and constricted—is seen when the bird is stimulated in some way, whether it is excited, curious, anxious, angry, or afraid. This behavior is often demonstrated in a clinical setting, which is stressful for most birds.

The avian **pupil** is circular, and a small degree of anisocoria (Gr. *anisos* unequal, uneven + Gr. *korē* pupil + *-ia* L. and Gr. state or condition = disparity in pupil size between left and right eyes in varying light) in uneven light is normal. As pupil size is under conscious control, the pupils may be used as a means of communication in birds.

Posterior segment

On fundoscopic examination, the **fundus** is gray or reddish in most avian species, and the **choroidal vessels** are not always visible. In many species of nocturnal birds, the **retina** is nonpigmented, allowing visualization of the underlying choroidal blood vessels, as seen in the normal owl in Fig. 25.6-4 (right).

IRIS COLOR

In birds, the color of the iris varies widely among species and includes black, brown, red, orange, yellow, blue, green, and even white.

In some species, iris color varies with gender, maturity, and/or breeding status (changing color in the breeding season). Thus, it is important to consider the species when determining whether iris color is normal in a particular bird.

OCULAR EVISCERATION

The relatively large ocular size in most birds makes surgical enucleation extremely dangerous, with high morbidity and mortality rates due to blood loss. Therefore, ocular evisceration (removal of the contents of the eye) is the preferred technique. The cornea is excised by cutting along the limbus, and the contents of the globe are carefully removed. The globe is then flushed with sterile saline solution and packed with a sterile gel foam. The nasolacrimal puncta are left intact to facilitate post-operative drainage. The margins of the eyelids are excised, and a complete permanent tarsorrhaphy is placed. This procedure results in a comfortable, cosmetically satisfactory appearing orbit.

Visual acuity in birds is estimated to be 2–8 times greater than that in mammals, and the structure of the retina plays a critical role. The avian retina is comprised of the same divisions as that in mammals: the anterior parts that extend over the ciliary body and posterior surface of the iris (**pars ciliaris retinae and pars iridica retinae**) are avisual, and the optical part (**pars optica retinae**) in the posterior region is the visual portion. However, the avian retina is avascular and lacks a tapetum (reflective layer often found in mammalian eyes).

Birds also have color vision, and most species have specialized cones in the retina that can detect ultraviolet light.

Visual acuity among avian species is largely determined by the presence/absence and location of the **fovea**, a depression in the retina where the cones are particularly dense and visual acuity is therefore greatest. Some birds, including many domestic species, lack a discrete fovea (they are afoveate). Many other birds have a single fovea (monofoveate) that is either centrally located (most common) or temporally located. For example, owls have a single, temporal fovea. Several birds that hunt during flight, including eagles, falcons, and various insectivores, are bifoveate, having a primary central fovea and a secondary temporal fovea.

The **optic disc** in birds is ovoid and elongated, but it is usually difficult to see because of the overlying **pecten oculus** (plural, **oculi**)—a highly vascular, typically pleated, darkly pigmented structure that protrudes from the retina into the vitreous fluid in the region of the optic disc (Fig. 25.6-4). The pecten oculus is involved in producing and supporting the vitreous fluid, so it nourishes the avascular retina by diffusion from the surrounding fluid.

The shape of the pecten varies by species. In most birds, it is pleated, as illustrated. However, in kiwis (small, flightless birds native to New Zealand), the pecten is small and conical; in ostriches and rheas (large, flightless birds native to Africa and South America, respectively), the pecten is "vaned" or fern-like. The conical and vaned pecten oculi arise from the optic disc, while the pleated type arises from the **choroidal fissure** (where the choroidal vessels enter the eye), even though the blood supply to the pecten arises separately from the choroidal vasculature.

Adnexa

Birds have upper and lower **eyelids**, but in most avian species the lower lid is larger and/or more mobile than the upper. Thus, when the eye is closed most of the eye is covered by the lower lid, which has a fibroelastic **tarsal plate** for support. The meeting of the upper and lower lids may form a curve when the eye is closed. When the eyelids are fully open, the **palpebral fissure** in birds forms a circle that is filled by the cornea; little, if any, sclera is visible (Fig. 25.6-5).

The palpebral response (closure of the eyelids when the lids or lashes are touched) is normally present in birds. However, the menace response (closure of the lids against perceived threat to the eye) is weak in birds and of limited clinical value in ocular and neurologic examinations.

Eyelashes are not universal in birds; however, when present, eyelashes are modified feathers, located near the palpebral margins of the eye. As with mammalian eyelashes, these modified feathers serve protective and tactile functions.

HATCHLINGS

In precocial avian species, in which the chicks are feathered and quite independent from the moment they hatch (e.g., chickens), the eyelids are open when the chicks hatch. In altricial species, in which the hatchlings are naked and unable to feed themselves (e.g., parrots), the eyelids are sealed closed, like those of neonatal puppies and kittens, and remain closed for 2–3 weeks after hatching. For example, the eyes of cockatoo chicks open at 10–17 days of age, and the eyes of macaw chicks open at 17–26 days of age.

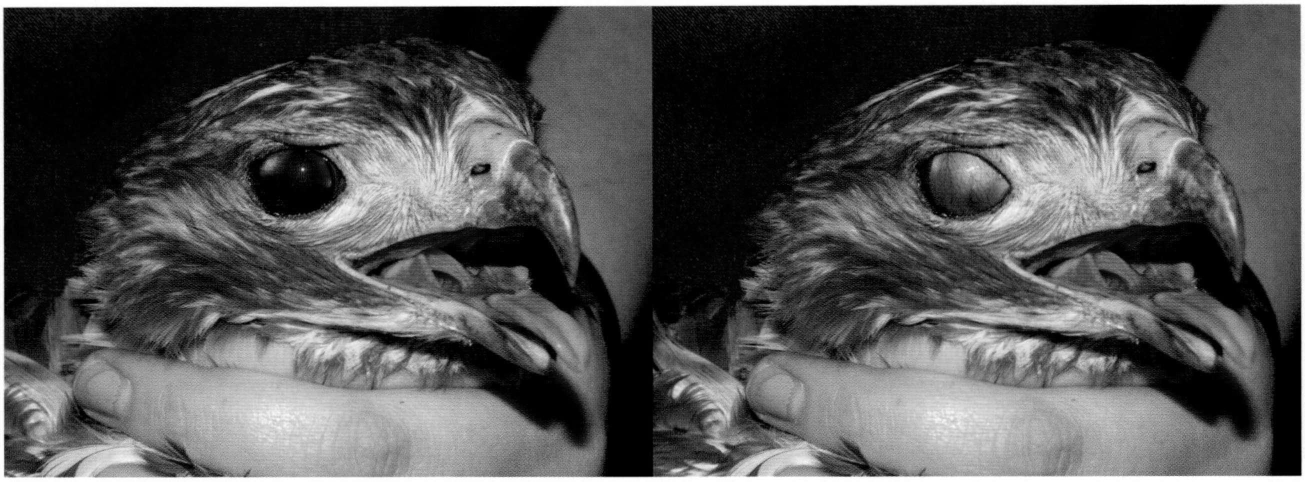

FIGURE 25.6-9 Red-tailed hawk (*Buteo jamaicensis*) with ocular trauma. Left: Hyphema (blood in the anterior chamber) is present in the right eye, obscuring the pupil and most of the iris (which in this species normally is gold or straw-colored). Right: The bird blinked when this photo was taken, so the cornea is completely covered by the nictitating membrane. The blood vessels within the nictitans can be seen coursing through this translucent membrane, which appears darker than normal because of the hyphema beneath.

The **nictitating membrane** (or nictitans; also called the "third eyelid") is well-developed and highly mobile in birds. It flicks across the surface of the cornea, covering it completely as it draws lacrimal fluids across the eye for lubrication and sweeps away surface debris. Most of the membrane is thin and translucent (Fig. 25.6-9), but its free edge is thickened into a marginal fold that may be pigmented. In birds, the nictitans moves over the cornea from dorsonasal to ventrotemporal, drawn by the **pyramidal muscle** that originates on the posterior aspect of the sclera and attaches to the lower nasal edge of the nictitans. The upper temporal edge of the nictitans is firmly attached to the adjacent conjunctiva and sclera.

The **Harderian gland**, which is located near the base of the nictitating membrane, is the main source of tear production in birds. Its duct opens into the conjunctival space between the nictitans and the cornea. In addition, a **lacrimal gland** is generally present in the lower temporal region of the conjunctival sac, although it is absent in owls and some other birds. As in mammals, ocular secretions are drained via **puncta** (small holes) in the **conjunctival sac** into the **nasolacrimal duct**, which empties into the nasal cavity.

Some avian species, such as budgerigars, also have a **nasal gland** or **salt gland** that lies dorsonasal to the globe within the orbit. The duct of this gland penetrates the frontal bone and enters the nasal cavity (see Case 25.1).

Selected references

[1] Stiles J, Buyukmihci NC, Farver TB. Tonometry of normal eyes in raptors. Am J Vet Res 1994;55(4):477–9.
[2] Murray M, Pizzirani S, Tseng F. A technique for evisceration as an alternative to enucleation in birds of prey: 19 cases. J Avian Med Surg 2013;27(2):120–7.
[3] McLelland J. Aves: sense organs and common integument. In: Getty R, editor. Sisson and Grossman's The anatomy of the domestic animals. 5th ed, volume II. Philadelphia: Saunders; 1975. p. 2063–6.
[4] Holmberg BJ. Exotic pet and avian ophthalmology. In: Maggs D, Miller P, Ofri R, editors. Slatter's fundamentals of veterinary ophthalmology. 6th ed. St Louis: Elsevier; 2018. p. 510–3.
[5] Walls GL. The vertebrate eye and its adaptive radiation. Michigan: Cranbrook Institute of Science, Bloomfield Hills; 1942. p. 641–62.

CHAPTER 26

THORACO-ABDOMINAL CAVITY

Cynthia M. Faux and Marcie L. Logsdon, Chapter editors

26.1 Ischemic stroke—*Cynthia M. Faux and Marcie L. Logsdon* 1316
26.2 Egg-yolk peritonitis—*Cynthia M. Faux and Marcie L. Logsdon* 1326
26.3 Air sacculitis—*Cynthia M. Faux and Marcie L. Logsdon* 1334
26.4 Ventricular foreign body—*Cynthia M. Faux and Marcie L. Logsdon* 1343
26.5 Sertoli cell tumor—*Cynthia M. Faux and Marcie L. Logsdon* 1349
26.6 Marek's disease—*Ricardo de Matos and James K. Morrisey* 1355

CHAPTER 26

CASE 26.1

Ischemic Stroke

Cynthia M. Faux[a] and Marcie L. Logsdon[b]
[a]University of Arizona College of Veterinary Medicine, Oro Valley, Arizona, US
[b]Exotics & Wildlife Department, Washington State University College of Veterinary Medicine, Pullman, Washington, US

Clinical case

History

A 17-year-old female African gray parrot (*Psittacus erithacus*; Fig. 26.1-1) was presented for lethargy, decreased appetite, and abnormal behavior of 1 day's duration. She spent all day sitting fluffed up on the floor of her cage, whereas she had previously been very active and vocal. Her diet primarily consisted of seeds and nuts with some fresh vegetables. The owner reported that the bird had recently stopped eating nuts. The other birds in the household were unaffected.

Physical examination findings

The patient was standing in her travel carrier either unaware of, or disinterested in, her surroundings. She was grooming with her left foot but was unable to switch to grooming with her right foot because she could not balance on her left foot without faltering. In addition, tremors were observed when she walked. When a bright light was shone into her left eye to assess its "dazzle" reflex, she showed no response, but when the light was moved to her right side, she was surprised and fell backward. (Assessing the dazzle reflex in birds is described in Case 25.6.)

FIGURE 26.1-1 An African gray parrot gently restrained in a towel for examination.

Comparative Veterinary Anatomy: A Clinical Approach. https://doi.org/10.1016/B978-0-323-91015-6.00131-X

The rest of the physical examination was performed with the bird wrapped in a towel for restraint. She was bradycardic with a heart rate of 320 bpm (normal in a stressed/restrained parrot of this size is > 400 bpm). The asymmetry in her dazzle reflex was confirmed on a closer examination. However, no abnormalities were found on fundic examination of either eye (see Case 25.6).

Following the trend in other animals, larger birds generally have slower heart rates than smaller birds. For example, a hummingbird may have a heart rate of > 1000 bpm, a budgerigar around 600 bpm, and a macaw around 350 bpm. In most avian species, the heart rate is so fast that only 1 heart sound per beat is heard on auscultation.

In addition, the bird had weak grip strength in her left foot compared with the right, and decreased strength in her beak.

A wooden tongue depressor is often used to gently encourage a bird to open its mouth for oral examination. In parrots, a tongue depressor can also serve as a useful distraction during physical examination by providing the bird something on which to chew. In this case, the tongue depressor also allows the assessment of masticatory muscle strength, or "crushing power" in the beak. Decreased crushing power may explain why this bird had stopped eating nuts.

Differential diagnoses

Stroke (ischemic event involving the CNS); primary neurologic disease, such as brain tumor or abscess

Diagnostics

No abnormalities were found on routine bloodwork or whole-body radiography. However, indirect measurement of blood pressure in the metatarsal artery (Fig. 26.1-2) revealed hypertension: the systolic blood pressure was 240 mmHg (normal, 120–180 mmHg). In addition, her peripheral pulses were weak.

It may seem counterintuitive that peripheral pulses would be weak in a patient with systemic hypertension ("high blood pressure"). However, hypertension is often accompanied by peripheral vascular disease, in which blood flow through one or more of the peripheral arteries is impeded by fatty plaques (atherosclerosis), blood clots (thromboembolism), or vasospasm (smooth muscle contraction in the arterial walls).

No further diagnostics were performed because this patient was too small for CT or MRI to be helpful in examining the brain for signs of ischemic injury.

Diagnosis

Ischemic CNS event secondary to hypertension (presumptive diagnosis based on the bird's signalment, history, and clinical signs)

African gray parrots are predisposed to cardiovascular disease, for reasons that are poorly understood.

MEASURING BLOOD PRESSURE IN BIRDS

Systemic blood pressure can be measured either directly or indirectly. Direct measurement is more precise and accurate, but it requires the placement of an arterial catheter, which in birds requires general anesthesia and technical skill. The most common site for catheter placement is either the deep radial artery or the superficial ulnar artery in the wing.

In conscious birds, blood pressure can be measured indirectly using Doppler ultrasound and a small sphygmomanometer (inflatable cuff with pressure gauge attached). Either the metatarsal artery or distal tibiotarsal artery is used. The distal ulnar artery can be used for indirectly monitoring blood pressure trends in anesthetized birds, but it is impractical in conscious patients as most birds resist having a wing stretched out and restrained.

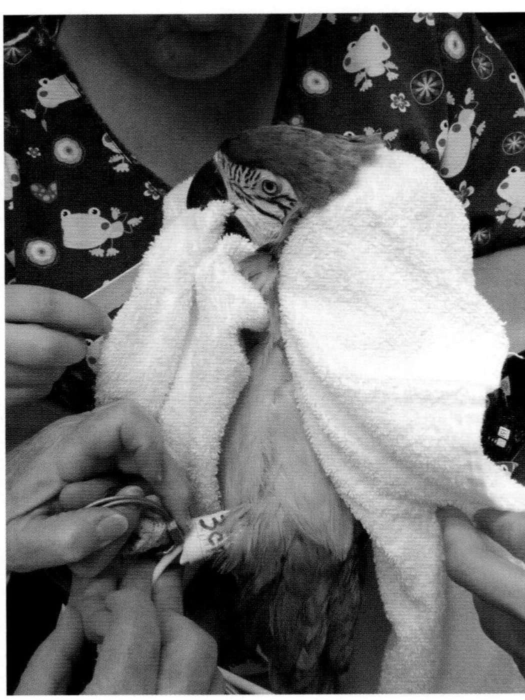

FIGURE 26.1-2 Indirect measurement of blood pressure in the metatarsal artery of a macaw. A small blood pressure cuff is applied to the tarsometatarsus, and a Doppler ultrasound probe is placed over the metatarsal artery distal to the cuff to monitor blood flow in the artery as the cuff is inflated and deflated. (Note the wooden tongue depressor being used to distract the bird during the procedure. The bird is also chewing on the towel that is used for restraint.)

In cases such as this, the prognosis for improvement in neurologic function generally is very good, with time and supportive care. However, if the underlying cause is not addressed, there is a higher likelihood of another ischemic event. In this case, vascular disease and hypertension resulting from a high-fat, unbalanced diet are likely the primary factors, along with an inherent species predisposition.

Treatment

The bird was treated with (1) a peripheral vasodilator to lower the blood pressure, (2) an omega-3 fatty acid supplement (flax oil) to reduce the risk of blood clots that could precipitate another ischemic event, (3) anti-inflammatory therapy to manage coexisting inflammation or arthritis that might be complicating the clinical picture, and (4) diet change to decrease fat intake and ensure a nutritionally balanced diet.

At recheck 2 weeks later, the bird's mental state was remarkably improved; ocular reflexes were normal in both eyes; and grip strength in the left foot and crushing power in the beak were improved, although there remained mild asymmetry between the left and right feet. Her peripheral pulses were stronger, and her indirect systolic blood pressure was 170 mmHg. The owner also reported a noticeable improvement in the bird's attitude and appetite.

Anatomical features in avian species

Introduction

This section covers the avian cardiovascular system. As the general structure and function are the same in birds and mammals, the features specific to birds are emphasized here.

Function

Despite the larger relative heart size in birds than in mammals, the tiny body size and high heart rate of many avian patients can make clinical examination of the heart difficult.

The avian cardiovascular system is highly adapted for flight, which requires a large energy expenditure. Thus, the avian heart is larger than that of a mammal of comparative size. Across the range of body sizes, the heart accounts for approximately 0.6% of the body mass in mammals. In comparison, the heart in a hummingbird weighs 2.4% of the bird's body mass. These small birds have the largest relative heart size of all avian species. In general, larger birds have smaller heart size relative to body mass than smaller birds. For example, the heart is relatively smaller in a goose than in a wren, although there is a variation among species.

External cardiac features

The basic structure of the **heart** in birds is like that in mammals—4 chambers with atrioventricular, aortic, and pulmonary valves—although its shape differs. In birds, the apex of the heart is narrower or pointed, so the heart appears much more conical than that of mammals (Figs. 26.1-3 and 26.1-4). As in dogs and cats, the avian heart

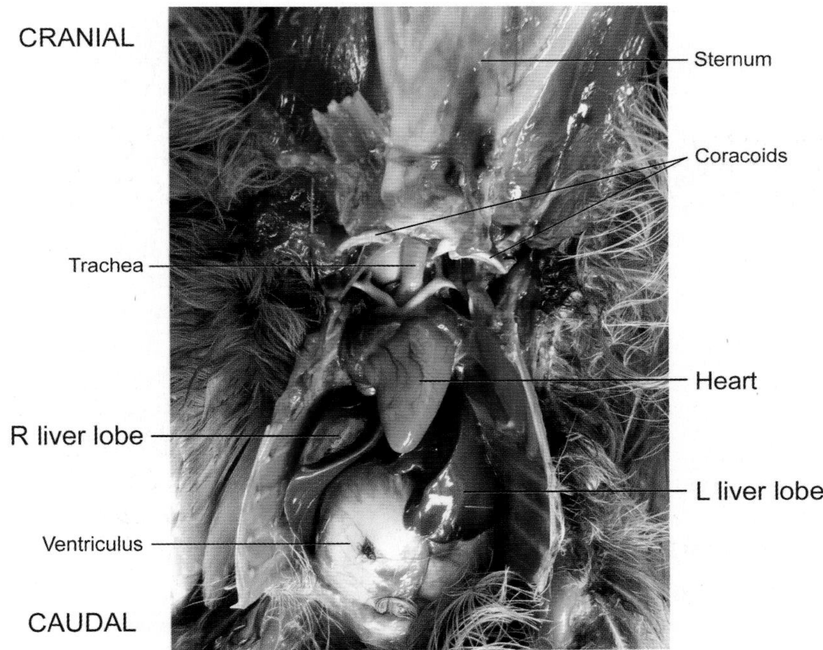

FIGURE 26.1-3 Ventral view of the coelomic contents of an American coot. The sternum has been separated from the ribs and reflected cranially to show the organs in the cranial and middle portions of the coelomic cavity, including the heart, liver, and ventriculus (gizzard). Importantly, birds lack a diaphragm and the lungs lie entirely dorsal to the heart. In this prosection, the pericardium and hepatopericardial ligament have been removed to expose the heart and great vessels, which are shown in a greater detail in Fig. 26.1-4. Key: *L*, left; *R*, right.

ADAPTATIONS FOR FLIGHT

Cardiac output volume of blood pumped by the heart per minute is higher in birds than in mammals. For example, cardiac output in a budgerigar during flight has been measured to be 7 times that of an exercising human. The average heart rate in birds is lower than in mammals of similar size, but birds have higher mean blood pressure (average pressure over the cardiac cycle) and lower peripheral vascular resistance (arterial resistance to flow). In this, birds meet the high energy and oxygen demands of flight with greater economy.

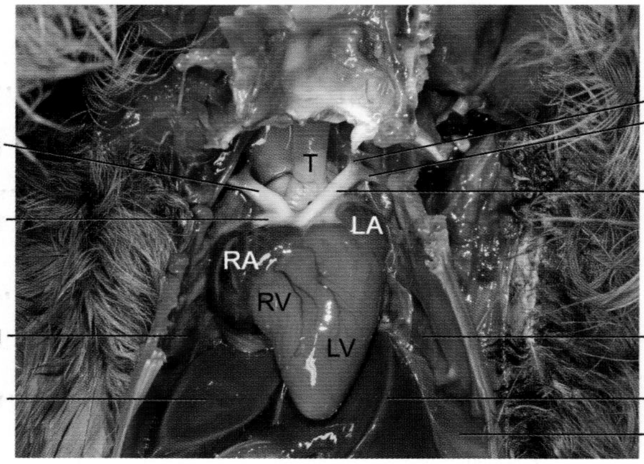

FIGURE 26.1-4 Close-up image of the prosection shown in Fig. 26.1-3, the ventral aspect of the heart of an American coot. Key: *a.*, artery; *A*, atrium; *L*, left; *R*, right; *V*, ventricle; *T*, Trachea.

is oriented obliquely, its apex directed caudoventrally in relation to its base. However, in birds, extensions of the pulmonary air sacs lie between the base of the heart and the sternum (see Case 26.3).

Perhaps the most noteworthy external difference in cardiovascular anatomy between birds and mammals is that birds do not have a diaphragm, so the heart lies in intimate contact with the liver (Fig. 26.1-3). The apex of the heart is situated between the left and right lobes of the **liver**, and the **lungs** are entirely dorsal to the heart and liver (Fig. 26.1-4). The continuation of the fibrous **pericardium** in birds is called the **hepatopericardial ligament**, as it anchors the pericardial sac to the ventral coelomic wall and liver rather than to the diaphragm as in dogs and cats.

Radiographically, the avian heart and liver form a contiguous structure, the cardiohepatic silhouette. On the dorsoventral/ventrodorsal view, this silhouette normally has an "hourglass" shape, highlighted by the surrounding air sacs (Figs. 26.1-5 and 26.1-6). Trauma to the body wall can cause the separation of the heart and liver by free fluid (e.g., blood) or escaped air (from injured air sacs) within the coelomic cavity (Fig. 26.1-7).

Atria (Fig. 26.1-8)

In birds, the **right atrium** is larger than the left (Fig. 26.1-4). Two **cranial** and one **caudal vena cavae** enter the right atrium, merging to form a distinct **sinus venosus** in most species. The walls of the sinus venosus invaginate deep into the atrial lumen, forming the muscular **sinoatrial valve** that extends most of the length of the dorsal atrial wall. Increased pressure within the right atrium during systole closes the valve, preventing the backflow of blood into the vena cavae. The right atrium is funnel-shaped as it leads to the **right atrioventricular (AV) orifice**.

In the **left atrium**, distinct **right** and **left pulmonary veins** usually enter the atrium separately and immediately form a single vessel, the **common pulmonary vein**, which invaginates into the atrium to the level of the **left AV orifice**. Sometimes, the pulmonary veins fuse into the common pulmonary vein before entering the atrium. This structure forms a **pulmonary chamber (camera pulmonalis)** which is separate from the atrium. Its left wall channels some of the blood from the pulmonary veins directly into the left ventricle. Like the sinoatrial valve in the right atrium, contraction of the left atrium pushes the wall of the pulmonary chamber against the **interatrial septum**, preventing the backflow of blood into the pulmonary veins during systole.

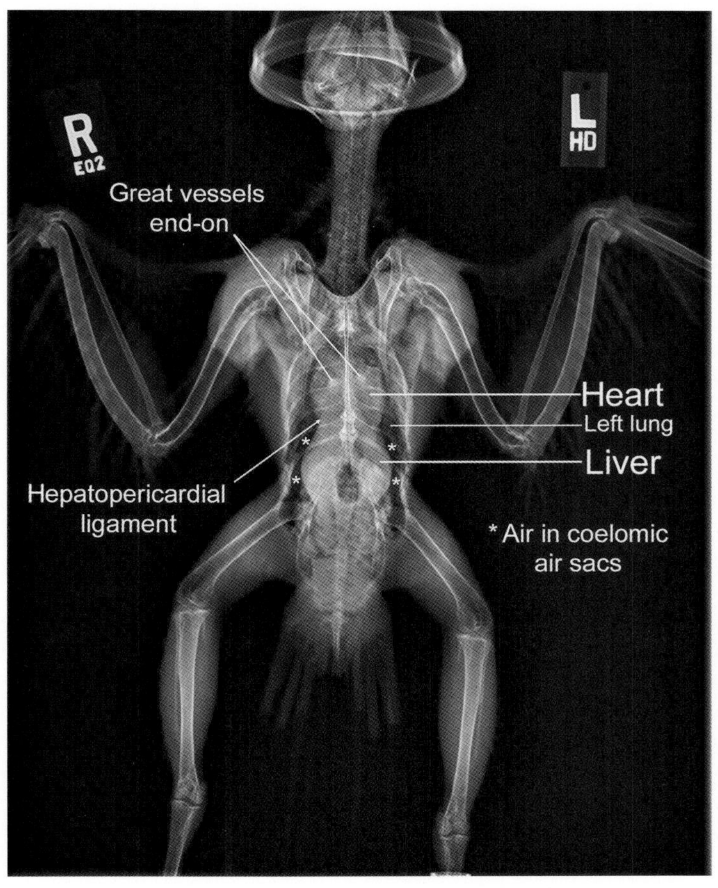

FIGURE 26.1-5 Ventrodorsal radiograph of a Peregrine falcon illustrating the normal "hourglass" cardiohepatic silhouette.

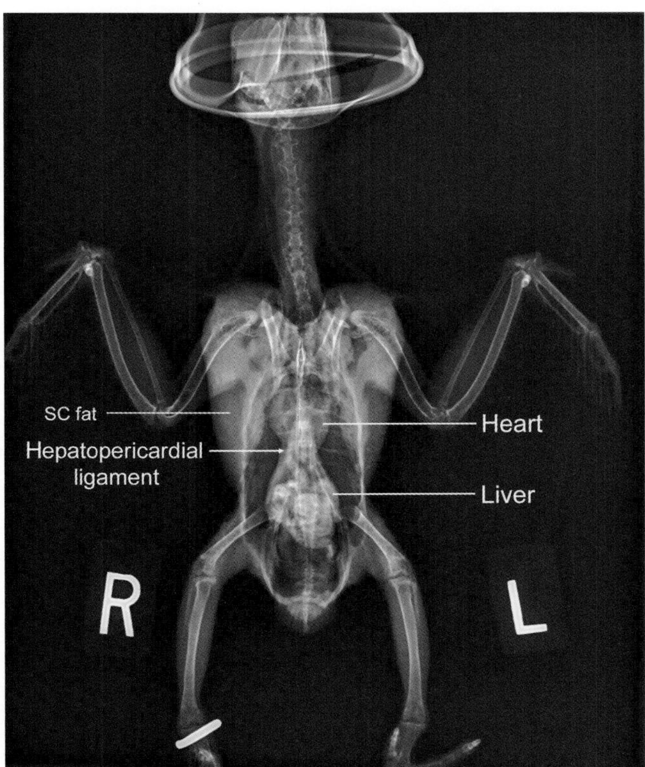

FIGURE 26.1-6 Ventrodorsal radiograph of an African gray parrot. This species typically has an exaggerated "hourglass" cardiohepatic silhouette. (Compare this parrot with the falcon in Fig. 26.1-5.) Also note that this pet parrot has less muscle mass and more superficial fat deposits than the raptor. Key: *SC,* subcutaneous.

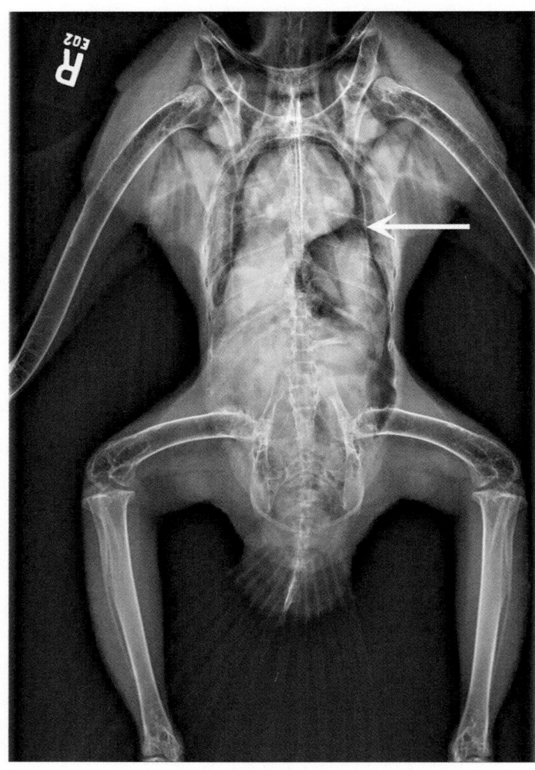

FIGURE 26.1-7 Ventrodorsal radiograph of the eagle from Case 27.1. Separation of the cardiohepatic silhouette is visible on the bird's left side (*arrow*). This separation is caused by free air in the coelomic cavity, likely the result of a ruptured coelomic air sac. In addition, there is decreased aeration of the caudal air sacs on the left side and a complete collapse on the right side, with displacement of the caudal coelomic organs to the right. Loss of normal serosal detail in these organs is likely attributable to hemorrhage. Compare to Fig. 26.1-5.

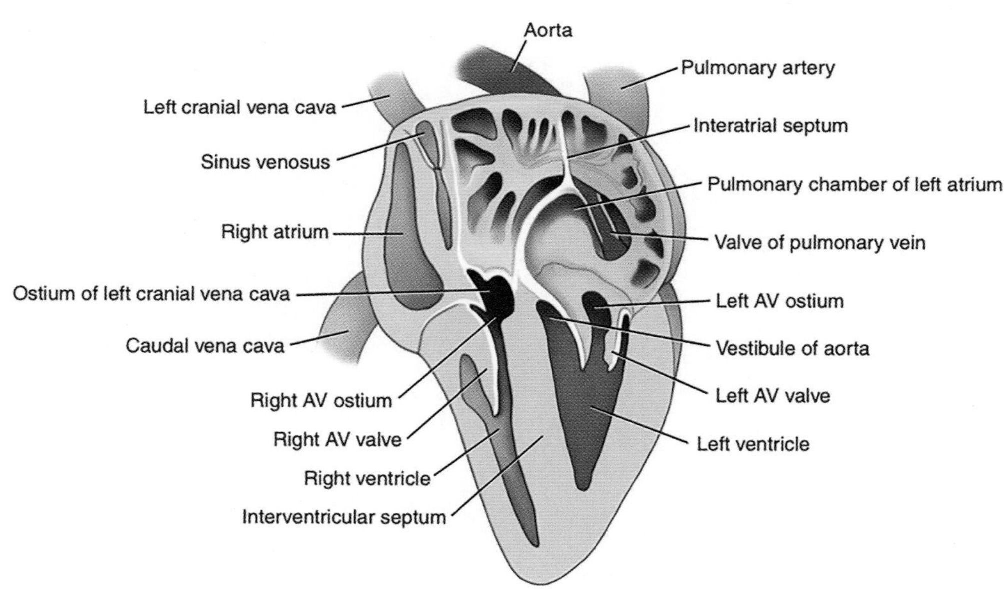

FIGURE 26.1-8 The avian heart, with an emphasis on the interior of the atrium, ventricle, and AV valve.

Ventricles (Fig. 26.1-8)

As in mammals, the **right ventricle** is smaller and has a thinner wall than the left, and it wraps part of the way around the left ventricle, so it has a crescent shape on cross-section. In birds, the right ventricle extends only about 2/3 of the distance from the base to the apex, so the cardiac apex in birds is formed solely of the myocardium of the left ventricle (Fig. 26.1-4).

In birds, the **right AV valve** is a distinctly muscular structure that, unlike the mammalian heart, does not have membranous valve cusps. Rather, it consists of a spiraling fold of myocardium that arises from the walls of the right ventricle and right atrium. During systole, ventricular contraction closes the right AV valve, directing blood through the **pulmonary trunk** and **pulmonary valve** into the pulmonary arteries. The pulmonary valve in birds is like that in mammals, having 3 membranous **cusps** or **valvules**.

Distinct from mammals, there are no chordae tendineae attached to the right AV valve in birds. Instead, a unique ring of **Purkinje fibers** (specialized cardiac conducting fibers) encircles the right AV orifice and supplies the myocardium of the right AV valve, which actively contracts during systole.

In birds, the wall of the **left ventricle** is 3–4 times thicker than the right, and the left ventricle has large **trabeculae carnae** on its inner wall (see Case 4.4). **Papillary muscles** support the **chordae tendineae** of the membranous **left AV**, or **mitral**, **valve**. Unlike mammals, the avian mitral valve has 3 cusps, the septal cusp being the largest.

In birds, the 3 membranous cusps of the **aortic valve** are completely encircled by myocardium, creating an outflow sphincter, which may allow the additional control of blood leaving the heart toward the general circulation.

Coronary circulation

The coronary circulation of the avian heart differs significantly from that in mammals. In birds, the myocardium is supplied by deep **coronary arteries** rather than superficial ones, and the **cardiac veins** typically do not travel with the arteries, so the surface veins are more prominent than the arteries (Fig. 26.1-4). The many small cardiac veins usually drain separately into the right atrium.

Aorta and its major branches

In birds, the **aorta** is the embryologic derivative of the right 4th aortic arch, rather than the left as in mammals, so in birds the ascending aorta curves to the right as it proceeds into the aortic arch and descending aorta (Fig. 26.1-4). The **coronary arteries** are the first branches of the aorta. The left and right **brachiocephalic trunks** arise next (Fig. 26.1-4). In "strong fliers," the brachiocephalic trunks are larger in diameter than the descending aorta.

> Radiographically, when the "great vessels" entering/exiting the heart are viewed in cross-section (e.g., Fig. 26.1-5), they may be mistaken for granulomas, such as the fungal granulomas seen in birds with respiratory aspergillosis. Bilateral circular structures of similar size, shape, and position in the region of the heart base are more likely to be the great vessels end-on, particularly if they are not evident on the orthogonal (right-angle) view, such as the lateral compared with the ventrodorsal (VD) projection. Typically, an abnormal mass would be evident on both the VD and lateral views, unless obscured by a superimposing structure. Additionally, increased opacity of the great vessels and cardiomegaly (enlargement of the heart) are common findings in parrots with peripheral vascular disease.

The paired **common carotid artery** arises from the brachiocephalic trunk (Fig. 26.1-4) and soon bifurcates into the **internal carotid artery** and the **vertebral-vagal trunk**. Like mammals, a baroreceptor of the carotid sinus is in the common carotid artery. Some birds have only one common carotid artery, but in most species the artery is paired. The left and right internal carotid arteries converge on midline at the base of the neck and travel cranially side-by-side within an osseous-fibrous canal—the **cervical carotid canal (subvertebral canal)**. The internal carotid arteries do not supply the tissues surrounding the canal; they simply travel along the canal. They exit at the 4th cervical vertebra and once again diverge to the left and right. The **external carotid artery** branches from the internal carotid artery before the latter enters the skull.

In the case described here, it is assumed that peripheral vascular disease involved one or more of these intracranial vessels. Although hypertension and peripheral vascular disease cause systemic abnormalities, the clinical signs were asymmetrical in this case, as is typical of "ischemic stroke" in humans.

Once inside the skull, each internal carotid artery divides into the **external ophthalmic** and **cerebral carotid arteries**. The avian equivalent of the Circle of Willis (circulus arteriosus cerebri, an anastomosis of several vessels suppling blood to the brain and surrounding tissues) is the anastomosis of the internal carotid arteries.

Birds possess numerous examples of **arteriovenous rete** or plexuses, which are thought to participate in regulating blood temperature—i.e., countercurrent heat exchange—in the brain, nasal passages, and cutaneous structures of the head such as the comb and wattles.

The right jugular vein is routinely used for venipuncture in birds (Fig. 26.1-9), especially in small patients such as budgerigars, because it is generally larger than the left. In larger birds, the basilic or ulnar vein may be used (Fig. 26.1-10). It is safe to take a sample that is 1% of bodyweight, or 10% of estimated blood volume.

Venous return

Venous drainage of the head is via paired **jugular veins**, of which the right is usually substantially larger than the left. Anastomosis between the left and right jugular veins may exist in some species. The jugular veins combine with the **subclavian veins** to form the left and right **cranial vena cavae**. The right cranial vena cava is also larger than the left.

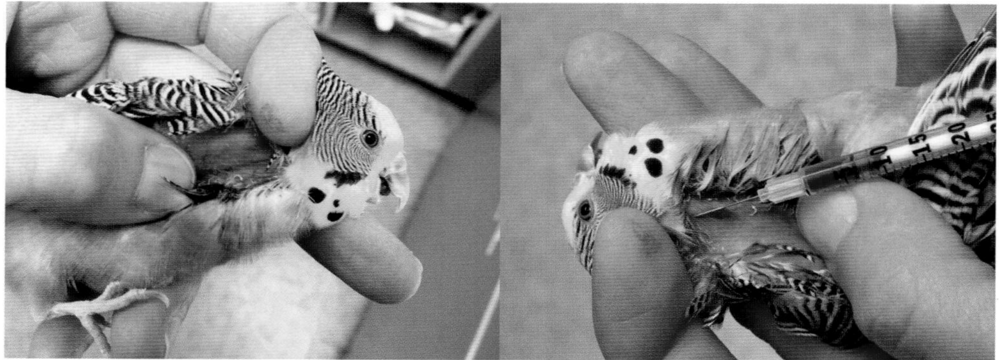

FIGURE 26.1-9 Jugular venipuncture in a budgerigar. Left: The apteric (featherless) area overlying the jugular veins makes the right jugular vein an ideal site for blood sample collection in small birds. Right: blood collection from the right jugular vein.

STROKE (CEREBRAL VASCULAR ACCIDENT [CVA]) IN HUMANS

"Stroke" is a term used to describe a medical condition due to inadequate blood flow to the brain resulting in cell death. There are 2 main types of stroke—ischemic, due to the lack of blood flow, and hemorrhagic, due to bleeding from the rupture of a blood vessel. Both result in brain injury and decreased function. Signs and symptoms in people vary and include an inability to move or sense a side or part of the body, decreased comprehension, difficulty speaking, dizziness, and loss of vision. If signs are transient, the stroke is termed a "transient ischemic attack" (TIA) or "mini-stroke." Time is critical in treating this emergency condition to minimize the long-term sequela (L. lesion or affection caused by a disease) from the lack of blood flow to the brain.

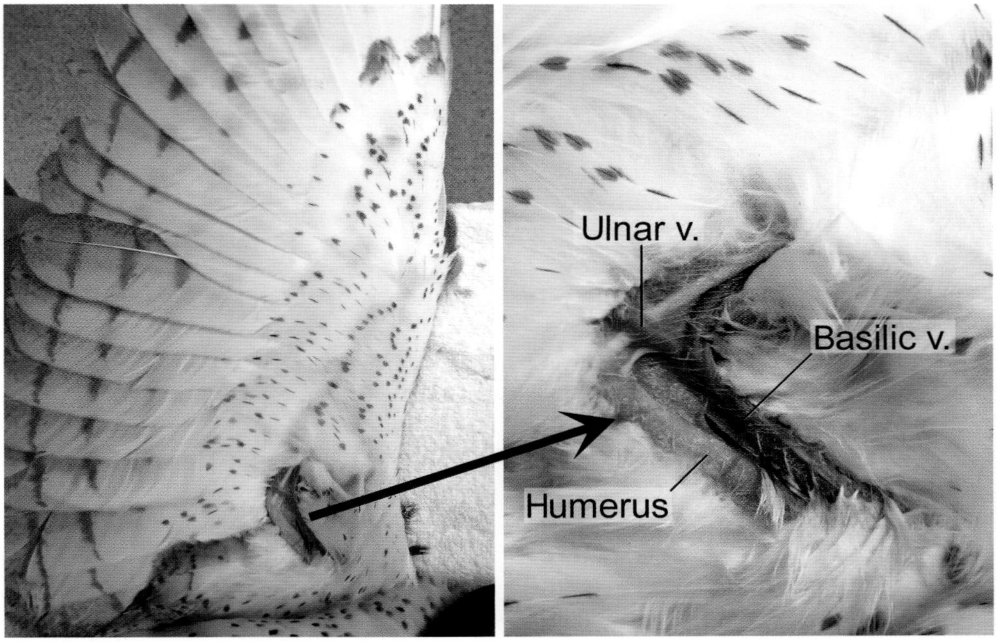

FIGURE 26.1-10 Right basilic/ulnar vein of a common barn owl. Left: The wing is extended to show the location of the apteric area overlying the vein on the ventral aspect of the elbow joint. Right: magnified view of the vein, with the elbow flexed. The feathers have been dampened with alcohol and swept aside to allow visualization of the superficial vein just beneath the skin. This vein can be used for blood sample collection in larger birds.

The **hepatic portal system** in birds is like that in mammals, in that venous return from the digestive tract—including the ventriculus ("gizzard"), part of the proventriculus, pancreas, spleen, and intestine—passes through the liver before entering the **caudal vena cava**. However, the **cranial proventricular vein** drains into the left cranial vena cava or sinus venosus of the right atrium, bypassing the hepatic portal system. In addition, the portal vein is paired in birds. The **right portal vein** corresponds to the mammalian portal vein, while the smaller **left portal vein** drains branches of the **celiac vein** in birds.

Other important highlights of the peripheral vasculature are discussed in specific cases. They include the **rete tibiotarsale** and **medial tibiotarsal vein** (Case 27.3) and the **renal portal system** (Case 26.5).

Selected references

[1] Exotic Animal formulary, 5th ed. Carpenter J ed. Table 5-48 Blood Pressure Values Reported in Birds' pp 330.
[2] Avian medicine 3rd ed. Samour J ed. 'Disorders of the cardiovascular system' pp 397–399.
[3] Baumel JJ, editor. Handbook of avian anatomy: Nomina Anatomica Avium. 2nd ed. Nuttall Ornithological Club (USA), World Association of Veterinary Anatomists (USA), International Committee on Avian Nomenclature; 1993.

CHAPTER 26

CASE 26.2

Egg-yolk Peritonitis

Cynthia M. Faux[a] and Marcie L. Logsdon[b]
[a]University of Arizona College of Veterinary Medicine, Oro Valley, Arizona, US
[b]Exotics & Wildlife Department, Washington State University College of Veterinary Medicine, Pullman, Washington, US

Clinical case

History

A 2-year-old laying hen (*Gallus gallus domesticus*) presented for lethargy, inappetence, and a fluffed appearance. She was housed with 5 other laying hens, so the owner was unsure when the hen last laid an egg. The remainder of the flock appeared healthy, and the owner had no birds other than the single flock of chickens. All the hens were acquired as chicks from a single source approximately 2 years before. Biosecurity measures primarily consisted of limited exposure to other facilities with chickens.

Physical examination findings

The hen was depressed and lethargic but responsive to stimulation. She ignored food, including live mealworms, which are a very appealing food for chickens. She was moderately dyspneic, displaying a slight head bob with each breath at rest that rapidly progressed to open-mouth breathing when stressed. There was a slight decrease in fat deposits over her keel, so she was given a body condition score of 2 (1 = emaciated, 5 = obese). Her oral mucous membranes were pink but tacky, indicating moderate dehydration. Her coelomic cavity, palpable between the caudal point of the keel and the ischia (Fig. 26.2-1), was distended and taut. There was fecal and urate staining around her vent. In addition, her crop was distended, and she passively regurgitated a small amount of greenish-brown, malodorous liquid when she was picked up. Digital palpation of the cloaca was unremarkable.

Although a distended crop can be a sign of "sour crop" or other GI disease, it is a common finding in any chicken with coelomic disease. Because birds lack a diaphragm, any buildup of pressure in the coelomic cavity can compress its contents, such as the air sacs (which accounted for the dyspnea seen in this patient) and the GI tract. In conjunction with general debilitation and dehydration, coelomic distention delays or halts emptying of the crop. This hen regurgitated because of the increased pressure on her coelomic cavity when she was lifted. (The crop is described in Case 25.2.)

Digital palpation of the cloaca is performed in birds of ample body size to check for an egg in the caudal reproductive tract or other palpable mass(es) in the caudal coelomic cavity. It involves gently inserting a gloved, lubricated finger through the vent and carefully palpating the cloaca while applying light pressure to the caudal coelomic cavity with the other hand (i.e., a bimanual examination).

Comparative Veterinary Anatomy: A Clinical Approach. https://doi.org/10.1016/B978-0-323-91015-6.00122-9

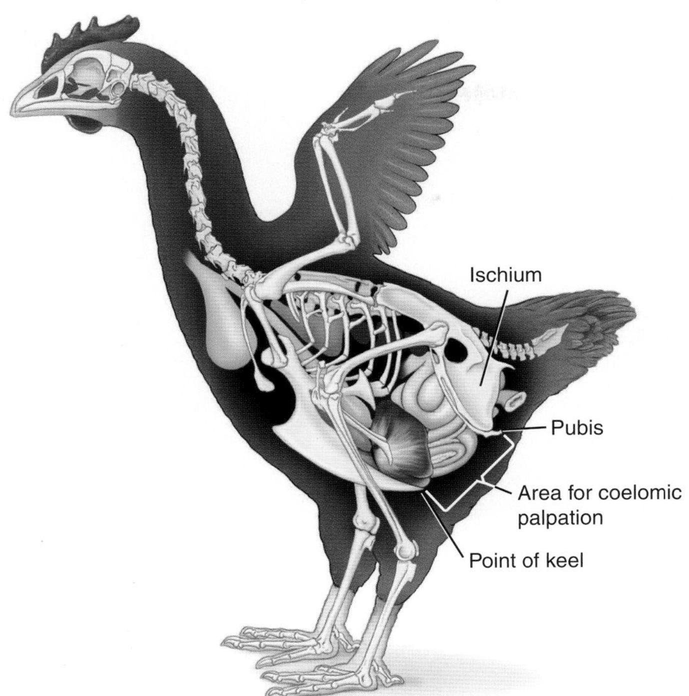

FIGURE 26.2-1 Coelomic cavity of a chicken (lateral view), showing the external landmarks for palpation (keel of the sternum and ischia), the vent, and the contents of the coelomic cavity.

Differential diagnoses

Egg binding (dystocia), egg-yolk peritonitis (coelomitis), oviductal/uterine rupture, ovarian neoplasia, systemic disease not involving the reproductive tract

This patient's signalment (female chicken of laying age) made primary reproductive disease most likely. Breeds of laying hens selected for high egg production are prone to reproductive problems such as egg-yolk peritonitis. Even so, systemic disease should not be ruled out until a definitive diagnosis is reached.

Diagnostics

Radiographs revealed severe fluid distension of the coelomic cavity which removed serosal detail and compressed the caudal thoracic and caudal abdominal air sacs (Figs. 26.2-2 and 26.2-3). There was mild hyperostosis of the long bones, indicating an active reproductive status.

Hyperostosis of the long bones is a normal physiologic response seen in reproductively active female birds, in which the bird lays down large stores of extra calcium in the bones at the onset of reproductive activity (Fig. 26.2-4). This process allows rapid mobilization of calcium for shell formation during egg production. It should not be confused with a pathologic condition; rather, its absence in reproductively active birds is a sign of inadequate dietary calcium. (Note: Avian bones normally have a much thinner cortex than mammalian bones, so hyperostosis may not be readily apparent if one is using mammals as the frame of reference.)

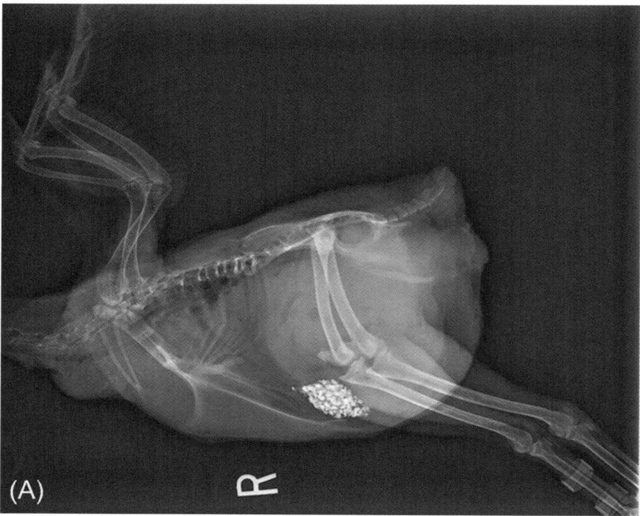

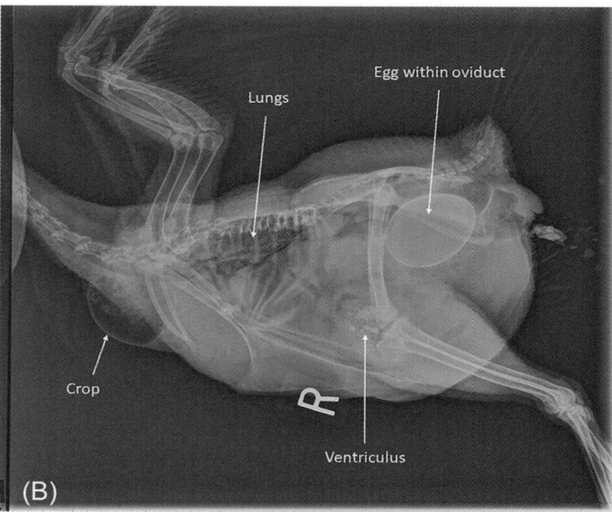

FIGURE 26.2-2 (A) Lateral radiograph of a laying hen with severe fluid distension of an identifier for the "coelomic cavity" caused by egg-yolk peritonitis (coelomitis). There is increased soft-tissue opacity and loss of serosal detail in the caudal coelomic cavity, along with compression of all coelomic organs, including the lungs and air sacs. The ventriculus (identifiable by the mineral-opacity grit it contains) is displaced ventrally. In addition, identify the crop (at the ventral thoracic inlet) is distended with granular material. (The two cylindrical bands on her right leg are identification bands.) (B) Compare with the lateral radiograph of a normal hen. The serosal detail within the coelomic cavity is much greater (not obscured by fluid). The lungs are also more clearly visualized. Although the presence of a large egg can also somewhat displace the organs within the abdominal region of the coelom, note that the ventriculus lies more dorsally in the coelomic cavity in the normal hen than in the hen with severe fluid distension.

Heterophils are avian granulocytes that fulfill a similar function to mammalian neutrophils (see Case 26.6).

Coelomic centesis yielded a large quantity of clear, yellow fluid with an elevated total protein and lipid content. Cytological examination revealed variably sized basophilic protein globules and a mixed inflammatory response consisting of heterophils and macrophages. No definitively neoplastic cells were seen.

OVULATION

The mechanics of ovulation are similar in birds and mammals, but the ovum is vastly different. Mammals ovulate a microscopic ovum, so if it misses the oviduct and enters the peritoneal cavity, it does not cause a problem. In contrast, the developing follicles on the avian ovary contain a developing yolk which comprises a large amount of organic material that is not readily absorbed by the body once separated from its vascular supply. Therefore, if an oviduct is injured and a bird ovulates into the coelomic cavity, it may cause peritonitis (coelomitis). Although initially a sterile peritonitis, secondary infection is common and may lead to sepsis.

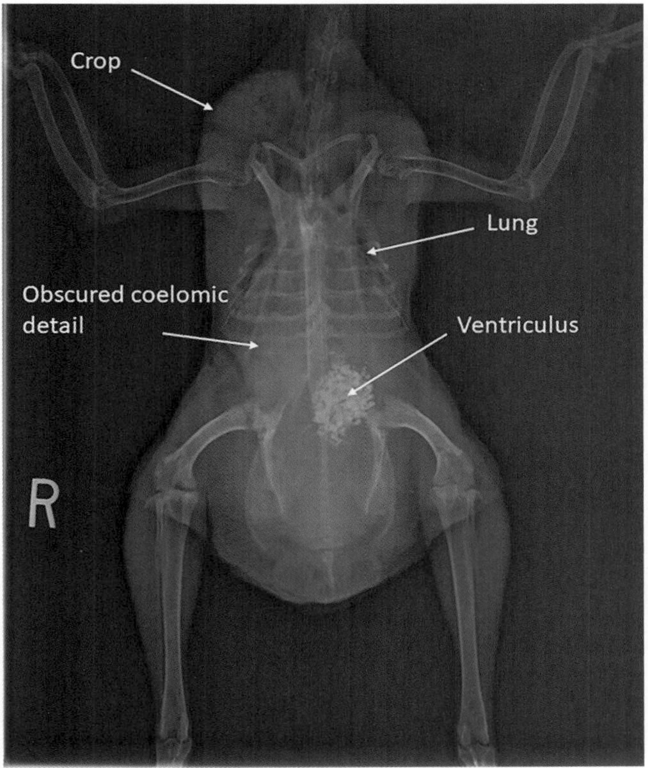

FIGURE 26.2-3 Ventrodorsal radiograph of the hen in Fig. 26.2-2. Severe compression of the coelomic airsacs is evident. There is no formed egg (calcified shell) in the caudal reproductive tract, so this hen is not "egg-bound." Distension of the crop is even more apparent on this view, at the base of the neck on the right side. Note the lack of serosal detail.

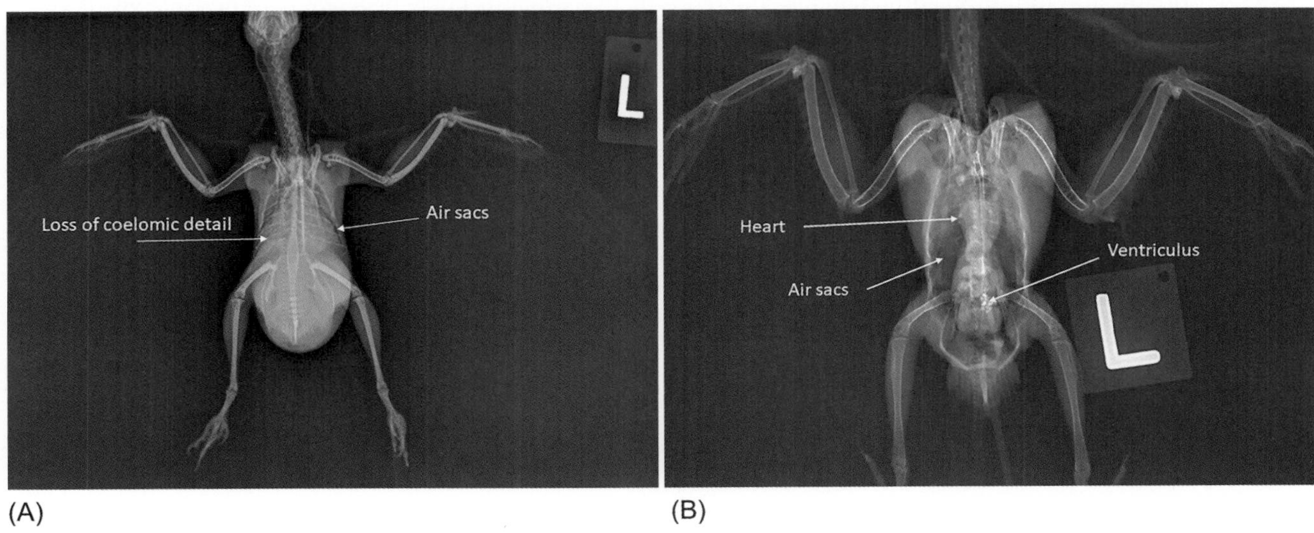

FIGURE 26.2-4 (A) Ventrodorsal radiograph of a female cockatiel with reproductive disease and coelomitis, showing the normal hyperostosis of reproductive activity more clearly. The humerus is a pneumatic bone, so it still has a radiolucent medullary cavity, but the other long bones are diffusely radiodense. Similar to Fig. 26.2-3, this bird has severe compression of the coelomic airsacs and loss of serosal detail. This bird would be expected to be dyspneic on exam. (B) A similar radiograph of a normal bird for comparison. Note how the airsacs help define the outline of the viscera within the coelomic cavity and serosal detail is clear.

Surgical treatment involves exploratory coelomotomy (Gr. *koilōma* the body cavity + Gr. *temnein* to cut) and salpingohysterectomy (surgical removal of the oviduct and uterus). Because the ovary is both difficult to access and is in close proximity to the kidneys, aorta, and caudal vena cava, the ovary itself is usually left intact. In most avian species, reproductive activity ceases once the oviduct is removed. With incomplete salpingohysterectomy, the ovary may continue to produce follicles, predisposing the bird to recurrent egg-yolk peritonitis unless hormonal therapy is used, and dietary/husbandry changes are made (see side box entitled "Medical management of egg-yolk peritonitis").

Diagnosis

Egg-yolk peritonitis (coelomitis)

Treatment

Treatment for this disease is strongly driven by owner preference. None of the current treatments can restore egg production, so if egg production is the bird's primary purpose, then humane euthanasia is the most practical option. If the hen is a pet, then either medical or surgical treatment may be indicated, depending on disease severity.

Anatomical features in avian species

Many species of raptors retain 2 functional ovaries but only one oviduct.

Introduction

The female reproductive tract of birds comprises the ovary and oviduct, the latter consisting of 5 regions: infundibulum, magnum, isthmus, uterus (shell gland), and vagina (Fig. 26.2-5). As embryos, female birds develop paired ovaries and oviducts; however, in most species, only the ovary and oviduct on one side—usually the left, become functional. In most species, development of the right ovary and oviduct ceases and these structures regress to a remnant at the cloaca (the common opening of the digestive and urogenital tracts in birds). Occasionally, the right ovary and oviduct do not completely diminish and larger vestiges can be identified grossly. Seasonally, and in nonbreeding females, the reproductive structures may degenerate to become virtually unidentifiable; this may be a general adaptation for flight (minimizing unnecessary body mass).

MEDICAL MANAGEMENT OF EGG-YOLK PERITONITIS

For mild or early cases, coelomic drainage, antimicrobial therapy, and hormonal down-regulation of ovulation may be effective. Drainage of the accumulated coelomic fluid provides immediate respiratory relief by reducing pressure on the lungs and air sacs. Antimicrobial therapy is used to prevent/treat secondary infection and any underlying salpingitis. Gonadotropin-releasing hormone agonists may be used to prevent ovulation. However, cases managed in this way commonly regress and require repeated treatment or surgery.

ENVIRONMENTAL CUES

Many avian species have specific environmental cues that trigger the reproductive activity. They vary by species but include seasonal, and particularly day-length, changes; increased carbohydrate or sugar content in the diet (indicating plentiful resources); physical stimulation (most commonly inappropriate petting behaviors by parrot owners); and increased rainfall (an important factor in species native to areas with definitive dry and rainy seasons, which can be artificially simulated in captivity by the sound of running water).

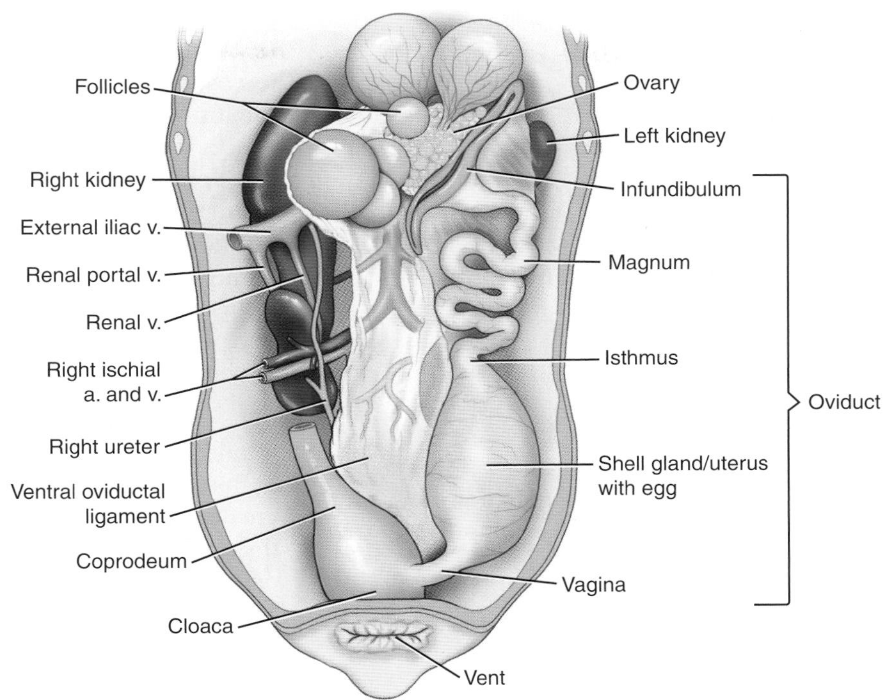

FIGURE 26.2-5 Female reproductive system of a chicken (ventral/dorsal view), showing the ovary (and its relation to the kidneys), oviduct (infundibulum, magnum, isthmus, uterus, vagina), cloaca, and vent.

Function

The avian female reproductive system produces (and, when sperm is available, allows for fertilization of) eggs, providing for future generations. Birds lay in a pattern of "one-a-day," which may be another adaptation for flight—having a reproductive tract laden with a large cluster of mature follicles and developing eggs could negatively impact the female's ability to fly.

Ovary (Fig. 26.2-5)

As the female bird matures and becomes reproductively active, the **left ovary** is completely encased in follicles, which mature sequentially. Birds generally lay one egg per day until the species-specific clutch size is attained (ranging from one egg to over a dozen). 

A clutch is a group of eggs incubated together in the nest.

The left ovary is generally located at the cranial pole of the left kidney. As the ovary begins to enlarge with follicles, the entire ovary is located on the dorsal midline of the coelomic cavity and extends to the right side. The ovary may be in contact with the left lung and is dorsal to the left abdominal air sac, which forms an **ovarian pocket** that helps the infundibulum guide the newly released follicle.

"INTERNAL LAYING"

Asynchrony of the infundibulum with ovulation may result in an ovum missing the infundibulum and entering the coelomic cavity. The ovum is simply resorbed in most of these instances of "internal laying." Sometimes, however, and for reasons unknown, it leads to egg-yolk peritonitis (coelomitis).

Sperm may be stored for an extended period of time within the oviduct or within specialized "sperm host" glands situated at the uterovaginal junction or within crypts located in the vaginal wall.

The chalazae (singular, chalaza) are a pair of thin, white, rope-like strands that suspend the yolk in the center of the "egg white" (the thick albumen that surrounds the yolk).

Not all eggs are fertilized. A female bird does not need to have mated in order to lay an egg. In addition, the presence of a "blood spot" (a small streak of blood or small mass of bloody or fleshy tissue) adjacent to the yolk results from small blood vessel rupture during ovulation or passage through the infundibulum and is not an indication of fertilization.

Oviduct (Fig. 26.2-5)

Once mature, the follicle is released from the ovary and enters the **oviduct** through the funnel-shaped **infundibulum**. Fertilization of the ovum, when it occurs, takes place in the infundibulum. The ovum does not simply drop into the infundibulum; rather, the infundibulum manipulates the ovary to guide the mature follicle into the oviduct during ovulation. The ovarian pocket (discussed in the previous section) assists this process.

In the chicken, the ovum resides within the infundibulum for approximately 15 minutes. At this stage, the **infundibular glands** cover the ovum with the chalaziferous layer of thick albumen. Thus, fertilization must occur before this point if the egg is to produce a chick.

The narrow neck of the infundibulum opens into the magnum. The **magnum** is the longest portion of the oviduct—up to two times the body length of the hen. It is coiled, with deep internal folds. Most of the egg's albumen is deposited around the yolk during its 3-hour passage along the magnum.

The **isthmus** is demarcated from the magnum by a narrow, translucent band of tissue. Additional albumen production and the initiation of shell membrane formation are believed to occur in the isthmus. The egg spends about 1 hour here. Psittacines (parrots) lack an isthmus.

In avian species or breeds that lay colored eggs, the pigments are laid down in the outer layers of the shell. Thus, shell color is merely superficial.

For the next 20 hours, the egg resides in the **uterus**, or **shell gland**. Additional fluid may be added before the egg is encased in the shell and its final protective cuticle.

EGG BINDING

Female birds may become "egg-bound" a condition in which the completely formed egg is delayed in the caudal oviduct (specifically, the uterus or vagina). Egg-binding (dystocia) is often a consequence of nutritional or metabolic issues, such as hypocalcemia or obesity. Certain species—including budgerigars, canaries, and love birds—seem more prone to the disorder. At the beginning of reproductive activity, the young pullet will usually lay eggs in a variety of sizes, both larger and smaller than breed normals. Sometimes these larger eggs result in dystocia or rupture of the reproductive tract.

The retained egg may be palpable in the caudal coelomic cavity. It is also identifiable radiographically. However, radiographs cannot distinguish an egg inside the oviduct from one that is free in the coelomic cavity following the rupture of the oviduct. Therefore, transcoelom (L. *trans-* through + Gr. *koilõma* body cavity) perforation and aspiration of the egg is contraindicated.

Transcloacal (L. through + L. drain by a common passage for fecal, urine, and reproductive evacuation) manual removal of the egg is the preferred method of treatment under general anesthesia. A speculum—e.g., otoscope—is inserted via the vent to view the vaginal opening into the cloaca, where the accessible part of the eggshell is punctured and aspirated. Surgery is needed if the egg cannot be seen or the oviduct is ruptured.

Once completely formed, the egg passes through the **vagina** into the **cloaca**, from which it is laid (blunt end first) into the nest. When the egg is being laid, the slit-like vaginal opening into the cloaca protrudes through the vent (the external opening of the cloaca), which helps minimize fecal contamination of the egg as it passes.

The pink protrusion of the vaginal opening through the vent during egg-laying can stimulate other hens in the flock to peck at it, potentially injuring the vaginal opening. This injury may not be apparent on digital examination of the cloaca but may be identified using a speculum. This can be prevented by avoiding overcrowding and providing private nesting boxes.

Brood patch

Female birds of many species pluck feathers to line the nest, resulting in a bald area on the ventral body surface known as the **brood patch**. The skin in this plucked area also thickens. The brood patch allows warm skin to contact the eggs and later the nestlings (chicks in the nest). The plucked feathers create a warm, comfortably lined nest bowl. In some species in which both parents share brooding duties, such as the Great horned owl, males also create a brood patch. Feathers generally start to regrow in these areas once the chicks have hatched.

The brood patch should not be mistaken for feather-destructive behavior ("feather picking" or "feather plucking") or other dermatologic abnormalities. (The avian skin and feathers are discussed in Case 28.1.) Some galliformes spend so much time on the nest incubating eggs that it is normal for the brooding parents to lose weight from this significantly decreased food and water intake.

Selected references

Summa NM, Guzman DS, Wils-Plotz EL, et al. Evaluation of the effects of a 4.7-mg deslorelin acetate implant on egg laying in cockatiels (*Nymphicus hollandicus*). Am J Vet Res 2017;78:745–51.
Caruso KJ, Cowell RL, Meinkoth JH, et al. Abdominal effusion in a bird. Vet Clin Path 2002;31:127–8.
King AS. Aves urogenital system. In: Sisson and Grossman's The anatomy of the domestic animals. 5th ed, vol. II. Philadelphia: WB Saunders; 1975. p. 1919–64.
Dyce KM, Sack WO, Wensing CJG. Textbook of veterinary anatomy. 4th ed. Philadelphia: Saunders; 2010.

CHAPTER 26

CASE 26.3

Air Sacculitis

Cynthia M. Faux[a] **and Marcie L. Logsdon**[b]
[a]University of Arizona College of Veterinary Medicine, Oro Valley, Arizona, US
[b]Exotics & Wildlife Department, Washington State University College of Veterinary Medicine, Pullman, Washington, US

Clinical case

History

A 7-year-old female sun conure (*Aratinga solstitialis*; Fig. 26.3-1) presented for an acute onset of lethargy and anorexia. She was fed a commercial seed-nut mix diet. The other 2 parrots in the household were acting normally.

FIGURE 26.3-1 The sun conure is a medium-sized South American parrot that is a popular pet species.

Comparative Veterinary Anatomy: A Clinical Approach. https://doi.org/10.1016/B978-0-323-91015-6.00123-0

Physical examination findings

The bird was responsive to handling but was quiet and weak. Her body condition score was 4 (overweight) (1 = emaciated, 5 = obese). She was sitting fluffed-up on the bottom of her travel cage, with a markedly increased respiratory effort.

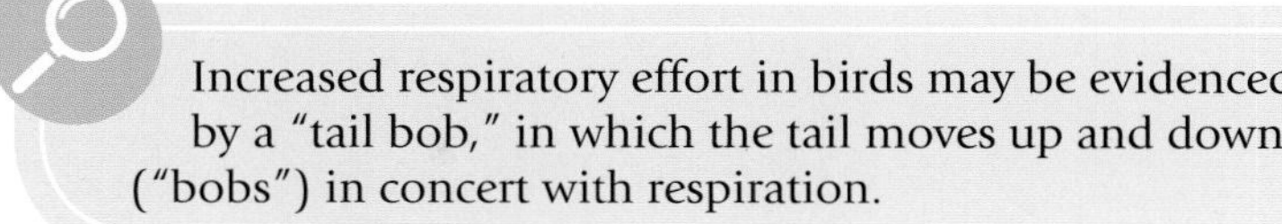

Increased respiratory effort in birds may be evidenced by a "tail bob," in which the tail moves up and down ("bobs") in concert with respiration.

Differential diagnoses

Ischemic event ("stroke," see Case 26.1); air sacculitis/pneumonia (fungal or bacterial infection of the air sacs/lungs); polytetrafluoroethylene (PTFE, or Teflon®) toxicity; other systemic disease causing generalized debilitation

Diagnostics

The bird was anesthetized for diagnostic evaluation. Whole-body radiography revealed asymmetry of the lateral clavicular air sac diverticula and a scalloped appearance to the coelomic air sacs. In addition, the cardiohepatic silhouette was mildly enlarged.

The clavicular air sac is located cranial to the heart, but its lateral diverticula extend around the heart and along the sternum, in addition to various extracoelomic extensions (described later). The coelomic air sacs include the cranial and caudal thoracic and the abdominal air sacs.

Complete blood count revealed a markedly elevated total white blood cell count (48,000 cells/µl; normal = 5000–13,000 cells/µl) and moderately elevated total protein concentration.

With the bird still under general anesthesia, coelomic endoscopy was performed to examine the coelomic air sacs. The membranes were thickened and opaque, with increased vascularity and multiple tan plaques. Cytological examination of the plaque material revealed branching fungal hyphae, consistent with *Aspergillus* spp. (Fig. 26.3-2).

Because the coelomic air sacs in birds contain air, and are further expanded with air during each inhalation (described later), endoscopy of these structures can be performed without insufflation.

Diagnosis

Aspergillosis/fungal air sacculitis

Treatment

The bird was promptly started on antifungal therapy and intensive supportive care. The antifungal medications were administered both systemically and via nebulization.

Healthy air sacs have very low vascularity, so it is difficult to achieve therapeutic concentrations of medication within them using systemic administration alone, even in cases of severe air sacculitis with increased vascularity. Nebulization allows fine particles of the medication to be inhaled directly into the air sacs, as well as into the lungs where systemic administration may also fail to achieve therapeutic levels.

Despite intensive care, the patient died shortly after initiating therapy. Necropsy revealed severe air sacculitis and hepatitis (inflammation of the liver). Fungal hyphae consistent with *Aspergillus* spp. (similar to those shown in Fig. 26.3-3) were found in the air sacs and liver.

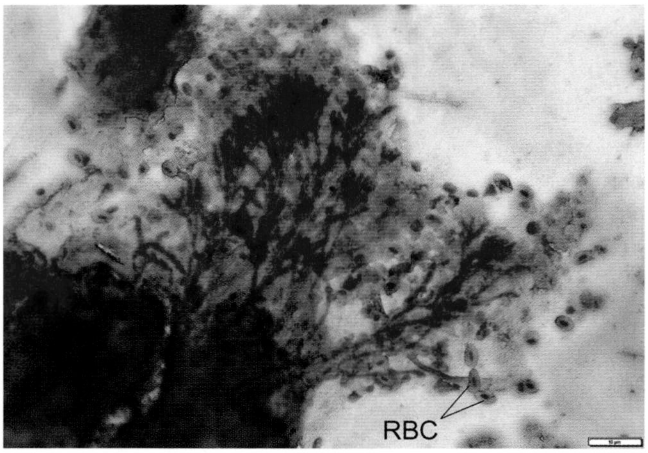

FIGURE 26.3-2 Photomicrograph of the plaque material from the coelomic air sacs. The numerous, branching fungal hyphae (stained dark blue) are consistent with *Aspergillus* spp. Note the nucleated avian red blood cells (*RBC*), which are also numerous in this sample. Scale bar: 10 μm.

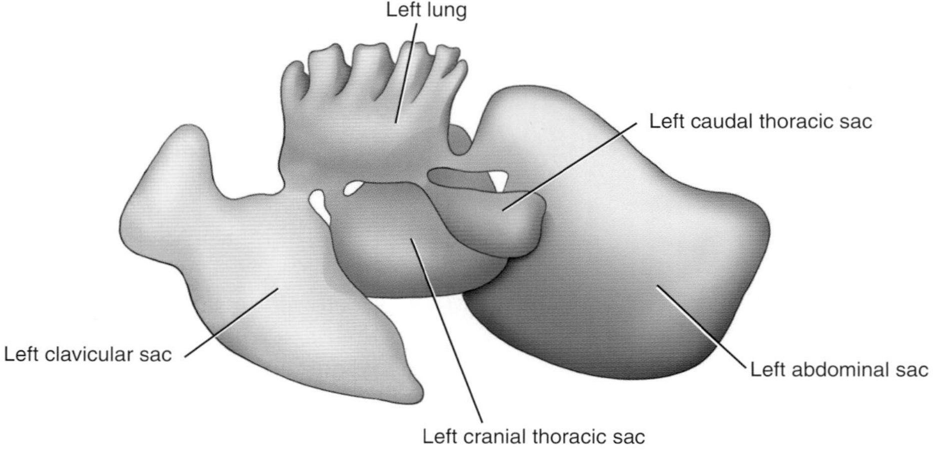

FIGURE 26.3-3 Main air sacs depicted in birds include the cervical, clavicular, cranial and caudal thoracic, and abdominal. There is variability in the number of air sacs in the avian species.

Anatomical features in avian species

Introduction

This section covers the lower respiratory tract, from the primary bronchi to the lungs and air sacs. The components of the upper respiratory tract, including the nares, nasal cavity, pharynx, larynx, trachea, and syrinx, are described in Case 25.3.

Function

The function of the lungs and air sacs is to deliver oxygen to the systemic circulation and remove carbon dioxide.

Overview of the avian respiratory system

The avian respiratory system differs in fundamental ways from the mammalian respiratory system, both in structure and in function. Although physiology is beyond the focus of this book, the physiology of the avian respiratory system and anatomy are intricately linked, and thus, one cannot be described without the other.

Relative to body weight, the avian lungs are similar in mass to those of mammals. However, by volume, the avian lung is approximately 25% smaller. In addition, for reasons that will become clear, avian lungs inflate or expand very little during inspiration. Up to 25% of the bird's lung volume is tucked between the ribs, so deep costal grooves mark the dorsolateral surfaces—an impossible arrangement in mammals, whose lungs require room to expand. Additionally, the lungs in birds lie entirely dorsal to the heart.

Probably the most significant difference between the mammalian and avian respiratory system is the one-way respiratory pathway in birds: Air passes in only one direction through the lungs (except for the neopulmo region in species that possess it; see side box entitled "Paleopulmo and neopulmo"). This one-way air flow over gas exchange tissue in avian lung means that fresh, oxygenated air is continuously presented to the air capillaries (roughly analogous to alveoli of mammalian lungs). How this airflow pattern is achieved is described later, following a description of all the components.

In mammals, the coelomic cavity is divided by a muscular diaphragm into separate thoracic and abdominal cavities. The avian coelom (G. *coelom* cavity) is not divided by a diaphragm, so although we may refer to the "thoracic" and "abdominal" regions of the coelom in birds, these are not separate compartments. As described in Case 26.1, the avian heart is not separated from the liver by a diaphragm; instead it is intimately associated with the liver, the ventricles nestled between the left and right lobes of the liver. Without a diaphragm, movement of the respiratory system involves the entire coelom in birds.

In many, if not most, avian species, the pleural space (between lung and chest wall) is elusive, as the parietal and visceral pleura fuse during embryonic development. However, a pleural space may be present, particularly in the dorsolateral region, in some species (e.g., chickens).

As described in Case 25.3, the syrinx (vocal organ in birds) is located at the tracheobronchial bifurcation in most species and thus forms the origins of the primary bronchi. Distal to the syrinx, the main components of the avian respiratory system include the primary bronchi, lungs (comprising secondary bronchi, parabronchi, and air capillaries, which are the sites of gas exchange), and an elaborate series of air sacs, all of which arise from the lungs.

Air sacs (see Landscape Figs. 6.0-5 and 6.0-6)

The **air sacs** consist of thin, translucent, relatively avascular membranes that hold air but do not take part in gas exchange (Fig. 26.3-3). Their specific role in respiration is described later. The number and configuration of the air sacs varies widely across species. Six pairs of air sacs are typically found in the embryo, with great variability in the number and communication among final numbers developing through different fusions. The major air sacs typically described are the clavicular, cervical, cranial thoracic, caudal thoracic, and abdominal.

Most species have a single (unpaired) **clavicular air sac**, also called the interclavicular air sac. It is located cranial to the heart; its cranial extent occupies the thoracic inlet. In chickens, this air sac has a large median chamber and a pair of small lateral chambers.

PALEOPULMO AND NEOPULMO

"Paleopulmo" and "neopulmo" are terms used to describe portions of the lung considered phylogenetically old or new, respectively. All birds are considered to have the paleopulmo region of the lung, where airflow is unidirectional.

In birds that possess it, the neopulmo consists of the lateroventral region of the lung and the connections to the caudal air sacs (caudal thoracic and abdominal). Airflow through this portion of the lung is bidirectional.

Some birds common to veterinary practice, such as owls and ducks, have a poorly developed neopulmo. Others, such as chickens, have a well-developed neopulmo, which can comprise up to 25% of the total lung. The physiological and clinical implications of the neopulmo region are not fully understood.

In many birds, including chickens, the **cervical air sac** is also unpaired. It lies immediately ventral to the vertebral column in the caudal cervical and cranial thoracic region, dorsal to the clavicular air sac. Fusion between the cervical and clavicular air sacs occurs in some species, forming a **cervicoclavicular air sac**.

The **cranial** and **caudal thoracic air sacs** are both paired, resulting in a set of 4 thoracic air sacs (left and right, cranial and caudal). Generally, all 4 are confined to the cranial or "thoracic" region of the coelom bounded by the ribs. Typically, the cranial sac is smaller in volume than the caudal sac, although the opposite is the case in chickens. The caudal thoracic air sac may be totally absent in turkeys. In some species, the cranial thoracic air sacs communicate with the clavicular air sac.

The paired **abdominal air sacs** lie dorsally in the "abdominal" region of the coelom. Their size varies by species. Although the abdominal air sacs are typically the largest of the air sacs, there are species (e.g., the kiwi) in which they are the smallest of the air sacs.

If a clinical procedure depends upon air sac anatomy, a careful review of the species is needed before proceeding.

Diverticula or outpouchings of the air sacs are common and add an appreciable variation among species. They occur primarily in the cervical, clavicular, and abdominal air sacs; the thoracic air sacs generally do not have diverticula. These pouches may extend between muscle layers (intermuscular), beneath the skin (subcutaneously), and into bones (via pneumatic foramina).

Vertebral diverticula can extend from the cervical air sac as far cranially as the atlas (1st cervical vertebra) and may enter—and thus "pneumatize"—the cervical vertebrae. Subcutaneous and intermuscular diverticula from the cervical air sac are described in several species.

Intra- and extrathoracic diverticula may extend from the clavicular air sac. The **intrathoracic diverticula** lie around the heart (**cardiac diverticula**) and along the sternum (**sternal diverticula**), while the **extrathoracic diverticula** pneumatize the humerus via a pneumatic foramen (see Case 27.1). Additional extrathoracic diverticula may be found stretching between the muscles of the pectoral girdle.

Intraosseous (IO) infusion involves the administration of fluid directly into the medullary cavity of a bone via a small catheter (small needle). This technique is used for a rapid resuscitation or maintenance fluid administration when IV access is not practical. Owing to the presence of extracoelomic air sac diverticula, IO administration of fluid in birds can result in drowning if one of the pneumatized bones such as the humerus or femur is used. A nonpneumatized bone such as the ulna or tibiotarsus must be used instead, as long as the bone is confirmed to be nonpneumatized in that species. For example, the pelican ulna is typically pneumatized.

Diverticula of the **abdominal air sacs** include the **perirenal diverticula**, which surround the kidneys and may have extensions that can pneumatize the vertebral column and pelvis, and the **femoral diverticula** that pneumatize the femur.

SUBCUTANEOUS EMPHYSEMA

Rupture of a diseased or injured air sac can lead to subcutaneous emphysema and/or free air in the coelomic cavity (Fig. 26.3-4).

Owing to the extension of the cervical and clavicular air sac diverticula in particular, SC emphysema is a common consequence of air sac disease in some species. The emphysema may be relatively localized, fairly widespread (Fig. 26.3-5), or generalized.

Once the air sac membrane has healed, the escaped air is slowly resorbed over a period of approximately 2 weeks. No specific treatment is needed, although antimicrobial therapy may be indicated because air sac rupture can be a consequence of air sacculitis.

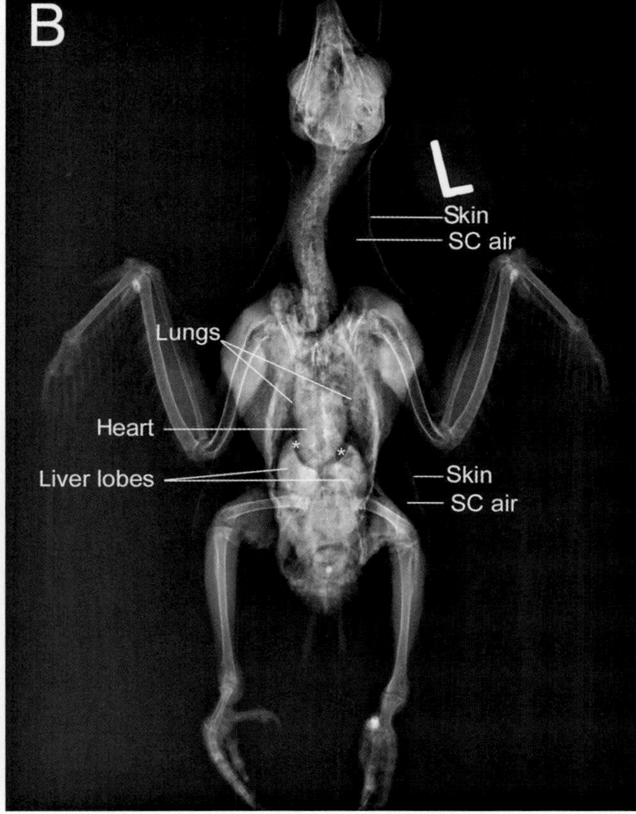

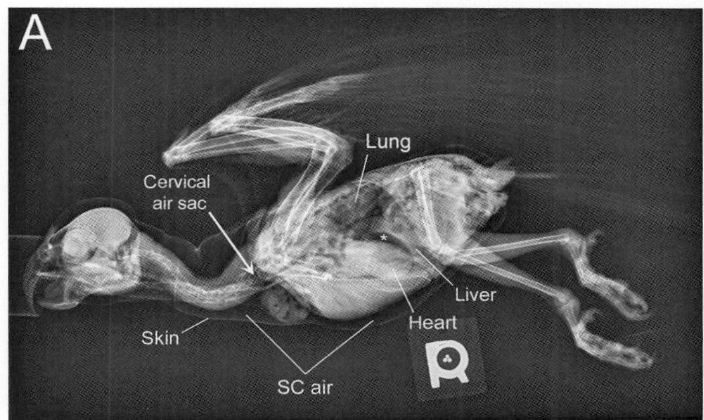

FIGURE 26.3-4 Lateral (A) and ventrodorsal (B) radiographs of a Goffin's cockatoo (*Cacatua goffiniana*) with diffuse air sacculitis. Note the asymmetrical aeration of the clavicular air sacs on the dorsoventral view. Rupture of the coelomic air sac has resulted in generalized subcutaneous emphysema; from head to tail, the skin is grossly distended with escaped air. In addition, free air in the coelomic cavity has caused separation of the cardiohepatic silhouette (*asterisks*) and, on the ventrodorsal view, even of the left and right lobes of the liver. Key: *L*, left; *R*, right; *SC*, subcutaneous.

Bronchi and lungs

The trachea bifurcates into the paired **primary bronchi** either at—or distal to—the syrinx, depending on the species. Each primary bronchus (left or right) enters the lung on that side, becoming progressively narrower in diameter as the secondary bronchi branch off. The primary bronchi eventually terminate at the **abdominal air sacs**.

There are 4 major groups of **secondary bronchi,** named for the region of lung they supply: medioventral, mediodorsal, lateroventral, and laterodorsal. For example, there are 4 or more medioventral secondary bronchi which branch from the primary bronchus and supply the medioventral portion of that lung.

The small (tertiary or greater order) bronchi that branch from the secondary bronchi are classified simply as **parabronchi**. The parabronchi of each group of secondary bronchi loop anastomose with parabronchi of the other groups of secondary bronchi, creating a network of parabronchi, which enables the one-way air flow through the lungs in birds.

Clinical use of the caudal air sacs

Gaseous anesthetics can be infused directly into the caudal air sacs via a cannula placed through the body wall. The anesthetic gas is then drawn through the lung and absorbed into the systemic circulation as the bird breathes.

Cannulation of the caudal air sacs is a useful technique for diagnostic or surgical procedures involving the head or upper respiratory tract, where an endotracheal tube would interfere in performing the surgery.

Caudal air sac cannulation can also be used in emergency situations when the upper airway is obstructed. Supplemental oxygen may also be supplied to the patient by this route.

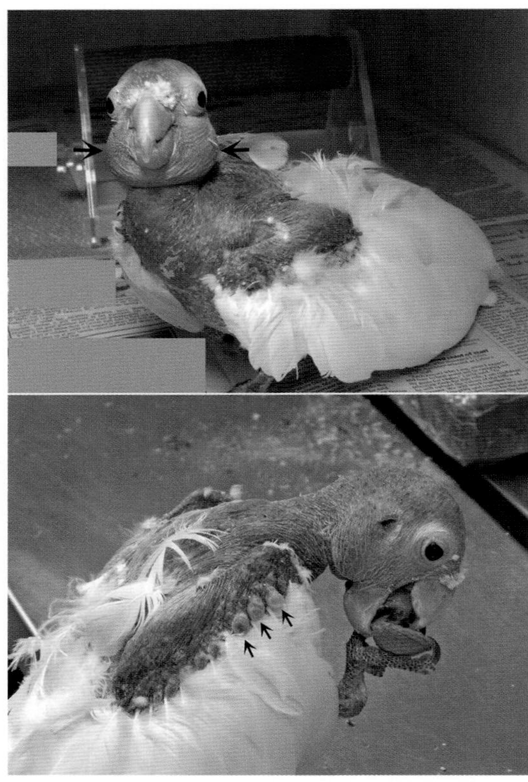

FIGURE 26.3-5 Images of the cockatoo from the radiograph in Fig. 26.4. Top: Initial presentation of the bird showing extreme, diffuse SC emphysema. (Note the unrelated chronic feather loss secondary to feather destructive behavior/plucking which reveals the full extent of the subcutaneous emphysema (*arrows*). Below: Clinical appearance after one week of antibiotic treatment. Note that the emphysema has almost resolved however there are still small pockets of SC emphysema around the feather follicles on her wing (*arrows*).

The gas exchange tissues in birds are termed **air capillaries**. The air capillaries are a network of fine tubules that extend from the parabronchi. They are lined with respiratory epithelium and are intimately entwined with blood capillaries where gas exchange occurs. The air capillaries in birds are much narrower than the alveoli in mammals—3 μm in a small passerine bird such as a finch vs. 35 μm in the smallest mammal. Each air capillary is preceded by a small "atrium" in the parabronchus. Smooth muscle at the **astral openings** controls airflow through air capillaries.

This aspect of avian pulmonary anatomy is important when considering nebulization for the treatment of respiratory infections in birds. Nebulized particles must be smaller than the diameter of the air capillaries in order to reach all parts of the avian respiratory system.

Mechanics of respiration

(Fig. 26.3-6) With the notable exception of the diaphragm, the muscular component of respiration is similar in birds and mammals. In birds, air is drawn into the respiratory tract when the thoracic muscles expand the rib cage, moving the ribs craniolaterally and drawing the sternum cranioventrally. This movement expands the coelom, creating a negative pressure within. Exhalation involves a coordinated contraction of the thoracic and abdominal muscles, much like in mammals.

As birds lack a diaphragm, restricting the movement of the sternum and body wall, whether by a hand, tightly wrapping the bird in a towel, or heavy draping during surgery, can restrict the bird's ability to spontaneously ventilate, potentially causing suffocation.

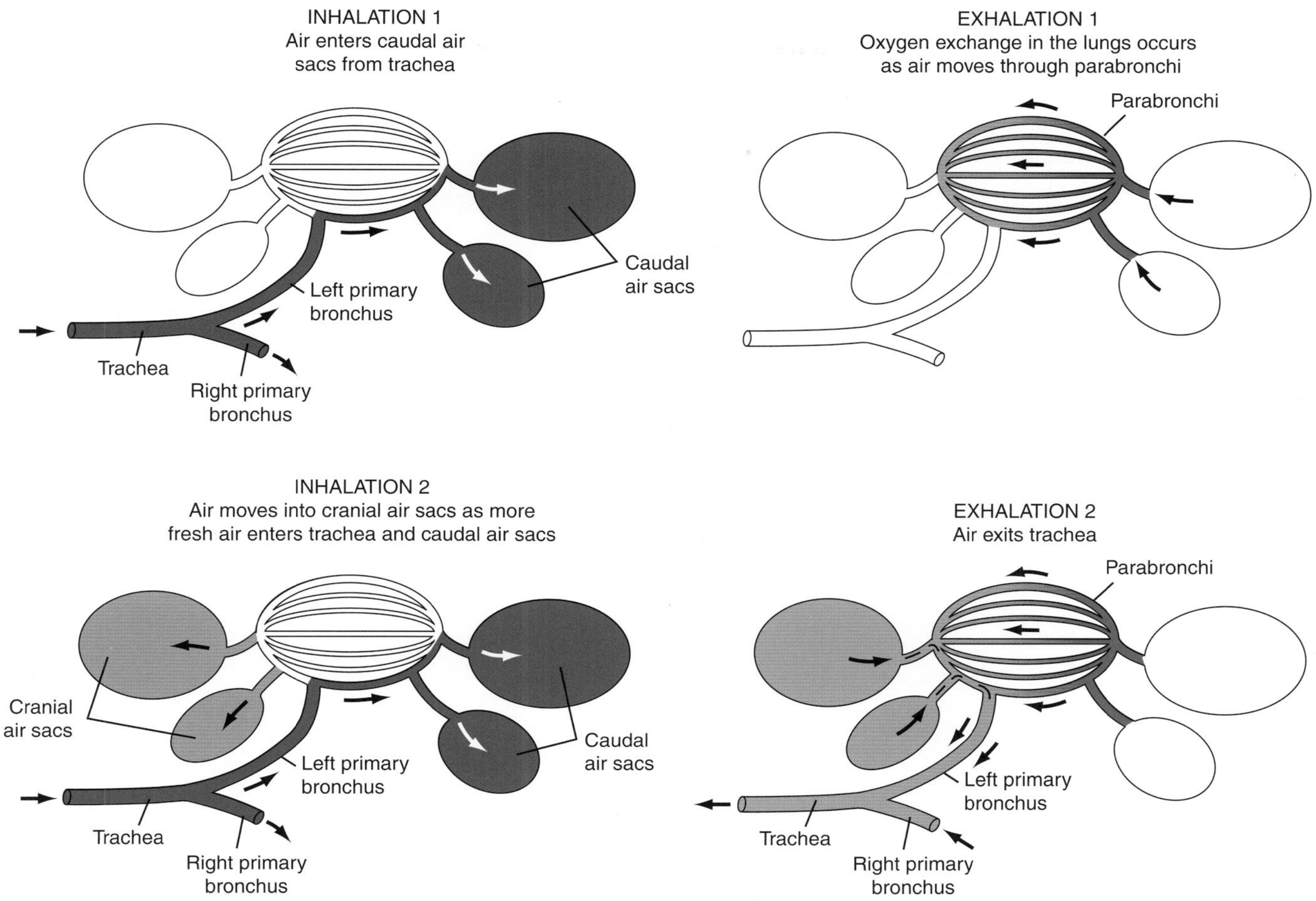

FIGURE 26.3-6 The pattern of air flow through the avian air sacs.

The one-way passage of air through the avian lungs occurs as a 4-phase process:

1. When the bird inhales, air enters the trachea and passes through the primary bronchi into the caudal thoracic and abdominal air sacs, collectively called the caudal air sacs. Although some air does enter the secondary bronchi and thus the lungs, most of this inhaled breath is deposited into the caudal air sacs.
2. When the bird exhales, this initial breath travels through the parabronchi and is thereby drawn across the air capillaries of the lungs.
3. With the next inhalation, the original breath travels from the lungs to the cervical, clavicular, and cranial thoracic air sacs, collectively called the cranial air sacs. At the same time, a second breath of fresh air is inhaled through the trachea and enters the caudal airsacs (and another step 1 begins).
4. On the next exhalation, the initial breath is finally exhaled through the trachea.

Thus, with each inhalation, the air sacs are inflated, and with each exhalation, the air within the air sacs is either drawn through the lung or exhaled through the trachea. Because of this one-way airflow system, respiratory "dead space" (the volume of inspired air that never participates in gas exchange) is limited to the trachea in birds. This highly efficient system allows some species, such as the trumpeter swan, to have a surprisingly long trachea (see Case 26.4).

Selected references

[1] Carpenter JW. Hematologic and biochemical values of selected psittaciformes. In: Exotic animal formulary. 5th ed. St Louis: Saunders Inc.; 2017. p. 287.
[2] Harris DJ. Clinical tests. In: Tully T, Dorrestein G, Jones A, editors. Handbook of avian medicine. 2nd ed. St. Louis: Saunders Inc.; 2009. p. 78.
[3] Kirchgessner M. Aspergillosis. In: Mayer J, Donnelly TM, editors. Clinical veterinary advisor: birds and exotic pets. St. Louis: Saunders Inc.; 2013. p. 155–7.
[4] Powers LV. Air sac tube placement. In: Mayer J, Donnelly TM, editors. Clinical veterinary advisor: birds and exotic pets. St. Louis: Saunders Inc.; 2013. p. 537.

CASE 26.4

Ventricular Foreign Body

Cynthia M. Faux[a] and Marcie L. Logsdon[b]
[a]University of Arizona College of Veterinary Medicine, Oro Valley, Arizona, US
[b]Exotics & Wildlife Department, Washington State University College of Veterinary Medicine, Pullman, Washington, US

Clinical case

History

A 7-year-old male Umbrella cockatoo (*Cacatua alba;* Fig. 26.4-1) was presented for anorexia of 3 days' duration. He was fed a balanced diet consisting mostly of formulated pellets and had free range of the house when his owners were home. Just before the onset of clinical signs, he was seen getting into things on his owner's dresser, including an open bottle of nail polish.

Physical examination findings

On presentation, the bird was quiet but alert and responsive. He had a body condition score of 3.5 (1 = emaciated, 5 = obese). The fecal droppings were scant and bright green; the urine and urate components were normal. The remainder of the physical examination findings were normal.

The color, consistency, and volume of the fecal component part of bird droppings are determined by the bird's diet and health. There are several possible causes of green feces, including lead toxicity and ingestion of food dyes. When found in conjunction with decreased fecal volume (scant fecal droppings), green discoloration is usually the result of excreted bile pigments and indicates the decreased passage of food through the gastrointestinal tract—in this case, as a result of anorexia.

Differential diagnoses

Ingestion of foreign material leading to toxicosis (most commonly zinc or lead) and/or gastrointestinal obstruction

BIRD DROPPINGS

Bird droppings consist of a mix of feces, urine, and urates. Fecal droppings are solid waste from the digestive tract and urine is liquid waste from the kidneys. The urate or uric acid component is produced by the kidneys and is typically a white semi-solid. Uric acid is the main by-product of protein metabolism in birds. Unlike urea, uric acid is not water-soluble, so its excretion requires less water.

Comparative Veterinary Anatomy: A Clinical Approach. https://doi.org/10.1016/B978-0-323-91015-6.00124-2

FIGURE 26.4-1 A male Umbrella cockatoo. When the crest feathers are raised for display, they fan out like an upturned umbrella.

Diagnostics

Under general anesthesia, an esophageal feeding tube (e-tube) was placed via an esophagostomy (surgically created stoma, or opening, in a wall of the esophagus) to facilitate supportive care while diagnostic tests were performed. Radiography confirmed the correct placement of this e-tube (i.e., ending in the ventriculus) and revealed the presence of several metallic foreign bodies in the ventriculus (Figs. 26.4-2 and 26.4-3). The owner indicated that the foreign bodies were most likely the teeth from a zipper the bird had destroyed. Bloodwork, including measurement of zinc and lead blood levels, revealed an increased serum zinc concentration.

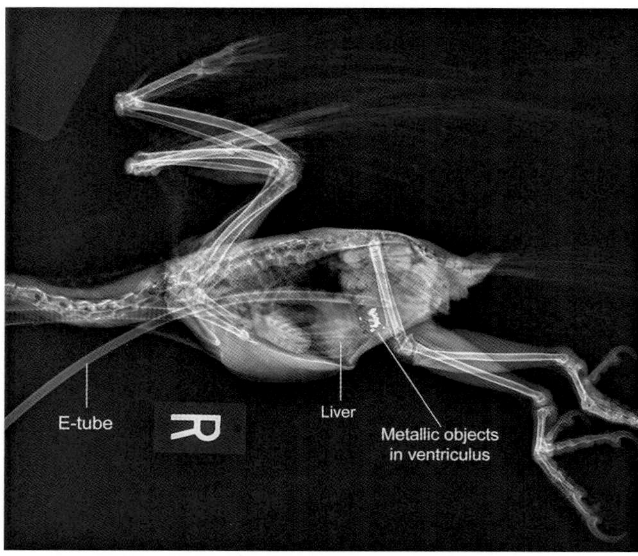

FIGURE 26.4-2 Lateral radiograph of the patient, showing several metallic foreign bodies (zipper teeth) in the ventriculus. The esophagostomy tube (labeled *E-tube*) is correctly placed in the esophagus and terminates within the ventriculus. The ventriculus is located caudal to the liver.

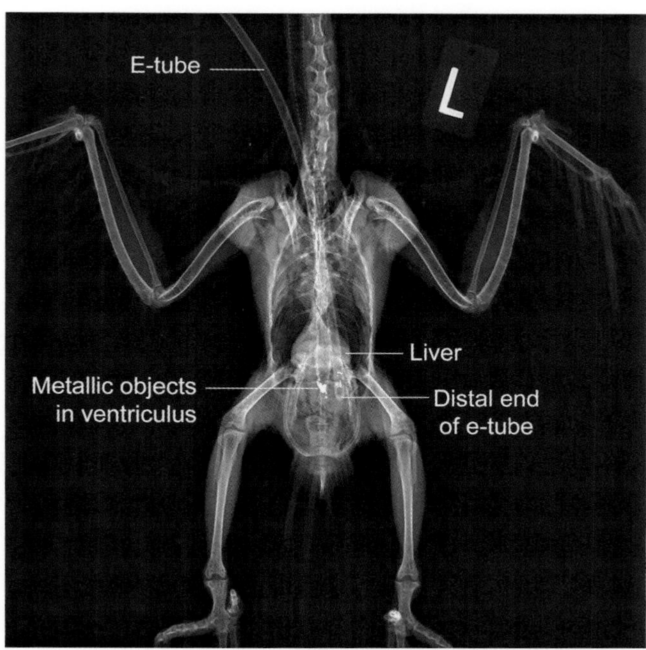

FIGURE 26.4-3 Ventrodorsal radiograph of the patient, showing the distal end of the e-tube within the ventriculus, alongside several metallic foreign bodies (zipper teeth). Note that the e-tube is present on the right side of the bird's neck, as the avian esophagus is located to the right of the trachea in the cervical region.

Diagnosis

Ventricular foreign bodies leading to acute zinc toxicity

Treatment

After discussing the treatment options (described later) , the owners elected medical management. Systemic chelation therapy to bind the excess zinc and allow excretion by the body was initiated. Psyllium (a bulk laxative) was administered via the e-tube to encourage the zipper pieces to pass through the digestive tract on their own. Within a few days, the bird began eating, at which point the e-tube was removed.

HEAVY METAL TOXICITY

In birds, heavy metal toxicity is a common sequela to foreign body ingestion. Pet parrots such as Cockatoos can be exposed to zinc from any number of metal alloys. Because they use their beaks to help maneuver around their environment, they can be exposed to heavy metals from chewing on galvanized (zinc-coated) cage wire or hardware cloth on homemade cages, and from lead-based paint in older houses. Poultry may ingest galvanized hardware such as bolts and nails in their search for grit. Raptors are more commonly exposed to lead from fragments of lead-based ammunition found in the bodies of animals they consume.

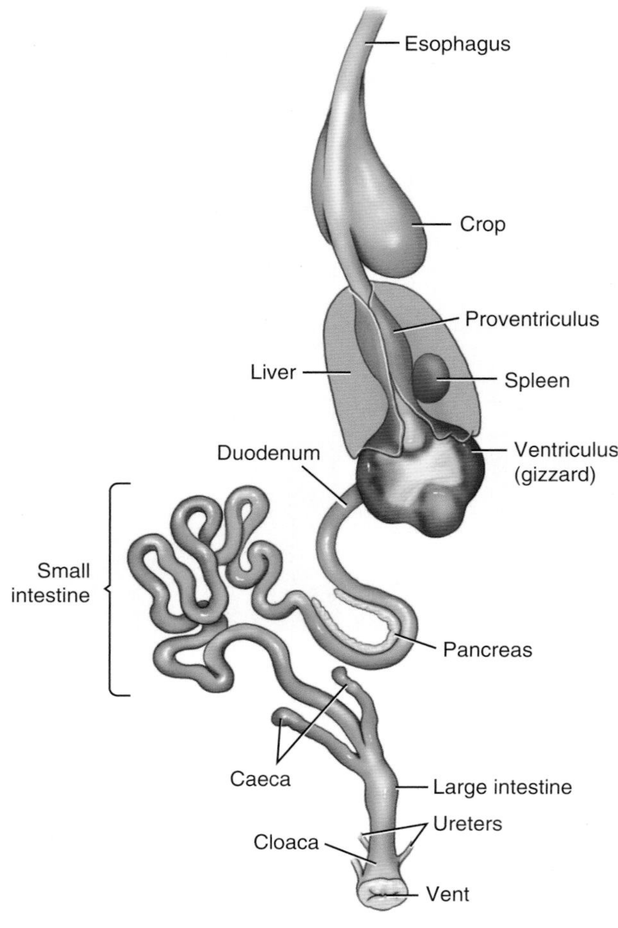

FIGURE 26.4-4 Gastrointestinal tract of a parrot.

Anatomical features in avian species

Introduction

The avian digestive tract (Fig. 26.4-4) varies across species, with most differences occurring in the shape and function of the beak, presence or absence of a crop, architecture of the stomach, length of the gastrointestinal tract, and presence of variations in the cecum. This case covers the anatomy of the avian gastrointestinal tract. The crop is discussed fully in Case 25.2 and the beak and oral cavity are discussed further in Case 25.4.

Function

The digestive system in birds functions in similar ways to other animals in the ingestion of food, secretion of fluids and digestive enzymes needed to break down the ingested foods, mixing and movement of food and by-products through the body, assimilation of food into smaller pieces, absorption of nutrients, and removal of indigestible food stuffs. There are modifications and adaptations in the digestive system including the beak and cloaca that provide different function in the avian species compared to its mammalian counterparts.

Gastrointestinal system

Birds lack a soft palate, so the avian pharynx is not divided into oral and nasal portions; it is simply referred to as the oropharynx. The oral cavity is usually considered to be the area rostral to the base of the tongue.

The **esophagus** is relatively thin-walled in birds. For most of its length, the cervical part lies to the right of the trachea until it reaches the crop (see below), at which point the esophagus becomes more midline in position.

> Care must be taken to avoid rupturing the esophagus when tube-feeding avian patients. When esophageal feeding tubes are surgically placed, the esophagostomy incision is made on the right side of the neck in most avian patients.

The **crop**, or **ingluvies**, is an expandable diverticulum of the esophagus that is present in most avian species common to veterinary medicine. Its primary function is to store food; in pigeons, it also provides nutritive secretions for feeding chicks. The crop is usually located just cranial to the thoracic inlet. It is palpable when filled with food, which is useful when determining if a meal was recently consumed. The crop is discussed in more detail in Case 25.2.

> An e-tube may also be surgically placed via an ingluviotomy (surgical incision into the crop).

The typical avian **stomach** consists of 2 separate compartments: a glandular portion (proventriculus) and a muscular portion (ventriculus). The **proventriculus** has a thinner wall and usually lies to the left of midline. Its function is like that of the stomach in monogastric mammals—it mixes ingesta with secreted gastric acid and proteolytic enzymes. The **ventriculus**, or **gizzard**, has a thick, muscular wall and a tough, koilin-lined inner surface in many species. As birds lack teeth, the ventriculus aids in mashing and macerating hard-coated food items such as seeds. Some species of birds ingest grit (small particles of rock or sand) to assist in the grinding process. Avian species that are primarily carnivorous (e.g., raptors) or masticate, shred, or hull food with their beak (e.g., parrots) have a less muscular ventriculus than species that consume whole seeds (e.g., chickens). Soft foods can bypass the ventriculus and move from the proventriculus directly into the duodenum.

> Grit in the ventriculus may be visible radiographically and should not be mistaken for foreign material.

> Ventricular foreign bodies requiring surgical removal are often retrieved via a proventriculotomy (a surgical incision into the proventriculus), because the proventriculus has a much thinner wall than the ventriculus. An alternative approach is through a thinner portion of the ventricular wall.

Ingesta passes from the stomach into the **small intestine**, which comprises the **duodenum**, **jejunum**, and **ileum**. The length of these sections varies among avian species, depending on diet. The **pancreas** lies in the mesentery within a loop of the duodenum (Fig. 26.4-4).

The **large intestine** variably consists of the **cecum** (or paired **ceca** in some species) and **colon**. The cecum is present in chickens and ducks but is generally reduced in parrots and raptors. The avian colon serves to recover water and electrolytes from the digesta.

The colon terminates as the **rectum**, which empties into the **cloaca** via the coprodeum, one of the regions of the cloaca. The external opening of the cloaca is called the **vent**. The cloaca is described further in Cases 26.2 and 26.5.

Birds do not have a diaphragm, so the **liver** normally lies in contact with the heart. The presence of a **gall bladder** is not universal among avian species; notably, most parrot species lack a gall bladder, so bile flows directly from the liver to the duodenum.

As the heart and liver have a similar radiodensity, their common borders are normally indistinct on radiographs, creating a cardiohepatic silhouette on ventrodorsal views.

The **spleen** lies next to the stomach and varies in shape, from round in parrots to elongated in other avian species.

TREATMENT OPTIONS FOR VENTRICULAR FOREIGN BODIES

Depending on the size, shape, and composition of the material, there are several options for removing foreign bodies from the ventriculus:

- Surgical removal. Surgical approaches through the thinner walled proventriculus (a proventriculotomy) or through the thinner walled caudal ventral portion of the ventriculus (a ventriculotomy) are used to retrieve foreign bodies. Incising through the muscular lateral walls of the ventriculus is contraindicated. These procedures involve general anesthesia and a coelomotomy (incision through the body wall into the coelomic cavity), so are usually reserved for large or dangerous foreign bodies that cannot be safely passed on their own
- Endoscopic removal. Under general anesthesia, a flexible endoscope is passed either orally or via an ingluviotomy into the ventriculus to allow the visualization and removal of the foreign body(ies); care must be taken to avoid traumatizing the thin-walled esophagus during removal
- Gavage (gastric lavage). The material is flushed (lavaged) from the ventriculus using a rubber catheter, passed as described above for endoscopic removal
- Elimination. Feeding a high-fiber or mucilaginous material such as psyllium (which forms a bulky gel when mixed with water) encourages normograde (aboral) passage of the foreign body(ies); this approach should not be attempted unless the particles are small enough that they will not cause an obstruction as they pass through the rest of the gut
- Magnet retrieval. Ferrous (iron-containing) metallic foreign bodies can sometimes be retrieved using a magnet glued or otherwise firmly attached to a tube and passed into the ventriculus as described above for endoscopic removal

In certain species, such as owls and other raptors, no intervention may be needed as the foreign body(ies) will usually be incorporated into the “pellet” of indigestible material that is regurgitated after most meals.

Selected references

[1] Harris DJ. Basic anatomy, physiology and nutrition. In: Tully T, Dorrestein G, Jones A, editors. Handbook of avian medicine. 2nd ed. St. Louis, MO: Saunders Inc., Elsevier; 2009.

[2] McLelland J. Aves digestive system. In: Sisson and Grossman's The Anatomy of the Domestic Animals. 5th ed, vol. II. Philadelphia: WB Saunders; 1975. p. 1857–82.

[3] Singh B. Dyce, sack, Wensing - Textbook of Veterinary Anatomy. 5th ed. St. Louis, MO: Elsevier; 2018.

CHAPTER 26

CASE 26.5

Sertoli Cell Tumor

Cynthia M. Faux[a] and Marcie L. Logsdon[b]
[a]University of Arizona, College of Veterinary Medicine, Oro Valley, Arizona, US
[b]Exotics & Wildlife Department, Washington State University, College of Veterinary Medicine, Pullman, Washington, US

Clinical case

History

A 7-year-old male Budgerigar (budgie) presented for left leg lameness of 3 days' duration. The owner also reported that the bird's cere had gradually changed color from blue to brown over the last few weeks.

> The cere is the fleshy area around the nostrils in some avian species (Fig. 26.5-1). In healthy, adult male budgies, the cere is bright blue (sometimes referred to as "royal blue"), fading to pale blue when the bird is not reproductively active. In healthy, adult females, the cere is pink or brown, fading to pale pink, light tan, or white when the bird is not reproductively active. Normally, the cere thickens and may become crusty when the bird is reproductively active, but it retains the color appropriate for the bird's gender. In immature budgies (less than 4 months of age), the cere is pink in both males and females.

Physical examination findings

The patient was bright, alert, and responsive. He was unable to stand on his left leg. His body condition score was 3.5 (1 = emaciated, 5 = obese). A small amount of droppings were caked to the feathers around his vent (external opening of the cloaca). His left leg exhibited weakness and an incomplete withdrawal reflex. On palpation of the coelomic cavity, the ventriculus was prominent and unable to be retropulsed. The rest of the physical examination was unremarkable.

> Inability to manipulate the position of the ventriculus on coelomic palpation often indicates the presence of an abnormal mass within the coelomic cavity.

BUDGERIGARS

Budgerigars ("budgies") are commonly called parakeets in the United States. However, the term "parakeet" actually refers to a large number of species of small psittacines (parrots) that are characterized by their long tails, including—but not limited to—budgerigars.

Comparative Veterinary Anatomy: A Clinical Approach. https://doi.org/10.1016/B978-0-323-91015-6.00125-4

FIGURE 26.5-1 A pair of Budgerigars. The bird on the left (green and yellow) is male, and the bird on the right (blue and white) is female. In adult budgies, the cere (the fleshy area around the nostrils) is blue in males and brown in females.

A testicular tumor is suggested by the "feminization" of the cere in this male bird (brown instead of blue). While the lameness and neurologic deficits suggest a primary musculoskeletal or neurological disorder, these signs are also consistent with a renal or testicular tumor in birds, owing to the intimate association between the kidneys/testes and the spinal nerves that supply the pelvic limbs (discussed later).

Differential diagnoses

Limb fracture or other musculoskeletal injury, renal or testicular tumor (e.g., Sertoli cell tumor)

Diagnostics

Whole-body radiography revealed a mass effect in the caudodorsal coelomic cavity. (Radiographs of a cockatiel with a similar coelomic mass are shown in Figs. 26.5-2 and 26.5-3). No musculoskeletal abnormalities were found. As the mass was most likely a renal or testicular tumor, and as both have a poor prognosis and are progressive with poor quality of life (inability to ambulate), the owners elected humane euthanasia. Necropsy and histopathology revealed Sertoli cell tumor in the left testis.

Diagnosis

Sertoli cell tumor

Treatment

Even if benign, renal and testicular tumors are difficult to treat in avian patients. Surgical resection carries a high risk of fatal hemorrhage because of the large blood vessels in the surrounding area (described later). Laparoscopic orchidectomy (surgical removal of the testis) may be an option in larger avian species. However, when the tumor is already impacting locomotion, quality of life declines rapidly and humane euthanasia is usually indicated.

FEMINIZATION SYNDROME

Sertoli cell tumors often cause feminization in males due to the increased secretion of estrogens. In Budgerigars, this is most readily observed as a change in cere color from blue to pink or brown. Behavioral changes that may be apparent in mammals, such as attraction of other males and a sudden willingness of a male dog to be mounted by another male dog, are not often observed by bird owners, particularly if there is only one male bird in the enclosure.

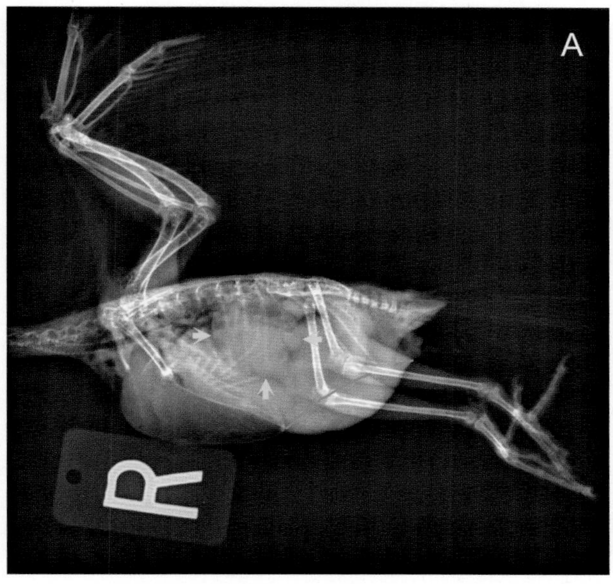

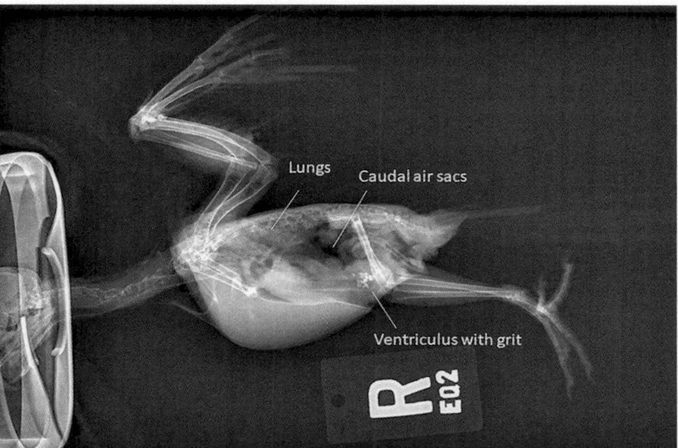

FIGURE 26.5-2 (A) Lateral radiograph of a Cockatiel with a large soft-tissue mass in the caudodorsal coelomic cavity (*arrows*). In addition, there is a general loss of serosal detail and air sac space and gross distension of the ventral body wall (normal contour shown in *green*). (B) Cockatiel with a normal coelomic cavity.

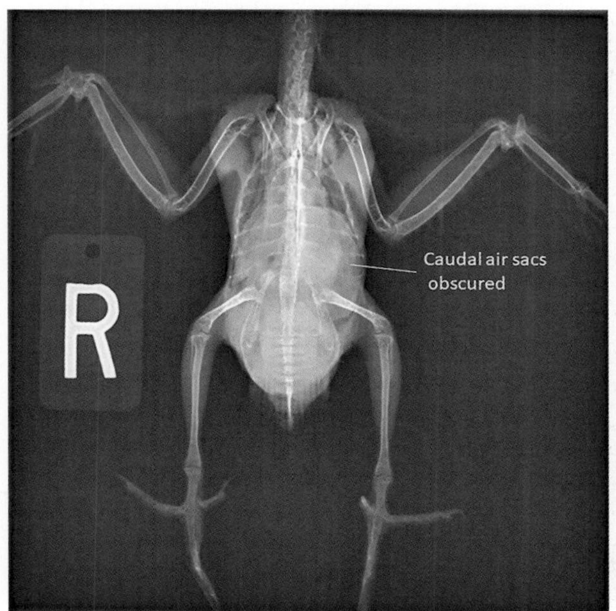

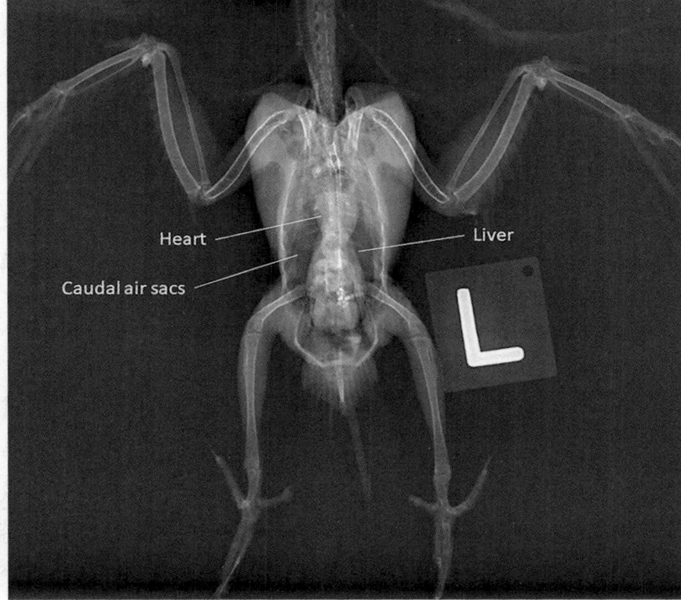

FIGURE 26.5-3 (A) Ventrodorsal radiograph of the Cockatiel shown in Fig. 26.5-2A. There is a large "mass effect" in the left side of the coelomic cavity. The cardiohepatic silhouette is greatly widened, and the margins of the right liver lobe are displaced laterally; the margins of the left liver lobe are difficult to distinguish from the mass. (B) Cockatiel with a normal coelomic cavity.

Anatomical features in avian species

Introduction

The clinical anatomy illustrated by this case includes the urinary system and the male reproductive system. Each is discussed separately below.

Function

The urogenital system functions in the formation, storage, and elimination of urine and for reproductive purposes. Many of the functions are controlled by autonomic and somatic efferent pathways of the nervous system and originate in the lumbosacral spinal cord. In general, "urogenital" refers to the urinary and reproductive systems.

The synsacrum is an avian skeletal structure formed by fusion of the sacrum with a variable number of lumbar and caudal (coccygeal) vertebrae (see Landscape Fig. 6.0-7). This fusion of vertebral elements is an adaptation for flight and locomotion in birds. The location of the kidneys immediately ventral to the pelvis and synsacrum makes their radiographic assessment challenging (see Landscape Figs. 6.0-5 and 6.0-6). On the lateral projection, the associated abdominal air sacs may allow some assessment of kidney size: Loss of this air density dorsal to the kidneys may be indicative of renal enlargement. Large renal tumors may also result in ventral displacement of other coelomic organs.

Urinary system (Fig. 26.5-4)

In birds, the right and left **kidneys** are retroperitoneal and lie immediately lateral to the spine, within fossae ventral to the pelvis and synsacrum. Diverticula of the abdominal air sacs lie between the kidneys and the synsacrum.

The avian kidneys are relatively larger than the kidneys in mammals of similar size. In birds, each kidney is roughly rectangular in shape with three segments or "divisions": cranial, middle, and caudal. Each division is supplied by its own **renal artery** (**the cranial, middle,** and **caudal** renal arteries), which are branches of the aorta (cranial renal artery) or external iliac or ischiatic arteries (middle and caudal renal arteries).

Avian patients with abnormal enlargement or displacement of the kidney (e.g., renal tumor, renal gout, testicular tumor) may initially present with lameness as the primary clinical sign. Surgical removal of such tumors is difficult and risky because of the intimate association of the kidneys and testes with the great vessels (aorta, caudal vena cava) and these neural plexuses.

The avian kidneys are closely associated with the **lumbar** and **sacral plexuses**, and the **sciatic nerve** passes through the kidney parenchyma.

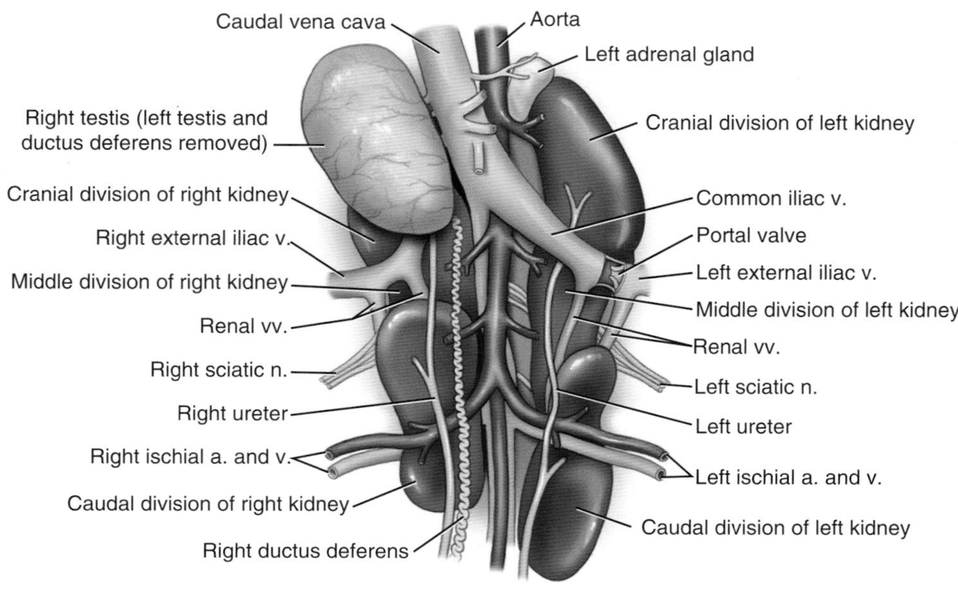

FIGURE 26.5-4 Anatomy of the avian male reproductive and urinary systems (the right testis and deferent duct are illustrated; the left testis and deferent duct have been removed to better visualize the left kidney). Note the position of the sciatic nerve passing through the middle division of the kidneys.

The avian kidneys have two separate venous systems. The first is similar to that in mammals: The **cranial** and **caudal renal veins** return blood from the kidneys to the heart via the caudal vena cava. The caudal renal vein is the larger of the two, as it drains both the middle and caudal divisions of the kidney.

The second venous drainage pathway is via is the **renal portal system**, in which blood returning from the caudal regions of the body is diverted through the kidneys before entering the caudal vena cava. The **renal portal veins** (**cranial** and **caudal**) are branches of the **external iliac vein**, and they deliver blood to the capillary network surrounding the renal tubules along with the renal arteries. Thereafter, the blood is returned to the systemic circulation via the renal veins.

This portal system is enabled by the **portal valve**, a conical or cylindrical ring of smooth muscle within the external iliac vein (Fig. 26.5-4) which, when closed, shunts blood to the kidneys instead of proceeding directly to the caudal vena cava. Opening and closing of this valve is under autonomic control: parasympathetic stimulation closes the valve, so birds at rest shunt blood to the kidneys via the renal portal system; sympathetic stimulation opens the valve, allowing blood from the caudal regions of the body to return directly to the caudal vena cava. This mechanism is likely an adaptation for flight, in which the kidneys, with their high energy demands, are partially bypassed.

Birds lack a urinary bladder. Each kidney is drained by a **ureter**, both of which (left and right) deliver urine and urates directly into the **urodeum**, a shallow vestibule within the **cloaca** that is located just caudal to the **coprodeum** (the vestibule into which the rectum empties; see Fig. 26.2-5 and Landscape Fig. 6.0-6). The ureteral openings are located on the dorsolateral walls of the urodeum, dorsal to the genital opening(s): the deferent ducts (left and right) in males and the left oviduct in females. (The female reproductive tract is described in Case 26.2.)

Birds excrete the nitrogenous waste products of protein metabolism predominantly as uric acid, or urates (whereas urea is the principal excretion product in mammals). Uric acid is excreted as a colloidal suspension that may be refluxed from the urodeum into the coprodeum, and thus into the large colon, for further water resorption.

Because the cloaca is common to the digestive, urinary, and genital tracts, collection of urine that is free of fecal contamination is difficult or impossible in birds. Furthermore, urine samples are of limited diagnostic value because post-renal fluid and electrolyte resorption occurs in birds.

The **adrenal glands** are not part of the urinary system, but given their close proximity to the kidneys, they are described here. The avian adrenal glands are paired (left and right) and lie at the cranial pole of their respective kidney; however, the left and right adrenal glands may meet at midline. Avian adrenal glands are irregular in shape, the shape varying with species. In male birds, the adrenal gland overlies the testis on that side. In mature female birds, the left ovary covers the left adrenal gland. The **cranial renal arteries** supply the adrenal glands.

Male reproductive system (Fig. 26.5-4 and Landscape Fig. 6.0-6)

The avian **testes** (left and right) are bean-shaped and located internally, cranioventral to the cranial division of each kidney. Like the kidneys, the testes are surrounded by the abdominal air sacs. As another example of

RENAL PORTAL SYSTEM

In birds, the renal portal system must be taken into consideration when administering drugs that undergo significant renal clearance. Such drugs administered by injection into the pelvic limb or other caudal regions of the body may be delivered to the kidneys before entering the systemic circulation. Thus, the drug may be cleared from the body before reaching a therapeutic concentration systemically.

In addition, drugs with nephrotoxic potential (e.g., aminoglycoside antibiotics, nonsteroidal anti-inflammatory drugs) may be more likely to cause renal impairment in birds when administered by injection into the pelvic limb. The large pectoral muscles are often the preferred site for the intramuscular injection of such drugs in birds.

Seasonal testicular enlargement is so noticeable in some species, radiographic determination of sex is possible base on visualization of the gonads. During the nonbreeding season, gonads are not apparent radiographically. When evaluating a patient for testicular enlargement, it is important to take into account not only the species of bird, but also the bird's age, physical appearance, time of year, environmental cues, and other species-specific indicators of reproductive status.

conservation of mass for flight, testicular size in many avian species greatly increases during seasonal breeding activity and decreases once the breeding season is over.

The testis and epididymis are supplied by branches of the **cranial renal artery**. In birds, the **deferent duct** (ductus deferens) leaves the epididymis and follows the ureter to the urodeum, into which it empties via a **papilla** (plural, **papillae)** in the dorsolateral wall, adjacent to its respective ureter.

Male birds do not have well-defined accessory sex glands. However, the paired **vascular body** (left and right) in the ventrolateral wall of the urodeum between the deferent duct and the phallic body (see below) may serve a similar function to the bulbourethral gland in mammals.

Birds do not have a penis, as such, but rather a **copulatory organ**. It is located in the caudoventral region of the cloaca, just cranial to the vent. This part of the cloaca in males is called the **proctodeum**. In its resting state, the copulatory organ is not visible externally; it is everted through the vent only during sexual activity.

In general, the copulatory organ consists of a bilateral set of 4 structures: papilla of the deferent duct, vascular body, phallic body, and lymphatic fold (see Landscape Fig. 6.0-6). The pair of phallic bodies, together forming the **phallus**, is broadly equivalent to the mammalian penis, in that it facilitates the delivery of semen into the female reproductive tract. However, birds do not have a urethra, so the avian phallus does not serve any urinary function. Furthermore, even the presence of a phallus varies among avian species, from absent (e.g., raptors, parrots), to not extensible (e.g., chickens), to extensible (e.g., ducks, geese, ratites such as ostriches). Identifying a bird's gender based on the presence or absence of a phallus is thus species-dependent. In species without a phallus, sperm is transferred into the vagina simply through cloacal contact between male and female.

GENDER DETERMINATION IN BIRDS

Sexual dimorphism is not universal in birds, and physical examination may not be adequate for gender determination, especially for many species of parrots, pigeons, and raptors. Relative body size is usually reliable in raptors—females tend to be larger than males—although large males and small females can be found within the spectrum of body size. The advent of DNA analysis of blood or feather samples has decreased the need for surgical or endoscopic gender determination in birds.

Selected references

[1] Lierz M. Avian renal disease: pathogenesis, diagnosis, and therapy. Vet Clin North Am Exot Anim Pract 2003;6:29–55.
[2] Burgos-Rodrıguez AG. Avian renal system: clinical implications. Vet Clin North Am Exot Anim Pract 2010;13:393–411.
[3] King AS. Aves urogenital system. In: Getty R, editor. Sisson and Grossman's, The anatomy of the domestic animals. 5th ed, vol. II. Philadelphia: WB Saunders; 1975. p. 1919–64.
[4] Dyce KM, Sack WO, Wensing CJG. Textbook of veterinary anatomy. 4th ed. Philadelphia: Saunders; 2010.

CASE 26.6

Marek's Disease

Ricardo de Matos and James K. Morrisey
Department of Clinical Sciences, Cornell University College of Veterinary Medicine, Ithaca, New York, US

Clinical case

History

A 3-month-old Wyandotte pullet (a young hen that is not yet laying eggs) was presented for progressive lethargy, decreased appetite, and weight loss. The bird was part of a mid-sized flock of laying breed pullets with indoor/outdoor access. Older birds were present at the same facility but housed in a different area. No other birds in this flock were showing similar clinical signs at the time of presentation. The housing, bedding, fixtures, and feed were considered adequate for the species and current age of the hens.

Physical exam findings

On presentation, the bird was quiet, depressed, and minimally responsive to handling. The mucous membranes were pale and dry. The bird was thin with a body condition score of 1.5 (1 = emaciated, 5 = obese).

Differential diagnoses

Toxicity (heavy metals, aflatoxin), parasites (including histomoniasis), Marek's disease, avian leukosis, reticuloendotheliosis, Newcastle disease, *Mycobacterium* spp. infection, septicemia (*Escherichia coli, Salmonella* spp.), metabolic disease, nutritional disease

Diagnostics

The results of a CBC revealed a moderate elevation of the total white blood cell count ($56.3 \times 10^3/\mu L$; reference interval $9–32 \times 10^3/\mu L$) with mostly lymphocytes ($46.7 \times 10^3/\mu L$ or 83% of the total white blood cell count). Except for an elevated total protein (8.1 g/dL; reference interval 3.3–5.5 g/dL), the results of the plasma biochemistry panel

MAREK'S DISEASE

Marek's disease (MD), a common viral disease of chickens, is a transmissible neoplastic disease caused by a herpesvirus. The virus affects mostly chickens, although quail, turkeys, and pheasants are susceptible to infection and disease. The virus causes the development of lymphocyte-based tumors in several tissues. Many strains of Marek's disease virus (MDV) have been identified, with variable pathogenicity. Depending on the strain of the virus, individual immunity, and organ affected, clinical presentation can vary from asymptomatic infection to neurologic disease (progressive or persistent), lameness due to ischiatic nerve involvement, skin disease, ocular disease, and/or nonspecific clinical signs such as weight loss and crop stasis/impaction. Lymphoma (neoplastic disease of lymphoid tissue) associated with MD is the most prevalent pathologic syndrome. Clinical disease is more common in birds between 4 and 14 weeks of age. The prognosis for affected birds is poor, especially for younger birds where the disease is associated with a high mortality rate. Mortality in adult birds is usually sporadic.

Comparative Veterinary Anatomy: A Clinical Approach. https://doi.org/10.1016/B978-0-323-91015-6.00126-6

were within normal limits. Due to the relatively poor prognosis for recovery, and the suspicion of a diagnosis that would have health implications for the entire flock, the bird was humanely euthanized and a necropsy performed in order to try to make a definitive diagnosis.

On the gross pathologic examination, all thymic lobes were noted to be enlarged and pale (Fig. 26.6-1). Spleno- and hepatomegaly were present. The proventriculus was firm and white to pale tan, with a thick wall (Fig. 26.6-2). Several white to tan, poorly demarcated masses were noted in the pancreas (Fig. 26.6-3). The ovary consisted of a single poorly demarcated, white to tan mass.

The most relevant histopathologic finding was the presence of large infiltrative or mass-forming aggregates of neoplastic lymphocytes within several tissues, including the thymus, pancreas, proventriculus, skin, ischiatic nerves, lung, skeletal muscle, ovary, heart, liver, cecal tonsils, duodenum, and eye ganglia (Figs. 26.6-4 and 26.6-5). Gallid herpersvirus-2, also known as Marek's disease virus, was identified in affected tissues using PCR.

Diagnosis

Marek's disease

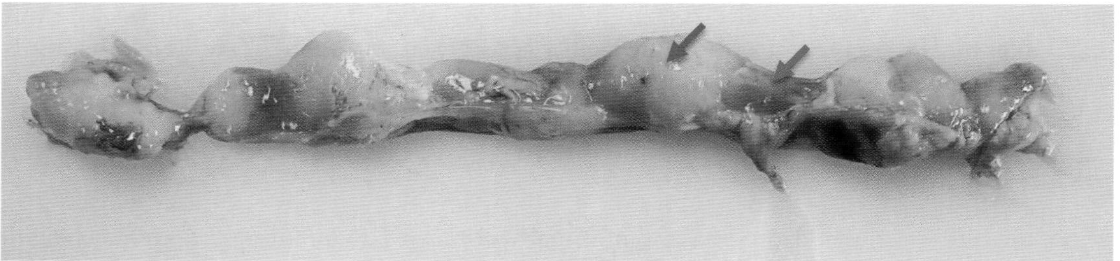

FIGURE 26.6-1 Enlarged and pale thymic lobes (*red arrow*) in a chicken with Marek's disease. Normal thymus (*blue arrow*) can also be seen. (Image courtesy of Dr. Elena Demeter.)

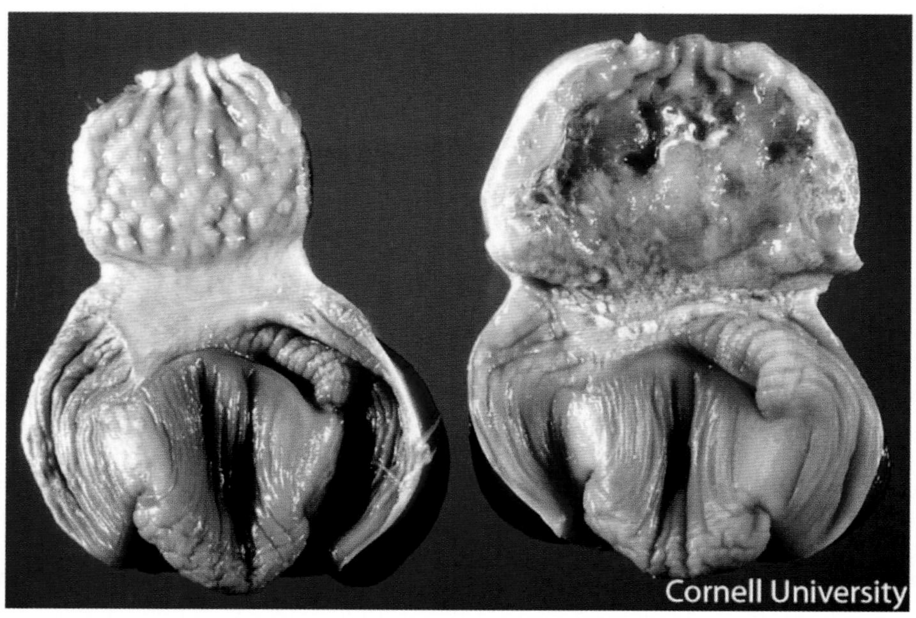

FIGURE 26.6-2 Cross section of a firm, white, irregular, and thickened proventriculus from a chicken with Marek's disease. A normal proventriculus is shown on the left for comparison.

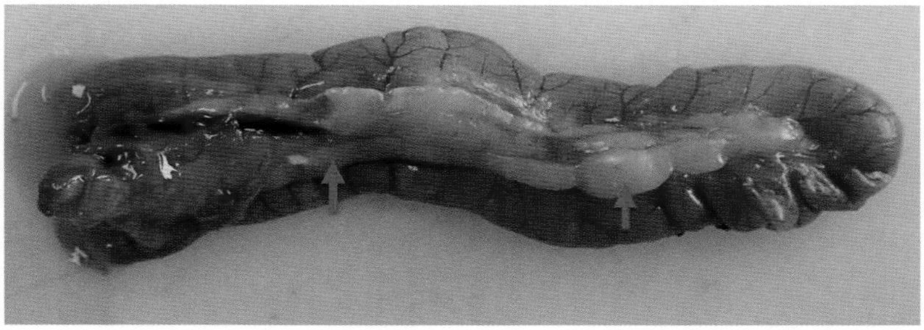

FIGURE 26.6-3 Pancreas (with adjacent duodenum) with several poorly demarcated masses (*red arrow*) in a chicken with Marek's disease. The *blue arrow* highlights an area of normal pancreas. (Image courtesy of Dr. Elena Demeter.)

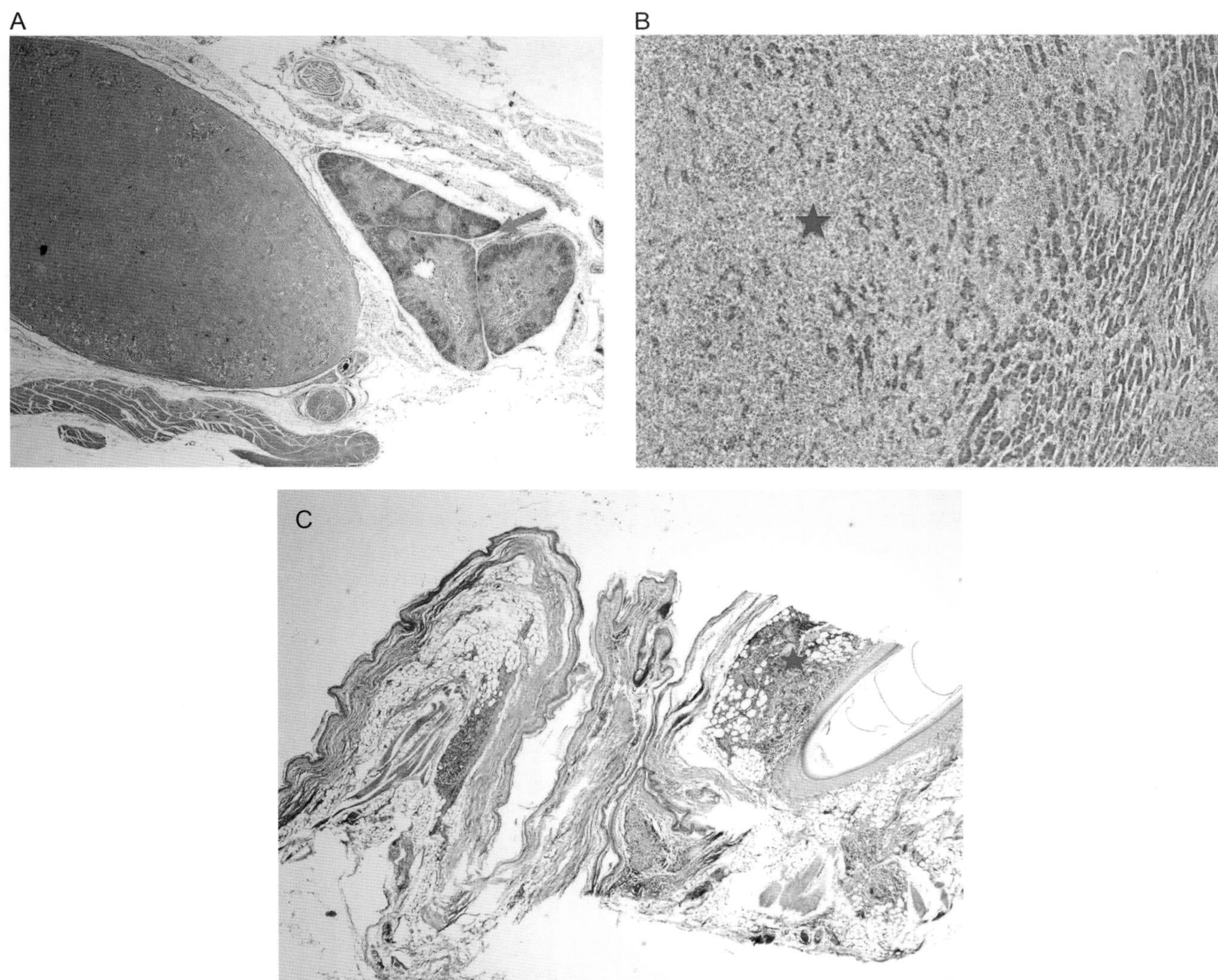

FIGURE 26.6-4 Histologic images of the thymus (A), pancreas (B), and skin (C) demonstrating the presence of large aggregates of neoplastic lymphocytes (*red star*). Areas of normal tissue can also be seen (*blue arrow*). (Images courtesy of Dr. Elena Demeter.)

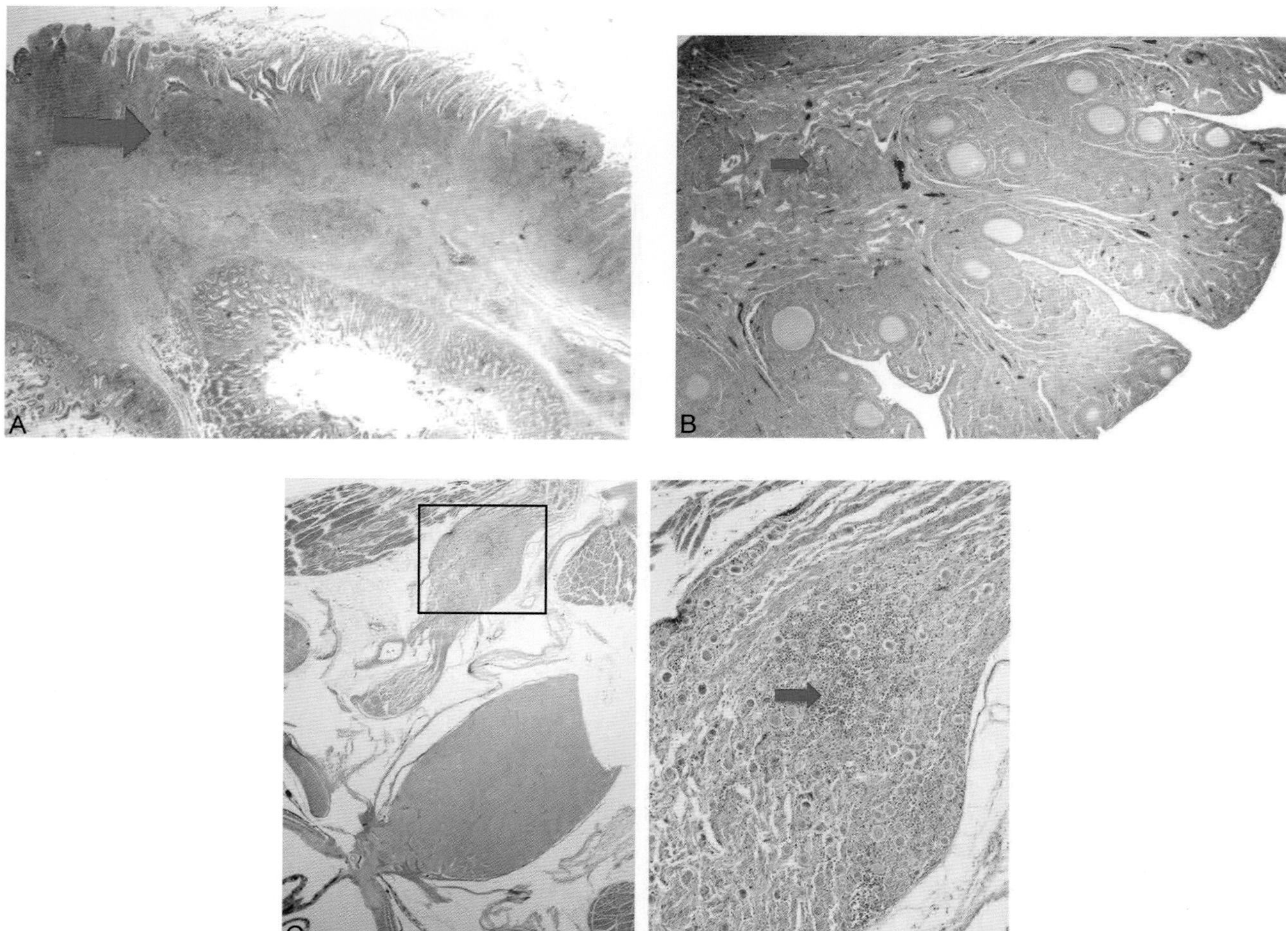

FIGURE 26.6-5 Histologic images of the proventriculus (A), ovary (B), and eye (C) demonstrating the presence of large aggregates of neoplastic lymphocytes (*red arrow*). (Images courtesy of Dr. Elena Demeter.)

Treatment

Recommendations were made to monitor the rest of the flock for the development of clinical signs like this bird or consistent with any of the different syndromes of the disease (see side box entitled "Marek's disease"). There is no effective treatment for birds affected with Marek's disease. Control, biosecurity, and disinfection are difficult to achieve in infected flocks due to the physical characteristics of the virus and its long-term persistence in feathers and dander. As most vaccination protocols are done *in ovo* (L. in the egg) or soon after hatch, the efficacy of vaccinating the remaining birds in the flock is unknown.

Anatomical features in avian species

Introduction

The avian immune system is phylogenetically intermediate between reptiles and mammals, bearing features of both. The system is composed of primary and secondary lymphoid organs. The primary lymphoid organs include the bursa of Fabricius, thymus, bone marrow, and yolk sac (in the embryo). The secondary lymphoid organs are the spleen and disseminated lymphoid tissues. Birds generally lack "true" lymph nodes, but have lymph aggregates within various organs, as well as lymph vessels and lymph hearts to help move lymphatic fluid around the body. These tissues are depicted in Fig. 26.6-6.

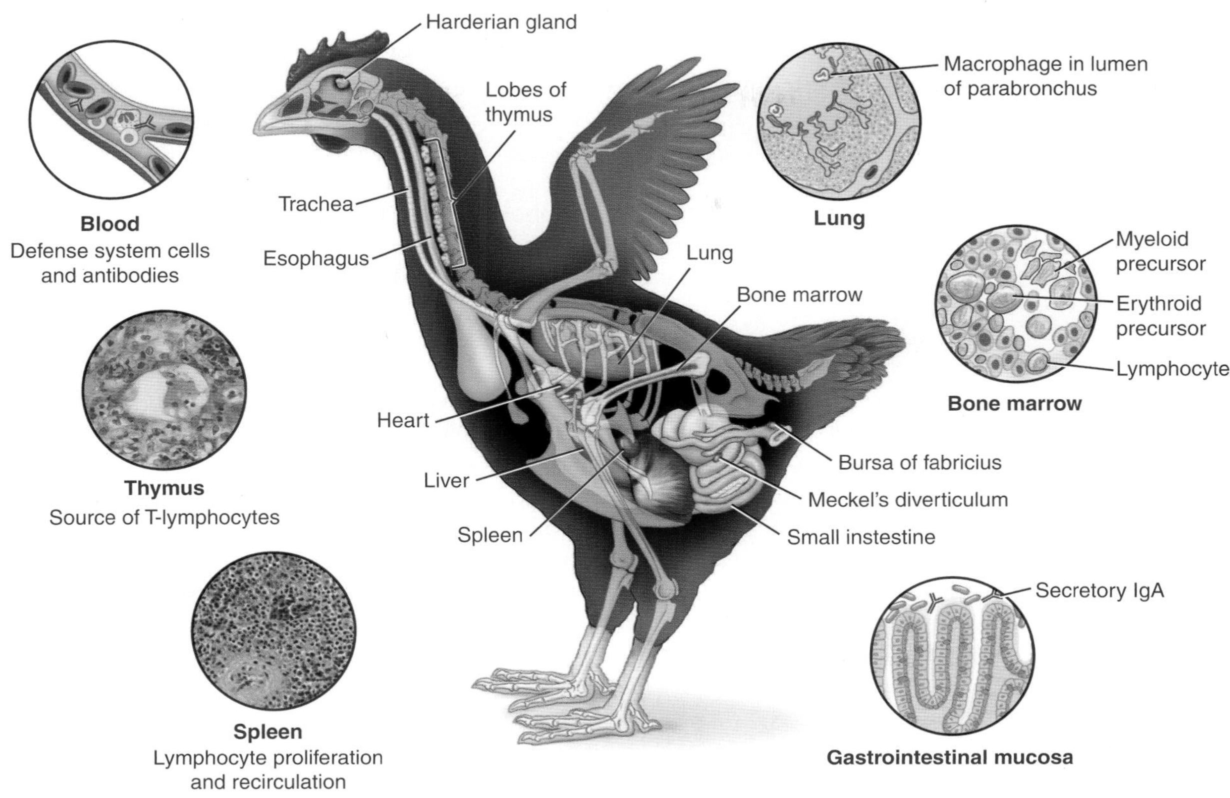

FIGURE 26.6-6 Lymphoid tissue and other immunologic defense mechanisms in the bird.

Function

The immune system functions to protect the body against infectious organisms and foreign substances. The specific function of each structure along with its anatomy is discussed in the following sections.

Bursa of Fabricius

The **bursa of Fabricius**, or cloacal bursa, is the site of B-lymphocyte (B-cell) maturation and is unique to birds. It is a dorsal outpouching of the proctodeum or most distal portion of the cloaca. The bursa develops before hatching and reaches its maximum size by several weeks of age. It then involutes and basically disappears by sexual maturity. In ratites (flightless, large, long-legged birds—e.g., ostrich, rhea, emu, cassowary, and kiwi), the remnant of the bursa is incorporated into the wall of the proctodeum and functions as a urinary bladder.

> The ostrich (a ratite) is the only bird that has a structure which functions as a vesicular urine storage structure. This bladder is found in the wall of the proctodeum in the remnants of the bursa of Fabricius. Unlike mammals, the ureters do not empty into this "bladder."

B- AND T-CELLS—WHAT IS THE DIFFERENCE?

T-cells (T-lymphocytes or thymus-derived cells) and B-cells (B-lymphocytes or bone marrow- or bursa of Fabricius-derived cells) are a distinct line of defense in the adaptive or acquired immune system. T-cells are involved in cell-mediated immunity—i.e., cytotoxic/killer T-cells and T-helper cells—and produce cytokines that boost the immune system, whereas B-cells are primarily responsible for humoral immunity—i.e., antibody production. Besides their function in the acquired immune system, these cells are playing an important part in the treatment of cancer in immunotherapy—i.e., checkpoint inhibitors—for multiple cancers, and CAR T-cell therapy—i.e., chimeric antigen receptor T-cell therapy—in cancers of the blood—e.g., leukemia and lymphoma.

Evaluation of the bursa is important in clinicopathologic investigation in poultry for diseases such as infectious bursitis.

The bursa consists of lymphocytes embedded in folds of epithelial tissue. Scattered throughout the folds are lymphoid follicles. These follicles have a cortex and medulla. The cortex contains lymphocytes, plasma cells, and macrophages. The medullary epithelial cells are replaced by lymphoblasts and lymphocytes at the center of the follicle. This epithelium produces a mucin-like substance and takes up antigen to present to the bursal lumen. Lymphoid progenitor cells of the B-cell line mature within the bursa to reach humoral immunocompetence.

Thymus

The **thymus** is divided into 7–8 lobes and is located on each side of the neck adjacent to the jugular vein from the 3rd cervical vertebra to the thoracic inlet. Each lobe is 10–12 mm (0.4–0.5 in.) in diameter and surrounded by thin fibrous connective tissue and embedded in adipose tissue. The thymus enlarges for the first few weeks of life and then involutes at around 1–2 months of age, but not to the extent that it does in mammals. Lymphocytic stem cells migrate here during development and become the main source of T-lymphocytes (T-cells) for the body. Each lobe has a **cortex** and **medulla**. The medulla contains **Hassall's corpuscles,** which are bundles of concentrically arranged reticuloendothelial cells that are shielded from antigenic stimulation, while the cells in the rest of the medulla can empty into circulation for antigen production.

Spleen

The spleen can easily be seen on radiographs and may enlarge with infectious diseases such as chlamydiosis, polyomavirus, septicemia, and systemic protozoal diseases (e.g., *Atoxoplasma* spp., *Sarcocystis* spp., *Plasmodium* spp.).

The **spleen** can be round or flat in shape and lies to the left of the junction of the proventriculus and ventriculus. It contains both red and white pulp, so it is a part of both the hematologic and immune systems. The **red pulp** contains lymphoid and non-lymphoid cells. The **white pulp** is comprised of **lymphatic sheaths** (T-lymphocyte zones) and **splenic nodules** (B-lymphocyte zones). The spleen is the largest lymphoid organ and receives antigen from the circulating blood. It is part of the immune response of lymphocyte proliferation and recirculation.

Surface lymphoid tissues

The **Harderian glands** are located medioventrally in the orbit behind the globe and have secretory ducts that open onto the 3rd eyelid to lubricate the nictitating membrane. They—along with the lymphoid tissue within the conjunctiva—are responsible for mounting responses to aerosolized antigens, producing antibodies and cytokines such as interferon. The glands are divided into a **head** and **body**, each of which contains a variety of B- and T-lymphocytes.

Peyer's patches, also known as gut-associated lymphoid tissue (GALT), are small masses of lymphoid tissue distributed throughout the epithelial tissues of the ileum in mammals.

The **cecal tonsils** are lymphoid aggregates found at the base of each arm of the cecum. The general structure is like that of a Peyer's patch with a lymphoepthelium, germinal centers, and interfollicular areas (see Case 5.6). They appear before hatching, and the lymphocytes increase in number to adulthood.

Lymph nodes

Actual lymph nodes are rare in birds (seen in some waterfowl), but many birds have mural lymph formations within the walls of the lymphatic vessels. These thickenings are visible grossly and can be considered a modified version of the avian lymph node. They contain areas of B- and T-lymphocytes. These structures have full immune potential and fulfill the role of mammalian lymph nodes.

True lymph nodes found in water birds are limited to paired cervicothoracic and paired lumbar lymph nodes. The **cervicothoracic lymph nodes** are located at the end of the jugular lymphatic vessel next to the thyroid gland. The **lumbar lymph nodes** are modifications of the wall of the thoraco-abdominal vessel (trunk). The nodes are composed of lymph sinuses and lymphoreticular cords between the sinuses. The sinuses are branches of the end of the lymphatic vessel and are loosely arranged. The cords are made up of T- and B-lymphocytes and reticular fibers. Antigen processing takes place between the lumen of the sinuses and cords.

Avian hematopoietic system (Fig. 26.6-7)

Although beyond the scope of clinical anatomical description, the fluids that flow within the described vessels deserve description, in particular because of the differences between the avian and mammalian systems. Avian white blood cells are grouped into mononuclear cells and granulocytic cells. The mononuclear cells include the lymphocyte and monocyte. The granulocytic cells include the heterophil, eosinophil, and basophil. The avian white cells function similarly to their mammalian counterparts for the most part, but a few notable differences exist.

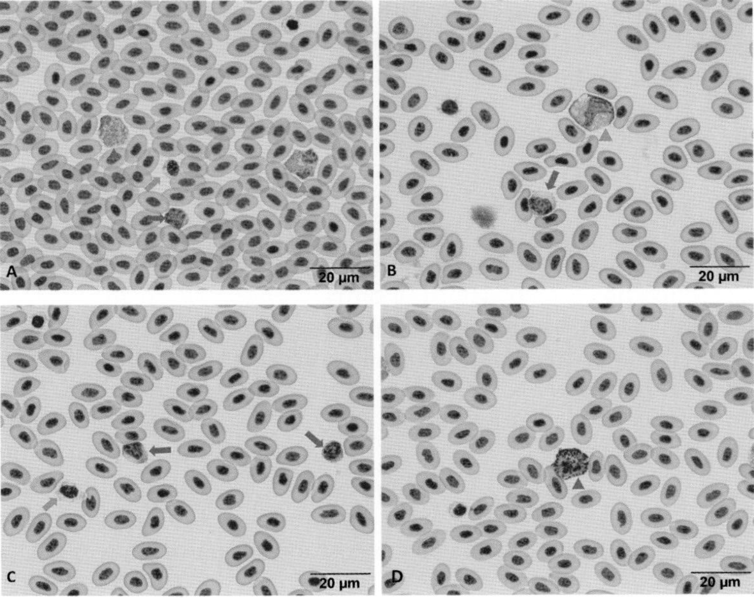

FIGURE 26.6-7 Chicken blood smears showing the different types of blood cells with erythrocytes. (A) Heterophil (*red arrowhead*), thrombocyte (*blue arrow*), lymphocyte (*red arrow*), and eosinophil (*blue arrowhead*); (B) lymphocyte (*red arrow*) and monocyte (*blue arrowhead*); (C) thrombocyte (*blue arrow*) and lymphocytes (*red arrows*); (D) basophil (*red arrowhead*); Wright-Giemsa stain. (Images courtesy of Dr. Ashleigh Newman, Cornell University.)

Erythrocytes

Avian **erythrocytes** are oval and nucleated and considerably larger than mammalian erythrocytes. The avian red cell is typically 10–15 μm in length, depending on the species; however, the typical mammalian erythrocyte is biconcave and 6–7 μm in diameter. Because the erythrocytes are larger, red blood cell (RBC) numbers are lower in birds compared with mammals. Numbers of RBCs also tend to be lower in young birds and females. The lifespan of the avian erythrocyte is 28–45 days, compared with the human erythrocyte lifespan of 120 days. This shorter lifespan is hypothesized to be a function of the increased metabolism and higher body temperature of birds. Being nucleated, avian erythrocytes do not rely on anaerobic glycolysis for cellular energy as mammalian cells do, but instead use aerobic metabolism of fat and protein. Additionally, avian erythrocytes contain several types of hemoglobin and are capable of synthesizing hemoglobin in situ.

Heterophils (Fig. 26.6-7A)

The **heterophil** is the avian equivalent to the mammalian neutrophil and has many of the same functions. These are the most common granulocytes involved in the acute inflammatory response in birds and work to control bacterial, viral, and parasitic infections. Avian heterophils lack myeloperoxidase, and thus mount a weak respiratory burst, relying more heavily on non-oxidative killing mechanisms.

> Elevations in heterophils are seen with infections, inflammation, stress, trauma, and some toxins. Avian abscesses are typically filled with caseous pus and surrounded by a thick wall. Heterophils accumulate—and are surrounded by—fibrin and epithelioid macrophages. This is thought to be related to a lack of myeloperoxidase or the presence of anti-tryptic enzymes that inhibit the protein breakdown.

Immature heterophils can be enlisted from the bone marrow in response to cytokines and other inflammatory modulators. The presence of these band cells are seen in the face of severe infections and are present in the first 12–24 h after an acute inflammatory event; this response lasts about 7 days.

Eosinophils (Fig. 26.6-7A) and basophils

The avian **eosinophil** has round, pink granules within the cytoplasm. The granules contain lysozymes, peroxidase, and high concentrations of arginine. These cells play a significantly different function from that of their mammalian counterparts. Experiments that trigger eosinophilia in mammals have produced inconsistent results in birds. This makes eosinophils an unreliable indicator for intestinal parasitism and hypersensitivity reactions in birds. Experimental findings suggest that avian eosinophils participate in delayed rather than immediate hypersensitivity reactions.

> An empirical correlation between eosinophilia and respiratory tract injury has been observed in birds. Conditions such as smoke inhalation, asthma, feather-picking, and drug interactions may cause an increased eosinophil count. Tissue trauma such as self-mutilation, predator-induced trauma, and flying accidents can also occasionally cause these elevations.

The avian **basophil** often has a single oblong nucleus and contains large, round, intensely basophilic granules within the cytoplasm. The granules contain histamine, making basophils important for the avian immune response. The degranulation of these cells plays a role in the early phase of acute inflammation in birds. Basophils also appear to play a role in anaphylaxis and the immediate hypersensitivity reaction in birds.

Lymphocytes (Fig. 26.6-7A, B, and C)

The avian **lymphocyte** originates in the yolk sac and migrates to either the bursa of Fabricius or thymus, creating B- and T-cells, respectively. B-cells are responsible for humoral immunity, while T-cells work through cellular immunity. In addition to the bursa and thymus, lymphocytes are also found in large numbers in the cecal tonsils and the skin.

Thrombocytes (Fig. 26.6-7C)

Avian **thrombocytes** are similar in function to mammalian platelets. They are round to square cells with dark blue cytoplasm. These cells are derived from stem cell precursors rather than megakaryocytes (as is the case in mammals). They function in hemostasis and coagulation by producing thromboplastin (a substance with procoagulant activity). They also have a phagocytic activity, especially against bacteria, and play a minor role in removing foreign material from the blood.

Selected references

[1] Pendle H, Tizard I. Immunology. In: Speer B, editor. Current therapy in avian medicine and surgery. St. Louis, MO: Elsevier; 2016. p. 400–32.
[2] Kaiser P, Balic A. The avian immune system. In: Scanes C, editor. Sturkie's avian physiology. 6th ed. St. Louis, Missouri: Elsevier; 2015. p. 403–18.
[3] Wakenell P. Management and medicine of backyard poultry. In: Speer B, editor. Current therapy in avian medicine and surgery. St. Louis, Missouri: Elsevier; 2016. p. 550–65.
[4] Greenacre C. Musculoskeletal diseases. In: Greenacre C, Morishita T, editors. Backyard poultry medicine and surgery a guide for veterinary practitioners. Ames, IA: Wiley Blackwell; 2015. p. 145–59.
[5] Schat K, Nair V. Marek's disease. In: Saif YM, Fadly AM, Glisson JR, LR MD, Nolan LK, Swayne DE, editors. Diseases of poultry. 12th ed. Ames, IA: Blackwell Publishing; 2008. p. 452–514.

CHAPTER 27

THORACIC AND PELVIC LIMB

Cynthia M. Faux and Marcie L. Logsdon, Chapter editors

27.1 Humeral fracture—*Cynthia M. Faux and Marcie L. Logsdon* 1366
27.2 Vertebral column trauma—*Cynthia M. Faux and Marcie L. Logsdon* 1377
27.3 Pododermatitis (bumblefoot)—*Cynthia M. Faux and Marcie L. Logsdon* 1384

CHAPTER 27

CASE 27.1

Humeral Fracture

Cynthia M. Faux[a] and Marcie L. Logsdon[b]
[a]University of Arizona College of Veterinary Medicine, Oro Valley, Arizona, US
[b]Exotics & Wildlife Department, Washington State University College of Veterinary Medicine, Pullman, Washington, US

Clinical case

American kestrels are a small species of falcon native to North America. By 2 months of age, they have left the nest and are of mature size but are still being supported by their parents.

Birds normally breathe through the nares (nostrils), except when stressed or over-heated. Handling and examination alone are stressful for wild birds; having a painful injury adds to the bird's level of stress.

History

A 2-month-old female American kestrel (*Falco sparvarius;* Fig. 27.1-1) was found on the side of the road unable to fly, presumably having been hit by a car.

Physical examination findings

The bird was visibly stressed and was open-mouth breathing. There was a noticeable droop to her left wing and palpable instability proximal to the elbow. Her body condition score was 3 (1 = emaciated, 5 = obese). The rest of the physical examination was normal.

Differential diagnoses

Fracture of the humerus or part of the pectoral girdle (scapula, coracoid, or clavicle)

Diagnostics

Whole-body radiography revealed an oblique fracture of the mid-diaphysis of the left humerus (Fig. 27.1-2). Separation of the cardiohepatic silhouette was also noted, consistent with blunt-force trauma which injured the surrounding air sacs, allowing free air to escape into the coelomic cavity and separate the apex of the heart from the cranial margins of the liver. The cardiohepatic silhouette is discussed in Case 26.1 and the air sacs in Case 26.3.

Comparative Veterinary Anatomy: A Clinical Approach. https://doi.org/10.1016/B978-0-323-91015-6.00127-8

Diagnosis

Closed, midshaft, oblique humeral fracture and blunt-force trauma to the coelomic cavity

Treatment

The bird was placed under general anesthesia; the fractured humerus was surgically reduced (realigned) and stabilized using an intramedullary (IM) pin attached

Fractures may be described as "open" or "closed." With a closed fracture, the skin overlying the fracture remains intact, whereas in an open fracture the skin has been perforated either by the fractured bone or by whatever external force fractured it. The distinction between an open and closed fracture is clinically important, as open fractures carry a much greater risk of infection which compromises bone healing—and in birds, they may lead to respiratory infection if a "pneumatic" bone such as the humerus is fractured (discussed later).

FIGURE 27.1-1 The young female kestrel described in this case. This photo was taken post-operatively; the distal portion of the external skeletal fixator is apparent on the left wing.

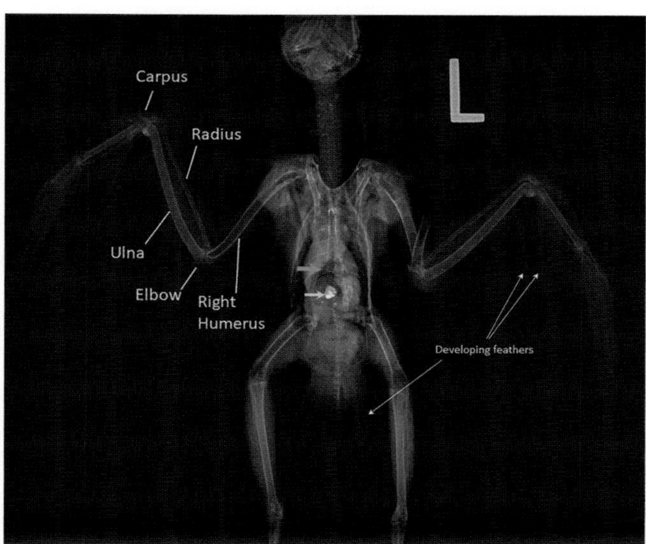

FIGURE 27.1-2 Pre-operative radiograph, ventrodorsal (VD) view, showing an oblique fracture of the mid-diaphysis of the left humerus. Also note the gap (*green arrow*) between the V-shaped apex of the heart and the broad, M-shaped cranial margin of the liver—separation of the cardiohepatic silhouette—consistent with blunt-force trauma to the body, resulting in free air in the coelomic cavity. The metallic foreign material in the ventriculus (*yellow arrow*) is an incidental finding. Of interest, the blood-filled shafts of developing flight feathers (blood feathers) are more radio-dense and thus visible on the extended wings and the tail of this young bird.

to an external skeletal fixator (ESF). Through a small skin incision over the dorsal aspect of the fracture site, the IM pin was introduced into the proximal medullary cavity of the humerus and directed proximally ("retrograde") until it exited the proximal humerus and the overlying muscle and skin adjacent to the shoulder joint. The pin was then advanced distally (normograde)—bridging the fracture—and into the distal diaphysis. The external, proximal part of the pin was bent at a 90° angle to serve as the attachment for the ESF. Two short bone pins were placed transversely across the shaft of the humerus, proximal and distal to the fracture. These pins and the IM pin were then connected by an acrylic side bar. Post-operative radiographs showed good alignment of the humerus (Figs. 27.1-3 and 27.1-4).

This fracture repair technique is commonly known as an "IM-ESF tie-in" (intramedullary-external skeletal fixator). The IM pin counteracts bending forces on the bone, while the ESF protects the bone from collapse and from shearing and torsional forces. Once the bone has healed, the surgical implants are removed.

HUMERAL FRACTURES IN BIRDS

The humerus is a common site for fractures in birds. The species and life history of the bird greatly influence the treatment and prognosis. Wild birds have a requirement for return to full function, whereas captive/pet birds may be able to manage with some degree of impairment.

Pet birds, such as parrots, often have decreased bone density due to poor diet, excessive egg-laying (which depletes calcium stores), and lack of exercise, so they may not have a sufficient bone density to support orthopedic pin placement. These birds often have a poor prognosis for return to flight and are sometimes best treated conservatively with cage rest, pain management, diet correction, and calcium supplementation.

Wild birds, in contrast, often have good bone density due to a balanced diet and the physiologic loads that flying places on the skeleton. The prognosis for return to full function can be good if the periosteum is not devitalized and the fracture is closed and repaired promptly, followed by appropriate physical rehabilitation.

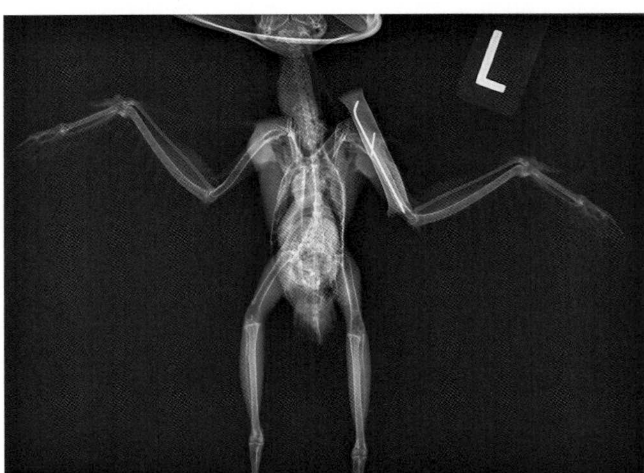

FIGURE 27.1-3 Post-operative radiograph, VD view. The intramedullary (IM) pin extends the length of the medullary cavity of the left humerus. In this view, the external skeletal fixator (ESF) is superimposed on the IM pin. The proximal part of the IM pin that protrudes from the skin near the shoulder was bent at a 90° angle to serve as the proximal anchor for the ESF. Only the proximal of the two transverse pins stabilizing the humerus is readily visible on this view. (The metallic material in the ventriculus, seen in Fig. 27.1-2, was passed in the feces or regurgitated in a pellet, as is typical with indigestible material in raptors.)

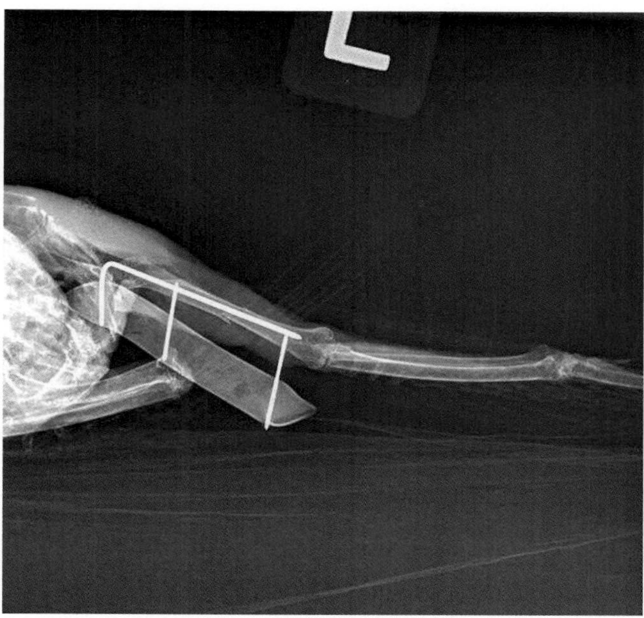

FIGURE 27.1-4 Post-operative radiograph of the left wing, craniocaudal view. The relationship between the intramedullary pin and the external skeletal fixator is best seen on this view. Alignment of the humerus (fracture reduction) is good, and return to full function (i.e., flight) can be anticipated once the fracture has healed.

The bird made an uncomplicated recovery and the coelomic trauma resolved with rest. The sidebar and cross pins were removed 2 weeks after surgery, and the IM pin was removed 1 week later. The bird was released to the wild after an additional 2 weeks of reconditioning.

In birds, orthopedic fractures that have been appropriately stabilized and treated heal surprisingly quickly compared with those in mammals. Clinical fracture stabilization, based on palpation of the fracture site, is often achieved before radiographic union is apparent.

Anatomical features in avian species

Introduction

Although some avian species are flightless, flight is one of the distinctive characteristics of birds. The various structural and functional modifications birds have made for flight are summarized in Chapter 24, and the flight feathers are described and illustrated in Case 28.1. Following is a discussion of the anatomy of the wing and pectoral girdle.

Function

The function of the pectoral girdle is to support the thoracic limb (i.e., the wing). In most avian species, the thoracic limb enables flight. In addition, some species may use their wings for swimming, display, defense, or to protect and cover the nest and nestlings.

Bones of the pectoral girdle (Figs. 27.1-5–27.1-7)

The bones of the wing include the humerus, radius, ulna, carpal and metacarpal bones, and digits. Proximally, the **humerus** articulates with the axial skeleton via the **pectoral girdle**, which in birds includes the paired scapulas, coracoids, and clavicles. Cranioventrally, the left and right clavicles unite, thus forming a single U- or V-shaped structure, the **furcula** (L. *furca* fork).

The bones of the thoracic limb and pectoral girdle in birds are homologous to the mammalian thoracic limb, except for the coracoid, which has no mammalian equivalent. The coracoid process of the scapula in mammals

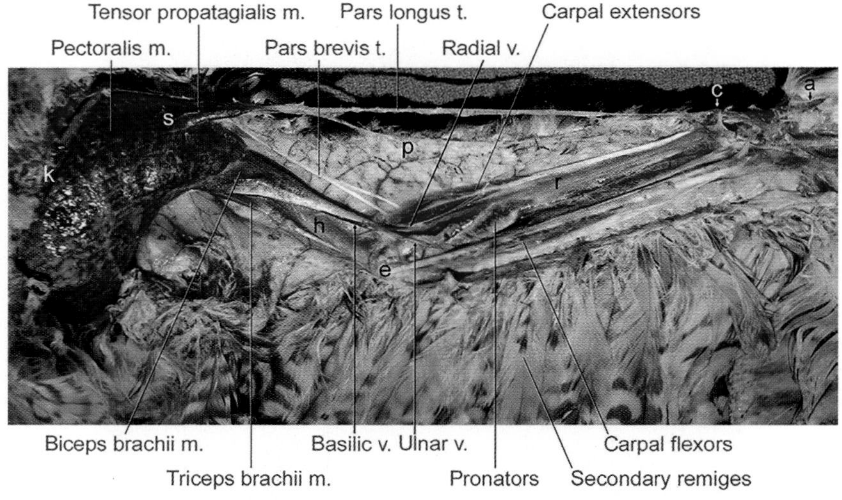

FIGURE 27.1-5 Left wing of a Red-tailed hawk (*Buteo jamaicensis*), ventral aspect. The skin and superficial fascia have been removed from the ventral surface to show the major muscles and vessels of the wing. Pars brevis is the short tendon and pars longus is the long tendon of the tensor propatagialis muscle. Key: *m.*, muscle; *t.*, tendon; *v.*, vein. Key: *a*, alula; *c*, carpal joint; *e*, elbow joint; *h*, humerus; *k*, keel of the sternum; *p*, propatagium (inner surface); *r*, radius (the ulna is obscured by muscles and tendons); *s*, location of the shoulder joint.

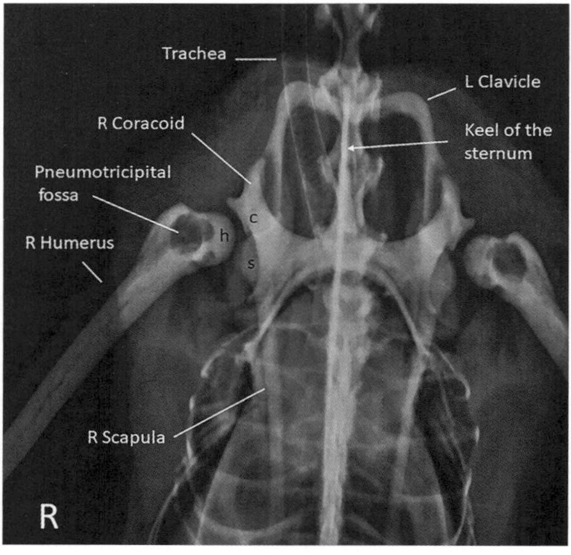

FIGURE 27.1-6 Standard VD radiograph of an adult female domestic duck, with the elements of the pectoral girdle labeled. To obtain this view, the bird was positioned in dorsal recumbency and the X-ray beam was directed 90° to the X-ray plate beneath the bird's body. Compare this image with Fig. 27.1-7. Key: *L*, left; *R*, right.

is not homologous to the coracoid bone in birds. As illustrated, the avian coracoid, and in some species the furcula as well, provides a bony connection between the wing and the axial skeleton (specifically, the sternum), whereas in many mammalian species, the connection between thoracic limb and axial skeleton is composed entirely of soft tissues.

In most avian species, the **scapula** is flat, long, and narrow. It extends caudally from its articulation with the **humerus**, typically lying parallel to the thoracic spine and overlying the dorsal aspect of the ribs. However, the scapulohumeral joint in birds is more complex than that in mammals, as the **glenoid fossa** in birds is formed by the articulation between the **scapula** and the **coracoid**. The **head of the humerus** thus articulates with both bones,

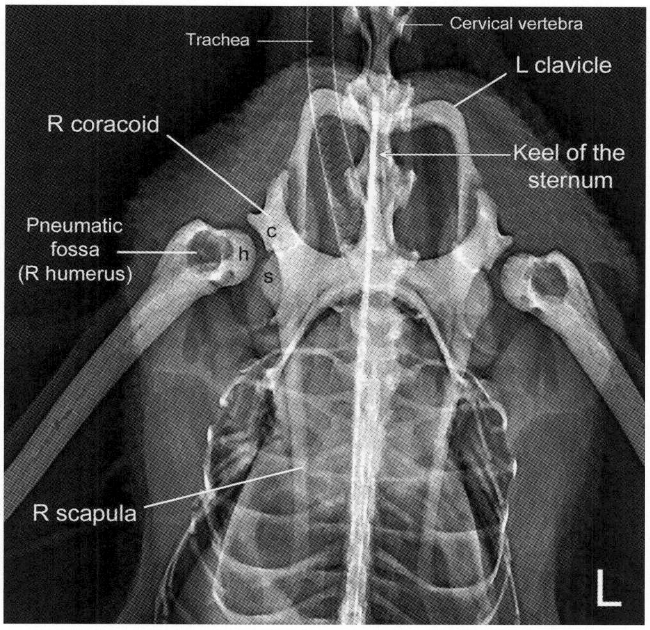

FIGURE 27.1-7 Radiograph of the duck in Fig. 27.1-6. This 45° caudoventral-craniodorsal oblique, commonly called the "H view" or "H shot," is used to evaluate the pectoral girdle because it eliminates much of the superimposition that occurs with the standard VD view. The 3 bony components of the avian shoulder joint are also clearly seen on this view: scapula *(s)*, coracoid *(c)*, and head of the humerus *(h)*. Note that this duck is exhibiting the increased bone density that normally occurs in reproductively active adult females. Also note the large pneumotricipital fossa within which lies the pneumatic foramen, in the proximal humerus.

"H VIEW" OF THE PECTORAL GIRDLE

The individual components of the pectoral girdle are visible on a standard VD radiographic view (Fig. 27.1-6), in which the anesthetized bird is positioned in dorsal recumbency and the vertical X-ray beam is at 90° to the horizontal X-ray plate beneath the bird. However, superimposition makes it difficult to assess each component in full. The 45° caudoventral-craniodorsal oblique ("H view" or "H shot"; Fig. 27.1-7) eliminates much of this superimposition. The X-ray beam is directed at a 45° angle to the table (and thus to the X-ray plate) from the patient's ventral abdomen cranially toward the base of the neck. This view is particularly useful when unilateral fracture of some component of the pectoral girdle is suspected.

making this multifaceted joint a "scapulocoracohumeral" or "coracoscapulohumeral" joint. Generally, though, the term **shoulder joint** is used for succinctness.

The **coracoid** extends caudoventrally from the shoulder joint to articulate distally with the craniolateral aspect of the **sternum**. Its size, shape, and position allow it to act as a rigid support, providing extra stability to the wing. In some species, the coracoid is a pneumatic bone, invested with diverticula of the clavicular air sac, like the humerus (described later).

The scapulae and coracoids also articulate with the proximal ends of the furcula. Typically, the furcula consists of a pair of rod-like **clavicles**, left and right, which originate at the shoulder (as described below) and unite on the ventral midline at the base of the neck, thus forming the eponymous "fork"—and defining the skeletal component of the thoracic inlet. At its midpoint, the furcula may be attached to the cranial aspect of the sternum by a short, strong ligament; in some species, this is a bony attachment. However, there is considerable species variation in this component of the pectoral girdle: The 2 clavicles may fuse completely or form a merely fibrous union; in some parrot species and ratites (ostrich, emu, rhea, etc.), the clavicles may be reduced or even absent.

Immediately dorsal to the shoulder joint, the scapula, coracoid, and clavicle converge to form a channel, the **triosseal canal** (literally, "3-bone" canal), through which the **tendon of the supracoracoideus muscle** passes from the sternum to the humerus. This flight muscle originates on the **keel** of the sternum and inserts on the dorsal aspect of the proximal humerus so, when it contracts, it raises the wing. The triosseal canal acts as a pulley to relay the contraction of the muscle, via the tendon, from the ventrum of the bird to the dorsum of the wing. In this way, the muscles that raise the wing and those that lower the wing (the pectoralis muscles; described later) both originate on the sternum, below the bird's center of gravity. In some species, the clavicle does not contribute to the canal.

A notable exception is the normal increase in bone density that occurs in adult females in advance of breeding season, their bones serving as a calcium reservoir for egg shell production (Fig. 27.1-7). The bones of domestic poultry that are bred for relatively continuous egg production may be increased in density year-round, unless the bird's diet is lacking in calcium.

Bones of the wing (Fig. 27.1-5)

Avian bones, including those of the wing, tend to be hollow and thin-walled, with strut-like trabeculae as internal reinforcements, creating a lightweight yet strong bone. These features are evident radiographically in the long bones of the kestrel in this case (see Fig. 27.1-2).

In addition to its lightweight structure, the avian **humerus** is usually "pneumatic": Diverticula from the **clavicular air sac** extend into the medullary cavity of the humerus as a direct extension of the respiratory system (see Case 26.3). This anatomic feature presents some unique clinical challenges (see the side box entitled **Fractures of pneumatic bones**). The **pneumatic foramen** through which the air sac diverticulum passes is found within a cavitary region, the **pneumotricipital fossa** (Fig. 27.1-7). The size and shape of this fossa are quite variable among species. Some species, such as penguins, lack a pneumatic foramen.

FRACTURES OF PNEUMATIC BONES

In birds, fractures of pneumatic bones such as the humerus may lead to respiratory infection, particularly with open fractures. Conversely, because anesthetic gases may leak from the fracture site during surgical repair, electrocautery cannot be used for hemostasis (control of bleeding) because many anesthetic gases and the presence of high concentrations of oxygen are potentially combustible. Fractures of pneumatic bones can also lead to subcutaneous emphysema, although usually self-limiting. Lastly, excessive lavage of the fracture site may flood the air sacs literally drowning the patient.

PECTORAL MUSCLE MASS

Assessment of pectoral muscle mass and its overlying layer of subcutaneous fat is an important part of physical examination in birds. Reduction in muscle mass (atrophy) for the bird's age and life history is important, whether unilateral or bilateral. Common causes of bilateral pectoral muscle atrophy include:

- Malnutrition (starvation or specific nutritional deficiencies)
- Parasitism, especially coccidiosis in young, immune-compromised birds
- Improper housing (too little space)
- Chronic flightlessness in flighted species, caused by illness or injury

Unilateral pectoral muscle atrophy should direct a thorough examination of the associated wing and pectoral girdle, as it may be indicative of an injury or other impairment.

Note: In obese birds, pectoral muscle mass may be difficult to assess.

The radius and ulna are not typically pneumatic bones, although the ulna may be in some species. In birds, the **radius** is thinner and shorter than the ulna (Fig. 27.1-2). In contrast to mammals, the **ulna** is the principal load-bearing bone in the avian antebrachium. Movements of the radius and ulna are relative and reciprocal—the parallel radius and ulna move in opposite directions (proximal-distal movement) as the wing is extended and flexed. This dynamic allows synchronized extension and flexion of the joints of the carpus and elbow. As in mammals, there is also some limited rotation around the long axis of the avian antebrachium.

A flight-limiting synostosis (abnormal fusion of 2 bones) can develop between the radius and ulna if both are fractured or if a misaligned fracture of one is not corrected. This condition is evident clinically by the inability to extend the carpus. Because of this possibility, surgical stabilization of radial and ulnar fractures is likely to provide the best outcome. Physical rehabilitation, including passive range-of-motion exercises, is also very beneficial during healing.

The secondary flight feathers, or **secondary remiges**, attach directly to the periosteum of the **ulna** and may form raised bumps called "quill knobs" along the caudal edge of the ulna. These feathers are responsible for lift in flight (see Case 28.1).

Plucking of the secondary remiges in preparation for surgery to repair an ulnar fracture can traumatize the periosteum and inhibit bone healing. Therefore, these feathers should be preserved.

In birds, the proximal row of carpal bones is reduced to 2: the **radial carpal bone** and the **ulnar carpal bone**. The distal row of carpal bones is fused with the metacarpal bones, creating the **carpometacarpus**.

The avian metacarpus is generally reduced to 3 bones, with corresponding digits. The homology of the metacarpus and digits of the avian manus has been debated, as to whether these bones correspond to mammalian metacarpal (MC) bones I, II, and III or MC II, III, and IV. Avian clinicians typically avoid confusion by referring to these bones as the major, minor, and alular metacarpal bones. The **major (carpo)metacarpal bone** is the largest. Many of the long primary flight feathers, the **primary remiges**, attach to this bone via its periosteum. The **minor metacarpal bone** and the small **alular metacarpal bone** are fused to the major metacarpal bone, forming a rigid structure.

The major metacarpal bone is typically associated with two **phalanges**, which bear the remainder of the primary remiges. The minor metacarpal bone usually articulates with a single phalanx, which bears none of the primary remiges. There are typically 2 phalanges associated with the **alula**, which forms the "thumb" of the avian manus. The alula, and the feathers attached to it, performs a distinct aerodynamic function—when extended, it changes the curvature of the airfoil of the wing. This action is typically deployed at low airspeeds, the alula acting in the same fashion as the wing flaps on an airplane.

Muscles of the wing (Fig. 27.1-5)

The primary depressor of the wing is the **pectoralis muscle**, the muscle that provides the main "power stroke" in flight. It is the largest muscle of the typical bird. The pectoralis muscle originates on the broad lateral surface of the sternum and **keel** (ventral midline of the sternum), and its tendon inserts on the deltoid crest of the humerus, distal to the humeral head.

INTRAOSSEOUS INFUSION IN BIRDS

The distal ulna and the proximal tibiotarsus are safe and convenient sites for intraosseous (IO) fluid administration in most birds. This procedure involves placing a needle through the cortex of the bone so that fluid or medication may be administered directly into the highly vascular medullary cavity, where it is rapidly absorbed into the systemic circulation. An alternative to oral and IV fluid administration in most animal species, IO administration is particularly useful in birds—although care must be taken to avoid IO infusion into pneumatic bones such as the humerus.

The smaller, spindle-shaped **supracoracoideus** muscle lies deep to the pectoralis muscle on the **sternum**. As mentioned above, the tendon of the supracoracoideus muscle passes through the **triosseal canal** and attaches to the proximal **humerus**, thereby raising the wing. Many smaller muscles of the shoulder and wing also help flight and allow the bird to control its movements in the air and during takeoff and landing.

The **propatagium** (often referred to simply as the "patagium") is a thin, feathered fold of skin that spans the angle between the proximal humerus (at the shoulder) and the carpal bones on the cranial aspect of the wing (Fig. 27.1-8). It serves as a smooth leading edge for the wing, contributing to the airfoil shape of the wing, which provides the aerodynamic lift necessary for flight. The **tensor propatagialis** muscle has two parts, the **pars longus** and **pars brevis**, and it tenses the propatagium when the wing is fully extended. The tendon of the pars brevis, often simply called the **pars brevis tendon**, runs parallel to the humerus within the propatagium. The **pars longus tendon** runs along the leading edge of the propatagium.

Injury to, or contracture of, the propatagium can be catastrophic as it may lead to permanent flight loss. Specifically, injury to the pars longus tendon can be a negative prognostic indicator in cases of propatagial trauma.

PARS BREVIS TENDON ENTRAPMENT

When the humerus is fractured, the proximal end of the distal bone fragment is often displaced laterally, toward the radius and ulna (Fig. 27.1-2), by the pull of the carpal extensor muscles. The distal fragment of the humerus may then impinge on the pars brevis tendon within the propatagium. If the fracture is not quickly reduced and surgically repaired, the pars brevis tendon may become entrapped by fibrous tissue or bony callus, preventing full extension of the wing and therefore flight (Fig. 27.1-9).

Good surgical repair is necessary to prevent the entrapment of the pars brevis tendon and thus preserve the bird's ability to fly. Regular passive range-of-motion exercises (usually performed under anesthesia) also help prevent this consequential complication.

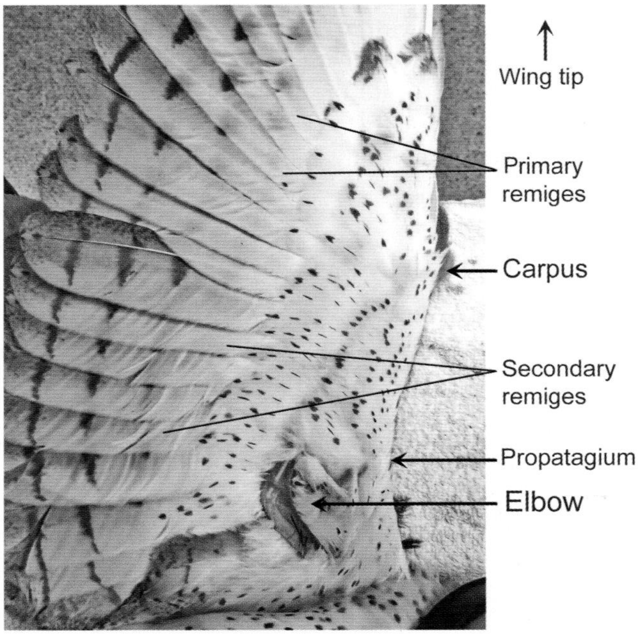

FIGURE 27.1-8 Ventral aspect of the right wing of a barn owl (*Tyto alba*). The wing is extended to show the propatagium, the primary and secondary flight feathers (remiges). Note, alcohol has been applied to the featherless area at the elbow to aid in the visualization of the veins in this area (see Fig. 27.1-10).

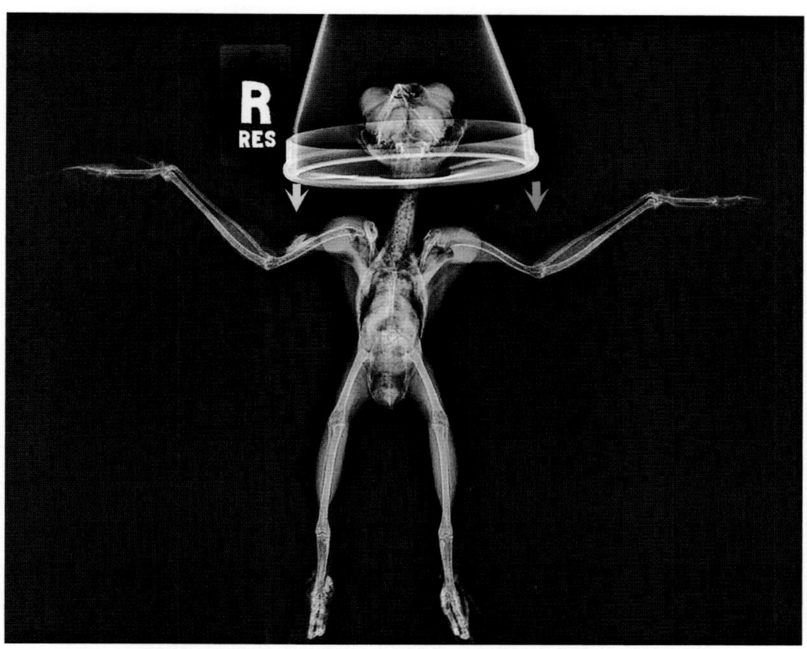

FIGURE 27.1-9 Radiograph of a small owl with entrapment of the tensor propatagialis pars brevis tendon following fracture of the right humerus. Note the bony callus on the cranial aspect of the right humerus, just distal to midshaft. Whereas the leading edge of the propatagium on the left wing forms a straight line between shoulder and carpus (*green arrow*), it dips toward the callus on the right wing (*yellow arrow*).

Blood supply and innervation (Fig. 27.1-5)

Arterial supply to the wing is via the **axillary artery**, a branch of the **subclavian artery**, which arises from the **brachiocephalic trunk** (see Case 26.1). In many avian species, the paired brachiocephalic trunks, which arise from the ascending aorta, are larger in diameter than the descending aorta, highlighting the importance of the arterial supply to the wing in birds.

Venous drainage generally follows the arterial pattern. The veins of clinical importance are visible just beneath the skin on the ventral aspect of the elbow, in a relatively featherless area (Fig. 27.1-10). The **deep ulnar vein** and the larger and more superficial **ulnar vein** converge in the proximal antebrachium, and the ulnar vein may be seen crossing the proximal end of the ulna just distal to the elbow joint. The **basilic vein** is the primary superficial vein in the brachium. It may be seen tracking parallel to the humerus just proximal to the elbow joint in the same featherless area and may also be seen continuing toward the muscles of the shoulder. The **radial nerve** coils around the humerus and must be protected when performing surgery in this area.

The ulnar and basilic veins are often used for venipuncture in birds. Which vein is chosen depends on the size of the bird.

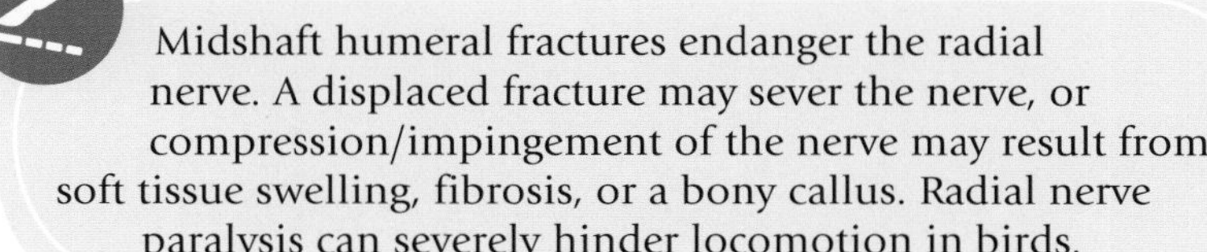

Midshaft humeral fractures endanger the radial nerve. A displaced fracture may sever the nerve, or compression/impingement of the nerve may result from soft tissue swelling, fibrosis, or a bony callus. Radial nerve paralysis can severely hinder locomotion in birds.

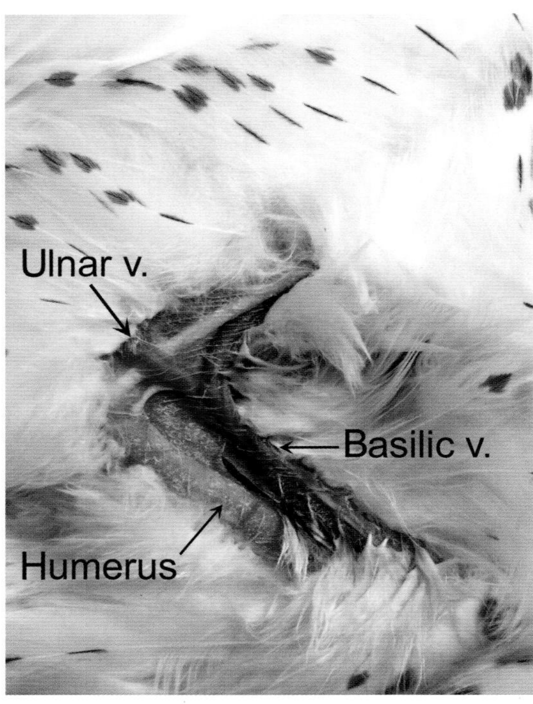

FIGURE 27.1-10 Ventral aspect of the right elbow of the barn owl shown in Fig. 27.1-8. In this image, the elbow is in greater flexion. The feathers have been dampened with alcohol and swept aside to show the ulnar and basilic veins just beneath the skin in the featherless area overlying the elbow. The skin in this area is so thin that the veins are easily identified, allowing them to be used for blood sample collection. Key: *v.* vein.

Selected references

[1] Redig P, Ponder J. Orthopedic surgery. In: Samour J, editor. Avian medicine. 3rd ed. St. Louis, MO: Mosby Ltd; 2016. p. 312–58.
[2] Getty RA. Sisson and Grossman's The anatomy of domestic animals. 5th ed. Philadelphia: W.B. Saunders Company; 1975. p. 1786–2095.
[3] Baumel JJ, editor. Handbook of avian anatomy: Nomina Anatomica Avium. 2nd ed. Nuttall Ornithological Club (USA), World Association of Veterinary Anatomists (USA), International Committee on Avian Nomenclature; 1993.
[4] King AS, McLelland J. Birds: Their structure and function. 2nd ed. London; Philadelphia: Baillière Tindall; 1984.
[5] Visser M, Hespel AM, de Swarte M, et al. Use of a caudoventral-craniodorsal oblique radiographic view made at 45° to the frontal plane to evaluate the pectoral girdle in raptors. J Am Vet Med Assoc 2015;247(9):1037–41.

CHAPTER 27

CASE 27.2

Vertebral Column Trauma

Cynthia M. Faux[a] and Marcie L. Logsdon[b]
[a]University of Arizona College of Veterinary Medicine, Oro Valley, Arizona, US
[b]Exotics & Wildlife Department, Washington State University College of Veterinary Medicine, Pullman, Washington, US

Clinical case

History

A juvenile Bald eagle (*Haliaeetus leucocephalus;* Fig. 27.2-1) was found down and unable to fly in a field about 20 yards from a road. A passerby brought the bird to a wildlife rehabilitation center for evaluation, where it was observed that the young eagle seemed to have trouble using his feet.

The feather patterns in immature bald eagles change dramatically over the first 4 or 5 molts. Bald eagles do not have the characteristic white head (for which they are named) and tail feathers until they are approximately 5 years of age.

Physical examination findings

The bird was bright, alert, and responsive. There were no palpable fractures of the wings or legs, but the bird had difficulty standing. His tail was dirty and unkempt, likely from sitting against it for some time. In addition, there was fecal staining around the vent (the shared external opening of the digestive and urogenital tracts in birds), tail, and legs. Faint bruising was visible overlying the spine.

Neurologic examination revealed flaccid paresis (weakness) in both legs. A withdrawal reflex was present bilaterally, but it was weak, as was the deep pain response. Vent and tail tone were present, but there was no tail-flare response.

The deep pain response is often difficult to assess in wild animals, as their fear may mask any response to pain. In this case, movement of the nictitating membranes was the only consistent response to the assessment of deep pain when pinching the toe with a hemostat). The tail-flare response is evaluated by holding the bird and rocking it from a "tail-up" to "tail-down" position. A normal bird "flares" or spreads its tail feathers in response to this sudden change in posture.

LEAD TOXICITY IN RAPTORS

Lead toxicity is very common in raptors. Presenting signs include generalized weakness or pelvic limb paresis, clenched feet, and green feces. Raptorial species that feed largely on carrion are most often affected. Viscera (gut piles) left by hunters are a common source of exposure, as fragments of lead ammunition often remain in these discarded tissues.

Comparative Veterinary Anatomy: A Clinical Approach. https://doi.org/10.1016/B978-0-323-91015-6.00128-X

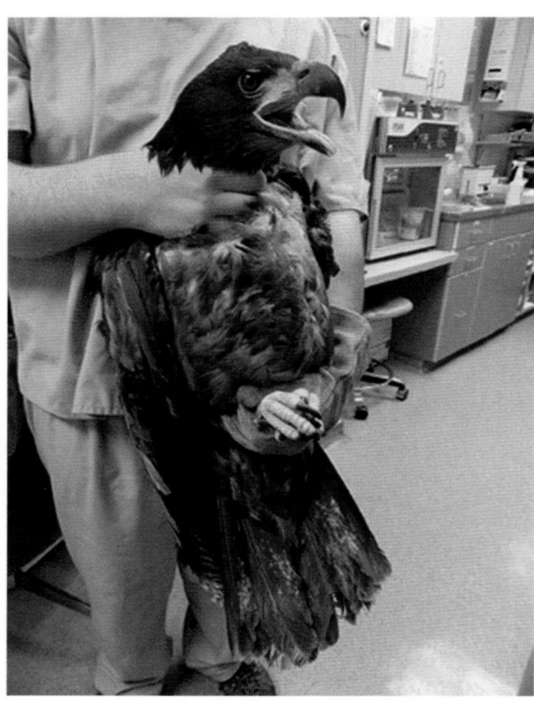

FIGURE 27.2-1 A juvenile bald eagle.

Differential diagnoses

Lead toxicity, vertebral column trauma

Diagnostics

Whole-body radiography showed evidence of coelomic trauma; there were no apparent bony lesions (Figs. 27.2-2 and 27.2-3). Fig. 27.2-4 depicts the normal appearance of the notarium, synsacrum, and pygostyle for comparison. The bird's blood lead concentration was 0.12 mg/dL, which is below the diagnostic threshold for lead toxicity in birds (0.20 mg/dL).

VERTEBRAL COLUMN TRAUMA IN BIRDS

The thoracic intervertebral articulations between the notarium and synsacrum are movable.

These "free" caudal thoracic vertebrae are the most common site of vertebral column trauma. It is often the result of a "rigid" impact, such as being hit by a car, in which the notarium and synsacrum briefly shift in relation to one other, traumatizing the spinal cord in between. Often, these structures return to their normal position immediately afterward, making radiographic diagnosis of vertebral column trauma difficult or impossible unless, and until, bone remodeling occurs. The presence of a vertebral column lesion on initial radiographs is a negative prognostic indicator.

While MRI is useful for identifying radiographically silent vertebral column trauma, it is often cost-prohibitive. Instead, a presumptive diagnosis of vertebral column trauma is generally based on the history and physical examination findings and ruling out the other causes for the bird's clinical signs.

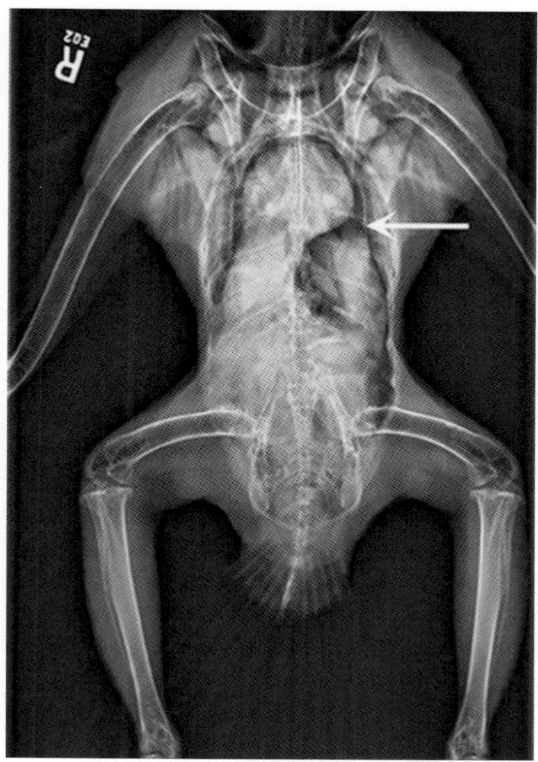

FIGURE 27.2-2 Ventrodorsal radiograph of the eagle at the initial presentation. Separation of the cardiohepatic silhouette is seen on the bird's left side (*arrow*). This separation is caused by free air in the coelomic cavity, likely the result of a ruptured coelomic air sac. In addition, there is decreased aeration of the caudal air sacs on the left side and a complete collapse on the right side, with displacement of the caudal coelomic organs to the right. Loss of normal serosal detail in these organs is likely attributable to bleeding. (Compare this radiograph with the post-recovery one in Fig. 27.2-5.)

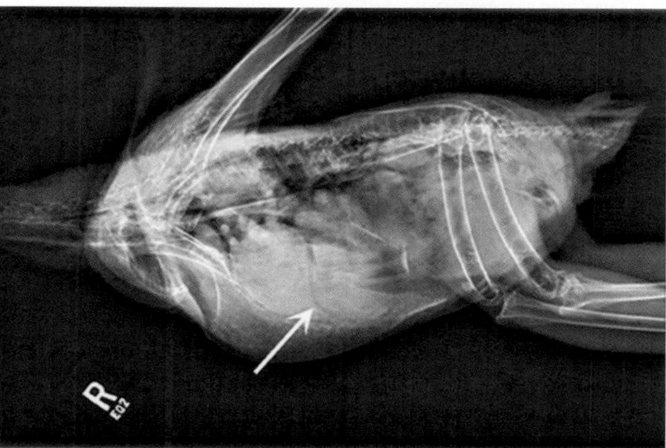

FIGURE 27.2-3 Lateral radiograph of the eagle at the initial presentation. Separation of the cardiohepatic silhouette is also seen on this view (*arrow*). However, there is no apparent bony injury to the vertebral column.

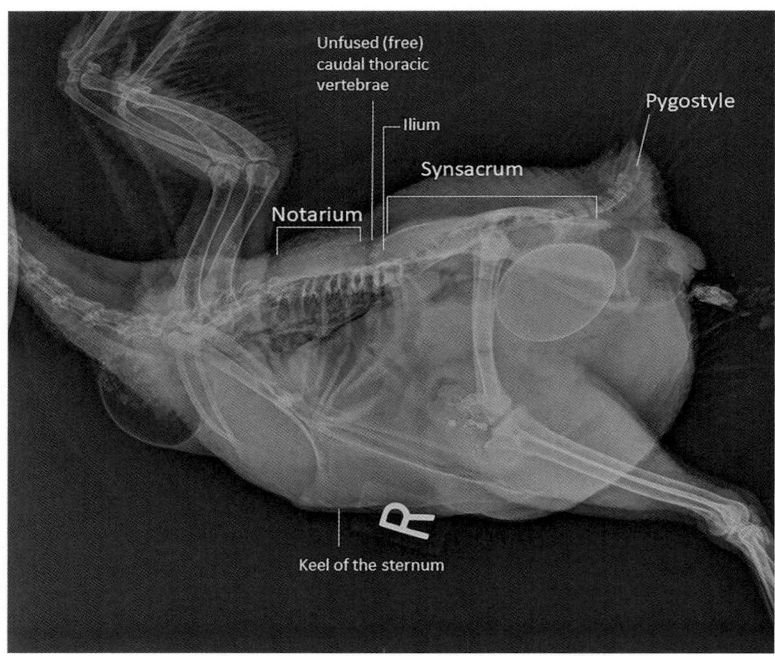

FIGURE 27.2-4 Lateral radiograph of a chicken, showing the normal appearance of the notarium, synsacrum, and pygostyle in this species.

Diagnosis

Vertebral column trauma (presumptive)

> In wild birds, outcome is strongly correlated with the severity of neurologic deficits on the initial presentation. In this case, the bird retained some sensory and motor functions in the legs and caudal region, so the prognosis was considered "fair" for complete recovery.

Treatment

With anti-inflammatory therapy and supportive care, the eagle slowly regained normal neurologic function and was able to be released 3 months after presentation. Follow-up radiography before the bird's release showed resolution of coelomic trauma and focal hyperostosis/bone remodeling of the caudal thoracic spine (Figs. 27.2-5 and 27.2-6).

Anatomical features in avian species

Introduction

There is considerable variation in the vertebral formula among birds; the number of vertebrae varies among species, both in total and within vertebral regions. In addition, anatomical fusion of a variable number of vertebrae is a universal feature. The 3 vertebral fusions normally and commonly found in birds are the notarium (thoracic), synsacrum (lumbosacral), and pygostyle (caudal). Each is described in its associated section below.

CSF FLUID ASPIRATION

The atlanto-occipital (AO) space is generally the only option for the aspiration of CSF fluid because of the fusion of much of the vertebral column; additionally, the free spaces are very narrow, making the AO space the only viable site. The spinal cord terminates as a small extension into the pygostyle after the last spinal nerve has exited.

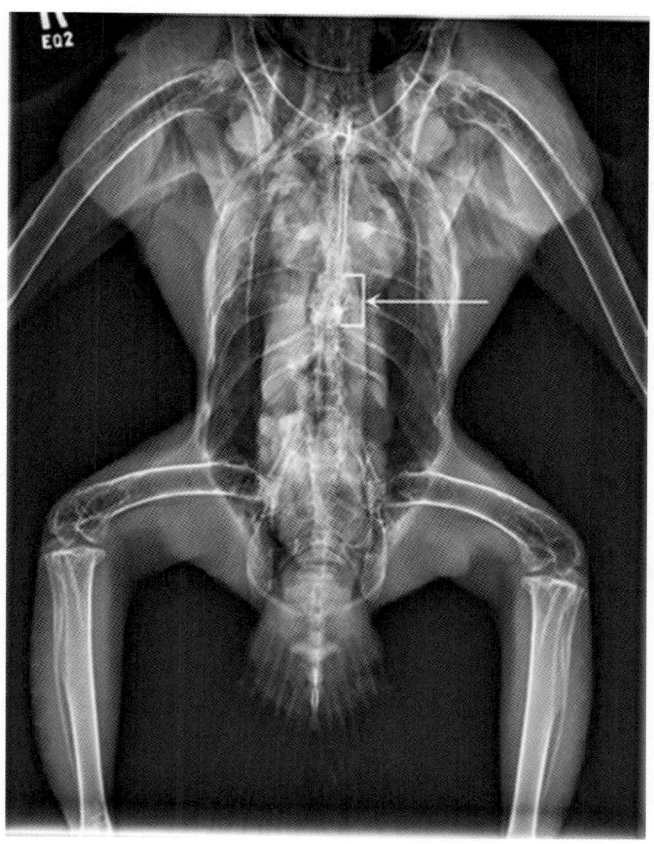

FIGURE 27.2-5 Ventrodorsal radiograph of the eagle 3 months after presentation. The coelomic trauma evident at presentation (Fig. 27.2-2) has resolved, and there is now an area of bone remodeling and hyperostosis in the caudal thoracic vertebral column (*arrow and bracket*).

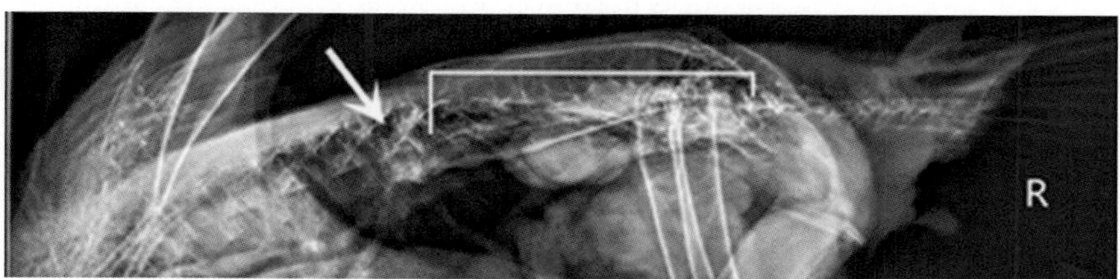

FIGURE 27.2-6 Lateral radiograph of the eagle 3 months after presentation, cropped and magnified to highlight the vertebral column. Note the bone remodeling and hyperostosis in the caudal thoracic vertebral column (*arrow*), just cranial to the synsacrum (*bracket*).

Function

The major function of the axial skeleton is to protect the spinal cord, nerve roots and, with the ribs, to provide protection to internal organs. It also provides the structural support and balance for posture, enabling flexibility and mobility in motion while affording attachment sites for the pectoral and pelvic muscles.

Cervical vertebral column

The neck is the most flexible portion of the avian vertebral column. The **cervical** region of the vertebral column is the most variable among, and even within, avian species. In mammals, lengthening of the neck occurs by lengthening of the 7 cervical vertebrae themselves, whereas in birds it is accomplished by increasing the number of cervical vertebrae. For example, chickens typically have 16 cervical vertebrae, while swans may have as many as 25.

Unlike in mammals, the avian skull articulates with the 1st cervical vertebra (the atlas) via a **single occipital condyle**, so the avian atlas has a single condylar surface. Thereafter, the articulations between adjacent vertebral bodies (**centra**) in most birds are saddle-shaped (heterocoelous).

The vertebral body is also known as the centrum (plural, centra).

In many birds, the intervertebral joints of the neck have thin **menisci** which attach to the joint capsule. In contrast to mammalian intervertebral disks, these menisci are thinner in the center and may be open. Some species, such as chickens and pigeons, have less intervertebral cushioning, with just a ring of thickened tissue (**annulus fibrosus**) between the articular surfaces.

Thoracic vertebral column and ribs (see Landscape Fig. 6.0-7)

The **thoracic** region of the vertebral column is generally considered to begin with the first vertebra that is attached to the **sternum** via a pair of **ribs**—i.e., the first vertebra with a pair of sternal or 'true' ribs. The true ribs in birds are those that have both dorsal (or vertebral) and ventral (or sternal) parts. Rib number varies among, and even within, species. For example, chickens have 6 or 7 pairs of true ribs (Fig. 27.2-4).

However, the transition from cervical to thoracic regions can be indistinct in birds, because the first 1 or 2 pairs of ribs may be asternal, simply having a short vertebral component. The associated vertebrae are usually considered to be cervical, even though they have attached ribs. Sometimes these vertebrae are called 'transitional,' as they have features of both cervical and thoracic vertebrae.

Except for the first and last pair, adjacent true ribs bear an **uncinate process**, which is a short caudal projection of the vertebral portion of the rib that overlaps the cranial portion of the rib immediately caudal to it. These processes help stabilize the lateral aspect of the rib cage, creating a more rigid structure.

In a variety of avian species, 2 or more of thoracic vertebrae may fuse to form a rigid structure known as the **notarium**, or **os dorsale** (Fig. 27.2-4). A notarium is present in chickens, pigeons, and some raptors, but not in ducks or parrots. Species without a notarium strenghten the thoracic vertebral column by ossification of epaxial muscle tendons or by buttressing the dorsal spinous processes. A notarium, and other reinforcing strategies, prevents the bending of the thoracic vertebral column during flight.

Sternum

The **sternum** in most birds is extensive. Unlike in mammals, the avian sternum is not segmented; rather, it is a large, single bone that acts as a ventral support for the coelomic organs. In avian species that fly, the sternum also provides a broad surface for attachment of the flight muscles, notably the pectoralis and supracoracoideus muscles (see Case 27.1). A ventrally directed, median **keel**—or **carina** (L. *carina* keel of a ship)—increases the surface area for muscle attachment in most flying species (Fig. 27.2-4). Flightless birds such as ostriches lack a keel on the sternum.

Palpation over the keel is an important component of physical examination in birds. A decrease in pectoral muscle mass may indicate that the bird has been malnourished or unable to fly for some time. Asymmetry indicates unilateral wing or nerve injury. Many species deposit subcutaneous fat over their pectoral muscles, so palpation over the keel is useful for body condition assessment.

The "ratites" (ostrich, emu, rhea, cassowary) are so-named because their sternum is shaped more like a raft (L. *ratis* raft).

Caudally directed lateral processes extend from the sternum in some species, creating notches or perforations. Rods of bone between these lateral processes are called **trabeculae**. A fibrous membrane occludes these perforations in the sternum. The shape of the caudal border of the sternum can be round or notched, squared off, or pointed.

The sternum in many birds is a pneumatized bone, in which diverticula of the thoracic air sacs enter the bone. In species with an elongated trachea that is contained within the sternum, cavities within the bone accommodate these loops or coils of extracoelomic trachea (see Case 25.3). In addition to articulating with the true ribs, the sternum articulates with the coracoid, which is described in Case 26.3.

Synsacrum

In birds, the most caudal thoracic vertebrae as well as the lumbar and sacral vertebrae are fused to form the **synsacrum**, which is often fused bilaterally with the **os coxae**, forming an inflexible structure (see Figs. 27.2-4 and 27.2-6). Unlike the notarium, the synsacrum is a feature that is common to all birds.

Cranially, the synsacrum articulates either with the notarium or with a free thoracic vertebra, forming an intervertebral joint that is potentially more vulnerable to traumatic displacement than the rest of the post-cervical vertebral column. Caudally, the synsacrum articulates with the 1st caudal vertebra. The kidneys lie well-protected tucked beneath the ventral surface of the synsacrum.

Caudal vertebrae

The number of **caudal** vertebrae in birds is quite variable and, typically, the last 5 or 6 caudal vertebrae are fused to form the **pygostyle**, to which the caudal (tail) muscles, fascia, and tail feathers (**rectrices**) are attached (Fig. 27.2-4). The shape of the pygostyle varies among species, and often it is laterally flattened into a vane or rudder-like structure, as the rectrices (L. *rector* ruler or governor) are used for steering in flight. In woodpeckers, the pygostyle is disk-shaped to accommodate the strong caudal muscles used by the bird to prop itself against a tree. Additionally, the paired **uropygial (preen) glands** lie on either side of the pygostyle in many avian species. (The uropygial glands are described in Case 28.1.)

Selected references

[1] Stauber E, Holmes S, DeGhetto DL, et al. Magnetic resonance imaging is superior to radiography in evaluating spinal cord trauma in three bald eagles (*Haliaeetus leucocephalus*). J Avian Med Surg 2007;21:196–200.
[2] Scott DE. A retrospective look at outcomes of raptors with spinal trauma. In: Proceedings of the Association of Avian Veterinarians Annual Conference; 2014. p. 43.
[3] Sisson S, Grossman JD, Getty R. Sisson and Grossman's the anatomy of the domestic animals. 5th ed. Vol. II. Philadelphia: Saunders; 1975.
[4] Dyce KM, Sack WO, Wensing CJG. Textbook of veterinary anatomy. 4th ed. Philadelphia: Saunders; 2010.
[5] Baumel JJ, editor. Handbook of Avian Anatomy: Nomina Anatomica Avium. 2nd ed. Nuttall Ornithological Club (USA), World Association of Veterinary Anatomists (USA), International Committee on Avian Nomenclature; 1993.
[6] Singh B. Dyce, Sack, and Wensing—Textbook of Veterinary Anatomy. 5th ed. St. Louis: Elsevier; 2018.

CHAPTER 27

CASE 27.3

Pododermatitis (Bumblefoot)

Cynthia M. Faux[a] and Marcie L. Logsdon[b]
[a]University of Arizona College of Veterinary Medicine, Oro Valley, Arizona, US
[b]Exotics & Wildlife Department, Washington State University College of Veterinary Medicine, Pullman, Washington, US

Clinical case

Ospreys are also known as sea, river, or fish hawks. They are diurnal raptors that primarily eat fish, so their nests are built near water.

History

An approximately 2-month-old Osprey (*Pandion haliaetus*) (Fig. 27.3-1) was presented for injuries sustained when his nest caught fire. The nest was located on a power line and caught fire during a storm when a wet piece of baling twine amalgamated the nest connected with

a power line and caused a short-circuit. The vanes of several of the flight feathers on the patient's left wing were burned and/or melted, and only the rachis (main shaft) of those feathers remained (Fig. 27.3-2). In addition, there were superficial wounds over the right tibiotarsus, with some associated soft tissue swelling. No musculoskeletal abnormalities were identified at that time.

Owing to the prospect of a long period of recovery and the young age of the patient (not self-sufficient or hunting on his own), the decision was made to place the bird at an environmental educational facility. While awaiting transfer, the bird was housed in a standard hospital enclosure with a smooth, sealed concrete floor and a single perch.

FIGURE 27.3-1 The young Osprey described in this case, adopting a defensive posture over his meal (a fish) in his enclosure. His feet have been bandaged following examination and treatment. A protective guard covers his tail feathers.

Comparative Veterinary Anatomy: A Clinical Approach. https://doi.org/10.1016/B978-0-323-91015-6.00129-1

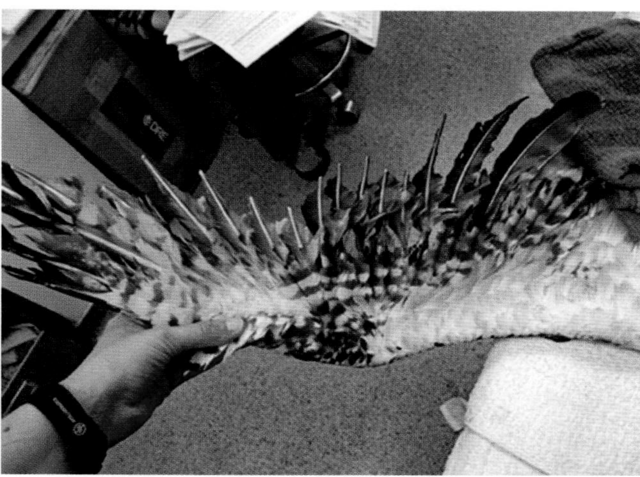

FIGURE 27.3-2 The patient's left wing, extended. The vanes of many of his flight feathers had been burned and/or melted by the electrical fire, and only the rachis (main shaft) of those feathers remained.

Approximately 2 weeks post injury hospitalization (see above), the talon sloughed from the hallux (digit I) of the left foot, presumably secondary to the underlying tissue trauma. The injury was managed with the topical antibiotic therapy. During a routine weight check approximately 2 weeks later, the caretakers noticed additional lesions on the bird's feet, which prompted referral for veterinary examination.

Physical examination findings

The bird was bright, alert, and responsive. The feather abnormalities were unchanged from first presentation, but the swelling and abrasions on the right leg had resolved. The bird's body condition score was 4 out of 5 (1 = emaciated, 5 = obese). There were notable fat deposits over the keel, but pectoral muscle mass was reduced compared with a healthy wild Osprey of similar age, a difference attributable to the patient's inability to fly. The plantar surfaces of both feet had multiple areas of abrasion, erythema and loss of normal skin texture (Fig. 27.3-3).

In addition, there were focal skin lesions on the distal aspect of the hallux bilaterally and a deep skin lesion between digits III and IV on the left foot. The distal hallux of the left foot, from which the talon had sloughed, was healing well, and there was no swelling or other signs of infection.

Compared to other raptor species, the plantar (solar) surfaces of the feet in Ospreys have a very rugged texture (Fig. 27.3-3). In addition to long, hooked talons, this feature is a species adaptation designed to facilitate gripping the fish Osprey's obligate prey. However, it also makes Ospreys very prone to developing pododermatitis in captivity.

Many types of birds nest on power poles and other electrical structures. This poses a hazard as it carries the potential for electrical burns, other electrocution injuries, and even death. These injuries are often characterized by blackened necrotic tissue and two wounds on the distal extremities—i.e., the entrance and exit path of the current. These wounds usually appear small on first presentation, but they continue to enlarge over the following 1–2 weeks as the full extent of tissue injury and subsequent necrosis becomes apparent. Electrical burns therefore carry a guarded to poor prognosis.

Differential diagnoses

Electrical burns, pododermatitis ("bumblefoot")

A diagnosis of bumblefoot is readily made on physical examination. However, often the true diagnostic challenge is determining what caused the lesions.

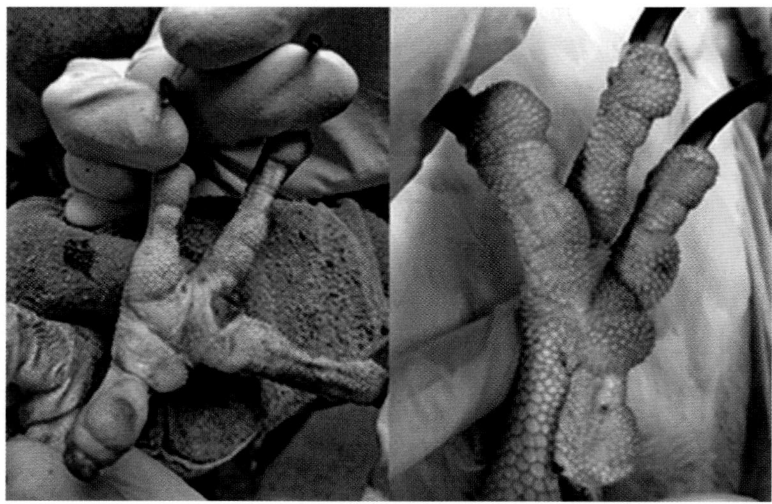

FIGURE 27.3-3 Left: plantar (solar) surface of the patient's left foot. Note the loss of normal texture on the plantar surface of the foot and the hallux (digit I) compared with the roughly textured pads of the other digits. In addition to multiple areas of abrasion, there is a full-thickness skin lesion between digits III and IV, which contains areas of necrotic tissue. Right: plantar surface of a normal Osprey's foot.

Diagnostics

Radiographs of both feet did not reveal any underlying injury or infection of the metatarsal bones or phalanges.

Treatment depends on lesion severity (graded I-V). Class I lesions usually are responsive to correction of the underlying management problems, although some warrant antimicrobial or anti-inflammatory therapy. Class II and III lesions usually are effectively managed with antimicrobial and anti-inflammatory therapy, bandaging designed to relieve pressure on the affected areas (e.g., "donut" bandages; Fig. 27.3-4), and mild surgical debridement (removal of devitalized tissue). Class IV lesions require intensive surgical debridement and staged surgical closure. Class V carries a poor prognosis, so humane euthanasia is usually recommended. As with any management-related problem, correction of the underlying issue is essential to prevent recurrence.

Diagnosis

Pododermatitis class II (see side box entitled "Pododermatitis (bumblefoot)")

Treatment

The bird was treated with antimicrobial and anti-inflammatory therapy, bandages designed to relieve pressure on the ulcerated areas of the feet, and a more species-appropriate perching material. He was provided a sand floor and a variety of perching materials in his enclosure to promote healing and prevent recurrence.

PODODERMATITIS (BUMBLEFOOT)

Bumblefoot is the common name for pododermatitis (inflammation of the skin on the foot) because the resulting foot pain causes the bird to stumble or "bumble" when walking. Most avian species have little soft tissue padding on the plantar surfaces of their feet, so they are very susceptible to developing pododermatitis and pressure sores when confined. Bumblefoot is seen most often in raptors, waterfowl, chickens, and—to a lesser extent—parrots.

Predisposing factors include inadequate or species-inappropriate enclosure size, flooring, and/or perches; obesity; injury to one leg or wing, resulting in excessive weight-bearing on the contralateral limb; vitamin A deficiency; overgrown talons; and unsanitary living conditions. All of these factors are encountered almost exclusively in captive birds. Bumblefoot is extremely rare in birds living in the wild, as they have species-appropriate and clean perching materials in abundant variety, are physically active, and are rarely obese or vitamin A deficient.

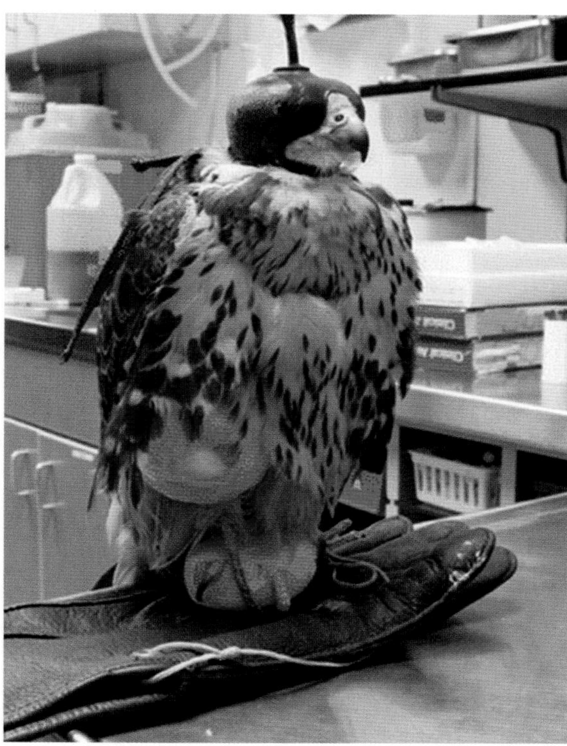

FIGURE 27.3-4 "Donut" bandages are used in this Prairie falcon with bilateral pododermatitis to relieve pressure the central plantar surfaces of the feet. The bird is standing on its left foot, with its right foot raised, revealing the underside of the bandage. The padding is made from a foam "pool noodle," cut to size and secured with a self-adhesive bandage. The hollow center allows access to the wounds for cleaning and dressing without having to replace the entire bandage.

Anatomical features in avian species

Introduction

Birds are bipedal (L. *bi* tow + *pes* foot = having 2 feet). When not flying or swimming, they support their weight on their legs and feet by perching or by standing, walking, or running. During these activities, the bird's entire body weight is concentrated on the plantar surfaces of the feet.

This section covers osteology and integument of the leg and foot, the digital and metatarsal pads, and the regional blood supply and innervation.

BUMBLEFOOT LESION CLASSIFICATION

Pododermatitis lesions are classified according to severity, from mild (superficial) to severe (deep):

- I: Loss of skin texture in affected areas, in some cases with associated erythema
- II: Hyperkeratosis with mild abrasion and scabbing, mild swelling, and an initial appearance of focal skin necrosis
- III: Inflammation extending into the subcutaneous tissues; often separation of the dermis from the subcutis in affected areas, and the beginnings of abscess formation
- IV: Extension to the digital flexor tendon(s), which may lead to ascending infection along the tendon sheath to the intertarsal joint
- V: Extension to the bones of the feet, in the form of osteolysis or osteomyelitis, and in some cases septic arthritis of the metatarsophalangeal or interphalangeal joints

Function

The avian foot is adapted to a bird's lifestyle perhaps as much as is the beak (Fig. 27.3-5). Birds use their feet for grasping, perching, hunting, and preening, as well as for locomotion (swimming, walking, or running, depending on the species). Some species such as hummingbirds do little or no walking, and they use their feet primarily for perching. In contrast, ostriches do not perch nor grasp with their feet, but they can run at great speed. Grebes can "walk" for a short distance on the surface of the water during their mating dances; however, they have a difficult time walking on land. Parrots can manipulate objects with their feet, while raptors catch and carry their prey with their feet.

> Injuries to the legs, and particularly to the feet, can greatly impact a wild bird's survival and a captive bird's quality of life.

> The synsacrum is a bony structure common to all birds, comprising the fused lumbar and sacral vertebrae. (see Case 27.2.)

Osteology of the leg (Fig. 27.3-6)

As in mammals, the **pelvis** in adult birds is composed of the fused **ilium**, **ischium**, and **pubis**. However, in birds, the pelvis is fused with the **synsacrum** at the ilia (left and right). This fusion forms a rigid framework such that the muscular activity is not required to maintain flight posture.

As in mammals, the **femur** articulates proximally with the pelvis at the **coxofemoral joint**, supported by a strong **ligament of the head of the femur**, which spans the short distance between the **acetabulum** and the **capitis fovea** on the head of the femur.

FIGURE 27.3-5 Top left: nighthawks have relatively small feet for their body size, as they spend much of their time in flight. Top right: plantar surface of a cockatoo's foot (with pododermatitis); cockatoos use their feet for perching as well as grasping and manipulating objects. Bottom left: webbed feet of a domestic goose; geese move awkwardly on land but are graceful and agile in the water. Bottom right: left foot of an Ostrich; these feet are well adapted for supporting the bird's large body mass as well as for walking and running over long distances.

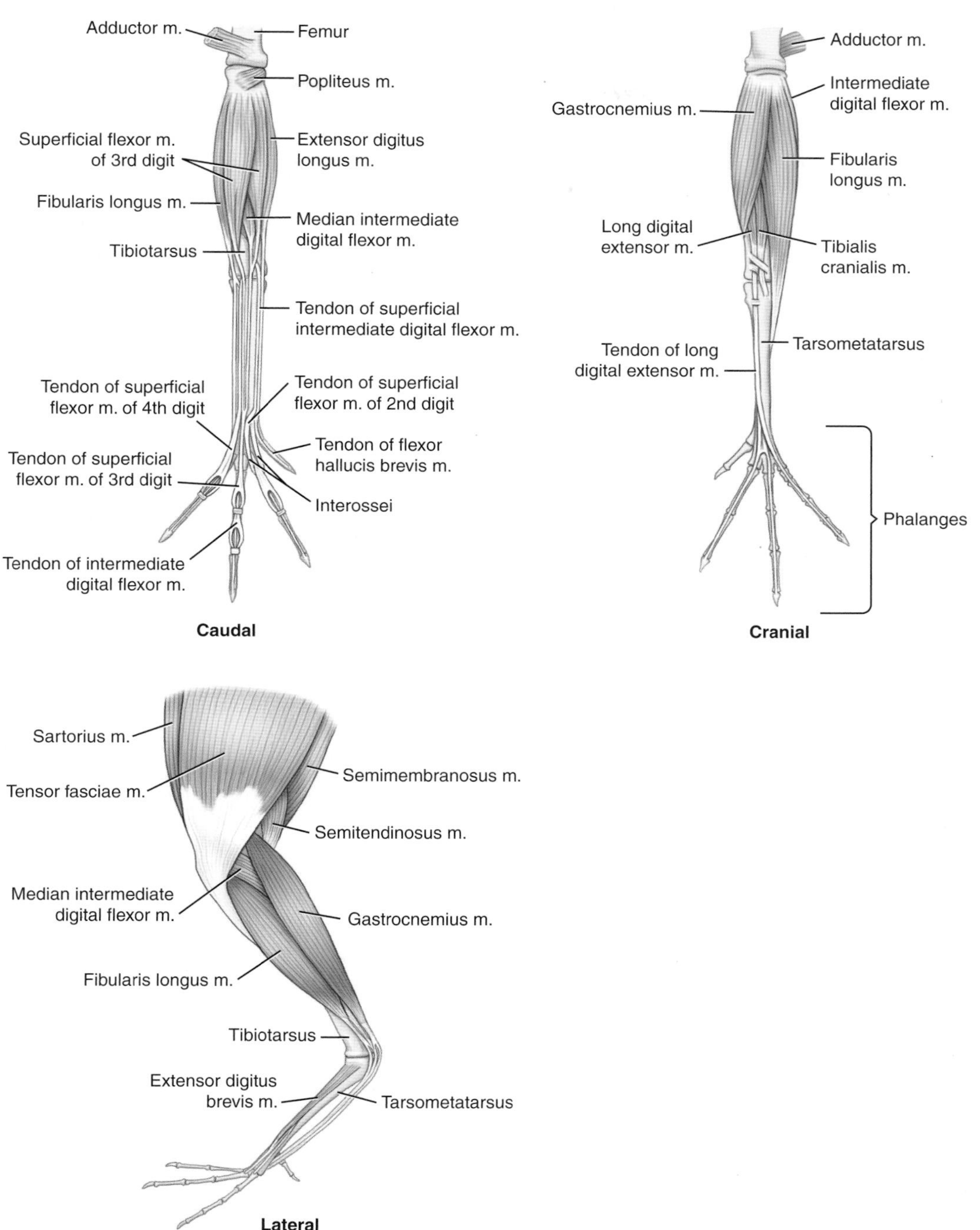

FIGURE 27.3-6 Caudal, cranial, and lateral views of the avian pelvic limb depicting tendons, muscles, and bones.

Distal to the femur, however, the avian limb differs significantly from the basic pattern found in mammals, being somewhat simplified yet highly adapted and species-specific. In birds, the femur articulates distally with the **tibiotarsus** (described below) and the **fibula**. In most avian species, a **patella** is present within the common tendon of the **femorotibiales** and **iliotibiales** muscles.

The **tibiotarsus** is a compound bone made up of the **tibia** and the fused **proximal row of tarsal bones**. The tibiotarsus is paired with the much smaller fibula, as are the tibia and fibula in mammals. The tibiotarsus articulates distally with the tarsometatarsus (described below) at the **intertarsal joint**, while the fibula does not in most species. The intertarsal joint is sometimes called the "hock."

The **tarsometatarsus** is also a compound bone, consisting of the fusion of the **distal row of tarsal bones** and the 3 long **metatarsal (MT) bones: MT II, III, and IV**, which correspond to mammalian metatarsal bones of the same number. **Metatarsal I** in birds is typically a short, separate bone that is attached by a small ligament to the distal third of the tarsometatarsus on its medial aspect.

At the distal end of the tarsometatarsus, and at the distal end of MT I, there is a separate condyle for articulation with each of the proximal phalanges. Thus, the number of condyles is determined by the number of digits, or toes, in that species (discussed in the next section). For example, in the chicken, the distal tarsometatarsus has 3 separate condyles, which, along with the single condyle on MT I, accommodate the 4 digits typically present in this species (digits I–IV).

Osteology of the foot (Fig. 27.3-7)

There is a considerable variation among avian species regarding the number and pattern of the **digits**. The number generally ranges from 2 to 4, with 4 digits being the most common. The ostrich has 2 toes (digits III and IV) (Fig. 27.3-5). Other ratites (e.g., emus and rheas) and many of the wading birds have 3 functional toes. Usually, it is digit I that is reduced or absent, but not always. For example, geese have 4 toes (digits I–IV), with webbing between the 3 larger digits (II, III, and IV), but digit I—on the plantaromedial aspect of the leg—is vestigial (see Fig. 27.3-5). Silkies, a highly selectively bred chicken breed, have 5 toes (Fig. 27.3-8), whereas most other chicken breeds have 4 toes. In silkies, the extra digit next to digit I is a genetic variant that is a form of polydactyly (extra digits).

When 4 toes are present, the arrangement varies with function. Perhaps the most common arrangement is 3 toes pointing forward and 1 toe pointing back, which allows for both perching and walking. This pattern is termed **anisodactyl** (G. *anisos* unequal + *daktulos* fingers). Parrots have a **zygodactyl** (G. *zugon* yoke or pair) pattern, where digits II and III are directed forward and digits I and IV are directed backward (Fig. 27.3-5). Parrots use this arrangement to precisely manipulate food and other objects, as well as to climb. Woodpeckers also use this variation to climb. Owls and Osprey have a zygodactyl foot as well; however, these species can move digit IV forward as necessary to grasp their prey.

Other variations exist. For example, swifts have a **pamprodactyl** foot (G. *pan-* all + *pro-* in front) in which all 4 toes point forward. Kingfishers have a **syndactyl** foot (G. *syn-* together) in which digits III and IV are fused. Cormorants have a **totipalmate** foot (L. *totus* whole + *palma* palm of the hand), where all 4 toes point forward and are webbed. Furthermore, huge adaptive variation exists within groups, including webbed toes (**palmate**) for swimming (Fig. 27.3-5), long talons for grasping prey (Fig. 27.3-9), feathered feet for insulation against cold (Fig. 27.3-9), and **lobate** toes for swimming that also allow walking with ease (Fig. 27.3-10).

The **phalanges** (the individual bones of the digit) also vary in number (see Fig. 27.3-8). When present, digit I (the **hallux**) has 2, digit II has 3, digit III has 4, and digit IV has 5 phalanges. As in mammals, the phalanges within a digit are numbered from proximal to distal, where the most proximal is phalanx (P) 1. For example, the most proximal phalanx of digit IV is P1 and the most distal is P5.

OVERGROWN NAILS OR TALONS

Overgrown nails or talons in captive birds can cause puncture wounds to the feet, inoculating the tissues with fecal bacteria and other environmental microbes, leading to pododermatitis or abscess formation. Failure to keep flooring, perches, and feeding stations clean increases the likelihood that even a slight breach in the integrity of the plantar skin results in pododermatitis or abscessation.

More commonly, however, overgrown nails or talons cause a shift in weight-bearing that results in abnormal pressure on the plantar surfaces of the foot, which can lead to pressure sores.

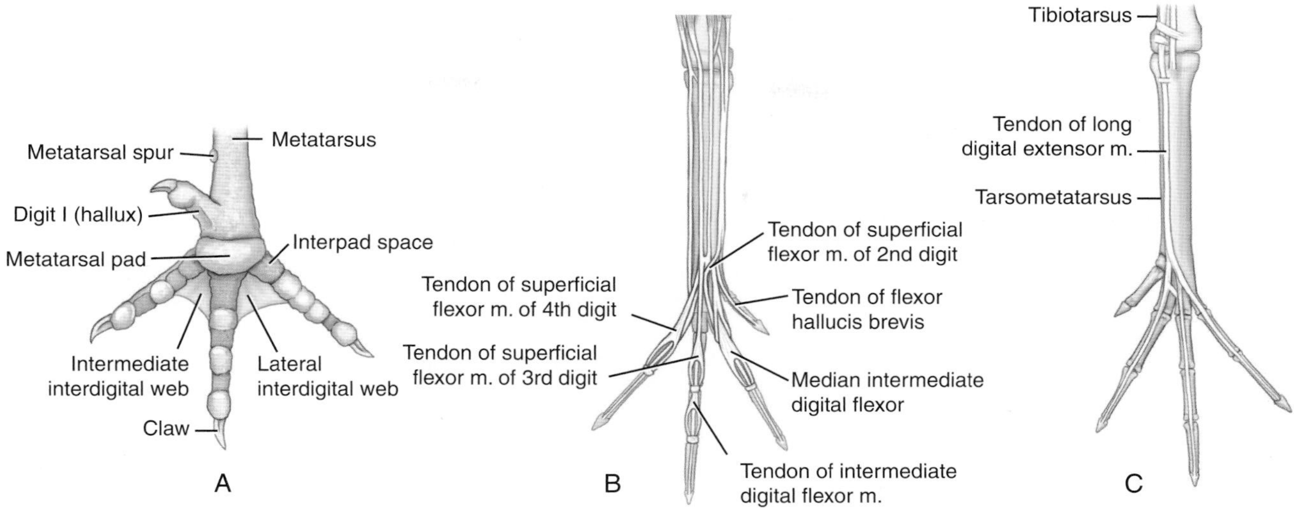

FIGURE 27.3-7 The avian foot in detail, showing the bones, joints, tendons, and metatarsal and digital pads. (A) Plantar view, showing pedal topography of the digital and metatarsal pads. (B) Plantar view of the flexor tendons. (C) Dorsal view, showing the digital extensor tendons.

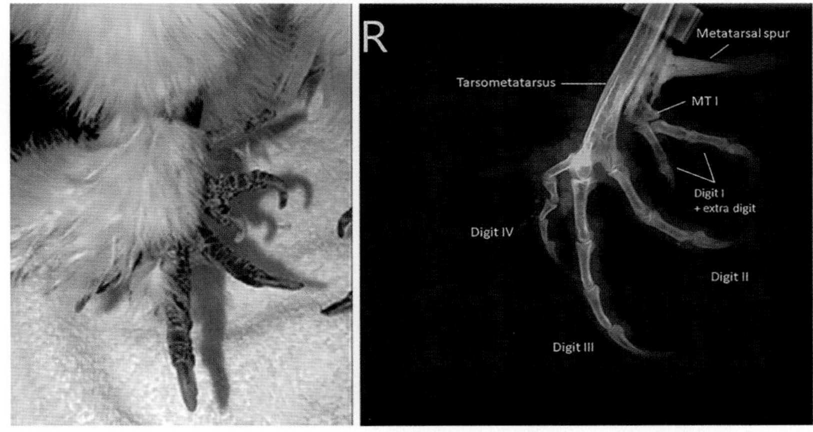

FIGURE 27.3-8 Silkie chickens typically have 5 toes instead of the 4 toes usually found in most other chicken breeds. Left: dorsal view of the right foot of a silkie rooster (note the large metatarsal spur just below the dark blue identification band). Right: radiograph of the right foot of a silkie rooster. Metatarsal (MT) I has a second condyle to accommodate the extra digit adjacent to digit I. Also visible on this radiograph is the extensive bony core of the metatarsal spur in this mature male and the feathering of the leg and foot that extends to the proximal phalanges in this breed.

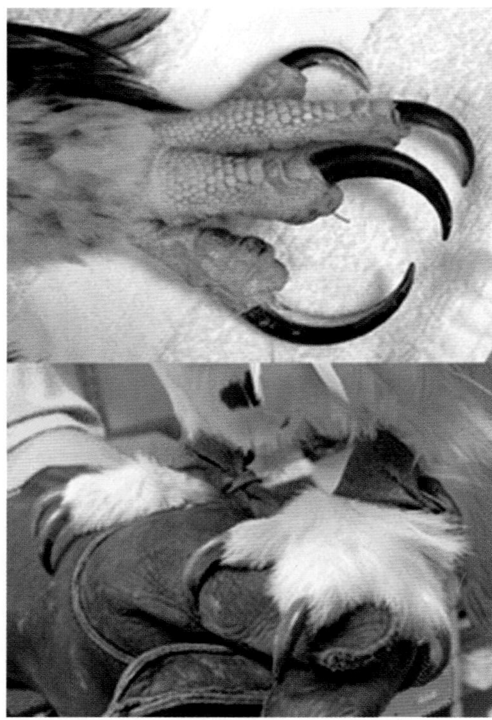

FIGURE 27.3-9 Left: Foot of a Golden eagle. Right: Foot of a Snowy owl.

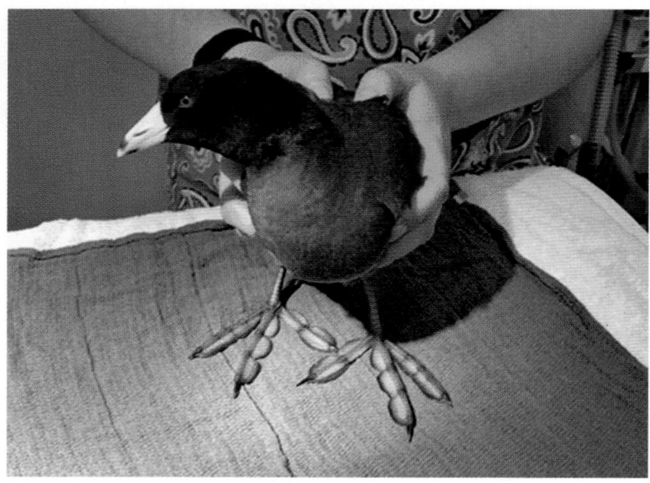

FIGURE 27.3-10 Lobate toes of the American coot.

In all digits, the **distal phalanx**, also called the **ungual phalanx**, supplies the bony core for the keratinized **nail** or **talon**. Nails may be vestigial (as in digit IV of the ostrich), short and adapted for walking and scratching (as in chickens), or long and sharp for grasping prey (as in raptors).

Integument of the leg and foot

Podotheca (L. *pod-* foot + *theca* box or case) is the term used for the highly specialized integument of the avian leg, generally extending from the intertarsal joint to the tips of the digits (Fig. 27.3-11). In some long-legged wading birds, it may continue proximally to the distal tibiotarsal area.

The podotheca primarily consists of highly keratinized **scales**, which range in size and shape in consistent patterns. **Scutes** are the larger scales that are generally found on the dorsal and plantar surfaces of the leg and on the dorsum of the digits. **Scutellae** are the smaller scales that are found on the lateral and medial surfaces of the leg and digits. **Reticulae** are smaller still and are found on the plantar surfaces of the digits and may also be found between the scutes and scutellae on the leg and digits. The smallest of all, **cancellate scutes**, are found on the toe webs of waterbirds.

In some avian species or breeds (e.g., Snowy owls and some chicken breeds), the entire leg and even the dorsal aspect of the digits are feathered (Figs. 27.3-8 and 27.3-9). The skin in these areas is soft and flexible, unlike the podotheca of the bare-legged species and breeds, although feathers and scutes may both be found in transitional areas.

Some galliforme males such as roosters and tom turkeys sport a **metatarsal spur** on the medial aspect of the lower leg that projects plantaromedially and somewhat proximally (Fig. 27.3-11). This **process calcaris** consists of dense bundles of connective tissue that—in the mature male—become calcified (Fig. 27.3-8). It is covered by a hard keratinous sheath, much like nails, which can become very long, and even overgrown, with age. These spurs are an effective weapon for offense and defense.

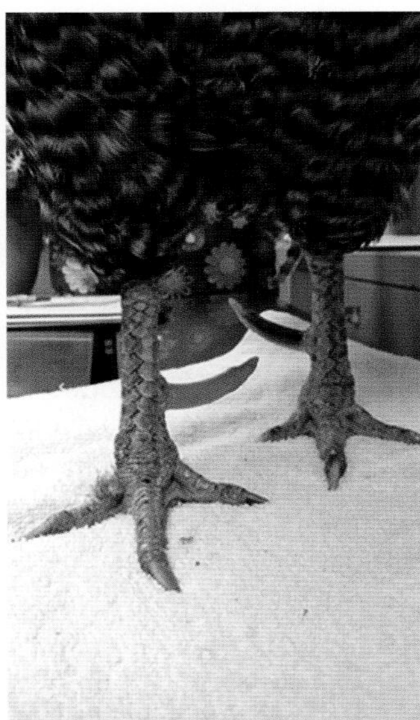

FIGURE 27.3-11 This mature rooster has a thick podotheca and long metatarsal spurs.

Digital and metatarsal pads

Heavy-bodied birds housed in small enclosures without adequate space to slow their flight can bruise the plantar surfaces of their feet by bouncing from perch to perch. Hard flooring such as river rock or concrete can also cause concussive trauma to the underside of the feet.

The plantar surfaces of the avian foot have only small amounts of subcutaneous tissue, but these tissues are important. As shown in Fig. 27.3-3, the plantar surfaces of the digits have a series of small, fleshy **digital pads**. A broader **metatarsal pad** covers the plantar aspect of the tarsometatarsal-phalangeal joints and proximal phalanges. These pads supply subcutaneous cushioning for the bones of the feet, and the digital pads provide additional grip when perching.

Flexor and extensor tendons (Fig. 27.3-6)

Erosion through the digital or metatarsal pads, as occurs with class III or IV pododermatitis, exposes the flexor tendons to direct trauma and infection, either of which can have serious long-term consequences for the bird's quality of life or survival in the wild.

The **digital flexor tendons** course down the caudal/plantar aspect of the bird's leg and foot to insert onto the plantar aspect of the digits. The associated flexor muscles are found on the caudal aspect of the tibiotarsus, and their primary role is to flex the digits. Each digit has its own flexor tendon, so the number of flexor tendons is dictated by the number of digits in that species. Within the foot, the flexor tendons run deep to the metatarsal and digital pads.

PROTECTING THE FEET

Species-appropriate perch shape and size is essential for preventing bumblefoot. Perches that are too wide or flat predispose to pressure sores on the digital pads (the fleshy pads under the toes), while perches that are too narrow can cause pressure sores on the metatarsal pad (in the center of the foot). Certain species such as Peregrine and Saker falcons are adapted to perch predominantly on rock ledges in the wild, so they require flat, shelf-like perches in captivity.

Appropriate perch texture is also essential: hard, smooth surfaces such as bare wood apply consistent, inappropriate pressure to the feet. Covering perches with sisal (rope), long-leaf astroturf, cork, or textured rubber can prevent this problem by varying the forces on the metatarsal and digital pads.

PASSIVE PERCHING MECHANISM?

Birds are purported to have a "passive perching mechanism" which causes the toes to flex (e.g., around a perch) when a bird squats, thus allowing birds to sleep on a perch without falling off. However, experiments with anesthetized starlings showed that these birds, at least, did not spontaneously grasp a perch when so positioned and would fall off if relying solely on this mechanism. The authors concluded that active mechanisms are also required for resting or sleeping while perching [4].

The **digital extensor tendons** generally pair with the flexor tendons, extending down the cranial/dorsal aspect of the leg and foot to insert onto the dorsal aspect of the digits. The associated extensor muscles are found on the cranial and lateral aspect of the tibiotarsus, and their primary role is to extend the digits. Each digit has its own extensor tendon, so the number of extensor tendons is dictated by the number of digits in that species.

Intratendinous ossification is a phenomenon that occurs in several avian species, including turkeys and owls, particularly in the tendons of the pelvic limb (Fig. 27.3-12).

This normal ossification must not be confused with tendinopathy or other pathology involving abnormal calcification.

Grasping stability in many species of birds is augmented by a digital tendon-locking mechanism. In mammals, tendons glide smoothly within their tendon sheaths, as both facing surfaces (the outer surface of the tendon and the inner surface of the sheath) are smooth. However, birds have roughened areas on the inner surface of the tendon sheath and the outer surface of the tendon, which enable the tendons to "lock" or engage with the sheath, thereby providing greater grasping stability, particularly in birds of prey and in birds that perch.

Blood supply (Fig. 27.3-14A)

Two aortic branches supply blood to the pelvic limb: the **femoral** and **ischiatic arteries**, the latter of which provides the major arterial supply in most avian species. The pattern of veins in the pelvic limb generally follows that of the corresponding arteries. The largest vein that drains the foot is the **medial metatarsal vein**. It is quite superficial, lying just beneath the podotheca on the medial aspect of the proximal tarsometatarsus and intertarsal joint.

The medial metatarsal vein is one of the 3 most common sites for blood sampling in birds (Fig. 27.3-13). (The other two are the jugular vein in the neck and the basilic or ulnar vein in the wing.) In some birds, the medial metatarsal vein may be seen through the skin over the medial aspect of the proximal tarsometatarsus or intertarsal joint. The needle is carefully inserted through the skin between the overlying scutellae or reticulae and into the vein.

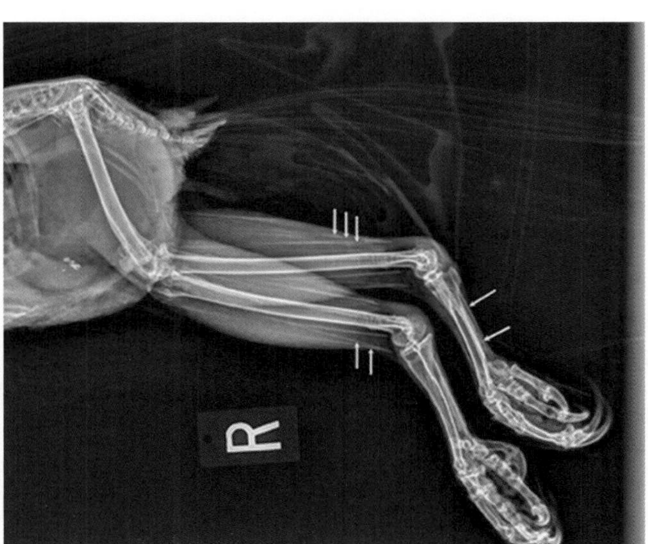

FIGURE 27.3-12 Intratendinous ossification in the pelvic limbs of a Great horned owl. The *arrows* are pointing to just some of the many calcified tendons/fascial bands in this mature bird. Both the flexors and extensors are involved.

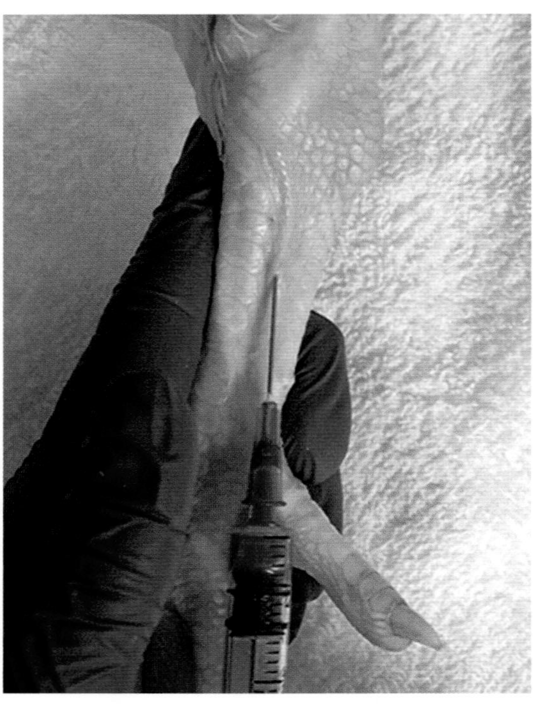

FIGURE 27.3-13 Drawing a blood sample from the medial metatarsal vein at a site just distal to the intertarsal joint (hock) of a broiler chicken. (Also, compare the thickness of the podotheca in this young, shed-raised bird with that of the mature rooster in Fig. 27.3-11.)

The **rete tibiotarsale** is a thermal countercurrent exchange network of arteries and veins along the cranial aspect of the tibiotarsus in many species, including ducks, pelicans, and galliformes (e.g., chickens). The complexity of the rete varies by species, but basically cool venous blood returning from the foot and lower leg is warmed by the adjacent arterial blood before it returns to the central circulation. In this way, body heat is conserved and less heat is lost to a cold environment.

Innervation supply (Fig. 27.3-14B)

The pelvic limb is supplied by **spinal nerves** from the lumbar and sacral plexuses. The nerves of the **lumbar plexus** exit the synsacrum and pass toward their target structures. The **femoral nerve** exits cranial to the acetabulum, while the **obturator nerve** exits the pelvis via the obturator foramen. As in mammals, the femoral nerve innervates the primary extensors of the stifle, and the obturator nerve innervates the adductors of the hip.

The spinal nerves comprising the **sacral plexus** exit the synsacrum near the cranial division of the kidney and pass through the kidney parenchyma, traveling caudolaterally to exit the pelvis through the **ilio-ischiatic** foramen, caudal to the hip. The **sciatic (ischiatic) nerve** is composed of 2 nerves, peroneal and tibial, which travel within a common epidural sheath. The **peroneal (fibular) nerve** branches innervate the extensors of the distal limb, while the **tibial nerve** branches innervate the flexors.

> As described in the case of Sertoli cell tumor (Case 26.5), the sciatic nerve in birds passes through the parenchyma kidney, so renal or testicular tumors may compress this nerve, resulting in lameness or paresis/paralysis in the affected limb.

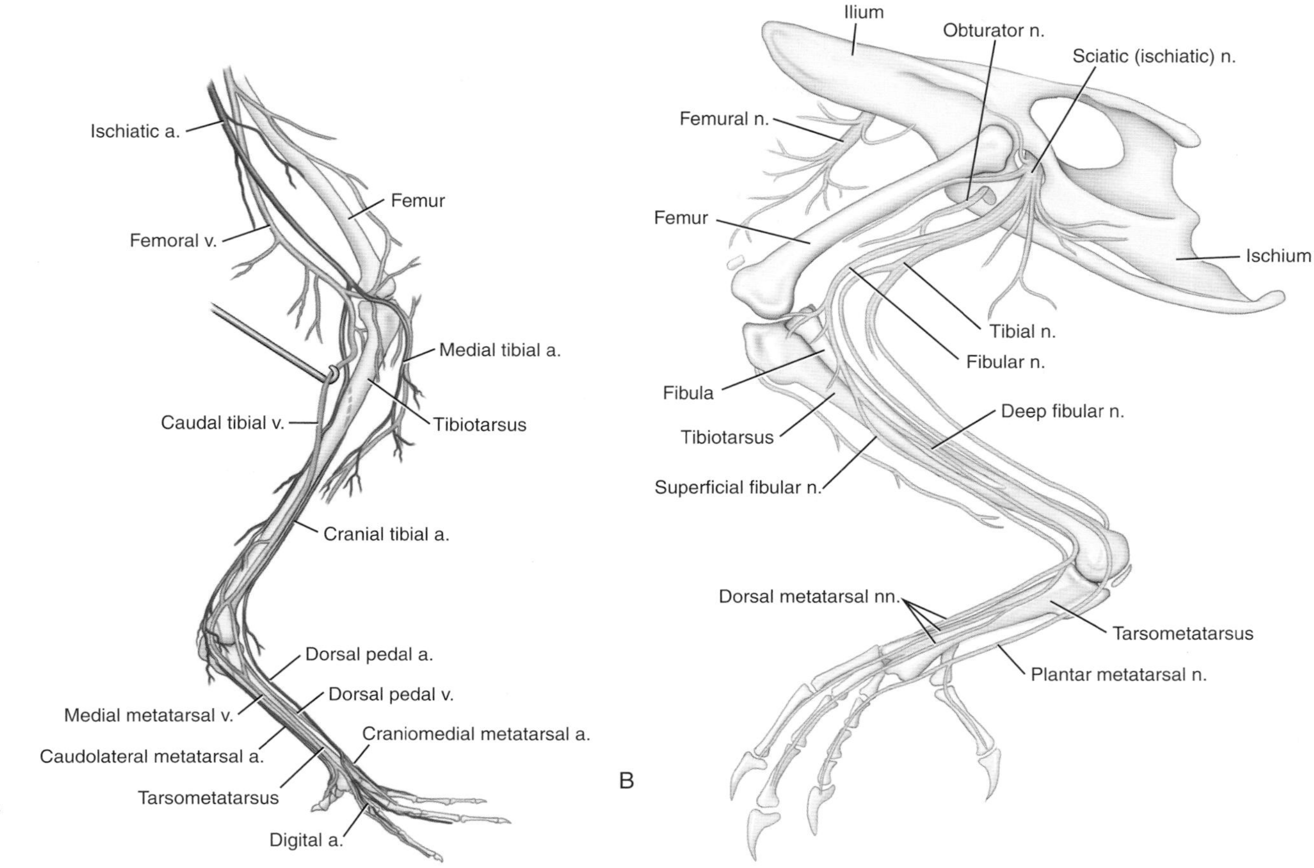

FIGURE 27.3-14 (A) Blood supply to the avian pelvic limb, medial view. (B) Nerve supply to the avian pelvic limb, lateral view.

Selected references

[1] Arent L. Raptors in captivity: Guidelines for care and management. Hancock House Publishers; 2018.

[2] Samour J. Avian medicine. 3rd ed. Elsevier; 2016.

[3] Redig PT, Cooper JE, Remple JD. Raptor biomedicine. University of Minnesota Press; 1993.

[4] Galton PM, Shepherd JD. 2012. Experimental analysis of perching in the European starling (Sturnus vulgaris: passeriformes; passeres), and the automatic perching mechanism of birds. J Exp Zool 317:205–15.

INTEGUMENT/FEATHERS

Cynthia M. Faux and Marcie L. Logsdon, Chapter editors

28.1 Impacted uropygial (preen) gland—*Cynthia M. Faux and Marcie L. Logsdon* 1400

CHAPTER 28

CASE 28.1

Impacted Uropygial (Preen) Gland

Cynthia M. Faux[a] and Marcie L. Logsdon[b]
[a]University of Arizona College of Veterinary Medicine, Oro Valley, Arizona, US
[b]Exotics and Wildlife Department, Washington State University College of Veterinary Medicine, Pullman, Washington, US

Clinical case

History

An 8-month-old male African gray parrot (*Psittacus erithacus*) was presented for a wellness examination and evaluation of poor feather quality. He had recently been acquired from an out-of-state breeder whose premises the owners had never visited, so the health status of the bird's parents and siblings was unknown. Since purchase, the bird had been fed a commercial seed-and-nut diet and frequent human food.

Although still a common practice by many breeders and pet store owners, trimming the flight feathers is now discouraged in young birds before they have learned how to fly. It is thought to hinder the bird, both physically and mentally, during an important developmental stage. In addition, poorly performed feather trims can lead to feather-destructive behaviors such as plucking or "barbering" (biting off part of the feather) and—in severe cases—self-trauma.

Physical examination findings

The bird was alert and responsive but nervous. He was grossly over-conditioned with a body condition score of 5 out of 5 (1 = emaciated, 5 = obese). Pectoral muscle mass was decreased, consistent with a nonflighted bird. Of note, the primary flight feathers had been trimmed.

The remaining feathers were dull, unkempt, and frayed. The rectrices (tail feathers) were broken, and the body coverts over the chest were injured. There were also some mild erythema (redness) and loss of texture on the plantar surfaces of both feet. (The integument of the avian leg is described in Case 27.3.)

Differential diagnoses

The common causes and contributing factors for poor feather quality and feather-destructive behavior are summarized in the side box entitled "Poor feather quality"; the most likely factors leading to the mild pododermatitis (inflammation of the skin on the feet) in this case included vitamin A deficiency, obesity, and improper housing (particularly perch size and surface)

The uropygial gland is a bilobed, ducted sebaceous gland at the base of the bird's tail, on the dorsum of the sacrocaudal region (Fig. 28.1-2). It secretes an oily substance the bird then distributes over the feathers and skin with the beak during preening—hence, the common name for this gland is the "oil" or "preen" gland. A normal uropygial gland releases a small amount of oily secretion when gently palpated. In this case, nothing could be expressed from the distended gland, indicating that the duct was blocked.

Diagnostics

The bird was anesthetized for diagnostic evaluation. Routine bloodwork was normal. Whole-body radiographs were within normal limits, except for a grossly enlarged uropygial gland (Fig. 28.1-1). On closer physical examination, the uropygial gland was distended and nonpatent.

Comparative Veterinary Anatomy: A Clinical Approach. https://doi.org/10.1016/B978-0-323-91015-6.00130-8

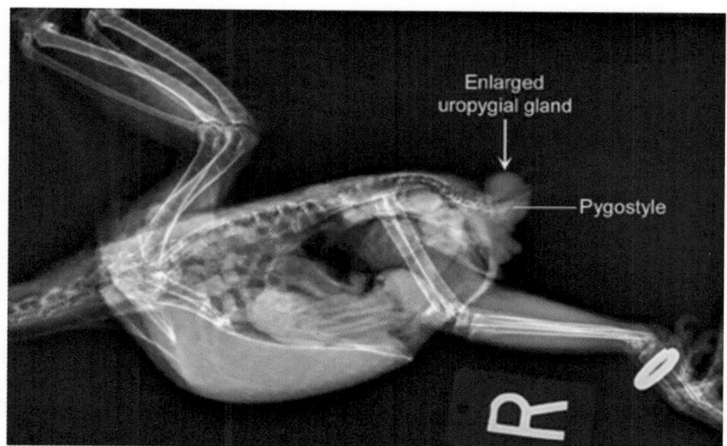

FIGURE 28.1-1 Lateral radiograph of the juvenile African gray parrot in this case, showing the grossly enlarged uropygial gland overlying the pygostyle (fused caudal vertebrae) at the base of the tail. Normally, the uropygial gland is not visible radiographically in this species.

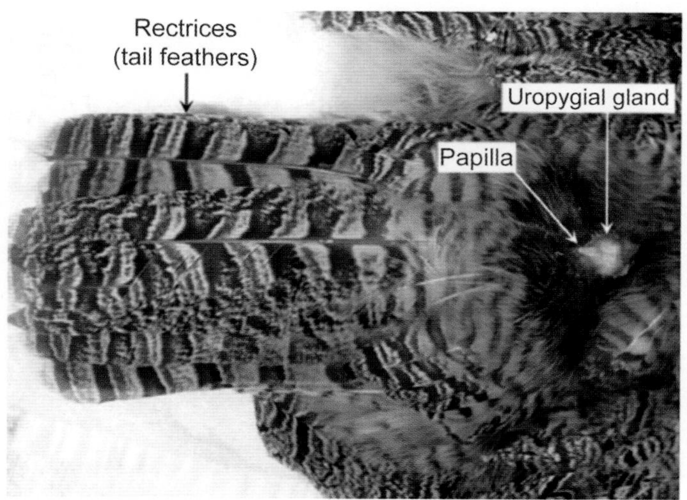

FIGURE 28.1-2 Dorsal view of the tail of a Great horned owl (*Bubo virginianus*), showing the normally large uropygial gland in this species. The covert feathers over the dorsum of the sacrocaudal area have been gently swept aside to reveal the bilobed gland and the single papilla through which the ducts of both lobes release their secretions when the gland is stimulated during preening. Small, downy feathers surround the gland and papilla in this species. Also shown in this photo are the dorsal surfaces of the rectrices (tail feathers used in flight).

Diagnosis

Impacted/abscessed uropygial gland

Treatment

With the bird still under general anesthesia, both lobes of the uropygial gland were cannulated and flushed to remove the debris obstructing the ducts and impacting the gland. The bird was also treated with systemic anti-inflammatory and antimicrobial medications, and switched to a balanced diet consisting of commercial pellets and fresh vegetables, with occasional fruits and nuts as training treats.

For the next 3 weeks, the bird was anesthetized 2 or 3 times a week for repeated flushing of the uropygial gland. For the first week, an "E-collar" was applied to prevent self-trauma to the gland, which was bruised and irritated after the

An Elizabethan (E-)collar is a flexible, plastic, cone-shaped collar used to prevent the patient from biting or chewing at an injured area while it heals. The E-collars used in birds serve the same purpose as those used in dogs and cats but are often made of different materials (e.g., neoprene) and placed upside-down (the broad end of the cone facing caudally), because the E-collars used in dogs and cats are not well tolerated by birds.

Feathers do not grow continuously, so broken or otherwise injured feathers are not replaced until the bird next molts, which may be several months or up to a year later.

first treatment. As the gland healed, the collar was removed to encourage normal preening behaviors and allow the bird to work on the gland himself.

By the end of the third week, the uropygial gland had healed and the bird was doing well on his new diet. The owners were also directed to several behavioral resources. (Regardless of the inciting cause, many skin or feather conditions in birds can lead to chronic feather-destructive behaviors if not fully addressed.) The bird molted normally and on-schedule the following year, after which his feathering was greatly improved (Fig. 28.1-3).

FIGURE 28.1-3 The young African gray parrot in this case, 1 year—and a full molt—after treatment.

POOR FEATHER QUALITY

Several different management factors and disease states may contribute to poor feather quality in pet or captive birds. The most common of these factors are:

- Improper diet
- Improper housing
- Lack of role model for basic grooming behaviors in young, hand-raised birds
- Uropygial ("preen") gland disease

Common causes of feather-destructive behavior include:

- Stress
- Improper diet
- Systemic infection
- Parasitism
- Pain or discomfort
- Reproductive dysfunction/hormonal imbalance (adult birds)

Anatomical features in avian species

Introduction

This section describes the avian integument, which includes the skin and feathers. The highly specialized integumentary structures such as the rhamphotheca—keratinous sheath—of the beak (Case 25.4) and the podotheca of the lower legs and feet, the digital and metatarsal pads, and the nails or talons (Case 27.3) are described elsewhere.

Function

The primary functions of the skin, often referred to as the largest organ of the body, are protection, thermoregulation, and sensation. These functions are similar in birds and mammals. Feathers, on the other hand, are the most distinctive characteristic of birds, and they serve several different functions. Feathers form the lift and control surfaces of flight, and they provide insulation and weatherproofing. Color variation in feathering allows for species recognition as well as display and mating behaviors within species. Modified feathers serve as eyelashes. Feathers can muffle the sound of a wing in flight, which is useful in raptors such as owls; and conversely feathers can create sound, as in the aerial displays of hummingbirds.

Skin

This lack of skin "redundancy" can result in excessive tension on skin sutures, making primary closure of some surgical or traumatic wounds difficult.

Although the primary functions of the **skin** are identical in birds and mammals, avian skin differs from mammalian skin in several important ways. In general, avian skin is thinner than that of a mammal of similar size, and in some regions, such as the head and dorsum, it is closely adhered to the underlying tissues.

The microanatomy of avian skin differs considerably from mammals both in organization and in terminology, although avian skin still has an epidermis, dermis, and subcutis. Generally, the avian **epidermis** is thin, merely 4–7 cells thick in most areas. It is thicker in some specific regions, such as the podotheca of the lower legs and feet and the rhamphotheca of the beak.

The avian **dermis** consists of several layers, from superficial (**stratum superficiale**) to deep (**stratum profundum**). Over much of the body, the dermis is attached to the underlying musculoskeletal structures by a subcutaneous layer of loose connective tissue. Two notable exceptions are the dorsum of the body and the ridge of the keel (ventral midline of the sternum); in these areas, the dermis is closely adhered to the underlying structures.

Within the dermis, small **feather muscles** attach via elastic tendons to the embedded feather follicles, permitting fine motor control of the feathers. Specific feather muscles act as erectors (lifting the feather from the body), depressors (bringing the feather closer to the body), and retractors (pulling the feathers closer together). Additionally, more specialized muscle patterns can allow the rotation of feathers. Similar muscles are found in the featherless regions of the skin (described later), forming a network of interconnected intradermal muscles.

SUTURING AVIAN SKIN

In most situations, primary closure (suturing) of surgical or traumatic wounds in birds can be achieved using a simple continuous suture pattern or other simple "through-and-through" pattern (e.g., simple interrupted). Use of absorbable suture material avoids the need for suture removal once the wound has healed.

Because the skin over the dorsum of the body is closely adhered to the underlying tissues, open wounds over the back are often left to heal by second intention—i.e., granulation tissue fills the defect, which is then gradually covered by re-epithelialization from the skin margins. For example, "raccoon vs. chicken" incidents may result in extensive skin loss or separation ("degloving") over the chicken's back. Primary closure of these wounds can be difficult, so they are often left open to heal by second intention.

Compared with mammals, avian skin is relatively aglandular. Three notable exceptions are the uropygial gland (described below), the sebaceous glands of the external ear canal, and the mucosal glands surrounding the vent (external opening of the cloaca; see Case 26.5). However, as the keratinocytes move from the basal layer to the cornified layer of the epithelium, they produce lipids which eventually arrive at the skin surface. In this way, the entire epidermis acts as an oil gland, which may explain why some birds lack a uropygial gland.

Uropygial gland

The **uropygial gland** is a bilobed, simple-tubular holocrine gland that is present in most avian species. It lies deep to the epidermis on the dorsal midline at the base of the tail, dorsal to the levator (elevating) muscles of the tail. The gland typically forms a distinctive bulge (see Fig. 28.1-2). Each **lobe** has 1 or 2 ducts which empty onto the skin surface caudal to the gland, usually via a shared **papilla**. A tuft of downy feathers may surround the gland and its papilla, or the area may be completely devoid of feathers.

Holocrine glands are exocrine (externally secreting) glands in which the secretory cells as well as their products (G. *holos* whole, all) are released into the lumen of the gland. Sebaceous glands such as the uropygial gland produce a thick secretion ("sebum"), which contains cell debris as well as a complex oily and waxy (ceruminous) material. These glands release their contents onto the skin surface via a narrow duct, which may become blocked by the thick secretions.

The composition of the uropygial gland secretion varies among avian species. The size, shape, and even presence of the gland are also species-specific. In most species, the 2 lobes lie in close apposition, separated by a median septum, and both lobes empty via a single, shared papilla. The uropygial gland is absent in ostriches, some psittacine species (e.g., Amazon parrots, Hyacinth macaws), some species of pigeon, and various other birds.

Feather overview

Feathers differ from mammalian hair in several ways. Avian feathers (as well as beaks and nails/talons) are made primarily of β-keratin, rather than the α-keratin that primarily composes mammalian hair (as well as claws/nails, hooves, and horns). The α-keratins are found in all vertebrates, but the β-keratins are exclusive to the reptile-bird lineage.

While most avian species may appear to be uniformly covered in feathers, they are not. Penguins (Fig. 28.1-4) are one of the rare exceptions in which feather follicle coverage is uniform. Generally, feathers grow in relatively discrete patches or tracts called **pterylae** (L. *pteryla* feathered part) (Fig. 28.1-5). Featherless regions between the pterylae are therefore referred to as **apteria** (L. *a-* absence of; Gr. *pteron* wing; wingless; singular, **apterium**; adjective, **apteric**). Small "down" feathers may be found in the apteria, as well as an occasional "semiplume," but "contour" and "flight" feathers are absent from apteria. (All of these feather types are described and illustrated below.)

SKIN AND FEATHER CARE

Birds use the oils and waxes expressed from the uropygial gland ("oil" or "preen" gland) to protect and waterproof their feathers and skin. This activity is particularly important in waterbirds such as ducks.

The secretions from this gland are also thought to contain a precursor of vitamin D, which is then converted by sunlight to its active form once spread over the feathers and skin. In this way, some preformed vitamin D is ingested when the bird preens.

Other mechanisms of feather care include preening behavior and dust- or water-bathing. Certain species also have "powder down" feathers which disintegrate into a fine powder that coats the skin and feathers.

FIGURE 28.1-4 Penguins are one of the rare exceptions in which feather follicle coverage is uniform, so they lack apteria.

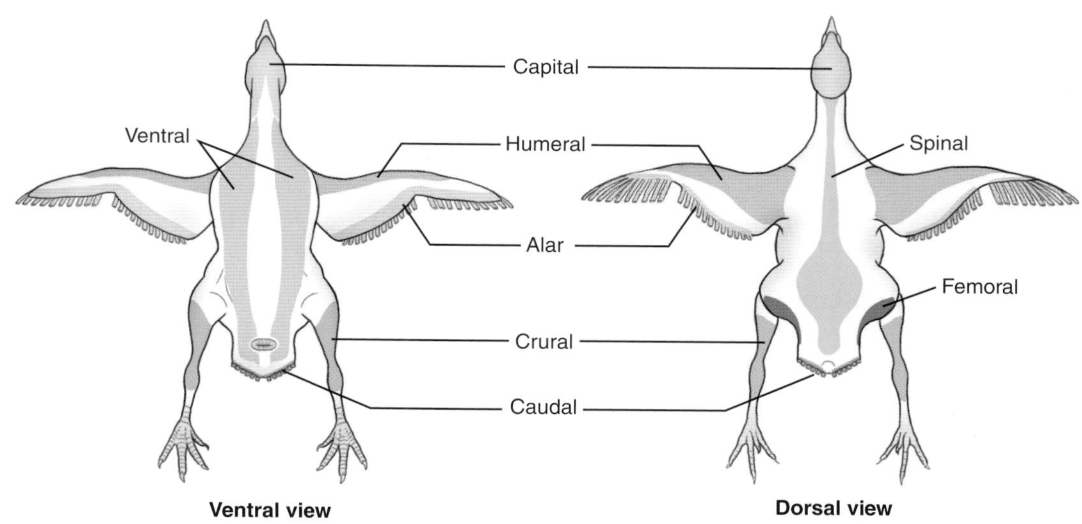

FIGURE 28.1-5 The principal pterylae (feather tracts) of the chicken.

Within each species, apteria separate the pterylae in a predictable pattern. The pterylae have been named by pterylologists; for example, the **pteryla capitalis** refers to the feathers on the head. However, the wide variation among species means that the study of this avian feature is itself a science.

The edges of the pterylae can serve as useful surgical landmarks. For example, one described approach to the proventriculus in the ostrich involves an incision through the apterium on the ventral abdomen, thus avoiding the need to pluck feathers when preparing the surgical site.

Feather morphology

The basic range of feather morphology is shown in Fig. 28.1-6. Contour feathers are those that cover the bird and provide a smooth, streamlined, and aerodynamic surface. **Body contour** feathers are symmetrical (both sides of the feather lateral to the central shaft are of similar size and shape) but the feathers themselves vary somewhat in

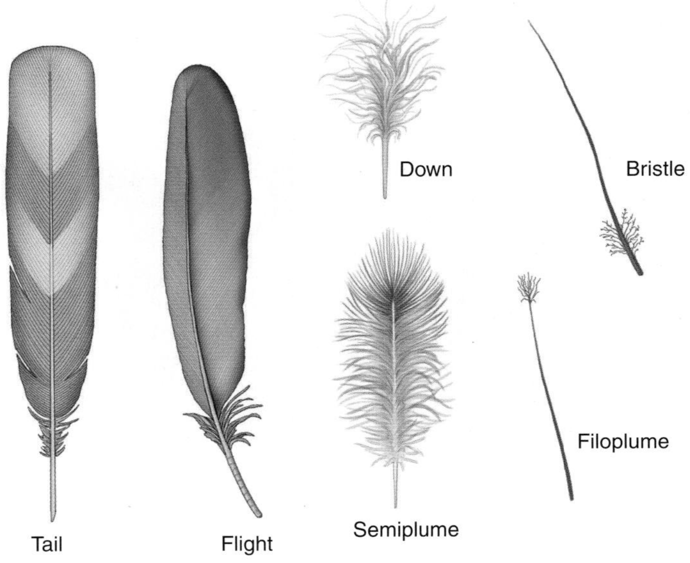

3

FIGURE 28.1-6 Basic range of feather morphology.

size and shape over the body to follow the form or contour of the bird. In contrast, **flight feathers** of the wing are asymmetrical, to meet the needs of flight (described later).

Most contour feathers are dorsoventrally flattened, so they have a dorsal/upper or outer surface which faces away from the body and a ventral/lower or inner surface which faces toward the body. Each has an axial **shaft** that extends from the **feather follicle** within the dermis to the **feather tip**. The **calamus** is the follicular part of the shaft, and the **rachis** is the part that extends distally from the skin surface toward the feather tip (Fig. 28.1-7). In large feathers, such as flight feathers, the rachis may have a groove on its ventral surface.

The **vanes** are the paired lateral portions of the feather. They are composed of multiple fine, linear **barbs**, which extend from the rachis laterally and distally (toward the feather tip). Depending on the feather type, the barbs may each be covered by tiny **barbules** which interlock adjacent barbs like zipper teeth or hook-and-loop fasteners (see Fig. 28.1-7). In body contour feathers, interlocking of the barbs creates a smooth and relatively impenetrable surface to the vane, which helps to facilitate the insulating and water-shedding properties of the body contour feathers. In flight feathers, interlocking of the barbs is essential for lift and thrust.

Much of the bird's preening activity involves straightening and "re-zipping" any barbs that may have become "unzipped."

BROOD PATCHES

In addition to the normal apteria, incubation or "brood" patches develop in many species for the incubation of eggs. Feather loss and an increase in skin vascularity over the ventral abdomen occur in whichever parent or parents are involved in incubating the clutch (nestful of eggs). In many species, both parents actively participate in incubating the eggs. The presence of a brood patch therefore cannot be used to distinguish males and females in those species. The feathers regrow in the brood patch once incubation is complete.

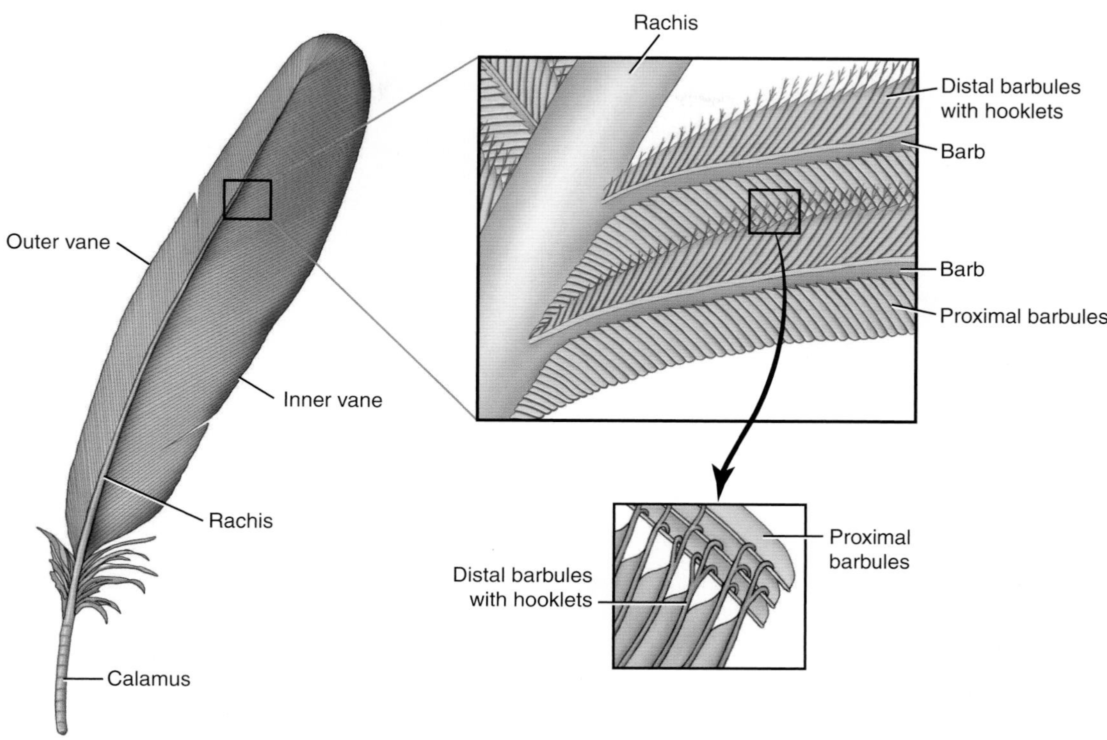

FIGURE 28.1-7 Details of a flight feather of the right wing.

If the vanes have interlocking barbs, the feather is described as **pennaceous**. The main portion of a flight feather is an example of a pennaceous vane. If the barbs are not normally interlocked, the feather has a fluffy appearance and is described as **plumaceous**. An ostrich feather is an example of a plumaceous vane. Some contour feathers, such as the flight feathers, have a short plumaceous portion near the skin before transitioning into a pennaceous type.

A small **afterfeather** or aftervane may be found on the ventral surface of the rachis, just above the skin. The afterfeather can be quite large in some species. Although it may appear to be a separate feather, it is considered part of the main feather to which it is attached, and from which it arises. Afterfeathers are not found on flight feathers.

Flight feathers

The **flight feathers** of the wing and tail are larger and stiffer than the body contour feathers, as they provide the lift required for flight. The flight feathers of the **wing** are termed the **remiges** (L. *remus* oar; *remiges* rowers). These feathers are long and asymmetrical, with a "leading" vane that is narrower than the "trailing" vane (see Figs. 28.1-6 and 28.1-7). Thus, a flight feather can be assigned to the right or left wing based on this asymmetry.

The remiges are classified as either primary or secondary, depending on whether they are attached to the carpus and digits (primaries) or the ulna (secondaries). The **primary remiges** are numbered distally from the carpus to the wing tip (Fig. 28.1-8), while the **secondary remiges** are numbered proximally from the carpus to the elbow. (As the

FEATHER TRIMMING

The flight feathers of the wing are sometimes trimmed to prevent flight in birds (pets, poultry, and wild birds in captivity). As the secondary flight feathers (between the carpus and elbow) provide the main lift in flight, it is harmful to trim these feathers, because the bird would be unable to control its descent if it does try to fly. As a result, the keel or beak may be injured if the bird falls to the ground. Trimming some or all of the primary flight feathers (between the carpus and wing tip) is sufficient to prevent a sustained flight.

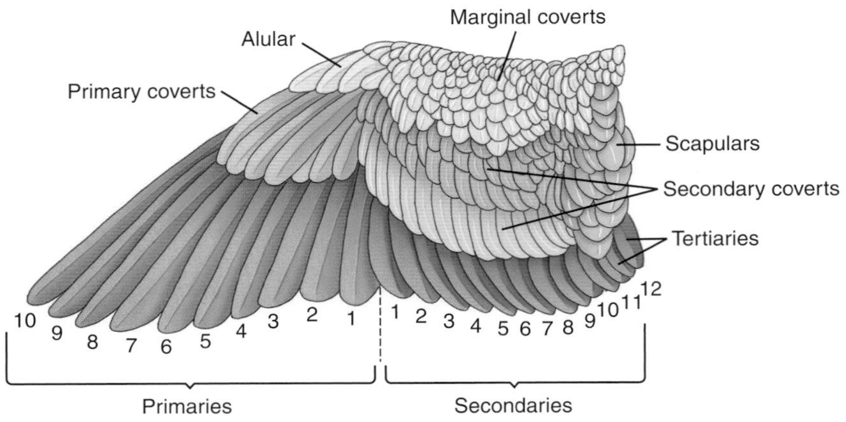

1

FIGURE 28.1-8 Dorsal surface of the left wing. The primary remiges (flight feathers of the wing) are numbered distally from the carpus (wrist) to the wing tip. The secondary remiges are numbered proximally from the carpus to the elbow.

most distal primary flight feather is small or absent in some species, numbering the primaries in this way allows for a variation in the most distal feather without affecting homologous feathers between closely related species.) Most species have 10 primary remiges, although 11 is not uncommon. Ostriches have the most, with 16 primary remiges. The number of secondary remiges is much more variable; hummingbirds are reported to have 6, while albatrosses may have as many as 40.

Flight feathers are firmly attached to the periosteum of their respective bones by short, strong ligaments. The follicular attachments of the **secondary remiges** can be observed on the trailing (caudal and ventral) edges of the **ulna** as evenly spaced "quill knobs" or **papillae remigales**.

The flight feathers of the **tail** are termed the **rectrices** (L. *rector* ruler). Depending on the species, they may be short or long. Generally, the rectrices are more symmetrical toward the midline of the tail, while the laterally located feathers are more asymmetrical. They are numbered from axial (inner) to abaxial (outer). The rectrices assist the bird during flight by providing additional lift but also by providing directional control and stability.

Additional functions of tail feathers include ornamental or auditory display (e.g., peacocks and hummingbirds, respectively), biomechanical assistance (e.g., woodpeckers use their tails to brace themselves against a tree), and balance when perching, hopping, and running.

In the clinical restraint of birds, the tail is good for neither catching nor for restraining a bird. Feather "sloughing" in response to capture is a predator-escape tactic that necessitates appropriate handling in the veterinary examination room in order to avoid this outcome.

Prey species such as the mourning dove may have "loosely" attached feathers, such that a predator may come away with a mouthful of tail feathers rather than the entire bird. While this strategy can be a life-saving maneuver, it leaves the bird flight-impaired until a new set of feathers replaces those that were lost. (Replacement of lost or damaged feathers is described later, in the section on molting.)

Other feather types

Wing and **tail coverts**, or **tectrices** (L. *tect-* covered, covering), are feathers that cover the base of the flight feathers on the wings and tail. As illustrated in Fig. 28.1-8, the wing coverts may include several rows, with the major coverts covering the base of the flight feathers. **Ear coverts** cover the opening of the external ear canal in many species (Fig. 28.1-9); these feathers allow air flow smoothly past the opening of the ear canal.

Semiplumes are feathers that have a long rachis but short, plumaceous barbs (Fig. 28.1-6). They differ from down feathers (see below), which have a short rachis, or no rachis at all, and longer plumaceous barbs. If the tip of the semiplume is pennaceous, like that shown in Fig. 28.1-6, then the feather is a contour feather. Semiplumes are located with the other contour feathers in the pterylae, while down feathers may also be distributed in the apteria.

Down feathers are short and fluffy, thereby creating air pockets near the skin. By lying deep to the contour feathers, down feathers provide additional insulation. There are wide species differences in where down is distributed on the body.

Chicks hatch with a variable amount of **natal down**. Precocial species, which are well-developed and quite independent from the time they hatch, are mostly covered in natal down (Fig. 28.1-10). In contrast, altricial species are entirely dependent on their parents for the first few weeks of life and hatch mostly naked (Fig. 28.1-11). Natal down is the precursor to either adult definitive down or contour feathers (Fig. 28.1-12). As with adult down feathers, natal down serves to insulate and protect the hatchling. Down insulates best when the air pockets formed are either deep to the contour feathers (adults) or deep to the protective cover of the nesting parent (young chicks).

Powder down is a special type of down feather that is designed to disintegrate into a very fine powder of keratin particles. This powder is believed to help protect the feathers and skin, as well as provide some waterproofing. Some species, including pigeons and parrots, produce far more powder than others; chickens, for example, produce little to none.

Powder down is implicated in human allergies to birds.

Bristles or setae (L. *seta* bristle) are feathers with barbs only near the base of the otherwise naked rachis (Fig. 28.1-6). Typically, these stiff feathers are found on the head or around the mouth or nostrils; they also form "eyelashes" in some species. **Semibristles** have barbs for a greater distance along the rachis than true bristle feathers. Classification is not exact, as intergrades between bristle, semibristle, and fully barbed feather exist. Bristle feathers likely have a sensory function; in addition to their strategic location around the mouth, nostrils, and/or eyes, the base of a bristle feather is usually surrounded by sensory corpuscles (sensory nerve endings).

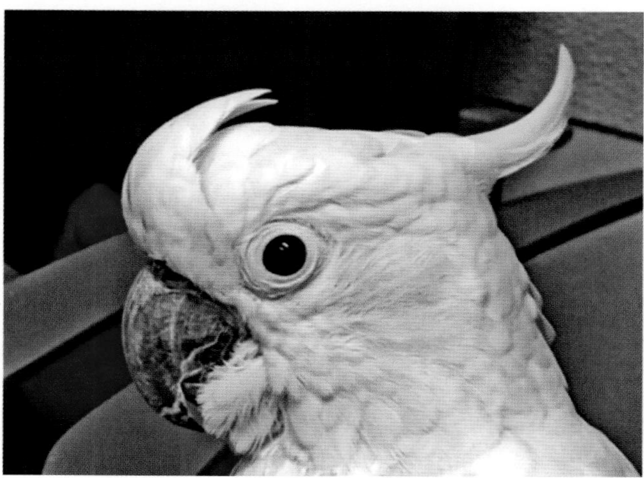

FIGURE 28.1-9 The ear coverts of the lesser sulfur-crested cockatoo (*Cacatua sulphurea*), located caudoventral to the eye, are pale yellow.

FIGURE 28.1-10 An Aplomado falcon (*Falco femoralis*) chick. The chick is covered in the preplumulae, or second down, which develops posthatching. The Aplomado falcon is so-named because of its "lead-colored" (blue-gray) adult plumage, which is suggested by the color of this nestling's down.

FIGURE 28.1-11 The 10-day-old Goffin's cockatoo (*Cacatua goffini*) with a crop impaction described in Case 25.2. This altricial species has barely any natal down, yet it already has pin feathers.

The "bristles" of a tom turkey's beard, located on the caudoventral aspect of the neck, do not grow from feather follicles but directly from a raised area of the epidermis. Like feathers, these beard "bristles" are composed of β-keratin, but they are not easily classified as a type of feather and are not a bristle feather as described above.

Filoplumes are feathers that have a long, thin, naked rachis with a tuft of barbs at the tip (see Fig. 28.1-6). They are always found alongside other feathers; often, multiple filoplumes are associated with each contour feather. Filoplumes can be tiny or the same length as the associated contour feather. They are believed to function in a proprioceptive fashion that allows the bird to discern the relative position of the contour feathers on its body.

FIGURE 28.1-12 A 5-week-old sun conure (*Aratinga solstitialis*) showing natal down and pin feathers that are well on their way to becoming covert and flight feathers.

Molting

Feathers are routinely molted or shed. The molt process replaces worn or damaged feathers, as well as shedding any attached feather parasites such as lice or mites. Breaking or cutting a feather does not trigger replacement, but plucking a feather often triggers replacement. In most small species, a plucked feather starts to regrow right away.

In general, birds usually molt once or twice a year on a schedule typical for that species, although a more frequent molt schedule can occur. There is considerable species variation in the frequency of molting. Most small birds molt once or twice a year, whereas larger birds (e.g., eagles) may do only a partial molt every year.

In birds that molt twice a year, molting often corresponds with the development of breeding plumage (nuptial plumage), which may be more colorful to help protect the nest. Particularly in males, the nonbreeding or basic plumage may be more drab, such as the "eclipse" plumage of male waterfowl.

The molting process typically is orderly and symmetrical. Feathers are molted in sequence, and flight feathers are lost in a manner that keeps the bird aerodynamically balanced. For example, the 6th primary remiges are lost at the same time on both wings. (In fact, primary and secondary remiges were originally numbered to reflect the order of molt.) Some ducks molt all of their flight feathers at once, resulting in temporary flightlessness.

> In captive birds, a balanced diet helps facilitate regrowth of feathers. In the wild, the large energy outputs are molting, breeding, and migration, so birds are less likely to do any of these activities simultaneously.

Pin feathers are incipient new feathers (Figs. 28.1-11 and 28.1-12). An **axial artery** feeds the developing feather as it grows. Often referred to as a **blood feather** (Fig. 28.1-13), this immature feather is vulnerable to injury, and the bird can suffer considerable blood loss if one is injured. Once the feather is mature, the artery regresses (Fig. 28.1-14). The keratin sheath of the emerging feather is preened away to reveal the fresh feather.

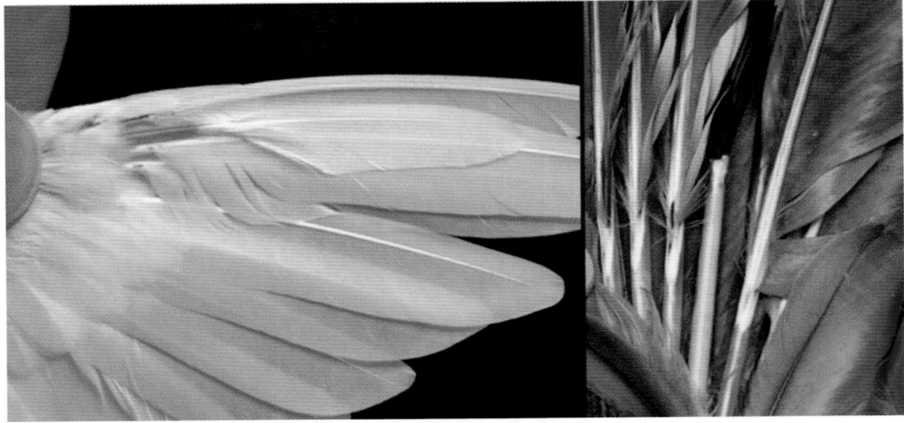

FIGURE 28.1-13 Examples of "blood feathers" in a Budgerigar (left) and a Scarlet macaw (right). Blood is clearly visible in the shaft of the white feathers, but the brightly colored feathers of the macaw make the shaft of its blood feather appear blue.

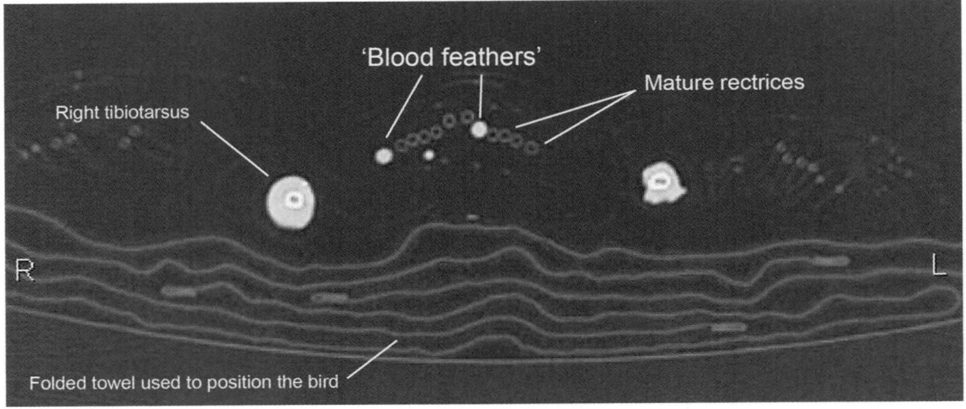

FIGURE 28.1-14 Computed tomographic image of the tail feathers of a cockatoo, transverse view, showing the difference in cross section between the new (blood) feathers and the mature rectrices (flight feathers of the tail).

Feather color

Feathers are more brightly colored on their dorsal/outer or exposed surface than on their underside. The great variety of colors is achieved through several means, including pigment content and feather microstructure. The pigments involved include melanins, carotenoids, and porphyrins.

> In some species, plumage colors vary noticeably between juveniles and adults, such that the identification of young birds of these species can be challenging.

Melanins are responsible for black, gray, some browns and yellows, and some reddish browns. Melanins are packaged in granules, the concentration of which affects the color of the feather. Melanin granules also contribute to the strength and durability of a feather, allowing the feather to better resist wear. A good example is the black feather tips of many white birds such as seagulls.

Melanin content can also be used as a mechanism for feather color change without the need to molt. For example, the European starling sports a speckled appearance during the nonbreeding season. The light-colored tips of the feathers wear away by the time the breeding season occurs to reveal shiny black feathers. The common house sparrow undergoes a similar change. Abrasion wears away the more cryptic (camouflaging) light ends of the feathers of its bib, such that by breeding season the male sparrow sports a large black bib without having to grow new feathers.

Carotenoids contribute to bright reds, yellows, and oranges. Carotenoids are obtained from food, so diet can impact the color of the feathers in certain species. Flamingoes are the classic example, as they require dietary carotenoids to maintain their pink color. The scarlet ibis (Fig. 28.1-15) is another.

Porphyrins are constructed from amino acids and result in reds, greens, and browns. Thus, depending on the species, both feather color and brightness can be an indication of the health of the bird and the adequacy of its diet.

In addition to these pigments, color may be achieved by structural means. Iridescence, the reflection of light from the feather microstructure, often leads to positional color changes. For example, the angle from which one views the gorget (throat piece) of a hummingbird affects the color perceived (Fig. 28.1-16).

Blue colors are commonly produced by light reflected from microscopic air pockets within the feather structure. The structural origin of this range of colors can be demonstrated by grinding the feather, thereby eliminating air pockets. Feather color may also be dramatically changed by saturating the feather (Fig. 28.1-17), which disrupts the combination of blue-reflecting air pockets and the underlying pigment colors; the normal feather color returns once the feather dries.

Lastly, birds can perceive light in the ultraviolet range, so it should come as no surprise that feathers we see as perhaps dull and lacking in detail may be much more vibrant and intricate when observed under ultraviolet light (Fig. 28.1-18)—and, we may assume, when viewed by another bird. The avian perception of each other, and of the world around them, is not necessarily our human reality.

FIGURE 28.1-15 The scarlet ibis (*Eudocimus ruber*) requires dietary carotenoids to show its striking red plumage to its full advantage.

FIGURE 28.1-16 The refractile feathers on the gorget, or throat piece, of this Costa's hummingbird (*Calypte costae*) appear to change color from black to purple (*arrow*) simply by a change in angle.

FIGURE 28.1-17 Symmetrical flight feathers from a Red lored Amazon parrot. The feather on the left has been soaked with water, which disrupts the reflectivity of the blue structural colors. When the feather dries, it returns to the same bright colors as the dry feather on the right.

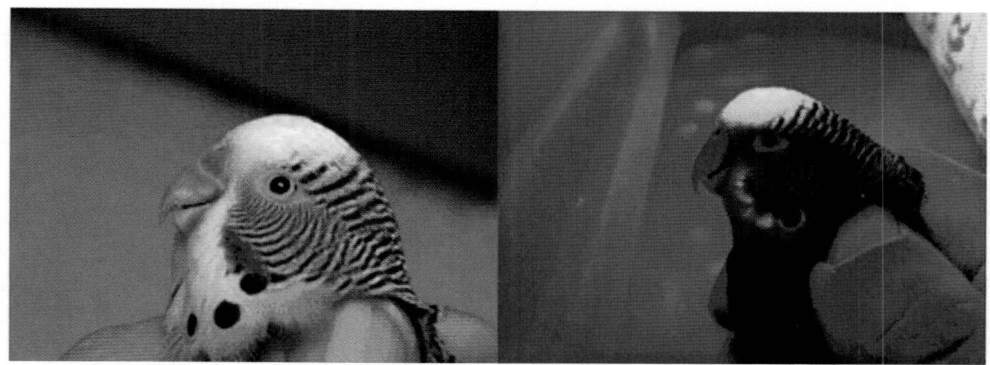

FIGURE 28.1-18 A Budgerigar under normal light (left) and ultraviolet light (right).

Selected references

[1] Baumel JJ, editor. Handbook of Avian Anatomy: Nomina Anatomica Avium. 2nd ed. Nuttall Ornithological Club (USA), World Association of Veterinary Anatomists (USA), International Committee on Avian Nomenclature; 1993.
[2] King AS, McLelland J. Birds: Their structure and function. 2nd ed. London: Baillière Tindall; 1984.
[3] Lucas AM, Stettenheim PR. Avian anatomy—Integument part I and II. Agricultural handbook. US Government Printing Office; 1972. p. 362.

SECTION VII

APPENDICES

APPENDIX 1: STANDARD ABBREVIATIONS 1419

APPENDIX 2: NORMAL RESPIRATORY RATE, HEART RATE, AND TEMPERATURE REFERENCE RANGES 1423

APPENDIX 3: HEMATOLOGY REFERENCE INTERVALS 1425

APPENDIX 4: BIOCHEMISTRY REFERENCE INTERVALS 1427

Standard Abbreviations

A

a. L.[a] Arteria (artery)
aa. L. Arteriae (arteries)
AFib Atrial fibrillation
ALP Alkaline phosphatase
ALT Alanine aminotransferase
Ao Aortic root
A.S. L. Auris sinistra (left ear)
AST Aspartate aminotransferase
AV Atrioventricular

B

BAL Bronchoalveolar lavage
BBB Blood-brain barrier, Bundle branch block
BID, bid L. Bis in die (twice a day)
BP Blood pressure
bpm Beats per minute
brpm Breaths per minute
BUN Blood urea nitrogen

C

°C Degree Celsius
CBC Complete blood count
Cl^- Chloride
CK Creatine kinase
CN Cranial nerve
CNS Central nervous system
CO Cardiac output
CPR Cardiopulmonary resuscitation
CRT Capillary refill time
CSF Cerebrospinal fluid, Colony-stimulating factor
CT Computed tomography
CV Cardiovascular
CVP Central venous pressure

D

DDSP Dorsal displacement of the soft palate
DNA Deoxyribonucleic acid

E

ECF Extracellular fluid
ECG Electrocardiogram or electrocardiographic
ECM Extracellular matrix
EDTA Ethylenediaminetetraacetic acid
EEG Electroencephalogram
e.g. Latin for *for example*; use only in parenthetical statements
EKG Electrocardiogram
ELISA Enzyme-linked immunosorbent assay
EMG Electromyogram
ERG Electroretinogram
ESWT Extracorporeal shock-wave therapy

F

°F Degree Fahrenheit

G

g Gram
GGT γ-Glutamyltransferase
GI Gastrointestinal

H

H^+ Hydrogen
Hb Hemoglobin
HCO_3^- Bicarbonate
HCT, Hct Hematocrit
H&E Hematoxylin and eosin
Hp Haptoglobin
hpf High-power field

I

ICU Intensive care unit
ICS Intercostal spaces
ID Intradermal
i.e. Latin for *that is*; use only in parenthetical statements

Ig Immunoglobulin
IM Intramuscular
IOP Intraocular pressure
IV Intravenous
IVS Interventricular septum

J and K

K^+ Potassium
kg Kilogram

L

L, l. Liter
L Lumbar vertebra
LA Left atrium
LDH Lactate dehydrogenase
Lig. Ligament
LRS Lactated Ringer's solution
LV Left ventricle

M

M Median, Meter
MAP Mean arterial pressure
mEq, meq Milliequivalent
mg Milligram
μg Microgram
mL Milliliter
μL Microliter
MM, mm Mucous membrane
Mol wt Molecular weight
mOsm Milliosmole
MR/MRI Magnetic resonance imaging
MW Molecular weight
MV Mitral valve

N

n. Nervous (nerve)
Na^+ Sodium
NA Nomina anatomica
NAV Nomina anatomica veterinaria
Nd:YAG Neodymium:yttrium-aluminum-garnet (laser)
ng Nanogram
nm Nanometer
nn. Nerves
NPO L. nil per os (nothing by mouth)
NSAID Nonsteroidal anti-inflammatory drug

O

O_2 Oxygen
OD L. Oculus dexter (right eye), Overdose
OL L. Oculus laevus (left eye)
OS L. Oculus sinister (left eye)
OTC Over-the-counter
OU L. Oculus uterque (both eyes)

P

P Phosphorus
PA Pulmonary artery, Posteroanterior
PABA Para-aminobenzoic acid
PCR Polymerase chain reaction
PCV Packed cell volume
PET Positron emission tomography
pg Picogram
PMI Point of maximal intensity
PMN Polymorphonuclear neutrophil
PO L. Per os (orally)
ppm Parts per million
prn Pro re nata (as circumstances require)
PT Prothrombin time

Q

QID, qid L. Quarter in die (four times a day)

R

RA Right atrium
RBC Red blood cell, Red blood count
RES Reticuloendothelial system
RNA Ribonucleic acid
rRNA Ribosomal RNA
RR Respiratory rate
RV Right ventricle

S

S Sacral vertebra, Sulfur
S_1 First heart sound
S_2 Second heart sound
S_3 Third heart sound
S_4 Fourth heart sound
s Second
s. L. Sinister (left)
š L. Sine (without)

SA Sinoatrial
SGOT Serum glutamic-oxaloacetic transaminase
SGPT Serum glutamate pyruvate transaminase
SID, sid L. Semel in die (once a day)
sp., gr Specific gravity
SQ Subcutaneous
SUN Serum urea nitrogen

T

T Thoracic vertebra
t Temperature
$T^{1/2,}$ $t^{1/2}$ Half-life
3D Three-dimensional
TID, tid L. Ter in die (three times a day)
TNF Tumor necrosis factor
TP Total protein
TPN Total parenteral nutrition
TSH Thyroid-stimulating hormone

U

U Units
US United States, Ultrasound

V

V Volume
v. L. Vena (vein)
vv. L. Venae (veins)

W–Z

WBC White blood count
wt Weight

[a] L., Latin.

APPENDIX 2

Normal Respiratory Rate, Heart Rate, and Temperature Reference Ranges

Resting respiratory rates (brpm)

	Ref. [1]	Ref. [2]	Ref. [3]
Cat	16–40		20–42
Dairy cow	26–50	12–46	
Dog	18–34		16–32
Goat		15–40	
Horse	10–14		
Pig	32–58		
Sheep	16–34	12–72	

Resting heart rates (bpm)

	Ref. [1]	Ref. [2]	Ref. [3]
Cat	120–140		140–220
Chicken	250–300		
Dairy cow	48–84	40–80	
Dog	80–120		60–160 (smaller up to 200)
Goat	70–80	70–110	
Horse	28–40		
Pig	70–120		
Sheep	70–80	60–120	

Normal rectal temperature

	Ref. [1]		*Ref. [2]*		*Ref. [3]*	
	°F	*°C*	*°F*	*°C*	*°F*	*°C*
Cat	100.5–102.5	38.1–39.2			100–102.2	37.8–39
Chicken	105–109.4	40.6–43				
Dairy cow	100.4–102.8	38–39.3	100.5–102.5	38–39		
Dog	100.2–103.8	37.9–39.9			100–102.2	37.8–39
Goat	101.3–103.5	38.5–39.7	101.5–103.5	38.5–39.5		
Horse	99–100.8	37–38.2				
Pig	101.6–103.6	38.7–39.8				
Sheep	100.9–103.8	38.3–39.9	102–103.5	39–40		

References

[1] Anon. The Merck veterinary manual. 11th ed; 2016.
[2] Smith B, Van Metre D, Pusterla N. Large animal internal medicine. 6th ed. St Louis: Elsevier; 2019.
[3] Bassert J, Beal AD, Samples OM. McCurnin's clinical textbook for veterinary technicians. 9th ed. St Louis: Elsevier; 2018.

APPENDIX 3

Hematology Reference Intervals

Test[a,b]	Units	Canine	Feline	Equine	Bovine	Alpaca	Caprine
Red blood cells							
Hematocrit (HCT)	%	41–58	31–48	34–46	25–33	26–45	28–44
Red blood cells (RBC)	$\times 10^6/\mu L$	5.7–8.5	6.9–10.1	6.6–9.7	5.0–7.2	10.6–18.4	13.6–23.7
Hemoglobin (Hb)	g/dL	14.1–20.1	10.9–15.7	11.8–15.9	8.7–12.4	11–19.3	8.9–13.8
Mean cell volume (MCV)	fL	64–76	40–52	43–55	38–51	22–28	16–22
Mean corpuscular hemoglobin (MCH)	pg	21–26	13–17	15–20	14–19	9–12	5–7
Mean corpuscular hemoglobin concentration (MCHC)	g/dL	33–36	32–35	34–37	34–38	42–49	32–34
Red cell distribution width (RDW)	%	10.6–14.3	13.2–17.5	16.3–19.3	15.0–19.4	ND	17.8–24.4
Nucleated red blood cell count (NRBC)	/100 WBC	0–1	0–1	0	0	0–3	0
Reticulocyte percentage	%	0.2–1.5	0.1–0.7	ND	ND	ND	ND
Reticulocyte count	$\times 10^9/L$	11–92	9–61	ND	ND	ND	ND
White blood cells							
White blood cells (WBC)	$\times 10^3/\mu L$	5.7–14.2	5.1–16.2	5.2–10.1	5.9–14.0	7.1–18.6	7.2–17.7
Neutrophils	$\times 10^3/\mu L$	2.7–9.4	2.3–11.6	2.7–6.6	1.8–7.2	3.5–11.7	1.9–9.5
Band neutrophils	$\times 10^3/\mu L$	0–0.1	0–0.1	0	0	0	0–0.1
Lymphocytes	$\times 10^3/\mu L$	0.9–4.7	0.9–6	1.2–4.9	1.7–7.5	1.1–5.5	2.6–11.7
Monocytes	$\times 10^3/\mu L$	0.1–1.3	0–0.7	0–0.6	0–0.9	0–1.0	0–0.9
Eosinophils	$\times 10^3/\mu L$	0.1–2.1	0.1–1.8	0–1.2	0–1.3	0.1–4.3	0–0.8
Basophils	$\times 10^3/\mu L$	0–0.1	0–0.2	0–0.2	0–0.3	0–0.4	0–0.3
Platelets							
Platelets (PLT)	$\times 10^3/\mu L$	186–545	195–624	94–232	252–724	220–817	247–912
Mean platelet volume (MPV)	fL	8.4–14.1	9.1–24.3	5.3–8.4	5.7–8.0	4.4–6.9	5.3–9.4
Total protein (TP)	g/dL	5.9–7.8	5.9–7.5	5.2–7.8	5.9–8.1	6.0–7.5	ND
Fibrinogen (by heat precipitation)	mg/dL	ND	ND	0–200	100–600	ND	0–200

[a] Reference values courtesy of Cornell's e-clinpath website: https://www.vet.cornell.edu/animal-health-diagnostic-center/laboratories/clinical-pathology/reference-intervals/hematology.

[b] There is a variation in reference intervals based on the laboratory and method of determination.

ND, not determined.

APPENDIX 4

Biochemistry Reference Intervals

Test[a,b]	Units	Canine	Feline	Equine	Bovine	Alpaca	Caprine
Sodium	mEq/L	143–150	149–158	134–142	134–144	149–156	143–154
Potassium	mEq/L	4.1–5.4	3.8–5.5	2.4–4.8	4.0–5.9	4.2–5.9	4.2–6.0
Chloride	mEq/L	106–144	111–124	95–104	92–99	105–116	101–116
Bicarbonate	mEq/L	14–24	14–20	24–31	22–30	20–32	16–26
Anion gap	mEq/L	17–27	18–29	12–19	19–26	16–22	19–27
Glucose	mg/dL	68–104	71–182	71–122	57–79	99–146	35–142
Urea nitrogen	mg/dL	9–26	17–35	10–22	7–19	11–30	10–35
Creatinine	mg/dL	0.6–1.4	0.8–2.1	0.8–1.5	0.4–0.9	1.0–2.3	0.3–0.8
Total bilirubin	mg/dL	0–0.2	0–0.1	0.5–2.1	0–0.1	0–0.1	0–0.1
Direct bilirubin	mg/dL	0–0.1	0	0.1–0.3	0	0	0
Indirect bilirubin	mg/dL	0–0.1	0–0.1	0.3–2.0	0–0.1	0–0.1	0–0.1
ALP	U/L	7–115	11–49	88–261	27–127	17–111	ND
GGT	U/L	0–8	0–2	8–33	17–54	9–35	24–64
ALT	U/L	17–95	28–109	ND	ND	ND	ND
AST	U/L	18–56	17–48	222–489	54–135	119–286	62–145
SDH	U/L	3–13	1–11	1–6	8–48	0–6	24–63
GLDH	U/L	1–20	0–12	2–10	14–141	4–19	ND
CK	U/L	64–314	74–386	171–567	88–292	35–132	116–464
LDH	U/L	24–388	71–406	218–555	725–1122	ND	116–443
Total protein	g/dL	5.5–7.2	6.6–8.4	5.4–7.0	6.7–8.8	5.6–7.0	6.2–8.0
Albumin	g/dL	3.2–4.1	3.2–4.3	2.9–3.6	3.3–4.3	2.8–4.2	2.9–4.0
Globulin	g/dL	1.9–3.7	2.9–4.7	2.3–3.8	2.8–5.4	2.1–3.1	3.0–4.7
Albumin/globulin ratio		0.9–1.9	08.–1.5	0.8–1.5	0.6–1.6	1.0–2.0	1.0–2.0
Calcium	mg/dL	9.4–11.1	9.0–11.3	10.8–12.9	8.9–10.9	8.1–9.9	8.3–10.3
Phosphate	mg/dL	2.7–5.4	2.6–5.5	2.1–4.7	4.1–7.3	3.3–7.7	4.5–7.8
Magnesium	mEq/L	1.5–2.1	1.7–2.2	1.2–1.9	1.6–2.5	1.7–2.5	1.5–2.2
Cholesterol	mg/dL	136–392	101–323	68–133	163–397	46–117	ND
Triglycerides	mg/dL	23–125	30–106	14–65	10–19	9–53	ND
Amylase	U/L	322–1310	603–2183	3–8	14–50	824–1686	ND

(Continued)

Biochemistry Reference Intervals

Test[a,b]	*Units*	*Canine*	*Feline*	*Equine*	*Bovine*	*Alpaca*	*Caprine*
Lipase	U/L	15–228	7–23	7–16	5–13	ND	ND
Iron	µg/dL	97–263	59–169	95–217	64–224	86–203	102–194
Total iron binding concentration (TIBC)	µg/dL	280–489	222–423	289–535	320–490	245–441	266–408
Saturation	%	27–66	20–56	27–56	18–54	30–59	28–59
Cholinesterase	U/L	1895–5240	736–3016	2684–5889	53–126	242–603	ND
Beta-hydroxybutyrate (BHB)	mg/dL	0–1.0	0–1.0	1.0–3.1	2.3–10	ND	ND
Non-esterified fatty acids (NEFA)	mEq/L	0.13–1.25	0.15–1.05	0.02–0.43	ND	ND	0.02–0.27
Fructosamine	Umol/L	266–381	219–365	284–387	ND	ND	ND
pH	ND	7.31–7.46	7.21–7.41	7.36–7.43	7.36–7.46	ND	ND

[a] Reference values courtesy of Cornell's e-clinpath website: https://www.vet.cornell.edu/animal-health-diagnostic-center/laboratories/clinical-pathology/reference-intervals/chemistry.

[b] There is a variation in reference intervals based on the laboratory and method of determination.

ND, not determined.

BIBLIOGRAPHY

Anatomy textbooks referenced for clinical cases and figures

[1] Budras KD, Habel RE, Müllerig CKW, Greenough PR, Weinche A, Budras S. Bovine anatomy. 2nd ed. Germany: Schlütersche; 2011.
[2] Budras KD, Sack WO, Röck S, Horwitz A, Berg R. Anatomy of the horse. 6th ed. Germany: Schlütersche; 2011.
[3] de Lahunta A. Applied veterinary anatomy. Philadelphia: W.B. Saunders; 1986.
[4] Anon. Dorland's illustrated medical dictionary. 32nd ed. Philadelphia: Elsevier/W.B. Saunders; 2012.
[5] Evans HE, de Lahunta A. Miller's dissection guide for the dog. 8th ed. Philadelphia: Elsevier/W.B. Saunders; 2017.
[6] Getty R. Sisson and Grossman's, the anatomy of the domestic animals. 5th ed. Philadelphia: W.B. Saunders; 1975.
[7] Hermanson JW, de Lahunta A, Evans HE. Miller and Evan's anatomy of the dog. 5th ed. Philadelphia: Elsevier/W.B. Saunders; 2020.
[8] Popesko P. Atlas of topographical anatomy of the domestic animals. vols. 1–3. Philadelphia: W.B. Saunders; 1978.
[9] Schummer A, Nickel R, Sack WO. In: Nickel R, Schummer A, Seiferle E, editors. The viscera of the domestic mammals. 2nd ed. New York: Springer-Verlag; 1979.
[10] Shively MJ. Veterinary anatomy basic, comparative, and clinical. College Station: Texas A&M Press; 1984.
[11] Dyce SB. Sack and Wensing's textbook of veterinary anatomy. 5th ed. St. Louis: Elsevier; 2018.
[12] McIlwraith CW, Nixon A, Wright IM. Diagnostic and surgical arthroscopy in the horse. 4th ed. St. Louis: Elsevier; 2015.
[13] Pollitt CC. The illustrated horse's foot: a comprehensive guide. St. Louis: Elsevier; 2016.
[14] Maggs D, Miller P, Ofri R. Slatter's fundamentals of veterinary opthamology. 4th ed. Philadelphia; W.B. Saunders; 2008.

List of Illustrations under Editor Copyright

The following figures are under the copyright of Nora S. Grenager and James A. Orsini and cannot be used without their permission. To request permission to reuse these figures, you can contact the editors at orsini@vet.upenn.edu and ngrenagervmd@gmail.com.

Note that the figures are listed in numerical order per illustrator. Please review each of the three lists of figures to locate an illustration. Figures not listed here are either under Elsevier copyright or the copyright of a third party. You can check the figure in the book to determine who owns the copyright. Those with a credit line in the figure legend will belong to a third party and you should contact that copyright owner for permission. Illustrations without a credit line and not listed in the Copyright Exceptions pages are under Elsevier copyright. See page IV for requesting permission to use these figures.

Figures illustrated by Jeanne Robertson: 1.1, 1.2, 1.3, 1.4, 1.5, 1.6, 2.3-2, 2.3-3, 2.3-9, 2.4-1AB, 2.4-2A, 2.4-3, 2.7-3, 3.1-3, 3.1-4, 3.1-5, 3.10-2, 3.10-3, 3.10-4, 3.10-5, 3.11-6, 3.11-7, 3.11-8, 3.12-3, 3.12-4, 3.12-5, 3.12-6, 3.12-7, 3.12-8, 3.2-3, 3.2-5, 3.2-7, 3.3-4, 3.3-5, 3.4-4, 3.5-1, 3.5-2, 3.5-3, 3.5-4, 3.5-5, 3.5-6, 3.5-8, 3.6-16, 3.7-2, 3.7-3, 3.8-8, 3.9-4, 3.9-5, 4.2-4, 4.3-4, 4.4-10, 4.4-6, 4.4-7, 4.4-8, 4.5-3, 4.5-4, 4.5-5, 4.5-6, 4.6-4, 4.7-1, 4.7-5, 4.7-6, 4.8-2, 4.8-3, 4.8-4, 5.5-2, 5.1-5, 5.1-7, 5.2-2, 5.2-3, 5.2-4, 5.3-1, 5.3-2, 5.3-3, 5.4-4, 5.4-5, 5.5-5, 5.5-6, 5.6-2, 5.6-3, 5.8-5, 5.8-6, 6.1-4, 6.2-3, 6.2-4, 6.2-6, 6.2-7, 6.3-7, 6.4-4, 6.4-5, 6.4-6, 6.5-4, 7.1-7, 7.1-8, 7.1-9, 7.5-3, 7.5-4, 7.5-5, 7.5-6, 8.3-4, 8.3-5, 8.3-6, 8.6-2, 8.6-3, 8.6-4, 8.6-5, 8.6-6, 8.6-7, 8.7-4, 8.7-5, 9.1-3, 9.1-4, 10.1-3, 10.1-5, 10.1-7, 10.10-7, 10.10-8, 10.12-3, 10.2-3, 10.2-4, 10.2-5, 10.2-6, 10.2-7, 10.3-4, 10.3-5, 10.3-6, 10.3-8, 10.3-9, 10.4-2, 10.4-6, 10.4-7, 10.4-8, 10.4-9, 10.5-10, 10.5-5, 10.5-6, 10.5-7, 10.6-10, 10.6-11, 10.6-14, 10.6-8, 10.6-9, 10.7-4, 10.7-7, 10.7-8, 10.8-5, 10.8-6, 10.8-7, 10.8-8, 10.9-3, 10.9-6, 10.9-7, 11.1-12, 11.1-13, 11.2-10, 11.2-7, 11.3-8, 11.3-9, 12.1-3, 12.1-4, 12.2-2, 12.2-3, 12.3-2, 12.3-3, 12.3-4, 12.4-1, 12.7-2, 12.7-4, 13.1-3, 13.1-4, 13.10-4, 13.10-5, 13.10-6, 13.11-3, 13.11-4, 13.11-5, 13.2-3, 13.2-5, 13.4-3, 13.4-4, 13.4-5, 13.4-6, 13.6-4, 13.6-5, 14.1-2, 14.1-2, 14.1-3, 14.1-3, 14.1-4, 14.1-4, 14.1-5, 14.1-5, 14.1-6, 14.1-6, 14.7-3, 14.9-5, 15.4-6, 15.5-2, 15.5-3, 16.1-2, 17.1-11, 17.1-7, 17.2-4, 17.3-10, 17.3-10, 17.3-8, 17.3-8, 17.3-9, 17.3-9, 18.1-6, 18.2-5, 19.1-3, 19.1-4, 19.5-2, 19.5-3, 19.5-5, 19.6-4, 19.7-4, 19.7-5, 19.7-7, 19.7-8, 20.1-1, 20.2-10, 20.2-12, 20.2-14, 20.2-2, 20.2-3, 20.2-5, 20.2-6, 20.2-7, 20.2-8, 20.2-9, 21.1-8, 21.4-7, 22.4-5, 23.1-10, 23.1-14, 25.1-6, 25.3-7, 25.3-8, 25.3-9, 25.6-7, 25.6-8, 26.1-8, 26.2-1, 26.2-5, 26.4-4, 26.5-4, 26.6-6, 27.3-6, 27.3-7, 28.1-5, 28.1-6, 28.1-7, 28.1-8
Canine Landscape: 3.0-1, 3.0-2, 3.0-3, 3.0-4, 3.0-5, 3.0-6, 3.0-7, 3.0-8, 3.0-9
Feline Landscape: 3.1-1, 3.1-2, 3.1-3, 3.1-4, 3.1-5, 3.1-6, 3.1-7, 3.1-8, 3.1-9
Equine Landscape: 4.0-1, 4.0-2, 4.0-3, 4.0-4, 4.0-5, 4.0-6, 4.0-7, 4.0-8, 4.0-9, 4.0-10, 4.0-11
Bovine Landscape: 5.0-1, 5.0-2, 5.0-3, 5.0-4, 5.0-5, 5.0-6, 5.0-7, 5.0-8, 5.0-9
Avian Landscape: 6.0-1, 6.0-2, 6.0-3, 6.0-4, 6.0-5, 6.0-6, 6.0-7

Figures illustrated by Libby Lamb Wagner; MPS NA, LLC: 12.6-4, 13.3-5, 16.2-3, 13.7-2A–F, 13.7-5A, 13.7-6, 13.7-7, 14.5-4, 14.3-2, 14.3-4, 14.3-5, 14.4-4, 14.4-5, 14.4-6, 14.2-3, 14.2-4, 14.2-5, 14.7-3, 14.8-2AB, 14.8-5, 14.9-2, 14.9-3, 15.3-3, 15.3-4, 14.6-3, 14.6-4, 14.6-5, 15.2-3, 15.2-6, 15.4-4, 15.4-5, 15.1-2, 15.1-3, 17.4-8, 19.4-2, 19.3-6, 19.3-7, 20.1-4, 20.4-6, 20.3-6, 21.1-11, 21.1-7, 21.3-2, 22.1-2, 22.3-3, 23.2-4

Figures illustrated by Stefan Németh (SciMed-Illustration): 10.11-7A, 7B, 10.11-5, 10.11-6A, 6B, 6C

Index

Note: Page numbers followed by *f* indicate figures, *t* indicate tables, and *b* indicate boxes.

A

Abdomen
avian, 1338–1339
female (left), 1256*f*
male (right), 1257*f*
canine, 31*f*, 53*f*
female (left), 1257*f*
male (right), 90*f*
ventral view, 91*f*
equine, 735*f*
female (left), 560*f*
male (right), 561*f*
ventral view, 562*f*
feline, 333
female (left), 99*f*
male (right), 100*f*
ventral view, 101*f*
ruminant, 1116*b*, 1117*f*, 1120*f*
female (left), 997*f*
male (right), 998*f*
ventral view, 999*f*
Abdominal wall
canine, 91*f*
equine, 562*f*
ruminant, 999*f*
Abducent nerve (CNVI)
canine, 88*f*
equine, 652
Abductor digiti I longus muscle, feline
Abomasum, ruminant, 1047*f*, 1064, 1064–1065*f*, 1066, 1067–1068*b*, 1067*f*
Accessory axillary lymph node, canine, 87*f*
Accessory, dorsal branch nerve, ruminant, 996*f*
Accessory carpal bone
canine, 84–85*f*
equine, 554*f*, 894–896
feline, 94*f*
ruminant, 992*f*
Accessory cephalic vein
equine, 555*f*
ruminant, 993*f*
Accessory gluteal muscle, equine, 938
Accessory head of scapula, ruminant, 1154
Accessory hepatic lymph node, ruminant, 1094
Accessory lobe of right lung
canine, 90*f*, 224*f*
equine, 560*f*, 721*f*
feline, 99–100*f*
ruminant, 997–998*f*, 1087–1088
Accessory nerve
equine, 652–653
ruminant, 996*f*
Accessory reticular artery, ruminant, 1060
Accessory parathyroid gland, canine/feline, 159
Accessory thyroid gland, canine, 159
Accessory vaginal vein, ruminant, 1122, 1124
Acetabulum
avian, 1388
canine, 92*f*, 459
equine, 563*f*
feline, 458–459
ruminant, 1034*f*, 1196
Acinus lobule, equine, 767, 768*f*
Acromion
canine, 85*f*, 92*f*
feline, 102*f*
ruminant, 1149
Adductor mandibulae ext. muscle, avian, 1255*f*
Adductor muscle
canine, 88*f*
equine, 956
feline (magnus and cruris caudalis), 477*f*
ruminant, 1036
Adnexa
of eye, avian, 1313–1314
of skin, canine, 546–548
Adrenal gland
avian, 1353
canine, 295
equine, 735*f*
feline, 296
ruminant, 1099
Adrenal plexus, canine, 296
Air capillaries, avian, 1340
Air sac, avian, 1335*b*, 1336–1339
Alar fold, equine, 589
Allantochorion/chorioallantois
canine, 367, 369
feline, 362
Allantoic duct, ruminant, 1103
Alula, avian, 1258*f*, 1373
Alveolar bone
canine, 124, 227
equine, 595
Alveoli, equine, 607, 986
Amnion
canine, 367
ruminant, 1107*b*
Ampulla of rectum, equine, 798, 812–813, 812*f*, 825, 826*f*, 829*f*, 831–832, 831*f*
Ampullar excretory ducts, equine, 831, 831*f*
Ampullary crests, feline, 143
Anal sac, canine, 89*f*
Anal sac gland, canine, 89*f*
Anal sphincter
avian, 1323
canine, 217
equine, 761
feline, 254
ruminant, 1080
Anconeal process of ulna, canine, 408
Angularis oculi muscle, canine, 85*f*
Angular process, canine, 92*f*
Anisodactyl, avian, 1390
Annular ligament, palmar/plantar, equine, 718
Annulus fibrosus, equine, 683
Anorectal lymph node, equine, 761–762, 778
Ansa subclavia, ruminant, 998*f*
Ansa trachealis, avian, 1282
Antebrachiocarpal joint
canine, 421*f*
equine, 894
ruminant, 1184–1185
Antebrachium
canine, 84*f*
equine, 554*f*
feline, 94*f*
ruminant, 1149–1157, 1153*f*, 1158*f*, 1159–1160

Anticlinal vertebra
canine, 85*f*
equine, 555*f*
feline, 95*f*
ruminant, 993*f*
Antimesenteric band, equine, 741, 759, 759*f*
Antimesenteric ileal branch, ruminant, 1080
Antitragus, canine, 136*f*
Antrum of stomach, equine, 726, 728*b*
Anus
canine, 321–323
equine, 760–761
feline, 321–323
ruminant, 997*f*
Aorta, ascending, descending, and thoracic
avian, 1253*f*, 1323–1324
canine, 226, 254–256, 271, 295
equine, 761–762, 833
feline, 99*f*
ruminant, 997*f*, 1047*f*, 1052, 1053*f*
Aortic arches, canine, 90*f*
Aortic valve
canine, 253
equine, 711
feline, 100*f*
ruminant, 1052
Apex
of cecum
equine, 562*f*
ruminant, 1079
of bladder, canine, 344
of heart
canine, 85*f*
equine, 711
ruminant, 1046
of teeth
canine, 124*f*
equine, 609*b*
Apical delta, canine, 124
Apocrine sweat gland
canine, 547*b*
equine, 980*f*, 983
Aponeurosis of external abdominal oblique muscle
canine, 88*f*
equine, 557*f*
feline, 102*f*
ruminant, 996*f*
Aponeurosis of internal abdominal oblique muscle
canine, 88*f*
equine, 557*f*
feline, 102*f*
ruminant, 996*f*
Appendix vesiculosa, equine, 798
Apteria, avian, 1404
Arcuate vessels, equine, 760, 760*f*
Arcus terminalis, ruminant, 1167
Arrector pili muscle, canine, 547
Arteriovenous rete, avian, 1324
Artery of the bulb, caprine, 1132–1133
Aryepiglottic fold
canine, 154
equine, 641, 641*b*
Arytenoid cartilage
avian, 1281
equine, 636–638, 637–639*f*, 640–642, 641*f*
ruminant, 1008*f*
Ascending colon
canine, 91*f*
feline, 101*f*
ruminant, 998*f*, 1079–1080, 1079*f*
Ascending duodenum
canine, 91*f*
ruminant, 1071–1073, 1079–1080
Ascending reticular activating system, canine, 191–192
Atlas
avian, 1258*f*
canine, 92*f*
equine, 563–564*f*
feline, 102*f*
ruminant, 1000*f*
Atrichial sweat gland, canine, 544, 547
Atrioventricular node
canine, 89*f*
equine, 710
ruminant, 1053
Atrioventricular valve
avian, 1323
canine, 89*f*
ruminant, 1052, 1053*f*
Atrium
left
avian, 1320
canine, 89*f*, 270
equine, 710
feline, 97*f*
ruminant, 1052, 1053*f*
right
avian, 1320
canine, 89*f*, 268–269
equine, 712
feline, 97*f*
ruminant, 1052, 1053*f*
Auditory ossicles
equine, 658–659, 660*f*
feline, 109
Auditory tube
avian, 1303
equine, 658
feline, 107*f*, 108, 110, 142
Auricle, left and right
canine, 84–85*f*
equine, 655, 656*f*, 710, 712
feline, 94*f*, 135–137
ruminant, 999*f*
Auricular artery, feline, 110–111
Auriculopalpebral nerve
canine, 88*f*
equine, 569*t*
feline, 141
Auriculotemporal nerve
equine, 556*f*
feline, 96*f*
ruminant, 994*f*
Autonomic nervous system
canine, 175
caprine, 1102
equine, 870
ruminant, 1102
Avian plumage, 1254*f*
Axial artery
avian, 1411
ruminant, 1167
Axial ligament
equine, 916
ruminant, 1166, 1175
Axial nerve
equine, 1165
feline, 172
ruminant, 993*f*, 1168
Axial palmar proper digital artery, ruminant, 1189
Axial proper plantar digital vein, ruminant, 1189
Axillary lymph node
canine, 85*f*
feline, 95*f*
Axillary vessels
avian, 1375
canine, 271, 399
equine, 880–881, 890
ruminant, 1158–1160, 1190
Axis of spine, skeleton
avian, 1258*f*
canine, 92*f*
equine, 563–564*f*
feline, 102*f*
ruminant, 1000*f*
Azygous vein
canine, 90*f*, 241, 274–275
equine, 561*f*
feline, 100*f*
ruminant, 997*f*

B

Barbs, avian, 1406
Barbules, avian, 1406
Bartholin's gland, equine, 776
Basal border of lung
canine, 84–85*f*, 89–90*f*
equine, 554–555*f*, 560–561*f*
feline, 94–95*f*, 99*f*
ruminant, 992–993*f*, 997*f*

Basihyoid bone
canine, 154–156, 176
equine, 614*f*
ruminant, 1008*f*
Basilic vein, avian, 1253*f*, 1255*f*, 1375
Basioccipital bone, equine, 615, 646–647, 648*f*
Basisphenoid bone, equine, 646–647, 648*f*, 650
Basophils, avian, 1362
B-cells, avian, 1359*b*
Beak, avian, 1297*b*, 1346
Biceps brachii muscle
avian, 1255*f*
canine, 399
equine, 559*f*
feline, 98*f*
ruminant, 996*f*, 1154–1155
Biceps femoris muscle
avian, 1255*f*
canine, 536
equine, 558*f*
feline, 478
Bicipital groove, ruminant, 1149–1151
Bijugular trunk, equine, 560*f*
Bill tip organ, avian, 1297
Biventer cervicis muscle, avian, 1255*f*
Blood vessels
artery
avian
aorta, 1253*f*
caudal femoral, 1253*f*
superficial ulnar, 1253*f*
canine
carotid, 85*f*
caudal femoral, 88*f*
collateral ulnar, 88*f*
digital, lateral and medial, 85*f*
dorsalis pedis, 85*f*
femoral, 88*f*
internal thoracic
median, 85*f*
palmar, lateral and medial, 85*f*
plantar, 85*f*
popliteal, 85*f*
saphenous, 85*f*
equine
aorta, 560*f*
carotid, 555*f*, 560*f*
coronary, 561*f*
cranial mesenteric, 561*f*
external pudendal, 561*f*
femoral, 555*f*
internal thoracic, 560*f*
median, 555*f*
palmar, 555*f*
palmar digital, 555*f*
plantar, 555*f*
plantar digital, 555*f*
popliteal, 555*f*
pudendal, external, 561*f*
saphenous, 555*f*
supreme intercostal, 560*f*
transverse facial, 559*f*
umbilical, 561*f*
feline
aorta, 95*f*
carotid, 95*f*
digital, lateral and medial, 95*f*
dorsal pedal, 95*f*
femoral, 95*f*
internal thoracic, 95*f*
maxillary, 95*f*
median, 95*f*
palmar, lateral and medial, 95*f*
plantar, 95*f*
popliteal, 95*f*
saphenous, 95*f*
ruminant
abaxial palmar/plantar, 993*f*
abaxial palmar/plantar digital III and IV, 993*f*
aorta, 997*f*
axial palmar, 993*f*
axial palmar digital, 993*f*
carotid, 993*f*
facial, 996*f*
femoral, 993*f*
mammary, caudal, 997*f*
median, 993*f*
palmar, 993*f*
palmar digital, lateral and medial, 993*f*
plantar, 993*f*
plantar common digital, 993*f*
popliteal, 993*f*
pudendal, 998*f*
saphenous, 993*f*
subclavian, 993*f*
thoracic, 997–998*f*
vein
avian
basilic, 1253*f*, 1255*f*
external iliacus, 1253*f*
jugular, 1253*f*, 1255*f*
left subclavian, 1324
ulnar, 1253*f*, 1255*f*
canine, 85*f*
cephalic, 85*f*
common digital, dorsal, 85*f*
digital, 85*f*
external jugular, 85*f*
facial, 87*f*
femoral, 87*f*
labial, inferior and superior, 88*f*
lateral saphenous, 87*f*
left subclavian, 87*f*
lingual, 87*f*
linguofacial, 87*f*
maxillary, 87*f*
median, 85*f*
nasal, dorsal, 87*f*
palmar, 85*f*
plantar, 85*f*
right azygos, 90*f*
saphenous, 85*f*
equine
accessory cephalic, 555*f*
azygos, 561*f*
bijugular trunk, 560*f*
brachiocephalic trunk, 560*f*
buccal, 559*f*
cephalic, 555*f*
caudal vena cava, 559*f*
costocervical, 559*f*
cranial vena cava, 561*f*
deep cervical, 561*f*
deep facial, 559*f*
digital, 555*f*
external jugular, 557*f*
external pudendal, 561*f*
femoral, 555*f*
lateral palmar/plantar, 559*f*
medial palmar/plantar, 559*f*
maxillary, 559*f*
median, 555*f*
medial saphenous, 559*f*
palmar, 555*f*
palmar digital, 559*f*
plantar, 555*f*
plantar digital, 559*f*
pudendal, external, 561*f*
subclavian, 561*f*
superficial thoracic, 555*f*
transverse facial, 555*f*
vertebral, 561*f*
feline, 97*f*, 100*f*
digital, 95*f*
external jugular, 95*f*
femoral, 95*f*
lateral saphenous, caudal and cranial branch of, 95*f*
left subclavian, 99*f*
linguofacial, 95*f*
maxillary, 95*f*
median, 95*f*
palmar, 95*f*
plantar, 95*f*
saphenous, 95*f*
ruminant
abaxial palmar/plantar digital, 993*f*
axial palmar/plantar digital, 993*f*
cephalic, 993*f*
common digital, dorsal, 993*f*
deep facial, 993*f*
external jugular, 993*f*, 996*f*
facial, 993*f*
femoral, 993*f*
intercostal, external, 993*f*

Blood vessels *(Continued)*
lateral saphenous, 993*f*
left azygos, 993*f*, 998*f*
linguofacial, 996*f*
maxillary, 993*f*
medial saphenous, 993*f*
median cubital, 993*f*
palmar, 993*f*
palmar digital, 993*f*
plantar, 993*f*
plantar digital, 993*f*
pudendal, 998*f*
subclavian, 998*f*
saphenous, 996*f*
subcutaneous abdominal (milk, mammary), 993*f*
tail, 993*f*
Bone
avian
alula, 1258*f*
atlas, 1258*f*
axis, 1258*f*
carpal - radial, ulnar, 1258*f*
clavicle, 1258*f*
coracoid, 1258*f*
digit - major and minor, 1258*f*
femur, 1258*f*
fibula, 1258*f*
frontal, 1258*f*
humerus, 1258*f*
hypotarsal crest, medial, 1258*f*
ilium, 1258*f*
ischium, 1258*f*
long digital extensor, major, 1255*f*
jugal arch, 1258*f*
keel, 1258*f*
lacrimal, 1258*f*
mandible, 1258*f*
manubrium sterni, 1258*f*
maxilla, 1258*f*
medial hypotarsal crest, 1258*f*
metacarpal - major and minor, 1258*f*
metatarsal - 1st, 1258*f*
nasal, 1258*f*
notarium (fused thoracic vertebrae), 1258*f*
occipital, 1258*f*
palatine, 1258*f*
parietal, 1258*f*
patella, 1258*f*
phalanges, middle, 1258*f*
phalanx, distal, extensor process, 1258*f*
phalanx - 1st, 4th, 1258*f*
phalanx - distal, middle, proximal, 1258*f*
premaxilla, 1258*f*
pubis, 1258*f*
pygostyle, 1258*f*
radial carpal, 1258*f*
radius, 1258*f*
scapula, 1258*f*
sternum, 1258*f*
synsacrum, 1258*f*
tarsometatarsus - 2nd, 3rd, and 4th, 1258*f*
temporal, 1258*f*
tibial tuberosity, 1258*f*
tibiotarsus, 1258*f*
ulna, 1258*f*
uncinate process, 1258*f*
vertebrae - cervical, thoracic, caudal, 1258*f*
vertebrae - free and fused, 1258*f*
xiphoid process of sternum, 1258*f*
zygomatic (jugal arch), 1258*f*
canine
acromion, 92*f*
atlas, 92*f*
axis, 92*f*
carpal - accessory, intermedioradial, ulnar, 2nd, 92*f*
condylar process of mandible, 92*f*
coronoid process of mandible, 92*f*
costochondral junction, 92*f*
fabellae (sesamoid), 92*f*
femur, 92*f*
fibula, 92*f*
fibula, head, 92*f*
greater trochanter, femur, 92*f*
humerus, 92*f*
ilium, shaft of, 92*f*
lateral epicondyle, 92*f*
malleolus, lateral, 92*f*
mandible, 92*f*
manus, 92*f*
maxilla, 92*f*
metacarpal - 1st, 5th, 92*f*
metatarsal - 1st, 2nd, 5th, 92*f*
parietal, 92*f*
patella, 92*f*
pes, 92*f*
phalanx - distal, middle, proximal, 92*f*
radius, 92*f*
sacrum, 92*f*
scapula, 92*f*
scapula, spine of, 92*f*
sesamoid, 92*f*
sesamoids (fabellae), 92*f*
spine of scapula, 92*f*
sternum, 92*f*
supraglenoid tubercle, 92*f*
symphysis pelvis, 90*f*
talus, 92*f*
tarsal - central, 1st, 2nd, 4th, 92*f*
tibia, 92*f*
tibial tuberosity, 92*f*
tuber calcanei, 92*f*
tuber ischiadicum, 92*f*
tuber olecrani, 92*f*
tympanic bulla, 92*f*
ulna, 92*f*
ungual crest, 92*f*
ungual process, 92*f*
vertebra - cervical, thoracic, lumbar, sacral, caudal, 92*f*
zygomatic, 92*f*
zygomatic arch, 92*f*
zygomatic process of temporal, 92*f*
equine
atlas, 563–564*f*
axis, 563–564*f*
carpal - accessory, intermediate, radial, ulnar, 2nd, 3rd, 4th, 563–564*f*
condylar process of mandible, 563–564*f*
coronoid process of mandible, 563–564*f*
costal arch, 563–564*f*
costochondral junction, 563–564*f*
deltoid tuberosity of humerus, 563–564*f*
external sagittal crest, 564*f*
fabellae (sesamoid), 563–564*f*
facial crest, 563*f*
femur, 563–564*f*
fibula, 563–564*f*
frontal, 563–564*f*
humerus, 563–564*f*
ilium, 563–564*f*
ilium, shaft of, 563*f*
incisive, 563–564*f*
infraspinous fossa, 563–564*f*
ischium, 563*f*
lateral epicondyle of femur, 564*f*
lateral malleolus, 564*f*
lateral styloid process, 564*f*
lateral supracondylar crest, 563*f*
lesser trochanter, 564*f*
malleolus of tibia, 563–564*f*
mandible, 563–564*f*
manubrium sterni, 563–564*f*
maxilla, 563–564*f*
metacarpal - 2nd, 3rd, 4th, 563–564*f*
metatarsal - 2nd, 3rd, 4th, 563–564*f*
nasal, 563–564*f*
navicular (distal sesamoid), 564*f*
parietal, 563–564*f*
patella, 563–564*f*
pes, 563–564*f*
phalanx - distal, middle, proximal, 563–564*f*
radial carpal, 563*f*
radius, 563–564*f*
rib, 563–564*f*
sacrum, 563–564*f*

scapula, 563–564*f*
scapula, spine of, 563*f*
sesamoid, distal (navicular), 563–564*f*
sesamoid, proximal, 563–564*f*
sternum, 563–564*f*
styloid process of ulna, 563–564*f*
supracondylar crest, 563–564*f*
supraglenoid tubercle, 563–564*f*
supraspinous fossa, 563–564*f*
symphysis pelvis, 561*f*
talus, 563–564*f*
tarsal - central, 1st, 2nd, 3rd, 4th, 563–564*f*
tibia, 563–564*f*
tibial tuberosity, 563–564*f*
trochanter of femur, greater, 3rd, 563–564*f*
tuber calcanei, 563–564*f*
tuber coxae, 563–564*f*
tuber ischiadicum, 563–564*f*
tuber olecrani, 563–564*f*
tuber sacrale, 563–564*f*
tuber of scapular spine, 563*f*
ulna, 563–564*f*
vertebra - cervical, thoracic, lumbar, sacral, caudal, 563–564*f*
withers, 563–564*f*
xiphoid, 563–564*f*
zygomatic, 564*f*
zygomatic arch, 563*f*
zygomatic process of frontal bone, 564*f*
feline
acromion, 102*f*
atlas, 102*f*
axis, 102*f*
carpal - accessory, distal carpals, intermedioradial, ulnar, 1st, 2nd, 102*f*
clavicle, 102*f*
condylar process, 102*f*
coronoid process, 102*f*
costochondral junction, 102*f*
digit(s), 95*f*
fabellae (sesamoid), 102*f*
femur, 102*f*
fibula, 102*f*
fibula, head, 95*f*
frontal, 102*f*
greater trochanter, 102*f*
humerus, 102*f*
humerus, epicondyloid, lateral, 102*f*
humerus, greater tubercle, 102*f*
ilium, 102*f*
ilium, shaft, 102*f*
lateral supracondylar tuberosity, 102*f*
lateral epicondyle, 102*f*
malleolus, lateral, 102*f*
mandible, 102*f*
manubrium sterni, 102*f*
manus, 102*f*
maxilla, 102*f*
metacarpal - 1st, 5th, 102*f*
metatarsal - 1st, 2nd, 5th, 102*f*
nasal, 102*f*
occipital condyle, 102*f*
parietal, 102*f*
patella, 102*f*
pes, 102*f*
phalanx - distal, middle, proximal, 102*f*
radius, 102*f*
sacrum, 99*f*
scapula, 102*f*
sesamoid - popliteal, 102*f*
sesamoids (fabellae), 102*f*
sternum, 102*f*
supraglenoid tubercle, 102*f*
symphysis pelvis, 102*f*
talus, 102*f*
tarsal - central, 1st, 4th, 102*f*
tibia, 102*f*
tibial tuberosity, 102*f*
tuber calcanei, 102*f*
tuber coxae, 95*f*
tuber ischiadicum, 102*f*
tuber olecrani, 102*f*
tuber sacrale, 95*f*
tympanic bulla, 102*f*
ulna, 102*f*
ungual crest, 102*f*
ungual process, 102*f*
vertebra - cervical, thoracic, lumbar, sacral, caudal, 102*f*
zygomatic, arch, process, 102*f*
ruminant
atlas, 1000*f*
axis, 1000*f*
calcaneus, 1000*f*
carpal - accessory, intermediate, radial, ulna, 2nd, 3rd, and 4th, 1000*f*
carpal, fused carpal and tarsal, 1000*f*
condylar process of mandible, 1000*f*
cornual process, 1000*f*
coronoid process of mandible, 1000*f*
costochondral junction, 1000*f*
deltoid tuberosity of humerus, 1000*f*
epicondyle, lateral of humerus, 1000*f*
facial tuberosity, 1000*f*
femur, 1000*f*
fibula, 1000*f*
frontal, 1000*f*
humerus, 1000*f*
humerus, major tubercle of, 1000*f*
ilium, 1000*f*
ilium, shaft of, 1000*f*
incisive, 1000*f*
ischial spine, 1000*f*
lateral malleolus, 1000*f*
malleolus of tibia - lateral, medial, 1000*f*
mandible, 1000*f*
manubrium sterni, 1000*f*
maxilla, 1000*f*
medial cuneiform, 1000*f*
metacarpal - 3rd, 4th, 5th, 1000*f*
metatarsal - 3rd, 4th, 1000*f*
nasal, 1000*f*
occipital condyle, 1000*f*
paracondylar process of skull, 1000*f*
patella, 1000*f*
phalanx - distal, middle, proximal, 1000*f*
radial carpal, 1000*f*
radius, 1000*f*
sacrum, 1000*f*
scapula, 1000*f*
scapular spine, 1000*f*
sesamoid, distal, 1000*f*
styloid process of ulna, 1000*f*
sustentaculum tali, 1000*f*
symphysis pelvis, 998*f*
talus, 1000*f*
tarsal - centroquartel, 2nd, 3rd, 4th, 1000*f*
tibia, 1000*f*
tibial tuberosity, 1000*f*
tuber calcanei, 1000*f*
tuber coxae, 1000*f*
tuber ischiadicum, 1000*f*
tuber olecrani (point of the elbow), 1000*f*
tuber sacrale, 1000*f*
tubercle, major, 1000*f*
ulna, 1000*f*
ulna, styloid process of
vertebra - cervical, thoracic, lumbar, sacral, caudal, 1000*f*
xiphoid process of sternum, 1000*f*
zygomatic arch, 1000*f*
Bony labyrinth, feline, 108, 142–143
Bony septum of bulla
avian, 1280–1281
canine, 110
equine, 594*f*
feline, 110
Brachial artery
canine, 88*f*
equine, 557*f*
ruminant, 1159
Brachialis muscle
canine, 88*f*
equine, 557*f*
ruminant, 1155

Brachial plexus
canine, 84*f*
equine, 876, 877*f*
ruminant, 1157, 1157*b*
Brachial vein
canine, 84*f*
equine, 889, 897
ruminant, 1159
Brachiocephalic trunk
avian, 1323, 1375
canine, 160, 254, 271
equine, 557*f*
feline, 99*f*
ruminant, 997*f*
Brachiocephalicus muscle
canine, 87*f*, 177
equine, 557*f*
feline, 98*f*
ruminant, 1152
Brachiocephalic vein
canine, 87*f*
equine, 559*f*
feline, 168
ruminant, 997*f*
Brachium
canine, 84*f*
equine, 554*f*
feline, 94*f*
ruminant, 996*f*, 1149–1157, 1153*f*, 1158*f*, 1159–1160
Brainstem
canine, 174
equine, 650–653, 653*f*
ruminant, 996*f*
Bristles, avian, 1409
Broad ligament
canine, 349–350
equine, 787–788, 795, 804–805
ruminant, 1119–1120, 1120*f*
Bronchi
avian, 1283, 1339
canine, 226
equine, 719, 719*f*
ruminant, 1086*b*
Bronchial artery
canine, 225, 268*f*
equine, 560*f*, 720–721
ruminant, 997*f*
Bronchial lymphocenters/lymph node
canine, 275–276
equine, 560*f*
ruminant, 1089
Bronchioles
canine, 222
equine, 719, 719*f*
ruminant, 998*f*
Bronchoesophageal artery, equine, 678
Brood patch, avian, 1333
Buccal branch, facial nerve
canine, 87*f*
equine, 555*f*
Buccal nerve
canine, 87*f*
equine, 581
ruminant, dorsal and ventral, 993*f*, 996*f*
Buccal salivary gland
equine, 582
feline, 98*f*
ruminant, 993*f*
Buccal vein, equine, 555*f*
Buccinator muscle
equine, 578
ruminant, 996*f*
Bulbospongiosus muscle
canine, 90*f*
equine, 822, 830, 830*b*, 858
Bulbourethral gland
canine, 90*f*
equine, 820, 825, 826*f*, 829*f*, 831*f*, 833
ruminant, 998*f*, 1102
Bulbourethralis muscle, equine, 833
Bulb segment, ruminant, 1178
Bulbus glandis, canine, 90*f*
Bursa of Fabricius, avian, 1359–1360
Buttocks, equine, 554–555*f*

C

Calamus, avian, 1406
Calcaneal tendon
canine, 535–536
equine, 960
Calices of kidneys, major and minor, ruminant, 1100
Cancellate scutes, avian, 1393
Canine teeth
canine, 5*f*
equine, 603, 607
feline, 5*f*
Caninus muscle
equine, 554*f*
ruminant, 995*f*
Capitis fovea, avian, 1388
Cardia, of stomach, equine, 726
Cardiac diverticula, avian, 1338
Cardiac impression, ruminant, 1087
Cardiac notch
canine, 223
equine, 563*f*
feline, 248
ruminant, 993*f*
Cardiac plexus, ruminant, 1053
Cardiac sphincter, equine, 677, 727
Cardiac vein, avian, 1323
Carina of trachea
avian, 1382
canine, 222
equine, 50*f*
ruminant, 993*f*
Carnassial teeth, canine, 116
Carotid artery
avian, 1274–1275, 1297, 1323
canine, 85*f*, 125
equine, 555*f*, 560*f*, 611–613, 612*f*, 618, 644, 678
feline, 95*f*, 110
ruminant, 993*f*
Carotid sheath
canine, 175
equine, 676
Carpal bone(s)
avian, 1373
canine, 84*f*
equine, 894, 894*f*
feline, 94*f*
ruminant, 996*f*, 1183–1185, 1183–1184*f*, 1184*b*
Carpal bone - accessory, distal carpals, intermedioradial, ulnar, 1st, 2nd, feline, 102*f*
Carpal bone - accessory, intermediate, radial, ulna, 2nd, 3rd, and 4th
equine, 563–564*f*
ruminant, 1000*f*
Carpal bone - accessory, intermedioradial, ulnar, 2nd, canine, 92*f*
Carpal bone - radial, ulnar, avian, 1258*f*
Carpal canal, equine, 897
Carpometacarpal joint, equine, 564*f*
Caudal accessory leaflets, equine, 710
Caudal auricular artery, equine, 658, 660
Caudal auricular vein, equine, 658
Caudal cartilage, avian, 1283
Caudal cerebellar artery, equine, 662
Caudal circumflex humeral artery, canine, 399
Caudal conchae, avian, 1280–1281
Caudal cutaneous antebrachial nerve
canine, 88*f*
feline, 95*f*
Caudal cutaneous sural nerve, equine, 950–951
Caudal epigastric lymph node, feline, 95*f*
Caudal deep cervical lymph node
canine, 85*f*
equine, 881, 890
ruminant, 1160*b*
Caudal femoral artery, equine, 938
Caudal femoral vein, equine, 939
Caudal flexure of duodenum
canine, 312–313
equine, 734
Caudal flocculonodular lobe, equine, 689
Caudal gluteal nerve
equine, 941
ruminant, 1037

Caudal grooves of cranium, ruminant, 1058
Caudal humeral head, canine, 92*f*
Caudal interosseous artery
canine, 87*f*
feline, 100*f*
Caudal laryngeal nerve, equine, 644, 644*f*
Caudal mammary vessels
equine, 776
ruminant, 993*f*
Caudal mesenteric artery
equine, 760
ruminant, 1080
Caudal oblique muscle of head, equine, 563*f*
Caudal pancreaticoduodenal artery, ruminant, 1075
Caudal rectal artery
equine, 761
ruminant, 1080, 1111
Caudal rectal nerve, equine, 761, 776, 791, 823
Caudal renal artery, avian, 1352
Caudal renal vein, avian, 1353
Caudal superficial epigastric artery, equine, 822
Caudal superficial epigastric vein, equine, 986
Caudal thoracic air sac, avian, 1338
Caudal vena cava, 53*f*
avian, 1320, 1325
canine, 226, 295
equine, 740, 765
feline, 97*f*
ruminant, 1047*f*, 1052, 1053*f*, 1086–1087, 1111
Caudate lobe of liver
canine, 282*b*
equine, 740*b*, 741–742
feline, 100*f*
ruminant, 998*f*
Caudodorsal blind sac of rumen, ruminant, 1058
Caudoventral blind sac of rumen, ruminant, 1058
Caval foramen, canine, 241
Ceca, avian, 1256*f*
Cecal lymph node, equine, 1076
Cecocolic band/fold, equine, 754
Cecocolic orifice
equine, 561*f*
ruminant, 1079
Cecum
avian, 1347
canine, 320
equine, 746–747, 746*f*
feline, 746*b*
ruminant, 1077–1079, 1078–1079*f*
Cecum, lateral band of, equine, 561*f*
Celiac artery, ruminant, 1060, 1066, 1075
Celiac ganglia, ruminant, 1060
Celiac lymph node
equine, 731
ruminant, 1080
Celiac trunk, equine, 736
Celiac vein, avian, 1325
Cementum
canine, 124
equine, 605, 605*f*
Central and 4th tarsal bone, canine, 509–515
Central flexure, ruminant, 1079–1080
Central tarsal bone
canine, 87*f*, 92*f*
equine, 564*f*, 960–961
feline, 94*f*
Central vestibular system
canine, 186–187
equine, 646
feline, 94*f*
Centrifugal gyri of spiral colon, ruminant, 1079–1080
Centripetal gyri of spiral colon, ruminant, 1079–1080
Centrum, avian, 1382, 1382*b*
Cephalic vein
canine, 85*f*, 413*f*
equine, 555*f*, 889
feline, 167*f*
ruminant, 1189
Ceratohyoid bone, canine, 176
Cere, avian, 1280, 1291, 1349*b*, 1350*f*
Cerebellopontine medullary angle, canine, 186
Cerebellum
canine, 449
equine, 652, 689
feline, 449
ruminant, 1189
Cerebral carotid artery, avian, 1324
Cervical air sac, avian, 1338
Cervical carotid canal, avian, 1323
Cervical intumescence
canine, 211, 450
equine, 863*f*
Cervical lymph node
canine, 158, 162–163
equine, 897
feline, 95*f*
ruminant, 996*f*
Cervical spine
avian, 1381–1382
canine, 172–175, 172*f*
equine, 866
feline, 171
ruminant, 1000*f*, 1033
Cervical sympathetic trunk, equine, 870–871
Cervicoauricularis superficialis muscle, equine, 656–658, 658*b*
Cervicocephalic diverticula, avian, 1281
Cervicoclavicular air sac, avian, 1338
Cervicomedullary cistern/junction, canine, 174
Cervicothoracic ganglion
canine, 176
equine, 870–871
Cervicothoracic lymph node, avian, 1361
Cervix
canine, 318
equine, 805–806
feline, 349
ruminant, 1119
Chalazae, avian, 1313
Cheek teeth, equine, 605, 606*f*, 607
Chestnut, equine, 554*f*
Chin
avian, 1254*f*
equine, 554*f*
ruminant, 992*f*
Choanal slit, avian, 1281*b*
Choanae
avian, 1280–1281
equine, 589
Chordae tendineae
avian, 1323
canine, 269
equine, 708*f*
Chorioallantois
canine, 367
feline, 362
Chorion
canine, 367
equine, 804*b*
feline, 362
Choroidal fissure, avian, 1313
Choroidal vessels, avian, 1312
Ciliary muscle
avian, 1312
canine, 131
equine, 570
Circumflex femoral artery, ruminant, 1197
Circumflex marginal artery, equine, 930
Cisterna chyli
canine, 53*f*, 296
equine, 731, 736, 748, 755, 761–762
ruminant, 1100
Clavicles
avian, 1258*f*, 1371
feline, 95*f*, 102*f*
Clavicular air sac, avian, 1335*b*, 1337, 1372
Clavicular intersection
canine, 87*f*
feline, 98*f*
Clavicular tendon, canine, 177
Claw
canine, 434
feline, 434
ruminant, 1173–1174, 1185

Claw capsule, ruminant, 1165
Cleidobrachialis muscle
canine, 87–88*f*
feline, 97–98*f*
Cleidocephalicus muscle, ruminant, 995–996*f*
Cleidomastoideus muscle
equine, 557–559*f*
ruminant, 996*f*
Clitoral sinus/fossa
canine, 89*f*
equine, 776, 782*b*
feline, 99*f*
Clitoris
canine, 364
equine, 775–776, 782–783, 782*f*
feline, 97*f*
ruminant, 1123
Cloaca, avian, 1348, 1353
Cnemial crest of tibiotarsus, avian
Coccygeus muscle
avian, 1255*f*
ruminant, 995–996*f*
Cochlea
avian, 1303
equine, 652, 960
feline, 144–145
Cochlear duct, equine, 661–662
Cochlear ganglia, feline, 143
Cochlear nerve
equine, 661–662
feline, 144
Cochlear window, feline, 109
Coelomic air sac, avian, 1335*b*
Coffin joint, equine, 921–922, 922*f*. *See also* Distal interphalangeal joint
Colic artery
equine, 754–755
ruminant, 1075, 1080
Colic branches
equine, 754–755
ruminant, 1080
Colic lymph node, ruminant, 1076, 1080
Collateral cartilage, equine, 919*f*
Collateral ligament, medial and lateral, equine, 948
Collateral sesamoidean ligament, ruminant, 1166*b*
Collateral ulnar artery, equine, 974
Colliculus seminalis, equine, 820, 825, 826*f*, 829–831*f*, 830–831
Collum glandis, equine, 820
Colon
avian, 1347
canine, 31*f*, 290
equine, 560*f*
feline, 318–321
ruminant, 1077–1080, 1078*f*
Colon, ascending
canine, 31*f*, 290
feline, 318–321, 561*f*
ruminant, 998*f*
Colon, descending
equine, 560*f*
canine, 31*f*, 290
feline, 318–321, 561*f*
ruminant, 998*f*
Colon, left and right ventral lateral free band of, 560*f*
Colon, left dorsal and ventral, equine, 560*f*
Colon, diaphragmatic/sternal flexure of, equine, 560*f*
Colon, small descending, equine, 560*f*
Colon, distal ansa/loop of spiral, ruminant
Columellar muscle, avian, 1303
Comb, avian, 1255*f*
Commissural fibers, feline, 202
Commissure
of the mouth, 775–776, 1109
of the vulva, 775–776, 1109
Common bile duct
canine, 289
equine, 734*f*, 735–736
Common carotid artery
avian, 1274–1275, 1297, 1323
canine, 125
equine, 611–613, 612*f*, 618, 644, 678
feline, 110
Common digital artery, ruminant, 1189
Common digital extensor muscle
canine, 410*f*
equine, 900, 916
feline, 410*f*
ruminant, 1189
Common iliac vein, ruminant, 1086–1087
Common interosseous artery, equine, 889
Common meatus, ruminant, 1006
Common fibular (peroneal) nerve
equine, 559*f*, 941
feline, 98*f*
ruminant, 996*f*
Common pulmonary vein, avian, 1320
Complexus muscle, avian, 1255*f*
Compound follicle, canine, 546
Concha auriculae, equine, 656
Conchae
avian, 1280–1281
equine, 586
ruminant, 1006, 1006*f*, 1008*f*
Conchal sinus, equine, 587*f*, 590*f*, 593–594, 594*f*, 596*f*
Conchofrontal sinus. *See* Frontal sinus
Conchomaxillary aperture, equine, 594
Condylar fossae, canine, 174
Condylar process of mandible
canine, 92*f*
equine, 563–564*f*
feline, 102*f*
ruminant, 1000*f*
Cone beam computed tomography, 77–78, 78*t*
Conjunctiva, canine, 131
Conjunctival ring, equine, 572
Conjunctival sac, avian, 1270, 1314
Constrictor vestibuli muscle, equine, 783–784
Constrictor vulvae muscle
equine, 776, 782*f*, 783–784, 783*b*
ruminant, 1109
Conus medullaris, ruminant, 1034, 1037
Conus terminalis, ruminant, 1037
Coprodeum, avian, 1348, 1353
Copulatory organ, avian, 1354
Coracobrachialis muscle, ruminant, 1155
Coracoid, avian, 1258*f*, 1370–1371
Corium, ruminant, 1174, 1174*f*
Corneoconjunctival region, equine, 572–574
Corneoscleral junction, equine, 567*b*
Cornual artery, ruminant, 1028
Cornual dermis, ruminant, 1025–1026
Cornual epidermis, ruminant, 1025–1026
Cornual nerve, ruminant, 1027
Cornual process, ruminant, 1000*f*
Cornual vein, ruminant, 1028
Corona glandis, equine, 820, 822*f*
Corona radiata, feline, 202
Coronary artery
avian, 1323
canine, 254
equine, 712–713
ruminant, 1052–1053
Coronary band, equine. *See* Coronet
Coronary cushions, ruminant, 1174–1175
Coronary dermis, equine, 927–928
Coronary epidermis, equine, 928–929
Coronary grooves, ruminant, 1058
Coronary ligament
equine, 765, 767*f*
ruminant, 1092
Coronary segment, ruminant, 1175
Coronary sinus
canine, 254
equine, 712–713
feline, 94*f*
ruminant, 1052–1053
Coronary vein, ruminant, 1052–1053
Coronet, equine, 554*f*
Coronoid process of ulna
canine, 92*f*
equine, 563–564*f*
feline, 102*f*

Corpus cavernosum of penis
canine, 90*f*, 365
caprine, 1132
equine, 563*f*, 822–823, 830
ruminant, 1137–1140, 1137*b*, 1142–1143*f*
Corpus luteum
canine, 362
equine, 801*b*
Corpus spongiosum of penis
canine, 90*f*
equine, 822–823, 830
Cortical ribbon, feline, 202
Corticospinal tract, feline, 206
Costal arch
canine, 90*f*
equine, 563–564*f*
feline, 95*f*
ruminant, 1000*f*
Costocervical vessel
canine, 90*f*, 160
equine, 561*f*
ruminant, 997*f*
Costochondral junction
canine, 85*f*, 92*f*
equine, 563–564*f*
feline, 102*f*
ruminant, 1000*f*
Covert feather, 1265*b*
Coxofemoral joint
avian, 1388
canine, 459–460
equine, 935, 937–938, 937*b*
ruminant, 1194
Cranial accessory leaflets, equine, 710
Cranial cartilage, avian, 1283
Cranial cervical ganglion
canine, 141–142
equine, 870–871
ruminant, 1080
Cranial gluteal nerve
equine, 941
ruminant, 1037
Cranial grooves, ruminant, 1058
Cranial laryngeal nerve, equine, 644, 644*f*
Cranial mammary artery, equine, 986
Cranial mediastinal lymph node
canine, 89*f*
equine, 678
Cranial mesenteric artery, 758
equine, 740
ruminant, 1075, 1080
Cranial mesenteric lymph node, ruminant, 1076, 1080
Cranial mesenteric vein
equine, 758, 760
ruminant, 1075
Cranial oblique muscle of head, equine, 557*f*
Cranial pancreaticoduodenal artery, equine, 736
Cranial pillar of rumen, ruminant, 997*f*
Cranial proventricular vein, avian, 1325
Cranial rectal artery, ruminant, 1080
Cranial renal artery, avian, 1352
Cranial renal vein, avian, 1353
Cranial sac of rumen, ruminant, 1047*f*
Cranial thoracic air sac, avian, 1338
Cranial thyroid artery
canine, 152, 158
equine, 644
feline, 168
Cranial tibial artery
avian, 1255*f*
equine, 967, 974–975
ruminant, 1189
Cranial tibial muscle
avian, 1255*f*
canine, 87–88*f*
equine, 557–559*f*
feline, 97–98*f*
ruminant, 995–996*f*
Cranial uterine artery, equine, 790–791
Cranial vena cava
avian, 1320, 1324
canine, 89–90*f*
equine, 561*f*
feline, 99*f*
ruminant, 997*f*, 1052, 1053*f*
Craniodorsal blind sac of rumen, ruminant, 1058
Craniofacial hinge, avian, 1295
Cremasteric artery, equine, 839
Cremaster muscle, equine, 841
Crest, equine, 563–564*f*
Cribriform plate
canine, 124, 124*b*
equine, 650–652
Cricoarytenoideus dorsalis muscle, equine, 637–638, 639*f*, 642*b*, 644*b*
Cricoarytenoideus lateralis muscle, equine, 636, 637*f*, 642*b*, 644*b*
Cricoid cartilage
avian, 1281
canine, 274–275
equine, 639*f*, 643, 676
ruminant, 1008*f*
Cricopharyngeus muscle
canine, 274–275
equine, 637–638, 639*f*, 643
Cricothyroid ligament
canine, 156
equine, 642
Cricotracheal ligament, equine, 643
Crop (ingluvies), avian, 1270, 1271*b*, 1274
Croup (sacral region)
canine, 84*f*
equine, 554*f*
feline, 94*f*
Crown, avian, 1254*f*
Crown of teeth, equine, 606, 606*f*
Cruciate ligament of stifle, cranial and caudal
canine, 94*f*
equine, 950
ruminant, 1209–1210
Cruciate sesamoidean ligament, equine, 908–909
Crura
of penis, ruminant, 1139–1140
of diaphragm, canine, 239
Crus of diaphram
canine, 503–504
equine, 676
feline, 301–302
Cubital lymph node, equine, 881, 890, 897, 974
Culmen, avian, 1292
Cunean bursa, equine, 966*b*
Cunean tendon, equine, 966
Cuneiform processes, equine, 641
Cuneocerebellar tract, feline, 204
Cupula of cecum, equine, 101*f*
Cusps of cardiac valve
avian, 1323
canine, 270
equine, 711
Cutaneous branch of the musculocutaneous nerve, ruminant, 1190
Cutaneous circulation, ruminant, 1234–1235
Cutaneous colli muscle
canine, 177
equine, 557–559*f*
ruminant, 995–996*f*
Cutaneous trunci muscle
equine, 557–559*f*
ruminant, 995–996*f*
Cuticle
canine, 546–547
ruminant, 1224*b*

D

Deep artery of the penis
caprine, 1132–1133
ruminant, 1141
Deep auricular artery, equine, 660
Deep branches of the fibular nerve, ruminant, 1190
Deep cervical lymph node, feline, 111
Deep digital flexor muscle
equine, 901, 919*f*, 922–923
ruminant, 1155, 1187–1189
Deep facial vein
equine, 597
ruminant, 993*f*
Deep fibular nerve, ruminant, 1168

Deep inguinal/iliofemoral lymph node, ruminant, 1122
Deep inguinal lymph node
  equine, 761–762, 939, 950, 975
  ruminant, 1143
Deep leaves of the greater omentum, ruminant, 1058
Deep pectoral muscle, ruminant, 1154
Deep perineal nerve
  equine, 823
  ruminant, 1111
Deep fibular (peroneal) nerve, equine, 973–974
Deep plantar arch, equine, 974–975
Deep plexus of skin, equine, 981
Deep popliteal lymph node, ruminant, 1167
Deep ulnar vein, avian, 1375
Deferent duct (Ductus deferens)
  avian, 1354
  canine, 381
  equine, 829*f*, 836
  ruminant, 1131–1132
Deltoideus muscle
  avian, 1255*f*
  canine, 87–88*f*
  equine, 557–559*f*
  ruminant, 1154
Deltoid tuberosity
  equine, 563–564*f*, 879–880
  ruminant, 1000*f*, 1151
Dentin
  canine, 123–124
  equine, 604–605, 605*f*
Depressor labii inferioris muscle, equine, 578
Dermal lamellae, equine, 1175*b*
Dermatomes
  canine, 86*f*, 544
  equine, 556*f*, 982
  feline, 96*f*
  ruminant, 994*f*
Dermis
  avian, 1403
    beak, 1294
  canine, 544–546
  equine, 980*f*, 981
  ruminant, 1024
Descending colon, ruminant, 1080
Descending palatine artery, ruminant, 1011
Desmosomes, canine, 542
Detrusor muscle, equine, 849
Dewclaw (vestigial digit)
  canine, 502*b*
  ruminant, 992*f*, 1174, 1185
Dewlap
  avian, 1254*f*
  ruminant, 1233
Diaphragm
  canine, 89*f*, 241–242
  equine, 560*f*
  feline, 247
  ruminant, 1047, 1047*f*
Diaphragmatic flexure of large colon, equine, 560–561*f*
Diencephalon, canine, 190
Digastricus muscle, equine, 617
Digital artery
  canine, 85*f*
  equine, 974–975
Digital cushion
  equine, 922–923
  ruminant, 1174–1175
Digital extensor muscle, ruminant, 1155
Digital extensor tendon, avian, 1395
Digital imaging and communications in medicine, 74
Digital nerve, ruminant, 1170*f*
Digital pads, avian, 1394
Digital radiography, 23, 24–25*t*, 29
Digit(s)
  avian, 1390
  canine, 84*f*, 434–435
  equine, 555*f*
  feline, 95*f*
  ruminant, 1173–1179, 1173*f*, 1176–1178*f*, 1185
Digit - major and minor, avian, 1258*f*
Distal annular ligament, ruminant, 1166
Distal colon, feline, 323*b*
Distal cruciate interdigital ligament, ruminant, 1173–1174
Distal curves of urethra, caprine, 1132
Distal cutaneous nerve, ruminant, 1111
Distal deep palmar branch, ruminant, 1189
Distal digital annular ligament, equine, 910–911
Distal intermediate ridge, equine, 960
Distal interphalangeal joint
  equine, 921–922, 922*f*
  ruminant, 1161–1167, 1163*b*, 1163–1164*f*, 1165–1166*b*
Distal intertarsal joint
  equine, 921–922, 922*f*
  ruminant, 1219, 1219*b*
Distal ligament of phalanx, ruminant, 1166
Distal loop of colon, ruminant, 1079–1080
Distal perforating artery, equine, 974–975
Distal phalanx, avian, 1393
Distal sesamoid bone
  equine, 921*f*, 923–924
  ruminant, 1166
Distal sesamoidean ligament, equine, 907–909, 909*f*
Distal sigmoid flexure of penis, ruminant, 1140
Diverticulum of air sac, avian, 1338
Dorsal apical ligament, ruminant, 1140
Dorsal proctodeal gland, avian, 1256*f*
Dorsal scalenus muscle, ruminant, 996*f*
Dorsocaudal sac of the rumen, ruminant, 1056–1057
Down feathers, avian, 1409
Dorsum (back)
  canine, 84*f*
  equine, 554*f*
  feline, 94*f*
  ruminant, 992*f*
Duct, thoracic
  canine, 53*f*, 89*f*
  equine, 560*f*
  feline, 99*f*
  ruminant, 997*f*
Duct, mandibular salivary, feline, 95*f*
Ductus arteriosis, canine, 256*b*
Ductus venosus
  canine, 262
  ruminant, 1087, 1087*b*
Duodenocolic fold
  canine, 91*f*
  equine, 734–735
  ruminant, 1080
Duodenocolic ligament, equine, 759, 759*b*
Duodenum, ascending, cranial sigmoid loop of
  canine, 91*f*
  ruminant, 1071–1073, 1079–1080
Duodenum, descending
  avian, 1347
  canine, 312–314
  equine, 735*f*, 736
  feline, 100*f*
  ruminant, 1071–1073
Duodenum, sigmoid flexure of
  canine, 290
  ruminant, 1092
Dura mater, ruminant, 1034

E

Ear
  avian, 1303
  canine, 84–85*f*
  equine
    external, 656–658, 657*f*, 659*f*
    inner, 661–662, 661*f*
    middle, 658–660, 660*f*
  feline, 108–111
  ruminant, 999*f*
Ear canal, avian, 1270
Ear covert, avian, 1301, 1409
Ear lobe, avian, 1303

Egg tooth, avian, 1293
Ejaculatory ducts, equine, 830, 831*f*, 832
Ejaculatory orifice, equine, 830, 830*f*
Elbow joint
 canine, 402
 equine, 555*f*
 feline, 94–95*f*, 402
 ruminant, 1151, 1156
Enamel, equine, 604–605, 605*f*
Endocardium
 canine, 249
 equine, 710
 ruminant, 1052*b*
Endometrial folds, equine, 804
Endometrium
 canine, 367
 equine, 805
Enteric nervous system, equine, 740
Eosinophils, avian, 1362
Epicardium, ruminant, 1046
Epiceras, ruminant, 1025–1026
Epicondyle, lateral of humerus, feline, 102*f*
Epidermal horn sheath, ruminant, 1028
Epidermic (horn) lamellae
 equine, 1175
 ruminant, 1174–1175
Epidermis
 avian, 1403
  of beak, 1294
 canine, 542–544, 543*f*
 equine, 980*f*, 981
 ruminant, 1174
Epididymis
 canine, 380–381, 380*b*
 equine, 841
 ruminant, 998*f*
Epigastric vessel, superficial, canine, 331*f*, 333*b*, 334*f*
Epiglottis/epiglottic cartilage
 canine, 117*f*, 151
 equine, 630, 637*f*, 640–641
 feline, 117
 ruminant, 1008*f*
Epihyoid bone, canine, 176
Epiploic foramen, equine, 728, 741–742, 741–742*f*, 769, 769*b*
Epithelial receptors, ruminant, 1060–1062
Epitrichial sweat gland, canine, 547
Epitympanic recess, feline, 109, 140
Epoophoron, equine, 789*b*, 798*b*
Ergot, equine, 554*f*
Erythrocytes, avian, 1362
Esophageal plexus, avian, 1274–1275
Esophagus
 avian, 1274, 1347
 canine, 85*f*, 152, 275–276
 equine, 678
 feline, 95*f*
 ruminant, 1047*f*
Ethmoidal artery
 feline, 108
 ruminant, 1011
Ethmoidal foramina, canine, 131–132
Ethmoidal labyrinth, equine, 591
Ethmoidal meatus, ruminant, 1007
Ethmoidal nerve, equine, 597
Ethmoid conchae/turbinate
 canine, 112*b*
 equine, 1008*f*
 ruminant, 1008*f*
Exocrine pancreas, equine, 735–736, 735*f*
Extensor bursa, equine, 916
Extensor carpi obliquus muscle
 canine, 87–88*f*
 equine, 557–559*f*
 ruminant, 996*f*, 1155*b*
Extensor carpi radialis muscle
 canine, 87–88*f*
 equine, 557–559*f*
 ruminant, 996*f*, 1155, 1189
Extensor retinaculum
 canine, 87–88*f*
 equine, 557–559*f*
 ruminant, 995–996*f*
Extensor tendon sheath, ruminant, 1186
External abdominal oblique muscle
 avian, 1255*f*
 canine, 87–88*f*
 equine, 557–559*f*
 feline, 97*f*
 ruminant, 1117*f*, 1118*b*, 1120*f*
External acoustic meatus
 avian, 1301
 equine, 656
 feline, 137–139
 ruminant, 995–996*f*
External carotid artery
 avian, 1297, 1323
 canine, 125
 equine, 618
 feline, 110
 ruminant, 993*f*
External ear
 canine, 84–85*f*
 equine, 657*f*, 658, 659*f*
 feline, 110
 ruminant, 999*f*
External ear canal, avian, 1301–1303
External ethmoidal artery, equine, 597
External iliac artery
 camelids, 1204
 equine, 806, 938
External iliac vein, avian, 1253*f*, 1353
External jugular vein
 canine, 85*f*, 161*b*, 179
 equine, 557*f*, 676, 889
 feline, 95*f*
 ruminant, 993*f*, 996*f*
External ophthalmic artery
 avian, 1324
 equine, 574–575
External pudendal artery
 caprine, 1132–1133
 equine, 822, 986
External pudendal vein
 caprine, 1132–1133
 equine, 986
External pudendal vessels, equine, 839
External sacral lymph node, ruminant, 1111
External urethral orifice
 caprine, 1132
 equine, 785
 ruminant, 1123
Extrahepatic bile duct, canine, 290
Extraocular muscle, avian, 1310
Extraperitoneal segments, equine, 760–761
Extrapyramidal tracts, feline, 206
Extrathoracic diverticula, avian, 1338
Eye
 avian, 1309
 canine, 3, 5*f*
 equine, 570, 574–575
 ruminant, 1002*f*
Eyelid
 avian, 1313
 canine, 130–131
 equine, 568*b*, 568*f*, 570*f*, 572, 574*b*, 574*f*
 ruminant, 1069

F

Fabellae (sesamoid)
 canine, 92*f*
 equine, 563–564*f*
 feline, 102*f*
Facets of vertebrae, equine, 683
Facial disc, avian, 1304
Facial nerve (CNVII)
 avian, 1298
 canine, 88*f*
 equine, 609, 652
 feline, 95*f*, 108, 110, 141, 202–203
 ruminant, 996*f*
Facial tubercle, caprine, 1011*f*
Falciform ligament
 feline, 99*f*
 equine, 100*f*
 ruminant, 1087, 1092
Fan beam computed tomography, 77, 78*t*
Fascia latae (tensor fascia latae)
 avian, 1255*f*
 canine, 87*f*
 equine, 558*f*, 948

Fasciculus cuneatus, canine, 192–193, 214, 214*f*
Fasciculus gracilis, canine, 192–193, 214, 214*f*
Feather, avian, 1260, 1402, 1404–1407, 1409–1410, 1412–1413
  afterfeather, 1407
  auricular feathers, 1254*f*
  blood feather, 1411
  body contour, 1405–1406
  covert feathers, 1265*b*
  down feathers, 1409
  flight feathers, 1405–1409
  follicle, 1406
  muscle, 1403
  pin feathers, 1411
  tip, 1406
Female reproductive tract
  canine, 370–371
  equine, 775, 775*f*, 790–791, 791*f*, 796–797*f*
  ruminant, 1119–1120
Femoral artery
  avian, 1395
  canine, 85*f*
  equine, 555*f*
  feline, 95*f*
  ruminant, 993*f*
Femoral head ligament, ruminant, 1196
Femoral nerve
  avian, 1396
  camelids, 1204, 1204*b*
  equine, 555*f*, 941
  ruminant, 1036, 1197
Femoral vein
  canine, 87*f*
  equine, 555*f*
  feline, 95*f*
  ruminant, 993*f*
Femoropatellar joint
  canine, 489
  equine, 946
  ruminant, 1207–1208
Femorotibialis, avian, 1389
Femorotibial joint, equine, 947
Femur
  avian, 1258*f*, 1388
  camelids, 1202
  canine, 84*f*, 92*f*, 460, 472–474
  equine, 563–564*f*, 945–946, 945*f*
  feline, 102*f*
  ruminant, 1000*f*
Fetal circulation
  canine, 262
  equine, 855*b*
  ruminant, 1083, 1087*b*
Fetlock (metacarpophalangeal and metatarsophalangeal) joint, equine, 545*b*
Fibroelastic, caprine, 1132
Fibrous tunic, equine, 570
Fibrovascular penis, ruminant, 1139–1140
Fibula
  avian, 1258*f*, 1389
  canine, 92*f*, 501–502
  equine, 563–564*f*, 946
  feline, 95*f*, 102*f*
  ruminant, 1000*f*
Fibula, head
  canine, 92*f*
  equine, 563*f*
  feline, 102*f*
Fibular nerve
  avian, 1396
  canine, 85*f*
  equine, 555*f*, 967
  feline, 95*f*
  ruminant, 1037, 1168, 1219, 1219*b*
Fibularis (peroneus) brevis muscle
  canine, 87–88*f*
  feline, 97–98*f*
Fibularis (peroneus) longus muscle
  canine, 87–88*f*
  feline, 97–98*f*
  ruminant, 996*f*
Fibularis (peroneus) tertius muscle
  equine, 557–559*f*
  ruminant, 995–996*f*
Fighting teeth, camelid, 1015
Filiform papillae, equine, 580
Filoplumes, avian, 1410
Fimbriae, equine, 812
First (proximal) phalanx, equine, 907
Fissures, canine, 131–132
Flexor carpi radialis muscle
  equine, 557–559*f*
  ruminant, 995–996*f*, 1155, 1187
Flexor carpi ulnaris muscle
  equine, 557–559*f*
  ruminant, 995–996*f*, 1155, 1187
Flexor hallucis brevis muscle, avian, 1255*f*
Flexor perforans et perforatus muscle, avian, 1255*f*
Foliate papillae, equine, 580
Follicle of hair, canine, 546
Foramen of the skull
  canine
    infraorbital, 92*f*
    mental, 92*f*
  equine
    infraorbital, 563–564*f*
    mandibular, 563–564*f*
    mental, 563–564*f*
    obturator, 563–564*f*
  feline
    infraorbital, 102*f*
    mental, 102*f*
  ruminant
    infraorbital, 1000*f*
    mental, 1000*f*
    supraorbital, 1000*f*
Foramen magnum
  canine, 174
  equine, 650
Foramen ovale, canine, 262
Foramen venae cava, ruminant, 1092
Forebrain, feline, 197*b*
Forestomach, ruminant, 1059*f*, 1060–1062, 1061*f*
Fornix vagina
  equine, 785, 805–806
  ruminant, 1110
Fossa, supraorbital, equine, 554*f*
Fossa, infraspinous and supraspinous, canine, 85*f*
Fossae of lacrimal sac, canine, 131–132
Fossa glandis, equine, 820
Fovea, avian, 1313
Frontal bone
  avian, 1258*f*
  canine, 129
  equine, 563–564*f*
  feline, 102*f*
  ruminant, 1000*f*
Frontalis muscle
  canine, 87*f*
  ruminant, 996*f*
Frontal sinus
  equine, 592–593, 592*f*, 599, 600*f*
  ruminant, 84*f*, 1005, 1007–1009, 1007*b*, 1007–1009*f*, 1009*b*
Fundus, eye, avian, 1306*f*
Fundus, stomach
  avian, 1312
  equine, 726
  ruminant, 1066
Furcula, avian, 1369

G

Gall bladder
  avian, 1348
  feline, 561*f*
  ruminant, 1092
Ganglions, feline, 204
Gastric artery
  equine, 678
  ruminant, 1060
Gastric ligament and omentum, canine, 309
Gastrocnemius muscle
  avian, 1255*f*
  canine, 536
  equine, 962
  ruminant, 996*f*

Gastroduodenal artery
canine, 308–309
equine, 736
Gastroduodenal vein, ruminant, 1075
Gastroepiploic artery, equine, 728–729
Gastroepiploic vessels, canine, 302–303
Gastropancreatic fold, equine, 742
Gastrophrenic ligament (gastrosplenic ligament), equine, 728–729
General somatic efferent neuron. *See* Lower motor neuron
Genioglossus muscle, equine, 580–581, 632, 632*b*
Geniohyoid muscle, equine, 581
Genital ducts, ruminant, 1108
Genitofemoral nerve
equine, 839, 987
ruminant, 1035
Germinal epidermis, equine, 928–929
Germinal epithelium, equine, 795–796
Gestational sac, canine, 369
Gingiva
canine, 125
equine, 607
Gizzard, avian, 1347
Glands
avian
dorsal proctodeal, 1256*f*
preen (uropygial), 1255*f*
uropygial, 1256–1257*f*
canine, 89*f*
bulbis glandis, 90*f*
mammary, 89*f*
mandibular, 89*f*
parotid, 89*f*
prostate, 89*f*
thymus, 89*f*
thyroid, 89*f*
equine, 555*f*, 557–558*f*
bulbourethral, 560*f*
mammary, 560*f*
parotid, 560*f*
prostate, 560*f*
thyroid, 560*f*
feline
bulbourethral, 98*f*
mandibular, 98*f*
parotid, 98*f*
prostate, 100*f*
thymus, 100*f*
thyroid, 100*f*
ruminant
bulbourethral, 998*f*
mammary (udder), 997*f*
mandibular salivary, 993*f*, 996*f*
parotid salivary, 993*f*, 996*f*
prostate, 998*f*
thyroid, 993*f*, 996*f*
vesicular, 998*f*
Glandular cistern
equine, 986
ruminant, 1238–1239
Glandular region of stomach, equine, 727
Glans clitoris
canine, 365
equine, 782
Glans penis
canine, 385
caprine, 1132
equine, 820
feline, 97*f*
ruminant, 1140–1141
Glenohumeral joint
canine, 396–399
ruminant, 1149, 1150*f*
Glenoid cavity, ruminant, 1149, 1150*f*
Glisson's capsule, canine, 284
Globe
avian, 1268, 1310–1311
canine, 126
equine, 570, 570*f*
ruminant, 1021*f*
Glossopharyngeal nerve (CNIX)
avian, 1274–1275
canine, 117, 152
equine, 581, 632, 652–653, 678
feline, 141–142
Glottis
canine, 151*b*
equine, 640
Gluteal muscle, superficial, middle, deep
canine, 87–88*f*
equine, 557–559*f*
ruminant, 995–996*f*
Gluteal nerve, ruminant, 1197
Gnathotheca, avian, 1287, 1297
Goblet cells, equine, 759
Gonydeal angle, avian, 1292
Gonys, avian, 1292
Gracilis muscle
canine, 87–88*f*
equine, 557–559*f*, 956
ruminant, 995–996*f*
Gray matter, feline, 202
Great auricular nerve, equine, 658, 658*b*, 659*f*
Great cardiac vein
equine, 713
ruminant, 1053
Greater ischiatic notch, ruminant, 1034*f*
Greater omentum
canine, 302
equine, 728
ruminant, 1066
Greater palatine artery
feline, 108
ruminant, 1011
Greater trochanter of femur
canine, 92*f*
equine, 563–564*f*
feline, 102*f*
ruminant, 992*f*
Greater tubercle of humerus
canine, 84*f*
equine, 563–564*f*
feline, 102*f*
ruminant, 992–993*f*, 1000*f*
Great vessels, ruminant, 1052, 1053*f*
Gut-associated lymphoid tissue, avian, 1360*b*
Guttural pouch, equine, 590, 658, 870–871

H

Hallux, avian, 1390
Harderian gland, avian, 1314, 1360
Hard palate
canine, 116, 150
ruminant, 1008*f*
Hassall's corpuscles, avian, 1360
Haustra of the cecum, equine, 746, 754, 758–759, 759*f*
Heart
avian, 1256*f*, 1261
equine, 709, 709*f*, 712–713
feline, 247–249
ruminant, 1046–1047, 1047*f*, 1052, 1053*f*
Heart border
canine, 84–85*f*
equine, 554*f*
feline, 94*f*
ruminant, 992–993*f*
Hematopoietic system, avian, 1361
Hepatic artery
canine, 282, 308–309
ruminant, 1085, 1085*f*, 1087*f*, 1094
Hepatic ligament
equine, 765, 767*f*
ruminant, 1092–1094, 1093*f*
Hepatic lobule, equine, 767
Hepatic lymph node
equine, 765
ruminant, 1066, 1094
Hepatic nervous plexus, ruminant, 1094
Hepatic node, ruminant, 1076
Hepatic portal system, avian, 1325
Hepatic sinuses, ruminant, 1085
Hepatic trunk, ruminant, 1094
Hepatic vein, ruminant, 1085–1086
Hepatic venules, ruminant, 1086
Hepatocytes
canine, 285*f*
equine, 763
ruminant, 1084, 1085*f*, 1091

Hepatoduodenal ligament, equine, 734–735, 740
Hepatoid gland, canine, 547
Hepatopericardial ligament, avian, 1320
Hepatorenal ligament
  canine, 285*f*
  equine, 765, 767*f*
  ruminant, 1092
Heterophils, avian, 1328*b*, 1362
Hilus
  canine, 222
  equine, 718, 765, 767*f*, 788
  ruminant, 1099–1100
Hilus pulmonis, ruminant, 1087
Holocrine gland, avian, 1404*b*
Hoof
  equine, 554*f*
  ruminant, 445
Horn, ruminant, 1023–1026, 1028–1029
Horn capsule, ruminant, 1174
Horner syndrome, feline, 109*b*
Horn sheath, ruminant, 1025–1026
Humeral head of the flexor carpi ulnaris muscle, ruminant, 1187
Humeroradial joint
  equine, 879*f*
  feline, 410
  ruminant, 1156
Humeroulnar joint
  equine, 879*f*
  feline, 410
  ruminant, 1156
Humerus
  avian, 1253*f*, 1258*f*, 1370–1372, 1374
  canine, 85*f*, 92*f*
  equine, 563–564*f*
  feline, 94*f*, 102*f*
  ruminant, 1000*f*, 1149–1151, 1150*f*
Hydrosalpinx, equine, 814*b*
Hymen
  equine, 777
  ruminant, 1123
Hyoepiglottic muscle
  canine, 154–156
  equine, 634–635
Hyoglossus, equine, 580–581
Hyoid apparatus
  canine, 176
  equine, 614–615*f*, 615, 621–622, 630–631
Hypogastric ganglia, ruminant, 1080
Hypogastric lymph node, canine, 344
Hypogastric nerve (CNXII)
  canine, 344
  caprine, 1133–1134
  equine, 791, 806–807, 850, 858
  ruminant, 1102, 1111, 1123
Hypoglossal nerve
  avian, 1274–1275
  equine, 581, 632–633, 652–653
Hypophyseal fossa, equine, 652
Hypotarsal crest, medial, avian, 1258*f*
Hypothalamus
  canine, 190, 293*b*
  ruminant, 1241*b*
Hypothalamic-pituitary-adrenal axis, canine, 293*b*
Hypsodont teeth, equine, 606*f*, 607

I

Ileal ostium
  canine/feline, 314*b*
  equine, 746*b*
Ileal papilla, equine, 747
Ileal sphincter, ruminant, 1079
Ileal vein, ruminant, 1075
Ileocecal
  artery, equine, 741
  fold
    equine, 741
    ruminant, 1079
  junction, ruminant, 1074, 1079
  orifice, equine, 741, 746–747, 746*f*
Ileocolic artery
  equine, 748
  ruminant, 1075, 1080
Ileocolic valve, canine/feline, 320*b*
Ileum
  avian, 1347
  canine, 314
  equine, 741
  feline, 101*f*
  ruminant, 1074, 1078–1079
Iliac node
  canine, 92*f*
  equine, 563*f*, 761–762, 778, 833, 849, 939, 967
  feline, 102*f*
  ruminant, 1111, 1122
Iliacus lateralis muscle, ruminant, 996*f*
Iliofemoral lymph node, ruminant, 1197
Iliohypogastric nerve, equine, 839
Ilioinguinal nerve, equine, 839
Ilioischiatic foramen, avian, 1396
Iliosacral lymph node, ruminant, 1197
Iliotibiales, avian, 1389
Ilium
  avian, 1258*f*, 1388
  equine, 563–564*f*, 935
  feline, 97*f*
  ruminant, 1000*f*, 1032, 1034*f*
Ilium, shaft
  canine, 92*f*
  feline, 102*f*
Impar ligament, equine, 920
Incisive bone
  canine, 116
  caprine, 1011*f*
  equine, 563–564*f*
  ruminant, 1000*f*
Incisors
  canine, 122
  equine, 603–605, 605–606*f*
  ruminant, 1015
Incus
  equine, 658, 660*f*
  feline, 109
Inferior alveolar artery
  alpaca, 1018
  canine, 125
Inferior alveolar nerve
  alpaca, 1018, 1019*f*
  equine, 581, 608, 608*b*
Inferior alveolar vein, alpaca, 1018
Inferior tarsus muscle, equine, 870
Infraorbital artery
  alpaca, 1018
  canine, 125
Infraorbital canal
  equine, 595
  ruminant, 1008*f*, 1011, 1011*f*
Infraorbital foramen
  canine, 85*f*
  caprine, 1011*f*
  feline, 95*f*
  ruminant, 1011, 1011*f*
Infraorbital fossa, avian, 1268
Infraorbital nerve
  alpaca, 1018
  avian, 1270
  canine, 125
  caprine, 1011*f*
  equine, 581, 608, 608*b*
  ruminant, 1011*b*
Infraorbital palatine artery, ruminant, 1011
Infraorbital sinus, avian, 1268–1270, 1268*f*
Infraorbital vein, canine, 125
Infraspinatus
  bursa
    equine, 879
    ruminant, 1156–1157
  fossa
    canine, 85*f*
    equine, 563–564*f*
  muscle
    canine, 399
    ruminant, 996*f*, 1154, 1156
  tendon
    equine, 879
    ruminant, 1156–1157
Infratrochlear nerve, ruminant, 1028, 1028*f*
Infundibular cement, equine, 605, 605*f*
Infundibular gland, avian, 1332
Infundibulum of uterine tube
  avian, 1332
  equine, 812, 812*f*

Ingluviales artery, avian, 1274–1275
Ingluvies/crop, avian, 1347
Inguinal canal, equine, 840*f*, 843–844
Inguinal lymph node
  canine, 85*f*
  equine, 555*f*, 823, 967
  ruminant, 993*f*
Inguinal ring, internal (deep) and external (superficial), equine, 377*f*
Inguinofemoral lymph node, ruminant, 1197
Inner ear
  avian, 1303–1304
  equine, 652, 662
Integument
  avian, 1393
  canine, 541
Interatrial septum, avian, 1320
Intercarpal ligament
  equine, 1185
  ruminant, 1185
Intercentral articulation, equine, 683
Interclavicular air sac, avian, 1284, 1337
Intercornual protuberance, ruminant, 993*f*
Interdigital artery, ruminant, 1189
Interdigital collateral ligament, ruminant, 1173–1174
Interdigital ligament, ruminant, 1173–1174
Interdigital pouch, ruminant, 1231
Interdigital space, ruminant, 1173–1174
Intermediate carpal bone, equine, 894–896
Intermediate cartilage, avian, 1283
Internal abdominal oblique muscle
  canine, 87–88*f*
  equine, 557–559*f*
  feline, 97*f*
  ruminant, 1118
Internal acoustic meatus
  equine, 652, 661–662
  feline, 108
Internal auricular branch of ear, equine, 658, 658*b*, 659*f*
Internal carotid artery
  avian, 1323
  equine, 611–613, 612*f*
Internal iliac artery
  caprine, 1133
  equine, 833
  ruminant, 1080, 1111, 1141
Internal jugular vein
  canine, 160, 175
  feline, 168
Internal ophthalmic artery, equine, 574–575
Internal pudendal artery
  caprine, 1132–1133
  equine, 778, 822, 833, 847, 858
  ruminant, 1111, 1141
Internal pudendal vein
  canine, 344
  caprine, 1132–1133
  equine, 823, 858, 986
Internal spermatic nerve, equine, 791
Internal thoracic artery
  feline, 235
  ruminant, 1048
Internal thoracic vein, ruminant, 1048
Internal urethral orifice, equine, 857
Internuncial neuron, equine, 868–870
Interosseous medius muscle, ruminant, 1166–1167
Interosseous muscle
  canine, 440–441
  equine (*see also* Suspensory ligament), 557*f*
  ruminant, 1188
Interosseous ventralis muscle, avian, 1255*f*
Intertarsal joint, avian, 1389–1390
Intertransverse ligament, equine, 936
Intertrochlear notch, ruminant, 1185
Intertubercular bursa
  equine, 1157*b*
  ruminant, 1156–1157, 1157*b*
Intertubular horn, ruminant, 1025–1026
Interventricular septum
  equine, 711
  feline, 250
  ruminant, 1053*f*
Intervertebral foramen, equine, 683, 683*b*, 684*f*
Intralobular duct, equine, 735
Intramuscular plexus, equine, 760
Intrathoracic diverticula, avian, 1338
Iris
  avian, 1311*f*, 1312*b*
  canine, 127
  equine, 570
  feline, 203*f*
Ischial arch, equine, 820, 830
Ischial spine, ruminant, 1000*f*
Ischiatic artery, avian, 1395
Ischiatic lymph node, ruminant, 1197
Ischiatic spine, ruminant, 1034*f*
Ischiocavernosus muscle, equine, 820–822
Ischiourethalis muscle, equine, 858
Ischium
  avian, 1258*f*, 1388
  equine, 820–822
  ruminant, 1032
Islet cells, canine, 289
Isthmus
  avian, 1332
  canine, 157
  equine, 812–813, 812*f*

J

Jejunal intestinal loops, ruminant, 1069, 1069*f*
Jejunal lymph node, ruminant, 1076
Jejunal vein, ruminant, 1075
Jejunoileal flange, ruminant, 1072*b*, 1072*f*, 1074
Jejunoileum, avian, 1256*f*
Jejunum
  avian, 1347
  canine, 314, 355*b*
  equine, 740–741
  feline, 97*f*
  ruminant, 1070–1073*f*, 1072, 1075*f*
Jugal arch, avian, 1258*f*
Jugal bar, avian, 1297
Jugal bone, avian, 1268, 1297
Jugal process, avian, 1297
Jugular foramina, equine, 652
Jugular groove
  equine, 557*f*
  ruminant, 993*f*, 996*f*
Jugular vein
  avian, 1253*f*, 1255*f*, 1274–1275, 1324
  canine, 85*f*
  equine, 897
  feline, 95*f*
  ruminant, 993*f*, 996*f*, 1049*f*
Jugular vein, external
  canine, 85*f*
  equine, 557*f*
  feline, 95*f*
  ruminant, 993*f*, 996*f*

K

Keel, avian, 1258*f*, 1372–1373, 1382, 1382*b*
Keratinocytes, canine, 542
Kidney
  avian, 1256–1257*f*, 1261, 1352
  canine, 295, 339*f*, 340–341
  equine, 560–562*f*
  feline, 99–100*f*
  ruminant, 1092, 1092*b*, 1098–1100, 1098*f*, 1100–1101*f*
Kinocilium, feline, 143
Kupffer cells, ruminant, 1085

L

Labia
  canine, 357, 364
  equine, 776
  ruminant, 1109
Labrum, canine, 459
Labyrinthine artery, equine, 662

Labyrinthine vein, equine, 662
Labyrinthine wall, equine, 659
Lacrimal bone
avian, 1268
canine, 129
caprine, 1011*f*
Lacrimal gland
avian, 1258*f*, 1314
canine, 130, 132
equine, 571, 571*f*, 593
Lacrimal nerve, cornual branch in ruminant, 1028, 1028*f*
Lacrimal puncta, canine, 131
Lacrimal sac
canine, 131
equine, 571
Lacrimojugale ligament, avain, 1255*f*
Lactiferous ducts
equine, 986, 986*f*
ruminant, 1238–1239
Laminae
equine, 643
ruminant, 1174
Lamina muscularis, equine, 739–740
Laminar dermis, equine, 927–928
Laminar epidermis, equine, 928–929
Langerhans cells
canine, 544
equine, 981
Lanolin, caprine, 1229, 1230*f*
Large colon, equine, 753–755
Laryngeal artery, equine, 644
Larynx
avian, 1281–1282
inlet, 1281
mound, 1281
canine, 153–157, 274–275
equine, 630–631, 633, 643–644, 644*f*
ruminant, 1011
Lateral cutaneous antebrachial nerve, equine, 868–870, 887–889
Lateral digital extensor muscle, ruminant, 996*f*
Lateral digital flexor muscle
equine, 900
ruminant, 996*f*, 1189
Lateral digital flexor tendon, equine, 901
Lateral dorsal metatarsal nerve, equine, 916
Lateral epicondyle of humerus
equine, 900
ruminant, 1151
Lateral femorotibial joint, equine, 947
Lateral ligament of the urinary bladder
canine, 340–341
equine, 856
Lateral longitudinal esophageal muscle, equine, 678
Lateral malleoli of the tibia, equine, 960
Lateral retropharyngeal lymph node, feline, 95*f*
Lateral sacral crest, ruminant, 1032
Lateral sublingual recess, equine, 577*f*, 581
Lateral suspensory ligament of the uterus, equine, 986
Lateral tympanic membrane, avian, 1283–1284
Latissimus dorsi muscle
avian, 1255*f*
ruminant, 1154
Leaflets of cardiac valves, equine, 710
Left view of the thoracic and abdominal cavities and pelvis (female)
avian, 1256*f*
canine, 89*f*
equine, 560*f*
feline, 99*f*
ruminant, 997*f*
Lesser ischiatic notch, ruminant, 1034*f*
Lesser omentum, ruminant, 1066
Levator anguli oculi medialis muscle, equine, 870
Levator ani muscle
avian, 1255*f*
canine, 321–323
Levator labii superioris muscle, equine, 578
Levator nasolabialis muscle, equine, 578
Levator palpebrae superioris muscle
canine, 131
feline, 202–203
Levator veli palatini muscle
canine, 117
equine, 631
Leydig cells, canine, 379*b*
Ligament of the tail of the epididymis
canine, 380–381, 380*b*
equine, 837*f*
ruminant, 998*f*
Ligamentum flavum, canine, 175
Limbus, avian, 1311
Linea alba
canine, 330*b*, 331*f*
equine, 844
feline, 327
ruminant, 1118
Lingual cartilage, equine, 580
Lingual frenulum, equine, 579, 581, 584–585
Lingual muscle, equine, 579
Lingual nerve, equine, 581, 608*b*
Lingual tonsil, equine, 580
Linguofacial vein
canine, 87*f*
equine, 582–583, 637–638, 639*f*
feline, 95*f*
ruminant, 996*f*
Lips, equine, 578, 580*f*
Liver
avian, 1320, 1348
canine, 282–286
equine, 765, 767*f*
feline, 99–100*f*
ruminant, 1082, 1083–1084*f*, 1084, 1086*f*, 1087, 1091–1092, 1091*f*, 1092*b*, 1093–1095*f*
Lobate toes, avian, 1390
Lobe of lung
canine, 89–90*f*
equine, 560–561*f*
feline, 99–100*f*
ruminant, 997–998*f*
Lobules
canine, 89–90*f*
of kidneys, ruminant, 1099–1100
of liver, equine, 767, 768*f*, 986
ruminant, 997–998*f*
Long digital extensor muscle
equine, 900
ruminant, 1189, 1214
Longissimus atlantis muscle, ruminant, 996*f*
Longissimus capitis muscle, ruminant, 996*f*
Longissimus cervicis muscle, avian, 1255*f*
Longissimus lumborum
feline, 97*f*
ruminant, 1033
Longissimus muscle, equine, 936
Longitudinal intermammary groove, equine, 986, 988*f*
Long plantar ligament, equine, 966
Longus atlantis muscle, ruminant, 996*f*
Longus capitis muscle
canine, 180
equine, 615, 650, 651*f*, 653–654
Longus colli muscle
avian, 1255*f*
canine, 180
equine, 676, 718
Lower motor neuron
canine, 210–211, 210*b*
equine, 863*b*
Lumbar artery, ruminant, 1037
Lumbar center, canine, 296
Lumbar intumescence, canine, 211, 522–523
Lumbar lymphatic trunk, ruminant, 1122
Lumbar lymph node
avian, 1361
caprine, 1133
equine, 858
Lumbar nerve, ruminant, 1035, 1036*f*
Lumbar plexus, avian, 1352, 1396
Lumbar trunk, equine, 761–762
Lumbar vein, ruminant, 1037

Lumbar vertebral column
equine, 71*f*
ruminant, 1032
Lumbo-aortic lymph node, ruminant, 1037
Lumbosacral cisterna, ruminant, 1034*b*
Lumbosacral dorsal root ganglia, equine, 758, 760
Lumbosacral interarcuate space
equine, 555*f*
ruminant, 993*f*
Lumbosacral plexus
canine, 536
equine, 833
Lumbosacral trunk, ruminant, 1037
Lung(s)
avian, 1320, 1336, 1339
canine, 222–225
equine, 719*f*, 720–722
feline, 248
ruminant, 1087–1089
Lung, basal border
canine, 89–90*f*
equine, 560–561*f*
feline, 99–100*f*
ruminant, 997–998*f*
Lung, cranial and caudal lobe of right
equine, 560–561*f*
ruminant, 997–998*f*
Lung, root of left and right
canine, 89–90*f*
equine, 560–561*f*
feline, 99–100*f*
ruminant, 997–998*f*
Lymphatic system
avian, sheaths, 1360
canine
axillary lymph node, 85*f*
cervical, superficial lymph node, 85*f*
inguinal, superficial lymph node, 85*f*
mandibular lymph node, 85*f*
parotid lymph node, 85*f*
popliteal lymph node, 85*f*
equine, 555*f*, 561*f*
cervical, superficial lymph node, 555*f*
hilar lymph node, 555*f*
inguinal, superficial lymph node, 555*f*
mandibular lymph node, 555*f*
mesenteric lymph node, 555*f*
superficial cervical lymph node, 555*f*
superficial inguinal lymph node, 555*f*
feline
axillary lymph node, 95*f*
caudal epigastric lymph node, 95*f*
lateral retropharyngeal lymph node, 95*f*
mandibular lymph node, 95*f*
popliteal lymph node, 95*f*
superficial cervical lymph node, 95*f*
superficial inguinal lymph node, 95*f*
ruminant
cervical, superficial lymph node, 993*f*
inguinal, superficial lymph node, 993*f*
lateral retropharyngeal lymph node, 993*f*
mandibular lymph node, 993*f*
mediastinal, caudal lymph node, 993*f*
parotid lymph node, 993*f*
retropharyngeal lymph node, 993*f*
subiliac lymph node, 993*f*
superficial cervical lymph node, 993*f*
Lymphocytes, avian, 1362

M

Maculae acousticae, equine, 662
Magnum of oviduct, avian, 1332
Major caruncle, avian, 1254*f*
Major duodenal papilla, equine, 734–736, 734*f*
Major jejunal artery, equine, 740
Major (carpo)metacarpal bone, avian, 1373
Major palatine artery, canine, 116
Major palatine foramina, canine, 116
Major salivary gland, equine, 582
Major splanchnic nerve, equine, 731
Malar artery, equine, 574–575
Malaris muscle, ruminant, 996*f*
Male reproductive system
avian, 1353–1354
canine, 374
equine, 833
feline, 375
ruminant, 1131*f*
Malleolus(i) of tibia, lateral and medial
canine, 92*f*
equine, 563–564*f*
feline, 102*f*
ruminant, 1000*f*
Malleus
canine, 134*f*
equine, 658, 660*f*
feline, 109*f*
Mammary gland
canine, 85*f*, 371–372
equine, 984–988, 984–986*f*
feline, 333–335
ruminant, 997*f*, 1077, 1238–1241, 1239*f*
Mandible
avian, 1258*f*
canine, 92*f*
equine, ramus, 563–564*f*
feline, 102*f*
ruminant, 1000*f*
Mandibular branch
of beak, avian, 1297–1298
of dentition, canine, 125, 162
of head, ruminant, 1011
of tooth, alpaca, 1018
of trigeminal nerve, equine, 575, 581, 584–585, 631, 654
Mandibular foramen, alpaca, 1018
Mandibular lymph node
canine, 85*f*
equine, 583*f*
feline, 95*f*
Mandibular salivary gland
canine, 85*f*
equine, 583*f*, 584
feline, 95*f*
ruminant, 993*f*, 996*f*
Manica flexoria, equine, 900–901, 909–910
Manubrium sterni
avian, 1258*f*
equine, 563–564*f*
feline, 102*f*
ruminant, 1000*f*
Manus
canine, 92*f*
equine, 563–564*f*
feline, 102*f*
Margo plicatus, equine, 726, 728*b*
Masseter muscle
canine, 130
equine, 609
ruminant, 996*f*
Maxilla
avian, 1258*f*, 1297
canine, 92*f*, 116, 129
caprine, 1011*f*
equine, 563–564*f*
feline, 95*f*, 102*f*
ruminant, 1000*f*
Maxillary septal bulla, equine, 594
Maxillary sinus
caprine, 1011*f*
equine, 591*f*, 592, 594–595, 596*f*, 599, 600*f*, 601, 602*f*, 607
caudal, 593
cranial, 589–590
ruminant, 1002, 1002*f*, 1004*f*, 1005, 1008*f*
Maxillary vein
canine, 85*f*, 118
equine, 658
feline, 95*f*
ruminant, 1011

Meatus, external acoustic
avian, 1301
equine, 656
feline, 137–139
ruminant, 995–996*f*, 1006, 1008*f*
Meatus, nasal
Nasal meatus
equine, 589–590
ruminant, 1006
Mechanoreceptors, canine, 548–549
Medial cuneiform bone, ruminant, 1000*f*
Median artery
canine, 85*f*
equine, 555*f*, 974
feline, 95*f*
ruminant, 1159, 1189
Median cubital vein, equine, 889, 897
Median ligament of the bladder, equine, 849
Median nerve
canine, 85*f*, 88*f*
equine, 555*f*
feline, 95*f*
ruminant, 1158–1159, 1167–1168
Median raphe, equine, 839
Median sacral crest, ruminant, 1032
Median septum of guttural pouch, equine, 614–615
Median vein
canine, 85*f*
equine, 555*f*
feline, 95*f*
ruminant, 1159
Mediastinal caudal lymph node, ruminant, 993*f*
Mediastinal lymph node
canine, 89*f*
equine, 678
ruminant, 1048, 1053, 1089
Mediastinal pleura, ruminant, 1087
Mediastinum
canine, 222, 271
equine, 676
feline, 233
ruminant, 1047–1048, 1047*f*
Medulla of kidney
avian, 1360
canine
equine, 652
ruminant, 1224*b*
Meibomian gland
canine, 131
equine, 571*f*
Meissner's corpuscles
canine, 548
caprine, 1235
equine, 982
Melanocytes
canine, 544
equine, 981
Membranous labyrinth
equine, 661
feline, 108, 143
Meniscus of stifle
avian, 1382
canine, 490–493
equine, 949*f*, 950
ruminant, 1206, 1207*f*, 1208–1209, 1209*b*
Mental nerve
canine, 85*f*
equine, 581
ruminant, 996*f*, 1018
Mentum
equine, 554*f*
ruminant, 992*f*
Merkel cells
canine, 544, 548–549
equine, 981
Mesenteric ganglia
equine, 736, 748, 755
ruminant, 1080
Mesenteric lymph node, equine, 759
Mesenteric vein, equine, 758, 760
Mesenteric artery, equine, 759, 759*f*
Mesentery
canine, 91*f*
equine, 734–735
ruminant, 1063*b*
Mesocolon
equine, 759, 759*f*
ruminant, 1080
Mesodiverticular band, equine, 743
Mesoduodenum, equine, 734–735, 740
Mesometrium
canine, 348*b*
equine, 788, 788–789*f*, 790, 805, 809, 810*f*
ruminant, 1119–1120, 1120*f*
Mesonephric remnant, equine, 798*b*
Mesorchium, equine, 841
Mesorectum
equine, 759
ruminant, 1080
Mesosalpinx
canine, 335
equine, 788–789, 789*f*, 795, 799*f*, 805, 810*f*, 815
ruminant, 1119–1120, 1120*f*
Mesovarium
canine, 349
equine, 788, 788*b*, 788–789*f*, 795, 797–798, 805, 815
ruminant, 1119–1120, 1120*f*
Metacarpal bone - 1st, 5th
canine, 92*f*
feline, 102*f*
Metacarpal bone - 2nd, 3rd, 4th, equine, 563–564*f*
Metacarpal bone - 3rd, 4th, 5th, ruminant, 1000*f*
Metacarpal bone - major and minor, avian, 1258*f*
Metacarpophalangeal joint
canine, 434–435
equine, 907–908, 908*f*
ruminant, 1185, 1186*f*
Metatarsal bone - 1st, avian, 1258*f*
Metatarsal bone - 1st, 2nd, 5th
canine, 92*f*
feline, 102*f*
Metatarsal bone - 2nd, 3rd, 4th, equine, 563–564*f*
Metatarsal bone - 3rd, 4th, ruminant, 1000*f*
Metatarsal bone, avian, 1390
Metatarsal nerve, ruminant, 1168*b*
Metatarsal pad, avian, 1394
Metatarsal spur, avian, 1393
Metatarsal vein, avian, 1395
Metatarsophalangeal (fetlock) joint
equine, 545*b*
ruminant, 1185
Metatarsus, equine, 973–975, 973*f*
Microvilli brush border, equine, 739–740
Middle carpal joint
equine, 894
ruminant, 1184–1185
Middle colic artery
equine, 561*f*
ruminant, 1075, 1080
Middle concha
avian, 1280–1281
ruminant, 1008*f*
Middle cranial fossa, equine, 652
Middle ear
avian, 1303
equine, 660
feline, 109, 109*f*
Middle nasal meatus
equine, 589–590
ruminant, 1007
Middle plexus, equine, 981
Middle rectal artery, ruminant, 1080
Middle renal artery, avian, 1352
Middle scutum, equine, 909
Milk cistern, equine, 986, 986*f*, 988*f*
Milk vein, ruminant, 1240
Minor calice, ruminant, 1100
Minor duodenal papilla, equine, 734–735, 734*f*
Minor metacarpal bone, avian, 1373
Mitral (left atrioventricular) valve
avian, 1323
equine, 710
ruminant, 1052

Moderator band, equine, 711
Molars
canine, 123
equine, 603–604, 607
ruminant, 1015
Mucogingival junction, canine, 125
Mucoperiosteum, canine, 116
Müller's muscle, equine, 870
Muscle
avian
abdominal oblique, external, 1255*f*
adductor mandibulae ext., 1255*f*
biceps femoris, 1255*f*
biventer cervicis, 1255*f*
coccygeus, 1255*f*
complexus, 1255*f*
cranial tibial, 1255*f*
deltoid, 1255*f*
external abdominal oblique, 1255*f*
fibularis longus, 1255*f*
flexor carpi ulnaris, 1255*f*
flexor, deep digital, 1255*f*
flexor hallucis brevis, 1255*f*
flexor perforans et perforatus, 1255*f*
gastrocnemius, 1255*f*
lateral coccygeus, 1255*f*
latissimus dorsi, 1255*f*
levator ani, 1255*f*
levator coccygeus, 1255*f*
long digital extensor, 1255*f*
longissimus cervicis, 1255*f*
longus colli, 1255*f*
sartorius with superficial gluteus, 1255*f*
scapulohumeralis caudalis, 1255*f*
semimembranosus, 1255*f*
semispinalis cervicis, 1255*f*
semitendinosus, 1255*f*
sphincter ani, 1255*f*
sternothyroideus, 1255*f*
superficial (major) pectoral, 1255*f*
tendon of flexor perforatus, 1255*f*
teres major, 1255*f*
trapezius, 1255*f*
canine, 88*f*
abdominal oblique, external and internal, 88*f*
bulbospongiosus, 88*f*
cranial tibial muscle and tendon, 88*f*
deltoideus, 88*f*
dorsal internal sacrocaudal, 88*f*
extensor carpi radialis, 88*f*
flexor carpi radialis, 88*f*
flexor carpi ulnaris, 88*f*
frontalis, 88*f*
gluteal, deep, 88*f*
intercostal, external, 88*f*
lateral digital extensor, 88*f*
latissimus dorsi, 88*f*
levator ani, 88*f*
levator nasolabialis, 88*f*
long digital extensor, 88*f*
longissimus dorsi, 88*f*
longus cervicis, 88*f*
mentalis, 88*f*
middle gluteal, 88*f*
quadratus femoris, 88*f*
rectus abdominis, 88*f*
rectus femoris, 88*f*
rectus sheath, external and internal lamina of, 88*f*
retractor penis, 88*f*
rhomboideus, 88*f*
sartorius, 88*f*
semimembranosus, 88*f*
semitendinosus, 88*f*
serratus, cranial and caudal dorsal, 88*f*
serratus ventralis, 88*f*
splenius, 88*f*
spinalis et semispinalis, 88*f*
sternocephalicus, 88*f*
sternocephalicus, pars mastoideus and occipitalis, 88*f*
sternohyoideus, 88*f*
supraspinatus, 88*f*
triceps brachii, lateral and long head of, 88*f*
zygomatic, 88*f*
equine, 557–559*f*
abdominal oblique, external and internal, 558–559*f*
anal sphincter, external, 558–559*f*
bulbospongiosus, 558–559*f*
deep digital flexor, 558–559*f*
deep digital flexor, medial head of, 558–559*f*
deltoideus, 558–559*f*
depressor anguli oris, 558–559*f*
depressor labii inferioris, 558–559*f*
extensor carpi radialis, 558–559*f*
flexor carpi radialis, 558–559*f*
flexor carpi ulnaris, 558–559*f*
gastrocnemius, lateral head of, 558–559*f*
gluteal, middle and superficial, 558–559*f*
gracilis, 558–559*f*
iliacus lateralis, 558–559*f*
iliocostalis thoracis, 558–559*f*
intercostal, external of, 558–559*f*
interosseous (suspensory ligament), 558–559*f*
interosseous (suspensory ligament), extensor branch of, 558–559*f*
interscutular, 558–559*f*
lateral digital extensor, 558–559*f*
lateral head of, deep digital flexor, 558–559*f*
latissimus dorsi, 558–559*f*
levator anguli oculi medialis, 558–559*f*
levator labii superioris, 558–559*f*
levator nasolabialis, 558–559*f*
long digital extensor, 558–559*f*
longissimus atlantis, 558–559*f*
longissimus capitis, 558–559*f*
longissimus cervicis, 558–559*f*
longissimus thoracis et lumborum, 558–559*f*
longus capitis, 558–559*f*
longus colli, 558–559*f*
malaris, 558–559*f*
masseter, 558–559*f*
middle gluteal, 558–559*f*
pectoral, descending, 558–559*f*
pectoral, deep, 558–559*f*
pectoral, superficial, 558–559*f*
rectus abdominis, 558–559*f*
rectus femoris, 558–559*f*
retractor penis, 558–559*f*
rhomboideus, 558–559*f*
sacrocaudalis, 558–559*f*
semimembranosus, 558–559*f*
semispinalis capitis, 558–559*f*
semitendinosus, 558–559*f*
serratus dorsalis caudalis, 558–559*f*
serratus, thoracic ventral, 558–559*f*
soleus, 558–559*f*
splenius capitis et cervicis, 558–559*f*
spinalis thoracis, 558–559*f*
sternocephalicus, 558–559*f*
sternohyoid, 558–559*f*
sternothyroid, 558–559*f*
stylohyoideus, 558–559*f*
subclavius, 558–559*f*
supraspinatus, 558–559*f*
triceps brachii, lateral and long head of, 558–559*f*
zygomatic, 558–559*f*
feline, 97–98*f*
abdominal oblique, external and internal, 97–98*f*
deltoideus, 97–98*f*
digastricus, 97–98*f*
extensor carpi radialis, 97–98*f*
extensor carpi ulnaris, 97–98*f*
flexor digitorum profundus, 97–98*f*
flexor digitorum superficialis, 97–98*f*
flexor carpi radialis, 97–98*f*
flexor carpi ulnaris, 97–98*f*
frontalis, 97–98*f*
infraspinatus, 97–98*f*
intercostal, internal, 97–98*f*
lateral digital extensor, 97–98*f*
latissimus dorsi, 97–98*f*
levator nasolabialis, 97–98*f*
longissimus lumborum, 97–98*f*
masseter, 97–98*f*

Muscle (Continued)
rectus abdominis, 97–98f
rectus femoris, 97–98f
rectus sheath, external and internal lamina of, 97–98f
rhomboideus, 97–98f
rhomboideus capitis, 97–98f
sacrocaudalis dorsalis lateralis, 97–98f
sartorius, 97–98f
scalenus, 97–98f
semimembranosus, 97–98f
semitendinosus, 97–98f
serratus, dorsal and ventral, 97–98f
soleus, 97–98f
splenius, 97–98f
sternocephalicus, pars occipitalis, 97–98f
sternohyoid, 97–98f
supinator, 97–98f
supraspinatus, 97–98f
triceps brachii, lateral and long head of, 97–98f
ruminant
abdominal oblique, external and internal, 996f
biceps brachii, 996f
buccal part of buccinator, 996f
cleidomastoideus, 996f
coccygeal, 996f
deltoideus, 996f
depressor labii superioris, 996f
dorsal scalenus, 996f
deep fibular (peroneal), equine
extensor carpi obliquus, 996f
extensor carpi radialis, 996f
extensor, common, lateral and long digital, 996f
fibularis longus, 996f
fibularis tertius, 996f
flexor, lateral digital, 996f
frontalis, 996f
frontoscutularis, 996f
flexor carpi radialis, 996f
flexor carpi ulnaris, 996f
flexor, deep digital, 996f
gastrocnemius, 996f
iliacus lateralis, 996f
infraspinatus, 996f
intercostal, external, 996f
lateral digital extensor, 996f
lateral digital flexor, 996f
lateral dorsal sacrococcygeal, 996f
latissimus dorsi, 996f
levator nasolabialis, 996f
long digital extensor, 996f
longissimus atlantis, 996f
longissimus capitis, 996f
longissimus lumborum, 996f
longus atlantis, 996f
levator ani, 996f
levator labii superioris, 996f
malaris, 996f
masseter, 996f
middle gluteal, 996f
pectoral, deep and superficial, 996f
rectus abdominis, 996f
rectus femoris, 996f
retractor penis, 996f
scalenus medius, 996f
scalenus medius, dorsal, 996f
sacrocaudalis dorsalis and lateralis, 996f
semimembranosus, 996f
semitendinosus, 996f
serratus muscle, caudal and cranial dorsal, ventral, 996f
serratus dorsalis caudalis, 996f
serratus ventralis cervicis, 996f
serratus ventralis thoracis, 996f
splenius, 996f
sternocephalic, 996f
sternomandibular, 996f
scutuloauricular, dorsal superficial, 996f
supraspinatus, 996f
thoracic rhomboid, 996f
triceps brachii, lateral head of, 996f
vastus lateralis, 996f
zygomatic, 996f
zygomaticoauricularis, 996f
zygomaticoscutularis, 996f
Musculocutaneous nerve
canine, 84f
equine, 554f
feline, 96f
ruminant, 994f, 1158–1159
Muzzle
canine, 84f
equine, 554f
Myenteric ganglion cells, equine, 678
Myenteric plexus
equine, 740
ruminant, 1080
Mylohyoid muscle, equine, 581
Myocardium
canine, 254
equine, 711
feline, 248
ruminant, 1052
Myometrium
canine, 348b, 355
equine, 805

N

Nail
avian, 1293, 1390b, 1393
canine, 442–445
feline, 435
Nape, avian, 1254f
Nares. *See also* Nostril
avian, 1279–1281, 1287
canine, 84f
equine, 554f, 589
feline, 94f
ruminant, 992f, 1003f
Nasal bone
avian, 1258f, 1268, 1295
caprine, 1011f
equine, 563–564f
feline, 95f, 102f
ruminant, 1000f
Nasal cavity
avian, 1268, 1279–1281
equine, 591
ruminant, 1006–1007, 1006b, 1006f
Nasal conchae
avian, 1269
equine, 591
ruminant, 1006
Nasal diverticulum, equine, 589
Nasal gland, avian, 1269b, 1314
Nasal plate
canine, 84f
feline, 94f
Nasal process, caprine, 1011f
Nasal septum, ruminant, 1006, 1008f
Nasal vestibule, equine, 589
Nasofrontal angle, canine, 84f
Nasofrontal joint, avian, 1295
Nasoincisive notch
equine, 563f, 589
ruminant, 993f
Nasolabial plate, ruminant, 992f
Nasolacrimal duct
avian, 1270, 1314
canine, 132
equine, 589, 593
ruminant, 1010f
Nasomaxillary aperture, equine, 589–590, 590f
Nasopharynx
canine, 116–117, 150
equine, 615, 616–617f, 624–626, 625f, 658
feline, 107–108, 107f
ruminant, 1008f
Natal down, avian, 1409
Navicular bone, equine, 921f, 923. *See also* Distal sesamoid bone
Navicular bursa, equine, 922f, 923
Nephrosplenic ligament, equine, 729f
Nephrosplenic space, equine, 730b, 730f
Nerve
avian, 1255f
canine
auriculopalpebral, 88f
axillary, 88f
caudal cutaneous antebrachial, 88f

digital, medial and lateral, 85*f*
dorsal buccal branch of facial, 88*f*
facial, 88*f*
frontal, 88*f*
infraorbital, 88*f*
infratrochlear, 88*f*
mandibular (CN-V), 88*f*
maxillary (CN-V), 88*f*
median, 85*f*, 88*f*
mental, 88*f*
musculocutaneous, 88*f*
palmar, medial and lateral, 85*f*
palpebral, 88*f*
fibular (peroneal), 85*f*
phrenic, 88*f*
plantar, medial and lateral, 85*f*
radial, 85*f*
saphenous, 88*f*
sciatic, 85*f*
superficial radial, 88*f*
suprascapular, 85*f*
tibial, 85*f*
ulnar, 88*f*
vagus, dorsal and ventral branches of, 88*f*
ventral buccal branch of facial, 88*f*
zygomaticofacial, 88*f*
zygomaticotemporal, 88*f*
equine, 555*f*, 559*f*
auriculotemporal, 556*f*
axillary, 556*f*
caudal cutaneous sural, 557*f*
common fibular (peroneal), 557*f*
digital, 555*f*
medial and lateral, palmar and plantar, 557*f*
deep fibular (peroneal), 557*f*
dorsal branch of lateral palmar digital, 557*f*
dorsal branch of ulnar, 557*f*
fibular, 555*f*, 557*f*
foramen, 555*f*
frontal, 557*f*
infraorbital, 557*f*
mandibular (CNV), 557*f*
median, 555*f*
mental, 557*f*
musculocutaneous, 557*f*
ophthalmic, 557*f*
palmar, lateral and medial, 555*f*
fibular (peroneal), superficial, 557*f*
phrenic, 557*f*
plantar, lateral and medial, 555*f*
popliteal, 555*f*
radial, 555*f*
radial, superficial branch of, 557*f*
saphenous, 557*f*
sciatic, 555*f*
superficial fibular (peroneal), 557*f*
suprascapular, 557*f*
tibial, 557*f*
ulnar, 557*f*
vagus, 557*f*
zygomatic, 557*f*
zygomaticotemporal, 557*f*
feline, 95*f*, 99–100*f*
auriculotemporal, 95*f*
caudal cutaneous antebrachial, 95*f*
digital, palmar and plantar, medial and lateral, 95*f*
facial, 95*f*
frontal, 95*f*
infraorbital, 95*f*
infratrochlear, 95*f*
mandibular (CNV), 95*f*
medial cutaneous antebrachial, 95*f*
median, 95*f*
mental, 95*f*
musculocutaneous, 95*f*
palmar, medial and lateral, 95*f*
fibular (peroneal), 95*f*
phrenic, 99*f*
plantar, medial and lateral, 95*f*
radial, 95*f*
saphenous, 99*f*
sciatic, 95*f*
suprascapular, 95*f*
tibial, 85*f*
ulnar, 95*f*
vagus, dorsal and ventral branches of, 99*f*
zygomaticofacial, 99*f*
zygomaticotemporal, 99*f*
ruminant
abaxial palmar and plantar digital III and IV, 996*f*
accessory, dorsal branch of, 996*f*
auriculotemporal, 996*f*
axillary, 996*f*
buccal, 996*f*
dorsal buccal, 996*f*
common digital IV, 996*f*
common fibular (peroneal), 996*f*
accessory, dorsal branch of, 996*f*
auriculotemporal, 996*f*
dorsal common digital III, 996*f*
facial, 996*f*
frontal, 996*f*
infraorbital, 996*f*
infratrochlear, 996*f*
lumbar spinal, 996*f*
mandibular (CNV), 996*f*
median, 996*f*
mental, 996*f*
musculocutaneous, 996*f*
ophthalmic, 996*f*
palmar, medial and lateral, 996*f*
fibular (peroneal), 996*f*
phrenic, 996*f*
radial, 996*f*
saphenous, 996*f*
sciatic, 996*f*
tibial, 996*f*
ulnar, 996*f*
vagus, 996*f*
zygomatic, 996*f*
zygomaticotemporal, 996*f*
Nerve sheath, canine, 449
Neurocranium, avian, 1297
Nictitating membrane
avian, 1314
canine, 130–131
equine, 572
Nostril. *See also* Nares
avian, 1287
canine, 84*f*
equine, 554*f*
feline, 94*f*
ruminant, 992*f*, 1003*f*
Notarium, avian, 1378, 1380, 1382
Nuchal bursa, equine, 687, 687*b*
Nuchal ligament
canine, 667*b*
equine, 680, 687
ruminant, 993*f*
Nucleus pulposus, equine, 683

O

Oblique sesamoidean ligament, equine, 908–909
Obturator artery
caprine, 1132–1133
equine, 822, 858
Obturator foramen
avian, 1258*f*
equine, 936
ruminant, 1034*f*
Obturator nerve
avian, 1396
equine, 941
ruminant, 1036, 1036*f*, 1038–1039*f*, 1197
Obturator vein
caprine, 1132–1133
equine, 986
Occipital artery
equine, 658
feline, 110
Occipital bone
avian, 1258*f*
equine, 650
ruminant, 1000*f*
Occipital condyle
canine, 174
equine, 650
feline, 102*f*
ruminant, 1000*f*

Occipitohyoid muscle, equine, 617–618
Occipitomandibularis muscle, equine, 617–618
Occlusal surface of tooth
canine, 120
equine, 605*b*
Oculi, avian, 1313
Oculomotor nerve (CNIII)
canine, 131–132
equine, 652
feline, 144
ruminant, 652
Odontoblast, canine, 123
Olecranon
canine, 421–422
equine, 887
ruminant, 1151
Olfactory nerve, equine, 650–652
Omasal groove, ruminant, 1058
Omaso-abomasal orifice, ruminant, 1058
Omasum, ruminant, 1058
Omental bursa
equine, 728, 769
ruminant, 1117*f*, 1119
Omental sling, ruminant, 1117*f*, 1119
Omentum, lesser and greater
canine, 89*f*
equine, 560*f*
feline, 99*f*, 101*f*
ruminant, 999*f*
Omentum, ruminal insertion of, ruminant, 999*f*
Omohyoideus muscle, equine, 633–634, 633*b*
Omotransversus muscle
canine, 87*f*
equine, 879
ruminant, 1152*b*
Operculum, avian, 1279–1280, 1301–1303
Ophthalmic branch of trigeminal nerve, equine, 652
Ophthalmic nerve
avian, 1298
canine, 131–132
equine, 652
feline, 203
ruminant, 996*f*
Ophthalmic plexus, equine, 597
Optic canal
canine, 130–132
equine, 650–652
Optic chiasm
canine, 202
equine, 651*f*
feline, 203
ruminant, 999*f*
Optic disc, avian, 1313
Optic nerve (CNII)
canine, 202
equine, 650–652
feline, 202
ruminant, 652
Oral cavity
avian, 1268, 1293–1294, 1347
canine, 105
equine, 578, 579*f*, 581–582, 582*f*
feline, 105
Orbicularis oculi muscle
canine, 87*f*, 131
equine, 559*f*
feline, 202–203
ruminant, 996*f*
Orbicularis oris muscle
canine, 87*f*, 131
equine, 559*f*
feline, 202–203
ruminant, 996*f*
Orbit
avian, 1309
canine, 129–130
equine, 570, 571*f*
feline, 95*f*
Orbital fascia, canine, 130
Orbital fissure
canine, 131–132
feline, 204
Orbitalis muscle, equine, 870
Orbital ligament, canine, 129
Orbital soft tissues, canine, 130
Orifice (meatus), external urethral
caprine, 1132
canine, 90*f*
equine, 785, 986, 986*f*
ruminant, 1123
Oropharynx
avian, 1347
canine, 116–117, 150
equine, 629
Os coxae
equine, 935
ruminant, 1196
Os dorsale, avian, 1382
Osseous labyrinth, equine, 661–662
Ostium uteri externum, equine, 805–806
Ostium uteri internum, equine, 805–806
Otoliths, feline, 143
Oval foramen, equine, 652, 701
Ovarian artery
canine, 363
equine, 790–791, 806, 814
ruminant, 1122
Ovarian bursa, equine, 769, 788, 797–798, 799*f*
Ovarian follicle
avian, 1256*f*
canine, 363
equine, 800*b*
Ovarian ligament, ruminant, 1119–1120
Ovarian pedicle, canine/feline, 351–352
Ovarian vein, ruminant, 1122, 1124
Ovary
avian, 1331–1332
canine, 350–352
equine, 788, 789*f*, 795–797, 799–800, 799*f*
feline, 97*f*
ruminant, 1119
Ovary, proper and suspensory ligament of
canine, 350–352
equine, 788, 795
feline, 97*f*
Oviduct. *See also* Uterine tube
avian, 1332–1333
canine, 352
equine, 788, 789*f*, 811–814, 812–813*f*
ruminant, 1119
Ovulation fossa, equine, 788, 789*f*, 795–796, 798*f*

P

Pacinian corpuscles
canine, 548
caprine, 1235
equine, 982
Palatine aponeurosis, equine, 631
Palatine artery, alpaca, 1018
Palatine bone
avian, 1258*f*, 1268
canine, 116, 129
equine, 630–631, 650
Palatine foramen, canine, 131–132
Palatine process, canine, 117
Palatine sinus, ruminant, 1008*f*, 1010
Palatine tonsil, canine, 151–152
Palatinus muscle
canine, 117
equine, 631
Palatoglossal arch, canine, 151–152
Palatopharyngeal arche, equine, 630, 630*f*
Palatopharyngeus muscle
canine, 118
equine, 631
Palmar artery
canine, 85*f*
feline, 95*f*
equine, 555*f*, 897, 974
ruminant, 993*f*, 1167
Palmar nerve
canine, 85*f*
equine, 555*f*, 889, 897, 971–972
feline, 95*f*
Palmar branch of ulnar nerve, ruminant, 1190
Palmar carpal ligament, equine, 897, 902
Palmar common digital artery, ruminant, 1189
Palmar digital nerve, equine, 973

Palmar metacarpal nerve, equine, 897, 971–972
Palmar/plantar annular ligament, equine, 910–911
Palmar/plantar digital artery, equine, 911, 916, 930
Palmar/plantar digital nerve, equine, 911, 916, 929
Palmar/plantar metacarpal nerve, equine, 911
Palmar proper digital artery, ruminant, 1189
Palmar vein
 canine, 85*f*
 equine, 555*f*
 feline, 95*f*
Palmate toe, avian, 1390
Palpebral conjunctiva
 canine, 131
 equine, 572
Palpebral fissure, avian, 1313
Palpebral nerve, canine, 87*f*
Pamprodactyl foot, avian, 1390
Pancreas
 canine, 290
 equine, 736
 feline, 561*f*
 ruminant, 1093*f*
Pancreatic duct, canine/feline, 290
Pancreaticoduodenal node, ruminant, 1076
Papilla
 avian, 1354, 1404
 ruminant, 1025–1026, 1079
Papillae remigales, avian, 1408
Papillary cistern, equine, 986
Papillary duct, equine, 986
Papillary muscle
 avian, 1323
 canine, 269
Parabronchi, avian, 1339
Paracondylar process
 equine, 617
 ruminant, 1000*f*
Paralumbar fossa, ruminant, 1097, 1115–1116*b*, 1116–1117*f*
Paramesonephric (Müllerian) ducts, equine, 841*b*
Paranasal sinus
 avian, 1268*b*
 equine, 597–599, 600*f*
 ruminant, 1002*b*, 1003*f*, 1007–1011, 1007*b*, 1008*f*
Parapatellar fibrocartilage, equine, 948
Parasympathetic trunk, ruminant, 1100
Parathyroid gland
 canine, 158–160
 feline, 168
 equine, 797
 ruminant, 993*f*
Parietal bone
 avian, 1258*f*
 canine, 84*f*
 equine, 554*f*
 feline, 102*f*
Parietal groove
 avian, 1258*f*
 canine, 92*f*
 equine, 563–564*f*, 930
 feline, 102*f*
Parietal leaflet, equine, 712
Parietal peritoneum, equine, 767–769, 768*f*
Parietal vaginal tunic, equine, 837
Paroophoron, equine, 789
Parotid duct
 canine, 87*f*
 equine, 609
 ruminant, 996*f*
Parotid lymph node
 alpaca, 1018
 canine, 85*f*, 162
 equine, 575
 feline, 111
 ruminant, 1011, 1027
Parotidoauricular muscle
 equine, 584, 656–658
 feline, 97*f*
 ruminant, 996*f*
Parotid salivary gland
 canine, 85*f*
 equine, 582–583, 583*f*, 617, 656–658
 feline, 95*f*
 ruminant, 993*f*, 996*f*
Pars brevis tendon, avian, 1374
Pars flaccida
 canine, 139
 feline, 109*f*
Pars longus, avian, 1374
Pars optica retina, avian, 1313
Patagii accessorius muscle, avian, 1255*f*
Patagii brevis muscle, avian, 1255*f*
Patella
 avian, 1258*f*, 1389
 canine, 84–85*f*
 camelids, 1202
 equine, 554*f*, 563–564*f*, 945*f*, 946
 feline, 94*f*, 102*f*
 ruminant, 992*f*, 1000*f*, 1207, 1207*b*
Patellar ligament
 canine, 475*f*
 equine, middle, medial, lateral, 944*b*, 948
 feline, 501*b*
 ruminant, 993*f*
Paw pads
 canine, 84*f*, 542
 feline, 94*f*
Pecten oculus, avian, 1313
Pectoral girdle
 avian, 1369
  bones of, 1369–1372
  H view of, 1371*b*
 ruminant, 1149–1155, 1153*f*
Pectoral muscle
 avian, 1372*b*
 deep
  equine, 557*f*, 559*f*
  ruminant, 1151, 1154
 descending, equine, 558–559*f*
 superficial
  avian, 1255*f*
  equine, 557*f*, 889*b*
  ruminant, 1154
 transverse, equine, 558*f*
Pedal artery, equine, 967, 974–975
Peduncles, equine, 689
Pelvic brim, equine, 760–761
Pelvic diaphragm, canine, 321–323
Pelvic flexure of large colon, equine, 560*f*
Pelvic nerve
 canine, 344
 caprine, 1133–1134
 equine, 761, 791, 850, 858
 ruminant, 1080, 1111
Pelvic plexus
 equine, 806–807
 ruminant, 1111, 1123
Pelvic sacral foramina, ruminant, 1037
Pelvic symphysis. *See* Symphysis pelvis
Pelvis
 avian, 1388
  female, 1256*f*
  male, 1257*f*
 canine, 458–459
  female, 1257*f*
  male, 90*f*
 equine, 935–936, 936*f*
  female, 560*f*
  male, 561*f*
 feline
  female, 99*f*
  male, 100*f*
 ruminant, 1032, 1034*f*
  female, 997*f*
  male, 998*f*
Penile artery, equine, 822
Penile vein, ruminant, 1141
Penis
 canine, 387–389, 388*b*
 equine, 822–823
 ruminant, 1137–1139, 1137*f*, 1140*b*, 1141, 1142*f*
Pennaceous, avian, 1407
Pericardium, parietal and visceral
 avian, 1320
 canine, 271
 feline, 248–249
 ruminant, 1046–1047, 1047*f*

Perimetrium, equine, 805
Perineum
  canine, 358
  caprine, 1131–1132
  equine, 761*b*, 782*f*, 783–784
  ruminant, 1111*f*
Perineal nerve, equine, 776, 839
Periodontal ligament
  canine, 124
  equine, 607
Periodontium, canine, 123–125
Periople
  equine, 927–929
  ruminant, 1175
Periorbita, canine, 130
Peripheral vestibular system, canine, 184–186
Peri-sinusoidal space, ruminant, 1086
Peritoneum, parietal and visceral
  canine, 240–241
  equine, 767–770, 768*f*
  feline, 328–330
Peroneal nerve. *See* Fibular (peroneal) nerve
Pes
  canine, 92*f*
  equine, 563–564*f*
  feline, 102*f*
Pessulus, avian, 1283
Petrous temporal bone, feline, 108
Peyer's patch
  avian, 1360*b*
  ruminant, 1076
Phalanx - distal (3rd), middle (2nd), proximal (1st)
  avian, 1258*f*
  canine, 92*f*
  equine, 563–564*f*
  feline, 102*f*
  ruminant, 1000*f*
Phallus, avian, 1354
Pharyngeal ostium, feline, 142
Pharyngoesophageal branch, canine, 276
Pharynx
  avian, 1281, 1347
  canine, 116–117
  equine, 629, 646, 647*f*
  ruminant, 1008*f*
Photoreceptors, feline, 202
Phrenic nerve
  canine, 89–90*f*
  equine, 560*f*
  feline, 99–100*f*
  ruminant, 997*f*
Phrenicoabdominal vein, 53*f*
Phrenicopericardiac ligament
  canine, 240–241
  equine, 560*f*
  feline, 247
Phrenicosplenic ligament, equine, 729
Philtrum
  canine, 84*f*
  feline, 94*f*
Pillars of the rumen, ruminant, 1058
Pilosebaceous unit, canine, 547
Pinna
  canine, 84–85*f*
  equine, 655, 656*f*, 710, 712
  feline, 94*f*, 135–137
  ruminant, 999*f*
Piriformis muscle, canine, 88*f*
Placenta
  canine, 262, 369*b*
  equine, 804
  ruminant, 984, 1063
  feline, 367
Plantar
  annular ligament, equine, 910–911
  artery
    canine, 85*f*
    equine, 555*f*, 974–975
    feline, 95*f*
    ruminant, 1167
  deep arch, equine, 974–975
  digital artery, equine, 911, 916, 930
  digital nerve, equine, 911, 916, 929
  digital vein, ruminant, 1189
  lateral artery, equine, 974–975
  lateral nerve
    canine, 85*f*
    equine, 555*f*, 973–974
    feline, 95*f*
    ruminant, 1190
  medial artery, equine, 974–975
  medial nerve, ruminant, 1190
  metacarpal nerve, equine, 911
  metatarsal nerve, equine, 973–974
  nerve, ruminant, 1190
Plantar vein
  canine, 85*f*
  equine, 555*f*
  feline, 95*f*
Pleura, parietal and visceral
  canine, 240–241
  equine, 720
  feline, 231*b*, 235
  ruminant, 1087
Pleural reflection
  canine, 84–85*f*, 89*f*
  equine, 554–555*f*, 560*f*
  feline, 94–95*f*, 99*f*
  ruminant, 992–993*f*, 997*f*
Plumaceous, avian, 1407
Plumage, avian, 1254*f*
Pneumatic bone, avian, 1372*b*
Pneumatic foramen, avian, 1372
Pneumotricipital fossa, avian, 1372
Podotheca, avian, 1393
Podotrochlear bursa, ruminant, 1166
Poll
  equine, 554*f*
  ruminant, 993*f*
Polystomatic sublingual gland, equine, 583*f*, 584–585
Pons, equine, 652
Pontine vestibular nuclei, feline, 144
Popliteal artery
  canine, 85*f*
  equine, 555*f*, 967
  feline, 95*f*
  ruminant, 993*f*
Popliteal lymph node
  camelids, 1204
  canine, 85*f*
  equine, 939, 967, 975
  feline, 95*f*
Popliteus muscle, equine, 558*f*
Popliteal nerve, equine, 555*f*
Portal lobule of liver, equine, 767, 768*f*
Portal tracts, equine, 767
Portal valve, avian, 1353
Portal vein, 53*f*
  avian, 1325
  canine, 282
  equine, 740, 755, 765
  ruminant, 1066, 1075, 1080, 1085, 1085–1087*f*, 1094
Portio vaginalis, equine, 805–806
Positron emission tomography (PET), 72, 78–80
Posterior chamber of globe, equine, 570
Posterior scleral ossicles, avian, 1311
Powder down, avian, 1409
Preen (uropygial) gland, avian, 1255*f*
Prefemoral (superficial inguinal) lymph node, ruminant, 1049–1050
Preganglionic ciliary ganglion, feline, 203
Premaxilla bone, avian, 1258*f*, 1295
Premolars
  canine, 122–123
  equine, 603–604, 607, 607*b*
Prepubic
  canine, 84*f*
  equine, 936
  feline, 94*f*
  ruminant, 1196
Prepubic tendon
  equine, 936
  feline, 97*f*
  ruminant, 1196
Prepuce
  canine, 388–389, 388*b*
  equine, 782
  feline, 333–335
  ruminant, 1141
Preputial orifice
  canine, 385
  ruminant, 1141

Prescapular (superficial cervical) lymph node
canine, 85*f*, 163
equine, 669, 897
feline, 95*f*
ruminant, 1049–1050
Presphenoid bone, equine, 650–652
Primary bronchi, avian, 1339
Primary epidermal lamellae, equine, 929
Primary hairs, canine, 546
Primary remige, avian, 1373, 1407–1408
Process calcaris, avian, 1393
Procricoid cartilage, avian, 1281
Proctodeum, avian, 1354
Promontory of eye, feline, 109
Pronator longus et brevis, avian, 1255*f*
Pronator quadratus muscle, canine, 424
Pronator teres muscle, canine, 428
Propatagium, avian, 1374
Proper axillary lymph node, ruminant, 1190
Proper ligament of ovary
canine, 89*f*
equine, 560*f*
feline, 351*f*
ruminant, 1120
Proprioceptors, canine, 192–193
Prosencephalon
canine, 190–191
feline, 201
Prostate gland
canine, 383*b*
equine, 829*f*, 832–833
feline, 97*f*
ruminant, 998*f*
Prostatic artery
canine, 341, 344
equine, 833
Prostatic ducts, equine, 833
Protractor prepuce, ruminant, 1141
Proventriculus, avian, 1347
Proximal annular ligament, ruminant, 1166
Proximal curves of urethra, caprine, 1132
Proximal digital annular ligament, equine, 910–911
Proximal interphalangeal (pastern) joint
equine, 915*f*, 916
ruminant, 1185
Proximal loop of ascending colon, ruminant, 1079–1080, 1079*f*
Proximal radioulnar joint, ruminant, 1156
Proximal sesamoid bone
equine, 907, 910*f*
ruminant, 1000*f*
Proximal tubercle of the talus, equine, 901
Psoas muscle, canine, 239
Pterygoid bone
avian, 1297
equine, 630–631
Pterygoid hamulus, feline, 142
Pterygoid muscle
canine, 130
equine, 617, 654
Pteryla capitalis, avian, 1405
Pterylae, avian, 1404, 1405*b*, 1405*f*
Pubis
avian, 1258*f*, 1388
equine, 936
ruminant, 1032, 1034*f*
Pudendal nerve
canine, 321–323
caprine, 1131–1134
equine, 761, 806–807, 823, 850, 858, 987
ruminant, 998*f*, 1102, 1111, 1143
Pudendal rectal nerve, equine, 791
Pulmonary artery
equine, 712
ruminant, 1052, 1089
Pulmonary chamber, avian, 1320
Pulmonary ligament
canine, 222
ruminant, 1087
Pulmonary pleura, ruminant, 1087
Pulmonary plexus, ruminant, 1089
Pulmonary trunk
avian, 1323
canine, 89*f*
equine, 559*f*
feline, 99*f*
ruminant, 1047*f*, 1052
Pulmonary vein
avian, 1320
canine, 90*f*
ruminant, 1052, 1089
Pulmonic valve
canine, 259*f*, 262*b*
equine, 694*b*
feline, 262*b*
ruminant, 1046*b*
Pulp of spleen
canine, 124
equine, 605, 605*f*
Puncta of nasal cavity, avian, 1314
Pupil
avian, 1306–1307*f*, 1312
canine, 182
equine, 570*f*
ruminant, 1002
Purkinje fibers
avian, 1323
equine, 690, 690*b*, 691*f*
ruminant, 1053
Pygostyle, avian, 1258*f*, 1380, 1383
Pyloric part of stomach
canine, 91*f*
feline, 101*f*
ruminant, 1066
Pylorus
canine, 307, 312–313
equine, 726, 728, 734
Pyramidal muscle, avian, 1314
Pyramidal system, feline, 206
Pyramidal tracts, equine, 652

Q

Quadrate bone, avian, 1297
Quadrate lobe of the liver
equine, 561*f*
feline, 101*f*
Quadratojugal bone, avian, 1297

R

Rachis, avian, 1406
Radial artery
canine, 85*f*
equine, 974
ruminant, 1189
Radial carpal bone
avian, 1373
equine, 894–896
Radial fossa, ruminant, 1151
Radial nerve
avian, 1375
canine, 85*f*
equine, 555*f*, 866
feline, 95*f*
ruminant, 1159*b*, 1167–1168
Radius
avian, 1253*f*, 1258*f*
canine, 92*f*, 422–423
equine, 563–564*f*
feline, 102*f*
ruminant, 1000*f*, 1151, 1152–1153*f*
Ramus of mandible, equine, 607
Reciprocal apparatus
equine, 952*f*
ruminant, 1213–1214, 1215–1216*f*
Rectal ampulla, equine, 760–761, 760*b*
Rectococcygeus muscle, equine, 760–761
Rectogenital pouch, ruminant, 1123
Rectovaginal septum
equine, 560*f*
ruminant, 1109*f*
Rectum
avian, 1256–1257*f*
canine, 89–90*f*
equine, 560–561*f*
feline, 99–100*f*
ruminant, 997–998*f*
Rectus capitis muscle, ventral
avian, 1255*f*
equine, 558*f*

Rectus femoris muscle
equine, 558*f*
ruminant, 996*f*
Rectrices, avian, 1348, 1383, 1408
Rectus abdominis muscle
canine, 89–90*f*
equine, 559*f*
feline, 99*f*
ruminant, 997*f*
Rectus capitis lateralis, equine, 654
Rectus capitis ventralis muscle, equine, 615, 650, 651*f*
Recurrent laryngeal nerve
canine, 175–176
equine, 643–644, 643–644*b*
Red pulp of spleen
avian, 1360
equine, 726
Regional anatomy
avian, 1258*f*
canine, 84*f*
equine, 554*f*
feline, 94*f*
ruminant, 992*f*
Reissner's vestibular membrane, 144
Remiges, avian, 1407
Renal artery
avian, 1352
canine, 341
ruminant, 1100, 1100*f*
Renal lymph node, ruminant, 1100
Renal pelvis, caprine, 1130
Renal plexus, ruminant, 1100
Renal portal system, avian, 1325, 1353, 1353*b*
Renal portal vein, avian, 1353
Renal vein, ruminant, 1100, 1100*f*
Renosplenic ligament, equine, 729. *See also* Nephrosplenic ligament
Rete tibiotarsale, avian, 1325, 1396
Reticulae, avian, 1393
Reticular artery, ruminant, 1060
Reticular (reticulo-omasal) groove, ruminant, 1047*f*, 1058
Reticulo-omasal orifice, ruminant, 1058
Reticulorumen, ruminant, 1060
Reticulum, ruminant, 1047, 1047*f*, 1058
Retina
avian, 1312
canine, 84*f*
equine, 554*f*
ruminant, 996*f*
Retinaculum, external of carpus, crural, extensor, flexor, metacarpal, metatarsal, tarsal
canine, 87*f*
equine, 557*f*
feline, 94*f*
ruminant, 996*f*
Retractor bulbi muscle, avian, 1310
Retractor penile muscle
caprine, 1132
equine, 822
ruminant, 1141
Retractor prepuce muscle, ruminant, 1141
Retrobulbar nerve block, equine, 574*b*
Retropharyngeal lymph node
alpaca, 1018
canine, 118, 125, 163
equine, 617–618, 644
ruminant, 993*f*
Rhamphotheca, avian, 1287
Rhinotheca, avian, 1285, 1287, 1295
Rhomboideus muscle, ruminant, 1152
Rib
avian, 1382
canine, 85*f*
equine, 563–564*f*
feline, 95*f*
ruminant, 993*f*
Rictus, avian, 1293–1294
Right view of the thoracic and abdominal cavities and pelvis (male)
avian, 1257*f*
canine, 90*f*
equine, 561*f*
feline, 100*f*
ruminant, 998*f*
Rima oris, equine, 578
Rosette of Fϋrstenberg, ruminant, 1243*f*
Rostral alar foramen, canine, 131–132
Rostral auricular artery, equine, 658
Rostral auricular vein, equine, 658
Rostral colliculus, feline, 203
Rostral conchae, avian, 1280–1281
Rostral cranial fossa, equine, 650–652
Rostral tympanic artery, equine, 660
Rostrum, avian, 1287
Round ligament of urinary bladder
equine, 765, 767*f*
ruminant, 1087, 1092
Ruffini corpuscles, canine, 548–549
Rumen, cranial, dorsal, and ventral sac of, ruminant, 1058, 1059*f*, 1060, 1062*f*
Ruminoreticular fold, ruminant, 1047*f*
Rump, avian, 1254*f*

S

Saccule, canine, 184
Saccus cecus, equine, 727
Sac of rumen, ruminant, 1047*f*
Sacral foramen, ruminant, 1034*f*, 1037
Sacral lymph node, ruminant, 1122
Sacral nerve, ruminant, 1036*f*, 1037
Sacral plexus
avian, 1352, 1396
ruminant, 1037
Sacroiliac joint, equine, 936
Sacrosciatic ligament, equine, 936
Sacrum
canine, 92*f*
equine, 563–564*f*
feline, 97*f*
ruminant, 1000*f*, 1032, 1033*f*
Salivary gland
canine, 85*f*, 149–150
equine, 555*f*, 582–585, 583*f*
feline, 95*f*
ruminant, 993*f*
Salt gland, avian, 1314
Saphenous artery
camelids, 1204, 1204*b*
canine, 85*f*
equine, 967
Saphenous nerve
canine, 85*f*, 88*f*, 493
equine, 941, 950–951, 966
ruminant, 1036
Saphenous tibial artery, ruminant, 1189
Saphenous vein
canine, 85*f*
feline, 95*f*
ruminant, 993*f*, 1189
Sartorius muscle
equine, 956
feline, 97*f*
Sartorius with superficial gluteal muscle, avian, 1255*f*
Scales, avian, 1393
Scalenus medius muscle, ruminant, 996*f*
Scapula
avian, 1258*f*, 1370–1371
canine, 92*f*
equine, 875–876
feline, 97*f*, 102*f*
ruminant, 1000*f*, 1148*f*, 1149, 1150*f*, 1154*b*
Scapula feathers, avian, 1254*f*
Scapular cartilage
equine, 875–876
ruminant, 1149
Scapular spine
canine, 85*f*
equine, 564*f*
ruminant, 1000*f*, 1149
Scapula, tuber, equine, 563–564*f*
Scapulohumeral (shoulder) joint
equine, 876
ruminant, 1148*f*, 1149
Sciatic nerve
avian, 1352, 1396
canine, 85*f*, 537
equine, 555*f*, 941

feline, 95*f*
ruminant, 1036*f*, 1039*f*, 1197, 1219
Scleral ossicle, avian, 1311
Scrotal ligament, equine, 839
Scrotal lymph node, caprine, 1133
Scrotum
canine, 375–379, 380*b*
equine, 836–840, 837–838*f*
ruminant, 998*f*
Scutellae, avian, 1393
Scutes, avian, 1393
Sebaceous gland
canine, 547
equine, 980*f*, 982
Sebum, canine, 547
Secondary bronchi, avian, 1339
Secondary epidermal lamellae, equine, 929
Secondary feathers, avian, 1254*f*
Secondary hairs, canine, 546
Secondary remiges, avian, 1373, 1407–1408
Secondary tympanic membrane, equine, 659
Second (middle) phalanx, equine, 913*f*, 915, 915*f*
Secretory tubules, equine, 833
Semibristles, avian, 1409
Semicircular canals
canine, 184
equine, 661–662
feline, 108
Semimembranosus muscle
avian, 1255*f*
canine, 88*f*
equine, 558*f*
feline, 98*f*
ruminant, 996*f*
Seminal vesicle (vesicular gland)
equine, 825, 826–828*f*, 827, 832
ruminant, 998*f*
Semiplumes, avian, 1409
Semispinalis cervicis muscle, avian, 1255*f*
Semitendinosus muscle
avian, 1255*f*
canine, 88*f*
equine, 558*f*
feline, 98*f*
ruminant, 996*f*
Sensory pretectal nucleus, feline, 203
Septal leaflet, equine, 712
Septomarginal trabecula. *See* Moderator band
Septum of eyes, avian, 1310
Serratus ventralis cervicis muscle, ruminant, 996*f*
Serratus ventralis muscle
canine, 88*f*
equine, 558*f*
feline, 98*f*
ruminant, 996*f*, 1154
Sertoli cells, canine, 379*b*
Sesamoid bone
canine, 92*f*, 437*b*, 474–475, 475*b*
equine - proximal, 563–564*f*
equine - distal (navicular), 563–564*f*
feline - popliteal, 95*f*, 102*f*
ruminant, 1000*f*
Sesamoidean ligament
equine, 907–909, 909*f*
ruminant, 1166–1167
Setae, avian, 1409
Shaft, avian, 1406
Sheath
equine, 561*f*
ruminant, 1141
Shell gland, avian, 1332
Short ligament of the tarsus, equine, 902
Short plantar ligament of the tarsus, equine, 902
Short sesamoidean ligament, equine, 908–909
Shoulder
alpaca, 1146–1147, 1147–1148*f*
avian, 1370–1371
canine, 392*b*, 399
equine, 876
Sigmoid colon, ruminant, 1080
Sigmoid flexure of pancreas
caprine, 1132
equine, 734
feline, 94*f*
ruminant, 1141
Sigmoid loop of duodenum, ruminant, 1071–1073
Single-slice scanners, 47, 49*f*
Sinoatrial node
canine, 84*f*
equine, 710
ruminant, 1053
Sinoatrial valve, avian, 1320
Sinus hairs, canine, 548–549
Sinus of Valsalva, equine, 711
Sinusoid, ruminant, 1085
Sinus venosus, avian, 1320
Skeleton - cranial and caudal view, equine, 564*f*
Skeleton - lateral view
avian, 1258*f*
canine, 92*f*
equine, 563*f*
feline, 102*f*
ruminant, 1000*f*
Skin, unfeathered, avian, 1254*f*
Skull
avian, 1253*f*
canine, 172*f*
equine, 646–647, 648*f*, 653
feline, 95*f*
ruminant, 993*f*
Small (descending) colon, equine, 760, 760*f*
Small intestine
avian, 1347
canine, 312
equine, 739–740, 740*f*
feline, 311–312
ruminant, 1071–1075
Snood, avian, 1254*f*
Soft palate
canine, 116–118, 117*b*, 150
equine, 625, 626*f*, 630, 630*b*
Solar dermis, equine, 927–928
Solar epidermis, equine, 928–929
Solar foramen, equine, 930
Sole
equine, 918
ruminant, 1178
Soleus muscle
equine, 962
feline, 97*f*
Spermatic cord
canine, 381
equine, 841–843
Sphenoid bone
canine, 129
equine, 596
Sphenopalatine artery
alpaca, 1018
equine, 597
feline, 108
ruminant, 1011
Sphenopalatine foramen, canine, 131–132
Sphenopalatine sinus
equine, 592, 594*f*, 595–597, 596*f*
ruminant, 1011
Sphenopalatine vein, equine, 597
Sphincter ani muscle, avian, 1255*f*, 1312
Spinal cord
avian, 1396
canine, 208*b*
equine, 563–564*f*, 652, 685–687
feline, 102*f*
ruminant, 1034–1035
Spinal nerve
avian, 1396
canine, 211, 241, 523–524
equine, 685–687, 866, 868*f*
ruminant, 1034–1035
Spinal segments, canine, 211–212
Spinous process, dorsal
canine, 85*f*
equine, 555*f*
feline, 993*f*
ruminant, 993*f*
Spiral colon, ruminant, 1079*b*
Spiral loop of colon, ruminant, 1079–1080

Spiral organ (organ of Corti)
equine, 662
feline, 145
Splanchnic nerve
canine, 296
equine, 758, 760
Spleen
avian, 1348, 1348*b*, 1360
canine, 301–302
lymph node, 303
vascularization, 302–303
equine, 726, 729*f*
feline, 98*f*, 301–302
Splenogastric artery, ruminant, 1060
Splenic hilus, equine, 730
Splenic lymph node, equine, 731
Splenic nodules (and accessory), avian, 1360
Splenic vein, ruminant, 1066
Splenius muscle, ruminant, 996*f*
Squamous (nonglandular) mucosa of stomach, equine, 726
Stapedius muscle, equine, 658–659
Stapes
equine, 658, 660*f*
feline, 109
Stellate ganglion, equine, 870–871
Sternal diverticula, avian, 1338
Sternal flexure of large colon, equine, 560*f*
Sternocephalicus muscle
canine, 177
ruminant, 996*f*
Sternohyoideus muscle
canine, 179
equine, 582–583, 583*f*, 628, 628–629*f*, 633–634, 633*b*
feline, 167
Sternomandibular muscle, ruminant, 996*f*
Sternopericardiac ligament, ruminant, 1046
Sternothyroideus muscle
avian, 1255*f*
canine, 179
equine, 628, 628*f*, 633–634, 633*b*
feline, 167
Sternum
avian, 1258*f*, 1371, 1374, 1382–1383
canine, 85*f*, 92*f*, 239
equine, 563–564*f*
feline, 97*f*, 102*f*
Stifle (knee) joint
canine, 85*f*, 490
equine, 944
feline, 95*f*
ruminant, 1207–1210
Stomach
avian, 1347
canine, 307–309
equine, 726–728, 726–727*f*
feline, 97*f*
Straight sesamoidean ligament, equine, 908–909
Stratum basale, equine, 805, 980*f*, 981
Stratum compactum, equine, 805
Stratum corneum
avian, 1294
canine, 542, 543*f*
equine, 980*f*, 981
Stratum externum, equine, 927
Stratum granulosum, equine, 980*f*, 981
Stratum internum, equine, 927, 929
Stratum lamellae, equine, 929
Stratum lucidum, equine, 980*f*, 981
Stratum medium, equine, 927
Stratum profundum, avian, 1403
Stratum spinosum, equine, 980*f*, 981
Stratum spongiosum, equine, 805
Stratum superficiale, avian, 1403
Styloglossus muscle, equine, 581, 633
Stylohyoid bone
canine, 176
equine, 614–617*f*, 615, 618, 654
Styloid process
of ulna, ruminant, 1000*f*
lateral, equine, 563–564*f*
Stylomastoid foramen
equine, 652, 658
feline, 110, 141, 202–203
Stylopharyngeus muscle, equine, 615, 615–617*f*, 632–633, 635*f*
Subclavian artery
avian, 1375
canine, 160, 254
equine, 880–881
Subclavian vein, avian, 1324
Subclavius muscle
equine, 554*f*
ruminant, 1154
Subcutaneous olecranon bursa
equine, 564*f*
ruminant, 1156–1157
Subcutis
canine, 549
equine, 980*f*, 981
ruminant, 1174–1175
Subiliac lymph node, equine, 967, 975
Sublingual caruncle, equine, 580*f*, 584–585
Sublingual fold, equine, 580*f*, 581, 584–585
Sublingual polystomatic salivary gland, equine, 581
Submental organ, canine, 547
Submucosal plexus
equine, 740, 760, 760*f*
ruminant, 1080
Subperitoneal node, equine, 814
Subpulmonary infundibulum, equine, 700–701
Subscapular artery, ruminant, 1159
Subscapularis muscle, ruminant, 1154
Subscapular nerve, ruminant, 1158–1159
Substantia propria, canine, 131
Suburethral diverticulum, ruminant, 1103*b*, 1110, 1124
Subvertebral canal, avian, 1323
Sulcus of frog/hoof, equine, 918, 918*f*
Sulcus of tooth, canine, 125
Supercilium, avian, 1254*f*
Superficial artery, feline, 110
Superficial branch of the fibular nerve, ruminant, 1190
Superficial branch of the radial nerve, ruminant, 1190
Superficial cervical artery, canine, 160
Superficial cervical lymph node
canine, 85*f*, 163
equine, 555*f*, 669, 897
feline, 95*f*
ruminant, 993*f*, 1160, 1160*b*, 1167
Superficial digital flexor muscle
canine, 536
equine, 900–901
ruminant, 1155, 1187–1189, 1214, 1216*f*
Superficial fibular nerve, ruminant, 1168
Superficial inguinal lymph node
canine, 85*f*
equine, 761–762, 839, 987
feline, 95*f*, 111
ruminant, 1111, 1143
Superficial radial nerve, canine, 88*f*
Superficial palmar arch, equine, 974
Superficial plexus, equine, 981
Superficial temporal artery, equine, 658
Superficial temporal vein, equine, 658
Superficial ulnar vein, avian, 1253*f*
Superficial vein, equine, 986
Superior mesenteric ganglion, ruminant, 1076
Superior tarsus muscle, equine, 870
Supernumerary digit, canine/feline, 434*b*
Supracondylar crest of humerus, lateral, equine, 563–564*f*
Supracondylar foramen, canine/feline, 405*b*
Supracondylar tuberositiy of humerus, equine, 962
Supracoracoideus muscle, avian, 1374
Supraglenoid tubercle of scapula
canine, 92*f*
equine, 563–564*f*, 875
feline, 95*f*
ruminant, 1149*b*
Supraomental recess (omental sling), ruminant, 1066, 1071–1073, 1119

Suprascapular artery, equine, 880–881
Suprascapular nerve
canine, 85*f*
equine, 563*f*
feline, 95*f*
ruminant, 1158–1159
Supraspinatus muscle
canine, 88*f*
equine, 878*t*
ruminant, 996*f*, 1154
Supraspinous bursa, equine, 687, 687*b*
Supraspinous fossa, equine, 563–564*f*
Supratrochlear foramen, canine/feline, 400
Surface lymphoid tissues, avian, 1360–1361
Suspensory ligament
equine, 902
ruminant, 1119–1120, 1166–1167, 1239–1240
Sustentaculum tali
equine, 960
ruminant, 1000*f*
Sympathetic trunk
canine, 176
equine, 678
ruminant, 1100
Symphyseal tendon
canine, 997*f*
ruminant, 1100
Symphysis of pelvis, ruminant, 1032
Syndactyl foot, avian, 1390
Synsacrum, avian, 1258*f*, 1352*b*, 1378, 1380, 1383, 1388
Syringeal bulla, avian, 1284
Syrinx, avian, 1277, 1282–1284

T

Tail
avian, 1408
canine, 530
equine, 554*f*
Tail coverts, avian, 1409
Tail gland, canine, 547
Tail head
equine, 554–555*f*
ruminant, 992*f*
Tail of epididymis
equine, 837*f*
ruminant, 998*f*
Tail vein, ruminant, 993*f*
Talocalcaneocentral joint, ruminant, 1217–1219
Talon, avian, 1393
Talus
canine, 92*f*
equine, 563–564*f*
feline, 102*f*
ruminant, 1000*f*
Tarsal bone
avian
distal row, 1390
proximal row, 1389–1390
equine, 960–962
ruminant, 1211*f*, 1217, 1218*f*
Tarsal bone - central, 1st, 2nd, 3rd, 4th, equine, 563–564*f*
Tarsal bone - central, 1st, 2nd, 4th, canine, 92*f*
Tarsal bone - central, 1st, 4th, feline, 102*f*
Tarsal bone - centroquartel, 2nd, 3rd, 4th, ruminant, 1000*f*
Tarsal joints
canine, 131, 515
equine, 962
Tarsal plate, avian, 1313
Tarsocrural joint
canine, 84*f*
equine, 960, 964*b*
ruminant, 1217
Tarsometatarsal joint
equine, 960
ruminant, 1217, 1219, 1219*b*
Tarsometatarsus bone - 2nd, 3rd, and 4th, avian, 1258*f*, 1390
Tarsus, canine, 131
T-cells, avian, 1359*b*
Teat
canine, 89*f*
equine, 560*f*
feline, 97*f*
ruminant, 1245–1246
supernumerary, 1247*b*
Tectorial membrane, feline, 145
Tectrices, avian, 1409
Teeth
canine, 123
equine, 608–609
feline, 95*f*
ruminant, 1015–1018
Telencephalon
canine, 190
feline, 201–202
Temporal artery
feline, 110
ruminant, 1028
Temporal bone
avian, 1258*f*
equine, 656
Temporalis muscle
canine, 130
equine, 557*f*
feline, 98*f*
Temporal vein, ruminant, 1028
Temporohyoid joint, equine, 621
Tendon of flexor perforatus muscle, avian, 1255*f*
Tenon's capsule, canine, 130
Tension receptors, ruminant, 1060–1062
Tensor fasciae latae muscle
canine, 527
equine, 557*f*
feline, 98*f*
ruminant, 1093*f*
Tensor patagii longus muscle, avian, 1255*f*
Tensor propatagialis muscle, avian, 1374
Tensor tympani muscle, equine, 658–659
Tensor veli palatini muscle
canine, 117
equine, 631, 634*f*
Tentorium cerebelli, equine, 689
Teres major muscle
avian, 1255*f*
ruminant, 1154
Teres major tuberosity
equine, 879–880
ruminant, 1151
Teres minor muscle
equine, 559*f*
ruminant, 1154
Terminal arch
equine, 554*f*
ruminant, 1189
Testis
avian, 1353–1354
canine, 379–380, 380*b*
equine, 836–840, 837–838*f*
feline, 100*f*
ruminant, 998*f*
Testicular bursa, equine, 769, 841
Thalamus, 63*f*
Thebesian vein, equine, 713
Third (distal) phalanx, equine, 920–921, 921*f*
Third trochanter of femur, equine, 563*f*
Throat latch, equine, 554*f*
Thoracic cavity
avian
female (left), 1256*f*
male (right), 1257*f*
canine
female (left), 1257*f*
male (right), 90*f*
equine, 718
female (left), 560*f*
male (right), 561*f*
feline
female (left), 99*f*
male (right), 100*f*
ruminant
female (left), 997*f*
male (right), 998*f*
Thoracic duct
canine, 53*f*, 89*f*
equine, 560*f*
feline, 99*f*
ruminant, 997*f*

Thoracic rhomboid muscle, ruminant, 996*f*
Thoracic sympathetic trunk, equine, 870–871
Thoracolumbar fascia, canine, 87*f*
Thrombocyte, avian, 1363
Thymus
  avian, 1360
  canine, 163
  feline, 99*f*
Thyroepiglottic ligament, equine, 641
Thyrohyoid bone
  canine, 176
  equine, 642
Thyroid cartilage
  canine, 156
  equine, 641–642
  ruminant, 1008*f*
Thyroid gland
  canine, 85*f*, 157–159
  equine, 555*f*
  feline, 95*f*, 167–168
  ruminant, 993*f*, 996*f*
Thyropharyngeus muscle, equine, 643
Tibia
  canine, 92*f*
  equine, 563–564*f*
  feline, 95*f*, 102*f*
  ruminant, 1000*f*
Tibial crest, camelids, 1202
Tibial artery
  camelids, 1204
  canine, 85*f*
  equine, 967
  feline, 85*f*
  ruminant, 996*f*
Tibial nerve
  avian, 1396
  canine, 85*f*, 537
  equine, 966
  ruminant, 996*f*, 1037, 1168, 1190
Tibial tuberosity
  avian, 1258*f*
  canine, 92*f*
  equine, 563–564*f*, 948
  feline, 102*f*
  ruminant, 1000*f*
Tibiotarsal vein, avian, 1258*f*, 1325
Tibiotarsus, avian, 1389–1390
T-ligament, equine, 909, 923
Tomia, avian, 1293
Tongue
  canine, 176
  equine, 579–581, 580*f*
  feline, 105*f*
  ruminant, 1233
Tonsil, feline, 108
Topographical anatomy
  avian, 1253*f*
  canine, 85*f*
  equine, 555*f*
  feline, 95*f*
  ruminant, 993*f*
Totipalmate foot, avian, 1390
Trabeculae, avian, 1383
Trabeculae carnae, avian, 1323
Trachea
  avian, 1282–1283
  canine, 85*f*, 153–157, 222
  feline, 95*f*
  ruminant, 1047*f*
Trachealis muscle, equine, 718
Tracheal loop, avian, 1282
Tracheal rings, avian, 1282
Tragus, canine, 136*f*
Transversus abdominis muscle
  canine, 87–88*f*
  equine, 559*f*
  feline, 98*f*
  ruminant, 996*f*, 1118
Transverse colon
  canine, 91*f*
  equine, 758
  ruminant, 998*f*, 1079–1080
Transverse facial artery and vein
  equine, 555*f*, 559*f*
  feline, 98*f*
Transverse fold of vestibule, equine, 785
Transverse (humeral) retinaculum, ruminant, 1155
Transverse process of vertebra
  equine, 559*f*
  ruminant, 993*f*
Transverse spinal fascia, ruminant, 996*f*
Trapezius muscle
  avian, 1255*f*
  canine, 87*f*
  equine, 557–558*f*
  feline, 98*f*
  ruminant, 996*f*, 1152
Triangular ligament
  equine, 765, 767*f*
  ruminant, 1092
Triceps brachii muscle
  avian, 1255*f*
  canine, 87*f*
  equine, 557*f*
  ruminant, 1154
Trichoglyph, ruminant, 1024
Tricuspid (right atrioventricular) valve
  canine, 269
  equine, 712
  feline, 268
  ruminant, 1052, 1052*b*
Trigeminal nerve (CNV)
  avian, 1270, 1298
  canine, 116
  equine, 581, 652, 660
  feline, 204
  ruminant, 1012, 1027
Trigone of urinary bladder
  canine, 341
  equine, 849
  ruminant, 1100–1101, 1101*f*
Triosseal canal, avian, 1372, 1374
Trochanter of femur
  canine, 84*f*
  equine, 554–555*f*, 559*f*
  feline, 94*f*
  ruminant, 992–993*f*
Trochanteric bursa, equine, 938
Trochanteric fossa, equine, 937–938
Trochlear groove
  equine, 563–564*f*, 945
  ruminant, 1207
Trochlear nerve (CNIV), equine, 652
Trochlear notch, ruminant, 1151, 1152*f*
Trochlear ridge of femur, medial and lateral, equine, 945
Trophoblast layer, canine, 369
T-shaped cartilage, equine, 572
Tubal membrane, equine, 789
Tubal pole, equine, 797
Tuber calcanei
  canine, 84*f*, 92*f*
  equine, 563–564*f*, 960, 962
  feline, 94*f*, 102*f*
  ruminant, 992*f*, 1000*f*
Tuber coxae
  canine, 458
  equine, 554*f*, 563–564*f*, 935
  feline, 94*f*
  ruminant, 992*f*, 1000*f*, 1034*f*
Tubercle of humerus, major, ruminant, 1000*f*
Tuber ischiadicum
  canine, 84*f*, 92*f*
  equine, 554*f*, 563–564*f*, 936
  feline, 94*f*, 102*f*
  ruminant, 992*f*, 1000*f*
Tuber of scapular spine, equine, 563–564*f*
Tuber olecrani
  canine, 92*f*
  equine, 554*f*, 563–564*f*
  feline, 102*f*
  ruminant, 1000*f*
Tuber sacrale
  canine, 92*f*
  equine, 563–564*f*, 935
  feline, 102*f*
  ruminant, 1000*f*, 1032, 1034*f*
Tunica adventitia, equine, 678
Tunica albuginea
  equine, 823, 837
  ruminant, 1137–1140, 1137*b*
Tunica dartos, equine, 839
Tunica mucosa, equine, 739–740
Tunica muscularis, equine, 678, 678*b*, 739–740

Tunica serosa, equine, 739–740
Tunica submucosa, equine, 739–740
Tunica vaginalis, equine, 656
Tylotrich hairs, canine, 548
Tympanic bulla
  canine, 92*f*
  equine, 656
  feline, 102*f*, 110, 140–142
Tympanic cavity
  avian, 1270
  equine, 658
  feline, 107*f*, 109, 140, 140*b*
Tympanic membrane
  avian, 1283–1284, 1303
  canine, 92*f*
  equine, 656
  feline, 139–140
Tympanic nerve, feline, 141–142
Tympanohyoid cartilage, canine, 176
Tympanum, avian, 1283

U

Udder
  equine, 554*f*
  ruminant, 1239*f*, 1240–1241
Ulna
  avian, 1258*f*, 1373, 1408
  canine, 92*f*, 421–422
  equine, 563–564*f*, 884, 884*f*
  feline, 102*f*
  ruminant, 1000*f*, 1151, 1152–1153*f*, 1189
Ulnar carpal bone
  avian, 1373
  canine, 92*f*
  equine, 894–896
  feline, 102*f*
  ruminant, 1000*f*
Ulnaris lateralis muscle
  canine, 87*f*
  equine, 557*f*
  feline, 410*f*
  ruminant, 1153*f*, 1155
Ulnar nerve
  avian, 1253*f*
  canine, 88*f*
  equine, 889, 897
  feline, 95*f*
  ruminant, 1159, 1167–1168
Ulnar vein, avian, 1375
Umbilical artery
  canine, 262
  caprine, 1132–1133
  equine, 847
Umbilical cord
  canine, 262
  equine, 847
  ruminant, 1087
Umbilical vein
  canine, 262
  equine, 847
  ruminant, 1087, 1087*b*, 1088*f*
Umbilicus
  equine, 847–849, 848*f*
  feline, 335
  ruminant, 1087
Uncinate process, avian, 1258*f*, 1382
Ungual crest
  canine, 92*f*
  feline, 102*f*
Ungual process
  canine, 92*f*
  feline, 102*f*
Ungual phalanx, avian, 1393
Upper motor neuron
  canine, 210–211, 210*b*, 527*b*
  equine, 867–868
  feline, 204
Upper tail covert feather, avian, 1265*b*
Urachus
  equine, 847
  ruminant, 1103–1104
Ureter
  avian, 1353
  canine, 340–344
  equine, 856
  feline, 99*f*
  ruminant, 1098, 1098*f*, 1100–1102, 1100–1101*f*, 1101*b*
Urethra
  canine, 89*f*
  caprine, 1129, 1130–1131*f*, 1131–1132
  equine, 820, 829*f*, 830, 849, 857–858
  feline, 97*f*
  ruminant, 1101*f*, 1102*b*, 1103*f*, 1124, 1132*b*, 1133*f*, 1140
    female, 1103
    male, 1100–1102
Urethral meatus, external (orifice)
  canine, 90*f*
  feline, 99–100*f*
Urethral process
  equine, 561*f*
  ruminant, 996*f*
Urethral recess, ruminant, 996*f*
Urethral tubercle, canine, 89*f*
Urethralis muscle
  canine, 88*f*
  caprine, 1131–1132
  equine, 822, 830, 833, 858
  ruminant, 1124
Urinary bladder
  canine, 342*f*, 344
  caprine, 1128, 1129*f*, 1130–1134
  equine, 849, 849*f*, 855–857, 855–856*f*
  feline, 97*f*
  ruminant, 1102
Urinary bladder, lateral ligament of, equine, 998*f*
Urodeum, avian, 1333, 1353
Urogenital artery, ruminant, 1111
Urogenital fold, equine, 829*f*, 831
Uropygial (preen) gland, avian, 1256–1257*f*, 1383, 1401, 1404
Uterine artery
  canine, 356
  equine, 786, 787*f*, 806
  ruminant, 1120–1122, 1124
Uterine branch of urogenital artery, ruminant, 1111, 1122
Uterine tube. *See also* Oviduct
  avian, 1256*f*
  canine, 352–353
  equine, 559*f*, 812–814, 812–813*f*
  feline, 99*f*
  ruminant, 997*f*
Uterine vein, ruminant, 1122
Utero-tubal junction, equine, 812–813, 812*f*
Uterus
  avian, 1332
  canine, 353–356
  equine, 788, 789*f*, 804–805, 830–831*f*, 831, 831*b*
  feline, 97*f*
  ruminant, 1113, 1115*f*, 1116, 1119, 1120*f*, 1121*b*
Uterus, round ligament of, equine, 560*f*
Utricle, canine, 184
Utriculus masculinus, equine, 831*f*
Uvea, equine, 570

V

Vagal nuclei, ruminant, 1060–1062
Vagina, 782–783*f*, 785
  avian, 1333
  canine, 89*f*, 335
  equine, 777–778, 782–783*f*, 785
  feline, 97*f*
  ruminant, 1106, 1108, 1110–1111, 1123–1126, 1124–1125*f*, 1125*b*
Vaginal artery
  canine, 341, 344
  equine, 778, 790–791, 806
  ruminant, 1124
Vaginal fornix
  equine, 785
  ruminant, 806
Vaginal vein
  equine, 1108
  ruminant, 1124
Vaginal vestibule
  canine, 340*f*
  equine, 851
  ruminant, 1103*b*

Vagosympathetic trunk
canine, 160, 175
equine, 644, 870–871
Vagus artery, avian, 1274–1275
Vagus nerve (CNX)
avian, 1274–1275
canine, 89*f*, 117, 152, 175, 290
equine, 652–653, 678, 731
feline, 324
ruminant, 996*f*, 1053, 1068, 1076
Vallate papillae, equine, 580
Valve cusp
canine, 269
equine, 711*f*
Valvules of ventricle, avian, 1323
Vanes, avian, 1406
Vascular notch
equine, 563*f*
ruminant, 1000*f*
Vastus medialis muscle, feline, 478
Vastus lateralis muscle
canine, 88*f*
equine, 558*f*
feline, 475*f*
ruminant, 996*f*
Venous plexus of penis, equine, 823, 986
Vent, avian, 1348
Ventral buccal branch of facial nerve
canine, 87*f*
equine, 610*f*
Ventral lateral sacrocaudal muscle, canine, 88*f*
Ventricle, left and right
avian, 1323, 1347
canine, 269–270
equine, 701, 711–712
feline, 249
ruminant, 1052, 1053*f*
Ventriculus (gizzard), avian, 1256*f*
Ventrodorsal view of the abdominal cavity
canine, 91*f*
equine, 562*f*
feline, 101*f*
ruminant, 999*f*
Vermis, equine, 689
Vertebral formula - cervical, thoracic, lumbar, sacral, caudal
canine, 92*f*
equine, 563–564*f*
feline, 102*f*
ruminant, 1000*f*
Vertebral column
avian, 1258*f*, 1383
canine, 92*f*
equine, 71*f*, 563*f*, 653–654, 683
feline, 102*f*
ruminant, 1031*f*, 1032, 1033*f*
Vertebral diverticula, avian, 1338
Vertebral-vagal trunk, avian, 1323
Vesical artery, canine, 344
Vesicourethral junction, canine, 340–341
Vesicular gland
equine, 825, 826–828*f*, 827, 832
ruminant, 998*f*
Vestibular apparatus, feline, 108
Vestibular bulb, equine, 784
Vestibular fold, equine, 641
Vestibular ganglia, feline, 143
Vestibular gland, ruminant, 1123
Vestibular labyrinth, feline, 143–144
Vestibular nuclei, canine, 186
Vestibular system, 782*f*, 784–785
avian, 1303
canine, 184, 364
equine, 649, 776–777, 782*f*, 784–785, 857
feline, 108
ruminant, 1108, 1110, 1123–1126, 1124–1125*f*
Vestibule
canine, 85*f*, 89*f*
equine, 560*f*
ruminant, 997*f*
Vestibulocochlear nerve (CNVIII)
canine, 183
equine, 652
feline, 108, 144–145
Vestibulovaginal ring/junction, equine, 777, 785
Villi of small intestine
canine, 312
equine, 739–740
ruminant, 1078–1079
Visceral peritoneum
canine, 314
equine, 739–740, 767–769, 768*f*
ruminant, 1131
Visceral vaginal tunic, equine, 837
Vocal cord
equine, 637–638*f*, 640–643, 645*f*
ruminant, 1008*f*
Vocalis muscle, equine, 643
Vocal ligament, equine, 643
Vocal process, equine, 641–642
Vomer bone, ruminant, 1006, 1008*f*
Vomeronasal organ, ruminant, 1006
Vulva
canine, 335, 363–364
equine, 775–776
feline, 99*f*
ruminant, 1063–1064, 1106, 1107*b*, 1108–1109, 1109*f*, 1112–1113, 1125*f*

W

Wall segment of lamellae, ruminant, 1175
Wattle
avian, 1254*f*
caprine, 1229–1230, 1231*f*
White line
equine, 929
ruminant, 1164*f*
White matter, feline, 202
White pulp of spleen
avian, 1360
equine, 726
Wing, avian, 1372–1373, 1375–1376, 1407, 1409
Wing of atlas
canine, 85*f*
equine, 563*f*
ruminant, 1000*f*
Wing of ilium
canine, 85*f*
equine, 563*f*
feline, 555*f*
ruminant, 1120*f*
Withers, equine, 563–564*f*

X

Xiphoid
avian, 1258*f*
canine, 84*f*
equine, 94*f*, 563–564*f*
feline, 561*f*
ruminant, 1000*f*, 1066
Xiphoid, cartilage of
canine, 85*f*, 91*f*
equine, 95*f*, 102*f*
feline, 555*f*, 561*f*
Xiphoid process of sternum
avian, 1258*f*
canine, 85*f*
equine, 563–564*f*, 747
feline, 555*f*, 561*f*
ruminant, 993*f*, 998*f*

Z

Zona fasciculata, canine, 295
Zona glomerulosa, canine, 295
Zona reticularis, canine, 295
Zygodactyl, avian, 1390
Zygomatic arch
canine, 92*f*
equine, 563–564*f*

feline, 102*f*
ruminant, 1000*f*
Zygomatic bone
avian (jugal arch), 1258*f*
canine, 129
caprine, 1011*f*
feline, 95*f*
Zygomatic muscle
canine, 87*f*
equine, 557*f*, 559*f*
ruminant, 996*f*
Zygomatic nerve
equine, 575*f*
ruminant, 1027
Zygomaticoauricularis muscle
equine, 657*f*
ruminant, 996*f*
Zygomaticofacial nerve
canine, 86*f*
feline, 96*f*
Zygomaticotemporal nerve
canine, 86*f*
equine, 556*f*
ruminant, 1027
Zygomatic process
canine, 92*f*
equine, 656–658
Zygomatic region
canine, 84*f*
equine, 554*f*
feline, 94*f*
Zygomatic salivary gland, canine, 130

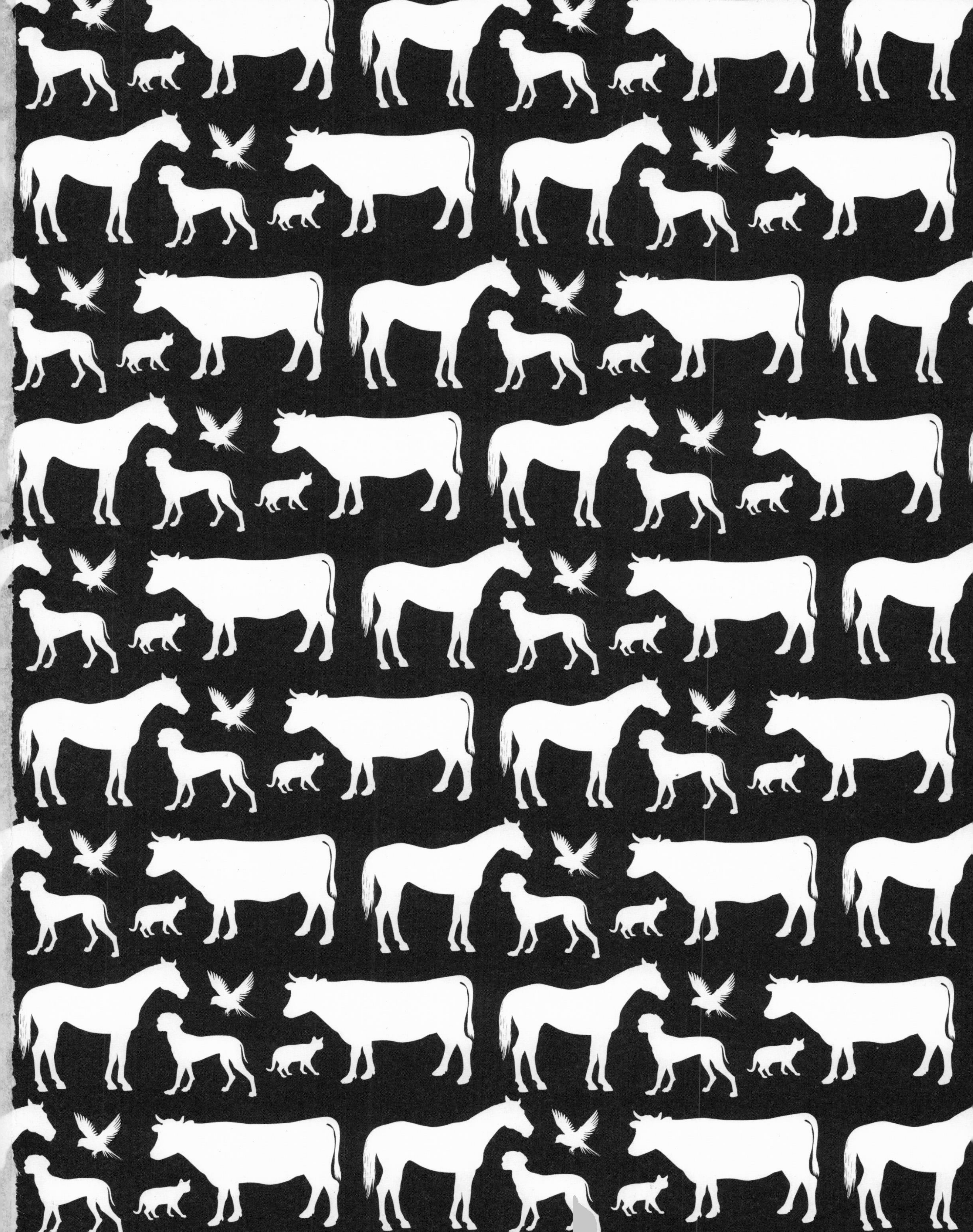

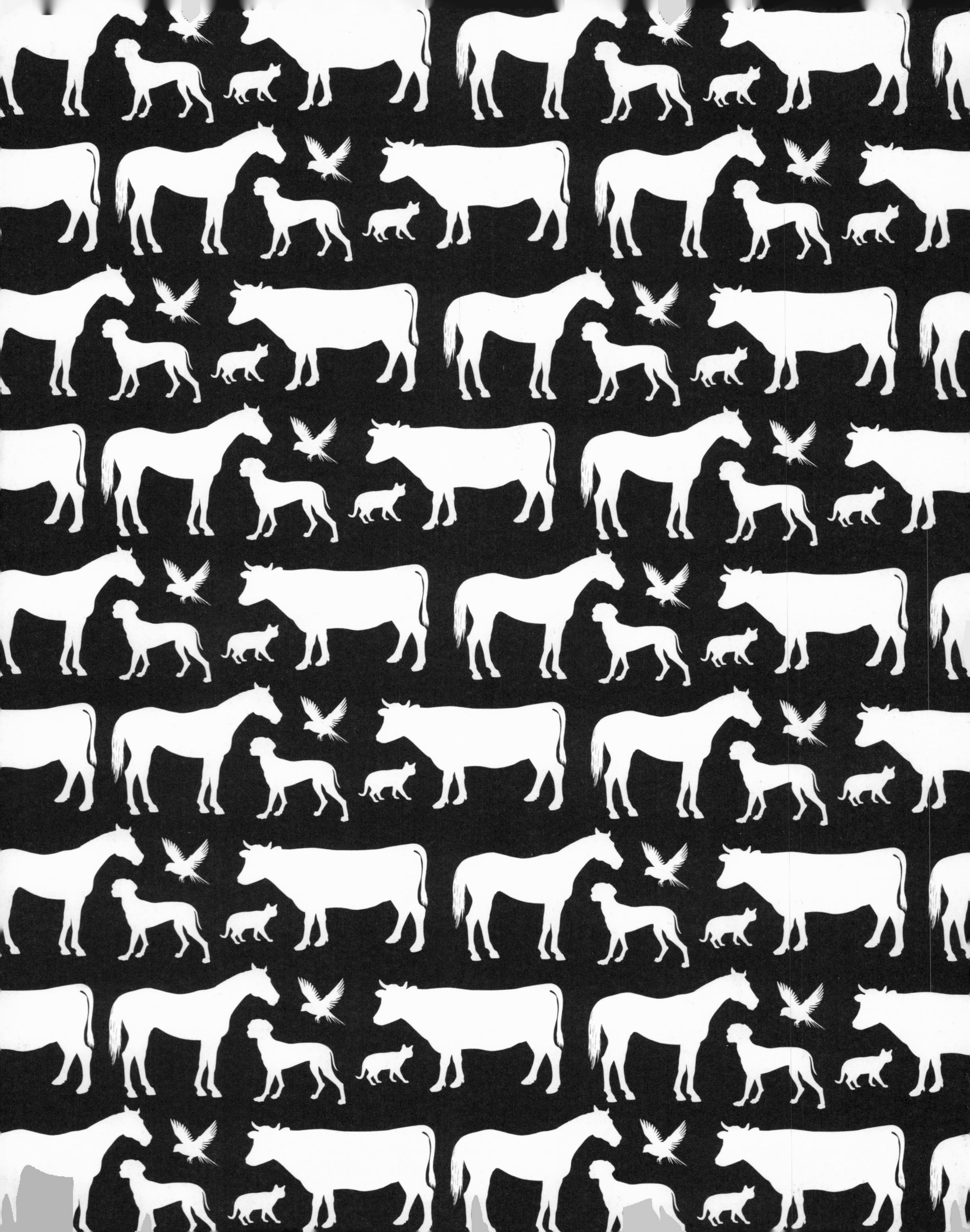